About the Editors:

David G. Seiler is the Chief of the Semiconductor Electronics Division, Electronics and Electrical Engineering Laboratory, National Institute of Standards and Technology. His Division provides technical leadership in research and development of the semiconductor measurement infrastructure essential to silicon and other advanced semiconductor technology needs. Dr. Seiler, Fellow of the American Physical Society, has worked extensively with the characterization of the electrical, optical, and nonlinear optical properties of numerous semiconductors and artificially structured materials. Prior to joining NIST in 1988, Dr. Seiler served as a Solid State Physics Program Director in the Materials Research Division at the National Science Foundation and was a physics professor at the University of North Texas. He obtained a Ph.D. in Physics from Purdue University in 1969.

Alain C. Diebold is manager of Metrology Coordination at SEMATECH and leader of the Analytical Laboratory Managers Working Group of the SEMATECH member companies. Through direct interaction between customers, government laboratories, and metrology supplies, these industry coordination activities foster the development of metrology and analytical technology that is required for the development and manufacture of future generations of integrated circuits. Dr. Diebold chaired the Metrology Technical Group that provided the Metrology Roadmap found in the 1997 National Technology Roadmap for Semiconductors. Prior to joining SEMATECH, he worked at Allied Signal Corporation in the area of molecular beam epitaxy of III-V compounds, as well as with a broad range of materials characterization issues. He obtained a Ph.D. in Chemistry from Purdue University in 1979 and was a post-doctoral fellow at Pennsylvania State University.

W. Murray Bullis is Vice President, International Standards for Semiconductor Equipment and Materials International (SEMI). Prior to joining SEMI in 1996, he served the industry for five years as a private consultant. Before that, he held positions at Siltec Silicon, the Fairchild Research Center, the National Bureau of Standards, and Texas Instruments. Dr. Bullis received his A.B. from Miami University, Oxford, Ohio, in 1951, and his Ph.D. in Physics from MIT in 1956. He is a senior member of IEEE, a fellow of ASTM, and a member of The Electrochemical Society and the American Physical Society.

Thomas J. Shaffner is Manager of Materials Characterization R&D within Silicon Technology Research at Texas Instruments Incorporated. He has directed a broad range of characterization programs over the past 20 years, with emphasis on advanced materials growth and techniques for microstructure and chemical analysis. Previously, he worked at E. I. duPont de Nemours in fiber surface research. He is an active short course instructor at Arizona State University and the International Society for Optical Engineering, and serves on committees of regional and national organizations, including SEMATECH, The Electrochemical Society, and the Board on Assessment of NIST Programs. He obtained a Ph.D. in Physics from Vanderbilt University in 1969, and was a post-doctoral fellow in electron microscopy at the University of Manchester, UK.

Robert McDonald is currently an independent consultant. Bob retired this year from Intel after serving in various management and technical capacities for twenty years. His efforts for Intel included startup of their first materials analysis labs and the formation of a worldwide lab network. He was responsible for the introduction of new methods to meet the demands of rapidly advancing semiconductor technology. This included the application of advanced materials analysis techniques for critical manufacturing measurement and metrology needs as well as for technology development. He served as Intel's representative to the SEMATECH Analytical Lab Managers Working Group and the SIA Metrology TWG. Prior to joining Intel, He was responsible for the Advanced Materials Analysis Lab at Fairchild R&D for 5 years. He received his Ph.D. degree in Materials Science from the UCLA School of Engineering in 1981. Bob currently serves on the Board of Directors for Balazs Analytical Lab.

E. Jane Walters is Lead Editor for the Semiconductor Electronics Division of the Electronics and Electrical Engineering Laboratory of the National Institute of Standards and Technology. For over 20 years, she has provided technical support and information to the semiconductor industry, publishing bibliographies and conference proceedings and compiling quarterly EEEL Technical Progress Bulletins. Major activities include representing NIST at the SEMICON/West show and developing world wide web materials. She is responsible for ensuring the quality of all technical documents and papers from the Division.

CHARACTERIZATION AND METROLOGY FOR ULSI TECHNOLOGY

Conference Sponsors:

NATIONAL INSTITUTE OF STANDARDS AND TECHNOLOGY
Semiconductor Electronics Division
National Semiconductor Metrology Program
Electronics & Electrical Engineering Laboratory

SEMATECH

SEMICONDUCTOR RESEARCH CORPORATION

AMERICAN VACUUM SOCIETY
Manufacturing Science and Technology Division

SEMICONDUCTOR EQUIPMENT AND MATERIALS INTERNATIONAL
(SEMI)

Wine and Cheese Poster Sessions Sponsors:

VLSI STANDARDS, INC.
EATON
KLA-TENCOR

CHARACTERIZATION AND METROLOGY FOR ULSI TECHNOLOGY

1998 International Conference

Gaithersburg, Maryland March 1998

EDITORS

David G. Seiler
NIST, Gaithersburg, Maryland

Alain C. Diebold
SEMATECH, Austin, Texas

W. Murray Bullis
SEMI, Mountain View, California

Thomas J. Shaffner
Texas Instruments, Dallas, Texas

Robert McDonald
Intel Corp., Santa Clara, California

E. Jane Walters
NIST, Gaithersburg, Maryland

CD-ROM INCLUDED

AIP CONFERENCE PROCEEDINGS 449

American Institute of Physics **Woodbury, New York**

Editors:

David G. Seiler, Chief, Semiconductor Electronics Division
National Institute of Standards and Technology
Room B344, Building 225
Gaithersburg, MD 20899

E-mail: seiler@nist.gov

Alain C. Diebold
SEMATECH
2706 Montopolis Drive
Austin, TX 78741

E-mail: alain.diebold@sematech.org

W. Murray Bullis, Vice President, International Standards
SEMI
805 Middlefield Road
Mountain View, CA 94043

E-mail: mbullis@semi.org

Thomas J. Shaffner
2016 Aliso Road
Plano, TX 75074

E-mail: txbeaker@prodigy.net

Robert McDonald
540 Roxbury Lane
Los Gatos, CA 95032

E-mail: hightechsearch@msn.com

E. Jane Walters
Semiconductor Electronics Division
National Institute of Standards and Technology
Room B344, Building 225
Gaithersburg, MD 20899

E-mail: walters@nist.gov

L.C. Catalog Card No. 98-87959
ISBN 1-56396-753-7 Set
ISBN 1-56396-867-3 Print
ISBN 1-56396-868-1 CD-ROM
ISSN 0094-243X
DOE CONF- 980364

Printed in the United States of America

CONTENTS

FRONT END PROCESSES—DOPING AND 300mm WAFER ISSUES

FRONT END PROCESSES—GENERAL

INTERCONNECT

INTERCONNECT—GENERAL

LITHOGRAPHY

LITHOGRAPHY—GENERAL

PACKAGING

PACKAGING—GENERAL

MATERIALS CHARACTERIZATION

MATERIALS CHARACTERIZATION—DOPANT PROFILING

MATERIALS CHARACTERIZATION—GENERAL

Preface

The 1998 International Conference on Characterization and Metrology for ULSI Technology was held at the National Institute of Standards and Technology from March 23 to March 27, 1998. The Conference was dedicated to summarizing major issues and giving critical reviews of important semiconductor techniques that are crucial to continue the advances in semiconductor technology. This Conference was the second in a series; the first, Semiconductor Characterization: Present Status and Future Needs, was held at NIST, on January 30 through February 2, 1995. Papers from that Workshop were published in a hardback Proceedings by the American Institute of Physics Press.

The 1998 Conference brought together leaders, scientists, and engineers concerned with all aspects of the technology and characterization techniques for silicon research, including development, manufacturing, and diagnostics. Key, knowledgeable people in the semiconductor field addressed the unique characterization requirements of silicon IC development and manufacturing. Sessions on silicon ICs were based on the technology drivers in the National Technology Roadmap for Semiconductors.

The editors believe that the invited talks, poster papers, and informal discussions provided a basis for stimulating practical perspectives and ideas for research and development, as well as a chance to explore collaborations and interactions. The editors feel that the conference and this book of collected papers provide a concise and effective portrayal of industry characterization needs and some of the problems that must be addressed by industry, academia, and government to continue the dramatic progress in semiconductor technology.

Characterization and metrology are key enablers for developing semiconductor technology in improving manufacturing. The worldwide semiconductor community faces increasingly difficult challenges as it moves into the manufacturing of chips with feature sizes approaching 100 nm. Many of these challenges are materials related, such as transistors with high k dielectrics and on-chip interconnects made from copper and low k dielectrics. The magnitude of these challenges demands special attention from those of us in the metrology and analytical measurements community. New paradigms must be found to better work together. Adequate research and development for new metrology concepts is sorely needed.

The Conference opened with introductory remarks by Ray Kammer, Director of the National Institute of Standards and Technology, on "NIST and the Semiconductor Industry: A Four-Decade Partnership Focuses on the Future." Mark Melliar-Smith, President and CEO of SEMATECH, opened the technical sessions with a challenging talk "Metrology Needs for the Semiconductor Industry over the Next Decade. At the Banquet on Monday evening, Paul Peercy, President of SEMI/-SEMATECH, spoke on "Semiconductor Materials Research for the 21st Century." These talks provided a larger context for the detailed discussions of semiconductor characterization issues in the Conference program which consisted of formal invited presentation sessions and poster sessions for contributed papers.

The invited papers provided up-to-date reviews of major issues and characterization techniques for semiconductor device research, development, and manufacturing. Poster papers, which were presented in three sessions, emphasized new developments and improvements in characterization technology.

The chairs of the invited paper sessions were: Ralph Cavin, SRC; Paul Peercy, SEMI/-SEMATECH, Rinn Cleavelin, Texas Instruments/SEMATECH, and Anne Testoni, DEC; Jeff Bindell, Lucent Technologies, and Tom Shaffner, Texas Instruments; Robert McDonald, Intel, and Bryan Tracy, Advanced Micro Devices, Inc.; Shane Palmer, Texas Instruments; George Harman, NIST; Bill Oosterhuis, DoE, and Stephen Knight, NIST; Alain Diebold, SEMATECH, and Ellen Williams, University of Maryland; and Lori Nye, MEMC, and Murray Bullis, SEMI.

An evaluation survey was conducted following the Conference. Respondents indicated that the Conference had been meaningful and relevant for them because it:

- was an important meeting for keeping abreast of the latest developments
- did a good job of including the breadth of the industry - this is the strength of the meeting.
- provided a good high level overview of industry and technology, including business trends
- included people from semiconductor manufacturing, suppliers of equipment, labs, government institutes, consortia
- presented state-of-the-art developments
- provided interaction with experts and co-workers, strong interaction among industry, academia, and NIST people
- touched on a variety of topics - no area was completely left out
- gave a broad overview of process issues

The respondees felt that this is an important meeting for keeping abreast of the latest developments and is definitely a needed forum for the semiconductor metrology community.

This proceedings volume is organized along the lines of the Conference program. The papers are grouped in major sections under the topics of the invited paper sessions: Challenges; Overview; Front End; Interconnect, Back End of Line; Lithography; Packaging; Review of Critical Analytical Techniques; and 300 mm and Beyond.

The Editors thank the members of the Program Committee, the Session chairs, and many of the invited speakers for their assistance in reviewing the manuscripts submitted for publication in this volume. They also thank the authors for their diligence in producing the camera-ready copy for their papers and in responding to reviewer comments and suggestions. Special thanks go to VLSI Standards, Inc., Eaton, and KLA-Tencor, sponsors of the wine and cheese poster sessions. Finally, they especially thank Brenda Main of the NIST staff, whose expert assistance greatly facilitated the planning and conduct of the Conference. Her tireless efforts helped contribute both to a successful Conference and to the timely publication of this volume.

David G. Seiler
Alain C. Diebold
W. Murray Bullis
Thomas Shaffner
Robert McDonald
E. Jane Walters
Editors

Committee Co-Chairs:

Committee Members:

OVERVIEW

Metrology Needs for the Semiconductor Industry Over the Next Decade

Mark Melliar-Smith

Alain C. Diebold

SEMATECH • 2706 Montopolis Drive • Austin, TX 78741

Metrology will continue to be a key enabler for the development and manufacture of future generations of integrated circuits. During 1997, the Semiconductor Industry Association renewed the National Technology Roadmap for Semiconductors (NTRS) through the 50 nm technology generation and for the first time included a Metrology Roadmap (1). Meeting the needs described in the Metrology Roadmap will be both a technological and financial challenge. In an ideal world, metrology capability would be available at the start of process and tool development, and silicon suppliers would have 450 mm wafer capable metrology tools in time for development of that wafer size. Unfortunately, a majority of the metrology suppliers are small companies that typically can't afford the additional two to three year wait for return on R&D investment. Therefore, the success of the semiconductor industry demands that we expand cooperation between NIST, SEMATECH, the National Labs, SRC, and the entire community.

In this paper, we will discuss several critical metrology topics including the role of sensor-based process control, in-line microscopy, focused measurements for transistor and interconnect fabrication, and development needs. Improvements in in-line microscopy must extend existing critical dimension measurements up to 100 nm generations and new methods may be required for sub 100 nm generations. Through development, existing metrology dielectric thickness and dopant dose and junction methods can be extended to 100 nm, but new and possibly in-situ methods are needed beyond 100 nm. Interconnect process control will undergo change before 100 nm due to the introduction of copper metallization, low dielectric constant interlevel dielectrics, and Damascene process flows.

INTRODUCTION

Integrated circuits (ICs) are the cornerstone of a multi-billion dollar electronics industry that is leading civilization into the 21st century. IC technology is being driven by consumer demands for improved applications, such as information processing and communications, with greater capability at a constant cost. Thus, consumers have forced the historical trend of a 25 to 30% cost reduction per unit function per year. More than 30 years ago, Gordon Moore observed that the number of transistors in a manufactured die increased by a factor of two every year (2). From this observation, it was postulated that the number of bits in memory chips would increase by a factor of four every three years. A 30% decrease in feature size every three years and a 1.5x increase in chip size are two of the factors that enable the continuation of Moore's law. The other factors that have kept the industry on the historical cost reduction trend are yield improvement, wafer size increase, and overall equipment effectiveness. The cost productivity curve is shown in Fig. 1.

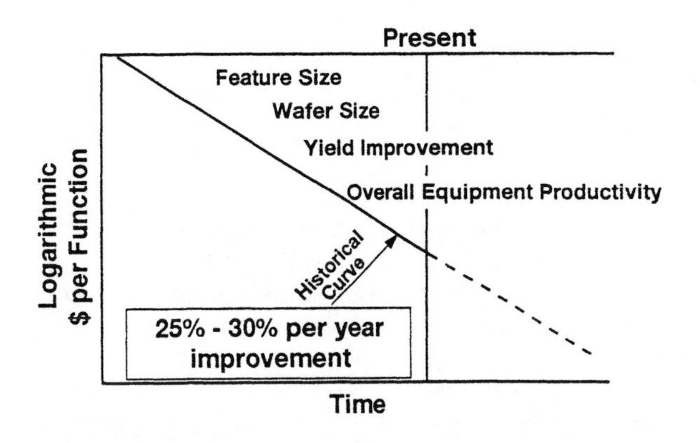

FIGURE 1. *Historical Trend of 25-30% Annual Cost Reduction per Unit Function for the Semiconductor Industry.*

The overwhelming cost of research and development including product development for each new technology generation has forced international cooperation for the semiconductor industry. For example, the total cost of developing 300 mm-wafer-capable manufacturing is estimated to be more than $10 billion, and the development of 193 nm lithography is estimated at around $1 billion (3). One example of cooperation is the National Technology Roadmap for Semiconductors (1). The 1992, 1994, and 1997 Roadmaps provide a consensus view of the most critical technology requirements for IC manufacture with a 15 year horizon. The 1997 NTRS projects these technology needs to the 50 nm technology generation. The NTRS allows R&D organizations such as SEMATECH, SRC, MARCO, and interested national labo-

CP449, *Characterization and Metrology for ULSI Technology: 1998 International Conference*
edited by D. G. Seiler, A. C. Diebold, W. M. Bullis, T. J. Shaffner, R. McDonald, and E. J. Walters

ratories to focus their resources on the most critical requirements and plan activities according to the industry's implementation timeline (1).

Metrology enables the lithography, wafer size, yield, and equipment effectiveness drivers of cost reduction, and therefore metrology technology needs to keep pace with process technology. In addition, metrology must also enable the rapid introduction of new products which require the "ramping within ramps" such as the manufacture of higher speed versions of a generation of microprocessor (4). For the first time, the 1997 version of the NTRS contains a Metrology Roadmap. In this paper, we will describe the industry's metrology needs over the next ten years using the 1997 NTRS for technical guidance.

The 1997 NTRS describes the key challenges that the industry faces if it is to keep on the historical cost-productivity curve (1). These critical industry challenges provide a means of prioritizing metrology requirements. The overall challenges are listed below with a brief description (1):

- *The ability to continue affordable scaling*
  Continue the roadmapped timeline for new technology generations
- *Affordable lithography at and below 100 nm*
  Move to 193 nm optical lithography and find affordable replacements for optical lithography
- *New materials and structures*
  Short term: replace aluminum/silicon dioxide interconnect materials with copper and low dielectric constant materials including appropriate cost saving Damascene processes; Long Term: find a solution for transistor design and materials for sub 100 nm technology generations
- *GHz frequency operation on- and off-chip*
  High-frequency on-chip interconnects and off-chip packaging based interconnects must allow full utilization of increases in chip speed
- *Metrology and test*
  This paper describes the required metrology, and electrical testing of chip functionality must remain cost-effective as chip functionality increases
- *The research and development challenge*
  Provide an affordable infrastructure that moves technology from basic research into manufacturing

Metrology needs must be prioritized and developments driven to enable the key processing technology that will allow us to maintain the cost reduction trend and profit margins that maintain an economically healthy industry. In a perfect world, metrology capability would be available when process tool suppliers initiate tool development for the next technology generation (1). Unfortunately, this forces the metrology suppliers to wait three years longer than process tool suppliers for a return on their investment in new tool development. The same principle applies to changing wafer size, and time for return on investment will be longer than three years for 300 mm wafer metrology tools. The entire community must cooperate to overcome this challenge.

The Metrology Roadmap describes the requirements and potential solutions for the off-line, in-line, and in-situ measurements. Other metrology needs are found in the Defect Reduction Technologies Roadmap. In Fig. 2, an overview of the multiple measurement requirements is depicted in terms of the fab process flow.

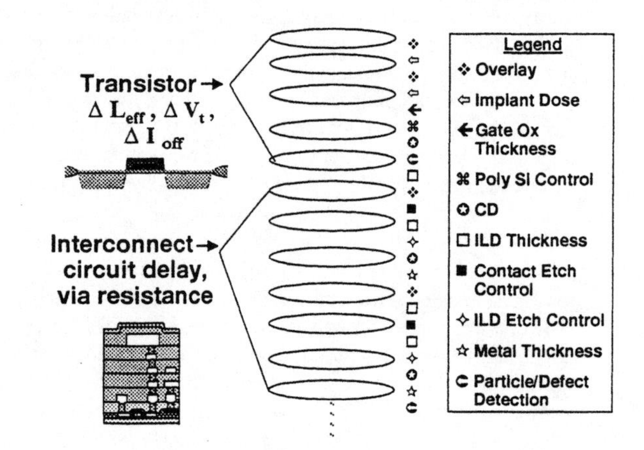

FIGURE 2. *In-FAB Metrology*
Metrology is used to control fabrication processes so that integrated circuit electrical performance falls within product specifications. At the 150 nm technology generation, there will be approximately 500 process steps requiring more than 20 overlay and 35 CD measurements.

In addition to fab process metrology, packaging metrology is covered in the 1997 NTRS. Metrology is expected to continue to migrate from off-line toward in-line and in-situ application. However, off-line measurements will continue to provide highly detailed information that requires special capabilities. For example, transmission electron microscopy (TEM) is often the only means of analyzing small features found in modern IC's. Evolutionary improvements of current in-line metrology methods are expected to enable control of implant and thin film processes. Currently, many transistor and interconnect processes are controlled using physical measurements on unpatterned (test) wafers. The need for nondestructive measurements on product wafers is an additional challenge for in-line metrology. Initial implementation of in-situ sensor-based process control has proven the usefulness of this approach through improved tool throughput and reduced wafer scrap. Many of these sensors measure something that allows control of wafer level properties, often indirectly. In addition to increasing their range of applications, sensors are expected to evolve to provide more detailed information on the wafer with increased spatial resolution (i.e., die to die uniformity). Micro-electromechanical systems (MEMS) are expected to become the smart sensors of the future (1). One visionary example is an etch chamber with a strategically located mass spectrometer on a chip with built-in circuitry for real time process control. Packaging metrology is not as mature as wafer fab metrology, and thus there are many opportunities to implement existing technology as well as to develop appropriate new methods.

The advantages of in-line measurements, such as improved analysis cycle time, have resulted in the "FAB-LAB" concept (5,6). The laboratories contain tools and data management/analysis systems capable of characterizing product wafers using information such as particle/defect location maps (5). The FAB-LAB is an integral part of pilot line as well as fab start-up and operation, and there are examples of FAB-LABs both inside and next to the clean room. It is interesting to note that some IC manufacturers have located traditionally out of the fab tools such as TEM inside clean rooms. This is one example of how the need to reduce the cycle time for product development and yield learning has driven the industry. The relationship between analysis cycle time requirements and the proximity of analysis tools to the fabrication line has been described elsewhere (5).

THE MEASUREMENT REQUIREMENTS AND FUTURE OF STATISTICAL PROCESS CONTROL

Statistical Process Control (SPC) methodology is used to control each IC manufacturing process so that a majority of the final product has a narrow range of electrical characteristics. In this section, we emphasize the need for process application based evaluation of metrology tools capability for use in SPC. Automated in-line and appropriate off-line metrology tools are evaluated for SPC by the measurement precision to process tolerance (process specification range) ratio criterion, P/T. The P/T should be $< 10\%$, but 30% is often tolerated. $P/T = 6\sigma/(UL-LL)$ where the measurement precision (variation), σ, includes the short-term repeatability and long-term reproducibility, and UL and LL are the upper and lower process limits, respectively. The P/T metric is well accepted by the semiconductor industry.

Determination of the true P/T ratio for the process range of interest requires careful implementation of the P/T methodology and reference materials with identical feature size, shape, and composition to the processed wafer being measured. Often, there are no certified reference materials that meet this requirement, and P/T ratio and measurement accuracy are determined using best available reference materials. One key example is the lack of an oxide thickness standard for sub 5 nm SiO_2 and nitrided oxides.

Another aspect of true determination of P/T capability is measurement linearity. If one determines the P/T ratio for a thickness or width measurement using a reference material that has larger (or smaller) dimensions than the features being measured during IC manufacture, lack of measurement linearity could result in insufficient resolution to distinguish changes over the entire process range of interest. For the sake of argument, we will call tool precision determined using inadequate reference materials as "estimated tool precision."

Often, the concept of measurement resolution is confused with "estimated tool precision." It is possible for a tool to appear to have small (i.e., good) precision and still have poor resolution. As transistor and interconnect features shrink, greater resolution is becoming more critical. The criteria for metrology tool applicability to SPC need to include both P/T and resolution.

For example, we need to be able to distinguish a 2.0 nm thick transistor gate dielectric from a film having 2.1 nm or 2.2 nm thickness. Another example is distinguishing a 100 nm gate electrode width (critical dimension, CD) from a 102 nm CD. SPC criteria are usually applied to physical measurements, but electrical measurement variability also needs to be suitably small.

Another example of the varied and thus confusing usage of the term resolution is in critical dimension (CD) measurement by scanning electron microscopy (SEM). The microscopy community has used two different methods to determine the resolution of a scanning electron microscope (SEM). One is the width of the electron beam, and the other is the ability to distinguish two closely spaced features in an image of a well characterized test sample. Fine gold particles on a carbon background is a well accepted test sample.

Despite the long history of these methods, there do not seem to be standardized (SEMI, ASTM, etc.) resolution procedures. Another measure of the resolution of an SEM dedicated to CD measurement is its ability to distinguish repeatedly differences in transistor gate linewidth (e.g., 100 nm from 102 nm). This process application based metric for resolution would facilitate evaluation of true SPC capability.

One way to assure that the metrology tool has adequate resolution is to determine the true P/T capability by using a series of standardized, accurate reference materials over the measurement range specified by the upper and lower process limits. In Fig. 3, we depict how the multiple reference materials approach might work. We note that some measurements may require one suitable reference material for P/T determination.

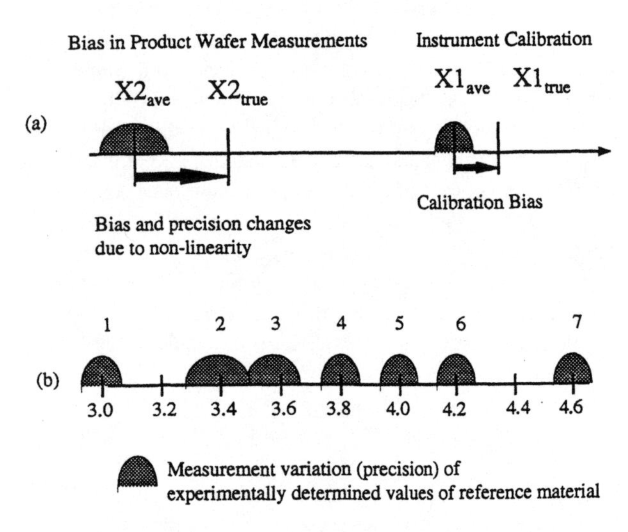

FIGURE 3. *Relationship between Measurement Resolution and Precision*

(a) Measurement non-linearities can result in bias (difference between true and measured value) changes between the calibrated value and values within the range of interest.

(b) For this example let us assume that the process tolerance (also called process specifications) is from 3.0 nm to 4.6 nm. The measurement precision at 3σ (variation) is shown for reference materials inside the process range. The experimental P/T capability observed using reference materials 4, 5, and 6 indicates that a single measurement of a 4.0 nm is different from one at 3.8 nm or 4.2 nm. Thus this fictitious metrology tool can resolve those values. The tool is not able to resolve 3.4 nm from 3.6 nm at 3σ.

A SEMI task force will be standardizing the implementation of P/T over the next several years. This type of activity can greatly improve the usefulness of the P/T metric, and perhaps it could be called the "Resolution P/T." Implementation of this methodology is hampered by the lack of reference materials with features (size, composition, etc.) inside the measurement range of interest.

The goal of sensor-based process control is to become real time. In-line measurement tools are (when appropriate) used for run to run control. The concept of P/T must be extended to allow evaluation of sensors. Sensors appear to have the interesting challenge of being required to measure process variation without the luxury of averaging over multiple measurements to improve precision.

One revolutionary improvement in SPC has been the use of cumulative data sets (from the same sample, across a wafer, and previous samples) to greatly improve estimation of measurement precision (6,7). A smaller estimated precision may mean that a metrology tool is capable of meeting P/T criteria for SPC without hardware improvements or it can allow reduction of process tolerances while maintaining P/T. Since this improvement was not associated with the measurement tool itself, it raises questions about how to evaluate improvements in metrology tools.

CORRELATION OF PHYSICAL AND ELECTRICAL MEASUREMENTS

IC performance and yield are ultimately evaluated by the electrical parametric and functional testing. It is possible to measure the relevant electrical properties of transistor and interconnect structures on product wafers during IC manufacture. Clearly, this requires that the wafers have been processed to the point where a complete transistor (or interconnect structure) is fabricated. When the manufacturing process is mature and robust and a majority of the process flow is transferred to the next product or technology generation, electrical metrology may provide a majority of process control needs. However, when excursions in process tool performance or process material quality occur, they need to be observed and controlled long before an electrical test can be performed. In this section, we describe several levels of correlating physical and electrical measurements. These levels can be described as follows: correlation of electrical and physical measurements of a specific physical feature such as gate dielectric thickness; correlation of a set of physical parameters with the electrical properties of the transistor or interconnect structure; and correlation of die, wafer, or lot level data with yield.

Since scrap prevention is one of the goals of physical, in-line metrology, physical measurement must correlate with electrical performance. The gate dielectric thickness, gate electrode critical dimension, and implant dose and profile all influence transistor electrical parameters such as threshold voltage, off current, and gate delay. Modeling of transistors provides insight into the nature of the correlation of physical parameters with electrical performance, including manufacturing sensitivities (8).One can model the change in variation of electrical parameters using device simulation and selected ranges of physical parameter variation. The variation of threshold voltage range with range of gate length and channel implant is shown for a fixed range of gate dielectric thickness and other physical parameters in Fig. 4. Zeitzoff and Tasch discuss manufacturing sensitivity in these proceedings (8).

Another issue is that integrated circuit development requires statistically significant information which can only come from electrical test. The need for early measurement capability during process tool development is highlighted by the push for more rapid ramping of pilot line yield. Delivery of well characterized process tools greatly facilitates rapid pilot line yield ramping. Bartelink is credited with initiating the field of Statistical Metrology which utilizes electrical test structures to determine across the die and across the wafer properties (9, 10). Statistical Metrology is a set of procedures designed to deconfound measurement error from true process variation. Statistical Metrology is discussed in the Process Integration, Devices, and Structures Roadmap and the Metrology Roadmap's one page contribution to that roadmap (1).

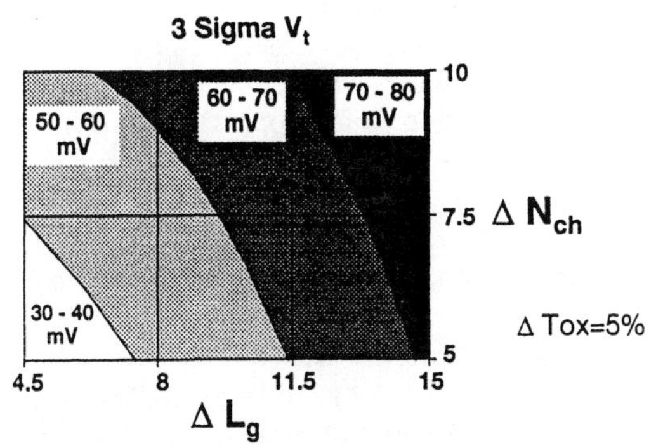

(effect of CD variation on 180 nm transistor)

FIGURE 4. *Modeling Manufacturing Sensitivity for a Transistor at the 180 nm Technology Generation*
The 3σ variation in threshold voltage is plotted as a function of the variation in gate length (critical dimension) and channel implant dose. The variation of oxide thickness was kept at 5%. Figure courtesy Tasch and Zeitzoff (8).

In its most general sense, "Defect to Fault" mapping is another aspect of correlating physical phenomena with electrical performance (11, 12). Defect to Fault mapping/modeling uses a hierarchical set of modeling tools that connects process and device modeling to circuit design/operation modeling. The concept is a generalization of the manufacturing sensitivity modeling discussed above. Defect to Fault modeling can be used to implement Statistical Metrology. We will use two examples to illustrate the different applications of "Defect to Fault" Modeling. The first type of application models the effect of parametric variations on circuit peformance. The effect of changes in average gate length on chip-level electrical performance such as circuit speed or the effect of an increased range of threshold voltages can be modeled. The second application is to yield limiting defects. A physical fault such as a short between metal lines can be placed in the circuit design, and the resulting distinctive electrical signature of this defect is calculated. Present generation software utilizes in-line metrology including defect detection and electrical test data from pilot line fabrication of memory test vehicle as an experimental database. Experimental data are used to verify and calibrate the models. The goal of this approach is to design circuits that are less susceptible to manufacturing variation and physical defects. Again, this methodology requires mature metrology capability for pilot line measurement of next generation process flows and tools. "Defect to Fault" modeling software is being developed through university-industry cooperation.

Pilot line and fab defect databases are another critical part of correlating physical and electrical measurements. Cumulative use of wafer maps often shows systematic process issues (13, 14). For example, adding the wafer maps from a single lot or from several lots can show clusters of defects in a particular area of the wafer (13, 14). This information is used to determine the source of process defects. The defect detection and silicon materials communities have provided the customer drive for development of software systems. These software systems are typically part of the total support provided by suppliers of defect detection equipment. Some of these systems have already been generalized to include data and analysis from all metrology capability. In the future, data management systems are expected to merge into Computer Integrated Manufacturing (CIM) systems where their full potential can be realized (2,15). Integration of sensor-based process control, off-line and in-line metrology, and defect metrology data management and analysis into CIM will allow true factory and electrical performance control (1, 15).

INFLUENCE OF MANUFACTURING STRATEGY ON METROLOGY IMPLEMENTATION

IC manufacturers have a variety of approaches to manufacturing that match their products. Some manufacture a very small number of similar products, such as microprocessors (or memories), with a small variety of process flows in very high volume fabs. Others manufacture a large variety of products (such as ASICs) with several process flows at small to medium volume. The goal of many manufacturers is to send a robust, high-yielding process flow from development (pilot line) to fab with little or no change. This is possible when a considerable effort is made to characterize process tools and flows in a statistically significant manner. Fabs having robust process flows seem to need less in-line metrology than fabs that must alter process flows to accommodate many different products.

Manufacturing strategy has a large impact on the type of metrology tool used during manufacturing. Some metrology tools are engineering intensive. These tools provide more detailed information. Examples include focused ion beam (FIB) systems, scanning electron microscopes equipped with energy dispersive x-ray detection (SEM/EDS), and unpatterned wafer defect detection (2). Defect autoclassification is also used in this manner (2, 15, 16). A manufacturing strategy that transfers a high yielding process flow from pilot line to volume FAB would minimize use of these tools after successful yield ramping. We also note that the need to ramp pilot line yield more rapidly is expected to push more of the detailed characterization back into process tool development (2, 15, 16).

The move to eliminate setup and monitor wafers implies the use of product wafer metrology. This trend is often given as a driver for in-situ sensor-based tool control because many in-fab metrology tools use monitor wafers.

CRITICAL CHALLENGES FACING METROLOGY

In the 1997 NTRS, the most difficult challenges for metrology centered around the changes associated with manufacturing before and after the 100 nm technology generation (1). Many of these challenges are due to the introduction of new materials or processes such as high-k transistor dielectric, copper metallization, low-k dielectrics, and Damascene (in-laid metal) processing. Other challenges are a result of shrinking device features. Indirectly this is also true for physical measurements of transistor processes which must improve so that they provide the same control as electrical measurement of transistor parametrics. The ten most critical challenges facing metrology are (1):

Five Difficult Challenges for > 100 nm before 2006

- Robust sensors, process controllers, and data management that allow integration of add-on sensors.
- Metrology for silicon wafers — impurity detection (particles, oxygen, and metallics) at levels of interest for starting materials.
- High-frequency dielectric constant for new ILD materials — measurement of frequency dependent dielectric constant of low-k interconnect materials at 5X to 10X base frequency.
- Metrology for new interconnect processes — control of new process such as Damascene (in-laid metal) and copper metallization.
- Reference materials and standard methods for gate dielectrics, thin films, and other process needs.

Five Difficult Challenges for < 100 nm beyond 2006

- In-line microscopy — non-destructive, manufacturing capable microscopy for critical dimension measurement, defect detection, and analysis
- Gate oxide reliability testing — standard electrical test method for reliability of ultra thin silicon dioxide and new gate dielectric materials
- Metrology tools for 450 mm wafers
- 3D dopant profiling
- Transistor fabrication metrology — manufacturing capable, physical in-line metrology for transistor process that provides statistical process control (SPC) required for electrical properties of the transistor.

Metrology Technology Requirements

Future measurement requirements are driven by the process, device, and structure requirements predicted by the focused technology roadmaps from lithography, transistor (front end processes) and interconnect, and process integration roadmaps. Many of these requirements cross the boundaries of the focused technology areas. Examples of this include microscopy, materials and contamination characterization, and dopant characterization. Microscopy will be applied to critical dimension, in-line particle and defect characterization, and defect detection. In-line and off-line dopant characterization requirements come from NTRS sections on implant process development, modeling and simulation, and process integration. Other requirements for integration of metrology tools into the pilot line and fab computer integrated manufacturing (CIM) systems for metrology and defect data management and analysis are described in the Factory Integration Roadmap and standards documents from SEMI. In Table 1, we show the Metrology Technology Requirements from the Metrology Roadmap, along with key technology requirements from the Front End Processes (transistor fabrication processes), and Interconnect Roadmaps. The Lithography Metrology Requirements are listed in the Lithography section of this paper. We included process technology requirements in this section to illustrate NTRS process development issues that will require advances in metrology. For example, the increase in the aspect ratio of interconnect structures will drive requirements for process control microscopy.

ADVANCED MICROSCOPY DEVELOPMENT (As applied to Critical Dimension, Defect Detection, and Materials and Contamination Characterization)

The Metrology Roadmap emphasizes the need to accelerate longer term research and development of microscopy for all areas. Microscopy is used in most metrology tools. Examples of the use of microscopy include critical dimension and overlay measurement, off-line characterization, defect detection and autoclassification, optical microscopes used for pattern recognition and focusing in both process and metrology tools, and failure analysis. New microscopy capability is typically first utilized for materials characterization and then it quickly migrates to lithography applications. One example of this migration is low voltage scanning electron microscopy. The low voltage SEMs moved from off-line characterization of insulating materials such as ceramics into critical dimension measurement. Low voltage operation required the development of new electron optics and field emission sources capable of high resolution at low electron beam energies. Defect detection applications of microscopy require higher throughput, and thus microscopy developments migrate more slowly into this area. Clearly, microscopy research and development has cross-functional applications and is found in the main section of the Metrology Roadmap. The Metrology roadmap lists microscopy for sub 100 nm technology generations as a critical metrology challenge. In this

TABLE 1. *Metrology Technology Requirements*

Year of First Product Shipment ***Technology Generation***	***1997*** ***250 nm***	***1999*** ***180 nm***	***2001*** ***150 nm***	***2003*** ***130 nm***	***2006*** ***100 nm***	***2009*** ***70 nm***	***2012*** ***50 nm***
Inline, nondestructive microscopy resolution (CD precision is different from resolution) (nm)	2	1.4	1.2	1	0.7	0.5	0.4
Particle analysis size (on patterned wafers) (nm)	75	60	50	45	35	25	15
Surface detection limits (Al, Ti, Zn)/ (Ni, Fe, Cu, Na, Ca) (atoms/cm^2)	5×10^9 5×10^8	2.5×10^9 4×10^8	2×10^9 3×10^8	1.5×10^9 2×10^8	1×10^9 1×10^8	5×10^8 $\leq 10^8$	$\leq 5 \times 10^8$ $\leq 10^8$
Front End (Transistor) Processes							
Wafer diameter (mm)	200	300	300	300	300	450	450
Oxygen (tolerance ± 1.5 ppma) (ASTM '79) *	20–31	19–31	18–31	18–31	18–31	18–31	18–31
Localized Light Scatterers (LLS) (includes particles) (nm) **	125	90	75	65	50	35	25
Composition and thickness gate dielectric (equivalent film thickness ± 3σ control) (nm)	4-5 ± 4%	3-4 ± 4%	2-3 ± 4%	2-3 ± 4–6%	1.5-2 ± 4–8 %	< 1.5 ± 4–8 %	< 1 ± 4–8 %
2- and 3-D dopant profile spatial resolution (nm)	5	3	3	2	1.5	1	0.8–0.6
Dopant concentration precision (across concentration range)***	5%	5%	4%	4%	3%	2%	2%
Interconnect Processes							
Planarity requirements within litho field for minimum interconnect CD (nm)	300	250	230	200	175	175	175
Minimum metal CD (nm)	250	180	150	130	100	70	50
Minimum contact/via CD (nm)	280/360	200/260	170/210	140/180	110/140	80/100	60/70
Metal height/width aspect ratio—logic (microprocessor)	1.8	1.8****	2.0****	2.1****	2.4****	2.7****	3.0****
Via aspect ratio—logic	2.2	2.2****	2.4****	2.5****	2.7****	2.9****	3.2****
Barrier layer film thickness measurement capability (nm)	50	23	20	16	11	8	6
Interlevel metal insulator—effective dielectric constant (k)	3.0–4.1	2.5–3.0	2.0–2.5	1.5–2.0	1.5–2.0	≤ 1.5	≤ 1.5

Solutions Exist | Solutions Being Pursued | No Known Solutions

* Range of center point value to provide IG (no IG) based on IC requirements. ± tolerance is min-max range about center point value; tolerance of ± 2 ppma appropriate @ 250 nm CD. Bulk Micro Defects (BMD) in IG (no IG) polished wafer > 1 ′ 10^8/cm^3 (< 1 ′ 10^7/cm^3) after IC processing. IOC '88 value obtained by multiplying ASTM value by 0.65

** Total front surface Localized Light Scatterers size (includes particles and COPs) = K_2 (CD); K_2 = 0.5. Less than 50 "visual" back-surface particles recommended

*** Accuracy for dopant profiling depends upon use of accurate reference materials (see Reference Materials Discussion).

**** Metal and via aspect ratios are additive for dual-Damascene process flow

paper, we chose to emphasize the synergy between lithography and microscopy, and we combine these topics in our discussion of long-term development needs.

The crosscut nature of long-term microscopy development can be illustrated by examples from critical dimension measurement, interconnect process control, and defect detection. In Fig. 5 a to c, we show these three different applications using a patterned, low dielectric constant layer before metal deposition. This type of interconnect fabrication is referred to as Damascene or in-laid metal processing. The Lithography and Metrology Roadmaps call for future critical dimension measurement to be independent of line shape, line materials, and line density (1). This implies the need to characterize linewidth and the angle of the sidewall which can be obtained from three dimensional images as shown in Fig. 5a. Metal deposition for Damascene processes can be adversely affected by "lip" structures and incompletely etched via/contact as shown in Fig. 5b. Particle and defect detection requirements also call for locating particles and incompletely etched via/contacts in patterned interlevel dielectric structures as shown in Fig. 5c. All of these microscopic measurements are increasing in difficulty as feature sizes shrink.

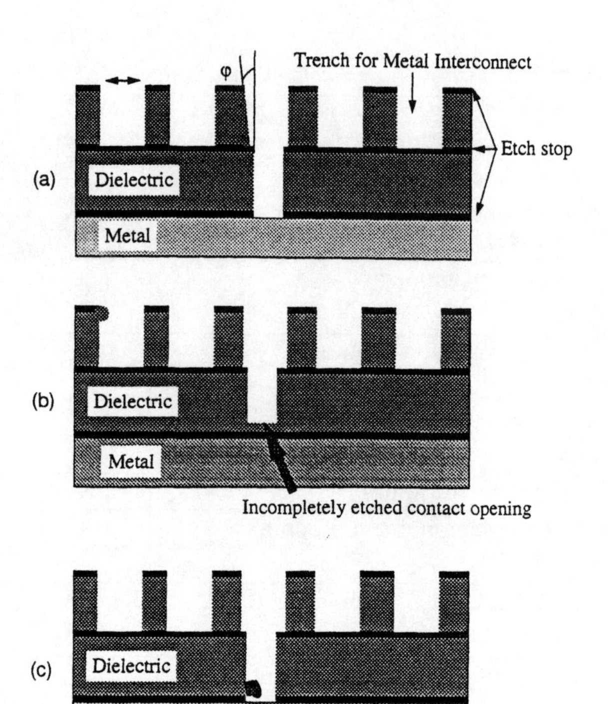

Figure 5. *Microscopy Requirements for Critical Dimension, Interconnect Process, and Defect Detection.*

A patterned dielectric layer is used to illustrate the crosscut nature of different applications of microscopy. Future 3D microscopy methods could be applied to all three applications.

(a) Critical dimension measurement must characterize linewidth and side wall features.

(b) Metal deposition processes can be adversely affected by lip structures at the top of dielectric sidewalls.

(c) The same microscopy capability used for critical dimension and interconnect process control can be applied to particle and defect detection.

LITHOGRAPHY METROLOGY

Development and application of new lithography technology remains the most critical and costly challenge facing the semiconductor industry. Therefore, research and development of lithography measurement tools and methods remains one of the highest priorities for the metrology community. According to the Metrology section in the Lithography Roadmap, critical dimension measurement is more challenged than overlay by the measurement precision requirements. Improved CD-SEM is listed as the potential solution for CD measurement until the 100 nm technology generation. At this juncture, other methods such as CD-scanned probe microscopy (CD-SPM, i.e., CD-AFM) or sensor-based control will potentially replace or supplement CD-SEM. The requirements for CD and overlay measurements by technology generation are presented in Table 2. Also listed are the microscopy resolution requirements from the Metrology Roadmap requirements table (1). This line describes the requirements for distinguishing differences in linewidth for each technology generation.

The nature of the challenge facing linewidth measurement is representative of the general issue of correlating physical and electrical measurements. Physical CD-SEM measurement is affected by process induced roughness and waviness in the sidewalls. This can result in the CD-SEM considering the linewidth out of tolerance while the electrical properties are in tolerance. CD-SEMs also respond differently to materials changes from gate electrode to metal line to resist. This requires level by level calibration (6).

Overlay metrology is largely challenged by new processes such as chemical mechanical polishing (CMP) and Damascene. The traditional methods of producing overlay registration features have been changed to accommodate CMP. In addition, the CMP process increases the level of uncertainty in feature location due to feature roughness.

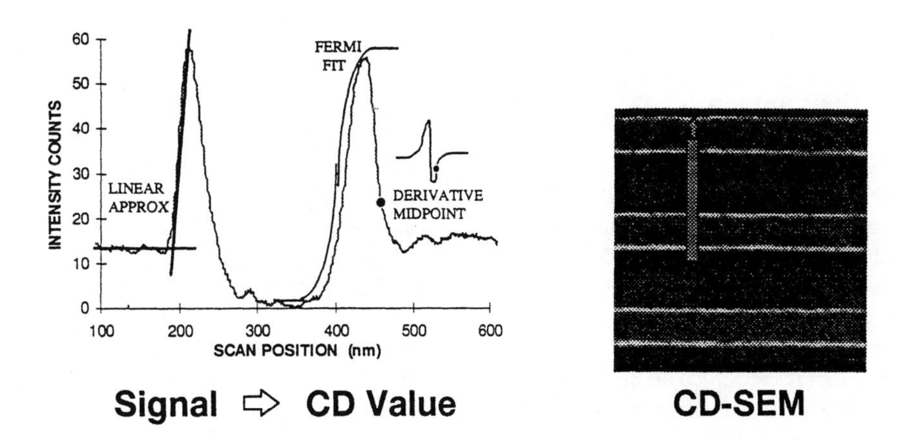

FIGURE 6. *Relationship of CD-SEM Signal to Feature Dimensions*
Present generation CD-SEMs use an empirical relationship to determine linewidth from the signal. A more fundamental model for this relationship would improve measurement. The inset showing the SEM signal is provided by H. Marchman and was first used in reference 17.

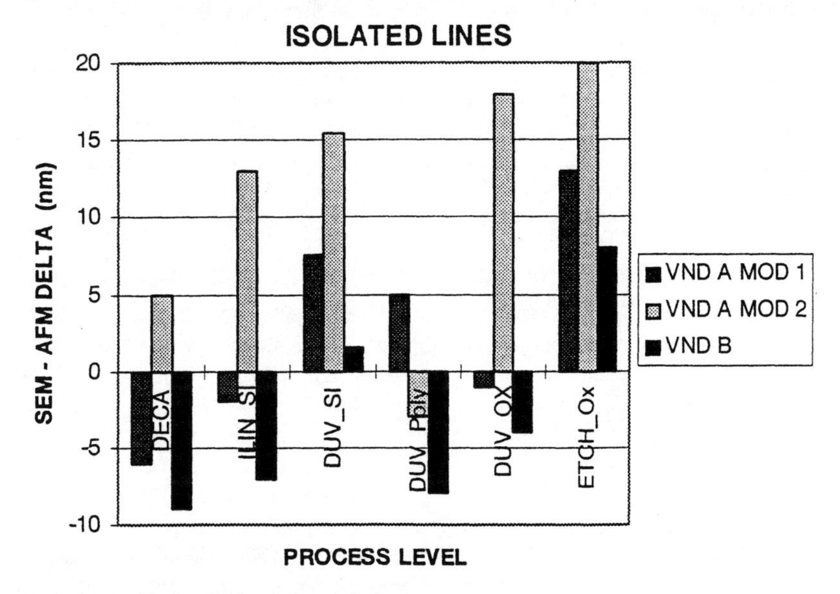

FIGURE 7. *Level Dependent CD-AFM Calibration of CD-SEM Measurements*
Difference between SEM and AFM CD measurements is shown for several different process levels. This information is plotted for two different suppliers, labeled VND A and VND B. Figure provided by H. Marchman was first used in reference 17.

TABLE 2. Lithography and Microscopy Metrology Requirements

Year of First Product Shipment *Technology Generation*	*1997* *250 nm*	*1999* *180 nm*	*2001* *150 nm*	*2003* *130 nm*	*2006* *100 nm*	*2009* *70 nm*	*2012* *50 nm*
Gate CD control (nm)	20	14	12	10	7	5	4
Final CD output metrology precision (nm, 3σ) *	4	3	2	2	1.4	1	0.8
Inline, nondestructive microscopy resolution (CD precision is different from resolution) (nm)	2	1.4	1.2	1	0.7	0.5	0.4
Overlay control (nm)	85	65	55	45	35	25	20
Overlay output metrology precision (nm, 3σ)*	9	7	6	5	4	3	2

Solutions Exist ☐ Solutions Being Pursued ▒ No Known Solutions ■

* Measurement tool performance needs to be independent of line shape, line materials, and density of lines

Near-Term Lithography and Microscopy Metrology Development

The Metrology roadmap team (TWG) considers CD-SEM as the most likely method of measuring CD until the 100 nm technology generation. This means that the industry expects that improvements in SEM tools will extend CD-SEM technology. One example frequently cited by practitioners of CD-SEM is the need to have a quantitative, fundamental model for relating the physical feature being measured to the signal inside the CD-SEM (1). This model would be incorporated inside the CD-SEM tool for routine measurement. The present status of this process is summarized by Fig. 6.

Calibration of CD measurement can be done by scanned probe microscopy (SPM), cross-sectional analysis, or electrical test structures. Atomic force microscopy (AFM) is the most commonly used SPM method. Process level calibration of CD-SEM measurements has been demonstrated by CD-AFM (17, 18). This is shown in Fig. 7.

The extension of these measurements to the 100 nm technology generation will require development of new probe tip technology and improved CD-SPM. Carbon nanotubes are being investigated for use as SPM probe tips (19). Marchman reviews developments in CD measurement in this volume (18).

Optical microscopy continues to be used for off-line and in-line particle detection and identification, pattern recognition, and product failure analysis. DUV microscopes are in the process of being introduced into the industry. The near-term need is for microscopes that have increased resolution without requiring liquid immersion. Frequently, lithography requirements drive development of new optics, and the opportunity for applying lithographic technology to microscopy should not be lost. Ager reviews optical microscopy in this volume (20).

Long-Term Lithography and Microscopy Metrology Research

Optical and electron holography and interferometry have been proposed as solutions for all in-line applications of microscopy (1). Holographic methods have the advantage of providing rapid, three dimensional measurements of sub 100 nm features. Ultra-low energy SEM has also been proposed as a method of overcoming charging and other artifacts (1). Therefore, electron beam based holography is expected to use sub 100 eV electron beams. Joy describes the potential of electron beam based holography in this volume (21).

FRONT END PROCESSES METROLOGY

Processing used to fabricate transistors is referred to as Front End Processes in the 97 NTRS. Thermal processes are used to grow the gate oxide and diffuse implants. Silicon dioxide and nitrided silicon dioxide are expected to extend thermal oxide type gate dielectrics to the 100 nm technology generation. At the 100 nm technology generation, the Front End Processes Roadmap indicates that the equivalent oxide thickness drops below 2 nm. The use of the term equivalent thickness means that a non-silicon dioxide layer will have dielectric properties (capacitance) equivalent to a perfect silicon dioxide layer of the specified thickness while maintaining low leakage currents. Alternate materials such as TiO_2 are being investigated for longer term replacement of thermal SiO_2. Implant dominates doping process technology, and low energy implants are expected to extend implant to at least the 100 nm technology generation.

Much of the existing metrology must be improved so that it can be extended to meet process needs to the 100 nm technology generation. The biggest challenge for transistor fabrication metrology is improving the correlation of physical methods with electrical methods. This is described below for gate dielectric thickness measurement. Another challenge is forming a consensus method for evaluating gate dielectric reliability as silicon dioxide thickness allows direct tunneling. Reliability of alternate dielectric materials must also be understood. The Metrology Roadmap has a one page section found in the Front End Processes Roadmap that describes appropriate metrology needs and potential solutions.

There are many aspects to correlating electrical and physical measurements for transistor processes. Each process tool must have a metrology that allows localized control. For example, thermal fabrication processes such as vertical furnaces or rapid thermal processors require temperature and oxide thickness measurement. Oxide thickness measurement must be correlated with electrical thickness measurements of thickness and thus gate capacitance. Implant requires dose and uniformity measurement. The physical measurements of dose must be correlated to the electrically active portion of the implant. There is an additional aspect to correlation of electrical and physical thickness measurements. The oxide thickness and implant dose must fall within a range of values so that the threshold voltage and leakage current variation across the die, across the wafer, and wafer to wafer is minimized. Yet, these electrical properties are also determined by the gate length (CD) as is shown in Fig. 4 from Tasch and Zeitzoff (8).

Near-Term Development for Transistor Processes Metrology

The extension of ellipsometric measurement of oxide thickness to 2 nm SiO_2 requires optical models, optical constants, and reference materials. Additional development is also required for extending electrical measurements such as capacitance-voltage and current-voltage to sub 2 nm SiO_2 and nitrided oxides. Correlation of electrical and physical thickness measurements will require the correct inclusion of the interfacial layer between Si and SiO_2 in optical and electrical models. Capacitance-voltage data for very thin gate dielectrics are shown in Fig. 8. These data illustrate the unusual shape of C-V data for thin oxides.

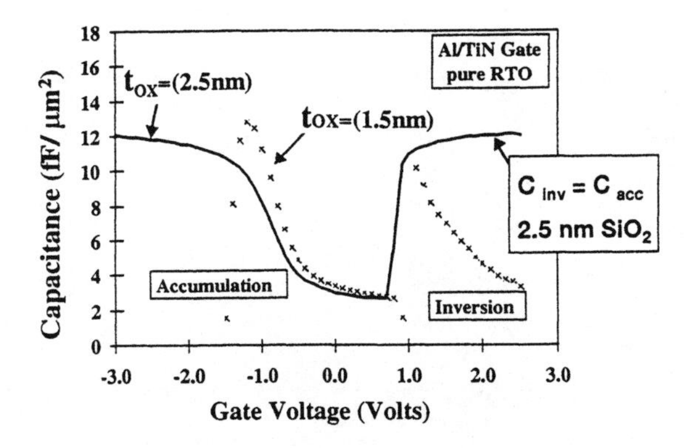

FIGURE 8. *Capacitance-Voltage Measurement of Oxide Thickness*

The behavior of C-V data changes dramatically when silicon dioxide thickness falls below 2.5 nm. The depletion of charge across doped poly silicon gate electrodes and quantum effects due to oxide thickness change the observed capacitance in both the depletion and accumulation region. Data provided by G. Brown.

Control of thermal processes is a very difficult challenge especially for rapid thermal processes. The wafer temperature needs to be controlled across the wafer and wafer to wafer to minimize stress induced damage and ensure uniform processing of gate dielectric thickness or implant anneal.

Larson and Current have reviewed metrology requirements for transistor doping processes, and Harris and Vandervorst have reviewed secondary ion mass spectroscopy (SIMS) and 2-dimensional dopant profiling (22, 23). The precision of four-point probe and optically modulated optical reflection methods must be improved for future technology generations (1). New methods based on acoustic response or optical density changes in a polymer film have the potential of providing near-term in-line implant control (1). New, wafer-capable SIMS have pushed this method toward more routine process control (5, 6). SIMS can be used to analyze test structures on product wafers before or after thermal activation of dopant electrical activity. In addition, SIMS can routinely quantify dose of several dopant species simultaneously (5, 23). Optically modulated optical reflectance can measure dopant dose on an unannealed (unactivated) monitor wafer. Four-point probe uses special monitor wafers, and is capable of rapid mapping of resistivity (resistivity is then related to dopant dose) across a wafer (6, 22). Optical density measurements also rapidly map dose uniformity, and this method is performed using a quartz wafer with a polymer coating (6, 22).

Long-Term Research for Transistor Processes Metrology

The Metrology roadmap lists several topics in the area of transistor process metrology as key challenges for the sub 100 nm technology generations. New gate dielectric materials, doping processes, and transistor designs will all be challenges.

The applicability of existing physical and electrical methods to alternate dielectric materials must be evaluated. Research into the appropriate methods for process control for the sub-100 nm technology generation must begin now. Optical constants and models should be ready before development of process tools begins.

The need to reduce monitor and setup wafers will drive doping metrology, and thus new product wafer capable, non-destructive methods seem appropriate. One example is that the acoustic measurement technology mentioned above may be capable of in-situ process control for implant or non-traditional doping methods.

INTERCONNECT METROLOGY

Traditional interconnect structures based on patterning of aluminum and silicon dioxide interlevel dielectrics are quickly being replaced by Damascene using copper and low k dielectric layers. A comparison of these process technologies is shown in Fig. 9a, and the dual Damascene process flow is shown in Fig. 9b.

The process tools for metal and low k dielectric materials are being developed, and IBM has already announced a commercial IC based on copper interconnects. At this point, there is no consensus about what changes to process control are required. Traditional measurement of film thickness for blanket layers (or large test structures) for either metal or dielectric films could provide process control. Contamination and defects in high aspect ratio features such as contact/via and trench structures continue to challenge metrology. Critical dimension measurements after etching metal patterning will be replaced by CD measurement of either patterned low-k dielectric or metal filled structures. CD-SEM measurements of developed photoresist features are considered difficult due to electron beam-resist interactions.

Near-Term Development for Interconnect Metrology

Measurement of the high-frequency dielectric properties of new low-k materials is a critical need and challenge for the metrology community. This information will be used to design interconnect structures and evaluate materials for use in high-frequency circuits. Electrical testing equipment exists and NIST has described test structures and methodology for high-frequency measurements (24). The challenge is to

(a)

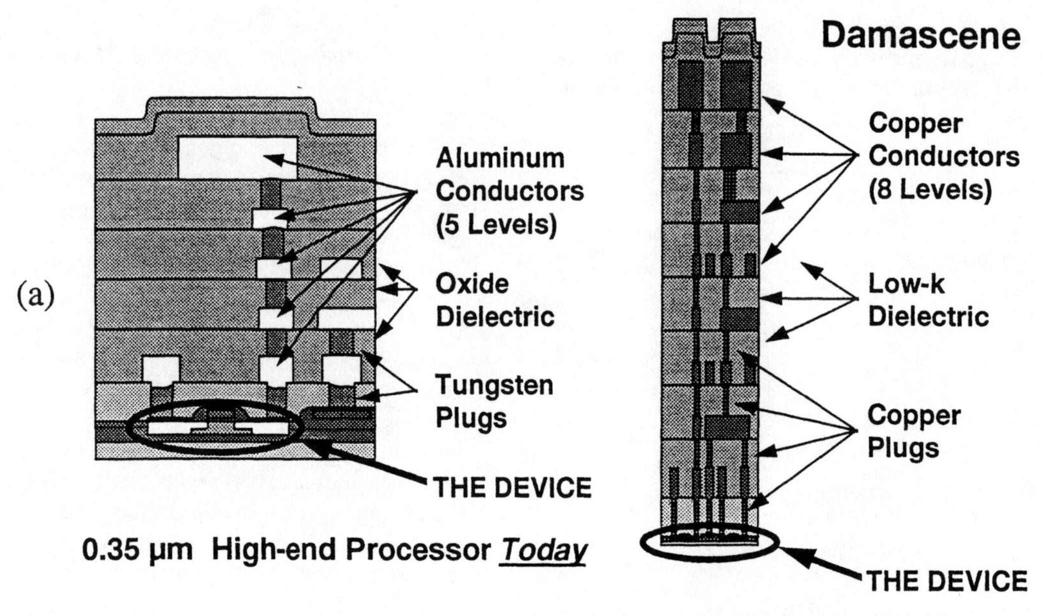

(b)

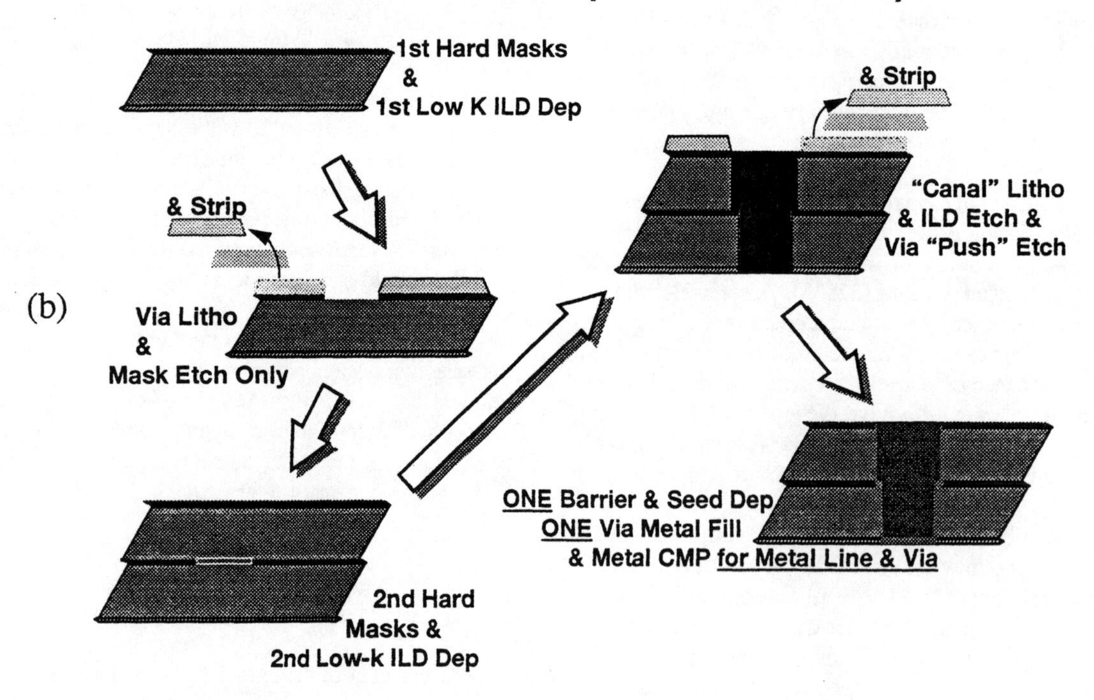

FIGURE 9. *Comparison of Traditional Aluminum-Silicon Dioxide Dielectric Interconnect Technology With Copper-Low-k Dielectric / Damascene Interconnect Process Technology.*
(a) New Interconnect Materials and Structures
(b) Dual Damascene Process Flow

evaluate new materials, design new test structures for specific applications, and provide this information in time for IC development.

The Metrology Roadmap also lists process control for new interconnect materials and processes as a key challenge before the 100 nm technology generation. New photo-acoustic measurement methods can provide rapid in-line or in-situ measurement of stacked, opaque structures (25, 26). Systems equipped with pattern recognition capability can measure large areas. These methods are being evaluated for copper and stacked copper-tantalum film control, and are being used for aluminum-titanium nitride stacks. In addition, acoustic methods have been shown to be capable of determining film density independently, and can be used to measure low k film properties (26). Metal film density is a function of deposition conditions and interlevel dielectric density is a function of process conditions (25, 26). Film thickness characterization methods such as x-ray fluorescence require film density (6). Acoustic, x-ray fluorescence, and 4-point probe are expected to continue to be used for control of metal film thickness.

Long-Term Research for Interconnect Metrology

Measurement of patterned features on product wafers remains an elusive goal. Several publications point to the possible development of the photo-acoustic methods described above (25, 26). The acoustic response is strongly affected by interfaces and boundary conditions. Therefore, film thickness measurement on Damascene patterned features should be investigated. Evaluation of metal adhesion may also be possible using this new method (25, 26).

SENSOR-BASED PROCESS CONTROL

The Metrology Roadmap describes the status and projects the future of Sensor-based Metrology for Integrated Manufacturing (1). The goal of sensor-based process control is to provide real time fault detection and model based operation of process tools. Achieving the full potential of this approach requires both hardware (process tool controllers) and software that are fully integrated into the factory control CIM system. In other words, the sub-system that receives sensor data and controls the process tool must comply with factory CIM data transfer protocols and standards. Presently, many sensors, process controllers, and data management systems are added to the process tool at pilot line and manufacturing sites. Examples of this include miniaturized residual gas analyzers and vacuum based particle sensors. Process controllers added to the process tool are often referred to as "piggy-back" controllers as shown in Fig. 10. Some process tools come equipped with embedded sensors and process controllers as shown in Fig. 11. Examples of this include fast ramp furnaces that have accelerated ramping based on temperature sensors and model based, embedded process controllers. Higher throughput provides a competitive advantage for these tools. Another example is lithography exposure tools that auto-focus. Presently, the conversion to 300 mm wafer capable tools is driving much of the factory control protocol and standardization activities for Advanced Process Control (APC)/Advanced Equipment Control (AEC). This community's assessment is that no sensor-based process control capability is fully compliant with these protocols and standards. Adoption of standards such as "Sensor Bus" will facilitate the process of achieving full compliance (1).

It is important to note that there are two customer bases having slightly different approaches involved in the evolution of sensor-based process control. One is the process engineering/fab production communities that typically do not wish to use add-on sensors. This group has utilized sensor controlled process tools when the capability is embedded and transparent to the user. Improvements in process tool effectiveness such as higher throughput, lower wafer scrap, and reduction of setup wafers justify any added expense of sensor-based control. A variety of different communities are adding sensors to process tools including early activities in the area of process tool development to inclusion in volume manufacturing. This community often requires more detailed information and hands-on optimization. The most effective way to implement sensor-based control is to design the process tool to optimally locate sensors inside the process chamber.

Near-Term Development of Sensor-Based Metrology

The Metrology Roadmap lists development of robust sensors, process controllers, and data management systems as one of the key challenges to metrology before the 100 nm technology generation. This includes improvement and utilization of existing sensor technology and accelerated movement toward compliance with AEC/APC standards.

Long-Term Research for Sensor-Based Metrology

The Metrology Roadmap predicts that MEMS (micro-electromechanical systems) will revolutionize sensor-based metrology. MEMS technology has the potential of providing new sensing capabilities, miniaturizing present in-line measurements, and self calibrating sensors. The concept of MEMS is revolutionary enough that it is difficult to categorize MEMS. Mass flow controllers with built-in calibration are MEMS technology that is an example of future sensor-based process control. Other examples of developmental MEMS technology include: arrayed infrared sensors for temperature measurement, and sensors for preventing particle flaking from

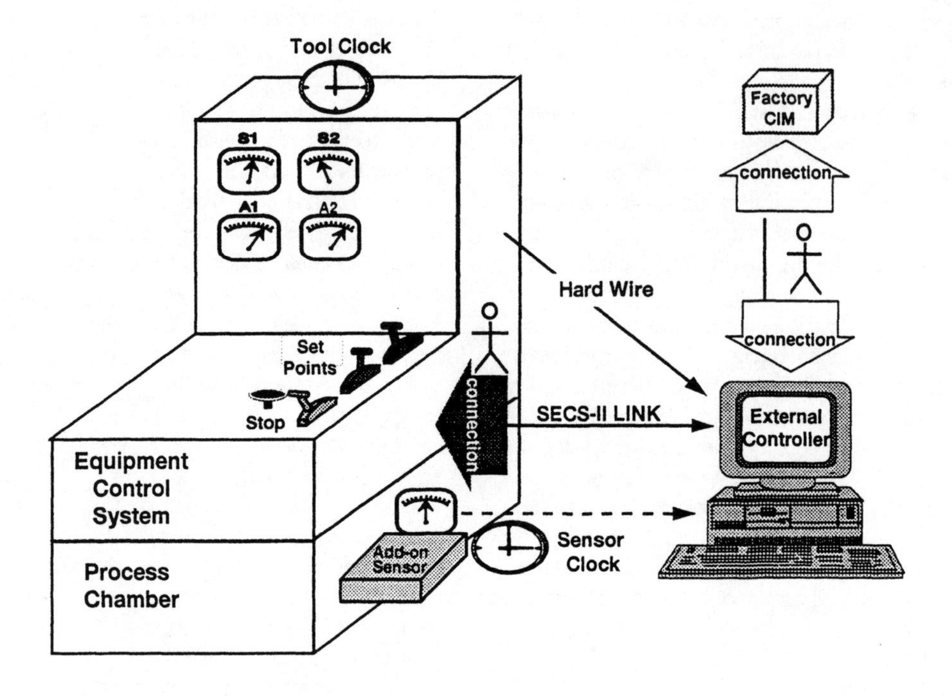

FIGURE 10. *Add-on Sensors and Piggyback Controller*
Presently, most sensors are added onto the process tool along with an extra control computer. This figure was first published in reference 27.

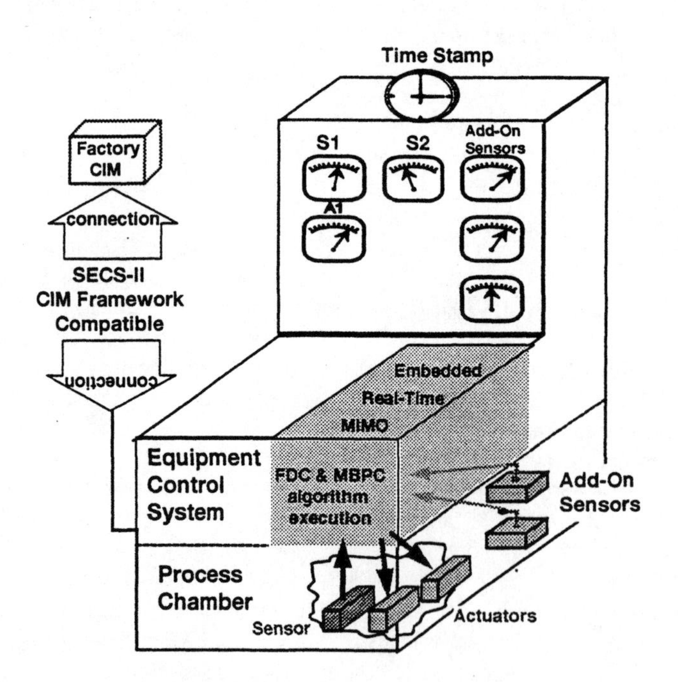

FIGURE 11. *Internal Sensors with an Embedded Process Controller*
Sensor-based process control is expected to evolve into an architecture that utilizes process chambers designed for sensor-based control to achieve real-time process control. This architecture includes an embedded process controller capable of running sophisticated control algorithms in addition to fault detection. The process controller should be capable of taking data from multiple sensors and outputting control information to multiple actuators. Thus the term MIMO (multiple input-multiple output) is used to describe the controller. The control system should be capable of using add-on sensors also. This figure was first published in reference 27.

film buildup on process tool chambers. Visionary examples of MEMS technology include scanning electron microscopes, mass spectrometers, and gas chromatographs on a chip.

Sensors need to evolve toward measurement of more localized properties. The Metrology roadmap describes this as 3-dimensional measurement of wafer level and process level properties such as etch uniformity and particle deposition.

PACKAGING METROLOGY

Metrology methods used during fabrication of chip packaging are often not as mature as those used during chip manufacture. Packaging is driven by the market application and the IC chip technology. According to the 1997 NTRS, the package pin count and the chip pad count (connection from chip to package) are increasing especially for high performance applications such as microprocessors. A Metrology Roadmap for Assembly and Packaging appears in the 1997 NTRS (1). Packaging metrology needs were summarized in a table form in the NTRS and are presented here in Table 3. Modeling, accelerated stress testing, and materials and interface models and parameters are highlighted as research and development requirements for packaging.

MATERIALS CHARACTERIZATION FOR R&D, PILOT LINE, AND FAB

Materials and contamination characterization was discussed in the Metrology Roadmap (1). This discussion covered several key topics such as improvements in electron and ion beam resolution, detector technology, surface sensitivity, and off-line dopant analysis. Here, we describe the trend in implementation of these tools and key development requirements in x-ray detector technology.

Over the past several years, a new approach to materials characterization has emerged. Many characterization tools have become capable of measuring whole samples using precision stages and software capable of using and storing information in the wafer map format used by electrical test and defect detection tools. The tools are typically clean room compatible and most can measure properties on patterned wafers. Data management tools that network the characterization laboratory to the FAB are an essential part of this trend (28). This trend has its origins in the pioneering development of in-line SEM by Hattori (29). SEMATECH's Analytical Laboratory Managers Working Group has played a key role in spreading this trend to many other tools including Auger microanalysis, secondary ion mass spectrometry (SIMS), focused ion beam (FIB), and total reflection x-ray fluorescence (30). The reason for this trend is the need to find small particles or specific features rapidly (e.g., failed bits) without contaminating the wafers. Wafer suppliers, process tool suppliers, and IC manufacturers all apply unpatterned defect detection and whole wafer characterization tools to development. These new product wafer tools are utilized by many different engineering groups with responsibilities from off-line product failure analysis to in-line product and process maintenance. Many IC manufacturers now locate a "FAB-LAB" next to or inside the clean room equipped with whole wafer tools and other essential methods such as transmission electron microscopy, spectroscopic ellipsometry, and Fourier transform infrared spectroscopy (30).

A key new x-ray detector technology developed at NIST and elsewhere is in the process of commercialization (30, 31, 32, 33). Martinis et al., and others have developed the microcalorimeter x-ray detector. The advantage of this detector is to obtain high energy resolution with high sensitivity to the widest possible range of x-ray energies. A comparison of traditional energy dispersive x-ray spectroscopy (EDS) detectors and microcalorimeter EDS is shown in Fig. 12.

SEMATECH member companies have pushed for the new detectors to be retrofitable in existing EDS applications so that these tools are extended to future technology generations. Microcalorimeter EDS will probably first be commercialized for use on SEM to improve small area analysis such as particle and defect identification. Traditional EDS detectors do not easily identify titanium when oxygen is present. Recent data from NIST (Martinis et al.) indicate that the latest detectors have energy resolution capable of determining chemical information such as distinguishing aluminum from aluminum oxide (31). *Other applications are also expected to be improved by microcalorimeter EDS.* Improved transition metal detection limits and extension of light element sensitivity are expected when the microcalorimeter EDS is applied to total reflection x-ray fluorescence. Wollman describes microcalorimeter EDS in this volume (32).

REFERENCE MATERIALS AND STANDARDS

Reference materials and standard procedures for measurement enable instrument calibration and precision determination. The 1997 Metrology Roadmap lists several types of traceable reference materials that can be commercially obtained: certified reference materials (CRM), consensus reference materials, NIST Traceable Reference Materials (NTRM®), and Standard Reference Materials (SRM®). Our discussion of SPC points to the need for reference materials in the process range of interest. Another class of reference materials sometimes referred to as internal standards is typically employed during semiconductor manufacture. Important examples include the CD reference structures that are produced using the true manufacturing process for each critical step (poly Si gate, metal level 1, etc.). There are many challenges that NIST and other suppliers of reference materials face. One issue is that each IC manufacturer uses different process flows and tools. Metrology that is sensitive to slight variations in materials properties can require specialized reference materials. The NIST or commercial reference

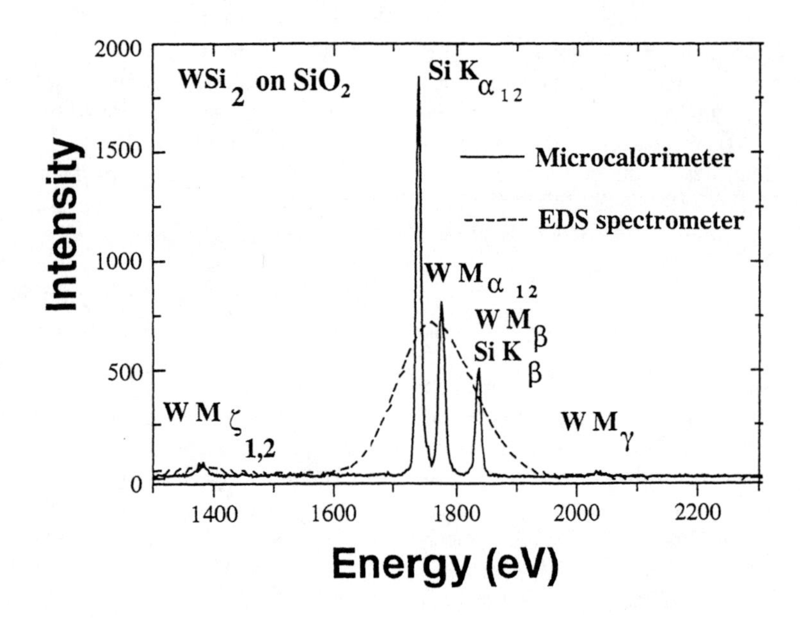

FIGURE 12. *Microcalorimeter X-ray Detector*
The new microcalorimeter EDS (energy dispersive spectrometer) detector system has considerably better energy resolution as shown in this x-ray spectrum of Wsi. Figure courtesy Dave Wollman, NIST, and was first published by the Royal Microscopical Society in J. Microscopy. Vol. 188, pp. 196-223, Figure 25.

TABLE 3. Assembly & Packaging Metrology Crosscut Issues

Assembly and Packaging Metrology Needs	Summary of Issues
Electrical simulation models of packages and systems (modeling system of chips pushing the practical limits of cost and time effectiveness)	• Improvement needed for coupling between components, mixed signal simulation, power disturbs, and EMI • Parameter extractions of 3-dimensional interconnect and power delivery structures • Integrated electrical (architecture), mechanical, thermal, and cost modeling tools needed for cost and cycle time reduction
Accelerated stress methods tests representative of application	• Improved accelerated stress test techniques needed to qualify manufacturing processes, and to improve the lifetime and successful operation of the product • Research needed on failure modes and mechanisms for finding accelerated stress techniques that mimic "real life"
Measurement and modeling of interfaces (thermal performance, reliability, yield, and cost are driven by understanding of interfaces)	• Need to measure, design, and control the basic mechanisms (physical, chemical, mechanical) for interface bond strength (adhesion)
Thermal and mechanical simulation models of packages and assemblies	• Comprehensive thermal and mechanical model tools fully supported by "real life" materials data • Measurement of *in situ* properties, location, and characterization of defects and failures
Material parameters	• Measurement, collection, and dissemination of materials properties of packaging materials for the sizes, thicknesses, and temperatures of interest
Material application and assembly process control	• Improvements in the online measurement of solder systems, solder alternatives, underfills, encapsulants, attachment materials, etc., in the manufacture of packages and bumped chips

material must be robust, i.e., the measured value must be stable in time and not easily damaged by the measurement. The properties of the material must be well defined. Often this requires non-standard processing to produce smooth features or interfaces. Therefore, reference materials are listed as one of the challenges for metrology of sub 100 nm technology generations.

HIGHEST PRIORITY FOR METROLOGY RESEARCH AND DEVELOPMENT

- Integration of off-line and in-line physical and electrical metrology, sensor-based process control, defect metrology, and final electrical test data into a complete unified electrical performance and factory control CIM system.
- Provide metrology, including reference materials, in time for process tool development
- Accelerate short- and long-term microscopy R&D
- Provide near-term high-frequency dielectric characterization of new low k dielectrics
- Streamline the commercialization of important new capabilities such as microcalorimeter x-ray detectors.
- Fund and push MEMS technology toward critical applications

ACKNOWLEDGMENTS

We acknowledge the important work of the metrology technical working group team that assembled the Metrology Roadmap for the 1997 *National Technology Roadmap for Semiconductors*. Anne Schmidt provided drawings of interconnect structures and processes. Pam Hanners very kindly reviewed this manuscript and Danny Lufkin provided expert editorial and graphics assistance. Paul Tobias reviewed our discussion of precision and resolution.

REFERENCES

1. Diebold, A.C., and Monahan, K.A., *National Technology Roadmap for Semiconductors*, Semiconductor Industry Association, 1997, pp. 179-186.
2. Moore, G.E., "Intel-Memories and the Microprocessor," *Daedelus* 125, No. 2, pp. 55-80 (1964).
3. Huff, H.R., Goodall, R.K., Nilson, R.H., and Griffiths, S.K., "Thermal Processing Issues for 300 mm Silicon Wafers: Challenges and Opportunities," *Symposium Proceedings of ULSI/97*, pp. 135-181 (1997).
4. Parker, G.H., "Intel's Manufacturing Strategy," *Future FAB International*, Volume 1, Issue 2, pp. 25-27 (1997).
5. Diebold, A.C., and McDonald, R., "The At-line Characterization Laboratory of the 90s: Characterization Laboratories (FAB-LABS) Used to Ramp-up new FABS and Maintain High Yield," *Future FAB International*, Volume 1, Issue 3, pp. 323-330 (1997).
6. Diebold, A.C., "In-Line Metrology," *Handbook of Semiconductor Manufacturing*, in progress.
7. Monahan, K.M., Forcier, R., Ng, W., Kudallur, S., Sewell, H., Marchman, H., and Schlesinger, J, "Application of Statistical Metrology to Reduce Uncertainty in the CD-SEM Measurement of Across-Chip Linewidth Variation," *Proceedings of SPIE* Vol 3050, pp. 54-67 (1997).
8. Tasch, A.F., and Zeitzoff, P., to appear in the *Proceedings of 1998 International Conference on Characterization and Metrology for ULSI Technology*, March 23-27, 1998.
9. Bartelink, D., "Statistical Metrology: At the Root of Manufacturing Control," *J. Vac. Sci. Technol.* B12, pp. 2785-2794 (1994)
10. Yu, C., Maung, T., Spanos, C.J., Bonning, D., Chung, J., Liu, H-Y, Chang, K-J, and Bartelink, D., "Use of Short-Loop Electrical Measurements for Yield Improvement," *IEEE Trans. on Semiconductor Manufacturing*, Vol. 8, pp. 150-159 (1995).
11. Gaitond, D., and Walker, D.M.H., "Test Quality and Yield Analysis Using DEFAM Defect to Fault Mapper," *Proceedings of the IEEE International Conference on CAD 1993*, pp.78-83 (1993).
12. Khare, J., and Maly, W., "Rapid Failure Analysis Using Contamination-Defect-Fault (CDF) Simulation," *Proceedings of the Fourth IEEE/UCS/SEMI International Symposium on Semiconductor Manufacturing 1995* (ISSM '95), IEEE Catalogue Number 95CH35841, p. 136 (1995).
13. Lee, F., Wang, P., and Goodner, R., "Factory Start-Up and Production Ramp: Yield Improvement through Signature Analysis and Visual/Electrical Correlation," *Proceedings of the 1995 Advanced Semiconductor Manufacturing Conference and Workshop*, IEEE Catalog #95CH35811, (1995).
14. Tobin, K.W., Gleason, S.S., Lakhani, F., and Bennett, M.H., "Automated Analysis for Rapid Defect Sourcing and Yield Learning," *Future FAB International*, Volume 1 Issue 4, pp. 313-319 (1998).

15. Diebold, A.C., "Overview of Metrology Requirements based on the 1994 National Technology Roadmap for Semiconductors," *Proceedings of the 1995 Advanced Semiconductor Manufacturing Conference*, pp. 50-62 (1995).

16. Diebold, A.C., "Critical Metrology and Analytical Technology Based on the Process and Materials Requirements of the 1994 National Technology Roadmap for Semiconductors," *Semiconductor Characterization: Present Status and Future Needs*, American Institute of Physics Press, New York, 1996, pp. 25-41.

17. Marchman, H., *Future FAB International*, Vol. 3, pp. 345-354 (1997).

18. Marchman, H.M., this volume

19. Dai, H., Hafner, J.H., Rinzler, A.G., Colbert, D.T., and Smalley, R.E., *Nature* 384, p. 147 (1996).

20. Ager, J., this volume

21. Joy, D., this volume.

22. Larson, L., and Current, M., this volume.

23. Harris, G., and Vandervorst, W., this volume.

24. Williams, D.F., Janezic, M.D., Ralston, A.R.K., List, R.S., "Quasi-TEM Model for Coplanar Waveguide on Silicon," 1997 IEEE 6th Topical Meeting on Electrical Performance of Electronic Packaging, pp. 225-228 (1997).

25. Banet, M., Fuchs, M., Yeung, H., Hanselman, J., Sola, H., Lam, H., "All-Optical, Noncontact Measurement of Blanket Copper and Tantalum Structures Deposited by PVD, ECD, and in Single Damascene Structures," this volume.

26. Stoner, R., this volume.

27. Butler, S., Hosch, J., Diebold, A.C., and Van Eck, B., "Sensor-based Process Control," *Future FAB International*, Volume 1, Issue 2 (1997).

28. Diebold, A.C., "Materials and Failure Analysis Methods and Systems Used to Develop and Manufacture Silicon Integrated Circuits," *Journal Vacuum Science and Technology* **B12**, p. 2768 (1994).

29. Hattori, T., *Microelectronic Manufacturing and Testing 11*, p. 31 (1988).

30. Diebold, A.C., and McDonald, R., "Evolution of Characterization Equipment for the Wafer FAB of the 90s," Materials Analysis and Defect Characterization for the Wafer FAB of the 90s, presented at the meeting of the Arizona Chapter of the American Vacuum Society, May 14, 1996.

31. Irwin, K.D., Hilton, G.C., Wollman, D.A., and Martinis, J.M., submitted to *Journal of Applied Physics*.

32. Wollman, D.A., this volume.

33. Wollman, D.A., *Journal of Microscopy*, Vol. 188, pp. 196-223 (1997).

Industry/University/Government Partnerships in Metrology: A New Paradigm for the Future

C. R. Helms

Texas Instruments Inc., Dallas, TX 75265

A business process is described where Industry/University/Government interactions are optimized for highest productivity across these three sectors. This cross-functional approach provides for the rapid development of differentiated products for competitive advantage in industry, best of class scholarship and academically free university research, and the assurance of U.S. economic and military strength. The major focus of this paper will be R&D. However, the above objectives will only be met if effective transition from R&D into final product marketing, design, and manufacturing are included as an additional required concurrent, cross-functional activity. Metrology will be shown as an area that meets all the requirements for the development of a broad cross-functional partnership between industry, academia, and the Government that creates significant value for each sector.

INTRODUCTION

The past few years has ushered in a new era of research & development in semiconductor technology. In industry new management paradigms, advanced cost allocation models, and so-called repositioning has changed the way R&D is done & how technology transfer is accomplished (1-6). A concerted effort to minimize fixed costs through the use of outsourcing is a common theme throughout industry. The pressure for higher added value at lower cost is complimented by the drive to shortened development cycle times and faster production ramps.

Academia is under pressure from higher costs and the changing landscape of Government funding. This is especially true for research in semiconductor technology where the high cost of facilities and equipment makes it difficult for academic research institutions to maintain state of the art facilities. Balancing the responsibility to educate students and create new knowledge with the pull from industry to solve today's problems is not a task many academics want to take on (6-8).

The Federal Government's research establishment, at least as it relates to semiconductor technology, is also going through a transition. Between Government funding of industry and universities as well as work in the National labs, enormous progress in semiconductor technology was enabled for over three decades. This provided the U.S. defense establishment with superior technology and led to significant commercial developments as well (9). As priorities have shifted, the level of effort supported by the Government in semiconductor technology is on the decline (6-8).

How do we continue or even accelerate the rate of semiconductor technology development in the face of these major changes? Higher productivity of our research endeavors will clearly be required. This leads me to the focus of this paper, the development of broad cross-functional teams and partnerships to guarantee that we make the most efficient use of our resources. We will also show that metrology is an area ripe for the development of such a broad cross-functional approach, involving the industrial, academic, and government sectors.

The paper will be divided into six additional sections. In the next section, some of the key factors necessary for a successful R&D partnership are outlined. The desired focus of one member of the partnership treating the others as "customers" is discussed in the next section. The third section outlines metrics for determining which technical areas in the semiconductor R&D process can lead to high payoff R&D partnerships. A successful business process for such partnerships is described in section four. In section five some examples of the application of these concepts is discussed. Finally in the last section, these models are applied to the possible development of a broad cross-functional partnership in metrology between Industry, Academia, and Government. We show that metrology has all the necessary elements for high payoff to all sectors with such a partnership.

PARTNERSHIP SUCCESS FACTORS

As the title suggests this paper is endorsing the formation of cross-functional partnerships between industry, government, and academia, especially in the area of metrology. As this will mainly take the form of R&D activity, it is first useful to clearly define what R&D means and how it relates to the three sectors involved (10). After that discussion, the metrics by which we can identify an opportunity where a cross-functional partnership has potential for success will be discussed. Indeed, partnerships and teams are not a cure-all for many problems we encounter in business, education, or government (11,12).

CP449, *Characterization and Metrology for ULSI Technology: 1998 International Conference*
edited by D. G. Seiler, A. C. Diebold, W. M. Bullis, T. J. Shaffner, R. McDonald, and E. J. Walters

Research & Development

R&D can be viewed as having three different functions enabled by three different R&D processes (10). First, from a functional point of view, R&D can support and expand existing capabilities. Second, it can drive new opportunities. Finally, R&D can broaden and deepen core competencies. From a business process perspective R&D can focus on incremental, innovative, exploitation of existing knowledge. It can focus on more radical creation of knowledge new to the organization. R&D can also focus on the creation of fundamental knowledge new to the world. Unfortunately the three different functions cannot map into the three different business processes effectively. This author's attempt to devise such a map is shown below.

Process / Purpose	Incremental, Innovative Exploitation of Existing Knowledge	Creation of Knowledge new to the Organization	Creation of Fundamental Knowledge new to the World
Support & Expand Existing Capabilities	High Payoff	Some Payoff	Low Payoff
Drive New Opportunities	Low Payoff	High Payoff	Some Payoff
Broaden & Deepen Core Competencies	Low Payoff	Some Payoff	High Payoff

FIGURE 1. Diagram of the Business Process of "R&D," illustrating metrics for high payoff.

High payoffs for industry will come from the two upper left diagonals; the NIST ATP program has achieved good results right in the middle of the matrix (9). Academia has historically played a major role at the lower right (6-8). There are many examples in the semiconductor industry where our inability to understand such a map as led to a low return on R&D investment over the years (6).

Metrics for Value Added Partnerships

The discussion of R&D purpose and process has already illustrated that the value creation proposition of industry, academia, and government are not the same. For a partnership to be successful there must be a significant opportunity for value creation for each sector participating. We will show later that the best opportunity for partnerships in the present context is in middle to lower right of Fig. 1. This provides each sector with value in and of itself and a reasonable expectation of what the other will provide that adds overall value to the partnership.

In the next section we will more specifically discuss our views on the value proposition of each of the sectors. Each will be viewed from a customer/supplier relationship as part of an overall vision for a successful partnership. Each supplier must understand how to create value for the customer.

At the same time knowledge created by one sector must be transferred rapidly to the other to be of maximum utility. This is part of the process to implement shorter development cycletimes. Before discussing the technology transfer process between one sector and another it is useful to look at the evolving process within one organization. An historical view is shown in fig. 2 below (4-7).

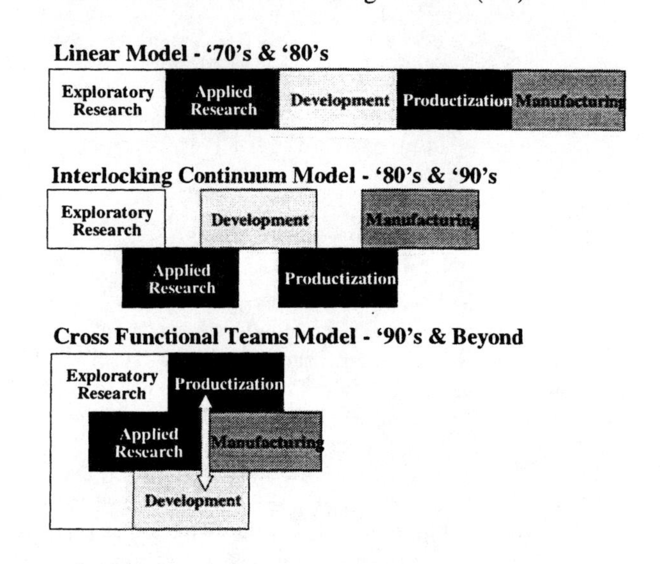

FIGURE 2. Historical view of relationships and effect on development cycletime (horizontal cycletime axis implied).

We are moving from a linear model (some call this "throwing technology over the wall") with relatively long development cycletimes to a cross-functional teams model. Even in the exploratory phase, there is significant participation from the other functions; note new product development, design, and marketing are also involved, but cannot easily be shown on a two-dimensional diagram. Focusing longer-term research on the proper space of Fig. 1 for a particular company's strategy is an additional benefit of this model.

How does a university research activity fit into a scheme as the one depicted in the lower panel of Fig. 2? A critical component is building the right bridges so that information that is generated in one sector can be rapidly transferred to another. A good part of this is people oriented, as we will discuss below. The worst medium for information transfer is, unfortunately, paper or hard copy. Today rapid and effective information transfer requires that the information generated in one sector be embedded in software to be readily used by the appropriate customer. This can go all the way from a spreadsheet database to a large TCAD program. This is a critical requirement for a successful broad, cross-functional partnership. All information must be captured in software and made easily accessible. Today's information technology is clearly up to this task.

CUSTOMER ORIENTATION

Partnership success can be assured if each partner treats the others as "customers." The value proposition for each partner can then be properly viewed from that partner's perspective and actions taken accordingly.

Industry as the Customer

In the context of metrology for semiconductor development and manufacturing there are two industry metrics. First is low development and manufacturing cost. The second, which also impacts the first, is short development cycletime and rapid yield learning (13,14).

It's clear that the company with the lowest cost structure for a particular product has the potential to be the most successful. However, in semiconductors, the health of the entire industry over the long term requires us to focus on cost from a broader perspective. If IC's are too expensive to produce, the mass markets we are targeting will be closed to us.

We will consider DRAM's as a timely example. Only the lowest cost producers are in a position today to be even close to profitable. Historically the cost per bit in a DRAM has decreased nearly 30% per year, while the number of bits per die has increased by 60% per year. This is projected to continue well into the future. Even with the cost per function decrease, the functionality is increasing faster driving the cost per die to increase by 12% per year. This is illustrated in Fig. 3 from industry-wide predictions, where the cost at product maturity is shown as a function of the year of first introduction (13). It should be noted that product maturity is about 5 years after first introduction. Over the 15-year period shown the DRAM capacity goes from 256 Megabits to 256 Gigabits, a factor of 1000. The cost of producing the mature technology goes from about $16 to $90.

This example illustrates the importance of keeping the manufacturing cost of semiconductors as low as possible and the reason why so much effort is going into reducing cost. Focusing on the semiconductor manufacturing industry as a customer therefore requires solutions that significantly reduce cost and are targeted correctly on the roadmap so they will be available when needed.

Academia as the Customer (6-8,15)

The value proposition of academia is, of course, very different from industry. Education is at its heart, providing the best education to the student population. The second critical goal for academia is to create new knowledge in an environment of free information exchange. The analog to cycletime in industry is rapid dissemination of research knowledge in academia.

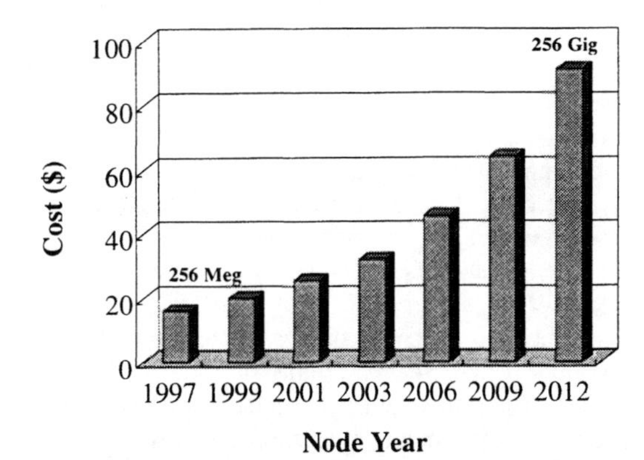

FIGURE 3. Mature DRAM cost per die projections from (9). Note the time at which mature production occurs is about 5 years after the "node year."

Just as high cost is threatening the health of the semiconductor industry, it is also threatening academia. This is especially true in high technology research where the cost of facilities and infrastructure are out of reach for many institutions.

Government as the Customer

The Federal Government as the customer is more difficult to quantify, being so pervasive in many aspects of our society. For the purpose of this discussion we will limit consideration to economic and military sectors. We would propose that the value proposition of government in this context would be to provide sustainable competitive advantage – or at least parity – for the U.S. in the economic and military sectors as well as overall quality of life across the board (8, 9, 15).

Cost is not an insignificant consideration here either. The high cost of achieving parity on both the military and economic fronts, for example, led to the end of the cold war.

CROSS-FUNCTIONAL PARTNERSHIP OPPORTUNITIES IN SEMICONDUCTOR TECHNOLOGY

To summarize the previous section, a successful partnership between industry, academia, and government must create significant value for each. For industry minimum time to market at lowest development and manufacturing cost is the overriding value proposition. High level scholarship, research, and academic freedom describe the value proposition for academia. For the Federal Government it is sustainable competitive advantage in the economic and military arenas.

Rationalizing these seemingly divergent goals is no simple matter. In many ways the situation may have seemed almost hopeless years ago when DARPA began focusing on Industry/University/Government partnerships (7). If we look a bit further, however, the situation was not so dire. For example, industry needs a well trained, agile workforce. Academia serves its own mission and can help achieve industry's goals at limited additional cost. Academia needs resources, both financial and intellectual. Both government and industry can provide these resources. Finally government needs to provide the infrastructure for education, information archiving, and standards ownership (9, 15). Both academia and industry can assist with these functions.

There are four business processes that can be applied for maximum value creation to all sectors. The first is the application of multidisciplinary approaches. Second is the formation of cross-functional teams. Third is the creation and use of a system for rapid and effective information transfer. Finally, the sum total of the financial and intellectual resources must be allocated in a highly coordinated manner.

If we turn back to the diagram of Fig. 1 we can determine the final set of factors that needs to be considered to assure success of a broad cross-functional partnership. We focus on the middle of the matrix so that the partnership " creates new knowledge" to "drive new capabilities" and those capabilities remove a major barrier on the industry roadmap, there is a high probability for success for all concerned.

Are there areas where such partnerships have a low probability for success? The upper left corner of the Fig. 1 matrix represents such activities. This focus will provide the company significant competitive advantage through new capabilities, lower costs, or faster cycletimes. In this case broad cross-functional partnerships cannot succeed.

We will now assess semiconductor technology areas based on the potential to limit the overall industry vs. the potential of the technology to provide significant competitive advantage. Areas we might consider include Design & Test, Front End Processes (FEP), Lithography, Interconnect, Factory Integration, Assembly & Packaging, ESH, Defect Reduction, Metrology, and Modeling & Simulation (13). We will only consider the semiconductor manufacturers; this leaves the role of the equipment and materials suppliers, which we will return to later.

Areas showing good potential for broad cross-functional partnerships are shown in Fig. 4. Unfortunately there will be no general agreement on a quantitative measurement of these two factors across all the technology areas. It is clear that lithography, metrology, ESH, and interconnect all have the ability to limit the overall industry. It is also clear that solutions for the first three need to be tackled industry-wide as the investment required by one company to achieve a solution is so large that a good ROI would be hard to achieve. These areas in the upper left of the diagram are just those where a broad cross-functional partnership would provide significant advantage.

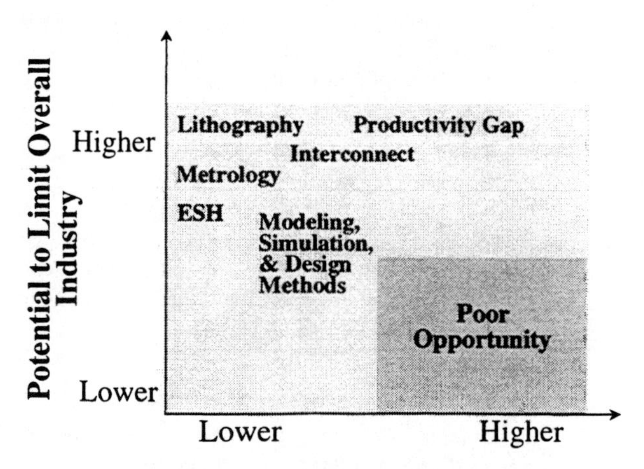

FIGURE 4. Diagram to assess the potential of broad cross-functional partnerships. Excellent potential exists at the upper left, poor potential at the lower right.

PARTNERSHIP BUSINESS PROCESS MODELS

We have now established that there are a number of possible areas in semiconductor R&D, including metrology, where there is significant potential for broad cross-functional partnerships. How can we structure such partnerships so that each sector can create maximum value? There are a number of approaches that have been developed over the years, including those developed by the DOD, specifically DARPA (6-8), the NSF, various "affiliates" programs in academia (3), the SRC and SEMATECH (16), National Lab CRADA's, NIST ATP's (9), and most recently SIA Focus Centers (17). Each of these has its advantages and disadvantages. In the space we have here we will not try to review each of these models, but suggest a model that is built from the best practices of each. We have had the good fortune to come to this model through about 20 years of trial and error.

It is based on effective cross-functional teams and a robust process, with appropriate contractual documentation. The team arrangement is illustrated in the Fig. 5 (7). It involves activity by the university professor for strategic planning, program definition, and project planning; a research associate for project planning and execution and knowledge and information transfer, and a student (or team of students, including at least one Ph.D. candidate) for knowledge transfer and program execution. Lest someone assume that we are proposing that the students and faculty have little contact in this model, to the contrary, it also includes the natural interaction between student and professor that is an integral par of academic life (8, 15).

The individual industrial contributor participates in project planning and execution, and knowledge and information transfer, the manager in program definition and project planning and technology transfer, and the "director" in strategic planning and project definition. The glue that holds this system together is the research associate in academia. This is an oft times overlooked part of most successful industry/university programs.

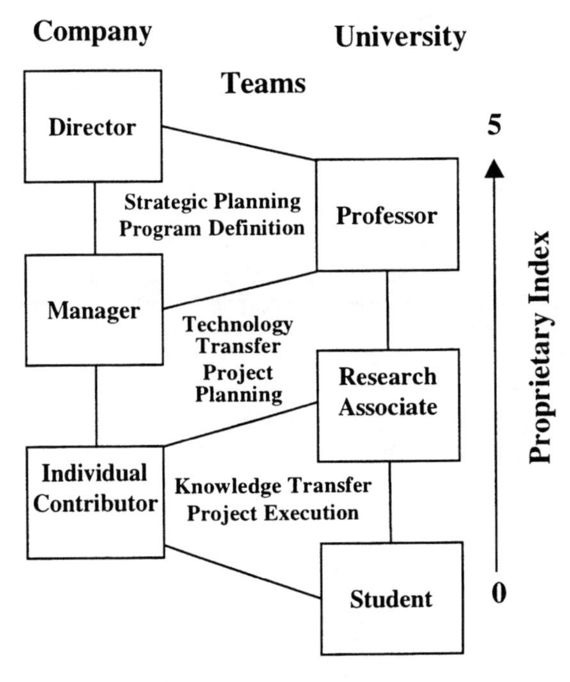

FIGURE 5. Diagram of the relationship of the cross-functional teams formed to provide optimum interactions between industry and academia; possible level of confidentiality is also indicated (7).

The diagram indicates the formation of three interlocking teams (systems of from two to four teams can be established with these functions). Each of these teams owns the functions or processes described above. Although this two-dimensional diagram does not indicate the role of government personnel, a three dimensional diagram with three functions similar to those indicated for industry can be constructed easily.

There are three major issues to be dealt with when constructing such relationships. First, from the academic perspective, the issue of student involvement in confidential or proprietary research needs attention. My experience suggests one simple model works best: students shall not participate in aspects of any research activity that involves confidential or proprietary information. This assures that the student's research is publishable in without restriction.

However, there must be a convenient mechanism for transfer of knowledge or technology from academia rapidly into the proprietary space of industry. The research associate or professor can serve this purpose so long as this does not lead to any issue of conflict of interest.

This leads to the issue of intellectual property, its ownership, and licenses thereto. This issue is complex and has consumed more pages than is in this entire volume. It is fair to say that there is no solution comprehended by all the communities. There is the concern on the part of academia that it receive appropriate compensation for IP generated; there is the concern on the part of industry that it not be blocked or charged an unfair amount for IP that it partially paid for in the first place. Finally the Government and National Lab personnel must pay attention to obtaining proper benefits to the U.S. economic infrastructure for IP generated through their funding mechanisms or in part by the actions of Government employees.

The model we propose has been used successfully over the past few years. It has the requirement that all the personnel, who are represented at the bottom of the diagram above, including but not limited to the student(s), industry individual contributor, and academic research associate, participate to a sufficient extent to be considered co-inventors of the IP generated. In industry this implies that management acknowledge the importance of the individual contributor's participation in the program and mentorship of the student. Each organization therefore has access to the IP through its personnel's co-inventorship.

In summary, the model of Fig. 5 represents a best practices approach to industry/university partnerships. It provides significant value creation potential for each sector and has built-in protections for publishable research for the student and IP access and ownership for industry. In the following section, examples where this model was developed and applied will be discussed.

EXAMPLES

In this section we will discuss two examples from personal experience where the above concepts were applied effectively. In one case knowledge was created for and applied to simulation of processes, device structures, and performance of HgCdTe infrared detector arrays. In the other the focus was on ultra-pure (de-ionized) water (UPW) use reduction in semiconductor manufacturing. These seemingly disparate activities had one thing in common: the formation of the correct broad cross-functional team to identify the correct problem and create a productive solution (minimum development time and lowest cost). I will discuss the second first.

The problem is the need for large quantities of UPW for semiconductor manufacturing and the cost of producing, distributing, and disposing of wastewater. Due to environmental concerns, the continued access to ever increasing supplies of feed water is a significant risk. Current use can be 2000 gallons per wafer; this corresponds to over 10 billion gallons per year for major semiconductor manufacturers. Developing processes (e.g. wafer rinsing) that use less water is one potential solution to the overall problem. Once the **problem is defined** (bold print indicates the sequence of steps in developing this successful partnership), the next step is to **form the right team** to investigate the problem and potential solution and verify that the problem is correctly specified and a high productivity solution is being pursued.

Since we are assuming that industry, academia, and government are involved in the partnership, we must also verify that there is **significant value creation** for the latter two sectors (the problem is industry driven and thus, if the problem is properly specified, value creation for industry is built-in). From the outset it was clear that a multi-disciplinary approach would be necessary and there was significant potential for a unique educational experience and the creation of new engineering knowledge. Thus significant value could be created in the academic sector. The whole area of environmental quality is a significant responsibility of government. Reduced water semiconductor manufacturing therefore has the potential for value creation for the government. Finally, an industry-wide solution is required; **little competitive advantage** will be derived for an individual semiconductor manufacturer. All of the conditions seem to be correct for a broad cross-functional approach.

The team formed needed a significant cross-functional membership from industry as well as academics and government lab personnel. An additional benefit of the participation of the industry outsiders was in pulling the internal industry functions into a closer working relationship. The solution required participation by process integration, process engineering, equipment engineering, facilities and ESH. The more of these functions missing from the team, the longer the delays in technology insertion and the higher the cost. In this case the teams were formed from Stanford University, Sandia National Lab, and industrial participants identified through SEMATECH, including those from Hewlett Packard, AMD, IBM, Motorola, and Texas Instruments.

The next two steps in the process relate to knowledge generation. The first is benchmarking, the second the generation of knowledge to be applied to the specific problem. For an area of on-going technology development within industry a competitive benchmark should be available as part of the overall industry core competency. **Benchmarking** across the whole industry serves to confirm that the correct problem is being addressed and that the correct team has been assembled. After benchmarking, the problem can be redefined and team modified to address the modified problem. In the present example, benchmarking showed that there was a significant difference (over a factor of ten) in water use for various seemingly similar processes; the potential for defining and **implementing best practices** across the industry was therefore significant.

Finally, work to **obtain the new knowledge** required to determine, design, and implement an appropriate solution is planned and performed. An appropriate solution does not imply the technically optimum solution. Overall optimization with respect to technical advancement, development cost, and development cycletime, in most cases, does not lead to the optimum technical solution – that solution would likely be too expensive and take too long to develop. In the present case, the team rapidly developed solutions, based on fundamental experimentation and physical models, that provided up to a factor of five improvement in point of use water consumption (18). The results were captured in computer databases and software for rapid portability industry-wide so that the **information is retained** and made available on a broad scale. The know-how generated was also transferred on a wide scale by university research associates who, through non-disclosure and personal service agreements, could rapidly insert the student generated published work into manufacturing facilities and equipment.

The success of this program and others like it led to the formation of an even broader partnership between industry, academia, and government, the NSF/SRC Engineering Research Center for Environmentally Benign Semiconductor Manufacturing (18).

In the other example, the drive to low manufacturing cost, especially for low volumes, led the Government and industry to consider flexible manufacturing and programmable factory approaches to HgCdTe infrared detector array production. A critical part of this capability is the availability of processes, device, performance, and equipment simulation tools. In the limit this provides for a virtual factory capability giving first pass success for the production of new designs.

This idea was not a new one, with technology computer aided design (TCAD) for Si and GaAs technologies being developed over the last 20 years. One of the distinguishing features of this program was that the team formed had participants all the way from manufacturing engineers at Texas Instruments to students at Stanford University working on fundamental understanding of process induced point defect mechanisms. In addition to a contractual relationship between Texas Instruments (prime contractor, Air Force/DARPA) and Stanford University, the Stanford principal investigator and later a university research associate were able to provide both a rapid feedback and feedforward mechanism through personal service and non-disclosure agreements. It was therefore possible to plan experimental work in industry using industrial processes and process flows to support calibration and validation of the models and simulators in nearly real time. On the other hand fundamental work performed by the students could not only be immediately published, but also captured in the models and simulators as well.

The development, use, and validation of the models and simulator is illustrated in Fig. 6. Rapid turn-around of the overall process led to a robust process flow and physically based simulation capability. The capture of university-based information in a computer program assured convenient access to the information even after the end of the program. The software has also been transferred to SRI International to provide longer term archiving. Although SRI is not a Government Laboratory per se, it is serving a similar function in this case, as an archive for information.

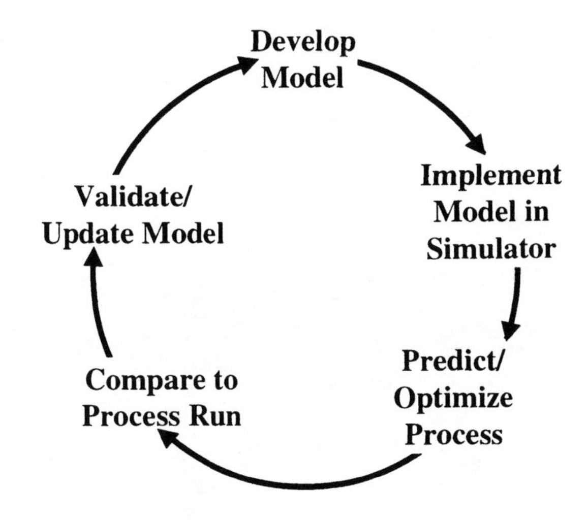

FIGURE 6. Illustration of the cycle used to development, implement, validate, and update HgCdTe process and performance models and simulators. In all cases quantitative models are validated early, with optimized parameter sets continually upgraded and improved (7, 19, 20).

To summarize this section, two examples were given where the models discussed in the previous sections were developed and tested over 20 years in programs based at Stanford University. They are far from perfect and require significant energy and commitment on the part of the individual contributors to make it all work. In addition it is clear that teams can't solve all problems and should not be considered a cure-all; value must flow to all sectors and individuals involved for a successful partnership (7, 11, 12, 18).

OPPORTUNITY FOR A CROSS-FUNCTIONAL PARTNERSHIP IN GSI METROLOGY

Metrology and test have been identified as one of the "grand challenges" the semiconductor industry faces as we move into production with feature sizes approaching 100nm (11). Yet the level of external funding applied to this area by either industry or government is surprisingly low. The specific areas of concern include 2/3-dimension dopant profiling, in-line microscopy, and effective thickness determination for barrier layers and gate insulators. This must be accomplished over 300mm wafers and down vias and contacts. These factors have the potential to limit the overall industry in the next few years (see Fig. 4). Even if indirect methods are found to keep manufacturing under control (i.e. statistical metrology), the lack of failure analysis tools and metrology will add significantly to development cycletimes (21). This in turn will make it impossible to achieve the productivity gains needed to keep the industry healthy (22).

Except for possibly Hitachi, none of the semiconductor manufacturers develops metrology methods or tools (23). Each is looking for a turnkey solution to metrology, not competitive advantage per se. Significant competitive advantage can, however, derive from the way metrology is performed and information collected and analyzed.

From a semiconductor manufacturer perspective metrology has the right features to be considered for a broad cross-functional approach. Specifically, given the need for metrology early in the development phase of a particular technology node, metrology for the 0.13 micron technology node (0.1 micron isolated lines) and 300 mm wafers and beyond is an excellent target.

Measurement fundamentals and methods have long been a focus of academic research. The near atomic level of tenth micron technologies provides an excellent testbed of academic ideas and education. In addition to the overall health of the semiconductor industry, the need for measurement standards and standard methodologies provides a significant driver for government participation.

If we look back at Fig. 4 some of the areas identified as good candidates for broad cross-functional partnerships, in particular lithography and ESH, have active broad cross-functional support and interconnect and design will soon have SIA sponsored Focus Centers. In addition the historical level of support in these areas has been significant – very large compared to metrology. The time is clearly right to emphasize the development of new metrology methods and tools, and their integration into the overall process and factory integration paradigm. Participation from the semiconductor manufacturing industry, academia, the U.S. government funding agencies, and the National Labs is clearly indicated.

Might there and should there be other participants in such a partnership? The answer is yes, specifically the process and metrology equipment supplier community. However, this leads to a dilemma that we have yet to discuss.

Competitive advantage, the bane of broad cross-functional partnerships, is exactly what the process and metrology tool suppliers are seeking in the same arena we are proposing such a partnership. In addition, for a semiconductor manufacturer/ university/government partnership in metrology to be of optimum benefit, the technology needs to be inserted seamlessly into the supplier community. We will not solve this problem here; it is clear that a first task of a team set up to plan this program will be to design a process for equitable inclusion of the supplier community. The solution might include participation by the suppliers in the form of internal resources directly applied, in-kind contributions (i.e.

equipment) to the universities and government labs, and additional financial support to the overall program.

The other issue that needs attention is the relationship of U.S. government participation in and support of programs which not only provide significant benefit to U.S. industry and the U.S. educational infrastructure, but also significant benefit to off-shore interests. Half (six of the top twelve) of the dominant technology insertion partners for metrology advancements, the equipment suppliers, are non-U.S. companies (22). These same companies are significant suppliers to the U.S. semiconductor manufacturing companies. For maximum benefit, technology transfer to the non-U.S. equipment suppliers must not only be possible, but encouraged.

Progress in dealing with this issue is fortunately being made. The best example is I300I and its expansion to include lithography and ESH in International SEMATECH. However, since many of the non-U.S. equipment suppliers are in Japan and significant customers for metrology equipment will also be in Japan, their participation in a broad cross-functional partnership in metrology is almost a must. This may be an ideal opportunity for the U.S., Japan, and other countries to develop a all-inclusive international program.

SUMMARY

We have presented and discussed a number of factors that lead to successful broad cross-functional partnerships. Quantifying and understanding the value proposition of all the possible participants is a first step in this process. Making sure that there is sufficient value creation for all "customers" is a must. For a program to have significant participation from an industrial sector, the best chance for success lies with issues that are overall industry limiters, but which are also not perceived to provide significant competitive advantage. This is not to say that the work is pre-competitive; rather, the industrial sector has made a conscious decision to forego competitive advantage in favor of the benefits of a cross-sector focus.

A successful partnership must not only be cross-functional from one sector to another, but also cross-functional within a given sector. Participation by the multiple functions within the semiconductor manufacturing sector (i.e. design, process integration, facilities, etc.) is necessary for minimum cost and time to market efforts. In academia participation by faculty, research associates, and students are necessary.

It is time to apply these concepts to future developments in metrology. This effort must include the process and metrology equipment supplier sector and be international in nature. The benefits of an international effort, including the supplier sector, far outweigh the potential political and business process barriers we might encounter.

REFERENCES

1. Savage, C. M., *5th Generation Management,* Burlington, Digital Press, 1990.
2. Hammer, M., *Beyond Reengineering,* New York, Harper, 1996.
3. Saxenian, S., *Regional Advantage,* Cambridge, Harvard University Press, 1994.
4. Jacob, R., The Struggle to Create an Organization for the 21st Century, *Fortune,* 90-99 (April 3, 1995).
5. Majchrzak, A., Wang, Q., Breaking the Functional Mind-Set in Process Organizations, *Harvard Business Review,* 92-99 (Sept.-Oct., 1996).
6. Vandendorpe, L., ed., Industry & Government Search for their Roles in Basic Research, *R&D Magazine,* 17-21 (October 1997) and references therein.
7. Helms, C. R., Optimized Model for University/Industry Interactions: Development of Designed Coupled Manufacturing Tools for Infrared Imagers, *Proc. 11th University Government Industry Microelectronics Symp.,* 12-15, IEEE, Piscataway, NJ (1995).
8. Vest, C. M., Stewards of the Future: The Evolving Roles of Academia, Industry, & Government, *MIT Report of the President 1995-96,* Cambridge, MIT (1997).
9. Powell, J. W., Development, Commercialization, & Diffusion of Enabling Technolgies: Progress Report for ATP Projects Funded 1993-1995, NISTR 6098, Gaithersburg, U.S. Dept. of Commerce NIST (1997).
10. Roussel, P. A., Saad, K. N., Erickson, T. J., *Third Generation R&D,* Cambridge, Harvard Business School Press, 1991.
11. Dumaine, B., The Trouble with Teams, *Fortune,* 86-90 (Sept. 5, 1994).
12. Lambert, D. M., Emmerlhainz, M. A., Gardner, J. T., So You Think You Want a Partner?, *Marketing Management,* **5**, 2, 25-41 (1996).
13. The National Technology Roadmap for Semiconductors, San Jose, Semiconductor Industry Association, 1997.
14. Ross, P. E., Moore's Second Law, *Forbes,* 116-117 (March 25, 1995).
15. Casper, G., Cares of the University: Five-Year Report to the Board of Trustees & the Academic Council of Stanford University, Stanford, Office of the President, Stanford University (1997).
16. Rea, D.G., Brooks, H., Burger, R. M., LaScala, R., The Semiconductor Industry-Model for Industry/University/ Government Cooperation, Research-Technology Management, 46-54, (July-August 1997).
17. Radack, D. J., Department of Defense Semiconductor Technology Council First Annual Report, Houston, Science Applications International (1996).
18. Helms, C. R., Partnerships in Environmentally Benign Semiconductor Manufacturing, *Solid State Technology,* 52-53 (March 1997).
19. Melendez, J. L., Helms, C. R., Process Models for HgCdTe, *J. Electron. Mater.,* **24**, 565-590, 1995.
20. Holander-Gleixner, S., Robinson, H. G., Williams, B. L., Helms, C. R., HgCdTe Process Model Boundary Conditions, *J. Electron. Mater.,* **26**, 628-636, 1997, and references therein.

21. Bartelink, D. J., Statistical Metrology: At the Root of Manufacturing Control, *J. Vac. Sci. Technol.*, **B 12**, 2785-2794, 1994.
22. Perloff, D. S., Gauging the Future: The Long Term Business Outlook for Metrology & Wafer Inspection Equipment, this volume.
23. Melliar-Smith, M., Diebold, A. C., Metrology Needs for the Semiconductor Industry Over the Next Decade, this volume.

Gauging the Future: The Long Term Business Outlook for Metrology and Wafer Inspection Equipment

David S. Perloff

Veeco Process Metrology Group, Veeco Instruments, San Jose, CA 95119

This paper describes a long term view for semiconductor metrology and inspection which, at $2.3 billion, was approximately 10% of the front end equipment market in 1997. Topics covered include consumption patterns for each of the major categories of measurement equipment, consolidation among capital equipment suppliers, and the business implications of integrated measurement and sensing in next generation process equipment.

INTRODUCTION

The traditional viewpoint of chip manufacturers has been that discrete measurement steps, even when performed directly on product wafers, represent non-value-added contributors to the cost of wafer fabrication. This thinking, however, has changed dramatically over the past several years with the need to meet the challenges presented by smaller device geometries and larger wafer diameters.

Metrology and inspection tools are now widely viewed as essential elements in rapidly achieving and sustaining high manufacturing yields and are among the first equipment sets brought on line in new facilities. In addition, many established manufacturing lines continue to upgrade contamination control and process diagnostic equipment as increased resolution and measurement performance becomes available.

This paper provides an overview of the process control capital equipment market, which was approximately 10% of the front end equipment market in 1997. With revenue projected to increase from $2.3 billion to $4.5 billion by 2002, the business aspects of designing, selling, and supporting metrology and inspection tools have, in many respects, become as important as the underlying measurement technology. Topics covered include consumption patterns for each of the major categories of measurement equipment, consolidation among capital equipment suppliers, and the business implications of integrated measurement and sensing in next generation process equipment.

METROLOGY AND INSPECTION

Metrology involves the accurate determination of a value and tolerance for key wafer parameters. Figure 1 depicts the manner in which in-line metrology is typically employed in modern manufacturing lines. Parameters such as film thickness (h), critical dimension (w), layer-to-layer overlay (d_x,d_y), step height (h) and resistivity (R_s) are measured at a limited number of sites (typically five), on a subset of wafers from each production lot, and then tracked using various statistical process control techniques.

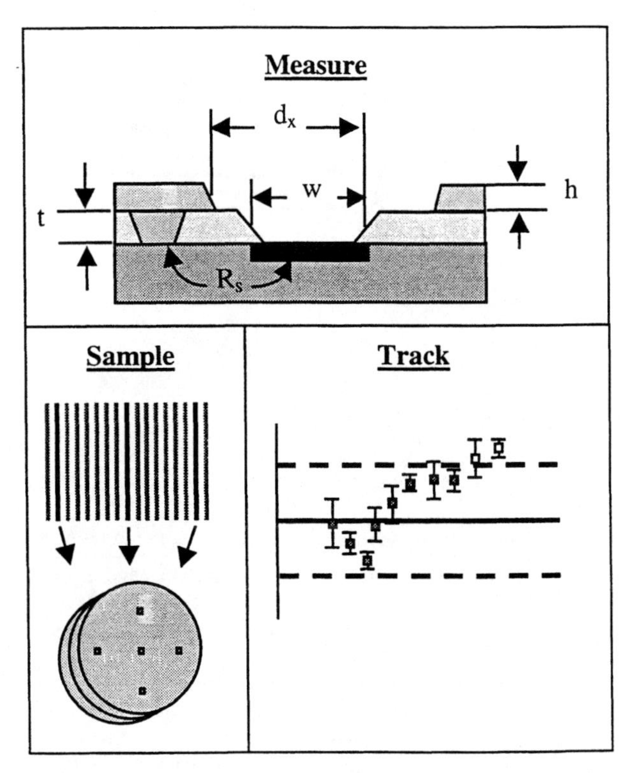

FIGURE 1. Metrology involves the accurate determination of a value and tolerance of key wafer parameters.

Inspection involves detecting, observing and categorizing process exceptions, such as particles and pattern flaws, which

CP449, *Characterization and Metrology for ULSI Technology: 1998 International Conference*
edited by D. G. Seiler, A. C. Diebold, W. M. Bullis, T. J. Shaffner, R. McDonald, and E. J. Walters

could adversely affect end-of-line product yield. Inspection generally is carried out on unpatterned monitor wafers using laser-scanning systems, and on patterned production wafers using a combination of direct imaging, laser-scanning and scanning electron microscope (SEM) tools. As shown in Fig. 2, inspection is followed by a separate operation in which a subset of the defect population is reviewed and classified, either by operators or using automatic defect classification (ADC) software.

Defects detected at different steps in the process may have differing correlation with the ultimate functionality of devices. Various tools are employed to identify root causes of yield loss, an example of which involves systematically partitioning the defects among the various process layers.

Finally, the defects that are directly correlated with yield reduction are assigned relative weightings by employing Pareto analysis, thereby establishing priorities for yield improvement efforts.

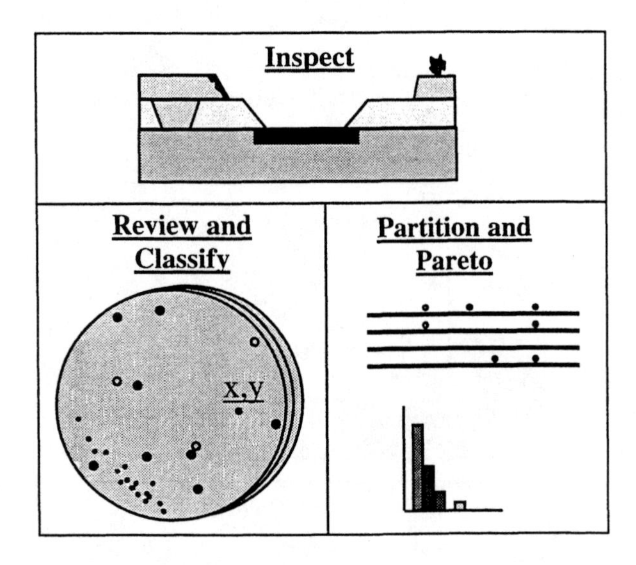

FIGURE 2. Inspection involves the detection, observation and categorizing of process exceptions.

Stand-Alone Metrology and Inspection Tools

Today's manufacturing lines invest primarily in stand-alone metrology and inspection tools which, for the purposes of this discussion, may be divided into two broad categories: *off-line* and *in-line* (Fig. 3).

Off-line tools generally address multiple process requirements and involve measurements primarily on single-use monitor wafers. Examples of off-line measurements include x-ray fluorescence, resistivity and unpatterned wafer defect detection.

In-line metrology and inspection tools are an integral part of the process flow, and must therefore be matched in throughput with the process equipment tool set. Examples of in-line measurements are film thickness, pattern-to-pattern overlay and critical dimensions.

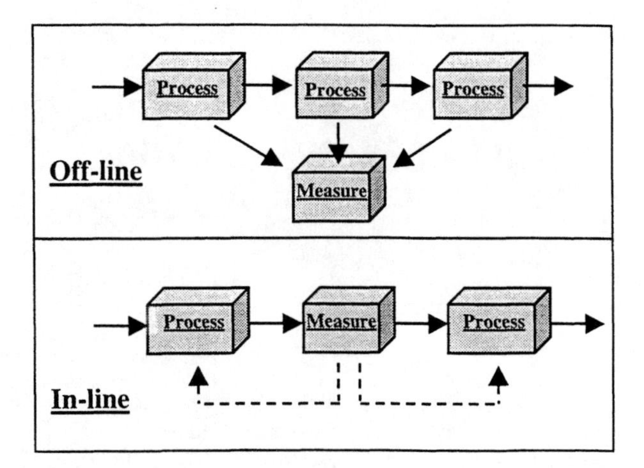

FIGURE 3. Manufacturing facilities invest primarily in stand-alone metrology and inspection tools.

Integrated Measurement and Sensing

The increasing use of metrology and inspection tools in manufacturing has been identified as an area of concern to device manufacturers. To illustrate this point, we cite the following comment by Rose (1) with emphasis added:

> *"One proposal that needs more than simple consideration is to eliminate the need for inspection. If process tools were capable of delivering and maintaining the technical capability by design rather than by nearly continuous inspection, we might see a reversal in this escalation of metrology equipment capital cost."*

The cost of such metrology and inspection tools has in fact escalated (as have deposition, lithography and etch equipment), so that less costly alternatives are gaining favor. In particular, integrated measurements and sensing, which may be either *on-line* or *in-situ*, offer the prospect of more tightly linking metrology with the wafer processing equipment, at significantly lower cost and complexity.

On-line measurements (Fig. 4) provide essential metrology capability directly on the process tool, allowing the process set-point to be adjusted wafer by wafer. For example, some manufacturers of chemical mechanical polishing (CMP) equipment have integrated film measurement modules that permit the thickness of films to be determined without first cleaning or drying the wafers. The information derived from such tools is used either to adjust the amount of polishing required to complete the operation on the wafer being measured, or used to "tune" the operating conditions for the next set of wafers to be processed.

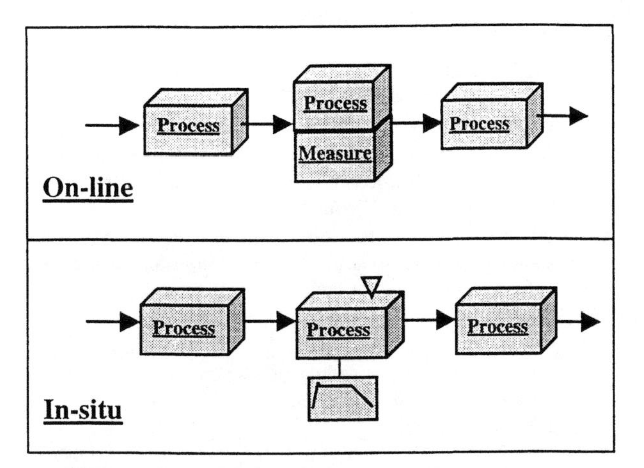

FIGURE 4. Integrated measurement and sensing are becoming increasingly important for next generation processes.

In-situ sensors are often considered an ideal solution for real-time equipment control, because they are conceived of as enabling the process tool to dynamically adjust its own performance (Fig. 4). In-situ sensors, and equipment models that tie the sensor data to the control of the equipment, are applicable to both single-wafer and batch processes. As shown in Fig. 5, in-situ sensing may be divided into three broad categories: *direct sensing*, *indirect sensing* and *virtual sensing*.

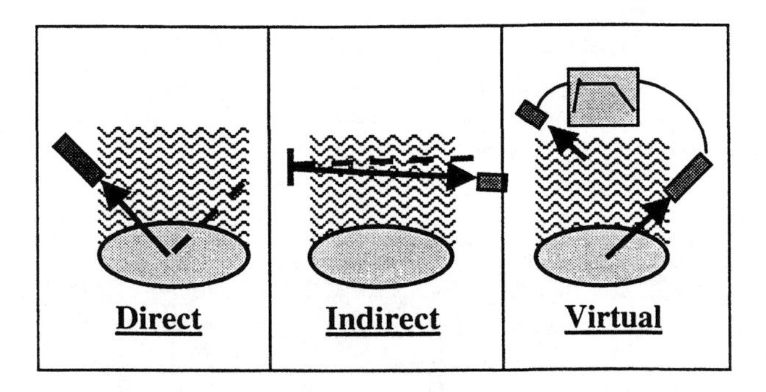

FIGURE 5. The three categories of in-situ sensing.

Direct sensing, also known as *wafer state sensing*, continuously monitors key parameters of interest, such as the thickness of an insulating film as material is removed during a plasma etching operation. Examples of direct sensing are emission spectroscopy to control etch uniformity, and spectroscopic ellipsometry to control film growth.

Indirect sensing, also known as *process state sensing*, continuously monitors the processing environment, tracking parameters such as temperature, pressure, flow and heater output. Examples are end-point detection using residual gas analysis (RGA) and particle detection using laser scattering.

Virtual sensing involves a combination of *wafer state* and *process state* measurements, and utilizes conventional sensors and real-time adaptive models to achieve improved control. The models employed are capable of describing the wafer state at multiple locations and over time.

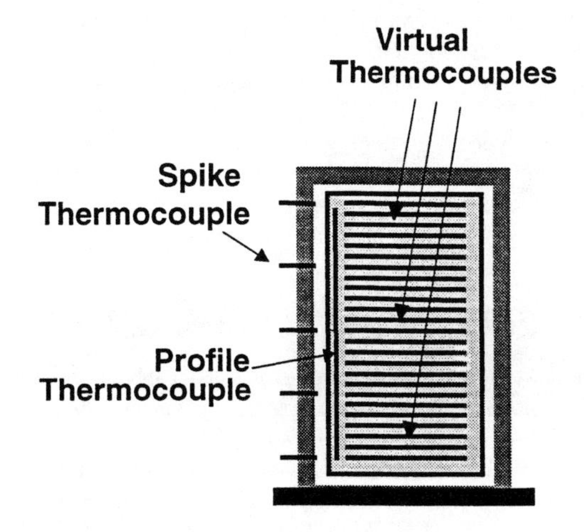

FIGURE 6. Example of *virtual* sensing used in advanced furnace control.

Figure 6 is an example of the application of virtual sensing for improving the performance of a fast ramp, multi-zone vertical thermal reactor. Instrumented wafers are used to develop a model that allows the furnace to be controlled using thermocouples "virtually" located on the wafers themselves. The results have been 20-30% higher throughput as a result of faster stabilization, 3-5X center-to-edge improvement in film uniformity and improved tool-to-tool matching. Rapid set-up and recipe transportability have proved to be an added advantage in production, resulting in a substantial reduction of monitor wafer use (2).

Impediments to the Adoption of Integrated Measurement and Sensing

In-situ sensors and add-on measurement modules are often viewed as non-value-added by customers of wafer processing equipment, who strive to minimize capital equipment costs and avoid reliance on third party suppliers for support. In addition, process equipment manufacturers are reluctant to absorb additional cost-of-goods to achieve improved process control. (However, equipment suppliers do use integrated measurement and sensing during the tool development phase for its informational and diagnostic benefits.)

As a result of these factors, small metrology suppliers are often at a disadvantage when dealing with large equipment companies, especially when attempting to preserve their intellectual property and marketing rights.

The Integrated Measurement Association (IMA) has been established to help deal with these concerns. It is a consortium of sensor developers, chip manufacturers, tool suppliers and software integrators. The IMA is attempting to foster a dialogue which will result in open architecture and interface standards, so that ultimately a third-party sensor may be added to the control loop of a process tool by the end user without fear of voiding warranties or requiring additional support from the tool supplier.

Industry initiatives to improve communication and establish standards are essential. Despite years of effort and devotion to this type of metrology, there are still few examples of companies specializing in this area that have been successful in terms of growth, profitability or wide scale adoption.

THE PROCESS CONTROL MARKET

According to Dataquest's 1998 forecast, wafer fabrication capital equipment revenue reached $22.3 billion in 1997, with $2.3 billion of that number represented by the category of Process Control equipment. As shown in Fig. 7, the Process Control market (as defined by Dataquest) includes patterned defect, CD-SEM, auto review and classification, thin film metrology, automated unpatterned wafer defect and optical metrology. The category "other" equipment represents tools such as resistivity probes, atomic force microscopes and x-ray fluorescence systems.

Figure 8 shows the relationship between the various metrology and inspection segments tracked by Dataquest and those of the overall front end equipment market.

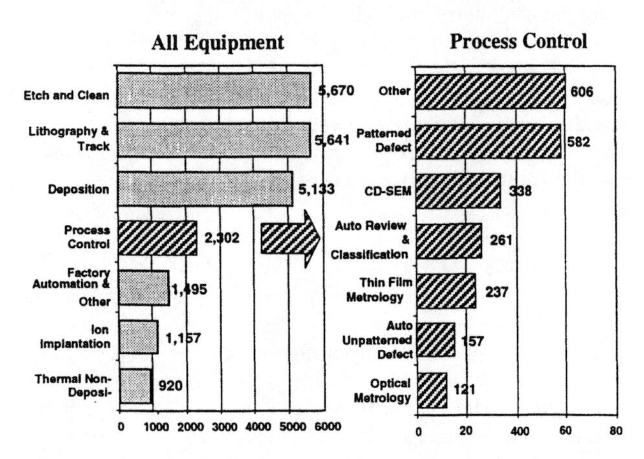

FIGURE 7. Wafer fab revenue reached $22.3 billion in 1997. Source: Dataquest.

Figure 9 depicts the top twelve wafer fab equipment companies in 1997. These companies largely control the front end equipment market, accounting for about 2/3 of the total revenue. However, only three of these companies supplied metrology and inspection equipment: Applied Materials, KLA-Tencor and Hitachi.

Two points are worth noting. First, KLA-Tencor placed number four on this list, despite the fact that just 10% of the market is represented by metrology and inspection equipment. This reflects the impact of the consolidation occurring in the Process Control market, a key example of which were the successive mergers of Prometrix, Tencor and KLA to create a mega-supplier of metrology and inspection equipment. Second, at this time only Applied Materials, Hitachi and Veeco Instruments supply both metrology and processing equipment. However, this strict demarcation between the two broad categories of equipment is likely to change as metrology and inspection products become recognized as being an essential part of a *total* process solution for semiconductor manufacturers.

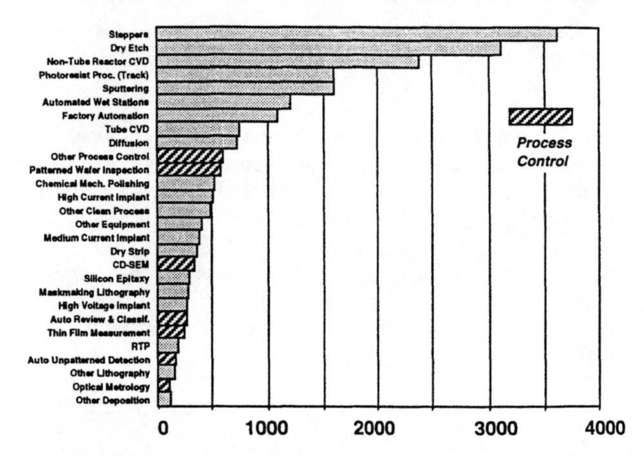

FIGURE 8. Metrology and inspection equipment represented $2.3 billion of the overall equipment market in 1997. Source: Dataquest.

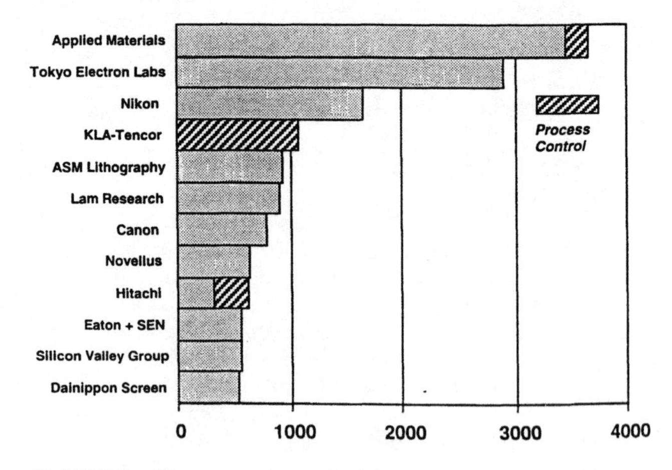

FIGURE 9. The top twelve wafer fab equipment companies in 1977 controlled 2/3 of the overall market. Source: Dataquest.

MARKET CONSOLIDATION

Recent acquisitions by KLA and Applied Materials have dramatically altered the competitive balance within the metrology and inspection market. As shown in Fig. 10, these two companies, along with Hitachi, dominate this marketplace, holding close to a 75% combined market share.

The fact that mergers and acquisitions are gradually reducing the number of metrology and inspection equipment suppliers is not surprising (Table 1). First, major metrology and inspection suppliers are attempting to keep pace in revenue and market share with their front end counterparts, which are typically in the >$500M range. Second, metrology and inspection companies in the $25M - $50M range are finding it too costly to invest in world-wide support and the new platforms required for 300mm wafers.

TABLE 1. Mergers and acquisitions since 1993

- *Phase Shift by ADE (1998)*
- *Amray and KLA-Tencor (1998)*
- *Digital Instruments and Veeco (1998)*
- *IVS and Schlumberger ATE (1998)*
- *Technical Instruments and Zygo (1997)*
- *Tencor and KLA (1997)*
- *Knights Technology and Electroglas (1997)*
- *WYKO Instruments and Veeco (1997)*
- *Opal and Orbot and Applied (1996)*
- *Metrologix and KLA (1994)*
- *Censor and Tencor (1993)*
- *Prometrix and Tencor (1993)*

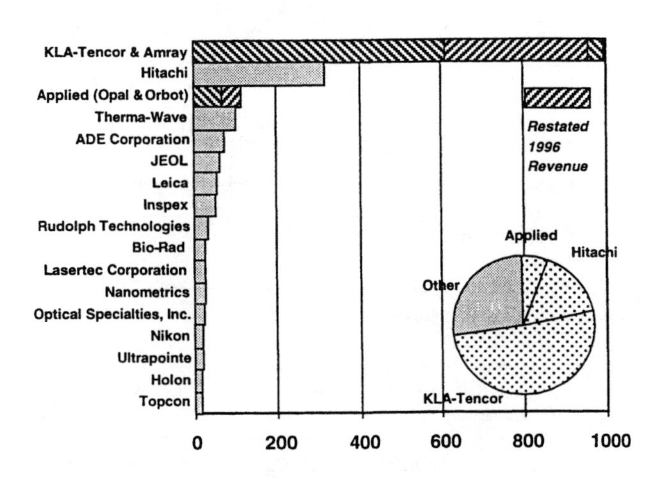

Figure 10. Acquisitions by KLA and Applied Materials alter the competitive balance in Process Control equipment. Source: Dataquest.

THE WORLD MARKET

The U.S. is the dominant supplier of metrology and inspection equipment. As shown in Fig. 11, the U.S. in 1996 supplied over 70% of the equipment in this category, with Japan at 24% and Europe at 5%. In comparison, in 1989, the U.S. supplied 58%, Japan 35% and Europe 7%. This trend may be expected to continue, as Hitachi gradually loses its dominant role as a CD-SEM supplier, a process which is being driven by the emphasis of Applied Materials and KLA-Tencor on this product area.

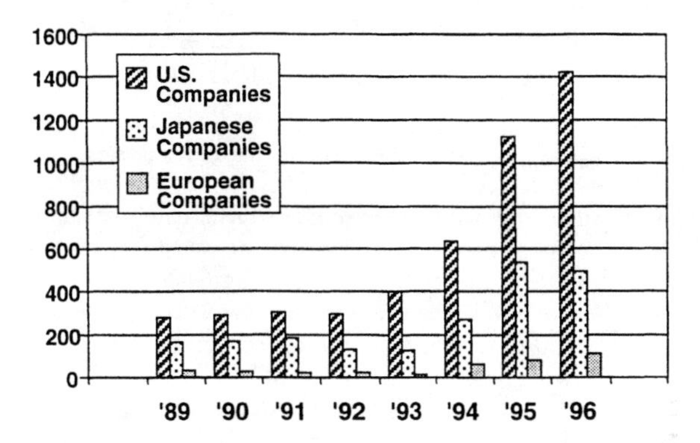

FIGURE 11. The U.S. has steadily increased its share of the world metrology and inspection equipment market. Source: Dataquest.

MARKET OUTLOOK

Figure 12 summarizes the overall metrology equipment market for the years 1990-2002. The rapid growth in the 1994-1995 time frame is not expected to be repeated between 1997 and 2002. Likewise, the compound annual growth rate will, according to Dataquest, diminish, from 23% between 1990 and 1996 to about 13% between 1996 and 2002.

The 5% year-on-year growth between 1997 and 1998 noted in Fig. 12 represents Dataquest's "optimistic" forecast, released in January 1997, in which it is assumed that a number of economic and technology factors combine to soften the impact of the Asia-Pacific crisis. This regional situation, which has resulted in a virtual shutdown of semiconductor plant expansion in Korea, is viewed in this scenario as being moderated in part by continued expansion of capacity in Taiwan.

Dataquest has provided an alternative "downside" scenario in which reductions in discretionary spending actually result in a down year for the industry as a whole, and for the Process Control market in particular, of about -7%.

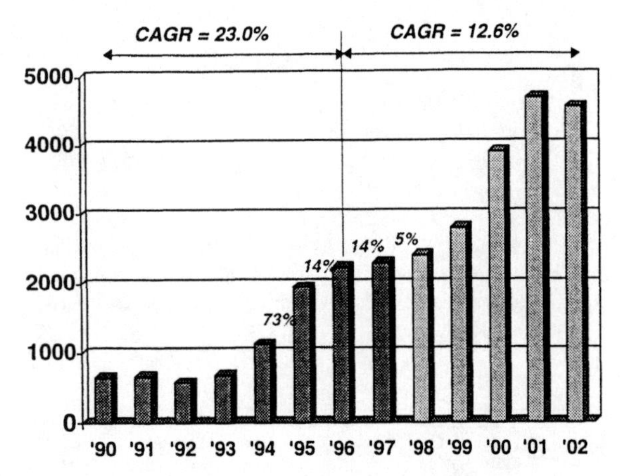

FIGURE 12. Dataquest forecasts a somewhat slower growth rate for metrology and inspection equipment over the next five years.

It is an interesting fact that the semiconductor chip market, which of course drives consumption of manufacturing equipment, has experienced no "average" growth year between 1978 and 1998. That is, although year-to-year growth averaged 17% during that time period, any given year fell into one of two categories: less than 10% and greater than 21%. Figure 13 depicts this behavior, and indicates that 1998 has been estimated as a low growth year.

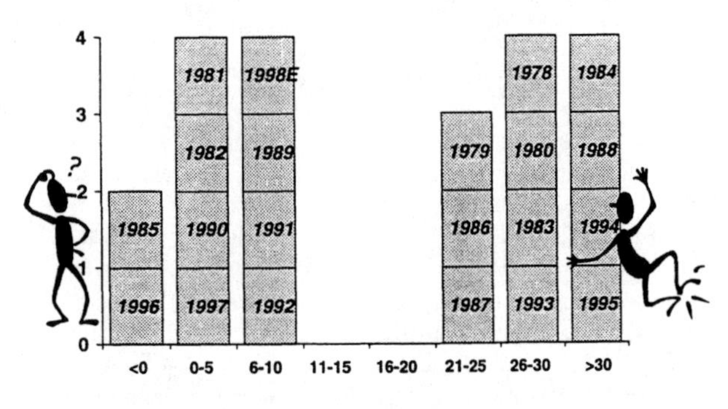

FIGURE 13. Worldwide semiconductor revenue growth "averaged" 17% between 1978 and 1998E. Source: WaferNews, IC Insights.

Not surprisingly, a similar situation exists with respect to the growth of semiconductor equipment sales. Figure 14 compares the year-on-year growth of chip revenue with that of equipment for the years 1990-1998E. Shown in the figure are both the optimistic and downside scenarios.

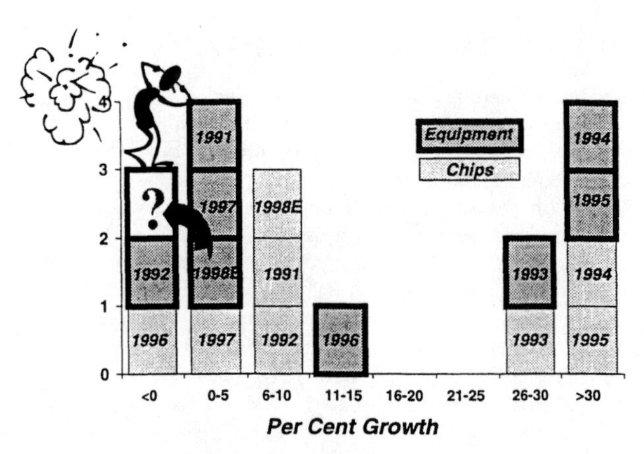

FIGURE 14. Similar behavior is reflected in the growth of semiconductor equipment sales (1990-1998). Source: WaferNews, SIA.

Despite the rather large year-on-year fluctuations of growth rate for the semiconductor capital equipment market, individual market components tend to track closely with the overall market. This is also true of the Process Control market segment, as shown in Table 2.

TABLE 2. Optimistic growth outlook for semiconductor equipment. Source: Dataquest.

Equipment Segment	1996	1997	1998	1999	2000	2001	2002
Total Process Control ($B)	2.2	2.3	2.4	2.8	3.9	4.7	4.5
% change	14.4%	3.1%	4.9%	15.9%	39.5%	20.0	-2.8%
Total Fab Equipment ($B)	21.7	22.3	22.7	26.6	37.5	44.5	42.8
% change	13.4%	2.9%	1.8%	17.2%	40.8%	18.7%	-3.7%
Share of Market	**10.3%**	**10.3%**	**10.6%**	**10.5%**	**10.4%**	**10.5%**	**10.6%**

The various factors which are likely to constrain the growth of metrology and inspection equipment to approximately 10% of the total market are summarized in Table 3.

TABLE 3. Factors which may inhibit the long-term growth of metrology and inspection equipment revenue.

- *The movement to multiple-step, single-wafer processing reduces the number of measurement and inspection opportunities.*
- *Front end equipment ASPs accelerate due to costs associated with providing turn-key process solutions and 7x24 support.*
- *Significant new process control requirements do not emerge, while newer front end segments (e.g. CMP) absorb additional equipment dollars.*
- *In-situ and integrated process control displaces the need for pre- and post-process measurements.*
- *Information management strategies lead to more efficient use of process control tools, requiring fewer tools to get the job done.*

TECHNOLOGY DRIVERS

The relentless drive toward smaller feature sizes should continue to fuel the demand for metrology and inspection tools over the next decade. Figure 15 summarizes the outlook for feature size reduction and the year in which 300 mm and 450 mm diameter wafers will first be used in volume production.

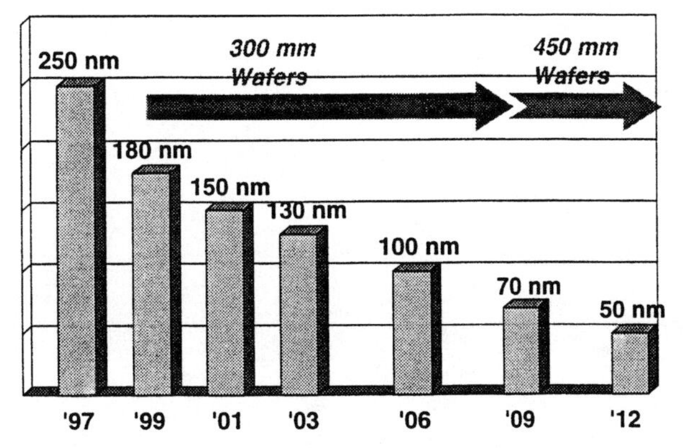

FIGURE 15. Year of first product shipment for successive technology generations. Source: SIA Roadmap (1998).

Figure 16 illustrates the projected rate of technology conversion of Intel Corporation's manufacturing capacity between 1996 and 2001. Again, the process control requirements associated with these successive technology generations should result in a demand for more advanced metrology and inspection tools, as well as the more rapid adoption of the integrated metrology solutions (*on-line* and *in-situ*) described above.

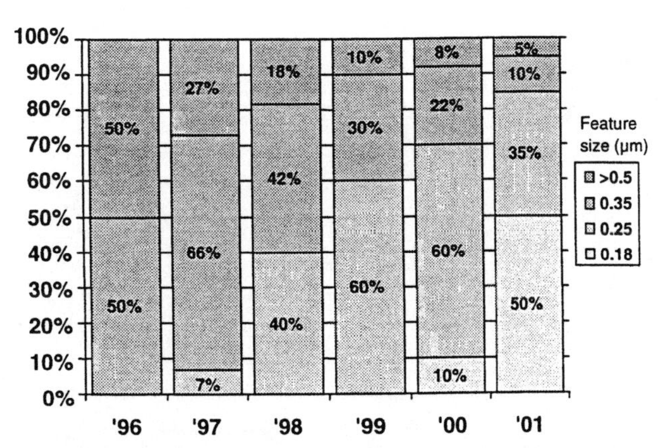

FIGURE 16. Intel's forecasted capacity by feature size suggests no slackening in the demand for advanced metrology and inspection equipment. Source: WaferNews.

The movement to 300 mm wafers has been slowed by the acceleration of feature size "shrinks" which has extended the life cycle of existing 200 mm lines. This situation has challenged the equipment industry, since the investment in new platforms must be absorbed over a longer time period before any significant sales revenue can be realized.

For metrology and inspection tools, this has perhaps been less of a concern than for deposition, etch and lithography systems which require substantial modification in order to achieve the throughput, uniformity, repeatability and cost-of-ownership objectives established by semiconductor chip manufacturers.

Figure 17 indicates that 300 mm wafers will still represent less than 10% of the silicon area devoted to chip manufacturing in 2002. Figure 18 shows the corresponding number of 300 mm wafers associated with this utilization, indicating that there will be on the order of just fifteen fully operational 300 mm fabs in 2002 (assuming 20,000 wafer starts per month).

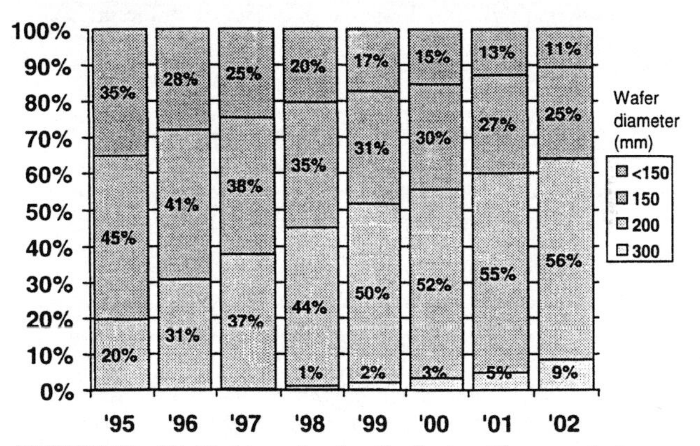

FIGURE 17. Worldwide wafer size distribution (% square inches by diameter) shows gradual shift to 300 mm. Source: Dataquest.

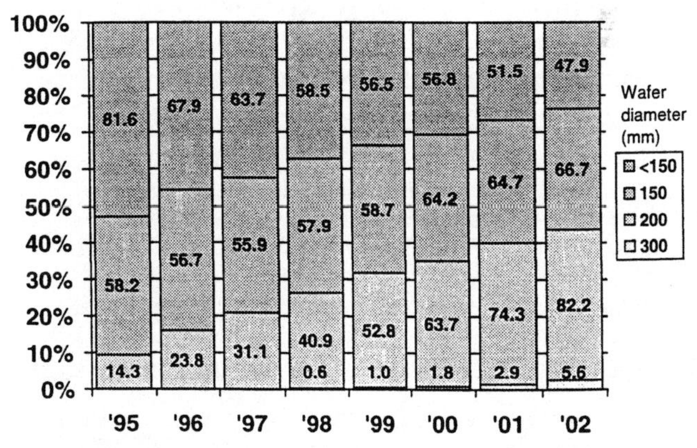

FIGURE 18. Worldwide wafer usage (millions of wafer starts/yr) shows 300 mm wafer consumption still small in 2002. Source: Dataquest.

CONCLUSION

The metrology and inspection capital equipment market is forecasted to grow to $4.5 billion in the year 2002, providing a significant opportunity for technologically and financially strong companies. At the same time, market consolidation will continue due to the inability of small suppliers to absorb the costs of world-wide support, the capital-intensive conversion to 300 mm wafers, and the inclination of chip manufacturers to reduce their supplier base.

The author believes that innovative metrology solutions will continue to emerge over the next five years. Many of these will come from small companies and research laboratories, whose efforts must continue to be encouraged through joint development programs and capital infusions to assure their timely passage from the laboratory to the production line.

In particular, in-situ solutions will become a more critical factor in maintaining the historical level of manufacturing productivity that has been the basis for decreases in the price of DRAM and microprocessor chips. However, there is considerable commercial complexity involved with a transition to integrated metrology, which can only be overcome by a high degree of cooperation among suppliers, integrators and customers. Lack of cooperation, in the interests of maintaining propriety advantages, will slow the rate of adoption and may ultimately limit the growth potential of the entire industry.

ACKNOWLEDGEMENTS

The author wishes to acknowledges the support of Klaus Rinnen and Clark Fuhs of Dataquest, as well as the encouragement of Alain Diebold of Sematech and James Greed of VLSI Standards in updating and extending the current presentation beyond those given previously at SEMI's ISS Meeting (January 1996) and the SIA's Metrology TWG Roadmap Meeting (Summer 1997).

REFERENCES

1. Rose, D., "Cost Benefit of Process Metrology," *Semiconductor Characterization: Present Status and Future Needs*, W. M. Bullis, D. G. Seiler and A. C. Diebold, eds. (American Institute of Physics Press, 1996).
2. Porter, C., Laser, A., Hertle, J., Pandey, P., Erickson, M., and Shah, S., "Improving Furnaces with Model-Based Temperature Control," *Solid State Technology*, November, 1996, p. 119.

Effect of Technology Scaling on MOS Electrical Characterization

Mohsen Alavi , Rafael Rios

Intel Corporation, Component Technology Development, 5200 NE Elam Young Parkway, Hillsboro OR 97124

Some shortcomings of the standard MOS device models and algorithms used for parametric characterization of the device in deep sub-micron technologies are discussed. These include field and channel length dependent mobility, un-pinned surface potential due to operation in weak inversion, Poly-depletion, quantum mechanical (QM) effects, and carrier velocity saturation. Enhancements to the basic model are proposed to account for these higher order effects to improve the accuracy of extracted parameters. Issues and challenges for evaluation of gate oxide parameters are also discussed. The effect of direct tunneling induced gate leakage is the most significant hurdle to overcome there. This leakage impacts many characterization methods as well as reliability evaluation methodologies. Scaling related issues in interconnect evaluation are also discussed. There, the main issues are in coping with narrow, high aspect ratio lines and spaces with varying thickness due to CMP.

1. INTRODUCTION

Electrical testing and characterization (E-test) is one of the most effective means of evaluating an integrated circuit fabrication process. It is widely used during process development in conjunction with test chips for evaluating intrinsic and yield performance of various process modules and during manufacturing for wafer disposition using scribe-line test structures. It lends itself well to automation and high volume data collection and is generally non-destructive. Therefore, development and improvement of electrical test structures and algorithms receives a lot of effort and attention in integrated circuit fabrication.

As technology scaling results in smaller device dimensions and increased level of density, new physical phenomenon become significant which impact E-test in many areas and require new solutions both in terms of structures and algorithms. These problems are not unique to any section of the process and can be seen in both the front half (transistor) and the back half (interconnect) areas. This paper points out some of these problem areas and offers potential solutions.

2. TRANSISTOR CHARACTERIZATION

MOS testing is one of the areas which encounters numerous difficulties in the deep sub-micron regime. Usually, the desired outcome from MOS evaluation is extraction of transistor parameters which provide physical insight about the device design and behavior. Such insight is subsequently used to optimize the device for improved performance.

Effective channel length L_e, effective channel width W_e, external source-drain series resistance R_{ext}, effective channel mobility μ, and threshold voltage V_t, are of special interest since they are key parameters for modeling, predicting, and bench-marking device performance. Readily measured electrical responses from the transistor are its I-V and C-V characteristics in various bias conditions and the challenge in all e-test algorithms is to come up with consistent and robust ways of inferring and evaluating the above parameters from these characteristics. The success of such algorithms, therefore, is strongly dependent upon the accuracy and simplicity of the model used to relate the device I-V and C-V to its parameters.

Almost every method for MOSFET characterization proposed in the literature [1] is based on the well known strong inversion drain current expression in the linear region of operation as given below:

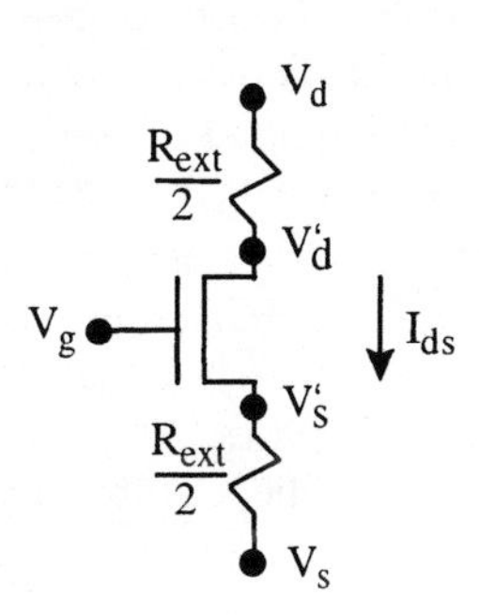

CP449, *Characterization and Metrology for ULSI Technology: 1998 International Conference*
edited by D. G. Seiler, A. C. Diebold, W. M. Bullis, T. J. Shaffner, R. McDonald, and E. J. Walters

$$I_{ds} = \mu \cdot C_{ox} \cdot \frac{W_e}{L_e} \cdot \left[\left(V'_{gs} - V_t \right) - \frac{V'_{ds}}{2} \right] \cdot V'_{ds}$$
$$V'_{ds} = V_{ds} - R_{ext} I_{ds} \qquad (1)$$
$$V'_{gs} = V_{gs} - \frac{R_{ext}}{2} I_{ds}$$

Here, C_{ox} is the gate capacitance per unit area, R_{ext} is total external resistance in series with drain and source (1/2 per side for a symmetric device) and the voltages V'_d and V'_s refer to the drain and source of the intrinsic device, respectively.

To extract the parameters in (1), algebraic manipulation is used to show linearity of some quantity with respect to applied biases or transistor sizes. For example, if (1) is valid, then plots of the total resistance, $R_t = V_{ds}/I_{ds}$, vs. gate length for constant gate drive, $V_{gt} = V_{gs} - V_t$, should fall on straight lines intercepting at one point. This intercept thus yields R_{ext} and ΔL, the difference between the drawn and effective channel length. Based on this assumption, L_e and R_{ext} are extracted by fitting a least square line to the measured R_t vs. L_{drawn} data at constant V_{gt}.

Unfortunately, as technologies advance and transistor dimensions are made smaller, the validity of Eq. (1) is broken. A well known shortcoming of (1) is the assumption of a constant, bias independent R_{ext}, which is violated for technologies with lightly doped drain extensions, LDD. Several methods have been proposed to deal with this problem [2,3].

Recent technologies result in even more severe limitations of (1), where the quantities μ, C_{ox}, and V_t can no longer be assumed to be constants. In the next sections, we shall analyze some of the shortcomings of assumptions leading to expression (1) and the impact on the extracted parameters.

2.1 The Constant Mobility Assumption.

The channel mobility μ is assumed to be a constant in most extraction methods. However, it is very well known that the linear region channel mobility has a strong dependency on the vertical field, or equivalently, on the applied gate bias. Some techniques have been proposed to take this dependency into account by assuming that μ is a known function of the applied gate drive $V_{gt} = V_{gs} - V_t$ [4, 5]. These methods continue to assume that μ is independent of device size at a given gate drive.

In deep sub-micron devices, in addition to a gate bias dependency, effective mobility of carriers in the channel is no longer the same between short and long channel devices due to channel length dependent doping concentration caused by defect enhanced diffusion near source and drain. Use of halo implants further increases this variation by modulating dopant concentration along the length of the channel, as illustrated in Fig. 1. This phenomenon manifests itself in a 'hump' in the relation between V_t and channel length which is sometimes referred to as the 'reverse short channel effect' (see Fig. 2).

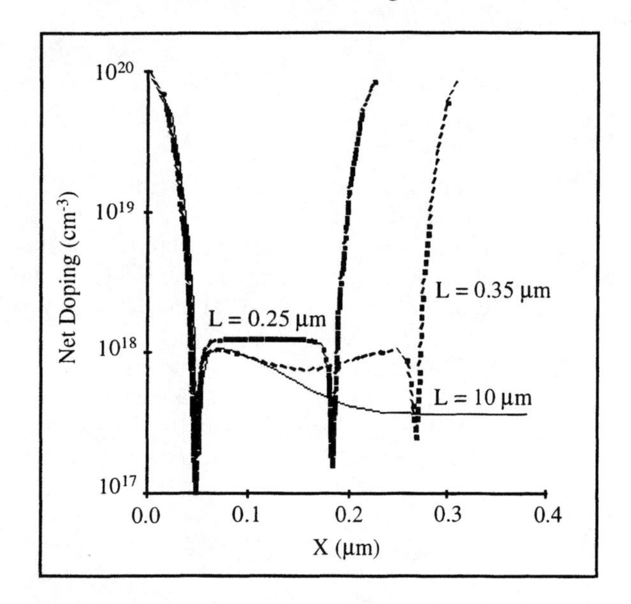

FIGURE 1. Simulated surface doping profiles vs. depth under the channel for devices with halo implants. The channel doping increases significantly as the devices get shorter.

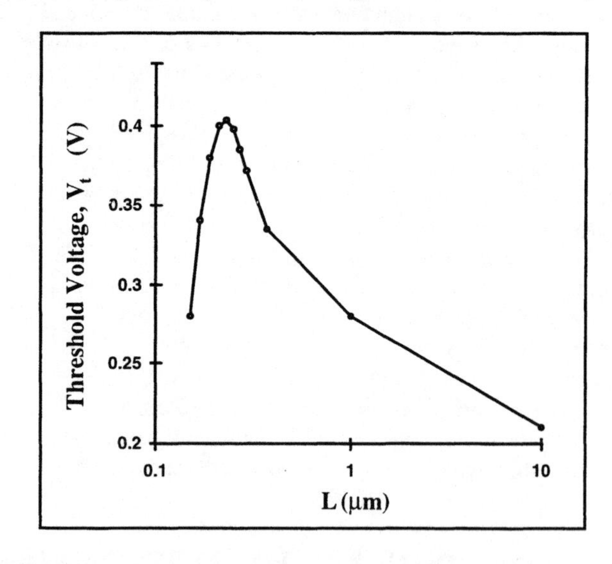

FIGURE 2. V_t vs. channel length for NMOS devices from a process flow which incorporates a channel halo implant showing a 'hump' as channel length becomes small.

In addition, the required channel doping for modern technologies is now in the 10^{18} cm^{-3} range. At this doping level the contribution of impurity scattering to μ cannot be neglected. Moreover, power supply scaling coupled with the maintenance of a relatively high V_t to ensure adequately low off state leakage is forcing the device to operate near the weak inversion regime [6], where the mobility degradation due to impurity scattering tends to dominate.

Fig. 3 illustrates the effect of increased channel doping on channel mobility. Note that assuming that all devices have a channel mobility equal to that of the long channel device leads to over-prediction of short channel mobility. If the extraction procedure does not take this effect into account, the error in μ would cause an error in the extracted L_e, making it appear larger than the actual value.

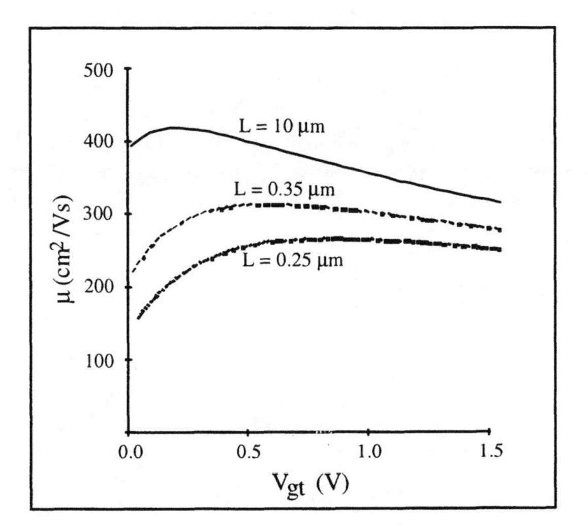

FIGURE 3. Channel mobility as a function of gate drive for devices with halo implants showing that channel mobility is reduced with decreasing channel length.

2.2 The Pinned Surface Potential Assumption

The derivation of (1) assumes that in the strong inversion, the surface potential ϕ_s is pinned to a constant value (see Fig. 4). This leads to the appearance of a constant quantity, the threshold voltage V_t, independent of applied gate bias. Historically, the concept of a threshold voltage and of the gate drive V_{gt} have been tightly coupled to the understanding of MOSFET operation. As discussed above, however, modern devices operate near the weak inversion regime and equation (1) looses accuracy.
Derivation of the drain current equation without the pinned surface potential assumption leads to [7]:

$$I_{ds} = \mu \cdot C_{ox} \cdot \frac{W_e}{L_e} \cdot \left[\left(V_{gs}^{'} - V_0 \right) - \alpha \frac{V_{ds}^{'}}{2} \right] \cdot V_{ds}^{'} \qquad (2)$$

which is valid in weak and strong inversion regimes. The quantity V_0 is dependent on the applied biases and is approximated by:

$$V_0 = V_{fb} + \phi_s + V_{bs} + \gamma \sqrt{\phi_s} \qquad (3)$$

Here, α is the bulk charge factor, V_{fb} is the flat-band voltage and γ is the body factor which is dependent on channel doping and gate oxide thickness [8]. The surface potential ϕ_s is determined by solving the implicit equation:

$$\frac{\left(V_{gb} - V_{fb} - \phi_s \right)^2}{\gamma^2} = \phi_s + \frac{kT}{q} \exp\left(-q \frac{2\phi_f}{kT} \right) \cdot \left\{ \exp\left(q \frac{\phi_s}{kT} \right) - 1 \right\} \qquad (4)$$

which is valid through the depletion, weak inversion and strong inversion regimes. An evaluation of the effect of non-pinned surface potential on gate drive for a modern device will be shown after the following discussion on two other important gate bias dependent effects.

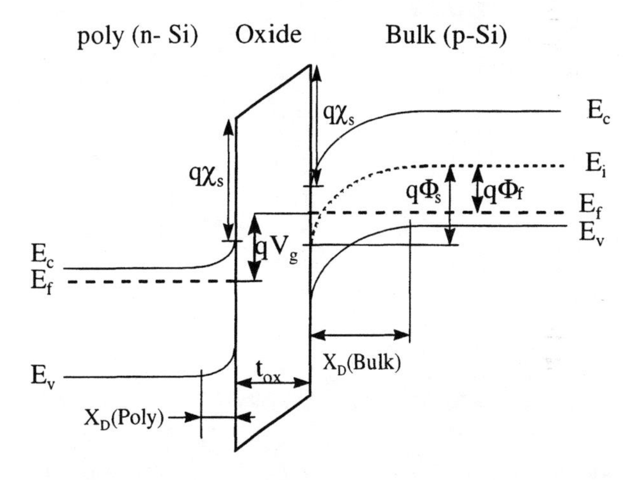

FIGURE 4. Energy band diagram of an NMOS structure biased in inversion.

2.3 Poly-Depletion and Quantum Mechanical (QM) Effects

Another assumption used in the derivation of (1) is that the MOS device is a perfect parallel plate capacitor, where the induced gate and channel inversion charges are infinitely thin sheets right at the two oxide interfaces. However, when the device is operating in inversion, the polysilicon side is depleted and the gate charge is distributed in a finite volume (see Fig. 4). Similarly, due to quantum mechanical effects, the channel inversion charge occupies a finite volume and its centroid is displaced from the interface. In modern devices, the thickness of both the poly-depletion charge and the channel inversion charge are of the same order as the gate oxide thickness, which violates the assumption of infinitely thin sheets of charges at the interfaces.

Fig. 5 depicts the simulated charge distribution in a modern device biased in inversion, both with the classical solution and with QM corrections [9]. The effective or "electrical" oxide thickness is larger than the physical one and is given by the sum of the poly and channel charge centroids multiplied by the ratio of oxide and Si dielectric constants:

$$t_{oxe} = t_{ox} + \frac{\varepsilon_{ox}}{\varepsilon_{Si}}\left(c_P + c_{Si}\right) \quad (5)$$

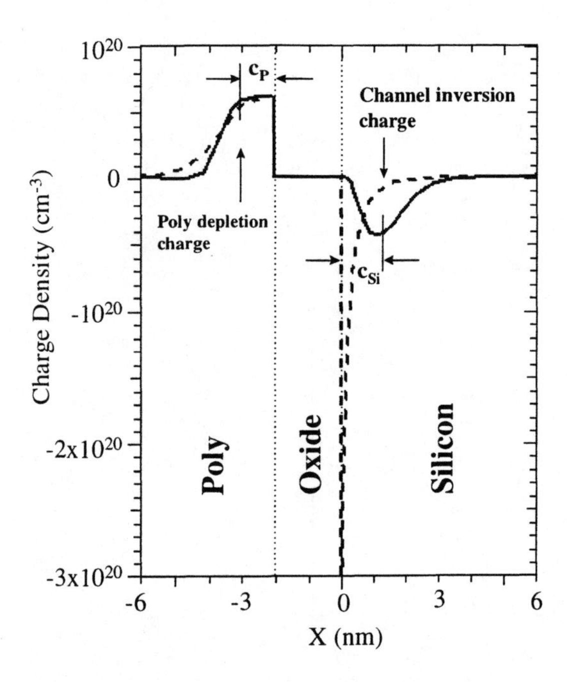

FIGURE 5. Simulated charge distribution in a MOS system with classical solution (dashed line) and QM solution (solid line). The charge centroid displacement from the interface is of the same order as the oxide thickness on both sides.

Since these charge centroids vary with applied bias, gate capacitance per unit area C_{ox} (= ε_{ox} / t_{oxe}) in (1) and (2) is dependent not only on the device physical oxide thickness, but also on the poly layer doping level, and the applied gate bias.

Note that Eq. (2) automatically removes the charge sheet assumption on the channel side, since expression (4) comes from a direct integration of the Poisson equation. QM effects, that tend to displace the centroid of inversion charge away from the interface, can be included in the solution of Eq. (4) [10]. If Eq. (4) is corrected to take into account QM effects, it has been shown that the surface potential in the inversion regime is larger than that given by the classical solution [10], and that, again, the effect is dependent on the gate bias. Therefore, in addition to decreasing C_{ox}, QM effects also introduce a gate bias dependent loss in gate drive.

Similar to the above, in addition to affecting C_{ox}, poly-depletion charge also causes a drop in gate potential which is yet another gate bias dependent effect on gate drive [11]. The loss in gate drive becomes more pronounced as the gate bias is increased. Fig. 6 illustrates the contributions of all the effects mentioned above on the device gate drive for a modern device. The dashed line corresponds to the standard gate drive of Eq. (1) with a constant V_t. Note that after a non-pinned surface potential solution, and accounting for poly-depletion and QM effects, in inversion, gate drive at any given gate bias is significantly smaller than that given by the standard V_{gt}. Therefore, Eq. (1) uses an over-estimate of the gate drive and any measurement method based on (1), in turn, would over-estimate the value of L_e.

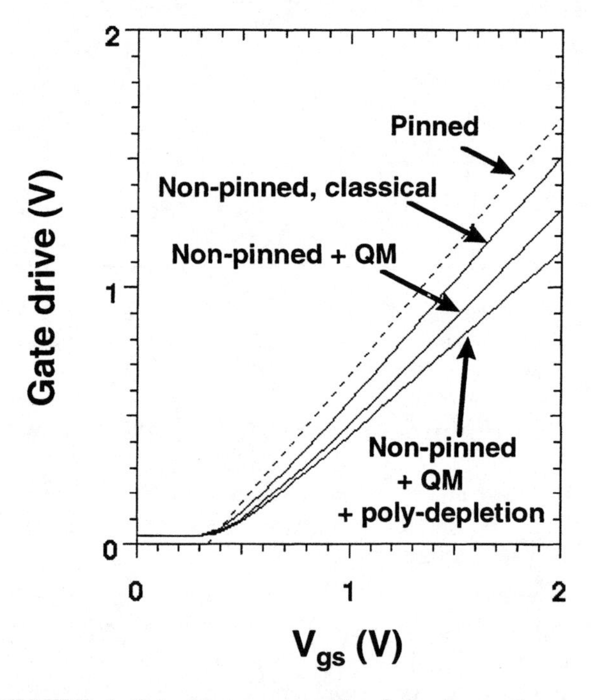

FIGURE 6. Gate drive vs. gate bias for various phenomena incorporated in the I-V model. The dashed line corresponds to the classical V_{gs} - V_t, that results from the pinned surface potential assumption.

2.4 The Velocity Saturation Effect

As pointed out previously, normally, L_e, R_{ext}, and μ extraction methods rely on measured I-V characteristics in the linear region of operation. Therefore, carrier velocity is assumed to be well below its saturation value. However, the velocity-field relation [12] shows that as the electric field is increased, carrier velocity starts to deviate significantly from what is predicted by a linear relation. Therefore, for very small channel lengths, mobility degrades not only due to the vertical field previously discussed, but also, due to the lateral field in spite of the small value of drain bias. Accounting for this phenomenon, the effective channel mobility is approximated by:

$$\mu_e = \frac{2\mu}{1+\left[1+\left(\frac{2\mu V_{ds}}{v_{sat} L_e}\right)^n\right]^{1/n}} \qquad (6)$$

where v_{sat} is the saturation velocity, and n is 2 for electrons and 1 for holes [13].

Expression (5) indicates that for PMOS devices, the velocity saturation effect can no longer be ignored in linear region characterization. For example, assuming an L_e of 0.1 μm (representative of technologies under development), a low field mobility of 120 cm^2/Vs, a saturation velocity of $6x10^6$ cm/s, and a drain bias of 50 mV (the usual value used for measurements), it is observed that μ_e of PMOS devices is reduced by about 10% due to velocity saturation which is the source of significant error if not accounted for.

2.5 Characterization Algorithms

The discussion in previous sections indicated that MOS device characterization becomes increasingly more complex with technology scaling. This is the result of increased significance of previously negligible physical phenomena (poly-depletion, QM effects, etc.) as well as power supply constrains resulting in device operation near the weak inversion regime. Consequently, validity of the simple drain current expression (1) diminishes and parametric extraction methods based on this equation tend to give rise to invalid and sometimes misleading results.

Some of the required model corrections to (1) have been described above. Unfortunately, these models lead to more complex expressions and the simplicity of Eq. (1) is lost. Therefore, simple linear regression methods enabled by the linear relationships based on (1) are no longer possible.

To use Eq. (2), which incorporates additional physical phenomena, the implicit surface potential Eq. (4) must be solved iteratively. Thus, algorithms for extracting device parameters with Eq. (2) result in lengthy and fairly complicated computer code. This, however, may not pose a serious limitation since, given the power of today's computers, it is possible to run more complex algorithms based on more accurate models simultaneously with data acquisition by the measurement equipment.

Alternatively, in cases where changes in the fabrication process are limited (such as scribe-line testing during manufacturing for process control), empirical corrections to the simple methods may be considered. Such corrections can be calibrated against the more elaborate algorithms or against physical measurements.

3. GATE CHARACTERIZATION

Gate characterization problems arise due to aggressive scaling of the gate oxide thickness and failure of conventional theory to properly model charge separation and distribution in the MOS structure. The main effects, namely, QM effects and poly depletion have been discussed in detail above.

Parameters of interest in the MOS stack include dielectric thickness, dielectric constant, poly depletion, oxide charge, bulk mobility, and bulk carrier lifetime. Most of these parameters are evaluated from the C-V characteristics [14]. Additionally, there are a number of diagnostic methods for evaluating these parameters such as charge pumping (interface states) [15], split CV (mobility) [16], and Ct (carrier lifetime) [17].

Gate oxide scaling impacts these measurements in two ways. First, ultra-thin oxides can no longer withstand a sufficiently high bias to reach deep accumulation. This is the preferred region for evaluation of gate oxide thickness where the effect of poly depletion is eliminated and QM effects are minimized. Fig. 7 shows an example of NMOS (p-well) C-V characteristics where even at -3V of gate bias, there is considerable slope in the curve. As the gate thickness scales, it becomes increasingly difficult to perform measurements at that point without inflicting device degradation or damage.

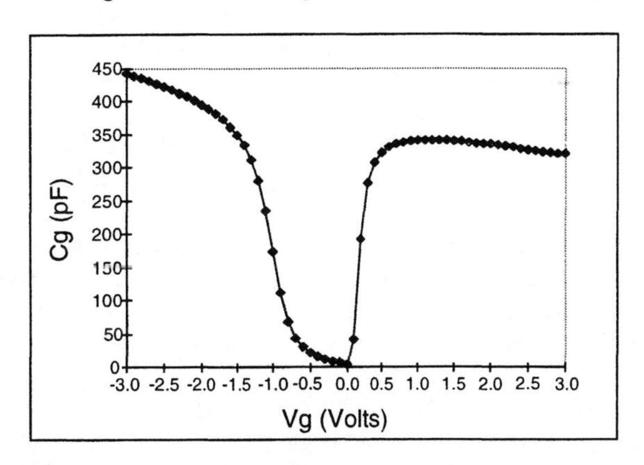

FIGURE 7. Typical C-V characteristics of an NMOS (P-well) capacitor with ultra-thin (<45A electrical) oxide showing a significant slope in the curve in the accumulation region.

The second problem encountered from oxide scaling is increased gate leakage due to direct tunneling. This problem happens later in the course of scaling. However, it has far reaching impact and compromises many of the characterization algorithms. Figure 8 shows typical J-V characteristics of ultra-thin oxides starting from an electrical thickness of 60A to increasingly thinner oxides. As is evident from this figure, the increase in leakage is large and far outweighs any reduction in operating voltage.

Any measurement method relying on a well insulating gate will suffer from this phenomenon. For example, the charge pumping method which relies on measurement of very small amounts of bulk current (pA range) and the Ct method which relies on variation of capacitance in deep depletion due to carrier generation are first to fail.

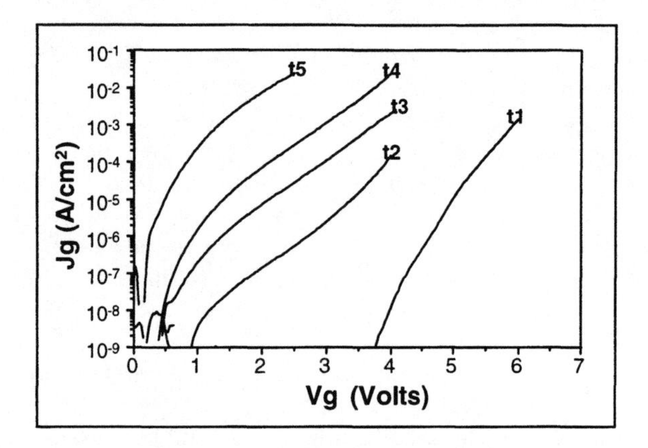

FIGURE 8. Typical J-V characteristics of ultra-thin oxides. Curve t1 corresponds to an electrical oxide thickness of 60A. Curves t2 to t5 correspond to 45A and thinner oxides.

With sufficiently high leakage, even the standard CV characterization falls victim. Fig. 9 shows an example of C-V characteristics of NMOS (P-well) capacitors with different oxide thicknesses. A grounded substrate here gives only the component of capacitance due to inversion (referred to as split C-V). As seen in this figure, for very thin oxides in which leakage is sufficiently large, the measured capacitance starts to decrease in inversion with increasing gate bias. This problem vanishes on the same material in smaller area capacitors with decreased source-drain spacing.

The need for measuring capacitance vs. bias accurately in a very leaky structure (i.e. high dissipation factor) is a new difficulty set forth for the measurement instrument. However, in addition to that, it is not hard to imagine a scenario in which the leakage is high enough and the source-drain diffusions (source of carriers for inversion charge) are placed far enough that charge in the inversion layer is not completely maintained at increased gate bias, thereby, resulting in the behavior shown in Fig. 9.

Characterization of reliability of the gate insulator is another area which is significantly impacted by scaling. As direct tunneling becomes a more significant portion of the oxide current, charge to breakdown (QBD) tests are affected and test results become more difficult to interpret. The problem is twofold. First, since direct tunneling current does not generate damage in the same manner as the Fowler-Nordheim current, QBD results for thinner oxide are more optimistic [18]. Therefore, correlation between such tests and constant voltage stress seen by the device during normal operation may be different.

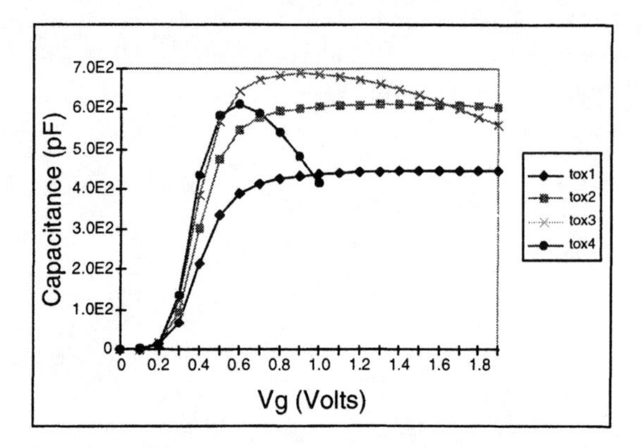

FIGURE 9. Split C-V (grounded body) characteristics of NMOS (P-well) capacitors with different oxide thicknesses. Curve tox1 corresponds to an electrical oxide thickness of 45A. Curves tox2 to tox4 correspond to increasingly thinner oxides

The second and more significant problem is that thinner oxides do not exhibit the 'hard breakdown' behavior which has been traditionally observed after current stress [19]. An example of this phenomenon is shown in Fig. 10. The 'soft breakdown' observed in the scaled oxide is more difficult to detect during test and also more difficult to relate to any possible functional failure of a real circuit.

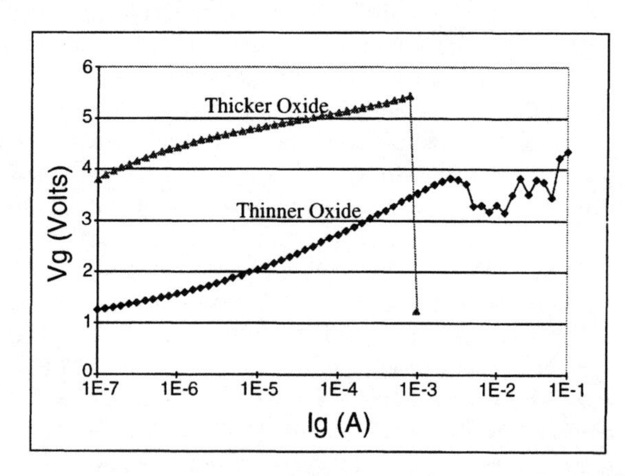

FIGURE 10. Breakdown characteristics from a QBD test with stepped current. The graph denoted by 'thicker oxide' corresponds to an electrical oxide thickness of 45A. As evident here, thinner oxide does not exhibit 'hard breakdown'.

4. INTERCONNECT CHARACTERIZATION

Scaling related problems also exist in interconnect testing, but, to a lesser complexity due to simpler physical properties. An example is algorithms to measure interconnect sheet resistance and line width. The classical approach is to use two equations obtained from I-V of lines with different dimensions. Simultaneous solution of these equations, in turn, provide line width and sheet resistance.

Narrow lines with high aspect ratios which are typical in advanced processes, however, don't necessarily have the same sheet resistance or print/etch bias as wide lines. So, erroneous results arise. Alternative single structure test do exist [20] but are often prone to noise when small values of sheet resistance are being measured.

Another challenge in interconnect characterization is evaluation of the dielectric constant of insulating layers between interconnect, since adopting dielectric materials with reduced dielectric constant compared to SiO_2 is being widely pursued in the industry. While interconnect capacitance can be readily measured at the desired frequency, difficulty arises since the use of CMP makes the value of dielectric thickness uncertain. The challenge, therefore, is to find electrical means of separating the value of dielectric thickness from that of dielectric constant.

5. SUMMARY

Some shortcomings of the basic standard device model and algorithms used for parametric characterization of MOS devices in the deep sub-micron region were discussed. These include field and channel length dependent mobility (specially in halo doped devices), an un-pinned surface potential due to operation in weak inversion, Poly-depletion, quantum mechanical (QM) effects, and carrier velocity saturation. Without proper modeling of these phenomena, extraction of parameters such as effective channel length L_e, external source-drain series resistance R_{ext}, and effective channel mobility μ may give unphysical results. Models and methods for inclusion these phenomena have been proposed.

Similar issues and challenges for evaluation of the gate oxide parameters such as thickness and capacitance have been discussed. The effect of direct tunneling induced gate leakage is the most significant hurdle to overcome there. This leakage impacts many gate oxide characterization methods as well as reliability characterization methodologies. Scaling related issues in interconnect evaluation have also been discussed. There, the main issues are measurements of interconnect resistance and capacitance from multiple structures while coping with narrow high aspect ratio lines and spaces with varying thickness due to CMP.

6. REFERENCES

1. K.K. Ng and J.R. Brews, "Measuring the effective channel length of MOSFETs," *IEEE Circuits and Devices*, Nov. 1990, pp. 33.
2. S. Biesemans, et al., "Accurate determination of Channel Length, Series Resistance and Junction Doping Profile for MOSFET Optimisation in Deep Submicron Technologies," VLSI Symposium Digest, 1996, pp. 166.
3. J. Otten, F. Klaasen, "Determination of Series Resistance Using One Single MOSFET," ESSDERC 1995, pp. 483.
4. F.H. De La Moneda, H.N. Kotecha, M. Shatzkes, "Measurement of MOSFET Constants", *IEEE Electron Device Letters*, Vol. EDL-3, No. 1, Jan. 1982, pp. 10.
5. Y. Taur, et al., "A New "Shift and Ratio Method for MOSFET Channel-Length Extraction," *IEEE Electron Device Letters*, Vol. EDL-13, pp. 267, May 1992.
6. M. Bohr and Y. El-Mansy, "Technology for Advanced High-Performance Microprocessors," *IEEE Transactions on Electron Devices", Vol.* ED-45, pp. 620, March 1993.
7. H. Park, et al., "A Charge Sheet Capacitance Model of Short Channel MOSFETs for SPICE," *Transactions on Computer Aided Design,* Vol. CAD-10, pp. 376, March 1991.
8. R.S. Muller, T.I. Kamins, *Device Electronics for Integrated Circuits 2nd Edition*, New York: Wiley, 1986, ch. 9, pp. 437.
9. R. Rios and N. Arora, "Determination of Ultra-Thin Gate Oxide Thicknesses for CMOS Structures Using Quantum Effects," IEDM Tech. Digest, pp. 613, Dec. 1994.
10. R. Rios et al., "A Physical Compact MOSFET Model, Including Quantum Mechanical Effects, for Statistical Circuit Design Applications," IEDM Tech. Digest, pp. 937, Dec. 1995.
11. R. Rios, et al., "An Analytical Polysilicon Depletion Effect Model for MOSFETs," *IEEE Electron Device Letters,* Vol. EDL-15, pp. 129, April 1994.
12. R. Coen et al., "Velocity of surface carriers in inversion layers of silicon," *Solid-State Electronics*, Vol. 23, pp. 35, 1980.
13. C. Canali, et al., "Electron and Hole Drift Velocity Measurements in Si and Their Empirical Relation to Electric Field and Temperature," *IEEE Transation on Electron Devices*, Vol. ED-22, pp. 1045, 1975.
14. E.H. Nicollian, J.R. Brews, *MOS (Metal Oxide Semiconductor) Physics and Technology,* New York: Wiley, 1982, ch. 12, pp. 581.
15. G. Groeseneken, et al, "A Reliable Approach to Charge-Pumping measurements in MOS transistors", *IEEE Transactions on Electron Devices", Vol.* ED-31, No. 1, pp. 42, Jan 1984.
16. J. Koomen, "Investigation of The MOST Channel Conductance in Weak Inversion", *Solid-State Electronics*, Vol. 16, pp. 801, 1973.
17. D.K. Schroder, J. Guldbery, "Interpretation of Surface and Bulk Effects Using the Pulsed MIS Capacitor", *Solid-State Electronics*, Vol. 14, pp. 1285, Dec. 1971.
18. M. Alavi, et al, "Effect of MOS Device Scaling on Process Induced Gate Charging", International Symp. on Plasma Process-Induced Damage P2ID Tech. Digest. pp.7, May 1997.
19. B.E. Weir et al, "Ultra-Thin Gate Dielectrics: They Break Down, but do they Fall", IEDM Tech. Digest, pp. 73, Dec. 1997.
20. M.G. Buehler, S.D. Grant, W.R. Thurber, "Bridge and van der Pauw Sheet Resistance Resistors for Characterizing the Line Width of Conducting Layers", *J. Electrochem. Soc.*, Vol. 125, No. 4, pp.650, April 1978.

Elements for Successful Sensor-Based Process Control {Integrated Metrology}

Stephanie Watts Butler

Texas Instruments, Silicon Technology Development, Dallas, TX 75265

Current productivity needs have stimulated development of alternative metrology, control, and equipment maintenance methods. Specifically, sensor applications provide the opportunity to increase productivity, tighten control, reduce scrap, and improve maintenance schedules and procedures. Past experience indicates a complete integrated solution must be provided for sensor-based control to be used successfully in production. In this paper, Integrated Metrology is proposed as the term for an integrated solution that will result in a successful application of sensors for process control. This paper defines and explores the perceived four elements of successful sensor applications: business needs, integration, components, and form. Based upon analysis of existing successful commercially available controllers, the necessary business factors have been determined to be strong, measurable industry-wide business needs whose solution is profitable and feasible. This paper examines why the key aspect of integration is the decision making process. A detailed discussion is provided of the components of most importance to sensor based control: decision-making methods, the 3R's of sensors, and connectivity. A metric for one of the R's (resolution) is proposed to allow focus on this important aspect of measurement. A form for these integrated components which synergistically partitions various aspects of control at the equipment and MES levels to efficiently achieve desired benefits is recommended.

ELEMENTS FOR SUCCESS

In the past few years, several semiconductor manufacturing companies have attempted to install sensor-based applications to improve productivity and control. In addition, several companies have formed to meet the needs of the industry by providing unique sensor-based solutions and applications assistance. The applications have varied in their level of success. An examination of the applications reveals that 4 elements are required for success. These elements are listed in Table 1. The term "Integrated Metrology" is proposed to signify the integration of all these components into a particular form that meets a business need. In other words, Integrated Metrology is a successful sensor-based controller.

TABLE 1. Elements of Successful Sensor Applications

- All Needed Components Addressed
- Integration of components
- Form of integrated components Which Meets Goal
- Required Business Factors Met

In this paper, each of the 4 elements will be presented. All the necessary components, such as hardware and software, will be noted with emphasis on the key ones relevant for sensor based control. To be able to explain the importance of certain components, a clearer definition of integration to meet business needs must be given first. The paper concludes with a proposal for the form that these integrated components must take and the benefits that will then be achieved. However, because of the importance of business factors for driving integration, business issues will be reviewed first.

REQUIRED BUSINESS FACTORS

The easiest route for integration of a sensor is for the process equipment vendors to do the integration. In order for this integration to occur, 4 business factors must be present. The business factors are described in Table 2, with examples for each factor. Only recently have all of these factors become present in the semiconductor industry. Having a strong industry-wide drive (business factor #1), such as the drive to shrink the die size, will provide the critical mass of effort required to solve a difficult control need. Identifying the need (business factor #2) is explicitly stated because if the goal is not understood by all it is unlikely that the sensor solution will fulfill that goal. In addition, an understanding of how the process improvement made will impact the final device must already exist if the desired goal is a final device result, such as yield or speed. Finally, it is important that the impact upon the goal is measurable so that progress toward success can be tracked. The importance of measurable and tractable metrics will be discussed again in the components section. The importance of feasibility of integration (business factor #3) becomes more obvious as some of the more intricate controllers are incorporated. The need for profitability (business factor #4) is obvious.

CP449, *Characterization and Metrology for ULSI Technology: 1998 International Conference*
edited by D. G. Seiler, A. C. Diebold, W. M. Bullis, T. J. Shaffner, R. McDonald, and E. J. Walters

TABLE 2. Business Factors for Integration to Occur

1) Strong Industry-wide Business Needs (Drivers) of
   - IC manufacturer or Equipment supplier
   - 300mm
   - Shrinking dimensions (e.g., linewidth)
   - Shrinking die size
   - Environmental, safety, and health (ESH)
2) The Business Need (Driver) is Identified & Measurable
   - Decreased Scrap
   - Reduced Pilot Usage
   - Improved Yield
3) Integration must be feasible
   - Hardware/software changes spawned by 300mm
   - Availability of commercial sensors & sensor bus
   - Viability of software suppliers
   - Emergence of controller consultants/trainers
4) Integration must be profitable
   - Increased productivity
   - Creates a competitive edge

As viewed from a supplier's standpoint, business drivers can also be grouped into the three categories of process, cost of ownership (CoO), and market driven. A process business driver is the inability of the process to be capable without the controller. A cost of ownership business driver is to create a competitive advantage by making the process more productive, such as by increasing the throughput, through the use of a controller. A market driver is an opportunity filled by a third party control or sensor supplier. These markets are generally related to CoO opportunities, although they can meet process capability needs. To demonstrate the importance of the business driver and profit aspects, Table 3 lists current controllers available on the market and their corresponding drivers. All these controllers also contain the other 3 elements listed in Table 1.

TABLE 3. Controllers Available Now & Their Drivers

1) Process needs driven
   - Pressure control
   - Uniformity control for RTP & litho bake plate
   - Etch endpointing for small open area
   - RGA on sputter etch for contamination control
2) Cost of Ownership (CoO) advantage
   - Small batch fast ramp furnace control for reduced cycle time
   - Retrofit model based controllers for older furnaces
   - CMP On-Line Thickness Metrology*
3) Market driven
   - Equipment monitoring systems, such as Brookside; Triant; Semy; others
   - CMP On-Line Thickness Metrology*

*While CMP On-line Thickness Metrology is not truly "integrated metrology" as defined by this paper, it is believed that it will be soon

Except for the equipment monitoring systems, the other controllers listed meet the business needs of the process equipment vendor. While some of these controllers might have originally been driven by IC makers or control/sensor suppliers, their widespread use only occurred after they were marketed by the equipment supplier. Note that for On-line CMP thickness metrology, both routes are currently occurring, by a third party supplier and by the process vendors. The equipment monitoring systems still continue to be marketed directly by the software supplier, but many process vendors are now looking into these systems due to the strength of the market.

Another driver for sensor based control is the expense of the traditional equipment vendor approach to maintenance. This approach assumes a single process, but a single process does not exist in the production environment. The aging of the machine and the impact of maintenance are ignored, i.e., the chamber state is assumed not to change. In addition, the wafer state itself may impact the process. Thus, if the equipment is used to manufacture more than one device or for more than one step in the flow, even with the same recipe, the equipment will function differently. Several of the controllers listed in Table 3 are able to assist the vendor in maintaining their processes more cost effectively.

Besides the control drivers listed above, there are several opportunities for improvement for a process vendor by using control. Table 4 lists these opportunities. Some of these opportunities alone may be drivers enough for the vendor to add control, or they may be included as the vendor adds control for other reasons, such as process needs or field support issues.

Table 4. Future Opportunities & Drivers

- Measurement between steps in multiple-step single wafer processing
  - Only Chance for Measurement is With Sensors
- Start-up/shut-down control
  - Importance for plasma damage just beginning to be addressed
- Improved steady state control
  - Correlation of yield/faults to noisy behavior suggest tighter real-time steady state control may result in tighter wafer results
- Integrate process,hardware,and control design
  - Controllability and observability determined by hardware and process not by control system
- Better sensor for control & monitoring
  - Current sensor used in real-time control loop may not be best variable to control.
- Machine process checks without a wafer
  - Characterize machine and process not including wafer to allow checks during idle times

As the control needs of the processes and field service increase, and market opportunities become apparent, the process equipment vendor will continue to expand the sensor based control applications on their equipment. Thus, many of the future suppliers of sensor applications will also be the future users.

INTEGRATED METROLOGY

There are several types of measurement tools. To contrast sensors with stand-alone metrology, Table 5 classifies various measurement tools.

TABLE 5. Measurement Tool Classification in 2 Categories: Stand-Alone and Sensor (? = No Common Term)

Stand-Alone Measurement Tool	
Off-Line:	Outside of the Fab, Usually Destructive or Contaminating
At-Line:	In the Fab, Monitor Wafers Since Unable to Measure Pattern Wafers, Destructive, or Contaminating
In-Line:	Is or Can Be Part of Routing, Can Measure Pattern Wafers
Sensor (On Process Equipment Measurement Tool)	
On-Line:	Measurement Tool Integrated with Process Equipment, But Not Able to Measure During Wafer Processing
In Situ:	Measures During Processing (Wafer, Process, or Equipment State)

Table 5 only defines where the metrology is physically located and what it may measure, which is different from how the data may be used. Table 6 provides a list of 3 items that prove that Integrated Metrology has been achieved as defined in this paper. Table 6 shows the emphasis is on how the data is used. For sensors, this requires heavy software and hardware integration.

TABLE 6. Proof of Integrated Metrology

- Equipment Is Not Used Unless Sensor is Functioning
- Sensor Hardware, Software, And Computational Hardware Integrated
- Operational Procedure Determined By Analysis Of Sensor Data

Thus, the contents of Table 6 define Integrated Metrology as the use of measurement data, especially sensor data, to decide how to operate the equipment. Unless the operational procedure is dependent upon the sensor, the equipment will be allowed to operate without the sensor functioning. Ironically, the operational procedure will never become dependent upon the sensor until it is integrated, but the sensor will never become completely and reliably integrated until it is used to determine the operating procedure of the processing equipment. Thus, a conscious decision must be made to create a successful integrated solution. It is a business need that will cause this decision to be made. Some of the business needs that drive this decision were presented in the first section. More business needs will be presented in the last section as well.

Decision Making

To provide a clearer idea of what it takes to have the operational procedure determined by analysis of the sensor data, i.e., make a decision using sensor data, Figure 1 provides a graphical explanation of the steps involved in decision making. Note the emphasis of converting data into information and the use of this information to decide upon an action. Examples of data would be an optical emission spectrum trace from an etcher or the lamp power trace from a rapid thermal processor. Examples of information are "the etch rate is 2Å/second less than desired" or "the temperature is unusually high". Examples of actions are "increase the time by 2 seconds" or "fix the power supply to lamp number 2."

The consequences of only converting data to information without providing an action to take will now be examined to show how it will result in a sensor application failure. If the sensor based controller simply alerts the engineer and operator to a process failure without telling them how to fix the problem, then the engineer or operator is likely to view the controller as an efficient shrew which alerts everyone else to how bad the engineer and operator are. Due to the number of machines for which a given operator or engineer is responsible, the number of alarms will be astronomically high. This scenario of frequent alarms with little assistance increases the complexity of the engineer's or operator's job. In addition, most people will not be willing or able to work through improving the sensor and controller reliability as will be initially needed if the final result is increased highlighting of problems with no assistance to solve them. The final result will be that people will "fix" their problem by turning off the controller.

For sensors, due to the volumes and volumes of data usually measured, extracting information from the signals which can then be used to decide what actions to take is one of the greatest challenges. In addition to volume, sensor data may be confounded by extreme aging of sensor components, such as viewport windows, or be influenced by factors unrelated to the process. The result is difficulty in creating high quality information that is effective for use in deciding what action to take.

COMPONENTS

To achieve the integration stressed above, many components need to be considered. The Ishikawa

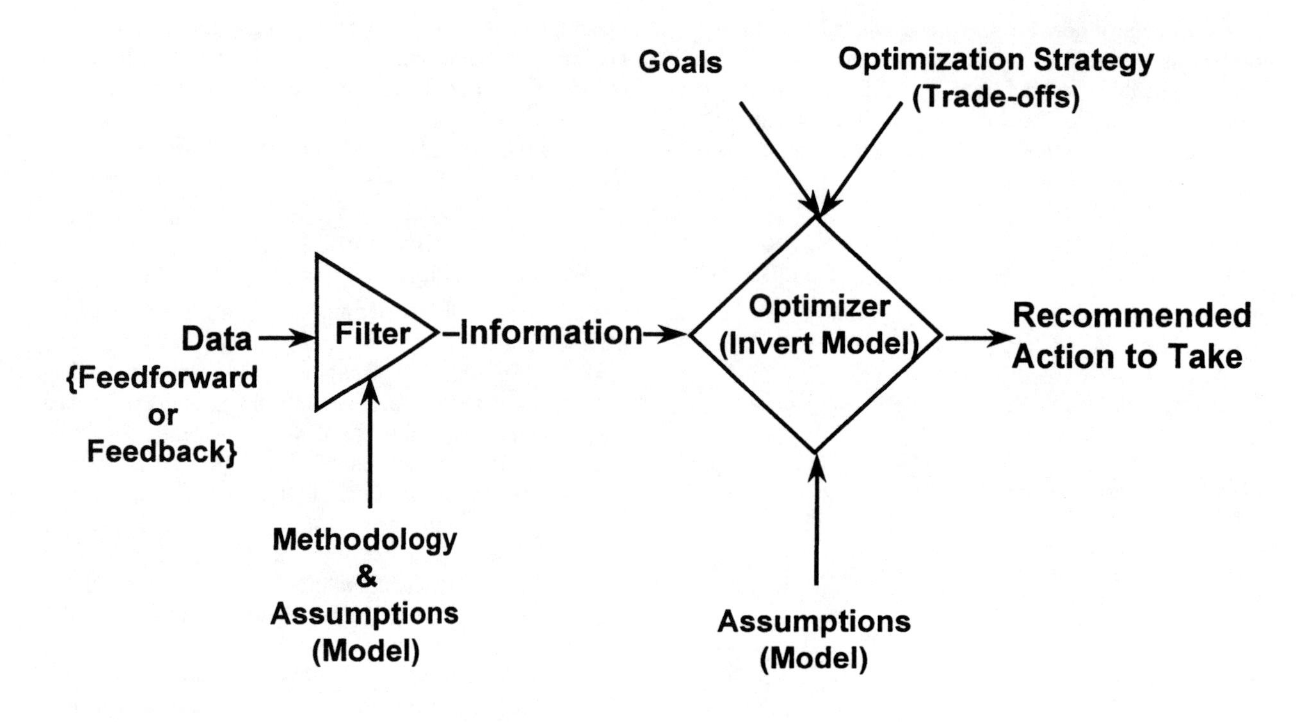

FIGURE 1: Steps In Decision Making

Fishbone Diagram for analyzing sources of variation has 4 components: Man, Machine, Material, and Methods. These sources can also be considered the components present in any system and which contribute to a system functionally correctly. These 4 components are often known as 4M. Table 7 provides an expanded list of components involved in a sensor application system.

TABLE 7. Components (The "M's)

People
Skills
Money (funding)
Material (wafers, gases, etc.)
Manufacturing
Manufacturability (Margin, Robustness)
Management
Metrics for Monitoring
Methods
Machine (Hardware)
Software

People, skills, and money are self-explanatory. Material refers to the incoming wafers, how the wafers will change during processing, gases and liquids used during processing, and all the other consumables involved in processing. Materials need to be considered to ensure that the sensor will function adequately in the intended processing environment and that needed consumables are adequately supplied. Manufacturing refers to the organizational environment in which the sensor application will be used, i.e., is the application compatible with the way the fab operates and its culture. Manufacturability indicates how robust a process are to typical variation, i.e., will the sensor and controller function correctly and achieve yield under all expected variations in wafer state and chamber state. Management and metrics are linked. Successful projects should consider how and when management will be updated on progress and final results. Most managers prefer data rather than opinions. Thus, what metrics will be needed must be considered to ensure the data is gathered, especially "before" data if improvements are to be monitored. Management of the project itself is important too, i.e., basic project management tools will help drive a project to success and prevent unnecessary problems. Some aspects of the last 3 components will be described in the sections below.

Methods

Methods include the mathematics, practices, and techniques used. Based upon the need to produce a decision as illustrated in Figure 1, the methods of key interest are those involved in decision making. Table 8 highlights that models are needed not only for how the process behaves, i.e., how it responds to a change in pressure, but also how it ages. Models alone are not sufficient, algorithms must be generated which use those algorithms to create information from the data and to decide upon an action based upon that data.

TABLE 8. Key Methods for Decision Making

- Models (capture assumptions)
  - Process & disturbance models
- Algorithms (filters, optimizer)
  - Use models
  - Suggest compensation for wafer state & disturbances
  - Detect faults & assign causes

In the previous section, the issue of the volume of machines for which each operator is responsible and the impact of too many alarms was mentioned. Understanding the false positives (Type I error) and false negatives (type II error) of the methods is a necessity. If the false positive rate of the controller for manual actions is too high than manufacturing personnel will become frustrated at trying to fix what isn't broke and turn off the controller. Conversely, if the false negative rate is too high, then yield loss will go uncorrected and the controller will be considered useless. While the false positives are the most visible, it is the long-term false negative rate which determines the improvement that the controller will achieve. Thus, the right method, and the tuning of that method, to achieve acceptable Type I and Type II errors is a very import component of any control system.

Data & Information Links

An obvious data source in Figure 1 and used in the methods of Table 8, are data from sensors. Sensors reduce measurement cycle time and eliminate wafers at risk. However, another data source will still be traditional stand-alone (predominantly in-line and at-line) metrology. To use this data efficiently and in conjunction with the sensor data, links (hardware/software connections) between this data and the controller must exist. The data and information of Figure 1 also refer to data and information about the wafer, such as which process to perform, goals, optimization strategy, previous controller results, etc. All this data is distributed throughout the fab in a variety of databases and machines. Thus, connecting all of these databases including the Computer Integrated Manufacturing (CIM)/Manufacturing Enterprise System (MES) and machines, to the controller is required for the controller to have the necessary data and information to make a good decision. Conversely, the controller needs to connect to the relevant databases and machines to allow accurate wafer tracking and action execution. Fortuitously, the change to 300mm platforms is also causing equipment suppliers to change their software and address this issue of connectivity.

The 3R's of Sensors

Another important aspect of hardware is the repeatability, reproducibility, and resolution of the sensor. The importance of repeatability and reproducibility is generally understood. As reproducibility/repeatability become worse, the controller must be detuned to prevent decisions being made based upon noise. Thus, the controller is also less able to recognize and respond to small changes and is slower to respond to big changes.

Resolution, or sensitivity, is the 3rd R. Resolution is the smallest difference that the measurement device can see. The new Semiconductor Industry Association (SIA) National Technology Roadmap stresses the emerging importance of resolution. Resolution is important because it determines how tightly a system can be controlled. As the device dimensions continue to shrink, measurement tools must not only have less noise, but they must also have better resolution. As noted in the SIA roadmap, a metric is needed for resolution so that a gauge exists to compare different metrology and determine if it is good enough for the intended application. A metric is proposed which is similar to the common metric used for capability of a measurement tool, Cp:

$$Cp = \frac{USL - LSL}{6\sigma} \quad (1)$$

where

USL = Upper Specification Limit
LSL = Lower Specification Limit
σ = Over-all repeatability and reproducibility of measurement tool

Let us define

$$\text{Resolution} = \frac{\sigma_T}{\Delta} \quad (2)$$

where

Δ = Smallest Change Can Measure
σ_T = Targeted σ (see Equation (3))

and

$$\sigma_T = \frac{USL - LSL}{6Cp_T} \quad (3)$$

where

Cp_T = Targeted Cp for Measurement tool

While a metric for resolution is important, the industry agreed target value for this metric is even more important. For example, the industry accepts a measurement Cp of less than 3 unacceptable, a value of 3-10 only marginal, and a Cp value greater than or equal to 10 acceptable. These industry standards help the metrology developers understand what capability they must achieve.

Another issue with resolution is how to determine Δ, the smallest change measurable, because the only way to make this determination is to know the size of features being measured. This requirement drives the need for a calibration source, which is not required for gauge repeatability and reproducibility studies. This need for calibration may be the reason resolution is not examined. However, the lack of understanding of the importance of resolution to control may be also responsible for the lack of focus on resolution.

FORM

There are a variety of forms that the final solution can take. However, some forms are more compatible with the fab environment than others. A list of proposed solution characteristics is given in Table 9. The top of the list is reliable software because regardless of how nice the features, the sensor and its associated software must not impede the throughput of the fab.

TABLE 9. Some Proposed Solution Characteristics

- RELIABLE software
- Ability to align & correlate events to data
  - Factory mean time clock
- Wafer/die tracking
  - Randomization capability
- Intelligent data collection, reduction, storage, and transmission
  - Event and test result driven
- Run tests during idle time with no wafer
- Ability to add sensors (windows, space, ...)

In order to find correlation among different data sources and identify root causes of faults, different sources of data and events must be aligned to one another. To achieve this alignment, a single clock must be used for every piece of equipment and data collection in the fab. Also to identify correlations, wafer positional tracking is needed. Tracking through cluster tools by the cluster tool is key to this effort. Due to the lack of randomization during fab processing, wafers are likely to remain in the same position through many machines, thereby causing confounding when trying to identify which machine or position is causing faults. In other words, low yielding wafers are likely to occur in the same run position across many lots on several different machines. Randomization, either within the tool itself or when placing wafers in the boat, breaks this confounding because low yielding wafers will most likely occur only in a given position across many lots on only 1 machine. However, a single agreed upon standard for randomization is necessary since equipment can not use a single boat if randomization into an outgoing boat is required. Some equipment today unload and load back into the same boat.

To send all the data up to the CIM/MES system for analysis would swamp both data connection lines, as well as the CIM/MES host computer. Thus, data needs to be analyzed at the equipment and only summaries sent to the CIM/MES. However, storage of more (all) of the information on the equipment for shorter period of time, like 3 months, would be needed to reconcile major problems. What data is sent should depend upon what happened during processing. More data should be sent if there is a failure or less data should be sent if the process appears to be normal. Similarly, data collection needs to be event driven and test driven so that more data is collected when needed, such as after a test failure or during ramps, and less data is collected when appropriate, such as during normal operation or steady-state processing periods. The desired behavior may also be different for different processes performed on the same equipment.

One of the basic performance issues with current equipment is their lack of self-checking during idle time. For example, vacuum problems on Chamber C of a cluster tool could be detected by trying to pull a vacuum when a machine has no queue of wafers. However, today, this vacuum problem would be discovered when the first wafer arrived at Chamber C (after processing through A and B), when the machine has problems pumping down Chamber C. Another example is that parts have hard fails that go undetected until a wafer arrives at that chamber. Merely doing a "are you there" electronic ping of the equipment parts would detect many hard fails.

Finally, in order to integrate sensors onto processing equipment, space and possibly view ports must be present. As equipment has gotten large, space has become a premium. In addition, to simplify chambers, view ports have decreased in size and quantity. Consideration of lack of sensors needs to be considered when designing the chamber. Retrofit kits for adding sensors need to also be made available for adding sensors. Process equipment vendors are showing increased understanding of the value of these kits. For example, kits are available from CMP vendors for adding on-line thickness sensors and from etch vendors for adding full-wafer interferometers.

The above characteristics in a Integrated Metrology solution will result in equipment that is able to comprehend it's own processing capability, able to adjust itself to achieve desired results, and able to perform diagnosis and prognosis to identify possible causes of current and future failures. A control system will still exist at the CIM/MES level, but it will focus more on module control and work in synergy with the equipment's control system by specifying the more advanced equipment control strategy requirements and targets. In other words, the CIM/MES control system will transform from a "wrapper" on equipment which is not aware of nor capable of utilizing the wrapper, to a CIM/MES system which works in harmony with an equipment control system that is aware and relies on its presence.

SUMMARY & BENEFITS DERIVED

In this paper, Integrated Metrology was defined and examined. This examination may be summarized as:

> *A strong business need is required to drive a complete integrated solution, and only a complete integrated solution will provide benefits to the business.*

The business drivers were illustrated by highlighting currently available control systems. The future business opportunities for control were also suggested. The necessary components were reviewed, with emphasis on decision making methods, data connectivity, and the 3R's of sensors. To facilitate the focus on the 3rd R (resolution), a metric was proposed for consideration by the industry. Several solution characteristics, such as intelligent data collection, were proposed. The benefits that Integrated Metrology can achieve are summarized in Table 10. The benefits are broken into 2 categories: those related to how the process will function in the fab and those related to the equipment suppliers' responsibility for maintaining the equipment and process. By improving how the process and the field service personnel perform, the process equipment vendor creates a competitive advantage for their machine.

TABLE 10. Benefits that Integrated Metrology Can Produce

For Integrated Circuit Maker; Process Itself:

- Increase productivity
  - X
  - People
  - Equipment
  - Fab
  - Technology transfer & ramp-up
  - Sales (customer relations)
- Eliminate outliers
  - Points not belonging to typical distribution
- Improve typical distribution
  - Drive average to target
  - Reduce variance

For Equipment Supplier:

- Reduced engineering support
  - Increases effective skill set of field service personnel
  - Automatic re-tunings of process
  - Diagnosis/prognosis to reduce down-time
  - Quicker time to recover (repair & condition & tune)
- Intelligent maintenance
  - Maintenance on need basis scheduled when convenient
  - Parts available at site when needed

FRONT END PROCESSES

Characterization and Metrology Implications of the 1997 NTRS

W. Class

Eaton SEO, Beverly MA 01915

J. J. Wortman

NC State University, Raleigh NC 27695

In the Front-end (transistor forming) area of silicon CMOS device processing, several NTRS difficult challenges have been identified including; scaled and alternate gate dielectric materials, new DRAM dielectric materials, alternate gate materials, elevated contact structures, engineered channels, and large-area cost-effective silicon substrates. This paper deals with some of the characterization and metrology challenges facing the industry if it is to meet the projected needs identified in the NTRS. In the areas of gate and DRAM dielectric, scaling requires that existing material layers be thinned to maximize capacitance. For the current gate dielectric, SiO_2 and its nitrided derivatives, direct tunneling will limit scaling to approximately 1.5nm for logic applications before power losses become unacceptable. Low power logic and memory applications may limit scaling to the 2.0-2.2nm range. Beyond these limits, dielectric materials having higher dielectric constant, will permit continued capacitance increases while allowing for the use of thicker dielectric layers, where tunneling may be minimized. In the near term silicon nitride is a promising SiO_2 substitute material while in the longer term "high-k" materials such as tantalum pentoxide and barium strontium titanate (BST) will be required. For these latter materials, it is likely that a multilayer dielectric stack will be needed, consisting of an ultra-thin (1-2 atom layer) interfacial SiO_2 layer and a high-k overlayer. Silicon wafer surface preparation control, as well as the control of composition, crystal structure, and thickness for such stacks pose significant characterization and metrology challenges. In addition to the need for new gate dielectric materials, new gate materials will be required to overcome the limitations of the current doped polysilicon gate materials. Such a change has broad ramifications on device electrical performance and manufacturing process robustness which again implies a broad range of new characterization and metrology requirements. Finally, the doped structure of the MOS transistor must scale to very small lateral and depth dimensions, and thermal budgets must be reduced to permit the retention of very abrupt highly doped drain and channel engineered structures. Eventually, the NTRS forecasts the need for an elevated contact structure. Here, there are significant challanges associated with three-dimensional dopant profiling, measurement of dopant activity in ultra-shallow device regions, as well as point defect metrology and characterization.

INTRODUCTION

The Front End Process section of the 1997 NTRS (1) is divided into sections devoted to Starting Materials, Surface Preparation, Thermal/Thin Films & Doping, and Front End Plasma Etch. This paper focuses on metrology requirements which arise out of the Thermal/Thin Films & Doping section, as well as pertinent portions of the Surface Preparation section which specifically relate. Each section contains a set of requirements which are based upon evolutionary scaling of the MOS transistor or DRAM storage cell. This exercise results in the identification of areas where current technology fails or must be fundamentally changed in order the allow further device miniaturization. Many such areas have been identified, several of which pose difficult challenges, which if not overcome could fundamentally change the economic drivers underlying the growth of the semiconductor industry. Table 1 identifies a collection of such requirements vs. the respective technology nodes abstracted from the NTRS, which are pertinent to the subject of this paper. Figure 1 is a cross-sectional view of the MOS transistor which will be used to clarify the meaning of the various parameters identified in Table 1.

GATE STACK ISSUES

Gate Dielectric

Depicted in Figure 1 is a structure, known as the gate stack, which overlies the silicon transistor material. It is comprised of a gate dielectric, a doped polysilicon gate electrode and a titanium silicide over-layer, frequently called a shunt. The titanium silicide has a much lower resistivity than the polysilicon and serves the purpose of lowering the overall resistivity of the gate. A potential applied to the gate results in a potential appearing in the silicon region which underlies the gate stack (called the channel). A channel potential of appropriate sign can cause depletion and inversion of the silicon (2). Inversion charge carriers in the channel will result in a current flow from the source to the drain of the transistor, when the appropriate potential is applied to the drain. Fundamental to the performance of such a transistor is the effectiveness with which a potential applied to the gate will result in the generation of inversion charge carriers in the channel. The full potential applied to the gate does not appear as a potential in the channel because of potential drops associated with the capacitance of the gate dielectric, and other capacitances associated with depletion of the gate polysilicon material, and the

CP449, *Characterization and Metrology for ULSI Technology: 1998 International Conference*
edited by D. G. Seiler, A. C. Diebold, W. M. Bullis, T. J. Shaffner, R. McDonald, and E. J. Walters

Table 1 1997 NTRS Requirements.

Year of First Product Shipment	**1997**	**1999**	**2001**	**2003**	**2006**	**2009**	**2012**
Technology Generation (nm)	250	180	150	130	100	70	50
Operating Voltage, Vdd (V)	2.5-1.8	1.8-1.5	1.5-1.2	1.5-1.2	1.2-0.9	0.9-0.6	0-6-0.5
Equivalent Oxide Thickness (nm)	4-5	3-4	2-3	2-3	1.5-2	<1.5	<1.0
Gate Dielectric Thickness Control (%3σ)	±4	±4	±4	±4-6	±4-8	±4-8	±4-8
Maximum Electric Field (MV/cm)	4-5	5	5	5	>5	>5	>5
DRAM GOI (per cm2)	0.06	0.03	0.026	0.014	0.006	0.003	0.001
Logic GOI (per cm2)	0.15	0.15	0.11	0.08	0.05	0.04	0.03
DRAM Particle Defects (per cm2)	0.3	0.15	0.1	0.075	0.03	0.015	0.01
Logic Particle Defects (per cm2)	0.75	0.5	0.45	0.4	0.25	0.2	0.15
Particle Size (nm)	125	90	75	65	50	35	25
Critical Metal (atoms/cm2)	5E(9)	4E(9)	3E(9)	2E(9)	1E(9)	<1E(9)	<1E(9)
Oxide Residue (O atoms/cm2)	1E(14)	7E(13)	6E(13)	5E(13)	3.5E(13)	2.5E(13)	1.8E(13)
Drain Extension Jn Depth (nm)	50-100	70-140	60-150	50-100	40-80	15-30	10-20
Contact Jn Depth (nm)	100-200	70-140	60-120	50-100	40-80	15-30	10-20
Drain Ext. Dopant Conc. (Atoms/cm3)	1E(18)	1E(19)	1E(19)	1E(19)	1E(20)	1E(20)	1E(20)
Contact Silicide Rs(Ω/Sq.)	2	2.7	3.3	3.8	2	2	2
Si/Silicide Max Resistivity (Ω-cm2)	<1E(-6)	<6E(-7)	<4E(-7)	<3E(-7)	<2E(-7)	<8E(-8)	<3E(-8)
Drain Structure	Drain Extension				Elevated Contact + single Dr.		
Sidewall Spacer Thickness (nm)	100-200	72-144	60-120	52-104	20-40	7.5-15	5-10

No Known Solutions

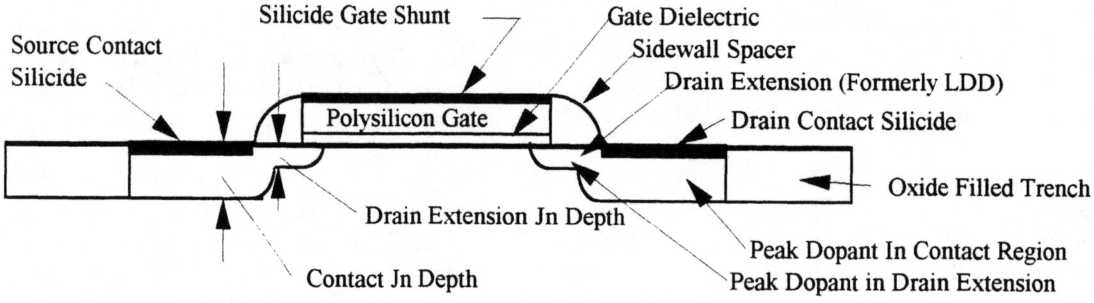

Figure 1 MOS Transistor with Drain and Contact Terminology

depletion of the channel region itself. These undesirable parasitic potential drops become increasingly onerous as the total voltage available to operate the transistor becomes smaller. This voltage (V_{dd}), as shown in Table 1, must scale with each device generation in order to manage the overall power consumption of the integrated circuit. The achievement of optimum transistor performance in this scaled voltage environment requires that all these capacitance values be maximized in order that the greatest fraction of gate voltage be available for inversion of the channel. This mandates the thinnest possible gate dielectric, and the elimination of depletion in the gate polysilicon material. Channel depletion capacitance cannot be arbitrarily minimized because of other scaled device performance requirements such as off-state current, threshold voltage, and drive current (3)

Table 1 shows the anticipated scaling of the gate dielectric, T_{ox} in units of equivalent oxide thickness. For a dielectric other than silicon dioxide, the equivalent thickness is a function of the respective dielectric constants, ϵ_d, and ϵ_{ox}, of the other dielectric and SiO_2, and is given by the equation:

$$T_{ox}(\text{equivalent}) = \frac{\epsilon_{ox}}{\epsilon_d} T_d \qquad (1)$$

Thus, a material with higher dielectric constant can be made thicker in direct proportion to the ratio of its dielectric constant to that of silicon dioxide, and still maintain the same capacitance. Table 1 also defines the maximum electric field which appears across the gate dielectric, which is $\cong$ 5 MV/cm for most of the technology nodes. At these electric fields it becomes possible for charge carriers to tunnel through the thin gate dielectric layers, giving rise to undesirable leakage currents, with attendant power dissipation (4, 5, 6). Thus, while MOS transistors have been realized with silicon dioxide gates as thin as 1.5nm, direct tunneling currents become a major obstacle. Another factor to be considered is that when the dielectric film is very thin, then a small variation in film thickness will give rise to significant variations in tunneling current (5). Also, since the capacitance associated with the dielectric layer is proportional to the inverse of the thickness, the capacitance variance associated with a thickness variation is proportional to the inverse square of the thickness. For these reasons, control of transistor performance requires that film thickness variations be held to a minimum. Table 1 gives the NTRS requirements. These pose a very significant metrology problem.

Current thinking is that SiO_2-based gate dielectric materials will be extended to approximately the 100nm technology node at which point some logic applications will make use of 1.5nm thick SiO_2 layers. Getting to this point is in itself a task of significant proportions, and implies some very significant measurement and control problems. Firstly, the migration of boron doping from the doped polysilicon gate (for P-channel tranistors) must be blocked. Here it has been found that nitrogen doping of the gate dielectric can be used to suppress boron penetration (7,8) but if present at the dielectric/channel interface, can give rise to interface traps which degrade low voltge device performance. Therefore the requirement is for a SiO_2 film, of $\approx$ 1.5nm thickness with a graded nitrogen profile such that the nitrogen is locallized at the interface with the doped polysilicon gate. In addition, if such films are to be produced by oxidation/nitridation of the substrate material, and as well exhibit reproducible growth rates and reliable dielectric breakdown behavior, then utmost quality of the silicon surface in terms of surface smoothness, absence of particle defects, and surface metal and organic contamination, must be maintained immediately prior to, and during film growth. Table 1 gives the NTRS surface preparation requirements driven by these needs. These present some very significant measurement and characterization problems.

A change to new materials, with higher dielectric constants will ultimately be required if MOS device scaling is to continue. These materials will be required for both the gate dielectric and in the DRAM storage capacitor. The DRAM material conversion will most likely occur earlier because of the greater sensitivity of these devices to leakage currents. This presents many challenges, especially with the MOS transistor because of the electrically active nature of the dielectric-semiconductor interface, its sensitivity to contamination and the high applied electric fields.

Although many materials can be identified having high dielectric constant, the above requirements significantly narrow the choices. For example, the tunneling resistance of a dielectric depends upon the barrier height which is exhibited by the semiconductor-insulator pair (9). This barrier height depends on the band gap of the insulator, the semiconductor band-gap, and the semiconductor Fermi level. Unfortunately, many dielectric materials which exhibit a high dielectric constant, also have relatively small band gaps in comparison with SiO_2. Table 2 gives some representative examples of potential candidate materials.

Table 2. Band-gap, barrier height and dielectric constant of several candidate gate dielectric materials.

Dielectric Material	Band Gap (eV)	Barrier Height (eV)	Dielectric Constant
Silicon Dioxide	9	3.2	3.9
Silicon Nitride	5	2.1	7.5
Titanium Dioxide	3.2	1	40-80
Tantalum Pentoxide	5.2	1-1.5	25

Silicon nitride has attractive properties as a near-term dielectric material choice. However before it can be used it must demonstrate electrical stability and suitable semiconductor interface properties. Since the cation from which it is comprised is silicon, it will not contaminate the device channel. Silicon nitride however cannot be produced by in-situ nitriding of silicon since the thermal budget for such a process would be unacceptable. Therefore unless some unanticipated developments take place, industry conversion to a silicon nitride gate dielectric would require that the gate forming process change from a in-situ oxidation-nitriding process to deposition. The metrology and characterization issues associated with such a conversion are considerable. One candidate low temperature silicon nitride deposition process, PECVD, yields films which are electrically unstable. Therefore, characterization is required in order to understand the origin of this electrical instability.

Materials other than silicon nitride, while attractive from the dielectric constant viewpoint pose added problems. Cations from these materials can contaminate the semiconductor device requiring therefore that some barrier be placed between the high-k material and the underlying semiconductor. In addition, these materials may be crystalline or amorphous when deposited and may as well be deposited in a range of stoichiometries, which could influence electrical stability as well as tunneling resistance.

One solution to the above problems may be to use a thin layer of SiO_2 or silicon nitride as a interface between the high-k material and the semiconductor (10, 11). However, this places the high-k and lower-k materials in series. The total capacitance of such a combination, C_{tot}, is related to the capacitance of the high-k dielectric and the oxide, C_d and C_{ox} respectively by the equation:

$$C_{tot} = \frac{C_d C_{ox}}{(C_d + C_{ox})} \qquad (2)$$

which requires that the thickness of the low dielectric layer must be very, very small to achieve an acceptable overall capacitance for the dielectric stack. Looking into the future therefore, one can envision the possibility of a production process for a deposited multi-layer dielectric stack having a total 2-3 nm thickness, with a <1nm silicon oxide or nitride under-layer, and high-k over-layer of controlled crystal quality and composition. In addition, this stack would require reproducible characteristics of all interfaces and be able to withstand the ≅ 5MV/cm electric fields encountered with device operation. This represents a metrology and characterization task of very significant complexity and scope.

GATE STACK ISSUES

Gate Metal

CMOS technology, which represents >80% of all silicon integrated circuits (12), requires that MOS transistors be configured in N-channel and P-channel pairs, with matched threshold voltages. The low power consumption inherent to CMOS depends upon such matching. Since the threshold voltage of an MOS transistor depends, among other factors, upon the work function of the gate material, modern CMOS technology uses dual doped polysilicon to achieve such threshold matching, consistent with sub-micron transistor performance. Here, p-doped polysilicon is used as the gate material for the P-channel MOS transistor, and n-doped polysilicon is used as the gate material for the N-channel MOS transistor. The turn-on of the P-channel device requires that a negative voltage be applied to the p-doped gate, and N-channel turn-on requires that a positive voltage be applied to the n-doped gate. Thus both gate materials can become locally depleted by the application of the respective gate voltage. This depletion occurs in the region near the gate/gate dielectric interface. The voltage fall associated with such depletion limits the drive current of the MOSFET, and limits as well the sharpness with which the transistor turns on. This latter characteristic is usually described by a parameter known as the sub-threshold swing S, (in mV/decade) the inverse slope of the of the transistor sub-threshold voltage-current (V-I) characteristic. A minimum value of S implies abrupt transistor turn-on. The sub-threshold swing is given by the relation (13):

$$S(\text{mV/decade}) = \left(\frac{kT}{q}\right) \times (\ln 10) \times \left(1 + \frac{C_D}{C_{ox}}\right) \qquad (3)$$

Where k = Boltzmann's constant, T is the absolute temperature, and q is the electronic charge, ln(10) is the natural log of the number ten, C_D is the total capacitance associated with the depletion layers in the gate as well as the silicon channel which underlies the gate dielectric, and C_{ox} is the capacitance associated with the gate dielectric. KT/q has a value of 25.9mV at room temperature (300°K) and ln10 has the value 2.303, giving the smallest possible value for S = 59.6 mV per decade of current. Reference to equation 3 shows that increasing the capacitance of both the channel depletion region and the gate depletion region has the effect of increasing S, which is undesirable for low voltage device operation. Similarly, increasing the capacitance of the gate dielectric layer has the desired effect of minimizing the swing. Since the capacitance associated with channel depletion cannot be arbitrarily reduced (short channel effect & DIBL), management of the total depletion capacitance requires the absolute minimization of the capacitance associated with depletion of the doped polysilicon gate material.

In the near term, this requires that the process used for gate doping achieve maximum uniform dopant concentration throughout the thickness of the gate. This can be difficult when using low energy ion implantation to simultaneously dope the gate as well as the source and drain regions. Here, the shallow doping requirements of the source/drain regions may result in a dopant profile in the polysilicon where the dopant concentration is below optimum at the gate/dielectric interface, which will exacerbate the gate depletion effect. A solution here is to use separate resist masked ion implantation steps to dope the polysilicon layer before the etching the gate structure. An alternate approach, in-situ doping, requires that a means be found to efficiently create separate adjacent regions of p+ and n+ polysilicon doping required for CMOS manufacture. The metrology and characterization issues here are the determination of active dopant concentration as a function of depth in the polysilicon layer.

Near term solutions to the depletion problem may involve the optimization of the doping process the achieve the maximum degree of dopant activity at the gate-dielectric interface. In the long term, the solution of this problem may require that metal gate structures be developed. This transition will involve many characterization and metrology issues, which arise out of the need to maintain CMOS process compatibility. Of greatest importance is the need to control the threshold voltage of both N-channel and P-channel transistors of the CMOS pair. Since threshold voltage is determined in part by the Fermi level of the gate material relative to that of the channel, the task of controlling threshold voltage with a single metal in conjunction with channels having intrinsically different Fermi levels becomes a difficult task. Dual metal gate approaches may be hampered by manufacturing efficiency or other CMOS compatibility problems This limits the choice of gate electrode materials. Also, many good electrical conductors i.e. copper are highly deleterious silicon contaminants and can significantly degrade device performance. Therefore reliability issues associated with gate material diffusion through the gate dielectric and into the channel become important.

Finally, in the near term the issue of gate material conductivity gives rise to several characterization and metrology issues. Referring to Figure 1 it is noted that current technology requires that the gate material be shunted with a low resistivity, CMOS compatible material. The current choice for logic applications is titanium silicide which has a relatively low resistivity (15μ-ohm-cm), excellent CMOS process flow compatibility, and readily forms a stable silicide on both n-and p-doped single and polycrystalline materials. These features allow the use of self-aligned silicide formation (salicide) to shunt both the p- and n-doped polycrystalline regions, as well as the p- and n-doped source/drain regions of the CMOS device. The salicide process requires that a relatively low temperature (≈750°C) be used to control the selective reaction between the uncovered substrate silicon and the deposited titanium metal. This results in titanium silicide phase with a C49 crystal structure which has an inherently high resistivity. The desired C54 low resistivity phase is achieved by a second higher temperature anneal (≈850°C) which is carried out after the remaining unreacted titanium metal has been chemically removed. Problems associated with scaling this titanium salicide process creates added characterization and metrology issues. Firstly, scaling requires that the self-aligned silicide be thinner to satisfy requirements for ultra-shallow source/drain regions, and be reliably produced over very narrow polysilicon gates lines. The salicide gate shunting process becomes difficult to scale because of two problems- agglomeration of the silicide, and unreliable conversion from the low temperature C49 phase to the high temperature C54 phase. Both phemonena lead to high resistivity of the shunted gate and local interconnnect structures. In addition both phenomena are inherently stochastic making difficult the problems of failure detection and failure mode identification.

CMOS DOPING ISSUES

Source/Drain Doping

MOS transistor scaling results in a reduction of the separation between the source-channel and drain-channel junctions of the device. As a consequence, the potential applied to the drain may not be completely screened by the immobile charges (ionized doners or acceptors) in the channel allowing some of the drain electric field to be present near the source-channel junction. The result is a lowering of the potential barrier at the source - channel pn junction. Since source and channel are normally at the same applied potential, only the built-in potential barrier of this junction prevents charge injection from the source into the channel. The unscreened potential from the nearby drain, lowers this barrier and thereby allows for carrier injection. This phenomenon, known as DIBL (drain induced barrier lowering) leads to many undesirable phenomena such as a dependence of threshold voltage on channel length, a dependence of threshold voltage on drain voltage and high off-state current. In a worst case, this barrier lowering may result in source-drain current flow even when no potential has been applied to the gate (punch-through).

Another consequence of reduced source-drain separation is that the drain potential asserts a relatively greater degree of influence on the inversion charge in the channel, which is ideally determined solely by the potential applied to the gate. This phenomenon is known as charge sharing, also gives to a drain induced lowering of the threshold voltage.

These above phenomena are collectively referred to as "short channel effects" One way of minimizing short channel effects is to reduce the depth of the source and drain junction. Therefore, the scaling of MOS devices requires that the junction depths at the edge of the channel be reduced in proportion with the degree that the channel length is reduced. A current solution is the use of a duplex drain structure, made up of a drain extension, and a deeper drain contact region, as depicted in Figure 1. The drain contact region, which must be deeper to allow for contact silicide formation, must nevertheless also scale to minimize short channel effects. Usually, a "drain engineering" dopant implant is performed near the edge of the contact junction to minimize local depletion lengths near the edge of the deeper contact junction, thereby allowing for better screening of the drain potential. In addition to the drain engineering dopants, many channel doping strategies must be developed to maintain scaled device performance. These are beyond the scope of this paper and are discussed elswhere (14).

Table 1 gives requirements for the drain extension and contact junction depths. Reference to Figure 1 shows that as the contact junction depth is reduced, less silicon is available to form the contact silicide. Therefore, a thinner silicide must be formed, which creates problems of manufacturing control, since both candidate silicide materials, titanium and cobalt silicide tend to agglomerate when made very thin. Manufacturing control is made more difficult by the previously referenced need to simultaneously form these silicides on the doped polysilicon gates of the CMOS device.

Since, as previously discussed, device operating voltage falls with future device generations, parasitic resistance in the drive current path must also be reduced in order to maintain optimum low voltage performance. This mandates that the dopant concentration in the drain extension be increased as shown in Table I. Ultimately, there is a limit to the concentration of activated dopant which can be achieved in the very shallow drain extension. Therefore the drain extension length must be decreased, leading to the evolution of "single drain" structures, where an explicit drain extension is no longer produced. Such single drain structures cannot be used in conjunction conventional device structures, since these yield unacceptable short channel behavior.

Also, as device dimensions scale, the source and drain contact areas proportionately decrease, and overall contact resistivity must therefore also decrease to preserve drive durrent. This places a lower limit on the allowed thickness for the contact silicide, and also requires that the specific contact resistivity of the silicide/silicon interface be reduced. These requirements are also shown in Table I. At the present time, it is not know how to decrease the contact resistivity of the silicide/silicon junction to maintain conformance with the roadmap requirements. This arises because the resistivity of this interface is dependent upon the barrier height, specific to the metal-semiconductor pair, and the concentration of activated dopant in the semiconductor (14). Some contact engineering will be required to solve this problem.

A common solution to all these contract scaling problems is a device where the contact structure is elevated above the plane of the channel. Table 1 shows the emergence of such devices with the 100nm technology node. There are numerous characterization and metrology problems associated with the scaling of conventional devices to this 100nm technology node, and a whole host of new problems associated with the emergence of the elevated contact structure.

The preferred contact doping process for present generation devices of the type depicted in Figure 1 is ion implantation. This process has many attractive features including; precise control of doping dose and depth, compatibility with the use of resist masking to produce adjacent regions of n- and p- doping as required for CMOS manufacture, and compatibility with the use of existing device topographical features to mask the doping process (self-aligned doping). A major drawback of the process is the collateral crystal damage which is produced. A second drawback is the limited dose rate achievable at the very low ion energies required to form ultra- shallow junctions. These could impact the continued cost effectiveness of the implant process.

The crystal damage associated with the doping process introduces a non-equilibrium excess of point defects (interstitials and vacancies) which for a transient period, can give rise to very rapid diffusion (15, 16, 17, 18). For Boron, the most challenging dopant, the transient enhanced diffusivity is characterized by the equation (15):

$$D_A^{enh} = D_A^* \frac{[I]}{[I^*]} \tag{4}$$

Where D_A^{enh} denotes the enhanced diffusivity, D_A^*, the equilibrium diffusivity, $[I]$, the time dependent concentration of excess silicon self-interstitials, and $[I^*]$, the equilibrium self-interstitial concentration. The implant process introduces the self interstitials which are thought to cluster on the time scale of the implant damage cascade (19). Post-implant heating then results in a ripening process during which time the cluster configuration changes into progressively more stable interstitial arrays. During this time the excess interstitial concentration may be many orders of magnitude greater than the equilibrium concentration. Upon completion of this transient period, thermally stable dislocation arrays or extrinsic stacking faults remain. These latter can give rise to leakage currents when located within the depletion region of a reverse biased p-n juction such as is the case with the drain junction of the MOS device. Another consequence of these point defects and defect arrays is the ability to capture dopant atoms into configurations which are difficult to activate using any reasonable thermal budget (20). Therefore, a central issue surrounding the extension of ion implantation to future generation devices is the understanding of dopant-defect, defect-defect, defect-contaminant, and defect-surface interactions in silicon, and the influence of these on three dimensional dopant diffusion, and associated activation/deactivation. This problem is made more difficult by the fact that the kinetics of point defect migration and clustering remains poorly understood, although significant progress is being made using ab initio calculations (19). Further understanding of these issues is hampered by the lack of suitable means of measuring three-dimensional dopant profiles, and the collateral inability to measure active vs. inactive dopant distributions. It is noted, that many of these issues will remain if means other than ion implantation are found to accomplish the doping process. It has been found that non-damaging means of dopant introduction, such as molecular beam epitaxy can yield a non-equilibrium excess of point defects and associated transient enhanced diffusion (21).

NTRS suggested long term solutions to the formation of elevated contact structures involve the use of selective CVD deposition of epitaxial silicon, titanium silicide, or sacrificial polysilicon. It is recognized that CMOS compatible solutions remain elusive, and that alternate solutions may emerge. The referenced processes all have advantages of self-alignment, a virtue which has significant manufacturing cost and yield implications. However, CMOS compatibility requires that these selective CVD depositions be accomplished in a robust manner on single- and polycrystalline silicon, having a variety of dopants. The achievement of these aims requires much greater understanding of nucleation processes, as well as better characterization of the chemistry whereby dopants are incorporated into the selectively deposited films.

It is also noted that the development of elevated contact structures will add significantly to the reliability requirments of the side-wall spacer, which acts as the insulator separating the contact and the gate. Here, as referenced in Table I, spacer thicknesses equivalent to current generation gate dielectric thickness are anticipated in order to achieve single drain devices. The electrical and physical characterization of sidewall spacer material, as well as the characterization of the gate sidewall on which these spacers are deposited, present formidable characterization and metrology problems.

CONCLUSIONS

The 1997 NTRS makes clear that there are many characterization and metrology challenges associated with the extension of current technology to evolutionary future generation devices in the areas of MOS gate stack and contact formation. Included are the needs for the characterization and control of composition profiled ultra-thin gate dielectric films, and the silicon surfaces upon which these are produced. Included as well are characterization needs associated with contact doping and associated three dimensional dopant redistribution and defect interactions. New gate stack and contact structures which may emerge in response to difficult challenges, will require characterization and control of multi-layer, ultra-thin high-k dielectric structures, metal gates, and elevated contact structures.

REFERENCES

1. 1998 National Technology Roadmap for Semiconductors, available from the Semiconductor Industry Association, 181 Metro Drive, Suite 450, San Jose CA 95110, pp. 59-81

2. Tsividis, Y.P., *Operation and Modeling of the MOS Transistor,* New York, McGraw-Hill, 1987, pp.35-62

3. Tsividis, V.P., *Operation and Modeling of the MOS Transistor,* New York, McGraw-Hill, 1987, pp. 168-214

4. Depas, M., Vermeire, B., Mertens, P.W., Meirheaghe, R.L., and Heyns, M.M., *Sol.St. Elecs.,* **38**, 1465-1471 (1995)

5. Rana, F., Tiwari, S., and Buchanan, D.A., *Appl. Phys. Lett.,* **69**, 1104-1106 (1996)

6. Momose, H.S., Ono, M., Yoshitomi, T., Ohguro, T., Nakamura, S-I., Saito, M., and Iwai, H., *IEEE Trans. Elec.Dev.,* **43**, 1233-1242(1996)

7. Momose, H., Morimoto, T., Ozawa, Y., Yamabe, K., and Iwai, H., *IEEE Trans. Elec.Dev.,* **41**, 546-552(1994)

8. Hasegawa, E., Kawata, M., Ando, K., Makabe, M., Ishitani, A., Manchanda, L., Green, M., Krisch, K., and Feldman, L., *IEDM Tech. Dig.,* 327-330 (1995)

9. Duffy, J.A., *Bonding Energy Levels and Bands in Inorganic Solids,* New York, Wiley, 1990

10. Yang, H., Niimi, H., and Lucovsky, G., *J. Appl. Phys.,* **83**, 2327-2337 (1998)

11. Vogel, E., Ahmed, K., Hornung, B., Henson, W., McLarty, P., Lucovsky, G., Hauser, J., and Wortman, J., To be Published, *IEEE Trans. Elec.Dev.,* May 1998

12. McClean, B., Ed., *Status 1997,* Scottsdale, Az., Integrated Circuit Engineering Corp., (1997) pp. 4-1 to 4-7

13. Tsividis, Y.P., *Operation and Modeling of the MOS Transistor,* New York, McGraw-Hill, 1987, pp.102-160

14. Hauser, J.R., and Lynch, W.T., To be published *IEEE Trans.Elec.Dev.,* 1998

15. Stolk, P.A., Gossmann, H.J., Eaglesham, D.J., Jacobson, D.C., Rafferty, C.S., Gilmer, G.H., Jaraiz, M., Poate, J.M., Luftman, H.S., and Hayes, T.E., *J. Appl. Phys.* **81**, 6031-6080 (1997)

16. Jones, K.S., Elliman, R.G., Petravic, M.M., and Kringhoj, P., *Appl. Phys. Lett.,* **68**, 3111-3280 (1996)

17. Privitera, V., Coffa, S., Priolo, F., Larsen, K.K., and Mannino, G., *Appl. Phys. Lett.,* **68**, 3422-3424 (1996)

18. Chao, H.S., Griffin, P.B., Plummer, J.D., and Rafferty, C.S., *Appl. Phys. Lett.,* **69**, 2113-2115 (1996)

19. Bedrossian, P.J., Caturla, M.-J., and Diaz de la Rubia, T., *Appl. Phys. Lett.,* **70**, 176-178 (1997)

20. Chao, H.S., Griffin, P.G., and Plummer, J.D., *Appl. Phys. Lett.,* **68**, 3570-3572 (1996)

21. Agarwal, A., Eaglesham, D.J., Gossmann, H.-J., Pelaz, L., Herner, S.B., Jacobson, D.C., Haynes, T.E., Erokhin, Y., and Simonton, R., *IEDM Tech. Dig.,* 467-470 (1997)

Characterization of Ultrathin Gate Oxides for Advanced MOSFETs

M. Hirose, W. Mizubayashi, M. Fukuda and S. Miyazaki

Department of Electrical Engineering, Hiroshima University,Higashi-Hiroshima 739-8527, Japan

This paper reviews recent progress in structural and electronic characterizations of ultrathin SiO_2 thermally grown on Si(100) surfaces. Based on the accurate energy band profile determined for the n+poly-Si/SiO_2/Si(100) system, the tunnel current through nanometer thick gate oxides is quantitatively examined by multiple scattering theory. Metrological issues for determining the real oxide thickness in the ultrathin regime are also discussed. It is further shown that nanometer thick gate oxides maintain the sufficient reliability for scaled MOSFETs.

INTRODUCTION

The high performance nMOSFETs with 1.3-1.5nm thick gate oxides has been demonstrated [1,2]. Also, a large transconductance(gm=1.07S/mm) for a 90nm gate length nMOSFET with an acceptable level of the gate leakage has been demonstrated[2]. The energy bandgap of 1.7-5.2nm thick SiO_2 is maintained at 8.95eV[3], indicating that the intrinsic nature of ultrathin SiO_2 is basically identical to that of bulk oxide although the density of the oxide layer near the SiO_2/Si interface is higher than that of the bulk SiO_2[4,5]. The thickness uniformity for 1-5nm oxides on Si wafers is extremely good because the thermal oxidation of silicon proceeds through a layer-by-layer process on an atomic scale as directly demonstrated by scanning reflection electron microscopy [6]. Wearout behavior of sub-5nm gate oxides is significantly different from that of thicker oxides in which hard breakdown accompanied with a dramatic increase of the leakage current is a typical failure mode. In contrast thinner(sub-5nm) oxides under constant current or constant voltage stressing exhibit quasi-breakdown or soft breakdown [7-10]. A possible mechanism of triggering the soft breakdown is believed to be formation of local conductive paths in SiO_2 when the percolation path is generated by stress-induced electron traps. The leakage current is considered to be controlled by variable range hopping via localized states [11], by direct tunneling through locally thin oxide in contact with conductive paths[8,10] or by trap-assisted tunneling conduction [9,12,13]. Possible mechanisms of defect generation in SiO_2 under stressing have been argued in connection with hydrogen release by high energy electron injection into the oxide and anode [14, 15], Si-O bond dissociation due to interaction between injected electrons and strained Si-O-Si bonds[16], or precursor oxide defects existing in as grown oxides[11]. The reciprocal of the charge-to-breakdown (Q_{BD}) or the defect generation probability measured as a function of stressing voltage yields a threshold near 5V which corresponds to the excess electron energy of about 2eV with respect to the oxide conduction band at the SiO_2/Si interface. This energy is believed to be related to hydrogen release threshold near the oxide/anode interface and/or chemical reaction with existing precursor sites [14,15]. It is also important to note that the reciprocal of Q_{BD} is proportional to oxide thickness beyond about 5nm [15], where hard breakdown becomes predominant. This is consistent with the fact that the soft breakdown is the major failure mode for sub-5nm oxides.

Quantitative modeling of tunnel leakage current through sub-3nm oxides is particularly important for future circuit design in which the gate oxide resistance due to direct tunneling should be taken into account [17]. In the analysis of tunnel current we encounter three crucial problems as follows: (1) An appropriate theoretical framework must be chosen on the basis of WKB approximation [10], multiple scattering theory [18], or transverse-resonant method [19,20]. (2) The effective mass of tunneling electron must be properly given although the reported values are in the broad range from $0.29m_0$ to $0.5m_0$. (3) Relevant oxide thickness must be determined by ellipsometry, C-V analysis or x-ray photoelectron spectroscopy (XPS). The oxide thickness determined by high resolution electron microscope (HREM) is higher by 0.2-0.5nm than the ellipsometric value for 1-2nm SiO_2 [2]. It is also known that the ellipsometric value is higher by 0.1-0.2nm than the XPS value. In determining the oxide thickness from C-V curves, poly-Si depletion and quantization effect in the inversion or accumulation layer must be carefully taken into account [19,20].

GROWTH AND STRUCTURE OF ULTRATHIN OXIDES

The measured oxide thickness on a silicon wafer is extremely uniform as reported in a previous paper, where 64 capacitors on a 200mm wafer with 3nm thick, thermally grown oxides yielded the thickness variation of ±0.014nm [20]. This is basically due to the fact that atomic-scale, layer-by-layer oxidation controls the growth process. Scanning reflection electron microscopy (SREM) has been employed to reveal the initial oxidation of Si(100) surfaces. The periodic reversal of the SREM contrast during oxidation directly showed the layer-by-layer oxidation while interfacial step

CP449, *Characterization and Metrology for ULSI Technology: 1998 International Conference*
edited by D. G. Seiler, A. C. Diebold, W. M. Bullis, T. J. Shaffner, R. McDonald, and E. J. Walters

structure was maintained [6].

The existence of a dense (~2.4gm/cm^3), thin(~1nm) layer at the SiO_2/Si(100) interface has been demonstrated by a high-accuracy difference x-ray reflectivity method [4]. Note that the oxide density near the interface is higher than that of bulk SiO_2(2.35-2.36gm/cm^3) as well as bulk Si(2.33gm/cm^3). Recently consistent results have also been reported by another group [5]. Si-O-Si bonds in the interfacial layer (1-1.5nm thick) with the higher density should be compressively strained. The extent of built-in strain can be detected by an infrared absorption band due to the longitudinal optical (LO) phonon mode originating from Si-O-Si lattice vibration. A distinct redshift of the LO phonon peak from 1250 to ~1210cm^{-1} is observed within 2nm from the interface. According to a central force model, the observed redshift of about 30cm^{-1} for the LO phonon peak corresponds to about 7.7° reduction of the Si-O-Si average bond angle which is considered to be 144° for the relaxed SiO_2 network [21]. The presence of different levels of compressive, built-in stress in the interface is also suggested by the fact that the etch rate for 800°C SiO_2 on Si(100) and 1000°C SiO_2 on Si(111) by dilute HF is significantly reduced in the thickness range below 1.5nm, while the rate for 1000°C SiO_2 on Si(100), which shows the smaller redshift of the LO phonon peak, is nearly constant as shown in Fig.1.

ELECTRONIC STATES IN ULTRATHIN OXIDES AND THE SiO_2/Si INTERFACE

For the purpose of calculating the tunnel current the electron barrier height at the SiO_2/Si interface has to be known and the tunnel electron effective mass must be given. Figure 2 represents the energy band profile for an n$^+$poly-Si gate/ ultrathin SiO_2/p-Si(100) structure as determined by recent

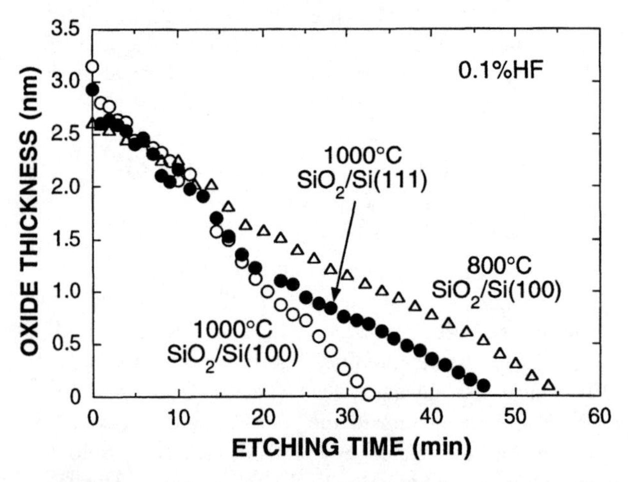

FIGURE 1. Oxide thickness as a function of etching time in a 0.1% HF solution. The oxides grown on Si(111) and Si(100) were simultaneously etched in the same HF solution. The oxide thickness uniformity was confirmed by AFM at representative steps of oxide stripping.

work. Note that the barrier height at the poly-Si/SiO_2 interface is tentatively set equal to the value at the SiO_2/Si(100) interface. The energy bandgap for ultrathin (5.2 to 1.7nm) SiO_2 has been recently determined to be 8.95eV[3] and compared with the valence band alignment or hole barrier height at the SiO_2/Si interface to obtain the electron barrier height [22]. The oxide bandgap has been obtained from the energy loss spectrum of the O_{1s} core level photoelectron because the photoexcited electrons in SiO_2 suffer inelastic losses such as plasmons and band-to-band excitation whose onset appears at 8.95eV for oxides thicker than 1.7nm. For oxides thinner than 1.2nm the onset of the bandgap excitation becomes less clear and a significant energy loss yield emerges in the subgap energy region mainly because the energy loss due to the interface suboxide states contributes to the 5-9eV signal. Further, the integrated intensity of enhanced O_{1s} energy loss signal over 4 to 10 eV is not proportional to the integrated Si_{2p} suboxide signal intensity for oxides thinner than 0.8nm. This might arise from the compressive, built-in stress in the interfacial layer because the stress changes the local density of states in SiO_2 and induces the localized band tail states below the conduction band [23]. The barrier height indicated in Fig.2 is slightly higher than those previously reported. For obtaining the barrier height we have determined simultaneously the oxide bandgap and the valence band alignment at the SiO_2/Si interface for 0.8-5.2nm thick oxides by using high resolution x-ray photoelectron spectroscopy. This is self-consistent determination of the barrier height for both electron and hole, being different from the case of previous work.

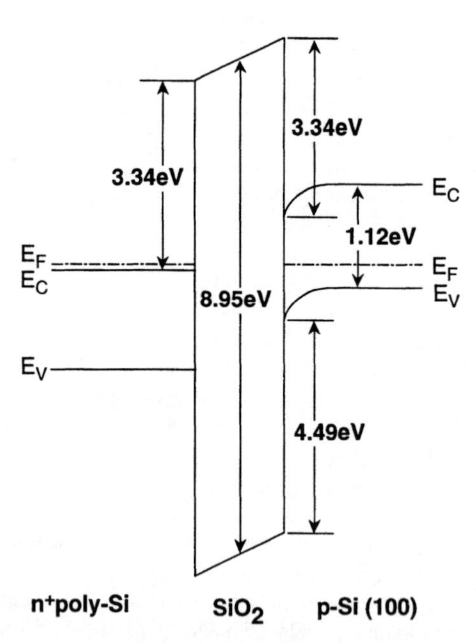

FIGURE 2. Energy band profile of an n$^+$poly-Si/SiO_2/p-Si(100) structure for dry oxides in the thickness range 1.6 to 40nm. The wet oxides provide a slightly larger (0.05eV) electron barrier height. Also, the electron barrier height at the SiO_2/Si(111) interface is larger by 0.13eV than the value indicated in the figure.

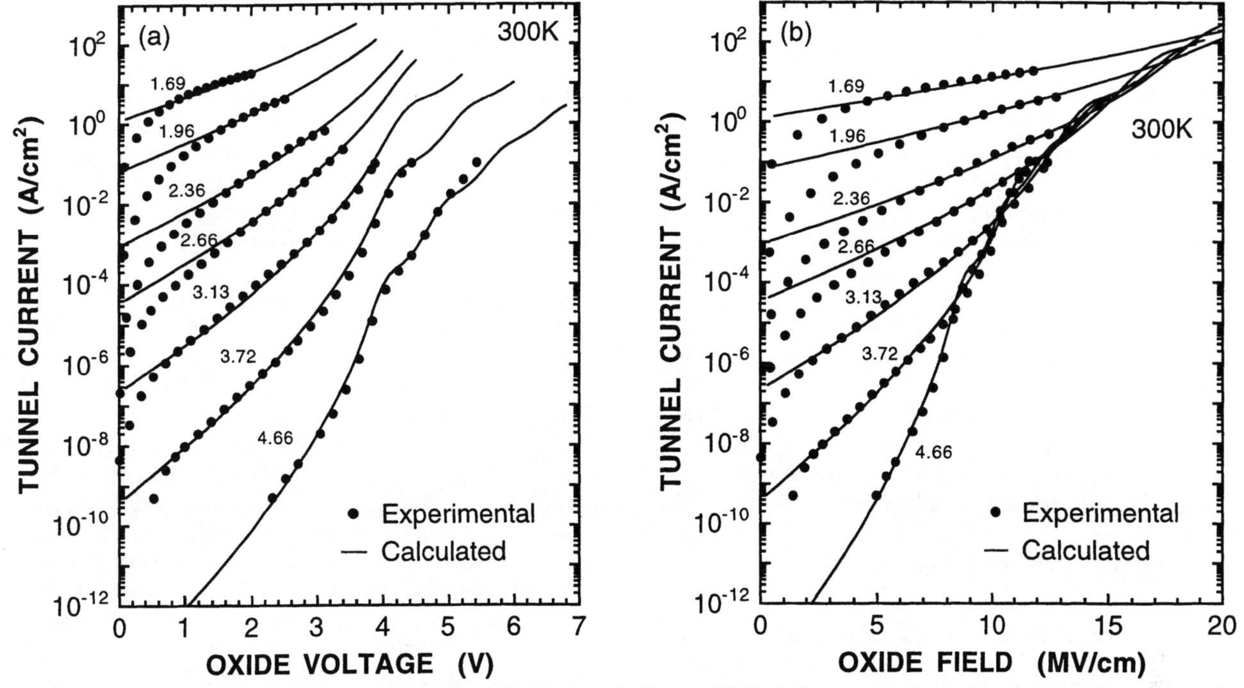

FIGURE 3. Tunneling current through ultrathin oxides for n+poly-Si gate MOS diodes measured as a function of voltage across the oxide(a). The poly-Si gate is negatively biased for all measurements. The oxide thicknesses indicated in the figure are obtained by fitting theoretical I-V curves to measured ones, being in fair agreement with the ellipsometric values. Also, tunnel current versus oxide field plot is given in (b).

MODELING OF TUNNEL CURRENT

The effective mass for tunneling electron has been empirically determined as a kind of fitting parameter based in most cases on the WKB approximation. The reported values of electron effective mass for the direct tunnel current regime are in the range $0.29m_0$ to $0.32m_0$ for the WKB approximation with a parabolic E-k relationship [10,24-26]. On the other hand, the tunnel current more accurately calculated by the multiple scattering theory has been fitted to measured current with the effective mass of $0.36m_0$ for oxide thicker than 3nm[18]. We have also compared measured tunnel current with the multiple scattering theory for sub-5nm oxides, and the best fit has been obtained for $m^*=0.35m_0$ as shown in Fig.3(a) [27]. In the calculation of tunnel current by the multiple scattering theory both forward and backward tunneling are taken into account, whereas there is a significant discrepancy between the measured and calculated current at the oxide voltages below 1 volt. This arises mainly from difficulty to predict exact electron density at n+poly-Si/SiO_2 interface at low negative gate voltages. For oxides with thicknesses above 3nm, the thickness determined by ellipsometry is consistent with convensional CV data although the CV thickness is lager by about 0.3nm than the ellipsometric value as shown in Fig. 4, while for sub-3nm oxides, the discrepancy becomes more significant because of the enhanced contribution of poly-Si depletion and inversion layer quantization effects [19,28] and reduced carrier concentration in the n+poly-Si gate doped at 850°C. The ellipsometric thickness obtained by using the refractive index of 1.460 was larger by

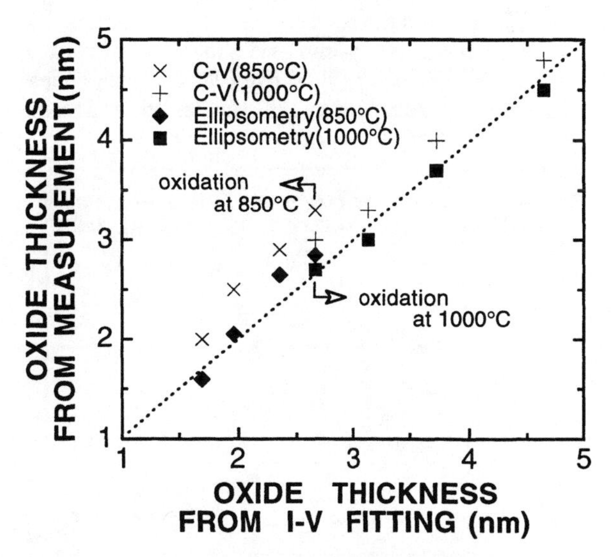

FIGURE 4. Ellipsometric and CV oxide thickness versus corresponding thickness obtained from I-V fitting. Dry oxidation and $POCl_3$ diffusion in poly-Si were performed at 1000°C for thicker oxides, while at 850°C for thinner ones.

about 0.2nm than the XPS value in which the Si_{2p} photoelectron escape depths for SiO_2 and Si were assumed to be 3.4 and 2.7nm, respectively[27]. The result of Fig. 4 indicates that the ellipsometric oxide thickness agrees with that obtained from the best theoretical fitting of the measured tunnel current. It should be noted that at the oxide voltages above 3.34V, which is equal to the electron barrier height, Fowler-Nordheim tunneling current oscillation is also quantitatively reproduced as understood from I-V curves for 3.13 to 4.66nm oxides in Fig. 3(a), where the oxide conduction band effective mass of $0.60m_0$ is used together with the tunneling effective mass of $0.35m_0$. As is well known the direct tunnel current is very sensitive to the barrier height Φ_B, the electron effective mass m* and the oxide thickness T_{OX}. For extracting T_{OX} from I-V fitting, we fixed Φ_B=3.34eV and $m^*=0.35m_0$. Consistency between ellipsometric T_{OX} and extracted T_{OX} implies that Φ_B and m* employed here are valid to characterize the n^+poly-Si/ultrathin SiO_2/Si(100) system.

DIELECTRIC DEGRADATION

The charge-to-breakdown Q_{BD} under constant current stress has been employed as one of methods to test oxide reliability. The current level has been set mostly at 0.1A/cm^2. However, for the oxide thicknesses below 2.36nm, the initial oxide field strength yielding a current level of 0.1A/cm^2 becomes smaller than 10MV/cm and the further decrease will occur during the stress as understood from Fig.3(b). By increasing the stress current up to 1A/cm^2 for the 2.36nm oxide, the initial oxide field becomes 13.5MV/cm which is maintained at the same value even after the soft breakdown, and hence the Q_{BD} value will make sense. This indicates that Q_{BD} under constant current stress is not always a useful measure for testing the oxide reliability in the thickness range below 2.36nm unless the stress current density is properly chosen. Therefore, for sub-3nm oxides the constant voltage stress at electric field strengths above 10MV/cm can offer a useful measure of the reliability [29]. On the other hand the stress-induced leakage current or the soft breakdown current for oxides thicker than 3nm always remains smaller than 0.1A/cm^2 and the gate voltage or oxide field under the constant current stress condition remains almost unchanged. In fact the soft breakdown current is controlled by the direct tunneling current component whose current level is significantly lower than 0.1A/cm^2 [10]. Thus both the constant current and constant voltage stressing can be used for characterizing the gate oxide lifetime.

Next, influence of the absolute value of gate voltage on oxide wearout mechanism should be discussed in particular for sub-3nm oxides. As shown in Fig.3(a) and (b) the maximum oxide voltage which corresponds to 10MV/cm for 1.69 to 2.36nm oxides is smaller than the barrier height of 3.34eV. This means that tunneling electron at $V_{OX}<3.34V$ is no longer enter to the oxide conduction band and the stress induced degradation related to hydrogen release from SiO_2 near the anode/oxide interface, which needs a threshold electron energy of about 2eV in the oxide conduction band at the interface and corresponding gate voltages $V_G>5V$ [30,31], might be less and less important mechanism in determining the soft breakdown process. Also, it is interesting to note that the time-to-breakdown t_{BD} measured as a function of oxide voltage can be uniquely explained by anode hole injection model. This model shows that the oxide lifetime for a given oxide thickness is determined by oxide voltage or oxide electric filed strength during stressing [32]. Since the anode hole injection occurs with a threshold electron energy of about 5eV or $V_G>8V$[33], it is not clear if this mechanism plays an important role at low oxide voltages.

Reliability of ultrathin gate oxides has been evaluated by the time-dependent dielectric breakdown(TDDB) under constant voltage stressing. As shown in Fig.5(a) the Weibull plots for 2.17nm oxides give 50% time-to-breakdown t_{BD} at different oxide electric fields. Figure 5(b) represents 50%t_{BD} as a function of oxide electric field E_{OX}. The results shows that the oxide integrity apparently depends on the oxide electric field as proposed for the anode hole injec-

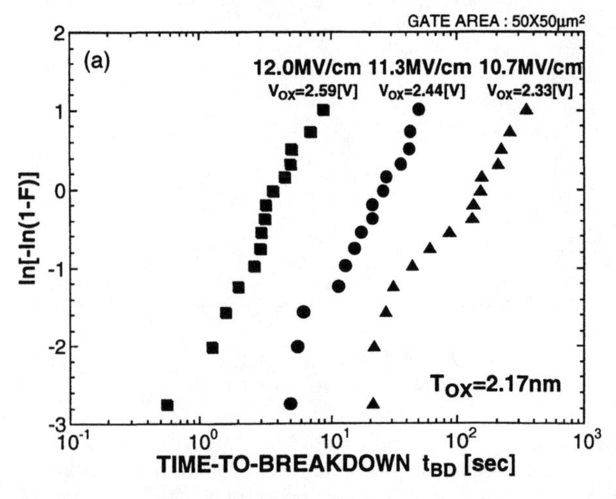

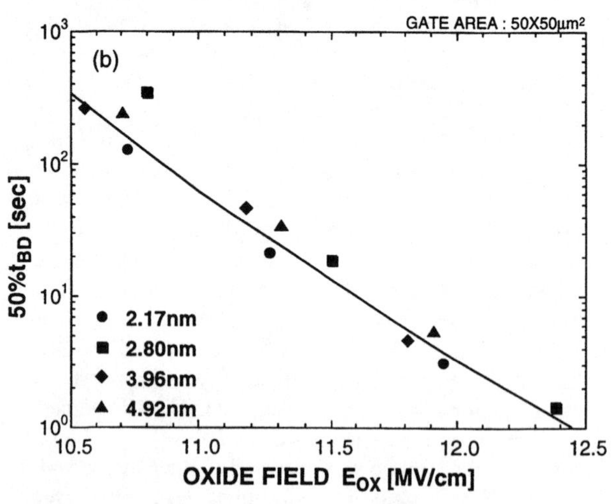

FIGURE 5. The Weibull plots for 2.17nm oxides give 50% time-to-breakdown t_{BD} at different oxide electric field(a), and 50%t_{BD} as a function of oxide electric field E_{OX}(b).

tion model[32,33], although it is not clear that such t_{BD} vs E_{OX} relationship can hold over the wide range of E_{OX}.

SUMMARY

Electronic density of states in ultrathin (1.7-5.2nm) gate oxides is basically identical to those of bulk SiO_2. Existence of localized band tail states below the oxide conduction band is suggested for oxides thinner than 0.8nm. Tunneling current measured for ultrathin oxides is well fitted to the multiple scattering theory and the extracted oxide thickness is in fair agreement with the ellipsometric value. The electron effective mass for direct tunneling was obtained to be $0.35m_0$ in agreement with the previous work. The oxide conductin band effective mass in the Fowler-Nordheim tunneling regime is found to be $0.60m_0$. Wearout behavior of sub-5nm oxides is well described by oxide electric field.

REFERENCES

1. H. S. Momose, M. Ono, T. Yoshitomi, T. Ohguro, S. Nakamura, M. Saito and H. Iwai, *Tech. Dig. International Electron Device Meeting* , 593 (1994).
2. G. Timp, A. Agarwal, F.H. Baumann, T. Boone, M. Buonanno, R. Cirelli, V. Donnely, M. Foad, D. Rrant, M. Green, H. Gossmann, S. Hillenius, J. Jackson, D. Jacobson, R. Kleiman, A. Kornblit, F. Klemens, J. T-C. Lee, W. Mansfield, S. Moccio, A. Murrell, M. O'Malley, J. Rosamilia, J. Sapjeta, P. Silverman, T. Sorsch, W. W. Tai, D. Tennant, H. Vuong and B. Weir, *Tech. Dig. International Electron Device Meeting* , 930 (1997).
3. S. Miyazaki, H. Nishimura, M. Fukuda, L. Ley, and J. Ristein, *Appl. Surf. Sci.* **113/114**, 585 (1997).
4. N. Awaji, S. Ohkubo, T. Nakanishi, Y. Sugita, K. Takasaki and S. Komiya, *Jpn. J. Appl. Phys.* **35**, L67 (1996).
5. S. D. Kosowsky, P. S. Pershan, K. S. Krisch, J. Bevk, M. L. Green, D. Brasen, L. C. Feldman and P. K. Roy, *Appl. Phys. Lett.* **70**, 3119 (1997).
6. H. Watanabe, K. Fujita, T. Kawamura and M. Ichikawa, *Extended Abstracts of 1997 Intern. Conf. on Solid State Devices and Materials*, 538 (1997).
7. K. Okada, Symposium on VLSI Technol., *Digest of Technical Papers*, 143 (1997).
8. S. H. Lee, B. J. Cho, J. C. Lo and S. H. Choi, *Tech. Dig. International Electron Device Meeting*, 605 (1994).
9. M. Depas, M. M. Heyns, T. Nigam, K. Kenis, H. Sprey, R. Wilhelm, A. Crossy, C. J. Sofield and D. Graef, ed. H. Z. Massoud, E. H. Poindexter and C. R. Helms, *The Physics and Chemistry of SiO_2 and Si-SiO_2 Interface-3*, The Electrochmical Society, pp.352 (1996).
10. T. Yoshida, S. Miyazaki and H. Hirose, *Ext. Abstracts of 1996 Intern. Conf. on Solid State Devices and Materials*, 539 (1996).
11. K. Okada and S. Kawasaki, *Ext. Abstracts of 1995 Intern. Conf. on Solid State Devices and Materials*, 473 (1995).
12. R. Degraeve, G. Groeseneken, R. Bellens, M. Depas and H. Meas, *Tech. Dig. International Electron Device Meeting*, 863 (1995).
13. T. Tanamoto and A. Toriumi, *Extended Abstracts of 1996 Intern. Conf. on Solid State Devices and Materials*, 368 (1996).
14. D. J. DiMaria, *Appl. Phys. Lett.* **68**, 3004 (1995).
15. A. Toriumi, J. Koga, H. Satake and A. Ohta, *Tech. Dig. International Electron Device Meeting*, 847 (1995).
16. E. Hasegawa, A. Ishitani, K. Akimoto, M. Tsukiji and N. Ohta, *J. Electrochem. Soc.* **142**, 273 (1995).
17. B. E. Weir, P. J. Silvermann, D. Monroe, K. S. Krisch, M. A. Alam, G. B. Alers, T. W. Sorsch, G. L. Timp, F. Baumann, C. T. Liu, Y. Ma and D. Hwang, *Tech.Dig. International Electron Device Meeting*, 73 (1997).
18. S. Nagano, M. Tsukiji, K. Ando, E. Hasegawa and A. Ishitani, *J. Appl. Phys.* **75** ,3530 (1994).
19. S.-H. Lo, D. A. Buchanan, Y. Taur, L.-K. Han and E. Wu, Symposium on VLSI Technol., *Digest of Technical Papers*, 149 (1997).
20. D. A. Buchanan, and S.-H. Lo, ed. H. Z. Massoud, E. H. Poindexter and C. R. Helms, *The Physics and Chemistry of SiO_2 and Si-SiO_2 Interface-3*, The Electrochmical Society, 3 (1996).
21. T. Yamazaki, S. Miyazaki, C. H. Bjorkman, M. Fukuda and M. Hirose, *Mat. Res. Soc. Symp. Proc.* Vol. **318**, 419 (1994).
22. J. L. Alay and M. Hirose, *J. Appl. Phys.* **81**, 1606 (1997).
23. R. B. Laughlin, J. D. Jannopoulos and D. J. Chandi, ed. S. T. Pantelides, *The Physics of SiO_2 and Its Interfaces*, Pergamon Press, pp.321 (1978).
24. M. Depas, R. L. Van Meirhaeghe, W. H. Laflere and F. Cardon, *Microelectronic Engineering* **22**, 61 (1993).
25. M. Hiroshima, T. Yasaka, S. Miyazaki and M. Hirose, *Jpn. J. Appl. Phys* **33**, 395 (1994).
26. B. Brar, G. D. Wilk and A. C. Seabough, *Appl. Phys. Lett.* **69**, 2728 (1996).
27. M. Fukuda, W. Mizubayashi, S. Miyazaki and M. Hirose, unpublished.
28. Khairurrijal, S. Miyazaki and M. Hirose, *Jpn. J. Appl. Phys.* **36**, L1541 (1997).
29. T. Nigam, M. Depas, R. Degrave, M. Heyns and G. Groeseneken, *Ext. Abstracts of 1997 Intern. Conf. on Solid State Device and Materuals*, 90 (1997).
30. D. J. DiMaria and E. Cartier, *J. Appl. Phys.* **78**, 3883 (1995).
31. D. J. DiMaria and J. H. Stathis, Appl. Phys. Lett. **70**, 2078 (1997).
32. J. C. Lee, I-C. Chem and C. Hu, *IEEE Trans. on Electron Devices* **35**, 2268 (1988).
33. K. F. Schuegraf and C. Hu, *IEEE Trans. on Electron Devices* **41**, 761 (1994).

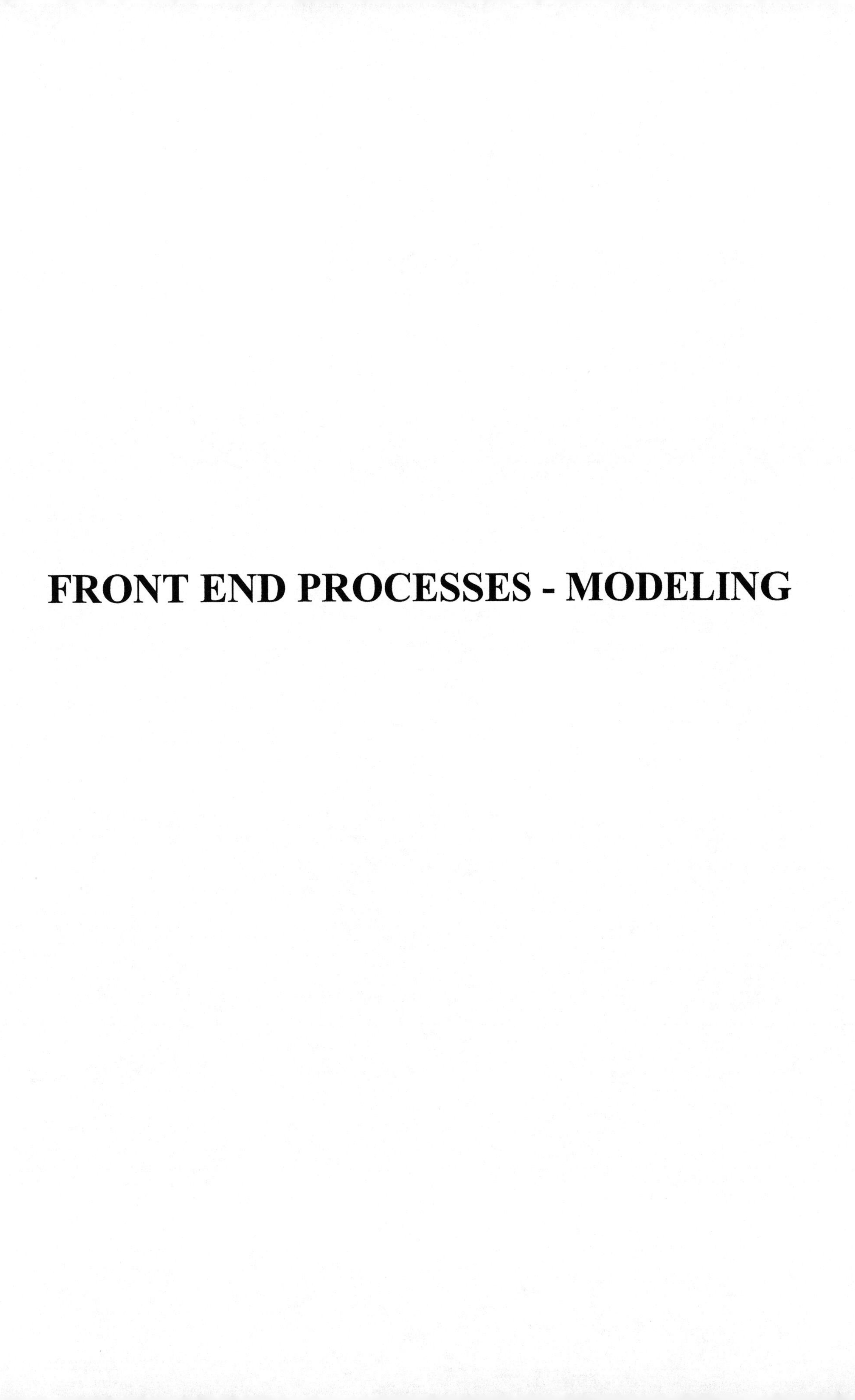

FRONT END PROCESSES - MODELING

Modeling of Manufacturing Sensitivity and of Statistically Based Process Control Requirements for a 0.18 μm NMOS Device

P. M. Zeitzoff*, A. F. Tasch**, W. E. Moore*#, S. A. Khan**, and D. Angelo**&

*SEMATECH, Austin, TX 78741; **University of Texas, Austin, TX; #Currently at AMD, Inc., Austin, TX; &Currently at Applied Materials, Inc

Random statistical variations during the IC manufacturing process have an important influence on yield and performance, particularly as technology is scaled into the deep submicron regime. A simulation-based approach to modeling the impact of these variations on a 0.18μm NMOSFET is presented. The result of this modeling is a special Monte Carlo simulation code that can be used to predict the statistical variation of key device electrical characteristics and to determine the reduction in these variations resulting from improved process control. In addition, the level of process control needed to satisfy specified statistical targets for the NMOSFET electrical performance was analyzed. Meeting these targets requires tight control of five key parameters: the gate length (optimal statistical variation is 9% or less), the gate oxide thickness (optimal statistical variation is 5% or less), the shallow source/drain extension junction depth (optimal statistical variation is 5% or less), the channel dose (optimal statistical variation is 7.5% or less), and the spacer width (optimal statistical variation is 8% or less).

INTRODUCTION

Random variations during the IC manufacturing process cause corresponding variations in device electrical characteristics. These latter variations can result in substantial reductions in yield and performance, particularly as IC technology is scaled into the deep submicron regime. In an effort to understand and deal with this problem, simulation-based modeling of the manufacturing sensitivity of a representative 0.18μm NMOSFET has been carried out, and the results have been used to analyze the process control requirements.

In the sensitivity analysis, eight key device electrical characteristics (responses) were modeled as second order polynomial functions of nine key structural and doping parameters (input parameters). The polynomial models were embedded into a special Monte Carlo code that was used to do statistical simulations. In these simulations, the mean values and the statistical variations of the nine structural and doping parameters were the inputs, and the resulting mean values and statistical variations of the responses were the outputs. Through numerous Monte Carlo simulations with a variety of values for the input parameter variations, the level of process control of the input parameters required to meet specified targets for the device electrical characteristics was analyzed, including the allowable tradeoffs.

DESIGN AND OPTIMIZATION OF A NOMINAL DEVICE

The design and optimization of a 0.18μm NMOSFET for use as the nominal device in the sensitivity analysis involved targeting a set of device structural parameters and their values anticipated in 0.18μm technology. Figure 1 illustrates many of the structural parameters important in the specification of the final device structure. The basic structural parameters of the device such as gate oxide thickness, poly re-oxidation thickness and deep junction depth were chosen to be consistent with the values listed in the 1994 SIA National Technology Roadmap for Semiconductors [1]. Since the completion of this design analysis, the 1997 SIA Roadmap has been published [2]. Some of the structural parameters for the 0.18μm technology node have been changed slightly in the 1997 version; however, the structural parameters used in the analysis are still representative of what is currently planned for use by the industry.

When determining the nominal device structure for this analysis, the eight key electrical parameters shown in Table I were chosen to characterize the device electrical performance. The primary goal was to design a device that showed maximum drive current while still satisfying specific requirements for off-state leakage, short-channel effects, and peak substrate current (to ensure hot carrier reliability). This device would be used in high performance desktop applications. Due to the short gate length of the device, it was necessary to include a boron 'halo' implant as part of the device structure, in order to obtain an optimal combination of turn-off and saturation current performance for the device. The effectiveness of the halo implant in suppressing short-channel effects as well as maintaining hot carrier reliability has been previously reported [3,4]. This implant, along with a V_T adjust channel implant, is found to improve both the V_T rolloff with decreasing channel length and the device reliability, while maintaining acceptable I_{dsat} vs. I_{leak} characteristics of the device. Because of the 1.8V power supply assumed for this technology, consideration had to be given to ensuring hot carrier reliability. This was done through the use of a device in which shallow source-drain extensions were doped with a peak concentration of $4\times10^{19}cm^{-3}$ and were self-aligned to the edge of the oxide grown on the polysilicon gate (15nm from the poly edge).

CP449, *Characterization and Metrology for ULSI Technology: 1998 International Conference*
edited by D. G. Seiler, A. C. Diebold, W. M. Bullis, T. J. Shaffner, R. McDonald, and E. J. Walters

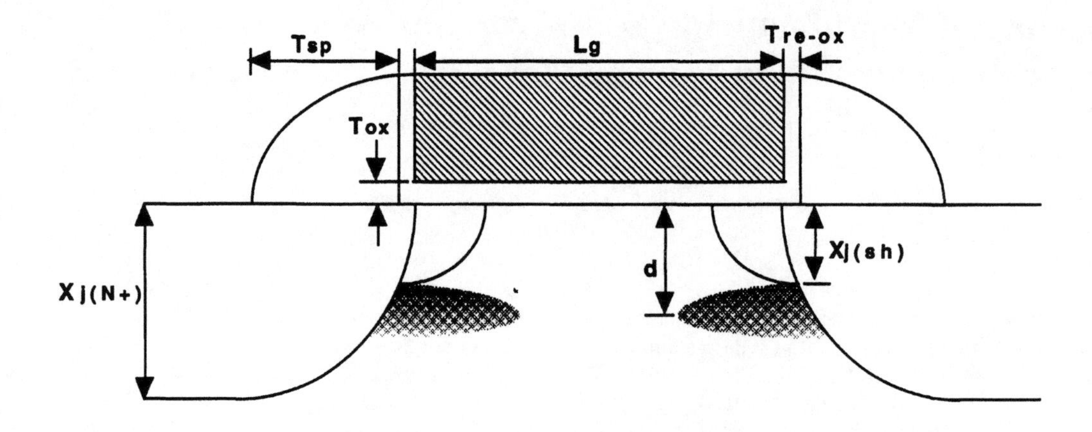

FIGURE 1. *Cross-section of the 0.18μm NMOSFET structure. The nominal values of the structure parameters and the variations that were used in the sensitivity analysis are shown in Table II. The poly re-oxidation thickness, $t_{re\text{-}ox}$, was fixed at 15nm for all simulations.*

The deep source/drain regions were self-aligned to the spacer oxide edge and had a junction depth of 150nm, which was held constant throughout the analysis.

The nominal device structure was determined by examining a large number of simulated devices with different halo peak depths and doses. The only other parameter that was varied was the boron V_T adjust implant, in order to satisfy the requirement for the maximum allowable off-state leakage current. The dose of the V_T adjust implant was adjusted in order to meet an off-state leakage current requirement of $\leq$ 50pA/μm at room temperature. A number of simulations were performed to examine these ranges of variation, and I_{dsat} and ΔV_T due to drain induced barrier lowering (DIBL) were examined for an optimal combination of current drive and turn-off characteristics. The results of these simulations lead to the selection of a boron halo implant dose of $1.5\times10^{13}cm^{-2}$ with a peak doping profile depth of 80nm, in order to obtain maximum drive current while maintaining acceptable turn-off characteristics.

The one-dimensional process simulator SUPREM-3 and the two-dimensional process simulator TSUPREM-4 (both from Avant! Corp.) were used to generate the doping profiles for the various regions of the device. Due to the uncertain accuracy of two-dimensional diffusion models for arsenic implanted junctions with short thermal cycles, the one-dimensional vertical profile of both the shallow and deep S/D junctions was simulated using SUPREM-3. For each junction, the two-dimensional profile was then generated by extending the vertical profile laterally using a complementary error function with a characteristic length corresponding to 65% of the vertical junction depth. Conversely, the two-dimensional halo implant profile was directly simulated using TSUPREM-4. The V_T adjust implant vertical profile was simulated using SUPREM-3, and was then extended laterally without change over the entire device structure. A composite profile containing all the above individual profiles was generated and imported to the device simulator, UT-MiniMOS[5]. UT-MiniMOS was chosen to simulate the device's electrical characteristics because it has both the UT hydrodynamic (HD) transport model based on non-parabolic energy bands and the UT models for substrate current[6], quantum mechanical effects [7,8,9] and mobility in the inversion layer [10,11,12]. Also, UT-MiniMOS has adaptive gridding capability, and this capability was used to adapt the grid to the potential gradient and the carrier concentration gradients during the simulations. Table II lists the final structural and doping parameters for the nominal device along with the amount of statistical variation used for each of the nine parameters that were varied in the sensitivity analysis.

SENSITIVITY ANALYSIS

A primary aim of this analysis was to obtain a set of complete, second-order empirical equations relating the random manufacturing variations in the structural and doping parameters of the representative 0.18μm NMOSFET to the key device electrical parameters listed in Table I. A three-level Box-Behnken design[13] was performed in order to obtain the responses of the output parameters to the input parameters. Besides the center point where all factors were maintained at their nominal values, the other data points were obtained by taking three factors at a time and developing a 2^3 factorial design for them, with all other factors maintained at their nominal values. The advantage of this design is that fewer runs are required to obtain a quadratic equation as compared to other designs. A total of 97 (96 variations plus the one nominal device simulation) runs were required for this analysis for the case of nine input factors. One drawback of this design, however, is that all of the runs must be performed prior to obtaining any equation, and it is not amenable to two-stage analyses. Hence, there is

TABLE I. *Key device electrical characteristics and target values for the nominal device.*

Electrical Characteristic	Target Value
Threshold Voltage (from extrapolated linear I-V, @V_d = 0.05V), $\mathbf{V_T}$ (Volts)	≤0.5
Drive current (@V_g = V_d = V_{dd}), $\mathbf{I_{dsat}}$ (µA/µm of device width)	450-500
Peak off-state leakage current (@V_d = 2V, V_g = 0, T = 300°K), $\mathbf{I_{leak}}$ (pA/µm of device width)	≤50
D.I.B.L. (V_t @ V_d = 0.05V - V_t @ V_d = V_{dd}), $\mathbf{\Delta V_T}$ (mV)	≤100
Peak substrate current (@V_d = V_{dd}), $\mathbf{I_{sub}}$ (nA/µm of device width)	<200
Subthreshold swing (@V_d =0.05V), **S** (mV/decade of Id)	≤90
Peak transconductance (@V_d =2.0V); $\mathbf{g_m}^s$ (mS/mm of device width)	300-400
Peak transconductance (@V_d =0.05V); $\mathbf{g_m}^l$ (mS/mm of device width)	30-60

TABLE II. *Nominal value and statistical variation for each of the input parameters*

	Input Parameter	Nominal	3σ Variation
1	Gate length, $\mathbf{L_g}$ (µm)	0.18	±15%
2	Gate oxide thickness, $\mathbf{T_{ox}}$ (nm)	4.50	±10%
3	Spacer oxide width, $\mathbf{T_{sp}}$ (nm)	82.50	±15%
4	Shallow junction doping profile depth, $\mathbf{X_{j(sh)}}$ (nm), @N_D =4.36×10^{18}cm^{-3}. (This is where N_D =N_A for the nominal device)	50	±10%
5	Peak shallow junction doping, $\mathbf{N_{sh}}$ (cm^{-3})	4×10^{19}	±10%
6	Channel dose, $\mathbf{N_{ch}}$ (cm^{-2})	5.65×10^{12}	±10%
7	Halo dose, $\mathbf{N_{halo}}$ (cm^{-2})	1.5×10^{13}	±10%
8	Halo peak depth, **d** (nm)	80	±10%
9	Series resistance (external), $\mathbf{R_s}$ (Ω-µm)	400	±15%

no indication of the level of factor influence until the entire experiment has been conducted.

In addition to the nine input parameters that were varied (shown in Table II), several device parameters, such as the deep junction profile and its peak doping used in the nominal device, were held constant throughout all of the simulations. In addition, a background substrate doping of $5\times10^{15}cm^{-3}$ and an interface charge of $3\times10^{10}cm^{-2}$ were also uniformly applied. The eight key device electrical characteristics listed in Table I were chosen as the response variables. After the completion of the 97 simulations, two sets of model equations were generated for each response; one in terms of the actual values of the input parameters, and the other in terms of their normalized values. The normalized values are calculated using the following equation

$$\Delta x_i = 2\frac{\left(\xi_i - \bar{\xi}_i\right)}{d_i} \tag{1}$$

where $\bar{\xi}_i$ is the nominal value of the ith input parameter, ξ_i is the actual value of the ith input parameter for any given run, and d_i is the magnitude of the difference between the two extreme values for x_i used in the sensitivity analysis. The equations that use the Δx_i's for their variables will be called "normalized model equations."

After each of the 97 simulations was performed, the eight electrical responses were extracted for each device, and were then entered into a design matrix. Analysis of variance (ANOVA) methods were used to estimate the coefficients of the second order model equations, and to test for the significance of each term in the model. For the model as a whole, information such as the coefficient of determination, R^2, and coefficient of variation was generated using the data from the ANOVA. A diagnosis of the model was performed by using normal probability plots to evaluate the normality of the residuals. A transformation of the response was made if necessary. The ANOVA was performed again if a transformation was made, and comparisons of the normal probability plots and coefficients of determination were made to decide whether to use the transformed data or not. In addition, plots of Cook's distance vs. run order and Outlier-t vs. run order were generated in order to check for the occurrence of any extraneous data (outliers). The corresponding model was then used to generate a set of reduced equations, i.e. equations including only those terms with at least a 95% level of significance (terms with a lower significance value were discarded from the ANOVA). For input parameters outside of the range of values used in this analysis, the model equations are not guaranteed to be accurate, and they are expected to become less and less accurate as the values move further and further outside the range.

The final resulting normalized model equations are shown in the Appendix for the eight key electrical responses. In these equations, all of the input parameters will take on a value between -1 and +1, as determined by equation (1) above. The relative importance of any term is determined solely by the relative magnitude of the coefficient of that term. For example, in the model equation for the saturation drive current, I_{dsat} is most sensitive to the gate length (ΔL_g), followed by the oxide thickness (ΔT_{ox}), the shallow junction depth (ΔX_{sh}), and the spacer oxide thickness (ΔT_{sp}). In some cases, the output responses were transformed into a form that yielded a more reliable model equation. Typically, if the residual analysis of the data suggests that, contrary to assumption, the standard deviation (or the variance) is a function of the mean, then there may be a convenient data transformation, Y=f(y), that has a constant variance. If the variance is a function of the mean, the dependence exhibited by the variance typically has either a logarithmic or a power dependence and an appropriate variance stabilizing transformation can be performed on the data. Once the response is transformed, the model calculations and coefficients are all in terms of the transformed response, and the resulting empirical equation will also be in terms of the transformation. In this study, both a logarithmic and an inverse square root dependence were used in transforming three of the output responses so that a better model equation fit was obtained (see the Appendix).

Once the model equations were obtained from the ANOVA, Monte Carlo techniques were used to extract the statistical distribution of each response for a specified set of statistical variations of the nine structural and doping parameters. A special Monte Carlo code was written that treats each of these parameters as a random input variable in the normalized model equations. This is a good approximation, since any correlations amongst these parameters is weak and second order. Each of these random variables comes from a normal distribution with a specified mean and standard deviation, σ. Using a random number generator that returns a series of non-uniform deviates chosen randomly from a normal distribution that correctly predicts the specified mean and standard deviation, the code selects individual values that are representative of the random manufacturing variations in each of the nine input parameters. These values are used as inputs to the second-order normalized model equations in order to calculate the final values of each of the eight electrical characteristics. A large number of such trials (typically ~5000) is run to generate the probability distribution of the characteristics. The resulting distribution is then analyzed to obtain the mean and standard deviation. The results of the Monte Carlo simulations (with the 3σ input parameter variations listed in Table II) are listed in Table III, and are explained in the next section.

MONTE CARLO-BASED DETERMINATION OF PROCESS CONTROL REQUIREMENTS

The Monte Carlo tool was utilized for several purposes: (1) to determine the impact on the statistical variations of the device electrical parameters of changing the input parameter statistical variations by arbitrary amounts; (2) to find an optimal set of reductions in the input parameter variations to meet device electrical statistical targets; (3) to analyze process, device, equipment, and metrology statistical control requirements, including those necessary to meet National Technology Roadmap for Semiconductors [14] (NTRS) requirements; and (4) to analyze whether device electrical targets (for example, the maximum leakage current specification) are reasonable. In this section, we will focus on Items (1) and (2) above.

In Table III, the statistical target value for each response is listed, as well as the Monte Carlo results from the previous section, where relatively large input parameter statistical variations were assumed. (In the "Parameter of Interest" column, the "Maximum Value" is calculated as the mean value + [3σ statistical variation], while the "Minimum Value" is calculated as the mean value - [3σ statistical variation].) The targets for the first four responses in the table (V_T, ΔV_T due to DIBL, I_{dsat}, and I_{leak}) are not met, but the targets for the last four parameters in the table (I_{sub}, S, g_m^s, and g_m^l) are met. To bracket the problem, and to determine whether the targets for the first four responses are realistic, the first step was to reduce all the input parameter variations in two stages, first to a set of more "realistic" values and second to a set of "aggressive" values. These sets are listed in Table IV, and reflect the judgement of several SEMATECH experts[15]. A Monte Carlo simulation was performed for each of these sets of variations, and the simulation results are listed in Table V. The targets for the third and fourth responses, I_{dsat} and I_{leak}, are satisfied with the more realistic input variations, but the targets for V_T and ΔV_T are only satisfied with the aggressive input variations. The conclusion is that the targets for all the responses can probably be met, but that it will be especially difficult to meet them for V_T and ΔV_T.

Next, Monte Carlo simulation was used to meet the targets for the output parameters with an optimal set of reductions in the input parameter variations. Each input parameter variation was reduced in steps, as listed in Table VI. (Note that, for each input parameter, the maximum is the variation used in the previous section, and the minimum is half or less than half of the maximum.) The straightforward approach is to run a series of Monte Carlo simulations covering the entire range of possible combinations for the input parameter variations. However, the number of simulations is 46,656 for each response (see Table VI), an unreasonably high number. In order to reduce the number of simulations to a more manageable total, the following procedure was used. For each response, the normalized model equation was examined to select those input parameters that are either missing from the normalized model equation or are included only in terms with very small coefficients. Since the variations of those inputs are unimportant in influencing the response, the variation was held fixed at its maximum value for each of these selected parameters. As shown in Table VII, following this procedure, two parameters were selected for each response, and hence the number of Monte Carlo simulations was reduced to a more manageable 3888 or 5184. (Note that, based on physical grounds, the selected input parameters would be expected to be relatively unimportant in determining the value of the response.) Since each Monte Carlo simulation took about three seconds to run on a Hewlett-Packard work station, the total simulation time was about three to four hours for each response. (Table VII does not include listings for I_{sub}, S, g_m^l, and g_m^s since those responses are within specification with the maximum values for the variation of all the input parameters, as shown in Table III.) The outputs from the Monte Carlo simulations were imported to a spreadsheet program for analysis and display. By utilizing the spreadsheet capabilities, the input variations were then iteratively reduced from their maximum values to meet the targets for the responses.

As discussed above, it is most difficult to meet the targets for V_T and for ΔV_T due to DIBL. Hence, these two were dealt with first. From the size of the coefficients in the normalized model equation for V_T, the terms containing ΔL_g, ΔT_{ox}, ΔX_{sh}, and ΔN_{ch} are the most significant. Thus, the 3σ statistical variations of only these parameters were reduced to meet the V_T target, while the variations of the other input parameters were held at their maximum values. Contour plots of constant 3σ variation in V_T were determined using the spreadsheet program. The results are shown in Figure 2, where the 3σ variations of ΔT_{ox} and ΔX_{sh} were fixed at their realistic values of 5% each, and the variations of ΔL_g and ΔN_{ch} were varied. Along Contour 1, the 3σ variation in V_T is 50 mV, and the variations of both ΔL_g and ΔN_{ch} are less than 7.5%. Since these variations are quite low (see Table IV), the 50 mV target will be very difficult to meet. Along Contour 2, the 3σ variation in V_T is 60 mV. This target is realistic because the variations of both ΔL_g and ΔN_{ch} on the contour are achievable, particularly in the vicinity of the point where the variations are about 9.5% for ΔL_g and 7.5% for ΔN_{ch} (see Table IV). Figure 3 also shows contours of constant 3σ variation in V_T; the only difference from Figure 2 is that the 3σ variation of ΔX_{sh} is 7.5%, not 5% as in Figure 2. The 60 mV contour here, labeled Contour 3, is shifted significantly to the left from the 60 mV contour in Figure 2, and hence is much more difficult to achieve. For the case where the variation of ΔT_{ox} is 7.5%, while that of ΔX_{sh} is 5%, the 60 mV contour is shifted even further to the left than Contour 3. The contour plots can be utilized to understand quantitatively the impact of the statistical variations of the key input parameters, and how they can be

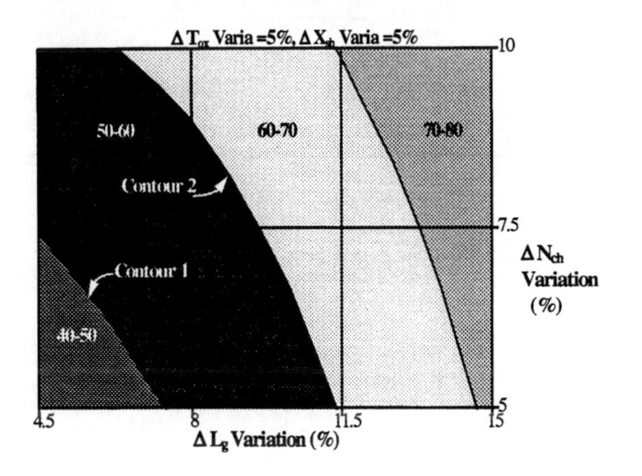

FIGURE 2. *Contours of constant 3σ V_T variation (mV)*

TABLE III. *Monte Carlo Results with Maximum Input Parameter Statistical Variations*

Responses (key device elec. characteristics)	Parameter of Interest	Target Value	Monte Carlo Simulated Value for Parameter of Interest
V_T (mV), Extrapolated	3σ variation	**<50-60**	**97.1**
ΔV_T (mV), DIBL	Maximum Value	**<100**	**121.8**
I_{dsat} (μA/μm)	Minimum Value	**>450**	**430.2**
I_{leak} (pA/μm)	Maximum Value	**<50**	**95.5**
I_{sub} (nA/μm)	Maximum Value	<200	55.6
S (mV/decade)	Maximum Value	≤90	86.8
g_m^s (mS/mm)	Minimum Value	>300	397.8
g_m^l (mS/mm)	Minimum Value	>30	48.0

TABLE IV. *Sets of input parameter statistical variations*

Input Parameter	Maximum 3σ Variation	Realistic 3σ Variation	Aggressive 3σ Variation
ΔL_g	±15%	±10%	±7.5%
ΔT_{ox}	±10%	± 5%	±2.5%
ΔT_{sp}	±15%	±10%	±5%
ΔX_{sh}	±10%	± 5%	±2.5%
ΔN_{sh}	±10%	±10%	±7.5%
Δd	±10%	± 5%	±2.5%
ΔN_{halo}	±10%	±7.5%	±5%
ΔN_{ch}	±10%	±10%	±7.5%
ΔR_s	±15%	±10%	±7.5%

TABLE V. *Monte Carlo results for different levels of input parameter statistical variations*

Responses (key device electrical parameters)	Parameter of Interest	Target Value	Monte Carlo Simulation Results: Maximum Input Parameter Variation	Realistic Input Parameter Variation	Aggressive Input Parameter Variation
V_T (mV), Extrapolated	3σ variation	**<50-60**	**±97.1**	**±66.1**	±46.0
ΔV_T (mV), DIBL	Maximum Value	**<100**	**121.8**	**104.8**	95.8
I_{dsat} (mA/μm)	Minimum Value	**>450**	**430.2**	457.8	482.6
I_{leak} (pA/μm)	Maximum Value	**<50**	**95.5**	30.3	15.3
I_{sub} (nA/μm)	Maximum Value	<200	55.6	49.4	46.2
S (mV/decade)	Maximum Value	≤90	86.8	85.7	85.1
g_m^s (mS/mm)	Minimum Value	Maximize	397.8	427.8	442.0
g_m^l (mS/mm)	Minimum Value	Maximize	48.0	52.1	54.1

TABLE VI. *Steps in input parameter statistical variation*

Parameters	Maximum (%)	Minimum (%)	Step Size	# of Steps	# of combinations
$\Delta L_g, \Delta X_{sh}, \Delta N_{sh}, \Delta d, \Delta N_{halo}, \Delta N_{ch}$	10	5	2.5	3	3^6=729
$\Delta Lg, \Delta Tsp, \Delta Rs$	15	4.5	3.5	4	4^3=64
Total # of combinations			64 x 729 =		**46,656**

TABLE VII. *Reduced number of steps in input parameter variation*

Response	ΔL_g	ΔT_{ox}	ΔT_{sp}	ΔX_{sh}	ΔN_{sh}	Δd	ΔN_{halo}	ΔN_{ch}	ΔR_s	**# of comb's.**
V_T	4	3	4	3	3	**1**	3	3	**1**	3888
ΔV_T	4	3	4	3	3	3	3	**1**	**1**	3888
I_{dsat}	4	3	4	3	**1**	**1**	3	3	4	5184
I_{leak}	4	3	4	3	3	**1**	3	3	**1**	3888

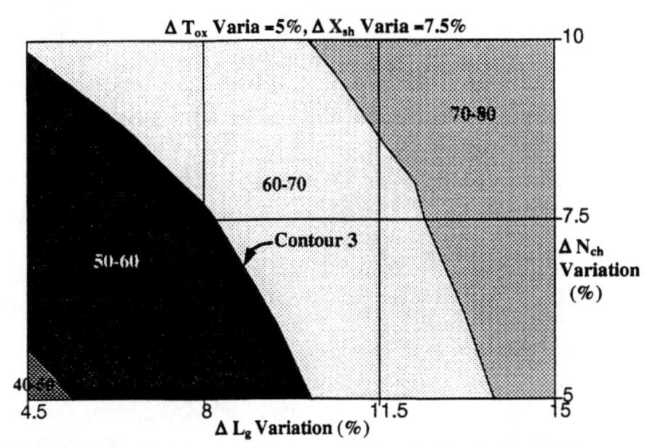

FIGURE 3. *Contours of constant 3σ variation in V_T (mV)*

traded off to reach a specific target for V_T variation. Looking particularly at Contour 2 in Figure 2, and utilizing "realistic" values of the variations as much as possible (see Table IV), an optimal choice for the variations is 5% for ΔT_{ox} and ΔX_{sh}, 7.5% for ΔN_{ch}, and 9.5% for ΔL_g.

Next, the requirements to meet the target for ΔV_T due to DIBL were explored. From the size of the coefficients in the normalized model equation for ΔV_T, the terms containing ΔL_g, ΔX_{sh}, ΔT_{sp}, and ΔT_{ox} are the most significant. Thus, the variations of only these parameters were reduced to meet the ΔV_T target, while the variations of the other input parameters were held at their maximum values. Figure 4 shows contours of constant maximum value of ΔV_T, where the variations of both ΔT_{ox} and ΔX_{sh} were held at 5%, as in Figure 2. Along Contour 4, the value is 100 mV, the target value. The variations of ΔL_g and ΔT_{sp} on this contour are realizable, particularly in the vicinity of the point where the variations are about 9% for ΔL_G and 8% for ΔT_{sp}. Utilizing the same reasoning as discussed above for meeting the V_T target, this point is about optimal.

Finally, the requirements to meet the targets for I_{dsat} and I_{leak} were explored. From the size of the coefficients in the normalized model equations, the terms containing ΔL_g, ΔT_{ox}, , and ΔX_{sh} are the most significant for I_{dsat}, while the terms containing ΔL_g, ΔX_{sh}, ΔN_{ch}, and ΔT_{ox} are most

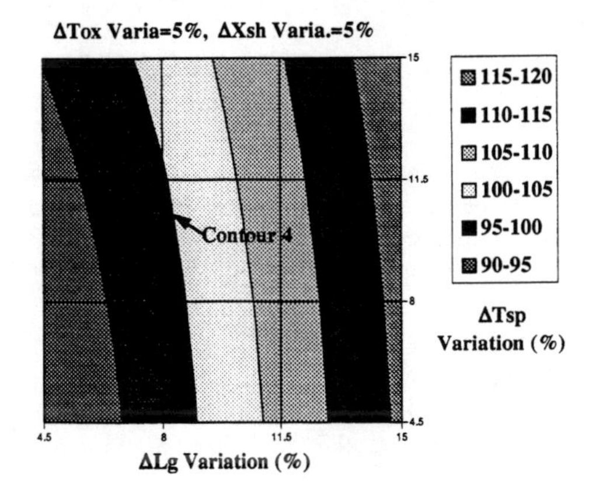

FIGURE 4. *Contours of constant maximum ΔV_T [DIBL] (mV)*

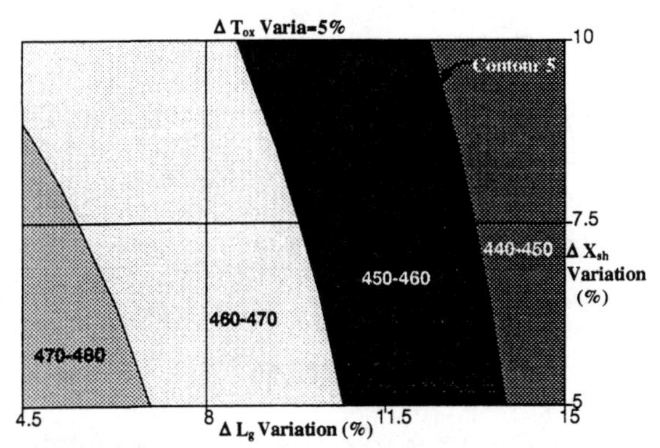

FIGURE 5. *Contours of constant minimum I_{dsat} (μA/μm)*

significant for I_{leak}. Figure 5 shows the contours of constant minimum value of I_{dsat}, with the variation of ΔT_{ox} held at 5% as in Figures 2, 3, and 4. Along Contour 5, the minimum I_{dsat} is 450μa/μm, the target value. Similarly, Figure 6 shows the contours of constant maximum value of I_{leak}, with the variations of all input parameters except ΔX_{sh} and ΔL_g held at their maximum variations. Along Contour 6, the maximum I_{leak} is 50pA/μm, the target value. The input parameter variations along both Contours 5 and 6 are significantly larger than those required to meet the targets for V_T and ΔV_T due to DIBL (see Contour 2 in Figure 2 and Contour 4 in Figure 4). Hence, if the V_T and ΔV_T targets are met, then the targets for I_{leak} and I_{dsat} are also automatically met.

Tying all the above results and discussion together, only five input parameters, ΔL_g, ΔT_{ox}, ΔX_{sh}, ΔN_{ch}, and ΔT_{sp}, need to be tightly controlled (i.e., the variation of each of them must be notably less than the "Maximum 3σ Variation" in Table IV) to meet the targets in Table III for the device electrical characteristics. The other four input parameters, Δd, ΔN_{halo}, ΔN_{sh}, and ΔR_s, can be relatively loosely controlled (i.e., the variation of each of them can be equal to or possibly larger than the maximum variation in Table IV), and the targets will still be met. An optimal set of choices that satisfies all the output response targets is: variation of $\Delta L_g \leq$ 9%, variation of $\Delta T_{ox} \leq$ 5%, variation of $\Delta X_{sh} \leq$ 5%, variation of $\Delta N_{ch} \leq$ 7.5%, and variation of $\Delta T_{sp} \leq$ 8%. The

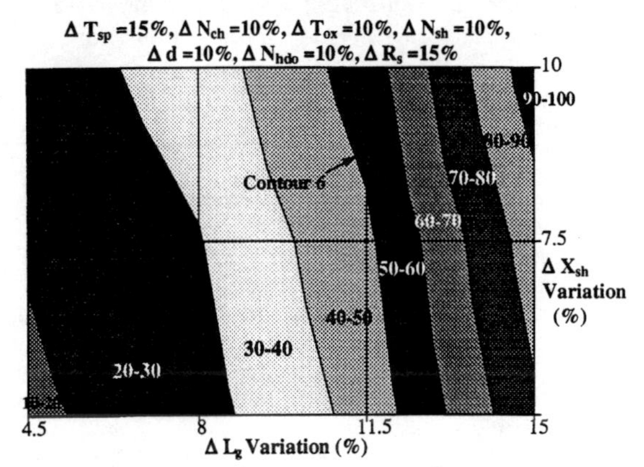

FIGURE 6. *Contours of constant maximum I_{leak} (pA/μm)*

9% requirement on gate length control will be pushing the limits, since, according to the 1997 SIA Roadmap [16], the gate CD (critical dimension) control is 10%. This approach can also be used to determine tradeoffs. If, for example, the process variation of ΔT_{sp} can only be controlled to 12%, it is evident from Contour 4 of Figure 4 that the control of ΔL_g would have to be tightened so that its process variation is 8% or less.

CONCLUSIONS

A 0.18 μm NMOSFET device was designed and optimized to satisfy a specified set of electrical characteristics. This optimized device was the nominal design center for a simulated sensitivity analysis in which normalized second-order polynomial model equations were embedded within a special Monte Carlo code. Monte Carlo simulations with the code were used to correlate the random statistical variations in key electrical device characteristics to the random variations in the key structural and doping parameters. Using these simulations, process control tradeoffs amongst the different structural and doping parameters were explored, and the level of process control required to meet specified statistical targets for the device electrical characteristics was analyzed. Meeting these targets requires tight control of five key parameters: the gate length (optimal statistical variation is 9% or less), the gate oxide thickness (optimal statistical variation is 5% or less), the shallow source/drain extension junction depth (optimal statistical variation is 5% or less), the channel dose (optimal statistical variation is 7.5% or less), and the spacer width (optimal statistical variation is 8% or less).

ACKNOWLEDGMENTS

This work was supported by SEMATECH, the Semiconductor Research Corporation, the Texas Advanced Technology Program, Motorola, Advanced Micro Devices, Micron, Rockwell International, and SGS Thomson.

We are grateful to Terri Moore of SEMATECH's Statistical Methods Department for providing valuable critiques of the statistical modeling aspects of this work.

REFERENCES

[1] Semiconductor Industry Association, "The National Technology Roadmap for Semiconductors," 1994.

[2] Semiconductor Industry Association, "The National Technology Roadmap for Semiconductors," 1997.

[3] A. Hori, A. Hiroki, H. Nakaoka, M. Segawa and T. Hori, "Quarter-Micrometer SPI (Self-Aligned Pocket Implantation) MOSFET's and Its Application for Low Supply Voltage Operation," *IEEE Trans. on Electron Devices*, Vol. 42, No. 1, Jan. 1995.

[4] A. Chatterjee, J. Liu, S. Aur, P.K. Mozumder, M. Rodder and I.-C. Chen, "Pass Transistor Designs using Pocket Implant to Improve Manufacturability for 256Mbit DRAM and Beyond," p. 87-90, *IEDM* 1994.

[5] UT-MiniMOS 5.2-3.0 Information Package, Microelectronics Research Center, The University of Texas at Austin, 1994.

[6] V. Martin Agostinelli, T. James Bordelon, Xiaolin Wang, Khaled Hasnat, Choh-Fei Yeap, D.B. Lemersal, Al F. Tasch, and Christine M. Maziar, "Two-Dimensional Energy-Dependent Models for the Simulation of Substrate Current in Submicron MOSFET's," *IEEE Trans. on Electron Devices*, vol. 41, no. 10, Oct. 1994.

[7] M. .J. van Dort, P.H. Woerlee, and A.J. Walker, "A Simple Model for Quantisation Effects in Heavily-Doped Silicon MOSFETs at Inversion Conditions," *Solid-State Electronics*, vol. 37, no. 3, pp. 411-414, 1994.

[8] S. A. Hareland, S. Krishnamurthy, S. Jallepalli, C.-F. Yeap, K. Hasnat, A.F. Tasch, Jr., and C.M. Maziar, "A Computationally Efficient Model for Inversion Layer Quantization Effects in Deep Submicron N-Channel MOSFETs," *IEDM*, Washington, D.C., pp. 933-936, December 1995.

[9] S.A. Hareland, S. Krishnamurthy, S. Jallepalli, C.-F. Yeap, K. Hasnat, A.F. Tasch, Jr., and C.M. Maziar, "A Computationally Efficient Model for Inversion Layer Quantization Effects in Deep Submicron N-Channel MOSFETs," *IEEE Trans. on Elec. Dev.*, vol. 43, no. 1, pp. 90-96, Jan. 1996.

[10] H. Shin, G.M. Yeric, A.F. Tasch, and C.M. Maziar, "Physically-Based Models for Effective Mobilityy and Local-Field Mobility of Electrons in MOS Inversion Layers," *Solid-State Electronics*, vol. 34, no. 6, pp. 545-552, 1991.

[11] V.M. Agostinelli, H. Shin, A.F. Tasch, "A Comprehensive Model for Inversion Layer Hole Mobility for Simulation of Submicrometer MOSFET's," *IEEE Trans. on Electron Devices*, vol. 38, no. 1, pp. 151-159, 1991.

[12] S.A. Khan, K. Hasnat, A.F. Tasch, and C.M. Maziar, "Detailed Evaluation of Different Inversion Layer Electron and Hole Mobility Models," *Proc. of the 11th Biennial University/Government/Industry Microelectronics Symposium*, Austin, TX, p. 187, May 16-17, 1995.

[13] G.E.P. Box and D.W. Behnken, "Some New Three Level Designs for the Study of Quantitative Variables," Technometrics, Vol. 2, No. 4, Nov. 1960.

[14] Semiconductor Industry Association, "The National Technology Roadmap for Semiconductors," 1997.

[15] Private conversation between P.M. Zeitzoff and L. Larson.

[16] Semiconductor Industry Association, "The National Technology Roadmap for Semiconductors," 1997, Table 24, p. 85.

APPENDIX: NORMALIZED MODEL EQUATIONS

The equations given below for each of the eight output responses are a function of the normalized values from the normalized set of input factors. Each of these input factors, for example ΔL_g, will have a range of values between -1 and +1. Referring to Table II, a -1 value for ΔL_g would correspond to a gate length, L_g , of 0.18μm-15%=0.153μm while a +1 value for ΔL_g would indicate an L_g =0.18μm+15%=0.207μm. For the nominal case, each normalized input variable would have a value of 0. Each input parameter value in this table must be correlated to the appropriate range of values as shown in Table II.

$\mathbf{V_T}$ (mV) = 454.9 + 56.1 ΔL_g + 52.4ΔT_{ox} + 41.4ΔN_{ch} - 33.6ΔX_{sh} + 9.9ΔN_{sh} + 9.7ΔT_{sp} + 7.1ΔN_{halo} - 1.6Δd - 19.9ΔL_g^2 - 5.9ΔT_{ox}^2 - 3.8ΔT_{sp}^2 + 3.8Δd^2 + 21.9ΔL_g x ΔX_{sh} - 6.9ΔL_g x ΔT_{sp} - 6.2ΔL_g x ΔN_{sh}

$\mathbf{\Delta V_T}$ (mV) = 72.3 - 39.4ΔL_g + 17.8ΔX_{sh} - 12.7ΔT_{sp} + 8.4ΔT_{ox} - 5.9ΔN_{halo}- 4.7ΔN_{ch} - 3.0ΔN_{sh} - 1.5Δd + 12.4ΔL_g^2 + 7.8ΔT_{sp}^2 - 2.3ΔN_{sh}^2 - 11.9ΔL_g x ΔX_{sh} + 5.9ΔL_g x ΔT_{sp} - 5.3ΔL_g x ΔT_{ox} + 4.4ΔT_{sp} x Δd + 2.4ΔL_g x ΔN_{halo} + 2.3ΔT_{sp} x ΔN_{halo}

[a]$\mathbf{I_{dsaT}}$ (log[μA/μm]) = 2.721 - 0.060ΔL_g - 0.052ΔT_{ox} + 0.028ΔX_{sh} - 0.019ΔT_{sp} - 0.016ΔN_{ch} - 0.008ΔN_{halo} - 0.007ΔR_s - 0.005ΔN_{sh} + 0.015ΔL_g^2 + 0.009ΔT_{ox}^2 + 0.009ΔT_{sp}^2 + 0.013ΔX_{sh} x ΔN_{ch} - 0.011ΔT_{sp} x ΔNch- 0.008ΔL_g x ΔT_{ox} - 0.006ΔL_g x ΔX_{sh}

[a]$\mathbf{I_{leak}}$ (log[pA/μm]) = 0.385 - 1.189ΔL_g + 0.571ΔX_{sh} - 0.508ΔN_{ch} - 0.417ΔT_{ox} - 0.241ΔT_{sp} - 0.144ΔN_{sh} - 0.127ΔN_{halo} + 0.011Δd - 0.424ΔL_g^2 + 0.104ΔT_{sp}^2 - 0.080Δd^2 - 0.063ΔN_{sh}^2 + 0.055ΔT_{ox}^2 - 0.449ΔL_g x ΔX_{sh} + 0.156ΔL_g x ΔT_{sp} + 0.112ΔL_g x ΔN_{sh} - 0.088ΔL_g x ΔT_{ox}

$\mathbf{I_{sub}}$ (nA/μm) = 35.4 - 17.6ΔL_g - 1.9ΔT_{sp} - 1.7ΔN_{ch} - 1.7ΔR_s + 1.5ΔN_{sh} - 1.5ΔT_{ox} + 1.2ΔX_{sh} + 6.2ΔL_g^2 + 2.0ΔT_{ox}^2 + 1.7ΔT_{sp}^2 + 1.7ΔX_{sh}^2 + 1.5ΔN_{ch}^2 + 3.2$\Delta L_g \times \Delta N_{ch}$

[b]$\mathbf{S^{-1/2}}$ (mV/decade)$^{-1/2}$ = 0.10916 - 0.00137ΔT_{ox} - 0.00085ΔL_g - 0.00050ΔN_{ch} - 0.00018ΔN_{halo} - 0.00016ΔT_{sp} + 0.00016ΔX_{sh} - 0.00030ΔX_{sh}^2 - 0.00027ΔL_g^2 - 0.00022ΔN_{halo}^2 + 0.00047ΔL_g x ΔX_{sh} - 0.00028ΔL_g x ΔT_{sp}

$\mathbf{g_m^s}$ (mS/mm) = 470.6 - 50.4ΔT_{ox} - 41.7ΔL_g + 14.4ΔX_{sh} - 6.0ΔN_{halo} + 2.9ΔN_{ch} - 14.3ΔX_{sh}^2 - 12.5ΔN_{halo}^2 - 20.2ΔT_{ox} x ΔN_{ch}

$\mathbf{g_m^1}$ (mS/mm) = 58.63 - 9.44ΔL_g - 3.93ΔT_{ox} + 2.74ΔX_{sh} - 1.69ΔT_{sp} - 1.53ΔR_s - 0.85ΔN_{ch} - 0.47ΔN_{sh} + 0.43Δd - 0.34ΔN_{halo} + 1.68ΔL_g^2 + 1.36ΔT_{ox}^2 + 0.94ΔN_{halo}^2 + 0.69ΔN_{ch}^2 + 0.64ΔT_{sp}^2 - 1.33ΔL_g x ΔX_{sh} + 0.78ΔL_g x ΔT_{sp}

[a] A logarithmic transformation was used to achieve better normality and fit.
[b] An inverse square root transformation was used to achieve better normality and fit.

An Analytical Framework for First-Order CMOS Device Design

J. A. del Alamo[1], W. T. Lynch[2], and D. A. Antoniadis[1]

[1] MIT, Cambridge, MA 02139; [2] SRC, Research Triangle Park, NC 27709

We have constructed a simple physics-based analytical framework for MOSFET device simulation. The suite of models is optimized for "well designed" devices that exhibit sufficient electrostatic integrity. This is assessed by means of an analytical model for drain-induced barrier lowering. Using the developed framework, we have simulated 0.1 μm n-channel MOSFETs fabricated at MIT. The agreement obtained between simulations and experiments suggests the viability of this approach in the sub-0.1 μm regime. This analytical framework could form the basis of a compact and efficient first-order CMOS device design environment that could be used to explore key trade-offs and constraints involved in device design several generations ahead. Such a framework could find applications in roadmap planning, sensitivity analysis, and microelectronics education.

INTRODUCTION

Sophisticated two- and three-dimensional device modeling is routinely utilized in CMOS device design. After proper calibration, this simulation environment is extremely effective in guiding device and process technology. Unfortunately, the large number of input parameters that need to be specified, the complexity of grid management and the cost of acquiring and running the tools prevent all but the most sophisticated users from utilizing professional device simulators.

There is a need for a simple, physics-based simulation tool capable of first-order device analysis. Such a tool could be used to map out the range of required values for various CMOS structural parameters several generations ahead. It could also be instrumental in understanding the key constraints and tradeoffs facing a given technology generation depending on the choice of various parameters, such as voltage or oxide thickness. Additionally, a tool of this kind would help to explain past developments and predict future trends in CMOS design. A direct application is its utility for roadmap planning, such as for the National Technology Roadmap for Semiconductor (NTRS) (1). It would be particularly valuable in predicting design and operating points that can then be subjected to the more extensive robustness and parameter sensitivity Monte Carlo analyses that are required for manufacturability studies (2,3). Finally, a simulation tool as envisioned here would be an unprecedented educational aid for CMOS device physics and design.

We have constructed an analytical framework that could form the basis for a compact and efficient physics-based device simulation environment for CMOS. Over the years, a wealth of analytical models have been developed to describe a variety of MOSFET device physics aspects. The range of applicability of these models is often limited. As a consequence, it is essentially impossible to develop a simple analytical formulation that can be reasonably accurate for any choice of device parameters. To overcome this difficulty, we have restricted ourselves to integrating a set of models that is reasonably accurate for "well designed" devices, since these are the only ones that eventually reach the manufacturing stage. A well designed device is a device with sufficient electrostatic integrity, that is, good isolation between the output and the input, or a high aspect ratio of vertical to lateral electric fields. The demand of good electrostatic integrity substantially narrows down the range of required models and simplifies their development. This suite of models needs to be complemented by a judicious choice for a "flag" that indicates violation of electrostatic integrity. In our work, we have selected Drain-Induced Barrier Lowering (DIBL) as such a flag.

This paper describes the set of physics-based models utilized in the analytical framework and its implementation in a commercial Spreadsheet software package. In order to evaluate its validity, we have simulated n-channel MOSFETs that feature a Super-Steep-Retrograde well doping and halos around the source and drain. These devices were fabricated at MIT and have effective channel lengths down to 0.1 μm (4,5). The agreement that has been obtained between simulations and measurements suggests the viability of an improved version of our analytical approach in the sub-0.1 μm regime.

ANALYTICAL MODELS

Fig. 1 shows a cross section of a prototypical n-channel MOSFET. In the simple implementation explored here, there are only five input parameters that describe the technology: the polysilicon gate length, L_g, the effective channel length, L_{eff}, the oxide thickness, x_{ox}, the junction depth, x_j, and the acceptor doping level in the well, N_A, which is assumed to be uniform

CP449, *Characterization and Metrology for ULSI Technology: 1998 International Conference*
edited by D. G. Seiler, A. C. Diebold, W. M. Bullis, T. J. Shaffner, R. McDonald, and E. J. Walters

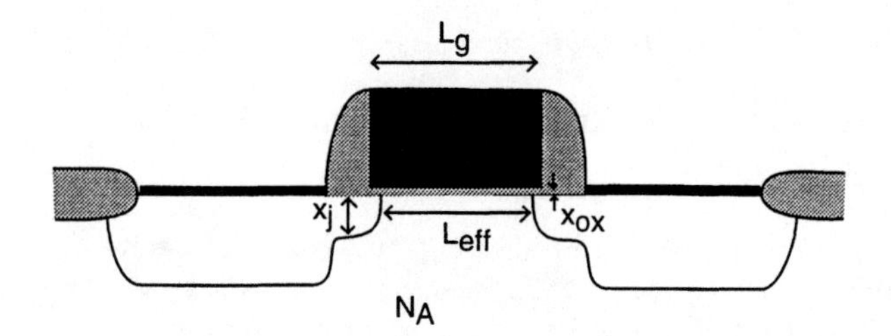

Figure 1: Cross-section of a prototypical MOSFET indicating the five input parameters to the analytical framework.

in the immediate vicinity of the channel.

There are several modules in our model framework. Three are described below: the threshold voltage, the *on* current, and the *off* current. Additional modules compute peak electric fields, device delay, and inverter switching delay and power-delay product. The models are described below for an n-channel device. A parallel set of equations describes the p-MOSFET.

Threshold Voltage Model

The threshold voltage model consists of a geometry-independent portion, the "long" V_T, plus a term that accounts for all shifts due to the small geometry and finite V_{DS}:

$$V_T = V_T(long) + \Delta V_T \tag{1}$$

$V_T(long)$ follows the standard form:

$$V_T(long) = V_{FB} + \phi_{sth} + \frac{1}{C_{ox}}\sqrt{2\epsilon_s q N_A \phi_{sth}} \tag{2}$$

where ϕ_{sth} is the surface potential at threshold and the rest of the symbols have their usual meaning (6).

The proper definition of "threshold" is a critical issue in any MOSFET model. In this work, we have defined threshold when:

$$\phi_{sth} = 2\phi_f + \frac{kT}{q} \tag{3}$$

with ϕ_f given by:

$$\phi_f = \frac{kT}{q}\ln\frac{N_A}{n_i} \tag{4}$$

The traditional definition of $V_T(long)$, without the factor kT/q in Eq. 3, yields too low a threshold voltage. In our experience, the addition of a thermal voltage to $2\phi_f$ in Eq. 3 brings $V_T(long)$ much closer to the measured values.

The flat-band voltage, V_{FB}, is given by:

$$V_{FB} = -\phi_{MS} = -(\frac{E_g}{2} + \phi_f) \tag{5}$$

This assumes that the n$^+$-polySi gate is sufficiently doped so that its Fermi level is located at the conduction band edge.

In an actual small MOSFET under typical operating conditions, V_T differs from $V_T(long)$ due to a number of effects. In our model, we have captured the shift in V_T due to the short gate length and the drain-induced barrier lowering (DIBL) that accompanies the application of a significant V_{DS}. The two effects are handled simultaneously by means of the model of Liu *et al.* (7):

$$\begin{aligned}\Delta V_T \;=\; & -[3(V_{bi} - \phi_{sth}) + V_{ds}]e^{-L_{eff}/\lambda} - \\ & 2\sqrt{(V_{bi} - \phi_{sth})(V_{bi} - \phi_{sth} + V_{ds})}e^{-L_{eff}/2\lambda}\end{aligned} \tag{6}$$

In this equation, V_{bi} is the built-in potential of the source-substrate and drain-substrate junctions which is approximately equal to ϕ_{MS} of the MOS structure given in Eq. 5. λ is the characteristic length of the vertical electrostatics of the channel and is given by (7):

$$\lambda = \sqrt{\frac{\epsilon_s}{\epsilon_{ox}} x_{ox} x_{dmax}} \tag{7}$$

x_{dmax} is the thickness of the depletion region under the source end of the channel:

$$x_{dmax} = \sqrt{\frac{2\epsilon_s \phi_{sth}}{qN_A}} \tag{8}$$

This simple model for ΔV_T has been shown to compare well with 2D simulations for devices that have sufficient electrostatic integrity, that is, if $L_{eff} \gg \lambda$ (7).

The model for V_T used in this work does not include the narrow-width effect nor the so-called "inverse short-channel effect." Additionally, and unlike other models, this model for V_T does not depend on x_j. The relative insensitivity of V_T on x_j has been established in (7). We have nevertheless kept x_j as an input to the model because it impacts other modules not described here, such as the one that computes the maximum channel electric field.

On Current Model

The on current model used here is a conventional one that provides a continuous transition between the mobility and velocity saturation regimes (8). In this model, in the saturation regime of operation, the drain current is given by:

$$I_{dsat} = W_g v_{sat} C_{ox}(V_{gs} - V_T - V_{dsat}) \quad (9)$$

with

$$V_{dsat} = \sqrt{V_c^2 + 2(V_{gs} - V_T)V_c} - V_c \quad (10)$$

and

$$V_c = \frac{v_{sat}}{\mu_{eff}} L_{eff} \quad (11)$$

μ_{eff} is the inversion layer mobility which is affected by the vertical electric field. A first-order empirical description of this effect is:

$$\mu_{eff} = \frac{\mu_{n0}}{1 + (\frac{\mathcal{E}_{av}}{\mathcal{E}_0})^\alpha} \quad (12)$$

where $\mathcal{E}_{av}$ is the average vertical field which can be approximated by (9):

$$\mathcal{E}_{av} = \frac{V_{gs} + V_T}{6x_{ox}} \quad (13)$$

In all our work, we have used the values suggested by Ko in (9) for the above physical constants: $\mu_{n0} = 670\ cm^2/V.s$, $\alpha = 1.6$, $\mathcal{E}_0 = 6.7 \times 10^5\ V/cm$, and $v_{sat} = 10^7\ cm/s$.

The drive or *on* current is defined as:

$$I_{on} = I_{dsat}(V_{gs} = V_{ds} = V_{dd}) \quad (14)$$

where V_{dd} is the supply voltage.

Off Current Model

The subthreshold current of the MOSFET gives rise to a trickle current when the device is nominally off. This is the so-called *off* current. We use a simple model for the subthreshold current that is given by:

$$I_{subth} = \frac{W_g}{L_{eff}} \mu_{n0} (\frac{kT}{q})^2 C_{sth} \exp \frac{q(V_{gs} - V_T)}{nkT} \quad (15)$$

where n is given by:

$$n = 1 + \frac{C_{sth}}{C_{ox}} \quad (16)$$

and C_{sth} is the capacitance associated with the depletion region underneath the gate on the source side of the device at threshold:

$$C_{sth} = \frac{\epsilon_s}{x_{dmax}} \quad (17)$$

The off current is the subthreshold current when $V_{gs} = 0$ and $V_{ds} = V_{dd}$, that is:

$$I_{off} = I_{subth}(V_{gs} = 0, V_{ds} = V_{dd}) \quad (18)$$

Eq. 18 depends on V_{ds} through the V_T term in Eq. 15.

This simple model for the *off* current uses the same expression for V_T as the one used for the on current. As we will see below, this tends to result in slightly lower *off* currents than experimentally observed. Two different expressions for V_T should probably be used in a more advanced model (10).

COMPARISON WITH EXPERIMENTS

The set of analytical models described in the preceding section has been programmed in a commercial Spreadsheet package. We have investigated the suitability of this simple model by simulating MIT's SSR III CMOS technology. This technology features an n-channel MOSFET with a super-steep-retrograde well doping (made by means of a deep B implant and a shallow In implant) and In halos around the source and drain. These devices have effective channel lengths down to 0.1 μm (4,5).

The inputs to our analytical model that correspond to MIT's SSR III technology were determined in (4) and

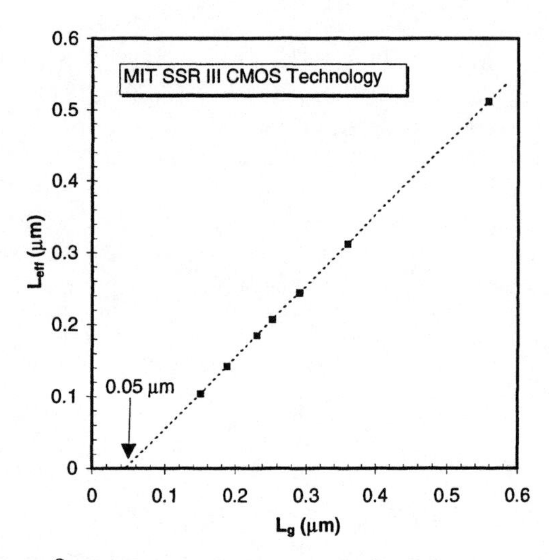

Figure 2: Experimentally determined values of L_{eff} vs. L_g for MIT's SSR-III technology.

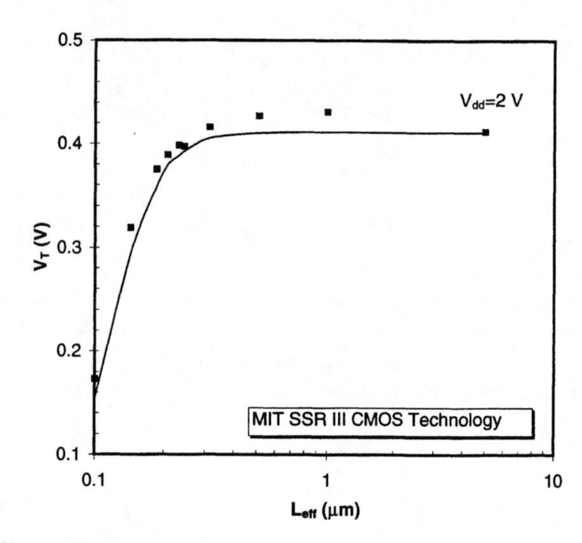

Figure 3: Measured (symbols) vs. simulated V_T as a function of L_{eff} for $V_{dd} = 2\ V$.

(5) following a very detailed procedure that is summarized next.

- L_g and L_{eff} were obtained using a new method that involves the comparison of measured and simulated gate-to-source/drain capacitances (5). The extracted L_g and L_{eff} were confirmed by SEM and experimentally obtained 2D source/drain profiles. Fig. 2 plots the experimentally determined values of L_g and L_{eff}.
- For oxide thickness, an effective value of $x_{ox} = 5.3\ nm$ was used as obtained from capacitance measurements in inversion (4).
- The well doping level was obtained from a combination of SUPREM and MINIMOS-4 simulations (4). An average value of $N_A = 2.6 \times 10^{17}\ cm^{-3}$ was found to provide a good description of the device. As detailed below, this value had to be raised to $N_A = 3.3 \times 10^{17}\ cm^{-3}$ to obtain good agreement with the long-channel V_T. The small discrepancy is related to our definition of threshold made in Eq. 2 and is nevertheless within the uncertainty of the experimental determination of N_A.
- For completeness and although not relevant for the modules presented in this paper, a value of x_j of 35 nm was determined by means of reverse modeling (4). This value is consistent with the data in Fig. 2 if we assume a lateral diffusion of the source and drain of 70% of the junction depth.

Figs. 3 through 12 graph the experimental measurements in the MIT SSR III n-channel MOSFETs at room temperature and the simulation results. We discuss these graphs in detail below.

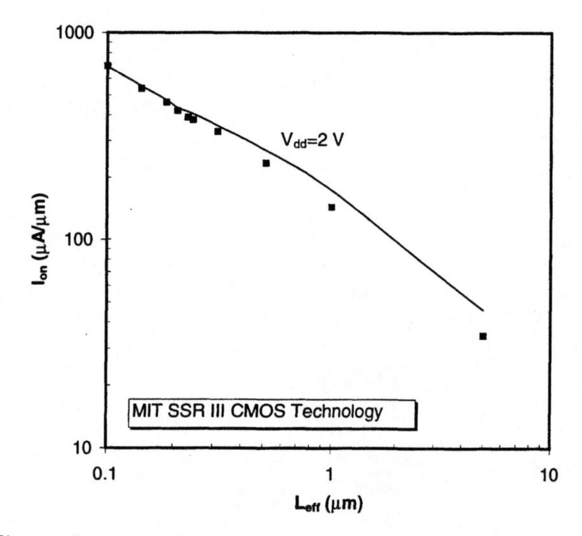

Figure 4: Measured (symbols) vs. simulated I_{on} as a function of L_{eff} for $V_{dd} = 2\ V$.

Fig. 3 shows V_T vs. L_{eff} for $V_{dd} = 2\ V$. Experimentally, V_T was extracted as the value of V_{gs} that corresponds to a drain current of $10^{-7}\frac{W}{L}\ A$, at the appropriate V_{ds}. In the simulations, as mentioned above, N_A has been slightly adjusted to match V_T for the $L_{eff} = 5\ \mu m$ device. Once this is done, the simulations closely follow the experimental measurements down to $L_{eff} = 0.1\ \mu m$. We did not attempt to model the slight reverse short-channel effects that are seen in Fig. 3.

Fig. 4 graphs I_{on} vs. L_{eff} for $V_{dd} = 2\ V$. The agreement is fairly good throughout. For the long channel devices, the simulated current is slightly higher than

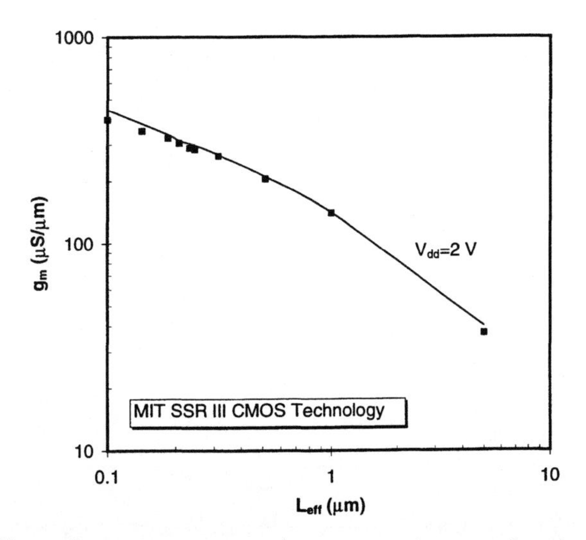

Figure 5: Measured (symbols) vs. simulated g_m as a function of L_{eff} for $V_{dd} = 2\ V$.

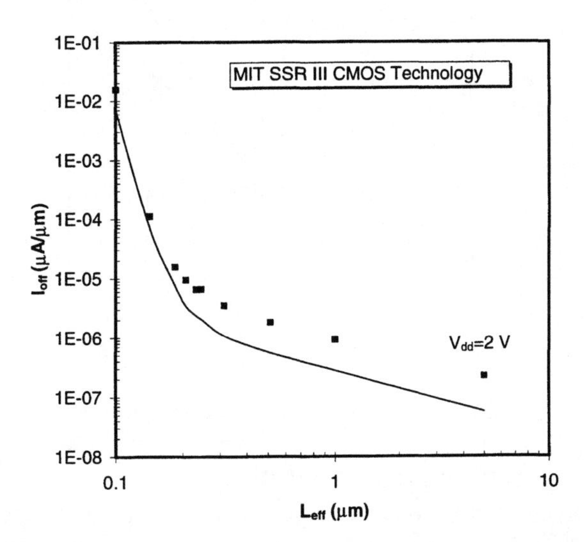

Figure 6: Measured (symbols) vs. simulated I_{off} as a function of L_{eff} for $V_{dd} = 2\ V$.

the measured one. This is probably due to the fact that our I_{on} model lacks the body effect. This does not affect short-channel devices in a significant way since they operate close to velocity saturation.

Fig. 5 graphs g_m vs. L_{eff} for $V_{dd} = 2\ V$. In the simulations, g_m was obtained by calculating the values of I_{on} for V_{dd} and for $V_{dd} + 0.01\ V$ and computing $g_m = \Delta I_{on}/\Delta V_{dd}$. The agreement between simulations and measurements is excellent throughout. For the longest device, a small discrepancy arises, probably due to the absence of the body effect. For short devices, the discrepancy that is seen is probably due to the absence of series resistance effects in the simulation. This is easy to correct.

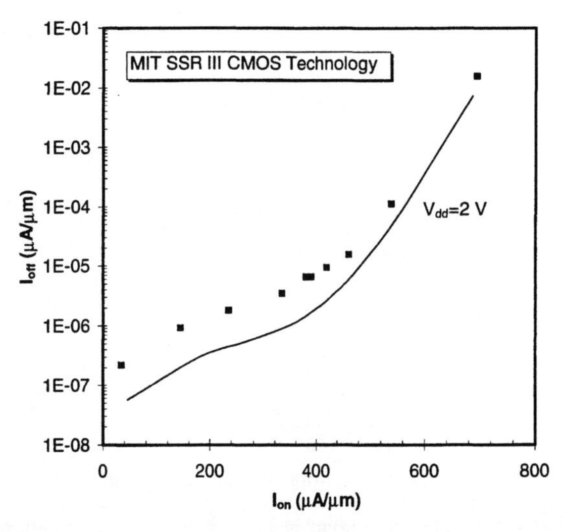

Figure 7: Measured (symbols) vs. simulated I_{off} as a function of I_{on} for $V_{dd} = 2\ V$. The symbols graph data for different values of L_{eff}.

Fig. 6 graphs I_{off} vs. L_{eff} for $V_{dd} = 2\ V$. The model calculations exhibit a similar dependence to the data. The model shows a gentle rise in I_{off} as L_{eff} is reduced in the long regime followed by a much steeper rise for short channels. This is exactly what the data do. However, the model predicts too small a value of I_{off}, particularly for long channels. This might be due to the definition of V_T, which is the same for I_{on} and I_{off}, as mentioned above. The discrepancy can be resolved by introducing a different threshold voltage for the *off* current (10). A rough calculation indicates that for SSR III technology, V_T for I_{off} should be about 40 mV smaller than the value used for I_{on}. This difference is consistent with results of the analytical work of Wann et al. (10).

Fig. 7 shows I_{off} vs. I_{on} for $V_{dd} = 2\ V$. The data points are obtained for different values of L_{eff}. This type of plot is favored by experimentalists because both I_{off} and I_{on} can be measured unambiguously at any bias point. There is no need to extract L_{eff}, as is the case of the previous plots. In Fig. 7, the simulation exhibits a similar behavior as the measurements. However, due to the discrepancies noted above on I_{on} and I_{off}, the quantitative agreement is only fair.

Fig. 8 graphs g_m vs. $DIBL$ for $V_{dd} = 2\ V$. The various data points are obtained for different values of L_{eff}. DIBL is a figure of merit that quantifies the change in V_T as a result of the application of V_{ds}. Here, $DIBL$ is defined as:

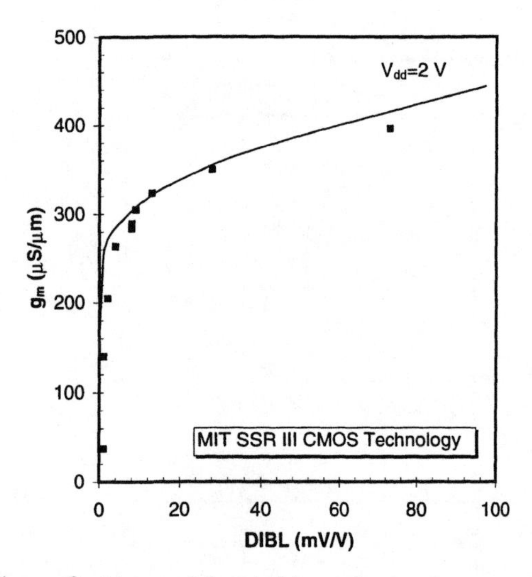

Figure 8: Measured (symbols) vs. simulated g_m as a function of $DIBL$ for $V_{dd} = 2\ V$. The symbols graph data for different values of L_{eff}.

$$DIBL = \left|\frac{V_T(V_{ds} = 2.01\ V) - V_T(V_{ds} = 0.21\ V)}{1.8}\right| \quad (19)$$

This definition has been used so as to match the available experimental data. As it is customary in the field, $DIBL$ is given in units of mV/V.

A plot of g_m vs. $DIBL$, such as the one in Fig. 8, emphasizes a key trade-off that device designers face in their quest for higher g_m. $DIBL$ is one of the costs that are incurred. As L_{eff} is reduced, g_m increases but so does $DIBL$. As Fig. 8 shows, the simulations fairly accurately capture this trade-off all the way up to $DIBL$ values of 100 mV/V which is often considered an upper limit to what can be tolerated in a manufacturable process.

Figs. 9 through 12 compare in more detail the measurements obtained in the shortest device ($L_{eff} = 0.1\ \mu m$) with the predictions of the models. In these simulations, the set of input parameters remains unchanged.

Fig. 9 shows V_T vs. V_{dd} for $L_{eff} = 0.1\ \mu m$. The agreement is rather good throughout. Fig. 10 shows I_{on} vs. V_{dd} for $L_{eff} = 0.1\ \mu m$. The agreement is excellent too.

Fig. 11 shows g_m vs. V_{dd} for $L_{eff} = 0.1\ \mu m$. The discrepancy at high values of V_{dd} is probably due to the absence of series resistance in the model. The discrepancy at intermediate values of V_{dd}, between 0.4 and 0.6 V, probably arises from the simplistic model of in-

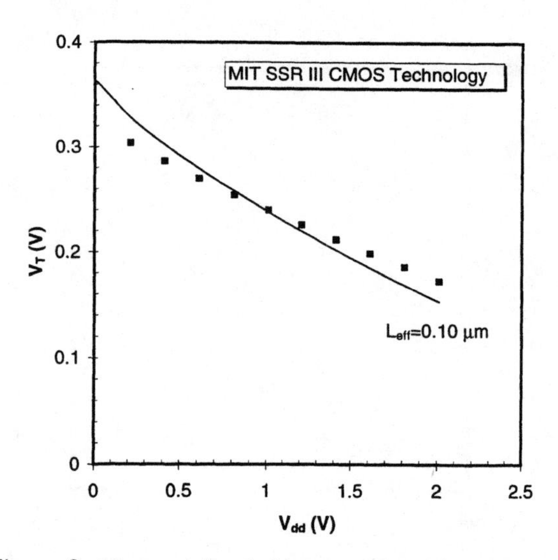

Figure 9: Measured (symbols) vs. simulated V_T as a function of V_{dd} for $L_{eff} = 0.1\ \mu m$.

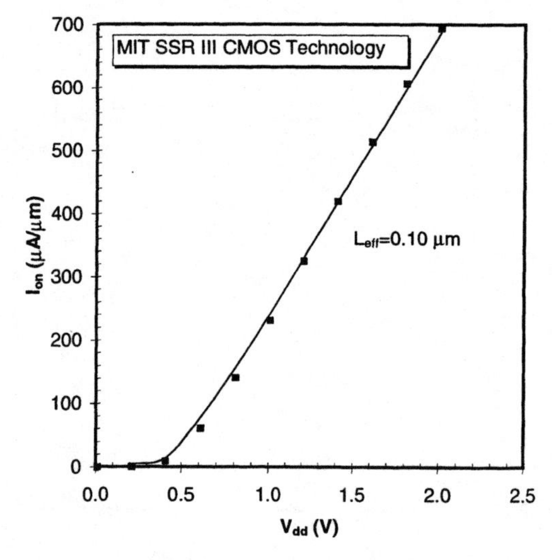

Figure 10: Measured (symbols) vs. simulated I_{on} as a function of V_{dd} for $L_{eff} = 0.1\ \mu m$.

version capacitance used in Eq. 9.

Finally, Fig. 12 shows I_{off} vs. V_{dd} for $L_{eff} = 0.1\ \mu m$. The qualitative behavior of I_{off} is well captured by the model. There is, however, a quantitative disagreement of as much as an order of magnitude particularly at low values of V_{dd}. As mentioned above, this probably arises from the unified V_T definition used in this work.

DISCUSSION

The comparative study described in the previous section between the output of the model and the results of well characterized hardware suggests the considerable merit of the analytical framework presented here

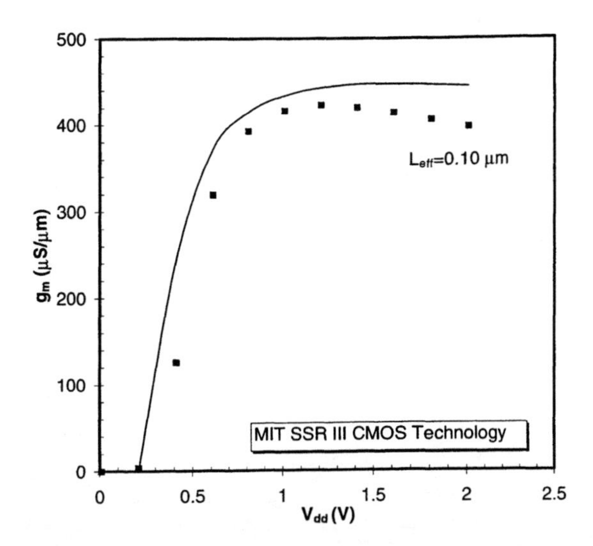

Figure 11: Measured (symbols) vs. simulated g_m as a function of V_{dd} for $L_{eff} = 0.1\ \mu m$.

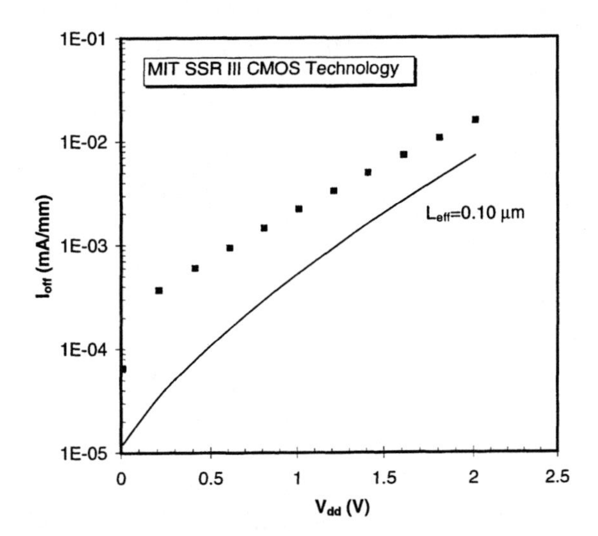

Figure 12: Measured (symbols) vs. simulated I_{off} as a function of V_{dd} for $L_{eff} = 0.1\ \mu m$.

to describe MOSFETs down to 0.1 μm effective channel lengths. At the heart of this modeling suite is an analytical model for threshold voltage that is remarkably accurate for well designed devices. These are devices with sufficient electrostatic integrity (7). In fact, as Fig. 8 suggests, the set of models holds well up to a value of $DIBL$ of 100 mV/V, which is considered the upper limit for manufacturable devices. We can use the value of $DIBL$ as a flag to bound the limits of applicability of the framework proposed here. If $DIBL \leq 100\ mV/V$, the set of models presented here should be reasonably accurate.

In order to further explore the applicability of this analytical framework to sub-0.1 μm effective channel length MOSFETs, we have attempted to simulate re-

L_{eff}=0.07 µm
V_{dd}=1.8 V

	Su et al.	simulations
V_T(long) (V)	0.38	0.38
V_T (V)	0.20	0.18
DIBL (mV/V)	100	110
I_{on} (µA/µm)	800	874
g_m (µS/µm)	625	623
I_{off} (µA/µm)	$6x10^{-3}$	$5x10^{-3}$

Table 1: Measurements vs. simulations of $L_{eff} = 0.07\ \mu m$ n-MOSFETs fabricated by IBM ($V_{dd} = 1.8\ V$) (11).

cently reported $L_{eff} = 0.07\ \mu m$ nMOSFETs fabricated at IBM (11). In this paper, values for L_g (0.1 μm), L_{eff} (0.07 μm), and x_{ox} (3.5 nm) are given. Furthermore, among a number of electrical characteristics, $V_T(long)$ (0.38 V) is given. This can be used to determine an effective uniform well doping level, N_A ($6 \times 10^{17}\ cm^{-3}$). With this set of inputs, Table I compares the outputs of the model with the values of electrical parameters given in the paper. The V_T for $L_{eff} = 0.07\ \mu m$ device is closely matched. Our simulations predict a value of 0.18 V against a measured one of 0.2 V. This arises from a discrepancy of about 10% in $DIBL$. This is most surely because of our uniform doping level against the super-steep retrograde actual one. In spite of this, I_{on} and g_m are within 10%, while I_{off} is underestimated by about 20%. These encouraging results give us hope that the analytical framework explored in this paper has applicability in the sub-0.1 μm channel length regime.

There are several improvements that can be made to this model while preserving it physically meaningful, simple, and analytical. First, as noted above, the body effect and series resistance should be added. Additionally, a better definition for the threshold voltage for the *off*-current has to be identified and implemented. For deep submicron devices, inversion-layer quantization should also be introduced. Also for very small devices, a simple way to handle the super-steep-retrograde well doping has to be found. In addition to this, useful and probably viable enhancements to the model are incorporation of gate current, substrate current, the impact of back bias, gate-induced drain lowering, and dynamic threshold-voltage switching.

Before a simple model such as the one proposed here is widely used, rigorous testing should be done against well-characterized technology and two-dimensional simulations. It is unclear whether a single model can tackle the multiplicity of advanced device designs currently under consideration (10). The suitability of the model to different design strategies has to be established.

CONCLUSIONS

We have shown that a reasonably accurate physics-based analytical framework for MOSFET device simulation can be constructed that has the potential of being applicable in the sub-0.1 μm regime. Good agreement has been obtained between simulations and experiments down to $L_{eff} = 0.1\ \mu m$. The set of models could form the basis for a compact and efficient first-order device design environment for CMOS of potential use in roadmap planning, sensitivity analysis, and microelectronics education.

ACKNOWLEDGEMENTS

This work has been partially funded by the Semiconductor Research Corporation and by the National Science Foundation through the ECSEL coalition. The authors thank Keith Jackson for valuable discussions and for making available the experimental data from MIT used in this work.

REFERENCES

1. *The National Technology Roadmap for Semiconductors*, Semiconductor Industry Association, 1994.

2. P. M. Zeitzoff, A. F. Tasch, W. E. Moore, S. A. Khan, and D. Angelo. *This Conference.*

3. K. Hasnat, S. Murtaza, and A. F. Tasch, *IEEE Trans. Semic. Manuf.* **7**, 53 (1994).

4. H. Hu, J. B. Jacobs, L. T. Su, D. A. Antoniadis, *IEEE Trans. Electr. Devices*, **42**, 669 (1995).

5. C.-L. Huang, J. V. Faricelli, D. A. Antoniadis, N. A. Khalil, and R. A. Rios, *IEEE Trans. Electr. Devices*, **43**, 958 (1996).

6. Y. P. Tsividis, *Operation and Modeling of the MOS Transistor*, McGraw Hill, New York, 1987.

7. Z.-H. Liu, C. Hu, J.-H. Huang, T.-Y. Chan, M.-C. Jeng, P. K. Ko, and Y. C. Cheng, *IEEE Trans. Electr. Devices*, **40**, 86 (1993).

8. R. S. Muller and T. I. Kamins, *Device Electronics for Integrated Circuits*, 2nd ed., Wiley, New York, 1986.

9. P. K. Ko, in *VLSI Electronics: Microstructure Science*, Vol. 18, Ch. 1, Academic Press, 1988.

10. C. H. Wann, K. Noda, T. Tanaka, M. Yoshida, and C. Hu, *IEEE Trans. Electr. Devices*, **43**, 1742 (1996).

11. L. Su, S. Subbanna, E. Crabbe, P. Agnello, E. Nowak, R. Schulz, S. Rauch, H. Ng, T. Newman, A. Ray, M. Hargrove, A. Acovic, J. Snare, S. Crowder, B. Chen, J. Sun, and B. Davari, *1996 Symp. VLSI Techn.*, p. 12.

Physical Modeling of Shallow/Deep Junctions

Conor Rafferty
Bell Laboratories, Lucent Technologies
600 Mountain Ave, Murray Hill, NJ 07974

ABSTRACT

Predicting impurity profiles after diffusion is a key step in process design for deep submicron integrated circuits. Recent technology trends give rise to strong anomalous diffusion effects arising from implantation damage. Present models describe such effects at high energy but require improvement for shallow implants.

Introduction

Process and device modeling is widely used throughout the IC design cycle, from exploration of novel device concepts, to process design, to performance optimization. A key part of process modeling is predicting the impurity profile after implant and anneal. Recent technology trends are extending the range of ion implantation to both higher and lower energy, and lowering process temperatures. These conditions give rise to strong anomalous diffusion effects, known as transient enhanced diffusion. The models developed for medium energy implants and the subsequent diffusion have worked well for high energy implants and can be used to predict and explain a number of surprising effects. At low energies, a number of new physical phenomena has greatly complicated the picture and a full understanding is still in development.

Technology Trends

The constant drive to increase integrated circuit packing density by shrinking lateral dimensions requires a concomittal reduction in vertical length scales. The National Technology Roadmap for Semiconductors(1) calls for junctions as shallow as 300-600Å by 2001. Very low energy implants are required to achieve such junctions depths; in the case of B implants, energies below 1keV are anticipated.

At the same time, high energy implantation is finding a number of applications in modern technology, particularly in the formation of profiled tubs and buried conducting layers. Profiled tubs have a number of advantages, including reduced front end processing steps, better latchup immunity, and the possibility of replacing expensive epitaxial substrates with implanted substrates.

Lower processing temperatures are another feature of modern technologies. Lower temperatures are favorable to reduce the equilibrium component of dopant diffusion, allowing shallower junctions and retrograde profiles for high device performance, and to prevent dopant penetration through ever-thinner gate oxides.

Transient Enhanced Diffusion (TED)

It has been known for over a decade that during the first anneals following implantation, impurity diffusivity can be greatly enhanced relative to equilibrium diffusion; enhancements of 10,000 or more are common. The temporary increase in diffusion is known as transient enhanced diffusion (TED). At temperatures where no measurable equilibrium diffusion is expected, substantial displacements have been observed via TED. The displacement increases with dose and energy, and the implantation of one species can induce the rapid diffusion of another already present in the wafer.

The technology trends mentioned above have brought TED to the forefront of impurity diffusion modeling. It is not uncommon for the majority of total profile broadening to arise from TED. Lower temperatures reduce the equilibrium component of diffusion, but actually exacerbate TED. As the temperature is reduced for a fixed-length anneal, the amount of diffusion will first decrease,

CP449, *Characterization and Metrology for ULSI Technology: 1998 International Conference*
edited by D. G. Seiler, A. C. Diebold, W. M. Bullis, T. J. Shaffner, R. McDonald, and E. J. Walters
 1-56396-753-7/98/$15.00

then rise again, before finally falling down an Arrhenius curve. The amount of diffusion in the TED displacement can be substantial, up to several tenths of a micron. High energy implantation increases damage and hence transient diffusion, so significant anomalous tub diffusion can be expected. Fortunately, such effects are much reduced in the device area, as is shown later. Shallow junctions introduce a host of complicated physical phenomena affecting impurity diffusion, related to the proximity of the surface and the high concentrations present. Modeling TED under all these conditions is a challenging but important step in modern technology development.

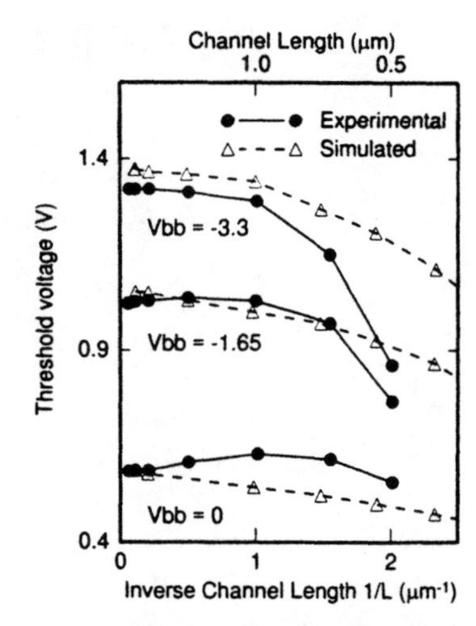

Figure 1. RSCE caused by TED. The threshold voltage of short devices increases before rolling off; predictions based on equilibrium diffusion miss the trend. This technology had a flat channel profile, excluding most alternative possible mechanisms of RSCE.

Device Effects of TED

An interesting example of the effect that TED can have on devices is given by reverse short channel effect (RSCE). It is sometimes observed that the expected decrease of device threshold voltage at smaller device size is reversed; there is an unexpected increase before the final threshold collapse. As shown in Fig.1, the threshold can increase substantially before finally rolling over as the ends of the device approach one another. A frequent mechanism by which RSCE can occur is transient diffusion of the channel profile induced by implants in the source/drain area(2).

Even more serious effects can occur in heterostructure bipolar devices. The base boron profile, carefully grown to lie inside a SiGe layer to form a heterojunction, can be induced to broaden to ten times its original width due to TED induced by the emitter implant. Even if the emitter formation is constrained to reduce TED, extrinsic base implants can produce the same result in narrow stripe devices. The outdiffusion from the SiGe layer transforms the device from a heterojunction to a homojunction device and makes it unusable(3).

Modeling Deep Junctions

The current understanding of TED was developed for medium energy implants and applies well at high energies. Ion implantation creates a large number of displaced lattice atoms and lattice vacancies (Frenkel pairs) as well as one interstitial for each implanted ion. The majority of the Frenkel pairs rapidly recombine, leaving in many cases as the majority component of damage the extra interstitials (the "plus-one" model(4)). These interstitials aggregate into extended defects (mostly rod-shaped {311} defects(5)). The defects gradually dissolve during thermal processing and locally support a steady supersaturation of interstitials above their equilibrium value, which enhances the diffusion of dopants. Interstitials diffuse from the defects to the surface and are recombined there(6), eventually depleting the reservoir of interstitials stored in extended defects and bringing the transient enhancement to an end.

Several intriguing physical phenomena arise from this picture. In the first place, enhancements should be relatively constant and independent of implant dose or energy; this is in fact observed(7)(8). Secondly, the flux of interstitials to the surface gives rise to a "defect wind" which drives dopants to the surface and piles them up there. Surface piles present metrology challenges and have only been observed in a few instances(9); however their effects manifest themselves electrically in a number of ways(2)(10).

An important feature of transient diffusion after high energy implantation is that the enhancement varies with depth and is deepest around the high energy peak. A high dose tub implant may experience substantial self-TED; however it may have a modest effect on the surface profiles which dictate device behavior (Fig.2). This fortuitous result, arising from the surface recombination of defects, limits the impact of high energy TED on devices(12).

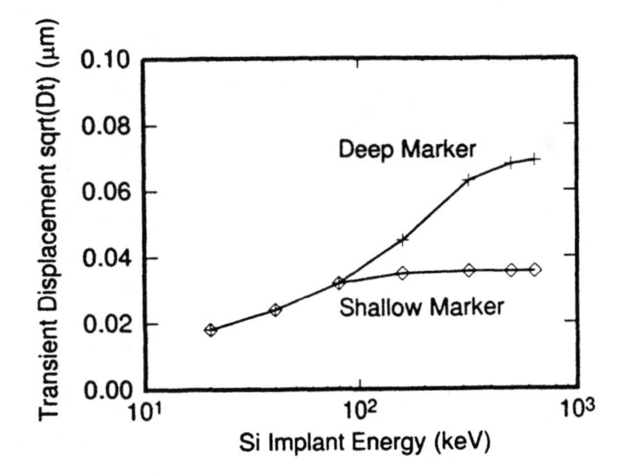

Figure 2. TED induced by high energy implants is more marked in the deep tub profile than in the shallow profiles which dominate device behavior.

High energy implants can also help prevent reverse short channel effect, paradoxically by the presence of extra TED. The TED caused by the high energy implant gives rise to laterally uniform dopant profiles which are already piled up towards the surface. When a subsequent source/drain implant induces nonuniform pileups, the later TED is working against a pre-existing concentration gradient and is relatively ineffective(13).

Modeling Shallow Junctions

For shallow junctions, accurate implant modeling including the effect of channeling is very important. Detailed Monte Carlo simulations including the effect of damage can be predictive down to the lowest energies (Fig.3)(14). For diffusion modeling, several pieces of the previous TED picture must be reconsidered. The "plus-one" model of damage is no longer accurate, as

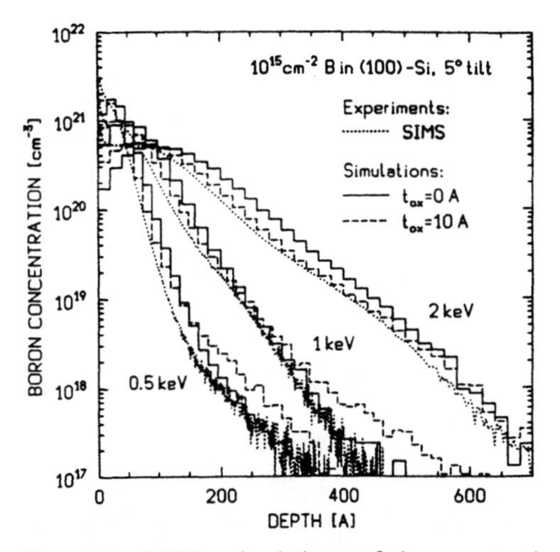

Figure 3. IMSIL simulations of low-energy boron implantation compared to SIMS. 100 channeling is more important at 2keV than 0.5keV, where the presence of a native oxide actually increases scattering into 110 chan-

the effect of unrecombined Frenkel pairs becomes more important at low energies. Figure 4 shows that already at 20keV, the net interstitial population after Frenkel pair recombination is equivalent to about half the implanted dose, so that the total effect in diffusion can be roughly characterized as "+1.5". Lower energies and heavier atoms exacerbate the trend. Since Monte Carlo simulation is rather expensive to build into routine process flow simulation, one method of efficiently modeling such effects is to use a Monte Carlo simulator to build a prepared table of "plusfactors" characterizing the total effect of the damage relative to the simple plus-one model, in much the same way as prepared moment models of implanted impurity profiles are built from detailed MC simulation.

Shallow implants make greater demands on the surface as a recombination site, since the source is brought into close proximity and the flux of defects is large. However, it is found experimentally that down to the lowest energies, the amount of TED induced by silicon implants decreases steadily, showing that the surface recombination remains sufficiently fast to absorb all the necessary interstitials(15).

The surface has another important effect on impurity profiles in shallow junctions; a substantial fraction of the implanted dose can be lost at the surface(16)(17). The loss is exacerbated by

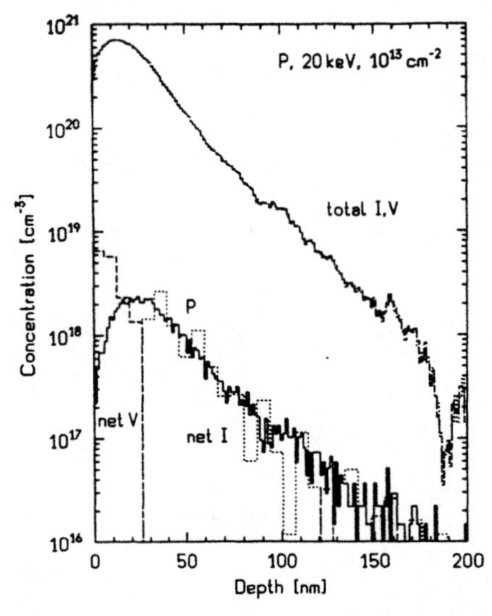

Figure 4. IMSIL simulation of net Frenkel pairs and dopant profiles for 20keV phosphorus. The Frenkel interstitial contribution matches the plus-one contribution over half its range.

the interstitial gradient effect which brings impurity to the surface. The amount of the loss is proportional to the amount of TED (Fig.5)(18). This dose loss to the surface is quite different from the normal segregation which occurs at a growing oxide; impurities such as arsenic which preferentially segregate to silicon during oxidation show substantial dose loss effects during inert annealing(16)(17). The present understanding is that the lost dose resides at the interface, borne out by the observation that re-annealing at high temperature causes the dose in silicon to partially recover(17).

Several complications arise due to high dopant concentrations in the shallow junctions. Competition between dopant clusters and interstitial clusters causes other as-yet uncharacterized defects to form(19). Conversely, precipitation of dopants during anneal can cause large numbers of point defects to be released, with consequences as significant as implant damage(11). Dislocation loops introduce another stage of damage recovery with different time constant and enhancement level from {311}-driven TED.

Summary

Transient diffusion leads to significant effects in modern devices. TED is reasonably well modeled at medium and high energy. Low-energy shallow junctions require more characterization and modeling before simulations tools will be truly predictive.

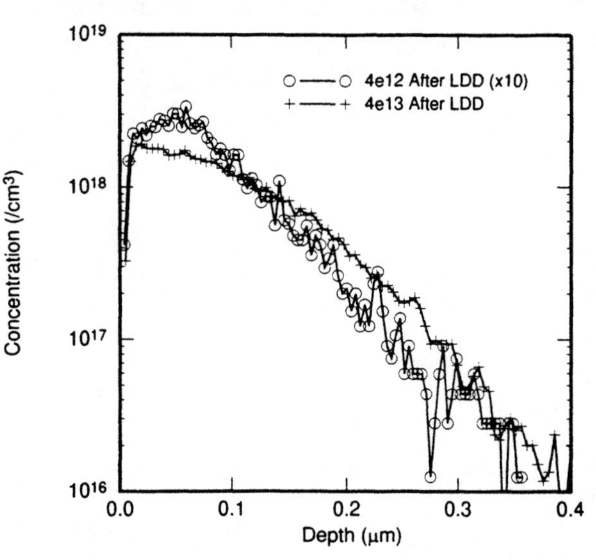

Figure 5. Dose loss is related to amount of TED. Boron was implanted at $4\times10^{12}/\text{cm}^2$ and $4\times10^{13}/\text{cm}^2$ and annealed. The high dose implant diffused more and lost more at the surface. (The doses have been scaled to match for graphing).

References

(1) National Technology Roadmap for Semiconductors, Semiconductor Industry Association (1997)
(2) C.S. Rafferty *et al.*, IEDM p.311 (1993)
(3) L. Lanzerotti *et al.*, IEDM p.249 (1996)
(4) M.D. Giles, J. Electrochem. Soc. 138 p.1160 (1991)
(5) D.J. Eaglesham *et al.*, Appl. Phys. Lett. 65 p.2305 (1994)
(6) D. Lim *et al.*, Appl. Phys. Lett. 67 p.2302 (1995)
(7) P.Packan *et al.*, Appl. Phys. Lett. 56 p.1787 (1990)
(8) H. Chao *et al.*, Appl. Phys. Lett 69 p.2113 (1996)
(9) M.D. Giles, Appl. Phys. Lett. 62 p.1940 (1993)
(10) D. Esseni *et al.*, Elec. Dev. Lett. 19 p.131 (1998)
(11) S. Crowder *et al.*, IEDM p.427 (1995)
(12) C.S. Rafferty *et al.*, IEDM p.91 (1996)
(13) S. Chaudhry *et al.*, IEDM p.679 (1997)
(14) G. Hobler, private communication
(15) A. Agarwal *et al.*, IEDM p.467 (1997)
(16) G.A. Sai-Halasz *et al.*, Elec. Dev. Lett. 6 p.285 (1985)
(17) P.B. Griffin *et al.*, Appl. Phys. Lett. 67 p.482 (1995)
(18) H.-H. Vuong, private communication
(19) L.H. Zhang *et al.*, Appl. Phys. Lett 67 p.2025 (1995)

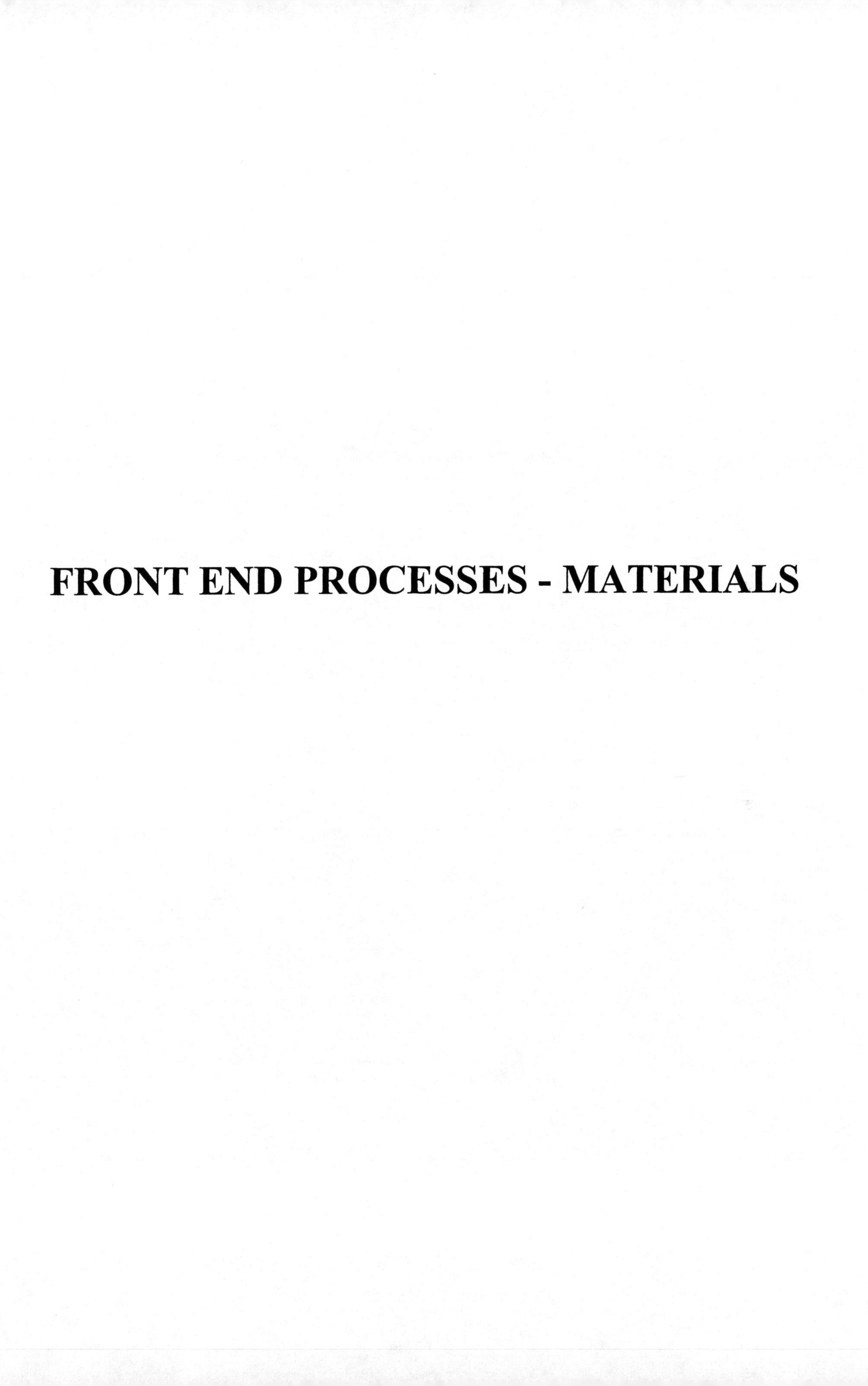

FRONT END PROCESSES - MATERIALS

MODEL-BASED SILICON WAFER CRITERIA FOR OPTIMAL INTEGRATED CIRCUIT PERFORMANCE

Howard R. Huff [(1)], Randal K. Goodall [(2)], W. Murray Bullis [(3)],
James A. Moreland [(4)], Fritz G. Kirscht [(5)], Syd R. Wilson [(6)] and
The NTRS Starting Materials Team

[(1)] SEMATECH, Austin, TX 78741
[(2)] I300I, Austin, TX 78741
[(3)] SEMI, Mountain View, CA 94043
[(4)] Wacker Siltronic AG, Burghausen, Germany
[(5)] Mitsubishi Silicon America, Salem, OR 97303
[(6)] Motorola, Inc., Tempe, AZ 85284

The Starting Materials requirements in the 1997 Semiconductor Industry Association (SIA) National Technology Roadmap for Semiconductors (NTRS) were developed by a fifty-nine member team comprised of industrial (e.g., silicon suppliers, equipment and IC manufacturers) and university personnel. The silicon wafer parameter values as-received by the IC manufacturers are generally derived from model-based analyses based on the technology generation critical dimension (CD), bits, wafer diameter, etc. The values represent the perceived critical material characteristics required to ensure that silicon materials support the continued growth of the IC industry while being cognizant of cost-of-ownership (CoO) considerations. The characteristics were developed via a modular approach of a core set of general characteristics applicable to all product wafers, plus specific recommendations for polished, epitaxial and silicon-on-insulator (SOI) wafers. The formulae utilized, assumptions made and the issues involved, as well as opportunities for improvements in the modeling process, are discussed.

INTRODUCTION

The fabrication of a state-of-the-art DRAM, microprocessor or application specific integrated circuit (ASIC) has become a truly complex manufacturing process with over 500 significant process steps. A paradigm shift has concurrently occurred during the '90's with an increased emphasis towards the scientific understanding of the physico-chemical processes and the selective application of design-of-experiment methodologies in the improved fabrication of ICs. Indeed, model-based experiments and simulation procedures have become *de rigeur* in the IC industry with technology advancements based on scientific principles rather than exclusively experience per se. This approach has especially focused on front-end-of-the-line (FEOL) core processes associated with transistor formation (e.g., dielectric, electrode and plasma etching) in conjunction with component isolation, shallow p-n junction formation and contact structures with the impending sub-100 nm era. The FEOL section of the NTRS includes Starting Materials, Surface Preparation, Thermal/Thin Film and Doping, and related plasma etching phenomena (1). Although the Starting Materials requirements must be supportive in achieving these device structures, it is especially important to balance the "best wafer possible" against the cost-of-ownership (CoO) opportunity of not driving silicon requirements to the detection or ultimate resolution limit but to some less stringent and optimized value (2,3). For example, the incoming Starting Material wafer contamination values are greater than the analogous requirements for pre-gate Surface Preparation cleans due to CoO considerations (1).

The Starting Material wafer parameter values as-received by the IC manufacturers are generally derived from model-based analyses based on the technology generation critical dimension (CD), bits, wafer diameter, etc. The understanding of these underlying models of IC performance-characteristic relationships is, in fact, more critical than the specific numerical values. Although empirical models are employed utilizing extrapolated

CP449, *Characterization and Metrology for ULSI Technology: 1998 International Conference*
edited by D. G. Seiler, A. C. Diebold, W. M. Bullis, T. J. Shaffner, R. McDonald, and E. J. Walters

trends, as appropriate, the use of anecdotal opinions was minimized. A fifty-nine member team comprised of industrial (e.g., silicon suppliers, equipment and IC manufacturers) and university personnel was formed to identify the required metrics, including the formulae, assumptions and issues involved (see Table 1). The Starting Materials characteristics were developed via a modular approach of a core set of general characteristics applicable to all product wafers as-received by the IC manufacturers plus specific recommendations for polished, epitaxial and silicon-on-insulator (SOI) wafers. The Starting Materials values represent the perceived critical material characteristics required to ensure that silicon materials support the continued productivity growth of the IC industry as described by Moore's law (4-7).

Table 1
Starting Materials Sub-Teams and Personnel for NTRS

Model-based Structured Approach	Statistical Distributions/ Metrology	Polished Wafer Trends	Epitaxy Wafer Trends	SOI	Corresponding Consultants
Randy Goodall	Larry Beckwith	Steve Bay	Chi Au	Mike Alles	Olli Anttila
Kim Kimerling	**Murray Bullis**	Jeff Butterbaugh	Michael Brohl	George Cellar	Werner Bergholz
Harold Korb	Alain Diebold	John Crabtree	Jeff Epstein	Harry Hovel	Cor Claeys
David Jensen	Worth Henley	Bob Graupner	Dinesh Gupta	Pat O'Hagan	Laszlo Fabry
Lubek Jastrzebski	Paul Langer	Dick Hockett	Bob Helms	George Rozgonyi	Dieter Huber*
Harold (Skip) G. Parks	Larry Larson	Bob Johnston	**Fritz Kirscht**	Witek Maszara	Bill Lynch
K.V. Ravi	Don McCormack	Bob Kunesh	Wen Lin	Dieter Schroder	Mike Mendicino
Eicke Weber	Richard Novak	**Jim Moreland**	Fred Meyer	Paul Smith	J-G Park
	Chris Sparks	Jagdish Prasad	David Myers	**Syd Wilson**	Paul Patruno
		Larry Shive	P.K. Vasudev		Jon Rossi
		Mike Walden			Shin Takasu
					Masaharu Watanabe

* Deceased

WAFER AREA GENERATION MODEL

The phenomenal growth of the IC industry, evidenced by a 25-30% compound annual growth rate (CAGR)—achieved by staying on the "productivity learning curve"—continues to be the gauge by which the industry is measured (6,7). Concurrently, the number of transistors per chip has increased with a CAGR of approximately 40% such that the number of transistors is over 10 million on a state-of-the-art microprocessor (1). This enables a 25-30% per year cost reduction per function for nearly the past thirty years. This growth has been fueled by four factors, (a) shrinking lithographic design rules; (b) yield improvements; (c) increased equipment utilization and (d) larger wafer diameter (8). The largest opportunity growth factor to maintain the IC productivity engine is increased equipment effectiveness; that is, the percentage of time the equipment is adding value to the wafer (8). This is especially important inasmuch as the largest challenge to maintaining the productivity curve may be the enormous financial infrastructure required, rather than technological limits, to increase chip density (6,9,10). Indeed, departures from Moore's law have already been noted (6). Nevertheless, increased wafer diameter has been an historical method by which the number of IC chips per wafer has been increased. The conversion to 300 mm diameter wafers, beginning about 1998-1999, with initiation of volume manufacturing expected about 2000-2001, is necessary to maintain the required economy of scale for large volume IC manufacturing. Business issues are the primary migration concern as it appears the engineering issues associated with the cost-effective crystal growth (11) and wafer gravitational stresses (10) can be addressed. It should be noted, however, that new design rule super-shrinks, resulting in increased numbers of chips per wafer, might delay the onset of volume conversion by a year from earlier projections. Figure 1 illustrates the historical area demand for silicon per year versus year for wafer diameter from 38/51 mm to the present era and the prognosis for future diameters (12). As a given diameter nears the peak of its utilization, initiation of the next diameter must begin. Projections of the wafer diameter beyond 300 mm suggest that 450 mm (followed by 675 mm) may be the appropriate next sizes to maintain the historical productivity enhancement growth rate (12), modeled by a doubling of the wafer area for the next generation wafer size. These future wafer diameters, however, will require the most severe business and economic global discussions before their projected implementation becomes a reality (see the Summary section).

Concurrently, the edge exclusion is projected to decrease from 3 to 1 mm (see Table 2). The decreasing edge exclusion has a second order effect on increasing the number of chips per wafer due to the more effective layout of rectangular chips on a larger diameter (e.g. circular) wafer. Indeed, in some cases, wafer specifications are denoted over the whole wafer albeit there may be temporary limitations in productivity increase due to metrology issues and the fact that nearly all IC processes have significant edge exclusions.

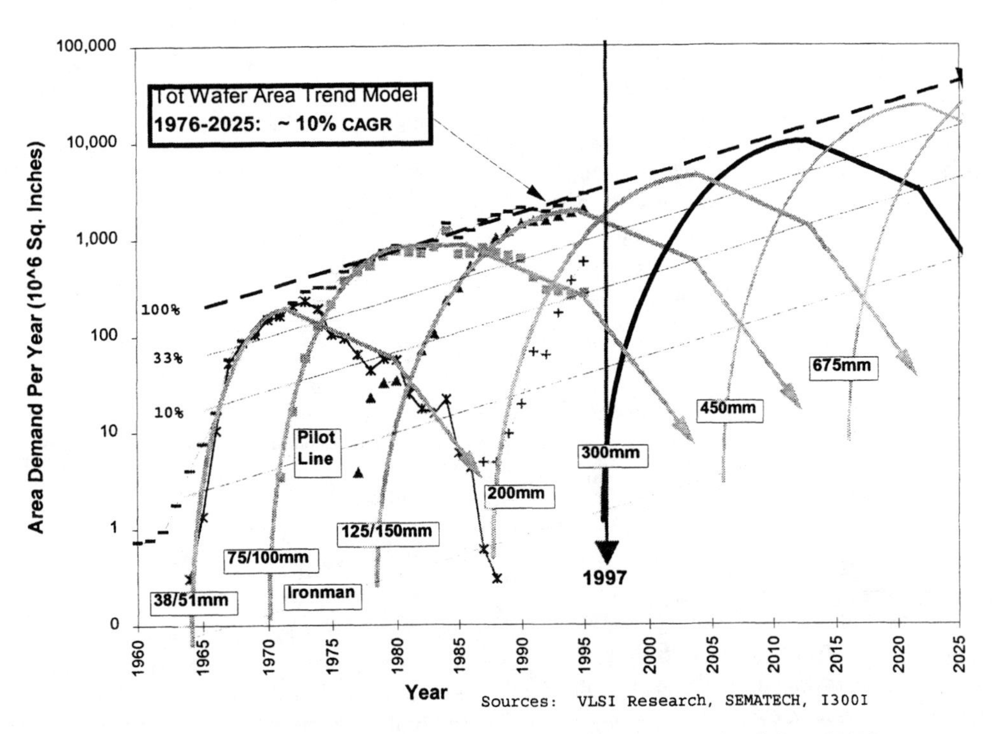

Figure 1 - Modeled Wafer Area Demand Per Year (updated from [12]).

Table 2
Wafer Diameter and Edge Exclusion With Technology Generation

Year of First Product Shipment Technology Generation	1997 250 nm	1999 180 nm	2001 150 nm	2003 130 nm	2006 100 nm	2009 70 nm	2012 50 nm
Wafer Diameter (mm)	200	300	300	300	300	450	450
Edge Exclusion (mm)	3	2	2	2	1	1	1

Solutions Exist | Solutions Being Pursued | No Known Solution

YIELD-DEFECT DENSITY MODEL

The lack of effective, experimentally based data or models at the present time precludes mutually constraining several parameters by partitioning their yield impact (i.e., assuming they are statistically dependent) and, therefore, the parameters are taken as statistically independent. The numerical values of the parameters are selected as upper limits corresponding to 99% yield for each characteristic parameter. It is assumed that the median value of the typically manufactured distribution for each parameter is much better (e.g. usually smaller) than the listed value and that the upper limit values would rarely coincide for more than one parameter at a time on a silicon wafer, thereby ensuring the total yield for all parameters is about 99% (1). This approach is expected to provide the most cost-effective solution to wafer quality since it is essential to balance the "best wafer possible" against the CoO opportunity of not driving silicon requirements to the detection or ultimate resolution limit but to some less stringent and optimal value(2,3).

The Poisson probability distribution (13) is generally expressed:

$$Y = \exp\left[-D_e A_{active}\right] \qquad (1)$$

where D_e is the electrical defect density per cm^2 and A_{active} is the active chip area (cm^2). More significantly, D_e is related to the physical defect density per cm^2 through the kill ratio, R_i, for defect type i and A_{active} is replaced by the total chip area, A_{chip}, through the critical area ratio, α, expressed, respectively, by:

$$D_e = \sum_i R_i D_i \qquad (2a)$$

$$A_{active} = \alpha A_{chip} \qquad (2b)$$

Equation (1) may, therefore, be replaced by:

$$Y = \exp\left[-\sum_i D_i R_i (\alpha A_{chip})\right] \qquad (3)$$

The critical area ratio, α, is expressed as the total number of transistors per chip, T, for an MPU or the number of bits per chip for a DRAM, divided by the chip area, multiplied by three terms which take into account the area per transistor or active device component as expressed by eq. 4. These terms are, (a) the square of the CD of the basic "unit cell" making up the device component; β, the width to length ratio of the unit cell; and, δ, the number of unit cells required to fabricate the device component of interest.

$$\alpha = (T/A_{chip}) (CD)^2 \beta \delta \qquad (4)$$

For a DRAM, $\beta_{DRAM} = 1$ whereas for logic ICs, β_{LOGIC} is modeled as 10, although values of β_{LOGIC} can range from approximately five to 20. Delta, δ, varies from unity for a gate oxide integrity (GOI) test to six for a buried oxide, *BOX*, structure in the case of SOI. Replacing eq. 4 in eq. 3, one obtains the basic, modified Poisson probability distribution relating the yield to the defect density:

$$Y = \exp\left[-\sum_i D_i R_i \{T (CD)^2 \beta \delta\}\right] \qquad (5)$$

A major assumption in the current model, clearly requiring rectification, is the representation of D_i as a fixed defect size with fixed kill ratio, R_i, for all defect sizes, rather than determining an effective value by integration over the particle size and kill ratio distribution. Table 3 summarizes the relevant CDs, transistors per chip, bits per chip and the MPU and DRAM chip area, critical area and critical area ratios.

Table 3
Critical Dimension, Transistors Per Chip, Bits Per Chip, MPU and DRAM Areas, Critical Areas and Critical Area Ratios With Technology Generation

Microprocessor

Year	CD (cm)	Transistors / Chip	MPU Area (cm^2)	MPU Critical Area (cm^2)	MPU Critical Area Ratio
1997	2.50×10^{-5}	$1.1 \times 10^{+7}$	3.0	6.88×10^{-2}	2.29×10^{-2}
1999	1.8×10^{-5}	$2.1 \times 10^{+7}$	3.4	6.80×10^{-2}	2.00×10^{-2}
2001	1.5×10^{-5}	$4.0 \times 10^{+7}$	3.85	9.0×10^{-2}	2.34×10^{-2}
2003	1.3×10^{-5}	$7.6 \times 10^{+7}$	4.3	1.28×10^{-1}	2.98×10^{-2}
2006	1.0×10^{-5}	$2.0 \times 10^{+8}$	5.2	2.00×10^{-1}	3.85×10^{-2}
2009	7.0×10^{-6}	$5.2 \times 10^{+8}$	6.2	2.55×10^{-1}	4.11×10^{-2}
2012	5.0×10^{-6}	$1.4 \times 10^{+9}$	7.5	3.5×10^{-1}	4.67×10^{-2}

DRAM

Year	CD (cm)	Bits / Chip	DRAM Area (cm^2)	DRAM Critical Area (cm^2)	DRAM Critical Area Ratio
1997	2.50×10^{-5}	$2.67 \times 10^{+8}$	2.8	0.167	5.96×10^{-2}
1999	1.8×10^{-5}	$1.07 \times 10^{+9}$	4.0	0.347	8.68×10^{-2}
2001	1.5×10^{-5}	$1.7 \times 10^{+9}$	4.5	0.383	8.61×10^{-2}
2003	1.3×10^{-5}	$4.29 \times 10^{+9}$	5.6	0.725	1.29×10^{-1}
2006	1.0×10^{-5}	$1.72 \times 10^{+10}$	7.9	1.72	2.18×10^{-1}
2009	7.0×10^{-6}	$6.87 \times 10^{+10}$	11.2	3.36	3.0×10^{-1}
2012	5.0×10^{-6}	$2.75 \times 10^{+11}$	15.8	6.88	4.35×10^{-1}

STARTING MATERIALS PARAMETERS

The starting materials parameters discussed below are highlighted in terms of the formulae, assumptions and issues requiring further clarification. The various physical phenomena, however, are not discussed per se, but referenced to the current literature as appropriate for more detailed discussions.

Localized Light Scatterer (LLS)

The localized light scatterer, *LLS*, size is modeled as:

$$\text{Size} = K_1 \times (\text{CD}) \text{ (nm)} \qquad (6)$$

with $K_1 = 0.5$ and CD in nm. The LLS defect density (which includes particles; crystal originated pits, COPs; residual polishing micro-damage; surface microroughness; surface chemical residues including organics; and structural defects such as epitaxial stacking faults) (14) is modeled by the modified Poisson probability distribution of Eq. 5 with the yield Y_{LLS} = 99%, LLS kill ratio $R_{LLS} = 0.1$, $\beta = 1$ and $\delta = 1$ (see Table 4). The particle density is empirically modeled (15) as:

$$\text{Particle Density} = K_2 \times (\text{CD})^{1.42} \text{ (cm}^{-2}) \quad (7)$$

with $K_2 = 5.50 \times 10^{-5}$ and CD is in nm. A major assumption in this empirical model is the extrapolation of the particle density from the regime where COPs is insignificant (15) to the less than 150 nm technology generation, where COPs become significant for polished wafers. The value of R_{LLS} for Starting Materials, R_{LLS} = 0.1, is one-half the value utilized for Surface Preparation with its emphasis on pre-gate cleans, due to CoO considerations (see Table 4). That is, the wafer supplier performs a final clean while the IC manufacturer often performs an incoming clean as well as a number of cleans prior to gate oxidation; accordingly, the incoming surface particles and metals requirements are less stringent than the pre-gate oxidation requirements. Nevertheless, a more rigorous determination of the kill ratio, R_{LLS}, its functional dependence on size, and the development of compatible metrology tools must also be addressed.

Both the chemical structure of the wafer surface produced and delivered by the wafer supplier (hydrophobic versus hydrophilic) and the wafer-shipping box interaction are critical issues in controlling the subsequent adsorption of metals, ionics, organics and particles on the wafer surface. The physical structure of the silicon surface has also emerged as a critical concern. Both polished and epitaxial surfaces exhibit defects that must be controlled to achieve high yielding ICs. The detection, counting, compositional analysis, morphology, removal and prevention of surface defects are state-of-the-art challenges for both metrology and silicon wafer technology as is the development of laser scanning and other instrumentation utilized to monitor these defects.

Table 4
LLS and Particle Densities for Starting Materials and Surface Preparation With Technology Generation

Year of First Product Shipment Technology Generation	1997 250 nm	1999 180 nm	2001 150 nm	2003 130 nm	2006 100 nm	2009 70 nm	2012 50 nm
LOCALIZED LIGHT SCATTERERS (LLS) SIZE (includes particles) (nm)	125	90	75	65	50	35	25
LLS: STARTING MATERIALS Total (cm^{-2})	$\leq$ 0.60	$\leq$ 0.29	$\leq$ 0.26	$\leq$ 0.14	$\leq$ 0.06	$\leq$ 0.03	$\leq$ 0.015
Particles (#/cm^2)	$\leq$ 0.14	$\leq$ 0.088	$\leq$ 0.068	$\leq$ 0.055	$\leq$ 0.038	$\leq$ 0.023	$\leq$ 0.014
LLS: SURFACE PREPARATION Total (cm^{-2})	$\leq$ 0.30	$\leq$ 0.15	$\leq$ 0.13	$\leq$ 0.07	$\leq$ 0.03	$\leq$ 0.015	$\leq$ 0.007

Additionally, the wafer back-surface is increasingly being prepared with a shiny or polished finish to improve flatness, the uniformity of monitoring the wafer temperature during IC processes and, where applicable, distinguishing particles and crystal micro-defects such as COPs from microroughness for ULSI cleanliness requirements. The improvement in back-surface cleanliness, however, more readily reveals microscopic contamination and handling scratch damage from robotic handling systems. Standards for robotic handlers may be necessary to ensure conformance of the surface to the implicit quality requirements.

Organics/Polymers and Surface Microroughness

Organics and polymers are assumed to be sufficiently innocuous if modeled as approximately 0.1 of a monolayer (e.g. $\leq 1 \times 10^{14}/cm^2$). Oxygen-free ambients can vaporize organics (16). Other studies indicate that increased numbers of residual surface carbon atoms can reduce the charge-to-breakdown (Q_{bd}) for gate oxide integrity (GOI) tests (17). Nevertheless, it is expected that the wafer container will tend to saturate the surface with organics and it may be appropriate to only specify organics for the Surface Preparation roadmap. Clearly, this is an area requiring further clarification (18).

The front-surface microroughness has been taken as $\leq$ 0.10 nm for all technology generations except $\leq$ 0.15 nm at the 250 nm technology node. The instrumentation choice, target values and spatial frequency range (scan size) selected should be based on the application (19). The power spectral density analysis is recommended to fully utilize the currently accessible spatial frequency range of 0.01-50 μm^{-1} using atomic force microscopy (AFM). The effects of different surface termination and the wafer-carrier as well as storage ambient on surface microroughness are critical issues that are not yet sufficiently comprehended for optimal device/IC performance.

Surface Metals - Critical Metals

The critical surface metals are modeled by the modified Poisson probability distribution Eq. 5 with the yield Y_M = 99%, metal kill ratio $R_M = 1$, $\beta = 1$ and $\delta = 1$. The metal kill ratio R_M is taken as unity, which implies that every metal is an equally effective generation-recombination (g-r) defect center. This, of course, is inappropriate inasmuch as the effectiveness of a g-r center depends on its energy level relative to mid-gap, its capture cross-section, etc. Furthermore, metallics are most detrimental when they decorate structural irregularities in space-charge regions (20,21) rather than as g-r centers per se. Nevertheless, we have utilized this zeroeth order approximation, in conjunction with an empirically derived relation between the metal (Fe) defect density (D_M) and critical metal (Fe) surface concentration [M] noted by Eq. 8, from which [M] is determined in conjunction with Eq. 5:

$$D_M = K_3 [M]^3 \exp(-T_o/0.7 \text{ nm}) (\text{cm}^{-2}) \quad (8)$$

with $K_3 = 1.854 \times 10^{-29}$ cm^4 and T_o is the equivalent oxide thickness in nm for each technology generation. The experimental data in this model (see Table 5) are based on experimentation which extends the precursor publication of oxide films to the 8 nm regime (22,23). Nevertheless, eq. 8 is not based on experiments in the sub 4 nm regime and, therefore, the mechanism and dependence of the oxide breakdown with surface metallics, such as Fe, may not be adequately represented by extrapolation of the experimental data. The experimental data for Fe, furthermore, has been extended to include critical metals such as Ca, Co, Cu, Cr, K, Mo, Mn, Na and Ni, although all these metals redistribute differently between the silicon surface and the growing oxide (22-24). The model analysis explicitly assumes that the listed value corresponds to the highest of the several possible candidate metals, with the other metals more likely near the mean of their distribution and, thereby, exhibiting a yield much greater than 99%. Clearly, it appears that herein is an opportunity for a partitioning scheme. The Starting Materials critical metals are approximately 4x the Surface Preparation pre-gate values, consistent with the CoO approach. If the IC manufacturer uses an initial cleaning process, one may group K and Na with the other surface metals (see below). The present analysis assumes the presence of a wafer gettering mechanism which, if not present, would decrease the Starting Materials critical metal concentration by a factor of 2x.

Table 5
Critical Surface Metal Densities for Starting Materials and Surface Preparation With Technology Generation

Year of First Product Shipment Technology Generation	1997 250 nm	1999 180 nm	2001 150 nm	2003 130 nm	2006 100 nm	2009 70 nm	2012 50 nm
CRITICAL SURFACE METALS Starting Materials (at/cm^2)	$\leq 2.5 \times 10^{10}$	$\leq 1.3 \times 10^{10}$	$\leq 1 \times 10^{10}$	$\leq 7.5 \times 10^9$	$\leq 5 \times 10^9$	$\leq 2.5 \times 10^9$	$\leq 2.5 \times 10^9$
CRITICAL SURFACE METALS Surface Preparation (at/cm^2)	$\leq 5 \times 10^9$	$\leq 4 \times 10^9$	$\leq 3 \times 10^9$	$\leq 2 \times 10^9$	$\leq 1 \times 10^9$	$\leq 10^9$	$\leq 10^9$

Surface Metals - Other Metals

Other surface metals include Al, Ti, V and Zn, which were assigned a value of 10^{11}/cm^2 for all technology generations; here also, the presence of a wafer gettering mechanism is assumed. The Al value is the maximum surface concentration to prevent modification of silicon's oxidation rate (25); Al also forms charged defects in the oxide which may require consideration (26).

Site Flatness

The site front-surface least squares reference plane total indicator range flatness (SFQR) (27) is modeled by equating it to the critical dimension (CD) for dense lines (DRAM half pitch) for each technology generation, including partial sites, by:

$$SFQR = K_4 \times (CD) \quad (9)$$

with $K_4 = 1$. For isolated lines (MPU Gates), the CD is approximately 80% of the dense line values for technology generations $\geq$ 130 nm and approximately 70% for technology generations < 130 nm (see Table 6), although this does not impact the SFQR requirements. Recent epitaxial wafer SFQR values for a number of suppliers are consistent with 250 nm and 180 nm DRAM requirements and correspond to approximately 25% and 20% of the allowable depth-of-focus for the 250 nm and 180 nm technology generations, respectively (see Table 6) (28). Double-side polishing utilized on 300 mm wafers (as well as for 200 mm wafers) is expected to result in significantly improved site flatness. Scanning steppers are expected to be implemented, in some cases, for technology generations $\leq$ 180 nm and utilization of the new site flatness metric, SFSR (29,30) (site front-surface scanning total indicator range) will be required. Non-optical lithography may be required for technology generations $\leq$ 100 nm in which case the site flatness may be de-coupled from the CD. Finally, warp is selected to be less than 50 μm for all CD generations, taking into consideration different back-surface films. However, significantly improved bow and warp performance for both 200 mm and 300 mm diameter wafers is expected in the case of wire-sawing and the relevance of these parameters is expected to decrease (28,31).

Table 6
Several Site Flatness and Related Parameters With Technology Generation

Year of First Product Shipment Technology Generation	1997 250 nm	1999 180 nm	2001 150 nm	2003 130 nm	2006 100 nm	2009 70 nm	2012 50 nm
Dense Lines (DRAM Half Pitch) *	250	180	150	130	100	70	50
Isolated lines (MPU Gates) *	200	140	120	100	70	50	35
Site Flatness (SFQR) (nm)	≤ 250	≤ 180	≤ 150	≤ 130	≤ 100	≤ 100	≤ 100
Site Size (mm x mm) Site Area (mm^2)	22 x 22 484	25 x 32 800	25 x 34 850	25 x 36 900	25 x 40 1000	25 x 44 1100 **	25 x 52 1300 **
Depth of focus (μm, usable @ full field with ± 10% exposure)	0.8	0.7	0.6	0.6	0.5	0.5	0.5

* Requirements scale with resolution for shrinks
** Site size requirements are based on Year 2 chip sizes, the year demanding the full site size for high volume production

Oxygen

The values for oxygen and bulk micro-defects (BMD) are based on extensive anecdotal evidence. The oxygen values represent the range of the center point requirement to provide internal gettering (IG) for generally high oxygen values or minimal IG for lower oxygen values, based on experimentally derived IC requirements. The tolerance of ± 1.5 parts per million atomic (ppma) represents the min-max range about the center point value; a tolerance of ± 2 ppma is appropriate for the 250 nm technology generation (see Table 7). Bulk micro-defects (BMD) in IG (no IG) polished wafers are > 1 x 10^8/cm^3 (< 1 x 10^7/cm^3) after IC processing. The oxygen, BMD value and the relationship between them providing effective IG requires a more rigorous examination inasmuch as the formation of uncontrolled BMD (e.g., SiO_x precipitates) in both polished and epitaxial wafers, may result in excessive device leakage current, necessitating the optimization of IG (32). The magnitude and uniformity requirements for oxygen, as well as dopants, may require the utilization of magnetic Czochralski (MCZ) silicon, especially for 300 mm and larger diameter wafers (33).

The determination of the BMD density (cm^{-3}) by preferential etching requires accurately monitoring the silicon etch rate in order to convert the area density to a volume density. The measurement of interstitial oxygen and the accompanying acceleration of BMD formation in heavily doped silicon continues to remain an area of active research (34).

Table 7
Oxygen Concentration and Tolerance With Technology Generation

Year of First Product Shipment Technology Generation	1997 250 nm	1999 180 nm	2001 150 nm	2003 130 nm	2006 100 nm	2009 70 nm	2012 50 nm
Oxygen (Tolerance ± 1.5 ppma) (ASTM '79)	20 - 31	19 - 31	18 - 31	18 - 31	18 -31	18 - 31	18 - 31

Recently, the characterization of the time dependence of oxygen precipitation has been studied in 300 mm sectioned polished wafers using the Avrami relation (28,31,35) as described by:

$$\xi = 1 - \exp[-kt^n] \qquad (10)$$

where ξ is the fraction transformed (e.g., precipitated) material and n and k are parameters characteristic of the precipitation kinetics. The n parameter relates to the dimension (3D, 2D, fractal) of the system participating in the polymorphic change and is also influenced by the growth mechanism of an individual island (28,31). The k parameter is proportional to the number of nucleation sites and is governed by the mechanism of individual island growth at the phase boundary, e.g., diffusion-limited, reaction-limited, capture at the perimeter, etc. (36). A detailed discussion of the Avrami analysis and the kinetic parameters *n* and *k* (determined by the slope and intercept, respectively), with O_i^o have been presented (28,31). The results appear consistent with the present understanding of oxygen in silicon (37,38) and may be useful to enhance our understanding of the precipitation kinetics of oxygen in silicon. It has been noted, however, that *n* and *k* cannot uniquely determine the rate limiting processes inasmuch as the different mechanisms noted above sometimes yield similar values for *n* and *k* (28). These topics continue to remain at the forefront of silicon technology (39).

The above analysis does not explicitly address the coupling of the thermal history of the crystal, the variation of O_i^o along the crystal axis and the BMD formation during subsequent wafer thermal processes and IC fabrication. The decoupling of the oxygen content and the IC process details from the BMD density and the denuded zone (DZ) depth, using vacancy concentration engineering techniques, may be required (40-42). The utilization of an appropriate thermal anneal requires a fundamental understanding of the critical temperature range for vacancy agglomeration (COPs, D-defects) and oxygen precipitation control (40-45). For example, the dependence of ΔO_i versus O_i^o has been shown to exhibit a rather weak dependence with O_i^o, compared to the previous cases (41,46). This technique offers the opportunity to tune the BMD depending only upon the O_i^o value inasmuch as the thermal history has been decoupled from the BMD formation (41). Quantitative formulation of this methodology will facilitate model-based BMD formation by tuning the crystal growth parameters to minimize the formation of undesired point-defect clusters. The IC DRAM process flow, however, may also need to be re-engineered, in partnership with the silicon supplier, to ensure mutual compatibility.

Carrier Lifetime

The carrier diffusion length, *L*, is related to the bulk carrier recombination lifetime, τ_r, by:

$$L = \sqrt{D_n \tau_r} \qquad (11)$$

where D_n is the minority-carrier diffusion coefficient (47-49). The latter is taken as representative of lightly doped p-type material at 27C (34.5 cm^2/s). The carrier diffusion length is modeled as equal to the wafer thickness by:

$$L \equiv \text{wafer thickness} \qquad (12)$$

and is therefore dependent on the technology generation since the wafer thickness increases as the wafer diameter increases. Equation (11) is accordingly written as:

$$\tau_r = 2\ (\text{wafer thickness})^2 / D_n \qquad (13)$$

where the factor two enters in because τ_r has been implicitly partitioned evenly between the silicon supplier and the IC manufacturer. Appropriate techniques to control, passivate or correct for surface effects are mandatory when measuring τ_r. The bulk carrier recombination lifetime is a measure of the cumulative effect of all the metals. Iron, however, is a common contaminant and the bulk carrier recombination lifetime listed for the 250 nm technology generation in Table 8 is consistent with recent state-of-the-art experimental Fe concentrations (50). Iron concentrations $\leq 1 \times 10^{10}\ /cm^3$, however, may not impact a proportionally significant carrier lifetime improvement because of the influence of residual crystalline flaws in the silicon material (51,52), although one might expect larger values for τ_r even for today's technology. Relating the carrier diffusion length to the wafer thickness as is presently the case, however, is rather artificial and precludes accountability to the material physics issues; clearly, rectification of this model analysis is required. In any case, it should be noted that the bulk Fe concentration (cm^{-3}) cannot be converted to its surface value (cm^{-2}) via the wafer thickness due to thermodynamic distribution effects (2,24).

The SOI effective recombination lifetime, $\tau_{r,eff}$, is represented by:

$$(\tau_{r,\,eff})^{-1} = (\text{SOI layer thickness}/2s)^{-1} + (\tau_{r,\,bulk})^{-1} \qquad (14)$$

where *s*, the surface recombination velocity, is taken as 1-10 cm/s in this model calculation. Here also, proper control, passivation or correction for surface effects at the upper silicon surface is required. Future model improvements should properly account for the surface recombination velocity at the silicon/BOX interface. Since $\tau_{r,\,bulk}$ is several orders of magnitude larger than the reciprocal of the first term in Eq. (14), the value for $\tau_{r,\,eff}$ is approximately represented as (2s/SOI layer thickness) (see Table 8).

Table 8
Bulk Carrier Recombination Lifetime, Bulk Fe Concentration and Effective SOI Recombination Lifetime With Technology Generation

Year of First Product Shipment Technology Generation	1997 250 nm	1999 180 nm	2001 150 nm	2003 130 nm	2006 100 nm	2009 70 nm	2012 50 nm
Recombination Lifetime (μs)	≥ 300	≥ 325	≥ 325	≥ 325	≥ 325	≥ 450	≥ 450
Total bulk Fe (at/cm³)	3×10^{10}	1×10^{10}	1×10^{10}	$< 1 \times 10^{10}$	$< 1 \times 10^{10}$	$< 1 \times 10^{10}$	$< 1 \times 10^{10}$
Effective SOI Recombination Lifetime (μs)	0.3	0.4	0.5	0.6	1	1	1

The generation lifetime, τ_g, is a harbinger of device performance in, for example, a DRAM refresh operation. This lifetime parameter has been modeled by considering only space-charge region leakage currents. We have neglected contributions due to the device sub-threshold, gate dielectric, and diffusion leakage currents, which represents a shortcoming of the current model for the first two components. Accordingly, τ_g is modeled utilizing the definition of τ_g by Eq. 15 (47), where the generation rate, g, is considered to represent only space-charge region leakage:

$$\tau_g = n_i/g \qquad (15)$$

Accordingly, τ_g is represented by:

$$\tau_g = (qWn_i)\,(I_{limit} / A_{space\text{-}charge})^{-1} \qquad (16)$$

where q is the absolute value of the electron charge (1.602×10^{-19} coulomb), W is the space-charge region width (taken as 0.5 μm) and n_i is approximately $1.02 \times 10^{10}/cm^3$ at 27 C and $1.4 \times 10^{12}/cm^3$ at 100C (53,54). The value of $(I_{limit}/A_{space\text{-}charge})$ is modeled by taking the leakage current, I_{limit}, $\leq 10^{-16}$ A/bit at 27 C ($\leq 10^{-13}$ A/bit at 100 C) (55-58) and the total space-charge area of a trench storage cell for the 256 MDRAM is taken as 2.5 μm^2 (59). These parameters result in a room-temperature value of τ_g, approximately equal to 20 μs, which ensures the DRAM storage cell will not be upset due to the space-charge leakage current. The critical current is modeled to scale with the DRAM technology generation (e.g., with decreasing $A_{DRAM\ cell}$ due to decreasing CD).

In view of the assumptions made in developing the above models for τ_r and τ_g, it is not surprising that these parameters do not relate with each other as previously indicated (47,60). Comprehending the periphery, area and volume generation/recombination processes will become more important with the continued miniaturization of silicon device components (61).

Gate Oxide Integrity

Gate oxide integrity (GOI) is modeled by the modified Poisson probability distribution of Eq. 5 with the yield $Y_{Do} = 99\%$, $R_O = 1$, $\delta = 1$ and $\beta = 1$ and 10 for the DRAM and MPU, respectively (see Table 9). The GOI is measured at 10 MV/cm for 100 s under accumulation and D_O is determined using the charge-to-breakdown, Q_{bd}, criteria. The GOI is indicative of potential IC performance, although significant reassessment of the relevance of this methodology is required for sub-4 nm dielectrics (62,63), especially < 2 nm, where direct tunneling current predominates and surface metals may be more detrimental than COPs (64). While other IC electrical parameters such as leakage current, etc. are also important, these parameters were not explicitly addressed in the Starting Materials specifications as they are generally IC design and process dependent.

Table 9
DRAM and MPU Defect Density With Technology Generation

Year of First Product Shipment Technology Generation	1997 250 nm	1999 180 nm	2001 150 nm	2003 130 nm	2006 100 nm	2009 70 nm	2012 50 nm
$D_{O,\,DRAM}$ (cm^{-2})	≤ 0.06	≤ 0.029	≤ 0.026	≤ 0.014	≤ 0.006	≤ 0.003	≤ 0.001
$D_{O,\,MPU}$ (cm^{-2})	≤ 0.15	≤ 0.15	≤ 0.11	≤ 0.08	≤ 0.05	≤ 0.04	≤ 0.03

Oxidation Stacking Faults

Oxidation stacking faults (OSF) are empirically modeled by:

$$OSF = K_5 \times (CD)^{1.42}\ (cm^{-2}) \qquad (17)$$

with $K_5 = 2.75 \times 10^{-3}$ and CD in nm (15) (see Table 10). The OSF density depends on a number of material and IC process conditions (7,65,66); the recommended test is steam oxidation at 1100 C for 1 hr and preferential etch after stripping the oxide. Oxidation stacking faults generated during IC fabrication result in competitive and deleterious sinks for metallics compared to bulk SiO_x precipitates, due to their proximity to the active device regions. Control of OSF is more difficult in n-type material.

Table 10
Oxidation Stacking Faults With Technology Generation

Year of First Product Shipment Technology Generation	1997 250 nm	1999 180 nm	2001 150 nm	2003 130 nm	2006 100 nm	2009 70 nm	2012 50 nm
Oxidation Stacking Faults (OSF) (cm^{-2})	≤ 7	≤ 4	≤ 3.5	≤ 3	≤ 2	≤ 1	≤ 1

Epitaxial Layer Properties

The epitaxial layer thickness represents the range of the center point value with the tolerance expressed as the min-max % range about the selected center point value (see Table 11). The epitaxial layer thickness is empirically modeled from the epitaxial flat-zone required by IC designers by:

$$\text{Epi Layer Thickness} = K_6 \times (\text{Epi Flat-Zone}) \qquad (18)$$

with $K_6 = 1.25$. The epitaxial growth induced structural defects such as mounds and stacking faults are modeled by the modified Poisson probability distribution of Eq. 3 with the yield $Y_{EPI} = 99\%$, $R_{EPI} = 1$ and $\alpha = 1$ (see Table 11). The representation of α by unity assumes that every defect within the MPU's chip area (which scales with technology generation) is detrimental. Further clarification is expected from ongoing studies of epitaxial defects and defect-specific studies of their impact on GOI and/or junction leakage tests.

Table 11
Epitaxial Layer Thickness and Defect Density With Technology Generation

Year of First Product Shipment Technology Generation	1997 250 nm	1999 180 nm	2001 150 nm	2003 130 nm	2006 100 nm	2009 70 nm	2012 50 nm
Layer Thickness (μm) (± % Tolerance)	2 - 5 (± 5%)	2 - 4 (± 4%)	2 - 4 (± 4%)	2 - 4 (± 4%)	1 - 3 (± 3%)	1 -3 (± 3%)	1 - 3 (± 3%)
Layer Structural Defects (cm^{-2})	≤ 0.0033	≤ 0.0029	≤ 0.0026	≤ 0.0023	≤ 0.0019	≤ 0.0016	≤ 0.0013

The improved GOI in epitaxial material is due to the improved structural perfection of epitaxial material as compared to residual polishing micro-damage and grown-in micro-defects in polished wafers. The utilization of polished wafers as well as p/p^- material with hydrogen or argon annealed substrates for advanced ICs is also receiving attention (67). The benefit of both epitaxial surface quality and reduced system capacitance (due to a larger space-charge region width as a result of the lightly doped substrate) may offset the lack of enhanced solubility gettering of iron in p^-, compared with conventional p^+ epitaxial, substrates. The oxygen content, however, may have to be re-assessed since oxygen precipitates slower in p^-, compared to p^+, material (34,68), at least for conventionally prepared materials. Back-surface polysilicon gettering may become more fully utilized, in conjunction with the general reduction of the oxygen content, in both epitaxial substrates and polished wafers (69). Selection of the gettering system, however, is very dependent on the IC thermal process sequence. The role of MeV implantation and associated annealing procedures in polished wafers as a replacement for epitaxial structures continues to receive attention, although the control of COPs and related defects is aggravated in polished wafers. The use of MeV implant/thermal anneal procedures, in conjunction with epitaxial wafers, has also been proposed to achieve enhanced device architectures.

Silicon-On-Insulator

The silicon layer thickness represents the range of the center point value with the tolerance expressed as the min-max % range about the selected center point value (see Table 12). The buried oxide (BOX) thickness represents the range of the center point value with the tolerance expressed as the min-max % range about the selected center point value. A fully depleted silicon layer is anticipated for CD < 100 nm, dependent on the IC application and is reflected in the silicon thickness (in conjunction with the BOX thickness) required to withstand the device breakdown voltage. The interface charge at the SOI/BOX interface is specified to be < 1 x $10^{11}/cm^2$.

The three types of SOI defects examined are BOX defects, inclusions and threading dislocations (70); these defects have been calculated independently at the 99% level for the case of an MPU. The BOX defect density, D_{BOX}, is modeled by the modified Poisson probability distribution Eq. 5 with the yield Y_{BOX} = 99%, BOX kill ratio $R_{BOX} = 0.2$, $\beta = 10$ and $\delta = 6$. The six CD units for δ are represented as the gate, the sidewall spacers, and the salicided source and drain. The inclusion density, D_{INC}, is modeled with the yield Y_{INC} = 99%, inclusion defect kill ratio $R_{INC} = 1$, $\beta = 10$ and $\delta = 1$ (gate). The threading dislocation density, D_{TD}, is modeled with the yield Y_{TD} = 99%, threading dislocation kill ratio $R_{TD} = 1 \times 10^{-6}$, $\beta = 10$ and $\delta = 2$. The 2 CD units arise from the gate and the sidewall spacers, where the dislocations could impact the gate oxide or p-n junctions, respectively. The selection of the kill ratios is expected to require re-assessment. It also appears that a two-way partitioning of the BOX defects and inclusions (e.g., to be expanded to reflect pits/COPs as well as inclusions) may be appropriate while threading dislocations may not be particularly harmful (71).

Table 12
SOI Layer Thickness and Defects With Technology Generation

Year of First Product Shipment Technology Generation	1997 250 nm	1999 180 nm	2001 150 nm	2003 130 nm	2006 100 nm	2009 70 nm	2012 50 nm
Silicon Layer Thickness (Tolerance ± 5%) (nm)	50 - 200	50 - 200	50 - 200	50 -200	50 - 100	30 - 100	20 - 100
Buried Oxide Thickness (Tolerance ± 5%) (nm)	≤ 400	≤ 200	≤ 200	≤ 200	≤ 100	≤ 70	≤ 50
D_{BOX}, MPU (Box Defects) (cm^{-2})	≤ 0.12	≤ 0.12	≤ 0.09	≤ 0.06	≤ 0.04	≤ 0.03	≤ 0.02
D_{INC}, MPU (Inclusions) (cm^{-2})	≤ 0.15	≤ 0.15	≤ 0.11	≤ 0.08	≤ 0.05	≤ 0.04	≤ 0.03
D_{TD}, MPU (Threading Dislocations) (cm^{-2})	≤ 7.2 x 10^4	≤ 7.4 x 10^4	≤ 5.6 x 10^4	≤ 3.9 x 10^4	≤ 2.6 x 10^4	≤ 2.0 x 10^4	≤ 1.4 x 10^4

Evaluation of the various SOI wafer fabrication techniques by material characterization and identification of the relationships between defects and SOI properties such as gettering and the effective SOI recombination lifetime (see Table 8) on subsequent device properties is essential (71). The small SOI supplier base impacts the cost structure although different SOI approaches may be necessary to service different IC applications. Some bulk IC designs can be directly transferred to SOI. Process and mask redesign may further improve performance and chip size. However, SOI applications may be limited until conventional silicon materials reach a technological or economic wall, perhaps at 130 nm and beyond. Nevertheless, SOI obviates the concern about latch-up while offering the potential for improved device performance, soft-error immunity, low-power applications, fewer process steps and, presumably, smaller chip size with the associated opportunity of utilizing the previous generation's factory equipment to achieve the required number of chips per wafer.

SUMMARY

For technology generations down to 100 nm, the CoO of silicon material quality, both polished and epitaxial wafers, is of paramount importance. This includes distinguishing particles, microroughness and silicon micro-defects for the front-surface (and, where the requisite instrumentation is available, for the back-surface) with improved understanding and control of both grown-in micro-defects and the wafer-carrier interaction, and their subsequent impact on relevant device characteristics. The critical challenge for technology generations below 100 nm is the CoO for the production of 450 mm, 675 mm and SOI starting materials. The methodology for fabricating silicon for the 64 Gbit era (450 mm and 675 mm wafers) as compared to the effective introduction of SOI and compatibility with IC processing needs to be established. The engineering issues associated with these larger diameters, however, appear to be enormous and a paradigm shift in approaching the fabrication of cost-effective silicon materials, including the cost-effective introduction of SOI, is required. The manufacturing cost for the 450 mm and larger diameter wafers, for example, is projected to become both an increasing percentage and increase the total device manufacturing cost (72). The fabrication of silicon materials on an appropriate substrate or re-engineering the IC package could help mitigate the escalating costs associated with grinding off approximately 50% of the wafer volume, as is currently done with conventional silicon materials and IC packages.

In addition to the effective partitioning of relevant material parameters, it will be essential to incorporate the defect size distribution and kill ratio in the Poisson probability distribution as well as identify improved kill factors for the various defects. The de-convolution of particles, COPs and microroughness for the front-surface remains critical as the CD is approaching the COPs grown-in dimensions. Back-surface particle detection may be limited to ≥ 200 nm, even for smaller CD, as a result of the back-surface finish. The improvement in back-surface finish compared to current standards, however, will more readily exhibit microscopic contamination and handling scratch damage from robotic handling systems. Standards for robotic handlers, therefore, may be necessary to ensure conformance of the surface to the implicit cleanliness requirements. The characterization of the surface microroughness is essential inasmuch as the excursion of microroughness may be comparable to the gate dielectric thickness, which may be only several unit cells thick, at least for the silicon dioxide, oxynitride and silicon nitride dielectric systems. Finally, determination of material parameters over the whole wafer surface (taking cognizance of the edge exclusion) is a most critical metrology issue.

FUTURE OPPORTUNITIES

To reduce the CoO of wafers, the silicon supplier parameter distributions should be utilized as materials acceptance criteria at IC companies. This method would alleviate the challenge of separating measurement variability from the true parameter variability that is becoming more difficult as the values for many parameters are approaching detection or resolution limits. Use of parameter distributions established by wafer suppliers for their processes to demonstrate that an ensemble of wafers will be satisfactory for the intended purpose will require improvements in both process capabilities and a significant paradigm shift in IC materials acceptance practices. Currently available models, however, cannot sufficiently establish the real requirements for parameter uniformity or the effects of parameter variability on IC properties. The requirements presented in the 1997 edition of the Starting Materials Roadmap do not adequately address either of these issues and development of such models will be essential to enhance the utility of future Roadmaps.

ACKNOWLEDGMENTS

The authors appreciate discussions with Tibor Pavelka, Al Tasch and Peter Zeitzoff.

REFERENCES

1. SIA 1997 Roadmap, Semiconductor Indusry Association, 181 Metro Drive, Suite 450, San Jose, CA 95110
2. H. R. Huff and R.K. Goodall, Silicon Materials and Metrology: Critical Concepts for Optimal IC Performance in the Gigabit Era, *Semiconductor Characterization* (edited by W.M. Bullis, D.G. Seiler and A.C. Diebold), 67-96 (1996)
3. R.K. Goodall, Silicon Wafer Specifications and Characterization Capabilities ...Another Perspective, *Flatness, Roughness and Discrete Defect Characterization for Computer Disks, Wafers and Flat Panel Displays*, SPIE 2862, (edited by J.C. Stover), 2-9 (1996)
4. G.E. Moore, Cramming More Components Onto Integrated Circuits, Electronics, 38, No. 8, 114-117 (1965)
5. G.E. Moore, Progress in Digital Integrated Electronics, *IEDM*, 11-13 (1975)
6. G. E. Moore, Lithography and The Future of Moore's Law, *Integrated Cicuit Metrology, Inspection and Process Control XI* (edited by M. H. Bennett), SPIE 2439, 2-17 (1995)
7. H.R. Huff, Twentieth Century Silicon Microelectronics, *ULSI Science and Technology/1997*, (edited by H.Z. Massoud, H. Iwai, C. Claeys and R.B. Fair), 53-117 (1997)
8. D. Rose, Future of 200 mm Wafers, *SEMI-JEIDA Joint Technical Symposium: What's Ahead for 200 mm Wafers?*, SEMICON/West, July 22, 1993
9. H.R. Huff and R.K. Goodall, Growing Pains, *Interface*, 31-35 (1996)

10. H.R. Huff, R.K. Goodall, R.H. Nilson and S.K. Griffiths, Thermal Processing Issues for 300 mm Silicon Wafers: Challenges and Opportunities, *ULSI Science and Technology/97*, (edited by H.Z. Massoud, H. Iwai, C. Claeys and R.B. Fair), 135-181 (1997)
11. S. Yamagata, 300 mm Wafer Processing, *SEMICON/KOREA 96*, 117-126 (1996)
12. F. Robertson and A. Allan, 300 mm Conversion Timing, *Future Fab International*, 1, No.3, 27-33 (1997)
13. W. Maly, H.T. Heineken and F. Agricola, A Simple New Yield Model, *Semiconductor International*, No.7, 148-154 (1994)
14. H.R. Huff, R.K. Goodall, E. Williams, K.-S Woo, B.Y.H. Liu, T. Warner, D. Hirleman, K. Gildersleeve, W.M. Bullis, B.W. Scheer and J. Stover, Measurement of Silicon Particles by Laser Surface Scanning and Angle-Resolved Light Scattering, J.Electrochem. Soc., 144, 243-250 (1997)
15. M. Kamoshida, Trends of Silicon Wafer Specifications vs. Design Rules in ULSI Device Fabrication. Particles, Flatness and Impurity Distribution Deviations, DENKA KAGAKU, No. 3, 194-204 (1995)
16. K. Saga and T. Hattori, Influence of Silicon-Wafer Loading Ambients in an Oxidation Furnace on the Gate Oxide Degradation Due to Organic Contamination, Appl. Phys. Letts., 71, 3670-3672 (1997)
17. F. Tardif, G. Quagliotti, T. Baffert and L. Secourgeon, Hydrocarbons Impact on Thin Gate Oxides, UCPSS '96, 309-312 (1996)
18. P.J. Smith and P. Lindley, Analysis and Reduction of Organic Contamination in Semiconductor Processing, this volume
19. W.M. Bullis, Microroughness of Silicon Wafers, *Semiconductor Silicon/1994*, (edited by H.R. Huff, W. Bergholz and K. Sumino), 1156-1169 (1994)
20. W. Shockley, Problems Related to p-n Junctions In Silicon, Solid-State Electronics, 2, 35-67 (1961)
21. Y. Ichida, T. Yanada and S. Kawado, Influence of Impurity-Decorated Stacking Faults on The Transient Response of Metal Oxide Seimconductor Capacitors, *Lifetime Factors in Silicon* (edited by R.D. Westbrook), ASTM STP 712, 107-118 (1980)
22. W.B. Henley, L. Jastrzebski and N.F. Haddad, The Effects of Iron Contamination on Thin Oxide Breakdown—Experimental and Modeling, MRS 262, (edited by S. Ashok, J. Chevallier, K. Sumino and E. Weber) 993-998 (1992)
23. C.R. Helms and W.B. Henley (unpublished data)
24. C.R. Helms, Silicon Surface Preparation and Wafer Cleaning: Role, Status, and Needs for Advanced Characterization, *Semiconductor Characterization* (edited by W.M. Bullis, D.G. Seiler and A.C. Diebold), 110-117 (1996)
25. J.M. deLarious, D.B. Kao, C.R. Helms and B.E. Deal, Effect of SiO_2 Surface Chemisty on The Oxidation of Silicon, Appl. Phys.Letts., 54, 715-717 (1989)
26. H. Shimizu, Alison Shull and C. Munakata, Metal-Induced Charge and Growth Rate of Ultra-Thin Thermal Oxide Films Grown on Al or Fe Contaminated Si (100) Surfaces, *Optical Characterization Techniques for High-Performance Microelectronic Device Manufacturing III* (edited by D. DeBusk and R.T. Chen), 174-185 (1996)
27. *SEMI Int'l Stds*. 1994 - Materials M1-96, 20 (1996)
28. H.R. Huff, D.W. McCormack, Jr., C. Au, T. Messina, K. Chang and R.K. Goodall, Current Status of 200 mm and 300 mm Silicon Wafers, *ISSDM'97*, 456-457,575 (1997), (to be published, Japanese J. Appl. Phys.)
29. R.K. Goodall and H.R. Huff, Wafer Flatness Modeling for Scanning Steppers, *Metrology, Inspection and Process Control for Microlithography*, SPIE 2725, 76-85 (1996)
30. *SEMI Int'l Stds*. - Materials M1 - (rev. 98, to be published)
31. C. Au, T. Messina, K. Chan, R.K. Goodall and H.R. Huff, Characterization of 300 mm Polished Wafers, *Semiconductor Silicon/1998* (edited by H.R Huff, U. Gosele and H. Tsuya), 641-659 (1998)
32. H. Fujimori, Y. Ushiku, T. Ihnuma, Y. Kirino and Y. Matsushita, The Interrelation Between The Morphology of The Oxygen Precipitates and The Junction Leakage Current in Czochralski Silicon Crystals, *Semiconductor Silicon/1998*, (edited by H.R. Huff, U. Gosele and H. Tsuya), 1033-1044 (1998)
33. S. Kumai, 300 mm Wafer Development - Current Status and Future Challenges, *SEMI Technology Symposium*, SEMICON/Japan, 1-45 -- 1-50 (1996)
34. T. Ono, E. Asayama, H. Horie, M. Hourai, K. Sueoka, H. Tsuya and G.A. Rozgonyi, Effect of Heavy Boron Doping on Oxide Precipitate Growth in Czochralski Silicon, *Semiconductor Silicon/1998*, (edited by H.R. Huff, U. Gosele and H. Tsuya), 1113-1125 (1998)
35. J.W. Christian, *The Theory of Transformations in Metals and Alloys*, Pergamon Press, Oxford, England, (1965)
36. P.H. Holloway and J.B. Hudson, Kinetics of The Reaction of Oxygen With Clean Nickel Single Crystal Surfaces, I. Ni (100) Surface, Surface Science, 43, 123 (1974)
37. T.Y. Tan and C.Y. Kung, Oxygen Precipitation Retardation and Recovery Phenomena in Czochralski Silicon: Experimental Observations, Nuclei Dissolution Model, and Relevancy With Nucleation Issues, J. Appl. Phys., 59, 917-931 (1986)
38. F. Shimura, *Oxygen in Silicon*, Academic Press, Inc., 1994

39. R.C. Newman, Oxygen Aggregation and Interactions With Carbon and Hydrogen, *Semiconductor Silicon/1998*, (edited by H.R. Huff, U. Gosele and H. Tsuya), 257-271 (1998)
40. R. Falster, D. Gambaro, M. Olmo, M. Cornara and H. Korb, The Engineering of Silicon Wafer Material Properties Through Vacancy Concentration Profile Control and The Achievement of Ideal Oxygen Precipitation Behavior. MRS Proceedings, Spring, 1998
41. J. Binns, S. Pirooz and T.A. McKenna, Silicon Material Issues and Requirements for Device Design Rules of $\leq$ 0.35 µm, *Future Fab International*, 1, No. 4, 263-277 (1998)
42. R. Falster, V.V. Voronkov, J.C. Holzer, S. Markgrafh, S.S. McQuaid and L. Mule'Stagno, Intrinsic Point Defects and Recations in The Growth of Large Silicon Crystals, *Semiconductor Silicon/1998,* (edited by H.R Huff, U. Gosele and H. Tsuya), 468-489 (1998)
43. M. Hourai, H. Nishikawa, T. Tanaka, S. Umeno, E. Asayama, T. Nomachi and G. Kelly, Nature and Generation of Grown-In Defects in Czochralski Silicon Crystals, *Semiconductor Silicon/1998*, (edited by H.R. Huff, U. Gosele and H. Tsuya), 453-467 (1998)
44. E. Dornberger, J. Esfandyari, J. Vanhellemont, D. Graf, U. Lambert, F. Dupret and W. von Ammon, Simulation of Non-Uniform Grown-In Void Distributions in Czochralski Silicon Crystals, *Semiconductor Silicon/1998*, (edited by H.R. Huff, U. Gosele and H. Tsuya), 490-502 (1998)
45. T. Sinno and R.A. Brown, Modeling and Microdefect Formation in Czochralski Silicon, *Semiconductor Silicon/1998*, (edited by H.R. Huff, U. Gosele and H. Tsuya), 529-545 (1998)
46. W. Bergholz, J.L. Hutchison and G.R. Booker, Order-Related Defects in CZ-Silicon After Annealing at 650C, *Semiconductor Silicon/1986*, (edited by H.R. Huff, T.Abe and B. Kolbesen), 874-888 (1986)
47. D.K. Schroder, *Semiconductor Material and Device Characterization*, New York; John Wiley & Sons, 1990
48. W.M. Bullis and H.R. Huff, Interpretation of Carrier Recombination Lifetime and Diffusion Length Measurements in Silicon, J. Electrochem. Soc., 143, 1399-1405 (1996)
49. D. Schroder, Carrier Lifetimes in Silicon , IEEE Trans. Elec. Devices, 44, 160-170 (1997)
50. Y. Kitagawara, T. Yoshida, T. Hamaguchi and T. Takneaka, Evaluation of Oxygen-Related Carrier Recombination Centers in High-Purity Czochralski-Grown Si Crystals by the Bulk Lifetime Measurements, J. Electrochem. Soc., 142, 3505-3509 (1995)
51. A. Buczkowski, F.G. Kirscht and H. Koya, Sample Preparation Impact on MetalContamination Evaluated by Surface Photovoltage and Photoconductance Decay, *Crystalline Defects and Contamination: Their Impact and Control in Device Manufacturing: II*, (edited by B.O.Kolbesen, P. Stallhofer, C. Claeys and F. Tardiff), 376-384 (1997)
52. A. Buczkowski, Comparative Analysis of Photoconductance Decay and Surface Photovoltage Techniques: Theoretical Perspective and Experimental Evidence, *Silicon Recombination Lifetime Characterization Methods*, (edited by D.C. Gupta, F. Bacher and W.H. Hughes) ASTM (to be published) (1998)
53. M.A. Green, Intrinsic Concentration, Effective Densities of States, and Effective Mass in Silicon, J. Appl. Phys., 67, 2944-2954 (1990)
54. A.B. Sproul and M.A. Green, Improved Value For The Silicon Intrinsic Carrier Concentration From 275 to 375 K, J. Appl. Phys., 70, 846-854 (1991)
55. J.S. Schmid, R. Craigin, C. Damianou, J. Hohl, H. Schrimpf, H. Parks, J. Ramberg, N. Brown and R. Jones, A Simple Model For Estimating Allowable Transition Metal Contamination Levels in DRAMs, *TECHON '90*, 263-266 (1990)
56. K. Koyama, DRAM Technology Development From The Standpoint of Material Technology, *SEMI Technology Symposium*, SEMICON/Japan, 3-9 -- 3-12 (1997)
57. C.S. Hwang, B.T. Lee, H. Horii, K.H. Lee, W.-d Kim, H.-J Cho, C.S. Kang, S.I. Lee and M.Y. Lee, A Process Integration of (Ba, Sr) TiO_3 Capacitor into 256M DRAM, *ISSDM'97*, 272-273 (1997)
58. K. Itoh, H. Sunami, K. Nakazato and M. Horiguchi, Pathways to DRAM Design and Technology For The 21st Century, *Semiconductor Silicon/1998*, (edited by H.R. Huff, U. Gosele and H. Tsuya), 350-369 (1998)
59. A.F. Tasch and L.H. Parker, Memory Cell and Technology Issues for 64- and 256-Mbit One-Transistor Cell MOS DRAMs, Proceedings of IEEE, 77, 374-388 (1989)
60. D.E. Schroder, The Concept of Generation and Recombination Lifetimes in Semiconductors, IEEE Trans. Electron Devices, ED-29, 1336-1338 (1982)
61. E. Simoen, A. Poyai and C. Claeys, Optimised Diode Assessment of The Surface and Bulk Generation/ Recombination Properties of Silicon Substrates, *Semiconductor Silicon/1998*, (edited by H.R. Huff, U. Gosele and H. Tsuya), 1576-1592 (1998)
62. M. Hirose, W. Mizubayashi, M. Fukuda and S. Miyazaki, Tunnel Current and Wearout Phenomena in Sub-5 nm Gate Oxides, *Semiconductor Silicon/1998* (edited by H.R Huff, U. Gosele and H. Tsuya), 730-744 (1998)
63. J.R. Hauser and K. Ahmed, Characterization of Ultra-Thin Oxides Using Electrical C-V and I-V Measurements, this volume

64. T. Miera, J. Jablonski, M. Danbata, K. Nagai and M. Watanabe, Structure of The Defects Responsible For B-Mode Breakdown of Gate Oxide Grown on The Surface of Silicon Wafers, *Defects in Electronic Materials II*, MRS 442, 107-112 (1997)
65. S.M. Hu, Defects and Device Processing: Achievements and Limitations, *Semiconductor Silicon/1986*, (edited by H.R. Huff, T.Abe and B. Kolbesen), 722-750 (1986)
66. S.M. Hu, Silicon Defects in Silicon Technology, *Semiconductor Silicon/1998* (edited by H.R Huff, U. Gosele and H. Tsuya), 220-256 (1998)
67. Y. Matsushita, M. Sanada, A. Tanabe, R. Takeda, N. Shimaoi and K. Kobayashi, Hydrogen Anneal of Silicon Wafer Formation of High Quality Device Active Layer, *Semiconductor Silicon/1998* (edited by H.R Huff, U. Gosele and H. Tsuya), 683-697 (1998)
68. M. Schrems, Simulation of Oxygen Precipitation, 391-447, Oxygen in Silicon (edited by F. Shimura) (1994)
69. W.M. Bullis and W.C. O'Mara, Large Diameter Silicon Wafer Trends, *Solid State Technology*, No. 4, 59-65 (1993)
70. H.J. Hovel, Status and Prospects for Silicon-on-Insulator Materials, *Future Fab International*, 1, No. 2, 225-230 (1997)
71. W.P. Maszara, R. Dockerty, C. Gondran and P.K. Vasudev, SOI Materials for Mainstream CMOS Technology, *Silicon-on-Insulator Technology and Devices* (edited by S. Cristoloveanu, P.L.F. Hemment, K. Izumi and S. Wilson) 15-26 (1997)
72. M. Kikuchi and K. Koyama, A Cost Model Analysis and Expectations For Large Diameter Wafers, *SEMI Technology Symposium*, SEMICON/Japan, 1-15--1-20 ((1996)

FRONT END PROCESSES - GATE DIELECTRICS

Reliability Characterization of Ultra-Thin Film Dielectrics

J. S. Suehle

Semiconductor Electronics Division, National Institute of Standards and Technology, Gaithersburg, MD 20899

The reliability of gate oxides is becoming a critical concern as oxide thickness is scaled below 4 nm in advanced CMOS technologies. Traditional reliability characterization techniques must be modified for very thin gate oxides that exhibit excessive tunneling currents and soft breakdown. As intrinsic reliability limits are approached by increasing chip temperature and electric fields, it becomes essential to fully understand the physical mechanism(s) responsible for gate oxide wear-out and eventual breakdown. Issues relating to the reliability testing of ultra-thin oxides are discussed with examples.

INTRODUCTION

As the semiconductor industry continues to scale device dimensions to achieve channel lengths below 100 nm, SiO_2 gate dielectrics must be scaled to have an equivalent thickness between 1 and 2 nm. The reliability of gate oxides in this thickness range is becoming a critical concern since devices will operate with higher gate electric fields and direct tunneling currents passing through the gate dielectric. The physics of failure and traditional reliability testing techniques must be reexamined for ultra-thin gate oxides that exhibit excessive tunneling currents and soft breakdown.

Highly accelerated ramped voltage or current tests are traditionally used in production monitoring. Common metrics used to assess and compare oxide quality during manufacture are charge-to-breakdown (Q_{bd}) and electric-field-at breakdown (E_{bd}). Q_{bd} is obtained by integrating the current that flows through a dielectric film before it fails in an accelerated stress test. It becomes very difficult to detect hard breakdown (a surge in current for ramped /constant voltage stress or a collapse in voltage for ramped /constant current stress) in dielectrics less than 4 nm thick.

Section II discusses soft breakdown and shows that a more sophisticated failure criterion must be used to detect breakdowns in ultra-thin films. Section III discusses time-dependent-dielectric-breakdown (TDDB) and the importance of choosing the correct electric field acceleration model for extrapolating oxide reliability to operating conditions. Section IV presents conclusions.

II. GATE OXIDE INTEGRITY TESTS

The primary problem in characterizing the dielectric integrity of ultra-thin oxides is determining the breakdown event. Unlike thicker films, ultra-thin films (films with a thickness less than 4 nm) exhibit soft breakdown. An overestimation of the reliability of the film can result if the initial breakdown event is not detected.

Figure 1 illustrates a JEDEC standard voltage ramp procedure used to determine the breakdown electric field for thin gate oxides. The bottom figure shows the voltage versus time applied to the oxide during the test, and the y-axis of the top figure shows the current measured through a 3, 5, and 10 nm thick film. Oxide breakdown is exhibited as a sharp increase in current for the 5 and 10 nm thick film; however, a sharp breakdown cannot be readily observed in the 3 nm film.

A similar observation for another common gate oxide reliability test, the bounded current ramp test, is shown in Fig. 2. In this case current is ramped up to a constant value and is held until breakdown occurs. Breakdown is usually shown as a large sharp decrease in the measured gate voltage. Note that a large sharp decrease is difficult to observe in the 3 nm thick film.

Constant voltage or long-term reliability tests can also be affected by soft breakdown as shown in Fig. 3. The figure shows the current versus time measured through

CP449, *Characterization and Metrology for ULSI Technology: 1998 International Conference*
edited by D. G. Seiler, A. C. Diebold, W. M. Bullis, T. J. Shaffner, R. McDonald, and E. J. Walters

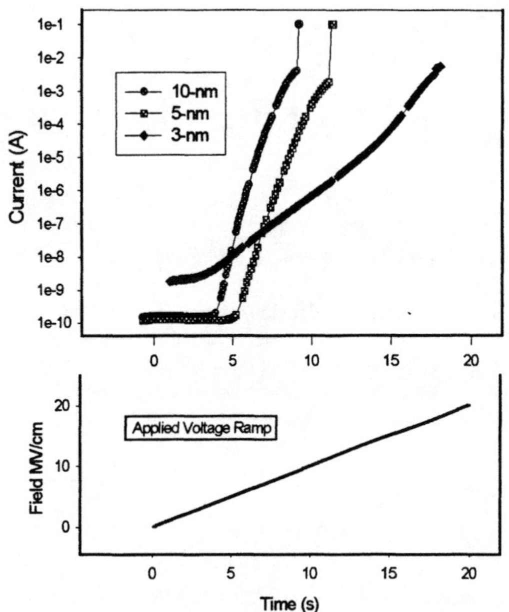

FIGURE 1. Current versus time characteristics obtained during a JEDEC standard voltage ramp test for 3, 5, and 10 nm thick SiO_2 films. The electric field versus time is shown at the bottom. Note that there are clear breakdown events for the 5 and 10 nm thick films. The breakdown is not obvious in the 3 nm thick film

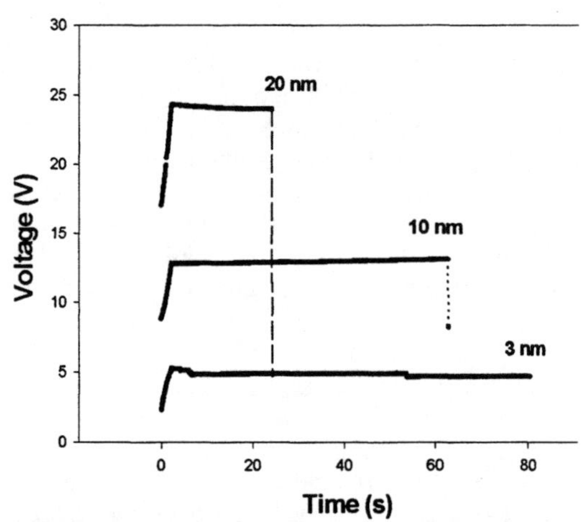

FIGURE 2. Voltage versus time characteristics obtained during a bounded current ramp test for a 3, 10, and 20 nm thick gate oxide sample. The failure criterion is a sudden large decrease in the measured gate voltage. Note that this is not observed in the 3 nm sample.

a 3, 5, and 10 nm thick oxide. Again, the breakdown event is clearly observed as a large surge in current for the 5 and 10 nm thick samples; however in the case of the 3 nm film several "steps" or breakdown events are exhibited. One explanation for soft breakdown is that the energy stored on the gate of the ultra-thin capacitor (CV^2) is not large enough to cause a thermal destructive breakdown.

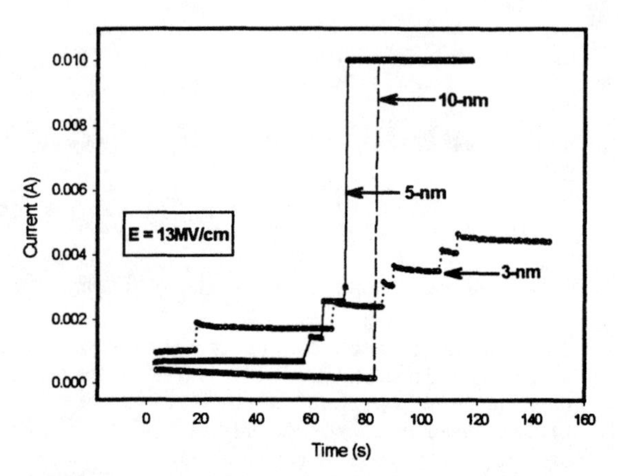

FIGURE 3. Current versus time characteristics measured during constant voltage breakdown tests for a 3, 5, and 10 nm film. Note the multiple breakdown events observed by the 3 nm film.

Soft or "quasi" breakdown has also been explained as charge trapping and de-trapping in a physically damaged region near the Si/SiO_2 interface [1] or multiple tunneling paths caused by generated electron traps [2].

It has also been observed that very thin films with a thickness less than 2 nm do not exhibit any hard breakdown event [3]. The presence of current or voltage noise appears to be the only way of detecting that a breakdown event has occurred [4].

Standards groups such as ASTM (American Society for Testing and Materials) and JEDEC (Joint Electron Devices Engineering Council) [5] are modifying older voltage/current ramp test standards and changing failure criterion to accommodate the testing of ultra-thin dielectrics. Examples of these changes include modifying the tenfold increase in the expected current failure criterion in the voltage ramp test to a more sophisticated but better failure criterion such as current versus voltage slope change [5].

This modification is expected to work for testing gate oxides down to the 3 nm thickness range. For thinner oxides it has also been shown that a change in current noise can be used to detect breakdown [6].

III. OXIDE WEAR-OUT

Time-dependent-dielectric-breakdown (TDDB) or "wear-out" has been studied for thin SiO_2 films for many years. There have been a number of physical models developed for predicting breakdown under normal operating conditions from results obtained from accelerated life tests.

One of the most popular models includes the linear "E" model observed by Crook [7], Berman [8], and by McPherson and Baglee [9]. This model assumes that oxide wear-out is a thermodynamic process and predicts that the log of the median-time-to-fail $\log(t_{50})$ is linearly dependent on the stress electric field.

A second model developed by Chen [10], Lee [11] and Moazzami [12] is based on anode hole injection. Electrons tunneling through the oxide allow holes to be injected and trapped in the oxide. The trapped holes enhance localized electric fields in the oxide which cause an increase in the tunneling current. Breakdown occurs via a positive feedback between hole trapping and enhanced electron injection. The model predicts that the $\log(t_{50})$ is proportional to the reciprocal electric field.

Care must be used in the choice of this model because oxide lifetime at normal operating conditions is extrapolated over many orders of time from testing conditions. An example extrapolation of the linear E and reciprocal E mode is shown in Fig. 4.

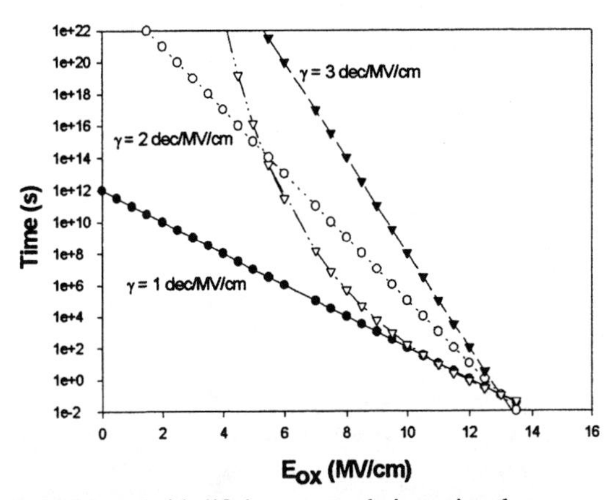

FIGURE 4. Oxide lifetime extrapolation using the reciprocal E model (open triangles) and the linear E model (γ =1 dec/MV/cm, solid circles), (γ =2 dec/MV/cm, open circles), and (γ =3dec/MV/cm, solid triangles).

The figure shows the extrapolated lifetime for three different values of the field acceleration parameter (also referred to as "γ" and defined as the slope of the $\log(t_{50})$ versus E plot) used in the linear E model and for the reciprocal E model. Note that the extrapolated lifetime of the oxide at lower electric fields can vary by many orders of magnitude in time depending on the model used and the parameters used in the model.

Early studies reported varying field and temperature dependencies for TDDB. Also, a large variation in the value of the field acceleration parameter and the thermal activation energy was reported. Most of these studies were conducted over limited test conditions on oxides of varying quality from different processes.

A comprehensive TDDB study performed by NIST [13] demonstrated that intrinsic breakdown of thin oxides can be described by a linear electric field dependence. The study was conducted over a wide range of stress electric field, temperature, and oxide thickness. The results are shown in Fig. 5.

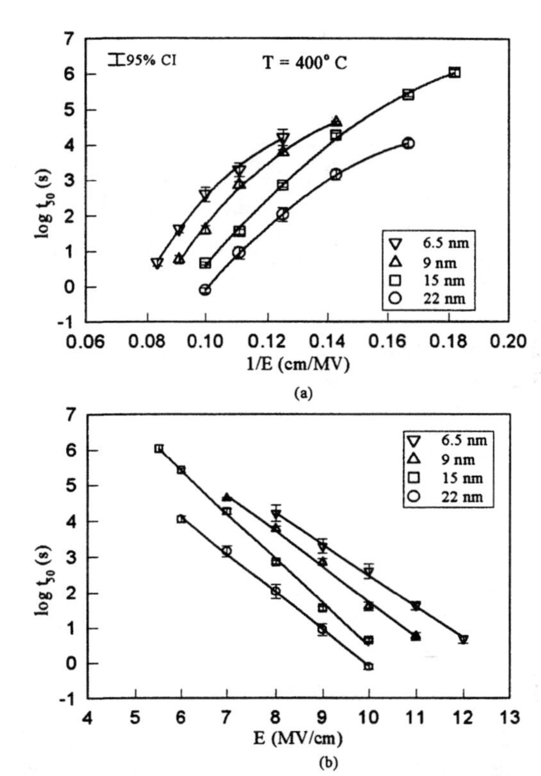

FIGURE 5. The plot of ln (t_{50}) versus reciprocal electric field (a) and linear electric field (b) for a 6.5, 9, 15, and 22 nm thick oxide at stress temperatures of 125 °C, 250 °C,

The TDDB data exhibit a linear dependence with electric field over the entire range of electric fields studied, as shown in Fig 5b. Note the deviation of the 1/E dependence is only observed at lower electric fields as shown in Fig. 5a and a good fit to both models can be obtained at higher electric fields

Figures 6 and 7 show that the TDDB acceleration parameters exhibit consistent dependencies with temperature and stress electric field and do not vary appreciably with oxide thickness.

The field acceleration parameter plotted as a function of temperature for different oxide thicknesses is shown in Fig. 6. Note that the value ranges between 0.8 and 1.4 dec/MV/cm and is independent of temperature.

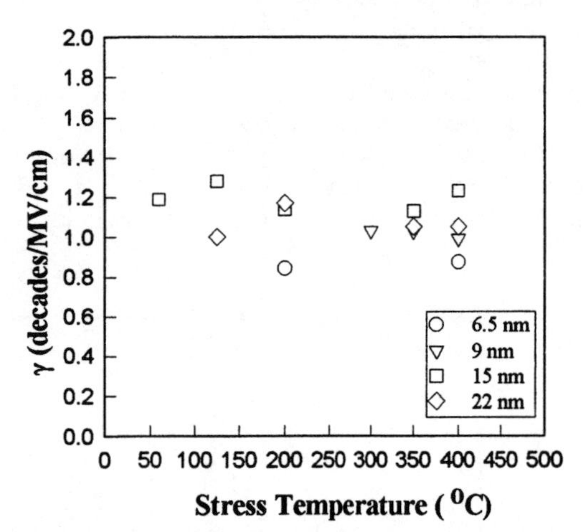

FIGURE 6. Electric field acceleration parameter (γ) plotted as a function of temperature for 6.5, 9, 15, and 22 nm thick oxides.

Figure 7 shows a plot of the thermal activation energy as a function of electric field. Note that the value remains constant and has a value between 0.7 and 1 eV for fields above 10 MV/cm.

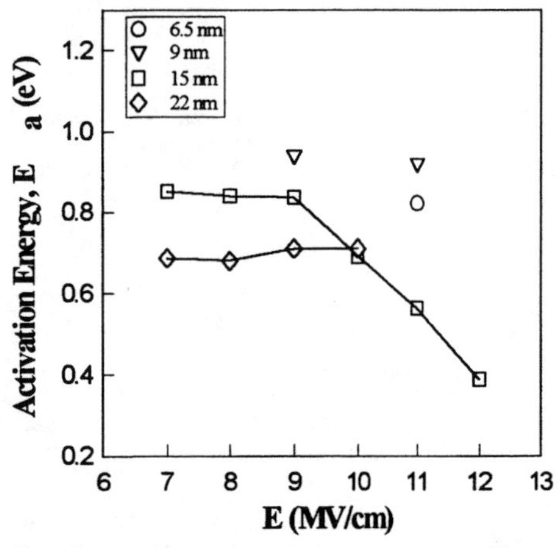

FIGURE 7. Plot of TDDB thermal activation as a function of stress electric field for the 6.5, 9, 15, and 22 nm thick films.

The results presented in Figs. 5 through 7 have recently been explained and predicted by a theoretical model [14]. The model is based on field-induced breakdown unlike carrier-induced breakdown. Breakdown is initiated by the thermal breakage of the relatively weak Si-Si bonds in pre-existing oxygen vacancy defects, also known as the E' centers [15].

A better understanding of oxide wear-out allows long-term reliability parameters to be extracted from highly accelerated oxide breakdown tests and how thickness scaling impacts dielectric reliability. Also, it is important to understand what effect the excessive leakage current observed in ultra-thin dielectrics due to direct tunneling current has on the long-term reliability of the film.

IV. CONCLUSIONS

Characterizing the reliability of ultra-thin gate oxides presents a new challenge because breakdown is difficult to detect due to soft or "quasi" breakdown. The breakdown criteria in standard gate oxide integrity tests must be modified to detect the breakdown event. These modifications should work for ultra-thin oxides as thin as 3 nm. Thinner oxides will require new characterization techniques such as monitoring current or voltage noise.

Recent studies [14] indicate that oxide wear-out and breakdown is induced by the electric field. Dipole interaction of oxygen vacancy defects with the applied field is the most likely mechanism of oxide wear-out at lower electric fields.

REFERENCES

[1] S-H Lee, B-J Cho, J-C Kim, and S-H Choi, "Quasi-Breakdown of Ultra-Thin Gate Oxide Under High Field Stress," IEEE IEDM Technical Digest, p. 605, 1994.

[2] M. Depas, T. Nigam, and M. H. Heyns, "Soft Breakdown of Ultra-Thin Gate Oxide Layers," *IEEE Trans. on Electron Devices* **43**. No. 9, p. 1499, (1996).

[3] K. R. Farmer, R. Saletti, and R. A. Burhman, "Current Fluctuations and Silicon Oxide Wear-Out in Metal-Oxide-Semiconductor Tunnel Diodes," *Appl. Phys. Lett.* **52** (20), p. 1749, (1988).

[4] B. E. Weir, P. J. Silverman, D. Monroe, K.S. Krisch, M. Alam, G.B. Alers, T. W. Sorsch, G. L. Timp, F. Baumann, C. T. Liu, Y. Ma, and D. Hwang, "Ultra-Thin Gate Dielectrics: They Break Down, But Do They Fail?," IEEE IEDM Technical Digest, p. 73, 1997.

[5] EIA/JEDEC Standard 35 "Procedure for the Wafer-Level Testing of Thin Dielectrics," JESD35,

Electronics Industries Association, Washington, D.C., July 1992.

[6] G. B. Alers, B. E. Weir, M. A. Alam, G. L. Timp, and T. Sorch, "Trap Assisted Tunneling as a Mechanism of Degradation and Noise in 2-5 nm Oxides," Proc of the IRPS, 36, p. 76, 1998.

[7] D. L. Crook, "Method of Determining Reliability Screens for Time Dependent Dielectric Breakdown," Proc. Rel. Phys. Symp., 17, p. 1 1979.

[8] A. Berman, "Time Zero Dielectric Reliability Test by a Ramp Method," Proc. Rel. Phys. Symp., 19, p. 204 1981.

[9] J. W. McPherson and D. A. Baglee, "Acceleration Factors for Thin Gate Oxide Stressing," Proc. Rel. Phys. Symp., 23, p. 1 1985.

[10] I. C. Chen, S. E. Holland, and C. Hu, "Electrical Breakdown in Thin Gate and Tunneling Oxides," *IEEE Trans. on Electron Devices* **32**, No. 2, p. 413 (1985).

[11] J. C. Lee, I. C. Chen, and C. Hu, "Statistical Modeling of Silicon Dioxide Reliability," Proc. Rel. Phys. Symp., 26, p. 131 1988.

[12] R. Moazzami, J. C. Lee, and C. Hu, "Temperature Acceleration of Time-Dependent Dielectric Breakdown," *IEEE Trans. on Electron Devices* **36**, No. 11, p. 2462 (1989).

[13] J. Suehle and P. Chaparala, "Low Electric Field Breakdown of Thin SiO_2 Films Under Static and Dynamic Stress", *IEEE Trans. on Elec. Dev.* **44**, (5), p. 801 (1997).

[14] J. W. McPherson, V. K. Reddy, and H. C. Mogul, "Field-Enhanced Si-Si Bond-Breakage Mechanism for Time-Dependent Dielectric Breakdown in Thin-Film SiO_2 Dielectrics," *Appl. Phys. Lett.* **71**, No. 8, p. 1101, (1997).

[15] T. R. Oldham and A. J. Lelis, "New Insights into Oxide Trapped Holes and Other Defects: Implications for Reliability Studies," Proc. of the Mat. Res. Soc., 428, p. 329, 1996.

Thin Film Ellipsometry Metrology

Prabha Durgapal

VLSI Standards, Inc.
San Jose, CA 95134-2006

James R. Ehrstein and Nhan V. Nguyen

National Institute of Standards and Technology
Gaithersburg, MD 20899-0001

A wide variety of commercial ellipsometers are available in the market today. They all measure the change in the state of polarization of light on reflection, but the measurement techniques adopted vary from instrument to instrument. Further, the models used to evaluate the thickness and refractive index of the oxide film during analysis of measurement data vary in complexity. The two main techniques of measurement are single wavelength ellipsometry and spectroscopic ellipsometry. The NIST Standard Reference Materials available today are based on conventional single wavelength ellipsometry. We discuss the challenges encountered in providing reference materials by using spectroscopic ellipsometry. First, the limits of conventional single wavelength ellipsometric determination of film thickness are investigated and then possible new technologies are explored. We present a discussion comparing the different types of instruments available and how their unique designs affect the accuracy of thickness determinations. Manufacturing, and accurate determination, of thickness of films this thin (<10 nm) is a challenging task. Results from independent ellipsometric measurements on two different types of instruments are compared for the case of ultra thin thermally grown silicon dioxide films on silicon crystal substrates. Stability curves for the thickness of thin dielectric films over a period of two years are also presented.

INTRODUCTION

The subject of measuring thin films dates back to the seventeenth century, but what is meant by a "thin film" has changed considerably over time and is still evolving. In the early days, thin films were used for optical filters (1), and film thickness was of the order of the wavelength of visible light (~600 nm). Thickness values of this order of magnitude could not be measured directly; therefore, various types of optical phenomena were used for the measurement of film thickness. The practice still continues today. Modern instruments use a spectrum ranging from ultraviolet to infrared wavelengths. Today, thickness determination is very important in the semiconductor industry. It is one of the first measurements that are made in the process leading to an integrated circuit.

With the advancement of integrated circuit technology, both lateral and vertical dimensions are decreasing (the gate oxide film thickness is becoming smaller and smaller). The gate dielectric is one of the most difficult challenges for future device scaling. The 1994 edition of the National Technology Roadmap for Semiconductors (NTRS) (2) had envisioned the need for 4 to 5 nm gate oxides with a process tolerance of 4% at the 0.18 µm technology node in 2001. In the new edition of NTRS (3) for 1997, these figures have been revised, reflecting the more rapid pace of technology that has actually been achieved. The above target was set for the year 1997 and the subsequent technology nodes of 0.18 µm and 0.15 µm are now spaced apart by only 2 years, predicting gate oxides of 3 to 4 nm in 1999 and 2 to 3 nm for the year 2001 with a process tolerance of 4% at the 150 nm technology node. Further, an uncertainty of 1% in the value is required at 95% or better confidence level! An oxide equivalent thickness of less than 1 nm is projected for the year 2012. Tunneling currents preclude the use of SiO_2 dielectric films below about 2 nm. These goals are putting pressing demands on thin film metrology. How ever it is done, thickness determination with subatomic (statistical) uncertainties and measurement with small spot sizes are a must.

In order to meet the challenge of the 1997 NTRS for 2 to 5 nm gates, not only the enabling process technologies must be developed, but, concurrently, improved experimental techniques for thin film thickness determination must also be developed. The results from a recent survey (4) of optical characterization methods for materials indicate that for thickness evaluation an ellipsometer is the instrument of choice. This survey represents a broad view of the semiconductor industry and includes material suppliers, device manufacturers, and makers of characterization equipment. Ellipsometers are very accurate instruments for measuring Δ and Ψ (see eq. 1), and can be very precise for determining thickness and index of a stable sample. However, even with these instruments, achieving accurate

CP449, *Characterization and Metrology for ULSI Technology: 1998 International Conference*
edited by D. G. Seiler, A. C. Diebold, W. M. Bullis, T. J. Shaffner, R. McDonald, and E. J. Walters

measurements of very thin dielectric films (oxides, oxynitrides, nitrides, or layer stacks) is becoming increasingly difficult, both in manufacturing and in the world of instrument calibration and standards. Part of this is due to limits on instrumentation and data analysis procedures, and part is due to the reality that for process control, measurements need to be made to the level of a few hundredths of a nanometer, i.e., below atomic dimensions. At the same time, there is significant evidence that interfaces from silicon to SiO_2 or from SiO_2 to polysilicon are nonuniform at a level of perhaps 0.1 to 1.0 nm. Thus, correctly modeling the structure and composition of the films being measured is also extremely important for achieving the required accuracy. Assuming that the modeling and data analysis challenges are met, it is likely that the best of the commercial ellipsometric-based tools may be capable of the precision needed to meet the NTRS Roadmap p/t (precision to tolerance) requirements for current and near-term technology nodes for gate dielectrics. However, precision alone is no longer sufficient for efficient process transfer, for support of predictive modeling, or even for telling whether the current process run will yield class A or class B devices. Although precision is enough for process control, the need for accuracy of measurements is now recognized, and it is not uncommon for an IC manufacturer to require that optical monitoring tools be matched at, or near, the 0.01 nm level. While a common expedient is to establish the "relative accuracy" of the optical tools through correlation with TEM or final capacitance measurements, the "accuracy" so obtained is generally specific to a given company and to a given process.

The preferred approach to establishing and maintaining accuracy of measurement tool performance is through the use of reference materials, but the requirements and impediments to establishing accurate reference materials suitable for current IC manufacturing needs are stringent. Among the requirements are (5):

A) Materials - fully understanding the structure of the films (oxide, or similar) being used for the reference materials, including interfaces and roughness; knowing, or being able to extract, accurate optical index values for the film, the interface, and the substrate; and having a method for stabilizing or cleaning the sample (the latter through a procedure that can be accomplished in both the lab and the fab);
B) Instrumentation - accurately establishing the ellipsometer angle of incidence on the sample and the polarization azimuth angle, selecting an angle of incidence (generally very close to the wavelength-dependent Brewster angle) with maximum sensitivity to small changes in the film being measured, use of a collimated beam with narrow wavelength bandwidth, and various other requirements on the electronics and optical components; and
C) Data analysis - using the correct model of the material structure, using the correct goodness-of-fit criteria, and examining the extent of correlation of the material parameters extracted from the fit to the data.

While the uncertainty of ellipsometric component alignment is often seen as setting the lower limit on the accuracy with which reference material films can be calibrated and transferred, several of the other less well appreciated considerations related to materials modeling and numerical analysis of the ellipsometry also result in potential, or real, limitations on achievable accuracy.

Separate from the issue of using the correct structural model of the film being measured, there are complexities in the numerical fitting of model to measurement data that can limit the accuracy of results. Jellison (6) points out that an unbiased estimator, or unweighted mean square error, is commonly used as a goodness-of-fit criterion for analysis of SE data, but that it is not as good, or as objective, as using a biased estimator, or reduced χ^2. The latter relies on a weighted uncertainty of the measured Δ and ψ values vs. wavelength, with the weighting determined by an estimate of experimental errors; this places increased emphasis on the more accurately determined measurement points. Jellison (7) also points out that when it is necessary to extract both thickness and index of the dielectric film, one is confronted with the fact that for very thin films, the thickness and index parameters are highly correlated in the numerical fits to the measurement data. The result is that the confidence limits on the values of the extracted thickness and index are dominated by errors due to correlation. These correlated errors may, in fact, be many times larger than the noncorrelated errors. This problem remains, but the description becomes more complex if additional parameters, such as those describing the interface layer, are simultaneously extracted from the data analysis.

As explained above, the demands on thin film metrology are numerous, but the payoffs from properly executed metrology are also large. Diebold and Monahan (8) contend that metrology reduces the cost of manufacturing by bringing in more robust processes, preventing scrap, and helping in ramping and maintaining yield. They list the need for reference materials and standard methods for gate dielectrics and thin films among the most difficult challenges facing metrology. These reference materials and calibration standards have been a key factor in assuring the measurement quality of the instruments. The thin film community desires that thin dielectric film (<10 nm) standards should be available and certified for thickness, refractive index, and the parameters Δ and Ψ. These standards should be stable and yield accurate and repeatable certified values. In addition, cleaning and handling techniques that will not significantly alter the certified values are needed. In summary, the standard cleaning procedure, instrumentation factors, and data analysis all need to be defined accurately and simultaneously. It is not possible to discuss all these points here for thin film ellipsometry metrology. The measurement and mathematical modeling techniques have been discussed in detail by a number of authors (7, 9-15). The focus of this paper is to concentrate on present measurement and modeling capabilities, currently available reference materials, and to discuss what needs to be done to keep up with the requirements of the NTRS 1997. We also limit our discussion to thermally grown thin silicon dioxide films on silicon substrates.

In the next section, a concise overview of ellipsometry is provided. In subsequent sections, the single wave and spectroscopic ellipsometers are summarized and discussed. Results are presented for thin oxide films measured at NIST and VLSI Standards Inc. The final section includes discussions of polarization modulation and other modern ellipsometers that are being used to achieve faster speed and higher accuracy of the measurements.

GENERAL ELLIPSOMETRY

Ellipsometry is based on the principle of polarization of light, and relates film thickness and other optical parameters to the change in the state of known polarization on reflection. In the general scheme of ellipsometry, a polarized light-wave is allowed to interact with the thin film sample by means of reflection, refraction, and transmission of light, thereby changing the state of polarization, along with some other optical properties of the incident light beam. Figure 1 illustrates the essential elements of a conventional null ellipsometer.

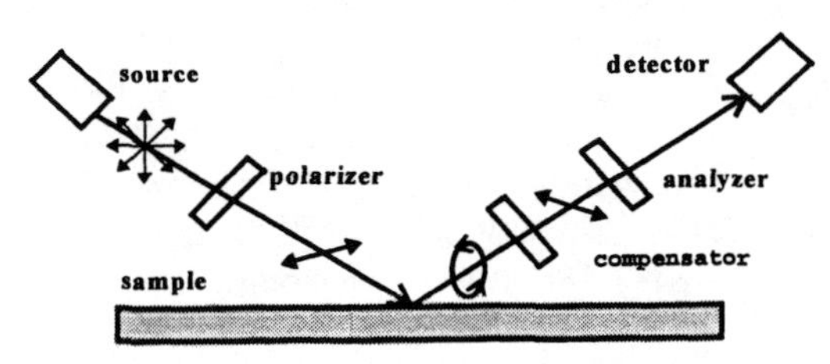

FIGURE 1. Ellipsometry Concept

An unpolarized light beam starts from a light source. On passing through the polarizer, the beam is linearly polarized. On reflection from the surface of the sample, the light becomes elliptically polarized. Another polarizer is used as an analyzer to determine the null and hence the angle of polarization. To convert the elliptically polarized light reflected from the surface to linearly polarized light, a compensator or quarter wave plate (QWP) is used. This is the basic concept. In reality, various alterations to this arrangement are used.

If one assumes that the interaction of light with the air/film/substrate system is linear, non-depolarizing, and frequency-conserving, then simple relationships between the film thickness and the change in the state of polarization can be derived. Polarized light consists of two components: one oscillating perpendicular (s component) and one oscillating parallel (p component) to the plane of incidence. The equation,

$$\rho = \frac{R_p}{R_s} = \tan \Psi e^{i\Delta}, \tag{1}$$

relates the measured ellipsometric parameters, Δ and Ψ, to the ratio of the amplitude reflectivities, R_p and R_s, for p and s polarized light. This in general is a complex number since there is a relative phase shift between the p and s reflected waves, and so ellipsometry at one angle of incidence yields two known quantities The beauty of the method is that by taking a ratio of reflectivities, the method does not need intensity normalization which direct measurements of the intensity reflectivities do. In general the ratio ρ is a function of several parameters,

$$\rho = \rho\ (d, N_0, N_1, N_2, \phi, \lambda), \tag{2}$$

where d is the thickness of the film, N_0, N_1, and N_2 are the refractive indices of the ambient, film, and substrate, respectively, ϕ is the angle of incidence, and λ is the wavelength of the source. In general, if the refractive indices of the ambient and the substrate, the angle of incidence, and the wavelength of the source are known, then upon measuring Δ and Ψ, the film thickness and film refractive index may be deduced. Multiple-Angle-of-Incidence can be used to deduce more parameters using basic ellipsometric measurement data, provided the associated set of equations passes the independence-of-correlation test.

In most IC fabrication facilities, single wavelength ellipsometry (SWE) (9) is the standard measurement technique employed for thickness and index determination. Generally, the measurements are conducted at a single wavelength (mostly 632.8 nm, sometimes at

546.1 nm). A single layer model is used for analysis, and the effects of the interfacial layer between the film and the substrate are neglected. This is a necessity in single wave ellipsometry as only two parameters, Δ and Ψ, are obtained from the measurement. The inclusion of the interlayer would require more than two measured parameters per measurement. Several studies have shown that many of these simplifying assumptions in the interpretation of single wavelength ellipsometry data are not valid. Taft and Cordes (11) have demonstrated the existence of an interfacial sublayer with considerable compressive strain between the SiO_2 film and the Si substrate. They have reported the thickness of this layer to be ~0.6 nm, with a refractive index of 2.8. In the SRM 2530 series (16) (nominal thickness values 50 nm, 100 nm, and 200 nm), NIST has determined a common thickness value of (1.0 ± 0.4) nm for the interlayer. In the recent additions, SRM 2535 and 2536, NIST has reported this value to be (1.1 ± 0.4) nm. For all SRMs, a refractive index of 2.8 is used for the interlayer. In his 1991 paper Jellison (14) has concluded that the interfacial layer thickness decreases with decreasing film thickness, and the nature of the interfacial layer is consistent with a microroughness layer. In a more recent study, Nguyen et al. (17) have reported that for a 10 nm oxide sample, the thickness of the interface region is 2.2 nm (0.7 nm roughness plus 1.5 nm strained silicon). It should be mentioned that the interlayer thickness may depend on the oxide growth process explaining the differences in the interlayer thickness reported by different authors. But clearly these results do indicate that a lot more work is required to accurately characterize the interlayer, and for very thin films, it must be included in the model. However, it is not possible to determine interface layer values from single wavelength ellipsometry on individual samples.

In attempts to improve the conventional null ellipsometry (CNE) technique and provide multiple measurements, photometric ellipsometers were developed (9,18). In the rotating analyzer ellipsometer (RAE), photometric measurements at various angular positions of the analyzer are made, in contrast to the CNE technique where only one reading at the null point is taken. The detector signal is then Fourier-analyzed.

For the earlier generations of devices where the gate oxide was fairly thick (>10 nm), single wave ellipsometry yielded satisfactory results. Over the last decade spectroscopic ellipsometry (SE) has become an important tool for the study (7, 10, 17-20) of very thin films. The advent of computer control and multi-channel detectors has made it possible to develop spectroscopic ellipsometers that can scan an entire range of wavelengths using an entirely automatic data acquisition and processing system. Most commercial spectroscopic ellipsometers measure values of $\tan\Psi$ and $\cos\Delta$ at about a hundred wavelengths which can then be used to determine multiple useful parameters through data regression. This includes thickness and optical properties of multilayer thin film stacks. Coupled with effective medium theories, spectroscopic ellipsometry enables the study of microstructures of deposited thin films. This is of importance for both semiconductor and optical coating technologies.

SINGLE WAVE ELLIPSOMETRY

Between the years 1988 and 1994, NIST provided Standard Reference Materials for thermally grown silicon dioxide films on silicon with nominal thickness values of 10 nm, 14 nm, 25 nm, 50 nm, 100 nm, and 200 nm. The instrument used for certifying these SRMs is a custom-designed, high accuracy rotating analyzer ellipsometer. All measurements are done at the single wavelength of 632.8 nm. Aside from the ellipsometric parameters Δ and Ψ, oxide thickness based on a two layer model and batch analysis of multiple oxide thickness values at the principal angle of incidence are certified. In the semiconductor industry, single wavelength ellipsometry has been the most widely used technique, and a single layer model is used for evaluating the thickness of dielectric films on a silicon substrate. The inclusion of an interlayer in the modeling process will require more known parameters in addition to the Δ and Ψ currently measured. Consequently, single layer analysis is a more practical mode from the industry's point of view.

In 1996-97, a Cooperative Research and Development Agreement (CRADA), CN-1364 (21) between the Semiconductor Electronics Division at NIST and VLSI Standards, Inc. was completed. One of the objectives of this collaboration was to develop and test artifacts with SiO_2 film thickness values less than 10 nm and devise a method by which transferability and traceability to NIST could be established in a straightforward way.

Spectroscopic ellipsometry and CNE were used by VLSI Standards, Inc. to qualify wafer uniformity. The formal comparison was based on single wavelength data only: 70 ° CNE at VLSI Standards and 70° RAE and principal angle RAE at NIST. For conventional null ellipsometry ,

$$I_{CNE} = \frac{I_o\left(|R_p|^2+|R_s|^2\right)\left[1-\cos2\Psi\cos2A+\sin2\Psi\sin(\Delta-2P)\sin2A\right]}{4} \qquad (3)$$

TABLE I: Comparison of the ellipsometric angles Δ and Ψ measured at VLSI Standards, Inc. and NIST.

Sample	Nominal	Δ in degrees		Ψ in degrees		Diff. VLSI - NIST	
I.D.	thickness, nm	VLSI	NIST	VLSI	NIST	Δ°	Ψ°
3723-001	4.5	164.578	164.856	10.713	10.792	-0.2781	-0.079
3723-002	4.5	164.465	164.699	10.720	10.801	-0.2337	-0.081
3722-001	7.5	157.745	157.883	11.013	11.089	-0.1377	-0.075
3722-002	7.5	157.706	157.857	11.017	11.093	-0.1514	-0.076

where I_0 is the intensity of light incident on the sample surface, and P and A refer to the polarizer and analyzer azimuth angles. When the quarter wave plate is set at a fixed angle of ±45°, the intensity near the null point is expressed as:

$$I_{null} = \frac{I_0(|R_p|^2 + |R_s|^2)[(\delta A)^2 + (\delta P)^2 \sin^2 2\psi]}{2}. \quad (4)$$

At the null point:

$$P_{null} = \frac{\Delta}{2} \pm \frac{\pi}{4}, \qquad A_{null} = \Psi. \quad (5)$$

Thus in null ellipsometry, the angles Δ and Ψ are measured directly from the orientation of the optical elements.

The rotating analyzer ellipsometer at NIST is based on a photometric method and does not require a null measurement. The intensity varies with the analyzer azimuth angle A,

$$I_{RAE} = \frac{I_0(|R_p|^2 \cos^2 P + |R_s|^2 \sin^2 P)[1 + \alpha \cos 2A + \beta \sin 2A]}{2} \quad (6)$$

where

$$\alpha = \frac{\tan^2 \psi - \tan^2 P}{\tan^2 \psi + \tan^2 P} \quad (7)$$

and

$$\beta = \frac{2 \tan \psi \tan P \cos \Delta}{\tan^2 \psi + \tan^2 P}. \quad (8)$$

The parameters α and β are obtained by Fourier analysis of the detector signal.

The samples used for this study were manufactured at VLSI Standards, Inc.. During a 6 month period, the artifacts were exchanged four times between the two labs. Table I compares the ellipsometric angles Δ and Ψ measured at the two facilities. The data demonstrate that the measured parameters show excellent agreement between the two laboratories: the Δ values differ by less than 0.2% and Ψ by less than 0.8%. As pointed out earlier, the measured parameters do not convey any information regarding the film properties. In order to arrive at the desired parameter viz. thickness of the film, both laboratories used their own algorithms initially. The program used at NIST was MAIN1 (22). In both cases the silicon dioxide film is modeled as a single, transparent

TABLE II: Comparison of thickness values obtained by using different models.

		Single layer model					Two layer model		
Sample	Nominal	VLSI thickness in nm		NIST thickness	Difference (VLSI-NIST)		Thickness in nm using MAIN1		Difference
I.D.	thickness,	using own	using NIST	in nm using	in nm using				(VLSI-NIST)
	nm	algorithm	MAIN1	MAIN1	own algorithm	MAIN1	VLSI	NIST	in nm
3723-001	4.5	5.184	5.200	5.103	0.080	0.097	5.28	5.15	0.13
3723-002	4.5	5.226	5.240	5.159	0.067	0.081	5.32	5.21	0.11
3722-001	7.5	7.711	7.740	7.691	0.020	0.049	7.8	7.72	0.08
3722-002	7.5	7.726	7.760	7.701	0.026	0.059	7.82	7.73	0.09

homogeneous material layer. The refractive index of the oxide film is taken to be 1.46. The thickness evaluated is given in columns three and five of Table II. In all cases the thickness values evaluated by the two algorithms differ by less than 0.1 nm. In order to determine how much difference is caused in thickness values by use of the different algorithms, the thickness values for the raw Δ and Ψ data from VLSI Standards were evaluated using the NIST algorithm MAIN1 for the single layer model. These values are listed in column four of Table II. Columns three and four of Table II are thickness values calculated using the same raw data but different algorithms. A comparison of these two columns indicates that algorithm differences do not contribute substantially to thickness differences. However, in this case the differences in thickness values between VLSI Standards and NIST are slightly higher but still less than 0.1 nm. The NIST program MAIN1 was used to make another set of calculations assuming a two layer model. In the two layer model, aside from the transparent homogeneous oxide layer, a transparent isotropic interlayer between the oxide layer and silicon substrate is assumed. The two layer thickness is taken to be the sum of the interlayer and the oxide layer. The details of this calculation (and higher thickness values) are given in Ref. (21). Columns 8 through 10 of Table II illustrate the results of two layer modeling. The refractive indices of the oxide film and the interlayer are found to be 1.464 and 2.8 from the data analyses of each lab. The interlayer thickness was determined to be 0.91 nm for NIST data and 0.95 nm for VLSI Standards' data. Although physically more accurate, the two layer model did not produce results very different from the single layer model. However, the thinner films undergo a much larger change in the thickness values when switching from the single layer model to the two layer model. This is so because in reality the interlayer and strain effects for a very thin film do contribute more to the measurement parameters Δ and Ψ then they do for thicker films.

The difference between thickness values calculated from VLSI Standards' and NIST data is less than the expanded uncertainty of difference calculated at the 95% confidence level (21). This is true for all models and algorithims used in Table II. In both the models used here (single layer and two layer), there are inherent model errors. These arise because we have assigned ideal properties to the surface which it does not possess. There is surface roughness and the layers are not homogeneous.

For very thin films, the ability to determine both thickness and index is affected by the wavelength used. Figure 2 shows the Δ-Ψ trajectories for different materials characterized by their refractive index, n, at a wavelength of 632.8 nm. The angle of incidence is 70°. The point marked "no film" corresponds to the bare substrate. It is seen that for very thin films ($t \leq 5$ nm) the Δ - Ψ trajectories belonging to different materials are clustered together. If we plot the Δ-Ψ trajectories for the same materials as shown in Fig. 2 but at a lower

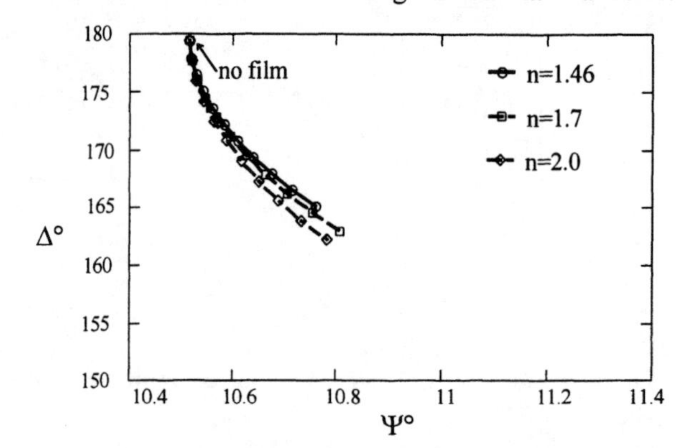

FIGURE 2. Δ–Ψ Trajectory for different refractive indices. Angle of incidence $\phi = 70°$. Wavelength λ=632.8 nm.

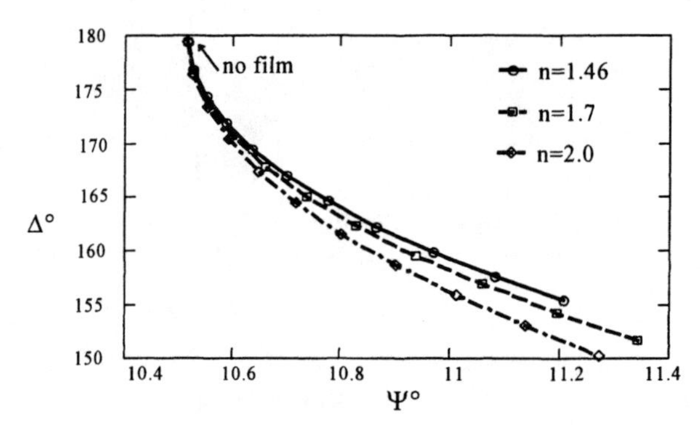

FIGURE 3. Δ–Ψ Trajectory for different refractive indices. Angle of incidence $\phi = 70°$. Wavelength λ=365.0 nm.

wavelength, 365 nm (mercury vapor i-line), then the resolution is better. This is illustrated in Fig. 3. Both these figures are plotted for thickness values ranging from 0.0 to 5.0 nm, with a thickness step of 0.5 nm. At the wavelength 632.8 nm, the resolution in Δ is from 162° to 180°, while the Ψ values range from 10.5° to 10.8°. Under the same conditions at the source wavelength of 365 nm, the resolution in Δ is from 150° to 180°, while the Ψ values range from 10.5° to 11.4°. This shows that for very thin films, shorter source wavelengths will yield better measurement results.

Thus the metrology needs for advanced gate dielectric films are confronted by the limitations of what can be resolved with single wavelength ellipsometry at the commonly used wavelength of 632.8 nm.

SPECTROSCOPIC ELLIPSOMETRY

The optical constants for silicon and silicon dioxide have been quite well established for SWE ellipsometry at 632.8 nm, but such measurements can only extract two parameters for a single specimen, and are thus limited to sharp-interface slab models of the material structure; hence they cannot readily deal with interface roughness or strain effects. SE measurements, on the other hand, provide more flexibility, particularly for modeling the interface using an effective medium approximation. The need to account for a transition, or interface, layer between a silicon substrate and an SiO_2 film has been demonstrated by numerous authors (12, 17, 23), although the exact structure of this interface remains a subject of active ongoing research. Failure to account for this interface layer results in a thickness or an index of the bulk oxide that is erroneously high for thin oxides. Thus the use of SE for the measurements of very thin films becomes rather a necessity. VLSI Standards has provided reference materials for SE technology for the past four years, but because of traceability issues and the undetermined nature of the interlayer, the data analysis is based on a single layer model. This is a significant limitation on the ability to respond to the semiconductor industry's need for advanced reference materials to support atomic layer metrology with uncertainties < 1/4 of process tolerance, or to provide reference materials for 2 or 3 film stacks with the thickness determination of the buried layers being as accurate as that for a single layer.

As mentioned earlier, spectroscopic ellipsometry is a monolayer sensitive optical characterization technique used widely to study thin films and properties of semiconductor materials. We have used this technique to further study the two samples with nominal thickness of ~4.5 nm. The SE employed for measurements at VLSI Standards is a rotating polarizer type and is equipped with a photodetector. The intensity varies as a function of the polarizer angle and is given by

$$I = I_0 \frac{\cos^2 A}{\tan^2 \psi + \tan^2 A}[1 + \alpha \cos 2P + \beta \sin 2P], \tag{9}$$

where the parameters α and β are obtained from the detector signal by a Hadamart transform and are given by

$$\alpha = \frac{\tan^2 \psi - \tan^2 A}{\tan^2 \psi + \tan^2 A} \tag{10}$$

and

$$\beta = \frac{2 \tan \psi \tan P \cos \Delta}{\tan^2 \psi + \tan^2 P}. \tag{11}$$

The ellipsometric parameters are obtained directly from the detector signal as $\tan\psi$ and $\cos\Delta$,

$$\tan \psi = \tan A \sqrt{\frac{1+\alpha}{1-\alpha}}, \tag{12}$$

$$\cos \Delta = \frac{\beta}{\sqrt{1-\alpha^2}}. \tag{13}$$

Figures 4 and 5 illustrate the precision and repeatability curves for the two thin samples with nominal thickness value of 4.5 nm. These data were taken over a period of nearly two years. GA refers to grand average of thickness for all the 11 runs, and σ is the standard deviation for the same. In order to achieve high sensitivity, all spectra were recorded at an incident angle in the vicinity of 75°, which is close to both the Brewster angle and the principal angle for the silicon substrate. At the principal angle, the ellipsometric parameter Δ equals 90° and Ψ equals the polarizer azimuth angle. Under these conditions, the reflected light is circularly polarized, there is greater sensitivity to small changes in the sample, and highly accurate data are obtained. A wavelength window ranging from 250 to 650 nm was chosen. A regression technique based on the Levenberg-Marquardt-Fletcher algorithm is used for the calculation of thickness, and the spectrum of refractive indices for silicon is taken from Ref. (24) and that for SiO_2 from Ref. (25, 32). The charts exhibit no apparent film growth, and the standard deviations produced support the contention that we can have uncertainties in the 0.05 nm range for 4 to 5 nm silicon dioxide films.

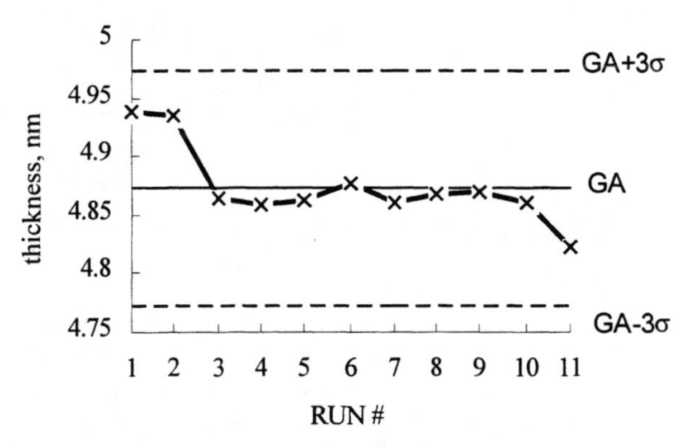

FIGURE 4: Precision and repeatability data for sample SN 3723-001.

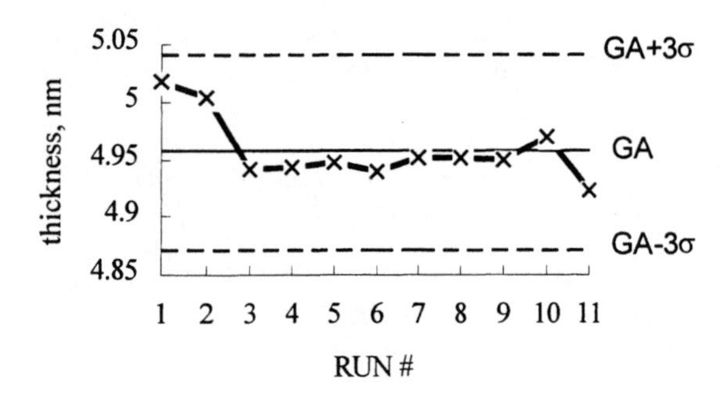

FIGURE 5: Precision and repeatability data for sample SN 3723-002.

Offsets between TEM thicknesses and those from one-layer analyses of SE data typically show ellipsometric derived thicknesses to be 0.4 to 0.5 nm larger, but this has been shown to be reducible to the order of 0.1 nm when the interface layer is accounted for in a two layer model of the film (26). The two samples, above, of nominal thickness value of 4.5 nm were used for this study, and they were measured at NIST and VLSI Standards. They were measured on a custom-built high-accuracy rotating-analyzer spectroscopic ellipsometer at NIST (27). The spectral range covers energies from (1.5 to 6.0) eV (equivalent wavelength range 206.6 to 826.6 nm) in steps of 0.05 eV. The angle of incidence was set at 75.0°. To determine the SiO_2 film thickness, the single layer model consisting of silicon substrate, SiO_2 film, and ambient was used in fitting the experimental data. For each wafer, the SiO_2 film was modeled as bulk fused silica and also as effective medium approximation of bulk fused silica and voids. The latter was employed in the modeling because the thin SiO_2 film is believed to be more dense than bulk oxide. For these two wafers, the introduction of voids to the film appears to only slightly improve the goodness-of-fits. The results are summarized in Tables III and IV.

TABLE III: NIST Results, film thickness, nm.
Angle of incidence: 75°, Measurement range:1.5-6.0 eV

	as Bulk	as EMA
3723-001	5.39 ± 0.05	5.19 ± 0.3
3723-002	5.48 ± 0.05	5.26 ± 0.3

In the samples, the void fraction resulting from the EMA fit is -7.1% (3723-001) and -7.5% (3723-002); similar values were obtained for the VLSI Standards data. The negative void fraction in effect indicates that the films are more dense than bulk oxide. In other words, they have a higher index of refraction.

TABLE IV: VLSI Standards Results, film thickness, nm.
Angle of incidence: 75.03°, Spectral range:250-650 nm

	as Bulk	as EMA
3723-001	5.12 ± 0.01	5.06 ± 0.2
3723-002	5.18 ± 0.01	5.09 ± 0.2

Comparing results presented in Figs. 4 and 5 with Tables III and IV, one might wonder why the thickness values for the two samples differ in the two cases. The silicon material used for the substrate is boron doped, p-type (100) surface prime silicon. In the analysis done for Tables III and IV, refractive indices for silicon (111) from Ref. (28) are used, whereas values for (100) silicon, (24) were used for Figs. 4 and 5.

In a recent study (15), it has been shown that for thin SiO_2 films in the regime of less than 10 nm, different models such as Cauchy or Sellmeier dispersion can be used to fit the experimental data equally well. However, these different models combined with the simple models used in this study can result in a large discrepancy in the determined thicknesses which can be as much as 13%. These models are not detailed here since in this paper, the focus is to show the variation of thickness, for the same set of wafers, measured and determined by different measurement instruments and modeling methods. Also, unlike the situation for SWE measurements at 632.8 nm, there is some disagreement in the literature regarding the dielectric function of silicon over the visible spectral range used for SE measurements. Nguyen (15) has shown that various choices for the silicon dielectric function and for parametrically modeling the optical index of the oxide film can lead to an unacceptably large variation in the derived thickness and index of the oxide, with all modeling choices having comparable goodness-of-fit to the measured Δ and ψ values. Improvements in the values for silicon substrate dielectric function are possible through measurements on specially prepared hydrogen-terminated samples. Unfortunately, while there are established chemical treatments for achieving this for (111) silicon, treatments for (100) silicon have been less successful.

Spectroscopic ellipsometers provide more flexibility than the single wavelength ellipsometer and allow for simultaneous determination of multiple parameters: for example, thickness, refractive index and composition of multilayer thin film stacks. However, quantitative results will be affected by the choice of structure models used and by small residual uncertainties about available optical constant values over the wavelength band being used.

MODERN ELLIPSOMETERS

Many advances have been made in improving ellipsometric techniques. One of the disadvantages of the single wave and spectroscopic ellipsometers described above is that they are both slow. This could lead to surface contamination even in a good vacuum. Several modern techniques have been developed to counter this difficulty.

One of the techniques (29) consists of performing spectroscopic measurements on evolving surfaces using rapid photon energy scanning such that all energies are collected on a time scale faster than that of the surface evolution.

Another ellipsometry technique that is currently in use for characterizing very thin films is based on polarization modulation. The polarization modulated ellipsometer has the same series of elements as the manual null system, but with the compensator replaced by a birefringent phase modulator. Jasperson and Schnatterly (30) were one of the first to construct a polarization modulation ellipsometer using the photoelastic modulator. In their device, a time dependent phase retardation is imparted to the light beam using the photoelastic effect. The polarization of the light beam is continuously and periodically varied in such a way that information regarding the reflecting surface is obtained synchronously, detecting and recording the magnitudes of the subsequent intensity modulations of the reflected and suitably analyzed beam. Their system was fast and obtained data continuous in wavelength. In addition to the ellipsometric angle Δ, the quantity $(1-\rho^2)/(1+\rho^2)$ is directly measured instead of ρ or ρ^2. The differential quantity leads to highly sensitive measurements even in the high reflectivity limit. More recently a two-channel polarization modulation ellipsometer has been developed by Jellison and Modine (31). They show that the instrument is very sensitive to Δ when Δ is close to 0° or 180°. Jellison (14) has used this technique for the determination of thin SiO_2 films on a silicon substrate. The range of thickness values used for the oxide were from 3 to 325 nm. Since $\cos(\Delta)$ is an even function of Δ, the RA (RP) type spectroscopic ellipsometers cannot measure the sign of Δ which can sometimes lead to an ambiguity of data interpretation. On the other hand, PM ellipsometers measure quantities $\sin(2\Psi)\sin(\Delta)$ and $\sin(2\Psi)\cos(\Delta)$, or $\sin(2\Psi)\sin(\Delta)$ and $\cos(2\Psi)$, depending on the two different orientations of the modulator. When equipped with a Wollaston prism instead of an analyzer, a PME can simultaneously measure three quantities, i.e., $\sin(2\Psi)\sin(\Delta)$, $\sin(2\Psi)\cos(\Delta)$, and $\cos(2\Psi)$. This results in an accurate measurement of angle Δ in all regions. It is claimed that thickness determination is accurate to 0.05 nm.

We have briefly discussed the instrumentation, precision, and accuracy of ellipsometers. An extensive review would be necessary in order to have a detailed evaluation of the performance of different ellipsometer designs with respect to thin film measurement and in-situ applications. A phase modulation ellipsometer does not suffer degraded precision when Δ is near 0° or 180° as occurs in the case of the more popular rotating-polarizer or analyzer designs, and hence is advantageous for analysis of thin films. However, the phase modulation frequency is incompatible with wavelength fast scanning using a multichannel analyzer with a linear photo diode array. Therefore, its extension to spectroscopic fast-scanning capability might be limited due to the slow wavelength scanning of a monochromator. Another solution in this case is to improve on the popular and simple system of the rotating-polarizer or rotating-analyzer. This approach has been taken by various research groups by adding a compensator as the rotating element. (33-39)

DISCUSSIONS AND CONCLUSIONS

Characterization of very thin films is critically important but extremely difficult to do with assured accuracy. With the decrease in film thickness, the value of the film thickness is approaching that of the interlayer itself. The NTRS (1997) projects a 2 to 3 nm gate oxide for the year 2001, with a process tolerance of 4% at the 150 nm technology node. Furthermore, precision is no longer sufficient for efficient process transfer; the need for accuracy is equally important. The preferred approach for establishing and maintaining accuracy of measurement tool performance is through use of reference materials. However, the requirements and impediments to establishing accurate reference materials are stringent. The requirements are 1) better understanding of material structure and improved material maintenance, 2) superior instrumentation, and 3) realistic data analysis. An improved knowledge of the interface layer and its inclusion in the data analysis will greatly improve the accuracy of present ellipsometry techniques.

Single wavelength ellipsometers (SWE), particularly those operating at 632.8 nm, continue to be used with acceptable measurement precision to monitor gate oxide film thickness. Analysis of SWE measurements at 632.8 nm benefit from extensive research and well defined optical index values for the silicon substrate, which are needed for data reduction. However, it cannot extract enough material parameters to deal with the silicon-to-SiO_2 interface layer (with its higher than oxide index of refraction). The result is that either an effective layer thickness with normal oxide index or an effective index with approximately correct total film thickness must be calculated from the measurements. But one cannot expect correct thickness and a normal oxide index simultaneously to result from SWE measurements of individual wafers.

Spectroscopic ellipsometry (SE) offers sufficient increased measurement information for extraction of additional film parameters, e.g., the interface layer, from the analysis. While advances in instrumentation may reduce the uncertainty of measured values, limitations on accuracy and uncertainty of the calculated film parameters still come from 1) choice of starting assumptions during data analysis, e.g., there are different available dielectric functions in the literature for the silicon substrate or dispersion relations for the dielectric film, none of which is yet taken to be the standard; and 2) the inevitable correlation of variables when extracting multiple parameters in the thin film regime increases the uncertainty values for the parameters that are evaluated. These latter issues are of lesser concern in the fab environment where a fixed choice for the silicon dielectric function can be locally standardized on, and the main interest is detecting small changes from baseline values. However, these issues do impact the certification of reference materials. It is not yet possible with SE measurements alone to extract values of all the parameters for a thin SiO_2 film with its interface layer and to provide certified values with 1% uncertainty at a 95% confidence level.

A centralized and consistent database needs to be established for universal use in thin film metrology. There should be universally agreed-upon spectra of refractive indices of silicon and silicon dioxide. In order to reduce final uncertainty, it becomes necessary for a provider of reference materials to stay with simple models and to state clearly what choices were made in the calculation of thickness and index of the oxide film. The NIST multiple-thickness batch-analysis of SRM wafers, when using a common interlayer model, results in a derived value of 1.461 for the refractive index of SiO_2 at 632.8 nm. However, if the same wafers are measured on a spectroscopic ellipsometer, and the refractive indices from Ref. (32) are used for SiO_2, the value at 632.8 nm is then 1.457. At first, these demands on thin film metrology may seem enormous and insurmountable; however, as Diebold and Monahan point out, the payoffs from implementing these improvements can also be large. This is what drives the need for better reference materials.

ACKNOWLEDGMENTS

The first author would like to thank Kenneth Nguyen for assisting with ellipsometric data acquisition, and Jim Greed for many helpful discussions. The NIST authors would like to acknowledge the support of the NIST Office of Microelectronics Programs.

REFERENCES

1. K. L. Chopra, *Thin Film Phenomena* (McGraw-Hill Book Company, New York, 1969).
2. The National Technology Roadmap for Semiconductors, Semiconductor Industry Association, 1994, San Jose, California.
3. The National Technology Roadmap for Semiconductors, Semiconductor Industry Association, 1997, San Jose, California.
4. W. M. Bullis, S. Perkowitz and D. G. Seiler, *Semiconductor Measurement Technology:* Survey of Optical Characterization Methods for Materials, Processing, and Manufacturing in the Semiconductor Industry, NIST Special Publication 400-98 (1995).
5. J. A. Woollam, at NIST Workshop on Thin Film Metrology, Gaithersburg, Md., Oct. 30-31, 1997 (unpublished).
6. G. E. Jellison, Jr., *Thin Solid Films*, **234**, 416 (1993).
7. G. E. Jellison, Jr., *Thin Solid Films* **290-291**, 40-45 (1996).
8. A. C. Diebold and K. Monahan, Next-Generation Metrology Must Meet Challenges, *Solid State Technology* **41**, 50 (Feb. 1998).
9. R. M. A. Azzam, and N. M. Bashara, *Ellipsometry and Polarized Light*, third printing (Elsevier Science B.V., Amsterdam, 1992).
10. R. M. A. Azzam, Editor, Selected Papers on Ellipsometry, *SPIE Milestone Series*, Vol. **MS27** (The International Society for Optical Engineering, Bellingham, Washington, 1991).
11. E. Taft and L. Cordes, *J. Electrochem. Soc.* **126**, 131 (1979).
12. D. E Aspnes and J. B. Theeten, *J. Electrochem. Soc.* **127**, 1359 (1980).
13. F. L. McCrackin, E. Passaglia, R. R. Stromberg, and H. L. Steinberg, *J. Res. Natl. Bur. Standards-A. Physics and Chemistry* **67A** (4), 363 (1963).
14. G. E. Jellison, Jr., *J. Appl. Phys.* **69** (11), 7627 (1991).
15. N. V. Nguyen and C. A. Richter, in *Silicon Nitride and Silicon Dioxide Thin Insulating Films*, Proceedings of the Conference, Vol. 97-10, M. J. Deen, W. D. Brown, K. B. Sundaram, and S. I. Raider, Editors, (The Electrochemical Society Inc., New Jersey, 1997), pp. 183-193.
16. G. A. Candela et al., *Standard Reference Materials:* Preparation and Certification of SRM-2530, Ellipsometric

Parameters Δ and Ψ and Derived Thickness and Refractive Index of a Silicon Dioxide Layer on Silicon, NIST Special Publication 260-109 (1988).
17. N. V. Nguyen, D. Chandler-Horowitz, P. M. Amirtharaj, and J. G. Pellegrino, *Appl. Phys. Lett.* **64**, 2688 (1994).
18. D. E. Aspnes, *J. Opt. Soc. Am.* **64**, 812 (1974).
19. D. E. Aspnes and A. A. Studna, Appl. Opt. **14**, 221 (1975).
20. J. L. Stehle, J. P. Piel, J. H. Lecat, C. Pickering, and L. C. Hammond, *Mat. Res. Soc. Symp. Proc.* **159**, 459 (1990).
21. B. J. Belzer et al., *Semiconductor Measurement Technology:* Thin Film Reference Materials Development: Final Report for CRADA CN-1364, NIST Special Publication 400-100 (1998).
22. J. F. Marchiando, *Semiconductor Measurement Technology:* A Software Program for Aiding the Analysis of Ellipsometric Measurements, Simple Models, NIST Special Publication 400-83 (1989).
23. V. Nayar, C. Pickering, and A.M. Hodge, *Thin Solid Films* **195**, 185 (1991).
24. G. E. Jellison, Jr., *Opt. Mater.* **1**, 41-47 (1992).
25. E. D. Palik, Editor, *Handbook of Optical Constants of Solids* (Academic Press, Inc., Orlando, Florida, 1985).
26. S. J. Fang, W. Chen, T. Yamanaka, and C. R. Helms, *J. Electrochem. Soc.* **144**, L231 (1997).
27. N. V. Nguyen, D. Chandler-Horowitz, J. G. Pellegrino, and P. M. Amirtharaj, in *Semiconductor Characterization: Present Status and Future Needs*, W.M. Bullis, D. G. Seiler, and A. C. Diebold, Editors (AIP Press, Woodbury, New York, 1996), pp. 438-442.
28. D. E Aspnes, J. B. Theeten, and F. Hottier, *Phys. Rev. B.* **710**, 24 (1986).
29. R. W. Collins, *Rev. Sci. Instrum.* **61** (8), 2029-2062 (1990).
30. S. N. Jasperson and S. E. Schnatterly, *Rev. Sci. Instrum.* **40** (6), 761 (1969).
31. G. E. Jellison, Jr., and F. A. Modine, *SPIE Proceedings* **1166**, 231 (The International Society for Optical Engineering, Bellingham, Washington, 1989).
32. I. H. Malitson, *J. Opt. Soc. Am.*, **55**, 1205 (1965).
33. P.S. Hauge, *Surf. Sci.* **56**, 148 (1976).
34. D.E. Aspnes, *Surf. Sci.* **56**, 161 (1976).
35. S.A. Henck *J. Vac. Sci. Technol. A* **10**, 934 (1992).
36. R. Kleim, L. Kuntzler, and A. Elghemmaz, *J. Opt. Soc. Am. A*, **11**, 2550, (1994).
37. D.A. Ramsey and K.C. Ludema, *Rev. Sci. Instrum.* **65**, 2874 (1994).
38. J. Lee, P.I. Rovira, I. An, and R.W. Collins *Rev. Sci. Instrum.* **69**, 1800, (1998).
39. J. Opsal, J. Fanton, J. Chen, J. Leng, L. Wei, C. Urich, M. Senko, C. Zaiser, and D.E. Aspens, *Thin Solid Films,* **313**, 58 (1998).

Analysis of Organic Contamination In Semiconductor Processing

Patrick J. Smith[1] and Patricia M. Lindley[2]

(1) IBM, 1580 Route 52, Hopewell Jct., NY 12533
(2) Charles Evans and Associates, 301 Chesapeake Dr., Redwood City, CA 94063

The cleanroom ambient contains numerous sources of volatile organic contaminants which can deposit on wafers. The effects of these contaminants on various aspects of device processing are reviewed. Possible sources of the contaminants, along with analytical techniques that can be used to identify and measure organic compounds are described.

INTRODUCTION

"Water insoluble organic compounds tend to make semiconductor and oxide surfaces hydrophobic, thus preventing the effective removal of adsorbed ionic or metallic species... Typical molecular contaminants... include... greasy films that are deposited when surfaces are exposed to room air or stored in plastic containers" (1). In their 1970 paper introducing the RCA clean, Kern and Puotinen identified two of the main features of organic contamination: airborne contamination can be deposited from processing ambients, and the effects of organic contamination can be magnified by their influence on the effectiveness of inorganic cleans. In this paper, we will review recent studies of the effects of organic contamination on device processing and briefly discuss some sources of the contaminants. We will also give overviews of cleaning methods being used to remove organics and of analytical techniques used to characterize molecular contaminants.

EFFECTS OF ORGANIC CONTAMINATION ON DEVICES

Oxide Quality

Kasi, *et al.*, showed, by exposing cleaned wafer surfaces to controlled amounts of known organic molecules, that significant amounts (10E15 atoms/cm^2) of hydrocarbon can cause serious degradation of metal-oxide-semiconductor (MOS) devices grown on hydrogen passivated, HF cleaned, (100) Si (2). Measurements of the amount of bulk charge in the oxide and of the number of defect sites at the Si-SiO_2 interface indicated that carbon was distributed throughout the oxide. Organic contaminants that had high sticking probabilities, desorption temperatures higher than the temperature where hydrogen passivation is lost, and that fragment easily during thermal processing had stronger effects on performance. The amount of organic material on the wafer surface was determined from the area under the carbon 1s peak measured by X-ray Photoelectron Spectroscopy (XPS). The C, Si, and O XPS peaks gave information on the bonding between carbon and Si.

Differences in the behavior of gate oxides were observed by Iwamoto and co-workers to depend on whether a wafer was processed with or without exposure to the clean room ambient (3). Attenuated Total Reflection Fourier Transform Infrared Spectroscopy (FTIR-ATR) measurements showed that the amount of contamination on the oxide increased with exposure time in the clean room. After deposition of a phosphorus-doped polysilicon, carbon was found by SIMS to be present at the polysilicon/ SiO_2 interface.

Saga and Hattori eliminated the deleterious effects of contamination on oxide quality by a suitable choice of ambients as the wafers were being loaded into the oxidation furnace (4). Electrical measurements detected degradation of gate oxide integrity if wafers were intentionally contaminated with the anti-oxidant butylated hydroxytoluene (BHT) and loaded into the gate oxidation furnace in the presence of a nitrogen atmosphere; SIMS detected carbon in the silicon dioxide. Addition of a small amount of oxygen to the nitrogen caused degradation and evaporation of the BHT, as seen by improved electrical parameters, and lower levels of carbon in the SiO_2.

Oxide Growth Rates

Licciardello and co-workers found that the initial room temperature oxide growth rate of HF-etched (hydrophobic) Si wafers was retarded by the presence of surface organics (5). The growth kinetics were consistent with an oxidation which takes place by the diffusion of

CP449, *Characterization and Metrology for ULSI Technology: 1998 International Conference*
edited by D. G. Seiler, A. C. Diebold, W. M. Bullis, T. J. Shaffner, R. McDonald, and E. J. Walters

oxygen through an overlayer on the silicon surface. Hossain, *et al.*, showed that high temperature oxide growth on hydrophobic wafers was affected by a layer of contamination; while no effect was found for hydrophilic wafers (6).

CVD of Silicon Nitride

As discussed above, organic contamination affects processing in non-oxidizing, ambient atmospheres. Saga and Hattori exposed wafers to the clean room atmosphere or stored the wafers in plastic boxes and found that initial nitride growth in a low-pressure chemical vapor deposition (CVD) was retarded (7). After exposure times of one month, an incubation time of several minutes was observed for CVD nitride growth, while clean wafers showed no incubation time. Thermal desorption gas chromatography mass spectrometry (GC-MS) found BHT, associated with antioxidants, and dibutylphthalate (DBP), a plasticizer, on wafers stored in a plastic box. DBP, tributyl tricarboxylate, and dioctylphthalate (DOP), also a plasticizer, were found on wafers exposed to the clean room air. The wafers exposed to the clean room contained higher levels of contamination than those stored in boxes.

Silicon Epitaxial Growth

Rapid thermal chemical vapor deposition (RTCVD) combines rapid thermal annealing and CVD to produce a thin, high quality silicon epitaxial layer with abrupt profiles. Kim, *et al.*, etched wafers in HF, and, after a H_2 bake at 1000 C for 45 seconds to remove the native oxide, grew a RTCVD epitaxial layer at 1000 C (8). They identified beta SiC precipitates at the substrate-epi interface by SIMS and transmission electron microscopy. By growing successive layers, and by only observing the precipitates at the interface between the substrate and the first epitaxial layer, they showed that the carbon originated at the initial surface, not in the process gas or from the vacuum system.

Wafer Doping

Clean room materials have caused unintentional doping of Si wafers by both boron and phosphorus. Stevie, *et al.*, improved the quantification of small amounts of boron in Si by SIMS by creating a sample where the matrix was identical on both sides of the interface, eliminating the problem of equilibrium of the primary beam species and secondary ion yield in the region before the implant range of the analysis beam is reached (9). A layer, such as polysilicon, was deposited on Si, the layer was then exposed to atmosphere and then covered with a layer identical to the first layer. After exposure to the clean room air, boron was found on the wafer. The source of the contamination was borosilicate glass, which was present as sub-micron glass fibers in HEPA filters. Electron microprobe and inductively coupled plasma atmospheric analyses found that boron could constitute 10-20% of the weight of HEPA filters. Lebens and co-workers found that an organophosphate flame retardant in polyurethane potting material used to secure and seal filter material to the module in some HEPA filters caused phosphorus doping (10).

Wafer Bonding

Si wafer bonding, one method for producing silicon-on-insulator (SOI) structures, requires homogeneous bonding across the joined surfaces. Unbonded areas, or bubbles, were shown by Mitani, *et al.*, to be due to hydrocarbon contamination originating in storage in plastic boxes and from exposure to controlled amounts of organic vapors (11).

Effect on Photoresists

Several groups have shown that photoresists are affected by various organic contaminants, particularly organic amines (12, 13).

SOURCES OF ORGANIC CONTAMINATION

Clean rooms have been designed to remove particulate contamination, with less thought given to the outgassing properties of the materials used to construct the rooms. Wall materials, window materials, floor and ceiling materials, and air filters all have organic components. Delivery systems for liquids and gases, bottles, clean room garments, and especially the plastic boxes used to hold wafers are all potential sources of volatile contaminants.

Evaluation of the contamination potential of materials can be carried out by measuring the outgassing of the material itself or by measuring the amount of contamination deposited on a wafer surface after exposure to the material. The importance of measuring the wafer surface was demonstrated by Saga and Hattori in their analysis of the effects of storage in plastic boxes (14). Although unpolymerized monomers and oligomers outgassed in large quantities, the major species found on the wafer surfaces originated from the small quantity of additives in the materials, such as antioxidants, plasticizers, and cross-linking agents. In general, the lower the vapor pressure and the lower the molecular weight the more easily the materials adsorbed onto the wafer surface. The material's polarity also affected its adsorption.

PROCESSING TO MINIMIZE EFFECTS OF ORGANIC CONTAMINATION

In situ processing, in which the wafer surface is not exposed to the clean room atmosphere between cleaning and processing, and cleaning procedures using ultraviolet light (UV), ozone, and, more recently, high temperature water have been developed to supplement the traditional cleaning procedures, such as the RCA clean. The desire to reduce the large volume of contaminated liquid waste from the cleaning processes is another incentive to develop alternate cleaning procedures.

Vig reviewed the use of UV/ozone cleaning procedures to remove organic contamination (15). The removal mechanism was seen to be primarily a photosensitized oxidation process. Short wavelength UV light excites and/or dissociates the contaminant molecules and dissociates molecular oxygen and ozone to atomic oxygen. The excited contaminant molecules and the free radicals produced by the dissociation of the contaminants react with atomic oxygen to form simpler volatile molecules, such as CO_2, H_2O, N_2, *etc.* The rate of contaminant removal for the UV/ozone combination was greater than two orders of magnitude faster than when either was used alone.

Hydrocarbon removal by a UV/O_2 combination was studied by Kasi and Liehr by exposing wafers to known gases (16). After HF etching to produce a hydrogen-terminated, hydrophobic surface, wafers were contaminated with a drop of a specific hydrocarbon, spin-dried in a nitrogen atmosphere, and moved without exposure to air through a vacuum load lock for *in situ* cleaning and analysis. The rate of hydrocarbon removal in O_2 gas in the presence of UV depended on the nature of the deposited hydrocarbon.

Ohmi showed that ozone-injected ultrapure water was effective in removing organic impurities and promoting native oxide growth on Si (17). Bakker and Hess have developed a cleaning process based on water and water/CO_2 mixtures at elevated temperatures (100 to 700 C) (18). Below 100 C, water dissolves inorganic (ionic) species. Between 100 C and the critical point (374.4 C) water's properties change from polar to nonpolar. At the critical point, hydrocarbons become very soluble, while ionic salts no longer dissolve. The cleaning properties of water can thus be adjusted by controlling its temperature. Bakker and Hess also describe how water at high temperature can dissolve SiO_2, thus removing any organic contaminants in native oxide films.

ANALYTICAL TECHNIQUES FOR ORGANIC CONTAMINATION

In selecting the analytical techniques to be covered in this review, two basic criteria have been used. First, the relevant techniques can be used to analyze contamination that deposits on wafers, *i.e.*, methods used only for air sampling are not discussed. Second, appropriate techniques for evaluating organic contamination must provide information that allows individual organic compounds or classes of compounds to be identified. In other words, these techniques must provide either direct or indirect chemical identification and/or bonding information.

The review of the known effects of organic contamination presented earlier highlights the importance of both of these points. Saga and Hattori noted significant differences between the outgassing products from plastic boxes and the compounds that ultimately deposit on wafers (14). At the same time, identifying a contaminant as simply containing certain elements is not sufficient, because specific compounds or classes of compounds will adhere to surfaces preferentially and will produce different effects depending on subsequent processing. In addition, more specific information can be valuable for tracing contaminants to their sources in order to reduce or eliminate them.

There are a number of analytical techniques that can meet these requirements, ranging from traditional wet chemical analysis to advanced surface analysis. However, the techniques most commonly used to evaluate organic contamination on semiconductor surfaces are Gas Chromatography-Mass Spectrometry (GC-MS), Ion Mobility Spectrometry (IMS), Time-of-flight Secondary Ion Mass Spectrometry (TOF-SIMS), X-ray Photoelectron Spectroscopy (XPS; also known as Electron Spectroscopy for Chemical Analysis or ESCA), and Fourier Transform Infrared Spectroscopy (FTIR).

These techniques can be classified into two groups. The first, covering GC-MS and IMS, includes methods that detect contaminants after desorption from a surface. In the second group, which includes TOF-SIMS, XPS and FTIR, the surface of the wafer is analyzed directly. Each of these techniques is discussed below, and a summary of some advantages and limitations for each is in Table 1.

Desorption techniques

Many common types of organic contamination are relatively volatile materials that deposit on wafer surfaces from the air. IMS and GC-MS, the two main techniques that rely on desorption of contaminants, take advantage of this volatility by measuring material outgassing from a wafer under controlled conditions. For both techniques, the desorption is typically induced thermally.

TABLE 1. Summary of Techniques Used to Analyze Molecular Contaminants on Wafers

Technique	Advantages	Limitations
GC-MS	-Good sensitivity for volatile contaminants -Direct chemical ID possible (MS) -Separates contaminants before identification -Semi-quantitative results can be obtained	-Poor sensitivity for ionic, nonvolatile or thermally unstable contaminants -Long analysis times -Detection limits poorer than IMS
IMS	-Excellent sensitivity for volatile contaminants -Separates contaminants -Quantification is possible	-Poor sensitivity for ionic, nonvolatile or thermally unstable contaminants -Sensitive to overloading -No direct chemical information
TOF-SIMS	-Surface of wafer analyzed directly -Excellent sensitivity for nonvolatile/ ionic contaminants -High resolution mass spectrum for chemical ID	-Vacuum technique; requires cold stage to detect volatile contaminants -Absolute quantification is difficult because it requires standards
ESCA/XPS	-Surface of wafer analyzed directly -Quantitative for elements and bonding states	-Not as sensitive to low levels of contamination -Vacuum technique; requires cold stage to detect volatile contaminants -Chemical identification ability is limited
FTIR	-Surface of wafer analyzed directly -Atmospheric pressure technique, for both volatile and nonvolatile contaminants -Extensive libraries available to ID contaminants	-Not as sensitive to low levels of contamination -No separation of contaminants, which complicates data interpretation

IMS has the advantage of very high sensitivity, which makes it possible to detect the small amounts of material that can affect device performance. However, this high sensitivity also makes IMS susceptible to overloading by more abundant, but relatively harmless species. GC-MS is not as susceptible to overloading, but special procedures may be required to detect low levels of contaminants. A practical advantage of GC-MS is its presence in many labs.

Gas Chromatography-Mass Spectrometry

GC-MS combines the separation capabilities of a gas chromatograph with the chemical identification abilities of a mass spectrometer. To evaluate contaminants present on wafers, GC-MS is typically combined with some type of thermal desorption apparatus that is used to volatilize the contaminants. This combined technique is also referred to as Thermal Desorption GC-MS or TD-GC-MS.

While commercially available TD-GC-MS instruments designed specifically for use with wafers are now available, many laboratories have developed their own in-house apparatus that can be attached to available GC-MS instruments. The time, temperature and other desorption conditions can be varied in the desorption step, though some efforts are being made to standardize protocols for specific types of analyses (19).

After the contaminants are desorbed thermally into a carrier gas, they are typically trapped on a cold finger before entering the GC. When the thermal cycle is completed, the cold finger is heated quickly to re-volatilize the contaminants. An inert carrier gas is used to take the contaminant mixture through the column, where the various compounds are partitioned into separate components based on their interaction with the stationary phase of the column. At the end of the column, each component is detected and tracked as a function of time.

In GC-MS, the gas chromatographic separation, which separates chemical species but does not provide chemical information, is followed by a mass spectrometric analysis. The mass spectrum obtained for each component separated by the GC provides direct chemical information that often can be used to identify the species. In practice, analysis of all components can be a time-consuming process, so the species reported for a given analysis may be limited to those above a selected minimum peak intensity or to specific compounds that are of interest in the analysis. A combination of standards, available reference compounds, mass spectrometry and, less frequently, retention times, is used to identify and semiquantitatively report data for most species observed in the analysis.

GC-MS has several advantages in analyzing organic contamination. It is an atmospheric technique, so no species are lost due to vacuum conditions. The use of standards and reference materials make it possible to get semiquantitative results fairly easily. As noted above, instruments are also available in many laboratories.

Some limitations of the technique are its detection sensitivity, which may require special procedures to detect low levels of contaminants; analysis times, which can be fairly long; possible cross-contamination that may require purging after very dirty samples; and insensitivity to non-volatile or easily condensable contaminants.

Ion Mobility Spectrometry

IMS separates ions based on their differences in travel time in an applied electric field. A silicon wafer is typically placed in a quartz sample chamber. The temperature of the sample is increased, and the volatile components are vaporized and mixed with a heated carrier gas, such as nitrogen. The carrier gas and organic volatiles are carried into a reaction chamber, where a foil of ^{63}Ni emits 60 keV beta rays and ionizes the gas. As the gas is carried through the reaction chamber, sample molecules are ionized through a series of charge transfer and energy transfer reactions between the carrier ions, the electronically excited species, and neutral sample molecules (20).

An applied electric field moves the ions through the reaction region until they reach a shutter grid. The grid is opened repeatedly to admit pulses of ions into a drift region. Separation of the ions occurs in the drift region due to the differences in mobility of the various ions in the presence of the inert carrier gas and the applied electric field. In general, smaller, lighter ions have higher mobilities.

Under normal operating conditions, ions migrate at velocities between 1 and 10 m/s, with arrival times on the order of 3-30 ms. After monitoring the collector current for approximately 30 ms, the grid is reopened and the measurement repeated to improve the signal to noise ratio.

IMS peaks are typically broad compared with the range of possible drift times, and identification of unknown species by mobility data alone is difficult. In practice, species are usually identified using a combination of standards and supplemental mass spectrometric analysis.

IMS has several significant advantages in measuring the small quantities of organic contamination that can affect device performance. Since it can measure ion currents below 10^{-12}, IMS has very high sensitivity; picograms (10^{-12}) and even femtograms (10^{-15}) of contaminants have been detected. As noted for GC-MS, IMS is an atmospheric technique so no species are lost due to vacuum conditions. By collecting mobility data as a function of sample temperature, additional separation data can be obtained.

The sensitivity of the system to overloading, and the possibly long cleaning periods required to purge the system preclude characterizing the outgassing properties of polymers or highly outgassing samples directly. As noted for GC-MS, the technique is also insensitive to contaminants that are nonvolatile or have only low volatility. Extreme care is even needed in handling silicon wafers to avoid introducing extraneous contamination. Budde, *et al.*, provide a good review of the application of IMS to device analyses (21).

Surface Analysis Techniques

Unlike the desorption methods discussed above, surface analysis techniques are used to analyze the wafer surface directly. The advantage of this approach is a higher sensitivity to less volatile polar or high molecular weight contaminants. This can also be important when reactions occur on a surface, for example when salts form from volatile organic amines and anions such as sulfates. A disadvantage of direct analysis is that there is no separation step, so spectra obtained from the wafer contain data from all of the compounds detected.

Time-of-flight Secondary Ion Mass Spectrometry

In TOF-SIMS, mass spectra are obtained by detecting positive or negative ions generated from a surface. A wafer or piece of wafer is placed in a sample chamber that is typically at a pressure of 10^{-8} to 10^{-10} torr (ultrahigh vacuum, UHV, or near-UHV). A pulsed primary ion source, for example $^{69}Ga^+$, Cs^+ or Ar^+, is used to bombard the surface of the sample, causing the emission of molecular and atomic secondary ions. The secondary ions are electrostatically accelerated into a field-free drift region. At the end of the drift region, the mass of each ion is determined from its flight time (22).

Secondary ions are generated and recorded with each primary ion pulse. A complete spectrum is generated by accumulating secondary ions detected in successive pulses for extended periods, typically 1-15 minutes. Typical analytical areas for the detection of organic contamination are on the order of 80 μm x 80 μm.

There are several advantages of TOF-SIMS for the measurement of organic contamination. It has very good sensitivity for low levels of contaminants and for low volatility or thermally unstable contaminants that can be difficult to analyze by desorption techniques. The high mass resolution of the mass spectrometer can provide unambiguous identification of chemical species. In addition, because the analytical areas are localized and the ions detected are coming from specific, identifiable regions of a wafer, it is possible to compare the levels of contamination from different regions on the same wafer (23).

The most important limitations of the technique are related to difficulties with quantification and to the effects of the UHV conditions of the experiment. Quantification is complicated by the large differences in relative ion yields among organic ions. This means that, at present, it is only possible to make relative comparisons of the intensities of a particular species among a series of wafers. The standard UHV conditions of the TOF-SIMS analysis mean that more volatile compounds present on wafers are often removed from the sample before it is analyzed. This problem can be mitigated by using a cold stage to perform the analysis (24).

X-ray Photoelectron Spectroscopy

In XPS, also known as Electron Spectroscopy for Chemical Analysis (ESCA), X-rays excite photoelectrons from a sample surface. The analysis is performed by placing a wafer or piece of wafer in a UHV sample chamber. Soft X-rays, usually either Al-kα or Mg-kα, excite photoelectrons in the near surface region (10-100 Å) of a sample. The photoelectrons are collected by an electrostatic lens and then are passed through a high resolution analyzer before detection. The photoelectron binding energy is characteristic of the element and chemical bonding environment from which the photoelectron originated. Information about oxidation states is obtained by comparing lineshapes and peak positions with published literature values. Quantitation is performed using sensitivity factors (25).

The advantages of XPS for the analysis of organic contamination lie in its quantitative capabilities and its ability to detect compounds that are nonvolatile or have low volatility. It is also a valuable tool for measuring the amount and type of carbon present after subsequent processing or when species such as silicon carbides may be present (2).

The limitations of the technique for organic applications are related to the relatively low sensitivity of the method and the limited chemical identification that is possible. Typical detection limits are on the order of 0.01 atomic percent, significantly worse than those obtainable with TOF-SIMS or IMS. The chemical state identification possible with XPS provides general information about the type of bonding present, but it is generally not possible to identify specific organic contaminants from XPS data. Finally, as with TOF-SIMS, the UHV analysis conditions can introduce problems in the analysis of volatile contaminants.

Fourier Transform Infrared Spectrometry

In FTIR analyses, the absorption of infrared radiation by a sample is measured. The IR radiation is absorbed at wavelengths corresponding to natural vibrational frequencies of covalent (molecular) bonds. The portion of the infrared spectrum that is of interest lies between 4000 and 400 cm^{-1}. In Fourier Transform spectrometers, analyses are performed using a radiation source that emits an infrared beam covering all of these wavelengths simultaneously. The waveform that is measured after interaction with the sample is deconvoluted to produce an infrared spectrum. Spectra are typically presented as either absorption or percent transmittance versus infrared frequency.

For wafer contamination analysis, samples generally must be analyzed using either Attenuated Total Reflectance (ATR) or grazing angle adaptations of the technique that enhance the surface sensitivity of the measurement.

Since most organic compounds contain several types of covalent bonds that absorb different frequencies of IR radiation, the complete spectrum for individual compounds can provide a relatively unique fingerprint for the material. Large databases of IR spectra are available that can make identification of unknown materials very practical.

The major advantage of FTIR in the analysis of organic contamination is this ability to make use of extensive databases to identify compounds. Another advantage of the technique is that it can be used to look directly at wafer surfaces without requiring vacuum conditions. This minimizes difficulties in the analysis of volatile contaminants.

The chief disadvantage of FTIR is its relatively poor detection sensitivity. This is especially a problem when a contaminant needs to be identified at low concentrations. A second disadvantage is the difficulty of identifying a particular contaminant when, as is often the case, more than one compound is present on the sample. As noted in the general discussion of the direct surface analysis techniques, these methods do not separate the contaminants prior to identification. For TOF-SIMS, the high mass resolution and specificity of individual mass spectral peaks minimizes the difficulties caused by this, but it is a problem for FTIR.

Summary of Analytical Techniques

The analytical techniques discussed above all meet the two criteria that were previously defined. Each can be used to analyze contaminants that deposit on wafers, and each provides molecular or bonding information, either directly or indirectly.

In the explanations of each technique, other factors are discussed that can be important in determining the correct tool for a particular analysis. These include: (a) detection sensitivity, since levels of individual contaminants are generally very low; (b) consideration of the volatility and surface affinity of the contaminants; (c) the type of chemical information provided by the technique; (d) the ability to provide quantitative or semi-quantitative results; (e) an awareness that contamination is often chemically altered and redistributed throughout a material during subsequent processing. Beyond these factors, others, such as availability and cost, are often important in the selection of a technique for a particular application.

In general, no single analytical technique can provide the answer to every contamination problem. The specific requirements of individual situations will often dictate the choice of technique that is most appropriate. In the analysis of organic contamination, this choice can be complicated because the effects of different contaminants are only beginning to be understood.

ACKNOWLEDGMENTS

We thank William Fil, Nancy Klymko, Steve Molis, and Al Passano of IBM and David Nehrkorn of Surface Science Laboratory for valuable discussions, and William Fil for reviewing the manuscript.

REFERENCES

1. Kern, W. and Puotinen, D. A., *RCA Rev.*, **31**, 187-206, 1970
2. Kasi, S. R., Liehr, M., Thiry, P.A., Dallaporta, H., and Offenberg, M., *Appl. Phys. Lett.*, **59**, 108-110, (1991)
3. Iwamoto, T., Miyake, T., and Ohmi, T., *Proc. Of SPIE*, **2875**, 207-215, (1996)
4. Saga, K., and Hattori, T., *Appl. Phys. Lett.*, **71**, 3670-3672, (1997)
5. Licciardello, A., Puglisi, O., and Pignataro, S., *Appl. Phys. Lett.*, **48**, 41-43, (1986)
6. Hossain, S. D., Pantano, C.G., and Ruzyllo, J., *J. Electrochem. Soc.*, **137**, 3287-3291, (1990)
7. Saga, K., and Hattori, T., *J. Electrochem. Soc.*, **144**, L253—L255, (1997)
8. Kim, K-B, Maillot, P., Morgan, A. E., Kermani, A., and Ku, Y., *J. Appl. Phys.*, **67**, 2176-2179, (1990)
9. Stevie, F. A., Martin, Jr., E. P., Kahora, P. M., Cargo, J. T., Nanda, A. K., Harrus, A. S., Muller, A. J., and Krautter, H. W., *J. Vac. Sci. Technol.*, **A 9**, 2813-2816, (1991)
10. Lebens, J. A., McColgin, W. C., Russell, J. B., Mori, E. J., and Shive, L. W., J. Electrochem. Soc., **143**, 2906-2909, (1996)
11. Mitani, K., Lehmann, V., Stengl, R., Feifoo, D., Goesele, U. M., and Massoud, H. Z., *Jpn. J. Appl. Phys.*, **4**, 615-622, 1991
12. MacDonald, S. A., Hinsberg, W. D., Wendt, H. R., Clecak, N. J., and Wilson, C. G., *Chem. Mater.*, **5**, 348-356, (1993)
13. Czech, G., Lehner, N., Schlueter, C., Lepuschitz, T., Beauchemin, B., Schulz, R., Daraktchiev, I., and Sinkwitz, S., *Microelectr. Eng.*, **23**, 331-335, (1994)
14. Saga, K., and Hattori, T., J. *Electrochem. Soc.*, **143**, 3279-3284, (1996)
15. Vig, J., *J. Vac. Sci. Technol.*, **A 3**, 1027-1034, (1985)
16. Kasi, S. R., and Liehr, M., *J. Vac. Sci. Technol.*, **A 10**, 795-801, (1992)
17. Ohmi, T., Isagawa, T., Kogure, M., and Imaoka, T., *J. Electrochem. Soc.*, **140**, 804-810, (1993)
18. Bakker, G.L. and Hess, D. W., *J. Electrochem. Soc.*, **145**, 284-291, (1998)
19. Camenzind, M. and Kumar, A., *IES 1997 Proceedings, Contamination Control*, Mount Prospect, IL: IES, 1997, pp. 211-226.
20. Carr, T. W., *Thin Solid Films*, **45**, 115-122, (1977)
21. Budde, K. J., Holzapfel, W. J., Beyer, M.M., *J. Electrochem Soc.*, **142,** 888-897,(1995)
22. Schueler, B. W., *Microsc. Microanal. Microstruct.*, **3**, 119-139, (1992).
23. Lindley, P., Radicati, F., Mowat, I., McCaig, L., and Kendall, M., *Secondary Ion Mass Spectrometry (SIMS XI)*, Chichester, England: John Wiley & Sons, 1998, pp. 175-178.
24. Strossman, G., Lindley, P., and Bowers, W., *Secondary Ion Mass Spectrometry (SIMS XI)*, Chichester, England: John Wiley & Sons, 1998, pp. 699-702.
25. Wagner, C. D., Riggs, W. M., Davis, L. E., Moulder, J. F., and Mullenberg, G. E., *Handbook of X-ray Photoelectron Spectroscopy*, Eden Prairie, MN: Perkin-Elmer Corp., 1978.

FRONT END PROCESSES – DOPING AND 300mm WAFER ISSUES

In-line Characterization of Doping Technologies for ULSI: Requirements and Capabilities

Larry Larson
SEMATECH, Austin, TX 78741, Larry.Larson@SEMATECH.org
Michael Current
Applied Materials, Santa Clara, CA 95051, current_michael@amat.com

Doping requirements for CMOS devices have been defined out to 50 nm gate size devices [1] and provide a context for the evaluation of in-line characterization needs. The wide range of doping processes used in CMOS fabrication, ranging in energies from ≤1 keV to over 1 MeV and in dose from $\leq 10^{11}$ to $\geq 10^{16}$ ions/cm^2 presents a significant challenge to provide adequate coverage in a high-accuracy, fast-response metrology. An added challenge is to provide characterization tools that function in-line with fab operations or, even better in-situ with the process chamber. Fortunately, a wide variety of materials characterization techniques have been adapted to the needs of doping technology processes [2]. However, the rapid advance of IC process requirements, a fundamental characteristic of the IC business, has pushed most of these techniques to their limits of performance.

The list of critical areas for in-line monitoring starts with improved dosimetry for both shallow (source/drain and channel doping) and deep (CMOS well) junctions. Added to the dosimetry requirements is an increased need to monitor dopant profile shape and junction depth, especially for shallow junction's [3]. Monitoring of elemental and particulate contamination levels is critical for adequate yield and cost-effective operations. Because the ion ranges for ultra-shallow junctions are ≤10 nm and the smallest particle size that can be imaged with routine techniques is ≥180 nm, new technology is required to detect small size "killer particles" that can mask low-energy (≤1 keV) ion beams. Control of wafer charging for gate dielectrics shrinking below 5 nm and increased antenna factors for large-area logic chips will drive the need for increased monitoring of net current flow from the ion beam-charge control plasma leading to development of in-situ sensing of plasma j-V characteristics.

INTRODUCTION

Doping requirements for CMOS devices have been defined out to 50 nm gate size devices [1], providing a context for the evaluation of in-line characterization needs. The incumbent technology for doping of CMOS transistors is accelerator-based ion implantation combined with furnace annealing or RTP in advanced processes. The characterization of these processes relies on process control methods using bare-Si monitor wafers evaluated for dose by 4-point probe and/or Thermawave, with occasional use of machine-based SPC, in-situ and in-line particle monitoring and charging monitoring wafers. Even though the use of accelerator-based ion implantation followed by thermal annealing has been very successful and has a good chance to continue well past the year 2000, significant opportunities exist for emerging technologies such as Plasma Immersion Ion Implantation (PIII) and Gas Immersion Laser Doping (GILD). These new technologies provide challenges to traditional metrology methods as the characteristics of the processes and the opportunity for unobtrusive measurement vary dependent on the process characteristics of the candidate technology. The wide range of doping processes used in CMOS fabrication, ranging in energies from ≤1 keV to over 1 MeV and in doses from $\leq 10^{11}$ to $\geq 10^{16}$ ions/cm^2 presents a significant challenge to provide adequate coverage in a high-accuracy, fast-response metrology. Fortunately, a wide variety of materials characterization techniques have been adapted to the needs of doping technology processes [2]. However, the rapid advance of IC process requirements, a fundamental characteristic of the IC business, has pushed most of these techniques to their limits of performance.

Advanced CMOS transistors call for both deep and shallow doping implants (Fig. 1). The deep implants extend to beyond 1 μm and form the CMOS wells with retrograde profiles. These are used to suppress latch-up effects along with vertical anti-punch-through ("mid-well") doping and deep "triple wells" for SRAM and FLASH devices. The shallow implants form the source/drain and channel structures. These structures include: source/drain extensions, channel doping, and deeper junctions for the source/drain contacts. Alternatives to threshold control through direct channel doping include super-steep retrograde profiles to confine the channel conduction path. In addition, control of short-channel effects calls for tailoring of the lateral doping profiles with combinations of HALO and Pocket implants to suppress off-state leakage and lateral punch-through.

CP449, *Characterization and Metrology for ULSI Technology: 1998 International Conference*
edited by D. G. Seiler, A. C. Diebold, W. M. Bullis, T. J. Shaffner, R. McDonald, and E. J. Walters
 1-56396-753-7/98/$15.00

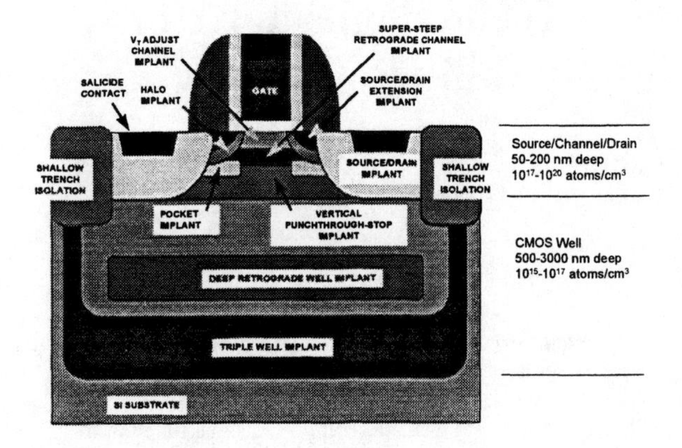

Figure 1. Doped regions for CMOS transistors.

Figure 2 shows the variation of critical shallow junction depths (source/drain extensions, channel doping and contacts) as predicted in the NTRS'97 [1] as a function of year and of the production node that is expected to be introduced at that time. This shows the expected trend in vertical profiles that will need to be measured in-line for the shallowest critical layers. A trend is shown from 100 nm junctions for the present 250 nm production (these junctions are presently difficult to measure) to 20-40 nm junctions in 2003. These junction depths are comparable with the standard deviation (straggling) of a typical implant and will be difficult to produce and to measure.

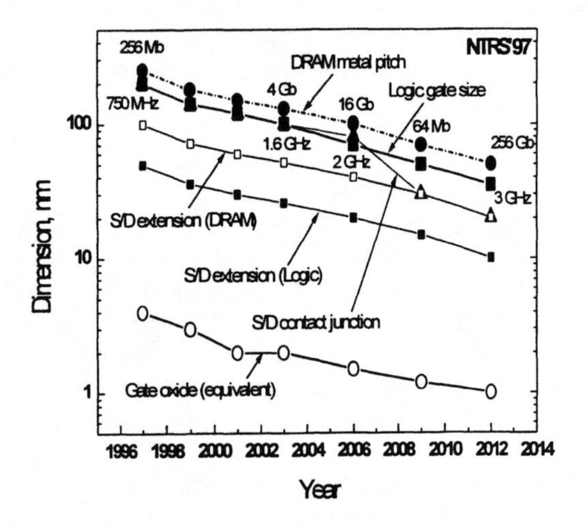

Figure 2. Expected evolution of shallow junction depth and gate oxide (equivalent) thickness according to the SIA National Technology Roadmap for Semiconductors, NTRS'97 [1].

REQUIREMENTS AND CAPABILITIES FOR IN-LINE DOPING CONTROL

The requirements for dosimetry and depth control are a trade-off of the characteristics of the transistor performance versus what profiles can be obtained and what may be measured to control those profiles. Because of this performance trade-off, which occurs between several attributes of the transistor, the individual control needs vary slightly between devices and products and are generally hard to nail down. One of the more open analyses of these issues is the paper in this volume by Zeitsoff and Tasch [4]. They found that the drain extension implant was one of the most critical implants in the 180 nm device process and that the threshold voltage and several other electrical parameters depend on both the depth and the dose of this implant. The process limits for this implant were ± 5% for depth (3-sigma, all sources) and ± 7.5% for dose.

Given this, note that the metrology technique for that process needs to also have significant process capability for the measurement of the dose and depth for the implant. This is generally expressed as a precision / tolerance ration (P/T ratio). The tolerance, in this case, is the process limits. These limits were established above as 5% for depth and 7.5% for dose. The P/T ratio is desired to be 10. Working backwards, the metrology tools need to be able to resolve a difference of 0.75% in dose. For the drain extension implant in the 180 nm process this is a variance of 7.5 X10^{13} /cm^2 in a dose of 1 X10^{15} /cm^2. In addition they need to be able to resolve 0.5% in depth. In the 180 nm process the extension layer is 50 nm deep, which indicates that the metrology technique needs to be sensitive to changes in the junction of 0.25 nm. This is comparable to the width of an atom!

In-line dosimetry measurement for implant has been traditionally done with either 4-point probe or Thermawave, or both. Some characteristics of these techniques, along with the more analytical profile measurement techniques are shown in Table 1.

The 4-point probe has good sensitivity for doses roughly above 10^{13} ions/cm^2 and continuing past 10^{16} ions/cm^2. As a very general statement, it has been considered capable of the required in-line dosimetry measurement as measured by the P/T ratios defined above. Issues in contacting the layer arise when the layer is buried or when it is too shallow. Noise and signal issues due to these contacting problems will challenge this assessment of the capability of the 4-point probe technique. Resistivity measurement also requires that the doped layer be annealed before measurement. This is an extra step - adding extra time and the risk of process variation in the anneal step. Thermawave has sensitivity down to 10^{12} doses but looses sensitivity in the 10^{14} region. This makes it an excellent complimentary tool to the higher dose 4-point probe measurement. The Thermawave also is capable of reading product wafers, with its small spot size, but has issues with signal stability so that a standard has to be routinely used.

Table 1. Principal Techniques for In-line Monitoring.

Technique	Dose Range (ions/cm^2)	Dose	Profile	Damage (in Si)
4-point probe	10^{13}-10^{16}	yes	no	no
Thermawave	10^{11}-10^{14}	yes	no	yes
SIMS	10^{12}-10^{16}	yes	1-D/(2-D)	no
Optical Density Map	10^{10}-10^{13}	yes	no	no
Spreading Resistance	10^{10}-10^{16}	yes	1-D/(2-D)	some
AFM/CPM	10^{10}-10^{13}	some	2-D	no
TEM (x-section)	10^{14}-10^{16}	no	2-D	some
TEM (plane view)	10^{11}-10^{16}	no	no	yes
CHARM-2 (EEPROM sensors)	10^{11}-10^{16}	no	no	(charging)
Contactless C-V	10^{11}-10^{16}	some	some	(charging)

The principal in-line tools for dose measurement are 4-Point probes for higher doses and Thermawave, or similar damage measurement, for lower doses. Shallow junctions present difficulties for 4-point probe because physical contact is used to measure the resistivity of the layer. Typical contact pins penetrate into the Silicon surface of the order of 100nm and can punch through the shallow junctions. Romig et al.[5] described the difficulties of measuring shallow junctions with a 4-point probe and the issues related to probe penetration. Recent work by Boyd et al. [6] report successful measurement of shallow junction sheet resistance with large-diameter (20 mil) probes.

Thermawave and the OMS technique [7] also have issues in the measurement of shallow junctions. The most aggressive of shallow junction implants have energies of a few hundred eV [8]. At these energies, especially for low mass ions such as Boron, the implant damage level is so low that it is a severe challenge to detect a sufficient amount of damage to provide a reliable signal for damage sensitive techniques such as Thermawave.

Several other techniques are useful for in-line dosimetry measurement and have their supporters. OMS is an optical damage measurement that has shown good sensitivity in the standard dose range and promise in extreme implants [7]. SIMS has been primary a profile analytical technique. Recent work has enabled SIMS to be used in fab lines as an in-line metrology technique [9]. The key message in doing this is that the methodology must be very well controlled for this to succeed, as the measurement (in this case) has as many variables (if not more) than the process. CV measurement has primarily been used for oxide control in the fabs to date. It is sensitive, however, to dopant dose and profile [10]. The limitation has been to be able to measure the capacitance of the whole layer. With the planned decrease in junction depth, carefully executed CV measurement may be competitive for in-line measurement of shallow junctions [10]. SRP has been a standby for electrical profile measurement and is sensitive to the same issues as discussed above for 4-point probe. SRP has been shown capable of shallow profile measurement under very careful control [11,12]. Issues include reducing surface roughness to a minimum and the correct deconvolution of the doping in very shallow (<50 nm) depleted layers.

An increasingly critical issue is the measurement of profile *shapes*, as well as the total dose in that profile. Doping profiles which approximate a uniform, "box" shape can be used to form shallower junctions than Gaussian or exponential profiles, which are limited by direct tunnel leakage currents [13]. Detailed information on doping profile shape after implantation and annealing is a critical component of a successful process development and TCAD modeling program [14]. The use of profile measurement tools, such as SIMS, continues into the new area of in-line measurement in production fab lines to assure process control at the precision required for high-yield, high performance fabrication.

An additional challenge to profile shape metrology is that the most critical profile control direction is the *lateral* dopant profiles at the source boundary through the channel to the drain, a much more difficult task than the vertical profile. For CMOS devices of the order of 100 nm and smaller, the junction abruptness at the source is the limiting factor for attaining high drain currents [13]. Given the criticality of getting these profile shapes right, one can expect increasingly urgent pressure to evolve practical and accurate 2- and 3-D metrology methods [15].

ADVANCED SHALLOW JUNCTION DOPING TECHNIQUES, CHARACTERIZATION FEATURES AND PROCESS RISKS:

The fabrication of advanced CMOS transistors with gate sizes at an smaller than 100 nm will almost certainly drive the invention of new techniques for doping, especially for shallow junctions [1]. The new technologies include: (1) evolution of beamline accelerator technology to transport sub-keV ions at high beam current, (2) use of high-mass molecular ions containing dopant atoms (such as $B_{10}H_{14}$ [16]), (3) plasma immersion chambers, (4) laser-melting in combination with dopant gases and (5) thermal growth of doped films by various CVD methods and (6) growth of doped layers by MBE techniques [13, 17]. Each of these techniques produces different characteristic dopant profile characteristics. Differences in the configuration and physical environments of the various process chambers also complicate the choice of measurement tools for in-situ monitors.

The ion-beam techniques (1 through 3) require a follow-on thermal step to repair the lattice damage from the ion beam and to facilitate activation of the dopant atoms. The choice of methods for the activation step will strongly effect the resulting dopant profile depth and shape. The general nature of these effects is illustrated in Fig. 3.

Not only the depth but the *shape* of the dopant profile is a strong function of ion energy for the ion beam techniques. In "conventional" ion implantation, with ion energies ranging up to several hundred keV, a typical dopant profile is strongly skewed towards the deeper regions. At low energies (< 10 keV for B) and for heavy elements (such as As, Sb and In) the profile is strongly skewed towards the surface. For plasma immersion implants, where the wafer is repeatedly pulsed to high negative potentials for short (<10 μsec) intervals, the spread in the ion energy at the wafer surface produces an exponential-like profile.

For thermal anneals following implantation, the longer times needed to operate in a batch furnace environment results in deeper dopant junctions than RTP or laser annealing. Recent tests of shockwave techniques result in no net diffusion but incomplete dopant activation [18].

Thermal doping techniques are in two groups, one that incorporate dopants from a gas at the wafer surface and one that utilizes diffusion from a solid film or the buildup of a solid film containing dopants. The gas-source techniques include use of a pulsed thermal source to facilitate the diffusion of dopants absorbed on the wafer surface, resulting an exponential dopant profile. If a laser is used to melt the wafer surface (in selected regions if the laser is shown through a mask array) the dopants are uniformly incorporated into the melt up to the depth of the melted region. This results in an abrupt dopant profile fall off at the bottom "box-like" junction. When dopants are diffused from a deposited solid layer, the profile is exponential in shape. However when CVD and MBE techniques are used to deposit doped and un-doped Si layers, then abrupt buried or surface doped layers can be formed.

The message in this section is that due to the decreasing thermal budget and required junction depth, several doping techniques that produce radically different profiles are under development. Measuring and controlling these profiles is the key challenge for in-line metrology of the future. This will need to be done with similar precision as the discussion in the previous section but in-line for process control.

The challenge for development of *in-situ* metrology for dose and profile shape is that these various doping techniques utilize various physical methods and a range of physical environments. In the case of accelerator beamlines, even though the ion beam is confined to a directed beam a few cm in diameter in a larger vacuum chamber, dopant gases are present as well as such noise-generating factors as uv-radiation and electrical arcing. There are also time-scale issues for conventional implantation. In batch-type ion implanters, using a spinning wheel or disk containing 13 to 25 wafers (depending on the wafer size), the speeds of the wafers passing through beam range up to 90 m/s. For serial-type implanters, with single wafer irradiation, the beam is often scanned across the wafer at frequencies of several kHz and the wafer is often mechanically scanned at frequencies of a few Hz. These various scanning mechanisms complicate the task of in-situ metrology techniques to monitor the wafer surface while the implantation is going on.

For plasma immersion, direct measurements from the wafer surface are complicated by the presence of the dopant containing plasma, which surrounds the wafer surface and pulsed bias on the wafer.

For thermal anneals following implantation, the longer times needed to operate in a batch furnace environment results in deeper dopant junctions than RTP or laser annealing. Recent tests of shockwave techniques result in no net diffusion but incomplete dopant activation [18].

In-situ monitors for the gas-based doping methods would need to distinguish signals from the dopant gas or dopant containing film and the dopant profile below the Si surface. Molecular beam techniques, which need UHV environments, provide the cleanest operating environment. But the challenge here would be to accurately monitor sub-monolayer dopant concentrations without compromising the extreme cleanliness needed for these methods.

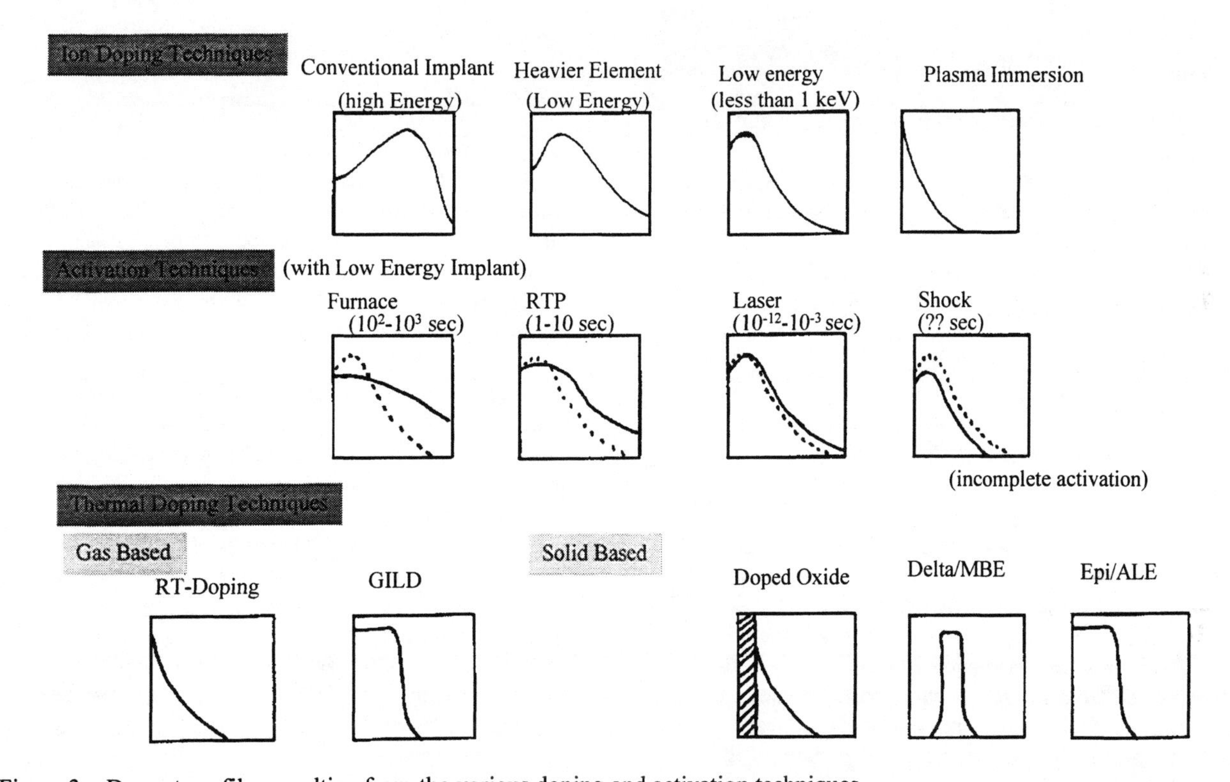

Figure 3. Dopant profiles resulting from the various doping and activation techniques.

CONTAMINATION MONITORS: TOOLS AND CHALLENGES - SENSORS AND LIMITS:

In-line characterization for the implant area implies somewhat more than just monitoring the process for ion dose and energy. The next three sections consider areas that are not unique to implant but have a significant impact on the doping process, thus are critical areas to cover for the in-line metrology needs of the implant area. Within the topic of contamination - there are two primary areas of concern: elemental contamination and particulate contamination.

Monitoring of elemental and particulate contamination levels is critical for adequate yield and cost-effective operations. Figure 4 demonstrates the effect of particles on the implant process. The simplest of these effects is blockage of the implant so that the device layer under the particle is not doped by the implant. Since ultra-shallow junctions are ≤100 nm and the smallest particle size that can be imaged with routine techniques is ≥180 nm, new technology is required to detect small size 'killer particles" that can mask low-energy (≤1 keV) ion beams.

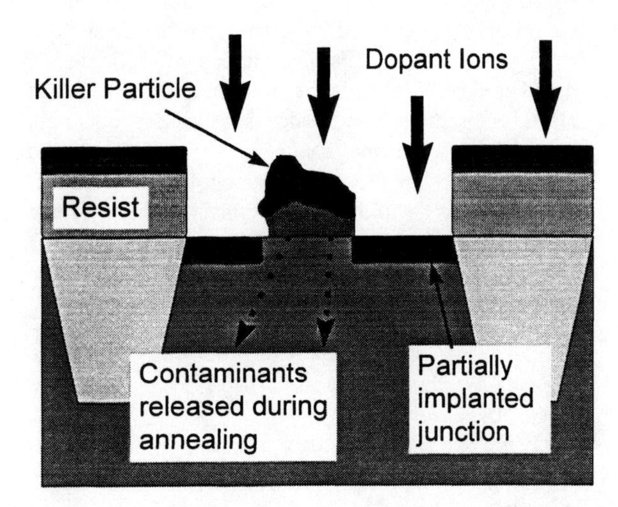

Figure 4: The effect of a particle on an implanted junction

There are two device effects that can occur when a particle blocks an implant. The least serious of these is when the area being doped should be of the opposite type as the implant or when it forms a Schottkey barrier. In this case, the device has reduced gain and/or maximum drive current, but little else occurs to disrupt performance. In the other case, the area can be shorted out by the lack of the implant. This would severely limit performance and likely kill the circuit.

Elemental contamination from particles is also illustrated in Fig. 4. Material can be sputter deposited from the particle and directly deposited into the silicon. In addition to the particle acting as a sputter and diffusion source, elemental contamination can be transported from the resist, and from any other part of the wafer that is in proximity to the ion beam. The implant itself can act to drive the contaminants, by recoil collisions, deep into the wafer surface.

A special concern for the future for this area is the new device structures and the new materials that will be used to make those devices. In particular, High K gate dielectrics and perhaps metal gates are planned for the 100-70 nm nodes and early development for that need is starting in development areas today. Candidate systems contain Ba, Ti, Ta, Sr and Zirconium. Each of these is potentially disruptive to device performance and could easily be sputtered onto, then driven into, the silicon active areas. Works by Hobler [19, 20] nicely describes the injection of dopants at the very edge of the gate by a sputtering mechanism. The same mechanism can sputter advanced gate materials directly into the most critical part of the active device.

One of the issues with classical particle measurement is the use of test wafers. This has, to some extent, been addressed by the introduction of in-situ particle counters. These enable constant monitoring of the tool performance while eliminating the use of special test wafers. At the same time, there are several issues to be considered. Placement: The sensor head is most effectively placed with a direct view of the wafer and of the ion beam and preferably as close as possible. The reality is that the space inside the wafer chamber is limited and port access may not be convenient for proper head placement. Another issue is sensitivity: It was noted previously that particles of the size quoted in the he roadmap are significantly larger that the shallow junction goals, yet the detection capability of today's in-situ sensors is significantly less than either of those sizes.

A worrisome mechanism of contamination is the beamline transport of implanter materials, first formulated by D. Brown [22]. In this case, implanter construction materials can be broken loose by a physical shock and/or and shorts, and the particle transported down the beamline to the wafer. The beamline potentials and the shape of those potentials are appropriate to suspend small particles once they have been charged by the ion beam. These suspended particles can be literally pushed by the ion "flow" down towards and onto the wafer. This appears to be one of the major mechanisms of contamination in the high current family of implanters. Measurement of these particles is a serious challenge to in-situ particle counters. Obviously, one would want the particle counter located directly in the beamline and accelerated particles may not have a long residence time in the detection area. The particles could be expected to be small - and - the particle counter would need to be able to distinguish between the plasma "glow" and the signal from true particles.

Despite all these issues, we are positive about the development and use of in-situ monitors as the reduction of test wafers and the efficient use of the implanter are the correct direction for process control tools. Another prime opportunity for in-situ metrology is the use of the data offered by the tool itself characterizing its operation. By carefully collecting machine data and analyzing it via SPC-like awareness, the operating characteristics of the tool can be documented and protected. Several authors have published contributions based on this concept [22, 23, 24].

CHARGING: WAFER-LEVEL MONITORS AND NEW GATE MATERIALS CHARACTERIZATION

Control of wafer charging for gate dielectrics shrinking below 5 nm and increased antenna factors for large-area logic chips will drive the need for increased monitoring of net current flow from the ion beam-charge control plasma leading to development of in-situ sensing of plasma current-voltage (j-V) characteristics. In-line monitors, such as the use of EEPROM-based sense and memory devices [25], provide direct measures of the beam plasma j-V characteristics. Contact-less CV probes can provide in-line measures of net trapped charge and trap densities in thick oxide layers. Both methods provide the ability to detect gross malfunction of charge level control of a plasma environment (such as an error in setup following a maintenance cycle) for beamline and plasma immersion doping. The EEPROM sensors, which provide data that can be straightforwardly linked to machine operating and wafer surface characteristics [26], has already proved itself as a key tool for machine and process development.

The challenge for in-situ charge sensors is the complexity of the plasma environment [26] and noise from other factors such as UV-radiation and arcing. Some results have been reported with Langmuir probe arrays [27] and instrumented wafer disks [28], but the challenge of obtaining real-time, in-situ charging data with the directness and sophistication of the in-line CHARM monitors remains.

In-line monitoring of charging damage in sensitive dielectric layers will jump up to new levels of complication when the familiar characteristics of SiO2 are replaced by those of the new materials under consideration as advanced gate dielectrics, such as oxide-nitride multilayers, Ta_2O_5 and $BaSrTiO_3$ [1]. Measurements of such straightforward characteristics as leakage current and trap densities will require extensions of existing electrical test methods and new models for data analysis to account for the increased complexity of these materials.

INTEGRATION TO SIMULATION AND PROBLEM ANALYSIS ENGINES:

The application of TCAD to in-line characterization of doping technologies may not be obvious to the casual observer. Simulation is uniquely valuable for the evaluation of process sensitivity and process control issues. Simulation of the effects of process variation on complex yet fundamental parameters, such as MOS threshold voltage, provides a way to evaluate the requirements for process controls on individual process steps in a fashion that could only be obtained by exhaustive and exquisitely performed experimentation [15, 29].

The relationship of process control to transistor electrical properties was emphasized in the keynote paper of this volume. [30] "The physical measurements of dose must be correlated to the electrically active portion of the implant. There is an additional aspect to correlation of electrical and physical thickness measurements. The oxide thickness and implant dose must fall within a range of values so that the threshold voltage and leakage current variation across the die, across the wafer, and wafer to wafer is minimized." TCAD simulations offer the linkage needed to correlate in-line characterization of dosimetry and uniformity to the electrical requirements stated as threshold voltage and various drive and leakage currents.

Accuracy requirements for profile measurements to the needs of TCAD simulation have been estimated by several sets of authors [4,15,31,32]. These need to be stated in terms of a depth accuracy and a concentration accuracy. For junction depth, these range from a relatively conservative 20% effective estimate [15] to a 5% value. An interesting analysis has been done by Zeitsoff & Tasch [4]. They used a 10% assumption for both depth control and concentration for the simulation of manufacturing variances as they apply to electrical performance of transistors. (In the case of this report the transistors were 180nm design rule.) With that assumption, the depth control of the drain extension was the third most important parameter following CD and gate oxide thickness. These had similar assumptions of manufacturing performance and tolerance. The evaluation parameters were threshold voltage and drive current. Dopant concentration was significantly lower in importance in their multivariate experiment.

Using this analysis and the fact that the numbers are somewhat mid-range in the several other estimates of the needs for profile and concentration accuracy, one can adopt the 10% requirement as a consensus goal. Unfortunately, this is a consensus through being equally unsuitable to all parties. The concentration requirement is only somewhat pressing as it implies a 10% accuracy from roughly the mid 10^{19} range (for peak drain concentration) to the mid 10^{17} range (where the junction is formed with the channel). More difficult is the depth requirement. For 1997, the roadmap has the largest junction depth for the drain extension as 100nm. By the 10% requirement, this then needs to be measured to a 10nm accuracy (all variances included). By 2006 at the 100nm node, this would decrease to measuring a 40nm junction to an accuracy of 4nm. This is significantly beyond the known capabilities of today's measurements!

The National Technology Roadmap for Semiconductors has a single line describing the needs for 2-D dopant profile measurement [1]. It places the requirement for 2-D measurement at a 5 nm resolution for the present 250 nm process and this requirement decreases in a regular fashion to a 1 nm requirement for the 70 nm processes expected in 2012. All of these requirements are colored red, which indicates that there is no known method to achieve this need. One could expect that this requirement line is based around the needs of models describing the active regions. This model need tends to ask for at least 10 well defined points to be measured to establish the shallowest profile.

Fair requires 10nm depth accuracy in roughly 100nm profiles and states "Truly useful 2D process simulations require high accuracy of 10nm in the location of lateral diffusion contours. However, none of the simulation cases we studied gave anything near this accuracy over multiple concentration contours, especially in the low concentration ranges where such accuracy is most critical. We find that both 2D ion implantation models and 2D diffusion models are unsuitable in current simulation programs." [31]

SUMMARY

Doping requirements for CMOS devices have been defined out to 50 nm gate size devices [1] and provides a context for evaluation of in-line characterization needs. The wide range of doping processes used in CMOS fabrication, ranging in energies from ≤1 keV to over 1 MeV and in dose from $\leq 10^{11}$ to $\geq 10^{16}$ ions/cm^2, presents a significant challenge to provide adequate coverage in a high-accuracy, fast-response metrology. These processes were broadly described and the characteristics of these processes were described within the context of their challenges to the metrology techniques used to monitor them. A list of these techniques was presented along with some commentary as to characteristics of each technique. An added challenge is to provide characterization tools that function in-line with fab operations or, even better, in-situ with the process chamber, some of the challenges to in-situ metrology were discussed along with several leading applications where this may be key to future process control. Fortunately, a wide variety of materials characterization techniques have been adapted to the needs of doping technology process, these were included in the list of techniques under discussion. However, the rapid advance of IC process requirements, a fundamental characteristic of the IC business, has pushed most of these techniques to their limits of performance.

The list of critical areas for in-line monitoring starts with improved dosimetry for both shallow (source/drain and channel doping) and deep (CMOS well) junctions. Added to the dosimetry requirements is an increased need to monitor dopant profile shape and junction depth, especially for shallow junctions [3]. Monitoring of elemental and particulate contamination levels is critical for adequate yield and cost-effective operations. Because the ion ranges for ultra-shallow junctions are ≤10 nm and the smallest particle size that can be imaged with routine techniques is ≥180 nm, new technology is required to detect small size 'killer particles" that can mask low-energy (≤1 keV) ion beams. In-situ particle measurement will be challenged by both the particle size and the plasma environment of the implanter due to the need to measure particles that are transported down the beamline.

Control of wafer charging for gate dielectrics shrinking below 5 nm and increased antenna factors for large-area logic chips will drive the need for increased monitoring of net current flow from the ion beam-charge control plasma leading to development of in-situ sensing of plasma j-V characteristics.

The application of TCAD process simulation is uniquely valuable for the evaluation of process sensitivity and process control issues. Simulation of the effects of process variation on complex yet fundamental parameters, such as MOS threshold voltage, provides a way to evaluate the requirements for process controls on individual process steps in a fashion that could only be obtained by exhaustive and exquisitely performed experimentation Measurement of 2-dimensional doping profiles will be a key need for the successful use of models to predict processes and to actively evaluate deviations from process control. The present capability to do this lags far behind the stated needs found both in the NTRS and by concerned authors.

REFERENCES

1. National Technology Roadmap for Semiconductors (NTRS97), Semiconductor Industry Association, San Jose, CA, (Dec. 1997).

2. Materials and Process Characterization for Ion Implantation, eds. M.I. Current and C.B. Yarling, Ion Beam Press, Austin, TX, (1997).

3. Proceedings of 4th Inter. Symp. on Ultra-Shallow Junctions (usj-97), J. Vac. Sci and Technol. B 16 (1) (1998) 255-480.

4. Modeling of Manufacturing Sensitivity and of Statistically Based Process Control Requirements for a 0.18 micrometer NMOS Device, P.M. Zeitsoff, A.F. Tasch, et al., - This volume

5. T. Romig, . Materials and Process Characterization for Ion Implantation, eds. M.I. Current and C.B. Yarling, Ion Beam Press, Austin, TX, (1997).

6. W. Boyd, M. Lee, D. Wagner, T. Romig, J. Bennett, L.A. Larson, W. Johnson, L. Zhou, J. Vac. Sci and Technol. B 16 (1) (1998) 447-452.

7. Optical Densitometry Applications for Ion Implantation, J.P. Estives and M.J. Rendon, - This volume

8. M.I. Current, D. Lopes, M.A. Foad, J.G. England, C. Jones, D. Su, J. Vac. Sci and Technol. B 16 (1) (1998) 327-333.

9. Metrology Aspects of SIMS Depth Profiling for Advanced ULSI Processes, J.L. Hunter, A. Budrevich and S. Corcoran - This volume

10. Threshold Voltage Control using both Deposited and Mercury Gate MOS Capacitors, R.J. Hillard et al., - This volume

11. Verification of Carrier Density Profiles Derived from Spreading Resistance Measurements by Comparing Measured and Calculated Sheet Resistance Values, R.G. Mazur et al., - This volume

12 SRP, . Proceedings of 4th Inter. Symp. on Ultra-Shallow Junctions (usj-97), J. Vac. Sci and Technol. B 16 (1) (1998) 255-480

13. M.I. Current, E. Ishida, L.A. Larson, E.C. Jones, in Silicon Materials Science and Technology, Si-98, Electrochemical Society, San Diego, May, 1998.

14. M.D. Giles, S. Yu, H.W. Kennel, P.A. Packen, Solid State Technology, 41(2) (Feb. 1998) 97-104.

15. Process Simulation Challenges for ULSI Devices: A users Perspective, M.I. Current et al., Nucl. Instr. and Meth. in Phys. Res. B 102 (1995) 198-201

16. K. Goto, J. Matsuo, Y. Taka, T. Tanaka, Y. Momiyama, T. Sugii, I. Yamada, IEDM-97, 471-474.

17. E.C. Jones, E. Ishida, to appear in Mater. Sci. Eng. Rpts. (1998).

18. J. Grun, et al., Phys. Rev. Letters 78, pg. 1584 (1997), D.W. Donnelly, et al., Appl. Phys. Lett. 71, pg. 680 (1997)

19. G. Hobler, S. Selberherr, IEEE Trans. Comput.Aided Des. 8 (1989) 450.

20. R. von Criegen, F. Jahnel, R. Lange-Geisler, P. Pearson, G. Hobler, A. Simonescu, J. Vac. Sci and Technol. B 16 (1) (1998) 386-393.

21. P. Sferlazzo, D. A. Brown, S.E. Beck, and J. O'Hanlon , Ion Implantation Technology-92, Elsevier (1993) 565

22. J. Sedgewick, R. Hertel, G. Rizzo, Ion Implantation Technology-94, Elsevier (1995) 583

23. J.E. Sedgewick, A.J.Franceschelli, J. Hazelton, J. Malenfant, Ion Implantation Technology-96, IEEE (1997) 501.

24. B. Axon, L. Pivin, J. Irle, Ion Implantation Technology-96, IEEE (1997) 505.

25 W. Lukaszek, "The Fundamentals of CHARM-2", Tech. Note 1, Wafer Charge Monitors, 127 Marine Road, Woodside, CA 94926.

26 M.I. Current, M.C. Vella, W. Lukaszek, Ion Implantation Technology-96, IEEE (1997) 53-56.

27. P. Kellerman, V. Benveniste, Ion Implantation Technology-96, IEEE (1997) 360-363.

28. M.E. Mack, P. Barschall, P. Corey, S. Satoh, S. Walther, Nuclr. Instr. Meth. B74 (1993) 287-290.

29. H. Glawischnig, in: Ion Implantation Technology Handbook, ed. J.F. Ziegler, (Elsevier, 1994) pp. 223-269

30. Metrology Needs for the Semiconductor Industry Over the Next Decade, M. Melliar-Smith and A.C. Diebold, this volume

31. Ultra-Shallow 2D Dopant Profile Simulation Versus Experimental Measurement in the Low Thermal Budget Regime
R. B. Fair and M. Shen, J. Vac. Sci and Technol. B 16 (1) (1998)

32. Duane, Langler & Larson in 2D Dopant Profiling Workshop Notes

METROLOGY OF 300 MM SILICON WAFERS: CHALLENGES AND RESULTS

P. Wagner

Wacker Siltronic AG, P.O. Box 1140, 84479 Burghausen, Germany

Challenging requirements have to be met by metrology tools for 300 mm wafers and technology generations ≤0.25 µm in near future. Measurement equipment for some specific wafer parameters presently operates already at its limits and will not be able to meet the future requirements. New tools therefore were or are currently developed. The future requirements are outlined and examples for new approaches are presented.. New geometry measurement tools and scanning surface inspection systems are available now and their performance is encouraging. Repeatability of less than 10 nm (1 sigma) for flatness measurement is obtained by a transmitted light interferometer. The sensitivity of surface inspection tools has been improved to detect surface flaws <100 nmLSE and these tools allow to distinguish between particles and other surface defects. The results presented provide also an impression about the state of development of 300 mm wafers.

1 INTRODUCTION

The conversion to 300 mm Si wafers certainly imposes particular technological and financial challenges on the entire industry involved. Equipment and manufacturing processes have to be developed in short time and tuned to the specific applications. These tasks can be completed successfully only if appropriate measurement equipment is available in time. This infers that measurement tools have to be developed, tested and manufactured well in advance of device or wafer production and specifications for these tools have to be anticipated. The measurement tools also have to match the increasingly tighter demands regarding performance of wafers for design rules ≤ 0.25 µm occurring in parallel with the change in wafer size.

The level of challenge regarding upgrading, modifying or developing equipment can vary considerably for the different tasks to be met by the measurement tools. The spectrum of measurement and analytical techniques used for Si wafer development and manufacturing includes invasive and non-invasive, local and global, highly spatially resolving and entire wafer averaging techniques which have to provide electrical, chemical, topological, morphological parameters of wafer surface and bulk.

In the present paper the requirements for Si wafers and corresponding measurement equipment are outlined. Recent results obtained with new measurement tools are reported and compared with the requirements. Finally, future requirements are discussed and anticipated.

2 SI WAFERS FOR FUTURE TECHNOLOGY GENERATIONS

The evolving requirements for future Si wafers are anticipated and summarized in the National Technology Roadmap for Semiconductors (NTRS) /1/ of the Semiconductor Industry Association (SIA). The new version of this document now extends until the year 2012 and the design rule of 0.05 µm. It is assembled from contributions of the device manufacturers as well as the Si wafer suppliers and attempts to anticipate future requirements for materials and technologies. The model-based future requirements for Si wafers are mainly driven by yield considerations of the device manufacturers which are related to the perfection of the Si wafers. In general the number of "defects" is assumed to decrease during the years lying ahead. The surface contamination – metallic, organic, and particulate – is required to decrease as well as the number of surface and bulk defects. In parallel the allowed deviations of the wafer geometry from a perfectly flat disc with constant thickness are shrinking. This development is driven mainly by the decreasing depth of focus budget of steppers and the increasing number of CMP process steps (chemo-mechanical polishing) during device manufacturing. The most important parameters are listed in TABLE 1 for the time period 1997-2003. Afterwards the diameter of Si wafers is assumed to increase to 450 mm and the critical parameters decrease further.

CP449, *Characterization and Metrology for ULSI Technology: 1998 International Conference*
edited by D. G. Seiler, A. C. Diebold, W. M. Bullis, T. J. Shaffner, R. McDonald, and E. J. Walters

TABLE 1: Extract from NTRS of SIA regarding specifications for the starting material for the technology generations ≥130 nm..

Year of first product shipment	1997	1999	2001	2003
Technology Generation	250 nm	180 nm	150 nm	130 nm
Wafer diameter / mm	200	300	300	300
Edge exclusion/ mm	3	2	2	2
Critical surface metals / cm^{-2}	≤2.5E10	≤1.3E10	≤1E10	≤7.5E9
SFQR / nm	≤250	≤180	≤150	≤130
Site size / mm^2	22*22	25*32	25*34	25*36
Threshold for Localized Light Scatterers (LLS) / nm	125	90	75	65
Number of LLS / #/wfr	≤172	≤192	≤178	≤96
Number of particles / #/wfr	≤41	≤60	≤47	≤38
Gate oxide defects DRAM / cm^{-2}	≤0.06	≤0.029	≤0.026	≤0.014
Structural defects in epi layers / #/wfr	≤1	≤2	≤1.8	≤1.6
Frontside roughness / nm	≤0.15	≤0.1	≤0.1	≤0.1

The abstract numbers of TABLE 1 are illustrated by comparing a 300 mm wafer with a hypothetical wafer with a diameter corresponding to the diameter of the earth, which is called equatorial wafer in this example. This equatorial wafer would have a diameter of 12900 km and an area of 130698108 km^2. Two adjacent Si atoms would then be away from each other by 10.1 mm (~0.4 inch). Expressing the numbers in TABLE 1 in the frame of this equatorial wafer is performed in TABLE 2 and results in quite surprising numbers. Defect densities are more easily visualized when viewed on this length scale.

3 REQUIREMENTS FOR METROLOGY TOOLS

Requirements for metrology tools usually are derived from the critical wafer parameters assuming that three times of the standard deviation of the measured values is

TABLE 2: Parameters of an "equatorial" wafer

Technology Generation	0.25 µm	0.18 µm	0.15 µm	0.13 µm
Year of first product shipment	1997	1999	2001	2003
Wafer diameter / km	12900	12900	12900	12900
Edge exclusion/ km	129	86	86	86
Mean distance of metal atoms / m	2.72	3.77	4.30	4.97
SFQR / m	10.75	7.74	6.45	5.59
Site size / km	946.00	1216.22	1253.65	1290.00
Threshold for Localized Light Scatterers (LLS) / m	5.4	3.9	3.2	2.8
Area per LLS	Finland	Texas	~ Turkey	~ Tchad
Area per particle	Indonesia	Greenland	Argentina	~ India
Area per GO defect	0.5*USA	0.5*Arctic Sea	~ Australia	~ Antarctica
Roughness / mm	6.5	4.3	4.3	4.3

less than 10% of the allowed tolerance (specification range) corresponding to a precision to tolerance P/T ≤0.1 (3 sigma). Such numbers as derived from the critical parameters in TABLE 1 are listed as a guideline in TABLE 3 and obviously they are extremely challenging. These values need to be supplemented by requirements important for the operation of tools in a production ambient but not specified in the NTRS, like throughput, reliability, robustness, operator interface etc. These requirements depend on the specific mode of operation of the measurement tools and are not discussed here.

Comparing these requirements with some basic physical parameters of Si wafer helps to illustrate them. A value of 6 nm as is required as repeatability for SFQR for the 0.18 µm technology generation corresponds to about 25 atomic distances of the Si lattice or about 44 atomic step heights of a 1-0-0-Si surface. It is also one hundredth of the light wavelength of the HeNe-laser which is often used for interferometric applications. A temperature increase of 1°C results in a thickness change of a 300 mm wafer of about 1.8 nm, approximately one third of the required repeatability of 6 nm. The thickness of the native oxide layer of a hydrophilic Si surface is also about 1-2 nm. Inhomogeneous temperature of a wafer or variations in the native oxide layer therefore can contribute to flatness deviations.

Tools with capacitive sensors are widely used for measuring shape and flatness of Si wafers. The distance between the sensor and the wafer surface is approximately 0.5 mm in such tools. A variation of 6 nm then corresponds to a capacitance change of about 10 ppm. The capacitance of the sensor/wafer pair of a few tenths of 1 pF therefore has to be measured with a repeatability – or noise – of a few aF! Ten ppm of an assumed capacitance of 0.3 pF correspond to 3 aF. Such a small change in capacitance then results in a change of capacitor charge of only 20 electrons per Volt!

The assumptions regarding sensor capacitance are based on a capacitor area of 4x4 mm^2 and a distance of 0.5 mm between sensor and wafer surface. Higher spatial resolution is not demanded explicitly by the NTRS but implicitly by anticipating increasingly smaller edge exclusion zones. However, reducing sensor area also diminishes the capacitance of the sensor/wafer system which can be compensated only partly by reducing the distance between sensor and wafer surface as long as wafer handling is not inhibited.

The response of a geometry measurement tool depends very much on its spatial frequency bandwidth – or lateral resolution -- similarly to roughness measurement or surface inspection. Higher lateral resolution allows to see finer features of a surface and usually results in larger numbers for flatness or thickness variation. Achieving more compatibility between instruments of different suppliers would be a major progress and would help to stimulate competition and improvements.

For SSIS (Scanning Surface Inspection Systems)

TABLE 3: Specifications required for metrology tools to be met, derived from the critical parameters in TABLE 1. The values for repeatability refer to 1 sigma and are derived with a P/T ≤0.1 (3 sigma).

Technology generation	250 nm	180 nm	150 nm	130 nm
Wafer diameter / mm	200/300	300	300	300
Edge exclusion / mm	≤ 3	≤ 2	≤ 2	≤ 2
Laserrmark exclusion	yes	yes	yes	yes
Flat/notch detection	yes	yes	yes	yes
Repeatability critical surface metals / cm^{-2}	≤ 0.8E9	≤ 0.44E9	≤ 0.3E9	≤ 0.25E9
Repeatability SFQR / nm	≤ 8	≤ 6	≤ 5	≤ 4
LLS Threshold (50% capture rate)[1] / nm	110	80	65	60
Repeatability LLS counts / #/wfr	≤ 6	≤ 6	≤ 6	≤ 3.3
Repeatability particle counts / #/wfr	≤ 1.3	≤ 2	≤ 1.7	≤ 1.3
Repeatability Gate Oxide Defects / cm^{-2}	≤ 0.002	≤ 0.001	≤ 0.0008	≤ 0.0005
Repeatability Structural Defects in Epi Layers / #/wfr	≤ 0.03	≤ 0.07	≤ 0.06	≤ 0.05
Repeatability roughness / nm	≤ 0.05	≤ 0.03	≤ 0.03	≤ 0.03

[1] a standard deviation of 3% is assumed for peaks in the counts histogram of SSIS for PSL-spheres.

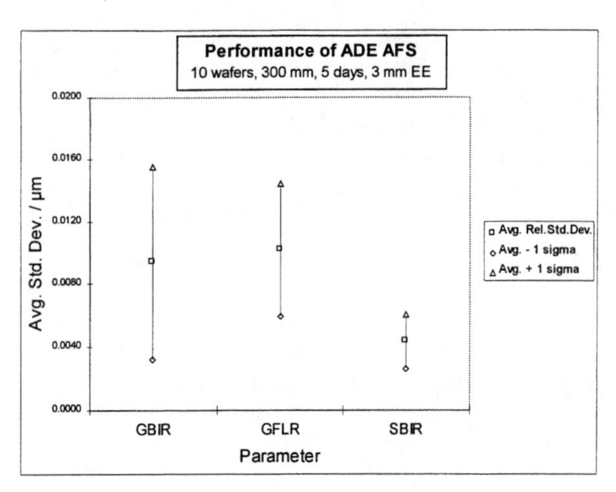

FIGURE 1: Performance of ADE AFS flatness measurement tool for 300 mm wafers (EE = edge exclusion).

sensitivity, discrimination between LLS and haze, coordinate accuracy and repeatability are the challenges to be met.. Several thousand photons are collected by the detector if a small LLS is hit completely by the laser spot during scanning assuming realistic laser power, laser efficiency, laser spot size, scanning time and scattering cross section of LLS. Much less photons, however, are scattered if the LLS is not hit completely but only with the tail of the intensity distribution within the laser spot, which is the general case, or if the LLS threshold becomes smaller. Repeatability and reproducibility are significantly influenced when LLS are not hit in precisely the same way during successive scans. The accuracy of hitting the same coordinates therefore becomes extremely important and imposes quite severe requirements regarding positioning accuracy of the wafer and the laser spot, respectively, in the SSIS during scanning. Repeatability and reproducibility, respectively, did not differ very much from an almost statistical behavior for SSIS of previous generations. This is discussed in more detail in sect. 4.1.2.

Another aspect to be considered by future SSIS is the discrimination of the various surface flaws. Presently available tools are capable to distinguish between LLS, extended defects or XLS (Extended Light Scatterers, sometimes called "area defects") and scratches. The characteristic defects of epitaxial layers, like stacking faults, hillocks, etc., are not distinguished by SSIS. The tools report these defects to be either LLS in general or large LLS. More detailed information has to be obtained then by using review stations with appropriate digital image processing. (see section 4.3).

4 RESULTS

Different approaches were used in solving the metrology problems for 300 mm Si wafers. Methods where the wafers are broken or cut in pieces anyhow need not to be adjusted to 300 mm wafers. Such methods include e.g. Deep Level Transient Spectroscopy, Low Temperature Photoluminescence, Transmission Electron Microscopy. Other techniques need only to be upgraded regarding the handling of 300 mm wafers, in particular techniques used in the production process of wafers. These techniques include e.g. oxygen measurement with FTIR, film and epi thickness measurement, electrical resistivity, microscopy. Such measurements can be performed with pieces of 300 mm wafers, at least for a transition period. Finally, there are the techniques for which the new diameter is a new qualitative aspect and not only a quantitative one and where the entire wafer has to be measured. This includes certainly the geometry and shape measurement of wafers, including the wafer edge profile, and surface contamination and defects. Examples for new measurement tools developed for 300 mm wafers and for the increased requirements are reported in the next sections.

4.1 Geometry Measurement

Particularly severe requirements are specified for the wafer geometry – thickness variation and flatness. The values as given in TABLE 1 refer traditionally to measurement tools equipped with capacitive sensors. Such a tools (ADE AFS) is now available for 300 mm and displays excellent repeatability. The capacitive sensors, however, have a very limited spatial resolution of only a few mm and therefore average characteristically over an area of 10-20 mm^2. Therefore these tools cover the spatial frequency range $<\sim 0.25\text{-}0.5\ mm^{-1}$. The entire wafer is scanned by such tools, however, with limited spatial resolution. Tools providing information about the spatial frequency range $> 0.1\ mm^{-1}$ and measuring the entire wafer were not available until recently. E.g. profilers (mechanical or optical) or standard phase shift interferometers had to be used to close the gap in the

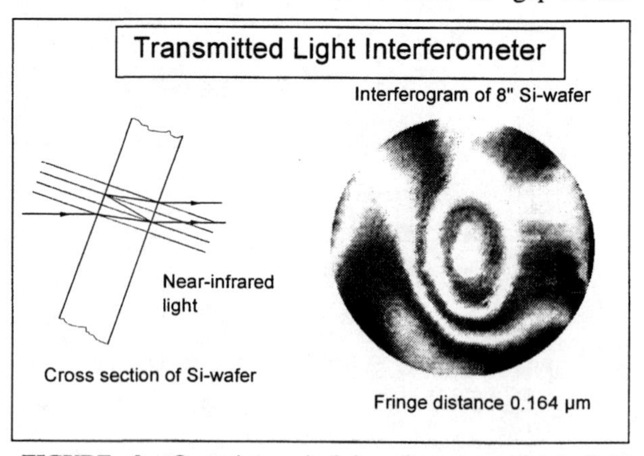

FIGURE 2: Operating principle of a transmitted light interferometer

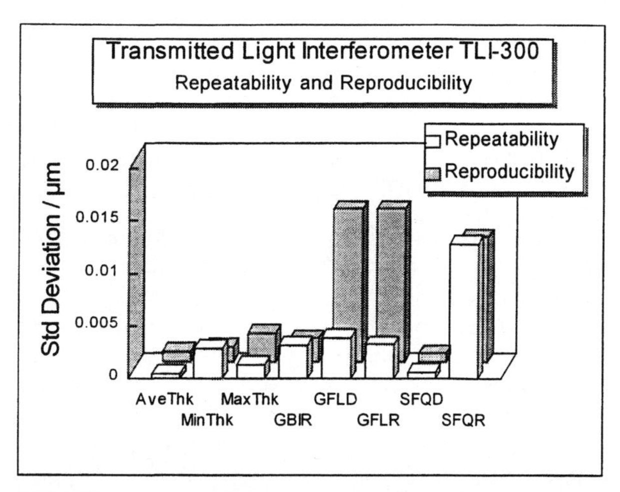

FIGURE 3: Repeatability and reproducibility of the transmitted light interferometer TLI-300. One 300 mm wafer was measured ten times with an edge exclusion zone of 1 mm. The site size was 25x25 mm^2.

spatial frequency spectrum between flatness measurement and tools like Atomic Force Microscopes. A few tens of nm SFQR are expected to be specified in this spatial frequency range. This corresponds to a precision of a few nm only allowed for the measurement tools. This certainly is a real challenge remembering that the distance of Si atoms in the crystal lattice is only 0.235 nm. A repeatability of about e.g. 2 nm is equivalent to about only eight to nine atomic distances (see sect. 3).

New tools for measuring the wafer geometry based on optical interferometry were developed recently to overcome the drawbacks of the capacitive sensors. Their operating principles, basic specifications and results obtained with them are reported in the following sections. The already impressing results were so far obtained with

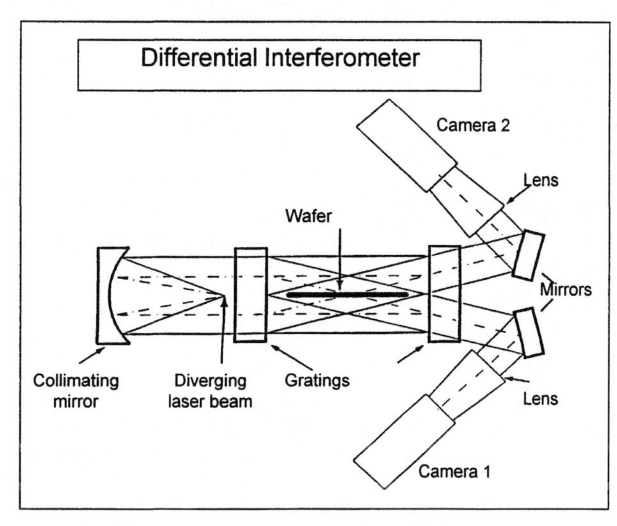

FIGURE 4: Schematics of a differential interferometer

prototypes or first-of-a kind tool. It is easy to imagine that there is a lot of room for further development of these tools e.g. with respect to spatial resolution, throughput, reliability. The results presented in the next sections also outline the current status of development of 300 mm wafers.

4.1.1 Transmitted Light interferometer (TLI)

The worldwide acceptance of a polished or highly glossy wafer backsurface was an important paradigm change which took place in the conversion to 300 mm wafers. This allows a new concept for a new transmitted light interferometer (TLI)/2, 3/. Not too highly doped Si is transparent in the near infrared wavelength range. Interferograms can easily be generated by superposing the directly transmitted light and the twice internally reflected light of a coherent light source (FIGURE 2). The high refractive index of Si in the near infrared (approximately 3.4) results in a high sensitivity of this technique. A pair of fringes corresponds to a thickness variation of only 0.164 µm when a HeNe-laser is used as light source.

A very high sensitivity can be obtained by applying a kind of phase-shift interferometric technique – recording e.g. five successive interferograms where the angle of the wafer has been changed very slightly. A commercial version of such an interferometer – Nanopro TLI-300 - is available since about one year /4/. The data sheet claims an accuracy of 20 nm for local flatness and a corresponding repeatability of 10 nm (3 sigma) at a spatial resolution of 0.6 mm. This is much better than is required in TABLE 3. Repeatability and reproducibility for the standard geometry measures – terms as defined by SEMATECH /5/ -- are also found to lie in this range (FIGURE 3). This is particularly remarkable as the measurements were performed with an edge exclusion zone of only 1 mm.

Unfortunately, TLIs can be used only for doubleside polished wafers which are not too highly doped. Wafers with rougher surfaces or highly doped wafers cannot be measured with this tool. Another approach was used for a new kind of interferometer for such wafers. This is outlined in the next section.

4.1.2 Differential Interferometer (DI)

The reflectivity of rough surfaces increases with increasing angle of incidence. This well know fact is utilized by the differential interferometer /6/ where both sides of a wafer are measured in parallel by grazing incidence interferometry (FIGURE 4). The effective wavelength of the light used is increased considerably by the large incidence angle and the required sensitivity is obtained by using a phase-shift technique. The data sheet of the tool claims an accuracy of ±30 nm for local flatness

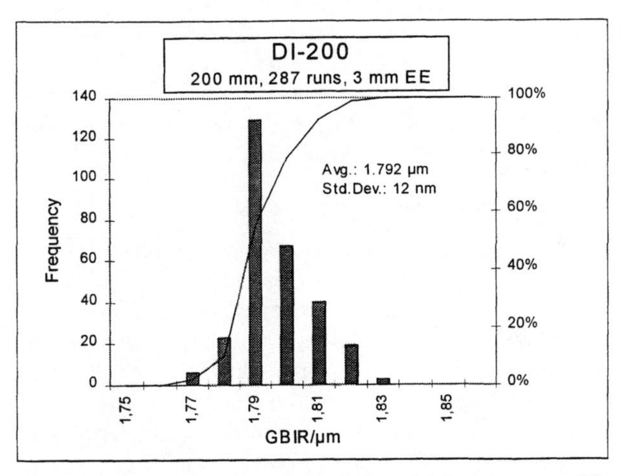

FIGURE 5: Repeatability of the differential interferometer DI-200. The solid line refers to the relative cumulative frequency as given on the axis on the right (EE = edge exclusion).

and a corresponding repeatability of 30 nm (10 meas., 3 sigma) combined with an impressive spatial resolution of 0.4 mm. Preliminary results obtained with a prototype of such a tool are very encouraging (FIGURE 5) and agree reasonably well with the claimed values.

4.2 Scanning Surface Inspection Systems (SSIS)

All major suppliers of SSIS presently market tools for 300 mm wafers featuring threshold sensitivities ≤ 0.1 µm LSE. All systems available also provide multiple dark field channels for diffusely scattered light and a bright field channel for recording slight variations of the specularly reflected light or a technique providing essentially the same information. The multiple dark field channel may be utilized to distinguish particles from other LLS (Localized Light Scatterers). Examples for the performance of one of these advanced SSIS are presented in the next paragraphs.

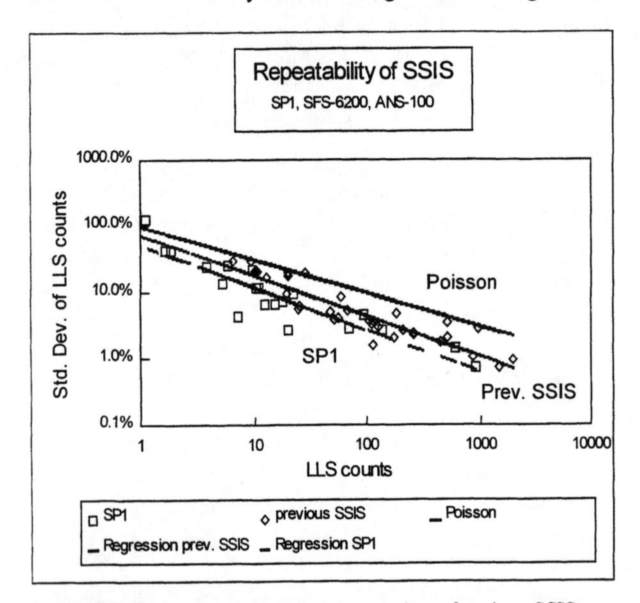

FIGURE 6: Repeatability of LLS-counting of various SSIS as obtained from 10 measurements. DFN- and DFW-data of the SP1 are combined. The lower regression line refers to SP1, the upper solid line indicates a purely statistical behavior.

A major concern of the older SSIS was their limited repeatability and reproducibility not to mention their mutual incompatibility with respect to reported LLS sizes and haze /2/. Repeatability and reproducibility of a Tencor SP1 are in the ≤ 10% range with the exception of bins with very low counts (FIGURE 6). The SP1 features two dark field channels – DFN (dark field narrow) and DFW (dark field wide), denoting the respective solid angles of the two channel – and a DIC-channel (differential interference contrast), designed to detect long wavelength surface irregularities, similar to so-called bright field channels of other systems. Correspondingly, the performance of both the DFN- and DFW-channels is reported in FIGURE 6.

The capability of present SSIS to distinguish particles and pits – or so-called COP (Crystal Originated Particles /7/) – is again demonstrated for the SP1. LLS of a 300 mm wafer were investigated regarding their stationary behavior during two successive measurements with an intermediate particle cleaning treatment. Stationary LLS (S-LLS) were identified as COP, non-stationary LLS were identified to be particles (N-LLS). In a next step the LSE-sizes (Latex Sphere Equivalent) as reported by the DFN- and the DFW-channels, respectively, were compared for each LLS. It turned out that both channels reported about the same size for particles whereas for COP the DFW reported a smaller size as compared to the DFN (FIGURE 7) /8/. The size difference for COP is about 20 nm in the average for

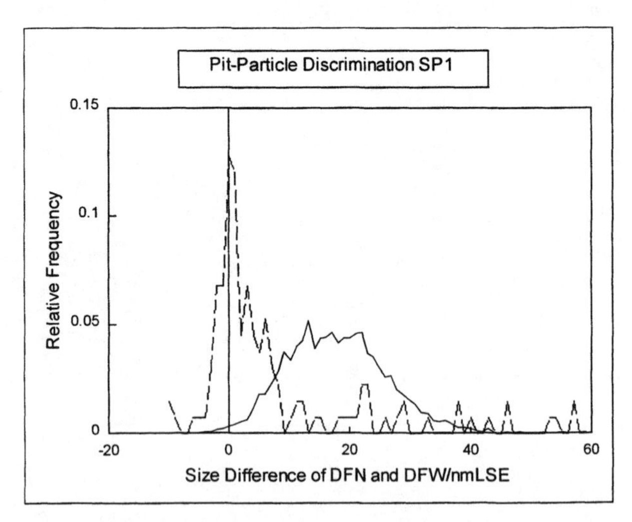

FIGURE 7: Relative frequency of particles and pits as identified by subsequent measurements with intermediate cleaning. The broken line refers to particles (non-stationary LLS), the solid line to pits (stationary LLS).

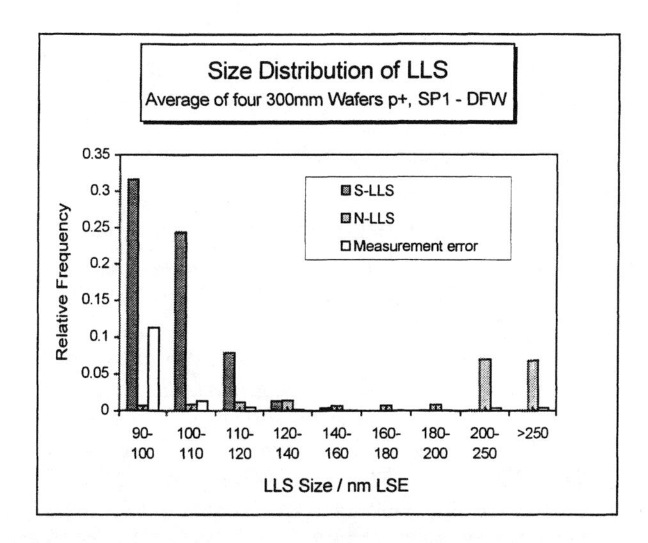

FIGURE 8: Frequency of particles (N-LLS) and stationary LLS (S-LLS) averaged over four 300 mm wafers (p+) relative to the total amount of LLS.

LLS≥0.12 µm LSE with the DFW reporting the smaller size. The system therefore is capable to distinguish COP and particles with a high probability.

This effects results from the still sufficiently large size of COP and particles. Raleigh scattering – or dipole scattering – of light occurs only for LLS smaller than about one tenth of the wavelength of the laser light. Most of the commercially available SSIS utilize Ar-ion laser with a wavelength around 500 nm. The limit for Raleigh scattering therefore is around a LLS size 50 nm. Particles or COP investigated currently are in the size range around or above 100 nm and information about their shape is contained in the angular distribution of the light scattered by them. COP appear to scatter light slightly more preferentially closer to the specularly reflected beam as compared to particles /7/. The ratio of light collected by the DFN- vs. the DFW-channel of the SP1, or comparable tools, is therefore slightly higher for COP as compared to particles and a larger size is reported by DFN-channel. SSIS are calibrated with specific particles, PSL-spheres (Polystyrene Latex) with a well-defined diameter. Therefore the different dark field channels of the SP1 have to report by definition the same size for particles.

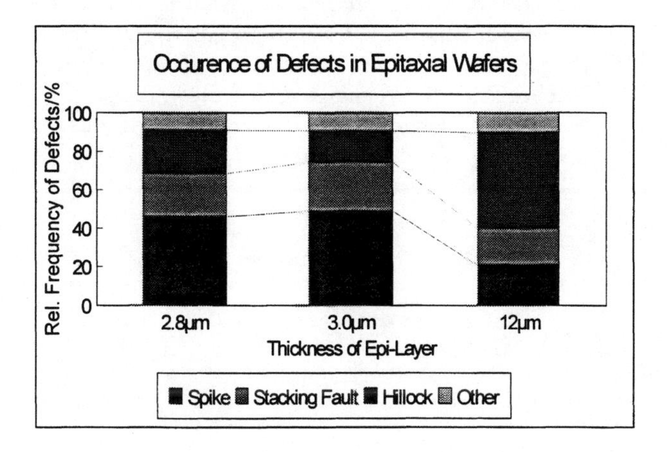

FIGURE 9: Relative frequency of occurence of defects in epitaxial wafers reported by SSIS to be large LLS.

Another, quite significant, information resulted from these investigations. The identification of N-LLS (particles) and S-LLS allows to sort them separately in the various size bins (FIGURE 8). The measurement errors indicated in this figure refer to the limited repeatability found in successive measurements without intermediate cleaning of the wafers. Most of the very small LLS turn out to be due to surface damage in this specific case and only very few particles are found on the wafers in the size range 0.09-0.2 µm LSE. Otherwise only very few COP are observed on p^+-wafers /9/.

A similar result with respect to distinguishing N-LLS and S-LLS is obtained when p^--wafers are investigated. However, the overwhelming number of S-LLS then turns out to be COP. Most particles found on wafers are larger than 0.2 µm LSE. This demonstrates that even the present cleaning technologies are capable to remove small particles from Si wafers, provided they occur at all.

4.3 Review Stations

Present SSIS do not provide information about the nature or shape of the surface flaw reported as LLS with the exception of the above mentioned discrimination of COP and particles, scratch detection and identifying large or extended LLS. The quite extended, characteristic defects of epitaxial layers (several µm in diameter, depending on the thickness of the epitaxial layer) are usually reported as LLS or XLS but are not discriminated from COP or from each other. Therefore review stations have been used for this task which are able to distinguish

Total detection rate for structural epi defects as large LLS

Thickness/µm	2.8	3	12
DFN	79%	82%	62%
DFW	56%	60%	32%
SS6200	57%	62%	31%

FIGURE 10: Rate of detection for structural epi defects as large LLS by SP1 and Surfscan 6200 for defects in epitaxial layers of various thickness on 200 mm wafers.

the various kinds of structural epi defects. /10/. The relative frequency of occurrence for specific epitaxial defects was evaluated with such a tool (Jenoptik DefectFinder 2000) for 200 mm wafers where the thickness of the epitaxial layer was varied between 2.8 and 12 μm (FIGURE 9). So-called "spikes" appear to be the most frequent defects on wafers with epitaxial layers with a thickness of about 3 μm followed by stacking faults.

The detection rate of SSIS for epitaxial defects also can be determined by using such a review station as reference tool. This is outlined in FIGURE 10 where the detection rate as XLS is listed for the DFN- and DFW-channels of the SP1 and the Tencor Surfscan 6200 as compared to the DefectFinder 2000. The DFW-channel of SP1 appears to be comparable to the Surfscan 6200, whereas the DFN-channel detects the epitaxial defects with a higher reliability.

5 CONCLUSIONS

A wide range of measurement equipment for 300 mm wafers was developed within the past two years and is available now. Challenging goals were met in those cases where improvement in quality was required in addition to the blow-up in wafer diameter. Most, if not all, measurement tools utilized in wafer manufacturing are available even if in some cases performance and reliability might needed to be improved in particular regarding <0.18 μm design rules. This includes in particular wafer handling without vacuum endeffectors but with edge gripping devices, identifying surface flaws beyond the particle/COP discrimination, increasing sensitivity, repeatability and spatial resolution (for geometry measurement equipment). Ease of operation, high throughput and low cost-of-ownership is always a concern of the users of measurement equipment. TABLE 1 and TABLE 3 may serve as guidelines for the requirements of future metrology tools. More detailed lists – although not completely up to date – regarding future requirements including the operational issues of measurement tools not discussed here can be found in /2/. Most of the values listed in TABLE 1and TABLE 3 are extremely challenging and require certainly focussed joint efforts of manufacturers of such tools, Si vendors and - last but not least - standardization organizations. Standards are the basis for a common understanding and language for material specifications and exchange of measurement data.

ACKNOWLEDGEMENTS

The willingness of H.A. Gerber, Ch. Maithert, F. Passek and R. Velten to provide unpublished data and their assistance is highly appreciated. This work has been supported by the Federal Department of Education, Science, Research, and Technology of The Federal Republic of Germany under contract number 01 M 2973 A. The author alone is responsible for the contents of this publication.

REFERENCES

1. National Technology Roadmap for Semiconductors (NTRS) of Semiconductor Industry Association (SIA) 1997
2. P. Wagner, H.A. Gerber, R. Schmolke, R. Velten, *Int. Symposium on Optical Science, Engineering, and Instrumentation, Conference on Flatness, Roughness, and Discrete Defect Characterization for Computer Disks, Wafers, and Flat Panel Displays*, Denver, August 96, SPIE Proceedings Vol. 2862, p. 152-162
3. Patent No. DE 39 31 213 C2
4. NANOPRO Transmitted Light Interferometer TLI-300
5. *Glossary of Terms Related to Particle Counting and Surface Microroughness of Silicon Wafers*, Technology Transfer 9508291A-TR, SEMATECH, Inc., Austin, 1995
6. Patent No. DE 196 02 445 A1
7. P. Wagner, H.A. Gerber, D. Gräf, R. Schmolke, M. Suhren, *Int. Symposium on Optical Science, Engineering, and Instrumentation, Conference on Flatness, Roughness, and Discrete Defect Characterization for Computer Disks, Wafers, and Flat Panel Displays*, Denver, August 96, SPIE Proceedings Vol. 2862, p. 18-27
8. F. Passek et al., to be published
9. M. Suhren, D. Gräf, U. Lambert, P. Wagner, *High Purity Silicon IV*, The Electrochem. Soc. Proc.Vol. PV96-13, The Electrochem. Soc., Pennington, N.J., 1996, C.L. Claeys, P. Rai-Choudhury, P. Stallhofer, J.E. Maurits, Eds., p.132
10. F. Passek, R. Schmolke, U. Lambert, G. Puppe, P. Wagner, *Crystalline Defects and Contamination: their Impact and Control in Device Manufacturing II*, The Electrochem. Soc. Proc. Vol. PV97-22, The Electrochem. Soc., Pennington, N.J., 1997, B.O. Kolbesen, P. Stallhofer, C.L. Claeys, F. Tardiff, Eds., p. 438

Metrology Issues for Processing of 300 mm Wafers

Kenji Watanabe and Atsuyoshi Koike

Semiconductor Manufacturing Technology Center, Semiconductor & IC Division, Hitachi, Ltd.
6-16-3 Imai, Omi-shi, Tokyo 198-8512 Japan

Construction of IC lines using 300 mm wafers will begin in 1998-99 for pilot lines, with construction of mass production lines in 2000 or later. These lines will most likely be introduced with design rules in the range from 250 to 180 nm. Difficult challenges for these feature sizes include critical dimensions, registration, thin film measurement, and defect inspection. At the same time, metrology is expected to play a major role in controlling costs, increasing manufacturing flexibility, and decreasing cycle time in addition to assisting in the reduced emission of global warming gases.

INTRODUCTION

For many decades, the bit price of DRAM has decreased while the density has increased in such a way as to maintain the price per chip virtually constant. As a result, the development of new markets has fueled the growth of the industry during this period. Although the rate of decrease in bit price is slowing, the price of each generation is still significantly less than the price of combining chips from the previous generation to obtain the same amount of memory.

To maintain this growth, future fabs must continuously improve productivity, reduce time to market, and become environmentally benign, while process complexity is increasing. This will require the introduction of 300 mm wafers, minienvironments, and stringent exhaust gas and chemical treatments.

One major problem in current semiconductor production is low productivity. Actual operating time of production equipment is only about 50% of the total time while the ratio of actual wafer processing time to total turn around time (TAT) is less than 30%. In addition, the usage ratio of materials including silicon, photoresist, chemicals and the like is less than 1%.

This provides a tremendous opportunity for produc-tivity increases as semiconductor processes move from the laboratory to true manufacturing. Future manufacturing fabs using large diameter wafers will feature high flexibility, advanced minienvironments, super short cycle time, quick TAT, and material recyclability.

300 mm FAB STRATEGY

The fundamental 300 mm fab strategy is based on the assumption that the per chip cost of equipment to produce chips on a 300 mm wafer will be 70% of that to produce chips on a 200 mm wafer. The investment cost for new fabs is very large. If the current value of the critical attributes is taken as 1 for a 200 mm fab, our suppliers tell us that the value for a 300 mm fab will be that listed in the third column of Table 1. This results in an investment efficiency for the 300 mm fab that is only a modest 18 to 37% larger than that of a 200 mm fab. To achieve an investment efficiency 100% more than that of a 200 mm fab, the critical attributes will have to have the more aggressive target values listed in the last column of Table 1.

TABLE 1. 300 mm Investment Efficiency Target

Attribute	Current		Target
	ϕ 200 mm	ϕ 300 mm	ϕ 300 mm
Equipment Cost	1	1.3-1.5	1.1-1.2
Throughput	1	0.8	1.0
Foot Print	1	1.5	1.0
Chips/Wafer	1	2.5	2.5
Investment Efficiency	1	1.18-1.37	2.08-2.27

It is well known that the relative investment efficiency increases to a saturation value with the quantity of wafers per day processed through the fab. Calculations show (1) that the slope of this curve decreases as the throughput and equipment costs increase. Achievement of the target values of the critical attributes listed in Table 1 could allow a 200 wafer per day fab to operate with the same relative investment efficiency as an 800 wafer per day fab with the "current" values of these critical attributes.

Another critical issue facing 300 mm fabs is the threat of global warming due to the emission of greenhouse gases CO_2, CH_4, N_2O, SF_6, HFC (such as CHF_3, $C_2H_2F_4$, etc.) and PFC (such as CF_4, C_2F_6, C_4F_8, etc.). The final plan of the Kyoto Protocol issued December 11, 1997, calls for reduction in usage of these gases from 1990 levels by 6% in Japan, 7% in the USA, and 8% in the EU. Consequently,

CP449, *Characterization and Metrology for ULSI Technology: 1998 International Conference*
edited by D. G. Seiler, A. C. Diebold, W. M. Bullis, T. J. Shaffner, R. McDonald, and E. J. Walters

TABLE 2. Measurement Precision Necessary to Meet Device Roadmap Projections

Year	1997	1998	1999	2000	2001	2002	2003	2004	2005	2006	2007 ...
DRAM half pitch (nm)	250		180		150		130			100	
DRAM generation	256 Mbit		1 Gbit		Note a		4 Gbit			16 Gbit	
Wafer diameter (mm)	200		300		300		300			300	
CD control/3σ precision (nm, nm)	20/4		14/3		12/2		10/2			7/1.4	
Overlay control/3σ precision (nm, nm)	85/9		65/7		55/6		45/5			35/4	
Critical particle size (nm)	125		90		75		65			50	
Defect size (monitor) (nm)	200		150		120		100			80	
Defect size (analysis) (nm)	150		120		100		80			60	

a. This node was inserted for trend purposes only.

the new 300 mm fabs will be expected to reduce the emission of these gases. Laws such as the one being submitted to the Japanese Diet by the Environment Agency of Japan are enacted will legally mandate these reductions.

METROLOGY ISSUES

The trends projected in the 1997 edition of the National Technology Roadmap for Semiconductors (1) are based on continuation of the traditional scaling of feature size, chip area, wafer diameter, film thickness, and other related parameters to maintain the exponentially declining price per bit. These trends pose significant metrology challenges, especially as related to reduced feature size, cost, and turn around time and increased wafer diameter.

Feature Size Reduction

Table 2 summarizes some of these challenges related to the reduction in feature size over the next several device generations. For example, even at 250 nm, the patterns are not resolvable by typical optical microscope as illustrated in Fig. 1. Fig. 2 illustrates the fact that modern scanning surface inspection systems can separate 90-nm (LSE) particles from 90-nm crystal originated pits (COPs) by comparing the wafer map obtained with vertical incidence (which displays both the particles and pits) with one obtained at oblique incidence (which minimizes the contribution from COPs. The situation is less favorable for detection of particles or other defects on a patterned product wafer. Here the practical sensitivity limit attainable with modern detection systems on typical patterned DRAM product wafers is 300–500 nm while it is desired to be able to detect particles and other defects in the size range from 120 to 150 nm.

Another issue related to the feature size reduction is the definition of flatness of the wafer. For 200 mm wafers, for which the lithographic pattern is generated by a instrument that steps from site to site on the wafer, the site flatness is defined as the maximum deviation within the site from a reference plane fitted to the front surface within the site by least squares regression analysis (verify definition). This parameter is identified as SFQR in the SEMI specification for polished silicon wafers (2); it is conventionally determined by mapping with a capacitance sensor, as shown in Fig. 3(A). In scanning steppers, the field exposed is a narrow stripe with a width equal to that of the site. This is shown in Fig. 3(B) for a site size approximately equal to the lithographic field size stipulated for the 150 nm generation in the Roadmap (3). It is suggested to define the flatness in the stripe as SFSR to indicate that it is determined only over a narrow stripe instead of the whole site. Then SFQR would be defined as the maximum value of SFSR within the site. However, there is as yet no agreement on these various issues. Additional issues with the stripes at the edges of the sites and with the values of the site dimensions and the stripe width remain under discussion in the standards community.

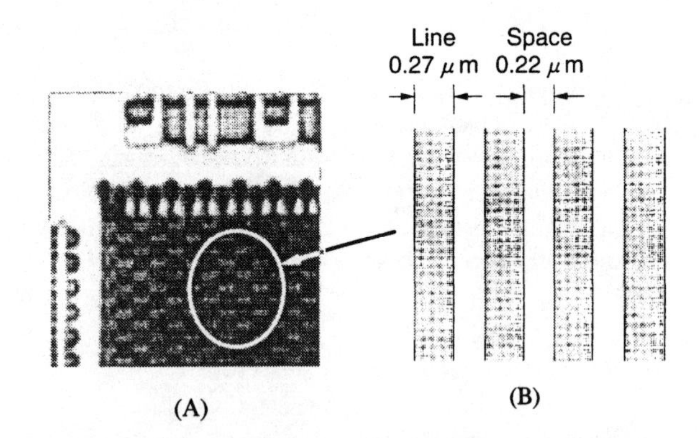

FIGURE 1. Gate pattern of a memory cell as viewed by an optical microscope with 100× magnification and an NA=0.90 (A) and schematic view of a portion of the pattern (B).

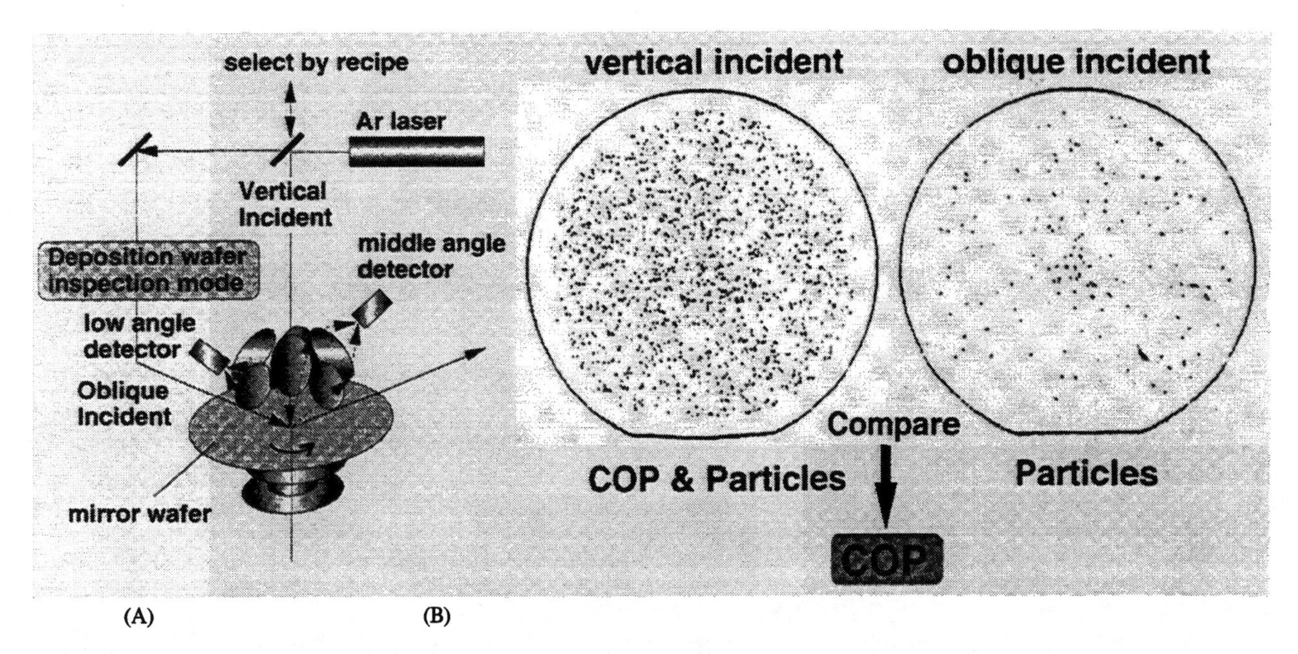

FIGURE 2. Arrangement of the optical system of a scanning surface inspection system (A) and the resulting wafer maps (B) showing the separation of particles and COPs on a polished Czochralski wafer.

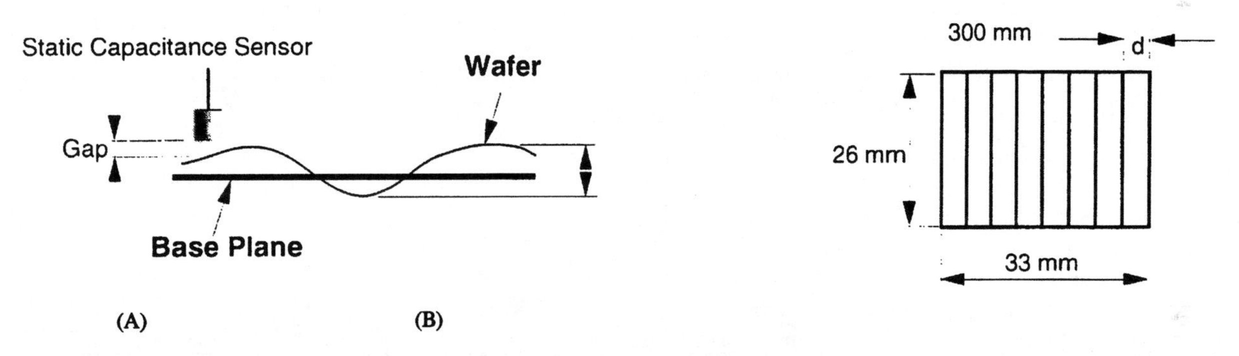

FIGURE 3. Schematic illustration of the definition of SFQR on a 200 mm wafer (A) and the complications in defining the site geometry arising from the introduction of scanning stepping lithographic systems for 300 mm wafers (B).

Cost Reduction

As noted in the Introduction, cost reduction is a critical aspect of the transition to 300 mm wafer fabs. However, these fabs are expected to use exclusively automated wafer handling systems; many will use a type of minienvironment known as front opening unified pod (FOUP). In-line metrology equipment as well as process equipment will be expected to provide standardized interfaces for FOUPs transported to the equipment from overhead hoist transfer systems or from floor guided vehicles. this situation is illustrated schematically in Fig. 4. This requirement is likely to increase, not to decrease, the initial cost of the equipment.

Increasing the overall equipment effectiveness (OEE) is a factor in factory cost reduction. For example, the I300I/Selete equipment metrics for 250 nm technology (4) cite a target mean time between failure (MTBF) of 1000 h and a mean time to repair (MTTR) of 1 h for metrology equipment including CD-SEMs, and overlay, film thickness, and patterned wafer defect detection instruments.

Another factor in controlling costs is reduction of wafer cost. At present, the final price of *p*-type Czochralski polished 300 mm wafers is projected to be about 3.2 times the price of a 200 mm wafer. Device manufacturers consider this to be too expensive. The desired multiplier is between 1.7 and 2.1. Costs can be reduced further by reducing the number of monitor wafers employed from 1.4–1.5 times the number of product wafers (as is the case for 200 mm lines) to 30–40% of the number of product wafers for 300 mm lines. This means that quality control based on use of monitor wafers will have to be replaced by increased in-line or in-situ monitoring of product wafers.

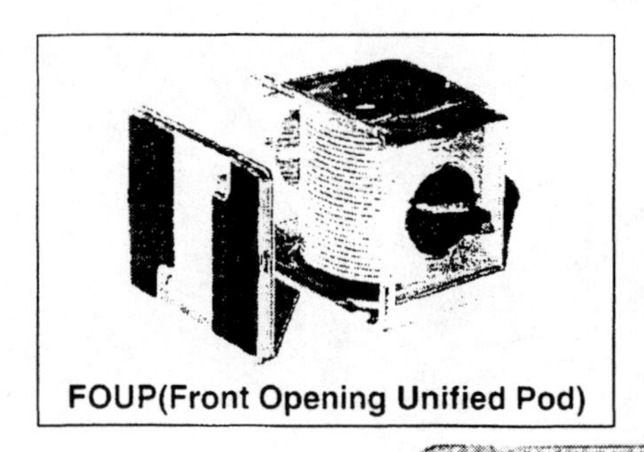

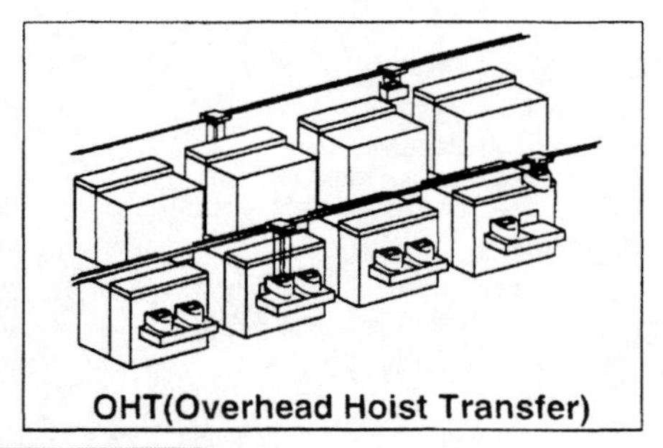

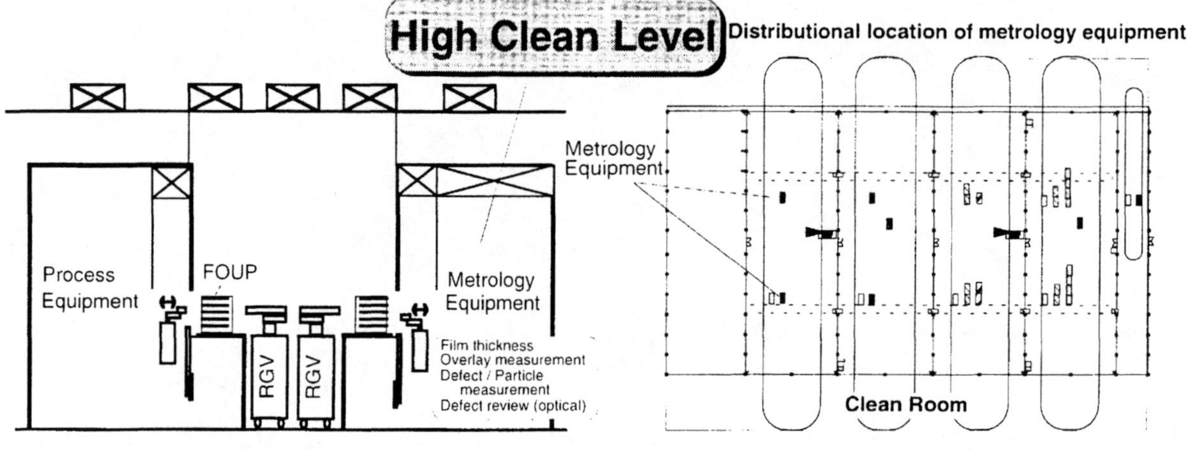

FIGURE 4. Schematic illustration of automated wafer handling for metrology equipment.

Short Cycle Time

Current trends require shorter times to market for logic circuits, especially for ASICs. For memory, time to maturity of the process, even in the face of new technologies, increased process complexity and number of process steps, and reduced process margins are critical for success. This translates to a need for faster TAT in the factory. Metrology tools are expected to contribute to this factor.

The area of a 300 mm wafer is about 2.25 times the area of a 200 mm wafer. Full wafer defect detection requires 112 22 mm × 22 mm shots on a 300 mm wafer compared with 44 shots on a 200 mm wafer. Instead of 5 shots for CD and overlay measurement, 13 shots are required. Therefore, to maintain the same wafer throughput, inspection time per shot must be reduced.

Simply speeding up the measuring and scanning time is not sufficient to achieve this reduction. In addition, error factor analysis algorithms must be improved for CD and overlay measurements, and more effective auto defect review and classification sampling algorithms must be developed for particle and pattern defect detection and analysis.

The time to identify the source of defects can be reduced by increasing the efficiency of observation and material analysis. This might include reducing the time for the cross sectioning and analysis procedures by more efficient sampling and on-wafer analysis in the review cycle, shown schematically in Fig. 5.

Another way of increasing the metrology-related throughput in the factory is to reduce the use of pre-process quality control (QC) procedures, especially in photo-lithographic and CMP steps. By replacing processing time and QC steps before the main process, the unit processing time can be significantly decreased. This "preceding QC-less" procedure involves application of in-line monitoring and requires the effective use of both equipment QC data and product wafer QC data on immediately prior runs. It is also expected to facilitate the improvement of process equipment reliability. An example of a combination of in-line and in-situ measurements during and after CMP processing is shown schematically in Fig. 6.

An example where feed-forward techniques can be usefully employed is the determination of etching times following film deposition. By providing film thickness data from the previous process to the etching controller, the etching time can be more accurately controlled and the accuracy of end point detection in the etching process can be enhanced.

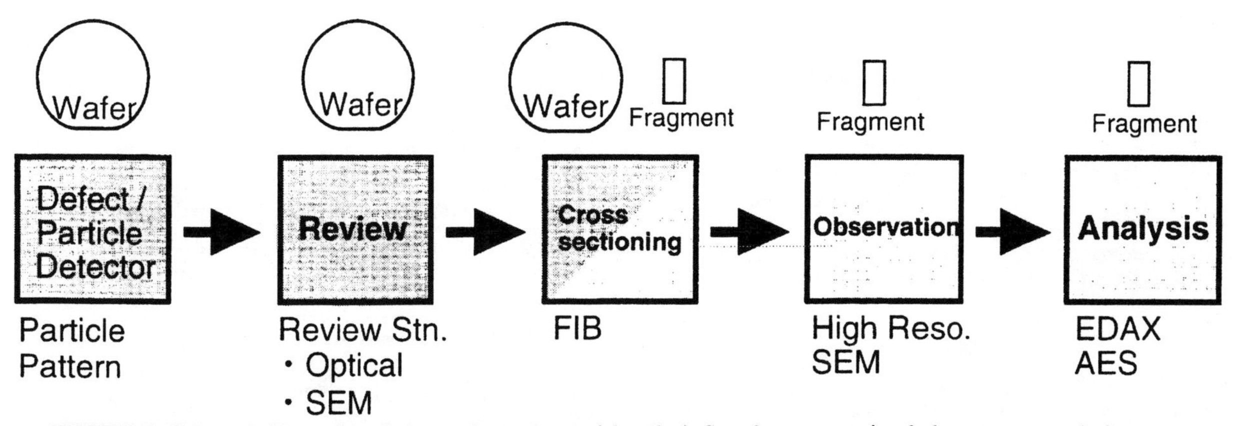

FIGURE 5. Schematic illustration of observation and material analysis flow from conventional clean room to analysis room.

Effective application of in-situ monitoring and feed-forward technology requires clarification of the relationships between control parameters and the measuring parameters through improved process models as outlined in Fig. 7. In many cases, currently available process models do not provide reliable information about these relationships. If suitable improvements in in-situ monitoring are implemented, these techniques can be effectively implemented for process optimization during development, for reducing differences in process results between multiple process machines using the same recipe or process conditions, and for improving process repeatability during deployment of volume production.

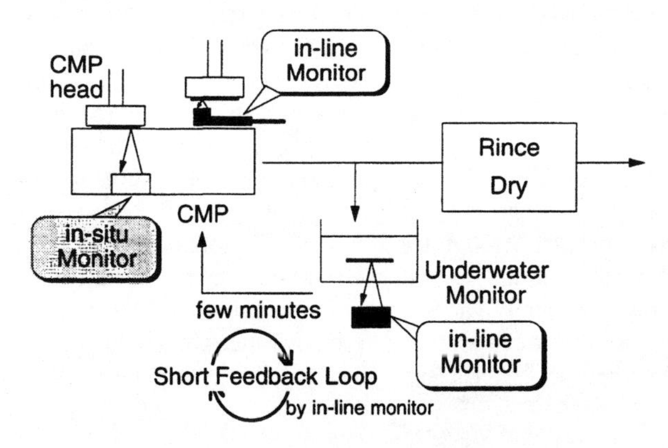

FIGURE 6. Schematic illustration of in-situ or in-line monitoring of film thickness after chem-mechanical planarization.

Examples of in-situ monitoring purposes and the object monitored are given in Tables 3 through 5 for PVD, etching, and CVD processes. These examples illustrate the power of in-situ monitoring. For effective deployment of in-situ monitoring, it is necessary to establish first the process model and appropriate control algorithm. Reduction of related expenses by means of hardware and software modification requires close cooperation between equipment and sensor suppliers. Standardization should proceed after the technology has been established. In-situ monitoring is applicable to all phases of production from process development to mass production, but implementation should be adjusted to correspond to the production phase.

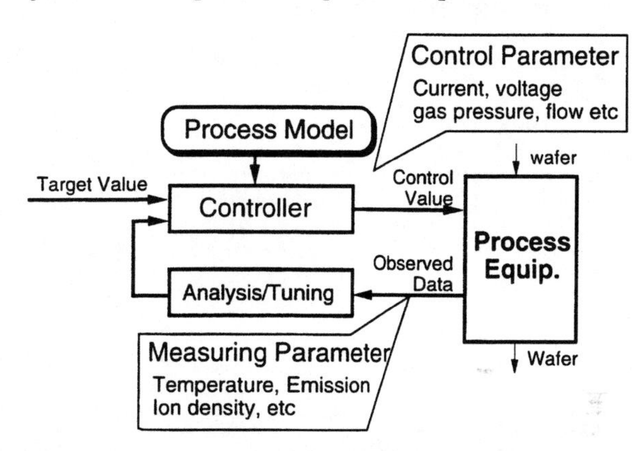

FIGURE 7. Schematic of in-situ monitoring showing examples of control parameters.

Environmental Issues

Another effective application of in-situ metrology is in optimizing process conditions so as to eliminate waste of resources. For example, in the gas cleaning example, pictured schematically in Fig. 8, the ion intensity is monitored by a quadrupole mass spectrometer (Q-mass). By observing the relative signals from the cleaning gas and the generated gas, process time can be reduced and both excess gas usage and excess emissions from the process can be eliminated. This not only improves the environmental aspects of the process but results also in a savings of consumable materials and hence costs.

TABLE 3. In-situ Monitoring of PVD Processes

Purpose	Monitor Object
Equipment Control	
Target change period	Film thickness on target
Collimator change period	Accumulated operation hours
Cleaning period	Particle
Abnormality Detection	
Leak detection	Left gas
Abnormal discharge	RF power waveform
Abnormal wafer	Wafer temperature
temperature	Particle (wafer)
Particle	Particle (chamber)
Process Control	
Film thickness (sheet resistivity)	Vortex current
RF power	RF power
Film morphology	Surface reflectivity

TABLE 4. In-situ Monitoring of Etching Processes

Purpose	Monitor Object
Equipment Control	
Shield ring change period	Accumulated gas flow
Focus ring change period	Accumulated RF power
Cleaning period	Particle
Abnormality Detection	
Gas temperature	Gas density
Abnormal discharge	RF power waveform
	Plasma emission
Abnormal wafer temperature	Wafer temperature
	Susceptor temperature
Particle	Particle (wafer)
	Particle (chamber)
Process Control	
Cleaning end point detection	Emission, Q-mass
RF power	RF power
Film morphology	Surface reflectivity

TABLE 5. In-situ Monitoring of CVD Processes

Purpose	Monitor Object
Equipment Control	
Part change period	Accumulated gas flow
Wafer ring change period	Accumulated RF power
Cleaning period	Particle
Abnormality Detection	
Leak detection	Left gas
Abnormal discharge	RF power waveform
	Plasma emission
Abnormal water temperature	Wafer temperature
	Susceptor temperature
Particle	Particle (wafer)
	Particle (chamber)
Process Control	
Cleaning end point detection	Emission, Q-mass
RF power	RF power
Film morphology	Surface reflectivity

SUMMARY

Wafer fabs for future generation will employ ever decreasing feature sizes and 300 mm wafers to achieve the desired increases in productivity and reduced time to market. The fundamental strategy in introducing such fabs is targeted at obtaining reduced manufacturing cost per chip. Metrology will play a key role in obtaining the desired goals.

The feature size reductions will challenge both the sensitivity and reliability of most metrology systems used in semiconductor manufacture. In addition, new definitions of parameters such as flatness will be required.

Control of costs is particularly critical. To achieve the desired targets, 300 mm equipment prices should be only 1.1 to 1.2 times that of 200 mm equipment, while maintaining the same wafer throughput and footprint. Widespread application of minienvironments is expected to reduce overall costs. Significant reductions can also be achieved by increasing overall equipment effectiveness. Targets for metrology equipment include a mean time between failure (MTBF) of 1000 h or more and a mean time to repair (MTTR) not in excess of 1 h. Finally, control of wafer cost involves both initial price and reduction in the use of monitor wafers.

Metrology equipment can contribute to the achievement of short cycle times in a number of ways. Inspection equipment throughput (per site) must increase by 2.5× in order to maintain the same wafer throughput as can now be achieved in 200 mm factories. Improvements in error fac-tor analysis algorithms and use of effective sampling algorithms for automatic defect review and classification are essential. Efficient observation and material analysis procedures can reduce the time to determine the source of defects in the line. Increased use of in-situ monitoring can reduce the overall turn around time through the elimination of QC procedures based on pre-process steps on non-product wafers. Process optizimation through the use of in-situ monitoring can also control consumable costs while at the same time reducing environmentally dangerous emissions from the factory.

Thus it can readily be seen that metrology will play a key role in the efficient operation of large diameter wafer fabs.

REFERENCES

1. National Technology Roadmap for Semiconductors, 1997 edition, San Jose, CA: Semiconductor Industry Association, 1997.
2. SEMI M1, Specification for Polished Monocrystalline Silicon Wafers, in *Book of SEMI Standards*, Mountain View, CA: Semiconductor Equipment and Materials International (http://www.semi.org).
3. Reference 1, Table 2.
4. Unified Equipment Performance Metrics for 0.25 μm Technology, I300I/Selete, November 1997, available from http://www.i300i.org.

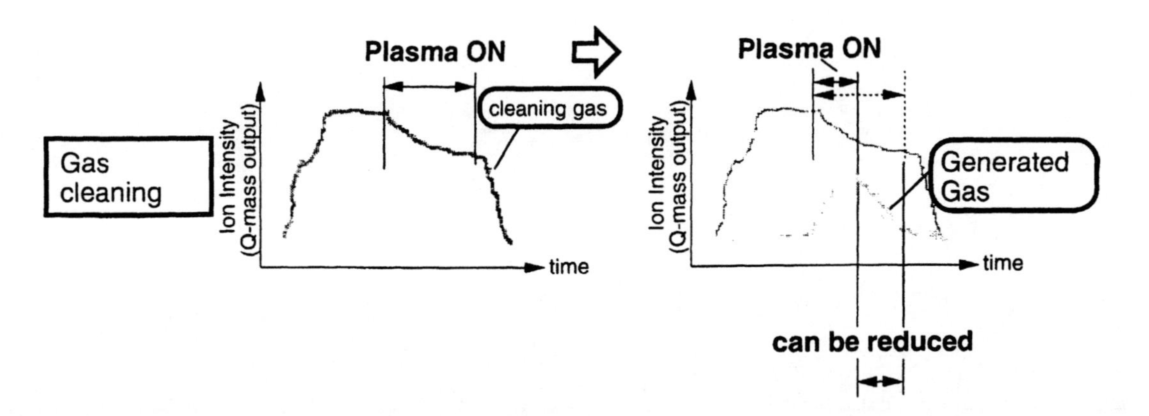

FIGURE 8. Application of in-situ monitoring with a quadrupole mass spectrometer (Q-mass) to reduce both consumption of consumables and emission of environmentally degrading gases.

Metrology Aspects of SIMS Depth Profiling for Advanced ULSI Processes

Andre Budrevich* and Jerry Hunter

Intel Corporation, Materials Technology Department, Santa Clara, CA 95051
*Intel Corporation, Fab 18 Yield Department, Kiryat Gat, Israel

As the semiconductor industry roadmap passes through the 0.1 μm technology node, the junction depth of the transistor source/drain extension will be required to be less than 20 nm and the well doping will be near 1.0 μm in depth. The development of advanced ULSI processing techniques requires the evolution of new metrology tools to ensure process capability. High sensitivity (ppb) coupled with excellent depth resolution (1 nm) makes SIMS the technique of choice for measuring the in-depth chemical distribution of these dopants with high precision and accuracy. This paper will discuss the issues, which impact the accuracy and precision of SIMS measurements of ion implants (both shallow and deep). First this paper will discuss common uses of the SIMS technique in the technology development and manufacturing of advanced ULSI processes. In the second part of this paper the ability of SIMS to make high precision measurements of ion implant depth profiles will be studied.

INTRODUCTION

Secondary Ion Mass Spectrometry (SIMS) has been widely used to support technology development and high volume manufacturing of ULSI devices. The SIMS technique is capable of giving quantitative information about contamination levels (atmospheric and metallic), in-depth dopant profiles and chemical composition information. Additionally, SIMS has been used for manufacturing tool performance control and in-fab equipment troubleshooting (1).

To keep pace with Moore's law scaling (2), a requirement is placed on the analytical methods used for process characterization to provide better control of dopant composition and finer control of contamination levels. SIMS is uniquely positioned to meet the requirements of future ULSI processes due to its high sensitivity, excellent depth, wide range of elements/molecules detectable and high precision and accuracy.

The ultimate precision achievable by SIMS has been previously demonstrated by Adriaens and Adams (3) for isotopic ratio measurements for geological applications in the range of 0.01%-.001% relative standard deviation (RSD). In this paper we will show data that a precision of 0.1%-1.0% RSD is achievable for SIMS dosimetry of ion implantation. We also will present a short review of critical applications and metrology capabilities of the SIMS technique for ULSI technology development and manufacturing.

1. SIMS INSTRUMENTATION AND CAPABILITIES

The SIMS technique is based on extraction and mass separation of secondary ions formed during the sputtering process.

Mass separation

Mass separation is accomplished by three major methods each with associated advantages and limitations as detailed below:

Magnetic sector (Cameca IMS) – is a high mass resolution double focusing sector field instrument. This type of SIMS tool provides both high transmission (~40%) and high mass resolution (~10,000 m/Δm). The transmission of the magnetic sector instrument is mass independent. The high transmission is possible due to the high secondary extraction voltage (kV range). The magnetic sector SIMS tool is able to operate in either ion microprobe or ion microscope mode by using electronic or optical gating. The primary limitations of the magnetic sector tool is that the primary beam incidence angle (variable between 21°-65°) is dependant on the primary beam energy, the initial capital cost is high and the tool is instrumentally complicated, requiring an expert operator.

Quadrupole (Atomika and PHI) – mass separation is accomplished by use of a quadrupole mass filter. The main advantages of the quadrupole SIMS tool results from the use

CP449, *Characterization and Metrology for ULSI Technology: 1998 International Conference*
edited by D. G. Seiler, A. C. Diebold, W. M. Bullis, T. J. Shaffner, R. McDonald, and E. J. Walters

of a low secondary extraction potential (between 10-100 V). This low extraction potential allows for the independent adjustment of primary beam energy and angle of incidence. Additionally, very low energy primary beams are achievable (<100eV) resulting in the excellent depth resolution achievable with such tools. Other key advantages include easy to use electron beam charge compensation for insulator analysis, low initial capital outlay and lower level of expertise required for operation. The key disadvantages of the quadrupole SIMS tool type is low mass resolution (~200-300m/Δm), mass dependent transmission and low transmission (0.1%) which makes small area analysis difficult.

Time-of-Flight (Cameca, and PHI) – separates ions based on their time of flight after sputtering with a pulsed primary ion beam. The key advantages of this type of tool are the high transmission and mass resolution (>10,000), excellent lateral resolution (ion probe imaging), and full 200 mm wafer navigation. The key disadvantage of the TOF-SIMS tool is that very little work has been done with depth profiling since it has been used primarily as a static SIMS tool for molecular identification.

Importance of primary beam parameters

Primary ion beam sources are a very important part of SIMS instrumentation and are especially important for achieving high depth resolution SIMS measurements. The typical requirements for primary ion beams are: primary ion species of Cs^+, O_2^+, O^-, Ga^+, primary beam stability better than 1% during a typical analysis time (10 min.), beam current 0.1 – 1000 nA, beam spot sizes of 0.05-10 μm and energies from 100eV to 20keV.

The capabilities of SIMS instruments that are relevant to ULSI applications are summarized in table 1 below.

TABLE 1. Summary of SIMS capability.

Dynamic range	10^6
Detection limits, (at/cm^2)	10^7-10^9
Depth resolution, (nm)	<1
Lateral resolution, (nm)	100
Elements	H –U
Accuracy, (RSD)	(1-3)×10^{-1}
Precision, (RSD)	(1-2)×10^{-2}

SIMS quantification

The most common method of SIMS Quantification is the Relative Sensitive Factor (RSF) method (4,5). The method is based on calculation of a scaling factor using SIMS measurements of a known and well-characterized elemental standard. The formula for RSF calculation is given as:

$$RSF = \frac{\phi\, C\, I_m\, t}{d\, \Sigma I_i - d\, I_b C}\left(\frac{EM}{FC}\right) \qquad (1)$$

Where ϕ is the dose (in atoms/cm^2) from a known standard, C is the number of data points in the profile, (EM/FC) is the ratio of electron multiplier to Faraday cup counting efficiency, d is the crater depth, ΣI_i is the sum of the impurity isotope secondary ion counts, I_b is the background ion intensity, and t is the analysis time. If the RSF value is known from the measurement of a SIMS standard, the SIMS raw data for an unknown sample can be calibrated by using formula (2) below:

$$\rho_i = RSF\frac{I_i}{I_m} \qquad (2)$$

where ρ_i is the impurity atom density in an unknown sample, I_i is the impurity isotope secondary ion intensity, and I_m is the matrix isotope secondary ion intensity .

The accuracy of the SIMS measurement is directly dependent on quality and availability of standards. A boron implant standard certified by NIST (SRM #2137, ±6% absolute accuracy) is available now (6). Arsenic (SRM #2134) and Phosphorus Consensus Reference materials are currently under development (7). Other SIMS standards are available commercially with a stated accuracy of ±15-20% (8,9).

2. SIMS APPLICATIONS FOR ULSI TECHNOLOGY DEVELOPMENT AND MANUFACTURING

SIMS data are widely used in the semiconductor industry for technology development, process control or quality assurance, failure analysis and equipment performance control. The main applications of SIMS for the semiconductor industry divide as follows: implantation support, surface/interface contamination analysis, thin films analysis and 2D/3D imaging. Table 2 lists the typical SIMS applications for ULSI process characterization and the current level of precision achievable as defined below:

Metrology (M) – SIMS precision is currently better than technology requirements.

Analytical (A) – SIMS precision is currently equal to technology requirements.

Unique (U) – SIMS precision is unknown.

2.1 Implantation support

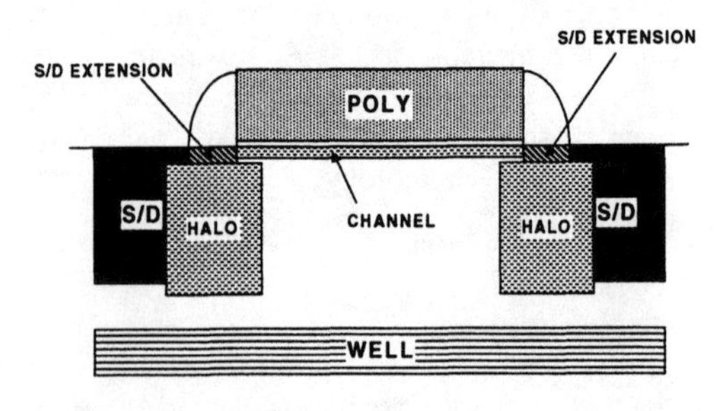

FIGURE 1. Typical implant structure for CMOS transistor.

TABLE 2. Main SIMS applications for ULSI support.

APPLICATION	TASK	SIMS MODE	(*)
Implantation	-dosimetry -profile shape -Shallow junction profiling -junction depth scaling	Dynamic SIMS	M M A M
Surface/Interface contamination	-trace metals -organic contamination	TOF/ Surface SIMS	A A
Thin films	-mobile metal atoms -nitrogen in gate oxide	Dynamic/ TOF SIMS	M A
Imaging	-particles on surface	TOF SIMS	A
	-lateral depth profiling	Dynamic SIMS	U
	-failure analysis	TOF SIMS	A

(*) U is unique application, A is analytical application, M is metrology application.

The most common SIMS application for ULSI process characterization is the depth profiling of implanted species. The high level of precision achievable with the SIMS technique (10, 11) makes it possible to use SIMS as a metrology and monitoring tool for ULSI processes (12, 13). A typical CMOS transistor implantation structure is shown in Fig.1 (14). The dopant in-depth profile shape and dose have a large effect on a CMOS transistor's performance. The various implantation steps include: wells, source/drain (S/D), S/D extension (LDD), channel doping (Vt adjust), anti-punch-through S/D halo formation and doping of poly-gate structures. Figure 2 gives typical process windows for both energy and doses for these implantation steps (14).

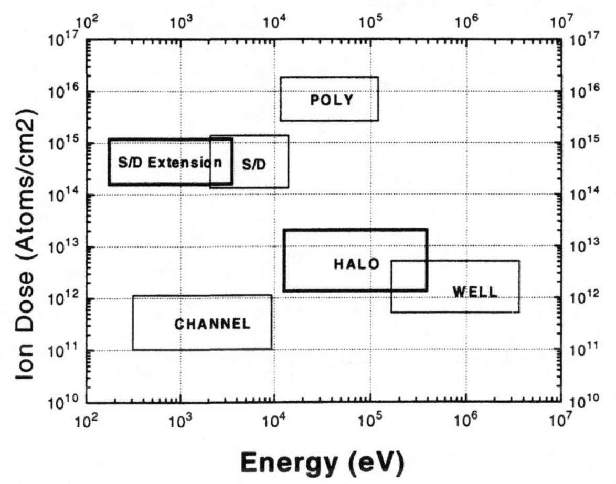

FIGURE 2. Typical implantation dose/energy process windows.

a) Dose and profile shape control

Typical energies of implanted ions are in the range of 10^2-10^6 eV with typical doses in the 10^{11}-10^{16} ions/cm^2 range. Common dopant species are As, P, and Sb (N-type) and B (P-type). In this paper we will refer to general implant types as given below:

High Energy Low Dose (HELD) - Dose e12 - e13 range, energy >100KeV.
Low Energy High Dose (LEHD) - Dose E14 - E15 range, energy < 10KeV.
Medium Energy High Dose (MEHD) - Dose E14 - E15 range, energy 10KeV < X < 100KeV.
Medium Energy Low Dose (MELD) -- Dose E12 - E13 range, energy 10KeV < X < 100KeV.

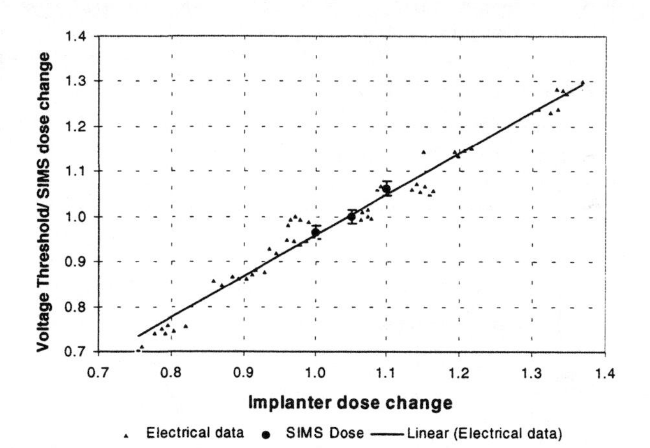

FIGURE 3. Correlation of V_{th} and SIMS dosimetry with implanter dose variation.

Device performance characteristics are very dependent on the implanted dose and resulting profile shape. As an example fig. 3 illustrates the dependence of transistor threshold voltage (V_{th}) on implanted dose (15), also shown in figure 3 are normalized SIMS dose results for a MELD Boron implant. The SIMS results in figure 3 demonstrate that for this particular implant step the SIMS and electrical test data have the same level of sensitivity to dose variations (slope ~ 1.0).

SIMS depth profiling has previously demonstrated the ability to achieve a precision of 1 – 5% RSD (16, 17), however, in many cases this precision is insufficient for implanter performance monitoring. The total variability of a SIMS dose measurement can be calculated as given in formula 3:

$$\sigma = \sqrt{\sigma_{impl}^{2} + \sigma_{SIMS}^{2}} \qquad (3)$$

Where σ_{impl} is the RSD for the implanter variability and σ_{SIMS} is the RSD for SIMS variability. Reliable dosing information can only be obtained in the case where $\sigma_{impl} >> \sigma_{SIMS}$. For example, if $\sigma_{impl} \approx 0.03$ and $\sigma_{SIMS} \approx 0.02$ the variability in the SIMS measurement will result in 17% of the total variability of dose measurement. Implanter variability tolerances for submicron technology require that the SIMS variability be on the order of 0.5% RSD. 300 mm implantation program will require a dose uniformity /repeatability of ~0.1% (18), and, as a result, a significant improvement of SIMS repeatability.

In addition to dose control, the other key parameter for device performance is in-depth profile shape control.

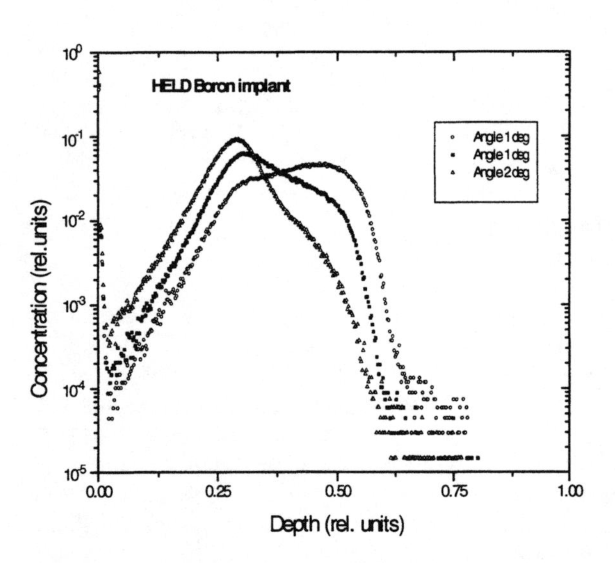

FIGURE 4. SIMS depth profile of HELD Boron implant. Profile shape strongly depends on implantation angle.

Energy contamination in the implant primary beam, angular variations, and wafer preparation prior to implantation can have a profound effect on the dopant profile shape. Due to its ability to monitor the in-depth distribution of dopants with ppm sensitivity, SIMS is the primary analytical method of choice for the precise monitoring of as-implanted profile shapes.

Other workers have shown significant differences in profile shapes based on angular variations in ion implanters (19). Figure 4 gives an example of the change in the dopant depth profile shape with variation of implant incidence angle. The 0° implant shows a strong channeling effect and the channeling effect nearly disappears with a slight deviation of incidence angle from 0° - 2°. The dose determined with SIMS is the same for all profiles shown in Fig. 4, however, the slight angle change results in large profile shape differences that could degrade device performance. Due to its ability to provide in-depth dopant distributions SIMS is ideally suited to detect these types of problems.

b) Shallow Junction depth profiling

One of the major current challenges for SIMS depth profiling is the characterization of ultra-shallow junctions (USJ) in Si (20). Table 3 shows the National Technology Roadmap for Semiconductors (NTRS) (21) for dopant junction depth scaling along with the peak concentrations that need to be characterized and the estimated depth resolution required for both as implanted and annealed junctions. Ultra shallow junction depth profiling, especially for source/drain extensions requires the ultimate depth resolution of the SIMS technique. The currently achievable depth resolution using the SIMS technique (Quadrupole or TOF) is ~ 0.5-0.6 nm meaning that SIMS capable of meeting the technology requirements up to 0.1μ technology generation.

TABLE 3. National Technology roadmap for dopant junction depth scaling and requirements for SIMS depth resolution.

TECHNOLOGY GENERATION	**0.25μ**	**0.18μ**	**0.15μ**	**0.13μ**	**0.1μ**	**0.07μ**	**0.05μ**
Year (1st product ship.)	1997	1999	2001	2003	2006	2009	2012
X_J at Channel (nm)	50-100	36-72	30-60	26-52	20-40	15-30	10-20
Drain extension concentration (cm^3)	1×10^{18}	1×10^{19}	1×10^{19}	1×10^{19}	1×10^{20}	1×10^{20}	1×10^{20}
Contact X_J (nm)	100-200	70-140	60-120	50-100	40-80	15-30	10-20
Required depth resolution for as-implanted profiles (nm)	2-4	1-2	0.8-1.8	0.7-1.5	0.4-0.9	0.3-0.7	0.2-0.4
Required depth resolution for annealed profiles (nm)	6-11	3-6	2.5-5	2.2-4.3	1.3-2.7	1-2	0.7-1.3

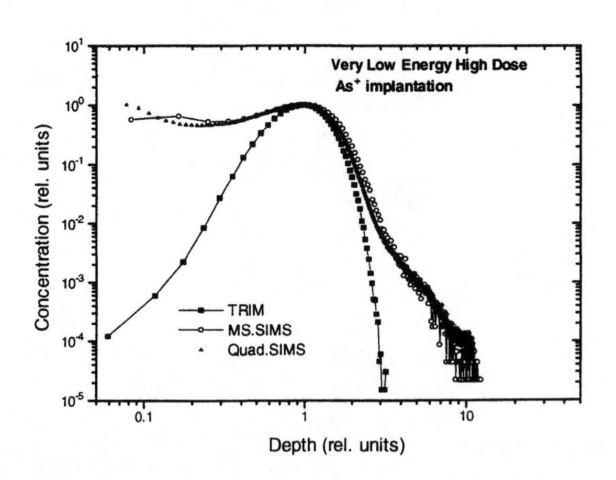

FIGURE 5. SIMS depth profiles for LEHD As implant. Quad. SIMS shows better depth resolution than Magnetic sector tool. Peak position is ~60 Angstrom.

Figure 5 shows SIMS depth profile measurements for a shallow As implant. The data in fig. 5 were acquired with Quadrupole and Magnetic sector SIMS tools. The data in fig. 5 shows that the quadrupole SIMS tool demonstrates superior depth resolution.

Some limitations are associated with the measurement of shallow junction depth profiles using SIMS as given below:

1) *concentration scale quantification* (22) - for high dose and low energy S/D extension implants the depth profile peak concentration is typically 10^{21}-10^{22} (atoms/cm^3). The high concentration of implanted atoms affects the secondary ion yield and, as a result, can lead to a depth dependent matrix effect. The RSF method is based on the assumption of unchanging secondary ion yield and is invalid if there is a depth dependent secondary ion yield. The utilization of cesium attachment SIMS, whereby the analyte of interest (M) is measured in molecular form as an MCs^+ ion (references) can minimize this concern for certain sample matrices. Unfortunately the MCs+ method has not proven very applicable to Si sample matrices (23).

2) *depth scale calibration* – typical sputtered depths for shallow junction profiles are in the range of 20-70 nm, the accurate measurement of these shallow craters is a difficult problem. Modern profilometers (both stylus and optical) have a vertical resolution 0.02-0.05 nm and can meet the requirements for measurement of these shallow craters. The more serious problem is that typical wafer roughness for test wafers used for ion implant studies is ~10-20 nm. This roughness can cause significant error in the crater depth determination. Preparation of deep craters ~100-250 nm can improve the confidence in the depth measurements but significantly increases the throughput time of SIMS measurements. Another alternative is utilization of epi-wafers (roughness of epi wafers is below 10 nm) for sputtering rate measurement (24).

3) *transient region* - presence of a native oxide (thickness ~2-3 nm) can significantly change the profile shape especially in the dopant peak area for shallow implants. As a result surface and peak concentrations of the dopant will not be properly estimated from the SIMS data. Moreover, sputter differences between native SiO_2 and Si can also sputter rate to vary with depth resulting in significant distortions to the dopant profile shape in the transient region. Encapsulation of the sample surface with poly-Si (25), flooding of the sample surface with a directed oxygen jet (26) and very low energy depth profiling (27) have been used to minimize this transient region effect.

2.2 Gate oxide composition/contamination control

Characterization of sub 5 nm gate oxides requires high depth resolution and in some cases the high depth resolution must be combined with high mass resolution. Currently quadrupole SIMS tools are achieving the highest depth resolution, however, they can not combine high depth resolution with the high mass resolution required for typical oxide contaminants (Fe, Ni, Cu, etc.) for the reasons previously mentioned. Typical gate characterization experiments include the analysis of metallic contamination (either in the gate oxide or at the SiO_2/Si interface), analysis of F in the gate oxide (reference) and the analysis of N at the SiO_2/Si interface. A discussion of three of these typical analyses, along with the driving force behind the need for the analyses, is given below.

SIMS detection limits in silicon dioxide films for critical metal contaminants are given in Table 4.

TABLE 4. SIMS Detection limits of critical metal contaminants in silicon dioxide (8).

Contaminants	Detection Limit (10^{10} at/cm^2)
Fe	0.7
Ni	1.6
Cr	1.3
Cu	1.6

Gate oxide composition control is important to prevent boron diffusion from the gate into the channel region. The presence of nitrogen in the gate oxide can significantly reduce boron penetration into the channel regions from the gate.

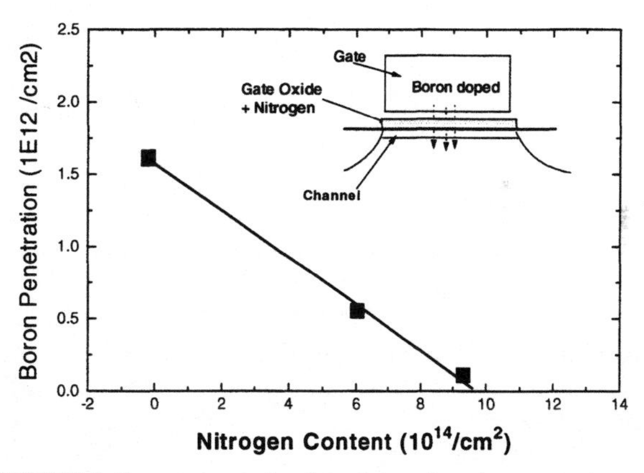

FIGURE 6. Boron penetration from boron doped gate vs. nitrogen composition in gate oxide (28).

Figure 6 shows data indicating that the presence of nitrogen in the gate oxide at the ~10^{15} (atoms/cm^2) level significantly reduces boron penetration into the channel region. The SIMS technique has been extensively applied for the monitoring of the nitrogen content in gate oxide (28). Precision has been demonstrated to be on the level of ±3%,with depth resolution of ~2 nm (29, 30) and nitrogen detection limits of ~0.001 atomic % (31). Figure 7 shows overlays of three nitrogen depth profiles in a typical gate oxide demonstrating the excellent reproducibility of the SIMS technique for this analysis. All data were acquired on magnetic sector SIMS tool using low impact energy Cs primary ions.

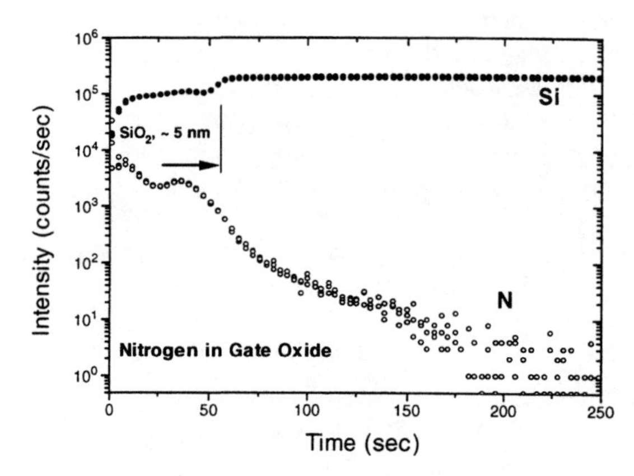

FIGURE 7. Nitrogen depth profile in a thin (~ 5 nm) gate oxide. 3 depth profiles are overlaid.

2.3 Surface contamination analysis

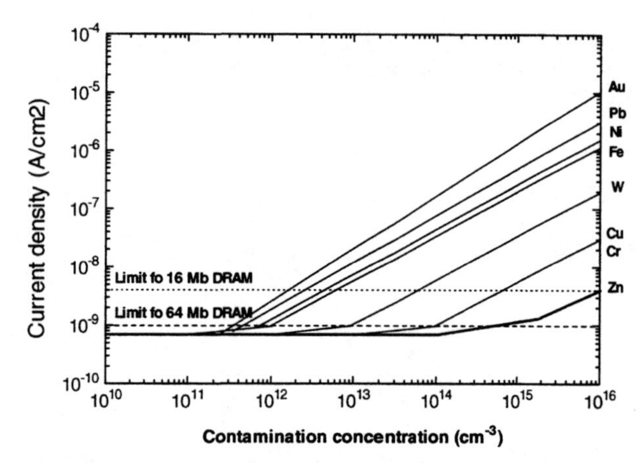

FIGURE 8. Junction leakage current density at 2V reverse bias vs. contaminant concentration in bulk extrinsic Silicon with N_A= 1.1 10^{17} (cm^{-3}) (32).

Surface contamination control includes analysis of metal ions and organic compounds accumulated on the wafer surface during processing. Trace metallic contamination on the wafer surface can detrimentally affect device performance by increasing leakage current and reducing carrier lifetime.Figure 8 (32) shows electrical data of junction leakage measurements vs. concentration of metallic contaminants. The data in fig. 8 shows that metallic concentrations below 10^{11} (atoms/cm^3) have little effect on device performance, however, device leakage current is especially sensitive to Au, Pb, Ni and Fe contaminants above the 10^{11} range.

Figure 9 (32) shows carrier diffusion length L_{diff} degradation with increasing Fe concentration along with the Fe detection limits per technology generation.

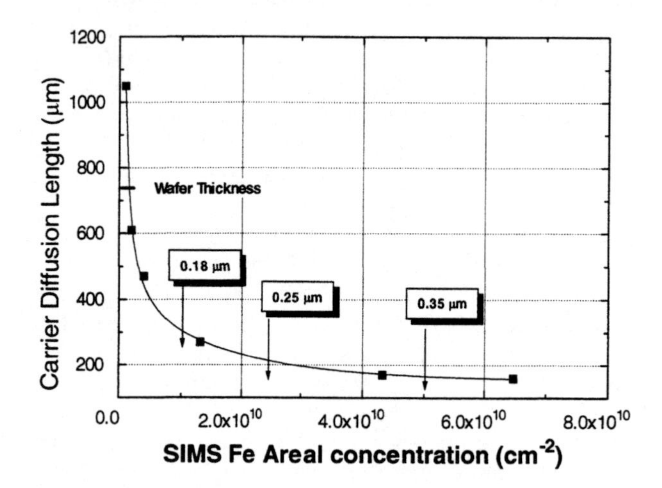

FIGURE 9. Carriers diffusion length degradation with the increasing iron concentration and iron concentration limits by technology generation (30).

The SIMS technique has been successfully utilized for surface contamination analysis. Surface SIMS (33) and TOF SIMS (34) have previously demonstrated excellent results for trace metals analysis. The National Technology Roadmap for Semiconductors (NTRS) for critical contaminants and the associated SIMS detection limits are given in Table 5 below (8,21).

TABLE 5. National Technology roadmap for critical contaminants and achievable SIMS detection limits.

CONTAMINANT	UNITS	0.35 µ	0.25 µ	0.18 µ	0.13 µ	SIMS DETECTION LIMIT
Al	10^{10}, at./cm^2	10	5	2.5	1	0.05
Ca	10^{10}, at./cm^2	10	5	2.5	1	0.1
Fe	10^{10}, at./cm^2	5	2.5	1	0.5	0.08
Ni	10^{10}, at./cm^2	5	2.5	1	0.5	0.2
Cu	10^{10}, at./cm^2	5	2.5	1	0.5	1
Zn	10^{10}, at./cm^2	5	2.5	1	0.5	0.5
Na	10^{10}, at./cm^2	5	2.5	1	0.5	0.1
C	10^{14}, at./cm^2	5	3	1	0.5	0.01

2.4 Lateral SIMS depth profiling;

Lateral dopant distribution is very important for prediction and control of device performance. The normal dynamic SIMS technique is unable to give reliable information on lateral dopant distribution due to:

i) the low concentration of dopant atoms in the analyzed area in combination with the instrumental transmission and secondary ion yield, can not give a high *S/N* ratio. The S/N ratio can be calculated as below

$$\frac{S}{N} \approx \frac{\alpha \, T(CD)^2 \Phi}{I_b \, t} \qquad (4)$$

where α is the dopant secondary ion yield (typically 10^{-3}-10^{-7}), T is instrumental transmission factor, *(CD)* is technology Critical Dimensions, Φ is implantation dose, I_b is background signal (counts/sec), t is acquisition time. For example, CD = 0.25μ, α =0.001, T =0.1, Φ =1×10^{15}(cm^{-2}), I_b=0.1 (counts/sec), t = 300 (sec) will give *(S/N)* ~ 2.

ii) beam spot size. To achieve good lateral resolution the primary beam spot size should be << *(CD)*. Current achievable oxygen and Cs beam spot sizes on commercially available SIMS tools are ~0.1-0.2 μ. The smallest beam spot sizes currently achievable are ~20-50 nm with Ga liquid metal ion guns (LMIG). However, Ga has little secondary ion yield enhancement resulting in lower signal levels and higher detection limits over oxygen or Cs bombardment.

iii) Von Criegen et. al. (35) have developed a technique allowing for the measurement of lateral dopant distributions using a standard dynamic SIMS setup. The technique is based on depth profiling of a wafer cross-section. Cross-sectioning is done in the direction parallel to the mask line. TEM/SEM micrographs are used to evaluate roughness and alignment of the cross- section. The obtained lateral dopant profile is convoluted with the normal dopant distribution. Deconvolution of the two profiles can be accomplished if the normal in-depth dopant distribution has been previously measured. This method utilizes the advantages of the dynamic SIMS method (very good depth resolution, high dynamic range and precision) in the depth dimension and capitalizes on them in the lateral dimension. The method has good profile shape repeatability (~5 nm) and the dynamic range for As and Boron profiles is reported as ~3 decades (35), detection limits are ~10^{17} (at/cm^3) and 2D-junction depth mapping has been demonstrated. The lateral SIMS depth profiling results can be used to check the validity of 2D simulation results and to predict device performance. Additional methods for lateral dopant profiling are discussed in papers by Dowsett (36) and Ray (37),.but both of these methods require specialized structures and data processing methods.

3. PRECISION SIMS METROLOGY MEASUREMENTS

To gain an increased level of understanding into the limitations of high precision SIMS depth profiling a number of experiments were conducted as follows:

Short term repeatability -- the capability of the SIMS tool to deliver the same dose on a single setup during a single analytical session.

SIMS linearity – the capability of SIMS to measure small dose differences.

Multiple setup reproducibility – the capability of SIMS to measure the same sample dose across a number of different instrumental setups over a long (3-6 month) time frame.

Cross tool matching – the capability of different SIMS tools to produce the same dose and profile shape.

3.1 Experimental

Measurements were performed on both magnetic sector and Atomika 4100 quadrupole SIMS tools. Boron profiles were acquired with O_2^+ primary ion beams and As and P profiles were acquired with Cs^+ primary ion beams. To study the ability of SIMS to make high precision measurements samples were prepared using ion implantation with a range of different implant species, energies and doses as outlined in table 6 below.

TABLE 6. Dose and energy ranges for various implants used in this study

Implant	Dose (atoms/cm2)	Energy
HELD	E12 – E13	>100keV
LEHD	E14 – E15	<10 KeV
MEHD	E14 – E15	10 keV < X < 100keV

3.2 Short term repeatability measurements

The ability of the SIMS technique to repeatably measure doses for a single setup was studied for implants of HELD B, HELD P, LEHD As, MELD As, (fig. 10 a-d, respectively). All short-term repeatability measurements were conducted on a single SIMS tool setup during a single session (<24 hours). An analysis cycle consisted of 3 depth profiles on a single sample, a loading and unloading cycle (including removal of the sample from the sample holder) between each analysis and a crater depth measurement. Analysis for each setup was performed by 3-4 different

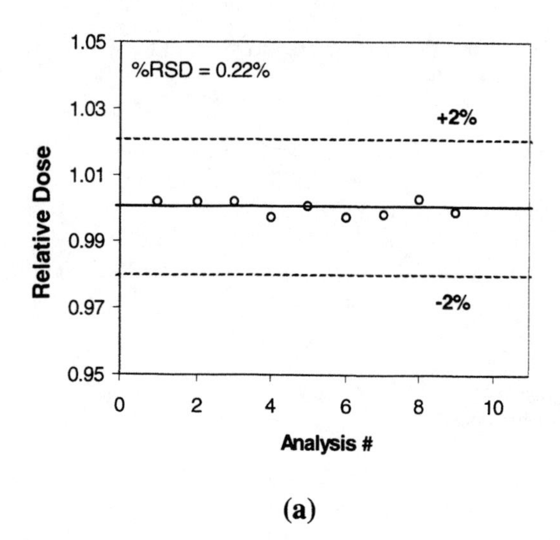

(a)

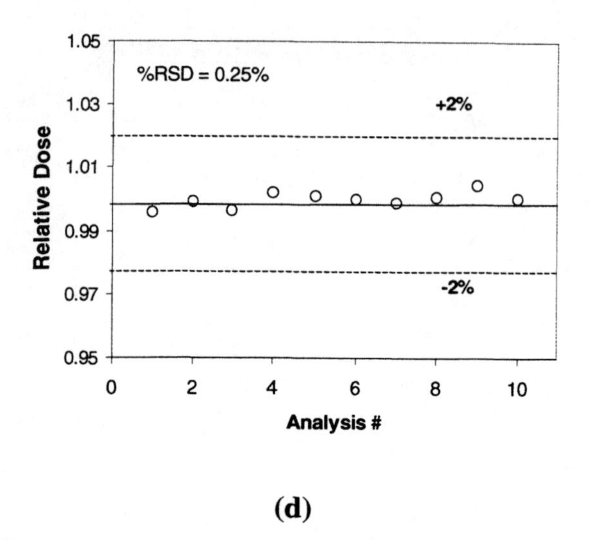

(d)

FIGURE 10. Short term repeatability for different implant recipes: (a) HELD boron; (b) HELD phosphorus; (c) LEHD arsenic (d) MELD arsenic. All implants were performed in Si substrate

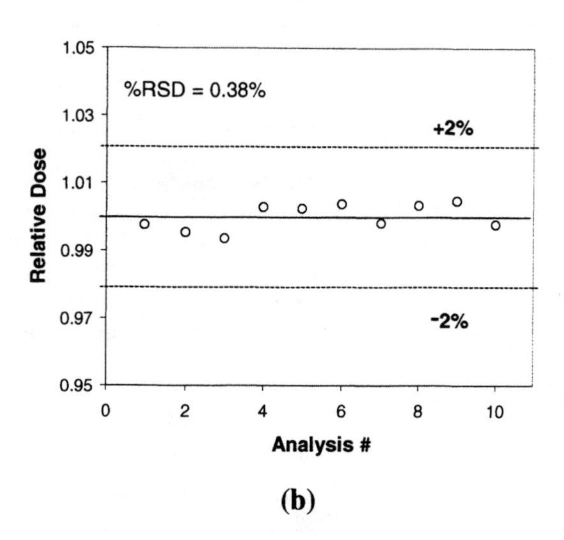

(b)

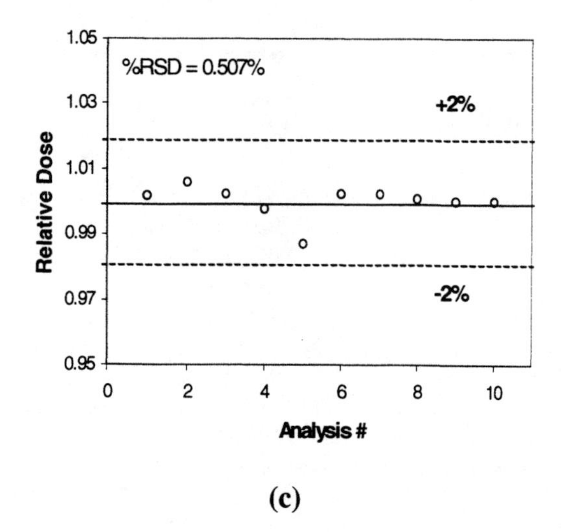

(c)

operators. The analytical session consisted of ten analysis cycles for each implant studied performed over a period of 12-24 hours depending on the implant being investigated.

a) Dose measurements

This experiment shows the ability of SIMS to make high precision measurements under real world conditions (depth and concentration measurements). Figure 10 a-d plots dose versus analysis #, all doses are scaled to 1.0 and the relative dose number shown in the plot is an average of three measurements taken for the particular analysis.

Figure 10 (a-d) shows that the highest precision is achievable on a HELD Boron implant (0.22% RSD one sigma) and the lowest is on LEHD As implant (0.507% RSD). In all cases the achievable precision is excellent and shows that SIMS is very capable of achieving precision needed for metrology measurements for ULSI processing. The short-term repeatability precision is a pre-requisite for the ability of SIMS to compare performance of two different implant tools (see section 3.4).

b) Profile shape comparison

Overlays of the 30 depth profile measurements acquired on the HELD Boron implants are shown in figures 11a (linear) and 11b (log). Note the excellent profile shape repeatability for all profiles achievable with the SIMS technique.

c) SIMS linearity response

To test the ability of SIMS to measure small dose differences, implants were generated for HELD B and Phos and MEHD B and As at a nominal dose (100%), nominal dose –5% (95%) and nominal dose +5% (105%). Additional sets were generated for LEHD B and As as above plus additional implants at nominal dose –3% (97%) and nominal dose +3% (103%) (38).

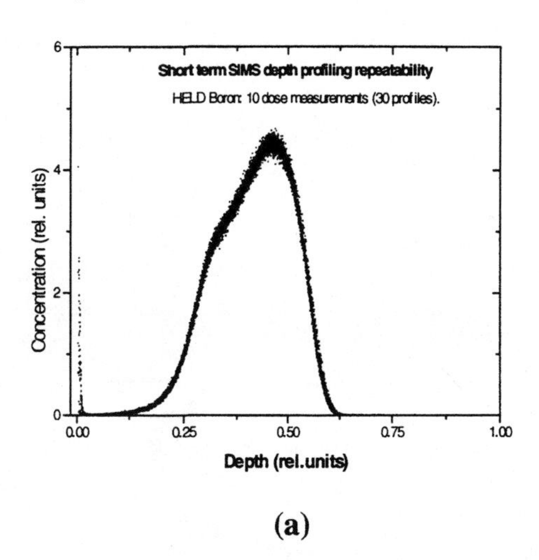

(a)

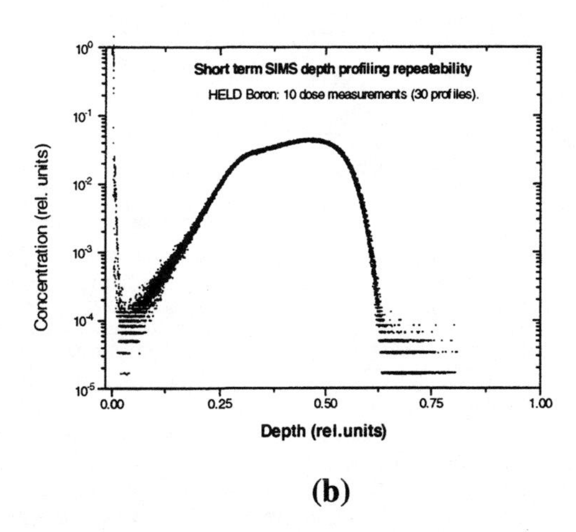

(b)

FIGURE 11. Short term repeatability of the profile shape for HELD B implant. (a) Linear scale; (b) logarithmic scale.

Figure 12 a-c show the linear SIMS response curves for selected implants (LEHD B (a), HELD As (b) and MEHD B (c))along with the 3σ error bars (based on the short term repeatability obtained above. Table 7 summarizes the results for all implants generated for this test, it includes three figures of merit for each curve: slope, intercept and correlation coefficient (R2). Slopes are in the range of 0.90 to 1.15, most likely due to the error associated with the measurement. Note that in all cases SIMS shows a measurable difference between different implanted doses, demonstrating SIMS ability to measure fine implant dose differences with high precision.

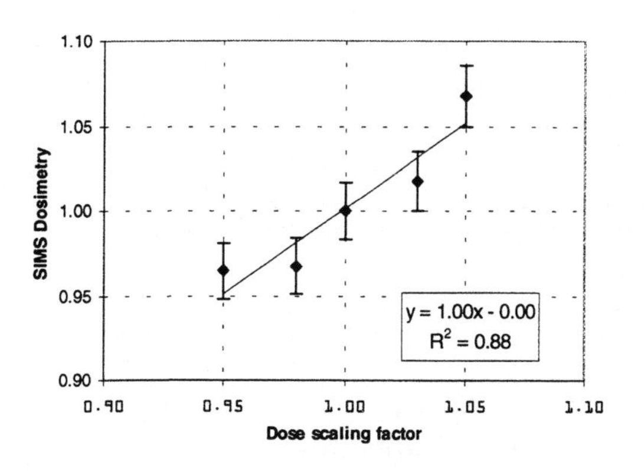

(a)

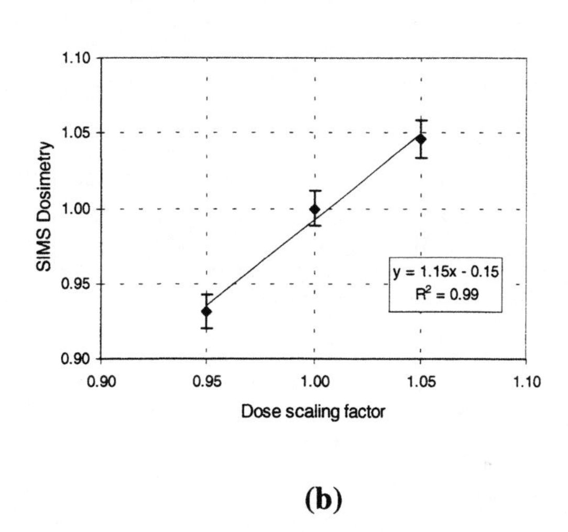

(b)

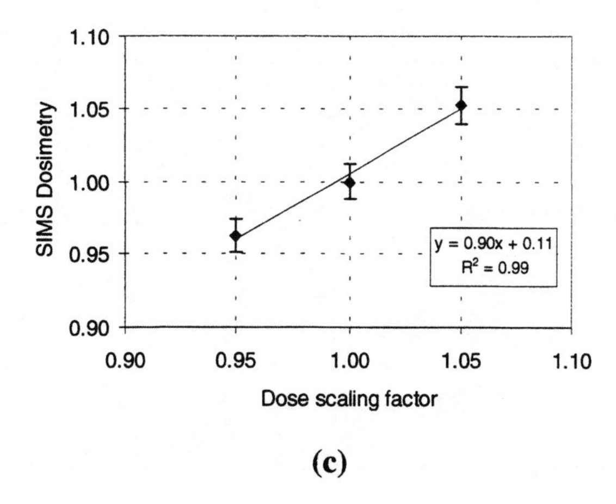

(c)

FIGURE 12. SIMS linearity response with the implanter dose scaling factor variation. (a) LEHD Boron; (b) HELD Arsenic; (c) MEHD Boron.

TABLE 7. Summary of SIMS linearity response curves

Implant	Slope	Intercept	R2
HELD P	1.15	0.15	0.99
HELD B	1.09	0.08	0.97
MEHD As	1.02	0.02	1.00
MEHD B	0.90	0.11	0.99
LEHD B	1.00	0.00	0.88
LEHD As	0.95	0.06	0.98

3.3 Multiple setup reproducibility

a) Relative Sensitivity Factors and dosimetry precision

The multiple setup variation in the relative sensitivity factors (RSF) used for profile quantification for an As standard is shown in figure 13. The figure shows that the RSF can vary by as much as 70 %, however, this variation does not prevent the high precision quantification of an As sample (note the minimal variation in the measured dose). The source of the RSF variation is traceable to detector degradation over time and the difference in tuning from various operators, but it has little effect on the ability of SIMS to reproducibly achieve the measurement of a particular dose over an extended period of time.

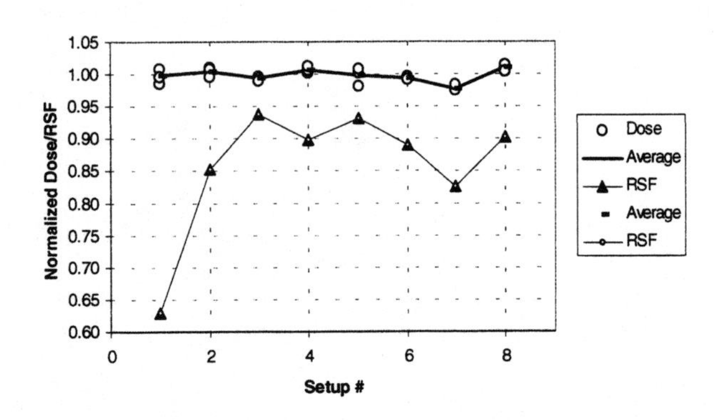

FIGURE 13. Long-term SIMS dosimetry and RSF variations for MEHD As implant

b) Dose reproducibility

The ability of SIMS to measure the same dose on the same sample over a period of time is a critical sanity check of the SIMS tool setup reproducibility. Dose reproducibility was studied on a single SIMS tool across a number of SIMS setups over a 4-6 month period. Table 8 summarizes the results from this experiment

TABLE 8. Multiple setup reproducibility data

Implant	Min	Max	Spread	%RSD
HELD B	0.985	1.015	3.0%	1.14%
MEHD B	0.988	1.008	2.0%	0.72%
HELD P	0.970	1.039	6.7%	2.19%
HELD As	0.986	1.022	3.6%	1.05%
MEHD As	0.978	1.009	3.1%	0.98%

Figure 14 a-b show the multiple setup reproducibility data along with 3σ error bars (based on the short-term repeatability obtained in section 3.2).

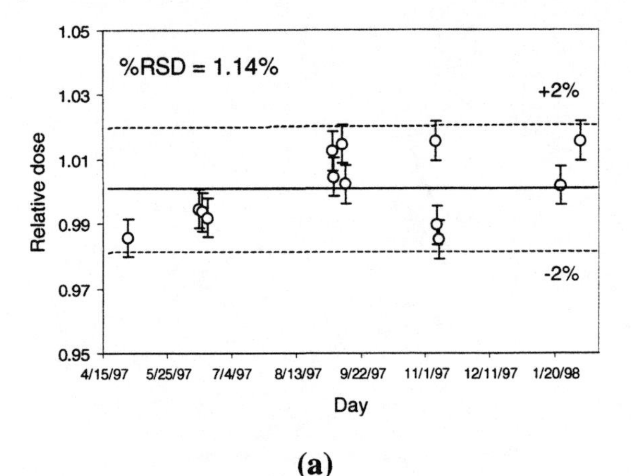

(a)

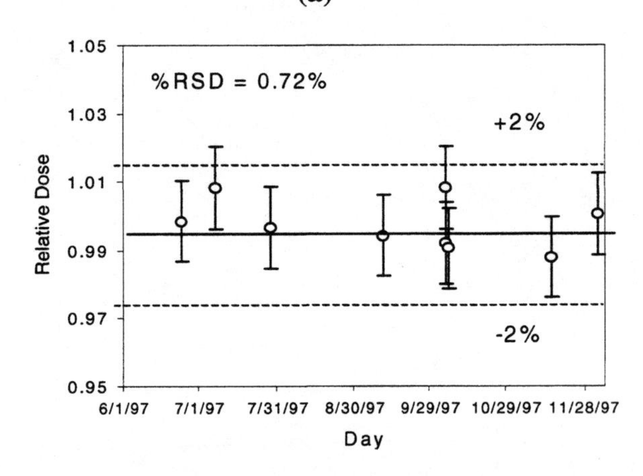

(b)

FIGURE 14. Multiple setup reproducibility of SIMS dosimetry for (a) LEHD boron; (b) MEHD boron implants

c) Profile shape reproducibility

The ability of the SIMS technique to reproduce a particular profile shape on the same sample over an extended period of time is shown in figures 15a (MEHD As) and 15b (HELD As) implants. These figures show overlays of data from > 10 different SIMS setups over a 4-6 month period. Excellent profile shape overlays are shown, demonstrating that the SIMS technique can very reproducibly measure a given profile shape.

As and B. Table 9 summarizes the results from the various laboratories and figures 16 a-c show the statistical response curves for three of the implants.

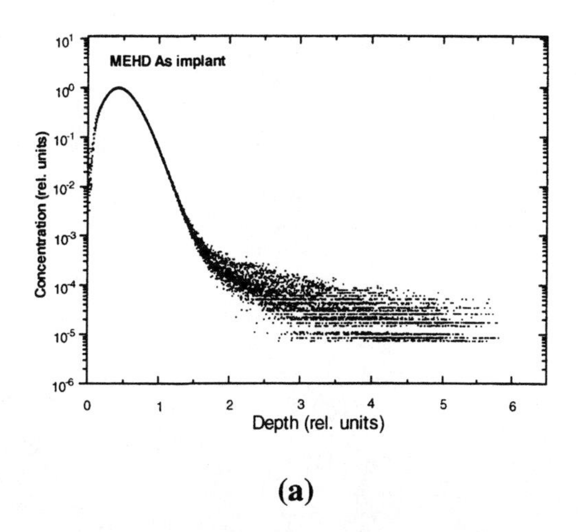

(a)

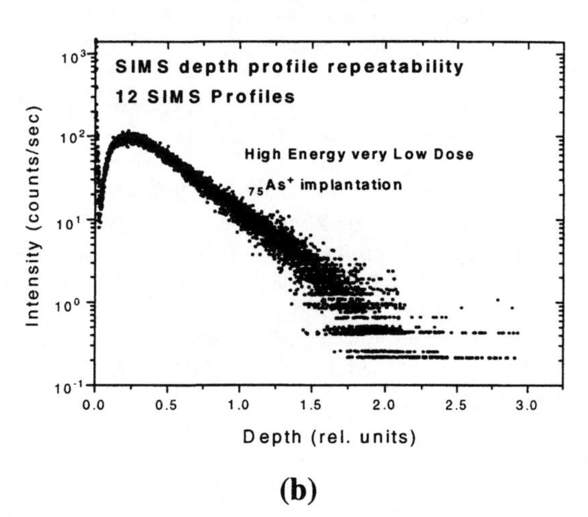

(b)

FIGURE 15. Multiple day SIMS reproducibility of dopant profile shape (a) 21 MEHD Arsenic profiles; (b) 12 HELD arsenic profiles.

3.4 Cross tool dose comparisons

To study the ability of the SIMS technique to obtain identical output from several tools samples were distributed to up to 5 different SIMS laboratories containing various generations of magnetic sector SIMS tools.

TABLE 9. Multiple SIMS tools repeatability data.

Implant	Min	Max	Spread	%RSD
HELD B	0.993	1.004	1.04%	0.43%
HELD P	0.994	1.006	1.15%	0.67%
MEHD B	0.968	1.017	4.86%	2.26%
MEHD As	0.992	1.001	1.39%	0.58%

The samples were analyzed by the various laboratories following the methodology for repeatability measurements described in section 3.2. Concentration vs. depth profiles were generated for HELD B and Phosphorus, and MEHD

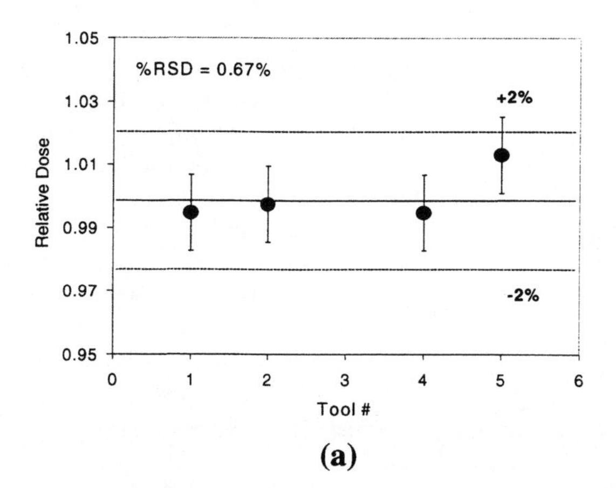

(a)

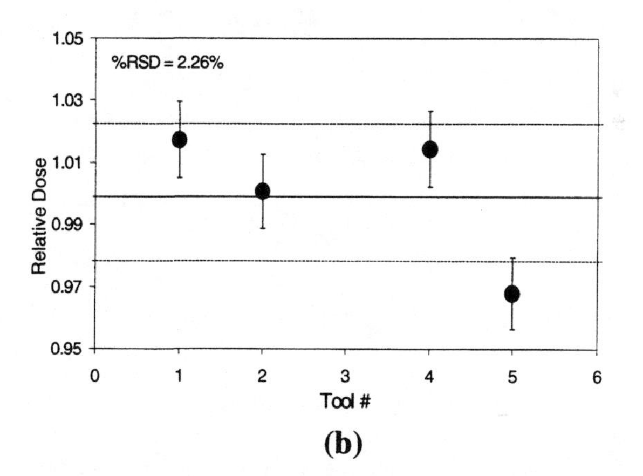

(b)

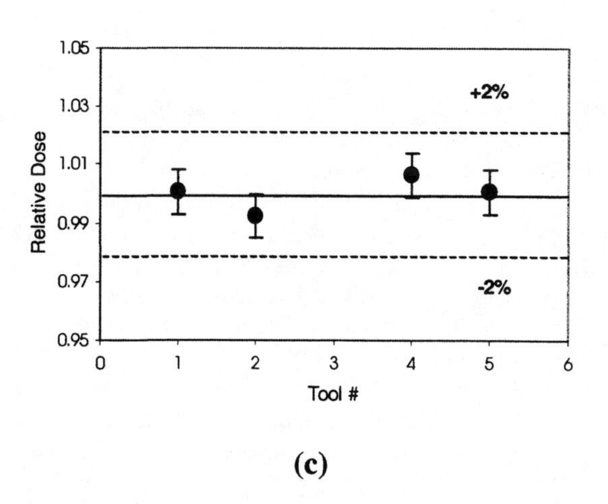

(c)

FIGURE 16. Cross SIMS tools matching for implant dosimetry. (a) HELD phosphorus; (b) MEHD boron; (c) MEHD arsenic implants.

3.5 Crater Depth repeatability

The depth scale on a SIMS depth profile is normally determined by measurement of the crater depth resulting from the SIMS sputtering process. The accuracy and precision of the crater depth measurement is equally important to the precision and accuracy of the SIMS dose measurement as the measurement of the impurity profile. To that end the ability of stylus profilometers to reproducibly measure the same depth was studied. A Si sample was sputtered with Oxygen bombardment to an approximate depth of 4000 angstroms, the resulting crater was measured with several generations of stylus profilometers. The results are shown in fig. 17.

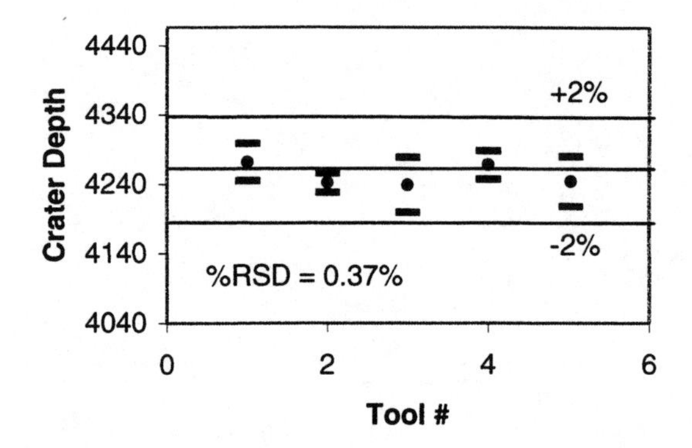

FIGURE 17. Stylus profilometer crater depth repeatability determination

The results in fig. 17 demonstrate that a high level of precision can be obtained from one profilometer to the next, allowing very precise depth scales to be placed on SIMS depth profile data from various laboratories.

Conclusions

SIMS has been used extensively for supporting technology development and manufacturing of advanced ULSI processes. The most common applications are depth profiling of dopant distributions and contamination control. SIMS has demonstrated a depth resolution and, when properly practiced, also a precision sufficient to satisfy requirements of the current and future 2-3 generations of ULSI processes. The authors have shown short term repeatability to be better than 0.5% for all implants studied and long term reproducibility of better than 2.0% RSD. SIMS has demonstrated an excellent linearity of response to small changes in implanter dose. When strict protocols are followed different SIMS tools can be setup to deliver the same dose information within better than 2% RSD.

Acknowledgements

The authors would like to acknowledge the following people for aid in gathering the results presented in this paper: Leonard Kulig, Brad Bates, Mike Frost, Jim Reinhardt, Roman Khait, Glen Martin, Jim Hauke, Jeanette DeBar, Jean Moran and Greg Scott.

References

1. Corcoran, S.F., "Challenges for SIMS in microelectronics", in *Secondary Ion Mass Spectrometry: SIMS XI*, John Wiley & Sons Inc., 1997, pp.107-112
2. Moore, G., *Electron. Aust.*, **42**, 14 (1983)
3. Adriaens, A. and Adams, F., "Reproducibility aspects of SIMS isotope ratio measurements: A statistical point of view", *Secondary Ion Mass Spectrometry: SIMS XI*, John Wiley & Sons Inc., 1997, pp.35-38
4. Benninghoven, A., Rudenauer, F.G., and Werner, W.H., *SIMS - Basic concept, instrumental aspects, applications and trends*, John Wiley & Sons Inc., 1987
5. Wilson, R.G., Stevie, F.A. and Magee, Ch.W., *Secondary Ion Mass Spectrometry*, John Wiley & Sons Inc., 1989
6. Trahey, N.M. *Standard Reference Materials Catalog 1998-1999*, NIST Special publications 260, US Department of Commerce, 1998, p.118
7. Simons, D., "Development of SRM 2134, an As implant in Si standard for SIMS", in *11 Annual SIMS Workshop*, Austin, Texas, May 1998
8. CEA seminars Series: *Working smarter with Surface Analysis Techniques*, Charles Evans & Associates, 1997
9. Simons, D., in *Secondary Ion Mass Spectrometry: SIMS IX.*, 1993, p.167
10. Simonton, R.B., Magee, C.W., and Tasch, A.F., *Nucl. Instr. & Meth.*, **B74**, 142 (1993)
11. Corcoran, S.F., and Simons, D.F., "Quantitative aspects of SIMS: precision accuracy and reproducibility", in *Int. Workshop for Measurements, Characterization and Modeling of Ultra-Shallow Doping Profiles*, RTP, MCNC, 1995,
12. Parks, C. C., "SIMS - a key tool in advanced DRAM Development", *Secondary Ion Mass Spectrometry: SIMS X*, John Wiley & Sons Inc., 1997, pp.55-63
13. Lu, S., Golonka, L., Schenk, R., and Evans, K., "Implantation/Diffusion process matching characterization via SIMS", in *Ion Implantation Technology-96*, Austin, Texas, IEEE Press, 1997, pp.222-225
14. Current, M., Ishida, M., Larson, L., and Jones, E., "Challenges of doping shallow junctions for giga-scale devices", in *Challenges for 2000 and beyond*, GSVIUG Symp., NCCAVS, 1997
15. Hemment, P.L.F., and Roxburgh, C., "Device Level measurements", in Materials and Process Characterization of Ion Implantation, ed. By M.I. Current and C.B. Yarling, Austin, Texas, Ion Beam Press, 1997, pp.74-98
16. Liu, G.L., Uchida, H., Aikawa, I., Kuroda, S., and Narashita, N., *J. Vac. Sci. Technol.*, **B14(1)**, 324 (1996)
17. Wilson, R.G., and Novak, S.W., *J. Appl. Phys.*, **69**, 466 (1991)

18. Current, M., et.al in Ion Implantation Technology-94, Amsterdam, Elsevier, 1995, pp.622-625
19. Jones, M.A., and Sinclair, F., "A cross wafer channeling variations on batch implanters: a graphical technique to analyze spinning disk systems", in *Ion Implantation Technology-96*, Austin, Texas, IEEE Press, 1997, pp.264-267
20. Dowsett, M.G., "SIMS depth profiling of ultrashallow implants and junction in silicon - present performance and future potential", *Secondary Ion Mass Spectrometry: SIMS XI*, John Wiley & Sons Inc., 1998, pp.260-264
21. National Technology Roadmap for Semiconductors, Semiconductor Industry Association, 1997, p. 70.
22. Canteri, R., in *Secondary Ion Mass Spectrometry: SIMS VII*, John Wiley & Sons Inc., 1992, p.135
23. Smith, H., and Harris, W., "SIMS Quantification of AsCs+ at CoSi2/Si Interfaces, in *Secondary Ion Mass Spectrometry: SIMS VIII*, John Wiley & Sons Inc., 1991, p. 99.
24. Hunter, J., and Budrevich, A., in preparation
25. Itgen, K., Benninghoven, A., and Niehuis, E., in 4th Int. Workshop - *Measurements, Characterization and Modeling of Ultra-Shallow Doping Profiles*, MCNC, 1997, p.10.1
26. Corcoran, S., and Felch, S., "Evaluation of Polyencapsulation, Oxygen Leak and Low Energy Ion Bombardment in the Elimination of SIMS Surface Ion Yield Transients", in J. Vac. Sci. Tech., B10, 1992, p. 342.
27. Smith, N.S., Dowsett, M.G., McGregor, B., and Phillips, P., in *Secondary Ion Mass Spectrometry: SIMS X*, John Wiley & Sons Inc., 1997, p.363
28. Manchanda, L., "Gate dielectrics for high performance ULSI: practical and fundamental limits", in Semiconductors Characterization, ed. by W.M. Bullis, D.G.Seiler and A.C.Diebold, NY, AIP Press, 1996, pp.123-130
29. Wu, L., Neil, T., Sieloff, D., Lee, J.J., and Hedge, R.I., "SIMS characterization of Interfacial Nitrogen for Oxynitride dielectric films", in *Secondary Ion Mass Spectrometry: SIMS XI*, John Wiley & Sons Inc., 1998, pp.260-264
30. Pundenzi, M.A.A., Diniz, J.A., Tatsch, P.J., Herion, J.K., and Muck A., "SIMS analysis of Silicon oxynitride films formed by N2+ and NO+ implantation", *ibid.*, pp.249-252
31. Oakey, P.R., Schauer, S.N., Cosway, R.G., and Griswold, M.D., "Characterization of thin oxynitride film", *ibid.*, pp.253-256
32. Variam, N., "Defect reduction strategies in ion implantation", in *Challenges for 2000 and beyond*, GSVIUG Symp., NCCAVS, 1997
33. Smith, S.R., Chia, V.F., and Ming Hong Yang, "Ion implanter diagnostics using Surface SIMS", in *Ion Implantation Technology-96*, Austin, Texas, IEEE Press, 1997, pp.512-515
34. Gates, J.K., and Molis, S., in *Secondary Ion Mass Spectrometry: SIMS X*, John Wiley & Sons Inc., 1996, p.511
35. Von Criegen, R., Jahnel, F., Lange-Giesler, R., Pearson, P., Hobler, G., and Simionescu, A., "Verification of Lateral SIMS as a method for measuring lateral dopant dose distributions in microelectronics test structures", in 4th Int. Workshop - *Measurements, Characterization and Modeling of Ultra-Shallow Doping Profiles*, MCNC, 1997, p.22.1-22.11
36. Cooke, G.A., Pearson, P., Gibbons, R., Dowsett, M.G., and Hill, C., J. Vac Sci. Technol., B14(1), 312 (1996)
37. Goodwin-Johansson, S., Xuefeng Liu, and Ray, M., "Two-dimensional doping profiles from experimentally measured one-dimensional Secondary Ion Mass Spectroscopy data", in J. Vac. Sci. Technol., B12(1), 1994, p. 247.
38. Stevie, F., Simons, D., McKinley, J., McMacken, J., Santiesteban, R., Flatch, P., and Becerro, J., "Dose calibration of ion implanters for semiconductor production", in *Secondary Ion Mass Spectrometry: SIMS XI*, John Wiley & Sons Inc., 1998, p. 1007.

FRONT END PROCESSES - GENERAL

Characterization of Thin SiO_2 on Si by Spectroscopic Ellipsometry, Neutron Reflectometry, and X-Ray Reflectometry

C.A. Richter, N.V. Nguyen, J.A.Dura[‡], and C.F. Majkrzak[‡]

Semiconductor Electronics Division and [‡]NIST-Center for Neutron Research
National Institute of Standards & Technology
Gaithersburg, MD 20899 USA

We compare the results of neutron reflectometry, x-ray reflectometry, and spectroscopic ellipsometry measurements of a thin oxide film (≈10 nm). These methods, which arise from three physically different scattering mechanisms, each determine physical properties of the film, and each has its distinctive strengths. This comparison of the extracted depth profiles of the physical properties gives multiple perspectives on the thickness and interfacial characteristics of an SiO_2 film on Si. This information improves our understanding of the material system and is helpful for refining the models used to analyze similar structures. The extracted thickness of the SiO_2 film is in agreement for these three methods.

INTRODUCTION

Improved metrology methods are critically needed for characterization and process control of the ultrathin gate dielectrics of today's and tomorrow's silicon technology. Gate dielectrics in advanced commercial products have been aggressively scaled to effective thicknesses of 4 nm and less. ULSI scaling laws predict that leading-edge gate dielectrics will be effectively 2 nm thick by 2001 (1). In order to meet these goals, film thickness measurement techniques and appropriate reference artifacts need to be improved for process monitoring and to expedite process development (2).

Optical techniques such as ellipsometry are generally used to monitor gate dielectric thickness in manufacturing. Historically, these methods have been effective and robust. Unfortunately, for very thin films, the accuracy of these optical methods degrades because of their model dependence (3). Improved models must be developed to confidently extract accurate optical properties and film thickness. In an effort to gather data needed to develop better physical models and subsequently improve the overall accuracy of the measurements, we have performed a comparison of neutron reflectometry, x-ray reflectometry, and spectroscopic ellipsometry (SE) on a thin oxide film. We are able to gain multiple perspectives on the thickness and interfacial structure of this thin SiO_2 film on Si from this comparison of the structural profiles obtained by these three different techniques.

Neutron and x-ray reflectometry are nondestructive, short wavelength (λ = 0.475 nm, and λ = 0.154 nm, respectively, in this experiment) techniques. X-ray reflectometry is a relatively common analytical tool used for structural determination due to its high resolution and wide availability. It is not well suited for the case of SiO_2 on Si because the x-ray contrast (or difference in scattering length densities [SLDs]) is relatively low, 7.6% (4). Neutron reflectometry is better suited for the study of the SiO_2/Si system because there is a large contrast, 65%, between the nuclear scattering length densities of the two materials. Its principal disadvantage is that measurements must be made at a facility for neutron scattering. In contrast, SE is a quick, powerful, and commonly used nondestructive method to determine the dielectric functions and thicknesses of thin films. All three of these methods have submonolayer sensitivity to changes in film thickness.

The fundamental physics that gives rise to the scattering mechanisms differs for each of these characterization techniques; therefore, they are independent measures of the film thicknesses and film properties. The x-ray SLD is primarily determined by the total electron density. For x-rays, larger atomic numbers lead to larger scattering lengths. Neutron scattering lengths result from a nuclear reaction between a neutron and a nucleus; the strength of this interaction specifically depends upon the nuclei type and there is no simple trend such as is found for x-rays. Thus, x-ray and neutron scattering depend solely upon the average scattering lengths of the constituent elements weighted by the appropriate number density. Ellipsometry is dependent upon the optical properties and film thickness(es) of the material(s) being measured. Because ellipsometry measures relative changes in the phase and intensity of light polarization upon reflection from a sample, it is highly sensitive to these quantities, i.e., thickness and optical properties.

In both x-ray and neutron reflectometry, the reflected intensity, rather than amplitude, is measured, and thus phase information is lost. The reflectivity is the square of the complex amplitude reflected from the sample, and includes refractive and multiple scattering events. The interference of the waves scattered from the interfaces causes

CP449, *Characterization and Metrology for ULSI Technology: 1998 International Conference*
edited by D. G. Seiler, A. C. Diebold, W. M. Bullis, T. J. Shaffner, R. McDonald, and E. J. Walters

oscillations in the reflected intensity. At relatively higher values of momentum transfer ($Q = (4\pi/\lambda)sin\theta$, where θ is the angle of incidence of the neutrons), where the refractive effects are negligible, the period of these oscillations is inversely proportional to the separation of the interfaces. The data in Figures 1 and 3 show examples of these oscillations.

EXPERIMENTAL

The sample used in this comparison is a nominally 10.0 nm thick SiO_2 film thermally grown on a <100> Si substrate (5). This moderate thickness was chosen to increase our confidence in the results for each of the methods while remaining thin enough to be technically relevant. An aggressive clean (6) was initially used in an effort to remove any organic contamination on the SiO_2 surface before the comparison began. Ellipsometry was used throughout the comparison to monitor the wafer. The specific history of the sample plays an important role in this comparison. Ellipsometry performed before and after the neutron scattering indicated that a contamination layer was deposited on the sample some time during the transit, handling, loading, and/or unloading of the sample for the neutron studies. X-ray scattering, and subsequently SE, were performed on the sample while this contamination was present on the sample's surface. The sample was again cleaned. SE and then x-ray reflectometry were performed and indicated the removal of the contamination layer. Ellipsometry data taken after this second x-ray measurement indicate that minimal contamination occurred during the x-ray measurement process.

RESULTS

Neutron Reflectometry

The neutron reflectivity data and fits are shown in Figure 1. The period of the strong oscillations gives an accurate indication of the thickness of the SiO_2 film. Modeling of the data is necessary in order to properly fit the data and to derive more detailed structural information. Details of this fitting are described elsewhere (7). This modeling indicates the SiO_2 film is 10.27 nm thick with a 0.24 nm transition region at the SiO_2/Si interface and 0.69 nm of contamination at the film's surface. In the analysis of both the neutron and x-ray reflectivity data, interfaces are modeled by an error function transition between the two materials; therefore, unlike the discrete slabs used below in the SE analysis, the film thicknesses quoted are defined as the center-to-center distance of the two interfaces. In addition to determining the thickness of the various films, the analysis indicates that the SiO_2 film has a uniform density (of 2.21 $g{\cdot}cm^{-2}$) with respect to depth.

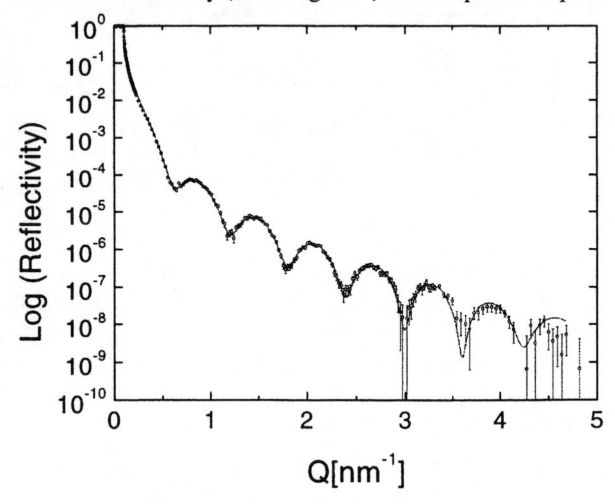

Figure 1. Neutron reflectivity results: Points are experimental data. Solid line is the best fit model.

Spectroscopic Ellipsometry

A well-characterized rotating-analyzer ellipsometer (8) was used to acquire data in the energy range (1.5 to 6.0) eV. Ellipsometry measures relative changes in amplitude (Ψ) and phase (Δ) of polarized light upon reflection from a sample (9). Figure 2 shows Ψ and Δ for the sample, both immediately after an aggressive clean and after neutron reflectivity. Qualitatively, the additional surface contamination is readily observed as a change in Δ. These data can be fit moderately well by using a single-slab model consisting of uniform SiO_2 on a Si substrate. When this is done, it is found that the sample is effectively 11.04 nm before and 11.97 nm after the neutron reflectivity measurements were made. We attribute this increase in the effective thickness of the sample to a contamination layer on the top surface.

A multi-layer model was used to fully interpret the SE data and extract the film thickness. This discrete-slab model consists of: silicon substrate, a thin "inter-layer" between the substrate and the SiO_2 film, SiO_2, and a contamination layer on top of the SiO_2. This model is physically realistic and was chosen to be consistent with those used in the interpretation of the neutron and x-ray reflectivity data. To simplify the analysis, literature values of dielectric function of the silicon substrate (10) were used, and the refractive index of bulk fused silica was used for the SiO_2 layer (11). An effective medium approximation consisting of a mixture of Si and fused silica was chosen for the interlayer. Because the properties of the surface contamination are unknown, the dielectric function of this layer was extracted via a Sellmeier dispersion model. This complete multi-layer model gives an improved goodness of fit when compared to a single-slab model, and it yields excellent fits as illustrated in Figure 2.

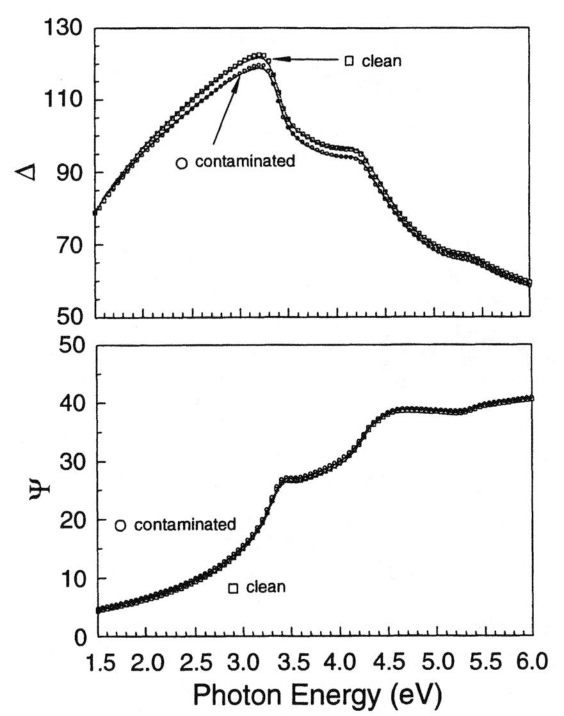

Figure 2. Spectroscopic ellipsometry results for the contaminated and cleaned sample. Open points are experimental data. Solid lines represent the multi-layer model.

The multi-layer model yields an SiO_2 film thickness of 10.21 nm and 10.37 nm for the clean and contaminated sample, respectively, and a 0.20 nm interlayer for both. The results indicate 1.02 nm of contamination on the sample after the neutron reflectivity and 0.43 nm of contamination on the SiO_2 surface after the cleaning procedure. The optical constants used in these models are shown in Figure 4 for one photon energy (1.96 eV).

X-ray Reflectometry

Figure 3 shows the x-ray reflectometry results. There are two periods observed in the data taken while the sample was contaminated: a high-frequency period corresponding to the SiO_2 film, and a low-frequency modulation due to the contamination layer. When these data are modeled (7), it is found that there is a 0.18 nm SiO_2/Si interface region, 10.15 nm of SiO_2 (with a density of 2.24 gm/cm^3), and 1.03 nm of surface contamination. After cleaning, only one oscillation period is easily observed corresponding to a 10.17 nm SiO_2 film. However, the best fits indicate that, even after cleaning, a small layer of less dense surface contamination (0.54 nm) exists. As was found in the neutron results, the x-ray analysis indicates that the SiO_2 film has a uniform density with respect to depth.

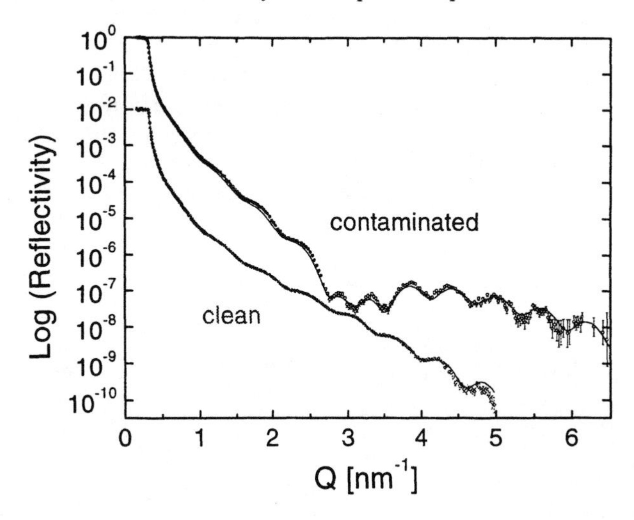

Figure 3. X-ray reflectivity results for the contaminated and cleaned sample. The data for the clean sample is offset. Open points are experimental data. Solid line represents the multi-layer model.

COMPARISONS

Neutron and x-ray SLD depth profiles and the dielectric function for the specific models used to fit the data are shown in Figure 4. The derived thicknesses are in good agreement for the three methods. The interface between the Si substrate and the SiO_2 is found to be relatively abrupt (0.18 nm to 0.24 nm) by all the methods. The methods are also in good agreement for the thickness of the SiO_2 layer. By treating the SE and x-ray reflectivity on the clean and contaminated sample as individual measures of the film properties, we have five independent measures of the film thickness. The mean of these five separate measurements is 10.27 nm $\pm$ 0.13 nm (a 1σ standard deviation of 1%). For this numerical comparison of the abrupt slab model used in the SE interpretation with the graded interface description used to analyze the reflectivity results, half of the SE interlayer (0.1 nm) is added to the thickness of the SiO_2 film. Note that there is no large variation in the density of the SiO_2 at either the top or bottom interface. The agreement between the three techniques is weakest for the thickness of the contamination layer, as expected. All properties of this layer are derived from the measurements, because we have no information *a priori* for this layer. The agreement between SE and x-ray reflectivity is surprisingly good for the thickness of the contamination on both the clean and dirty sample. There is some disagreement between the neutron scattering results and the values obtained for the thickness of the contamination layer on the sample after neutron scattering as measured by SE and x-ray reflectometry. This discrepancy is most likely due to a true difference in the amount of surface contamination on the sample. One can easily hypothesize that the sample was partially contaminated before/during the neutron reflectivity measurements and partially contaminated during the unloading of the sample and transportation to the SE and x-ray equipment. This hypothesis is consistent with the observation that the thickness of the contamination layer is less as measured by the neutrons than when measured by SE and x-ray.

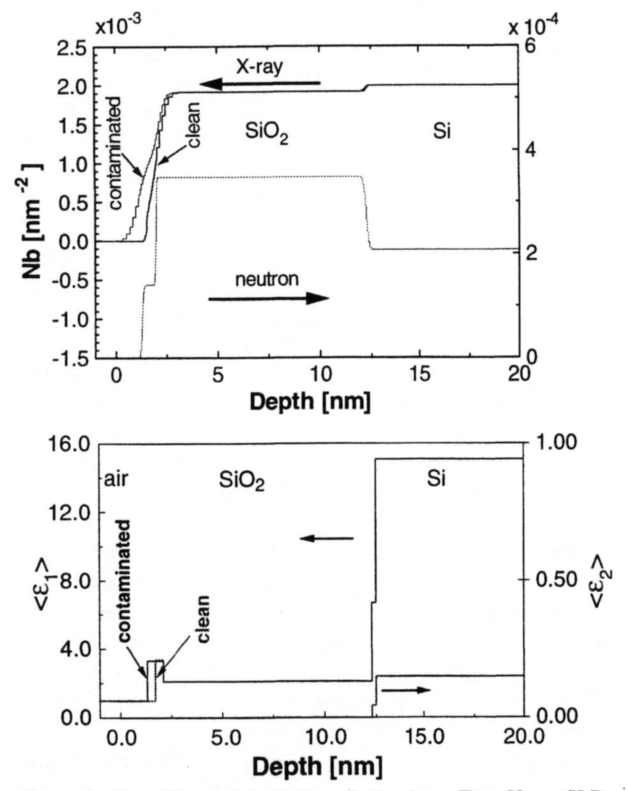

Figure 4. Specific models used to fit the data. Top: X-ray SLD depth profiles for the sample after neutron scattering (contaminated) and after cleaning (clean). Also, SLD extracted from neutron reflectivity. Bottom: The optical constants to fit the SE data for the sample after neutron scattering (contaminated) and immediately after cleaning (clean).

This set of experiments points out how difficult such multi-method comparisons can be due to handling and contamination issues. In this particular instance, if the contamination had not been explicitly identified, the results would not be in such good agreement, even for this moderately thick film. This issue will only become more critical for ultrathin films in the 1.0 nm to 3.0 nm thickness range. A second difficulty arises due to the models used to extract the results. When the models are not identical, it is conceptually difficult to compare the results with the sub-0.1 nm resolution demanded in this type of comparison. For example, in the work presented here, the SiO_2/Si interface was modeled by a discrete slab of mixed media to interpret the SE data. In the analysis of the reflectivity data, this same interface was modeled as a graded (error function shaped) transition between Si and SiO_2.

The results are model dependent for all three techniques. It should be noted that these models are not uniquely determined by the experimental data, and other models may also produce good fits. For example, an ellipsometry model based upon a different choice for the dielectric function or a reflectivity model based upon different SLDs may lead to acceptable fits and produce different results. However, the particular models used here were guided by knowledge of the structure of the system. Furthermore, because the data from three diverse, independent characterization probes can all be interpreted by using nominally the same model, we can be confident that this is a realistic model for this sample.

Each of the three probes has its own distinctive advantages. X-ray reflectometry, for example, had the ability to clearly resolve the contamination layer. Neutron reflectometry, due to the high contrast at the Si/SiO_2 interface, should have the best ability to resolve structural changes (if present) at that interface. SE is readily available and has high sensitivities. This allowed it to be used as a monitor for sample modification during the course of the experiment, and the contamination layer was quickly observed and identified as contamination by this monitoring process.

Looking forward to ultrathin dielectrics (with film thicknesses on the order of 2 nm), one can ask in what ways will these methods continue to be useful. It is known that all three techniques have the capability to resolve sub-monolayer changes in layer thickness. Simulations based upon realistic material and interface parameters indicate that SE and neutron reflectometry (at $\lambda = 0.475$ nm) can easily measure SiO_2 layers down to 2.0 nm thick (7). Thus, they will remain useful measurement methods for ultrathin dielectrics. Due to the lack of contrast between Si and SiO_2, comparable simulations for x-ray reflectivity (at $\lambda = 0.154$ nm) (7) show very weak oscillations for SiO_2 layers 6 nm and thinner, indicating that x-ray reflectometry at this wavelength is not an ideal choice for measuring the thickness of ultrathin SiO_2 films.

CONCLUSIONS

We have used three nondestructive metrology methods to experimentally determine the physical characteristics of a thin SiO_2 film. For the moderate thickness (10 nm) of the film in this comparison, each of the methods – SE, x-ray and neutron reflectometry – effectively extracts the SiO_2 thickness. There is good agreement among the results obtained by these techniques. Because the fundamental basis of these techniques differs, it strengthens our confidence in the models used to interpret the data. Since ellipsometry will remain an essential tool to monitor and control the production of gate dielectrics for future integrated circuit generations, multi-method comparisons such as the one presented here are expected to assist in the identification of a correct model to be used for ellipsometry and thus improve metrology accuracy. We are actively pursuing further multi-method comparisons and plan to expand to thinner and more diverse gate dielectric materials as well as more measurement techniques. However, the sample handling and contamination issues need to be addressed to ensure the success of such comparisons.

ACKNOWLEDGMENTS

This work was funded in part by the National Semiconductor Metrology Program at NIST. We would like to thank D.L. Blackburn and J.R. Ehrstein for insightful discussions and B.J. Belzer for technical assistance.

REFERENCES

1 The 1997 National Technology Roadmap for Semiconductors.
2 Durgapal, P., Ehrstein, J.R., and Nguyen, N.V., *Thin Film Ellipsometry Metrology,* this volume.
3 Nguyen, N.V., and Richter, C.A., Thickness Determination of Ultra-Thin SiO_2 Films on Si by Spectroscopic Ellipsometry, Silicon Nitride and Silicon Dioxide, Thin Insulating Films IV, M.J. Deen, W.D. Brown, S.I. Raider and K. Sundaram, Editors, PV 97-10, Montreal, Canada - May 1997.
4 Dura, J.A., and Majkrzak, C.F., in Semiconductor Characterization: Present Status and Future Needs, W.M. Bullis, D.G. Seiler, and A.C. Diebold, Editors (AIP Press, Woodbury, NY, 1996), pp. 549-554.
5 The sample was grown in O_2 at 900 °C followed by an annealing step in N_2 at 1050 °C for 30 min.
6 As an aggressive clean, the sample was immersed in a solution of H_2SO_4 with hydrogen persulate (1 mg/l) at 80 °C for 10 min, followed by a deionized H_2O rinse.
7 Dura, J.A., Richter, C.A. Majkrzak, C.F., and Nguyen, N.V. submitted to Appl. Phys. Lett.
8 Nguyen, N.V., Chandler-Horowitz, D., Pellegrino, J.G., and Amirtharaj, P.M., in Semiconductor Characterization: Present Status and Future Needs, W.M. Bullis, D.G. Seiler, and A.C. Diebold, Editors (AIP Press, Woodbury, NY, 1996) pp. 438-442.
9 Azzam, R.M.A. and Bashara, N.M., Ellipsometry and Polarized Light (Elsevier, Amsterdam, 1986).
10 Yasuda, J. and Aspnes, D.E., Appl. Optics, 33, 7435 (1994).
11 Malitson, H., J. Opt. Soc. Am., 55, 1205 (1965).

Structure, Composition and Strain Profiling of Si/SiO_2 Interfaces

G. Duscher[1,2], S. J. Pennycook[1], N. D. Browning[2], R. Rupangudi[3], C. Takoudis[3], H-J Gao[1,4], and R. Singh[4]

[1]*Solid State Division, Oak Ridge National Laboratory, PO Box 2008, Oak Ridge, TN 37831-6030, USA*

[2]*Department of Physics, University of Illinois at Chicago, Chicago, IL 60607-7059*

[3]*Department of Chemical Engineering, University of Illinois at Chicago, Chicago, IL 60607-7000*

[4]*Department of Materials Science and Engineering, University of Florida, Gainesville FL 32611*

Recently, the scanning transmission electron microscope has become capable of forming electron probes of atomic dimensions. This makes possible the technique of Z-contrast imaging, a method of forming incoherent images at atomic resolution having high compositional sensitivity. An incoherent image of this nature also allows the positions of atomic columns in a crystal to be directly determined, without the need for model structures and image simulations. Furthermore, atomic resolution chemical analysis can be performed by locating the probe over particular columns or planes seen in the image while electron energy loss spectra are collected. We present images of the Si/SiO_2 interface showing no crystalline oxide, compositional profiles at 2.5 Å resolution across a sample formed by Si(100) oxidation with an oxygen/nitrogen containing gaseous source showing an extended sub-stoichiometric zone, and strain profiles at a rough interface showing static rms displacements of ~ 0.1 Å extending 10 Å into the crystalline Si.

INTRODUCTION

A scanning transmission electron microscope (STEM) is designed to form a small electron probe which is scanned across a thin specimen as shown in the schematic of Fig. 1. With a crystalline specimen aligned parallel to the beam, the probe preferentially channels along the atomic columns, one by one as it is scanned across the specimen. A Z-contrast image results from mapping the intensity of electrons scattered through relatively large angles, using the annular detector. Because of the large size of the detector, phase contrast effects between different Bragg reflections are integrated to give an image based on total scattered intensity. As high angle scattering comes predominantly from the atomic nuclei, bright features in the image correspond directly to columns of atoms, with the brightness determined by their relative atomic number Z. Unlike conventional high resolution electron microscopy (HREM), this technique allows the positions of atomic columns to be determined directly and uniquely from the image to a high accuracy, without the need for extensive simulations of model structures. It provides an incoherent image of the atomic structure of materials and effectively bypasses the phase problem of HREM (1-4).

The world's highest-resolution Z-contrast microscope is located at ORNL, a VG Microscopes HB603U with a 300 kV accelerating voltage, with a Scherzer resolution of 1.26 Å. This is sufficient to resolve and distinguish the

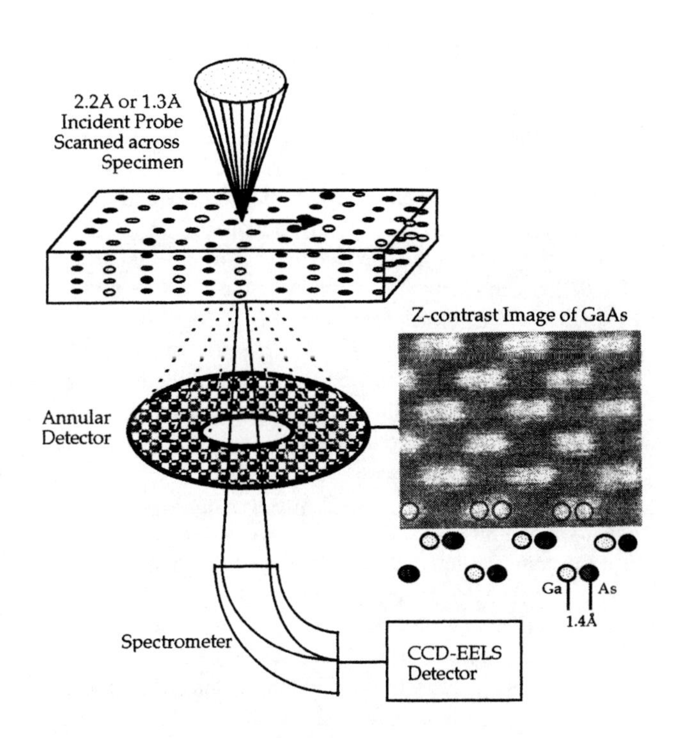

FIGURE 1. Schematic showing Z-contrast imaging and atomic resolution electron energy loss spectroscopy on a STEM. The image is of GaAs taken on our 300 kV STEM which directly resolves and distinguishes the sublattice.

CP449, *Characterization and Metrology for ULSI Technology: 1998 International Conference*
edited by D. G. Seiler, A. C. Diebold, W. M. Bullis, T. J. Shaffner, R. McDonald, and E. J. Walters
1998 The American Institute of Physics 1-56396-753-7/98/$15.00

sublattice in compound semiconductors for the first time, as seen in the image of GaAs shown in the schematic. For example, several different dislocation core structures at a CdTe/GaAs interface have been distinguished, including one structure not previously proposed (5). Electron energy loss spectroscopy (EELS) can be performed simultaneously with the Z-contrast image, allowing atomic resolution compositional profiling to be achieved (6). Furthermore, the fine structure on the core loss ionization edges maps the conduction band structure at the atomic columns illuminated by the probe. Since the probe can illuminate a single plane of atoms at an interface, local changes in band structure, valence and bond angles can be detected. At present the EELS capability is only installed on our older 100 kV STEM, which has a probe size of 2.2 Å, although a spectrometer is currently under construction for the 300 kV STEM.

To eliminate the possibility of beam damage affecting the energy loss results, a time sequence of spectra are taken from each region. This is particularly important at an interface, which typically damages before the perfect crystal adjacent to it. The spectra may then be compared, and if no evolution is seen through the sequence they can be added to improve statistics. Similarly, it is also simple to assess the contribution of ion milling surface damage to the shape of the edge by comparing spectra taken from different thicknesses.

THE STRUCTURE OF THE Si/SiO_2 INTERFACE

Conventional HREM images of amorphous materials always show a speckle pattern due to random interferences which tends to obscure the image at the interface and makes it difficult to infer atomic column positions. In addition, HREM images from the interface may show Fresnel fringe effects that can change the contrast of the last one or two Si layers. The direct nature of the Z-contrast image, however, makes it particularly simple to determine the structure at the Si/SiO_2 interface. Because there are no interference phenomena in an incoherent image, there are no Fresnel fringes or speckle pattern, and the last atomic columns of the crystal may be clearly seen.

The sample shown in Fig. 2 was prepared by thermal oxidation of a Si(111) wafer, which was subsequently used as a substrate for liquid phase epitaxy of Si. In this way a single microscope sample contained two Si/SiO_2 interfaces grown in different ways (7). Both interfaces were found to be atomically smooth and abrupt, as evident in Fig. 2. The oxide appears dark in the Z-contrast image because of the reduced atomic number (Z) and scattering power of the O, and also because the atoms are not aligned in well ordered atomic columns and no channelling of the electron probe occurs.

Superimposed on the image is the structure proposed by Ourmazd et al. (8) from HREM studies, based on a transition from crystal Si through crystal SiO_2 to amorphous SiO_2. Clearly the positions of the Si atoms are inconsistent with even a single monolayer of crystalline oxide at this particular interface. An alternative interpretation of the same phase contrast pattern has been made by Akatsu and Ohdomari (9). These authors proposed that interface roughness was responsible for the phase contrast effects. Clearly interface roughness will vary significantly from sample to sample. However, a sample having a rough interface is shown later for strain profiling (Fig. 6) and again there is no evidence for the presence of crystalline oxide at the interface.

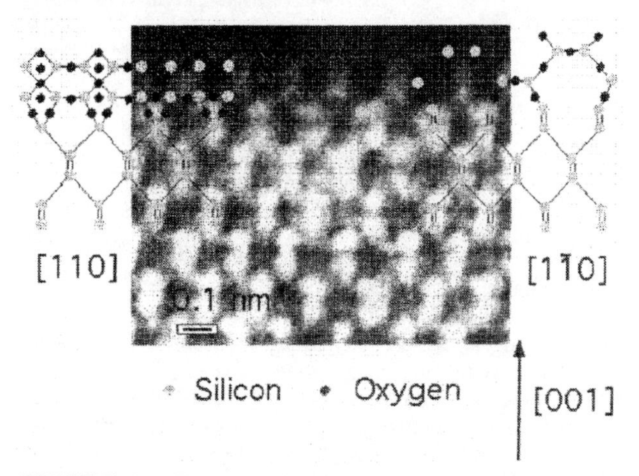

FIGURE 2. Z-contrast image of a Si/SiO_2 interface with superimposed structure of crystalline oxide. The "dumbbells" characteristic of the Si [110] projection are still present at the interface ruling out the presence of any crystalline oxide.

COMPOSITIONAL PROFILING AT THE Si/SiO_2 INTERFACE

There is no information in the Z-contrast image from the amorphous oxide. Its atomic structure is unresolved because of the limited resolution and detection sensitivity. Columns of atoms are visible in a crystal, but images from individual Si atoms are below the detection limit. However, information on composition and bonding can be readily obtained with EELS. Figure 3 shows a series of Si-$L_{2,3}$ ionization edges from a Si/SiO_2 interface produced by oxidation with a gaseous oxygen/nitrogen source. Each spectrum is obtained with the probe located at a different distance from the interface, as determined from the Z-contrast image. Due to the large band gap of the SiO_2, the onset of the ionization edge is 104 eV in the SiO_2 compared to 99 eV in the Si. It is clear that the edge profile evolves from Si to SiO_2 over a surprisingly extended region, greater than 2 nm.

It is worth stressing that this large interface width is more than an order of magnitude greater than the resolution of the measurement. Allowing for beam broadening in the amorphous oxide and the intrinsic delocalization of the inelastic scattering process (4,10) we calculate an instrumental resolution of 2.7 Å for this measurement.

Figure 4 shows the corresponding set of spectra at the O-K edge. Again, the shape of the edge evolves over a significant distance from the interface, only becoming characteristic of stoichiometric SiO_2 after 24 Å. The ratio of the first to second peak following the ionization threshold may be used to quantify the stoichiometry of the oxide (11,12). The resulting quantitative profile in Fig. 5 provides an independent verification of an extended sub-stoichiometric zone with an interface width comparable to that estimated from the Si-K edge spectra.

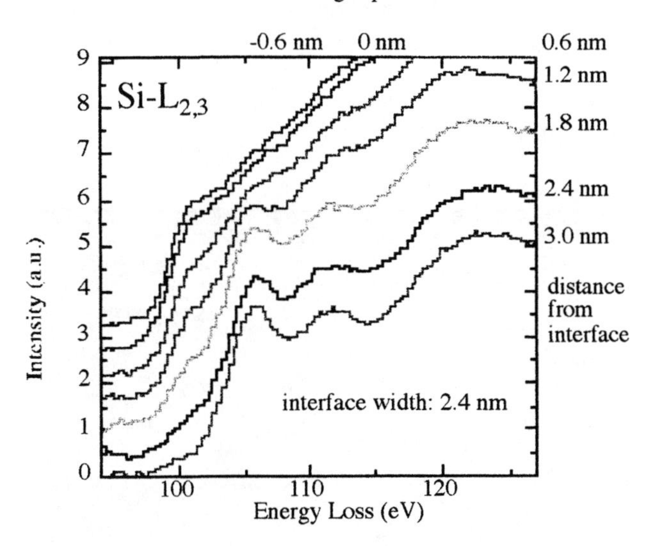

FIGURE 3. Si-$L_{2,3}$ spectra across a Si/SiO_2 interface showing evolution of the SiO_2 band gap. The full gap is not established until 2.4 nm into the oxide.

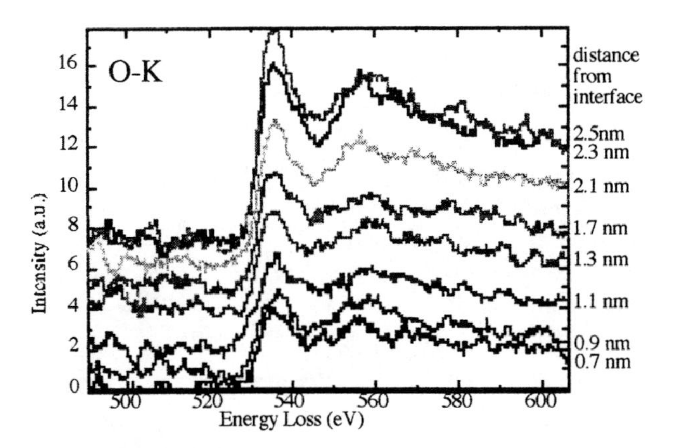

FIGURE 4. O-K spectra across a Si/SiO_2 interface. No further changes in the edge shape were seen closer to the interface than 0.7 nm.

STRAIN PROFILING ACROSS THE Si/SiO_2 INTERFACE

With a crystalline sample in a zone axis orientation, the intensity reaching the annular detector is predominantly thermal diffuse scattering. The cross section σ per atom depends on scattering angle and temperature through the atomic form factor f and the Debye-Waller factor M^T, as

$$\sigma = f^2 \left(1 - \exp(-2 M^T s^2) \right), \quad (1)$$

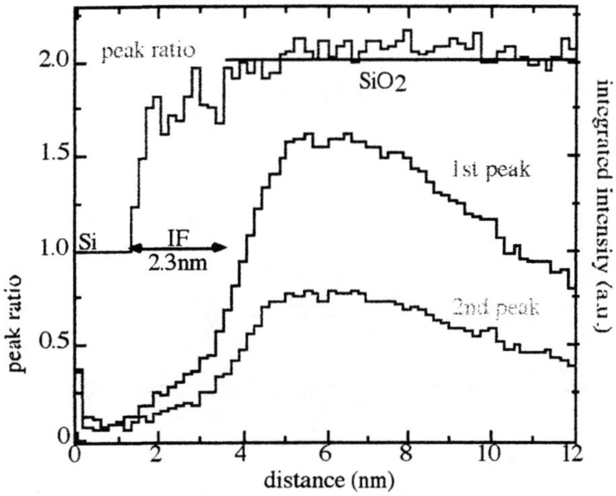

FIGURE 5. Profile of oxide stoichiometry obtained from the ratio of first to second peaks in the O-K edge spectra shown in Fig. 4 providing independent verification of an extended sub-stoichiometric zone.

where s = scattering angle, $M^T = 8\pi^2 \overline{u_T^2}$, and $\overline{u_T^2}$ is the mean square thermal vibration amplitude of the atom. In the presence of static random atomic displacements due for example to the proximity of an incoherent oxide interface, and assuming a Gaussian distribution of strain with a mean square static displacement of $\overline{u_S^2}$, the atomic scattering cross section will be modified to

$$\sigma^S = f^2 \left(1 - \exp(-2 (M^T + M^S) s^2) \right). \quad (2)$$

where $M^S = 8\pi^2 \overline{u_S^2}$. It is clear from the form of these expressions that both tend to the full atomic scattering cross section f^2 at a sufficiently high scattering angle. At lower angles where the Debye-Waller factor is significant, static strains comparable to the thermal vibration amplitude may lead to a significantly enhanced scattering cross section.

Strain contrast is seen at the thermally grown interface in Fig. 6. Here we show a bright field phase contrast image (a), showing dark contrast that could be due to a number of effects, such as strain, thickness variation, bending of the crystal or a combination of these mechanisms. Figure 6(b) shows a Z-contrast image collected simultaneously using a low (25 mrad) inner radius for the annular detector. Now there is a bright band near the interface indicating additional scattering. With this image alone, this additional scattering could be due either to strain or to the presence of some heavy impurity atoms. However, when the inner detector angle is increased further to 45 mrad the bright line disappears, showing that the

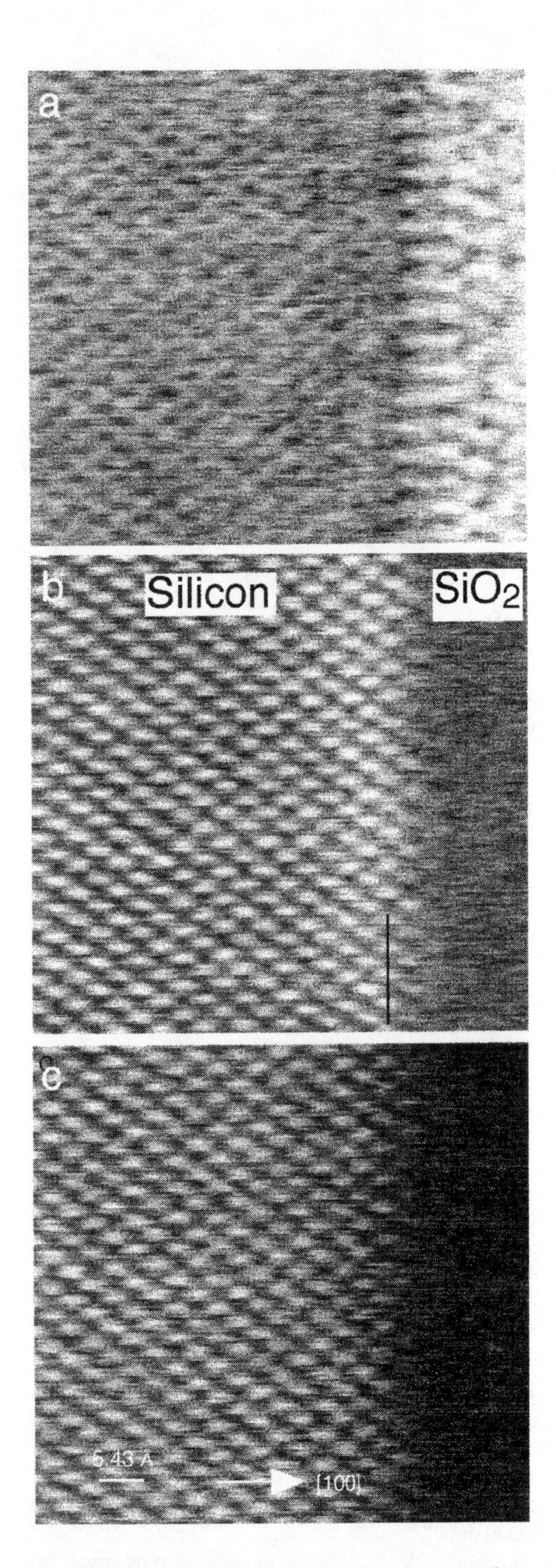

FIGURE 6 . Z-contrast image of a rough Si/SiO_2 interface; (a) bright field phase contrast image, (b) Z-contrast image with 25 mrad inner detector angle showing strain contrast, The vertical line marks the last Si plane used for strain profiling. (c) Z-contrast image with 45 mrad inner detector angle.

contrast can not be due to the presence of heavy impurity atoms. It must therefore be due to a strain effect.

In Fig 7 we show a higher magnification *Z*-contrast image from this interface which can now be seen to have significant roughness, much more than the interface shown in Fig. 2.

Even 10 Å into the oxide, local intensity maxima are seen at exactly the positions expected for a continuation of the Si lattice, although at a much reduced intensity. This implies that small crystalline regions must be protruding into the oxide, still connected to the Si substrate and so maintaining their correct relative positions. The interface width is therefore several monolayers, and we presume that this extended interface width is the source of the significant strain contrast in this specimen.

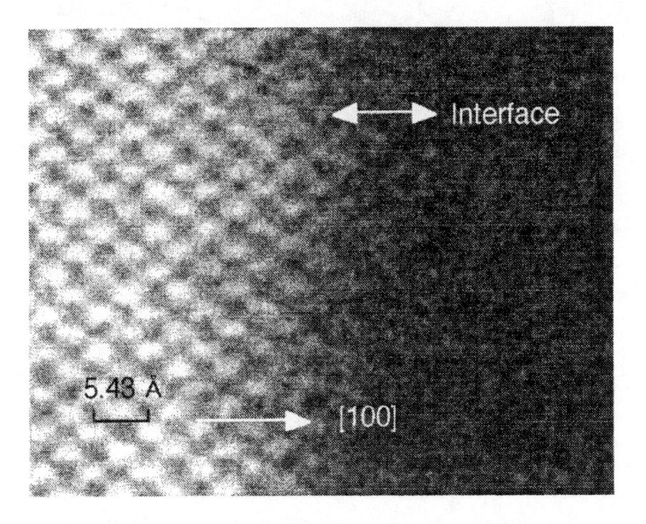

FIGURE 7 . High magnification Z-contrast image of a rough Si/SiO_2 interface showing a gradual transition into the oxide.

The image intensity is not a direct measure of atomic scattering cross section. In a perfect zone-axis crystal, the electron flux is concentrated into the vicinity of the atomic nuclei, increasing the effective columnar cross section. This channelling effect means that the intensity *I* scattered by a column in a zone-axis crystal is given by

$$I = \phi^{1S}\,\sigma\,, \qquad (3)$$

where ϕ^{1S} is the thickness integrated flux at the nuclei (given by the thickness integrated 1s Bloch state intensity at the column in question). Since this channelling effect is not dependent on detector angle (provided scattering angles remain sufficiently high) then images taken at a high angle (where the exponential factor in Eq. 2 is zero) are a direct measure of the channelling effect, and can be used to normalize images taken at lower detector angles.

For example, to quantify the strains seen in Fig. 6, intensity profiles are taken across the Z-contrast images and normalized to the same intensity far from the interface, as shown in Fig. 8. As the interface is approached, the intensity in the high angle image reduces because the columns are distorted and channel less effectively. The image at 25 mrad detector angle shows increased intensity due to the increased scattering cross section. Dividing this

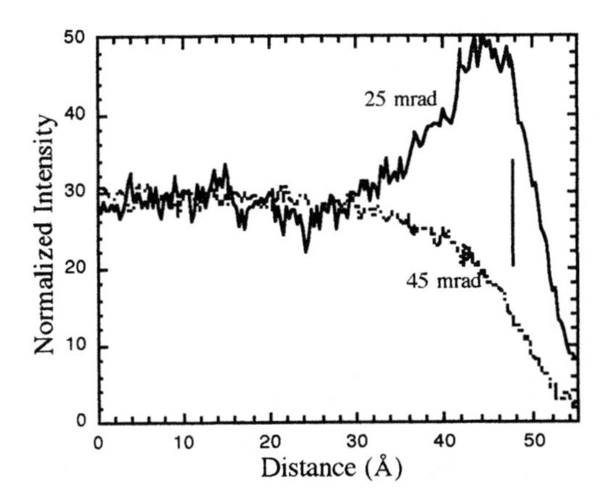

FIGURE 8. Intensity profiles across the Z-contrast images with 25 and 45 mrad inner detector angles.

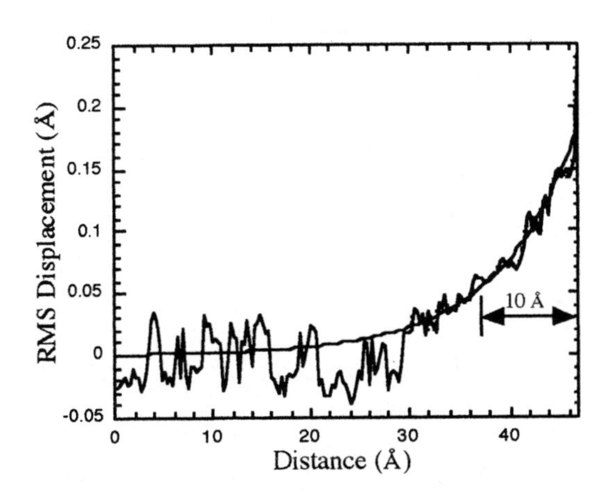

FIGURE 9. RMS atomic displacement due to static strain induced by the Si/SiO_2 interface and an exponential decay fitted to the data. Profile ends at the vertical lines shown in Figs. 6 and 8.

trace by the high angle profile normalizes out the channelling effect to give the relative scattering cross section at each position in the crystal. This in turn can be simply converted into a mean square atomic displacement as shown in Fig. 9. It is interesting to note that this displacement falls exponentially from the interface as would be expected for a planar periodic array of misfit dislocation cores. The 1/e decay length is 10 Å. Note that this simple normalization procedure assumes that the column is entirely Si, and therefore is not valid across the diffuse oxide interface itself.

SUMMARY

The STEM techniques presented in this paper show how the atomic resolution Z-contrast image combined with EELS can give insights into the atomic structure, composition, and strain in the vicinity of a Si/SiO_2 interface, at a spatial resolution limited primarily by the size of the electron probe. Plans are underway to install an EELS spectrometer on the ORNL 300 kV STEM, which would increase the available spatial resolution from ~2.5 Å to ~1.5 Å. With shrinking device dimensions, these techniques would seem to be ideally suited for the characterization of gate dielectrics and correlation with electrical properties.

ACKNOWLEDGMENTS

We are grateful to V. Kaushik for provision of samples for Figs. 6-9. This research was supported under DOE Contract Nos. DE-AC05-96OR22464 and DE-GF02-96ER45610 and by an appointment to the ORNL Postdoctoral Research Associates Program administered jointly by ORNL and ORISE. H-J Gao is supported in part by the NSF Engineering Research Center at the University of Florida.

REFERENCES

1. S. J. Pennycook and D. E. Jesson, *Phys. Rev. Lett.* **64**, 938-941 (1990).
2. S. J. Pennycook, et al., *Appl. Phys. A.* **57**, 385-391 (1993).
3. S. J. Pennycook, "Z-Contrast Electron Microscopy for Materials Science" in *Encyclopedia of Advanced Materials,* Pergamon Press, Oxford, (1995) pp. 2961-2965.
4. S. J. Pennycook, "STEM: Z-contrast," in *Handbook of Microscopy*, ed. by S. Amelinckx, G. Van Tendeloo, D. Van Dyck, J. Van Landuyt, VCH Publishers, Weinheim, Germany (1997) pp. 595-620.
5. A. J. McGibbon, S. J. Pennycook, and J. E. Angelo, *Science* **269**, 519-521 (1995).
6. N. D. Browning, M. F. Chisholm, and S. J. Pennycook, *Nature* **366**, 143-146 (1993).
7. R. Bergmann et al., Appl. Phys. Lett., **57**, 351-353 (1990).
8. A. Ourmazd, D. W. Taylor, J. A. Rentschler and J. Bevk, *Phys. Rev. Lett.*, **59**, 213-216 (1987).
9. H. Akatsu and I. Ohdomari, *Appl. Surf. Sci.* **41-42**, 357-365, (1989).
10. G. Duscher, N. D. Browning and S. J. Pennycook, *Phys. Stat. Sol.* (1998 in press).
11. D. J. Wallis, P. H. Gaskell and R. Brydson, *J. Microscopy*, **180**, 307-312 (1995).
12. G. Duscher, F. Banhart, H. Müllejans, S. J. Pennycook and M. Rühle, in *Proceedings: Microscopy and Microanalysis 1997*, San Francisco Press (1997) pp. 459-460.

Thickness Evaluation of Ultrathin Gate Oxides at the Limit

Dae Won Moon, Hyun Kyong Kim, Hwack Joo Lee,
Yong Jai Cho, and Hyun Mo Cho

*Materials Evaluation Center, Korea Research Institute of Standards and Science,
Yusoung P.O.102, Taejon 305-606, KOREA*

The thickness of ultrathin gate oxides thinner than 10nm were analyzed with Medium Energy Ion Scattering Spectroscopy (MEIS), High Resolution Transmission Electron Microscopy (HRTEM) and Spectroscopic Ellipsometry (SE). MEIS showed the presence of a 1.3nm transition interlayer between SiO_2 and Si substrate. For an ultrathin gate oxide, the thickness values estimated with MEIS, TEM, and SE were 6.2nm, 7.3nm and 6.7nm, respectively. The discrepancies and the complimentary features were discussed.

INTRODUCTION

As the thickness of gate oxides for high density DRAM shrinks thinner beyond 10nm, the thickness measured by conventional thickness measurement methods like spectroscopic ellipsometry (SE) and High Resolution Transmission Electron Microscopy (HRTEM) is subject to not negligible uncertainties. The uncertainties mainly come from the surfaces and interfaces of ultrathin gate oxide films. The surface layers formed due to physisorption of water and hydrocarbon molecules from the atmosphere and the interfacial transition layer between the gate oxide and the substrate should be considered for the analysis of ultrathin gate oxides. The commonly used SE is a very sensitive technique to the thickness of the analyzed thin films but it requires optical parameters to fit the layered structure of thin films, which are not always known. Anyhow, SE can be used for in-situ monitoring and has been most widely used for SiO_2 thickness determination in semiconductor industries. However, SE results on the thickness evaluation of SiO_2 gate oxide thinner than 10nm are subject to uncertainty due to the lack of appropriate models applicable to complex ultrathin layered structures. Since HRTEM has the atomic resolution, the sharp interface between SiO_2-Si in HRTEM images suggested that the transition from the amorphous SiO_2 to the crystalline Si might be very abrupt. However, it should be pointed out here that HRTEM images are based on interference patterns from the regular structures. Therefore HRTEM may not detect very local defects like kinks in the interface quantitatively. In addition, the thickness determination of the amorphous SiO_2 layer is based on visual inspection, compared to that of a crystalline layer based on the counting of the atomic layers. In HRTEM analysis, only a very small fraction of the sample is

CP449, *Characterization and Metrology for ULSI Technology: 1998 International Conference*
edited by D. G. Seiler, A. C. Diebold, W. M. Bullis, T. J. Shaffner, R. McDonald, and E. J. Walters

used to measure the thickness, which may deteriorate the statistical significance.

In this work, medium energy ion scattering spectroscopy (MEIS) was used to analyze a few nm SiO_2 on Si(001) and compared with SE and TEM results. In MEIS, the kinematic binary scattering energy analysis gives the elemental information and the precisely measured electronic energy loss of 100 keV proton ions with a electrostatic energy analyzer is used to estimate the thickness of thin films with almost single atomic layer depth resolution. MEIS also provides atomic structural information from the channeling and blocking analysis.

EXPERIMENTS AND RESULTS

MEIS energy spectra with 97.8 keV H^+ in the double alignment condition are shown in Figure 1 for thermal oxides of nominal thickness of 5.5nm and 3.0nm. The incidence ions were along <111> and the scattered ions were along <001> with a scattering angle of 125.3°.

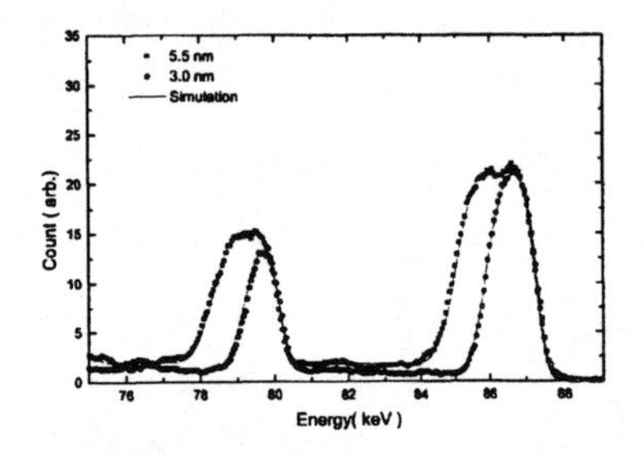

Fig1: MEIS energy spectrum of nominally 5.5nm and 3.0nm thermal oxides on Si(001)

MEIS analysis showed that for a nominally 5.5nm SiO_2 grown by thermal oxidation, there is an amorphous 4.9nm SiO_2 and a 1.3nm interface transition layer between the amorphous SiO_2 and the Si substrate. The interface transition layer consisted of amorphous SiO_2 and crystalline Si. The depth distribution of Si and O calculated from ion scattering simulation is shown in Figure 2. The crystalline Si was found to be vertically strained to 1% for thermal oxides and 2.8% for ion beam oxides.(1)

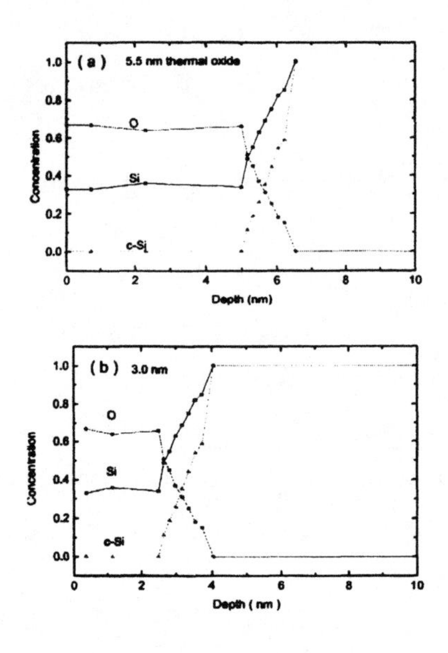

Fig2: Depth distribution for further development of Si and O calculated from ion scattering simulation

The HRTEM image in Figure 3 showed a 6.6 nm amorphous SiO_2 with a 0.7 nm interlayer. The dark for further development layer at the interface was considered as the interlayer. The dark layer in the interface could be due to the roughness of the interface caused by steps present in the HRTEM specimen of the thickness around several 10nm. Therefore the HRTEM image gives average information on the interfacial structure over the specimen thickness. If the specimen is thin enough, the dark layer may not appear. A study on the effect of the specimen thickness is under progress. The specimens for HRTEM observations were prepared by mechanical polishing, followed by Ar ion milling with an acceleration voltage of 4 keV. Microstructural

observation was
carried out by H9000-NAR working at 300 kV with

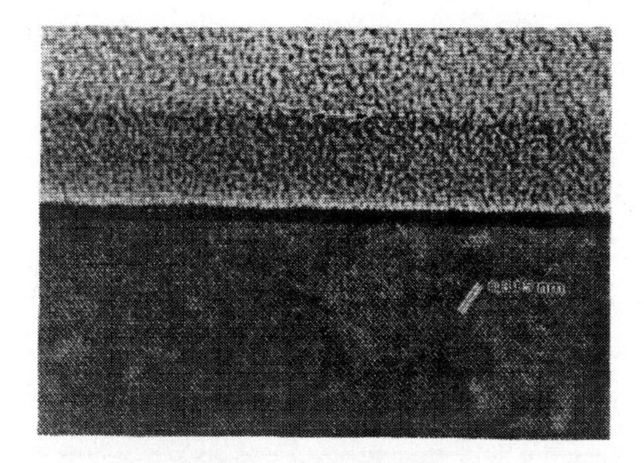

Fig 3: Cross-sectional TEM micrograph of a nominal 5.5 nm SiO_2 on Si(001)

a point resolution of 0.18 nm. The TEM image was taken in a high-resolution TEM mode with the <110> zone axis.

Spectroscopic ellipsometric measurements are summarized as following. The experimental values of $\tan\psi$ and $\cos\Delta$ for a 6-nm-thick sample were obtained over 1.5~5 eV range of a spectroscopic phase-modulated ellipsometer, where $\tan\psi$ is the ratio of the amplitude attenuation, Δ is the total phase shift. To simulate the possibility of a non-abrupt interface, an interlayer between the SiO_2 layer and the Si substrate was assumed. The dielectric function of the interlayer was done mathematically by evaluating the effective dielectric function of a physical mixture of SiO_2 and a-Si in the Bruggeman effective medium approximation. To calculate the model spectra, we used the literature values [2] of the dielectric functions of SiO_2, Si, and a-Si. When the oxide thickness d_{ox} and the volume fraction of a-Si in the interlayer are used as the adjustable parameter and the interlayer thickness d_{int} is fixed, the best-fit results are obtained as shown in Fig. 1. Here the curve of the mean-square deviation σ shows its minimum value at the interlayer thickness around 0.45 nm. As d_{int} increases from 0.2 to 1.3 nm the volume fraction of a-Si in the interlayer decreases from 20 to 3.2 % and d_{ox} from 6.60 to 5.48 nm. From these results, we put a starting point as d_{int}=0.45 nm, d_{ox}=6.3 nm, and (volume fraction of a-Si)=16.5 % in the best-fit model calculation and obtained the best-fit parameters of $d_{int}=0.46\pm0.18$ nm, $d_{ox}=6.26\pm0.25$ nm, and (volume fraction of a-Si)=16.4 ± 5.8 %. Finally, the 0.46-nm thickness and the index of refraction of $n_{int}=1.8\pm0.5$ at 2.27 eV are smaller than the previous ellipsometric results of d_{int}~0.7 nm and n_{int}~2.8 obtained by Aspnes et al. [3].

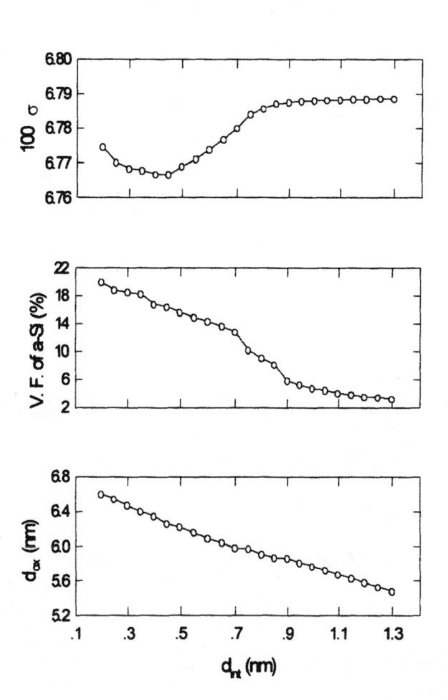

Fig 4: Optimization of the interlayer thickness and the volume fraction of a-Si in the interlayer.

Discussions

All the information from SE, HRTEM and MEIS was combined together in Table 1 to compare the results and to get more detailed information on ultrathin gate oxides from various aspects. It is quite clear that each technique gives different information. First of all, MEIS gives the lowest thickness. It could be due to errors in the electronic stopping power taken

Technique	SiO_2 (nm)	Interlayer (nm)	Total (nm)
MEIS	4.9	1.3	6.2
TEM	6.6	0.7	7.3
SE	6.3	0.4	6.7

Table 1: Summary of the thickness evaluated by MEIS, TEM and SE.

from the empirical formula by Anderson and Ziegler (4), of which the uncertainty can be 10-20%. In addition, the SiO_2 stopping power was estimated from elemental Si and oxygen values, where any chemical effect was not considered at all. Accurate MEIS measurement of the electronic stopping power of 100 keV H^+ in SiO_2 is required with a 20-30 nm SiO_2 analyzed, of which the thickness determined with TEM and SE is reliable. To minimize the uncertainty due to the visual determination of the interface position between the SiO_2 and epoxy capping layer in TEM analysis, capping with Si or metal should be useful. To assess the surface roughness of the SiO_2 layer, AFM analysis should be added.

The largest interlayer thickness, 1.3 nm given by MEIS could be partly due to the wide area analysis over 1x0.5mm^2 determined by the probing ion beam size. The effect of roughness over the wide area should be reflected in the MEIS spectrum. The effect of surface and interface roughness on MEIS spectrum is being studied with stepped Si surfaces. The high sensitivity of MEIS to local composition and structure could be additional factor for the largest interlayer thickness.

Finally it should be pointed out that electrical measurement of the thickness of ultrathin gate oxides will be compared with those from the above techniques to make the results practically meaningful.

Conclusions

The thickness of ultrathin gate oxides with nominal thickness of 5.5nm and 3.0 nm was evaluated with MEIS, HRTEM and SE. From the discrepancies between the three techniques, it was discussed how to provide reliable and meaningful information on the thickness of a few nm ultrathin gate oxides through complimentary use of the three techniques

References

1) Y. P. Kim, S. K. Choi, H. K. Kim and D. W. Moon, 71,3504, Appl. Phys. Lett.,(1997)

2) E. D. Palik, *Handbook of Optical Constants of Solids*, Academic Press, New York (1985).

3) D. E. Aspnes and J. B. Theeten, Phys. Rev. Lett. 43, 1046 (1979).

4) H.H.Anderson and J.F.Ziegler, *Hydrogen Stopping Powers and Ranges in All Elements*, vol.3 of *The Stopping and Ranges of Ions in Matter*, Pergamon, New York (1977).

Determination of Shallow Dopants in Silicon by Low-Temperature FTIR Spectroscopy

H. Ch. Alt

Physikalische Technik, Fachhochschule München, D-80335 München, Germany

M. Gellon, M. G. Pretto and R. Scala

MEMC Electronic Materials, I-39012 Merano, Italy

F. Bittersberger, K. Hesse and A. Kempf

Wacker, D-84489 Burghausen, Germany

The shallow dopants B, P, Al, As, and Sb in Si have been investigated by low-temperature Fourier-transform infrared absorption measurements. Experimental procedures are described to obtain quantitative results for the B and P concentration with improved precision. New calibration factors, based on a correlation with four point probe resistivity measurements, are given which can be used in the concentration range from 10^{10} to 10^{16} cm^{-3}. Preliminary results are reported for Al, As, and Sb.

INTRODUCTION

The quantitative assessment of electrically active dopants in Si is of crucial importance for the development of crystal growth technology and the use of the material in electronic industry. Group III and group V impurities like Al, As, B, P and Sb form hydrogen-like states which can be detected at low temperatures by sharp lines in the far-infrared absorption spectra (1,2). State-of-the-art Fourier-transform infrared absorption (FTIR) spectrometers can detect impurity levels down to 10^{10} cm^{-3}.

However, despite of considerable progress in calculating the absorption strength of the electronic transitions (2,3), a calibration of the FTIR method by some reference method is still necessary to obtain the required precision. Previous attempts to calibrate the line intensity (4,5) need to be improved, as either experimental conditions are not adequate or the applicability of the method is restricted.

EXPERIMENTAL PROCEDURE

The majority of the FTIR samples was prepared from single-crystal silicon ingots grown by the float-zone (FZ) technique. For comparative purposes also some refined polycrystalline samples were investigated. The typical sample geometry was a slab with a diameter of about 12 mm and a thickness between 0.3 and 10 mm according to impurity concentration (see below). Surfaces received a wedge angle to avoid interference fringes and were mechanically or chemically polished.

Infrared absorption spectra were taken with a high-resolution FTIR spectrometer (Bruker IFS 113v) equipped with a helium cryostat and Si bolometer as well as DTGS detectors. The cryostat used was of the exchange gas type with CsI and/or Si infrared windows. Strain-free mounting of the samples was accomplished by using tape. 'White-light' illumination was performed by focusing light from a quartz-halogen lamp (100 W) onto one sample surface using a side window of the cryostat. 'Dark' conditions were obtained by placing a long pass filter ($\lambda > 10$ µm) in front of the sample. Typically, 30 to 100 scans were necessary to obtain spectra with a reasonable signal to noise ratio. The use of the Si bolometer, which gives a 3-5 times better signal to noise ratio as compared to the DTGS, was of advantage in particular for samples with impurity concentrations $< 10^{13}$ cm^{-3}.

Four point probe resistivity measurements were done according to the standard test method ASTM F 84-93. Conversion to dopant concentration was performed by using ASTM F 723-88.

EXPERIMENTS: GENERAL REMARKS

A typical FTIR absorption spectrum taken at a temperature of 10 K is shown in Fig. 1. The sample contains predominantly phosphorus and boron, however, also small amounts of arsenic and aluminum. Due to

CP449, *Characterization and Metrology for ULSI Technology: 1998 International Conference*
edited by D. G. Seiler, A. C. Diebold, W. M. Bullis, T. J. Shaffner, R. McDonald, and E. J. Walters

above-band gap illumination ('white light'), both donor and acceptor species are visible. The absorption bands used for the analysis of the impurity concentration are indicated. This study is focused on B and P. Preliminary results on Al (473 cm^{-1}), As (382 cm^{-1}), and Sb (293 cm^{-1}, not shown in Fig. 1) are reported. The B bands at 245, 278, and 320 cm^{-1} and the P bands at 275, 316, and 323 cm^{-1}, respectively, are investigated in detail.

The bands at 320 and 316 cm^{-1} are usually taken for the quantitative determination of B and P, respectively (4,6). However, in some samples these bands are overlapping, together with the P band at 323 cm^{-1}. Also, the band at 320 cm^{-1} actually consists of three lines (5), making line fitting more difficult. Therefore, the bands at 275 and 278 cm^{-1}, which are both strong and consist of a single line each, may be used alternatively. The smaller B and P lines at 245 cm^{-1} and at 323 cm^{-1}, respectively, are suitable for samples containing higher concentrations of dopants. In this case, the other bands often are too strong.

The thickness of the sample has to be adapted to the expected impurity concentration range. In terms of the measured resistivity ρ, it is useful to have samples with a thickness of about 1 cm for $\rho \geq 500$ Ωcm, of 0.3 cm for 50 Ωcm $\leq \rho < 500$ Ωcm, of 0.1 cm for 10 Ωcm $\leq \rho < 50$ Ωcm, and of 0.03 cm for $\rho < 10$ Ωcm.

In contrast to the case shown in Fig. 1, the calibration of the absorption bands of B and P was performed on samples having a low degree of compensation. In addition, long-pass filtering ($\lambda > 10$ μm) was used to avoid any band-to-band absorption. The spectra thus obtained reflect the *net impurity* concentration and may be compared to the free carrier concentration measured by the four point probe technique at room temperature.

On the other hand, by using white-light illumination a large number of electron-hole pairs is generated, which are captured by ionized compensating impurity centers. If the illumination intensity is large enough, it can be expected that all centers are neutralized. The FTIR method is therefore capable of determining also the *total impurity* concentration. It is found that at 10 K a illumination intensity of 10 mW/cm^2 in the spectral range from 0.4 to 1.0 μm is sufficient to obtain saturation of the band intensity.

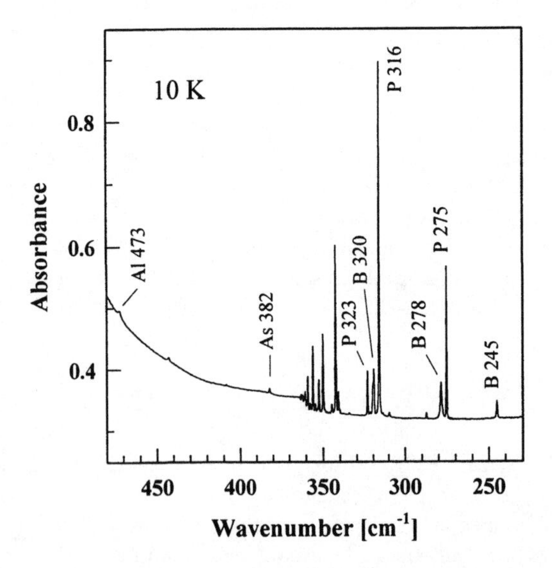

FIGURE 1. FTIR absorption spectrum taken at 10 K under white-light illumination. Sample contains phosphorus, boron, arsenic, and aluminum.

INFLUENCE OF EXPERIMENTAL PARAMETERS

Spectral resolution

Only the area of an absorption band is a reliable measure for the impurity concentration, as the width of the band and, as a consequence, the peak absorption is strongly dependent on parameters such as sample temperature, impurity concentration, and built-in or externally applied strain. Experimentally, always the convolution of the absorption band with the instrument function (given by the spectral resolution and the apodization function) is obtained. Therefore, the apparent area of a band decreases with decreasing spectral resolution. The effect is shown in Fig. 2 for the phosphorus line at 316 cm^{-1} in a sample containing 1.4×10^{13} cm^{-3} P atoms. The situation is extremely critical for all donor lines investigated, which have a FWHM of 0.2 cm^{-1} or even less. Furthermore, the FWHM is dependent on sample temperature. For example, for the sample shown in Fig. 1, the FWHMs of the P lines at 275 and 316 cm^{-1} increase between 7 to 22 K from 0.15 to 0.29 cm^{-1} and from 0.19 to 0.23 cm^{-1}, respectively.

As a compromise between experimental efforts and precision achievable, P doped samples were measured in general with a resolution of 0.1 cm^{-1}. For the boron bands the situation is less critical. The bands at 320 cm^{-1} (which is actually composed of 3 lines) and 278 cm^{-1} have a FWHM of not less than 1.0 cm^{-1}, the 245-cm^{-1} line of 0.5 cm^{-1}, respectively. Therefore, the spectral resolution of 0.5 cm^{-1} is sufficient for the quantitative evaluation of B.

Measurement temperature

Another important experimental parameter is the sample temperature. As not all cryostat systems allow the accurate determination of the actual sample temperature, for practical purposes it is necessary to have a temperature range where the FTIR method can be used. The carrier freezeout in a semiconductor is governed by Fermi statistics. Assuming that only one impurity with concentration N has to be considered, the fraction of

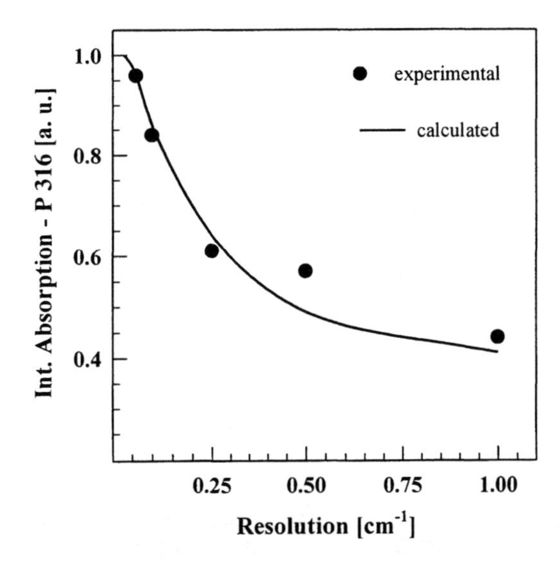

FIGURE 2. Influence of the spectral resolution on the P line at 316 cm^{-1}. Dots, experimental data; full curve, calculation. Absorption band is approximated by a Lorentzian line shape with a FWHM of 0.19 cm^{-1}.

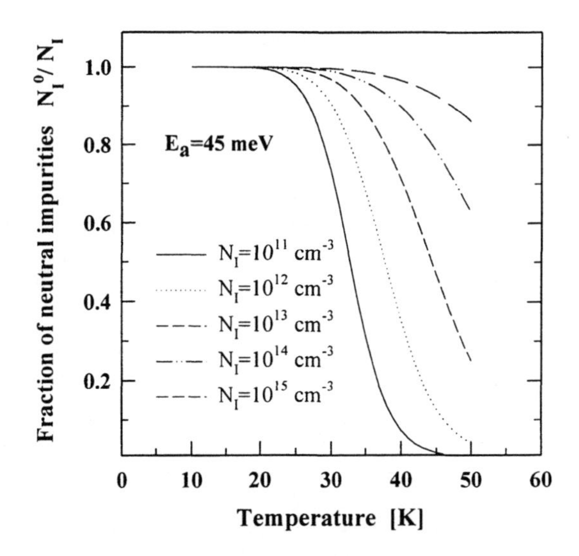

FIGURE 3. Model calculation of the fraction of occupied impurity sites as a function of temperature.

occupied dopant sites N_I^0/N_I can be expressed by the equation (7)

$$N_I^0/N_I = 1 - 2 / [1 + (1 + 4\, g\, N_I/N_{c/v} \exp(E_a/k_BT))^{1/2}]. \quad (1)$$

$N_{c/v}$ is the effective density of states in the conduction or valence band, g the degeneracy of the impurity ground state, and E_a the ionization energy. An example is shown in Fig. 3 for an ionization energy of 45 meV. This value holds for both boron and phosphorus in silicon (the difference of the degeneracy factors is only a minor correction). It is evident that the ionization depends on the concentration of the dopant itself. However, as a general rule, it can be concluded that in the cases of interest a temperature of less than 20 K is sufficient to avoid thermal ionization.

A further aspect is the influence of increasing temperature on the absorption line itself. As mentioned above, the width of the lines changes considerably, as well as the peak absorption. This may be attributed to a decreasing lifetime of the excited state due to phonon scattering. The optical matrix element, however, which is responsible for the strength of an optical transition, should remain unaffected in first order. Therefore, it can be assumed that the integrated absorption remains constant at least up to 20 K. This is in agreement with experimental observations.

Strain

Strain in the samples arises from crystal imperfections or from externally applied stress. The effect of uniaxial stress on the donor lines in Si has been studied in detail by Aggarwal and Ramdas (8). Dependent on the crystallographic direction of the applied stress relative to the orientation of the sample, the lines undergo a splitting into one, two or three lines. The reason is the lifting of the multivalley structure of the conduction band minimum. Normally, unintentionally introduced strain is not uniaxial.

For example, polycrystalline material contains internal strain due to the large density of growth defects at the grain boundaries. Clamping of the samples by screws, which is necessary in some cryostat constructions to obtain good thermal contact with the cold finger, leads to irregular strain fields in the sample. In those cases, usually a broadening of the band is observed as compared to the result on good FZ material. From the practical point of view, it is of interest whether this broadening will affect the area under the band, i.e. the integrated absorption, or not.

We investigated the influence of strain by placing a well-characterized rod-like FZ-sample in a copper frame. Due to the large difference in the thermal expansion coefficient between copper an silicon, a large compressive stress (1-2 GPa) is applied to the sample at 10 K. Despite of an extreme distortion of the bands (in this case B bands), the integrated absorption did not change by more than 10 %. It is concluded that the much smaller strain fields occurring under normal measurement conditions do not influence the integrated absorption of a band.

EXPERIMENTAL RESULTS

B- and P-doped samples

More than 30 boron and phosphorous doped single crystal Fz samples have been investigated. The dopant concentration ranges from 10^{12} to 10^{16} cm^{-3}. Areas of absorption bands were determined by using computer algorithms. In the first step a Lorentzian/Gaussian line and a baseline is fitted to the measured band using the usual absorbance representation of the spectrum. This procedure is believed to be more reliable than a direct evaluation of the area, as the extension of the tails of the bands is often underestimated. From the fitted line an apparent absorption coefficient α' for each wavenumber is deduced using the relation

$$\alpha' = 1/d \ln(10) A , \quad (2)$$

where d is the thickness of the sample and A the absorbance.

Because of the high refraction index of Si (n=3.41) a correction for multiple internal reflections is necessary. Typically, the apparent absorption coefficient α' is too large by about 10 %. The correction is done with the following equation

$$\alpha = 1/d \ln\{(1-R^2)/2 \times y \times$$
$$\times [1 + (1 + 4R^2/((1-R^2)^2 y^2))^{1/2}]\} , \quad (3)$$

with

$$y = \exp(\alpha' d) .$$

The reflectivity R is 0.30 for Si.

The final step is the numerical integration over the band given by α (3).

The correlation of the results with the free carrier concentration is shown in Figs. 4 and 5 for B and P, respectively. A good linearity with a slope very close to 1 is obtained for all bands investigated. The calibration factor is derived from the linear regression analysis. The results are given in Table 1. We estimate that the precision of the calibration factors derived is ± 5 %. No quantitative estimate is possible for bias.

Other dopants

Only a few samples were available with predominant doping of Al, As, and Sb. Al-doped samples were available in the concentration range from 10^{13} to 10^{14} cm^{-3}. Strong striation patterns observed by resistivity measurements make the correlation with the strongest Al FTIR lines at 473 and 443 cm^{-1} less reliable. The calibration of the As line at 382 cm^{-1} is based on one sample which contained only a negligible amount of P. The Sb-doped samples were also contaminated considerably with P, therefore the P concentration has been subtracted first, using the calibration given in Table 1. The resulting calibration factors for the lines at 473 (Al), 382 (As), and 293 cm^{-1} (Sb) are given in Table 2. The precision is less than in the case of B and P. A rough estimate gives ± 20 %.

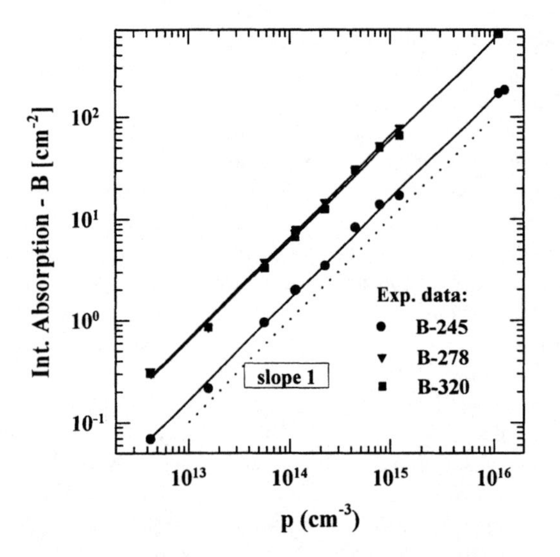

FIGURE 4. Correlation between the integrated absorption of the 3 boron bands at 245, 278, and 320 cm^{-1} and the free hole concentration.

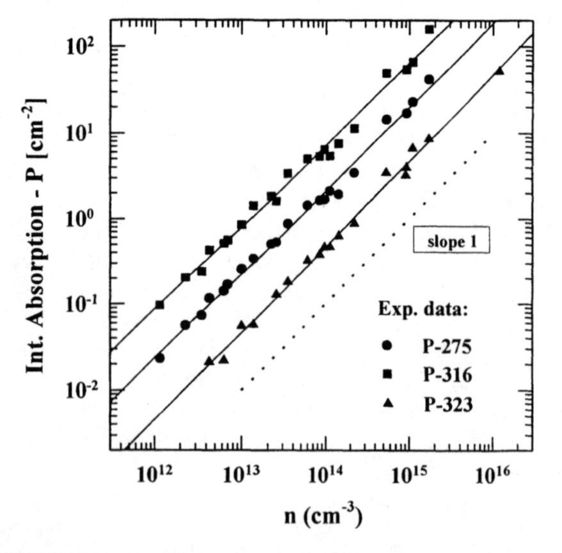

FIGURE 5. Correlation between the integrated absorption of the 3 phosphorus bands at 275, 316, and 323 cm^{-1} and the free electron concentration.

TABLE 1. Calibration factors for absorption bands of B and P.

Band	Calibration factor (cm^{-1}) This work	Theory*
B-245	6.85×10^{13}	5.12×10^{13}
B-278	1.55×10^{13}	1.42×10^{13}
B-320	1.75×10^{13}	1.38×10^{13}
P-275	4.36×10^{13}	3.15×10^{13}
P-316	1.22×10^{13}	7.42×10^{12}
P-323	2.29×10^{14}	2.87×10^{14}

*Theoretical values are taken from Ref. 5.

TABLE 2. Calibration factors for absorption bands of Al, As and Sb.

Band	Calibration factor (cm^{-1}) This work	Theory*
Al-473	7.4×10^{13}	3.3×10^{13}
As-382	1.6×10^{13}	
Sb-293	1.0×10^{13}	

*Theoretical values are taken from Ref. 5.

Polycrystalline samples

A limited number of refined polycrystalline samples were investigated for comparative purposes. Polycrystalline samples usually have a much larger line width than single crystal samples (typical is a FWHM of 1-2 cm^{-1} for both donor and acceptor bands). Polycrystalline samples show also larger scattering in correlation plots such as Figs. 4 and 5. However, no significant shift from the calibration given in Table 1 was found for the area of the bands. The larger scattering of the data is attributed to spatial inhomogeneities of impurity concentration as well as electrical properties within the sample. Therefore, the method can be used also for polycrystalline samples, with the restriction that the results are less precise.

DISCUSSION AND CONCLUSION

Comparing the calibration factors for boron and phosphorus with theoretical calculations (see Table 1), it was found that in the case of B the calculated values are lower by about 10-20 %, whereas in the case of P no systematic trend is observable. The discrepancy reaches 40 % for the 316-cm^{-1} line. It is beyond the scope of this study to discuss possible reasons for these discrepancies. From the practical point of view, it is obvious that a calibration of the FTIR method with an established reference method is still necessary.

In conclusion, it is believed that the experimental procedures defined above and the new calibration factors given provide the means to make the FTIR method to a highly sensitive analytical tool which meets the higher requirements for the quantitative determination of shallow dopants in silicon in the future.

REFERENCES

1. Burstein, E., Davisson, J. W., Bell, E. E., Turner, W. J., and Lipson, H. G., A. K., *Phys. Rev.* **93**, 65-74 (1954).
2. Beinikhes, I. L., and Kogan, Sh. M., *Sov. Phys. JETP* **66**, 164 (1987).
3. Pajot, B., Beinikhes, I. L., Kogan, Sh. M., Novak, M. G., Polupanov, A. F., and Song, C., *Semicond. Sci. Technol.* **7**, 1162-1169 (1992).
4. Baber, S. C., *Thin Solid Films* **72**, 201-210 (1980).
5. Andreev, B. A., Kozlov, E. B., and Lifshits, T. M., *Materials Science Forum,* Vols. **196-201**, Switzerland: Trans Tech Publications, 1995, pp. 121-126.
6. Kolbesen, B. O., *Appl. Phys. Lett.* **27**, 353-355 (1975).
7. Blakemore, J. S., *Semiconductor Statistics*, New York: Dover Publications, 1987, ch. 3.
8. Aggarwal, R. L., and Ramdas, A. K., *Phys. Rev.* **137**, A602-A612 (1964).

High-Resolution, High-Accuracy, Mid-IR ($450\ cm^{-1} \le \omega \le 4000\ cm^{-1}$) Refractive Index Measurements in Silicon

Deane Chandler-Horowitz and Paul M. Amirtharaj
Semiconductor Electronics Division, NIST, Gaithersburg, MD 20899
and
John R. Stoup
Precision Engineering Division, NIST, Gaithersburg, MD 20899

The real and imaginary parts of the refractive index of silicon, $n(\omega)$ and $k(\omega)$, have been measured by using Fourier Transform Infrared (FTIR) transmission spectral data from a double-sided-polished grade Si wafer. An accurate mechanical measurement of the wafer thickness, t, was required and two FTIR spectra were used: a high-resolution ($\Delta\omega = 0.5\ cm^{-1}$) yielding a typical channel spectrum dependent mainly on t and $n(\omega)$, and a low resolution ($\Delta\omega = 4.0\ cm^{-1}$) yielding an absorption spectrum dependent mainly on t and $k(\omega)$. Independent analysis of each spectrum gave initial $n(\omega)$ and $k(\omega)$ estimates which were then used together as starting point values for an iterative fit of the high- and low-resolution spectra successively. The accuracy of $n(\omega)$ and $k(\omega)$ values determined using this procedure is dependent upon the measurement error in the sample thickness, the absolute transmission values obtained from a sample-in and sample-out method, and the modeling of the influence of wafer thickness nonuniformity and the degree of incident light beam collimation. Our results are compared with previously published values for $n(\omega)$ and $k(\omega)$ in the 450 cm^{-1} to 4000 cm^{-1} spectral region. The error in $n(\omega)$ is < 1 part in the 10^4, a factor of 10 better than published values. The $k(\omega)$ values are in good agreement with previous measurements except in the vicinity of the 610 cm^{-1} peak feature where our values are ~ 20% lower.

INTRODUCTION

Infrared (IR) analysis is a well-developed spectroscopic method for nondestructive materials characterization. The current appeal and use of the technique notwithstanding, the probe has not yet fully met the needs of the rapidly advancing semiconductor industry for quantitative measurements. The critical limitation is the accuracy of the values of the intrinsic optical constants, $n(\omega)$ and $k(\omega)$. The accuracy with which small quantities of impurities in a wafer or a thin overlayer on the wafer may be determined is closely related to the accuracy with which the optical properties of the intrinsic wafer are known.

The published values of $n(\omega)$ possess uncertainties of about one part in 10^3, whereas the reported values of $k(\omega)$ vary by a factor of 2 or even greater in some cases. Edwards (1) has compiled the values for $n(\omega)$, measured by two techniques, the minimum angular deviation and channel spectra methods, and $k(\omega)$. Values for $k(\omega)$ were also measured by Humlicek and Vojtechovsky (2). Spectroscopic ellipsometry is a powerful technique, but it is most effective in the part of the spectra where the specimen is opaque extending from the near IR region with ω >10,000 cm^{-1}.

We describe in this paper a procedure that allows us to determine both $n(\omega)$ and $k(\omega)$ to a greater accuracy than has been achieved in the past, a factor of 10 improvement over the published values. The crux of the technique (3-5) is the measurement of a low- ($\Delta\omega = 4\ cm^{-1}$) and a high-resolution ($\Delta\omega = 0.5\ cm^{-1}$) spectrum and fitting both to predicted theory using a realistic model for the wafer configuration and properties of the probe beam (3).

EXPERIMENTAL DETAILS

Room temperature FTIR measurements were performed using a Bomem DA-8* interferometer fitted with mid-IR optics, source, and detector. A collimator was used to greatly reduce the divergence in the probe beam. p-type Si wafers were used as the specimen source. Careful attention was paid to use only the samples with the most parallel faces. Parallelism measurements were carried out using an ADE Corp. Model 6033 non-contact gauging station. Additionally, relative thickness measurements were performed by measuring the transmission spectrum at five equally spaced points on the 2.5 cm x 2.5 cm sample. The measurement spot size was 5 mm in diameter. A typical high-resolution scan is shown in Fig. 1. It can be seen that the fringe intensity maximum is near unity but falls off slightly at high wavenumbers due to the slight sample wedge. Low-resolution transmission measurements were

---

* Certain commercial products are identified in the manuscript to adequately specify the procedure. This does not imply recommendation or endorsement by NIST, nor does it imply that those commercial products are the best available for the purpose.

CP449, *Characterization and Metrology for ULSI Technology: 1998 International Conference*
edited by D. G. Seiler, A. C. Diebold, W. M. Bullis, T. J. Shaffner, R. McDonald, and E. J. Walters
1998 The American Institute of Physics 1-56396-753-7/98/$15.00

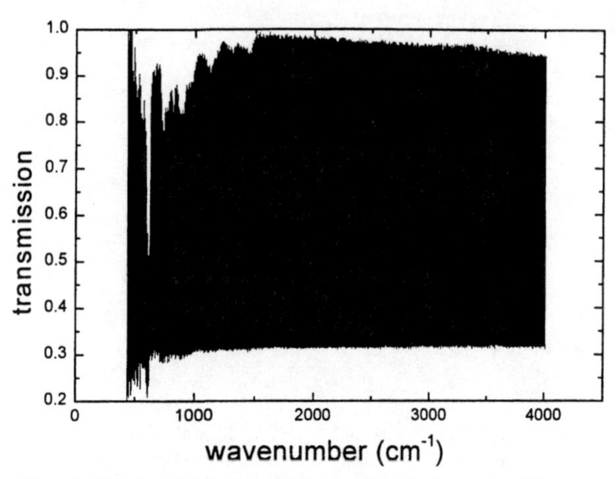

Fig. 1: Typical high-resolution transmission scan ($\Delta\omega$ = 0.5 cm^{-1}). Fringe spacing is about 3 cm^{-1} and unresolved in this view.

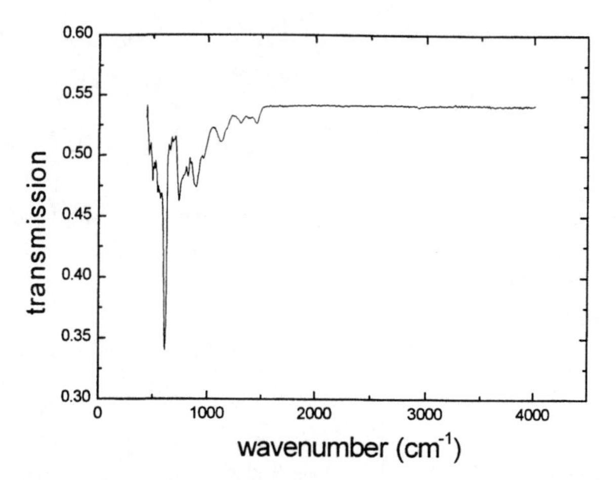

Fig. 2: Typical low-resolution transmission ($\Delta\omega$ = 4 cm^{-1}) regenerated from the same interferogram used to produce the spectrum in Fig. 1 and averaged over the five positions as described in the paper.

performed with and without the beam collimator in an effort to improve the signal-to-noise ratio. Figure 2 shows the results for the transmission regenerated from the same interferogram used for the high-resolution spectrum shown in Fig. 1 and averaged over the five wafer positions.

The wafer thickness, t, was measured by the comparison method using an electro-mechanical comparator. Two master gauge blocks, originally measured by static interferometry that utilizes a stabilized HeNe laser as the length standard, are used for the transfer comparison to the wafer. Corrections for the elastic deformation of the comparator and wafer contacts are calculated. The analysis of the length data and the process information yielded a combined standard uncertainty of ±18 nm. The thickness of the sample was nominally 467 μm.

THEORY

The transmission through a substrate of thickness t and with refractive index, $n + ik$, immersed in a medium of air with refractive index, n_a, is well known from classical electromagnetic theory. For the case of high-resolution transmission measurements, the following expression is valid (5) when $k << (n-n_a)$:

$$T(\omega) = \frac{16 n_a^2 n^2 \beta}{C_1^2 + C_2^2 \beta^2 - 2C_1 C_2 \beta \cos(4\pi n t \omega)} \tag{1}$$

where $C_1 = (n+n_a)^2$, $C_2 = (n-n_a)^2$, $\beta = \exp(-4\pi k t \omega)$, and ω is the photon frequency expressed in wavenumbers (cm^{-1}).

The variation of the wafer thickness, Δt, across the extent of the incident beam on the transmission spectrum, T, was accounted for by suitable integration over t, neglecting the small variation in β as shown below:

$$T_W = \frac{1}{\Delta t} \int T(t) dt \,. \tag{2}$$

From eq. 1 and eq. 2 we get:

$$T_W = \frac{32 n_a^2 n^2 \beta}{\Delta t (C_1^2 - C_2^2 \beta^2)} \tan^{-1}\left[\frac{C_1 + C_2 \beta}{C_1 - C_2 \beta} \tan\left(\frac{4\pi n t \omega}{2} \right) \right]_{t-\Delta t/2}^{t+\Delta t/2} . \tag{3}$$

For low-resolution and $k << (n-n_a)$, the following expression (2) is valid:

$$T_L = \frac{(1-R)^2 \exp(-4\pi k t \omega)}{1 - R^2 \exp(-8\pi k t \omega)} \tag{4}$$

where

$$R = \frac{(n-n_a)^2 + k^2}{(n+n_a)^2 + k^2} . \tag{5}$$

The dispersion of $n(\omega)$ may be described using the Sellmeier formula shown below.

$$n(\omega) = \varepsilon + A\omega^2 + \frac{B\omega^2}{\omega_1^2 - \omega^2} . \tag{6}$$

$\omega_1 = 9.0236 \times 10^3$ cm^{-1} for silicon (1).

DISCUSSION

The approach we used is initially similar to that employed in the past to obtain $n(\omega)$ from the wavenumber spacing of the interference fringe maxima in the transmission spectrum, $T(\omega)$, i.e., to determine the $n(\omega)t$ product, where t is the known wafer thickness (1,4). However, in addition, we have used both the complete high-resolution transmission spectrum, which includes a large number of interference fringes, and the low-resolution transmission spectrum, the shape of which is determined mostly by variations in $k(\omega)$. The advantage of this new procedure is that it helps separate the effects of $n(\omega)$ and t on the one hand and $k(\omega)$ and spectral drift errors on the other. An iterative procedure that fits the high- and the low-resolution spectrum successively yields values of both $n(\omega)$ and $k(\omega)$ to greater accuracies than can been obtained using each spectrum separately.

The details of the sample and beam geometry had to be taken into account to calculate the high-resolution spectrum. The slightly nonparallel faces of a wedge-shaped sample or slightly convergent probe beam geometry have a distinct influence on the transmission values at the maxima and minima of each interference fringe. Because of the good collimation obtained, only an integration of transmission across the wafer surface extending to the width of the probe beam was necessary to realistically account for the nonideal sample configuration.

The large number of oscillatory features in the high-resolution transmission spectrum, combined with the complex absorption structure, required good initial values for the parameters for the computer fit to converge and yield meaningful results. Good initial values of $n(\omega)$ were obtained using the method discussed in ref. (4) that uses the variation of the fringe order with the spectral position of the maxima to extract the values of $n(\omega)$. Using these values of $n(\omega)$, the low-resolution spectrum was utilized to determine initial values of $k(\omega)$. $n(\omega)$ and interpolated values of $k(\omega)$ were then used as the initial values to fit the high-resolution spectrum. Unlike the low-resolution transmission spectrum, variations in $k(\omega)$, and the associated absorption $\alpha(\omega)$, affect both the average transmission, as well as the modulation intensity. Hence, the high-resolution spectrum is qualitatively superior to the low-resolution spectrum. In this experiment the low-resolution spectrum was regenerated directly from the original high-resolution interferogram.

The dominant source of error in $n(\omega)$ is the accuracy to which the thickness, t, is known. The primary source of error in $k(\omega)$, obtained from the low-resolution transmission, is the systematic error in measuring the absolute value of the beam intensity and resultant errors introduced by the sample-in-sample-out ratio method. Accurate mechanical measurements and detailed modeling addressed the former, and the use of a sample manipulator adapted for the evacuated FTIR instrument sample chamber mitigates the latter.

Fitting the high-resolution spectra required good initial values for $n(\omega)$. The following simple procedure was used to get those initial values. First, the fringe intensity maxima spectral positions were located. The fringe order was then plotted as a function of their maxima positions and fit to a third-order polynomial expression in ω. $n(\omega)$ was then calculated from its known relationship to the fringe order. For silicon, which has a broad transparent region from 450 cm^{-1} to 4000 cm^{-1}, the index of refraction, n, can be fit to the following quadratic expression

$$n(\omega) = n_0 + n_1\omega + n_2\omega^2 . \quad (7)$$

Following the procedure by Edwards and Ochoa (5), the fringe orders, M, of the channel spectrum may be expressed as follows:

$$M = M_0 + a_1\omega + a_2\omega^2 + a_3\omega^3 . \quad (8)$$

The quadratic coefficients for $n(\omega)$ may then be derived from the relationship between fringe order M and the index $n(\omega)$ as follows:

$$M = 2tn(\omega)\omega = 2t(n_0\omega + n_1\omega^2 + n_2\omega^3) . \quad (9)$$

By comparing eqs. 8 and 9 we get the following expressions which are independent of M_0

$$n_0 = a_1 / 2t , \; n_1 = a_2 / 2t , \text{and } n_2 = a_3 / 2t . \quad (10)$$

It was later found that in spectral regions with appreciable absorption, anomalous jumps in M versus ω were introduced from absorption related structure. These were subsequently eliminated. This problem of fringe counting was fully circumvented during the final fit of data to eq. 3. Such an artifact could have been a source of error in previous studies involving similar calculations for $n(\omega)$.

Finally, the low-resolution data was fit to eq. 4 to solve for $k(\omega)$ using the above initial values of $n(\omega)$. $k(\omega)$ values for ω >2000 cm^{-1} are negligibley small and hence they were set to zero. A final normalization regression parameter, close to unity (typically ~1.000 ± 0.004), was invoked to further account for any small changes in the probe beam intensity in the sample-in and sample-out ratioing procedure. The resulting values of $k(\omega)$ were then interpolated into a grid of wavenumber values exactly matching the high-resolution ω sampling points. No functional form for $k(\omega)$ was employed in this work because of the complexity in spectral structure.

The values of $n(\omega)$ and $k(\omega)$ determined above were then used as starting parameters for a nonlinear regression fitting of the high-resolution transmission spectrum. Figure 3 shows a typical portion of the spectra fit to the high-resolution transmission. The fitting used a physically realistic Selmeier dispersion as shown in eq. 6 for $n(\omega)$. The starting values for the Sellmeier coefficients ε, A, and B were determined from the values of the coefficients in eq. 10. The regression used the fixed wafer thickness, t, and four varied parameters, the three Sellmeier coefficients, and

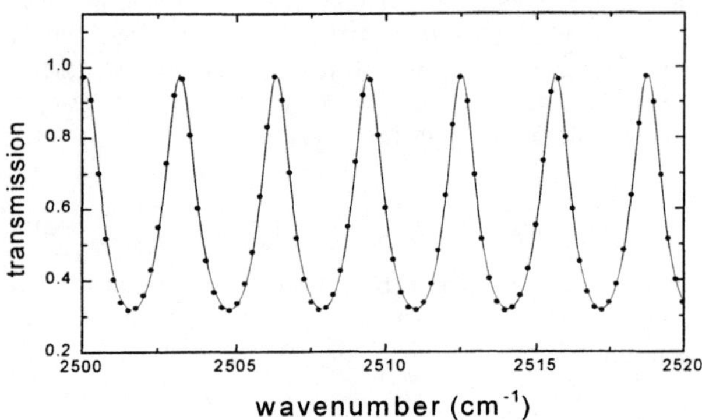

Fig. 3: The measured data (points) and the fitted spectrum (line) are shown.

the wafer wedge thickness variation, Δt.

The values of $n(\omega)$ determined from a complete fit of the high- and low-resolution spectrum are presented in Fig. 4. Data from published references (1,2,7) are also displayed for comparison. The data from Humlicek and Vojtechovsky (2) are an extrapolated version of that measured by Salzberg and Villa (7). The coefficient values determined by the nonlinear regression are $\varepsilon = 11.67300 \pm 0.00005$, $A = (0.92904 \pm 0.00005) \times 10^{-8}$ cm^2, and $B = 0.04097 \pm 0.00002$ where ω is expressed in wavenumbers (cm^{-1}). The uncertainties were determined from the 95% confidence intervals in the nonlinear regression. The uncertainty analysis shown above accounts only for the noise in the measured spectrum and amounts to about 1 part in 10^5. It does not include the uncertainty in the thickness value, which is the dominant overall source of error. Since the interference fringe structure depends mostly on the product $n(\omega)t$, the final error in the reported values of $n(\omega)$ is approximately equal to that in t, namely 1 part in 10^4. The values of $n(\omega)$ range from 3.4169 ± 0.0004 at 4000 cm^{-1} ($\lambda = 2.5$ μm) to 3.4397 ± 0.0004 at 450 cm^{-1} ($\lambda = 22.222$ μm). The Sellmeier form shown in eq. 6 gives the dispersion and the values for points in between. The uncertainties reported here are a factor of 10 better than those presented in ref. (1).

Our results for $n(\omega)$ are in good agreement with the measurements of Salzberg and Villa (7) but are lower than those reported by Edwards (1). The source of the discrepancy is being investigated. Salzberg and Villa used the minimum deviation method that relies solely on precise geometrical measurements and is a direct measurement that avoids the complexities of the approach in measuring and computer fitting the interference fringes used by us and Edwards (1). However, the minimum deviation method is limited to the near infrared where dispersive spectroscopy is effective. The disagreement between our data and ref. (1) may be due to the differences in the electrical properties of the samples used, i.e., 3 Ω.cm to 4 Ω.cm, n-type in ref. (1) versus 32 Ω.cm, p-type in our analysis or in differences in the models and computer fitting approaches. Our analysis fits the entire spectrum, unlike ref. (1) that used only the positions of the interference fringe extrema.

The final values of $k(\omega)$ obtained from the high- and low-resolution spectra are presented in Fig. 5. The spectrum shows the expected structure due to two-phonon absorption (8) bands below ~1050 cm^{-1} and possible oxygen band around 1107 cm^{-1} (9). The dominant source of error is the accuracy with which the absolute transmission values can be measured, especially in the semitransparent region extending from ~1200 cm^{-1} to 4000 cm^{-1}. As explained earlier, we have minimized the effect by reducing the possibility of probe intensity drift and adding a near-unity normalization factor. In addition, the fit to both high- and low-resolution spectra and the highly accurate values of $n(\omega)$ already presented lead us to believe that the values of $k(\omega)$ are reasonable.

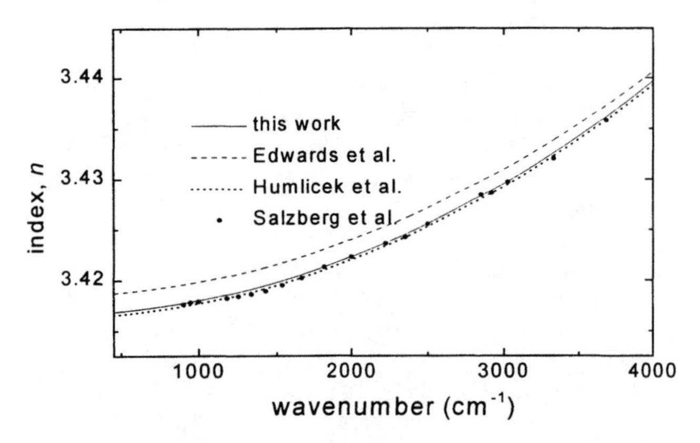

Fig. 4: Plot of n(ω) versus ω. Our results are compared with earlier data from refs. (5), (7) and (8).

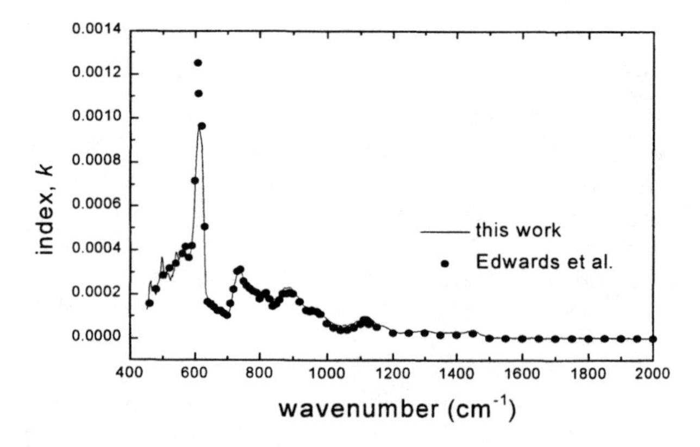

Fig. 5: Plot of k(ω) versus ω the wavenumber, showing the comparison with the values from ref. (1).

The difficulties of obtaining an accurate measure of $k(\omega)$ are reflected by the limited number of published values. In addition, unlike $n(\omega)$ that follows a simple dispersion relationship, as shown in eq. 6, $k(\omega)$ in the IR region contains complex structure that reflects the multiphonon density-of-states, and this factor adds to the difficulty of extracting reliable values. We have compared our results to those compiled by Edwards (1) and absorption measurements performed earlier by Johnson (10) and Bendow et al. (11) as shown in Fig. 5. The agreement seems reasonable in the whole region except for the peak near 610 cm^{-1} where our values are ~20% lower those available from ref. (1). We believe that the source of this disagreement may be in the inaccurate measurement of the reference and transmitted beam intensities in the earlier measurements. The spectral regions with the largest error should clearly be those with the largest absorption. We have taken care to avoid or minimize these errors. The earlier measurements focused mainly on understanding the spectral nature of the variations with less emphasis on photometric accuracy. The ability of our procedure to correct for normalization errors in the low-resolution spectrum and the superior $n(\omega)$ values yielded by the analyses suggest that the our values of $k(\omega)$ may be more accurate than those reported in the past. We are examining this point further and will publish details, along with the numerical values of $k(\omega)$, in the near future.

CONCLUSION

We have presented the methodology to extract accurate values of the optical constants in the mid-IR region that relies on the use of both high- and low-resolution transmission spectra. Armed with a realistic model that accounts for slightly nonparallel sample faces, and small deviations from ideal collimation, we can fit the measured spectra to extract the optical constants $n(\omega)$ and $k(\omega)$ in the 450 to 4000 cm^{-1} range. The low- and high-resolution spectra were fit iteratively to extract $n(\omega)$ and $k(\omega)$ for a Si wafer. We have demonstrated a factor of 10 improvement in the uncertainties in $n(\omega)$ over results available in the published literature.

REFERENCES

1. Edwards, D.F., *Handbook of Optical Constants of Solids*, Ed. Palik, E.D., Academic Press, 547 (1985).
2. Humlicek, J. and Vojtechovsky, K., Phys. Stat. Sol. **(A)** **92**, 249 (1985).
3. Randall, C.M. and Rawcliffe, R.D., Appl. Opt. **6**, 1889 (1967).
4. Loewenstein, E.V., Smith, D.R., and Morgan, R.L., Appl. Opt. **12**, 398 (1973).
5. Edwards, D.F. and Ochoa, E., Appl. Opt. **19**, 4130 (1980).
6. Manifacier, J.C., Gasiot, J., and Fillard, J.P., J. Phys. E: Sci. Instr. **9**, 1002 (1976).
7. Salzberg, C.D. and Villa, J.J., J. Opt. Soc. Am. **47**, 244 (1957)
8. Birman, J.L., *Theory of Crystal Space Groups and Lattice Dynamics: Infrared and Raman Optical Processes of Insulating Crystals* (Springer-Verlag, Berlin, 1984), p. 398.
9. Newman, R.C., Advances in Physics **18**, 545 (1969).
10. Johnson, F.A., Proc. Phys. Soc. London **73**, 265 (1959).
11. Bendow, B., Lipson, H.G., and Yukon, S.P., Appl. Opt. **16**, 2909 (1977).

INFRARED SPECTROSCOPY FOR PROCESS CONTROL AND FAULT DETECTION OF ADVANCED SEMICONDUCTOR PROCESSES

P. Rosenthal[a], W. Aarts[b], A. Bonanno[a], D. Boning[c], S. Charpenay[a], A. Gower[c], M. Richter[a], T. Smith[c], P. Solomon[a], M. Spartz[a], C. Nelson[a], A. Waldhauer[d], J. Xu[a], V. Yakovlev[a], W. Zhang[a], L. Allen[e], B. Cordts[e], M. Brandt[e], R. Mundt[f], A. Perry[f]

[a] On-Line Technologies, Inc., East Hartford, CT 06108
[b] Wacker Siltronic Corporation, Portland, OR 97283
[c] Massachusetts Institute of Technology, Cambridge MA 02139
[d] Applied Materials, Inc., Santa Clara, CA 95054
[e] Ibis Technology Corp., Danvers, MA
[f] Lam Research Corp., Fremont, CA

Fourier transform infrared (FTIR) spectroscopy has emerged as an attractive sensor for in-situ monitoring and control of semiconductor fabrication processes. New applications are being enabled by advances in FTIR hardware and software that provide for: compact size, fast measurements with exceptional stability and signal to noise, and intelligent model based algorithms for thin film and gas analysis. In previously reported work, FTIR instrumentation with automated spectral analysis software was demonstrated as a novel sensor for monitoring layer properties such as thickness, composition and temperature. Recent work has emphasized applications to practical problems in modern semiconductor manufacturing. In this paper we will report pioneering results on: 1) Run-to-run closed loop control of a single wafer epitaxial silicon process using integrated infrared thickness and doping profiling metrology, 2) Fault detection during cluster tool plasma etching using real-time infrared exhaust gas analysis, and 3) oxygen implantation process monitoring during the formation of silicon on insulator (SOI) wafers using infrared reflectometry.

1. INTRODUCTION

With the ever increasing demands on equipment capabilities required for cutting edge semiconductor device manufacturing along with the rapidly increasing cost of the required new capital equipment, new technologies must be implemented in order to increase the overall equipment effectiveness (OEE) of specific tools and the entire fab in general. In this paper, we will present recent results obtained using On-Line Technologies' (OLT) Fourier-transform infrared (FTIR) spectrometer for three specific applications: Run-to-run thickness control of the epitaxial silicon (epi) deposition process; Oxygen implantation for SOI wafer fabrication; and real time fault detection during plasma etch using exhaust gas analysis. In all three cases, we have demonstrated that the use of Advance Process Control (APC) will lead to increased OEE by producing wafers with tighter product specifications, decreasing the number of required test wafers, reducing scrap by instantly detecting process fault conditions well as reducing the time required for process development and yield learning.

2. RUN-TO-RUN CONTROL OF EPI SILICON PROCESSES

In current practice, fixed set-point control is used for epi growth. In this procedure, reactor recipes are set so that parameter values (thickness and resistivity) match target specifications. The process is run until the parameter value (as determined by measurements on 1 in 25 wafers, typically) is outside preset limits. Then the process is stopped and the reactor is re-tuned. This practice results in down time for process re-tuning, scrapped material which is produced before the process is stopped, and wider variations in the product specifications than necessary. Higher quality products with tighter specifications, produced at lower cost, with reduced scrap can be achieved through closed loop process control, where measurements are made on every wafer and recipe optimization is made continuously to keep the parameter on target.(1)

Wafer-by-wafer process control is the most logical and practical next step in improving semiconductor fabrication. Thin film metrology (TFM) tools can be installed for unobtrusive measurements on most fabrication tools, and most fabrication tools are configured to accept recipe changes via a SECS/GEM interface. Run-to-Run i.e. wafer-by-wafer control (in a single wafer process tool) can be implemented with only minor modification of the basic semiconductor process tool. What is required is a computer cell controller (2) which can bring the following required individual components together: 1) the fabrication tool; 2) wafer state sensors; 3) a communications and control package on a computer that can manage operating and coordinating the sensor data and the run-to-run control algorithm; and 4) the process tool.

We have demonstrated these components on a single wafer process tool, specifically, the Applied Materials Epi Centura HT single wafer epitaxial silicon cluster deposition tool. Wafer state data was provided by an OLT Epi On-Line film thickness monitor integrated onto the cooldown chamber of the Centura reactor. The cell controller software was implemented on a PC running Windows NT.

A series of experiments were conducted to evaluate the performance of the system. First, to evaluate the robustness of the sensor and software, a long baseline run

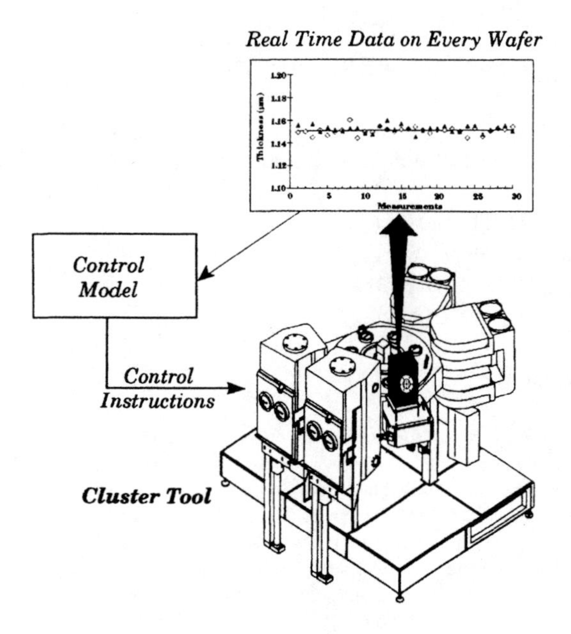

Figure 1. Illustration of the Applied Materials Centura cluster tool and the OLT integrated film thickness monitor.

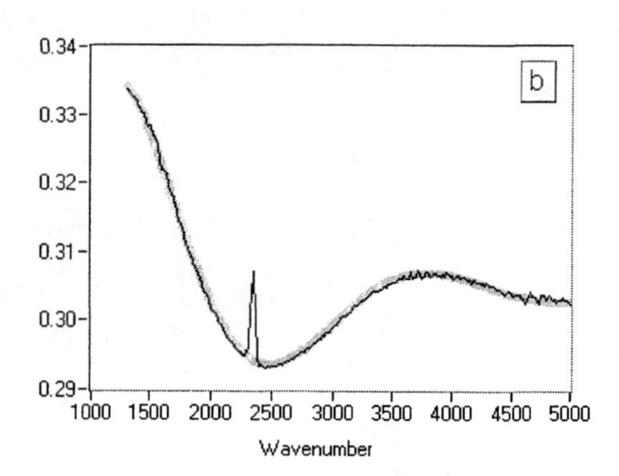

Figure 2. Measured reflectance (black line), and model based fit (gray line) for a 0.6 micron epi sample.

without control was performed on a Centura with three epi chambers. Film thickness and other data were collected from the sensor on over 800 consecutive wafers. To evaluate the control algorithm, wafers were processed under conditions that simulated several process upsets. A comparison was made to compare the wafer-to-wafer variability under open loop and closed loop control.

2.1 Epi System Architecture

The physical layout of the system is shown in Fig. 1. The figure illustrates the Applied Materials Centura cluster tool, showing two loadlocks, a central robotic transfer chamber, several process chambers and a cooldown chamber. The film thickness monitor is shown mounted on top of the cooldown chamber.

The Epi Centura reactor is a production tool used primarily for the fabrication of lightly doped epitaxial silicon layers, grown on heavily doped or ion implanted substrate wafers. A typical deposition takes around 100 seconds, and the overall throughput of a system can be in the range of 25 wafers per hour. Process variables include gas flows, dilution with hydrogen, temperature history and deposition time.

The Epi Centura is an extremely stable reactor. Drifts in deposition conditions can be largely ignored within a single deposition run, and generally speaking, the wafer-to-wafer reproducibility of the process is extremely high. Because of this, the process is ideally suited to a run-to-run control system. The capability of tracking material flow within the cluster tool as well as to read and write recipes to the Centura allow the implementation of a run-to-run control scheme with minimal modifications to the tool, mainly to allow physical and optical integration of the sensor. Because the sensor can be mounted on the cooldown chamber, the system can be implemented without affecting the design or operation of the deposition process chambers at all, a key practical advantage of this technology. Similarly, because the measurements are performed during the normal cooldown process of the tool, the metrology and control are implemented with no effect on the tool throughput.

The film thickness monitor consists of an FTIR reflectometer, a reference mirror mounted in the cooldown chamber of the Centura, and a controller with software to perform data acquisition and analysis. The analysis algorithms employ a model based analysis of the reflectance that fits a simulated reflectance spectrum to the measured data. The fitting parameters, which include variables such as film thickness, epi/substrate doping transition width and substrate carrier concentration, are selected to provide an accurate and relatively complete description of the optical properties of the wafers using only a few free parameters. Fig. 2 shows a model based fit for a submicron epi sample. In the figures, the layer thickness, the substrate carrier density and the doping transition width were varied for the fits. The peak at 2350 wavenumbers is due to atmospheric CO_2 in the optical path. The excellent quality of the fit illustrates how accurately the On-Line Technologies reflectance model accounts for the quantitative aspects of the epi silicon optical properties. Correlation studies comparing the model based FTIR analysis with destructive measurements (Secondary Ion Mass Spectroscopy (SIMS) and spreading resistance profiling (SRP)) show excellent agreement between the techniques, including the details of the shape of the doping transition. Fig. 3 shows a comparison between SRP and model based FTIR analyses of two epi wafers fabricated at slightly different deposition temperatures, illustrating the excellent agreement between the two techniques.

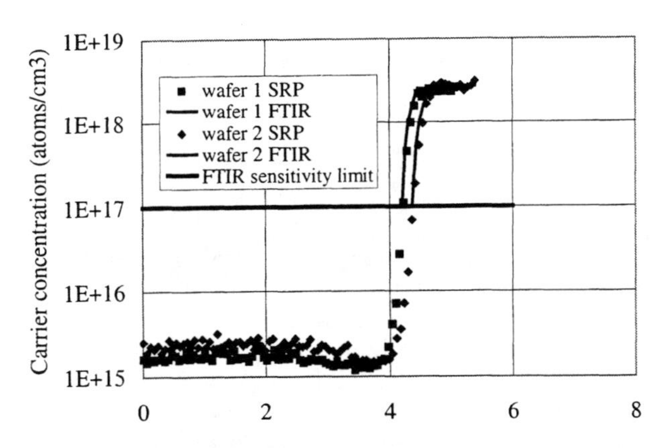

Figure 3 Comparison of doping profile results using model based FTIR and SRP analysis for two epi samples.

The epitaxial thickness sensor is fully automated. Data is collected, analyzed and archived without operator interaction. The cluster tool robot provides wafer handling to load and unload wafers. Calibration is automated as well, and employs a reference mirror mounted within the cooldown chamber, allowing for frequent corrections for spectrometer drift. Details describing the Run-to-Run control agorithm can be found in Refs. 3-6.

2.2 Experimental Results

2.2.1 Baseline Tests

A set of wafers was processed on a production three chamber Epi Centura cluster tool installed at Wacker Siltronic. Integrated in-line thickness measurements were performed on all the wafers spanning 89 hours of production of 809 commercial epitaxial silicon wafers. Out of the measurement set, a total of 13 FTIR measurements failed to provide thickness results (i.e., 1.6 %, primarily due to electrical noise or glitches). The thickness results for the remaining 796 wafers are shown as a scatter plot in Fig. 4. As can be seen from the figure, several batches of wafers of different epi thickness were processed during the period of investigation, the first batch of wafers having an epi thickness of ~2.15 μm, the second a thickness of ~2.65 μm, the third a thickness of ~2.16 μm, and the fourth a thickness of ~2.6 μm.

The data also show that drifts in the thickness over time were detected. Figure 5 shows details for the first batch (wafers 0 to 74). The data show that 1 out of every 3 data points is different from the two others. This trend is caused by one of the chambers (the second) being slightly out of tune from the other two. The repetitive pattern (i.e., thickness equal to ~ 2.16 μm, ~ 2.14 μm, ~ 2.16 μm and repeat) can be clearly observed after wafer #41, while for lower wafer numbers the pattern is sometimes slightly out

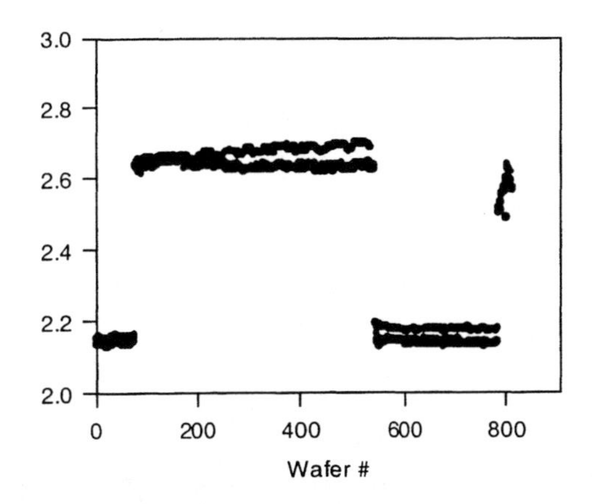

Figure 4. In-Line thickness measurement from a 89 hour run of 809 wafer produced on a three chamber Epi Centura cluster tool.

of sequence. The thickness sequence correlates to the chamber sequence.

The baseline experiment showed that: 1) the sensor provided reliable thickness results for nearly all of the wafers during an extended process run, and 2) drifts within the different process chambers occurred and were identifiable from the sensor data. In particular, during these runs, the second chamber behaved slightly differently than the two others and produced a noticeable drift in thickness with time. After completion of this experiment, it was found that the drifting chamber actually had a small air leak which eventually required a shut down for repair.

2.2.2 Control Runs

A series of wafers was processed in a single chamber system at Applied Materials while under run-to-run control. The run-to-run controller estimated the current deposition rate by perfoming an exponentially weighted moving

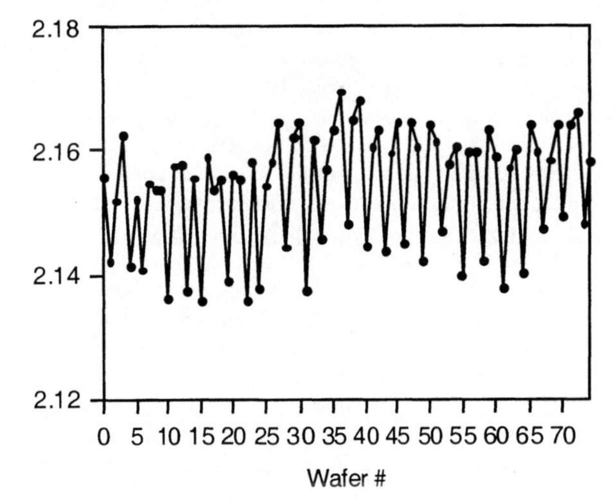

Figure 5. Detail View of in-line thickness measurements from the first batch of the baseline test.

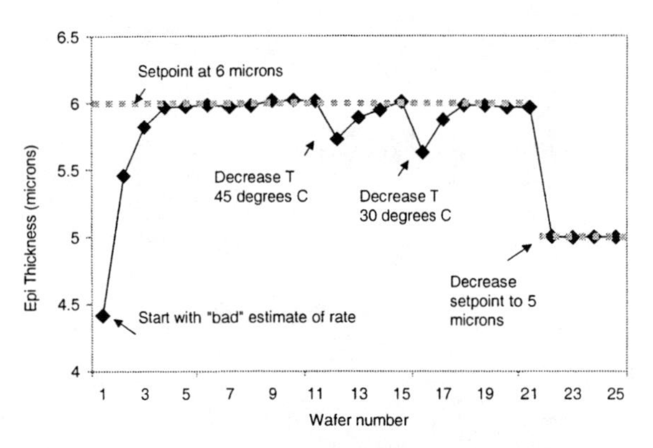

Figure 6. Run-to-Run control during epi process upsets.

average (EWMA) of the measured thickness on the previously processed wafers. The deposition time was updated after each wafer was processed and measured. The new deposition time was calculated as the target thickness divided by the EWMA estimated rate. The recipe changes were automatically downloaded to the cluster tool via the SECS port after each measurement. The response to process upsets was tested by deliberately modifying the temperature in the recipe several times during the lot.

The performance of the system under run-to-run control is shown in Fig. 6. Within 4 wafers of startup, the controller reached steady state at the target thickness with a deposition time of 100 seconds. Next, a small process upset was simulated at wafer 9. For the upset, the recipe was modified so as to increase the growth temperature by 15 °C. This resulted in a change in growth rate of less than 1%. Because the correction to the recipe was smaller than the 1 second time granularity imposed by the cluster tool settings, round off resulted in no modification to the recipe and a small increase in thickness was observed in wafers 9-11. Next, the temperature was decreased by 45 °C. This larger upset resulted in a significant decrease in growth rate for wafer 12. During the next four wafers, the controller adjusted the deposition time to bring the epi thickness back to the target value. The temperature was again decreased starting at wafer 16, and again, the controller brought the film thickness back to the target value during the next 4 wafers. At wafer 22, the target thickness of the controller was reduced to 5 microns, but the growth conditions were not modified. The run-to-run controller correctly modified the deposition time and brought the thickness to the new target value in the next wafer.

The primary benefits of closed loop control for the epi process are in improved overall equipment effectiveness (OEE), tighter product specifications, and faster yield learning. The reduction of monitor wafer usage alone is a powerful driver for this technology. A modern 200 mm epi line producing 50,000 wafers per month typically allocates 4% of its production capacity to monitor wafers which are not sold. At a typical selling price of $200 per epi wafer, a reduction of monitor wafer usage in half represents increased profits of $200,000 per month and the payback period for the inline monitors would be between two and three months, using monitor wafer savings alone.

3. SILICON ON INSULATOR DIAGNOSTICS

The use of silicon-on-insulator (SOI) material as substrate for integrated circuits has several advantages over conventional Si wafers, including high device speed, reduced circuit size, and lower power consumption. One of the main techniques for the fabrication of SOI wafers is the separation by implantation of oxygen (SIMOX) process in which a high dose of oxygen ions is implanted into a Si wafer to form a layer of buried oxide (BOX) followed by high-temperature annealing. A thin layer of device quality Si is formed on the top of the BOX layer after the processing. The BOX layer has to be a good dielectric, i.e., with a high breakdown field and a low density of defects. The BOX properties are very sensitive to the processing conditions, and SOI fabrication requires accurate control of the major processing parameters: wafer temperatures, ion energy and dose. The dose, (the amount of the implanted oxygen ions), is proportional to the ion beam current and implantation time and ultimately determines the thickness of the buried oxide layer.

Currently, the implant dose is estimated via post-process, *ex-situ* ellipsometry measurements of the BOX thickness of annealed wafers. The long delay between the process and the metrology reduces the effectiveness of the metrology as a process control tool. Monitoring the implantation beam current does not provide direct information on the wafer properties, and does not satisfy the demanding manufacturing requirements of 1-2% implant reproducibility. As we report in this work, the FTIR Reflectance Spectroscopy measurements provided rapid characterization of the dose deviation from the target

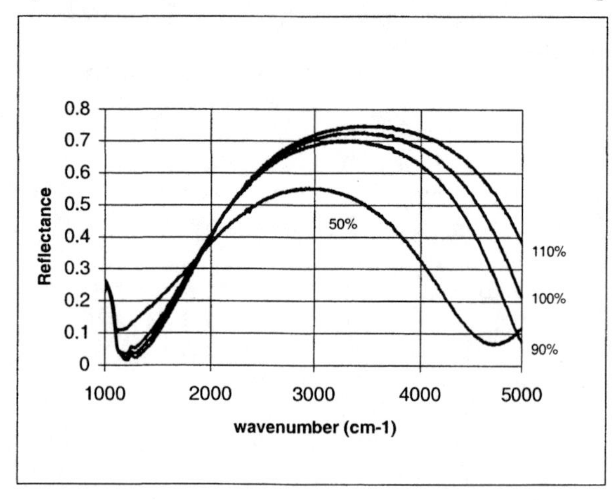

Figure 7. The reflectivity as a function of oxygen dose from 1000 to 5000 cm^{-1}.

levels with precision better than 1%.

FTIR reflectometry measurements were performed with a spectrometer equipped with confocal optics to eliminate reflections from the backside of the (ir-transparent) wafers under investigation.

The wafers were prepared with full, nominal doses of the implanted oxygen, and with doses deviated from the norm. Wafers from lot 9433 were processed with high dose deviation, -50%, -10%, and +10%. Wafers from the lot 9611 were processed with low dose deviation, +3%, +6%, and +9%. These dose deviations are of practical relevance to the fabrication process and provide a direct way to estimate sensitivity of the FTIR reflectance measurements.

Fig. 7 demonstrates typical mid-IR reflectance spectra of the SOI structures. The spectra include two main areas: low-wavenumber, 800 – 1500 cm^{-1}, and high-wavenumber, 1500 – 5000 cm^{-1}, one. The spectral features at low wavenumbers are influenced mainly by the optical and structural properties of the buried oxide layer. Silicon oxide has a strong absorption in this range, which is influenced by the material compositional and structural properties. The high wavenumber spectral range is influenced mainly by the interference effects in the multilayer structure. Both lightly doped Si (which is used for the SOI wafers), and SiO_2 are transparent in the high-wavenumber range, and the bulk refractive index of these materials can be used for the multilayered spectral analysis. Based on this consideration, the spectral range of 2000 - 5000 cm^{-1}, (only weakly affected by the compositional properties of the buried oxide layer), was chosen to deduce the thicknesses of top Si layer, D_{si}, and the BOX layer, D_{BOX}.

To analyze the SOI reflectance spectra, The following film stack model was employed: transparent Si semi-infinite substrate – BOX layer – top Si layer – (2 nm native oxide layer). Thin (on the order of a few nm) interfaces of the BOX layer were included in the effective thickness, D_{BOX}, of the layer. The multi-layer reflectance analysis minimized the difference between the experimental and simulated spectra, varying thicknesses of the top Si, D_{si}, and, the BOX layer, D_{BOX}.

Table II. Si layer thickness, BOX layer thickness and deviation as a function of ion dose.

Lot 9611 Wafer #	Dose	D_{si}, Å	D_{BOX}, Å	BOX thickness deviation, Å
12	Full	2611	1623	0
11	+3%	2601	1675	52
10	+6%	2584	1718	95
9	+9%	2570	1768	145

The spectra of the wafers, processed at dose deviations smaller than 10%, showed a nearly linear dependence between the dose level and reflectance in the spectral ranges of 3000 – 4500 cm^{-1}, and 1300 –1700 cm^{-1} (Fig. 8). Thicknesses of the top Si, D_{si}, and, the BOX layer,

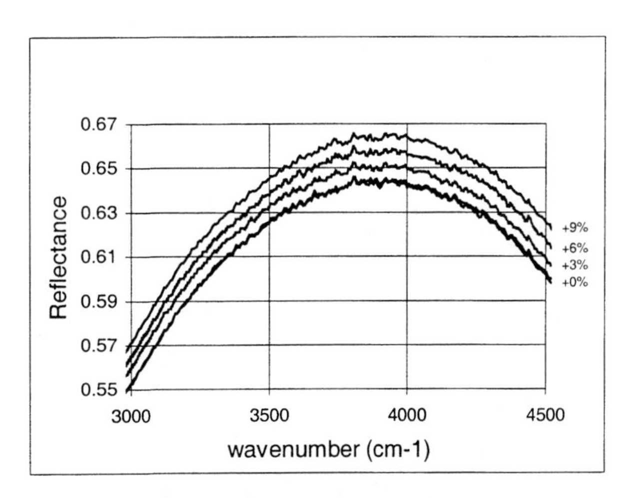

Figure 8. The region that shows a linear behavior between oxygen dose and reflectivity lies between 3000 and 4500 cm-1. Note that two traces overlap for the full dose (+0%). Sensitivity to dose is better than 1%.

D_{BOX} were deduced from the spectra. The results are given in Table II.

The table confirms that the BOX thickness deviation is linearly proportional the small dose deviation. Increased doses of oxygen ions leads to the consumption of Si from the top layer (see the table). A 3% dose deviation results in about 47 Å BOX thickness deviation (for a typical dose level of $1.6x10^{18}$), i.e. **1% dose variation would result in more that 15 A thickness change.** Given the repeatability and sensitivity of FTIR reflectance measurements to provide reproducibility of the BOX thickness measurements in a level a few angstroms, the precision on the the dose was of the order of 0.2-0.3%.

4. GAS ANALYSIS BY FTIR

Fourier transform infrared (FTIR) spectrometry has been shown to be an ideal technique for quantitative analysis of complex mixtures of gases. FTIR spectrometers characteristically have large optical throughputs and obtain the entire spectrum simultaneously. These two features allow for full spectral coverage while maintaining high sensitivity. One other significant advantage inherent to FTIRs is their frequency precision. This allows data to be collected day-after-day with no drifts in the frequency scale. Historically, however the complexity of the IR data set required a trained user to interpret and quantify the species present. Automated spectral analysis software now available greatly simplifies the interpretation and quantification of the complex data set.

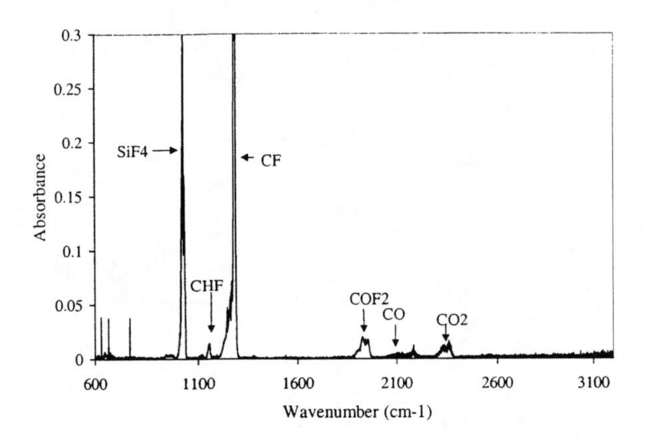

Figure 9. A typical spectrum from one of the CF_4 etch processes shows 6 IR-absorbing gases.

4.1 EXPERIMENTAL

Experiments were performed on a Lam Research Corporation, Research and Development Test Bench (RDTB). This experimental system is a large, high-flow, high density reaction chamber with good diagnostic access, which provides a flexible, high performance test bench for reactor testing. The reactor incorporates one 2000 l/s turbornolecular pump and a transformer coupled plasma (TCP) source with a 2.0 kW rf power supply (1 3.56 MHz). Provisions were made to collect and store parameters such as reactor set points (flows, pressure, plasma power, etc.) plus other sensor data (coolant temperature, Langmuir probe, Vi probes (for rf characterizations, RTSPC and fault detection), and optical emission spectroscopy for plasma emission). The spectrometer was positioned above the exhaust duct of the etch chamber. The IR beam was directed to a set of focusing and steering mirrors and into a multi-pass mirror assembly enclosed in an exhaust line tee.

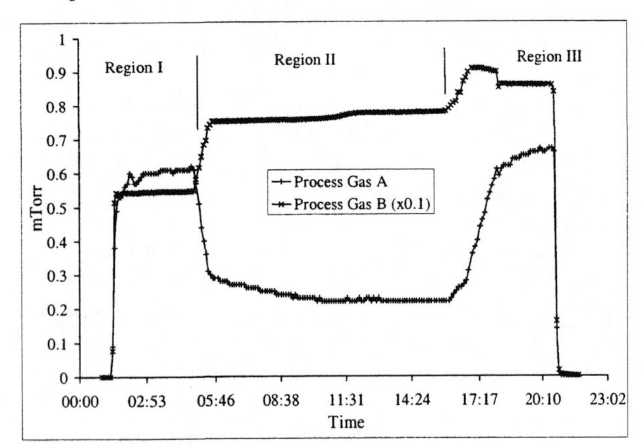

Figure 10. The gas concentration evolution of a CF_4 etch. The film stack is photoresist on oxide on silicon.

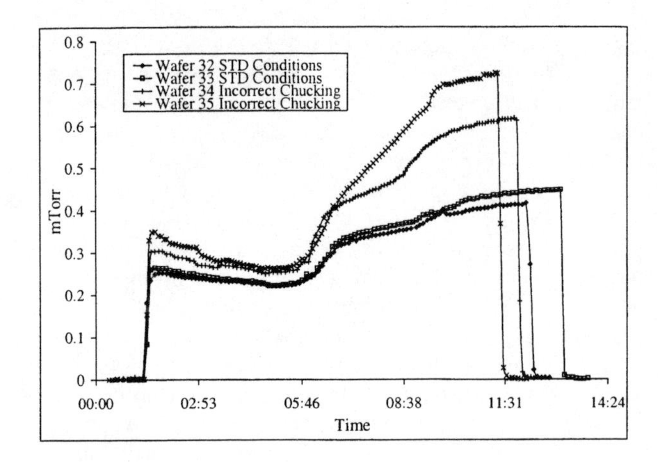

Figure 11. Simulated faults during etch due to low backing pressure show up as an increase in intensity of SiF4 gas concentration.

This tee was placed between the turbo and the mechanical pumps. The multi-pass cell folded 20 passes through the tee to provide a 5 meter path length.

Wafers were etched by a CF_4 plasma (100 sccm CF4, 1000 Watts TCP, 200 Watts bias, 15 Torr He chucking back pressure, and 10 mTorr chamber pressure). The wafers were etched using either standard conditions or simulated and actual fault conditions. The faults included air leaks and varying the chucking back pressure, plasma power, bias power, wafer moisture, and load lock position.

The exhaust gas passed through the in-line gas cell and spectra were collected at 1 cm^{-1} resolution. Sixteen scans were co-added over 5 sec. to provide approximately 11 data points/min. Figure 9 shows a typical spectrum obtained during a standard CF_4 etch. Peaks for SiF_4, CF_3H, CF_4, COF_2, CO, H_2O and CO_2 are identified. The spectrometer transfer optics were open to the atmosphere and the CO_2 variation may have been due to atmospheric fluctuations. The peak-to-peak noise level across the

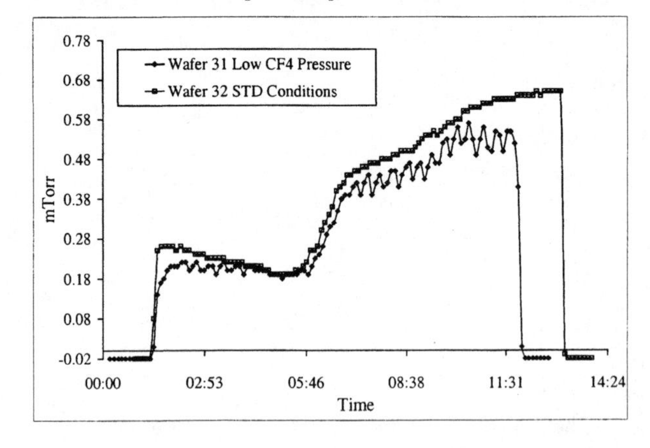

Figure 12. Insufficient back-pressure on a mass flow controller resulted in the oscillations observed during the etch.

fingerprint region was less than 0.002 absorbance units.

Quantitative analysis was performed using a classical least squares analysis routine. The algorithm uses selected spectral regions to minimize interferences and accounts for species with Non-Beer's Law behavior. For these processes, up to 6 gas intensities (no other IR active gases were detected) were tracked as a function of time during the etch process. Figure 10 shows the time evolution of two process gases during the etch of a wafer composed of photoresist (98% coverage) on 2.4 um SiO_2 on Si. The evolution of the gases is interpreted as follows: In Region 1, both photoresist and oxide are etched (there is 2% open area). For this specific film stack, the photoresist is consumed before the oxide is completely removed from the open areas (this would not be the case during an actual production run). In region 2, oxide is being removed, initially over the entire wafer. In the middle of region two, the small increase in signal in the SiF_4 trace corresponds to the clearing of the 2% oxide originally not covered by photoresist. At the end of region two, the 98% remaining coverage of oxide clears, leaving the etch products charecteristic of bare silicon in region 3.

Figure 11 presents the effect of an incorrectly chucked wafer on the generated exhaust gas composition. Under conditions of incorrect chucking, the wafer is hotter than normal, which modifies the etch rate and plasma chemistry. Two standard etches are compared to two simulated chucking faults by varying the helium back pressure. The two trend lines from the simulated faults show significant variation in the exhaust gas concentration for this compound.

During the simulated faults, an oscillation in the CF_4 intensity was observed, as shown in Fig. 12. This effect was traced back to a real fault caused by an incorrect back pressure on the CF_4 mass flow controller. When the backpressure was corrected, the oscillations disappeared. Other faults, including plasma power variations, air leaks and load lock states, we re investigated as well. Details on these results are described in Ref. 7.

5. CONCLUSIONS

We have described applications of FTIR for the charecterization, and control of semiconductor processes. Exhaust gas monitoring provided a reproducible and rapid measurement of a rich variety of compounds produced during the etch process. This information can be used for process control (Endpoint) as well as fault detection and classification. For silicon on insulator fabrication, oxygen implantation doses were correlated to infrared reflectance meaasurements of the BOX layer. The thickness of the top Si layer as well as the thickness of the buried oxide film were measured with greater than 1% accuracy. Closed loop run-to-run control of the epitaxial silicon process was demonstrated using a fully automated integrated metrology system.

6. ACKNOWLEGMENTS

We gratefully acknowledge the support of the National Science Foundation SBIR Grant No. DMI 9660643.

---

1 J. Hosch, SEMATECH, Private Communication

2 Moyne, J., and McAfee, L., Jr., IEEE Trans. Semi. Manuf.,. **5**, (2), 77-87, 1992.

3 Smith, T., D. Boning, Moyne, J., Hurwitz, A., and Curry, J., "Compensating for CMP Pad Wear Using Run by Run Feedback Control," VLSI Multilevel Interconnect Conference, pp. 437-440, Santa Clara, CA, June 1996.

4 Boning, D., Hurwitz, A., Moyne, J., Moyne, W., Shellman, S., Smith, T., Taylor, J., and Telfeyan, R., ''Run by Run Control of Chemical Mechanical Polishing," IEEE Trans. on Components, Packaging, and Manufacturing Technology, Vol. **19**, No. 4, 307-314, Oct. 1996.

5 Boning, D., Moyne, W., Smith, T., Moyne, J., and Hurwitz, A., "Practical Issues in Run by Run Process Control," 1995 SEMI/IEEE Advanced Semiconductor Manufacturing Conference & Workshop, pp. 201-208, Cambridge, MA, Nov. 1995.

6 Sachs, E., Hu, A., and Ingolfsson, A., "Run by Run Process Control: Combining SPC and Feedback Control," IEEE Trans. Semi. Manuf., vol. **8**, no. 1, pp. 26-43, Feb. 1995.

7 Solomon, P. R., Rosenthal, P. A., Nelson, C. M., Spartz, M. L., Mott, J., Mundt, R. and Perry, A., Proc. of the SEMATECH AEC/APC Workshop IX, SEMATECH Advanced Process & Equipment Control Program, Lake Tahoe, NV, Sep. 1997

Measurement of Silicon Doping Profiles using Infrared Ellipsometry Combined With Anodic Oxidation Sectioning

Thomas E. Tiwald, Aaron D. Miller and John A. Woollam

Center for Microelectronic and Optical Materials Research, and the Department of Electrical Engineering, University of Nebraska, Lincoln, NE 68588-0511

Silicon carrier concentration profiles were determined by resistivity measurements acquired optically using infrared ellipsometry combined with anodic oxidation sectioning. The method is based upon a technique developed in the early 1960's, except that the resistivity is measured optically using infrared ellipsometry instead of direct electrical contact to the sample. Infrared ellipsometry is a non-contact optical technique that can simultaneously determine the oxide thickness and the resistivity profile of the underlying substrate (via the free-carrier Drude effect). This eliminates the need for special sample geometries, implanted ohmic contacts, repeated hydrofluoric acid etches and separate measurements of oxide thicknesses. In this study the optical constants of anodic oxides are determined and a carrier profile of an As implanted sample is characterized using infrared ellipsometry.

INTRODUCTION

In the early 1960's several researchers (1-3) developed techniques to determine carrier concentration profiles that combined differential resistivity and/or Hall measurements with the removal of thin, uniform layers of silicon. Tannenbaum (2), appears to be the first to section silicon using anodic oxidation followed by a HF strip, rather than a planar chemical etch (1) or precision polishing (3). Anodic oxides grow at room temperature and consume uniformly thin layers of silicon over large surface areas, both requisites for differential electrical measurements. Since the 1960's, this technique has been further developed (e.g., see Refs. 4-10). In the 1980's, Galloni and Sardo (11) developed a fully automated differential Hall measurement system.

The most difficult aspects of this technique are the differential Hall and sheet resistivity measurements. They require special sample geometries (e.g., Van der Pauw), and deep, reliable ohmic contacts, usually by ion implantation and annealing followed by deposition of metal contacts (10,11).

Additional difficulties arise from the HF strip of the oxide after each oxide growth cycle, the need for an independent determination of oxide thickness, and growth rate calibration.

Here we describe a method where the resistivity and thickness of each silicon layer is measured by optical probe instead of direct electrical contact, eliminating the need for a Van der Pauw geometry and implanted ohmic contacts. The contact required for electrolysis is made by a bit of silver paint in one corner of the sample.

Infrared spectroscopic ellipsometry (IR-SE) is a non-contact optical technique that can measure the resistivity profiles of multiple-layered samples via the free-carrier Drude effect (12-14). Standard ellipsometric analysis techniques allow the simultaneous determination of the oxide thickness and resistivity of the underlying layers after each anodization cycle. Since the thickness of the consumed silicon layer can be calculated from the oxide thickness, there is no need to calibrate oxide growth or remove the oxide following each oxidation cycle. From knowledge of the resistivities and thicknesses of the consumed silicon layers, one can resolve the carrier (active dopant) concentration profile of the sample.

In this study a carrier profile is characterized and the optical constants of the anodic oxides are determined using IR-SE.

EXPERIMENTAL PROCEDURE

After the initial sample preparation, the procedure consists of a repeated cycle consisting of an anodic oxidation followed by an IR-SE measurement. When ellipsometric spectra show no evidence of a change in substrate optical properties, all the doped surface layers have been removed and only the substrate remains.

CP449, *Characterization and Metrology for ULSI Technology: 1998 International Conference*
edited by D. G. Seiler, A. C. Diebold, W. M. Bullis, T. J. Shaffner, R. McDonald, and E. J. Walters

Sample preparation

For this non-contact method, the size or shape of the sample is not critical, although it is preferable that the sample be wider than the diameter of the infrared beam. Prior to oxidation, the samples are cleaned, dipped in 49% hydrofluoric acid for one minute, and then thoroughly rinsed in deionized (DI) water and dried. Finally, one corner of the unpolished backside of the sample is coated with silver paint and dried for ten minutes at 125°C. Since even heavily doped silicon is partially transparent at near infrared wavelengths, it is important to avoid completely covering the back surface with silver paint, which can strongly reflect infrared light and complicate analysis of the ellipsometric results.

The size, electrical resistance, and ohmic or rectifying nature of the contact are not critical, since it only provides a path for anode current. These factors do effect the total voltage drop across the cell (see next section).

Electrolysis

The electrolysis cell consists of a 500 ml beaker above which two alligator clips are suspended (see schematic in Figure 1). One clip leads from positive output of a constant current source and is attached to the silicon sample (anode), the other is attached to platinum foil (cathode). Both are immersed in the electrolyte. Since evolved gas bubbles will stick to the sample, causing non-uniform oxide growth, it is important to agitate the solution. This is accomplished with a magnetic stirrer and stir bar.

During anodization, the current density is influenced by sheet resistance and the presence of P-N junctions, both of which vary during oxide growth. On the other hand, the forming voltage, V_f, is primarily a function of anodic oxide thickness (6). Therefore we operated our power supply in a constant-current mode and monitored the V_f to estimate oxide thickness.

The compliance voltage at the source (see Figure 1) equals the sum $V_f + V_{cell} + V_{si}$. V_{cell} is the voltage dropped across the platinum cathode and electrolyte solution, and V_{Si} is the voltage dropped across the silicon contact and the series resistance of the bulk. Assuming V_{cell} and V_{Si} are constant, any change in the compliance voltage represents a change in V_f. For our deposition conditions, roughly 1 nm of oxide growth occurred for every 1 volt increase in V_f. Since we determine the oxide thickness using IR-SE after each oxidation cycle, it is not necessary to precisely determine V_f.

Because our power source had a maximum compliance voltage of 300 V, our maximum oxide thicknesses were limited to approximately 230 nm ($V_f \cong 230V$ and $V_{cell}+V_{si} \cong 70V$). When additional silicon sectioning was needed, the oxide was removed with HF.

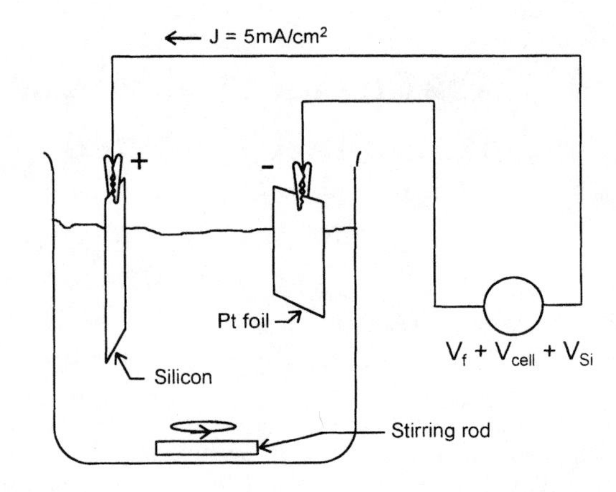

Figure 1. Electrolytic cell. SiO_2 grows on both sides of silicon sample.

Previous researchers have used current densities ranging from 3 to 11 mA/cm^2 (4-9,11). Barber et al. found current densities in the range of 2-8 mA/cm^2 resulted in similar oxide properties (8). For this study, we chose 5 mA/cm^2. The total current is determined by multiplying 5 mA/cm^2 times the total immersed surface area, including both the front and back surfaces of sample. It is likely that more current actually flows through the unpolished back surface, since it has a larger surface area.

A number of electrolyte solutions have been used for anodic oxidation (see (6) for list of references). Many have chosen a mixture of ethylene glycol, water and potassium nitrate (4-6,7,8,11). The proposed anodic reactions for this solution are (4)

$$\begin{matrix} CH_2 - CH_2 \\ | \quad\quad | \\ OH \quad OH \end{matrix} \Leftrightarrow [\text{alcohol}] \Leftrightarrow 2HCHO + 2H^+ + 2e^- \tag{1}$$

$$Si + 2H_2O \rightarrow SiO_2 + 4e^- + 4H^+ \tag{2}$$

As Eqn. (2) indicates, the production of SiO_2 depends on the presence of water. Thus the rate of anodization and oxide quality depend on the amount of water in solution (4-6). We chose an electrolyte solution consisting of 90% ethylene glycol, 10% DI water and 0.05 M of potassium nitrate.

Infrared ellipsometry

The infrared variable angle spectroscopic ellipsometer is described elsewhere (13,14). For this study, the spectral range was 700 to 5500 cm^{-1} (1.8 to 14.2 μm, 0.089 to 0.681 eV) at a resolution of 16 or 32 cm^{-1}. The angle of incidence φ was 72°, which yielded ellipsometric spectra with reasonable sensitivity and signal-to-noise ratios.

Ellipsometry measures the change in polarization state of light as it reflects from a sample surface. That change is expressed as the ratio ρ of the complex Fresnel reflection coefficients r_p and r_s (for light polarized parallel and perpendicular to the plane of incidence, respectively). That ratio is (15)

$$\rho = \frac{r_p}{r_s} = \tan(\Psi)e^{i\Delta} \quad (3)$$

Specifically, tan(Ψ) is the ratio of the magnitudes of r_p and r_s, and Δ is the phase difference between the coefficients.

Values for ψ and Δ at each wavelength comprise the ellipsometric spectrum This spectrum – when measured at oblique angles of incidence – is an extremely sensitive function of the various layers and microstructure of a sample. Unfortunately, for all but the simplest samples, this function cannot be inverted, therefore information must be extracted by optimizing the parameters of a appropriate optical model to fit the data. The numerical regression procedure used here (described by Herzinger et al. (16)) adjusts the various parameters until the mean squared error between the calculated and measured ellipsometric values is minimized.

A related formula is called the pseudo dielectric function. It is defined as (15)

$$\langle\varepsilon\rangle = \langle\varepsilon_1\rangle + i\langle\varepsilon_2\rangle = \sin^2\varphi\left[1 + \tan^2\varphi\left(\frac{1-\rho}{1+\rho}\right)^2\right] \quad (4)$$

Provided that a bulk sample is optically thick with a perfectly smooth, abrupt surface, the pseudo dielectric function will be identical to the bulk dielectric function. Although that is not the case here, we use the pseudo dielectric function to define a common-pseudosubstrate (17) which incorporates all the silicon layers into a single substrate. This approximation, which can be applied to silicon and other high dielectric constant materials, simplifies analysis and reduces the accumulation and propagation of errors during analysis.

Drude optical model

The optical properties of each silicon layer are defined by the classical Drude equation (18)

$$\varepsilon_j = \varepsilon_\infty - \frac{1}{\rho_{dc_j}} \frac{4\pi\hbar^2}{(E^2\tau + i\hbar E)}, \quad (5a)$$

$$\rho_{dc_j} = \frac{m^*}{N_j e^2 \tau} = \frac{1}{N_j e \mu} \quad (5b)$$

where ε_j is the complex dielectric function of the jth layer, ε_∞ is residual dielectric response from the interband transitions, E is the energy of the incident photons, and τ is the mean scattering time of the free carriers. The quantity ρ_{dcj} is the dc resistivity of the jth layer. It is inversely proportional to the electronic charge e, the carrier concentration N_j, and the carrier mobility μ. ($\mu = e\tau/m^*$ where m^* is the carrier effective mass.) For this study, we assume that τ is independent of photon energy.

Ultimately, we wish to determine N_j; unfortunately, N_j is not independent of mobility, since $1/\rho = e\mu N_j$. This is problematic because $\mu = e\tau/m^*$, and both m^* and τ are strong functions of doping in heavily doped samples. Rather than fix m^* at an appropriate value as was done in our first study (13), we use a method based on ASTM standard 723-88 (19), which yielded good results in a later IR-SE carrier distribution study (14). This method converts silicon resistivities directly into carrier densities, using empirical equations which were developed from a set of carefully characterized samples.

DATA ANALYSIS

Data are analyzed in reverse order; that is, the carrier concentration of the deepest silicon layers are determined first and the shallowest layers are determined last. Accurate results depend on the determination of the thicknesses of the silicon sections, which, in turn requires accurate optical constants for the anodic oxide.

Oxide Optical Constants

The oxide dielectric function is determined from the last oxide growth for two reasons: first, it is the thickest oxide layer; and second, the substrate can be treated as an undoped, optically-thick silicon layer, which simplifies analysis. We first model the oxide as a Lorentz oscillator in order to obtain the oxide thickness and ε_∞ of the silicon substrate; then determine the exact dielectric function using a wavelength by wavelength fit.

The results, shown in Figure 2, reveal two important characteristics of these anodic oxide films. First, they contain water, as indicated by the broad absorption peak centered around 3400 cm^{-1} (νOH stretching modes), and the scissor deformation mode at 1630 cm^{-1} (δH_2O). Second, they are less dense than quartz glass, as indicated by the reduced size of the νSiO stretching modes between 1000 and 1100 cm^{-1}, and the lower ε_1 background seen in the transparent region from 1600 to 4000 cm^{-1}.

When the data are modeled using an effective medium approximation layer (EMA) (20),volume fractions of 65% for SiO_2, 15% for water and 20% for void provide the best fit. This suggests that the ratio of silicon consumed to oxide growth would be 65% of 0.44 or 0.29 (0.44 is the ratio for a dense thermal oxide (4,8)). In fact, Duffek et al. (4) measured a ratio of 1/3.43 or 0.292 for an oxide grown in a similar solution. For these reasons, a ratio of 0.29 was

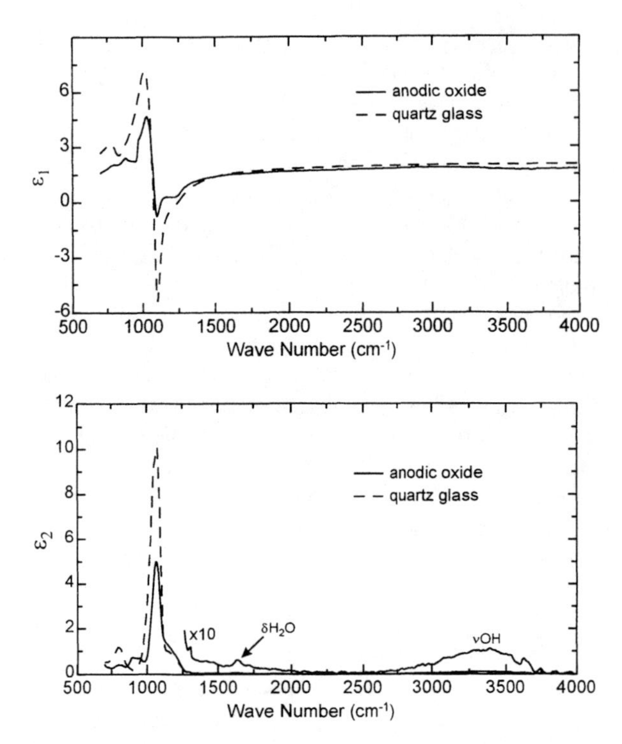

Figure 2. Real (ε_1) and imaginary (ε_2) parts of the dielectric function for anodic oxide from boron doped substrate, with quartz glass from Philipp (21) included for comparison.

used to calculate the amount of silicon removed in the carrier profiling described in the next section.

These dielectric constants fit the ellipsometric data for film thicknesses ranging from 7 to 220 nm.

Carrier Concentration Profiling

We modeled the IR-SE data taken after each anodic oxidation step as three layers consisting of a common-pseudosubstrate (20), a doped silicon layer and an anodic oxide (see Figure 3). The last anodic oxidation cycle associated with the deepest silicon layer is analyzed first. As the analysis proceeds to shallower layers, the optical properties of the deeper silicon layers are integrated into the common-pseudosubstrate. In Figure 3 the integer j indicates the jth silicon layer removed by the jth anodic oxidation, thus the thickest oxide would have the largest j. The thickness of the jth silicon layer is

$$t_{Si}(j) = r \times \left[t_{ox}(j) - t_{ox}(j-1) \right]. \qquad (6)$$

(j - 1) anodic oxide	tox(j -1)
jth n-type Si Drude layer	t_{Si}(j) = r x [tox(j) - tox(j - 1)]
common-psuedosubstrate	Optically thick

Figure 3. Optical model used for each anodic oxidation step.

The factor r is the ratio of silicon consumed to oxide grown. Here, r = 0.29, as discussed previously.

To analyze the jth silicon layer, the parameters N_j and τ_j from Eqn. (5) as well as oxide thickness t_{ox}(j-1) are determined using numerical regression analysis. The parameter ε_∞ is fixed at the value (usually between 11.6 to 12) found for the silicon bulk during the first analysis. The parameter t_{Si}(j) is calculated from Eqn. (6) using t_{ox}(j), which is known from analysis of the j+1 silicon layer and an initial estimate of t_{ox}(j-1). A first regression provides a better estimate for t_{ox}(j-1) and t_{Si}(j), which are applied to additional regressions until t_{ox}(j-1) converges, which it invariably does after just a few iterations. The values of t_{ox}(j-1), t_{Si}(j), N_j and τ_j are then recorded.

Before proceeding to the next data set, the silicon Drude layer is deleted from the model and – with the oxide thickness t_{ox}(j-1) fixed at the value just determined – the common-pseudosubstrate dielectric function is calculated using wavelength-by-wavelength regression.

With the values of t_{ox}(j-1), t_{Si}(j) and N_j determined, the next set of data (from the previous anodic oxidation) can be analyzed using the new common-pseudosubstrate.

The carrier concentration profile results for a 10^{15} cm^{-2}, 80 keV arsenic ion implant are presented in Figure 4 along with Spreading Resistance Probe (SRP) data (measured by Solid State Measurements, Inc. of Pittsburgh PA.) for comparison. The 20 oxide growth steps ranged from 11 to 60 nm, yielding silicon sections 3.3 to 18 nm thick.

The two profiles are similar. Some variability in the anodic oxide data could be the result of an inability to precisely measure at the same spot after each anodic oxidation cycle. Also, the IR-SE is currently insensitive to carrier concentrations less than 8×10^{18} cm^{-3}. This could be improved by increasing the wavelength range beyond 14

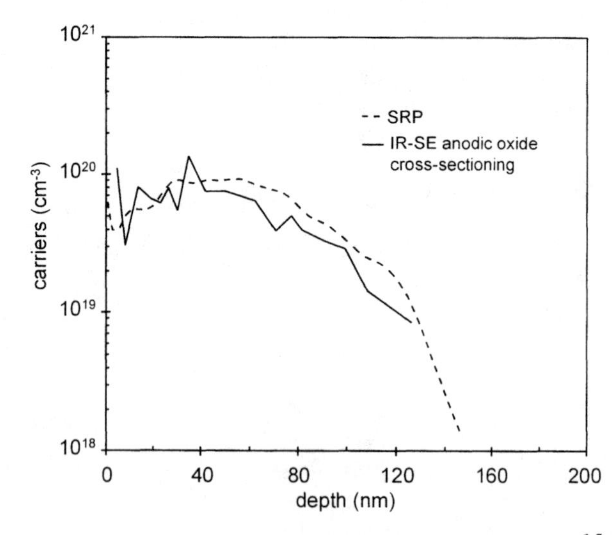

Figure 4. Results of the IR-SE anodic oxide method on a 10^{15} cm^{-2}, 80 keV As implant sample with a rapid thermal anneal. The SRP data provided for comparison.

μm (700 cm^{-1}), since there is a λ^2 in Eqn. (5a). Although the minimum section thickness here was about 3 nm, steps sizes of 1 to 2 nm should be possible.

CONCLUSIONS

We have developed a method to determine carrier (activated dopant) concentration profiles using anodic oxidation sectioning combined with infrared spectroscopic ellipsometry. Sample geometry, sample preparation and anodization parameters are simpler and less restrictive than differential resistivity or Hall effect measurements.

Future work should include automation of the anodization, data acquisition and data analysis processes, as well as measurements of technologically relevant samples. This would aid in the determination of minimum possible step size, reproducibility, and measurement accuracy.

ACKNOWLEDGMENTS

This research was funded by NSF grant # DMI-9761473, Sumitomo Sitix Silicon Inc., and the Center for Microelectronic and Optical Materials Research.

REFERENCES

1. M.F. Lamorte, *Solid-State Electron.* **1**, , 164-171 (1960).
2. E. Tannenbaum, *Solid-State Electron.* **2**,.123-132 (1961).
3. P.A. Iles and B. Leibenhaut, *Solid-State Electron.* **5**, 331-339 (1962).
4. E.F. Duffeck, E.A. Benjamini, and C. Mylroie, *Electrochem. Technol.* **3**, (3-4)., 75-80 (1965).
5. K.M. Busen and R. Linzey, *Trans. Metall. Soc. AIME* **236**, 306-309 (1966).
6. A. Manara, A. Ostidich, G. Pedroli, and G. Restelli, *Thin Solid Films* **8**, 359-375 (1971).
7. H. Jaskólsia L. Walis, and H. Golkowska, *Thin Solid Films* **33**, 281-286 (1976).
8. H.D. Barber, H.B. Lo, and J.E. Jones, *J. Electrochem. Soc.* **123** (9), 1404-1408 (1976).
9. M. Finetti, P. Negrini, S. Solmi, and D. Nobili, , *J. Electrochem. Soc.: Solid-State Sci. Tech.* **128** (6), 1313-1316 (1981).
10. F. Cembali, R. Galloni and F. Zignani, *J. Phys. E* 7, 698-700 (1974).
11. R. Galloni and A. Sardo, *Rev. Sci. Instrum.* **54** (3), 369-373 (1983).
12. J. Humlícek, R. Henn, and M. Cardona, *Appl. Phys. Lett.* **69**, 2581-2583 (1996).
13. T.E. Tiwald, D.W. Thompson, J.A. Woollam, *J. Vac. Sci. Tech. B.* **16** (1), 312-315 (1998).
14. T. E. Tiwald, D.W. Thompson, J.A. Woollam, W. Paulson, and R. Hance, *Thin Solid Films*, accepted for publication (1998).
15. R.M.A. Azzam and N.M. Bashara, *Ellipsometry and Polarized Light*, (North-Holland, New York, 1977), Chap. 4, p. 274.
16 C.M. Herzinger, P.G. Snyder, B. Johs, and J.A. Woollam, J. Appl. Phys. **77** (4), 1715-1724 (1995).
17. D.E. Aspnes, *J. Opt. Soc. Am. A* **10** (5), 974-983 (1993).
18. C.R. Pidgeon, "Free Carrier Optical Properties of Semiconductors." In T.S. Moss and M. Balkanski (ed.), *Handbook on Semiconductors, Vol. 2*, North Holland, Amsterdam, 1980, pp. 227-230.
19. Committee F-1 on Electronics, ASTM F 723-88, *1996 Annual Book of ASTM Standards: Electrical Insulation and Electronics Vol. 10.05 Electronics (II)*, (American Society for Testing and Materials, West Conshohocken, PA, 1995), p. 339-353.
20. D.E. Aspnes, J.B. Theeten and F. Hottier, *Phys. Rev. B* **20** (8), 3292-3302 (1979).
21. H.R. Philipp, "Silicon Dioxide (SiO_2) (Glass)," editor, E.D. Palik, *Handbook Of Optical Constants In Solids*, Academic Press, Boston, 1985, p. 749.

Verification of Carrier Density Profiles Derived from Spreading Resistance Measurements by Comparing Measured and Calculated Sheet Resistance Values

R. G. Mazur, S. M. Ramey, C. L. Hartford, E. J. Hartford

Solid State Measurements, Inc., 110 Technology Drive, Pittsburgh, PA 15275 USA

M. Kouno

SSM Japan K.K., 2-3-1 Haramachida, Machida-Shi, Tokyo, Japan

L. S. Tan

National University of Singapore, 10 Kent Ridge Crescent, Singapore 119260

In the past, resistivity and carrier density profiles derived from spreading resistance measurements have been satisfactorily verified by comparing a doped layer's measured sheet resistance, ρ_s, with a sheet resistance value calculated from the on-bevel resistivity profile. However, this verification process has been found to fail in the case of ultrashallow (<100nm) source/drain implants. This paper examines all possible sources of the discrepancy between ρ_s-measured and ρ_s-calculated, and concludes that the lack of a sheet resistance edge correction in the standard multilayer SRP analysis is the dominant factor. The paper also discusses the role played by bevel surface damage, on-bevel carrier diffusion, and variations in carrier mobility in high dopant density regions.

INTRODUCTION

For many years, the resistivity and carrier density profiles produced by the spreading resistance technique on junction isolated structures have been verified by comparing the sheet resistance (ρ_s) calculated from the SRP-derived resistivity profile to the ρ_s value measured on the test sample. While this verification process works well on thicker layers (typically yielding agreement to within about ±10%), a problem has been observed in ultra-shallow (<100nm) source/drain structures. In this case, the calculated ρ_s value is too high – often by a factor of about two – leading to a too-low estimate of the peak carrier density. Two recent publications (1,2) have suggested that surface damage incurred during sample beveling raises the measured resistance (R_m) values by reducing the conductance through the doped layer. This work will show that, in fact, surface damage has a minor effect on measured resistance values, at least when samples are beveled with 0.05µm diamond abrasive and optimal sample preparation techniques are used. We show that the ρ_s – calculated to ρ_s – measured discrepancy is generated by an edge-of-layer boundary effect on the sheet resistance component included in R_m values.

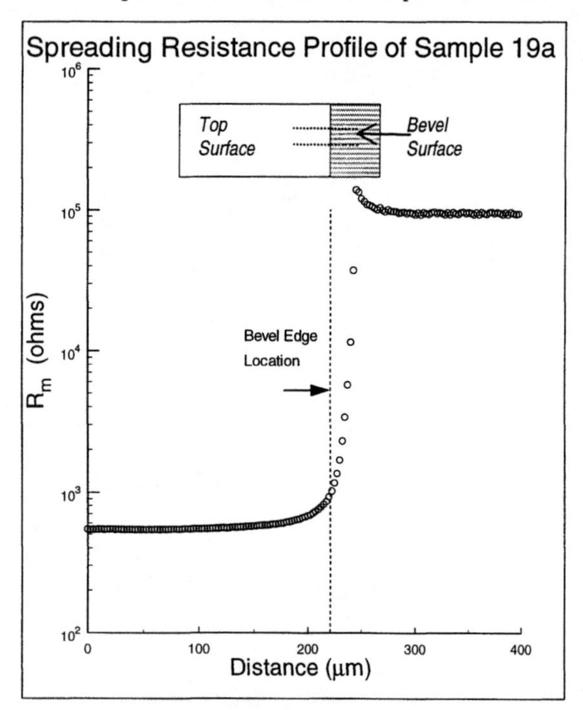

FIGURE 1. Spreading resistance profile starting on the top surface of a typical ultra-shallow source/drain structure. The probe spacing used was 100µm.

The paper includes both experimental data and theoretical considerations that support our contention. We also discuss (a) the effects of surface damage and carrier diffusion on calculated ρ_s values for a group of boron-doped ultrashallow source/drain structures and (b) the effect of mobility variations on the carrier density profiles in these same structures.

EXPERIMENTAL SUPPORT

Figure 1 shows an example of the type of spreading resistance profile that seems to support the idea that surface damage done in beveling affects top surface R_m values near the bevel edge. This profile was run on a boron implant sample with a junction depth of about 73nm (Sample 19a). In the Figure 1 measurements, the probes were set to begin the run on the top surface of the sample some distance away from the bevel edge (see the inset sketch). Note that the measured resistance values begin to increase before

CP449, *Characterization and Metrology for ULSI Technology: 1998 International Conference*
edited by D. G. Seiler, A. C. Diebold, W. M. Bullis, T. J. Shaffner, R. McDonald, and E. J. Walters

reaching the bevel edge and then continue to rise on the bevel surface itself to the position corresponding to the on-bevel source/drain junction. The rise in top surface R_m values has been used (1,2) as support for the idea that damage done during beveling increases the R_m values by introducing a less conductive (damaged) region under the surface being probed. In fact, Reference 2 suggests that the rise in R_m values as the probe approaches the bevel edge results from top surface damage that is not visible in an optical microscope.

A similar run on the same sample with a magnified scale is shown in Figure 2 (Run A). Figure 2 also contains data from another top surface run on the same sample (Run B); these Run B measurements start at a point on the sample's top surface well away from the bevel edge and run toward the cleaved (side) edge of the sample. Note that the Run B resistance values begin to rise as the probes approach the sample edge, similar to the Run A data. Since there is no damage on the cleaved edge of the sample and also no damage on the top surface near the Run B data, the rise in R_m in this case must be attributed to something other than surface damage.

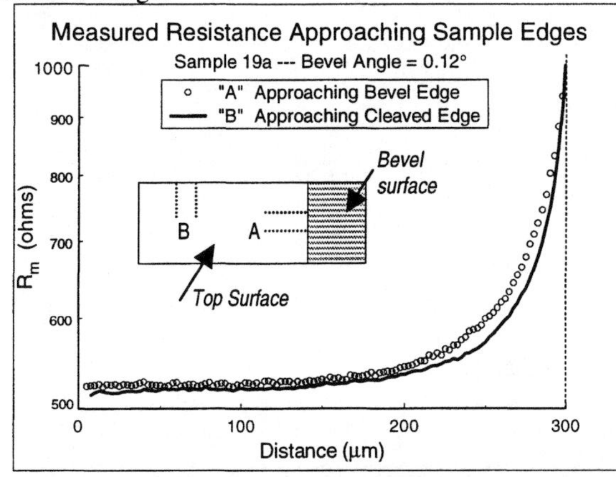

FIGURE 2. Top surface spreading resistance measurements approaching the bevel and cleaved side edge of Sample 19a. The cleaved and bevel edges are located at 300μm as denoted by the dashed line. The probe spacing used was 100μm.

We suggest that the rise in R_m in the Run B data is directly related to a boundary effect on the sheet resistance of the layer. It is well known that a four-point-probe measurement yields twice the voltage drop for a given current (that is, twice the resistance) when placed at the very edge of a conducting layer as opposed to a position many probe spacings away from the sample edges; this is because the current must now flow through a half sheet instead of the whole conducting sheet.

We believe that it is also obvious that the two-probe resistance values (R_m) on the top surface of a doped layer must contain a component that is directly related to the sheet resistance of the layer. In fact, this is the basis for the variable probe spacing (VPS) technique that is commonly used to measure a layer's sheet resistance with a spreading resistance system.(3)

In the case of ultrashallow source/drain structures, the sheet resistance component dominates the measured resistance R_m. This is because the high carrier density at the surface makes the contact barrier resistance part of R_m very small while the thinness of the layer yields a large value of sheet resistance. Therefore, as the probes approach the cleaved edge, R_m approximately doubles in value.

Furthermore, we believe that the major portion of the increase in R_m as spreading resistance probes approach a bevel edge is also due to the "sheet resistance edge effect". This is because of the similarity between the rise in R_m as the probes approach the same sample's cleaved side edge together with the fact that the position of the onset of the increase in R_m is related to the probe spacing, S.

Further empirical evidence for the ρ_s edge effect comes from a series of top surface measurements on samples with various bevel angles. For example, Figure 3 shows data obtained from a boron implant with a junction depth of about 105nm (Sample 7). Note that the magnitude of the R_m increase approaching the bevel edge increases with increasing bevel angle. Also note that the measured resistance at the bevel edge increases from 279Ω for a 0.03° bevel to 464Ω for a 1.17° bevel; this latter value is approximately twice the value of R_m in the center of the top surface well away from any edges. This near doubling is consistent with the notion of the bevel behaving more and more like a cleaved edge as the bevel angle increases.

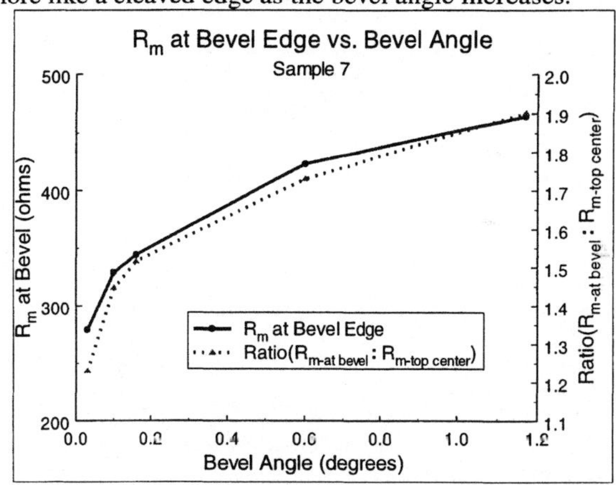

FIGURE 3. Effect of bevel angle on the measured resistance at the bevel edge.

Figure 3 also provides additional direct evidence against the surface damage model. Beveling causes a small amount of damage to the top surface of the sample as a result of diamond slurry brushing against the top surface of the sample near the bevel edge. This becomes more pronounced at shallower angles, since the top surface of the sample is then closer to the grinding plate and thus the slurry more easily brushes the surface. An implicit assumption in the surface damage model is that increased levels of damage will increase the measured resistance. However, the data in Figure 3 shows an opposite trend; the test chips with smaller bevel angles (and thus, presumably, greater top surface damage) have lower R_m values at the bevel edge.

THEORETICAL SUPPORT

Additional support for the concept of a "sheet resistance edge effect" comes from consideration of the multilayer model for "correcting" spreading resistance profiles.(4) A key assumption underlying the multilayer calculation is that the sub-layers comprising the multilayer model are infinite in lateral extent (all side boundaries are far away compared to the probe spacing), as shown in Figure 4a. In practice, spreading resistance measurements are made on a beveled sample as shown in Figure 4b. Note, however, because all multilayer analyses are based on Schumann and Gardner (4), all spreading resistance data is, in fact, processed as if the sub-layers extended to infinity which includes the missing portions of the sub-layers to the right of the bevel line in Figure 4b.

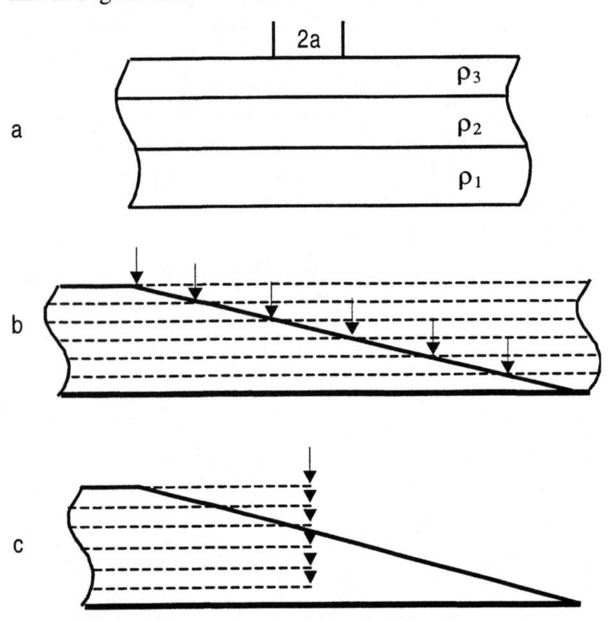

FIGURE 4. Detail of the multilayer analysis method used to convert spreading resistance to resistivity.

This situation has not caused a significant problem in the past. However, for high surface concentration ultrashallow junction-isolated structures (such as source/drain implants), the lateral distance to the on-bevel (insulating) junction is small relative to the probe spacing. Therefore, the real situation is much more like that shown in Figure 4c; that is, the probes are essentially on the edge of the stack of sub-layers.

In a Schumann and Gardner multilayer analysis, the spreading resistance values are calculated with an integral of the form,

$$R = \frac{\Delta V}{I} = \frac{\rho}{2a}\left[8\int_0^\infty dx\left\{\frac{J_1(x)}{x} - \frac{J_0(xS)}{2}\right\}I(x)F(xt)\right] \quad (1)$$

where ΔV is the voltage between two probe contacts to a semiconductor layer, I is the current flowing between the probes, I(x) is a function that depends on the current distribution through the contacts, F(xt) is determined by fitting boundaries between sub-layers, and S is the spacing between probes.

Schumann and Gardner chose to refer to the value of the integral as a "correction factor." That is

$$R = \frac{\rho}{2a} \times C \quad (2)$$

where ρ/2a is the spreading resistance for a bulk sample of resistivity ρ and C is the complete expression in the brackets in Equation 1. This "correction factor" view made some sense in 1969, as the layers being profiled were then relatively thick; but the tendency to continue calling the result of a multilayer calculation a "correction factor" has caused considerable confusion.

This is particularly true for ultrashallow layers where this so-called "correction factor" is often in the range of 100-1000. The reason why this "correction factor" is so large is that the integral value in Equation 1 actually contains a component related to the layer's sheet resistance. For ultrashallow layers, the sheet resistance becomes very large. In particular, in the near-surface region of a source/drain profile, the sheet resistance component dominates the value of R_m, the resistance measured with a pair of spreading resistance probes.

For ultrashallow source/drain structures, this means that the "missing half" of the sub-layers (shown in Figures 4b and 4c) must be accounted for in a spreading resistance analysis. This amounts to factoring in a resistance related to the sheet resistance of the "missing half" of the sub-layer stack. Ultimately, this will require a multi-dimensional solution of the Laplace and/or Poisson equations. In order to provide a near-term improvement to the accuracy of spreading resistance profiles, we propose the following:

Our standard spreading resistance analysis uses a working equation of the form,

$$R_n = \frac{\rho_n}{2a}C_n + R_B \quad (3)$$

where R_n is the calculated resistance for the nth sub-layer and is calculated from Equation 1. R_B is obtained from a calibration curve relating bulk R_m values to ρ and $C_n(\rho_n, \rho_{n-1}, \ldots)$ is calculated using the variational analysis published by Choo, *et al.*(5)

In practice, as the multilayer calculations proceed, starting with the first sub-layer above the substrate having a resistivity value of ρ_1, the resistivity of each sub-layer is derived via an iterative procedure, using Equation 3. In this procedure, an estimate of ρ_n is made; then the corresponding value of $C_n(\rho_n, \rho_{n-1}, \ldots)$ is calculated and a value of R_B is determined from the appropriate calibration curve; Equation 3 is then solved for R_n.

R_n is then compared to the corresponding measured value R_m and the estimated value of ρ_n is adjusted in successive iterations until the calculated value of R_n equals the measured spreading resistance value, R_m.

What we now propose is to add to Equation 3 the resistance associated with the "missing half" of the stack of sub-layers. This is the sheet resistance of the sub-layer stack consisting of the sub-layer currently being analyzed and all the sub-layers below it; that is,

$$R_s = \frac{1}{\frac{\pi}{\ln(S/a)} \sum_{i=1}^{n} \frac{t}{\rho_i}} \tag{4}$$

changing the working equation to:

$$R_n = \frac{\rho_n}{2a} C_n + R_B + R_s \tag{5}$$

The results of applying the new algorithm to typical source/drain structures are given in Figure 5 and Table 1. The dashed line carrier density profile in Figure 5 uses the standard SSM 150 algorithm with a fixed radius *a* which is determined by optical microscopy, a barrier resistance determined from the applicable calibration curve, and calculations based on Equation 3. The solid line profile uses the new algorithm based on Equation 5. Note that the carrier density peak using the new algorithm is close to a factor of two greater than that obtained with the old algorithm.

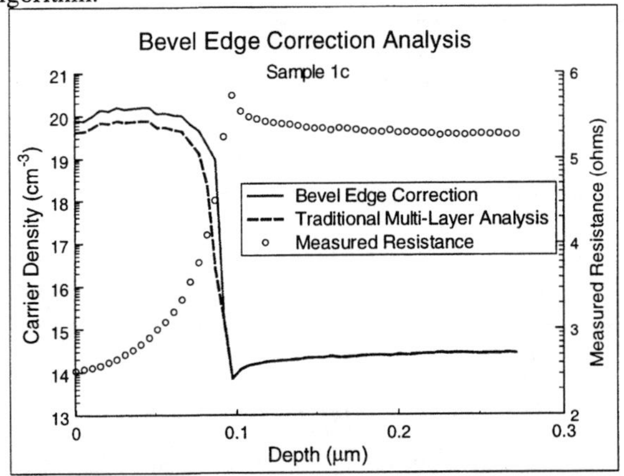

FIGURE 5. Results obtained from a multilayer spreading resistance analysis of sample 1c with ρ_s edge correction (Equation 5) and without ρ_s edge correction (Equation 3).

Table 1 shows that for sample 1c the value of ρ_s calculated with the edge correction agrees with the measured ρ_s better than the Schumann and Gardner value of ρ_s. The table also shows that, as the bevel angle increases, the approximation of Figure 4c becomes more accurate and the ρ_s values calculated with the edge correction agree more closely with the measured ρ_s values.

TABLE 1. Comparison of calculated sheet resistances using Equation 3 (conventional Schumann & Gardner multilayer algorithm) and Equation 5 (multilayer algorithm with Edge Correction, E.C.)

Sample	Angle (°)	X_j (nm)	VPS ρ_s (Ω)	SRP-S&G ρ_s (Ω)	SRP –E.C. ρ_s (Ω)
1a	0.10	96.	136.	202.	104.
1b	0.15	97.	136.	214.	111.
1c	0.29	95.	136.	257.	126.
19a	0.12	68.	394.	679.	336.
19b	0.31	73.	394.	829.	387.
7a	0.10	106.	163.	231.	118.
7b	0.33	121.	163.	273.	140.
7c	1.17	123.	163.	354.	179.

DAMAGE EFFECTS

This paper has established that the direct effect of subsurface damage from beveling plays a relatively minor role in the increase in R_m values as probes approach a bevel. However, we note that these results may be particular to the type of source/drain structures measured in our work and to the particular sample preparation techniques employed in our laboratory.

In normal practice we prepare samples for ultrashallow layer measurements with 0.05µm diamond paste. As a test of the possible effects of preparing samples in a different way, we profiled several of the samples tested here with 0.1µm diamond abrasive. The results show that, for a 100nm source/drain structure, the 0.1µm diamond profile is very similar to the 0.05µm diamond profile and the calculated ρ_s values closely agree. On the other hand, a 50nm sample prepared with 0.1µm diamond gives a profile that differs significantly from the 0.05µm diamond profile and a ρ_s (calculated) that differs from the 0.05µm diamond value by approximately a factor of four. Therefore, caution must be used in quantifying spreading resistance profiling results on ultrashallow layers unless bevel surface quality has been thoroughly assessed.

CARRIER DISTRIBUTION EFFECTS

A possible concern in comparing measured and calculated sheet resistances arises from the fact that the measured ρ_s depends on the vertical resistivity profile while the calculated ρ_s comes from the on-bevel resistivity profile. In some cases the vertical and on-bevel resistivity profiles will differ because of carrier diffusion arising from material removed in beveling.(6) To examine possible differences between the on-bevel and vertical profiles in this current work, we solved the Poisson equation for a Gaussian implant profile typical of ultrashallow source/drain structures. Figure 6 gives the results of these calculations in the form of vertical and on-bevel carrier density profiles, along with the net dopant profile. These results show that the vertical profile matches the on-bevel profile until carrier densities drop below the 10^{18} cm^{-3} range. Converting the carrier density profiles to resistivity and calculating ρ_s values results in an on-bevel ρ_s of 293Ω and a vertical ρ_s of 301Ω. Therefore, carrier redistribution from beveling can be neglected as a source of a significant discrepancy between measured and calculated sheet resistances for the ultra-shallow source/drain structures covered in this work.

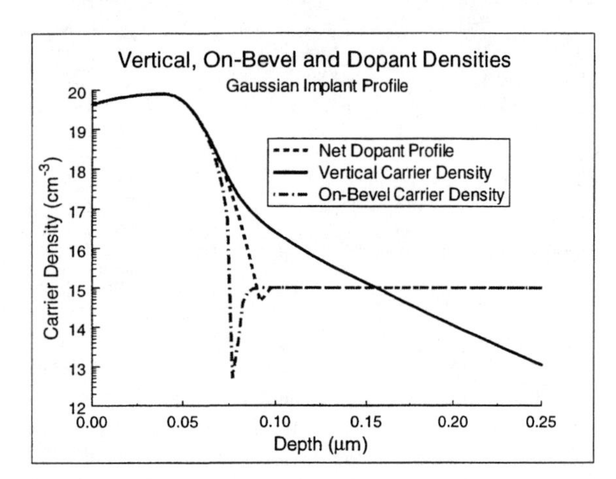

Figure 6: Density profiles comparing the on-bevel to the vertical profile, as well as the dopant profile.

MOBILITY EFFECTS

Spreading resistance measurements provide resistivity profiles that are normally converted to carrier density profiles using an ASTM standard (ASTM F723).(7) This method is an empirical inter-conversion between dopant density and resistivity for bulk silicon. A publication by Masetti (8) provides an alternative method to convert resistivity to carrier density using an empirical relation between carrier density and mobility based on Hall effect measurements on highly doped implanted samples. Since the two methods begin to differ at carrier densities above 10^{19} cm^{-3}, several source/drain profiles were analyzed using the two techniques. Table 2 indicates that the greatest discrepancy occurs for samples with peak carrier densities near $10^{20} cm^{-3}$, which results, for example, in about a 5% difference in the dose calculated for sample 7a. For samples with lower peak carrier densities, the difference becomes much smaller, such as the 2% difference in calculated dose for sample 7c. Therefore, though uncertainty in the mobility may contribute some inaccuracy in carrier density calculations, the magnitude of the effect is less than 5% for these boron-doped ultra-shallow source/drain structures; this is clearly not as important as the bevel edge effect on sheet resistance.

TABLE 2. Influence of mobility on the carrier density profiles of Source/Drain structures.

Sample	Dose ($x10^{14}$ cm^{-2}) [Peak ($x10^{19}$ cm^{-3})]	
	ASTM F-723 Dose [Peak]	Masetti Dose [Peak]
7a	4.85 [9.02]	5.09 [9.72]
7d	4.67 [8.10]	4.88 [8.66]
7b	4.00 [6.09]	4.13 [6.40]
7e	3.59 [4.79]	3.67 [4.97]
7c	2.98 [5.19]	3.03 [5.40]

CONCLUSION

The discrepancy between the measured sheet resistance of source/drain implants and the sheet resistance calculated from a spreading resistance-derived on-bevel resistivity profile is due to the lack of a sheet resistance edge correction in the conventional SRP multilayer analysis. This is shown to especially affect results in the case of source/drain structures with junction depths less than 100nm. The effects of bevel surface damage and carrier diffusion on calculated sheet resistance values are shown to be relatively minor in these structures. Carrier mobility variation is also shown to have a minor effect on the accuracy of these profiles – up to a level of about 1×10^{20} cm^{-3} for the boron-doped implants discussed in this paper.

In light of the data in Table 1 and the associated discussion, a conclusion can be drawn relative to practical use of the sheet resistance edge correction algorithm in SRP measurements that source/drain implants should be profiled with the largest possible bevel angle, consistent with the depth resolution required.

ACKNOWLEDGEMENTS

Some of the samples used in this paper were supplied by Dr. Susan Felch of the Varian Research Center. The authors would also like to acknowledge both past and present help in SRP analysis by Dr. S. C. Choo of the Nanyang Technological University of Singapore and Dr. M. S. Leong of the National University of Singapore.

REFERENCES

1. Osburn, C.M., Berkowitz, H.L., Heddleson, J.M., Hillard, R.J., Mazur, R.G., and Rai-Choudhury, P., J. *Vac. Sci. Technol. B*, **10** (B), p. 533, 1992.
2. Pawlik, M., "Profiling of Ultra-Shallow Layers Using Spreading Resistance," 4th International Workshop on the Measurement, Characterization, and Modelling of Ultra-Shallow Doping Profiles, Research Triangle Park, North Carolina, April 1997.
3. Dickey, D.H., and Ehrstein, J.R., NBS Special Publication 400-48, pp. 15-18, 1979.
4. Schumann, P.A., and Gardner, E.E., *J. Electrochem. Soc.* **116** (1), p., 87, 1969.
5. Choo, S. C., Leong, M. S., and Sim, J. H., *Solid-State Electronics*, **26**, pp. 723-730, 1983.
6. Mazur, R., *J. Vac. Sci. Tech. B* **10**, No. 1, pp. 397-407, Jan/Feb 1992.
7. ASTM F233-88, ASTM Annual Standards, 10.05, 1996.
8. Masetti, G., Severi, M., and Sandro, S., IEEE Trans. E.D., ED-30, No.7, pp. 764-769, July 1983.

Leakage Compensated Charge Method for Determining Static C-V Characteristics of Ultra-thin MOS Capacitors

Hui Song, Edwin Dons, Xi Qing Sun and K. R. Farmer

Microelectronics Research Center, New Jersey Institute of Technology, University Heights, Newark, NJ 07102

We describe a charge measurement method to determine static capacitance versus voltage characteristics of ultra-thin silicon oxide dielectric (<~3.5 nm), metal-oxide-silicon device structures in which significant leakage current is present. The leakage is accounted for using current compensation circuitry in combination with elementary charge integration electronics. For the thinnest oxides where direct tunneling is significant (<~2.5 nm), the circuit is combined with a straightforward numerical compensation technique. This measurement method extends the range of application of low frequency capacitance characterization of silicon oxides well into the direct tunnel thickness regime, and additionally is expected to be useful for studying ultra-thin *alternate* dielectric materials.

INTRODUCTION

For more than two decades quasistatic capacitance-voltage (C-V) measurement of metal-oxide-silicon (MOS) capacitors has been used extensively to study the electrical properties of MOS devices and to monitor integrated circuit fabrication. However, as CMOS technologies beyond the 0.1 μm regime have evolved with gate dielectric thickness less than ~3.5 nm, the usefulness of the quasistatic C-V method has been brought into question. In this thickness regime a leakage current is unavoidable due to the presence of direct tunneling. As shown in Figure 1, the direct tunnel current through a 6.25×10^{-4} cm^2, 3.5 nm thick oxide can reach several pA at bias voltages where capacitance would be measured for the purposes of characterization, while in a 2.4 nm oxide of the same area, the leakage is nearly six orders of magnitude higher.

Figure 2(a) demonstrates the effect of this tunnel leakage on a conventional quasistatic C-V curve. It can be observed that even for the thicker 3.5 nm oxide a leakage current as low as 1 pA per 100 pF capacitance, using an HP 4140B system, is enough to cause severe distortion in the C-V. The distortion is obvious even at the relatively high sweep rate of 50 mV/sec where the capacitor's displacement current is enhanced relative to the leakage.

In the present work, we overcome this difficulty using a leakage current compensated charge measurement

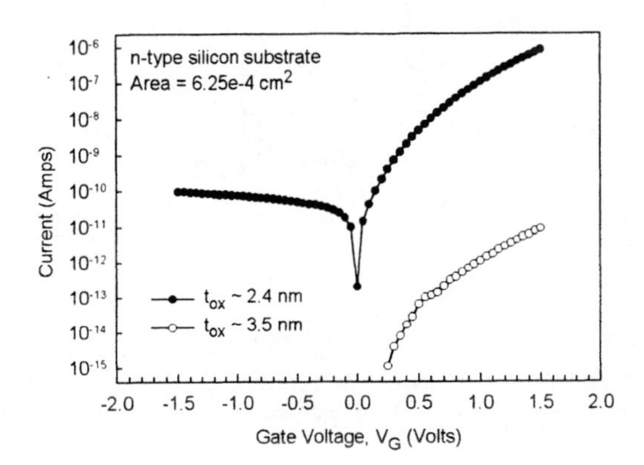

Figure 1. Current-voltage curves for 2.4 nm and 3.5 nm oxides on n-type silicon substrates. The gate electrodes are n-type polycrystalline silicon doped approximately to degeneracy.

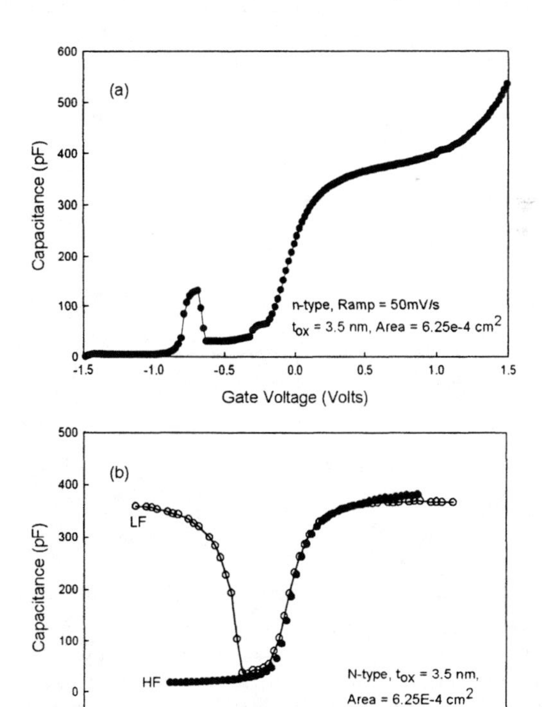

Figure 2. C-V curves of a 3.5 nm thick gate dielectric MOS capacitor: (a) Quasistatic method, (b) Leakage compensated charge method.

CP449, *Characterization and Metrology for ULSI Technology: 1998 International Conference*
edited by D. G. Seiler, A. C. Diebold, W. M. Bullis, T. J. Shaffner, R. McDonald, and E. J. Walters

method which is capable of producing true static C-V curves with high precision for samples even when the leakage current is greater than 10^{-7} A, such as in the 2.4 nm oxide of Figure 1. Figure 2(b) shows both static and high frequency C-V curves measured by the leakage compensated charge method using the same 3.5 nm thick oxide that was used for the quasistatic measurement in Figure 2(a). The leakage compensated static curve, in contrast to the quasistatic one, is not distorted, and thus can be used for device and defect characterization. For the thinnest oxides where direct tunneling is significant (<~2.5 nm), leakage compensation circuitry is combined with a straightforward numerical data reduction technique to extend this measurement method well into the direct tunnel thickness regime.

MEASUREMENT METHOD

The principle of the leakage compensated charge measurement method is illustrated in Figure 3. A small amplitude square wave signal $V_g(t)$ is applied across a MOS capacitor at DC bias level V_G. The total device current contains both a displacement current component responding to the small signal and a DC leakage current component. The current is integrated, with the DC component being eliminated using a feedback leakage compensation scheme. The output of the circuit, V_O is thus a transient charge waveform, Q(t) generated by the applied small signal. We note that this approach eliminates the leakage current using a single circuit, despite the fact that the leakage is a strong function of the applied voltage.

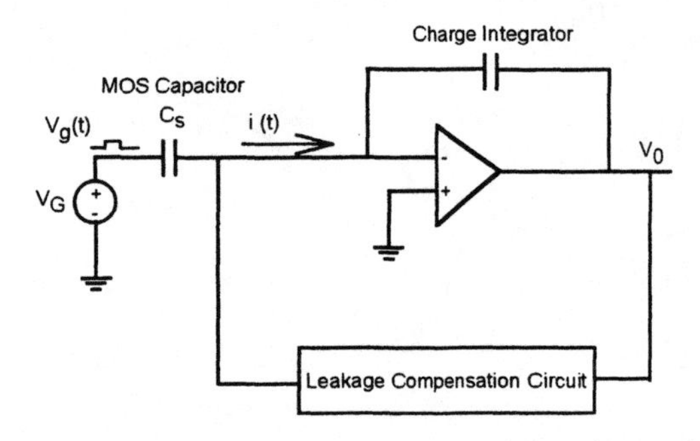

Figure 3. Principle of leakage compensated charge method.

RESULTS AND DISCUSSION

As illustrated in Figures 4 and 5, the shape and amplitude of the output waveform vary with V_G, for example displaying a slow rise time in inversion where the minority carrier response time is long, and conversely a fast rise time in accumulation where majority carriers can easily follow the applied square wave. The amplitude of the response waveform is proportional to the device capacitance, that is $C(V) = Q_{max}(V)/V_g$, where V_g is the amplitude of the excitation square wave, and $Q_{max}(V)$ is the maximum value of the transient charge response curve.

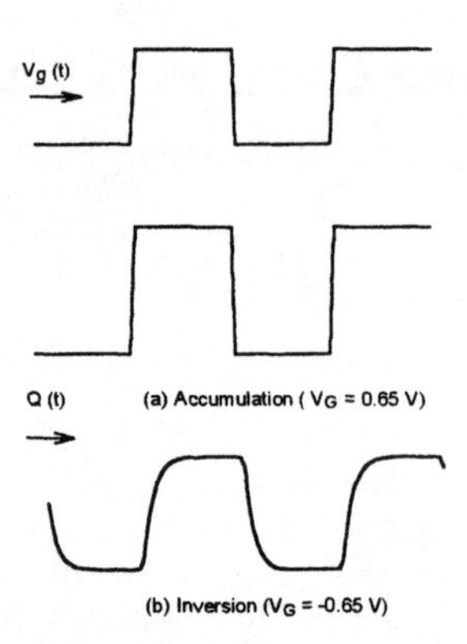

Figure 4. 2 Hz square excitation pulse $V_g(t)$, and transient charge waveforms Q(t), measured by the leakage compensated charge method using a 3.5 nm oxide. (a) Accumulation bias, and (b) Inversion bias.

Thus in Figure 5, the dashed curve traces out the shape of the static C-V curve that would be extracted from $Q_{max}(V)$ measured at each bias voltage. In accumulation and inversion, the measured capacitance tends toward a saturation value, nominally the oxide capacitance, but also including the series addition of capacitance due to, for example, quantum effects or the finite width of the accumulation or inversion layers. When the silicon is swept through depletion, the measured capacitance goes through the usual minimum. Longer time traces confirm that the inversion response curve amplitudes in Figure 5 are at the onset of saturation.

While it is beyond the scope of this introductory presentation to explore the details of the response waveforms, we note that minority carrier dynamics and associated quantities such as minority carrier lifetime and depletion layer width can be investigated using this approach since the response waveform in inversion is

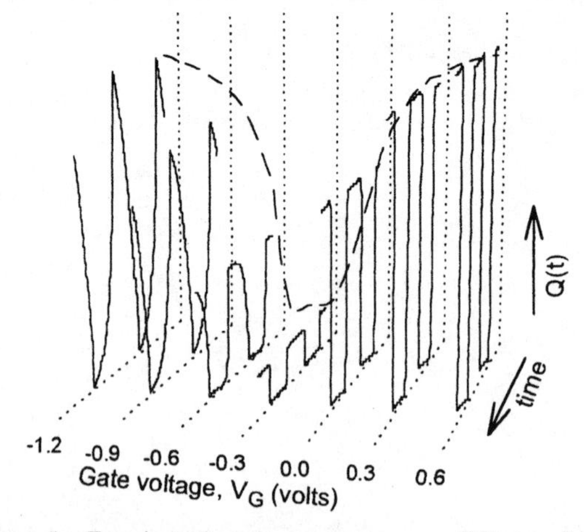

Figure 5. Transient charge response curves, Q(t) in arbitrary units, measured at different gate bias levels V_G using the 3.5 nm oxide device of Figure 2(b). The applied square wave excitation signal has amplitude V_g = 50 mV and frequency f = 2 Hz. The dashed curve, $Q_{max}(V)$ is proportional to the device capacitance, C(V).

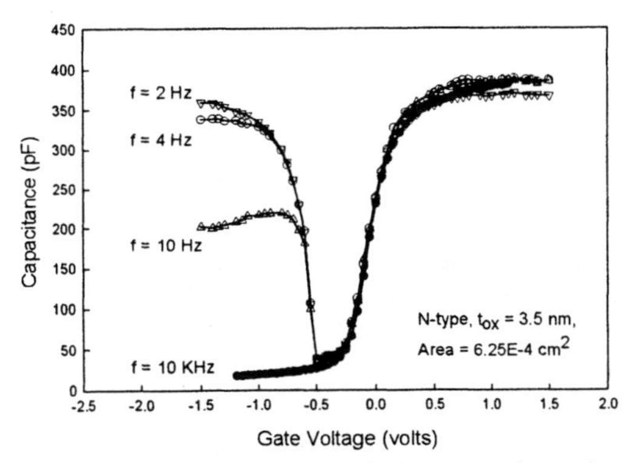

Figure 6. C-V curves extracted from transient charge response curves at different excitation signal square wave frequencies using the 3.5 nm oxide device of Figure 2(b). The amplitude of the excitation signal is 50 mV.

acquired as part of the measurement. The saturation of minority carrier response can easily be confirmed by the flatness at the end of each section of the charge waveform and thus assures that the measured curve is a static C-V curve, instead of "quasistatic" as in the conventional approach. The well-known bias sweep direction and rate effects of the quasistatic approach are avoided in this method, because the Fermi level E_F is always allowed to equilibrate at each measurement point.

The expected transition from a low frequency C-V characteristic to high frequency behavior can be demonstrated by varying the frequency of the excitation square wave signal. If the frequency is too high, then the minority carrier response in inversion does not saturate, and thus the capacitance extracted from the the values of Q_{max} in this bias region will not be the saturation value. This effect is shown in Figure 6, where the minority carriers in the inversion layer are unable to reach equilibrium at

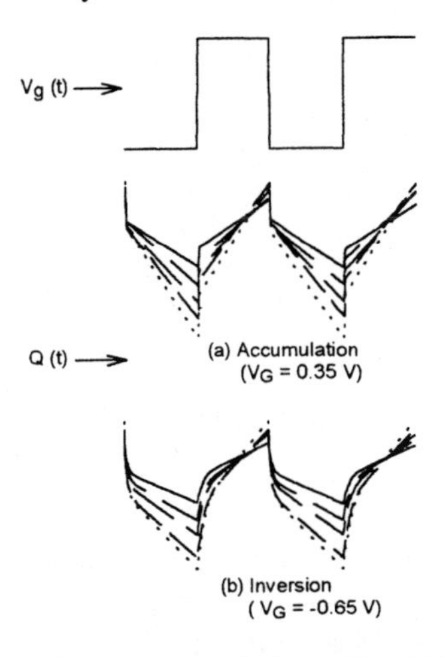

Figure 7. 2 Hz square excitation pulse $V_g(t)$, and transient charge waveforms Q(t), measured by the leakage compensated charge method using a 2.4 nm oxide. (a) Accumulation bias, and (b) Inversion bias. In both cases, response curves are shown for five different excitation signal amplitudes ranging from 30 mV (solid curves) to 70 mV (dotted curves) in 10 mV increments.

frequencies less than ~2 Hz, and do not respond at all above ~10 kHz.

We find that for the thinnest oxides, the circuitry alone is not sufficient to compensate all of the device leakage. In this situation, the transient response curves do not saturate as in Figure 4, but rather they display the behavior presented in Figure 7, where the charge transient amplitude for a 2.4 nm oxide increases indefinitely, varying linearly with time once "saturation" is achieved. The slope of this variation, dQ/dt is found to increase linearly with the excitation pulse amplitude, as shown in Figure 8. We term dQ/dt an "excess" leakage current since it is not accounted for by our circuitry.

Because of the straightforward relationship with the excitation pulse amplitude, the excess leakage is easy to

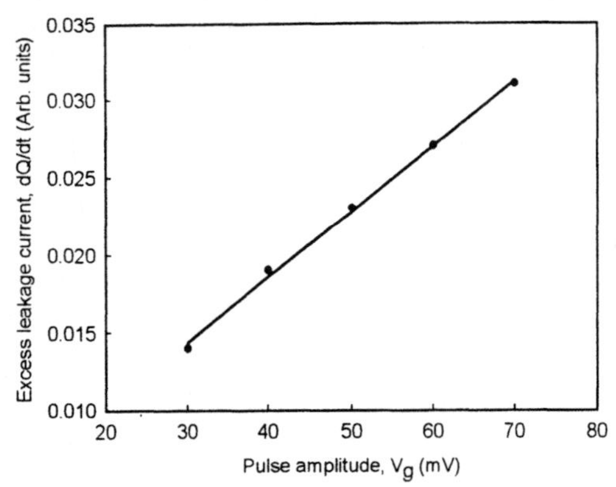

Figure 8. Transient response curve "saturation" slope versus excitation pulse amplitude.

account for numerically. Independent of our choice of excitation pulse height, subtracting the linearly increasing component of the response curve leads to recovery of the "normal" behavior of Figure 4. It is important to note that the total tunnel leakage through this oxide can be as high as 10^{-7} A, while a quasistatic capacitance measurement would have to detect a displacement current on the order of ~1 pA. Thus our charge compensation circuitry has eliminated most of the leakage current, and left only a small "excess" leakage to be dealt with numerically.

In Figure 9 we present a high frequency C-V for the 2.4 nm oxide as well as the static C-V's obtained both without and with numerically compensating the "excess" leakage. The fully compensated static curve (solid triangles) shows that even at the measurement frequency of 2 Hz, which was sufficient to achieve minority carrier equilibrium in 3.5 nm oxides, the minority carriers in the thinner oxide device cannot yet achieve equilibrium. This might be expected because of the higher fields present in the thinner system. More importantly, the fully compensated data show that it is possible to measure static capacitance even in oxides as thin as 2.4 nm. Clearly from the high and low frequency curves presented for this device it will be possible to use the leakage compensated charge method to extract defect and interface state information for devices having such a thin oxide.

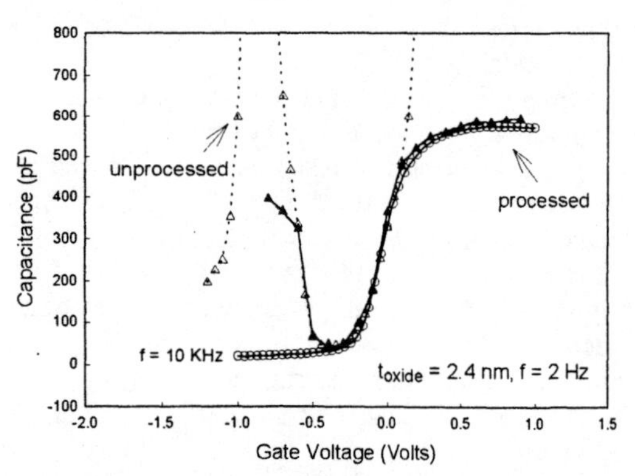

Figure 9. High frequency and static C-V's obtained by the leakage compensated charge method using a 2.4 nm oxide. The static curves are determined both without and with numerical compensation of the "excess" leakage current, represented by the open and closed triangles, respectively.

SUMMARY

A charge measurement method has been described to determine static C-V characteristics of ultra-thin silicon oxide dielectric (<~3.5 nm), MOS device structures in which significant leakage current is present. The leakage is accounted for using a current compensation circuitry. In the thinnest oxides, additional numerical compensation is required. This measurement method extends the range of application of low frequency capacitance characterization of silicon oxides well into the direct tunnel thickness regime, and additionally is expected to be useful for studying ultra-thin *alternate* dielectric materials.

ACKNOWLEDGEMENT

This work acknowledges partial support from the National Science Foundation through awards No. ECS-9530984 and ECS-9624798.

Characterization of Ultra-Thin Oxides Using Electrical C-V and I-V Measurements

J.R. Hauser and K. Ahmed

Department of Electrical and Computer Engineering, North Carolina State University, Raleigh, NC 27695

The measurement of electrical parameters from capacitance-voltage (C-V) and current-voltage (I-V) curves provides a fast means of characterizing oxides in MOS capacitors or transistor structures. For ultra-thin oxides (< 2 nm), conventional, well-established techniques must be reconsidered and modified due to several increasingly important physical effects including polysilicon depletion and surface quantum mechanical effects. In this work these effects have been incorporated into a rapid analysis program for extracting ultra-thin oxide parameters from measured C-V and I-V data. The technique uses a physically based model of structure charge and potential combined with a non-linear least squares fitting technique to extract device parameters.

INTRODUCTION

Gate oxide thickness used in the manufacture of silicon ICs is rapidly decreasing with each new technology generation. The recent National Technology Roadmap for Semiconductors (1997 Edition) projects future IC scaling at essentially a constant field which means that both operating gate voltage and oxide thickness scale linearly with future generations. An oxide thickness of below 2 nm is projected for the 100 nm technology node for the year 2006 and below 1 nm is projected for the 50 nm technology node at about 2012. To meet these goals, research today must be exploring oxide thicknesses (or equivalent oxide thicknesses for high K dielectrics) of 2 nm and below, and test devices have already been reported at about 1.5 nm. Conventional electrical analysis techniques for oxides and MOS devices have included C-V and I-V measurements which can provide very important information about oxide thickness, oxide charges, surface doping density, interface states and surface carrier mobility. However, it becomes increasingly difficult below 2 nm to apply conventional electrical characterization techniques because of the increasing importance of such physical effects as gate oxide tunneling and carrier confinement (quantum mechanical) effects. This paper discusses these effects and the limitations of applying C-V and I-V techniques to ultra-thin oxides for the purpose of modeling the electrical characteristics and extracting thin oxide parameters from electrically measured data. While gate tunneling I-V characteristics become of increasing importance for ultra-thin oxides, this paper considers only transistor channel I-V characteristics.

C-V MODELS AND CHARACTERIZATION

In order to accurately model the capacitance of a MOS capacitor or transistor, one needs an accurate model of semiconductor and gate charge as a function of gate-to-substrate voltage. Classically this is obtained by solving Poisson's equation near the surface with the substrate doping profile and applying appropriate boundary conditions. For ultra-thin oxides, additional factors which must be considered are: (a) the quantum mechanical (QM) confinement effects near the surface and (b) a finite voltage drop in the polysilicon gate contact material. Quantum confinement effects have been known for many years [1-3] but it is only recently that they have become of major importance.

In terms of device operation there are two major QM physical effects: (a) additional band bending occurs because the surface electrons (or holes) are located in localized energy levels above the edge of the conduction band and (b) the charge centroid is located further from the surface than predicted by the classical analysis. Both of these effects are illustrated in Figure 1. An exact analysis of these effects requires a self consistent solution of the coupled QM equations and Poisson's equation, and several researchers have reported such results [4-6]. In this work a simpler approach has been used to obtain a model containing the appropriate first order physics but which can rapidly be calculated for use in a non-linear least squares package to rapidly extract device parameters. The approximate analysis is based upon the following approximations [7]: 1) The classical relationship between potential and charge density is assumed to apply between the bulk of the semiconductor and the surface quantized energy state. 2) The additional band bending $\Delta\varepsilon$ is added to the classical results to obtain the total surface potential and 3) additional bulk charge is included corresponding to the distance Δz by which the charge centroid exceeds that of the classical results. A simple model predicts:

$$\Delta\varepsilon = \left(\hbar^2 / 2m^*\right)^{1/3}\left(\frac{9}{8}\pi q E_s\right)^{2/3} \qquad (1)$$

for the lowest quantized energy level, where E_s is the surface electric field and m^* is the electric effective mass [2]. More complete models predict a more complex dependence on depletion layer charge and inversion layer charge. All the models, however, predict the 2/3 power dependence on surface field at large fields where QM affects are most important. By adjusting m^* in the simple equation a good first order model can be achieved. The additional charge displacement Δz has been shown to be relatively constant at about 1.2 nm over most of the range of surface fields of interest in silicon devices [8].

CP449, *Characterization and Metrology for ULSI Technology: 1998 International Conference*
edited by D. G. Seiler, A. C. Diebold, W. M. Bullis, T. J. Shaffner, R. McDonald, and E. J. Walters

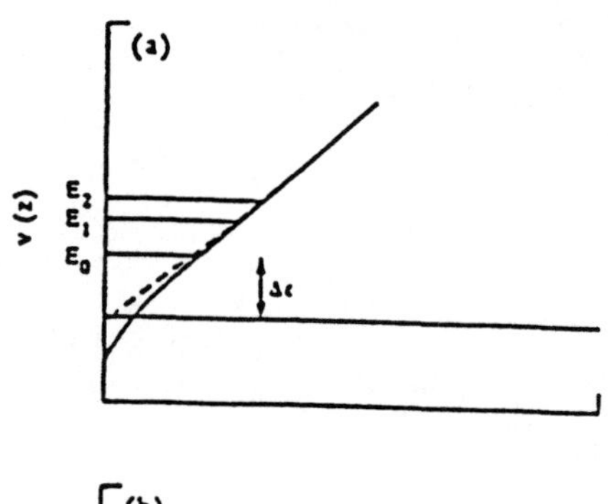

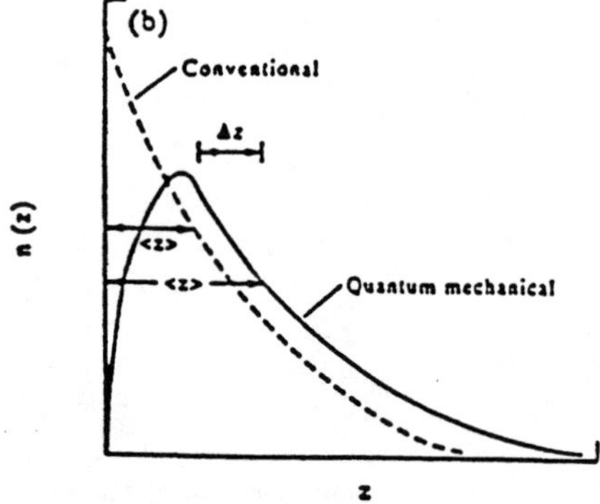

FIGURE 1. Major QM effects on surface energy levels and charge densities. (a) Increase in quantized energy level, Δz is difference in charge centroid [4].

Polysilicon depletion effects are included through solving the classical Poisson's equation in the polysilicon gate. The projected NTRS scaling is essentially at constant fields. This means that as oxide thickness and device voltage scales for future technology generations, the oxide and semiconductor field remains essentially constant. Then for a maximum doping density in the polysilicon, the voltage drop in the polysilicon remains constant and becomes a larger percentage of the device voltage with future scaling.

Figure 2 shows theoretical QM and polysilicon depletion effects for an oxide thickness of 2 nm. In general, the QM effects cause a reduced maximum capacitance resulting in an increase in the "effective" oxide thickness. The polysilicon depletion effect is most pronounced under depletion conditions in the polysilicon. For n^+ poly on p bulk (or p^+

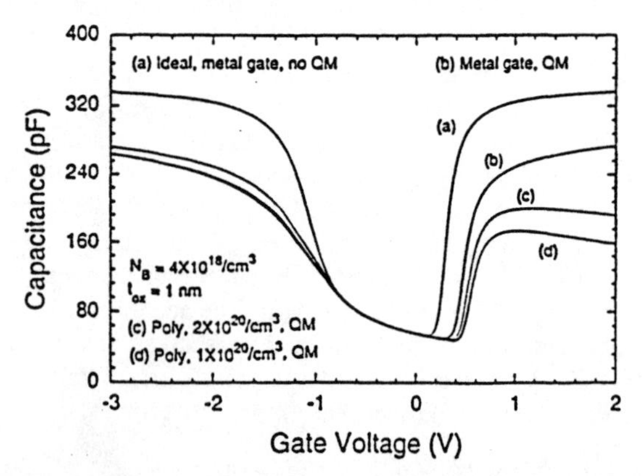

FIGURE 2. Illustration of QM and polysilicon depletion effects on C-V. Theoretical effects at 1 nm oxide thickness.

poly on bulk) this occurs in the accumulation region resulting in an asymmetry in the maximum capacitance under inversion and accumulation.

With the model presented above, the device parameters determining the C-V characteristics are: 1) Flat band voltage (or fixed oxide and interface charges), 2) Substrate doping density, 3) Oxide thickness and 4) Polysilicon doping density. The theoretical C-V model has been incorporated into a non-linear least squares curve fitting package to obtain values of these device parameters. Figure 3 shows examples of the agreement between the model and experimental data for a p-channel device at 3.5 nm and two n-channel devices at about 1.8 nm. In order to obtain C-V data, large area device structures were used with the source, drain and substrate grounded and gate capacitance measured. This gives a complete C-V characteristic which is needed to extract polysilicon doping density. For this example, the parameters extracted from the model fit to the C-V data were (a) oxide thickness, (b) flat band voltage, (c) substrate doping density and (d) polysilicon doping density.

Figure 4 shows the agreement between the model and high frequency C-V data for oxides of thicknesses from about 5.5 nm to about 2.7 nm. In all cases the agreement between model and data is very good. Also, given in the figure for comparison are oxide thickness values as measured by an ellipsometric optical technique (Opti-Probe instrument from Therma-Wave).

A more detailed comparison of oxide thicknesses obtained by several techniques is shown in Figure 5. The techniques are (1) high resolution TEM (HTEM), (2) our C-V model, (3) Quantum interface oscillations in I-V [9], (4) the Optiprobe measurement and (5) Masserjian's C-V techniques [10]. The various techniques are plotted in terms of the Optiprobe results. All the techniques (except for the maximum capacitance) show reasonably good results. The NCSU C-V model and HRTEM results are very close and are slightly larger than the Optiprobe results by about 0.2 - 0.3 nm. The dashed curve is the oxide thickness calculated from the capacitance at -2.5V neglecting QM and polysilicon deplection effects.

LIMITATIONS OF C-V FOR ULTRA-THIN OXIDES

The C-V technique can be applied in a relatively straight-forward manner to oxides with thicknesses down to about 2.0 nm. Below this thickness it becomes more difficult to obtain good C-V data due to increased leakage through the oxide, and a degraded C-V characteristic is observed in both directions. The least squares technique described here does not require a complete C-V curve, so data can still be extracted over a limited voltage range. However, the technique can be extended to thinner oxides if the limitations for ultra-thin oxides are properly understood. The first limitation can arise from the dissipation factor D. For many C-V meters, D must be less than 10 and with the exponential increase in tunneling

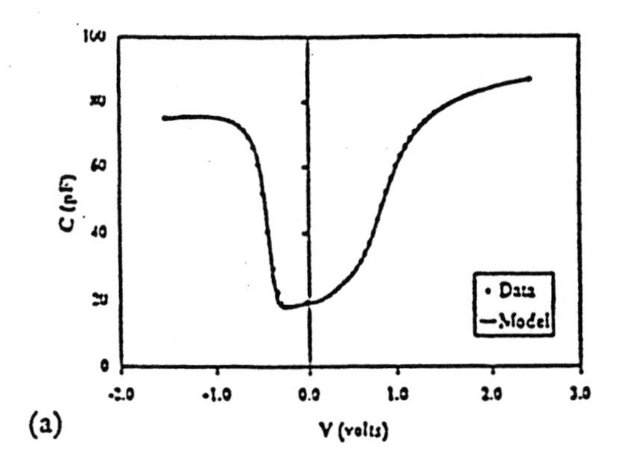

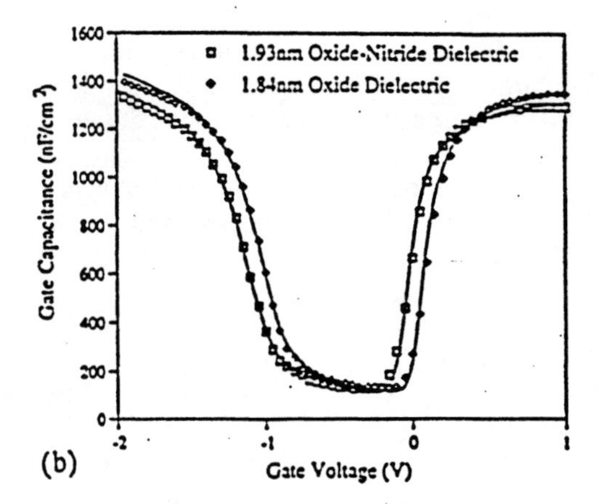

FIGURE 3. Model fits to experimental data for parameter extraction. (a) p-channel MOS device with a 3.5 nm oxide (b) n-channel MOS devices with i) 1.84 nm oxide and 1.93 nm oxide/ nitride gate stack.

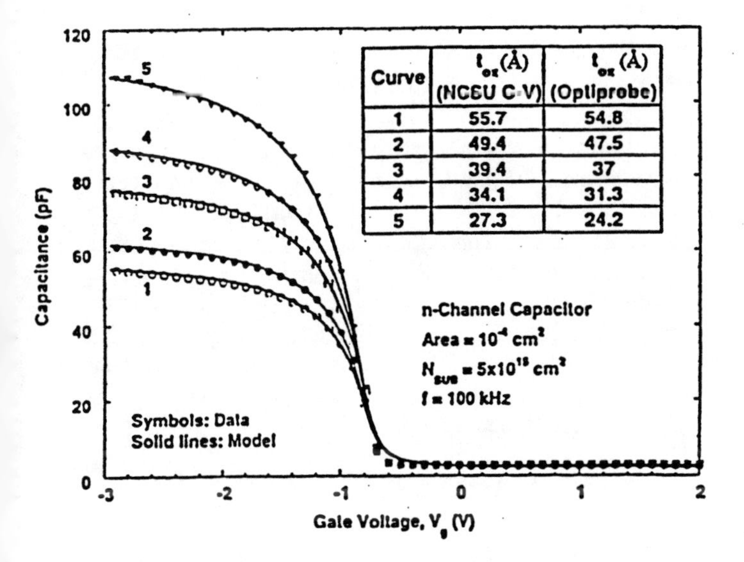

Curve	t_{ox}(Å) (NCSU C-V)	t_{ox}(Å) (Optiprobe)
1	55.7	54.8
2	49.4	47.5
3	39.4	37
4	34.1	31.3
5	27.3	24.2

FIGURE 4. Agreement between C-V model and high frequency C-V data for oxides of thickness 5.6 nm to 2.7 nm.

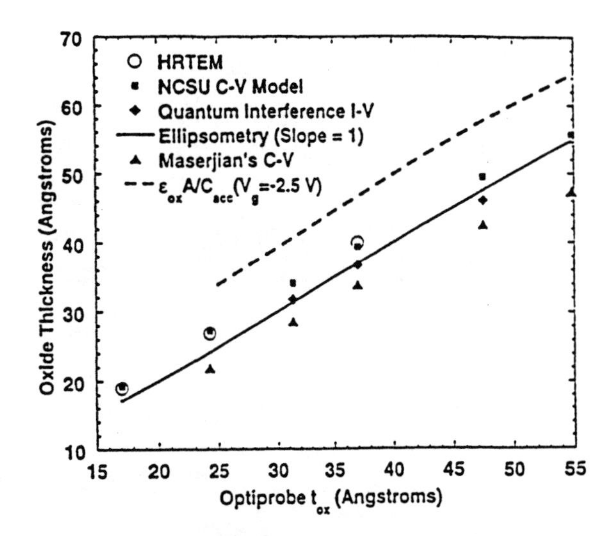

FIGURE 5. Comparison of extracted oxide thickness values by several experimental techniques.

current for thin oxides, this limit can easily be exceeded. Since D depends on frequency, the use of higher frequencies such as 1 MHz is required for thicknesses below 2 nm.

If a transistor structure is used to obtain C-V data in inversion, R-C drops along the channel can become especially important for thin oxides. This is shown in Figure 6 which shows a physical RC model [11] with gate leakage and simulated capacitance results for a 1.5 nm oxide thickness at 100 kHz. The simulations indicate that for a 1.5 nm oxide, the measurement structure must have a channel length of 10 μm or less. Since large areas are required to obtain sufficient capacitance, this requires a test structure with a large W/L ratio [12].

Another important limitation in accumulation with high frequencies and gate leakage is series resistance. For thin oxide this must be kept sufficiently low or errors will occur as illustrated in Figure 7. For this simulation at a measurement frequency of 1 MHz, an area of 10^{-4}cm² and a 1.5 nm oxide, it is seen that series resistance must be below 10Ω. The important points for ultra-thin oxides are:

(a) Use high measurement frequencies
(b) Use test structures with large W/L
(c) Minimize series resistance

By careful attention to these limitations, we believe that C-V characterizations can be extended down to about 1.5 nm with existing C-V meters.

Two major sources of error in extracting parameters from C-V measurements are: (1) approximations made in the theoretical C-V model and (2) measurement errors in obtaining an exact C-V for thin oxides. Both of these are manifest by errors between the modeled C-V and the measured data as can be seen in Figure 3. The larger the errors between the model and the data, the more certainty exists in the extracted parameter values. The least squares fitting technique provides an estimate of the uncertainty in the extracted parameter values. For oxide thickness, typical confidence limits (1 signa) are within the range of .01 nm to 0.05 nm.

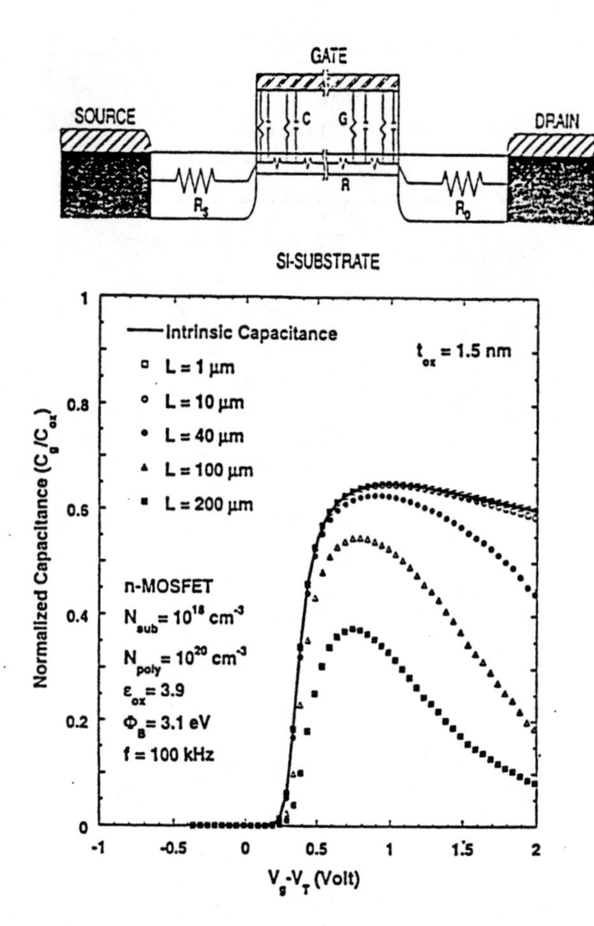

FIGURE 6. R-C-G limitation for C-V measurements in inversion using a transistor structure (a) R-C-G line model [11] (b) Calculated results for transistors of varying lengths.

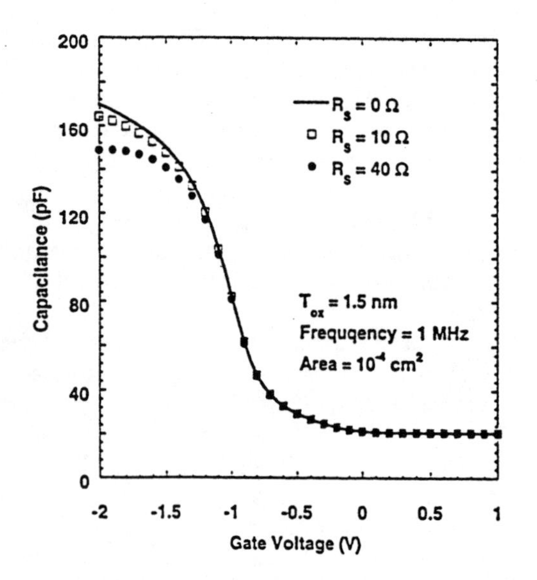

FIGURE 7. Theoretical effects of series resistance on measured inversion layer capacitance ($A=10^{-4}cm^2$).

I-V MODELS AND CHARACTERIZATION

To accurately model the I-V characterization of a MOS test structure, one needs an accurate model of inversion layer charge, as discussed in the previous section, and an accurate model of surface mobility. I-V data provide additional information on the properties of ultra-thin oxides and especially on the oxide-silicon interface properties. The NCSU I-V characterization approach combines the charge model of the previous section with a physical model of surface mobility. The surface mobility model is a 2-D mobility model [13] including (a) bulk ionized impurity scattering, (b) surface phonon scattering, (c) charged interface state scattering and (d) surface roughness scattering.

In order to extract effective surface mobility, I-V measurements are typically taken on relatively large (20 µm by 20 µm or larger) transistor structures at low V_{ds} values and at varying gate voltages. This varies the effective surface electric field and provides data such as shown in Figure 8a for a p-channel device. The rise in effective mobility with effective field is due to the increased carrier screening effects of channel charges which reduce the scattering effects of bulk ionized impurities and interface charges. The fall-off of mobility at high effective surface fields is due to surface roughness scattering. As described elsewhere [14], the NCSU analysis provides a least-square fit of the mobility model to experimental data.

Additional parameters provided by the I-V characterization are: (a) Interface charge scattering density and (b) Interface roughness. Figure 8b shows data for two devices with slightly different values of interface roughness. The I-V characterization is a very sensitive measurement technique for characterizing interface charge scattering density – usually in the range of $10^{10}/cm^2$, and interface roughness characterized by the product of roughness height and lateral period.

LIMITATION OF I-V FOR ULTRA-THIN OXIDES

Limitations on I-V measurements for ultra-thin oxides arise primarily from the gate leakage. To obtain accurate drain current data, the gate current must be much less than the drain current. This requires a device with a short channel length and the thinner the oxide the shorter the required channel length. For 1.5 nm oxides this will require 1-10 µm channel lengths. The only additional difficulty is that series source and drain resistance can become more of a problem in obtaining accurate data and corrections should be made for such resistance.

SUMMARY AND CONCLUSIONS

The characterizations of oxides using C-V and I-V measurements can provide very important information about physical parameters such as oxide thickness and electrical parameters such as interface charge scattering centers. These techniques have been successfully used for oxide thicknesses

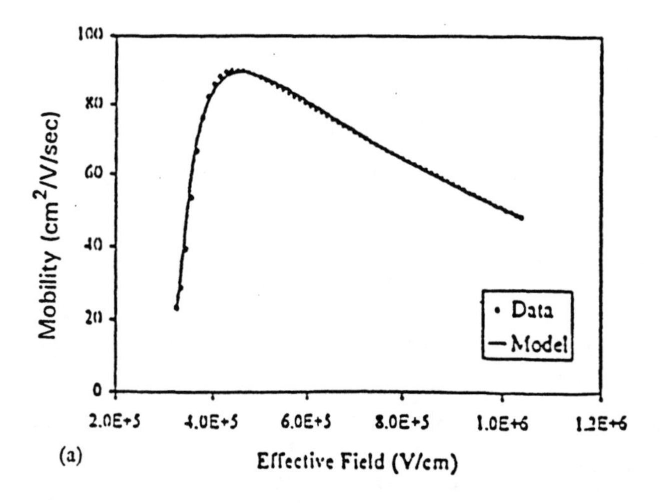

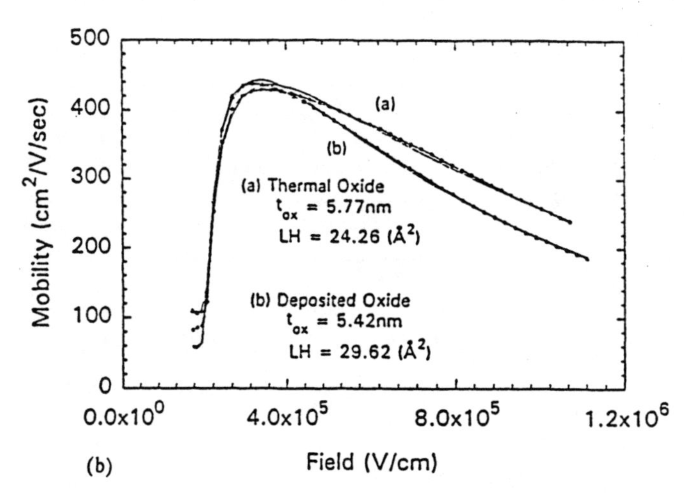

FIGURE 8. Extension of mobility parameters from I-V data (a) p-channel device with 3.5 nm oxide (b) n-channel devices with slightly different values of interface roughness.

down to about 1.8 nm. With more careful attention to instrumentation problems and test structure design, these techniques should be useful down to about 1.5 nm. As alternative gate dielectrics and gate stacks are considered, electrical interface charges are likely to become of more importance. The combination of first order physical models of interface charges and surface mobility combined with least squares curve fitting provides a rapid technique for extracting fundamental parameters from C-V and I-V data.

ACKNOWLEDGMENT

This work was supported in part by the Semiconductor Research Corporation, SRC Contract BP-132 and by the NSF Engineering Research Centers Program through the Center for Advanced Electronic Materials Processing.

REFERENCES

1. Stern, F. and Howard, W.E., *Phys. Rev.* **163**, 816 (1967).
2. Stern, F., *Phys. Rev.* **B5**, 4891 (1972).
3. Ando, T., Fowler, A.B., and Stern, F., *Rev. Mod. Phys.* **54**, 437 (1982).
4. Lopez-Villanueva, J.A., Melchov, I., Gamiz, F., Banqueri J., and Jimeneg-Tejoda, J.A., *Solid-State Elec.* **38**, 203 (1995).
5. Jallepalli, S., Bude, J., Shih, W.-K., Pinto, M.R., Maziar C.M., and Tasch, A.F., *IEEE Trans. on ED* **44**, 297 (1997).
6. Hu, C.-Y., Banerjee, S., Sadra, K., Streetman B.G., and Sivan, R., *IEEE Elec. Dev. Lett.* **17**, 276 (1996).
7. VanDort, M.J., Woerlee P.H., and Walker, J.A., *Solid-State Elec.* **37**, 411 (1994).
8. Ohkura, Y., *Solid-State Elec.* **12**, 1581 (1990).
9. Zafaz, S., Conrad, K.A., Liu, Q., Irene, E.A., Hames, G., Kuehna R., and Wortman, J.J., *Appl. Phys. Lett* **67**, 1031 (1995).
10. Maserjian, J., Petersson G., and Svensson, C., *Solid-State Elec.* **17**, 335 (1974).
11. Ernst, A.N., Somerville, M.H., and del Alamo, J.A., *IEEE Trans on ED* **45**, 9 (1998).
12. K. Ahmed, to be published.
13. Bayoumi, A.M., Ph.D. Thesis, NCSU (1995).
14. Hauser, J.R., *IEEE Trans. on ED* **43**, 1981 (1996).

Threshold Voltage (V_T) Control of sub-0.25μm Processes Using Mercury Gate MOS Capacitors

R. J. Hillard, R. G. Mazur, J.C. Sherbondy, L. Peitersen, M. Wilson, and R. Herlocher

Solid State Measurements, Inc. Pittsburgh, PA

MOSFET threshold voltage (V_T) is the most important parameter governing sub-0.25μm MOS device operation. Critical device performance issues such as speed, off state current and active power depend on V_T (1). The primary components that influence V_T are oxide thickness (W_{ox}), oxide charge (Q_{ox}), interface trap charge (Q_{it}), and channel dopant profile. Although conventional based CV methods have led many to believe that CV cannot be used for ultra-thin oxides, accurate methods for measuring V_T and its components using both a hard probe and a high repeatability mercury probe are described in this paper.

INTRODUCTION

One of the most important parameters controlling the operation of sub-micron devices is the threshold voltage (V_T). As device technology goes beyond 0.25μm design, the acceptable ranges for critical device and process parameters decrease. For example, the SIA roadmap (2) calls for an allowable V_T range of 60 mV for 0.25μm designs and 40 mV for 0.1 μm designs. Since V_T is composed of many process dependent parameters, an even tighter control of these V_T parameters is required. One example is gate oxide thickness. It is projected that a gate oxide thickness of 15 to 20Å will be needed with a 3σ thickness variation of 4 to 8% (2). It will also be required to control the channel doping level to 2 – 3 x 10^{18} cm^{-3} with a 3σ control of 3% (2). Maintaining tight control on V_T and its components places considerable demands on the accuracy and precision of metrology methods and equipment.

Conventional methods of controlling V_T have relied on controlling diffusion and oxidation processes for mobile ion charge and fixed charge. These methods utilize metal gate MOS capacitors (MOSCAPs) formed by short loop processes that evaporate or sputter aluminum on MOS monitor wafers. The primary parameter measured with these methods is the flatband voltage (V_{FB}). The V_{FB} is determined with high frequency capacitance-voltage (HFCV). Although these conventional methods are adequate for larger geometry designs, they are inadequate for use in sub-0.25μm geometry processes. First, controlling more than just V_{FB} is required in order to control V_T. As mentioned previously, gate oxide thickness, channel dopant distribution and interface trap density must also be controlled. Secondly, the accuracy and precision obtained with conventional HFCV methods degrade as oxide thickness decreases due to the lack of proper equivalent circuit and material considerations. Several other important considerations include the influence of processing effects, such as boron penetration from polysilicon gates, which can change V_T and the time and expense required for the short loop processing of monitor wafers. The effects of these complexities have led many to conclude that CV-based methods cannot be used for controlling sub-micron processes.

This paper describes a V_T management methodology that provides rapid and highly accurate production mode measurements of V_T and the components of V_T immediately after gate oxidation or polysilicon patterning of small-area (10^{-4} to 10^{-3} cm^2) MOSCAPs. These are located within the scribe lines of product wafers and measured with a fully automated hard probe system. We show that precise measurements can be made on ultra-thin gate oxides through a comprehensive consideration of the MOSCAP equivalent circuit as well as material related effects. Accuracy and precision are demonstrated for oxides as thin as 16Å. We can also demonstrate the use of a high repeatability mercury probe for process development on unpatterned wafers. The mercury probe can make HFCV, charge-voltage (QV), low frequency CV (LFCV), and current-voltage (IV) measurements (3,4) immediately after critical process steps such as gate oxidation.

EXPERIMENTAL AND PROBE DESIGN

The probes used in this work feature a kinematic bearing system to reduce damage to the MOSCAP test structure by the probe. The hard probe system utilizes pattern recognition so that the mapping of product wafers is fully automatic. The mercury probe used for process development features a one sigma area repeatability better than 0.1%. The advantages of this design over conventional mercury probe systems are described elsewhere (4). To improve measurements on thin oxides, the effective electrical area of the mercury gate is determined with a 35Å MOS reference wafer. Mapping of the blanket wafer with the mercury probe is also fully automatic.

The polysilicon gate MOSCAPs used for this work consist of both n- and p-type silicon substrates and have oxide thicknesses that range from 28Å to 300Å. Typical polysilicon gate areas ranged from 10^{-4} to 10^{-2} cm^2. Unpatterned MOS wafers used for mercury gate MOS CV, QV, and IV were also both n and p-type and had oxide thicknesses ranging from 16Å to 200Å. The mercury gate measurements were made using a 0.5 mm or 1.0 mm diameter capillary. The capacitance measurements for both the hard and mercury probes were made using a frequency of 0.1 MHz. A Keithley model 236/237 source-measure unit (SMU) was used for the IV measurements.

CP449, *Characterization and Metrology for ULSI Technology: 1998 International Conference*
edited by D. G. Seiler, A. C. Diebold, W. M. Bullis, T. J. Shaffner, R. McDonald, and E. J. Walters

V_T EQUATION FOR SUB-0.25μM TECHNOLOGY

SPICE (1) can be used to simulate submicron threshold voltages (here defined as $V_{T,MOSFET}$) as:

$$V_{T,MOSFET} = V_{T,MOSCAP} + \gamma(\Psi_{s,inv} + V_{BS})^{1/2} - K_T(\Psi_{s,inv} - V_{BS}) - \eta V_{DS} \quad (1)$$

where,

$V_{T,MOSFET}$ is the threshold voltage of the MOS field effect transistor (MOSFET)

and,

$V_{T,MOSCAP}$ is the threshold voltage of an MOS capacitor (MOSCAP).

and,

$\gamma = (2K_s qN_B)^{1/2}/C_{ox}$ ($V^{1/2}$)

K_T = charge sharing coefficient

η = drain induced barrier lowering (DIBL) coefficient

$\Psi_{s,inv}$ = silicon band bending in inversion (V)

V_{BS} = substrate bias (V)

V_{DS} = drain-to-source bias (V)

The second term in the $V_{T,MOSFET}$ Equation 1 accounts for body or substrate effects. The third and fourth terms account for short channel effects and are dependent on MOSFET effective channel length (L_{eff}); as L_{eff} decreases, K_T and η increase causing a decrease in $V_{T,MOSFET}$.

Note that, although $V_{T,MOSFET}$ is strongly affected by short channel effects, there remains a linear dependence on $V_{T,MOSCAP}$. Therefore, it is possible to obtain the necessary process control by measuring $V_{T,MOSCAP}$:

$$V_{T,MOSCAP} = V_{FB} \pm (Q_B + Q_I)/C_{ox} \pm \Psi_{s,inv} \quad (2)$$

where the flatband voltage V_{FB} is:

$$V_{FB} = -q(Q_f + Q_{it} + Q_{ot} + Q_m)/C_{ox} + \Phi_{ms} \quad (3)$$

and,

Q_f = fixed oxide charge (cm^{-2})

Q_{it} = interface trap charge (cm^{-2})

Q_{ot} = oxide trap charge (cm^{-2})

Q_m = mobile ion charge (cm^{-2})

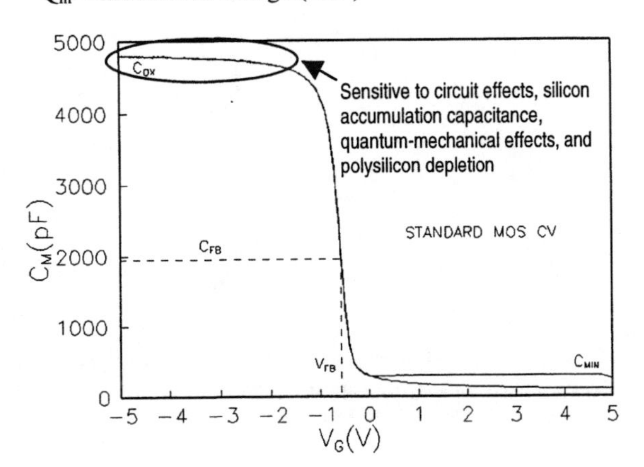

FIGURE 1. Illustration of a typical MOS HFCV curve.

Q_B = bulk charge (cm^{-2})

Q_I = threshold adjust partial implant dose (cm^{-2})

Φ_{ms} = metal-semiconductor work function difference (V)

C_{ox} = oxide capacitance (F/cm^2) = $\varepsilon_{ox}K/W_{ox}$

The primary parameters that influence $V_{T,MOSCAP}$ are oxide thickness, oxide thickness uniformity (σ/W_{ox}), V_{FB}, and Partial Implant Dose (PID) and substrate bulk dopant density. Interface trap charge (Q_{it}) must also be controlled because it affects V_{FB}, MOSFET subthreshold behavior, and noise.

ACCURATE DETERMINATION OF $V_{T,MOSCAP}$ AND COMPONENTS FOR ULTRA-THIN OXIDES

Oxide Thickness and Flatband Voltage

Conventionally, $V_{T,MOSCAP}$ is determined with HFCV data by measuring the parameters given in Equations 2 and 3. The four primary parameters of interest obtained from the HFCV curve are C_{ox}, C_{min}, flatband capacitance (C_{FB}), and V_{FB} as shown in Figure 1. The C_{ox} value depends heavily on the oxide thickness. The C_{min} value is determined by the bulk dopant density, threshold adjust implant, and temperature. In addition, experimental conditions such as measurement setup, moisture, and oxide charge influence C_{min}. After determining C_{ox} and C_{min}, the C_{FB} is calculated and V_{FB} determined from the intersection of C_{FB} with the HFCV curve. When the oxide thickness is greater than 200Å, HFCV data easily yields accurate values of the $V_{T,MOSCAP}$ components. For the thinner oxides used in 0.25μm technology and below, CV measurements get significantly more complex. The source of these complexities is twofold: First there are a number of effects that can only be understood and accounted for by considering the MOS equivalent circuit. Second, a number of materials-related issues need to be considered, specifically the effects of quantum-mechanical confinement of the silicon density of states, the silicon accumulation capacitance, and polysilicon depletion.

Equivalent Circuit Effects

The part of the HFCV that is affected most by equivalent circuit effects is the accumulation region. The origin of these circuit effects may be understood from the MOS equivalent circuit shown in Figure 2.

The stray capacitance and inductance are represented as C_{stray} and L_{stray}, respectively. The bulk series resistance, backside contact resistance, and backside capacitance are shown as R_s, R_{BS}, and C_{BS} respectively. Oxide leakage is represented by the parallel resistor, R_{ox} (linear leakage) and by the indirect and direct tunneling (1,5) diodes (nonlinear leakage). R_{ox} is used to account for surface moisture effects and for ohmic conduction due to the use of highly nonstoichiometric deposited insulators.

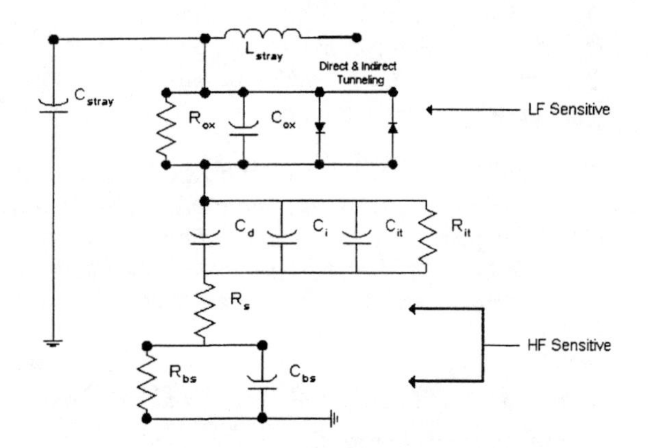

FIGURE 2. A schematic representation of the MOS equivalent circuit. The primary parameters that affect HFCV are R_{BS}, R_s, C_{BS}, and L_{stray}. LFCV data are influenced by R_{ox} and C_{it}.

The equivalent circuit components that involve the backside contact to a test wafer are of particular importance to HFCV. When R_S and R_{BS} increase, the dissipation factor increases causing a decrease in measured capacitance. The measured capacitance also decreases when C_{BS} decreases since C_{BS} is in series with C_{ox}. The effect of L_{stray} is to increase the measured capacitance. Details of these circuit effects are discussed elsewhere (6). The errors in measured capacitance that occur when the effects of these circuit elements are ignored are shown in Figure 3.

Reduction of Circuit Effects

Errors in measured capacitance due to L_{stray} are best reduced through the use of a lower frequency (0.1 MHz) and by applying empirical cable corrections.

The real problems associated with making ultra-thin oxide HFCV measurements are caused by the backside contact. The errors due to R_s and R_{BS} are reduced or eliminated through the use of a small gate area (about 10^{-3} cm^2 or less), the use of an operating frequency of 0.1 MHz or lower, and by correcting the measured capacitance for the effects of high dissipation factors (6). The effects of too low a value of C_{BS} introduce the largest errors as shown in Figure 3. This effect is frequency independent so that using a lower frequency does not reduce the error. For HFCV, the best approach is to use as small a gate area as possible and to ensure a high quality backside contact. Metal on the backside is the best solution. Since this is usually not possible in a production environment, the chuck vacuum should be as high as possible. For best results, the effective backside contact area should be at least 100 times the gate area. A high chuck vacuum coupled with the use of a 0.5mm diameter gate is usually sufficient for measuring ultra-thin oxides.

Another alternative is to use charge-voltage (QV) LFCV (3). In QV LFCV, the low frequency capacitance is determined from displacement current. Accurate capacitance values will be obtained as long as $C_{BS} \geq C_{ox}$, which is a relatively easy criterion to meet. Although these LFCV values are unaffected by the backside contact or series pn junctions, they are highly sensitive to leakage current through the oxide which mandates the use of high quality oxides. For such oxides, the limitation of QV LFCV is determined by the onset of excessive direct tunneling. An example of backside contact effects on HF and LFCV for a 45Å oxide is illustrated in Figure 4. With appropriate leakage corrections and backside compensation are applied, the results shown in Figure 5 are obtained for the worst case shown in Figure 4, that is for an oxide on the backside.

In addition to the circuit considerations just described, the accumulation capacitance associated with HFCV data obtained on ultra-thin oxides yields an electrical thickness that is higher than the physical thickness of the oxide. There are three primary causes for this discrepancy. One is due to the fact that the silicon capacitance in accumulation can no

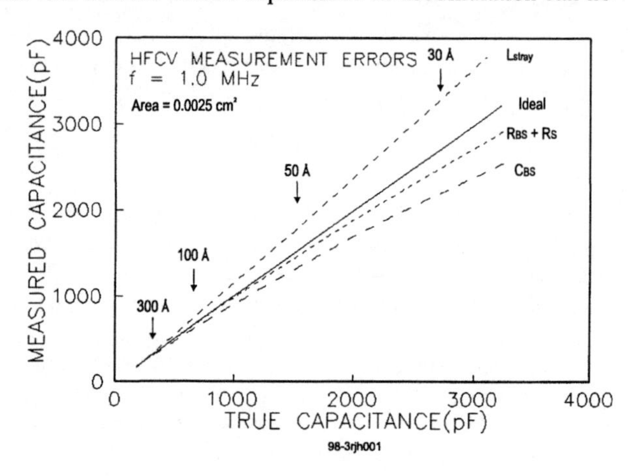

FIGURE 3. Measured HFCV versus true capacitance for conventional CV showing the effects of C_{BS}, L_{stray}, R_s, and R_{BS} for an ac frequency of 1.0 MHz. The magnitude of these errors for W_{ox} of 30Å, 50Å, 100Å, and 300Å is indicated.

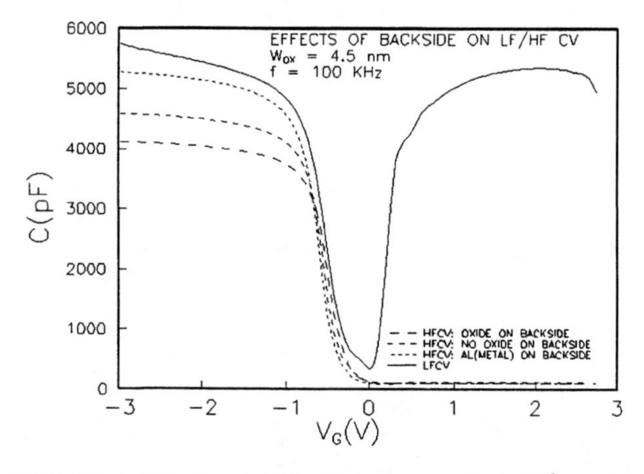

FIGURE 4. QV LF and HF CV data obtained on 45Å p-type MOS wafers with varying backside conditions. A 1.0mm diameter mercury gate was used. Note that the largest errors are obtained for the case of an oxide on the backside. The most accurate C_{ox} is obtained from the LFCV curve in inversion where the effects of leakage and silicon capacitance are minimized.

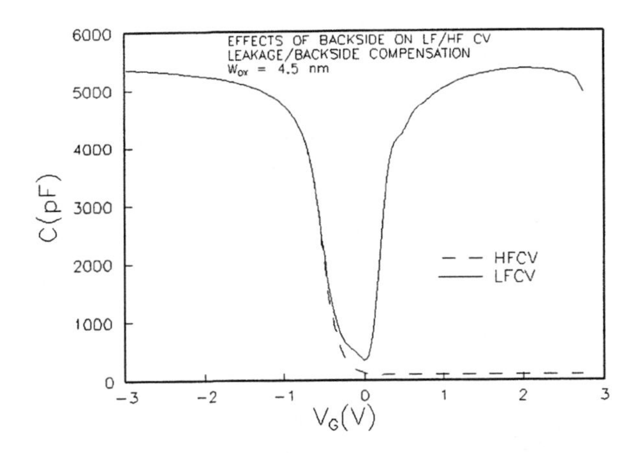

FIGURE 5. QV LF and HF CV data for the 45Å MOS wafer shown in Figure 4. The LFCV data is corrected for leakage current and the HFCV data are compensated for C_{BS}.

longer be neglected compared to the oxide capacitance. There are a number of methods available that correct for the silicon accumulation capacitance (7). A second consideration is that the silicon band bending in accumulation confines the silicon density of states so that they behave as discrete energy levels rather than a continuum. This causes a further increase in the electrical oxide thickness. The third is the effect of polysilicon depletion. To overcome these effects, the polysilicon doping level must be at least five orders of magnitude above the silicon substrate doping level. While this is a fundamental limitation associated with polysilicon, the use of a metal gate such as mercury overcomes this effect.

The W_{ox} accuracy and precision obtained by applying the methods discussed in this section are demonstrated in Table 1.

This table is a comparison of nine site maps made on oxides with optically measured thicknesses from 32Å down to 16Å. The mercury probe W_{ox} values were obtained with HFCV using a 0.5mm diameter capillary.

The repeatability of ultra-thin W_{ox} for mercury gate MOS CV was determined from ten measurements made at the same site; raising and lowering the probe prior to each measurement. The three sigma values for W_{ox} repeatability are 0.32Å and 0.39Å for oxide thicknesses of 32Å and 16Å respectively.

Polysilicon depletion has a major effect in determining V_{FB}. Conventional methods for measuring V_{FB} require accurate determination of C_{ox} and C_{min} from the HFCV curve. For ultra-thin oxides, where polysilicon depletion effects dominate, it is impossible to determine an accurate C_{ox}. However, an alternative method for determining V_{FB} exits. It is based on the second derivative of the $1/C^2$ versus V_G plot (3). The peak in the second derivative occurs at the flatband voltage. This method is unique in that no knowledge of the MOSCAP being measured is required.

Interface Trap Density

The interface trap density is determined from the difference between the QV LFCV and HFCV data that has been accurately corrected for leakage and circuit effects as discussed in the previous section. Once these data are obtained, curves similar to those shown in Figure 5 are used to generate a D_{it} versus energy plot. The average and minimum D_{it} and the D_{it} at two user-defined energy levels are calculated from this plot. The energy limits over which the D_{it} data are valid are also calculated and displayed. The QV method described (9,10) is sensitive to D_{it} values as low as about 6 x 10^9 cm^{-2} eV^{-1} for lightly doped substrates. The one sigma repeatability for mercury gate MOS QV D_{it} is typically less than 2%.

Partial Implant Dose (PID)

Threshold adjust implants are the primary means for establishing V_T for final devices. This requires implanting a dopant species such as boron, boron diflouride, phosphorus, or arsenic into the channel region of MOSFETs. The implant dose is typically in the 10^{12} cm^{-2} range and increases as the design geometry decreases (2). Monitoring and controlling this dose is critical. The best method for measuring the dose and resultant carrier profile in the channel is a method based on HFCV measurements which is called the partial implant dose (PID) method (11). High degrees of accuracy and precision are obtained by acquiring two deep depletion HFCV curves. The first curve is a low resolution

TABLE 1. Optical thickness versus electrical thickness obtained with 0.5mm-diameter mercury gate HFCV.

Measurement Method*	Site 1	Site 2	Site 3	Site 4	Site 5	Site 6	Site 7	Site 8	Site 9	Mean W_{ox} (Å)	σ (%)
Optical	32.67	32.75	32.85	32.94	33.35	32.70	32.84	32.89	33.24	32.91	0.711
SSM Hg Probe	32.45	33.4	32.69	32.29	32.47	32.82	31.94	32.57	33.69	32.7	1.660
Optical	21.91	22.00	22.14	21.92	22.17	22.26	21.97	21.93	22.04	22.04	0.568
SSM Hg Probe	24.43	25.34	24.75	24.53	24.87	25.06	24.08	24.44	25.39	24.76	1.780
Optical	22.02	21.94	22.01	21.74	22.00	22.25	21.84	21.87	21.94	21.96	0.651
SSM Hg Probe	24.8	25.35	25	24.71	25.14	25.41	24.64	25.05	25.64	25.09	1.320
Optical	16.62	15.85	15.80	15.87	16.85	16.84	15.90	15.66	15.97	16.15	2.940
SSM Hg Probe	20.06	20.94	20.43	20.43	21.06	21.07	20.17	20.22	21.24	20.62	2.180
Optical	16.72	15.89	16.02	16.03	16.68	16.76	16.12	15.85	16.07	16.24	2.290
SSM Hg Probe	19.52	20.12	19.76	19.70	20.09	20.01	19.43	19.64	20.25	19.84	1.460

* SSM Hg Probe – SiO_2/N_2O oxides

CV curve that yields C_{ox}, $V_{FB,}$ and the silicon substrate carrier density profile. The second curve is a high resolution CV curve obtained near the peak of the carrier density profile. This method measures the PID up to about 10^{13} cm^{-2} with a three sigma repeatability is better than $\pm 2\%$. (11)

The PID method has been shown to correlate well with final device V_T and is capable of controlling V_T to well within the requirements for 0.1μm designs. An example of the correlation between measured PID and final device V_T is shown in Figure 6. The total variation of device V_T shown in this figure is about 8 mV.

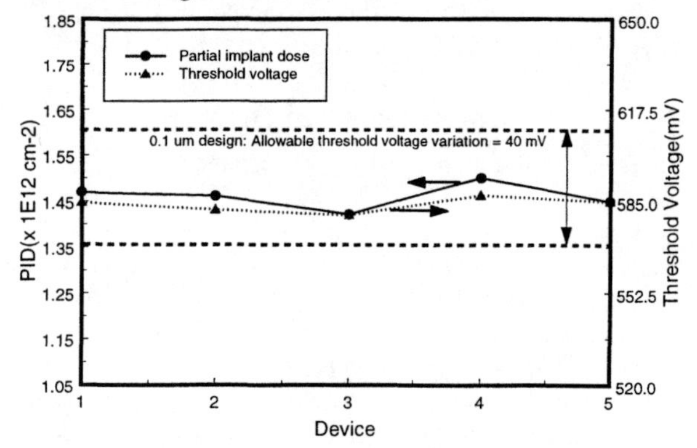

FIGURE 6. A plot of measured PID versus device V_T measured on five devices. These data demonstrate that measured PID correlates well with device V_T and is capable of monitoring V_T changes within 8 mV.

Oxide Quality

Final device V_T stability strongly depends on the quality and reliability of the gate oxide. For oxides less than 40Å, large direct tunneling currents can flow. These currents can affect V_T and other important device parameters such as transconductance and off-state leakage. Current stress over time may cause changes in oxide and interface trapped charge resulting in shifts in V_T. Therefore, it is desirable to monitor oxide reliability. The best approach is to use current-voltage (IV) time zero dependent breakdown (TZDB) and time dependent dielectric breakdown (TDDB) methods. An example of stepped voltage IV measurements made on oxides with thicknesses ranging from about 37Å down to 16Å is shown in Figure 7.

SUMMARY

V_T for sub-0.25μm geometry devices has been shown to be dependent on many process and geometry dependent parameters. The process dependent parameters include oxide thickness, oxide charge, interface trapped charge and threshold adjust implant dose. Although the device V_T becomes increasingly dependent on MOSFET geometry as design rules and channel lengths decrease, a strong dependence on these process parameters remains. In this paper methods for accurately determining and monitoring each of these process dependent parameters in ultra-thin gate oxides have been described. This monitoring can either be done with small area scribe line MOSCAPs for in-line production control or with a high repeatability mercury probe for process development. Sensitivity to changes in V_T of less than 8 mV have been demonstrated. Also, the monitoring of oxide quality for gate oxides as thin as 16Å has been shown.

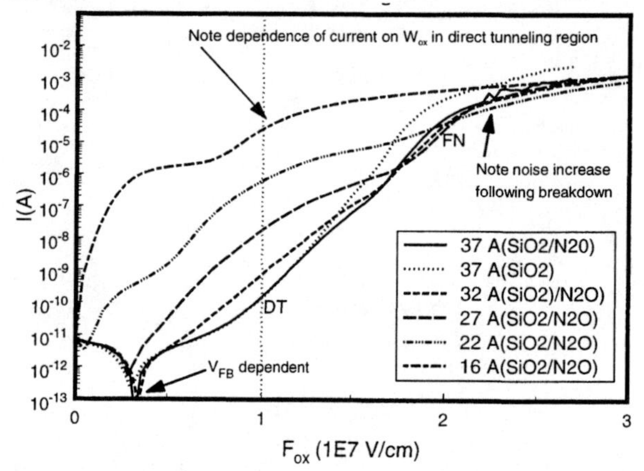

FIGURE 7. Current versus oxide electric field plots obtained with the stepped voltage method on oxides with thicknesses from about 37Å down to 16Å. Note the sensitivity of the direct tunneling current to oxide thickness. A sensitivity of 3.5Å/decade was determined from these data. Also, breakdown for these ultra-thin oxides is best determined by the field where an increase in noise is observed.

REFERENCES

1. Wolf, S., *Silicon Processing for the VLSI Era*, **3**, Lattice Press, 1995.
2. Diebold, A.C., "Critical Metrology and Analytical Technology Based on the Process and Materials Requirements of the 1994 National Technology Roadmap for Semiconductors", *AIP* 1996.
3. Solid State Measurements *CV/QV/IV Technical Seminar Book.*
4. Hillard, R.J., Mazur, R.G., Gruber, G.A., and Sherbondy, J.C., "Diagnostic Techniques for S Semiconductor Materials and Devices", *ECS PV97-12*, 1997, p. 310.
5. Sze, S.M., *Physics of Semiconductor Devices*, Second Edition, Wiley and Sons, 1981, p.456.
6. Schroder, D.K., *Semiconductor Material and Device Characterization*, Wiley and Sons, 1990.
7. Walstra, S.V., and Sah, C., *IEEE TED*, **44**, No. 7, 1997, p.1136.
8. Sune, J., Olivo, P., and Ricco, B., *IEEE NED*, **39**, No. 7, 1992, p. 1732.
9. Gruber, G.A., Heddleson, J.M., Hillard, R.J., and Weinzierl, S.R., SSM Internal Note: "Rapid n Interface Trap Density Measurements for MOS Production Control", Ref. 11.087, 1996.
10. Nicollian, E.H., and Brews, J.R., *MOS (Metal Oxide Semiconductor) Physics and Technology*, Wiley and Sons, 1982.
11. Ledudal, R., Hillard, R., Heddleson, J., Weinzierl, S., Rai-Choudhury, P., and Mazur, *R.G., Vac. Sci. Tech.*, 12, 1994, p. 336.

Advances in Surface Photovoltage Technique for Monitoring of the IC Processes.

K. Nauka [1], J. Lagowski [2]

1) Hewlett-Packard Laboratories, Palo Alto, CA 94304; 2) Semiconductor Diagnostics, Inc., Tampa, FL 33612

Surface Photovoltage has been employed for the in-line, contactless, and real-time monitoring of a variety of integrated circuit manufacturing processes. Recently demonstrated applications of Surface Photovoltage include monitoring of front-end cluster tool operations, measurement of the properties of ultra-thin gate dielectrics, and detection of interface states in Silicon-on-Insulator devices. A combination of Surface Photovoltage and Contact Potential Difference techniques has been employed to obtain detailed information about dielectric charges, plasma processing-related damage, and impurities introduced during device processing. Extensions of this technique satisfying future characterization requirements of ULSI manufacturing are also discussed.

INTRODUCTION

Successful manufacture of integrated circuits (IC) requires introduction of in-line monitoring techniques capable of real-time detection of misprocessed wafers and the enabling of immediate corrective action. Thus, further processing of already defective wafers can be avoided and subsequent wafers entering the manufacturing cycle can avoid faulty processing steps. These monitoring techniques must be clean-room compatible, fast, reliable, and integrated into the processing cycle. In order to minimize cost and accelerate monitoring they should not require any additional wafer processing beyond the routine IC manufacturing steps.

Among the many monitoring techniques recently proposed, Surface Photovoltage (SPV) appears to offer the largest number of demonstrated applications. Some of the early applications of the SPV technique for IC process monitoring have been previously discussed (1). This paper presents recent advances of SPV monitoring into new areas of IC processing. Additional monitoring applications have been obtained by combining SPV and Contact Potential Difference (CPD) measurements.

EXPERIMENTAL TECHNIQUES

The SPV signal is obtained by illuminating the front side of a Si wafer with a monochromatic photon flux (photon energy > Si bandgap), while the back side of the wafer is not illuminated. Non-equilibrium carriers generated by absorbed photons decrease the surface potential barrier Ψ_s (Fig.1) giving rise to the SPV signal V_{SPV}. At low light intensities (excess carrier density << equilibrium carrier density), V_{SPV} is related to the minority carrier diffusion length and surface recombination velocity. This relationship facilitates application of the SPV technique to detect metallic impurities related to minority carrier recombination centers in bulk Si (2). SPV-measured minority carrier diffusion lengths can then be used for qualitative, and sometimes also quantitative, evaluation of metallic impurity concentrations (2). Increasing the photon flux further decreases the surface barrier until it completely disappears (excess carrier density >> equilibrium carrier density) and the SPV signal reaches a maximum (Fig. 1.B). The maximum V_{SPV} is a measure of the equilibrium surface barrier height. Standard space charge expressions can then be used to calculate the surface charge density or, in the case of an oxidized silicon surface, it can be used to calculate the oxide – silicon interfacial charge density (3,4).

The CPD signal, V_{CPD}, represents the contact potential difference between the measured Si wafer and the reference metal electrode (5). When the Si surface is free of a dielectric, V_{CPD} is simply the work function difference between the Si and the reference metal. When the Si wafer is oxidized, V_{CPD} consists of the oxide barrier, V_{ox}, and the Si barrier, V_s (Fig.2). V_{ox} and V_s are determined by different electrical charges present in the Si space charge region, at the Si – oxide interface, within the oxide bulk, and at the oxide surface. Since high intensity illumination, as in the case of SPV measurement, can suppress the semiconductor surface barrier V_s, V_{CPD} measured under this illumination represents the oxide barrier V_{ox} and V_s is equal to the difference between the CPD signal measured in the dark and under illumination. When V_s and V_{ox} are known, the respective oxide charge components can be calculated.

This monitoring procedure can be further extended by application of a corona discharge. Corona discharge is used to deposit controlled amounts of positive or negative charges on the oxide surface. Deposited charges introduce an electric field causing drift of mobile ions that can be further enhanced when the wafer is resting on a hot chuck. Corona deposited dielectric surface charges can also provide an electric field used for the stress oxide measurements (5).

CP449, *Characterization and Metrology for ULSI Technology: 1998 International Conference*
edited by D. G. Seiler, A. C. Diebold, W. M. Bullis, T. J. Shaffner, R. McDonald, and E. J. Walters

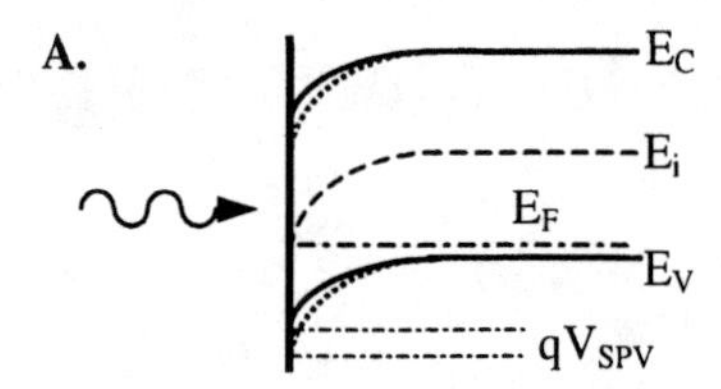

$$\Phi_{eff} / V_{SPV} = \text{const} * (s + D/L)(1 + \alpha/L)$$

Φ_{eff} = effective photon flux, s = surface recombination rate, D and L are minority carrier diffusivity and diffusion length, α = absorption coefficient

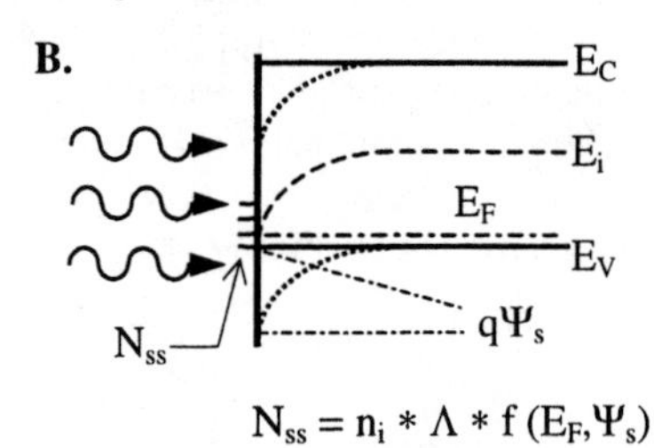

$$N_{ss} = n_i * \Lambda * f(E_F, \Psi_s)$$

N_{ss} = surface (or interface) charge density, Λ = screening factor, $f(E_F,\Psi_s)$ = function of Fermi level and SPV measured surface barrier

FIGURE 1. A schematic diagram of the illuminated silicon surface. A: low photon flux, only a small change of the surface barrier height occurs; B: high photon flux, the surface barrier is almost completely flattened.

SPV and CPD have been integrated into a commercial monitoring tool satisfying the requirement of clean-room compatibility. Both measurements are conducted in a "contactless" mode without need for additional wafer processing. They are accomplished by using an ac signal for capacitive coupling between the wafer and the SPV / CPD probe placed near the wafer surface. The SPV signal is obtained using either a filtered, incandescent light source for low light intensity applications, or a high intensity diode laser providing complete suppression of the Si surface barrier. The CPD signal is generated with a transparent reference probe vibrating in front of the Si wafer.

RESULTS AND DISCUSSION

Some of the recently demonstrated applications of SPV and CPD are described in this section. They include: monitoring of the Si surface barrier during the early stages of device manufacturing, measurement of interface charges in Silicon-on-Insulator devices, and a demonstration showing that the combination of SPV and CPD can provide information about all types of defects introduced by plasma processing. Surface barrier monitoring was used to provide detailed information about the degree of hydrogen termination of clean silicon surfaces. It was also employed to gain better understanding of the oxidation processes leading to gate oxides with thicknesses below 2 nm.

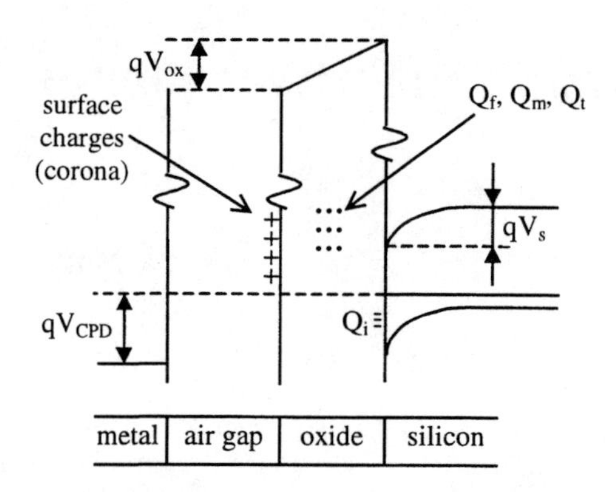

FIGURE 2. A schematic diagram of the silicon–oxide-air-metal probe system demonstrating electrical charges at the oxide surface, within oxide bulk, and at the silicon-oxide interface.

Monitoring of silicon surface in front-end processes

Hydrogen terminated silicon surface

Understanding and controlling the electronic properties of silicon surfaces during the early stages of wafer processing are critically important in IC manufacturing. Frequently, clean silicon surfaces are intentionally terminated with hydrogen (6,7) to assure their cleanliness, stability, and reproducibility for further processing. Hydrogen termination can be accomplished by either annealing in hydrogen ambient or by exposure to HF solutions. It is often assumed that both methods provide roughly equivalent hydrogen surface coverage that gradually degrades when exposed to air at room temperature. Recently, SPV has been employed (8) to study surface barriers obtained by these hydrogenation processes and their evolution when surfaces were exposed to air.

Fig.3 compares surface barriers obtained by both hydrogenation methods. Annealing in hydrogen ambient was a two step process; annealing at 1150°C was followed by heat treatment at temperature between 300°C and 700°C. Both annealing steps were conducted in H_2 ambient without breaking vacuum, in a load-locked, single wafer, lamp heated epitaxial reactor. Two groups of wafers were exposed to HF: as-received prime test wafers, and the same type of wafers previously annealed at elevated temperature. Pre-annealing step introduced well organized surface terrace structure, as observed with the Atomic Force Microscopy (AFM) and reduced surface roughness.

SPV results demonstrate that these two hydrogenation methods are not equivalent. H_2 annealing provides relatively stable hydrogen coverage that reduces barrier height. Hydrogen coverage is retained for days at room temperature. HF exposed wafers lose most of the surface hydrogen within first few minutes after removing from the HF solution. The same behavior was observed regardless whether wafers were or were not water rinsed after the HF exposure. Ordered surface terraces allow formation of unstable hydroxides within first few minutes after HF exposure which, then, disappear. Preference to form one type of hydride bonding rather than variety of hydrides is responsible for different steady-state barriers of ordered and unordered surfaces.

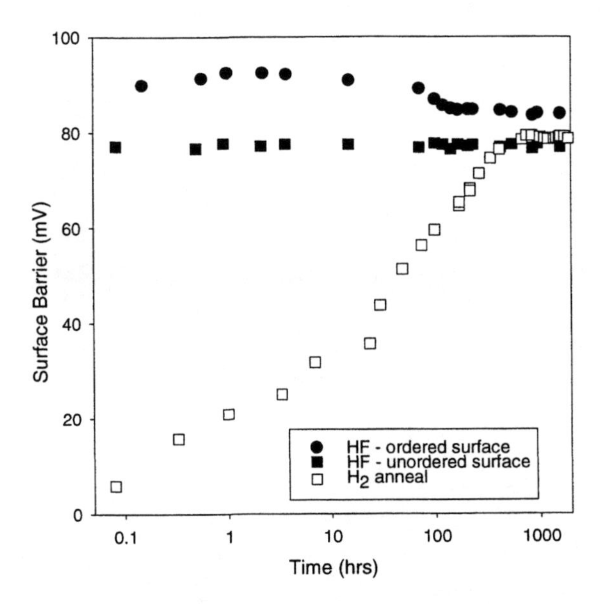

FIGURE 3. Evolution of surface barrier height during room temperature air exposure of Si (001). P-type (10 – 25 Ω-cm) prime test wafers were used in this experiment.

Thin gate dielectrics

Advanced IC devices require gate dielectrics with thicknesses below 2 nm (7). However, their manufacturing can provide a significant challenge and, frequently, they can only be fabricated in a cluster tool rapid thermal annealing (RTO) system. Very thin oxides tend to grow non-stoichiometric with increased number of oxide defects. RTO conditions need to be adjusted to maintain Si:O ratio in these oxides as close to 1:2 as possible. In addition, in order to improve oxide electrical strength and reduce boron diffusion, thin oxides frequently need to be doped with nitrogen introduced by adding N_2O during oxidation or by annealing in NH_3. Sometimes, a uniform dielectric is replaced with a stack of ultrathin dielectric layers, some of them containing nitrogen, with a total thickness below 2 nm. Because of these technological challenges, a monitoring technique is needed that provides information regarding nitrogen content and oxide quality. SPV has been employed to address both needs (8).

Fig.4 demonstrates the relationship between the nitrogen content and SPV measured Si surface barrier (nitrogen content has been independently measured with Auger spectroscopy). Good correlation has been observed in the case of nitrogen introduced by N_2O nitridation. Weaker correlation is observed in the case of NH_3 nitridation. It is probably due to residual hydrogen remaining in the film.

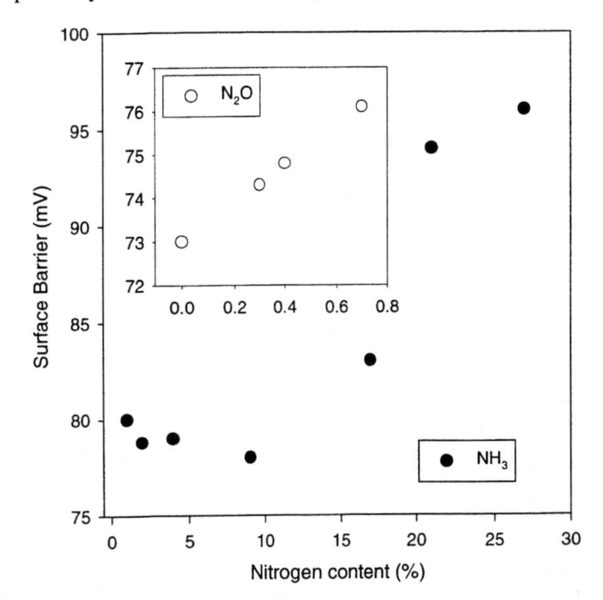

FIGURE 4. Correlation between silicon barrier height and nitrogen content. Nitridation process: N_2O – N_2O added during oxidation, NH_3 – post-oxidation annealing in ammonia.

Fig.5 presents correlation between the barrier height and the oxide thickness. In this experiment pure oxides without nitrogen doping were grown by RTO under the similar conditions. Barrier increase, and corresponding increase of the number of interface charges, is likely due to a gradual loss of the oxide stoichiometry when the oxide thickness is less than 1.8 nm.

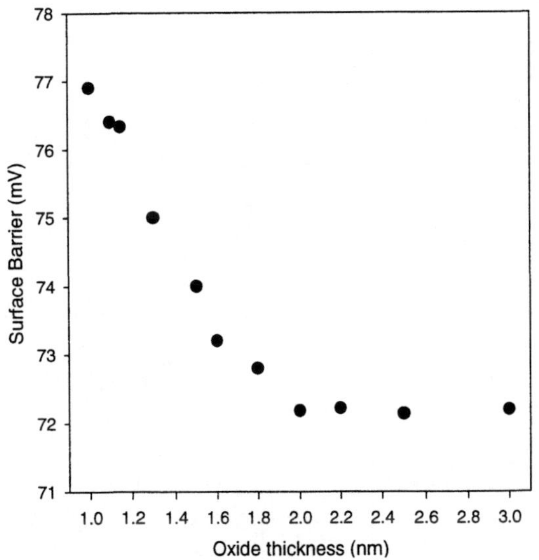

FIGURE 5. Correlation between the silicon barrier height and the oxide thickness.

Interfacial charges in Silicon-on-Insulator substrates.

Silicon-on-Insulator substrates offer opportunities for a rapid extension of the performance of CMOS IC devices (9,10). SOI MOSFETs consume less power and switch faster than the corresponding bulk Si CMOS transistors. In addition, presence of a buried dielectric can simplify insulation between PMOS and NMOS devices. SOI wafers ca be fabricated by oxygen implantation (SIMOX) followed by annealing that causes recrystallization of damaged silicon and forms buried oxide layer, or by wafer bonding (BSOI). Advanced SOI MOSFETs require substrates with thin silicon active layers and silicon - buried oxide (BOX) interfaces that remain in close proximity of the device active regions or even become part of devices. Electrically active interfacial defects can significantly degrade device performance, as shown in Figure 6. Therefore, understanding and control of the electrical properties of BOX and BOX – silicon interface are among the major challenges facing the SOI technology. Monitoring of the interface defects frequently relies on dedicated device structures that incorporate the interfacial region. Defects data are obtained at the expense of an additional processing of test structures and their analysis. In addition, these results are frequently convoluted with the data related to other properties of the SOI test structures. Recent advancement in the SPV (11) addresses these problems offering quick and reliable monitoring of the silicon-BOX interfacial charges.

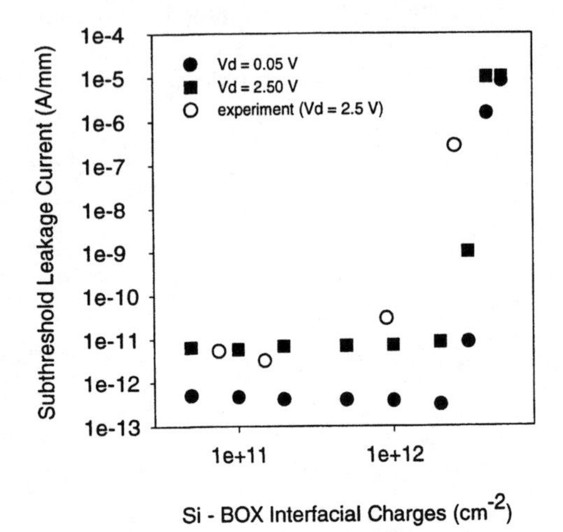

FIGURE 6. Subthreshold leakage as a function of the silicon – BOX interface charge density. Black squares and circles represent modeling data, white circles are experimental results. Interfacial charge densities were measured with the SPV (11).

SPV surface barrier measurements of SOI wafers differ from the previously described measurements because the SPV-SOI signal has three components: surface silicon active layer barrier, silicon active layer – BOX interfacial barrier, and BOX – silicon substrate interfacial barrier. However, it has been shown (11) that the last component has negligible value and the first component is a constant factor that can be easily eliminated from the analysis. SPV measurements demonstrated a strong correlation between the SOI substrate manufacturing process and the density and distribution of the interface charges. It was also observed that thermal cycle encountered during the CMOS manufacturing had no effect on the interface charges. Figure 7 compares interface charge distributions observed in the SIMOX and in the BSOI wafers.

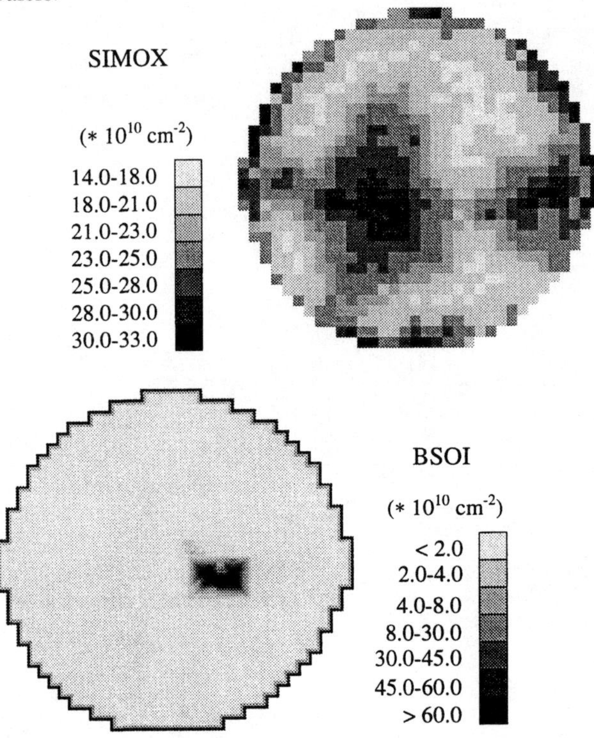

FIGURE 7. Distribution of interface charges in SOI substrates. SIMOX: Si active layer = 100 nm, BOX = 380 nm; BSOI: Si active layer = 160 nm, BOX = 400 nm.

A complete approach to the plasma damage monitoring in MOS structures.

Fabrication of modern IC involves large number of plasma processes. Plasma processing is often assisted by deleterious interactions between the plasma and materials causing undesirable modifications of device properties. Two frequently identified degradation mechanisms are: "antennae damage" caused by an unbalanced charge transfer between the dielectrics surface and the plasma sheath, and "radiation damage" introduced by highly energetic ions, neutral particles and photons bombarding device structure. Previous reports (1,12) demonstrated detection of particular types of plasma damage with CPD and SPV that was linked with device performance. However, device degradation frequently results from a sum of various damage mechanisms, additionally compounded by their interactions

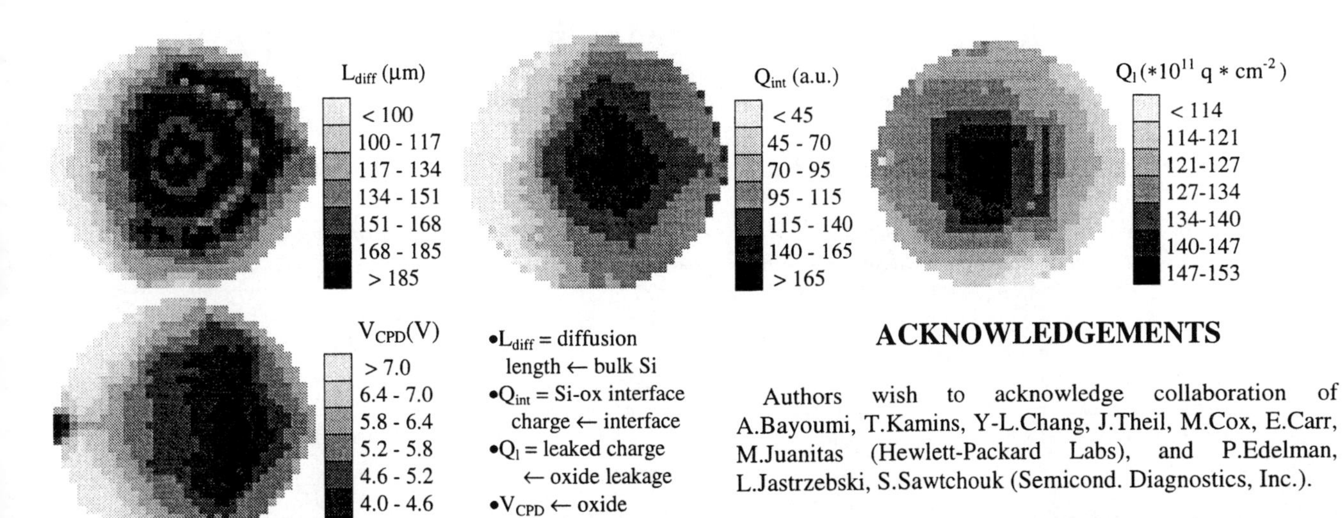

FIGURE 8. Maps of four types of defects introduced by plasma metal etch (13).

during the subsequent processing steps and topographic complexity of processed devices. Recent report (13) demonstrated how combination of SPV and CPD can provide a detailed information about all types of defects introduced by the various damage mechanisms during the plasma processing of submicron MOS structures.

Fig.8 shows distribution of defects induced by metal etching of submicron MOS structures. Four types of defects were considered: bulk Si recombination centers, Si-oxide interface charges, bulk oxide defects causing oxide leakage, and plasma induced oxide surface charges. Oxide leakage maps were obtained by corona charging of the oxide surfaces with uniformly distributed positive or negative charges, followed by a hot chuck anneal (experiment equivalent to the accelerated oxide stressing). Subsequent CPD scans showed locations on the wafer surface where, due to oxide defects, corona charges leaked through the oxide. Even more detailed information about the oxide defects can be obtained with a recently developed advanced CPD analysis of the oxide properties (14).

SUMMARY AND FUTURE WORK

Recent advancements in application of SPV and CPD for the monitoring of IC processes have been described. They include: monitoring of the hydrogen termination of Si surfaces, measurements of the properties of oxides with thicknesses below 2 nm, analysis of the interface charges in SOI substrates, and detection of various types of defects introduced by plasma processing. The future work will be primarily focused on the new applications of the SPV Si surface barrier measurements. This analytical approach could be applied to measure the reproducibility of wet and dry etches. Another potential application might include in-situ SPV monitoring of the front-end cluster tool operations.

ACKNOWLEDGEMENTS

Authors wish to acknowledge collaboration of A.Bayoumi, T.Kamins, Y-L.Chang, J.Theil, M.Cox, E.Carr, M.Juanitas (Hewlett-Packard Labs), and P.Edelman, L.Jastrzebski, S.Sawtchouk (Semicond. Diagnostics, Inc.).

REFERENCES

1. Nauka, K., *Semiconductor Characterization: Present Status and Future Needs*, AIP Press, 1996, pp.231 – 236.
2. Lagowski, J., Edelman, P., Dexter M., Henley W., *Semicond. Sci. Technol.* **7**, A185 – A192 (1992).
3. Monch, W., *Semiconductor Surfaces and Interfaces*, Springer, 1993, ch.5.
4. Edelman, P., Lagowski, J., Jastrzebski, L., "Surface Charge Imaging in Semiconductor Wafers by Surface Photovoltage", presented at the Mater. Res. Soc. Meet., San Francisco, CA, April 1992.
5. Lagowski, J., Edelman, P., "Contact Potential Difference Method for Full Wafer Characterization of Oxidized Silicon", presented at DRIP VII – 7th Internat. Conf. on Defect Recognition and Image Processing in Semicond., Templin, Germany, Sept. 1997.
6. Aoyama, T., Goto, K., Yamazaki, T., Ito, T., *J.Vac.Sci.Technol.* **A14**, 2909 (1996).
7. Vatel, O., Verhaverbeke, S., Bender, H., Caymax, M., Chollet, F., Vermeire, B., Mertens, P., Andre, E. Heyns, M., *Jpn.J.Appl. Phys.* **32**, L1489 (1993).
6. Nauka, K., Kamins, T.I., "Surface Photovoltage Measurement of Hydrogen-Treated Silicon Surfaces", submitted to J.Electrochem.Soc.
7. Maiti, B., Tobin, P.J., Misra, V., Hedge, R., Reid, K.G., Gelatos, C., *1997 Internat. Electron. Dev. Meet. Tech. Digest*, IEEE 1997, pp. 651 – 653.
8. Nauka, K., Bayoumi, A., Chang, Y-L., unpublished.
9. Shadini, G., 1993 *Internat. Electron. Dev. Meet. Tech. Digest*, IEEE 1993, pp.813 – 815.
10. Yamaguchi, Y., *IEEE Trans. Electron. Dev.*, **40**, 179 (1993).
11. Nauka, K., *Microelectron. Eng.*, **36**, 351 – 357 (1997).
12. Hoff, A.M., Esry, T.C., Nauka, K. *Sol.St.Technol.*, **39**, 79 (1996).
13. Nauka, K., Theil, J., Lagowski, J., Jastrzebski, L., Sawtchouk, S., "A complete approach to the in-line monitorng of materials defects introduced by plasma etching", presented at the 2nd Internat. Symp on Plasma Process-Induced Damage, Monterey, CA, May 1997.
14. Lagowski, J., Jastrzebski, L., Wilson, M., Edelman, P., Hoff, A.M., "COCOS Metrology – application to advanced gate dielectrics", presented at he 1997 SPIE Microelectronics Manufacturing Conf., Austin, TX, Oct. 1997.

Non-Contact Monitoring of Electrical Characteristics of Silicon Surface and Near-Surface Region

P. Roman, M. Brubaker, J. Staffa, E. Kamieniecki *, and J. Ruzyllo

Electronic Materials and Processing Research Laboratory, Department of Electrical Engineering, Penn State University, University Park, PA 16802
** QC Solutions, Inc., 150-U New Boston St., Woburn, MA 01801*

The SPV-based method of Surface Charge Profiling (SCP) is discussed, and its applications in silicon surface monitoring in IC manufacturing are reviewed. The SCP method shows high sensitivity to changes in the condition of the Si surface (e.g. surface cleaning operations) and a very thin near-surface region (e.g. variations of active dopant concentration near the surface).

INTRODUCTION

The increased size of wafers and decreasing device dimensions in today's ULSI manufacturing have resulted in a well recognized need for a broader and more rigidly executed in-line, real time monitoring of IC manufacturing processes. Particularly, as devices are becoming extremely shallow, the need to monitor the surface and near-surface region of Si wafers is becoming increasingly important.

In this work, the Surface Charge Profiling (SCP) method is studied in applications involving monitoring of the condition of the silicon surface and near-surface region. The tool used in this study is a commercial system that uses a probe in which an ac-Surface Photo Voltage (ac-SPV) induced by a collimated beam of chopped blue light is capacitively coupled through an air gap and amplified. Electrical characteristics of the silicon surface and near-surface region, namely the total surface charge, the recombination lifetime at the surface and the active dopant concentration in the surface space charge region are determined from analysis of the in-phase and quadrature components of the ac-SPV signal.

This paper discusses the principles of operation of the SCP method and the determination of these electrical characteristics. Experimental results demonstrating the response of these parameters to the condition of the silicon surface in various process monitoring applications, including surface cleaning and conditioning, as well as monitoring of variations in the active dopant concentration near the surface are presented.

PRINCIPLES OF OPERATION

When there is no electrical contact made to the silicon wafer, its surface remains in equilibrium with the bulk of the material, and the surface charge is neutralized by a charge collected in the surface space-charge region. This is illustrated in Fig. 1 which shows charge distribution in a p-type semiconductor under depletion conditions at thermal equilibrium, where Q_s represents surface charge and Q_{sc} surface space-charge (depletion layer charge). Because of the neutrality requirement, $Q_s = -Q_{sc}$.

In the SCP system, the electronic properties of the surface are determined by measurements of the ac-SPV (1-4). The schematic depicting the SCP measurement technique is shown in Fig. 2. The ac-SPV signal is generated with a collimated beam of chopped light of photon energy greater than the semiconductor bandgap. Because of the short wavelength (450 nm), the

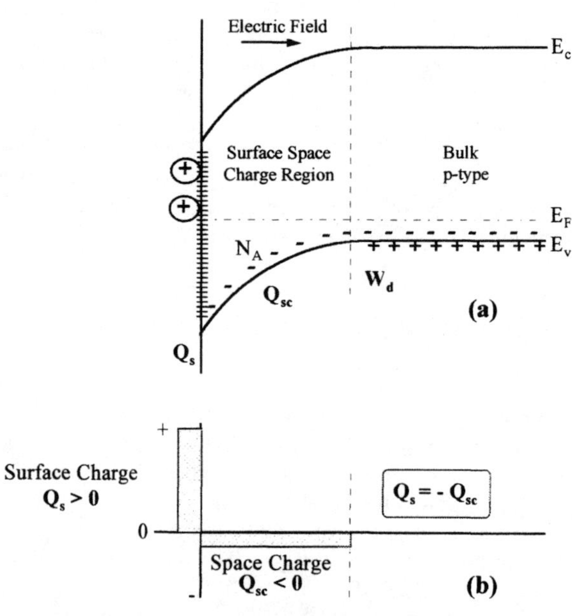

FIGURE 1. Surface of p-type silicon under depletion conditions at thermal equilibrium (no illumination), (a) energy band diagram, (b) charge distribution at the surface.

CP449, *Characterization and Metrology for ULSI Technology: 1998 International Conference*
edited by D. G. Seiler, A. C. Diebold, W. M. Bullis, T. J. Shaffner, R. McDonald, and E. J. Walters

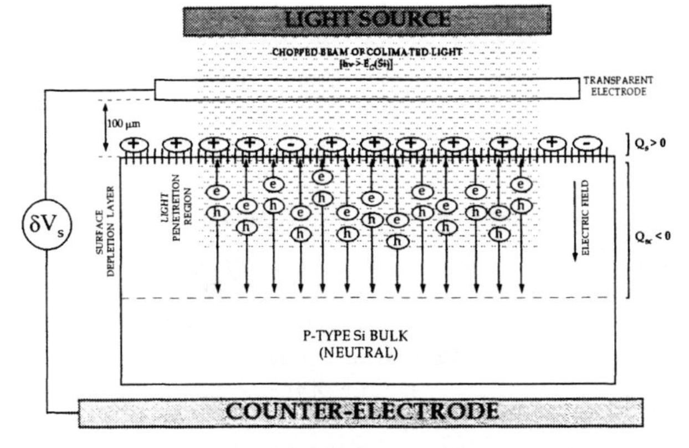

FIGURE 2. Schematic of SCP measurement technique, Qs-Surface Charge, Qsc-Semiconductor Space Charge.

light is absorbed close to the surface (light penetration depth in Si is 0.4 µm). Under depletion conditions the light is absorbed within the depletion layer generating electron-hole pairs. The strong electrical field that is present in the depletion layer separates electron-hole pairs in such a way that minority carriers flow toward the surface and majority carriers flow away from the surface toward the edge of the depletion layer (Fig. 2). Because of the influx of these additional minority carriers, both the width of the depletion layer and the barrier height at the surface decrease. The decrease of the barrier height causes an increase in the surface recombination so that when the light is off, the barrier height and depletion layer width are restored. Variations in the barrier height due to chopped illumination, known as the ac-SPV, are detected by the measurement electrode. Via capacitive coupling the voltage change induces a corresponding change in the charge on the electrode. The rate of change of the charge is the current which can be measured and amplified. The measurement electrode is separated from the front of the wafer by an air gap, and a counter-electrode is separated from the back of the wafer by an insulator. The basic theory of ac-SPV generation at low intensity, short wavelength illumination, is discussed in detail in Refs. (3) and (4). Here, we outline only those theoretical aspects that are essential for understanding the operation of the SCP method.

In the case of a uniformly-doped wafer under depletion or inversion conditions being illuminated with a low intensity light modulated sinusoidally at angular frequency ω, the real and imaginary parts of the ac surface photovoltage (δV_s) are described by the equations (3, 4):

$$Re(\delta V_s) = (q\Phi_o/\omega\varepsilon_s)\,(1\text{-}R)\,[\omega\tau/(1+\omega^2\tau^2)]\,W_d \qquad (1)$$

and

$$Im(\delta V_s) = -(q\Phi_o/\,\omega\varepsilon_s)\,(1\text{-}R)\,[\omega^2\tau^2/(1+\omega^2\tau^2)]\,W_d \quad (2)$$

where q is an elementary charge, Φ_o is the incident photon flux, ε_s is the permittivity of the semiconductor, R the reflection coefficient of the wafer surface, τ the surface recombination lifetime, and W_d the depletion layer width.

As indicated earlier, the SCP tool used in this study uses short wavelength illumination that is absorbed close to the silicon surface. Because of this, terms associated with the diffusion of minority carriers in the bulk of the silicon have been neglected in Eqs. (1) and (2).

The two main surface parameters determined using the SCP method are the depletion layer width, W_d and the surface recombination lifetime, τ. The depletion layer width is determined from the imaginary part of the ac surface photovoltage corrected for the surface recombination lifetime:

$$W_d = -[\omega\varepsilon_s/q(1\text{-}R)\Phi_o]\,Im(\delta V_s)(1+1/\omega^2\tau^2), \qquad (3)$$

and the surface recombination lifetime, τ, is defined by the equation (3, 4):

$$Re(\delta V_s)/Im(\delta V_s) = -1/\omega\tau \qquad (4)$$

Among other methods based on similar principles the SCP method distinguishes itself by features which make it specifically suitable for in-line real-time process monitoring (5). While other methods either require contact to the surface, or use high voltage, corona charging or high intensity illumination, each of which affects the height of the potential barrier on the measured surface, the SCP is a contactless method, and perturbations of the surface barrier during measurement are limited to a few tenths of the thermal energy, kT. Therefore, in the case of SCP the surface under measurement is not being affected by the measurement. Moreover, the light penetration depth in the case of SCP is very shallow, and hence, the information obtained is dominated by characteristics of the surface and near-surface region.

RESPONSE OF MEASURED PARAMETERS TO THE VARIED CONDITIONS OF THE SURFACE

As indicated earlier, the parameters measured by SCP include the surface charge density, the surface recombination lifetime, and the near-surface active dopant concentration. In this section examples showing how each of these can be used to extract information regarding the condition of the Si surface and sub-surface region are discussed.

Surface Charge Density

The depletion layer width, W_d is a direct measure of the semiconductor space-charge, Q_{sc}. For the net doping concentration in the space charge region, $N_{sc} = |N_D - N_A|$ the semiconductor space-charge is given by:

$$Q_{sc} = \pm qN_{sc}W_d \tag{5}$$

which is positive in n-type and negative in p-type semiconductors. Since during the SCP measurements the semiconductor remains neutral, the surface charge (Q_s) is compensated by the semiconductor space-charge (Q_{sc}), so that $Q_s = - Q_{sc}$, as shown in Fig. 1.

The maximum magnitude of the surface charge that can be determined using the SCP is reached when the Si surface becomes inverted and the depletion layer width reaches its maximum, W_{dmax}. Therefore, the maximum magnitude of the surface charge that can be determined using the SCP, $|Q_{s, inv}|$ depends on N_{sc}. At the opposite polarity the measurements of the surface charge are limited by the surface reaching accumulation conditions. For p-type Si the SCP can resolve positive surface charge values up to $+|Q_{s, inv}|$, whereas for n-type Si the SCP can resolve negative surface charge values down to $-|Q_{s, inv}|$.

It has been demonstrated that the surface charge density measured using SCP is very sensitive to changes of the condition of the Si surface during cleaning operations. For example, the surface charge responds to changes in composition of both standard cleaning solutions (6) as well as gas-phase cleaning chemistries (7). It has also been shown to respond to surface Fe, Al, and Mg (8-11). It has also been successfully used to detect noble metal (Ag, Au, Cu, Pt) surface contamination deposited during immersion in dHF with a Low Limit Detection (LLD) < 1ppb. Surface charge density measurements have also been shown to be correlated with surface roughness (9).

SCP's non-invasive quality makes it possible to take successive measurements of surface charge density without the measurement itself affecting this parameter and thereby monitor its evolution as a function of time (12). This approach is illustrated in Fig. 3 in which the stability of the surface charge density on a wafer etched in 1 % dHF solution reflects the more complete etching and hydrogen passivation of this surface and has been used to gauge the surface termination resulting from different gas-phase processes (13, 14).

Besides its use in bare surface monitoring the SCP method shows good potential in monitoring oxidized surfaces. The surface charge density has been shown to be sensitive to Na and Al contamination, and levels below $6x10^{10}$ at/cm^2 were detected (9, 10). Figure 4 shows experimental results of surface charge measurements using SCP on p-type Si wafers with thermally grown oxides annealed for various times in nitrogen following the oxidation. Such annealing is known to reduce the positive fixed oxide charge. The same behavior of SCP measured charge was also observed for 60 Å thermal oxides (15).

Surface Recombination Lifetime

The surface recombination lifetime, τ determined using the SCP method is defined by Eq. (4) and is determined from the real and imaginary components of the ac-SPV, or the decay time of the ac-SPV once the light is turned off. Since the recombination of the minority carriers at the surface proceeds mainly through the surface/interface states, N_{ss}, this parameter is sensitive to the chemical/physical state of the Si surface or Si/SiO_2 interface.

Previously, a good correlation has been observed between this recombination lifetime and wetting angle and XPS O_{1s} count following oxide etching in dilute hydrofluoric acid solution (dHF) (11, 16, 17). The experimental results shown in Fig. 5 demonstrate a response of the surface recombination lifetime to the presence of a chemical oxide remaining on the Si surface following a failed gas-phase oxide removal process (18). As shown, surfaces having low wetting angles indicating the presence of a remaining chemical oxide all displayed low values of recombination lifetime whereas surfaces with high wetting angles displayed high values of lifetime. The surface recombination lifetime has also been shown to be responsive to surface contamination resulting from immersion in standard SC1 solutions contaminated with Al (LLD < 10 ppt), Fe (LLD < 50 ppt) or Ca (LLD < 100

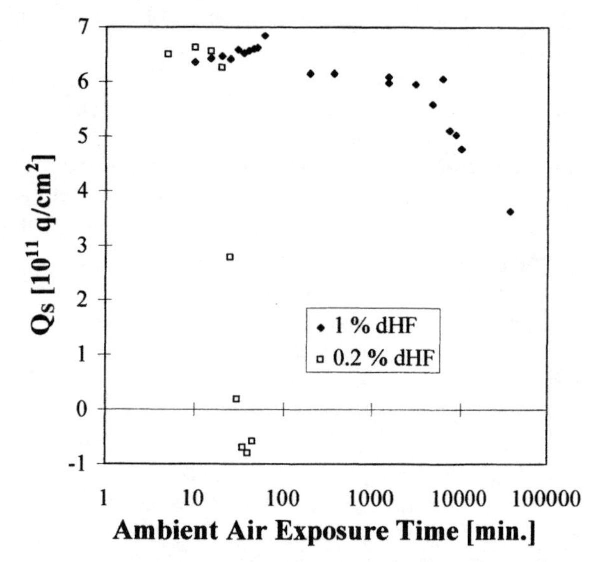

FIGURE 3. Surface charge stability following oxide etching in 1 % and 0.2 % dHF solutions for 30 seconds.

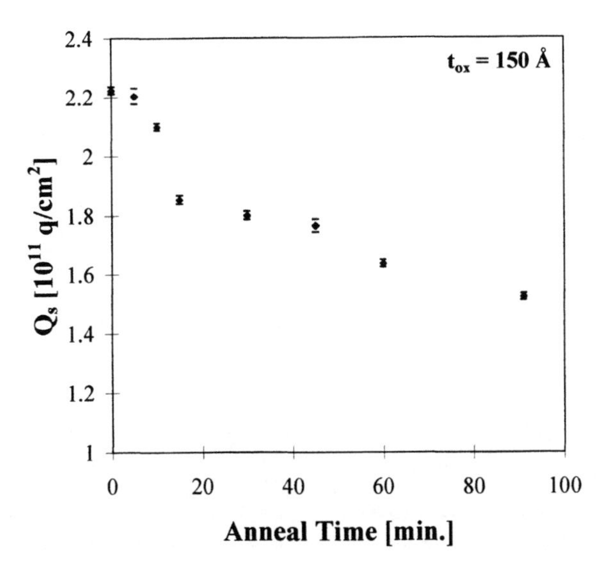

FIGURE 4. Surface charge density vs. post-oxidation anneal time on p-type wafers with thermally grown oxides.

ppt) and DI water contaminated with Fe (LLD < 50 ppt) (9).

Near-Surface Active Dopant Concentration

Under steady-state inversion conditions the inversion layer formed at the surface shields the bulk of the semiconductor, and the depletion layer width reaches its maximum value, W_{dmax}. The inversion layer can be formed at p-type silicon surfaces by dipping in dHF solution (5). Under this condition W_{dmax} is determined by SCP measurements using Eq. 3, and the active dopant concentration in the space charge region, N_{sc} is determined from the relationship (19):

$$W_{dmax} = [2\varepsilon_s kT\ln(N_{sc}/n_i)/(q^2 N_{sc})]^{1/2} \qquad (6)$$

It is known that shallow acceptors like boron can be rendered electrically inactive by pairing with hydrogen, or metals such as copper or iron. Hydrogen can be introduced into the near-surface region during device manufacturing processes such as reactive ion etching (RIE), and both hydrogen and copper can be introduced during chemo-mechanical polishing of the wafer causing uncontrolled variations in active dopant concentration near the surface which would not be seen by techniques such as four point probe and C-V dopant profiling.

Figure 6 shows the results of an experiment in which following our observations that the surface dopant concentration in polished p-type wafers from the manufacturer was initially several times less than the bulk value and that a brief low temperature (250 °C) anneal was sufficient to completely restore the surface value, the temperature dependence of the surface dopant concentration was studied using SCP (20). From these results an activation energy was determined that enabled the identification of the contaminant responsible for the initially observed boron deactivation as atomic hydrogen. Measurements of the surface dopant concentration made on epi layers are consistent with values given by Hg probe measurements (9, 11). This approach can be extended with very good results to measurements of dopant concentration and distribution in thin epi layers.

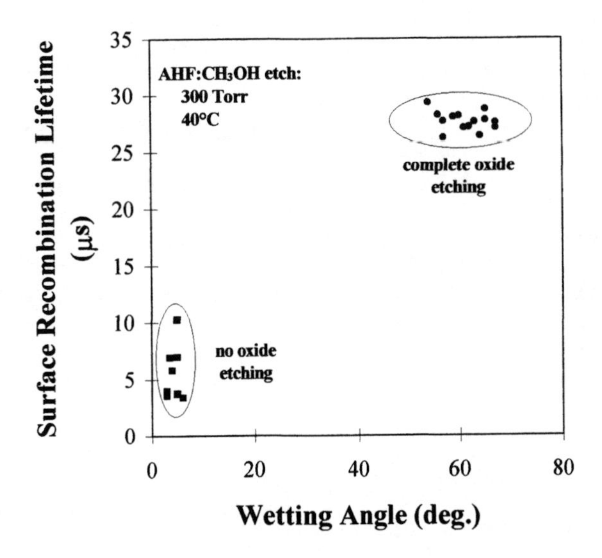

FIGURE 5. Surface recombination lifetime vs. wetting angle on Si wafers following failed and complete gas-phase oxide removal processes.

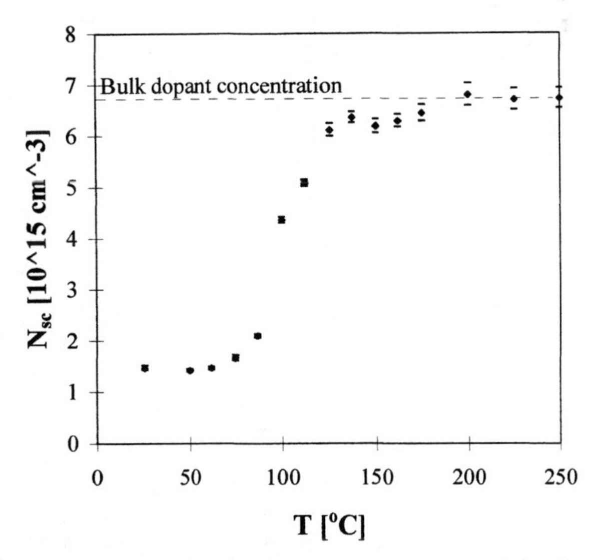

FIGURE 6. Surface dopant concentration vs. anneal time in polished boron-doped Si wafers.

SUMMARY

The SCP method discussed in this paper shows high sensitivity to changes in the condition of the Si surface and a very thin near-surface region. SCP measurements are fast and entirely non-invasive. Hence, this method is well suited for applications in in-line monitoring of surface treatments in IC manufacturing. In addition, experience shows that the SCP method is very useful in process development and equipment qualification. The SCP system can be integrated with cluster tools or used as a stand alone system. In the latter case, in order to obtain meaningful results, measures should be taken to assure reproducible ambient conditions; e.g. moisture in the clean room air as well as wafer storage conditions.

ACKNOWLEDGMENTS

This study contains work performed under a research grant from the Semiconductor Research Corporation. An equipment donation from QC Solutions, Inc. is also acknowledged.

REFERENCES

1. Kamieniecki, E., "Noninvasive Method and Apparatus for Characterization of Semiconductors", U.S. Patent 4,827,212, May 1989.
2. Kamieniecki, E., *J. Vac. Sci. Technol.*, **20**, 811-814 (1982).
3. Kamieniecki, E., *J. Appl. Phys.*, **54**, 6481-6487 (1983).
4. Kamieniecki, E. and Foggiato, G., "Analysis and Control of Electrically Active Contaminants by Surface Charge Analysis," in *Handbook of Semiconductor Cleaning Technology*, Ed. W. Kern, Noyes Publ., 1993, pp. 497-536.
5. Danel, A., Straube, U., Kamarinos, G., Kamieniecki, E., and Tardif, F., "Monitoring of Noble Metals in HF Based Chemistries by m-PCD, SPV, SCI and SCP," *Proc. of the Fifth International Symposium on Cleaning Technology in Semiconductor Device Manufacturing*, Eds. J. Ruzyllo and R. Novak, Pennington, NJ, The Electrochemical Society, Inc., 1998, pp. 400-407.
6. Kamieniecki, E., Roman, P., Hwang, D., and Ruzyllo, J., "A New Method for In-Line, Real-Time Monitoring of Wafer Cleaning Operations," *Proc. of the Second International Symposium on Ultra-Clean Processing of Silicon Surfaces*, Eds. M. Heyns, M. Meuris, and P. Mertens, Leuven, Belgium, Acco, 1994, pp. 189-192.
7. Torek, K., Ph.D. Thesis, Pennsylvania State University, 1995.
8. Roman, P., Kashkoush, I., Novak, R., Kamieniecki, E. and Ruzyllo, J., "Monitoring of Fe Contamination on Si Surfaces Using Non-Contact Surface Charge Profiler," *Proc. of the Fourth International Symposium on Cleaning Technology in Semiconductor Device Manufacturing*, Eds. R. Novak and J. Ruzyllo, Pennington, NJ, The Electrochemical Society, Inc., 1996, pp. 344-349.
9. Danel, A., Lardin, T., Kamarinos, G. and Tardif, F., "Sensitivity of the Surface Charge Profiler (SCP) Method for Some Monitoring Applications," *Proc. of the Symposium on Crystalline Defects and Contamination: Their Impact and Control in Device Manufacturing II*, Eds. B. Kolbesen, C. Claeys, P. Stallhofer and F. Tardif, Pennington, NJ, The Electrochemical Society, Inc., 1998, pp. 394-403.
10. Torek, K., Lee, W., Palsulich, D., Weston, L. and Gonzales, F., "Applications of Non-Invasive AC-Surface Photovoltage Monitoring in Integrated Circuit Cleaning," Same as in Ref. 5, pp. 392-399.
11. Trauwaert, M., Kenis, K., Caymax, M., Mertens, P., Heyns, M., Vanhellemont, J., Graf, D. and Wagner, P., "Evaluation of Si Surface Conditions by the Use of a Surface Photovoltage Technique," Same as in Ref. 5, pp. 455-462.
12. Torek, K., Mieckowski, A., and Ruzyllo, J., "Evolution of Si Surfaces After Anhydrous HF:CH_3OH Etching," Same as in Ref. 8, pp. 208-213.
13. Staffa, J., Fakhouri, S., Brubaker, M., Roman, P. and Ruzyllo, J., "Effects Controlling Initiation and Termination of Gas-Phase Cleaning Reactions," Same as in Ref. 5, pp. 315-321.
14. Staffa, J., Ph.D. Thesis, Pennsylvania State University, 1998.
15. Ruzyllo, J., Roman, P., Staffa, J., Kashkoush, I., and Kamieniecki, E., "Process Monitoring Using Surface Charge Profiling (SCP) Method," *Process, Equipment, and Materials Control in Integrated Circuit Manufacturing II*, Austin, TX, The International Society for Optical Engineering, 1996, pp. 162-173.
16. Roman, P., Hwang, D., Torek, K., Ruzyllo, J. and Kamieniecki, E., "Monitoring of HF/H_2O Treated Silicon Surfaces Using Non-Contact Surface Charge Measurements," *Ultraclean Semiconductor Processing Technology and Surface Chemical Cleaning and Passivation*, Eds. M. Liehr, M. Heyns, M. Hirose and H. Parks, Pittsburgh, Materials Research Society, 1995, pp. 404-405.
17. Kondoh, E., Trauwaert, M., Heyns, M. and Maex, K., "In-Line Monitoring of HF-Last Cleaning of Implanted and Non-Implanted Silicon Surfaces by Non-Contact Surface Charge Measurements," Same as in Ref. 5, pp. 221-228.
18. Brubaker, M., Staffa, J., Roman, P., Fakhouri, S. and Ruzyllo, J., "Determination of Native Oxide Removal in a Cluster Compatible Dry Cleaning System Using a Surface Photovoltage Technique," Same as in Ref. 5, pp. 409-410.
19. Sze, S.M., *Physics of Semiconductor Devices*, New York, Wiley, 1981, Ch. 7.
20. Roman, P., Staffa, J., Fakhouri, S., Brubaker, M., Ruzyllo, J., Torek, K. and Kamieniecki, E., *J. Appl. Phys.*, **83**, 2297-2300 (1998).

Contactless Transient Spectroscopy for The Measurement of Localized States in Semiconductors

S. Kishino and H. Yoshida

Department of Electronics, Faculty of Engineering, Himeji Institute of Technology, 2167 Shosha, Himeji 671-2201, Japan

We have succeeded in obtaining first-ever signals of COntactLess Transient Spectroscopy (COLTS) using an MIOS structure which is composed of metal, insulating helium gas, oxide, and silicon wafer. Feasibility test was carried out using a gold-doped CZ silicon wafer. The insulating air (helium gas) gap was controlled to be less than one micrometer using piezo-electric actuators with the help of a capacitance measurement in which three parallelism electrodes were prepared besides a reference electrode for COLTS measurements. It is plausible that COLTS measurements are indispensable for the wafer characterization and in-line monitoring of electrically active impurities in the advanced ULSI age. This is because electrodes might degrade surface properties, the oxide film, and interface properties during the fabrication and the characterization of the device.

INTRODUCTION

ULSI devices have been scaled down year after year to obtain a highly integrated LSI chip (1). Resultant MOS devices require higher reliability during their fabrication and operation in which the degradation of signal to noise (S/N) ratio must be avoided by all means.

This is a very severe requirement for the scaled-down devices with small signal charges because a faint noise charge could degrade S/N ratio in the situation. Moreover, there are many sources of noise charges in the scaled-down devices partly because of short channel effects including hot carrier effects (2).

In this situation, precise characterization of localized states such as bulk traps and interface traps is very important because they are primary causes of noise charges. The localized states have been characterized so far with the use of a sample device such as a Schottky diode or an MOS diode. Fabrication of a sample device could distort the intrinsic characteristic of the surface and interface.

This is very serious in the age of scaled-down MOS devices partly because a very thin gate oxide less than 4 nm in thickness will be used (3). Moreover, electrical active impurities and interface traps might be induced during the fabrication of electrodes. In order to eliminate most problems of conventional measurement techniques, it is preferable to use contactless measurement which does not require making a sample device. In line with the requirement, Sakai et al. (4) have already reported contactless C-V and C-t measurements using an MAIS (Metal-Air-Insulator-Semiconductor) structure. However, their technique is restricted to the measurements at room temperature.

Based on the background mentioned above, we have proposed and developed a novel technique for the inspection of localized states in semiconductors. This is a COLTS (COntactLess Transient Spectroscopy) in which an MIOS (Metal-Insulating helium gas-Oxide-Semiconductor) structure is used instead of an MOS diode. Namely, the semiconductor wafer itself is used as a sample in the COLTS technique without the fabrication of a sample device.

Regarding the signal analysis of the transient spectroscopy, the principle of COLTS is the same as that of ICTS (Isothermal Capacitance Transient Spectroscopy) (5), the principle of which is the same as that of DLTS (Deep Level Transient Spectroscopy) (6) which has been well known in the semiconductor world.

In this paper we report preliminary results of a newly developed COLTS system and a first-ever observation of a signal of contactless transient spectroscopy using an MIOS structure without fabricating a sample device.

PRINCIPLE

Comparison of Techniques among DLTS, ICTS and COLTS

Regarding transient spectroscopy, the most familiar technique is DLTS which has been prevailing in the semiconductor world. Following DLTS, ICTS was developed by Okushi and Tokumaru (5), the principle of which is the same as that of DLTS.

CP449, *Characterization and Metrology for ULSI Technology: 1998 International Conference*
edited by D. G. Seiler, A. C. Diebold, W. M. Bullis, T. J. Shaffner, R. McDonald, and E. J. Walters

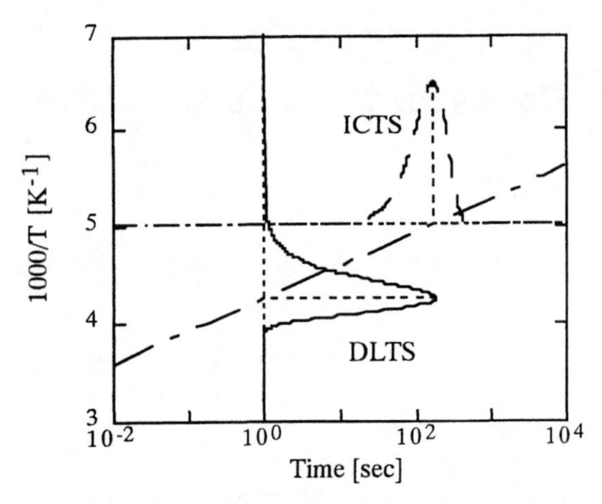

FIGURE 1. Comparison of DLTS and ICTS techniques.

However, there is a difference with respect to signal analysis of transient spectroscopy between DLTS and ICTS. Namely, the emission time from traps is measured in ICTS at a constant temperature while signals of DLTS are obtained from a sample device, the temperature of which is continuously varied from a low to a high temperature as shown in Fig. 1 for comparison.

A newly developed COLTS (COntactLess Transient Spectroscopy) operates according to the principle of ICTS. However, mere semiconductor wafers are used for the measurements instead of fabricating a sample device such as a Schottky diode or an MOS diode, the use of which is essential in the case of DLTS and ICTS measurements.

Although the principle of transient spectroscopy is the same in both DLTS and ICTS, the COLTS technique is only possible when the procedure of ICTS is used. This is because the isothermal measurements are essential for COLTS; otherwise the capacitance of an MIOS structure varies depending on the change of temperature during the measurements even when there are no localized states in the semiconductor wafers.

This is because the piezo actuator changes at a rate of 0.9 μm/K owing to thermal expansion. This value is huge in comparison with the detection sensitivity of COLTS as described later.

Tool for Making MIOS Structure

MIOS structure used for COLTS measurements is shown in Fig. 2(a). As is shown in Fig. 2(a), the maintenance of both the distance and parallelism between the upper electrode and the surface of a sample wafer is important for the capacitance measurement. The maintenance is achieved using both a micrometer and a piezo-electric actuator as shown in Fig. 2(a).

The upper electrode is composed of three parallelism electrodes besides one reference electrode as shown in Fig. 2(b). Each electrode area is about 6.3 mm^2. These parallelism electrodes are used for controlling the parallelism of a sample wafer during the measurements. The detailed structure and operation of the upper electrode will be reported in a separate paper. Both the parallelism and the distance are controlled with the use of capacitance measurements.

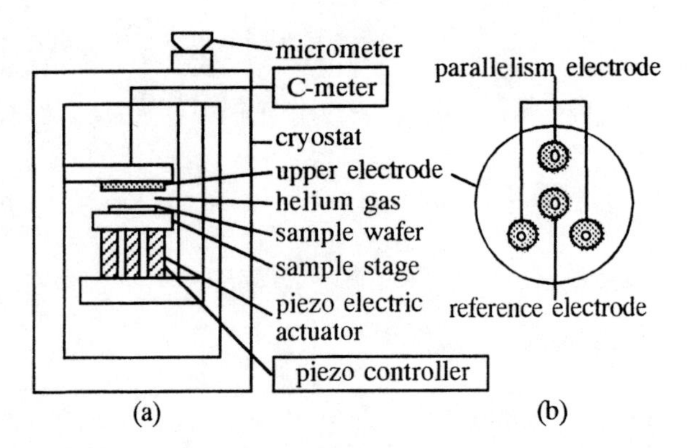

FIGURE 2. Schematic diagram of MIOS structure (a) and upper electrode (b).

For cooling a sample wafer liquid nitrogen was used. At the same time the atmosphere around the sample is filled with thermally conductive helium gas as shown in Fig. 2(a).

NEEDS AND RANGE OF APPLICATION

In the ULSI age using advanced scaled-down MOS devices, the intrinsic characteristic of the surface and interface of the semiconductor must be fully understood in order to make the ultimate use of semiconductor materials. In this case, it is electrical and not structural properties that effect great significance to the device properties.

In addition, bulk traps induced by a low density of heavy metals are also serious. Therefore, in-line monitoring of electrically active impurities is necessary during the wafer fabrication processes.

Moreover, in the scaled-down MOS devices in which a very thin gate oxide of less than 4 nm in thickness is used, it is thought that the interface characteristic becomes fragile as compared before. Namely, the increase of interface trap density might occur during the fabrication processes or device operation. Hence, contactless measurements of semiconductor surface and interface properties are urgently required to characterize material properties precisely.

Summarized range of application will be given in an itemized form as follows.

1. Study of the transition mechanism from surface states to interface traps, the knowledge of which is very

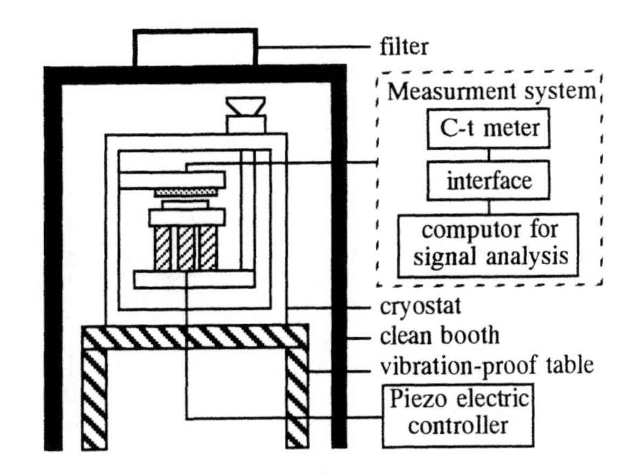

FIGURE 3. Schematic diagram of apparatus for COLTS equipment.

important to make ultimate use of the surface and interface of semiconductors.

2. Precise electrical characterization of the interface of silicon and SiO_2 in the case of a thin oxide film of less than one nanometer in thickness.
3. Selection of best passivation films, which cover semiconductor surface, from a view point of surface electrical charges.
4. Wafer inspection of as-grown and as-received state regarding localized states and carrier density.
5. In-line monitoring of electrically active impurities and interface traps during the wafer fabrication of LSI processes.

APPARATUS

The diagram of COLTS equipment is shown in Fig. 3. As shown in Fig. 3, a cryostat and a clean booth must be provided with COLTS. This is because a low temperature atmosphere is essential for the COLTS measurements just as it is for DLTS and ICTS measurements. The clean booth is also requisite because the dusts with electric charges disturb MOS C-V and C-t curves because the charged dusts change their position between the upper electrode and the surface of a sample wafer, depending on the polarity of the bias voltage applied to the upper electrode as shown in Fig. 4. The resultant C-t curve gives us misleading information of localized states as is discussed

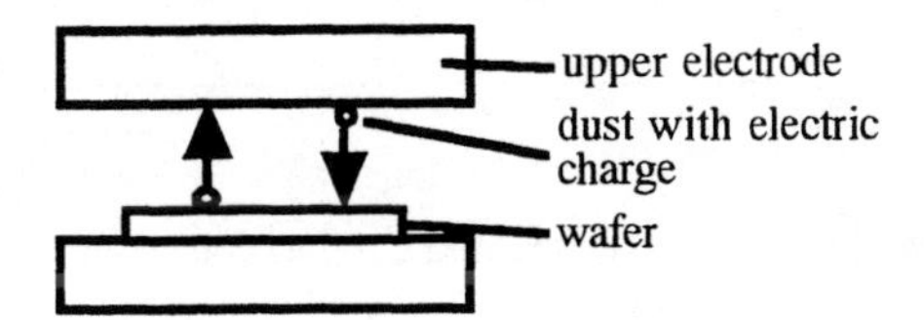

FIGURE 4. Schematic explanation for movement of dusts with electric charges.

later. The system for the signal analysis of COLTS is the same as that of ICTS. It is composed of a C-t meter, a pulse generator, a temperature controller and a personal computer as shown in Fig. 3. A piezo-electric controller and a capacitance meter for the parallelism and the air gap control must be added in the COLTS measurement compared with the conventional ICTS system.

DETECTION SENSITIVITY

The value of capacitance in the case of COLTS, C(colts), is given by

$$C(colts) = \left(\frac{d_s}{\varepsilon_s} + \frac{d_{air}}{\varepsilon_{air}} + \frac{d_{ox}}{\varepsilon_{ox}} \right)^{-1} \cdot A \quad (1)$$

where d_s, d_{air} and d_{ox} are the spacings of depletion layer of silicon, air gap and oxide thickness, respectively. And ε_s, ε_{air} and ε_{ox} are the dielectric constants of silicon, air and oxide, respectively. Detailed derivation of the equation (1) will be reported in a separate paper. Now, if Δd_{air} is induced as a variation of d_{air}, C(colts) is changed to C'(colts) where

$$C'(colts) = \left(\frac{d_s}{\varepsilon_s} + \frac{d_{air} + \Delta d_{air}}{\varepsilon_{air}} + \frac{d_{ox}}{\varepsilon_{ox}} \right)^{-1} \cdot A. \quad (2)$$

Therefore, delta C(colts) is obtainable as follows,

$$\Delta C(colts) = C(colts) - C'(colts). \quad (3)$$

The change of COLTS's capacitance, ΔC(colts), is shown in Fig. 5 as a function of the variation of air gap for three air gap spacings. The lateral straight broken line shows the minimum tolerable variation (0.05pF) of transient capacitance signal of COLTS. From Fig. 5 it is estimated that the variation of around one nanometer could be tolerable in the measurement. Actual stability of the

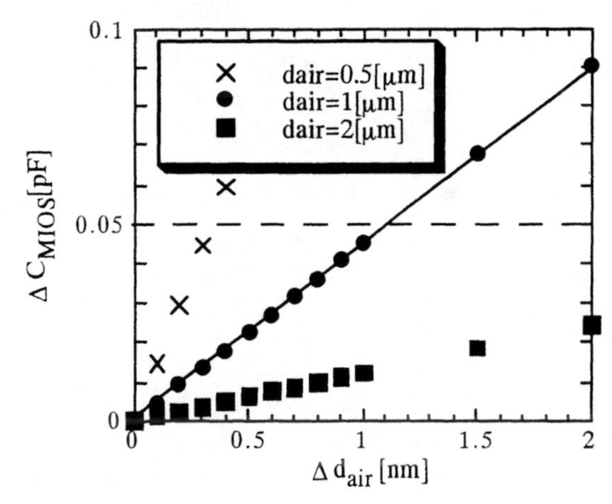

FIGURE 5. Deviation of COLTS capacitance with change in air gap.

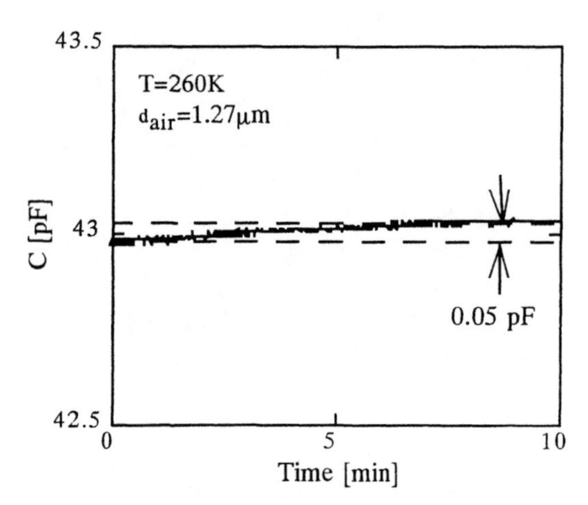

FIGURE 6. Fluctuation of COLTS's capacitance for a duration of 10 minutes.

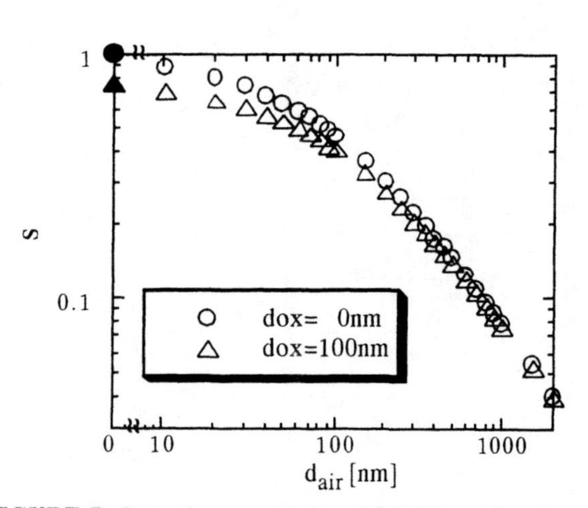

FIGURE 7. Detection sensitivity of COLTS as a function of air gap.

apparatus is shown in Fig. 6, and is seen to be acceptable for practical use.

Next, the detection sensitivity in the case of an MIOS structure was compared with those of a Schottky diode and an MOS diode. In this case the quantity s is used as detection sensitivity. The s value is given by the following equation,

$$s = \left(\frac{\frac{d_s}{\varepsilon_s}}{\frac{d_s}{\varepsilon_s} + \frac{d_{air}}{\varepsilon_{air}} + \frac{d_{ox}}{\varepsilon_{ox}}} \right). \qquad (4)$$

The detailed derivation and discussion of the s value will be reported separately. At any rate, s value is shown in Fig. 7 as a function of air gap. It is to be noted that s values of a Schottky diode and an MOS diode are marked on the longitudinal axis with solid circle and triangle, respectively. Reasonably, the detection sensitivity is decreased as the air gap is increased.

Figure 8 shows minimum measurable D_{it} values as a function of air gap where D_{it} is the interface trap density. In the present experiment, in which the air gap is from 500 nm to 1 μm, the detection sensitivity is around 10^{13}cm^{-3} and 10^{10}cm^{-2}eV^{-1} for bulk traps and interface traps, respectively, although the experimental data are not shown here.

EXPERIMENTAL

As a starting material a silicon wafer doped with phosphorous of 10 ohm cm was used. The sample wafer was additionally doped with gold impurity. The additional doping procedure is as follows (7). A silicon wafer was

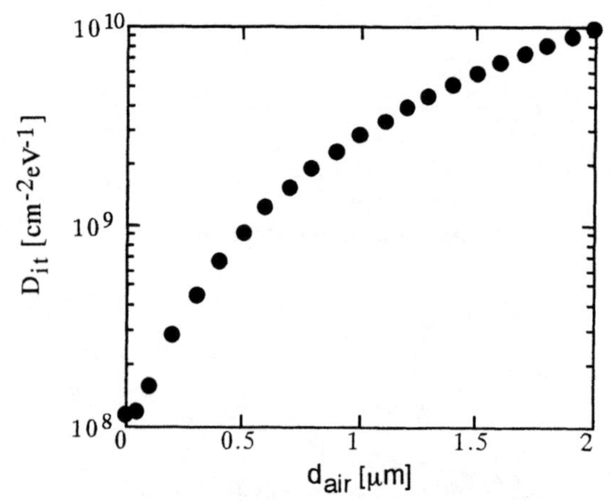

FIGURE 8. Detection sensitivity of interface trap density D_{it} as a function of air gap where depletion width is 1 μm.

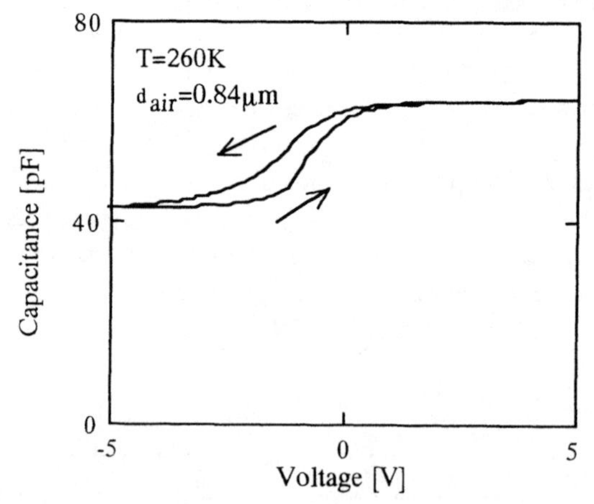

FIGURE 9. C-V curve obtained by COLTS technique using a gold-doped oxide Si wafer with dusts of electric charges.

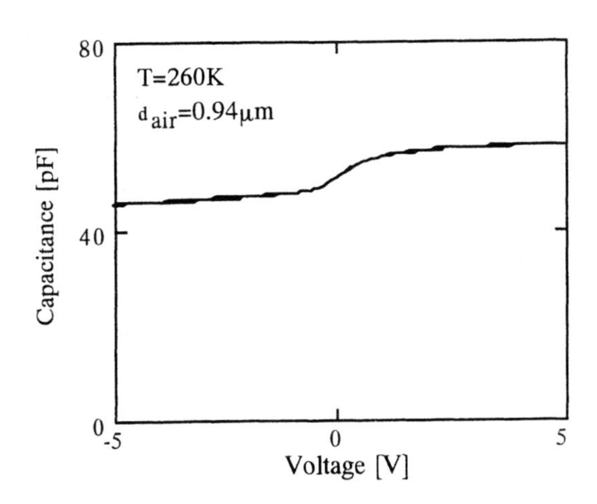

FIGURE 10. C-V curve of COLTS in clean atmosphere using the same wafer as that of Fig. 9.

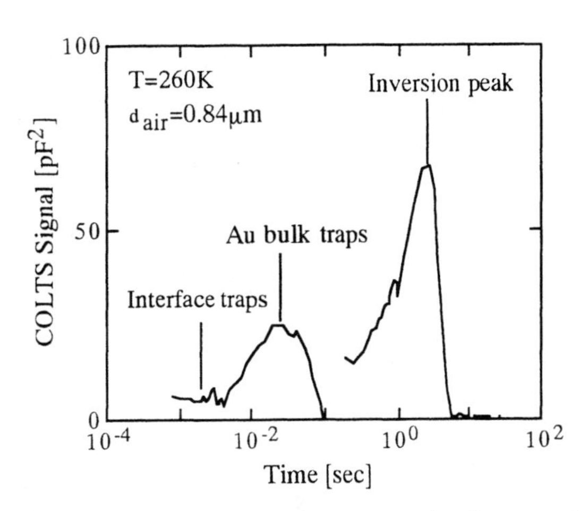

FIGURE 11. COLTS signal of interface and bulk traps of a gold-doped oxide Si wafer.

first oxidized and an oxide film of 100 nm in thickness was grown. After removing the oxide film on the rear surface, gold was diffused from that surface, followed by polishing and chemical etching.

A representative C-V curve is shown in Fig. 9 in which above mentioned voltage hysteresis is seen. The hysteresis is due to the dusts with electric charges (8). This was verified by the remeasurement in the clean atmosphere as shown in Fig. 10. Similarly, we obtained a C-t curve from the sample wafer at 260 K. Using the C-t curve we were able to obtain COLTS signals (the same as that of ICTS) as shown in Fig. 11 in which both interface traps and bulk traps due to gold impurity (9) are seen as is explained in the figure.

CONCLUSIONS

1. We have succeeded in obtaining first-ever signals of transient spectroscopy from an MIOS structure using a gold-doped silicon wafer without fabricating sample devices such as MOS or Schottky diodes.
2. The MIOS structure is formed by controlling an air (helium gas) gap using a piezo-electric actuator in which the gap distance is below one micrometer. The control of the gap is achieved by capacitance measurement using three parallelism electrodes which surround the reference electrode for the COLTS measurements.
3. Sensitivity of COLTS measurements is acceptable for the practical use of semiconductor wafers used in ULSI technology.
4. It is plausible that various promising versions of COLTS will be developed in the near future. One might be a scanning version of COLTS for the distribution measurements of traps and carrier density. In addition, an optical version of COLTS will be usable for the measurements of localized states in wide band-gap semiconductors.

ACKNOWLEDGMENTS

The authors are grateful to K. Kobayashi, M. Yoshida and R. Nakanishi for technical support, to Dr. H. Yamamoto of ULSI Laboratory of Mitsubishi Corp. for providing oxide silicon wafers and M. Adachi of Iwatani Industrial Gases Corp. for the design of a cryostat. This work has been partly supported by Hyogo Science and Technology Association and partly supported by the Japan Society for Promotion of Science.

REFERENCES

1. Shimatani, T., Pidin, S., and Koyanagi, M., Jpn. J. Appl. Phys. 36, pp.1659-1662(1997).
2. Sze, S. M., Physics of Semiconductor Devices, New York:John Wiley & Sons, 1981, ch.8, pp.467-486.
3. Ogawa, S., Kobayashi, T., Nakayama, S., and Sakakibara, Y., Jpn. J. Appl. Phys. 36, pp.1398-1406(1997).
4. Sakai, T., Kohno, M., Hirae, S., Nakatani, I., and Kusuda, T., Jpn. J. Appl. Phys. 32, pp.4005-4011(1993).
5. Okushi, H., and Tokumaru, Y., Jpn. J. Appl. Phys 28, L335-L338(1980).
6. Lang, D. V., J. Appl. Phys. 45, pp.3023-3032(1974).
7. Yoshida, H., Niu, H., and Kishino, S., J. Appl. Phys. 73, pp.4457-4461(1993).
8. Nojiri, K.,(Semiconductor & Integrated Circuits Div., Hitachi LTd.); private communication.
9. Yoshida,H., Niu, H., Matsuda, T., and Kishino, S., Semiconductor Characterization, Present Status and Future Needs, New York:AIP Press, 1996, pp.237-240.

Tunneling Spectroscopy of the Silicon Metal-Oxide-Semiconductor System

Whye-Kei Lye, Tso-Ping Ma and Richard C. Barker

Center for Microelectronic Materials and Structures, Department of Electrical Engineering, Yale University, New Haven, CT 06520-8284

Eiji Hasegawa

NEC Corporation, ULSI Device Development Labs, Crystal Technology Development Lab, 1120 Shomokuzawa, Sagamihara, Kanagawa 229-11, Japan

Yin Hu, John Kuehne and David Frystak

Semiconductor Process and Device Center, Texas Instruments, Dallas, TX 75265

In this work we demonstrate the application of tunneling spectroscopy to the silicon Metal-Oxide-Semiconductor system. As an electrical characterization method, this technique allows for the direct study of the structure of ultra-thin gate oxides in standard as-fabricated test structures, and their dependence upon processing and electrical stress.

INTRODUCTION

The properties of thin insulators and their interfaces between semiconductors and conducting electrodes are critical to the semiconductor microelectronics industry, and have been under intensive study for at least three decades. For the most part, the presence of charge trapped within the insulator and at its interface with the semiconductor, and the changes that take place with materials and process technology, with the device itself as the preeminent instrument, has answered the 'what' questions - the 'how' questions are more difficult to deduce from this evidence. Analysis methods such as infrared, Raman, and XPS spectroscopy have been helpful in understanding some of the structural properties of these insulators, although they are not compatible with as-fabricated test structures.

Inelastic Electron Tunneling Spectroscopy (IETS), on the other hand, was demonstrated as an in-situ, non-destructive spectroscopic technique in the 1970s for thermally grown oxides on silicon in a Metal-Oxide-Semiconductor (MOS) configuration. Even though credible MOS transistors were demonstrated with oxide thicknesses of between 25 and 40 Å, little interest was shown because few could imagine that oxides of thickness less than 100 Å would ever be of practical interest, and IETS requires insulators in the order of 20 Å. Designs today are approaching this thickness range, motivating this work. The results demonstrate the capabilities of IETS in determining changes in the insulator structure with process parameters and with electric-field and current-fluence stress.

TUNNELING SPECTROSCOPY

Direct electron tunneling takes place through a thin insulating barrier between two conductors when the properties of the insulating material support an evanescent electron wave in one conductor that reaches through the insulator to a conductor on the other side and finds an empty state at the same energy and the same wave vector **k** normal to the tunneling direction on the other side (1).

The total electron current at a given voltage separating the two Fermi energies is obtained by integrating over all the occupied electron states on the transmitting side connecting to a matching unoccupied state on the receiving side, weighted by the corresponding transition rate, i.e. tunneling probability. The models and calulations for this band-to-band tunneling current are only tangentially important to the subject of this paper. Suffice it to say that the current increases approximately exponentially with applied voltage, which is nearly linear for 'small' voltages.

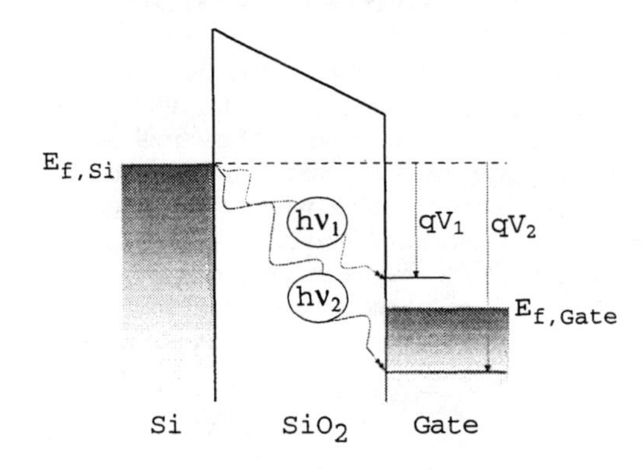

FIGURE 1. Principle of operation of IETS. Onset of inelastic tunneling conduction occurs at voltage biases $V_{bias}=h\nu/q$, where ν is the characteristic frequency of the vibrational excitation.

Based on the assumption of statistically independent processes, one models other channels of conductance as

CP449, *Characterization and Metrology for ULSI Technology: 1998 International Conference*
edited by D. G. Seiler, A. C. Diebold, W. M. Bullis, T. J. Shaffner, R. McDonald, and E. J. Walters

being simply additive. Thus, if an electron at the Fermi level on the transmitting side loses energy to a phonon with energy hν, the energy of the electron arriving at the receiver electrode will be at energy E_F-hν. Hence as the voltage applied to the receiver electrode passes the level hν, a tunneling channel is opened, discontinuously if T=0. The process is illustrated in Fig. 1. This conductance discontinuity, ideally, produces a sharp, nearly infinite peak in the second derivative of the current with respect to the applied voltage, which is drastically broadened by temperature, requiring measurements at liquid helium temperatures. Each time the applied voltage reaches one of these energy-loss values, a sharp peak is encountered. This method of measuring vibrational theshold energies is the underlying operating principle of IETS (2,3).

A special advantage of IETS is the use of tunneling spectra taken with opposite polarities of applied voltage. Parallel tunneling channels amounts to independent transition rates modeled as independent matrix elements between the two evanescent waves attached to the transmitter and receiver sides. A localized perturbation operator (dipole due to a vibrational degree of freedom) representing the energy absorption process corresponding to a spectral feature, enhances the transition rate due to that process if it is located closer to the receiver electrode where it couples more with the dipole images in and evanescent wave attached to the receiver. Thus, by comparing a given feature with spectra taken with both polarities, one should be able to localize a mode within the tunneling barrier.

To reveal these resonant peaks in the desirable spectral form, sophisticated, low-noise instrumentation is required, in which the direct band to band tunneling and other broad background tunneling is removed, and computer techniques must be employed to resolve and identify the spectral modes. A summary of the methodology can be found in a recent publication. (4)

EXPERIMENTAL METHOD

Experimentally the second derivative is measured using a modulation technique. Here, the voltage applied across the tunnel junction consists of a small AC signal (V_m) superimposed onto a DC bias (V_b). The current flow through the device is measured, with the signal component at the second harmonic of the AC modulation being of primary interest. The Taylor expansion in Eq. 1, shows that the magnitude of the second harmonic provides a direct measure of the differential conductance, ΔG.

$$I(V_b + V_m \cos\omega t) = I(V_b) + G_{V_b} V_m \cos\omega t + \frac{1}{4}\Delta G_{V_b} V_m^2 (1+\cos 2\omega t)\cdots \quad (1)$$

Two factors affect the resolution of tunnel spectra: the temperature at which the sample is measured and the modulation voltage used. These affect the measured peak width according to the relation

$$\Delta V_{fwhm}^2 = \left(5.4\frac{kT}{q}\right)^2 + \left(1.7V_m\right)^2. \quad (2)$$

Reducing modulation voltage and temperature can increase resolution, but in order to maintain the same signal-to-noise ratio, a decrease in modulation voltage by a factor 1/r must be accompanied by a r^4 increase in integration time.

Instrumentation and Electronics

A schematic of the instrumentation electronics is shown in Fig. 2. Passive components used are 1 %, 1/4 W metal film resistors, 50 V polypropylene signal capacitors and 16 V tantalum power supply bypass capacitors. The circuit is point-to-point soldered on an perforated epoxy-glass board, using 24 gauge single core copper wire. The ± 9 V dual rail power supply is provided by lithium batteries.

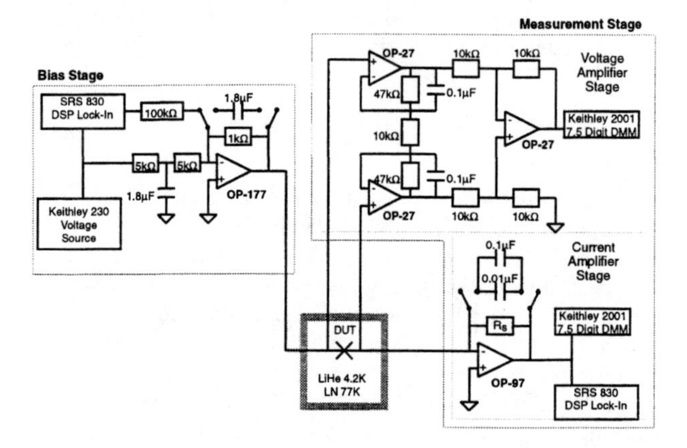

FIGURE 2. Circuit schematic of IETS electronics. Offset adjust potentiometers and power supply omitted for clarity. Switch positions are shown for measurement in AC mode.

A Stanford Research SR-830 DSP lock-in provides both the DC bias and AC modulation signals, which are combined in a scaling inverting adder bias stage. The output of this voltage bias stage is applied across the device under test (DUT) in a four terminal configuration. The voltage across the device is monitored via an instrumentation amplifier by a Keithley 2001 7-1/2 digit DMM. The current through the junction flows into a transresistance amplifier whose voltage output is applied to the SR-830 for synchronous detection. All instruments are controlled and data acquired by via a GPIB bus. The data presented was acquired at both 4.2 K and 77 K, with an integration time of 1 sec., a voltage step rate of 200 μV/sec., and a modulation voltage of 2 mV.

In addition to operating in the AC measurement mode just described, the circuit can also function in a DC measurement mode. In this instance, a high precision IV is taken using a Keithley 230 voltage source to provide the DC bias. The second derivative is then obtained by mathematical differentiation of the data. This mode of measurement has a lower signal-to-noise ratio compared to the lock-in technique, and is useful only for system checks.

Test Devices

The MOS devices studied were fabricated on 1-10 Ωcm *N*-type, (100) silicon. A high dose implantation of phosphorus, 5×10^{15} ions/cm^2 at 80 keV, was used to form the degenerate layer. After definition of the 0.01 cm^2 active device area by local oxidation of silicon (LOCOS), the 1.5 nm tunnel oxide was formed by rapid thermal oxidation (RTO), followed immediately by deposition of the 250 nm *in-situ* doped (ISD) *N*-type polycrystalline silicon gate. Following patterning and a dopant activation anneal at 800 °C for 20 min, 600 nm of tetra-ethyl ortho-silicate (TEOS) was deposited for device passivation. Contact openings were then lithographically defined, followed by metallization with 50 nm TiN/400 nm AlCu(0.5%)/50 nm TiN. A schematic cross-section of a completed device is shown in Fig. 3.

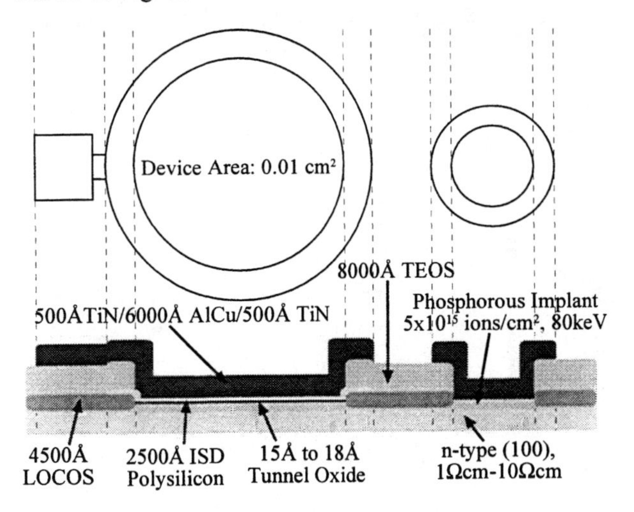

FIGURE 3. Cross-section of tunnel junctions studied.

The samples are mounted onto a 16 pin dual in-line packages (DIP) and placed in a socket at the end of a cryogenic probe. During measurement the devices are immersed in either a reservoir of liquid helium (4.2 K) or liquid nitrogen (77 K), as required.

Data Analysis

In applying IETS to the silicon MOS system, the contributions to the second derivative by the band structure of silicon have to be accounted for. This has previously been cited (5) as a major impediment to the application of this technique to this material system. Previous work (4), reported that by expressing experimental data as $G=G_e+G_i$ and $\Delta G=\Delta G_e+\Delta G_i$, where the subscripts e and i indicate elastic and inelastic tunneling contributions respectively, it can be shown that

$$\Delta G_i\big|_{4.2K} = \frac{(\Delta G_{4.2K}-\varepsilon)G_{77K}-\Delta G_{77K}G_{4.2K}}{G_{77K}-\gamma G_{4.2K}}. \qquad (3)$$

where γ is a dimensionless constant equal to 0.0545 (=4.2K/77K), and ε is an error term. This normalization technique is utilized on all data presented.

RESULTS

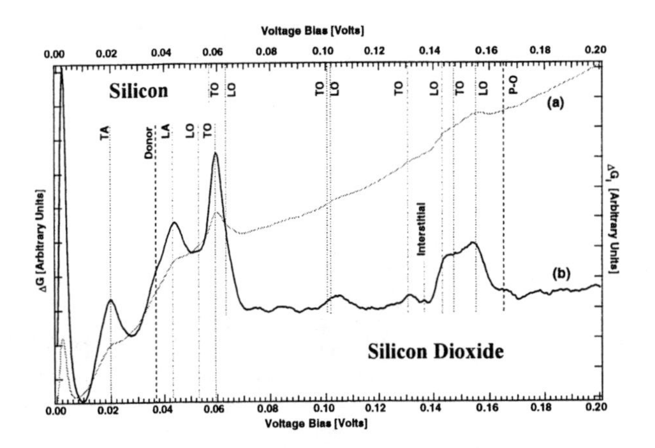

FIGURE 4. Tunnel spectra of an MOS junction, before (a) and after (b) elastic background subtraction using Eq. 3. The two curves shown are not on the same scale.

A representative forward bias (positive on gate) spectrum, is shown in Fig. 4. Superposed is a processed spectrum obtained by the background subtraction method discussed in reference (4). This process enhances features in the data between 0 and 200 mV (1 mV ≈ 8.065 cm^{-1}) which, over a series of runs on this and similar samples, can be unambiguously distinguished from noise. With the assistance of IR and neutron spectroscopy, we know the number, type and approximate location of phonon modes in both the silicon (6) and the oxide (7,8), but subsequent analysis offers some surprises. The locations of modes ultimately found, are indicated in Fig. 4 to summarize the results of analysis on one comprehensive figure. In addition to the four expected modes in the silicon range between 0 and 65 meV, we find evidence of one oxide mode and what appears to be a donor ionization mode. In the range above the silicon phonons, where we expect to find most of the oxide modes, we find six of the expected eight oxide modes, a P-O (9) mode, and a very small mode that corresponds to oxygen in the silicon substrate (8).

Deconvolution of the spectra by least-squares Gaussian fits (Lorentzians do not converge), utilizing commercial data analysis software, have been made over limited ranges of expected activity. The analysis for silicon phonons is limited to the range 35 to 75 meV, since the 20 meV mode stands alone and clear. For SiO_2, the range of 125 to 170 meV is studied.

Below 75 meV, four of the constituent mode positions coincide to the four branches of the Si phonon spectrum (Fig. 5) corresponding to the conduction band minimum in the [100] direction, located at 0.85 of a wave vector to the the Brillouin Zone boundary - the largest modes corresponding to intersections closest to the phonon critical points.

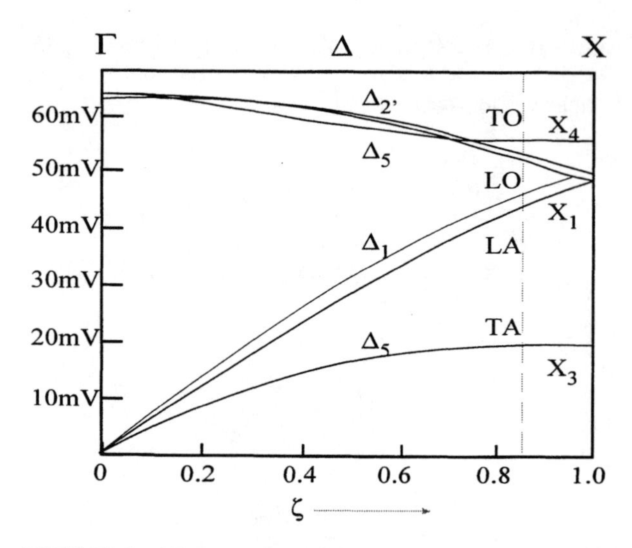

FIGURE 5. Silicon phonon dispersion curves in the [100] direction. The dashed line corresponds to the location of the conduction band minimum. After Asche (6).

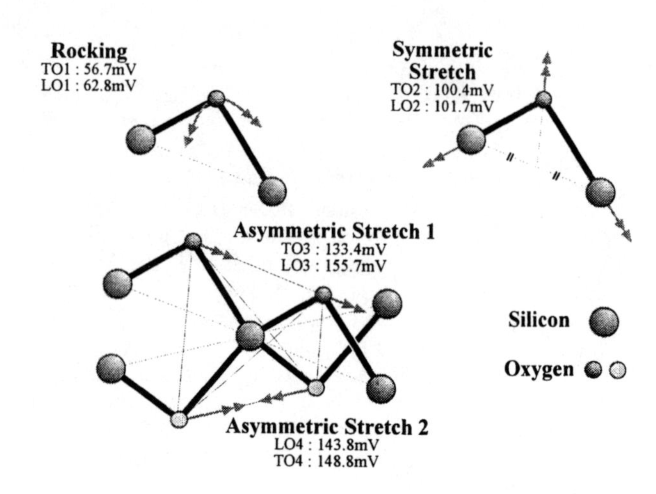

FIGURE 7. Schematic representation of oxide vibrational modes. After Kirk (7).

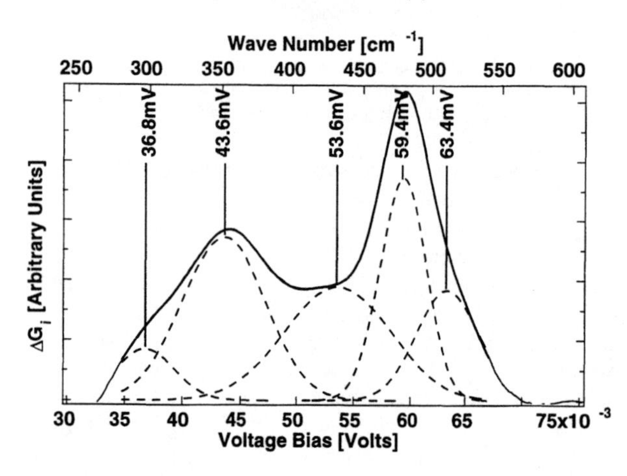

FIGURE 6. Deconvolution of silicon phonon modes.

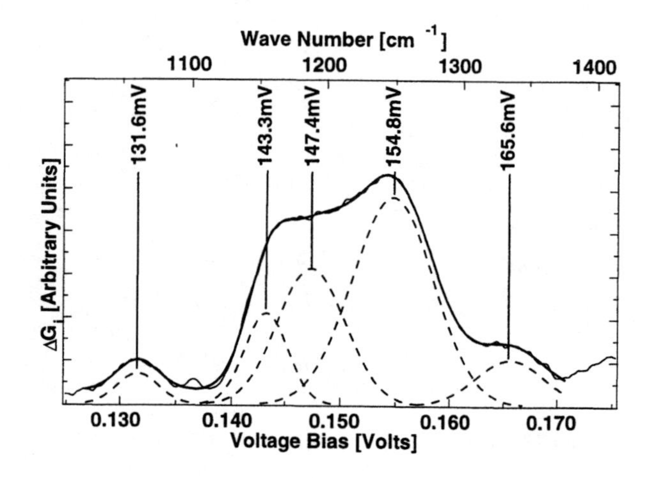

FIGURE 8. Deconvolution of silicon dioxide modes.

Deconvolving the feature in the range of 30 to 70 meV, Fig. 6, reveals five overlapping modes. They are identified as follows (6,7,8): a strong mode at about 19 meV, corresponding to the high energy-density of Δ_5 TA phonon modes (separate from the deconvolved states of Fig. 6); a mode that corresponds to the phosphorous donor ionization energy at 36.8 meV (10); a $\Sigma_{1'}$ or Δ_1 LA mode at 43.6 meV; a $\Delta_{2'}$ LO mode at 53.6 meV; a $\Sigma_{1'}$ or $\Sigma_{2'}$ TO mode at 59.4 meV; and a mode that corresponds to a LO rocking mode in SiO_2 at 63.4 meV. A similar SiO_2 TO mode at 56.7 meV, expected to be weak, does not seem to appear. The solid curve in this figure is an actual superposition of the original data by the sum of the extracted modes shown. There is no perceptable difference.

The 59.4 meV TO mode is about 2 meV higher than values found from IR data (6) and the early work by neutron scattering (11). However refinements of the early work on tunnel diodes (12) and many years of experimental study based on degenerate n-type silicon (13) suggest that this is the most accurate energy assignment for this mode.

The broad peak between 100 and 110 meV in Fig. 4, is consistent with IR active symmetric stretch TO2 and LO2 modes of SiO_2. The configurations of these and other oxide vibrational modes are illustrated Fig. 7 Deconvolved modes in the range between 125 meV and 170 meV, shown in Fig. 8, are consistent with the following SiO_2 asymmetric stretch vibrational modes: a TO3 mode at 131.6 meV, an LO4 mode at 143.3 meV, a TO4 mode at 147.4 meV, and an LO3 mode at 154.8 meV. The origin of the feature at 165.6 meV is consistent with a P-O vibrational mode resulting from phosphorus in the oxide (9). Again, the deconvolution best-fit is a remarkable fit to the corrected experimental spectrum.

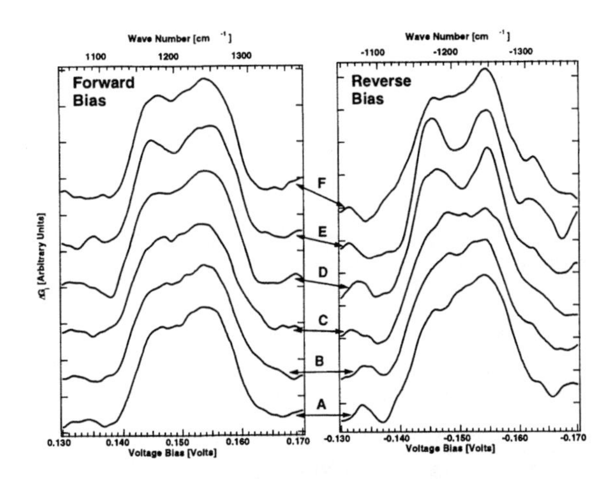

FIGURE 9. Modification of oxide vibrational mode structure by electrical stress. The curves labeled are: A - Initial; B, C and D - After 10^3 sec., 10^4 sec. and 10^5 sec., respectively, of ±0.5 V, 1kHz square wave stress; E - After 10^3 sec. of +2 V, 5μs pulses at 1kHz; F - After 10^3 sec of -2 V, 5μs pulses at 1kHz. The stress and measurements shown are on the same sample with the electrical stress being cumulative.

The large feature in the deconvolution of Fig. 8 is made up from distortions of the tetrahedral structure, the right side being made up of the LO AS1 mode derived from TO3 (131.6 meV), and the left side being made up of an entierly different LO/TO pair called AS2. Evolution of these modes under the effect of electrical stress can followed in Fig. 9. The silicon dioxide mode re-configuration appears preferentially in the reverse bias (negative on gate) spectra, suggesting localization of restructuring at the oxide-gate interface. It remains to be determined how these distortions affect the properties of the oxide in a device structure.

DISCUSSION AND CONCLUSIONS

Despite its technological importance, the structure of ultra-thin gate oxides within the silicon metal-oxide-semiconductor (MOS) system and its dependence upon processing and electrical stress has hitherto been difficult to study. An analysis tool capable of probing standard electrical test structures would be of great utility towards this end.

Issues regarding the application of this technique still need to be resolved. Immutable parameters such as the necessity for cryogenic temperatures have to be balanced against real-world non-degenerately doped substrates. Freeze out must be addressed, tunneling from an inversion layer must be tested. Also the limiting factor in instrumentation is the current through the device (not current density). The test samples used here had very large areas and instrumentation must be improved to apply this technique to smaller gate devices. Even if these problems are not solved, at the very least, this method allows us to study oxide structure and dependencies upon processing and stress - furthering a fundamental understanding of the behaviour of this material system and the associated correltation with device performance and characteristics.

With the advent of sub-20 Å gate transistors (14), the ability to directly probe the structure of ultra-thin oxides in as-fabricated devices, makes IETS a powerful aid for study of the technologically important MOS system.

ACKNOWLEDGEMENTS

The guidance and contribution of P. J. Kindlmann on instrumentation design is gratefully acknowledged. This work was supported by an NEC Graduate Fellowship and by the Semiconductor Research Corporation.

REFERENCES

1. Bohm, D., *Quantum Theory*, New Jersey: Prentice-Hall, 1951
2. Hansma, P. K., *Tunneling Spectroscopy*, New York: Plenum, 1982.
3. Wolf, E. L., *Principles of Electron Tunneling Spectroscopy*, New York: Oxford University Press, 1985.
4. Lye, W., Hasegawa, E., Ma, T. P., Barker, R. C., Hu, Y., Kuehne, J., Frystak, D., *Appl. Phys. Lett.* **71**, 2523-2525 (1997).
5. Salace, G. and Patat, J. M., *Thin Solid Films* **207**, 213-219 (1992).
6. Asche, M., and Sarbei, O. G., *Phys. Stat. Sol. B* **103**, 11-50 (1981).
7. Kirk, C. T., *Phys. Rev. B* **38**, 1255-1273 (1988).
8. Lange, P., *J. Appl. Phys.* **66**, 201-204 (1989)
9. Pliskin, W. A., *Appl. Phys. Lett.* **7**, 158-159 (1965).
10. Fistul, V. I., *Heavily Doped Semiconductors*, New York: Plenum Press, 1969.
11. Brockhouse, B. N., *Phys. Rev. Lett.* **2**, 256-258 (1959).
12. Logan, R. A., Rowell, J. M., and Trumbore, F. A., *Phys. Rev.* **136**. A1751-A1755 (1964).
13. Bencuya, I., Ph.D. thesis, Yale University, 1984.
14. Momose, H. S., Ono, M., Yoshitomi, T., Ohguro, T., Nakamura, S., Saito, M., and Iwai, H., *IEEE Trans. Electron Devices* **43**, 1233-1242 (1996).

Characterization of Ultra-Shallow Junctions with Tapered Groove Profilometry and Other Techniques

S. Prussin and Christiaan A. Bil

Laboratory for Point Defect Engineering, Department of Electrical Engineering
University of California, Los Angeles, California, 90095-1594

Daniel F. Downey and M.L. Meloni

Varian Ion Implant Systems, Gloucester, Massachusetts 09130

Carlton M. Osburn

North Carolina State University, Raleigh, North Carolina 27695

The tapered groove profilometry (TGP) technique was applied to a series of activated ultra-shallow junctions produced by implants of 1×10^{15} B cm^{-2} at 0.25, 0.50 and 1.0 keV, by implants of 1×10^{15} As cm^{-2} at 2.0 keV, and by implants of 1×10^{15} BF_2 cm^{-2} at 2.0, 3.5, 4.25 and 5.0 keV. The values obtained were compared to SIMS and SRP measurements made on the identical wafer. It was found that the TGP measurements tracked the SIMS values but at a lower level. For the B implants this was approximately 200Å, for the As and BF_2 implants it was approximately 100Å. The TGP measurements tracked the SRP values much more closely.

BACKGROUND

From its beginnings in the 1950's, the semiconductor industry recognized the need to precisely measure the depths of the pn junctions which were generated, first in discrete transistors and diodes and then in integrated circuits. It was found that pn junction depths could be easily and accurately measured by the angle lapping technique. This consisted of mounting the impurity diffused silicon surface on a 1 to 5° angle lap and abrading away the surface, thus exposing the junction. Junction delineation was accomplished by treating the Si with concentrated HF containing a small amount of HNO_3. This solution stains the p-type regions darker than the n-type. The use of an intense light improves the contrast. With passing years, as device geometries became ever smaller, the pn junctions were proportionately scaled down, and it became increasingly difficult to obtain an accurate measurement of pn junction depth by angle-lapping and staining. In 1983, Prussin [1] introduced the tapered groove profilometry (TGP) technique for measuring junction depths. The tapered groove exposed the junction in a manner similar to that of the angle-lap method so that the junction could be indicated by a staining technique, but now the junction depth was determined by a direct profilometric measurement. State-of-the-art bipolar junctions with junction depths of 2,500 Å, 4,000 Å and 9,200 Å were measured to a precision of 200 Å.

Since then our junctions have become much shallower but fortuitously a new generation of profilometers has emerged with capabilities of greater precision.

TECHNIQUES

The TGP method was developed with the use of two instruments, then frequently found in semiconductor processing laboratories, a cylindrical wafer groover and a profilometer. Wafer groovers were used to measure junction depths, but, having limitations similar to that of the angle lap method, were abandoned as junctions were significantly reduced below 1 μm depths. Profilometers, on the other hand, have found great use for the study of microelectronic structures and have been greatly improved over the years. The application of a modern profilometer to the measurement of ultra shallow junctions and the validation of these measurements by comparison with high resolution cross-sectional transmission electron microscopy has been described [2].

This method is based on the use of a tapered rather than a uniform groove. This was simplified by the fact that the cylindrical wafer groover used in this study, Model 2015C, Philtec Instrument Co., Philadelphia, PA, is designed with set screws to level the wafer holding chuck.

CP449, *Characterization and Metrology for ULSI Technology: 1998 International Conference*
edited by D. G. Seiler, A. C. Diebold, W. M. Bullis, T. J. Shaffner, R. McDonald, and E. J. Walters

These can be easily adjusted to introduce a slight fixed taper in the grooves produced. For our shallow junction studies, we set our angle at about 0.1-degree. Figure 1 represents the appearance of the tapered groove, illustrating how the pn junction is exposed and measured.

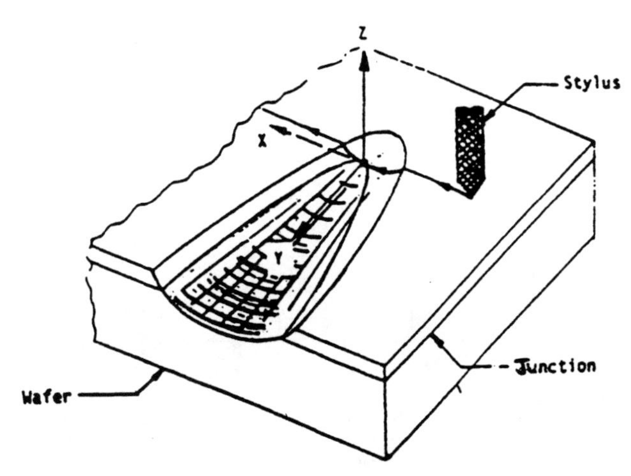

Figure 1: HF Stained TGP Groove for Ultrashallow Junction.

The depth of the groove corresponding to the peak of the contour of the junction corresponds to the junction depth. This can be easily measured by aligning the stylus of the profilometer so as to pass through the peak of the contour of the junction. Since this measurement takes only a minute or so, we normally make a series of 10 successive measurements from which we can extract a mean and a standard deviation.

The profilometer used in this study was an Alpha Stepper 200 Å, which is now obsolete. State-of-the-art profilometers, such as Tencor P-series and HRP-series have features which contribute to improved precision of the measurements. These features are image enhancement, which brings out the junction much more sharply and a cursor, which can be brought to the peak of the junction contour, and which subsequently directs the stylus to pass through this position.

MEASUREMENTS

In this study we evaluated several series of implants which were prepared with the same implant parameters but were subjected to different annealing times and temperatures for activation. Implants were performed on a Varian VIIS ion - 80 PLUS ion implanter. The TGP values were compared with SIMS and SRP measurements made on the same wafers. For the SIMS measurements, made by Evans East, the junction depth was chosen to coincide with the 1×10^{17} cm^{-2} impurity level. The SRP measurements were carried out by Semiconductor Assessment Services, Limited.

Table 1 represents a typical data array for eight 1 keV 1×10^{15} B cm^{-2} implantations subjected to various anneals. The values of the ten successive measurements yield a mean and a standard deviation. This data is plotted in Fig. 2 together with the corresponding SIMS values. These TGP measurements track the SIMS values at a level approximately 200Å lower than the SIMS values. Figure 3 represents a study for 1×10^{15} As cm^{-2} 2keV implantation for a series of anneals. The TGP measurements were carried out by two different individuals and their results compared together with the corresponding SIMS values measured on the same wafers. Figure 4 represents a study of wafers implanted with 1×10^{15} B cm^{-2} with various energies, 0.25, 0.5 and 1.0 keV. Figure 5 represents the results of two studies, 1×10^{15} As cm^{-2} at 2.0 keV and 1×10^{15} BF_2 at 2.2 keV. Figure 6 represents the results of two studies, 1×10^{15} BF_2 cm^{-2} at 3.5 keV and 4.25 keV. Figure 7 represents a study of wafers implanted with 1×10^{15} BF_2 at 5.0 keV. Figure 8 illustrates the relative junction depths for SIMS, SRP and TGP measurements. We note that the TGP values track the SRP values more closely than the SIMS values or than the SIMS and SRP values track each other.

For the As and BF_2, the TGP tracks the SIMS measurement with values approximately 100Å lower.

Table 1: Comparison of TGP data vs. SIMS data (all units are in Angstroms) for 1 keV 1×10^{15} B cm^{-2} implanted specimens at various anneal temperatures.

	PN-Junction data													
	1	2	3	4	5	6	7	8	9	10	TGP-value	Std. Dev	SIMS	SIMS-TGP
O1B0-2	690	675	685	680	685	680	675	685	705	700	686	10	930	244
O1B0-7	735	605	675	625	660	650	665	685	680	680	666	35	870	204
O1B0-11	500	535	495	510	505	515	510	495	525	510	510	13	740	230
O1B0-12	615	600	590	605	585	605	610	595	615	600	602	10	830	228
O1B0-13	595	605	625	610	605	615	620	600	595	625	610	11	800	190
MB-20	505	520	515	530	500	580	565	520	525	580	534	30	740	206
MB-21	680	670	660	685	690	695	660	675	685	660	676	13	780	104
MB-22	655	725	740	650	665	730	725	720	745	660	702	39	860	159

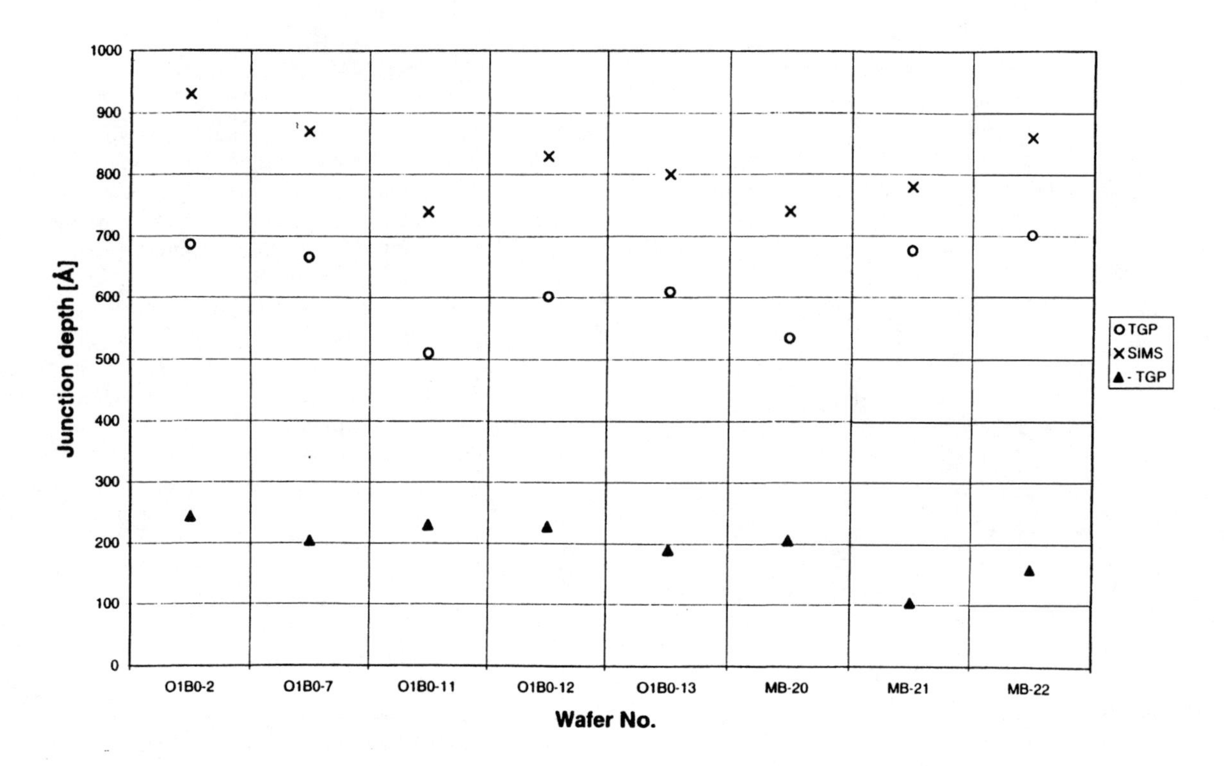

Figure 2: TGP vs. SIMS data for 1×10^{15} B cm^{-2} implanted at 1 keV with various anneal treatments.

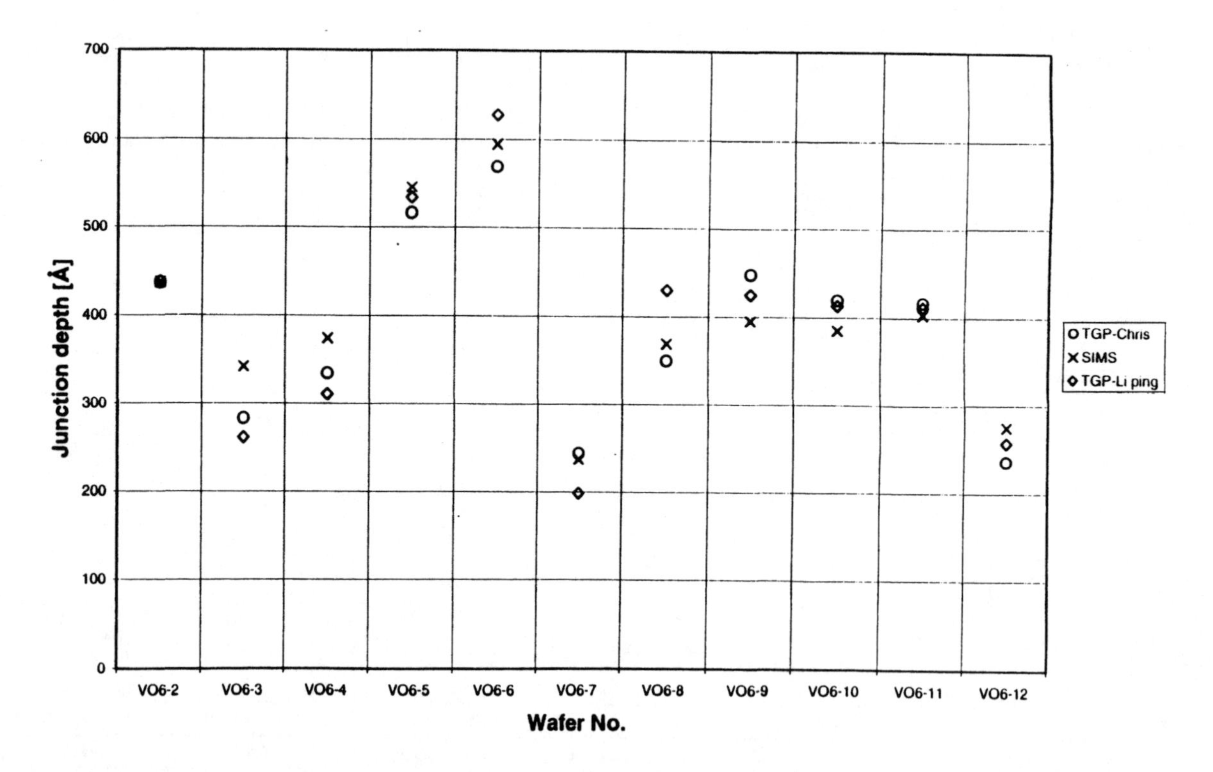

Figure 3: Comparison of TGP data for two individuals vs. SIMS data for 2 keV 1×10^{15} As cm^{-2} implanted specimens at various anneal temperatures.

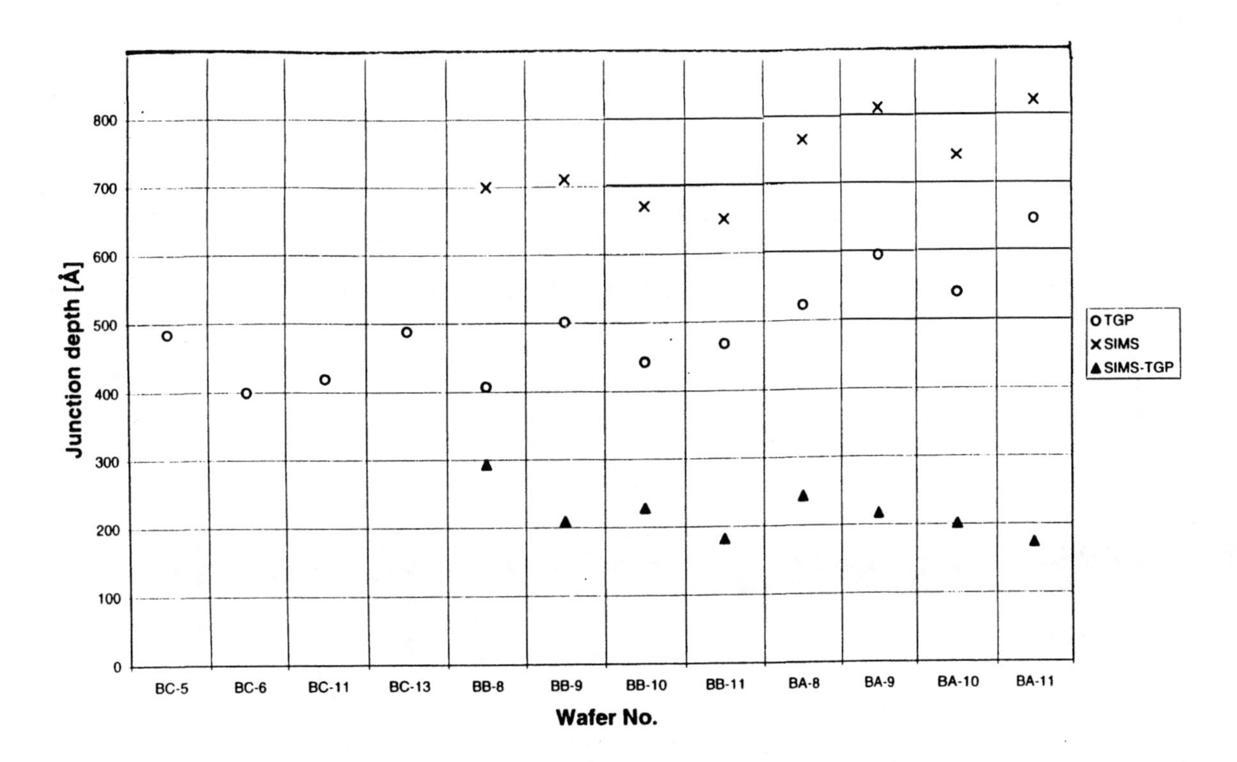

Figure 4: TGP vs. SIMS data for 1×10^{15} B cm^{-2} implanted specimens with various implant energies: BC-5, ..., BC-13, at 0.25 keV; BB-8, ..., BB-11 at 0.5 keV; and BA-8, ..., BA-11 at 1.0 keV.

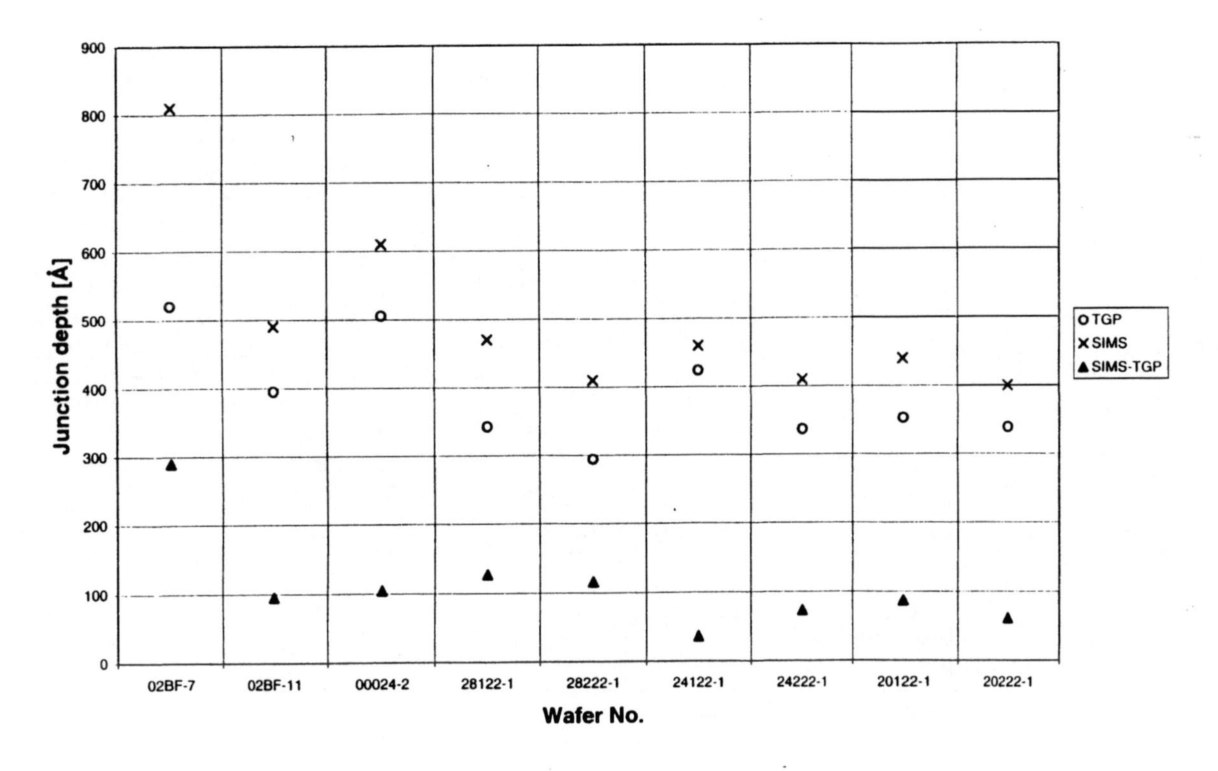

Figure 5: TGP vs. SIMS data for 1×10^{15} As cm^{-2} implanted at 2.0 keV (specimens 02BF-7 and 02BF-11) and 1×10^{15} BF_2 cm^{-2} (specimens 00024-2, ..., 20222-1) at 2.2 keV.

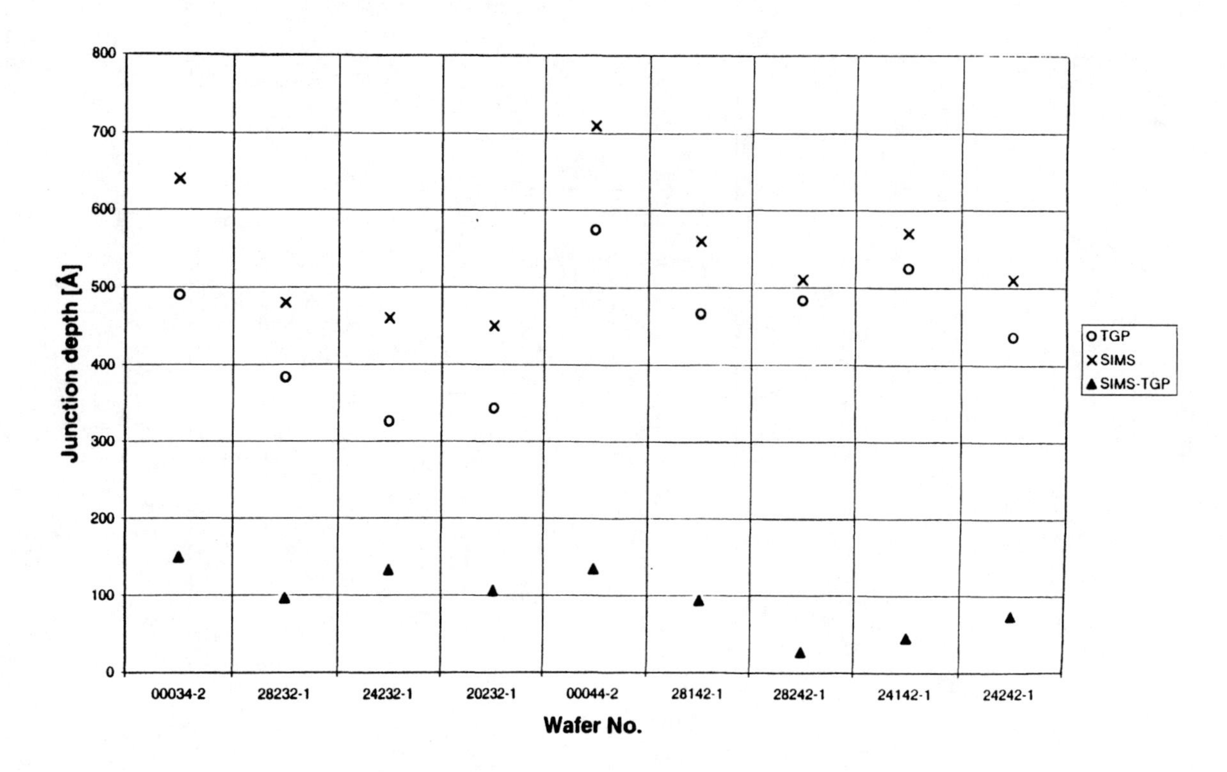

Figure 6. TGP vs. SIMS data for 1 × 10^{15} BF_2 implanted at 3.5 keV (specimens 00034-2, ..., 20232-1) and 1 × 10^{15} BF_2 implanted at 4.25 keV (specimens 00044-2, ..., 24242-1).

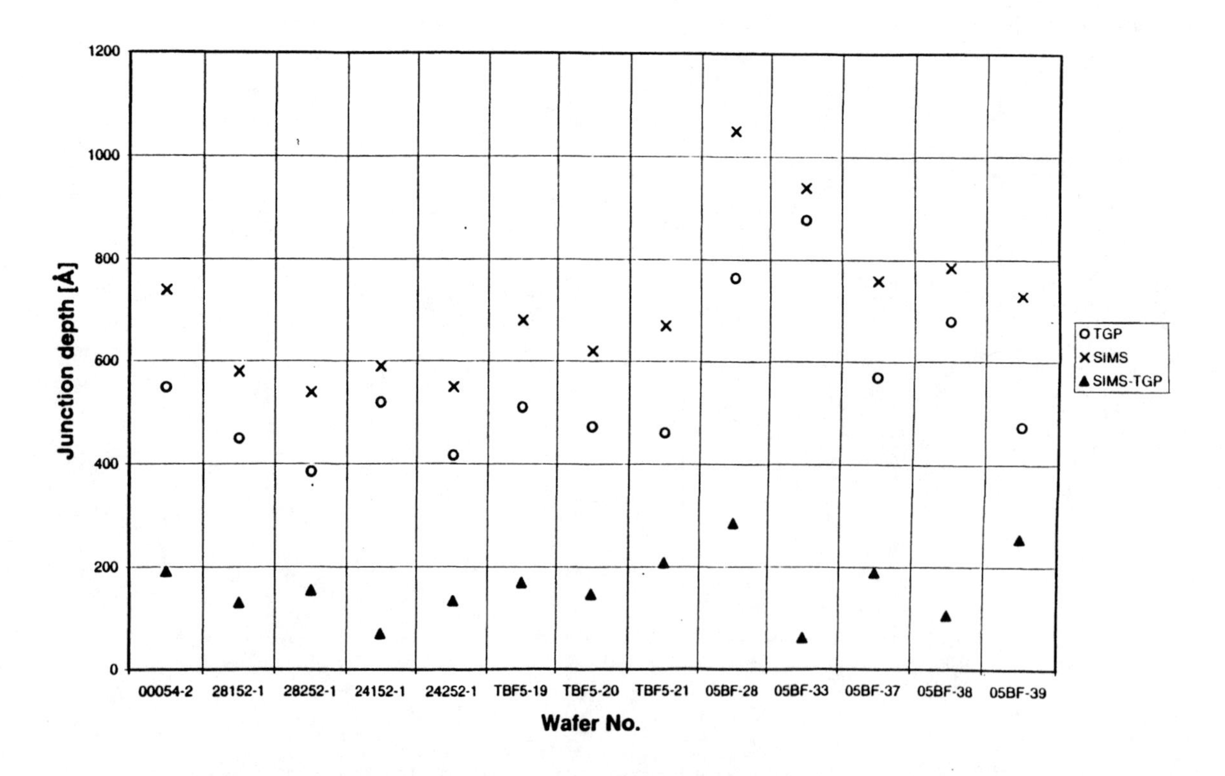

Figure 7: TGP vs. SIMS data for 1 × 10^{15} BF_2 implanted at 5.0 keV.

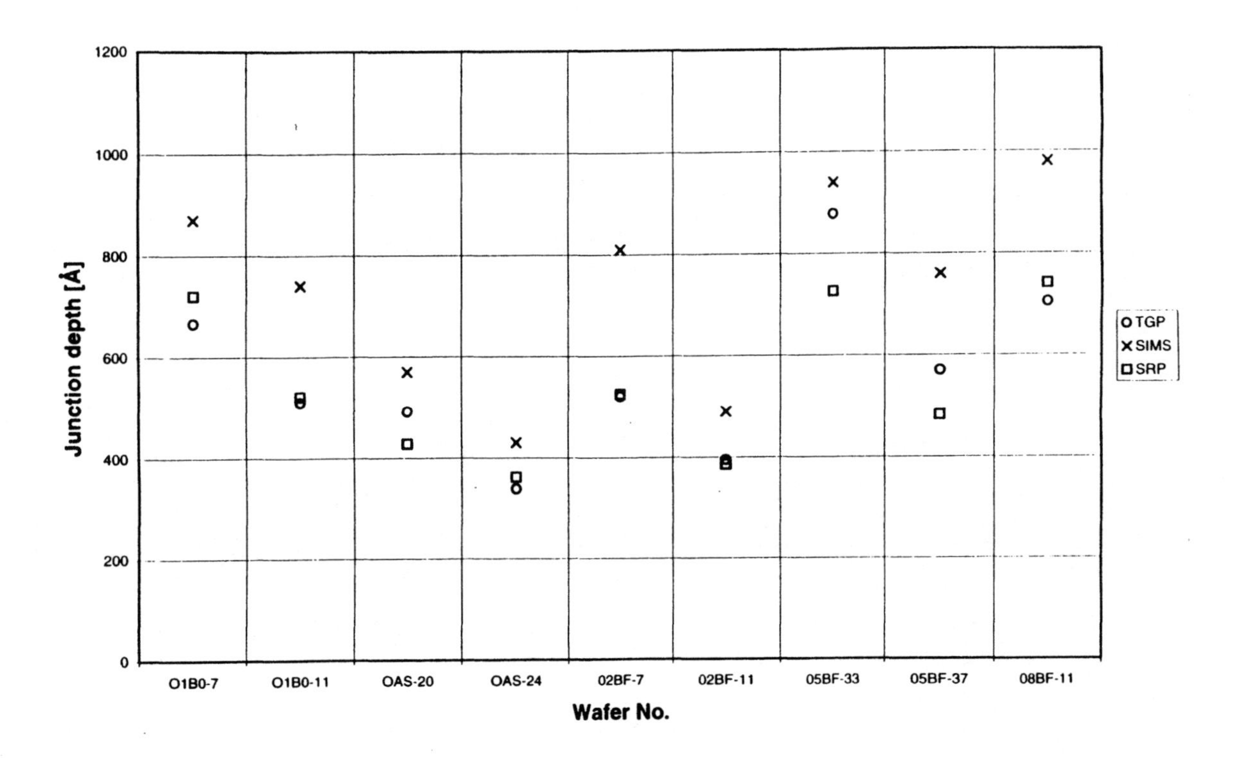

Figure 8: Comparison of TGP, SRP, and SIMS data for different species and implant energies.

DISCUSSION

With requirements of ever-shallower junctions it was found that the SIMS and SRP techniques exhibited greater relative discrepancies. This has been attributed to the role of ion mixing for SIMS which is expected to extend the measured depth of the junction and to carrier spilling which is believed to result in measured junction depths which appear to be too shallow. The TGP technique like SRP responds to carrier concentration and therefore understandably tracks the SRP values closely.

CONCLUSIONS

The TGP technology represents a simple, inexpensive means of evaluating ultra shallow junctions. Comparisons with the established techniques of SRP and SIMS indicate that TPG tracks the SRP measurements closely and tracks the SIMS measurements but at a value lower by 200Å for B and 100Å for As and BF_2 implants.

REFERENCES

1. S. Prussin, Jrl. Electrochem. Soc., 130, 184 (1983).
2. S. Prussin and L.P. Ren, Fourth Int'l. Workshop, on Measurement, Characterization and Modeling of Ultra Shallow Doping Profiles in Semiconductors (1997), Paper 35.

A New Low Thermal Budget Approach to Interface Nitridation for Ultra-Thin Silicon Dioxide Gate Dielectrics by Combined Plasma-Assisted and Rapid Thermal Processing

H. Niimi,[a] H.Y. Yang,[b] and G. Lucovsky [a][b][c]

Departments of Materials Science & Engineering,[a] Electrical & Computer Engineering,[b] and Physics,[c] North Carolina State University, Raleigh, North Carolina 27695, U.S.A.

This paper reports a new low thermal budget approach to interface nitridation for ultra-thin gate dielectrics. Monolayer level nitrogen incorporation at Si-SiO_2 interfaces is accomplished by combining a 300 °C remote plasma-assisted oxidation step that forms ~ 0.5 to 0.6 nm of SiO_2 with a post-oxidation remote plasma assisted nitridation. Incorporation of nitrogen is studied by a combination of secondary ion mass spectrometry (SIMS) and Auger electron spectroscopy (AES), from which it is shown that interfacial nitrogen atom incorporation is linearly proportional to plasma exposure time. This process sequence therefore can control (i) the oxide thickness and (ii) the degree of nitridation for ultra-thin oxides independently. Device results are presented.

INTRODUCTION

Interface nitridation of ultra-thin gate oxides has been shown to yield increased reliability in metal-oxide-semiconductor field effect transistor (MOSFET) devices without sacrificing device performance. Interface nitridation has been accomplished by several techniques (1-10); (i) high-temperature nitridation of thermally-grown oxides in ammonia (NH_4), nitrous oxide (N_2O), or nitric oxide (NO) gasses, (ii) high-temperature direct thermal oxidation/nitridation of silicon (Si) surface in N_2O or NO source gasses, and (iii) low-temperature plasma-assisted oxidation/nitridation using N_2O. For the processes labeled (ii) and (iii), the oxide growth and interfacial nitridation proceed concurrently and as such it may not be possible to control the oxide thickness (t_{ox}) and the degree of nitridation for ultra-thin oxide ($t_{ox} < 3.0$ nm) simultaneously as required for devices in advanced ultra-large scale integrated (ULSI) circuits. This is especially true for high-temperature thermal processing because of potential differences in thermal activation energies for oxide growth and interfacial nitridation.

We have developed a complementary integrated process (8,9,11) that includes remote plasma-assisted interface formation, bulk oxide deposition, and post oxide deposition rapid thermal anneal (RTA) in an ultra-high vacuum (UHV) multi-chamber system: (i) O_2 or N_2O 300 °C plasma oxidation/nitridation of Si surface to form 0.5-0.6 nm of SiO_2 and a device quality Si-SiO_2 interface, (ii) 300 °C bulk SiO_2 deposition from silane (SiH_4) and O_2 or N_2O gasses, and finally (iii) post oxide deposition rapid thermal annealing at 900 °C for chemical and structural relaxations in the bulk dielectric films. Plasma-assisted interface nitridation in N_2O gas has demonstrated improvements in device reliability in MOSFETs, in particular in increased resistance to hot carrier induced defect generation as manifested in decreases in transconductance, and increases in threshold voltage for n-channel MOSFET devices (8).

In this paper, we present an alternative approach to interface nitridation in which oxide thickness and interface nitridation are separately and independently controlled in a low-temperature low-thermal budget two-step interface formation: O_2 plasma oxidation of Si surface followed by N_2 plasma nitridation (12) to incorporate nitrogen at the Si-SiO_2 interface. This process sequence described below is readily extendible to stacked O-N or O-N-O structures which are needed when oxide equivalent thickness of less than 2.0 nm are required to scale with decreases in channel lengths to 100 nm and below. Thus the work reported here has important implications for the future generations of advanced microelectronic devices.

EXPERIMENTS

The processing for incorporation nitrogen at Si-SiO_2 interface is two-step interface formation: (i) low-temperature (300 °C) remote plasma-assisted oxidation of Si surface to form a superficial oxide layer, ~0.5-0.6 nm thick, and (ii) followed by a remotely activated N_2/He plasma nitridation to incorporate nitrogen at the Si-SiO_2 interface. For the oxidation process, the O_2/He discharge used flows of 200 sccm He and 20 sccm O_2, a pressure of 0.3 Torr, and an RF power at 13.56 MHz of 30 W. For the nitridation process, an N_2/He discharge was initiated using flows of 160 sccm He and 60 sccm N_2, a pressure of 0.3 Torr, and an RF power of 30 W. The degree of nitrogen incorporation at the Si-SiO_2 interface was controlled by varying the N_2 plasma exposure

CP449, *Characterization and Metrology for ULSI Technology: 1998 International Conference*
edited by D. G. Seiler, A. C. Diebold, W. M. Bullis, T. J. Shaffner, R. McDonald, and E. J. Walters

time. For the bulk SiO_2 deposition, a remote plasma enhanced chemical vapor deposition (RPECVD) with SiH_4 and He/O_2 gasses was employed. He and O_2 gasses are excited by the RF plasma but the SiH_4 is injected outside of the plasma excitation region of the chamber. Rapid thermal annealing was performed in He at 0.3 Torr at 900 °C for 30 s in on-line RTA chamber. The multi-chamber system, which is shown in Fig. 1, contains; (i) a load-lock chamber, (ii) a remote plasma processing chamber for oxidation, nitridation and deposition, (iii) a rapid thermal annealing chamber, (iv) a surface analysis chamber equipped with AES, and (v) a transfer chamber.

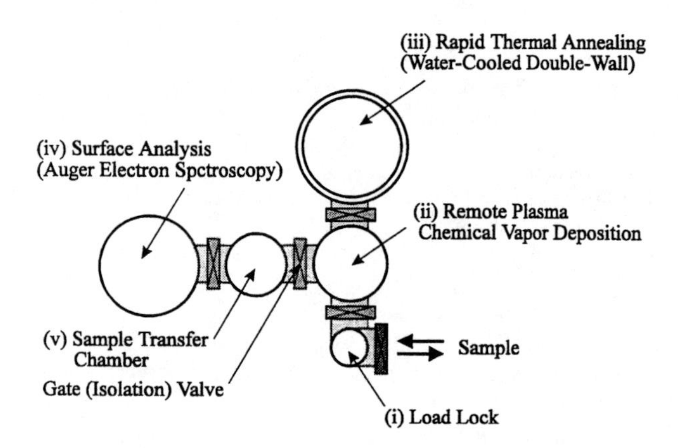

FIGURE 1. Schematic of ultra-high vacuum multi-chamber system.

The substrates used in this research were 50 mm n-type (phosphorous doped) Si(100) with a resistivity of 0.02-0.045 ohm-cm and p-type (boron doped) Si(100) with a resistivity of 5.0-10.0 ohm-cm. MOS capacitors were fabricated with field oxide isolation using aluminum (Al) or n^+ polycrystalline silicon (poly-Si) gate electrode formed by phosphorus implantation (35 keV, $5x10^{15}$ cm^{-2}). A post-implantation activation anneal was done at 1,000 °C for 30 s in argon ambient and post metallization anneal was performed in forming gas (H_2/N_2) for 30 minutes at 400 °C.

On-line AES was performed in the analytical chamber to quantify the initial stage of oxidation of the Si surface and nitridation of the oxide under vacuum. SIMS analyses were done at EVANS EAST, NJ, using both CsN^+ and SiN^- ions for depth profiling of the interfacial nitrogen.

High frequency (100 kHz) capacitance-voltage (C-V) traces were used to determine the oxide thickness.

RESULTS AND DISCUSSIONS

Interfacial confinement of nitrogen atoms has been confirmed using on-line AES to monitor the nitrogen features as a function of exposure time to the N_2/He plasma post-oxidation treatment. Figure 2 shows on-line AES data from (i) 15 s O_2 plasma exposure (this oxidation process give us ~0.5-0.6 nm thick superficial oxide on Si surface) followed by N_2/He plasma nitridation (ii) 30 s, (iii) 60 s, (iv) 90 s, and (v) 120 s. The intensity of the nitrogen KLL peak (N_{KLL}) increases with increasing exposure time to the nitrogen plasma, suggesting that longer exposures result in an increasing nitrogen incorporation either at the Si-SiO_2 interface, or in the bulk of the 0.5 to 0.6 nm oxide layer.

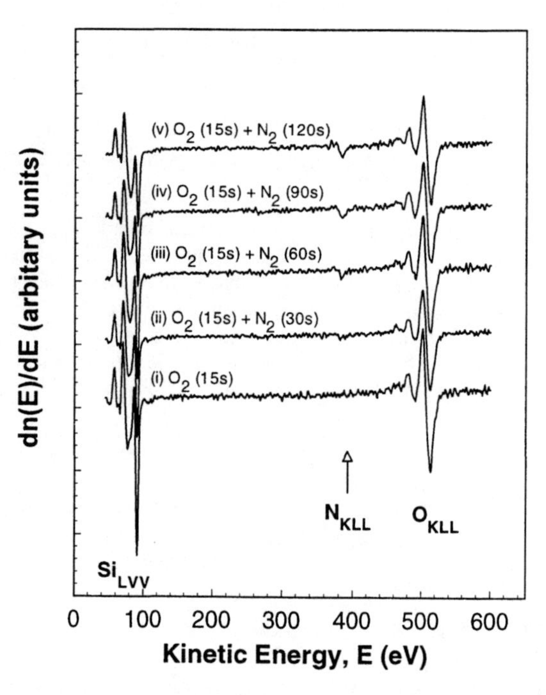

FIGURE 2. Differential AES spectra for (i) O_2 plasma oxidation of Si surface, and followed by N_2 nitridation varying exposure time: (ii) 30 s, (iii) 60 s, (iv) 90s, and (v) 120 s. Growth of N_{KLL} AES feature indicates nitrogen incorporation increased with increasing the plasma exposure time.

The following experiment suggests that this nitrogen incorporation occurs at Si-SiO_2 interface. The following sequence was used to obtain the data in Fig. 3. The bottom trace is for an O_2 plasma oxidation for 15 s and forms ~0.62 nm of SiO_2, as determined by the ratio of chemically-shifted (78 eV, Si-O bonding) and unshifted (91 eV, Si-Si bonding) Si_{LVV} AES (13), with no detectable N_{KLL} feature. And the following 60 s N_2 plasma nitridation produces a weak N_{KLL} AES feature (see middle trace in Fig. 3). The sample is then removed from the vacuum environment and 'dipped' in a very dilute hydrofluoric acid (0.5 at.% HF) for a few seconds to remove a portion of the thin oxide. The sample is then reintroduced into the vacuum system, and annealed at 900 °C for 30 s to reduce the suboxide bonding at the Si-SiO_2 interface (14,15) (see upper trace in Fig. 3). As seen in this trace, by the increased relative intensity of the Si_{LVV} feature at ~91 eV, the oxide thickness is reduced considerably; however, the weak nitride feature remains almost unchanged. In addition, there are other changes in the Si_{LVV} feature at ~ 83 eV, that also support retention of nitrogen after the HF dip and RTA. Therefore the data in Fig. 3 suggests that post-oxidation nitridation incorporates nitrogen in the immediate vicinity of the Si-SiO_2 interface.

The nitrogen content at the interface was quantified by SIMS using both CsN^+ and SiN^- ions to monitor the nitrogen. Since the data are essentially the same, we show only the SiN^- ion data. Figure 4 indicates the SIMS depth profile analysis, and Fig. 5 presents normalized area densities as function of time. The data in Fig. 5, obtained by integration of the nitrogen peak, indicate that the interfacial nitrogen concentration varies linearly with the nitrogen exposure time. In addition, comparison of the SIMS data in this experiment with that in Ref. (8) indicates approximately one monolayer is achieved at an exposure time of 90 s.

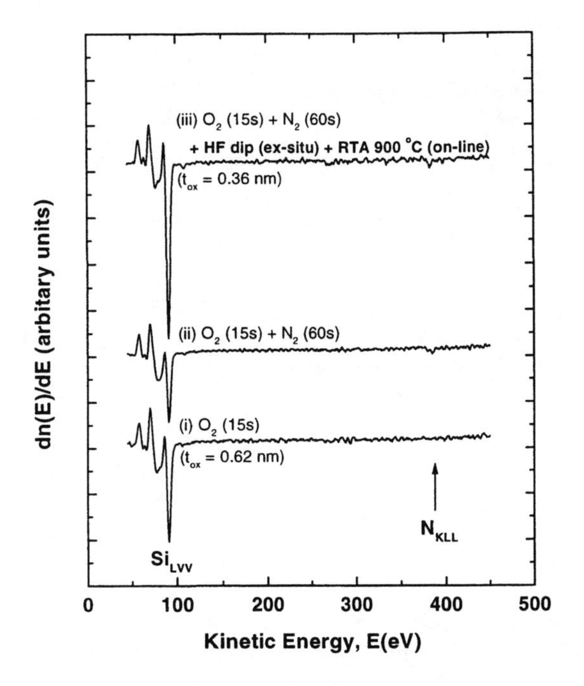

FIGURE 3. Differential AES spectra for (i) 15 s O_2 plasma oxidation of Si surface, (ii) followed by 60 s N_2 nitridation, and (iii) followed by ex-situ HF 'dip' to remove a portion of the thin oxide annealed at 900 °C in on-line RTA chamber.

This is consistent with electrical data presented below since exposure times of 90 and 120 s yield essentially the same current voltage characteristics, whereas shorter exposure times show reduced systematic shifts with respect to threshold voltages for Fowler-Nordheim (F-N) tunneling in structures with non-nitrided interfaces (see Fig. 6).

This nitridation process has been used in the fabrication of MOS capacitors with thickness ranging from 5.0 nm down to 2.0 nm on n-Si(100) and p-Si(100) using both Al and n^+ poly-Si gate electrodes. Current density and gate voltage (J-V) plots have been obtained. Figure 6 includes a series of J-V traces for a 5.0 nm-thick oxide.

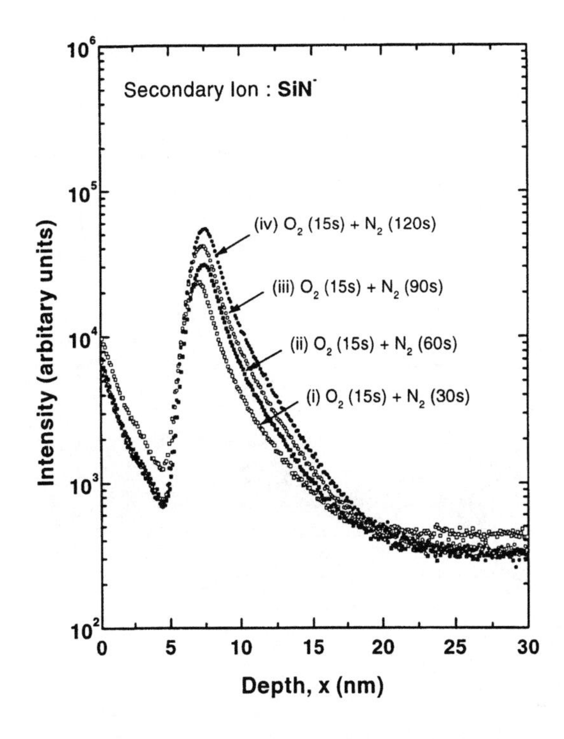

FIGURE 4. Nitrogen depth profile using SIMS for nitrogen plasma exposures of (i) 30 s, (ii) 60 s, (iii) 90 s, and (iv) 120 s. Bulk oxide thickness is 8 nm. Detection ions are SiN^-.

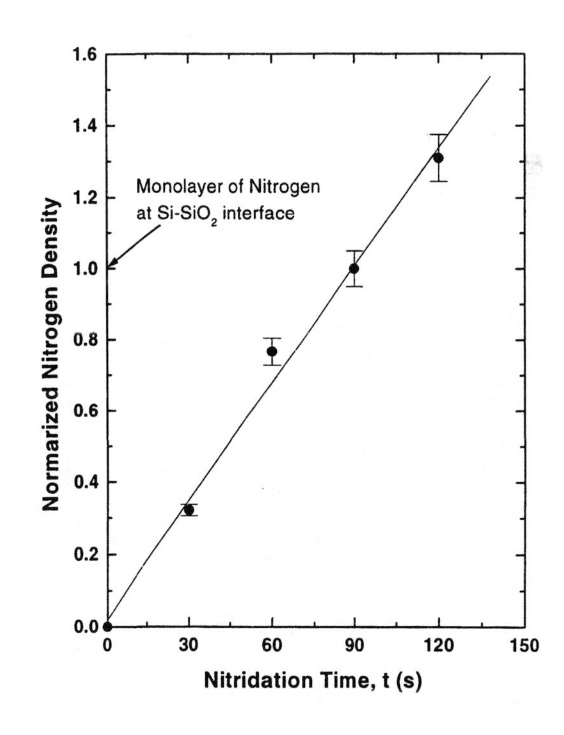

FIGURE 5. Normalized integrated nitrogen areal densities from SIMS (Figure 4) as a function of nitrogen exposure time. One monolayer is achieved for an exposure time of 90 s.

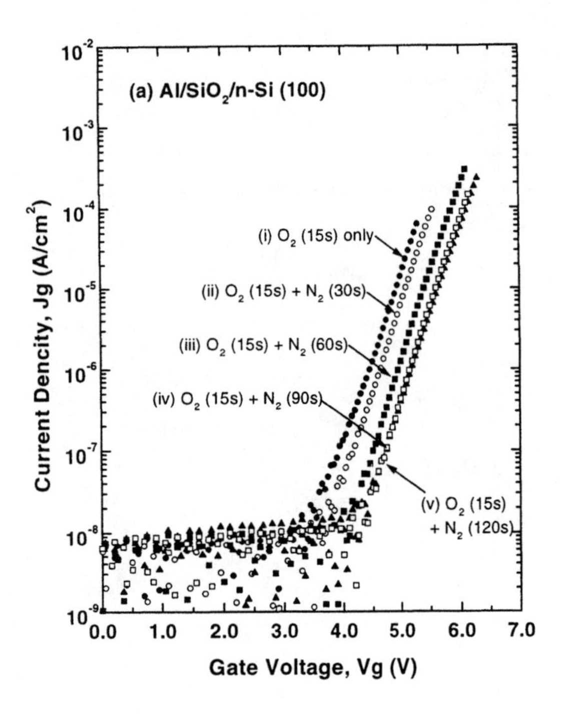

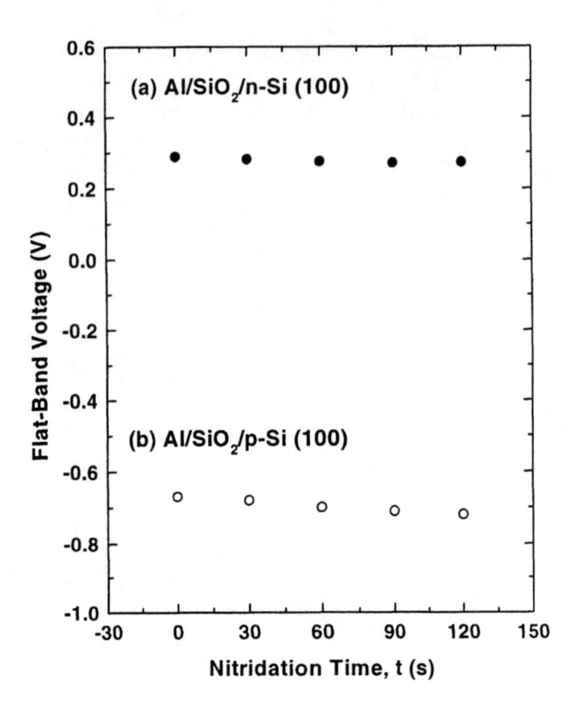

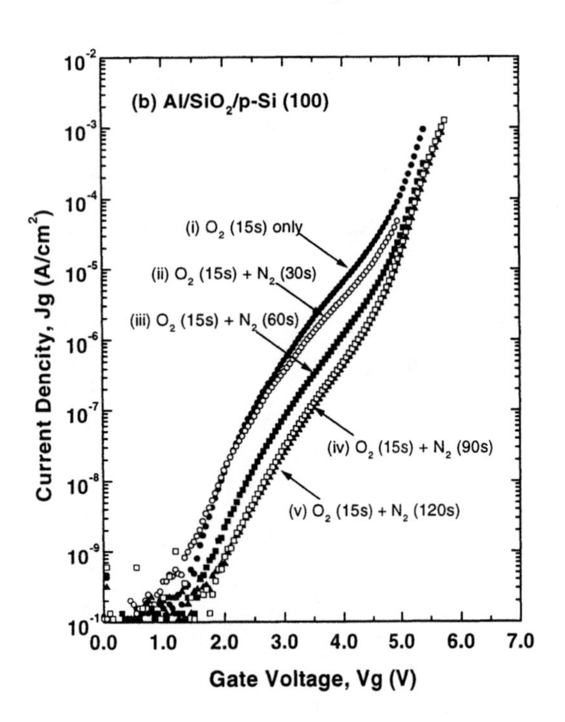

FIGURE 6. Current density-voltage plots in the F-N tunneling regime for 5 nm thick oxide as a function of nitrogen exposure time. Note that the F-N current is decreased as the nitrogen incorporation increased. (a) n-type substrate and substrates injection and (b) p-type substrates and gate injection.

FIGURE 7. Flat-band voltage plots as a function of nitrogen exposure time. Zero second nitridation means there is no nitrogen at interface. Note that the variation of flat band voltage is very small for both NMOS and PMOS.

These traces demonstrate the effect of interfacial nitrogen in reducing tunneling current in the F-N regime. This reduction is not due to a flat band voltage shift that results from incorporation of nitrogen at Si-SiO_2 interface generating fixed charge as that interface because as the amount of nitrogen at the interface is increased since the flat band voltage is essentially constant with increasing interface nitridation. (See Figure 7).

The decrease in tunneling current with increasing interfacial nitridation is systematic and significant in magnitude. Similar decreases in tunneling current have been observed using p-type substrates and gate injection (see Fig 6 (b)). Finally, this reduction of the tunneling current for substrate injection with increasing interface nitridation has also been observed for oxide thicknesses of 3.0, and 2.0 nm with an n^+ poly-Si gate electrode as shown in Figure 8. The range of oxide thicknesses spans a range from direct tunneling in the 2.0 nm oxide to F-N tunneling in the 5.0 nm oxide, with similar reductions in current with increasing nitridation.

Hao, et al. (16) reported that ultra smooth Si-SiO_2 interfaces reduce the tunneling current. Hegde, et al. (17) reported that incorporation N at the Si-SiO_2 smoothes the interface. Therefore, this suggests that interfacial smoothing with increased nitridation is responsible for the decreased tunneling currents. Barrier lowering with increased nitridation is not consistent with the constancy of the flat band voltages.

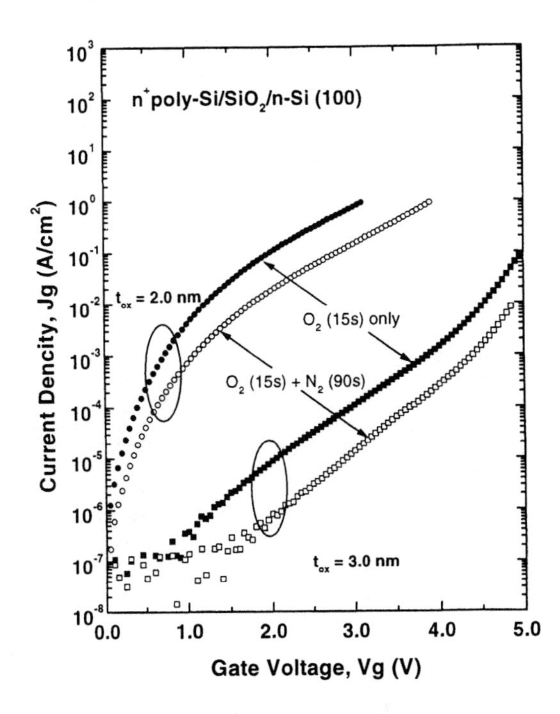

FIGURE 8. Current density-voltage plots for 2 and 3 nm thick oxide of non-nitrided and nitrided interface. Substrate injection.

CONCLUSIONS

We have demonstrated controlled incorporation of nitrogen at the Si-SiO_2 interface by two-step interface formation using remote plasma-assisted processing: O_2 oxidation of Si surface followed by N_2 nitridation. We have confirmed that nitrogen is at the Si-SiO_2 interface and not in the 0.5 to 0.6 nm oxide film. This process sequence therefore can control (i) the oxide thickness and (ii) the degree of nitridation for ultra-thin oxide independently as required in the advanced ULSI devices. Incorporation of a monolayer of nitrogen at Si-SiO_2 interface reduces both direct and F-N tunneling drastically for the substrate injection. Interfacial smoothing with increased nitridation might be the mechanism for these reductions. This low-thermal budget remote plasma-assisted processing, therefore, represents a promising alternative approach for fabricating devices with ultra-thin gate dielectrics as required by deep sub-micron lateral dimensions.

ACKNOWLEDGMENTS

This work is founded by partially, Office of Naval Research (ONR), National Science Foundation – Engineering Research Corporation (NSF-ERC), and Semiconductor Research Center (SRC).

REFERENCES

1. H.S. Momose, T. Morimoto, Y. Ozawa, K. Yamabe, and H. Iwai, IEEE Trans. Electron Devices, **ED-41**, 546 (1994).
2. U. Sharma, M. Moazzami, P. Tobin, Y. Okada, S.K. Cheng, and J. Yeargain, IEDM, **92**, 461 (1992).
3. M. Bhat, J. Kim, J. Yan, G.W. Yoon, L.K. Han, and D.-L. Kwong, IEEE Electron Device Letters, **EDL-15**, 421 (1994).
4. Z.Q. Yao, H.B. Harrison, S. Dimitrijev, and Y.T. Yeow, IEEE Electron Device Letters, **EDL-15**, 516 (1994).
5. N.S. Saks, D.I. Ma, and W.B. Fowler, Appl. Phys. Lett., **67**, 374 (1995).
6. M.L. Green, D. Brasen, K.W. Evans-Lutterodt, L.C. Feldman, K. Krisch, W. Lennard, H.T. Tang, L. Manchanda, and M.T. Tang, Appl. Phys. Lett., **65**, 848 (1994).
7. L.K. Han, *Extended Abstracts of the 1995 Int. Conf. on Solid State Devices and Materials, Osaka, 1995*, 261 (1995).
8. D.R. Lee, G. Lucovsky, M.S. Denker, and C. Magee, J. Vac. Sci. Technol., **A13**, 1671 (1995).
9. D.R. Lee, C.G. Parker, J. Hauser, and G. Lucovsky, J. Vac. Sci. Technol., **B13**, 1788 (1995).
10. S.R. Kaluri and D.W. Hess, J. Electrochem. Soc., **145**, 662 (1998).
11. T. Yasuda, Y. Ma, S. Habermehl, and G. Lucovsky, Appl. Phys. Lett., **60**, 434 (1992).
12. S. Hattangady, H. Niimi, and G. Lucovsky, Appl. Phys. Lett., **66**, 3495 (1995).
13. H. Niimi, K. Koh, and G. Lucovsky, Proc. 11th International Symp. on Plasma Processing, **96-12**, 623 (1996).
14. G. Lucovsky, A. Banerjee, B. Hinds, B. Claflin, K. Koh, and H. Yang, J. Vac. Sci. Technol., **B15**, 1074 (1997).
15. X. Chen and J.M. Gibson, Appl. Phys. Lett., **70**, 1462 (1997).
16. M.-Y. Hao, K. Lai, W.-M. Chen, and J.C. Lee, Appl. Phys. Lett., **65**, 1133 (1994).
17. R.I. Hegde, B. Maiti, R.S. Rai, K.G. Reid, and P.J. Tobin, J. Electrochem. Soc., **145**, L13 (1998).

Development of a Metrology Method for Composition and Thickness of Barium Strontium Titanate Thin Films

Thomas Remmel, Dennis Werho, Ran Liu, and Peir Chu*

Semiconductor Technology, Semiconductor Products Sector, Motorola Inc.
*Mesa, Arizona, 85202 and Austin, Texas, 78721**

Thin films of barium strontium titanate (BST) are being investigated as the charge storage dielectric in advanced memory devices, due to their promise for high dielectric constant. Since the capacitance of BST films is a function of both stoichiometry and thickness, implementation into manufacturing requires precise metrology methods to monitor both of these properties. This is no small challenge, considering the BST film thicknesses are 60 nm or less. A metrology method was developed based on X-ray Fluorescence and applied to the measurement of stoichiometry and thickness of BST thin films in a variety of applications.

INTRODUCTION

The continually increasing demand for more memory and the trend toward higher operating speeds of integrated circuits dictates the need for high density memory accessible at rapid rates. The time-proven bit cell design of DRAMs (dynamic random access memories) relies upon the storage of charge in an integrated capacitor. To date, the dielectric used in this capacitor has been SiO_2; the challenge of maintaining sufficient charge storage density as the bit cell size decreases has historically been met by decreasing the SiO_2 thickness. However, this approach is reaching its limit and although nitridation of SiO_2 affords some increase in dielectric constant, radically higher dielectric constant films are required to replace SiO_2. Barium strontium titanate (BST) is particularly attractive because of its very large dielectric constant (as high as 2500 in bulk form), low leakage behavior, thermal stability, low dissipation factor and promising fatigue behavior (1-4).

The dielectric constant of $Ba_{1-x}Sr_xTiO_3$ is dependent upon both the Ba/Sr ratio and the A/B [(Ba+Sr)/Ti] site ratio (5-8). Since the capacitance per unit area is a function of both the dielectric constant and the film thickness, implementation of BST into manufacturing requires precise metrology methods to monitor both the stoichiometry and thickness of the films. This is no small challenge, considering the BST film thicknesses are 60 nm or less and control of their stoichiometry to better than 1% (relative) will be required. In addition, the metrology method needs to be rapid, non-destructive, non-invasive, wafer-fab compatible and capable of handling wafers up to 300 mm in diameter. The ability to measure across-wafer variations is also desirable.

Several measurement methods were considered as potential metrology options, including x-ray fluorescence (XRF), Rutherford Backscattering Spectrometry (RBS), Fourier Transform Infrared Spectroscopy (FTIR), Spectroscopic Ellipsometry (SE), Inductively Coupled Plasma-Atomic Emission Spectroscopy (ICP-AES), Glow Discharge-Optical Emission Spectroscopy (GD-OES), Glow Discharge Mass Spectrscopy (GD-MS), and small spot XRF. The utility of SE for film thickness measurement is well accepted; however composition measurement via SE is not likely. FTIR shows some promise as an indirect "fingerprinting" method for BST film process monitoring; however, the ability of FTIR to directly measure BST film stoichiometry has not yet been proven. RBS was ruled out due to its high cost and lack of wafer fab compatibility; ICP was not considered due to its destructive nature, lenghty analysis time, poor spatial resolution and lack of fab compatibility. GD-OES, GD-MS and small spot XRF were also not considered viable options for various technical reasons. XRF was chosen as the preferred method because it satisfied most of the criteria, is already an accepted, fab-compatible tool and offers a high degree of maturity in quantitative analysis of thin films. The weakest attribute of XRF is a rather large analysis area, 35mm on our instrument; although not very small, it still affords the ability to measure film composition and thickness at several locations across the wafer.

X-RAY FLUORESCENCE OF BST

X-ray fluorescence is based on the principle of x-ray excitation of the sample using continuous radiation and the subsequent measurement of excited, characteristic x-rays associated with the atoms present. The characteristic x-ray intensity is proportional to the number of atoms in the irradiated volume. This intensity is measured using x-ray spectrometers, of which two principle types exist: energy dispersive x-ray spectrometers (EDS) and wavelength dispersive spectrometers (WDS). For BST analysis on Si, several of the elements have characteristic x-ray lines with

CP449, *Characterization and Metrology for ULSI Technology: 1998 International Conference*
edited by D. G. Seiler, A. C. Diebold, W. M. Bullis, T. J. Shaffner, R. McDonald, and E. J. Walters
 1-56396-753-7/98/$15.00

wavelengths in close proximity to those of other elements; to ensure unambiguous intensity measurement, high wavelength resolution is required, such as that afforded by WDS.

Wavelength Dispersive XRF Method

Wavelength dispersive spectrometers consist of an analyzing crystal and an x-ray detector selected for their sensitivity to a particular x-ray wavelength. Several of these channels are mounted on the same instrument to achieve simultaneous and optimized detection of a number of different x-ray lines. For BST films, Ti-Kα intensity was measured using a fixed channel, optimized for Ti-Kα, whereas Ba-Lβ and Sr-Kα x-rays were measured using a scanning channel. Oxygen was not measured due to the unavailability of an oxygen channel on the instrument. (The recent acquisition of fixed oxygen channel has allowed for its analysis; however those results are not reported here).

To quantify stoichiometry and thickness from x-ray intensity, calibration is required through the use of appropriate standards. We created a series of MOD (metal-organic-decomposition) films made from mixtures of strontium titanate (STO) and barium titanate (BTO) precursors, spun onto oxidized Si <100> wafers, and subsequently baked to drive off the volatiles. These standard films ranged from 50 to 200 nm in thickness and covered the range of $Ba_{1-x}Sr_xTiO_3$ from STO to BTO. The thicknesses of the BST standard films were determined using SE.

Stoichiometry of the standards was independently determined using three methods: (1) weight ratios of the STO and BTO precursors; (2) RBS; and (3) ICP-AES. The results from RBS and ICP validated the weight ratio values, therefore those values were taken to be the film stoichiometries. Empirical relationships were then derived between the Ba/Ti and Sr/Ti x-ray intensity ratios and their respective film stoichiometries. Shown in Fig. 1 is this relationship for Sr stoichiometry versus Sr/Ti x-ray intensity ratio; the linear fit is excellent. After calculating the Ba and Sr stoichiometries, Ba+Sr was normalized to 1.0 to yield the Ti stoichiometry in $Ba_{1-x}Sr_xTi_yO_3$.

An empirical relationship for film thickness based on Ba, Sr and Ti x-ray intensities was also derived. The results are shown in Fig. 2 plotted against film thickness determined using SE. The agreement between the two methods is very good. Note that consideration (and further calibration) is required when analyzing BST films that are deposited using processing techniques other than MOD, such as MOCVD and sputtering. These processing methods result in films that are typically more dense than MOD films, and thus the thickness results will be in error. (XRF effectively measures mass-density, not true thickness; therefore a density must be assumed to calculate film thickness).

Errors Due to Substrate Diffraction

There are several sources of error in the analysis of films using XRF, including the accuracy of the standards, the quality of the calibration process, counting statistics, instrumentation stability, accuracy of the calculation algorithm. In addition, particularly problematic for wavelength dispersive XRF (WDXRF) analysis of BST films is the issue of substrate diffraction (9).

Substrate diffraction refers to diffraction of the incident radiation (Rh tube operated at 40 kV and 70 mA) by the single crystal substrate. This diffracted intensity, which is dependent upon the geometrical arrangment of the radiation source, substrate and the x-ray detector, will vary as a function of crystallographic orientation of the substrate and its angular position. The diffracted intensity manifests itself as additional background x-ray intensity which is detected by the various x-ray detectors.

In WDXRF, with fixed position detectors, it is not possible to scan the spectrometer wavelength to measure

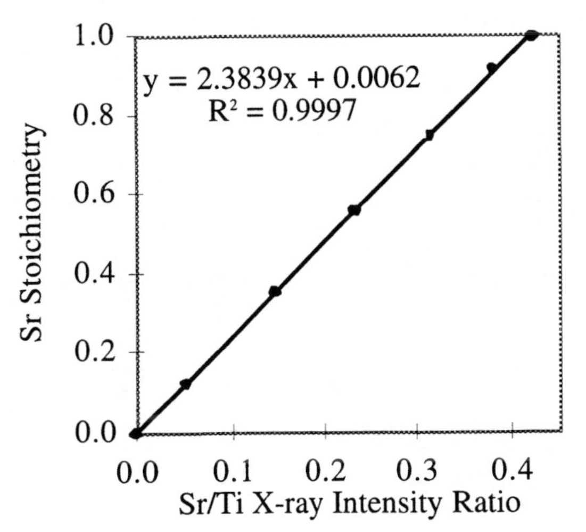

FIGURE 1. Sr stoichiometry calibration for MOD BST films.

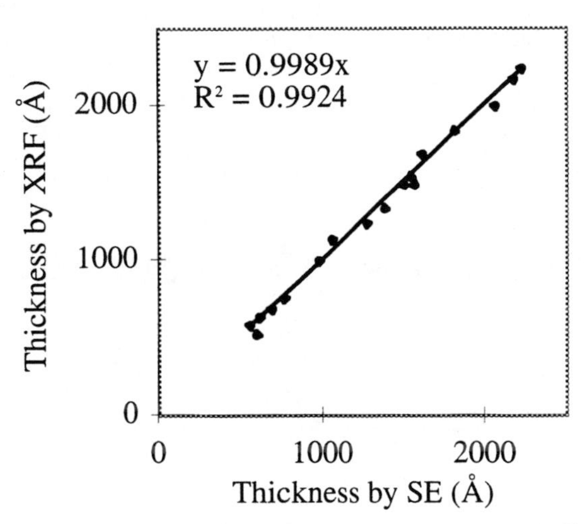

FIGURE 2. Film thickness calibration for MOD BST films (XRF thickness derived from Ba, Sr & Ti x-ray intensities).

backgrounds levels. Therefore background levels were measured on BST film-free wafers (for Ti) or measured for Ba in STO films and for Sr in BTO. These "element free" background levels are then assumed to be applicable to the unknown films. However, substrate diffraction can contribute to these background level measurements and therefore must be well characterized from wafer to wafer. As an added complication, to achieve full wafer coverage, many XRF tools use an "r-θ" stage to reach all locations on a wafer, rather than an x-y stage. This introduces the added variable of angular geometry (θ) for each measurement location.

Shown in Fig. 3 is a plot of the x-ray intensity collected by the Ti-Kα channel measured for a bare Si <100> wafer (no BST), as a function of angular rotation (θ). Note the strong diffraction effect; although there is no Ti present in the wafer surface, the background x-ray levels vary by as much as an order of magnitude as a function of θ. The maximum background levels are as high as 10% of the maximum Ti intensity measured in our BST films. The repeating pattern with angular rotation is due to the symmetry of the Si <100> wafer.

The preferred method to address substrate diffraction is to measure the effect for the particular substrate orientation and tool/detector geometry, and then select an angular rotation where the intensity of the diffraction contribution is minimal and non-varying as a function of θ. Significant substrate diffraction contributions to the background levels of Ba and Sr were also observed, with diffraction being a potentially large source of error for Sr.

XRF MEASUREMENT RESULTS

Repeatability

The repeatability of the XRF method for measuring BST film stoichiometry and thickness was determined by repetitive analysis of a nominal $Ba_{0.25}Sr_{0.75}TiO_3$ film over a nine month time frame. Shown in Figs. 4-6 are the results of the gauge capability study for Ba stoichiometry, A/B ratio and film thickness, respectively. The statistics associated with the gauge capability study are shown in Table 1.

One challenging aspect of developing metrology methods is the transfer of the method to other labs and the ability to achieve inter-lab agreement in results. This has been accomplished with the BST XRF metrology method within Motorola. Shown in Fig. 7 is a comparison of A/B

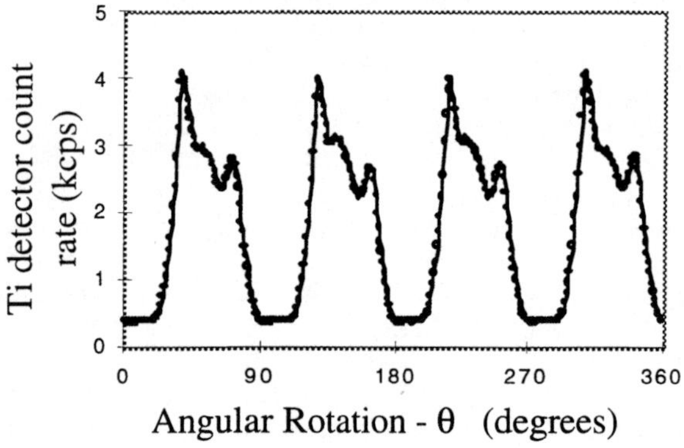

FIGURE 3. Diffraction effect for Ti-Kα on bare Si (100) wafer. This shows background levels measured by Ti spectrometer as a function of angular rotation of Si substrate.

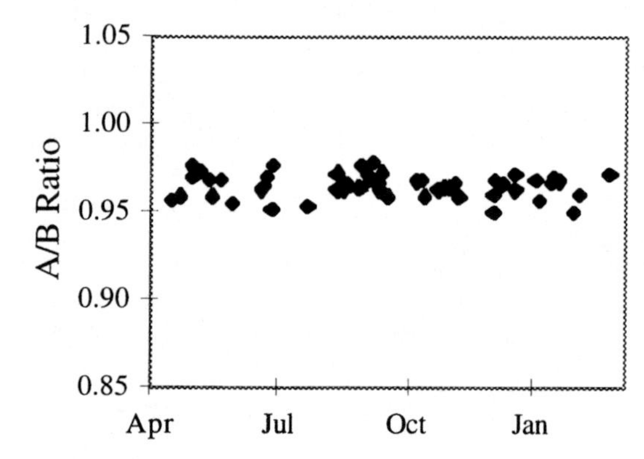

FIGURE 5. XRF measurements of A/B ratio on the same $Ba_{0.25}Sr_{0.75}TiO_3$ film over a 9 month period.

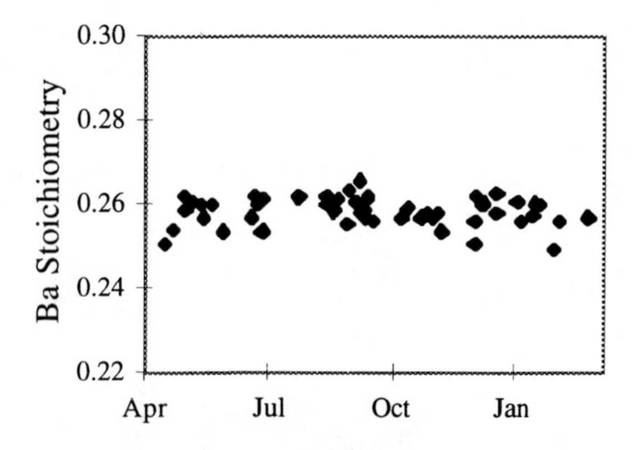

FIGURE 4. XRF measurements of Ba on the same $Ba_{0.25}Sr_{0.75}TiO_3$ film over a 9 month period.

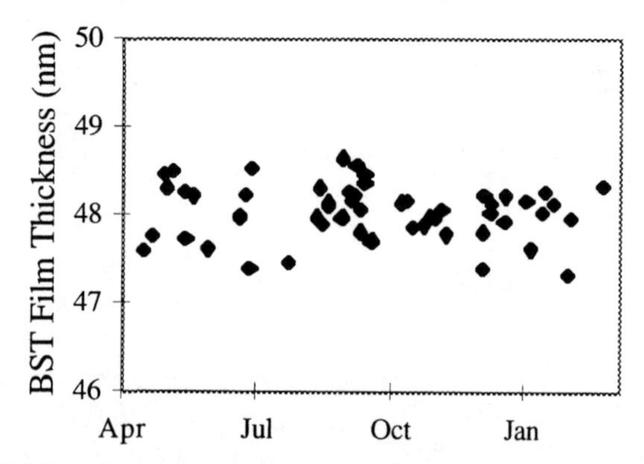

FIGURE 6. XRF measurements of BST film thickness on the same $Ba_{0.25}Sr_{0.75}TiO_3$ film over a 9 month period.

TABLE 1. Repeatability Results for XRF Analysis of the Same $Ba_{0.25}Sr_{0.75}TiO_3$ Film over Nine-month Period.

Measurement	Mean	Std Dev	RSD (%)*
Ba Stoichiometry	0.258	0.003	1.3
A/B Ratio	0.965	0.007	0.7
BST Film Thickness (Å)	480	3	0.6

* RSD - relative standard deviation (std dev/mean)

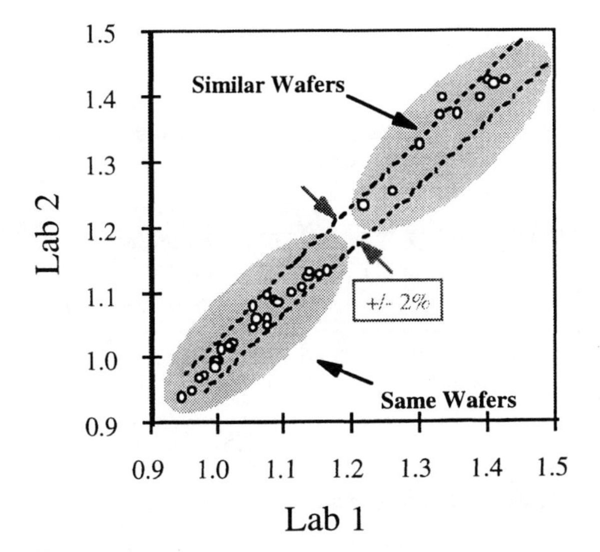

FIGURE 7. Inter-laboratory comparison of A/B ratio measurements of BST films on similar and identical wafers.

ratios between two different labs on a combination of similar and identical BST films. The agreement between the two labs is within the ± 2% precision of the method for A/B ratio measurements.

The XRF method was used to monitor a BST film deposition process over an extended period of time; the technique was sensitive enough to detect changes after a target change (different A/B ratio in new target) and to identify a change in stoichiometry due to a deposition tool problem.

Across Wafer Variation

As noted previously, the XRF method, with its 35 mm diameter analysis area, is capable of measuring across wafer variations in film stoichiometry and thickness. This has been used extensively during the development process for process optimization, since uniformity of BST film properties across the wafer is critical to yield functional and reliable devices. It can also readily be used as a process monitoring tool, to ensure process control.

Shown in Fig. 8 is an example of the results from measurement of the A/B ratio across a 150 mm wafer. In this case, "overlapping" measurements were taken every 10 mm, from the wafer center to within a few mm of the edge,

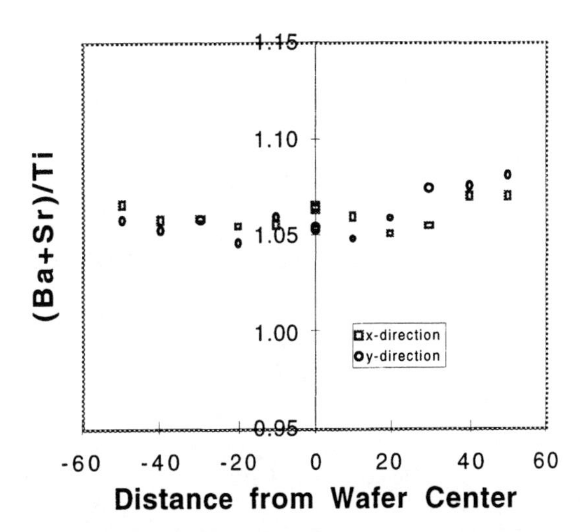

FIGURE 8. Across-wafer measurements of the A/B ratio of a BST film on a 150 mm wafer.

along orthogonal x- and y- directions. The results show a relatively uniform A/B ratio for the BST film, with an indication of Ti deficiency near the wafer edge. The analysis time for across-wafer measurements such as those shown in Fig. 8 is on the order of 1.5 hours/wafer, and of course, also yields results for Ba/Sr and BST film thickness variations.

FUTURE ENHANCEMENTS

Although XRF has been shown to be a viable and useful metrology method for BST films, there remains a need for higher spatial resolution and even more rapid analysis. To meet those requirements, optical metrology methods based on FTIR are under investigation. Some success has been achieved in measuring a correlation between the frequency of the FTIR phonon absorption and the Sr stoichiometry of BST films. Figure 9 shows a series of FTIR absorption spectra, acquired in the transmission mode, for the TO

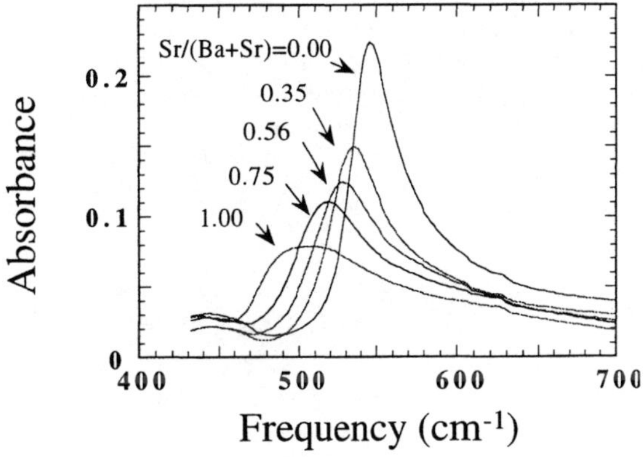

FIGURE 9. FTIR TO phonon absorption spectra for various in BST films (transmission mode).

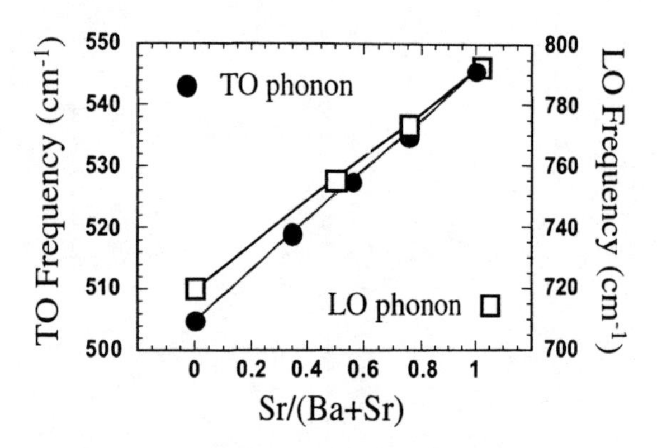

FIGURE 10. FTIR phonon peak frequency peak as function of Sr stoichiometry in BST films.

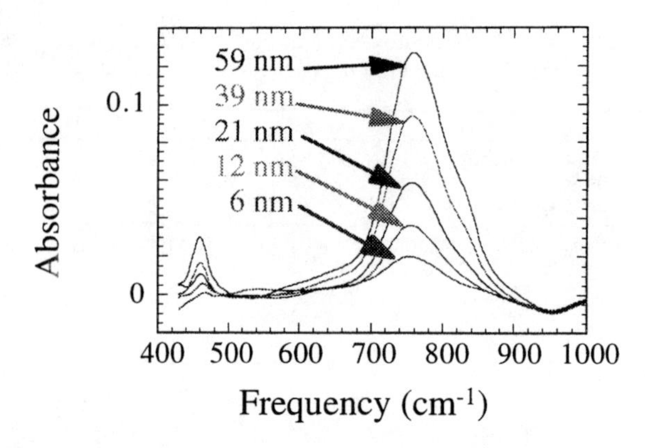

FIGURE 11. FTIR LO phonon absorption spectra as a function BST film thickness (reflection mode).

vibration of BST films ranging in composition from STO (x=1) to BTO (x=0). The frequency and shape of the absorption peaks vary with BST stoichiometry. The frequency dependence of the TO and LO peaks as a function of Sr/Ba stoichiometry are plotted in Fig. 10, revealing a linear relationship between phonon peak frequency and Ba/Sr ratio. FTIR has also shown promise for measurement of BST film thickness. Shown in Fig. 11 are a series of reflection mode FTIR absorption spectra for the LO phonon vibration in BST films ranging from 6 to 59 nm in thickness; the intensity of the absorption peak increases with increasing film thickness.

DISCUSSION AND CONCLUSIONS

WDXRF has been shown to be a viable metrology method for the composition and thickness of BST films. Accuracy and precision have been measured through comparison with other measurement methods and repeatability studies. Current limitations to the WDXRF method include substrate diffraction effects, counting statistics and accuracy of the independent measurement techniques for validation of the standards. Improvements can be achieved through high efficiency x-ray detectors for Ba and Sr, more thorough understanding of the substrate-to-substrate variability of the diffraction contribution and improved standard validation. Optical methods based on FTIR have shown early promise for process monitoring of BST films in a "fingerprinting" mode. However, further development is necessary to demonstrate that FTIR is capable of measuring A/B ratios and to validate the precision and accuracy of FTIR as a BST film metrology method.

ACKNOWLEDGMENTS

The authors want to acknowledge the efforts of many process technologists within Motorola's Semiconductor Technology group who enabled the performance of this work. Special thanks for the individual contributions from Beth Baumert, Gary Mote, Steve Primrose and Dan Sullivan. Motivation for this work came from the Materials Research and Strategic Technologies group; many thanks to the support of Bradley Melnick, Robert E. Jones, Jr., Peter Zurcher and Sherry Gillespie.

REFERENCES

1. Peng, C-J. and Krupanidhi, S. B., *J. Mater. Res.*, **10**, 708-726 (1995).
2. Hwang, C. S., Park, S. O., Cho, H.-J., Kang, C. S., Kang, H.-K, Lee, S. I. and Lee, M. Y., *Appl. Phys. Lett.* **67** (1995).
3. Lee, W.-J., Park, I.-K., Jang, G.-E. and Kim, H.-G., *Jpn. J. Appl. Phys.* **34**, 196-199 (1995).
4. Cho., H.J., Kang, C. S., Hwang, C. S., Kim., J.-W., Horii, H., Lee, B. T., Lee, S. I. and Lee, M. Y., *Jpn. J. Appl. Phys.* **36**, L874-L876 (1997).
5. Miyaska, Y. and Matsubara, S., in *Proceedings of 1990 IEEE Seventh Int. Symp. on Application of Ferroelectrics*, edited by S. B. Krupanidhi and S. K. Kurts, pp. 121-124.
6. Kim, T. S., Kim, C. H. and Oh, M. H., *J. Appl. Phys.*, **75**, 7998-8003 (1994).
7. Yamamichi, S., Yabuta, H., Sakuma, T. and Miyasaka, Y., *Appl. Phys. Lett.*, **64**, 1644-1646 (1994).
8. Tsu, R., Liu, H-Y, Hsu, W-Y., Summerfelt, S., Aoki, K. and Gnade, B., *Mat. Res. Soc. Symp. Proc.*, **361**, 275-280 (1995).
9. Kohno, H., Kobayashi, H., Kuraoka, M. and Wilson, R., "Next Generation Semiconductor Thin Films: Specific Analysis Problems and Solutions," presented at the 45th Annual Denver X-Ray Conference, Denver, Colorado, August 3-8, 1996.

Physical and Chemical Characterization of Barium Strontium Titanate Thin Films

Thomas Remmel, Wei Chen, Ran Liu, Mike Kottke, Richard Gregory, Peter Fejes, Beth Baumert and Peir Chu*

*Semiconductor Technology, Semiconductor Products Sector, Motorola Inc. Mesa, Arizona, 85202 and Austin, Texas, 78721**

Thin films of barium strontium titanate (BST) are under consideration as a replacement for silicon dioxide as the dielectric in advanced memory devices. The major attraction of BST is its very high dielectric constant compared to SiO_2. Integral to the development and implementation of a new materials system such as BST is physical and chemical characterization. To achieve optimal electrical performance with a robust process requires an in-depth understanding of the relationships between materials/process variables, physical/chemical properties and the electrical behavior of the BST films. This paper covers specifics of many of these inter-relationships.

INTRODUCTION

Increasing demand for more functionality of integrated circuits dictates the need for higher density memories. To achieve the higher packing density, the evolutionary path of dynamic random access memories (DRAMs) has been to decrease bit cell size. To maintain sufficient charge storage, DRAMs have relied upon shrinkage of the capacitor dielectric (SiO_2) thickness. However, this approach is reaching its limit and although nitridation of SiO_2 affords some increase in dielectric constant, radically higher dielectric constant films are needed to replace SiO_2. Barium strontium titanate (BST) is particularly attractive because of its very large dielectric constant, up to 20,000 in bulk form (1). It also shows potential with respect to low leakage behavior, thermal stability, low dissipation factor and promising fatigue behavior (2-4). $Ba_{1-x}Sr_xTiO_3$ is a perovskite that exhibits complete solid solubility over all compositions with a cubic structure at room temperature for $0.3 \leq x \leq 1$, becoming tetragonal for $x \leq 0.3$ (5). The stoichiometry of the films discussed in this paper is $Ba_{0.5}Sr_{0.5}TiO_3$.

The electrical behavior of BST films is heavily dependent upon the physical and chemical properties of the films, which in turn, are a function of the material and process variables used to create the films. In RF magnetron sputtering, these materials/process input variables include the deposition process (sputtering gas, deposition temperature, rate, power, pressure, etc.), the target material (composition, density, etc.) and the deposition tool. Physical/chemical properties of the BST films include thickness, microstructure, grain size, composition, impurities, crystal structure, crystallinity, surface/interface chemistry, refractive index and defects. Key electrical properties include dielectric constant, leakage behavior, relaxation and reliability.

Development of a process that produces high quality BST films which exhibit the desired electrical performance requires a thorough understanding of the inter-relationships between material/process variables, physical/chemical properties and electrical performance. To obtain this understanding numerous analytical methods were used to characterize BST films, including Scanning Electron Microscopy (SEM), Transmission Electron Microscopy (TEM), Atomic Force Microscopy (AFM), X-ray Diffraction (XRD), X-ray Fluorescence (XRF), Rutherford Backscattering Spectrometry (RBS), Fourier Transform Infrared Spectroscopy (FTIR), Auger Electron Spectroscopy (AES), Inductively Coupled Plasma-Atomic Emission Spectroscopy (ICP-AES) and Spectroscopic Ellipsometry (SE).

CHARACTERIZATION RESULTS

Grain Size

A direct relationship between BST grain size and dielectric constant has been reported by several authors (6-8), therefore BST film grain size was the focus of many studies. Shown in Fig. 1 are TEM images of BST films deposited at two different substrate temperatures; as

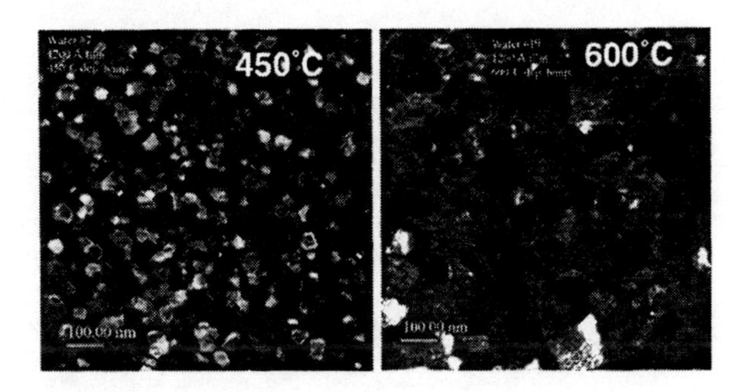

FIGURE 1. Plan view TEM images of BST films sputter-deposited at 450°C and 600°C.

CP449, *Characterization and Metrology for ULSI Technology: 1998 International Conference* edited by D. G. Seiler, A. C. Diebold, W. M. Bullis, T. J. Shaffner, R. McDonald, and E. J. Walters

expected, the grain size increases with deposition temperature. XRD analyses verified the linear relationship between grain size and deposition temperature (Fig. 2).

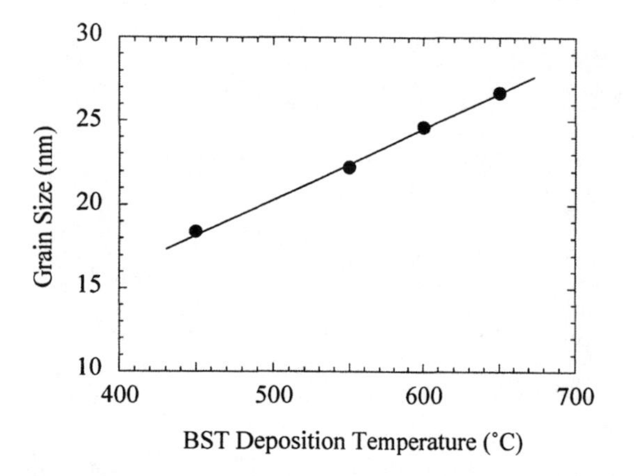

FIGURE 2. Grain size in growth direction of 600Å BST films determined using XRD.

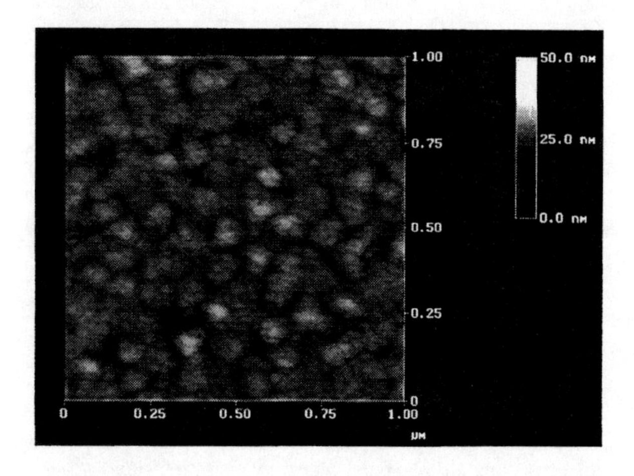

FIGURE 3. AFM image of the grain structure of a 600Å BST film, sputter-deposited on an underlying Pt electrode. Grain size measurement of both BST and Pt were achieved through Fourier analysis of the images.

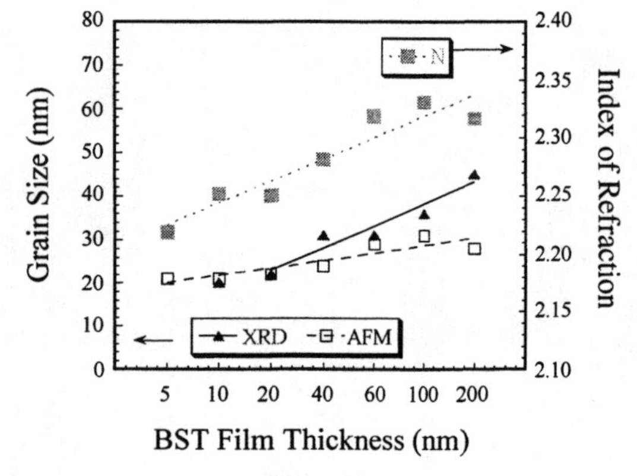

FIGURE 4. Grain size in film thickness for BST films determined using AFM and XRD.

The relationship between grain size and film thickness was determined using XRD and AFM (Figs. 3-4). Grain size, derived from x-ray diffraction peak broadening, represents the grain size in the growth direction of the film, whereas plan-view AFM measures grain size normal to the growth direction. Grain size was determined from AFM images through Fourier analysis of the digital image and interpretation of the Power Spectral Density function (Fig. 5). Using this method, it is possible to deconvolute the underlying Pt grain structure from that of the BST film. In studying BST film grain size, a correlation was found between grain size and index of refraction, determined by SE (Fig. 4). This is consistent with the relationship between grain size (measured by AFM and XRD) and dielectric constant, as shown in Fig. 6.

Film Crystallinity

XRD and FTIR were used to evaluate film crystallinity, since there is a direct relationship between BST film crystallinity and dielectric constant. Shown in Fig. 7 is a plot of integrated XRD peak intensity for the (110) BST peak as a function of deposition temperature, indicating an increase in film crystallinity with temperature. The FTIR

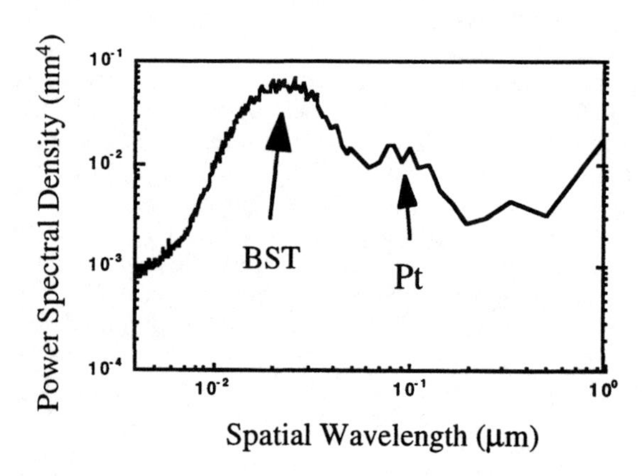

FIGURE 5. Power spectra density of image in Fig. 3, showing frequency resolved grain sizes of BST and Pt.

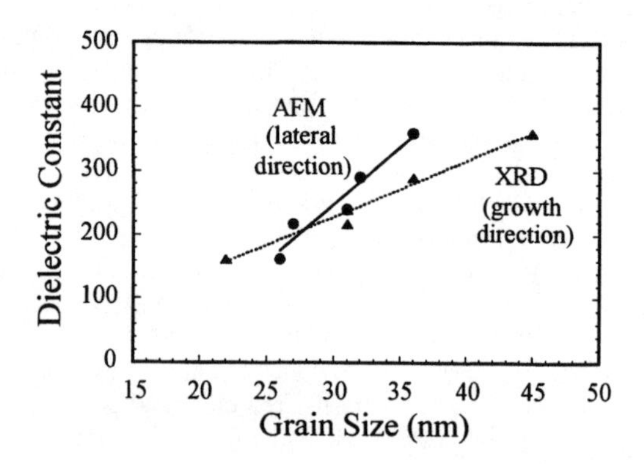

FIGURE 6. Dielectric constant vs. grain size measured by AFM and XRD methods.

absorption spectra in Fig. 8 confirm this relationship, with the absorption peak width decreasing with increasing deposition temperature. The narrower width and the higher absorption peak intensity are indicative of greater crystallinity in the BST films.

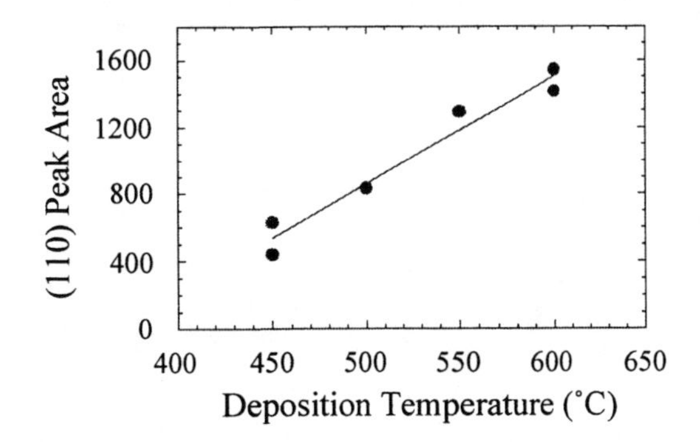

FIGURE 7. Plot of XRD (110) BST peak intensity vs. deposition temperature.

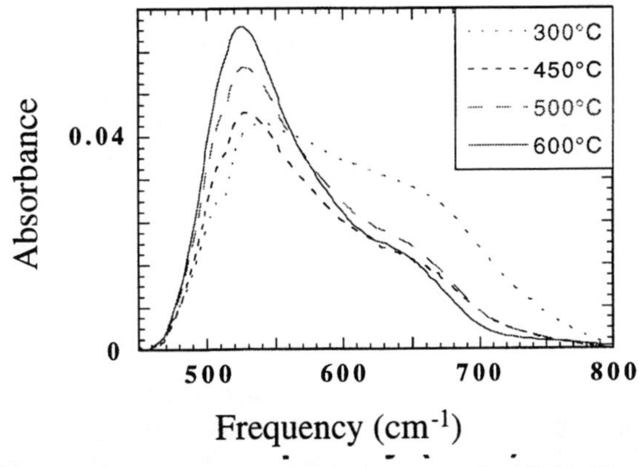

FIGURE 8. FTIR absorption spectra for BST films deposited at various temperatures, indicating increasing crystallinity with increasing deposition temperature.

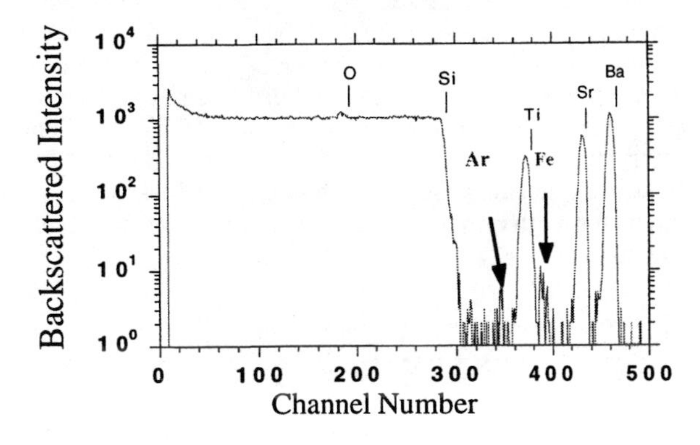

FIGURE 9. RBS spectrum of BST thin film, showing major constituents and impurities from the sputtering process.

Film Chemistry

The electrical behavior of BST films is a strong function of the film chemistry. We developed methods based on XRF, AES, RBS and ICP-AES to measure the bulk stoichiometry and trace levels in thin BST films. Shown in Fig. 9 is a RBS spectrum of a BST film, where both bulk chemistry and trace constituents (Fe and Ar) are evident. The source of the Ar is the sputtering gas, whereas the Fe was traced to stainless steel shielding within the sputter chamber. Quantification of the Fe, Ni and Cr levels in this contaminant was accomplished using ICP-AES; reduction of these impurity levels in the BST films was achieved by application of a DC bias during sputtering (Fig. 10). The DC bias appears to move the plasma away from the stainless steel source.

The A/B site ratio, defined as (Ba+Sr)/Ti, was measured using XRF. Shown in Fig. 11 is the relationship between A/B ratio and sputtering gas ratio (O_2:Ar). The A/B ratio was shown to decrease dramatically with the introduction of small amounts of oxygen. This is significant, since small deviations in A/B ratio from 1.0 adversely affect the dielectric constant (9-10).

Auger depth profiling methods were developed to measure the distribution of major constituents through the

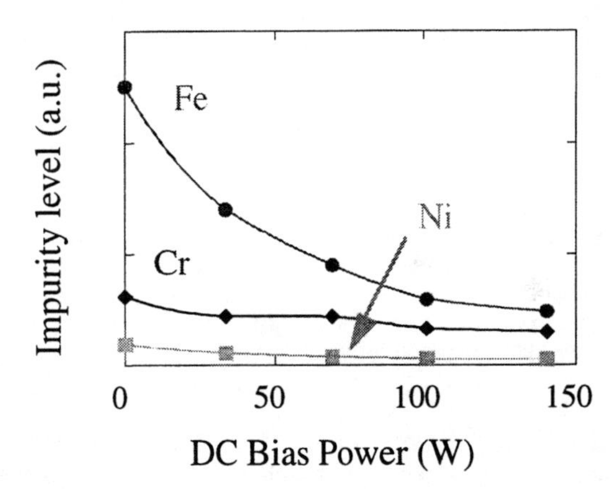

FIGURE 10. Reduction of stainless steel impurities in BST films through increased DC bias during sputtering.

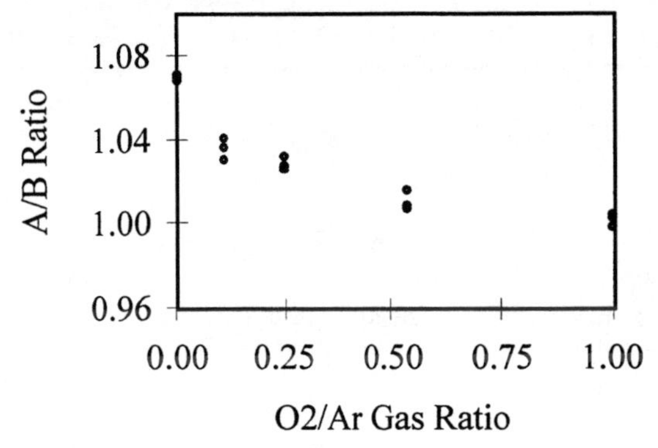

FIGURE 11. A/B site ratio measured as a function of O_2:Ar sputtering gas ratio.

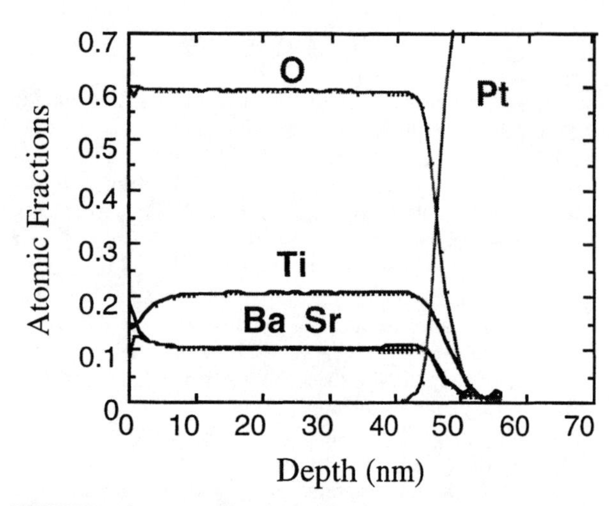

FIGURE 12. Auger depth profile through the thickness of a BST film sputter deposited on a Pt bottom electrode.

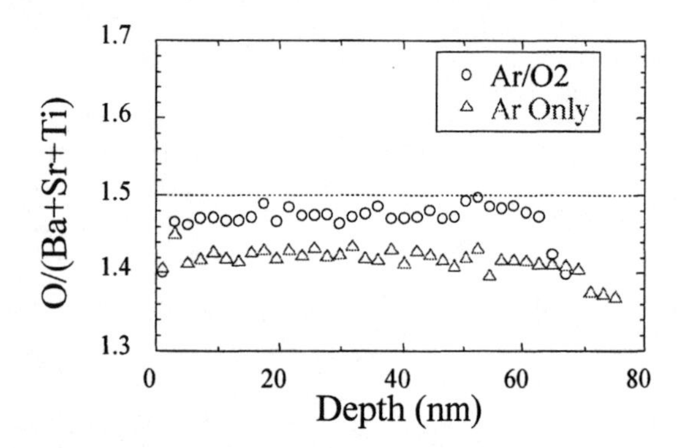

FIGURE 13. Comparison of the relative oxygen content through the thickness of two BST films; one film was deposited using an argon-only plasma; the other using an argon-oxygen plasma. These depth profiles were measured using Auger electron spectroscopy.

thickness of the BST films. A typical depth profile is shown in Fig. 12. Through appropriate calibration, profiles such as this can be converted into quantitative elemental distributions through the BST film thickness. Figure 13 shows an example of quantified Auger depth profiles comparing the relative oxygen content of BST films deposited using two different sputtering gases, 100% Ar and an argon-oxygen mixture. The lower levels of oxygen in the Ar-only sputtered film are indicative of substoichiometric oxygen content.

Surface Roughness

Film breakdown under electric field can be affected by variations in film roughness. The surface roughness of various thickness BST film deposited on Pt electrodes is shown in Figs. 14 and 15. The roughness, measured using AFM, decreases with increasing film thickness; this

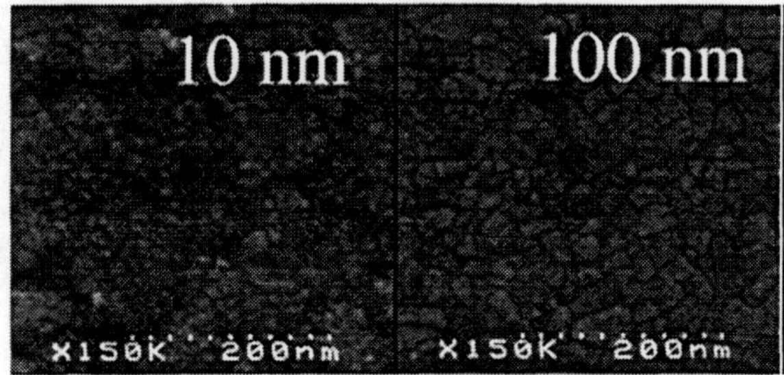

FIGURE 14. AFM images of the surfaces of 10 and 100nm thick BST films on Pt electrodes, showing the change in surface morphology with BST thickness.

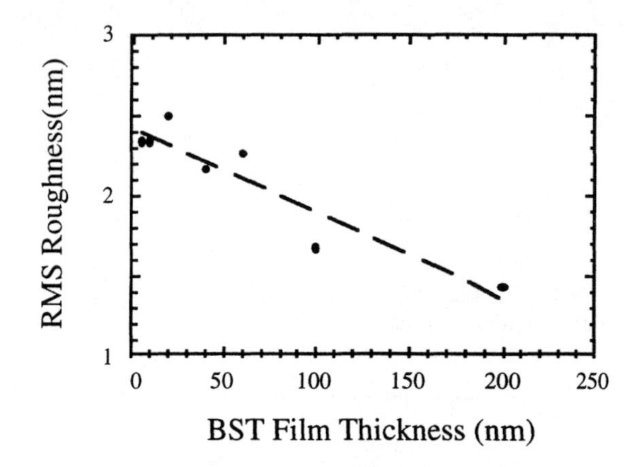

FIGURE 15. Plot of AFM surface roughness as a function of BST film thickness.

is because the BST has a smoothing effect on the rougher, underlying Pt.

Film Microstructure

BST film microstructure was also characterized with a view toward understanding structure/property relationships. The TEM image in Fig. 16 shows the columnar structure of a BST film deposited on an iridium electrode. The grains extend through the entire thickness of this particular film. Film crystallinity, preferred orientation and lattice parameter were measured using XRD. The lattice parameter of the sputtered films was found to be significantly larger than that expected for bulk across all compositions (Fig. 17); the lattice parameter decreased with increasing deposition temperature (Fig. 18), both observations consistent with other authors (11-14). Post deposition annealing in air was found to decrease the lattice parameter, with bulk values nearly achieved at 850°C after five minutes.

CONCLUSIONS

Numerous analytical methods were applied to the characterization of BST thin films to achieve a greater understanding of the relationships between material/process

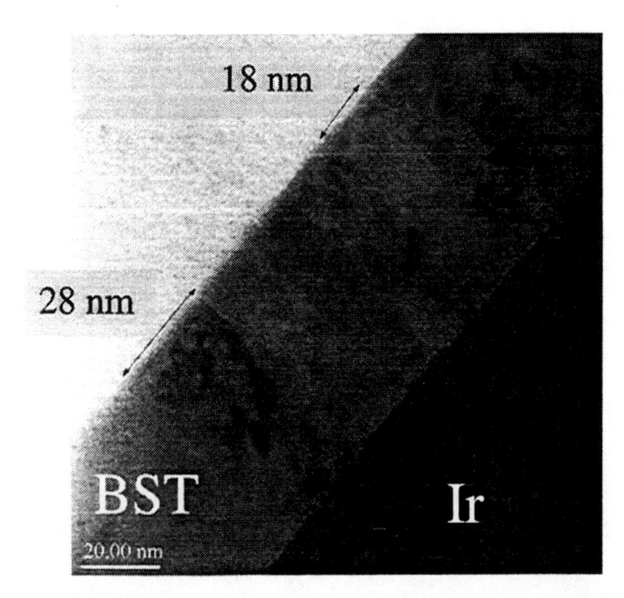

FIGURE 16. Cross-sectional TEM image of a BST film sputter deposited onto an Ir electrode. Lateral grain diameters are 15-30 nm.

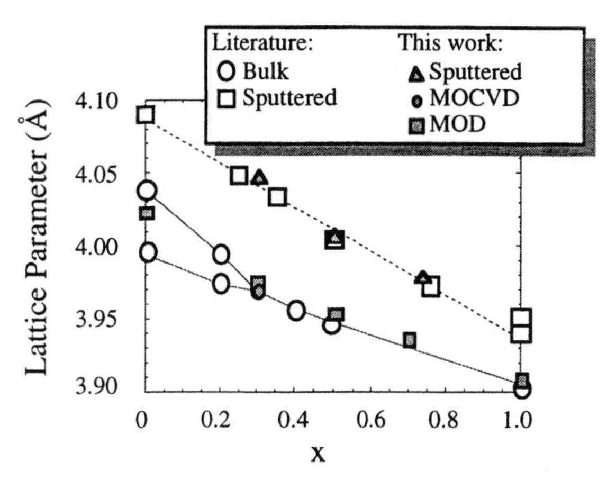

FIGURE 17. Lattice parameter of sputtered BST films as a function of stoichiometry. For comparison, bulk values (5) and sputtered films (11-13) are also shown. The sputtered films were deposited at 550°C.

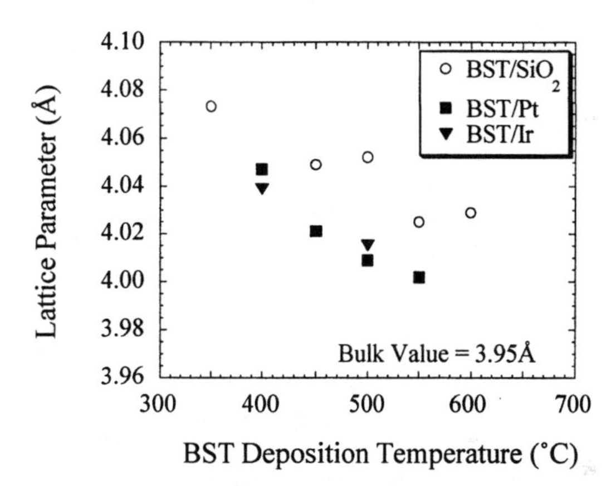

FIGURE 18. Lattice parameter vs. BST deposition temperature on several different substrate films.

variables, physical/chemical properties and electrical performance. The application of several, complementary analysis techniques was essential to achieve optimal insight into the various structure/property relationships. This learning is accelerating the introduction of high dielectric constant materials into advanced memory circuits.

ACKNOWLEDGMENTS

The authors wish to acknowledge the efforts of many process technologists within Motorola's Semiconductor Technology group which enabled the performance of this work. Special thanks for the individual contributions of Ray Doyle, Joe Hooker, Kathy Monarch, Gary Mote and Dan Sullivan. The Materials Research and Strategic Technology organization provided the motivation for this effort; many thanks to the support of Bradley Melnick, Robert E. Jones, Jr., Peter Zurcher and Sherry Gillespie.

REFERENCES

1. Fielding Jr., J. T., Jang, S. J. and Shrout, T. R., *Proceedings of the Ninth IEEE International Symosium on Applications of Ferroelectrics*, 363 (1994).
2. Nowotny, J. and Rekas, M. in *Key Engineering Materials* (Trans. Tech. Publications, Switzerland, 1992), p. 45.
3. Hwang, C. S., Park, S. O., Cho, H.-J., Kang, C. S., Kang, H.-K, Lee, S. I. and Lee, M. Y., *Appl. Phys. Lett.* **67** (1995).
4. Lee, W.-J., Park, I.-K., Jang, G.-E. and Kim, H.-G., *Jpn. J. Appl. Phys.* **34**, 196-199 (1995).
5. Basjamin, A. and DeVries, R.C., *J. Am. Ceram. Soc.* **40**, 373 (1957).
6. Arlt, G., Hennings, D. and de With, G., *J. Appl. Phys.* **58**, 1619-1625 (1985).
7. Horikawa, T., Mikami, N., Makita, T., Tanimura, J., Kataoka, M., Sato, K. and Nunoshita, M., *Jpn. J. Appl. Phys.* **32**, 4126-4130 (1993).
8. Kuroiwa, T., Tsunemine, Y., Horikawa, T., Makita, T., Tanimura, J., Mikami, N., Sato, K., *Jpn. J. Appl. Phys.* **33**, 5187-5191 (1994).
9. Yamamichi, S., Yabuta, H., Sakuma, T. and Miyasaka, Y., *Appl. Phys. Lett.* **64**, 1644-1646 (1994).
10. Tsu, R., Liu, H.-Y, Hsu, W.-Y., Summerfelt, S., Aoki, K. and Gnade, B., *Mat. Res. Soc. Symp. Proc.* **361**, 275-280 (1995).
11. Abe, K. and Komatsu, S., *J. Appl. Phys.* **77**, 6461-6465 (1995).
12. Kim, T. S., Kim, C. H. and Oh, M.H., *J. Appl. Phys.* **75**, 7998-8003 (1994).
13. Fujimoto, K., Kobayashi, Y. and Kubota, K., *Thin Solid Films* **169**, 249-256 (1989)
14. Ichinose, N. and Ogiwara, T., *Jpn. J. Appl. Phys.* **32**, 4114-4117 (1993).

Suppression of Boron Penetration for P+ Polysilicon Gate Electrodes by Ultra-Thin RPECVD Nitride Films in Composite Oxide-Nitride Dielectrics

Yider Wu [a] and Gerald Lucovsky [b]

Departments of Electrical & Computer Engineering [a], and Physics [b]
North Carolina State University, Raleigh, North Carolina 27695-7911, U.S.A.

Ultra-thin (0.4~0.8 nm) silicon nitride films were deposited by RPECVD (Remote Plasma Enhanced Chemical Vapor Deposition) onto thin oxides for the p+-polysilicon gated PMOS devices. Boron penetration is suppressed when the top nitride layer thickness increased to at least 0.8 nm. Suppression of boron diffusion is monitored by flat band voltage shifts as well as polysilicon depletion. Compared to oxides, improved charge to breakdown (Q_{bd}) is obtained for nitride/oxide dual layers, while interface state density remains at a level of $3x10^{10}$ $eV^{-1}cm^{-2}$.

INTRODUCTION

One of the metrology concerns of gate dielectric material for dual gate CMOS technology is the suppression of boron penetration. Surface-channel P-MOSFET with p+ polysilicon gate was recommended to suppress the short channel effect associated with buried-channel devices and improve turn-off characteristics. Unfortunately, diffusion of boron from heavily doped p+ polysilicon gate into underlying silicon substrate becomes a critical issue since it increases flatband voltage instability and decreases the sub-threshold slope. The boron diffusivity is even enhanced by the decrease of oxide thickness or the presence of fluorine, which is accompanied with BF_2 implantation. Oxynitride gate dielectrics has been suggested as an alternative for sub-0.25 micron technologies because of lower boron diffusivity. Previous study shows that nitridation process, such as with N_2O and NO, results in nitrogen incorporation at the SiO_2/Si interface with N amount less than 3% typically [1, 2]. The increased nitrogen concentration improves reliability and reduces diffusivity of boron [3], but increases fixed charge and interface trap density [4].

Furthermore, boron penetrated into the bulk oxide remains a problem as the diffused dopant saturated inside the gate dielectrics, which degrades the oxide reliability. Thus, it is ideal to have the diffusion barrier at the top polysilicon/dielectric interface. Because the boron diffusivity in oxynitride is decreased for increasing nitrogen concentration, ultra-thin layer of nitride on the top of gate oxide should be the optimal gate dielectric of PMOSFET. In this work, we report the use of Remote Plasma Enhanced Chemical Vapor Deposited thin nitride layer as boron diffusion barrier. The effects of boron penetration on flatband voltage, polysilicon depletion, and charge to breakdown are examined for p+ poly gated MOS devices. It is known that boron penetration inside the oxide increases the electron trapping, which was observed for high temperature NO or N_2O nitrided SiO_2 [5].

EXPERIMENTAL

MOS capacitors (100μm x 100μm) were fabricated on 1~10 ohm-cm <100> n-type silicon substrates. Gate oxide layers ranging from 4 to 6 nm were thermally grown in dry oxygen at 800°C, followed by RPECVD nitride depositions of 0.2 ~ 0.8 nm[6]. Active species from a remote excited He/N_2 mixture are combined with downstream injected SiH_4 to deposit the nitride films. The substrate temperature is 300°C, the process pressure is 0.2 Torr and an rf (13.56MHz) power input of 30W is used. The thickness of top nitride layer was estimated from the deposition rate, and confirmed by Auger electron spectroscopy [7] and spectroscopic ellipsometry. Post-nitride-deposition annealing of the dual layer was performed at 900°C for 30 sec in an inert ambient to drive out excess hydrogen, and minimize bonding defects in the nitride layer. Polysilicon was implanted by boron to a dose of $5\times10^{15}/cm^2$. A 140nm LTO film was then deposited onto the polysilicon to prevent boron out diffusion during activation annealing. Rapid thermal annealing in Argon at 1000°C for 1 to 4 minutes was used to activate the implanted boron. After metallization, the wafers were subjected to a post-metal-anneal in forming gas at 400°C for 30 minutes. High frequency and quasi-static C-V curves were measured. All sweeps were made from inversion to accumulation. The equivalent oxide thickness of N/O dual layers was determined by the accumulation capacitance. A PMOS capacitor with thermal oxide was fabricated and used as a control device.

CP449, *Characterization and Metrology for ULSI Technology: 1998 International Conference*
edited by D. G. Seiler, A. C. Diebold, W. M. Bullis, T. J. Shaffner, R. McDonald, and E. J. Walters

RESULTS AND DISCUSSION

Figure 1 shows AES spectra from the top surface of oxide, nitride/oxide dual layer, and nitrogen plasma nitrided gate dielectrics. Spectrum (a) shows the as loaded surface featuring the dominant Si-LVV and O-KLL signals. Spectra (b) and (c) show that the nitrogen signal strength increases with increasing the nitride deposition time. To compare the efficiency of incorporation of nitrogen atoms onto the oxide top surface, a sample was prepared with top nitridation process at the same plasma power (30W) and pressure (0.2torr), as shown in spectrum (d) [4]. The nitrogen peak strength of 1 minute plasma nitride deposition process is stronger than that of 10 minute plasma top nitridation process. Therefore a higher nitrogen areal density can be obtained at a reduced plasma energy for the nitride film deposition. This increased areal density is critical for reduction of boron penetration from p+ polysilicon gate electrode.

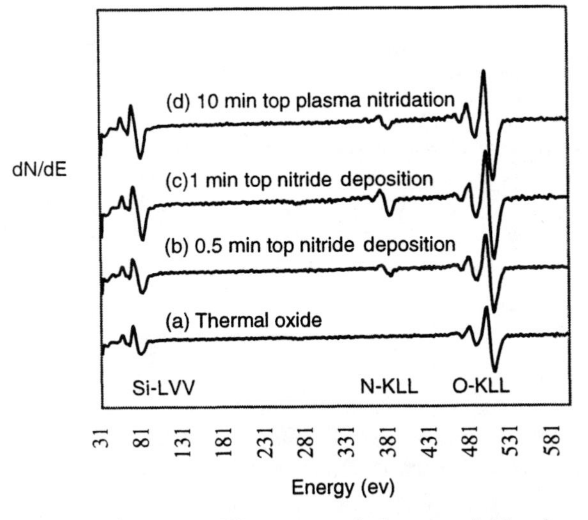

Figure 1. AES spectra of the top nitride deposition process and the top plasma nitridation process on the thermal oxide.

To quantity the nitrogen areal density at the oxide top surface, the thickness of top nitride layer was estimated from the nitride deposition rate, and then confirmed by Auger electron spectroscopy (AES) and spectroscopic ellipsometry (SE), as shown in Figure 2. The 0.8 nm of nitride is equivalent to an areal density of nitrogen atoms of approximately $4.3 \times 10^{15} cm^{-2}$. This is obtained directly from an assumed density of 3.1 gm/cm^3 for the plasma-deposited nitride after the 900°C anneal. If it is assumed that the effectiveness of blocking boron atom diffusion is related to the total boron cross-section defined as the areal density times the effective area of a single boron atom, then this normalized cross section is ~0.7 for the 0.8 nm nitride film. Experiments are currently underway using dual layer oxide-oxynitride alloy $[(SiO_2)_x(Si_3N_4)_{1-x}]$ dielectrics to control the effective areal nitrogen atom density, and to see how dilution of the nitrogen atom volume density affects the blocking power to boron atoms.

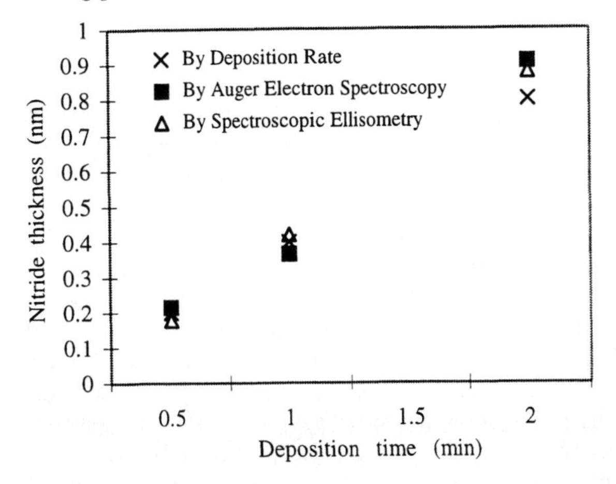

Figure 2. Determination of top nitride layer thickness by Auger Electron Spectroscopy and Spectroscopic Ellipsometry: comparison with the thickness as determined by the deposition rate.

In the nitride deposition process, the oxide film is exposed to the nitrogen plasma. Nitrogen atoms generated by rf plasma from the nitrogen source gas may diffuse into the oxide bulk. A post nitride deposition annealing at 900°C for 30sec is performed for chemical and structural relaxation, as well as to reduce the interface state density [8]. To understand the effect of plasma deposition and annealing on the diffusion of nitrogen in the oxide bulk, SIMS analysis of a nitride/oxide ~ 0.8nm/4.0nm dual layer structure was performed. As shown in Figure 3, after nitride deposition, the nitrogen concentration decays exponentially into the oxide due to the SIMS ion bombardment.

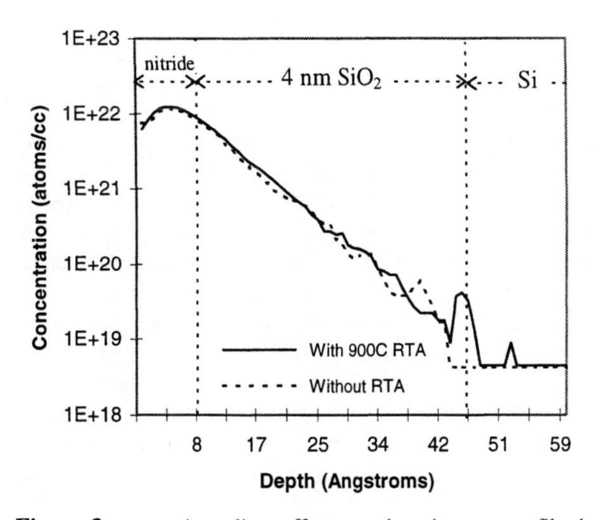

Figure 3. Annealing effect on the nitrogen profile by SIMS analysis for a nitride/oxide (~0.8nm/4.0nm) dual layer gate dielectrics. The N-atom profile is possibly broaden by the SIMS ion bombardment technique.

However, after an annealing at 900°C for 30sec, a nitrogen concentration peak at the oxide/silicon interface was found, showing that nitrogen atoms pile up at the oxide/silicon interface during annealing. The Si-N bonds at the interface replace Si-O bonds and have been assumed to relieve interface strain with the smaller effective size of the nitrogen atoms, and additionally provide a smoother interface[9]. This enhanced smoothness of the oxide/silicon interface may be responsible for increased high field mobility reported for N_2O oxides[10].

It is of interest to determine if fixed charge is introduced into oxide during the plasma nitride deposition process. The flatband voltage of a capacitor is determined by the work function difference between the gate electrode and the substrate doping, and modified by fixed oxide charge. By using the same gate electrode and substrate doping, the fixed oxide charge can be monitored by the flatband voltage shift as a function of nitrogen exposure. Al gate NMOS capacitors were fabricated with various top nitride thickness to compare the dependence of nitride deposition time on the flatband voltage. As shown in Figure 4, compared to thermal oxide, no flatband voltage shift was found by deposition of 0.2nm, 0.4nm, and 0.8nm of plasma nitride onto 6.1nm thermal oxide. No flatband voltage shift between thermal oxide and nitride/oxide dual layer gate dielectrics means: 1). effectively no charge is introduced into the bulk oxide during top nitride deposition and 2). no charge fixed at oxide/nitride interface. In addition, after forming gas annealing at 400°C for 30minutes, no significant difference in mid-bandgap D_{it} was observed between control oxide and top nitrided-oxide films for Al-gate capacitors, with the value ~ 2 - 5 x 10^{10} $eV^{-1}cm^{-2}$. This means that no measurable defects were introduced at the oxide/silicon interface by the remote PRCVD top nitride deposition process.

Figure 5 shows the superposition of three quasi-static C-V curves for capacitors with oxide and N/O dual layers. The C-V curve for control oxide was shifted dramatically to the right as compared to top-nitride-deposited devices. Based on the anticipated value of the flat band voltage as determined solely by substrate and gate electrode doping, this large shift of the reference oxide compared to the ON dielectric with the 0.8 nm nitride indicates significant boron penetration into the Si substrate. The flat band voltage shift of N/O dual layer with the 0.4 nm top nitride is intermediate, which means the amount of boron penetration is controllable by varying the top nitride thickness. Additionally, with this aggressively annealing at 1000°C for 4minutes, strong poly-depletion was observed for the control oxide, while almost no inversion capacitance reduction is observed for the device with the 0.8nm top nitride film, indicating that boron is depleted near the oxide/poly interface for control oxide device due to boron diffusion into the oxide.

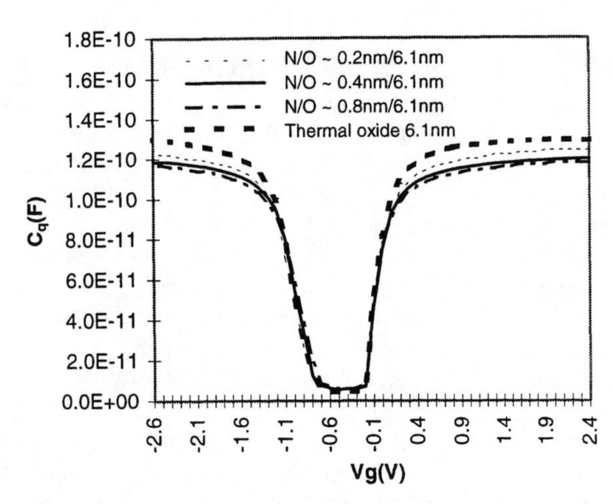

Figure 4. Superposition of CV curves for Al gate NMOS capacitors with various thickness top nitride on 6.1nm oxide.

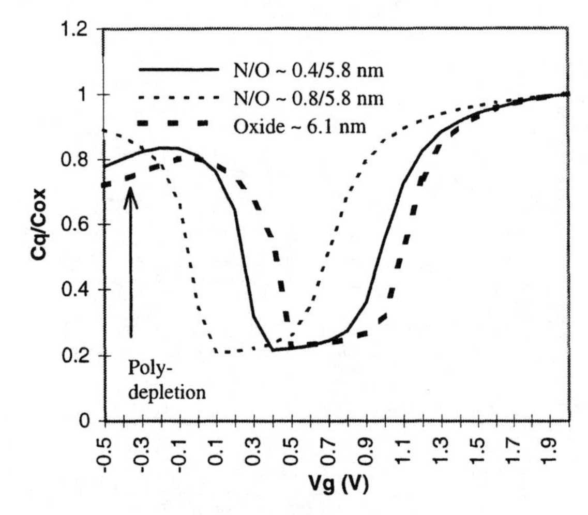

Figure 5. Normalized quasi-static C-V curve for thermal oxide, and 0.4 nm and 0.8 nm top nitride deposited onto thermal oxides. Curves are shifted due to boron penetration through thin gate material. The dopant activation annealing is 1000°C for 4min.

Oxide reliability was examined by charge to breakdown under constant current stressing (100 mA/cm^2). The Q_{bd} value (50% failure) of 0.8 nm top nitride N/O film is 15.8 C/cm^2 while that of control oxide is smaller, 9.1 C/cm^2. The improvement of Q_{bd} for N/O dual layer is associated with preventing boron diffusion into oxide bulk. In contrast, nitrided oxides fabricated by oxidation, or high temperature annealing in NO or N_2O, with N-rich layers at oxide/Si interface, show accumulation of boron at the SiO_2-Si interface and oxide bulk, resulting in poorer Q_{bd} reliability than for a non-nitrided thermal oxide[5].

An enhanced boron blocking capability by Post-Nitride-Deposition Annealing is observed by monitoring the CV curve shift. Annealing at 900°C for 30 sec drives out excess hydrogen, densifies the film, and minimizes bonding defects in the nitride layer, and hence further retards the diffusion of boron through the top nitride layer for the additional high temperature activation anneal of polysilicon. As shown in Figure 6, the CV curve with nitride/oxide dual layer without annealing shows a small shift to the positive voltage compared to that of N/O with RTA, which means a little amount of boron can migrate through the unannealed nitride films to reach the oxide/silicon interface. Also, the distortion of CV curve at the onset of inversion region for the N/O dual layer without annealing is possibly due to a p-type layer under oxide/Substrate interface.

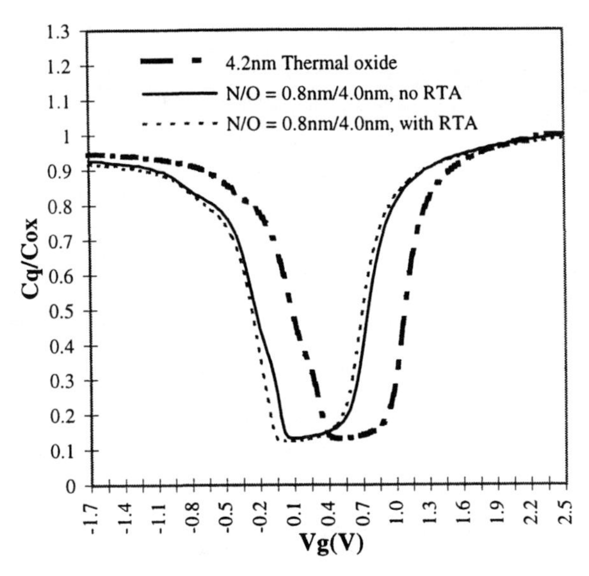

Figure 6. Normalized quasi-static C-V curve for oxide and N/O stack with and without post-deposition treatment. The RTA condition is at 900°C for 30 sec in the He ambient. And the dopant activation annealing is 1000°C for 60sec.

CONCLUSION

We have confirmed that the deposition of 0.4 to 0.8 nm Remote PECVD nitride on a thin gate thermal oxide provides an effective barrier against the diffusion of boron into the PMOS channel region, resulting in higher flatband voltage stability and lower polysilicon depletion. The integrity of dielectric films was examined by Q_{bd} and D_{it}, showing the nitride barrier improved gate dielectric properties. By preventing boron diffusion into the oxide bulk, the reliability degradation by boron accumulation in oxide, which is observed in thermally nitride oxides, does not exist in the plasma-nitride/oxide dual layer films. As compared to other high temperature NH_3, N_2O, and NO nitridation processes, which only form the nitrogen rich layer at the oxide/silicon interface, the N/O dual layers can obtain high nitrogen concentration at oxide/polysilicon interface as well as interfacial nitridation. Thus the dual-layer ON structure is a promising gate dielectric material for the aggressively scaled sub-micron CMOS technology.

REFERENCES

1. L. K. Han, D. Wristers, J. Yan, M. Bhat, and D. L. Kwong, "Highly Suppressed Boron Penetration in NO-nitrided SiO_2 for p+-Polysilicon Gated MOS Device Application," *IEEE Electron Device Letter,* vol. 16, p. 319, 1995.
2. Z. J. Ma, J. C. Chen, Z. H. Liu, J. T. Krick, Y. C. Cheng, C. Hu, and P. K. Ko, "Suppression of Boron Penetration in p+ Polysilicon gate P-MOSFET's using Low-Temperature Gate-Oxide N_2O Anneal," *IEEE Electron Device Letter,* vol. 15, p. 109, 1994.
3. R. B. Fair, "Modeling Boron Diffusion in Ultra-thin Nitrided Oxide p+ Si Gate Technology," *IEEE Electron Device Letter,* vol. 18, p. 244, 1997.
4. S. Hattangady, R. Kraft, D. Grider, M. Douglas, G. Brown, P. Tiner, J. Kuehne, P. Nicollian, and M. Pas, "Ultra-thin Nitrogen-Profile Engineered Gate Dielectric Films," *IEDM Tech. Dig.,* p.495, 1996.
5. D. Wristers, L. K. Han, T. Chen, H. H. Wang, and D. L. Kwong, "Degradation of Oxynitride Gate Dielectric Reliability due to Boron Diffusion," *Applied Physics Letter,* vol. 68, p. 2094, 1996.
6. S. Hattangady *et al.*, *J. Vac. Sci. Technol.*, vol. A14, 3017, 1996.
7. C. Chang, *Surface Science*, vol. 48, 8, 1975.
8. Y. Ma, T. Yasuda, and G. Lucovsky, "Ultrathin Device Quality Oxide-Nitride-oxide Heterostructure formed by Remote Plasma Enhanced Chemical Vapor Deposition," *Applied Physics Letter,* vol. 64, p. 2226, 1994.
9. M. L. Green *et al.*, "Rapid Thermal Oxidation of Silicon in N_2O between 800 and 1200°C: Incorporated Nitrogen and Interfacial Roughness." *Applied Physics Letter,* vol. 65, p. 848, 1994.
10. Z. H. Liu, J. Krick, H. Wann, P. Ho, C. Hu, and Y. Chang, *IEDM Tech. Dig.*, p. 625, 1992.

Luminescence Measurements of Sub-Oxide Band-Tail and Si Dangling Bond States at Ultrathin Si-SiO2 Interfaces

A.P. Young[1], J. Schäfer[2], G. H. Jessen[1], R. Bandhu[3], L. J. Brillson[1,2,3]

[1]Department of Electrical Engineering, [2]Center for Materials Research, [3]Department of Physics
The Ohio State University, 2015 Neil Ave., Columbus OH, 43210-1272, USA

G. Lucovsky and H. Niimi

Department of Material Science and Engineering, North Carolina State University,
Raleigh, NC 27695-8202, USA

We have directly observed cathodoluminescence (CLS) in ultrahigh vacuum (UHV) over a broad spectral range (0.7-4.0 eV) from ultrathin 5 nm layers of remote plasma enhanced CVD grown a-SiO_2:H deposited on silicon substrates. In the infrared regime, luminescence is observed at 0.8 eV consistent with the presence in the as-deposited dielectric of Si dangling bond localized states, as well as at 1.1 eV due to band edge emission. In the optical regime, three peaks are observed showing evidence for band tail state emission from an amorphous silicon-oxygen bonded sub-oxide region in the film, together with smaller contributions from either substrate related c-Si or defect containing, stoichiometric SiO_2. CLS spectra at varying excitation indicates that the stoichiometric SiO_2 is very close to the surface of the film, possibly due to oxidation of the air-exposed wafer or due to a nonuniformity in the film. When the films are annealed *in situ* in stages up to 500 °C, we observe no change in the shape of the a-SiO_x:H associated peak indicating the stability of this sub-oxide to such temperatures. These observations are consistent with CLS measurements of thicker films of a-SiO_2:H and a-SiO_x:H, demonstrating the utility of CL spectroscopy for the study of ultrathin dielectric studies.

INTRODUCTION

Si-SiO_2 interfaces are a subject of both fundamental interest as well as being extremely important to the performance of electronic devices, notably metal-oxide-semiconductor (MOS) type structures. Recently gate linewidths have shrunk deep into the submicron regime, and if the scaling laws of electronic devices are to continue to hold, oxide thicknesses must shrink as well. For amorphous SiO_2, the current oxide used in MOS structures to remain as a viable dielectric for future generation ULSI circuitry, the thickness of this layer must decrease to 3 nm and less.

At the Si-SiO_2 interface, it is well known that a silicon-rich sub-oxide exists between the stoichiometric, amorphous SiO_2 (a-SiO_2) and the crystalline Si (c-Si) substrate(1). As the thickness required of the oxide shrinks to the point where a significant fraction of the sub-oxide interface is part of the dielelctric, detailed knowledge correlating the chemical and electrical properties of the interface becomes of increasing importance to the performance of any future device.

Because the thickness of the transition layer is so thin, it is difficult to easily deconvolute the electronic properties of the suboxide from the rest of the "bulk" stoichiometric oxide. Photo-induced luminescence is difficult with conventional sources, given the large bandgap of SiO_2 (9.1 eV) and other dielectrics(2) . Cathodoluminescence and photoluminescence spectroscopy (CLS and PL) have been used to study the electronic properties of bulk (~1 μm) SiO_2 films(3) , thick (~1 μm) SiO_x films(4,5,6), as well as oxidized porous silicon structures (7). While CLS has been extensively used for the study of bulk SiO_2, to the authors' knowledge, we know of no positive CL results for extremely thin, silicon dioxide films. In this work, we present what we believe is the first spectroscopic CLS measurements of ultrathin (5 nm) SiO_2 films on Si substrates.

EXPERIMENT

The 5 nm thick hydrogenated SiO_2 (a- SiO_2 :H) films were deposited onto p-type silicon wafers by remote plasma

CP449, *Characterization and Metrology for ULSI Technology: 1998 International Conference*
edited by D. G. Seiler, A. C. Diebold, W. M. Bullis, T. J. Shaffner, R. McDonald, and E. J. Walters

enhanced chemical vapor deposition (RPECVD) onto ultrathin oxide surfaces (~0.5 nm) prepared by in-situ plasma-assisted oxidation in a multi-chamber UHV system for both oxides and dielectrics. Additional details of the growth chamber apparatus and the growth process have been described elsewhere (8,9,10). All the films were then transported through air to a separate UHV chamber with a base pressure of ~1-2 x 10^{-10} Torr for CLS analysis.

The Si-SiO_2 films were measured at room temperature and to a temperature of T~90K. The experimental setup employs an electron gun using voltages and currents in the range of 0.6-4.5 keV, and ~1.0-4.0 μA, respectively for the excitation of the luminescence. The beam spot size was ~200 - 500 μm. A LN_2 cooled North Coast p-i-n Ge detector measured luminescence in the energy range 0.7-2.0 eV, while a thermoelectrically cooled S-20 photomultiplier (PMT) was used to measure luminescence in the spectral range 1.4 - 4.0 eV.

The Leiss monochromator employed contains a flint glass prism with nominally 50 meV resolution over the measurement range from the near infrared through the near UV part of the spectrum. For the in-situ annealing experiments, the specimens were rapidly brought up to temperature (~1 min.), held for 5 min. at the specified annealing temperature, then brought back to room temperature for analysis. During the annealing process, the specimen temperature was measured using an infrared pyrometer. The CLS spectra presented here have not been corrected for the spectral responsivity dependence of the detector or the transmission of the optical train.

RESULTS

Even though the a-SiO_2:H film was only 5 nm thick, luminescence was strong enough to be faintly visible to the naked eye at room temperature. Subjectively, the light appeared predominately white, with some blue and red emission noticeable. Besides providing quantitative luminescence spectra for the wide bandgap materials, CLS has another distinct advantage over optical excitation techniques, namely a large variable excitation depth controllable via the incident electron beam energy. Fig 1, shows spectra from the as-deposited specimen for various excitation voltages (0.6 - 4.5 keV) with discrete peaks apparent in the visible and near UV region of the electromagnetic spectrum. There is an obvious change in spectral features as a function of beam energy. The low energy spectra at 0.6 and 1.0 keV show broad luminescence peak features centered at 2.7 eV and 1.9 eV. When the beam intensity is increased from 1.5 keV up to 4.5 keV, the features change dramatically, with a much broader band centered at 1.9 eV, and a weak peak developing at 3.4 eV, which increases with increasing beam voltages.

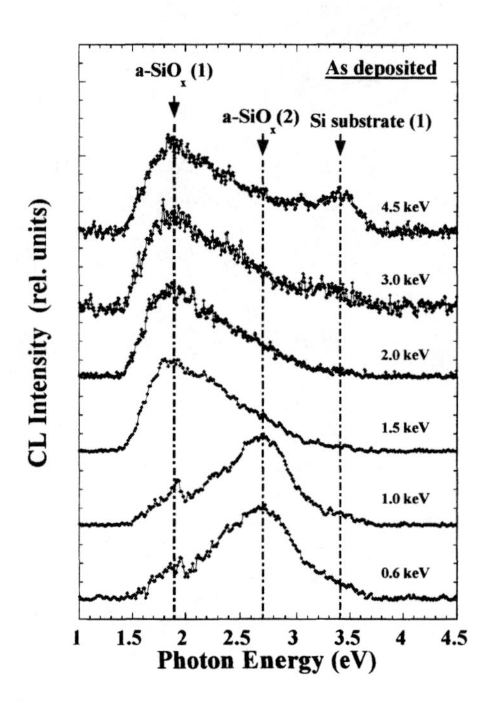

FIGURE 1. Voltage-dependent CLS spectroscopy of the as-deposited a-SiO_2:H film in the visible and near UV spectrum using an S-20 PMT. The spectra were taken with the substrate at room temperature.

The temperature dependence of the luminescence can help clarify the stoichiometry and bonding arrangements of the various forms of Si and SiO_x in the film, since the luminescence from a-Si:H is known to be strongly quenched, while a-SiO_x:H shows a significantly smaller quenching effect. As can be seen in Fig. 2, after lowering the temperature of the as-deposited film to 90 K, the overall spectrum shape did not change significantly, with only a minor intensity increase observed.

In addition to the spectra in the visible and UV, the infrared spectrum is shown in Fig. 3 for the as-deposited film at an excitation voltage of E=2 keV. Two peaks were clearly resolved at ~0.8 eV and ~1.0 eV. At other voltages, the spectra was qualitatively similar to the 2 keV spectrum with only minor changes observed. For voltages between 0.6 keV and 3.0 keV, the only significant change observed was an increase in the relative intensity of the

1.0 eV band compared to the 0.8 eV band as the excitation voltage increased. Also shown in Fig. 3 is the CLS spectrum after the specimen was annealed in 5 min. stages *in-situ* up to 500°C. In the figure, it is clear that there is dramatically enhanced intensity at 0.8 eV after the annealing, compared to the as-deposited case.

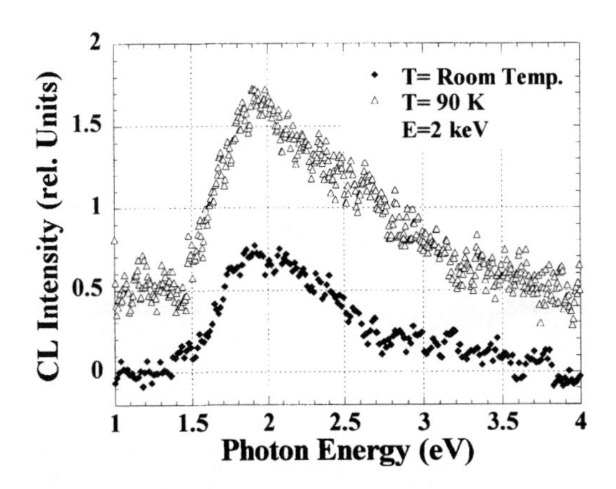

FIGURE 2. Temperature-dependent CLS spectra of the as-deposited a-SiO_2:H. Notice the lack of change in the spectra intensity consistent with the picture of an amorphous sub-oxide film.

Fig. 4 shows the corresponding visible/UV spectrum as a function of annealing temperature for the same excitation voltage. In contrast with the strong increase in luminescence in the infrared, there is relatively little change in intensity observed in the visible region of the spectrum. The only significant change that we observed is the slight strengthening of the 3.4 eV band as the annealing progresses.

We also addressed the possibility of electron-beam induced damage as there is always the question of the stability of the material. For a 2.0 keV beam with a current of ~1 μA, the total exposure in on the order of 10^{17} electrons / cm^2, two orders of magnitude below the reported damage threshold for a thicker SiO_2 film(11). For our experiment, a typical CLS spectrum was measured, then the beam was allowed to remain on one spot for two hours, and the spectra was then remeasured. As can be seen in the Fig. 5, there is only a minor change in the intensity of the 1.9 eV peak, within the experimental error of apparatus. By contrast, typical CL spectra are obtained with over an order of magnitude lower exposure (flux times time), so that damage is not deemed a significant problem for this system with the short (~12 min.) exposure times and currents.

DISCUSSION

The high energy of the incident electron beam relative to the insulator band gap (i.e. 1-2 keV vs. 9.1 eV) is a distinct advantage over traditional optical excitation for the generation of large densities of electron-hole pair recombination in wide band-gap materials such as SiO_2. With such a high energy excitation source, the initial electron generates a cascade of electrons and holes, multiplying the generation of free electron-hole pairs. Compared to the 15-20 keV normally used by a scanning electron microscope (SEM) for CLS, the relatively low energy 0.5-4.5 keV beam we use has a significantly shallower penetration depth compared to typical SEM based CL and PL spectroscopies(12,13). Therefore, low energy CLS is particularly well suited for probing thin, wide bandgap layers and can explain why CL has been successful in producing detectable luminescence from 5 nm of SiO_2, while laser excitation has not been successful.

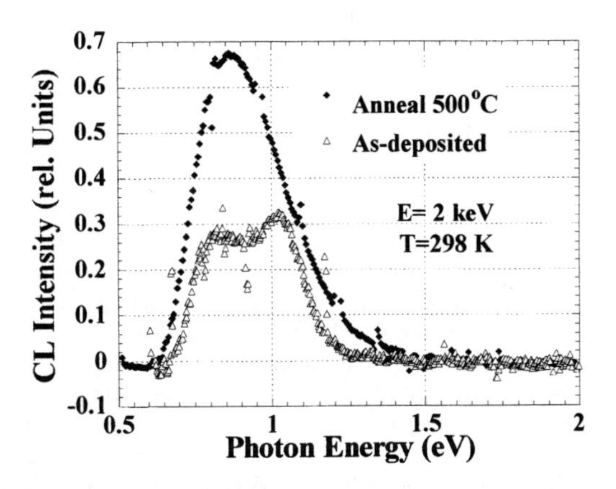

FIGURE 3. Room temperature infrared CLS spectra of the a-SiO_2:H film before and after a five minute anneal at 500°C. The spectra were obtained with a Ge detector and with the electron beam voltage at 2 keV. The 0.8 eV peak increases significantly with annealing, indicating an increase in luminescence from Si dangling bonds, correlating with the removal of H from the film.

From the data presented in Fig. 3, we can immediately identify the presence of stoichiometric SiO_2 at the very top surface layer. The combination of both 2.7 eV and 1.9 eV emission is very similar to that observed in bulk layers of a-SiO_2(3). Furthermore, for the higher excitation energies (1.5 - 4.5 keV), the 2.7 eV peak diminishes significantly, and in its place is a single, broad peak at 1.9 eV (FWHM= 0.9 eV), very suggestive of band tail transitions in an amorphous, hydrogenated, Si rich sub-oxide(14,15,16). Additional evidence supporting the presence of this local excess silicon bonding is provided by the data in the infrared. The 0.80 eV peak has been observed in thick, a-Si:H films corresponding to an optical transition from amorphous Si band tail to mid-gap dangling bond states (17). In addition to the 0.80 eV peak, the luminescence band at 0.95-1.05 eV can be assigned to the crystalline silicon substate, since the band edge (E_g=1.12 eV) emission increased as the electron beam penetrated deep into the substrate. Therefore, both the infrared and visible spectrum support the presence of sub-oxide and silicon-silicon bonding features in the film.

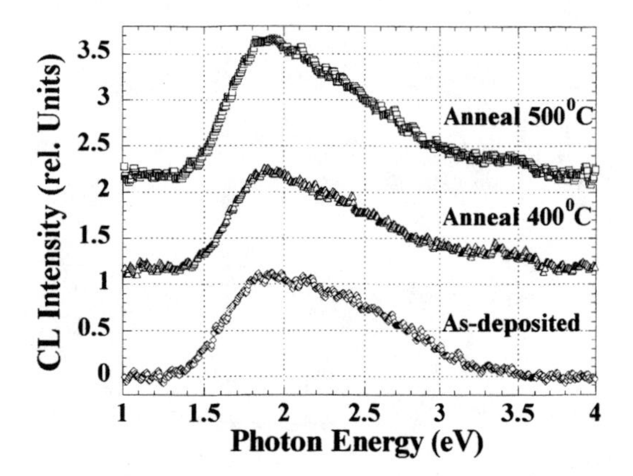

FIGURE 4. Room temperature CLS spectroscopy of the a-SiO_2:H film after five minute consecutive anneals between room temperature and 500°C using the S-20 PMT detector. All spectra were obtained with the electron beam voltage at 2 keV. There is no change in the overall spectra with annealing indicating no change in the stoichiometry of the film up to 500°C. The spectra are offset for clarity.

The final peak feature observed is a weak ~3.4 eV shoulder at higher incident beam energies in the as-deposited specimen. A clue to its origin is the increasing intensity of the shoulder as the electron beam penetrates deeper into the film. From this observation and the lack of a strong 2.7 eV peak, the luminescence band might be due to the Γ-Γ transition (~3.5 eV) of bulk-crystalline silicon, most likely coming from the substrate, not the SiO_2 film itself. On the other hand, this band could also be due to a defect in SiO_2 where peaks have been observed before by PL(18). This latter interpretation is less likely however, since most luminescence measurements of SiO_2 do not have this defect feature, and when present, it is only in combination with luminescence from other defect bands between 2.2 -2.7 eV which are not observed in these spectra. Therefore, without additional data, we cannot definitively correlate this feature with the properties of bulk films.

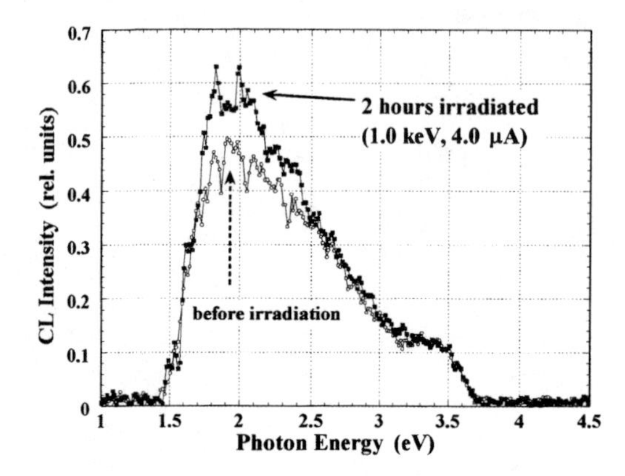

FIGURE 5. Test for the possibility of e-beam damage during the measurement of a single CLS spectrum. Only a minor change within experimental error in the 1.9 eV peak is observed after 2 hours of continuous irradiation on a single spot.

Temperature-dependent studies provide further information concerning the origin of the luminescence peaks. In Fig. 2, we detected only a minor increase in the 1.9 eV luminescence intensity for the as-deposited film at T= 90 K, compared to the room temperature spectra. This observation further supports the conclusion that the 1.9 eV luminescence is associated with SiO_2 or a sub-oxide rather than a-Si:H, whose PL spectra is known to be strongly quenched at room temperature (19). In addition to the *in-situ* annealing experiment points out the stability of the main bonding arrangement, and it emphasizes the role of H in passivating Si dangling bonds in the film. While the visible and UV range spectra in Fig. 4 show little change, the infrared spectrum shows a dramatic increase in intensity of the 0.8 eV band consistent with the evolution of H from the film, and the consequent increase of Si dangling bond states. Simultaneously, we observed a small intensity increase at 3.4 eV. Either this intensity increase infers an increase in c-Si if the 3.4 eV peak is associated with silicon, or to SiO_2, if the same 3.4 eV peak is associated with a defect in a-SiO_2. Annealing to higher temperatures such as 900°C, where it is known any residual sub-oxide would completely separate into c-Si and a-SiO_2(15) would help to clarify the role of H and the excess Si.

CONCLUSION

In summary, we have directly observed cathodoluminescence from ultrathin 5 nm SiO_2 dielectric layers. From our measurements, we can clearly differentiate between Si bonding arrangements in these films using a relatively simple, yet surface-sensitive characterization technique. While the layers are thin, they are nevertheless thick enough to show characteristic luminescence spectrum similar to what is observed in thicker films. The 5 nm SiO_2 grown by RPECVD is shown to initially have silicon-rich regions in addition to stoichiometric SiO_2 stable up to at least 500°C. Simultaneously, the passivating role of H on Si dangling bonds was also clearly observed in the luminescence spectra. From these measurements, it is clear CLS is a powerful tool for probing the electronic structure at these ultrathin Si-SiO_2 interfaces and how they change with thermal annealing.

ACKNOWLEDGMENTS

This work is supported in part under NSF contract # DMR-9711851, the NSF Division of Materials Research, the NSF ERC for Advanced Electronic Materials Processing, and the Office of Naval Research.

REFERENCES

1. F.J. Grunthaner and P.J. Grunthaner, Mater. Sci. Reports 1, 65, (1986).
2. W.C. Choi, M.S. Lee, C.K. Kim, S.K. Min, C.Y. Park, and J.Y. Lee, *Appl. Phys. Lett.* **69**, 3402 (1996).
3. M. A. Kalceff and M. R. Phillips, *Phys. Rev. B* **52**, 3122 (1995).
4. B. J. Hinds, F. Wang, D. M. Wolfe, C.L. Hinkle, and G. Lucovsky, *J. Non-Cryst. Solids,* (in press).
5. F. Koch and V. Petrova-Koch, *J. Non-Cryst. Solids* **198-200**, 840, (1996).
6. Y. Nakayama, M. Uecha, and T. Ikeda, *J. Non-Cryst. Solids* **198-200**, 915 (1996).
7. A.G. Cullis, L.T. Canham, P.D.J. Calcott, and references therein, *J. Appl. Phys.***82**, 909 (1997).
8. G. Lucovsky, A. Banerjee, B. Hinds, B. Claflin, K. Koh, and H. Yang, *J. Vac. Sci. Technol. B* **15**, 1074 (1997).
9. D.R. Lee, G. Lucovsky, M.S. Denker, and C. Magee, *J. Vac. Sci. Technol. A* **13**, 1671 (1995).
10. T. Yasuda, Y. Ma, S. Habermehl, G. Lucovsky, *Appl. Phys. Lett.* **60**, 434 (1992).
11. J. S. Johannessen, W.E. Spicer, and Y.E. Strausser, *J. Appl. Phys.* **47**, 3028 (1976).
12. B.G. Yacobi and D.B. Holt, Cathodoluminescence Microscopy of Inorganic Solids, (Plenum Press, New York, New York, 1990), pg.151.
13. L.J.Brillson and R.E. Viturro, *Scan. Electron Microscope* **2**, 789 (1988).
14. M.A. Paesler, D.A. Anderson, E.C. Freeman, G. Moddel, and W. Paul, *Phys. Rev. Lett.* **41**, 1492 (1978).
15. J. C. Knights, R.A. Street, and G. Lucovsky, *J. Non-Cryst. Solids* **35-36**, 279 (1980).
16. R. Carius, R. Fischer, E. Holzenkampfer, and J. Stuke, *J. Appl. Phys.* **52**, 4241 (1981).
17. R. A. Street, *Advances in Physics* **30**, 593 (1981).
18. Keunjoo Kim, M.S. Suh, T.S. Kim, C. J. Youn, E.K. Suh, Y.J. Shin, K.B. Lee, H.J. Lee, M.H. An, H.J. Lee, and H. Ryu, *Appl. Phys. Lett.* **69**, 3908 (1996).
19. B. Hinds, A. Banerjee, R.S. Johnson, and G. Lucovsky, *Mat. Res. Society Proceedings*, 1997 (in press).

Optical Studies of Phosphorus-doped Poly-Si Films

Stefan Zollner*, Ran Liu, Jim Christiansen, Wei Chen, Kathy Monarch, Tan-Chen Lee

Motorola Semiconductor Technology, Arizona Technology Laboratories, Technology Test and Analysis Laboratory, MD M360, 2200 West Broadway Road, Mesa, AZ 85202

Rana Singh, Jane Yater, Wayne M. Paulson, Chris Feng

Motorola Semiconductor Product Sector, MD K21, Austin, TX 78721

We have performed a detailed optical study of phosphorus-doped polycrystalline silicon films (prepared by amorphous deposition with *in situ* doping and annealing by thermal oxidation up to 1000°C) using Raman spectroscopy and variable-angle spectroscopic ellipsometry from 0.7 to 5 eV. We focus on the determination of strain and grain size using optical techniques and compare with results of a structural analysis using plan-view transmission electron microscopy and atomic force microscopy. We analyze the derivatives of our ellipsometry data (peak shifts and broadenings) using analytical lineshapes, which are affected by grain size, film thickness, doping, and inhomogeneity, but only minimally by macroscopic biaxial strain.

INTRODUCTION

Dielectric breakdown of thin gate oxides is a concern for ULSI circuit failure. Because of the difference in molar volumes between Si (20 Å^3 per atom) and its oxide (40 to 45 Å^3 per molecule), a large biaxial stress is expected at the interface. The shear forces act along, but not normal to the interface, since the oxide can expand vertically, but is bound by the Si bonds at the interface. While most of this stress is released by a plastic viscoelastic flow of the amorphous oxide at high temperatures (typically present during deposition or thermal oxidation), some of it remains and leads to a warping of the Si wafer, which can be measured using a FLEXUS tool. This stress has been identified as the culprit for stress-induced leakage currents which lead to dielectric breakdown (1).

Two optical studies (2,3) using Fourier-transform infrared (FTIR) spectroscopy, photoreflectance, and spectroscopic ellipsometry (SE) claim to have observed this stress by measuring the energy of the Si-O stretching vibration in SiO_2 and of the E_1 direct band gap in Si. However, two similar studies (4,5) using the same techniques (FTIR, SE) interpret the data in an alternate way without invoking stress. Therefore, the existence and magnitude of stress at the Si/SiO_2 interface is still doubtful.

The situation is different for poly-Si on SiO_2, where the existence of stress can be deduced from the Si-Si optical vibration energy measured by Raman spectroscopy (1). It was found that the stress in poly-Si is less tensile (and the dielectric breakdown improved) by high arsenic doping concentrations. Apparently, the As atoms pile up at grain boundaries and at the interface and thus release the stress. In order to test this hypothesis, we have performed a similar study using phosphorus-doped poly-Si films, where a similar stress relief is expected.

EXPERIMENTAL RESULTS

We have prepared 13 high-quality poly-Si films (up to 3000 Å thick) on Si (001) with 90 Å of tunnel oxide by amorphous deposition at 550°C using silane-based CVD at a pressure of 400 mT (growth rate 20 Å/min) and thermal processing (up to 1000°C) as described elsewhere (6). The films were *in situ* doped with phosphorus by adding phosphine to the gas flow. Secondary ion mass spectrometry measurements found P concentrations up to 3×10^{20} cm^{-3}. The sample parameters are given in Table 1.

The rms surface roughness (measured by atomic force microscopy) was between 12 and 17 Å for all films. The maximum contrast was about ten times larger. The roughness increases slightly by annealing, doping, and with increasing film thickness. No obvious lateral features (due to grain structure) were observed in the images. Images for amorphous and polycrystalline samples looked similar. The lateral features are about 200 to 500 Å in size, much smaller than the grain size, as discussed below.

Using plan-view transmission electron microscopy, it was possible to observe individual grains, which range in size from 1000 to 10000 Å. Most grains were about 3000 to 5000 Å large. There was no obvious dependence of the grain size on doping or annealing temperature. However, we

* Electronic Mail: R37595@email.sps.mot.com.

CP449, *Characterization and Metrology for ULSI Technology: 1998 International Conference*
edited by D. G. Seiler, A. C. Diebold, W. M. Bullis, T. J. Shaffner, R. McDonald, and E. J. Walters

noticed a difference between samples annealed at 800°C and 1000°C: At 800°C, the grains were shaped irregularly and showed small-scale internal structure, possible due to inhomogeneous strain fields. At 1000°C, the grains were shaped like simple polygons with less internal structure except for an occasional straight line (dislocation) across the grain. Apparently, annealing at 1000°C does not significantly increase the grain size, but reduces the inhomogeneous strain fields present in the grains produced by the 800°C anneal.

Raman spectra of the annealed samples were acquired at 300 K with the 457.9 nm Ar^+-ion laser line (penetration depth 2800 Å in Si) and a power of 5 mW (to avoid laser heating) using a Dilor XY800 triple-spectrometer (equipped with an LN_2 cooled Spectrum One CCD) in high-dispersion additive mode. A Raman microscope was used to focus the laser to a spot with less than 1 μm diameter and to collect the scattered light. The integration time was about 5 min. For bulk Si, the Raman shift (optical phonon energy) was found as ω_0=520.4 cm^{-1}.

All samples exhibited narrow symmetric Raman lines similar to those of bulk Si. No additional (amorphous) peaks were observed. After removing spikes from the CCD detector, the spectra were fitted using a combination of a Gaussian and Lorentzian lineshape. Peak energies could be determined with an accuracy of about 0.01 cm^{-1}. The changes of the Raman shift (relative to bulk Si) and the linewidths for the annealed samples are given in Table 2.

The ellipsometric angles (tanψ and cosΔ) were determined from 0.7 to 5.4 eV at three angles of incidence (65, 75, and 85°) using a variable angle-of-incidence spectroscopic ellipsometer with rotating analyser (J.A. Woollam, Co., Inc.). The experimental data for wafer #26 are given in Fig. 1. Using a commercial software package, the thicknesses of the tunnel oxide, the poly-Si (or a-Si), and the thermal (or native) oxide were determined. The thicknesses were close (±5-10%) to the nominal values in Table 1. On the as-deposited films, the native oxide was about 20–30 Å thick. The thickness of the thermal oxide on the annealed films was between 130 and 170 Å.

In order to model the optical constants for a-Si and poly-Si, we first use the parametrization of McGahan (6) (included in the software), which has the crystallinity as the only parameter (3 to 6% for as-deposited films, 98% for annealed films). In a second step, we model them with the parametric oscillator model, which has many parameters (7). This barely changes the film thicknesses, indicating that the first step was a good approximation. Finally, we fix the film thicknesses determined in the previous steps and determine the optical constants for the Si films at each photon energy from the measured ellipsometric angles and the known dielectric function of the substrate and the oxide. For the as-deposited film, we find a broad absorption centered at 3.3 eV as expected for amorphous Si. The dielectric functions for the films annealed at 800°C and 1000°C are shown in Figs. 2–3.

TABLE 1. Nominal thickness, phosphorus concentration, and annealing conditions for poly-Si wafers studied here.

wafer	poly thickness	P content	annealed at
#2	NONE	N/A	NONE
#4	1500 Å	3×10^{20} cm^{-3}	NONE
#6	1500 Å	3×10^{20} cm^{-3}	800° C/3 hrs
#8	1500 Å	3×10^{20} cm^{-3}	1000° C/12 min
#10	3000 Å	3×10^{20} cm^{-3}	NONE
#12	3000 Å	3×10^{20} cm^{-3}	800° C/3 hrs
#14	3000 Å	3×10^{20} cm^{-3}	1000° C/12 min
#7	300 Å	1×10^{20} cm^{-3}	800° C/3 hrs
#16	1500 Å	1×10^{20} cm^{-3}	NONE
#18	1500 Å	1×10^{20} cm^{-3}	800° C/3 hrs
#30	1500 Å	1×10^{20} cm^{-3}	1000° C/12 min
#22	3000 Å	1×10^{20} cm^{-3}	NONE
#24	3000 Å	1×10^{20} cm^{-3}	800° C/3 hrs
#26	3000 Å	1×10^{20} cm^{-3}	1000° C/12 min

DISCUSSION

For the analysis of the Raman results, we assume that the stress is biaxial in the (001) plane, i.e., that the shear forces only act along the interface, and can be described by the stress tensor (8-9)

$$\vec{X}=\begin{pmatrix} X & 0 & 0 \\ 0 & X & 0 \\ 0 & 0 & 0 \end{pmatrix}. \qquad (1)$$

TABLE 2. Raman shift Δω (compared to bulk Si), deduced biaxial stress *X* (ignoring free-carrier effects), and peak width Γ (full width at half maximum). The negative sign for *X* corresponds to a compressive stress.

wafer	Δω (cm^{-1})	*X* (GPa)	Γ (cm^{-1})
#2	0.01	0.00	2.80
#6	–0.30	0.06	3.35
#8	–0.01	0.00	3.24
#12	–0.65	0.14	3.72
#14	–0.18	0.04	3.69
#18	–0.24	0.05	3.47
#30	0.24	–0.05	3.24
#24	0.00	0.00	3.93
#26	0.06	–0.01	3.35

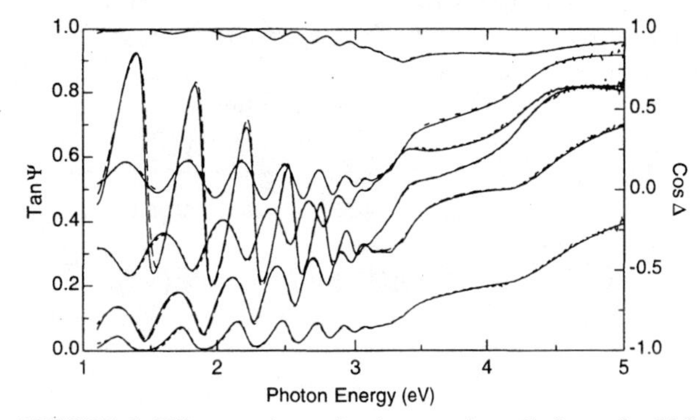

FIGURE 1. Ellipsometric angles (tanψ and cosΔ) for wafer #26 from 1 to 5 eV acquired at three angles of incidence (dashed) and best fit using a four-layer model containing the thicknesses and crystallinity as parameters (solid lines).

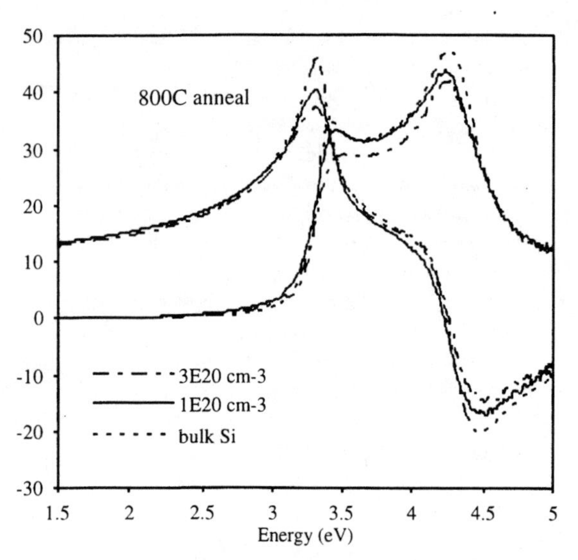

FIGURE 2. Dielectric function of polycrystalline Si annealed at 800°C at different doping densities in comparison with bulk Si from the literature.

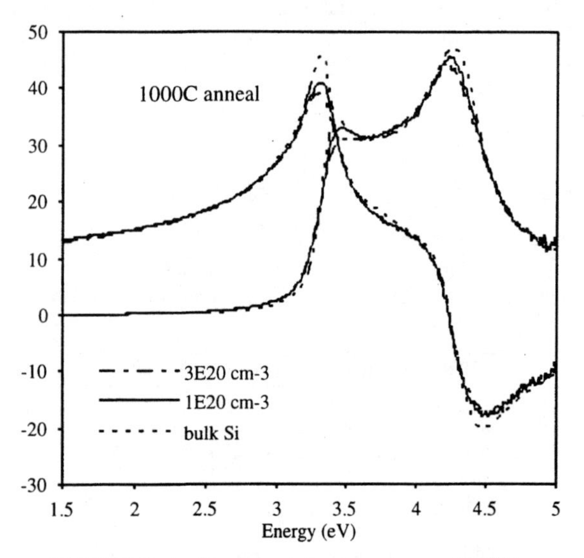

FIGURE 3. As Fig. 2, but after annealing at 1000°C.

A compressive stress reduces the lattice constant a' along the interface and is therefore considered negative, whereas a tensile stress is positive.

A biaxial stress not only causes a strain (relative lattice constant change) $\varepsilon=(a'/a_{Si})-1=\varepsilon_{xx}=\varepsilon_{yy}$ along the interface (a_{Si} is the bulk Si lattice constant), but the crystal responds elastically by changing its lattice constant perpendicular to the interface with a vertical strain $\varepsilon_{\perp}=-2\nu\varepsilon/(1-\nu)=\varepsilon_{zz}$, where ν is Poisson's ratio. The resulting strain tensor (8-9)

$$\vec{\varepsilon} = \begin{pmatrix} (S_{11}+S_{12})X & 0 & 0 \\ 0 & (S_{11}+S_{12})X & 0 \\ 0 & 0 & 2S_{12}X \end{pmatrix} \quad (2)$$

is related to the stress X through the compliance tensor elements S_{11} and S_{12}. It is convenient to decompose the biaxial strain into a hydrostatic strain $\varepsilon_h=2(S_{11}+S_{12})X/3$ and a pure shear strain $\varepsilon_s=(S_{12}-S_{11})X/3$.

The hydrostatic strain causes a shift of the three-fold degenerate LTO phonon, whereas the shear strain splits the triplet into a singlet and a doublet. In the backscattering geometry used here, only the singlet can be observed. Calculating the energy of the singlet is quite involved (10–13), but the final result is the following: Under a (001) biaxial stress, the change in the Raman shift of the singlet is $\Delta\omega=(p\varepsilon_{zz}/2\omega_0)+q\varepsilon_{xx}/\omega_0=b\varepsilon$, where p and q were determined in (11) resulting in b=–831 cm^{-1}. The stress is thus X=–$\Delta\omega\times$0.217 GPa/cm^{-1}, since X=$\varepsilon\times$180 GPa. Since our grain sizes and film thicknesses are large, the position of the Raman peak is not affected by confinement (14-15).

We have calculated the stresses from the Raman peak positions for all annealed samples and list them in Table 2. First, we note that the stress is small for most samples and comparable in magnitude to results for As-doped poly-Si reported in (1). Both negative (compressive) and positive (tensile) stresses occur, in contrast to As-doped poly-Si, where the stress is always tensile. Because of different interactions between the film and the substrate (14), only data for the same film thickness should be compared. The broadenings are between 3.2 and 3.8 cm^{-1}, indicating a very high film quality in comparison with the literature (14-16). Two trends can be noted: (i) The stress is smaller (more compressive) and the linewidth reduced after annealing to 1000°C. (ii) The stress is smaller (more compressive, less tensile) for the samples with lower doping if all other sample parameters are kept the same (i.e., when comparing #6 with #18, #8 with #30, etc.). This result is in contrast with As-doped samples (1), where the stress is more tensile at lower doping.

The reduction of the tensile stress in poly-Si due to As doping observed in (1) was explained with the accumulation of dopants at grain boundaries and interfaces. Since this explanation should also apply to P-doped films, it is probably not correct for our films. Instead, we suggest that the stress is also affected by the different ionic radii of Si, P, and As. Since As is larger than Si and P is smaller, we expect that higher As doping leads to a more compressive (less tensile) stress, whereas P doping leads to a more tensile stress compared to undoped poly-Si films. In other words, the stress dependence on doping is (at least mostly) a volume and not a surface effect.

It is also possible (maybe even more plausible) that the observed softening of the Si optical phonon energy (Raman redshift) is not due to stress, but due to an interference of the Raman-active inter-conduction-band transitions with the zone-center optical phonon (17).

Next, we discuss our ellipsometry data. First, we note that the parametrization of McGahan (J.A. Woollam Co., Inc.) gives an excellent description of the optical constants for our polycrystalline Si films, see Fig. 1. Second, we see from Figs. 2-3 that the dielectric function of our films is very close to that of bulk Si, except near the critical points, i.e., the peaks at 3.3 eV (E_1) and 4.2 eV (E_2). These differences are important and will be discussed in the following sections. Qualitatively, we can say that grain size, film thickness, doping, and strain all affect the position and/or broadening of these peaks. Therefore, we analyze the structure of the peaks in order to find information about the films.

First, we enhance the critical-points by calculating the second derivative of the dielectric function as described in (14), see Fig. 4 (symbols). Next, we fit the derivative in the E_1 region with the simple 2-D analytical lineshape (14-16)

$$\varepsilon(\omega) = A\ln\left(E_1 - \hbar\omega + i\Gamma\right)\exp(i\Phi) \tag{3}$$

where A is the amplitude, E_1 the energy, Γ the broadening, and Φ the phase angle of the critical point and ω the angular frequency of the photon. The best fit to the data is shown by the lines in Fig. 4. The parameters determined from this fit are given in Table 3. Values for a high-resistivity bulk Si (111) wafer are given for comparison.

The E_1 peak is caused by optical interband transitions from the highest valence band to the lowest conduction band in a region along the (111) direction and near the center of the Brillouin zone. We can simply call it the direct gap of Si. The exact microscopic nature of this critical point is quite complicated, since at least four different transitions contribute to this peak, which cannot be resolved with our methods. Therefore, we describe this peak as a single critical point.

In microcrystalline Si or thin Si films, the energy and broadening of this peak is different from bulk Si. According to (18), the E_1 energy is equal to that of bulk Si (3.37 eV) for grain sizes larger than 500 Å. Only for very small grains, the E_1 energy increases (to 3.41 eV for 50 Å grains) due to confinement. This was also found in (19). The broadening parameter increases with decreasing grain size L as $\Delta\Gamma = Q/L$, where Q is either 18 or 39 eV/Å depending on whether a 2-D (14) or excitonic (18) lineshape is chosen for E_1. For a film with 1500 Å thickness we expect a broadening increase of about 12 meV, for a 3000 Å thick film 6 meV. The observed broadenings for wafers #30 and #26 are equal and about 20 meV larger than in the bulk, see Table 3. We conclude that for our films, confinement effects (or scattering at the grain boundaries) do not change the E_1 position, but increase the broadening by a measureable amount.

Next, we discuss the effects of doping. The spectra in Fig. (3) resemble those of phosphorus-implanted bulk Si with similar carrier concentrations (20). The optical constants change only a little below the direct gap due to free carrier and heavy doping effects. Near the peaks, however, the influence of doping on the critical point parameters is quite obvious, see Fig. 3. As explained by Viña and Cardona (21), the electron Bloch wave is scattered by the difference between the Si and P potentials, which

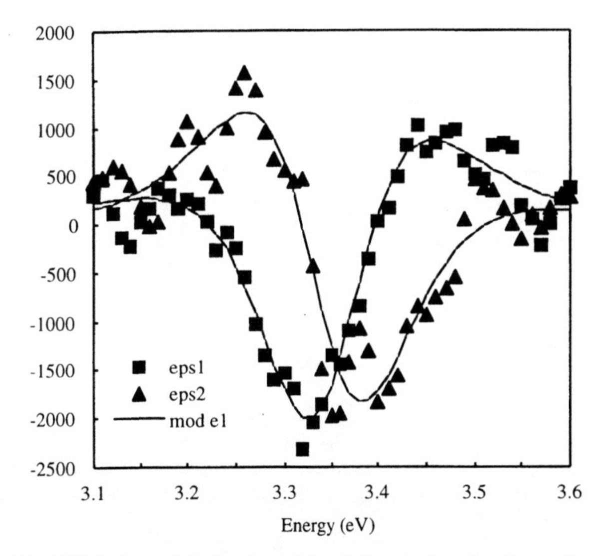

FIGURE 4. Second derivative of the dielectric function for wafer #24 in the E_1 critical-point range (symbols) along with a fit to the data using analytical lineshapes (lines).

leads to decrease of E_1 and an increase of Γ with increasing doping concentration. For a P density $n=10^{20}$ cm^{-3}, the increase of Γ is 10 meV and E_1 decreases by about 20 meV. For $n=3\times10^{20}$ cm^{-3}, Γ increases by 20 meV and E_1 decreases by about 35 meV (21). The phase angle Φ also changes because of the screening of the excitonic effects at higher carrier densities. Φ is expected to decrease by about 20° as n increases from 0 to 10^{21} cm^{-3}. All observations (decrease of E_1 and Φ, increase of Γ with increasing doping levels) are consistent with the data. It is peculiar, however, that Φ is larger for poly-Si with high doping (3×10^{20} cm^{-3}) than at lower doping (10^{20} cm^{-3}).

Finally, we discuss the influence of strain on the E_1 critical point. As explained in (9), a biaxial stress does not affect the E_1 energy, since the hydrostatic and uniaxial strain contributions to the E_1 shift cancel almost exactly. Therefore, in contrast to Raman scattering, ellipsometry cannot be used to determine stress in poly-Si or at the Si/SiO_2 interface, in contradiction to the conclusions reached in (3). However, a biaxial stress will increase the value of the spin-orbit splitting Δ_1, which cannot be observed in ellipsometry at 300 K, since Δ_1 (30 meV) is much smaller than Γ. Therefore, we may expect a small increase of the broadening with increasing stress. But even for a stress of X=0.2 GPa (larger than for any film studied here), this increase is only 2 meV (9). Therefore, the biaxial stress in our poly-Si films does not lead to observable changes in E_1 or Γ.

In addition to a macroscopic (average) stress, there may be fluctuations of the lattice constant from one grain to the next, which should lead to an additional broadening of the Raman and ellipsometry lines. It is likely that the decrease of the Raman and ellipsometry broadenings (see Tables 2 and 3) with increasing annealing temperatures is due to the relaxation of such inhomogeneous strain fields. This is con-

TABLE 3. Energy E_1, broadening Γ, amplitude A, and phase angle Φ of the as-deposited (top) and annealed films (bottom) determined from a lineshape analysis of ellipsometry spectra (assuming a two-dimensional critical point).

wafer	E_1 (eV)	Γ	A (1)	Φ (°)	dim
#4	3.387	721	37.6	242	2-D
#10	3.366	719	37.9	238	2-D
#16	3.368	708	39.4	241	2-D
#22	3.370	695	39.0	239	2-D
#6	3.344	107	139	207	2-D
#8	3.321	109	157	195	2-D
#12	3.333	97	147	204	2-D
#14	3.314	103	138	194	2-D
#18	3.357	93	185	219	2-D
#30	3.351	88	186	211	2-D
#24	3.344	88	204	210	2-D
#26	3.337	87	185	204	2-D
bulk	3.373	68	330	230	2-D

sistent with TEM results. We note that dislocations can also cause an increase of the broadenings (9), but the exact microscopic nature of this effect is not clear. It is likely that dislocations cause local strain fields, but they may also affect the local band structure.

CONCLUSIONS

We conclude that Raman scattering is a powerful optical tool to study biaxial stress at Si/SiO_2 interfaces and is less ambiguous than some other techniques, such as FTIR spectroscopy of the SiO_2 vibrational modes (since it is not affected by compositional variations at the interface) and spectroscopic ellipsometry (which has no sensitivity to biaxial stress at all), although the position of the Raman peak is affected by doping as well as stress. Unlike previous work on As-doped poly-Si (1), the phonons soften at higher doping levels in our P-doped films, possibly indicating a less tensile stress at higher doping levels. Therefore, the influence of doping on the stress in poly-Si is most likely a volume effect related to the different ionic radii of Si, As, and P or a pure interference effect due to doping rather than caused by the accumulation of doping atoms near grain boundaries and interfaces (where we would expect similar effect for As and P).

The derivative of the dielectric function near interband critical points (peak positions and broadenings) contains valuable information about uniformity, grain size, doping, and film thickness. For our films, the observed increase in broadening and decrease of the direct gap E_1 are mostly due to doping, but there are also contributions to the broadening due to scattering by grain boundaries and inhomogeneous strain fields. Since the data for our poly-Si films are very close to those of bulk Si, they can be used as benchmarks for poly-Si films deposited elsewhere.

REFERENCES

1. Wristers, D., Wang, H. H., de Wolf, I., Han, L. K., Kwong, D. L., and Fulford, J., "Ultra thin oxide reliability: Effects of gate doping concentration and poly-Si/SiO2 interface stress relaxation", in 1996 IEEE International Reliability Physics Proceedings, (IEEE, New York, 1996), pp. 77-83.
2. Fitch, J. T., Bjorkman, C. H., Lucovsky, G., Pollak, F. H., and Yin, X., *J. Vac. Sci. Technol. B* **7**, 775-781 (1989).
3. Nguyen, N. V., Chandler-Horowitz, D., Amirtharaj, P. M., and Pellegrino, J. G., *Appl. Phys. Lett.* **64**, 2688-2690 (1994).
4. Devine, R. A. B., *Appl. Phys. Lett.* **68**, 3108-3110 (1996).
5. Herzinger, C. M., Johs, B., McGahan, W. A., Woollam, J. A., and Paulson, W., *J. Appl. Phys.* **83**, 3323-3326 (1998).
6. Hegde, R. I., Paulson, W. M., and Tobin, P. J., *J. Vac. Sci. Technol. B* **13**, 1434-1441 (1995).
7. Herzinger, C. M., Johs, B., and Woollam, J. A., (unpublished).
8. Yu, P.Y., and Cardona, M., *Fundamentals of Semiconductors*, Berlin: Springer, 1996.
9. Lange, R., Junge, K. E., Zollner, S., Iyer, S. S., Powell, A. P., and Eberl, K., *J. Appl. Phys.* **80**, 4578-4586 (1996).
10. Anastassakis, E., *Acta Physica Hungarica* **74**, 83-105 (1994).
11. Anastassakis, E., Cantarero, A., and Cardona, M., *Phys. Rev. B* **41**, 7529-7535 (1990).
12. Englert, T., Abstreiter, G., and Pontcharra, J., *Solid-State Electron.* **23**, 31-33 (1980).
13. Lockwood, D. J., and Baribeau, J.-M., *Phys. Rev. B* **45**, 8565-8571 (1992).
14. Boultadakis, S., Logothetidis, S., and Ves, S., *J. Appl. Phys.* **72**, 3648-3658 (1992).
15. Boultadakis, S., Logothetidis, S., Ves, S., and Kircher, J., *J. Appl. Phys.* **73**, 914-925 (1993).
16. Ingels, M., Stutzmann, M., and Zollner, S., *Mat. Res. Soc. Symp. Proc.* **164**, 229-233 (1990).
17. Chandrasekhar, M., Renucci, J. B., and Cardona, M., *Phys. Rev. B* **17**, 1623-1633 (1978).
18. Logothetidis, S., Polatoglou, H. M., and Ves, S., *Solid State Commun.* **68**, 1075-1079 (1988).
19. Nguyen, H. V., and Collins, R. W., *Phys. Rev. B* **47**, 1911-1917 (1993).
20. Jellison, G.E., Withrow, S. P., McCamy, J. W., Budai, J. D., Lubben, D., and Godbole, M. J., *Phys. Rev. B* **52**, 14607-14614 (1995).
21. Viña, L., and Cardona, M., *Phys. Rev. B* **29**, 6739-6751 (1984).

Calibration Wafer for Temperature Measurements in RTP Tools

K. G. Kreider*, D. P. DeWitt**, B. K. Tsai**, F. J. Lovas**, and D. W. Allen*

*Chemical Science and Technology Laboratory and **Physics Laboratory

National Institute of Standards and Technology, Gaithersburg, MD, 20899

Rapid thermal processing (RTP) is a key technology that is used to produce integrated circuits at lower cost and reduced thermal budgets. One of the limiting factors in expanding the use of RTP is the accuracy of temperature measurements of the wafer during processing. We are developing a wafer for calibrating radiometric temperature measurements in RTP tools. The calibration wafer incorporates thin-film thermocouples with platinum / palladium (Pt/Pd) wire thermocouples welded to thin- film pads at the periphery of the 200 mm wafers. We have reduced the uncertainty of the temperature measurements up to 1200 K with this system. This has been accomplished by reducing the uncertainty due to the thermocouple itself and due to reduction of heat transfer near the junction.

We report results of NIST calibrations of radiometers using Pt/Pd wire thermocouples welded to the thin films on the wafer and of calibrated type K thermocouples. The thin-film thermocouples were sputter deposited from high purity Pt, Pd and Rh. These thin-film thermocouples were calibrated by comparison with Pt/Pd wire thermocouples in a specially designed test cell at temperatures up to 1150 K. Radiometric temperature measurements were made on the calibration wafer in the NIST RTP sensor test bed, using a commercial radiometer, and compared to those obtained from the thermocouple measurements. A model is presented to account for errors in the radiometric measurements due to stray radiation from the heating lamps, reflection of wafer emission from the chamber walls, and wafer emissivity. The calibrated type K thermocouples indicated temperature measurements within 4 K of both the Rh/Pt and Pt/Pd thermocouples on the 200 mm calibration wafer between 1000 K and 1150 K. The Pt/Pd thin films proved less durable than the Rh/Pt thin films and the limitations of these systems are discussed.

INTRODUCTION

Rapid thermal processing (RTP) is a rapidly developing technology which is gaining market share in the ULSI industry. RTP has advantages in reproducibility, short turnaround times for development, lower risk per batch, and adaptability for large wafers and vacuum-integrated processing. One of the limiting factors in expanding the use of RTP is the accuracy of temperature measurement of the wafer during processing. The Technical Working Group on Manufacturing of the 1997 National Technology Roadmap for Semiconductors established a 2 °C total uncertainty requirement in temperature measurements in RTP tools derived from the 1999 requirement for 0.18 μm line widths.

A major interest of the RTP community for the last seven years has related to temperature measurement of the wafer during processing. Moslehi et al. (1,2) recognized the problems related to optical temperature measurement of the wafer such as knowing the emissivity, the effects of reflections in the chamber and of wafer doping, and the characteristics of the optical path of the radiometer. There were also many problems with lift pin thermocouples related to heat transfer which made them less desirable than optical thermometry. It was recognized by Barna et al. (3) that the accuracy of the sensors in RTP is a significant limitation in semiconductor manufacturing. In the Third International Symposium on ULSI, 1991, Wortman et al. (4) explained the advantages of calibrating radiation thermometers in the RTP chamber with a test wafer. This approach mitigates the problem of the uncertainty related to moving the radiometer

CP449, *Characterization and Metrology for ULSI Technology: 1998 International Conference*
edited by D. G. Seiler, A. C. Diebold, W. M. Bullis, T. J. Shaffner, R. McDonald, and E. J. Walters
1998 The American Institute of Physics 1-56396-753-7/98/$15.00

from the blackbody calibrator to the RTP chamber.

Several approaches for temperature measurements in the chamber have been developed, including *in situ* resistance measurements (5); diffraction analysis to measure thermal expansion (6); laser speckle interferometry (7); and calibrated narrow band radiation sources in the RTP chamber (8). The "ripple" technique has been developed (9,10,11) which enables a separation of the dc thermal emission of the wafer from the ac emission from the lamps in the RTP chamber thus reducing the uncertainty related to emitted and reflected power. The complexity of the measurement problem using radiation thermometers in RTP has been well treated by DeWitt et al. (12), Timans (13), and Rogner et al. (14). The general conclusion is that in order to have confidence in the temperature measurement with a traceability to the absolute temperature scale and a total uncertainty of less than 2 °C, further efforts are required.

Our approach is to provide for the RTP radiation thermometers a calibration wafer the temperature of which is traceable to the International Temperature Scale of 1990 and has a well defined uncertainty as a result of the thermocouple temperature measurement of the silicon wafer surface. In order to achieve the highest accuracy, we are using the new Pt/Pd wire thermocouple system, which has an expanded uncertainty (k =2) and drift of less than 0.2 °C at 1000 °C, and a thin-film thermocouple (TFTC), the junction of which has less than 10^{-4} mm^3 volume. The very small thin film thermocouple causes minimal disturbance to the wafer surface temperature during the measurement.

EXPERIMENTAL PROCEDURE

This investigation was intended to measure the temperature of the wafer inside the RTP tool during thermal cycling. We calibrated and used the new Pt/Pd wire thermocouples (TCs) (15,16) to establish the temperature of its measuring junction with the expanded uncertainty (k=2) of less than 0.2 °C at the periphery of the wafer and calibrated and used sputter deposited Pt/Pd (17) and Rh/Pt TFTCs (18) to measure the temperature difference from the Pt/Pd junction to the radiometer target and other locations on the wafer. In addition we used a dual instrumented 200 mm wafer which also had type K wire thermocouples mounted at corresponding locations on the wafer according to commercial practice.

Pt/Pd Wire Thermocouple

NIST has developed improved wire TCs consisting of high purity Pt and Pd that minimize errors associated with instability and inhomogeneities (15). Such TCs constructed using 0.5 mm diameter wire have uncertainties of less than 0.03 °C below 960 °C. In collaboration with the Istituto di Metrologia "G. Colonnetti", the Italian national standards laboratory, NIST has produced an accurate Pt/Pd reference function for the range 0 °C to 1500 °C (16,19) to facilitate the use of this new TC in industrial and scientific applications. Its accuracy is approximately one order of magnitude better than PtRh alloy thermocouples. We calibrated the 0.25 mm diameter wire Pt/Pd thermocouples used for our calibration wafer, and found them to agree to within 0.2 °C with the reference function up to 1000 °C without calibration.

The Silicon Calibration Wafer

The silicon calibration wafer was instrumented with conventional fine-wire (0.08 mm diameter) type K TCs, the new thin-film Pt/Pd TCs and the new wire Pt/Pd TCs. The region at the center of the wafer in close proximity to the TC junctions is intended as the primary target for sighting of a radiation thermometer.

The 200 mm diameter silicon wafers were obtained with a 310 nm thermal oxide layer. These wafers were alcohol rinsed, dried under nitrogen gas, and ultraviolet/ozone irradiated to remove adsorbed hydrocarbons. The thin films were sputter deposited using physical masks prepared from 3 mm thick aluminum alloy plates with beveled 0.5 mm wide patterns for the 0.5 μm thick metal films. The purity of the targets was 99.99 % Pt, 99.97 % Pd, and 99.95 % Rh and the films were bonded to the SiO_2 with a 8 nm thick sputter-deposited bond coat of Ti. This procedure is described in more detail by Kreider and DiMeo (17). The thin-film pattern included welding pads 10 mm from the edge of the wafer and the Pd pad was covered with Pt to improve its welding characteristics. The 0.25 mm diameter Pt and Pd TC wires were welded to 1 cm long, 0.1 mm diameter Pt and Pd wires, respectively, which were welded to the weld pads with a parallel gap welder. These wires were anchored to the wafer with alumina cement in an alumina tube mounted in the wafer. This instrumented wafer permited a comparison of the temperature measurements from the radiation thermometer, wire TCs (type K and Pt/Pd), and thin-film thermocouples (TFTCs).

Experimental Apparatus and Instrumentation

At NIST the RTP Temperature Sensor Test Bed has been developed to evaluate the calibration wafer and a variety of radiation thermometers. The major component of the test bed is a chamber with many of the features of a production RTP tool which radiatively heats the wafer from the top side. The lower portion of the chamber was constructed so that the radiation environment of the wafer's lower surface can be configured to meet experimental design objectives. The system, shown schematically in Fig. 1, was originally designed as a rapid thermal anneal unit for atmospheric pressure operation with a cold-wall chamber and single-sided wafer heating provided by a linear array of tungsten-halogen lamps (48 kW). This test bed is described in detail by Tsai et al. (21).

During the calibration wafer evaluation in the RTP test bed, the thin-film and wire thermocouple emf values were acquired with a SensArray thermal mapping system [20] and a precision HP 3458a digital voltmeter combined with a Keithley 7001 scanner and a PC. The expanded uncertainty of the voltage readings on the voltmeter was 0.3 μV and we averaged four samples and corrected for drift of the DVM by measuring a short circuit. Radiometric measurements were made using an optical fiber thermometer with a pyrometer head having a 60 cm focal length. A Na heatpipe blackbody and a Au/Pt reference thermometer were used to calibrate the optical fiber thermometer with an expanded uncertainty of 2 °C. (22).

RESULTS

Calibration of the Thermocouples

The wire thermocouples were calibrated by comparison with a Au/Pt wire TC which had an expanded uncertainty (k=2) of 10 mK at 950 °C in a Na heatpipe oven. A flow of 15 cm^3/min of N_2 through a 5 mm diameter quartz tube supporting the three TCs (Pt/Pd-0.25 mm, type K

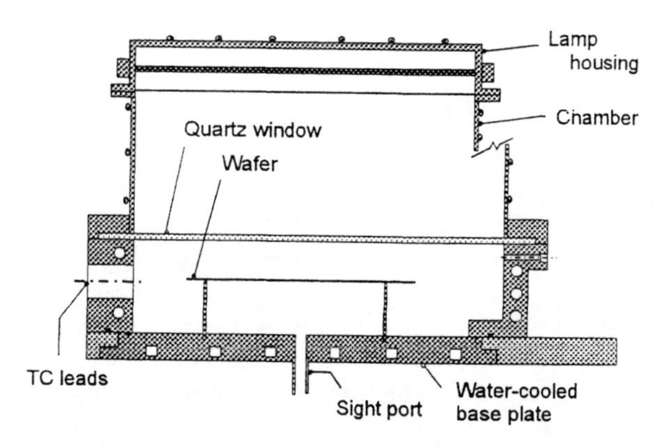

FIGURE 1. Schematic drawing of the NIST RTP sensor test bed showing the heater and wafer placement.

(0.08 mm), and type K (0.25 mm diameter) was used. The immersion depth of the thermocouples was over 200 mm. The results of the tests are presented in Table I with the differences indicated by Δ. Note that between 700 °C and 900 °C the Pt/Pd TC agreed to within 0.1 °C of the Au/Pt thermocouple. We also tested the type K TCs for drift in air at 797 °C and measured a drift of 1 °C in 4 h and 2 °C in 16 h.

TABLE I

COMPARISON OF AU/PT, PT/PD AND TYPE K THERMOCOUPLES

Au/Pt	Pt/Pd	type K (0.08 mm)	type K (0.25 mm)
°C	°C	°C	°C
694.2	694.1	694.6	698.3
Δ	-0.1	0.4	4.1
800.3	800.2	799.9	804.9
Δ	-0.1	-0.4	4.6
900.3	900.2	898.9	905.4
Δ	-0.1	-1.4	5.1

Calibration of the TFTCs has been described in detail by Kreider et al. (18). The thin-film thermoelement test coupon is sputter deposited simultaneously with the calibration wafer and the 10 mm x 50 mm silicon wafer coupon is calibrated by comparison with calibrated Pt/Pd or Pt 10%Rh/Pt TCs. The reference TCs were clamped at the water-cooled end and at the high temperature end of the thin film and provided both the emf of the thin film with respect to Pt and also enabled the temperature measurement at the position of the junctions (18). The Seebeck coefficients were determined for the calibration wafer thin films versus a Pt reference wire at 800 °C and were 1.4 μV/K for the Pt thin films; -14.3 μV/K for the Pd thin films; and 14.0 μV/K for the Rh thin films. The expanded uncertainty of these Seebeck coefficients was 3 % (18) . This uncertainty in the TFTC calibration is the primary uncertainty in our measurements since the Pt/Pd TC uncertainty is less than the equivalent of 0.2 °C and the instrumentation uncertainty is less than 0.1 °C. The uncertainty in the Seebeck coefficient of 3 % for the TFTCs would lead to an equivalent uncertainty of 0.6 °C over the 20 °C maximum temperature differential between the reference junction and the measuring junction of the TFTC. Considering the above we are estimating an expanded uncertainty of 1 °C for the temperature of the silicon wafer at the TFTC junctions.

Testing of the Calibration Wafer

The 200 mm calibration wafer with TFTCs and wire Pt/Pd and type K TCs was inserted in the test bed and four power levels were used to heat the wafer to temperatures of approximately 700 °C, 750 °C, 800 °C, and 850 °C. Heating took approximately 12 min to the first level followed by a 3 min hold for data acquisition, and 8 min heating or cooling between 3 min holds at the next temperatures. Observations were made manually from the radiometer display.

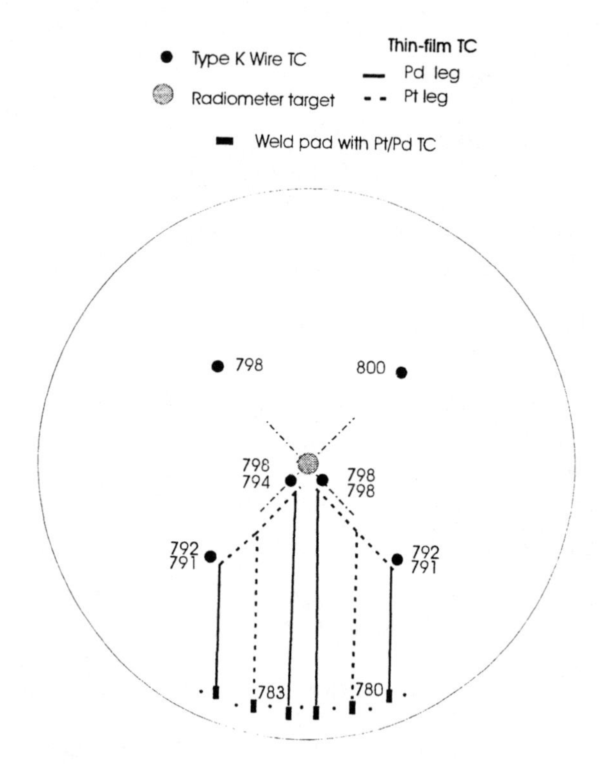

FIGURE 2. Schematic drawing of 200 mm wafer with temperature measurements of radiometer, type K thermocouples, Pt/Pd wire thermocouple junctions on weld pads, and thin-film thermocouple junctions compared (nominally at 800 °C).

Results of the comparison evaluation in the RTP test bed are shown in Fig. 2. In this figure, the locations of the six type K TCs are schematically presented as dots with the temperatures in °C beside the dots. In addition, the TFTC junctions have temperature readings given below the type K values and the Pt/Pd wire TCs have the temperatures displayed next to the appropriate welding pad. Fig. 2 shows typical results which agreed with similar comparison values for lower temperatures and repeated tests. Previous tests for thermal mapping of a commercial wafer also indicated high thermal gradients of up to 20 °C on the wafer, with the coolest part of the wafer being near the water-cooled access door. The temperatures of the Pt/Pd wire TCs were determined from the inverse function derived from the investigation of Ripple et al. (19) and those of the type K TCs were obtained from the SensArray (20) software. The temperatures of the thin film junctions were derived by adding the Pt/Pd TC temperatures at the welding pads to the voltage output of the TFTCs divided by their Seebeck coefficients as described above.

Unfortunately, the Pd thin films on the calibration wafer started to peel off at the welding pads with repeated thermal cycling and further tests with this wafer were not fruitful. Although improvements in the bond coat have been achieved using 10 nm of Ta, other problems with Pd thin films on Si/SiO_2 may preclude its use at these temperatures in the test bed. As was reported earlier (17) Pd recrystallizes and develops large pores at 900 °C after 2 h. The degradation severely limits the high temperature application of Pd thin films on Si wafers.

In order to avoid these problems with diffusion and coalescence of the Pd films on Si/SiO_2, we investigated the more refractory Rh as the thin-film thermoelement opposing Pt. Rh has a higher melting point (1966 °C vs 1554 °C for Pd), much smaller self diffusion coefficients, and lower

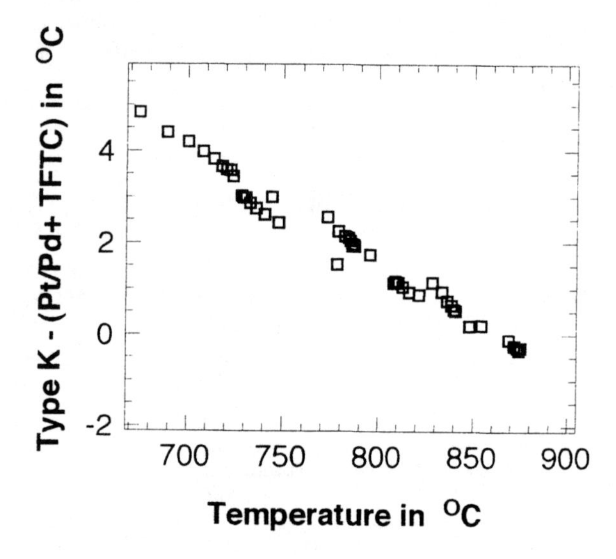

FIGURE 3. Plot of differences in temperature as indicated by type K TCs and Rh/Pt thin-film plus Pt/Pd wire TCs versus temperature for a location near the radiometric target.

diffusion rates in SiO_2. Its oxide is stable in air at higher temperatures (23) (≈1100 °C vs. 840 °C for Pd) and both the elastic constants and yield stresses are much higher. Our calibration of the Rh thin films versus Pt on 0.8 mm Si wafers yielded a Seebeck coefficient of 14.0 μV/K at 800 °C and nearly identical thermoelectric output to previous results on bulk Rh as reported by Roeser and Wensel [24]. The evaluation of the Rh/Pt TFTC calibration wafer in the test bed confirmed the superior qualities of these Rh films. No visual change of the TFTCs was observed after six cycles with peak temperatures reaching 875 °C.

Figure 3 shows the results of the comparison between the type K TCs and the Rh/Pt TFTCs combined with the Pt/Pd wire thermocouples between 650 °C and 880 °C. The location for this comparison is near the radiometer target, slightly (5 mm) to the right . These results are typical of all six cycles and the four comparison locations, in that the type K TCs read higher at 700 °C and lower at 850 °C and the range of deviation is 4 °C to 6 °C. There are several potential reasons for the differences. First, the TC comparison (Table I) would account for approximately half this difference and for the slope. Similar comparison data were reported by Vandenabeele and Renken (25). Second, the ceramic cement used to secure the type K TCs may be absorbing more heat than the Si wafer at the lower temperatures because of emissivity differences with the silicon which transmits radiation below 700 °C. Third, the TC wires may be conducting away more heat at the higher temperatures since they pass near the water cooled bottom plate.

The Temperature Measurement Equation – Radiation Mode

In order to estimate the test wafer temperature from the spectral radiance temperature, T_λ, it is necessary to construct a model of the wafer and its radiation environment based upon the temperature measurement equation [12]. Referring to Fig. 1 illustrating the test bed configuration, the backside of the wafer experiences irradiation from the lower chamber surfaces due to emission and reflection from the wafer, as well as due to irradiation from the heating lamps. The measurement equation can be expressed as

$$L_{\lambda,b}(\lambda,T_\lambda) = \varepsilon_\lambda \cdot L_{\lambda,b}(\lambda,T) + f \cdot L_{\lambda,b}(\lambda,T) + g \cdot L_{\lambda,\mathrm{ref}} \qquad (1)$$

where the term on the left-hand side of the equation is the blackbody spectral radiance corresponding to the spectral radiance temperature indicated by the radiometer, T_λ. From Planck's law, the spectral radiance is

$$L_{\lambda,b}(\lambda,T) = c_{1,L} \cdot \lambda^{-5} \cdot [\, 1 + \exp(-c_2 / \lambda T_\lambda) \,]^{-1} \qquad (2)$$

where $c_{1,L}$ = 1.1911 x 10^8 W·μm^4·m^{-2}·sr^{-1} and c_2 = 1.4388 x 10^4 μm·K

The first term on the right-hand side of Eq. (1), representing the wafer self-emission, is the product of the spectral emissivity, ε_λ, and the blackbody spectral radiance corresponding to the (true) temperature. The second term, the enclosure effect, represents the spectral radiance emitted from the test wafer which, after multiple reflections between the wafer and chamber, reaches the radiometer. The f-coefficient is a function of the wafer emissivity, chamber wall emissivity (reflectivity), and the geometry of the enclosure formed by the wafer-chamber walls. The third term, the lamp irradiation effect, represents the spectral radiance from the heating lamps which, after multiple reflections reaches the radiometer. The g-coefficient contribution is a function of the radiative properties of the chamber walls and wafer as well as, chamber geometry.

The spectral emissivity of undoped smooth, polished silicon at the effective wavelength of the radiometer (0.95 μm) at 700 °C is 0.65 and decreases nearly linearly to 0.64 at 900 °C (26). Although the test wafer backside is not polished, the spectral emissivity is not likely to be substantially higher. Because the chamber walls are cold relative to the test wafer, the enclosure effect can be included in the first term on the right-hand side of Equation (1) by replacing the emissivity, ε_λ, with the effective emissivity, $\varepsilon_{\mathrm{eff}}$. For a two-surface (parallel planes), diffuse-gray enclosure, the effective emissivity has the form

$$\varepsilon_{\mathrm{eff}} = \varepsilon_w \,/\, \{\varepsilon_{cw}\,(1 - \varepsilon_w) + \varepsilon_w\} \qquad (3)$$

where ε_w is the spectral emissivity of the wafer and ε_{cw} is the spectral emissivity of the cold chamber walls. As can be seen in Figure 1, the lower wafer surface (A_w) and the surrounding chamber walls (A_{cw}) only grossly approximate the parallel-plane configuration required of Eq. (3). In the limit, if $A_{cw} >> A_w$, we would expect ε_{cw} =1; and for $A_{cw} \approx A_w$, we would expect ε_{cw} = 0.2, the typical value for polished stainless steel. Recognizing that the apparatus configuration is much closer to the former, rather than the latter, we chose to use ε_{cw} = 0.8, resulting in $\varepsilon_{\mathrm{eff}}$ = 0.70 for the subsequent analysis.

There is no direct way to estimate the lamp irradiation effect from first principles because of the complexity of the geometry. Referring to Fig. 1 illustrating the arrangement of the heating lamps relative to the radiometer target on the backside of the test wafer, it is evident that stray radiation should be low compared to wafer emission effects, but should not be neglected. In our analysis, we chose to evaluate the g-coefficient by matching the model and measured temperatures, T_λ and T_{tc}, at a selected temperature, T_o = 800 °C. We postulated that the reflected lamp irradiance, $L_{\lambda,\mathrm{ref}}$, is proportional to the lamp spectral radiance at the operating temperature of the lamp,

$$L_{\lambda,\mathrm{ref}} = L_{\lambda,\mathrm{ref},\,o} \cdot L_{\lambda,b}(\lambda, T_{\mathrm{lamp}}) \,/\, L_{\lambda,b}(\lambda, T_{\mathrm{lamp},o}) \qquad (4)$$

where the variables subscripted with o represent the match point condition. We further postulate that the lamp power, P, is proportional to the blackbody exitance, $M_b = \sigma T^4$, where the Stefan-Boltzmann constant is σ = 5.670 x 10^{-8} W/m^2·K^4,

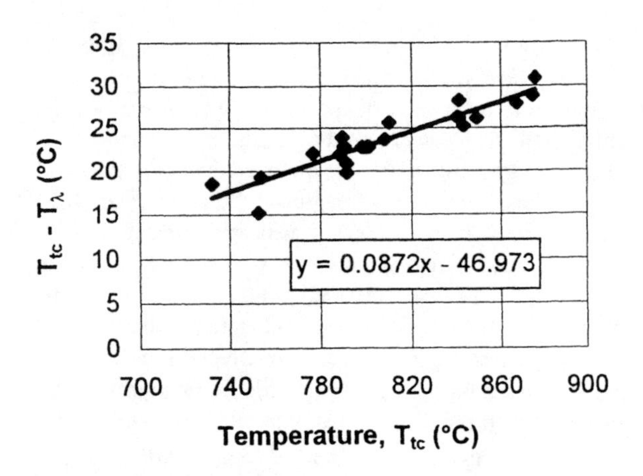

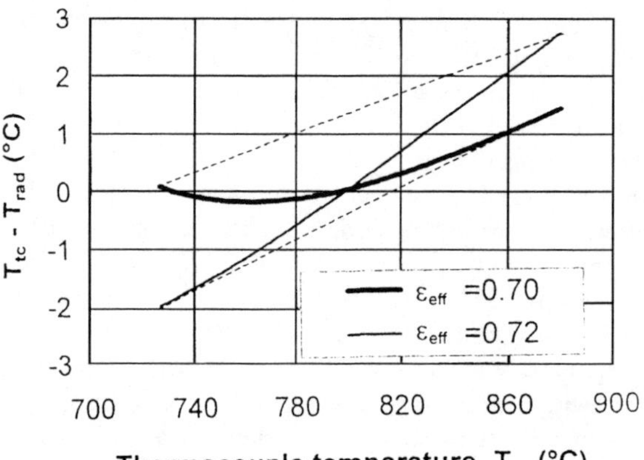

FIGURE 4. Differences between temperatures indicated by the Rh/Pt thin-film and wire Pt/Pd thermocouples, T_{tc}, and the radiation thermometer, T_λ, for six data sets taken in air and nitrogen atmospheres.

FIGURE 5. Differences between temperatures indicated by the thin-film and Pt/Pd wire TC, T_{tc}, and the temperature predicted by the measurement equation model of the radiation thermometer, T_{rad}. Accuracy of the predicted temperature is about 2 °C.

$$T_{lamp} = T_{lamp,o} \cdot (P_{lamp} / P_{lamp,o})^{1/4} \quad (5)$$

To select a single match point that would suffice for modeling the entire data set determining the proportionality factor g through Eqs. (4) and (5), least-squares fits were performed on plots of (T_{tc}- T_λ) vs. T_{tc} (see Figure 4) and P_{lamp} vs. T_{tc} (not shown). We assume that for the full-power condition, P_{lamp} = 100 %, that the lamp operating temperature is 3300 K.The results of the analysis with the model are shown in Fig. 5, where (T_{tc}- T_{rad}), the difference between the experimentally determined wafer temperature from thermocouple readings, T_{tc}, and the temperature predicted by the model based upon the observed spectral radiance temperature, T_{rad}, is plotted against T_{tc}. Results are shown for two assumed effective emissivities, ε_{eff} = 0.70 and 0.72. With these values, the reflected lamp irradiance terms are approximately 3.5 % and 1.5 %, respectively, of the wafer self emission term. If ε_{eff} were assumed a larger value, like 0.74, the contribution of the reflected lamp irradiation would vanish, a condition not possible with the RTP test bed. We assume that the precision of our estimate for the wafer temperature from the radiometer observations can be represented as the envelope of the two curves.

CONCLUSIONS

We have fabricated and evaluated a novel thin-film and wire thermocouple calibration wafer for RTP optical temperature measurements. This wafer has the advantage of reducing the uncertainty of the measurement of surface temperature of 200 mm wafers in the RTP tool. The reduced uncertainty stems from the use of the more accurate Pt/Pd wire thermocouple and thin-film thermocouples which provide a minimal disturbance of the heat transfer in the vicinity of the thermocouple junction. Our estimate of the expanded uncertainty (k =2) of the surface temperature measurement is less than 2°C. This could be reduced by reducing the uncertainty of the thin-film thermocouple calibration, which indicates an objective of our future activities.

Pt/Pd thin-film thermocouples can be used on the wafers, but are limited to temperatures below 900 °C and have problems with adhesion. Rh/Pt thin-film thermocouples performed well over six cycles in the RTP tool up to 875 °C and their usefulness and durability will be further evaluated.

We have developed a radiation model of the wafer-chamber to estimate wafer temperature from radiometer observations of the spectral radiance temperature. At 800 °C, the model analysis shows that the enclosure and reflected lamp effects influence the true temperature estimate by amounts of 5 °C to 8 °C and 2 °C to 4 °C, respectively. From Fig. 5, we conclude that the precision of the radiometer estimate for the true temperature is about 2 °C. Full confidence in the estimated wafer temperature from the radiometer observation in any RTP tool environment can be assured only if the stray radiation is made negligible through the chamber design configuration (using, for example, shields and baffling) and if the radiative surroundings can be well characterized. By well characterized surroundings, we mean that the geometry is amenable to representation as an enclosure with regular surfaces and that the radiative properties of the surfaces are known.

REFERENCES

1. Moslehi, M.M., Najm, H., Velo, L., Yeakley, R., Kuehn, J., Dostalik, B., and Yin, D., Mat. Res. Symp. Proceedings, Vol.224, Mat. Res. Soc., Pittsburgh, PA, 1991, 143-157
2. Moslehi, M.M., Paranjpe, A., Velo, L.A., and Kuehn, J., Solid State Technology, May 1994, 37-45
3. Barna, G.G., Moslehi, M.M., and Yong, J.L., Solid State Technology, Apr. 1994, 57-61
4. Wortman, J.J., Houser, M.C., Ozturk, J.R., and Sorrell, F.Y., Proceedings of Third International Symposium on ULSI, 1991, 528-540
5. Colgan, E.G.,Cabral, C., Clevenger, L.A.,and Harper,J.M., Mat. Res. Soc. Proc.V 387, 1995, 49-55
6. Brueck, S.R.J., Zaida, S.H., and Lang, M.K., Mat. Res. Soc. Proc., V 303, 1993, 117-125
7. Brueck, S.R.J., Zaida, S.H., and Lang, M.K., Mat. Res. Soc. Proc.,V 387, 1995, 125-131
8. Peuse, B., Yam, M., Bahl, S. And Elia, C., Proceedings of 5th Int. Conf. On Advanced Thermal Processing of Semiconductors, RTP '97, Round Rock, TX, 358-365
9. Shietinger, C., Adams,B.,and Yarling, C., Mat. Res. Soc. Proc.,V 224,1991, 23-28
10. Fiory, A.T., Shietinger, C.S., Adams, B.A., and Yarling, C., Mat. Res. Soc. Proc., V 303, 1993, 139-145
11. Xing, G.C., Wang, Z.H., Shietinger, C.S., Wasserman, Y., and Sun, M.H., Mat. Res. Soc. Proc.,V 338, 1995, 119-125
12. DeWitt, D.P., Sorrell, F.Y., and Elliot, J.K., Mat. Res. Soc. Proc.,V 470, 1997, 3-15
13. Timans, P.J., J. Appl. Phys., 74, 1993, 6353-6360
14. Rogner, H., Timans, P.J., and Amed, H., Appl. Phys. Lett., 69, 1996, 2190-2
15. Burns, G.W., and Ripple, D.C., Proceedings of 6th Int'l. Symp. On Temperature and Thermal Measurements in Industry, TEMPMEKO '96, Turin, Italy, 1996,
16. Battuello, M., Ali, K., and Girard, F., Proceedings of 6th Int'l Symp. On Temperature and Thermal Measurements in Industry, TEMPMEKO '96, Turin, Italy, 1996
17. Kreider, K.G., and DiMeo, Proc. Of Fifth International Conf. On Advanced Thermal Processing of Semiconductors, RTP '97, Round Rock, TX, 1997, 105-113
18. Kreider, K.G., Ripple, D.C., and DeWitt, D.P., Proceedings of 44th International Instrumentation Symposium, Reno, NV, ISA, Research Triangle Park, NC, May, 1998
19. Ripple, D.C., Burns, G.W., and Battuello, M., Proceedings of NCSL 1997 Workshop and Symposium, Atlanta, GA,1997, 481-488
20. Identification of commercial equipment and materials in this paper does not imply recommendation of endorsement by the National Institute of Standards and Technology, nor does it imply that the equipment and materials are necessarily the best available for the purpose.
21. Tsai, B.K., Lovas, F.J., DeWitt, D.P., Kreider, K.G., Burns, G.W., and Allen, D.W., Proc. of Fifth International Conf. On Advanced Thermal Processing of Semiconductors, RTP'97, Round Rock, TX, 1997, 340-346
22. Lovas, F.J., Tsai, B.K., and Gibson, C.E. Private communications
23. Barin, I., *Thermochemical Data of Pure Substances*, VCH Verlagsgesellscraft mbH, Weinheim, FRG, 1989, 1168
24. Roeser, W.F. and Wensel, H.T., in *Temperature; Its Measurement and Control in Science and Industry*, _Reinhold Pub, New York, NY, 1941, 1293
25. Vandenabeele, P., Renken,W., Mat. Res. Soc. Symp. Proc.,V 470, 1997, 17-28
26. DeWitt, D.P. and Nutter, G.D. (Eds.), *Theory and Practice of Radiation Thermometry*, John Wiley & Sons, 1987, pp 41.

ACKNOWLEDGEMENTS

The authors would like to acknowledge the valuable contributions of George Burns, and Dean Ripple of NIST for their help and advice.

Rapid Non-invasive Temperature Measurement of Complex Si Structures Using In-Situ Spectroscopic Ellipsometry.

R T Carline, J Russell, C Pickering and D A O Hope

Defence Evaluation and Research Agency, St Andrews Road, Malvern, Worcs WR14 3PS, U.K.

Use of spectroscopic ellipsometry for temperature monitoring of complex Si structures is discussed. System calibration is demonstrated for Si epitaxy between 610°C and 850°C on our growth system and shown to be in reasonable agreement with reference data in the literature. Use of virtual interface analysis of spectroscopic ellipsometry spectra is shown to allow an emittance-free determination of wafer temperature during growth on complex silicon-on-insulator substrates and highlights the large errors obtained from the corresponding pyrometer measurement. Measurement of emittance using spectrosopic ellipsometry to correct the pyrometer values is shown to be feasible. Accurate realisation of epitaxial designs is obtained using spectroscopic ellipsometry to end-point thicknesses of epitaxial layers and differential techniques are introduced as a potentially more accurate and absolute source of temperature values.

INTRODUCTION

Wafer temperature is important in determining the rates of many semiconductor processing steps. In some cases, such as rapid thermal processing (RTP) and chemical vapour deposition (CVD), temperature control is critical if reproducible and accurate results are to be achieved. As the complexity and density of semiconductor circuits and devices increases the demands on the necessary temperature monitors can only increase. Present solutions based on application of closed-loop control have run into problems through limitations of the existing temperature monitors. This is especially true for processes using RTP units (1,2) where the high ramp speeds and contact/contamination problems limit the use of thermocouple probes (3), and the intense radiative background and non-temperature related emittance changes introduce significant errors for single wavelength pyrometers. Such units are predicted to have increased importance in future semiconductor manufacturing and are considered essential in realising a viable single-wafer device process (2). Thus, there is a requirement to find temperature monitors with which real-time closed-loop control of temperature in RTP systems can be achieved.

SPECTROSCOPIC ELLIPSOMETRY

A wide range of temperature measurement techniques are being considered for use in process control (4,5). Being non-invasive and rapid, optical probes have received much of the attention (4). One such technique is real-time spectroscopic ellipsometry (RTSE) (4,6,7). Spectroscopic ellipsometry (SE) measures the change in polarisation induced by specular reflection from a sample surface. Spectra representing the amplitude ratio of the r_s and r_p reflection coefficients and the phase difference between them are generated (termed $\tan(\psi)$ and $\cos(\Delta)$ respectively) from which spectra of the refractive indices (n,k) or dielectric function ($\varepsilon=\varepsilon_1+i\varepsilon_2$) can be calculated using a model. If a bulk model is assumed for a multilayer sample these spectra are referred to as pseudo-dielectric function spectra. Temperature is usually obtained as a fit parameter in a comparison between the measured refractive index spectra of a layer and reference spectra for the same material at different temperatures. The acquisition of Si reference dielectric function spectra and their subsequent use for evaluating temperature from RTSE spectra obtained during Si epitaxy is the topic of this paper.

Standard methods to acquire the refractive index of a layer from the measured spectra involve complex multilayer modelling. The refractive index of a surface layer can, however, be obtained by modelling the spectra from a current RTSE measurement using the spectra of a previous RTSE measurement as a pseudo-substrate on which a single 'overlayer' has been grown (8). By defining the substrate/overlayer interface (termed the virtual interface) at a fixed depth below the moving (e.g. growing) surface the pseudo-substrate is continually up-dated avoiding problems caused by variations in substrate temperature or sample emittance. To the best of our knowledge this technique has not been applied to obtain temperature information from RTSE data recorded during epitaxial growth. The bulk of this paper discusses the application of virtual interface analysis to evaluate the wafer temperature during growth of Si epitaxy in situations where standard pyrometry fails. The use of RTSE to provide emittance as a correction to pyrometry readings is also discussed.

Virtual interface analysis requires appropriate reference dielectric function spectra. Where optical activities of elements in the path of the SE probe beam are small and a bulk is measured which exhibits no surface layers the measured SE spectra can be assumed to provide

CP449, *Characterization and Metrology for ULSI Technology: 1998 International Conference*
edited by D. G. Seiler, A. C. Diebold, W. M. Bullis, T. J. Shaffner, R. McDonald, and E. J. Walters
1998 The American Institute of Physics 1-56396-753-7/98/$15.00

system and material specific reference spectra. The application of this 'system-calibration' approach in our epitaxial growth system is discussed in the next section.

Temperature calibration

In this section we describe the calibration of our low pressure chemical vapour deposition (LPCVD) system to allow SE temperature measurement of Si during epitaxy. SE is sensitive to temperature through variations in refractive indices and also shifts in the energies of direct critical point (CP) transitions seen as peaks in the SE spectra. This is illustrated in Fig 1 which shows the pseudo-dielectric function spectra measured using SE at several temperatures, determined by pyrometry and assumed accurate, from a Si (100) substrate in our LPCVD Si/SiGe reactor after a 900°C oxide removal procedure.

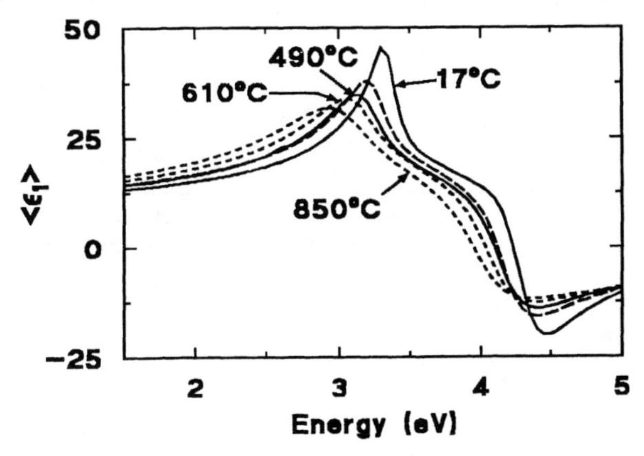

FIGURE 1. Comparison of reference dielectric function spectra of Si (——) at 17°C (9) and 490°C (10) with spectra obtained on a Si (100) substrate in our growth reactor using RTSE at an intermediate temperature (— —) and at 610°C and 850°C (----).

The origin of the observable peaks is the E_0' (and E_1) CP in the range 3-3.4eV and E_2 above 4eV. Measurements were made with a SOPRA rotating polariser RTSE instrument which uses a UV-enhanced optical multichannel analyser detector (OMA) to obtain data at 256 points simultaneously in the range 1.5-5.0eV (signals from 512 pixels are paired to give improved signal/noise). The measurements were made through Bomco viewports designed to minimise errors in the SE measurement due to residual strain in the windows. Signals were averaged over 10s. The angle of incidence was 70° giving a 5mmx~13mm spot incident at the wafer centre where the temperature is uniform. The magnitude and peak position of these spectra compare well with reference Si spectra at room temperature (9) and 490°C (10) from the literature, also shown in Fig 1, and this demonstrates their validity as reference dielectric function spectra for use on our system. Spectra at an arbitrary temperature are necessary for temperature analysis and were generated from these reference spectra using an interpolation procedure (11).

Figure 2 shows the shift in CP energy with temperature of the E_o' CP between 610°C and 850°C calculated from the numerical second differential of measured Si reference spectra. The inset shows the second differential of the ε_2 spectrum at 610 °C.

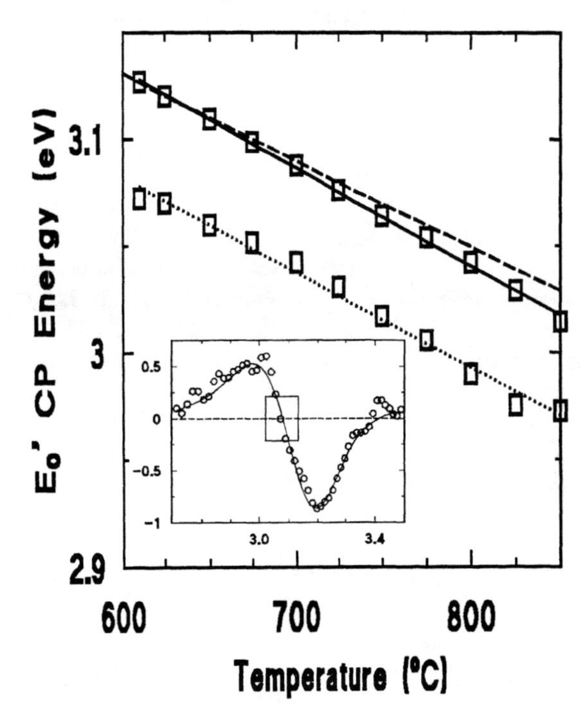

FIGURE 2. Temperature dependence of the E_0' CP transition in Si calculated from SE spectra measured on our system (—□—) compared with extrapolated literature data from lower temperature (12) (— —). Also shown is the variation in the $d^2\varepsilon_1/dE^2$ zero crossing point (- -□- -), which is highlighted in the inset which also shows a Lorentzian fit to $d^2\varepsilon_2/dE^2$ at 610 °C.

This is compared to a Lorentzian function, $Ae^{i\theta}/(E_{cp}-E+i\Gamma)^2$, fitted to the second differential, $\partial^2\varepsilon_2/\partial E^2$, to obtain the CP energies, E_{cp}. Other parameters fitted in this equation include amplitude A, phase θ and a broadness parameter Γ. The CP energies calculated from our reference spectra compare well with values extrapolated from literature data at <550°C (12) and shown in Fig 1.

The sensitivity of the SE measurement to temperature through shifts in CP energy position and changes in magnitude of dielectric function spectra is different. Sub-pixel resolution of CP energies can be obtained using the spectroscopic fitting of RTSE spectra. Hence, at 3eV, where 1 pixel corresponds to 8meV, Fig 2 suggests a resolution of ±10°C (1 pixel) or below. The noise level observed for Si spectra measured at 3eV for 1s integration time is approximately ±0.1% giving an observable temperature shift of approximately ±2°C. Unfortunately the magnitude is affected by both the sample surface and the RTSE measurement calibration making absolute values difficult to determine. Thus, though use of spectrosopic data should give a temperature precision of <±2°C, an *accuracy* of only <±10°C (CP energy) is expected.

IN-SITU GROWTH AND MONITORING OF TEMPERATURE USING RTSE

A range of Si/SiGe device structures have been grown successfully on silicon (100) substrates using our LPCVD reactor. Silane, germane and diborane (dopant) source gases were mixed with a hydrogen carrier gas and passed over the heated substrate. Growth rates for Si epitaxy varied between ~6nm/min at 610°C and ~80nm/min at 850°C with the temperature closed-loop controlled using a pyrometer sensitive to radiation in the 0.7-1.0μm range. Growth of the alloy is restricted to 610-650°C to maintain material quality. The growth rate calibration at a given wafer temperature is normally used to achieve a specified epitaxial thickness by timing of the growth. RTSE is used to monitor composition and thickness during growth. Here, we discuss its application to temperature monitoring.

Growth of SiGe/Si MQW longwave infrared photodetectors on BESOI

Recently epitaxy has been attempted on bond-and-etch-back silicon-on-insulator (BESOI) substrates. The buried oxide was required to act as one reflector of the resonant cavity used to enhance the absorption in SiGe/Si multi-quantum well (MQW) infrared detectors operating in the 8-14μm range (13). As the second reflector was defined at the sample surface, correct device operation required that the total thickness of epitaxy gave a cavity which resonated at the design wavelength, λ_T, and that the position of the MQW coincided with a peak in the resonant electric field at this wavelength.

Errors in the pyrometer measurement of temperature

Early attempts to realise epitaxial designs on BESOI wafers failed. Post-growth characterisation using cross-sectional transmission electron microscopy (XTEM) and Fourier transform infrared reflectance (FTIR) measurements showed this to be a problem with the Si epitaxy which produced layers which varied by as much as 40% from their target values. The origin of this problem was identified as an error in the pyrometer measurement of wafer temperature by comparison with temperatures calculated from the corresponding RTSE spectra*. This is illustrated in Fig 3 which compares the power supplied by the pyrometer temperature control system with the temperature calculated from RTSE spectra during the growth at 650°C of nominally 1.5μm Si on a [1.3μm / 2μm SiO_2 / Si] BESOI substrate. Although the shapes of the two curves are similar a monotonic power-to-temperature relation is not seen as the wafer temperature is a function of both wafer emittance and heater power. The spectra were recorded on an earlier version of the SOPRA RTSE instrument and analysis used a modified version of the virtual interface analysis+ which fitted all the spectra recorded during the time corresponding to the growth of the virtual layer. The technique gave improved signal/noise and was particularly effective where the virual layer was defined to be thick enough to encompase 4 or more measurements.(4 being typical).

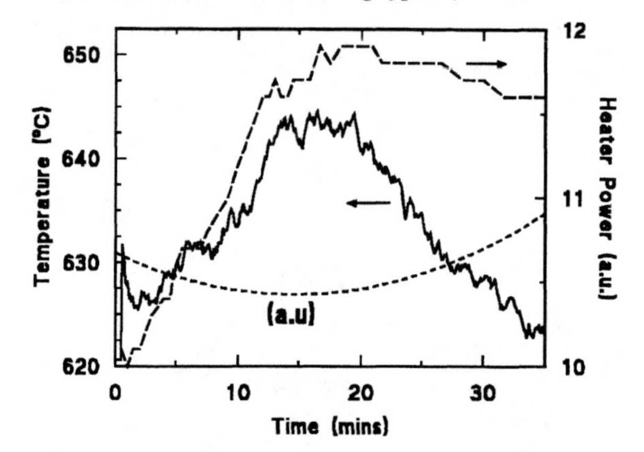

FIGURE 3. Temperature variation (——) and the fit to the integrated (0.9-1μm) emittance change (- - - -) calculated from RTSE compared to heater power (— — —) supplied during growth on a BESOI substrate.

To obtain adequate signal/noise during the RTSE measurement pixels were grouped giving 128 spectral points This gives a corresponding decrease in spectral resolution. The results in Fig 3 indicate a temperature precision of ±1.5°C. This is only achieved through use of a relatively thick overlayer (20nm) and with the noise reduction provided by the modified virtual interface procedure. Analysis using an overlayer defined by the penultimate measurement (i.e. the minimum) gave a precision of >±3°C.

RTSE corrections to pyrometer errors

In principle an emittance calculated from the RTSE measurement can be used to correct errors in the pyrometer measurement of temperature (14). An integrated emittance calculated over the pyrometer range (0.7-1μm) from the individual RTSE spectra of Fig 3 showed significant scatter.An integrated emittance calculated over the pyrometer range (0.7-1μm) from the individual RTSE spectra of Fig 3 showed significant

+ fastdyn method by J A Woollam Co., Inc. 650 J Street, Suite 39, Lincoln, NE, USA 68508

scatter. The scatter is caused by the RTSE measurement which, with this pixel grouping did not resolve short period fringes in the emitance spectra. This is illustrated in Fig. 4 which compares a room temperature RTSE spectrum measured in-situ from a BESOI substrate with a simulation of the same substrate. A quadratic fit to the emittance data integrated over a reduced range of 0.9-1μm, where most change is observed in the RTSE, is shown in Fig 3.

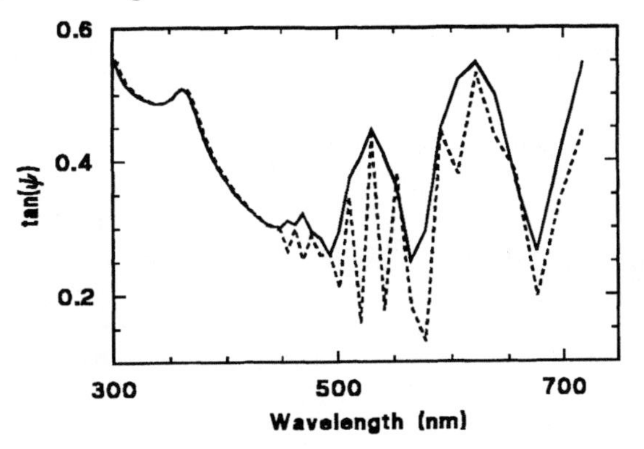

FIGURE 4. Measured (——) and simulated (- - - -) spectra for a room temperature in-situ SE measurement of a [0.96μm Si / 1.52μm SiO2 / Si] BESOI substrate.

The trend shown by this curve agrees with that expected from the temperature variation calculated from the same spectra. This suggests that such a correction is feasible. The improved signal/noise characteristics of the current RTSE instrument allow a 2-4 times improvement in spectral resolution. However, the spectral range is limited to <826nm. With either instrument operation of the pyrometer over a reduced spectral range to match the SE is necessary.

RTSE monitoring of growth on BESOI substrates

The correct cavity width and MQW position have been achieved for our resonant cavity SiGe/Si MQW infrared detectors by using fringes in tan(ψ) measured by RTSE during growth to end-point individual layer thicknesses. This is despite simultaneous use of pyrometer-based closed-loop temperature control. Figure 5a shows the variation in RTSE signal, tan(ψ), at 814nm recorded during growth of the first p+ Si contact layer of one such device. The epitaxial thickness represented by each fringe is available from the reference dielectric function spectra at the growth temperature. Figure 5b shows the responsivity of one such device produced from a design aimed at a λ_r=8.8μm resonance. Radiation is incident through the Si substrate and resonance occurs between the buried BESOI oxide and a metal deposited on the surface of the epitaxial structure. Epitaxial Si grown on either side of the MQW was made p+ to provide contact whilst the MQW consisted of 16 periods, each being

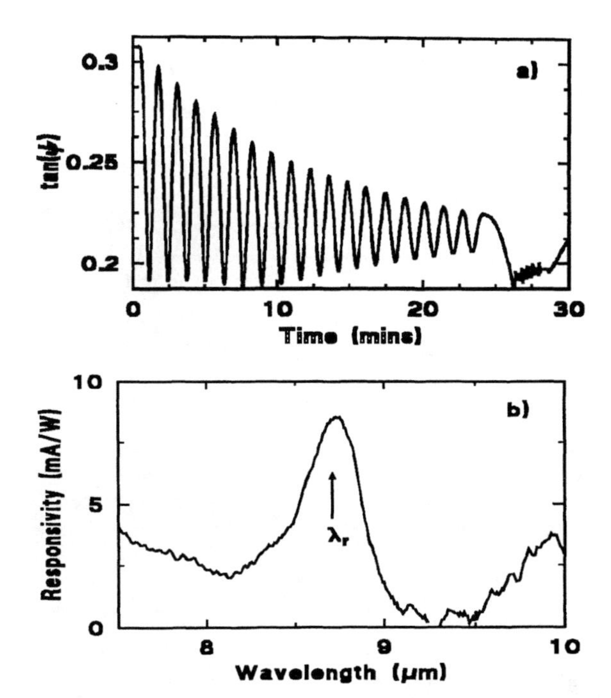

FIGURE 5. a) Interference oscillations in the RTSE data recorded at 814nm of the first p$^+$ Si contact layer of a resonant cavity SiGe/Si MQW infrared detector and b) the relative responsivity of the resulting detector with a peak at the design wavelength, λ_r.

[$Si_{0.845}Ge_{0.155}$ (5nm with the central 20% p-type 1.7x10^{12}cm^{-2}) / Si(50nm)].

Fringes of the type seen in Fig 5a are not seen at shorter wavelengths which penetrate only the top few hundred nm of the sample. Figure 6 shows the variation in tan(ψ) at 413nm and the temperature calculated from virtual interface analysis of the same RTSE spectra that gave the data in Fig 5a but using the region of the E_0' CP. The relation between the magnitude of tan(ψ) and temperature is obvious. Again, large variations in a nominally constant wafer temperature are seen. The magnitude of these variations reduce as the growth proceeds due to increased absorption in the Si. The rapid temperature drop between 24-26 minutes is required prior to growth of the SiGe/Si MQW and the same variation is calculated when the RTSE spectra are analysed using the minimum overlayer of 13nm (within the error of 2°C). The ~1.4°C/s temporal resolution is promising but insufficient for many RTP applications. Variations equivalent to ramp rates of >30°C/s have been calculated from RTSE spectra. With the superior signal/noise characteristics of the new RTSE instrument allowing acquisition times to be reduced by a factor of 10 (with a corresponding reduction in minimum overlayer thickness) ramp rates of the type used in RTP (e.g. 50°C/s) should easily be resolved.

Due to a non-uniformity in wafer heating in our system wafer bowing exists at high temperatures which

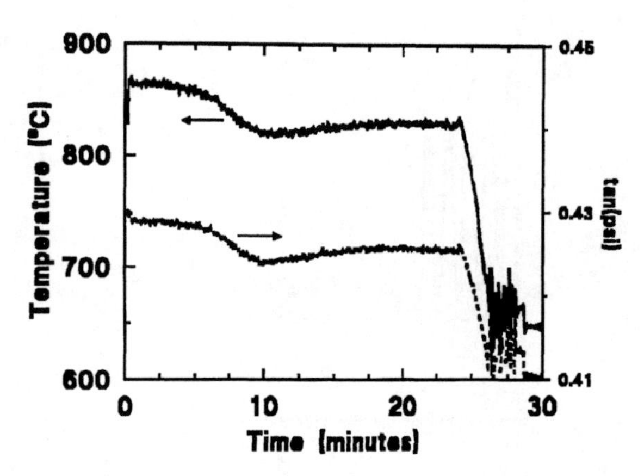

FIGURE 6. tan(ψ) at 413nm recorded using RTSE during the growth of a p+ contact layer on a BESOI substrate (- - - -) and the temperature calculated from the same RTSE spectra between 2.5 and 3.5eV (——).

changes as the wafer is cooled. As a result, the wafer requires re-alignment for the RTSE measurement after it experiences large temperature changes. This process is the origin of the noise seen between 24 and 26 minutes and the 18°C discontinuity in temperature at 28.5 minutes. Mis-alignment introduces errors in the magnitude of spectra measured by RTSE and errors as large as 30°C have been observed due to wafer movement during growth. Though this represents a large source of potential error in the RTSE measurement of temperature on our reactor the temperature uniformity necessary for commercial wafer processing means this problem should be greatly reduced.

Differential techniques

Temperature errors usually correspond to errors in the magnitude of either the measured or reference SE spectra. Differential spectra, of the type shown in the inset of Fig 2, are less sensitive to changes in spectral magnitude and so may be unaffected by the sources of such errors (e.g. mis-alignment or poor calibration). An absolute temperature comparison may also be possible between systems despite differences in the RTSE 'system calibration'. Differentiation of the pseudo-dielectric function spectra provided by the virtual interface analysis provides surface sensitivity and prevents errors caused by a difference in the CP structure in underlying layers (e.g. SiGe). For a given material this means that a feature of the differential structure may be calibrated against temperature rather than the CP transition energy calculated from equation 1. This is illustrated in Fig 2 where the crossing point of the E_0' differential structure in $d^2\varepsilon_2/dE^2$ is plotted against temperature. Such a feature can be located with a high degree of accuracy allowing sub-pixel resolution in energy and associated temperature resolutions lower than ±10°C. The similarity in gradient between this curve and the E_0' CP curve is due to the similarity in phase (ϑ in equation 1) of the differential structure at different temperatures.

CONCLUSIONS

In conclusion, the potential of SE to provide real-time temperature measurements for RTP applications has been discussed. A system calibration for Si has been demonstrated using SE spectra measured in-situ on a bulk Si wafer as reference spectra at temperatures measured using a single wavelength pyrometer. Virtual interface analysis using these spectra was shown to provide temperature information from SE spectra recorded during the growth of SiGe/Si MQW detectors on BESOI substrates as well as real-time end point detection for epitaxial thickness. Results show the technique has potential for application to RTP systems with high ramp rates. Though an observed temperature precision of <±2°C, is adequate for RTP applications the precision of the standard measurement is only <±10°C The use of differential techniques have been presented as potential methods to improve both these figures.

REFERENCES

1. Roozeboom, F., *Mat. Res. Soc. Symp. Proc.* **303**, 149-164 1993.
2. Moslehi, M.M., Chapman, R.A., Wong, M., Paranjpe, A., Najm, H.N., Kuehne, J., Yeakley, R.L. and Davis, C.J., *I.E.E.E. Trans. Elect. Dev.*, **39**, 4-32, 1992.
3. Wang, Z., Kwan, S.L., Pearsall, T.P., Booth, J.L., Beard, B.T. and Johnson, S.R., *J. Vac. Sci. Technol.* B. **15**, 116-121, 1997.
4. Herman, I.P., *Optical diagnostics for thin film processing,* Academic Press, 1996.
5. Peyton, D., Kinoshita, H., Lo, G.Q. and Kwong, D.L., SPIE **1393**, 295-308, 1990.
6. Wakagi, M., Hong, B., Nguyen, H.V., Collins, R.W., Drawl,W. and Messier, R., *J. Vac. Sci. Technol.* A **13,** 1917-1923, 1995.
7. Maracas, G.N., Kuo, C.H., Anand, S., Droopad, R., Sohie, G.R.L. and Levola, T., *J. Vac. Sci. Technol.* A **13**, 727-732, 1995.
8. Aspnes, D.E., *J. Opt. Soc. Am.* A **10**, 974-983, 1993.
9. Palik, A.D., (Ed), *Handbook of Optical Constants of Solids,* New York, Academic, 1985.
10. Jellison, G.E. and Modine, F.A., *Phys. Rev.* B 27, 7466, 1983.
11. Snyder, P.G., Woollam, J.A., Alterovitz, S.A. and Johs, B., *J. Appl. Phys.* **68**, 5925-5926, 1990.
12. Lautenschlager, P., Garriga, M., Vina, L. and Cardona, M., *Phys. Rev.* B **36**, 4821-4830, 1987.
13. Carline, R.T., Robbins, D.J., Stanaway, M.B. and Leong, W.Y., *Appl. Phys. Lett*, **68**, 544-546,0 1996.
14. Hansen, G.P., Krishnan, S., Hauge, R.H., and Margrave, J.L. *Appl. Opt.* **28**, 1885, 1989

Fabrication of SiGe and SiGeC epitaxial layers by ion implantation and excimer laser annealing

P.Boher, J.L.Stehle and J.P.Piel

SOPRA S.A., 26 rue Pierre Joigneaux, 92270 Bois Colombes, France

E.Fogarassy,

Laboratoire PHASE, BP20, 23 rue du Loess, 67037 Strasbourg Cedex, France.

Epitaxial $Si_{1-y}C_y$ and $Si_{1-x-y}Ge_xC_y$ substitutional alloy layers are prepared on monocrystalline Si crystal by carbon and germanium multiple energy ion implantation followed by pulsed excimer laser induced epitaxy. The properties of the alloy layers obtained by this technique were demonstrated to depend strongly both on ion implantation and laser processing conditions. The growing of pseudomorphic epitaxial layers from group IV semiconductor alloys was successfully achieved by using high energy and large area beam (Up to 2J/cm^2 per pulse over 20cm^2) excimer laser developed by SOPRA for industrial applications.

INTRODUCTION

The $Si_{1-x}Ge_x$ system has been studied intensively for applications in the field of optoelectronics and high speed heterojunction bipolar transistors. These applications require a very good control of composition and stress of the epitaxial layers. To reduce the layer stress, inclusion of carbon has been studied by different techniques like molecular beam epitaxy, chemical vapor deposition (1), implantation (2) or laser annealing after implantation (3-4). However, due to the formation of carbide precipitates, only two ways of preparation can be applied; the growth can be made at low temperature (<850°C) to prevent carbide formation, or the process can be faster than the carbide formation using for example rapid thermal annealing (5) or liquid phase epitaxy induced by pulsed laser heating (3-4,6). In this work, the latter method is used to recrystallize silicon layers amorphized by implantation of carbon, germanium or both elements. Very high Energy excimer Laser system (VEL) developped by SOPRA which allows homogeneous annealing on large surfaces is used. Non destructive structural characterization after implantation and after laser annealing are made using spectroscopic ellipsometry, x-ray diffraction and Rutherford backscattering techniques.

EXPERIMENTAL DETAILS

Ion implantation

Float zone <100> monocrystalline silicon wafers are used. Two kinds of implantation have been realized. First series of samples have been implanted with variable carbon content and four successive energies (15, 25, 40 and 50keV) to achieve flat carbon profiles over about 150nm. Three wafers have been implanted with different doses to get 0.67, 1.1 and 1.32% for the mean carbon content. A second series of samples has also been prepared in the same way but with only one implantation energy for the germanium (50keV for a content of ~ 12%) and for the carbon (25keV with contents of 0.2, 1.1 and 2.2%).

Excimer laser annealing

Annealing has been performed using a SOPRA VEL excimer laser with maximum total energy on the sample surface of 45J/pulse. This is a XeCl excimer laser working at 308nm with x-ray preionization to initiate the discharge (7-8). Three different laser cavities have been associated to provide 45J/pulse on the sample surface. The pulse duration is around 200ns and the frequency can be one pulse every 6s. All the treatments of the study have been made with only one laser shot in air at room temperature.

CP449, *Characterization and Metrology for ULSI Technology: 1998 International Conference*
edited by D. G. Seiler, A. C. Diebold, W. M. Bullis, T. J. Shaffner, R. McDonald, and E. J. Walters

Spectroscopic ellipsometry

Spectroscopic Ellipsometry measurements (SE) were made with a commercially available SOPRA GESP5 rotating polarizer instrument with a high resolution double monochromator and photon counting detector. The angle of incidence can be changed in the range 6-90° but was fixed in this study to 75° close to the pseudo Brewster angle of silicon for optimum sensitivity. A 3mm diameter beam with <0.5mrad divergence is produced by a mirror system. The polarisation state of the light is modulated by a rotating quartz Rochon polarizer and analyzed by another fixed Rochon after reflection from the sample. The standard wavelength range is 0.25 to 0.85μm but can be extended from 0.192 to 2.05μm (9).

Other techniques

X-Ray diffraction (XRD) measurements were made with a conventional Philips HR1 diffractometer especially adapted to the grazing measurements. A cobalt x-ray tube with fine focus is setup with a monochromator consisting of four Ge crystals to select the Co K-α line. After reflection the beam is detected by a proportional counter. Details on the setup can be found elsewhere (10). Depth resolution of Rutherford backscattering was enhanced using 2MeV He^+ and a glancing angle detection geometry. The tetragonal expansion (or contraction) of the films was estimated from the measurement of the angular shift between oblique incidence channeling dips respectively recorded on distorted layers and substrate as reported in reference (11).

EXPERIMENTAL RESULTS

SE analysis of $Si_{1-y}C_y$ layers

SE has been applied systematically on all the samples before and after laser annealing. As reported in Figure 1 in the case of the 1.32% C implanted sample, SE measurements show a great variation with the energy of the laser beam. The as-implanted sample shows a smooth shape both for the tan Ψ and cos Δ (Cf. Figure 1) ellipsometric parameters. On the contrary the sample annealed at 1.2J/cm2 shows very well defined interference fringes. In the UV range the measurement changes drastically compared to the as-implanted sample and the shape of the curves are also very similar for all the annealing energies. Since in the UV range, the light absorption allows to see only the top surface of the sample (penetration depth < 200Å), we can conclude that the samples are crystallized almost at the top surface even for the lower laser energy. For the higher laser energies (2.1 and 2.3J/cm^2), there are no more interference fringes and the sample seems completely crystallized.

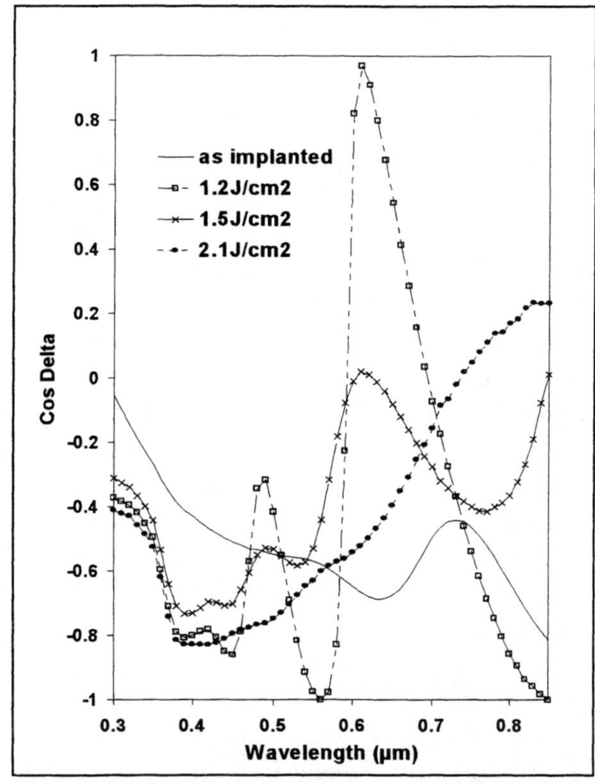

Figure 1. CosΔ parameter measured by SE on the 1.32% C implanted silicon sample before and after laser annealing.

The SE measurements reported in Figure 1 have been simulated using graded multilayer structures including Gaussian distribution of amorphous silicon inside crystalline silicon. Optical indices of both materials have been extracted from our database (crystalline silicon indices have been extracted from measurement on monocrystalline silicon wafers, amorphous silicon has been extracted from measurement on LPCVD hydrogen free amorphous silicon layers). More precisely, the sample structure is split into 30 layers with the same thickness and the total thickness is fitted during the regression with the amorphous content of each layer. In each case, an equivalent SiO_2 top layer of 40-60Å in thickness, is added to the structure to take into account the oxide layer detected by Rutherfor backscattering at the surface of the layers after laser annealing. We have used a fixed composition on a top layer and a Gaussian shape for the amorphous content decrease at the bottom interface to account for the measurement on the as-implanted samples. For the samples annealed at low density of energy we have used a double asymetric Gaussian profile for the remaining amorphous region located in depth of the sample surface.

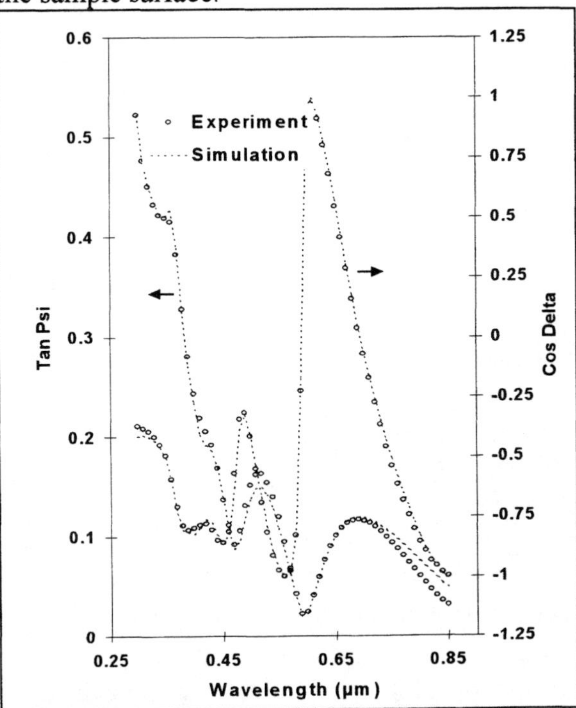

Figure 2. Experimental and simulated SE curves obtained on the 1.32% C implanted silicon sample after annealing at $1.2 J/cm^2$.

The last model applied to the sample annealed at $1.2 J/cm^2$ gives the simulation reported in Figure 2 with the experimental measurement. We can see that, even if the model is quite complicated, the agreement between simulation and experiment is very good in the entire wavelength range. Both the UV part which includes the crystalline top layer and the IR part with the interferences due to the amorphous bottom region are taken into account with the model. We found that a 500Å width zone at the bottom of the implanted region stays amorphous after annealing at $1.2 J/cm^2$. After irradiation at higher density of energy ($1.5 J/cm^2$), the amount of amorphous material is reduced in the perturbed region. After irradiation at $2.1 J/cm^2$ or more, SE cannot detect any remaining amorphous region and the sample seems completely crystallized in the entire depth but with a slightly different optical index.

RBS and XRD analysis of $Si_{1-y}C_y$ layers

As shown in Figure 3, the SE model is confirmed by RBS analysis of the same samples. Indeed, an amorphous region is clearly detected in depth of the sample whereas the top surface is crystallized. The depth of the perturbated region can be evaluated around 1500Å from the RBS measurement in agreement with the SE measurements. After annealing at $2.1 J/cm^2$ the RBS spectrum is very similar to bare silicon showing that the layer is really epitaxial and monocrystalline with a low quantity of defects.

Examples of XRD spectra obtained after annealing of the 1.1% C implanted silicon sample are reported in Figure 4. In addition to the Bragg peak of Si some additional contributions appear at higher angles. These contributions are more or less well defined depending on the annealing conditions. For energy densities not sufficient to get complete recrystallization of the implanted zone (1.2 and 1.5J/cm2 from SE measurements), the contribution is very broad. On the contrary for energy densities sufficient to melt the entire layer (2.1 and $2.4 J/cm^2$), the contribution becomes very well defined. In the case of 1.1% C implanted sample, at $2.1 J/cm^2$ an additional well defined Bragg peak can be detected (Cf. figure 4). The additional Keissig fringes can be used to determine the thickness of the $Si_{1-y}C_y$ epitaxial layer in this case (~1500Å in agreement with SE measurements). The position of the peak can be used to extract the deviation of the lattice parameter from silicon. As expected, the strain of the layer increases with the carbon content from 0.67% to 1.10%. For the highest carbon content, the stress is reduced due to either some silicon carbide precipitation in agreement with other studies (4-11), or partial relaxation of the layer. For the highest density of energy ($2.4 J/cm^2$) used in the study, the structural quality seems also slightly reduced compared to $2.1 J/cm^2$. The occurance of two contributions for the 1.1% C implanted sample at this energy (Cf. figure 4) confirms the partial relaxation of the layer and the occurance of dislocations. Further studies are needed to clear this point. Oblique incidence channeling dips around the <111> plane have been also performed to provide the lattice deviation of the $Si_{1-y}C_y$ layer.

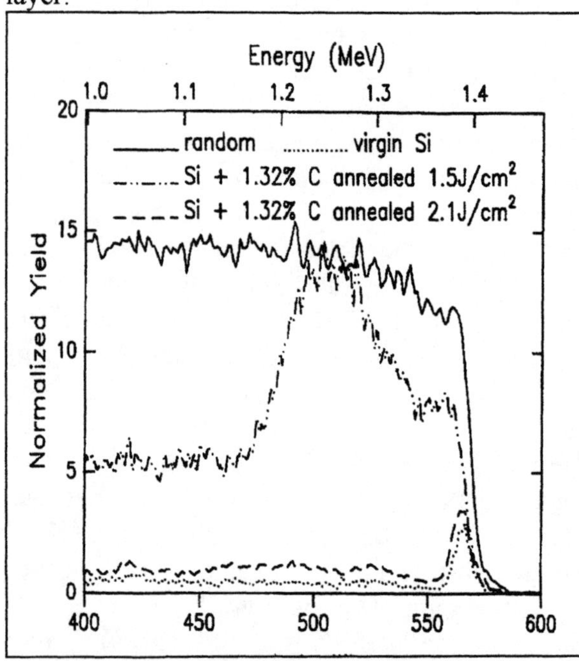

Figure 3: Experimental RBS spectrum of the 1.32% C implanted silicon sample annealed at $1.5 J/cm^2$ and $2.1 J/cm^2$ compared to random and channeling Si spectra.

Similar results as those deduced from XRD are found except for the highest carbon concentration (1.32%), for which the lack of any angular shift indicates that the layer is fully relaxed

First results on $Si_{1-x-y}Ge_xC_y$ layers

First experiments have been realized on the second series of samples ($Si_{1-x-y}Ge_xC_y$ with x ~12% and y variable). We have fixed the annealing conditions to $2.1 J/cm^2$ following the results on the first series of samples reported above. XRD spectra measured for three different carbon contents are reported in Figure 5. Even if the diffraction peaks due to the top epitaxial layers is not so well defined than for the first series of samples (mainly because of the non flat profile of Ge and C after implantation in this case), the peak position is clearly in agreement with a reduction of the stress of the layer when increasing the carbon content of the layer. The tensile stress due to the Ge element is partly reduced by the compressive strain of the C element.

CONCLUSION

In this study we have shown that excimer laser annealing of carbon implanted silicon can provide high quality epitaxial layers when the implantation profile is flat and when the energy density is optimized accurately. The crystalline quality is very good as observed by RBS and XRD techniques. Moreover, the lattice mismatch observed in these films increases with the carbon content showing that the carbon inclusion is probably free of carbide formation up to 1.1%. First results on $Si_{(1-x-y)}Ge_xC_y$ structures show the expected effect of the carbon element on the residual stress of the layers. Further studies will be necessary to get optimized $Si_{(1-x-y)}Ge_xC_y$ epilayers with optimized structural properties but the possibility up to now to anneal large surfaces (up to 20 cm^2 in one laser shot) with the SOPRA VEL laser is a key point for further industrial applications.

ACKNOWLEDGMENTS

Mr. **L. Hennet** from University Henry Poincaré (Nancy, France) is kindly acknowledged for the XRD measurements.

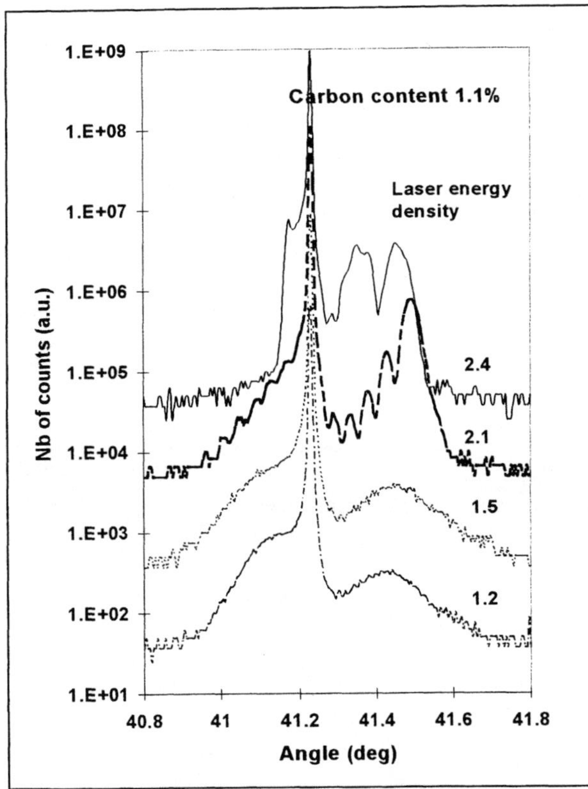

Figure 4: Experimental XRD spectra of the 1.1% C implanted silicon samples after annealing conditions. For figure clarity the curves have been shifted.

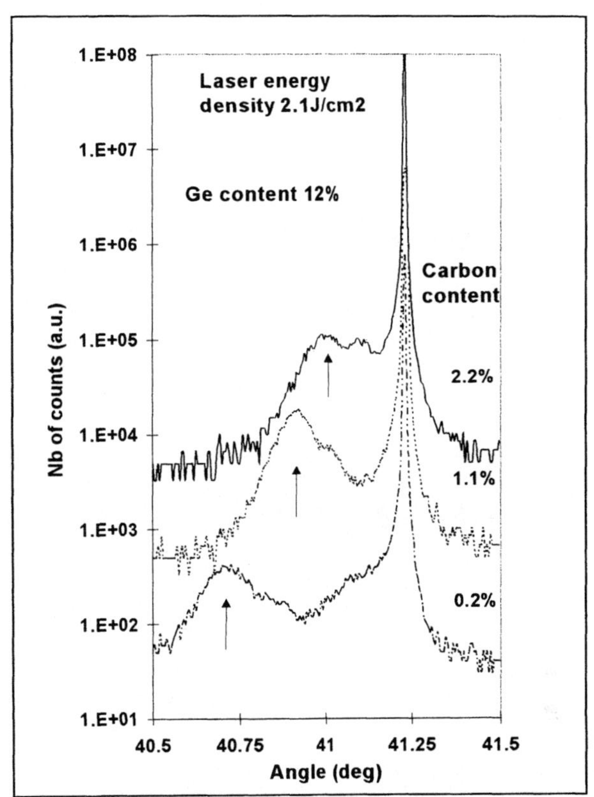

Figure 5: Experimental XRD spectra of the 12% Ge implanted silicon samples after annealing. For figure clarity the curves have been shifted.

REFERENCES

1) S.C. Jain, H.J. Osten, B. Dietrich, H. Richter, Semicond. Sci. Technol., 10, p. 1289 (1995)

2) J.W. Strane, H.J. Stein, S.R. Lee, B.L. Doyle, S.T. Picraux, J.W. Mayer, Appl. Phys. Lett., 63, N°20, p. 2786 (1993)

3) E. Fogarassy, D. Dentel, J.J. Grob, B. Prevot, J.P. Stoquert, R. Stuck, MRS Symp. Proceed., V, 354, p. 585 (1995)

4) Z. Kantor, E. Fogarassy, A. Grob, J.J. Grob, D. Muller, B. Prevot, R. Stuck, Appl. Phys. Lett., 69, N°7, p. 969 (1996)

5) J.W. Strane, S.R. Lee, H.J. Stein, S.T. Picraux, J.K. Watanabe, J.W. Mayer, J. Appl. Phys., 79, N°2, p. 637 (1996)

6) A.G. Cullis, R. Series, H.C. Weber, N.G. Chew, Semiconductor silicon 1981, Edited by R.F. Huff (Electrochemical Society, Pennington, N.J.), p. 518 (1981)

7) B. Godard, P.Murer, M. Stehle, J. Bonnet, D. Pigache, GLC 92; Heraklion, Greece, September (1992)

8) M. Stehle, Laser Focus World, june, p. 135 (1993)

9) P.Boher, J.P.Piel, C.Defranoux, J.L. Stehle, L.Hennet, SPIE'1996 International Symposium of Microlithography, 10-15 March, Santa Clara, California (1996).

10) J.Bobo, B.Baylac, L.Hennet, O.Lenoche, M.Piecuch, B.Raqet, J.Oussel, V.Viel, E.Snoeck, J. of Magn. And Magn.Mat., 121, 291, 1993

11) A. Grob, J.J. Grob, D. Muller, B. Prevot, R. Stuck, E. Fogarassy, EMRS Spring Meeting, Symposium D, paper D-P4 (1996)

Application of Electrical Step Resistance Measurement Technique for ULSI/VLSI Process Characterization

Walter H. Johnson
KLA-Tencor, FMD, three Technology Dr. Milpitas, Ca 95035

Charlie Hong and Bill Glover
Motorola Inc., 3501 Ed Bluestein Blvd., Austin, TX 78721

The electrical step resistance measurement technique is used to determine the step coverage performance of an aluminum deposition process. The results show that the electrical step resistance results are comparable to the traditional SEM cross section method. This technique is sensitive enough to detect the across the wafer step coverage variation caused by sputter target erosion. Other applications, such as measurement of poly silicon side wall slope after etch process, are also possible.

INTRODUCTION

One of the most commonly used techniques for determining the step coverage of a metal deposition process and the side wall profile of an etch process is SEM (scanning electron microscopy) cross section. The SEM technique requires a significant amount of time and resources for sample preparation, image acquisition and measurement; it is also subject to influencing artifacts from sample preparation. The step resistance measurement technique uses a pre-manufactured test wafer and an automated measurement system; therefore, the post-experiment sample preparation is eliminated and the data acquisition efficiency improves significantly (1, 2). Since the step resistance measurement is performed on an automatic stage, it is also possible to acquire a high resolution diameter scan or wafer map that is very useful for studying the across-the-wafer uniformity of a process.

EXPERIMENTAL

Principle of Measurement Technique

The step resistance technique is based on the non-contact electrical eddy current measurement of a conducting metal film. The measurement technique used for this study is based on the well known mutual inductance method. This technique is heavily used in the aerospace and nuclear power industries for metal thickness or defect detection. But, in those fields, the metal is typically hundreds of microns to several millimeters thick. For semiconductor applications, the small signals obtained from thin films require improvements to these existing techniques.

The principle of the mutual inductance method is described below:

If an alternating current is flowing in the drive coil, a magnetic field is developed, Fig. 1; this is called the primary field (3).

The impedance of the drive coil, Z, is expressed by Equation (1).

$$Z^2=R^2+X_L^2. \quad (1)$$

where R is the AC resistance of the coil and X_L is the inductive reactance.

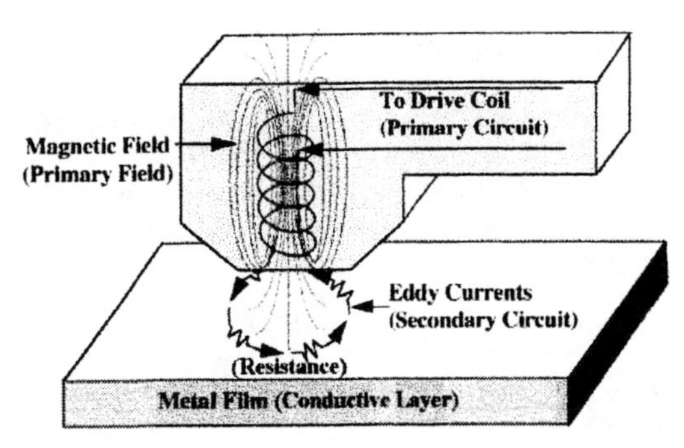

FIGURE 1. Mutual Inductance Measurement Setup.

The inductive reactance is a function of the frequency used and the coil inductance and is calculated from Equation (2).

$$X_L = 2\pi f L_0. \quad (2)$$

where f = frequency in Hertz, and L_0 = coil inductance in Henrys.

If this coil is placed in the vicinity of a conductive film the magnetic field created by the alternating current will interact with the film, inducing a current in the film. These currents are commonly called eddy currents and are dependent on the electrical and magnetic properties of the film as well as the distance from the coil. These eddy currents create a secondary magnetic field which interacts with the primary field (mutual induction) altering the resistance and inductive reactance of the coil.

The eddy currents produced in the film (secondary circuit) are dependent on the film resistance (sheet resistance) and magnetic permeability. Because the films that are normally used in the manufacturing of semiconductor devices have a relative permeability of 1, we will ignore that parameter for the purposes of this paper. Figure 1 shows a simple circular circuit with the film's resistance designated as a series of resistors. The magnetic field created by these eddy currents and their effect on the primary circuit will depend on the sheet resistance of the film.

The density of eddy currents will vary with depth, depending on the standard skin depth of the material as indicated in Fig. 2.

CP449, *Characterization and Metrology for ULSI Technology: 1998 International Conference*
edited by D. G. Seiler, A. C. Diebold, W. M. Bullis, T. J. Shaffner, R. McDonald, and E. J. Walters

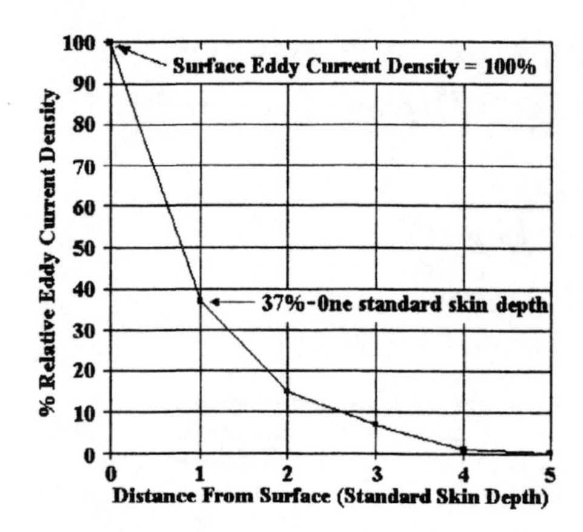

FIGURE 2. Eddy Current Density Variation Throughout the Depth of a Non-Magnetic Film.

The depth at which the eddy current density has decreased to 37% of its value at the surface is called the standard skin depth or standard depth of penetration. The standard skin depth for non-magnetic films can be approximated by the following equation:

$$S = 5.03E7\ (\rho/f)^{1/2} \tag{3}$$

where S is the standard skin depth in µm, ρ is the resistivity in Ω-cm and f is the frequency in Hertz.

The aluminum standard skin depth curve is charted in Fig. 3 as a function of frequency. For aluminum film sheet resistance measurements, the standard skin depth should be much larger then the thickness of metal. In this study, 10 MHz was used, which puts the aluminum standard skin depth at approximately 27 µm.

In the case of a metal thin film on a conductive substrate, such as an aluminum film on a doped silicon substrate, eddy currents will also form in the silicon substrate and this must be taken into account. However, if the metal film resistivity is significantly lower than the substrate resistivity, the effect of the substrate can be ignored.

In this study, the resistivity (2.8µΩ-cm) of aluminum is far lower than the resistivity of the silicon (in the Ω-cm range). Therefore, it is possible to ignore any eddy currents formed in the silicon substrate.

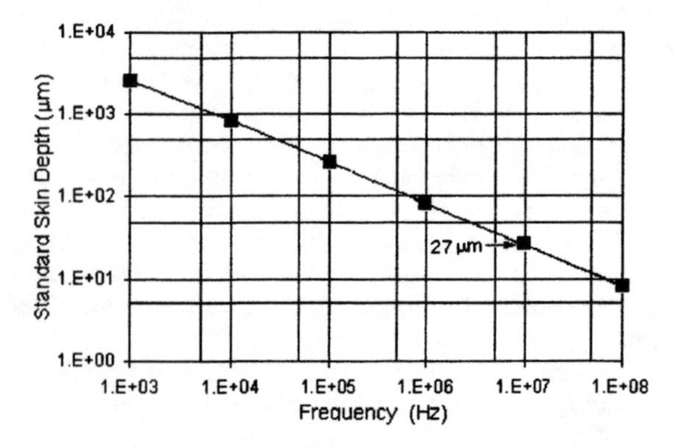

FIGURE 3. Aluminum Standard Skin Depth Curve.

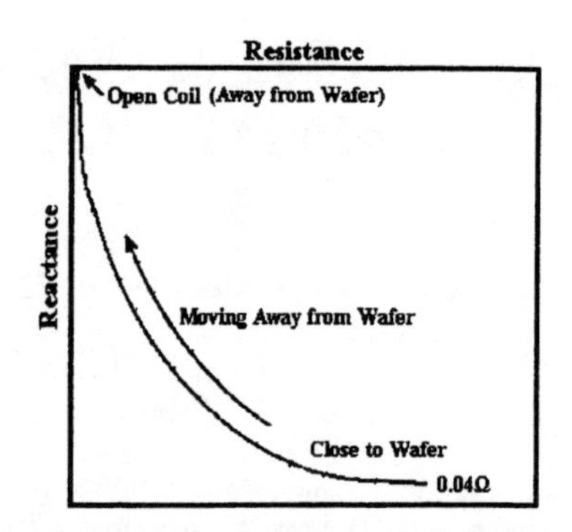

FIGURE 4. Measurement of Liftoff Curve

In this eddy current measurement technique, the change in coil values are typically measured rather than the absolute coil parameters. This greatly simplifies the procedure. To obtain a reference point, the coil is placed sufficiently far away from the film as to have no significant effect, and the coil resistance and reactance are measured. This is the reference value and is commonly called the open coil value. As the coil is brought close to the film the resistance and reactance will gradually change. To prevent the probe head from crashing into the substrate, the measurement is started at or near the surface and the probe pulled away or lifted off the surface; this is commonly called a liftoff. Plotting these resistance and reactance values creates a curve (liftoff curve) which is illustrated in Figure 4.

Typically, these liftoff curves are collected on standard samples which have been measured on a calibrated four-point probe, with the values of the respective centers assigned to each liftoff curve, as shown in Fig. 5.

If the reactance and resistance values obtained from each of the set of standard samples at a specified constant height are connected, they form a continuous curve called an iso distance curve. A series of these iso distance curves can be overlaid on the liftoff curves as shown in Figure 5. The sheet resistance of an unknown test wafer can then be interpolated from the sheet resistance values of the two adjacent calibrated liftoff curves at any given height.

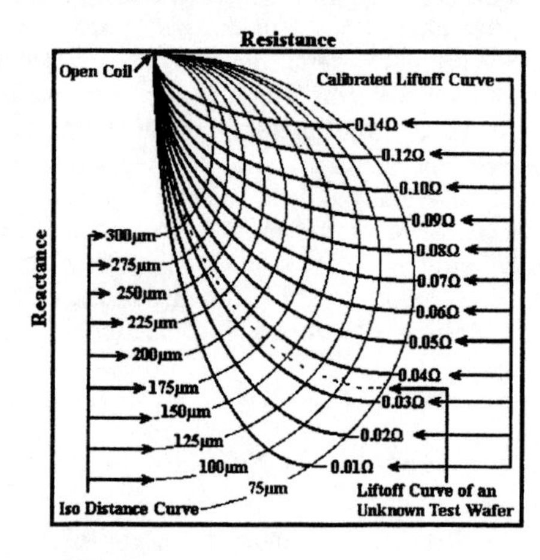

FIGURE 5. Determination of Sheet Resistance from a Series of Calibrated Liftoff Curves.

Step Resistance Measurement

A convenient unit was derived for this measurement which we call step resistance, it is defined in Equation (4).

$$\text{Step Resistance}=\frac{Rs_{TS}}{Rs_B}=\frac{\rho_{Al}\cdot Thk_B}{\rho_{Al}\cdot Thk_{TS}}=\frac{Thk_B}{Thk_{TS}} \quad (4)$$

Where Rs_{TS} and Rs_B are the sheet resistance of the patterned test structure wafer and blank control wafer, respectively. Thk_B is the metal thickness on the blank control wafer and Thk_{TS} is the effective metal thickness on the test structure. ρ_{Al} is the resistivity of aluminum for the specific process conditions used. The effect of ρ_{Al} cancels out in the equation because all wafers were processed with the same process conditions and have the same ρ_{Al}.

The step resistance is an arbitrary unit and is highly dependent on the test structure. Therefore, it is important to maintain the consistency of the test structure within an experiment.

Test Wafer Preparation

Silicon wafers were coated with a specified thickness of oxide, then patterned. The test pattern can consist of repeating line/space pairs, or densely packed repeating contact holes. In this study the test pattern with repeating line/space pairs was used. Figure 6 shows the layout of the test pattern and the test structure on the wafer.

The oxide films were etched and then aluminum films were deposited on the matrix of oxide line/space test structure. Table 1 shows the 3 factor full factorial experimental design. The variables used for this experiment are the metal thickness, t_m, the oxide step height, t_{ox}, and the pitch of the line/space pairs.

TABLE 1. Experimental Design

	Group 1				Group 2			
Wafer ID	A	B	C	D	E	F	G	H
t_m*	1.35	1.35	1.35	1.35	0.88	0.88	0.88	0.88
t_{ox}*	0.79	1.16	0.79	1.16	0.81	1.17	0.79	1.17
Pitch	3.0	3.0	2.0	2.0	3.0	3.0	2.0	2.0
Aspect ratio	0.54	0.86	0.71	1.10	0.55	0.87	0.72	1.09

* All measurements were in μm and obtained from an SEM cross section of the test wafer.

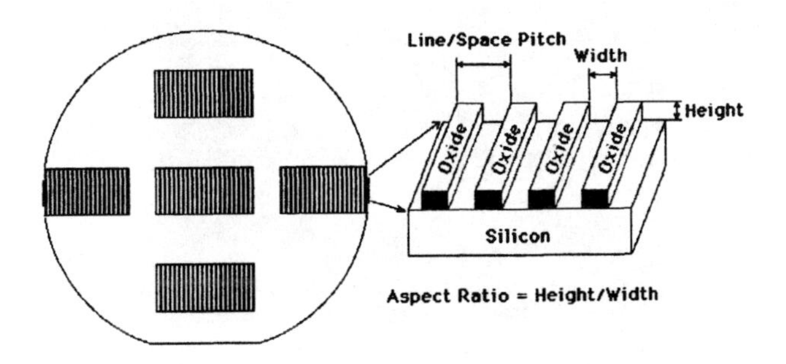

FIGURE 6. Test Pattern Layout and Test Structure.

It is extremely important to note that, in each group, a blank control wafer was processed with the patterned test wafers to produce a metal film with identical microstructure and electrical properties. The sheet resistance of this blank control wafer was measured and used in the step resistance calculations.

RESULTS AND DISCUSSION

Diameter scans of the sheet resistance on the step coverage test wafers were collected and the step resistance values were calculated from Equation (4). As an example, Figure 7 shows the step resistance of the Group 1 wafers (A, B, C and D in Table 1). Only the center values of the structure were used for comparisons in this study. This is because the side test patterns were affected by unexpected photo lithography problems.

In Fig. 8, the step resistance acquired from the center of the wafer was plotted against the relative aspect ratio. The relative aspect ratio is defined as the aspect ratio divided by the pitch ratio, where the pitch ratio is the test structure pitch divided by the reference pitch. The reference pitch used here is 2μm. The relative aspect ratio is introduced to compensate for the line/space pair density difference caused by different line/space pair pitch.

Figure 8 shows that the step resistance is a function of metal thickness, aspect ratio and line/space pair pitch suggesting that the step resistance measurement results are an overall measurement of deposited metal conformity on the test structure.

FIGURE 7. Step Resistance Diameter Scan on Group 1 Test Wafers

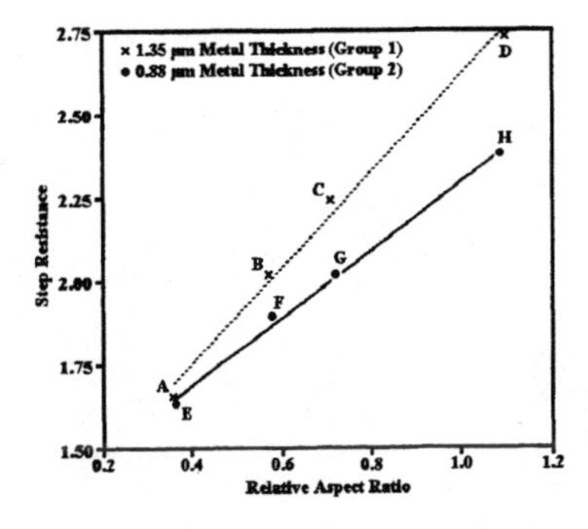

FIGURE 8. Correlation between Step Resistance and Relative Aspect Ratio.

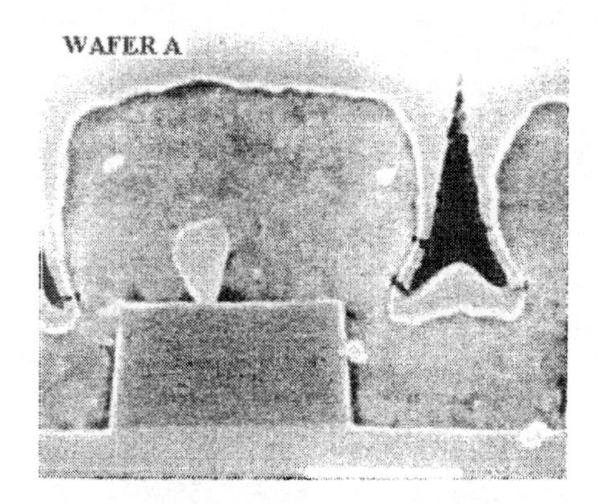

FIGURE 9A. Test Wafer A

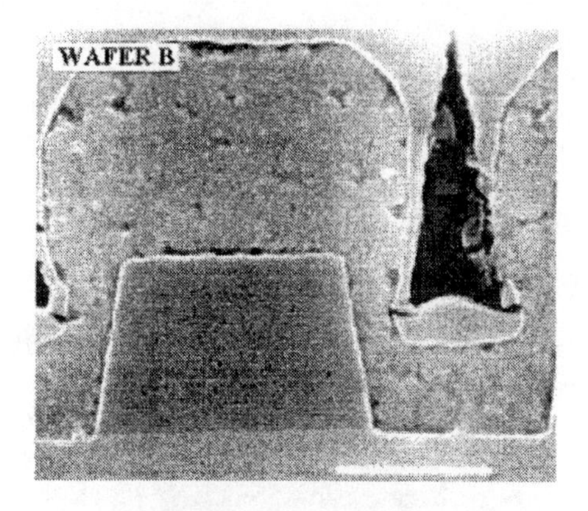

FIGURE 9B. Test Wafer B.

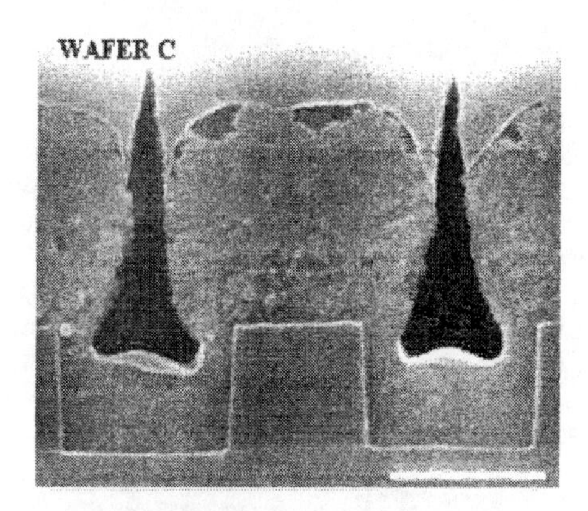

FIGURE 9C. Test Wafer C.

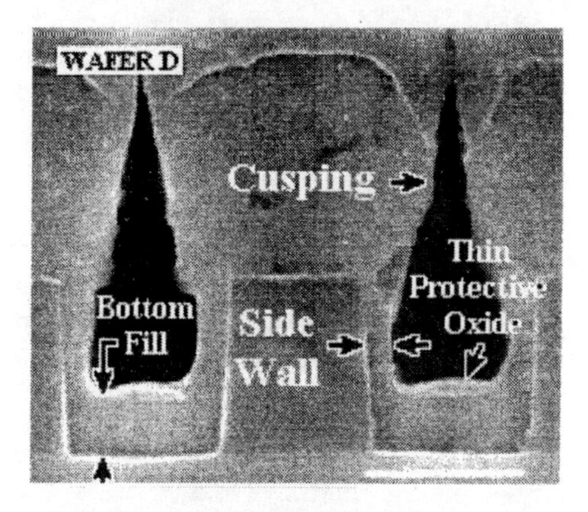

FIGURE 9D. Test Wafer D. Critical Elements of Step Coverage.

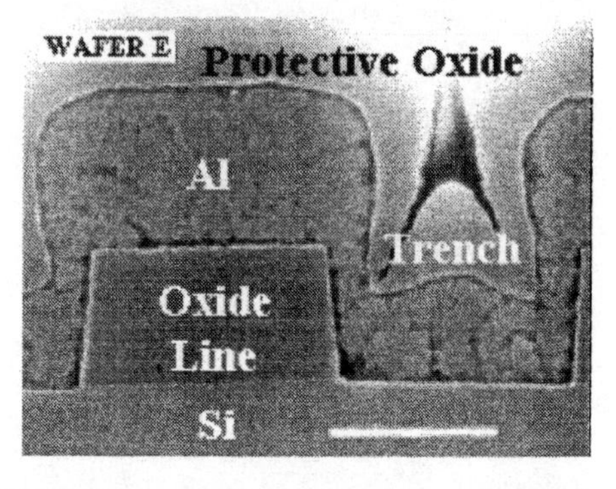

FIGURE 9E. Test Wafer E. Components of Test Structure.

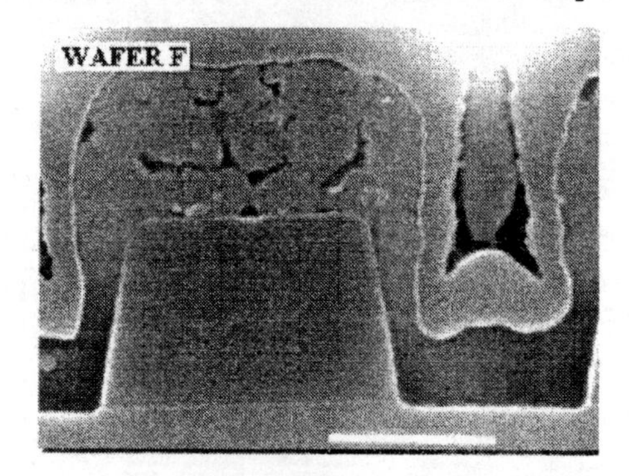

FIGURE 9F. Test Wafer F.

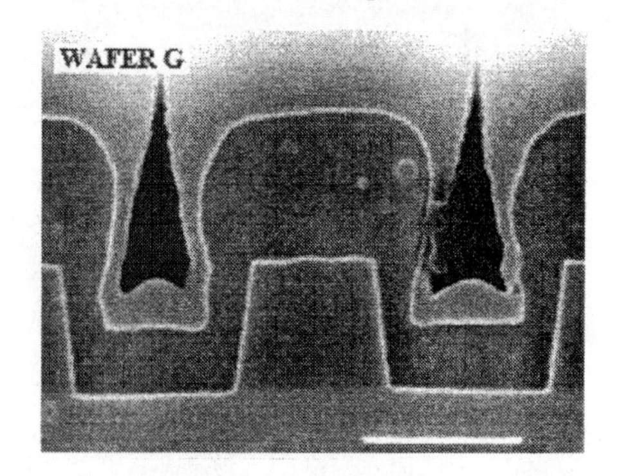

FIGURE 9G. Test Wafer G.

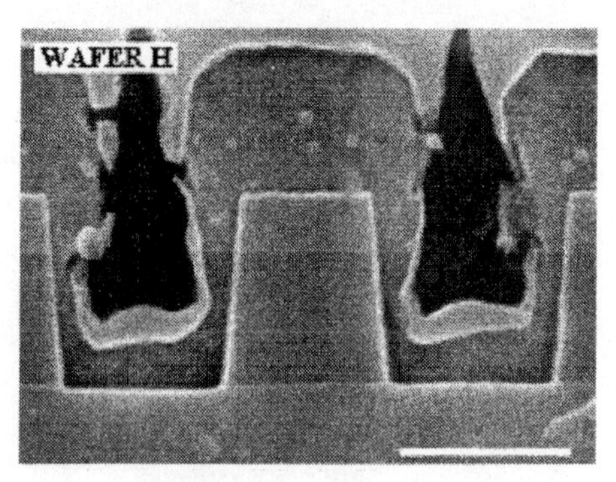

FIGURE 9H. Test Wafer H.

It is very difficult to describe the exact relationship between the step resistance and critical elements of the metal step coverage, such as side wall, bottom fill and cusping. Figure 9D illustrates these elements.

Figure 9A to Fig. 9H shows the SEM cross section of each test wafer to provide a visual correlation between the step resistance and the metal step coverage. The letter assigned to each SEM cross section corresponds to the wafer ID given in Table 1. Right before the SEM sample preparation, these wafers were deposited with approximately 10 μm of protective oxide on top of the metal film to protect the metal and to minimize the distortion caused by sample preparation. Figure 9E illustrates the key components of the test structure.

The test wafers with better metal side wall coverage and bottom fill (Figure 9A and 9B) have lower step resistance (Figure 8, point A and point B). In contrast, the wafers with highest aspect ratio (Figure 9D and 9H), have the highest step resistance within their groups (Figure 8, point D and point H). The cross section in Fig. 9D and Fig. 9H shows that the longer deposition time (thicker metal film Fig. 9D) provided very limited improvement in side wall thickness and bottom fill thickness, due to the closing of the gap between two metal cuspings, which virtually prevented metal deposition inside the trench and resulted in higher step resistance. The very thin protective oxide inside the trench illustrates the effect of the gap size, Fig. 9D.

APPLICATION OF STEP RESISTANCE MEASUREMENT

Across the Wafer Step Coverage

Figure 10 illustrates the relative location of a used sputter target erosion profile and the step resistance of a test wafer processed prior to the target change. The blips on the step resistance curve (approximately 15 mm and 45 mm from the center) coincide with the seam between two stepper patterns. The step resistance variation across the wafer is less than 0.1. This example demonstrates that even a subtle across the wafer metal step coverage variation can be measured with this technique.

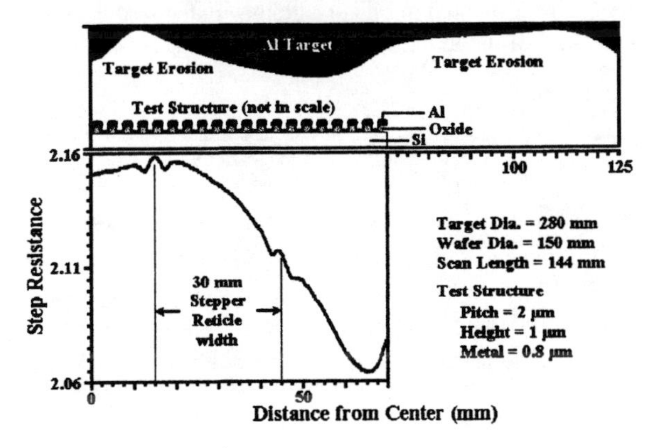

FIGURE 10. Correlation between Sputter Target Erosion Profile and Step Resistance Diameter Scan

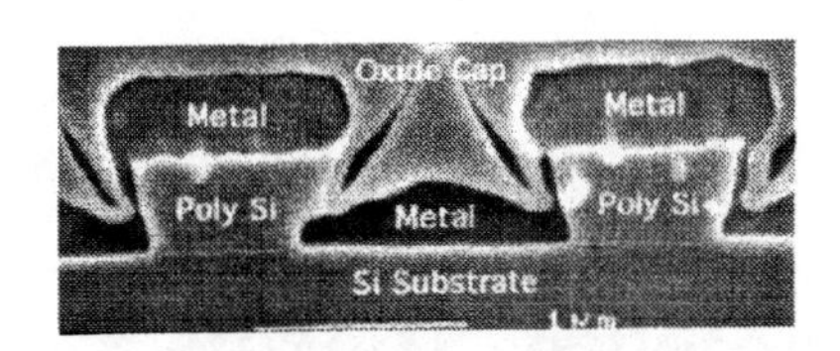

FIGURE 11A. Poor metal step coverage over the retrograde poly silicon side wall resulted in higher step resistance. Step Resistance = 1.746.

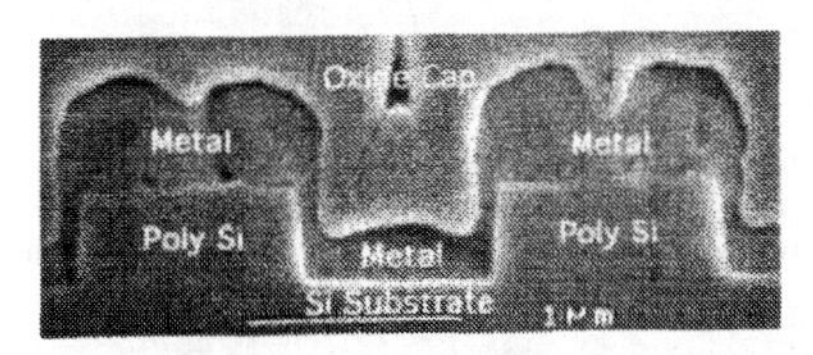

FIGURE 11B. Better metal step coverage over the normal poly silicon side wall resulted in lower step resistance. Step Resistance = 1.573.

Poly Silicon Etch Profile Study

In poly silicon etch, the poly silicon side wall slope is highly dependent on the etch process variation. The change in side wall slope cannot be observed with an optical microscope and, therefore, the most common method is using the SEM cross section. However, if the test wafers are deposited with aluminum using the same metalization process, it is possible to compare the after-etch poly silicon side wall slope by a step resistance measurement. This is because the test wafers with poor side wall profile tend to have higher step resistance caused by poor metal step coverage.

Figure 11A and 11B shows the cross section of these poly silicon test wafers. The test wafer with retrograde side wall (Fig. 11A) shows discontinuous metal under the side wall. The test wafer with normal poly silicon sidewall has better metal step coverage.

CONCLUSION

The step resistance measurement is a very useful technique for metal step coverage study and side wall profile study. This technique complements the present SEM cross section technique to provide timely experimental results. The small probe head and automatic stage also enable the instrument to perform high resolution diameter scans and wafer maps for an across-wafer process uniformity study.

REFERENCES

1. Johnson, W.H., Hong, C., and Becnel, V., "Step Coverage Measurements Using A Non-Contact Sheet Resistance Probe," 1997 Proceedings, 4th VMIC Conference, Santa Clara, Ca., pp. 198-200.
2. Hong, C., Becnel, V., Foster, S., and Johnson, W.H., "The Effect of Deposition Time, Temperature and Power on Step Resistance of Aluminum Films," 1997 Proceedings, 4th VMIC Conference, Santa Clara, Ca., pp. 213.
3. *Metals Hand Book,* ASM Int., May 1992, Vol. 17, pp. 165-170.

In-situ Surface and Interface Characterization by Optical Second Harmonic Generation (SHG) of Silicon Dioxide Fabrication with High Purity Ozone

Ken Nakamura, Akira Kurokawa, and Shingo Ichimura

Electrotechnical Laboratory, Tsukuba, Ibaraki 305-8568, Japan

Second harmonic generation (SHG) on silicon surfaces, giving signals only from surfaces and interfaces, was applied for in-situ analysis of ultra-thin oxide fabrication by high purity ozone. In addition to its capability of in-situ measurement, the characteristics of SHG analysis by use of fundamental Nd:YAG laser beam on silicon surfaces is highlighted with such merit as continuous observation of surface and film growth processes and information depth limited to outermost surface layer. However, such limitation as angular condition of optical path must be also considered due to the symmetry of surface structure.

INTRODUCTION

In-situ characterization of surfaces and interfaces is indispensable while developing novel materials processing technique because the results of diagnosis during materials processing will give feedback to optimize processing condition. Although many surface analysis methods such as photoelectron spectroscopy, ion scattering spectroscopy and secondary ion mass spectrometry have already become conventional analysis technique and offer significant data for analysis, in-situ measurement is difficult to perform with them because of their need of high vacuum to obtain enough mean free path of electron or ion before their detection. However, optical measurements such as infrared spectroscopy (IR) and Raman spectroscopy require no such condition and that makes in-situ characterization easily possible with their high energy resolution, while their sensitivity is not so high as is required and their surface-sensitivity is not always guaranteed. Therefore, another optical characterization technique, in addition to its possibility of in-situ characterization, possessing high sensitivity to surfaces and interfaces is strongly required.

Nonlinear optical techniques such as second harmonic generation (SHG) which is highly sensitive to surfaces and interfaces (1-3) are becoming a versatile tool among surface analysis methods. Generally, SHG is a coherent light generation with a frequency as double as, or with a wavelength as half as that of intense incident laser beam in a nonlinear optical medium(4-6). Within the electric dipole approximation, this phenomena requires a condition that nonlinear optical medium has its structure without inversion symmetry. This condition is always satisfied at surfaces and interfaces of any materials so that surfaces and interfaces always have such nonlinear optical effect with intense laser beam given. Especially, in a centrosymmetric medium such as silicon or in an isotropic material such as amorphous silicon oxide in the bulk of which SHG is forbidden, signals are only originated on surfaces and interfaces so that SHG is a relevant characterization tool to analyze those of silicon by a coherent laser beam with a fixed as well as a tunable wavelength.

The purpose of this article is to summarize phenomena and principle of SHG at surfaces and interfaces and to indicate the suitability of its application, as well as its limitation to be taken into consideration, to surface and interface analysis of silicon materials system through our series of study on silicon oxidation processes (7-14) by high purity ozone(15,16). Ozone is expected to be an alternative oxidant of silicon because of its strong activity (17) which is expected to lower substrate temperatures during oxidation, an indispensable condition for device fabrication in the near future. Several studies have been reported so far to utilize ozone for silicon oxidation by use of conventional ozonizer (18-21) or UV ozone (22) with their results indicating that ozone is a promising oxidant. Therefore, ozone in high concentration is expected to enhance oxidation rate and to upgrade quality of synthesized oxide. Our work on adsorption / oxidation of Si(111) and Si(100) surfaces with high purity ozone has been performed with various analysis methods including surface SHG(10-12,14). Method and characteristics of surface SHG analysis are presented in this article through our application to the study of initial oxidation mainly of Si(111)7x7.

PRINCIPLE OF SHG

SHG is a second-order nonlinear optical effect(4-6). It is, quantum mechanically, a multi photon process: two-photon absorption from initial state to final state via virtual intermediate state and one photon emission from final state to initial state. Therefore, second harmonics (SH) intensity indicates second-order linearity to the intensity of incident light such as intense pulsed laser beam. This can be also represented in

CP449, *Characterization and Metrology for ULSI Technology: 1998 International Conference*
edited by D. G. Seiler, A. C. Diebold, W. M. Bullis, T. J. Shaffner, R. McDonald, and E. J. Walters

principle by electromagnetics as one of electric dipole radiation phenomena(23): incident light first induces polarization in the media and then this polarization radiates electromagnetic waves. In the case of SHG, nonlinear polarizations oscillating with the double frequency of the incident light are sources of SHG. This effect is widely used in the wavelength conversion technique of laser beam from its fixed fundamental wavelength to its second harmonics.

Nonlinear polarization for SHG in silicon is limited to surfaces and interfaces under electric dipole approximation because inversion symmetry is only broken at surfaces and interfaces of silicon with centrosymmetric crystalline structure. Interaction of light with matter is basically described by oscillating electric field of light and induced polarization. Second-order nonlinear polarization, ***P***, can be represented in the Cartesian coordinate with nonlinear optical susceptibility tensor, χ and incident light, ***E***, as follows:

$$P_i = \Sigma \chi_{ijk} E_j E_k \quad (i, j, k = x, y, z) \qquad (1).$$

Inversion of this Cartesian coordinate turns E_i into $-E_i$ and P_i into $-P_i$ so that, in this case, (1) must turn into

$$-P_i = \Sigma \chi'_{ijk} (-E_j)(-E_k) \quad (i, j, k = x, y, z) \qquad (2).$$

If this media has inversion symmetry, then

$$\chi_{ijk} = \chi'_{ijk} \qquad (3)$$

so that, comparing (1) with (2) after inserting (3) into (2), it must be

$$\chi_{ijk} = -\chi'_{ijk} \qquad (4).$$

Therefore, it is deduced from (3) and (4) that in the media with inversion symmetry

$$\chi_{ijk} = 0 \qquad (5).$$

That is, SHG is forbidden in the media with inversion symmetry.

These tensor elements are reduced in number and related to each other based upon the symmetry of the medium. For example, in the case of C_{3v} (Si(111)7x7) and C_{2v} (Si(100)2x1) symmetry, tensor elements are, for C_{3v} with the mirror plane of x-z

$$\chi_{xxx} = -\chi_{xyy} = -\chi_{yyx} = -\chi_{yxy}, \ \chi_{zxx} = \chi_{zyy}, \\ \chi_{xxz} = \chi_{yyz} = \chi_{xzx} = \chi_{yzy}, \ \chi_{zzz} \qquad (6)$$

and for C_{2v} with mirror planes of x-z and y-z

$$\chi_{xxz} = \chi_{xzx}, \ \chi_{yyz} = \chi_{yzy}, \ \chi_{zxx}, \ \chi_{zyy}, \ \chi_{zzz} \qquad (7)$$

Hence, by observing dependence of SH intensity on angular condition of incident light, information on symmetry of surfaces and interfaces can be obtained because surface symmetry is reflected in the nonlinear susceptibility, that is, SH intensity. This phenomena can be used as structural analysis of flat surfaces(24)as well as vicinal surfaces(25) and interfaces(26,27).

These elements of nonlinear optical susceptibility show their dependence on incident and SH wavelength due to the resonance of specific state localized at surfaces or interfaces. Second-order perturbation theory indicates that dipole transition probability of SHG is proportional to the product of each dipole transition probability of two photon absorption and one photon emission (1) so that resonance of either incident or SH light to each transition enhances the overall transition probability which can be used as spectroscopy. Recent advancement of tunable coherent light source has enabled to perform such SHG spectroscopy(28-31). Characteristics of high energy resolution obtained from optical probe are made highly use of to obtain spectroscopic SHG data and information on energy states localized in the surface and interface layers. Use of resonance is more appropriately applied for another second order nonlinear optical effect, sum-frequency generation (SFG)(1-3), as vibrational spectroscopy.

MEASUREMENT SYSTEM

Our measurement system of SHG consists of a laser as a light source for SHG, an ultra-high vacuum chamber for processing and diagnostics, a monochromator and a photomultiplier as a detector (Figure 1). Nd:YAG laser was used as the light source with a series of optical devices for the rotation of the plane of polarization of incident light. The laser wavelength was fixed at 1.064μm, with a pulse repetition rate of 10Hz and a pulse width of 10ns. The p-polarized light was

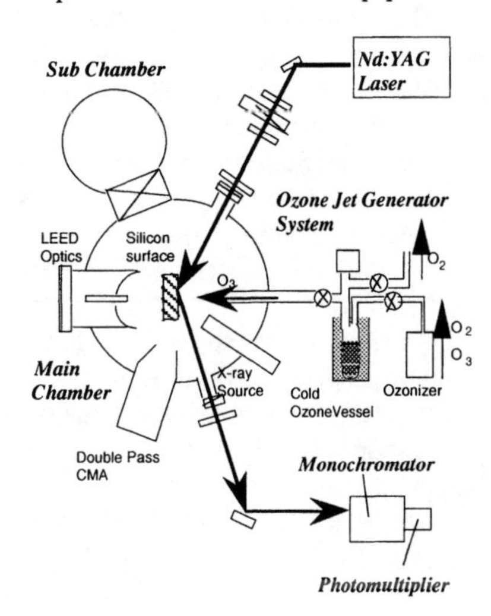

FIGURE 1. Experimental setup for SHG analysis of ozone oxidation.

incident onto the surface along the $[2\bar{1}\bar{1}]$ axis with the incidence angle of 55° to normal to the surface and was focused in a spot of 2.5mmϕ on the surface. This angle of incidence is a critical factor for setting up an optical path of SHG measurement especially on Si(100) because of its C_{2v} symmetry which will be described later. The intensity of the incident laser was $1.4x10^8$ W/cm^2 at the sample, lower than the threshold for optical breakdown(32) but high enough to generate nonlinear optical effect. P-polarized SH light of 532 nm was detected in the direction of the reflectance angle by a photomultiplier after separating only SH light from a fundamental laser beam by a grating-type monochromator. SH signals were averaged by boxcar for 100 laser shots and recorded.

For oxidation procedure, we used high purity ozone jet generator system, which is described in detail elsewhere(15). The system consists of a conventional ozonizer of electric discharge type and an ozone vessel. After a production of ozone in the ozonizer, a mixture gas of ozone and oxygen (ozone concentration ~5%) was introduced from the ozonizer into the ozone vessel at 90K. During this procedure, the vessel was evacuated by a pumping system. After only ozone was liquefied in the vessel (purity of ozone >98mol%) while oxygen was evacuated, the vessel was isolated with its temperature controlled within accuracy of 0.1K so that the precise control of saturate vapor pressure of ozone enables its flux introduced into the chamber to be constant. At the sample position, the concentration of ozone was estimated >80at% (16).

APPLICATION OF SHG

The purpose of our application of SHG is to analyze adsorption kinetics and interface formation in-situ on silicon surfaces and interfaces during oxidation processing by high purity ozone at lower temperatures than by conventional technique. In the R&D of novel processing technique to fabricate thin film in the semiconductor technology, kinetics of adsorption and initial film growth is to be analyzed in-situ to explore and control condition of fabrication to its best. However, during such processing as our oxidation with ozone, the pressure inside the main processing chamber is up to 10^{-3}Pa and in the sub-processing chamber up to 10^{-1} Pa so that only optical in-situ analysis can be utilized. Although UV light is contributing to the photodissociation of ozone(33), infrared light of Nd:YAG laser has shown no such effect on oxidation kinetics that in-situ surface adsorption analysis by SHG pumped by Nd;YAG laser is possible(34,35). Also possible is specific characterization only of Si/SiO_2 interfaces (36). Therefore, surface and interface analysis of initial adsorption and oxide formation, respectively, by high purity ozone can be monitored in-situ by SHG because all the condition for characterization technique to be highly specific only to surfaces and interfaces are satisfied.

Two types of 532nm light appeared during the course of

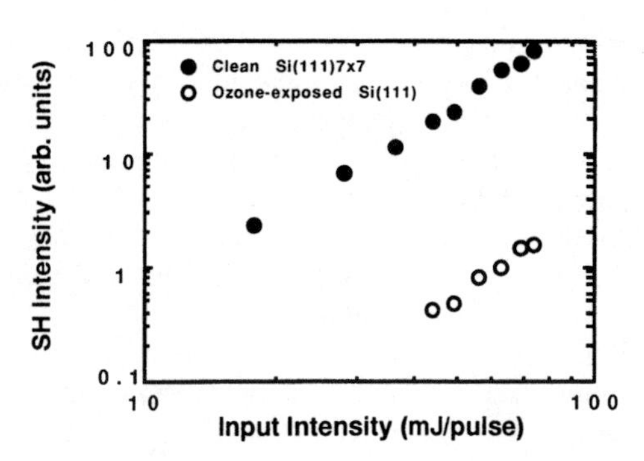

FIGURE 2. Dependence of second harmonic intensity of (●)clean and (○)ozone-exposed Si(111)7x7 on input laser intensity.

initial oxidation by ozone: one is large optical intensity on original clean Si(111)7x7 surface and another is weak intensity which appears in the course of exposure. Both intensities indicate their second-order linearity to that of incident light (Figure 2). This means that not only clean surfaces with vacant dangling bonds but also adsorbed oxide states are contributing to SHG. The latter can be utilized to analyze interface formation. This weak SH intensity appeared on any surfaces at temperatures ranging from room temperature to 400°C after they once decreased during the initial exposure. Especially at higher substrate temperatures than 260°C, this latter intensity becomes more intense than that at RT as is indicated

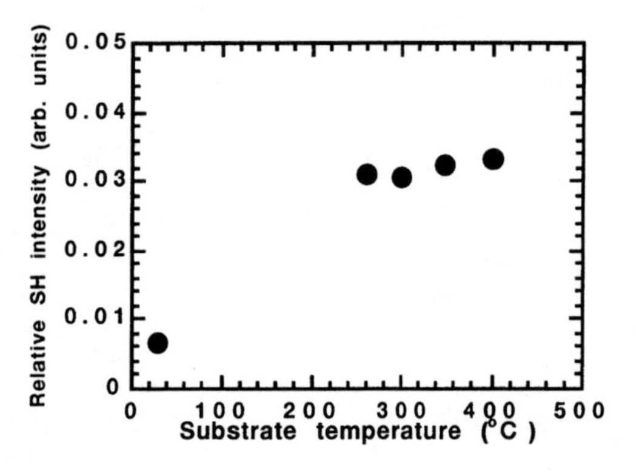

FIGURE 3. SH intensity after exposure of Si(111) at room-temperature (RT), 260°C, 300°C, 350°C and 400°C to 20L ozone, relative to that of the clean 7x7 surface. SHG measurements were performed at the same temperature as the ozone exposure.

in Figure 3. Due to the different characteristics between that appeared at RT and that at 260°C and 380°C of SH intensity recovery by desorption of adsorbed species which was described elsewhere(12), we expect that this latter intensity is due to the formation of more stable initial oxide states, although physical mechanism of this weak SHG remained unclear whether this is only due to the electric dipole contribution or also due to quadrupolar and third-order (electric-field induced SHG) contributions (25,36).

Origins of SHG on silicon surfaces have been investigated with spectroscopic SHG data by recent literature. Pedersen *et al.*(28) observed dependence of SH intensity of Si(111)7x7 on the different wavelength of pumping laser. They also observed, during the exposure of Si(111)7x7 at room temperature to molecular oxygen, that SHG by fundamental Nd:YAG laser at 1.17eV first decreased to its minimum and then started increasing again to some extent(29). The first decrease was reported due to the decrease of 1.15 and 1.3eV SHG peaks attributed to transition between surface-states by terminating dangling bonds. The second increase or appearance of weak intensity is due to the broadening of higher peak at 1.7eV by, according also to Daum *et al.*(30), strain induced by the reconstruction. This broadening increases the SH intensity pumped by, for example, 1.17eV fundamental Nd:YAG laser. They have concluded, therefore, that addition of these two factors due to different origins indicated both decrease and increase of SH intensity with its minimum during the same course of oxygen adsorption(29).

The contribution of this more stable oxygen adsorbates to SHG in our measurement was confirmed directly by XPS analysis. After the exposure of Si(111)7x7 at RT and 400°C to 20L ozone gas, both of the samples were increased to the temperatures shown and decreased down to take XPS spectra. Figure 4 indicates the change of O1s intensity during this procedure. Apparently, this result supports that the exposure of Si(111)7x7 at 400°C forms more stable oxide species than at RT because oxygen atoms rest on the sample at higher temperature. Therefore, appearance of weak SH intensity can be related to the formation of more stable oxide species. The existence of different oxide species with different thermal stability was shown also clearly by SH intensity recovery(12), although desorption temperatures are higher when observed by XPS than by SHG.

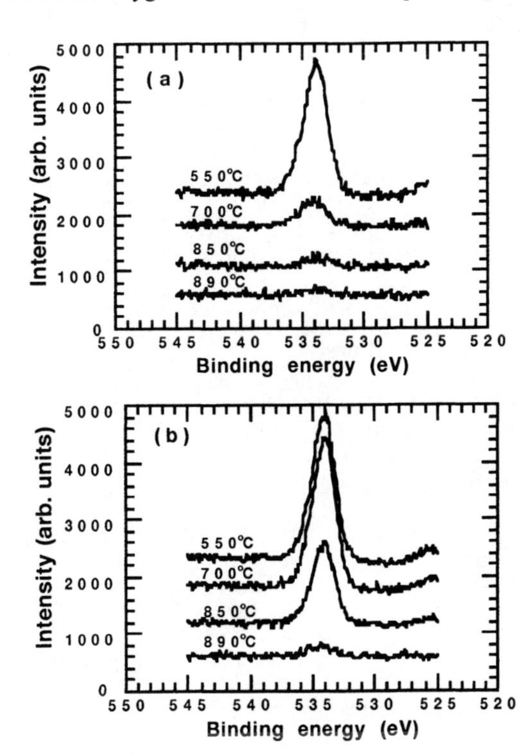

FIGURE 4. XPS O1s analysis of desorption of oxygen species formed by exposing Si(111)7x7 at (a)room-temperature and (b)400°C to 20L ozone.

One of the possible reason for the difference of desorption temperatures is the difference between information depth of SHG at 1.064μm pump beam on clean Si(111)7x7 and that of XPS. That is, lower desorption temperatures by SHG indicate that desorption first starts on dangling bonds. In the case of XPS, Tanuma *et al.* estimated that the escape depth for O1s photoelectron emission is approximately 2nm(37) in Si. Therefore, the total amount of photoelectron emitted from oxygen atoms within this information depth, i.e., not only those

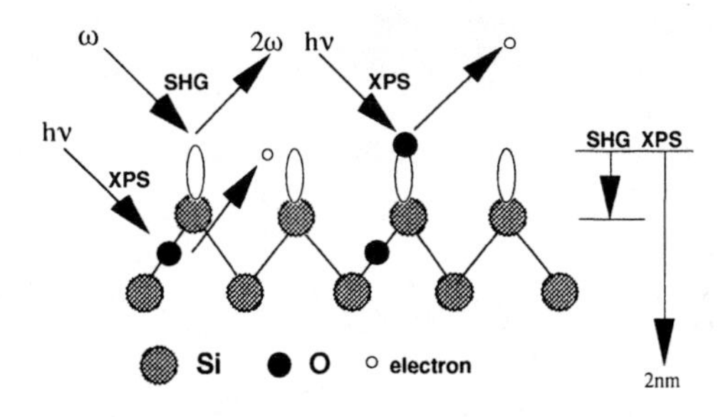

FIGURE 5. Schematic diagram of information depth of SHG on clean silicon surface with vacant dangling bonds and XPS.

on the dangling bonds but also in the backbonds in the subsurface layers, can be detected by O1s XPS intensity. However, SH intensity by 1.064μm Nd:YAG laser on clean silicon surfaces is sensitive to the surface state originated from vacant dangling bonds on the surfaces(28). Hence, change in the SH intensity on clean surface must be attributed mainly to the dangling bonds. (Figure 5) One of the example which is affected by this difference in the information depth is our previous observation by XPS and SHG of apparently different kinetics during the same ozone adsorption processes on Si(111)7x7(11). This is because adsorption on dangling bonds can be detected by both SHG and XPS while backbonds insertion can only be detected by XPS but scarcely by SHG(14).

Appropriate angular condition of laser incidence in this setup enables SHG analysis of Si(100) surfaces, more important for ULSI technology. Structure of Si(111)7x7 surface has C_{3v} symmetry while that of Si(100)2x1 has C_{2v} symmetry. This C_{2v} symmetry limits the condition of incidence / scattering angles to detect SH intensity effectively. If the normal or near-normal incidence and normal detection angles of laser beam are employed for SHG on Si(100)2x1, SHG is not detected because nonlinear polarization parallel to the surface giving SH intensity cannot be induced due to its surface symmetry, and polarization perpendicular to the surface is not contributing to SH detection in this setup (Figure 6) . Our previous condition of optical setup with incidence and detection angles of 20° prohibited intense SHG on Si(100)2x1, although SH

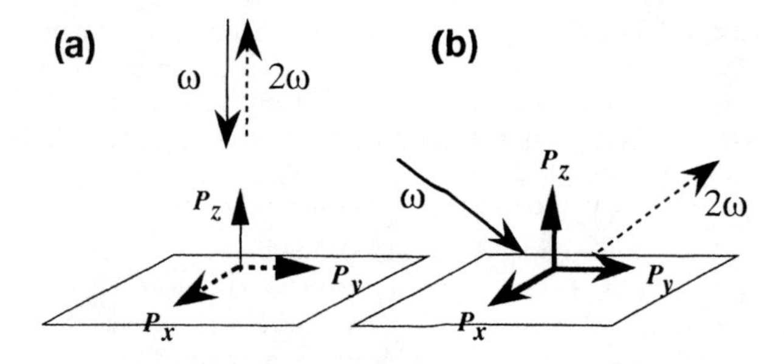

FIGURE 6. Nonlinear polarization on Si(100)2x1 by (a) normal and (b) off-normal laser incidence.

intensity on Si(111)7x7 was clearly detected(38). But with off-normal incidence of laser, nonlinear polarization perpendicular to the surface can radiate second harmonics in the direction of reflection for detection. Successful SHG measurement on Si(100) (39,40) including our measurement(10) is always carried out by using inclining incident laser beam. SHG analysis of oxide growth by ozone on Si(100) (13) is under preparation.

CONCLUSION

Application of SHG at R&D stage of semiconductor technology as well as its principle was presented and discussed through our example of its application for developing oxidation technique. Relevant application of SHG to the analysis of initial oxidation kinetics is by use of fundamental beam of Nd:YAG laser with 1.17eV pump energy, i.e., 1.064μm wavelength. At this energy, 532nm second harmonics are generated mainly due to surface states induced by vacant dangling bonds and also oxygen adsorbates during exposing Si(111)7x7 to ozone: the latter may be related to the strain on the surfaces. Because SHG on clean silicon surfaces is dependent upon surface states induced by dangling bonds, adsorption on different sites other than dangling bonds must be considered by complementary use of other surface characterization technique with larger information depth such as XPS. In addition to Si(111), Si(100) is also contributive to SHG. In that case, however, angles of incident and scattering light must be taken large enough to induce polarization perpendicular to the surface and to detect radiated SH intensity because of C_{2v} symmetry of Si(100) surfaces.

REFERENCES

1. Shen, Y. R., *Annu.Rev.Phys.Chem.* **40**, 327-350 (1989).
2. Shen, Y. R., *Nature* **337**, 519-524 (1989).
3. McGilp, J. F., *J. Phys. D*, **29**, 1812-1821 (1996).
4. Bloembergen, N., *Nonlinear Optics*, Reading: Addison-Wesley, 1992.
5. Shen, Y. R., *The Principles of Nonlinear Optics*, New York: John Wiley and Sons, 1984.
6. Boyd, R. W., *Nonlinear Optics*, San Diego: Academic Press, 1992.
7. Kurokawa, A. and Ichimura, S., *Jpn. J. Appl. Phys.* **34**, L1606-1608 (1995).
8. Kurokawa, A.and Ichimura, S., *Appl. Surf. Sci.* **100/101**, 436-439 (1996).
9. Kurokawa, A., Ichimura, S., Kang, H. J. and Moon, D. W., *Mat. Res. Soc. Proc.* **429**, 269-274 (1996).
10. Nakamura, K., Kurokawa, A., and Ichimura, S., *Proc. 3rd Int'l Symp. Ultra Clean Processing of Silicon Surfaces,* Antwerp, Belgium, p.279-282, (1996).
11. Nakamura, K., Kurokawa, A., and Ichimura, S., *J. Vac. Sci. Technol. A* **15**, 2441-2445 (1997).
12. Nakamura, K., Kurokawa, A., and Ichimura, S., *Surf. Interface Anal.* **25**, 88-93 (1997).
13. Kurokawa, A., Ichimura, S., and Moon, D. W., *Mat. Res. Soc. Proc.* **477**, 359-364 (1997).
14. Nakamura, K., Kurokawa, A., and Ichimura, S., *Surf. Sci.* in press.
15. Hosokawa, S. and Ichimura, S., *Rev. Sci. Instrum.* **62**, 1614-1619 (1991).
16. Ichimura, S., Hosokawa, S., Nonaka, H. and Arai, K., *J. Vac. Sci. Technol. A* **9**, 2369-2373 (1991).
17. Horvath, M., Bilitzky, L., and Hüttner, J., *Ozone,* New York: Elsevier, 1985, p30.
18. Chao, S.C., Pitchai, R. and Lee, Y.H., *J. Electrochem. Soc.* **136**, 2751-2752 (1989).
19. Kazor, A. and Boyd, I. W., *Electron. Lett.* **29**, 115-116 (1991).
20. Kazor, A. and Boyd, I. W., *Appl.Phys. Lett.* **63**, 2517-2519 (1993).
21. Kazor, A., Gwilliam, R. and Boyd, I. W., *Appl. Phys. Lett.* **65**, 412-414 (1994).
22. Kazor, A. and Boyd, I. W., *Appl.Surf. Sci.* **54**, 460-464 (1992).
23. Panofsky, W. K. H., and Phillips, M., *Classical Electricity and Magnetism,* Reading: Addison-Wesley, 1962, p.257.
24. Heinz, T. F., Loy, M. M., and Thompson, W. A., *Phys. Rev. Lett.* **54**, 63-66 (1985).
25. van Hasselt, C. W., Verheijen, M. A., and Rasing, Th., *Phys. Rev. B* **42**, 9263-9266 (1990).
26. Bjorkman, C. H., Shearon, Jr., C. E., Ma, Y., Yasuda, T., Lucovsky, G., Emmerichs, U., Meyer, C., Leo, K., and Kurz, H., *J.Vac.Sci.Technol.A* **11**, 964-970 (1993).
27. Bjorkman, C. H., Yasuda, T., Shearon, Jr., C. E., Ma, Y., Lucovsky, G., Emmerichs, U., Meyer, C., Leo, K., and Kurz, H., *J. Vac. Sci. Technol. B* **11**, 1521-1527 (1993).
28. Pedersen, K. and Morgen, P. , *Phys. Rev. B* **52**, R2277-2280 (1995).
29. Pedersen, K. and Morgen, P. , *Phys. Rev. B* **53**, 9544-9547 (1996).
30. Daum, W., Krause, H. J., Reichel, U., and Ibach, H., *Phys. Rev. Lett.* **71**, 1234-1237 (1993).
31. Xu, Z., Hu, X. F., Lim, D., Ekerdt, J. G., Downer, M. C., *J. Vac. Sci. Technol.* B15, 1059-1064 (1997).
32. Nakamura, K., Ichimura, S. and Shimizu, H., *Jpn. J. Appl. Phys.* **33**, L1035-1037 (1994).
33. Okabe, H., *Photochemistry of Small Molecules,* New York: John Wiley and Sons, 1978, p.237.
34. Bratu, P., Kompa, K. L., and Höfer, U., *Phys. Rev. B* **49**, 14070-14073 (1994).
35. Shklyaev, A. A. and Suzuki, T., *Surf. Sci.* **351**, 64-74 (1996).
36. Dadap, J. I., Doris, B., Deng, Q., Downer, M. C., Lowell, J. K., and Diebold, A. C., *Appl. Phys. Lett.* **64**, 2139-2141 (1994).
37. Tanuma, S., Powell, C. J., and Penn, D. R., *Surf. Interface Anal.* **11**, 577-589 (1988).
38. Nakamura, K., Ichimura, S. and Shimizu, H., *Appl. Surf. Sci.* **100/101**, 444-448 (1996).
39. Hollering, R.W., Hoeven, A. J., and Lenssinck, J. M., *J. Vac. Sci. Technol. A* 8, 3194-3197 (1990).
40. Höfer, U., Li, L., and Heinz, T. F., *Phys. Rev. B* **45**, 9485-9488, (1992).

Analysis of Reflectometry and Ellipsometry Data from Patterned Structures

M.E. Lee, C. Galarza, W. Kong, W. Sun and F. L. Terry, Jr.
Department of Electrical Engineering and Computer Science, University of Michigan, Ann Arbor, MI. 48109-2122

Specular reflected light techniques, including both single wavelength and spectroscopic versions of ellipsometry and reflectometry, have been used for both etch and growth rate control. However, use of these techniques for process control on products has been limited due to the problems inherent in the analysis of reflected light from patterned structures. In this paper, we examine techniques for the quantitative analysis of data from both highly regular grating structures and from patterns with low local order. We find good quantitative agreement of vector diffraction theory to specular reflection data. We conclude that there is significant promise for the use of specular techniques for in situ monitoring of topography provided that computational speed issues can be improved.

Introduction

High accuracy, high speed, non-invasive wafer state monitors are important for advanced semiconductor process development and control. Reflected light measurements have proven to be very successful for *in situ* monitoring, endpoint detection, and feedback control of vacuum processes on unpatterned substrates. Single wavelength and spectroscopic ellipsometry (1), laser and spectral reflectometry (2,3), and related methods have all been used to demonstrate monitoring accuracies in the ~1nm or better range. However, applications of these methods in actual production have been very limited due to the problems inherent in monitoring patterned structures. Laser reflectometry (4) and laser ellipsometry (5) have been used in some patterned structure monitoring and endpoint detection applications. Recent success in the use of multi-wavelength ellipsometry to monitor and control high density reactive ion etching of digital circuit structures with long pattern repeat distances has been achieved using a scalar theory for modeling the polarization dependent reflection problem (6). However, as expected, these scalar methods are inaccurate on periodic patterns due to strong diffraction effects. Some initial efforts have been made to experimentally and theoretically examine ellipsometry data from gratings (7,8,9). However, quantitative analyses have been limited.

Ex situ applications of diffraction-based measurements (scatterometry) have shown that accurate critical dimensions, thickness, and sidewall shapes can be obtained by analyzing the specular and higher order diffraction intensities vs. angle of incidence (10) or by analyzing the relative intensity of several diffracted orders at a single angle of incidence (11). However, for *in situ* applications, it is often only possible to obtain specular reflection information at a single angle of incidence. In this paper, we will present results on applications of scalar models and vector diffraction theory to the analysis of both spectroscopic ellipsometry (SE), and spectral reflectometry (SR) data from patterned structures. Strong potential of specular measurements for topographic monitoring is indicated.

Experiments

We fabricated two types of experimental samples for tests of reflection models.

The first was a set of relief grating etched in (100) orientation single crystal Si wafers. Gratings with nominally equal lines and spaces were fabricated with periods of 2, 4, and 10 μm. The grating areas were 2.5 x 2.5 mm. They were etched by first patterning approximately 1.3 μm of Shipley 1813 photoresist using contact photolithography. The patterns were then etched using a Lam 9400 SE TCP high-density plasma etch tool. The etch conditions were: 10 mTorr pressure, an etch gas mixture of CL_2 (100 sccm) and HBr (100 sccm), a TCP power of 350 W, and a bias power of 160W. With this recipe, the etch rate is approximately 0.4 μm/min. Six wafers were etched to different depths (approximately100 to 600 nm in 100 nm increments) by changing the etching time. Due to time limitations, we concentrated our analysis efforts on the 500nm deep grating. We used the other wafers to simulate *in situ* data from the etching process. Post-etch cross-section scanning electron microscopy (SEM) was used to evaluate the line structure profile on the 500 nm sample (shown in Figure 1). There is no visible surface roughness and the lineshapes can be modeled as trapezoidal. There appears to be some concave-down rounding of the bottoms of the grating. This may have some influence on the results of our attempts to fit optical data from these structures.

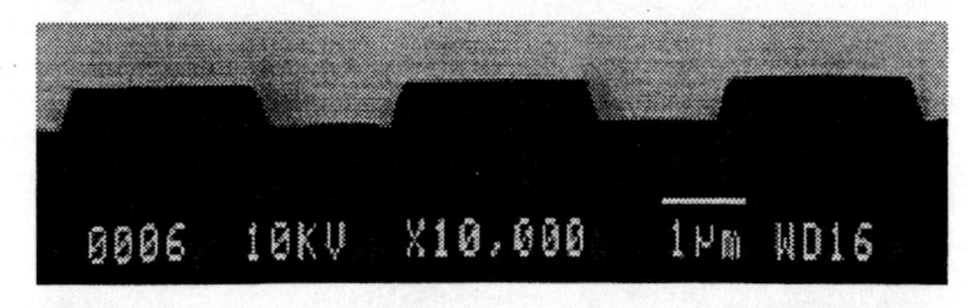

Figure 1. Cross-sectional view of 4 μm period Si relief grating. We estimate the structure as: period=3.96 μm, top linewidth = 2.2 μm, depth = 0.52 μm, and wall angle = 73.9°.

CP449, *Characterization and Metrology for ULSI Technology: 1998 International Conference*
edited by D. G. Seiler, A. C. Diebold, W. M. Bullis, T. J. Shaffner, R. McDonald, and E. J. Walters

The second set of samples was a set of aSi:H on Cr on glass samples from our flat panel display efforts. The samples contain no array structures within each die. The die are square and have a repeat distance of about 360 μm. Approximately 100 nm of Cr was deposited by magnetron sputtering on Corning 1737F glass substrates. The aSi:H was deposited using a PlasmaTherm Clusterlock 7000 PECVD chamber. The aSi:H patterns were etched using conventional photolithography and selective wet chemical etching.

All SE and SR measurements were performed on a Sopra GESP-5 spectroscopic ellipsometer/photometer system. The rotating polarizer ellipsometric measurements were performed in the tracking analyzer mode. For grating measurements at an angle of incidence of 75°, focusing optics were used to reduce the spot size so that only the grating was measured. For near-normal ellipsometry and reflectometry, an aperture was used to reduce the spot size. For measurements of the aSi:H samples, the full spot size of the instrument was used. Polarization-dependent reflectometry measurements were conducted in a double beam mode with a straight-through calibration of the instrument before each measurement.

Modeling and Simulation

For scalar analysis, we are using the modified Heimann approach (4) of the Lucent group (6). For diffraction analysis, we are using both commercial grating simulation software (12) which uses the rigorous coupled wave analysis (RCWA) method (13) and our own software which uses a surface integral equation (SIE) method (14).

The scalar model assumes that the profile of a patterned wafer is divided into separate uniform thin film regions. The total reflection coefficient of the structure is calculated as a complex combination of the individual reflection coefficients from the different regions, i.e.,

$$R_p = \sum_{i=1}^{2} af_i \exp\left(2j\frac{2\pi}{\lambda}\delta_i \cos(\phi_0)\right) R_{p_i} \qquad (1)$$

$$R_s = \sum_{i=1}^{2} af_i \exp\left(2j\frac{2\pi}{\lambda}\delta_i \cos(\phi_0)\right) R_{s_i}$$

where R_{p_i} and R_{s_i} are the reflectances in the p and s polarization corresponding to the i-th region. The nonnegative real number af_i is the area fraction of the i-th region. In Eq. (1), δ_i is the thickness of a layer of vacuum added on top of the i-th region to consider the phase lag due to the different heights of the stacks. Now, as usual, the ratio of R_p and R_s determines $\rho = \tan(\Psi)e^{j\Delta}$. The coefficients R_{p_i} and R_{s_i} are well-known nonlinear functions of the thickness of the layers in each stack (15). The computational load of this model is equivalent to the requirements of four different models for blanket wafers and is negligible compared with the vector diffraction models.

The RCWA algorithm finds a set of inhomogeneous plane waves which approximate the exact solution of the Maxwell's equation boundary value problem defined by the grating. The set of differential equations is solved with standard difference equation - eigenmatrix methods. Arbitrarily complex grating line profiles are approximated with discrete slices. The number of slices, s, and the number of plane waves (the order of algorithm, N) determines the computation time/wavelength point and simulation precision. A 2(N+1)s x 2(N+1)s matrix problem is created, but can be solved relatively efficiently due to the 4s x 2s block diagonal character of the matrix. For ellipsometry simulations, the complex *s*- and *p*-polarization fields were first computed and the $\tan(\psi)$ and $\cos(\Delta)$ quantities were computed from these fields. In all our RCWA simulations, we used 10 slices to approximate the grating profiles. One limitation of the current software is that it will not use pointwise optical refraction indexes. To overcome this difficulty, we fitted silicon's n and k by a 10th order polynomial (the maximum the software allowed). This introduced about 3% error in the worst region. The order, N, for our simulations was determined on the basis of energy conservation, and was typically in the range of 45-65. Our typical run times were on the order of 2.1-5.4 minutes/wavelength on a 300 MHz Pentium II™ system (so that a 100 wavelength SE simulation for a given structure could take up to 9 hours). Parallel computation can offer great time advantages but was not pursued in this initial effort.

The SIE algorithm uses a number of surface current filaments placed at the boundary between two media (air and Si in our case). By utilizing the equivalence theorem (a generalization of Huygen's Principle) and a two-dimensional periodic Green's function as a kernel, the diffracted field above and below the Si surface can be convoluted to form a single integral equation. By matching boundary conditions, the continuity of tangential components of both the E and H fields, the amplitude and phase of each surface current filament can be solved. The field of each diffraction mode can then be evaluated by the interference from these secondary filament sources. The typical execution times for this algorithm were 1.3 minutes/wavelength using 16 filament/(wavelength spacing on the grating) on a Sun Ultra Sparc 1 workstation.

In comparison with RCWA, the SIE approach uses fewer unknowns, and thus reduces the size of the variable matrix. Also, the most time-consuming part in the SIE algorithm, the evaluation of the periodic Green's function, must only be calculate once for both s and p polarization. Therefore, this method is much more computationally efficient than RCWA. To date, we have only been successful in applying the SIE model to the normal incidence case. A program including oblique incidence is under development and will be reported in a future publication.

Experimental Results

For the scalar analysis of the aSi:H/Cr/glass samples, the structure was divided into two thin film regions. We have included a native SiO_2 layer on top of the wafer. In addition, based on SE measurements of unpatterned aSi:H films on these substrates, we have inserted a thin surface roughness layer between the aSi:H and the Cr substrate. The refractive index of this layer was calculated with Bruggeman's effective media approximation using 50% of amorphous silicon and 50% of Cr.

We have measured tan(Ψ) and cos(Δ) at an angle of incidence of 75°. The cost function used to fit the model was as follows:

$$\sum_{\lambda}\left[\begin{array}{l}(\tan(\Psi)-\tan(\hat{\Psi}))^2+((\cos(\Delta)-\cos(\hat{\Delta}))^2+ \\ (\frac{\partial(\tan(\Psi)-\tan(\hat{\Psi}))}{\partial\lambda})^2+(\frac{\partial(\cos(\Delta)-\cos(\hat{\Delta}))}{\partial\lambda})^2\end{array}\right] \quad (2)$$

where $\tan(\hat{\Psi})$ and $\cos(\hat{\Delta})$ are the estimates given by the model. The behavior of the patterned sample in the short wavelength range is very similar to bare Cr. Hence, we needed to include in Eq. (2) the error in the derivatives to force the model to fit the oscillations observed in the long wavelengths. These oscillations are not present in the bare Cr, and they are a distinctive characteristic of the patterned structure. Figure 2 shows the results of this procedure.

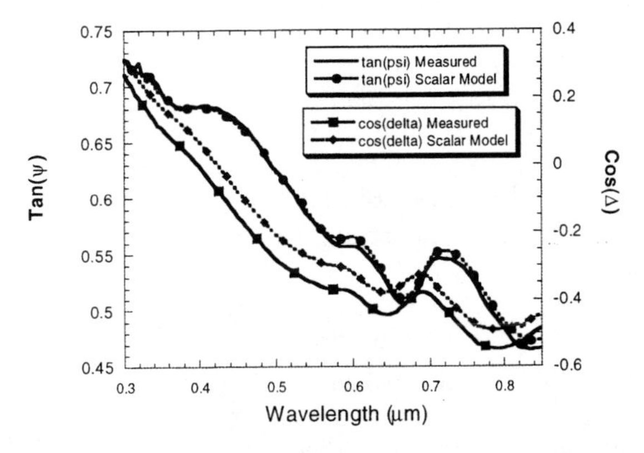

Figure 2. Measured SE data and fit of the scalar model for the aSi:H test structure. Model parameters: region 1- 10.96% (94.52Å SiO2 / 2888.42Å a-Si / 197.09Å roughness / Cr); region2- 89.04% (8.89Å roughness / Cr).

The model is in near-perfect agreement with the experimental data for tan(Ψ). Moreover, the area fraction of region 1 obtained by the model is very close to the nominal area fraction deduced from SEM measurements (about 9.3%). However, the scalar model shows an almost constant offset in cos(Δ). This effect is evident in the data presented by the authors of (6) as well. This difference may be due to diffraction effects even though the pattern repeat distance is very long for these samples.

On the grating structures, SE data were collected from several die on the 500 nm depth sample at 75°. Typical data are illustrated in Figure 3. Data were collected both with the plane of incidence normal to the grating and parallel to the grating. As expected, we observed the strongest grating induced structure in the SE data for the case normal to the grating direction. We observed significant differences in the SE data curves between different die which we believe to due to small differences in the etch depth across the sample. As very long times are required for the RCWA simulation, we concentrated our data analysis efforts on an arbitrarily selected single 4 μm period grating the normal-to-the-grating geometry.

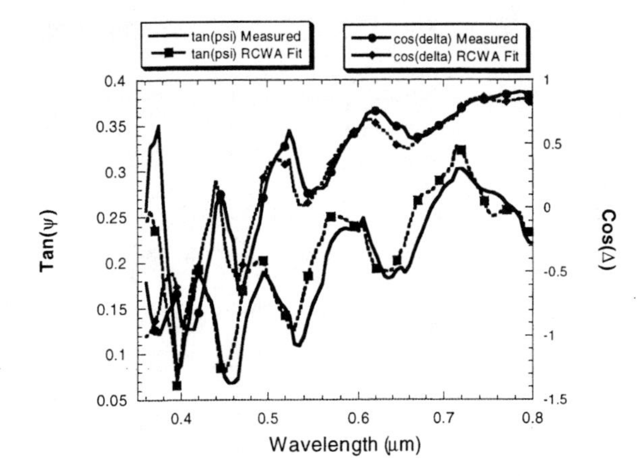

Figure 3. Measured SE data from the 500 nm depth, 4 μm period sample. The plane of incidence was normal to the grating. The RCWA simulation yielded a period of 4.0 μm, a top linewidth of 2.22 μm, a sidewall angle of 72.5°, and a depth of 480 nm.

Our RCWA simulations show several general trends. The positions of peaks and valleys in the oscillations in tan(ψ) and cos(Δ) vs. λ are very sensitive to the period of the grating, while the magnitudes of both ellipsometric parameters were very sensitive to the structure depth. More subtle but still strong structures in the curves are related to the details of the lineshapes. However, these effects are not mutually orthogonal, so it is not a straight-forward exercise to extract the topography information from even these simple test structures. Samples with additional thin film structure would present even stronger challenges.

Therefore, we used a hybrid procedure for finding approximate fits to the grating topography. We measured both near-normal (6°) s- and p-polarized SR and SE data, and 75° SE data. Simple scale theory allow the thickness to be estimated from ¼ wave interference between waves reflected from the top and bottom of the grating:

$$\frac{1}{4d} = \left(\frac{1}{\lambda_{peak}} - \frac{1}{\lambda_{valley}} \right) \quad (3)$$

We used this estimate on the p-polarized data as it is expected that these data will be less strongly influenced by coupled mode effects (16). We observed that the scalar-area fraction model could be applied to the p-polarized reflectance data to achieve very good fits, but that the s-polarized data showed stronger differences vs. the scalar model. Thus, for more complex samples, the scalar approach could be used on p-polarized normal-incidence reflectance data to establish initial guesses. These thickness estimates can then be refined, and linewidth and sidewall slope estimates can be added by iteratively fitting using the SIE approach. Finally, these estimates can be further refined using the RCWA algorithm.

The best fits we have achieved to the near-normal incidence data are shown in Figure 4 and Figure 5. The 75° SE data and RCWA simulation are illustrated in Figure 3. The agreement between the SEM-measured quantities and the ones from fits to optical data is good. The differences may be due to the differences between the two die, incomplete optimization of the RCWA analysis, magnification calibration errors in the SEM, or human error in measuring the SEM photo.

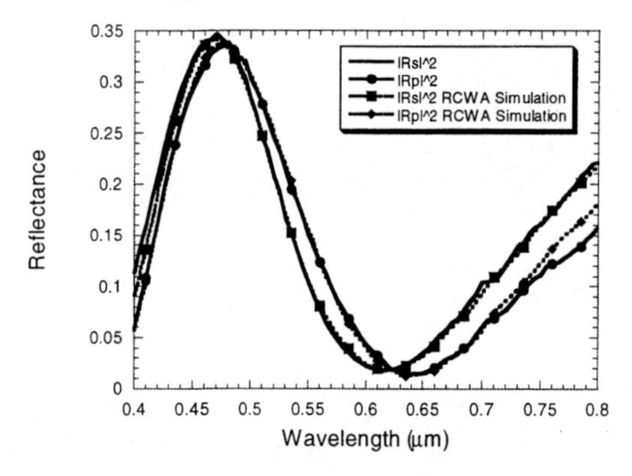

Figure 4. Measured and fitted s- and p-polarized reflectances at 6° for the nominal 500 nm deep, 4 μm period structure. The RCWA simulation yielded a period of 4.0 μm, a top linewidth of 2.22 μm, a sidewall angle of 72°, and a depth of 480 nm.

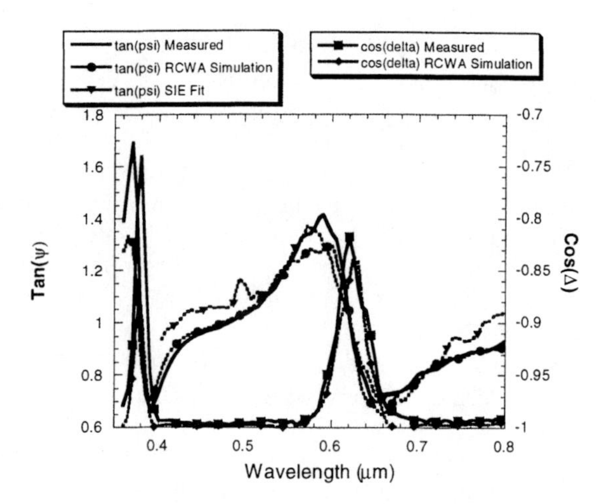

Figure 5. Near-normal (6°) SE data, SIE fit to tan(ψ) data, and RCWA simulation.

The potential of SE to measure evolving grating topographies during an etch was simulated by measuring the samples etched for varying times. We have not yet fully optimized the RCWA simulations for these experiments, but as can be seen in Figures 6 and 7, the basic trends in the experimental data are reflected in the modeled results. Careful examination of the side-by-side simulations and experiments reveals similar non-monotonic behavior in both tan(ψ) and cos(Δ) as the etch depth increases. In this simulation the top and bottom linewidths were held constant at 2.2 and 2.5 μm, respectively, and the period was held constant at 3.96 μm. This leads to a varying wall angle. A constant wall angle simulation may lead to more comparable results to the experimental data. By more accurately fitting theory to experimental data, it should be possible to accurate extract etch depth and wall vs. etch time from in situ data. Other RCWA simulations which we have made for 0.1 μm line/space gratings indicate that this technique will be applicable for monitoring the etching of deep submicron structures.

Conclusions

Use of SE and/or SR yields quantitatively accurate critical dimension and wall angle data on patterned semiconductor structures. These methods can be employed for *in situ* monitoring and process control. However, accurate analysis requires both good prior knowledge of the approximate structure and computationally-intensive vector diffraction calculations. Use of successively more complex but accurate approximation techniques can reduce the need for prior knowledge. Scalar simulation techniques show promise for locally irregular structures, but some method to account for polarization-dependent scattering appears to be necessary even in these cases.

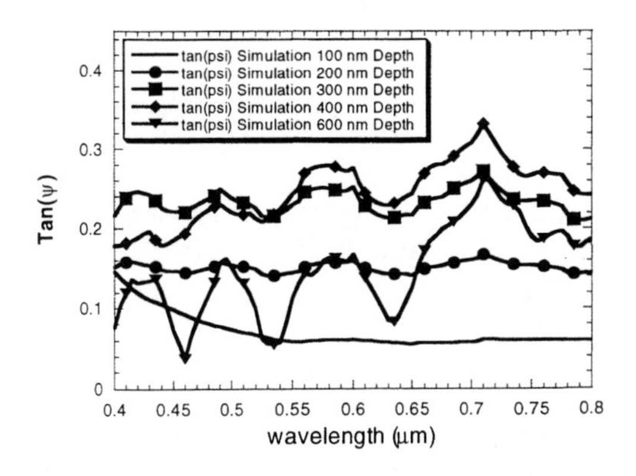

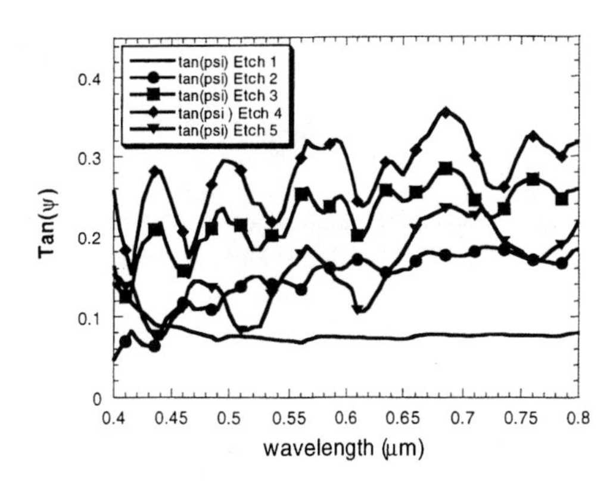

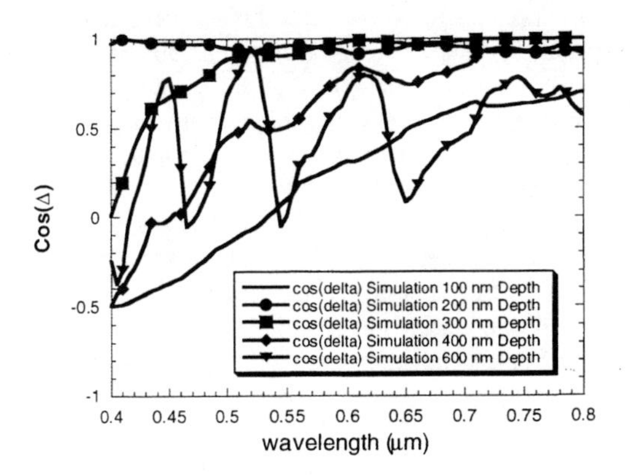

Figure 6. Simulated tan(ψ) (top) and cos(Δ) (bottom) curves at 75° angle of incidence for a series of etched 4 μm period Si gratings.

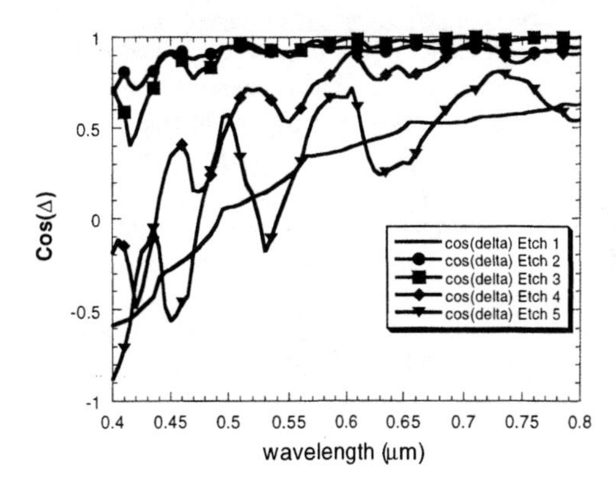

Figure 7. Experimental tan(ψ) (top) and cos(Δ) (bottom) curves at 75° angle of incidence for a series of etched 4 μm period Si gratings.

Acknowledgements

This work was supported in part by the Semiconductor Research Corporation (contract 97-FC085), AFOSR/DARPA MURI Center for Intelligent Electronics Manufacturing (AFOSR F49620-95-1-0524), and the State of Michigan Center for Display Technology and Manufacturing. The authors would also like to thank Dr. D. S. Grimard and Ms. M. Gulari for assistance with sample fabrication.

References

1. D.E. Aspnes, Solid State Communications, 101, pp.85-9 (1997)
2. T. L. Vincent, P.P. Khargonekar, and F.L. Terry, Jr., J. Electrochem. Soc., 144, pp.2467-72 (1997)
3. T. B. Benson, L.I. Kamlet, P. Klimecky, and F.L. Terry, Jr., J. Elec. Mat., 25, pp955-64 (1996)
4. P. A. Heimann and R. J. Schutz, J. Electrochem. Soc., 131, pp. 881-5 1984).
5. M. Haverlag and G. S. Oehrlein, J. Vac. Sci. Techn., B10, pp. 2412-8 (1992).
6. H. L. Maynard, A. N. Laydi, J. T. C. Lee, J. Vac. Sci. Tech., B15, pp109-15 (1997)
7. N. Blayo, R. A. Cirelli, F. P. Klemens, and J. T. C. Lee, J. Opt. Soc. Am., A 12, pp. 591-9 (1995).
8. D. W. Mills, R. L. Allen, and W. M. Duncan, Proc. SPIE, 2637, pp. 194-203 (1995).
9. H. Arimoto, Jpn. J. Appl. Phys., 36,pp. L173-5 (1997).
10. C.J. Raymond, M.R. Murnane, S.L. Prins, S.S.H. Naqvi, J.R. McNeil, J.W. Hosch, Proc. SPIE, 2725, pp.698-709 (1996)
11. T.M. Morris, D.S. Grimard, C.F. Shu, F. L. Terry, M.E. Elta, R.C. Jain, Proc.SPIE, 1926,pp.27-32 (1993)
12. Grating Solver Development Co., Allen, TX..
13. M. G. Moharam and T. K. Gaylord, J. Opt. Soc. Am., pp. 1385-1392 (1982).
14. R. Petit, ed., *Electromagnetic Theory of Gratings*, Springer-Verlag (Berlin, 1980).
15. Azzam, R.M.A. and Bashara , N.M., *Ellipsometry and Polarized Light*, Elsevier (Amsterdam, 1986).
16. D.A.Gremaux and N.C.Gallagher, Appl. Opt., 32, pp.1948-1953 (1993).

In situ Layer Characterization by Spectroscopic Ellipsometry at High Temperatures

W. Lehnert[*], P. Petrik[***], C. Schneider[*], L. Pfitzner[*] and H. Ryssel[*,**]

[*] Fraunhofer Institute of Integrated Circuits IIS-B, Schottkystrasse 10, 91058 Erlangen, Germany
[**] University of Erlangen-Nuremberg, Chair of Electron Devices, Cauerstrasse 6, 91058 Erlangen, Germany
[***] Research Institute for Material Science, P.O.B. 49, H-1525 Budapest, Hungary

Abstract

The demand for increased cost-effectiveness in semiconductor manufacturing is the driving force for the development of *in situ* and in-line measurement tools. Some of the most critical manufacturing steps are high-temperature processes such as thermal oxidation and chemical layer deposition. Solutions for accessing batch furnace processes by high-temperature single wavelength and spectroscopic ellipsometry for layer thickness and composition control have been proposed and studied intensively in the past. These techniques require comprehensive knowledge of the optical parameters at high temperatures. Therefore, a systematical study has been started to determine the optical high-temperature data (refractive index, extinction coefficient) of relevant semiconductor materials. Moreover, optical data of amorphous and polycrystalline silicon at high temperature are under investigation. All measurements were performed with a spectroscopic ellipsometer integrated in a vertical LPCVD-batch furnace. Optical access is provided by a special beam-guiding system. The established data are used to develop models for ellipsometric *in situ* monitoring of layer structure such as thickness, roughness, crystallinity, or density. Accurate and reliable optical models are particularly required for *in situ* monitoring of polycrystalline silicon because the optical properties vary considerably, depending on the deposition conditions. A Bruggeman-Effective Medium Approximation (B-EMA) is used to calculate the dielectric function of the layer. This method allows to characterize multilayer structures and to obtain all layer thicknesses and optical layer characteristics from one SE measurement. These models are implemented into the measurement programs and they are already used in a commercial spectroscopic ellipsometer for use in industrial applications.

INTRODUCTION

Today, *in situ* and in-line metrology is often regarded as the key to achieve increased cost-effectiveness in semiconductor manufacturing. The main advantages of integrated metrology in processing equipment (*in situ* and in-line) are a reduced number of monitor wafers and direct process control instead of off-line monitoring (1). Especially with respect to the processing of 300 mm wafers, the demand for process control without monitor wafers wherever possible is extremely high. Moreover, integrated metrology provides higher tool utilization through optimized maintenance instead of preventive maintenance and shorter ramp-up cycles of new processes.

In semiconductor processing, the formation of thin films in high temperature processes is one of the key techniques. Thermal oxidation and chemical layer deposition comprise some of the most decisive manufacturing steps such as gate oxide formation, deposition of stacked dielectrics, and deposition of conductive layers like polysilicon or amorphous silicon. The process control (layer thickness and quality) is completely carried out off-line. *In situ* monitoring of layer growth and composition as well as end-point detection support new control techniques to provide a better process stability without the use of monitor wafers and without additional handling for process and quality control. Solutions for accessing batch furnace processes by high-temperature single wavelength and spectroscopic ellipsometry have been proposed and studied intensively in the past (1)(2). The major requirement for the use of these techniques is to be in the exact knowledge of the optical material properties at high temperatures up to 1000°C. As these data are in most cases not available, a systematical study has been started to determine the optical high-temperature data (refractive index, extinction coefficient) of relevant semiconductor materials. In the first step, the optical high-temperature data up to 900°C of crystalline silicon, silicon dioxide, and silicon nitride were established (1). Now, these data are used to develop models for ellipsometric *in situ* monitoring of layer structure such as thickness, roughness, crystallinity, or density. An important layer material in semiconductor manufacturing is polycrystalline silicon because of its application as e.g. gate material or interconnects. Especially for this material accurate and reliable optical models are a precondition for *in*

CP449, *Characterization and Metrology for ULSI Technology: 1998 International Conference*
edited by D. G. Seiler, A. C. Diebold, W. M. Bullis, T. J. Shaffner, R. McDonald, and E. J. Walters
 1-56396-753-7/98/$15.00

situ monitoring because the optical properties vary considerably depending on the process conditions. Therefore, the change of the polycrystalline silicon structure depending on the temperature is presently under investigation. In order to calculate the dielectric function of the polycrystalline layer, a Bruggeman-Effective Medium Approximation (B-EMA) is used by employing a mixture of materials with dielectric functions. These functions can be determined independently. As studies showed in the past, this method allows to characterize multilayer structures and to obtain all layer thicknesses and optical layer characteristics from one SE measurement (3)(4).

EXPERIMENTAL

The off-line reference measurements were carried out with a SOPRA ES4G spectroscopic ellipsometer. For the *in situ* measurements a spectroscopic ellipsometer (SOPRA MOSS OMA) integrated into a vertical batch furnace for thermal oxidation and low pressure chemical vapour deposition (LPCVD) was used. Spectroscopic ellipsometry was chosen because of the possibility to measure on absorbing layers like polycrystalline silicon or in multilayer structures (5). As vertical furnaces in semiconductor manufacturing do not provide any optical access for a ready-to-use integration of metrology tools, a special beam guiding system is employed. This system keeps the modifications of the furnace geometry at a minimum. In Fig. 1, the ellipsometer arrangement and the beam path with the beam guiding system are shown. The two ellipsometer heads are placed side by side and are mechanically coupled to the base plate of the furnace. Together with the base plate and the wafer carrier, they form a mechanical unit that moves vertically with the boat loader. The light beam of the ellipsometer is guided through the base plate into the furnace tube and directed onto the wafer by quartz glass prisms operating in total internal reflection (TIR) mode (1).

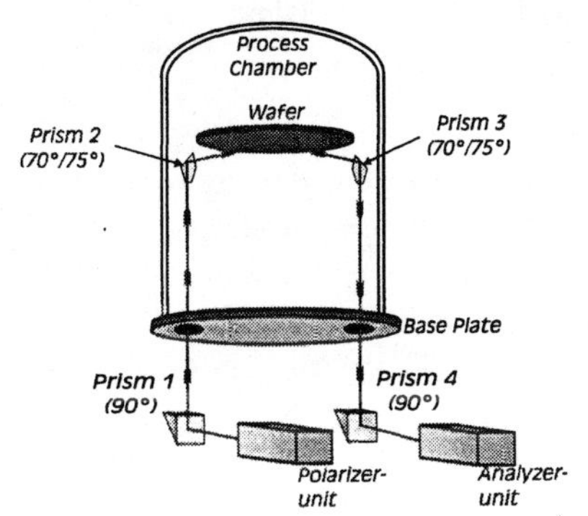

FIGURE 1: The *in situ* ellipsometer arrangement with the beam guiding system (1)

In this setup the calibration of the offsets and angle of incidence or an ellipsometric measurement can be carried out without boat insertion. In addition, measurements of the wafers´ "starting conditions" and *post process* measurements inside as well as outside the tube are possible. No modifications of the process tube and the heating cassette are necessary. In this configuration the wafers are loaded into the boat with the device side down. Conventional loading with the device side up is possible by using two additional beam deflections for the *in situ* metrology. In all measurements, the selected prisms provide an angle-of-incidence of the measuring beam of 75° onto the wafer (1).

As shown in Fig. 1, the ellipsometer beam must be deflected four times. The two reflections of 90° are only necessary if the overall height of the vertical furnace with integrated ellipsometer heads would otherwise exceed the usual cleanroom height. The TIRs lead to a well-defined additional phase shift in the polarization state of the light that can be calculated or measured and subtracted from the measured phase shift according to eq. (1).

$$\Delta = \Delta_m - (2\delta_{p90^\circ} + 2\delta_{p75^\circ}) \quad (1)$$

Δ_m is the phase shift as measured with the *in situ* SE setup, Δ is the phase shift caused by reflection on the silicon wafer, δ_{p90° and δ_{p75° are the phase shifts resulting from the TIRs in the prisms. A measurable change of the amplitude ratio $\tan\Psi$ does not occur. The phase shift at a constant angle of reflection depends only on the index of refraction of the prism material. In practice, the phase shift of the prisms can be detected by measuring a completely transparent sample (quartz glass wafer) in the vertical furnace. Additionally, the angle of incidence can be calculated from the same measurement (sensitive to $\tan\Psi$). The ellipsometer software has been adapted to support measurements with the additional optical components (prisms). With this system the optical material properties of crystalline silicon, silicon dioxide, and silicon nitride as a function of temperature were determined.

In order to enable *in situ* monitoring during the deposition of complex layer structures like polycrystalline silicon, the optical models of this material and the change of the layer structure and composition at deposition condition are presently under investigation. The poly samples were prepared by using LPCVD at a pressure of 250 mTorr and a gas flow of 50 sccm at 560°C. The bulk material was single crystalline (111) oriented, 15-20 Ωcm, p-type CZ-silicon with a thermal oxide of approximately 100 nm thickness.

The spectroscopic ellipsometric monitoring of the recristallization was carried out during the annealing process in the vertical furnace with the beam guiding system as described above. This batch furnace is not designed for rapid heating (max. ramp-rate 20 K/min). Therefore, the insertion of the samples into the process chamber with a pure nitrogen atmosphere was performed at 530°C where no changes of the layer structure occure, as test measurements showed. After the stabilization of the temperature, the monitoring started and the furnace was ramped to 600°C. The temperature of the furnace was measured with thermocouples. A special designed fuzzy controller provides an accuracy of less than 0.5 K. It is inevitable that the temperature of the thermocouples may differ from the temperature of the samples. Parallel measurements with a thermocouple wafer were executed to determine the dependence of the thermocouple temperature and sample temperature. With this known dependence it is possible to calculate the real sample temperature out of the course of the profile temperature with an accuracy of better than 2°C.

RESULTS AND DISCUSSION

The capability of the beam guiding system in the vertical furnace was proved by measuring the optical material properties of crystalline silicon, silicon dioxide, and silicon nitride as a function of temperature. These data are now available in the wavelength range between 250 nm and 900 nm. They are already used for growth monitoring and end-point detection during thermal oxidation. Examples of the measured spectroscopic refractive indices of crystalline silicon and silicon dioxide for different temperatures from 450°C to 900°C are shown in Fig. 2 and Fig. 3 (1).

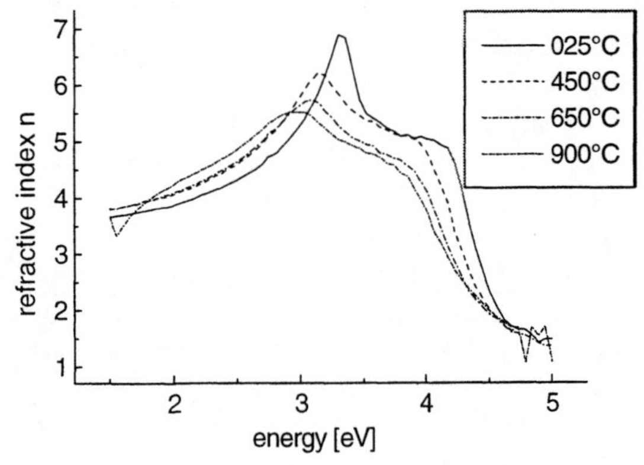

FIGURE 2: Spectra of refractive index of crystalline silicon at different temperatures

These data are used for calculating the dielectric functions of the poly layers during the anneal process. In order to get the initial conditions before the annealing, the layer structure was measured *ex situ* with a SOPRA ES4G spectroscopic ellipsometer (table device). The results are shown in table 1.

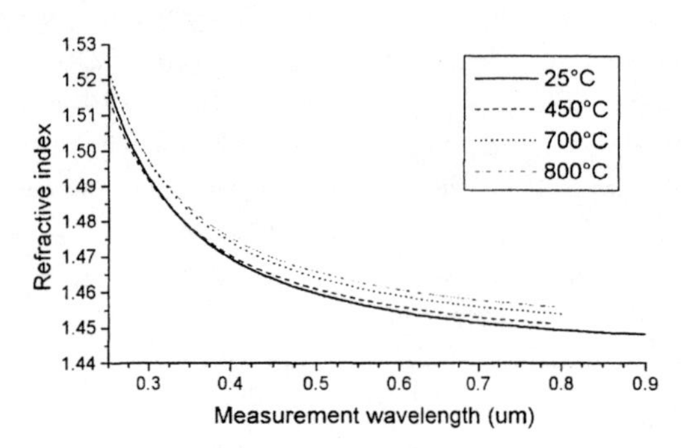

FIGURE 3: Spectra of refractive index of silicon oxide for different temperatures

TABLE 1. Poly-layer structure before annealing (correlation error: σ=0.014)

Layer-number	material 1. comp.	material 2. comp.	conc. of 2. comp. [%]	thickness [nm]
1. Layer	SiO2	--	--	2.0
2. Layer	a-Si	c-Si	5.42	477.9
3. Layer	SiO2	--	--	110.6
Bulk	c-Si	--	--	∞

The evaluation of the measurement was done by fitting theoretical spectra of the ellipsometric angles (tan Ψ and cos Δ) describing the polarization state of the reflected beam to the measured spectra. In this method appropriate models are assumed and the fit is done by varying the wavelength-independent model parameters using Linear Regression Analysis (LRA). The error in the correlation of measured and fitted spectra is given by the unbiased estimator (σ) of the mean square deviation:

$$\sigma=\sqrt{\frac{\sum[(\cos\Delta_{j,m}-\cos\Delta_{j,c})^2+(\tan\Psi_{j,m}-\tan\Psi_{j,c})^2]}{(2n-p-1)}} \quad (2)$$

where n is number of independent measurement values (corresponding to the different wavelengths), p the number of independent model parameters, m the means measured data, and c means the calculated data (6).

The Bruggeman-Effective Medium Approximation (B-EMA) is used to calculate the dielectric functions of the layers. This method allows the characterization of multi-layer structures (thickness and composition) by one measurement (3)(4). The poly layer is described by a mixture of amorphous silicon (a-Si) and single crystalline silicon (c-Si) as shown in Table 1.

Measurements after the annealing process (ES4G, *ex situ*) show that there remains no amorphous part in the sample. The best fit with a correlation error σ better than 0.02 could be obtained by using a mixture of c-Si and a fine-grain polycrystalline silicon (p-Si) without an amorphous component.

An example of a measured spectra (tanΦ and cosΔ) of a polycrystalline sample is given in Fig. 4. The change in the spectra depending on temperature and time is evident. Before a calculation of the dielectric functions is started, the additional phase shift caused by the prisms must be corrected.

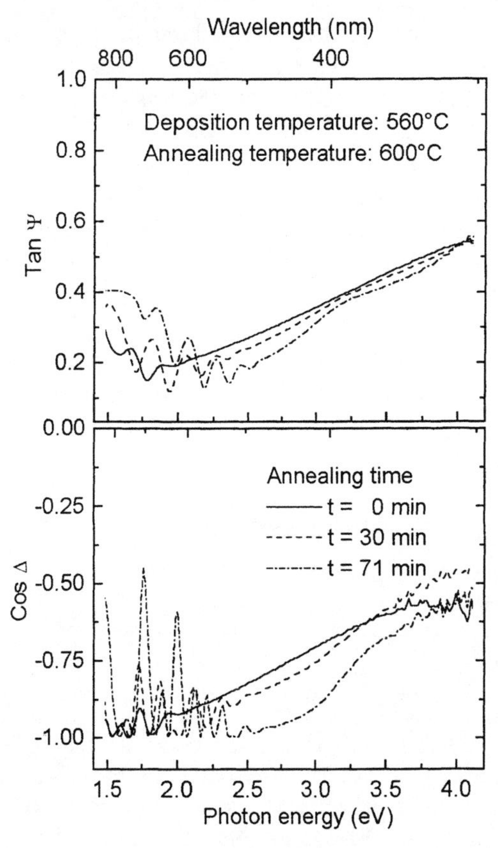

FIGURE 4: Measured spectra at different annealing times

Unfortunately, this correction is only accurate enough in the ultra violet region (2.5 eV to 3.5 eV), because there is no change in the sign of cosΔ for the correction. With thin layers it is possible to determine the points where the sign changes. But in this case there are so many changes in the sign of cosΔ that the error caused by the correction is too high for calculating the dielectric functions. The disadvantage of using only the region from 2.5 eV to 3.5 eV is a much smaller sensitivity to the structural change of the material. To circumvent this problem, in the *in situ* setup an additional compensator must be used to determine the signs of the cosΔ values in the spectrum during the measurement itself. Then, the whole spectral range of the measurement can be used for the calculation of the dielectric functions.

A first calculation of the dielectric functions in the UV-region is shown in Table 2. Here a one-layer model is used, because the polycrystalline layer is not transparent in this region. Therefore, the polycrystalline layer is assumed to be a bulk with a thin oxide layer on its top. The bulk material is assumed to be a mixture of amorphous and fine-grain polycrystalline silicon for the whole annealing time. For the oxide layer, the established data of silicon dioxide at 600°C (SiO2-600) is used. The sample temperature was calculated from the profile temperature and the measured temperature dependence of the profile thermocouple and the thermocouple wafer.

TABLE 2. Recrystallization of the polycrystalline sample

annealing time [min]	sample temp [°C]	a-Si 600°C [%]	p-Si 600°C [%]	SiO2 600°C [nm]	error (σ)
0	525	100	0	1.9	0
11	595	99.9	0.1	2.0	0.003
21	600	93.1	6.9	2.3	0.006
27	600	68	32	3.4	0.007
32	600	51	49	3.4	0.006
34	600	40	60	3.3	0.005
37	600	33	67	3.6	0.005
42	600	16	84	3.3	0.006
47	600	8	92	3.2	0.006
53	600	3	97	3.0	0.005
63	600	-1	101	2.9	0.005
71	600	-4	104	2.6	0.007

The dielectric function of amorphous silicon at 600°C (a-Si-600) is calculated from the first measurement at 530°C assuming that 70°C deviation in the temperature will cause only a small impact on the dielectric function. The data of the polycrystalline silicon is calculated from the last measurement at 600°C after 71 min annealing. To get a information about the reliability of the regression, the thickness of the top oxide was also calculated for each measurement.

The results show that the change in the structure of the polycrystalline silicon is very fast (Fig. 5). After 50 min the recrystallization process is finished. The proportion of polycrystalline silicon in the sample is 0% in the beginning and about 95% after 50 min. After this time the sample was for about 35 min at 600°C because approximately 15 min were needed for heating up to 600°C.

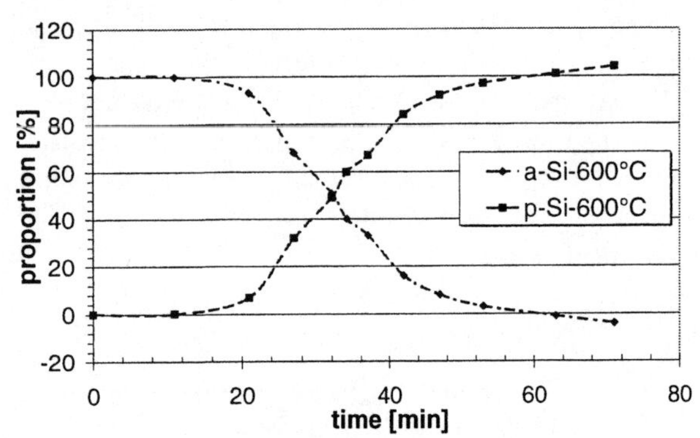

FIGURE 5: Proportion of amorphous and polycrystalline silicon during the annealing process

The quality of the fit is very well during the whole annealing time and the results of the top oxide are reproducable. This means that the model works reliable in this wavelength range (2.5 eV - 3.5 eV). But the error of the proportion values at the end of the annealing process (104% and -4%) and the little variation of the top oxide thickness results indicates that there is a small error in the calculation of the high temperature refractive indices of a-Si-600 and p-SI-600.

CONCLUSION

The capability of the beam guiding system for ellipsometric *in situ* measurement was demonstrated. With the established optical high temperature data of crystalline silicon, silicon dioxide, and silicon nitride the possibility of having *in situ* monitoring of layer growth, layer thickness, and end-point detection in production equipment like vertical batch furnaces was shown. It is very important that these data can be directly measured in the real processing environment, because the optical characteristics of some materials depend on the specific process conditions. The arrangement of the integrated ellipsometer enables *in situ* process control as well as high temperature material research. It could be proved that by the use of the B-EMA to describe the optical material properties the monitoring of layer structure and composition at high temperature is possible. The example of *in situ* recrystallization monitoring of polycrystalline silicon using the beam guiding system reveals a wide variety of applications for which *in situ* spectroscopic ellipsometry can be the key in establishing optical high temperature data. Investigations of the temperature dependence, process time, and changes in material structures can be carried out under real process conditions. Further work has to be done in order to calculate and prove the high temperature dielectric functions of amorphous and polycrystalline silicon in the whole wavelength range from 1.5 eV to 4.5 eV (280 nm to 840 nm). Adaptations of the ellipsometer software to enable measurements with the additional optics (prisms) for *in situ* monitoring of thin oxide and nitride layers are already completed and tested. Automatic measurements of multi-layers or complex layer materials will require the use of a compensator. This will enable automatic correction of the phase shift in the measured signal which is caused by the prisms operating in the total internal reflection mode. Additionally, for layer deposition processes, an especially constructed gas shielding system has been developed to prevent the prisms from coating. Coating of the prisms during a deposition process would cause an continuously increasing error which is difficult to correct. At present the capability of this prism protection system is under investigations. In addition to the established optical high-temperature data this will enable *in situ* layer deposition control during LP-CVD-processing in batch systems.

REFERENCES

1. Berger, R., Schneider, C., Lehnert, W., Pfitzner, L., Ryssel, H., "Advanced process control in vertical furnaces", SPIE Vol. 2876, 1996, pp. 16-26
2. Schneider, C., Berger, R., Pfitzner, L., Ryssel, H., *Applied Surface Science* **63**, 135-142 (1993)
3. Petrik, P., Lohner, T., Fried, M., Berger. R., Biro, L. P., Schneider, C., Gyulai, J., Ryssel, H., "Comparative Study of Polysilicon-On-Oxide Using Spectroscopic Ellipsometry, Atomic Force Microscopy, and Transmition Electron Microscopy", presented at the ICSE II conference in Charleston, South Carolia, USA, 1997, to be published in special volumes of Thin Solid Films
4. Aspnes, D. E., Thin Solid Films, **89**, 249 (1982)
5. Neumann, W., and Gardavsky, J., *Jahrbuch für Optik und Feinmechanik,* Berlin: Fachverlag Schiele&Schön GmbH, 1995, pp.51-86
6. SOPRA manual for ES4G and MOOS-OMA spectroscopic ellipsometer

Instrumental and Computational Advances for Real-time Process Control Using Spectroscopic Ellipsometry

C. Pickering, J. Russell, D.A.O. Hope, R.T. Carline, A.D. Marrs, D.J. Robbins and A. Dann

Defence Evaluation & Research Agency (DERA), St Andrews Road, Malvern, Worcs WR14 3PS, UK

An improved real-time spectroscopic ellipsometer (RTSE) operating over a wide energy range (UV-near-IR) with high resolution and precision, capable of producing 256-wavelength spectra at a rate of 5 s^{-1}, has been interfaced to a $Si_{1-x}Ge_x$ CVD reactor. Reference spectra for strained $Si_{1-x}Ge_x$ with x=0-0.2 have been generated at the growth temperature. Data analysis techniques for real-time determination of composition and growth rate are reviewed. Developments in novel algorithms based on Principal Component Analysis and Artificial Neural Networks to overcome correlation problems in very thin surface layers are discussed.

INTRODUCTION

The trend in the silicon industry towards large wafer sizes and single-wafer processing tools is increasing the need for improved equipment effectiveness and a reduction in the use of expensive monitor wafers. This can be achieved with in-situ process control (1,2). Real-time Spectroscopic Ellipsometry (RTSE) is a preferred technique since it is non-invasive and can provide information on wafer temperature, layer thickness/growth rate, crystallinity, composition and roughness. Before RTSE can be applied routinely as a process control sensor in silicon manufacturing, developments are required in (i) instrumentation and implementation in the growth system, and (ii) real-time data analysis and software. Taking the deposition process as an example, and building on our earlier work on in-situ monitoring in Si/SiGe epitaxy (3,4), this paper reports recent advances in both these areas. The ultimate aim is to integrate the sensor performance characteristics with a process model to provide real-time control of a deposition system based on a commercial cluster tool.

INSTRUMENTATION AND REACTOR INTERFACING

An improved version of the SOPRA RTSE system (5) has been interfaced to a Si/SiGe Chemical Vapour Deposition (CVD) growth reactor. Details of the earlier version and the CVD reactor have been discussed previously (3) and only significant differences are noted here. The more compact polarizer and analyzer arms are attached directly to the UHV flanges and both quartz Rochon prisms are stepper-motor controlled for improved reproducibility. The polarizer rotation speed is 11 Hz and an Intensified Photo-Diode Array detector is used which provides up to 512-pixel tan Ψ, cos Δ spectra over the range 1.5-5 eV at a rate of up to 5 s^{-1}. A Digital Signal Processor (DSP) allows calculation of tan ψ and cos Δ, or the pseudo-optical functions, $\langle\varepsilon_1\rangle$ and $\langle\varepsilon_2\rangle$, before the data are passed to the computer.

Instrumental noise has been evaluated by recording 100 spectra with a 1 s integration time using pixel grouping in pairs, i.e. 256 energies. Over most of the spectrum, up to 4 eV, the standard deviations, σ, depend on the magnitudes of ψ and Δ and are consistent with expected variations, i.e. σ increases as Δ approaches 180°. At higher energies, further increases of σ are observed due to the reduced source intensity. Typical noise results are shown in Fig.1 and a diagram of the equipment is shown in Fig.2.

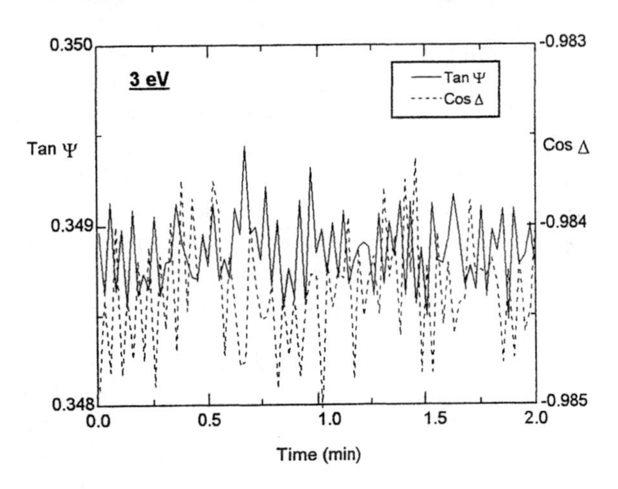

FIGURE 1. RTSE repeatability at 3 eV with 1 s integration time

The initial spectra of Si wafers with oxide overlayers produced by chemical pre-cleaning are reproducible from run to run within the precision reported above. Even after removing the arms, baking the UHV system, re-assembly and re-aligning, maximum observed differences were <0.005 in tan Ψ and cos Δ for energies below 4.5 eV. The calibration parameters, i.e. the polarizer and analyzer azimuths with respect to the plane of incidence, were highly reproducible from run to run, and also as the temperature was increased to ~850°C. Effects of the Bomco 'strain-free' windows, and non-idealities in the

CP449, *Characterization and Metrology for ULSI Technology: 1998 International Conference*
edited by D. G. Seiler, A. C. Diebold, W. M. Bullis, T. J. Shaffner, R. McDonald, and E. J. Walters
1998 The American Institute of Physics 1-56396-753-7/98/$15.00

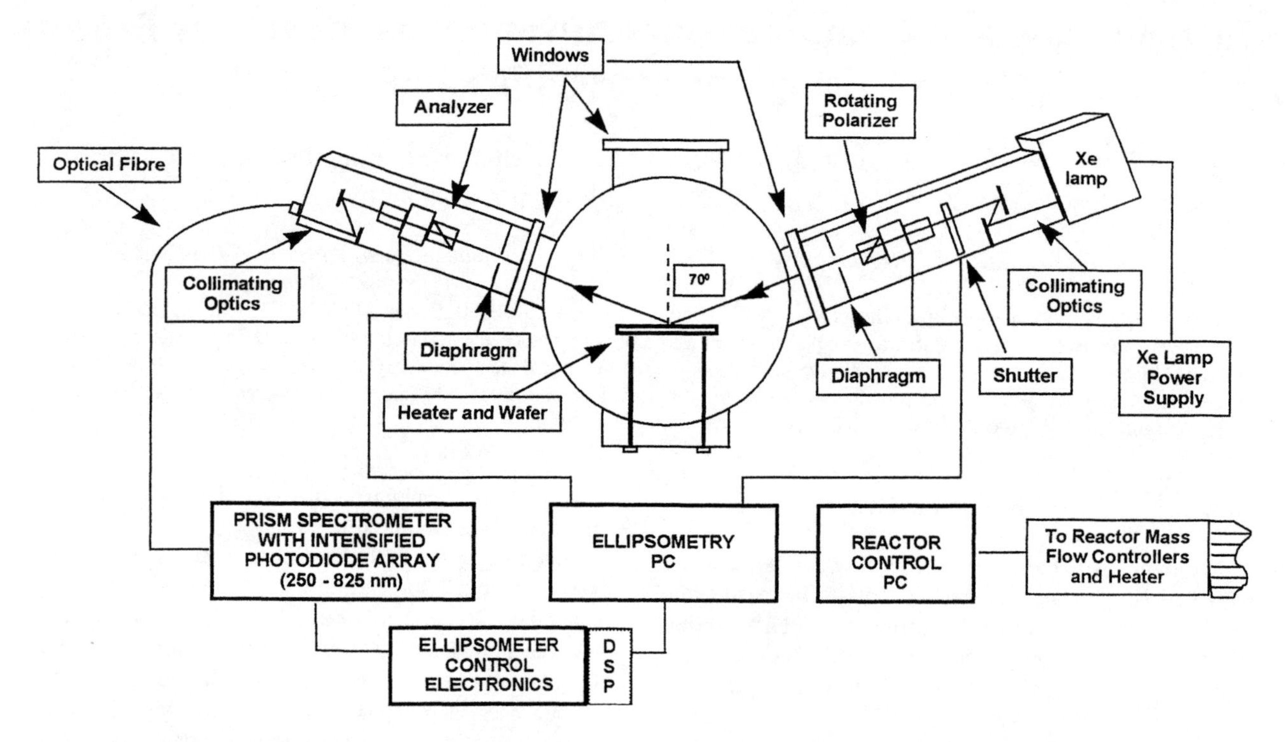

FIGURE 2. Schematic diagram of real-time spectroscopic ellipsometry equipment interfaced to Si/SiGe CVD system.

instrumental function, were assessed by comparison of spectra obtained from a Si wafer with a ~10 Å native oxide with and without windows present, and by comparison with reference data. The windows had no effect on ψ but caused an energy-dependent reduction in Δ. This reduction was ~2° at 4 eV and followed an approximately λ^{-1} dependence, which was used to correct the data. An effective attenuation factor (6) of the measured sine wave, $\eta = 0.994$, was used to correct for cross-talk in the detector array, which mainly affects Δ values near 180°. More details of these data corrections will be given elsewhere. The validity of the corrections was assessed by comparison of the spectra from a bare Si wafer at room temperature with reference spectra. This surface was prepared by desorbing the oxide overlayer from a cleaned Si wafer at ~880°C and leaving the wafer at room temperature inside the CVD system overnight. Figure 3 shows the raw and corrected data compared with calculated spectra for bare Si, the latter being a compilation of data from references (7) (>2.75 eV) and (8) (<2.75 eV).

DATA ANALYSIS AND COMPUTATION

Conventional Analysis

Improved, corrected reference dielectric function spectra of strained $Si_{1-x}Ge_x$ ($0 \le x \le 0.2$), deposited and analyzed as described previously (3), have been obtained at the growth temperature of 610°C. These are shown in Fig. 4 and form the basis of the data analysis procedures for fitting real-time data obtained during subsequent growths.

Closed-loop real-time control requires measurement, analysis and feedback all within the timescale of monolayer growth. Since SiGe growth rates are ~1Å/s, this must be achieved within 1-2 s. Table 1 shows the main approaches to analysis of data obtained in situ during layer deposition and their potential for real-time analysis. Normal spectral regression as used for ex-situ fitting would only provide the composition averaged over the grown layer depth and the total thickness. This is not useful where the composition is changing with time since

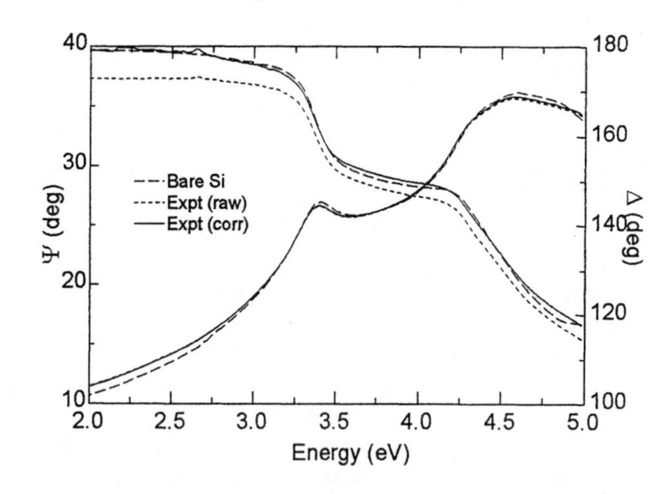

FIGURE 3. In-situ spectra at room temperature of Si wafer after oxide desorption.

TABLE 1. Analysis of real-time multi-wavelength data

Method	Composition (x)	Thickness/Growth Rate (R)	Comments
Normal spectral regression (9)	Average x over total thickness	Total accumulated thickness	Fits may be poor and parameters inaccurate if composition is non-uniform.
Wavelength-by-wavelength analysis (1,3)	Determines dielectric function, not composition	Average growth rate	Needs regions of uniform composition and growth rate. Useful for determining optical constants but not for real-time use.
Exponential spiral approximation (10) (single-wavelength)	Surface x as f(time)	Not possible	Needs known growth rate as a function of x.
Spectral regression with virtual interface (1,3,11)	Surface x as f(time)	Instantaneous growth rate or accumulated thickness	Works best when one parameter is fixed. x may be inaccurate if R is assumed constant. Simultaneous fitting of x and R needs x uniform and/or thick surface layer.
Novel algorithms (1,12)	Surface x as f(time) independent of R	Instantaneous growth rate or accumulated thickness	See text for discussion

the composition of the last-grown part of the layer is required for control. Wavelength-by-wavelength analysis, i.e. treating each wavelength independently, requires blocks of data with time-independent optical constants and is difficult to implement in real time. Methods based on the virtual interface or pseudo-substrate approximation, such as use of exponential spirals or spectral regression have difficulties in obtaining values of both growth rate (R) and composition, particularly when both are changing rapidly, since these parameters are highly correlated when the surface region for analysis is chosen to be very thin, as required for real-time control. This is illustrated in Fig. 5, which shows data from our implementation of virtual-interface spectral regression for real-time analysis. The RTSE data, shown for 3 selected wavelengths in Fig. 6, were obtained during growth of a $Si_{0.81}Ge_{0.19}$ layer. In Fig. 5 data have been determined for ~7.5 Å surface regions.

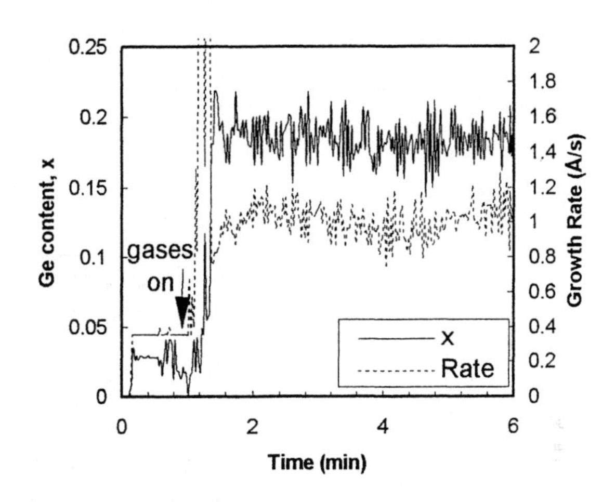

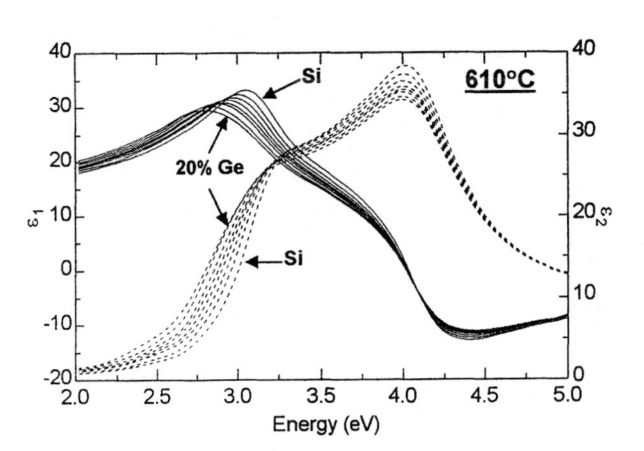

FIGURE 4. Real (ε_1) (——) and imaginary (ε_2) (·····) parts of the dielectric function of strained $Si_{1-x}Ge_x$ at 610°C for x = 0 (Si), 0.04, 0.06, 0.09, 0.11, 0.14, 0.17, 0.20.

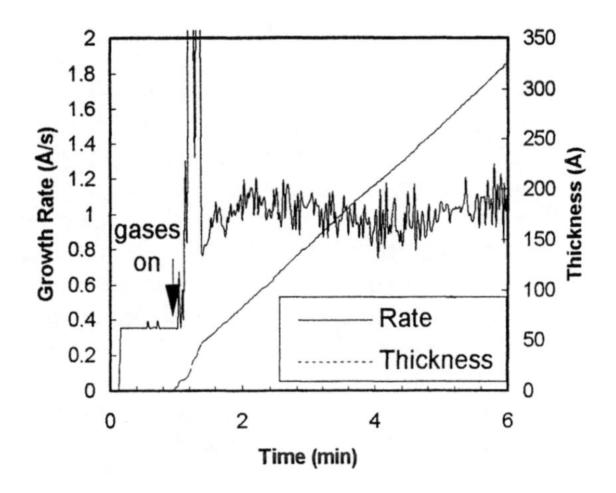

FIGURE 5. Ge fraction, growth rate and accumulated thickness obtained during growth of $Si_{0.81}Ge_{0.19}$ layer on Si at 610°C.

As discussed previously (3), there is a ~30 Å interfacial graded region and meaningful fitted results are only obtained when the composition approaches its target value after about 30 s. The growth rate of $Si_{1-x}Ge_x$ is a non-linear function of x (1), initially increasing with x up to x=0.07 followed by a decrease. The data of Fig. 5 show that beyond the initial region, x and R are both approximately uniform. The results for thickness were obtained by accumulating the individual computed depth regions, and again after ~30 s a straight line is obtained indicating a constant growth rate of 1.05 Å/s.

Principal Component Analysis

In addition to the correlation problems, significant computation time is required for the large data sets. We have applied Principal Component (P/C) Analysis (PCA) (12,13) for the first time to RTSE data to reduce the number of data input channels. Table 2 shows that >95% of the variance of Ψ and Δ can be accounted for by the first two P/Cs, reducing dimensionality of the data to 4.

TABLE 2. Percentage variance in principal components

Principal Component	% Total Ψ Variance	% Total Δ Variance
1	63.9	87.9
2	32.4	10.3
3	2.7	1.5

The PCA has the effect of removing regions of redundant data and efficiently encodes the significant information. Although it is difficult to relate P/Cs directly to regions of the RTSE data, some indication can be obtained from a comparison with the raw data. Figure 6(b) shows the time dependence of the first four P/Cs obtained from RTSE data determined during growth of a layer with a stepped Ge profile The raw data shown in Fig. 6(a) should be compared to the first region of the P/C data (12% GeH_4 flow). It would appear that the high energy data, where there is strong absorption and the layer soon becomes optically thick, are encoded in the first two P/Cs, while the other P/Cs contain information on the lower energy data which exhibit more oscillations.

The P/C data of Fig. 6(b) may also be plotted in the form of trajectories, as shown in Fig. 7 for the first two P/Cs. These are remarkably similar to single wavelength Δ-Ψ or ε_1-ε_2 trajectories (10). Further work is in progress to investigate this connection. These spirals are approximately exponential for constant growth rate and composition. This may provide a basis for process control by tracking a pre-determined trajectory from a specified growth recipe. Preliminary work has also indicated that the spirals may be used to obtain an estimate of growth rate which is independent of composition.

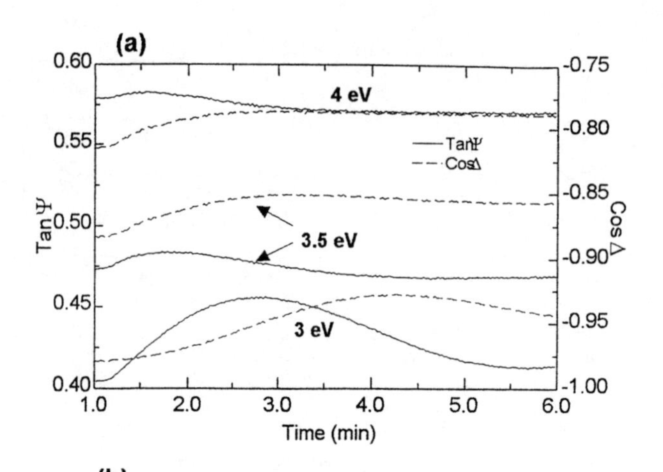

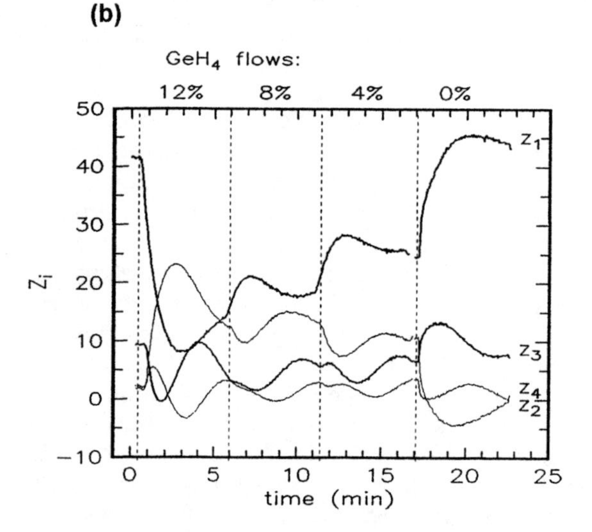

FIGURE 6. (a) RTSE data from $Si_{0.81}Ge_{0.19}$ layer growth; (b) First 4 principal components from stepped Ge layer.

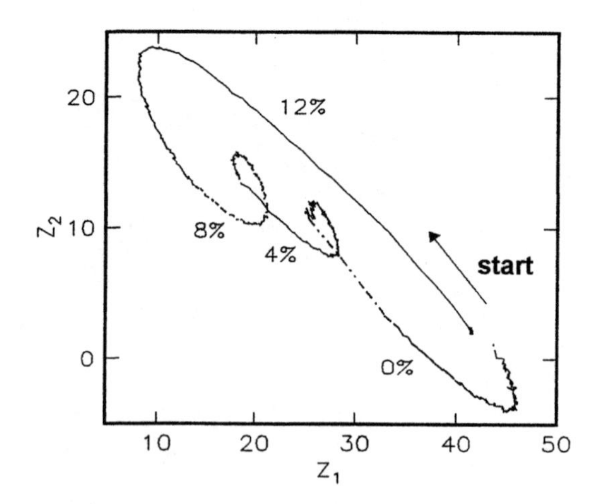

FIGURE 7. Trajectory in Z_1-Z_2 phase space from stepped Ge layer growth as in Fig. 6(b).

Neural Network Analysis

A large set of Ψ-Δ simulations was performed to generate training data for an artificial neural network (ANN). PCA was subsequently performed on the training data to reduce dimensionality. A 3d plot from this data set of the dependence of x on the two largest P/Cs is shown in Fig. 8(a).

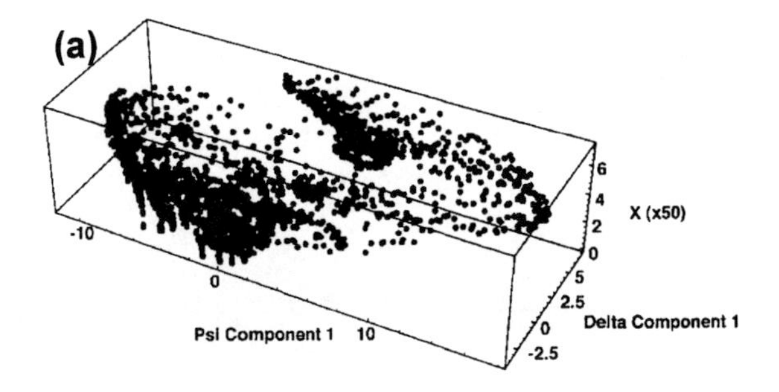

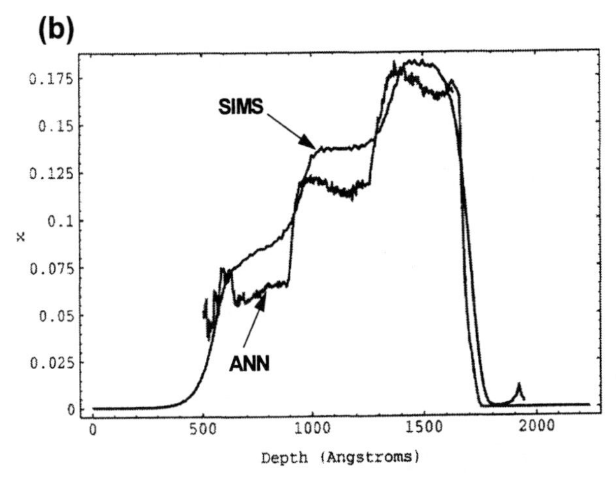

FIGURE 8. (a) Main principal component for Ψ and Δ plotted against Ge fraction, x; (b) Artificial neural network (ANN) estimate of x and SIMS profile of stepped Ge layer.

The spread of the data in Fig. 8(a) indicates a one-to-one mapping between a given measurement and its associated surface layer composition. A similar mapping for associated depth data showed that no direct relationship exists. Thus, the development of an ANN to determine x independently of depth, and hence growth rate, should be possible. A radial basis function network was developed and tested against simulated profiles. The composition returned by the ANN is related directly to the surface layer thickness sampled in the simulated training data. In this first representation 10Å was chosen as the sampling depth.

The ANN analysis has been validated by analysis of real-time data obtained during growth of the stepped composition layer discussed earlier, which had three composition regions (x~0.08, 0.13, 0.18). The results are compared with a post-growth SIMS profile in Fig. 8(b). Since no growth rate information is available from the ANN analysis, the result has been scaled to fit the SIMS result. A layer depth of ~3 Å between data points was required for scaling. Although, as noted earlier, the growth rate is composition-dependent, this corresponds well to an average rate of ~1.15 Å/s with the time interval between measurements of 2.6 s. The values of x obtained are within 0.02 of the SIMS data, which is within experimental error. Further improvements are expected in the near future by using more accurate reference data generated with the new RTSE instrument. A combination of algorithms is envisaged to provide simultaneous real-time values of composition and growth rate. Work is now in progress to incorporate these algorithms in optimised software code and implement them in a feedback loop for real-time process control.

CONCLUSIONS

Instrumentation capable of producing real-time SE spectra over a wide energy range with sufficient resolution and precision to monitor semiconductor alloy growth on the monolayer scale is now available. Progress has been made in the development of novel rapid analysis algorithms which can overcome the composition/growth rate correlation problem required for control of very thin near-surface layers.

REFERENCES

1. Pickering, C., "Complementary in-situ and post-deposition diagnostics of thin film semiconductor structures," presented at the 2nd Int Conf on Spectroscopic Ellipsometry, *Thin Solid Films* (in press) (1998).
2. Pickering, C., "In-situ optical studies of epitaxial growth," in *Handbook of Crystal Growth* (ed. D. T. J. Hurle), Amsterdam: Elsevier, 1994, vol **3**, ch. 19, p. 817.
3. Pickering, C., Hope, D.A.O., Carline, R.T. and Robbins, D.J., *J. Vac. Sci. Technol. A,* **13** 740 (1995).
4. Pickering, C., Carline, R.T., Hope, D.A.O. and Robbins, D.J., *Semiconductor Characterization: Present Status and Future Needs* (eds Bullis, W.M., Seiler, D.G. and Diebold, A.C.), Woodbury, NY: AIP Press, 1996, pp. 532-536.
5. Boher, P. and Stehlé, J.-L., *Mat. Res. Soc. Symp. Proc.,* **452** (1997).
6. Collins, R.W., *Rev. Sci. Instrum.,* **61** 2029 (1990).
7. Nayar, V., Leong, W.Y., Pickering, C., Pidduck, A.J., Carline, R.T. and Robbins, D.J., *Appl. Phys. Lett.,* **61** 1304 (1992).
8. Jellison, G.E., *Opt. Mater.,* **1** 41 (1992).
9. Pickering, C., *J. Min. Met. Mater. Soc.,* **46** 60 (1994).
10. Aspnes, D.E., *J. Opt. Soc. Am. A,* **10** 974 (1993).
11. Herzinger, C., Johs, B., Chow, P., Reich, D., Carpenter, G., Croswell, D. and Van Hove, J., *Mat. Res. Soc. Symp. Proc.,* **406** 347 (1996).
12. Marrs, A.D., "In-situ ellipsometry solutions using a radial basis function network," presented at the 5th Int. Conf. on Artificial Neural Networks, Cambridge, UK (1997).
13. Borggaard, C., *Spectroscopy Europe,* **6** 21 (1994).

Evaluation of an Automated Spectroscopic Ellipsometer for In-line Process Control

C. Pickering, J. Russell, V. Nayar

Defence Evaluation & Research Agency (DERA), St Andrews Road, Malvern, Worcs, WR14 3PS, UK

J. Imschweiler

TEMIC Semiconductor GmbH, Theresienstrasse 2, D-74025 Heilbronn, Germany

H. Wille

Austria Mikro Systeme International A.G., Tobelbader Strasse 30, A-8141 Unterpremstatten, Austria

S. Harrington, C. Wiggins

GEC-Plessey Semiconductors, Cheney Manor, Swindon, Wilts, SN2 2QW, UK

J.-L. Stehlé, J.-P. Piel

SOPRA S.A., 26 rue Pierre Joigneaux, F-92270 Bois-Colombes, France

J Bruchez

SITEC, Hazelwood, Damery Lane, Woodford, Glos, GL13 9JU, UK

The performance of an automatic spectroscopic ellipsometer, the SOPRA Multi-Layer Monitor (MLM) has been evaluated in the Defence Evaluation and Reasearch Agency's Cleanroom over a one-year period on representative CMOS/Bipolar wafer structures supplied by IC manufacturers. Precision, stability, throughput and accuracy have been assessed for ONO, OPO, oxide, Si/SiGe and PECVD nitride/metal wafers. Examples are given to illustrate the ability of the MLM to discriminate variations due to intrinsic and deliberately-induced process changes, as well as cross-wafer uniformity. Both monitor and patterned product wafers have been measured and the inadequacy of relying on monitor wafers demonstrated

INTRODUCTION

Metrology in integrated circuit production is currently perceived as a non-value added process since monitor wafers are expensive and reduce the effective yield and throughput. With the trend to larger wafer sizes and single-wafer processing, it is becoming more important to reduce the reliance on monitor wafers, which are not always representative of the product wafers. In-line monitoring, which can be applied at various process stages on actual product wafers, must be automatic, reliable, reproducible, non-contaminating, and have high throughput. If this can be achieved, metrology will be perceived to add value and provide a return on investment, since expensive test wafers and wasted product and processing time can be eliminated. This paper reports the overall results and conclusions from the European Union Semiconductor Equipment Assessment Initiative Project 'IMPROVE' (In-Line Monitor for Process Optimization and Verification). The project's aim was to evaluate the capabilities of automatic spectroscopic ellipsometry (SE) and, in particular, to assess the performance of the SOPRA Multi-Layer Monitor (MLM) over a period of one year (August 1996-July 1997) in cleanroom processing conditions. The MLM (1) has robot wafer handling and pattern recognition software with a small spot (~100 μm) to allow SE wafer-mapping measurements to be made in scribe lines or test pads on patterned product wafers. Representative key CMOS/Bipolar composite layers, supplied by the IC manufacturer partners, were assessed in both unpatterned and product wafers forms. Machine performance was evaluated in terms of precision, stability, accuracy, throughput, mean time between failures (MTBF) and mean time to repair (MTTR).

MEASUREMENT TRIALS

A representative set of wafer structures was selected to reflect mainstream CMOS/Bipolar key processes and the interests of the partners. These are summarised in the first column of Table 1. For one wafer from each type, detailed trials were made to assess performance in terms of the following parameters:

CP449, *Characterization and Metrology for ULSI Technology: 1998 International Conference*
edited by D. G. Seiler, A. C. Diebold, W. M. Bullis, T. J. Shaffner, R. McDonald, and E. J. Walters
1998 The American Institute of Physics 1-56396-753-7/98/$15.00

TABLE 1. MLM Performance on Representative Wafer Structures

Wafer Structure		Precision (3σ)	Stability (3σ)	Confidence (95%)	Throughput (mode)
Oxide	45Å	3.1Å	2.8Å	2.5Å	
Nitride	55Å	1.7Å	1.3Å	1.5Å	2.1w/hr
Oxide	50Å	2.3Å	2.5Å	2.5Å	(scan)
Oxide	20Å	0.25Å	1Å	2Å	
Poly-Si	1700-2400Å	0.2Å	1.7Å	5Å	37w/hr
Oxide	1100-1500Å	1.7Å	4Å	30Å	(OMA)
Oxide	1000Å	0.35Å	1.5Å	2Å	37w/hr
	n=1.457	0.001(n)	0.001	0.003	(OMA)
Si	200Å	1.2Å	1.1Å	3Å	
SiGe	480Å	3Å	2.4Å	5Å	2.3w/hr
	x=0.26	0.007(x)	0.006	0.01	(scan)
Si	1250Å	5.6Å	4.3Å	10Å	
SiGe	770Å	9Å	13.5Å	10Å	2.3w/hr
	x=0.12	0.01(x)	0.01	0.01	(scan)
PECVD	1000Å	3.7Å	2.7Å	8Å	3w/hr
Oxide	n=1.55	0.2%(n)	0.1%	0.5%	(scan)
(metal)	<n>=2.6	(0.5%<n>)	(1%)	(1%)	

Precision -- 3σ from 30 measurements made at the same site without unloading
Stability -- 3σ from 30 measurements made twice a day for 5 days, unloading and reloading between each measurement
Confidence -- 95% confidence limits from the SE model fit
Accuracy -- Difference between the mean value and a corroborative characterization result
Throughput -- Rate in wafers/hr for a 25-wafer batch, with 5 sites per wafer.

Before the trials were made, a comparison was made between results obtained using the two measurement modes of the MLM to establish the most appropriate measurement conditions for each structure (2). SE measurements may be made in two modes, either using the optical multi-channel analyzer (OMA) for high throughput measurements (full 512-wavelength spectra measured in 1 sec) or the slower, more accurate, Scan mode for accurate determination of reference spectra and for structures requiring deep-UV measurements (2,3). In our system the OMA measurement range is 1.1-4 eV while that of the Scan mode is 1.5-5.6 eV.

The overall results of the trials are summarised in Table 1 and indicate the power and flexibility of SE. Such results for the complex multilayer structures would not be achievable with single/dual wavelength ellipsometers. The wafer throughput was evaluated in both the OMA and Scan modes. More complex structures such as ONO require the more accurate Scan mode with consequently reduced throughput, although there is scope for reducing the number of wavelengths used. Optimization of data/throughput trade-offs could increase the throughput in this mode to 7 wafers/hr. During the 12-month evaluation period an uptime of >97% with MTBF >3000 hours was observed, with negligible maintenance and MTTR.

The 95% confidence data in Table 1, which indicates the sensitivity of SE to the parameters, together with the precision and stability data, establish the minimum meaningful variations that can be measured across wafers and from wafer to wafer. The accuracy has been established by comparison of selected structures with direct corroborative techniques. These results are shown in Table 2 and the agreement can be seen to be excellent.

PROCESS MONITORING

The ability of the MLM to discriminate small changes across wafers, within batches and from batch to batch was assessed on several wafer batches to quantify its usefulness for in-line control. The results are summarised in Table 3.

Oxide-Nitride-Oxide (ONO)

Figure 1(a) shows the variability of the individual layer thicknesses in very thin ONO stacks through a 25-wafer standard process batch (Batch 1). These results and those of Table 3 show that this process is very reproducible and uniform. Figure 1(b) shows the nitride layer thicknesses from 5 batches, two with small changes from the standard process (Batches 2 and 4) and two processed with non-standard recipes (Batches 3 and 5). The nitride layer thicknesses are increased in the latter and also vary systematically through these batches. These changes were subsequently traced to errors in the furnace control recipe which allowed deposition to take place during temperature ramping. Similar results have also been obtained on patterned wafers with the ONO stack produced on poly-Si.

TABLE 2. Comparison of MLM with Corroborative Techniques

Layer	MLM	SIMS	XTEM	XRD*/elec†
ONO				
Top Oxide	55(±2.5)Å	-	-	-
Nitride	61.5(±1.4)Å	65(±10)Å	-	-
Bottom Oxide	43.7(±2.4)Å	-	-	-
Total thickness	160(±7)Å	145(±20)Å	-	-
OPO				
Top Oxide	2054(±6)Å	2017(±25)Å	2071(±30)Å	-
Poly-Si	3106(±3)Å	3097(±25)Å	3370(±180)Å	-
Bottom Oxide	1030(±16)Å	-	1037(±15)Å	-
Si/SiGe				
Si Cap	840(±4)Å	850Å	840(±20)Å	850(±5)Å*
SiGe	508(±5)Å	500Å	490(±10)Å	505(±5)Å*
(Ge content, x)	0.25(±0.015)	0.25	-	0.244(±0.001)*
PECVD Nitride				
Nitride (metal)	1300(±10)Å	-	1320(±20)Å	1290Å†

TABLE 3. MLM Measurement of Standard Process Variations

Wafer Structure	Wafers in Batch	Stability (3σ)	Process Variation (3σ)	Cross-wafer Variation (3σ)
Oxide 45Å		2.8Å	2.7Å	3.5Å
Nitride 55Å	3 lots of 25	1.3Å	2.1Å	2.9Å
Oxide 50Å		2.5Å	2.8Å	2.9Å
Oxide 20Å		1Å	7.5Å	8.1Å
Poly-Si 2350Å	2 lots of 5	1.7Å	3.9Å	10.8Å
Oxide 1500Å		4Å	3.9Å	4.5Å
Si 200Å [thin cap		1.1Å	1.8Å	11.1Å
SiGe 480Å variant]	1 lot of 25	2.4Å	7.8Å	39.3Å
(x=0.26)		0.006(x)	<0.01	<0.01
Si 800Å [standard		3Å	15.3Å	24.6Å
SiGe 480Å process	1 lot of 25	8Å	6.3Å	25.2Å
(x=0.24) monitor]		0.01(x)	<0.01	<0.01
Si 800Å [standard		3Å	61.8Å	64.2Å
SiGe 480Å process	8 wafers	8Å	9.6Å	17.4Å
(x=0.24) product]				
Nitride 1300Å (metal)	2 lots of 11	2Å	25Å	45Å

Oxide-Poly-Oxide (OPO)

Figure 2(a) shows thicknesses for five 5-wafer OPO batches, produced with changes to the process conditions. These results are compared with Nanospec measurements made on monitor wafers processed in the same batches. The Nanospec can be seen to overestimate the poly-Si thickness significantly (4). This is because the poly-Si is amorphous in these samples and the Nanospec assumes a refractive index more appropriate to crystallised poly-Si. Since α-Si has a higher refractive index than poly-Si in the visible region of the spectrum this will cause the observed error. Nanospec measurements were also made on the oxide on Si wafers used for the poly-Si deposition and these can be seen to be in good agreement with the MLM measurements of the buried oxide thickness. Shown in Fig. 2(b) is a 25-site wafer map of the poly-Si thickness made with the MLM, showing a non-radial variation, as expected from the gas-flow configuration.

SiGe/Si

Two thickness configurations have been studied in the Si/SiGe system, one with a 200 Å Si cap and the other with a 800 Å cap (see Table 3), the latter being more representative of device product structures. The results for the thin cap wafers were validated by comparing with X-ray diffraction (XRD) data taken at 7 sites on the wafer. The MLM was programmed to measure the wafer at the same x-y co-ordinates as used for the XRD measurements. Figure 3 shows that the same variation in the SiGe layer, decreasing towards the wafer edges, is determined by both techniques. There is a small difference (~2%) in the SiGe thickness values, which is within the combined experimental error. At the sites compared, there is no significant variation of the Si cap as indicated by both techniques. However, higher resolution MLM maps (2), with points nearer the wafer edge, indicated a small variation in cap thickness, as shown in Table 3.

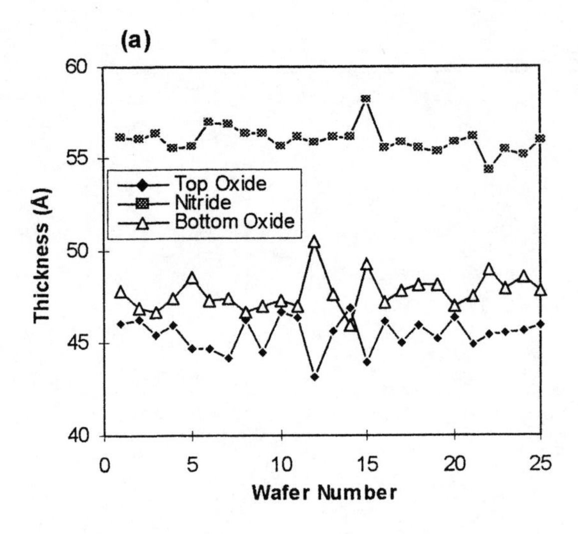

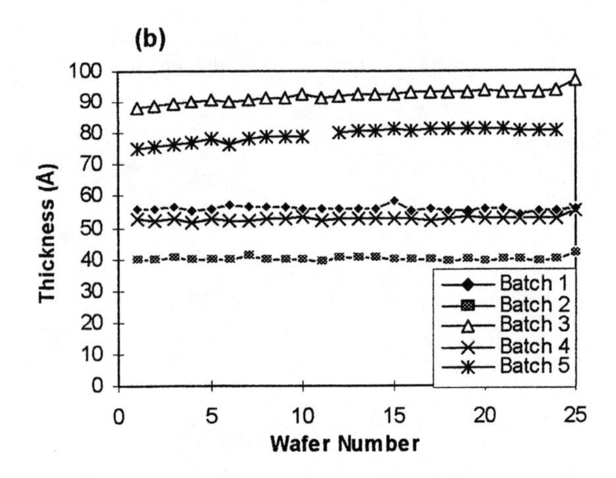

FIGURE 1. (a) Layer thicknesses in standard ONO batch (b) Nitride layer thicknesses in ONO batches.

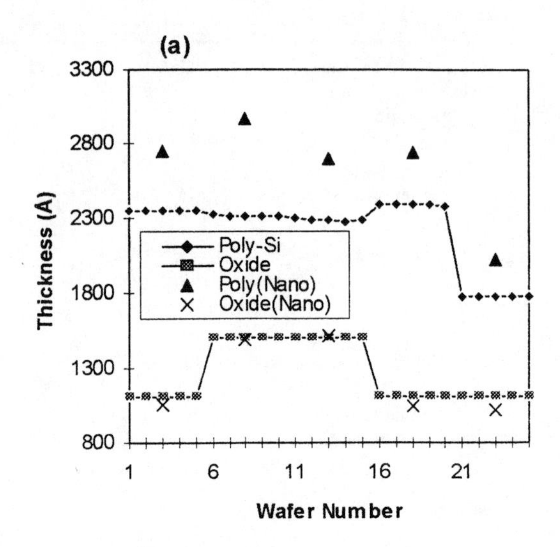

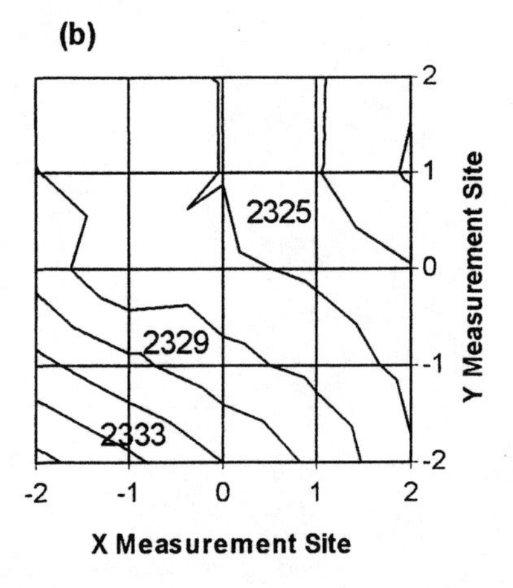

FIGURE 2. (a) Poly-Si and buried oxide layer thickness for five 5-wafer batches, compared to Nanospec measurements on monitor wafers; (b) Wafer map of poly-Si layer thickness (thicknesses are shown for selected contour regions in Å).

The Ge content, x, of the layers may also be measured by SE, using reference data for strained SiGe (5,6). Best results are obtained for uncapped layers, or layers with thin (~200 Å) caps. Layers with thicker caps give reduced confidence (e.g. $\pm$0.015-0.02) in x.

Wafer maps were also produced of Si/SiGe product wafers. Measurements were made at 13 sites in a cross configuration, pattern recognition being used to locate the positions of the 1x2 mm sites in the dies. These results showed unexpected thickness differences compared to unpatterned wafers, with large increases ~80 Å in the Si cap layer towards the edges of the wafer (see Fig. 4 and Table 3), which were subsequently confirmed by XRD. The differences are possibly due to dopant segregation effects caused by the presence of an additional high-doped buried Si epitaxial layer in the product wafers and highlight the inadequacy of the use of monitor wafers.

PECVD Nitride/metal

Measurements were made of 115 x 115 μm PECVD nitride on TiW/AlCu capacitor structures arranged in 5 x 10 matrices in each die. The MLM measurements were in good agreement with XTEM data and variations across the wafer correlated well with measurements on unpatterned monitor wafers. After processing, thicknesses were also calculated from the electrical data and compared with the MLM data as shown in Fig. 5 (see also Table 2). Also included are data measured with a Tencor UV-1250SE instrument using reference data generated with the MLM. Although the absolute values are slightly different, possibly due to the assumed value of dielectric constant, the data sets show the same distribution. Thus, the MLM can be used as a non-destructive measurement of device capacitance.

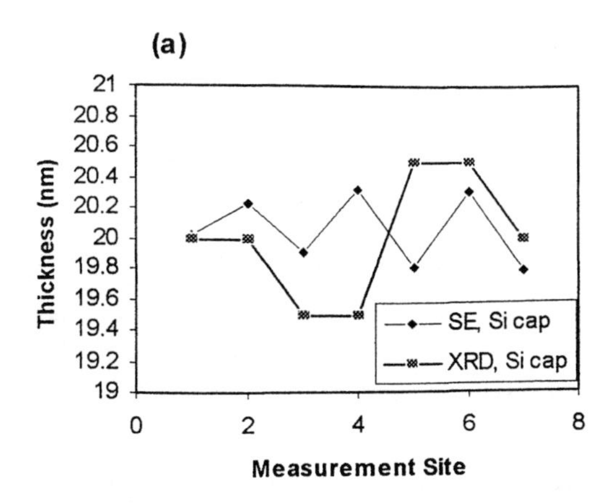

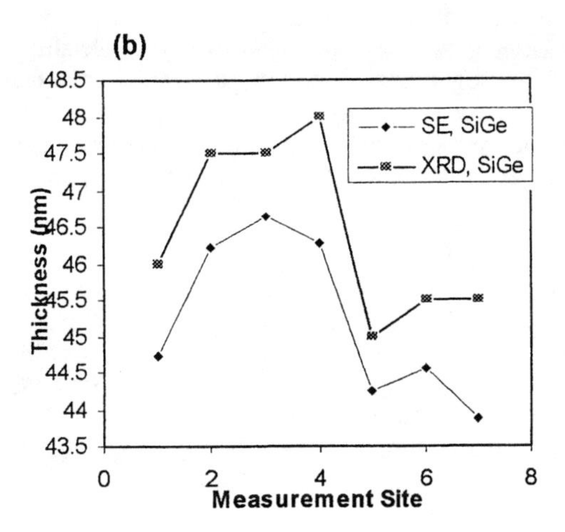

FIGURE 3. (a) Si cap and (b) SiGe layer thickness measured at same site co-ordinates by MLM and XRD.

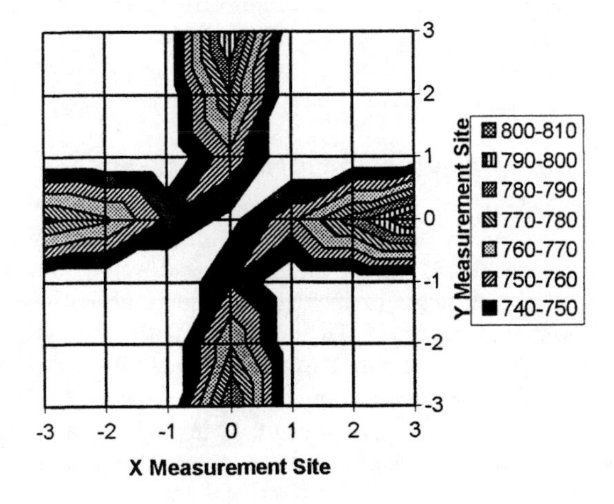

FIGURE 4. 2d wafer map of Si cap thickness of Si/SiGe patterned product wafer, measured at 13 sites in cross configuration.

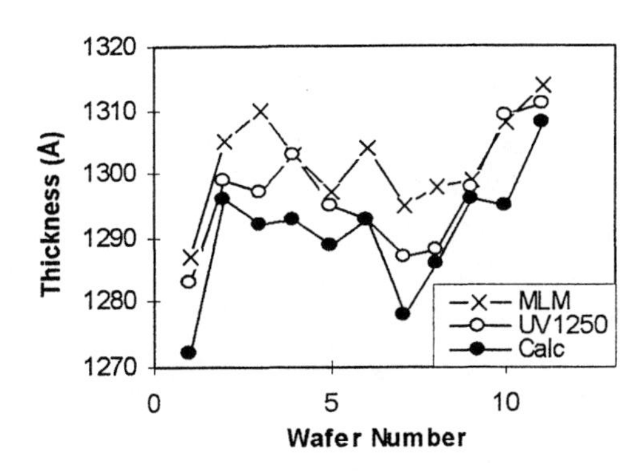

FIGURE 5. Nitride layer thicknesses measured on capacitor structures by MLM and UV-1250SE compared with values calculated from electrical data.

CONCLUSIONS

The IMPROVE project has demonstrated the accuracy and flexibility of SE for the non-destructive in-line measurement of several important multilayer wafer structures. The limits of process-induced variations that can be reliably measured have been assessed. Process-induced perturbations have been identified, which could not be reliably and quickly characterized by conventional monitoring equipment. The inadequacy of relying on monitor wafers has also been demonstrated.

REFERENCES

1. Zahorski, D., Mariani, J.L., Escadafals, L.and Gilles, J., *Thin Solid Films*, **234** 412 (1993).

2. Pickering, C. et al, "Evaluation of automated spectroscopic ellipsometry for in-line control -- ESPRIT SEA Project 'IMPROVE'" , presented at 2nd Int. Conf. SE, Charleston, SC, May, 1997, *Thin Solid Films* (in press) (1998).

3. Pickering, C., *J Min. Met. Mater. Soc.*, **46** 60 (1994).

4. Pickering, C., Sharma, S., Morpeth, A.G., Keen, J.M. and Hodge, A.M., *ECS Proc. 4th Int. Symp. SOI Technology and Devices*, **90-6** 175 (1990).

5. Carline, R.T., Pickering, C., Robbins, D.J., Leong, W.Y., Pitt, A.D. and Cullis, A.G., *Appl. Phys. Lett.*, **64** 1114 (1994).

6. Pickering, C., "Complementary in-situ and post-deposition diagnostics of thin film semiconductor structures", presented at 2nd Int. Conf. SE, Charleston, SC, May, 1997, *Thin Solid Films* (in press) (1998).

Metrology Standards with Ellipsometers

John A.Woollam, James N. Hilfiker, C. M. Herzinger, Ron A. Synowicki, and M. Liphardt, J.A.Woollam Co., Inc, 645 M Street, Lincoln, NE 68508

ABSTRACT

At the October 1997 NIST "Workshop for Thin Dielectric Film Metrology", uses of ellipsometers, and the role of standard materials for semiconductor metrology were two of several important topics discussed.[1] Much of the workshop discussion centered on metrology for thin gate oxides. This paper is an extension of some of the topics discussed at that workshop.

INTRODUCTION

Ellipsometers and reflectometers are traditionally used for semiconductor metrology.[2] Metrology normally includes layer thickness determinations, and sometimes optical constant and/or microstructure studies. Tools used in the Fab also often include robot loading, area mapping, and automated data acquisition and analysis. As dimensions shrink laterally they also shrink in thickness. Thus ellipsometers are increasingly important metrology tools.

Ellipsometers use polarized light at oblique angles, and have significantly greater sensitivity to determining the thickness of thin films, as compared to reflectometers. Metrology using reflectometers has an advantage in that the light beam is perpendicular to the sample so accurately knowing the angle of incidence is not a problem. However, the layer thickness sensitivity of reflectometers drops dramatically with decreasing thickness.[2] Reflectometers pick up sensitivity by using shorter wavelength light, but then UV optical systems are more difficult to deal with. For example, due to strong dispersion of the index of refraction of lenses the focal length depends strongly on wavelength. Also, UV light source intensity seriously degrades in time, and the UV light itself degrades other optical materials with which it interacts.

Ellipsometers are highly sensitive to properties of ultra-thin films, including thickness. This sensitivity occurs over wide spectral ranges, and can be enhanced for any materials system by wise selection of the angle of incidence.[3]

RESULTS AND DISCUSSION

In this paper we make seven points which we feel should be addressed by users of optical instruments for metrology: 1) Spectroscopic ellipsometry is recommended over one wavelength ellipsometry; 2) Angle of incidence accuracy is extremely important, and at the high parameter sensitivity angle, incidence angle accuracy needs to be greater; 3) The wavelength of light must be accurately known; 4) Standard reference materials are needed, but better control of the environment is needed; 5) Good optical constants of materials, and realistic optical models for the microstructure are needed; 6) Full regression calibration of the ellipsometer is recommended; and finally, 7) Details of the regression algorithms used to determine unknown parameters are an important part of metrology, and differences can result in different parameter solution values. We urge that all of the above be addressed simultaneously.

Spectroscopic Ellipsometry

We recommend spectroscopic ellipsometry, as opposed to single wavelength ellipsometry, for metrology.[4,5] This is especially true for materials such as oxides, nitrides, oxynitrides, polysilicon, photoresists, and ONOPO multilayer stacks.[6,7] There are a number of reasons why spectroscopic data are highly desirable. These include: a) finding the correct "order" for thick films for which there are interference oscillations, b) better precision, due to much larger acquired data sets, c) greater likelihood that at least part of the spectral range will provide high sensitivity to the desired unknown parameter, d) opportunity to introduce more realistic physical models for materials, and e) opportunity to analyze microstructurally complex multilayered materials such as ONO stacks. f) Opportunity to solve for angle of incidence in the regression, which reduces a possible source of systematic error.

Angle of Incidence Accuracy

In making ellipsometric measurements, the angle of incidence needs to be accurately measured, requiring accurate beam and sample alignment, and well-collimated light. Errors in angle of incidence result in errors in film thickness measurements. To illustrate, Figure 1 shows the ellipsometric delta spectra expected for 4 nm thick SiO_2 on Si at angles of incidence from 74 to 75 in steps of 0.1°. For example, at 632.8 nm an error of 0.1° in angle of incidence results in a delta error of about 3 degrees. This results in a thickness error of about 0.1 nm out of 4 nm.

CP449, *Characterization and Metrology for ULSI Technology: 1998 International Conference*
edited by D. G. Seiler, A. C. Diebold, W. M. Bullis, T. J. Shaffner, R. McDonald, and E. J. Walters
 1-56396-753-7/98/$15.00

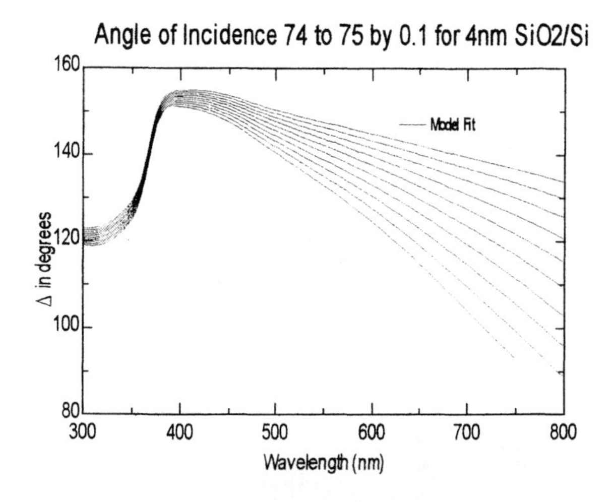

Figure 1. Ellipsometric delta for 4 nm thick SiO_2/Si for 0.1° angle of incidence steps from 74° to 75°.

Also, the sensitivity to thickness or optical constant measurement is strongly dependent on the angle of incidence used. The optimum angle of incidence to use depends on the particular materials, their thickness, and the wavelengths of light used. These points are illustrated in Figure 2, for the case of 5nm thick SiO_2 on Si. Notice the strong sensitivity to measuring the ellipsometric psi parameter for angles of incidence very near the Brewster Angle. This angle depends on wavelength, as seen in the figure. In the case shown, the high sensitivity region is near 75° angle of incidence at long wavelength (for example, at 632.8 nm), and shifts to higher angles of incidence for shorter wavelengths. This sensitivity becomes a sharper function of angle for thinner SiO_2. Gate oxides are prime examples of where this is important.

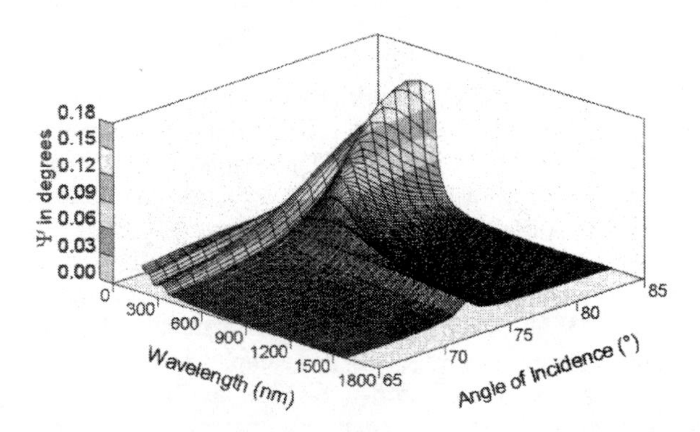

Figure 2. Sensitivity of Psi to 0.1 nm change of SiO_2 thickness for 5 nm thick SiO_2 on Si.

Every parameter of interest has an angle of incidence sensitivity, that is, a region of angle of incidence for which there is higher or reduced sensitivity to accurately measuring thickness. Along with the high sensitivity there is an increased danger of parameter error if the angle of incidence is not set accurately. These two points are shown in Figures 3 and 4. Figure 3 shows the sensitivity to measuring SiO_2 thickness on Si using wavelengths between 400 and 750 nm. ("Sensitivity" is defined here as the square root of the sum of the squares of shifts in psi and delta, divided by twice the number of data points, resulting from a 1% change in thickness.) Sensitivity peaks near Brewster Angles. This covers a smaller range of angles for thinner oxide layers, as stated above. This fact is well known, but we believe this is the first three-dimensional visualization.

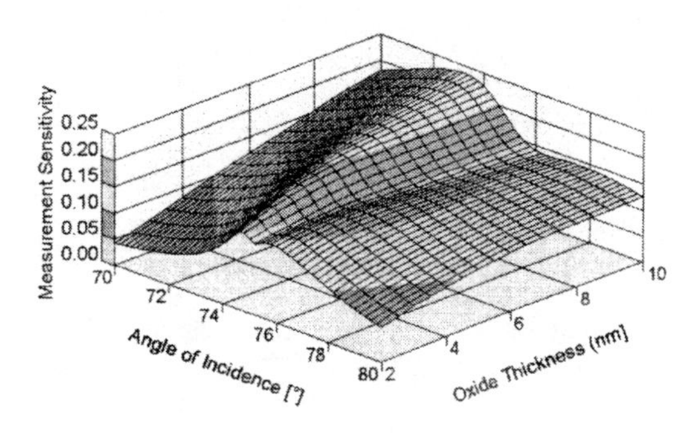

Figure 3. Sensitivity to measuring thickness of SiO_2 on Si, using wavelengths between 400 and 750 nm.

Likewise, Figure 4 shows the thickness error that would be introduced due to an angle of incidence error of 0.01°. Notice that in the regions of angle of incidence and thickness where the sensitivity to thickness increases (See Figure 3), the effect of angle of incidence error on thickness error increases. This means accurate metrology requires good absolute angle of incidence accuracy, not just good repeatability (precision). For experts in ellipsometry this is not new information. Our purpose here is to inform users in ways easy to observe. We believe these are the first three- dimensional graphs showing "desired parameter" (Not just psi or delta) error due to incidence angle error.

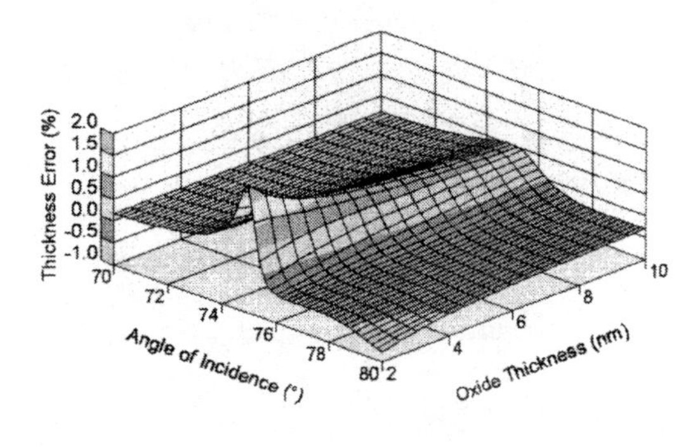

Figure 4. Thickness error that would occur should the ellipsometer have an angle of incidence error of 0.01°, using wavelengths between 400 and 750 nm.

As said, with increased angle of incidence sensitivity to a desired unknown parameter is increased need for accurate angle of incidence settings. How does the user know the angle? Obviously, the instrument and sample need proper accurate alignment. There are several instrumental aspects involved in angle of incidence accuracy. These include alignment relative to the axis of rotation of the optics-arms, as well as alignment of the sample relative to the ellipsometer.[8,9] With spectroscopic ellipsometers the angle of incidence can be solved for in the regression, if needed. For single wavelength data, there is too little information content to solve for the angle: too few measured quantities.

We believe that angle of incidence error is a major source of disagreement between laboratories in round robin sample exchanges or other comparative measurements.

Wavelength Accuracy

Third, optical metrology (including both ellipsometry and reflectometry) requires accuracy in setting and knowing the wavelength of light in spectroscopic systems. Wide spectral range, especially including the ultra-violet is highly desirable for both thick and thin films.[10] Ellipsometers taking accurate and precise data to 190 nm are now available. (See the companion paper by Hilfiker, et. al. on "Metrology Applications in Lithography with Variable Angle Spectroscopic Ellipsometry", in this book.) Figure 5 shows experimental data and model-dependent fits for silicon-oxynitride films on silicon. Notice the sharp peaks in the data. Good metrology for such materials requires a narrow wavelength bandwidth, and accurate wavelength value. Otherwise, peaks are washed out and shifted as a function of wavelength, resulting in poor regression fits and erroneous results. This requires accurate monochromators, with narrow bandwidth and good optical throughput.

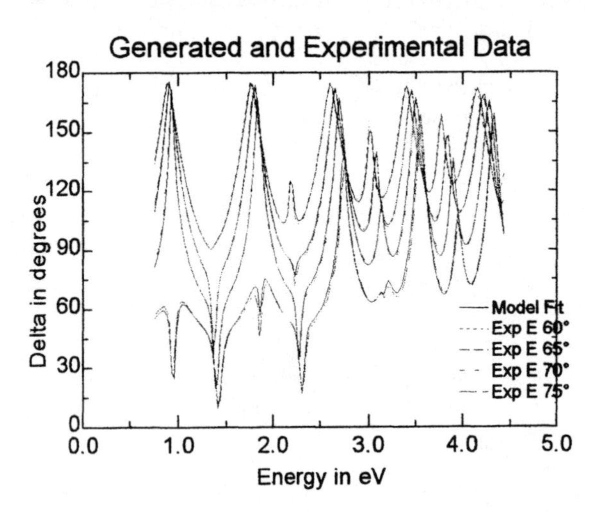

Figure 5 Ellipsometric delta parameter for 522nm thick index graded and spatially non-uniform silicon-oxynitride on silicon.

Figure 6 shows the error in thickness of SiO_2 on Si should there be an error of 0.5 nm in setting wavelengths in the spectral range 400 to 750 nm, plotted vs angle of incidence and thickness of the oxide. This is a quite small effect, for most angles smaller than about 78° for this spectral range. At shorter wavelengths the critical point optical constants of the underlying Si cause increased need for spectral accuracy. In conclusion, monochromator accuracy is less of a concern than is angle of incidence accuracy under conditions of no interference effects present in the spectral range. As seen in Figure 5, wavelength accuracy is quite important when spectra are sharp.

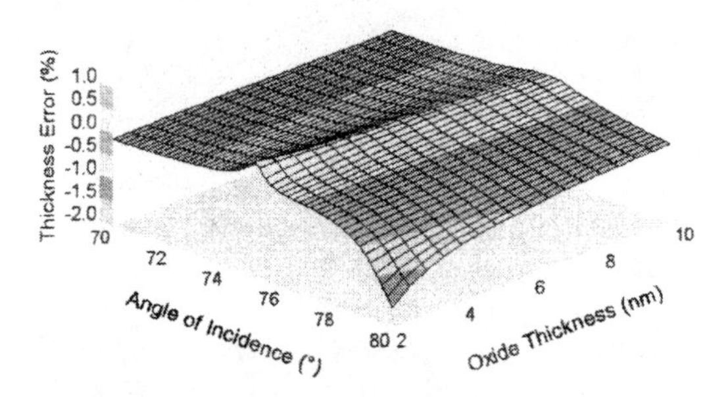

Figure 6. Error in thickness of SiO_2 on Si should there be an error of 0.5 nm in wavelength, using wavelengths of 400 to 750 nm.

Standard Reference Materials

Standard reference materials are definitely needed. However, HF cleaned (111) Si is probably not a practical standard for the semiconductor industry, due to the lack of (111) wafer availability, especially 300-mm diameter, and also due to lack of stability of hydrogen termination. Fused silica is an interesting possible standard for science, but providing robot loading, and compatibility with standard Fab equipment is difficult. We believe that several samples (differing thicknesses) of thermally grown SiO_2 on Si wafers is still viable for thin film metrology of gate oxides and ULSI materials.[8] Preparation of oxides with reproducible properties is well documented. Also, it is a material of direct interest; that is, it is desirable to use the same material for a standard. Furthermore, oxides on silicon are compatible with existing Fab tools, and optical constants of both materials are known (Discussed below).

There are other materials including dielectrics and metals, for which ellipsometers will become increasingly important for ULSI technology. For new materials, as well as some existing materials, standard samples in addition to the proposed thermally grown SiO_2/Si are likely to be needed. A major requirement is that the material can be made with reproducible properties that are stable in time.

However, reproducible surface cleaning methods for standard materials should be further investigated, as contamination becomes increasingly important in thin films, and can cause standard reference materials to change over time and during shipping.[8,1] Fabs can be clean with respect to particle contamination, yet hydrocarbons can easily condense from the air. The relative humidity may not be constant. Both hydrocarbons and moisture can condense on surfaces from the atmosphere. Another environmental factor is temperature, as optical constants are a function of temperature, and what is called "room temperature" can differ between measurement sites.

Optical Constants and Optical Models

Good optical constants and models for the geometry and microstructure are important integral aspects of proper metrology. Even for SiO_2 on Si, the physical model has important effects on thickness values. It is now well established that a very thin interface between Si and SiO_2 exists, with intermediate index values.[11-18] It is also known that there are small but significant differences between published values of the optical constants of Si. Further, the spectral range of published optical constant values is often limited. For example, Aspnes and Jellison Si data are for wavelengths 206 nm to 826 nm, and 234 nm to 850 nm, respectively, yet commercial metrology tools are available to 190 nm. Herzinger, et. al. have recently extended the spectral range for optical constants of both Si and SiO_2 to cover 190 nm to 1700 nm.[16]

For other materials, such as nitrides, oxynitrides, and oxide-nitride stacks, the modeling and use of optical constants becomes more complex. For example, effective medium theory is sometimes used to "mix" SiO_2 with Si_3N_4 optical constants to represent dispersion in the optical constants of oxynitrides.[6] This works reasonably well for optical constant parameterization, even though the chemistry may not be physically correct. Other conventional dispersion models also work but give small differences in parameter values determined. Thus differences between results in different laboratories or Fabs can depend on the model used for analysis as well as the optical constants of the constituent materials.

Even within a single layer of material such as silicon-oxynitride, there is often vertical index grading, as illustrated in Figure 7. The results in Figure 7 come from the data fits shown in Figure 5. The uncorrelated best fits to the data were found to also include 1.8% lateral inhomogeniety. Surface and interfacial roughness is another factor that needs to be accounted for. In optical modeling of the material, either the "apparent" or "pseudo" optical constants are found (for purposes of studying their dependence on material processing parameters), or microstructure aspects (for purposes of studying the material microstructure) are modeled. The choice of which way to optically model depends on users needs. In the Fab, parameter reproducibility may be of highest priority, and microstructural modeling not needed. For materials such as poly-silicon or poly-silicon with ONO stacks, the situation is even more complex, as the apparent optical constants are strongly related to the microcrystallinity, which is strongly dependent on temperature during deposition and anneal conditions.[19] With full spectroscopic and variable angle of incidence data, combined with realistic modeling, ellipsometers can be used for quite detailed materials analysis. In addition, once models and procedures are established, automated ellipsometric analysis can be done, including area mapping.

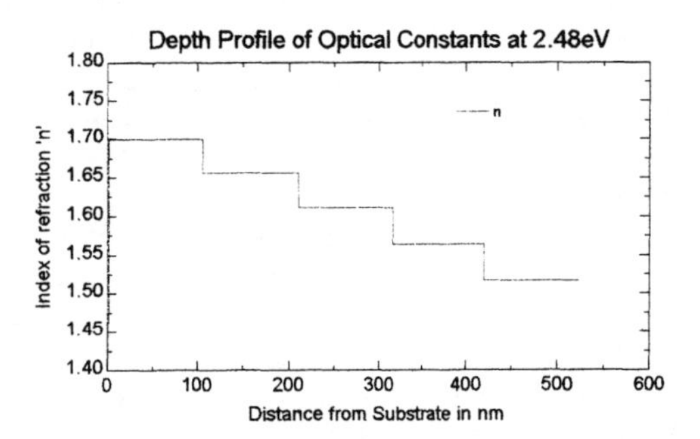

Figure 7. Example graded index depth profile: oxynitride with data shown in Figure 5. In addition to vertical index grading, there was 1.8% lateral thickness non-uniformity.

Instrument Calibration

Ellipsometers need to be calibrated. One part of this is to establish the azimuthal angles of the polarizers relative to the plane of incidence (defined as the plane containing the incident light, the reflected light, and the normal to the sample). The Aspnes method has been used by many groups since the 1970s.[20] However, the Johs "regression calibration method", improves accuracy significantly.[21]

Regression and Statistics of Fits

With any optical metrology technique the desired parameters are not directly measured. For ellipsometry and reflectometry the electromagnetic boundary value problem is solved, resulting in nested sets of Fresnel coefficients for reflection and transmission in and between constituent layers. These are too algebraically complex to be able to write an equation expressing desired parameter (such as layer thickness) in terms of measured quantity. Instead, guesses are made for the parameter value, and the expected measurable quantity is compared with the actual measured value. This "inverse problem" generally has solution in regression analysis. Most people use the Levenberg-Marquardt algorithm for ellipsometry regression; yet there are sometimes differences between parameter values found by different workers using the same original measured ellipsometric parameters but different regression software.[22] For example, the comparison functions can differ: The Mean Squared Error (MSE) is often used, and different software calculates different functions. Some normalize the fits by the random noise measured at each wavelength, calculate confidence limits, and calculate the correlation of parameters; others do not.

CONCLUSIONS

In summary, we have made seven major points: 1) Spectroscopic ellipsometry is recommended over one wavelength ellipsometry for semiconductor metrology, 2) Angle of incidence accuracy is extremely important, and at the preferred angle for high parameter sensitivity the need for angle of incidence accuracy is greater. 3) For conditions of strong dispersion in the data, the wavelength of light must be accurately known. 4) Thermal oxides on Si are good reference materials, but additional standard reference materials are needed. Better environment and cleaning control is needed. 5) Good optical constants of component materials and realistic models for the microstructure are needed. 6) Full regression calibration of the ellipsometer is recommended; and finally, 7) Details of the regression algorithms used to determine unknown parameters are an important part of metrology. We urge that all of the above points be simultaneously addressed. That is, that the semiconductor community looks not simply at standard samples as the solution for metrology needs. Rather, that all the aspects of these interdependent metrology components should be addressed simultaneously.

In round robin studies, we suspect that differences in results by different workers are most often due to: a) angle of incidence differences, b) different optical models assumed for the physical structure, c) differing details of how regression is carried out, as well as d) actual sample changes due to contamination during transit between measurement locations or due to differing local environments, such as temperature and/or humidity.

The next time a round-robin study is suggested, we feel it important that participants address and answer: What is the measuring temperature and humidity? What is the angle of incidence? Is it a high sensitivity angle? How accurate is the angle known? How was this angle measured, and what effect could a systematic angle of incidence error have on the reported result? What is the optical model used, and are other participants in the round robin study using the same model? Is it universally agreed to be a valid model? From where did the optical constants come, and how would errors in the optical constants affect the reported results? What calibration algorithm was used, and how would calibration errors affect reported results? These questions can be asked whether the instrument or software is home-built or commercial, and should lead to better agreement and thus better metrology.

Keywords: spectroscopic ellipsometry, thin film metrology, angle of incidence accuracy.

REFERENCES

1. "Speaker Presentations", Thin Dielectric Film Metrology Workshop, October 30-31, 1997, NIST, Gaithersburg, Maryland, Barbara Belzer, editor.
2. Hilfiker, J.N., Synowicki, R.A., "Spectroscopic Ellipsometery for Process Applications", Solid state Technology, October, 1996.
3. Snyder, P., Rost, M.C., BuAbbud, G.H., Woollam, J.A., and Alterovitz, S.A., J. Appl. Phys. **60**, 3293 (1986).
4. Tompkins, H., Users Guide to Ellipsometry, Academic Press, San Diego, CA, 1993.
5. Woollam, J.A., and Snyder, P.G., "Variable Angle Spectroscopic Ellipsometry", in Encyclopedia of Materials Characterization, Butterworth-Heinemann, Publishers, Boston, 401 (1992).
6. Xiong, Y.M., Snyder, P.G., Woollam , J.A., Strausser, Y., and Kroche, E.R., J. Vac. Sci. and Tech. **A10**, 950 (1992).
7. Snyder, P.G., Xiong, Y.M., Woollam, J.A., Al-Jumaily, G.A., and Gagliardi, F.J., J. Vac. Sci. Tech. **A10**, 1462 (1992).
8. Candela, G.A., Chandler-Horowitz, D., Marchiando, J.F., Novotny, D.B., Belzer, B.J., and Croarkin, M.C., NIST Special Publication **260-109**, October 1988, Library of Congress Catalog Number: 88-600591.
9. Azzam, R.M.A., and Bashara, N.M. Ellipsometry and Polarized Light, North-Holland Press, New York, 1977.
10. Hilfiker, J., and Synowicki, R.A., "Employing Spectroscopic Ellipsometry for Lithography Applications", Semiconductor Fabtech, Fifth Edition, 189 (1997).
11. Yakovlev, V.A., and Irene, E.A., J. Electrochem.Soc. **139**, 1450 (1992).
12. Irene, E.A., Thin Solid Films **223**,96 (1993).
13. Nguyen, N.V. Chandler-Horowitz, Amirtharaj, P.M. and Pellegrino, J.G., Appl. Phys. Lett. **64**, 2688 (1994).
14. Taft, E.A., and Cordes, L., J. Electrochem. Soc. **126**, 131 (1979).
15. Aspnes, D.E., and Theeten, J.B., Phys. Rev. Letts. **43**, 1046 (1979).
16. Herzinger, C.M., Johs, B. McGahan, W.A., and Woollam, J.A., and Paulson, W. J. Appl. Phys. **83**, 3323 (1998).
17. Herzinger, C.M., Johs, B., McGahan, W.A., and Paulson, W., Thin Solid Films **313-314**, 281 (1998).; and Proceedings of the Second International Conference on Spectroscopic Ellipsometry.
18. Nguyen, N.V., and Richter, C.A., Electrochemical Society Proceedings, **97-10**, M.J. Deen, W.D.Brown, K.B. Sundaram, and S.I. Raider, Eds. (The Electrochemical Society, Inc., Pennington, New Jersey, 1997), pp 183-193.
19. Snyder, P.G., Xiong, Y.M., Woollam, J.A., and Krosche, E.R., Mater. Res. Soc. **238**, 605 (1998).
20. Aspnes, D., and Studna, A.A., Appl. Opt., 14, 220 (1975).
21. Johs, B., Thin Solid Films, **234**, 395 (1993).
22. Belzer, B., and Blackburn, D.L. "The Results of An Interlaboratory Study of Ellipsometric Studies of SiO2 on Si", NIST Special Publication 400-99, May 1997.

Evaluation of Surface Depletion Effects in Single-Crystal Test Structures for Reference Materials Applications†

Richard A. Allen and Rathindra N. Ghoshtagore

Semiconductor Electronics Division
National Institute of Standards and Technology
Gaithersburg, Maryland 20899

Monocrystalline silicon test structures are being investigated for critical dimension (CD) reference materials applications. The goal of this work is to produce samples which do not exhibit the phenomenon of "methods divergence," where the measurement of a single sample, fabricated in an electrical conductor, by multiple techniques leads to results that differ from one another by more than the total known error budgets of the measurements. In this paper, measurements are described to determine the sources of differences observed between electrical and other measurements.

1. Background

Critical dimension (CD), or linewidth, measurements made on features patterned for semiconductor applications have tended to exhibit differences related to the type of metrology technique or instrument used to perform the measurement(1),(2). This effect, known as methods divergence, is primarily due to differences in the way each measurement instrument interacts with the edges of the feature being measured. These effects combine to create differences up to 100 nm between measurements of a single feature using different metrology techniques. Understanding these differences becomes critical for sub-half-micrometer processes where such methods divergence can equal *20% or more* of the design width.

Recent work at NIST and Sandia National Laboratories has focused on developing artifacts which can be used for instrument cross-calibration and which will give consistent and meaningful CD results no matter which metrology technique is used. The test structure in Figure 1 shows an example of such an artifact. This structure is based on the cross-bridge resistor test structure(3),(4) commonly used for measuring the electrical CD (ECD) of a feature. The horizontal line in the structure shown in Figure 1 is designed to be measured by current CD metrology techniques. This CD structure is produced to be free of the random sources of methods-divergence observed in conventionally produced materials (these random effects in conventional processes generally cannot be corrected for without introducing unacceptably high uncertainties). In other words, CDs measured using these new structures should be equivalent *independent of metrology tool* (e.g., electrical probe, scanning electron microscopy (SEM), transmitted-light optical microscopy, or scanning probe microscopy (SPM)). The features are patterned onto the surface film of (110) silicon-on-insulator (SOI) wafers which provide the isolation needed for electrical probing; further they are defined using lattice-plane selective etch techniques which align the features' sidewalls to the (111) crystallographic planes of the silicon film(5) and thus the features should have vertical, atomically smooth sidewalls with well-defined intersections between features.

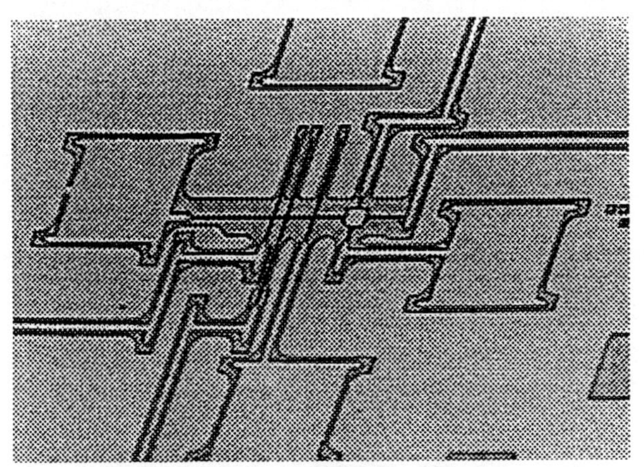

Figure 1 The basic single-crystal CD Test Structure.

Since the feature sidewalls of the cross-bridge resistor are bounded by the (111) planes of the SOI film, which are in <$11\bar{2}$> directions, the cross-bridge resistor structure has non-orthogonal intersecting features (since <112> vectors intersect each other at 70.529° in the (110) plane), thus the sheet resistance analysis procedure had to be modified.(6) That is, since the van der Pauw structure is made by superimposing a box on a non-orthogonal intersection, the sheet resistance R_s must be calculated using the generic van der Pauw equation for planar structures

†Contribution of the National Institute of Standards and Technology (NIST). Not subject to copyright.

CP449, *Characterization and Metrology for ULSI Technology: 1998 International Conference*
edited by D. G. Seiler, A. C. Diebold, W. M. Bullis, T. J. Shaffner, R. McDonald, and E. J. Walters

$$\exp\left(\frac{-\pi V_1}{R_s I}\right) + \exp\left(\frac{-\pi V_2}{R_s I}\right) = 1\,, \tag{1}$$

rather than the commonly used analytical expression, $R_s = (\pi/\ln(2))(V/I)$, which is valid only for van der Pauw structures which exhibit four-fold rotational symmetry. In Eq. (1), V_1 is the average voltage measured across the acute angle for a current I and its complement forced across the opposite acute angle. V_2 is the same measurement across the obtuse angles. R_s is the resistance, in ohms, of an arbitrary size square of material. Determination of R_s using Eq. (1) is accomplished numerically. Since the etching process leaves facets in the acute intersection regions of the features, the sheet resistance is also subject to correction for the three dimensional nature of the structure.(6)

Using the sheet resistance R_s extracted from Eq. (1), the linewidth w is calculated using the modified cross-bridge resistor equation

$$w = \frac{R_s(L - \delta L)}{R_b} \tag{2}$$

where L is the length of the bridge, δL is the length shortening correction factor due to the presence of the non-point-contact voltage taps,(7) and R_b is the V/I drop of the bridge segment being measured. In this process, misregistration-induced rotation of the photomask relative to the wafer at the time the image is transferred to the chip will cause the bridges to exhibit narrowing in proportion to the bridge length and angle of rotation. An algorithm was developed which allows for the simultaneous solution of the angle of rotation as well as the bridge width of each length using Eq. (2).(6)

Materials and Fabrication

The samples used in the first phases of this work were fabricated on SIMOX (Separation by IMplantation of OXygen) and BESOI (Bonded and Etched-back Silicon-On-Insulator) SOI wafers. The features of these prototype reference materials were patterned directly into the surface layer. This leads to CMOS-non-compliant requirements for the resistivity and doping of the starting material. That is, in a typical CMOS SOI process, the substrate is lightly doped and thus has a relatively high resistivity, approximately 10 Ω cm to 20 Ω cm for a 1 μm layer, which is too high for this application. Only in regions that must be highly conductive, e.g., transistor sources and drains, is the resistivity lowered by implanting donor or acceptor atoms. Implantation processes, however, do not provide acceptably defect-free starting materials in which to fabricate the reference material prototypes. Therefore, the SOI material used in this project were fabricated in a different manner than is done for CMOS/SOI processing to provide substantially lower resistivity, 0.003 Ω cm to 0.03 Ω cm.

As mentioned above, there were two types of SOI materials used in this experiment: SIMOX and BESOI. SIMOX is a mature technology and comparatively straightforward to acquire for a proof-of-concept experiment. Compared to BESOI, however, SIMOX processing techniques allow only limited variation in surface layer thickness and doping level while BESOI also provides a much more abrupt interface between the silicon and silicon dioxide layers. The device layers of the SIMOX wafers used were ~200 nm thick and doped with Sb to a level of 5×10^{17} cm^{-3} (0.031 Ω cm). The device layers of the BESOI wafers used were ~1000 nm thick and doped with As to a level of 2.3×10^{19} cm^{-3} (0.003 Ω cm).

Two surface effects also must be considered in the electrical measurement of the linewidth. The first is the non-conductive, native oxide that forms on the silicon surface. This oxide grows on each surface to a thickness of approximately 3 nm. Second, if a fixed charge remains on the silicon surface after processing, a surface depletion layer may form. The maximum thickness of this layer is(8)

$$d = \sqrt{\frac{4\epsilon_S kT\ln(N_D/n_i)}{q^2 N_D}}\,, \tag{3}$$

where ϵ_s is the permittivity of silicon, k is Boltzmann's constant, T is the temperature, N_D is the donor dopant concentration, n_i is the intrinsic carrier concentration in silicon at T, and q is the elementary charge. Figure 2 shows the predicted temperature dependence of the depletion depth for several doping levels, including the doping level

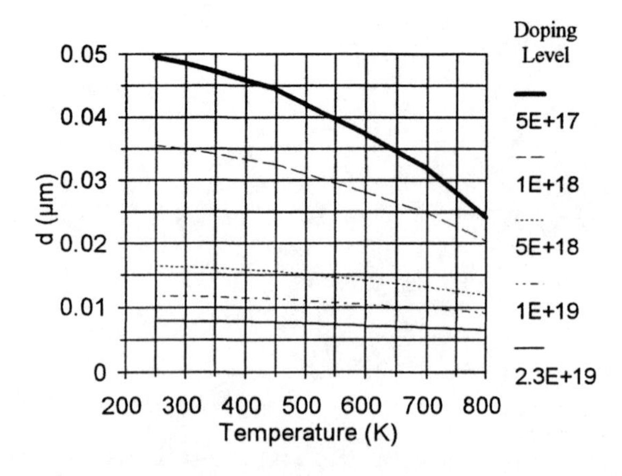

Figure 2 The maximum depletion depth as a function of temperature for different doping levels. Note that the electrical width of a feature will be decreased by twice this amount for two sidewalls.

of the SIMOX and BESOI samples (5 x 10^{17} cm^{-3} and 2.3 x 10^{19} cm^{-3}).

Results

The first artifacts were fabricated on a SIMOX wafer. Initial *V/I* measurements of these structures indicated that there is a very limited range of valid applied currents that may be used when measuring the van der Pauw structures fabricated using the SIMOX wafers. This is partly due to the relatively high resistivity of the SIMOX wafers and partly due to the non-orthogonality of the intersection between the lines. That is, a current forced between adjacent taps of the obtuse corner produces a voltage V_2, measured at the opposite corner, that is much higher than the voltage V_1, which is measured between adjacent taps of the acute corner with current forced at the opposite corner. Even though this difference was expected, its magnitude, over a factor of ten, was surprising. The difference in magnitude was ascribed to the presence of facets, i.e., shorts, across the acute angle intersections.

The lower limit of valid applied currents occurs due to the noise floor of the test system. That is, when attempting to measure extremely low voltages, the uncertainty due to the instrumentation is unacceptably high. At the other end, when the current is too high, Joule heating causes the resistance of the material to change as the measurement proceeds. Thus the measured voltage for a given current also changes. For the SIMOX structures, the range of valid currents was observed to be less than one decade.

Since there were a number of different design width devices, each with its own optimal current, a measurement algorithm was developed to ensure that the voltages used to extract the sheet resistance were measured at the optimal value. The van der Pauw resistors were measured over a number of different currents (approximately three orders of magnitude), and R_s is extracted where the standard deviation of repeated *V/I* measurements is minimum (both effects described above introduce noise into the measurements). Using this procedure, the sheet resistance is determined with a repeatability of 0.8% and an expanded uncertainty of 5%. A detailed discussion of the statistical method used to extract the R_s values and evaluate the uncertainty has been presented elsewhere.(6) Figures 3 and 4 show the sheet resistance versus the design parameters for the SIMOX and BESOI chips, respectively. These figures also reveal the importance, when producing cross-bridge resistors from these SOI starting materials, of using a van der Pauw resistor with a cross superimposed on the intersection. The van der Pauw resistors with the superimposed box (also known as the "box cross") convey sheet resistance independent of tap width. In contrast, the Greek cross van der Pauw resistors reveal a sheet resistance inversely dependent on the arm width of the device being measured. This unexpected effect, which is clearly visible on devices fabricated on both the SIMOX and BESOI wafers, has not been seen in measurements of cross-bridge resistors fabricated with orthogonal geometries using conventional materials (the sheet resistance is typicially independent of arm width in the Greek cross van der Pauw, although there may be a component of random uncertainty introduced in the measurement due to limitations of the instrumentation(9)).

An additional effect observed measuring the SIMOX wafers was a weak rectification between the voltages measured across a single resistor for a current and its complement. This effect, whose magnitude was not consistent from structure to structure, did not occur with the BESOI structures and is believed to be due to the non-planar bottom surface of the SIMOX conductive layer.

There were two or three bridge resistors in each structure from which the ECD was determined. The structure shown

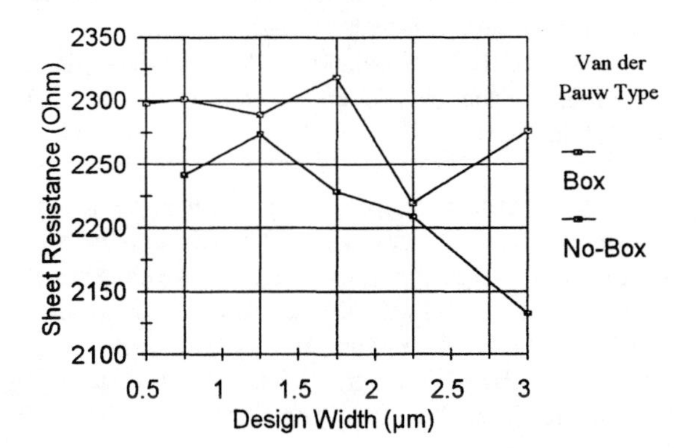

Figure 3 The measured sheet resistance as a function of design width for a SIMOX-processed structure.

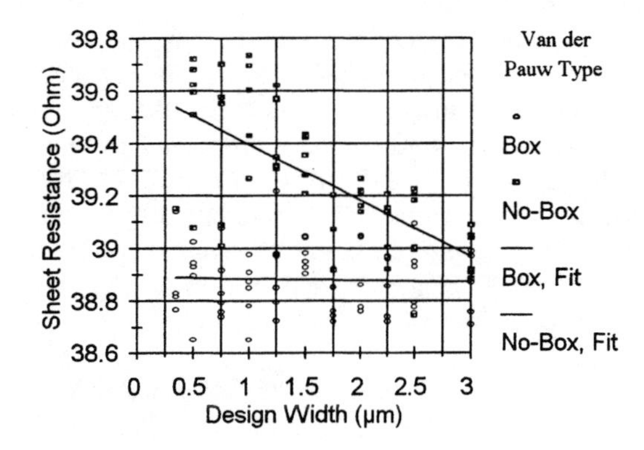

Figure 4 The measured sheet resistance as a function of design width for a BESOI-processed structure.

in Figure 1 has a box cross van der Pauw; therefore, there are only two bridges available for measurement. Similar devices with a Greek cross van der Pauw incorporate three bridges.

The bridge resistors were measured using the following procedure. Each of the three bridges was measured as well as the bridge comprising all three bridges; this sum measurement provides a check on the instrumentation by verifying that the *V/I* drop of the overall bridge was equal to the sum of the *V/I* drops of the individual bridges. The optimal measurement current was determined using a procedure similar to that of the van der Pauw resistors. Determination of the optimal current is not as critical for the bridges as it is for the van der Pauw resistors. This is because the voltage measured for a bridge is much higher than that measured for a van der Pauw resistor for the same applied current density. Thus, the primary concern was ensuring that the current was low enough to avoid Joule heating.

ECD Measurement and Extraction

From the V/I measurements, the ECD is determined using modified version of Eq. (2) to allow for simultaneous extraction of the widths of the two or three sequential lines as well as the line-shortening effect and the rotational misalignment of the photomask (which causes the linewidth of longer lines to be narrower than that of shorter lines). The ECD correction procedure and the numeric modeling developed to extract the ECD is described elsewhere.(6)

Initial measurements of these structures, using electrical probing, scanning electron microscopy (SEM), and atomic force microscopy (AFM), suggested that the electrical critical dimension (ECD) measurement of the features underestimated the actual physical width by approximately 100 nm. This difference was attributed to the electrical measurement since the amount of overetch needed to produce this difference was not consistent with the process used to fabricate the device. Additionally, the observed difference is consistent with twice the calculated maximum thickness of the surface depletion layer (the value from Figure 2 is doubled since there are surface depletion layers at each sidewall). Note that there also may be a surface depletion at the top surface of the material; this will have no effect on the measured ECD since it will change both the measured sheet resistance and the measured bridge resistance proportionally.

Temperature-Induced Changes in the Extracted ECD

To verify that the source of the observed difference between the expected and extracted ECDs was due to a surface depletion layer narrowing the width of the bridge, measurements were made to see if the ECD varied with temperature. This work, reported in detail elsewhere,(10) measured the ECD of SIMOX features as a function of temperature to determine if the change followed the curve shown in Figure 2. The temperature range used in this experiment was 300 K to 425 K. From Figure 2 it can be seen that a change in ECD of approximately 8 nm, that is, about 4 nm per surface, should occur over this range of temperatures for the SIMOX wafer's dopant concentration of 5×10^{17} cm^{-3}. Also note that over this range of temperatures there will be a change in the ECD due simply to thermal expansion of the line. Thermal expansion affects the CD extracted using Eq. (2) in two ways. The first is direct expansion of w, the second in lengthening of L. The sum of these two effects was calculated to be less than 2 nm for the nominally 1.0 μm to 3.0 μm lines used in this experiment.

The results of the measurements at different temperatures generally supported the depletion layer hypothesis. However, since only a small change in the ECD was expected over this limited temperature range, the results of the ECD measurements at different temperatures were not considered conclusive. Consequently, two other approaches to demonstrating the existence of a surface depletion layer, by injecting carriers into the edge regions of the feature, were investigated. The first was photon bombardment which creates free carriers in this surface layer; the second was exposing the artifact to environmental ozone which neutralizes the surface charge which produces the depletion layer. As with the temperature measurement, an increase in the ECD under either of these measurement conditions would support the depletion layer hypothesis.

Environmental-Induced Changes in the Extracted ECD

The sample was measured in all permutations of illumination and ozone on and off, beginning and ending with both off to check the repeatability. Upon analysis of the data, it became clear that the presence of the ozone had no effect on either the sheet resistance or the ECD for either the SIMOX or BESOI samples. This was likely due to limitations in the experimental setup that need to be explored further.

The presence or lack of strong illumination on the sample, in contrast, had a significant effect on both the sheet resistance as well bridge resistances. This change in the conductive area or volume will manifest as a change in the width of the features. For the structures fabricated on the highly doped BESOI substrates with minimal depletion width, there should be virtually no change in the ECD; for the structures fabricated on the SIMOX substrates with a large depletion width, a more substantial change is expected.

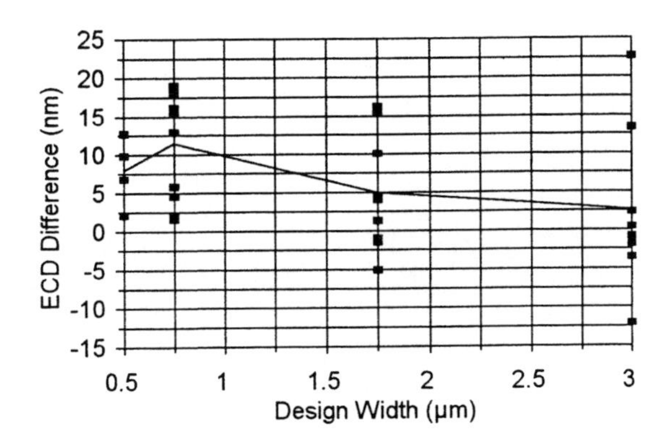

Figure 5 The calculated difference in ECD, from darkness to full illumination, for SIMOX-processed structures.

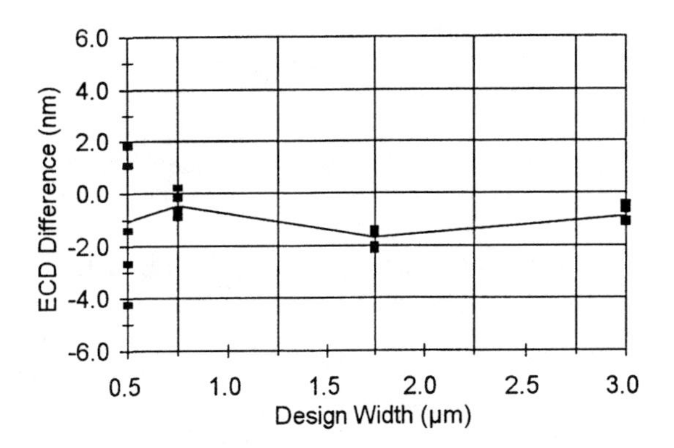

Figure 6 The calculated difference in ECD, from darkness to full illumination, for BESOI-processed structures.

Samples of both the SIMOX and BESOI were measured in this way. Figures 5 and 6 show the difference in measured ECD for the devices under conditions of darkness and strong illumination. The scatter observed in the measured ECD for the SIMOX (Figure 5) is high but is consistent with other ECD measurements for this high-resistivity material. As expected, there was no change for the BESOI wafer, but there was a slight increase (~7 nm) in the ECDs of the SIMOX chip. This change is similar to that observed during temperature variation (described earlier).

3. Summary

In this paper we have described our work on the investigation of the hypothesis that surface depletion on the sidewalls of the SIMOX artifacts was leading to underestimation of the ECD. The observations of the increase in ECD observed under conditions of strong illumination for the lightly doped SIMOX artifacts, in comparison with the observation of no change in ECD for the more highly doped BESOI devices, while necessarily qualitative, strongly support this hypothesis. Thus, at this time, it appears that the BESOI materials are most likely to provide method-divergence-free reference materials. For the proposed reference material any observed difference in CD, as measured by different metrology tools, must be readily quantifiable for the reference material to be of any value. Ideally, of course, there should be no difference between different metrology instruments. Measurements of the BESOI wafer showing no evidence of a significant depletion layer are encouraging since the electrical probe is the only metrology tool that relies on the doping level. This validates our approach of using the BESOI artifacts in this reference material application.

4. Acknowledgments

The authors would like to thank W. Robert Thurber, John Villarrubia, Michael W. Cresswell, Loren W. Linholm, Jim Potzick, Michael Postek, and Ron Dixson of NIST for technical discussions, James C. Owen III of NIST for performing measurements of these devices, and Jeffry Sniegowski and Sarah Everist of the Microelectronics Development Laboratory at Sandia National Laboratories, Albuquerque, New Mexico for producing the single-crystal test structures.

4. References

1. E.E. Chain and M. Griswold, "In-Line Electrical Probe for CD Metrology," Proc. SPIE 2876, 135 (1996).

2. R.A. Allen, P. Troccolo, J.C. Owen III, J.E. Potzick, and L.W. Linholm, "Comparisons of Measured Linewidths of Sub-Micrometer Lines Using Optical, Electrical, and SEM metrologies," Proc. SPIE 1926, 34 (1993).

3. M.G. Buehler, S.D. Grant, and W.R. Thurber, "Bridge and van der Pauw Sheet Resistors for Characterizing the Line Width of Conducting Layers," Journal of the Electrochemical Society 125, 650-654 (1978).

4. SEMI Standard P19-92, SEMI International Standards 1994, Microlithography Volume, 83 (1994).

5. M.W. Cresswell, J.J. Sniegowski, R.N. Ghoshtagore, R.A. Allen, W.F. Guthrie, A.W. Gurnell, L.W. Linholm, R.G. Dixson, and E.C. Teague, "Recent Developments in Electrical Linewidth and Overlay Metrology for Integrated Circuit Fabrication Processes," Japan Journal of Applied Physics 35, 6597-6609 (1996).

6. M.W. Cresswell, N.M.P. Guillaume, R.A. Allen, W.F. Guthrie, R.N. Ghoshtagore, J.C. Owen III, Z. Osborne, N. Sullivan, L.W. Linholm, "Extraction of Sheet Resistance from Four Terminal Sheet Resistors Replicated in Monocrystalline Films Having Non-Planar Geometries," Proc. 1998 International Conference on Microelectronic

Test Structures, Kanazawa, Japan, 29-38 (1998).

7. R.A. Allen, M.W. Cresswell, and L.M. Buck, "A New Test Structure for the Electrical Measurement of the Widths of Short Features with Arbitrarily Wide Voltage Taps," IEEE Electron Device Letters, 13 (6), 322-324 (1992).

8. S.M Sze, *Physics of Semiconductor Devices*, Second Edition, John Wiley and Sons, New York, 372-374 (1981).

9. P. Troccolo, L. Mantalas, R. Allen, and L. Linholm, "Extending Electrical Measurements to the 0.5 μm Regime" Proc. SPIE 1464, Integrated Circuit Metrology, Inspection, and Process Control V, 90-103 (1991).

10. R.A. Allen, O. Oyebanjo, M.W. Cresswell, and L.W. Linholm, "Surface Effect on Electrical Measurements of Single-Crystal Critical Dimension Artifacts," Proc. 1998 International Conference on Microelectronic Test Structures, Kanazawa, Japan, 57-60 (1998).

In-Situ Real-time Mass Spectroscopic Sensing and Mass Balance Modeling of Selective Area Silicon PECVD

Ashfaqul I. Chowdhury, Tonya M. Klein, and Gregory N. Parsons

Engineering Research Center for Advanced Electronic Materials Processing and Department of Chemical Engineering North Carolina State University, Raleigh, NC 27695-7905.

In this paper, we show that real-time in-situ mass spectroscopy can quantify deposition and etching reaction rates in plasma processes. Mass spectroscopy is a useful process sensor for SiH_4 processes because it is particularly sensitive to changes in silane concentrations in the sampled gas. We demonstrate rate sensitivity using a cyclic deposition/etch process that leads to selective area microcrystalline silicon PECVD. The procedure involves repeated cycles of a $SiH_4/He/H_2$ plasma followed by a He/H_2 plasma. During $SiH_4/He/H_2$ flow, when the plasma is initiated, we observe a decrease in the silane signal that is correlated to film deposition. During the He/H_2 plasma the silane signal is larger when the plasma is on, and the change is a quantitative indicator of silane produced by etching. Moreover, the transition from selective to non-selective conditions can be detected in real-time. A sharp change in slope of the silane signal is observed during etching, which is consistent with complete removal of stray nuclei from the non-receptive surfaces. Analyzing mass spectrum data with a mass balance model gives quantified deposition and etch rates.

INTRODUCTION

Some of the critical steps in processes with complex reactions cannot be observed without real-time process state sensors. Thin film processes in microelectronics fabrication often have intermediate steps that need to be quantified for optimization purposes. These intermediate steps, including concurrent etching and deposition cannot be quantified using only final state analysis. In such a situation incorporation of one or more real-time sensors such as in-situ mass spectroscopy, to reactor systems becomes essential. Mass spectroscopy is a very effective real-time sensor for gas phase process analysis and characterization because it is a relatively simple and robust technique for sensing neutral and ion species, and its application to residual gas analysis is well known (1). Mass spectroscopy has been used as a real-time sensor for various deposition (2,3,4,5) and etch (6,7) processes. In this paper, we outline a methodology of using real-time in-situ mass spectroscopic sensing to quantifying critical process parameters such as etch rates on various types of surfaces during selective area silicon PECVD, in the context of a mass balance model. Selective area silicon deposition from modulated flow of silane in a hydrogen plasma proceeds through repeated alternate steps of deposition and etching. Selectivity is attained because Si is etched at a lower rate on receptive surfaces such as Mo and c-Si, while etching on non-receptive surfaces such as SiO_2 and SiN_x is much faster. We have previously reported observation of deposition and etching reactions during selective area deposition of silicon using in-situ mass spectroscopy (4). The combination of deposition and etch chemistry involved in selective area silicon PECVD makes it a prime candidate for using mass spectroscopy as an in-situ process state sensor. Low temperature plasma enhanced selective area silicon deposition offers advantages in thin film transistor fabrication and performance (8,9,10,11), but is a difficult process to optimize because of sensitivity to process fluctuations and chamber conditions. Plasma enhanced selective area silicon deposition may also be useful in processing of silicon integrated circuitry using elevated source and drain structures. Monitoring this process in real-time will lead to process improvements, and enable optimization of process conditions.

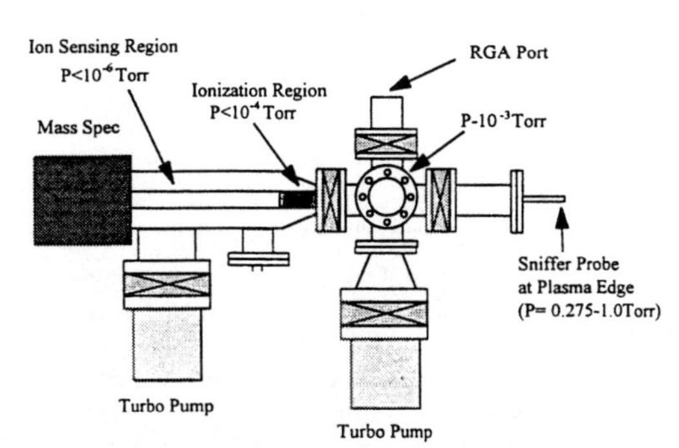

Fig. 1: Schematic of the double differentially pumped mass spectrometer sensing apparatus. The system samples neutral gas species exiting the plasma edge. Typical pressures in the reactor, pressure reduction stage, ionization region, and ion sensing region during operation are indicated.

EXPERIMENTAL PROCEDURE

A schematic of the two stage differentially pumped mass spectrometric process sensing system used is shown in Fig. 1. The sensing system consists of a changeable sampling aperture (0.305 mm) attached to a stainless steel tube connected through a conflat port on the reactor. This allows the aperture to be at the edge of the plasma zone of

CP449, *Characterization and Metrology for ULSI Technology: 1998 International Conference*
edited by D. G. Seiler, A. C. Diebold, W. M. Bullis, T. J. Shaffner, R. McDonald, and E. J. Walters

the parallel plate capacitively coupled PECVD reactor. A detailed description of the reactor is available elsewhere (3). The sampling aperture is made of sapphire embedded in a titanium disk. The sniffer probe is located at the plasma edge, but the ionization region of the quadrupole mass spectrometer is not in line of sight with the aperture. Hence, the mass spectrometer can only detect the stable gas phase species in the reactor. The first pumping stage consists of the sniffer tube, connected with a larger diameter tube to a chamber, with a 56 l/s turbomolecular pump to reduce the pressure to the low 10^{-3} Torr regime. For the apertures used, and a reactor pressure of 0.5-1.0 Torr, the flow in the thin sniffer tube is in the transition zone between molecular and viscous flow regimes, and the flow is molecular in the larger tube and the chamber. The gas composition in the sampling tube adequately describes the gas composition in the reactor (4). The quadrupole mass spectrometer is a Leybold Inficon CIS300 unit with a 0-300 amu range. It is configured with a partial enclosure around the ionization region, allowing the ionizer to be maintained at a higher pressure (~10^{-4} Torr) than at the filament in the housing manifold

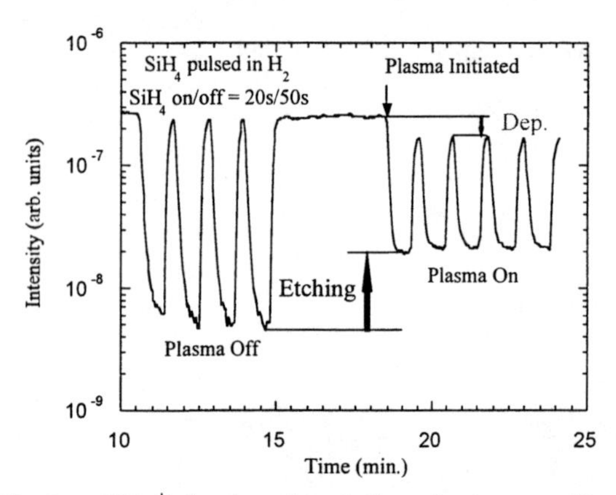

Fig. 2: SiH_2^+ signal vs. time during selective area silicon deposition with modulated gas flow. The time dependence is consistent with the gas residence time in the chamber. Note that when the plasma is initiated, the maximum silane signal is reduced, corresponding to deposition, and the minimum silane signal is enhanced, corresponding t etching (4).

(10^{-9} - 10^{-6} Torr). The signal is enhanced by the electron multiplier set at 1390 volts. The pressure in the manifold is maintained with a 56 l/s (N_2) turbomolecular pump.

Real-time data were collected using the mass spectrometer both before and after plasma initiation. In order to monitor reactant concentrations in real time, process trace data were collected at 30 amu corresponding to SiH_2^+, the principal SiH_4 related signal generated in the ionization region of the mass spectrometer. Higher mass signals in the 59-61 amu range were also monitored to obtain an indication about the robustness and sensitivity of the sensor.

Pulsed gas PECVD of substrate selective microcrystalline silicon is achieved at low temperature by using time modulated flow of silane into a hydrogen plasma. The process involves modulating the flow of 2% SiH_4/He into the reactor during the deposition cycle and into the exhaust, bypassing the reactor, during the H_2 etching cycle. To minimize pressure fluctuations in the chamber, the total flow rate into the reactor was kept invariant by substituting the flow of 2% SiH_4/He with an equivalent amount of He flow, as we pulsed the gases. So when 2% SiH_4/He was flowing into the exhaust an equivalent amount of He was flowing into the reactor and vice versa. The gas pulsing was set at 20-22 seconds 'on' time and the 'off' time was varied between 20-40 seconds, where 'on' and 'off' times refer to the cycle times when 2% SiH_4/He flows into or bypasses the reactor respectively. Selective area deposition experiments were performed using 500 sccm 2% SiH_4 /He (on), 250 sccm H_2 (on and off), and 500 sccm He (off), where 'on' denotes flow rates during the deposition cycle and 'off' denotes the flow rates during the etching cycle. The reactor was operated at 250 °C, 1 Torr, and 5 W rf power. Films were selectively deposited on 1''x3'' Corning 7059 microscope slides with a 1''x1'' area covered with ~100 nm sputtered molybdenum film. Films were also deposited on 0.5''x2'' sections of double side polished c-Si pieces. Film thicknesses were measured using a Tencor Alphastep 5000 profilometer.

RESULTS AND DISCUSSION

Selective deposition allows silicon to be deposited only on receptive areas of a patterned substrate, and be eliminated from other areas. Selective deposition of microcrystalline Si was achieved on Mo and c-Si while no deposition was observed on SiO_2 surfaces. Fig. 2 shows the intensity of the 30 amu silane signal plotted vs. time for both the plasma off and plasma on conditions (4). With the plasma off, the sensor clearly shows the time dependence of the SiH_2^+ signal corresponding to the partial pressure modulation. The asymmetric shape, with a gradual tail when silane flow is off, is due to the gas residence time in the chamber. When the plasma is on, the silane signal in Fig. 2 shows a similar time dependence as in the plasma off condition. However, when the plasma is on, the silane signal during the time that silane flows into the reactor (~19 min. point in Fig. 2) is smaller than measured when the plasma is off. During the time when silane is off and only hydrogen and helium flows into the reactor, the plasma on condition results in a larger silane signal than the plasma off condition. The larger silane signal when the plasma is on results from etching of silicon by atomic hydrogen in the plasma. The real time traces of the SiH_2^+ signal at 30 amu during the etch cycle for three different pulsing and plasma conditions are superimposed on each other as shown in Fig. 3. The plot shows signals for SiH_4 on/off = 22s/25s and 22s/30s with plasma on and 22s/40s with plasma off. The change in slope of the silane signal when the plasma in on for the 22s/25s and the 22s/30s conditions is also pointed out. It is seen that between 290 and 305 seconds in Fig. 3 the silane signal undergoes a distinct change in slope during the plasma on (i.e. etching) condition, while we don't see any slope change with the plasma off (i.e. no etching) condition. The first derivatives of the signals shown in Fig. 3 are plotted against time in Fig. 4. The change in the value of the slope is more prominent in Fig. 4. These signals were observed to be repeatable from run to run. In order to gauge the sensitivity of the mass spectrometer we observed the changes in the 60 amu signal which showed a similar behavior with time at an intensity level 2 orders of magnitude below that of the 30 amu signal. The change in slope of the 30 amu signal as shown in Fig. 3 and Fig. 4 is associated with a primary etch endpoint corresponding to complete removal of stray nuclei from non-receptive surfaces such as glass. The ability of in-situ mass spectroscopic sensing to give quantified deposition and etch rate information can be illustrated by developing a mass balance model for selective deposition.

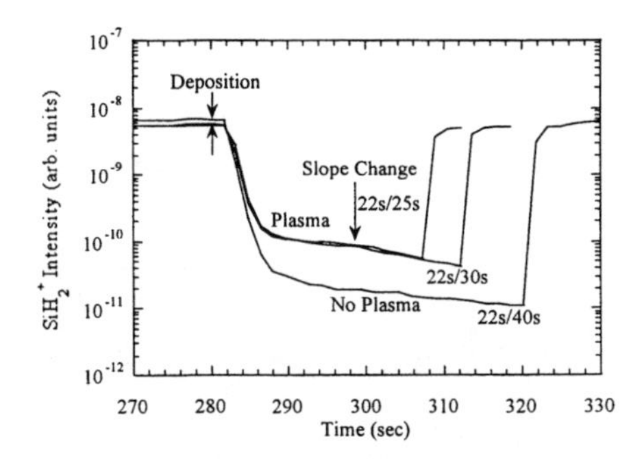

Fig. 3: SiH_2^+ signal intensity vs. time during selective area silicon deposition with SiH_4 on/off = 22s/25s and 22s/30s with plasma on and 22s/40s with plasma off. The slope change shown is indicative of a change in etch rate.

Mass Balance Model for Selective Deposition

Selective area silicon deposition from modulated flow of silane is attained because Si is etched at a lower rate on receptive surfaces, while etching on non-receptive surfaces is much faster. Process conditions resulting in selectivity depend on several factors such as substrate temperature, reactor operating pressure, rf power, SiH_4/H_2 ratio in the feed stream, and the silane modulation time. The modulation time is divided in two parts, t_{on} and t_{off}, where t_{on} is the time when SiH_4 flows into the reactor and t_{off} is the time when SiH_4 bypasses the reactor, flowing to the exhaust. In such an arrangement, t_{on} is the time period when net blanket deposition takes place and t_{off} is the time when selective etching takes place. During the deposition cycle (t_{on}) both silane and hydrogen flow into the reactor. With the plasma on, the intensity of the SiH_2^+ signal at amu 30 is smaller than when the plasma is off (Fig. 2). This difference in signal intensity during deposition is denoted by ΔI_d^{SiH2+}. Conversely, during the etch cycle (t_{off}) when only hydrogen flows into the reactor the intensity of the SiH_2^+ signal at amu 30 is larger when the plasma is on, relative to its value when the plasma is off (Fig. 2). This difference in signal intensity during etch is denoted by ΔI_e^{SiH2+}. In order to model film deposition

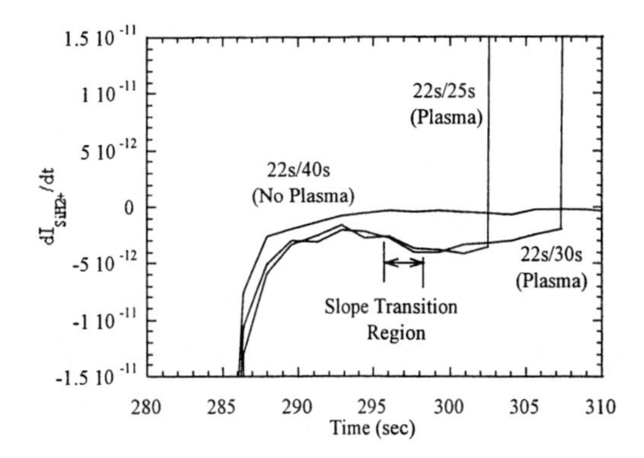

Fig. 4: First derivative of the SiH_2^+ signals shown in Fig. 3. The slope transition region is shown more clearly.

during t_{on} we can write,

$$\Delta I_d^{SiH_2^+} = K_d \cdot A^{Pl} \cdot r_d \qquad (1)$$

Where A^{Pl} is the surface area of the plasma volume, r_d is the blanket deposition rate (nm/min) during the deposition cycle time t_{on}, and K_d is a proportionality factor, which is a lumped parameter consisting of the mass spectrometer's partial pressure response and conversion factors. We can similarly write an expression for the etching process during t_{off} as,

$$\Delta I_e^{SiH_2^+} = K_e(A^{Re} \cdot r_e^{Re} + A^{Nr} \cdot r_e^{Nr}) \qquad (2)$$

Where the A's and the r_e's are the areas and etch rates on the receptive (Re) and non-receptive (Nr) surfaces, and K_e is a proportionality factor similar to K_d. The difference in integrated signal intensities during deposition and etching (denoted by I) is proportional to the amount of material deposited during t_{on} minus the amount of material etched during t_{off}. Based on this, we can write an overall mass balance combining both deposition and etching as follows:

$$I = \int_{t_{on}} \Delta I_d^{SiH_2^+} dt - \int_{t_{off}} \Delta I_e^{SiH_2^+} dt$$

$$= K_d \cdot A^{Pl} \cdot r_d \cdot t_{on}$$

$$- K_e(A^{Re} \cdot r_e^{Re} + A^{Nr} \cdot r_e^{Nr}) \cdot t_{off} \qquad (3)$$

The conditions that need to be satisfied for selective deposition include:

1) All material deposited on the non-receptive surfaces during t_{on} must be removed during t_{off} i.e.,

$$r_d \cdot t_{on} \leq r_e^{Nr} \cdot t_{off} \quad (4)$$

2) Some of the material deposited on the receptive surfaces during t_{on} must remain after the etching cycle t_{off} i.e.,

$$r_d \cdot t_{on} - r_e \cdot t_{off} > 0 \quad (5)$$

The second condition leads to the net mass balance for deposition on receptive surfaces over the total cycle time $(t_{on} + t_{off})$,

$$A^{Re} \cdot r_d \cdot t_{on} - A^{Re} \cdot r_e^{Re} \cdot t_{off} = A^{Re} \cdot r_{net}(t_{on} + t_{off}) \quad (6)$$

Where r_{net} is the net film deposition rate per cycle

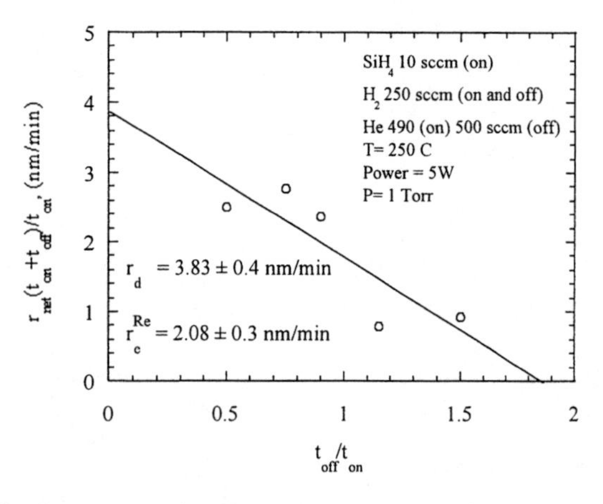

Fig. 5 Linear plot of equation 7. The blanket deposition rate r_d is obtained from intercept and the etch rate on receptive surfaces r_e^{Re} is obtained from the slope.

determined from profilometry. The above equation can be simplified to get

$$r_{net}\left(\frac{t_{on}+t_{off}}{t_{on}}\right) = -r_e^{Re} \cdot \left(\frac{t_{off}}{t_{on}}\right) + r_d \quad (7)$$

The etch rate on receptive surfaces r_e^{Re} and the blanket deposition rate r_d can now be extracted from a linear plot of the equation (7). This is shown in Fig. 3. From Figure 3 we find that the blanket deposition rate is 3.83±0.4 nm/min and the etch rate on receptive surfaces r_e^{Re} is 2.08±0.3 nm/min. The uncertainties are obtained from linear regression analysis. Assuming $K_d \cong K_e = K$ (since the response of the mass spectrometer is almost linear), equation 3 simplifies to give

$$I = \int_{t_{on}} \Delta I_d^{SiH_2^+} dt - \int_{t_{off}} \Delta I_e^{SiH_2^+} dt$$

$$= K \cdot (A^{Pl} \cdot r_d \cdot t_{on} - (A^{Re} \cdot r_e^{Re} + A^{Nr} \cdot r_e^{Nr}) \cdot t_{off}) \quad (9)$$

Which simplifies to

$$\left(\frac{A^{Pl} \cdot r_d \cdot t_{on} - A^{Re} \cdot r_e^{Re} \cdot t_{off}}{A^{Nr} \cdot t_{off}}\right) = \frac{I}{K} \cdot \left(\frac{1}{A^{Nr} \cdot t_{off}}\right) + r_e^{Nr} \quad (10)$$

The etch rate on non-receptive surfaces r_e^{Nr} can be obtained from a linear plot of equation (10) as shown in Fig 6. We find that r_e^{Nr} is 4.5± 0.5 nm/min. Comparing the values of etch rates we see that the etch rate on the

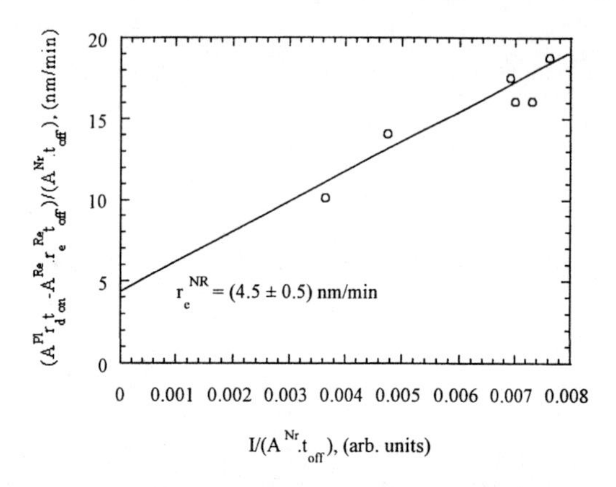

Fig. 6 Linear plot of equation 10. The etch rate on non-receptive surfaces is obtained from the intercept.

non-receptive surfaces is more than twice that of the receptive surfaces. Therefore, the model allows us to obtain quantitative estimates of etch rates on both receptive and non-receptive surfaces. The model can also be used in conjunction with other sensors such as an optical emission spectroscope or a plasma impedance analyzer. The information extracted from a model analysis of the process can be used to optimize the process in order to get maximum attainable growth rates while operating within the selective window. The analysis reveals that the etch rate on non-receptive surfaces can only be obtained if we have real-time process state data such as that collected by an in-situ mass spectroscopic sensor. The usefulness of process models for complex processes depends exclusively on the availability real time sensor data.

CONCLUSIONS

In this article, we have demonstrated that in situ mass spectrometry is a useful real-time process state sensor that can be used to understand and quantify concurrent deposition and etching processes. The sensor is particularly useful when analyzing SiH_4 containing processes due to its acceptable sensitivity and resolution in detecting changes in SiH_4 concentrations. We have used selective area PECVD of silicon as an example process. We have outlined a model that can be used in conjunction with real-time sensor data, such as mass spectroscopy, to quantify deposition and etching rates in selective deposition. We used the model and real-time mass spectroscopic data to obtain the etching rates on selective and non-selective surfaces during selective deposition. Real-time in-situ process sensing such as mass spectroscopy can help us understand processes where multiple chemical steps such as deposition and etching are occurring concurrently. The sensor provides us with some of the necessary tools to optimize the performance of these processes. Some of the problems associated with implementation of in-situ mass spectroscopy are sensor drift and loss of sensitivity due to contamination. Further hardware and software development along with investigation on industrially significant processes are required to make in-situ mass spectroscopy a commercially viable real-time process sensing tool.

ACKNOWLEDGEMENTS

The authors acknowledge support from the NSF Engineering Research Center Program through the Center for Advanced Electronics Materials Processing (Grant CDR 8721505) and the DARPA High Definition Systems Program, Contract No. DABT 63-94-C-0004.

REFERENCES

[1] L.Peters, Solid State Technol. **20 (12)** , 94 (1997).

[2] D.W. Greve, T.J Knight, X. Cheng, B.H. Krogh, M.A. Gibson, and J. LaBrosse, J. Vac. Sci. Technol. B **14** , 489 (1996).

[3] A. I. Chowdhury, T. M. Klein, T. M. Anderson, and G.N. Parsons, accepted for publication in J. Vac. Sci. Technol. B (May/Jun 1998).

[4] A. I. Chowdhury, W. W. Read, G.W. Rubloff, L. L Tedder, and G.N. Parsons, J. Vac. Sci. Technol. B **15**, 127 (1997).

[5] L.L Tedder, G.W. Rubloff, I. Shareef, M. Anderle, D.-H. Kim, and G.N. Parsons, J. Vac. Sci. Technol. B **13** , 1924 (1995).

[6] K.-T. Sung and S. Pang, Jpn. J. Appl. Phys. **33**, 7112 (1994).

[7] J.J. Chambers, K. Min, and G.N. Parsons, submitted for publication in J. Vac. Sci. Technol. B.

[8] L.L Smith, W.W. Read, C. S. Yang, E. Srinivasan, C. H. Courtney, H. H. Lamb and G.N. Parsons, accepted for publication in J. Vac. Sci. Technol. B (May/Jun 1998).

[9] P. Roca i Cabarrocas, N. Layadi, T. Heitz, and B. Drévillon, Appl. Phys. Lett **66**, 3609 (1995).

[10] G. N. Parsons, Appl. Phys. Lett. **59**, 2546 (1991).

[11] G. N. Parsons, IEEE Elec. Dev. Lett. **13**, 80 (1992).

Optical Densitometry Applications For Ion Implantation

John P. Esteves
Michael J. Rendon

Motorola MOS3, Austin TX 78721, USA
Motorola APRDL/Sematech, Austin TX 78741, USA

Ion Implantation, which has provided semiconductor manufacturers with the precise control over dopant concentration and junction profile needed for several device generations, is expected to remain a vital step in semiconductor processing in the future. In-line metrology gauges for ion implantation have had problems measuring the critical low dose range of $1x10^{11}$ to $1x10^{13}$ ions/cm^2. The two current industry standards for ion implant metrology are the four point probe and thermal wave probing. Four point probes have good sensitivity, but without additional processing are only capable of measuring the $1x10^{13}$ to $1x10^{15}$ ions/cm^2 dose range. Thermal wave probing is used for a wide range of doses, but suffers from low sensitivity and high gauge variation. However, recent developments have made available an optical mapping system (OMS) that exhibits very high sensitivity to both implant dose and energy in the critical low dose range of $1x10^{11}$ to $1x10^{13}$ ions/cm^2. Using this optical densitometry gauge, engineering has solved many issues such as dosimetry, across wafer uniformity, and equipment matching that were never as clearly revealed with other metrology systems.

INTRODUCTION

Today's most common forms of implant process monitoring are limited to two methods. However, these methods may not be optimal for monitoring all implant processes. There has long been a need for a tool with the capability to monitor the implant process with the sensitivity and accuracy required to control threshold voltage and other critical implants in the low dose regime of $1x10^{11}$ to $1x10^{13}$ ions/cm^2.

EXISTING METROLOGY METHODS

Resistivity mapping using a four point probe requires post processing of the implanted wafer for activation of the dopant. To ensure proper process control, the annealing equipment as well as the four point probe must be monitored and kept calibrated. Each additional process step adds another source of variation and one more item to cause an unwarranted, process control violation. Since resistivity mapping is a physical contact tool, the probe itself is a source of variation. Throughout the life of the probe, proper conditioning of the probe tips is required to remove contamination that may otherwise cause variation in resistivity measurements. Another disadvantage of the four point probe is its inability to monitor electrically inactive implants in silicon. The four point probe has problems measuring doses lower than the $1x10^{13}$ ions/cm^2 range without further processing or using a double implant technique (1).

Modulated-optical reflectance systems measure a change in the reflectance properties of silicon as a function implant induced damage (2) . In this technique, two lasers are used. The first is an intensity modulated laser to thermally pump the silicon surface and induce thermal waves. The second laser is used to probe the changes in reflectance of the substrate.

Since this is not an electrical measurement, it is capable of monitoring implants that are not electrically active in silicon. Although this method has become widely accepted to monitor implant, it is not without issues. Sources of variation include laser drift, post implant damage relaxation, and batch to batch variation of silicon. These issues cause process control violations that are not related to the implant process. Modulated-optical reflectance and four point probe systems have dose ranges with low sensitivity which may include a dose that must be monitored.

An optimal metrology gauge would be one that is flexible enough to allow implant engineers to customize its sensitivity curve for a particular application. Until such a gauge is available, it will be important for process engineers to understand the metrology tool that they are working with.

OPTICAL DENSITOMETRY

Optical densitometry is not a new technology for

CP449, *Characterization and Metrology for ULSI Technology: 1998 International Conference*
edited by D. G. Seiler, A. C. Diebold, W. M. Bullis, T. J. Shaffner, R. McDonald, and E. J. Walters

monitoring implant processes. In 1983, a gauge using a 390nm light source was used to monitor the darkening effects of implant into photoresist caused by the graphitization of long hydrocarbon chains that result from the breaking of hydrogen carbon bonds (3).

Optical densitometry systems used today to monitor ion implant equipment consist of two main parts, the substrate and a gauge.

The substrate is coated with a thin film that consists of a polymer carrier and an implant sensitive radiochromic dye. During implant, the dye molecule undergoes heterolytic cleavage resulting in a positive ion that has a peak light absorption at 600nm and a small negative ion (4). The film's ability to absorb 600nm light increases as more dye molecules undergo heterolytic cleavage. This effect will continue until saturation occurs, where sensitivity of the film is significantly reduced. There are several process parameters that will change the rate at which the film darkens including ion size, depth of penetration and the number of ions implanted resulting in a sensitivity to implant species, energy, and dose. This process also has the capability to measure implants that are not electrically active, making it an attractive alternative for GaAs manufacturers.

After implant, the substrate is placed in the gauge and used as a filter between the lamp and the sensor. It is rotated and translated past the sensor with the light incident as light intensity variation and position data are collected (Fig. 1). This database can then be used to generate high resolution uniformity maps. Currently the Optical Mapping System (OMS) OMS-3000™ is the only implant optical densitometry system available.

TABLE 1. Typical Dose and Energy Sensitivities

Specie	Dose Sensitivity	Energy Sensitivity
Arsenic	0.71	0.94
Boron	0.85	0.78
Phosphorus	0.78	0.87

OMS CHARACTERISTICS

Implant metrology gauges are of extreme importance in ensuring that the implanters are dosing production wafers with the correct dopant concentrations and energies. Production wafers are critically sensitive to low dose implants ($1x10^{11}$-$1x10^{13}$ ions/cm^2), and must be monitored and controlled within tight process limits.

Unfortunately, the critical low dose implants are the most difficult to measure accurately and precisely. As discussed above, most implant metrology gauges perform poorly at the low dose regime.

The optical densitometry technique provides a capability that did not previously exist in monitoring these very low doses. Its greatest characteristic is its high sensitivity (%Δ in measurement / %Δ in dose) for low dose implants.

Table 1 lists the typical OMS dose and energy sensitivities found for a $1x10^{12}$ ions/cm^2 100keV implant monitor. Sensitivity to beam current was found to be negligible at less than 0.07. Figure 2 demonstrates the sensitivity of the OMS versus a thermal wave measuring technique as compared to actual voltage threshold parametrics.

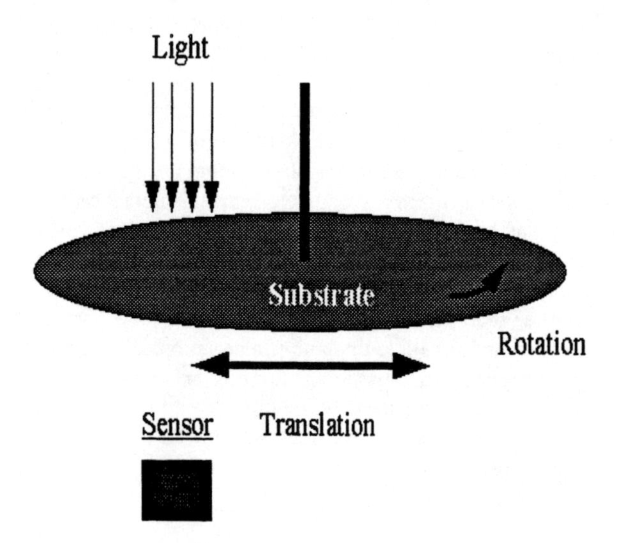

FIGURE 1. A diagram of the OMS-3000™ implant optical densitometry gauge.

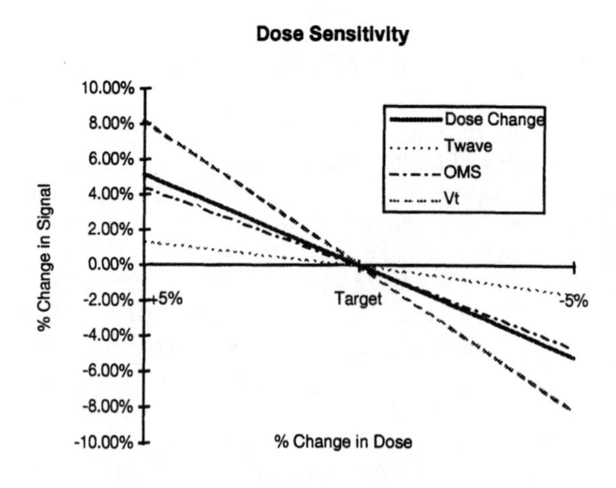

FIGURE 2. A sensitivity comparison of different metrology techniques. The solid line represents the actual change in dose.

Figures 3 and 4 display the OMS mean value per implanted dose and energy for boron and phosphorus. The OMS mean represents the delta density between the pre and post implant scan. The delta correlates directly to the implanted dose and energy.

The stability of the OMS process is its second greatest characteristic. Figure 5 displays the stability of the gauge and the implanted OMS ultra sensitive film for a period of 220 days. With a gauge repeatability and reproducibility of 1.43% and virtually no significant drift, the OMS has demonstrated to be an excellent low dose implant monitor.

Along with improved sensitivity and stability, the OMS has high resolution and is much faster than other metrology techniques. A wafer can be scanned quickly with high resolution using a 1mm radius beam of 600nm light. Table 2 lists the number of measurement sites and time of measurement for several wafer sizes.

OPTICAL DENSITOMETRY DRAWBACKS

The gauge's benefits have far exceeded any of its disadvantages. The single largest drawback is the backing plate. The backing plate is a thin aluminum plate with adhesive on the edges. The plate is adhered to the back of the clear glass monitor wafer after the background or pre implant measurement, but before the implant. The application and removal of the backing plates require operators to physically handle the OMS wafers.

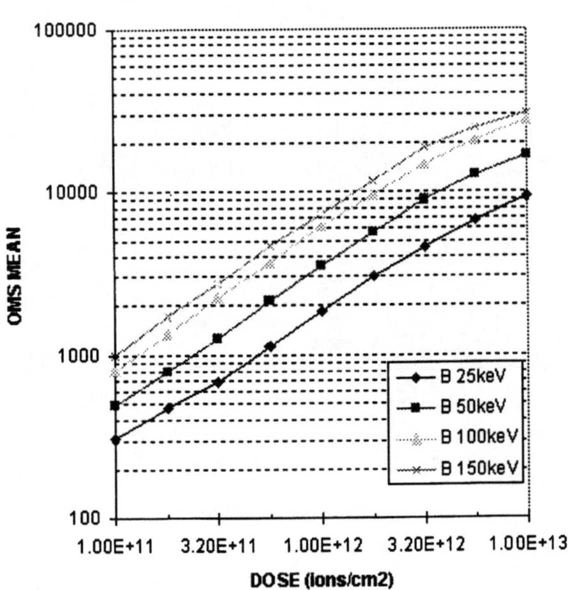

FIGURE 3. Expected OMS counts by boron dose and energy.

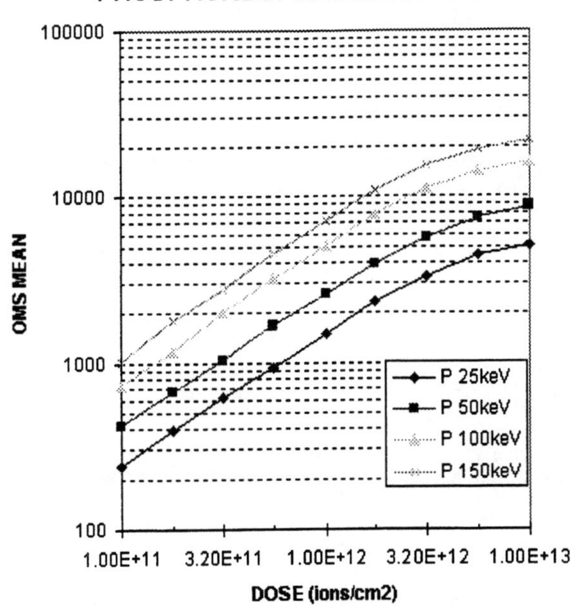

FIGURE 4. Expected OMS counts by phosphorus dose and energy.

The backing plate provides the implanter's optical sensors a surface to sense on the transparent monitor wafer. Trying to load this wafer without a backing plate will generate a loading error on most implant equipment. The plate must be removed after implant for the post implant measurement. These plates have a limited lifetime of approximately 30 uses before they no longer adhere or become badly kinked.

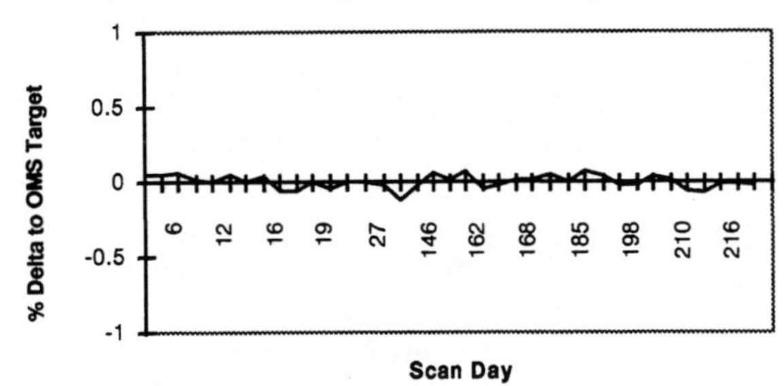

FIGURE 5. A long term trend for implanted ultra sensitive film.

TABLE 2. Measured Sites and Times Per Wafer

Wafer size (mm)	Measured sites	Time (s)
100	6,300	65
125	10,800	70
150	13,500	80
200	27,000	115

A smaller drawback involves the ultra sensitive film which is spun on the current optical densitometry monitor wafers. The current film is highly sensitive to ultra violet (UV) light and must be treated like undeveloped photoresist. Exposure of the monitors to standard fluorescent lights results in false monitor readings. The fab's implant area lights can easily be fitted with inexpensive sleeves to prevent UV light from affecting the monitor wafers.

The final drawback also involves the ultra sensitive film. The application of the film is typically done by the vendor. Used monitors are shipped to the vendor for strip and recoat and then are transported back to the fab. Accurate inventory is required so that new monitors are available and all monitors are accounted for. The film is available in canisters from the vendor for purchase. However, the dedication of equipment and personnel to maintaining the film strip and recoat process may not be cost effective.

PROBLEMS RESOLVED USING OMS

There are many implant equipment issues which occur on a daily basis in high volume wafer fabs. Most of them are easily identified by experienced engineers or technicians and are resolved quickly. However, the identification of implant problems such as poor uniformity, incorrect dosage, and equipment matching relies exclusively on the implant metrology gauge. The high sensitivity and resolution of the optical densitometry gauge allows for a quick and accurate identification of the problem. Several examples of using OMS in this manner are discussed in the remainder of this section.

Upon measuring an implanted OMS monitor wafer, an operator notified engineering of the bizarre pattern displayed by the gauge (Fig. 6). After seeing the high resolution display, the implant engineer easily concluded that one of the ion beam scan amplifiers had become faulty. The maintenance technician who repaired the implanter reported that a loose scan amp connector was the cause of the problem.

Another serial implanter with a history of intermittent microstriping had to be shutdown for immediate maintenance whenever the microstriping occurred. The severity of the microstriping could not be determined with the other metrology gauges being used at that time. Today, the OMS has had a chance to reveal the intermittent microstriping problem (Fig.7). Due to the sensitivity and resolution of the OMS gauge, engineering has the opportunity to precisely determine the severity of the microstriping.

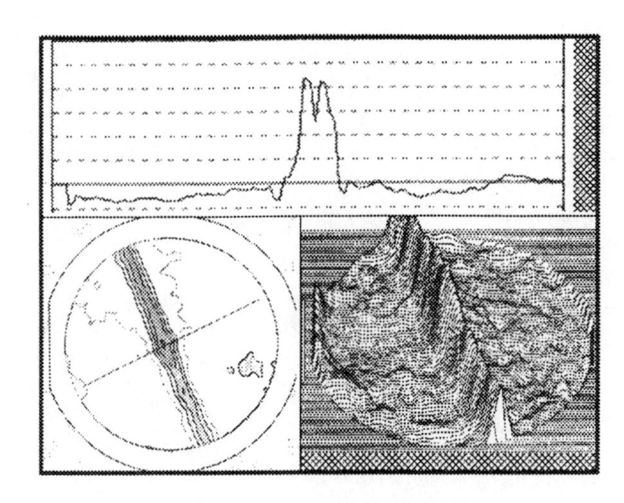

FIGURE 6. Three different OMS-3000™ displays representing the same improperly scanned implant.

Engineering can now assess the problem and decide whether to shut the implanter down immediately or wait for the next scheduled preventative maintenance (PM).

Another fab had just finished an intensive refurbishing of one of its implanters. They had finally run their dose monitors with successful results and were about to run production split lots in order to qualify the implanter as production worthy.

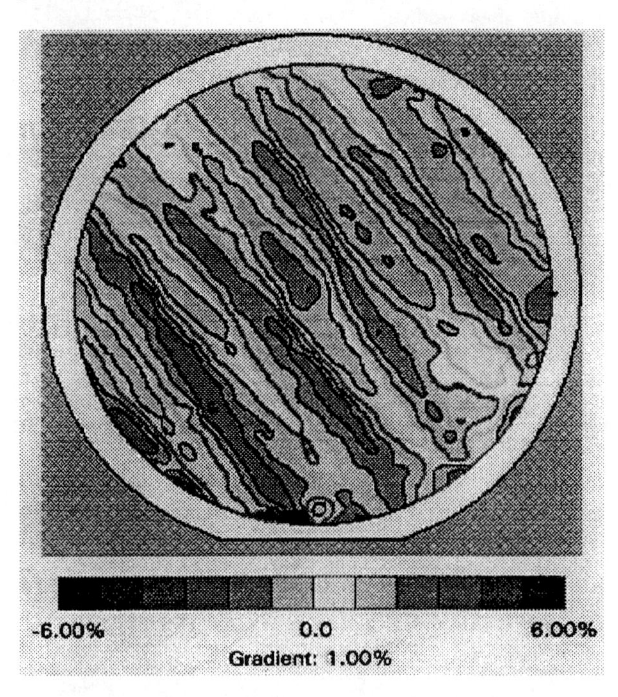

FIGURE 7. Evidence of microstriping as shown by the OMS-3000™ optical densitometry technique.

Before running the splits, they decided to use the OMS as a final monitor check. The OMS monitors demonstrated a surprising 7.4% overdose as compared to their control implanter. Their metrology gauges were not set up to monitor the $1x10^{12}$ ions/cm^2 dose due to their insensitivity at that range. Based on the OMS results, the splits were postponed until the implanter was properly matched. It turned out that the flood ring was not replaced correctly, and there was a dose linearity problem across dosing decades.

Along with the issues discussed above many other implant problems have been revealed using an optical densitometry gauge. Operator error is one of them. The gauge has served as an excellent means of demonstrating the operator's manual tuning ability. Operators can now witness in high resolution the effects of a poorly tuned ion beam. The gauge has served not only as a metrology tool, but as a learning instrument as well.

CONCLUSION

The OMS has demonstrated that optical densitometry is an excellent implant monitoring method for doses in the range of $1x10^{11}$ to $1x10^{13}$ ions/cm^2. Advantages over other techniques include sensitivity, stability after implant, resolution, and speed.

Optical densitometry applications are now just beginning to emerge. Evaluations of thicker films need to be conducted so that monitoring of MeV energy implants will be possible. With additional development, the film could be optimized for use on high current implanters to monitor doses at source/drain levels, providing excellent high resolution maps that are capable of detecting minute levels of microstriping (5). With higher penetration of implanted ions into this film as compared to silicon, ultra shallow implants could also be easier to monitor using this technique.

ACKNOWLEDGMENTS

The authors would like to thank Bill Burton, Jim McMillen, Tom Gruhn, Chad White, Paul Rodriguez, David Dyer and Irma Perez for their contributions and assistance.

REFERENCES

1. Omnimap Technical Brief #07 March 1987.

2. Yarling, C., Keenan, W., and Larson, L., "The History of Uniformity Mapping in Ion Implantation," Presented at the IEEE Ion Implantation Technology Conference, Surrey, UK 1990.

3. Chang, J., and Tripp, G., *Solid State Technology*, **Vol 26 11**, 143-152, 1983.

4. McMillen, J., Current, M., and Yarling, C., *Materials and Process Characterization of Ion Implantation*, Austin, TX 1997, pp. 58-73.

5. Heden, C., Heckman, V., Weisenberger, W., and McMillen, J., "Micro Uniformity in MeV Implantation," Presented at the IEEE Ion Implantation Technology Conference, Austin, TX, pp. 226-227, June 16-21, 1996.

INTERCONNECT

Dimensional Metrology Challenges for ULSI Interconnects

R. Havemann, H. Marchman, G. Dixit, M. Jain, E. Zielinski, A. Ralston, Y. Hsu, C. Jin, A. Singh and J. Schlesinger

Silicon Technology Development
Texas Instruments Incorporated, 13570 North Central Expressway, Dallas, Texas 75243

Manufacturing of Ultra-Large-Scale-Integration (ULSI) semiconductor devices will likely require feature sizes as small as 150nm by the year 2001, with an overlay accuracy of 55nm, a critical dimension (CD) control of 12nm and a precision of 2nm. While this level of overlay control is currently achievable, the forecast CD control/precision present formidable metrology challenges for multilevel interconnects due to the high aspect ratios of scaled features as well as the introduction of new materials such as low-k dielectrics and copper. New techniques for edge detection, profile definition and film thickness measurement will be needed to meet future metrology needs.

INTRODUCTION

The era of Ultra-Large-Scale-Integration (ULSI) of semiconductor devices has spurred an ever-increasing level of functional integration on-chip, driving a need for greater circuit density and higher performance. The demand for improved functional density and transistor performance has led to an approximately 70% dimensional scaling for each successive generation of integrated circuits. In addition, more interconnect levels are required to service the transistors and optimize speed and power distribution paths, escalating both process complexity and cost.

While scaling has served to reduce the device parasitics which plague transistor performance, it has caused a dramatic increase in interconnect RC delay due to narrower leads (increasing resistance, R) and tighter lead spacing (increasing capacitance, C) as shown in Figure 1 (1).

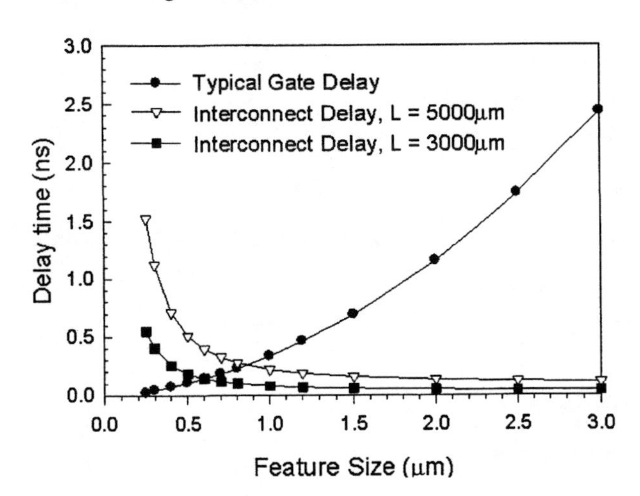

FIGURE 1. Gate and interconnect delay vs. feature size. Interconnect delay is shown for repeater spacing (L) of 3000μm and 5000μm.

In response to the interconnect scaling crisis, new materials (such as copper and low-k dielectrics) and processes (such as damascene) are being implemented to reduce interconnect parasitics. A successful strategy for implementation of scaled interconnects will be critically dependent upon concomitant advances in metrology, since the dimensional tolerances of thin films and patterned features play a dominant role in determining both performance and reliability of the interconnect system, while particle detection is a crucial part of yield improvement.

In contrast with older technology generations where interconnect feature sizes tended to lag far behind the minimum feature of transistor gate length, ULSI interconnects will strongly leverage lithographic pattern capability in order to maximize interconnect density. In addition, tighter tolerances will be required to control circuit RC delay as interconnect linewidth and spacing shrink. Interconnect linewidth control is particularly important for critical speed paths in high performance circuits, since designs must compensate, at the cost of circuit density, for any process marginality to insure performance goals are met.

Current and projected interconnect requirements for logic devices are summarized in Table 1 from the 1997 SIA National Technology Roadmap for Semiconductors (NTRS) (2). According to the NTRS roadmap, lithographic feature sizes as small as 150nm will be required in manufacturing by the year 2001, with an overlay accuracy of 55nm, a critical dimension (CD) control of 12nm and a precision of 2nm. While this level of overlay control is currently achievable, the forecast CD control/precision present formidable metrology challenges for multilevel interconnects due to the high aspect ratios of scaled features as well as the introduction of new materials and architectures.

CP449, *Characterization and Metrology for ULSI Technology: 1998 International Conference*
edited by D. G. Seiler, A. C. Diebold, W. M. Bullis, T. J. Shaffner, R. McDonald, and E. J. Walters

TABLE 1. Key Parameters from NTRS Roadmap for Interconnects (Logic).

SIA Roadmap for Logic

Year of First Product Shipment Technology Generation	1997 250 nm	1999 180 nm	2001 150 nm	2003 130 nm	2006 100 nm
Number of metal levels	6	6-7	7	7	7-8
Minimum contacted/non-contacted interconnect pitch	640/590	460/420	400/360	340/300	260/240
Minimum metal CD (nm)	250	180	150	130	100
Minimum contact/via CD (nm)	280/360	200/260	170/210	140/180	110/140
Metal height/width aspect ratio	1.8	1.8*	2.0*	2.1*	2.4*
Via aspect ratio	2.2	2.2*	2.4*	2.5*	2.7*
Metal effective resistivity (μΩ-cm)	3.3	2.2	2.2	2.2	2.2
Barrier/cladding thickness (nm)	100	23	20	16	11
Interlevel metal insulator - effective dielectric constant (K)	3.0-4.1	2.5-3.0	2.0-2.5	1.5-2.0	1.5-2.0

**Metal and via aspect ratios are additive for dual-Damascene process flow*

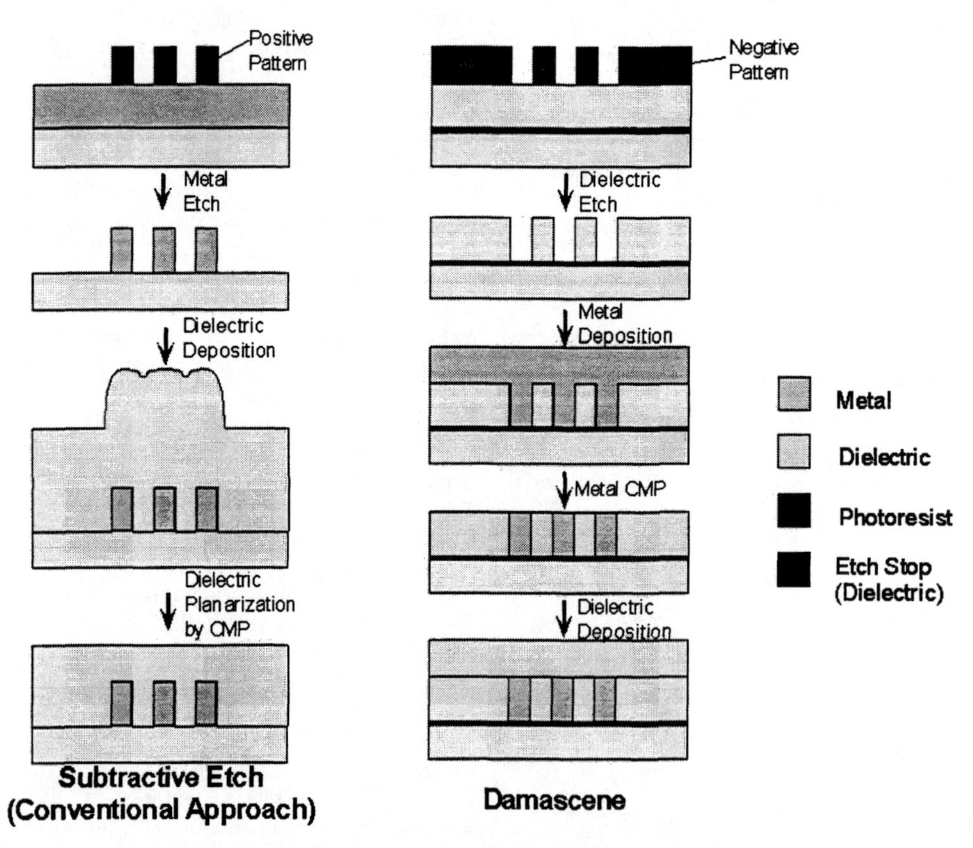

FIGURE 2. Interconnect fabrication options.

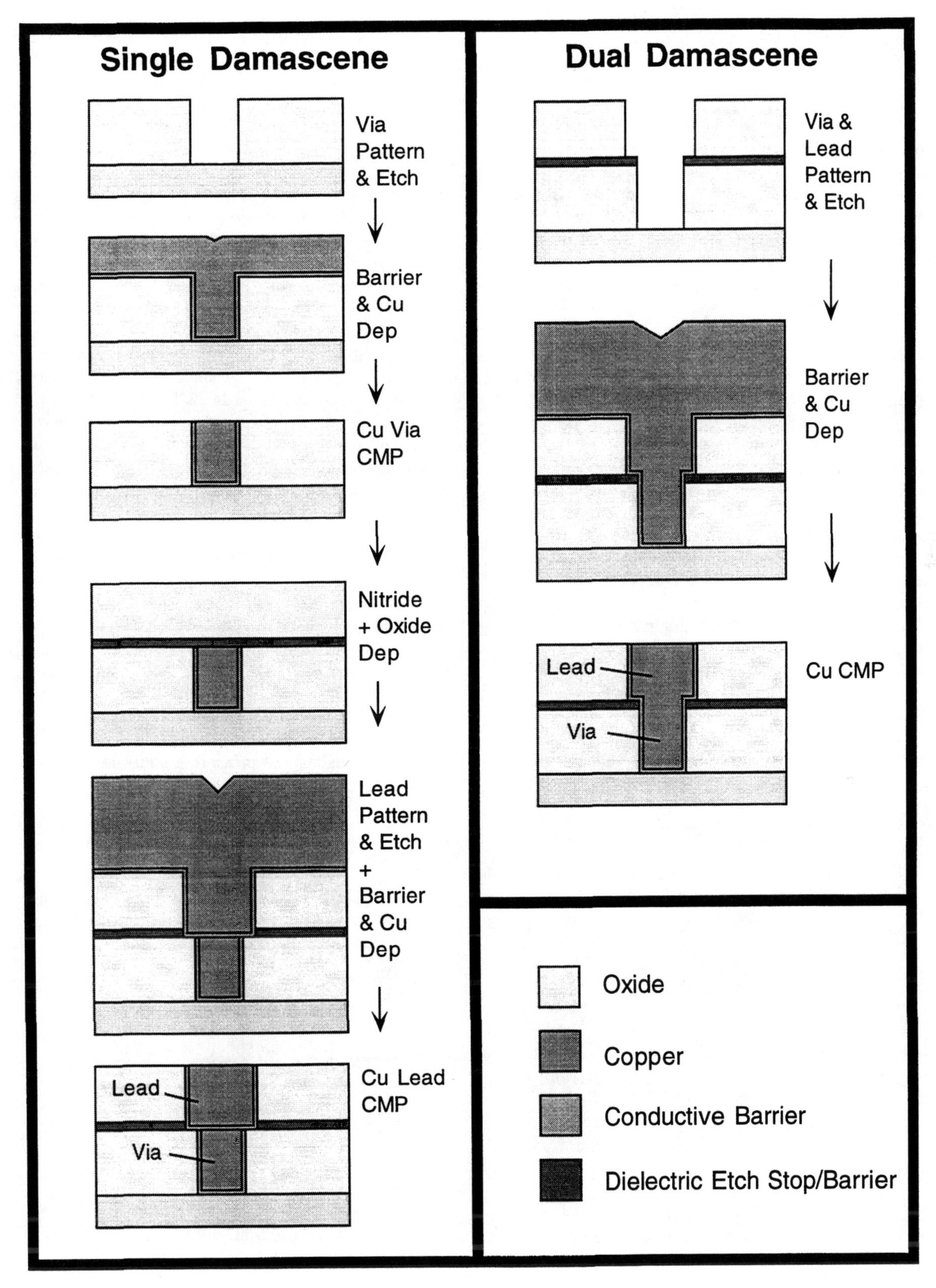

FIGURE 3. Copper damascene process flow options.

INTERCONNECT ARCHITECTURES AND FABRICATION OPTIONS

Fundamentally there are two interconnect fabrication options (Figure 2): subtractive etch of the metal or the damascene approach. The subtractive etch method has been traditionally used with aluminum metallization, and employs pattern/etch of the metal to form the conductors followed by dielectric deposition/planarization for electrical isolation. In the alternative damascene approach, holes and/or grooves are formed in the dielectric and then filled with metal, which is subsequently planarized by chemical mechanical polish (CMP).

State-of-the-art manufacturing processes today typically use a combination of damascene and subtractive etch, whereby different layers of etched aluminum conductors are connected through tungsten-filled vias planarized by the CMP damascene process. Future interconnects utilizing copper metallization will likely be fabricated strictly by damascene methods due to the difficulty in dry etching copper and the density-driven need for borderless vias. A dual damascene structure (shown in Figure 3) is preferred from a cost perspective but drives higher aspect ratio features and multilayer dielectrics which challenge both metrology and process.

Yet another trend is the growing need to integrate low-k dielectrics into the interconnect structure in order to reduce capacitance. Multilayer dielectrics confound film thickness measurements due to differences in density and absorption of each layer. Etch stop and barrier layers needed for the damascene process (Figure 3) further complicate the dielectric stack. Typical interconnect architectures for insertion of low-k dielectrics are shown in Figure 4. These include use of a homogeneous low-k dielectric and the so-called embedded approach, where the low-k dielectric is inserted primarily between the metal leads. As shown in Figure 5, sidewall capacitance tends to dominate overall capacitance as interconnects scale, so it is important to maintain low permittivity between the metal leads. Figure 6 illustrates the relative performance impact of low-k dielectrics for both the embedded and homogeneous cases; note most of the performance gain is due to the use of low-k between the metal leads.

While use of a homogeneous dielectric may simplify the process, the chief advantage of the embedded approach is the flexibility of choosing two different dielectrics. Thus, the dielectric system can be tailored for optimum performance by using the lowest k dielectric between metal leads for reduced sidewall capacitance while utilizing a different dielectric between metal levels for improved mechanical strength, heat transfer or ease of process integration.

Several examples of the embedded approach are highlighted in Figures 7, 8, 9 and 10. Figure 7 shows a five level aluminum metallization system with an embedded porous oxide (hydrogen silsesquioxane, k~3) that has been used in production (3). Figures 8 and 9 respectively show experimental aluminum metallization systems with embedded ultra-low-k dielectrics which include organics (AF4 or parylene-F, k~2.3) (4) and nanoporous oxide (xerogel, k~1.8) (5). In Figure 10, xerogel is integrated into a copper damascene architecture (6). These example embedded multilayer dielectric structures underscore the metrology challenge of measuring both film thickness and feature size over a wide range of materials.

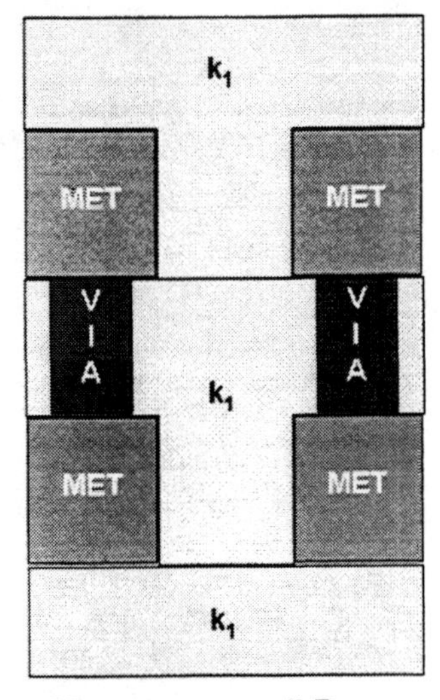

Homogeneous ILD

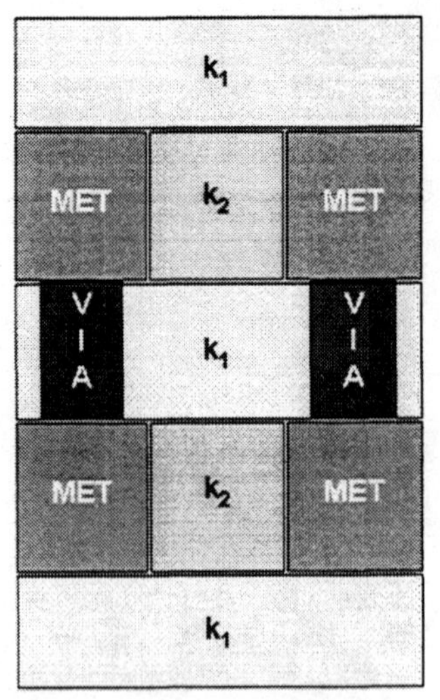

Embedded Low-k ILD
(Typically, $k_1 > k_2$)

FIGURE 4. Typical interconnect architectures for low-k options.

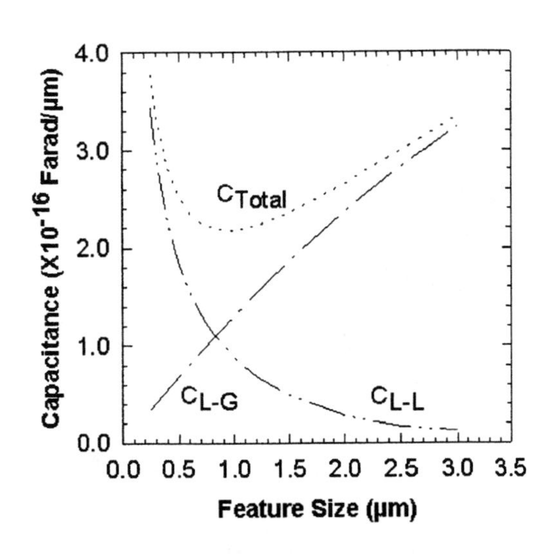

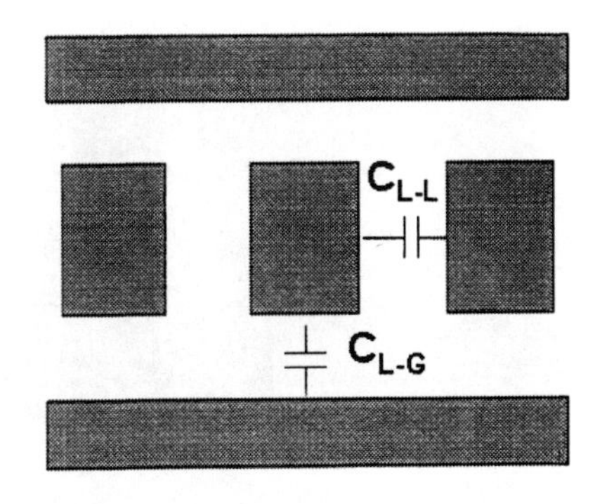

Line-to-line Capacitance = C_{L-L}
Line-to-ground Capacitance = C_{L-G}

FIGURE 5. Interconnect capacitance as a function of feature size.

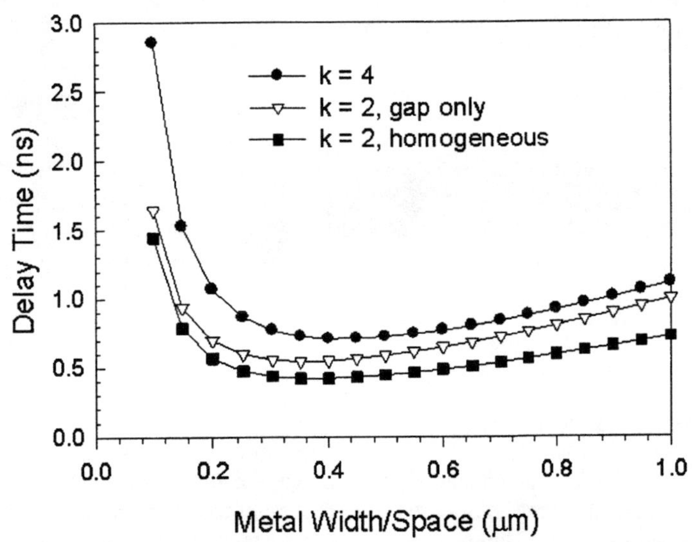

FIGURE 6. Interconnect delay as a function of feature size and dielectric.

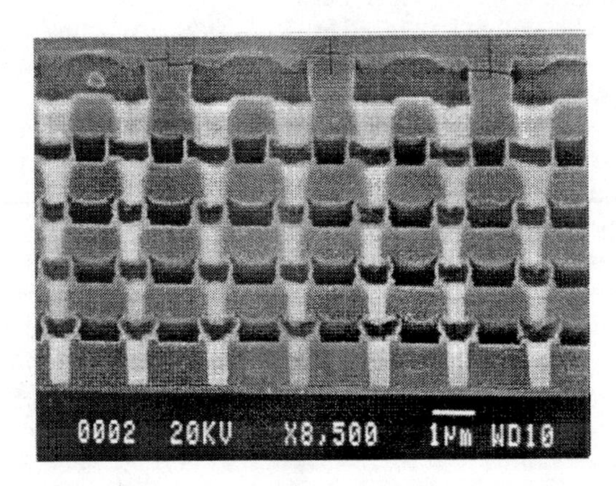

FIGURE 7. Example multilevel metallization with integrated low-k porous oxide dielectric (HSQ, k~3).

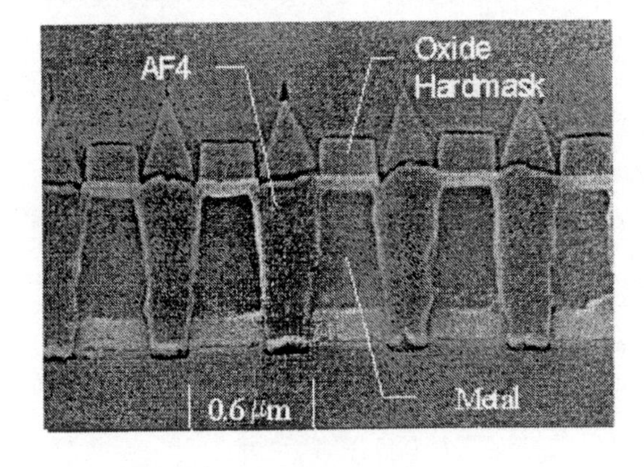

FIGURE 8. Example of aluminum metallization with integrated polymer dielectric (AF4, k~2.3).

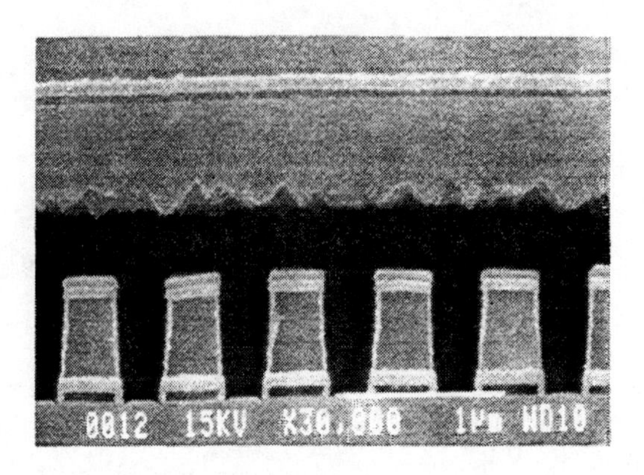

FIGURE 9. Example of aluminum metallization with integrated ultra-low-k dielectric (xerogel, k~1.8).

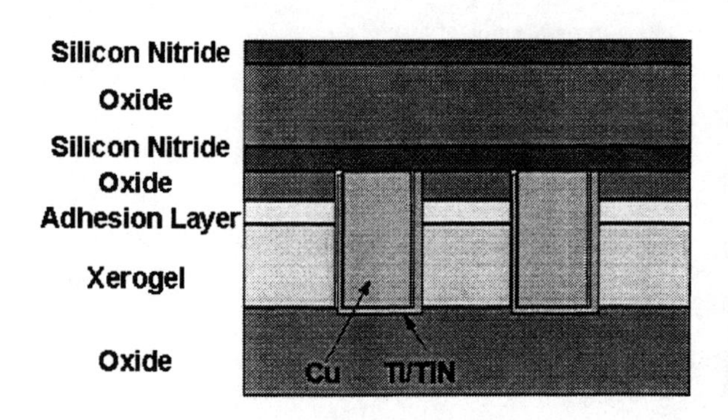

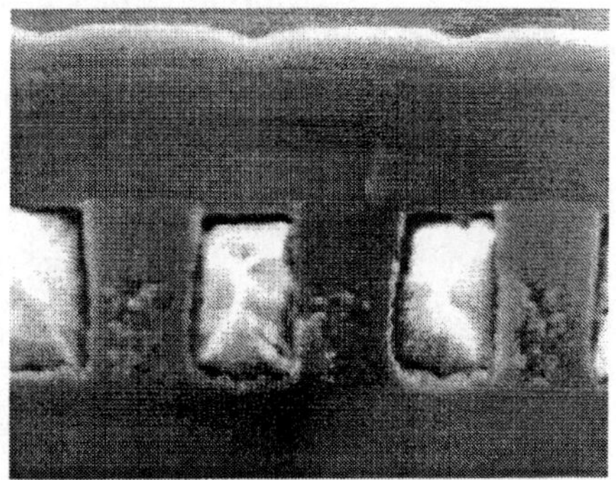

FIGURE 10. Example integration of xerogel into a copper damascene structure.

INTERCONNECT METROLOGY ISSUES

Edge and profile definition is the fundamental root of all CD metrology issues, including calibration, metrology tool matching, precision, etc. The process-dependent three dimensional geometric variation seen in typical interconnect structures (Figure 11) confounds correlation and calibration procedures. While SEM measurement tools are available to provide images of backscattered electrons from submicrometer geometries, the image intensity does not directly correspond to feature topography, nor does the peak intensity occur exactly at the feature edge (7). Thus, extracting the actual edge position from the detected signal presents a formidable challenge.

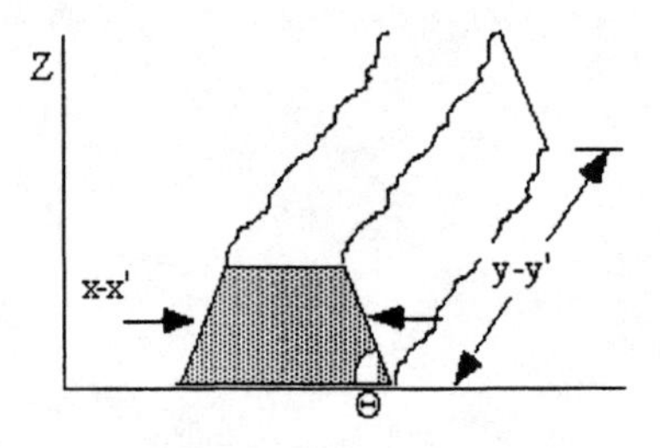

FIGURE 11. Illustration of three dimensional variation in interconnect features.

Several measurement algorithms (see Figure 12) have been used to approximate the intensity transition with a function such as a tangent line (linear approximation), nonlinear step (Fermi) function or a threshold in the first derivative of the detected intensity signal. Unfortunately the extracted linewidth can vary between algorithms depending on geometric shape, material and pattern density as shown in Figure 13, and variations of as much as 100nm have been noted (8). Therefore, other methods such as AFM calibration must be used to correlate actual feature size and shape with SEM measurement in order to choose the best algorithm for a particular combination of substrate and geometric configuration.

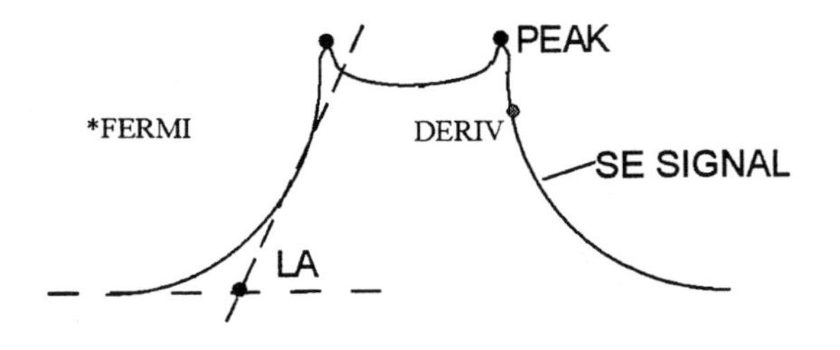

FIGURE 12. Linewidth measurement algorithms for secondary electron signals.

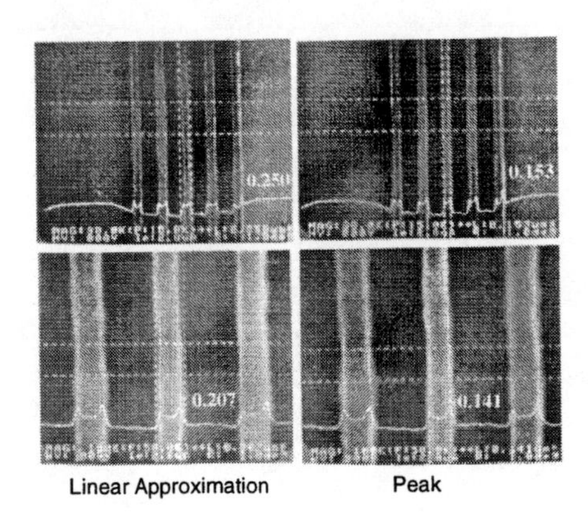

FIGURE 13. Example of algorithm-dependent linewidth measurement for dense and isolated features.

The AFM contour plot shown in Figure 14 illustrates how feature profiles can be mapped for correlation to SEM signal intensity. While AFM provides a useful means of metrology calibration, its use in real-time applications is limited by low throughput. An additional looming limitation as devices continue to scale is the AFM probe tip size and length. For feature sizes below 180nm and aspect ratios above 5:1, as shown in Figure 15, new approaches such as ion microprobes are needed to provide correlation and reference for future SEM metrology.

CONCLUSIONS

The performance- and density-driven scaling of feature size, film thickness and alignment tolerance for Ultra-Large-Scale-Integration of semiconductor devices presents difficult metrology challenges. Metrology requirements for multilevel interconnects are particularly problematic due to high aspect ratio features and the introduction of new multilayer dielectric materials.

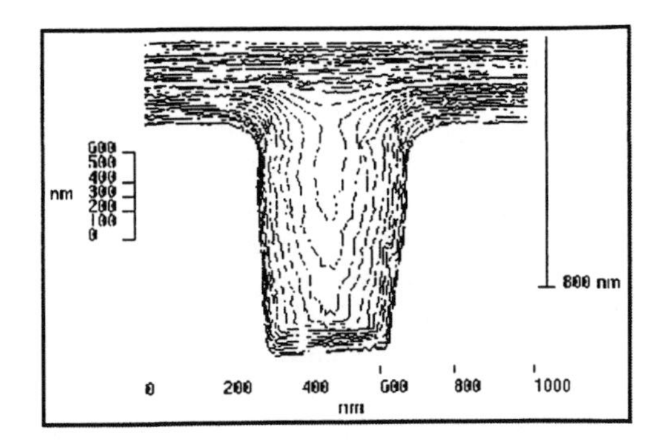

FIGURE 14. AFM profiles for interconnect via hole.

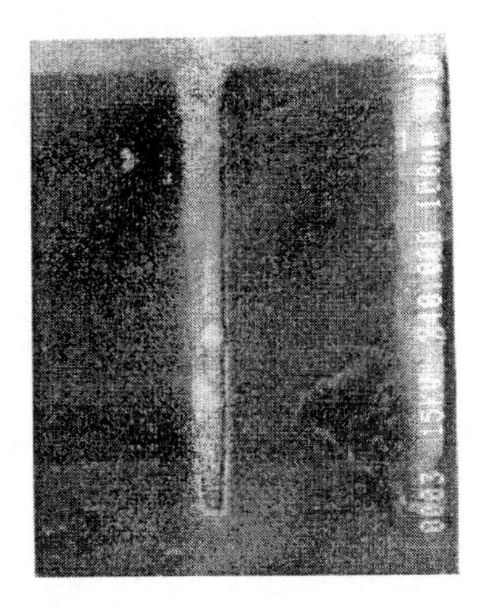

FIGURE 15. Example tungsten-filled contact with high aspect ratio (~10:1).

Correlation of SEM measurement with actual feature size/profile is critical for control of device performance and reliability. Current profile measurement techniques such as AFM provide valuable reference tools which meet today's needs. As scaling continues, new techniques for edge detection, profile definition and film thickness measurement will be needed to meet future metrology needs.

ACKNOWLEDGMENTS

The technology examples discussed in this paper involved the efforts of numerous people in the Silicon Technology Development group of Texas Instruments, and their help is gratefully acknowledged. The authors would like to especially thank the Kilby Fab personnel for providing process and metrology support.

REFERENCES

1. S.P. Jeng, M-C. Chang, R.H. Havemann, Adv. Met. for Devices and Circuits MRS Symp. Proc. (1994), pp. 25-31.
2. Semiconductor Industry Association, 1997 National Technology Roadmap for Semiconductors.
3. G. Dixit, M. K. Jain, S. P. Jeng, P. S. McAnally, W. L. Krisa, C. M. Garza, Ken Brennan, Ajit Paranjpe, S. Nag, and R. Havemann, Proc. on Advanced Metallization and Interconnect Systems for ULSI Applications in 1995. (Materials Research Society, Montreal, 1995), pp. 53 -58.
4. A. Ralston, J. Gaynor, A. Singh, L. Le, R. Havemann, M. Plano, T. Cleary, J. Wing and J. Kelly, VLSI Tech. Symp. Dig. (1997), pp. 81-82.
4. R. List, C. Jin, S. Russell, S. Yamanaka, L. Olsen, L. Le, L. Ting and R. Havemann, VLSI Tech. Symp. Dig. (1997), pp. 77-78.
6. E. Zielinski, S. Russell, R. List, A. Wilson, C. Jin, K. Newton, J. Lu, T. Hurd, W. Hsu, V. Cordasco, M. Gopikanth, V. Korthuis, W. Lee, G. Cerny, N. Russell, P. Smith, S. O'Brien and R. Havemann, Proc. IEEE IEDM (1997), pp. 936-938.
7. H. Marchman, J. Griffith, J. Guo and C. Celler, J.Vac.Sci.Technol. B 12(6), 1994, p. 3585.
8. ibid.

Picosecond Ultrasonics: A New Approach for Control of Thin Metal Processes

Robert J. Stoner[a], Christopher J. Morath[b], Guray Tas[b], George Antonelli[a], Humphrey J. Maris[a]

[a]*Division of Physics and Engineering, Brown University, Providence RI 02912*
[b]*Rudolph Technologies, Flanders NJ 07836*

The Picosecond Ultrasonics technique is described. The application of the technique to modern ULSI process metrology and characterization is illustrated with a number of examples. Specific applications discussed include measurement of thickness and density of PVD TiN and CVD WN_x, CVD W edge thickness profiles, 5 metal layers with in a single interconnect stack, ultrathin Ti, Ti/TiN adhesion stacks, as well as PVD and ECD copper.

INTRODUCTION

Metal processes have taken on steadily increasing importance with each succeeding LSI process generation. They have also increased greatly in complexity; it is now the exception for a metal process to comprise just a single film deposition. The great majority of modern metal processes now include at least two depositions: many interconnect processes call for as many as six sequential metal depositions within a single vacuum system. In such multilayer processes individual sublayers may play several roles, serving as conductors, adhesion promoters, nucleation or wetting surfaces, diffusion barriers, getters or as sources of material for chemical reactions. The electrical functionality and reliability of an integrated circuit depends ultimately on the chemical and mechanical properties of the dozens of metal layers deposited in the overall manufacturing process, and also on the interactions which take place at the interfaces between them. This places a considerable burden on process monitoring techniques since they must have the capability to measure the thicknesses of several metal films simultaneously, as well as other film and interface properties such as density, roughness and uniformity and adhesion.

With few exceptions[1] optical techniques are not used for monitoring processes using metal films for the obvious reason that such films are usually opaque. However, even for thin (i.e. semitransparent) metal films the strong dependence of the optical constants on surface conditions usually makes routine ellipsometry and reflectometry impracticable. In the absence of an alternative, metal film thicknesses are presently almost universally estimated from sheet resistance measurements made using a four-point probe. Since such measurements involve touching the wafer on which the film is deposited, it is necessary to make them on sacrificial witness wafers (rather than on the patterned wafers on which the chips are created). This is costly and inefficient compared to optical measurements used for transparent films which, because they can be carried out without contact using focussed light beams within microscopic areas, can usually be made directly on patterned wafers. Film thicknesses estimated from sheet resistance are also susceptible to errors associated with geometrical factors (for example, close to the edge of a wafer), fluctuations in film composition, and boundary scattering (which plays a dominant role for many films thinner than a few hundred Angstroms). The impact of these errors has grown in significance with the increasing use of CVD, CMP and ultrathin metal layers.

Recently x ray measurements (XRF and XRR) have been used with some success to overcome the inherent failings of sheet resistance as a means of thickness measurement. In this paper we describe an optical characterization technique for metals known as Picosecond Ultrasonics. This technique can be used to measure the metal film thicknesses ranging from tens of Angstroms to several microns. It can be applied to films deposited singly or within multilayers. Along with thickness, Picosecond Ultrasonics may be used under some conditions to measure intrinsic material properties such as density. We show how Picosecond Ultrasonics can be applied in a factory setting to make routine film measurements.

CP449, *Characterization and Metrology for ULSI Technology: 1998 International Conference*
edited by D. G. Seiler, A. C. Diebold, W. M. Bullis, T. J. Shaffner, R. McDonald, and E. J. Walters

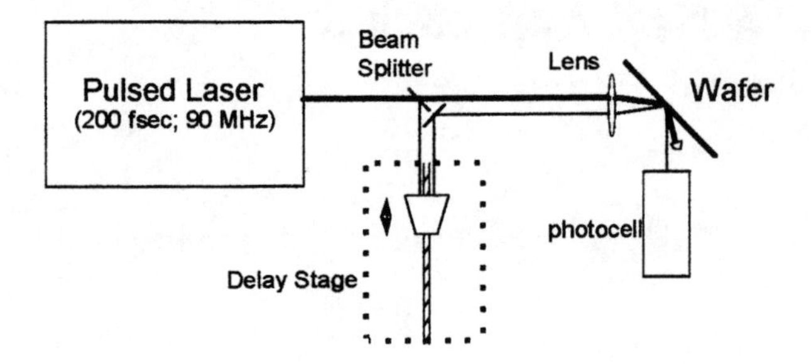

Figure 1. Optical schematic for picosecond ultrasonic measurement system.

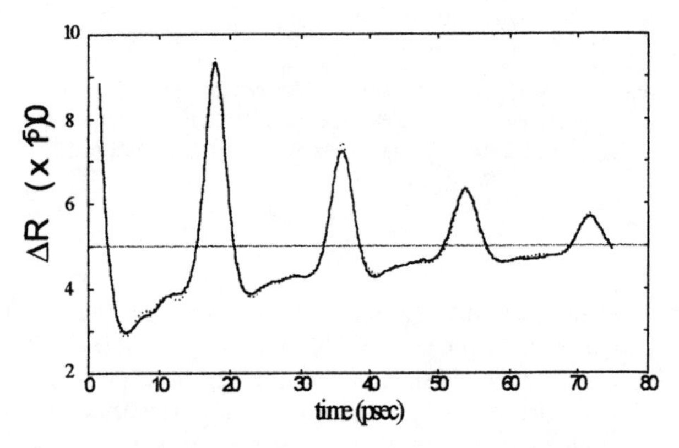

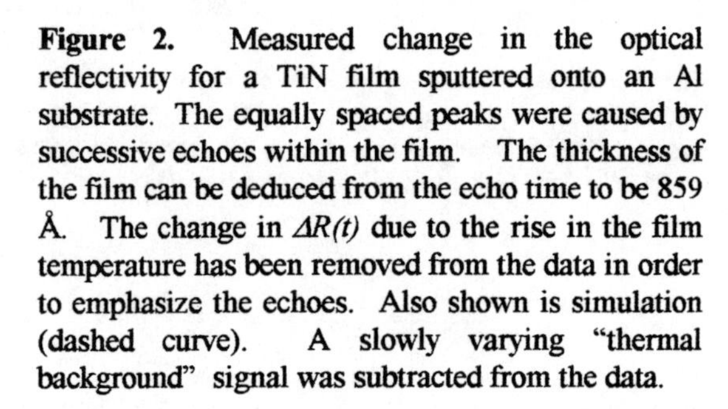

Figure 2. Measured change in the optical reflectivity for a TiN film sputtered onto an Al substrate. The equally spaced peaks were caused by successive echoes within the film. The thickness of the film can be deduced from the echo time to be 859 Å. The change in $\Delta R(t)$ due to the rise in the film temperature has been removed from the data in order to emphasize the echoes. Also shown is simulation (dashed curve). A slowly varying "thermal background" signal was subtracted from the data.

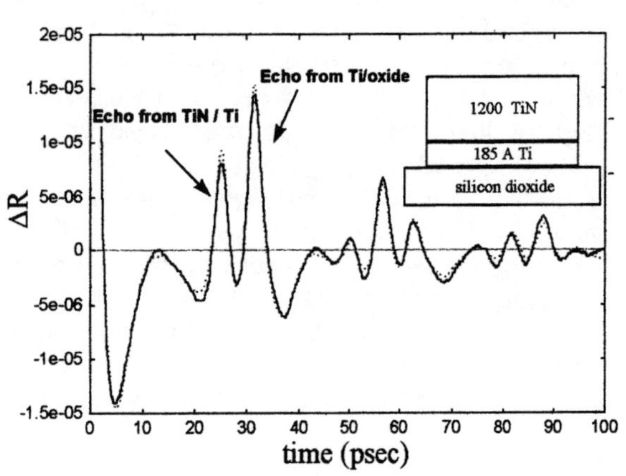

Figure 3. Measured change in the optical reflectivity for a TiN film sputtered on top of a thin sputtered Ti film on a thick film of thermal SiO_2. The echo pattern is a result of sound propagating within both metal layers. This pattern can be directly compared to the pattern of Fig. 1 obtained for a sample having a similar TiN thickness. Also shown is simulation (dashed curve).

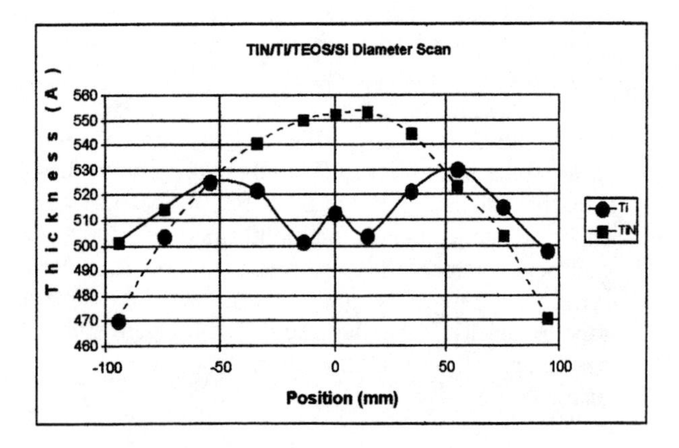

Figure 4. Thickness of TiN and Ti layers deposited in a cluster tool without a vacuum break. The measurements were made using Picosecond Ultrasonics along a diameter with position -100 corresponding to the notch position.

PICOSECOND ULTRASONICS

In a picosecond ultrasonic measurement, a 100 fsec laser pulse (the *pump*) is focussed onto a small spot (less than 10 microns in diameter) on the surface of a metal film. This pulse is partially absorbed within a volume close to the surface of the film, and the deposited energy gives rise to a sudden increase in the temperature within this volume of a few degrees (based on a typical pulse energy of 0.1 to 1 nJ). Rapid thermal expansion within this heated volume launches a sound pulse from the surface toward the interior of the film which travels at the velocity of sound for the material of which the film is composed. For a simple film deposited on a substrate this sound pulse travels until it reaches the substrate, whereupon it is partially reflected. The amplitude of the reflected pulse is proportional to the acoustic reflection coefficient r, where

$$r = \frac{Z_{substrate} - Z_{film}}{Z_{substrate} + Z_{film}}. \quad (1)$$

In this expression Z is the acoustic impedance which is defined to be the product of the density and sound velocity (which in general have different values for the film and substrate). The sound pulse returning to the surface of the film causes a change in the optical reflectivity ΔR of the film of order of 1 part in 10^5, depending on the film material, and on the pulse energy. This change is detected by means of a second laser pulse (the *probe*) focussed onto the same spot on the film as the pump. The two beams are derived from the same laser by means of a beam splitter. The arrival of this pulse can be controllably delayed by a time t relative to the pump pulse by adjusting the length of its path through the optical system using a mechanical stage. The reflected part of the probe beam is directed onto a photodetector, and its intensity is recorded as a function of pump-probe delay time over a suitable time range (less than 1000 psec for most practical film thicknesses). The arrangement of the optics in the system is illustrated in Fig. 1. The result of a measurement of the reflectivity change versus time $\Delta R(t)$ for a TiN film sputtered onto aluminum is shown in Fig. 2 (a slowly varying "thermal background" component has been subtracted from the data). For the example shown in Fig. 2 the film thickness can be determined from the product of one half of echo time (approximately 18.1 psec) and the sound velocity in the film (95 Å/psec) which gives 859 Å. To obtain a precise thickness it is usually preferable to carry out a simulation based on a model in which the thickness of the film is adjusted iteratively and compared with the data. In such simulations the ultrasound generation, propagation and corresponding time-resolved reflectivity signal are computed. In addition to film thicknesses, other optical and mechanical properties of the films and interfaces may be varied in the physical models underlying the simulations. Like the thicknesses, these properties may be treated as parameters, and their values adjusted iteratively in order to obtain agreement with the measured response. As a result it is usually possible to determine the thickness and density of a film simultaneously, and to report these quantities to an operator. Misprocessing events (for example, a vacuum leak during reactive sputtering) often give rise to films or interfaces with highly non-ideal characteristics. In these situations it may be impossible to obtain close agreement between the model simulation and measured response by adjusting a reasonable subset of layer properties. This gives rise to a large fit uncertainty which may be used as an indication that the process has deviated from normal.

For the sample of Fig. 1, the density of the TiN layer was determined using the relationships described in relation to Eq. 1. The sound velocity and density for the Al layer on which the TiN was deposited were assumed to take their nominal bulk values to give a value for $Z_{substrate}$, and the TiN film was assumed to have a sound velocity of 95 Å/psec. The density was accordingly found to be 5.25 g/cc. It has been shown in a series of experiments[2] in which films of TiN having widely differing densities were prepared and measured that the sound velocity for this material depends only very weakly on film composition and structure. Later in this paper we describe a similar study for CVD WN.

METAL BILAYERS

TiN is most often deposited as part of a stack made up of two or more different layers. These layers are typically deposited sequentially in one or more chambers forming part of a cluster tool (i.e. a tool having a common wafer handling system and vacuum, and several deposition of processing chambers). From a process control standpoint it is desirable to measure all of the films in the stack simultaneously: This is more efficient and less costly than making measurements on a series of individually deposited films, and it also allows for making such measurements on product wafers Product wafer-based measurements are desirable not only because of the cost savings they bring, but also because they broaden the scope of process monitoring. As a trivial illustration, for

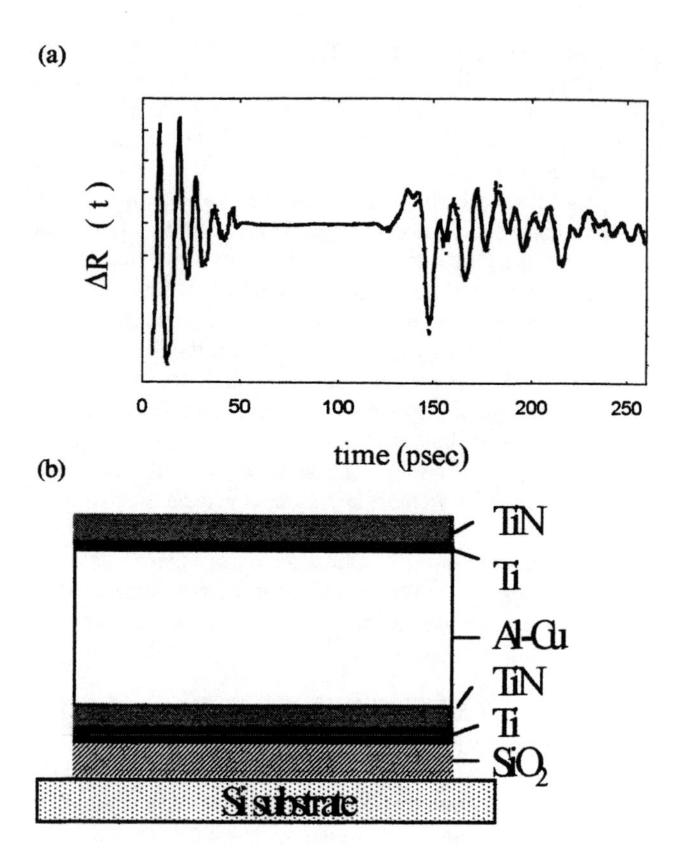

Figure 5. (a) Data and fit for a multilayered interconnect stack having layers as illustrated in Fig. 5a. The "ringing" at times less than 50 psec comes from sound reverberating within the thin surface layers. The complex echo beginning near 130 psec is due to sound generated in the top layers which has traversed the thick Al-Cu layer and reflected from the bottom layers before returning to the surface.(b) Schematic cross section of a multilayered interconnect stack deposited on a thin layer of silicon dioxide on silicon.

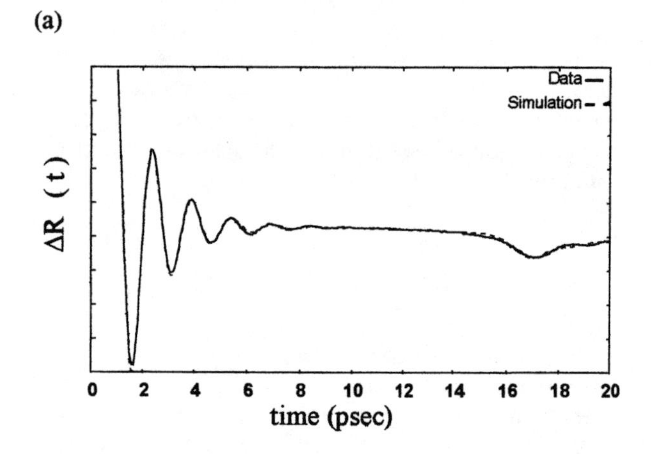

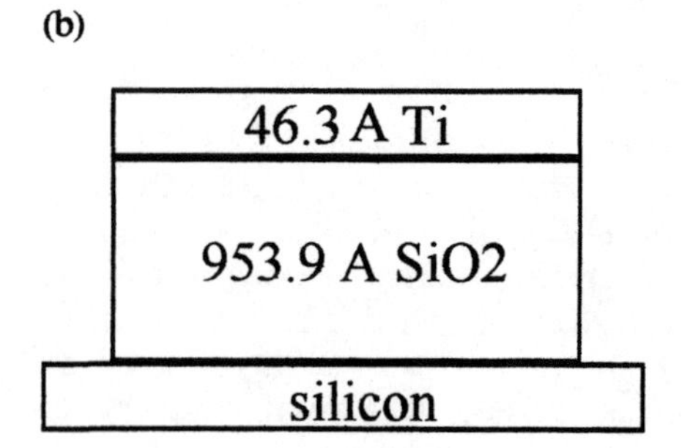

Figure 6. (a) Measured and simulated echo pattern for a 46.3 Å Ti film deposited on 953.9 Å SiO_2. The oscillations at times less than about 10 psec are due to vibrations of the Ti layer, while the downward echo at about 17 psec is due to sound generated in the Ti layer entering the underlying silicon after traversing the oxide layer.

example, based on measurements of single films a process may be deemed to be in control, but this does nothing to ensure that product wafers subsequently introduced to a deposition tool are loaded right-side up.

Echo patterns become more complex as the number of films is increased. To illustrate this, Fig. 3 shows a measurement obtained from a bilayer formed by depositing a thin PVD Ti layer, followed by a thicker PVD TiN layer (Fig. 3). The TiN layer is similar in thickness to the bare TiN film which gave the echo pattern of Fig.1. The bilayer gives echoes from both the top and bottom surfaces of the Ti layer, at times of approximately 25 and 30 psec respectively. Subsequently, repeated trips through the two layers give rise to the higher order echoes beginning at about 50 psec. The echoes shown in Fig. 3, which contains the measured pattern (solid line) as well as a simulation (dotted line - barely visible) span an interval corresponding to three round trips through the combined stack. Because the measured and simulated patterns remain in phase over this entire interval, the uncertainty in the fitted thicknesses is less than 1 Å for both films. Similar measurements have been made for bilayers comprised of IMP (Ionized Metal Plasma) Ti, deposited on CVD TiN with film thicknesses of less than 50 Å for both layers.

The measurements shown in Fig.4 show the variation across the diameter of a wafer in the thicknesses of Ti and TiN for a similar sample with different film thickness. The obvious absence of a correlation between the curves serves to illustrate that the measurements for the two layers are independent.

MULTILAYERED INTERCONNECTS

As a further illustration of measurements of stacks deposited in cluster tools, Fig. 5a shows a pattern of echoes measured for the multilayered interconnect stack depicted schematically in Fig. 5b. In a chip, the thick Al-Cu layer in this stack carries the majority of the electrical current. The much thinner cladding layers of TiN and Ti are designed to act together to serve many purposes, for example, to promote adhesion between the interconnect layer and surrounding insulators, to suppress diffusion and improve patternability. In a picosecond ultrasonic measurement, the thick Al layer has the effect of dividing the time axis into two regions: one at *short times* due to sound propagating in the top layers of the stack; and a second at *long times* corresponding to the interval sound originating in the upper layers to propagate to the bottom of the Al and return to the surface. In Fig. 5a the short time region ends at about 50 psec. The long time region begins after about 125 psec. The top TiN and Ti thicknesses can be determined from an analysis of the echoes in the short time region. This analysis can be carried out independent of the thicknesses of the underlying layers since they do not influence the short time echoes. The Al thickness can be determined from the position of the onset of the long time echo, and bottom Ti and TiN layer thicknesses can be determined from its amplitude and periodicity. The detailed shape of the long time echo is influenced by the thicknesses of the upper layers, which, however, are known independently from the short time echoes. Since it can be broken into parts in this manner, the overall analysis of this 5 layer structure, therefore, requires only two thicknesses at most to be determined simultaneously. This makes the analysis far more stable than it would be if all layer thicknesses had to be determined at once.

ULTRATHIN LAYERS

Much simpler applications that those considered in the preceding two sections also require new measurement techniques. In this section we consider single films whose thicknesses are less than about 100A. The electronic mean free path in films in this range becomes comparable to the film thickness, and as a result boundary scattering of electrons becomes a significant factor determining the sheet resistance. The usual assumption made for thicker films, that the sheet resistance is proportional to the bulk resistivity and inversely proportional to the thickness, therefore cannot be made for ultrathin films. It is possible in principal to apply correction formulae which account for boundary scattering, but these require many assumptions to be made about the character of the surfaces of the film, and their validity for any particular film may be questionable. Therefore, even for blanket metal depositions sheet resistance measurements become unreliable for very thin films.

The minimum measurable film thickness for picosecond ultrasonics depends on the effective generation and detection bandwidth of the measuring system, which can be up to about 1THz, depending on the metal. As an example, Fig. 6a shows a measurement for a Ti film less than 50Å thick deposited on top of a 954Å SiO_2 film (Fig. 6b). For a metal film whose thickness is less than or comparable to the corresponding optical absorption length, the echoes observed for thicker films take on the appearance of damped sinusoidal oscillations. In Fig. 6a, the frequency of these oscillations is

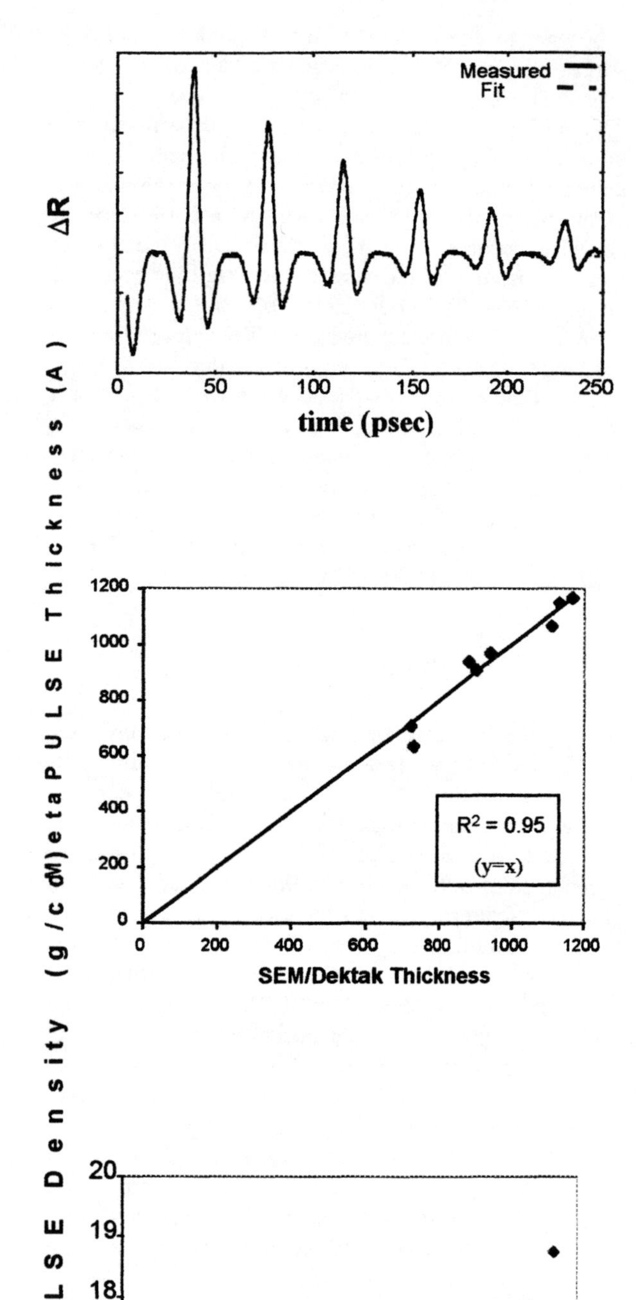

Figure 7. Series of six successive echoes observed in a CVD WN_x film of thickness 1025Å. The density was determined from the damping rate of the echoes to be 14.1 g/cm^3. For this sample the value of x was estimated to be 0.65.

Figure 8. Graph showing correlation of WN_x thickness measured by picosecond ultrasonics and secondary destructive measurement (SEM or profilometer). The samples used to compile this graph had x in the range 0.27 to 0.79. The sound velocity used to determine the thickness from picosecond ultrasonics was the same for all samples.

Figure 9. A graph showing on the vertical axis the density measured by picosecond ultrasonics for a series of WN_x samples with x in the range 0.27 to 0.79. The horizontal axis shows an value of density computed from x based on a simple model in which N atoms were assumed to occupy W sites on a W lattice substitutionally.

approximately 500GHz, corresponding to a film thickness of 46.3 Å. The minimum thickness measurable for Ti is less than 25Å.

METAL CVD FILM COMPOSITION

An increasing number of metals are deposited using chemical vapor deposition (CVD) because they can have much better conformality than films deposited by conventional methods such as sputtering (or physical vapor deposition, PVD). This is important when thin films must be deposited onto surfaces with a large amount of topography, for example, ILD patterned and etched to create openings for vias, or interconnects. CVD films may composed of one element (for example, Al or Cu) or several (for example, WN_x, WSi_8). In the latter case, the composition may depend strongly on the deposition conditions, and may be intentionally or unintentionally varied within wafer or from run to run. It is currently of interest to measure both the composition and thickness of these films in order to monitor the uniformity of a deposition process or to achieve and maintain a particular composition.

The accuracy of a picosecond ultrasonic measurement of thickness depends on the accuracy with which the sound velocity in the film material is known. In principle this may vary as a function of composition, and the particular functional dependence can be determined for any given material system. This is a topic of ongoing research. However, experiments have shown that for many common CVD and PVD materials the sound velocity varies by only a small amount when the composition is varied over an achievable range. The density, however, may be varied widely.

Figs. 7 through 9 show results for the case of CVD WN_x. Figure 7 is a measurement showing measured and simulated echo patterns for a sample where x=.65. Using an assumed value for the sound velocity based on independent thickness measurements, the density of the film was determined to be 14.1 g/cm^3 from the rate of decay of the echo amplitudes as explained above using Eq. 1. Results obtained in this way for a series of samples with x varied from 0.27 to 0.8 are plotted in Figs. 8 and 9. In Fig. 8 the thicknesses of the samples measured by picosecond ultrasonics (using a single value for the sound velocity, independent of composition) are plotted versus thicknesses measured using profilometry. Despite the wide variation in x for the samples used to compile this graph, the thicknesses obtained by the two techniques are in close agreement and give a correlation coefficient of 0.95 assuming straight line correlation through the origin. In Fig. 9 the film density obtained from picosecond ultrasonics is plotted versus an approximate value for the density based on a combination of x-ray and RBS measurements of x. To make contact in an illustrative way between these measurements and the densities obtained via picosecond ultrasonics, densities were computed from the measured x-values assuming that the nitrogen atoms replace tungsten atoms substitutionally on a W lattice; thus, density on the lower axis is computed from the formula $19.3\{(1-x)+M(N)/M(W)x\}$. Despite this obvious over simplification, the values obtained in this way show a clear relationship to each other, and the densities measured via picosecond ultrasonics depend on x in a reasonable way (i.e. approximately linearly).

CMP METALS: W

Metals which are deposited onto patterned substrates and subsequently polished by CMP (chemical-mechanical polishing) in order to create filled recesses (i.e. vias or wires embedded in ILD) present an unusual measurement challenge. Metal deposited close to the beveled wafer edge tends to break off during polishing and this can lead to severe scratching of the wafer surface. To prevent this, metals such as W (and more often, Cu) which are commonly polished must be deposited with techniques designed to prevent deposition within about 1 mm of the wafer edge, and yet achieve uniform deposition elsewhere on the wafer. A typical thickness profile for CVD W within the transition region measured by picosecond ultrasonics, from uniform thickness to zero, is shown in Fig.10.

The radius at which the thickness drops below the accepted value for the deposition process must be maintained around the entire circumference of the wafer. This requires control not only of the process gas flows, but also of the wafer positioning mechanisms within the deposition chamber. Failure to center a wafer relative to the chuck within the chamber results in a film which is miscentered with respect to the wafer. Film thickness profiles measured by picosecond ultrasonics for a properly positioned wafer, and a wafer which was not properly centered in the deposition chamber are shown in Figs. 11a and 11b.

To be able to make measurements with adequate spatial resolution it is important to have a measurement site size much smaller than the width of the transition region, generally about 50 microns or less. Since visible laser beams can easily be focussed to spots as small as 1 micron, this is easily achievable in a picosecond ultrasonics system. On the other hand, it is not currently

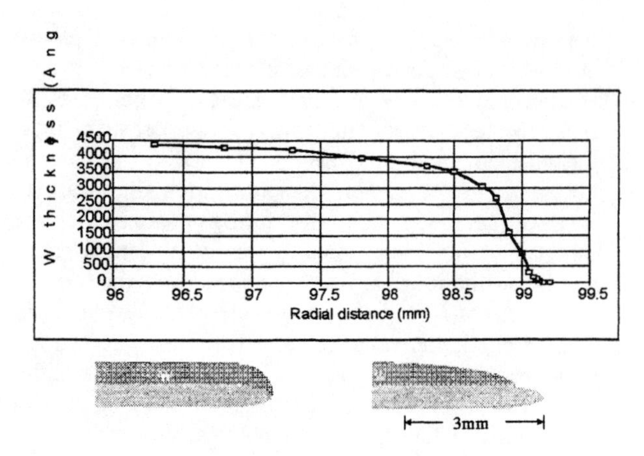

Figure 10. Graph of W thickness versus distance of the center for a 200mm diameter wafer. The inset figures show an undesirable deposition with W overlapping the beveled wafer edge (at left), and a desirable deposition for which the thickness goes rapidly to zero within about 1mm of the edge.

(a)

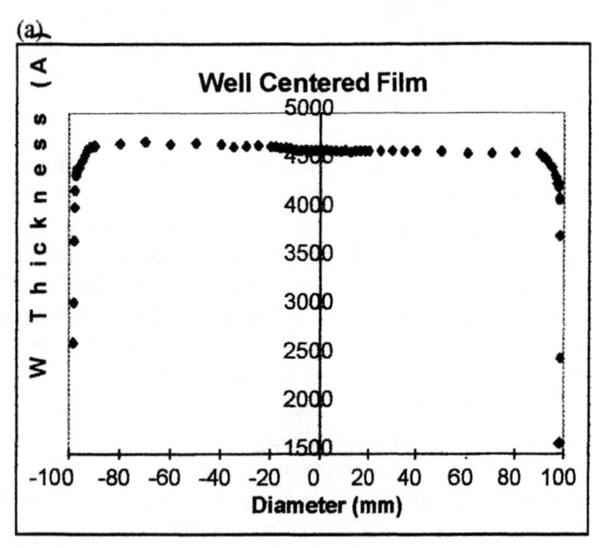

(b)

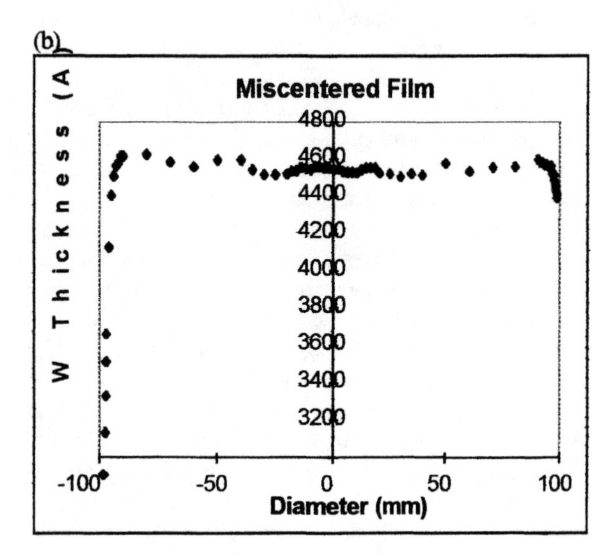

Figure 11. Picosecond ultrasonic measurements of the thickness of two tungsten films measured along a 200mm wafer diameter. The measurements in (a) are for a film which was well-centered on the wafer. For the wafer measured in (b) the wafer was offset relative to the correct position within the deposition chamber, and the film is poorly centered as a result.

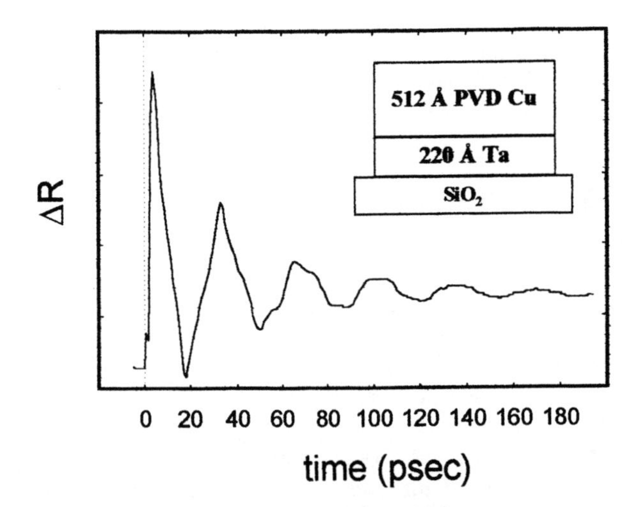

Figure 12. Echoes within a 512Å PVD copper film deposited on top of a 220Å Ta film.

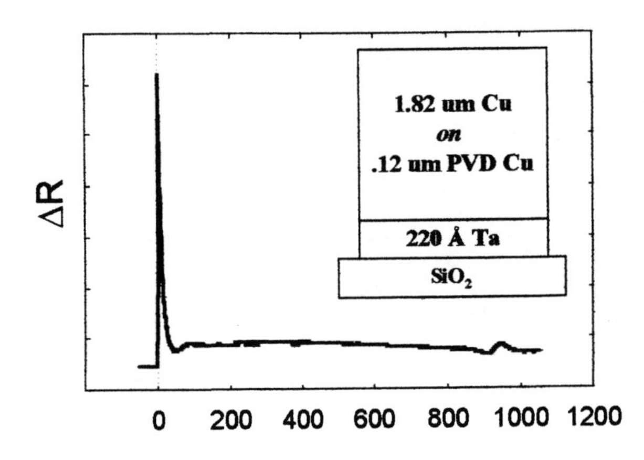

Figure 13. Echo from a thick film of plated copper deposited on a thinner film of PVD copper on 220 Å of Ta.

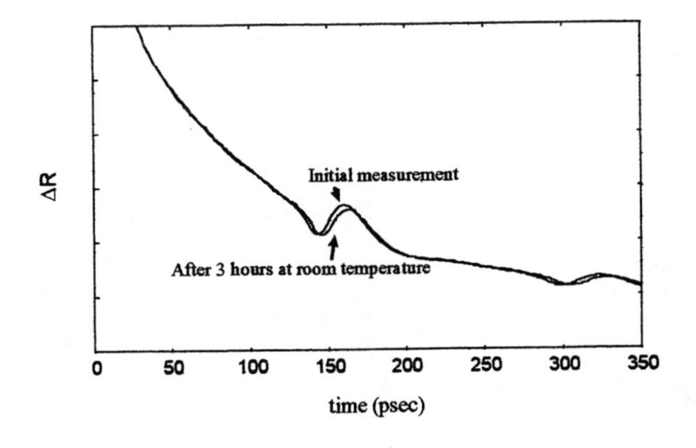

Figure 14. Graph showing echoes within a plated copper sample made 3 hours apart. The initial measurement was made immediately after removing the sample from dry ice. The second measurement was made after the sample was kept at room temperature for three hours. The echo was observed to shift to a longer time in the second measurement.

practical to make a four point probe on such a small scale, and so to make a measurement of film thickness based on sheet resistance within the transition region is not feasible.

COPPER MEASUREMENTS

Copper films may be deposited by electroplating, PVD, or CVD. These films (and the barrier films they are deposited on top of) are likely to be polished as part of a damascene or dual-damascene process, and so edge thickness control is important for all deposition techniques. These may be carried out using picosecond ultrasonics for films less than 100Å to more than several microns thick. Figure 12 shows a measurement for a 512 Å PVD film deposited on 220 Å of Ta on SiO_2 with six successive echoes well resolved. Figure 13 shows a measurement for a much thicker (1.82 micron) electroplated film deposited on top of a thinner PVD Cu film on Ta with a single echo at time of about 900 psec.

Figure 14 shows an overlay of two picosecond ultrasonics measurements made at the same point on an electroplated sample at two times separated by 3 hours. Based on previous observations indicating that similar films spontaneously anneal to some extent at room temperature, the sample was immediately packed in dry ice after plating. The first measurement in Fig. 14 ("initial measurement") was made within a few minutes after the sample was removed from the dry ice. The second measurement was made three hours later and the laser beam was blocked between measurements. It can be seen from the figure that the second measurement gave an echo shifted in time relative to the initial echo by a small amount (2.5 psec). We assume that this is due to a spontaneous change in the grain structure of the film in the interval between measurements. Further experiments to understand this change are underway.

CONCLUSION

Copper serves as a microcosm and a driver for many aspects of the trend toward ever increasing sophistication in metal processing. Processes which seemed inherently problematic or marginal until recently such as cluster tool sputter depositions, CVD metal depositions, intermetallic diffusion barrier formation, ultra-thin film growth, wafer scale plating and reactive sputtering are now rapidly becoming commonplace. The same processes pose characterization challenges that are beyond conventional measurement techniques. In this paper we have described a new approach to process monitoring based Picosecond Ultrasonics. This technique makes possible non-contact measurement of film thickness on product wafers within a measurement diameter of less than 10 microns. We have given examples of its applicability for measuring multilayer interconnect stacks, adhesion stacks, ultra-thin layers, CVD layers of one and two metal species, and thick plated layers.

ACKNOWLEDGMENT

Work at Brown University was supported in part by the US Department of Energy through grant number DE-FG02-86ER45267. The authors are grateful to Sailesh Merchant of Lucent Technologies who provided samples and helpful insights regarding interactions between metals in cluster tools, Tom Ritzdorf of Semitool who supplied the ECD copper samples, and also Applied Materials and Novellus Systems who supplied many of the other samples used in the course of the work presented in this paper.

Statistical Metrology - Measurement and Modeling of Variation for Advanced Process Development and Design Rule Generation

Duane S. Boning and James E. Chung

Microsystems Technology Laboratories, Dept. of Electrical Engineering and Computer Science, Massachusetts Institute of Technology, Cambridge, MA 02139

Advanced process technology will require more detailed understanding and tighter control of variation in devices and interconnects. The purpose of statistical metrology is to provide methods to measure and characterize variation, to model systematic and random components of that variation, and to understand the impact of variation on both yield and performance of advanced circuits. Of particular concern are spatial or pattern-dependencies within individual chips; such systematic variation within the chip can have a much larger impact on performance than wafer-level random variation.

Statistical metrology methods will play an important role in the creation of design rules for advanced technologies. For example, a key issue in multilayer interconnect is the uniformity of interlevel dielectric (ILD) thickness within the chip. For the case of ILD thickness, we describe phases of statistical metrology development and application to understanding and modeling thickness variation arising from chemical-mechanical polishing (CMP). These phases include screening experiments including design of test structures and test masks to gather electrical or optical data, techniques for statistical decomposition and analysis of the data, and approaches to calibrating empirical and physical variation models. These models can be integrated with circuit CAD tools to evaluate different process integration or design rule strategies. One focus for the generation of interconnect design rules are guidelines for the use of "dummy fill" or "metal fill" to improve the uniformity of underlying metal density and thus improve the uniformity of oxide thickness within the die. Trade-offs that can be evaluated via statistical metrology include the improvements to uniformity possible versus the effect of increased capacitance due to additional metal.

INTRODUCTION

Statistical metrology is the body of methods for understanding variation in microfabricated structures, devices, and circuits. The goal of this paper is to describe key features of statistical metrology, to review the tools and methods developed to date, and present an application of statistical metrology to advanced technology and design rule development.

A running application example will be used to illustrate the concepts and experience in statistical metrology. In this section, we present the variation problem being addressed, namely the variation of interlevel dielectric (ILD) thickness across the wafer and chip resulting from chemical mechanical polishing (CMP). The progression of our development of statistical metrology concepts and methods will then be presented. These phases are (I) variation assessment, (II) variation modeling, (III) semi-physical model calibration, (IV) circuit impact modeling, and (V) design rule generation. Finally, future work and a summary will be presented.

In essence, this paper serves as a review of the MIT work on statistical metrology. While the references are drawn almost entirely from the MIT research, it is important to note that much work on statistical metrology has been contributed elsewhere (e.g. Spanos et al. (6) and many others) which is not reviewed or addressed in the present paper.

Statistical Metrology

Statistical modelling and optimization have long been a concern in manufacturing. Formal methods for experimental design and optimization, for example, have been developed and presented by Box et al. (1) and Taguchi (2), and application to semiconductor manufacturing by Phadke (3). More recently, these methods are seeing renewed development in "statistical metrology" research. Bartelink introduces statistical metrology in (4), emphasizing the importance of statistical metrology as a "bridge" between manufacturing and design. In (14, 15) an early review of the defining elements of statistical metrology is presented. These include an emphasis on characterization of variation, not only temporal (e.g. lot-to-lot or wafer-to-wafer drift), but also spatial variation (e.g. within-wafer, and particularly within-chip or within-die). A second important defining element is a key goal of statistical metrology: to identify the systematic elements of variation that otherwise must be dealt with as a large "random" component through worst-case or other design methodologies. The intent is to isolate the systematic, repeatable, or deterministic contributions to the variation from sets of deeply confounded measurements. Such variation identification is critical to focus technology devel-

CP449, *Characterization and Metrology for ULSI Technology: 1998 International Conference*
edited by D. G. Seiler, A. C. Diebold, W. M. Bullis, T. J. Shaffner, R. McDonald, and E. J. Walters

opment or variation reduction efforts, as well as to help in development of device or circuit design rules to minimize or compensate for the variation.

In the next section we describe one process area which is confronted by such intertwined and important variation.

Problem: Oxide Thickness Variation and CMP

The planarization of dielectric layers between multi-level metal layers is critical for present and future interconnect technologies. Variation in ILD thickness remaining after CMP, however, is present at multiple length or time scales. For example, Fig. 1 illustrates example measured wafer-level oxide thickness nonuniformity after CMP, as well as the measured ILD thickness within a sample die. We see that the within-die variation can often be much larger than the wafer-level variation.

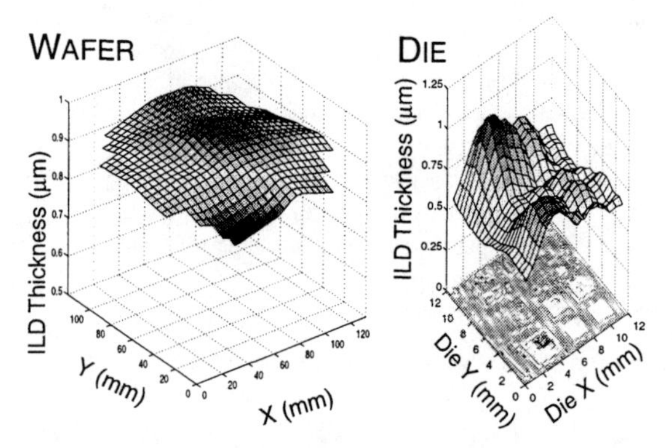

Figure 1: Typical within-wafer and within-die ILD thickness variation. Note that the range within the die is larger than the entire wafer-level trend.

After deposition of oxide over patterned metal lines, local steps are present which one wishes to remove, as illustrated in Fig. 2. While good local planarity can be achieved, different regions across the chip may not polish in a uniform fashion. As also shown in Fig. 2, the reality of oxide CMP is that the process transforms local step height nonuniformity into global nonplanarity: those regions of the chip with higher density of raised topography essentially polish more slowly, so that the final oxide thickness over metal lines in these regions is thicker than that over low density regions.

Progression of Statistical Metrology & ILD Thickness Variation

In order to attack the problem of ILD (oxide) thickness variation described in the previous section, a series of developments has taken place. These are summarized in Fig. 3. In Phase I, variation identification methods were employed. Screening experiments and variation decomposition methods were developed in order to separate and identify the components of oxide thickness variation. Once key elements

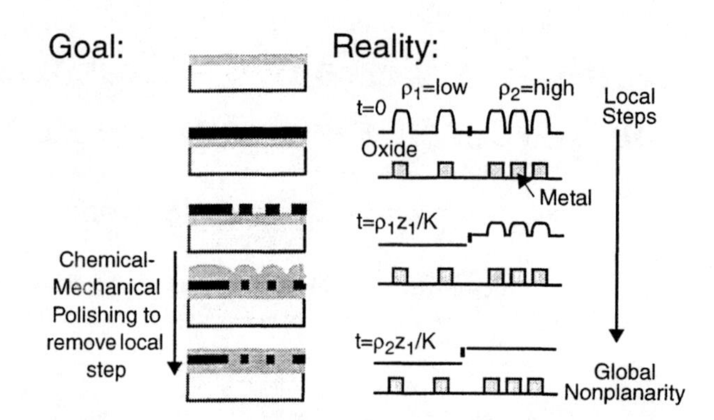

Figure 2: Oxide thickness variation and effect of CMP: the goal (removal of local steps) is shown on the left, while the reality (creation of global non-planarity during step height reduction) is shown on the right.

were identified (i.e. die-level variation), Phase II focused on methods to develop models of that variation. These methods include more focused factor experiments, as well as the generation of empirical models. These experiments result in models with a functional dependency on particular layout practices (e.g. the density, pitch, or area of layout structures). In the oxide CMP case, density was found to be the primary explanatory factor, enabling the development in Phase III of a semi-physical model for oxide polishing. In this phase, a key element is the creation of tightly coupled characterization and calibration methods for extraction of model parameters such as blanket removal rate and planarization length. The resulting CMP model can be applied to predict the topographical variation of oxide thickness across the entire die for a new arbitrary layout. In Phase IV, this CMP variation model is integrated with technology CAD and electronic CAD tools in order to understand the impact of such variation on circuit performance. Finally, in Phase V, trade-offs between structural variation and circuit impact can be evaluated, and design rules guiding the process technology or circuit layout practices generated. In particular, the use of a "dummy" metal fill to reduce the range of density across the chip is considered against the cost of additional capacitance.

In the following sections, each of these five phases will be discussed in more detail and illustrated with historical work in understanding the oxide CMP variation problem.

I. VARIATION IDENTIFICATION

The key purpose of variation identification is to understand what are the potential sources of ILD thickness variation. An experimental approach is undertaken, including determination of (A) measurement strategy and test structures; (B) definition of short flow experiments; and (C) experimental design methods in a search for important lay-

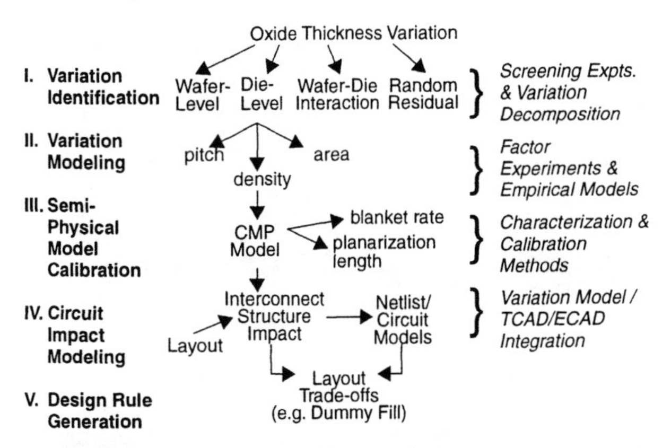

Figure 3: Phases of statistical metrology development as applied to oxide thickness and CMP variation.

out factors.

A. Test Structure Design

Initial investigations of oxide thickness included development of a fingered electrical capacitive test structure, as illustrated in Fig. 4, from which the oxide thickness between two metal layers could be inferred using TCAD tools in conjunction with a large volume of electrical probe measurements (5-8, 11).

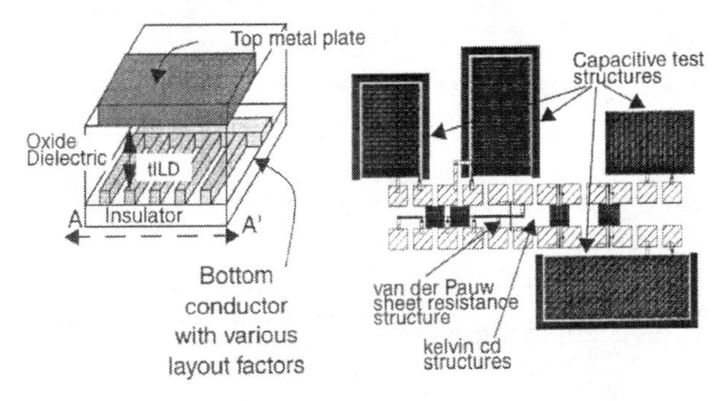

Figure 4: Electrical (capacitive) test structure used to infer the thickness of oxide between metal layers.

B. Short Flow Experiments

The statistical metrology methodology emphasizes short process flows. First, these enable fast feedback and data to be gathered. Second, a shorter flow ensures that the variation being studied is not confounded with that generated by subsequent processing steps. Indeed, substantial care was taken in the test structure of Fig. 4 in order to deconvolve linewidth variation from dielectric thickness variation in the experiment. In later work, optical test masks have been used with the advantage of even shorter process flows and complement the fine line electrical test structures.

C. Design of Experiment Phases

As discussed in (8), three stages of experimental design have been pursued in understanding variation. In the first stage, "screening" experiments seeks to explore a large space of possible layout or process factors that might influence the parameter of interest. Second, "environmental" experiments have been pursued which study how critical parameters depend on factors in a realistic circuit environment (e.g. that of particular product families such as microprocessors, memory, or ASICs). Finally, modeling experiments focus on very specific factors or their interactions, and probe multiple levels of those factors to provide the data needed for empirical models that capture key causal dependencies.

D. Variation Decomposition

Based on electrical or optical measurements taken on many sites within the die, and many or all die on the wafer, a key issue is the assignment of variation in that measure to different sources. In particular, spatial characteristics of the variation can provide substantial insight into the physical causes of the variation. Variation decomposition methods (captured in the "VarDAP" tool) have been developed which successively serve to extract key components from measurement data (16). First, wafer level trends are found, and are themselves often of key concern for process control. The trend is removed from the data, and methods employed to extract the "die-level" variation (that is, the component of variation that is a clear signature of the die layout) using either 2D fourier techniques, or modified analysis of variance approaches (6, 16). This is especially important in order to enable further detailed analysis of the causal feature- or layout-dependent effects on a clean set of data. Third, examination of wafer-die interaction is needed. For example, if the pattern dependencies and wafer level variation are highly coupled, then process control decisions (usually made based on wafer-level objectives) must also be aware of this die-level impact (20). For an example set of raw data (measurements based on the electrical test die (8)), this decomposition is pictured in Fig. 5.

In the case of oxide CMP, we found that the die-level variation is much larger than the wafer level, and is a key cause for concern in the process. ANOVA methods were used to identify both statistically significant effects (care must be taken even here, as spatial location within the die destroys some apparent replications), as well as to estimate the magnitude of effects. A second important aspect of the statistical analysis was the merger of different combinations of the original screening factors into more "natural" or meaningful combinations. For example, the original design was performed in terms of linewidths, line spaces, and number of lines; we found, however, that structure "density" and in some cases "area" provided clearer explanations. At this point, further detailed models of the sources and causes of

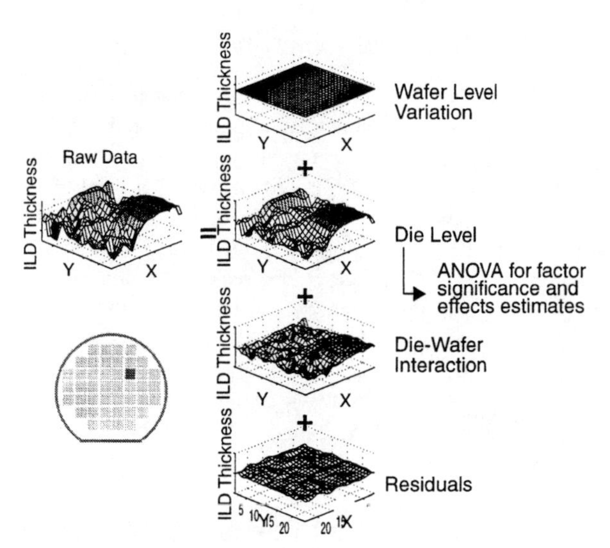

Figure 5: Variation decomposition of ILD thickness.

oxide thickness variation were desired.

II. VARIATION MODELING

As shown in the statistical metrology progression of Fig. 3, the second phase has the goal of more clearly understanding the sources of die-level variation. The key tools are further experimental methods, together with empirical modeling of results (28, 21).

In the case of oxide CMP, a new set of masks were designed which focus on a smaller number of factors (identified in the screening phase), but allowing many more levels of those factors to be explored. A CMP characterization mask set was designed with substantial input from the CMP community, as shown in Fig. 6. In this case, the masks examine: (a) total structure area or size; (b) structure density (e.g. density of patterned lines divided by total area) for a constant structure pitch; (c) structure pitch (sum of line-width and line space) for a constant 50% density; and (d) structure perimeter over area to explore feature edge effects.

The die-level signatures resulting from oxide polish experiments enable both comparison of the importance of these layout factors in determining polish performance, and effective empirical modeling of these effects. As shown in Fig. 7, quantitative measures of the corresponding layout effects can be estimated. In the case of oxide CMP, we find that pitch and perimeter/area have very little impact, while ILD thickness shows a very strong (and linear) relationship with density. After correction for the density variation, structure area also shows very little clear dependency.

The clear relationship between density and ILD thickness leads directly to the next phase of statistical metrology: development of physical or semi-physical models that can be calibrated to data for analysis of causal variation dependencies.

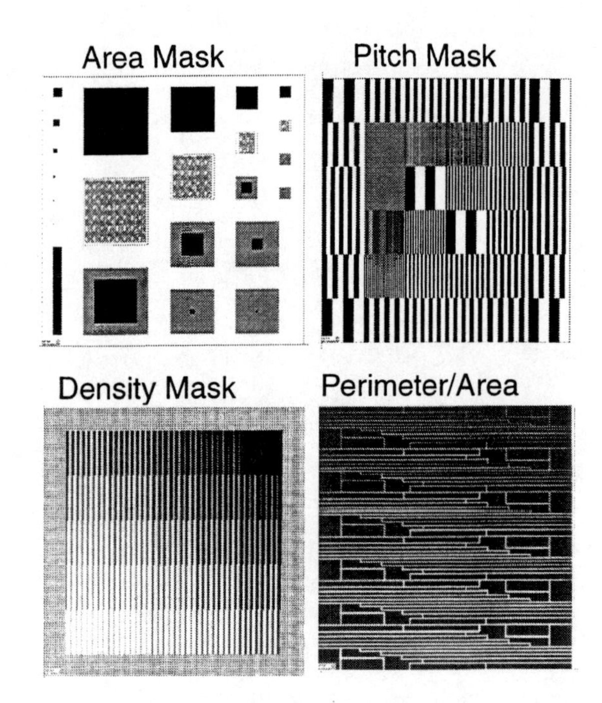

Figure 6: CMP Characterization mask set.

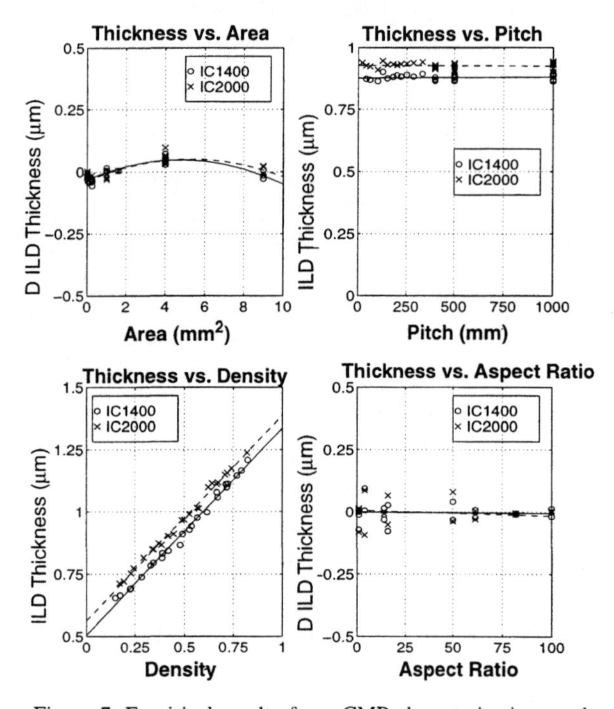

Figure 7: Empirical results from CMP characterization mask set.

III. SEMI-PHYSICAL MODEL CALIBRATION

Based on the characterization mask set, a clear relationship between pattern density and ILD thickness was found. A simple modification to Preston's equation (stating that the bulk removal rate is proportional to the product of pressure and velocity) is used, whereby the overall down-force is distributed only across the "raised" topography (oxide over metal features), leading to a rate dependence on density (18). This relationship is captured as:

$$z = \begin{cases} z_0 - \left(\dfrac{Kt}{\rho_0(x,y)}\right) & Kt < \rho_0 z_1 \\ z_0 - z_1 - Kt + \rho_0(x,y) z_1 & Kt > \rho_0 z_1 \end{cases} \quad \text{(Eq. 1)}$$

where K is the bulk or blanket removal rate, t is the polish time, z_0 is the initial oxide thickness, and z_1 is the initial step height (i.e. the difference in height of oxide between metal features and those over metal features). A key remaining issue, however, is exactly what is meant by "density" (ρ_0 in (Eq. 1) above). As illustrated in Fig. 8, the density is calculated as the raised area to total area in some square (or circular) window determined by a "window size" or planarization length of L. In the case of very fine linewidths, biasing to account for lateral deposition modifies the effective density (17). The size of the window can have a tremendous impact on the effective density calculated at a point on the layout, as well as the range of density found on a chip or along a cut line (such as the B-B' line in Fig. 8). For example, a relatively small window of length 2.5 mm might result in a density range of 57% along BB', while a large window size of 10.0 mm will "average" the local densities much more thoroughly, and result in only a 9% density range across BB'. Through the relationship of (Eq. 1), this density relates directly to the final oxide thickness measured after CMP.

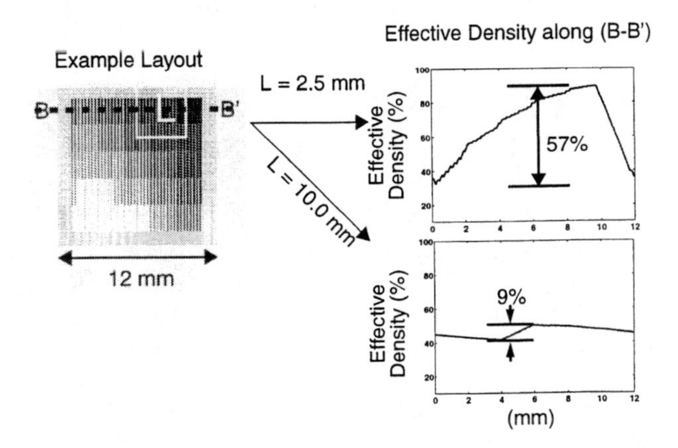

Figure 8: Semi-physical CMP model, illustrating the importance of planarization length (density calculation window size).

The determination of planarization length L, however, cannot as yet be determined based on purely first-principals physical understanding or external measurement of the CMP pad and process. Instead, as part of the third phase of statistical metrology, specific test structures and experimental methods are used to calibrate the model to a given CMP pad and process. As shown in Fig. 9, the planarization length, together with a wafer-level model of blanket polish rate $K(x,y)$, provide the information necessary to predict ILD thickness for any die on a new arbitrary layout (30).

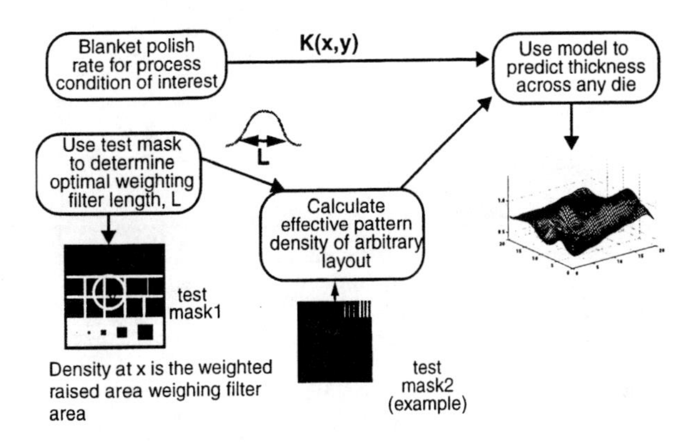

Figure 9: CMP model calibration and use.

Fig. 10 shows an example verifying a calibrated CMP model. In this case, a more sophisticated window shape extraction procedure is employed (30) based on one test mask and experiment. A second, different test mask is shown in Fig. 10, with the corresponding predictions (solid lines) and measurements (circles) for several cut lines across the layout. In particular, cut lines L1 and L2 are drawn along regions of gradual density increase, while L3 is drawn along 4mm blocks of "step density" where the underlying metal pattern density is changed dramatically from one block to the next. Finally, L4 is drawn along constant 50% density blocks where the pitch changes substantially.

Further modeling of the CMP variation can account for the variation of oxide thickness across the wafer (24), as well as other second order polishing effects (26). An important result for multilevel metal modeling is that if the oxide is polished through to "local planarity" then each level of oxide (e.g. between M2 and M3) is affected only by the density on the immediately preceding metal layer; that is, the oxide thicknesses are additive in a multilevel polishing system (25).

IV. CIRCUIT IMPACT MODELING

Based on the statistical calibration and modeling of the previous section, a prediction of the oxide thickness across the chip resulting from CMP is now possible. The fourth key phase of statistical metrology is now undertaken, to under-

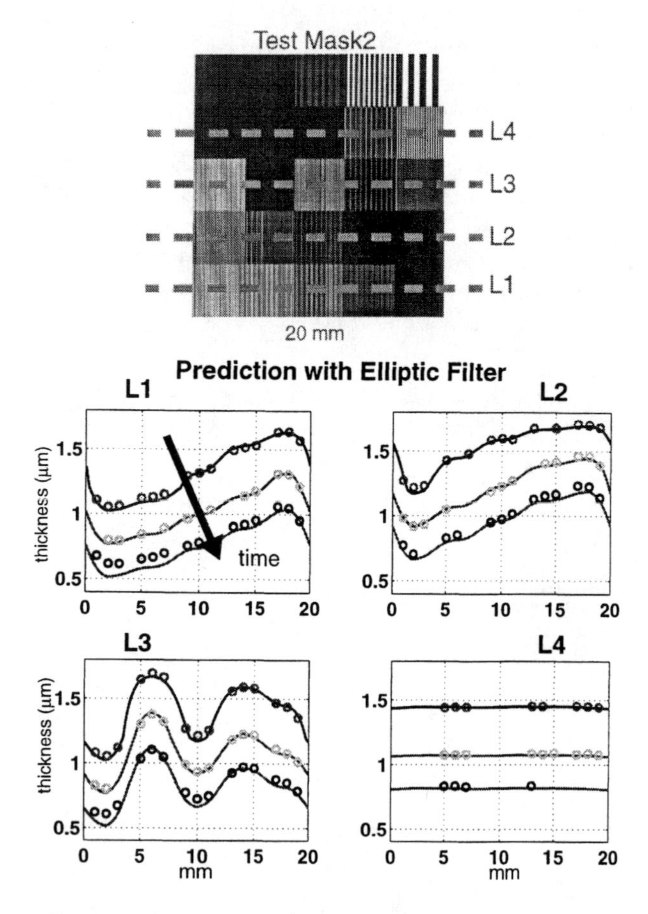

Figure 10: CMP model verification example.

stand the impact that such variation may have on circuit performance. In this case, the strategy is to integrate the variation models with process and device simulation (TCAD) tools, as well as with electronic CAD (ECAD) tools in order to evaluate circuit performance (27).

An H-bar balanced clock tree design case study illustrates this approach (27). A hypothetical chip floorplan is shown in Fig. 11, where the goal is to carry the clock signal generated at the center of the chip along the four different paths in such a way that no clock skew results. Unfortunately, the metal clock paths lie over regions of different underlying circuit or metal density. For example, path 1 runs over random logic (with perhaps a 30% density), and then over embedded memory with a higher density of perhaps 50%. Path 3, on the other hand, lies entirely over 30% local density.

As discussed in the earlier sections of this paper, this density difference will result in non-ideal polishing of the interlevel dielectric between the clock line metal layer and the circuit metal layers, as shown in the lower part of Fig. 11. As a result, the ILD thickness along the clock paths, as well as the layer to layer capacitance those paths experience, will be substantially different, as shown in Fig. 12. This disparity can be expected to have an impact on the

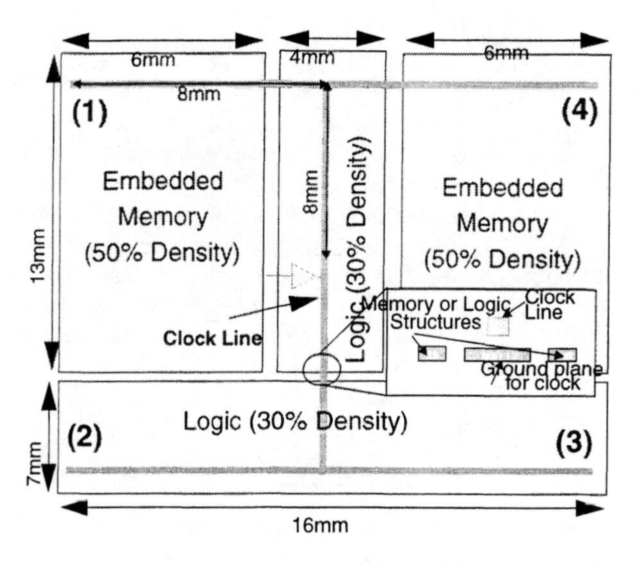

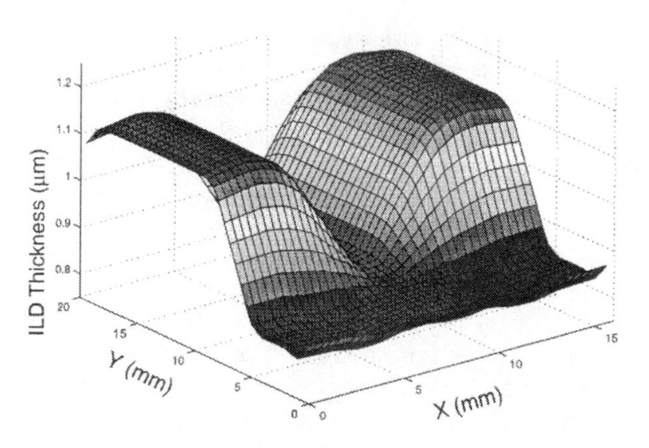

Figure 11: Hypothetical balanced H-bar clock tree floor plan (above); corresponding predicted ILD thickness map (bottom).

resulting clock skews; as shown in the lower part of Fig. 12, for example, the difference between the desired and "actual" timing for path 3 is about 17% in this example.

V. VARIATION MINIMIZATION: DESIGN RULE GENERATION

While the fourth phase of statistical metrology discussed above helps to understand the impact of variation, we have not as yet directly addressed a key goal: how does one minimize that variation or its impact? In the fifth phase of Fig. 3, we consider the development of design rules which attack the ILD thickness variation problem.

First, we note that the key variation component is due to the effective density variation or range across the chip. If the density were completely uniform (e.g. 50% everywhere on the chip), then the entire chip would polish at a uniform rate

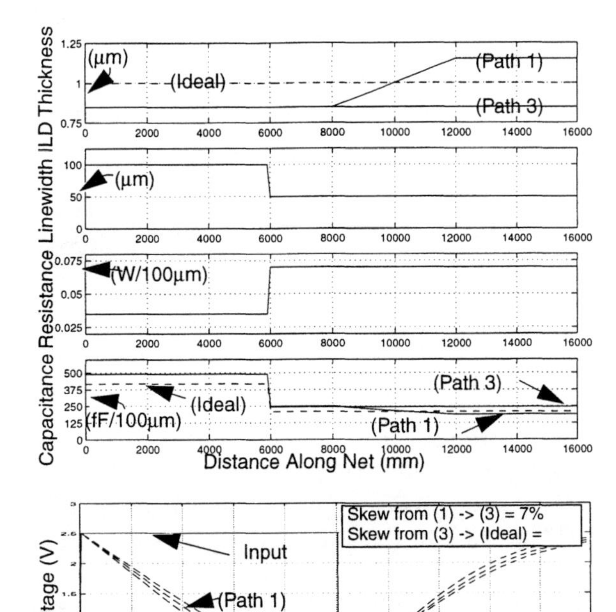

Figure 12: Resulting ILD thickness and capacitance along clock paths (top); resulting clock skews (below).

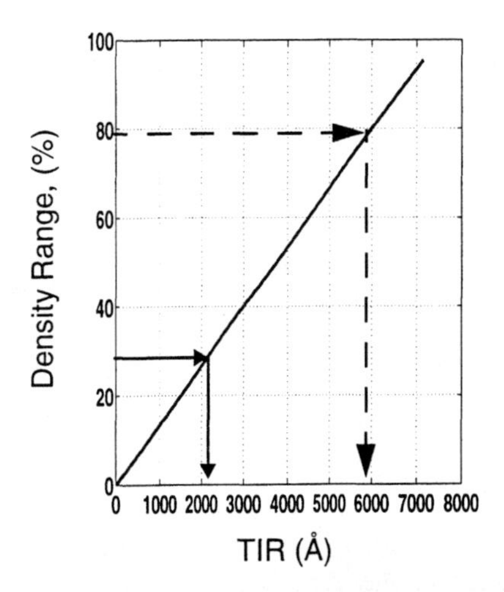

Figure 13: Range of density variation across the chip, and the corresponding total indicated range of oxide thickness across the chip (for initial step height of 7500A).

and a high degree of global planarity could be achieved. As pictured in Fig. 13, however, the range in oxide thickness across the chip directly contributes to the "total indicated range" (TIR) in oxide thickness across the chip:

$$TIR = \Delta z = \Delta\rho \bullet z_1 \qquad \text{(Eq. 2)}$$

where $\Delta\rho = (\rho_{max} - \rho_{min})$, and z_1 is the initial step height. For an initial step height of 7500Å and a density range of 80% across the chip, the TIR would be approximated 6000Å. If the density range could be reduced to 33%, then the TIR would be only 2500Å. One approach, then, is to simply mandate to the circuit designer that density may only vary within a relatively narrow range. Of course, CAD tools are needed to efficiently extract the effective density for a given layer, such as those described in Phase III.

One question remains, however: how is one to achieve a target density? One approach with much potential is "dummy fill" of metal lines at the target metal layer. As discussed in (9, 19), different dummy fill practices may be appropriate in different circuit design situations. For example, in the case of high speed microprocessor design, dummy lines (or even plates) may be the most effective; grounding these lines increases the capacitance but to a well-known value. For ASIC design, on the other hand, added capacitance may be undesirable but cross-coupling of lines a concern, so that arrays of floating blocks may be a better choice.

The feasibility of achieving a target density is also limited by pattern dependencies, as well as the availability of open space into which the dummy lines (or pedestals) may be placed. As shown in Fig. 14, for example, only linewidths or spaces above particular values my be considered achievable (19). Furthermore, the size of a "buffer distance" between the dummy fill and active lines (which may be substantial in order to minimize the opportunity for lateral capacitive coupling and defect-based shorting) may also impact the achievable density for a given layout.

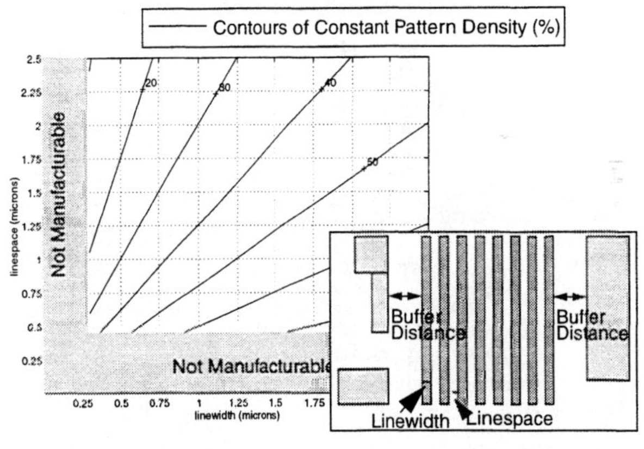

Figure 14: Feasibility of dummy fill densities for dummy line approach.

With even relatively conservative implementations of dummy fill, however, substantial improvements in within-die ILD thickness variation are achievable. For example, in Fig. 15, the distribution of oxide thickness across the die without (left) and with (right) dummy fill structures are shown for a development test chip (9). In this case, the variation was decreased by nearly 20% with the addition of dummy fill.

In creating more clear design rules, however, the trade-offs between the density and ILD thickness improvement

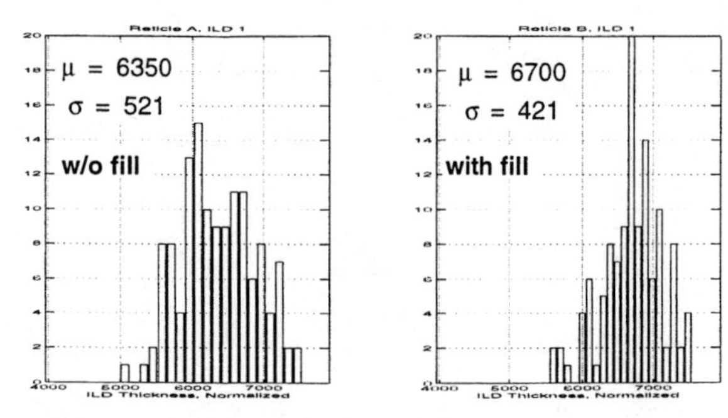

Figure 15: Improvement in ILD thickness distribution observed after addition of dummy fill.

possible with the effect of the added capacitance due to the additional dummy metal must be considered carefully. As shown in Fig. 16, the capacitance for dummy line (or dummy pedestal) can be evaluated for different buffer distances, line widths, and line spaces using capacitance simulation in conjunction with the oxide thickness variation model. A design rule can then be developed based on the trade-off, as shown in Fig. 17. In this case, for example, a 50% density was desired, while essentially 0% increase in the effective capacitance was desired, leading to a particular choice of line space and linewidth for the dummy fill to be added to the layout.

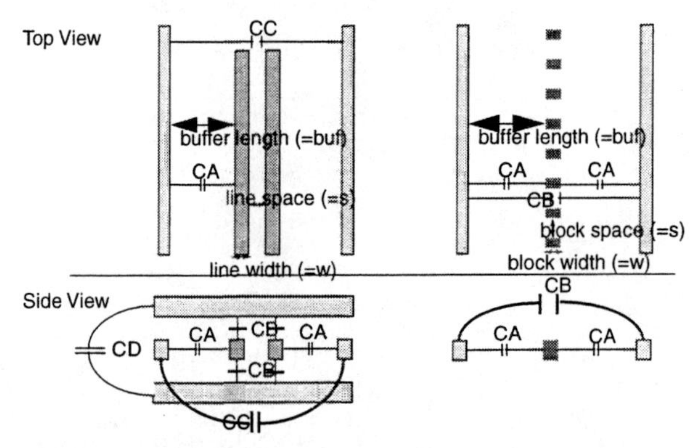

Figure 16: Capacitance impact of added dummy lines. The above capacitance components can be evaluated with capacitance simulation tools.

SUMMARY AND FUTURE WORK

Spatial variation, particularly that with systematic elements, will become an increasingly important concern in future scaled technologies. Statistical metrology methods are needed to gather data and analyze spatial variation, both to decompose that variation and to model functional dependencies in the variation. Methods are also needed to understand and minimize the impact of such variation.

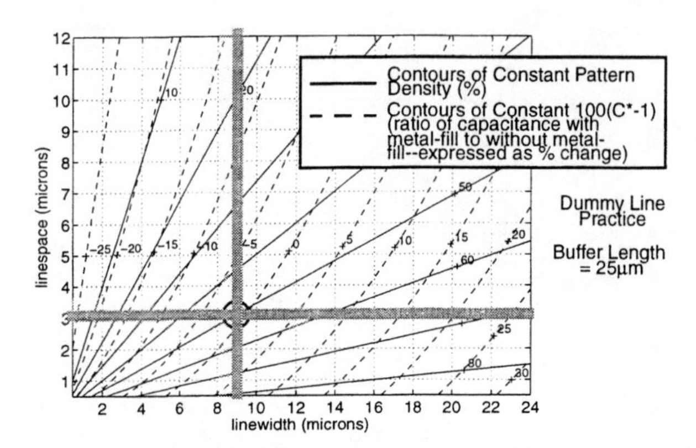

Figure 17: Trade-off between achievable dummy fill densities (solid line), and the net percentage change in capacitance.

In this paper, we have summarized the development and application of a statistical metrology methodology to understanding variation in ILD thickness arising in CMP. The methodology has developed through five successive phases: variation identification, variation modeling, semi-empirical modeling and calibration, circuit impact modeling, and design rule generation.

While the methodology has been developed for the case of oxide CMP, current work is underway to apply and further develop the methods for other pattern and spatial dependent issues in CMP. In particular, shallow trench isolation (STI) also suffers from pattern dependencies (e.g. dishing into large oxide trenches) (31). Even more important for future interconnect technologies, copper damascene suffers from substantial spatial variation, including both dishing into inlaid copper lines, and the erosion of supporting oxide regions in areas of high metal density (32).

We believe that the applications of statistical metrology will be widespread. While we discussed the connection to circuit performance evaluation in this paper, we believe that statistical metrology methods are also extremely powerful in developing and optimizing advanced process technology (e.g. the development of optimal CMP processes (10, 11, 12, 23, 29) or pads that minimize pattern dependencies), as well as in equipment selection and evaluation, and in process control. For example, CMP models are enabling more time and resource efficient processes which minimize the production of waste by-products in CMP. Statistical metrology methods are also seeing application in other process areas, particularly the study of within-chip variations in device and interconnect lines (e.g. MOS channel length) (13, 22).

ACKNOWLEDGMENTS

This work has been based on the contributions of numerous MIT students, including B. Stine, D. Ouma, T. Park, T. Tugbawa, C. Oji, V. Mehrotra, R. Divecha, E.

Chang, D. Maung, S. Kim, and B. Lee. The work has been performed in close collaboration with industry, including D. Bartelink, S. Nakagawa, and others at HP; D. Hetherington and others at Sandia National Labs; A. Kapoor, S. Prasad, B. Loh and others at LSI Logic; T. Equi and others at Digital Semiconductor; G. Shinn and others at TI, J. Kibarian, D. Ciplickas, and others at PDF Solutions; and the fabrication staffs at these locations. This work has been supported in part by DARPA under contract #DABT-63-95-C-0088, and AASERT grant #DAAHA04-95-1-0459, and by the NSF/SRC Engineering Research Center for Environmentally Benign Semiconductor Manufacturing.

REFERENCES

(1) G. E. P. Box, W. G. Hunter, and J. S. Hunter, *Statistics for Experimenters - An Introduction to Design, Data Analysis and Model Building*. John Wiley and Sons, New York, 1978.

(2) G. Taguchi, *Introduction to Quality Engineering*. Asian Productivity Organization, 1986. (Distributed by American Supplier Institute, Inc., Dearborn, MI).

(3) M. S. Phadke, *Quality Engineering Using Robust Design*. Prentice Hall, Englewood Cliffs, NJ, 1989.

(4) D. Bartelink, "Statistical metrology: At the root of manufacturing control," *J. Vac. Sci. Tech. B*, Vol. 12, pp. 2785-2794, 1994.

(5) D. S. Boning, T. Maung, J. Chung, K.-J. Chang, S.-Y. Oh, and D. Bartelink, "Statistical metrology of interlevel dielectric thickness variation," *Proceedings of the SPIE Symposium on Microelectronics Manufacturing*, SPIE Vol. 2334, pp. 316-327, Austin, TX, Oct. 1994.

(6) C. Yu, T. Maung, C. Spanos, D. Boning, J. Chung, H.-Y. Liu, K.-J. Chang, and D. Bartelink, "Use of Short-Loop Electrical Measurements for Yield Improvement," *IEEE Trans. Semi. Manuf.*, pp. 150-159, May 1995.

(7) E. Chang, B. Stine, T. Maung, R. Divecha, D. Boning, J. Chung, K. Chang, G. Ray, D. Bradbury, S. Oh, and D. Bartelink, "Using a Statistical Metrology Framework to Identify Random and Systematic Sources of Intra-Die ILD Thickness Variation for CMP Processes," *1995 International Electron Devices Meeting*, pp. 499-502, Wash. D.C., Dec. 1995.

(8) R. R. Divecha, B. E. Stine, E. C. Chang, D. O. Ouma, D. S. Boning, J. E. Chung, O. S. Nakagawa, S.-Y. Oh, S. Prasad, W. Loh, and A. Kapoor, "Assessing and Characterizing Inter- and Intra-die Variation Using a Statistical Metrology Framework: A CMP Case Study," *First International Workshop on Statistical Metrology*, pp. 9-12, Honolulu, HI, June 1996.

(9) B. Stine, D. Boning, J. Chung, L. Camilletti, E. Equi, S. Prasad, W. Loh, and A. Kapoor, "The Role of Dummy Fill Patterning Practices on Intra-Die ILD Thickness Variation in CMP Processes," *VLSI Multilevel Interconnect Conference*, pp. 421-423, Santa Clara, CA, June 1996.

(10) R. R. Divecha, B. E. Stine, D. O. Ouma, D. Boning, J. Chung, O. S. Nakagawa, S.-Y. Oh, and D. L. Hetherington, "Comparison of Oxide Planarization Pattern Dependencies between Two Different CMP Tools Using Statistical Metrology," *VLSI Multilevel Interconnect Conference*, pp. 427-430, Santa Clara, CA, June 1996.

(11) S. Prasad, W. Loh, A. Kapoor, E. Chang, B. Stine, D. Boning, and J. Chung, "Statistical Metrology for Characterizing CMP Processes," *European Materials Research Society Spring Meeting*, Strasbourg, France, June 4-7, 1996.

(12) D. Ouma, B. Stine, R. Divecha, D. Boning, J. Chung, I. Ali, and M. Islamraja, "Using Variation Decomposition Analysis to Determine the Effect of Process on Wafer and Die-Level Uniformities in CMP," *First International Symposium on Chemical Mechanical Planarization (CMP) in IC Device Manufacturing*, Vol. 96-22, pp. 164-175, 190th Electrochemical Society Meeting, San Antonio, TX, Oct.6-11, 1996.

(13) B. Stine, D. Boning, J. Chung, D. Bell, and E. Equi, "Inter- and Intra-die Polysilicon Critical Dimension Variation," Manufacturing Yield, Reliability, and Failure Analysis session, *SPIE 1996 Symposium on Microelectronic Manufacturing*, SPIE Vol. 2874, Austin TX, Oct. 1996.

(14) D. Boning and J. Chung, "Statistical Metrology: Understanding Spatial Variation in Semiconductor Manufacturing," Manufacturing Yield, Reliability, and Failure Analysis session, *SPIE 1996 Symposium on Microelectronic Manufacturing*, Austin TX, Oct. 1996.

(15) D. Boning and J. Chung, "Statistical Metrology: Tools for Understanding Variation," *Future Fab Int'l.*, December, 1996.

(16) B. Stine, D. Boning, and J. Chung, "Analysis and Decomposition of Spatial Variation in Integrated Circuit Processes and Devices," *IEEE Trans. Semi. Manuf.*, February 1997.

(17) R. Divecha, B. Stine, D. Ouma, J. Yoon, D. Boning, J. Chung, O.S. Nakagawa, S-Y Oh, "Effect of Fine-Line Density and Pitch on Interconnect ILD Thickness Variation in Oxide CMP Processes," *1997 Chemical Mechanical Polish for ULSI Multilevel Interconnection Conference (CMP-MIC)*, p. 29, Santa Clara, February, 1997.

(18) B. Stine, D. Ouma, R. Divecha, D. Boning, J. Chung, D. Hetherington, I. Ali, G. Shinn, J. Clark, O. Nakagawa, S-Y Oh, "A Closed-Form Analytic Model for ILD Thickness Variation in CMP Processes," *1997 Chemical Mechanical Polish for ULSI Multilevel Interconnection Conference (CMP-MIC)*, p. 266, Santa Clara, February, 1997.

(19) B. E. Stine, D. S. Boning, J. E. Chung, L. Camilletti1, F. Kruppa, E. R. Equi, W. Loh, S. Prasad, M. Muthukrishnan, D. Towery, M. Berman, and A. Kapoor, "The Physical and Electrical Effects of Metal Fill Patterning Practices for Oxide Chemical Mechanical Polishing Processes," *IEEE Trans. on Electron Devices*, Vol. 45, No. 3, pp. 665-679, March 1998.

(20) D. Boning, J. Chung, D. Ouma, and R. Divecha, "Spatial Variation in Semiconductor Processes: Modeling for Control," *Electrochem. Society Meeting*, Montreal, CA, May 1997.

(21) B. E. Stine, D. S. Boning, and J. E. Chung, "Rapid Characterization and Modeling of Spatial Variation: A CMP Case Study," *KLA-Tencor Yield Management Seminar*, CMP Metrology Session, Semicon West'97, July 1997.

(22) B. E. Stine, D. S. Boning, J. E. Chung, D. Ciplickas, J. K.

Kibarian, "Simulating the Impact of Poly-CD Wafer-Level and Die-Level Variation On Circuit Performance," *Second International Workshop on Statistical Metrology*, Kyoto, Japan, June 1997.

(23) N. M. Muthukrishnan, S. Prasad, B. E. Stine, W. Loh, R. Nagahara, J. E. Chung, D. S. Boning, "Evaluation of pad life in chemical mechanical polishing process using statistical metrology," Manufacturing Yield, Reliability, and Failure Analysis session, *SPIE 1997 Symposium on Microelectronic Manufacturing*, Austin TX, Oct. 1997.

(24) D. Ouma, B. Stine, R. Divecha, D. Boning, J. Chung, G. Shinn, I. Ali, and J. Clark, "Wafer-Scale Modeling of Pattern Effect in Oxide Chemical Mechanical Polishing," Manufacturing Yield, Reliability, and Failure Analysis session, *SPIE 1997 Symposium on Microelectronic Manufacturing*, Austin TX, Oct. 1997.

(25) O. S. Nakagawa, S.-Y. Oh, F. Eschbach, G. Ray, P. Nikkel, R. R. Divecha, B. E. Stine, D. O. Ouma, D. S. Boning, and J. E. Chung, "Modeling of CMP-Induced Pattern-Dependent ILD Thickness Variation in Multilevel Metallization Systems," *Advanced Metalization Conference*, San Diego, CA, Oct. 1997.

(26) A. Maury, D. Ouma, D. Boning, and J. Chung, "A Modification to Preston's Equation and Impact on Pattern Density Effect Modeling," *Advanced Metalization Conference*, San Diego, CA, Oct. 1997.

(27) B. E. Stine, V. Mehrotra, D. S. Boning, J. E. Chung, and D. J. Ciplickas, "A Methodology for Assessing the Impact of Spatial/Pattern Dependent Interconnect Parameter Variation on Circuit Performance," *1997 International Electron Devices Meeting*, pp. 133-136, Wash. DC, Dec. 1997.

(28) B. Stine, D. Ouma, R. Divecha, D. Boning, J. Chung, "Rapid Characterization and Modeling of Pattern Dependent Variation in Chemical Mechanical Polishing," *IEEE Trans. Semi. Manuf.*, Feb. 1998.

(29) D. Ouma, C. Oji, D. Boning, J. Chung, D. Hetherington, and P. Merkle, "Effect of High Relative Speed on Planarization Length in Oxide Chemical Mechanical Polishing," *1998 Chemical Mechanical Polish for ULSI Multilevel Interconnection Conference (CMP-MIC)*, Santa Clara, Feb. 1998.

(30) D. Ouma, D. Boning, J. Chung, G. Shinn, L. Olsen, and J. Clark, "An Integrated Characterization and Modeling Methodology for CMP Dielectric Planarization," to be presented, *International Interconnect Technology Conference*, Burlingame, CA, June 1998.

(31) J. T. Pan, P. Li, F. Redeker, J. Whitby, D. Ouma, D. Boning, and J. Chung, "Planarization and Integration of Shallow Trench Isolation," to be presented, *VLSI Multilevel Interconnect Conference*, Santa Clara, CA, June 1998.

(32) T. Park, T. Tugbawa, J. Yoon, D. Boning, R. Muralidhar, S. Hymes, S. Alamgir, Y. Gotkis, R. Walesa, L. Shumway, G. Wu, F. Zhang, R. Kistler, and J. Hawkins, "Pattern and Process Dependencies in Copper Damascene Chemical Mechanical Polishing Processes," to be presented, *VLSI Multilevel Interconnect Conference*, Santa Clara, CA, June 1998.

Wafer Inspection Technology Challenges for ULSI Manufacturing

Stan Stokowski and Mehdi Vaez-Iravani

KLA-Tencor, One Technology Drive, Milpitas, CA 95035

The use of wafer inspection systems in managing semiconductor manufacturing yields is described. These systems now detect defects of size as small as 40 nm. Some high-speed systems have achieved 200-mm diameter wafer throughputs of 150 wafers per hour. The particular technologies involved are presented. Extensions of these technologies to meet the requirements of manufacturing integrated circuits with smaller structures on larger wafers are discussed.

INTRODUCTION

Wafer inspection systems help semiconductor manufacturers increase and maintain integrated circuit (IC) chip yields. The manufacturers buy these systems at a rate of about $700 million per year. This capital investment attests to the value of these systems in manufacturing IC chips.

The IC industry employs inspection systems to detect defects that occur during the manufacturing process. Their main purpose is to monitor whether the process is under control. If it isn't, the system should indicate the source of the problem, which the manager of the IC fabrication process (fab manager) can fix. The important inspection system characteristics here are defect detection sensitivity and wafer throughput. As we discuss later, sensitivity and throughput are coupled such that greater sensitivity usually means lower throughput. There are both physical and economic reasons for this relationship.

The relative value of sensitivity and throughput depends on the function of the inspection system. There are three general functional requirements for these systems: first, detecting and classifying defects in process development, second, in monitoring a process line, and third, in monitoring a station. In process development one is willing to have low throughput in order to capture smaller defects and a greater range of defect types. However, in monitoring a production line or a station, cost-of-ownership (COO), thus throughput, becomes relatively more important. In this case, of course, the sensitivity must be adequate to capture the yield-limiting defects.

Evolution of the semiconductor manufacturing industry is placing ever greater demands on yield management and in particular, on metrology and inspection systems. Critical dimensions are shrinking to 0.13 μm and 0.10 μm in the near future. Wafer size is increasing from 200 mm to 300 mm. Economics is driving the industry to decrease the time for achieving high-yield, high-value production. Thus, minimizing the total time from detecting a yield problem to fixing it determines the return-on-investment for the semiconductor fabricator.

Thus, inspection systems are evolving from stand-alone "tools" that just found defects to a part of a more complete solution where detecting defects, classifying them, analyzing these results and recommending corrective action are their functions. We do not have space here to consider and review all aspects of yield management systems, which are nevertheless important. Thus, we concentrate on the front end of this process, which is the detection phase, and briefly on classification. In this paper we describe the physical laws and engineering system constraints that determine what an inspection system can do.

HISTORICAL PERSPECTIVE

In the 1970's and early 1980's manual inspection was the norm. Critical dimensions were micrometers and yield limiting defects were easily seen visually. However, production yields were low and their rate of increase slow. Furthermore, the variance in the inspection process was considerable, thus, the desire for automated inspection.

Two types of inspection tools appeared in the early 1980's. One type was an automated microscope that captured bright-field images of patterned wafers and looked for defects by comparing die images (die-to-die comparison). The inspection rate was about 0.1 100-mm diameter wafers per hour (wph). About the same time a tool that detected particles on bare silicon wafers became available. This tool found particles by detecting laser light scattered from the particle, commonly designated in microscopy as dark-field. Its inspection rate was quite rapid at the time, equivalent to about 30 100-mm wph. Its minimum detectable dielectric particle diameter was about 3 μm.

A large difference between the bright field and dark field approaches is that in the latter, one typically has a situation in which the average size of the detected defect is much

CP449, *Characterization and Metrology for ULSI Technology: 1998 International Conference*
edited by D. G. Seiler, A. C. Diebold, W. M. Bullis, T. J. Shaffner, R. McDonald, and E. J. Walters

smaller than the resolution of the optical system. This is why dark-field systems have higher throughput.

Over the ensuing years these inspection tools became more sensitive and faster. In the mid 1980's several companies introduced dark-field tools that measured defects and particles on patterned wafers. Inspection rates for 200-mm wafers reached 100 wph for unpatterned wafers and 30 wph for patterned wafers. Sensitivity for detecting particles and defects was now sub-micron.

In the early 1990's the bright field tools became more sensitive with higher resolution microscope objectives, and used both cell-to-cell comparisons for array regions of the die, and die-to-die comparisons for logic regions. Throughput increased dramatically with the use of time-delay-integration (TDI) detectors. The detector pixel rate was 100 million pixels per sec (Mpps) and the image processing computer ran at the same rate. For a pixel size of 0.39 μm the tool ran at a 200-mm wafer inspection rate of 1 wph. Sensitivity to defects was about the same as the pixel size.

The year 1995 saw the appearance of a fast scanning-electron-beam microscope (SEM) for inspection. It had a spot diameter of 100 nm and a pixel rate of 12.5 Mpps, which is equivalent to a throughput of 0.05 wph. It also had the capability for inspecting 300-mm wafers, the first system to accomplish this.

This very brief and extremely condensed history brings us to the present capabilities of inspection systems. Bright field systems now use broad-band visible light illumination and high magnification objectives (resolution of 0.4 μm). The pixel rate is 400-600 Mpps with throughputs in the range of 1 to 8 wph, depending on pixel size.

Dark-field systems have achieved a sensitivity of 60 nm (polystyrene spheres on bare silicon) at 50 wph and 150 nm defects on patterned wafers at 30 wph.

Defect review, classification, and analysis are now being integrated to decrease the time for correcting yield problems.

INSPECTION SYSTEM BASICS

To understand how inspection systems will meet industry requirements in the future, we now consider some of the basic physics, engineering, and economic constraints imposed on these systems.

Physics

To inspect an object we look at it via some interrogating means, which are usually photons or electrons scattered by the object. The detected scattered photons or electrons as a function of position (an image) hopefully contains the information needed to determine whether a defect is present. An image processing system then decides if there is a defect. Thus, defect detection naturally consists of three main steps: first, obtaining the image, second, processing the image, and third, applying criteria to this processed image to detect defects.

We find it interesting to compare inspection technology with that of lithography, in particular, the exposure process. Lithography is almost exclusively optical (I-line [365 nm wavelength], deep ultraviolet [DUV, 248 nm] , 193 nm, and eventually extreme ultraviolet [EUV, 13 nm]). The print rate of optical lithography is about 10^{10} resolution elements per sec now and is increasing to 10^{11} over the next few years. This high print rate, of course, is a consequence of the massive parallelism of optical techniques. On the other hand, the highest inspection rate currently is 6 x 10^8 pixels per sec. However, optical lithography has an easier task in that it does not have to process and analyze an image.

The challenge for inspection tools is then to detect small defects with a system resolution spot size much larger than the defect size. Fortunately, as we shall see, one does not have to "resolve" a defect in order to detect it. Resolution, appropriately, does impact defect classification and identification. However, again even for performing these functions, we can sometimes obtain sufficient information without necessarily "resolving" the defect.

Even the first step of obtaining the image, by its nature, includes an optical processing step. How we illuminate and collect the resultant scattered light determines the contrast between a defect and the background in which it resides (surface or pattern scattering). Ideally one wants to maximize this contrast by judicially choosing the optical arrangement. Fortunately there are tools and techniques for accomplishing this choice. We describe some of them in this paper. In addition, for periodic array cells optical spatial filtering is effective. Finally light polarization plays an extremely important role in enhancing sensitivity.

This paper concentrates on optical techniques for wafer inspection because they are most commonly used. However, before we describe these techniques in more detail, it is worth looking at the differences between an optical, or photon, system and an e-beam system for inspection. These differences are in the scattering process itself and in the image contrast.

Electron scattering occurs near the surface of the inspected object and thus, a SEM inspector looks at the surface morphology. Also because electric fields affect electron trajectories a SEM inspector can look for electrical defects.

Next, the throughput and sensitivity of the SEM system depends on its resolution and detected electron current. Typically the detected electron current is in the range of 25 nanoamps or more importantly, 10^{11} electrons per second. In a typical dark-field optical system the electron current at the photocathode of a photo-multiplier tube is in the range of 10^8 to 10^{14} electrons per second The contrast in a SEM system is in the range of 10%-50% whereas in a dark-field optical system the contrast varies from 10% to 10^6%. Thus, a dark field optical system has an inherent contrast advantage that translates into a throughput advantage.

A SEM inspector, however, does have very high sensitivity because of its 100 nm resolution and it can detect electrical defects, which an optical system can not do.

Bright-field and dark-field systems

All optical inspection systems depend on photon scattering from the inspected object. Bright-field systems collect both the scattered and reflected light through the same aperture to obtain an image. In addition, one illuminates the object through the objective aperture. Basically these systems are a high-speed microscope. Dark-field systems, on the other hand, only collect the scattered light; no part of the reflected light falls within the collection angle. They can have a multiplicity of configurations, depending on the angle and type of illumination, collection angles, and detector type.

We categorize dark field systems into two main groups: single dark field and double dark field. Single dark field occurs when either the illumination or the collection angle is greater than 45°; double dark field, when both the illumination and collection angle are >45°. As we shall see later, these definitions have their origin in the angular dependence of surface scattering. Figure 1 shows schematically single dark field optics and double dark field optics. This figure also shows two different methods of obtaining an image: the single dark field system using a TDI detector and the double dark field system using a scanner, such as an acousto-optic deflector (AOD), coupled with a PMT detector.

All these systems have their advantages and disadvantages for detecting different defect types. In general, dark field systems are particularly useful when the defect has some high-spatial-frequency topography, whereas bright field is good at finding planar defects. In most cases dark field systems find defects much smaller than the system resolution or spot size; whereas, in bright-field systems the detected defects are about the same size as the system resolution. This fact has important implications for system throughput. However, of particular importance for system sensitivity is the fact that *no one optical arrangement is optimum for detecting all possible defect types*.

Particle scattering

Particles or their effects are the source of a majority of defects in IC chips. Thus, understanding particle scattering

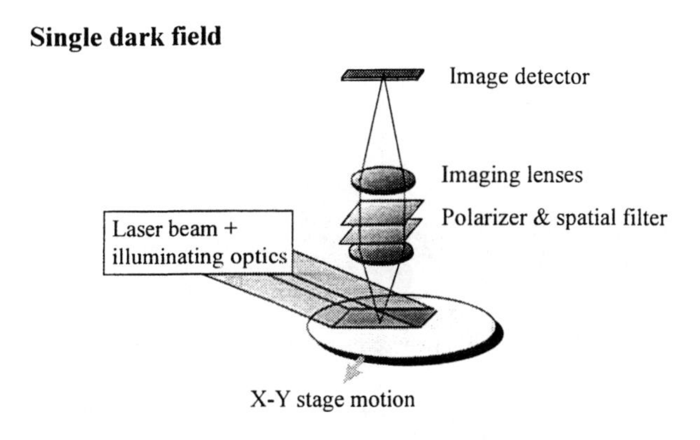

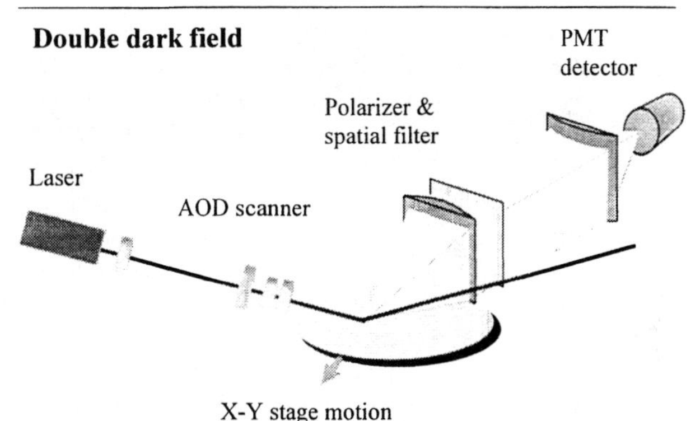

FIGURE 1. Schematics of a single dark-field configuration and a double dark-field configuration.

helps to design sensitive inspection tools. We have a program that calculates scattering from a sphere on a layered substrate when illuminated by a plane wave. It uses a formalism developed by Bobbert and Vleiger (1). Assi (2) and Stokowski et al.(3) developed application software that calculates the polarized scattering into the 2π hemisphere above the substrate. We used this software tool to calculate the scattering patterns shown in this paper, unless otherwise stated.

In Fig. 2 we show our definition of the spherical coordinates that we use in discussing scattering from particles, surfaces and defects. The polar angle is defined from the surface normal and the azimuthal angle counter-clockwise from the reflected beam projected on to the surface plane. The illumination polarization definitions are *s* (E perpendicular to the incidence plane) and *p* (E parallel to the incidence plane). The scattered field polarization is s or p relative to the plane containing the surface normal and the scattered light direction. When talking about polarized scattering, we use subscripts to refer first to the incident polarization and second to the scattered polarization; for example, E_{sp} refers to the p-polarized scattered far field with s-polarized light incident.

To understand some of the basic scattering rules we start with a polystyrene latex sphere (PSL) on silicon. Although PSL spheres are not found in IC fabs, they are convenient

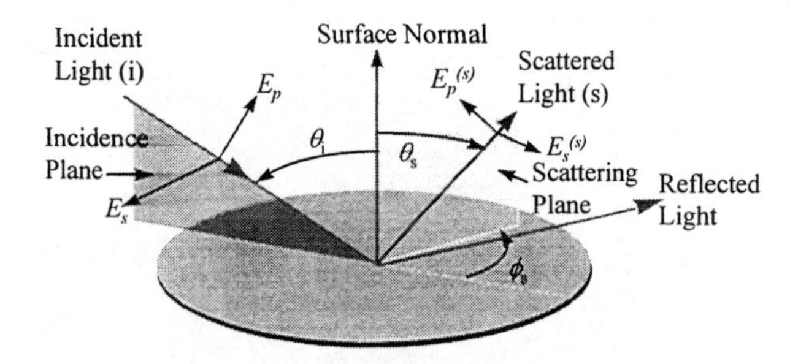

FIGURE 2. Diagram for coordinate and polarization definitions.

for calibrating inspection systems because they are spheres of known diameter.

Figure 3 shows scattering from PSL spheres as a function of diameter, polarization, and angle of incidence. Of particular interest is the advantage of using p-polarized light for detecting small particles. The total scattering cross section for a 60 nm PSL sphere with p-polarized illumination at 70° incident angle is 86 times that with s-polarized light and 42 times that with normal incidence. Also note that most of the scattered light under oblique p-polarized illumination is in the polar angular range of 20° to 70°. Thus, an optimum system for detecting small particles uses obliquely incident p-polarized light and collects the scattered light over a large solid angle, 20° to 70° in polar angle and almost 360° in azimuth.

The advantage of p polarization for small particle detection is a consequence of the standing E-M wave fields above a surface. We can best describe this effect by realizing that for a small enough particle, the particle acts as a probe of the near field because the far-field scattering depends on the E-M field present at the particle. If the particle is small enough, it does not perturb substantially the field that would be present in the absence of the particle.

For 70° incidence Fig. 4 shows the electric field as a function of distance above a silicon surface for s and p polarization. Note that s-polarized light has a low field at the surface ("dark fringe"), whereas, p-polarized light is at a maximum. It follows then that small particle scattering is greatest for p polarization.

Experimental results confirm the utility of the scattering model we use. For example, Figs. 5 and 6 show the agreement between measurements and modeling results for two types of collectors as a function of incidence angle and polarization. One collector covers polar angles from about 25° to 72° and the other from about 6° to 20°, with almost 360° in azimuth.

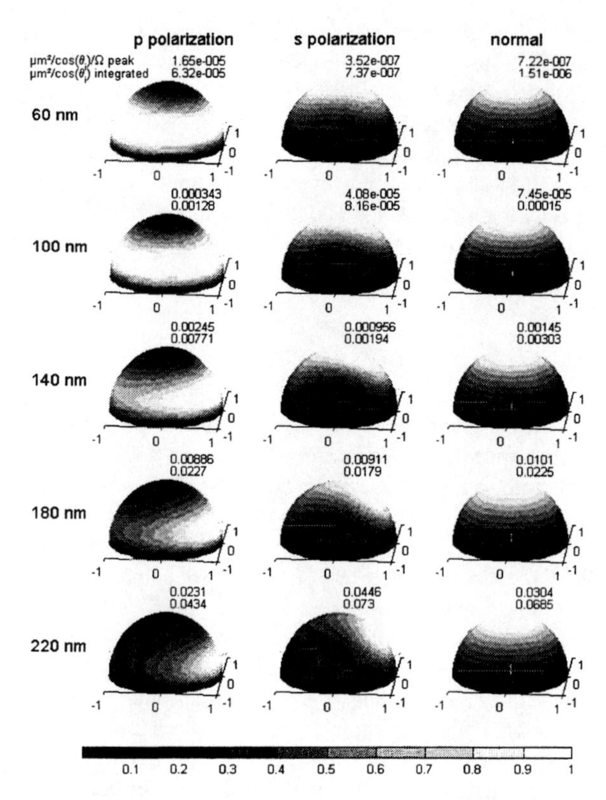

FIGURE 3. Scattered intensity patterns of PSL spheres on silicon as a function of diameter, polarization, and incidence angle. The 488-nm plane wave comes in from the left at 70° incidence and view is about -90° in azimuth from the incidence plane. The last column of images is for normal incidence, circular polarization. The numbers correspond to the peak differential cross sections and the total integrated cross sections in μm² divided by the cosine of the incidence angle.

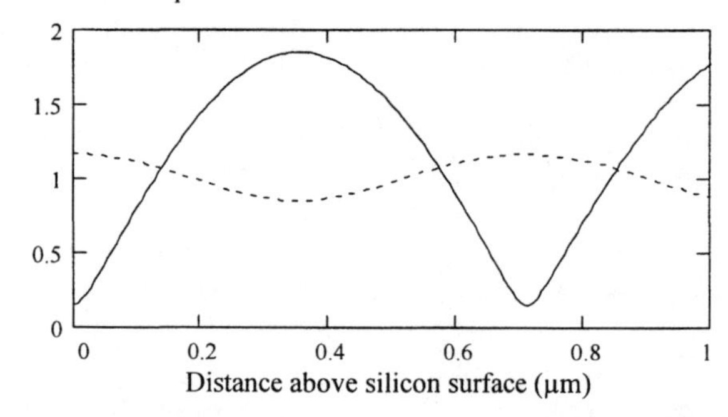

FIGURE 4. Magnitude of the electric field as a function of distance above a silicon surface for s polarization (solid line) and p polarization (dashed line) for 70° incidence and a input field amplitude of 1.

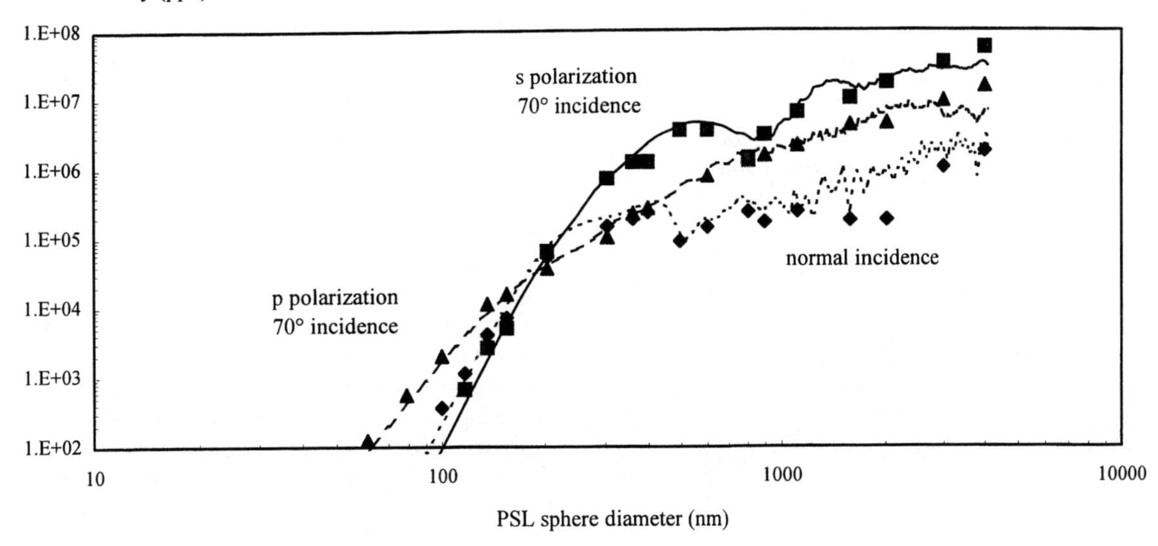

FIGURE 5. Scattering model calculations agree with measurements for PSL spheres on silicon with 70° incidence angle, s polarization (squares) and p polarization (triangles), and normal incidence (diamonds). The collector covers the polar angles from about 25° to 72° and nearly 360° in azimuth.

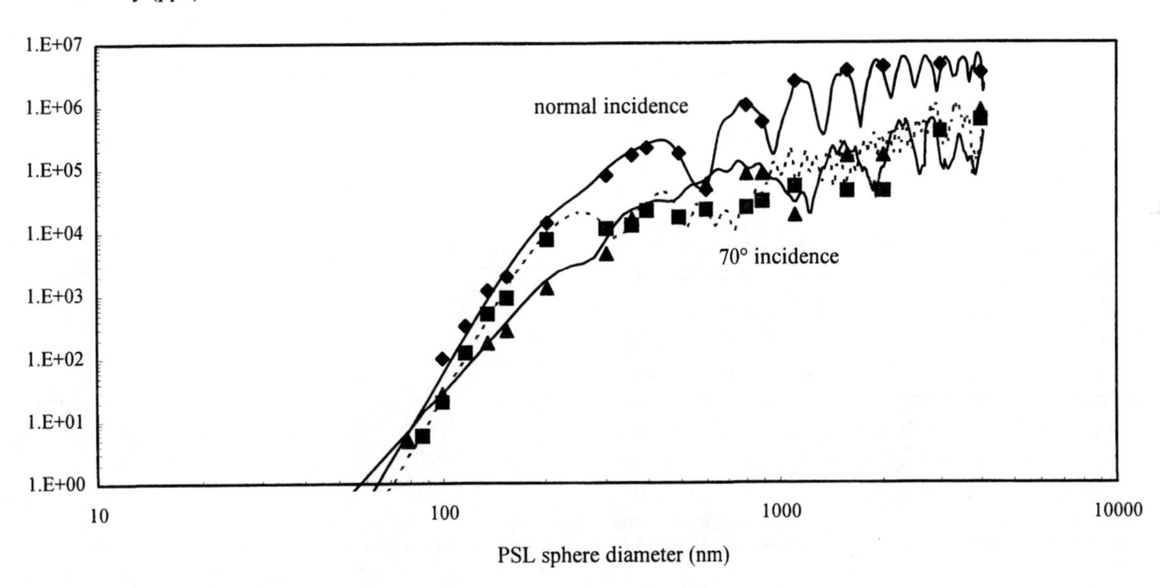

FIGURE 6 Scattering model calculations agree with measurements for PSL spheres on silicon with 70° incidence angle, s (squares) and p (triangles) polarization, and normal (diamonds) incidence. The collector covers the polar angles from about 6° to 20° and 360° in azimuth.

The sensitivity of an inspection system for small particle detection depends on the particle material. In the Rayleigh limit the total integrated scattering of a sphere in a medium depends on

$$\frac{d^6}{\lambda^4} \cdot \left| \frac{(n^2 - 1)}{(n^2 + 2)} \right|^2 \cdot |E|^2 \tag{1}$$

where d is the sphere diameter, λ is the illuminating wavelength, n is the refractive index of the sphere and E is the electric field at the sphere (4). Thus, higher refractive index materials, such as semiconductors and metals, scatter more light. Figure 7 compares the total integrated scattering (TIS) for PSL, silicon, and aluminum spheres on silicon. As a consequence, if a system can detect 60 nm PSL spheres on silicon, it can detect 40 nm aluminum spheres on silicon.

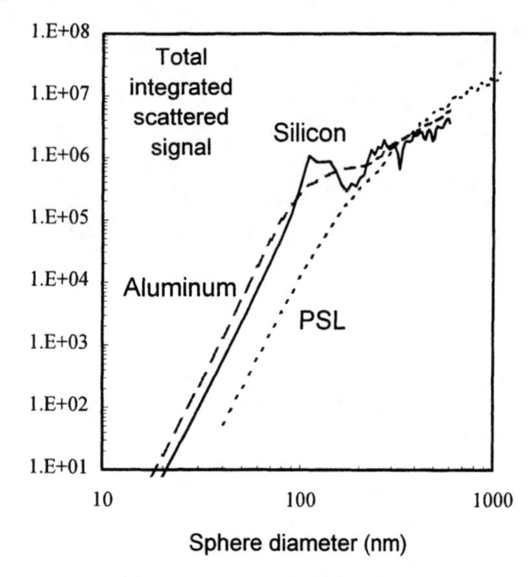

FIGURE 7. Total integrated scattering cross sections for PSL (dotted line), silicon (solid line), and aluminum (dashed line) spheres on silicon.

Particle sizing is always of interest. Typically the industry uses PSL spheres as a calibration standard. If an inspection system uses the total scattered light intensity as an indication of particle size, Figs. 5 and 6 reveal a problem: the intensity is not a monotonic function of the sphere diameter. (Oblique incidence has less of a problem than normal incidence.) Furthermore, the scattered intensity from spheres of other materials obviously do not relate in a simple fashion to the PSL sphere response unless one compares the curves for sphere diameters less than 100 nm. To obtain better sizing one needs to use more than one configuration or mode as suggested, for example, by the responses shown in Figs. 5 and 6.

Surface Scattering

For unpatterned wafers the background noise comes from surface scattering. There are several references that describe surface scattering in great detail; for example, Church et al (5) and Stover (6). We will only describe here the key parameters that determine surface scattering.

For surfaces that are rough, but with height variations much less than the light wavelength, the scattered power per unit solid angle as a function of the polar angle and azimuth is:

$$\frac{dP}{d\Omega} = P_i \cdot \frac{16\pi^2}{\lambda^4} \cdot \left[\cos(\theta_i) \cdot \cos^2(\theta_s) \cdot Q_{p,q}(\theta_i, \theta_s, \phi_s)\right] \cdot PSD(f_x, f_y) \tag{2}$$

where P_i is the input power, $d\Omega$ is the differential solid angle, θ_i, θ_s, ϕ_s are defined in Fig. 2, $Q_{ij}(\theta_i, \theta_s, \phi_s)$ is the polarization factor and $PSD(f_x, f_y)$ is the power spectral density of the surface height variation as a function of the x and y components of the surface spatial frequency. The frequency components, f_x and f_y, are, in turn, related to the scattering angles through the diffraction equations.

Cutting through all this detail, the "bottom line" is: once the surface PSD characteristic is known, we can calculate, to a good approximation, the angular distribution of the light scattered by the surface.

Obviously one tries to minimize surface scattering if one is to obtain good defect sensitivity on rough surfaces. Of particular importance is the polarization factor of Eqn. 2. Figure 8 shows the variation of this factor for silicon over the full scattering hemisphere for the four combinations of input and scattered polarizations and 70° incidence. To minimize surface scattering using the ss polarization combination and collecting light in the vicinity of 90° and 270° azimuth is very effective. In addition, depending on the underlying material the pp polarization combination and collecting scattered light in the forward direction is useful.

In eqn. 2 we can also see that the $\cos(\theta_i)$ and $\cos^2(\theta_s)$ terms also imply that greater sensitivity to detecting particles is obtained in the double dark field configuration, where both angles are >45°.

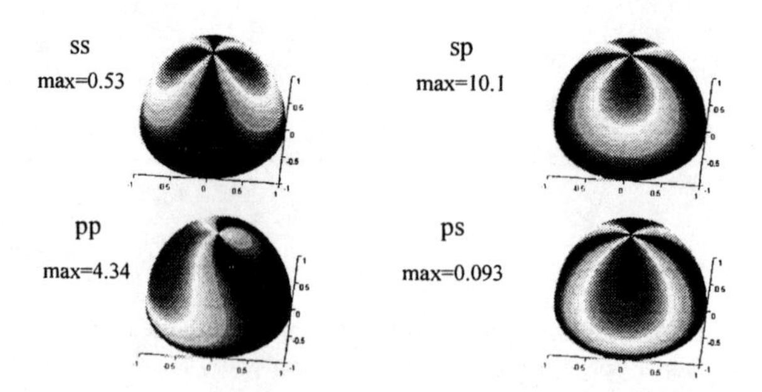

FIGURE 8. Relative magnitudes of the optical polarization factors $Q_{i,j} \cdot \cos^2(\theta)$ for 70° incidence on silicon: ss, pp, sp, and ps polarization combinations over the scattering hemisphere. (Gray scale converted from color: bright band contour is 0.5 of the maximum.) The 488-nm plane wave comes in from the left; view is near -90° in azimuth from the incidence plane.

Unpatterned inspection systems measure the background scattering level, which the industry refers to as "haze". The measured haze value obviously depends on where in the hemisphere we collect the scattered light. Obviously haze is related to the PSD characteristics of a surface, but the relationship is not necessarily a simple one.

Pit scattering

Pits are of great interest to the silicon wafer manufacturers. Pits have been a problem for inspection systems because they also scatter light and are indistinguishable from particles in a single channel detection system. The wafer manufacturers need to classify pits and particles on silicon wafers. Pits are octahedral voids in Czochralski-grown silicon that have been exposed at the surface by the polishing process. They are also known as "crystal-originated particles" (COP), obviously a misnomer. They sometimes are a single pit and, in a large number of cases, partially overlapping double pits.

Pits and scratches are "surface-breaking" defects; i.e., they are into the surface. The scattering characteristics, therefore, of pits and particles are different and as a consequence, we can classify detected defects as pits or particles if we have information from multiple channels or modes.

The first difference between pits and particles comes from their responses to normal and oblique illumination. Figure 9 is a simple cartoon of this difference. Part a shows the normal illumination with a sphere on a surface intercepting a cross section of the beam. Part b is the condition for an oblique beam where the illuminated area on the surface is the same as Part a. Note that in this plane the same-sized sphere intercepts a larger fraction of the incident beam cross section. Thus, a sphere will scatter significantly more with oblique incidence (see Fig. 3). However, Part c shows that in oblique incidence a pit is at a significant disadvantage relative to a sphere on the surface for scattering light. Thus, comparing the scattered light in normal and oblique incidence can classify pits and particles.

A more important difference between pits and particles is the angular pattern of the scattering. Both theoretical calculations and experimental results show that particles scatter light principally into the polar angle range from 20° to 70° when illuminated with p-polarized light. In contrast, pits scatter primarily toward the normal; therefore, comparing the light scattered into higher angles with those toward the normal will also classify pits and particles. Even for normal incidence this separation works; however, oblique incidence works best. We show experimental results for the p-polarized oblique incidence case in Fig. 10

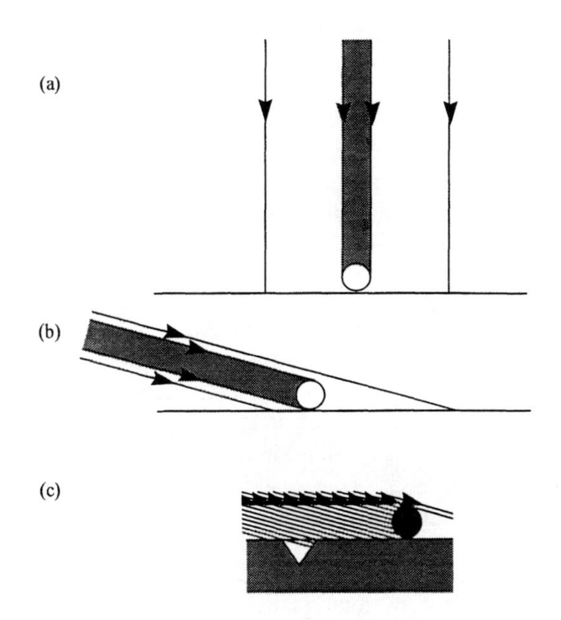

FIGURE 9. Schematic illustrating the difference between particles and pits relative to the illumination incidence angle.

Scratch scattering

Scratches are important in CMP processes and may be yield-limiting. Scratches preferentially scatter perpendicular to their long dimension. Of course, real scratches are not perfect linear defects; in many cases they have cross-sectional variations along the scratch, may have particulate debris nearby, and commonly are "chatter marks." These "chatter marks" or "micro-scratches" actually are a series of short small scratches along a line perpendicular to the long dimension of the scratches.

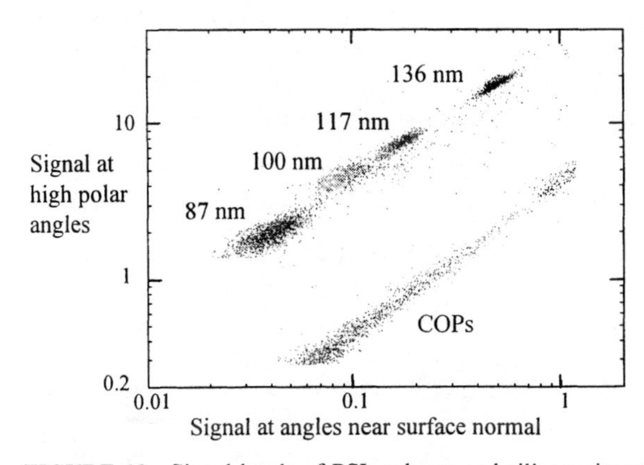

FIGURE 10. Signal levels of PSL spheres and silicon pits with about 25° to 72° collection vs. about 6° to 20° collection, showing the ease of classification.

Scratches also scatter primarily toward the normal, similar to pits. Furthermore, for uniformly detecting scratches of any orientation, normal incidence is preferred.

Dielectric film effects

On patterned wafers dielectric films are present. These lead to a couple of complications. One is the interference effect that produces color under broad band illumination and contrast variation under monochromatic illumination. These effects are particularly troublesome if the film thickness is not uniform and one is trying to do a die-to-die comparison. In bright field systems broad band illumination has helped. In dark field systems circularly polarized light is extremely useful in minimizing the film effect.

As a simple example of this we show the scattering cross section of a PSL sphere on silicon dioxide on aluminum as a function of the oxide thickness in Fig. 11. Note the substantial variations of total scattered light with film thickness with both s and p polarization with oblique incidence. However, because the s and p scattering are out-of-phase with respect to each other, scattering with circular polarization, which has both, is much less affected by film non-uniformity. For normal incidence, of course, s, p, and circular are all equivalent and the film effect is worse than that seen with oblique incidence.

Digs and scratch detection will also be affected by dielectric film thickness and polarization, but their variation in scattering is not in phase with the particle scattered intensity.

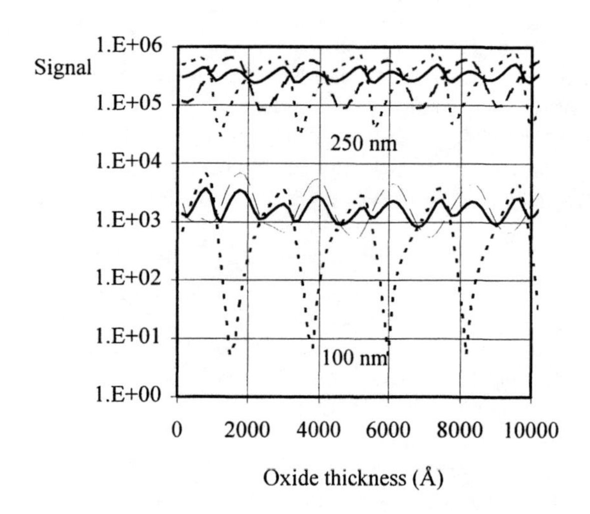

FIGURE 11. Scattered intensity for a 100 nm PSL sphere and a 250 nm PSL sphere on silicon dioxide on aluminum as a function of oxide thickness and input polarization, p polarization (long dash), s polarization (short dash), and circular polarization (solid). Scattered light collected from about 25° to 72° polar angle.

Previous layer defects

One may want to see or not see previous layer defects, depending on the system application. Usually for monitoring equipment one does not want to see down into the previous layers. Oblique illumination with s polarization has much less penetration of energy through transparent dielectrics than normal illumination and thus is preferred for detecting current layer defects.

System considerations

An inspection system obtains an image (electron or photon), then processes it to determine if a defect is present, classifies it according to some criteria, and finally passes the information on to a yield management system. Each of these steps may have certain limitations and we describe briefly some of the system considerations.

Ideally an inspection system should have high sensitivity, high throughput, and low cost of ownership (COO). However, all these desired system characteristics are coupled and one must do trade-offs to achieve the optimum system.

The semiconductor industry is shrinking the area density of devices by 40% per year. The challenge for the companies developing inspection systems is to maintain image acquisition time and COO constant while going to higher and higher image resolution. We consider how image acquisition, image processing, and defect classification might meet this challenge.

Obtaining the image

Image acquisition is the first step in the inspection process. It consists of illuminating the wafer with a source (lamp or laser), imaging or collecting the scattered light, and detecting this light with a photo-detector (PMT, TDI, or CCD). The source has to be bright enough to provide sufficient photo-electrons from the detector to obtain a reasonable signal-to-noise ratio (S/N). In the case of unpatterned wafers S/N should be about 8 to 10 for 95% capture probability and one false count per 200-mm wafer.

The source in bright field systems is usually a high-pressure mercury (mercury-argon) arc lamp, whereas dark field systems use lasers. The recent development of reliable solid-state, diode-pumped lasers of > 1 Watt power has provided inspection systems with sufficient power for most inspection tasks.

Image acquisition by existing inspection systems fall into one of two main categories: imaging systems or scanner systems. In imaging systems the source optics illuminates the inspected area, which microscope optics then image on to a TDI or CCD camera. In a scanning system a focused laser beam "paints" the inspected area and a single element detector (usually a PMT) detects the collected scattered light (ref. Fig. 1).

These two types of systems have their own advantages. An imager is basically a fast optical microscope; thus, the optical system design is straightforward. A TDI or CCD camera obtains the image elements in a parallel fashion. A scanner-type system has no constraints on the angles over which one collects the scattered light because it is a non-imaging system. It obtains the image in a serial fashion. An imaging system is useable in a bright-field or single dark field configuration, but not with double dark field. A scanner can have all three configurations. Its disadvantage is the high speed required for the scanner, the detector and its electronics.

At this time we need to describe the relationship between various terms commonly used in inspection systems, such as pixel size, spot size, and system spot size. In a camera-based system pixel size, as referenced to the wafer surface, is the detector element size divided by the magnification of the collection optics from the wafer to the detector. Note that this definition has nothing to do with the resolution of the objective lens. In a scanner system the focused Gaussian spot size is the full width between the e^{-2} points. Obviously in this case the resolution of the focusing optics determines the spot size. For these systems the pixel size is the spot size divided by the number of electronic samples per e^{-2} width.

The sensitivity of an inspection system is related to the system spot size, which includes the resolution, or modulation transfer function, of the optics, the detector element size, the front-end electronic bandwidth, and convolutions done after digitization. In addition, noise limits sensitivity; noise sources include photo-electron shot-noise, detector noise, electronic noise, noise from the analog-to-digital converter (ADC), aliasing noise, and spatial quantitization noise. These last two noise sources depend on the spatial sampling frequency relative to the system spot size. Generally one increases sensitivity by decreasing the system spot size.

Note that system spot size is a governing factor in sensitivity, *not* pixel size.

Throughput of an inspection system on the other hand is inversely related to the *square* of the pixel size. Thus, the time for actually inspecting the wafer is determined by the pixel rate of an inspection system, given its pixel size. Additional time affecting throughput is taken by operations such as wafer loading and unloading, alignment and registration, and data processing.

Fig. 12 shows the relationship between pixel size and inspection time for different pixel rates. Clearly one tries to use as large a pixel as one can while achieving a given sensitivity. Here is where dark field systems have a great advantage over bright field systems; the system spot size to defect size ratio is considerably greater than 1 in dark field systems, whereas bright field system have a ratio closer to 1. For example, albeit, a particularly advantageous situation, a dark field system exists that can detect small PSL spheres on bare silicon with a defect-to-spot area ratio of 3×10^5.

The detector or scanner is limited in speed. For imaging systems the fastest ones employ TDI detectors with 400-600 Mpps. The fastest scanners use AOD technology, currently running at an equivalent pixel rate of about 50 Mpps. However, the slower pixel rate in a scanning system is more than compensated by the larger defect to pixel size ratio in a dark field configuration.

Processing the image

After obtaining the image, the image processor has to determine the presence of a defect and to accomplish this function at a rate almost as large as that for the front-end detection. In a simple unpatterned wafer inspection system a simple threshold scheme works well. However, die-to-die and/or cell-to-cell comparisons are required for patterned wafers.

For DRAM chips with their highly periodic structures some inspection systems use optical spatial filtering to eliminate the light scattered from the periodic structure before it reaches the detector. Thus, only light scattered by

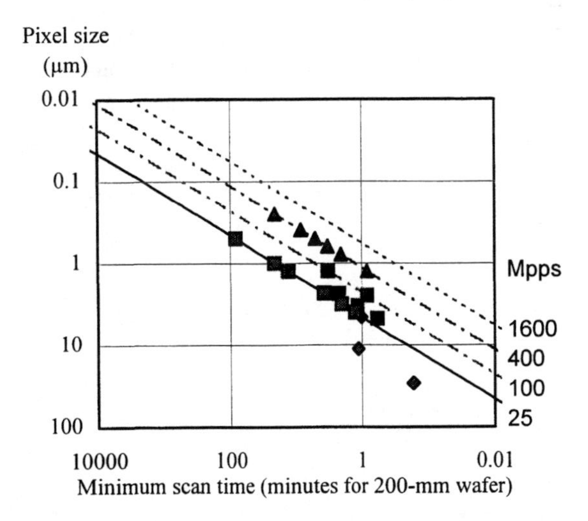

FIGURE 12. Actual inspection time (no overhead) for a 200-mm wafer as a function of pixel size, with pixel rate (Mpps) as a parameter. Points indicate range of some existing systems.

the non-periodic defects is detected. This technique only works with coherent laser illumination. Optical filtering typically lowers the background scattering from the array by 100 times or greater.

The speed and cost of the image processor for patterned wafer inspection is critical. Fortunately, inspection systems can leverage off of the improvements in the microprocessor industry. In a sense, we benefit from developments in the same industry we are helping to improve. Computer speeds have improved by ~30% per year over the last three decades (7) and the cost per MIP has fallen by ~65% per year. However, the semiconductor manufacturing industry is increasing the area density of IC devices by 40% per year. Thus, the time it takes for doing the image processing of a wafer should remain approximately constant, even as the required pixel rate needs to increase by 40% per year to maintain throughput. Processing cost should also fall, except for the fact that processing is becoming more complex (more MIPS!).

Classifying defects

In early days of wafer inspection systems classification consisted of reporting a defect size. High resolution, bright-field systems could resolve the defect and determine its area. Dark-field systems, because they detected defects much smaller than the system spot size, measured only the scattering light signal in a single channel. Defect sizing came from comparing this signal against a calibration curve for PSL spheres on the substrate.

For extended defects post-processing algorithms in current systems can classify clusters, scratches, and random defects. For defects smaller than the scanner spot size or the imager optical resolution, however, real time classification requires multiple views or channels. As mentioned in the section on pits vs. particles in this paper, multiple angles of incidence or multiple collection channels provide classification capability. Of course, all this comes at a price because each channel needs support, particularly in image processing.

FUTURE NEEDS AND DEVELOPMENTS

"Predictions are always difficult ....especially about the future." Niels Bohr

Fortunately here we do not have to predict the weather, the stock market, or the economy, all rather chaotic systems. We can predict some evolutionary changes, but revolution is in the minds of inventors. In light of this we'll confine ourselves to some evolutionary changes that will occur.

Smaller critical dimensions, larger wafers and more integrated inspection systems are part of our future. Inspection systems will follow the lead of lithography and migrate to ultraviolet wavelengths. We will also see an even closer coupling of inspection with process equipment, review stations, and yield management systems. However, before briefly discussing these coming developments, we talk about a current gap in inspection, that of inspecting contacts and vias or high aspect ratio structures.

Inspecting contacts and vias

Both optical and SEM inspectors are effective in helping to develop and control IC manufacturing processes. However, there is one major gap in the performance of current systems. It is seeing small defects or residue at the bottom of high aspect ratio structures.

Optically one can detect partially filled or missing contacts in high resolution systems. However, if a residue of 5 nm is at the bottom of a 250 nm diameter by 1000 nm deep via, we are asking the optical system to detect a volume difference equivalent to a 75 nm diameter sphere at the bottom of the hole, a very difficult task (8). Thus, if contact/vias must be checked individually, we are not going to do it optically on real wafers. However, if all the contact/vias within a local area are incompletely etched, then optical means can detect it.

In a SEM system a voltage contrast mode can detect a residue at the bottom of a via or contact. However, SEM inspectors are not fast; thus, to inspect contacts/vias in a reasonable time, we must resort to sampling small areas. Therefore, here again, as with optical techniques, we can observe incomplete etching if this fraction is on the order of roughly 10^{-4}, but finding 5 nm of residue in one contact/via out of 10^{10} of them is beyond practical consideration.

Using UV in inspection systems

For detecting smaller defects bright field systems need the higher resolution of shorter wavelengths. However, in dark field systems the system spot sizes currently employed are not limited by the visible wavelength. Thus, it is not imperative that these systems use UV immediately.

In dark field systems the shorter wavelength of a UV laser leads to a greater scattering cross-section from particles on bare silicon surfaces. That is clear from the Rayleigh "blue sky" factor of λ^{-4}. Therefore, UV systems will be able to detect particles in the range of 20 nm diameter on smooth surfaces

In terms of patterned wafers, however, using UV has the following issues. In dark-field scattering mode operations, one ultimately relies on the phase associated with the interaction of light with the structures. Patterned and unpatterned wafers with films on them will both see a more rapid thin-film effect fluctuation. Thus, process variations across the wafer will have a greater effect with UV illumination. It is, therefore, not obvious that one necessarily gains in detecting defects on dense structures where the amount of scattered power is not an issue. The shorter wavelength will result in the generation of more diffraction orders in the Fourier space to filter out. For larger cell sizes, this also means that the orders are closer together, causing difficulty in removing them. UV optics and lasers, of course, must be developed and available. For non-PMT detection, UV necessitates tricks such as back-thinning of TDI/CCD detector arrays. UV light also can cause photochemical deposition of air-borne contaminants on the optical surfaces, thus necessitating e.g. a constant nitrogen purge

Even with all of these issues, which can be solved, UV systems will be available in the not too distant future.

Integrated inspection systems

"Time-to-results" is always an important driver in the industry. Thus, we will see more and more integration of inspection hardware units into an overall system that can find the defects, review them, and determine the source of the problem.

The industry has a great incentive to "shorten the loop". As a result there is considerable investigation into bringing metrology and inspection within the process chamber ("in-situ") or into a port on the process equipment. However, both technical and economic barriers exist that make accomplishing this difficult. High performance (sensitivity and throughput) inspection has engineering constraints that make it not easily compatible with process equipment. In addition, the cost of a metrology/inspection module has to be relatively low compared to present-day systems to make it cost-effective. On the other hand, we will see some development of integrated inspection units that are tuned to the specific defects generated by process tools and are sensitive to relatively large defects.

SUMMARY

Wafer inspection system performances have kept up with semiconductor manufacturing industry requirements. Both dark field and bright field systems continue to increase in sensitivity and throughput. To meet future needs these systems will go to higher resolution with faster image acquisition and processing. Real time classification will improve, with better coupling to review, data management, yield learning, and yield management. Ultraviolet wavelength systems will provide an additional increase in capability.

ACKNOWLEGMENTS

We thank Mustafa Akbulut, Kurt Haller, Ning Yin, and Guoheng Zhao for providing us with data and data analyses of pits and particles on silicon wafers.

REFERENCES

1. Bobbert, P. A. and Vlieger, J., *Physica* **137A**, 209-241 (1986)
2. Assi, Fadi Ismail, "Electromagnetic Wave Scattering by a Sphere on a Layered Substrate," Master of Science thesis submitted to the Faculty of the Dept. of Electrical and Computer Engineering, Univ. of Arizona (1990).
3. Stokowski, S. E., Assi, F. I., Cangellaris, A., and Yin, Ning, "Confirming Experimentally the Utility of the Mie-Weyl Formalism for Calculating E-M Scattering by a Sphere on a Layered Substrate", *Proc. 2nd Workshop on Electromagnetic and Light Scattering: Theory and Applications*, Moscow: Moscow Lomonosov State University (1997).
4. Hulst, H. C. van der, *Light Scattering by Small Particles*, New York: Dover Publications, Inc., (1981), p. 432
5. Church, E. L., Jenkinson, H. A., Zavada, J. M., *Opt. Eng.* **18**, 125-138 (1979).
6. Stover, J. C., *Optical Scattering: Measurement and Analysis, second edition*, Bellingham: SPIE Press, 1995.
7. Brenner, A., *Physics Today* **49**, 25 (1996).
8. Socha, Robert J., Neureuther, Andrew R., *J. Vac. Sci. Technol. B* **15**, 2718-2724 (1997).

INTERCONNECT - GENERAL

All-Optical, Non-Contact Measurement of Copper and Tantalum Films Deposited by PVD and ECD in Blanket Films and Single Damascene Structures

Matthew Banet, Henry Yeung, John Hanselman, Hank Sola and Martin Fuchs

Active Impulse Systems, Inc.
Natick, Massachusetts

Hieu "Robbie" Lam

SEMATECH
Austin, TX

This paper describes use of a non-contact, all-optical measurement technology, called impulsive stimulated thermal scattering (ISTS), to monitor the thickness of copper and tantalum films contained in a variety of microelectronic structures. ISTS measurements were performed using an instrument designed to automate metal film thickness determination for integrated circuit fabrication. The instrument measured: 1) thickness contour maps of PVD and electrochemically deposited (ECD) copper films; 2) center-point thickness of copper, tantalum and copper/tantalum structures with films ranging from 20 to 200 nm; 3) thickness profile measurements of copper/tantalum structures from the wafer edge, through the exclusion zone, and into the copper film; and 4) single-point thickness of copper damascene structures. Whenever possible, the ISTS measurements were compared to those made by conventional instrumentation, such as a 4-point probe. The ISTS system measured the above-mentioned samples rapidly (typically about 1 second per point) and with high spatial resolution (as low as 25 microns per spot) without contacting the film. The repeatability of the measurements for both copper and tantalum was about 1A.

INTRODUCTION

Copper is an attractive alternative to aluminum for interconnect metalization in integrated circuits due to its high electrical conductivity and desirable electromigration properties (1). When used in combination with low-k dielectric materials, copper offers the semiconductor industry a means to address interconnection requirements for next-generation integrated circuits characterized by smaller feature size and higher operating speed. A number of deposition methodologies are currently under investigation as the industry seeks a reliable process for copper interconnect metalization. A particularly successful process involves deposition of a thin (approximately 1000Å) copper seed layer by physical vapor deposition (PVD), followed by electrochemical deposition (ECD) of a thicker (typically 1.5 microns or more) layer.

The thickness of copper and other metals can be measured over a small area (as low as 25 microns) using a rapid, all-optical instrument (called the InSite 300) that has recently been introduced (2-4). This metrology instrument measures metal-film thickness directly on product wafers without contacting the sample, and has the potential to greatly increase process tool effectiveness by: i) reducing reliance on unproductive (and often unrepresentative) monitor wafers; and ii) increasing the time that process tools are fabricating actual product wafers.

When used with copper/tantalum structures, the metrology instrument performs detailed profiling of "edge-exclusion" zones near the copper film's edge, measures a "usable" copper film diameter, and measures high-resolution contour maps of both small and large areas of the copper film. This paper reports on the use of this instrument to measure the above-mentioned properties from copper films deposited by both PVD and ECD.

CP449, *Characterization and Metrology for ULSI Technology: 1998 International Conference*
edited by D. G. Seiler, A. C. Diebold, W. M. Bullis, T. J. Shaffner, R. McDonald, and E. J. Walters

ALL-OPTICAL MEASUREMENT TECHNIQUE

The above-described metrology instrument employs impulsive stimulated thermal scattering (ISTS), a laser-based photoacoustic measurement technique, to measure the thickness of the copper films (3,4). In ISTS two excitation laser pulses having a duration of about 500 picoseconds are overlapped at the sample to form an optical interference pattern containing alternating "light" (constructive interference) and "dark" (destructive interference) regions. Optical absorption of radiation in the light regions leads to localized rapid thermal expansion. This launches counter-propagating acoustic waves whose wavelength and orientation match those of the interference pattern. A probe laser beam irradiates the acoustic waves on the film's surface and is diffracted to form a signal beam that is modulated at the acoustic waves' frequency. The signal beam is detected and digitized in real time, resulting in a signal waveform similar to that shown in Fig. 1 (these data were measured from a 5200A Cu/250A Ta:N/1000A Oxide/Si structure). The acoustic frequency is rapidly determined with high signal-to-noise ratio, and then used to calculate the film's thickness as described below. The entire process is very rapid: the data shown in Figure 1 were obtained in about 1 second.

The acoustic wave excited in these measurements has a velocity that is a sensitive function of the film thickness. Thickness is calculated by considering the measured acoustic frequency, the spacing of adjacent light regions in the optical interference pattern, and the density and sound velocity of the sample. The acoustic wavelength (set by the spacing of the regions in the interference pattern) that is excited in the film can be varied in an automated fashion. In this way, data collected at multiple acoustic wavelengths can be analyzed film properties other than thickness, such as the film's density.

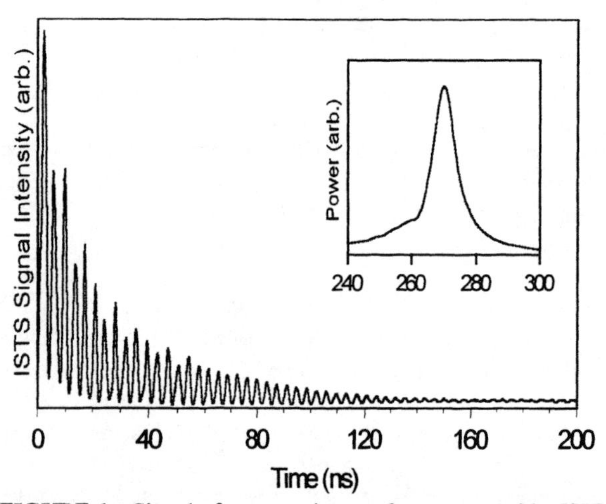

FIGURE 1. Signal of an acoustic waveform generated by ISTS in a copper film (5200A Cu/ 250A Ta:N/ 1000A Oxide/ Silicon). The FFT of the waveform is shown in the inset.

RESULTS AND DISCUSSION

Copper Thickness Measurement

Using the metrology instrument described above, center-point thickness measurements were performed on a series of wafers on which PVD copper and tantalum were deposited with systematically varying thickness. To verify the ISTS data, four-point probe measurements were performed on the same wafer sets. The results, shown in Figure 2, indicate strong correlation between the two techniques. We believe that small but systematic deviations from the line in the figure (drawn as a guide to the eye) may be due to complications in the four-point probe measurement, such as the fact that film resistivity used to calculate thickness may vary with film thickness.

Uniformity Determination

An important consideration for a metalization process is the uniformity of the film deposition. In this study, we investigated the uniformity of copper films deposited by both PVD and ECD processes by measuring thickness contour maps from copper films deposited by the different processes. Figures 3 and 4 show, respectively, three-dimensional contour maps from 121-point PVD copper film and 225-point ECD copper film. The maps clearly indicate physical differences in the two films. For example, the PVD copper film shown in Figure 3 has a well-defined inverse "bull's-eye" pattern, i.e., the film is relatively thick near its outer perimeter, and thinner near the film's center. In general, the PVD film is quite uniform (1.89% uniformity) and shows smooth, systematic thickness variations. In contrast, the ECD film shown in Figure 3 has a "crown" type shape where the copper film is relatively thick near the areas used to support the electrodes (distributed along the film's perimeter). The film build-up in these regions decreases the uniformity of the copper film considerably (4.7% uniformity). The ECD film is extremely uniform near the center region.

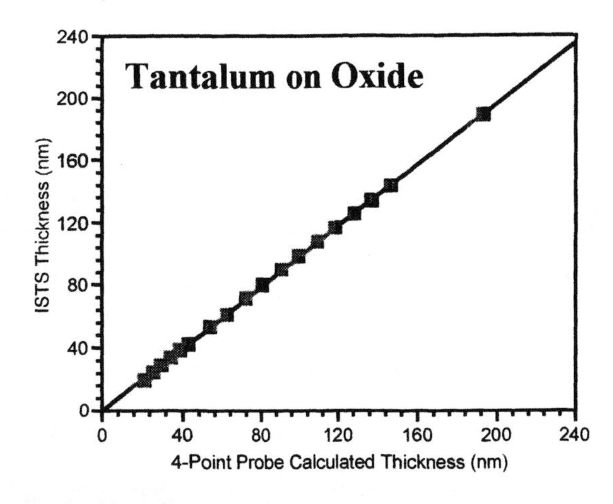

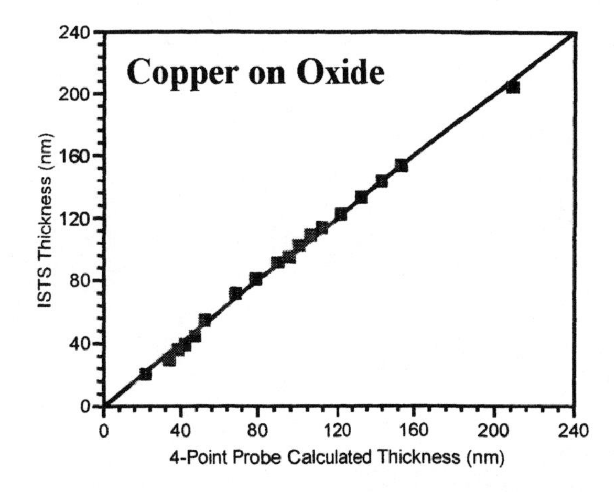

FIGURE 2. Tantalum and copper correlation data between ISTS measurements and 4-Point Probe calculated thickness. Each wafer set spanned the thickness range from 20 nm to 200 nm.

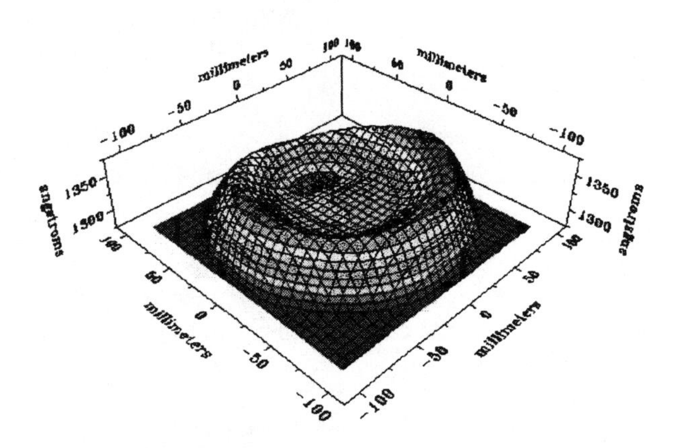

FIGURE 3. 121-Point contour map of a PVD blanket copper on oxide, with a test diameter of 192 mm.

FIGURE 4. 225-Point contour map of an ECD blanket copper on oxide, with a test diameter of 192mm.

Edge and Diameter Profiles

In addition to uniformity determination, an important application for film thickness measurement is the determination of the width of the exclusion zone at the edge of the wafer and the characterization of thickness properties near the film's edge. Tight control of these parameters is important to achieve maximum die yield on the wafer, as devices bordering the edge-exclusion zone can represent close to 10% of the total number of devices on a 200-mm wafer. The small spot size used in the ISTS measurement makes it possible to obtain a detailed thickness profile along the transition from the edge-exclusion zone to the film's center-point. Automated measurement of the diameter of the film can also be performed to determine a "usable" film diameter, i.e., one that is above a specified threshold thickness.

Figures 5A and 5B show edge-profile measurements for the PVD and ECD copper films described above. As is clear from the data in Figure 5A, the ECD copper film systematically increases in thickness near the edge. After a maximum thickness is reached, the film thickness drops dramatically just before the tantalum:nitride exclusion zone. In contrast, data from the PVD film shown in Figure 5B show a smooth, systematic decrease in film thickness that occurs over a relatively long distance when compared to the ECD film. The copper film ends at about 1 mm from the wafer's edge, leaving the tantalum exclusion zone.

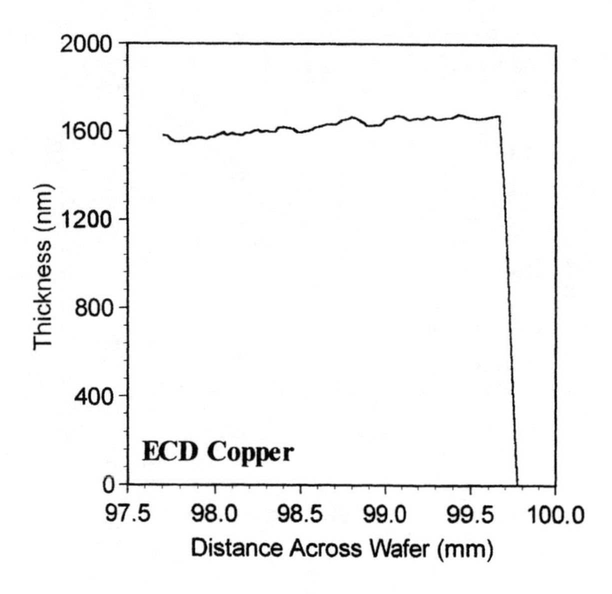

FIGURE 5A. An edge-profile of an ECD blanket copper film. The sharp transition in thickness starting at a radial distance of 99.6 mm represents the edge of the film

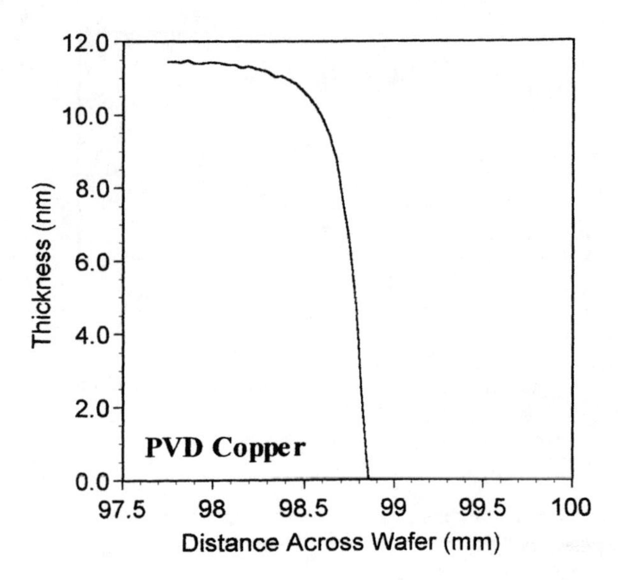

FIGURE 5B. An edge-profile of a PVD blanket copper film. The smooth transition in thickness starting at a radial distance of 98.2 mm represents the edge of the film

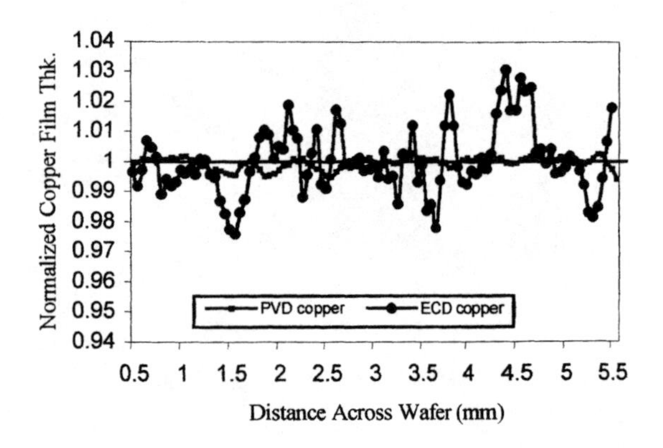

FIGURE 6. PVD versus ECD surface roughness comparison over a 5mm line scan, originating near the wafer's center-point.

Measurements along a linear sequence of points can also be performed away from the film's edge as a way of determining properties such as surface roughness. Figure 6 shows, for example, normalized thickness measurements made across a 5mm line (100 micron spacing) for both the PVD (dark squares) and ECD (dark circles) copper. The data for these experiments were highly repeatable, and attributed to thickness variations in the film over the 5mm distance. As is clear from the normalized data, the roughness of the PVD copper is significantly less than that of the ECD copper. These results were verified using a number of PVD and ECD films having different thickness.

Measurement of Copper Damascene

Due to a lack of suitable methods for etching copper, it is anticipated that copper patterning will be performed by a "damascene" process. In this process, copper is deposited over a dielectric material patterned to include a series of trenches and vias; excess copper is removed using chemical-mechanical polishing (CMP) to leave a series of copper bars in the previously formed trenches and vias.

Using the instrument described above a copper damascene structure having 1-micron wide trenches was measured with roughly the same frequency repeatability achieved for blanket copper films. Since the spot size for this measurement was approximately 25 microns, the ISTS technique produces data from more than one damascene trench. The thickness determined from these represents an average over the irradiated trenches

CONCLUSIONS

In summary, a novel, all-optical metrology technique called ISTS is shown to measure a variety of properties of copper/tantalum structures deposited using PVD and ECD. Measurements were successfully performed on both blanket copper films and copper damascene structures.

ACKNOWLEDGEMENTS

The authors would like to thank Alain Diebold for helpful discussions and Jim Sbrogna, Scott Whitman and Sean MacKinnon for their help with construction of the system.

REFERENCES

1. Linda Geppert, *IEEE Spectrum,* **January 1998**, pp. 23-28.

2. M. Banet, M. Fuchs, R. Belanger, J. B. Hanselman, J. A. Rogers and K. A. Nelson, *Future Fab International*, **4**, pp. 297-300 (1998).

3. M. Banet, et. Al., submitted to *Appl.. Phys. Lett.* for publication (02/98)

4. J. A. Rogers, M. Fuchs, M. J. Banet, J. B. Hanselman, R. Logan and K. A. Nelson, Appl. Phys. Lett., **71** (2), 225-227 (1997).

Grain Orientation Mapping of Passivated Aluminum Interconnect Lines by X-ray Micro-Diffraction

C.H. Chang[1,2], A.A. MacDowell[1], A.C. Thompson[3], H.A. Padmore[1], and J.R. Patel[1,2]

[1] *Advanced Light Source, Lawrence Berkeley National Laboratory, Berkeley, CA 94720*
[2] *SSRL/SLAC, Stanford University, Stanford, CA 94309*
[3] *Center for X-ray Optics, Lawrence Berkeley National Laboratory, Berkeley, CA 94720*

A micro x-ray diffraction facility is under development at the Advanced Light Source. Spot sizes are typically about 1-μm size generated by means of grazing incidence Kirkpatrick-Baez focusing mirrors. Photon energy is either white of energy range 6-14 keV or monochromatic generated from a pair of channel cut crystals. Laue diffraction pattern from a single grain in a passivated 2-μm wide bamboo structured Aluminum interconnect line has been recorded. Acquisition times are of the order of seconds. The Laue pattern has allowed the determination of the crystallographic orientation of individual grains along the line length. The experimental and analysis procedure used is described, as is the latest grain orientation result. The impact of x-ray micro-diffraction and its possible future direction are discussed in the context of other developments in the area of electromigration, and other technological problems.

INTRODUCTION

Electromigration is the physical movement of atoms in metallic interconnect lines passing current at high electron density (typically in the range of 10^5 amp/cm^2). Significant material movement results in voids that consequently leads to breakage and circuit failure in the metal lines. This problem gets more severe as the line dimensions continue to shrink on integrated circuits. In spite of much effort in this field (1,2), electromigration is not understood in any depth or detail, but is strongly associated with the physical material properties (stress and strain) within the interconnect material. Throughout this century x-rays have been a powerful tool to measure such material properties, but the ability to make such measurements on the micron scale required by the semiconductor industry has only come into realization with the advent of the latest generation of high brightness synchrotron sources. In this paper we describe the beginnings of a program to carry out various x-ray diffraction measurements on the micron scale. It is presumed that the electromigration properties of a metal line will be dependent, as well as stress/strain effect, on the grain orientation of adjacent grains in the line. This paper describes the experimental and analysis techniques that allow the grain orientation and indexing of individual micron sized grains along the length of aluminum interconnect line.

X-rays are quite well suited to such measurements as they are able to penetrate several microns into matter. In general, interconnect lines are encased in the insulator silicon dioxide (passivation). X-rays are able to penetrate and study such buried samples.

EXPERIMENTAL

Figure 1 shows the experimental setup. The synchrotron

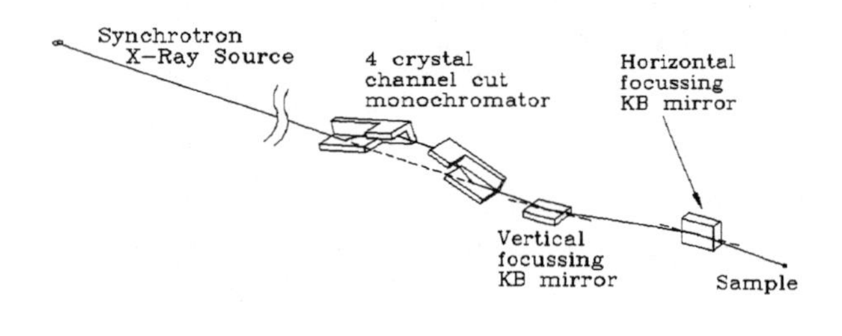

Figure 1. Schematic layout of the K-B mirrors and four crystal channel-cut monochromator.

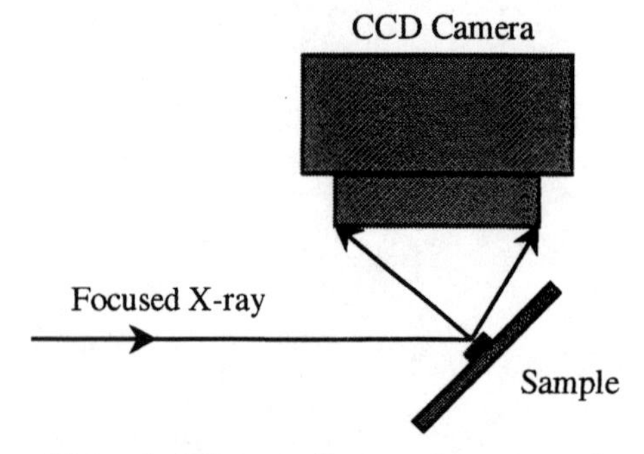

Figure 2. Schematic diagram of experimental arrangement.

CP449, *Characterization and Metrology for ULSI Technology: 1998 International Conference*
edited by D. G. Seiler, A. C. Diebold, W. M. Bullis, T. J. Shaffner, R. McDonald, and E. J. Walters

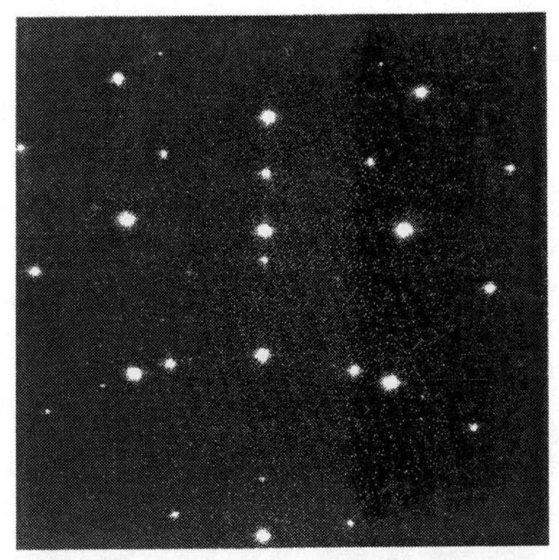

Figure 3. Laue pattern of silicon substrate.

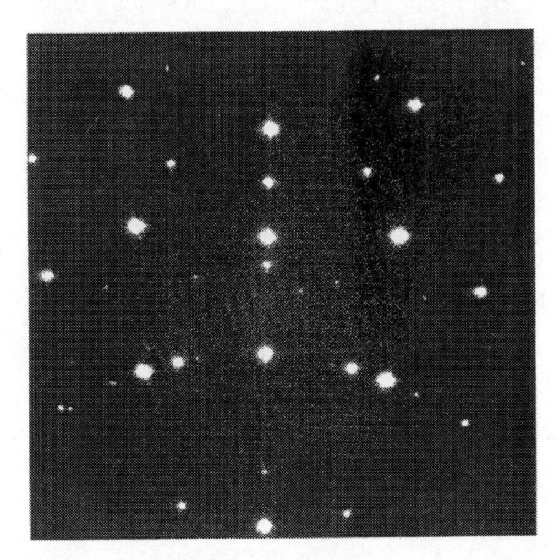

Figure 4. Laue pattern of a single grain in aluminum line as well as the silicon substrate.

a

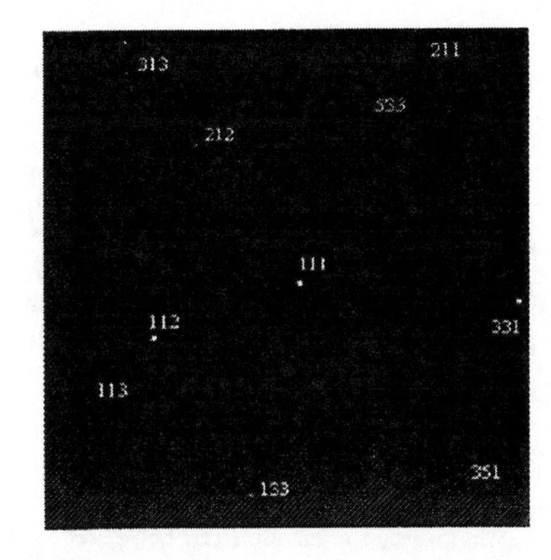

b

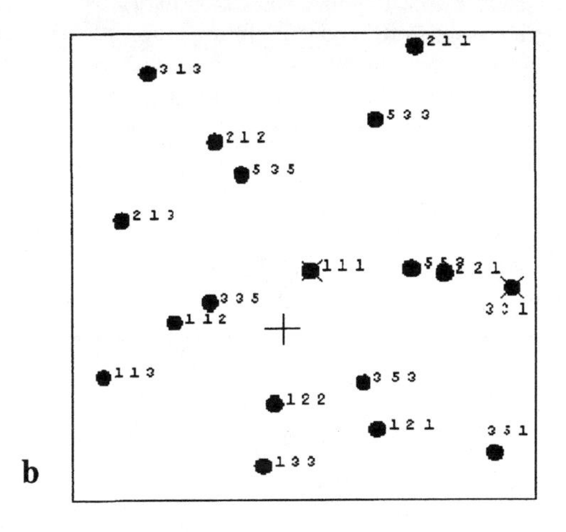

Figure 5. (a) Spot pattern of aluminum grain obtained by subtracting the silicon pattern (Fig. 3) from the aluminum pattern (Fig. 4). (b) Simulated pattern with the aluminum spot pattern indexed.

source of size typically 300 × 30 µm FWHM (horizontal and vertical) is imaged with demagnifications of 300 and 60 respectively by a set of grazing incidence platinum-coated elliptically bent Kirkpatrick-Baez (K-B) focusing mirrors (3). Imaged spot sizes on the sample are about a micron in size. Photon energy is either white of energy range 6-14 keV or monochromatic generated by inserting a pair of Si(111) channel-cut monochromator crystals into the beam path. A property of the four crystal monochromator is its ability to direct the monochromatic primary beam along the same direction as the white radiation. Thus, the sample can be irradiated with either white or monochromatic radiation. White radiation is chosen for Laue experiments for the orientation determination and monochromatic radiation for d-spacing measurements in stress/strain determination of single grains in the metal line.

The sample was a deposited aluminum line, 0.5-µm thick and 2-µm wide, on an oxidized silicon substrate. The line was passivated with a plasma-enhanced chemical vapor deposition (PECVD) nitride at 300°C to 0.3-µm thickness. Laue patterns were collected using white radiation and an x-ray CCD camera. The exposure time was 0.5 sec and the measured photon flux of the micro beam was about 10^7 photons/sec/µm^2/0.1% band in the energy range of 6-14 keV and sample-to-CCD distance was 16.4 mm. Fig. 2 shows the arrangement of the sample and CCD detector.

RESULTS

Figure 3 shows the Laue pattern from the silicon substrate. Figure 4 shows the Laue patterns from the silicon substrate and the fainter diffraction spots from a single grain in the aluminum line. Fig. 5a shows the aluminum

Laue pattern obtained following digital subtraction of the silicon pattern (Fig. 3) from the silicon and aluminum pattern (Fig. 4).

The origin on the CCD detector array was determined by moving the CCD camera radially from the sample and recording the silicon Laue patterns at various distances from the sample. The origin was determined at the CCD where the lines drawn through the succession of the same Laue spots intersected. All aluminum spot positions were coordinated to the origin and indexed using an indexing software package - LaueX (4). Fig. 5b shows the simulated pattern with reflections indexed. For confirmation of the indexation the 4 crystal monochromator was inserted into the beam and scanned in energy to determine the d-spacing of the Al (111) spot.

The aluminum grain orientation can be referenced to the silicon substrate based on the orientation matrix $\mathbf{R}_{Si}$ and $\mathbf{R}_{Al}$ in the silicon substrate and aluminum grain, respectively. The matrix $\mathbf{R}$ relates the crystal system S with axes parallel to the basic crystallographic axes in the crystal to the reference system S^R related to the primary beam direction:

$$S^R = \mathbf{R}S.$$

The aluminum grain orientation measured is then referenced to the silicon substrate as the following orientation matrix:

$$\mathbf{M} = \mathbf{R}_{Si}^{-1}\mathbf{R}_{Al}$$

$$= \begin{pmatrix} 0.707 & 0.475 & -0.524 \\ 0.003 & 0.737 & 0.676 \\ 0.707 & -0.481 & 0.518 \end{pmatrix}^{-1} \begin{pmatrix} 0.916 & 0.039 & 0.399 \\ 0.308 & 0.571 & -0.761 \\ -0.257 & 0.822 & 0.509 \end{pmatrix}$$

$$= \begin{pmatrix} 0.468 & 0.612 & 0.637 \\ 0.787 & 0.045 & -0.615 \\ -0.404 & 0.792 & -0.458 \end{pmatrix}.$$

The experimental accuracy in the determination of the orientation matrix $\mathbf{M}$ depends mainly on the angular resolution of the Laue camera system, because the Laue diffraction pattern of the aluminum grains always accompanies that of the silicon substrate that is the reference of the aluminum orientation. In the present case with the CCD of 23.5-μm pixel size and sample-to-CCD distance of 16.4 mm, the magnitude of the misorientation angle is determined within a precision of several minutes of arc.

CONCLUSION AND FUTURE DEVELOPMENT

We have demonstrated that x-ray micro-diffraction is capable of determining the crystallographic orientation of individual grains in passivated interconnect lines. The orientation mapping can be done by collecting the Laue patterns from individual grains along the length of the lines. A computerized indexing code to automate this is under development. Beyond this the requirement is to measure the d-spacing of various aluminum planes to determine the stress and strain state of individual grains along the length of the aluminum interconnect line.

ACKNOWLEDGEMENTS

This work was supported by the Director, Office of Basic Energy Sciences, Materials Sciences Division of the US Department of Energy, under Contract no. DE-AC03-76SF00098. Samples from T. Marieb and equipment support from Intel Corporation, Santa Clara, CA.

REFERENCES

1. Marieb T., Flinn P., Bravman J.C., Gardner D., and et al. *J. App Phys.* **78**, 1026-1032 (1995).
2. Wang P.C., Cargill G.S., and Noyan I.C., *MRS Proceedings* **375**, 247-252 (1995).
3. MacDowell A.A., Celestre R., Chang C.H., Franck K., Howells M.R., Locklin S., Padmore H.A., Patel J.R., and Sandler R., *SPIE Proceedings* (1998).
4. Soyer A., *J. App. Cryst.* **29**, 509 (1996); http://www.lmcp.jussieu.fr/sincris/logiciel/laueX/en/laueX_en.html.

In-Situ Plasma Chamber Monitoring for Feedforward Process Control

Jinsoo Kim and Kensall D. Wise

Center for Integrated Sensors and Circuits, Department of Electrical Engineering and Computer Science
University of Michigan, Ann Arbor, MI 48109-2122, USA

This paper examines the effects of polymer buildup in plasma etching systems and describes a micromachined sensor for in-situ polymer thickness measurement. Using gas flows of 45sccm CHF_3 and 15sccm CF_4 at 50mTorr and 1000W, the oxide:polysilicon selectivity ranges from 2.6 to 8.5 as the polymer thickness on the tool walls varies from 0 to 240nm. The polymer sensor is based on an electrothermal oscillator that measures the thermal mass change as polymer builds up on a stress-compensated dielectric window. The change in the thermal mass of the window can be detected as a variation in the pulse width (cooling time) of the oscillation. The device operates with a typical cooling time of 2.7msec and has a measurement resolution of better than 1nm. The device is flush-mounted in the chamber wall with the exposed window area protected by a thin film of iridium against damage by the plasma.

INTRODUCTION

In order to achieve the dry etching performance needed for next-generation ULSI wafer production, future etching equipment must employ sensor-driven adaptive control algorithms so that changes in the tool characteristics can be compensated for. In particular, the condition of the plasma etching chamber is important and today is rarely monitored as part of tool control. In the plasma etching process, the polymers used to enhance etch anisotropy and selectivity also deposit on the various parts of the chamber (1), changing the chemical and electrical properties of the discharge. The reactor surface, which serves as both a source and a sink for reactive gas species, not only strongly affects the concentration of reactants (eventually changing the etching characteristics) but also can produce particulates which lower yield. This paper reports the effects of chamber conditions on etching characteristics and describes a micromachined sensor for polymer buildup monitoring that is allowing new studies of etch chemistry and provides a means for improved run-to-run etch control in production reactors. While several approaches to polymer monitoring were considered, including infrared absorption spectroscopy (2), ultraviolet absorption spectroscopy (3), and the plasma emission spectrum (4), these techniques typically measure some radical (such as CF_2) concentration in the plasma rather than the chamber condition. The correlation between radical concentration and polymer buildup is not well understood and must be readjusted whenever there is change in the etch recipe. Besides this, changes in the chamber condition produced by sputtered masking and substrate materials and the inability to extract the information needed to optimally schedule chamber cleans require the use of in-situ measurements for feedforward process control. Using MEMS (MicroElectroMechanical Systems) technology, a micromachined electrothermal oscillator was successfully designed, fabricated, and tested in an Applied Materials 8300 hexode type plasma etcher outfitted for both real-time and run-to-run control.

ETCH CHARACTERISTIC VARIATIONS

Within the plasma etching process, a glow discharge is formed by exposing either fluorine-based (e.g., CF_4, CHF_3, SF_6, NF_3) or chlorine-based (Cl_2) gaseous monomers at low pressure to an electric field. In the case of CF_4 or CHF_3 (primarily used in dielectric etching), radicals such as CF_2 created by electron-neutral dissociation in the plasma discharge not only coat the pattern sidewalls, producing anisotropy, but also adsorb on the various parts of chamber. In addition, there are also deposits formed on the chamber by sputtering of the photoresist masking material and the etched substrate material. The polymer buildup process is a strong function of processing parameters such as power, base pressure, and flow rate (5) and also shows a large dependence on the types of surface materials used, temperature, and the hydrogen/oxygen concentrations in the plasma discharge. These changes in the chamber condition due to polymerization affect the chemical and electrical properties of the discharge, which alters the concentration of reactants (eventually changing the etching characteristics). Etch rate and selectivity variations for certain oxide and nitride etching recipes have been explored in an Applied Materials 8300 hexode etcher. While the oxide recipe used 45sccm of polymer-producing CHF_3 and 15sccm of CF_4 to achieve high selectivity, the nitride recipe employed 42sccm of CF_4, 15sccm of CHF_3 and 3sccm of polymer-suppressing O_2 at 50mTorr and 1000W. The chamber was wet cleaned first, and after each etching run, a 10min polymerization run producing about 50nm of polymer on the chamber wall surface was performed at 50sccm CHF_3, 50mTorr, and 1000W without processing wafers. Each etching run included one oxide, one nitride, and one polysilicon wafer. As expected, due to the polymerization-suppressing capability of the oxide, the oxide etch rate remained constant (± 1nm/min) as the polymer-producing CF_x radical concentration increased due to changes in the chamber as 0 to 240nm of polymer built up on the tool walls.

CP449, *Characterization and Metrology for ULSI Technology: 1998 International Conference*
edited by D. G. Seiler, A. C. Diebold, W. M. Bullis, T. J. Shaffner, R. McDonald, and E. J. Walters

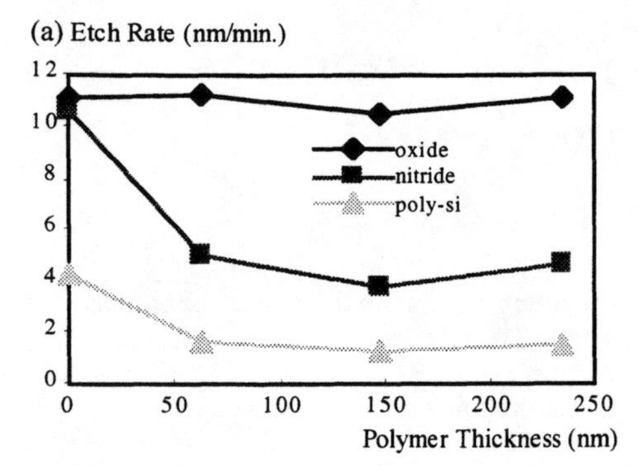

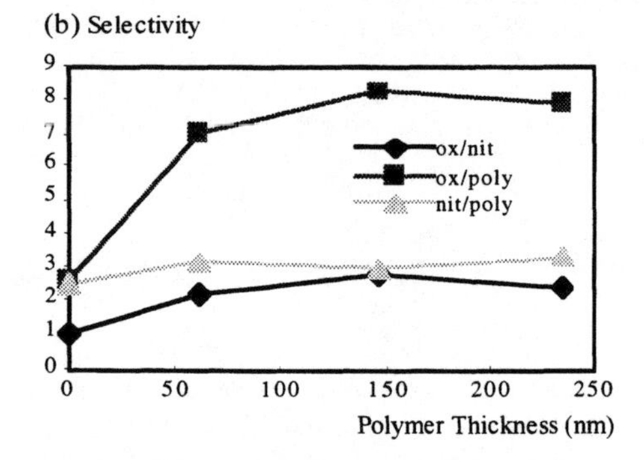

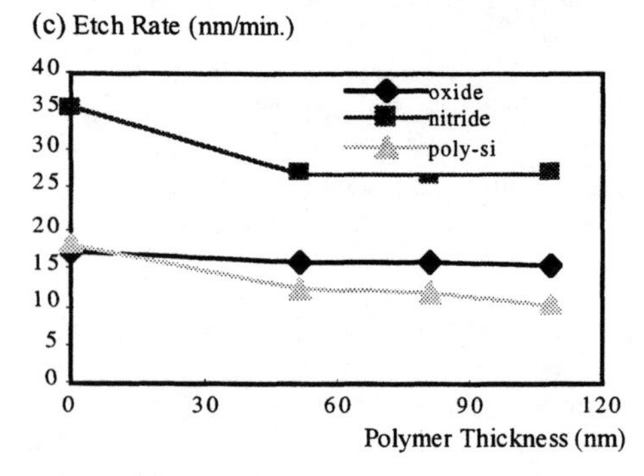

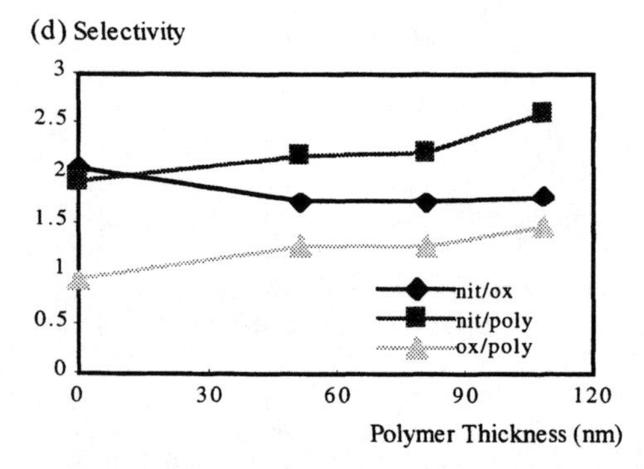

FIGURE 1. (a,c) Etch rate, and (b,d) Selectivity vs. Polymerization on the chamber wall. Experiment conditions for (a, b): 45sccm CHF_3, 15sccm CF_4, 50mTorr, 1000W, 15min etch each time. Experiment conditions for (c, d): 42sccm CF_4, 15sccm CHF_3, 3sccm O_2, 50mTorr, 1000W, 5min etch each time. The decreases in etch rates of silicon nitride and poly-Si may be due to an increase in the CF_x radical concentration in the plasma discharge which increases the polymerization on the wafers.

SENSOR DESIGN

Placing a micromachined sensor in the unfamiliar and rather hard-to-predict environment of a plasma chamber requires a careful examination of several important design requirements. Some of the important considerations are 1) protecting the device from the etching environment, 2) not disturbing the RIE operation with the sensor, 3) resolving polymer thickness to less than one nanometer, 4) maintaining stable device performance in spite of changes in other chamber characteristics such temperature and pressure, and 5) maintaining a known correlation between the amount of polymer deposited on the sensor and that deposited on the chamber wall. Several different approaches to meeting these requirements have been examined, including the use of an electrothermal oscillator, electrostatically-driven mechanical microbridge resonators, and acoustic wave devices (e.g., flexural-plate, surface acoustic and bulk crystal wave devices). Based on a detailed review of these sensors, a device based on an electrothermal oscillator was selected for its higher sensitivity, ease of fabrication, and reliable operation.

Operation of the Device

The electrothermal sensor can be operated by interfacing the heater and the temperature sensor to an amplifier, Schmitt trigger, and voltage-to-current converter as in Figure 2. During the first half cycle (heating), the Schmitt trigger is on and the V/I converter drives power to the heating resistor. As power is supplied, the window temperature rises and causes the temperature sensing film resistance to increase as well. The amplifier output voltage increases to reflect this change. When the amplifier output reaches the upper threshold of the Schmitt, the power from the heater is turned off and the sensing resistance starts to drop due to cooling of the window, resulting in a decreasing amplifier output voltage until it reaches the lower threshold of the Schmitt trigger. Then, the Schmitt turns on again, repeating the heating and cooling cycle. The electrothermal cooling time (t_c) can be derived from

$$[P_e - A(T - T_f)]dt = MCdT \qquad (1)$$

by integration and can be written as

$$t_c = \frac{MC}{A}\ln\left[\frac{T_H - T_f}{T_L - T_f}\right] \qquad (2)$$

where the input power $P_e = 0$ during the second half cycle (t_c), A is the no-flow thermal conductance, T is the diaphragm temperature, T_f is the ambient temperature, M is the diaphragm mass, and C is the diaphragm specific heat, T_H and T_L are the high and low switching temperature levels corresponding to the upper and lower thresholds of the Schmitt trigger, respectively. The thermal mass of the windows can be written as

$$MC = \Sigma m_i c_i = m_{SiO2}c_{SiO2} + m_{SiN}c_{SiN} + m_{Ti}c_{Ti} + m_{Ir}c_{Ir} + m_{polymer}c_{polymer.} \quad (3)$$

For these polymer buildup measurements, only the cooling time is monitored to allow simpler interpretation of the data. This cooling time changes with polymer deposition (i.e., with the thermal mass (MC) change) and has been monitored using the pulse width measuring function on a Motorola MC68HC11 evaluation board at an internal clock frequency of 2MHz. Figure 3 shows a typical output waveform from the sensor.

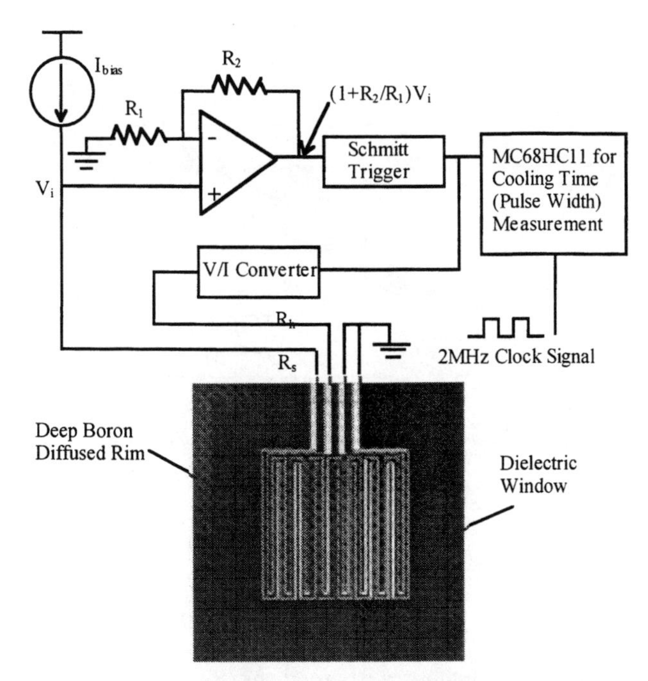

FIGURE 2. The electrothermal sensor measures changes in the electrothermal characteristics of a stress-compensated dielectric window ($SiO_2/Si_3N_4/SiO_2$) containing interleaved heating and sensing resistors (R_h, R_s). Polymer deposition on the window alters its mass and associated heat capacitance.

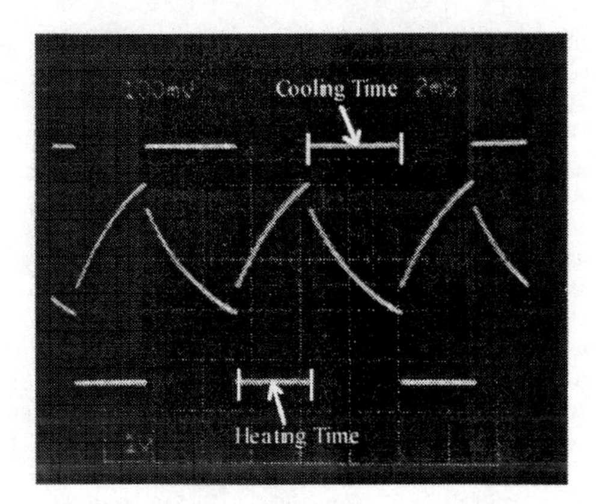

FIGURE 3. Output waveform of the electrothermal oscillator. The pulse width is a function of the polymer thickness.

Device Structure, Packaging, and Placement

The cross-section of the batch-fabricated polymer sensor and a packaged device are shown in Figure 4. A cavity-down pin-grid-array (PGA) package has been adapted for this device due to the ease of wire bonding and durability of the ceramic package inside the vacuum chamber (lack of outgassing). First, to expose the backside of the device window to the inside of the chamber (plasma), a hole was cut ultrasonically in the package base. Then, the device was attached over the hole, followed by wire bonding and sealing of the bonding area by a metal lid having a tiny hole to ensure adequate pressure balance. A packaged device was placed in an Applied Materials 8300 hexode RIE using a specially prepared mounting structure. A 10-pin electrical feedthrough and spot-welded device mounting socket were used in place of one of the normal ports on the tool. Figure 5 shows the mounted device.

TEST RESULTS

The sensor has been tested in an Applied Materials 8300 multiwafer RIE outfitted for both real-time and run-to-run control. The test results shown in Figure 7 represents twelve 10-minute etches using CHF_3 and a final 20-minute O_2/Ar plasma clean-up etch to remove the polymer film from the chamber surface. The device functions as expected. These tests employed a 30sccm flow of CHF_3, a base pressure of 20mTorr, and 1000W for the etching runs and 10sccm of O_2, 10sccm of Ar, a base pressure of 15mTorr, and 1000W for cleanup.

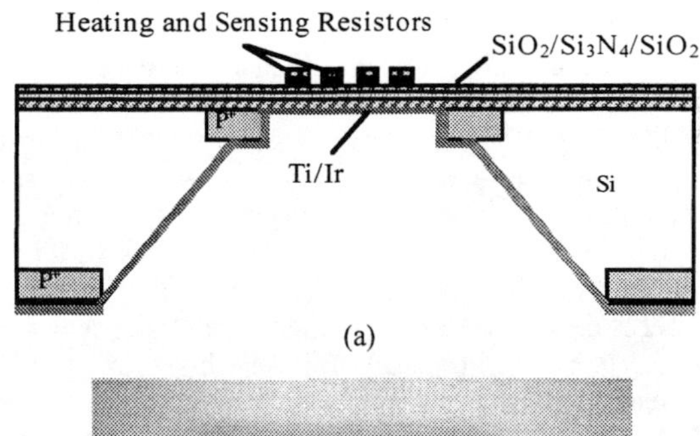

(a)

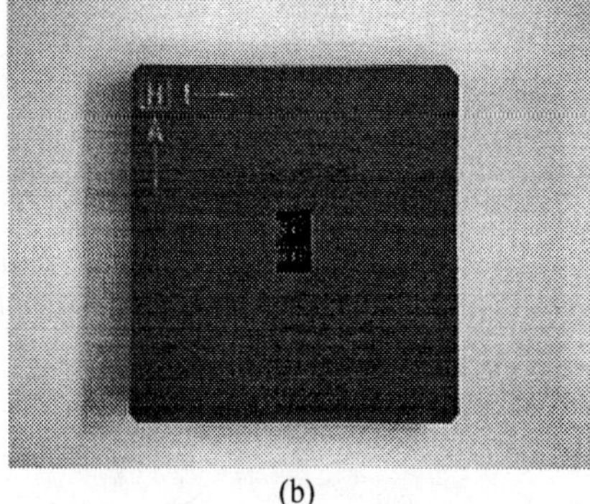

(b)

FIGURE 4. (a) Device cross-section showing the stress compensated dielectric window, heating and sensing resistors, backside Si etching by EDP and Ti/Ir layer for protection; (b) Packaged device.

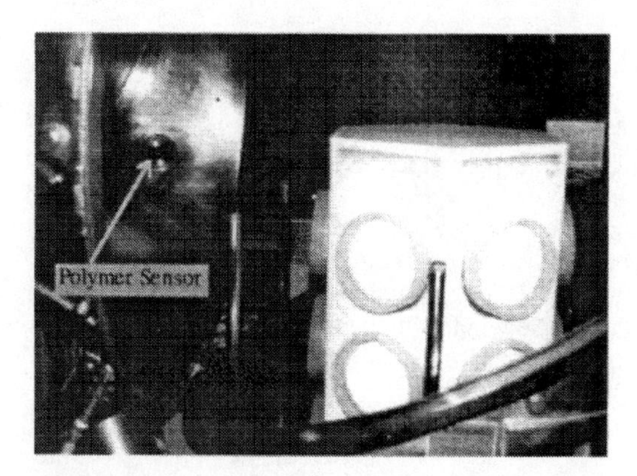

FIGURE 5. Picture showing the packaged polymer sensor mounted in the Applied 8300 hexode RIE chamber wall.

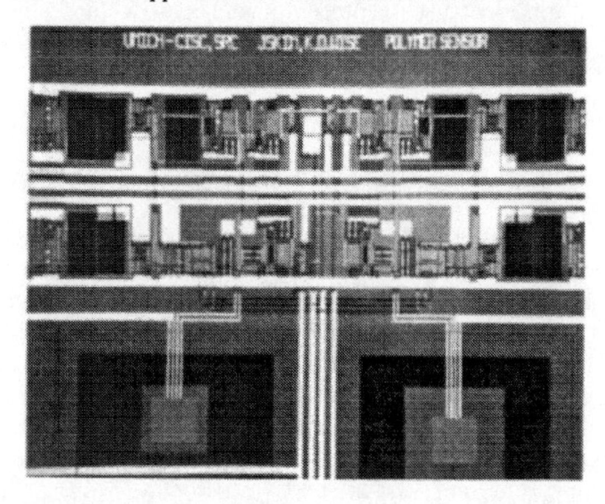

FIGURE 6. Picture of an active polymer sensor with on-chip temperature compensation and control circuitry.

The chamber was maintained at the cleanroom temperature of 23°C. The polymer film removal rate is more than 30 times greater when Ar is present during the cleanup cycle. The sensor response was verified by placing a silicon monitoring wafer on the chamber wall; after each 10-minute etch, the polymer thickness on this wafer was measured using an ellipsometer. About 3nm/minute of polymer is deposited by this particular etch. The sensor has a measurement resolution of <1nm. Each sensor employed four separate windows and the responses from these wndows track closely. The cooling time increases with polymer thickness in agreement with electrothermal models and decreases during the cleanup etch to a level similar to the starting condition. This device is now being used for detailed studies of polymer buildup in RIE and as the basis for advanced control algorithms aimed at improving the precision achieved in advanced pattern transfer tools.

CONCLUSIONS

A simple bulk-micromachined electrothermal oscillator with a dielectric window and metal film resistors has been designed and tested for the direct in-situ measurement of polymer buildup in plasma etching systems. Its high sensitivity, ease of fabrication, reliable operation, and simple interface are advantages over other approaches to polymer measurement. Test results show increases in the device cooling time during etching runs as polymer is deposited and decreases during cleanup cycles as it is removed. The resolution in measuring polymer thickness is <1nm. After successful design and testing of the prototype polymer sensor, a new active device with on-chip temperature compensation and control circuitry has been designed and fabricated (Figure 6); this device is currently under test. These polymer monitoring sensors are allowing new studies of dry etch chemistry and provide a means for improved run-to-run etch control in production reactors in concert with advanced feedforward control algorithms.

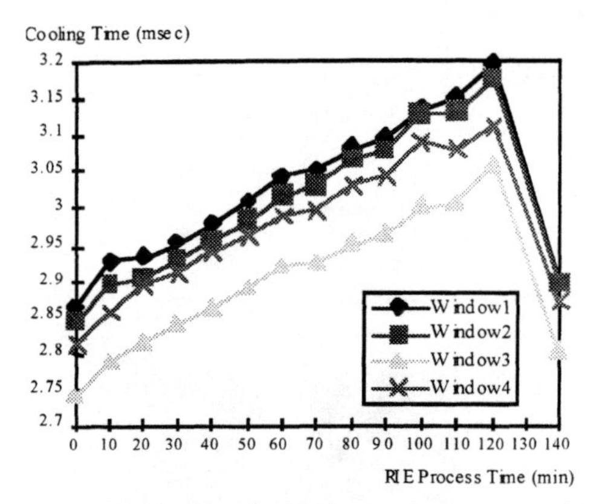

FIGURE 7. Test results showing increases in pulse width with polymer buildup and decreases with O_2/Ar plasma cleaning.

ACKNOWLEDGMENTS

The authors wish to thank the Semiconductor Research Corporation for supporting this work under contract 97-FC-085.

REFERENCES

1. Watanabe, S., "Plasma Cleaning by Use of Hollow-Cathode Discharge in a CHF_3-SiO_2 Dry-Etching System," *Japan Journal of Applied Physics*, Vol. 31, pp. 1491-1498, May 1992.
2. Haverlag, M., Stoffels, W., Stoffels, E., Kroesen, G., de Hoog, F., "Production and Destruction of CF_x Radicals in Radio-Frequency Fluorocarbon Plasmas," *Journal of Vacuum Science and Technology, A*, Vol. 14, No. 2, pp. 384-390, 1996.
3. O'Neill, J., Singh, J., "Role of the Chamber Wall in Low-Pressure High-Density Etching Plasmas," *Journal of Applied Physics*, Vol. 77, No. 2, pp. 497-504, 1995.
4. Nitta, T., "Significance of Ultra Clean Technology in the Era of ULSIs," *IEICE Transactions on Electron*, Vol. E79-C, No. 3, pp. 256-263, 1996.
5. Kim, J., Wise, K. D., Grizzle, J., "In-Situ Polymer Buildup Monitoring in Plasma Etching Systems," *1997 ISSM Conference*, pp. B37-B40, October, 1997.

Optical Computer Aided Tomography Measurements of Plasma Uniformity in an Inductively Coupled Discharge

E. C. Benck and J. R. Roberts

National Institute of Standards and Technology, Gaithersburg, MD 20899

Optical computer aided tomography (CAT) is being investigated as a potential *in situ* diagnostic for measuring plasma uniformity without making assumptions concerning the plasma symmetry. The presence of an opaque vacuum chamber wall severely limits the different directions from which optical emission measurements can be made of the plasma. The tomographic inversion problem with restricted optical access is being solved using Tikhonov regularization. The accuracy of this inversion process is investigated for several different observation geometries using theoretical test data generated from known distributions. Optical CAT is applied to an ICP-GEC plasma source, with all the measurements made through a single large 152 mm diameter window. Axially asymmetric plasma distributions are demonstrated as a function of gas flowrate and gas composition.

INTRODUCTION

With the increasing diameter of wafers used in semiconductor manufacturing, understanding and controlling plasma uniformity is becoming increasingly important. Variations of the etchant concentrations in a plasma will have a significant impact on the final etching uniformity of a wafer. Being able to predict and correct etching nonuniformities created by the plasma should reduce the cost of wafer production and improve control over the critical dimension on the wafer.

Optical computer aided tomography (CAT) may be applicable an *in situ* diagnostic of plasma uniformity without having to make assumptions concerning plasma symmetry. Optical emission measurements of a discharge do not provide a direct measure of the plasma uniformity. Each emission measurement is actually equal to the plasma emissivity integrated along the line of sight of the detector. The measured light intensities must be mathematically inverted to obtain the actual spatial distribution of the plasma. In many plasmas axial symmetry can be assumed and the radial plasma distribution can be obtained using Abel inversions. We have previously shown that in the ICP-GEC plasma source there are plasma conditions that do not have axial symmetry(1). To study these discharges, many additional optical measurements and the much more complicated CAT are necessary.

CAT has proven highly successful in the field of medical x-ray imaging. X-ray CAT scanners generally take measurements from a full 360° around an object with the number of x-ray measurements approaching 1 million per image. The resulting images are of very high quality with signal to noise ratios on the order of 1000(2). Unfortunately, the mathematical inversion techniques often applied in x-ray CAT are often not appropriate with optical CAT. This is because there is usually a vacuum chamber or some other opaque wall that significantly limits the directions from which the plasma can be measured.

THEORY

We are solving the optical CAT inversion using Tikhonov regularization. This type of inversion processes has been previously applied to microwave interferometry measurements used to determine electron densities in the Joint European Torus(3,4). The inversion region is divided into N pixels and it is assumed that the plasma emissivity outside of the inversion region is zero. There are M different line-integrated optical measurements, I_j, of the inversion region. These measured intensities are related to the emissivity of each pixel, ε_i, by the equation:

$$I_j \cong \sum_{i=1}^{N} A_{ji}\varepsilon_i \qquad (1)$$

or in its matrix form:

$$\underline{I} \cong \underline{A\varepsilon} \qquad (2)$$

where the coefficient A_{ji} is the weighting factor of pixel emissivity, ε_i, to the optical measurement I_j. If the maximum diameter of the light collection cone within the inversion region is small compared to the dimensions of a pixel, A_{ji} can be approximated by the length of the line segment along the line of sight which is included in the pixel.

CP449, *Characterization and Metrology for ULSI Technology: 1998 International Conference*
edited by D. G. Seiler, A. C. Diebold, W. M. Bullis, T. J. Shaffner, R. McDonald, and E. J. Walters
1998 The American Institute of Physics 1-56396-753-7/98/$15.00

The solution to the least squares minimization problem:

$$\min_{\varepsilon}\left\|(\underline{A\varepsilon}-\underline{I})^2\right\| \quad (3)$$

is generally very noisy(5). To overcome this problem, it is necessary to incorporate some prior knowledge about the general behavior of the plasma emissions. This is done by introducing a seminorm, defined as $\left\|(\underline{B\varepsilon})^2\right\|$, to create the Tikhonov regularized least squares problem:

$$\min_{\varepsilon}\left\{\left\|(\underline{A\varepsilon}-\underline{I})^2\right\|+\beta\left\|(\underline{B\varepsilon})^2\right\|\right\} \quad . (4)$$

The seminorm generally involves a discrete derivative equation, which represents the expected behavior of the inversion solution(5). Previous Abel inversions of axially symmetric plasma conditions showed that there is a radial inflection point in the plasma emissivity(1). Therefore, the following discrete matrix equation was used

$$\underline{B\varepsilon}=\left(\underline{\nabla}^2+(C\underline{1}-\underline{R})K^2\right)\underline{\varepsilon} \quad (5)$$

where $\underline{1}$ is the identity matrix, $\underline{R}$ is a diagonal matrix containing the distance each pixel is from the center of the discharge and the combined constant CK^2 is the smallest eigenvalue of the matrix

$$\left(\underline{\nabla}^2-K^2\underline{R}\right) \quad . (6)$$

The general shape of the solutions to the seminorm can be easily obtained by performing the Tikhonov regularization using a single measurement ($M = 1$). The resulting function for $C = 0.3$ is shown in Fig. 1.

Proper choice of the regularization parameter, β, is an important problem. If β is too small, the inversion will be plagued with excessive noise and if β is too large there will be excessive smoothing and loss of information. To determine the optimum value for β, test data was generated from several known distributions. The test data was then inverted and compared against the initial distributions as a function of β. The minimum value of the root mean square (RMS) error and the maximum pixel error occur around the range $0.000015 \leq \beta \leq 0.00002$. A better choice for the optimum value of β can be obtained from the L-curve criteria, which state that the optimum value of β will occur at the "corner" of the curve defined by ($\|(\underline{A\varepsilon_\beta}-I)^2\|$, $\|(\underline{B\varepsilon_\beta})^2\|$), where ε_β is the solution to the Tikhonov regularization for a particular value of β. This value of β will tend to balance the regularization and perturbation errors. These types of curves are shown in Fig. 2. The optimum value utilizing this technique for the experimental observation geometry, which will be described later in the paper, occurs in the range $0.00005 \leq \beta \leq 0.0001$. For most of the CAT presented in this paper, a regularization parameter of $\beta = 0.00005$ was used.

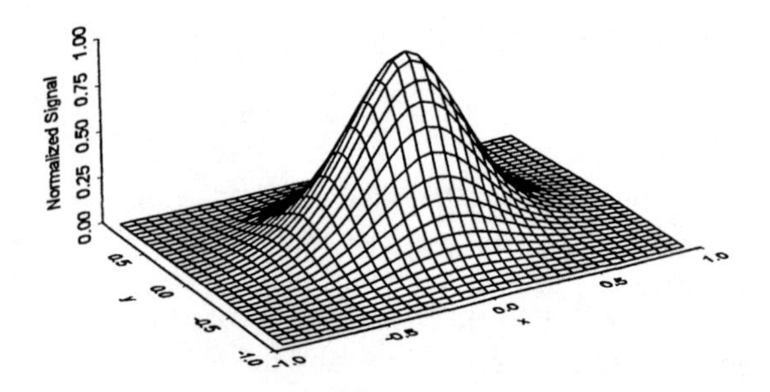

FIGURE 1. The solution to the seminorm with C=0.3 obtained from the CAT with a single measured value and then normalized with respect to the inverted distribution maximum.

Because there often is noise and uncertainty in the measurements, additional weighting factors, $w_j = 1/\text{var}(I_j)$, can also be incorporated into the problem. The resulting minimization equation becomes

$$\min_{\varepsilon}\left\{\left\|(\underline{A\varepsilon}-\underline{I})\underline{w}(\underline{A\varepsilon}-\underline{I})\right\|+\beta\left\|(\underline{Bw\varepsilon})^2\right\|\right\} \quad . (7)$$

Setting each second derivative of the previous equation with respect to the pixel emissivity equal to zero will now solve this problem. The solution found in this fashion can be written in terms of matrix equations. A matrix $\underline{Q}$ is defined

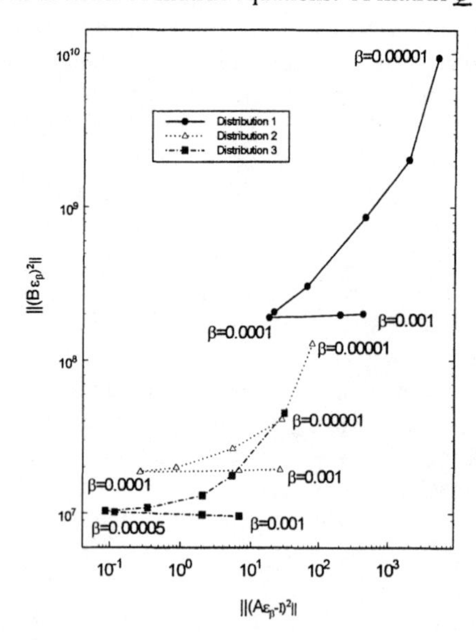

FIGURE 2. L-curves for three different distributions using the experimental observation geometry. The second derivative of distribution 1 is always negative and therefore the seminorm is not well matched to the distributions general behavior. Distribution 3 is shown in Fig. 8 and distribution 2 is of a similar form.

as:

$$\underline{Q} = \underline{A}^T \underline{wA} \quad . (8)$$

The solution to the problem is then given by

$$\underline{\varepsilon} = (\underline{Q}^{-1} \underline{A}^T \underline{w}) \underline{I} = \underline{M} \; \underline{I} \quad . (9)$$

Since the resulting N x N matrix $\underline{Q}$ is in general non-sparse, the matrix inversion can be computationally intensive. However, this matrix computation needs to be done only once and the matrix $\underline{M}$ reused in subsequent inversions, as long as the observation geometry and pixel assignments are kept fixed.

OBSERVATIONAL GEOMETRY EFFECTS

In order to study the effects of observation geometry, several known distributions were used to create test data based on different observation geometries. The different types of geometries investigated are shown in Fig. 3. The number of observations used was varied for each type of observation geometry and the results are presented in Fig. 4 and Fig. 5. After a rapid decrease in the inversion error with increasing number of observations, the error levels off. This type of behavior is easy to understand in the case of perpendicular arrays, since increasing the number of observations through the same row or column of pixels does not provide any new information to the inversion process. In the case of the fan arrays, there still is a slight reduction with increasing number of observations. This is probably due to the existence of regions of pixels which do not intersect with the observations in this geometry.

Because of the leveling off of the inversion error with respect to the number of observations, the type of observation geometry is very important to the inversion accuracy. Figure 4 shows that the lowest maximum pixel inversion error occurs with the perpendicular fans with a value of approximately 9% of the distribution maximum. The accuracy of the 7 fan array with large number of observations appears to be approaching a value comparable to that of the perpendicular fans.

The accuracy of the CAT also depends on the initial distribution as shown in Fig 5. The second derivative of distribution 1 is always positive and as expected the inversion error is quite large. The other two distributions have RMS errors very close to one another, with distribution 2 having a slightly lower RMS error. Increasing the number of pixels to 1089 causes the RMS error to decrease slightly to about 2.5% and with distribution 3 having the lowest error.

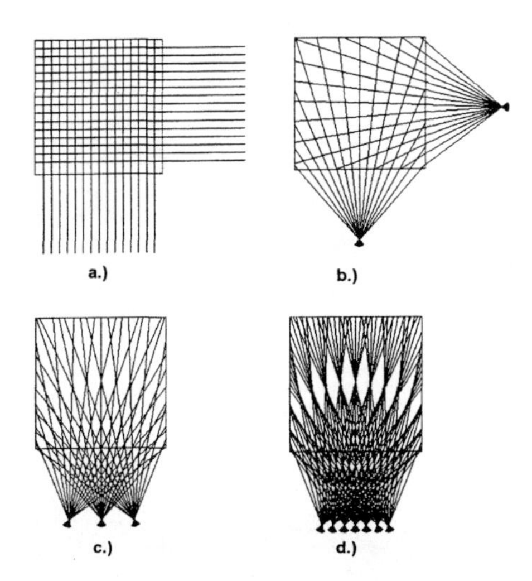

FIGURE 3. Test observation geometries: a.) perpendicular arrays, b.) perpendicular fans, c.) 3 fan array, d.) 7 fan array. The examples shown here have 30, 30, 45 and 105 observations respectively.

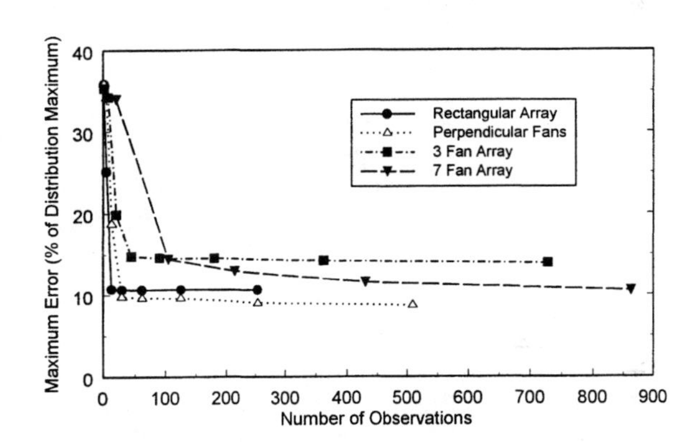

FIGURE 4. Maximum pixel inversion error versus the number of observations for several observation geometries. The test data was generated from distribution 3 with 729 pixels.

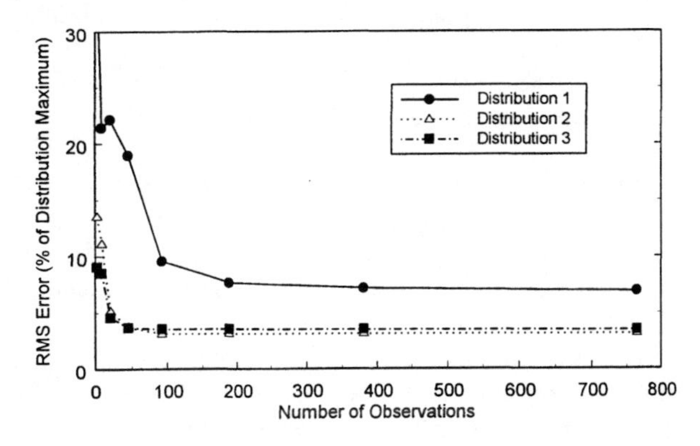

FIGURE 5. RMS inversion error versus the number of observations for the 3 fan array geometry for several test distributions. The inversion area was divided into 729 pixels.

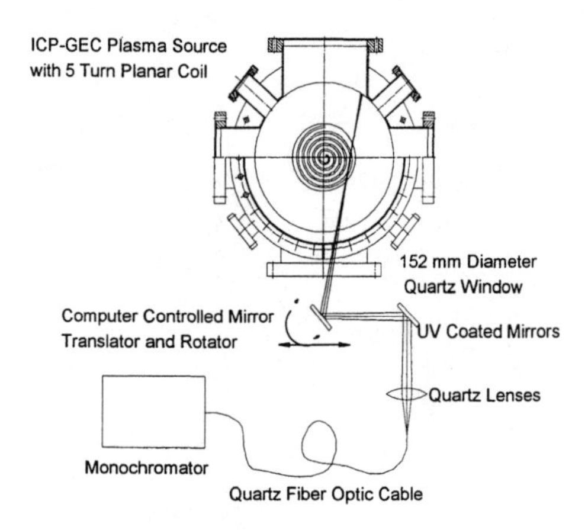

FIGURE 6. Experimental Apparatus.

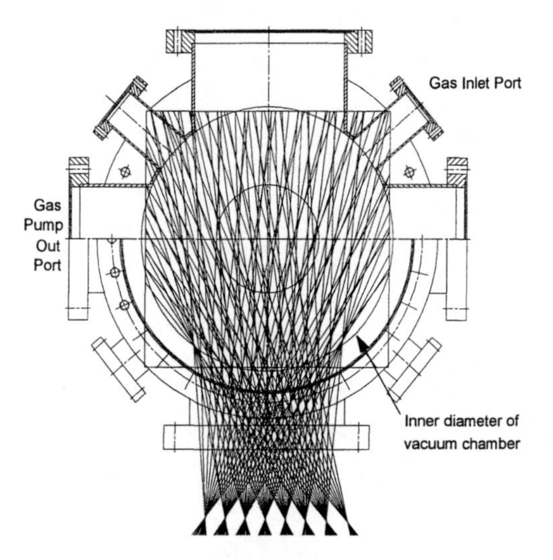

FIGURE 7. Experimental observation geometry. Approximately 1/10 of the total number of measurement directions is shown.

EXPERIMENTAL APPARATUS AND MEASUREMENTS

CAT has been applied to an inductively coupled plasma (ICP) source. A 5 turn, 100 mm diameter planar inductive coil with a dielectric vacuum interface has been installed in place of the powered electrode of a Gaseous Electronics Conference RF Reference Cell(6). In addition, an electro-static shield has been inserted between the coil and dielectric vacuum interface. A 165 mm diameter, grounded stainless steel lower electrode is 40.5 mm from the dielectric vacuum interface. The GEC RF Reference Cell vacuum chamber has many ports for diagnostics. The optical emission measurements were made through a single 152 mm diameter quartz window. Taking the data through a single window greatly simplifies the data acquisition process, although data could be taken through windows on four additional ports. A schematic of the experimental apparatus is shown in Fig. 6. UV coated optics are used so that optical measurements can be made from the near infrared to the UV. A ¼ m monochromator with a photomultiplier are used detect the desired optical emissions from the plasma. A computer controlled translator and rotator determined the detector line of sight through the plasma. The experimental observation geometry is shown in Fig. 7. This observation geometry is similar to an 8 fan array, but it has been slightly modified due to the refraction of the quartz window. The corners of the inversion geometry currently extend beyond the inner wall of the vacuum chamber.

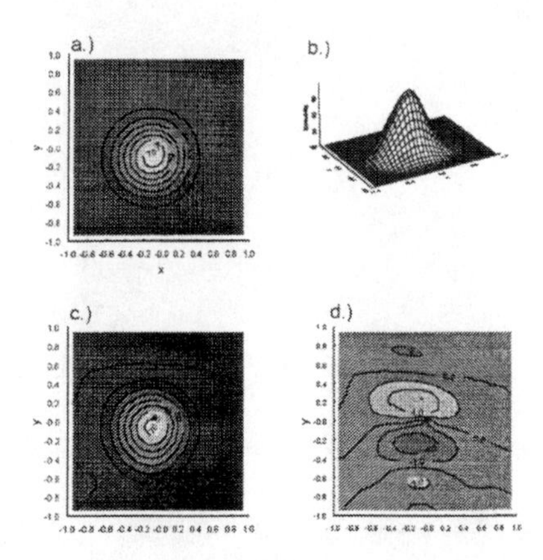

FIGURE 8. CAT inversion using test distribution 3 with 1089 pixels: a.) and b.) original distribution, c.) CAT image, d.) error in the CAT image.

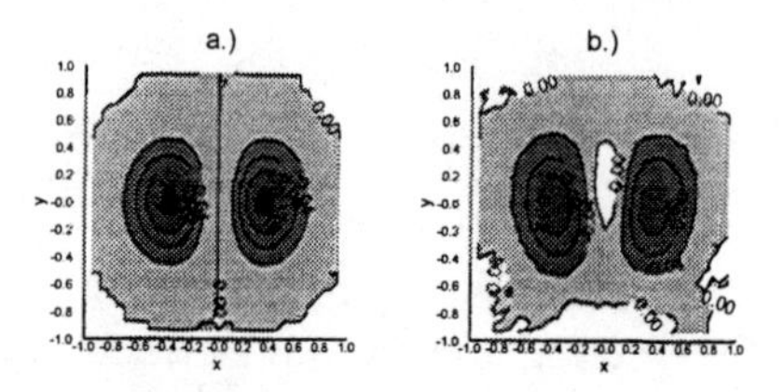

FIGURE 9. Differential CAT imaging. The two original distributions have maximum values of approximately 92 at the origin. a.) is the difference between the two original distributions and b.) is the difference between the inverted images.

The accuracy of the experimental observation geometry has also been tested using the same test distributions used previously. The result of one of these tests is shown in Fig. 8. The inversion region was a 240 mm x 240 mm square divided into 1089 pixels with a resulting spatial resolution of 7.1 mm x 7.1 mm. Unlike the previous tests, experimental weighting factors are now included. The maximum pixel inversion error is still on the order of 10 %, with the inverted distributions tending to be distorted perpendicular to the viewing window.

Figure 9 demonstrates the potential use of the CAT for monitoring small distribution changes, instead of concentrating on the absolute accuracy of a single image.

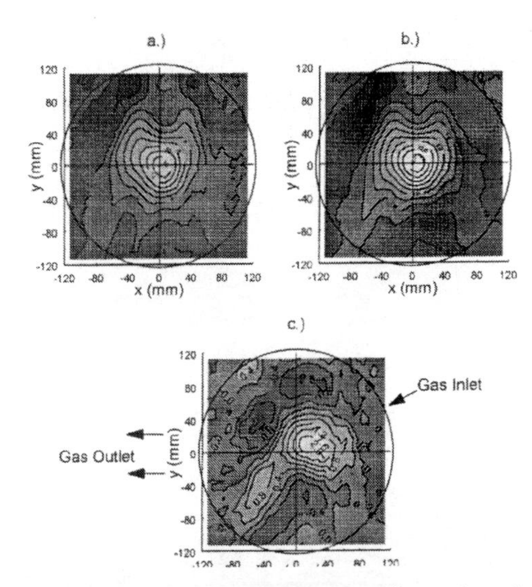

FIGURE 10. CAT images showing the effects of gas flowrate in an Ar discharge with 70 W applied RF power and 1.0 Pa pressure. The wavelength observed is $\lambda = 750.4$ nm. The flowrates were: a) 3.7 μmol/s (5 sccm) and b) 48 μmol/s (65 sccm). c.) is the difference between the two CAT images. The circles represent the inner diameter of the vacuum chamber wall.

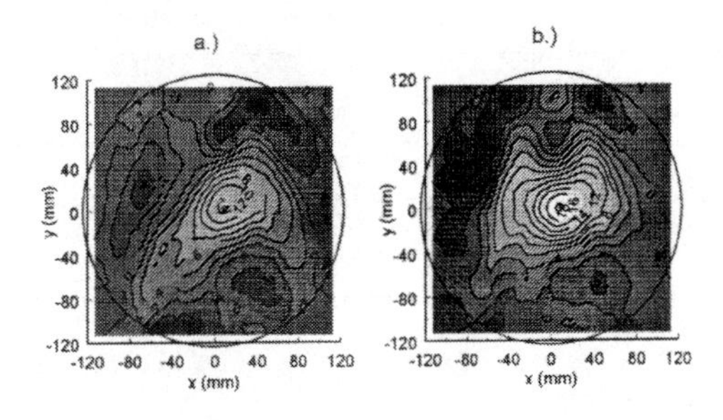

FIGURE 11. CAT images from a 50% Ar/O_2 discharge with plasma conditions of 100 W, 2.7 Pa and 3.7 μmol/s (5 sccm). a.) O I, $\lambda = 777.1$ nm and b.) Ar I, $\lambda = 750.4$ nm.

To create this figure, two distributions that differed only slightly from each other were inverted. The resulting CAT results are subtracted from each other and compared to the original distribution differences. The relative CAT image is qualitatively quite good, considering the differences in the original distribution were less than 0.1%.

Unlike the test distributions, the real experimental data will also be perturbed by reflections off the back wall of the vacuum chamber. The resulting systematic error can be many times larger than the statistical fluctuations of the optical measurements. The reflections are particularly large when there are sharp corners on the back wall. The resulting optical signals have sudden increases and decreases occurring over a small spatial length, which is contrary to what would be expected from a relatively smooth, nonnegative plasma emission distribution. These reflections are removed from the original data using a reflection matrix that is normalized to the maximum plasma intensity. The reflection matrix is generated so that the physically unrealistic jumps in the optical signals, which correlate with the vacuum chamber geometry, are removed for a variety of different plasma conditions. The uncertainty of the measurements perturbed by reflections is increased to account for the reflection matrix correction. Therefore, measurements perturbed by strong background reflections will receive very small weighting factors in the CAT inversion.

Previous optical emission measurements of this discharge source had shown that the plasma became axially asymmetric at high flow rates. Figure 10 shows the difference between a.) low (3.7 μmol/s) and b.) high (48 μmol/s) flow rates. Although it is not particularly obvious in the actual CAT images, the difference image c.) shows that the plasma emission maximum of Fig. 10 b.) has shifted towards the gas inlet port. This may be indicating that there is a neutral gas density variation within the vacuum chamber at these higher flow rates.

Plasmas with molecular gases, such as O_2 or N_2, also seemed to have axially asymmetric discharges. The CAT image of a 50% Ar/O_2 discharge is shown in Fig. 11. Although it is difficult to make to make quantitative statements with the 10% maximum pixel inversion error, it is apparent that the O I and Ar I optical emission distributions differ from one another in the same discharge and they both differ from the pure argon distribution of Fig. 10. It is possible that wall recombination effects may be altering the plasma distribution since the vacuum chamber walls are not axially symmetric.

CONCLUSION

CAT can be used to obtain information on plasma uniformity. The accuracy of the technique depends on several parameters including the normalization parameter, the seminorm, and the observation geometry. The experimental observation geometry could obtain an absolute accuracy of approximately 10% of the distribution maximum, although differential measurements seem suitable for monitoring small changes in plasma distributions. Asymmetric plasma distributions have been observed in high flow rate and molecular gas discharges.

REFERENCES

1. Schwabedissen, A., Benck, E. C., and Roberts, J. R., *Phys. Rev. E* **56**, 5866-5875.
2. Kak, A. C., and Slaney, M., *Principles of Computerized Tomographic Imaging*, New York: IEEE Press, 1988, ch. 4.
3. Williamson, J., H. and Evans, D. E., *IEEE Trans. on Plasma Sci.* **PS-10**, 82-93 (1982).
4. Williamson, J., H., Culham Rep., CLM R-210 (1982).
5. Hansen, P. C., *SIAM Review* **34**, 561-580 (1992).
6. Hargis, P. J. *et al.*, *Rev. Sci. Instrum.* **65**, 140-154 (1994).

Applications of Electron-Interaction Reference Data to the Semiconductor Industry

L. G. Christophorou and J. K. Olthoff

Electricity Division, Electronics and Electrical Engineering Laboratory
National Institute of Standards and Technology, Gaithersburg, MD 20899

Many advanced plasma diagnostic techniques and plasma models require fundamental physical data in order to make accurate measurements or predictions. Fundamental electron-interaction data are among the most critical since electron collisions are the primary source of ions, reactive radicals, and excited species present in etching, cleaning, and deposition plasmas. In this paper we present (*i*) a description of some semiconductor applications that require fundamental electron-interaction data, (*ii*) an explanation of the NIST program to provide these data, (*iii*) a summary of the available fundamental electron-interaction data for plasma processing gases, and (*iv*) a discussion of general data needs.

INTRODUCTION

As the push for smaller feature sizes and higher quality devices in the semiconductor industry has increased, so has the need for sophisticated models with predictive capabilities that can guide the technology, and for advanced diagnostics to probe the details of the plasmas used to etch features, deposit materials, or clean reactor chambers. Additionally, environmental concerns have fostered the demand for the more efficient use of global warming gases used in plasma processes. Advancement in each of these areas inherently requires detailed understanding of the physics and chemistry occurring within the discharge, which itself requires knowledge of the basic collision processes occurring between the species existing in the plasma. The most fundamental of the discharge processes are collisions between electrons and atoms or molecules. These collisions are the precursors of the ions and radicals which drive the etch, cleaning, or deposition processes.

The National Institute of Standards and Technology (NIST) started a program in 1996 to provide the semiconductor community with a concise source of complete, reliable electron-interaction reference data for plasma processing gases. To date, this program has assessed the available data for 6 major processing gases: CF_4, C_2F_6, C_3F_8, CHF_3, CCl_2F_2, and Cl_2. Details of the recommended data for these gases appear in Refs. (1–6), respectively, and downloadable tables of the recommended data are available via the World Wide Web (WWW) at *http://www.eeel.nist.gov/811/refdata*. In this paper we examine areas of the semiconductor industry that require these data for reliable operation or for continued development, along with a tutorial highlighting the industry's need for these data. Additionally, we review the present state of available electron-interaction data, and in some cases provide recently updated, and previously unpublished, data. In so doing, it is our intention to clearly inform the semiconductor community about the availability or unavailability of essential data of this type.

APPLICATIONS

Numerous applications exist in the semiconductor industry that require the use of electron-interaction data, and we briefly discuss five such applications here. They are: (*i*) plasma models designed to emulate discharge conditions and plasma processes; (*ii*) new diagnostic techniques being developed for the detection of gas-phase radicals utilizing negative ion mass spectrometry or threshold ionization mass spectrometry; (*iii*) diagnostic methods based on the detection of optical emission from the discharge; (*iv*) optimization of plasma-chamber cleaning processes utilizing global warming gases, and of post-processing emission abatement techniques; and (*v*) basic research into the identity and role of transient species in reactive plasmas.

(1) Plasma Models – Many models have been applied to simulating various aspects of reactive plasmas [see Refs. (7–10), for example]. The fundamental parameters required for the accurate modeling of reactor plasmas are electron-energy distributions, electron densities, positive ion fluxes and energies, negative ion densities, and reactive radical densities. Knowledge of these parameters is the precursor to the calculation of other more industrially-significant parameters, such as etch and deposition rates, etch profiles, and plasma uniformity. However, calculation of the fundamental parameters relies heavily on knowledge of the magnitude of electron-interaction cross sections, since virtually all physical processes in the discharge are initiated by electron motion through the gas.

Electron-energy distributions are determined by elastic and inelastic electron-scattering processes as electrons are

CP449, *Characterization and Metrology for ULSI Technology: 1998 International Conference*
edited by D. G. Seiler, A. C. Diebold, W. M. Bullis, T. J. Shaffner, R. McDonald, and E. J. Walters
1998 The American Institute of Physics 1-56396-753-7/98/$15.00

accelerated through the plasma gas by the applied electric field. Determination of the energies of the electrons in the discharge is the initial calculation in models based upon Boltzmann or Monte Carlo techniques (11). These models rely primarily upon cross sections for momentum transfer and vibrational excitation for determination of these distributions.

The electron density in a plasma is primarily determined by electron-impact ionization processes (which produce free electrons), electron attachment processes (which remove electrons by creating negative ions), and by secondary electron emission from surfaces. The first two processes are described by ionization and attachment cross sections, respectively, and the third process, while important and worthy of investigation, is a surface process and is not discussed here.

Positive ion bombardment is one of the main drivers of plasma surface reactions, and the ion flux is a direct result of electron-impact ionization. Although the final identity and magnitude of the positive ion flux may be dependent upon secondary reactions, such as ion-molecule reactions occurring as the ion travels through the discharge, the initial ion-formation process is driven by electron-impact processes. Partial ionization cross sections are required to determine the identity and quantity of the initial ions created in the plasma.

Negative ion production and the resulting densities are determined by direct electron attachment and dissociative attachment processes. Dissociative attachment processes can be a very effective way to produce both negative ions and radicals due to the large cross sections at low electron energies exhibited by some molecules.

Reactive radicals are produced by three basic processes: electron-impact dissociative ionization, where a positive ion and a radical are produced; electron-impact dissociative attachment, where a negative ion and a radical are produced; and neutral dissociation, where two neutral fragments are produced by electron impact. Knowledge of the cross sections related to these reactions is necessary for calculating the density of these highly reactive species that are critical to etching and deposition processes.

Electron transport parameters and rate coefficients are often the first parameters calculated by plasma models. Accurate measurements of these values are essential to validate the model calculations. The transport parameters commonly calculated in some models include: electron drift velocity, transverse electron diffusion coefficient to electron mobility ratio, density reduced ionization coefficient, density reduced electron attachment coefficient, and total electron attachment rate constant.

(2) Mass Spectrometric Diagnostics – The need to measure the identity and density of gas-phase plasma products, including reactive radicals, in industrial plasmas has led to the development of two mass spectrometric-based diagnostics: negative ion mass spectrometry and threshold ion mass spectrometry. Negative ion mass spectrometry detects gas-phase plasma products by monitoring negative ions formed by electron attachment to radicals, excited species, and molecules formed by reactions in the plasma (12,13). Cross sections for electron attachment and dissociative attachment are required for the feed gas and for the species to be detected in order for this technique to be effective.

Threshold ionization mass spectrometry detects radicals by monitoring positive ions generated by collisions of radicals with electrons whose kinetic energy is above the ionization threshold of the radical, but below the ionization threshold of the feed gas (14–16). This allows detection of radicals even when the mass spectra of the radicals are similar to those of the feed gas. This technique, too, requires detailed electron-impact ionization cross sections for the feed gas and the radicals at energies near threshold. This same technique can be used to detect excited species.

(3) Optical Emission Diagnostics – The light emission from a processing plasma is often used to monitor plasma uniformity, excited species densities, and electron-energy distributions. The later two applications require electron-impact excitation cross sections for the feed gas and for the gas-phase products produced in the discharge.

(4) Environmental Applications – Environmental concerns over the use of fluorinated compounds, that are often global warming gases, have prompted significant interest in increasing the efficiency of plasma-assisted cleaning techniques (17), and in the development of post-processing emission abatement techniques (i.e., the destruction or conversion of any remaining feed gas being exhausted from the reactor). Both of these processes, cleaning and abatement, require the dissociation of the feed gas into reactive radicals, which is nearly entirely dependent upon electron-impact ionization and dissociative collisions. Knowledge of the cross sections for these collisions is necessary to optimize such industrial processes.

(5) Basic Research – Fundamental electron-interaction data are applicable to a wide range of basic research being preformed in support of the semiconductor industry. Even a cursory discussion of all of these areas is obviously beyond the scope of this article. However, research into the detection and role of transient species in a plasma is of particular interest and importance.

As mentioned previously, the detection of transient species, such as radicals and excited species (vibrationally and electronically excited), often requires knowledge of the electron-interaction cross sections with these species. Many of these cross sections are currently unknown. Of perhaps greater importance, is the determination of the effect that these species have on the plasma itself. Early data indicate that electron interactions with transient species often have cross sections that are orders of magnitude greater than for similar interactions with the feed gas.

An example of this is shown in Fig. 1, where it can be seen that the cross section for electron scattering from metastable argon is significantly greater than from ground state argon (18,19). Similar results have been observed for molecules (20,21). This implies that the presence of these species may exert significant influence on the behavior of electrons in a plasma. The data for argon in Fig. 1 are especially noteworthy because argon is used as a buffer gas in many plasma reactors.

DATA ANALYSIS

The goal of the NIST effort in this area is to provide complete and accurate fundamental information on low-energy electron-molecule interactions with plasma processing gases used in the semiconductor industry. These data include electron-collision cross sections, and electron transport and rate coefficients. The gases currently being investigated include etching and cleaning gases (CF_4, CHF_3, C_2F_6, C_3F_8, Cl_2, NF_3, HBr, and SF_6), buffer gases (Ar and He), and common impurities or additives to plasma reactors (O_2, H_2O, and N_2). This list is modified and expanded as the needs of the industry change.

Data for each gas are accumulated, assessed, and evaluated, and then a NIST-recommended set of data is derived and placed in a database on the World Wide Web. The method for data review, assessment, and dissemination is as follows:

- perform thorough literature searches,
- review and assess available data,
- determine most reliable data,
- derive "recommended" data set,
- determine gaps in available data,
- perform measurement or calculation of needed data at NIST if possible,
- encourage the performance of required measurements or calculations elsewhere,
- write comprehensive archival paper, and
- present recommended data on the WWW (*http://www.eeel.nist.gov/811/refdata*).

The website allows easy access to the data for users in industry, academia, and government laboratories, including direct downloading of data. The site is maintained with the most up-to-date data available.

The assessment process is illustrated in Fig. 2 for the momentum transfer cross section of CF_4. A thorough evaluation of the literature reveals momentum transfer cross section data that differ by more than two orders of magnitude. However, assessment of the data allows the determination of the reliable cross section shown by the solid black line.

To date, analyses of the plasma processing gases CF_4, C_2F_6, and CHF_3 have been published (1,2,4), and the data for C_3F_8 have been accepted for publication (3). The assessment of electron-interaction data is nearly complete for Cl_2 (6). A similar analysis has been performed for CCl_2F_2, due to its importance in atmospheric chemistry and ozone depletion (5).

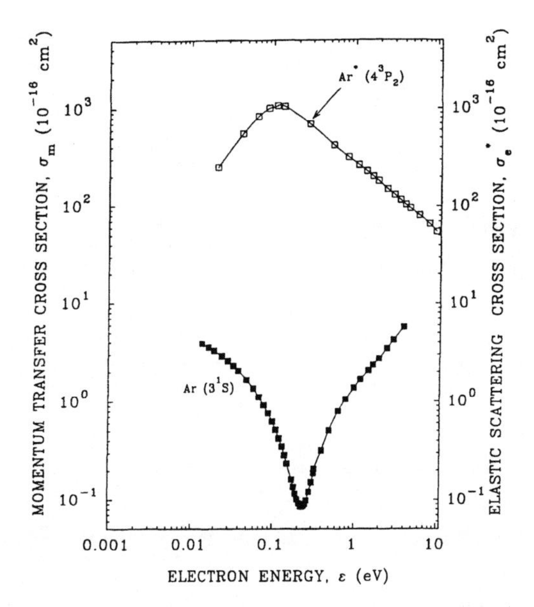

FIGURE 1. Momentum transfer cross section from ground state argon and elastic scattering cross section from metastable argon from Refs. (18) and (19), respectively.

REFERENCE DATA

In this section we review the completed NIST assessments of electron-interactions with the gases of greatest interest to the semiconductor industry: CF_4, CHF_3, C_2F_6, and C_3F_8. The data presented here reflect the most current data available,

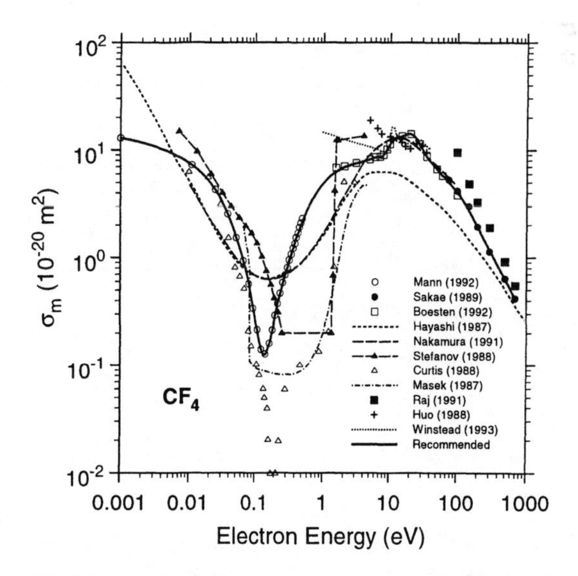

FIGURE 2. Momentum transfer cross sections for CF_4 from the archival literature. The solid line is the NIST recommended cross section. Details of the method to determine the recommended cross section, and of the references for the data shown in this figure may be found in Ref. (1).

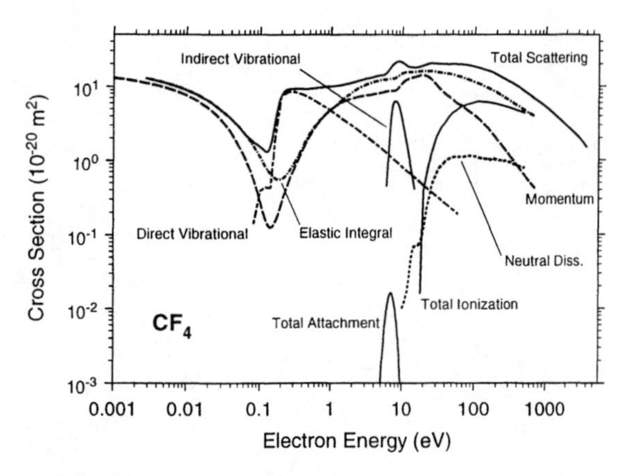

FIGURE 3. Electron-interaction cross sections for CF_4 (1).

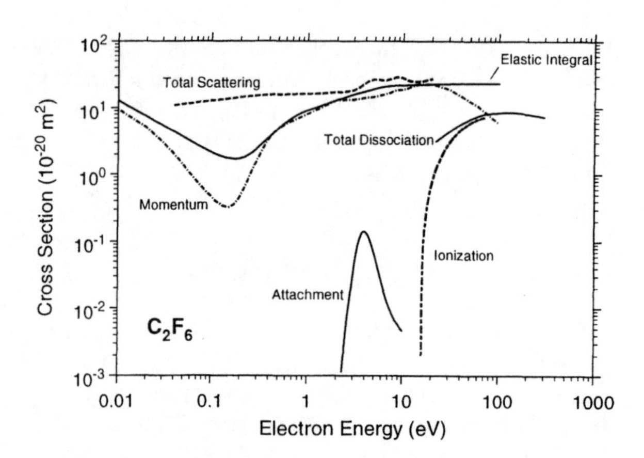

FIGURE 5. Electron-interaction cross sections for C_2F_6 (2).

and in some cases represent an update of the recommended cross section sets published previously.

Carbon Tetrafluoride (CF_4) – The data on electron-scattering cross sections, and attachment, ionization, and transport coefficients are reasonably complete for CF_4 (1). Figure 3 shows the independently assessed electron-interaction cross sections. The sum of the elastic, vibrational, ionization, attachment, and neutral dissociation cross sections has been shown to agree well with the recommended total electron-scattering cross section (1), thus demonstrating the overall consistency of the data.

The cross section for dissociation into neutrals shown here in Fig. 3 is updated from our previously published recommended set (1). Our previous set of recommended values for dissociation into neutrals was based upon the only directly-measured experimental data existing at the time (22). However, new direct measurements over a limited range of electron energies (23) confirm our earlier concern that the previous experimental measurements (22) of this cross section were too low. The new values presented here for dissociation into neutrals are based upon the recent measurements of Motlagh and Moore (24). Confirmation of this cross section over an extended range of electron energies is a significant data need for CF_4.

Additional data needs include experimental determination of direct and indirect vibrational excitation (the values shown here are derived from theoretical calculations), and resolution of some discrepancies in measurements of the total ionization and elastic integral cross sections.

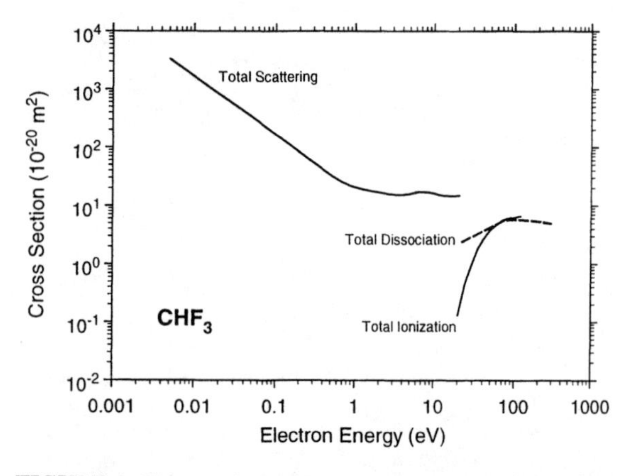

FIGURE 4. Electron-interaction cross sections for CHF_3 (4).

Trifluoromethane (CHF_3) – Figure 4 shows the meager electron-scattering cross section data for this important plasma processing gas (4). This data set serves as an illuminating contrast to the much more complete set of data available for CF_4. Basic measurements and calculations are needed for virtually all elastic and inelastic electron scattering processes, including momentum transfer, vibrational excitation, elastic scattering, differential scattering, and electron transport, attachment and ionization coefficients. Confirmation is required of the measured cross sections for total electron scattering and for total ionization. The previously recommended data for dissociation into neutrals are not shown here since these data are expected to suffer from the same shortcomings as those for CF_4 discussed above.

Perfluoroethane (C_2F_6) – Figure 5 shows the moderately complete cross section set for C_2F_6 (2). Many of the data are still preliminary (e.g., the cross section for momentum transfer at low energies). New cross section data are needed, especially for dissociation into neutrals, and vibrational and electronic excitation. Additional data are needed for momentum transfer, elastic integral, total ionization, and total scattering cross sections, in order to confirm existing data. There exist reasonably accurate data on attachment, ionization and transport coefficients.

Perfluoropropane (C_3F_8) – Figure 6 shows the presently available electron-scattering cross sections for C_3F_8 (3). Like C_2F_6, many of these data are preliminary and previously unpublished. The total electron-scattering cross section data are recent unpublished measurements, and the momentum transfer and elastic integral cross sections are from unpublished data derived from differential scattering cross

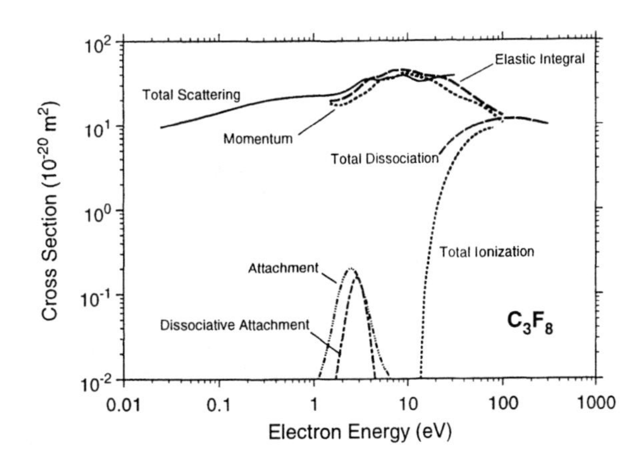

FIGURE 6. Electron-interaction cross sections with C_3F_8 (3).

section measurements. The preliminary values for elastic integral and momentum transfer cross sections exceed the measured total scattering cross section near 10 eV, but they are not incompatible within the combined estimated uncertainties. The total ionization, dissociative attachment (T≅300 K), and total dissociation cross sections are assessments based on the published experimental data in the literature. No measurements exist for electron-impact dissociation into neutral fragments and vibrational and electronic excitation. Much work is needed for virtually all cross sections and the electron transport coefficients for this molecule.

DATA NEEDS

At this time a reasonably complete set of electron-interaction cross sections exists only for CF_4. The other gases assessed in this project all have significant gaps in the known cross sections. Two nearly universal needs for these plasma processing gases are experimentally derived cross sections for vibrational excitation and for dissociation into neutrals. Both of these processes play critical roles in industrial plasmas, and a minimal amount of data are currently available for CF_4, with no data available for the other gases.

More generally, electron-interaction cross sections are needed for radicals and excited species commonly produced in industrial plasmas. Electron-impact ionization cross sections have been measured for some radicals produced in CF_4 plasmas (25), but no measurements are available for any other electron-collision interaction or any other plasma processing gases.

Existing data for electron transport and rate coefficients are reliable for most of the plasma processing gases, with a few major exceptions. No transport data are available for CHF_3, except for some recent measurements of drift velocity (26).

REFERENCES

1. Christophorou, L. G., Olthoff, J. K., and Rao, M. V. V. S., *J. Phys. Chem. Ref. Data*, **25**, 1342–1388, 1996.
2. Christophorou, L. G., and Olthoff, J. K., *J. Phys. Chem. Ref. Data*, **27**, 1–29, 1998.
3. Christophorou, L. G., and Olthoff, J. K., *J. Phys. Chem. Ref. Data*, in press.
4. Christophorou, L. G., Olthoff, J. K., and Rao, M. V. V. S., *J. Phys. Chem. Ref. Data*, **26**, 1–15, 1997.
5. Christophorou, L. G., Olthoff, J. K., and Wang, Y., *J. Phys. Chem. Ref. Data*, **26**, 1205–1237, 1997.
6. Christophorou, L. G., and Olthoff, J. K., *J. Phys. Chem. Ref. Data*, in preparation.
7. Bukowski, J. D., Graves, D. B., and Vitello, P., *J. Appl. Phys.* **80**, 2614–2623, 1996.
8. Lymberopoulos, D. P., and Economou, D. J., *IEEE Trans. Plasma Sci.* **23**, 573–580, 1995.
9. Meyyappan, M., and Govindan, T. R., *J. Appl. Phys.* **80**, 1345–1351, 1996.
10. Ventzek, P. L. G., Grapperhaus, M., and Kushner, M. J., *J. Vac. Sci. Technol. B* **12**, 3118–3137, 1994.
11. Bordage, M. C., Segur, P., and Chouki, A., *J. Appl. Phys.* **80**, 1325–1336, 1996.
12. Stoffels, W. W., Stoffels, E., and Tachibana, K., *Jpn. J. Appl. Phys.* **36**, 4638–4543, 1997.
13. Rees, J. A., Seymour, D. L., Greenwood, C. L., and Scott, A., *Nuc. Instrum. Methods B* **134**, 73–76, 1998.
14. Sugai, H., Nakamura, K., Hikosaka, Y., and Nakamura, M., *J. Vac. Sci. Technol. A* **13**, 887–893, 1995.
15. Nakamura, K., Segi, K., and Sugai, H., *Jpn. J. Appl. Phys.* **36**, L439–L442, 1997.
16. Schwarzenback, W., Tserepi, A., Derouard, J., and Sadeghi, N., *Jpn. J. Appl. Phys.* **36**, 4644–4647, 1997.
17. Sobolewski, M., Langan, J. G., and Felker, B. S., *J. Vac. Sci. Technol. B* **16**, 173–182, 1998.
18. Milloy, H. B., Crompton, R. W., Rees, J. A., and Robertson, A. G., *Aust. J. Phys.* **30**, 61, 1977.
19. Robinson, E. J., *Phys. Rev.* **182**, 196, 1969.
20. Christophorou, L. G., Pinnaduwage, L. A., and Datskos, P. G., in *"Linking the Gaseous and the Condensed Phases of Matter, the Behavior of Slow Electrons,"* L. G. Christophorou, E. Illenberger, and W. F. Schmidt (Eds.), Plenum Press, New York, 1994, pp. 415–442.
21. Christophorou, L. G., Van Brunt, R. J., and Olthoff, J. K., *Proceedings of XIth International Conf. Gas Discharges and Their Applications*, Tokyo, Japan, Sept. 10–15, 1995, Vol. 1, pp. 536–548.
22. Sugai, H., Toyoda, H., Nakano, T., and Goto, M., *Contrib. Plasma Phys.* **35**, 415, 1995.
23. Mi, L., and Bonham, R. A., *J. Chem. Phys.* **108**, 1910–1914, 1998.
24. S. Motlagh and J. H. Moore, *J. Chem. Phys.*, in press.
25. Tarnovsky, V., Kurunczi, P. Rogozhnikov, D., and Becker, K., *Int. J. Mass Spectrom. Ion Processes* **128**, 181, 1993.
26. Wang, Y., Christophorou, L. G., Olthoff, J. K., and Verbrugge, J., in *"Gaseous Dielectrics VIII,"* L. G. Christophorou and J. K. Olthoff (Eds.), Plenum Press, New York, 1998, in press.

RF Sensing for Real-Time Monitoring of Plasma Processing

Craig Garvin and Jessy W. Grizzle

University of Michigan Electronics Manufacturing Laboratory, 3300 Plymouth Road, Ann Arbor, MI 48105-2551
garv@eecs.umich.edu, grizzle@eecs.umich.edu

A novel sensing system based on the microwave resonance probe is compared to standard RF metrology. The system uses an antenna in the glow discharge to excite the bulk plasma at a frequency range of $30MHz$ to $1GHz$. Standard RF metrology is implemented by measuring the fundamental and five harmonics of the RF power signal. An experiment varying power, pressure, CF_4 and O_2 is constructed. Using a subset of the data to regress a model, standard RF sensing reconstructs the experimental variables with a best average R^2 of 0.586 at a high model coefficient variance (σ_b^2), whereas the novel sensing system results in a best average R^2 of 0.804 and an order of magnitude lower σ_b^2.

INTRODUCTION

Ever shrinking geometries and larger substrates are mandating improvements in sensor systems for control and diagnostics of plasma processing. Experts in industry and academia have recognized the RF signal ($13.56MHz$) and its harmonics as a potential source of process information. A number of researchers have used RF metrology as a tool for plasma diagnostics. Maynard *et al.* (1) have used RF metrology for end-pointing of an industrial etch process. Spanos *et al.* (2, 3) have extensively used RF metrology in their plasma diagnostic and control work. A number of researchers (4–7) have used ion flux information obtained via RF metrology to characterize etching. Researchers at the Adolph Slaby Institute (8, 9) have developed a diagnostic system that uses plasma physics models to infer process - relevant information from the fundamental frequency exciting the discharge and multiple harmonics resulting from plasma non linearities. Attempts to model the relationship between measured RF parameters and plasma physics have been presented in (10–12). This list is by no means complete. The common element is the attempt to relate measurement of the $13.56MHz$ signal to plasma, wafer, and tool conditions.

The ionospheric plasma community has followed a radically different approach to extracting information from plasmas, namely the use of resonance probes (13). A resonance probe is like a Langmuir probe in that it is an active probe. However, unlike the Langmuir probe, the resonance probe operates at a frequency far above the standard RF signal. In its standard mode of operation, an antenna is inserted into the plasma and driven over a frequency range centered about the plasma frequency in order to infer electron concentration.

This article investigates the use of a resonance probe in a micro-electronics processing plasma. The primary goal is to evaluate the relative observability of the state of the plasma under standard RF metrology and resonance probe techniques. More precisely, the aim of most plasma diagnostics is to *detect* a change in the plasma state and to *isolate* the source of this change. Accordingly, in this first study, a very simple experiment is conducted to compare the abilities of the two measurement techniques to detect and isolate plasma changes due to the variation of generator power, chamber pressure, and gas chemistry.

In contrast to the ionospheric plasma community's use of only plasma resonance frequency information, analysis of the resonance probe data presented in this article employs all frequency information. To underscore the frequency range of the modified resonance probe monitoring, it is referred to as *broad band sensing* whereas methods relying on the plasma's RF signal and its harmonics are referred to as *narrow band sensing*. It could be anticipated that a signal containing a broader range of data may carry with it more information about the plasma's condition. Initial results indicate that broad band sensing exhibits a sensitivity to plasma parameters that is significantly stronger than that achieved through conventional narrow band sensing.

EXPERIMENTAL SETUP

An experiment was designed in order to simultaneously collect broad band and narrow band data as a function of widely varying plasma conditions, as shown in Figure 1. Experiments were performed on a GEC research reactor, described in Ref. (14). Power is delivered using an ENI generator and matching network, and generator power is measured using the built in ENI power meter. Pressure is measured using an MKS barratron, and gas flow rates are measured with MKS flow meters with MKS gas correction factors.

Narrow band sensing is implemented using a Werlatone C1373 $1.5MHz$ to $80MHz$ directional coupler rated at 750 Watts power with a nominal $-30dB$ coupling between main line and sensor ports received by a Tektronix TDS 420 digital storage oscilloscope. Groups of 1000 points of both forward and reverse waves are sampled at 200 Ms/sec, resulting in a total of about 65 periods of each wave. Each waveform is then digitally filtered using separate 6th order discrete Chebyshev bandpass filters centered at the fundamental frequency and first five harmonic components respectively. The magnitude and relative phase of forward and reverse wave at each frequency component is then calculated.

CP449, *Characterization and Metrology for ULSI Technology: 1998 International Conference*
edited by D. G. Seiler, A. C. Diebold, W. M. Bullis, T. J. Shaffner, R. McDonald, and E. J. Walters

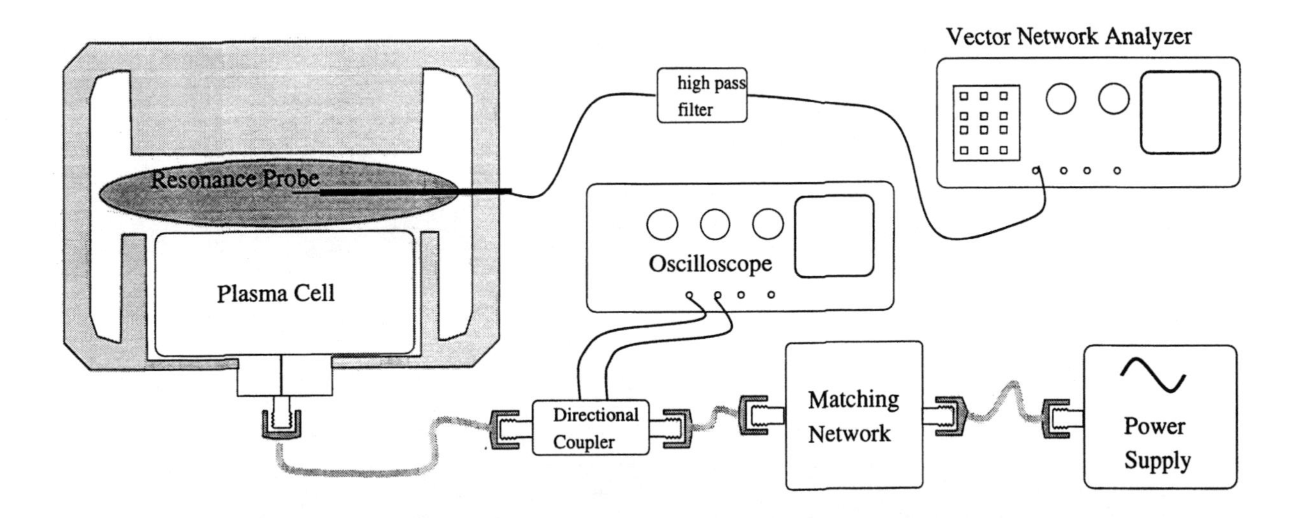

FIGURE 1. Experimental Setup for Comparing Broad Band and Narrow Band Sensing

TABLE I. Experimental Variables and Levels

Variable	Level 1	Level 2	Level 3
Power	$60W$	$70W$	$80W$
Pressure	$40mTorr$	$50mTorr$	$60mTorr$
CF_4 flow	$40sccm$	$50sccm$	-
O_2 flow	$0sccm$	$2sccm$	-

Broad band sensing is achieved by a resonance probe constructed of a length of RG402u stainless steel rigid coaxial cable. Approximately one inch of center conductor is exposed to the plasma to act as an antenna. The probe is inserted in the bulk plasma using an O-ring compression sealed vacuum port. A Hewlett Packard 8712B vector network analyzer drives the resonance probe over a range of $30MHZ$ to $1GHz$ at a power level of 0 dBM. A Mini Circuits $25MHz$ high pass filter is used to isolate the vector network analyzer from the discharge. After calibration, the complex reflection coefficient (Γ) is recorded at 801 frequency points linearly uniformly spaced between $30MHz$ and $1GHz$. To maximize accuracy, the data is collected using the analyzer's lowest IF bandwidth. The set-points of power, pressure and flow rate for the GEC as well as data acquisition are controlled with a Macintosh Quadra 950 running LabVIEW data acquisition and control software. All data is logged and written to file automatically.

EXPERIMENT AND INITIAL RESULTS

The goal of the experiment is to evaluate the ability of broad band and narrow band sensing to isolate basic plasma perturbations due to changes in power [1], pressure and chemistry. Accordingly, a full factorial experiment is performed, as summarized in Table I. Standard statistical methods are then used to construct a linear regression model to predict the variables listed in Table I based on measured broad band and narrow band data, respectively.

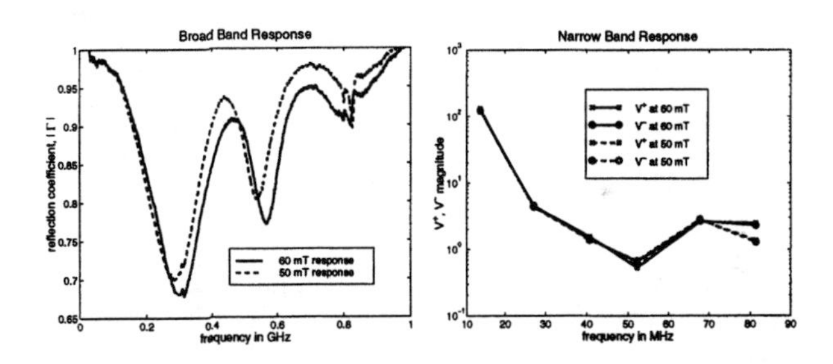

FIGURE 2. Broad Band and Narrow Band Response for 50 and 60 mTorr Plasma Pressures

It is informative to consider the response of the two sensing systems to a variation in pressure. Figure 2 shows both the broad band and narrow band magnitude response to a change in pressure from 50 $mTorr$ to 60 $mTorr$ with power at a constant 60 W, and a chemistry of pure CF_4 at 50 $sccm$ flow rate. In the broad band signal, there is a clear and distinct trend differentiating the two pressure conditions. Although both plasmas show absorptive peaks near 300, 600 and $800MHz$, the precise magnitude and location of these

[1] Power level refers to a generator set point and not a calculated delivered power to the plasma.

peaks changes visibly with pressure. Though it remains to be seen whether this qualitative observation will be elicited in the statistical analysis, the structure of the response using the broad band sensor suggests that substantial information about fundamental plasma physics may be embedded in the sensor data. The narrow band signal is also reported in Figure 2. A log scale is used due to the wide dynamic range of the signal[2]. It is more difficult to discern a pattern in the narrow band response, but this by no means indicates that structure is lacking.

EXPERIMENTAL ANALYSIS

The experiment described in Table I results in 36 experimental points, each consisting of a specific level of the independent variables. For each experiment, 12 complex points of narrow band and 801 complex points of broad band data are collected. The goal of the experimental analysis is to generate a linear regression model from each of the sensor systems to the plasma state as represented by power, pressure and chemistry. It is assumed that if the same regression methods are used in both cases, differences in fit can be attributed to fundamental differences in observability between the two systems. To present a direct comparison of methods, no additional transformations (such as using impedance or standing wave ratio representations) are used. The narrow band data is considered as $\mathbf{V}^+$, $\mathbf{V}^-$, and phase. The broad band data is considered as $|\,\mathbf{\Gamma}\,|$, and phase.

Two statistical parameters are used in the regression. The first is the standard R^2 evaluation of residuals as a function of output, defined as

$$R^2 = 1 - \frac{(y-\widehat{y})^T(y-\widehat{y})}{y^Ty}, \tag{1}$$

where y is the zero mean variable being estimated and $\widehat{y}$ is the estimate. The second is $\sigma_{\widehat{b}}^2$, the variance in the model parameters, which is estimated [3] by

$$\sigma_{\widehat{b}}^2 \approx s^2 \cdot \|(x^Tx)^{-1}\|, \tag{2}$$

$$s^2 = \sum_{i=1}^{n} \frac{(y_i-\widehat{y}_i)^2}{n}. \tag{3}$$

In the above, the induced 2-norm ($\|\cdot\|$) is used to reduce the covariance matrix of model coefficients to a single parameter.

The model variance, $\sigma_{\widehat{b}}^2$, essentially quantifies the relationship of the conditioning of the pseudo inverse ($(x^Tx)^{-1}$) to the residual ($(y_i-\widehat{y}_i)^2$). It is clear that as more sensor data is used to estimate y, the fit will improve. However, more data means poorer conditioning of $(x^Tx)^{-1}$. If the negative effects of poor conditioning outweigh improved fit, model variance increases. Physically, this suggests a model with poor robustness. It can be expected that a model with a high $\sigma_{\widehat{b}}^2$ would fit a new data set very poorly due to the extreme sensitivity of the model to sensor noise.

[2] The 6th harmonic is above the directional coupler frequency response of $80MHz$, slightly affecting the coupling.

[3] Precise conditions for the convergence of Eq. (2) to actual variance can be found in Sen (15) or other statistics texts.

TABLE II. Experimental Variables and Levels Used for Modeling Data Set

Variable	Level 1	Level 2
Power	$60W$	$80W$
Pressure	$40mTorr$	$60mTorr$
CF_4 flow	$40sccm$	$50sccm$
O_2 flow	$0sccm$	$2sccm$

The regression model is developed using a subset of the experimental data set, referred to as the *modeling set*, and tested on the full data set. The modeling set is composed of the 16 extremal experiments, shown in Table II. This modeling set is used for evaluating the relationship between principle components and $\sigma_{\widehat{b}}^2$. The fit (R^2) is then evaluated on the entire 36 experiment data set. This approach to evaluating model performance confounds two factors. By testing the model with new data, robustness is evaluated. Additionally, because the new data is at different experimental levels (the midpoints), the linearity of the underlying physics plays a major factor in fit performance. Because such a small data set is used for developing the model, it is expected that the fit to the entire 36 experiment data set may be relatively poor. Since the goal of this initial investigation is to compare the two measurement systems under similar conditions, only relative performance is of interest.

In order to study $\sigma_{\widehat{b}}^2$, principle component regression as described in (15–19) is used. Subsets of the original measurement data are formed using an increasing number of principle components (PC's). A plot of $\sigma_{\widehat{b}}^2$ as a function of PC's guides the selection of an optimum number of PC's. Principal component regression can be directly implemented on the narrow band data. It is straight forward to take the singular value decomposition of the 16×18 matrix composed of rows of experimental levels and columns of centered and scaled $\mathbf{V}^+$, $\mathbf{V}^-$, and phase. The same direct approach is not computationally practical for the broad band data where singular values of a 16×801 complex matrix are required. Accordingly, a preliminary reduction by FFT is performed on the data. A discrete Fourier transform of the data is taken, and the frequency components sorted by magnitude. The data is then reconstructed with increasing numbers of components until a compromise between data reduction and fit is found. For our data set, using the 300 most significant frequency components results in an R^2 *of the FFT reduced broad band sensor data to the unprocessed sensor data of* 0.999, and yields a tractable data set upon which to perform a principle component regression.

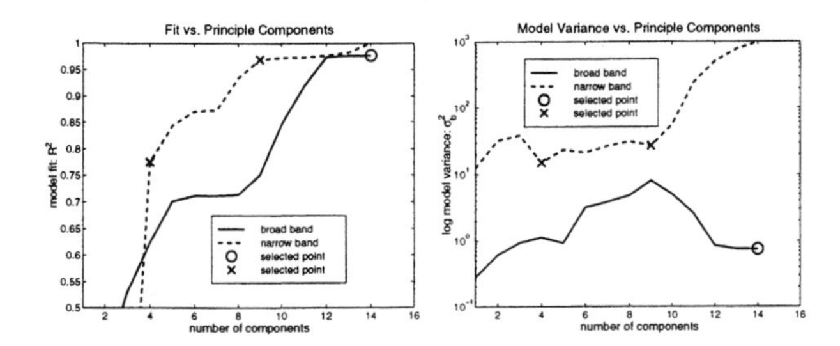

FIGURE 3. R^2 and σ_b^2 as a Function of Number of Principle Components on the 16 Experiment Modeling Data Set

Regression models fitting the four parameters of power, pressure, CF_4 flow and O_2 flow are generated using the PC's of the sensor data corresponding to the n strongest singular values. The procedure is performed for both broad band and narrow band data *on the 16 experiment modeling set.* Figure 3 shows both R^2 and σ_b^2 as a function of the number of PC's used in the regression model. It should be noted that Eq.(2) becomes less accurate as the number of measured data points approaches the number of experiments. A sharp drop in σ_b^2 serves as an indicator that the variance estimate is no longer accurate. In both the broad and narrow band cases, σ_b^2 is no longer accurate at 15 PC's and accordingly, data is only plotted to 14 PC's.

Choosing an optimal number of PC's for the broadband data is relatively straightforward. The variance estimate, σ_b^2, decreases monotonically above 9 PC's. Clearly, a 14 principle component model is best for the broad band data. Determining the optimal number of PC's in the narrow band case is more difficult. The variance estimate, σ_b^2, dips to a clear local minimum at 4 PC's. Another slightly higher local minimum is seen at 9 PC's. The model fit at 4 PC's is much poorer than that at 9 PC's. This may offset the better conditioning suggested by low σ_b^2. It makes sense to to test both these models against the complete 36 experiment data set.

TABLE III. R^2 and σ_b^2 of Broad and Narrow Band Sensor vs Experimental Variables: Entire 36 Experiment Data Set

	Broad Band		Narrow Band			
	14 PC's		9 PC's		4 PC's	
Variable	R^2	σ_b^2	R^2	σ_b^2	R^2	σ_b^2
Power	0.807	3.85	0.836	88.6	0.791	8.67
Pressure	0.885	2.46	0.667	192	0.416	25.9
CF_4	0.501	3.78	0.303	143	0.0570	14.9
O_2	0.997	0.0010	0.539	4.05	0.602	0.269

Table III summarizes R^2 and σ_b^2 data regression models using both the broad and narrow band sensor data to estimate experimental variables *in the complete 36 experiment data set.* With the exception of power, the fit with the broad band data is substantially better than that of the narrow band data. Even the fit to the power levels reflects favorably on the broad band sensor. Using only impedance information, the broad band sensor is able to determine power levels almost as accurately as the narrow band sensor, which has access to both forward and reverse voltage magnitudes. Of particular interest is the lower model variance achieved by broad band sensing. The low σ_b^2 promises a more robust parameter estimation under realistic process conditions, and will require additional experiments to confirm.

The R^2 and σ_b^2 data for 4 and 9 PC regression models using narrow band sensor show the trade offs that appear to be present narrow band sensing. Results of more than doubling the model size for the narrow band sensor are mixed. Average R^2 does improve from 0.466 to 0.586 but the variance estimate degrades by more than an order of magnitude. Additionally, R^2 performance is mixed. The CF_4 flow estimate improves dramatically with more PC's, but the O_2 flow estimate is actually worse. Clearly, there is a tradeoff in the narrow band data between improved predictive ability and model robustness.

FUTURE WORK

The promise shown by this initial experiment motivates several areas of future work. The range of narrow band sensing must be extended to provide an effective comparison. In this experiment, the coupler bandwidth limited data collection to 6 harmonics. Work at the Adolph Slaby Institute indicates that valuable information may be found at far higher harmonics. It appears that the main difference between broad and narrow band sensors is in model robustness. Accordingly, experiments should be designed to draw out these differences.

Although results using a resonance probe are promising, the potential of an intrusive diagnostic is clearly limited. In order to be practical, a non-intrusive application of broad band sensing is needed, as proposed in Figure 4. A combination of broad band combiner and coupler are used to add the broad band signal to the standard $13.56MHz$ power wave. The combiner has very strong directivity. As a result, power from the $13.56MHz$ power supply does not damage the analyzer. The broad band coupler is used to make a reflection measurement using the standard 'external test kit' mode of the network analyzer.

Finally, the plasma frequency response shows very interesting characteristics. An attempt to relate these to physical parameters through more sophisticated plasma models might be informative.

CONCLUSION

Power, pressure, CF_4 and O_2 levels relevant to microelectronics processing were varied using a full factorial experiment performed on a GEC reference cell. Standard RF sensing (referred to as 'narrow band sensing') was compared

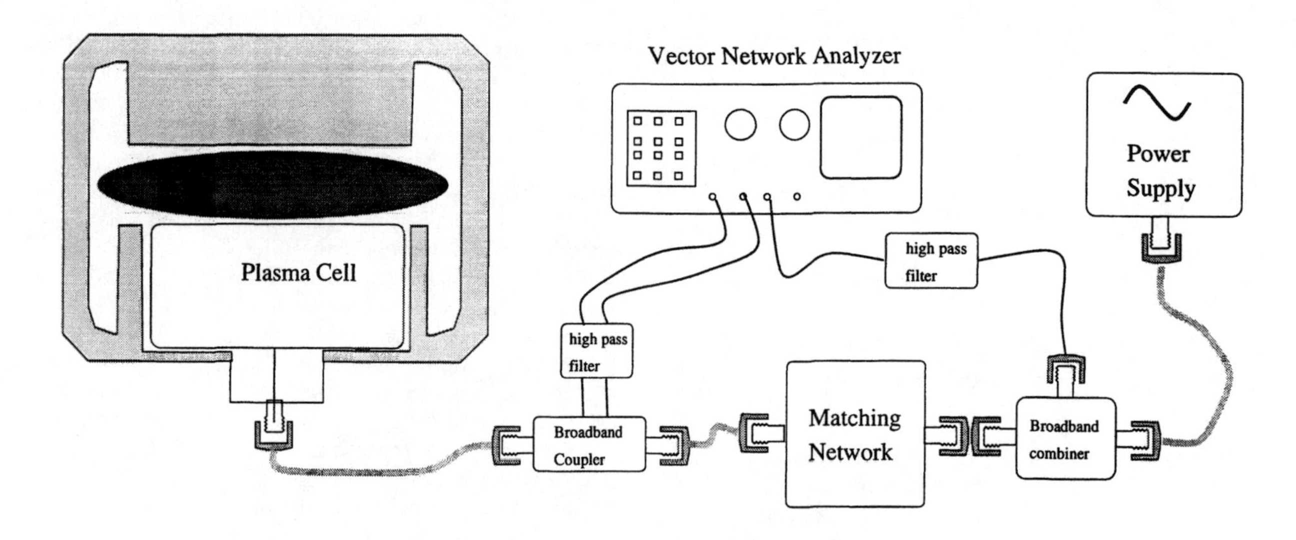

FIGURE 4. Potential Non-intrusive Setup for Broad Band Sensing

to a novel sensing technique based on resonance probes used in ionospheric research (referred to as 'broad band sensing'). Standard statistical techniques were used to regress a linear model against narrow band and broad band data respectively. A much better fit to the data was obtained using broad band sensing. Suggestions for further work include using more harmonics for the narrow band data and methods for designing a non intrusive broad band system.

ACKNOWLEDGMENTS

The authors sincerely thank Steven C. Shannon, and Professor Mary L. Brake for use of and assistance with the GEC cell. Sven G. Bilén constructed the resonance probe used in this research. Hyun-Mog Park designed and wrote the automated data collection algorithm. Dr. Dennis S. Grimard, Dr. Helen Maynard and Oliver Patterson provided many useful discussions.

This work was supported in part by AFOSR/ARPA MURI Center under grant # F49620-95-1-0524 and The Semiconductor Research Company under contract #97-FC-085.

---

[1] H. L. Maynard, E. A. Reitman, J. T. C. Lee, and D. E. Ibbotson, J. Echem. Soc. **143**, 2029 (1996).

[2] A. J. Miranda and C. J. Spanos, J. Vac. Sci. Technol. A **13**, 1888 (1996).

[3] A. Ison, W. Li, and C. J. Spanos, in *1997 IEEE International Symposium on Semiconductor Manufacturing* (The Institute for Electrical and Electronics Engineers, San Francisco, California, 1997), Vol. 1997.

[4] A. J. van Roosmalen, W. G. M. van den Hoek, and H. Kalter, J. Appl. Phys. **58**, 653 (1985).

[5] N. Hershkowitz, J. Ding, and J. Jenk, in *Proceedings of the IEEE International Conference on Plasma Science* (The Institute for Electrical and Electronics Engineers, Santa Fe, New Mexico, 1994), Vol. 1994, p. 110.

[6] S. Bushman, T. F. Edgar, I. Trachtenberg, and N. Williams, in *Proceedings of SPIE* (International Society for Optical Engineering, Bellingham, Washington, 1994), Vol. 2336.

[7] M. A. Sobolewski, IEEE Transactions on Plasma Sciences **23**, 1006 (1995).

[8] M. Klick, J. Appl. Phys. **79**, 3445 (1996).

[9] M. Klick, W. Rehak, and M. Kammeyer, Jpn. J. Appl. Phys. **36**, 4625 (1997).

[10] P. A. Miller, L. A. Romero, and P. D. Pochan, Phys. Rev. Letts. **71**, 863 (1993).

[11] V. A. Godyak, R. B. Piejak, and B. M. Alexandrovich, J. Appl. Phys. **69**, 3455 (1991).

[12] N. Hershkowitz, IEEE Transactions on Plasma Sciences **22**, 11 (1994).

[13] K. Takayama, H. Ikegami, and S. Miyasake, Phys. Rev. Letters **5**, 238 (1960).

[14] J. Hargis, P. J. *et al.*, Rev. Sci. Inst. **65**, 140 (1994).

[15] A. Sen and M. Srivastava, *Regression Analysis: Theory, Methods, and Applications* (Springer-Verlag, New York, New York, 1990).

[16] A. Ison and C. J. Spanos, in *1996 IEEE International Symposium on Semiconductor Manufacturing* (The Institute for Electrical and Electronics Engineers, Tokyo, Japan, 1996), Vol. 1996.

[17] R. Chen and C. J. Spanos, IEEE Transactions on Semiconductor Manufacturing **10**, 307 (1997).

[18] R. Chen, H. Huang, C. J. Spanos, and M. Gatto, J. Vac. Sci. Technol. A **14**, 901 (1996).

[19] S. Leang and C. J. Spanos, IEEE Transactions on Semiconductor Manufacturing **9**, 101 (1996).

Rapid Assessment of Plasma Damage Effects

L. E. Peitersen, G. A. Gruber, R. J. Hillard, R. G. Mazur, and R. H. Herlocher

Solid State Measurements, Inc., 110 Technology Drive, Pittsburgh, PA 15275

Richard Conti

IBM, Inc., IBM Microelectronics Division, IBM/Siemens DRAM Development Alliance
1580 Route 52 Hopewell Junction, NY 12533

This paper reports the use of a mercury probe with a highly repeatable contact area to monitor the effects of plasma damage on a gate oxide resulting from a PECVD process. The advantage of the mercury probe is that no processing is required to form the gate, allowing measurements to be made on plain oxidized wafers immediately before and after plasma exposure. The Hg probe provides a sensitive and rapid method for monitoring plasma damage without the need for antenna structures.

PLASMA DAMAGE MONITORING

Monitoring of plasma damage effects in MOS structures is critical to the reliability of ULSI circuits. Plasma damage causes an increase in both bulk oxide trapped charge (Q_{ot}) and in interface trap density (D_{it}). MOS device oxides are further degraded due to antenna effects (1) that cause locally high currents through the active gate oxide regions of the device.

Conventional CV and IV measurements made on simple MOS capacitor structures are not sufficiently sensitive to accurately monitor plasma damage. To date, only antenna structures that amplify the effects of plasma damage have proven to have sufficient sensitivity. However, antenna structures are costly and time consuming to prepare and test. This paper describes a technique for rapid and sensitive measurement of plasma damage without the need for antenna structures.

EXPERIMENTAL

To illustrate application of the technique, oxidized wafers (>2000Å SiO_2) were subjected to plasma enhanced phosphosilicate glass (PEPSG) deposition under conditions that resulted in either low or high plasma damage. The deposition time was 5 seconds, and the resultant average glass thickness was 300-400 Å.

All measurements were made with an SSM 495 Mercury Probe System with an area repeatability <0.1%. Details of this mercury probe are described elsewhere (2). Standard CV measurements were first attempted using a 1.0mm capillary and found to be insensitive to any differences in plasma damage. Similar results have been reported previously (3). Stepped voltage, stepped current and constant current IV measurements were made with a 4.0mm capillary.

RESULTS

Several IV based measurement methods were used to determine sensitivity to the difference in plasma charge in two oxide layers exposed to low and high plasma damage settings. The results are summarized in Table 1.

Values of Q_{BD} were obtained from stepped current IV measurements and show only small differences between low and high plasma charging.

Low-level oxide leakage measurements were made with the stepped voltage method. The leakage was measured at –2 volts and is higher for the high charging deposition.

The rate of change of voltage with time (dV/dt) was obtained during the application of a constant current stress at .2 nA. The dV/dt values at 50% cumulative probability for the two depositions show the highest sensitivity of the three methods reported here.

To confirm the Hg gate IV results, measurements were made on polysilicon gate sister wafers provided with antenna structures. Figure 1 shows leakage current measurements for the polysilicon gate antenna structures for both low and high plasma damage settings.

Figure 2 is a cumulative probability plot of the dV/dt values obtained during the Hg gate constant current stress measurements summarized in Table 1. The higher dV/dt values in the high charge cumulative probability plot indicate increased trapping within the oxide (4).

CP449, *Characterization and Metrology for ULSI Technology: 1998 International Conference*
edited by D. G. Seiler, A. C. Diebold, W. M. Bullis, T. J. Shaffner, R. McDonald, and E. J. Walters

TABLE 1. Summary of Plasma Damage Results Obtained by Hg Gate MOS IV

Wafer	PEPSG Conditions	Q_{BD} (50% Cumulative Probability) (C/cm^2)	I_{LK} (-2V) (nA)	dV/dt (50% Cumulative Probability) (mV/Sec)
KICG	Low Plasma Damage Settings, 5 sec. dep. time, 315 Å PEPSG	2.51E-3	0.150	1.2
KH5G	High Plasma Damage Settings, 5 sec. dep. Time, 404 Å, PEPSG	1.95 E-3	0.205	2.4

CONCLUSIONS

A number of Hg gate IV methods were used to evaluate plasma charging effects during a PEPSG deposition. It has been demonstrated that constant current dV/dt measurements are highly sensitive to charging from plasma enhanced processes. The use of a highly repeatable Hg probe system provides for rapid and highly sensitive assessment of plasma damage effects without the need for antenna structures.

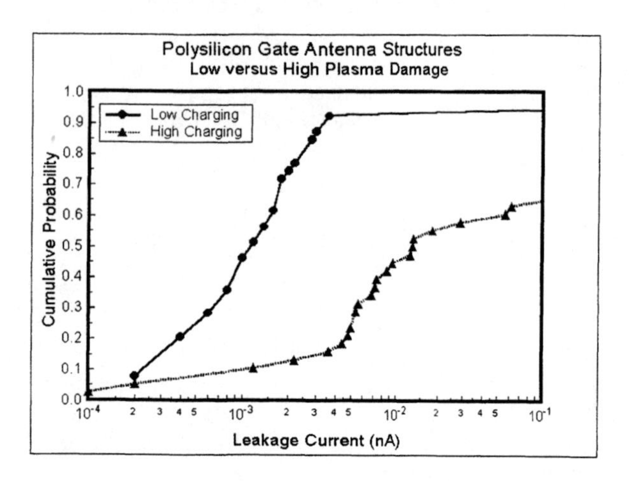

FIGURE 1. Leakage current cumulative probability obtained with polysilicon gate antenna structures.

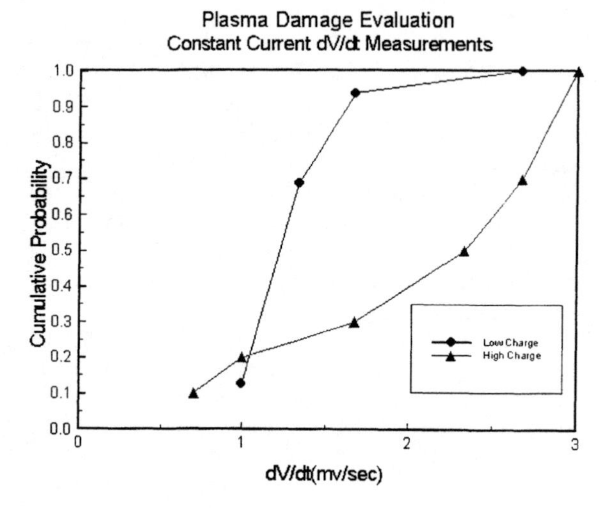

FIGURE 2. Constant current dV/dt cumulative probability obtained with Hg gate IV.

REFERENCES

1. Nariani, S.R., Gabriel, C.T. *"Gate Antenna Structures for Monitoring Oxide Quality and Reliability", Semi*, 1997 .
2. Hillard, R.J., Mazur, R.G., Gruber, G.A., and Sherbondy, J.C., *"Monitoring of Dielectric Quality with Mercury (Hg) Gate MOS Current-Voltage (Hg-MOSIV)," May Meeting of the ECS, 1997.*
3. Lin, H.C., Chen, C. Chen, C. H., Hsein, S. K. Wang, M. F., Chao, T.S. Huang, T. Y. Chang, C. Y., "Evaluation of Plasma Charging Damage in Ultrathin Gate Oxides", *IEEE Electron Device Letters,* **19** (3), 1998.
4. Ma, S., Adbel-Ati, L.N., McVittie, J.P., *"Limitations of Plasma Charging Damage Measurements using MOS Capacitors Structures". MRS Meeting,* 1997.

Novel Ion Current Sensor for Real-time, In-situ Monitoring and Control of Plasma Processing

Mark A. Sobolewski

Process Measurements Division, National Institute of Standards and Technology, Gaithersburg, MD 20899

This paper describes a technique for measuring the ion current at a semiconductor wafer while it is being processed in a high-density plasma reactor. The technique does not require the application of any voltage other than the rf bias voltage which is normally applied to wafers, thus avoiding any possible perturbations to the plasma. It relies on external measurements of the radio-frequency (rf) current and voltage at the wafer electrode. There is no need for any probe to be inserted into the plasma. To test the technique, comparisons were made with dc measurements of ion current at a bare aluminum electrode, for argon discharges at 1.33 Pa, ion current densities of 1.3-13 mA/cm^2, rf bias frequencies of 0.1-10 MHz, and rf bias voltages from 1 V to 200 V. Additional tests showed that ion current measurements could be obtained by the rf technique in electronegative gases and when electrically insulating layers or wafers were on the electrode.

INTRODUCTION

Maintaining control of plasma etching and deposition processes is often difficult. Plasma reactors are subject to many types of drift, so that settings of the control parameters that initially produced optimal results may no longer produce acceptable results at later times. If sensors were available to measure the relevant properties of ions and neutrals striking the wafer, such sensors could be used to detect process drift, diagnose its origin, and take corrective action, if needed.

One important parameter to monitor is the total ion current at the wafer. Etch rates, etch profiles, deposition rates and damage rates all depend on the total ion current.

Many methods have been used to measure ion current, but they are usually not suitable for monitoring commercial processes in industrial reactors. For example, Langmuir probes are often used in research, but they measure the ion current *within* the plasma, not at the wafer. Furthermore, materials sputtered from Langmuir probes may contaminate wafers. Probe measurements may be upset by the large rf currents circulating in plasma reactors. Conversely, the voltages applied to the probe may perturb the plasma. Finally, in many industrial applications, insulating layers are formed on the probes, causing them to fail.

The ion current may also be measured by sampling the ions through an orifice, as in mass spectrometry. Mass spectrometers, however, are difficult to calibrate and are not compatible with many industrial processes. Although a calibrated ion-sampling orifice can be incorporated into an electrode to measure the ion current there (1), wafers placed on the electrode will cover the orifice, interrupting the flow of ions and preventing the ion current measurement. To avoid this problem, devices for measuring ion current have been fabricated on specially-designed test wafers (2). But the difficulty of transferring the signal from the wafer makes such devices too impractical for use with actual wafers being processed. Furthermore, all of these techniques, including mass spectrometry (3) suffer from problems caused by the deposition of insulating layers.

The ion current at an electrode has been measured by a non-invasive technique (4, 5) in which a large, negative, dc voltage is imposed on the electrode itself, and the resulting dc current at the electrode is measured. If an insulating layer or insulating wafer is on the electrode, however, no dc current will be measured, and this technique will fail. Also, this "dc-biased electrode" technique strongly perturbs the plasma sheath adjacent to the electrode. This may alter the ion energies at the electrode, and even the ion current itself (4).

Recently, a new method (6) for measuring the ion current, which is well-suited for use in semiconductor manufacturing, has been developed. The technique is designed to measure the ion current at a wafer in real-time, while the wafer is undergoing high-density plasma processing. It does not require the application of any voltage other than the rf bias voltage which is normally applied to wafers, thus avoiding any possible perturbations to the process. The technique relies on measurements of the radio-frequency (rf) voltage and current at the wafer electrode. These measurements are made externally; there is no need for any probe to be inserted into the plasma. Because rf signals are used, ion current measurements can be obtained on electrically insulating as well as conducting wafers or electrodes.

EXPERIMENTAL APPARATUS

Figure 1 shows a diagram of the measurement system and the reactor in which it is installed, a GEC (Gaseous Electronics Conference) Reference Cell in which the standard upper electrode was replaced by an inductive, high-density plasma source (7, 8). The source is always powered at 13.56 MHz. Variable frequency "rf bias" power is supplied to the lower electrode by a sinusoidal signal generator and a power amplifier. On the lead that powers the lower electrode, a Pearson model 2877 current probe* and a Lecroy model PP002 voltage probe* are mounted.

* Certain commercial equipment, instruments or materials are identified in this paper to foster understanding. Such identification does not imply recommendation or endorsement by the National Institute of Standards and Technology, nor does it imply that the materials or equipment identified are necessarily the best available.

CP449, *Characterization and Metrology for ULSI Technology: 1998 International Conference*
edited by D. G. Seiler, A. C. Diebold, W. M. Bullis, T. J. Shaffner, R. McDonald, and E. J. Walters
1998 The American Institute of Physics 1-56396-753-7/98/$15.00

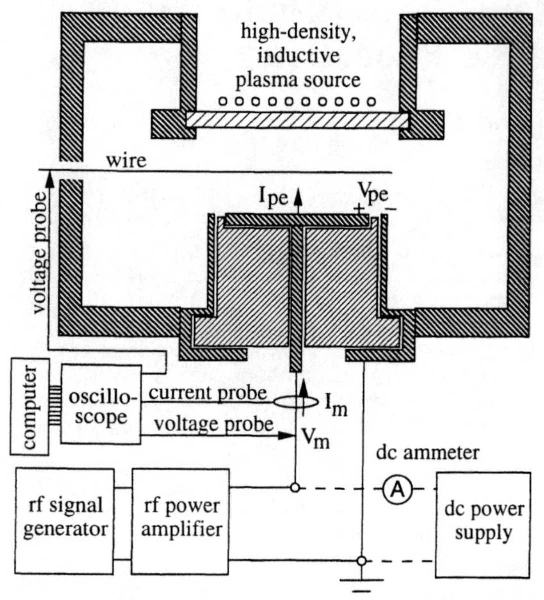

FIGURE 1. Diagram of the high-density plasma reactor equipped with the ion current measurement system. Measurements made by the probes connected to the lower electrode are used to determine the rf current and voltage at that electrode (I_{pe} and V_{pe}) and the ion current at that electrode. A wire probe was installed to measure sheath voltages, but it is not necessary for ion current measurements. The dc power supply and the dc ammeter used for independent, dc measurements of ion current are also shown.

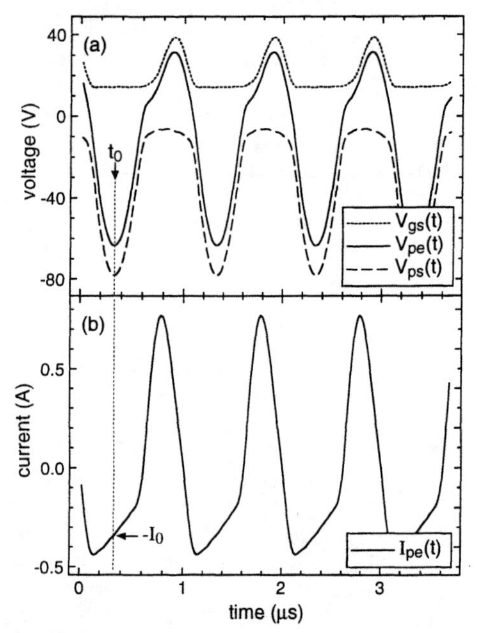

FIGURE 2. Waveforms of (a) the voltage on the powered electrode, $V_{pe}(t)$, the voltage across the sheath at the powered electrode, $V_{ps}(t)$, the voltage across the ground sheath, $V_{gs}(t)$, and (b) the current at the powered electrode, $I_{pe}(t)$. At time t_0, when $V_{pe}(t)$ and $V_{ps}(t)$ are minimized, $I_{pe}(t)$ is equal to the ion current, $-I_0$. Data were obtained for an argon discharge at 1.3 Pa, at a bias frequency of 1 MHz, an inductive source power of 120 W, and an ion current (determined independently from dc measurements) of 0.32 A.

Signals acquired by the probes are digitized by an oscilloscope and transferred to a computer for Fourier analysis. Procedures described previously (9) account for phase errors in the probes and for the stray impedance between the probes and the electrode, allowing one to determine $I_{pe}(t)$ and $V_{pe}(t)$, the current and voltage waveforms at the electrode itself. Also, the rf components of $V_{ps}(t)$, the voltage drop across the sheath at the rf-powered electrode, and $V_{gs}(t)$, the voltage drop across the sheath at opposing, grounded surfaces were measured, using a wire probe inserted into the plasma and techniques described previously (10, 11). Measurements of $V_{ps}(t)$ and $V_{gs}(t)$ are not required by the ion current sensor, however. Examples of the waveforms are shown in Fig 2.

METHOD

The $I_{pe}(t)$ waveform is the sum of several currents, which can be expressed as

$$I_{pe}(t) = -I_0 + I_e \exp\left(V_{ps}(t)/T_e\right) + C_s(t)\, dV_{ps}/dt. \quad (1)$$

The first term is the ion current. It is negative, corresponding to a flow of positive ions from the plasma to the electrode. The second term is the electron current, for a Maxwell-Boltzmann distribution of plasma electrons at temperature T_e (in volts). The final term is the sheath displacement current. The displacement current arises because the sheath contains a net positive charge, $Q(t)$, which is compensated by electrons on the electrode surface. As the sheath charge varies during an rf cycle, the surface charge must also vary, causing a current dQ/dt to flow in the electrode's electrical connections. In Eq. (1), this current has been expressed using the instantaneous sheath capacitance, $C_s(t) = dQ/dV_{ps}$.

When the voltage $V_{ps}(t)$ is strongly negative, electrons in the plasma are strongly repelled from the electrode, and the electron current in Eq. (1) will be negligibly small. Furthermore, when $dV_{ps}/dt = 0$, the displacement current is zero. Therefore, at t_0, the time when $V_{ps}(t)$ reaches its minimum value, provided that this voltage is sufficiently negative, both the electron current and the displacement current are negligible. The value of the current waveform at that time is therefore equal to the ion current,

$$I_{pe}(t_0) = -I_0, \quad (2)$$

as indicated in Fig. 2. Thus, the ion current can be determined using very general arguments, with no need for a detailed model of the displacement current or the electron current. Furthermore, in Fig. 2a (and throughout the experimental ranges given below) the minimum in the electrode voltage, $V_{pe}(t)$, occurs at the same time as the minimum in $V_{ps}(t)$. Thus t_0 (and I_0) can be determined from $V_{pe}(t)$ and $I_{pe}(t)$ alone. Wire probe measurements of the individual sheath voltages are not necessary.

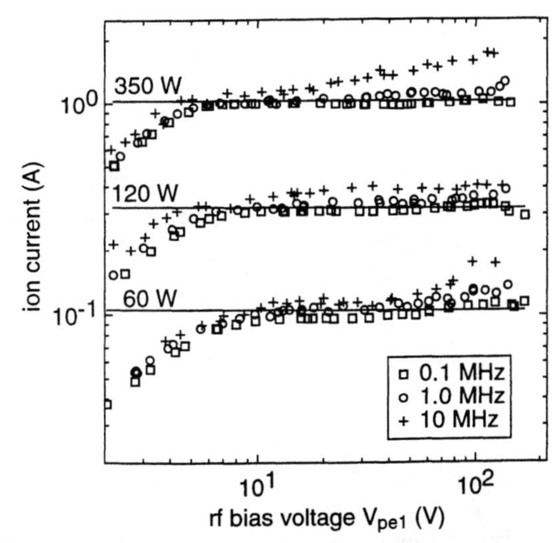

FIGURE 3. Comparison of rf measurements of ion current obtained from Eq. (2) (symbols) and dc measurements of ion current (lines). Measurements were performed for argon discharges at 1.33 Pa (10 mTorr) at varying rf bias frequencies (0.1-10 MHz) and varying rf bias voltages, given on the x-axis as V_{pe1}, the amplitude of the fundamental component of $V_{pe}(t)$ (i.e., the component at the rf bias frequency). The power supplied to the inductive plasma source was (from bottom to top) 60 W, 120 W, and 350 W, corresponding to dc ion currents of 0.105 A, 0.32 A, and 1.05 A, and dc ion current densities of 1.3 mA/cm^2, 4.0 mA/cm^2, and 13 mA/cm^2.

RESULTS

Values of $I_{pe}(t_0)$ were obtained for argon discharges at 1.33 Pa (10 mTorr), rf bias amplitudes from 1 V to 200 V, rf bias frequencies of 100 kHz, 1 MHz and 10 MHz, and inductive source powers of 60 W, 120 W and 350 W. At each source power, the rf bias power supply was temporarily disconnected and a dc power supply was connected to perform independent measurements of the ion current, using the dc method (4, 5) described above. Results from both methods are shown in Fig. 3. Good agreement was obtained over most of the experimental range, but not at high rf bias frequencies or at low rf bias voltages. At low voltages, electrons are not repelled strongly enough by the electrode, and some electrons are collected even at t_0. The collected electrons cancel part of the ion current, forcing $I_{pe}(t_0)$ to zero as the voltage approaches zero. Nevertheless, such low voltages are not used by industry. To accelerate ions to desired energies, tens or hundreds of electron volts, the rf bias in industrial reactors must supply tens or hundreds of volts of rf bias.

The deviation at high frequencies is more difficult to explain. It could in part result from the limited bandwidth of the current and voltage probes. Their accuracy starts to degrade around 25-30 MHz, but to faithfully reproduce the waveforms requires measurement of signals at and above 60 MHz, the sixth harmonic of the 10 MHz fundamental frequency. Alternatively, because the dc and rf measurements are not performed simultaneously, it is possible that the increase in ion current measured by the rf technique is a real effect, indicating that the high-frequency rf bias produces an increase in plasma density and ion current. Finally, the discrepancy might be explained by a time-modulation of the ion current. Simulations (12) predict that the ion current should vary with time when the rf bias frequency is comparable to the ion transit frequency (the reciprocal of the time required for ions to cross the sheath). Under these circumstances, $I_{pe}(t_0)$ would still be equal to the ion current at time t_0, but it would not necessarily agree with the time-averaged value of the ion current provided by the dc measurement.

Other methods have previously been proposed for estimating the ion current at an electrode from external electrical measurements (13-16), but those methods rely on untested or invalid assumptions about the power dissipation mechanisms within the discharge and the symmetry of the discharge. Therefore, estimates obtained by those methods do not agree as well with the dc measurements in Fig. 3. For example, the method of Ra and Chen (16) obtains the ion current from the equation

$$I_0 = P_{pe}/(V_{pe0} - V_{pef}), \tag{3}$$

where P_{pe} is the rf bias power, V_{pe0} is the dc voltage on the powered electrode when P_{pe} is applied, and V_{pef} is the dc voltage on the powered electrode when no rf bias power is applied. Estimates of I_0 obtained from Eq. (3), plotted in Fig. 4, deviate strongly from the dc measurements. Equation (3) assumes that all of the power P_{pe} is spent accelerating ions into the powered electrode. For most conditions, however, this assumption is invalid, so the agreement with dc measurements is poor. The method of Ra and Chen also suffers because it requires two measurements that cannot be performed simultaneously: V_{pe0} and V_{pef}. Furthermore, if an insulating layer is present on the wafer or on the electrode, V_{pe0} cannot be measured at all, and so Eq. (3) cannot be used.

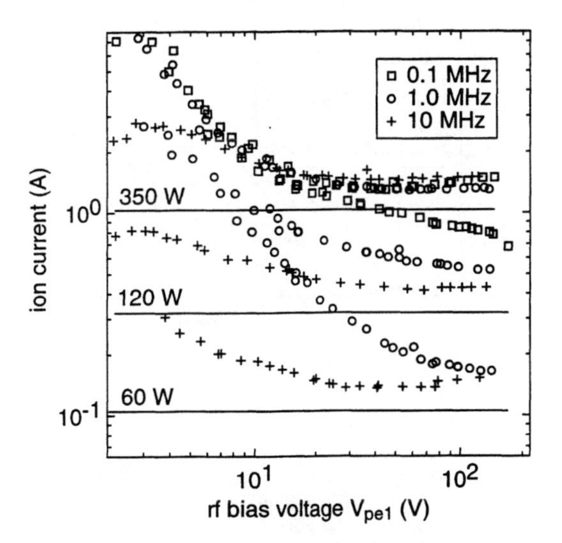

FIGURE 4. Comparison of ion current values obtained using the method of Ra and Chen (symbols) and dc measurements of ion current (lines). Experimental conditions are given in Fig. 3. For clarity, data from 0.1 MHz at 60W are not shown.

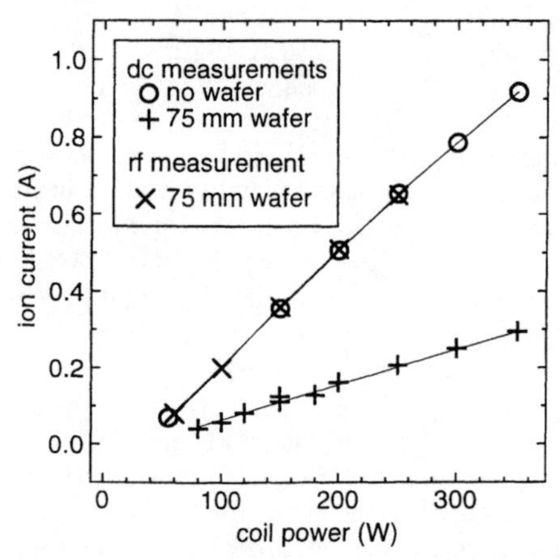

FIGURE 5. Values of the ion current obtained by dc measurements and by the novel, rf technique, Eq. (2), for argon discharges at 1.33 Pa, with and without a wafer placed on the rf biased electrode. The wafer diameter was 75 mm and the electrode diameter was 102 mm.

Figure 5 shows results obtained with a silicon wafer placed on the electrode. The wafer acts as an electrical insulator: no dc current is collected from the plasma over the area covered by the wafer. Therefore, the current measured by the dc-biased electrode technique becomes smaller when the wafer is placed on the electrode. The rf measurement technique, Eq. (2), is unaffected; it measures the same ion current with a wafer present as was measured by the dc technique when no wafer was present.

High-density plasma etching is usually performed using mixtures of electronegative gases, which supply fluorine or chlorine, and electropositive gases, which help sustain the discharge. The rf measurement technique succeeds in such mixtures. Waveforms measured in mixtures of argon with SF_6 were quite similar to those shown in Fig. 2. Ion current results from Ar/SF_6 plasmas are shown in Fig. 6,

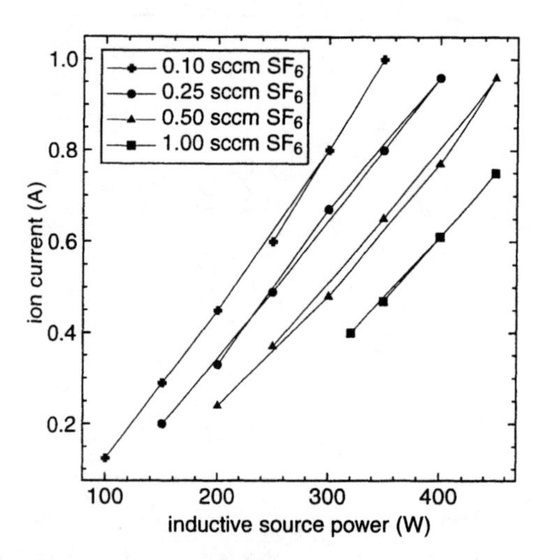

FIGURE 6. Ion current measured by the novel rf method at an aluminum electrode covered by an electrically insulating layer, in an Ar/SF_6 high-density plasma, as a function of inductive source power and SF_6 flow. The argon flow was 5 sccm, and the pressure was 2.9 Pa (22 mTorr).

which plots the dependence of the ion current on the inductive source power and the SF_6 flow. The ability to rapidly determine the ion current over a wide parameter space, as shown in Fig. 6, could be of use in process development.

Ar/SF_6 plasmas react with the aluminum electrode, forming an insulating layer on its surface. The data in Fig. 6 were obtained after several minutes of exposure of the electrode to Ar/SF_6 plasmas, during which a completely insulating layer had formed. This layer made it impossible to perform comparisons with dc ion current measurements such as those shown in Figs. 3-4.

The data in Fig. 6 show a slight degree of hysteresis. This is not necessarily a measurement error; the hysteresis may result from true, time-dependent changes in the ion current. Presumably, the ion current varies with time because surface conditions—and the gas-phase conditions that are coupled to them—require some time to respond to changes in the inductive source power or SF_6 flow. When experiments similar to Fig. 6 were performed on a clean, aluminum electrode being exposed to an Ar/SF_6 plasma for the first time, a much larger degree of hysteresis was observed, indicating that much larger changes in the density of gas phase reactant or product species were occurring during the initial formation of the insulating layer. The changes in the ion current during this "conditioning" of reactor surfaces is an example of the sort of uncontrolled processes that the ion current measurement technique is designed to monitor.

CONCLUSIONS

A new method for measuring the ion current at a wafer undergoing high-density plasma processing has been developed. The technique is well-suited for use in semiconductor manufacturing. It does not require any hardware to be inserted into the plasma reactor, thus avoiding any possibility of contaminating wafers. Instead, it relies on measurements of the radio-frequency voltage and current made externally. It does not require the application of any voltage other than the rf bias voltage which is normally applied to wafers during processing, thus avoiding any possible perturbations to the plasma. The technique has been demonstrated in both electronegative (Ar/SF_6) and electropositive (Ar) plasmas, for bare aluminum electrodes, electrodes covered by wafers, and electrodes covered by electrically insulating layers. Comparisons with independent, dc measurements of ion current in argon plasmas show that the new technique is more accurate than methods previously proposed for determining the ion current from external rf measurements. Future efforts include further investigation of the accuracy of ion current measurements at high rf bias frequencies and the development of modeling and analysis techniques that would enable external rf measurements to be used for real-time monitoring of other parameters such as sheath voltages and ion bombardment energies.

REFERENCES

1. A. Manenschijn, G. C. A. M. Janssen, E. van der Drift and S. Radelaar, *J. Appl. Phys.* **69**, 1253 (1991).
2. S. Pan, "Real-time plasma intensity and uniformity monitoring with WaferMonitor™", Applied Materials, *Etch Tech* **1** (2) p. 5, July 1996.
3. J. K. Olthoff, R. J. Van Brunt and S. B. Radovanov, *Appl. Phys. Lett.* **67**, 473 (1995).
4. V. A. Godyak, R. B. Piejak and B. M. Alexandrovich, *J. Appl. Phys.* **69**, 3455 (1991).
5. M. A. Sobolewski, *Appl. Phys. Lett.* **70**, 1049 (1997).
6. M. A. Sobolewski, *Appl. Phys. Lett.* **72**, 1146 (1998).
7. A. Schwabedissen, E. C. Benck and J. R. Roberts, *Phys. Rev. E* **55**, 3450 (1997).
8. P. A. Miller, G. A. Hebner, K. E. Greenberg, P. D. Pochan and B. P. Aragon, *J. Res. Natl. Inst. Stand. Technol.* **100**, 427 (1995).
9. M. A. Sobolewski, *J. Vac. Sci. Technol.* **A 10**, 3550 (1992).
10. V. A. Godyak and R. B. Piejak, *J. Appl. Phys.* **68**, 3157 (1990).
11. M. A. Sobolewski, *IEEE Trans. Plasma Sci.* **23**, 1006 (1995).
12. D. Martin and H. Oechsner, *Vacuum* **47**, 1017 (1996).
13. A. J. Van Roosmalen, *Appl. Phys. Lett.* **42**, 416 (1983).
14. B. Andries, G. Ravel and L. Peccoud, *J. Vac. Sci. Technol.* **A 7**, 2774 (1989).
15. T. Yamashita, S. Hasaki, I. Natori, H. Fukui and T. Ohmi, *IEEE Transactions on Semiconductor Manufacturing* **5**, 223 (1992).
16. Y. Ra and C.-H. Chen, *J. Vac. Sci. Technol. A* **11**, 2911 (1993).

Spatial Uniformity in Chamber-Cleaning Plasmas Measured using Planar Laser-Induced Fluorescence

Kristen L. Steffens, Mark A. Sobolewski

Process Measurements Division, National Institute of Standards and Technology, Gaithersburg, MD 20899

Planar laser-induced fluorescence (PLIF) measurements were made to determine 2-D spatial maps of CF_2 density as an indicator of chemical uniformity in 92 % CF_4/O_2 and 50 % C_2F_6/O_2 chamber-cleaning plasmas. Measurements were also made of broadband optical emission and of discharge current and voltage. All measurements were made in the Gaseous Electronics Conference (GEC) reference cell, a capacitively-coupled, parallel-plate platform designed to facilitate comparison of results among laboratories. The PLIF and emission results were found to correlate with discharge current and voltage measurements. Together, these optical and electrical measurements provide insight into the optimization of chamber-cleaning processes and reactors and suggest new methods of monitoring plasma uniformity.

INTRODUCTION

Fluorocarbon plasmas are widely used for both etching and chamber-cleaning in the semiconductor industry. Optimization of these plasmas is critical to assure high etch rates, high gas utilization efficiency, and the desired plasma spatial characteristics. (1) Non-uniformities in these plasmas cause etch or cleaning rates to vary across wafer or reactor surfaces. Etch rates on some reactor surfaces are difficult to measure directly; however, spatial variations in etch rate are likely to be related to spatial variations in reactive species densities, which can be measured by optical techniques. Planar laser-induced fluorescence (PLIF) is a sensitive optical diagnostic used for the spatially-resolved detection of gas-phase free radical reactive intermediates, such as CF_2. (2) In PLIF, the entire 2-D map of the species density is obtained simultaneously, eliminating the need for multiple point measurements.

The CF_2 radical is an important species in the chemical reactions that control the concentration of the fluorine radical, (3) which is believed to be the active chemical etchant. (4) As such, CF_2 is a useful marker of chemical uniformity in the plasma. In this work, PLIF was used to determine 2-D spatial maps of CF_2 density as an indicator of chemical uniformity in chamber-cleaning plasmas as a function of power, pressure, and composition. In addition, 2-D broadband optical emission measurements were made to learn where excited species are formed in the plasma. Measurements of discharge electrical characteristics were also performed. Correlations between the PLIF, emission, and electrical measurements were observed which help explain the observed variations in the plasma uniformity. These results suggest new methods of monitoring and optimizing plasma uniformity.

EXPERIMENTAL

Measurements were made in 92% CF_4/O_2 and 50% C_2F_6/O_2 chamber-cleaning plasmas at pressures varying from 13.3 Pa to 133.3 Pa (100 mTorr to 1000 mTorr) and a total gas flow rate of 9.5 standard cubic centimeters per minute (sccm) in the Gaseous Electronics Conference (GEC) Reference Cell, a standard platform designed to facilitate comparison of results among laboratories. (5) The cell, shown in Figure 1, is a capacitively-coupled, parallel-plate platform with 10.2 cm diameter electrodes spaced by 2.25 cm. The lower electrode is powered at 13.56 MHz, and the upper electrode and chamber are grounded. On the lead that powers the lower electrode, current and voltage probes were mounted. An identical current probe was mounted on the wire that grounds the upper electrode to the exterior of the cell. Using these probes and procedures described previously, (6,7) the stray impedance of the cell was characterized. This characterization enables the current and voltage waveforms at the electrodes themselves to be determined. The plasma power, which is the power delivered to the discharge itself, excluding all external power losses, was also determined. Measurements were made at plasma powers of 10 W and 30 W, corresponding to power densities of 0.12 W/cm^2 and 0.37 W/cm^2. These power densities are comparable to industrial reactors operated at hundreds of watts, because of the larger size of industrial reactors and the fact that external losses are usually not accounted for.

CP449, *Characterization and Metrology for ULSI Technology: 1998 International Conference*
edited by D. G. Seiler, A. C. Diebold, W. M. Bullis, T. J. Shaffner, R. McDonald, and E. J. Walters
1998 The American Institute of Physics 1-56396-753-7/98/$15.00

The PLIF experimental apparatus (2) is shown schematically in Figure 1. Using cylindrical optics, the 266 nm laser beam from a quadrupled Nd:YAG laser was expanded into a vertical laser sheet 2.5 cm tall and 0.5 cm thick. When the laser sheet passes through the plasma, it excites the CF_2 radicals from the X (0,1,0) ground electronic state to the A (0,2,0) excited electronic state. (8) CF_2 then fluoresces down to the X (0,0-20,0) states with a lifetime (8) of 61 ns, emitting light between 250 nm and 400 nm. Fluorescence between 300 nm and 400 nm is collected at a right angle to the direction of propagation of the laser beam, using an intensified charge-coupled device (ICCD) camera. Colored glass filters are used to block the 266 nm scattered laser light and to reduce the broadband plasma emission. For each image, 1000 laser shots were averaged. The spatial resolution was determined by the 5.0 mm laser sheet thickness and the 0.2 mm × 0.2 mm imaged dimensions of the camera pixels. Broadband, spontaneous plasma emission between 300 nm and 400 nm is measured over the same total time as a PLIF image, but with the laser blocked. The emission is then subtracted from the PLIF image before normalizing to spatial variations and drift in the laser power. A uniform field correction is also applied to the image to normalize to variations in collection efficiency across the ICCD. Because electronic quenching is expected to be negligible, (2) the resulting PLIF fluorescence intensity is directly proportional to the ground state density of the CF_2 radical.

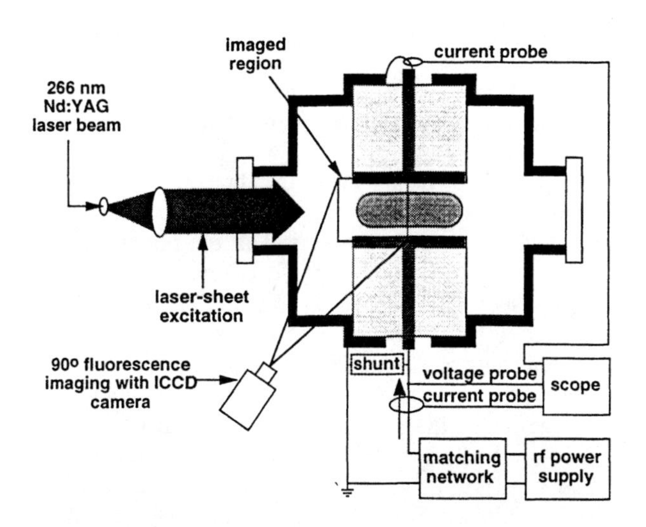

FIGURE 1. Experimental schematic of GEC reference cell, including electrical probes and PLIF apparatus.

RESULTS AND DISCUSSION

PLIF Measurements

Figures 2a-c and 3a-c show 2-D images of CF_2 density as a function of pressure for 92% CF_4/O_2 at 30 W and 50% C_2F_6/O_2 at 30 W, respectively. In both figures, three different regimes of behavior are observed. First, at low pressures of 13.3 Pa to 26.7 Pa (100 mTorr to 200 mTorr), the CF_2 density is diffusely distributed near the outer edge of the electrodes, closer to the powered (lower) electrode. As pressure is increased, the CF_2 density increases and shifts closer to the center of the electrodes. At intermediate pressures of 53.3 Pa to 80 Pa (400 mTorr to 600 mTorr), CF_2 reaches its maximum density and is radially uniform across the electrode surface. As pressure increases still higher, the CF_2 density decreases in the region near the electrode center, but begins to increase in a localized region at the edge of the powered electrode. At the highest observed pressure of 133.3 Pa (1000 mTorr), the CF_2 density has collapsed to a ring around the powered electrode edge. The transitions between regimes are gradual and occur at lower pressures in C_2F_6/O_2 than in CF_4/O_2. At 30 W, the

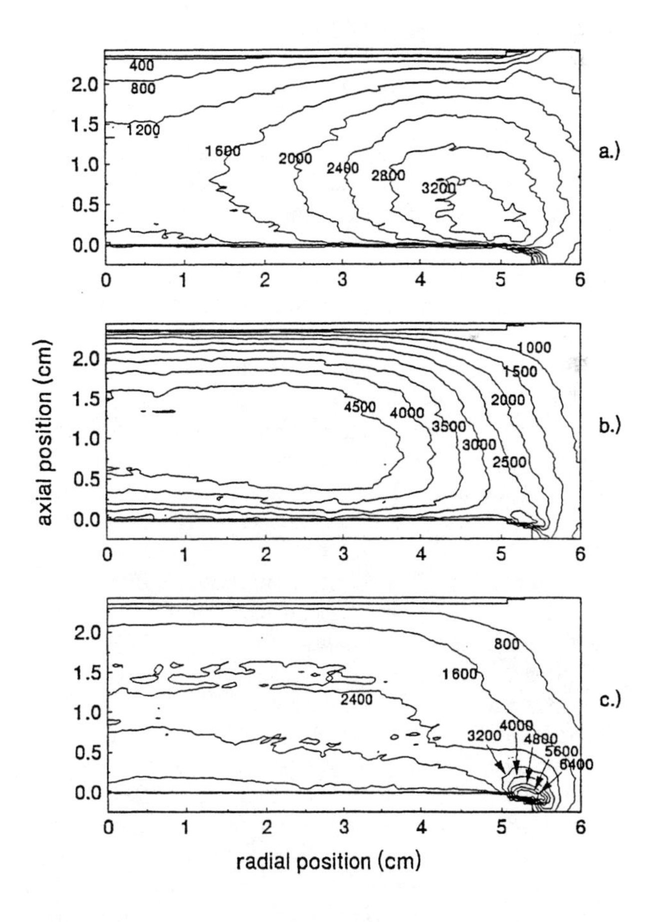

FIGURE 2. CF_2 PLIF images in a 30 W, 92 % CF_4/O_2 plasma at a) 26.7 Pa (200 mTorr), b) 80.0 Pa (600 mTorr), and c) 133.3 Pa (1000 mTorr). In each image, the right half of the plasma is shown. Upper and lower electrodes are outlined.

overall density in 92% CF_4/O_2 is roughly three to five times greater than that in 50% C_2F_6/O_2. For both 92% CF_4/O_2 and 50% C_2F_6/O_2, the overall CF_2 density at 10 W (not shown) is lower by about a factor of two than that at 30 W.

The CF_2 PLIF data provide insight into the relative cleaning rates at different cell surfaces. For example, one would expect the cleaning rate at the center of the powered electrode to be largest at the intermediate pressures shown in Figures 2b and 3b, where the CF_2 density at R = 0 and, presumably, the density of the active etchant species at R = 0, are maximized. In contrast, at R = 6 cm, a radial position far from the discharge center, the CF_2 density peaks around 13.3 Pa to 26.6 Pa (100 mTorr to 200 mTorr), as shown in Figures 2a and 3a. Thus, the etch rate at radially remote surfaces should be highest in the low pressure regime. Similar behavior is observed in commercial reactors, where chamber-cleaning is often accomplished with two steps: one at lower pressure to clean the outer regions of the reactor, and one at somewhat higher pressure to clean the showerhead and susceptor area. (1)

Emission Measurements

Broadband plasma emission occurs due to the electron impact dissociation and excitation of molecules in the plasma, which result in electronically excited molecular fragments and atoms which subsequently emit. Thus, the emission measurements give different information than the CF_2 PLIF measurements, which detect CF_2 in its electronic ground state. The lifetimes of many of the created excited state fragments are very short (61 ± 3 ns (8) for $CF_2(A\ ^1B_1\ (0,0,0))$ and 26.7 ± 1.8 ns (9) for $CF(A^2\Sigma^+, v'=0)$) such that the fragments emit before they have a chance to travel away from the region in which they were created. Thus, the spatial intensity of the emission indicates where reactive species, including CF_2, are created in the plasma. Once CF_2 emits and drops back down to the ground electronic state, it lives long enough (10,11) in the plasma to diffuse into other regions, where it is detected by PLIF.

Figure 4 shows emission images for 50% C_2F_6/O_2 at 30 W. Emission occurs primarily above the powered electrode and near the grounded electrode, with less emission occurring in the bulk regions. The emission intensities near the grounded electrode and in the bulk of the plasma are maximized at intermediate pressures. The radial distribution of emission near the powered electrode varies with pressure: at low pressures (Figure 4a) and high pressures (Figure 4c) it is highest near the edge of the powered electrode, but at intermediate pressures (Figure 4b) it is highest near R = 0. Because emission is a "line-of-sight" measurement, the observed emission comes not only from the focal plane of the camera, but also from in front of and behind the focal plane. Therefore, the emission at R = 0 in Figure 4 is not necessarily from the radial center; rather, it includes contributions from all the radial positions along the line-of-sight. Emission images at 10 W and in 92% CF_4/O_2 show similar spatial behavior with change in pressure. Overall emission intensity increases strongly with power but varies little with changing composition.

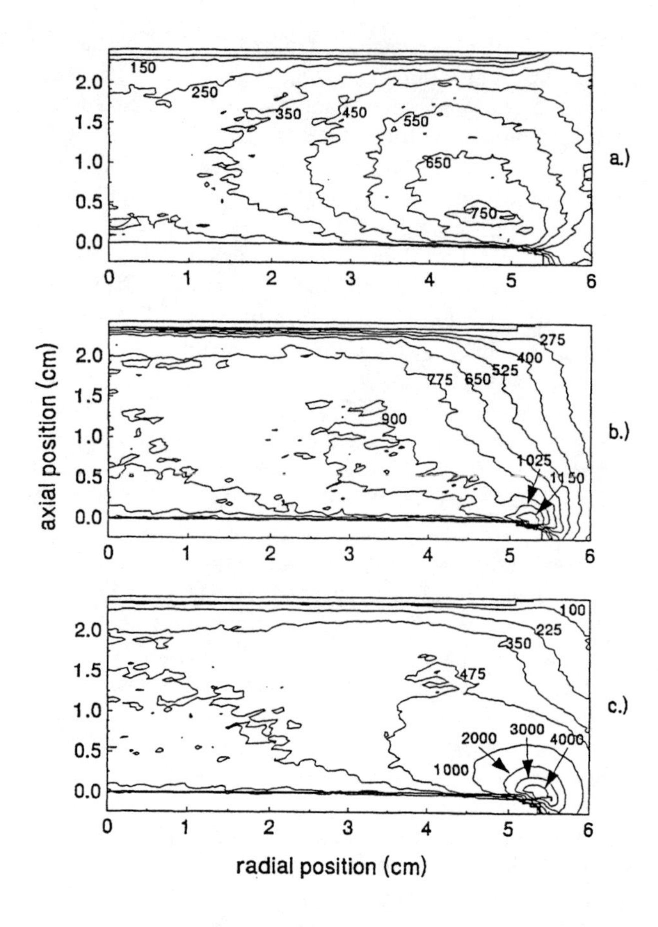

FIGURE 3. CF_2 PLIF images in a 30 W, 50 % C_2F_6/O_2 plasma at a) 13.3 Pa (100 mTorr), b) 53.3 Pa (400 mTorr), and c) 133.3 Pa (1000 mTorr).

Electrical Measurements

In previous studies, (12,13) the intensity of the optical emission from NF_3/Ar, $CF_4/O_2/Ar$, and $C_2F_6/O_2/Ar$ chamber-cleaning plasmas was found to correlate with several electrical parameters, and possible explanations were proposed to explain these correlations. The spatially-resolved optical measurements presented above have allowed us to further investigate such correlations. One particularly important electrical parameter identified in Ref. (13) is I_{ge}/I_{pe}: the ratio of the fundamental amplitude of the current at the ground electrode to the

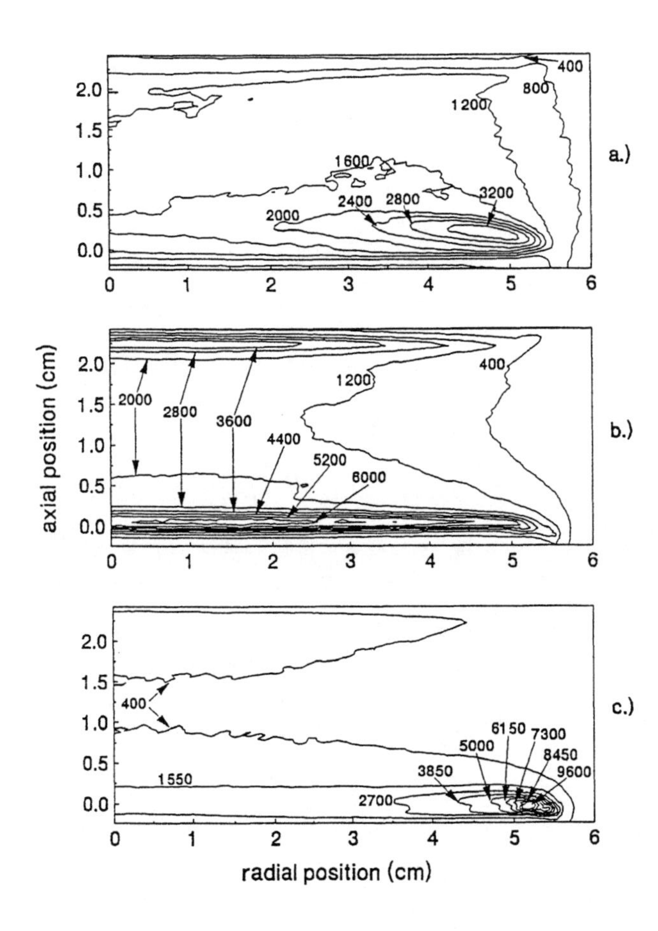

FIGURE 4. Broadband emission images in a 30 W, 50 % C_2F_6/O_2 plasma at a) 13.3 Pa (100 mTorr), b) 53.3 Pa (400 mTorr), and c) 133.3 Pa (1000 mTorr).

fundamental amplitude of the current at the powered electrode. In Figure 5a, a plot of I_{ge}/I_{pe} versus pressure exhibits three different pressure regimes which parallel the three pressure regimes apparent in the emission and CF_2 PLIF measurements. First, there is a regime at low pressures where I_{ge}/I_{pe} is small, because in this regime the current flows predominantly to the ground shields of both electrodes, and possibly out to the walls of the vacuum chamber. This regime coincides with the pressures at which CF_2 PLIF images show a broad maximum near the edge of the electrodes. In contrast, at intermediate pressures, the more remote current paths are less favored, and most of the current I_{pe} flows to the grounded electrode, so that I_{ge}/I_{pe} peaks, reaching a maximum of about 80%. This regime coincides with the pressure range at which radially uniform CF_2 PLIF images were observed. Finally, in the high pressure regime, where the discharge collapses to a region very close to the edge of the powered electrode, most of the current presumably takes the very short path to the ground shield of the powered electrode, so that I_{ge}/I_{pe} becomes small again. The current I_{ge}, shown in Figure 5b, also reaches a

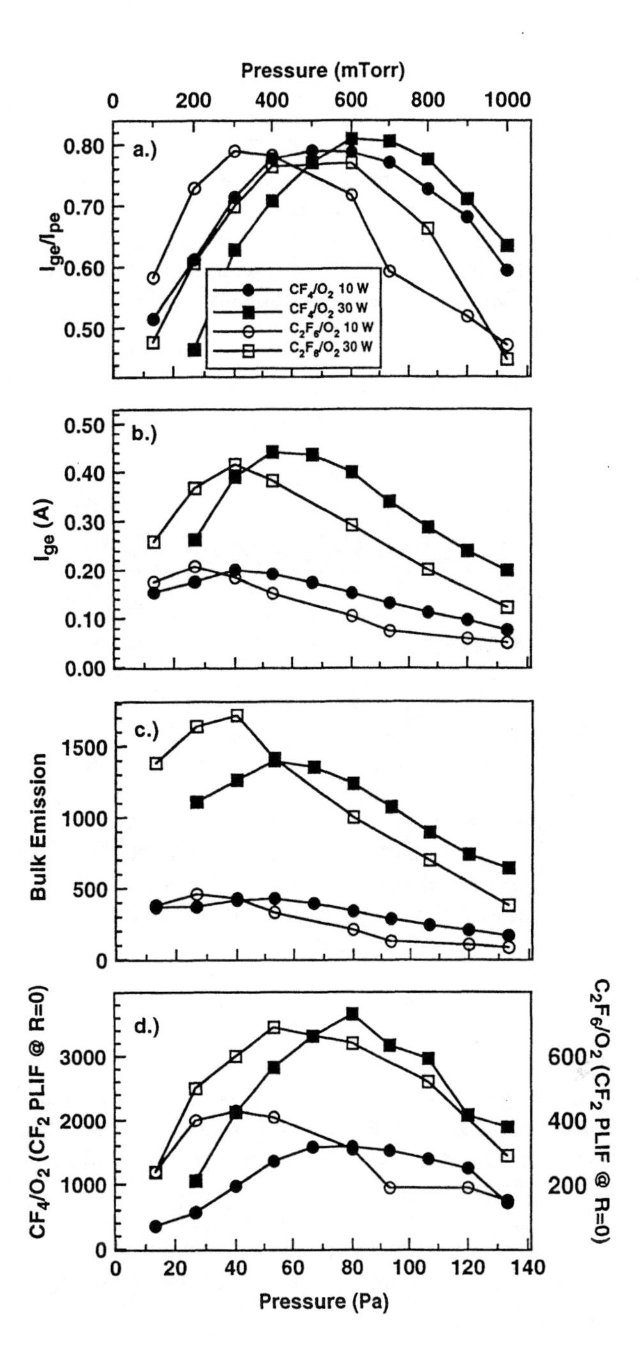

FIGURE 5. Plots vs. pressure for 92% CF_4/O_2 and 50% C_2F_6/O_2 at 10 W and 30 W: a.) I_{ge}/I_{pe}, b.) I_{ge}, c.) emission at the radial and axial center, d.) CF_2 PLIF integrated along the axial centerline.

maximum at intermediate pressures, but at lower pressures than I_{ge}/I_{pe}.

Variations in the rf current density at a surface, or along a surface, should result in variations in electron heating (14) near the surface and concomitant variations in optical emission intensity. For example, the bulk emission intensity (the emission intensity at a small area centered axially at R = 0) plotted versus pressure in Figure 5c correlates well with I_{ge} in Figure 5b. Because the hot electrons that make emission possible are also responsible for the creation of CF_2, one might expect PLIF intensities to be correlated with emission and electrical parameters. For example, the CF_2 PLIF at the center line, shown in Figure 5d, is correlated to the I_{ge}/I_{pe} data in Figure 5a, and to the plasma impedance (not shown), which reaches a minimum at intermediate pressures, as observed previously. (12,13,15) The pressure dependence of the PLIF and emission are correlated; however, some differences in behavior are observed because CF_2 is long-lived and can diffuse away from the regions where it is created.

CONCLUSIONS

The results presented here provide insight into the chamber-cleaning process and help explain previous measurements of the cleaning rate at the center of the powered electrode. (12,15) Prior studies have indicated the importance of changes in the power coupling efficiency with pressure (12,13,15) and of global changes in the mechanisms of power deposition in the plasma with pressure. (13) The spatially resolved optical measurements presented here show the importance of changes in the spatial uniformity of chemical species with pressure. At the radial center, the densities of CF_2, and presumably of the active etching species, peak at intermediate pressures, where the etch rate at the center of the powered electrode is maximized. However, the densities of CF_2, and presumably of the active etching species, vary widely throughout the plasma reactor. Thus, we would expect etch rates to be different at other reactor surfaces, where the cleaning rate has not been measured. Correlations observed between the PLIF, the optical emission and the electrical parameters suggest that emission or electrical measurements could be of use in optimizing and monitoring chamber-cleaning processes at both the center of the electrodes and at more remote surfaces.

REFERENCES

1. Streif, T., DePinto, G., Dunnigan, S., and Atherton, A., *Semiconductor International* **20**, 129-134 (1997).
2. McMillin, B. K. and Zachariah, M. R., *J. Vac. Sci. Technol. A* **15**, 230-237 (1997).
3. Pang, S. and Brueck, S. R. J., in *Laser-Induced Fluorescence Diagnostics of $CF_4/O_2/H_2$ Plasma Etching*, 1983 (Elsevier Science Publishing Co., Inc.), p. 161-167.
4. Flamm, D. L., *Solid State Technology* **22**, 109-116 (1979).
5. Hargis, P. J., Jr., Greenberg, K. E., Miller, P. A., Gerardo, J. B., Torczynski, J. R., Riley, M. E., Hebner, G. A., Roberts, J. R., Olthoff, J. K., Whetstone, J. R., Van Brunt, R. J., Sobolewski, M. A., Anderson, H. M., Splichal, M. P., Mock, J. L., Bletzinger, P., Garscadden, A., Gottscho, R. A., Selwyn, G., Dalvie, M., Heidenreich, J. E., Betterbaugh, J. W., Brake, M. L., Passow, M. L., Pender, J., Lujan, A., Elta, M. E., Graves, D. B., Sawin, H. H., Kushner, M. J., Verdeyen, J. T., Horwath, R., and Turner, T. R., *Rev. Sci. Instrum.* **65**, 140-154 (1994).
6. Sobolewski, M. A., *J. Vac. Sci. Technol. A* **10**, 3550-3562 (1992).
7. Sobolewski, M. A., *IEEE Transactions on Plasma Science* **23**, 1006-1022 (1995).
8. King, D. S., Schenck, P. K., and Stephenson, J. C., *J. Mol. Spec.* **78**, 1-15 (1979).
9. Booth, J. P. and Hancock, G., *Chem. Phys. Lett.* **150**, 457-460 (1988).
10. Booth, J. P., Hancock, G., Perry, N. D., and Toogood, M. J., *J. Appl. Phys.* **66**, 5251-5257 (1989).
11. Hansen, S. G., Luckman, G., and Colson, S. D., *Appl. Phys. Lett.* **53**, 1588-1590 (1988).
12. Langan, J. G., Rynders, S. W., Felker, B. S., and Beck, S. E., *J. Vac. Sci. Technol. A, in press* (1998).
13. Sobolewski, M. A., Langan, J. G., and Felker, B. S., *J. Vac. Sci. Technol. B* **16**, 173-181 (1998).
14. Lieberman, M. A. and Lichtenberg, A. J., *Principles of Plasma Discharges and Materials Processing,* New York: John Wiley & Sons, Inc., 1994.
15. Langan, J. G., Beck, S. E., Felker, B. S., and Rynders, S. W., *J. Appl. Phys.* **79**, 3886-3894 (1996).

In Situ Mid-Infrared Analyses of Reactive Gas-Phase Intermediates in TEOS/Ozone SAPCVD

Thomas K. Whidden and Sarah Doiron
Xylaur enterprises, Fredericton, New Brunswick, Canada E3B 6C2

In this report, we present *in situ* characterizations of chemical vapour deposition (CVD) reactors used in silicon dioxide thin film depositions. The characterizations are based on Fourier transform infrared spectroscopy. The infrared absorption data are interpreted within the context of process and thin film properties and the bearing of the spectroscopic data upon the chemical mechanisms extant in the deposition reaction. The relevance of the interpretations to real-time process control is discussed. The process under study in this work is TEOS/ozone-based deposition of silicon dioxide thin films at subatmospheric pressures. This process exhibits many desirable properties but has fundamental problems that may be solvable by reaction control based on *in situ* analyses and the real-time manipulation of reagent concentrations and process conditions. Herein we discuss our preliminary data on characterizations of TEOS/ozone chemistries in commercial reactor configurations. Reaction products and reactive intermediate species are detected and identified. Quantitative *in situ* measurements of the reagent materials are demonstrated. Preliminary correlations of these data with process and thin film properties are discussed.

INTRODUCTION

Real-time control of semiconductor process reactors is acknowledged as a requirement for ULSI implementation (1-5). We are developing a form of such control that uses real-time data on the internal reactor chemistry in CVD reactors for feedback control of the process. The initial test process for this development is the TEOS suite of CVD reactions. Atmospheric pressure (APCVD) and low-pressure (LPCVD) pyrolitic CVD TEOS processes have a history of successful use in device fabrication (6-8). These processes yield silicon dioxide thin films with excellent properties for applications within ULSI device fabrication. The conventional, thermal CVD processes, however, require temperatures much too high for use as intermetal dielectrics in multilevel metal structures. The temperature necessary for low pressure pyrolysis of TEOS, for example, is approximately 725 °C. Plasma-enhanced CVD (PECVD) has been employed for intermetal applications, but remains problematic in terms of film purity and substrate damage issues. Consequently, our choice of an initial CVD process target has therefore focused on the TEOS/ozone reaction. This process fulfills the low thermal budget requirements of multilevel metal processing while simultaneously retaining many of the advantageous properties of TEOS-based chemistries. These chemistries have not been definitively characterized and thus present an excellent opportunity to evaluate our approaches. The process and thin films have sufficient advantages over conventional TEOS or silane-based routes therefore our data will have an immediate relevance for device manufacturers.

The CVD reaction of TEOS with ozone yields silicon dioxide thin films with excellent conformal fill characteristics at unusually low process temperatures (350 °C to 400 °C) (9-11). Two critical issues exist in the process and film properties that have impeded implementation of this process in ULSI fabrication schemes. Firstly, unless the process is carried out at the high end of its temperature "window", the as-deposited films tend to be hygroscopic, with significant residual Si-OH and water. Depending upon the quality of metal under the oxide, the process temperature may be at the limit of acceptable conditions. The high temperature processing may cause defects in the conductor lines where metal of marginal quality exists. Annealing of the films to remove the residual hydroxyl moieties is not acceptable in that it negates the process advantage of low thermal budget. Secondly, the TEOS/ozone process tends to exhibit a narrow window for low particulate processing. Small process upsets lead to unacceptable particle levels. This latter problem correlates with the need for higher temperature processing, since parasitic reactions resulting in reduced deposition rates and leading to gas phase particle growth are accelerated at increased temperatures (12).

There is strong mass spectral (13) and other, more inferential evidence (14-16) for the role of gas phase intermediates in the characteristics of TEOS CVD processes. Oligomeric siloxanes and hydroxy-siloxanes have been observed in GC-MS studies of He-diluted APCVD depositions (17). Researchers at the University of Osaka (14) have analyzed particle growth in the gas phase of TEOS reactions. Their mass spectral data support the growth of particles in the process via a step-wise condensation polymerization initiated and sustained by hydroxylated siloxanes. More inferential data based on film growth characteristics have been reported by Sorita et al (18). Their results support the presence and active participation of at least two (probably siloxane) intermediate species in the process.

CP449, *Characterization and Metrology for ULSI Technology: 1998 International Conference*
edited by D. G. Seiler, A. C. Diebold, W. M. Bullis, T. J. Shaffner, R. McDonald, and E. J. Walters

The literature shows that TEOS and its derivatives exhibit well-behaved IR absorption characteristics under near-industrial process conditions (19). We have therefore selected this technique for *in situ* characterizations that will provide base data for our real-time process control in this CVD process.

Figure 1. Experimental reactor showing the chamber, the load lock and the infrared spectrometer with its interface to the CVD chamber.

EXPERIMENTAL

Reactions are performed in a system of our designs constructed by Intelvac of Georgetown, Ontario, Canada (Figures 1 and 2). The loadlock and chamber are configured for substrates of up to 125-mm diameter. The chamber operates in both high vacuum and at sub-atmospheric pressure modes. It is pumped by a combination of mechanical and turbomolecular pumps and normally achieves base pressures between 1 x 10^{-7} torr and 1 x 10^{-6} torr, depending upon the chamber history. High vacuum is required between runs to remove the residual moisture due to reaction products and ambient adsorption on the substrates. Pressure control at subatmospheric conditions uses a combination of an inline throttle valve and pumping speed reductions by the addition of nitrogen at the mechanical pump.

Gasses are added to the chamber through a water-cooled showerhead (Figure 2). Temperatures of the showerhead do not exceed 150 °C during process. Calibration tests indicated no reaction between TEOS and ozone at this temperature (as determined by substrate oxide measurements and by spectra of the gas mixtures). The distance from the showerhead to the substrate is adjustable and is typically ca. 3.0 cm. Substrates are heated using tungsten-halogen lamps. Temperatures are measured using J-type thermocouples in direct physical contact with the substrate and controlled by a Eurotherm 2216 controller.

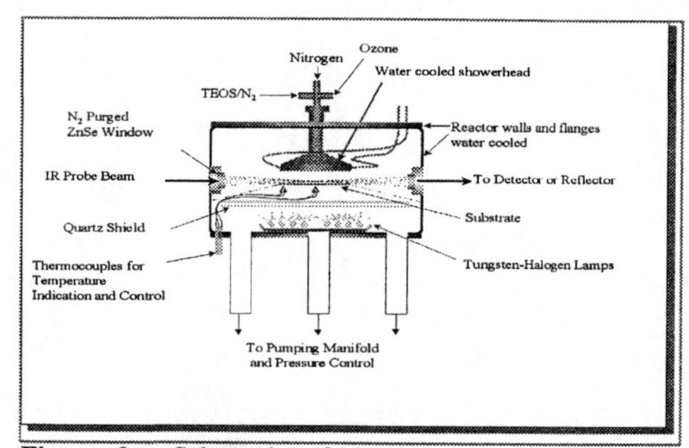

Figure 2. Schematic of the interior of the reaction chamber showing the arrangement of the showerhead, substrate and the probe IR beam.

A quartz plate between the lamp housing and the substrate minimizes deposition on the lamp surfaces. Spectra are obtained using a customized Bomem MB series FT-IR equipped with transfer optics between the spectrometer and the reaction chamber. The IR beam enters and exits the chamber through nitrogen-purged ZnSe windows and focuses directly over the substrate surface. The beam is detected with a liquid nitrogen-cooled mercury-cadmium-telluride detector and either one or two beam passes through the reactor. Double pass experiments utilize a ZnSe beam splitter and corner cube arrangement.

In a typical deposition, a 100-mm substrate was loaded and the reactor evacuated to base pressure. A background spectrum was obtained, the turbopump isolated and the pumping re-directed to a rotary vacuum pump (Alcatel 2004A, 90 slm). A 3 slm chamber purge of ultra-high purity (UHP) nitrogen was established and the pressure was set to 500 Torr using a Tylan General throttle valve and controller. Auxiliary nitrogen to pump was adjusted to maintain the throttle valve position at 15-30% of its control range. After the system stabilized, the temperature was adjusted to 375°C and a new background spectrum determined. Oxygen (1.5 slm) and TEOS (300 sccm of N_2 through a TEOS-filled bubbler maintained at 50°C) were added to the reactor and the pressure stabilized. The O_3 generator (PCI, model GL-1) was set to 8% O_3/O_2. Ozonator conditions for an 8% feed had been previously determined using a commercial monitor (PCI model HC-NEMA-12). After two minutes, an *in situ* infrared spectrum was obtained. Spectra were determined at 16 scans and 4 cm^{-1} resolution. The reaction was stopped by first reducing the temperature to ambient, then stopping the TEOS flow, followed by turning off the O_3 generator. O_3 lines were purged with oxygen for one minute, then the oxygen flow was turned off. The chamber was purged with 3 slm of UHP nitrogen for an additional 3 minutes, then the nitrogen flow stopped and the chamber pumped to base pressure by first the mechanical pump followed by the turbopump.

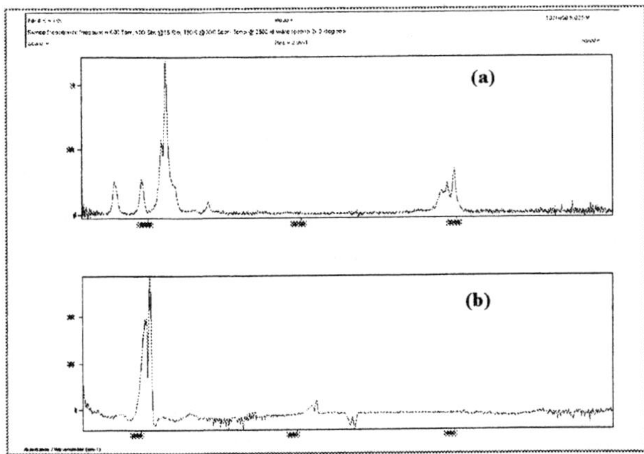

Figure 3. Individual *in situ* FT-IR spectra (600-4000 cm^{-1}) of TEOS (a) and O_3 (b) at 250°C in a production style CVD process chamber. The TEOS spectrum was obtained without diluent, while that of O_3 was obtained from a 2.7% mixture in oxygen and nitrogen. Process conditions: P=500 Torr, N_2 diluent flow = 3 slm, O_2 flow (ozone only) = 1.5 slm.

RESULTS AND DISCUSSION

The *in situ* spectrum of TEOS vapour at 250 °C is shown in Figure 3(a). The spectrum is consistent with reports by Van der Vis et al (19) and others (20,21). Table 1 shows a comparison of the spectra from this work with those of Van der Vis et al. The spectrum of ozone was compared with those determined by Kawahara et al (20) and by O'Neill et al (21). It is consistent with their results. Band assignments for the ozone spectrum are also presented in Table 1.

Signals that vary in a well-behaved linear fashion with concentration are important for fast and simple real-time control algorithms. Our work has examined the behaviour of the different IR bands in the TEOS and ozone spectra to determine their linearity. All major peaks in the spectra were analyzed. Distinct non-linearities were observed in correlations of the strongest peaks of the TEOS spectrum (1115 and 1088 cm^{-1}) with reagent concentration. Correlation of either peak height or area against the partial pressure of TEOS showed saturation effects in these bands at partial pressures typical for deposition processes. Other workers (21) have used these bands to measure TEOS concentrations in the feed lines of a PECVD reactor. Their experimental configuration used an infrared gas cell inserted in the feed line. The pathlength was only 5 cm and minimal saturation, if any, was observed. Our reactor has a diameter of 40 cm and has an effective pathlength of 15 cm. or more, allowing for partial pressure gradients across the chamber. Our experimental configuration thus more closely mimics those of production reactors that are processing 200 mm or larger wafers. Given the non-linearities observed in our data, alternative bands of the TEOS spectrum that can provide a reliable measure of the concentration in production environments need to be chosen as input for control algorithms.

Assignment	**Van der Vis et al**	**This work**
TEOS		
CH3asym stretch	2981	2981
CH3 asym def.	1447	1446
CH2 wag	1394	1395
CH3 sym def.	1374	1372
CH2 twist	1304	1305
CH3 rock	1176	1175
CO asym str.	1115	1115
CO asym str.	1089	1088
CH3 rock	965	965
SiO4 asym str.	794	794
Ozone	**Kawahara et al**	**This work**
		1021.4
	1043	1037
v_1	1055	1057
v_2	2105	2102/2123.9

Table 1. Infrared band assignments for TEOS and ozone.

We have correlated band area with concentration for both 100% TEOS and TEOS/N_2 vapour mixtures. The observed areas of the 794 cm^{-1} SiO_4 asymmetric stretch

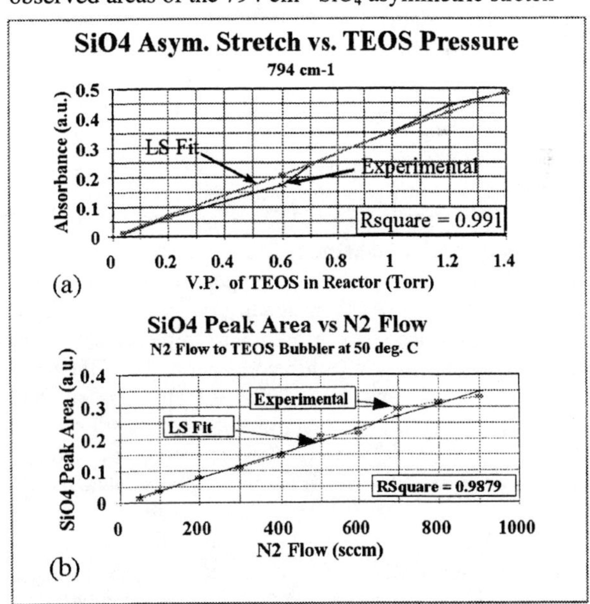

Figure 4. The variation of the 794 cm^{-1} absorption with (a) partial pressure of TEOS and (b) N_2 flow through a TEOS bubbler.

varied linearly with the partial pressure of 100% TEOS with no evidence of saturation over the range evaluated (Figure 4a). Figure 4b shows similar linearity for area vs.

nitrogen flow. Cross correlation of the data in Figures 4a and 4b permits a direct readout of TEOS partial pressures within the reactor when it is supplied from a nitrogen-bubbler (i.e. N_2 to bubbler = 900 sccm, P_{TEOS} = 0.9 torr). Such *in situ* concentration monitors for CVD reactors have not, to our knowledge, been reported in the literature. This is surprising as the information has value in setting reagent ratios and in solving process upsets. This data point can be critical in resolving production upsets. Confirmation of the reagent concentrations in a reactor cannot be determined from the standard instrument read-backs that are common to most CVD equipment configurations.

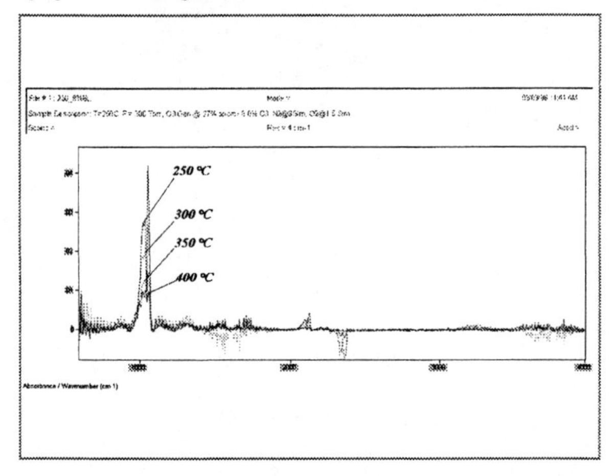

Figure 5. The changes in the infrared spectrum of ozone with increasing temperature in the reaction chamber.

We have examined the variations in the spectrum of O_3 upon changes in ozonator generation rates and process conditions. The 1055 cm^{-1} peak in the spectrum varies linearly with ozonator voltages and feed rates as reported by O'Neill et al (21). Our data differ from O'Neill's in that our determinations were made within the CVD chamber at process temperatures where O_3 undergoes thermal reactions leading to a loss in signal relative to measurements at ambient temperatures. This loss in band intensity is significant for our understanding of TEOS/O_3 chemistry. The literature does not distinguish whether the reaction occurs via direct O_3 insertion or via O_3 dissociation followed by atomic oxygen insertion into the alkoxide bond. Knowledge of the correct mechanism is necessary to form accurate models for advanced control. The data in Figure 5 support the dissociation of O_3 as the initiating step in the reaction. This is consistent with the observed trends for deposition rate in the process. Our data indicates that, at 400 °C, the O_3 peak area is reduced by ca. 70% over that observed at 250 °C. Rate studies in the TEOS/O_3 process show the deposition rate to peak at 400 °C, then fall off. It is likely that, near 400 °C, O_3 is nearly completely dissociated and further increases in temperature only promote parasitic side reactions. These side reactions consume TEOS, reduce deposition rates and induce gas phase nucleation. IR monitoring of the reaction thus provides a means to ensure optimal levels of O_3 dissociation and to minimize parasitic reactions.

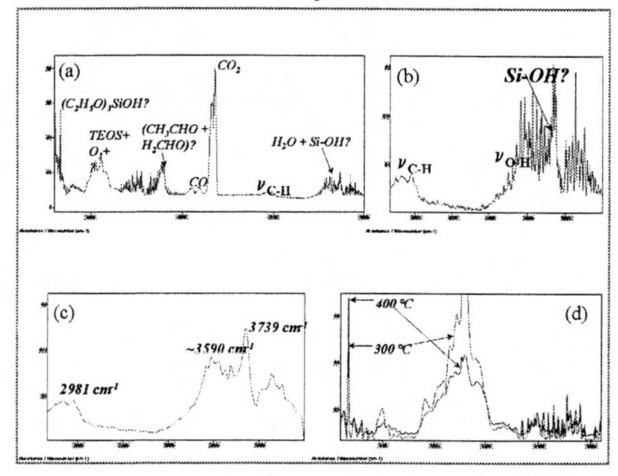

Figure 6. (a) The infrared spectrum of a TEOS/ozone mixture under simulated CVD process conditions, 600-4000 cm^{-1}; (b) C-H and O-H stretching region of the spectrum, ~2800-4000 cm^{-1}; (c) smoothed spectrum, O-H stretching region ~3400-4000 cm^{-1}; (d) spectral variations in the 670 cm^{-1} and 1000-1200 cm^{-1} regions with variations in process temperature

Figure 6 shows *in situ* spectra of TEOS/O_3 under CVD process conditions. Absorptions that may be ascribed to the starting materials, reaction products and intermediate species are present. Bands at 1055, 1115 and 1085 cm^{-1} are due to TEOS and O_3. Additional features in this region may be due to the presence of Si-O-Si oligomers in the mixture, but the data is not definitive. The intense band at 2300 cm^{-1} is due to the CO_2 product. CO is also present as indicated by the band at 2140 cm^{-1}. Absorptions between 1700-1800 cm^{-1} may be due to either less oxidized products or intermediate peroxide species. Kawahara et al (20), has assigned these bands to CH_3CHO and/or H_2CO, the expected products of hydride elimination (CH_3CHO), and subsequent oxidation of the organic fragments (H_2CO). Intensity ratios of the absorptions due to CO_2, CO and aldehydes may provide a useful measure of the relative degree of oxidation in the chemical system. This has consequences for the stoichiometry of the final oxide film, which in turn is related to its device quality. Correlation studies in this area are planned.

The spectral views in Figures 6(b) and (c) show features that may be due to the presence Si-OH intermediates. Figure 6(c) has been smoothed to eliminate the sharp, overlapping water bands. 6(b) shows peaks at ca. 3580 and 3735-3740 cm^{-1}, superimposed on the absorptions due to water. These bands do not appear to be due to CO + SiO combination bands, as the latter should occur at frequencies greater than 3800 cm^{-1}.

Spectra of very dilute solutions of polydimethylsiloxane (PDMS) (22) show Si-OH absorptions at ca. 3700 cm^{-1}, with an accompanying, broad, weak band at 3580 cm^{-1}. The OH group in TEOS derivatives may be more basic than that in PDMS, consistent with a frequency shift (~+35-40 cm^{-1}) for the SiO-H vibration. Our spectra also show a broad absorption at ca. 3580 cm^{-1}. Figure 6(d) shows a band at 670 cm^{-1} that may be attributable to triethoxysilanol. Theoretical calculations for the latter predict a peak at this position (23). Peak intensity varies with temperature and reagent flows in a manner consistent with the above assignment. Further work is being performed to determine whether this band is actually due to a silanol intermediate or to thermal artifacts within the system. We are currently correlating the absorption intensities in our spectra with characteristics of the process and oxide films. Figure 7 shows spectra of CVD oxides formed in the reactions at 300 and 400 °C. The spectra show increased water content in the films produced at lower temperatures, consistent with the literature. The results of these studies will be presented in a separate report (24).

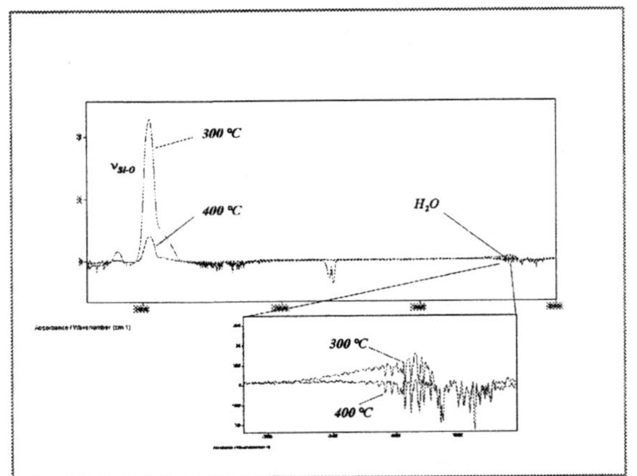

Figure 7. Infrared spectra of CVD oxides formed by TEOS/ozone reactions at 300 and 400 °C.

CONCLUSIONS

The data in this report demonstrate the feasibility of IR analyses for *in situ* determinations of reagent concentrations. The correlation of O_3 band intensities with temperature provides support for a deposition mechanism invoking O_3 dissociation as the initiating step of the reactions in TEOS/O_3 processes. The relative degree of oxidation of the deposition mixture is reflected in the infrared data, information that is of importance for process control. We have detected the presence, within the reaction mixture, of intermediate species (silanols) that influence particle generation and stoichiometries in the thin films. The use of this detection methodology in monitoring and manipulating concentrations of reactive intermediates has immediate consequences for the maintenance of optimal conditions within CVD processes.

REFERENCES

1. 1994 *National Technology Roadmap for Semiconductors.*
2. S. W. Butler, *J. Vac. Sci. Technol.* **B13**, 1917 (1995).
3. I. Nir, J. Winniczek, M. P. Splichal, H. M. Anderson and A. Stanton, *Photonics Spectra*, **27**, 87 (1993).
4. Gabrial G. Barna et al, *Solid State Technology*, **37**, 47 (1994).
5. "Process Control, Diagnostics and Modeling in Semiconductor Manufacturing", M. Meyyappan, D. J. Economou and S. W. Butler, Editors, Proceedings of The Electrochemical Society, 97-9.
6. A. C. Adams and C. D. Capio, *J. Electrochem. Soc.*, **126**, 1042 (1979).
7. E. J. Kim and W. N. Gill, *J. Electrochem. Soc.*, **142**, 676 (1995).
8. A. Chatterjee, I. Ali, K. Joyner, D. Mercer, J. Kuehne, M. Mason, A. Esquivel, D. Rogers, S. O'Brien, P. Mei, S. Murtaza, S. P. Kwok, K. Taylor, S. Nag, G. Hames, M. Hanratty, H. Marchman, S. Ashburn and I.-C. Chen, *J. Vac. Sci. Technol.* **B15**, 1936 (1997).
9. K. Fujino, Y. Nishimoto, N. Tokumasu and K. Maeda, *J. Electrochem. Soc.*, **140**, 2922 (1993).
10. A. Kubo, T. Homma and Y. Murao, *J. Electrochem. Soc.*, **143**, 1769 (1996).
11. T. Homma, M. Suzuki and Y. Murao, *J. Electrochem. Soc.*, **140**, 3591 (1993).
12. I. A. Shareef, G. W. Rubloff and W. N. Gill, *J. Vac. Sci. Technol.*, **B14**, 772 (1996).
13. O. Sanogo and M. R. Zachariah, *J. Electrochem. Soc.*, **144**, 2919 (1997).
14. M. Adachi, K. Okuyama, N. Tohge, M. Shimada, J. Sato and M. Muroyama, *Jpn. J. Appl. Phys.*, **33**, L447 (1994).
15. D. M. Dobkin, S. Mokhtari, M. Schmidt, A. Pant, L. Robinson and A. Sherman, *J. Electrochem. Soc.*, **142**, 2332 (1995).
16. K. Okuyama, T. Fujimoto and T. Hayashi, *Ceramics Processing*, **43**, 2688 (1997).
17. T. Satake, T. Sorita, H. Fujioka, H. Adachi and H. Nakajima, *Jpn. J. Appl. Phys.*, **33**, 3339 (1994).
18. T. Sorita, S. Shiga, K. Ikuta, Y. Egashiba and H. Komiyama, *J. Electrochem. Soc.*, **140**, 2952 (1993).
19. M. G. M. van der Vis, R. J. M. Konings, A. Oskam and T. L. Snoeck, *J. Mol. Struct.*, **274**, 47 (1992).
20. T. Kawahara, A. Yuuki and Y. Matsui, *Jpn. J. Appl. Phys.*, **31**, 2925 (1992).
21. J. A. O'Neill, M. L. Passow and T. J. Cotler, *J. Vac. Sci. Technol.*, **A12**, 839 (1994).
22. E. D. Lipp, *Applied Spectroscopy*, **45**, 477 (1991).
23. P. Ho, *private communication.*
24. T. K. Whidden, Abs. # 218, 193rd Meeting of the Electrochemical Society, San Diego, May 3-8, 1998.

X-Ray Scanning Photoemission Microscopy Of Titanium Silicides And Al-Cu Interconnects

G. F. Lorusso, H. Solak, S. Singh, F. Cerrina

Center of X-ray Lithography, University of Wisconsin, Madison, Wisconsin

P. J. Batson, J. H. Underwood

Center of X-ray Optics, Lawrence National Berkeley Laboratory, Berkeley, California

We use x-ray spectromicroscopy in order to investigate current problems in microelectronics. In particular, we report here the results obtained by MAXIMUM, a scanning photoemission microscope recently installed at the Advanced Light Source (ALS), at the Lawrence Berkeley National Laboratory. MAXIMUM is a spectromicroscope based on multilayer optics at 130 eV, where the images are achieved by collecting the photoelectron emitted from the sample by means of an electron energy analyzer. The spatial resolution is 100 nm, and the spectral resolution is 200 meV. In order to show the capabilities of this instrument, we will discuss the formation of $TiSi_2$ in patterned structures and the electromigration in AlCu lines. In the case of $TiSi_2$, we imaged lateral variations in the local chemistry of patterned samples with feature from 100 to 0.1 mm, which are attributed to the formation of different phases (C54 and C49). We studied the effect of electromigration in Al-Cu lines with different Cu content and width in situ, in order to distinguish different phases of Cu and follow its evolution electromigration proceeded.

INTRODUCTION

The term x-rays spectromicroscopy identifies a collection of experimental techniques combining conventional spectroscopy methods and high spatial resolution (1). In particular, photoemission spectromicroscopy is a relatively new experimental field, based on the analysis of soft x-rays photoelectrons. Since it is the spatial resolution that opens up new areas of physical investigation, much effort has been devoted to improve it. However, the development of the photoemission spectromicroscopy approach has been strongly limited by the low signal-to-noise level, making difficult to achieve at the same time high lateral and high energy resolution. This limitation has been only recently overcome by the development of new high-brightness synchrotron radiation sources.

One of the new x-ray photoemission spectromicroscopes developed in the last decade is MAXIMUM (Multiple Application X-ray Imaging Undulator Microscope) (2). In this paper we report some recent results obtained by MAXIMUM investigating two open problems in microelectronics: electromigration and interconnects.

In 1987, The University of Wisconsin, in collaboration with the Advanced Light Source, begun the development of an x-ray microscope system of the scanning type with the goal of reaching a spatial resolution better than 0.1 μm, and a spectral resolution better than 200 meV, working with a base pressure better than 10^{-10} torr. All these goals were achieved in 1992, after a period of development of about 5 years. Several breakthroughs had to be achieved in order to deliver the required performances. In particular, MAXIMUM mounted the first UHV-compatible Schwarzschild objective with in-situ alignment, using for the first time at-wavelength Foucalt test in extreme ultraviolet (EUV), demonstrating the high efficiency of multilayer optics near the silicon edge.

The photoemission microscope MAXIMUM was originally installed at the Synchrotron Radiation Center (SRC) in Madison, WI. During the period from 1992 to 1995, the microscope was successfully used to study semiconductor surfaces, interfaces, biological samples and organic particles. However, during the operation of the microscope, it become apparent that the microscope performance was severely hampered by the relatively low brightness of Aladdin, which limited the available flux at the microscope's focus and, consequently, the achievable spatial resolution because of signal-to-noise consideration. As a consequence, the experiments at SRC were often forced to operate at a reduced resolution in order to obtain realistic counting rates.

The installation of the microscope on a third-generation light source clearly appeared the only possible solution to

CP449, *Characterization and Metrology for ULSI Technology: 1998 International Conference*
edited by D. G. Seiler, A. C. Diebold, W. M. Bullis, T. J. Shaffner, R. McDonald, and E. J. Walters

this problem. A Participating Research Team (PRT) was formed in order to move maximum from Wisconsin to the Advanced Light Source (ALS). The microscope was moved to the ALS in April 1995, and it was preliminary installed on the bend magnet beamline 6.3.2, were the preliminary tests were performed. In the present paper we report on the successful installation of MAXIMUM on its final location (beamline 12.0 at the ALS) in 1997.

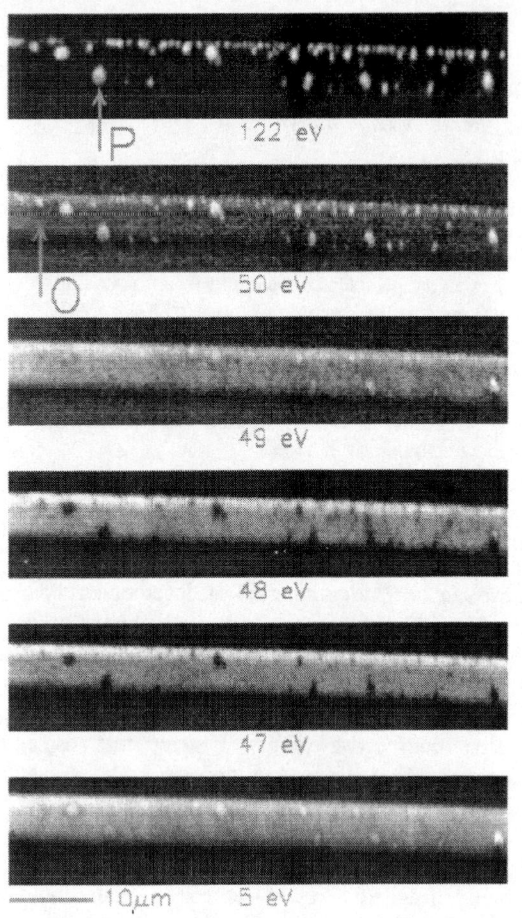

Figure 1. 60 μm X 10 μm images of 5 μm wide Al-4%Cu line before electromigration stressing acquired using photoelectrons of indicated energies.

Radiation from the synchrotron source is monochromatized and focused by a Kirkpatrick-Baez (KB) system to illuminate a pinhole, which serve as source for the microscope optics. A multilayer coated Schwarzschild Objective (SO) produce an image of the pinhole with a 20x demagnification. When a sample is placed at the focus, photoelectrons are collected by a cylindrical mirror analyzed (CMA) electron spectrometer. The sample is mounted on a scanning stage, and by rastering the sample it is possible to produce a 2-d photoemission image.

Al-Cu INTERCONNECTS

Electromigration (EM) is the movement of atoms in a conductor under the influence of an electric current. This process is one of the major reliability concerns in microelectronics industry because of its damaging effects on metal interconnect lines. Al, which has been the industrial choice for interconnect metallization, is especially susceptible to this damage mechanism (3). Addition of small amounts of Cu (0.5-4%) to Al has been found to increase the lifetimes of interconnect lines against electromigration damage significantly and it is practiced commonly in industry (4).

In order to investigate EM, it is essential to obtain information on the Cu content of grain boundaries, its chemical state and the dynamics of Cu distribution between grain boundaries, grains and Al_2Cu precipitates during process, operation and test conditions. MAXIMUM is a excellent tool to study this problem (5). It can map the distribution of an element and can differentiate chemical states of the same element (like Al or Cu). Furthermore, it is especially suitable in examining the dynamics of the surface electromigration process thanks to its surface sensitivity. In order to test EM *in situ* to preserve the chemical states of newly EM created surfaces, a special stage and a sample holder were constructed for in-situ testing of in the UHV chamber of the microscope. It allows heating of the sample above 300°C and four separate electrical contact.

For the EM test samples, 600nm of Al-Cu alloy was sputter deposited onto thermally oxidized Si wafers. Three different Cu concentrations, 0.5, 2 and 4% (weight) were chosen to observe the effect of Cu content. Lines were patterned using photolithography and wet etching processes. After patterning the wafers were annealed at 450°C for 30 minutes in forming gas to set the microstructure. No passivation was deposited onto the samples. Samples were slightly etched before introducing into the microscope in a solution of Ammonium-oxalate-monohydrate in ammonium hydroxide to reveal the Cu precipitates (6).

Samples were characterized before and after EM testing using spectromicroscopy. This is done by acquiring images of the sample at interesting photoelectron energies, as well as photoelectron spectra (PES, also called Energy Distribution Curves, EDC) on interesting spatial features on the sample. In this way we were able to compare the images and PES before and after the EM process and evaluate the changes that has occurred.

Figure 1 shows a typical set of images acquired in an area of sample at six different photoelectron energies before testing for EM. The sample in that case is a 5μm wide Al-4%Cu line. A reference image is acquired by collecting the secondary electrons of 5eV kinetic energy (KE); contrast at this energy arises from sample topography as well as elemental inhomogeneity, and the images are very similar to those obtained by SEM. Grain boundaries are resolved in that image and they appear darker due to topography. The 122 eV image is acquired at the Cu 3d core level and shows the distribution of Cu in the sample. Bright spots on that image correspond to the Cu-rich precipitates. Figure 3 shows two EDC's acquired at points P (precipitate) and O (ordinary) that are marked in Figure 1. Presence of Cu 3d signal is clearly evident as a shoulder around 121 eV. Images at 47, 48, 49 and 50 eV are acquired around the Al 2p core level. To understand the contrast mechanism in those images let us look at the Al 2p core level EDC's. The Al 2p core level exhibits a shift of 0.45eV towards higher kinetic energy in the Cu rich area. This core level shift causes Cu rich precipitates to appear bright in the 50 eV image, and dark in the 47 and 48 eV images.

Figure 2 shows a set of images taken after electromigration stressing. The images show the *cathode* end of the Al-Cu line. Extensive void formation is apparent in those images which appear as dark regions in all images except the one taken at 42 eV kinetic energy. The micro EDC's acquired at point V show several shifted Al 2p core level components are present near 45 eV. We attribute this energy shift in the void areas to the charging of Al that is left behind on the substrate. The charging is due to the electrical isolation of the left over Al in the voids from the rest of the line. Therefore the void areas appear bright in the 42 eV image. Images acquired at 47 and 48 eV in this set show some grain boundaries as dark areas. Some of the same grain boundaries appear as bright regions in the 50 eV and 122 eV images. This suggests an increased presence of Cu in the grain boundaries after EM.

Furthermore, the micro-EDC analysis on the labeled hillock acquired at point H marked in Figure 2 indicated a narrow peak at 51.6 eV, due to metallic Al 2p core level. This is fresh Al that was carried here by the EM flux. Oxidation of Al in that hillock is slowed down by the UHV environment of the microscope chamber. This area (H) appears bright in images acquired at all phototelectron energies due to high emission of clean Al surface. The fact that we observe metallic Al only in the hillocks and not anywhere else on the line indicates that surface electromigration did not take place in our experiment. In other words the EM was strictly limited to the grain boundary network.

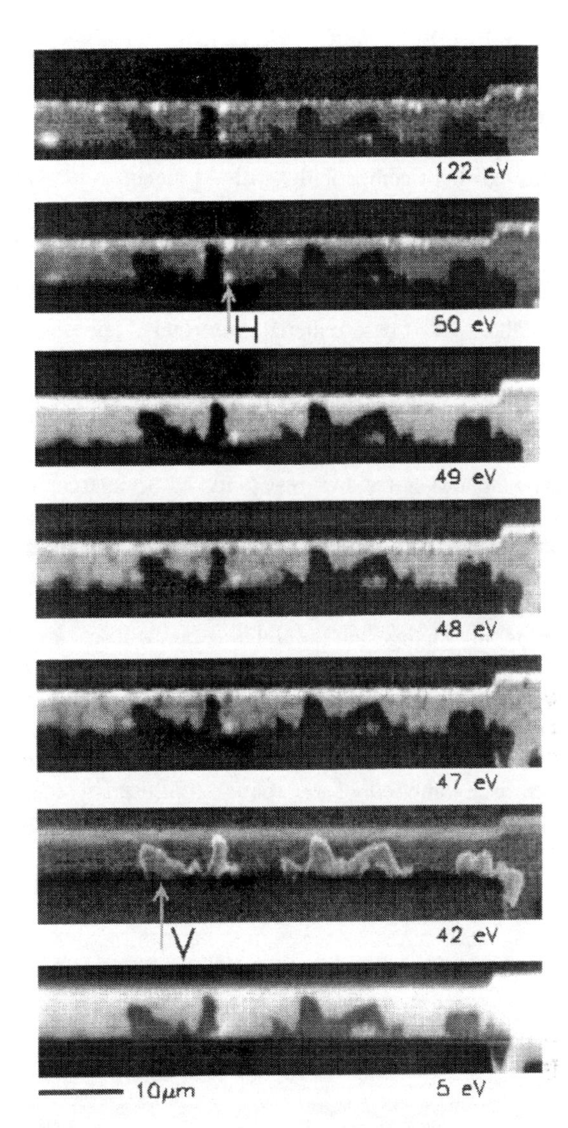

Figure 2. 60 μm X 10 μm images of 5 μm wide Al-4%Cu line after electromigration stressing acquired using photoelectrons of indicated energies

In conclusion we reported the first photoemission spectromicroscopy study of electromigration in Al-Cu lines. Our preliminary results indicate that we are able to resolve Cu in grain boundaries and precipitates using the Cu 3d emission and the chemical shift of Al 2p core level in the Cu rich phase. We found no evidence of surface electromigration even under favorable conditions of the UHV environment. Finally, MAXIMUM which was installed recently on beamline 12.0 as a permanent endstation is routinely operating with better than 0.1μm spatial resolution which made this experiment possible.

SILICIDES

Titanium silicide has the lowest resistivity of all the refractory metal silicides, and has good thermal stability as well as excellent compatibility with Al metallization (7). It is used as an intermediate buffer layer between vias and the Si substrate to provide good electrical contact in ultralarge scale integrated processes, namely a self-aligned silicide (salicide) formation (8). $TiSi_2$ exist in two phases: a metastable C49 base-centered ortorombic phase with specific resistivity of 60-90 μΩ cm that is formed at a lower temperature, and a stable 12-15 μΩ cm resistivity face-centered orthorombic C54 phase into which C49 is transformed with a higher temperature step (9,10). C54 is clearly the target for low resistivity ULSI interconnects. However, it has been observed that when the dimension shrink below 1 um the transformation of C49 into C54 is inhibited and agglomeration often occurs in fine lines at high temperature, resulting in a rise in resistivity. Since there is no change in the relative amount of species, spectromicroscopy should be an appropriate tool to study the evolution of the process. In fact, photoemission spectromicroscopy can directly image a patterned wafer and obtain information on the distribution of the different phases as a function of the spatial confinement and the local chemical environment for each of the phases. Because the nearest neighbor configuration differs in C49 and C54 phases (10), a shift in the core electron binding energy between the two phases should be apparent in

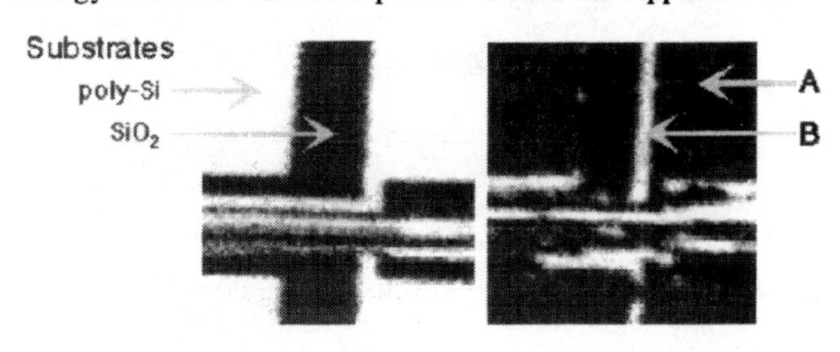

Figure 3. 100 X 100 μm x-ray microhtaph of a SRAM chip at 25.75 eV and 20.25.

Figure 3 (100x100 microns each with 1-micron steps), taken with the MAXIMUM spectromicroscope, show a small section of an SRAM chip that has just been through a siliciding process. This process begins with depositing a blanket layer of titanium (Ti) over the polycrystalline silicon (poly-Si) and silicon dioxide (SiO_2) substrates. Annealing (heating) to 700 degrees C causes the poly-Si to react with the Ti and form $TiSi_2$, which initially exists as a phase called C49. A second anneal at ~1000 degrees C transforms the C49 into a lower-resistivity phase called C54. (The sample above stops there, with unreacted Ti over the SiO_2 areas, but industrial production continues with an etching process to remove the unreacted Ti.) During the annealing process, the silicide spreads out laterally a short distance onto the SiO_2 areas. This lateral growth can cause problems in chips with submicron feature sizes. Both images were taken using an incoming photon energy of 130 eV. The left-hand image shows electrons with kinetic energies of 25.75 eV (unshifted Si 2p core level) and is bright in the silicided poly-Si areas. The right-hand image shows electrons with kinetic energies of 20.25 eV (oxide-shifted Si 2p) and is bright in areas with a strong oxide (from SiO_2) signal. The stronger oxide signal from the areas where the silicide has grown laterally could arise because the laterally grown silicide has agglomerated, exposing regions of the underlying SiO_2. This agglomeration, characteristic of edges andThe Si 2p photoemission spectra at left were taken from the locations marked A ($TiSi_2$ over poly-Si) and B (laterally grown $TiSi_2$ over SiO_2) on the image above. The lower-kinetic-energy peak comes from oxide-shifted Si in SiO_2 and corresponds to the image at right above. The higher-kinetic-energy peak comes from Si in $TiSi_2$ an[illegible] esponds to the image at left abo of spatially confined areas such as narrow (less than one micron wide) lines, can prevent the formation of continuous lines of C54 and disrupt circuit function.

REFERENCES

1. G. Margaritondo and F. Cerrina, Nucl. Instr. and Meth. A291, 26 (1990).
2. C. Capasso, A. K. Ray-Chaudhuri, W. Ng, S. Liang, R. K. Cole, J. Wallace, F. Cerrina, G. Margaritondo, J. H. Underwood, J. K. Kortright, and R. C. C. Perera, J. Vac. Sci. Technol. A9, 1248 (1991).
3. Tom Seidel and Bin Zhao, "0.1μm Interconnect technology challenges and the SIA roadmap," Mat. Res. Soc. Sym. Proc. 427, 3 (1996)
4. Ames, F.M. d'Heurle, and R. Horstman, IBM J. of Res. Develop. 4, 461 (1970)
5. W. Ng, A.Ray-Chaudhuri, S. Liang, S. Singh, H. Solak, J. Welnak, F.Cerrina, G. Margaritondo, J. Underwood, J. Kortright, and R. Perera, "High resolution spectromicroscopy reaches the 1000 angstrom scale," Nucl. Instr. and Meth. A347, 422 (1994).
6. E.G. Colgan, and K.P. Rodbell, "The role of Cu distribution and Al_2Cu precipitation on the electromigration reliability of submicrometer Al(Cu) lines," J. Appl. Phys., 75, 3423, (1994).
7. S.P. Murarka, Silicides for VLSI Applications (Academic, New York, 1983).
8. J. F. Jongste, F. E. Prints, and G. C. A. M. Janssen, Mater. Lett. 8, 273 (1989).
9. R. Beyers, and R. Sinclair, J. Appl. Phys. 57, 5240 (1985).
10. H. Jeon, C. A. Sukow, J. W. Honeycutt, G. A. Rozgonyi, and R. J. Nemanich, J. Appl. Phys. 71, 4269 (1992).

Thin-Film Metrology by Rapid X-Ray Reflectometry

L.N. Koppel, L. Parobek

AXIC, Inc., 493 Gianni Street, Santa Clara, CA 95054

Grazing-incidence X-ray Reflectometry (XRR) is emerging as a powerful thin-film and substrate metrology technique for the semiconductor industry. XRR measurements allow the thickness, density, and surface and interface microroughness of thin-film structures to be characterized non-destructively and without reference to standards. The density and microroughness of smooth substrates can also be accurately measured. We will report on the performance and range of application of a new type of reflectometer which uses a proprietary x-ray optical system to focus a converging fan of x rays onto a sample, and an x-ray sensitive electro-optic sensor to detect the reflected x-ray pattern all at once. This configuration allows very rapid analysis that supports multi-point mapping of thin-film and substrate properties. Percent-scale thickness measurement accuracy has been confirmed using titanium, titanium nitride, TiN-on-Ti, and tantalum pentoxide thin-film samples and correlated XRF and RBS data. The ability of the XRR technique to "optically" measure the density of as-built films has been confirmed using silica aerogel-on-silicon samples and RBS correlation. Silicon wafer frontside/backside measurements, correlated to AFM data, have confirmed the technique's ability to characterize angstrom-scale microroughness. Due to the penetrability and short wavelength of x rays, we believe that Rapid XRR Metrology will be particularly important for the monitoring and control of opaque metal barrier and adhesion films and low-k dielectric films used in advanced ULSI interconnect structures.

INTRODUCTION

While the phenomena of total external reflection and the formation of interference fringes in the x-ray regime have been understood since the 1930s, it is only in the last decade that the field of grazing-incidence x-ray reflectometry (XRR) has seriously undertaken the measurement of the structural properties of thin films (film thickness, density, and microroughness) and of smooth surfaces (density and microroughness) for the purposes of process development and control (1,2). Fortunately, the x-ray reflectometry technique has reached a state of maturity just at the time that the semiconductor industry is relying increasingly on new materials for the fabrication of deep-submicron devices. Some of these new materials, such as barrier films used in copper metallization schemes, low-k dielectrics used as advanced interlayer dielectric (ILD) films, and high-k dielectrics used in dense DRAM memories, are not well served by conventional thin-film metrology instruments such as four-point probe, scanned-microtip, and ellipsometer/spectrophotometer measurement tools. The measurement of the density of CVD-formed thin films, which can dramatically affect the functionality of a film (such as the permeability of a barrier film) but which can vary to a high degree across a wafer, is a particular challenge in this regard. A parallel need is emerging for a rapid, long scale-length "optically-based" surface microroughness characterization technique that may support the use of double-polished 300mm-diameter wafers in microelectronics fabrication. The impressive body of experimental work reported in the last decade demonstrates that x-ray reflectometry can satisfy a large number of the structural metrology challenges now confronting the semiconductor industry. In this paper, we will describe initial results which we have obtained in the last year using a method which we call Rapid-XRR. This technique uses a special x-ray optical configuration to perform measurements in seconds for each sampled point, rather than the hundreds of seconds required by conventional XRR instrumentation. We believe that these results support an expectation that the technique can contribute to the timely multipoint mapping of thin-film and substrate structural properties within the semiconductor industry.

X-RAY REFLECTOMETRY

XRR measurements most commonly involve observing the interaction of a collimated, monochromatic x-ray beam with a thin film structure or smooth surface, as illustrated in Figure 1. At a grazing angle of incidence less than a value known as the "critical angle", the x rays are specularly reflected with very high efficiency, representing the phenomenon of total external reflection. The critical angle ϑ_C scales as the square root of the sample's surface density ρ and linearly as the wavelength λ:

CP449, *Characterization and Metrology for ULSI Technology: 1998 International Conference*
edited by D. G. Seiler, A. C. Diebold, W. M. Bullis, T. J. Shaffner, R. McDonald, and E. J. Walters
 1-56396-753-7/98/$15.00

$$\vartheta_C = C \cdot \lambda \cdot \mathrm{sqrt}(\rho) \quad (1)$$

where C is a constant. The critical angle is in the range from about 0.25° to 0.75° for materials commonly encountered in the semiconductor industry, at an x-ray wavelength of 2Å. As the angle of incidence ϑ is increased through the critical angle, the x-ray reflection efficiency decreases rapidly as the x-ray beam begins to refract into the sample. The rate of decrease is determined by the microroughness of the sample, being proportional to ϑ^{-4} for a perfectly smooth sample, and decreasing even faster for a roughened sample. Finally, for an angle of incidence several times the critical angle, the beam is wholly refracted into the sample, and the externally observable reflection is extinguished. In the middle range of angles defined above ($\vartheta_C < \vartheta <$ several times ϑ_C), both reflection and refraction can occur, and for a thin-film sample, interference fringes can be formed by the constructive interference of x rays reflected from the sample's top surface and from density-contrast interfaces within the sample (at the thin-film/substrate interface, for example).

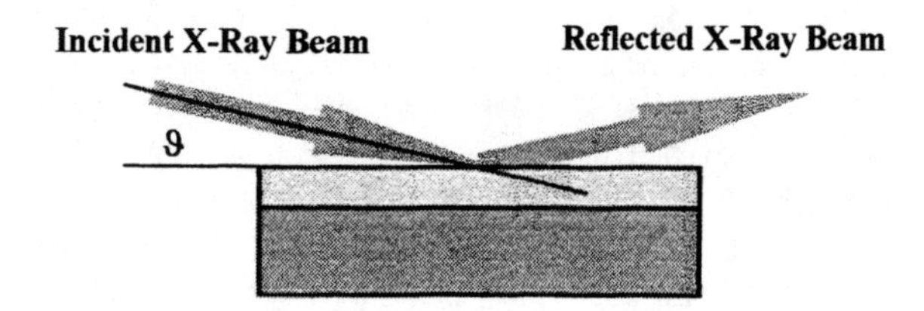

FIGURE 1. Interaction of a collimated, monochromatic x-ray beam with a thin film structure, at a grazing angle of incidence ϑ less than the critical angle.

The behavior of the specular reflection efficiency described above, illustrated in Figure 2 for a thin-film structure and in Figure 3 for a smooth substrate, provides the basis for understanding the structure of a sample. For example, the experimental data for a tantalum film deposited on a silicon substrate shows a series of interference fringes, the maxima of which are located at angles ϑ_n given by the relation:

$$\sin^2(\vartheta_n) = \sin^2(\vartheta_C) + [n+½]^2\,[\lambda/2T]^2 \quad (2)$$

where n is the fringe order number (n = 1,2,...) and T is the thickness of the thin film (3). Hence, the thin-film's thickness and density (through the density dependence of ϑ_C) can be deduced by straightforward observation of the fringe maxima locations, without reference to the amplitude of the fringes. However, the sample's surface and interface roughness values (expressed as Å-rms) affect the amplitude of the fringes, and can be quantified by recursive solution of the Fresnel Equations that predict the behavior of electromagnetic radiation at boundaries within a sample. A theoretical model of this type has been fitted to the data shown in Figure 2 to determine both the top surface and thin-film/substrate microroughness of the tantalum-on-silicon dioxide sample.

Modeling of the reflection efficiency for a smooth silicon substrate, shown in Figure 3, demonstrates the sensitivity of the XRR technique to surface microroughness on the angstrom scale, particularly for short x-ray wavelengths.

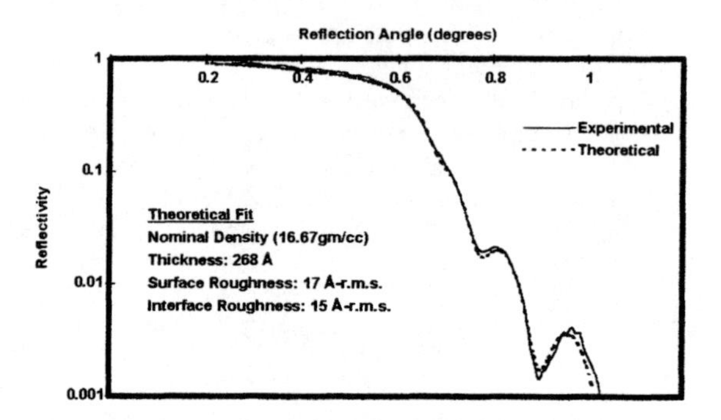

FIGURE 2. Experimental x-ray reflectivity measured at a wavelength of 1.936Å (iron $K\alpha_1$ characteristic x rays) for a tantalum film deposited on a silicon dioxide substrate. The critical angle for this film is 0.65°. The theoretical fit is based on recursive solution of the Fresnel Equations.

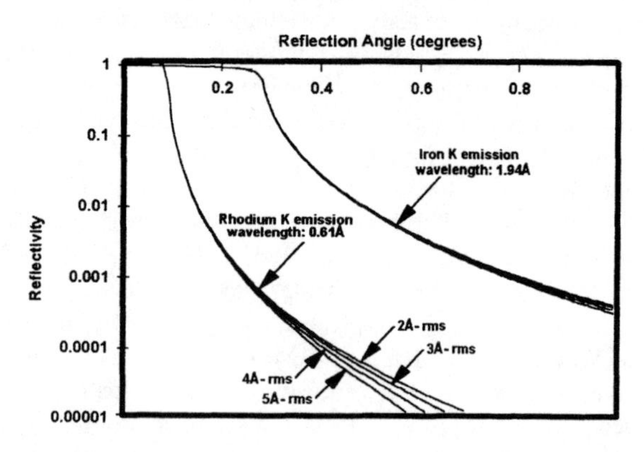

FIGURE 3. Calculated x-ray reflectivity of a bare silicon substrate at two wavelengths, for microroughness values of 2, 3, 4, and 5Å-rms.

Thin-film thickness measurement accuracies of the order of ±1% are routinely reported in the literature, as are density measurement accuracies of the order of ±3%. The accuracy with which XRR measures microroughness is an issue under active study, although several investigators claim performance that rivals or exceeds that of the atomic force microscope (4,5). For most XRR instruments, the practical upper limit on thin-film thickness measurement range is in the neighborhood of 2,000Å due to finite fringe resolution and x-ray absorption in the sample. Additionally, the specular reflectivity is reduced to virtually undetectable levels for microroughness exceeding about 50Å-rms.

THE RAPID-XRR TECHNIQUE

In the last year, we have demonstrated a technique for performing XRR measurements which is substantially faster than traditional methods, which we believe supports the use of XRR as a practical multipoint mapping technique for the first time. In the traditional method, a finely collimated x-ray beam is allowed to strike the sample at a particular value of angle of incidence ϑ while an x-ray counter records the intensity of reflection for that angle. A complete reflection efficiency curve is then built up by stepping and repeating the measurement over the required angular range, using an x-ray diffractometer or similar apparatus to step from one value of angle to the next. Due to limitations on maximum count-rate, a period of several hundred seconds or greater is required to record the reflection efficiency curve by the traditional method.

The Rapid-XRR technique relies on the novel x-ray optical configuration shown in Figure 4 to measure every point of the reflection efficiency curve simultaneously, by exposing the sample to a converging fan of monochromatic x rays incident over the required angular range, and recording all reflected x rays all at once (7). This configuration makes use of x-ray intensity which would otherwise be lost in collimation in the traditional apparatus, greatly reducing the data accumulation time. The two key components of the technique are a point-to-line x-ray focusing element known as a Johansson diffraction monochromator (8), and a spatially resolving electro-optic x-ray detector such as a self-scanning photodiode array (SSPA) (9). The source region of an x-ray tube, the Johansson monochromator, and the sample are carefully co-aligned on the monochromator's focusing circle so that a line image of the emission region, rendered at a characteristic emission wavelength associated with the x-ray tube target element, is projected onto the sample surface at grazing incidence. The fan of x rays that form this image is then reflected by the sample over a range of angles and detected by the SSPA. By aligning the sample with respect to the optical axis of the instrument in tilt and elevation, a one-to-one correlation between position on the SSPA and reflection angle is obtained, so that signal recorded at a pixel can be associated with a particular value of reflection angle. The distribution of intensity in the incident fan beam can be measured by removing the sample and allowing the beam to fall directly on the lower part of the sensor.

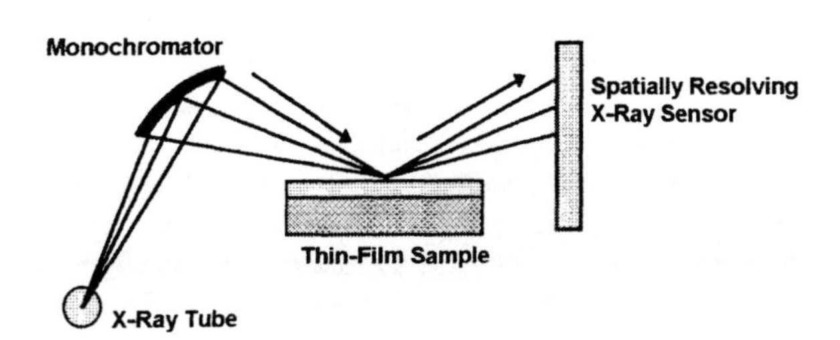

FIGURE 4. A schematic of the Rapid-XRR x-ray optical configuration, containing an x-ray tube source, a Johannson monochromator, and a sample co-aligned on the monochromator's focusing circle, and a spatially resolving sensor to record the fan of reflected x rays.

Since the beginning of 1997, we have assembled and tested an XRR analysis system based on the technique described above. The key components of this feasibility-demonstration system are the following:

- a low-power, iron-target x-ray tube operating at 40kV potential and 40watts beam power, with a 0.050mm x 0.300mm focal spot distribution
- a silicon (111) Johansson monochromator crystal formed to a 300mm focal circle (Rowland) radius
- a sample stage comprised of a 6-axis positioner, and an autocollimator required during the alignment of a sample with respect to the axis of the reflectometer
- a 1024-element, x-ray sensitive SSPA detector fitted with a thin beryllium-foil window and comprised of .025mm x 2.54mm rectangular sensing elements (pixels).

These components were designed so as to achieve a 30arcsecond reflection-angle resolution at the detector. Parameters that affect the resolution include the width of the x-ray tube focal spot in the plane of dispersion, the diffraction-profile width of the monochromator crystal, and the center-to-center spacing of the detector's sensing elements in the plane of dispersion. The critical components were supported and precisely co-aligned on a stable, vibration-isolated baseplate by a set of commercial and custom-made positioners and mounting fixtures. The most critical of these was the Johansson monochromator gimbal mount, which allowed rotation of the crystal about three axes and translation along two. To permit the monochromator to select between the $K\alpha_1$ and $K\alpha_2$ characteristic emissions of the iron target (at wavelengths of 1.936Å and 1.940Å, respectively) the pitch-rotation accuracy achieved by the gimbal was better than 6arcseconds, representing the most stringent requirement encountered during the assembly of the apparatus.

The feasibility-demonstration system is routinely capable of measuring reflection efficiency curves over the range from unit reflection efficiency to somewhat less than 0.001 in 10 to 30 seconds signal-integration time, using the low-power x-ray tube and for a sampled spot size measuring 1.5mm x 8.5mm. In the near future, we will upgrade the apparatus with a more intense x-ray source, drawing beam powers of hundreds of watts rather than tens of watts, allowing us to accumulate reflection efficiency data over a wider dynamic range and smaller spot size.

PERFORMANCE OF THE RAPID-XRR DEMONSTRATION SYSTEM

At this time the only Standard Reference Materials (SRM) available from NIST for the certification of thin-film metrology tools consist of silicon dioxide films deposited on silicon. These standards, developed for conventional optical tools such as ellipsometers, are not well suited for the testing of x-ray reflectometers, since the low density contrast between SiO_2 and silicon produces interference fringes having subtle amplitude changes from peak to valley. A standard composed of a titanium nitride thin film would be preferable in this regard. Since a suitable SRM was not available, measurements undertaken with the Rapid-XRR System in the last year have focused on demonstrating measurement performance comparable to that achieved by the conventional step-and-repeat x-ray reflectometry technique, and on correlating our XRR data with that obtained by other thin-film and surface structural metrology methods, such as the optical reflectometry and Rutherford Back Scatter (RBS) thin-film analysis, and Atomic Force Microscope (AFM) microroughness characterization. In the course of this work, we have examined a large number of samples consisting of single-layer and mulitlayer thin-film structures that spanned a thickness range from 100Å to 1,200Å, and thick-film and substrate samples having varying surface density and microroughness.

Our first measurements characterized the thickness of a set of titanium thin films loaned to us by Dr. Craig England of Digital Semiconductor (Hudson, MA). The thickness of these samples were measured by Dr. England using the conventional step-and-repeat XRR technique and XRF. Our thickness values, obtained by fitting reflection efficiency curves such as that shown in Figure 5 for the thinnest film, are compared to the calibrated values in Table 1. The error ranges listed for the present work represent the standard deviation at 1-sigma confidence obtained for 25 independent measurements of film thickness. The close correlation of the data sets demonstrates that measurements obtained by the Rapid-XRR technique are equivalent to those produced by the traditional XRR method.

TABLE 1. XRR Measurement Correlation for Ti Samples

Titanium Sample	Calibrated Thickness	This Work
Sample Number 1	133Å ±2Å	138Å ±4Å
Sample Number 2	232Å ±3Å	231Å ±4Å
Sample Number 3	338Å ±4Å	334Å ±6Å

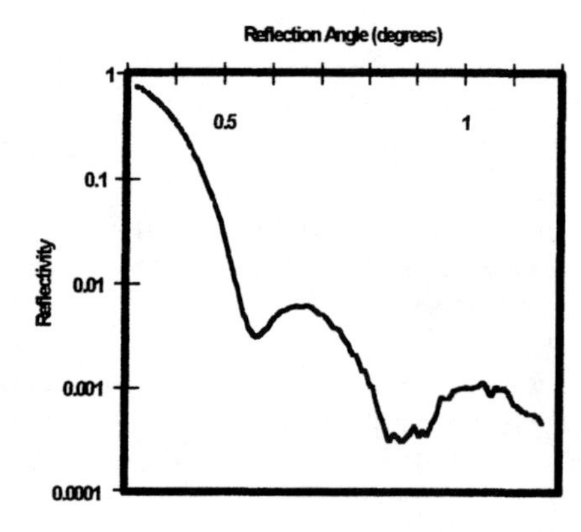

FIGURE 5. Experimental x-ray reflectivity measured at a wavelength of 1.936Å (iron $K\alpha_1$ characteristic x rays) for a titanium film, 133Å calibrated thickness, deposited on a silicon substrate.

To date, our most thorough cross-correlation with other thin-film thickness measurement techniques has been obtained with a series of tantalum pentoxide films deposited on silicon dioxide undercoats. As shown in Table 2 for three representative samples, these films were characterized by an AXIC Tyger multiple-angle optical reflectometer (OR) and by RBS, as well as by XRR. In Table 2, we have calculated the $\rho \cdot T$ values for XRR by multiplying the independently measured density (ρ) and thickness (T) values, to allow comparison to the RBS data. We find that the cross-correlation among the three methods is quite good, supporting confidence in the XRR metrology technique.

TABLE 2. XRR/UVE/RBS Measurement Correlation for Ta_2O_5 Samples

	XRR			OR	RBS
Sample	**Thickness**	**Density**	**ρ · T** (derived)	**Thickness**	**ρ · T**
Sample A	365Å	8.50g/cc	31.0μg/cm²	367Å	30.3μg/cm²
Sample B	490Å	8.55g/cc	41.9μg/cm²	497Å	39.7μg/cm²
Sample C	640Å	8.60g/cc	55.0μg/cm²	639Å	54.9μg/cm²

We believe that XRR provides an important, previously unavailable capability with respect to the measurement of surface density of layers formed by chemical and physical vapor deposition techniques. As an interesting demonstration of this capability, we have examined samples of a revolutionary low-k dielectric material consisting of a very low density silicon dioxide aerogel, proposed for use as an interlayer dielectric in device interconnect structures. Representative experimental data obtained by the XRR Demonstration System is shown in Figure 6. The reflectivity of a bare silicon substrate is also shown for comparison. The data for the aerogel is unique in that two critical angles are observed, one at low angle for the aerogel and a second for the silicon substrate that supports the relatively thick aerogel layer (for which no interference fringes can be resolved). X rays refracted through the aerogel at angles immediately above its critical angle but in the region of total external reflection for the silicon constitute the signal between the two critical angles. The location of the aerogel layer implies a density of about 0.3grams/cc, close to the value of 0.4grams/cc derived from RBS and profilometry measurements. The accuracy of this measurement could be substantially improved by using longer-wavelength radiation, to move the aerogel critical angle out to higher angles.

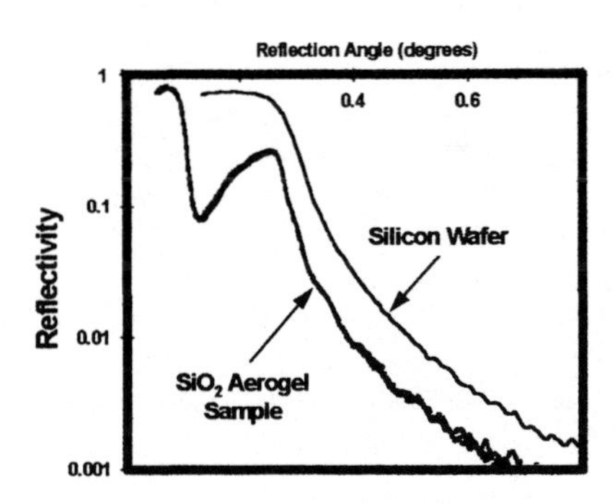

FIGURE 6. Experimental reflectivity measured for a low-density silicon dioxide aerogel sample and for a bare silicon wafer.

Finally, we have compared microroughness measurements obtained by the XRR Demonstration System for bare silicon wafers and thick tungsten and aluminum films to those produced by AFM examination. These measurements agree in relative magnitude, although the XRR microroughness values are consistently higher than those obtained by the AFM, possibly due to the finite tip radius of the AFM. Further work will be performed in this area.

FUTURE DIRECTIONS

We are now developing a Rapid-XRR Prototype System based on the technique described in this paper. This System's performance will be enhanced by the use of a more intense x-ray tube source and more efficient monochromator, allowing us to measure in several seconds reflectivity profiles down to the 10^{-4} - 10^{-5} decades over sampled spot areas having sub-millimeter dimensions. This performance is expected to provide to the semiconductor industry an important, new thin-film and surface structural metrology capability.

ACKNOWLEDGMENTS

The authors greatly appreciate the collaborative efforts of Dr. C. England of Digital Semiconductor and Dr. L. Rolly of Hewlett-Packard Company, both of whom provided samples and corroborative measurements used in this work.

REFERENCES

1. Toney M., Brennan S., *J.Appl.Phys.* **66**, 1861-1863 (1989).
2. England C., *Adv. X-Ray Anal.* **40**, to be published.
3. Huang T., Parrish W., ***Adv. X-Ray Anal.*** **35**, 137-142 (1992).
4. Lendeler B., *Adv. X-Ray Anal.* **35**, 127-136 (1992)
5. Stommer R., et al, *Adv. X-Ray Anal.* **41**, to be published.
6. U.S.Patent Number 5,619,548.
7. Johansson T., *Z. Physik* **82**, 507-511 (1933).
8. Koppel L., *Rev.Sci.Instrum.* **47**, 1109-1112 (1976).

Energy-Dispersive X-ray Reflectivity and the Measurement of Thin Film Density for Interlevel Dielectric Optimization

W.E. Wallace, C.K. Chiang, W.L. Wu

Polymers Division, Electronics Applications Group,
National Institute of Standards and Technology, Gaithersburg, MD 20899

A novel method for determining thin film electron density by energy-dispersive x-ray reflectivity is demonstrated for a commercial spin-on-glass dielectric. The effects of sample misalignment limit the accuracy of x-ray reflectivity as typically practiced. These effects may be properly accounted for by measuring the critical angle for reflection at many different x-ray wavelengths simultaneously. From this measurement, the thin film electron density can be ascertained with much improved accuracy. The electron density can be converted to a mass density with knowledge of the film composition.

INTRODUCTION

The search for low-dielectric-constant materials to serve as interlayer dielectrics in ULSI technology continues at an ever-increasing pace. Candidates from almost every broad class of insulating material have been considered (1). These include fluorine-modified silica, organically-modified silica, organic polymers, diamond-like carbon films, polymer foams, xerogels and aerogels, to name a few. The candidate materials are typically amorphous (i.e. non-crystalline) and are usually deposited at conditions far from equilibrium (e.g. by spin coating, or by chemical vapor deposition). For these reasons, film properties are very sensitive to the deposition conditions which must be closely controlled to insure reproducible properties and high process yields. One property that is very sensitive to process history in amorphous materials is density. Thus, rapid and accurate measurement of density can serve as a useful process diagnostic.

Until recently, measuring the mass density of thin films less than a 1000 nm thick has not been an easy task. In practice, thin-film density is often determined indirectly by making separate measurements of mass per unit area and of thickness. Methods of determining mass per unit area include Rutherford backscattering spectrometry, and x-ray fluorescence. Methods of thickness determination include step profilometry, optical interferometry, and ellipsometry. Ellipsometry and other optical methods are sometimes used to provide a measure of film density indirectly through the index of refraction. Changes in bonding configuration and molecular orientation that may accompany changes in density can be a significant source of error (2), especially in thin films. This effect is often found in confined films under stress (3).

Energy-dispersive X-ray reflectivity, performed on a modified x-ray powder diffractometer, has been applied to a variety of candidate interlevel dielectric materials. The electron density can be measured to ±1% in a few minutes. Changes in density can be followed as a function of annealing time and temperature. By adding information about film composition, the mass density can be calculated easily. Results demonstrated here were performed on a commercially prepared spin-on-glass dielectric: Allied-Signal AS418 (4). A wide array of other materials have also been investigated in our laboratory.

EXPERIMENTAL RESULTS

Currently, the most direct and precise technique to measure thin film density is x-ray reflectivity (5), where the critical angle for reflection, θ_c , is determined, typically at a single wavelength. The critical angle is defined as the grazing-incidence angle at which a well-collimated beam of x-rays is no longer totally reflected off the free surface, but starts penetrating into the sample. A critical angle exists when the index of refraction for x-rays is less than unity as given by the expression $n = 1 - \delta - i\beta$, where δ is proportional to the electron density in the sample, and β is the imaginary part of the index describing the absorption of x-rays. From the critical angle, the critical momentum-transfer vector normal to the surface is given by: $q_c = (4\pi/\lambda)\sin(\theta_c)$, where λ is the x-ray wavelength. The square of the critical momentum-transfer vector, $(q_c)^2$, is proportional to δ and, thus, to the electron density.

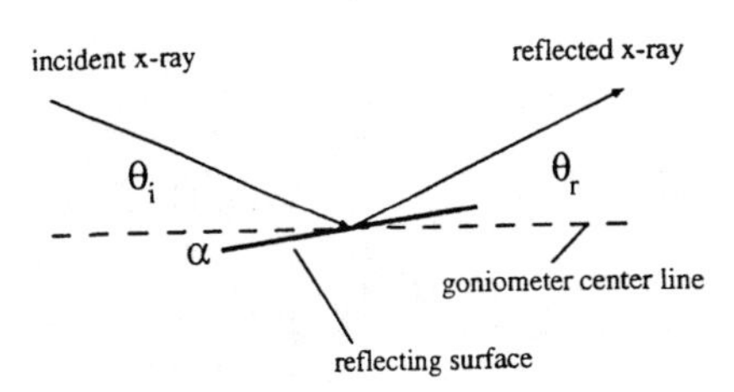

FIGURE 1. Diagram showing the angles of incidence and of reflection, as well as the sample misalignment α. Energy-dispersive x-ray reflectivity corrects for the effects of sample misalignment and yields a more accurate value for thin film density.

CP449, *Characterization and Metrology for ULSI Technology: 1998 International Conference*
edited by D. G. Seiler, A. C. Diebold, W. M. Bullis, T. J. Shaffner, R. McDonald, and E. J. Walters
1998 The American Institute of Physics 1-56396-753-7/98/$15.00

In a typical single-wavelength x-ray reflectivity experiment, instrument alignment and, to a lesser degree, beam collimation, are the limiting factors in determining the critical angle and, therefore, the density. Misalignments of the sample as small as 0.005° in sample tilt, α, shown in Fig. 1, with respect to the incoming x-ray beam can change the final value for the density by 5%. This is because now $q_c = (4\pi/\lambda)\sin(\theta_c+\alpha)$, and, thus, q_c is incorrectly determined. Improved methods of sample alignment could be made, but would be tedious and difficult to implement, especially in a production environment where measurement time and equipment cost are major considerations.

To overcome the sample-alignment problem, x-ray reflectivity can be performed at many different wavelengths simultaneously. This is done by using not just the characteristic intensities of the x-ray tube (e.g. the Kα line of a Cu tube like the one used in these experiments), but also the broad wavelength spectrum known as the bremsstrahlung. An energy-dispersive silicon(lithium) x-ray detector with 200 eV resolution is employed and its output directed through amplifiers to a multichannel analyzer (MCA). Regions-of-interest (ROI) are defined on the MCA at various, distinct wavelengths. Each ROI will result in a different reflectivity curve. Data is collected by incrementally varying the angle of incidence of the x-ray beam to the surface under study and measuring the intensity of the reflected x-rays on the multichannel analyzer in each ROI at each increment. For a full description of the method, its error analysis, and its demonstration of standard materials see reference (6).

Figure 2 shows the output of the MCA for reflection at a given low-incidence angle. The labeled areas give the six ROIs, A through F, used in a typical experiment. Figure 3 shows how these integrated areas change as a function of incidence angle. Each curves defines a critical edge for reflection at a different x-ray energy. The data is plotted as a function of momentum transfer normal to the surface of the sample by applying the equation $q = (4\pi/\lambda)\sin(\theta)$ for each x-ray wavelength at each angle. The reflected intensities are also all normalized by the intensity coming from the x-ray tube which varies as a function of x-ray energy as seen in Fig. 2. The data in Fig. 3 took about 5 minutes to generate.

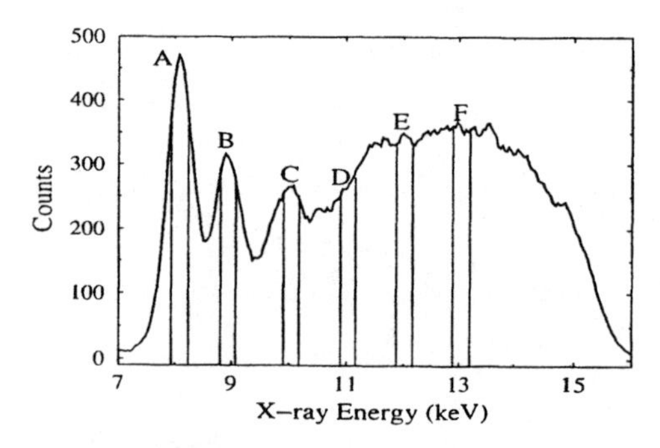

FIGURE 2. Output of the MCA at a low incidence angle. The number of counts in ROIs A through F are followed as a function of angle of reflection. ROI A corresponds to Cu Kα , ROI B to Cu Kβ and ROI C to the tungsten L lines due to evaporation of the filament onto the copper anode.

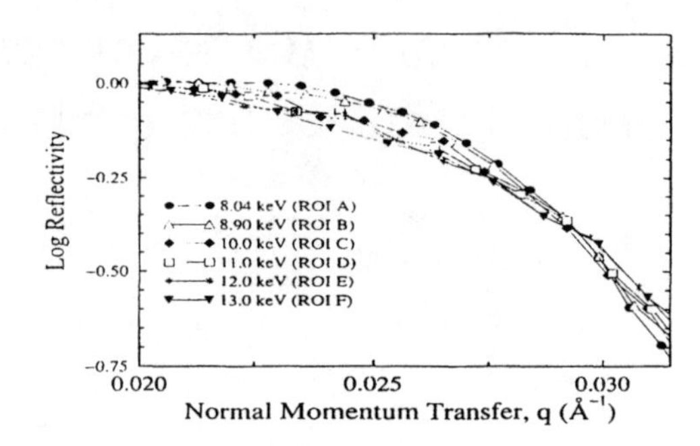

FIGURE 3. Logarithm of x-ray reflection vs. q near the critical edge for reflection. The six ROIs of Fig. 2 yield six reflectivity curves.

For the case of perfect sample alignment, all the reflectivity curves in Fig. 3 should be identical in momentum-transfer space regardless of the wavelength used because they are controlled solely by the electron density of the thin film which is invariant with x-ray energy (in this energy range). However, in practice the curves will be off-set in momentum transfer (i.e., horizontally) due to sample misalignment. In the particular case shown the curves shift to a smaller critical angle with increasing x-ray energy. If the density (and absorption) are calculated for each curve (7), a set of different apparent densities will result. Each particular (single-wavelength) measurement gives an ***incorrect*** value for the density. (The relative standard deviation of each of these data points is on the order of ±5%.) However, a plot of these apparent densities versus the reciprocal of the incident x-ray wavelength, as shown in Fig. 4, can be extrapolated to infinite wavelength at which point the true thin film density is recovered. This is true because at infinite incident wavelength, where θ_c would be also infinite, the errors due to sample misalignment vanish. As the misalignment of the sample varies, the slope of the fitted line will vary; however, extrapolation to infinite wavelength always recovers the true density to about ±1%. In the particular case shown the correct electron density of the thin film was 4.11×10^{23} e^-/cm^3 and the tilt misalignment of the sample was 0.0086 degrees.

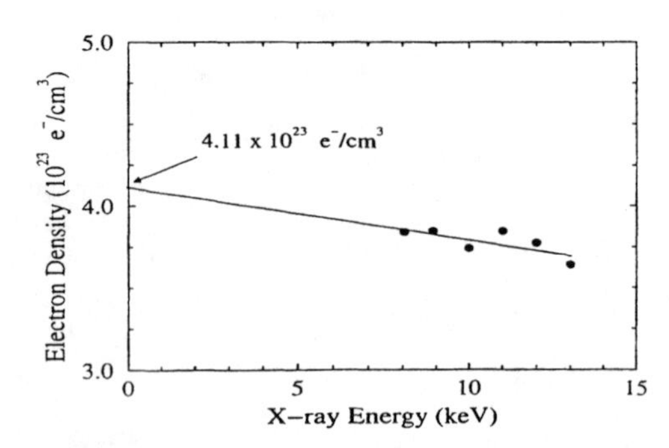

FIGURE 4. Electron density as measured from the critical angle for x-ray reflection vs. x-ray energy. Extrapolation to zero energy recovers the correct electron density for the thin film, and thus corrects for sample tilt. The extrapolation line is a least-squares fit to the data.

For the case shown, Allied-Signal AS418 (4), a methyl silsesquioxane oligomer processed according to manufacture's instructions, combining the measured electron density with the ideal film composition, $SiO_{3/2}CH_3$, yields a mass density of 1.31 g/cm^3. This value lies between that of a typical organic polymer, slightly more than 1.0 g/cm^3, and that of a typical thermal oxide, slightly more than 2.0 g/cm^3.

DISCUSSION

Work to date with energy-dispersive x-ray reflectivity in our laboratory has included studies of polymer thin film density (polystyrene and polyimide), diamond-like carbon films, and a variety of organically-modified silica films. Extensive work has been done on a novel β-choloethylsilsesquioxane (8) provided by Gelest, Inc. (4). This latter material has the advantage of being not only thermally processable but also processable by exposure to ultraviolet radiation. Exposure times as little as 20 minutes were shown to produce films with compositions and densities equivalent to heating 350 °C for four hours (9). Film composition was measured by Rutherford backscattering spectrometry, with residual hydrogen content measured by elastic recoil detection. Ultraviolet processing also opens the possibility of using this material in damascene processing. Experiments on using ultraviolet laser radiation to pattern fine silica lines for optoelectronic applications have also been performed (10).

Future work will center on measuring the density of organic and inorganic foams. Foams with small carefully-controlled pore sizes derived from a variety of methods are currently under investigation by many groups. Gas filled pores have the benefit of greatly reducing thin film dielectric constant. If mechanical integrity can be preserved, rigid foams may be the material of choice for next generation ULSI circuitry. At this time accurate methods to measure thin film foam density are not available. Energy-dispersive x-ray reflectivity it is believed offers such a method.

ACKNOWLEDGEMENTS

The authors would like to thank Dr. David Feiler and Dr. Wei Xia of Rockwell Semiconductors Systems for providing the AS418 sample.

REFERENCES

1. Lu, T.M., Murarka, S.P., Kuan, T.S., and Ting, C.H., *Low-Dielectric Constant Materials—Synthesis and Applications in Microelectronics*, Pittsburgh: Materials Research Society, 1995. Proceedings Volume 381.
2. Takahasi, H., Kataoka, H., and Nagata, H., *J. Mater. Res.* **12**, 1722-1726, (1997).
3. Schwan, J., Ulrich, S., Theel, T., Roth, H., Ehrhardt, H, Becker, P., Silva, S.R.P., *J. Appl. Phys.* **82**, 6024-6030 (1997).
4. Certain commercial materials are identified in this paper in order to specify adequately the experimental procedure. In no case does such identification imply recommendation or endorsement by the National Institute of Standards and Technology.
5. Chason, E. and Mayer, T.M., *Critical Reviews in Solid State and Materials Sciences*, **22**(1), 1-67, (1997).
6. Wallace, W.E. and Wu, W.L., *Appl. Phys. Lett.* **67**(9), 1203-1205 (1995).
7. Ankner, J.F. and Majkrzak, C.J., *Proc. SPIE* 1738, 260-269 (1992).
8. Arkles, B., Berry, D.H., Figge, L.K., Composto, R.J., Chiou, T., Colazzo, H., and Wallace, W.E., *J. Sol-Gel. Sci. Tech.* **8**, 465-469 (1997).
9. Composto, R.J., Arkles, B., Berry, D.H., Figge, L.K., Gonzales, G.B., and Wallace, W.E., (in preparation).
10. Sharma, J. and Dai, H.L., (unpublished).

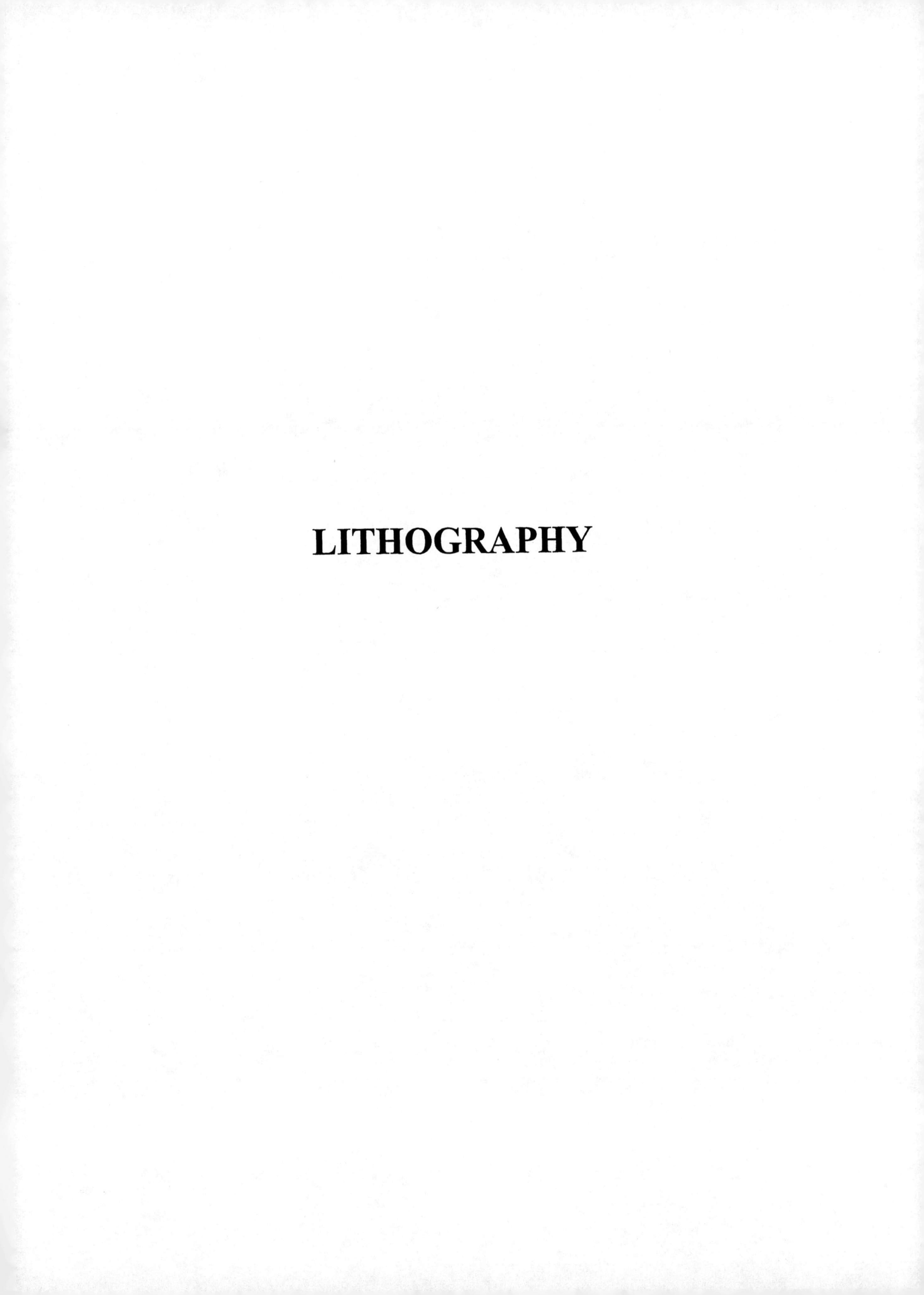

LITHOGRAPHY

Next Generation Lithography - The Real Challenge

Karen H. Brown

International SEMATECH, 2706 Montopolis Drive, Austin, TX 78741

Abstract

The semiconductor industry continues to move forward aggressively to make more productive wafers and lower cost per function in its products. The National Technology Roadmap for Semiconductors (NTRS) follows Moore's Law of four times the number of bits per semiconductor memory chip every three years. Additionally, this productivity will be based to a large degree on the continued ability to decrease the minimum feature size year by year. In the past three years the industry has actually accelerated the introduction of small feature sizes.

193nm lithography is expected to be introduced in manufacturing by 2001, for dimensions of 0.15μm, extending down to 0.13μm technology. Currently, SEMATECH is working to define and implement a plan for the next generation lithography (NGL). X-ray, SCALPEL (scanning e-beam projection), ion projection lithography (IPL and extreme ultraviolet (EUV), and e-beam direct write are all potential solutions.

It is expected that the next generation lithography will be introduced for the 100nm generation and be extended to feature sizes below 50nm. These technologies all offer a key advantage in much larger depth of focus than 193nm, and smaller imaging capability. However, the lack of a transparent pellicle material combined with the large depth of focus will mean that current techniques for insuring against repeating defects will be ineffective. Tools, reticle storage, handling and transport will all have to be designed to be ultraclean, with zero defects per pass. At 0.25μm pellicles are required to obtain adequate yield in manufacturing. There are many challenges ahead. It may be that cost and chip yield are the real limiters to the next generation lithography.

The semiconductor industry continues to move forward aggressively to make more productive wafers and lower cost per function in its products. The National Technology Roadmap for Semiconductors (NTRS) follows Moore's Law of four times the number of bits per semiconductor memory chip every three years. Additionally, this continuing productivity improvement will be based to a large degree on the continued ability to decrease the minimum feature size year by year. In the past three years the industry has actually accelerated the introduction of smaller feature sizes.

Today, optical lithography continues to be the mainstream technology for the industry and is being used in production by leading–edge high–volume factories at 0.25μm design rules at the deep ultraviolet (DUV) wavelength of 248nm. Exposure tool enhancements such as off-axis illumination (OAI) in the optics design and optical proximity correction (OPC) of the mask for optics distortions in the exposure tool and process effects in transfer of the aerial image into the resist result in better image fidelity to the design parameters. These optical extensions are being used to support 0.18μm product and process development. 193nm lithography is expected to be introduced into manufacturing by 2001, for dimensions of 0.15μm, and is expected to extend at least to 0.10μm. The industry momentum is behind optical extensions to 0.10μm and perhaps even to 70nm. The key challenge will be maintaining an adequate and affordable process latitude (depth of focus/exposure window) necessary for the 10% post–etch CD control required to meet the device/chip characteristics.

The key elements of the lithography infrastructure include

- Exposure equipment
- Resist materials and processing equipment
- Mask making, mask equipment and materials
- Metrology equipment for critical dimension (CD) and overlay

The NTRS defines the requirements for each of these elements for each of the technology nodes. Figure 1 and Table 1 show the critical elements for the exposure tool and for the mask /reticle. In optical technology today the area that presents the most difficulty is the mask infrastructure. With the move from 1x exposure tools to 5x steppers in the late 1980's it became possible to continue to make the latest generation masks without significant investment or development of new equipment and processes. Equipment with the capability of making 1.5μm images at 1x was readily used with modest improvements even for images of 0.30μm at wafer, or 1.5μm on the mask. At the same time many semiconductor manufacturers, in an effort to cut costs, sold their internal mask shops to commercial suppliers. As mask production costs rose and semiconductor manufacturers focused on cost issues, there was a major effort to reduce the number of masks used. The

CP449, *Characterization and Metrology for ULSI Technology: 1998 International Conference*
edited by D. G. Seiler, A. C. Diebold, W. M. Bullis, T. J. Shaffner, R. McDonald, and E. J. Walters

consolidation of internal and external mask manufacturing combined with the focus on mask usage resulted in an over-capacity for mask production. In the past two years this over-capacity has mostly been eliminated. In addition, as feature sizes have moved below 0.30μm the technology has required significant investment in both research and development. New mask writers, inspection and repair equipment are all in development. The capability to produce the leading edge masks lags that which is needed by the industry. As the roadmap moves to ever-smaller feature sizes the cost of the equipment, production times, and yield have all lead to escalating mask costs. As will be seen later, costs both of masks and wafer costs are one of the real challenges of the next century.

As critical dimensions are expected to decrease below 0.10μm, it is expected that the industry will move for the first time to a non-optical lithography technique. It is expected that the "next generation lithography" will be introduced for the 100nm generation and be extended to feature sizes below 50nm. SEMATECH is working to define and implement a plan for consensus on and development of the next generation lithography (NGL). X-ray, SCALPEL (scanning e-beam projection), ion projection lithography (IPL), extreme ultraviolet (EUV) and e-beam direct write are all potential solutions. These technologies all offer a key advantage in much larger depth of focus than 193nm, and smaller imaging capability. However, the lack of a transparent pellicle material combined with the large depth of focus will mean that current techniques for insuring against repeating defects will be ineffective. Even today with leading edge manufacturing at 0.25μm, pellicles are required to obtain adequate yield in manufacturing. Tools, reticle storage, handling and transport will all have to be designed to be ultraclean, with zero defects per pass. At dimensions less than 100nm, the defect size that can impact yields is less than 30nm. Even with a precision/tolerance ratio of 20% this means a measurement capability of less than 6nm. The NTRS indicates that measurement capability for these dimensions and full wafer characterization for defects at these dimensions will be required between 2003 and 2006. Today there are no known solutions.

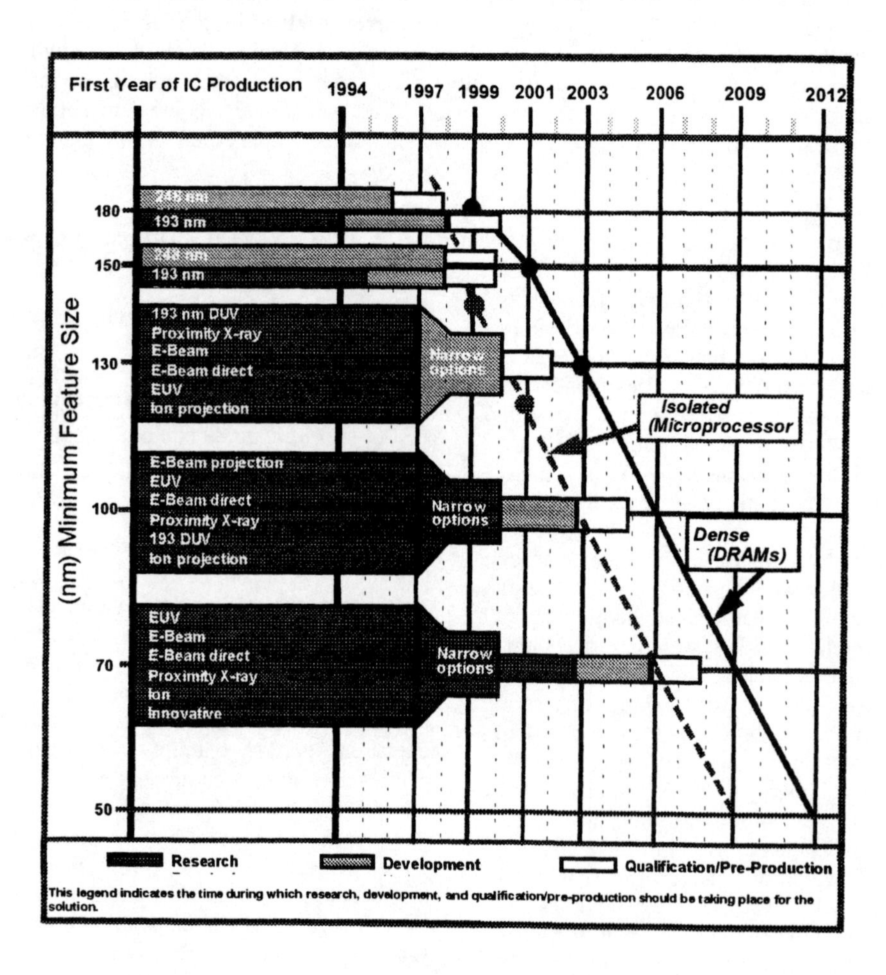

Figure 1. Critical Level Exposure Technology Potential Solutions

Year of First Shipment Technology Generation (nm)	1997 250	1999 180	2001 150	2003 130		2006 100	
Wafer Minimum Feature Size (nm)	200	140	120	100		70	
Magnification	4X	4X	4X	4X	1X	4X	1X
CD Uniformity (nm)	26	18	16	13	8	9	6
CD Mean to Target (nm)	20	12	10	8	4	6	3
Image Placement (Multi-point, nm)	52	36	32	28	14	20	10
Defect Size (nm)	200	150	125	100	26	80	20

Table 1. Mask Requirements

The Semiconductor Roadmap is a blueprint for technology development required to maintain the productivity of the semiconductor industry. It does this by targeting research for leading edge technology. It has become ever clearer as the industry moves to smaller feature sizes and new materials such as copper metallization that there is significant cost associated with these difficult technological changes. The cost of development, the cost of implementation become barriers to success. Mask cost is increasing dramatically as the cost of the mask writers, mask inspection and repair tools increase. Mask write times have increased dramatically while mask yields have decreased. Figure 2 shows the relationship between mask cost, mask cost per wafer and affordable lithography cost per wafer level exposure. There are many challenges ahead. It may be that cost and chip yield are the real limiters to the next generation lithography.

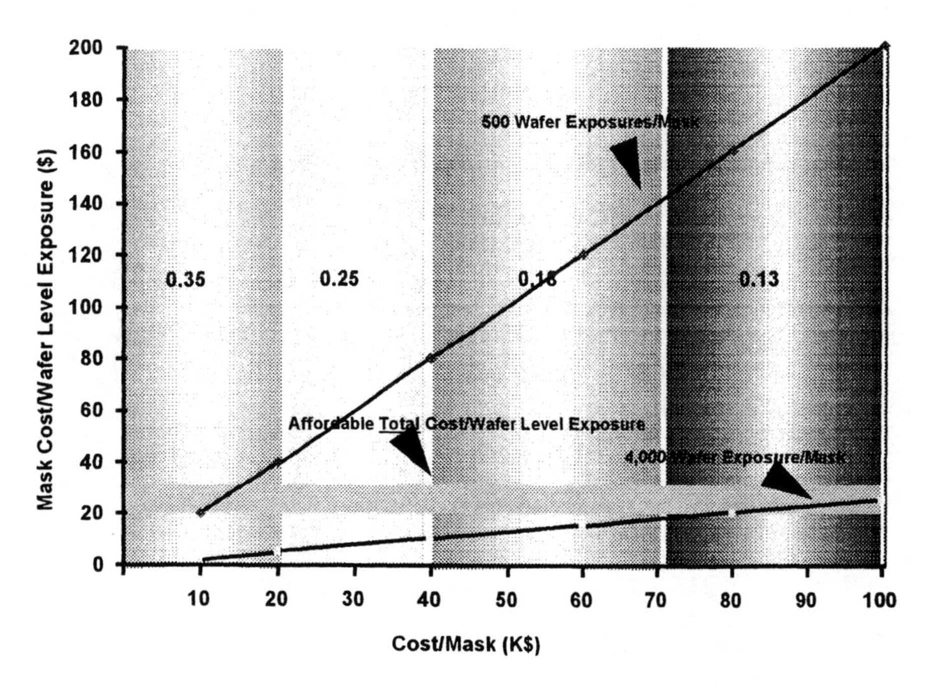

Figure 2. Mask Costs

A Survey of Advanced Excimer Optical Imaging and Lithography

Koichi Matsumoto and Kyoichi Suwa

2nd Optical Designing Department, Nikon
1-6-3, Nishi-Ohi, Shinagawa-ku, Tokyo 140-8601, JAPAN

The first item discussed in this paper is to estimate the future trend regarding minimum geometry and the optical parameters, such as NA and wavelength. Simulations based on aerial images are performed for the estimation. The resolution limit is defined as a minimum feature size which retains practical depth of focus (DOF). Pattern geometry is classified into two categories, which are dense lines and isolated lines. Available wavelengths are assumed to be KrF excimer laser (λ =248nm), ArF excimer laser (λ =193nm) and F_2 excimer laser (λ =157nm). Based upon the simulation results, the resolution limit is estimated for each geometry and each wavelength.

The second item is to survey ArF optics. At present, the ArF excimer laser is regarded as one of the most promising candidates as a next-generation light source. Discussions are ranging over some critical issues. The lifetime of ArF optics supposedly limited by the radiation compaction of silica glass is estimated in comparison with KrF optics. Availability of calcium fluoride (CaF_2) is also discussed. As a designing issue, a comparative study is made about the optical configuration, dioptric or catadioptric. In the end, our resist-based performance is shown.

SIMULATION ANALYSIS

Approach to the Estimation

The simulation is performed to estimate the limit of optical lithography. Patterns are assumed to be dense lines or an isolated line, where the isolated line is optically isolated dark line in the bright background, because positive photoresist is assumed. In the analysis, the "ED-tree" method (1) is applied. In order to estimate minimum feature size, it is defined that a pattern is resolvable when DOF is more than 400nm, anticipating technology development, such as chemical mechanical polishing (CMP) (2). Details of the simulation conditions are given in Table-1.

Items	Conditions
Allowance of linewidth	±10%
Allowance of dose	±3%
Mask manufacturing error	±3%
Wavelength	248nm, 193nm, 157nm
NA of projection optics	0.50~0.88

TABLE 1. ED-tree simulation conditions

Analysis of Dense Lines

As a variety of masks, firstly, attenuated Phase Shift Mask (PSM) (3) is considered and the transmission rate at Cr-covered part is assumed to be 8%. In the simulation, following illumination systems are considered:

[1] Conventional illumination: the coherence factor, σ , of which ranges from 0.45 to 0.90.

[2] 2/3 Annular illumination: 2/3 is the ratio of radius comprising the doughnut shape of the source.

[3] SHRINC: which stands for "Super High Resolution IllumiNation Control" and is one of the quadruple illuminations (4),(5).

The results are that the modified illuminations including 2/3 Annular and SHRINC are superior to the conventional one and that no significant differences are there between 2/3 Annular and SHRINC. Therefore, precise simulations are performed only for 2/3 Annular illumination. Figure-1, Figure-2 and Figure-3 are the simulation results. They show the achievable maximum DOF vs. NA of projection optics for KrF, ArF and F_2, respectively. As is observed in Figure-1, minimum feature size which still has more than 400nm DOF is 120nm for KrF. In the same manner, we see the minimum feature size is 100nm for ArF and 90nm for F_2.

As another choice of masks, alternative PSM, or so-called Shibuya-Levenson type PSM (6),(7), is considered. Since it is obvious that small σ is suitable for alternative PSM, σ is fixed to 0.3 which seems lower limit to retain practical illumination power. Though the details are not shown here, it is safe to say generally that the minimum feature sizes will be smaller than those of attenuated PSM by 30nm. The results are summarized in Table-2.

CP449, *Characterization and Metrology for ULSI Technology: 1998 International Conference*
edited by D. G. Seiler, A. C. Diebold, W. M. Bullis, T. J. Shaffner, R. McDonald, and E. J. Walters

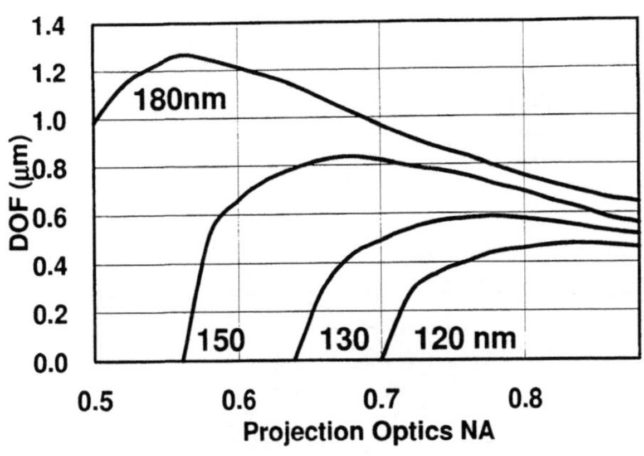

FIGURE 1. Simulated DOF vs. NA (λ =248nm)

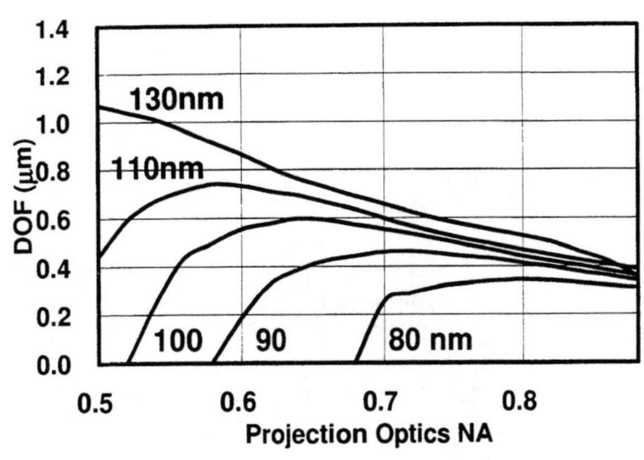

FIGURE 3. Simulated DOF vs. NA (λ =157nm)

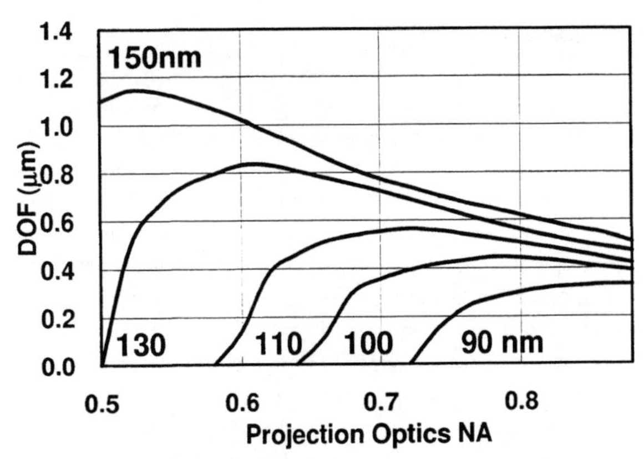

FIGURE 2. Simulated DOF vs. NA (λ =193nm)

Mask + Illumination	KrF (λ =248nm)	ArF (λ =193nm)	F_2 (λ =157nm)
attenuated PSM + 2/3Annuar illumination	120nm (NA=0.76)	100nm (NA=0.76)	90nm (NA=0.66)
alternative PSM + 0.3 σ illumination	90nm (NA=0.78)	70nm (NA=0.78)	60nm (NA=0.78)

TABLE 2. Achievable feature size for dense lines

Mask + Illumination	KrF (λ =248nm)	ArF (λ =193nm)	F_2 (λ =157nm)
attenuated PSM + SHRINC	130nm	110nm	100nm

TABLE 3. Achievable feature size for an isolated line

Analysis of an Isolated Line

Since alternative PSM is obviously not applicable to isolated patterns, only an attenuated PSM is considered as a choice of masks. A rough survey among the various illumination systems indicates that SHRINC has a little more DOF than the conventional and 2/3 Annular, though the difference is less significant than that of dense patterns. The simulation results are summarized in Table-3. Though not indicated in the table, minimum NA's required are lower than those of dense lines.

The feature sizes indicated in Table-2 and Table-3 should be development targets for optical lithography as well as the metrology field.

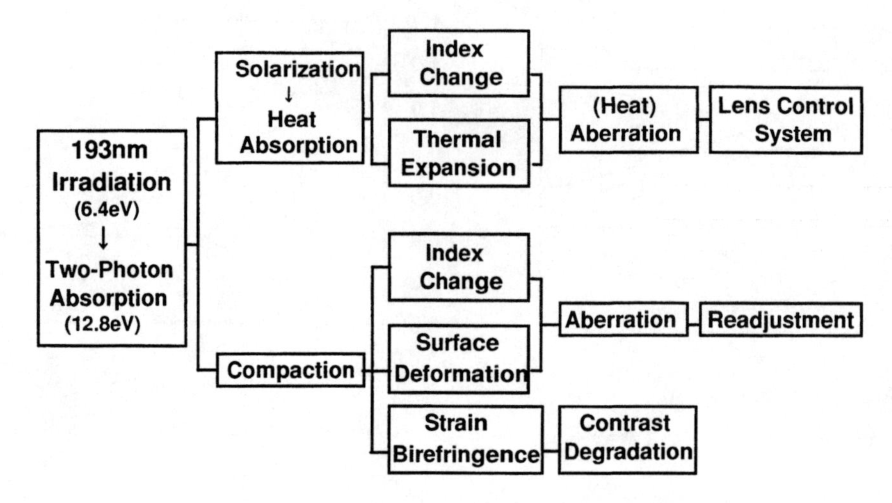

FIGURE 4. Schematic diagram of the damage of silica glass

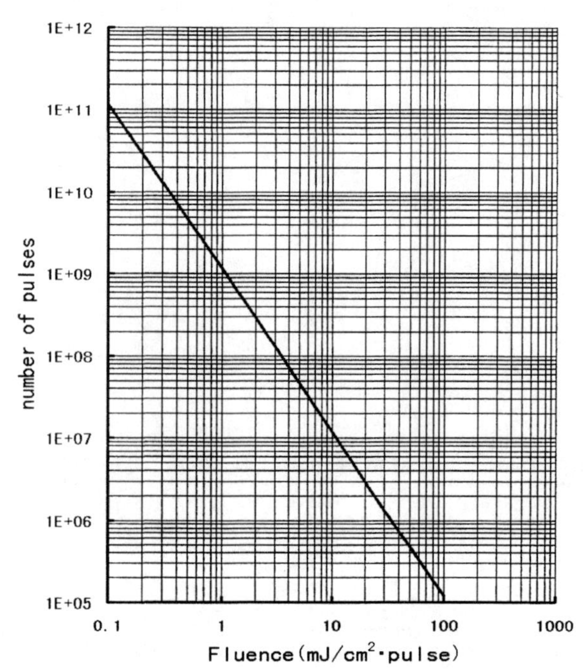

FIGURE 5. Lifetime of silica glasses

SURVEY OF ArF OPTICS

Material Study of Silica Glass (SiO_2)

In order to estimate the lifetime of ArF exposure equipment, optical materials are studied. Available materials at 193nm are almost limited to silica glass and calcium fluoride. The first discussion is related to silica glass. Figure-4 schematically shows the damage of silica glass caused by the irradiation of DUV light and the methods to compensate them. Two major phenomena relating to the damage is "solarization" and "compaction". The transmission rate of silica glass is decreased by the solarization. The compaction makes the material compacted locally and induces the increase of refractive index. It is known that these two phenomena are caused by a two-photon absorption process. Degradation of the transmission rate leads to increase of absorption, and the thermal aberration induced by the absorption may deteriorate the imaging performance. Though exposure tools generally have a mechanism and control system to compensate for the deterioration, it is the end of lifetime when the deterioration goes out of the compensation range. On the other hand, the compaction results in an increase of refractive index and it requires adjustment of optics at regular intervals. The optics will survive as long as the adjustment remains within an allowable domain.

Among currently available data of silica glass, the most excellent one in terms of lifetime is shown in Figure-5. It is assumed that the end of lifetime is represented by the number of pulses, with which internal absorption increases by 0.1%/cm. The figure shows the relation between the lifetime and the fluence or the power density per unit pulse. Accumulated pulse number will reach to as many as 10^{11}pulses when the tool is fully used for 10 years. Even so, the figure indicates that the silica glass have lifetime long enough to survive for 10 years, provided the fluence is retained under 0.1mJ/cm^2/pulse. Therefore, emphasis should be put on the compaction issue to study the lifetime.

Lifetime of ArF Optics

It will be mentioned later that calcium fluoride is almost free from the deterioration due to irradiation of DUV light. So, silica glass is a main item for the discussion. The solarization, as is mentioned, is not a critical issue in the discussion of lifetime, while the radiation compaction which is prominent at the wavelength of ArF should be an issue to be considered in detail. The compaction induces increase of the refractive index, which is empirically represented such that:

Laser (Name)	Laser (Spectral Bandwidth)	Optical Configuration
Super Narrow Band	FWHM＜0.3pm	Dioptric (Single material : SiO_2)
Narrow Band	FWHM＜1～3pm	Dioptric (SiO_2 and CaF_2)
Quasi-Narrow Band	FWHM＜5～100pm	Catadioptric

TABLE 4. Bandwidth of laser and optical configurations

$$\Delta n / n = \kappa \left(N \cdot I^2 \right)^\alpha \quad (1)$$

where n is the refractive index, Δ indicates its increase, N is the number of pulses, I is the fluence or the power density per unit pulse, κ is an empirical constant dependent on wavelength and grade of the material, and α is a universal constant. At present, κ of ArF is estimated 10 times larger than that of KrF. The experimental data are shown in Figure-6. The figure shows the relation between Δ n/n and pulse numbers. That is, when the wavelength is indicated by suffix, A for ArF and K for KrF:

$$\kappa_K / \kappa_A \approx 0.1 \quad (2)$$

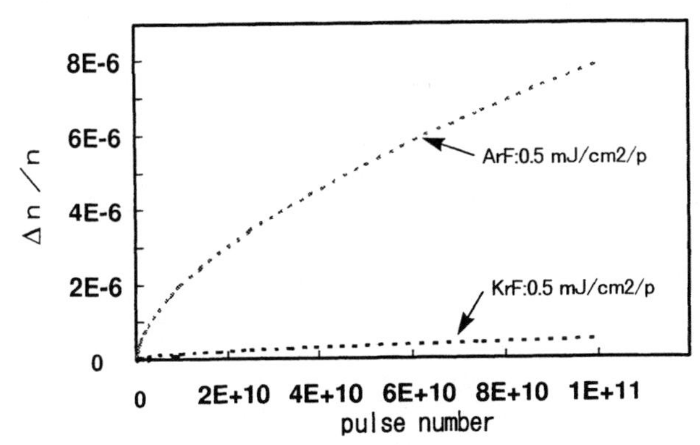

FIGURE 6. Radiation compaction of silica glasses

Hereafter, the lifetime of silica glass at the ArF wavelength is discussed in comparison with that of KrF. Assume "a shot" is exposed by the same number of pulses for both ArF and KrF. And let the number of pulses be N_{shot}, then the following equation holds:

$$N_{shot} = S_A / I_A = S_K / I_K \quad (3)$$

where S is the resist sensitivity. Here, the lifetime is defined as total number of pulses and they are denoted as N_A and N_K for ArF and KrF, respectively. At the end of lifetime, the material supposedly suffers from same index change, and then the situation can be expressed by the following equation:

$$\begin{aligned} \Delta n / n &= \kappa_A \left(N_A \cdot I_A{}^2 \right)^\alpha \\ &= \kappa_K \left(N_K \cdot I_K{}^2 \right)^\alpha . \end{aligned} \quad (4)$$

Substituting Equation-3 for Equation-4, the equation to estimate the lifetime is obtained as follows:

$$N_A / N_K = \left(\kappa_K / \kappa_A \right)^{1/\alpha} \cdot \left(S_K / S_A \right)^2 \quad (5)$$

The sensitivity of KrF resist, S_K, is typically 30mJ/cm^2. Though that of ArF resist, S_A, is not clear at present, it is widely accepted that its development target is around 5mJ/cm^2. The constant, α, is measured by many researchers. Among them, the smallest one is 0.53 proposed by Corning group (8) and the largest one is 0.7 suggested by U. C. Berkley group (9). Substituting these values into Equation-5, the lifetime of ArF optics is estimated 0.5×～1.3× than that of KrF. Although the estimation may not be accurate enough, it certainly gives us a sort of relief.

Material Study of Calcium Fluoride (CaF_2)

Calcium fluoride is worthy of notice for its laser durability. Synthesized calcium fluoride with high purity has substantially no absorption at 193nm. Moreover, the solarization is extremely small and the compaction as in silica glass is not observed. Therefore calcium fluoride has been used in the parts, such as a laser window, which are exposed to high energy density. Even so, calcium fluoride has been regarded as a material for illumination optics with less critical precision is required than projection optics. The reasons are these:

[1] It is difficult to achieve high uniformity and birefringence-free material.
[2] Since the calcium fluoride should be a single crystal, it is not easy to obtain large-diameter material.
[3] The polishing should be different than the conventional one because of its crystal structure.

Among the researchers, calcium fluoride has been one of the main topics these years. In a couple of years, the material will certainly be used in projection optics as well. What is more, the calcium fluoride is probably a main refractive material for F_2 excimer laser, because silica glass does not have enough transmission rate at this wavelength.

Optical Configuration

The optical configuration for ArF lithography should be considered with combination of spectral bandwidth of the laser. Candidates for ArF optics at present are summarized in Table-4.

The dioptric optics are highly indebted to the development of bandwidth narrowing technology and calcium fluoride manufacturing technology. The dioptric optics are, however, natural extension of conventional technology for most stepper manufacturers and less technical barriers are expected. Current dioptric projection optics are generally equipped with control system to stabilize its performance against the turbulence of temperature, air-pressure, irradiance and so forth. Because of the rotational symmetry of the optics, its controllability is considered to be accurate enough.

The catadioptric KrF exposure equipment is currently available on a commercial basis from a US-based company (10). In designing catadioptric optics, the key element must be a concave mirror. The mirror turns divergent wave into convergent wave. In this respect, the concave mirror functionally corresponds to positive power lens. Yet, some differences are there:

[1] The mirror is free from chromatic aberration, while the lens has the aberration.

[2] The mirror makes a negative contribution to Petzval sum, while the lens makes a positive one.

So the catadioptric optics, if its main power is allocated at the mirror, will have small chromatic aberration and can have wider acceptance towards the laser in term of spectral bandwidth. It is well known that positive refractive lens elements make positive contribution to Petzval sum. In order to reduce Petzval sum, or to achieve a flat image plane, various lens types have been historically developed. These include "Triplet type," "Gauss type" and so forth. Most of the current dioptric optics are supposedly evolved from "Gauss type." On the contrary, in the case of catadioptric optics, it is quite easy to reduce Petzval sum, because the mirror has negative contribution to Petzval sum, and this makes it possible to design the optics compactly. In the catadioptric system, a polarized beam splitter (PBS) and quarter wave plates (QWP) are equipped in order to physically separate the incident and reflected light on the concave mirror. So the imperfectness of the PBS and the QWP provides flares on the image plane and this immediately leads to the deterioration of image quality. Moreover, the silica glass for the PBS should be homogeneous as viewed from two orthogonal directions, because the PBS has two optical axes is the actual configuration. These are technically challenging items for catadioptric optics.

Actual Performance of ArF Optics

As has been mentioned, there are the pros and cons in the choice between dioptric and catadioptric. Should an emphasis be placed on lead time, the dioptric type is advantageous for us. We have been completing tentative projection optics, the specifications of which are 0.60NA, 25mm × 8mm field. The laser is a narrow band type and the optics is achromatized. Figure-7 shows its resolution capability. The resist is chemically amplified type with 0.41 micron thickness. Conventional binary mask is used and 2/3 Annular illumination is applied. Though the adjustment has not yet been completed, as is observed, the optics has capability enough to resolve down to around 130nm geometry and this meets the expectation of its wavelength and NA.

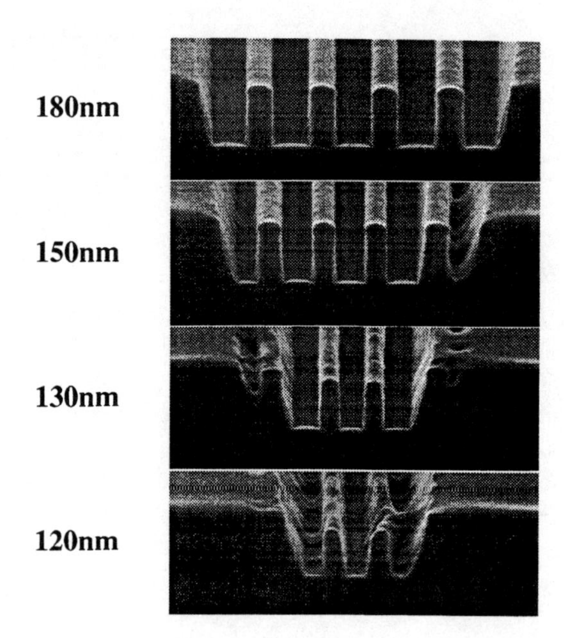

FIGURE 7. Cross-section SEM photos (ArF exposure)

SUMMARY

Firstly, the future trend of optical lithography is estimated. The resolution limit varies among a variety of masks and illumination method. In case of dense lines, for example, we obtain 120nm for KrF, 100nm for ArF and 90nm for F_2 excimer lasers, using attenuated PSM and 2/3 Annular illumination. These values gives us targets for the technological development. Secondly, ArF optics is surveyed, ranging over the lifetime of silica glass, availability of calcium fluoride, and comparative study between dioptric and catadioptric. The survey makes us confident in the realization of ArF optics.

ACKNOWLEDGMENTS

The authors would like to thank Mr. S. Hirukawa of Nikon for his simulation support and several engineers for their work in collecting ArF experimental data. Their thanks should be extended to Mr. T. Mori for his productive discussions.

REFERENCES

1. B.J.Lin : A Comparison of Projection and Proximity Printings -- From uv to x-ray, *Proc. of SPIE*, **1263**, pp.80 (1990)
2. M.E.Thomas et al. : The Mechanical Planarization of Interlevel Dielectrics for Multilevel Interconnect Applications, *Proc. of 1990 VMIC Conference*, pp.438 (1990)
3. T.Terasawa et al. : Imaging Characteristic of Multi-Phase-Shifting and Halftone Phase-Shifting Mask, *J. Jpn. Apl. Phys.*, **30**, pp.2991 (1991)
4. N.Shiraishi et al. : New Imaging Technique for 64M-DRAM, *Proc. of SPIE*, **1674**, pp.741 (1992)

5. M.Noguchi et al. : Subhalf Micron Lithography System with Phase-Shifting Effect, *Proc. of SPIE*, **1674**, pp.92 (1992)
6. M.Shibuya : *Japan Patent* 62-50811, No.1441789 (in Japanese)
7. M.D.Levenson et al. : Improving Resolution in Photolithography with a Phase-Shifting Mask, *IEEE Trans. Electron Dev.*, **ED-29**, pp.1828 (1982)
8. C.Smith et al. : Compaction of Fused Silica Under Low Fluence/Long Term 193nm Irradiation, *Proc. of SPIE*, **3051**, pp.116 (1997)
9. R.Schenker et al. : Durability of Experimental Fused Silicas to 193-nm-Induced Compaction, *Proc. of SPIE*, **3051**, pp.44 (1997)
10. D.Cote et al. : Micrascan^{TM}III-performance of a third generation, catadioptric step and scan lithographic tool, *Proc. of SPIE*, **3051**, pp.806 (1997)

An Overview of CD Metrology for Advanced CMOS Process Development

Herschel Marchman
Texas Instruments, Kilby Center, Dallas, TX

The scanning electron microscope (SEM) is currently poised as the instrument of choice for ULSI dimensional metrology because of its high resolution, throughput, and automation. Despite high resolution, several issues do exist which currently limit the metrological performance of the SEM. Image quality and measurement resolution is affected by such factors as the electron beam source design, lens properties, beam to sample interactions, and the detection system. Recent advances in electron optical lens and column design have provided improved image resolution and measurement precision. Improved automation has impacted throughput as well as measurement repeatability by establishing more consistent feature placement and better control of measurement conditions. A comparative study of the imaging and measurement performance of different SEM technologies and the effects of sample properties on precision as well as correlation of data with cross-sectional techniques will be presented.

INTRODUCTION

A normal, gaussian, distribution is typically obtained when repeated measurements having random error components are taken with all possible factors held constant. The mean ($\overline{X}$) is obtained when the average of these measurement values is taken such that

$$\overline{X} = \sum_{n=1}^{N} \frac{x_n}{N},$$

where x_n is the n th measurement and N is the total number of measurements. Repeatability (σ_{rpt}) is the short term variation of repeated measurements on the same feature under identical conditions. These are also referred to as ***static*** measurements. The variation in measurements should be random with normal distribution. An estimate of repeatability is found from standard deviation,

$$\sigma = \frac{\left[\sum_{n=1}^{N} (x_n - \overline{x})^2\right]^{1/2}}{N-1}.$$

This level of variation is considered to represent best case of instrument performance. Reproducibility (σ_{rpd}) is the fluctuation in the static mean values when the measurements are made under different conditions. The factors that are allowed to vary when determining reproducibility are those that affect the ability of an instrument to reproduce its own results after the sample has been completely removed from the instrument and reloaded. Reproducibility allows one to determine the effect of the systematic error components (i.e. those factors chosen to vary). Precision (σ) is the total variation in the measurements. This is obtained by,

$$\sigma = \sqrt{\sigma_{rpt}^2 + \sigma_{rpd}^2}\ .$$

Precision values are usually quoted in terms of an integral multiple of the standard deviations, such as 3σ or 6σ. These expressions are valid if the number of measurements is sufficiently large and the distribution is Gaussian, or normal. Normal distributions typically result when the errors are of a random nature, which can therefore be reduced by averaging. Systematic errors, on the other hand, skew the data and are more difficult to eliminate. In addition, obtaining a large number of measurements per site may be difficult in a production environment. Therefore, particular attention must be paid to experimental design and instrumentation so that valid results are obtained with a nonprohibitive number of measurements.

Accuracy refers to how closely the measurement conforms to an absolute standard, or truth, assuming there is a fundamental basis of comparison. The mean value **W** of a measurement set with precision σ has an offset **O** relative to an accepted standard. Calibration of an instrument involves subtracting the offset **O** between **W** and the reference value. In this case, calibration does not yield absolute accuracy because the reference standard has its own uncertainty. Even more importantly, imaging characteristics for most types of microscopes are not constant. Slight changes in feature wall angle, edge and surface roughness, or material properties can dramatically affect the precision and accuracy offset of the metrology instrument .[1,2,3] To achieve true accuracy, one would have to calibrate for all sample permutations. Of course it would be impossible to fabricate a reference standard for every type of sample, as this would require an *apriori* knowledge of the feature characteristics as well as an infinite number of reference artifacts. The minimization of this relative offset **M** between measurement tools is referred to as matching. It should be noted that one could use a member in the set of tools, but the question naturally arises of which one is most correct. Alternatively, the statistical mean of all the instruments could be used as a virtual matching

CP449, *Characterization and Metrology for ULSI Technology: 1998 International Conference*
edited by D. G. Seiler, A. C. Diebold, W. M. Bullis, T. J. Shaffner, R. McDonald, and E. J. Walters

source if the set is composed of a large number of very similar systems. The concept of defining a matching reference can become more obscured when the quantity to be measured is not well defined. The ideal structures for linewidth measurement would consist of flat features, or structures with uniform height and vertical sidewalls, with smooth edges. In reality, the line edges are usually not well defined. Structures typically encountered in IC technology may have edges that are and walls that are asymmetric or even re-entrant . A more appropriate term is line profile, which describes the surface height along the **X** direction at a **Y** location. Alternatively, the profile could be given by the width **W** at each height position **Z** of the line (line being a three-dimensional entity in this case). Edge roughness is presently a major factor in the uncertainty of linewidth standards. Sidewall profile also complicates the issue because linewidth must be defined at a particular height.

SEM METROLOGY

Initially, an electron probe is formed by thermionic or field emission gun sources. The smallest cross-section of the electron source, crossover, is focused by several electromagnetic lenses to form the primary beam of electrons (**Fig. 1a**). An interaction volume is formed where electrons from the primary beam enter the sample. These electrons are scattered by the atoms in the sample volume of interaction and, upon re-emission, are detected. This detected intensity level is then translated into an image pixel. The recorded signals from the electron detector are displayed on a cathode ray tube (CRT) rastered in synchronism with the primary beam to form, essentially, real-time images. Each gray level of the image corresponds to an detector intensity level. A single line cut of intensity versus horizontal scan position at a fixed vertical coordinate is super-imposed on the gray-scale image shown in **Fig. 1b**. The electron yield from the interaction volume is primarily a function of the change in surface height (topography) or atomic number (material transitions).[4] Peaks in the detected electron intensity typically occur at the feature edges. The dependence of electron yield on surface height rate of change is known as the *Secant Effect*.[5] Critical dimensional (CD) measurements are obtained from these intensity profiles. Unfortunately, SEM image intensity does not correspond directly to the surface height. This fact presents the main problem in CD-SEM metrology - where in the intensity transition is the feature edge actually located ? Studies have shown that edge position does not correspond to the peak value of intensity.[6] This presents ambiguity in the location of the feature edges relative to the SEM image

PRIMARY
BSE
SE
BSE
SE

(a)

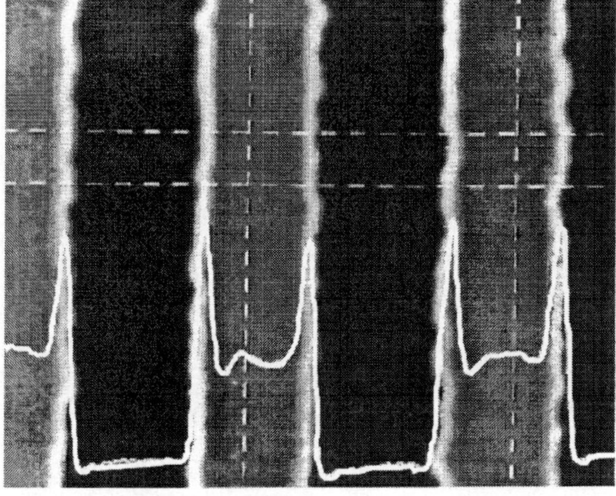

(b)

Figure 1 - (a) Cross-sectional view of primary beam and sample surface interaction and (b) SEM image created by rastering the e-beam and recording emitted electrons.

intensity profile. Techniques for determining the actual edge position from an intensity profile are still somewhat arbitrary. One generally attempts to approximate the intensity transition with some function (**Fig.2**), such as a tangent line (linear approximation), nonlinear step (Fermi) function, or a threshold in the first derivative of the detected intensity signal. The line width, or CD, measurement can vary by as much as 100 nm depending on what edge detection algorithm is used.[7] The data presented in this article will shown that differences in the edge detection algorithms are one of the primary factors in the offset between tools. This fact is finally being recognized throughout the semiconductor industry and efforts to standardize are currently being enacted.[8] In addition to edge detection algorithms, several other factors

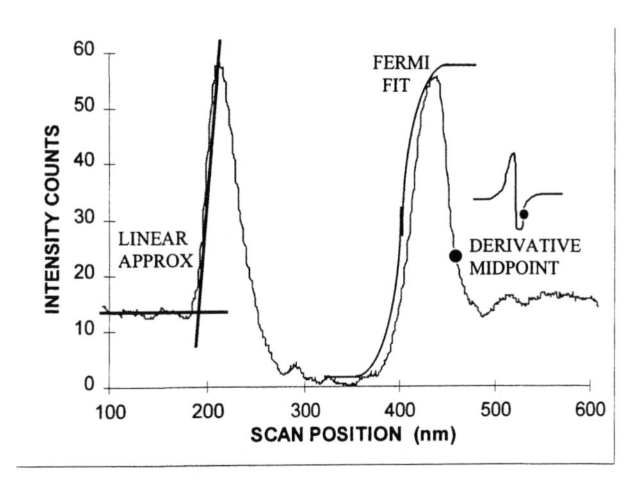

Figure 2 - An SEM image intensity line scan with associated edge detection algorithms.

exist which can cause a variation in the matching between CD-SEMs. These factors can be divided into those attributed to the either sample type or the metrology instrument itself.

Once the resolution, linearity, and precision are shown to be adequate for the particular feature size of interest, the most appropriate algorithm and threshold is determined by comparison to a suitable reference tool. In this study, the use atomic force microscopy (AFM) as a reference source for matching and calibration will be described. AFM offers an alternative to optical and electron microscopes for imaging submicrometer features. AFMs directly measure the topography of a surface by bringing a sharp probe very close (within angstroms) to the sample surface in order to detect small forces from the atoms of the surface. Contact mode scanning is said to occur when repulsive forces are used for regulating the tip to sample distance. If the tip is moved away from the surface, a small attractive force is encountered several nanometers back. Feedback control using these longer-range forces is employed with the non-contact, attractive, mode of imaging. Tip-to-sample separation in the attractive mode is usually an order of magnitude larger than in the repulsive mode, which helps to minimize unwanted tip-to-sample contact. Lateral forces from approaching sidewalls are also sensed in the attractive mode, which makes this type of AFM ideal for imaging high aspect features that are commonly found in semiconductor processing. The main advantage of non-contact operation is that tip shape stability and measurement repeatability are greatly enhanced. Accurate profiles of the surface can be extracted at any position in a completely nondestructive manner and a complete 3-dimensional rendering of the surface can be created. Problems do exist that currently limit the ability of most AFMs to provide precise topographic data. Tip shape effects are the main source of these concern in AFM metrology. For samples with topographies containing high aspect ratio features, such as those encountered with integrated circuits, one usually desires very sharp probe tips in order to scan areas having abrupt surface changes[9]. Most commercially available probes are conical or pyramidal in shape and have finite apex radii. When features having high aspect ratios are scanned, as in **Fig. 3a**, they appear to have sloped walls or curtains[10]. Even if we had known the exact shape of this probe, there is no way to recover the true shape of the feature side walls. Recently, an attractive mode AFM (2D-AFM) with separate feedback control in the vertical and horizontal directions has also been used to perform undercut side wall imaging.[11] This AFM is known as the SXM Workstation and is the product of more than fourteen years of research and development by the IBM corporation in the area of scanned-probe technology. The SXM workstation can be operated in a two-dimensional servo critical dimension (CD) AFM mode. In CD mode, the tip shape is somewhat cylindrical with a flared bottom end[12]. The boot shaped CD probe shown in **Fig. 3b** is wider at the bottom than in the

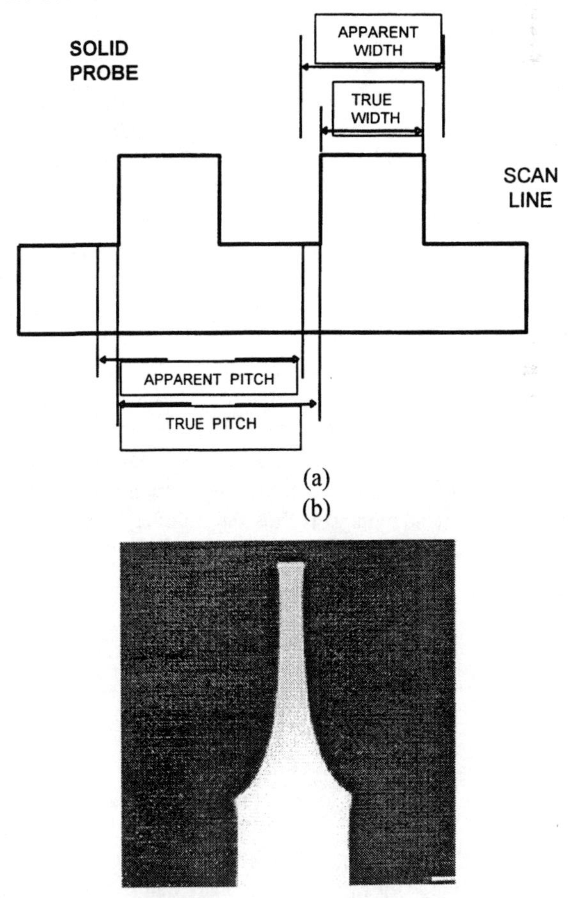

Figure 3 - (a) Effect of probe width on Pitch and Line Width measurements.(b) Image of flared CD AFM tip.

center, so that it is well suited to sense topography on vertical and undercut surfaces. The bottom corners of the boot shaped CD tip sense the side walls of a feature as the tip scans vertically along them. The position of the protrusions at the bottom corners of the tip is key for imaging the foot of a side wall, which can only be imaged if these tip protrusions are at the lowest part of the probe body and are not worn or rounded. Once the width of the probe is obtained, it can be subtracted from the raw data in order to obtain a CD value. A series of etched silicon ridges having sharp apexes can be used to determine the CD tip width. This structure is referred to as the Nanoedge[13]. If the tip had zero width, a triangular profile would appear in the AFM image. However, a set of trapezoidal lines are obtained whose top widths yield the size of the probe's bottom surface, after the width of the feature ridge (<2nm) is taken into account. Irregularities of the Nanoedge sample can produce variations in the reproducibility. It is very important to note at this point that the precision and accuracy of the CD AFM mode critically depend on, and are somewhat limited by, this calibration procedure. In addition to the CD tip, it is necessary to have a two-dimensional (2D) scanning algorithm that can provide servoed tip motion in both the lateral and vertical scan axis directions in order to image side walls. In the SXM CD mode AFM, servoing is achieved through the use of digitally controlled feedback and separate piezo-actuators in the x and z directions. When the tip encounters a side wall, the distance between tip and sample is adjusted through the lateral feedback servo. Instead of making evenly spaced steps in the X and Y scan directions and providing only a top-down view of the surface, this technique follows along the contour of the surface. The local scan direction is deduced from surface slope detection, described elsewhere[14], and is continually modified to stay parallel to the surface element. In this way, data are not acquired at regular intervals along x anymore, but at controlled intervals along the surface contour itself. A significant advantage is that the number of data points can be set to increase at feature side walls or abrupt changes in surface height. The CD-AFM mode offers a significant advantage in that the tip scans a path that follows the entire surface using the sharp bottom edges of the "boot" shaped tip. Real CD-width at each Z value can be measured.

CORRELATION

A flow chart to illustrate the proposed matching procedure is shown in **Fig. 4**. The objective is to define a process that permits different SEMs to be used interchangeably for line width measurements. The first step in the matching process involves optimizing the SEM operating point. It is usually desirable to minimize the number of factors, such as working distance, that can be adjusted on the CD-SEM in order to minimize the possibility of variation in that tool. Factors that typically can still be adjusted on commercial units are the beam acceleration voltage, beam current, magnification, and frame averaging. The optimum beam current and acceleration voltage are usually found by minimizing sample surface charging and optimizing signal-to-noise or contrast of the image. Recently, the use of ultra-low acceleration voltages[15] has become popular for minimization of the charging affects on deep-UV photoresist layers. The magnification is usually chosen for required field of view and pixel density.

Next, the scaling or "yardstick" at that magnification should be matched. Matching the systems for distance scale requires the use of a standard that has features spaced at known periodicity or "pitch". Since pitch involves measurement of the structure periodicity, the errors in width determination subtract out. Errors can creep into pitch measurements if the structures are not truly periodic or have irregular shape, edge roughness for example. It should also be noted that the actual periodicity of the test features is still susceptible to inaccuracies of the equipment used to pattern them. After establishing distance scale matching, the edge detection algorithm that best reproduces the measurement response (i.e. the linearity) of the reference tool should be found. A series of features having a trend in sizing or profile is needed. This can usually be accomplished by changing the design size or stepper focus/exposure conditions during lithography. Once the most appropriate algorithm has been found, the offset that is needed to make the two curves (hopefully the same shape) overlap must be found. AFM has been chosen as the reference tool. SEM cross-sectioning techniques were also tried but the results were too erratic due to lack of spatial averaging and sample preparation artifacts associated with that technique.

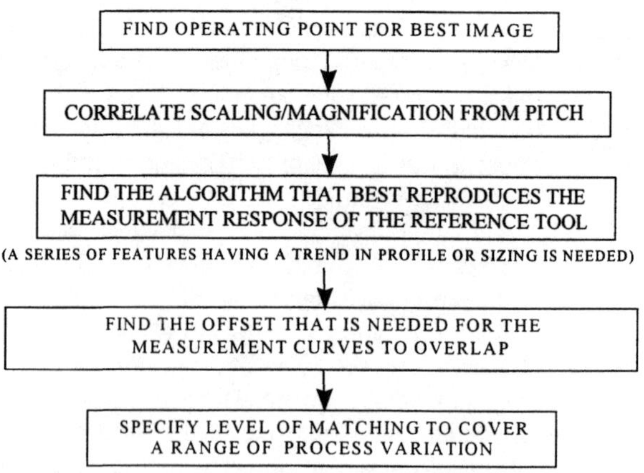

Figure 4 - Matching process flow diagram.

After correlating the magnification of each tool using a pitch reference, a set of etched poly-silicon lines (transistor gates) ranging in width from 1000 to 50 nm were used to look at measurement linearity. An AFM surface rendering of the 60 nm line of the set is shown in **Fig. 5a** . This image contains a three-dimensional volume of data. The effect of edge variation along the line was minimized by performing width measurements on each scan along the line and then averaging. One of these surface height profiles is plotted in **Fig. 5b** . The CD-AFM width measurement was taken at the feature bottom. The CD-AFM scans also verified that the edge wall profile remained fairly constant across the range of feature sizes. This was useful in trying to isolate the effects of size from those of wall profile on the SEM measurement linearity. By maintaining the edge profile and varying the feature width, we can evaluate the sizing linearity of different CD-SEMs. This trend in size will also allow us to find the most appropriate algorithm.

(a)

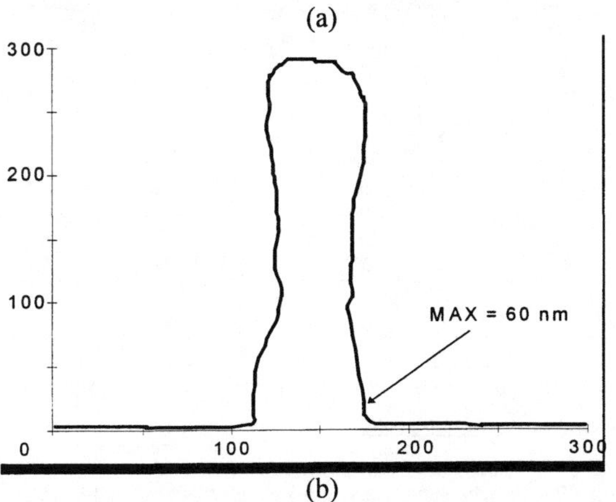

(b)

Figure 5 - Etched Poly-Si lines for nano-scale devices.

algorithm. Initially, each line was measured by the CD-AFM in order to obtain the reference line width. Then the features were measured on different SEMs. At first, several algorithms on each SEM were tried until the one that corresponded to the AFM in the most linear fashion was found. Line width measurements from three CD-SEMs are plotted versus the AFM values (representing actual feature size) in **Fig. 6a** . The oldest, most established SEM, was made by Vendor A and is denoted as Model 1. The next SEM was also from Vendor A, but was newer and had different imaging hardware (Model 2). The third and newest SEM was obtained from another manufacturer, Vendor B, and also differed in hardware configuration from the other two. All the SEMs continued to track changes in the line width down to line widths of 50 nm, but there were offsets between the three tools. The SEM-to-AFM measurement offsets are illustrated directly by the plot in **Fig. 6b**. The curves are fairly constant with respect to each other for feature sizes larger than 250 nm. To achieve matching for 300 nm lines, an offset of +6 nm would be added to the measurements from the first SEM (Vendor A , Model 1). Similarly, an offset of -6 nm is needed for the second SEM (Vendor A , Model 2) and +8 nm for the third SEM from Vendor B. It is interesting that the tool from Vendor B actually matched Model 1 of Vendor A better than Model 2 from Vendor A. This means that just because an SEM may be from the same vendor, doesn't mean that it will be any easier to match to the previous model than a tool from a another vendor. It will be shown later that matching is greatly improved and simplified if one maintains a homogeneous tool set of the same ***model*** (hardware configuration). The plots in **Fig. 6** also show that current CD-SEMs are capable of detecting changes in the width of an isolated poly-silicon line of less than 10 nm, in a linear fashion. The dip in all three measurement offset curves at the 200 nm line width is highly suggestive of an error (of about 5 nm) in the AFM tip width calibration at that site. This re-enforces the assertion made earlier that AFM performance is essentially limited by the ability to characterize and control the tip shape. The main aspect to point out is the increase in measurement noise for widths below the 250 nm node. As feature sizes decrease, it will be necessary to perform more measurements at each site in order to improve averaging. A more thorough estimation of the amount of averaging required for each technology node is given in the literature.[16] Unfortunately as fabrication processes are pushed harder to achieve smaller critical dimensions, the features will become rougher and introduce more sample variation effects. This problem could be minimized if spatial averaging along the feature is increased. However, SEMs provide only a square field of view so that when the magnification is increased enough to obtain sufficient pixel

determination method, the AFM is used to obtain the surface height topograph at each side wall profile (focus

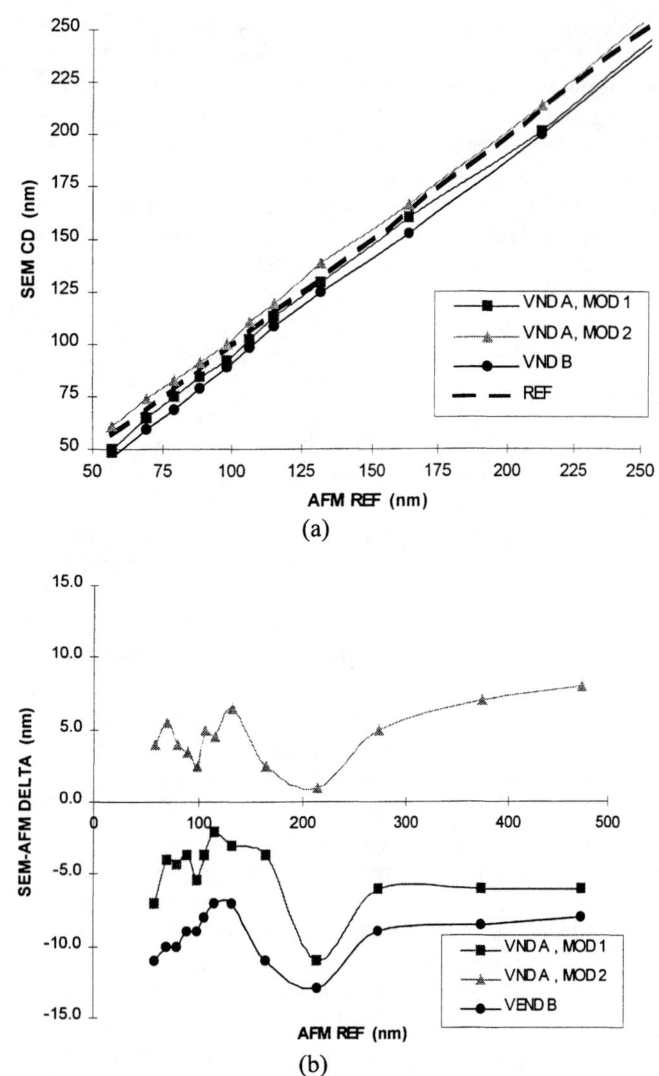

Figure 6 - SEM linearity for isolated poly-Silicon lines.

density, less averaging along the line in the normal direction occurs. The other way to create a trend in the feature series is to vary their profile. This can be achieved by varying the focus and/or exposure of the stepper during patterning of the photoresist. In addition to varying edge profile, this technique also provides even smaller increments in width variation than the previous method of design size coding. In this regard, measurement resolution (the ability to detect small changes in the feature size) is tested more thoroughly. By varying the lithographic conditions in this way, we are able to test the effects of both (the convolution) edge profile and size changes on matching simultaneously. This also provides a way to see how robust the matching between tools will be over a known range of litho-process variation. In this algorithm

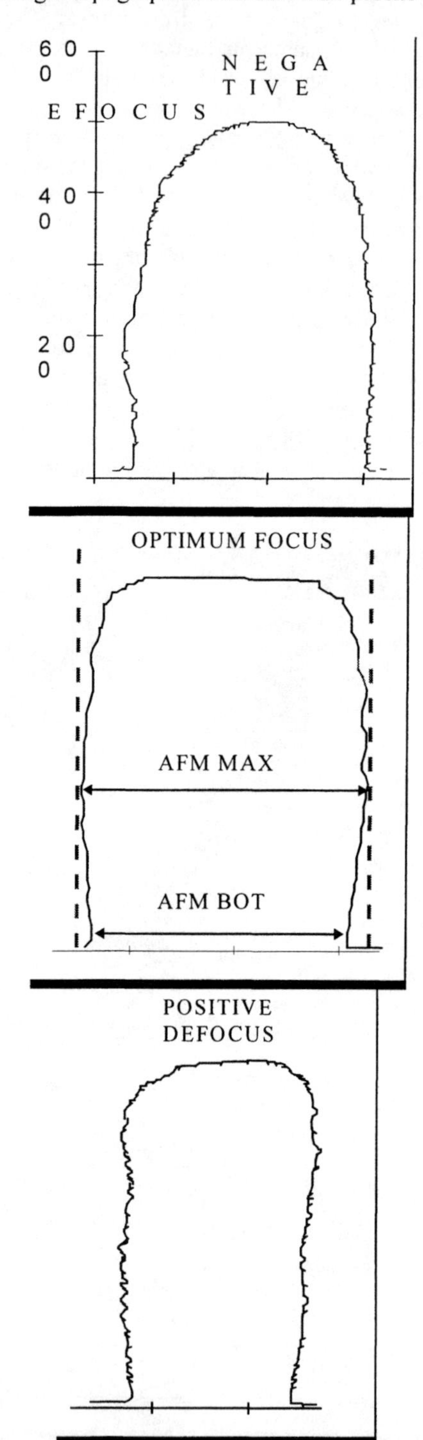

Figure 7 - Feature profile variation across stepper focus

level) condition. AFM topographs at the optimum focus level and at each opposite extreme are shown in **Fig. 7** .

The algorithm that best follows the reference topography over the combined variation in sizing and profile across the widest range can now be determined. This technique can also be implemented on etched layers when the resist has been removed by varying the etching conditions instead of the lithographic conditions. The surface topographs illustrate a wider range in the process variation than is ever (hopefully) encountered in the production line. Initially, there was great difficulty in finding an SEM edge algorithm that had the same shape as that of the AFM bottom CD curve. The middle profile in **Fig. 7** shows why!

The CD-SEM is only capable of providing a top-down image view of the feature so that the maximum portion indicated by the red lines is seen. This is the widest part of the feature and will be referred to as AFM MAX. As the side wall is varied, the threshold point in the SEM edge detection algorithm that corresponds desired width will also change. Since it would not be practical to try to change this threshold in production, a worst-case value in mismatch can be specified over a given range of process variation. SEM and AFM width measurements, before matching, are plotted verse stepper focus level (i.e. side wall profile) in **Fig. 8**. Indeed, it is possible to find SEM algorithms that follow AFM MAX in shape, but not for AFM BOT. This is because the bottoms of the features are being shadowed by the extreme widths. There is also a slight increase in the SEM-to-AFM offset as the focus level changes from negative to positive. The optimum focus that yielded the most desirable profile (**Fig. 7**) was found to be at focus level 5. This is mainly due to the effect of side wall profile variation alone on the beam-to-sample interaction volume and electron yield. The relatively large offset between AFM MAX and BOT (unrecoverable region for top-down SEM) brings up a rather philosophical issue : how does one control the pre-to-post etch bias with the CD-SEM ? Basically, the AFM MAX-to-BOT offset represents the worst case SEM error/bias due to metrology on pre-to-post etch offset. How bad this error is depends on the isotropy of the etch process. The more anisotropic the etch, the closer the worst case is for the SEM error component. This MAX-to-BOT offset will vary for different material layers, feature sizes, and pattern types (isolated, periodic, holes). Then the bias could falsely be attributed to material or pattern dependency of the SEM measurement, when it is actually a topographic effect.

In addition to feature size and profile, pattern type also has an effect on SEM offset variations. Isolated lines, dense grouped lines/spaces, and holes will be the patterns of consideration in this section. Pattern dependency of the SEM is mainly due to differences in charging. Three-dimensional AFM images of dense and isolated photoresist lines are shown in **Figs. 9 a and b** , respectively. The corresponding top-down CD SEM images of the same features are shown in **Figs. 9 c and d** .These images, clearly show differences in the SEM image behavior between dense and isolated lines. The dense lines appear opposite in tone (bright) from the isolated lines (dark) due to charge buildup. Charging seems to be more prevalent in

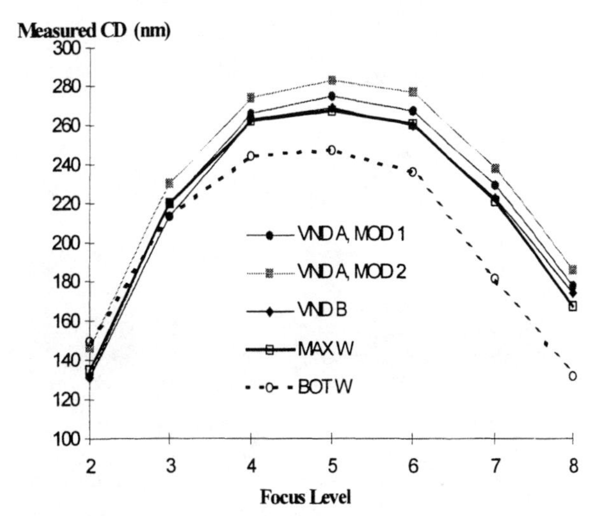

Figure 8 - SEM comparison across profile variation.

the dense line pattern, where intensity blooming is seen to occur, than for the isolated line case. Such large differences in the intensity profiles of dense and isolated lines necessitate separate matching parameters for each pattern type. In addition, if the dense lines are spaced close enough, charging can prevent electrons from the substrate region between lines from reaching the detector. The space between lines appears dark and less information about the feature is being collected. This will make it difficult to measure the feature bottoms and can cause aliasing in the measurements, since the minimum intensity level is a basis from which the algorithm fits are performed. It should be noted also that pattern type can affect the AFM's imaging ability if the probe is not narrow or long enough to reach the bottom of a trench or hole.[17] Differences in tip-to-sample regulation stability have also been noticed on occasion. This condition can quickly be detected by monitoring the amplitude of vibration of the AFM cantilever and tip width during operation.

Line width measurement from all three SEMs are plotted verses AFM width for dense periodic features, from the same sample as the isolated ones, in **Fig. 10**. The optimum focus was also at level 4 this time. The AFM MAX and BOT curves are very close in shape and value, which means the dense lines in this example had very little undercut. The amount of area that is unrecoverable from a top-down view is minimal (less than 3 nm) with this set of features. However, more charging effects exist than with the isolated line case (see **Fig. 8**) so that the SEM-to-AFM

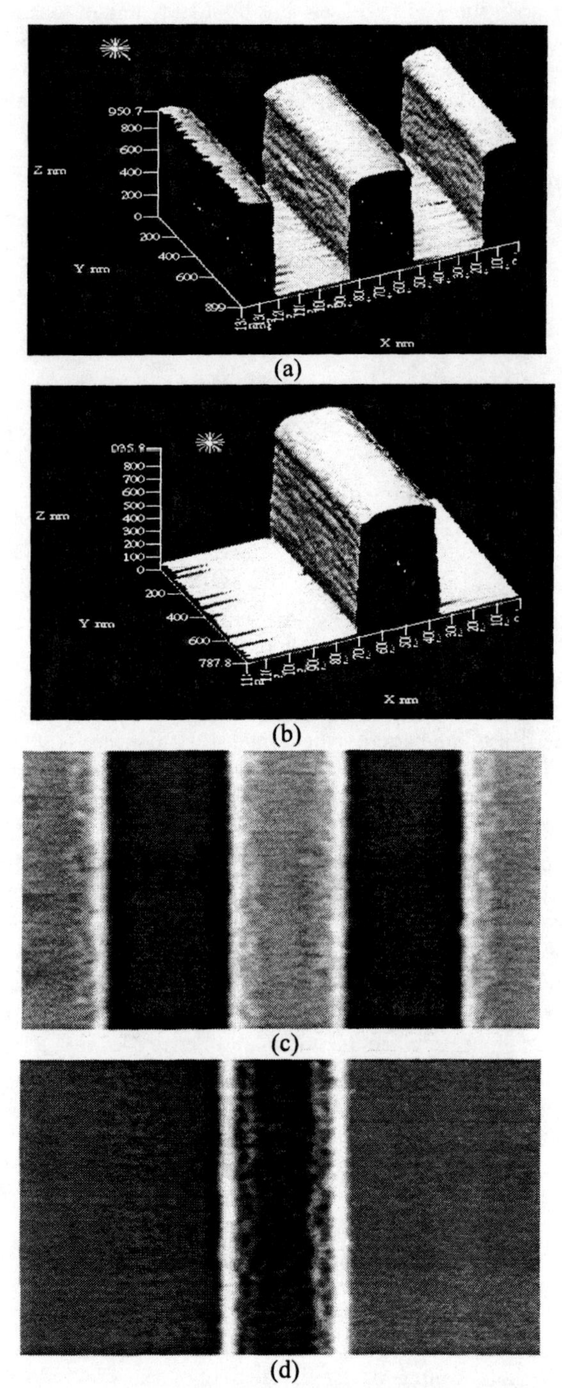

Figure 9 - I Line photoresist grouped-dense and isolated line (a + b) AFM topographies and (c + d) SEM images.

reference offset appears to have changed, or shifted, for both models from Vendor A. The first SEM (Vendor A, Model 1) measurement offset at the optimum feature profile site shifted down by about 20 nm from isolated to dense lines, with respect to the AFM reference values. Model 2 from Vendor A shifted down by a different amount of about 10 nm. Finally, the tool from Vendor B did not shift by any detectable amount. The main factor affecting the SEM measurement shifts was that of operating point, beam acceleration voltage to be specific.

Model 1 from Vendor A achieved optimum optical performance at 1000 V. Unfortunately, strong sample charging is seen to occur at that beam acceleration voltage (i.e. the region between features could not be imaged with adequate signal detection). Charging was less with Model 2 / Vendor A because it is optimized for performance at 800 V for the electron landing energy.[18] The tool from Vendor B is capable of imaging without apparent degradation in quality at an electron landing energy of 600 V. Evidently, the point of minimum charge accumulation for this material and pattern type combination, where electron yield is close to unity, is around 600 V. Sufficient signal level could be detected from the substrate between the features. As in the isolated line case, there is an increase in the SEM-to-AFM offset as the focus level changes from negative to positive. Again, this is most likely due to the variable of side wall angle. Closer comparison of the SEM-to-AFM offsets do yield that the systematic drift is essentially the same magnitude for dense and isolated lines.

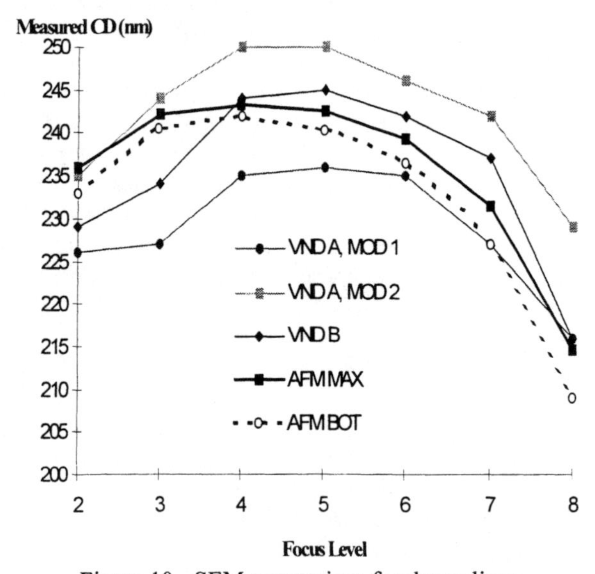

Figure 10 - SEM comparison for dense lines

One of the more difficult features to image and measure is the hole pattern, also referred to as a contact or via. SEM (left) and AFM (right) top-down images of a nominal 400 nm diameter hole patterned in deep-UV photoresist on an oxide substrate are shown in **Fig. 11a** . Cut-away views of the AFM data are also shown in **Fig. 11b** . For the SEM, holes are more prone to charging effects so that signal collection from the bottom is more difficult than with lines. Therefore, a larger size and material dependence of the matching parameters occurs with holes. The effect of edge detection algorithm also seems more pronounced for holes. On newer SEMs, an average diameter is computed by performing a radial average about all angles (see **Fig. 11a**). This is possible by either changing the beam scan direction (Vendor B) or the line cut direction from the final video image (Vendor A, Model 2). The hole structure also presents the most difficulty for the AFM reference tool to image. The key issues with AFM analysis of holes are measurement scan location and angular direction, sample induced variations, and tip shape/size effects. The usual method for imaging holes with CD mode AFM consists of first performing a quick standard mode overview scan of the hole. The location of the measurement scan is then set by placing the CD-mode image indicator (the long white rectangular box in **Fig. 11a**) at the desired position. Essentially, a small set of CD mode scans (denoted by 1,2,3 in the figure) are taken within a section of the top-down standard image. Only scanning in the horizontal image direction between left and right are possible due to the method of lateral tip-to-sample distance control used in CD mode AFM. The result is that we can only obtain CD imaging within a horizontal band through the hole center. Centering of this horizontal band through the hole turns out to be a major component of AFM hole diameter imprecision and error. Sensitivity to centering depends on the hole radius, of course. Radial averaging along different scan angles is also not possible, which can lead to sample variation (i.e. edge roughness or asymmetry) effecting the measurement quality. Although scanning in different directions is possible in the standard mode, diameter measurements can not be performed because the tip cross-section is elliptical (see the red trace in the AFM image of **Fig. 11a**), as seen facing the bottom of the probe.

A new method for improving hole width measurement with the CD mode AFM will be presented here. The new technique involves imaging the entire hole with the CD scan mode and NOT using the standard overview mode for measurement centering. CD mode width measurements are made at each image line scan, except at the hole walls parallel to the scan direction. The AFM MAX and AFM BOT width measurements are plotted for each scan line number in **Fig. 12a** . The parabolic curve shapes are from the change in contact width at each vertical scan position. A polynomial is then fit to the data points, whose maximum represents the hole diameter. This technique eliminates the issues of centering and spatial averaging, as well as improves the static measurement averaging. Residuals from the polynomial fit can also be used to estimate the combined precision components of sample variation and tool random error. The overall hole measurement reproducibility was reduced to 3 nm using this technique. It should be noted that this technique is still susceptible to errors caused by changes in

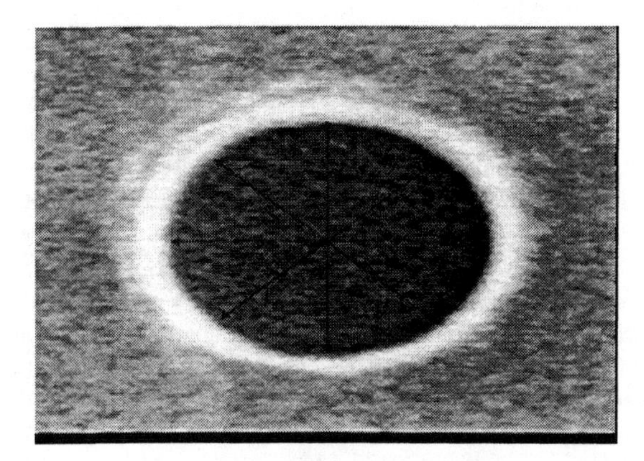

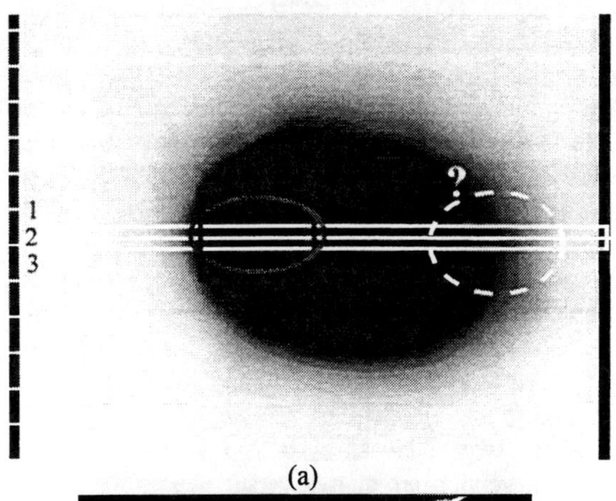

(a)

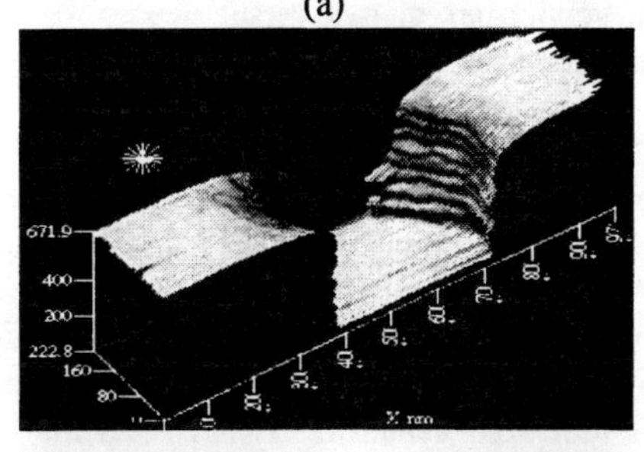

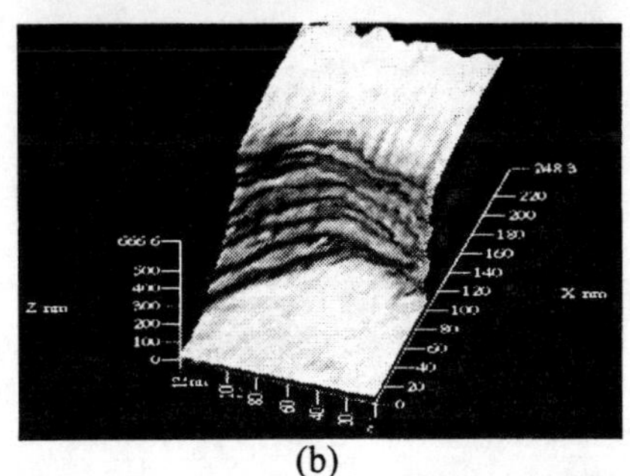

(b)

Figure 11 - Top-Down SEM and AFM views of DUV resist contact on Oxide substrate

the hole diameter along different angular directions through the hole. These errors can be corrected somewhat by correlating the horizontal diameter in the SEM image to the AFM value. Then relative changes in hole diameter at different angles can be found from the SEM image, assuming beam stigmation and rotational induced shifts (RIS) in the SEM have been corrected properly.[19] As noted earlier, eccentricity in the probe front will introduce errors in hole diameter measurements in the vertical direction. A more serious issue is starting to arise for sub-200 nm etched silicon CD probes - they develop a rectangular probe front, or footprint (see the blue trace in **Fig. 11a**). This causes the diameter measurement to be less than the actual hole diameter when the point of interaction switches from one probe bottom corner to another. A pointed probe shape characterization structure can then be used to measure the probe size in two dimensions to correct this problem. AFM measurements are relatively slow compared to the SEM. This is one reason why the SEM is still the tool of choice in the cleanroom fab. Low throughput of the AFM slows down the matching process and can create a potential for "bottle necks" in the measurement process. However, the linearity of the SEM (assuming the correct edge detection algorithms are chosen) has been established sufficiently such that only one AFM point is really necessary to determine the offset factor that is needed to make the curves overlap. Measurements from four SEMs, two "identical" systems of the same model type from each vendor, are plotted verses stepper focus in **Fig. 12b** . The diameter value found in **Fig. 12a** for the hole in deep-UV photoresist on oxide substrate was used to provide the offset necessary to match these four systems. The systems from Vendor B are more closely matched to each other than those from Vendor A, but Vendor A was closer to the AFM reference value. The measurement offset of Vendor B was due to its hole diameter measurement algorithm and was eliminated using the AFM reference value. The tool from Vendor B exhibited less charging and was able to detect the subtle change in feature size that occurred at focus 4 in **Fig. 12b** .

The last factor in CD SEM matching variation to consider at this time is that of material composition of the feature. The chart in **Fig. 13** shows the SEM-to-AFM offset of each tool for different sample types. Measurements where performed at the optimum feature profile on each material combination. This was done in order to isolated the effect on matching due to material type. The dependence of SEM measurement offset on material type is clearly different for all three SEMs. Material dependence of SEM measurement offset was also studied for dense lines and different behaviors were observed. Therefore, material induced variations in SEM-to-AFM offset can also introduce errors in the measurement of dense-to-isolated line bias of a process.

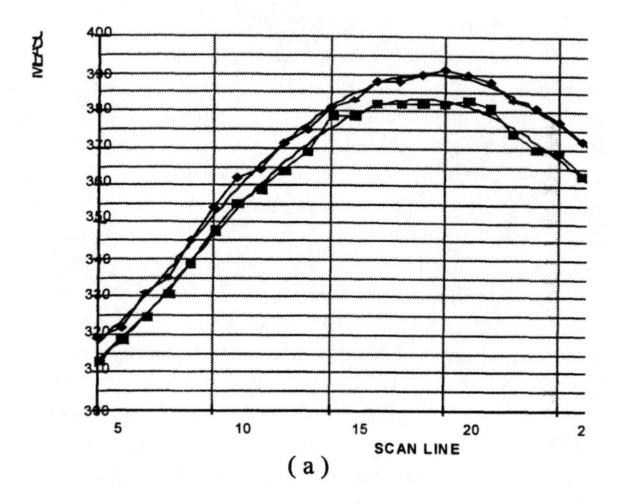

(a)

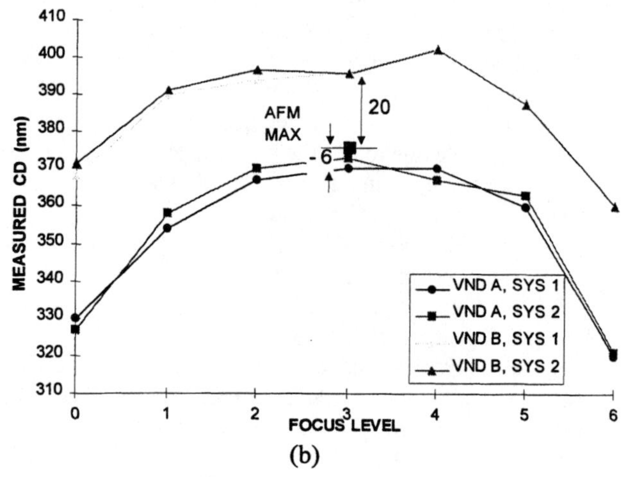

(b)

Figure 12 - (a) Polynomial fitting to determine the contact diameter (b) Matching data for tools of the same model .

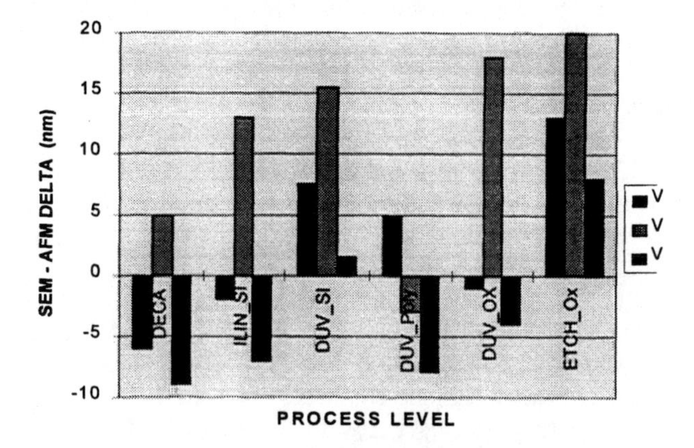

Figure 13 - Process Level material induced SEM bias.

CONCLUSIONS

The first property of a microscope to consider is resolution. Resolution is difficult to define quantitatively and there are many different context in which resolution is used throughout the industry. However, the one most preferred by this author is the related to determining the minimum change in feature size that can be detected. The ability to determine subtle changes in the feature topography encompasses the primary beam spot size, detection signal to noise and efficiency, as well as sample characteristics. This last point refers to the fact that SEM resolution is somewhat dependent on sample properties due to beam-to-sample interactions, such as surface charge accumulation. Distance and measurement linearity are the next characteristics that are required. These are related directly to, but not totally determined by, the instrument resolution. Measurement linearity is achieved when the image size corresponds to changes in the actual feature size in a direct, or linear fashion. The final aspect to consider when assessing how well the tool can be matched to another is precision. From **Fig. 1**, it is clear that the error in matching between two tools cannot be better than their precision. Therefore, precision is the first indication of what minimum difference between instruments is reasonable to expect. If the tool's precision is already outside of the specifications needed for cont rol of a particular process, further efforts to achieve matching within that specification will also be fruitless. If the level of matching cannot be made to meet the allowed metrology error budget for a given process, then sample measurement jobs should only be done on the same SEM each time.

REFERENCES

[1] D.Nyssonen and R. Larrabee, *NIST J. of Res.*, vol. 92, p. 195.

[2] M. Postek and R. Larrabee, *Solid-State electronics*, vol. 36, No. 5, pp. 673-684, 1993

[3] H. Marchman, J. Griffith, C. Pierrat, and S. Vadia, *J. Vac. Sci. & Technol.* **B** 11 (6), pp. 2482-2486.

[4] L. Reimer, Image Formation in Low-Voltage Electron Microscopy, SPIE Tutorials, Bellingham, Washington, 1993, p. 64.

[5] L. Reimer, Image Formation in Low-Voltage Electron Microscopy, SPIE Tutorials, Bellingham, Washington, 1993,

[6] H.M. Marchman, J.E. Griffith, J.Z.Y Guo, and C.K. Celler, *J.Vac. Sci. Technol.* B 12(6), 1994, p. 3585.

[7] H.M. Marchman, J.E. Griffith, J.Z.Y Guo, and C.K. Celler, *J.Vac. Sci. Technol.* B 12(6), 1994, p. 3585.

[8] SEMATECH SEM users group.

[9] H.M. Marchman, J.E. Griffith, and R.W. Filas, *Rev. Sci. Instrum.* 65 (8), 1994, pp. 2538-2541.

[10] J.E. Griffith, H.M. Marchman, and L.C. Hopkins, *J Vac. Sci. and Technol.* **B** 12 (6), 3580.

[11] Y. Martin and H. Wickramasinge, Appl. Phys. Lett. 64, 2498 (1994).

[12] O. Wolter, Th. Bayer, and J. Gresshner, J. Vac. Sci. Technol. **B 9**, 1353 (1991)

[13] Nanoedge sample is supplied with the Veeco SXM System and manufactured by IBM

MTC , FRG.

[14] Y. Martin and H. K.. Wickramasinghe, Appl. Phys. Lett. **64**, 2498 (1994)

[15] To be published, H.M. Marchman, Chapter on Metrology, Handbook of Microelectronics, Marcel Decker, J. Sheats, Ed., 1997.

[16] K. Monohan, H. Marchman, and H. Sewell, SPIE Proc. In Microlithography , 1997.

[17] J. Griffith, H. Marchman, M. Vasile, and L. Hopkins, J. Vac. Sci. & Technol. B 11, pp. 2473-2476.

[18] Landing energy is used instead of beam acceleration voltage because of the retardation technique used in the newer SEMs.

[19] K. Monohan, H. Marchman, and H. Sewell, SPIE Proc. In Microlithography , 1997.

Overlay Metrology: The systematic, the random and the ugly

Neal Sullivan
Jennifer Shin

Advanced Process Tool Development Group, Digital Semiconductor, Hudson, MA 01749

Typical advanced lithographic processes require an overlay tolerance that is approximately 30% of the minimum feature size. To achieve this metrology must be limited to an error budget of 3-5% of the minimum feature size. A discussion of the general use of overlay data for assessment of stepper performance is followed by an overview of key overlay metrology equipment performance parameters (Tool Induced Shift - TIS) and the interaction with the process (Wafer Induced Shift - WIS). The interactions between the target, the process and the equipment will also be reviewed. Finally, use of this information in a case study of overlay target design for Tungsten Chemical Mechanical Planarization (W CMP) will be presented. Relevant process data will demonstrate how such targets can be used to effectively monitor both stepper (systematic error) and metrology tool (random error) performance.

INTRODUCTION

Semiconductor pattern overlay is the measure of vector displacement from one process level (substrate) to another level (resist), usually separated by an intermediary (thin film) layer. A top down view of a typical overlay measurement target is shown schematically in Figure 1.

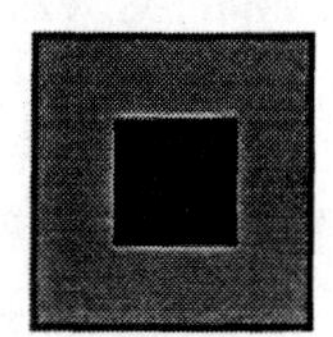

FIGURE 1. Overlay target schematic

Overlay error is defined as the planar distance from the center of the substrate target (gray box in figure 1) to the center of the resist defined target (black box in figure 1). Overlay measurement involves the determination of the centerline of each structure along both the X and Y axis. Centerline determination utilizes the symmetry around the structure center such that the error associated with edge determination will tend to cancel from each side of the structure. Conversely, in a linewidth determination the error associated with each edge combines in an additive fashion. Overlay measurement uncertainty is a complex function of tool, target and process interactions and is difficult to quantify. The device overlay tolerance specification is derived from the technology design rules. The circuit performance requirements are compared with established manufacturing tolerances. All error sources, from the mask to the measurement uncertainty are combined either linearly (most conservative) or are added in quadrature (for Gaussian error distributions) to calculate the overlay tolerance. Successful process control of 0.5 micron process technology was possible through initial stepper matching and subsequent control of layer to layer overlay through the use of x y offsets. Process tolerances for the 0.5 micron generation, typically 150 nm on critical levels, were generous enough that the non-normal (multi-modal) behavior and dependence on location within a field did not affect the ability to meet target performance specifications. A typical example of the non-normal nature of the raw overlay measurement results is shown in Figure 2. Product misregistration overlay budgets for current 0.25 micron production

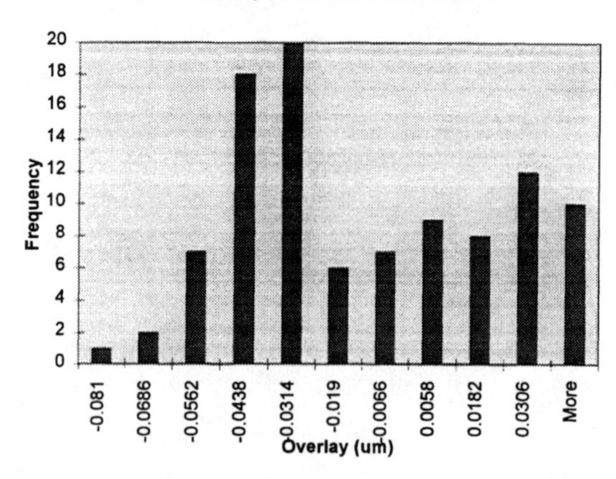

FIGURE 2. Non-Normal Overlay Distribution

process technologies however, are less than 100 nm. In order to consistently achieve these levels of performance every aspect of overlay error must be identified and

CP449, *Characterization and Metrology for ULSI Technology: 1998 International Conference*
edited by D. G. Seiler, A. C. Diebold, W. M. Bullis, T. J. Shaffner, R. McDonald, and E. J. Walters

corrected. To that end, modeling is employed more frequently to identify, and where possible, correct the systematic sources of overlay error above and beyond the zero order offset terms. Moreover, process control is more difficult unless these systematic errors can be accounted for since their presence will distort the observed nature of the variation.

THE SYSTEMATIC:

The sources of overlay error attributable to the wafer stepper, generally arise from two sources; the lens (intrafield) and the stage (Grid). Intrafield errors can be further broken out into errors which are intrinsic to the optics of the stepper (distortions) and errors which are derived from interactions of the reticle with the stepper's optical subsystem. The former set of errors are characteristic of a given lens and are typically addressed during lens manufacture (direct control over a wafer stepper to control intrinsic lens distortions is not available). The residual lens distortions are accommodated for, at the price of individual stepper overlay performance, in the process of stepper matching. It is essential therefore, that each new lens be thoroughly characterized with respect to distortion since these errors are an integral part of the performance of the stepper in manufacturing.

Equation 1[1] is a mathematical model for all x direction intrafield error terms associated with steppers (a similar equation applies for the y direction).

(1) $\delta x = \alpha x + (\delta M/M)x_0 - \Theta y_0 - t_1 x_0^2 - t_2 x_0 y_0 - E x_0(x_0^2+y_0^2) + F x_0(x_0^2+y_0^2)^2$ + Residuals

From equation 1, the terms up to first order in x; Offset (α), Magnification ($\delta M/M$) and Rotation (Θ) can typically be controlled from the stepper. The higher order terms; Trapezoid (t_1 and t_2), third (E) and fifth (F) order distortions are either not easily controlled (trapezoid) or are intrinsic to the stepper lens (third and fifth order).

Stage derived errors may also be divided into those over which the process engineer has control and those which are "intrinsic" to the mechanical subsystem. As shown in equation 2[2], the systematic Grid errors are very similar in mathematical form to the intrafield or lens errors shown in equation 1 but unlike the intrafield case the terms apply across the entire wafer.

(2) $\delta x = \alpha x + (\delta M_g/M_g)x_0 - \Theta_g y_0 + y_0^2 Dx$ + Residuals

$\delta M_g/M_g$ is the wafer scaling coefficient, Θ_g is the wafer rotation coefficient and D is the stage bow coefficient. Typically the stepper will allow direct control over only the scale and rotation terms. Accurate separation of the various error components shown in equations 1 and 2 above is heavily dependent upon both sample plan and model statistics. The primary statistical assessment for goodness of fit is the 3 sigma coefficient error. This estimate provides a check of the statistical significance of the estimated coefficient against the other modeled systematic errors. Without this it is possible to accept error terms which are not significant and may add to the overall process variability. The residual error from the fit is also an indicator of the goodness of the model and will contain all random metrology error as well. In addition to assessing model performance, these error terms can be used to identify sample plan weaknesses. When establishing the metrology sample plan it is important to maximize symmetry and spatial coverage. These concepts derive from an understanding of the systematic error components due to the lithography tool. To achieve symmetry, die and intrafield site positions should be chosen such that they are balanced by an opposite die or site. For good spatial coverage, measurement locations must be positioned such that they adequately cover the area of interest (wafer or die) in both the x and y directions. Poor symmetry can lead to incorrect systematic error assessment. In the extreme intrafield case - one site per field, magnification error will appear as a translation error. Correction of this through an offset term will result in a 2x error at opposing die locations. Poor spatial coverage will also lead to incorrect systematic error assessment. For example; sampling the field in four corners only will result in exaggerated field magnification terms. This is due to the inclusion of (non-correctable) higher order (3rd and 5th) distortion terms. In this instance, adding one or two sites along the die edge (at X=max, Y=0 or X=0, Y=max) will allow for differentiation of the higher order systematic errors from magnification terms.

THE RANDOM:

Overlay measurement must be sensitive enough to discriminate error components derived from all sources. It is critical, therefore, that the contribution of the measurement tool to the total be minimized and well controlled for all process levels. Figure 3 shows relative tool performance versus lithographic error budget requirements for the last several process generations. The bold line in Figure 3 represents Overlay metrology specifications as outlined in the 1994 and 1997 SIA National Technology Roadmap for Semiconductors. The dashed line in Figure 3 represents worst case performance of overlay metrology equipment at Digital Semiconductor fabrication facilities over the same time period. As can be seen, the raw system performance has been and will continue to be adequate for several process generations.

However, overlay measurement difficulties arise in manufacturing from an interaction of the measurement equipment with the instance of the measurement feature (process variations and target design); primarily due to low edge contrast. Low contrast edges often result from advanced planarization techniques, such as Chemical Mechanical Polishing (CMP), used to accommodate a decreasing photolithography process window. These problems require that implementation of overlay metrology be performed with a deeper understanding of the physics of the measurement instrument and the interaction of the process module with the overlay measurement target. The goal is to design effective targets that maximize edge contrast and allow equipment utilization in the most efficient modes of operation.

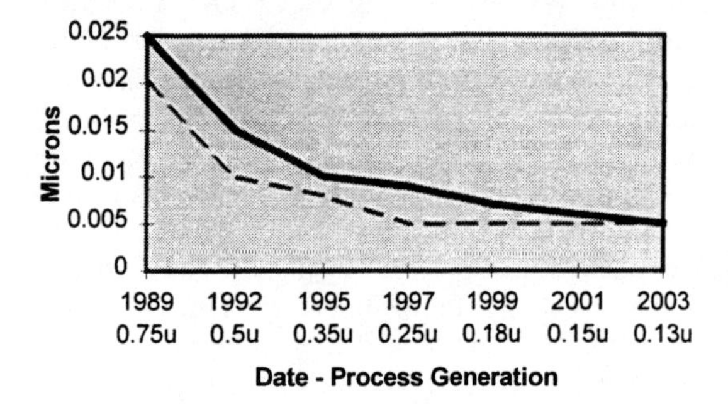

FIGURE 3. Performance of Overlay metrology equipment relative to SIA process error budget requirements.

The inherent, tool limited, measurement accuracy is commonly referred to as Tool Induced Shift[3] (TIS). TIS is quantified by measuring the same feature at 0 and 180 degree (wafer) rotations. TIS is one half the sum of the measurements from each orientation. The TIS error arises from; optical alignment, mode of illumination, and aberrations of the optics in the system. The second most common source of measurement error is termed Wafer Induced Shift (WIS). This error results primarily from physical changes to the measurement targets due to the manufacturing process (e.g. asymmetric film deposition / coat, severe noise due to grains or low contrast due to planarization). Unlike TIS, WIS is invariant with respect to rotation and, as a result is not as easy to identify. In many cases, it is not possible to quantify WIS until subsequent etch processing is performed. WIS must be addressed in both the measurement algorithms themselves and the overlay target design. The final component of the measurement contribution to the overlay error comes from the measurement precision of the tool. This component is easily quantified by performing repeated measurements of a sample, in a carefully designed Analysis of Variance (ANOVA) experiment. All of these sources of measurement error combine to constitute the total measurement uncertainty.

Sources of TIS and contributions to WIS within the measurement system are primarily found in; the stage, the illumination system and the imaging optics. The stage must be well behaved in the X-Y plane during Z motion (where Z is perpendicular to the wafer surface). Any translations in X or Y which result from Z motion will contribute to measurement error. The worst case measurement scenario for this Z axis induced transitional error is during a dual focus measurement. In this measurement mode, which is used for structures exhibiting either topographies larger than two microns in Z or poor contrast, centerline estimates from the substrate and resist levels are taken at two separate focal planes. Any translation in X and Y which takes place during the move between focal planes will show up as measurement error (TIS).

The effect of optical aberrations on TIS must be examined from the perspective of centerline determination. Assuming that the target is located within the center of the microscopes field of view, the effect of all symmetric aberrations (e.g. spherical), will cancel. This is true for all focal positions since, in the centerline determination, the edge error will be symmetric (equal and opposite in value) across a centered measurement target.[4] Asymmetric aberrations (e.g. coma), on the other hand, will result in an incorrect assessment of centerline for a symmetrically placed overlay measurement structure due to error asymmetry. Kirk[4] has shown that brightfield optical systems, in the presence of 0.6λ of coma, can have a range of error in edge determination as large as 100 nm, over 1 μm of defocus. From an overlay measurement perspective, the measurement instance most vulnerable to the effects of comatic aberrations is the single focus measurement on a dual layer sample; generally the standard measurement technique. In this case the two layers are at two different focus positions and each layer will experience different error magnitudes due to the comatic aberration.

Illumination will directly impact sensitivity to process variations and, if not properly aligned can result in large increases in TIS. The illumination bandwidth must be chosen such that it will; not expose photoresist, not be susceptible to diffraction phenomena, provide the required resolution, and maintain a stable output (intensity) over time. Typically broadband (450 - 650 nm) partially coherent[5] light is set up in Kohler[6] illumination in which the lamp filament is imaged on and completely fills the back focal plane of the objective. The aperture of the illuminator must be centered on the optical axis for symmetry. The importance of symmetric (telecentric) illumination for the objective is derived from the centerline estimation process. If the illumination axis is not perpendicular to the wafer surface a lateral image shift with focus[4,7,8] can result. Illumination can also be used to control system Numerical Aperture (NA) through coherence changes in the illumination subsystem. Initial work in this area[9] was on targets for which process induced asymmetry was present in conjunction with a high level of noise (e.g. grainy metals). The results demonstrated that the choice of illumination wavelength most severely impacted those targets in which thin film effects were expected. Illumination alone was not shown to improve the accuracy of measurement results for asymmetric targets. However, by varying the system NA in this way[10] it was shown that edge contrast enhancement is improved for certain grainy metal films. As always, trade-offs exist, and by increasing the system NA sensitivities to focus repeatability are accentuated.

Overlay measurement accuracy errors, if not recognized and accounted for can produce false systematic stepper errors. Zavecz[11] has reported that the introduction of systematic measurement errors into a simulated stepper set-up dataset results in the transposition of those errors to the stepper. TIS shows up in the translation term of the modeled stepper systematic errors and Pixel scale directly modifies the grid and field magnification (scale) coefficients. If these errors are not properly ascribed to the metrology tool they will end up as a stepper input correction which will further degrade product overlay performance. It is important to accurately quantify and assign the contribution of the measurement tool's error component to the measured value and to minimize its impact via a hardware modification (e.g. optical alignment / columnation) or software calibration.

THE UGLY:

TIS is observed to interact with process conditions, achieving different values for different; resist thickness', substrate surface roughness, substrate topographies, and structure designs.[8] Figure 4 depicts the TIS performance of two different overlay measurement systems over six different process layers. The substrates shown include; metals, poly-silicon, field oxide and multilayer dielectric; the resist thickness' shown range from 0.5 µm to greater than 2 µm. The "Nominal" wafer is a single layer, (1 µm) resist on field oxide, with designed overlay offset, frame-in-frame, targets covering a 125 nm total offset range. Calculated TIS for this sample shows, minimum values for both measurement instruments relative to the other process levels. This is due to the planar nature of the structure which permits measurements at a single focal plane. If TIS were independent of the process effects, one would expect to be able to predict a measurement instruments behavior on all process layers given an adequate understanding of the response in the ideal, minimum TIS instance. As can be seen from the figure, while it is generally true that minimum TIS results in best overall performance, it is also clear that there is a complex interaction between the various measurement schemes (optics, algorithms, illumination, etc.) represented by the different equipment and TIS performance on a given substrate. This data re-emphasizes the observations made during the discussion of the system optics TIS is a very complex function of; choice of focus method (e.g. single vs. Double grab), optical configuration, and substrate material. It is not possible to accurately predict the TIS response for a

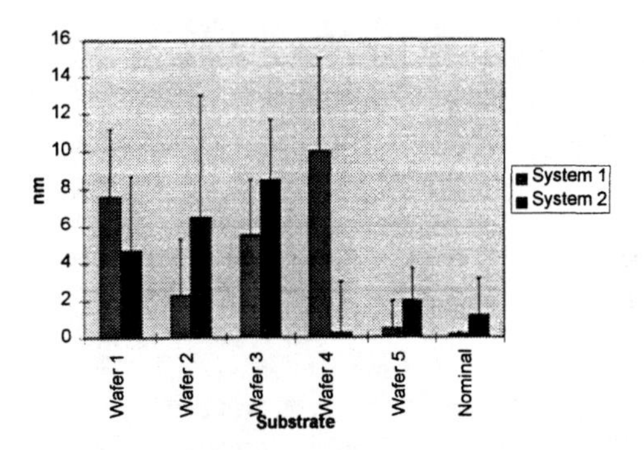

FIGURE 4. X TIS

given instrument on a new substrate based only upon an empirical understanding of the instruments performance on other materials. Figure 5 shows the interaction of focus and TIS. The data for each of the systems was collected in the single focus measurement mode using Wafer 3 from Figure 4 which has a resist thickness > 1.5 µm. The X axis of Figure 5 shows the deviation from best focus (0.0) in microns while the Y axis is the value

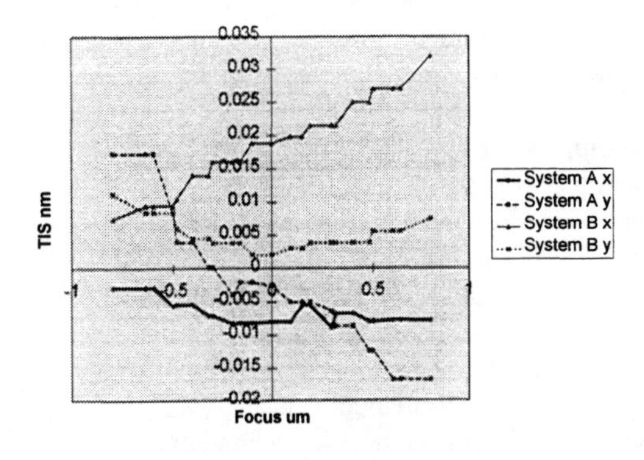

FIGURE 5. TIS vs. System focus

obtained for TIS. The behavior of the X TIS, for both systems, in the figure is typical of that which would be expected; a region of focus which corresponds to a stable operating point across the Depth of Focus (DOF) of the system optics. The Y TIS however is not so well behaved. Across the D.O.F. there does not appear to be a stable (TIS) point for either system. The anomalous behavior in Y TIS, corroborated by the two measurement tools, is believed to be due to an interaction with a target asymmetry.[12,13] In fact, it is conceivable that the variations in TIS as shown in Figure 4, both within and across wafers, are due to an interaction of the tool asymmetry (TIS) and the process asymmetry (WIS). Even across a single wafer, process conditions are interacting with the fundamental tool setup in different ways at different wafer locations. This is the cause of the measurement errors of differing magnitudes shown by the error bars in Figure 4. As the hardware component of the overlay tools TIS has been reduced, process asymmetries have become a significant fraction of the observed measurement error. Target asymmetry is usually the result of proximity effects or asymmetric film deposition conditions. For example, a substrate step, in close proximity to the measurement target, can result in a locally thicker resist layer due to the planarizing nature of spin coating.[14] There is also an effective increase in the exposure dose due to the greater reflectivity in the vicinity of the step. This effective dose change, coupled with the thicker resist across the step, results in an asymmetric target. Measurements taken on structures exhibiting this asymmetry by Starikov et al[12] demonstrate that a mean overlay difference of 70 nm, along the axis affected by the asymmetry was observed as compared to a structure which was not influenced by proximity effects. Further simulations by Starikov show that the introduction of even a 0.5% resist slope can result in a 23 nm measurement error. Tanaka[15] reports of a target asymmetry which results from uneven deposition of a film. In the instance sited a metal sputter deposition process produces film thickness' which are dependent upon topography and orientation with respect to the incident sputtered atoms. This results in a "snowdrift" effect within the box-in-box overlay target, whereby the metal is piled up in one side of the overlay target. From an overlay measurement perspective, the "snowdrift" effect results in an inaccurate measurement due to an incorrect estimate of the "true" target centerline. The measurement inaccuracy which results from this type of target asymmetry can be greater than 100 nm (3 sigma) when comparing measurements taken following etch processing with standard overlay measurements taken following the develop process.[15] The WIS component of inaccuracy must be addressed through measurement target design optimization and measurement algorithm development.

Modern overlay measurement algorithms are moving beyond the threshold algorithm. This departure is due to accuracy difficulties which result from the complex film stacks used in state of the art processes. Given the lack of symmetry within some targets, basic threshold algorithms are prone to systematic measurement error. Determination of a specified threshold on the image waveform, corresponding to a point on the edge of the overlay target will not match reality. In addition, asymmetry as a result of tool (e.g. minor illumination misalignment) or mark sources (grainy film or image contrast variation) causes different edge detection errors for the left and right edges. Errors greater than 50 nm can be expected to result. It is essential that more than a simple interpretation of the signal waveform for a percentage of the threshold be performed. The measurement algorithm can take advantage of "a priori" knowledge of the overlay target[13] and the physics of the subsequent interactions with the films and resist layers. All measurement algorithms, in one way or another, compare the centerline (or the intersection of two orthogonal centerlines) of a substrate layer to the photoresist layer. Certain "a priori" facts are known about the "ideal" target which can be used to interpret the manifestation of that target for a given substrate and target design. Use of this information in the measurement process can lead to improved measurement accuracy, even in the absence of certified standards. Starikov et al[13] describe an algorithm which takes advantage of the inherent symmetry and redundancies present in overlay target design. Since target designs follow standard dimension practices, such information can be used to assess any asymmetries resulting exclusively from tool or process. Another algorithm which has been proposed to address process variations is a modified intercorrelation algorithm.[16] This algorithm uses all of the data contained within a defined region encompassing the actual edge, rather than just a single threshold point along the maximum slope. Such a method provides added accuracy since using the entire waveform tends to average out any low contrast or noisy segments of the waveform. The intercorrelation data can be also used to judge the symmetry of the target. In a well-formed highly symmetric target, the output of the intercorrelation is expected to be well behaved. Based upon this, it is conceivable that a library of intercorrelation responses to varied target designs over all substrate and resist combinations could be used for target design optimization.

Another avenue used to address WIS is overlay target design. An example of typical overlay target design is shown in Figure 6. Typical target dimensions are for the inner box - a = 1-2μm b = 6 - 11 μm, c = 10 - 15μm and for the outer box: A= 1 - 4 μm, B = 16 - 26 μm C = 20 -

30 μm. (e.g. the classic box-in box structure of Fig 2 would have dimensions a=5, b=10, c=10 A=10, B=20, C=20). A number of possible combinations of segmented frame outer, solid box inner frame can be considered in relation to the amount of information derived from the structure, for a specific process layer. When evaluating target designs, it is also important to study different phases of the inner and outer boxes. Substrate trench structures will provide, in many instances vastly different contrast than mesa structures. In addition, the choice of contrast "tone " of the structure (e.g. resist inner box / substrate outer box vs. substrate inner / resist outer) will produce different measurement results. Typically the

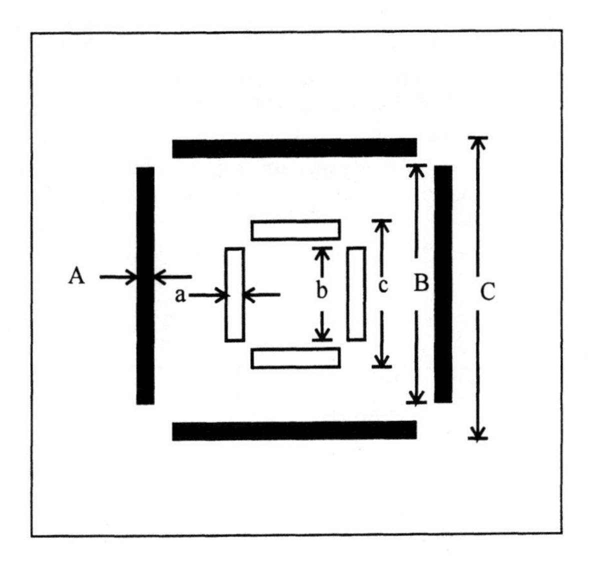

FIGURE 6. Typical automated overlay structure design

resist box provides well defined edges with little asymmetry (in the absence of proximity effects) whereas the substrate defined box is subject to process variations (grainy materials, reflectivity variations etc.). The best structures tend to be those which provide the greatest amount of redundant information for processing by the measurement algorithm. Although a detailed analysis of the target-to-process interaction must be performed for each process it is possible to define several generic overlay target design rules:

1. Make substrate level the outer "Box"
2. Maximize edge area for lowest contrast / highest noise edge
3. Use multiple edges for substrate and resist features
4. Select target profile to minimize z distance between levels.
5. Don't obscure substrate edges with resist overcoat
6. Strive to make targets comparable to reticle layout (e.g. contact and via levels % open area)
7. Avoid proximity effects near substrate edges (keep adjacent edges > 5 um from target edges).

OVERLAY METROLOGY CASE STUDY: W CMP

As process technology approached the 0.35 um generation, the use of Chemical Mechanical Polishing (CMP) became more pervasive. The increase in the number and thickness of layers being manufactured on the silicon wafer has resulted in large variations in the planarity of the surface. Such non-planarity's can often consume the entire Depth-of-Focus of the photolithography equipment. Further, such a high degree of non-planarity has been shown to impact device yield due to contamination following dry etch and poor step coverage of subsequent metallization layers. Initially CMP was relegated to polishing of interlayer dielectric materials (ILDs)[17] but as expertise in CMP grew, its application to other process modules followed. CMP has been implemented for shallow trench isolation; for polysilicon where improved planarization has resulted in reduced CD variation(2) and for metallizations (plug and inlaid) where reduced defectivity and improved recess control has been achieved. While very beneficial for process performance, CMP has resulted in extremely low contrast overlay (and alignment) targets which have significantly impacted metrology performance.

As the name implies, CMP is a process by which a sacrificial Oxide, metal, or polysilicon layer is removed through a combination of chemical and mechanical action. The wafer is placed on a mechanical carrier, circuit side down into a slurry (alkaline for oxide CMP and acidic for metal CMP). The carrier applies both downforce and rotational motion to the wafer which sits on a pad resting on the polishing table which is also in motion (rotational or linear) with respect to the carrier. The CMP process itself is very difficult to control. Planarization is a function of both wafer attributes (e.g. feature size, circuit density, and film type) and a host of CMP equipment parameters (time, downforce, slurry pH and particle size, pad hardness, pad condition, table rpm, wafer rpm, and etc.) many of which change over time. Additionally, the thickness uniformity of the planarized layer can vary significantly across both die and wafer. From an overlay measurement perspective this means that target behavior will vary across the wafer as well as from wafer-wafer and lot-lot. In oxide CMP, the material removal is heavily dependent on the mechanical aspects of the polishing. The goal of the oxide CMP process is to remove or minimize topography (perfect planarity) while maintaining uniformity of thickness for all affected films across a 200 mm wafer. Unfortunately for oxide CMP these two goals are mutually exclusive. This is in some ways fortunate for the overlay metrology function. The goal of overlay metrology target design (in direct opposition to the goal of the planarization process) is to maintain sufficient edge contrast (i.e. topography) for adequate measurement performance. The fact that perfect planarization can only be achieved at the (very high) cost of across-wafer non-uniformity, results in the possibility that some feature steps will remain. The design of the overlay measurement targets must take this fact into account and utilize additional understanding of the dependencies of planarization to achieve this end.

For Tungsten (W) CMP the slurry chemistry plays a more significant role than it does in dielectric CMP, enabling fairly high selectivity to oxide (as high as 150:1).[18] Additionally, due to grain size differences, different thickness of as deposited W will have different polish rates[19] further enhancing oxide selectivity. In theory, it should be possible, in this situation, to adequately control the CMP process to achieve the desired level of W removal. In reality however, the selectivity to oxide is greatly reduced as a function of pattern density[18] This phenomena, which is dependent on all aspects of the W CMP process (esp. pad hardness, slurry type, and thickness of as deposited W), is termed erosion and is shown schematically in figure 7 (right).

FIGURE 7. Dishing (left) and erosion (right) [18]

The dark areas in Figure 7 represent the oxide, the light areas in the figure represent the W. The impact of excessive erosion on the overlay target will be to completely planarize the edge step, rendering the target

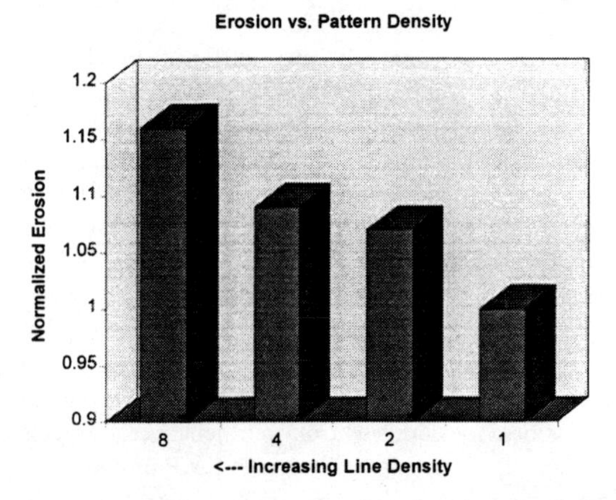

FIGURE 8: Measured Erosion as a Function of Pattern Density [18]

unmeasurable. This occurs because of higher polishing forces on the oxide edges. Since these dense areas clear first, the oxide between the W plugs is exposed to the polishing process for a longer duration. Higher local pressure exists in this case as well since less of the wafer oxide surface area is in contact with the polishing pad in these dense regions as compared to regions of large unpatterned oxide planes. Figure 8[18] depicts the magnitude of erosion to be expected for increasing line density. It is clear that for overlay target design, with respect to erosion, a relatively isolated feature would be expected to have minimal erosion vs. the rest of the die.

The other circuit dependent CMP phenomena is called dishing and is a function of feature width. This is shown schematically in figure 7 left. In this case, once the W has been removed from unpatterned areas (dark areas in figure 7) polishing of oxide will continue but at a much slower rate. However W patterned areas will continue to polish and those areas which are large enough for the pad to deflect into the W feature will produce non-uniform polishing from the center-out. Figure 9[18] depicts the magnitude of dishing to be expected as a function of increasing linewidth. For overlay target design, it is clear that the trenchwidth of the substrate box must not exceed a critical value (dependent upon the CMP pad hardness) in order to avoid dishing effects.

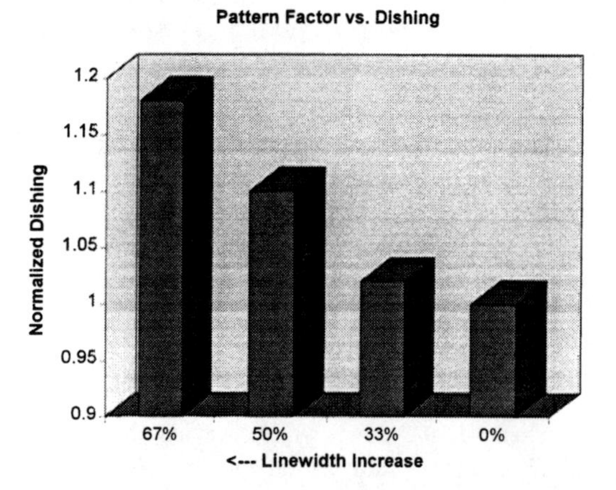

FIGURE 9. Measured Dishing as a Function of Linewidth.[18]

When excessive dishing occurs in an overlay measurement feature, the edge is effectively smoothed over the entire width of the trench (as opposed to having a clean sharp step transition from oxide to W). In summary, for overlay target design, the impact to edge contrast from planarity, uniformity, erosion and dishing must be given careful consideration. In some instances these phenomena can be used to promote edge contrast. Careful quantification of the dependence of erosion and dishing on circuit density and feature size (see figures 8 and 9), typically done during the development phase for the CMP process can be used to define design guidelines for overlay metrology targets.

As has been discussed, the primary effect of W CMP on the overlay targets is to significantly reduce the edge contrast by reducing and rounding the step height of the features used to define the edge. Prior to W CMP, typical edge definition for a standard W etchback process was on the order of 2000 Angstroms (measured by AFM to be the stepheight from oxide surround to the W surface). With W CMP this step is typically reduced to 200 Angstroms or smaller. This small step is then covered with metallization prior to the photo patterning. Such a feature poses significant challenges to the overlay measurement equipment. A treatment of the theoretical constraints of overlay measurement with respect to precise and accurate results has been performed by A. Shchemelinin et al[20] While the mathematics are beyond the scope of the present work, the general observations and conclusions are worth noting and will be related to system and target improvements for W CMP overlay. These workers found that: the measurement error is proportional to the square root of the noise. In this case the noise refers to all sources present in the measurement activity (electronic or Johnson noise and target noise). Another interesting finding, which supports the use of multiple edges, states that the measurement error increases with targets which contain intervals of constant signal (as would be found in the box-in-box target design). Thus increasing the number of edges and reducing their separation will result in improved measurement precision. Further support for the use of multiple edges was found in this analysis where the measurement precision is reduced by a factor between k and the square root of k (where k is equal to the number of lines present in the measurement target). Finally this theoretical analysis of measurement errors found that the measurement error is inversely proportional to the system resolution. Any system modifications which improve resolution (e.g. raise NA, lower illumination wavelength, use confocal or interferometric imaging) will improve measurement precision. Based upon considerations such as these, overlay equipment manufacturers have attempted to address the issue of low contrast edges through a number of novel approaches. These include; the use of phase information, Lens NA optimization, illumination optimization, algorithm optimization and signal processing / spatial filtering. Fundamental system design for W CMP overlay metrology examines the entire system as a whole.[21] In general, short wavelength illumination (for improved resolution) and a choice of bandwidth which minimizes both chromatic aberrations

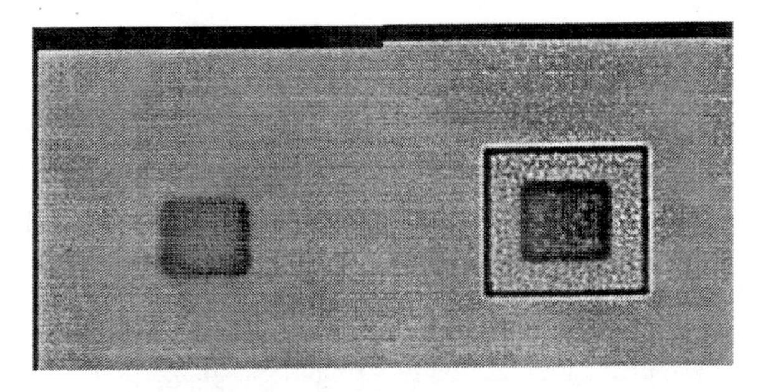

FIGURE 10. Contrast Improvements due to Image Enhancement [22]

and thin film effects[21]. Resolution, as discussed, contributes significantly to the performance of measurement algorithms. In many cases it is possible to exceed theoretical limits through spatial oversampling and Fourier based algorithms. However, for CMP, contrast not resolution, is the critical system parameter. Contrast enhancement can be accomplished by lowering the system NA. This is achieved through increasing the system response to lower spatial frequency elements by modifying the illumination towards coherence.[21] Significant contrast enhancement can be obtained with minimum impact to resolution (> 20% reduction). The danger with this approach comes as a result of increased diffraction effects. Finally elimination of all system induced noise (e.g. Johnson noise) must be accomplished to the point that the image noise is due solely to sample noise (e.g. background limited). Examples of the improvements from such system optimization are shown in Figure 10[22] (brightfield-left, enhanced-right).

A second system approach involves utilization of phase information. The optical phase change upon reflection from light waves traveling from the substrate as compared to those from the edge area is quite large and can be used to enhance the edge contrast. The commercial implementation of phase imaging[23] utilizes a Linnik interferometer to obtain the phase information from the overlay target. The Linnik interferometer utilizes an incoherent, broadband light source and relies on interference from the sample light path and the

FIGURE 11. Contrast Improvements due to Phase Imaging [24]

reference light path when the sample is within the coherence region. The resulting signal is produced from very small region in z (perpendicular to the plane of the sample) which is defined by the coherence length of the optics. As the sample is scanned in z, a coherence probe image (or "Cloud Plot" - refer to Figure 13) is constructed from the interference fringes within the coherence region. Contained within this image is phase information which provides the basis for the measurement algorithms. Figure 11 shows the comparison of a brightfield image (left) and a phase enhanced Coherence Probe image (right) of a typical, low contrast target at W CMP.

Measurement algorithms are another area where the overlay equipment has been significantly improved to accommodate the low contrast targets produced by CMP. One of the most straightforward improvements involves utilizing a priori knowledge of the overlay target (e.g. edge size and redundancy characteristics). This is an extension of some of the algorithmic methods discussed earlier. In practice the image enhancement attempts to utilize components known to be present in the image which the "noise factors" do not share. One such method[25] is to use the linear characteristics of the square shape of the typical overlay measurement target to enhance the image numerically. An example of this enhancement is shown in Figure 12[26] (brightfield-left, enhanced-right). Clearly the image contrast is greatly enhanced using this method. Comparison between stan-

FIGURE 12. Contrast Improvements due to a priori Target data[26]

dard brightfield measurements and the image enhanced measurements[25] has demonstrated significant improvements in precision. There was also an accompanying improvement in measurement accuracy, identified using stepper modeling. These same workers have also examined reducing the wavelength of the illumination and increasing the objective lens NA. This serves to reduce the depth of focus and improve the vertical resolution of the measurement instrument making it more sensitive to the low contrast edges of the CMP target.

EXPERIMENTAL:

AFM Characterization of the effect of the W etchback process on the overlay measurement targets (from figure 5: a=5 um, b=10, c=10; A=1, B=16, C=20 - resist box in substrate bars), which preceded the W CMP process, gave a step height into the substrate bar of > 2000 A. As the process was moved to W CMP, initially no modifications were done to the overlay targets. A dramatic degradation of the lot-lot variability of overlay measurement results was observed. AFM characterization of the W CMP targets showed a substrate edge step of 200 - 400 A. An attempt to utilize phase-based image enhancement techniques demonstrated success in reducing the measurement variability at the expense of throughput and ease of measurement setup. Figure 13 shows the brightfield image of the W CMP target. Very little edge contrast is available for measurement. Figure 14 shows a coherence probe generated cross-sectional image of the overlay target. The step height at the substrate edge, highlighted by the arrow, is quite small. As can be seen the phase enhancement technique can be quite effective in increasing the edge contrast. Short loop characterization of the W etchback process using overlay measurement performance and AFM step height characterization showed that a severe degradation in observed measurement variability occurred for step heights less than 1000A. An examination of the design variables available for the overlay target, in conjunction with available CMP characterization data led to target design

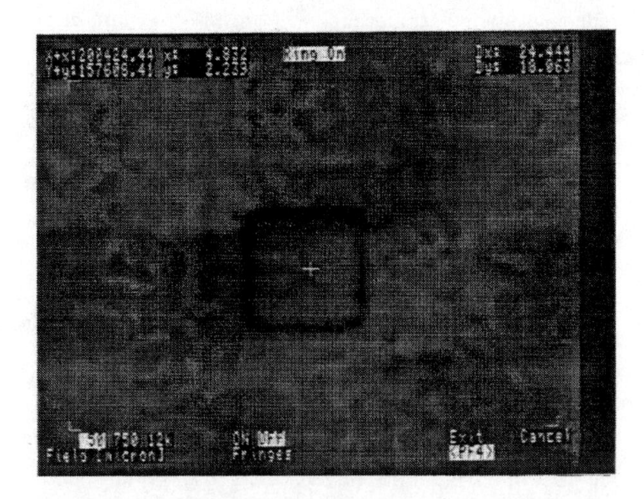

FIGURE 13. Brightfield View of W CMP Overlay Target.

modifications. For the experimental work: The substrate bar width (A in figure 5): A= 1, 2, 4 μm; The CMP parameters: slurry type, downforce, and overpolish; W thickness at two levels; and trench depth variations (by

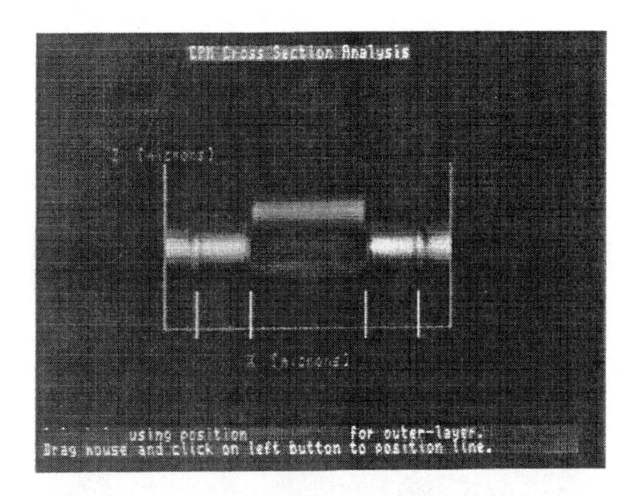

FIGURE 14. "Cloud Plot" of W CMP Overlay Target.

removing etch stops) were examined for isolated target structures. The primary output: measurement performance was supported by AFM characterization of the step height for each of the samples.

RESULTS:

Targets in which the W didn't completely fill the trench worked best. In the case of this experiment that includes targets of A= 2 um and 4 um. However edge contrast was lost in the larger targets due, for the most part, to

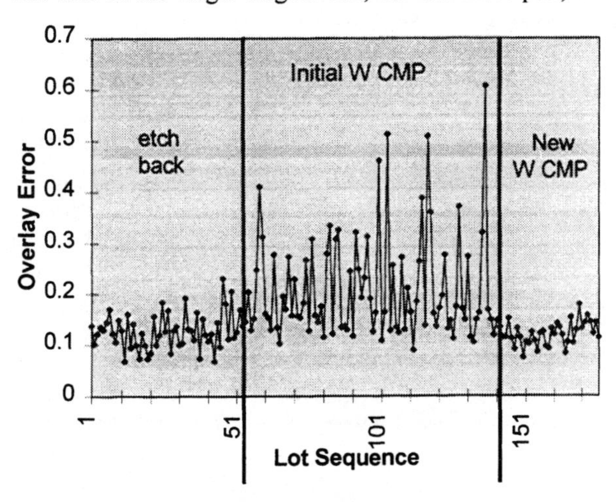

FIGURE 15. Manufacturing Data Demonstrating Overlay Target Performance.

dishing which smoothed the edge over the 4 um width, creating considerable uncertainty in the edge location for the metrology tool. From a CMP perspective; downforce had less impact on the metrology than overpolish and reduced W deposition created the best contrast targets. The 2 um target with the etchstop removed was designed into all product and test reticles. Figure 15 shows the resulting improvement on product material, due to the combination of the target modifications, the W thickness optimization and the trench depth increase. An improvement of nearly 3x from the "old" target with W CMP (Lot 52 - Lot 145) and the "new" target W CMP (Lot 146 - Lot 160) can be seen. Some small improvement is also visible from "new" W CMP as compared to the W etchback (Lot 1 - Lot 51).

DISCUSSION:

From the results achieved in the preceding study, in conjunction with some recent work by S. Hsu and co-workers at National Semiconductor Corp.,[27] it is now possible to articulate some overlay measurement target design rules for W CMP. In addition to the general rules listed earlier the following rules are proposed:

1) Substrate bar formed by isolated space to avoid erosion.
2) Maximize edge definition by maximizing post W CMP Step height:
    a) Minimized deposited W thickness
    b) Optimize trench width to 6 - 8 times minimum feature size (Max width determined by dishing).
    c) maximize trench depth (remove etch stops).

Work continues in this area to develop a single design rule which would account for the effects of erosion and dishing on the overlay target as a function of; target Critical Dimension, pre-W trench depth and W film thickness.

CONCLUSIONS:

Data has been presented which demonstrates the capability of current generation overlay metrology equipment for sub-0.25 micron process technologies, assuming moderate contrast overlay targets. It has also been shown that for certain process modules, most notably W CMP, serious effort must be devoted in the target design to maintain viable overlay metrology. Detailed examination of target response to CMP was used to design an experiment. From this, several broad design guidelines, both for generic overlay measurement targets and for target design specific to W CMP have been related. Improvements in overlay metrology have been shown to result, in an actual production application, from the use of these design guidelines.

As overlay metrology advances beyond the current focus on stable target design the next level of refinement with respect to CMP and its affect on these measurements must be undertaken. It has already been demonstrated that[28] physical changes to the substrate target (in a way which is analogous to the "snowdrift effect" discussed earlier) can be detected through the use of scanned probe measurements. These workers have discovered that an induced rotation, which is not the result of the lithography process, shows up in a systematic analysis of the target asymmetry.

ACKNOWLEDGEMENTS:

The authors wish to acknowledge Frank Krupa for his CMP expertise, his patience and his critical review of this work.

REFERENCES

1. Zych, L., Spadini, G., Hansan, T., Arden, B., "Electrical methods for precision stepper column optimization," *Proc. SPIE Optical Microlithography V*,Vol. 633-13, 98, 1986.
2. van den Brink, M., de Mol, C., George, R., "Matching performance fro multiple wafer steppers using an advanced metrology procedure," *Proc. SPIE* Vol. 921, 180, 1988.
3. Coleman, D., et al., "On the accuracy of overlay measurements: tool and mark asymmetry effects," *Proc. SPIE*: 1261, pp139-161, 1990.
4. Kirk, C., "A study of the instrumental errors in linewidth and registration measurements made with an optical microscope," *Proc. SPIE*: 775, pp 51-59 1987.
5. Hecht, E. and Zajac, A., *Optics*, Reading, MA, Addison-Wesley, 1979, pp 214ff and 425ff.
6. Born, M. and Wolf, E., *Principles of Optics*, 2nd Edition, New York, The Macmillan Co., 1964, pp522-526.
7. Troccolo, P., Smith, N. and Zantow, T., "Tool and mark design factors that influence optical overlay measurement errors," *Proc. SPIE*: 1673, pp148-156 1992.
8. Kawai, A., et al., "Dependence of offset error on overlay mark structures in overlay measurement," *Jpn. J. Appl. Phys*., **Vol 31**, pp385-390, 1992.
9. Han, J., Kim, H., Nam, J., Han, M., Lim, S., Yanowitz, S., Smith, N., Smout, A., "Effects of Illumination wavelength on the accuracy of optical overlay metrology," *Proc. SPIE* Vol. 2439, 1995.
10. Yanof, A., Windsor, W., Elais, R., Helbert, J., & Harker, C., "Improving metrology signal-to-noise on grainy overlay features," *Proc. SPIE* Vol. 2439, 1995.
11. Zavecz, T., "Lithographic overlay measurement precision and calibration and their effect on pattern registration optimization," *Proc SPIE*: 1673, pp191-202. 1992.
12. Starikov, A., et al., "Accuracy of overlay measurements: tool and mark asymmetry effects," *Optical Engineering*, **Vol. 31**, no. 6, pp1298ff 1992.
13. Starikov, A., et al., "Use of apriori information in centerline estimation," *Proc KTI Microlitography Seminar*, pp277-294, 1991.
14. Coleman, D. et al., "On the accuracy of overlay measurements: tool and mark asymmetry effects," *Proc. SPIE*: 1261, pp139-161, 1990.
15. Tanaka, Y. et al., "New methodology of optimizing optical overlay measurement," *Proc. SPIE*: 1926, pp429-439, 1993.
16. Hignette, O. et al., "A new signal processing method for overlay and grid characterization measurements," *Microelectronic Engineering*, **6**, pp637-643, 1987.
17. Ali, I., Roy, S., Shin, G., "Chemical Mechanical Polishing of Interlayer Dielectric: A review," *Solid State Technology*, Oct 1994, 63.
18. Rutten, M., Feeney, P., Cheek, R., Landers, R., "Pattern density effects in Tungsten CMP," *Semiconductor International*, 123, September 1995.
19. Kim, I., Murella, K., Schlueter, J., Nikkel, E., Traut, T., Castleman, G., "Optimized process developed for Tungsten CMP," *Semiconductor International*, 119,November 1996.
20. Shchemelinin, A., Shifrin, E., Zaslavsky, A., "Basic challenges of optical overlay measurement," *Proc. SPIE* Vol. 3050, 425, 1997.
21. Podlesny, J., Cusack, F., Redmond, S., "CMP overlay metrology: Robust performance through signal and noise improvements," *Proc. SPIE* Vol. 3050, 293, 1997.
22. Image courtsey of H. Vaule, Schlumberger Corp.
23. Davidson, M., Kaufman, K., Mazor, I., Cohen, F., "An Application of interference microscopy to integrated circuit inspection and metrology," *Proc. SPIE* Vol. 775, 233, 1987.
24. Image courtsey C. Harker, KLA Corp.
25. Yeo, J., Nam, J., Oh, S., Moon, J., Koh, Y., Smith, N., Smout, A., "Improved overlay measurement of CMP processed layers," *Proc. SPIE* Vol. 2725, 345, 1996.
26. Image courtsey of A. Gurnell, BioRad.
27. Hsu, S., Dusa, M., Vlassak, J., Harker, C., Zimmerman, M., "Overlay target design characterization and optimization for Tungsten CMP," *Proc. SPIE*, Vol 3050, 1998.
28. Mathai, A., Schneir, J., "High resolution profilometry for improved overlay measurements of CMP processed layers," *Proc. SPIE*, Vol 3050, 1998.

Metrology of Image Placement

Alexander Starikov *

Ultratech Stepper, Inc., San Jose, California 95134

Metrology of registration, overlay and alignment offset in microlithography are discussed. Requirements and limitations are traced to the device ground rules and the definitions of edge, linewidth and centerline. Precision, accuracy, system performance and metrology in applications are discussed. The impact of image acquisition and data handling on performance is elucidated. Much attention is given to the manufacturing environment and effects of processing. General new methods of metrology error diagnostics and technology characterization are introduced and illustrated. Applications of these diagnostics to tests of tool performance, error diagnostics and culling, as well as to process integration in manufacturing are described. Realistic overlay reference materials and results of accuracy evaluations are discussed. Requirements in primary standards and alternative metrology are explained. The role and capability of SEM based overlay metrology is described, along with applications to device overlay metrology.

INTRODUCTION

The microelectronics industry has had a long period of remarkable growth. From the time the integrated circuit (IC) was first introduced, the number of transistors per chip has steadily increased while both the size and cost of each device have decreased[1]. Rapid evolution of lithography and materials processing were among the primary technologies that enabled the increase of the device count to millions per substrate and the mass-production of computers. As customer expectation of system level reliability has increased, device reliability has improved and their failure rates have fallen[2] to below 10^{-10}.

Dimensional metrology of small device features on planar substrates has been a key technology supporting manufacture of ICs by microlithography. With decreasing device sizes, dimensional metrology is becoming ever more difficult. Nevertheless, the economics demand that the metrology sampling rates decrease. Even a rough estimate suggests that the overlay metrology sampling rates have already dropped below 10^{-8} measurements per device per layer.

Optical microlithography is already used for printing 250nm features. At this technology level, device ground rules[3] stipulate image linewidth control to <25nm and image placement to <75nm. By now, many aspects of microlithography and of dimensional control in IC manufacturing have become so advanced that they push the limits of the available materials, manufacturing tools and methods.

If the historical rates of change in device size, number, reliability etc. were extrapolated for another decade, many implications of a simple dimension scale-down[3] would be difficult to rationalize. Is there an end to the current trends in microlithography? Technology learning in metrology of image placement yields some insights.

What is metrology of image placement? What are its current practices and limitations? How can it be improved? When is an alternative required? These issues are reviewed in this paper.

DEFINITIONS

Device ground rules

Microlithography is used to print many layers of images that define the fine structure of thin film microelectronics devices. Design, manufacture and quality assurance of these devices involve a set of design rules. These design rules, in one way or another, affect every group involved in the microelectronics industry.

Metrology activities involve the use of the metrology systems on samples representative of the product, experiments, gathering and interpreting data. Metrology serves to control various manufacturing tools and processes, to meet design rules, as well as to assure the quality of a product. Technology characterization provided by metrology is used to extend the present technology, to forecast what may be manufacturable in the future and to anticipate technology obsolescence.

To define the common usage of the device ground rules[4] in microlithography, consider a convention illustrated in Fig. 1. The name *feature* denotes the simplest element in one *layer* of the design of a thin film device. Feature *linewidths* (LW, *critical dimension,* CD) in layer t (*reference* or *target*) and layer b (*resist, current* or *bullet*) are denoted LW_t and LW_b, respectively. If the edge coordinates are X_1, X_2, X_3 and X_4, then

$$LW_t = | X_2 - X_3 |$$

and

$$LW_b = | X_1 - X_4 |.$$

Likewise, feature *centerlines* CL_t and CL_b are

* Current address: Intel Corporation, 2200 Mission College Blvd., Santa Clara, California 95052-8119.

CP449, *Characterization and Metrology for ULSI Technology: 1998 International Conference*
edited by D. G. Seiler, A. C. Diebold, W. M. Bullis, T. J. Shaffner, R. McDonald, and E. J. Walters

$$CL_t = (X_2 + X_3) / 2$$

and

$$CL_b = (X_1 + X_4) / 2.$$

The centerlines are defined for two coordinate axes of each layer, typically in a Cartesian coordinate system with axes X and Y. Since microlithography involves patterning very many features in each layer, a set of all centerline coordinates or *registration*[5], is of primary interest. Registration vector field **R(x; y)** is made up of centerline vectors {x; y} of all features in one layer. In microlithography, devices are replicated with some periodicity and so registration is often referred to as *registration grid.* Registration is always defined with respect to some agreed on upon *reference,* $\mathbf{R_r(x; y)}$.

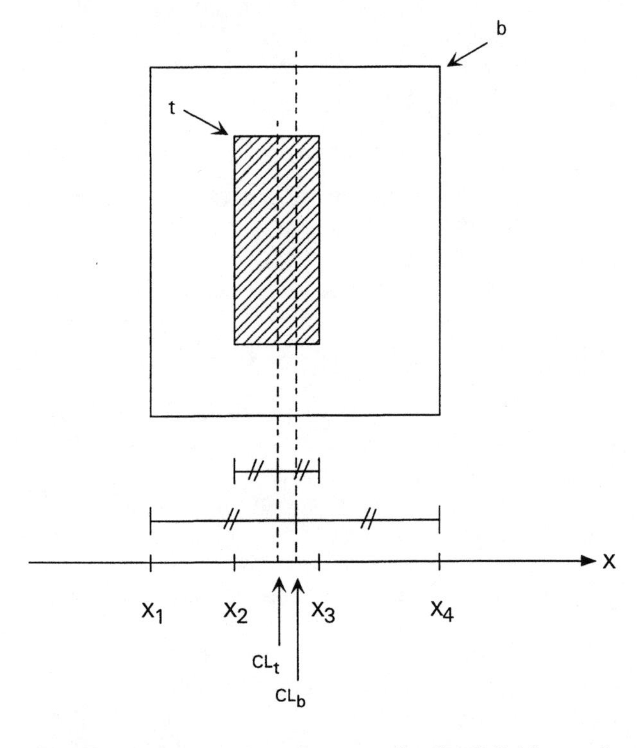

FIGURE 1. Illustrating the ground rule definitions of linewidth, centerline overlay and edge-to-edge overlay.

The difference in centerlines of one feature in one layer and one feature in another layer is called *centerline layer-to-layer overlay* or simply *overlay* (O/L). Referring again to Fig. 1,

$$O/L = CL_b - CL_t$$

Device ground rules require that CD and O/L be within specified tolerances. Dimensional metrology is conducted for both process control and quality assurance.

In addition, design rules specify requirements on *edge-to-edge overlay* for features of different image layers. Left and right edge-to-edge overlay in Fig. 1 is defined as

$$EE_L = X_2 - X_1$$

and

$$EE_R = X_4 - X_3.$$

Although edge-to-edge overlay may appear similar to linewidth, unlike linewidth, it involves two edges in two dissimilar materials in two different image layers. In the present metrology practice, edge-to-edge overlay is not measured directly. In order to control edge-to-edge overlay, CD measurements in each layer are made with CD metrology systems and centerline layer-to-layer overlay is measured with O/L metrology systems. The data are then mathematically combined to produce the estimates of edge-to-edge overlay.

Overlay budgeting

As in any manufacturing process, all of the groups involved in large scale manufacturing are governed by some kind of a social contract. In a mature industry with good control of manufacturing processes, such as manufacture of nuts and bolts, there are strict standards and the contract takes form of a *specification.* However, as microlithography pushes the limits of the available manufacturing capability, a more *flexible* social contract becomes the key enabling factor. It predicates both the manufacturability and quality of a product, as well as the efficiency of the whole manufacturing method. In the end, the optimal use of resources to manufacture a product is the key factor in the commercial success.

It is well recognized that overlay in thin film devices is a result of image placement errors incurred at many stages of the manufacturing process. Registration errors in reticles, at alignment and those due to manufacturing processes are among the contributors. The same factors that affect device O/L often also affect metrology of image placement. Based on this observation one may conjecture that a successful metrology for microlithography may only be achieved by a business process[6] similar to those used to assure quality[2, 7] in microelectronics manufacturing.

In order to achieve the required device O/L, and do so at a reasonable cost, it is necessary for all the parties affecting O/L to collaborate on O/L budgeting. Using a sensible model that describes components of device O/L, a portion of the O/L budget is allocated to each contributing error source. Each error source is then controlled, so that its total does not exceed the required tolerance. For example, for a 256Mb DRAM process with CD of 250nm, it is expected[3] that the O/L budget is <75nm. The error allocated to O/L metrology is 10% of the budget. That is, O/L metrology error must be <7.5nm.

What does an O/L metrology error of <7.5nm mean? Can this requirement be met? What is the impact of not meeting this requirement? Answers to these questions are needed in order to make sound decisions about the technology used to produce and assure image placement in devices.

The ultimate purpose of O/L metrology in microlithography is to limit the losses in device yield, performance and reliability that are due to O/L deviations from the nominal value of zero. Mathematical models estimating quantity of good fields address this objective[5, 8]. On the other hand, performance of O/L metrology may be stated as the mean error and three standard deviations:

$$O/L_x = |\langle O/L_x \rangle| + 3\sigma_{O/Lx.}$$

The mean error $|\langle O/L_x \rangle|$ of O/L metrology carries an especially heavy penalty in the O/L limited yield because it is introduced directly into device O/L by way of alignment offsets[9]. Consider its impact for a process with ±75nm O/L control limits. Assuming that the distribution of O/L data is Gaussian ($\sigma_{O/Lx}$ = 25nm and $\langle O/L_x \rangle$ = 0nm), it is expected that 0.27% of all O/L values are outside the control limits. A mean error of just 12.5nm would increase the failure rate to 0.62%. This is why the accuracy of O/L metrology is of much importance.

In evaluating accuracy of O/L metrology, we will make use of the concept of measurement uncertainty[10]. Unlike uncertainty in metrology on a single specimen, we are concerned with errors in a very large number of measurements on different metrology structures. We presume that O/L can be measured quite accurately by at least some means, no matter how slow and expensive. Since we cannot afford doing this in production and since the errors vary from one target to another, much larger O/L metrology errors are incurred in the applications. So, as a practical matter, our goal is to estimate those systematic metrology errors that cannot be removed in manufacturing. In our approach to evaluating measurement uncertainty, we first suppress the imprecision by averaging and apply all available calibrations. Then, for a population {E} of the residual measurement errors, measurement uncertainty is stated[11] as

$$U = |\langle E \rangle| + 3\sigma_E.$$

We also use this format in estimating the impact of errors from various specific error sources.

Linewidth and centerline

Metrology of linewidth and metrology of image placement for microlithography have much in common. Both involve the definition of an edge and estimation of edge coordinates.

What constitutes an edge and how to estimate the edge coordinates are among the most fundamental and difficult questions of dimensional metrology[12]. Consider a schematic presentation of a single feature in an image formed in a layer of thin film supported by a planar substrate, Fig. 2. The SEMI definition of linewidth[13], due to Nyyssonen and Larrabee[14], specifies the *linewidth* as the distance between two opposite line edges in the thin film material,

$$LW = |X_1 - X_2|,$$

where the *edge coordinates* X_1 and X_2. These coordinates are derived from the coordinated of the interface between materials, at an agreed upon height H from the substrate and in a single agreed upon cross-section of the feature.

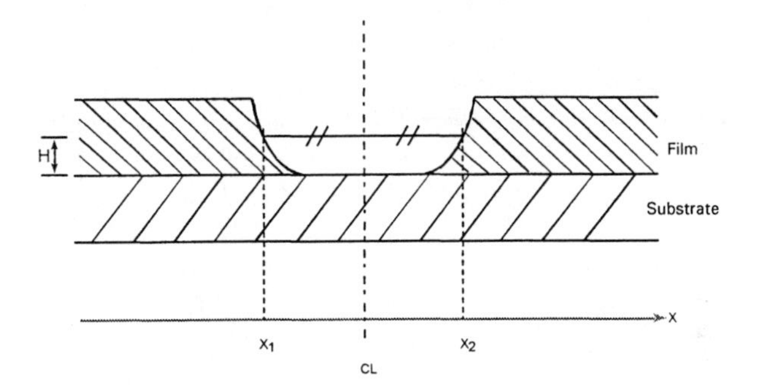

FIGURE 2. Definitions of linewidth and centerline in dimension metrology.

Clearly, a definition of centerline consistent with its use in microlithography must be linked to the definition of linewidth. Referring to Fig. 2, we define[6, 9] the *centerline* as an average coordinate of two opposing edge coordinates, that is,

$$CL = (X_1 + X_2) / 2.$$

Given the dimensions of interest and the realistic materials, these standard definitions are difficult to support with available metrology means [6, 9, 12 - 14]. Microscopic observations of real structures reveal a great deal of variability in the shape and material forming the sidewalls in the features of interest. As the device linewidth continues to shrink, variability becomes an ever larger fraction of the error budget.

Metrology of image placement

In order to manufacture devices with tight overlay, image placement is measured in three distinct metrology operations:

- image placement within one image layer or registration, is measured with X/Y metrology systems;
- alignment offset of a new image layer with respect to a prior image layer recorded on a substrate is measured with alignment systems of lithographic printers;

- layer-to-layer centerline overlay in images recorded on a substrate is measured with O/L metrology systems.

Each of these operations is made on special metrology structures, using different kinds of tools and supporting performance requirements of its own, so that the device O/L can be maintained.

Metrology of registration, alignment and overlay metrology are fundamentally related to each other, as well as to CD metrology. It is very helpful to view the tools, methods and limitations in metrology of image placement with this commonality in mind.

BASIC PROPERTIES

Reproducibility

Measurements must be *reproducible* [15]. By this we mean that repeated measurements of the same quantity should be somewhat similar. In testing reproducibility, a tool is used as expected and the variability of metrology is evaluated. Some elements of reproducibility relate to the random events in a measurement: photon noise, flicker and wander of illumination, electronics noise, conversion of analog signal to digital and so forth. In order to express the impact of random events, many measurements are made over a relatively short period of time, without changes in the state of the instrument. Sample or population standard deviation is then usually reported as the measure of variance. This is referred to as a *static precision* test.

In the course of making measurements on multiple targets and wafers, there will also be a small drift in focus, illumination, placement of a wafer on the instrument etc. By exercising many complete cycles of measurements, contribution to non-reproducibility of both random events and drift in sub-systems may be evaluated. This is referred to as a dynamic precision test. An inter-laboratory long term study of reproducibility may also be conducted along the same lines. Substitution of an operator, creation of a measurement sequence, or preventive maintenance may be a part of such a test.

Occasionally, an unexplained event may corrupt one of many repeated measurements, resulting in an *outlier*. Outlier removal [16, 17], an inherent part of metrology practices, is based on an *a priori* expectation of reproducibility. Culling of discrepant data, outliers, is made on the grounds of statistics applied to the repeated measurements of the same quantity. An appropriate statistical distribution is established from experience and the culling criteria are defined as best suited for the applications.

One property of the microlithographic environment is an expectation of sameness in the centerline estimates made on multiple samples selected along the length of a feature. Although reproducibility is a part of it, this expectation is based on more than just precision. It is expected from prior experience and before a new measurement is made, that is, it constitutes *a priori* information. Verification of this property is used in error diagnostics and data culling[11, 18, 19]. Unlike the use of statistics to reject an outlier from a set of many repeated measurements, these diagnostics are used in a single centerline estimate for one target in one layer.

Symmetry

The instruments used today in metrology of image placement can not readout the coordinates of the material-to-material interface in the feature sidewalls. The edge and/or centerline coordinates are estimated form an *image* of a structure. The accuracy of centerline estimation is supported by an *implied* assumption of symmetry[6, 9, 18, 20] of the target sidewalls and adjacent areas, of the imaging system and measurement procedure.

Symmetry, more than any other property, affects the accuracy of O/L metrology and alignment. A high degree of symmetry, and of tolerance to unavoidable asymmetries, are the essential attributes of an O/L metrology tool. On the other hand, it is the goal of process integration to assure the highest degree of symmetry in metrology targets. Much of this paper deals with improving symmetry, as a means to achieve accuracy.

Symmetry is the property we expect to be present in the O/L measurement targets and preserved by the metrology tools, that is, it is also considered to be *a priori* information. This form of *a priori* information enables diagnostics of metrology errors, culling and other forms of error reduction[6, 9, 18, 19, 21] in centerline estimation for one target in one layer.

Redundancy

A requirement of essence in microlithography is preservation of redundancy[6, 9]. Simply put, if two pairs of edges in one imaging layer are designed with a common (redundant) centerline, this relationship will be maintained at every stage. In fact, it must hold for all features in an imaging layer, after accounting for the built-in translation.

The reason why preservation of redundancy is so important for microlithography is because it is an extreme form of mass-production. A single imaging layer may contain many millions of individual features. Metrology of image placement on just a few strategically placed features assures image placement for all features linked by redundancy. Clearly, preservation of redundancy and tight control of image registration in one layer are closely related.

Redundancy is expected to be preserved at every step of the microlithographic process. Consider, a reticle written by a modern mask writer has registration errors <40nm (on the reticle) and its is maintained to within a fraction of a part per million (ppm or 10^{-6}). The imaging properties of a lithography system are designed and maintained so that when a print from a reticle is made, this level of registration error is achieved in the image (small distortion). Image recording and transfer processes are expected to maintain registration, as well.

In microlithography of integrated circuits, registration (and redundancy) is usually preserved very well. A consistent (and small) redundancy error is expected in the O/L measurement marks of one layer, even before a measurement is made. This form of *a priori* information about the measurement marks is also used for error diagnostics and culling[6, 9, 11, 18, 19] in the centerline estimation for one target in one layer.

Measurement structures

The users of metrology equipment expect some commonality across all available systems. It is in the user's best interests to have a choice among the substantially compatible metrology systems. In order to promote system-to-system compatibility, efficiency and competitiveness of our industry SEMI described[22] the designs of commonly used O/L metrology structures*.

An O/L metrology system is expected to handle the SEMI standard O/L metrology structures illustrated in Fig. 3. They are comprised of a set of the target and bullet level features. Both of these features are isotropic and mirror symmetric, with the same center. They are defined in a range of dimensions[22] to fit within a field of view of a typical O/L metrology system.

The simplest possible design is a box-in-box structure shown in Fig. 3a. It is comprised of a single solid square shape for both the target and bullet layer. For a given value of edge-to-edge distance between the target and bullet features, this is the most compact design. This structure can support the statistics and symmetry based error diagnostics and culling[6, 9, 11, 18 - 20].

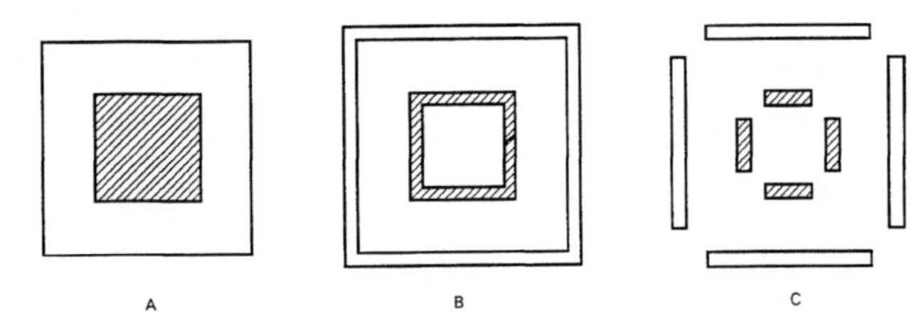

FIGURE 3. Typical O/L metrology structures.

Both the frame-in-frame design of Fig. 3b and the bars-in-bars design of Fig. 3c have two pairs of edges per axis per layer. These designs support the symmetry and statistics based diagnostics and culling. They can also support redundancy based diagnostics and culling and enable greater degree of measurement recovery[19, 20]. Built-in redundancy may also be used to improve measurement precision by averaging two redundant centerline estimates[11]. The differences between the frame-in-frame and bars-in-bars designs are small and process-specific. For example, a set of bars on the wafer which is coated with photoresist or chem-mech polished may lead to lesser asymmetry and thickness variation in the target and its vicinity.

Numerous studies of how the target design, polarity and linewidth affect robustness, precision and accuracy of alignment and O/L metrology can be found in literature. For example, some users reported[23, 24] that a box-in-box structure leads to exacerbated asymmetry and variation in the O/L metrology structures. They concluded that for their applications a frame-in-frame or bars-in-bars structure is better.

What is a better design of an alignment or O/L measurement structure? That can only be answered in the context of a specific application with a specific metrology system. To aid in this task, generally applicable in metrology of image placement diagnostics have been developed.

Metrology error diagnostics

Assumptions of statistical cohesiveness, symmetry and redundancy can be tested in the course of making an estimate of a target centerline. *A priori* information may be used to validate a measurement or to reject one. To do so, an *image* of the *target* is analyzed and measures of deviation from *a priori* properties are computed. If these deviations are acceptable, a measurement is validated. If they exceed some predetermined culling criteria, some image data used in a measurement or an entire measurement may be rejected.

Consider an example of the error diagnostics suite developed for SUSS XLS alignment system[19, 20]. This bright field optical polychromatic system produces a two-dimensional image of a three bar alignment target. Projections of the image along the bars is presented in Figs. 4. In order to estimate the centerline of the target, six edge coordinates are first produced using a maximum slope criterion. Their average is the centerline estimate. Quality of this estimate is evaluated from the data accumulated during intermediate computations.

Three redundant centerline estimates of the concentric pairs of edges are compared to produce the values of redundancy failure. When a redundancy failure exceeds a predetermined culling threshold, a centerline is culled. In the signal illustrated in Fig. 4a, redundancy failure exceeded the culling criterion of 125nm and one pair of edges was rejected from the computation of the target centerline. The remaining four edge coordinates were averaged, recovering alignment on this target. In this case, redundancy based diagnostics and culling corrected a gross 425nm error of alignment and recovered alignment on a partially corrupt target.

---

* Registration metrology marks on photomasks are also regulated by SEMI. This is not done in alignment, however, where there is much diversity in alignment targets and little brand-to-brand compatibility in alignment systems.

A measure of left-right asymmetry in each of three concentric pairs of edges (redundant centerlines) also can be estimated. For example, the ratio of two values of image intensity at edge coordinates serves as a measure of asymmetry. For the image illustrated in Fig. 4b, this ratio exceeded a culling criterion of 2:1 for two pairs of edges. Culling them and using the only symmetric pair of edges corrected a gross alignment error and recovered alignment.

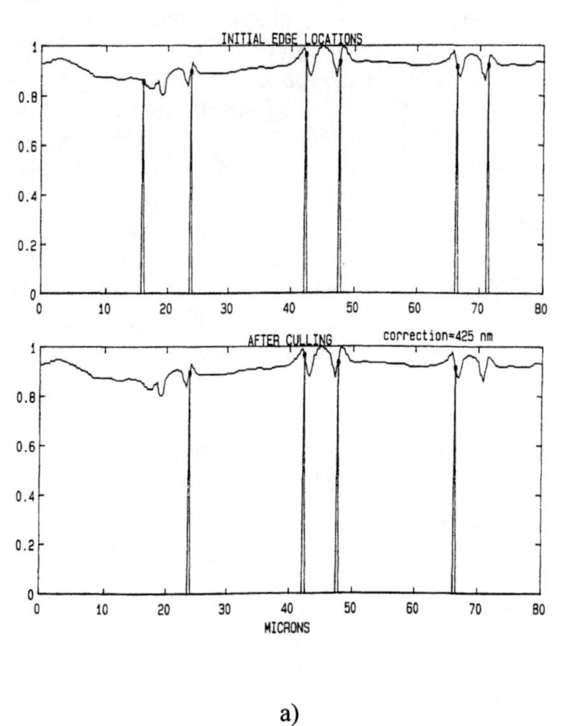

a)

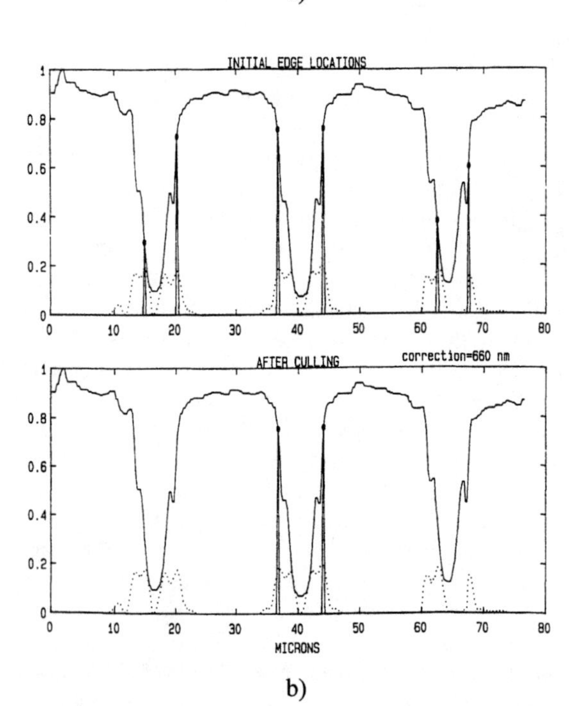

b)

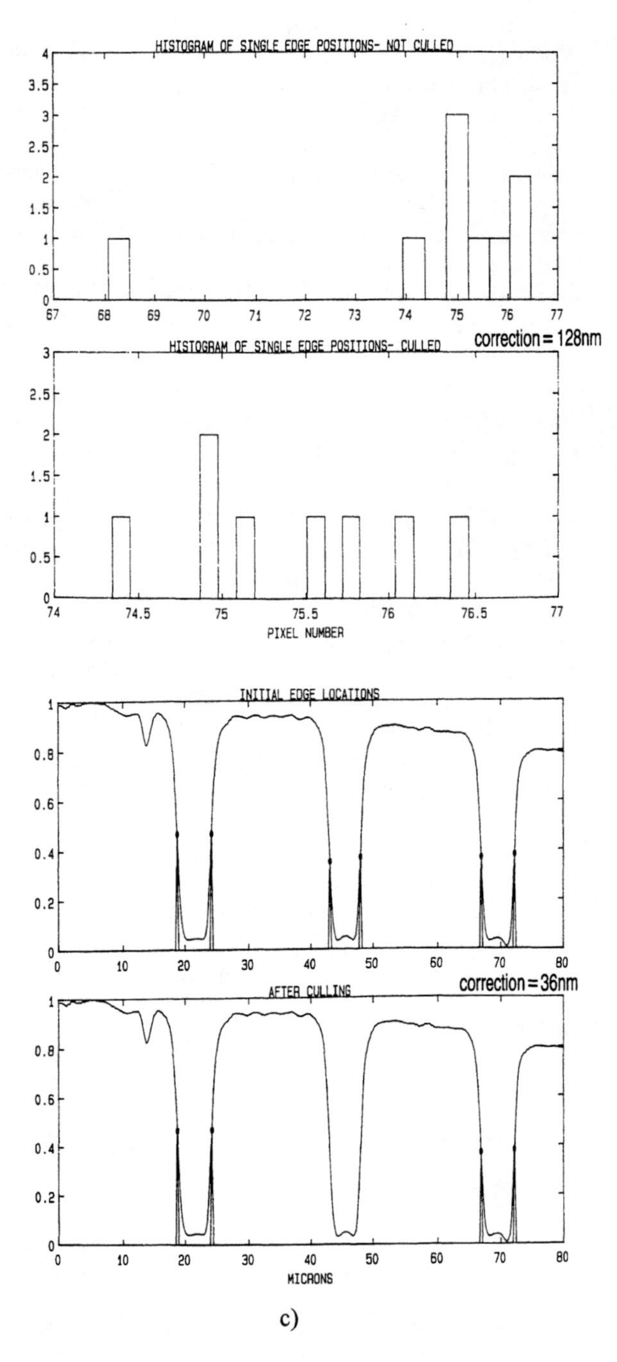

c)

FIGURE 4. Illustrating the image based metrology error diagnostics and culling by detecting: failure of redundancy (a), asymmetry (b) and spatially localized anomaly (c).

Since the two-dimensional image of the target is available, it is possible to estimate edge coordinates at multiple positions along the bars. One of the histograms of 9 estimates made in this implementation is shown in Fig. 4c. In this example, one outlier is rejected by a culling algorithm based on differences in pairs of estimates as a fraction of total range[16]. As a result, one edge coordinate was corrected by 128nm. The subsequent check of redundancy in the error diagnostics and culling algorithm resulted in rejection of one centerline. Alignment was recovered and corrected by 36nm.

Metrology error diagnostics and culling procedures illustrated here are generic, that is, they are widely applicable in metrology of image placement. Such error diagnostics, data culling and measurement recovery procedures have been implemented in other systems.

As the requirements for robust and accurate metrology of image placement tighten, the value of automated metrology error diagnostics rises. Not only are these diagnostics useful in data culling and measurement recovery, they are an efficient means of technology characterization. Metrology error diagnostics are becoming indispensable in process integration in manufacturing environment.

Optical O/L metrology systems

A typical O/L metrology system is a polychromatic optical bright field microscope. Illumination bandwidth is usually selected between 400nm and 700nm. Both single band or multiple user-defined illumination bands are available. Stable broad band W halogen and Xe arc long life light sources are common. These partially coherent optical systems use Köhler illumination with a filling ratio $\sigma > 0.5$. The primary measurement objective may have a numerical aperture between 0.5 to 0.95, depending on the model and system configuration.

These parameters of the imaging system may affect metrology performance. For example, it was shown[25, 26] that the illumination band strongly affects measurement accuracy and that a large filling ratio reduces[27] the sensitivity to asymmetry and granularity in the targets. It also was demonstrated[28] that a system with a higher NA yields better correlation of the O/L data taken in developed image to those in final image (after etch). Illumination band, numerical aperture, filling ratio and other parameters are subject to manufacturing tolerances. These variations affect both the job portability and tool-to-tool matching.

The early optical O/L metrology systems were equipped with multiple objectives and a conventional turret. That was required for wafer inspection, wafer alignment, navigation, O/L and CD metrology. The more recent instruments were designed as dedicated O/L metrology systems. They have a single O/L metrology objective that is either fixed or moved in a precision mechanism. Their simplified optical train provides a high degree of symmetry and temporal stability.

Focus and X/Y coordinate linkages

In the course of O/L metrology it is often necessary to acquire images of the target and bullet portions of an O/L measurement structure while focusing at two different planes. In doing so, three different types of inaccuracy may be introduced.

Dual focusing may result in a few nanometers of X and Y translation for every micron of movement in Z. In order to reduce this error, manufacturers of metrology systems can improve the hardware and introduce calibrations. Much improvement has been made in this area over the last few years. Correction of this error may also be made at the cost of lower throughput (see TIS). To enable further progress in stage technology, NIST is currently developing a stepped micro-cone structure[29] and test methods.

Digital signal processing (DSP)

A SEMI standard O/L measurement structure fits within a typical field of view of 30µmX30µm to 50µmX50µm. Unlike the earlier systems which used a scanning slit and a point detector or a full field video tube detector, the new systems use CCD cameras. Because these devices are spatially uniform, distortion-free and stable in time, introduction of CCD cameras improved accuracy and reproducibility.

The image of an O/L metrology target projected onto CCD is detected by a rectangular array of small detectors (pixels). The signal from each pixel is converted from analog form to digital (A/D conversion). Estimation of edge or centerline coordinates is carried out a variety of digital signal processing (DSP) algorithms. These algorithms may have a significant impact on performance. In order to properly use a metrology system, a metrologist must understand how a measurement is made. Although the algorithms may be closely guarded by a vendor for competitive reasons, a metrologist needs to know which algorithms are used and how they are implemented.

Some of the differences between CD and O/L metrology are evident in the their DSP and reliance on calibration procedures and standards. For example, in CD metrology a distance between positions of suitable transitions in an image of a sample is compared to that in an image of a certified reference material (standard). This way calibration of measurement can be made to arrive at an accurate CD estimate. In this sense, all CD metrology systems are comparators. CD metrology requires an essential similarity of all germane properties of a sample being measured and the standard, as well as similarity of the image of the sample and standard. Centerline estimation in conventional metrology of image placement, on the other hand, does not rely on such comparison, such calibration and such use of standards. Accuracy of O/L metrology is assured by maintaining symmetry in the sample and metrology system.

Differences between CD and O/L metrology, as practiced today, are also reflected in the DSP used to localize pairs of edges or centerlines. In centerline estimation, the image of a sample is compared with a symmetric replica of itself, not with an image of a standard. The image contrast, polarity, number of fringes and fringe separation may vary from one target to another, without necessarily impacting precision and accuracy of metrology of image placement. To achieve robust performance in applications, the algorithms used for centerline estimation rely heavily on *a priori* expectations of symmetry, statistical cohesiveness, redundancy, culling and various methods of error recovery.

Precision

Precision of an image placement metrology system is a function of numerical aperture, filling ratio, illumination band and bandwidth, of spatial sampling rate, electronics noise, digital representation and of the DSP algorithms used in centerline estimation. Hardware-limited precision of new O/L metrology systems has much improved from about $3\sigma = 30$nm in the late 1980's and is approaching $3\sigma = 1$nm. When the target quality is not the limiting factor, many commercial systems will perform at this level in both static and dynamic precision tests. High precision is often achieved on the O/L metrology targets formed in resist and many etched CVD films.

However, in addition to properties of the imaging optics, spatial sampling, A/D conversion and electronics noise[30], the signal-to-noise ratio (SNR) of an image of a real O/L metrology target is a strong function of the applications. In some cases, target or sample limited precision may be so poor as to render this metrology useless.

To improve precision on such targets, it is desirable to use as large as possible total length of target edges to boost the effective SNR. In this situation, a bars-in-bars target may be preferred over a box-in-box target. By using four edges in a bars-in-bars target, rather than two edges of a box-in-box target, it is possible to improve the target-limited precision by a factor of $\sqrt{2}$ or even better[11] (with culling).

In order to improve precision, it is also desirable to use an optical imaging system that produces an image with strong contrast (peak-to-valley signal). Various forms of interference microscopy are introduced to extend utility of the optical metrology systems in applications to levels with low amplitude contrast, e.g., chemical-mechanical planarization followed by metal deposition.

Accuracy

Accuracy of O/L metrology used to be taken for granted until a metrology crisis in the late 1980's, when many users reported systematic errors in excess of 100nm. Since then, a wide proliferation of new methods of performance testing and error diagnostics lead to significant improvement of metrology systems and increased the user awareness[31].

However, large errors are still reported in the applications on what is called "difficult" layers. These errors depend on the optical technology used in the tool. For a given tool type, these errors depend on the quality of tool manufacture and set up. They also vary due to the interactions with the adverse, but unavoidable, effects of processing on the O/L measurement structures. Complexity and variability of such effects lead to a situation where it is difficult to foresee all error mechanisms and to design or modify configuration of a metrology system, so that it is either insensitive or compensated for all error sources.

Overlay measurements may be made more accurate by characterization and removal of the error sources. The new methods performance testing and error diagnostics have been developed for this purpose. Although their scientific pedigree may not be obvious at the first encounter, these methods are intuitively understood and they have been widely confirmed as broadly applicable and extremely effective.

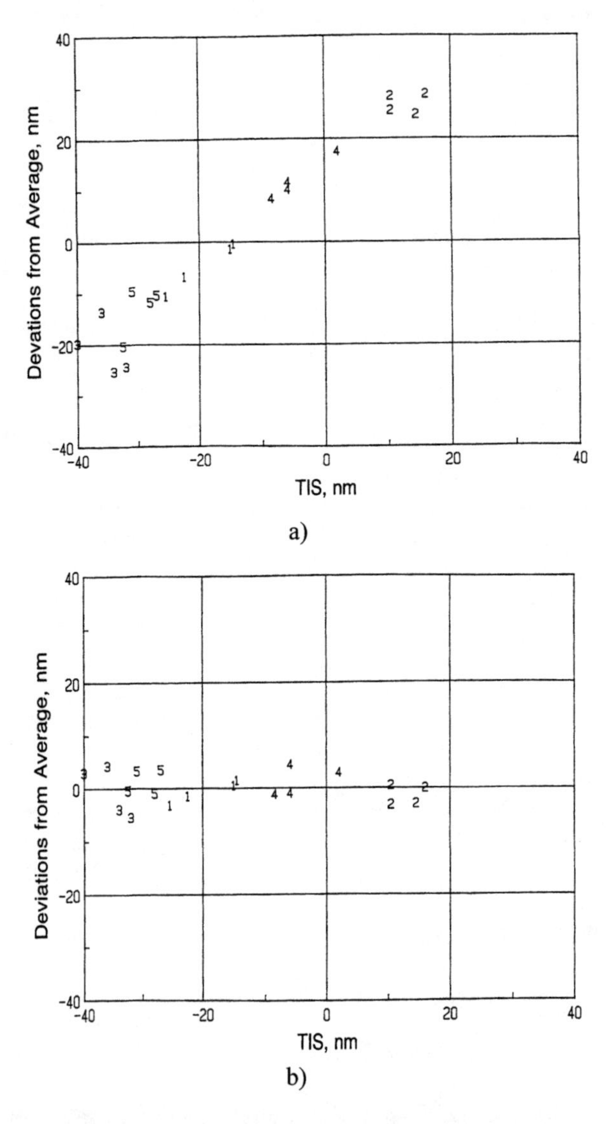

FIGURE 5. Tool-to-tool matching in a set of five systems of the same type from a population average on four sites on one wafer: a) using only the data taken at 0° and b) using an average of the data taken at 0° and 180° (that is, calibrated for TIS).

Tool-induced shift (TIS)

In order to estimate the impact of tool asymmetry on measurement error, a simple test was devised[20]. Two overlay measurements of the same target are made with sample at 0° and 180° orientation with respect to the tool. In the wafer coordinate system they should be the same, that is, $O/L_{0°} = O/L_{180°}$. A discrepancy between these two estimates is a measure* of O/L metrology error called tool-induced shift (TIS):

$$TIS = (O/L_{0°} - O/L_{180°})/2.$$

Using the estimates of TIS, it is possible to calibrate O/L measurements[9]. We define (in a wafer coordinate system) a TIS-corrected O/L estimate as an average of the measurements made at 0° and 180° orientation with respect to the tool,

$$AVE = (O/L_{0°} + O/L_{180°})/2 = O/L_{0°} - TIS.$$

Many metrologists have demonstrated that this calibration for TIS and reduction of sources of TIS are effective in improving system-to-system matching.

The example[9] illustrated in Fig. 5 shows that tool-to-tool matching between the systems of the same type may be improved from $3\sigma = 54.3$nm to $3\sigma = 8.7$nm by calibrating for TIS. Making measurements at 0° and 180° adds complexity and results in lower throughput. Another example illustrated in Fig. 6 shows that when TIS of these same systems is smaller, matching is better, $3\sigma = 11.4$nm, close to precision limit. TIS calibration only improves it to $3\sigma = 7.2$nm, about by $\sqrt{2}$.

Some sources of these errors are illustrated[9] in Figs 7a-c. Concepts of TIS as a measure of such error and compensation are simple and associated procedures are easy to apply in O/L metrology. Following the introduction of TIS as a measure of tool performance related to accuracy, users and manufacturers of optical O/L metrology equipment collaborated on reducing this form of error. Automated assessment of TIS and calibration for TIS became universally available. A number of experimental aides [32, 33, 34] were developed for TIS evaluation, tool set up and maintenance. Modeling of error sources and experiments brought about superior optical systems, designed and built exclusively for O/L metrology.

Since its introduction, TIS became an industry standard measure of performance for metrology equipment. Procedures for evaluation of TIS are available and fully automated. Enhancements have been developed that measure TIS over a range of some variables of interest, so as to assess errors in applications. Many systems also have the means to optimize set up parameters for best performance. From as much as 100nm observed[20] in 1989, TIS was reduced to <5nm. The same approach is applicable to other forms of metrology of image placement. Careful attention to TIS is already used to improve accuracy of alignment[35, 36].

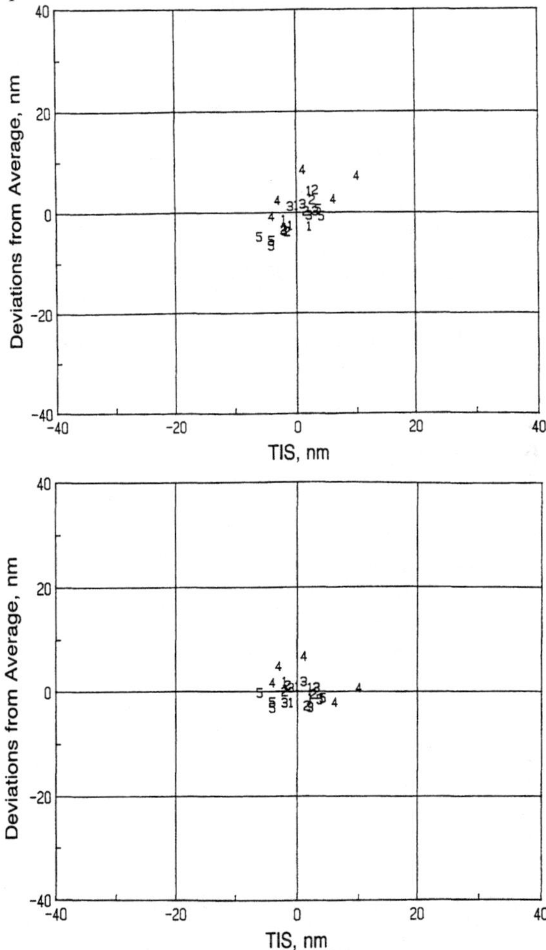

FIGURE 6. Tool-to-tool matching of the same tools as in Fig. 5, but when TIS is small.

An evaluation of TIS on a few samples does not yield an exhaustive measure of tool performance due to its asymmetries. In addition, as shown in Figs. 7d-e, there are some error mechanisms which affect the accuracy, even when the tool is symmetric or corrected for TIS.

Consider an example of tool-to-tool matching illustrated[9] in Fig. 8. Four bright field polychromatic O/L metrology systems of different configurations were used to measure the same sample that was used to produce data in Fig. 5. Even though corrected for TIS, the dissimilar systems did not match well because some other errors, different for each of them, were left uncompensated.

* This convention is intuitively obvious: the value of O/L is unique and it is linked with the O/L metrology structure on the wafer. In the tool coordinate system, the sign in equations should be reversed.

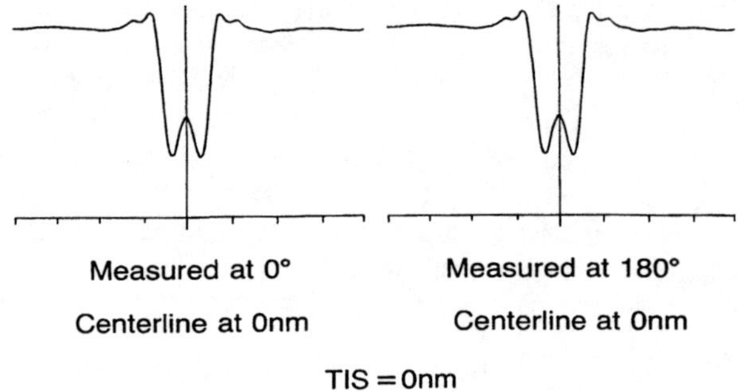

a) symmetric tool and mark

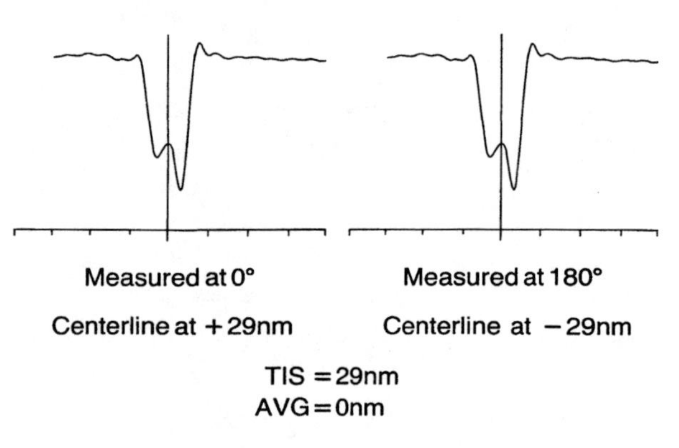

b) asymmetric illumination

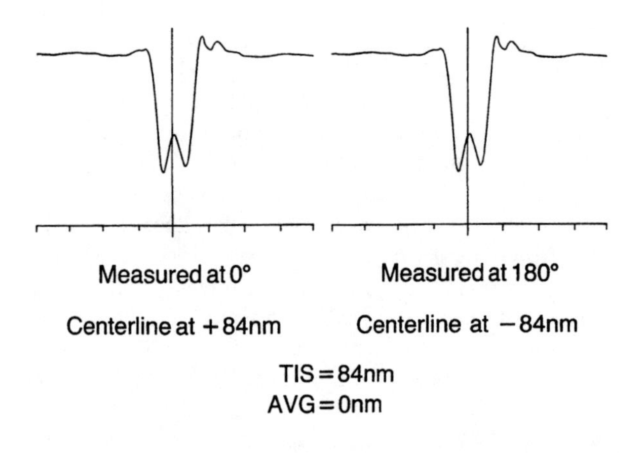

c) asymmetric imaging optics (coma)

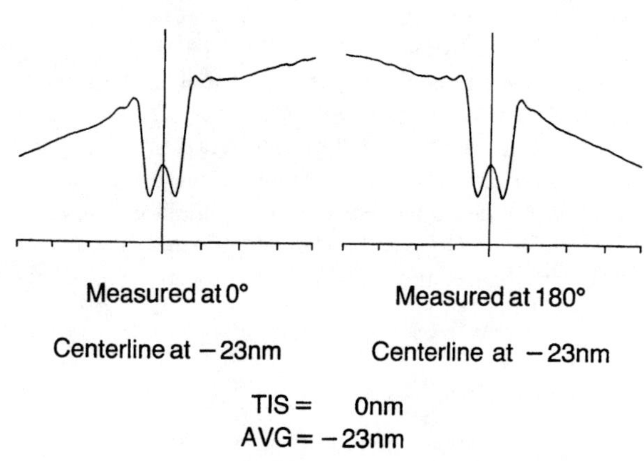

d) asymmetric photoresist

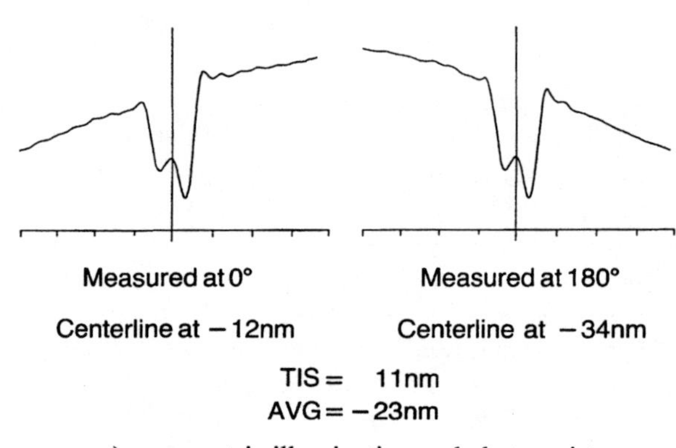

e) asymmetric illumination and photoresist

FIGURE 7. Examples of centerline estimation illustrating required symmetry (a) and the impact of asymmetries in the imaging system (b-c), measurement mark (d) and both (e).

Wafer-induced shift (WIS)

The notion of wafer-induced shift (WIS) was introduced[20] to account for the errors due to asymmetry in the properties of measurement structures. By WIS we mean an error that is attributable to asymmetry borne by the sample. This error would be incurred by a perfectly symmetric O/L metrology system. Such an error is exemplified by a symmetric tool on a symmetric target coated with asymmetric resist film, often resulting in an asymmetric signal, such as is illustrated[9] in Figs. 7d-e.It is now widely recognized that low TIS is required, but not sufficient for accurate metrology of image placement.

The magnitude of WIS is largely defined by technology and specific configuration of an instrument. For a particular type of sample asymmetry, one system type may be generally less liable to err than another. Selection of a detection criterion, focus and other parameters may also affect this type of error. In addition, for a fixed measure of sample asymmetry, WIS of a perfectly symmetric system varies as function of such parameters as film thickness, defocus, illumination band etc.

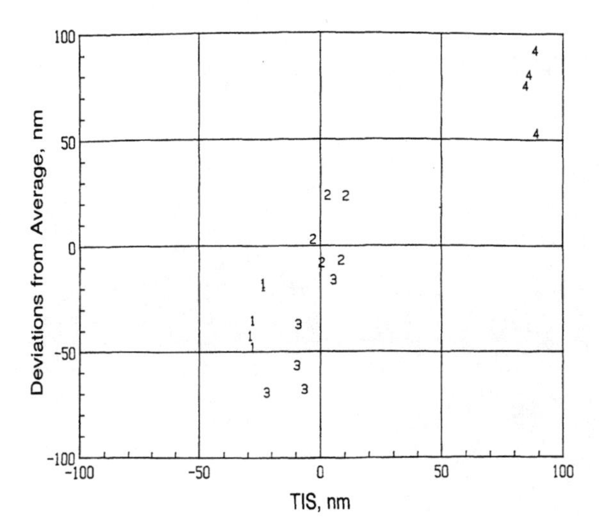

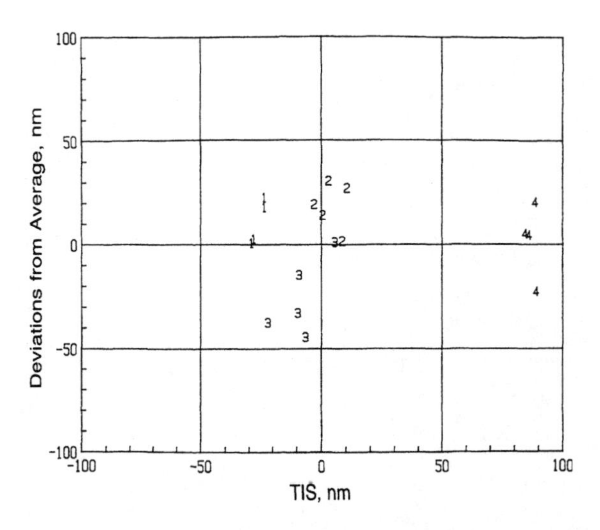

FIGURE 8. Tool-to-tool matching of four conventional optical bright field polychromatic systems with different optical configurations, focus and edge/centerline estimation criteria.

Unlike TIS, errors due to asymmetry in a wafer structure (WIS) are not that easy to estimate. However, physical properties of various forms of asymmetry in O/L measurement structures can be characterized[6]. Modeling the impact of asymmetry in a sample on viewing and resulting metrology errors[6, 9, 37, 38, 39, 40] is very useful in evaluating WIS. It is also possible to separate the tool related asymmetry from the sample related asymmetry[21]. Once the source and the impact of error are known, corrective actions may be taken.

METROLOGY IN MANUFACTURING

Device ground rules treat CD and O/L as if they were independent random variables. However, they are neither random nor independent. To illustrate this point, consider a metrology test site[6] whose photographs are presented in Fig. 9. This test site consists of device arrays, SEM and optical O/L measurement structures, isolated lines and gratings. The last image was printed in a 1 μm thick single layer resist (SLR) coated over the topographic features of the substrate. The photograph in Fig. 9a was taken on a bright field microscope with NA = 0.2 and with relatively coherent illumination with a 10nm band centered at 546.1nm, while that in Fig. 11b was made with NA = 0.9 and white light.

One can observe that the image size and image placement in the SLR images vary as a function of substrate topography and reflectivity variations. Exposure monitor structure[41] (EMS) was used to evaluate printing in a variety of environments. Consider the image of a narrow line and of an EMS (wider line) printed across an elevated area in the test site (bright square, 40μmX40μm, in the lower right portion of this image).

Reflectivity variations (Fig. 9a) clearly illustrate that this SLR is partially planarized and resist thickness is changing over distances comparable with 20μm. EMS, whose printed linewidth is about 5 times more sensitive than the narrow line to exposure dose, shows linewidth variations typical for reflective and edge notching in SLR. It is also apparent that both printed linewidth and centerline of SLR images vary as a function of position. The same effects are present, though hard to observe with the naked eye, in all SLR images illustrated here.

This example illustrates two important points. First, that the same physical phenomena which result in systematic CD variations in the thin film devices may also cause systematic O/L variations. Second, that both the device O/L and its metrology may be affected in similar ways.

Are the errors of image placement related to every aspect of printing and processing a part of the O/L metrology error budget? They should not be. However, such errors are first encountered in the course of O/L metrology. As the IC makers tighten the O/L control, these errors keep surfacing in the course of O/L metrology for the first time. Metrologists are the first to observe these problems. So, until they are attributed to their source and the proper owner is established, they may be considered a part of metrology error.

The real solution here is for the metrology community to interact with the other groups on O/L budgeting. In an efficient metrology business process, an error can be attributed and owner assigned, so that it be removed where is makes most sense. In this process, conventional optical O/L metrology and error diagnostics capabilities recently developed for O/L metrology are rapidly becoming the engine of technology characterization and catalyst of process integration. Many elements of the metrology business process[6] are typical to a continuous quality improvement program:

- use a comprehensive (systems) approach;
- analyze processes, sub-processes and points of hand-off;
- establish quantitative measures of metrology quality;
- automate gathering and analyses of quality feedback;
- assess quality of typical metrology (benchmarking);
- assess the failures of control (frequency and magnitude);
- establish the absolute values of error (to standard);
- account for technology limitations;
- rank the impact and cost to remove (Pareto analysis);
- remove the largest detractors first and re-assess.

This approach to metrology of image placement has proven extremely useful. In a sense, *this paper is as much about metrology of image placement as it is about process integration.*

Reticle

A detailed study of dimensional errors in binary conventional reticles is available[42]. It reports the butting errors at stripe boundaries >100nm. This work suggests that the image placement errors in reticles may be larger than would be expected, based on conventional sampling. This sampling and an assumption of Gaussian distribution of registration errors do not fully account for the systematic errors of the mask writer. Startling as these gross errors may seem, the extent of reticle butting errors is consistent with the magnitudes of stable redundancy failures observed in the O/L measurement structures[6].

The implications of image placement errors found in conventional reticles on device O/L and on O/L metrology are severe. Image placement error at a stripe boundary extended over an array of devices may result in a large number of devices exceeding the O/L control limits. Since the O/L metrology budget is only 10 % of permissible device O/L, the impact on O/L metrology may be much worse.

Centerline of a feature in a conventional reticle is determined by two edges in chrome on quartz[43]. In the less aggressive microlithography of the past, image placement of the reticle features correlated well with the placement of the printed image. As a result, maintaining tight control in relative placement of such centerlines in reticles was the only requirement related to reticle registration errors.

The advent of phase shift masks (PSM) and optical proximity correction (OPC) changed this paradigm. Image placement properties of a reticle with phase shifters are strongly affected by left-right symmetry of transmission, thickness, roughness, sidewall and, often, by the placement of phase shifter relative to chrome. Image placement properties of a chrome on quartz reticle with OPC correction also are affected by the symmetry of the sub-resolution correcting shapes adjacent to a feature. Assurance of registration error in such reticles requires control of multiple parameters, rather than just one.

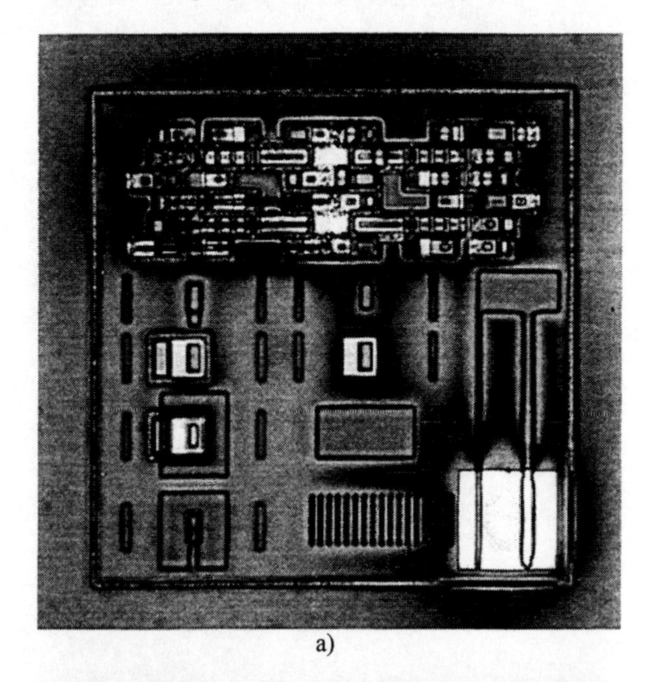

a)

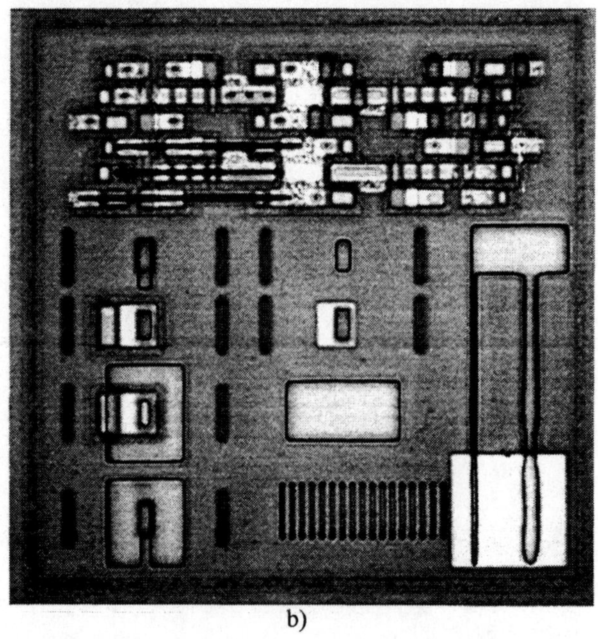

b)

FIGURE 9. Metrology test site at resist develop. Viewed at low NA and with narrow band light (a); high NA and white light (b).

In the new lithography paradigm[44, 45], transmission function (amplitude and phase) of a reticle is modified in order to improve performance of the entire manufacturing process. Properties of the OPC and PSM reticles are inextricably linked to the properties of the imaging systems, image recording and transfer process as well as to the measures of performance. The impact of the new lithography paradigm on the image placement errors and on metrology of image placement is likely to be profound.

Image formation

Registration errors found within a reticle will be modified in the course of image formation on a lithographic printer. Registration errors in aerial image differ from those in the reticle (at wafer scale). The simplest form of this error in image placement is a slowly varying in space local translation, known as distortion. Lens distortion is a well-known error of the optical projecting imaging systems and it is tightly controlled.

However, lens distortion alone is neither complete nor adequate to account for image placement errors at imaging. Indeed, these systems are used to print a wide range of images over a variety of exposure-defocus conditions. In such an environment, image placement error of devices varies as a function of feature type, size and polarity, as well as focus and exposure conditions.

It is well-known that the asymmetric aberrations of the imaging optics may result in considerable lateral shift [46, 47] of the image of a pinhole or a narrow slit, even though the image quality may remain high. It is also well-known that the aberrations and set up of the condenser[48, 49, 50, 51] may lead to considerable and variable errors of image placement[6]. The full impact of asymmetric aberrations and of illuminator misalignment on image placement and on its metrology is only beginning to be appreciated.

Image recording

As we have already observed in Fig. 9, the process of image recording may lead to errors of image placement when the SLR (or the substrate below it) is asymmetric. Accuracy of measurements made on O/L metrology structures located in proximity to substrate topography may be affected severely. However, *once* the errors are *identified* and *attributed*, the magnitude and/or frequency of errors *may be reduced.* To make this correction *efficiently*, metrology *error diagnostics* have been developed[6, 9] and the feedback of estimates *automated*[11, 19]. Numeric estimates of metrology errors were used as a key part of the O/L metrology *business process*[6]. This enabled a systematic review of quality in metrology of image placement and its process integration in order to achieve the highest performance at the lowest cost.

One example of process integration addresses the errors of image recording. Our initial observations[20] of these errors were up to 200nm. Having attempted to reduce these errors in the O/L metrology targets of a newly released scribe, we re-assessed them. Automated error diagnostics[11] provided the required quality feedback with a minimal loss of tool throughput. Figs. 10a and b present a summary of our observations for two O/ L metrology structures. The photographs of Fig. 10, made with a low NA bright field microscope and narrow band illumination, suggest that SLR standing wave effects due to adjacent

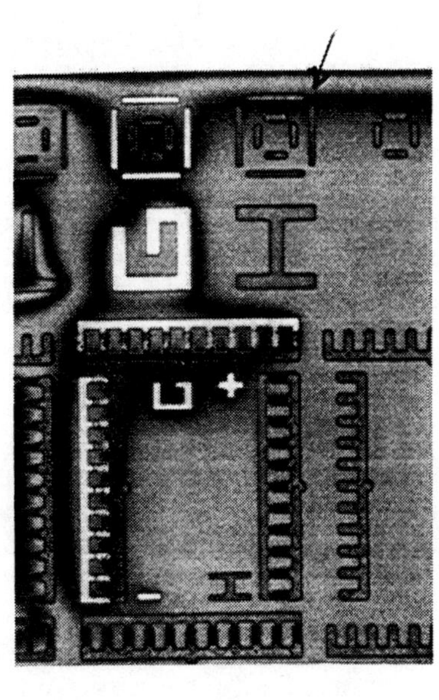

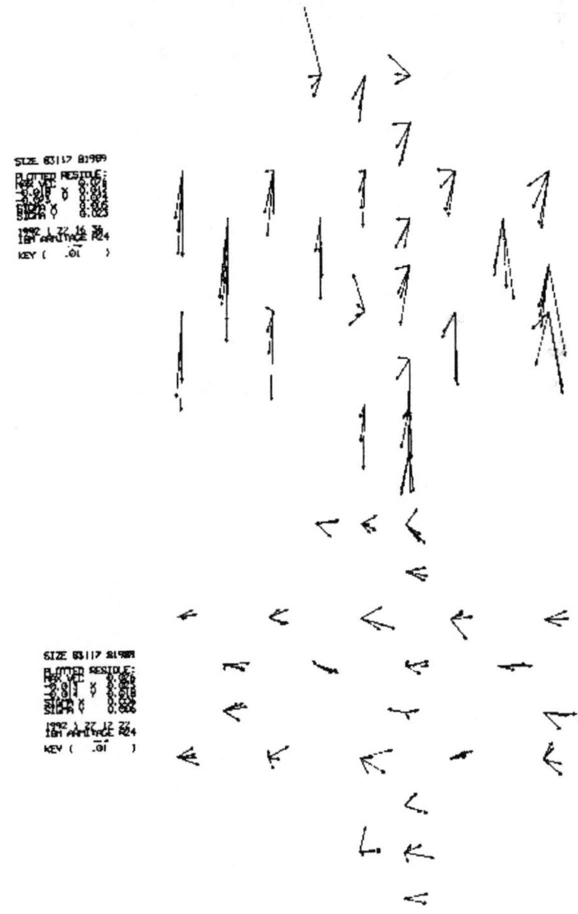

FIGURE 10a. Assessment of O/L metrology error in metrology process integration: large errors due to topography and SLR.

scribe and chip structures are readily observable for an O/L structure of Fig. 10a, but much less for that of Fig. 10b. The numeric estimates of the O/L metrology errors were reported as redundancy failure for the target (substrate) and bullet (resist) level bars. These errors, analyzed with KPLOT[52] data analysis package, are also shown in Fig. 10.

For the bullet portion of the O/L metrology structure in Fig. 10a, the estimates of measurement uncertainty solely due to redundancy failure are

$$U^X_{RED} = |< RED_X >| + 3\sigma_{RED} = 20nm;$$
$$U^Y_{RED} = |< RED_Y >| + 3\sigma_{RED} = 96nm.$$

These errors are due to the SLR effects at printing, but also due to viewing of the bars formed in asymmetric resist. These errors were triggered by the adjacent topography in the scribe (to the left from the target) and the chip (above the target).

A much better result is possible and it can be achieved through a further dialog with the scribe (kerf) and chip "neighbors". Indeed, by better placing the O/L metrology structure (or by eliminating the adjacent topography) these errors were reduced, see Fig. 10b:

$$U^X_{RED} = |< RED_X >| + 3\sigma_{RED} = 11nm;$$
$$U^Y_{RED} = |< RED_Y >| + 3\sigma_{RED} = 15nm.$$

The largest remaining component of redundancy error, $|<RED_Y>| = 9nm$, is likely due to mask making. Since this error is constant and the source is difficult to eliminate, an efficient solution here is calibration.

Estimates of measurement uncertainty solely due to redundancy failures in the target portion of the O/L structures in Figs. 10a and 10b are comparable. These errors appear to be unrelated to the adjacent topography. These observations are consistent with the process flow: the substrate structures were printed and polished well before the appearance of topography.

METROLOGY AND PROCESSING

Semiconductor processes are an endless source of the most challenging alignment and O/L metrology error mechanisms. Occasionally, the goals of processing and metrology may seem to be quite different. Consider an example of advanced back end of the line (BEOL). The best CD and electrical properties of metal interconnections are achieved when chemical-mechanical polishing (CMP) results in global planarization, that is, in a flat top surface. But, once the metal is sputtered and the resist is spun on, conventional alignment systems and O/L metrology systems *will* fail. After all, they are *optical* systems. A flat metal film is a mirror through which a substrate target *cannot* be observed.

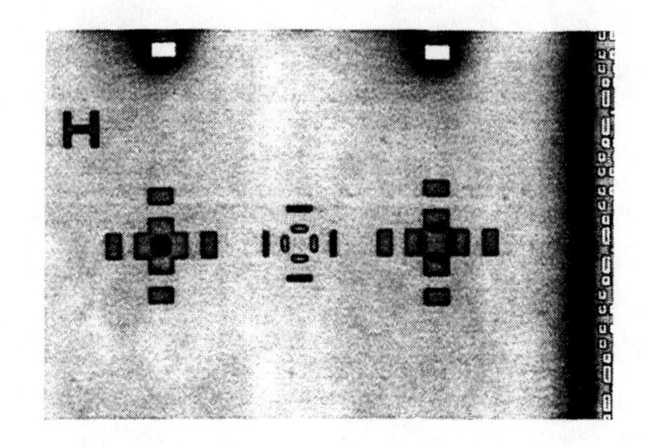

FIGURE 10b. Assessment of O/L metrology error in metrology process integration: large errors due to topography and SLR.

However, *the ultimate goal of IC manufacturing is to make high quality product at reasonable cost, so that the people in metrology and processing make their living.* When the social contract of the various groups involved in IC manufacture is seen this way, a solution is *always* found.

To arrive at the most efficient solution, a metrologist needs to know about processing and visa versa. Quantitative feedback of metrology error diagnostics provides the relevant measures of quality. It provides the subject and the language for the dialog, helps driving the corrective actions. The result is an effective business process, the outcome of which is the required performance achieved at the least cost.

Asymmetric resist coating

Characterization of the asymmetry in resist coverage over metrology structures was undertaken in an effort to better understand and solve the problem. A quantitative study[6] of resist flow across the alignment structures was carried out with a Tencor Alpha-Step 200 contact profilometer. A set of surface relief profiles formed by 1µm novolak resist spun across 370nm deep trenches is shown in Fig. 11. These trenches are located about 2 inches from the center of the wafer and trench widths range from 1µm to 15µm; direction of resist flow is from right to left. The test structures, data collection and processing afforded reduction of the asymmetries associated with the tip of the diamond stylus and direction of scanning. Measured resist profile was leveled, referenced to the top of the substrate and centered with respect to the trenches below. The left-right asymmetry readily seen in of the photoresist profiles of Fig. 11 is the trigger of inaccuracy in centerline estimation.

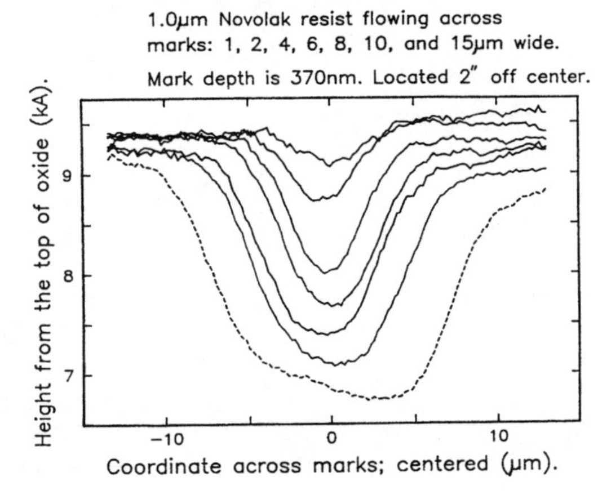

FIGURE 11. Surface profiles of spin coated novolak photoresist.

A qualitative assessment of the problem may be gained by review of microphotographs in Fig. 12. These observations were made with a low NA bright field microscope and narrow band illumination, an optical configuration somewhat similar to those of some early alignment and O/L metrology systems. Owing to interference of light in a thin film of resist, the image contrast is strongly dependent on resist thickness. Resist thickness and asymmetry vary across the wafer. Resulting variability of the image contrast, anti-reflection and left-right asymmetry in the images of alignment targets lead to algorithmic failures, poor signal-to-noise ratio and gross errors of centerline estimation.

A quantitative assessment of inaccuracy of centerline estimation is possible with a physical model that predicts an image produced by an optical metrology system on a sample of interest. This approach to optical CD metrology[12, 14, 43, 45, 53, 54] was pioneered in the 1980's by Larrabee, Nyyssonen, Kirk and Potzick at NBS (National Bureau of Standards, now NIST). The same approach was successfully applied in studies of alignment[37-40]. Modeling of alignment errors due to resist asymmetry (and other sources of error) proved extremely useful for understanding the causes of errors of centerline estimation in conventional optical alignment. Analyses were also done for conventional optical O/L metrology systems[6, 9, 20]. An example illustrating the notion of WIS, as error of measurement due to asymmetric resist coating, is in Figs. 7d and 7e.

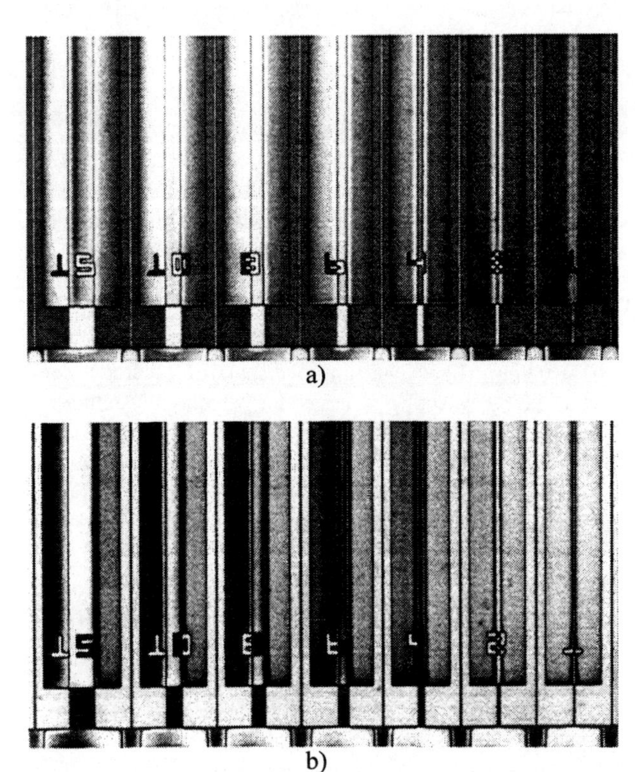

a)

b)

FIGURE 12. Illustration of the error of viewing through spun on photoresist over trenches (a) and islands (b).

Modeling of metrology error sources became indispensable in development of high performance alignment and O/L metrology systems. A majority of the recently introduced alignment and O/L metrology systems underwent exhaustive trials by performance modeling, which by now became an inherent part of design evaluation. As a result, vendors rapidly evolved their conventional optical systems and, more recently, those with interferometric or phase imaging.

Asymmetry in sputtered metals

Sputtering of metals is commonly used in IC manufacturing to deposit metals for device interconnection or metallization layers in BEOL[55, 56]. For example, a W plug may be formed by first etching a via hole in the inter-layer dielectric (ILD) and then sputtering metal until the via is filled. On those layers where via linewidth is large and aspect ratio (depth/CD) is small, bias sputtering is used. Given the size of a metal target and sputtering chamber with respect to the size of a wafer, metal incident upon the surface predominantly travels from the center. In this situation, a target perpendicular to radial direction is covered with an asymmetric[57] metal film.

An example of W sputtered over a 1μm via with an aspect ratio of 0.55 is illustrated* in Fig. 13. A physical model[57] of the sputtering process, SIMulation by BAllistic Deposition (SIMBAD), produces a vivid account of both the gross and fine structure of deposited tungsten, Fig. 13a. The V-shaped groove in the top surface of W is apparently shifted with respect to the center of the via, as defined by its sidewalls. The cross-section SEM microphotograph in Fig. 13b is consistent with the model.

This process has some similarity with spin-on deposition of resist. It is also associated with an asymmetry of a film over the target and results in inaccuracy of centerline estimation. Image placement errors in metal to via alignment were found[58] to be radial across the wafer with $3\sigma = 0.9\mu m$. Significant improvements of accuracy were achieved by selecting a target for which the sputtering process leads to the smallest error (and asymmetry). This is another example of a situation where process integration usually provides an efficient solution.

Granularity in sputtered metals

Sputtering may also result in granularity and voids in filling (see Fig. 13), spontaneously formed clusters of single metal (precipitate) and increased variability along the length of a feature. When such structures are used for O/L metrology, an increased variability of centerline estimates is the result.

Increasing the sampling size from within an image of an O/L metrology structure, and averaging, help to some extent. On the other hand, statistics based diagnostics within an image of a target is even more effective when dealing with small distance spatial variability in sputtered metal. Culling[11, 19, 20, 35] of the small and inconsistent with the rest of the image portions of an O/L measurement structure allows to recover a measurement. This practice is different from statistical methods of outlier rejection[16, 17]. Validation or culling of data within an O/L metrology structure, relying on an *a priori* expectation and known measure of sameness of the structure along its length, is entirely appropriate* within a metrology integration business process.

a)

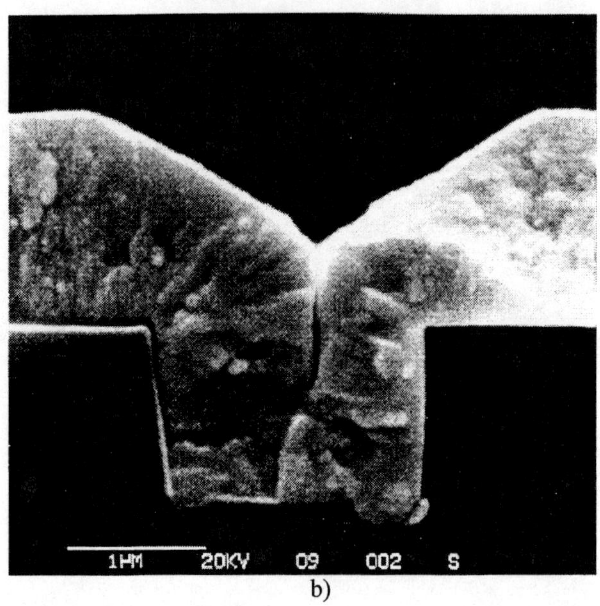

b)

FIGURE 13. Illustrating properties of W sputtered over a trench: SIMBAD model (a) and SEM cross-section (b).

Consider an example[6, 11] of O/L metrology on a BEOL metal layer with severe granularity and precipitate in AlCu metallurgy. Metrology sector in the manufacturing line reports poor O/L. The O/L related rework rate is high. There also is incidence of very large individual O/L measurements, but a visual review usually brings no explanation as to the cause - a metrology bottleneck.

* From S. K. Dew et al., *Modeling bias sputter planarization of metal films using a ballistic deposition simulation,* J. Vac. Sci. Technol. **A** 9(3); 1991. Reprinted with permission.

* One must be cautioned, however: indiscriminate culling outside the integration process may obscure the frequency and magnitude of anomalies in materials. When these properties are inconsistent with device ground rules, they may lead to device failures.

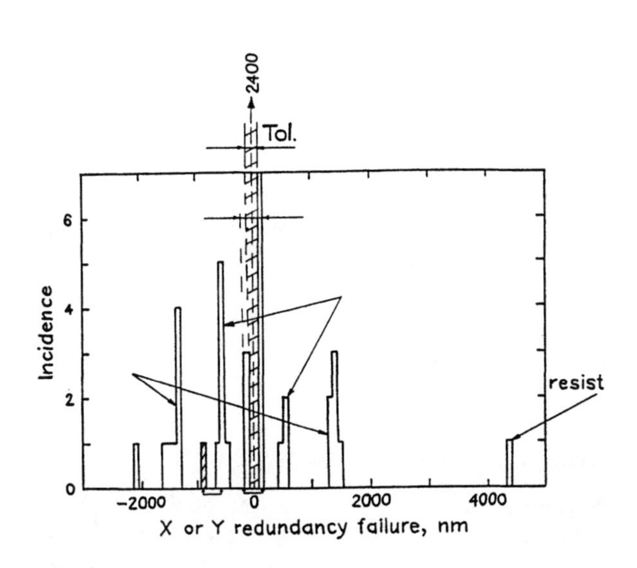

FIGURE 14. Redundancy failures in resist and metal bars-in-bars metrology structures.

Things changed with the introduction of O/L metrology error diagnostics and culling[11]. Redundancy errors in resist bars of the bars-in-bars targets are typically small. On the other hand, these errors are quite large in metal bars. To deal with that, when the two redundant centerline estimates in the metal bars were found discrepant by more than 70nm, an O/L measurement was automatically culled. The long term observation of the error diagnostics[6], shown in Fig. 14, revealed that the gross metrology errors occurred in only about 1.3% of measurement attempts. The primary culprit turned out to be the rough metal and Cu precipitate in the vicinity of the edges. For those measurements that passed culling, redundant centerline estimates were averaged. As a result, dynamic precision on this metal layer improved from 3σ = 16.3nm to 3σ = 8.5nm[11].

Implementation of O/L metrology error diagnostics and culling on metrology systems put an end to the apparently poor O/L performance and resulting re-work, metrology bottleneck and "bogus" data. Similar results were achieved with the error diagnostics implemented on a different metrology system[59].

Chemical-mechanical polishing (CMP)

To assess the impact of asymmetric ILD and film thickness variation on accuracy of alignment and O/L metrology, consider a target processed with W CMP illustrated in Fig. 15a.

Owing to the long distance effects of CMP and non-uniform distribution of W filled areas, the SiO_2 ILD may become asymmetric in the vicinity of the target. Viewing of the W filled target portion of the O/L metrology structure is affected by the asymmetry of the clear ILD film.

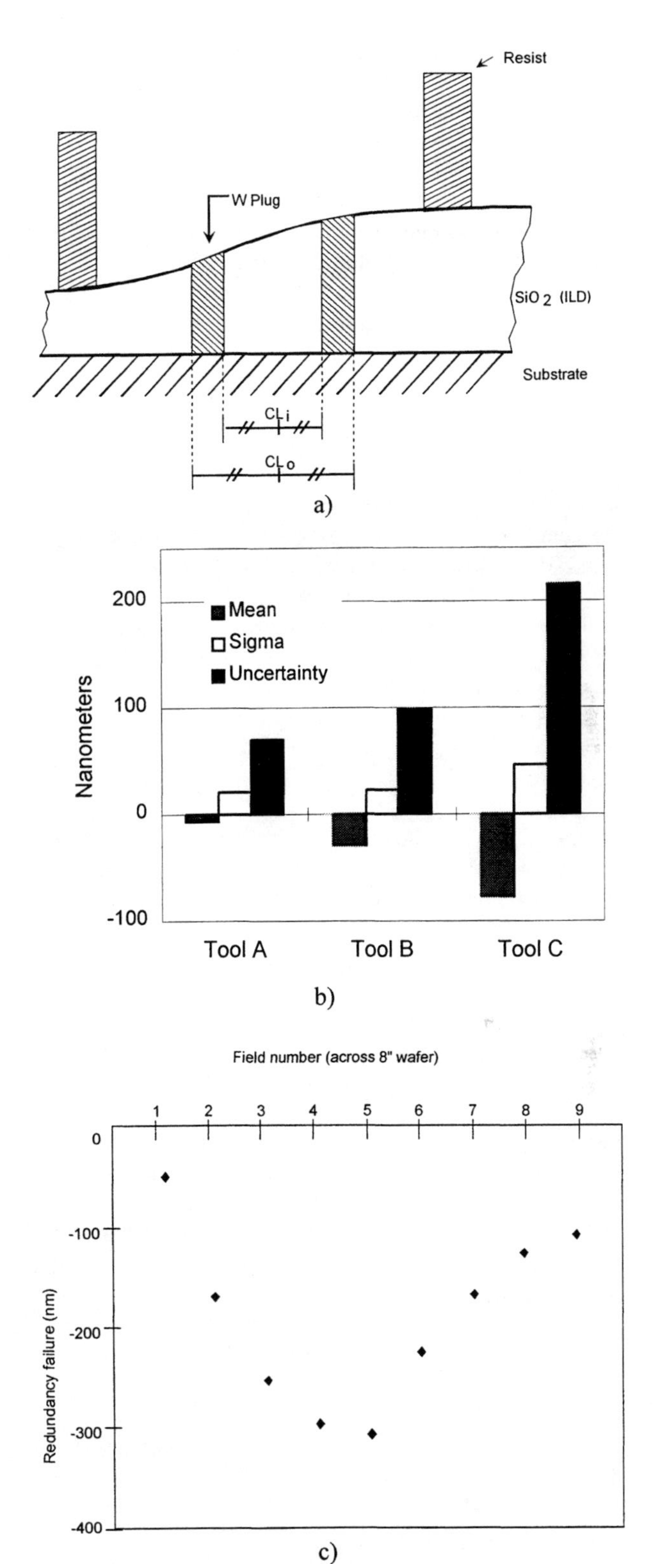

FIGURE 15. Illustrating a problem with viewing due to rounding of ILD in CMP (a), estimates of uncertainty associated with an apparent redundancy failure at viewing W bars in three metrology tools (b) and (c) site to site variation of apparent redundancy failure for tool C.

The potential impact of an asymmetric ILD on centerline estimation in W stud structures was assessed[60] using the O/L metrology structures in Fig. 15a. A single O/L measurement structure was measured in 9 fields across the diameter of an 8 inch wafer. Several repeated measurements were made on each field. The wafer was rotated and re-measured in the same way. After averaging and correcting for TIS, the estimated redundancy failure (RED) in the W target was computed from two measurements, one made with the inner edges of the bars and one with the outer edges. Redundancy error was estimated,

$$RED_W = CL_o - CL_i,$$

where CL_o is the centerline of the outer edges and CL_i is the centerline of the inner edges. These estimates were then corrected for the error of mask making and imaging, producing calibrated estimates RED'_W. The mean calibrated error $\langle RED'_W \rangle$ and standard deviation $\sigma_{RED'_W}$ for one set of 9 targets was produced. Measurement uncertainty solely due to apparent redundancy failure in W bars was estimated as

$$URED'_W = |\langle RED'_W \rangle| + 3\sigma_{RED'_W}.$$

Experiments were set up by the applications engineers of three vendor companies designated as A, B and C and run in comparable conditions at their respective demo facilities.

For each system, the target-limited precision and TIS were much higher than would be typical on resist and etched targets. However, the systematic redundancy failures incurred in imaging W targets were by far the largest errors. Unlike TIS and imprecision, they could not be removed by tool calibration and averaging. Results of this study are summarized in the chart shown in Fig. 15b. Although this is an evaluation of the worst case metrology, it is rather difficult to believe that the systems with hardware limited precision $3\sigma < 5$nm could produce metrology data with uncertainty >70nm. The error of viewing alone is $>12\sigma$!

However, consider the spatial behavior of redundancy failures for system C illustrated in Fig. 15c. This error slowly changes as a function of position in one wafer, reaching 300nm close to its center. This behavior is consistent with a slow and nearly radial variation of ILD thickness commonly observed in CMP. Yes, processing effects *can* lead to very large and systematic inaccuracies of centerline estimation.

Our example shows that the conventional optical O/L metrology on CMP layers may be very inaccurate. Can it support the ground rules of a 256 Mb manufacturing, with CMP on several critical layers? The answer is "yes" - in the course of O/L metrology business process. In fact, the first "all good" 256 Mb chips [61] were made with optical microlithography and using optical alignment and O/L metrology systems. CMP was used in both front end and back end critical layers. The key to success is integration.

All alignment and O/L metrology systems on the market today make at least some use of error diagnostics and culling. Manufacturers of metrology equipment also made many refinements of their hardware and software to reduce metrology errors. Automated error diagnostics and diagnostics feedback to the user are rapidly becoming available. These quantitative quality measures enabled practicing O/L metrology as a business process. The number of reported applications of these methods and of successful cases of process integration are many and growing [6, 9, 11, 19, 20, 23, 58 - 60, 62, 63, 64].

METROLOGY AS A BUSINESS PROCESS

Microlithography is an extreme form of mass production. This environment puts some very unique requirements on dimensional metrology. When metrology errors were frequent and gross, a metrologist could spot them. However, as the magnitude and frequency became small, the automated error diagnostics[6, 11, 19, 20, 59] become indispensable. Metrology of image placement for microlithography is not practical without the automated error diagnostics.

When alignment or O/L metrology error diagnostics detect a problem with a measurement, culling of data is made in response to a problem that has already occurred. This re-active use of the error diagnostics is very effective at reducing the impact of the problem, for example, in terms of error magnitude and frequency. Automated error diagnostics also provide a "paper trail" of metrology quality and play an important role in product quality assurance.

However, the most significant impact of metrology error diagnostics is achieved in pro-active uses for process integration. Here, the error diagnostics serve as a numeric quality feedback. This feedback fuels a metrology integration process, where the actions are taken to reduce the magnitude and incidence of the problem itself. These actions may be directed at changing something about the metrology system or the way it is used. They may also be directed outside the metrology itself, changing the environment in which metrology is conducted. The outcome is an efficient metrology that is optimal for its environment. That is the goal of the metrology business process.

It is a good practice of a metrology business process to address the fundamental technology limitations. It reduces the risk of some day being unable to support manufacturing. With some foresight, alternatives can be developed in steady business-like environment, rather than in a crisis.

Target-limited precision

Precision in metrology of image placement is strongly a

functions of the application. Examples of applications where the target quality may strongly impact precision are targets with small or slowly varying amplitude and phase (height). For grainy materials these difficulties are compounded by sampling considerations.

For example, conventional optical systems may produce poor signal on targets processed with CMP and covered by metal. The reason for poor signal is clear from physical characterization [6] of the alignment and O/L metrology structures: CMP may result in surface relief of <30nm and slopes of <2°! However, various forms of interference microscopy, including computerized interference microscopy used in those studies, are able to produce phase (related to profile) and amplitude image, as well as a conventional intensity image. Using phase information in the images they collect, the newer alignment and O/L metrology systems can better cope with such targets.

Although their applicability is extended, at a cost, these systems are still limited by the same basic causes and, in some cases, may still be rendered useless. These limitations are overcome, with certainty, by practicing metrology as a business process. In this approach, putting the increased demands and resources into new metrology tooling is seen as just one of many available options. The most efficient one is found through technology learning and optimization.

Target asymmetry

When a symmetric target is built on, coated with or surrounded by asymmetric films these films may make it either difficult or impossible to estimate the centerline.

When these films are transparent, accurate metrology may be (at least in principle) expected from an optical metrology system. Over the years, accuracy of O/L metrology in presence of asymmetric clear films has improved. A user may reasonably demand that a system used in O/L metrology for microlithography be insensitive to asymmetry in clear films, such as photoresist or inter-layer dielectric. Realistic calibrated O/L reference materials of this kind and performance evaluation of optical O/L metrology systems are described in the next section.

In order for a conventional measurement system to be accurate, the target must be fundamentally symmetric, that is, its sidewall have mirror symmetry. The same must be true for absorbing films above the target. If these are not symmetric, the state of symmetry (and accuracy) can be improved. For that goal, metrology error diagnostics related to symmetry in O/L metrology structures are used to either cull some data or modify the process and achieve greater symmetry of the image. As long as a metrology system can detect asymmetry, its utility can be extended with both re-active and pro-active corrections.

Failure to detect asymmetry

When an O/L metrology system is unable to detect asymmetry of the targets, technology employed in such a system is not suitable for standalone applications. The problem here may be the fundamental inability to resolve the feature sidewalls or simply the lack of error diagnostics. However, without the estimates of inaccuracy and systematic corrective measures that they fuel, O/L metrology could not be a quality process. For this metrology to be useful in manufacture of IC's, assurances of target symmetry must come from elsewhere.

Clearly, the same approaches used to manufacture and assure the quality of IC's may also be used to manufacture and assure the quality of the O/L metrology structures. Physical characterization, modeling and statistical process control are among them. In O/L metrology, as in IC manufacturing, the first step is to achieve and maintain high levels of process control in the O/L measurement structures. Next, to reduce those effects in manufacturing processes that result in asymmetry (and inaccuracy). Finally, the accuracy of O/L metrology is directly ascertained by using alternatives metrologies that overcome the limitations of the metrology systems used in production. Scanning electron microscopy[65, 66] supports one such alternative.

SEM based overlay metrology

It is often said that the high spatial resolution achievable with an SEM is not required for accurate optical O/L metrology. Strictly speaking, this is not true. SEM capabilities which made them so useful in CD metrology also make them useful in O/L metrology. A scanning electron microscope is a very attractive instrument, provided it has the basic attributes of a good O/L metrology system, such as symmetric imaging.

Asymmetry inherent to the left-to-right scanning and off-axis detectors was greatly reduced[65, 66] by retrofitting a Hitachi S-806 FE SEM with an annular microchannel plate (MCP) detector and using frame rate image acquisition. Symmetric images of specially designed O/L metrology structures and IC structures were acquired over a range of operating conditions. Various aspects of this metrology capability were assessed.

It was found[65] that, with these changes, effects of sample charging on the error of O/L measurement can be quite small. For example, sample charging in 20 repeated measurements of a 250nm feature lead to drifting CD measurements and 3σ = 44nm, variation of O/L measurements was only 6nm. When the O/L data from uncoated targets were compared to those after a 5nm Pd/Au coating, it was found that for an off-axis detector a 20nm apparent shift was observed, but only 4nm for the symmetric MCP detector. Precision and linearity of SEM based O/L metrology were found competitive with optical O/L metrology systems. With appropriate configuration and set up, TIS in SEM based O/L metrology can be very small.

This SEM was used for O/L metrology in devices, including gate to oxide in 200nm CMOS devices and emitter opening to base in advanced bipolar transistors. In the latter example[65], SEM O/L measurements were made on the transistors and SEM O/L metrology marks illustrated in the lower left of Fig. 9. Optical O/L measurements were made on the adjacent conventional sized O/L measurement structures. There data were acquired in developed resist and in final (after etch and strip) image. SEM O/L measurements in developed images of transistors were found to differ from those in SEM or optical O/L metrology marks by about 20nm. After etch, this level of correlation deteriorated to > 40nm. Unlike the SEM O/L measurement marks, this transistor is inherently asymmetric and the RIE may have effectively shifted centerlines of asymmetric resist openings. Other observations made with SEM based O/L metrology suggest the effects of linewidth and polarity on measured error, consistent with the more recent reports.

Capability to measure O/L in devices is very useful in both process control and quality assurance. In addition, it allows one to assess metrology errors due to the differences in imaging and image transfer between the device features and O/L measurement marks[6]. This way, the errors associated with the resolution limits and process sensitivities in optical O/L metrology systems may be addressed. In order to practice conventional optical O/L metrology, its accuracy must be ascertained. SEM based O/L metrology has been developed and proven useful for this purpose.

Different physics of image formation and of interaction with the sample also helps the application of SEM based O/L metrology for certification of accuracy. Using a single tool and edge detection criterion for both CD and O/L metrology permits linking these forms of metrology, so there is no ambiguity in estimating the edge-to-edge overlay required by the device ground rules. These are the basic considerations regarding the merits of SEM based O/L metrology, as reflected in the National Technology Roadmap for Semiconductors[3].

O/L reference materials and tests of accuracy

Accurate metrology of image placement can be achieved by assuring symmetry of the tools and targets. However, some forms of process related asymmetry are unavoidable and the tools are expected to tolerate them. It is worthwhile to evaluate the tool sensitivity to such effects. Although various forms of a priori information provide much data (error diagnostics) on how erroneous metrology is, accuracy is still unknown. In order to make statements that the error is no more than a certain value, the calibrated O/L reference materials are required.

Clearly, this reference material should be more than a derivative from the standard of length with which the tool distortion may be calibrated. What is required is a certified "difficult to measure" O/L metrology structure. Such structures may be used to evaluate how well this O/L type of measurement system will perform in the applications on "difficult" layers.

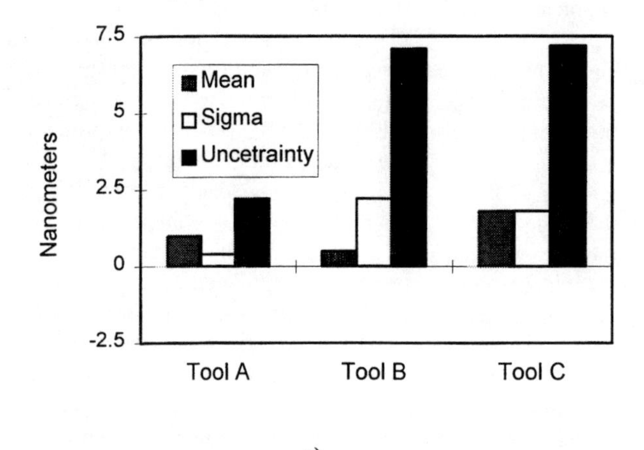

a)

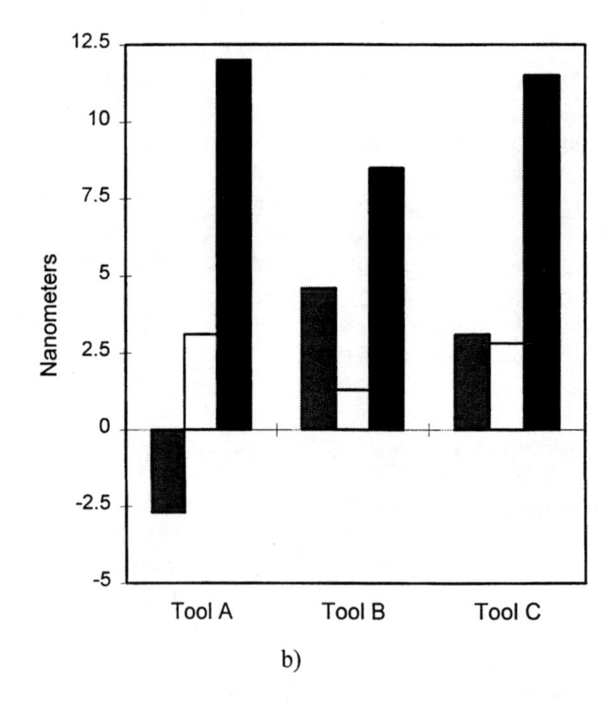

b)

FIGURE 16. Tests of accuracy of three O/L metrology systems on calibrated reference materials with asymmetric resist over marks: shallow trenches and high contrast (a); deep trenches and low contrast (b).

Reference materials of this kind [21, 60], known as TV-LITE, have been designed, built and used at IBM. In order to certify accuracy of these materials, X/Y metrology in reticles, as well as SEM and optical O/L metrology on wafers were used. A pair of critical and non-critical 5X reticles were built. The critical level reticle was measured on a Nikon-3i system with errors <5nm (at wafer scale).

Two values of SiO_2 thickness were selected, so that the images of etched structures have either high or low contrast[53]. Two sets of 5 inch ultra-flat Si wafers were oxidized, with multiple measurements made to assure that film thickness variation was small within each batch and across each wafer. These wafers were coated with novolak resists and exposed on a 5X I-line lithography system. Linewidth measurements were made to assure that the O/L metrology targets were printed as similar as possible. Etching was made with minimal over-etch into Si and with minimal radial asymmetry in SiO_2 sidewalls. SEM and optical characterization of sidewall asymmetry found it to be below measurable.

O/L metrology on multiple structures and wafers was conducted. Twenty measurements on each site were averaged to reduce imprecision. Measurements made at 0° and 180° were also made. The TIS corrected measurements of open etched targets were found to be within a total uncertainty (mean + 3σ deviation from calibration values) of 4.5nm for the high contrast wafers and 5.6nm for the low contrast wafers.

These wafers were then spin-coated with 1μm grade novolak resist and exposed again. This non-critical exposure cleared resist from the reference structures and left patches of resist over other structures. Resist asymmetry, a known source of measurement inaccuracy, was readily observable in coated structures. Now, these samples were both certified and (a little) difficult to measure.

In order to evaluate accuracy of the conventional optical O/L metrology, these samples were measured using three conventional O/L metrology systems designated as A, B and C. In each case, vendor's applications engineers set up the system according to their best judgment. Twenty measurements at 0° and 180° orientations at multiple sites and wafers were used to evaluate and suppress imprecision and TIS.

All systems easily measured the wafers with high amplitude contrast. After averaging, correcting for TIS and subtracting the calibration values, mean residual error and variance were produced, Fig. 16a. All three conventional optical systems demonstrate the required uncertainty of <7.5nm. However, for those applications where averaging and calibration for TIS on each site are not practical and one repeat and wafer orientation are used, a more realistic estimate of error would almost twice that large.

Metrology on the wafers with low contrast was conducted the same way, producing estimates of mean residual error and variance, Fig. 16b. On these samples, the combination of greater resist asymmetry over deeper trenches and low amplitude contrast resulted in higher imprecision, TIS and uncorrected residual error. Uncertainty of all three systems was better than 12nm. A more realistic estimate of metrology error, based on one repeat and orientation would exceed 20nm. Uncertainty due to TIS alone was > 13nm.

These experiments[60] demonstrate utility of certified O/L reference materials for evaluation of measurement capability. It was shown that an O/L metrology requirement of 7.5nm may be met at least in some cases. It also became apparent that to gain the required metrology capability across all process layers, further improvements in tools and process integration are required.

CONCLUSIONS

Over the last decade the optical O/L metrology systems have made a remarkable progress. The preferred metrology systems of today, their precision and TIS are often <5nm. The automated performance testing and error diagnostics enabled much faster technology learning and efficient process integration. Further utility of optical metrology systems in applications with faint targets is being extended by interference microscopy. The realistic calibrated O/L reference materials were built and used to certify accuracy. Many of the gains initially made in optical O/L metrology are already used to improve alignment. Metrology of registration may be improved, as well.

Mathematical models of processes, manufacturing tools and metrology systems were developed. More and more, the user and vendor dialog about metrology errors and process integration issues involves a broader spectrum of the IC manufacturing community.

SEM based O/L metrology was developed. Capabilities to measure O/L in devices, verify accuracy and complement the optical O/L metrology on difficult layers were demonstrated.

It appears that, as a whole, metrology of image placement can continue to advance at a fast pace and enable further use of microlithography in microelectronics manufacturing.

ACKNOWLEDGMENTS

This paper is based on a large body of original work produced at IBM East Fishkill, IBM Burlington and IBM Yorktown Heights Research Center between 1986 and 1994.

Alan Rosenbluth and Steve Knight supported the study of resist flow. Don Samuels, Bob Fair, Tim Brunner and Mike Rosenfield were instrumental in building and characterizing TV-LITE. Kurt Tallman helped me to appreciate the intricacies and usage of device ground rules. Howard Landis and Bill Guthrie shared their knowledge of materials and processes used in BEOL. Development of error diagnostics and process integration of alignment and O/L metrology in BEOL was conducted with ATX-4 and -4SX Task Forces. Process integration of optical O/L metrology in the 256Mb DRAM joint venture of IBM, Siemens and Toshiba was supported by Alois Gutmann of Siemens.

Development of optical O/L metrology, error diagnostics and culling and the evaluations of accuracy were done jointly with the tool vendors. I would like to thank BioRad Micromeasurements, Ltd., KLA Instruments, Inc. and Optical Specialties, Inc. for their collaboration. I would like to specifically acknowledge the assistance of Jackie Gueft of BioRad in preparation and initial runs of the studies of accuracy reported here for the first time.

The physical model and observation of W sputtering are contributions from SIMBAD Project at Alberta Microelectronic Centre, Edmonton, Canada; courtesy of Tomas Janacek.

I am very grateful to Dr. Kevin Monahan, Dr. Michael Postek and Mr. James Potzick for many recent stimulating dialogs on dimensional metrology, leading to a consolidated account of our work in this field.

I am also grateful to IBM Corporation for the permission to publish some material for the first time and to Dr. Timothy Brunner for assistance in making it possible.

REFERENCES

[1] G. E. Moore, *Lithography and the Future of Moore's Law*, SPIE Vol. 2439; 1995.

[2] D. L. Crook, *Evolution of VLSI reliability engineering,* 28-th Annual Proceedings on Reliability Physics, New Orleans, LA; IEEE Catalog No. 90CH2787-0; 1990.

[3] The National Technology Roadmap For Semiconductors, SIA; 1997.

[4] R. M. Booth, Jr. et al., *A statistical approach to quality control of non-normal lithography overlay distributions,* IBM J. of Res. and Dev., Vol. 36, No. 5; 1992.

[5] *Specification for overlay capabilities of wafer steppers,* SEMI P18-92.

[6] A. Starikov, *Overlay in Subhalf-Micron Optical Lithography,* Short Course, SEMICON/WEST '93.

[7] L. H. Breaux et al., *Pareto charts for defect analysis with correlation of in-line defects to failed bitmap data,* Handbook of Critical Dimension Metrology and Process Control, K. Monahan, Ed.; SPIE Volume CR52; 1993.

[8] W. H. Arnold, *Overlay Simulator for Wafer Steppers,* Proc. SPIE Vol. 922; 1988.

[9] A. Starikov et al., *Accuracy of overlay measurements: tool and mark asymmetry effects,* Optical Engineering, **31**; 1992.

[10] R. D. Larrabee, *Report on a Workshop for Improving Relationships Between Users and Suppliers of Microlithography Metrology Tools,* NISTR 5193, National Institute of Standards and Technology; 1993.

[11] N. H. Goodwin, A. Starikov, G. Robertson, *Application of mark diagnostics to overlay metrology,* Proc. SPIE, Vol. 1673; 1992.

[12] R. D. Larrabee, *Submicrometer optical linewidth metrology,* SPIE Vol. 775; 1987.

[13] *Specification for metrology pattern cells for integrated circuit manufacture,* SEMI P19-92.

[14] D. Nyyssonen, R. D. Larrabee, *Submicrometer Linewidth Metrology In the Optical Microscope,* J. of Res. of the National Bureau of Standards, Vol. 92, No. 3; 1987.

[15] *CD Metrology Procedures*, SEMI P24-94.

[16] F. Proschan, *Rejection of Outlying Observations,* Am. J. of Physics, Vol. 21, No. 7; 1953.

[17] J. R. Taylor, *An Introduction to Error Analysis (The Study of Uncertainties in Physical Measurements),* Oxford University Press; 1982.

[18] A. Starikov et al., *Use of a priori information in centerline estimation,* Proc. KTI Microlithography Seminar, 1991.

[19] C. J. Progler et al., *Alignment signal failure detection and recovery in real time,* J. Vac. Sci. Technol. **B** 11(6); 1993.

[20] D. J. Coleman et al., *On the accuracy of overlay measurements: tool and mark asymmetry effects,* Proc. SPIE, Vol. 1261; 1990.

[21] A. Starikov, *Beyond TIS: separability and attribution of asymmetry-related errors in centerline estimation,* talk at 1994 SPIE Microlithography Conference.

[22] *Specification for overlay-metrology test patterns for integrated-circuit manufacture*, SEMI P28-96.

[23] B. Plambeck et al., *Characterization of chemical-mechanical polished overlay targets using coherence probe microscopy,* Proc. SPIE, Vol. 2439; 1995.

[24] J.-H. Yeo et al., *Improved overlay reading on MLR structures,* Proc. SPIE, Vol. 2725; 1996.

[25] J.-S. Han et al., *Effects of Illumination Wavelength on the Accuracy of Optical Overlay Metrology,* Proc. SPIE, Vol. 2439; 1995.

[26] S. Kuniyoshi et al., *Contrast improvement of alignment signals from resist coated patterns,* J. Vac. Sci. Technol., **B** 5(2); 1987.

[27] Y. Oshida et al., *Relative Alignment by Direct Wafer Detection Utilizing Rocking Illumination of Ar Ion Laser,* Proc. SPIE, Vol. 633; 1986.

[28] A. Starikov, *On the Accuracy of Overlay Measurements: Mark Design,* IEEE/DARPA Microlithography Workshop, New Orleans, LA; Jan. 1990.

[29] R. Silver et al., *A Method to Characterize Overlay Tool Misalignments and Distortions,* Proc. SPIE, Vol. 3050; 1997.

[30] N. Bobroff, *Position Measurement with a Resolution and Noise Limited Instrument,* Rev. Sci. Instr., Vol. 57; 1986.

[31] J. Hutcheson, *Demand for Overlay Measurement Tools;* Executive Advisory, VLSI Research, Inc.; 1992.

[32] N. P. Smith and R. W. Gale, *Advances in optical metrology for the 1990's,* Proc. SPIE, Vol. 1261; 1990.
[33] A. Kawai et al., *Characterization of Automatic Overlay Measurement Technique for Sub-Half Micron Devices,* Proc. SPIE, Vol. 1464; 1991.
[34] P. M. Troccolo et al., *Tool and mark design factors that influence optical overlay measurement errors*, Proc. SPIE, Vol. 1673; 1992.
[35] C. J. Progler et al., *Overlay Performance of X-ray Steppers in IBM Advance Lithography Facility (ALF),* Publication No. 231, Karl Suss America, Inc.; 1993.
[36] T. Kanda, *Alignment Sensor Corrections for Tool Induced Shift (TIS),* Proc. SPIE, Vol. 3051; 1997.
[37] C. P. Kirk, *Theoretical Models for the Optical Alignment of Wafer Steppers,* Proc. SPIE, Vol. 772; 1987
[38] G. M. Gallatin et al., *Modeling the images of alignment marks under photoresist,* Proc. SPIE, Vol. 772; 1987.
[39] N. Bobroff and A. Rosenbluth, *Alignment errors from resist coating topography,* J. Vac. Sci. Technol. **B** 6(1); 1988.
[40] C.- M. Yuan and A. J. Strojwas, *Modeling of optical alignment and metrology schemes used in integrated circuit manufacturing,* Proc. SPIE, Vol. 1264; 1990.
[41] A. Starikov, *Exposure Monitor Structure,* Proc. SPIE, Vol. 1261; 1990.
[42] J. N. Wiley, L. S. Zubrick, S. J. Schuda, *Comprehensive detection of defects on reduction reticles,* Proc. SPIE, Vol. 2196; 1994.
[43] J. Potzick, *Automated Calibration of Optical Photomask Linewidth Standards at the National Institute of Standards and Technology,* Proc. SPIE, Vol. 1087; 1989.
[44] D. M. Levenson, *Extending the lifetime of optical lithography technologies with wavefront engineering,* Jpn. J. of Appl. Phys. **33**; 1994.
[45] J. Potzick, *NeoLithography,* 17-th Annual BACUS Symposium on Photomask Technology and Management, Proc. SPIE, Vol. 3236;1997.
[46] M. Born and E. Wolf, *Principles of Optics,* Pergamon Press.
[47] A. Starikov, *Structures for test of asymmetry in optical imaging systems*, IBM TDB, Vol. 33, No. 5; 1990.
[48] T. Brunner, *Impact of lens aberrations on optical lithography*; Olin Microlithography Seminar, INTERFACE '96; 1996.
[49] D. W. Peters, *The Effects of an Incorrect Condenser Lens Set-Up on Reduction Lens Printing Capabilities*; Kodak Microelectronics Seminar, INTERFACE '85; 1985.
[50] D. S. Goodman and A. E. Rosenbluth, *Condenser Aberrations in Köhler Illumination,* Proc. SPIE, Vol. 922; 1988.
[51] T. A. Brunner, *Pattern-Dependent Overlay Error in Optical Step and Repeat Projection Lithography,* Microelectronic Eng. **8**; 1988.
[52] J. D. Armitage, Jr. and J. P. Kirk, *Analysis of overlay distortion patterns*, SPIE Vol. 921; 1988.
[53] D. Nyyssonen, *Theory of Optical Edge Detection and Imaging of Thick Layers,* JOSA **72**; 1982.
[54] C. P. Kirk and D. Nyyssonen, *Modeling the Optical Microscope Images for the Purpose of Linewidth Measurement,* Proc. SPIE, Vol. 538; 1985.
[55] K. H. Brown et al., *Advancing the state of the art in high-performance logic and array technology,* IBM J. Res. Develop., Vol. 36, No. 5; 1992
[56] W. L. Guthrie et al., *A four-level VLSI bipolar metallization design with chemical-mechanical planarization,* ibid.
[57] S. K. Dew et al., *Modeling bias sputter planarization of metal films using a ballistic deposition simulation,* J. Vac. Sci. Technol. **A** 9(3); 1991.
[58] C. Lambson, A. Awtrey, *Alignment mark optimization for a multi-layer-metal process,* Proceedings of KTI Microlithography Seminar, INTERFACE '91; 1991.
[59] D. Meunier et al., *The implementation of coherence probe microscopy in a process using chemical-mechanical polishing,* OCG Microlithography Seminar, INTERFACE '95; 1995.
[60] A. Starikov, *SRDC/TRIAD overlay metrology development,* 1994. Material presented here for the first time was made available for this publication by IBM Corp.
[61] *IBM, Siemens and Toshiba alliance announces smallest fully-functional 256Mb DRAM chip*, Press Release; June 6, 1995.
[62] A. Yanof et al., *Improving metrology signal-to-noise on grainy overlay features,* Proc. SPIE, Vol. 3050; 1997.
[63] E. Rouchouze et al., *A CMP-compatible alignment strategy,* Proc. SPIE, Vol. 3050; 1997.
[64] P. R. Anderson and R. J. Monteverde, *Strategies for characterizing and optimizing overlay on extremely difficult layers,* Proc. SPIE, Vol. 2196; 1994.
[65] M. G. Rosenfield and A. Starikov, *Overlay measurement using the low voltage scanning electron microscope*, Microelectronic Engineering **17**; 1992.
[66] M. G. Rosenfield, *Overlay measurements using the low voltage scanning microscope: accuracy and precision,* SPIE Vol. 1673; 1992.

LITHOGRAPHY - GENERAL

Deep Ultraviolet Laser Metrology for Semiconductor Photolithography

M. L. Dowell, C. L. Cromer, R. W. Leonhardt, T. R. Scott

Sources and Detectors Group, Optoelectronics Division
National Institute of Standards and Technology, Boulder, Colorado 80303-3328

Recent improvements in calibration procedures have led to reductions in overall uncertainties of laser power and energy calibrations to ±1%. We plan to extend these services to include laser dose measurements and angular response. Deviations from cosine behavior for angular response can introduce measurement errors when dose meters, calibrated with parallel laser beams, are employed in stepper systems. These measurement errors will become important as variable numerical aperture systems become commonplace. Current and future laser measurement services at 193 and 248 nm will be reviewed.

INTRODUCTION

Several years ago, the Optoelectronics Division with Semiconductor Manufacturing Technology (SEMATECH) support, developed primary standard laser calorimeters for 248 nm excimer laser power and energy measurements [1].Our efforts are now focused on addressing the demands for 193 nm measurement capability as quickly as possible. At the same time, we are adding a laser dose, *i.e.*, energy density, capability to our measurement services. Accurate measurements of laser dose are crucial to the development of new mask and resist materials at 248 and 193 nm. Furthermore, transfer standards for dose measurement will be critical to ensure high accuracy to the end user. We will be working with industrial partners to improve transfer standard capabilities.

Since the first draft of the National Technology Roadmap for Semiconductors in 1992, the semiconductor industry has made an organized, concentrated effort to move toward smaller critical dimensions. As a result, there has been a shift toward shorter laser wavelengths in the optical lithography process. Ultraviolet (UV) lasers, specifically KrF (248 nm) and ArF (193 nm) excimer lasers, will be the preferred sources for high resolution optical lithography stepper systems until the future shift to nonoptical lithography for the fabrication of devices with features smaller than 100 nm. Therefore, UV laser metrology is crucial for the development of new optical lithography processes and tools.

LASER CALORIMETRY

Primary Standard Calorimeters

The Optoelectronics Division at NIST has designed and built several laser calorimeters for accurate measurements of laser power and energy [2].Current NIST laser measurement services are listed in Laser calorimetry. Briefly, a laser calorimeter operates as follows [3]: laser radiation is injected into a thermally isolated cavity. Upon absorption, the electromagnetic radiation is converted into thermal energy. The corresponding temperature rise is precisely measured with thermal sensors, such as thermocouples. An electrical heater is used to calibrate the calorimeter's temperature response by injecting equivalent, known electrical energy into the cavity.

Table 1. NIST laser measurement services.

	Laser Power and Energy Calibrations		
	Laser	**Wavelength**	**Range**
CW	Argon Ion	488 and 514 nm	1 μW - 1W
	HeNe	633 nm	1 μW - 20 mW
	Diode	830 nm	100 μW - 20 mW
	Nd:YAG	1064 nm	100 μW - 450 W
		1319 nm	100 μW - 10 mW
	HeNe	1523 nm	100 μW - 1 mW
	CO_2	10.6 μm	1 μW - 1 kW
Pulsed	KrF Excimer	248 nm	1 nJ - 200 mJ/pulse 50 μW - 9 W
	ArF Excimer	193 nm	*
	Nd:YAG	1064 nm	1 - 50 mJ/pulse 10^{-3} - 10 nJ/pulse
		peak power	10 nW - 100 μW

*Available September 1998

The localized temperature rise is related to the volume of absorbing material over which the laser energy is deposited. In order to avoid damage due to the short laser pulses, we use two volume absorbers in the calorimeter's cavity to safely absorb the incident radiation (see Figure 1). The volume absorber materials for both UV primary standards are UV absorbing glasses, chosen for durability for the wavelengths of interest.

CP449, *Characterization and Metrology for ULSI Technology: 1998 International Conference*
edited by D. G. Seiler, A. C. Diebold, W. M. Bullis, T. J. Shaffner, R. McDonald, and E. J. Walters
1998 The American Institute of Physics 1-56396-753-7/98/$15.00

Using this technique, NIST has developed a primary standard ultraviolet-laser calorimeter (QUV) for use at 248 nm, which can provide accurate measurements of UV laser power and energy within ±1%. Additional primary standards for use at 193 nm are now under construction and estimated to be completed in FY 98 (see Figure 1). After extensive testing, an appropriate volume absorber has been selected for the 193 nm standards.

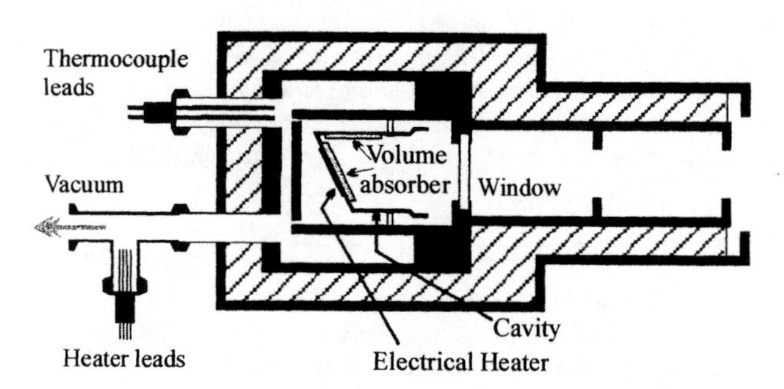

Figure 1. 193 nm laser calorimeter (under construction).

NIST Laser Measurement System

Laser Power and Energy Measurements

A schematic drawing of the NIST UV laser measurement system is shown in Figure 2. The incident laser beam is split with a 2° wedged fused silica beam splitter. The beam splitter ratio, B_Q, is determined from a series of energy measurements using the two primary standards, QUV-1 and QUV-2. The value for this ratio,

$$B = \text{(transmitted energy)/(reflected energy)}, \quad (1)$$

is approximately 25. Then, the detector under test is substituted for one of the standards and the beam splitter ratio B_{DUT} is measured. The detector calibration factor κ is determined from the ratio of these two beam splitter values,

$$\kappa = B_{DUT} / B_Q \,. \quad (2)$$

With judicious selection of transmitted and reflected beams, this system can span a four decade range in laser power and energy [4].

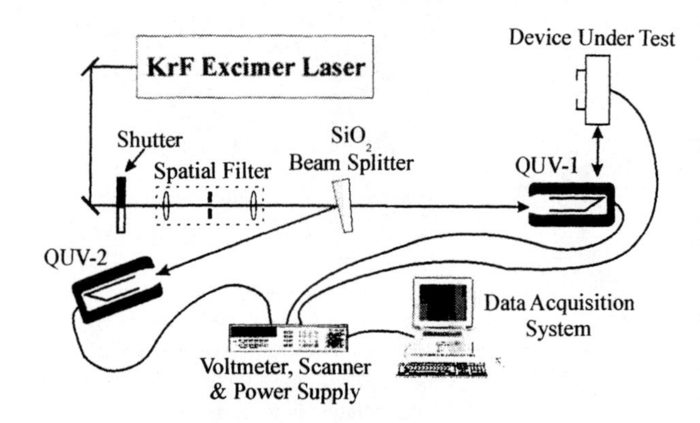

Figure 2. NIST UV measurement system for laser power and energy measurements at 248 nm. QUV-1 and QUV-2 are the primary standard calorimeters. A 2° wedged, fused silica (SiO_2) beam splitter is used as both a monitor and an attenuator for laser power and energy measurements.

Dose and Cosine Response Measurements

As the microelectronics industry moves toward smaller critical dimensions, process tolerances will shrink and accurate measurements of laser energy will play a crucial role in maintaining good quality control. Accurate dose measurements are important because small area photodiode detectors are widely used to monitor the laser energy that is deposited at the wafer plane. Furthermore, absolute dose measurements are important for the development of new mask and resist materials at 248 and 193 nm, since lower dose requirements can translate directly into higher throughput and extend the lifetime of a stepper's optical components as well. The selection and calibration of transfer and working standards for UV dose measurements, and ensuring high accuracy for the end user, will be difficult. In particular, detector damage issues related to long term UV exposure have yet to be fully investigated. We will be working with industrial partners to investigate new transfer and working standards.

One difficulty associated with dose calibrations is cosine, or angular response. Typically, dose meters are calibrated using a parallel laser beam with a uniform energy density. However, with the move toward high numerical aperture (N.A.) stepper systems, significant measurement errors can be introduced because the cosine response of dose meters can vary wildly (see Figure 3). The deviation from a true cosine response can be due to the angular response of a dose meter's constituent parts, the optical diffuser, the UV filter, and the UV photodiode detector (see Figure 4). An example is shown in Figure 3, where the difference in relative angular response between two *i*-line (365 nm) meters that are used in a stepper with off-axis illumination and N.A.=0.5 was approximately 40%.

The development of the calibration facility will evolve in two stages: (1) a 193 nm primary standard calorimeter will be developed and implemented for the calibration of laser energy meters, (2) a measurement system, including beam shaping optics with calibrated apertures, will be established for the direct calibration of dose meters that are used in wafer plane measurements. The techniques and instrumentation developed will also be used to add to the capability for dose meter calibrations to the existing 248 nm calibration facility which currently supports only large area detectors that capture the entire beam. In addition, a cosine response measurement capability will be added in order to ensure an accurate calibration transfer for meters that are used in high N.A. systems.

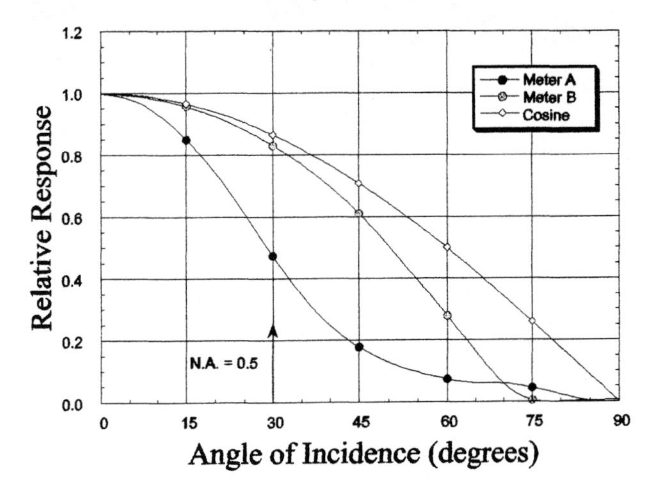

Figure 3. Angular dependence of *i*-line (365 nm) probes [5]. The arrow indicates the relative response at a numerical aperture (N.A.) of 0.5. As the plot indicates, a significant measurement error can be introduced due to the imperfect cosine response of a dose meter. This offset would be unaccounted for if the meters were calibrated using a parallel beam, but intended for use in a stepper with off-axis illumination.

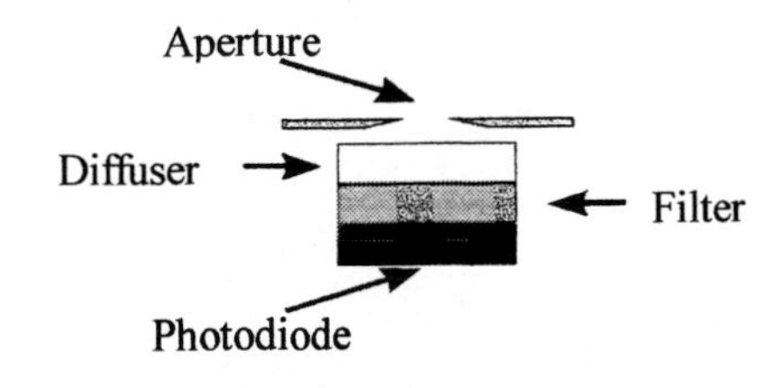

Figure 4. A typical UV detector for dose measurements at the wafer plane.

CONCLUSIONS

With the move towards critical dimensions of 0.2-µm or smaller, the dose exposure tolerances are expected to decrease precipitously. We anticipate that this will translate into a need for accurate, absolute dose measurements at the ±1-2% level [6]. Existing UV transfer standards and calibration methods are unable to meet this need. Therefore, improved transfer standards and techniques will be required to ensure accurate dissemination of the measurements to industry, and to facilitate accuracy improvement activities such as intercomparisons with calibration laboratories and industry round robins. NIST is working toward meeting these requirements by offering laser power and energy measurement services at 248 nm. Laser power and energy measurement services at 193 nm will be available to the general public during the latter part of 1998. Progress has been made towards establishing a calibration facility for dose meters. NIST solicits from end users and equipment manufacturers input that would aid in the identification of new transfer standards and measurement needs.

ACKNOWLEDGEMENTS

We thank SEMATECH for their ongoing support of this program.

Keywords: Laser metrology, dose control, transfer standards, ultraviolet detectors

REFERENCES

1. Leonhardt, R. W. and Scott, T. R., *Integrated Circuit Metrology, Inspection, and Process Control IX*, Proc. SPIE 2439, 448-459 (1995).
2. West, E. D., Case, W. E., Rasmussen, A. L., and Schmidt, L. B., *Journal of Research of the National Bureau of Standards* **76A**, (1972).
3. West, E. D., and Churney, K. L., *Journal of Applied Physics* **41**, 2705–2712 (1970).
4. Beers, Y., *National Bureau of Standards Monograph 146* (1974).
5. Cromer, C. L. and Bridges, J. M., "NIST Characterization of I-Line Exposure Meters," SEMATECH Technology Transfer Document, No. 91040516A-ENG (1991).
6. Cromer, C. L., Lucatorto, T. B., O'Brian, T. R., and Walhout, M, *Solid State Technology*, 75 (April 1996).

Metrology Applications in Lithography with Variable Angle Spectroscopic Ellipsometry

James N. Hilfiker*, Ron Carpio+, Ron A. Synowicki*, and John A. Woollam*

*J.A.Woollam Co., Inc., 645 M Street Suite 102, Lincoln, NE 68508
+Sematech, 2706 Montopolis Drive, Austin, TX 78741

Variable Angle Spectroscopic Ellipsometry is a common characterization technique for thin films used in lithography. These films include single and multiple layers of photoresists, anti-reflective coatings, and photomasks. The optical constants vary under different processing steps such as baking or UV exposure. Ellipsometry provides precise measurements of both the film thickness and optical constants, which are critical to ensure quality devices.

INTRODUCTION

Lithography requirements must keep up with future semiconductor processing demands. The current trend is toward shorter wavelength lithography to enable the shrinking features of 64, 256 kbit, and eventually 1 Gbit generations of DRAM production. Exposure wavelengths at 248nm and 193nm have introduced new challenges. Photoresists, anti-reflective coatings, and photomask materials are under development. Anti-reflective coatings become more important due to the increased reflectivity of the underlying silicon substrate at shorter wavelengths.

Both film thickness and optical constants are critical to the lithography process. Precise metrology tools are important for new materials research and process qualification. Variable angle spectroscopic ellipsometry (VASE®) is an optical metrology tool for thin films. It is a precise and non-destructive technique used to measure coating thickness and refractive index (n and k). Other material properties can be investigated with VASE; including surface or interfacial roughness, index grading, and optical anisotropy.

Ellipsometry measures two values, Ψ and Δ, which relate to the change in polarization due to reflection of light from the sample. These terms relate to the Fresnel reflection coefficients for p- and s- polarized light through the following relationship:

$$\tan(\Psi)e^{i\Delta} = \frac{R_p}{R_s} \quad (1)$$

Ellipsometry measures both the relative reflected intensities and the phase difference due to the reflection. The "phase" information contained in Δ provides high sensitivity to film properties; especially to the thickness of ultra-thin films (<10nm). Both Ψ and Δ are a ratio of parameters. This provides excellent accuracy, even under low or fluctuating light conditions.

METROLOGY CONSIDERATIONS

There are four major considerations for optical metrology of lithography applications: 1) reduced measurement light intensity near the exposure wavelength; 2) wide spectral range; 3) variable angle of incidence; and 4) fast measurements. Each is critical for certain applications.

Reduced Measurement Intensity

Photoresists are light sensitive by definition and strong UV light must be avoided during the optical measurement. VASE operates with a monochromator before the sample. Therefore, insignificant amounts of light reach the sample at any given time, as the wavelengths are scanned sequentially. In addition to protecting the sample from exposure during measurements, the monochromator protects the optics. Reduced light intensity through the lenses and polarizers reduces "solarization" effects, which would reduce the optical throughput over time.

The reduction of light in the ultraviolet protects the sample, but reduces the measurement signal-to-noise. Ellipsometer measurements are based on a ratio of intensities rather than an absolute value. This is beneficial with low or fluctuating intensities and allows the ellipsometer to maintain high accuracy in the deep ultraviolet (DUV).

Wide Spectral Range

A wide spectral range is critical to the thin films used in lithography. First, it is of interest to measure the film properties at the lithography wavelength. Most current processes use the g- or i-lines for lithography, but 248nm and 193nm will be important in the near future.

Second, many of the thin films are absorbing in the UV. Optical measurements at UV wavelengths lose information regarding film thickness. A single wavelength measurement in the absorbing region can determine the

CP449, *Characterization and Metrology for ULSI Technology: 1998 International Conference*
edited by D. G. Seiler, A. C. Diebold, W. M. Bullis, T. J. Shaffner, R. McDonald, and E. J. Walters
 1-56396-753-7/98/$15.00

"pseudo" optical constants. However, these values may not coincide with the actual film optical constants. Surface roughness, film non-uniformity, index grading, and other non-ideal properties affect the measurement. Therefore, a wide spectral range is required to determine both the thickness and the refractive index (*n* and *k*) accurately.

Third, spectroscopic data enhances multilayer measurements. Lithography applications often consist of optically similar materials. Typically the UV spectral range provides more optical contrast. Thus, the shorter wavelengths become very valuable to separate the thicknesses, while the longer wavelengths determine the thicknesses precisely.

Fourth, different spectral ranges contain unique information. As mentioned before, films are typically transparent in the visible and near infrared (NIR) which provides sensitivity to film thickness. Shorter wavelengths typically become absorbing due to the material properties. The optical properties will change with processing conditions, such as curing and UV exposure. Therefore, this region enables study of the development process. Spectroscopic ellipsometers from 2 to 30 microns (333 to 5000 cm^{-1}) are now commercially available. This mid-Infrared range provides new information that was not available in the UV, visible, and NIR. The mid-IR provides chemical bond specific information. The data shown in Fig. 1 are for a fluorinated polyimide coating on silicon. The significant number of features correspond to various bonding states in the polymer film.

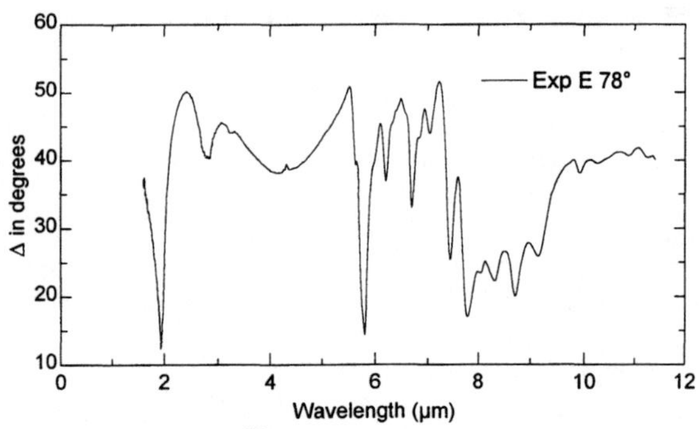

Figure 1. IR-VASE™ data taken on a thick polymer film.

Variable Angle of Incidence

The angle of incidence is an important consideration for ellipsometry measurements. The sensitivity to thin film parameters can be maximized with the proper choice of angle (See the companion paper by Woollam et. al. on "Metrology Standards with Ellipsometers").

Additional information can also be obtained by measuring and analyzing multiple angle data simultaneously. Multilayer samples can be precisely measured. As the angle changes, so does the path length through each of the films. These changes significantly improve the ability to precisely and accurately determine multiple film thicknesses.

Multiple angles of incidence are required for materials that exhibit optical anisotropy. Each angle introduces a different portion of light traveling along the in-plane and out-of-plane film direction. In the case of uniaxially anisotropic polymer films (typically oriented perpendicular to the sample), a single angle of incidence will lead to incorrect results. In fact, a different angle of incidence will provide a reasonable "fit" to the data, but with different results. Only when multiple angles of incidence are regressed simultaneously can the correct result be reached.

Fig. 2 shows the anisotropic optical constants for a fluorinated polyimide film that were determined before and after curing. This film is being considered as an interlayer dielectric. A few anti-reflective coatings under development have also exhibited similar anisotropy.

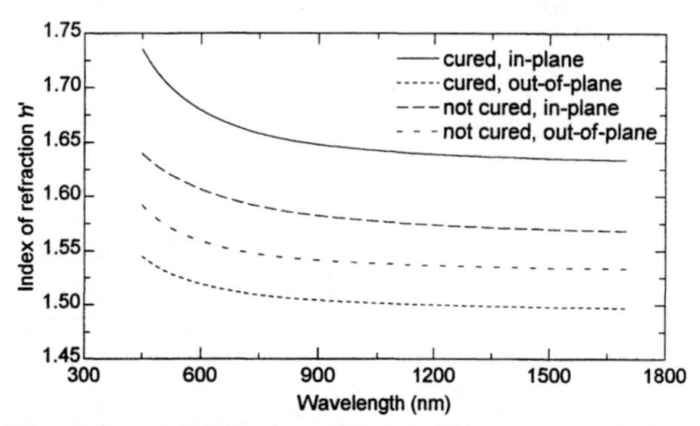

Figure 2. Anisotropic refractive index measured for a fluorinated polyimide film before and after curing. The curing process increases the anisotropy.

Fast Measurements

Previous considerations have focused on research and development applications. Similar measurements are required for process control, which demand faster measurement speeds. Current research VASE tools are limited to measuring a single wavelength at a time. This prohibits the fastest applications. However, reduced measurement sets can be developed to enhance high sensitivity to material properties with faster measurement speeds. This provides the user with another use for a research ellipsometer. This is commonly implemented when sample uniformity is measured. A reduced measurement is used to minimize mapping time. Fig. 3 shows the uniformity of an inorganic oxynitride antireflective coating measured on a VASE®.

New technologies allow spectroscopic ellipsometers to simultaneously measure multiple wavelengths. This is appropriate for applications where speed is a necessity. Precise thickness (1σ < 0.2Å) and refractive index values (1σ < 0.001 for n) can be determined in a matter of seconds.

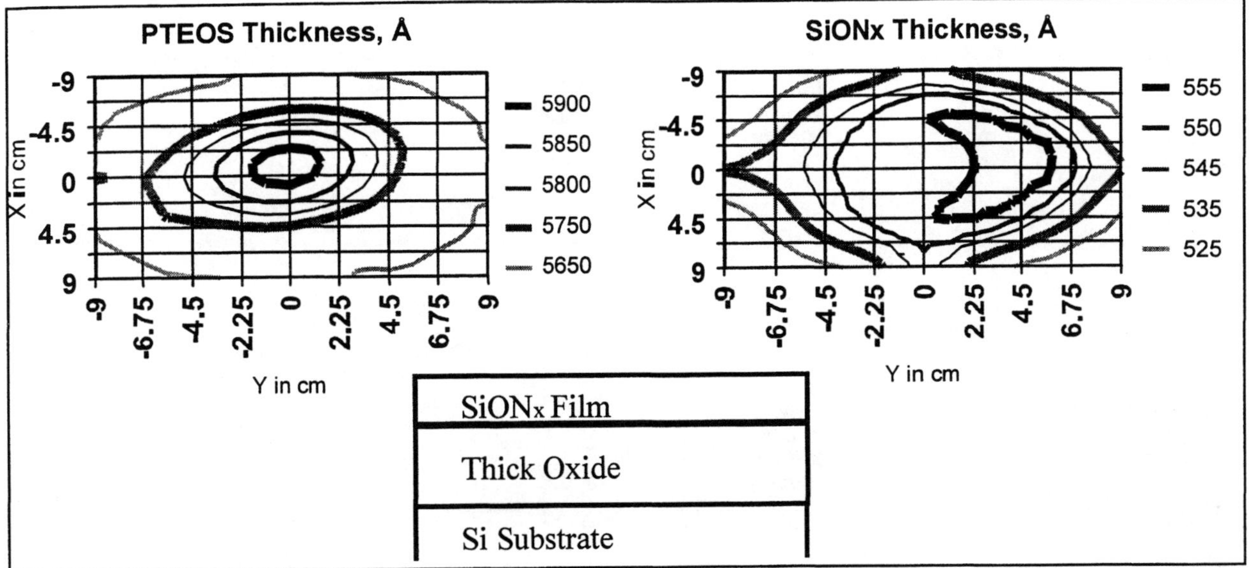

Figure 3. Wafer map of an inorganic oxynitride anti-reflective coating deposited on a thick oxide on silicon.

APPLICATIONS

The application of VASE® for lithography applications will be reviewed – including measurements of photoresists, anti-reflective coatings, photomasks, and multilayers of these materials.

Anti-Reflective Coatings

Anti-reflective coatings are becoming more popular in lithography because they enable better linewidth control by reducing reflective notching. Anti-reflective coatings also reduce (or eliminate) standing waves. Both inorganic and organic films are in use.

Inorganic thin films are of great interest for their favorable etch selectivity compared to organic films. Silicon oxynitride films are of particular interest, since these films can be used for lithography at 365, 248, and 193nm wavelengths; provided the appropriate thickness and refractive index (n and k) are obtained by process tuning. VASE measurements are valuable both in establishing and controlling the deposition process, since the above film parameters must be measured simultaneously. In many cases, the thickness of the underlying film must also be measured. This can be performed in an automated mode at multiple sites on a wafer. Results from a 49 point map of a silicon oxynitride film which was deposited on a PTEOS oxide film are shown in Figure 3. Notice, both the silicon oxynitride and PTEOS film thicknesses are obtained. The n and k values at the three lithographic wavelengths of interest were also determined, along with the Cauchy coefficients to plug into fast, on-line tools.

Organic anti-reflective coatings are in widespread use for lithography at 365nm, 248nm and 193nm. Effective use of such coatings requires that the appropriate thickness and

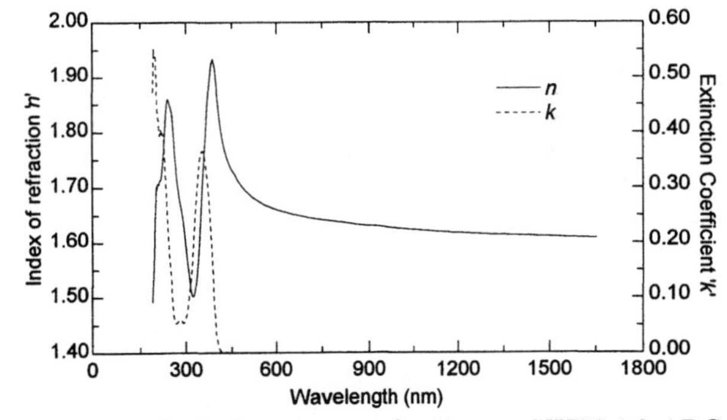

Figure 4. Optical constants of a Brewer XHRI-16 ARC designed for use at 365nm.

optical parameters be chosen and closely monitored. The end-user has less control over the optical constants compared to inorganic antireflective films. However, n and k can be modified by thermal or UV treatment processes. Figure 4 shows the optical constants for the Brewer XHRI-16 antireflective coating, which was designed for 365nm.

Photoresists

The optical constants of photoresists will vary with processing conditions. Because the monochromator is before the sample, VASE can measure the resist without exposing the resist. The resist optical constants can be measured as a function of the exposure dose. This is illustrated for a Shipley 510L i-line photoresist in Figure 5. Such measurements can have multiple applications, including both process development and process control.

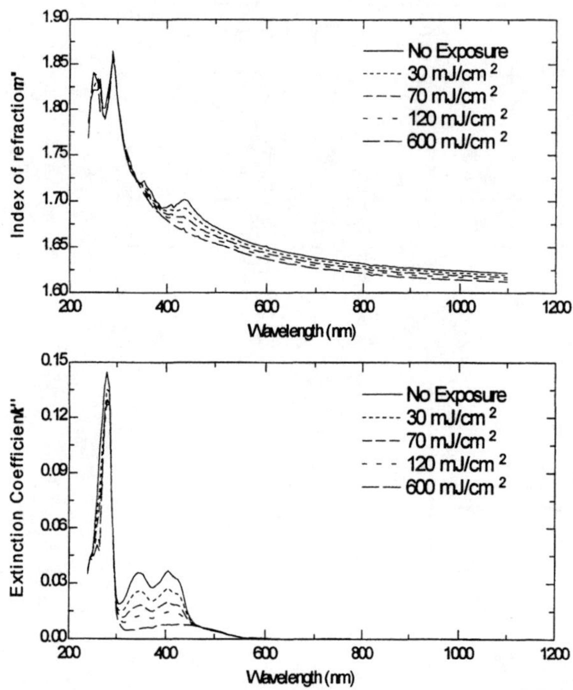

Figure 5. Index of refraction (a) and extinction coefficient (b) measured on a Shipley 510L i-line resist after different exposure conditions.

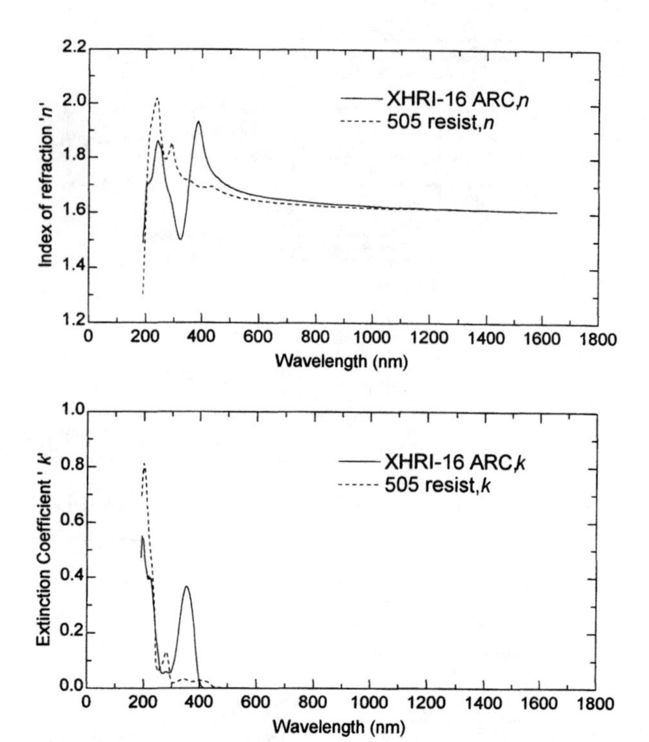

Figure 6. Index of refraction (a) and extinction coefficient (b) for both a Shipley SPR-505 resist and a Brewer XHRi-16 ARC.

Multilayers

One of the most important applications of VASE is measuring multilayer films. In lithography, it is very common to have multilayer films. An example multilayer would be photoresists on antireflective coatings. Most of the films used for lithography are optically similar. They are typically transparent in the visible and NIR, with absorption at shorter wavelengths (in the UV). When a film becomes absorptive, it will have anomolous dispersion of its refractive indices. In this region, there may be significant optical contrast between the layers, as they undergo significant anomolous dispersion. This fact can be utilized to unambiguously determine multiple layer thicknesses. The optical constants determined from a Shipley SPR-505 photoresist on a Brewer XHRI-16 ARC are shown in Figure 6. This figure illustrates the difference in optical constants that is most evident at shorter wavelengths. This fact was utilized to determine both layer thicknesses from a multilayer.

Photomasks

Photomasks are also very important to the lithography process. They can enable smaller features than considered possible if designed correctly. Complex material structures have been developed by DuPont(6) to accommodate a 180° phase shift in attenuated phase shifting masks. The composition of the Cromium was intentionally graded with oxygen, nitrogen, and carbon to produce complicated grading profile. An example profile is shown in Fig. 7. VASE measurements were able to measure the film thickness, refractive index, and grading profile by combining many different measurements(2).

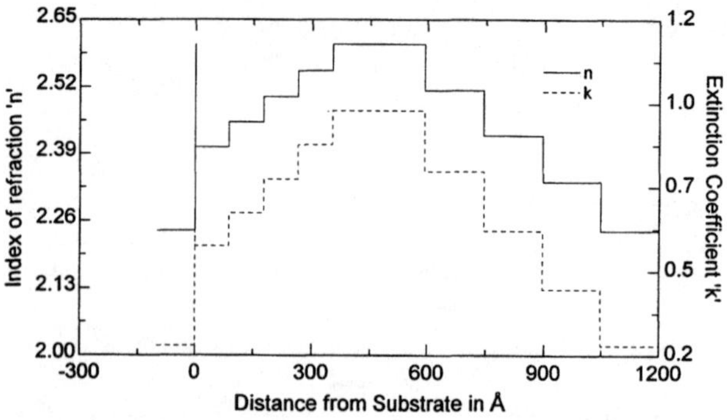

Figure 7. Complex index grading profile determined using VASE on a phase-shifting photomask.

CONCLUSIONS

Variable angle spectroscopic ellipsometry is a very powerful metrology technique that can be applied to all types of thin films in the lithography industry. It is non-destructive and very precise. By varying the angle of incidence and the measurement wavelength, extensive data can be collected for use in determining thin film properties – including thickness and refractive index (*n* and *k*). We have reviewed the merits of this technique for films in the lithography industry. We also report results from photoresists, anti-reflective coatings, photomasks, and multilayers. The high-accuracy and precision of this technique will ensure its future use for both process development and process control.

REFERENCES

1. Hilfiker, J. N., and Synowicki, R. A., *Solid State Technology* **10**, 157 (1996).
2. Hilfiker, J. N., and Synowicki, R. A., *Semiconductor Fabtech* 5th edition, 189 (1997).
3. Henderson, C. L., et. al., *SPIE Proc.* Vol. 3049, 585 (1997).
4. Dammel, R. R., et. al., *SPIE Proc.* Vol. 3049, 963 (1997).
5. Synowicki, R. A., et. al., "Refractive Index Measurements of Photoresist and Antireflective Coatings with Variable Angle Spectroscopic Ellipsometry", presented at SPIE Microlithography Conference, Santa Clara, CA, Feb. 98.
6. US Patents #5,393,465 and # 5,415,953 (DuPont).

Key Words: Variable Angle Spectroscopic Ellipsometry, Refractive Index, Thin Film Metrology, Lithography, DUV

Novel Metrology for the DUV Photolithography Sequence

Nickhil Jakatdar, Xinhui Niu, Haolin Zhang, Junwei Bao, Costas Spanos
University of California-Berkeley, CA 94720

In-situ metrology promises to provide effective manufacturing line operation, reduced cycle times and improved process quality for semiconductor processing. The challenges in developing this technology lie in identifying useful and relatively simple observables in the lithography sequence, relating these observables to the final quantity of interest, developing simple but effective control strategies and finally, integrating this with the production line equipment. This paper reviews the opportunities for metrology and control in the DUV lithography sequence. Metrology for the pre coat, post soft bake, post bake and post develop steps is presented here.

Keywords: DUV Photolithography, In-situ, Neural Network Algorithm, Deprotection Induced Thickness Loss

INTRODUCTION

Feature dimensions in semiconductor manufacturing are continually decreasing, while die and wafer sizes are increasing. As the critical feature size decreases below 0.3 µm, Deep Ultraviolet (DUV) lithography remains the key technology driver in the semiconductor industry, accounting for approximately 35% of processed wafer cost. However, submicron DUV photolithographic processes present significant manufacturing challenges due to the relatively narrow process windows often associated with these technologies. The sensitivity of the process to small upstream variations in incoming film reflectivity, photoresist coat and softbake steps as well as the bake plate temperature can result in the final critical dimension (CD) going out of specifications. Further, CD problems are usually not identified until the end of the lot. The high costs associated with the manufacture of Integrated Circuits necessitates higher yields and throughput, requiring a reduction in process variability. One approach to reducing process variability is to use a supervisory system that controls the process on a real time or run-to-run basis (1).

High end devices such as microprocessors require a considerable number of process steps. Therefore, it is becoming increasingly important to have an accurate, quantitative description of the submicron structure after each step. Currently, the lithography process is monitored before photoresist spin on (index, thickness and uniformity measurement of incoming stack) and after development (linewidth and profile measurement). Inspection at the initial and final stages of the process, however, provides only a measurement of the cumulative effects of all the upstream process steps. To isolate the effect of each process step, monitoring at each step is necessary. This need for wafer process monitoring requires in-line sensors and real time algorithms to facilitate real time analysis of sensor signals. In-line refers to processing steps or tests that "are done without moving the wafer" and are usually unobtrusive, non-contact and with little extra cost to the process. This is in contrast to off-line metrology, where the wafer needs to be removed from the processing environment to be measured. In-line metrology is preferred to off-line metrology due to increased throughput and possibly yield.

The need for in-situ and/or in-line process monitoring must however be balanced with critical manufacturing issues such as possible adverse effects on throughput, cost, sensor integration into an overall control strategy, possibly limited sensor reliability, etc. Most commercial metrology equipment is either too slow or too complex to be implemented in an in-line arrangement. An ideal in-line metrology sensor would be capable of making measurements that are sufficiently accurate, repeatable and rapid at a low cost. At present there is no single technique capable of meeting all of these demands. The following sections discuss some of the in-situ metrology developed for the lithography sequence that overcomes most of the above mentioned problems.

CP449, *Characterization and Metrology for ULSI Technology: 1998 International Conference*
edited by D. G. Seiler, A. C. Diebold, W. M. Bullis, T. J. Shaffner, R. McDonald, and E. J. Walters

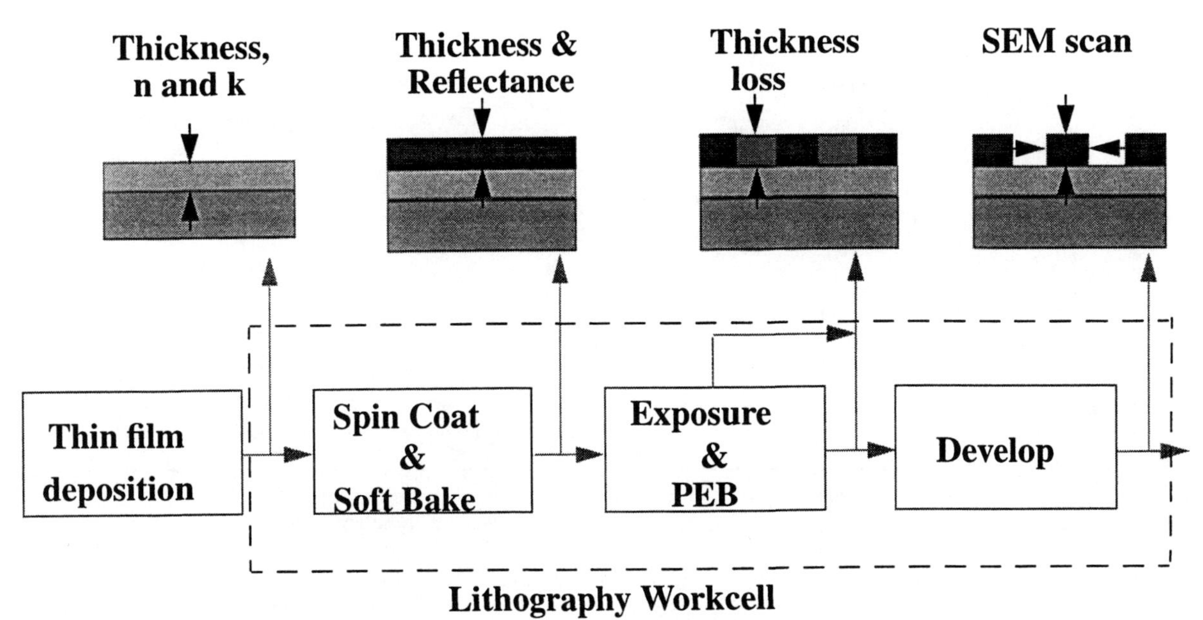

FIGURE 1. Opportunities for measurement in the lithography sequence

METROLOGY FOR DUV LITHOGRAPHY: THEORY & PRACTICE

The DUV lithography process provides the process engineer with numerous opportunities to monitor the process and wafer state (see FIGURE 1). It is important to identify quantities that are not only related to the photoresist critical dimension but also lend themselves to in-situ measurement.

The Neural Network-Adaptive Simulated Annealing Algorithm

The thickness and the optical constants (n and k) of the thin film are a good measure of the thin film reflectivity, both before and after the spin-coat step. The reflectivity of the thin film stack modulates the dose given by the stepper and hence changes the CD. In addition, the thickness and the optical constants also serve as statistical process control (SPC) variables to monitor the thin film deposition and the photoresist spin-coat steps. The measurement of these parameters should be done both accurately and in real time so as to be applicable for run to run (R2R) control applications. This problem required the use of an accurate dispersion relation as well as a fast global optimizer. The Forouhi-Bloomer (F-B) dispersion relation (2) was chosen in this example although other dispersion relations could also be used. For the optimization, we chose to use Adaptive Simulated Annealing (ASA) (3), which is a probabilistic optimization technique well suited to multi-modal, discrete, non-linear and non-differentiable functions. ASA's main strength is its statistical guarantee of global minimization, even in the presence of many local minima. However, simulated annealing methods are notoriously slow. To overcome the problem of speed, a Radial Basis Function Neural Network (NN) (4) was used for locally speeding up the optimization problem. The basic blocks of this algorithm are

1) Parameter Extraction using ASA - In this step, the optical constants and thickness of the wafers are extracted. This is done to determine the typical variation of the equipment.

2) Monte Carlo Simulation using the F-B formulation and Characteristic Matrix Approach - This step generates a large number of datasets based on experimental data, that are required for reliable training of the neural network.

3) Spectral Feature Selection - This step selects the features of the broadband signal as well as the range of wavelengths over which these features are to be selected. This is basically a physically based data reduction filter.

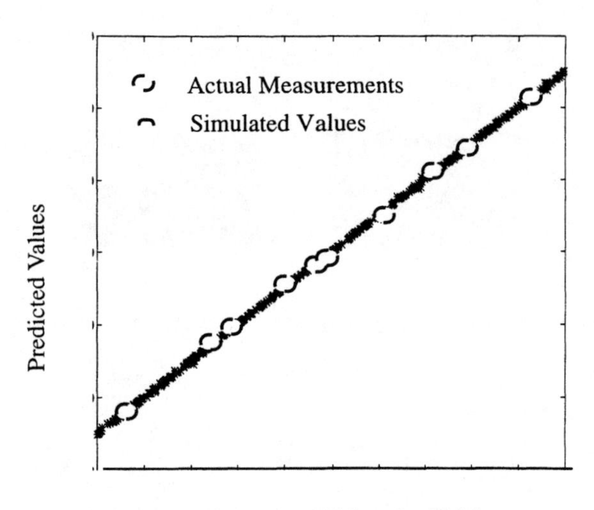

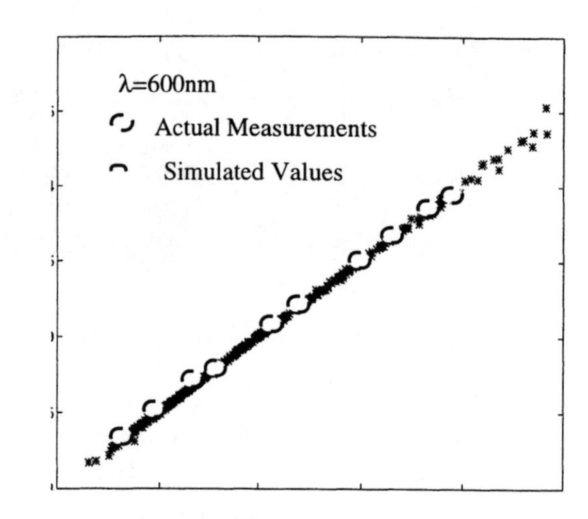

FIGURE 2. Performance of the NN-ASA algorithm for poly-Si on native oxide on Si stack.

Parameter	Range	Values	σ
Thickness	100 nm	350 - 450 nm	0.300 nm
n(600nm)	0.25	3.800 - 4.050	0.0012

4) Neural Network Training and Validation - This step is necessary to avoid either over training or under training the neural network.

5) Neural Network Feature Classification for Real Time Application - This step is a combination of some of the earlier steps and is executed in real time once the neural network has been trained.

The results of the extraction of the optical constants from the broadband reflectometry data is shown in FIGURE 2. The measured values were taken off Poly-Silicon wafers deposited at different temperatures, thus resulting in different thicknesses and optical constants.

Deprotection Induced Thickness Loss (DITL)

Chemically amplified photoresists do not bleach and hence a measurable other than the extinction coefficient was needed to characterize the state of the wafer after the PEB step (5). Deprotection, which is a quantitative measure of the photoresist state, can only be measured using Fourier Transform Infrared Spectroscopy (FTIR). We thus needed to identify an observable that reflected the deprotection while at the same time was easy to measure. Most chemically amplified resists undergo a thickness change during the post exposure bake step due to the volatility of the cleaved side-chains (6) as depicted in FIGURE 3(a). We found that the deprotection was a strong function of the exposure dose and the PEB temperature as was the thickness loss. The next step then was to measure the deprotection in the exposed areas and correlate this to the thickness loss. The deprotection was measured using FTIR which is a powerful analytical tool for characterizing and identifying organic molecules. We found that the thickness loss is very strongly correlated to the deprotection and could hence serve as a surrogate in-situ measurement for the deprotection. Since the final critical dimension is due to the deprotection and diffusion reactions taking place in the resist, the thickness loss provides a partial handle on the CD producing mechanism and could also be used for diagnosis and control of the exposure and PEB steps.

Deprotection was measured in flood exposed areas by following the ester bond for a Shipley UV-5 chemically amplified resist and was correlated to the thickness loss in the corresponding areas. Results of the linear fit are shown in FIGURE 3(b).

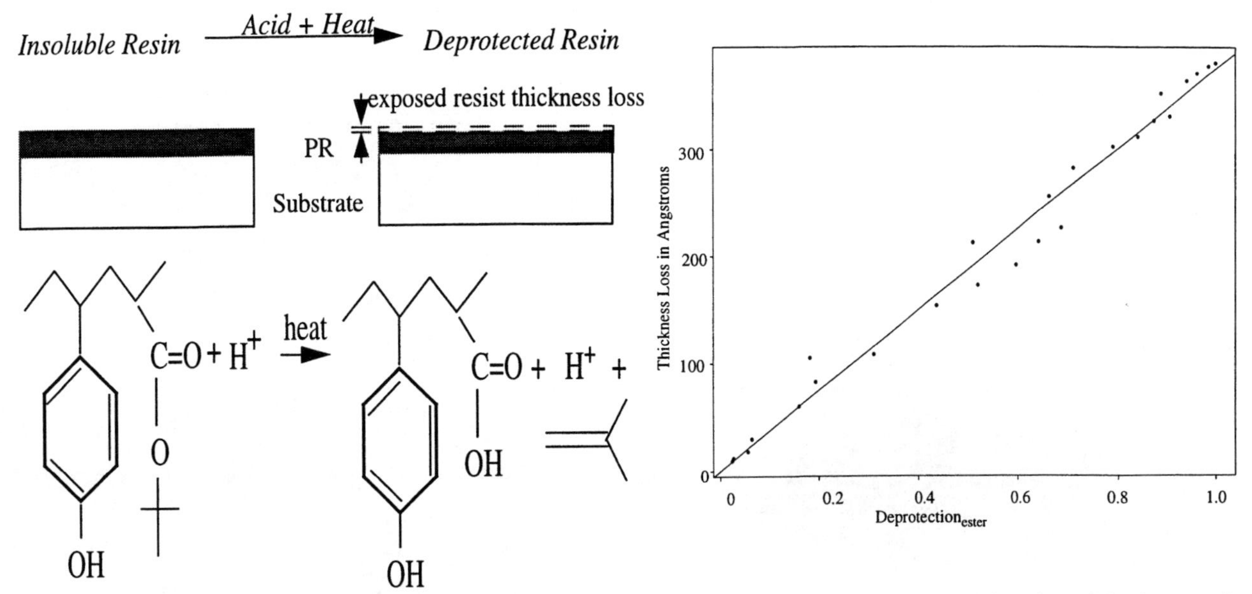

FIGURE 3(a). Resist thinning and deprotection during the PEB process

FIGURE 3(b). Thickness loss as a function of the deprotection measured by monitoring the ester absorbance

Focus Monitor

The CD is also a strong function of the focus. Scatterometry is a promising sensor for extraction of focus information from the post develop CD profiles. However, there are still some issues such as uniqueness and demonstration at sub-quarter micron feature sizes that need to be ironed out before the technology becomes mainstream. The small focus window available in most DUV processes along with the fact that the stepper focus tends to drift over the period requires running send ahead test wafers every few lots. This has motivated a sensor specifically for focus extraction using the broadband reflectometer sensor described. In this technique, we use a test structure populated with 2-dimensional post-like structures due to their high sensitivity to defocus. A focus-exposure matrix was carried out with this structure and broadband reflectometry signals were collected. A database was built for the combinations of focus and exposure. Having built the database, future signals from arbitrary focus exposure settings were compared to the database and the least squares criterion was used to pick the best fit. This provides an easy way to diagnose the focus setting used and could be used for feedback to the stepper in case of drifts. This also eliminates the need to have separate send ahead test wafers since each wafer now tests the focus setting. This technique does require the use of a test structure for the posts but this could be made on the areas of the wafer not being used for the product. Results of using this metrology are shown in FIGURE 4(a).

Smart CD-SEM

The CD-SEM is a routine inspection tool in today's IC fabrication lines. Even though relatively accurate critical dimensions can be obtained from a CD-SEM, much more information is hidden in the high resolution SEM images. We have implemented an easy to implement metrology by bringing the CD-SEM into the loop. In this work, we have examined the patterns with deep submicron feature size using the CD-SEM. Dimensionality reduction is used to summarize the information carried by the SEM scan, thus not only obtaining a reduced data set that is easier to handle but also extracting the significant "patterns" hidden in the original data. A Principal Component Analysis was used for the data reduction. A feedforward neural network was then trained by backpropagation to classify the different conditions of focus and exposure used. This technique used the first 30 principal components and used data from 169 combinations of focus exposure at 0.35 mJ/cm^2 features for the classification. Each measurement is the average of 32 scans to filter noise. The

FIGURE 4(a). Frequency of Predicted vs. Actual Focus Settings for the Focus Monitor

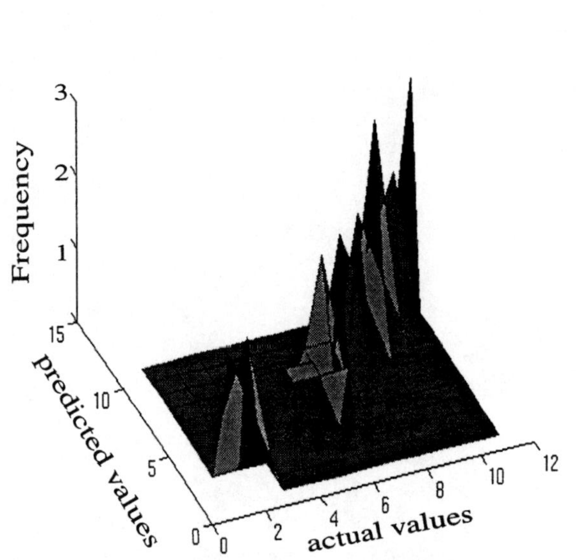

FIGURE 4(b). Frequencies of Predicted vs. Actual Exposure and Focus Settings

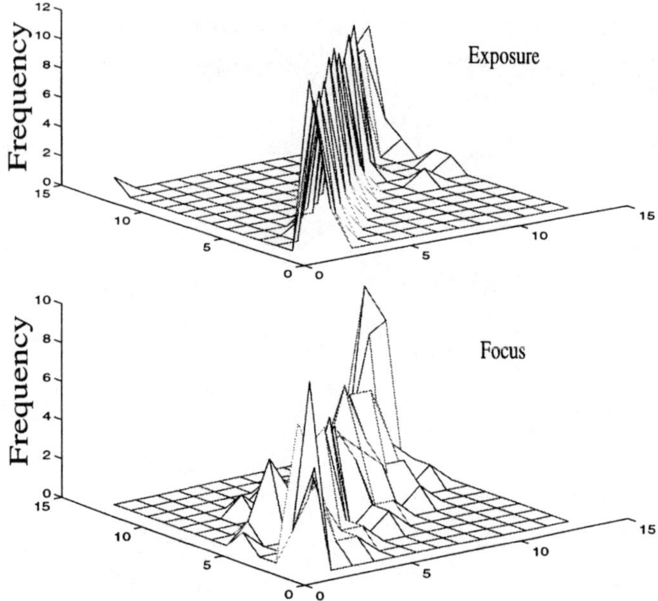

results of this metrology were very impressive and led to very good classification for the exposure settings and encouraging results for the focus setting. This work can be used for lithography process control by feedback to the stepper and also for diagnosis and monitoring of the stepper focus. This metrology is ex-situ; however it utilizes measurements that are already being made and hence does not add any complexity in implementation. Results of this metrology are shown above in FIGURE 4(b).

CONCLUSIONS

We have developed novel in-situ and ex-situ metrology for the DUV lithography sequence that can be used for process control and diagnosis. The in-situ metrology includes thickness and optical constants extraction for the incoming wafer and the pre-exposure steps, thickness loss for the exposure and PEB steps and focus monitoring for the stepper. The ex-situ metrology involves using the CD-SEM scan for additional information about the stepper focus and exposure settings.

This metrology has been shown to be accurate, repeatable, rapid and low cost. This is one of the first attempts to identify and measure intermediate lithography parameters in-situ and hence only comparison with ex-situ metrologies is possible at the moment. The NN-ASA technique for optical constant extraction has been verified with ellipsometry techniques for accuracy; the DITL methodology for deprotection has been verified with FTIR spectroscopy while the focus monitor and the smart CD-SEM have been cross validated with a CD-SEM. Repeatability has been shown to be very good in all cases. This metrology in conjunction with a simple broad-band reflectometer sensor is currently being tested with deep submicron lithography processes.

REFERENCES

(1) A. Iturralde, A Review of Sensing Technologies for Semiconductor Process Application, ISSM 1995

(2) A.R. Forouhi, I. Bloomer, Optical Properties of Crystalline Semiconductors and Dielectrics, Phys. Rev. B, vol38, 1865, 1988

(3) X. Niu, Timbre - An Adaptive Simulated Annealing Based Optimization Toolbox

(4) S. Haykin, Neural Networks - A Comprehensive Foundation, Macmillan, New York, 1994

(5) N. Jakatdar, In-situ Metrology for Deep Ultraviolet Photolithography Control, M.S. Thesis, 1998

(6) L. Thomson, C.G. Willson, M. Bowden, Introduction to Microlithography, ACS Professional, 1994

At-Wavelength Interferometry of Extreme Ultraviolet Lithographic Optics

Sang Hun Lee [a, b], Patrick Naulleau [a], Kenneth Goldberg [a], Edita Tejnil [c], Hector Medecki [a], Cynthia Bresloff [a], Chang Chang [a, b], David Attwood [a, b], and Jeffrey Bokor [a, b]

[a] Center for X-Ray Optics, Mail Stop 2-400, Lawrence Berkeley National Laboratory, Berkeley, CA 94720
[b] Department of Electrical Engineering and Computer Science, University of California, Berkeley, CA 94720
[c] Intel, Components Research, 2200 Mission College Blvd, SC-02, Santa Clara, CA 95052

A phase-shifting point diffraction interferometer (PS/PDI) has recently been developed to evaluate optics for extreme ultraviolet (EUV) projection lithography systems. The interferometer has been implemented at the Advanced Light Source at Lawrence Berkeley National Laboratory and is currently being used to test experimental EUV Schwarzschild objectives. Recent PS/PDI measurements indicate these experimental objectives to have wavefront errors on the order of 0.1 waves (~1 nm at a wavelength of 13.4 nm) rms. These at-wavelength measurements have also revealed the multilayer phase effects, demonstrating the sensitivity and importance of EUV characterization. The measurement precision of the PS/PDI has been experimentally determined to be better than 0.01 waves. Furthermore, a systematic-error-limited absolute measurement accuracy of 0.004 waves has been demonstrated.

INTRODUCTION

The fabrication of electronic devices with ever-smaller feature sizes is an ongoing challenge for the IC manufacturing community. Recently, the industrial standard critical dimension (CD) for IC devices has dropped from 0.35 μm to 0.25 μm. Further progressions to CDs of 0.18 μm and 0.13 μm are anticipated by extending conventional optical lithography methods to shorter wavelengths. For 0.1 μm and beyond, however, it is widely believed that new lithographic techniques are needed. One very promising candidate is extreme ultraviolet (EUV) lithography. This method continues on the path of projection optical systems but with a radical reduction in wavelength (10-15 nm) and conversion to lower numerical aperture (NA) all reflective system. EUV systems require unprecedented fabrication tolerances and hence unprecedented metrology accuracy (1). Here we describe an at-wavelength metrology technique expressly developed to meet this measurement challenge.

Achieving 0.1 μm resolution with EUV projection lithography places stringent tolerances on the fabrication and testing of these all-reflective multilayer-coated optical systems. In order to achieve a diffraction limited performance, the deviation from a perfect spherical wavefront at the exit pupil of the optical system must meet the Maréchal condition, $\lambda/14$ rms, at the operational wavelength of 13.4 nm (2). The interferometric characterization of these lithographic optical systems must provide measurement accuracy that is several times better than the Maréchal condition. Furthermore, since the properties of the multilayer reflective coatings strongly impact the optical performance, interferometry at the operational wavelength is desirable for alignment and qualification of EUV lithographic systems.

The required wavefront measurement accuracy is on the order of 0.01 waves (or 0.1nm). In order to meet this significant measurement challenge, a novel interferometer, called the phase-shifting point diffraction interferometer (PS/PDI), has recently been developed and implemented at the Center for X-Ray Optics, Lawrence Berkeley National Laboratory (3).

The accuracy of surface figure interferometers is limited by several factors; the two most important factors are systematic and random errors from the quality of the reference beam and systematic effects from the measurement geometry. A null test, required to identify these systematic and random errors, has recently been performed, demonstrating the PS/PDI to have a systematic error limited accuracy of 0.004 waves (0.05 nm or $\lambda/250$).

The PS/PDI is currently being used to characterize multilayer-coated 10X-demagnification Schwarzschild objectives. The 10X-reduction prototype lithographic optical system operates at a wavelength of 13.4 nm, with 0.1 μm resolution, a numerical aperture (NA) of 0.08, and a depth of focus of ~1 μm. Its general description is shown in Fig.1.

PS/PDI DESCRIPTION

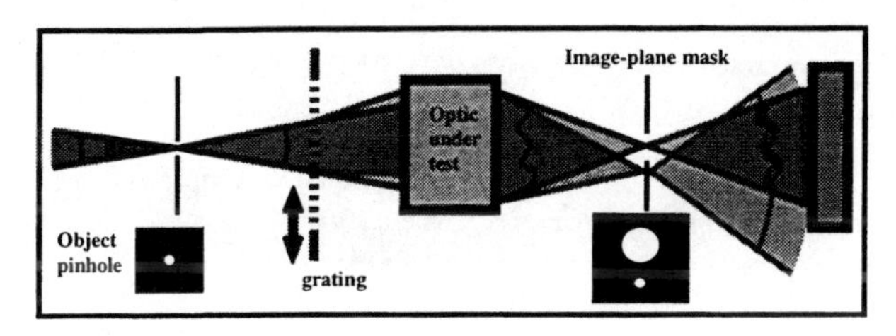

Figure1. Key elements of the PS/PDI

CP449, *Characterization and Metrology for ULSI Technology: 1998 International Conference*
edited by D. G. Seiler, A. C. Diebold, W. M. Bullis, T. J. Shaffner, R. McDonald, and E. J. Walters

The PS/PDI is briefly explained here; a more complete description can be found in the literature (3,5). The PS/PDI is a modified version of the conventional point diffraction interferometer with improved efficiency and phase-shifting capability. Because the PS/PDI is a nearly common path interferometer, it is extremely stable. The PS/PDI consists of an object pinhole, a coarse grating, the optic under test, an image-plane pinhole mask, and a CCD detector. Spatially coherent spherical wave illumination is provided by diffraction from the object pinhole. The transmissions grating serves as a beam-splitter providing the test and reference beams required for the interferometry. The optic under test focuses the various grating orders to the image-plane mask. The image plane mask generates the reference beam by way of pinhole diffraction whereas the test beam propagates, largely unaffected, through a large window (~4.5 um) in the mask. The rest of the grating orders are blocked by the mask. The test and reference beams propagate to the mixing (CCD) plane where they overlap to create an interference pattern. Phase-shifting interferometry can be achieved by translation of the transmission grating. The interferogram is a comparison of the test beam to the reference beam, which in the ideal case is a spherical wave.

EXPERIMENTAL RESULT

The present setup for EUV PS/PDI testing of 10X-Schwarzschild objectives is depicted in Fig. 2. EUV radiation is provided by an undulator beamline (beamline 12.0.1 at the Advance Light Source, Lawrence Berkeley National Lab). The undulator beamline provides continuously tunable illumination from 5 nm to 25 nm wavelength with a linewidth $\lambda/\Delta\lambda$ ~ 1000. The illumination is rendered spatially coherent by spatial filtering provided by the object plane pinhole, which is a commercially available 0.5 μm laser drilled pinhole.

The image plane mask is fabricated using electron beam lithography and reactive ion etching. The mask pattern is completely etched through a 1000 Å Si_3N_4 membrane and then 1000 Å of Ni is evaporated on each side to provide the required EUV attenuation. The test beam window is a 4.5 μm X 4.5 μm square open area, and the reference pinhole, typically 50 nm to 150 nm in diameter, is placed 4.5 μm from the center the test window. The detector used in the mixing plane is a back-illuminated, back-thinned, soft-X-Ray CCD detector.

Since the PS/PDI has been implemented, two different experimental EUV 10X-Schwarzschild objectives have been tested. These Schwarzschild systems utilize off-axis subapertures with each system containing three discrete subapertures, with NAs of 0.06, 0.07, and 0.08, respectively. These subapertures allow the system to be operated in three different NA modes. Measurements of the first Schwarzschild objective included characterization of three distinct subaperture regions, evaluation of the aberrations as a function of wavelength, and determination of the PS/PDI precision. E. Tejnil, et al, reports the detailed study of these experimental results (6).

Figure 3 shows a representative interferogram from the 0.07 NA subaperture in the first optic. Wavefront analysis is performed using the phase-shifting method (4). With the tilt and defocus removed, the wavefront errors of the 0.08, 0.07, and 0.06 NA subapertures were measured to be 0.26, 0.09, and 0.043 waves rms, respectively, at a wavelength of 13.4 nm.

EUV wavefront error is determined not only by the

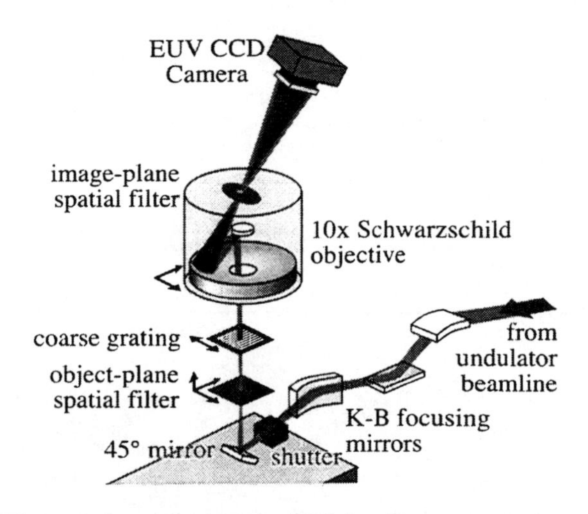

Figure 2. The set-up for the EUV interferometer to measure aberrations of the 10X-Schwarzchild optic.

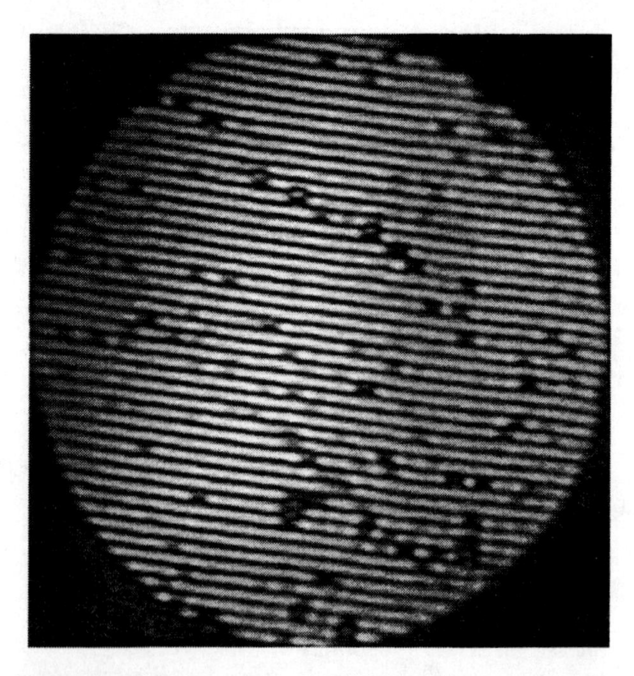

Figure 3. An interferogram from 0.07 NA subaperture in the first 10xSchwarschild optic.

quality of the surface figure of the test optic mirrors but also by the multilayer. Measuring subtle multilayer effects is not possible with visible light interferometry because reflection occurs at the top surface. These multilayer effects, however, can be directly observed with the PS/PDI at EUV wavelengths and isolated from substrate effects by varying the measurement wavelength, as has been reported in the literature (6). The precision of the interferometry was identified by way of repeatability measurements performed on the 0.07 NA subaperture of the first test optic. The measurements were decorelated in several ways: various pinholes were used, pinhole positions were changed relative to the illuminating beams, and the test optic was removed and replaced. The precision measurement was based on 23 separate phase-shifting measurement sets. The wavefront aberration map from the average of the 23 sets of data and the corresponding Zernike polynomial fit are shown in Fig. 4 and Fig. 5 respectively. A wavefront error of 0.090±0.008 waves (1.21±0.11nm) rms and 0.531±0.046 waves (7.11±0.61 nm) peak-to-valley was found. The most severe aberration in this optic is clearly astigmatism. The standard deviation from the 23 sets of data indicates the repeatability to be within 0.008 waves rms and 0.046 waves peak-to-valley.

One of the most significant values of the interferometry is to predict imaging performance of the optic. The performance of the first lithographic optic tested with PS/PDI has been verified using an experimental EUVL stepper at Sandia National Laboratory. Reference (6) shows the calculated images of a star resolution pattern and the actual printed star resolution patterns on the photoresist with different defocus regions. These results are in excellent agreement and verify that the PS/PDI is indeed an accurate wavefront measurement tool which can be used to predict the imaging performance of EUV lithographic systems.

Similar results were also obtained with the second Schwarzschild objective tested

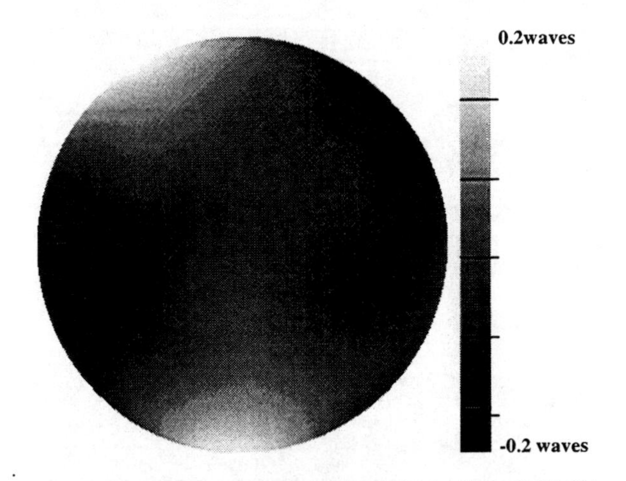

Figure 4. Reconstructed warfront aberrations of the 0.07 NA subaperture

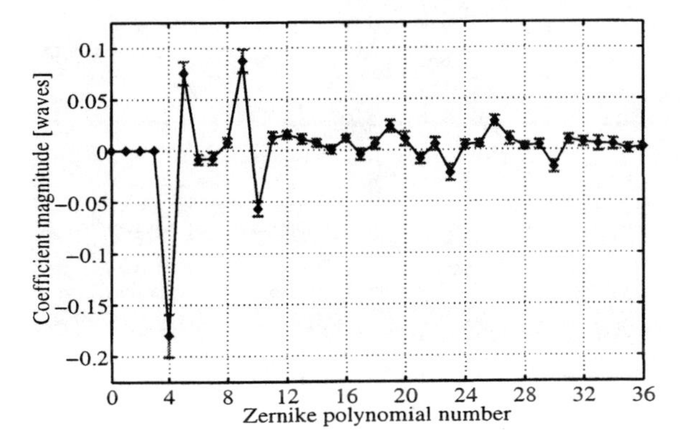

Figure 5. The Zernike polynomial fit coefficients of the Fig. 4.

ACCURACY OF THE PS/PDI

Above we described how repeatability measurements had been used to characterize the precision of the PS/PDI and how imaging tests had been performed to verify the accuracy of the interferometer. In order to quantify the absolute accuracy, however, a null test is required. In the null test, the standard PS/PDI mask is replaced by a two-pinhole mask where the test window is replaced by a second reference pinhole. A detailed discussion of the PS/PDI null test can be found in the literature (7).

There are two major sources of error in the PS/PDI: one is from systematic geometric effects, and the other is from the quality of the reference wave. The dominant geometrical effect is due to the lateral separation of the interfering, nominally spherical, test and reference waves. This situation leads to a hyperbolic fringe field which in turn leads to coma in the measurement. From the second order binomial expansion of the optical path length difference between the two waves, we can precisely predict the amount of measurement coma in the system. The geometrical coma effect can readily be removed from the reconstructed wavefront. For an image plane beam separation of 4.5 μm and a measurement NA of 0.08, the geometrical coma error is ~0.03 waves. Another possible source of geometrical error is astigmatism due to detector tilt misalignment (7). Proper alignment requires the CCD to be orthogonal to the optical system central ray. The detector tilt induced astigmatism can be shown to be about 0.019 waves per degree of detector tilt error. This error can be precisely controlled by careful mechanical design. Other geometrical effects have shown to be insignificant compared to the effects discussed here (8).

Error induced by reference wave imperfection is a much more subtle effect than the geometrical effects described above. Since the reference wave is generated by diffraction, its quality depends on many factors including pinhole size and shape, position of the pinhole relative to the image point, and the quality of the test optic.

In the two-pinhole null test, two high quality spherical waves generated by the two image plane pinholes create a fringe pattern (interferogram) on the CCD. Aberrations measured in this interferogram are indicative of errors in the PS/PDI and hence the accuracy of the PS/PDI. If the test optic was perfect or the pinholes were small enough, we would see only the predicted geometrical effects. In reality, however, we will measure other aberrations due to imperfections of the pinholes and the test optic.

Figure 6 shows an actual interferogram from a null test performed using 100 nm pinholes. The extremely straight fringe pattern is indicative of the accuracy of the PS/PDI. Fig. 7 and Fig. 8 show the reconstructed wavefront over a measurement NA of 0.08 before and after removing the geometrical coma effect respectively. The expected coma is 0.03 waves, whereas the measured coma is seen to be 0.031 waves. Removal of the geometrical coma leaves a residual wavefront error of 0.006 waves rms. This error includes systematic and random effects. The systematic-error-limited accuracy can be found by averaging many null-mask measurements, which eliminates the random errors. Performing this averaging shows the systematic errors to asymptotically approach 0.004 waves rms after removal of the geometrical coma.

It is also important to consider accuracy as a function of pinhole size. As expected, experiments have shown the accuracy to degrade as the pinhole size is increased.

Figure 6. The actual interferogram from the null test recorded on the 1" x 1" CCD. Two 100 nm pinholes have been used. The extremely straight fringes are observed.

The single measurement accuracy was $\lambda/74$, $\lambda/94$ and $\lambda/167$ for 140, 120, and 100 nm pinhole respectively.

Verification of the accuracy measurements using TEMPEST (8,9) simulation is currently under way.

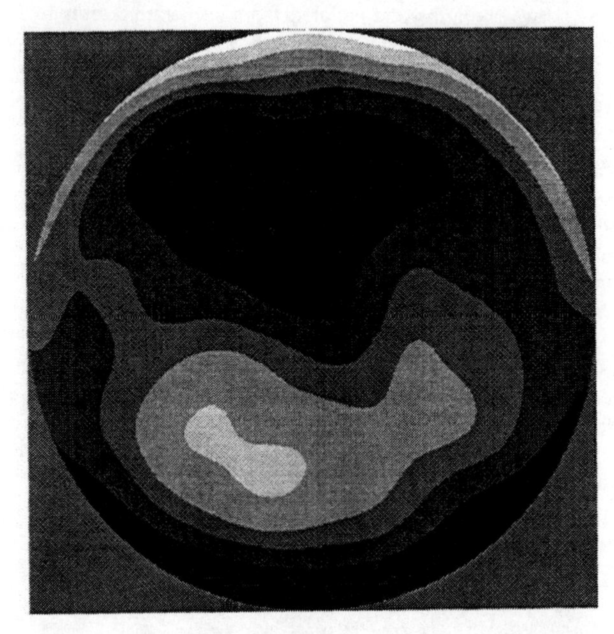

Figure 7. Average wavefront displayed as 8 level grayscale: prior to the predicted geometric coma removal. The full black to full white scale is from 0.5 nm to –0.5 nm

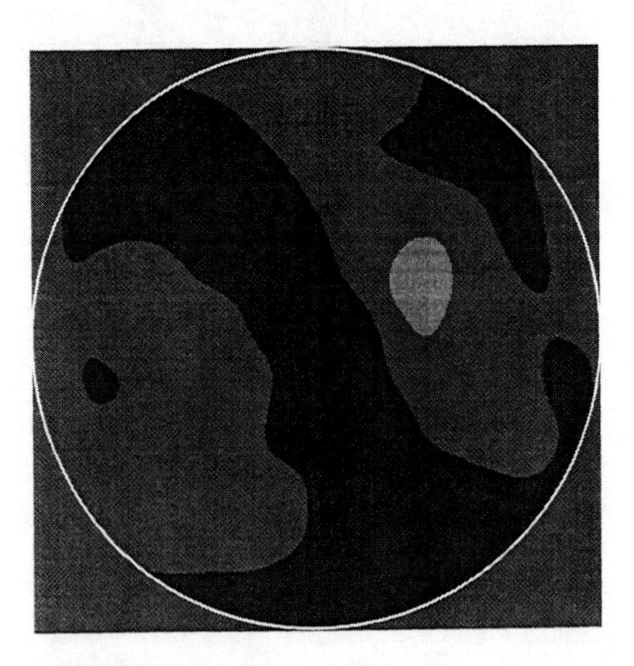

Figure 8. Average wavefront displayed as 8 level grayscale: after the predicted geometric coma removal. The full back to full white scale is from 0.5 nm to –0.5 nm.

CONCLUSION

We have successfully tested two different 10X-Schwarzchild lithographic objectives with the phase-shifting point diffraction interferometer at the operational wavelength of 13.4 nm. Aberrations measured were on the order of 0.1 waves. The ability of the PS/PDI to measure phase effects caused by the reflective multilayer coatings has been demonstrated, and the precision of the PS/PDI has been found to be better than 0.01 waves. Also the systematic errors in the PS/PDI have been shown to be limited to 0.004 waves.

ACKNOWLEDGMENT

The authors give special thanks to Erik Anderson for fabrication of the PS/PDI and Null masks. We also acknowledge the entire CXRO staff, including Phil Batson, Matthew Bjork, Joshua Cantrell, Shane Cantrell, Keith Jackson, and Drew Kemp for engineering and software support. Very special thanks to Paul Denham for expert assistance with experimental control systems. This research was supported by the EUV LLC, the Semiconductor Research Corporation, DARPA, and the DOE Office of Basic Science.

REFERENCES

1. "National Technology Roadmap for Semiconductors," (Semiconductor Industry Assoc., San Jose, CA, 1994).
2. Sommargren, G.E., "Phase shifting diffraction interferometry for measuring extreme ultraviolet optics," OSA Trends in Optics and Photons, Vol.4, 108 (1996).
3. Medecki, H., et al., "Phase shifting point diffraction interferometer," Optics Letter, 21, 19, pp. 1526-8, (1996).
4. Malacara, D., Ed. *Optical Shop Testing*, 2nd ed., (John Wiley & Sons, Inc, New York, 1992), Ch. 14.
5. Goldberg, K.A., Tejnil, E., Lee, S.H., Medecki, H., Attwood, D.T., Bokor, J., "Characterization of an EUV Schwarzschild objective using Phase-shifting point diffraction interferometry," SPIE Microlithography, 3048-24 , (1997).
6. Tejnil, E., et al., "At-wavelength interferometry for EUV lithography," J. Vacuum Science and Technology B, Vol.15, No.6, pp. 2455-61, (1997).
7. Naulleau, P., et al., "Characterization of EUV phase-shifting point diffraction interferometry," SPIE microlithography, 3331-13, to be published (1998).
8. Goldberg, K., "Extreme Ultraviolet Interferometry," doctoral dissertation, Department of Physics, University of California, Berkeley (1997).
9. Pistor, T., "TEMPEST version5.0," MS report at EECS, UC Berkeley (1997).

Assessing Polysilicon Linewidth Variation Using Statistical Metrology

N. Moorthy Muthukrishnan and Sharad Prasad

LSI Logic Corporation, M.S. A-100, Milpitas, CA 95035.

Assessment of wafer-level and die-level polysilicon linewidth variation is achieved by electrical linewidth measurement of a specially designed submicron linewidth measurement structures using an automated wafer prober connected to a parametric tester. The raw linewidth data is decomposed into wafer-level and die-level variations using statistical metrology techniques. The information thus obtained can be utilized for integrated circuit (IC) layout design and fabrication process improvements.

INTRODUCTION

As the minimum feature sizes continue to shrink and the wafer size continues to increase, variations in the device dimensions have a large impact on device performance and wafer yield. These variations can be broadly classified into lot-to-lot, wafer-to-wafer, and within wafer variations. The lot-to-lot variations are temporal, and can be attributed to process and equipment drift with time. The wafer-to-wafer variation can be temporal or spatial dependent on the type of processing equipment used. In batch processing equipments, the wafer-to-wafer variation is mainly due to the position of the wafer in the equipment during processing. Single wafer processing greatly reduces the spatial variation. However, temporal variations can still exist depending on the instability of the process and equipment parameters. In wafer-level, the inter-die and intra-die variations are spatial. The inter-die (or wafer-level) variations arise mainly due to equipment asymmetries and process imperfections and intra-die (or die-level) variations arise due to design parameters such as feature dimensions, density, and orientation. These variations can affect the die yield and device performance. The measurement techniques that are currently being used in the semiconductor industry are not necessarily capable of identifying the wafer-level and die-level components of variations. An insight into the nature of variations can be gained by performing large number of in-line measurements on each die of several wafers, which may be prohibitively expensive and time consuming. However, the availability of electrical parametric testers coupled with automatic wafer probers make electrical measurements on devices attractive and provide a means of gathering large amounts of data in a short time. Thus use of electrical characterization techniques, along with data analysis using statistical metrology, proves to be beneficial in good understanding of the variations of the feature sizes, such as polysilicon and metal linewidths, and dielectric thicknesses. This knowledge, if properly applied, would help to improve the design, fabrication and equipment aspects. The goal of this work described herein is to determine inter-die and intra-die polysilicon linewidth variations using electrical linewidth measurements and data analysis using statistical metrology techniques.

EXPERIMENT

The electrical linewidth measurement structure used in this work, as can be seen in Figure 1 consists of a van der Pauw structure for sheet resistance measurement and three Kelvin bridges for linewidth calculation. Two of the Kelvin bridges are of equal length, one of which contain dummy taps in order to estimate the effect of the voltage taps on the electrical length of the bridge. The use of this test structure for linewidth measurement of metal films was demonstrated by Allen et al. (1,2) and Cresswell et al. (3). The test structure was designed and fabricated on a silicon wafer using a standard silicon IC production process. The measurements were made on all the Kelvin bridge structures on 100% of the dice of each wafer. The sheet resistance of polysilicon film was measured from the van der Pauw structure in the same linewidth structure. A separate software program was written in PERL to calculate the electrical linewidth from the Kelvin resistance and sheet resistance measurements. A detailed discussion of the linewidth calculation can be found in (1). The linewidth data were analyzed using statistical metrology techniques. A statistical metrology program, VarDAP (Variation Decomposition Analysis Program), was used to separate the wafer-level and die-level variations from the raw

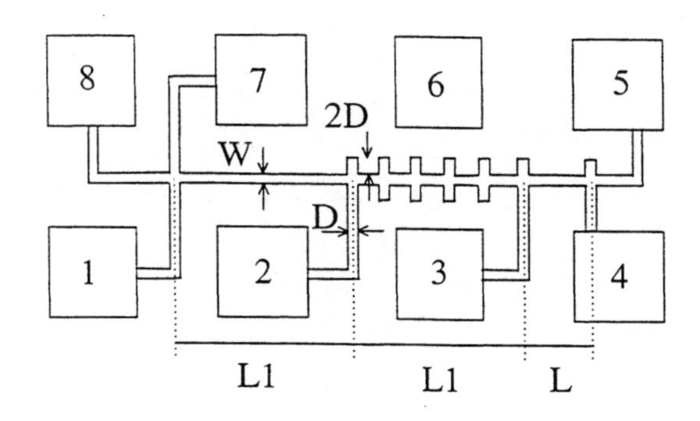

Fig. 1. Submicron linewidth measurement structure.

CP449, *Characterization and Metrology for ULSI Technology: 1998 International Conference*
edited by D. G. Seiler, A. C. Diebold, W. M. Bullis, T. J. Shaffner, R. McDonald, and E. J. Walters

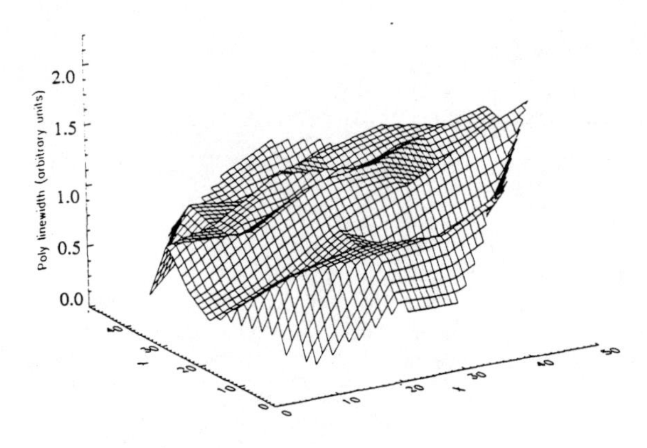

Fig. 2.a. Polysilicon linewidth - raw data.

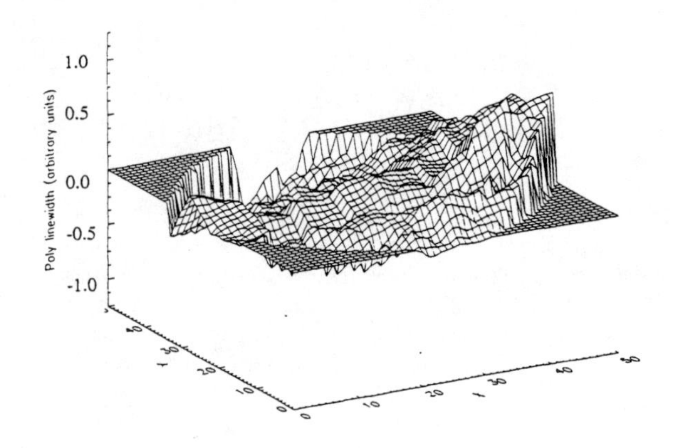

Fig. 2.d. Polysilicon linewidth - wafer-die interaction.

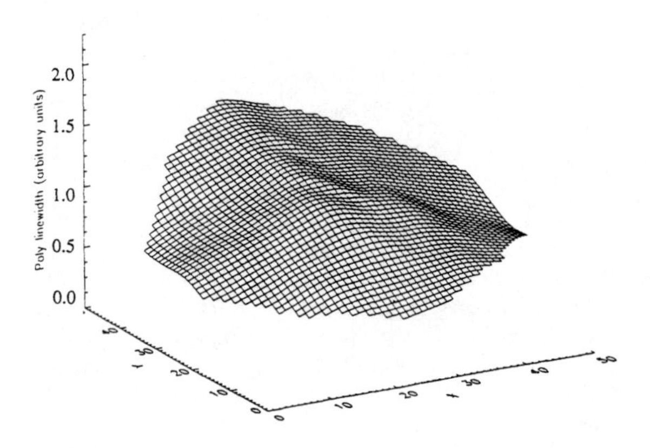

Fig. 2.b. Polysilicon linewidth - wafer-level variation.

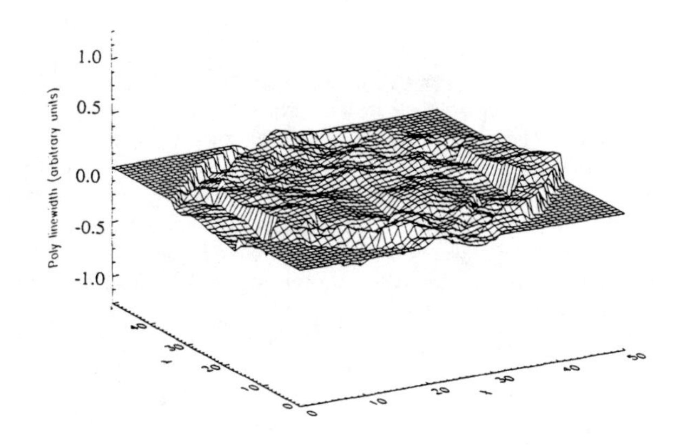

Fig. 2.e. Polysilicon linewidth - residuals.

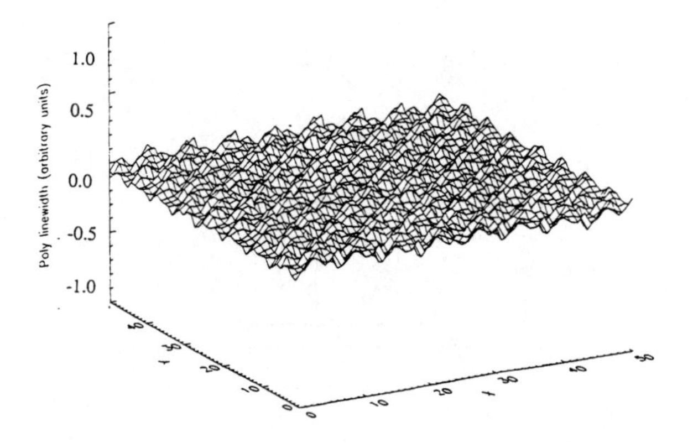

Fig. 2.c. Polysilicon linewidth - die-level variation.

data (4). The statistical metrology methodology works on an assumption that the parameter of interest, $f_{RAW}(x,y)$, can be modeled as the sum of wafer-level variation, $f_{WLV}(x,y)$, die-level variation, $f_{DLV}(x,y)$, wafer-die interaction, $f_{WLVxDLV}(x,y)$, and residuals, ε, as shown in equation 1.

$$f_{RAW}(x,y) = f_{WLV}(x,y) + f_{DLV}(x,y) + f_{WLVxDLV}(x,y) + \varepsilon \quad (1)$$

The wafer-level variation is estimated by down sampled moving average estimator, and the die-level variation is determined by two dimensional Fast-Fourier transformation (FFT) of the raw data, which is based on detecting the periodicity of the spatial repetition of the die across a wafer, and filtering the components near die frequency to capture the die-level variation components (4).

RESULTS AND DISCUSSIONS

From this work, the wafer-level and die-level variations were calculated using statistical techniques and plotted as shown in Fig. 2. The wafer-level variation, which is determined by taking the moving average of the critical dimension across the wafer, is mainly dependent on the process parameter settings and equipment configurations, making it useful for comparison of processes and equipments. The die-level variation, which is determined by calculating the critical dimension of identical structures across the die, is mainly dependent on the design parameters such as linewidth and density.

CONCLUSIONS

From this work, it is demonstrated that the subhalf micron electrical linewidth measurement structure, in combination with automated electrical measurements and data analysis using statistical metrology tools provide an efficient and powerful technique to study the critical dimension variations of poly silicon in a semiconductor process. The knowledge gained from this work can be applied to design modifications, fab process/ equipment development, qualification, and improvements.

ACKNOWLEDGMENTS

The authors would like to thank LSI Logic corporation for financial support and Richard Allen and Loren Linholm at NIST, and Jim Chung and Duane Boning at MIT for their technical inputs.

REFERENCES

1. Allen, R. A., Cresswell, M. W., and Buck, L. W., *IEEE Electron Device Letters* **13**, 322-324 (1992).
2. Allen, R. A., Cresswell, M. W., Ellenwood, C. H., and Linholm, L. W., *IEEE Transactions on Instrumentation and Measurement* **45**, 136-141 (1996).
3. Cresswell, M. W., Sneigowsky, J. J., Ghoshtagore, R. N., Allen, R. A., Linholm, L. W., and Villarubia, J. S., "Electrical test structures replicated in silicon-on-insulator material," SPIE Proceedings on Metrology, Inspection, and Process Control for Microlithography X, 659-675, Santa Clara, California, March 11-13, 1996.
4. Stine, B. E., Boning, D. S., and Chung, J. E., *IEEE Transactions on Semiconductor Manufacturing* **10**, 24-41 (1997).
5. Stine, B. E., Boning, D. S., Chung, J.E., Bell, D. A., and Equi, E., "Inter and Intra-die polysilicon critical dimension variation," presented at the SPIE Symposium on Microelectronic Manufacturing, Austin, Texas, October 16-17, 1996.

Nanometrology Using Scanning Probe Microscopy and Its Application to Resist Patterns

M. Nagase, K. Kurihara, H. Namatsu, and T. Makino

NTT Basic Research Labs., 3-1, Morinosato-Wakamiya, Atsugi-shi, Kanagawa, 243-0198, Japan

A new metrological method for critical dimension measurement of nano-structures had been developed. Two critical dimensions of a fabricated pattern - the width and the radius of an edge - were calculated from the height dependence of the appearent width measured by scanning probe microscopy. The feasibility of this method was confirmed by measuring sub-100 nm resist patterns. The modeled shape of ZEP resist patterns, fabricated by electron beam lithography, was presented along with the measured dimensions (width and radius). The distance accuracy and repeatability of the method were investigated for Si and resist patterns and were found to be good enough for nanometrology.

INTRODUCTION

At present, scanning electron microscopy (SEM) is the main tool for the measurement of critical dimension (CD) of sub-micron structures on Si LSI. The major advantage of SEM, compared with optical microscopy, is the high resolution that can be obtained. However, charge-up and contamination effects, which are problems with SEM-CD, cannot be completely eliminated despite the considerable efforts of many researchers (1,2). These problems place metrology for sub-100 nm resist patterns in a very serious status, because the degree of difficulty caused by these problems increases with magnification needed for observation of a smaller pattern. Scanning probe microscopy (SPM) is an alternative candidate as a tool for nano-structure metrology, because of the small sample-probe interaction and low noise level.

But, even though SPM provides atomic-scale resolution for smooth surfaces, SPM images of samples with a steep topography, such as LSI circuits, are distorted in the horizontal direction by the influence of the finite probe size. In the observation of a line pattern, a lateral range of distortion is of the same order as the height of the pattern. As a result of the distortion, the appearent resolution of an SPM image deteriorates as the height increases. This distortion can be removed using the reconstruction method described in many articles (3,4,5), if the shape of the probe is known. It is, however, very troublesome and difficult to evaluate the probe shape on a nanometer scale before and after taking measurements (6,7). And, the reconstruction method requires an enormous number of calculations. Since the most important point is to obtain a critical dimension of the sample structure, not a reconstructed image, a simplified and more practical method is required for SPM metrology.

With nano-metrology, of course, extremely high accuracy is required. The dimensional accuracy of a sub-100 nm fabricated structure is usually below 10 nm for 3σ. The accuracy of a measurement method must, by necessity, be higher and probably be below 1 nm for σ.

We have already reported on a metrological method that measures the critical dimensions of a nano-order structure with a simple symmetrical shape (8,9). In this paper, we will describe the results derived from our method as applied to sub-100 nm resist patterns. The details of our metrological method for resist patterns are described. The accuracy of SPM metrology for resist patterns is discussed also.

METROLOGICAL METHOD

The shape of microstructures for CD measurement on integrated circuits is symmetrical and has a flat part with rounded edges and almost vertical sidewalls ordinarily, as shown in Fig. 1. If the shape of the edges and sidewalls do not depend on the width of the structure, the width of the flat part is representative of the critical dimension of the structure. By taking into account the radius of the edges, the SPM-CD value will be approximately equal to the value of the real

(a)

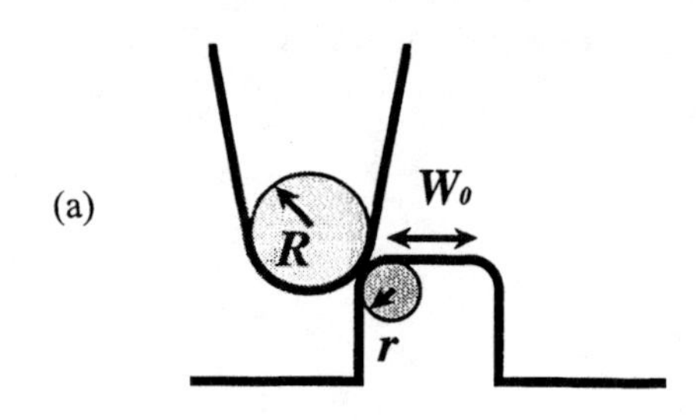

(b)

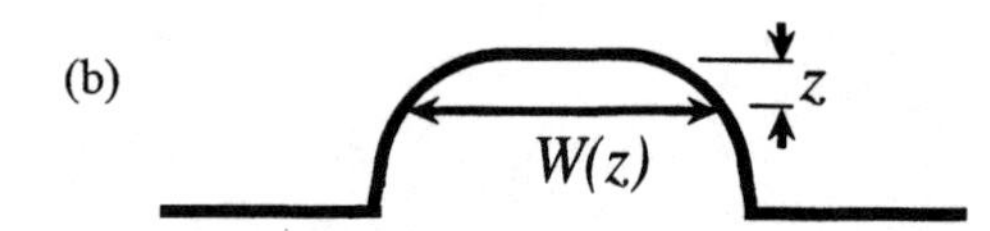

FIGURE 1. (a) Modeled shapes of a symmetric microstructure and an SPM probe. The line structure has a top flat part, rounded edges and vertical sidewalls. Critical dimensions of this system are the width of the flat part, W_0 and the edge radius, r. The tip of the SPM probe is spherical with radius, R. (b) $W(z)$ is the height dependence of the appearent width in the SPM image. z is the distance from the top surface in height.

CP449, *Characterization and Metrology for ULSI Technology: 1998 International Conference*
edited by D. G. Seiler, A. C. Diebold, W. M. Bullis, T. J. Shaffner, R. McDonald, and E. J. Walters

width. These are the basic concepts of the "top-CD" measurement by SPM.

In our previous work, the shape of the rounded edge of the pattern was confirmed to be cylindrical and that of the SPM probe tip was spherical (9). Figure 1 shows the modeling shapes of the sample and the probe. Critical dimensions of the shapes are the width of the flat part (W_0), the edge radius (r) and the probe radius (R). The height dependence of the appearent width ($W(z)$) is expressed as an analytical function as shown below.

$$W(z) = W_0 + 2\sqrt{(r+R)^2 - ((r+R)-z)^2}. \quad (1)$$

With a minor change in the above modeling equation, it is possible to adapt this method to trench structures.

$$W(z) = W_0 - 2\sqrt{(r+R)^2 - ((r+R)-z)^2}. \quad (2)$$

The critical dimensions (W_0 and r+R) are obtained from the modeling equation and the measured height dependence of the appearent width using a parameter fitting scheme, such as the least square method. The height dependence can be easily obtained from the SPM image. The modeling equation was derived by the mathematical scheme, 'the Legendre transform', which was used for the surface reconstruction method (5). This method is different from the published reconstruction method; the above equation only adapts to a limited part of the image (9). The adaptable area is the top portion of the structure and depends on the value of r+R.

The parameters, r and R cannot be separated using the above equations. Another measurement of the probe tip radius is necessary to determine the radius of the pattern edge. Since the edge radius depends on the fabrication process and does not depend on the pattern width, only one measurement of r is necessary for one process. In this sense, the only important value for the standard CD measurement is W_0. A major advantage of this method is that W_0 can be determined without defining the radius of the probe, R. An evaluation of the probe radius is not necessary to obtain the value of W_0.

EXPERIMENTAL CONDITIONS

The feasibility of this method has been confirmed by evaluating resist patterns with design widths of from 10 nm to 200 nm. A positive-tone resist, ZEP-520 with a thickness of 50 nm, was exposed to an electron beam (EB) and developed with hexylacetate. This combination of resist and developer is effective for improving resolution (10). A modified HL-700F system with a beam of less than a 10 nm in diameter was used for the pattern generation (11).

SPM measurements were performed with the SPI3700/SPA350 system (Seiko Instruments Inc.). A commercially available Si-probe with a typical spring constant of 20 N/m, a resonance frequency of 100-120 kHz, and a typical quality factor of 300 was used as a sensor of the dynamic force which is generated by cyclic contact between the probe and the sample surface. The average probe-sample distance was controlled by keeping the amplitude reduction of probe vibration at a constant value. The amplitude of the free vibration was 50 nm, and that of the controlled vibration was 40 nm. Because the dynamic force mode has several advantages, such as low damage and friction force free, over the contact-mode SPM, it is suitable for metrology of resist patterns. The scan area of a typical image is 1000 nm square. A typical scan rate is 1000 nm/s. A height dependence of apparent width was calculated using a standard subroutine of the SPI3700 system.

The radius of an SPM probe tip was determined by means of sampling inspection. The radius of several probes (about 10% of all probes) supplied from an SPM maker was measured using a tip characterizer made on an Si(110) substrate. The characterizer was fabricated by KOH anisotropic etching and had a strictly rectangular cross-section (8,12). As a result of our measurement of about sixty probes, the average of the radii (R) was evaluated to be about 11 nm and the standard deviation was about 3 nm. This value of the radius was used as the typical value of the probe radius.

RESULTS AND DISCUSSIONS

SPM Image of Resist Pattern

Figure 2 (a) shows an SPM image of a ZEP resist pattern

(a)

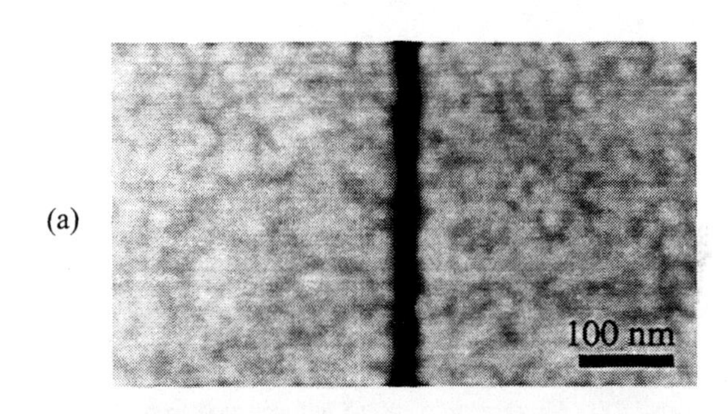

(b)

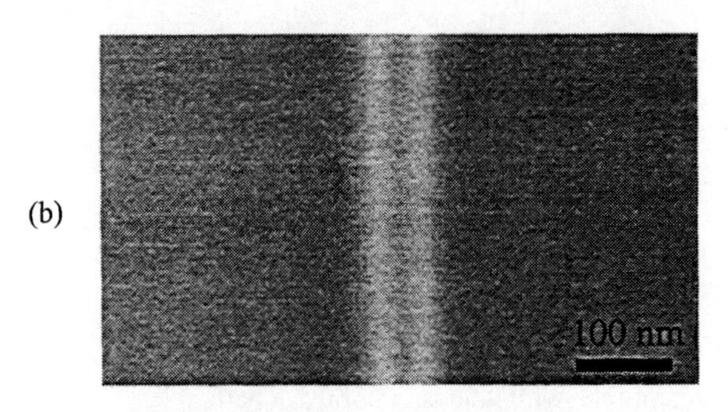

FIGURE 2. SPM and SEM images of the ZEP resist pattern with 10 nm design width. (a) SPM image as taken by a SPA350/SPI3700 system (Seiko Instrument Inc.) in dynamic force mode. Full scale of height of the gray SPM image is 10 nm. (b) SEM image as taken by a low voltage CD-SEM (S-7800; Hitachi). The diameter of the SEM probe is about 5 nm. The probe intensity is 3 pA at 1.5 kV for ZEP resist observation.

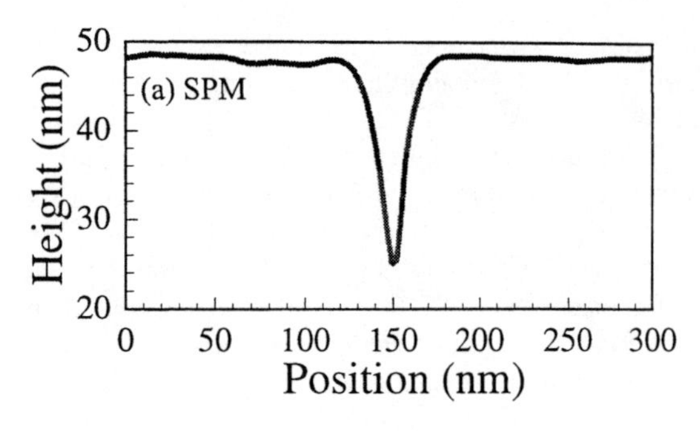

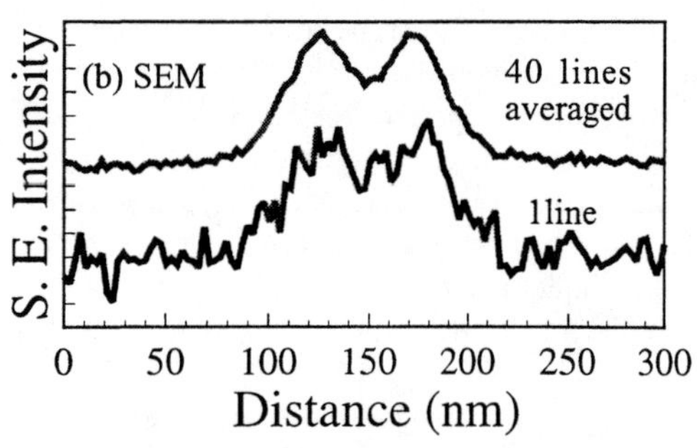

FIGURE 3. Cross-sectional profiles of the SPM and SEM resist images. (a) The SPM profile corresponds to one scan. Tilt compensation in a linear scheme is the only treatment for the image. (b) The SEM profile of one scan is very noisy. 40 lines are averaged to obtain a clear profile.

with a 10 nm designed width. The gray scale of the image was adjusted to be 10 nm peak-to-valley, because a part of the image below about 10 nm from the top surface is an image of the probe. Detailed morphologies of the pattern edge and resist surface are clearly observed in this SPM image. These morphologies cannot be distinguished in the SEM image shown in Fig. 2 (b), despite the fact that the diameter of the SEM beam (about 5 nm) is small enough to resolve the structures in the resist pattern. The SEM image was taken at the minimum probe current (3 pA) of a low voltage CD-SEM (S7800 ; Hitachi) in order to reduce contamination and charge-up effects. The SPM image gives us more information than the SEM image, because of the high sensitivity height imaging ability of the SPM.

Figure 3 shows the cross-sectional profiles of the resist images of the SPM and SEM. No additional treatment, such as smoothing or averaging, was applied to the SPM profile. On the other hand, with the SEM profile averaging was necessary to get a clear profile. Low noise level is one of the advantages of SPM metrology. The useful properties of clear imaging and low noise of the SPM have already been used in evaluating the width deviation of resist pattern.[13]

Critical Dimensions of Resist Pattern

How to get critical dimensions from an SPM profile is a technical issue for metrology. When taking only an SPM profile, it is impossible to decide where the pattern edge is and

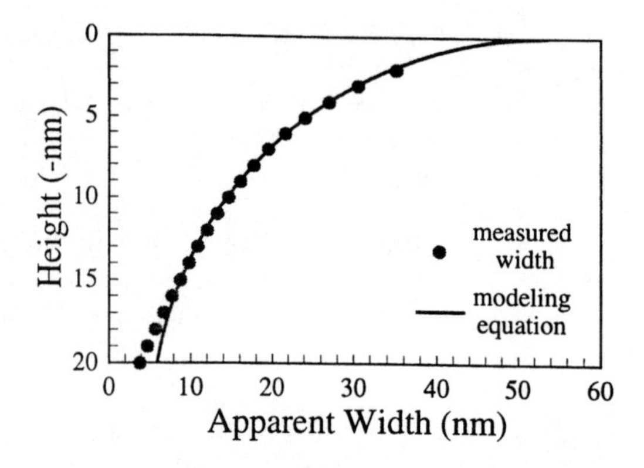

FIGURE 4. The height dependence of the appearent width of the resist pattern. Points are measured values from the SPM image shown in Fig. 2. The modeling equation (Eq. 2) fitted to the measured width is shown as the line. The critical dimensions, W_0 and $r+R$, extracted are 57 nm and 26 nm, respectively.

what are it's critical dimensions. The metrological method described above was applied to the profile shown in Fig. 3(a). Figure 4 shows the measured appearent width (points) and the modeling equation (line). Equation 2 for the trench structure was used. The critical dimensions, W_0 and $r+R$, extracted are 57 nm and 26 nm, respectively, as the fitting parameters. The range of the height for calculation was from 5 nm to 10 nm. Since the radius of probe (R) was already determined to be 11 nm, the radius of the pattern edge (r) can be estimated to be 15 nm. This method was applied to the other patterns with design width (W_d) of from 20 nm to 200 nm. Figure 5 shows the plots of W_0 and r for various design widths. The measured width (W_0) was 44 nm wider than the design width (W_d). By taking into account the edge radius (r), the effective width (W_{eff}) can be expressed as follows :

$$W_{eff} = W_0 - 2r. \quad (3)$$

Consequently, W_{eff} for the ZEP pattern was only 14 nm wider than the design width. This situation is summarized in Fig. 5(b). This result is consistent with the simulation results on a resist pattern (14).

Because the edge radius is not dependent on the width, the measured width (W_0) is a representative critical dimension of this resist system. The value of W_0 will be used as an indicator of the critical dimension control.

In this metrological method, the value of W_0 is not affected by changing the probe radius, R. This fact indicates that probe radius measurement is not necessary before nor after CD measurement.

Instant CD Measurement

If the values of r and R are known previously, a value of W_{eff} is instantly obtained by measuring the appearent width at the fixed height,

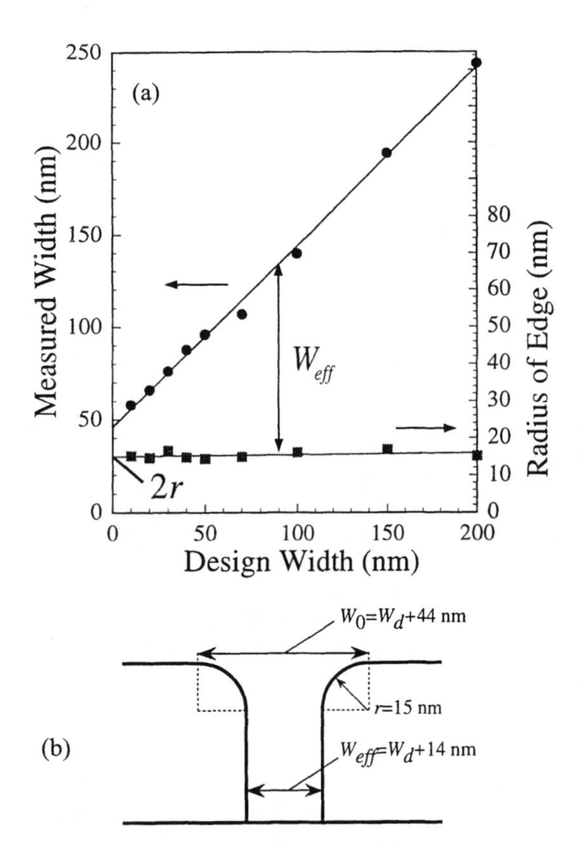

FIGURE 5. (a) Plots of W_0 and r for 10-200 nm design widths (W_d). The offset width from the design width is 44 nm. The average value of the edge radius is 15 nm. (b) The modeling shape of the ZEP resist pattern. The effective width, W_{eff}, is 14 nm wider than W_d.

$$z = r + R - \sqrt{(r+R)^2 - r^2}. \qquad (4)$$

For example, in the case described above (r=15 nm, R= 11 nm), the fixed height for W_{eff} was calculated to be 4.8 nm. In this case, only one point measurement of the appearent width , W(4.8 nm), was necessary to obtain a CD value.

Pattern Edge Radius

Two critical dimensions, W_0 and r, were obtained in our SPM-CD, while a conventional SEM-CD gives only one critical dimension. Of course, W_0 corresponds to the SEM-CD value. The value of the edge radius, r, depends on the fabrication process, such as contrast of resist, focus, and type of developer. For example, the edge radius of the ZEP resist pattern depended on the kind of developer. The edge radius of the pattern developed with hexylacetate was smaller than that developed with xylene (9). Since the hexylacetate developer gave a higher contrast (γ value) than xylene, resolution was improved (10). The edge radius, r, will be a useful parameter for improvement and optimization of a resist process.

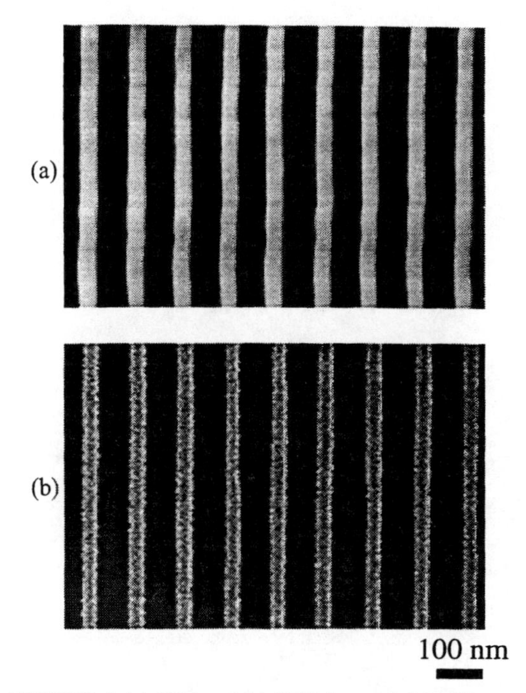

FIGURE 6. (a) SPM and (b) SEM images of the Si pattern with 100 nm pitch. The SEM image was taken by high resolution SEM (ERA-8000FE; Elionix). The diameter of SEM probe is about 2 nm at 10 kV, 1 pA.

Accuracy of SPM Metrology

In this section, two types of accuracy will be discussed. One is the distance accuracy in an image and the other is the repeatability of critical dimensions for the same pattern. The former was obtained to evaluate the Si structure with a well defined pitch. The latter was the measured results on a ZEP resist pattern.

The sample of Si lines with a 100 nm pitch was fabricated by the same process as that for the probe characterizer. The accuracy of the pitch, σ_{pitch} consists of the accuracy of the beam position in EB lithography(σ_{EB}), the deviation of line width(σ_{width}), and the distance accuracy of the instrument(σ_{inst}). The factors σ_{EB} and σ_{width} were evaluated to be 1.0 nm and 1.3 nm in Ref. 15, respectively. The accuracies of the SPM and SEM were evaluated from the images shown in Fig. 6. The pitch in the SPM image was measured from one profile. The averaged profile was used for SEM in order to reduce the noise. The deviation of 15 pitches measured was 1.9 nm for SPM and was 3.5 nm for SEM. The factor σ_{inst} was calculated as follows,

$$\sigma_{inst}{}^2 = \sigma_{pitch}{}^2 - \sigma_{EB}{}^2 - \sigma_{width}{}^2. \qquad (5)$$

The distance accuracy in the SPM image was 1.0 nm and that in the SEM image was 3.1 nm. The cause of accuracy deterioration in the SEM image was noise. This fact indicates that the SEM cannot be adapted to nano-metrol-

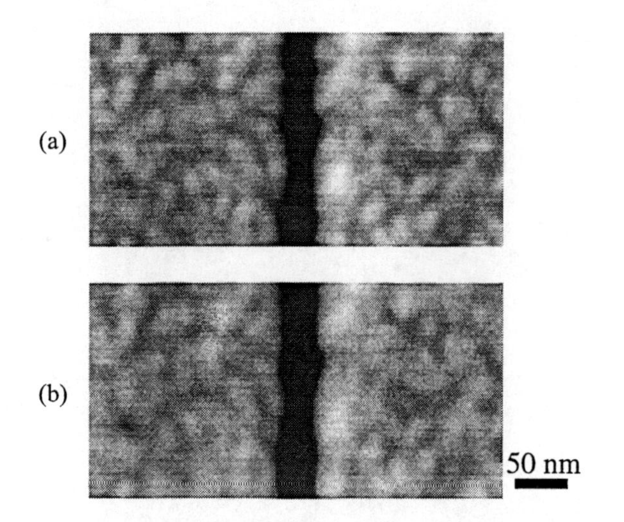

FIGURE 7. SPM images of the resist pattern with 10 nm design width for (a) 1st scan and (b) 16th scan. The surface morphology slightly changed by the SPM probe scanning.

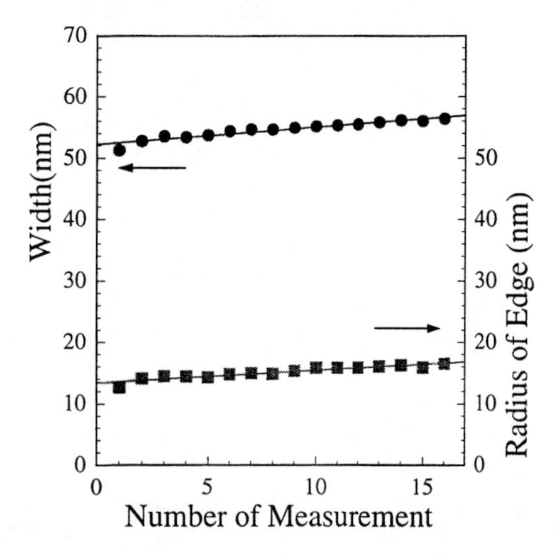

FIGURE 8. Repeatabilities of width and edge radius of the ZEP resist pattern as measured by SPM. Both values, W_0 and r, increased a little with an increasing number of measurements.

ogy.

The repeatability of SPM-CD measurements for Si structures was already reported and was below 1 nm in a standard deviation (9). The change of CD value, caused by wear or contamination, was negligibly small for the Si structure. In our present work, the repeatability for the ZEP resist pattern was investigated. Figure 7 shows the SPM image of a ZEP resist pattern with 10 nm design width, as taken with the first scan and with the 16th scan. A slight change is recognizable in the latter image. The CD values for each measurement were plotted in Fig. 8. Both values of width and edge radius increased a little with the number of measurements. The increasing rate for the width was 0.28 nm/scan and that for the radius was 0.20 nm/scan. The deformation by the scanning force of the SPM probe took place on the edge of the resist pattern. However, the deformation of the pattern was sufficiently small for CD measurement.

The accuracy of the SPM-CD is high enough for the CD measurement of nano-structures. There are, of course, other metrological problems, such as hysterisis of the piezo actuator or slow scan speed, but, they are not insurmountable and will be solved with an improved SPM system.

SUMMARY

In summary, a new metrological method for SPM-CD measurement was applied to sub-100 nm resist patterns. The critical dimensions, the width and edge radius of a ZEP resist were evaluated with high accuracy. The accuracies of the SPM-CD are already almost the same as or higher than that of SEM-CD. SPM technology will help to establish nanometrology.

ACKNOWLEDGMENTS

The authors are grateful to K. Inokuma for his contribution to this experiment.

REFERENCES

1. Hren, J. J., *Introduction to Analytical Electron Microscopy*, New York, : Plenum, 1979, pp. 481-505.
2. Bruenger, W. H., Kleinschmidt, H., Hassler-Grohne, W., and Bosse, H., J. Vac. Sci. Technol. **B15**, 2181-2184 (1997).
3. Chicon, R., Ortuno, M., and Abellan, J., Surf. Sci. **181**, 107-111 (1987).
4. Niedermann, P., and Fischer, O., J. Microsc. **152**, 93-101 (1988).
5. Keller, D., Surf. Sci. **253**, 353-364 (1991).
6. Montelius, L., and Tegenfeldt, J. O., Appl. Phys. Lett. **62**, 2628-2630 (1993).
7. Griffith, J. E., Marchman, H. M., Miller, G. L., and Hopkins, L. C., J. Vac. Sci. Technol. **B11**, 2473-2476 (1993).
8. Nagase, M., Namatsu, H., Kurihara, K., Iwadate, K., and Murase, K., Jpn. J. Appl. Phys. **34**, 3382-3387 (1995).
9. Nagase, M., Namatsu, H., Kurihara, K., and Makino, T., Jpn. J. Appl. Phys. **35**, 4166-4174 (1996).
10. Namatsu, H., Nagase, M., Kurihara, K., Iwadate, K., Furuta, T., and Murase, K., J. Vac. Sci. Technol. B13, 1473-1476 (1995).
11. Kurihara, K., Iwadate, K., Namatsu, H., Nagase, M., Takenaka, H., and Murase, K., Jpn. J. Appl. Phys. **34**, 6940-6946 (1995).
12. Kurihara, K., Iwadate, K., Namatsu, H., Nagase, and Murase, K., J. Vac. Sci. Technol. **B13**, 2170-2174 (1995).
13. Nagase, M., Namatsu, H., Kurihara, K., Iwadate, K., Murase, K., and Makino, T., Microelectron. Eng. **30**, 419-422 (1996).
14. Yamazaki, K., Kurihara, K., Namatsu, H., Jpn. J. Appl. Phys. **36**, 7552-7556 (1997).
15. Kurihara, K., Nagase, M., Namatsu, H., and Makino, T., Jpn. J. Appl. Phys. **35**, 6668-6672 (1996).

Intermittent-Contact Scanning Capacitance Microscopy Imaging and Modeling for Overlay Metrology

S. Mayo, J. J. Kopanski, and W. F. Guthrie[1]

Semiconductor Electronics Division
[1]Statistical Engineering Division
National Institute of Standards and Technology
Gaithersburg, MD 20899-0001 USA

Overlay measurements of the relative alignment between sequential layers are one of the most critical issues for integrated circuit (IC) lithography. We have implemented on an AFM platform a new intermittent-contact scanning capacitance microscopy (IC-SCM) mode that is sensitive to the tip proximity to an IC interconnect, thus making it possible to image conductive structures buried under planarized dielectric layers. Such measurements can be used to measure IC metal-to-resist lithography overlay. The AFM conductive cantilever probe oscillating in a vertical plane was driven at frequency ω, below resonance. By measuring the tip-to-sample capacitance, the SCM signal is obtained as the difference in capacitance, $\Delta C(\omega)$, at the amplitude extremes. Imaging of metallization structures was obtained with a bars-in-bars aluminum structure embedded in a planarized dielectric layer 1 µm thick. We have also modeled, with a two-dimensional (2D) electrostatic field simulator, IC-SCM overlay data of a metallization structure buried under a planarized dielectric having a patterned photoresist layer deposited on it. This structure, which simulates the metal-to-resist overlay between sequential IC levels, allows characterization of the technique sensitivity. The capacitance profile across identical size electrically isolated or grounded metal lines embedded in a dielectric was shown to be different. The floating line shows capacitance enhancement at the line edges, with a minimum at the line center. The grounded line shows a single capacitance maximum located at the line center, with no edge enhancement. For identical line dimensions, the capacitance is significantly larger for grounded lines making them easier to image. A nonlinear regression algorithm was developed to extract line center and overlay parameters with approximately 3 nm resolution at the 95% confidence level, showing the potential of this technique for sub-micrometer critical dimension metrology. Symmetric test structures contribute to facilitate overlay data extraction.

INTRODUCTION

Overlay metrology of integrated circuits (ICs) provides the most critical data for qualifying lithography. The National Technology Roadmap for Semiconductors calls for "development of new and improved alignment and overlay control methods independent of technology option" (1). While these measurements are mainly performed to sub-micrometer accuracy with optical techniques, scanning-probe microscopy-based techniques have been suggested to determine tool induced shifts (TIS) that limit the accuracy of current optical techniques. Previous work (2) reported on experimental contact-mode scanning capacitance microscopy (SCM) using an ac-voltage excitation to modulate a depletion volume in silicon and generate a differential capacitance signal for imaging junctions.

Recently, we have reported on the use of intermittent-contact scanning capacitance microscopy (IC-SCM) (3) to image metal components buried under a dielectric in IC structures. The imaging quality improves with lines electrically connected to the substrate. Moreover, due to its implementation on an atomic force microscope (AFM) platform, this technique conveniently yields simultaneous surface topography and capacitive data, which can be used to develop a micrometer-size test pattern structure for overlay metrology. Full characterization of the IC-SCM technique and development of an overlay test pattern structure will help advance the accuracy and precision of overlay metrology needed to qualify IC structures with sub-micrometer critical dimensions (4). This work is focused on the properties of this technique and critical elements in overlay test pattern design.

We have used a 2D electrostatic model to simulate differential capacitive data with structures built on a conductive substrate covered with a planarized dielectric layer, embedded in which are two identical metal lines, one electrically insulated from the substrate, and the other electrically connected to it. This simple model serves to evaluate the accuracy of this technique in determining a critical parameter such as the center of each line. By adding to the dielectric surface a properly patterned photoresist (PR) overlayer, the metal-to-resist alignment, i.e., the overlay, can also be measured.

EXPERIMENTAL

The IC-SCM was implemented with an AFM operated in intermittent contact (or tapping) mode and the capacitance sensor from a contact mode SCM. The conductive probe of the AFM is driven at a frequency ω, slightly below its natural resonance. The IC-SCM signal is generated by monitoring the capacitance sensor output at ω with a lock-in amplifier. The AFM produces an image of topography, while the IC-SCM simultaneously produces an image of differential capacitance, $\Delta C(\omega)$. More experimental details are given elsewhere (3).

CP449, *Characterization and Metrology for ULSI Technology: 1998 International Conference*
edited by D. G. Seiler, A. C. Diebold, W. M. Bullis, T. J. Shaffner, R. McDonald, and E. J. Walters
1998 The American Institute of Physics 1-56396-753-7/98/$15.00

The IC-SCM is sensitive to metal structures buried beneath an insulating layer and to variations in dielectric constant within a thin insulating film over a conductive substrate. Figure 1 shows an experimentally measured IC-SCM image of a "bars-in-bars" test structure (5), which is used for measurement of lithographic overlay between sequential metallization layers. The aluminum bars are fabricated on an oxidized silicon substrate and then covered with a 1 μm thick planarized oxide overlayer. The bars, which are thus electrically insulated from both the conductive silicon substrate and the SCM tip, are 2 μm wide with a 5 μm space between them. The AFM probes are micromachined from silicon and coated with Cr to make them highly conductive. The cantilever vibrational amplitude was 70 nm at a frequency of 75 kHz (3).

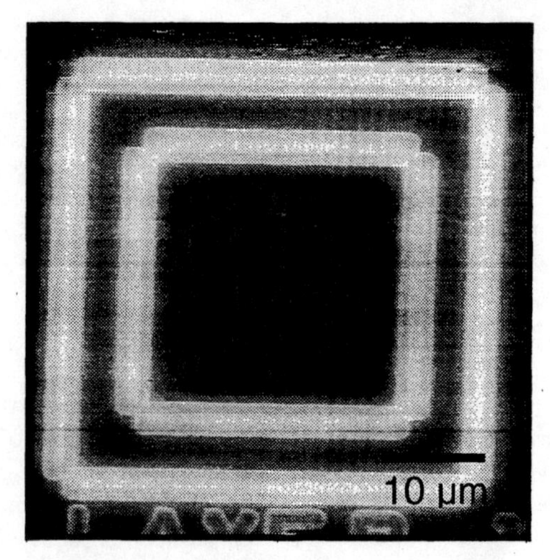

FIGURE 1. Experimental IC-SCM image of an aluminum bars-in-bars structure.

The bar edges in Fig. 1 are not sharply defined, but rather have an enhancement in signal near the line edges. Figure 2 shows two independent traces of IC-SCM data measured over the center of the test structure with probe vibration amplitudes of 53 nm and 120 nm. Because the response is symmetrical, data from just half the structure are shown. The signal plotted is the differential amplitude above the capacitance base level measured at the structure center. Determination of the line shape is limited by the overlayer oxide thickness and edge effects, but the well-defined peak shape allows centerline location to be accurately determined.

The controllable data acquisition parameters of IC-SCM are the probe vibration amplitude (t_{tip}), the tip vibration frequency (ω), and the phase of the capacitance signal with respect to the reference signal. As expected, the IC-SCM signal and image quality increase with t_{tip}. In Fig. 2, the base-level signal of the traces measured with the larger t_{tip} is approximately three times higher than the level measured with lesser amplitude. Also, when t_{tip} is less than the sample topography variations, the phase of the IC-SCM signal relative to the drive signal tends to shift, and this changes the apparent magnitude

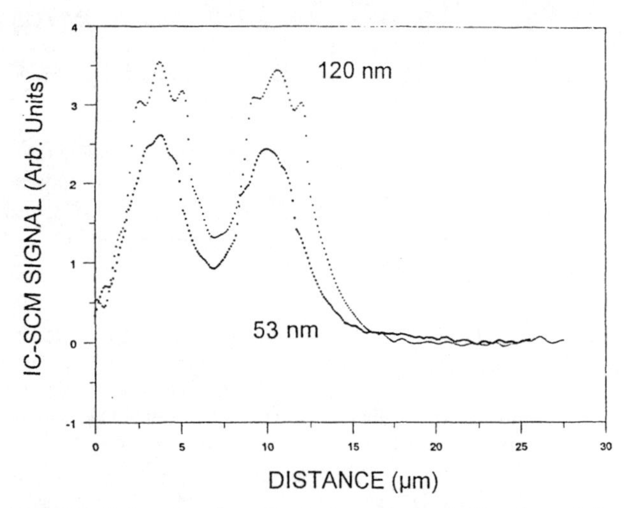

FIGURE 2. Raw data from Fig. 1 along a horizontal line through the structure center. The probe vertical-vibration amplitudes are 53 nm and 120 nm, respectively.

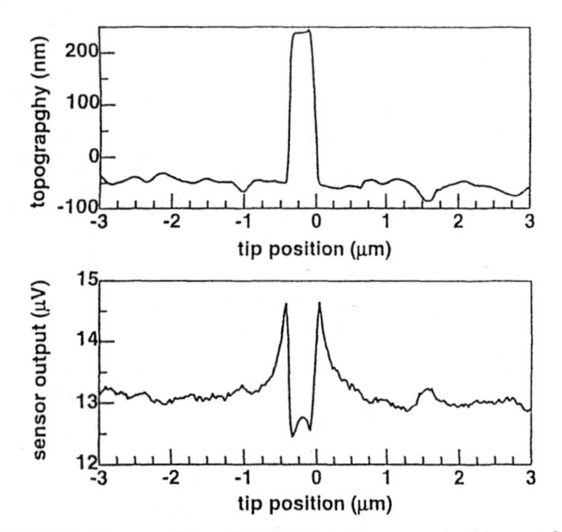

FIGURE 3. Topographic and IC-SCM data measured with an etched silicon line.

of the differential capacitance. The interaction of the tip sidewall with the IC-SCM is relatively more prominent than with the topography edge for AFM. Figure 3 shows topographic and IC-SCM data measured across a single conductive silicon line etched on a silicon substrate. The line is approximately 0.3 μm high and 0.35 μm wide, with no added overlayer except for its native oxide film. Here, the topographic image was acquired in intermittent contact, while the capacitive image was acquired in a separate "lift mode" scan with the tip offset from the surface by about 20 nm. The IC-SCM profile is dominated by the prominent probe-to-sample side wall interaction causing the large capacitance peaks at each line corner. The image illustrates the importance of considering the tip sidewalls in SCM measurements in any effort to physically model the SCM signal.

MODELING

IC-SCM data were modeled with a 2D electrostatic field simulator, based on a finite element method (6), applied to the structures described below. The probe was a conductive, knife-edge-shaped trapezoid 1 μm high, 0.5 μm wide, and with a 0.05 μm tip. In order to simulate the experimental intermittent contact technique, where the probe is vibrated in a vertical plane while being scanned over the model, the probe-to-sample total capacitance was calculated with a static probe set at two different levels: in contact with the dielectric surface and spaced by 0.1 μm, above the dielectric surface. The difference in probe-to-substrate capacitance at the two levels was calculated for each scan point. Line center locations and overlay parameters were extracted by fitting these calculated data with a nonlinear regression model. Details of the data analysis and comparison of results with the model design values are explained later.

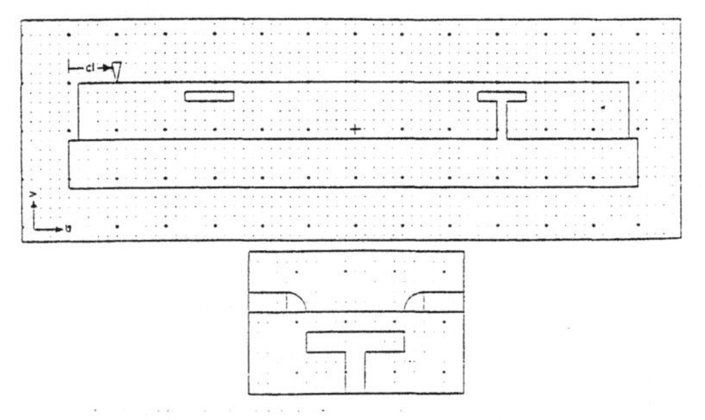

FIGURE 4. 2D model geometry used with IC-SCM simulations for line center and overlay extraction.

Figure 4 shows the model used to simulate some common situations encountered in IC fabrication. It consists of a conductive substrate covered with a thick dielectric layer with two embedded identical metal lines, 2.5 μm wide and 0.5 μm thick. One line is isolated and the other connected to the substrate, as mentioned above. The dielectric overlayer is uniform and 0.5 μm thick. Later, this model was first modified by adding a 0.5 μm thick PR overlayer on the planarized oxide with openings centered on the lines and identical to the line's width, as shown in the inset. Each PR opening edge over the lines was simulated with a 20-point quarter circle arc. In a second modification of this model, a defective lithography layer was simulated by shifting the PR toward the right side of the model by 50 nm with respect to the lines. This allowed us to evaluate the technique's sensitivity and the fitting procedures used to extract overlay data with three different configurations involving the same basic metal structure as shown in Fig. 5: a) no PR, b) centered PR, and c) shifted PR.

Figure 5 is a plot (arbitrary units) of the IC-SCM data versus scanning distance calculated with these three model configurations. The probe vibration amplitude was set at 0.1 μm, as mentioned above. A zero net charge value was assigned to the isolated line. The only difference between profiles b) and c) is the offset in the PR layer with respect to the metal lines. It is encouraging that this technique is sensitive enough to detect misalignments of a few nanometers in the PR. Although these are identical size lines, the isolated line yields a smaller signal than the grounded line due to probe-to-substrate field shielding.

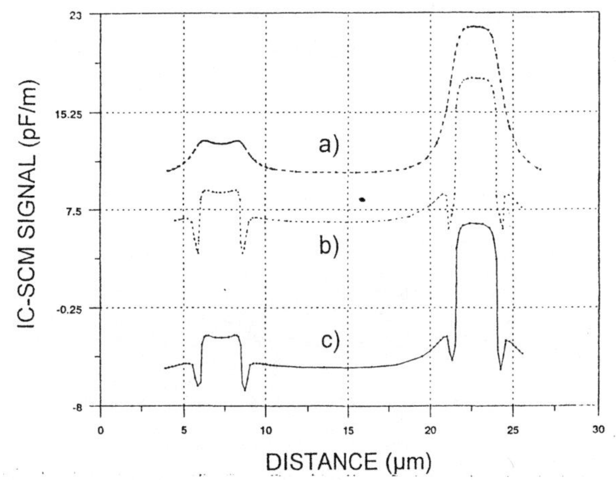

FIGURE 5. IC-SCM simulated data. a) no PR, b) centered PR, and c) shifted PR. Plots are shifted along the vertical axis. Capacitances at the lines center are about 9 pF/m and 18 pF/m, respectively.

Simulated capacitance data were obtained from the field energy, calculated with multiple iterations in order to reduce the energy error to about 0.02% by solving Laplace's equation

$$\nabla\cdot(\epsilon_r \epsilon_0 \nabla\Phi(x, y)) = 0 \qquad (1)$$

where $\Phi(x, y)$ is the electrical potential, ϵ_r is the relative dielectric permittivity, and ϵ_0 is the vacuum permittivity. The total electric charge in the dielectric is zero.

Proper modeling over the PR curved edges requires careful positioning of the probe in order to secure a minimum distance between the arc and the probe surface of about 1 nm in order to avoid creating excessively high-mesh density regions localized around the probe corners that would increase the iteration time. A typical energy calculation for this 2D model creates a mesh with about 3500 to 4000 triangles. Variations in probe-to-edge distance along the arcs cause data distortion and contribute to the TIS with this technique, thus altering line center and overlay data extraction. The capacitive profiles shown in Fig. 5 are slightly asymmetric due to the long-range probe sensitivity to neighboring objects when scanning over a particular line. Modeling done with symmetric structures shows symmetric IC-SCM profiles.

Figure 6 shows a nonlinear increase in IC-SCM data calculated versus probe vibration amplitude when the probe is located at the center of the isolated line, the connected line, or the model center shown in Fig 4a. At large vibration amplitude, a more robust and clear image is expected. Data measured with small amplitude are generally prone to pickup noise, due to the stronger relative influence of small amplitude variations on the data, as expected from the slope of the curves. Variations in probe vibration amplitude cause image distortion and contribute to the TIS with this technique.

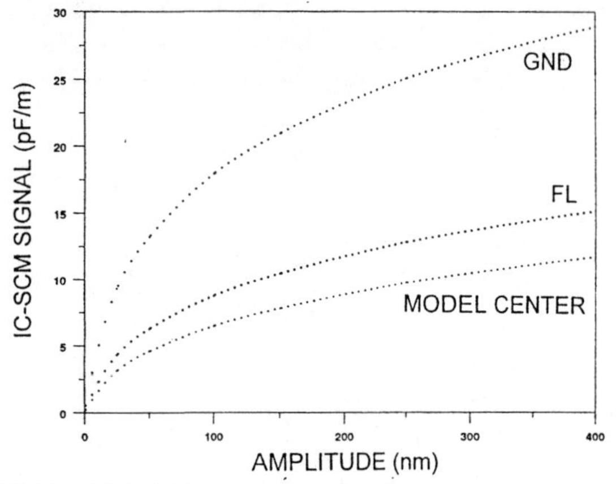

FIGURE 6. IC-SCM data versus probe vibration amplitude for the model shown in Fig. 5a.

DATA ANALYSIS and RESULTS

The primary tasks in analyzing the modeled IC-SCM data are the determination of the locations of conductive lines under the dielectric layer and openings in the PR. As discussed above, the line locations correspond to peaks in the capacitance data, so that the estimation of line centers can be done via estimation of capacitance peak locations. Estimation of the peak location can be done by either of the following approaches: 1) use of a differentiable function to describe the full peak, followed by standard, calculus-based computation of the line center location, or 2) use of a function to describe half of each peak combined with a "folding" of the data around an estimated point of symmetry (ps) in the peak. Our preliminary results indicate that the second approach provides estimates of line center location with smaller uncertainties.

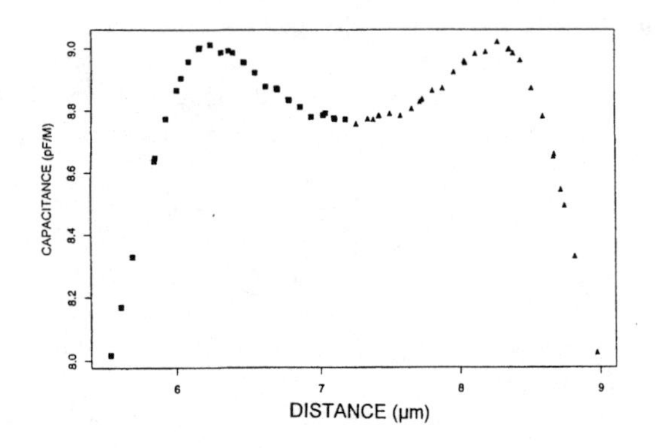

FIGURE 7. A subset of floating-line data calculated with the 2D electrostatic-field simulator from Fig. 5a. Squares and triangles correspond to data for x values less or greater than ps.

Figure 7 shows a subset of data from Fig. 5a used to determine the center of the capacitance peak corresponding to the floating line. These data were folded around a ps first approximation at x = 7.23 μm by plotting the data versus $(x-7.23)^2$. This plot showed how the two halves of the data begin to match when folded around the approximate ps. Next, we use a nonlinear regression to obtain an objective determination of ps. After testing a number of functions, a "quadratic over quadratic" rational function (2) in $(x\text{-ps})^2$, with a random error term, e, was selected:

$$y = (a_1 + a_2 (x\text{-ps})^2 + a_3 (x\text{-ps})^4) / (1 + b_1 (x\text{-ps})^2 + b_2 (x\text{-ps})^4) + e \quad (2)$$

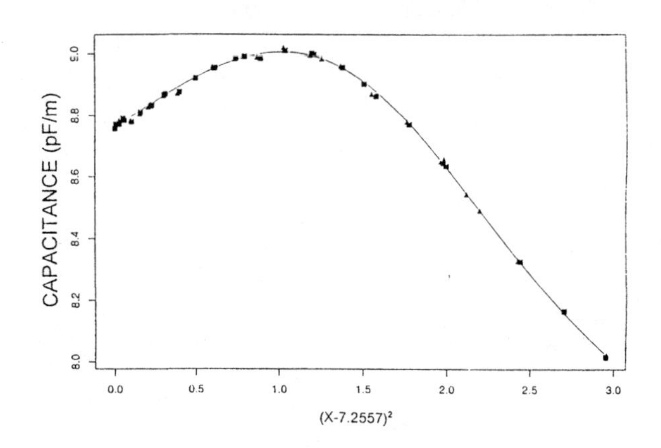

FIGURE 8. The data from Fig. 7 folded around the final estimate of ps = 7.2557 μm and the quadratic over quadratic fit.

Figure 8 shows both the data folded around ps = 7.2557 μm, which was extracted by fitting the data with Eq. (2) and the fitted quadratic over quadratic function. This figure shows that our model describes the data reasonably well in global terms. The model also seems quite reasonable for its specific purpose, estimating line center location, especially given our knowledge of the true model dimensions. This plot also allows good visualization of the peak asymmetry, part of which is due to the uneven influence of the neighboring grounded line on the line's capacitance profile.

Since the 2D electrostatic field simulator computes each data point to a specified error level, we also have additional information for calculation of the uncertainty of the estimated line centers via statistical simulation. In order to compute the uncertainty, the same analysis used to obtain line center estimates was repeated 1000 times with new random errors of the appropriate sizes added to the data. This method of computing the uncertainty offers the advantage of being free of some of the underlying assumptions that use of the regression data would require. With this approach, even if the regression model does not fit the data perfectly, our uncertainties will accurately reflect the variability in line center results due to random measurement error.

Table 1 shows the results, with 95% confidence intervals, from both the floating and the grounded lines. The estimated line centers are within a few nanometers of the model design values. Results without PR were calculated with a larger error at each data point, accounting for the larger uncertainties observed in those two cases. If the error bounds had been similar to those used for the cases with PR, the sizes of the uncertainties would have been commensurate. As mentioned above, there is a small bias in each line-center estimate due to the effects of neighboring structures. This results in slight overestimates for the floating line and slight underestimates for the grounded line.

The results with the off-center PR have larger biases than results for the other cases due to asymmetry in the capacitance data induced by the line-PR overlap. For the grounded line the bias is still not too large because the capacitance due to the line-PR overlap is small relative to the strength of the grounded line's signal, thus minimizing the asymmetry in the data. However, for the floating line the asymmetry has a relatively large effect, due to its weaker signal, which adversely impacts the line center estimate.

In order to estimate the center of each PR opening, our initial approach was to use a method similar to that used for the line centers. However, the asymmetries in the PR data were overwhelming since the signal produced by the PR is even less than the signal from the floating line. As an alternative, the edges of the PR openings were identified by averaging the pair of x values corresponding to the lowest capacitance values observed in the transition between PR and the embedded line. The results, which were excellent for these simulated data, are shown in Table 2, with uncertainties at the 95% confidence level. The success of this method depends on careful probe placement according to the surface topography. Capacitance calculations are made at a predefined series of points along the rounded edge of the PR. The primary disadvantage of this method is lack of a clear way to assess the uncertainty of the PR center estimates from the data. As a result, the uncertainties quoted here are based on engineering judgment of the repeatability of SCM probe placement capability.

With the results shown in Tables 1 and 2, overlay estimates can be computed by taking the difference of the line and PR locations and combining their individual uncertainties using the usual rules for propagation of uncertainty (7). For example, the centered-PR-to-floating-line overlay is given by the difference between the line center location, 7.2521 μm (Table 1), and the PR-opening location, 7.2500 μm (Table 2), 0.0021 μm. The results for the four overlay cases are given in Table 3. As expected from earlier results, the overlay estimates for three of the four cases are promising. The bias for the grounded line with off-center PR is large enough so that the confidence interval does not include the true overlay, but the answer is still fairly close to the design value. For the floating line with off-center PR, however, the failure to estimate the line center correctly ruins the overlay estimate. The regression model imposes symmetry on the asymmetric capacitance data by moving the line center estimate until the data are as symmetric as possible. This situation provides a good example of the importance of using symmetric capacitance profiles with this line center estimation method. It also supports using grounded line test structures because of their more robust, symmetric capacitance profile.

Future work to optimize test structure design and data analysis methods for this problem should alleviate the difficulties caused by PR-to-line overlap plus intrinsic capacitance biases caused by the influence of neighboring structures. Possible test-structure design alternatives include widening the PR openings to avoid PR-to-line overlap even for cases with misalignment. An alternative to refine the analysis is use of a weight function in the regression to weight data at the peak center over data at the peak edges, thus lessening asymmetry effects. Another potentially effective analysis alternative to reduce asymmetry-induced bias might be to average results that are biased in opposite directions.

TABLE 1. Line Center Estimates (μm)

Model Design	Floating Line Center Value	Floating Line Design Value	Grounded Line Center Value	Grounded Line Design Value
No PR	7.2557 ± 0.0082	7.2500	22.7489 ± 0.0037	22.7500
PR Centered	7.2521 ± 0.0010	7.2500	22.7490 ± 0.0008	22.7500
PR Off-Center	7.2958 ± 0.0009	7.2500	22.7562 ± 0.0022	22.7500

TABLE 2. PR Center Estimates (μm)

Model Design	Floating Line PR Center	Floating Line Design Value	Grounded Line PR Center	Grounded Line Design Value
PR Centered	7.2500 ± 0.0029	7.2500	22.7500 ± 0.0029	22.7500
PR Off-Center	7.3000 ± 0.0029	7.3000	22.8000 ± 0.0029	22.8000

TABLE 3. Overlay Estimates (μm)

Model Design	Floating Line Overlay	Floating Line Design Value	Grounded Line Overlay	Grounded Line Design Value
PR Centered	0.0021 ± 0.0031	0.0000	-0.0010 ± 0.0030	0.0000
PR Off-Center	-0.0040 ±0.0030	-0.0500	-0.0438 ± 0.0036	-0.0500

CONCLUSIONS

Line center and overlay parameters with micrometer size test structures can be determined with IC-SCM provided several requirements related to the probe properties and the sample design are met. The conductive probe needs to be small in order to minimize wall capacitive effects. No attempts have been made here to study the sample-probe electric deconvolution. The amplitude in the probe vibration should be stable and sufficiently large in order to generate robust IC-SCM signals with minimum capacitance data scattering, as shown in Fig. 6. This also contributes to keep a constant signal phase with respect to the reference signal. The dielectric film overlaying conductive components should be planarized and of uniform thickness in order to avoid distortion in capacitance data introduced by a wedge-shaped dielectric overlayer on metal lines. The metallic test structure should be electrically connected to the substrate in order to generate simple, stable IC-SCM imaging signals. The test structure should also be designed to yield symmetric capacitance profiles for optimal extraction of overlay parameters. Our data analysis shows encouraging results in extracting, with micrometer size test structures, line center and overlay parameters with approximately 3 nm resolution at the 95% confidence level. The model presented here shows that very simple and compact metal-resist structures could serve to qualify lithography. Once extensive evaluation of the factors causing TIS in IC-SCM overlay measurements is completed, imaging with this technique will offer an independent alternative to calibrate optics-based systems.

ACKNOWLEDGMENT

The authors are indebted to the NIST Office of Microelectronics Programs for partial support. We thank Lumdas Saraf, from SEMATECH, for supplying the bars-in-bars structure used in this work. Valuable discussions held at NIST with L. W. Linholm, J. R. Lowney, and M. W. Cresswell are recognized. This paper is a NIST contribution, not subject to copyright.

REFERENCES

1. The National Technology Roadmap for Semiconductors, Technology Needs, SIA (1997).

2. Kopanski, J. J., Marchiando, J. F., and Lowney, J. R., J. Vac. Sci. Technol. **B**14, 242 (1996).

3. Kopanski, J. J., and Mayo, S., "Intermittent-Contact Scanning Capacitance Microscope for Lithographic Overlay Measurements," to be published in Applied Physics Letters.

4. Rosenfield, M. G., "Overlay Measurements Using the Scanning Electron Microscope: Accuracy and Precision," Proceedings SPIE **1673**, 157-165 (1992).

5. Specification for Overlay-Metrology Test Patterns for Integrated-Circuit Manufacture, SEMI P28-96, Semiconductor Equipment and Materials, Mountain View, CA 94043.

6. ANSOFT Maxwell 2D Field Simulator. Certain experimental equipment, instruments, materials, or software are identified in this paper to specify the procedure involved. Such identification does not imply NIST recommendation or endorsement, nor does it imply that the mentioned elements are the best available.

7. Taylor, B. N., and Kuyatt, C. E., "Guidelines for Evaluating and Expressing the Uncertainty of NIST Measurement Results," NIST Tech Note 1297, 7-8 (1994).

Optimal Feedforward Recipe Adjustment for CD Control in Semiconductor Patterning

Steve Ruegsegger, Aaron Wagner, Jim Freudenberg, Dennis Grimard

Electrical Engineering and Computer Science, University of Michigan, Ann Arbor, Michigan 48109

Acceptable yields for nanofabrication will require significant improvement in CD control. One method to achieve better run-to-run CD control is through inter-process feedforward control. The potential benefits of feedforward control include reduced run-to-run post-etch CD variance, rework, and scrap. However, measurement noise poses a significant threat to the success of feedforward control. Since the stakes are high, an incorrect control action is unacceptable. To answer this concern, this paper will focus on how to properly use the available sensor measurement in a run-to-run feedforward recipe adjustment controller. We have developed a methodology based in probability theory that detunes the controller based on the confidence in the sensor's accuracy. Properly detuning the controller has the effect of filtering out the noise from the SEM. We will simulate this control strategy on industrial gate-etch data.

INTRODUCTION

The 1997 SIA Roadmap (7) suggests that 3σ CD control will need to be 10nm in 2003 (half of the current tolerance window) and 5nm by 2009. One method to achieve better run-to-run CD control is through inter-process feedforward control. Figure 1 shows a feedforward control system embedded into the patterning process. The lithography process has output X which is the input to the RIE process. The RIE has output Y. Disturbances D_{lith} and D_{rie} act on the outputs of the lithography and RIE, respectively. An in-line SEM is often employed in manufacturing systems for SPC on lithography CD. However, it may also be used for feedforward control. The measured photoresist (PR) CD is represented by M. The measurement also includes SEM disturbances, represented by D_{sem}. The feedforward controller adjusts the nominal RIE recipe in order to compensate for the estimated post-etch CD deviations $\hat{Y}$. The desired result of the RIE recipe adjustment is a reduction in the run-to-run variance of Y by rejecting D_{lith}. We call this strategy feedforward recipe adjustment (FFRA) control.

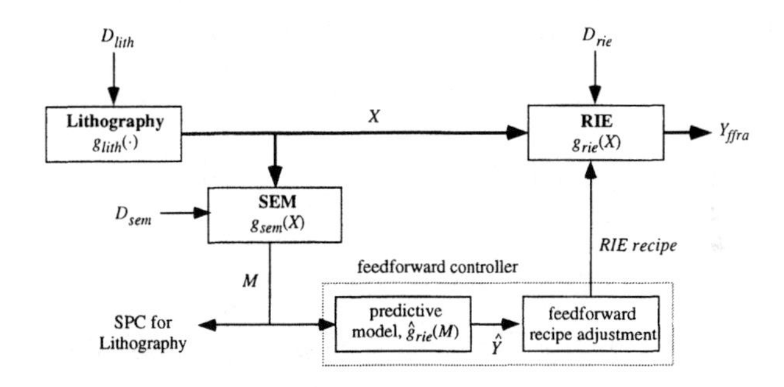

FIGURE 1. An inter-process feedforward controller implementation.

MOTIVATION

The potential benefits of feedforward control include reduced run-to-run post-etch CD variance, rework, and scrap. However, measurement noise D_{sem} poses a significant threat to the success of feedforward control. If the SEM noise is large enough, the measurement M will misrepresent the true PR CD X and the controller could command incorrect actions. Indeed, the variance of Y under feedforward control could actually increase! In most high-tech, high-cost manufacturing processes, incorrect control actions are unacceptable. The possibility of this scenario becoming reality is enough to prevent feedforward control from realization in manufacturing.

When a controller is subject to random measurement errors, compensation will only increase the variance of the process (4). Feedforward controllers can increase the output variance by making unnecessary or incorrect adjustments due to sensor noise. We will call this situation over-adjustment. The goal of this paper is to use variable estimation techniques to filter out the noise from the true underlying signal in order to avoid over-adjustment.

PREVIOUS WORK

Stoddard et al (10) implemented a feedforward and adaptive feedback controller in the manufacture of an on-chip capacitor. The goal was constant capacitance in the presence of varying dielectric thicknesses. Stefani et al (9) introduced a new eddy-current sensor and used it in a feedforward control strategy for controlling film thickness. They deliberately allowed a very wide input distribution in order to avoid over-adjustment. They state that the most important factor in their successful demonstration was having a very accurate and repeatable sensor measurement to feedforward. Leang et al (2) used a feedforward control strategy within a photolithographic stepper system. They recognize that feedforward control mechanisms are "... not well accepted in the semiconductor industry because of the high stakes involved.

CP449, *Characterization and Metrology for ULSI Technology: 1998 International Conference*
edited by D. G. Seiler, A. C. Diebold, W. M. Bullis, T. J. Shaffner, R. McDonald, and E. J. Walters

A corrective action that worsens a process is not tolerated." Their approach is to only activate the feedforward controller when a SPC alarm triggered. Rietman (6) discusses a pre-production demonstration of a neural network that is used to regulate the resistance of vias between the first and second metal layers.

Only one of the above examples took actions to avoid over-adjustment in their feedforward control strategy. The others mentioned the possibility of over-adjustment, but commented that the variance of the sensor was known to be much less than the variance of the manufacturing process. This gave them confidence, at some level, that over-adjustment was not going to be a problem. However, this is not always the case.

SYSTEM VARIABLE DEFINITIONS

In order to investigate over-adjustment in feedforward control, the variables of the system (Figure 1) need to be defined. First, it will be assumed that all the random variables (RV) are deviations from target. We will also assume that the variables are independent, identically distributed (i.i.d.) zero-mean Gaussian. Finally, we assume that the noise terms are uncorrelated.

The lithography process is represented by the function $g_{lith}(\cdot)$, the SEM by $g_{sem}(X)$, and the RIE by $g_{rie}(X)$. These represent the nominal manufacturing processes without any noise. The deviation terms are defined to add noise to the outputs of the processes. The variance of D_{lith} is σ^2_{lith}, the variance of D_{sem} is σ^2_{sem}, and the variance of D_{rie} is σ^2_{rie}.

Figure 1 shows the lithography process is defined without an input deviation RV. Any disturbance contributed by the incoming wafers to the lithography can be represented by the disturbance D_{lith} without loss of generality. The output of the lithography system is simply:

$$X = D_{lith} \ . \tag{1}$$

Due to the simple setup of this system,

$$\mathrm{Var}[X] = \sigma^2_X = \sigma^2_{lith} \ . \tag{2}$$

The output RV of the RIE is dependent upon the type of control system implemented. Under nominal recipe conditions (no feedforward control), the output is represented by Y_{nom}. In order to define Y_{nom} and its variance, a useful model of $g_{rie}(X)$ needs to be specified. Guided by industrial data sets, we are going to express $g_{rie}(X)$ as a linear model:

$$\begin{aligned} Y_{nom} &= g_{rie}(X) + D_{rie} \\ &= aX + D_{rie} \ . \end{aligned} \tag{3}$$

The variance of Y_{nom} is:

$$\mathrm{Var}[Y_{nom}] = a^2\sigma^2_{lith} + \sigma^2_{rie} \ . \tag{4}$$

The PR mask CD is measured by a SEM. SEMs are designed to be very accurate measurement tools and are regularly calibrated to give linear, unity gain outputs with no offset over their range of operation. Therefore, the measured lot-mean PR mask CD deviation is modeled as:

$$\begin{aligned} M &= g_{sem}(X) + D_{sem} \\ &= X + D_{sem} \ . \end{aligned} \tag{5}$$

The variance of the SEM output is:

$$\begin{aligned} \mathrm{Var}[M] = \sigma^2_M &= \mathrm{Var}[X] + \sigma^2_{sem} \\ &= \sigma^2_{lith} + \sigma^2_{sem} \ . \end{aligned} \tag{6}$$

Equation 3 modeled the RIE as a linear system. Therefore, the predictive model of the RIE will use the same structure,

$$\begin{aligned} \hat{Y} &= \hat{g}_{rie}(\hat{X}) \\ &= \hat{a}\hat{X} \ . \end{aligned} \tag{7}$$

The predicted RIE output deviation, $\hat{Y}$, becomes the input to the FFRA controller. The controller outputs an adjustment to the nominal recipe in order to compensate for $\hat{Y}$. The result of this control action can be represented as subtracting the *predicted* nominal output deviation from the *true* nominal output deviation. Therefore, the RIE output under feedforward control is

$$Y_{ffra} = Y_{nom} - \hat{Y} \ . \tag{8}$$

There are three sources of error between the predicted output $\hat{Y}$ and the true value Y_{nom}:

1. the estimate of the true RIE input $\hat{X}$ does not represent the true input X exactly,
2. the model parameter estimate $\hat{a}$ does not represent the true model parameters a exactly, and
3. the linear model structure of $\hat{g}_{rie}(\cdot)$ does not capture the process $g_{rie}(\cdot)$ exactly.

Each error source should be minimized in order to increase the accuracy of the controller. There is significant work in all three areas in the statistics literature. Variable estimation can be found in most any statistics book, (1) for example. Techniques for modeling and parameter estimation can be found in (5). This paper will focus on removing error source #1 by applying variable estimation to FFRA control systems.

For the sake of illustration, the next section will define a simple estimation method that can result in over-adjustment. We will then derive a better estimator and compare results.

A NAIVE ESTIMATION

One particularly naive estimate of $\hat{X}$ is equating it to the reported SEM measurement,

$$\hat{X}_M = M \ . \tag{9}$$

If this were so, then the prediction of the RIE output, based on Equation 7, becomes:

$$\hat{Y}_M = \hat{g}_{rie}(\hat{X}_M) = \hat{a}M \ . \tag{10}$$

Analogous to Equation 8, the FFRA output can be defined in terms of the nominal output and the control compensation $\hat{Y}_M$. Let Y_{naive} represent the FFRA output using the naive estimate,

$$\begin{aligned} Y_{naive} &= Y_{nom} - \hat{Y}_M \\ &= (aX + D_{rie}) - (\hat{a}M) \\ &= (a - \hat{a})X + D_{rie} - \hat{a}D_{sem} \ . \end{aligned} \tag{11}$$

For the sake of simplicity, let us assume that the model parameter estimate is accurate ($\hat{a} = a$). Then,

$$Y_{naive} = D_{rie} - aD_{sem} \ . \tag{12}$$

Since the noise terms are assumed independent, the post-etch variance becomes:

$$\text{Var}[Y_{naive}] = a^2\sigma^2_{sem} + \sigma^2_{rie} \ . \tag{13}$$

Compare this FFRA variance to the nominal variance in Equation 4. One can see that this naive implementation of FFRA rejects the lithography disturbances *in exchange for* the measurement disturbances. This may or may not be a desirable thing to do.

Figure 3 plots the nominal variance and $\text{Var}[Y_{naive}]$ as a function of the measurement noise. Notice that if the measurement tool is perfect ($\sigma^2_{sem} = 0$) then, under FFRA control, the input deviations would be compensated for exactly and only the RIE variance would remain. Also notice that under this naive implementation of feedforward control, $\text{Var}[Y_{naive}]$ can exceed the nominal variance. When the SEM variance is greater than the lithography variance, then feedforward control is *worse* than no control!

MMSE ESTIMATION

In contrast to the naive implementation described above, estimation theory can be used to define a better estimate of X. Classic signal processing techniques have a body of literature on estimating the value of an inaccessible RV in terms of the observation of an accessible RV. Since we are not ignorant about the RVs X or M, their expected behaviors can be used in the estimate of run-to-run x given m.

The problem statement is to find a $\hat{X}$ that minimizes the mean square error (MSE). The MSE is

$$\epsilon = E[(X - \hat{X})^2] \ . \tag{14}$$

The $\hat{X}$ that minimizes the MSE is the minimum mean square error (MMSE) estimator. The MMSE estimator of X based on observing the RV M is the conditional mean (8):

$$X_{mmse} = E[X \mid M] \ . \tag{15}$$

In general, solving the conditional expected value of the MMSE estimator is very difficult, except for the case of Gaussian RVs. Since we have assumed X and M are Gaussian deviations from target, the MMSE estimator can be calculated for our problem definition (3):

$$X_{mmse} = E[X] + \rho\frac{\sigma_X}{\sigma_M}(M - E[M]) \tag{16}$$

where ρ is the correlation coefficient. This is the "optimal linear estimator" of X given M (8).

Consider the effect of ρ on the MMSE estimate. If ρ is zero (i.e. the RVs are uncorrelated), then the best estimate of X is its mean $E[X]$ and the measurement provides no useful information. When $\rho \neq 0$, the measurement M is included in the estimate with appropriate scaling. The correlation coefficient is calculated by:

$$\rho = \frac{\text{Cov}[X, M]}{\sigma_X\sigma_M} \ . \tag{17}$$

The covariance of X and M is:

$$\begin{aligned} \text{Cov}[X, M] &= E[(X - \bar{X})(M - \bar{M})] \\ &= E[(X - \bar{X})(X + D_{sem} - \bar{X})] \\ &= E[(X - \bar{X})^2] \\ &= \text{Var}(X) = \sigma^2_X \ . \end{aligned} \tag{18}$$

Therefore, for our lithography and SEM setup,

$$\rho = \frac{\sigma_X}{\sigma_M} \ . \tag{19}$$

The MMSE estimator becomes:

$$X_{mmse} = M\,\frac{\sigma^2_X}{\sigma^2_X + \sigma^2_{sem}} \ . \tag{20}$$

Let S represent the ratio of variances,

$$S = \frac{\sigma^2_X}{\sigma^2_X + \sigma^2_{sem}} = \frac{\sigma^2_X}{\sigma^2_M} \ . \tag{21}$$

Note that S has the property

$$0 \leq S \leq 1 \ . \tag{22}$$

Consider the affect of SEM noise on X_{mmse}. If there is no SEM noise ($\sigma^2_{sem} = 0$), then $S = 1$ and $X_{mmse} = M$. That is to say, if the SEM is perfect, then the expected true input is, in fact, the measured value. When SEM noise exists, $S < 1$ and the expected true input X_{mmse} will be a fraction of the measured value M. Therefore, X_{mmse} will be closer to zero (target) than M. This is the de-tuning mechanism. As σ^2_{sem} increases, $S \rightarrow 0$, which essentially turns off all control actions since $\hat{Y} = X_{mmse} \rightarrow 0$. The *a priori* information of lithography and SEM process variances is what detunes the controller and avoids over-adjustment.

Figure 2 shows a graphical example of this estimator. The lithography output has a standard deviation of 10 and the

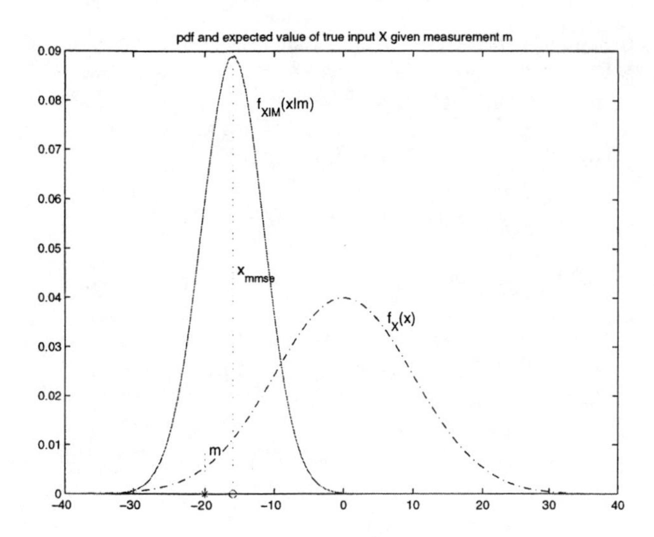

FIGURE 2. Distribution true input X given a measurement $M = m$.

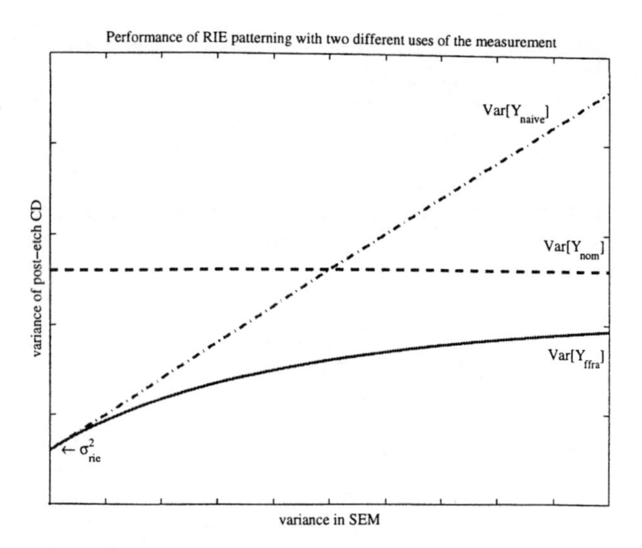

FIGURE 3. FFRA performance as a function of measurement noise.

SEM has a standard deviation of 5. A measured lot-mean CD deviation of $m = -20$ is given. The Gaussian distribution of the conditional RV $X|m$ is has expected value $E[X|m] = -16$ and variance of 20. Notice that x_{mmse} is closer to zero than the measurement m because of the disbelief in the measurement and *a priori* knowledge of the lithography distribution.

Using the MMSE estimator, the output of the RIE under FFRA becomes:

$$\begin{aligned} Y_{ffra} &= Y_{nom} - \hat{Y} \\ &= aX + D_{rie} - \hat{a}SM \ . \end{aligned} \tag{23}$$

The variance is calculated as:

$$\text{Var}[Y_{ffra}] = (a - \hat{a}\hat{S})^2\sigma_X^2 + \sigma_{rie}^2 + \hat{a}^2\hat{S}^2\sigma_{sem}^2 \ . \tag{24}$$

For simplicity, let us assume the model parameters are accurate ($\hat{a} = a$). A little algebra results in:

$$\begin{aligned} \text{Var}[Y_{ffra}] &= a^2 \frac{\sigma_X^2\sigma_{sem}^2}{\sigma_X^2 + \sigma_{sem}^2} + \sigma_{rie}^2 \\ &= a^2 S\sigma_{sem}^2 + \sigma_{rie}^2 \ . \end{aligned} \tag{25}$$

Compare Equation 25 to Equation 4 and Equation 13. They all contain an independent σ_{rie}^2 term. Feedforward control does not reduce the inherent variance of the RIE process. The first term for all equations contain an a^2 representing the effect of the RIE process. For the nominal RIE process, the output variance includes lithography deviations. For the naive FFRA implementation, the lithography variance is exchanged with SEM variance. When the MMSE estimator is used in the FFRA implementation, the lithography variance is exchanged for a scaled SEM variance, $S\sigma_{sem}$. Since $S < 1$, only a fraction of the SEM variance enters into the variance of Y_{ffra}.

Consider the limits of the SEM variance to understand this FFRA variance. If the SEM variance is large, using the FFRA strategy will result in an output variance of:

$$\lim_{\sigma_{sem}^2 \to \infty} \text{Var}[Y_{ffra}] = a^2\sigma_X^2 + \sigma_{rie}^2 \ . \tag{26}$$

This is just the nominal output variance (Equation 4). The FFRA design will *turn off* the controller if the measurement noise is too big. On the other hand, if the measurement noise is small:

$$\lim_{\sigma_{sem}^2 \to 0} \text{Var}[Y_{ffra}] = \sigma_{rie}^2 \ . \tag{27}$$

The input deviations are being compensated for perfectly, and only the random noise of the RIE process remains. Obviously, feedforward control cannot reduce the output variance beyond the inherent RIE variance.

The variance using FFRA can be plotted against increasing measurement noise. Figure 3 shows that the variance during FFRA will not increase above the nominal variance. As the SEM variance increases, the FFRA design detunes the controller gains. Knowledge of the increased measurement noise decreases the measurement tool's credibility and the MMSE estimate reduces the amount of control authority.

RESULTS

We have simulated this FFRA methodology on a 0.35μm gate etch data set obtained from an industrial fab. The data set contains a pair of SEM measurements for each lot. The first measurement is the pre-etch PR mask CD. This corresponds to M. The second measurement is the post-etch gate CD. This corresponds to Y_{nom}. We will use these two data points to simulate the RIE output as if FFRA control had been used.

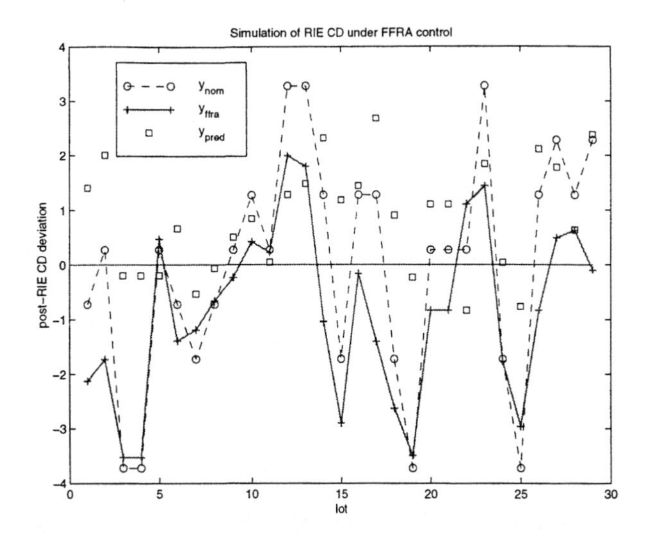

FIGURE 4. Simulated FFRA performances on normalized industrial data from a 0.35μm gate etch process.

First, we created the predictive linear model of M to Y_{nom} using linear regression (Equation 7). This calculated $\hat{a} = 0.73$. Next, we simulated the FFRA output by subtracting the predicted nominal output ($\hat{Y} = \hat{a}X_{mmse} = \hat{a}SM$) from the nominal output (Y_{nom}) as shown in Equation 23.

Figure 4 shows three data signals. The pre-etch measurements (M) and the RIE model were used to calculated the predicted nominal output ($Y_{pred} = \hat{a}X_{mmse}$) calculated from the pre-etch measurement and the RIE model. These are represented as '□'. The nominal etch deviations under no control (Y_{nom}) are shown by a dashed line and a 'o'. This is the second measurement from the data set directly. The simulated FFRA etch deviations (Y_{ffra}) are shown as a solid line and a '+'.

Notice that due to errors in the prediction, the controller does not always perform the proper action. For example, the measurement of the first lot predicted a CD above target, while the nominal CD was actually below target. Therefore, the FFRA simulated control action drove the output more negative. However, the MMSE estimation scaled the adjustment (by S) to avoid over-adjustment. Nonetheless, there are more corrections than improper adjustments. The standard deviation of the nominal output is $s[Y_{nom}] = 2.1$, while the standard deviation of the output with simulated FFRA control is $s[Y_{ffra}] = 1.6$. This is a reduction of 23%.

CONCLUSIONS

We have shown a methodology for proper integration of a sensor measurement into an inter-process feedforward controller. By using the MMSE estimator, over-adjustment is avoided and minimal variance is achieved.

Notice that this work is generic. It is applicable to many sensor and manufacturing processes. In fact, this work is currently being applied to lithography and RTP.

Future work will extend the use of MMSE estimation to another type of feedforward control strategy. Due to complexity and implementation issues, generating a unique recipe for each run may be undesirable. However, allowing the controller to select a recipe from within a pre-defined set of allowable, qualified recipes is sometimes acceptable. We call this control algorithm Feedforward Recipe Selection Control (FRSC). This approach will realize a portion of the FFRA benefits while minimizing the costs in complexity.

REFERENCES

1. E. Dougherty. *Probability and Statistics for the Engineering, Computing, and Physical Sciences.* Prentice Hall, 1990.
2. S. Leang, S. Ma, J. Thomson, B. Bombay, and C. Spanos. A control system for photolithographic sequences. *IEEE Transactions on Semiconductor Manfuacturing,* 9(2):191–207, May 1996.
3. A. Leon-Garcia. *Probability and Random Processes for Electrical Engineering.* Addison-Wesley Publishing Company, Inc., 2 edition, 1994.
4. J. MacGregor. A different view of the funnel experiment. *Journal of Quality Technology,* 22(4):255–259, Oct 1990.
5. J. Neter, W. Wasserman, and M. Kutner. *Applied Linear Statistical Models.* Richard D. Irwin, Inc., Homewood, IL, 2 edition, 1985. ch. 6.
6. E. Rietman. Multi step process yield control with large system models. In *Proceedings of the 1997 American Control Conference,* pages 1573–1574. American Control Conference, 1997.
7. The National Technology Roadmap for Semiconductors. Semiconductor Industry Association, 1997.
8. H. Stark and J. Woods. *Probability, Random Processes, and Estimation Theory for Engineers.* Prentice-Hall, Inc., 2 edition, 1994.
9. J. Stefani, K. Brankner, R. Jucha, W. Pu, and M. Graas. Open-loop predictive control of plasma etching of tungsten usin an in situ film thickness sensor. *SPIE Process Module Metrology, Control, and Clustering,* 1594:243–257, 1991.
10. K. Stoddard, P. Crouch, M. Kozicki, and K. Tsakalis. Application of feed-forward and adaptive feedback control to semiconductor device manufacturing. In *American Control Conference,* pages 892–896, June 1994.

PACKAGING

A Proposed Holistic Approach to On-Chip, Off-Chip, Test, and Package Interconnections

Dirk J. Bartelink

Hewlett-Packard Labs, Palo Alto, California 94304

The term interconnection has traditionally implied a 'robust' connection from a transistor or a group of transistors in an IC to the outside world, usually a PC board. Optimum system utilization is done from outside the IC. As an alternative, this paper addresses 'unimpeded' transistor-to-transistor interconnection aimed at reaching the high circuit densities and computational capabilities of neighboring IC's. In this view, interconnections are not made to some human-centric place outside the IC world requiring robustness—except for system input and output connections. This unimpeded interconnect style is currently available only through intra-chip signal traces in 'system-on-a-chip' implementations, as exemplified by embedded DRAMs. Because the traditional off-chip penalty in performance and wiring density is so large, a merging of complex process technologies is the only option today. It is suggested that, for system integration to move forward, the traditional robustness requirement inherited from conventional packaging interconnect and IC manufacturing test must be discarded.

Traditional system assembly from vendor parts requires robustness under shipping, inspection and assembly. The trend toward systems on a chip signifies willingness by semiconductor companies to design and fabricate whole systems in house, so that 'in-house' chip-to-chip assembly is not beyond reach. In this scenario, bare chips never leave the controlled environment of the IC fabricator while the two major contributors to off-chip signal penalty, ESD protection and the need to source a 50-ohm test head, are avoided. With in-house assembly, ESD protection can be eliminated with the precautions already familiar in plasma etching. Test interconnection impacts the fundamentals of IC manufacturing, particularly with clock speeds approaching 1GHz, and cannot be an afterthought. It should be an integral part of the chip-to-chip interconnection bandwidth optimization, because—as we must recognize—test is also performed using IC's.

A system interconnection is proposed using multiple chips fabricated with conventional silicon processes, including MEMS technology. The system resembles an MCM that can be joined without committing to final assembly to perform at-speed testing. 50-Ohm test probes never load the circuit; only intended neighboring chips are ever connected. A 'back-plane' chip provides the connection layers for both inter- and intra-chip signals and also serves as the probe card, in analogy with membrane probes now used for single-chip testing. Intra-chip connections, which require complicated connections during test that exactly match the product, are then properly made and all waveforms and loading conditions under test will be identical to those of the product.

The major benefit is that all front-end chip technologies can be merged—logic, memory, RF, even passives. ESD protection is required only on external system connections. Manufacturing test information will accurately characterize process faults and thus avoid the Known-Good-Die problem that has slowed the arrival of conventional MCM's.

INTRODUCTION

The mainstream semiconductor industry is slow to adopt new directions because its highly complex and expensive technology requires a broad consensus. Moore's Law and the National Technology Roadmap for Semiconductors (NTRS) are trusted tools used in planning the investment of available resources for new technology. To successfully guide the direction of mainstream technology around a newly recognized obstacle, it is necessary to gather a timely consensus by means of a clear vision of a recognized benefit that is based on a major market opportunity. In this paper, this market opportunity is the integration of various technologies into a functioning high-performance low-power combination, such as is now commonly called a system on a chip. The vision is of the far-reaching benefit of integrating arbitrary combinations of technologies through multi-chip assembly, in contrast with the limited options currently available through a monolithic merging of divergent technologies, such as memory and logic in embedded DRAMs. The goal of this paper is to present a clear understanding of the existing but surmountable barriers to integrating arbitrary technologies through advanced system assembly concepts in order to form a timely consensus on the need to act.

Major impediments to change are 1) the sunk cost of the installed manufacturing base and 2) an aversion to risk in shifting away from the tried-and-true consensus approach. Indeed, because only extensions of established approaches could be justified, the NTRS areas of Design & Test, Process Integration, Interconnect, Assembly & Packaging, and Factory Integration have been optimized separately along historical lines. Moreover, the success of technology scaling has entrenched the "same old way" for so many generations that there is no longer an overall architecture thrust that provides high-level optimization amongst the subdisciplines. What should be kept in perspective, however, is that the relative investment levels of various subdisciplines of the semiconductor industry are not the same and that the natural and understandable reluctance to change in one area must be balanced by the overall good of

CP449, *Characterization and Metrology for ULSI Technology: 1998 International Conference*
edited by D. G. Seiler, A. C. Diebold, W. M. Bullis, T. J. Shaffner, R. McDonald, and E. J. Walters
 1-56396-753-7/98/$15.00

the technology. In particular, chip-to-chip interconnection has been the domain of packaging and assembly where investments have traditionally lagged and where progress, including even that of multi-chip-modules (MCM's), has not kept pace with on-chip increases in density and speed.

In the funding of mainstream technology advancement, for example lithography, the institutional machinery is so well established that it is now difficult to grant a less established area equal access to resources even when that area would offer greater opportunities. Such is now the case with on-chip, off-chip, test, and package interconnections to which no holistic architectural vision has so far been applied. The traditional multi-vendor supply chain of the printed circuit board (PCB) business has dominated the mindset in this industry segment. The assumptions inherent in that business model still dominate the technology solution by requiring a high degree of robustness of the off-chip connections that permits multi-vendor relationships. Research investments have traditionally lagged in physical assembly of bare chips and the handling of whole wafers has been accepted, without any substantial investigation, as the most economical approach. Now that we have reached a cost basis of case-by-case merging of diverse technologies, such as embedded DRAM, it is time to re-evaluate the tradeoffs associated with multi-chip assembly because a case can be made that it provides an economical and flexible solution that has superior long-term prospects.

The vantagepoint of this paper is that, as the technology advances, the bulk of signals heading off-chip only go to other chips. Only a few signals need to venture beyond the chip-centric world to that of the printed circuit board and on to some human-centric system backplane. Monolithic integration is the only accepted means of remaining within the chip world of 'unimpeded' transistor-to-transistor interconnection because any chip-to-chip connection now suffers a penalty almost equal to that of reaching the backplane. This issue is widely recognized and many proposals exist, many of them centered on optical interconnects. The approach here is not to invent dramatically new communications solutions, but merely to bring off-chip electrical connection on par with on-chip through changes in chip handling procedures. Given that monolithic implementations are becoming common place and are completely 'in house,' it is reasonable that 'in house' multi-chip assembly can be implemented. Within this more protective environment, herein called a Composite IC, unimpeded interconnections can be designed to replace the disadvantageous practice of demanding only robust interconnections.

To change any status quo, there has to be a singular event that triggers the departure from incremental evolution of the incumbent scheme. In the present context, that event is the dramatic new importance of test interconnection. What is changing is that as speeds approach the GHz domain, at-speed test will have speed and bandwidth demands roughly of equal to those of the product. Moreover, the high cost of testers will require them in future to use high-density CMOS circuits located near the device under test. Thus, test access is fast becoming a chip-to-chip interconnection challenge on par with any required by the product.

Already, physical chip attachment is an integral part of manufacturing, but conventional wafer probing approaches use physical dimensions much larger than that optimum for GHz operation, particularly regarding the location of the decoupling capacitors. The electrical waveforms occurring on chip-to-test interconnection wires will affect the performance and thus yield verification of the chip. If the test and product wiring implementation are distinct, it will not be possible to truly qualify the chip, to say nothing of the engineering cost involved in characterizing the difference between the two. Coupled with this chip access issue is the fact that, at high circuit speeds, interconnect parasitics play a greater role than before and that manufacturing control of on-chip and off-chip interconnect demands more of our attention through high-speed testing. Perhaps the only way to truly learn about interconnections meeting specifications is by operating the chip at speed in the exact product environment. In a time critical way, test is the issue that forces the holistic approach to interconnect.

Although the goal of this paper is to give a clear statement of the scope of the problem and a vision of what needs to be accomplished, a convincing example is needed to give assurance that a solution is accessible. The Composite IC concept (1) is used to expose critical issues that come to light only when thinking through a specific physical implementation. To reiterate, the vision is to bring off-chip electrical connection on par with on-chip through changes in chip handling procedures. The Composite IC treats chip-to-chip interconnections in the product and during test on an identical footing by revising the fundamental design of testers. It introduces a demountable connection that acts as the probe card in place of test fixturing. A spring-like connection is fabricated using MEMS technology and is proposed as a universal standard interface layer that can be applied to any chip technology. This standard connection technology allows component pieces of the Composite IC to be assembled in the product with a permanently soldered connection or to serve as factory tooling during test.

TRADITIONAL INTERCONNECTIONS

Chip packaging was originally a means of protection and heat removal, not interconnection. In fact, not counting the lead frames found inside most packages, the only real interconnection technology that is part of packaging technology is in MCM's and in the recently emerging multi-chip packages. The progress of MCM's has been much slower than originally expected and multi-chip packages are taking their place. We now explore the slow growth in MCM's and the reasons for it: 1) lack of industry consensus on MCM technology, 2) the continued use of robust interconnections which leaves monolithic implementations as clearly superior and 3) difficulty in assuring fully functional parts without package insertion—the Known-Good-Die (KGD) problem. The last two are responsible for the first and have prevented MCM's from becoming a mainstream technology.

In order to understand the existing business constraints and the options within them for revised chip handling procedures, we must analyze the *food chain of the electronics industry*. Figure 1 shows a schematic view (2) of this food

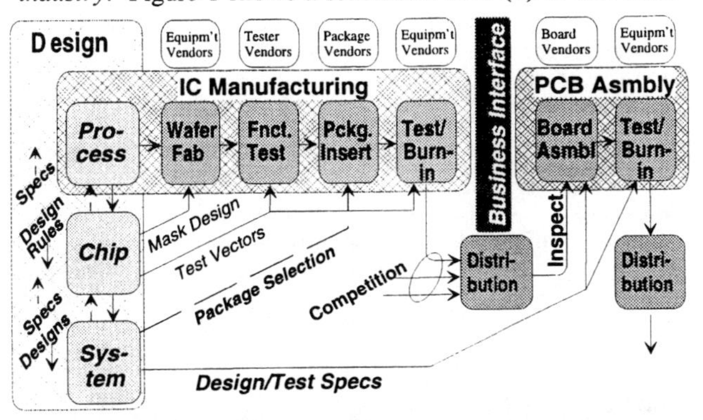

FIGURE 1. Food Chain of the Electronics Industry

chain. It presents three major segments: **Design**, **IC Manufacturing**, and **PCB Assembly**. In turn, each of these is made of several components. Design consists of **Process**, **Chip**, and **System** design sections that interact with each other through **Design Rules** and **Specifications**. Process Design is also part of IC Manufacturing, which further has **Wafer Fab**, **Functional Test**, **Package Insert**, and **Test/Burn-in**. Chip Design is shown to issue **Mask Designs** and **Test Vectors**, while System Design provides **Design/Test Specifications** to the PCB Assembly operation. The latter is closely aligned with the ultimate end product business and is free to purchase chips from the **Competition** through the **Distribution** channel.

Our focus in this figure is on the **Business Interface**, which separates PCB Assembly from IC Manufacturing. This interface is responsible for the additional specification issued by System Design, namely **Package Selection**. The reason is the need to protect the IC through the Business Interface, including the tests applied by PCB Assembly to **Inspect** incoming parts. With some chips costing thousands of dollars, the ownership of defective parts is a major factor in business-to-business relationships and the ability to determine who caused the defect fundamentally leads to the need for robust test interconnection to the interior of the chip. Any alternative chip handling procedure cannot violate these basic business considerations, including the issue of faults that can only be found by operating at speed, as can now be done only after package insertion. The convoluted path of package selection stems from the desire to optimize the PCB system for cost and speed using a package specially chosen for the application, while meeting the at-speed test. In doing so, it disrupts the orderly flow of all chips of the same design in IC Manufacturing, which now includes two tests: (die-sort) Functional Test and (at-speed) Final Test/Burn-in.

As clock speeds enter the GHz frequency regime, there is a basic shift in yield loss mechanisms that seriously impacts chip testing. Traditionally, the major load capacitance experienced by a critical-path transistor has been the gate capacitance of the following gate(s). With more of the energy of this transistor's switching transition going into parasitic wiring capacitance—which must be the case given all the attention given to low-K dielectrics—the critical-path speed becomes more dependent on the tolerances of the wiring as compared with the transistors. Transistor gate capacitance depends on gate-dielectric thickness control and gate-patterning accuracy. Both of these have been the foundation stones of the industry. Wiring capacitance control is a relatively new concern and at this time manufacturing-control procedures are not as developed as for transistor parameter control. The transient current passing through a gate and its load—which sets the transition speed—is equally controlled by the transistor tolerances and load tolerances, but the effects of the latter are only known with certainty in complex wiring meshes from yield losses due to race conditions. Thus at-speed testing will become the only sure way to learn the necessary yield improvement techniques. For those off-chip connections that run at similar speeds, as might be expected in MCM's, the only appropriate test for them is embed the chip in its final product environment. The current test approach of separate functional and final tests will become even more disruptive in IC manufacturing and is already unmanageable for MCM's. As noted, the driving force for changing the status quo within the food chain is this issue of test.

MCM technology Needs Industry Consensus

It can be stated without fear of contradiction that MCM's have thus far failed to become a mainstream technology despite their many obvious benefits. We need to understand the various causes of this phenomenon in order to tailor a solution that can become a winner in mainstream system integration. Figure 2 describes a sequence of reasoning about innovation and the lesson or rule of the mainstream—which is driven by Moore's Law and the NTRS—that has been a root cause in the MCM story. MCM's fall outside the established standards of mainstream ICs and without consensus no single company can afford to pioneer the needed technology, although it has been tried many times in the past. While mainstream experimentation with multi-chip packages (MCP) is gaining ground, in to-

The mainstream IC industry is very powerful, but. . . . Innovation is very difficult to introduce:
1. *Mainstream* power comes from low cost (but high capital)
2. *Low cost* comes from standardization (common learning curve, equipment infrastructure)
3. *Standardization* come from consensus (lowers risk)
4. *Consensus* allows no "end run" (no low-cost tooling)
5. *End run* has been traditional way to introduce. . . . Innovation
6. Everybody wants to be a fast follower
Then . . .
Who takes leadership when everybody wants to be #2?

FIGURE 2. Innovation and the 'Rule of the Mainstream'

day's environment there appear to be few takers for mainstream MCM's.

Figure 3 presents an interpretation of the end-system benefits and challenges as experienced historically by MCM's. Over time, there have been several examples of strong investments by a single company in a particular approach (left-most vertical dotted line). The end system derives immediate benefits from this investment in delivered system performance—taken here as the product of density and speed in the integrated system. As time progresses after the initial investment, however, the progress of the company by itself proves less than that of industry as a whole working on mainstream monolithic ICs, as shown by

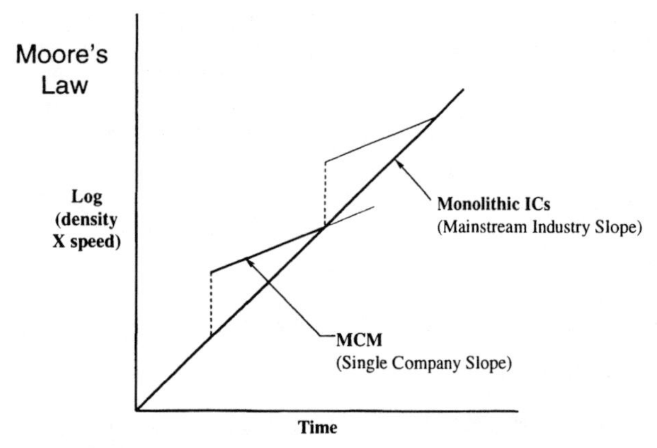

FIGURE 3. Rise and fall of single-company MCM technology

the lower slope of the single company learning curve. The inevitable intersection then poses a question for the company of whether to invest again (second vertical dotted line) or simply to wait for the mainstream to meet its needs. The negative impact of this mechanism on the introduction of MCM's is a matter of history.

To keep from repeating this history, it is necessary to identify a common need of the industry that is not being met by the current mainstream approach and to find a superior solution that embraces the major demands of the evolution of the electronics industry. While this clearly calls for opinion about the future, we can nevertheless analyze the important attributes of future technology scaling and system integration to gain most of the needed insight.

Why no mainstream support for MCM's?

There is a message in the fact that MCM's are losing the competition with system-on-a-chip implementations, such as embedded DRAMs. On the surface, the combination of memory and logic technologies on separate chips appears to be a perfect fit for MCM's.

Impact of robust interconnections

The current vision of MCM's has its origins some decades ago, well before clock speeds reached the high values of today. In the present market, embedded DRAMs and other system-on-a-chip implementations offer high-density unimpeded bandwidth, which is changing the nature of the memory-logic interaction. MCM's, with their robust interconnections and corresponding off-chip penalty, offer no substantial benefit over conventional packaging in speed or density. What lesson must be learned from this situation?

The Moore's Law learning-curve slope for mainstream ICs is well known, but the equivalent relation for the packaging industry is difficult to make out, despite the existence of the NTRS. A crude approximation of it is compared to the IC curve in Fig. 4. It may in fact not be a semi-log slope at all. Even with this limitation, drawing the two curves together, which serves to underscore their divergence, clearly illustrates that the penalty for leaving the

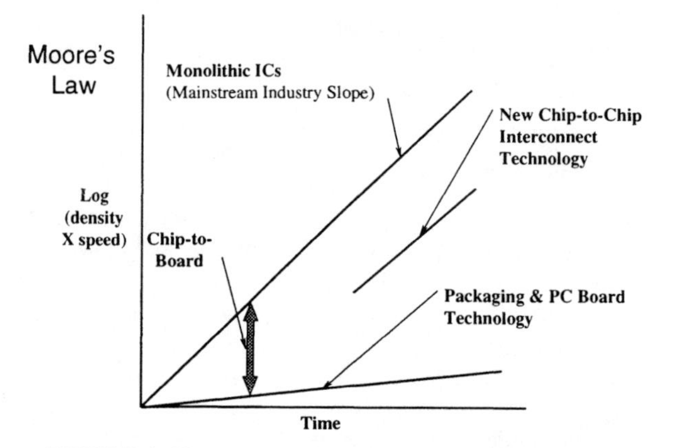

FIGURE 4. Needed: new chip-to-chip interconnect technology

monolithic IC world is growing exponentially with time. The double-headed arrow schematically summarizes the technology gap in making a round-trip chip-to-chip connection via the PC board. As the gap widens, there is growing need for a more direct technology that connects chips together, as MCM's have long forecast. That new technology is shown schematically in the figure as an intermediate locus progressing with its own slope.

In contrast with the end-system accounting of density times speed, as was done in Fig. 3, the vertical scale of Fig. 4 accounts for the benefits of interconnections on the basis of specific technology investments. It has been widely recognized that to sustain the economic miracle of ICs a new paradigm in interconnections is needed. Thus, a new accounting of interconnect performance is in order for the different connections between transistors, both on-chip and off. The steep upward sweep of the IC curve in Fig. 4 means that on-chip interconnect historically has progressed rapidly by lithographic scaling, although electrical performance limitations are now clearly at work. In keeping with our vision of bringing off-chip electrical connection on par with on-chip without inventing new technology, we can explore what would keep off-chip interconnects at a constant distance behind the main curve, i.e., what is the technological makeup of the double-headed arrow?

If a successful embedded DRAM chip or similar product were to be cut in half and reconnected with conventional technology, what would be changed? We can idealize that situation without losing the essential point by assuming that the two halves can be placed close enough together for short wire bonds to reattach the signal leads. In the limit, we can even assume that the added wiring capacitance and resistance of the bonds is negligible. In that case, have we recovered the original performance? The answer is categorically no because the two separate chips would be quite different from the two halves of the original chip. The key differences are the two major contributors to off-chip signal loading, ESD protection and the need to source a 50-ohm test head.

The IC manufacturing infrastructure that would produce the two halves is currently set up to handle chips only with a one-size-fits-all robust interconnection approach, even when all components are produced in house. To test each component part, for example, probes are brought down to the chip in the wafer. These probes present the on-chip drivers with 50 ohms loads as is necessary to source the high-speed data transfer over relatively long wires back to the tester. Also, ESD protection is mandatory because of the hostile environment on the tester. The added capacitance of the ESD circuitry completely swamps the wiring capacitance of our short wire bond.

We can now see what is in the technological makeup of the double-headed arrow: the low impedance of this off-chip circuit adversely loads the native high-impedance environment of most on-chip CMOS circuits, particularly short signal traces. While it is possible to adjust driver sizes to compensate for the reductions in speed, larger drivers consume power and chip area. This added area, even ignoring ESD circuit size, will then make it impossible to keep the on-pitch design of the original chip. Thus, as the IC locus of Moore's Law is being maintained with the advent of scaled low-power CMOS, it is leaving the low-impedance world of testers progressively further behind.

In the early history of MCM's, ECL was the dominant technology. Since ECL naturally operates near 50 ohms, off-chip driver scaling and tester sourcing were not considerations and MCM's provided a benefit through reduced wiring parasitics. In applying MCM technology to large CMOS chips, however, we have inherited the now faulty—but widely held—notion that the wiring capacitance of packages causes the major speed penalty. Since the actual loading comes from ESD protection and 50-ohm sourcing, the shorter package wires of MCM's are not a big benefit. Because of this hidden factor, multi-chip packages are a competitive alternative.

Balancing substrate wiring density and utilization

Beyond wiring-capacitance reduction, MCM's also have held the promise of tighter chip-to-chip packing densities by providing an intermediate interconnection level distinct from either the on-chip or board-level wiring technologies, as portrayed in Fig.4. The mainstream community, however, has not agreed on the nature or even the necessity of such a third wiring alternative since the latter must compete with the incumbent IC and PCB wiring technologies that over time have become highly cost competitive within their own applications. The long sought 'low-cost substrate' that would ignite the MCM business has never been found. This substrate must have higher wiring density than the PC board, but it has an inherent dilemma; namely, its wiring capability is often underutilized using conventional MCM chip assembly approaches. Not only has this substrate missed out on the benefit of cost reduction through mainstream use, it also has an inherent cost penalty from underutilization, as will now be examined.

Figure 5 depicts the issues of I/O escape and wiring utilization when a chip is connected to a substrate using area-array connections. I/O escape, Fig. 5a, refers to the need to provide sufficient wiring density when interior pads are routed beyond the footprint of the chip. Other chip attachment methods that connect only at the perimeter (e.g., fine-pitch quad flat pac—QFP) suffer from nearly identical limitations because in either case wires passing through this perimeter must carry all connections coming from the chip. This requirement is unavoidable and is most acute for low-cost substrates that have few metal levels and large feature sizes.

The companion issue, wiring utilization (Fig. 5b), looks at the balance between the signal-escape wiring density and the interconnect density needed for signals in the other areas of the substrate. With one bare microprocessor chip

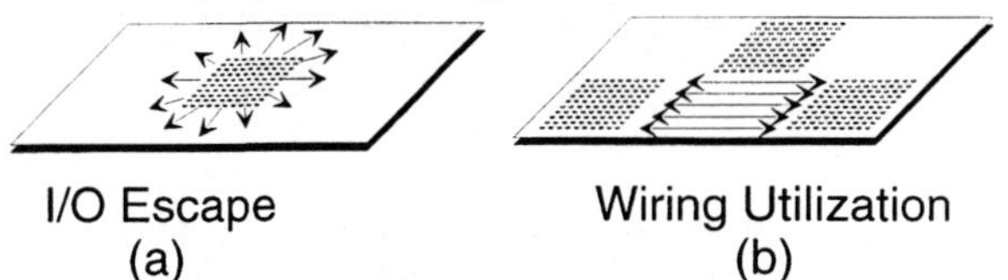

FIGURE 5. Issues in Substrate Utilization

having many I/O's, for example, other smaller components on the substrate may present too little need for its full wiring capability. Historically, MCM's have not been cost competitive in part because they have failed to balance I/O count and substrate wiring density. Conventional packages spread the I/O's over a larger footprint, while burdening only the one high-I/O-count component. In a fine-pitch QFP package, an interposer may match the pitch of the chip to that of the package perimeter leads. Other packaging approaches, such as chip-scale packages do not have that advantage. An alternative that is commonly used puts the microprocessor and its high-speed SRAM in a multi-chip package or on a PC board with an in-line connection (e.g., "Slot 1" technology) to ease the burden on the motherboard.

To take account of this issue in a revised MCM strategy, the solution must create a balance over a majority of the substrate area. Strangely, the escape problem that demands high substrate wiring density for a chip with high I/O density can be made cost competitive only by adding more chips—of comparable complexity and high I/O count. Before any intermediate substrate can become a mainstream player, we must develop a new architectural model and design procedure that ensures this balance as well as the resolution of other chip-to-substrate and chip-to-test interconnection issues yet to be discussed.

Figure 6 examines a third element in this interconnection riddle, namely hierarchy, which has delayed MCM's by confounding their base of support. In the two assembly hierarchies shown—direct chip-on-board and MCM—the

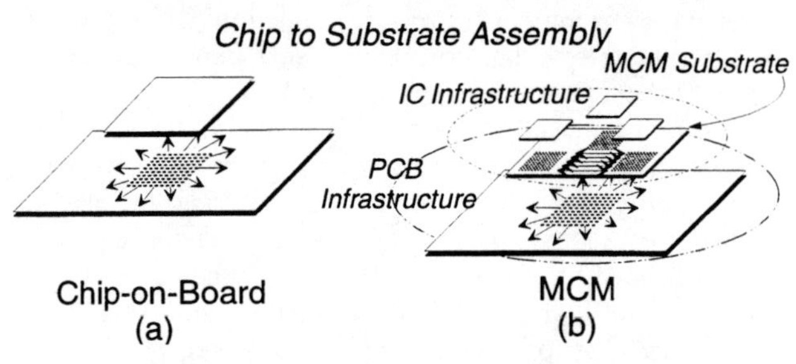

FIGURE 6 Chip-to-Substrate Interconnection Hierarchy

MCM substrate of Fig. 6b takes the place of the chip of Fig. 6a, thereby forming a repeating technology chain. A new relationship now opens up regarding who is the master of each technology. While for chip-on-board there is a simple relationship between the two parties, in MCM's two options occur, as shown in Fig. 6b: PCB infrastructure—the basis of all MCM's to date—and IC infrastructure.

MCM's—in common with the chip-on-board approach of Fig. 6a—require the handling of bare chips for assembly and test. The packaging community is set up for this, although it has only recently started developing solutions to the know-good-die (KGD) problem—such test issues are discussed more fully in a following subsection. The ultimate solution cannot be a 'post processing,' namely a test after the chip leaves the wafer-fab, because it intersects with the issue of yield management, which is the domain of IC manufacturing. This entanglement is characteristic of the factory floor where tough new issues are often first encountered. When different infrastructures are involved, finding a suitable technological solution can be particularly difficult. Since new technology development cannot easily be pushed upstream from the factory vantagepoint, a status quo seems to persist. The KGD issue is one of high-speed electrical chip access for testing, which in GHz-testing cannot be left as an afterthought and must be addressed.

In the evolving world of systems, MCM's—having had their early start in the more I/O-tolerant ECL world—have now cut across these CMOS manufacturing infrastructure constraints without providing a suitable solution. As a result the very name MCM has acquired a niche reputation and the mainstream momentum has swung to the costly system-on-a-chip paradigm.

Technology infrastructure to follow IC slope

The theme of this paper is that only by engaging the IC chip community and making the intermediate substrate part of the IC infrastructure (Fig. 6b) can there be a change from the status quo. The IC wafer manufacturing community, however, has no traditions for handling chips in any way other than in wafer form. Because of this history, it is totally unprepared for handling bare chips and has never considered any process technology or architecture beyond the bond pad to be their domain.

Yet, an urgent need exists for IC process designers, who in recent years have become well versed in interconnect issues, to become involved in off-chip interconnect technology beyond merely providing the bond pad. The packaging community has never been in a strategic position to influence the technology content of the IC except for the size, location and materials properties of the bond pads and the one-size-fits-all solution is no longer good enough. Continued CMOS scaling will move it further from the robust backplane and the 50-ohm test world. Without the IC community involvement in chip I/O, conventional IC-process architecture will fail to provide the needed direct means for communicating signals between adjacent chips, i.e., without suffering a large round trip penalty.

Off-chip technology for electrical signals needs only to be approximately as good as remaining on-chip. But, as Fig. 7 shows, unless the intermediate substrate and I/O technology track the IC learning curve, a new off-chip delay gap will open up—e.g., by following the packaging slope. This rapid progress of the new substrate technology pointedly implies that the wafer-fab community should take responsibility. History shows that the required resources

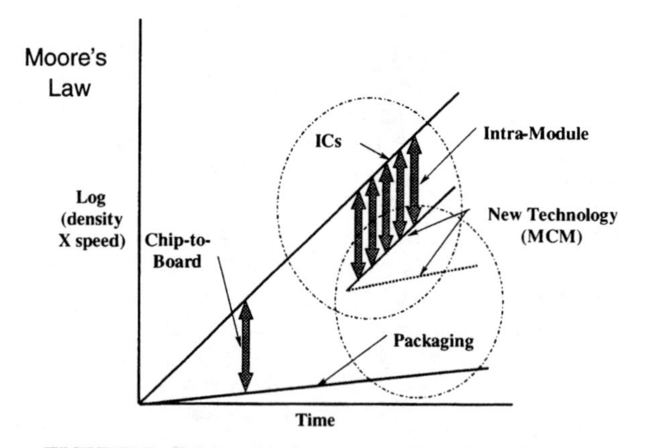

FIGURE 7. Chip-to-chip interconnect issues and strategy

and authorizations are more likely to be available in the wafer-fab community—it is experienced in maintaining the rapid pace. Of the two dotted circles in Fig. 7, which like those of Fig. 6b group the intermediate substrate with either IC or PCB technology, the former must be the choice.

Obstacles in Conventional MCM Testing

A third reason MCM's have made slow progress—along with a lack of mainstream consensus and the handicap of robust interconnections—relates to full functional test and the cost of repairing failed parts, the widely recognized KGD problem. The most obvious distinction in full functional verification of MCM's is that, upon discovery of a faulty chip, repair requires removal and replacement of permanently mounted chips—or their disposal along with the attached functioning chips, which is economically unacceptable. A second difference from conventional packaged parts is that burn-in is much more difficult for bare chips.

The most common and oldest approach to both problems has been to rework the solder connections between the chip and the substrate, but that has a heavy labor content compared with standard die sorting approaches. Another obvious solution, which is receiving considerable attention, is to perform the same tests on a bare chip as on a packaged chip by inserting the chip into a temporary package. A variant of this approach is to use a chip-scale package that does not use up much substrate space but provides a means of handling and contacting chips, although without the traditional package protection. In either case, there is a significant added complexity and cost. These obstacles may possibly be overcome for high-volume standard parts—processors, memory—where manufacturing procedures can be worked out, but in low-volume ASICs added engineering costs are not so easily absorbed.

The need for wafer-level burn-in is an outgrowth of MCM's and direct-chip-attach. It leverages the benefit of handling chips in the wafer, but then encounters a major hurdle in trying to power up all chips in the wafer at the same time. Burn-in is among those manufacturing technologies that are still an afterthought. In the manufacturing sequence, several of these steps require separate handling and electrical insertion operations, contrary to the well-known manufacturing metric of minimizing handling, ideally to a single insertion.

An important and often-overlooked architecture issue in chip interconnection is the need to provide yield improvement information to the fab from production testing. At-speed final testing of MCM's, which combine several chips from various sources, may provide the only source of feedback for certain critical parameters. The business implications of managing around the ownership of failed parts in this case presents an interesting challenge, as can be comprehended from Fig. 1.

IC chips are so complex that significant test bandwidth is needed to access the full complement of their interior circuits. For many of the critical combinations of key parameters, the only measure of the sensitivity of the manufacturing process comes from this information flow. In earlier technologies, active devices were the most poorly controlled items, mostly at the hands of lithography, but they benefited from having the single-crystal perfection of the silicon as the basis of their manufacturing control. Thus, other than lithography, few tools—for example, diffusion furnaces—had to be adjusted and maintained on the basis of functional test results. Interconnect parasitics become ever more important at high clock speeds, and interconnect parasitics—in contrast with active devices—are totally controlled by machine operations and settings, mainly deposition and etching steps. The performance of individual pieces of equipment cannot be known, nor adjusted, without this test feedback, in particular from at-speed testing. Also, as speeds increase in MCM's, this equipment sensitivity will even become acute in off-chip circuits.

Summary of Traditional Interconnections

The traditional large off-chip penalty in performance and wiring density is rooted in the entrenched multi-vendor business model for packaged parts and dates back to the beginning of the electronics industry. By retaining the off-chip penalty of ESD protection and 50-ohm test sourcing, MCM's have not delivered on their promise and have allowed the unimpeded interconnect style of monolithic system-on-a-chip technology to become the mainstream solution. MCM's are uniquely useful in combining diverse combinations of technologies with tight chip-to-chip packing densities, but they have become niche products because in they offer no real advantage over multi-chip packages.

The electronics industry is in urgent need of a generic approach for combining chips that can support GHz digital circuits speeds combined with mixed-signal waveforms —as promised by MCM's. The speed and density benefits of ICs remain the industry driving force and the sum of MCM versatility in combining chips from different process technologies and state-of-the-art performance can create a new mainstream solution. The multi-chip package is in many cases a single-vendor system integration approach that allows arbitrary combinations of chips, but it does so without the significant system-bandwidth benefits anticipated for unimpeded interconnections. To produce unimpeded-I/O chips within the current manufacturing infrastructure, component chips must be tested at speed with their neighbors, not only to select the working combinations, but also to identify which critical parameters are responsible for yield problems.

UNIMPEDED INTERCONNECTIONS

We now use the business model of the last section to understand the technology changes needed to achieve unimpeded interconnections while complying with the major constraints imposed by the entrenched infrastructure and resolving the electro-mechanical aspects of multi-chip assembly. In the following section, a proposal is presented of specific technology—the Composite IC— that implements the findings of this section.

Infrastructure Requirements

The food chain of the electronics industry cannot be altered for technological reasons alone, particularly the business interface of Fig.1. Because technology responds to market needs, complex technological structures have arisen that are not necessarily optimum, notably the convoluted path of package selection. Since we now want to add new technology to solve a particular need, we should not proceed without eliminating some of the unneeded structure.

Figure 8 presents a revised food chain that is in keeping with our 'in-house' chip fabrication concept. It introduces a new chip-to-chip interface within IC Manufacturing and consolidates functional and final test into one. Implicit in

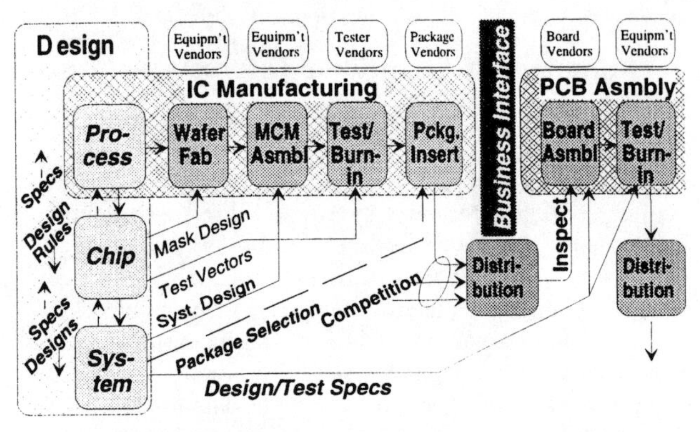

FIGURE 8. Revised Food Chain in 'IC Manufacturing'

this change is the involvement of Process Design in the architecture of the chip-to-chip interface. Some design functions are thereby interchanged with System Design and a corresponding revision is needed of the process for specification and design-rules.

To make these revisions in the food chain acceptable, the combination of chips that makes up a module is treated the same way as a single chip is normally passed through the business interface. This includes package selection according to the existing convoluted route; there is no conflict in this approach, it is just that the module's individual chips do not have to traverse it. System designers have to provide revised output, particularly if they want to participate at the chip level in the IC design process. Individual chip designs would not have to change, unless they want to take advantage of the new possibilities. It should be noted that significant differences might occur in intellectual property issues since complete chip designs and their processes could more easily be merged. Over the longer term, if an infrastructure has developed around this approach, test procedures might be found that permit module assembly from multiple suppliers, but that cannot be a major goal initially. A standard technology implementation throughout the industry would be a requirement for the chips to be able to communicate.

Architectural Requirements

The chip-module hierarchy of Fig. 6b places architectural requirements on any revised chip I/O technology. To improve chances for mainstream acceptance, the corresponding process and design changes must also offer positive benefits in the major roadmap challenges facing ICs.

Process technology and electrical performance

To balance the utilization of the intermediate substrate for low cost, as discussed in connection with Fig. 5, it is necessary to bring together several high-I/O density chips to neutralize the escape requirements of an individual high-I/O density chip. This seemingly singular requirement actually has a complement in mainstream technology that offers an ideal opportunity for synergy. As active devices and local interconnects (Front-End-of-Line—FEOL) are scaled, a technology gap is growing with the higher lying on-chip multi-level wiring planes (Back-End-of-Line—BEOL). Since the latter tend to be used for power/ground routing or long-distance signals that need fat wires they cannot make full use of the fine feature sizes available. Thus, this technology is effectively wasted on these levels, but in the normal process sequence there is no opportunity for change. With our chip-module approach, however, the upper wiring levels can be moved to the cheaper intermediate substrate.

The main system-level advantage of this architecture is that chip size is now a flexible parameter. While yield and stepper field size may limit FEOL chip size, the course-pitched BEOL technology allows much larger chips on the basis of yield. In conventional usage, BEOL chip size is constrained to be the same because they start in a common FEOL wafer. If they were separate chips—in the form of the intermediate substrate—their size could span several FEOL chips. Moreover, they can be made using projection aligners instead of steppers.

Regarding yield, we are used to seeing process technology being stretched at the introduction of large chip designs that try to encompass a whole system. This is not an issue with more flexible chip sizes, since they can be adjusted for optimum yield. Moreover, we commonly think only of chip area as the yield controlling parameter, but the shorter process sequences of separate chips also have great benefit.

A critical characteristic of mainstream technology for system implementations is that FEOL processes vary greatly by application—processors, memory, RF, etc.—and are therefore difficult to merge, while BEOL processes are becoming more alike because of technological limits. Thus, in establishing a universal system technology, a key feature of MCM's can be applied directly to our chip-module architecture, namely the merging of separate FEOL chip technologies with a common BEOL substrate.

Mechanical tolerances and physical connections

Three factors have traditionally presented barriers for participation by the IC process community in chip-to-chip assembly: 1) mechanical tolerances are much larger for package technology than ICs, 2) thermal expansion differentials introduce stresses and 3) product assembly approaches are unfamiliar to IC designers. The packaging community has shielded the whole IC industry from the complex product variety of the electronics industry by providing a standard interface, the universal bond pad. Our new chip-to-chip connection within IC manufacturing should not upset this external chip interface standard, but the system's internal physical connections do not have to adhere to it, which opens the opportunity for using tighter tolerances and increasing the I/O density.

In packaging terminology, an interposer is needed to connect IC dimensions to the packaging domain, in this case an interposer supporting several chips that replaces the usual on-chip BEOL technology. To complete this picture, the architecture must identify a mechanical assembly technology for joining FEOL chips to the intermediate substrate.

The most common MCM architecture employs flip-chip assembly using C4 technology that results in a rigid attachment between chip and substrate. This approach limits I/O density and chip size because of thermal expansion differentials and, therefore, does not offer the continued scaling capability discussed in connection with Fig. 7. Moreover, the only way to deal with faulty parts in test and burn-in of such MCM's is through reworking the solder connections, again not very promising for continued scaling.

The combination of tighter dimensions and thermal expansion in large chips argues that the flexible connection provided by the wire bond should be miniaturized in our method of joining chips. What is needed is a miniature spring-like structure that can be fabricated with IC technology—MEMS—that allows both ends to be permanently attached after testing. During test it functions as a probe tip.

Test Requirements

Conventional test interfacing, as done in the wafer using probe tips or flexible membranes, is part of the robust interconnection environment: it provides the temporary connection that has become central to the standard interface protocol of the bond pad. Complex test algorithms are executed using high-speed IC's. In current testers these are remotely located—by GHz standards—and need 50-ohm cabling.

Temporary test interfacing for unimpeded interconnections must be considered an integral part of the chip-to-chip interconnection bandwidth optimization because it must be capable of carrying high-speed intra-system signals that are not conditioned to go beyond their immediate neighborhood. This scenario calls for a rethinking of testers and test methodologies and discontinuing wafer-level tests.

Building on the earlier image of an interposer with multiple chips, the architectural vision here consists of augmenting the chip connection with temporary attachment capabilities and locating the critical test circuits on the interposer. This assembly forms a test fixture, namely a cross between an interposer and probe card. While it only accepts bare chips, it uses the same technology and design details as the product assembly. (The price now paid for testing chips in the wafer is that circuit details of the product must be matched in test, which is neither accurate nor easy. A key issue in test is the location of the decoupling capacitors for power and ground.) In the interposer, without added design overhead, this is the same for test as for the product. Bare-chips testing introduces its own set of problems, but once solved it offers better long-term options.

The revised test scenario, as previously presented (1), is configured as the 'farm' of test heads shown in Fig. 9. The tester distributes a set of accurate 'edges' over a 50-ohm network to each head. Signals that would normally return to the tester, but would be loaded or slowed by the traversal, are instead analyzed on each test head by comparitor circuits fabricated in the same technology as the product. These may be integrated with the product chips as BIST circuits or alternatively be collected together in a Built-Out-Self-Test (BOST) chip that is co-located on the interposer.

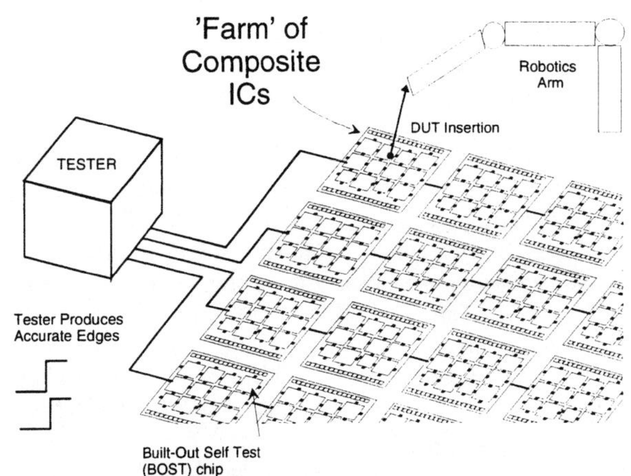

FIGURE 9. 'Farm' of test heads (1)

A key advantage is that clock speeds during test exactly equal product speeds. More importantly, current-generation CMOS is used in the tester-chip—even during technology development and prototyping. With conventional testers, it is becoming increasingly difficult to procure high-density, high-performance chips to test advanced chip designs because the custom chips used in testers are typically more than one generation behind. There is no alternative, given the lead-time for tester design and the low priority accorded to tester-chips at their low production quantities.

A major benefit using an 'in-house' test interposer populated with bare chips is a unified test/burn-in operation, which can handle the complete verification of the chip set with a single insertion. Because each test head in Fig. 9 represents a modest investment using mainly production parts, many can be duplicated to perform highly parallel testing. Time on the tester can then be extended to perform a thorough test, including burn-in. Such at-speed test information is fundamental to IC manufacturing control, particularly when clock speeds approach the GHz region and when production volumes are low (ASICs).

COMPOSITE IC

In the discussion so far, no mention has been made of how or even if this architecture can be implemented. We have examined architectural issues and made certain sweeping assumptions, such as bare chip testing, but we have left out the details of how an alternative interconnection technology can be introduced at this time, especially when previous attempts have failed, most notably MCM's.

The Composite IC (1) concept has been proposed for exactly this purpose, although more with an eye to truly wide-ranging integration of system components, such as sensors, analog circuits and passives—an 'integrated system.' This paper addresses mainstream digital systems, which need a greater degree of buy-in. As explained earlier, the IC wafer-fab community must first fully embrace this concept as the best alternative in moving mainstream technology along the Moore's-Law learning curve.

Figure 10 shows the major attributes of the Composite IC. There is the 'backplane' chip—called the intermediate (or MCM) substrate in our discussions so far—that supports several 'subsystem' chips. As seen from the second expanded view of this 6-level composite IC, these subsystem chips have only three levels of (local) interconnects in keeping with our earlier discussion of FEOL chips but are otherwise nearly conventional chips. The backplane chip is larger and has courser-pitch wiring than normal IC chips. It is fabricated with conventional three-level IC interconnect technology using projection aligners on 12" or 16" wafers.

Two processing operations based on MEMS (Micro-Electro-Mechanical System) technology are added to the normal IC complement. The first provides self-alignment features, while the second fabricates spring connections. A self-alignment chip is shown, somewhat schematically, in the first expanded view. By being permanently mounted on the backplane chip, it provides an inclined plane and vertical edge for registering the location of the subsystem chip.

A particular example of the MEMS-fabricated spring is shown in the second expanded view, with a fuller description being available in Ref. (1). This spring design is not immutable and other designs would work, provided they can be batch fabricated on the backplane chip in a manner that makes it convenient to use any subsystem chip in nearly unaltered form. It has two principal degrees of freedom, a large vertical travel and a small horizontal compliance. The vertical travel is incorporated to absorb manufacturing tolerances and ensure contact during test insertion. The horizontal compliance is needed to accommodate thermal expansion differentials once the spring is permanently soldered to the subsystem chip.

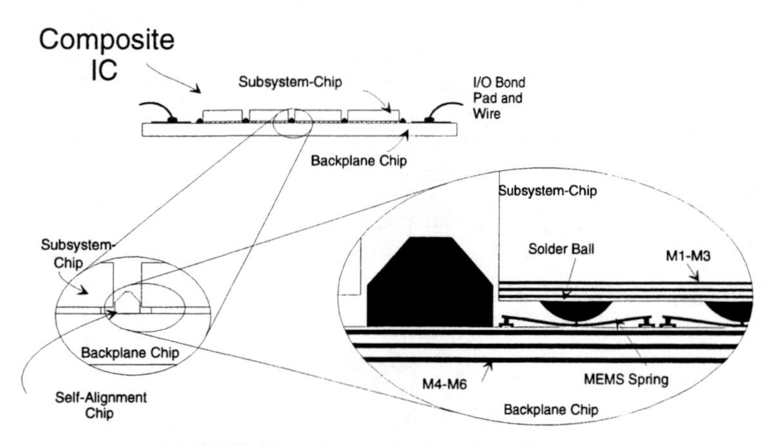

FIGURE 10. Major attributes of Composite IC (1)

The goal in spring design is to use only the 1 to 2 micron feature sizes of projection aligners, while still making an array of connections on a tight pitch. Ideally, they have enough force to break the oxide on the contact pad in test and enough vertical travel to press the chip against a heat sink. Another requirement is that, after chips finish being tested, a permanent solder connection is easy to achieve.

The basic plan of the composite IC is to provide enough intra-chip connections to the backplane chip to ensure its wiring density does not go underutilized. This increases the number of springs that must be provided, but with IC technology it can be assumed that these can be fabricated and soldered with high yield. During test, these intra-chip connections must be completed for the subsystem chip to function properly. The strategy of the composite IC is to test only the complete design, as illustrated in Fig. 9. In this way, the test operation only loads the CMOS drivers with the high-impedance circuit connections of the product and the corresponding exact electrical waveforms. By freeing test from the 50-ohm world of cabled interconnections, we can achieve our goal of unimpeded interconnections.

SUMMARY AND CONCLUSION

An in-depth analysis has identified the reasons for the large off-chip penalty of traditional interconnections and found them mainly rooted in the complex infrastructure of the IC and packaging businesses and not, as commonly thought, in the wiring capacitance of packages. Without upsetting the traditional channels of industry, a change has been proposed that only slightly alters the sequence of operations in IC manufacturing to disengage the delicate high-speed electrical connections between adjacent chips from the rigors of business-to-business protocols. This change is to remove this electrical connection from the robust low-impedance domain of conventional testers to the unimpeded signal-flow world of scaled CMOS. The change is made possible by modifying traditional MCM's to become composite-IC assemblies that are fabricated by the wafer-fab community. Packaging is reserved for the protection and heat removal of the whole assembly and interconnection of its perimeter circuits. Test has been given a premier position wherein the interconnect-bandwidth during test is the same as in the product. High-speed testing is an important requirement for manufacturing complex combinations of chips that must work together at GHz speeds. Compared with monolithically integrated circuits of the same complexity and diversity, multi-chip assembly greatly simplifies manufacturing, product design and test.

ACKNOWLEDGEMENTS

This investigation is one of long duration and the author thanks Dr. Hans Stork for supporting this latest segment. He is also indebted to Dr. William Lynch for his acceptance and encouragement of the fundamental concepts, for critical comments that have strengthened the arguments and for suggesting the name Chip-Scale Integration (CSI).

REFERENCES

1. D. J. Bartelink, "Integrated Systems," *IEEE Trans. Electron Devices,* vol. 43, pp.1678-1687, Oct. 1996.
2. D. J. Bartelink, "A unified approach to chip, test, assembly technologies for MCM's," in *Proc. '95 IEEE Multi-Chip Module Con.,* Santa Cruz, CA, Jan 1995.

Analytical Challenges in Next Generation Packaging/Assembly

Rajen Dias, Deepak Goyal, Shalabh Tandon and Gay Samuelson
Assembly Technology Development Quality and Reliability, Intel Corporation, Chandler Arizona

Next generation assembly/package development challenges are primarily thermal mechanical as interconnect levels increase and product performance drives the need for increased speed and power dissipation. The results of this trend present some distinct challenges for the analytical tool/technique set necessary to support this technical roadmap. The challenges in assembly analytical tool/technique development are in the areas of nondestructive imaging, board level fault isolation and materials property measurement to model and validate the thermal mechanical response of the assemblage. The purpose of this paper is to review the likely defects in next generation packages and the analytical approaches to identify these defects.

I. INTRODUCTION

Typical defects seen in C4 packages and multilayer organic boards are shown in Figure 1 and Figure 2 and consist of delamination, cracking and voids in interconnects and polymer composites. The analytical challenge is to expeditiously locate and understand the characteristics of these defects with a high success rate.

Some explicit examples of interconnect failures are presented in Figures 3-5. Figure 3 illustrates pad cratering, a typical defect of interest in printed circuit board (PCB) with metal defined pads. There is an inherent propensity to crack from the outer perimeter of corner BGA pads due to distance from neutral point stress effects. The crack can initiate as a result of delamination of the PCB dielectric from the copper pad and propagation occurs into the underlying dielectric as seen in Figure 3.

PCBs with solder mask defined pad lead to a re-distribution of stress that changes the fail mechanism from cratering to solder joint fatigue (1, 2). An example of such is provided in Figure 4 where the crack is seen to propagate cohesively across the diameter of the solder ball creating an electrical open.

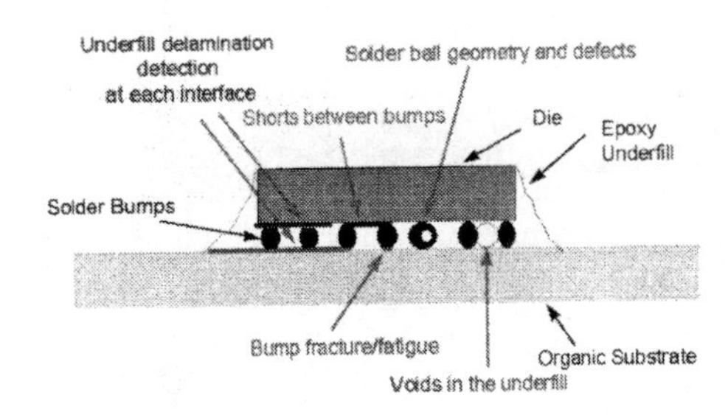

Figure 1: Typical defects in C4 interconnections that can be characterized by non destructive imaging techniques

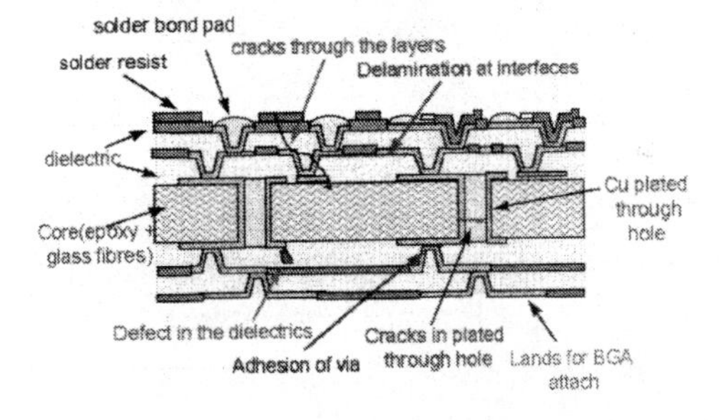

Figure 2: Typical defects in the organic board for which fault isolation and defect characterization techniques are required.

If there are board pad chemistry issues resulting in reduced adhesion, the stresses will manifest themselves as solder ball opens. This is yet a third fail mechanism of interest which is illustrated in Figure 5.

CP449, *Characterization and Metrology for ULSI Technology: 1998 International Conference*
edited by D. G. Seiler, A. C. Diebold, W. M. Bullis, T. J. Shaffner, R. McDonald, and E. J. Walters

Figure 3: Substrate insulator cracking around a BGA land resulting in a crater under the metal pad and subsequent electrical opens.

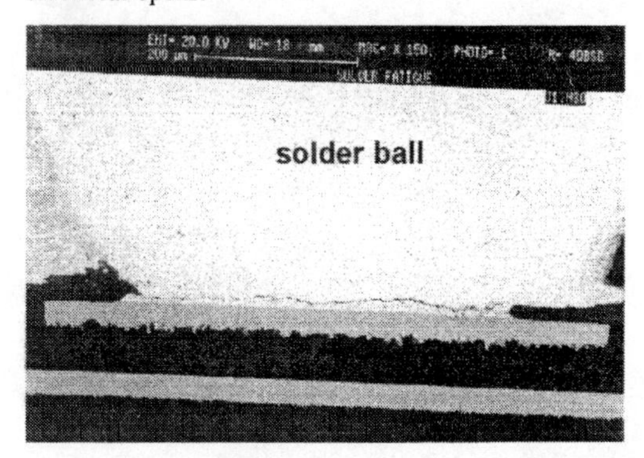

Figure 4: Solder Ball Fatigue in the solder mask defined land after thermo-mechanical stressing of the BGA resulting in electrical opens.

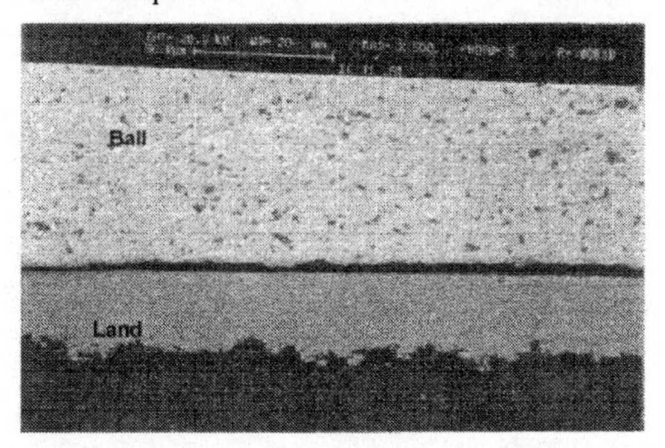

Figure 5: Solder Ball Lift from the metal land resulting in electrical opens.

Ideally, we would like to visualize all these defects non-destructively using next generation imaging tools.

II. NEXT GENERATION IMAGING

The driver in nondestructive imaging is the need for submicron (ideally <0.1 micron) resolution of defects in both x/y and z directions for multi-layer polymer/metal boards. The standard assembly imaging techniques of acoustics and x-ray radiography/tomography need to be extended into the next generation. Table 1 presents the current capabilities of the techniques extensively used while Table 2 discusses the capabilities of techniques applied in other industries but not currently used in microelectronic packaging failure analysis.

Driver	Current Capability				Required Capability
		Acoustics	Optical	X-ray	
Submicron defects in C4 & BGA; Multilayer Thick and polymer materials.	X/Y	>3 mils	1 μm	8 μm	<1 μm
	Z	<1 μm	1 μm	?	<1 μm
	Penetration	Good (?)	Poor	21 μm of Cu	Good
	Defect Location	Some	None	None	Good

TABLE 1 Currently used techniques, their capabilities and required capability to meet the imaging needs of the next generation packaging.

Driver	Technique	Available Capability		
		X/Y	Z	Penetration
Submicron defects in C4 & BGA;. Multilayer, thick and polymer materials	Laser Ultrasonics	>100 μm	?	- no issues
	X-ray Tomography	8 μm	?	21 mm of Cu
	Thermal Imaging	200μm	200μm	Poor
	Terahertz	100 μm	<100mils	- no issues
	MRI	10μm	10μm	Non-metallic materials

TABLE 2 Additional techniques and their capabilities as applied to microelectronic next generation packaging.

Current acoustic techniques featuring transducers of ~200MHz are able to resolve 10-15mils in spatial x/y and ~.1micron in the z direction. Unfortunately at these high frequencies, the penetration depth is limited to less than a millimeter by the highly attenuating organic materials in currently designed multi-layer interconnect structures. There are some indications that better understanding of transducer design for specific applications may be the key to improved resolution.

Transducers will need to be designed for optimum spot size and depth of focus at specific locations in the organic boards.

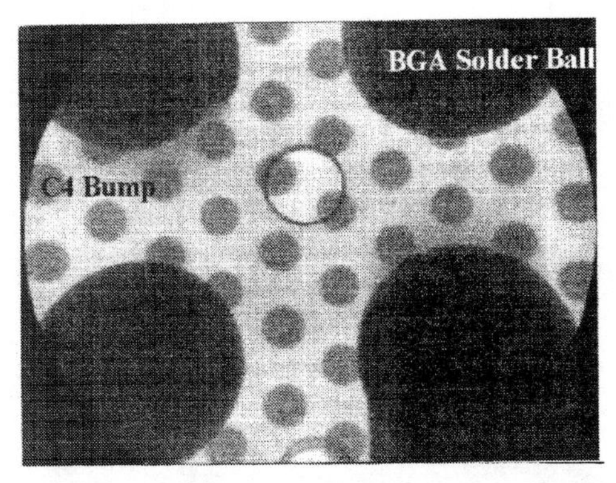

Figure 6: Low Magnification X-ray radiograph showing current X-ray radiography capability to image C4 bumps and BGA solder balls in a package mounted on a PCB.

For x-ray techniques, goals of 1 micron maximum spatial resolution and z resolution with geometric magnifications of >1000X appear to be attainable based on theoretical specifications from several well known vendors in the industry. The trade-off is reduced spatial resolution with increased penetration capability. Ideally the kV range should be 10-225 or more with maximum possible current to allow

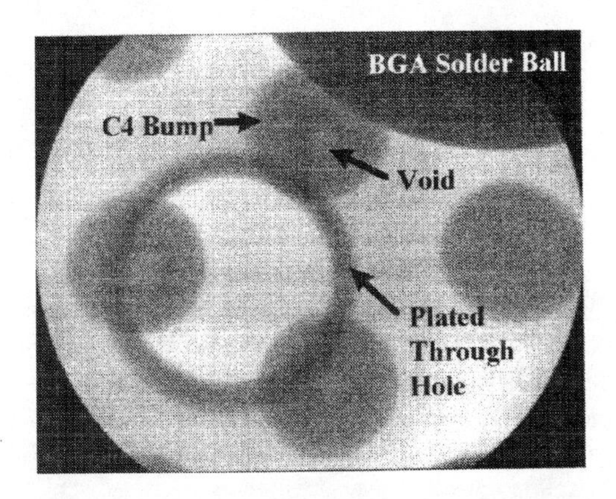

Figure 7: High Magnification x-ray radiograph showing current X-ray radiography capability to image C4 bumps and BGA solder balls in a package mounted on a PCB. Note the presence of voids in the solder balls. No other features could be resolved.

penetration through the thermal solutions attendant to high performance packaging. In addition, there should be good contrast resolution i.e. >256 grey/color scale and complete image manipulation capability such as filtering, density plots, multi-image layout, 2D/3D imaging and rotation. Figure 6 and Figure 7 show current x-ray radiographic capabilities for observation of the defects in the C4 bumps in a BGA package surface mounted to a PC board.

Another emerging need is for CAD navigation of a stage that will allow precise defect location that can then be translated into other tool stages to precisely characterize the defect physically once it has been identified non-destructively. The difficulty is providing the imaging industry with sufficient motivation to invest the necessary R&D funds to insure the next steps in assembly imaging can be taken.

This leads to an alternative avenue of needed investigation which is board level fault isolation and failure analysis.

III. BOARD LEVEL FAULT ISOLATION AND FAILURE ANALYSIS

Fault isolation and failure analysis of high performance packages and modules becomes increasingly challenging as the number of components on the module increases. A typical high performance module card could contain over 100 components on an organic board having over 10 layers of interconnects. Often, the throughput time and success rate of fault isolation and root cause failure analysis dictate the pace at which manufacturing and reliability concerns can be identified, addressed and validated. This section discusses six challenging areas in board level fault isolation and failure analysis : design of board level test structures, dielectric performance, time domain reflectometry, circuit edits, board disassembly and destructive physical analysis.

a. Design Of Board Level Test Structures

The key to expeditious development of a high performance board lies in understanding potential weaknesses in the thermal and thermo-mechanical performance of the multilayer board and components on the board. Finite element analysis modeling together with an in depth knowledge of the board and component material properties, component assembly processes and field reliability requirements are used to identify potential performance concerns that need to be evaluated. Once this is done, appropriate test structures can be designed and fabricated into the components and board at high stress locations.

Test structures can be designed specifically to understand the effects of variations in material properties, board and assembly process variations and changes or specific field reliability acceleration stresses. The test structures should be designed, located and routed to test points so as to simplify subsequent fault isolation and failure analysis. It is important that multiple failures in a variety of test structures can be sequentially analyzed on a single board because of the high cost and availability of the developmental boards.

An example of a typical test structure used for evaluating field reliability performance of ball grid array (BGA) interconnection between the component and the board is a daisy chain connection of BGA bumps. The high stress location would be at the maximum distance from the neutral stress point which is typically at the component corners. The bump chain should be in a "L" shape around the corner of the component and test points should be routed on the board surface from the ends of the bump chain as well as from the corner bump. If the bump chain is designed to be long then test points should be taken at frequent intervals along the chain. This will aid in destructive physical analysis of the failures in order to determine the root cause of the bump failure.

b. Time Domain Reflectometry

Time Domain Reflectometry (TDR) has been used for detecting location of opens and shorts of signal lines on printed circuit boards. The nondestructive technique involves sending a high frequency electrical pulse along a signal line and measuring the times of the reflected pulses. By comparing the signatures of the reflected pulses from a known good signal line to that of a defective line the location of the open or short defect can be measured and is illustrated in Figure 8 and Figure 9. The technique works well when the signal lines are long with few discontinuities. The technique becomes more challenging on advanced printed circuit boards that have multiple layers connected with through hole or via interconnects between the layers. The short signal lines and the numerous interconnects between layers makes analysis of the response signal signature difficult to interpret with or without the presence of defects. Multiple reflections from the vias and signal variability between units must be factored into the analysis. The long time and effort required to characterize each signal line makes this technique unappealing to the failure analyst.

The full potential of TDR can be realized by addressing these limitations. Elaborate software programs similar to those of speech recognition need to be developed to analyze the complex response signatures from each signal line, compare signatures of normal and defective signal lines and use statistical analysis to give a probabilistic estimate of the physical location of the defect on the line.

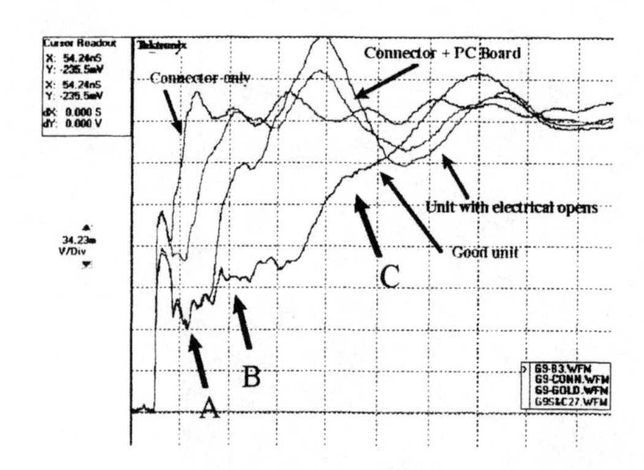

Figure 8: TDR signature of a good circuit and an open circuit at a point between location B and location C shown in the schematic on Figure 9

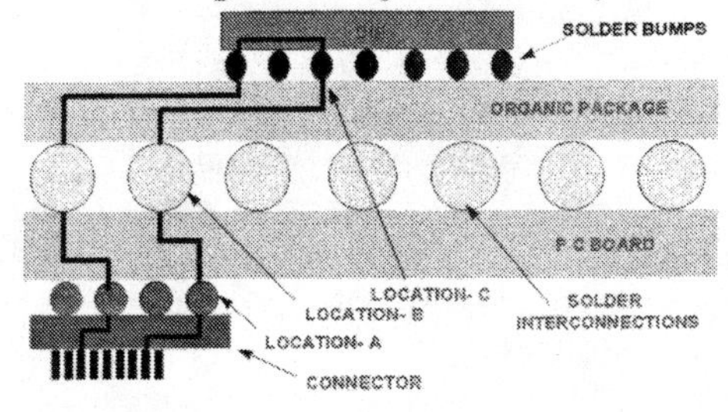

Figure 9: Schematic of a circuit loop that was tested on the TDR The open failure was detected between location B and location C.

c. Physical Analysis Capability

Once fault isolation has been successful, destructive physical analysis (DPA) follows to understand the characteristics of the defect in order to determine its root cause. Four main DPA methods; component removal, reactive ion etching, edit and repair capability and metallography are discussed below.

i. Component Removal

Defective components are normally removed by reflowing the solder joints that connect the component to the board. The thermal stress of the reflow operation can alter the defect or produce additional damage that make root cause analysis more complicated. For complex boards, non artifactual disassembly of components and testability of the components is critical. Tools and techniques need to be developed for non thermal removal of BGA components. Thin blade saws and wire cutters commercially available can result in damage or artifacts that complicate failure analysis. Benign methods to remove or cut the solder interconnections and still allow electrical test of the components need to be developed.

ii. Reactive Ion Etching (RIE)

RIE techniques used routinely for semiconductor device fault isolation and failure analysis are beginning to be used in package and board level analysis. With proper optimization of selected gases and radio frequency (RF) power, the organic insulation layers can be selectively removed revealing a 3D metal interconnect structure that can be optically inspected for defects as shown in Figure 10. Additional advantages of this technique are that the component does not need to be removed from the board, can still be electrically tested and sections of exposed interconnects can be electrically probed. One disadvantage of the RIE technique is that it is slow when etching boards that are greater than 1mm thick. Increasing the reaction rate by using higher RF power results in overheating artifacts. More research is needed in formulating the gaseous mixtures for rapid etching of the insulator materials.

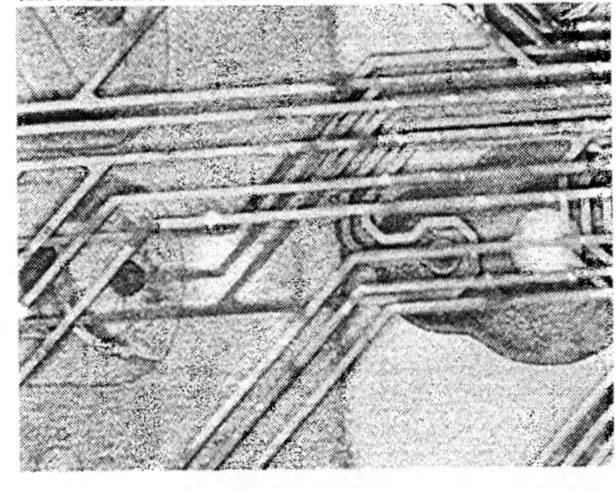

Figure 10: RIE etch of a multilayer organic board allows 3D viewing of fine metal interconnect structures which can be inspected for defects and probed for electrical continuity.

iii. Repair And Edit Capability

When defects are detected it is imperative to determine if the defect is the true cause of the failure. This can be achieved by developing a repair and edit capability that involves connecting a new metal line around the defect. Electrical test after the repair is used to confirm if the defect is responsible for the failure. Methods to connect a metal line are not trivial; they must make sure that the connected metal line has low contact resistance, does not short other interconnect lines and does no damage to the rest of the component or board. Chemical vapor deposition (CVD) and sputter deposition techniques are two approaches that should be investigated. The repair and edit capability can also be used to isolate the failure when the failure occurs in a region involving a large number of via interconnections between the layers on the board.

iv. Metallography

Standard metallographic techniques such as planar and x-section analysis are used for advanced board level analysis. The techniques are well established for single and dual level packages but can result in artifacts on complex multilayer packages and boards in which the residual stresses are high due to the sandwich of vastly dissimilar materials and the assembly processing effects. X-sectioning components or boards using traditional diamond impregnated saw blades can result in device and insulator cracking or delamination which makes defect analysis and root cause determination difficult. New approaches that minimize additional mechanical stress on the multilayer boards with high residual stresses will need to be investigated and optimized for addressing artifacts and reducing the analysis through put time.

The remaining area of major concern is metrology to determine materials properties that establish and validate accurate finite element analysis models.

III. MATERIAL PROPERTIES AND MODELING VALIDATION

a. Material Properties

There is widespread use of polymeric composites in packages. Although inexpensive and easily processable, polymeric materials still are not entirely reliable given their viscoelastic behavior. The state of stress of a viscoelastic body (polymer composites) changes with time and temperature, either via stress relaxation (constant strain) or creep (constant load). These materials are generally used as multilayer coatings (thickness range 1 - 30 μm) where the presence of intrinsic defects and electrical circuitry (metal traces, pads and vias) within the package increases the stress in the material leading to premature failure. Given the complexity of packages, it is increasingly important to rely on predictive modeling techniques to pinpoint potential problems that may be encountered in package technology. Furthermore, the modeling predictions have to be validated and correlated with actual failures observed.

Accurate predictive modeling requires that the constitutive [polymeric] properties such as the temperature dependent moduli (shear and tensile) and Poisson's ratio be accurately known. To further understand the thermo-mechanical response of the composite, its coefficient of thermal and humidity expansion, yield behavior, strength and fracture toughness should be known. Experience shows that the polymeric materials in packages usually fail well below their intrinsic strength. The presence of traces, foreign materials, vias and various defects lead to stress concentration effects that initiate or propagate cracks in the material. Although well defined metrologies to determine properties of bulk polymeric materials exist, reliable techniques need to be developed to measure these properties for thin, brittle polymeric composites that predominate in next generation packages.

Poisson's ratio is a case in point. It measures the ability of a material to change dimension in the lateral direction while the material is extended in an orthogonal direction. For isotropic materials, metrologies such as beam bending or the use of strain gauges (bulk material) allow evaluation of this parameter. However, for anisotropic thin films/coatings, this direction dependent parameter is not easily determined.

Tensile and shear moduli of brittle & thin polymeric composites also pose a challenge. Current techniques such as dynamic mechanical thermal analyzers (DMTA) can measure the stiffness of polymer composites as a function of temperature and frequency (strain rate) and convert that to a modulus value. The advantages of such techniques is the environmental control and the small sample size requirements. The limitations, however, are a small operational frequency range and the load transducers available that only allow extremely small strain ($\approx$0.02%) for high modulus materials. Thin, brittle polymeric composites pose even a greater challenge. Preparation of edge-defect free samples (typical dimensions: 12mm X 4mm X 0.02mm) becomes almost impossible and the gripping force of the instrument clamps usually cracks the material at the edges, leading to erroneous results. Hence, non contact techniques to measure moduli will be more appropriate for thin brittle films.

Coefficient of thermal expansion (CTE) is an important parameter for modeling of thermo-mechanical stresses. Current techniques, such as a thermomechanical analyzer (TMA) allow determination of linear CTE for bulk isotropic materials. For thin films of anisotropic composites, determination of the out-of-plane (through the thickness) CTE measurement is a challenge. The thickness dimensions are too small (10-30 μm) to accurately monitor the dimensional changes due to expansion. Alternate techniques exist to obtain the out-of-plane CTE, but they are indirect and laborious techniques (3). Given these limitations with the current techniques, there is a strong need for non-contact techniques that will allow characterization of thin brittle polymeric composites.

Predictive modeling based on accurate material properties allows understanding of potentially stressful areas in a package before empirical data can be collected. High stress regions in the package can be targeted and improved to withstand stresses (residual or externally generated) arising from CTE mismatches in the layered materials. To understand the margins available with the polymeric materials under applied stresses, properties such as strength, yield behavior, ductility and toughness become important. Knowledge of strength, toughness and ductility helps determine whether the polymer layer will fail under the applied stress.

Any cracks in the dielectric layer can lead to damage to the underlying metal traces and subsequent electrical failure, as shown in Figure 11 and Figure 12. The figures show the crack in the dielectric propagated through the thickness of the dielectric to damage the underlying metal trace.

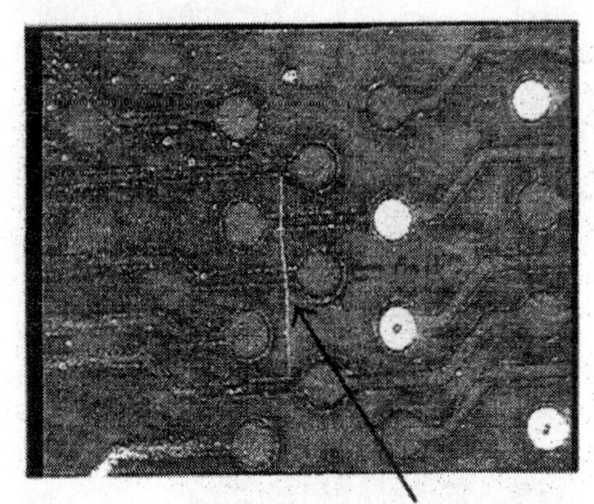

Figure 11: Optical image of a crack (white vertical line, magnification 100X) in the dielectric with underlying metal traces.

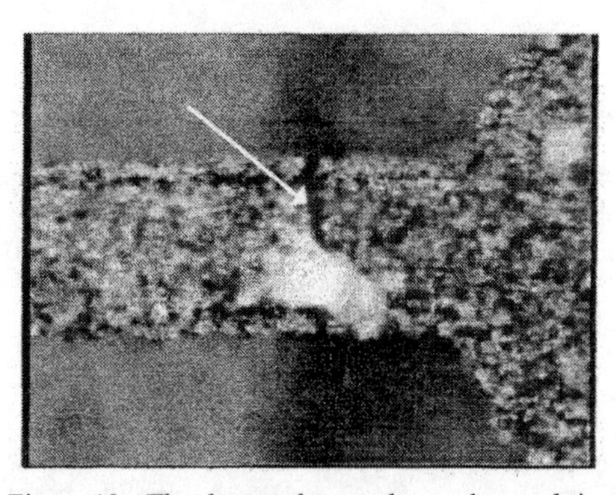

Figure 12: The damaged trace due to the crack in the dielectric material (magnification 1000X), as shown in Figure 11.

Existing metrologies such as micro-tensile testing allow characterization of strength and yield behavior for small samples. However, they are not very reliable for thin brittle films due to sample preparation limitations. Any defects in the form of nicks and scratches in the samples lead to artifactual cracks and yield compromised data. Reliable techniques are needed to address these material property gaps.

Even though the properties of individual polymeric materials may be known in a standalone mode, they may not reflect the true condition after the processing and stressing of the package is complete. Adhesion strength is a prime example. The adhesion strength of virgin materials may not be accurate when applied to composite packages. Processing conditions such as exposure to high temperatures and moisture will compromise the ability of materials to adhere since adhesion is susceptible to the presence of contaminants. In situ measurement techniques would be ideal to understand the post-processing and post- stress adhesion strength of polymeric materials. Unfortunately, metrology for this insitu adhesion is lacking at this time.

Predictive modeling is becoming crucial to understanding of weaknesses in packages prior to assembly and empirical data collection. Accuracy of these predictions is solely dependent upon use of accurate constitutive properties of polymer composites used in the packages. Hence, there is a dire need for reliable techniques to measure the aforementioned properties for polymer composite films and coatings.

b. Model Validation

Moire (4) and digital image correlation (5) are two key areas of focus for validating the strain fields predicted by finite element analysis modeling.

Digital image correlation is a technique that allows images to be taken pre and post loads which can be thermal, mechanical and/or chemical.

Digital image correlation software limits resolution to the digital imaging tool that is being used. Examples are optical microscopes or SEM. The benefit of the technique is that it can be done nondestructively, real time and with relatively large samples that are typically encountered in packaging failure analysis.

The other technique of choice is Moire where a grating of defined mesh is affixed at a known high temperature to a cross-sectional sample of interest. The distortion of the grating upon cool down is determined by two laser beams that strike the mesh grating at specific angles. The interference of the laser beams creates fringes whose spacing correlates to the strain field in the sample. Standard in plane Moire resolution is .42 microns/fringe whereas out of plane shadow IR Fizeau is 5.3 microns/fringe.

IV. CONCLUSIONS

It is clear that industry focus on development of acoustic and x-ray imaging tools and board level fault isolation tools needs to happen to prepare for next generation microprocessor and microcontroller technologies. Another avenue of need is correct materials properties for thermal mechanical predictive models. Attendant to this is the need to understand what stress level is problematic and so adhesion tools, fracture toughness tools all need to be evaluated. And lastly, there is a need for model validation tools beyond empirical studies (which are also necessary). Typical model validation tools of interest are Moire and digital image correlation.

IV. REFERENCES

1. S. Bolton, A. Mawer, and E. Mammo, "Influence of Plastic Ball Grid Array Design/Materials upon Solder Joint Reliability", The International Journal of Microcircuits and Electronic Packaging, Vol. 18, Number 2, P109, 1995

2. A. Mawer, S. Bolton and E. Mammo, "Plastic Ball Grid Array Solder Joint Reliability Considerations', Proceedings of the 1994 Surface Mount International Conference, P239, 1994.

3. Shalabh Tandon, "Modeling of Stresses in Cylindrically Wound Capacitors: Characterization and the Influence of Stress on Dielectric Breakdown of Polymeric Film", Ph.D. Dissertation, Department of Polymer Science & Engineering, University of Massachusetts, Amherst, 1997.

4. D.Post, B. Han, P. Ifju, "High Sensitivity Moire", Springer-Verlag.

5. H.A. Bruck, S.R. McNeill, M.A. Sutton, W.H. Peters III, "Digital Image Correlation Using Newton-Raphson Method of Partial Differential Correction", Expt. Mech. Vol 29, P. 261-267, 1989.

Trends in Nondestructive Imaging of IC Packages

T.M. Moore and C.D. Hartfield
DSPS Packaging Development
Texas Instruments, MS 921
13536 North Central Expressway
Dallas, TX 75265
moore@ti.com; hartfield@ti.com

Since the industry-wide conversion to surface mount packages in the mid-1980's, nondestructive imaging of moisture induced delaminations and cracks in plastic packaged ICs by scanning acoustic microscopy has been a critically important capability. Subsurface imaging and phase analysis of echoes has allowed scanning acoustic microscopy to become the primary nondestructive technique for component level inspection of packaged ICs and is sensitive to defects that are undetectable by real time x-ray inspection. It has become the preferred method for evaluating moisture sensitivity, and for many package processes, provides more reliable detection of wire bond degradation than physical cross sectioning or conventional electrical testing. However, the introduction of new technologies such as ball grid array (BGA) and flip chip packages demands improvements in acoustic inspection techniques. Echoes from the laminated substrates in BGA packages produce interference problems. Phase inversion detection is an important advantage of pulse-echo imaging of molded surface mount packages. However, phase inversion is not always helpful for delamination detection in these new packages, due to the properties of the materials involved. The requirement to nondestructively inspect flip chip interconnect bumps has arisen. Alternative approaches such as through-transmission screening of BGAs and high frequency (>200MHz) pulse-echo inspection of flip chip bumps are addressing these new issues. As the acoustic frequency approaches the limits dictated by attenuation, new methods of frequency-domain signal analysis will become important for routine inspection and may give acoustic microscopy a predictive capability.

INTRODUCTION

The semiconductor electronics market continues to demand an increase in circuit density and functionality per unit area. Greater numbers and types of personal electronic assemblies are being developed which are smaller, faster, lighter and more capable than their predecessors of only a few years ago. In this environment of low cost miniaturization, IC packaging is rapidly developing into new technologies such as BGAs, flip chip interconnect and direct chip attach.

The field of IC package characterization is swiftly evolving to satisfy the inspection needs of these new package technologies. The increase in the number of metal layers on the die in wire bonded product, as well as the growing importance of flip chip assembly, have encouraged advances in semi-destructive backside inspection techniques such as photoemission microscopy and infrared techniques, as well as nondestructive x-ray radiography and high frequency pulse-echo acoustic microscopy. Advances in x-ray imaging include finer resolution and better penetrating power, and improved post-processing of the images. Acoustic inspection offers several advantages over x-ray radiography including sensitivity to important package defects such as internal cracks and disbonds, which are practically invisible during x-ray inspection. A unique feature of pulse-echo acoustic inspection is the detection of delaminations at internal interfaces based on the inversion of the pressure pulse (180° phase shift) at the delamination (1). Scanning acoustic microscopy is a critical tool which enabled rapid surface mount package development, and continues to be an important tool for packaging process control and development.

BACKGROUND

Although acoustic absorption in particle-filled mold compounds has always been a challenge for acoustic inspection, the geometry of conventional leaded surface mount packages is ideal for pulse-echo inspection. The interfaces of interest lie along a common subsurface plane parallel to a featureless and flat package surface. The acoustic impedance transition from the mold compound to the die or metal leads was always such that delaminations at these interfaces were easily detected by the phenomenon of phase inversion of the echo pulse. Traditionally, the acoustic frequencies needed to detect critical package defects fell within the commercially available range of 15-150MHz.

IC manufacturers have depended on pulse-echo acoustic inspection for package development and process control since the beginning of an industry-wide conversion to surface mount packaging in the mid-1980's. The acoustic microscope offers sensitive imaging of moisture and thermal-induced damage. Defects such as cracks, delaminations, and voids can be identified rapidly and non-destructively. During surface mount package development, it was a widely held belief that wire bond degradation was associated with package cracking in temperature cycle failures of large die surface mount packages. Scanning acoustic microscopy identified die surface delamination, not package cracking, as the primary cause of wire bond degradation, which was shown to occur even in the absence of package cracks. For many package processes, the detection of wire bond degradation by acoustic microscopy proved more reliable than by other techniques. Acoustic microscopy does not induce new damage, as physical cross sectioning sometimes does. Delaminations are easily detected with acoustic microscopy, but practically invisible with real time x-ray. Also, acoustic microscopy is more sensitive to opens caused by delamination, which are not always detected by conventional electrical testing. Due to its non-destructive nature, sensitivity, and speed, scanning acoustic microscopy rapidly became the preferred method for determining moisture sensitivity (2-4).

In response to the introduction of new package styles, significant changes have occurred in the way acoustic microscopy is applied to the inspection of IC packages. The introduction of laminated substrate ball grid array (BGA) packages and flip chip assemblies has forced an evolution in both the transducer lens geometry and frequency used in acoustic inspection. The advent of finely layered BGA substrates has led to the use of through-transmission screening of BGAs (5). Also, the need for nondestructive inspection of flip chip interconnects has driven the acoustic frequency beyond 200MHz and toward the limits imposed by absorption in the coupling fluid and the sample. As the need for better depth resolution becomes more important, frequency domain analysis approaches will have to be implemented. These advances are discussed, and recommendations are made on how acoustic inspection must evolve to meet the needs of package designers through the next decade.

THROUGH-TRANSMISSION INSPECTION OF BGAs

Through-transmission acoustic inspection is a well known technique. Early in the application of acoustic microscopy to IC package inspection, pulse-echo acoustic analysis became the preferred method because of convenience, delamination detection by the phase inversion method, and the ability to correlate defects to specific interfaces. However, through-transmission acoustic inspection is advantageous for the evaluation of BGAs. These packages have a complex architecture which complicates the pulse-echo analysis setup and can lead to incorrect data interpretation. In some cases, even when set up properly, the pulse-echo technique will miss defects in BGAs that can be detected by the through-transmission technique (5).

The mechanical arrangements for pulse-echo and through-transmission acoustic inspection are diagramed in Fig. 1. Although similar transducers are used in both cases, in through-transmission inspection, two separate transducers are used for transmitting and receiving the acoustic pulse. When the receiving transducer is a focused transducer, its focus is set independently from that of the transmitter. The transmitter and receiver can be different types of transducers to optimize performance due to absorption and refraction in the sample and coupling fluid. Both transducers must be oriented and focused (assuming a focused receiver is used) for best results. In through-transmission imaging, depth information is lost, but only one scan is required to find a defect at any depth within the sample. Spatial resolution in the sample plane and penetration (signal/noise) are the most important characteristics of the transducers, and there is no need for very high bandwidth (short duration) pulses as exists for pulse-echo imaging.

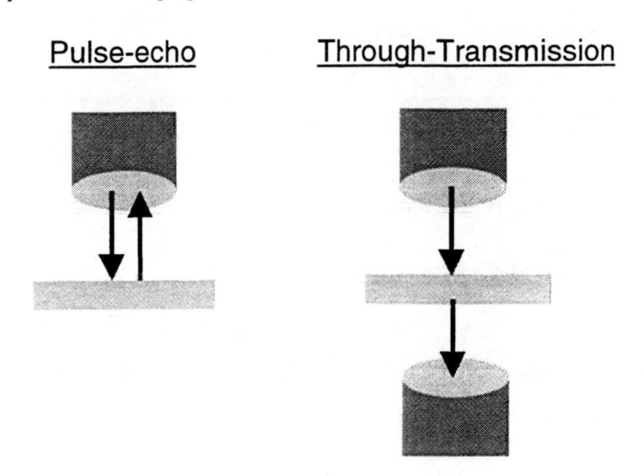

FIGURE 1. Sketch of pulse-echo and through-transmission inspection geometries.

Different BGA package types exist, such as wire-bonded over-molded BGAs and cavity-down BGAs. BGA packages are characterized by their laminated substrates and solder ball interconnection to the circuit board. Figure 2 better shows the multi-layered nature of a cavity-down BGA. The laminated layers of the BGA substrate include layers of patterned metal, fiber reinforced organic polymer, adhesive film, solder mask,

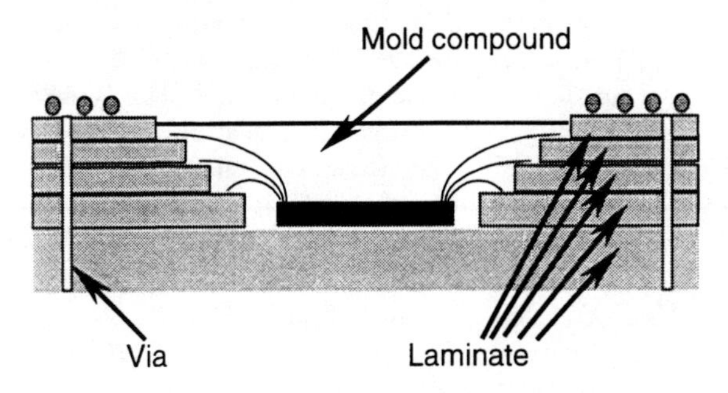

FIGURE 2. Sketch of the laminates in a cavity-down organic BGA package (inverted).

and metal interconnecting vias. Due to the velocities of sound and thicknesses of these materials, this complex layered structure produces acoustic resonance periods that range from 7ns for thin Cu traces to 160ns for individual polymer laminates. The multiplicity and thinness of these layers results in overlapping echoes that can constructively and destructively interfere with each other. Also, a continuum of reflections from the fibers in the laminate occurs. This can hinder proper pulse-echo acoustic microscopy setup and produce misleading data. Interface areas that have relatively few layers, such as die attach, are not excluded from these concerns. The die attach region revealed by the cross-section of a cavity-down BGA is shown by the SEM image in Fig. 3. Evident are the many reflecting layers beneath the die including the die attach layer, solder mask, die pad, and the fiber weave in the base laminate (substrate).

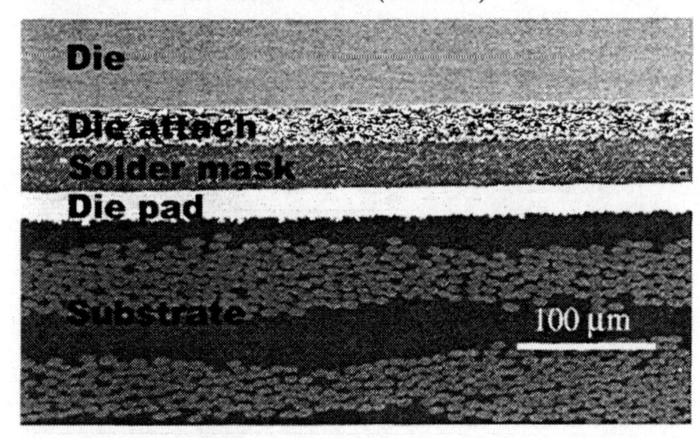

FIGURE 3. Scanning electron microscope image of the die attach region and base laminate in a cavity down BGA (inverted).

Only limited inspection opportunities by the pulse-echo technique exist in this laminated substrate package. Inspection from the base laminate side is not possible due to echo interference problems and signal attenuation and scattering. Even the apparently straightforward inspection of the mold compound interfaces in this package is complicated by the relative acoustic impedances of the mold compound and laminate. One of the primary advantages of pulse-echo acoustic imaging of plastic surface mount packages has been the phenomenon of phase inversion at delaminations. The acoustic pressure pulse echo inverts in phase at an interface where the acoustic impedance changes from higher on the incident side to lower on the transmitted side. This is the case in molded plastic packages at an internal delamination: the acoustic impedance of the air gap (essentially zero) is much lower than that of the plastic mold compound (Z_{MC}=6.8x10^{6}kg/m^{2}s). Interfaces such as the mold compound/die and mold compound/metal interfaces are low-to-high impedance transitions and produce no phase inversion at a bonded location (Z_{Si}=20x10^{6}kg/m^{2}s, Z_{Cu}=42x10^{6}kg/m^{2}s). As a result, phase inversion detection assists in the discrimination of delaminated interfaces from bonded ones.

However, the acoustic impedances of many organic laminate polymers are less than that of mold compound. For example, the acoustic impedance of the BT laminate in the package in Fig. 3 is roughly 4x10^{6}kg/m^{2}s for a 50% fiber loading (the acoustic impedance of the resin at the interface is less). So the transition in impedance at the mold compound/laminate interface is high-to-low (as in the case of a delamination) and the entire interface area will produce an inverted echo signal, delaminated or not.

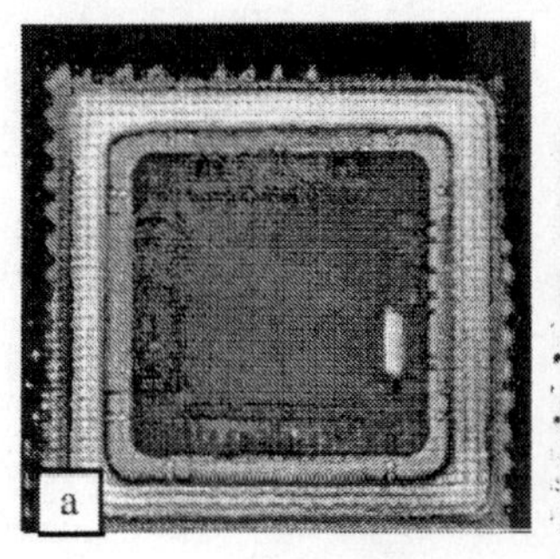

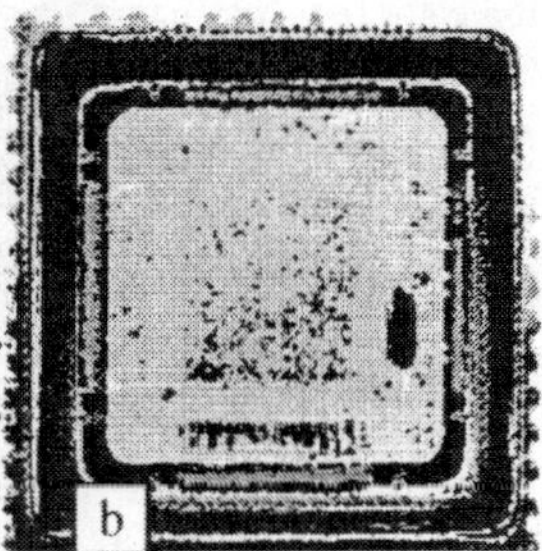

FIGURE 4. (a): pulse-echo intensity image at a mold compound/BT laminate interface, and (b): "delamination" image (with areas of phase inversion in black), and (c) the through-transmission image (true delaminations in black) (5).

An example of the confusion this can cause is shown in the acoustic study of a cavity down BGA in Fig. 4. Figure 4(a) shows the pulse-echo intensity image at the mold compound/BT laminate interface. Figure 4(b) shows the areas of phase inversion in black. The majority of the mold compound/BT laminate interface reports an inverted echo pulse. This can easily be misinterpreted as indicating delamination. However, the through-transmission image in Fig. 4(c) clarifies the situation and verifies the integrity of the mold compound/BT laminate interface. In addition, Fig. 4(c) reveals delaminations within the BT laminate that were not discovered in the pulse-echo images.

The echo interference problem described for the stacked laminates in the cavity-down BGA in Fig. 2 is also a problem for inspections beneath the die of any laminated substrate package, including the wire-bonded over-molded BGAs and flip chip BGAs. Constructive and destructive interferences can mislead the microscope operator to incorrectly select what appears to be a large internal reflection, or to overlook the correct reflection that has been almost completely canceled.

Figure 5 shows the pulse-echo image presumably of the die attach layer in a cavity-down BGA having the same structure as revealed by the cross section in Fig. 3. This image shows the characteristic wagon wheel shape of the die pad. Since delaminations reflect more signal than bonded areas (no transmission), they usually appear brighter in the pulse-echo image. Thus, Fig. 5 would likely be interpreted as showing that the majority of the die pad is delaminated.

FIGURE 5. "Die attach" image obtained when choosing the first noticeable subsurface echo below die surface (5).

However, the echo chosen to produce the image in Fig. 5 was not the correct die attach echo as intended, but was a partially overlapping echo from deeper in the package (solder mask layer) which was more prominent due in part to constructive interference. Through-transmission inspection clarifies the results of this pulse-echo study. The through-transmission image in Fig. 6 shows where delaminations, detected by the loss of transmitted intensity (dark areas in Fig. 6), truly occur.

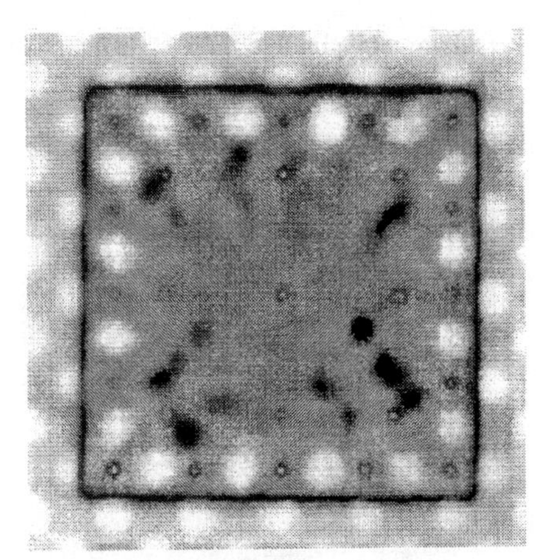

FIGURE 6. Delaminations appear black in this through-transmission image and should correlate with the correct die attach delamination pulse-echo image (5).

The pulse-echo inspection was revisited, and it was found that the correct echo was almost completely cancelled at bonded locations due to destructive interference. The correct die attach delamination image provided by this echo is shown in Fig 7.

FIGURE 7. Pulse-echo image of the die attach layer when the correct echo is chosen (5).

Pulse-echo and through-transmission acoustic inspections are complementary techniques. The above findings demonstrate that through-transmission inspection provides rapid and reliable screening of BGAs. In appropriate cases, pulse-echo inspection can still be used to acquire additional information.

HIGH FREQUENCY PULSE-ECHO IMAGING OF FLIP CHIP ASSEMBLIES

Process developers and failure analysts are demanding new methods and tools capable of analyzing IC's from the backside of the die. An increase in the number of metal layers and metal coverage, as well as the emergence of flip chip assembly, are making backside analysis more desirable. One of the main drawbacks associated with flip chip assembly is the difficulty in inspecting the device side of the die. Package related reliability issues include solder bump connection integrity and underfill defects. A schematic of a typical chip-scale flip chip BGA package is shown in Fig. 8 showing the die, underfill and multi-layer organic substrate.

Underfill delaminations, voids, density variations, and filler particle settling can all be imaged using pulse-echo acoustic microscopy and transducers that are commercially available for thin molded package inspection in the 50-110MHz range. However, imaging flip chip solder bumps and detection of solder bump defects has demanded an improvement in resolution. Transducer lens design, efficiency and center frequency are evolving to meet this need. Currently, transducers are available for flip chip bump inspection with center frequencies in the range of 110-250MHz. Both the pulser and receiver electronics must have sufficient bandwidth to take advantage of these very high frequency transducers. Frequency-dependent absorption of the broadband acoustic pulse, primarily in the coupling fluid (water), will determine the upper limit for transducer frequency. Preferential absorption of higher frequency components shifts the center frequency of the echo pulse to lower frequencies. Several

FIGURE 8. Schematic of a chip-scale flip chip BGA package.

conventions exist for labeling the transducer according to the frequency spectrum of its ideal output pulse. However, it is the frequency spectrum of the returning echo pulse that affects the image resolution (both depth and spatial). Spatial resolution in the image is a better metric of instrument performance than transducer frequency because many other system components significantly impact image resolution. This is similar to how image resolution is the metric for the performance of a scanning electron microscope, for example.

Figure 9 is an optical image from an unmounted, bumped flip chip die that has several bump defects. Excessive solder has coalesced and shorted three bumps, one bump has a reduced amount of solder, and a third shows thin film cracking at the pad (and no solder). Figure 10 shows the corresponding pulse-echo acoustic image acquired with a transducer that provided a 75MHz return echo center frequency, focused through the die to the bumps. Both the excessive solder and the missing solder bump defect sites were detected. The bump site with a reduced amount of solder appears normal because the footprint of the solder on the pad is similar to that of a normal bump.

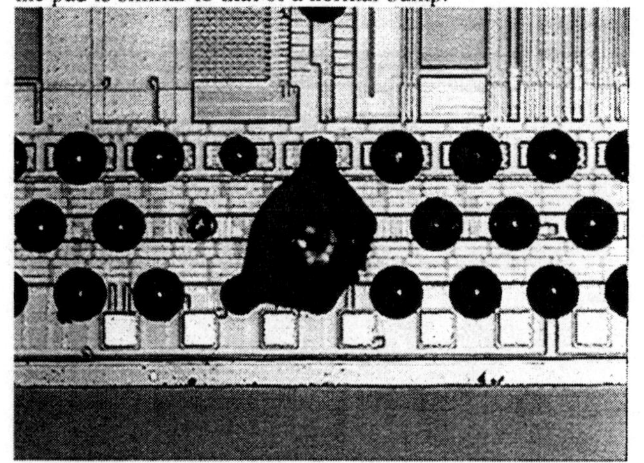

FIGURE 9. Optical image of a bumped flip chip die with solder bump defects (bump pitch = 225μm).

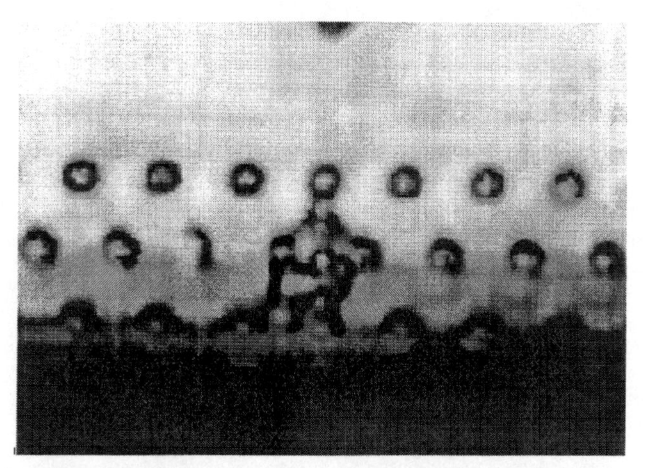

FIGURE 10. Pulse-echo acoustic image of the solder bump defects shown in Fig. 9 (flipped L-R for easier correlation to Fig. 9).

Despite the use of higher frequencies, distingishing echoes from both sides of a flip chip bump is not always possible. One can calculate the required temporal resolution for a material of a given thickness by the formula

$$(d/v) \times 2 = t \quad (1)$$

where d = thickness, v = velocity of sound, and t = roundtrip time of the echo. Figure 11 shows the frequency required for various pulse durations to resolve echoes for bumps of various heights. A solder bump with a post-assembly height of 100μm will require a temporal resolution of approximately 40ns to resolve the top and bottom echoes (t = 40ns). Assuming a pulse duration of 2 periods, this is equivalent to a return echo center frequency of 50MHz.

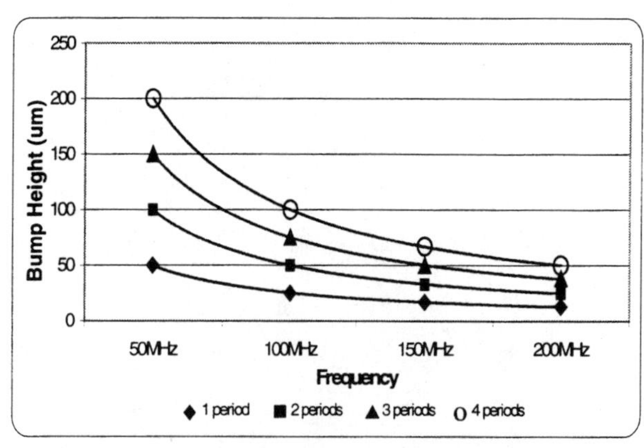

FIGURE 11. Bump height vs. pulse center frequency for several values of pulse duration.

However, in practice, transducers with much higher center frequencies are typically required to obtain this performance. There can be several reasons for this. The pulse produced by the transducer may have a duration of more than 2 periods. Multiple reflections and reverberation from thin film interfaces at the die surface may make temporal discrimination difficult and degrade signal strength. Spot size limitations may produce echoes from surrounding structures that overlap the desired echo. And, the lens of the transducer may not produce a well focused spot at the device side of the die.

Although a higher center frequency of the pulse enables better depth and spatial resolution, attenuation imposes a practical upper limit upon frequency. The attenuation of sound in most fluids at room temperature is proportional to the frequency squared, whereas in solids absorption is a linear function. Thus, the water path impacts attenuation to a higher degree than does the path through the silicon. The attenuation for a water path length (x) as a function of the attenuation coefficient (α) and the frequency (f), is given by:

$$\text{Attenuation (dB)} = \alpha f^2 x \qquad (2)$$

Figure 12 shows the decrease in maximum transducer-sample separation vs. pulse center frequency due to attenuation for different values of return signal strength ($\alpha = 0.217 \text{dB}/\mu\text{mGHz}^2$ at 20°C (6)). Assuming that 0.5mm is the minimum practical separation between the transducer and the sample for mechanical scanning of a flip chip die, the total path length for rays traveling from the surface of the lens cavity to the sample, and back, will be between approximately 1.0. – 2.0mm (depending on lens shape). Therefore, for a minimum acceptable return signal strength of 0.5% (99.5% loss), the cutoff for the return echo frequency lies between 230 - 326MHz. The attenuation can be reduced by approximately a factor of 2 by heating the water to 60°C (6,7). However, it is not practical to test IC packages nondestructively at this temperature. In practice, attenuation in the water may limit the radius of curvature of the transducer lens, which in turn will degrade the spatial resolution for flip chip bump inspection (smaller numerical aperture). Note that this calculation ignores all other losses such as interface reflections and attenuation in the sample.

Industry roadmaps forecast that the flip chip bump height will shrink from 100μm in 1998 to 20μm by 2012 (8). The temporal resolution required to separate the echoes from the top and bottom of a 20μm bump is 8ns, which translates to a return echo center frequency of 250MHz (assuming a 2 period pulse duration). This frequency is close to reaching the limit described above for practical water path lengths. In addition, multiple reflections from thin device layers will complicate the time domain echo signal. Because of these limitations, real-time frequency domain analysis of the echo signal may become necessary for routine inspection of flip chip bumps and underfill defects. This capability will be enabled by the availability of streamlined algorithms, fast and inexpensive PCs, and dedicated digital signal processor chips (DSPs).

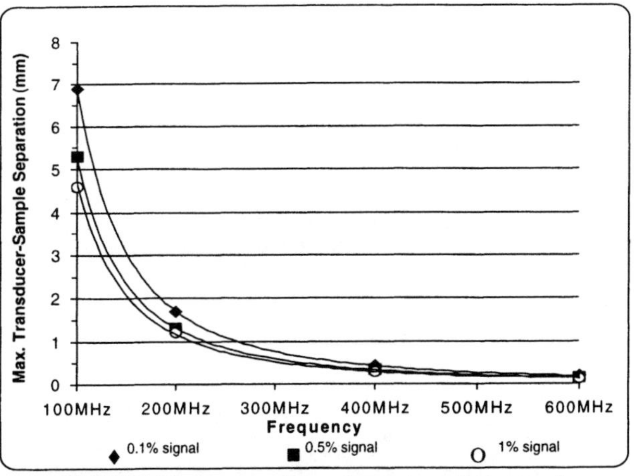

FIGURE 12. Decrease in maximum transducer-sample separation distance vs. frequency for different minimum return signal strength values.

SUMMARY

The application of scanning acoustic microscopy to IC package inspection has been an important factor in the successful development of surface mount packaging. Acoustic inspections were vital to demonstrating that the primary cause for electrical failure in large die surface mount packages during temperature cycling is delamination at the surface of the die, which can happen without the presence of "popcorn" package cracks. The introduction of innovative package styles, such as BGA and flip chip, has driven new developments in how nondestructive acoustic imaging is applied. Multi-layer substrate packages are inspected faster and more reliably with through-transmission acoustic inspection. Flip chip bump and underfill inspections require improvements in depth and spatial resolution. The frequency limit imposed by attenuation in water may mean that eventually, additional improvements in these areas will come from routine, real-time frequency domain analysis. The ability to perform routine frequency domain analysis will enable other innovative applications which may include nondestructive stress imaging. The ability to image stresses will depend on a correlation of phase and amplitude distortions observed in package inspection with localized stress fields (9).

For IC package inspection, it is unlikely that pulse-echo acoustic inspection will ever compete with real-time x-ray inspection for spatial resolution. Although real-time x-ray inspection cannot detect delaminations and open bumps as does acoustic inspection, it more easily images tiny solder bump voids. In this way acoustic and x-ray imaging are complementary for nondestructive flip chip inspection.

ACKNOWLEDGEMENTS

The authors wish to thank Eddie Moltz and Wayne Hambek for advice and technical support.

REFERENCES

1. Moore, T.M. "Identification of Package Defects in Plastic Packaged Surface Mount IC's by Scanning Acoustic Microscopy", *Proceedings 14th Int. Symposium for Testing and Failure Analysis* 61-67, 1989.
2. Moore, T.M., Kelsall, S.J. and McKenna, R.G. "Impact of Acoustic Microscopy on the Development of Surface Mount Packages", in "*Semiconductor Characterization: Present Status and Future Trends*", W.M. Bullis, D.G. Seiler and A.C. Diebold (eds.), (American Int. of Physics, Woodbury, N.Y.) 1996, pp. 202-209.
3. Moore, T.M., McKenna, R., and Kelsall, S.J. "Correlation of Surface Mount Plastic Package Reliability Testing to Nondestructive Inspection by Scanning acoustic Microscopy", *Proceedings Int. Reliability Physics Symposium*, 160-166, 1991.
4. Moore, T.M., Kelsall, S.J., and McKenna, R.G. *"Characterization of Electronic Packaging Materials"* New York, Butterworth/Manning 1992, Ch. 4.
5. Moore, T.M., and Hartfield, C.D. "Through-Transmission Acoustic Inspection of Ball Grid Array (BGA) Packages", *Proceedings 23rd Int. Symposium for Testing and Failure Analysis* 197-204, 1997.
6. Briggs, A. "*Acoustic Microscopy*" New York, Oxford University Press 1992, p. 81
7. Briggs, A. "*An Introduction to Scanning Acoustic Microscopy*" Royal Microscopy Society, Oxford University Press 1985, p. 3.
8. The National Technology Roadmap for Semiconductors, Semiconductor Industry Association (1998).
9. Drescher-Krasicka, E., Moore, T.M., and Hartfield, C.D. "Stress Measurements in Electronic Packaging", in "*Characterization and Metrology for ULSI Technology*", D.G. Seiler, W.M. Bullis, A.C. Diebold, R. McDonald, and T. Shaffner (eds.), (American Institute of Physics, Woodbury, N.Y.) December 1998.

PACKAGING - GENERAL

Interconnection Continuity Test for Packaged Functional Modules

Jan Obrzut

Polymers Division, National Institute of Standards and Technology, Gaithersburg, MD 20899

We developed an electrical test to evaluate interconnections in packaged, electrostatic-discharge (ESD) protected modules. The ESD protection circuit, which in modern integrated circuits is present at every I/O as an inherent part of the chip structure, can be employed to create an electrical path through the interconnection without powering ON the internal circuitry of the chip. The technique utilizes a chip-specific voltage–current characteristic, and a model of the interconnection, from which the interconnection ohmic resistance can be directly obtained. The technique is applicable to a broad range of functional packages. Its sensitivity is comparable or better than the sensitivity of standard test techniques that utilize specialized test chips. This technique can also be used to test interconnections in failed IC's, since integrity of the internal circuitry is not required in this test.

INTRODUCTION

The testing of interconnection of chip carriers prior to chip assembly has become an increasingly difficult and expensive task. The small size of the chip joints and their complex top-surface metallurgy makes them susceptible to contamination and damage during handling and electrical probing. This has become a growing problem for the direct flip-chip packaging technologies that utilize microvias and organic carriers. To minimize the possibility of damaging the chip bonding pads, as can occur during continuity testing, this test has often been replaced with procedure that involves optical tracing of the nets at each circuitry level, followed by a single-node charge deposition or an open ended time-domain reflectometry. However, neither the optical tracing nor the single-node electrical probing can detect nets with defects such as insufficient thickness of the conductor, microcracking, or a buried contact resistance. Once the functional chip is assembled to the carrier the closed interconnection loop then contains nonlinear elements such as diodes, bipolar transistors, or FET transistors. These elements limit the applicability of direct input testing, making the test for most part impractical to perform. This paper describes a new electrical test that can be used to evaluate the interconnections continuity resistance in packaged modules by utilizing the electrostatic discharge protection circuit.

INTERCONECTION MODEL

In modern integrated circuits the ESD is built as an inherent p-n junction in the I/O cells, located in close proximity to the bonding pads (1). Figure 1 illustrates the diagram of a single interconnection which consists of the ball grid array contacts (BGA), carrier circuitry, the chip solder joints (C4s) and the ESD protection circuit. During normal operation of the integrated circuit, the ESD is reverse biased. When the ESD is forward biased its dynamic resistance drops and the junction channels electric charge away from the internal circuit. Thus, the ESD junction can be employed to create an electrical

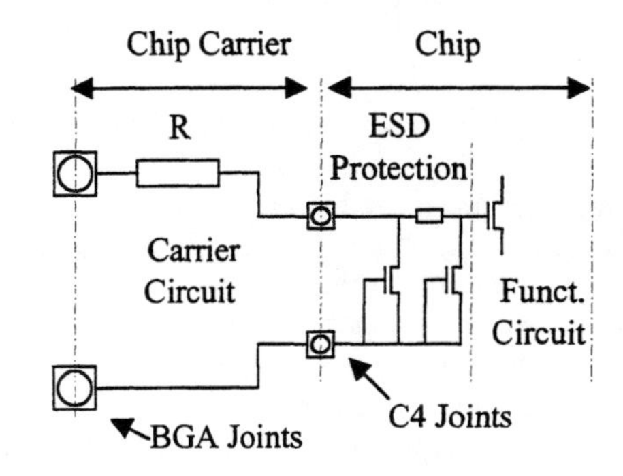

Fig. 1. Interconnection diagram.

path through the interconnection without powering ON the internal circuitry of the chip.

The relationship between the current drained from the forward biased junction, I_j, and the resulting voltage at the testing pads, V_p, can be described by the following expression (1) (see ref. (2)).

$$I_j = I_o\left[\exp\left(\frac{q(V_p - I_j R)}{\eta kT}\right) - 1\right] \qquad (1)$$

where, k is the Boltzman's constant, η is the junction ideality factor, I_o is the saturation current, and R represents the equivalent resistance of the interconnection consisting of a junction resistance and the resistance of the wiring. Application of a current wave digitized as a set of n data points ΣI_j, results in the corresponding voltage wave ΣV_p. These experimental data can be used to obtain the interconnection ohmic resistance, the saturation current, and the ideality factor

CP449, *Characterization and Metrology for ULSI Technology: 1998 International Conference*
edited by D. G. Seiler, A. C. Diebold, W. M. Bullis, T. J. Shaffner, R. McDonald, and E. J. Walters

from equations (2-5). Equations (2-5) have been derived from expression (1) by using a general linear least squares method (3).

$$0=\sum_{i=1}^{n}\frac{1}{\sigma_i^2}\left[\sum_{k=1}^{3}a_k X_k(I_{ji})-V_{pi}\right]X_l \quad l=1,2,3 \qquad (2)$$

where σ_i is the measurement error of the ith data point, and the parameters a_1 ... a_3 and the basic functions $X_1(I_j)...X_3(I_j)$ of the model are given as follows:

$a_1 = \eta kT/q \qquad X_1 = ln(I_j) \qquad (3)$
$a_2 = R \qquad X_2 = I_j \qquad (4)$
$a_3 = -\eta kT/q\ ln(Io) \qquad X_3 = 1 \qquad (5)$

Substituting equations (3-5) into equation (2) yields a set of linear equations **XA** = **V** which can be solved directly. **A** is a vector of the interconnection parameters, **X** is a square matrix, which elements represent combination of the basic functions, and **V** is a vector that represents combination of the voltage wave with the basic functions.

EXPERIMENT AND RESULTS

Interconnection testing has been performed on a flip chip, plastic, ball-grid array package that carried a CMOS-6S chip. During testing, a forward biased current I_j, was drained by probing the BGA I/O pads. A Keithley[1] Model 2400 SourceMeter™ was used as a current source and a voltmeter. The instrument was programmed to perform a current sweep operation during which discrete current values were applied to the testing pads while the corresponding voltage drop was measured. The current values, the instrument delay time, and the number of readings taken at each current level were programmed individually for every I/O to minimize the testing time while maintaining measurement errors at a acceptably low level. Typically, the current wave and the resulting voltage wave consisted of 20 data points. A computer was used to control the SourceMeter™ and to carry out the data acquisition and calculation of the interconnection parameters.

Figure 2 shows the experimental current – voltage data for a single interconnection compared with the least squares fit to the model. The experimental data agree very well with the predicted values confirming the validity of the model and the calculation procedure. The equivalent continuity resistance of this interconnection obtained from the model is 4.845 Ω. This value corresponds well to the circuit resistance estimated from the geometrical details of the interconnection. The two remaining parameters i.e. substrate reverse current and the ideality factor, which are global for a given integrated circuit, are 7.1 nA and 1.78, respectively. The

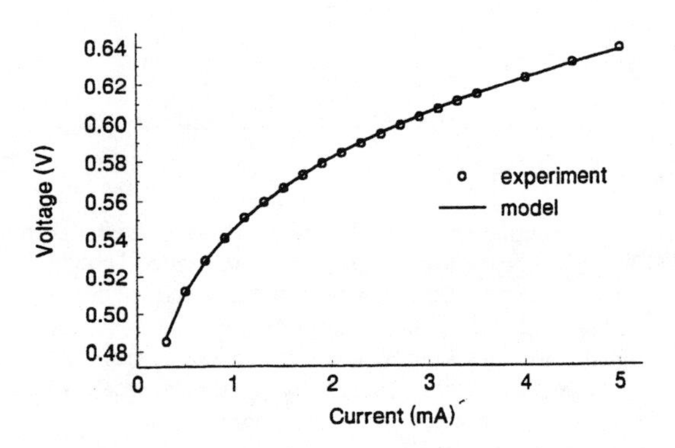

Fig. 2. The experimental and calculated current–voltage data for an interconnection path with ESD. R=4.845 Ω, η=1.78, Io=7.1 nA.

resolution of the resistance parameter, *R*, was determined experimentally to be about 5 mΩ by including a known standard, variable resistor in series with the interconnection.

There are several components, which contribute to the uncertainty of the estimated interconnection parameters *R*, η and *Io*. These include assumptions and simplifications of the model, computational as well as experimental errors. Experimental error includes error in making and maintaining reliable contacts as well as error introduced by variation of the junction temperature during the measurements. The interconnection model neglects contribution of the saturation current, *Io*, to the measured current. Since the saturation current is usually six orders of magnitude smaller than the measured current, this error was judged to be insignificant. Similarly, the computational errors were neglected since the number of significant figures in the computed solutions was >12. The number of significant figures in the computed solution was estimated by subtracting value of $\log(\|\mathbf{X}\|\ \|\mathbf{X}^{-1}\|)$ from the number of digits carried. From the instrument technical specification, the accuracy of voltage readings was better than 0.02 % with resolution of 50 μV, and the accuracy of the current readings was better than 0.05 % with resolution of 500 nA. Thus, the combined instrumental accuracy was expected to be in the range of about 0.07 %. However, the relative standard uncertainties of the interconnection parameters estimated as the diagonal elements of the covariance matrix were larger, in the range of 2.0 % for *R*, 1.5 % for η, and 1.7 % for *Io*. This corresponds to an error in *R* of about 100 mΩ. This error can be minimized further to about 80 mΩ by controlling and maintaining a constant temperature of the ESD junction.

1. Certain materials and equipment identified in his manuscript are solely for specifying the experimental procedures and do not imply endorsement by NIST or that they are necessary the best for these purposes.

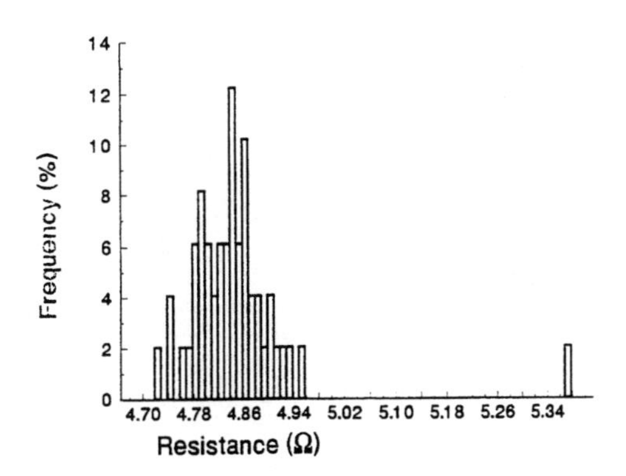

Fig.3. Histogram of the interconnection resistance.

Figure 3 shows a histogram of the interconnection resistance measurements for a batch of 36 modules. The distribution has a mean value of 4.836 Ω. The standard deviation of the distribution is 0.0514 Ω and the standard error of the mean is 8 mΩ. A measurement at 5.377 Ω falls outside of the distribution indicating an interconnection with questionable performance.

DISCUSSION AND CONCLUSION

It has been found experimentally that the test procedure described above can be used to determine the interconnection parameters, including the interconnection resistance, with considerable confidence and precision. For example, a current wave, digitized to 20 data points, is sufficient to determine the interconnection resistance, R, with resolution of 5 mΩ and relative standard uncertainty better than 2 %. A defective net will be distinctly separated from the entire population and easily detected on the histogram of the interconnection resistance. Likewise, statistical analysis of the other interconnection parameters, (the saturation current or the ideality factor) can be used to assess the quality and functionality of the chip. Unlike the results obtained by other techniques, which often produce unpredictable and unreliable results, the resistance values estimated with this method reflect the real continuity resistance of the interconnection circuit. This makes the method sensitive in the detection of defects caused by manufacture and environmental stressing. The computational algorithm is exceptionally stable and fast, while standard fitting procedures are usually computationally involved, subject to round off error affected by initial parameters, and often unable to meet the convergence criteria.

The testing time is a very important factor in assessing practical applicability of the test. A functional chip carrier having typically hundreds of input-output interconnections imposes rather demanding testing time requirements, which desirably, should not exceed a few tenths of a second per interconnection. In the presented laboratory set-up of the test, it takes about 200 ms to execute a suitable current wave for a single interconnection and to transfer the current-voltage data from the instrument buffer to the computing unit. In comparison, the computing time is negligible. The total testing time will also depend on the time it takes to make contacts to the test pads. Therefore, in a practical application, the testing time will most significantly depend on the speed of the hardware used and not on the model.

Although the above interconnection model emphasizes the electrical characteristic of the bipolar transistor, other types of the ESD protecting mechanisms, such as diode-connected FET transistors, can be employed and treated similarly. Thus, the presented test structure is of a general form, and therefore, is applicable to a broad range of functional packages. It can be used for process verification, non-destructive field failure analysis, and for interconnection reliability assessment with a comparable or better sensitivity than standard techniques that utilize specialized test chips. Even failed ICs can be tested using this technique since integrity of the internal circuitry is not required in this test.

REFERENCES

1. Amerasekera A. and Duvvury C., *ESD in Silicon Integrated Circuits*, New York: J. Wiley, 1995, pp. 100.

2. Luchies J. R. M., de Kort C. G. C. M. and Verweij J. F., *Journal of Electrostatics*, **36**, 81-92, 1995.

3. Press W. H., Flannery B. P., Teukolsky S. A. and Vettering W. T., Numerical Recipes, new York, Cambridge University Press, 1986, pp.509.

Scanning Acoustic Microscopy Stress Measurements in Electronic Packaging

Eva Drescher-Krasicka *, Thomas M. Moore** and Cheryl D. Hartfield**
*National Institute of Standards and Technology, Gaithersburg, Maryland 20899
dre@enh.nist.gov
**Texas Instruments, Inc., 13588 N. Central Expwy., MS-921, Dallas, Texas 75243
moore@ti.com, hartfield@ti.com

ABSTRACT

Scanning acoustic microscopy has been successfully implemented for the nondestructive detection of cracks and delaminations in integrated circuit packages. The incorporation of scanning acoustic microscope inspections in reliability tests of molded surface mount components has identified delamination at the mold compound/die interface as the primary cause of electrical failure during temperature cycling. The ability of the acoustic microscope to nondestructively image stress distributions prior to failure would greatly extend the impact of scanning acoustic microscopy on new package development and process control. This may be particularly important for the packaging of devices with fragile low-K (dielectric constant) layers in the device structure.

INTRODUCTION

Scanning acoustic microscope studies have had a significant impact on the successful development of molded surface mount devices (1,2). The acoustic microscope is sensitive to critically important package defects such as cracks and delaminations which are practically undetectable by x-ray radiography. The nondestructive nature of acoustic inspection enables smaller sample populations and more reliable results during reliability testing.

The scanning acoustic microscope uses a single piezoelectric transducer that is mechanically scanned over the surface of the package. The transducer focuses pulsed sound waves to a specific interface within the package. In pulse-echo mode, the same transducer then also receives the signals. The echoes are analyzed, and characteristics of the signal such as amplitude, phase, and interface depth, are used to form images of internal structures and defects. A water bath is used to couple the sound between the transducer and the package. An advantage of acoustic inspection is the ability to discriminate between bonded and delaminated interfaces. The pressure pulse echo from a delaminated portion of an interface has inverted polarity relative to that from a bonded portion.

The scanning acoustic microscope has been used to detect damage, such as delamination at the mold compound/die interface. Once the occurrence of a particular type of damage has been correlated to electrical failure of the component, the acoustic microscope can be used to predict the reliability of the component. The occurrence of cracks and delaminations results from the generation of thermal-mechanical stresses in the package. Mismatches in the coefficients of thermal expansion between the different materials in the package, absorbed moisture and mechanical stress raisers such as burrs, voids and corners all contribute to the generation of stress.

The phenomenon of phase inversion of the reflected longitudinal pressure pulse greatly assists in the discrimination of delaminated internal interfaces. Intermediate phase shifts in echo signals, although not permitted by the classical theory of reflectivity, are frequently detected. It has been suggested that these intermediate phase shifts provide additional information on the adhesion performance of the interface (3).

Recent studies of IC packages with the scanning acoustic microscope suggest that these intermediate phase shifts may be produced by localized stress fields and may offer the potential to image stresses nondestructively (4,5). The method described in these studies is based on the detection of distortions of the acoustic echoes due to localized stress fields. Physical properties of reflected polarized sound waves, such as phase and amplitude, are different for sound traveling through stressed and unstressed volumes of material.

A more detailed description of the principles behind this stress mapping method is available in the literature (6,7). More recent theoretical developments in stress induced anisotropy of longitudinal waves discuss the impact of residual stress concentration on phase measurements of the received signal (8). The ability of the acoustic microscope to nondestructively detect and image the distribution of stresses in the package would add a predictive capability and greatly extend the potential impact of this instrument for package development and improvement.

EXPERIMENTAL RESULTS

Figure 1 shows the amplitude image of the longitudinal wave reflection from the mold compound/die pad interface in a 132-pin QFP (quad flat pack) package that has been damaged during solder reflow exposure. The corners of the die pad have delaminated from the mold compound. The delaminated corners are outlined by a “dark line” (phase cancellation) boundary. Figure 2 shows the corresponding image of the mode-converted shear wave reflection from the mold compound/die pad interface.

The mode-converted shear wave is formed when the incident longitudinal wave reflects to form a longitudinal and

CP449, *Characterization and Metrology for ULSI Technology: 1998 International Conference*
edited by D. G. Seiler, A. C. Diebold, W. M. Bullis, T. J. Shaffner, R. McDonald, and E. J. Walters

a shear wave at the mold compound/die pad interface. The mode-converted shear reflection is distinguished due to its time-of-flight which is intermediate between that of the longitudinal and the shear reflections. The time-of-flight for the mode-converted shear reflection is identical to that of a reflected longitudinal wave that was formed by an incident shear wave. In some cases this may degrade the spatial resolution in this image because the focus is different for the shear and longitudinal waves. Figure 3 is a time-of-flight display (A-scan) showing the shear and longitudinal reflections from the mold compound/die pad interface (longitudinal focus).

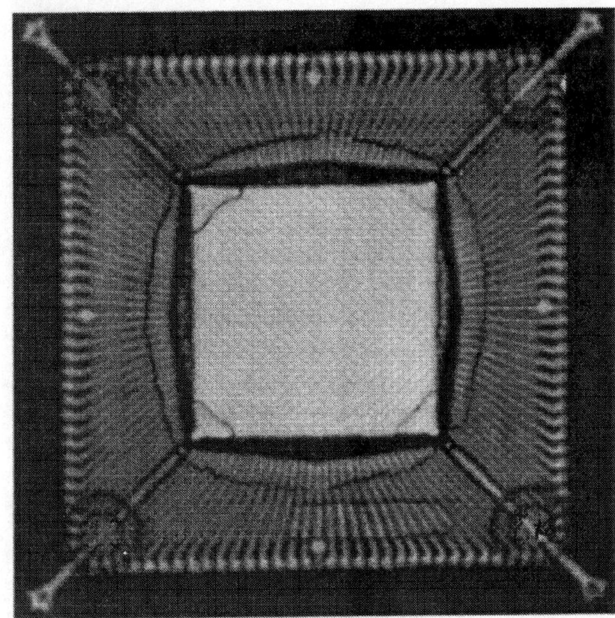

FIGURE 1. A longitudinal echo amplitude image of the mold compound/die pad interface in a 132PQFP.

The mode-converted shear wave image demonstrates additional contrast not seen in the longitudinal image. For example, the delaminated die pad corners are more clearly differentiated from the bonded central area of the die pad. And additional contrast is seen from the lead frame reflections. However, the arrival of the mode-converted shear wave after the longitudinal wave may be coincident with reflections and reverberations from deeper interfaces (die attach, die, etc.) at bonded areas of the mold compound/die pad interface. A similar caution exists for interpretation of the mode-converted shear image at other locations in the image where reflections and reverberations of the longitudinal wave exist and interfere with the mode-converted shear reflection. Also, contrast comparisons between the longitudinal and mode-converted shear images are valid only if the intensities in the two images are calibrated.

FIGURE 2. The corresponding mode-converted shear echo amplitude image of the same 132PQFP in Figure 1.

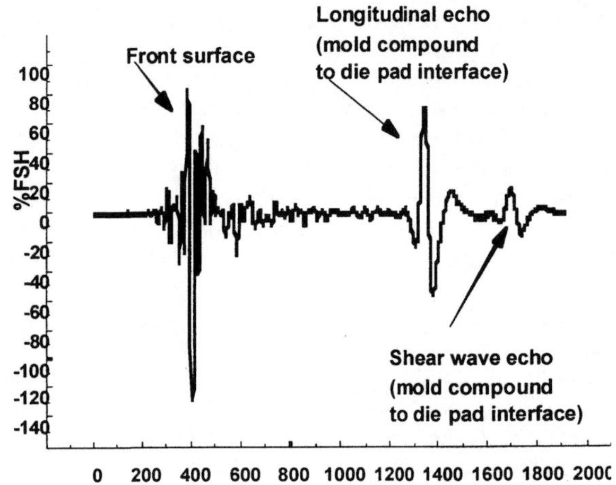

FIGURE 3. The time-of-flight relationship between the longitudinal and mode-converted shear echoes that formed the images in Figures 1 and 2.

In addition to the potential for imaging stress offered by the mode-converted shear wave, it may be possible to image stress with longitudinal waves. Figure 4 shows the longitudinal amplitude image of the back wall echo in a sample of 1mm-thick epoxy mold compound. The sample was uniaxially compressed during inspection, and a broadband acoustic pulse with a center frequency of 15MHz was transmitted through the sample and focused at the back surface. Localized stress is expected to distort the incident cone of longitudinal waves into a continuum of longitudinal waves having slightly different velocities. The interference of these waves may form amplitude variations in the image.

Theoretical efforts to describe this effect are underway(8). The contrast in Figure 4 suggests a sensitivity of the longitudinal echo amplitude to localized stress in this stressed sample.

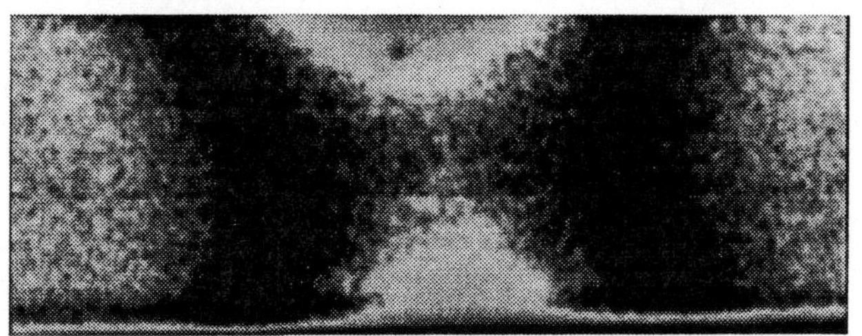

FIGURE 4. A longitudinal echo amplitude image of a uniaxially compressed strip of mold compound.

DISCUSSION

In order to quantify information from the acoustic images on the distribution and the state of stress within a biaxially deformed plane sample, one has to relate the measured amplitude or phase change caused by polarization and interference effects to the magnitude of stress. For example, if one can image a delaminated interface with the polarized shear wave (such as a mode-converted shear reflection) propagating perpendicularly or almost perpendicularly to the direction of acting principal stresses (in the plane of the die/mold compound interface), the amplitude image may indicate the distribution of maximal shear stresses in the material. (the difference of the principal stresses). In comparison, imaging with the longitudinal wave should result in an image of the pattern of the sums of principal stresses.

A recent effort in the quantitative evaluation of stress in IC components includes finite element modeling supported by in-situ piezoresistive measurements of mechanical stresses (9). This work was performed to assess the reduction of die surface stresses by the use of a polyimide die overcoat. The theoretical solutions for the intensity variations in the acoustic stress image can be generated for isotropic and elastic materials if the third order elastic constants are known. Calibration of the acoustic stress image should therefore be made possible by comparing with finite element modeling (once the finite element modeling is verified by the results of the in-situ piezoresistive stress sensor measurements).

CONCLUSIONS

Although immediate practical difficulties exist with isolating the mode-converted shear wave, the demonstrated sensitivity of the shear wave to local stress variations in other materials is promising for stress imaging in IC packages (7). Further, a correlation of the intermediate phase shifts seen in longitudinal reflections from IC packages with localized stress fields would give the acoustic microscope a predictive capability for IC package inspection.

ACKNOWLEDGEMENTS

We would like to express our thanks to Dr. James Sweet from Sandia National Laboratory for valuable suggestions.

REFERENCES

1. Moore, T.M., Kelsall, S.J., and McKenna, R.G. *"Characterization of Electronic Packaging Materials"* New York, Butterworth/Manning 1992.

2. Moore, T.M., Kelsall, S.J. and McKenna, R.G. "Impact of Acoustic Microscopy on the Development of Surface Mount Packages", in "Semiconductor Characterization: Present Status and Future Trends", W.M. Bullis, D.G. Seiler and A.C. Diebold (eds.), (American Int. of Physics, Woodbury, N.Y.) 202-209, 1996.

3. Moore, T.M., "Reliable Delamination Detection by Polarity Analysis of Reflected Acoustic Pulses", Proc. 17th Int'l. Symp. For Testing and Failure Analysis, 49-54, 1991.

4. Moore, T.M. and Drescher-Krasicka, E., "Comparison Between Images of Damage and Internal Stress in IC Packages by Acoustic Microscopy" *Proc. 6th Int'l. Workshop on Moisture in Microelectronics*, NIST, 202-209, 1996.

5. Moore, T.M. and Drescher-Krasicka, E. "Moisture/Thermal Induced Stress in Packaged IC's by Acoustic Imaging" *Proceedings of the Fourth International Conference on Composites Engineering,* ICCE, 35-38, 1997.

6. Drescher-Krasicka, E. "Scanning Acoustic Imaging of Stress in the Interior of Solid Materials" *J. Acoust. Soc. Am.* **94** (1), 453-464, 1993.

7. Drescher-Krasicka, E. and Willis, J.R. "Mapping Stress with Ultrasound" *Nature* **384**, pp. 52-55 November 1996.

8. Willis, J.R. and Drescher-Krasicka, E. "Theory of Acoustic Trirefringence for Acoustic Microscopy" - in preparation

9. Sweet, J.N., et al. "Piezoresistive Measurement and FEM Analysis of Mechanical Stresses in 160L Plastic Quad Flat Packs" *Proceedings of Inter*PACK'97. June 15-19, Mauna Lani, Hawaii, 1997.

MATERIALS CHARACTERIZATION

One- and two-dimensional dopant/carrier profiling for ULSI

W.Vandervorst[a], T.Clarysse, P.De Wolf, T.Trenkler, T.Hantschel, R.Stephenson and T.Janssens

Imec, Kapeldreef 75 B-3001 Belgium
[a also:] KULeuven, INSYS, Kard. Mercierlaan 92, B-3001 Belgium

Dopant/carrier profiles constitute the basis of the operation of a semiconductor device and thus play a decisive role in the performance of a transistor and are subjected to the same scaling laws as the other constituents of a modern semiconductor device and continuously evolve towards shallower and more complex configurations. This evolution has increased the demands on the profiling techniques in particular in terms of resolution and quantification such that a constant reevaluation and improvement of the tools is required. As no single technique provides all the necessary information (dopant distribution, electrical activation,..) with the requested spatial and depth resolution, the present paper attempts to provide an assessment of those tools which can be considered as the *main metrology* technologies for ULSI-applications. For 1D-dopant profiling secondary ion mass spectrometry (SIMS) has progressed towards a generally accepted tool meeting the requirements. For 1D-carrier profiling spreading resistance profiling and microwave surface impedance profiling are envisaged as the best choices but extra developments are required to promote them to routinely applicable methods. As no main metrology tool exist for 2D-dopant profiling, main emphasis is on 2D-carrier profiling tools based on scanning probe microscopy. Scanning spreading resistance (SSRM) and scanning capacitance microscopy (SCM) are the preferred methods although neither of them already meets all the requirements. Complementary information can be extracted from Nanopotentiometry which samples the device operation in more detail. Concurrent use of carrier profiling tools, Nanopotentiometry, analysis of device characteristics and simulations is required to provide a complete characterization of deep submicron devices.

INTRODUCTION

Dopant/carrier profiles constitute the basis of the operation of a semiconductor device and thus play a decisive role in the performance of a transistor. These dopant distributions are subjected to the same scaling laws as the other constituents of a modern semiconductor device and continuously evolve towards shallower and more complex configurations. Facing the increasing costs of the actual processing, new designs of process technology (and thus doping profiles) start off with an exploration of the various fabrication possibilities using process and device simulations (TCAD). If these simulations can be fine tuned to promising results, the experiment can be completed in silicon. Upon fabrication electrical charateristics can be measured and, in the ideal case, should match the expected performance.

Provided our understanding/modeling of doping processes and process control would be advanced to such a level that these simulations can be viewed as the perfect mimic of the actual processes, metrology (and more specifically dopant/carrier profiling) should be totally obsolete in a modern process environment. Unfortunately the latter is not yet achieved and process control as well as TCAD calibration hinges very strongly on feedback from dopant/carrier profiling tools. Moreover, since too many variables in the simulators are not exactly known, process splits during process development are a common approach during the first iteration as well. Again, assessment of the impact of these splits can only be performed when appropriate measurement tools are available. Finally, in order to prevent the recurrence of deficient processing and non-operational devices, failure analysis is required based on tools which provide detailed information on the structure under investigation.

As indicated in Fig. 1 a strong interplay exists between the process flow, process engineers, TCAD and experimental verification necessitating that metrology tools stay in line with the demands from processing technology. With the increasing dominance of two- (three-dimensional) effects in devices, it is obvious that the requirements placed on metrology have increased from simple 1D-dopant profiling for ancient technologies towards full 3D-chemical and electrical characterization. As simple 1D-models and experiments are no longer sufficient

CP449, *Characterization and Metrology for ULSI Technology: 1998 International Conference*
edited by D. G. Seiler, A. C. Diebold, W. M. Bullis, T. J. Shaffner, R. McDonald, and E. J. Walters

to explain the observed processing effects and device characteristics, availability of adequate metrology for ULSI technology is a very important issue and efforts should be directed towards making these available for routine applications.

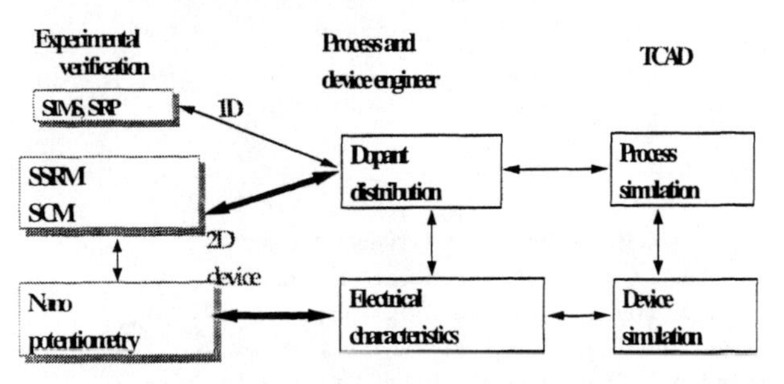

FIGURE 1. Interplay between experimental verification, process engineering and TCAD. Indicated are the main technologies for the experimental verification.

The changes in technologies have also increased the demands on the profiling techniques in particular in terms of resolution and quantification such that a constant reevaluation and improvement of the tools is required. Ideally one single technique should provide all the necessary information (dopant distribution, electrical activation,..) with the requested spatial and depth resolution. Since at present no single technique is available satisfying all the industrial needs, the present paper attempts to provide an assessment of those tools which can be considered as (candidates for) the *main metrology* technologies for ULSI-applications.

The latter implies that these tools do (or hold the prospect to) achieve all the requirements in terms of :

1. independent quantification,
2. quantification accuracy,
3. sensitivity and dynamic range
4. spatial resolution
5. depth resolution
6. direct applicability to any arbitrary device (i.e. no need for special test structures or extra processing)

Supporting metrology technologies can be viewed as those methods which fail on one of these criteria but may, within their limits, still provide substantial support to the main technologies. Where possible reference to these methods will be made but it is not the objective of this paper to provide a detailed overview of them also.

Facing the limitations of the techniques (dopant or carrier sensitive) and the various applications (homogeneous samples or device like structures) alternative methods can be used. Consequenty this paper is divided into one-dimensional and two-dimensional profiling, distinguishing as well between dopant and carrier profiles.

The outline of the paper is therefore based on the classification included in table 1.

TABLE 1. Classification of characterization methods available for dopant/carrier profiling.

	Main technologies	Supporting technologies
Dopant profiling		
1D-profiles	SIMS	Nuclear methods
2D-profiles device analysis	non-existent	2D-SIMS Lateral SIMS Tomography SIMS
Carrier profiling		
1D-profiles	SRP	C-V FPP (MicroSRP)[1] (Microwave Surface Impedance Profiler)[1]
2D-profiles device analysis	SSRM SCM Nanopotentiometry	Selective etching + TEM Selective etching + AFM Secondary electron emission Other scanning probe methods (SKM, EFM, STM,..)

[1]These could move to the main technologies if a commercial instrument were to be developed.

MEASUREMENT REQUIREMENTS

When relating the required measurement capabilities with the needs of the semiconductor industry as expressed in the SIA roadmap [1], one faces the requirement for highly repeatable and quantitative tools (< 5 % accuracy) for ever decreasing profile depths (< 50 nm) with a dynamic concentration range spanning 10^{15}-10^{20} at/cm^3 . Moreover in view of the increasing importance of the 2 (and 3)D-interactions between dopants and doping processes (among others leading to the Reverse Short Channel Effect [2,3]), a strong interest has surfaced in tools suited for 2D-characterization of doping profiles. As the intent is to

TABLE 2. Metrology requirements in relation to processing technology evolution according to NTRS for dopant profiling.

Year	1997	1999	2001	2003	2006	2009	2012
Junction depth contact (nm)	100-200	70-140	60-120	50-100	40-80	15-30	10-20
S/D extension (nm)	50-100	36-72	30-60	26-52	20-40	15-30	10-20
channel concentration (at/cm3)	4-6 10^{17}	6-10$10^{17}$	7-13 10^{17}	1-2$10^{18}$	2-3 10^{18}	> 3.5 10^{18}	> 7 10^{18}
specif. spatial resolution (nm)	5	3	3	2	1.5	1	0.8-0.6
average atom distance @channel conc.(nm)	13 nm	11 nm	10 nm	9 nm	7.5 nm	< 7 nm	< 5 nm

measure the dopant/carrier distribution within a transistor itself (for instance to probe the extent of lateral underdiffusion) the demands in terms of spatial resolution are extremely high. Their projections together with some of the relevant technological predictions are shown in table 2.

Referring to table 2 it is now important to consider two points. The specification of an extreme resolution is obviously meaningless if at the same time the concentration level is not considered. Indeed if one considers the average distance between two dopant atoms for a particular concentration level, one finds very quickly that a specification of 5 nm is less than the distance between two atoms at a level of $5x10^{18}$ at/cm^3. The implication of this observation is that as soon as the requested spatial resolution becomes less than the distance between the dopant atoms, the specification on profiling resolution must be viewed as the uncertainty on the determination of the position of the dopant atoms. Since the distance between the surrounding atoms determines the *local* dopant concentration, the specified resolution is nothing else than the uncertainty associated with the localization of the dopants atoms. The numbers mentioned above then refer to the apparent volume which must be assigned to these atoms as the limited resolving capabilities of the measurement technique do not allow to provide a more accurate determination of the atoms within that volume. The extreme resolution quoted for the year 2012 then calls for a method with near atomic resolution. For techniques probing the carrier distribution the situation is even more complex as extremely steep gradients do not exist but get reduced by carrier outdiffusion. The latter is clearly demonstrated in Fig.2. where we show the best possible resolution for a particular DOPANT step. These values were obtained by extracting the depth required for a 84-16 % transition in carrier concentration for a dopant concentration atomically abrupt varying from 10^N to 10^{N-1} at/cm^3. Calculations are based on simple solutions of 1D-Poisson equations.

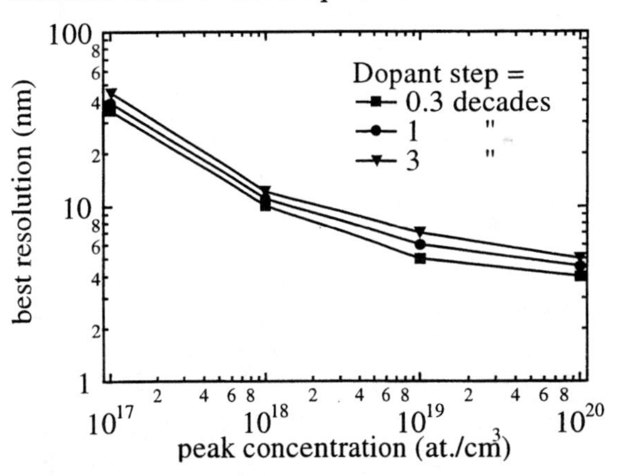

FIGURE 2. Theoretical resolution (84-16% transition) for a carrier profiling technique when an abrupt dopant step is encountered. Parameter is the dopant concentration step..

The results in Fig. 2 clearly show that the maximum attainable carrier resolution is strongly concentration dependent. Despite the rise in channel concentration (cfr table 2) for smaller geometries, this phenomena does impose a limit to all electrical techniques for satisfying the NTRS requirements. On the other hand one can argue that since the same carrier outdiffusion occurs in an ordinary device as well, providing a tool

with the capabilities to probe the latter carrier profiles with perfect accuracy should be sufficient to explain all the device characteristics observed.
Based on the above considerations one is tempted to conclude that eventually dopant observation is the only viable way to satisfy the NTSR needs. Unfortunately one must then also address the importance of the statistics involved. The continued shrinking of devices does imply that the number of atoms available for detection is decreasing rapidly.
For instance when calculating the number of dopant atoms in a (conservative) source/drain extension of 30 x 30 nm thereby assuming various concentration levels, one clearly sees that the number of atoms contained in that region becomes extremely limited (Fig.3).

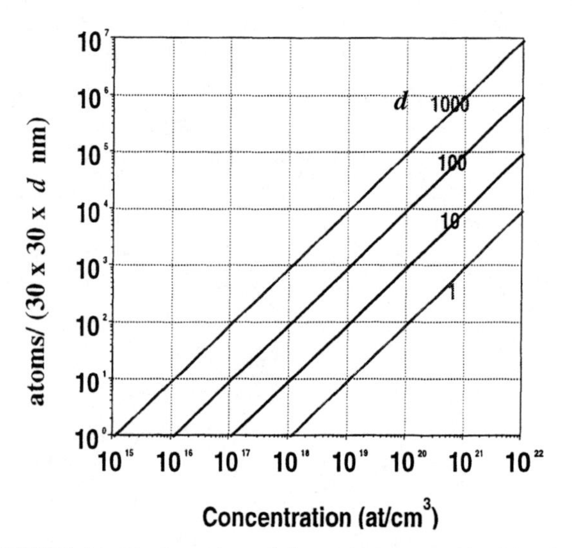

FIGURE 3. Total number of dopant atoms contained in the source/drain extensions (assumed to be doped homogeneously over 30x 30 nm) as viewed by a profiling technique with a specific information depth of *d* nm perpendicular to the cross section of the device

It is now immediately obvious that any technique solely observing the number of dopant atoms at the surface (information depth ~ 1 nm) are faced with serious problems in meeting the resolution, sensitivity and accuracy criteria due to the low counting statistics. Hence tools like STM-imaging of individual dopant atoms or direct SIMS analysis of the source/drain extensions will not be realistic concepts.

ONE DIMENSIONAL PROFILING

1D-dopant profiling

Historically achieving high depth resolution in combination with high sensitivity and quantification accuracy has been the driving force for using Secondary Ion Mass Spectrometry (SIMS) as the preferred method for determining dopant profiles. In a limited number of cases alternative methods based on high energy beams can be used to determine the total number of atoms contained in the sample such as As (RBS), Sb (RBS, NAA), B(p,α), P (p,γ) [4]. However, these methods usually do not provide sufficient sensitivity combined with high depth resolution to probe the dopant profile down to the detection limits which are of importance for the process engineers. The main explanation for this dominance of SIMS is the fact that as the technique is based on counting the number of atoms as they are emitted from the sample by the sputtering beam, a very large signal is generated. Compared to other techniqes such as RBS, AES, XPS,... the detected signal itself is not defined by a (limited) interaction cross section but the emitted atom itself constitutes the detected signal. Since the emission process in addition defines the depth scale, a low or high emission (sputter) probability does not translate into a low or high signal but only into a slow or fast acquisition of the profile. This process of sputter emission and detection of secondary ions is very efficient and well controlled such that SIMS has attained excellent qualifications as a dopant profiling tool primarily due to its unsurpassed sensitivity, multi-element detection and good depth resolution. Note that for instance detection limits down to the 10^{12}at/cm^3 level have been reported in favorable cases [5]. Refined analysis protocols [6,7] have led to increasing precision such that SIMS can now be used for implant dose control with an accuracy as good as 0.5 %.

Understanding the limitations of SIMS for one-dimensional profiling is inherently linked to our understanding of the mechanisms of the SIMS process. Basically SIMS is a method using continuous bombardment of the sample with an ion beam. The latter is equivalent with a high dose ion implantation and all mechanisms present during high dose implantation are operative as well. Hence during a SIMS experiment one must be aware of the effects of ion incorporation (eventually leading to compound formation) and energy deposition causing ion beam mixing and atomic displacements. Both mechanisms lead to a disturbance of the sample and might thus cause a discrepancy between the recorded profile and the original dopant distribution. Since the detected signal are ionized particles one must consider in addition to the emission process, the ionization process as well. The latter is a very complex phenomena which has been found to depend strongly on the oxygen concentration at the surface (for positive secondary ions) or cesium (for negative ions) as well as the details of the band structure of the bombarded surface. Compound formation (in particular the changes in band structure resulting from oxyde formation) is in that respect an important parameter influencing the ionization probability. The interplay between the various mechanisms is shown in Fig. 4.

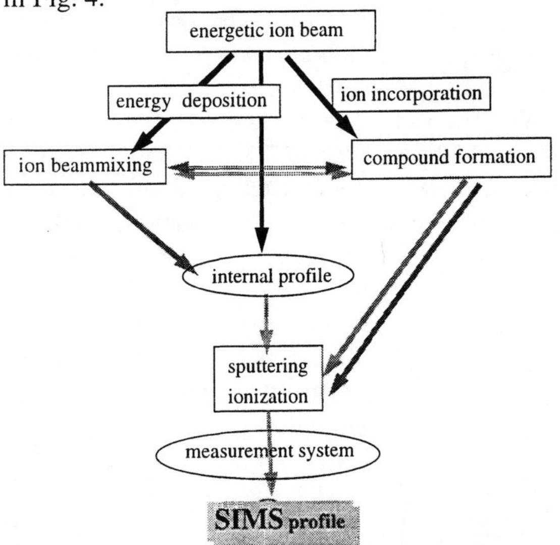

FIGURE 4. Basic mechanisms leading to the final SIMS profile

In order to understand the impact of these mechanisms and the quantification of SIMS it is now interesting to look at the basic SIMS equation relating the detected signal to the sample composition, eq (1).

$$I_M(t_i) = (I_p/q) .Y. \ \alpha_M . C^*_M (0,t_i) .\eta \qquad (1)$$

I_p : primary current

Y : sputter yield

α_M : ionization degree of element M

$z(t_i)$: depth sputtered at time t_i

$z^* = z - z(t_i)$: depth from the instantaneous surface

$C^*_M(z^*,t_i)$: depth distribution of M relative to instantaneous surface at time t_i

$C_M(z)$: original dopant profile of M

η : collection and transmission efficiency

The standard practice in interpreting a SIMS profile is to use a linear (constant) relation between bombardment time t and sputter depth z(t) and to assume a constant ionization degree. The difference between C^* and C is related to the ion beam mixing process which redistributes the dopants within the sample during the ion bombardment prior to their sputter emission. Normally one hopes that by a proper choice of the experimental conditions $C^*_M(z^*,t_i)$ is not too different from $C_M(z\text{-}z(t_i))$ such that $I_M(t)$ becomes proportional with $C_M(z)$.

Facing the requirements of shallow depth profiling (cfr table 2), it is important to reevaluate these assumptions.

The constant correlation between sputter time and sputter depth is based on the concept of a constant sputter yield Y. However in the case of Si bombarded with an oxygen beam it is widely accepted that an altered layer is formed (with a thickness commensurate with the range of the primary ion) whereby the original Si composition [11] is altered into an oxygen rich layer. For a particular set of bombardment conditions (angles of incidence less than ~ 30-35°) even a complete oxidized layer is formed. Disregarding possible variations in energy deposition and total sputter yield it is clear that a lot of the bombardment energy is now used to sputter oxygen atoms (which are replenished by the primary beam) such that the effective Si sputter yield is drastically reduced (with almost a factor three). The common practice of establishing the erosion rate by measuring a crater depth basically determines the stationary (reduced) sputter yield such that if the latter is used to quantify the depth scale of a profile, shallow features appear too close to the surface (this effect is usually termed "differential shift"). This phenomena has been extensively studied in the literature and depends on the energy and angle of incidence of the primary beam.[12,13] Determining factor is the "time (fluence)" it takes to form a stationary compound layer and the corresponding change in sputter yield. Stationary conditions are achieved when the amount of implanted oxygen atoms equals the number of sputtered oxygen atoms. At normal incidence the transient width is on the order of 2 nm/keV O_2, whereas the differential shift amounts to 1-1.5 nm/keV O_2. The increased sputter yield at non-normal incidence leads to a reduced oxygen incorporation level, causing a smaller differential shift despite the fact that the transient width increases quite drastically. A detailed description and modelling of this effect is based on the balance between the oxygen incorporation and sputter yield and was analysed qualitatively in [14] and more recently in detail in [15, 16]. Obviously a low primary beam energy is a requirement to arrive at the correct depth scales.

One of the main motivations to use an oxygen beam is the very strong enhancement of the ionization degree by the presence of oxygen, typically $\alpha \sim C_0^{3-4}$ with C_0 the near-surface oxygen concentration.[8] It is however clear that within the transient region towards the establishment of a stationary altered layer, a strong variation in the near surface oxygen concentration will occur such that the assumption of a constant ionization degree is not valid either. The latter can quite easily be observed when looking at the matrix signal within this region. Typically it increases very steeply over 2 orders of magnitude demonstrating the importance of this effect. Quantification of a dopant profile in this region will be extremely difficult as every element responds differently to this increase in oxygen concentration. Hence a simple ratioing of the impurity to matrix signals is not a valid procedure for quantification.[37]

A procedure suggested previously to eliminate this effect is the use of oxygen flooding. In this case sputtering is performed within a high pressure oxygen ambient speculating that the oxygen atoms from the gas phase will adsorb on the sample surface continously, thereby stabilizing the ionization degree from the start of the profile. Using the stability of the matrix signal as a (poor) indication for the stabilized ionization degree, flooding certainly does improve the quantification aspects of shallow profiles although even in this case a small transient (~ 1 nm) remains present [16,18]. The latter is linked with the fluence required to form this surface oxide layer and the (small) changes in sputter yield resulting from the still present primary ion incorporation and the formation a "thick" altered layer which does have a different energy deposition function and thus different sputter yield [16]. In particular the combination with non-normal bombardement appears attractive as the high incidence angle would reduce ion beam mixing effects and the flooding would eliminate the longer transients.

In the past this has been used quite successfully in many cases [17,19]. However when applied to very shallow profiles in combination with low energy bombardment (< 1kev), significant differences between profiles obtained at different angles and with and without flooding have been observed [20]. These differences are much larger than anticipated based on the above models and are now shown to be due to the rapid formation of surface roughness albeit at the nm-scale [21,22,23,24]. The data shown in Fig.5 illustrate the roughness as a function of eroded depth and the characteristic ripple structure. Apparently the well known phenomena of ripple formation [25,26] which for deep profiles (measured at higher energies) did occur at sputter depths of 2-3 μm (depending on the experimental conditions) and which could be eliminated by flooding as well, is much more effective at low energies and can not as easily be suppressed. (A complete detailed understanding of the formation of ripples is not yet available). The impact of this roughening is a reduction in erosion rate as a function of depth thereby causing a disturbance on the quantified profile when a constant erosion rate is used.

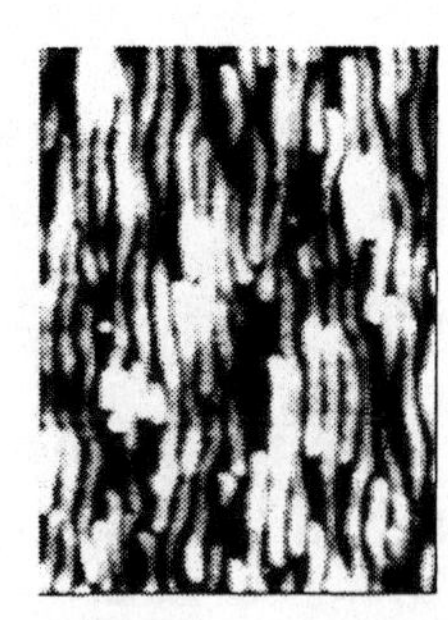

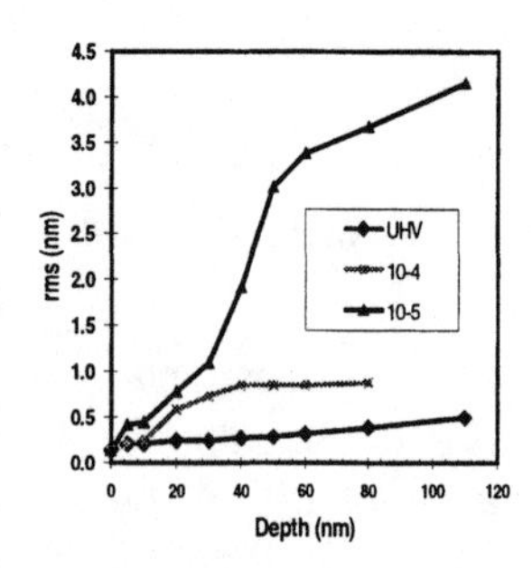

FIGURE 5. a) AFM image of crater bottom after erosion to 100 nm using 1 kev O_2 at 70°.and b) increase of crater bottom roughness as a function of eroded depth [23]. Parameter is the oxygen pressure during sputtering.

The trend towards shallower and steeper profiles has emphasized the need to reduce the basic mechanisms of SIMS inducing profile distortions i.e. the ion beam mixing process of the primary beam (which limits

the depth resolution) and the stabilization of the ionization and sputter yields (which determine the near surface quantification and the errors on the depth scale).
The latter has been achieved by recent developments in ion beam technology [27] enabling a strong reduction in primary beam energy (from the "standard" 2-8 keV down to the 100-200 eV range) and the introduction of cluster beams (SF6) [28]. The improvement in depth resolution (taken here as three times the decay length [29]) can be judged from the results recently reported by several groups (Fig 6) [28,30]. Assessment of the finer details of low energy implantation profiles and their behavior during subsequent anneal cycles can now be measured with high depth resolution.[32]

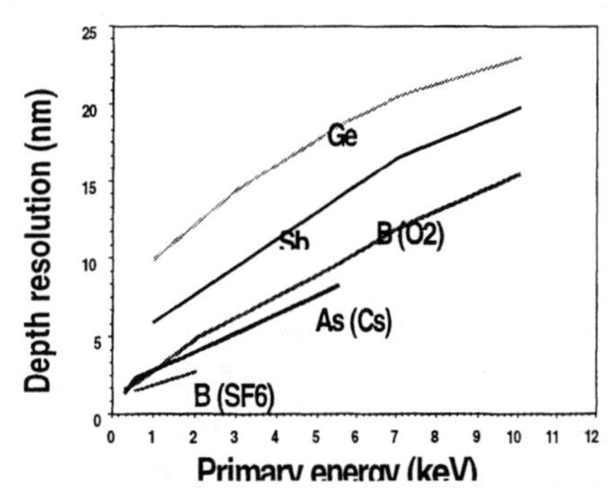

FIGURE 6. Depth resolution (defined as 3x decay length) for B, Sb, As and Ge in Si using an oxygen primary beam (B, Ge, Sb), an SF6-beam (B) or Cs (As) beam [28, 30].

Again low energy beams are of prime importance to achieve high depth resolution. In addition one must be conscious of the detrimental effect of surface roughness on the final dopant profile.
Such roughness can be initially present (for instance profiling a polycrystalline layer) or arise during the analysis. In that respect it is important to realize that the decay length in itself is not a sufficient criterium to assess the impact of surface roughness. The data shown in Fig. 7 demonstrate that whereas the FWHM of a delta layer increases proportionally with the surface roughness, the decay length is almost unaffected in line wtith theoretical predictions.

Surprisingly the impact of roughness is again element dependent (cfr the difference between B and Ge).

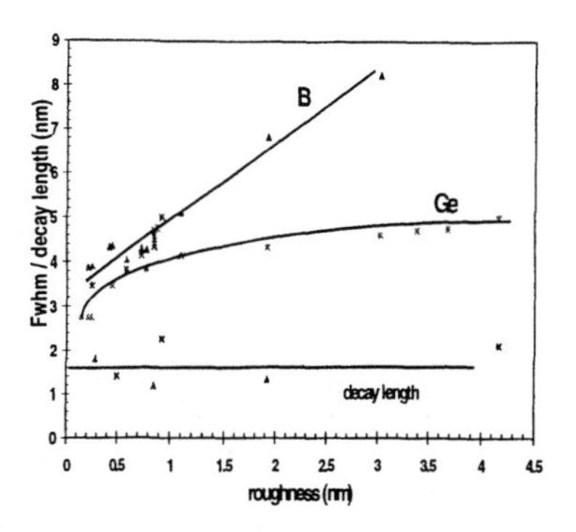

FIGURE 7. Influence of surface roughness on the decay length and FWHM of B (triangles) and Ge (squares)-delta layers in Si. Analysed using 1 kev O2 under various oxygen pressures.Extracted from [23].

The importance of surface roughness can also be seen on the typical dopant profile in a polygate-oxide silicon structure. The data in Fig. 8 were obtained under various flooding levels. In particular at intermediate oxidation levels roughness becomes prominent leading to a distortion on the As-profiles within the polylayer. The decay of the As into the oxide/Si is, however, almost independent of the analysis conditions and the degree of roughening.

As a summary one can conclude that the use of very low energy primary beams greatly has expanded our capability to probe shallow and very steep profiles. Errors and distortions induced relate to the build up of the altered layer, the non-stationary ionization yields and the development of surface roughness. Although these are still the subject of intense investigations, their magnitude (1-3 nm shifts) has been reduced so dramatically that for most cases of process assessment and evaluation they no longer form an impeding factor.[33]
When selecting the proper analysis conditions a depth resolution approaching the nm level can be achieved while still maintaining a high degree of quantification accuracy. Despite these appealing properties interpretational problems remain when dealing with more complex structures where, when going from one matrix to the next one (for instance SiO_2-Si), sputter yields and ionization yields may change and more complicated quantification schemes are required [34,35].

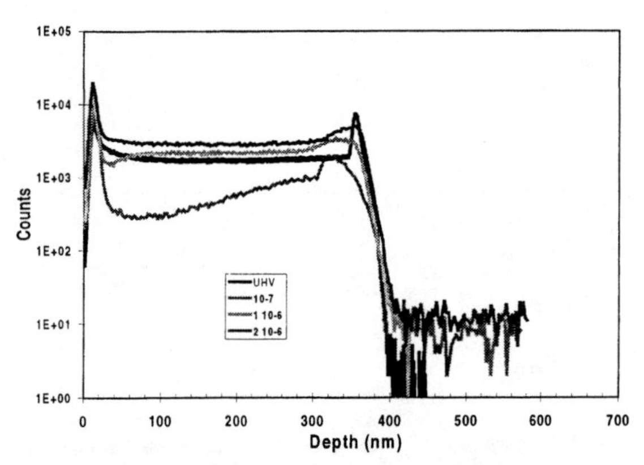

FIGURE 8. Distortions induced by roughnening on the As-profiles of a poly/ox/Si structure under various flooding conditions

1D-Carrier profiling

Spreading resistance profiling

Despite its very attractive features, one major limitation of SIMS is of course that it does not provide information on dopant activation. Therefore, alternative methods are necessary that sample the carrier distribution instead of the dopant distribution. The spreading resistance probe (SRP) is a technique which samples the electrical carrier distribution by measuring the (spreading) resistance between two metal probes which are stepping across a beveled sample surface [36].

SRP is in essence sensitive to the "local" resistivity and thus the net carrier concentration. Since a resistance is measured, very high dynamic range and low detection limits are feasible. The beveling process itself provides a geometrical magnification (100-1000x) such that with a horizontal step of 1-2.5 μm of the probes, a "geometrical" depth resolution of a few nm can be achieved. The advantages of probing the electrical distribution in addition to the SIMS profiles is the fact that one provides the information which is necessary to understand device operation[31]. Despite these interesting properties, SRP has not gained the same wide acceptance as SIMS due to poor results in terms of depth resolution, quantification accuracy. Results of a recent round robin on SRP [38] have demonstrated that the quality of the SRP profiles (depth resolution, reproducibility and quantification) can be correlated with the finer details of the probe tip and surface preparation. The ability to measure the narrow base of a bipolar transistor can for instance be directly traced back towards the "total" depth penetration in the measurement, i.e., the probe penetration plus the surface roughness (Fig 9).

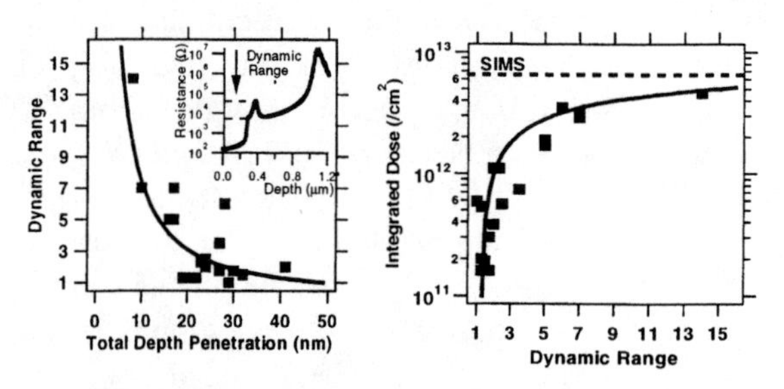

FIGURE 9. Correlation between dynamic range (as defined in the insert) in the raw SRP data for a thin base bipolar transistor and the total depth penetration of the measurement (left) and the calculated dose in the base region (right). Taken from the SRP-Round Robin [41].

It was for instance found that the delineation of the narrow base (characterized by the dynamic range within the base region) is strongly dependent on this penetration. The data shown in Fig.9 indicate a large difference in peak height which, after data conversion, leads to strong differences in calculated dose within that layer. Provided care is taken in these experimental procedures, profiles as shallow as 30-40 nm have been adequately measured [39] (Fig.10).

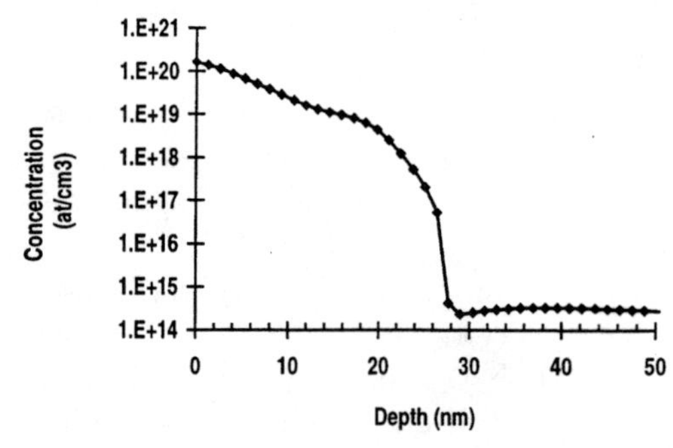

FIGURE 10. SRP profile of shallow implant showing a high depth resolution.[48]

Other very important parameters are the correct use of correction algorithms, the description of the probe-semiconductor contact and the stability of the (entire) calibration curve. The development of special staircase structures [40] for tracking the daily shift of the calibration has for instance enabled to highlight the importance of this calibration procedure. The recent efforts to develop detailed qualification procedure for SRP will enable users to fine tune their operation and to make a better use of this powerful technique [41]. The results of this round robin indicate that those labs meeting the criteria of surface preparation, probe conditioning, stability of the calibration curve and using the appropriate calibration and correction procedures, are able to measure the entire range of profiles present in the CMOS process with at least 10 % repeatability and reproducibility [41].

More fundamental limits in the SRP applications arise from the fact that the measurement not only samples the resistivity of the top surface layer but equally well contains a contribution from the underlying layers. Consequently the "electrical" resolution is not as good as the "geometrical" resolution and a complex deconvolution procedure is required to unravel the various contributions. Although adequate algorithms are used nowadays [42,49], the need for such a deconvolution implies that finer details will not be (easily) discernible. Another more important limitation is the enhanced redistribution of carriers (i.e their relocation with respect to the dopant atoms) as a result of the beveling procedure itself (usually termed "carrier spilling") [44]. This effect translates into a shift of the apparent junction position towards depths shallower than those obtained from SIMS profiles [45]. The magnitude of this effect is closely related with the gradient of the dopant profile and may become as large as 0.5-1 μm for slowly varying profiles like a well-structure.
For steeper profiles (like source/drain structures) the shifts are much less pronounced (20-60 nm) but of course in relative terms (these profiles only extend to 0.1-0.15 μm) are equally important. Whereas the modeling (and thus correction) of this effect can be performed by solving the Poisson equation, the detailed implementation requires the selection of boundary conditions properly describing the effect of the surface preparation (surface states !!) and the probe-semiconductor contact. Since pressures as high as 12 GPa are present under this contact, effects like bandgap narrowing, changes in dielectric constant and phase transitions need to be included in this kind of calculation [46]. A 1D-correction procedure using such a model has recently been developed to generate carrier profiles corrected for carrier spilling effects [47].
Although a significant improvement can be observed when compared to SIMS, this procedure is still not universally applicable (some of the model parameters depend on the profile under investigation) and can only be viewed as a first order iteration. One of its main limitations is that it is a pure one dimensional treatment whereas one really should consider the three-dimensional distortion of the carrier distribution (for instance the pressure effects will only be present under the tip and not in between the probes) in the SRP experiment. In particular the 3D-effects become more pronounced at very shallow profiles, necessitating the development of a full 3D-treatment of the carrier spilling problem [47].

Finally it needs to be mentioned that good agreement in particular for shallow profiles can only be obtained when the effect of the surface damage is taken into account as well. Indeed one can always observe a rise of the SRP-signal when stepping from polished surface towards the beveled surface even before the actual bevel edge is encountered. Moreover at the bevel edge itself a step in the resistance data can be seen as well. A tentative explanation for this effect is related to the surface damage introduced during the beveling process which does make part of the near-surface layer electrically non-active. RBS-channeling measurements have demonstrated this layer to be as thick as 10 nm. [48] When accounting for the presence of this layer as an extra barrier layer in the calculations [49], near surface concentrations for shallow profiles are increased and good agreement between sheet resistance values obtained from four probe measurements and SRP profiles can be obtained (cfr table 3).

TABLE 3. Comparison of sheet resistances from four point probe measurements, and calculated from SRP data using no correction and with correction for surface damage layer [49].

FPP (Ω/sq)	Standard SRP (Ω/sq)	Surface damage corrected SRP (Ω/sq)	Junction depth (nm)
353	695	348	127.6
696	1012	704	88.6
1373	3987	1421	42
3944	12300	3923	158
11100	71200	11920	32
455	765	462	49.6

Micro-SRP

One of the important conclusions of the 1D-model is that the pressure effects play a determining role in the extent of the carrier redistribution. An interesting approach to overcome (reduce) this limitation is to use a scaled down version of SRP i.e. MicroSRP [50]. In this approach the metal probes are replaced by a conductive Atomic Force Microscope (AFM) tip working at a much lower probe load (<10 mg versus 5 g) and with a much smaller contact radius (50-100 nm versus 1-2 μm). Obvious advantages are the reduction in "convolution" and thus correction factor (which scales with the probe radius) and the improved depth resolution. For instance for the shallow profile of Fig.11, the correction factors for MicroSRP go up to 5 whereas for SRP they go up to 100. Moreover since such a system can work with a radius of 50 nm and 50 nm horizontal steps, a theoretical depth resolution as small as 0.5 nm for a standard 34' bevel could be obtained. In order to achieve these numbers, extra efforts are required to arrive at smoother surfaces. But since the system now also provides in-situ height measurements (from the standard AFM operation), polishing procedures leading to surface rounding are no longer a problem. In fact with the AFM height information a depth dependent probing step can be introduced similar to the approach of D'Avanzo [51] . The data in Fig. 11 show that this is an important effect affecting the depth scale considerably. Note that this correction also has an impact on the calculated concentrations removing the typical too low surface concentrations in the SRP profile (Fig.11) [51].

Since the bevel is still present, carrier spilling will be operational as well. However the strong reduction in the contact radius does imply a significant reduction in the volume experiencing the high pressures, thereby reducing the importance of the pressure induced carrier spilling effects. This makes the entire problem more tractable and the technique more applicable for shallow profiling. Results reported in [50] demonstrate that for every profile measured with MicroSRP, the junction position is indeed deeper than compared to the standard SRP and much closer to the theoretical predictions solely accounting for the carrier spilling induced by the bevel geometry.
At present no commercial instrumentation is available (ideally one must be able to scan over distances larger than normally provided in a standard AFM-system) and one must still answer the questions regarding probe tip lifetime and reproducibility. Provided these can be resolved and a commercial instrument were to be developed, MicroSRP would move to main technology.

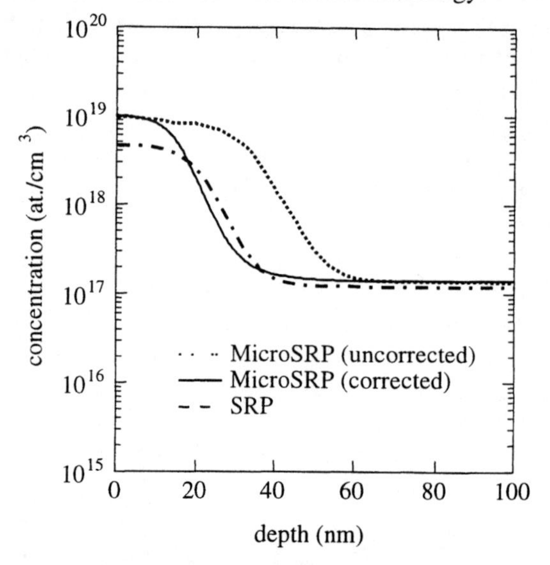

FIGURE 11. MicroSRP and SRP profile taken on a shallow profile. Corrections are made in MicroSRP to account for the effect of the bevel rounding.

Microwave Surface Impedance Profiling

One of the main problems associated with SRP and MicroSRP is the need to form a bevel. The material removal associated with this beveling procedure is the main cause for the carrier spilling problem. Note that the latter is not only present in techniques using beveling but equally well with techniques based on etching such as FPP combined with anodic oxidation and etching.
An interesting concept, Microwave Surface Impedance Profilometry (MSIP), was introduced several years ago by J.Martens et al. [52]. Their approach is to form a millimeter wave cavity between a spherical mirror and the sample surface. When the frequency of the applied mm-wave signal is such that the spacing between the mirror and the sample is an integral number of wavelengths, a resonance occurs. As the surface resistivity of the sample determines the frequency and the width of the resonance, their determination provides a way to probe the surface resistivity. By making measurements over a frequency range the skin depth

(which is basically the penetration of the electromagnetic field into the sample) can be varied allowing to probe the (integral) vertical variations in conductivity. A mathematical deconvolution is required to finally construct the carrier profile. When applied to shallow and steep profiles (with junction depths ranging from 45-100 nm), Ishida et al. [53] were able to demonstrate the excellent depth profiling capabilities of the technique. Surprisingly enough the profiles agreed with the SIMS profiles even in the tail region where one would expect the carrier profile to be deeper than the SIMS profile due to the carrier outdiffusion. Since the depth information is extracted from measurements averaging over increasing skin depths, the necessary deconvolution could result in some loss in depth resolution as a function of depth which might be the cause for this problem. Clearly more fundamental work is required to understand the limitations and properties of this technique but the appealing results demonstrated by Ishida et al. [53] justify a further investigation. The most obvious limitations are the relatively large sampling area (1 mm) and, at least until now, the absence of a commercial tool and more detailed knowledge. As this technique is totally non-destructive with good depth resolution and apparently highly quantifiable, it could however be the ideal method for shallow carrier profiling and its commercial implementation should be pursued further.

TWO-DIMENSIONAL PROFILING

The dopant/carrier profiling techniques described above relate to the analysis of relatively large, homogeneously doped (in the lateral direction) samples. Inevitably this implies that two-dimensional effects, such as the interaction between a dopant profile and the flux of vacancies or interstitials induced by a nearby implantation, can not be investigated. It is however true that many of the problems addressed today such as the reverse short channel effect, find their origin exactly in these spatially related interactions [54]. It is clear that in order to study these effects one needs to perform analysis on devices with the same geometry as the final structures. As it is important to determine not only the dopant distribution but equally well its registration with respect to the other structures such as gate material or spacers, it becomes essential to probe on the cross section of a device and to be able to image simultaneously these features. These considerations imply that a good 2D-profiling tool should not only be applicable to dedicated test structures but equally well to regular device structures. Of course the latter does imply that it will be of great value in failure analysis as well.
An important conclusion from these considerations is that spatial resolution is now becoming the most important criterion for the characterization tools. As one is facing the limitations put forward in Fig 2, one must concentrate on tools which are not directly counting atoms but merely probing a related property (for instance their electrical effect such as resistivity) since these can quite often be recorded with much better statistics and higher sensitivity. An alternative way to circumvent this statistical limitation is resort to some kind of multiple detection scheme or to increase the analyzed volume by some geometrical magnification. The latter may provide at the same time means to achieve the required spatial resolution. In particular the attempts to perform 2D-analysis with SIMS, all fall into this category.

2D-Dopant profiling

In SIMS the spatial resolution is eventually set by the spot size of the probing beam or the resolution of the imaging system. In either case the achievable resolution is much larger (0.5 μm - 50 nm) than the requirements put forward in table 1. Moreover, as pointed out in Fig. 2 direct application to a cross section will not be feasible in view of the low counting statistics. The consequence of these observations is that no technique can presently be promoted as the main technology for 2D-dopant profiling.

Solutions proposed to solve the spatial resolution problem all are based on some kind of magnification and/or averaging over larger areas. As such these methods can achieve (or hold the prospect to) the required spatial resolution at the expense of special sample preparation and routine applicability, justifying our classification as supporting technologies.

Among the solutions proposed the one of Cooke et al. [55] is quite interesting as use is made of a set of identical lines cut under a very small angle by some complex processing treatment. Details of this procedure have already extensively been described [55]. Suffices to say that when moving along the plane of the intersection with the ion beam, one basically gets now an expanded view of the 2D-distribution, the spatial resolution being set by the beam spot divided by the geometrical magnification (20-100). The use of multiple lines provides an increase in effective sampling volume thereby circumventing the limits related to the small number of atoms in one pixel. The main drawback is that the technique is only applicable to special test structures and fabrication of the angle cuts involves a very complex processing as well. Finally, spatial resolution is strongly subjective to the absence of any (small) disturbances such as roughness of the cuts, development of topography during the SIMS profiling etc. In particular the presence of these artifacts (and their impact) has kept the group from achieving until now, the expected geometrical resolution. Even if these problems can be resolved, the need for special structures and the complex extra processing, makes it not really suitable for routine applications nor as an failure analysis tool. Its role will be more in supplying for some dedicated 2D-dopant diffusion experiments, the dopant information complimentary to the electrical carrier profiles extracted from the tools described further on.

Along similar lines Goodwin-Johansson et al. [56] proposed a tomographic sectioning (i.e sputtering under different angles) thereby speculating that, if sufficient cuts could be made, sophisticated data treatment would allow to reconstruct the entire dopant distribution. As these experiments were performed in a standard SIMS instrument the number of cuts was fairly limited and the reconstruction conversely rather crude. It is possible that with the advent of more flexible FIB systems equipped with SIMS attachments better results can be obtained. However the features to be determined (a few nm of underdiffusion) are so small with respect to the rest of the source/drain profile that the latter will always dominate the collected signal and thus large errors on the lateral reconstruction will be observed.

An extreme case of tomography is the "Lateral SIMS" approach of von Criegern [57,58] whereby he sacrifices the depth resolution completely and performs a SIMS profile on the side of the sample. The results are somewhat ambiguous to interpret as now not a dopant distribution is determined but merely the integral over the entire depth at each spatial location. The advantage of this approach is that the "sidewall" sputtering implies that spatial resolution is set by the depth resolution of SIMS (a few nm) and no longer by the beam spot size (> 50-100 nm) . The requirements on the test structure are more relaxed, necessitating only a repetitive structure of a few 100 µm long coated with a Si-layer a few µm thick. (The latter turns out to be the most difficult item to obtain in a standard processing line). The volume sampled in one data point is again relatively small (i.e. depth of the profile 0.1-0.5 µm x 10 µm) implying that sensitivity is reduced. By running at low mass resolution (= full transmission) von Criegern [57] was able to show reasonable sensitivity for As and B. The sensitivity specification (dose > 10^{13} at/cm^2) implies that the finer details of the tails of the lateral extension profile are not really resolvable. Moreover interpretation of the profiles is closely dependent on the topographical details of the structure. Whereas predictions regarding the Lateral SIMS profile can be made using a particular 2D-profile + sample geometry, the inverse iteration (extraction of a 2D-profile) from the experimental data may lead to a non-unique solution. Due to the convoluted form of the information provided, detailed verification of samples in terms of channel length or underdiffusion will also be difficult to perform. Again the value of the technique primarily lies in the study of model experiments and it will probably not serve as a production control or failure analysis tool.

2D-Carrier profiling

Compared to the case of 2D-dopant profiling, the situation is largely different in case of carrier profiling. Here several techniques have been proposed and explored, all aiming at satisfying the requirements while probing on the cross section of an arbitrary sample. Consequently, a few techniques (SSRM, SCM, Nanopotentiometry) can be assigned as potential main technologies for this application.
Inevitably the need for extreme spatial resolution and applicability to a device has spurred the investigation of tools based on high resolution microscopy such as

Transmission Electron Microscopy or Scanning Probe Microscopy [59-61]. In principle all these methods have in their basic mode of operation a high inherent resolution (cfr. the near atomic resolution images obtained with TEM or STM). The key element deciding their applicability for ULSI-carrier profiling is however the concurrent generation of an appropriate signal related to the local carrier density. For that purpose these methods rely on (concentration dependent) selective etching (inducing a topography which then contributes to the image formation [60, 62-65]), or on the measurement of an electrical signal (resistance [66], capacitance [67], electric force [68]) which can be associated with the carrier concentration. It is clear that techniques sampling a property only weakly dependent on the carrier concentration are not very favorable as these will have a hard time satisfying the requirements in terms of concentration resolution or sensitivity. A good example in that context is for instance Scanning Tunneling Microscopy/Spectroscopy. The tunneling current is indeed dependent on the local Fermi level position and thus on the carrier concentration but since the latter varies only over a linear scale when the concentration varies over orders of magnitude, concentration resolution (i.e. the ability to detect small changes in carrier concentration) is compromised. The direct atom imaging capability of STM has only been demonstrated on III-V material [69, 70] but, even it would be feasible on Si, it would be of limited value in the present context as one will be faced directly with the poor statistics sketched out in Fig 2.

The limited concentration resolution is also a problem for the image formation based on secondary electron emission [71,72] since the observed contrast only varies from 0 to 15 % when the concentration ranges from 10^{14} to 10^{20} at/cm^3 [73].

One of the techniques reaching out almost to the main technology level is selective etching combined with TEM imaging. The preferential etch induces a concentration dependent topography which can be observed either using TEM or AFM. The use of AFM imaging originated from the idea to determine the topography in a quantitative manner as one then could convert the topography to concentration levels by the use of appropriate calibration curves [74].

Despite several attempts and the use of more controlled etching procedures (such as the etching within an electrolytical cell [75]) the latter concept has not materialized to a routine technique as the etching turns out to be fairly irreproducible. Variations in etch rates are observed related to different lighting conditions, size of the doped regions, dopant gradients and equally well local stress and defect levels in the sample [76]. Moreover tip convolution problems in the AFM imaging limit the final spatial resolution and in particular the determination of high dopant gradients.

Observation of the topography in the TEM leads to interference fringes which tentatively are interpreted as iso-concentration contour lines [61]. Quantification is always based on a cross calibration with a 1D-SIMS or SRP profile. Strengths of this technique are the simultaneous observation of the entire structure (highlighting fine details on overetching, surface recession, gate structure...) [32] and the apparent automatic generation of iso-concentration contour lines (Fig. 12).

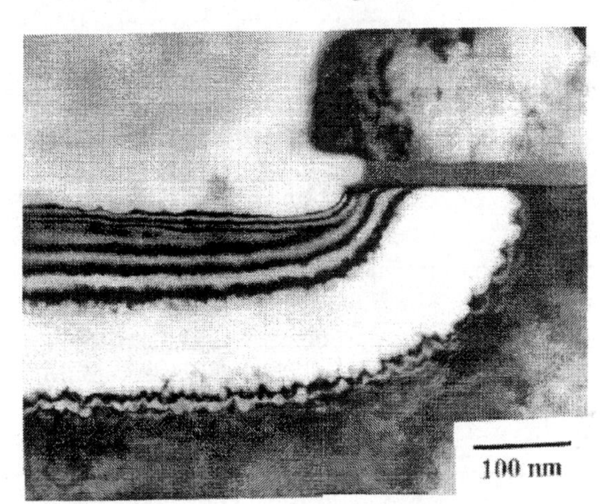

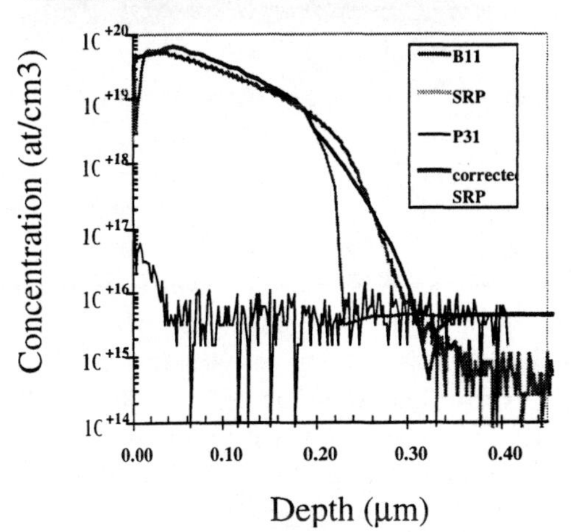

FIGURE 12: (top) TEM micrograph illustrating interference fringes obtained after selective etching. Quantification needs to be done by assigning a concentration level to the different fringes by correlating the depth of the fringe with the concentration level

at that depth derived from 1D- SIMS or SRP-profiles (bottom). Corrected SRP refers to the use of an algorithm for carrier spilling correction. [61]

The latter could however be an artifact from the technique as after all the surface will remain more or less continuous during the etching. The main limitation is the poor sensitivity limiting the practical application to concentration levels in excess of 10^{17} at/cm^3 . At present there is also a lack of information on the influence of gradients and stress fields on the etching mechanism and thus the accuracy of this quantification in the 2D-region.

The techniques discussed further in this section, map out the spatial distribution of a electrical property (capacitance, resistivity, potential...) which should be related to the local carrier concentration rather than the individual dopant atoms themselves. The difficulties in these approaches are the quantification, as one needs to establish a unique correlation between this property and the dopant/carrier concentration at the measurement point, the sensitivity or concentration range which can be covered, the concentration resolution and the final spatial resolution which is influenced by the intrinsic imaging resolution as well as by the response of the investigated property to concentration gradients and the sampling volume point (nearby layers could for instance contribute to the signal).

Scanning Capacitance Microscopy

The Scanning Capacitance Microscopy (SCM) technique is one of the first 2D-profiling tools to be provided as a commercial product and as such has gained considerable interest in recent years. The basic concept is a replica of the macroscopic C-V measurement on a MOS capacitor which allows to probe the carrier concentration at the edge of the depletion layer. In the present case a miniature MOS capacitor is formed between the tip (=metal) and the sample. The oxide layer normally present in the MOS capacitor is in this case the native oxide present on the tip or on the cross sectional surface or can originate from an extra oxide forming step (polishing, heat treatment, anodic oxidation, UV-treatment..).
The tip itself is connected to a highly sensitive capacitance sensor operating at a high frequency (915 MHz) such that the very small capacitance values still can be measured. The capacitance involved is typically in the order of 1 pF and almost entirely composed of stray capacitances. Hence spatial variations related to spatial changes in dopant concentrations will be very difficult to detect. Consequently, rather than looking at the total capacitance as a function of position along the cross section, the changes in capacitance are detected when an extra, slow AC voltage modulation is used. Depending on the amplitude of this AC-voltage the sample (below the tip) will be swept from accumulation into depletion and vice versa. The detected SCM signal (10^{-18}-10^{-20} F) can be extracted using appropriate lock-in techniques and basically provides the dC/dV-signal which is carrier concentration dependent. It needs to be mentioned that in addition to the AC-modulation voltage a DC-bias is used to position the AC-modulation on the steepest part of the C-V curve. A strong interplay exists between these voltages as the SCM-signal and the concentration contrast can be influenced by the bias and the AC-modulation voltage and even the amplitude of the high frequency signal [77]. Depending on the operational parameters the SCM-signal is not monotonically dependent on the concentration but a contrast reversal can occur at intermediate concentration levels (i.e low and high concentration levels may appear with the same SCM signal) [78]. Hence a careful selection of the operational voltages is required.

Two modes of operation exists for SCM : constant voltage and constant differential capacitance. In the first case a constant modulation voltage is used and the resulting changes in capacitance values form the basis for the image formation. The main problem in this mode arises from the fact that the extent of the depletion region (and thus the spatial resolution) depends on the concentration level being probed making a quantitative interpretation of the images very difficult. In the constant differential capacitance case this problem is circumvented by utilizing a feedback which automatically adjusts the modulation voltage such that a constant capacitance change (and thus constant change in depletion width) is maintained leading to a constant spatial resolution [79]. One of the main drawbacks of this mode of operation is its much slower speed for image acquisition (~ 20 min/image) making the whole measurement more susceptible to thermal drift problems.

Although very appealing SCM-images already have been produced [32], reproducibility of the measurement forms one of the major issues. Being a C-V measurement the technique is highly sensitive to the properties of the "oxide" layer and the underlying surface. Careful surface preparation and extra oxidation treatment (for instance the recently proposed UV-treatment of the cross sectioned surface) are essential to bring these problems to an acceptable level [80]. The (normally) quadratic dependence on the dopant density of the capacitance signal implies that the technique is highly sensitive at low concentrations and has a reduced signal at higher concentration levels [81]. Moreover when imaging complex transistors structures one must realize that neither oxide layers nor metal layers (or highly doped Si layers) lead to a measurable signal making the registration of the measured carrier profile with the details of the structure rather difficult (Fig. 13). For instance in Fig. 13 we compare the SCM and SSRM (see next section) image taken on the same DMOS transistor. Whereas in the SSRM-image, one can distinguish between the oxide and metal layers thereby providing insight in the sample structure, both layers show the same (zero) signal in the SCM image.

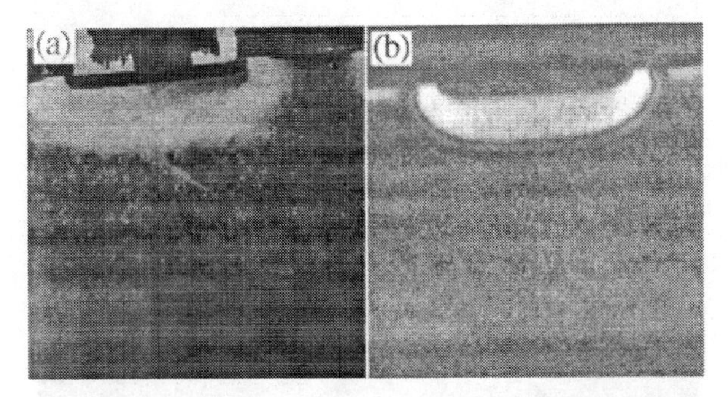

FIGURE 13: SSRM (a) and SCM (b) image of DMOS transistor.The white and dark layers in the top of the SSRM image correlate with the present oxide (white) and metal (dark) layers. The bands in the SCM image are linked to the junction positon.

The spatial resolution is determined by the extent of the depletion region which is correlated with the tip radius (including fringing fields coupling into the non-contacting part of the tip), the applied modulation voltage, the carrier concentration and the possible coupling towards nearby layers. Values in the order of 20-50 nm have been quoted in the literature but, as there exists at present no "standard" 2D-profiling tool or 2D-sample, it is hard to come up with very exact numbers for the spatial resolution. Judging from the resolution deduced from the measurements on the Sematech round robin structures (seethe discussion on Fig.17) the present resolution appears to be on the order of 30-40 nm.

Much of the work related to SCM is focused on the theoretical interpretation and quantification of the measured SCM images. In a simple MOS capacitor standard equations can be used to derive the carrier concentration [36]. However in the case of SCM the situation is more complex as one needs to consider the stray fields between the sample and the shaft of the tip, the laterally non-uniform carrier profile and the (unknown) properties of the surface. The most simple quantification procedures are based on a calibration using 1D-SIMS/SRP profiles (similar to TEM imaging) [67] or at least the use of 1(2) known concentration levels [82]. Implicit assumption is that there is no effect of the 2D-distribution on the SCM-signals such that a simple 1D-calibration can be applied. Huang et al. [83] developed a quasi-1D analytical model for the SCM signal by dividing the probe-sphere sample contact into annular rings and by calculating for each ring the corresponding capacitance. No effect of the profile gradient on the (individual) signal is included but the model already does allow to estimate the averaging resulting from the tip geometry [84]. As intuitively expected the results indicate that a flat contact has a much stronger averaging effect (i.e. leads to a stronger loss in profile resolution) than a perfect sphere (with of course only has a single point contact) with the same radius. The experimental SCM results obtained on the Sematech round robin samples do reflect a relatively strong averaging which can be simulated using a flat contact of 35 nm. The latter is not so surprising since in practice the outer end of a tip will probably look more like a flat contact than a perfect sphere. Deconvolution of this averaging effect is in principle possible [85] but does suffer from a limited degree of accuracy on the finer details of the profiles as the inevitable noise on the data may lead to the generation of unrealistic or biased solutions. Use of the maximum entropy deconvolution concept as used by Chu et al. [86] in SIMS profiling may represent a possible solution to improve the quality of the deconvoluted data. Despite the successes suggested

with this deconvolution method, prime objective should be to obtain directly data with higher resolution.
A fast algorithm has been developed by Kopanski et al.[87] using interpolation on the stored solutions of the three-dimensional Poisson equation [88] for various tip-sample bias voltages, oxide thicknesses, dopant concentrations and tip radii. The probe is modeled in this case as a cone with a hemispherical shape which does include the effect of stray fields. Moreover the calculations assume an infinitely small AC-signal (as well as HF-signal) whereas in practice this can be quite large. Including the effect of the amplitude (i.e averaging over part of the C-V curve) of these voltages does lead to slightly different results and can for instance explain some of the contrast reversals recently observed [77]. In the conversion of the SCM-signal towards carrier concentration, each point is treated individually and the effect of a profile gradient is ignored. Again the effect of the surface properties (although experimentally observed to be very important) is not included in the calculations.

A probably larger source of error at present is the perturbance caused by the SCM measurement voltage itself and the very broad delineation of a junction. Indeed the applied voltage influences the position of the mobile carriers and thus distorts the profile to be measured. In the case of a junction extra contour lines can be observed (depending on the bias conditions) which can be associated with the change in sign of the dC/dV signal although additional effects like flatband potential, band bending, and carrier generation may complicate the interpretation. Recent experiments (Fig. 14) and calculations also show that the sample-tip bias has a pronounced effect of the presence, the position and the extent of these bands making a one-to-one correlation with the junction position extracted from SIMS or SRP even more complicated. Recent modeling of these effects suggests that interpretation of SCM-images can only be achieved provided a full C-V curve at every point is obtained or that at least several images with different bias voltages are collected [89,90]. The fact that the shift of the apparent junction position can at present be predicted theoretically [89,91] is an encouraging step towards a full quantification of SCM. However a full 3D-treatment will be required before the objectives in terms of concentration resolution and accuracy can be met.

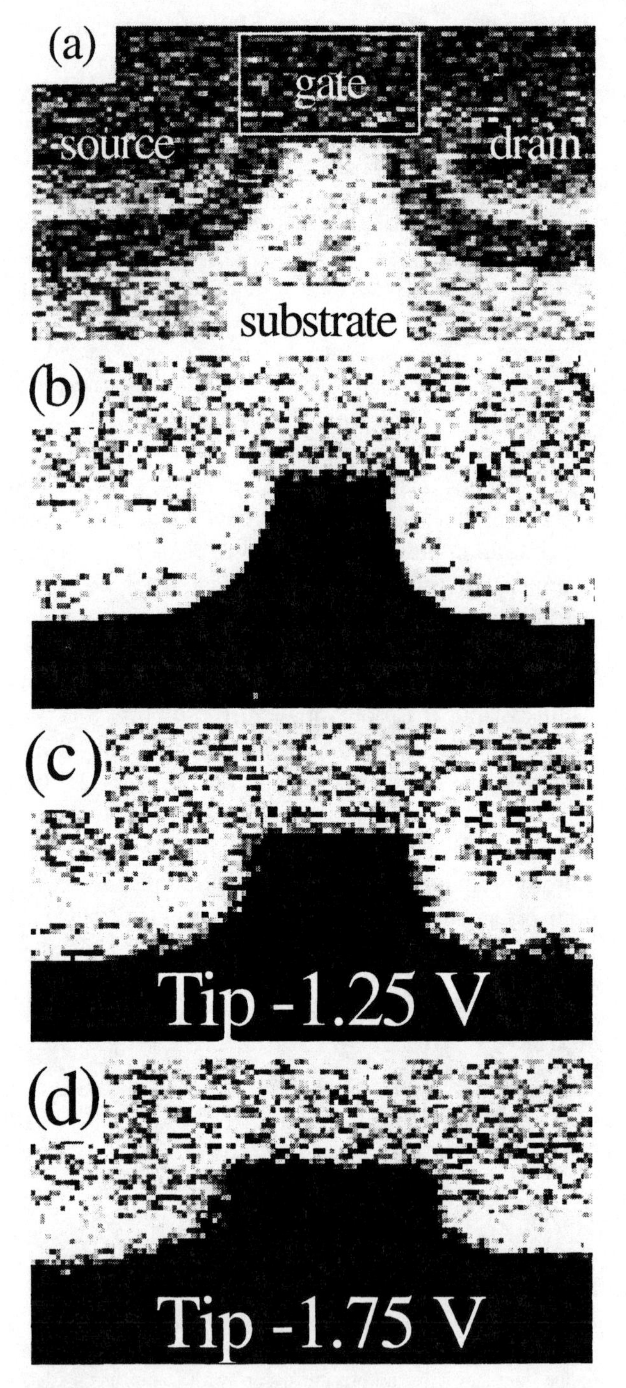

FIGURE 14. (a) Appearance of extra contour lines (magnitude of dC/dV-signal) in the junction region and (b) Displacement of apparent junction position (deduced from the phase of the dC/dV-signal, black-white transition) due to the bias voltage {b(0.25 V), c(-1.25V), d(-1.75V)} used in the SCM measurement.[90].

Scanning spreading resistance microscopy

Spreading resistance microscopy (SSRM) originates from the positive observation of the SR properties i.e the linear response towards the carrier concentration level and the relative ease of quantification. Compared to the standard SRP system which uses the two large metal probes and a bevel, a fine conductive AFM-tip is used which is scanned across the cross section of a device [92]. A large current collecting contact is placed on the other side of the cross section. The "spreading resistance" measured in such a configuration is a direct measure for the local resistivity (at least for uniformly doped samples) according to :

$$R=\rho/4a \qquad (2)$$

with ρ the local resistivity and a the tip radius [76, 92]. The interesting aspect of SSRM as exemplified by eq.(2) is its linear dependence on the resistivity (providing a constant concentration resolution over the entire concentration range) and the very large dynamic range. For instance when the resistivity varies from 10^{-3} Ω-cm (10^{20} at/cm^3) to 100 Ω-cm (10^{14} at/cm^3) the spreading resistance changes from 1000 Ω to more than 10^8 Ω [40]. The limits in measurable concentration are determined by the smallest current (for the lowest concentration) which still can be measured (as typically only a 5-50 mV bias is used, currents as small as 1 pA need to be measured) and the resistivity of the tip (for the highest concentration) as this has to be less than the resistivity to be measured. The recent development of tips based on highly doped diamond ($\rho < 10^{-3}$ Ω-cm) [93, 94] have moved this limit beyond the $3x10^{20}$ at/cm^3 level. Accurate determination of the resistance over such a wide range does represent no major problem in this technique.

In practice small deviations from this purely linear relation are observed probably due to non-linearities in the probe semiconductor contact but these are quite easily accounted for by using a calibration curve covering the entire range of interest. The recently developed staircase calibration samples [40] actually allow the acquisition of this curve in a very fast and elegant way.

The key element in establishing a reproducible contact with the SSRM method is the force used during the measurement. It has indeed been observed that a minimum force (for a particular tip) is required in order to penetrate through the native oxide and to form a (nearly) ohmic contact [95]. In fact when a pressure in the order of 10-12 GPA is achieved below the tip area, the semiconducting Si transforms into a metallic β-tin phase which drastically reduces the contact resistance. Typical values for this force are 1-10 μN which is much larger than the forces used in standard AFM or SCM imaging. A beneficial aspect of this high force is that the tip very slightly penetrates into the semiconductor surface making the measurement far less sensitive to the surface preparation as compared to SCM. Tip wear on the other hand is much more pronounced and high demands are placed on the mechanical properties of the tip and in particular its abrasion resistance. High aspect ratio tips (such as Si tapping tips or etched metal tips) are too fragile and deform/break very quickly under the high forces. Only tips based on solid diamond or Si tips coated with conductive diamond layers are found to withstand the measurement for a sufficiently long time. Note that since the diamond film is composed of fine microcrystallites the active tip radius is set by these fine crystallites and not by the global tip radius. Degradation of these tips is usually related to the loss of one those crystallites during the scanning.

When scanning across the cross section of the sample, the lateral distribution of the resistance can be correlated with the lateral distribution of the resistivity (or lateral carrier distribution). For that purpose it is necessary to modify eq. (2) in order to include the effect of profile gradients and nearby boundaries :

$$R= (\rho/4a)*Cf \qquad (3)$$

with a the contact radius of the tip and ρ the resistivity directly below the contact and Cf a correction factor accounting for the effect of nearby layers, doped to a different concentration level. For uniformly doped samples Cf is unity. The calculation of R (and thus the quantification of SSRM) is in principle based on the correct estimate of Cf and the determination of a. The latter can be extracted from the measurements on homogeneously doped samples (i.e. the calibration curve) whereas for Cf the solution of the three-dimensional Poisson equation needs to be sought. The latter takes into account the distorted current distribution resulting from the (unknown) 2D-carrier profile. It has been demonstrated that this calculation can be avoided through the use of parameterized solutions (based on the normalized distance to boundaries, profile gradient and contact radius) such that a simple iteration combined with look up procedure (for Cf) leads very rapidly towards a quantified profile [95]. Note that opposite to the SCM algorithms the present model does take profile gradients into account. One of the main reasons for the efficiency of this procedure is the relatively small corrections (typically less than a factor 2) which need to be applied to the data. As such a very good initial guess on the 2D-profile can be made simply by using eq. (2) (Cf=1) for the conversion. Final convergence only requires one or two extra iterations.

Spatial resolution is again set by the tip radius. By lack of a good 2D-calibration standard spatial resolution needs to be determined on cross sectional 1D-profiles. This can be done by determining the depth required for the 84-16 % decrease in measured carrier concentration on the (abrupt) dopant steps (cfr discussion on Fig.17) or from an analysis of the resistance variations when scanning across a series of very thin oxides [96]. In the latter case spatial resolution can be deduced from amount of resistance increase while scanning over a particular oxide (in our experiments ranging from 6-20 nm). Both experiments lead to a similar value of 15-20 nm for the tip radius.

Again several examples of 2D-images have been reported [32, 92, 97]. In Fig.15 the asymmetrical carrier concentration resulting from a 35° implant (As) is shown. One can clearly observe the different underdiffusion resulting from this asymmetrical implant and almost resolve the 5 nm oxide.

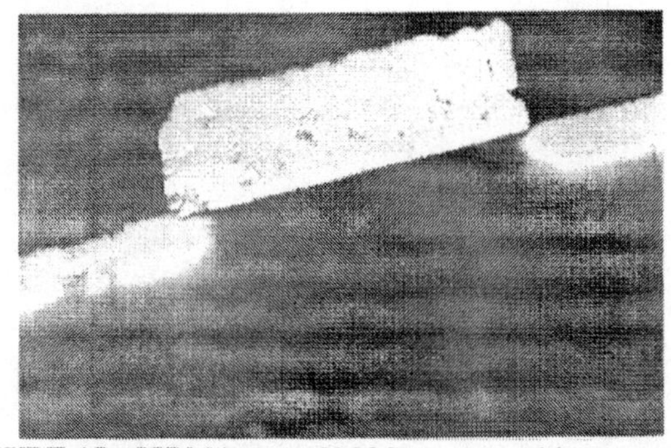

FIGURE 15 : SSRM image of MOS structure with asymmetrical (35 °) As-implant. Scan size 2.5 x 1.25 μm.

Advantage of the SSRM technique is that it does provide a strong signal for metal (small resistance) and oxide layers (very high resistance) such that details of the structure can quite easily be observed. The latter is exemplified in Fig.12 where the SSRM image clearly outlines the oxide and metal layers as compared to the SCM-data.

An even more important feature is that due to the small measurement voltages used (5-50 mV) virtually no carrier displacement occurs which complicates the SCM images so drastically (as there bias voltages in the volt range are used). Moreover the junction location shows up as a very well defined peak in the resistance profile (similar to the case in standard SRP) due to the low carrier density in the depletion region. The strong difference between SSRM and SCM imaging of junction isolated structures can be seen in Fig 16. (Although the extent of the bands in the SCM image could be influenced by the sample preparation and applied bias voltages, they never disappear and a crisp delineation as in the SSRM image is never obtained).

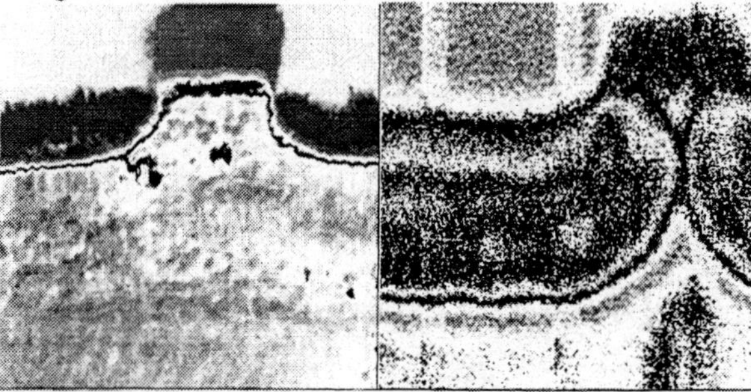

FIGURE 16. : SSRM (left) and SCM (right) image of NMOS transistor. Scan size 1 x 1 μm. The sharp black line (SSRM) and the broad band (SCM) delineate the junction position.

Disadvantage of the SSRM technique is the high force used during the analysis which is destructive for the tip as well as the sample surface. Subsequent images taken on the same spot usually lead to a serious loss in spatial resolution. Hence positioning of the tip in the area of interest requires the necessary precautions. We have found that using SCM for localizing the region of interest and SSRM for quantitative analysis appears to be an ideal combination of techniques.

In order to complement the above discussions, it is interesting to compare the 2D-profiling methods when applied to the same structure. As no calibrated 2D-standard exists yet, use can be made of a set of well characterized 1D-structures whereby the resolution of a particular technique can be tested by scanning (on the cross section) across a steep dopant profile. An interesting set in that respect are the Sematech round robin samples [98]. Structure #1 has

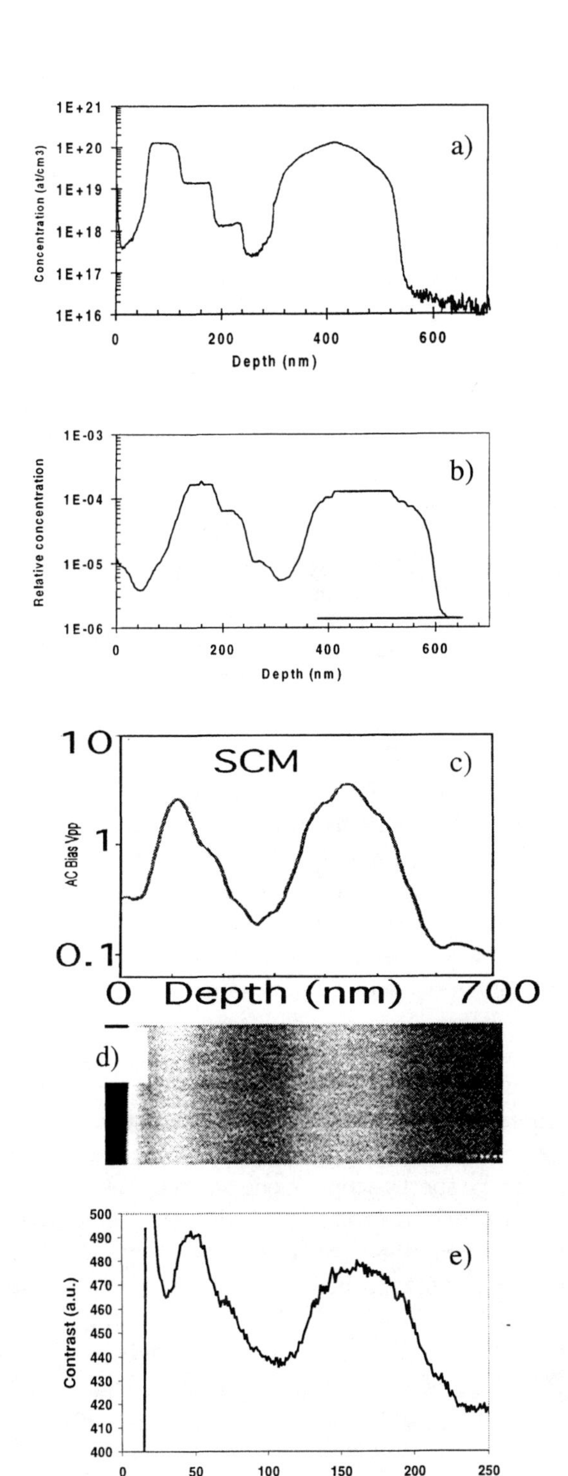

FIGURE 17 : Intercomparison of various 2D-profiling techniques using the Sematech structure #1: dopant profile measured with SIMS (a), carrier profile with SSRM (b), SCM (c) and secondary electron emission.(d)[72]. Fig.17e is a line scan through d).

abrupt boron doping steps between 10^{20}, 10^{19} and 10^{18} cm^{-3}. Each dopant concentration is confined to a thickness of roughly 50 nm (Fig. 17a). Low temperature growth was used in order to limit outdiffusion such that the dopant steps would be steeper than the resolution of the steps. Structure #2 contains an abrupt p-n junction and some narrow boron-doped plateaus.

The data in Fig.17b show the raw SSRM data but plotted as 1/R which is ~ρ (cfr eq.(2)) in order to facilitate comparison with the dopant profile. Raw SCM data are shown in Fig.17c [84]. The steps in the boron plateaus only show up in the SCM data as a change in gradient in the SCM profile whereas in the SSRM data a plateau in resistance values is observed corresponding to the various doped layers. From these results one can deduce that (at least in this case) the SSRM has a slightly higher intrinsic resolution as compared to the SCM data, since the raw data are already much closer to the actual profile shape. Using the 84-16% transition depth as a measure for the spatial resolution, we find an average resolution of 15 nm for SSRM and 25 nm for SCM. The shape of the SCM-profile and the derived value for the resolution are commensurate with the predictions made by McMurray et al. [84] based on the averaging effect of a flat 30 nm contact. They also showed that, at least theoretically, they can reconstruct from this data the actual dopant steps using their deconvolution algorithm. (The impact of noise on this deconvolution is however another problem). For comparison we also show the results obtained with secondary electron imaging. Whereas the dopant steps can be resolved, it is clear that the low dynamic range in signal contrast (ranging from 440 –490 for a change in concentration over 3 orders of magnitude) limits the accurate determination of concentration levels.

Our present assessment of the various techniques is summarized in table 4. In addition to the standard values as resolution and dynamic range, it is important to assess the concentration resolution (how does the signal vary with changes in concentration), the possibility to locate the carrier distributions with respect to the metal/oxide layers (structural information), junction delineation and of course prospect the (ease) of quantification based on first principles.

Table 4. Intercomparison of 2D-carrier profiling techniques.

	SSRM	SCM	Sel.etch + TEM	Sel. etch + AFM	Sec. Electron TEM
Resolution (nm)	10-30	20-30	25	25	25
Dyn.range	10^{15}-10^{20}	10^{15}-10^{20}	10^{17}-10^{20}	10^{17}-10^{20}	10^{17}-10^{20}
Concentration resolution	$\sim \rho$	~ 1/N	weak non-linear	weak non-linear	very weak
Quantification	relatively easy	difficult	1D-calib	1D-calib	1D-calib
Structural info	Metal Oxides	None	Excellent	None	excellent
Junction delineat.	resistance peak	broad bands	none	none	none
Problems	high forces tips	sample prep. reproducibility perturbance	2D-effects in etching	2D-effects in etching tip convolution	p-n junction necessary

Nanopotentiometry

Whereas the preceeding discussion was focussed on the direct measurement of dopant/carrier profiles, it is also interesting to look at a tool (Nanopotentiometry) which actually probes more directly the device operation. The basic concept of Nanopotentiometry is to analyze on the cross section of an active device (i.e. all bias voltages are applied) the potential distribution using a conductive AFM tip acting as a local voltage probe [99]. If a very small tip is used (<10 nm), a high spatial resolution analysis of the device becomes possible. In principle one thus obtains the potential distribution *inside* the device which reflects directly the device operation and which can be compared with the result from a device simulator (cfr Fig 2). Since the potential distribution is governed by the (two-dimensional) dopant distribution, it is also possible to extract from the potential measurements the dopant distribution as well. Although initially conceived to measure the static potential (and carrier) distribution, an extension using proper lock-in techniques could be to probe the AC-behavior of the device as well.

Since the potential measurement is a high impedance probe, one does not have to worry too much about the ohmic properties of the probe-semiconductor contact and lower forces (as compared to SSRM) can be used. Evidently this offers as important benefit reduced tip wear and smaller contact area (i.e higher spatial resolution). Moreover since this is a passive measurement no convolution effects need to be considered and data interpretation can be more straightforward. Another very important difference between SCM, SSRM and Nanopotentiometry is the fact that Nanopotentiometry looks at an active device whereas SCM and SSRM probe a passive one. The latter implies that Nanopotentiometry can also be used in a spectroscopic mode whereby the potential (re)distribution inside the device can be followed for changes in gate or source/drain bias conditions or even substrate bias. The information obtained from the potential distributions and these spectroscopic studies provides not only insight in dopant distributions (and as such enables TCAD calibration) but equally well on effects like punch through, ... bridging the gap between device characteristics and dopant distributions. The latter makes Nanopotentiometry an extremely powerful tool, exceeding and complementing SSRM and SCM.

Trenkler et.al. [99] have studied the potential distribution on a reverse biased p-n junction and showed that for instance the extent of the depletion layer (as derived from the Nanopotentiometry measurements) scaled directly with the square root of the applied voltages. Moreover they found an excellent agreement with the prediction of a device simulator and even demonstrated that one can extract from the data the electric field and space charge distribution in such a structure (Fig 18).

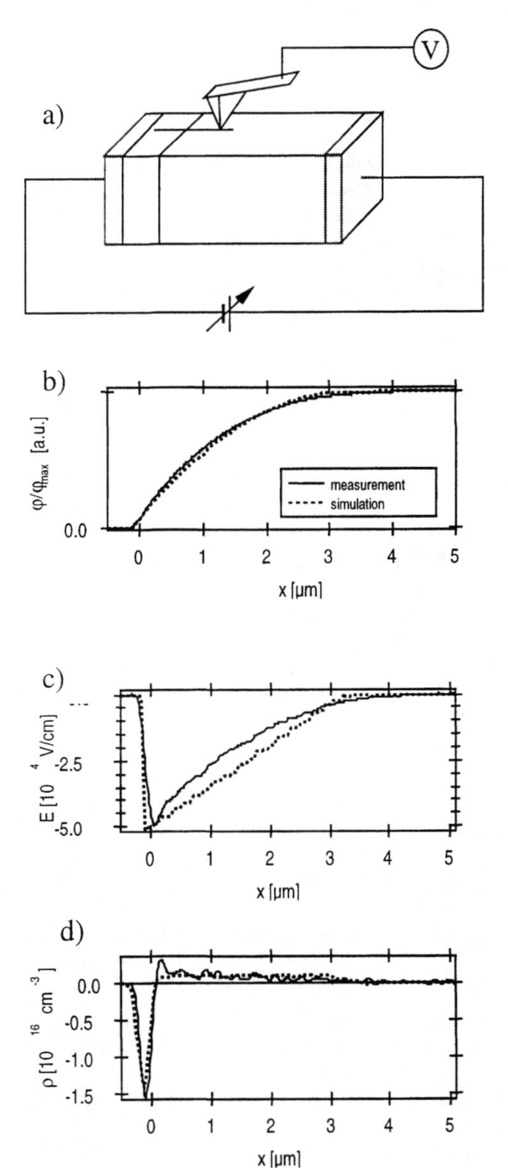

FIGURE 18. : Nanopotentiometry on a p^{+}-n junction. Fig 18a show the basic experimental set-up whereas in Fig.18(b-d) the measured potential distribution, electric field and space charge distribution are presented. Simulated distributions (assuming an abrupt dopant profile) are included as well.

With respect to its application to transistors, one must be conscient of the potential destruction of the transistor operation by the sectioning treatment or at least a strong deformation of the potential distribution by the surface damage, surface charge, interface states,...The results obtained by Trenkler et al. [100] however demonstrate that the transistors remain operational and that their drive characteristics scale with the reduction in transistor width. Depending on the surface treatment an increase in leakage current can however be observed. The quantitative nature of the technique can however be assessed from the data shown in Fig 19. In this case the potential distributions within a 0.25 µm are shown (V_{GS} = 0.5 V, V_{DS}= 1 V). The equipotential lines illustrate the detailed information which can be extracted from the technique. Switching of a transistor commensurate with the changes in applied gate voltages has also been observed [100].

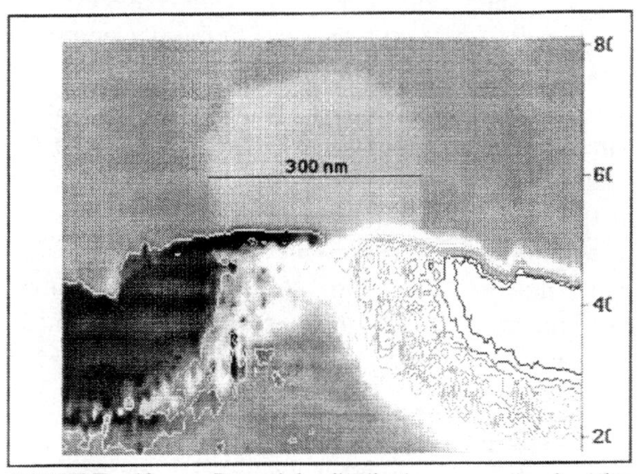

FIGURE 19. : Potential distributions measured using Nanopotentiomtry on a 300 nm transistor. Iso-potential contour lines represent 100 mV steps. The gate is biased to 0.5 V, drain-source voltage is 1 V.

Although the application of this technique is still in its early phase it does hold the promise of high spatial resolution due to the lower forces used in this measurement and the complementary information on device characteristics which can be obtained.

SUMMARY

The roadmap towards satisfying all the needs for dopant/carrier profiling for deep submicron technologies does still require the development and improvement of several techniques. With respect to 1D-dopant profiling high resolution SIMS has approached the levels requested. Much more work is still required in the field of carrier profiling using techniques based on SRP (MicroSRP) , in particular in removing the effects of the surface beveling and carrier outdiffusion. Non-destructive methods such as the MSIP might be an interesting alternative if developed commercially.

In the field of 2D-profiling or device characterization, tools based on AFM are the main technologies. SCM appears to be an attractive method but is strongly handicapped by reproducibility and quantification problems and its perturbant nature (causing large displacements in junction positions). SSRM on the hand is more quantitative and, at least at other achieves a slightly better resolution, but suffers from the large forces used making it more destructive for tip and sample. Junction positioning is however not a problem. Finally, Nanopotentiometry is an interesting complimentary tool probing the electrical characteristics of the device and establishing a (high resolution) link between device characteristics, carrier distributions and device simulation.

All the methods mentioned above do not achieve the resolution put forward in the NTRS (which might however be too stringent). Interesting prospects are contained in the recent developments of scanning probe concepts (SSRM, SCM, Nanopotentiometry). At the same time one should realize that characterization of devices (dopant/carrier profiling) using these tools should be integrated with the analysis of device characteristics and the results of process and device simulation as this represents the key route towards a more complete understanding and characterization of highly advanced technologies and devices.

ACKNOWLEDGEMENTS :

Support of the EEC for the SRP Round Robin project EEC (Standards and Materials Testing) and Nanopotentiometry (Long Term Research) is gratefully acknowledged. P.De Wolf , T.Trenkler, T.Hantschel and T.Janssens are indebted to the IWT for their Ph.D.fellowship and R.Stephenson to the EEC for a Marie Curie fellowship. Financial support from Digital Instruments, Intel, Digital Equipment Corporation and Motorola contributed to the development of the SSRM-concepts. P.Niedermann (CSEM) and O.Ohlsson (Nanosensors) are thanked for the supply of diamond coated tips.

REFERENCES

1. "The National Technology Roadmap for Semiconductors". Semiconductor Industry Association, San Jose, CA, (1997)
2. C.S.Rafferty, H.-H.Vuong, S.A.Esharadi, M.D.Giles, M.R.Pinto and S.J.Hillenius : IEDM Technical Digest, p311 (1993)
3. S.W.Crowder, P.M.Rousseau, J.P.Snyder, J.A.Scott, P.B.Griffin and J.D.Plummer : IEDM Technical Digest, p427 (1995)
4. B.Doyle, these proceedings
5. H.Gnaser, Surf. Int. Anal. 25, 737 (1997)
6. J.Hunter : these proceedings
7. R.S.Hockett, R.Bleiler and S.Smith : proc. SIMS-X, Eds.A.Benninghoven, B.Hagenhoff and H.W.Werner, Wiley,p 645, (1995)
8. K.Wittmaack, Appl. Surf. Sci 9, 315 (1981)
9. M.L.Yu, in "Sputtering by Particle Bombardment III", ed. R.Behrisch and K.Witmaack, Springer-Verlag (1991), p 91
10. J.Alay and W.Vandervorst, Phys. Rev B 50 (20), 15015, (1994)
11. W.Vandervorst, J.Alay, B.Brijs, W.De Coster and K.Elst : proc. SIMS-IX, Eds.A.Benninghoven, Y.Nihei, R.Shimizu and H.W.Werner, Wiley, p 599, (1995)
12. M.G.Dowsett, R.D.Barlow, H.S.Fox, R.A.A.Kubiak and R.Collins : J. Vac. Sci. Techn. B10, 336 (1990)
13. K.Wittmaack : Surf. Interface Anal. 24, 389 (1996)
14. W.Vandervorst and F.R.Shepherd : Appl. Surf. Sci. 21, 230 (1985)
15. K.Wittmaack : Phil. Trans. Roy. Soc. Lond. A 354, 2731 (1996)
16. H.De Witte, W.Vandervorst and R.Gijbels : proc. SIMS-XI, Eds.G.Gillen, R.Lareau, J.Bennet and F.Stevie, Wiley, 327 (1997)
17. W.Vandervorst and F.R.Shepherd : J. Vac. Sci. Techn. A5, 313 (1987)
18. K.Iltgen, O.Brox, A.Benninghoven and E.Niehuis : proc. SIMS-XI, Eds.G.Gillen, R.Lareau, J.Bennet and F.Stevie, Wiley, 305 (1997)
19. W.L.Harrington, C.W.Magee, M.Pawlik, D.F.Downey, C.M.Osburn and S.B.Felch : J. Vac. Sci. Techn. B16(1), 286 (1998)
20. K.Wittmaack and S.F.Corcoran : J. Vac. Sci. Techn. B16(1), 272 (1998)
21. Z.X.Jiang and P.F.A.Alkemade : proc. SIMS-XI, Eds.G.Gillen, R.Lareau, J.Bennet and F.Stevie, Wiley, 431 (1997)
22. Z.X.Jiang and P.F.A.Alkemade : Surf. Interface Anal. (in press, 1998)
23. K.Wittmaack : (submitted Surf. Interface Anal)
24. C.W.Magee, G.R.Mount, S.P.Smith, B.Herner and H-J.Gossmann : Appl. Phys. Lett. (to be published)
25. F.Stevie, P.M.Kahora, D.S.Simons and P.Chi : J. Vac. Sci. Techn. A6, 76 (1988)

26. K.Elst and W.Vandervorst : J. Vac. Sci. Techn. A 12, 3205 (1994)
27. M.G.Dowsett, N.S.Smith, R.Bridgeland, D.Richards, A.C.Lovejoy and P.Pedrick, proc SIMS-X, Eds.A.Benninghoven, B.Hagenhoff and H.W.Werner, Wiley, p367 (1995)
28. K.Iltgen, A.Benninghoven and E.Niehuis : proc. SIMS-XI, Eds.G.Gillen, R.Lareau, J.Bennet and F.Stevie, Wiley, 367 (1997)
29. M.G.Dowsett : formalism suggested at SIMS-XI (Orlando, 1997)
30. D.Kruger, K.Iltgen, B.Heinemann, R.Kurps and A.Benninghoven : J. Vac. Sci. Techn. B16(1), 292 (1998)
31. J.L.Maul and S.B.Patel : proc. SIMS-XI, Eds.G.Gillen, R.Lareau, J.Bennet and F.Stevie, Wiley, 707 (1997)
32. W.Vandervorst, T.Clarysse, P. De Wolf, T.Trenkler, T.Hantschlel, R.Stephenson : Future Fab International 4(1), 287 (1998)
33. M.I.Current, D.Lopes, M.A.Foad, J.G.England, C.Jones and D.Su : J. Vac. Sci. Techn. B16(1), 327 (1998)
34. K.Wittmaack : Surf. Interface Anal. 26, 290 (1998)
35. C.Tian, K.Elst, W.Vandervorst and K.Maex : proc SIMS-X, Eds.A.Benninghoven, B.Hagenhoff and H.W.Werner, Wiley, p367 (1995)
36. H.E.Maes, W.Vandervorst and R.Van Overstraeten in *"Impurity doping processes in Silicon"* eds F.F.Y.Wang (North Holland, 1981), p443
37. B.El-Kareh : J. Vac. Sci. Techn. B12(1), 172 (1994)
38. T.Clarysse and W.Vandervorst : J. Vac. Sci. Techn. B16(1), 260 (1998)
39. M.Pawlik : private communication
40. T.Clarysse, M.Caymax, P.De Wolf, T.Trenkler, W.Vandervorst, J.S.McMurray, J.Kim, C.C.Williams, J.G.Clark and G.Neubauer : J. Vac. Sci. Techn. B16(1), 394 (1998)
41. T.Clarysse and W.Vandervorst : (to be published)
42. H.L.Berkowitz and R.A.Lux : J. Electrochem. Soc., 128, 1137 (1981)
43. P.A.Schumann and E.E.Gardner : J. Electrochem. Soc. 116, 87 (1969)
44. T.Clarysse, W.Vandervorst and A.Casel : Appl. Phys. Lett. 57, 2856 (1990)
45. A.Casel and H.Jorke : Appl. Phys. Lett. 50, 989 (1987)
46. T.Clarysse, P. De Wolf, H.Bender and W.Vandervorst : J. Vac. Sci. Techn. B14(1), 358 (1996)
47. T.Clarysse and W.Vandervorst : J. Vac. Sci. Techn. B12(1), 290 (194)
48. W.Vandervorst and B.Brijs : (to be published)
49. M.Pawlik : proc. 4th Int. Workshop on the Measurement,Characterization and Modeling of Ultra Shallow Doping Profiles in Semiconductors (MCNC, 1997)
50. P.De Wolf, T.Clarysse, W.Vandervorst and L.Hellemans : J. Vac. Sci. Techn. B16(1), 401 (1998)
51. D.C.Avonzo, C.Clare and C.Dell'Oca : J.Electrocehm. Soc. 127, 2704 (1980)
52. J.S.Martens, S.M.Garrison and S.A.Sachtjen : Solid State Technol. 37, 51 (1994)
53. E.Ishida and S.B.Felch : J. Vac. Sci. Techn. B14(1), 397 (1996)
54. P.VandeVoorde, P.B.Griffin, Z.Yu, S-Y Oh and R.W.Dutton : Digest 1996 IEDM Electron Devices meeting, San Francisco, p 811
55. S.H.Goodwin-Johansson, M.Ray, Y.Kim and H.Z.Massoud : J. Vac. Sci. Techn. B10, 369 (1992)
56. R.von Criegern, F.Jahnel, M.Bianco and R.Lange-Gieseler : J. Vac. Sci.Techn. B12(1), 234 (1994)
57. R.von Criegern, F.Jahnel, R.Lange-Gieseler, P.Pearson, G.Hobler and A.Simionescu : J. Vac. Sci. Techn. B16(1), 386 (1998)
58. C.Spinella, V.Raineri, F.La Via and S.U.Campisano : J. Vac. Sci. Techn. B14(1), 414 (1996)
59. V.Raineri, V.Privitera, W.Vandervorst, L.Hellemans and J.Snauwaert : Appl. Phys. Lett. 64, 354 (1994)
60. A.Romano, J.Vanhellemont, J.R.Morante and W.Vandervorst : Micron and Microscopica Acta 20, 151 (1989)
61. S.S.Neogi, D.Venables, Z.Na and D.Maher : J. Vac. Sci. Techn. B16(1), 471 (1998)
62. D.Maher and B.Zhang : J. Vac. Sci. Techn. B12 (1), 347 (1992)
63. J.Liu, M.L.A.Dass and R.Gronsky : J. Vac. Sci. Techn. B12 (1), 353 (1992)
64. L.Gong, J.Lorenz and H.Ryssel : Proc. ESSDERC 90, eds. W.Eccleston and P.J.Rosser, Adam Hilger, p 93 (1990)
65. H.Cerva : J. Vac. Sci. Techn. B10, 491 (1992)
66. G.Neubauer, A.Erickson, C.C.Williams, J.J.Kopanski, M.Rodgers and D.Adderton : J. Vac. Sci. Techn. B14 (1), 426 (1996)
67. M.Nonnenmacher, M.P.O'Boyle and H.K.Wickramasinghe : Appl. Phys. Lett. 58, 2921 (1991) ; Ultramiscroscopy 42-44, 268 (1992)
68. M.B.Johnson, O.Albrektsen, R.M.Feenstra and H.W.Salemink : Appl. Phys. Lett. 64, 1454 (1994)
69. Ph.Ebert, M.Heinrich, M.Simon, C.Domke and K.Urban : Phys. Rev. B53, 4580 (1996)
70. K-J Chao, A.R.Smith, A.J.McDonald, D-L Kwong, B.G.Streetman and C-K Shih : J. Vac. Sci. Techn. B16(1), 453 (1998)
71. D.D.Perovic, M.R.Castell, A.Howie, C.Lavoie, T.Tiedje and J.S.W.Cole : Ultramiscroscopy, 58, 104 (1995)
72. D.Venables, H.Jain and D.C.Collins : J. Vac. Sci.Techn B16(1), 471 (1998)
73. T.T.Sheng and R.B.Marcus : Transmission Electron Microscopy : VLSI circuits and Structures, chap. 5 (Wiley, 1983)
74. M.Barret, M.Dennis, D.TIffin, Y.Li and C-K Shih : IEEE Electron Device Lett. 64, 118 (1995)
75. T.Trenkler, W.Vandervorst and L.Hellemans : J. Vac. Sci.Techn B16(1), 349 (1998)
76. R.Alvis, S.Luning, L.Thompson, R.Sinclair and P.Griffin : Journ. Vac. Techn. B14(1), 231 (1996)
77. Anne Verhulst, Thesis, KU-Leuven (1998), Belgium (UDC : 681.723 (043))
78. R.S.Stephenson, Anne Verhulst, P.De Wolf, W.Vandervorst and M.Caymax : (to be published)
79. Y.Huang, C.C.Williams and J.Slinkman : Appl. Phys. Lett. 66, 344 (1995)
80. V.V.Zavyalov, J.S.McMurray and C.C.Williams : presented at Materials Research Symposium, Symposium S : Nanoscale Materials Characterization using Scanning Probes, San Francisco Spring 1998.

81. C.C.Williams, J.Slinkman, W.P.Hough and H.K.Wickramasinghe : J.Vac. Sci.Techn. A 8, 895 (1990)
82. G.Neubauer, M.Lawrence, A.Dass and T.J.Thomson : Mat. Res. Soc. Symp. Proc. Vol 286, 283 (1992)
83. Y.Huang, C.C.Williams and H.Smith : J. Vac. Sci. Techn. B14(1), 433 (1996)
84. C.C.Williams : (private communication)
85. J.S.McMurray, J.Kim, C.C.Williams and J.Slinkman : presented at 4th Int. Workshop on the Measurement,Characterization and Modeling of Ultra Shallow Doping Profiles in Semiconductors (MCNC, 1997)
86. D.P.Chu and M.G.Dowsett : Phys. Rev. (1998)
87. P.De Wolf, T.Clarysse, W.Vandervorst, J.Snauwaert and L.Hellemans : J. Vac. Sci. Techn. B14(1), 380 (1996)
88. J.F.Marchiando, J.J.Kopanski and J.R.Lowney : J. Vac. Sci. Techn. B16(1), 463 (1998)
89. R.N.Kleiman, M.L.O'Malley, F.H.Baumann, J.P.Garno and G.L.Timp : Digest 1997 IEDM Electron Devices meeting, Washington, p 671
90. H.Edwards, R.MCGlothin, R.San Martin, E.U, M.Gribelyuk, R.Mahaffy, C-K Shih, R.Scott List and V.A.Ukraintsev : Appl. Phys. 72, 698 (1998)
91. J.J.Kopanski : private communication
92. P.De Wolf, J.Snauwaert, T.Clarysse, W.Vandervorst and L.Hellemans : Appl. Phys. Lett. 66, 1530 (1995)
93. Ph.Niedermann, W.Hanni, N.Blanc, R.Cristoph and J.Burger : J. Vac. Sci. Techn. A 14, 1233 (1996)
94. cfr product information *Nanosensors*
95. P.De Wolf, T.Clarysse and W.Vandervorst :: J. Vac. Sci. Techn. B16(1), 320 (1998)
96. P.De Wolf, T.Clarysse and W.Vandervorst :: J. Vac. Sci. Techn. B16(1), 463 (1998)
97. P.De Wolf and W.Vandervorst : (to be published)
98. M.Kump and A.C.Diebold : 1D/2D Dopant Metrology Round Robin Comparison Report, SEMATECH, Nov 1995.
99. T.Trenkler, P.De Wolf, J.Snauwaert, Z.Quamhieh, W.Vandervorst and L.Hellemans : Proc. ESSDERC-95, p 477
100. T.Trenkler, P.De Wolf, W.Vandervorst and L.Hellemans : J. Vac. Sci. Techn. B16(1), 367 (1998)

Overview of Optical Microscopy and Optical Microspectroscopy

Joel W. Ager III

Materials Sciences Division, Lawrence Berkeley National Laboratory,
University of California, Berkeley, CA 94720
ager@lbl.gov

Optical microscopy has historically been a major tool for semiconductor inspection. As the ULSI design rule continues to decline to 0.25 μm and below, standard optical microscopy methods will arrive at their resolution limit. In the first part of this paper an overview of currently used optical microscopy techniques will be given. The resolution limit for optical imaging will be discussed, and novel methods for increasing resolution, including deep UV microscopy and confocal laser microscopy, will be presented. The second part of the paper will discuss an emerging technology for contamination analysis in semiconductor processing, microspectroscopy. Three topics in this area will be discussed with an emphasis on applications to off-line defect identification in process development: (1) micro-Raman spectroscopy, (2) micro-fluorescence or micro-photoluminescence spectroscopy, and (3) micro-reflectivity. It will be shown that these microspectroscopy methods can provide composition information for defects down to 1 μm in size that is not accessible through the more commonly used methods such as scanning electron microscopy with energy dispersive spectroscopy (SEM/EDS) and scanning Auger microscopy. Classes of defects where optical micro-spectroscopy methods are useful include ceramic particles, thin films of organic material, and dielectric films.

INTRODUCTION

Optical microscopy has historically been a major tool for semiconductor inspection. As ULSI design rules continue to decline to 0.25 μm and below, traditional visible light optical microscopy will be unable to resolve the finest features on the integrated circuit. This is obviously a limitation for metrology applications. On the other hand, optical microscopy measurement and characterization techniques tend to be relatively fast and nondestructive. In addition, optical microscopy is not affected by the charging effects that occur when insulating features are examined by scanning electron microscopy (SEM). For these reasons, it is an advantage to optical microscopy methods when possible. The first part of this paper will present the resolution limits in optical microscopy and discuss some approaches for overcoming them, including deep UV microscopy, confocal laser microscopy and near-field scanning optical microscopy (NSOM).

In the second part of the paper, an emerging technology for contamination analysis in semiconductor processing, optical microspectroscopy, will be reviewed. As the smallest feature size on an IC continues to shrink, the size of the smallest defects that need to be located, characterized, and classified will also shrink. However, many defects, particularly those caused by contamination, will continue to be much larger than the minimum design rule size. Therefore, characterization techniques that can provide crucial information (e.g., chemical composition) from particles 1 μm or larger in size or from thin films will continue to be valuable in semiconductor manufacturing, regardless of the design rule size. Three topics in this area will be discussed with an emphasis on applications to off-line defect identification in Si IC processing: micro-Raman spectroscopy, micro-photoluminescence (micro-fluorescence) spectroscopy, and micro-reflectivity (micro-spectrophotometry). It will be shown that these microspectroscopy methods can provide composition information for defects down to 1 μm in size that is not accessible through more commonly used methods such as scanning electron microscopy with energy dispersive spectroscopy (SEM/EDS) and scanning Auger microscopy (1). Although the use of synchrotron light sources has improved the spatial resolution of infrared microscopy to 10 μm (2-3), making it more useful for semiconductor inspection, this technique will not be covered in this overview.

CP449, *Characterization and Metrology for ULSI Technology: 1998 International Conference*
edited by D. G. Seiler, A. C. Diebold, W. M. Bullis, T. J. Shaffner, R. McDonald, and E. J. Walters

TABLE 1. Lateral resolution for various imaging microscopy techniques. Values given are for imaging in air unless noted.

Objective (magnification/*NA*)	Illumination	Resolution (μm)	Comment
10x/0.25	white light	1.34	From Eq. (1)
20x/0.40	white light	0.84	From Eq. (1)
100x/0.95	white light	0.35	From Eq. (1)
100x/0.95	488 nm, confocal	0.22	Ref. 9. See also Fig. 5.
100x/1.35	248 nm (UV)	<0.16	glycerin immersion, from ref. 8.
Optical fiber	488 nm	0.05	Typical NSOM resolution. Values down to 10 nm have been reported, refs. 12 and 13.

OPTICAL MICROSCOPY AND THE DIFFRACTION LIMIT

A optical block diagram of semiconductor inspection microscope is shown in Fig. 1. The fundamental limit to resolution is determined by the numerical aperture (*NA*) of the objective lens (Fig. 2). For conventional brightfield illumination the resolution, defined as the minimum resolvable distance between two point light sources, is (4)

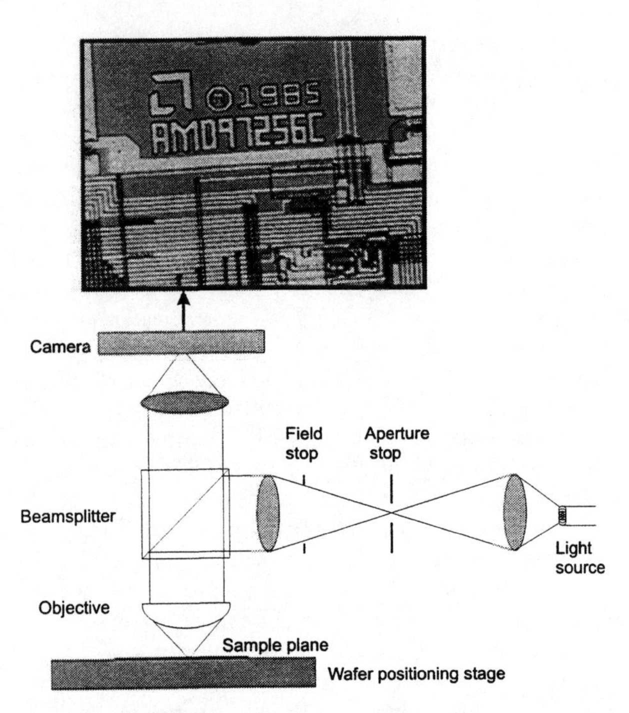

Fig. 1. Simplified block diagram of a semiconductor inspection microscope under brightfield illumination conditions. The image shown was taken with a 10x, NA = 0.25 microscope objective.

$$\frac{0.61\lambda}{NA} \tag{1}$$

where λ is the wavelength of light, NA is the numerical aperture,

$$NA = n\sin\alpha\,, \tag{2}$$

and n is the index of refraction of the medium between the microscope objective and the sample plane (n = 1 for air) and α is the collection half-angle as depicted in Fig. 2. Geometrical constraints limit α to 90° or less; the practical limit is on the order of 70° as shown in Fig. 2. For imaging with white light, an average wavelength of 550 nm is used (5) and Eq. (1) predicts a diffraction-limited resolution of 0.35 μm for an objective with NA = 0.95. Table 1 presents predicted and observed lateral resolutions for various microscope objectives and imaging conditions. It should be pointed out that objects smaller than the resolution limit can still be observed, but their outlines and shapes can no longer be clearly discerned.

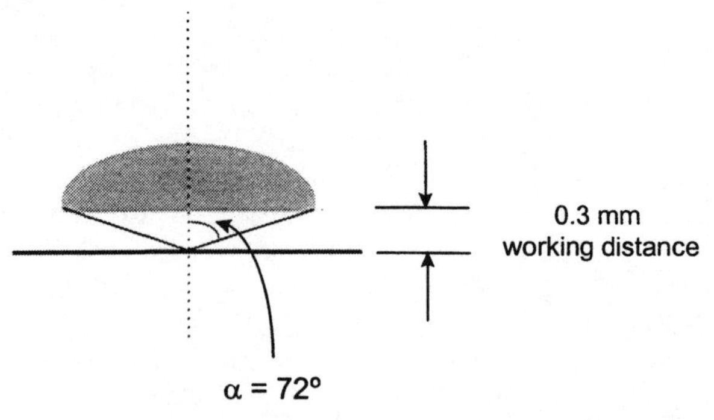

Fig. 2. Detail of 100x microscope objective with NA = 0.95.

OPTICAL MICROSCOPY TECHNIQUES WITH ENHANCED SPATIAL RESOLUTION

By examining Eq. (1), it can be seen that the spatial resolution of a microscope system can be improved by increasing the *NA* of the objective lens and/or using shorter wavelength light. The former is done in oil-immersion (n = 1.5 for mineral oil) objectives used in metallurgical inspection. *NA* values up to 1.4 are possible for this method and a corresponding increase in spatial resolution results. However, this technique has not been widely used in semiconductor inspection, presumably due to contamination concerns.

Ultraviolet Microscopy

The spatial resolution of optical microscopy has also been improved by using illuminating light with a shorter wavelength. Ultraviolet (UV) light microscopes were originally developed for biological applications (6) but are now being developed for high resolution semiconductor inspection (7-8). Fused silica lenses are used in the objectives and the spectral bandwidth of the source is restricted to limit chromic aberration. In recent work (8), objectives have been optimized for 436 nm (G-line), 365 nm (i-line) and 248 nm light generated by a Xe-Hg lamp equipped with bandpass filters. By immersing the objective lens in glycerin, a numerical aperture of 1.35 was achieved for the 248 nm objective, and a spatial resolution of better than 0.16 μm has been achieved in test structures. It should be mentioned that the increased spatial resolution achieved by brightfield UV microscopy is accompanied by a reduction in the depth of focus; this is apparent when comparing SEM images taken at the same magnification. A unique capability of imaging metal lines at sub-micron lateral resolution through dielectric layers (e.g. TEOS, oxide) of several μm in thickness has been demonstrated. This capability may be useful in characterizing buried defects.

Confocal Laser Microscopy

In confocal laser microscopy, the sample is illuminated by a focused laser beam that is raster scanned across the field of view of the microscopy objective (Fig. 3). The reflected laser light is refocused at the confocal pinhole (Fig. 4); light that is transmitted is measured by a detector and computer graphics software is used to create the image. The confocal pinhole acts to reject light from above and below the sample plane, substantially improving the axial resolution compared to brightfield imaging. As discussed in Wilson (9), the lateral resolution is improved by about 30% over the bright field imaging values in Eq. (1), such that the resolution of a confocal laser microscope operating with 488 nm laser light is 0.22 μm. This is demonstrated in Fig. 5 which compares images taken under

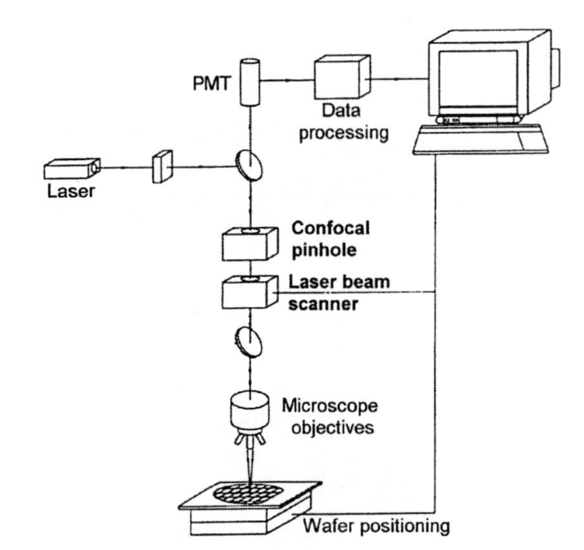

Fig. 3. Block diagram of a confocal laser microscope

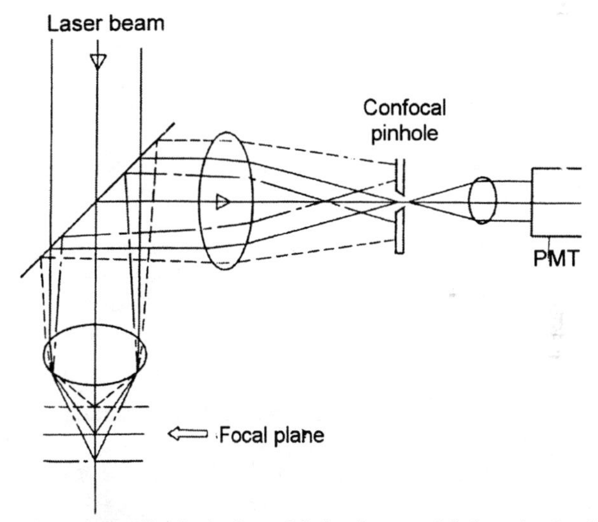

Fig. 4. Detail of rejection of light above and below the focal plane of the microscope objective. Dotted lines represent light emanating from above and below the focal plane that is blocked from reaching the detector by the confocal pinhole.

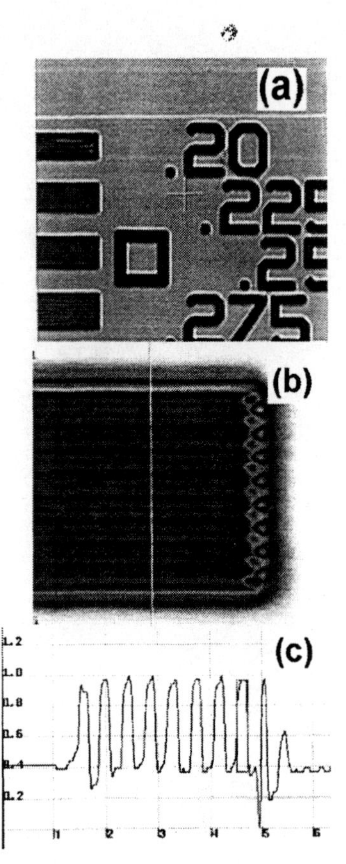

Fig. 5. Demonstration of the increase in lateral resolution in laser confocal microscopy: (a) brightfield image, 100x objective, little contrast observed for lines spaced 0.275 μm apart and below; (b) laser confocal image taken using same objective of lines spaced at 0.225 μm, good contrast is observed; (c) line scan of intensity along vertical line shown in (b) confirming the optical contrast. Unpublished data courtesy of Jim Xu, Ultrapointe Corp.

brightfield illumination and confocal laser scanning with the same objective.

The enhanced axial resolution also enables "optical sectioning" in thick samples. This has been used to image defects buried in transparent layers (10). Recently, an automatic defect classification scheme has been developed based on a combination of white light brightfield and confocal laser microscopy imaging (11). Image analysis software has been developed to classify defects based on "descriptors" such as brightness, contrast, size, etc. The system has been implemented in a full-wafer inspection tool, providing a high level of automation. Because it uses laser illumination, confocal laser microscopy is a convenient platform for implementing the optical spectroscopy methods (Raman, photoluminescence) discussed below.

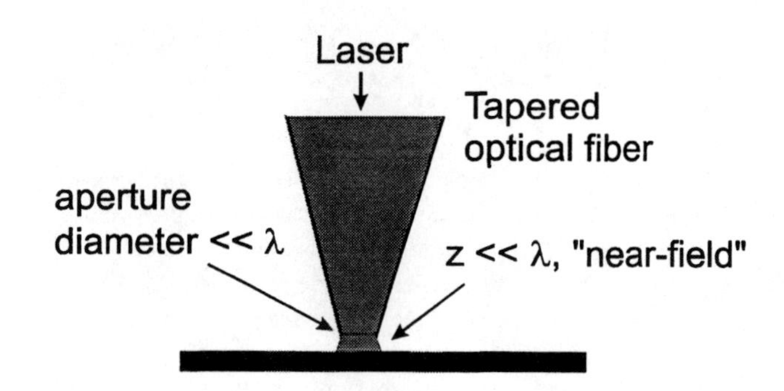

Fig. 6. Simplified picture of a near-field scanning optical microscope (NSOM). When tip to sample spacing is sufficiently small, lateral resolution is determined by the aperture diameter.

Near-field Scanning Optical Microscopy

In Near-field Scanning Optical Microscopy (NSOM), an objective lens is not used. Instead, an optical fiber with aperture diameter d is brought very close ($z < d/10$) to the sample (12). Typically, this is done with force feedback on the tip, similar to atomic force microscopy (AFM). In one NSOM implementation, a laser beam is propagated down the fiber (Fig. 6) and the image is formed by monitoring the intensity of the reflected laser light with far-field optics. In an alternative implementation, the sample is illuminated with a focused laser beam and reflected light is collected by the fiber. Typical aperture sizes are 50-150 nm with corresponding lateral resolution (13). Commercial NSOM instruments are available from several vendors. Although it has been demonstrated that NSOM can detect single molecules by fluorescence and write 100 nm lines in photoresist (14), a number of factors have limited the usage of NSOM for semiconductor defect evaluation (7). As in AFM, the effective field of view is limited by the range of the piezo tube scanners (up to 50 μm by 50 μm) and positioning the tip precisely near a defect is difficult. Because the near-field effect falls off rapidly with distance, only objects located at the sample surface can be imaged. The scanning process is relatively slow and only flat surfaces can be imaged at high spatial resolution (14). Although there are few examples of NSOM being used for defect identification, NSOM is used in research and development of optoelectronic components. Spectroscopic studies including photoluminescence (15-17) and Raman (18-19) have been reported.

OPTICAL MICROSCOPY AND THE SEMATECH ROADMAP

In the 1997 Metrology supplement (20) and the 1997 SEMATECH roadmap (21), characterization strategies for particle composition analysis are discussed. Manual inspection with optical microscopes and scanning electron microscopes with energy dispersive spectroscopy (SEM/EDS) are given as the baseline technologies for off-line defect composition analysis. Auger, time of flight secondary ion mass spectroscopy (ToF-SIMS), and NSOM are listed as potential near-term technologies for whole wafer particle/defect characterization. For auto particle classification for 0.18 μm design rule ICs, optical confocal microscopy and SEM are given as production tools, with NSOM following at the 0.13 μm design rule level. In accordance with this roadmap, auto-classification tools based on confocal laser microscopy are now commercially available. On the other hand, for the reasons discussed above, it will be difficult to implement NSOM in an automated classification tool. Furthermore, while the SEMATECH analysis captures the technical challenges found in imaging and autoclassifying the smallest defects on an IC, it probably underestimates the usefulness of optical techniques for performing defect characterization in general, especially in off-line analysis performed during process development.

In fact, many troublesome contaminants found in IC manufacturing are much larger than the minimum design rule size. In addition, standard bare wafer particle inspection tools (22) identify not only particles but also areas of the wafer with thin contaminant films. To eliminate such contaminants from the manufacturing tools, it is desirable to determine their chemical structure and process origin as quickly as possible. The SEMATECH roadmap accurately lists the techniques used most widely in the initial stages of contaminant identification (i.e. white light microscopy and SEM/EDS). However, depending on the contaminant or defect, the next stage of off-line analysis may involve other electron or X-ray based methods (e.g., Auger, XPS) and SIMS. It is at this stage that optical microspectroscopy methods have proven to be useful. It should also pointed out that in addition to characterizing contaminants on wafers, potential contaminant sources in the process tools themselves (e.g., chamber walls) and in the fab (e.g. wafer handling materials) can also require examination. Here, again, the scaling laws governing the design rules do not apply; rather, it is the ability of a characterization tools to provide a useful chemical or morphology signature is of interest.

TABLE 2. Raman scattering rates of selected materials for an excitation wavelength of 514.5 nm (24).

Material	Raman scattering rate 10^{-7} cm^{-1} Sr^{-1}
Ge	8600
graphite	650
GaAs	600
Si	85
diamond	6.5
ZnSe	2.2
benzene	0.72
quartz (SiO_2)	0.2
CaF_2	0.084

OPTICAL MICROSPECTROSCOPY

For the purposes of this paper, optical microspectrocopies are spectroscopic methods that can be performed with a spatial resolution close to that of the diffraction limit in brightfield microscope imaging. Perkowitz *et al.* (23) have reviewed recently the application of ellipsometry, IR spectroscopy, microspectrophotometry (i.e. micro-reflectance), modulation spectroscopy, photoluminescence (PL) spectroscopy, and Raman spectroscopy to characterization of properties in microelectronics manufacturing. In that work, both the characterization of bulk semiconductor properties (e.g. bandgap in compound semiconductors) and the analysis of defects were covered. The present paper will discuss the application of three of the above mentioned methods (Raman, PL, and micro-reflectivity) to the characterization of contaminant particles and films found in Si IC processing.

Micro-Raman spectroscopy

Raman spectroscopy is a light scattering technique that yields a vibrational spectrum from a laser illuminated spot. In the case of micro-Raman spectroscopy, the spot is formed by propagating a collimated laser beam through the microscope objective, exactly as in confocal laser microscopy. The Raman spectrum is obtained from a high-resolution spectrograph and high sensitivity detector (typically a CCD) coupled to the reflected light channel (Fig. 7). The technique is able to detect and characterize most covalently bonded materials and some ceramics and ionic species; metals are not detected. Raman scattering has been used extensively in semiconductor research (24-28). Micro-Raman spectroscopy has been applied to semiconductor inspection in the past, primarily for the identification of organic residues from semiconductor

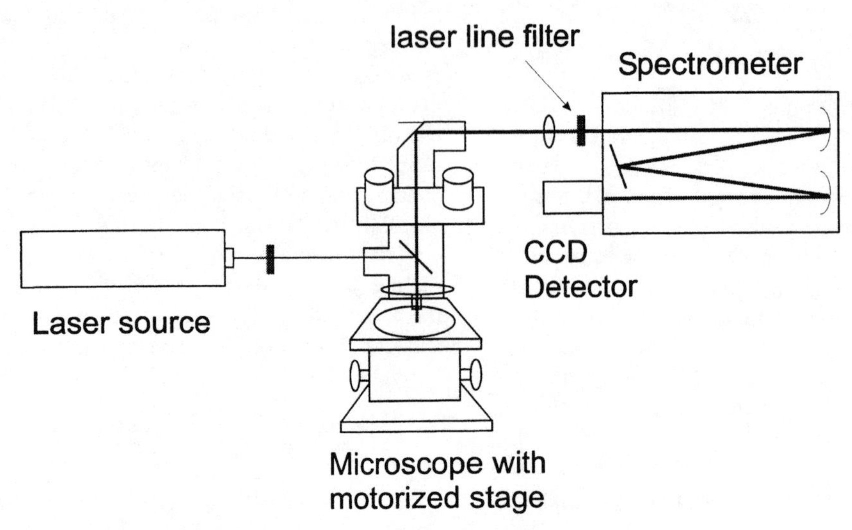

Fig. 7. Schematic of typical micro-Raman spectrometer for semiconductor defect identification.

handling materials (29). It has been limited from broader application to contaminant identification by the extreme weakness of the Raman scattering effect; in many cases the amount of material in the defect is insufficient to produce an observable signal from the defect. For example, although SiO_2 films have Raman-active vibrational modes, the signal from oxide films on Si is very difficult to observe against the strong background of scattering from the Si substrate. On the other hand, there are important classes of materials with strong Raman cross sections. Organic films, certain ceramics, and, in particular, graphitic particles are often observable and identifiable by their Raman spectrum. Table 2 lists Raman scattering cross sections for some materials (27). In general, covalent compounds (Ge, Si, diamond) tend to have larger scattering cross sections than more ionic compounds (CaF_2).

Recently, micro-Raman spectroscopy has been applied to defect identification in process tool development (30). Table 3 summarizes the signal strengths found in these initial studies. Incompletely cleaned organic films (e.g. residual photoresist) tend to graphitize and form thin amorphous carbon films in subsequent thermal processing. The sensitivity of Raman spectroscopy is relatively high for thin graphitic or amorphous carbon films; micro-Raman spectra with 1 μm spatial resolution are obtained routinely from the 2 nm (and thinner) amorphous carbon layers used as protective coatings on hard disk media and read/write heads. Figure 8 shows the Raman spectrum of a 200 nm graphite particle obtained with a prototype semiconductor inspection tool. Based on a minimum detectable count rate of 1 count per second (cps) at 10 mW

TABLE 3. Observed Raman scattering signal from different defects and defect standards (27). Count rates are adjusted to a laser power of 10 mW. In the absence of interfering signals from the Si substrate (cf. Si peak at 520 cm^{-1} in Fig. 11), in the absence of interference from fluorescence signals of 1 cps or greater should yield observable Raman spectra.

	Laser power (mW)	Counts	Collection time (s)	Count rate (cps)	Comment
Si substrate	6	4300	32	224	Probe depth using 488 nm laser is 570 nm.
200 nm graphite particle	18	20000	120	93	Sub-50-nm particles potentially detectable.
1 μm polystyrene sphere		1250	100	21	Sub-micron polymer particles and films potentially detectable.
1.5 μm quartz particle	6	100	32	5.2	Micron-sized particles detectable despite low Raman cross section.
NH_4Cl particles					
2.5 μm	20	5000	128	19	
1.0 μm	20	800	128	3.6	
0.5 μm	20	400	128	1.6	Micron-sized particles detectable.

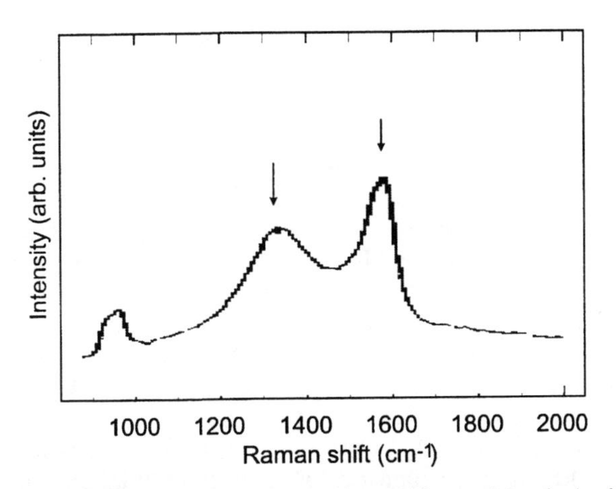

Fig. 8. Raman spectrum of 200 nm graphite particle obtained with prototype semiconductor inspection tool. Arrows mark the distinctive D-band (1360 cm^{-1}) and G-band (1580 cm^{-1}) features of microcrystalline graphite. The peak at 975 cm^{-1} is a 2nd-order feature from the Si substrate.

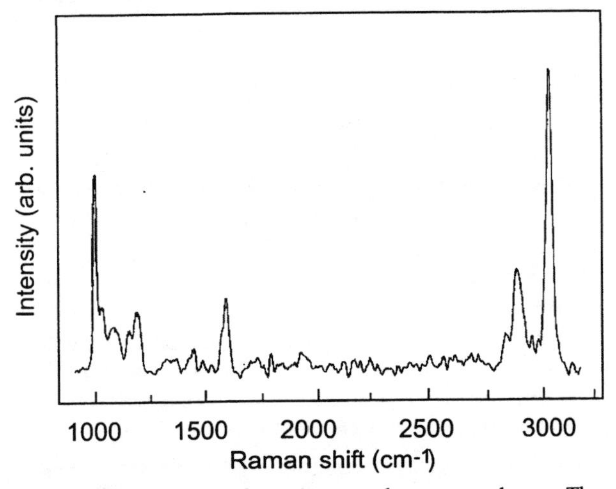

Fig. 9. Raman spectra from 1 μm polystyrene sphere. The features at 1000 cm^{-1} and 1600 cm^{-1} are due to C-H out of plane motion and aromatic ring breathing, respectively. The strong features at 2900 - 3100 cm^{-1} are due to C-H stretches.

laser power, a particle 100x smaller would have been detectable if it could be placed in the laser beam. The feasibility of detecting Raman scattering from thin organic films is simulated with a 1 μm diameter polystyrene sphere as shown in Fig. 9; the observed signal is 20x the detection limit. Raman spectra from a class of small particles generated in a metal etch tool are shown in Fig. 10. A detectable signal is obtained from a 1.5 μm sized particle. The combination of the peaks in the Raman spectrum and elemental analysis via SEM/EDS showed that the particles were NH_4Cl. The very weak Raman signal from a 1.5 μm quartz particle is shown in Fig. 11. In this case, the Raman

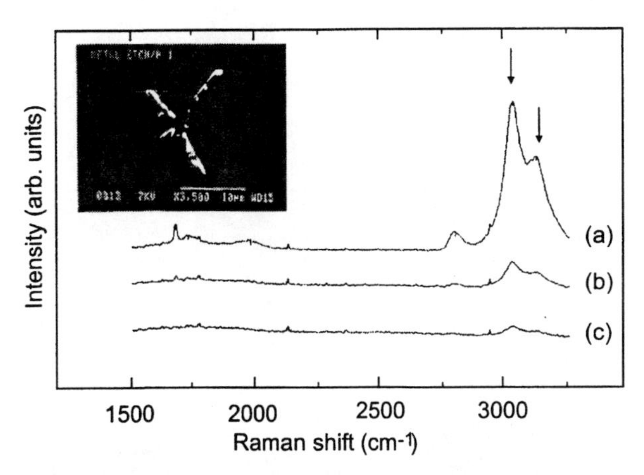

Fig. 10. Raman spectra from a series of NH_4Cl defects (example shown in inset) of decreasing size: (a) 2.5 μm; (b) 1.0 μm; (c) 0.5 μm. The peaks at 3040 and 3320 cm^{-1} (arrows) are due to the symmetric and asymmetric N-H stretches in the NH_4^+ cation. The position of these peaks established that the defect is NH_4Cl.

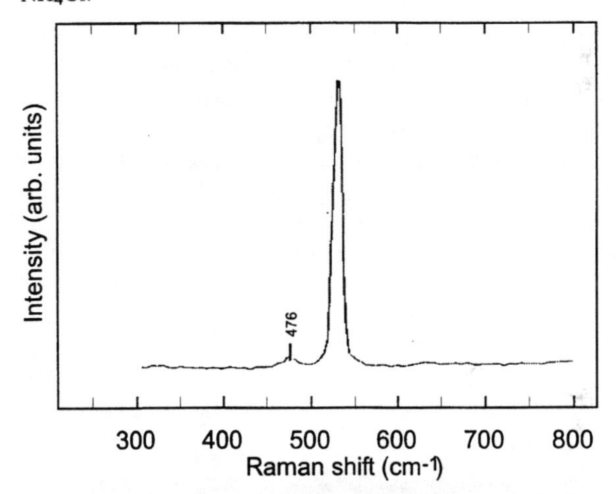

Fig. 11. Raman spectra from 1.5 μm quartz (SiO_2) particle on a Si wafer. The peak at 476 cm^{-1} is due to the quartz; the strong peak at 520 cm^{-1} is from the Si substrate which, because the quartz particle is transparent, is also in the field of view of the microscope objective.

information was used to distinguish the quartz particle from similarly sized glass particles.

Micro-photoluminescence spectroscopy

The same optical setup used to perform micro-Raman spectroscopy is used for micro-photoluminescence (micro-fluorescence) spectroscopy. Photoluminescence (PL) is used extensively in semiconductor research, particularly in

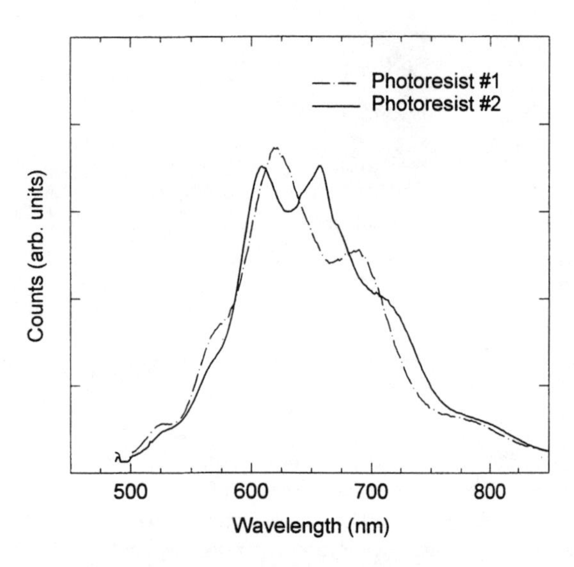

Fig. 12. PL spectra of two different photoresist films excited at 476.5 nm in a micro-PL system. Films are 630 nm thick.

the study of band and defect energy levels in compound semiconductors (28, 31-32). For defect identification, it can be used to distinguish defects by their PL signature. In contrast to Raman spectroscopy, for most materials found in semiconductor processing, it is not possible to deduce their structure from their PL spectrum. On the other hand, with the use of standards, micro-PL can be used to establish that the defect is the same as a known material or is of the same type as a previously identified defect.

Photoluminescence has been implemented in combination with SEM/EDS in an automated particle detection and identification scheme (33). In that work, a PL spectrometer coupled to a confocal microscope was shown to be valuable in distinguishing between residues of different photoresists. Berkeley Lab work shown in Fig. 12 shows that micro-PL can be used to distinguish two photoresist standards by their PL spectrum. Generally, the detector count rates in PL are much higher than in Raman scattering. In the case of the photoresist films shown in Fig. 11, no Raman spectrum could be observed due to interference from the PL, even though the films were sufficiently thick to expect a Raman signal (cf. Table 3).

There is a specific application of micro-PL spectroscopy for distinguishing between different forms of Al_2O_3. We have shown that strong PL at 692 nm ("ruby" fluorescence doublet) from the ubiquitous Cr impurity is observable from particles of α-Al_2O_3 (sapphire) down to 1 μm in size (34). Recently a combination of micro-PL, SEM/EDS, and Auger was used to establish that 1-5 μm-sized particles of nominal composition Al_2O_3 came from the anodized wall of an etch tool rather than from its ceramic components (34-35). In this case, the absence of the "ruby" PL signal established that the particles were not of ceramic origin (Fig. 13).

Micro-reflectance spectroscopy

Micro-reflectance spectroscopy or microspectrophotometry involves measuring the absolute reflectance of a sample as a function of wavelength. Similar optical techniques such as spectroscopic and fixed wavelength ellipsometry with mm scale spatial resolution (36) are used to measure the thickness of thin films in semiconductor processing. Usually, these instruments are not optimized for examining small defects. The implementation of microreflectance spectroscopy can be very simple. For

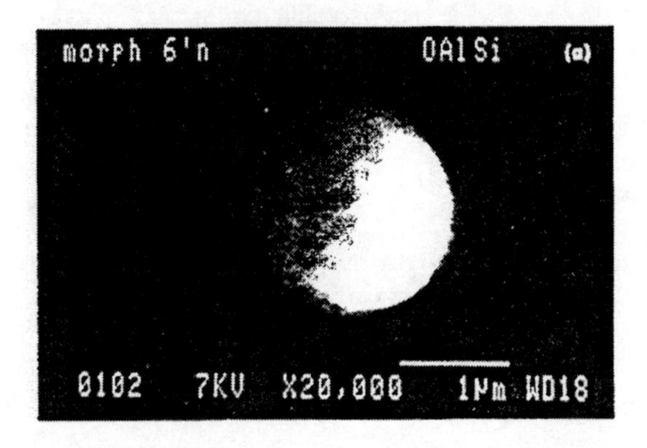

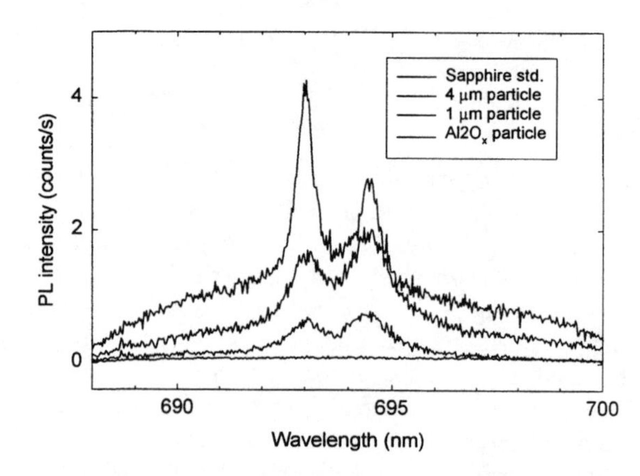

Fig. 13. SEM photo (left) of 1 μm Al_2O_x particle. PL spectra (right) in spectral region of Cr^{3+} (ruby) fluorescence of sapphire standard (top plot), 4 μm and 2 μm Al_2O_3 particles from a ceramic source (middle two plots), and 1 μm Al_2O_x particle (bottom plot). The absence of the distinctive ruby fluorescence peaks in the 1 μm Al_2O_x particle establishes that it is not of ceramic origin.

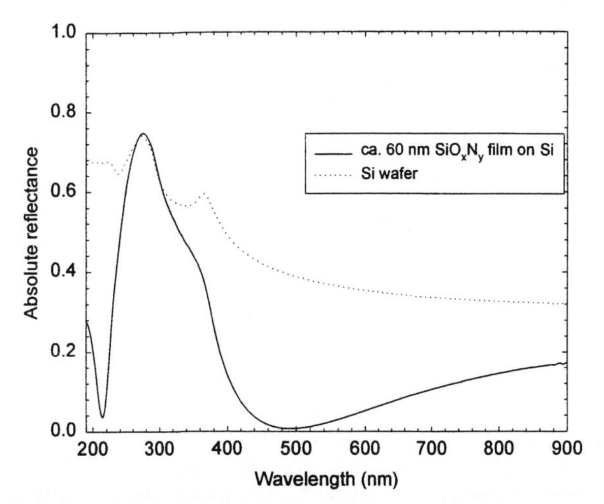

Fig. 14. Reflectivity spectrum of SiO_xN_y film deposited on Si. Minimum reflectivity is observed at ca. 490 nm. The absolute reflectance spectrum of Si is shown for comparison.

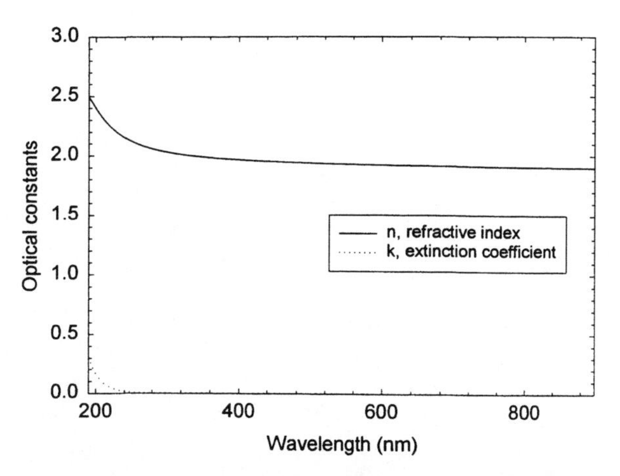

Fig. 15 Real (n, solid line) and imaginary (k, dotted line) parts of the refractive index of SiO_xN_y film deposited on Si. The measured film thickness is 624 Å.

example, Richards and Footner (37) provide a table for estimating oxide thickness based on the perceived color of reflected white light in a brightfield microscope. Differences in the thickness of transparent layers are often observable in the same way.

An example of a defect study performed using a more quantitative micro-reflectance method is given below. A SiO_xN_y coating (nominal thickness of 600 Å) is applied to Si wafers as an antireflection coating prior to photoresist application. The coated wafers appear brown in color. During process development in a prototype deposition tool, ca. 100 μm diameter "bright spots" were noted on the wafers. The spots are visible to the naked eye. Both SEM and AFM were employed to attempt to determine the structure and origin of the defects. SEM failed to observe any contrast between the defects and the rest of the wafer and AFM analysis found no evidence of any steps near their edges.

In the first part of the optical characterization, the optical constants of the bulk film were measured. The absolute reflectance spectrum (Fig. 14) was measured with reflectance spectrometer/analyzer (38). The reflectivity of bare Si is shown for comparison. Clearly, the SiO_xN_y coating is optimized for minimum reflectance at about 490 nm. The software in the commercial instrument was used to extract SiO_xN_y film thickness and optical constants from the reflectance data. The results are shown in Fig. 15. Although the film appears brown in color, the low value of k in the visible indicates that the film is actually transparent and that the film color is due to optical interference. The

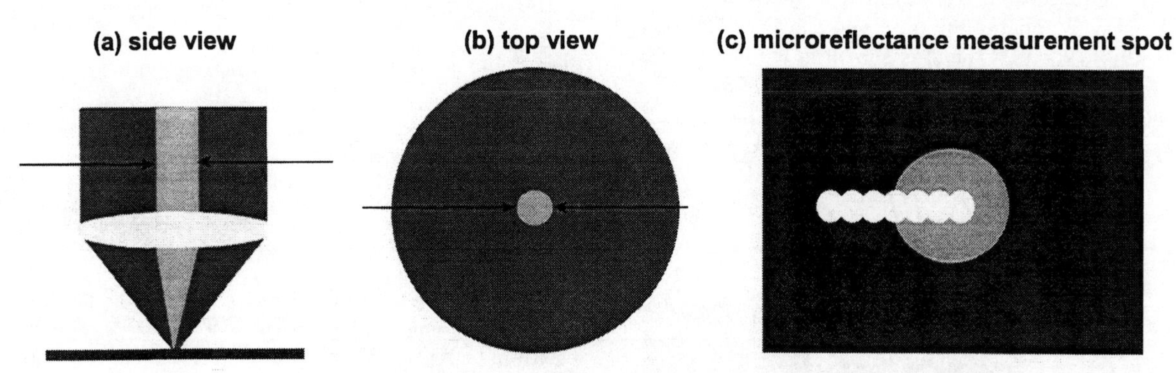

Fig. 16. Schematic modifications to optical microscope to perform micro-reflectivity measurements: (a) field stop is closed, reducing the effective NA of the 100x objective to 0.1; (b) 30 μm diameter illuminated spot is in center of field of view; (c) wafer is scanned under the illuminated spot. The intensity of the reflected light is monitored as a function of stage position to measure the changes in absolute reflectivity across the "bright spot" defect.

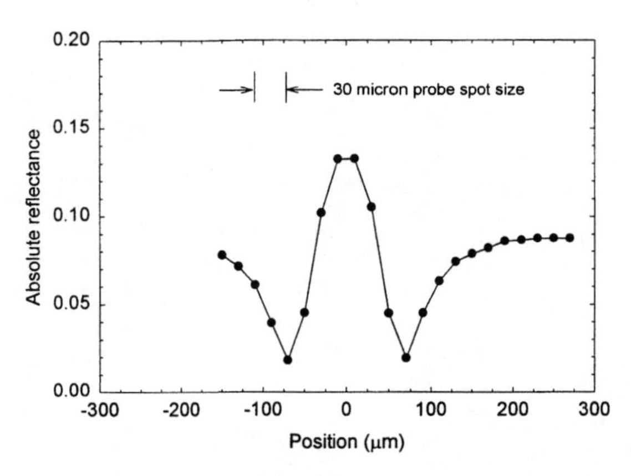

Fig. 17 Variation of the absolute reflectance of the SiO_xN_y at 632.8 nm as a function of distance. Zero position corresponds to the center of the 140 μm diameter "bright spot" defect.

analysis also extracts the SiO_xN_y film thickness, 624 Å, which is in good agreement with the nominal value of 600 Å.

Micro-reflectance measurements were performed in a modified Olympus semiconductor inspection microscope with a 100x objective. Figure 16 is a sketch of the measurement scheme. The field and aperture stops were closed completely to form a 30 μm spot on the sample surface with an effective *NA* of about 0.1, which allows near-normal incidence equations to be used in the subsequent reflectivity analysis. Although the lateral resolution used was 30 μm, it would have been possible to reduce this to 10 μm or less. The intensity of reflected light was monitored using a spectrometer and CCD detector.

The 30 μm spot of the micro-reflectance probe was scanned in 20 μm steps across a 140 μm diameter bright spot. The reflectance of the illuminated spot at 632.8 nm was obtained by ratioing the observed signal to that of a bare Si wafer (R = 0.347) and is shown in Figure 17. Far away from spot, the absolute reflectivity, 0.087, is in reasonable agreement with the value measured for the bulk film by the reflectance spectrometer, 0.070. As the probe is scanned over the bright spot, the reflectivity varies significantly. The high reflectivity (0.14) observed in the middle of the spot corresponds to the "bright" nature of the defect. The very low reflectivity near the edge of the spot is from an area where the film appears blue in color in the microscope.

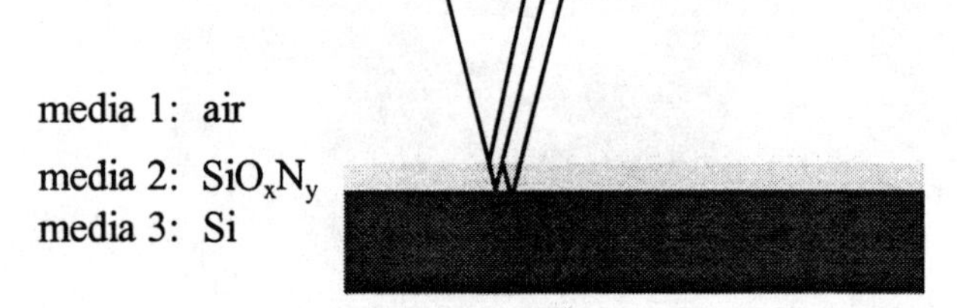

Fig. 18 Three media optical structure used in reflectance calculations. The surfaces are assumed to be sufficiently smooth to allow multiple interference.

The three media optical structure (air/SiO_xN_y/Si) used in modeling the reflectivity measurements is shown in Fig. 18. Standard multiple interference equations suitable for thin films are used to convert the observed reflectance to the refractive index of the SiO_xN_y. Normal incidence is assumed; this is a good approximation for both for reflectance spectrometer (angle of incidence = 5°) and for the microscope under conditions of low numerical aperture. We also assume that the material in the defect is non-absorbing at 632.8 nm, as is the bulk film (Fig. 15).

The results are shown in Fig. 19 in which the variation of the film refractive index is measured across the bright spot. Far from the bright spot (relative position > 200 microns), the film index is 1.9, in excellent agreement with the bulk measurement of 1.92. Sanchez *et al.* (39) have measured the refractive index of a series of nearly stoichiometric SiO_xN_y films produced by remote PECVD. Extrapolating the results of that work, a stoichiometry of x = 0.16 and y = 1.23 (N-rich) is estimated for the bulk film. The observed refractive index in the center of the spot is 1.74. Again, interpolating the results of ref. 36 yields x = 0.88 and y = 0.75 (O-rich) for the center of the spot. The refractive index observed at the edge of the spot is larger than that of Si_3N_4 (1.96); this indicates that the composition

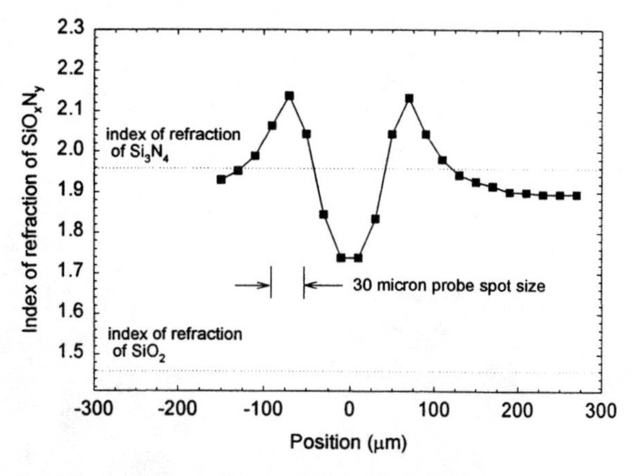

Fig. 19. Variation of the refractive index of the SiO_xN_y film across a 140 μm diameter "bright spot" defect. The position is taken to be zero at the center of the defect. The low refractive index area in the center of the plot corresponds to the bright center and is proposed to be O-rich. The high-index area at the edge of the spot is blue in color in brightfield microscopy and is proposed to be Si-rich.

at the edge of the "bright spot" is non-stoichiometric and probably Si-rich.

CONCLUSIONS

Although standard optical microscopy and optical micro-spectroscopy techniques cannot resolve extremely small features, they are fast and non-destructive and, for these reasons, they will continue to be used widely in semiconductor inspection. New methods to increase the resolution of optical microscopy include UV illumination, confocal laser microscopy, and NSOM. Confocal laser microscopy is already impacting automated defect inspection, and UV microscopy has demonstrated promising high-resolution capabilities with use of immersion objectives. The speed and positioning limitations of NSOM will restrict its application to defect classification and characterization, in spite of its sub-wavelength resolution capability. It was shown that the combination of optical microscopy (particularly confocal laser microscopy) with spectroscopy, can yield useful defect classification methods. Cases studies in which micro-Raman, micro-PL, and microreflectance spectroscopies yielded crucial defect information were presented. Classes of defects where optical micro-spectroscopy methods will useful include ceramic particles, thin films of organic material, and dielectric films.

ACKNOWLEDGMENTS

I thank Yuri Uritsky, Dick Brundle, and Cathy Cai of Applied Materials and Jim Xu of Ultrapointe for supporting the collaborations that made this work possible. Giuseppina Conti performed some of the reported Raman and PL microscopy work. I thank Rahim Forouhi and Iris Bloomer of n&k Technology for the long-term loan of an n&k Analyzer. Some of the this work was performed under a Cooperative Research and Development Agreement (CRADA) with Intel and a Personnel Exchange Agreement with Applied Materials and was supported by the Director, Office of Energy Research, Office of Computational and Technology Research, of the U.S. Department of Energy under Contract No. DE-AC02-76SF00098.

REFERENCES

1. U. Uritsky, V. Rana, S. Ghanayem, and S. Wu, *Microcontamination* **12**(5), 25-29 (1994).
2. J. A. Reffner, P. A. Martoglio, and G. P. Williams, *Rev. Sci. Instrum.* **66**, 1298-302. (1995).
3. G. L. Carr, D. DiMarzio, M. B. Lee, and D. J. Larson Jr., "Infrared Microscopy of Semiconductors at the Diffraction Limit," in *Semiconductor Characterization: Present Status and Future Needs*, eds. W. M. Bullis, D. G. Seiler, and A. C. Diebold, Woodbury, NY: AIP Press, 1996, pp. 418-421.
4. M. Born and E. Wolf, *Principles of Optics*, Oxford: Pergamon, 1980.
5. C. R. Brundle, C. A. Evans Jr., and S. Wilson, *Encyclopedia of Materials Characterization*, Boston: Butterworth-Heinemann, 1992, p. 62.
6. I. Pauluzzi, I. K. Lichtscheidl, and W. G. Url, *Proceedings of the Royal Microscopical Society*. **31**, 155 (1996).
7. J. Fitch, *EE Evaluation Engineering* **36**, 38 (1997).
8. T. Calahan and M. Bauerschmidt, "Imaging the 0.18 μm generation -- the UV challenge," presented at the Advances in Microscopy Workshop, SEMATECH, Austin, TX, November 19-20, 1997.
9. T. Wilson, *Confocal Microscopy*, London: Academic, 1990, pp. 1-39, 58.
10. Y. Uritsky, C. Cai, T. Francis, J. Xu, G. Lum, and B. Worster, *Solid State Technology* **38**, 61 (1995).
11. J. Xu, "Laser confocal microscopy for semiconductor defect review," presented at the Advances in Microscopy Workshop, SEMATECH, Austin, TX, November 19-20, 1997.
12. J. W. P. Hsu, *MRS Bull.* **22**, 27-30 (1997) provides an overview of NSOM techniques.
13. S. K. Buratto, *Curr. Opin. Solid State Mat. Sci.* **1**, 485-492 (1996).
14. H. F. Hess and E. Betzig, "Applications of Near-Field Scanning Optical Microscopy," in *Semiconductor Characterization: Present Status and Future Needs*, eds. W. M. Bullis, D. G. Seiler, and A. C. Diebold, Woodbury, NY: AIP Press, 1996, pp. 395-398.
15. J. Liu *et al.*, *Appl. Phys. Lett.* **69**, 3519-3521 (1996).
16. S.-K.Lee, W. Jhe, T. Saiki, and R. Ohtsu, *Opt. Rev.* **3**, 450-453 (1996).
17. J. K. Rogers *et al.*, *Appl. Phys. Lett. 66*, 3260-3262 (1995).
18. C. L. Jahncke, H. D. Hallen, and M. A. Praesler, *J. Raman Spectros.* **27**, 579-586 (1996).
19. C. L. Jahncke, M. A. Praesler, and H. D. Hallen, *J. Appl. Phys.* **82**, 5352-5359 (1997).
20. *Metrology Roadmap: A Supplement to the National Technology Roadmap for Semiconductors,* Technology Transfer Publication #94102578A–TR, Austin, TX: SEMATECH, 1995.
21. *The National Technology Roadmap for Semiconductors*, Austin, TX: Semiconductor Industry Association, 1997. URL: http://notes.sematech.org/97melec.htm
22. For example, the Tencor Surfscan wafer inspection tool.
23. S. Perkowitz, D. G. Seiler, and W. M. Duncan, *J. Res. Natl. Inst. Stand. Technol.* **99**, 605 (1994); S. Perkowitz, D. G.

Seiler, and W. M. Bullis, "Optical characterization of materials and devices for the semiconductor industry: trends and needs," in *Semiconductor Characterization: Present Status and Future Needs*, eds. W. M. Bullis, D. G. Seiler, and A. C. Diebold, Woodbury, NY: AIP Press, 1996, pp. 422-427.

24. *Light Scattering in Solids I*, ed. M. Cardona, Berlin: Springer-Verlag, 1975.
25. *Light Scattering in Solids II*, ed. M. Cardona and G. Güntherodt, Berlin: Springer-Verlag, 1982.
26. *Light Scattering in Solids IV*, ed. M. Cardona and G. Güntherodt, Berlin: Springer-Verlag1984.
27. *Light Scattering in Solids V*, ed. M. Cardona and G. Güntherodt, Berlin: Springer-Verlag, Berlin, 1989.
28. S. Perkowitz, *Optical Characterization of Semiconductors: Infrared, Raman, and Photoluminescence Spectroscopy*, London: Academic, 1993.
29. F. Adar and H. Schaffer, "Application of the Raman microprobe to identification of organic contaminants and to *in-situ* measurement of stress," in *Semiconductor Characterization: Present Status and Future Needs*, eds. W. M. Bullis, D. G. Seiler, and A. C. Diebold, Woodbury, NY: AIP Press, 1996, pp. 502-506.
30. Jim Xu of Ultrapointe provided the unpublished spectral data shown in this section.
31. G. D. Gilliland, *Mat. Sci. Eng.* **B 18**, 99-399 (1997).
32. L. Pavesi and M. Guzzi, *J. Appl. Phys.* **15**, 4779-4842 (1994).
33. T. Hattori and S. Koyata, *SPIE* **1464**, 367-376 (1991).
34. C. R. Brundle, Y. Uritsky, and J. T. Pan, *Micro* July/August, 43 (1995)
35. A. C. Diebold, Y. Uritsky, C. R. Brundle, T. Francis, H. Fatemi, K. Childs, and S. Clough, *Future Fab International*, 227-235 (1996)
36. W. A. McGahan and J. A. Woollam, "Ellipsometric characterization of thin oxides on silicon," in *Semiconductor Characterization: Present Status and Future Needs*, eds. W. M. Bullis, D. G. Seiler, and A. C. Diebold, Woodbury, NY: AIP Press, 1996, pp. 433-437.
37. B. P. Richards and P. K. Footner, *The Role of Microscopy in Semiconductor Failure Analysis*, London: Oxford University Press, 1992, p. 21.
38. n&k Technology, Santa Clara, CA.
39. O. Sanchez, M. A. Aguilar, C. Falcony, J. M. Martinez-Duarte, and J. M. Albella., *J. Vac. Sci. Technol.* **A. 14**, 2088 (1996).

Advanced SEM Imaging

D.C.Joy * and D.E.Newbury[+]

EM Facility, University of Tennessee, Knoxville, TN 37996-0810
**and Oak Ridge National Laboratory, Oak Ridge, TN 37831-6064*
[+]National Institute of Standards and Technology (N.I.S.T), Gaithersburg MD 20899

The scanning electron microscope (SEM) represents the most promising tool for metrology, defect review, and for the analysis of ULSI structures, but both fundamental problems such as electron-solid interactions, and practical considerations such as electron-optical constraints, are now setting a limit to performance. This paper examines the directions in which an advanced SEM might be developed to overcome these constraints. The SEM also offers considerable promise as a tool for the high spatial resolution X-ray microanalysis, especially for those situations where a thin cross-section is not practical and first surface analysis is required. The ways in which this capability can be incorporated in an advanced SEM are examined.

INTRODUCTION

The scanning electron microscope (SEM) has established itself in the semiconductor industry as the most widely used tool for metrology, wafer inspection, defect review and general characterization studies. It has been estimated that two out of every three SEMs manufactured go to some part or other of the semiconductor industry, and that these sales account for about three out of every four of the dollars spent on this type of instrumentation. The reasons for this dominance are that, compared to the alternative technologies that are available for the same tasks, the SEM offers a competitive level of spatial resolution, highly automated operation, a generally well understood methodology for data analysis, flexibility and extensibility, a unique combination of imaging and micro-analytical capabilities, and a familiar environment within which to work. However, over the next decade the SEM faces some severe challenges to its position because the continued extrapolation of Moore's law, as quantified by the SIA Roadmap (1), calls for instrument performance specifications that are at, or beyond, the boundaries that are currently accepted as fundamentally limiting. To meet these varied requirements advanced scanning electron microscopes, combining various new technologies as well as novel imaging techniques, will be required.

SPECIFYING AN ADVANCED SEM

Choosing a beam energy

The choice of the electron beam energy is crucial for an advanced SEM. In the 30 years that have elapsed since the commercial introduction of the SEM the typical operating energy has fallen from 20 or 30 keV to 5 keV, then to 1 or 2 keV, until now for many applications in the semiconductor field 0.6 to 0.8 keV is considered standard. These moves have been motivated by several different sets of considerations. Firstly, it is clear that a

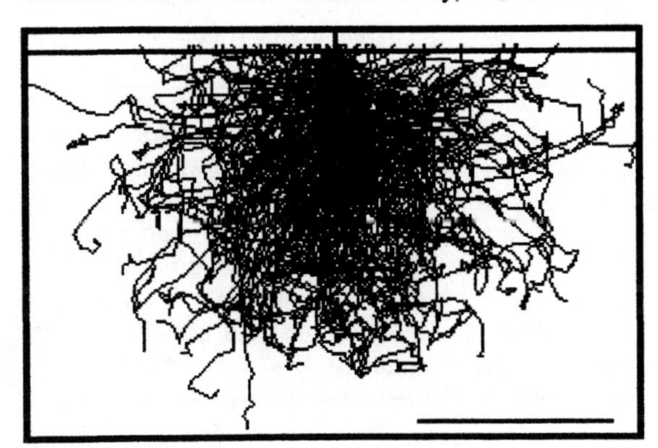

Figure 1. Monte Carlo simulation of electron trajectories in a bulk sample of silicon.

reduction of the incident energy minimizes the interaction volume of the electron beam within the sample. Figure (1) shows a Monte Carlo simulation of the electron interaction with silicon. The shape of this interaction volume does not change dramatically with energy, but its physical size does. Thus at 20 keV the

CP449, *Characterization and Metrology for ULSI Technology: 1998 International Conference*
edited by D. G. Seiler, A. C. Diebold, W. M. Bullis, T. J. Shaffner, R. McDonald, and E. J. Walters
1998 The American Institute of Physics 1-56396-753-7/98/$15.00

width of the scale marker shown would be about 5 micrometers, but at 5 keV it has fallen to 0.5 micrometers (500 nm), and by 1 keV it represents just 20 nm. The significance of this spreading associated with the incident beam it that it distorts signal profiles and lowers the image contrast. In general the signal profile from a feature is most readily interpreted when the scale of the beam interaction is either much larger or much smaller than the size of the feature itself. In the first case the point spread function of the probe is almost constant over the feature and so no distortion occurs although the contrast of the object is reduced. In the second case the point spread function is a delta function and again no distortion occurs but contrast is now maximized. The unacceptable case is when the interaction volume and the feature are of comparable sizes as then the profile becomes convoluted by the beam point spread function resulting in an image that is difficult to interpret, and poorly suited for measurement, because its exact form will depend on the details of the beam penetration. As the size of features in devices is ever decreasing there is thus good reason to minimize the beam interaction volume so as to maximize the contrast and avoid convolution effects. The depth penetrated by the beam also falls with a reduction in energy so the use of a low beam energy further enriches the information contributed by the important surface region to the image.

A second major reason for the continued reduction in beam energy has been the pervasive problem of charge build-up in samples irradiated by the electron beam. This leads to many problems including instabilities in the image intensities, random deflections of the incident beam, and even physical damage to the sample. Since the sample both receives charge, from the incident beam, and emits charge, as secondary (SE) or backscattered (BSE) electrons, the net charge accumulated in the specimen depends on the difference between the rates of arrival and departure of electrons. It can be shown that there are two energies, conventionally called E1 and E2, at which any given sample will be at dynamic charge balance in which on average every electron arriving is balanced by one electron leaving. At these energies charging is minimized or eliminated without the need to coat the sample to make it conductive. Figure (2) shows how E1 and E2 vary as a function of the atomic number of the target material (2). It can be seen that for the low atomic number elements that make up the majority of devices E1 is typically 100 eV while E2 is about 1 keV. Contemporary operational practice is aimed at matching the E2 energy for typical materials but if the beam energy is to be further reduced in future instruments then it is likely that it will be to a value close to E1 in order to maintain general charge neutrality. This would also have the merit of minimizing the depth at which any incident charge was implanted and so prevents damage to gate oxides and other similar regions which are at risk from radiation damage.

Preliminary experimental work on imaging at energies in the "ultra-low voltage" range from 20 to 200 eV is already in progress (3). In addition to confirming the benefits outlined above it has also been found that there is a simplification in the mechanisms of contrast formation at ultra-low energies which leads to images better suited for metrology and characterization than those at higher energies. It therefore seems probable that "ultra low" beam energies of from 50 to 150 eV could become standard practice for the new generation of SEMs.

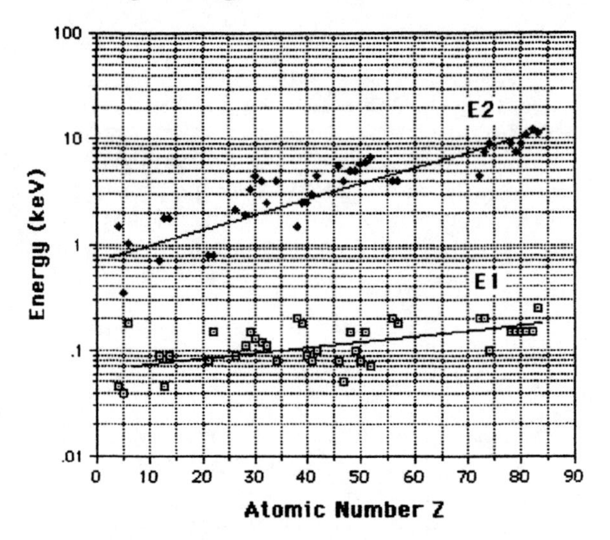

Figure 2. Variation of E1 and E2 energies as a function of the atomic number of the target material.

Specifying a resolution

The parameter which commands the most interest about any microscope is the spatial resolution that it can demonstrate. Based on the tasks to which an advanced SEM might be assigned, and using the data in the SIA Road Map as a guide, it can be assumed that spatial resolutions approaching one nanometer will be required for experimental purpose within the next five years, and on a routine basis within ten years. In the SEM the spatial resolution can be determined by one of a number of factors including the primary electron-solid interaction, the physics associated with the production of the signals used for imaging, the electron-optical performance of the tool, and the pixel density of the recorded image. Any one, or any combination of these factors might pose a limit to further progress, but the decision will depend on the manner in which the instrument is to be used.

If the secondary electron mode is to remain the preferred technique for imaging, then the target figure of one nanometer will be very hard to attain because the resolution of the SE image is inherently limited by the nature of SE production. The generation of secondaries by the incident electrons is a cascade process in which an initial scattering event between the incident electron and a free electron in the target produces a pair of secondary electrons each of which then initiates another stage of secondary production. Since the available energy is shared between the electrons at each stage, and because the mean free path (MFP) falls rapidly with energy, the cascade eventually stops multiplying leaving the cloud of electrons to diffuse through the material. The volume over which the cascade forms can be treated as a sphere of some effective radius λ_{SE}. As can be seen from Table 1, which lists some values for λ_{SE} deduced from SE yield measurements (2), the diffusion range of the SE is comparatively large and at low voltage is comparable to the range of the incident electrons. The resultant SE emission distribution from a surface at low beam energies is bell-shaped in form with a full width at half maximum of about λ_{SE}. When the incident beam approaches any edge in the sample (figure 3a) then the emitted SE yield will begin to rise when the beam is within about λ_{SE} of the edge because the SE can now escape through both the side and top surfaces. This produces the characteristic brightness enhancement which about doubles the signal intensity close to the edge. If the beam scans across a feature which is a few times greater than λ_{SE} in width then (figure 3b) the signal profile will show a corresponding signal rise at both edges. It is this edge contrast which "resolves" the feature and provides the information necessary to determine its shape and size. But if now (figure 3c) the size of the object becomes comparable to λ_{SE} then the two edge brightness regions overlap and merge to give a single feature. In this case the object appears bright but because it has no edges it has no definite size or shape, and so is not resolved. Thus λ_{SE} represents the practical resolution limit for SE imaging (4).

TABLE 1

Material	λ_{se}	Material	λ_{se}
Carbon	5.5nm	Silicon	3nm
Chromium	2.5nm	Copper	2.5nm
Silver	3.5nm	Gold	1nm
GaAs	5nm	SiO_2	5nm
Si_3N_4	4.5nm	PMMA	5nm

The data of Table 1 shows that this effect is most restrictive for precisely the low atomic number, low density, materials such as the polymer resists, oxides, metals encountered in semiconductor devices since λ_{SE} in such cases is typically 4 or 5nm. For the high resolution imaging of low atomic number materials (such biological tissue) this limit is circumvented by coating the sample with a very thin (2nm or less) film of a metal such as chromium. Because the SE production of the metal is much greater than that of the substrate, all the secondary generation is effectively localized within the metal film, and the resolution is now determined by the thickness of the coating (4). However this approach is clearly not feasible for most semiconductor applications so the advanced SEM must employ some other strategy to achieve its goal.

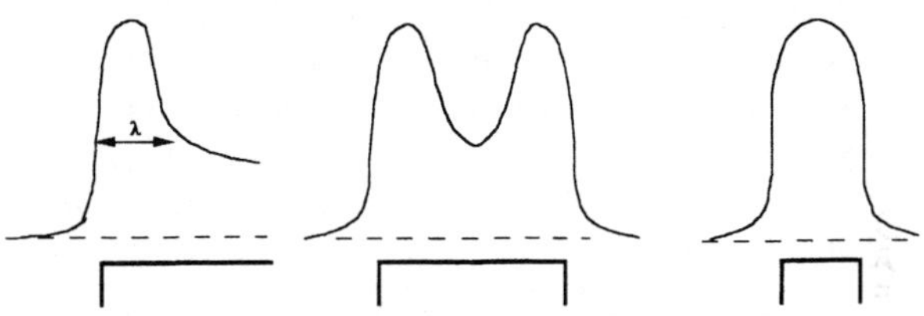

(a **Figure 3.** SE signal profiles (a) at an edge, (b) across a feature and (c) across a feature comparable in size with the SE mean free path λ_{SE}

The spatial resolution of the SEM is also potentially limited by its electron optical performance. The conventional assumption is that if the size of the electron beam at the specimen surface is d_0, then objects down to a size at least comparable with d_0 can be resolved. Since several production scanning microscopes claim to have the capability of producing probe diameters close to 1 nm this would seem to suggest that electron-optical considerations will not be a problem. However this conclusion is over-optimistic because it ignores the fact that the resolution of the SEM depends on other factors as well as the probe diameter.

This can be understood by considering the optical transfer function (OTF) of the SEM. The OTF(ω), which can be retrieved experimentally from a Fourier analysis of an image or computed given a knowledge of the properties of the probe forming lens, displays the response of the SEM to all the spatial frequencies ω passing through it. Figure (4) shows the OTF computed for the objective lens in a low voltage SEM designed to produce a nominal 1nm diameter beam of electrons. The shape of the profile depends on the current profile of the beam which in turn depends on the aberration coefficients, the energy, the energy spread, the focus, and the convergence angle of this lens. For the typical values

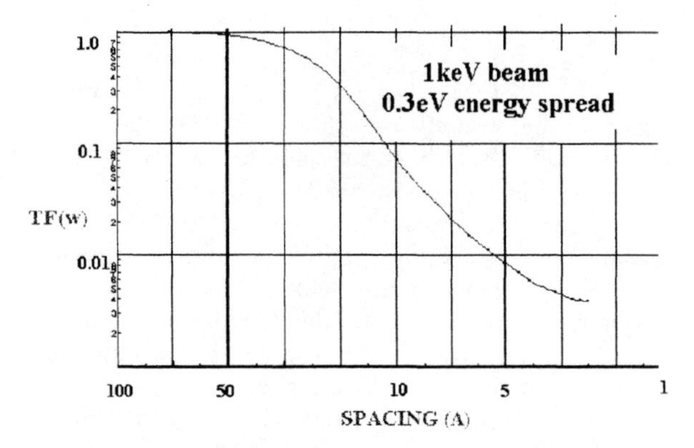

Figure 4. Computed optical transfer function for a high performance low voltage SEM system.

chosen here it can be seen that the transfer response falls rapidly with increasing spatial frequency, with the result that at 1 nm it is only about 0.06 of its value at 50 nm. This is significant because the signal-to-noise ratio of any frequency component ω in the image varies as $1/[DQE.OTF(\omega)]^{1/2}$, where DQE is the quantum efficiency of the SEM detector, so even though the incident beam current is sufficiently high to establish an acceptable S/N ratio at low frequencies it is much too poor to reproduce the higher frequency detail. To achieve the electron-optical goal of truly reaching 1 nm resolution it will be necessary to guarantee both a smaller probe diameter – to enhance the value of OTF(ω) at high frequencies – and an increase in the total beam current available to raise the signal to noise ratio at all frequencies to an acceptable value.

The aim of reducing the probe diameter while simultaneously increasing the numerical aperture of the lens to provide more current is daunting, especially at low beam energies, but it might now be possible because of recent advances in aberration corrected probe lenses. Although the principles of aberration correction have been known for fifty years it is only recently that viable, practical, systems have been demonstrated (5,6). For use in the SEM it is necessary to correct both of the major third order aberrations since, at low beam energies chromatic aberration is as much of a limiting effect as is spherical aberration. The systems now being discussed can reduce both spherical and chromatic coefficients to minimal values using arrangements of either electro-magnetic octopole and sextupole lenses, or equivalent electrostatic elements. Figure (5) shows the computed performance that could be obtained from a lens compensated by such a system. The graphs compare the 10%-90% diameter of a current high performance SEM lens at low energy, with the probe size from same lens after the spherical and chromatic aberration coefficients have each been reduced to 200 micrometers. While the uncorrected lens produces a minimum probe size of about 3.5 nm at a convergence angle of 7 milliradians, the corrected lens at the same energy could provide a 1.2 nm probe at a convergence angle of 35 milliradians, or a rather larger diffraction limited probe at smaller convergence angles. An enlargement of this scale in the numerical aperture of the lenses would enhance the signal-to-noise ratio by nearly an order of magnitude and, combined with the reduction in probe diameter and the consequent improvement in the OTF profile, would dramatically enhance the resolution. This technology would also remove the current need to position the specimen close to the lens in order to minimize the working distance, and hence the aberrations, and so could lead to improved sample access and simpler wafer handling in the SEM.

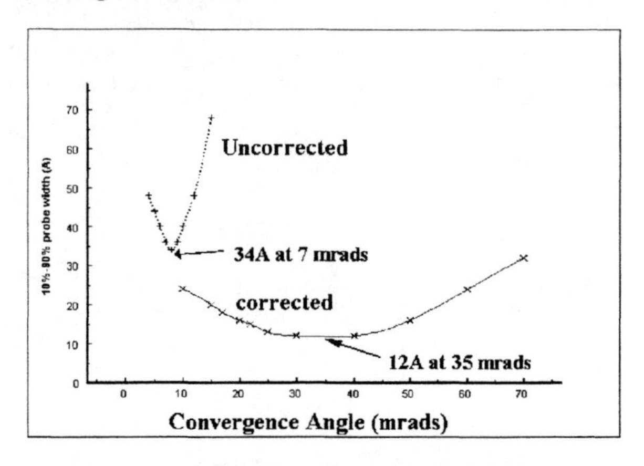

Figure 5. Comparison of the variation of beam probe size as a function of convergence angle for a standard lens and for the same lens corrected to reduce C_s and C_c to 200 micrometers. The beam energy is 1 keV and the energy spread is 0.3 eV

Aberration corrected lenses are thus likely to be a feature of an advanced SEM since they can enhance the performance by a very large factor compared to what is presently available. It must be noted, however, that an SEM combining nanometer-scale resolution and a high numerical aperture lens would have a depth of field of only a few tens of nanometers. Since current device design rules produce structures with high aspect ratios a minimal depth of field will result in most of the device being out of focus most of the time so appropriate corrections, or high speed automated focussing schemes, will have to be implemented to counter this difficulty.

Choosing an Electron Source

As noted above the signal-to-noise ratio is a major factor in setting the instrumental resolution limit of the SEM. Since the signal-to-noise ratio depends on the current in the incident beam, the brightness of the electron source will also be important in improving performance. All modern SEMs use field emission sources, since these provide an emission brightness which is as much as 100x greater than that of a conventional thermionic emitter at the same energy. However for the advanced SEM this level of performance may not be sufficient, since operation at low and ultra-low energies is contemplated and the brightness of all electron sources falls linearly with the energy. A possible solution for this problem is found in the 'nanotip" electron sources (7) in which the emitting tip has been shrunk to a single atom in size. These nanotips are closely related to the tips used in the Scanning Tunneling Microscope (STM) and are made in a similar way be ion etching a standard field emitter, by forming the tip in a gas atmosphere, or by a mechanical process. These tips possess a unique collection of desirable properties (8). Their brightness is a further factor of 100x above that of a standard field emitter, although the total emission current available is limited to a few tens of nanoamps. Because the tips are so sharp they directly emit at energies as low as 50 to 70eV, so they are well suited for the proposed ultra-low energy operation of the advanced SEM. The effective source size is of the order of 0.1 to 0.2 nm and so no demagnification is required to achieve a nanometer size probe. So far these nanotip emitters have only been demonstrated in a laboratory setting, and their performance in the more challenging environment of a fabrication line SEM is not easy to predict, but if even some fraction of their promise can be realized then one of the key instrumental limits to the desired level of performance will have been removed.

Detector Strategies

The secondary and backscattered electron detectors in the SEM have usually been regarded as so good that their effect on the instrument performance has been neglected but for an advanced SEM this optimistic assessment must be challenged. Firstly, recent measurements (9) have shown that the quantum efficiency of most SEM detectors is often far below the values usually assumed to apply. As noted above, this impacts directly on the signal-to-noise ratio of the image and so on the resolution that can be attained. A first step towards the advanced SEM will therefore be to design detector systems with the same care that is applied to lenses, scan coils and other components, and then to verify experimentally that the expected level of performance is indeed achieved. A second step will be to make detectors more selective to either the energy or the angular distribution of the electrons that they collect. For example, if the SE detector could be set to accept only those electrons with an energy of 40 ev or higher then the effective value of the SE mean free path λ_{SE} would be effectively reduced by a factor of two or more because the MFP reaches a minimum at about this energy. This would permit then a significant improvement in the resolution of the SE image by minimizing the SE diffusion zone discussed above. As an additional benefit these higher energy secondaries would also be much less susceptible to charge-induced artifacts. Similarly, the backscattered detector could be designed to collect only the "low loss" component - those BSE with an energy within a few percent of that of the incident beam - since these are the result of single high angle scattering events that occur very close to the beam impact point and can consequently offer a resolution approaching 1nm (10). The ultimate and ideal solution, in either imaging mode, for the advanced SEM would be a detector whose energy response was tunable so that any desired energy window could be chosen from which to form an image. While such a choice inevitably leads to a reduction in the total signal level this is more than compensated for by the increase in information and contrast. When the operator has the ability to choose the incident beam energy and the energy range of the emitted electrons, then the information in the image is optimized and the lateral and vertical localization of the region from which this data is obtained becomes highly specific..

Self Test, Calibration, Tool Matching

The SIA roadmap demands that all tools should be capable of testing and demonstrating their own level of performance. This is of importance for any microscope that is to be used in an automated, or remotely controlled, manner. It is particularly crucial for tools such as those used for metrology where significant production costs may be incurred if the instrument drifts and begins to report deviations from tolerable parameters in the devices that it is measuring. A goal for the ASEM is that it should be able to continuously monitor its own performance and produce an "audit trail" documenting its behavior so that deviations outside bounds can be identified and corrected. A simple example would be in the task of checking the resolution of the image. Until recently there has been no agreed way of experimentally determining SEM image resolution in contrast to the situation on the transmission electron microscope (TEM) where well established protocols are employed. Instead a variety of *ad hoc* methods were employed, such as measuring the size of the smallest spacing or object visible in an image, but these methods are subjective and give results that are both sample and observer dependent. Several techniques based on a Fourier analysis of the image are, however, now available (11,12). These attempt to identify the highest spatial frequency present in the image and equate this to the resolution. Figure (6)

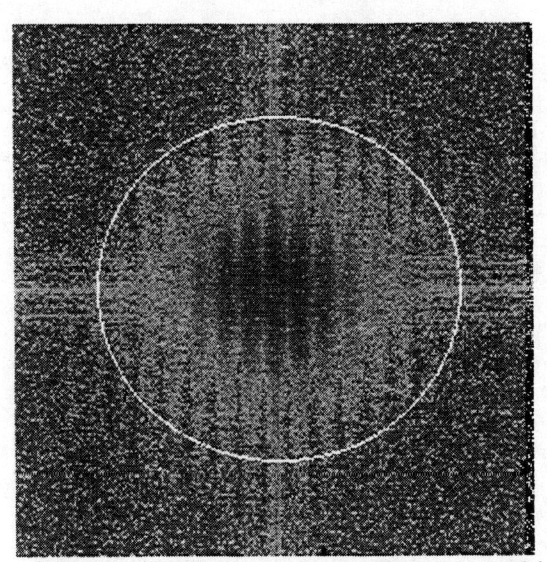

Figure 6. 2-D FFT power spectrum from two sequential images overlapped after a shift of 12 pixels. The line marks the image information cut-off at 10% of the peak intensity and defines the spatial resolution limit.

shows one such approach in which a pair of nominally identical digital images recorded successively from a sample have been overlaid with a small pixel shift and then Fourier transformed. Detail which is present in both images is coherently added in the transform but random noise or uncorrelated events are not. This results in an enhancement of the signal-to-noise ratio by a factor of 4x making it easier to assess the high frequency cut-off. As an additional bonus the fringe pattern, whose spacing is set by the pixel offset, provides a calibration scale in frequency space. The results of measurements such as this are, generally, pessimistic compared with more casual assessments and suggest that resolutions are not in fact as good as is often believed. Assessments of the image quality can also be made from these transforms by using the techniques of information theory (13) to generate parameters which describe not only the resolution of the image, but its signal-to-noise and its information content. For repetitive imaging tasks, monitoring parameters such as these would provide the information needed to track the variation of performance with time.

A key problem in metrology has been that of "tool matching" which attempts to compensate for the fact that nominally identical e-beam tools provide different results when used to measure the same device structure. The optical transfer function (OTF), which displays the response of the tool in the spatial domain, may provide a partial solution to this puzzle since the OTF of a particular machine is impressed like a fingerprint on all the data passing through the microscope. An experimental OTF profile, like that shown in figure (7) might then be compared to the corresponding OTF from other machines to identify how and why changes are occurring. This is not a complete solution, however, since the combination of detectors, amplifiers, and display and recording devices that make up an SEM also have a fingerprint of their own in the time domain, and work on appropriate ways to monitor and document this are still in progress.

Instrument configuration

The final question to consider is whether the scanning electron microscope as presently understood constitutes the best way to proceed for a more advanced instrument. Although the use of a focussed, scanned, electron probe produces a microscope of great flexibility and capability, this configuration also inherently produces limitations some of which have been noted above. For example the electron-optical performance is limited by lens aberrations or, if those are removed, by diffraction and the resolution of secondary electron imaging is limited by the mean free path of SE production. Focussed beams can also lead to charging artifacts and the high dose rate may produce irradiation damage in many samples. In a more general sense the production, and subsequent analysis, of an image a single pixel at a time

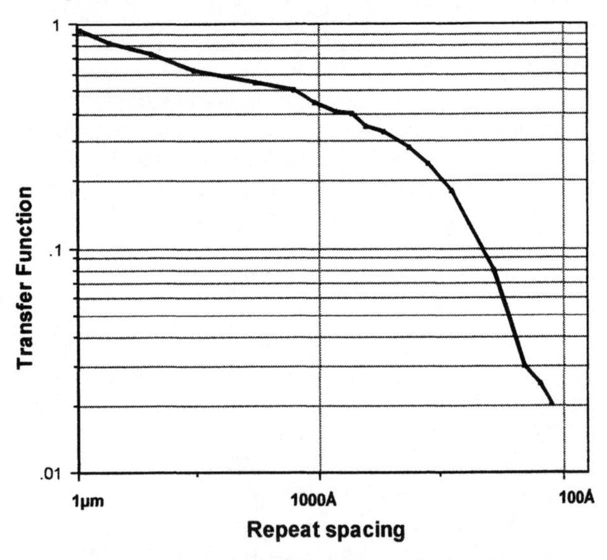

Figure 7. Experimental optical transfer function from a typical field emission gun SEM at 1keV

is a slow process and one that is not efficient because many of the pixels interrogated contain no information of any value. Thus in operations such as metrology only a small fraction of the pixel data acquired is actually utilized to make a measurement. Attempts to reduce the time involved can only be made by limiting the time spent per pixel, which then mandates higher beam current and places extreme requirements on the

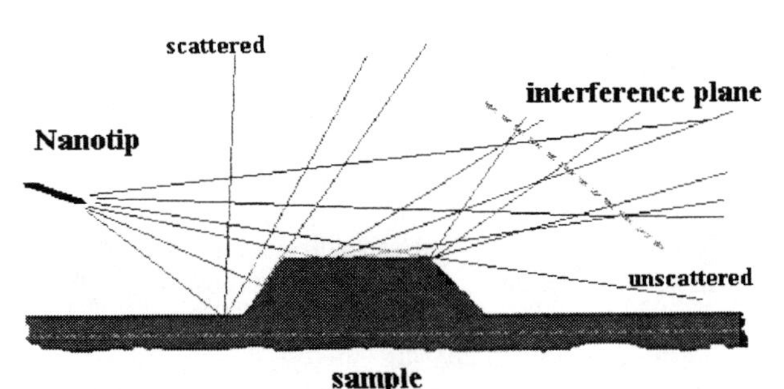

Figure 8. Schematic layout of an advanced imaging system using an ultra-low voltage nanotip and employing forward scatter holography.

performance of the video amplifiers, detectors, and display systems of the SEM.

Figure (8) shows the conceptual layout of a device for imaging which seeks to avoid the weaknesses of the conventional SEM while offering an equivalent level of convenience. The device uses a nano-tip to directly produce an electron beam with an energy in the range 50 -150 eV. No condensor or probe-forming lenses are employed and the beam is allowed diverge to produce a patch of illumination a few tens of micrometers in size on the sample. Because the nanotip produces highly coherent illumination some fraction of the signal is coherently scattered from the sample and interferes with unscattered radiation from the tip to produce a Fresnel hologram downstream from the specimen. This arrangement addresses many of the problems mentioned above. Since there are no lenses there are no lens aberrations, and because the beam is divergent there is no diffraction limit to the performance. The spatial resolution is ultimately limited to about three times the electron wavelength. The divergent beam also ensures a low electron dose at the specimen surface which will minimize charging and radiation damage. Holograms contain more information than an image because they contain both amplitude and phase information and they are also resistant to damage or corruption in transfer because they have a high degree of redundancy . An analysis of the hologram, performed by transforming it into Fourier space, then gives a statistical read-out of spatial information on the entire illuminated area simultaneously, so that only a relatively slow step and stop action is required to analyze a large sample. Instruments of this form have been demonstrated (14) and if they can be successfully developed they may well permanently transform the nature of all future advanced SEMs.

MICROANALYSIS IN THE ADVANCED SEM

The Current/Near Future Situation

X-ray Microanalysis with Conventional Semiconductor Energy Dispersive Spectrometry

Virtually all SEMs used for materials and device characterization, including failure analysis, are equipped with a semiconductor energy dispersive x-ray spectrometer (EDS). Most frequently the semiconductor is silicon with a limiting resolution performance (defined as the full peak width, half maximum (FWHM) peak intensity for Mn Kα, 5.89 keV) in the range 130 -150 eV, depending on the size of the detector. Additionally, some germanium detectors are also in use to obtain slightly better limiting spectral resolution (125 -130 eV, FWHM at Mn Kα). A few analytical SEMs are also equipped with the high resolution (10 eV or better) wavelength dispersive x-ray spectrometer (WDS) (15).

Throughout the history of electron-beam x-ray microanalysis, analysts have relied upon the strong dependence of the electron range R on the incident energy to control the spatial resolution of analysis, which is critical in such problems as characterizing inclusions in a matrix.

$$R = k E^n \qquad (1)$$

where k depends on matrix parameters (atomic number (Z), atomic weight (A), and density (ρ)) and the exponent n is in the range 1.5 - 1.7. The "conventional" energy range for quantitative electron beam x-ray microanalysis can be thought of as beginning at 10 keV and extending to the upper limit of the accelerating potential, typically 30 - 50 keV depending on the instrument. The lower limit of 10 keV for the conventional operating range is selected because this is the lowest incident beam energy for which there is at least one satisfactory analytical x-ray peak excited from the K-, L-, or M- shells for every element in the Periodic Table that is accessible to x-ray spectrometry, excluding only H, He, and Li. Electron-excited x-ray microanalysis thus extends from Be (E_K =0.116 keV) to the transuranic elements. This criterion is based upon establishing as the minimum acceptable overvoltage $U = E_0/E_c > 1.25$, where E_0 is the incident beam energy and E_c is the edge excitation energy ("critical ionization potential"). At E_0 = 10 keV, this overvoltage criterion would involve the use of K-lines for 4 (Be) < Z < 27 (Co); L-lines for 28 (Ni) < Z < 67 (Ho); and M-lines for

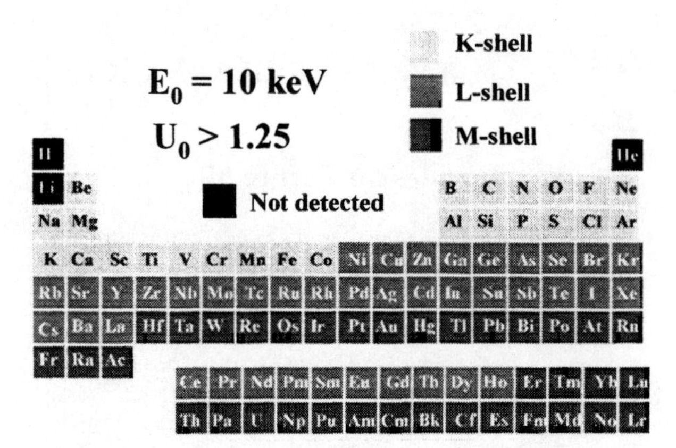

Figure 9. Selection of x-ray peaks available for analysis at a beam energy of 10 keV. Elements in black squares either do not emit x-rays or else are too low in energy for practical detection.

68 (Er) < Z to the transuranics, as illustrated by the Periodic Table in Figure (9) shaded according to the appropriate shell for analysis with $E_0 = 10$ keV and $U \geq 1.25$. This broad elemental coverage of electron beam x-ray microanalysis is one of its most important features.

Low electron beam energy microanalysis, arbitrarily defined as $E_0 < 5$ keV, has been possible for many years with SEMs based on tungsten and LaB_6 thermionic emitters, and has always been attractive because of the shallow excitation depth and the possibility of achieving high lateral resolution. Figure (10) shows the excitation range for several characteristic x-rays in a silicon matrix as a function of incident beam energy, as calculated with the Kanaya-Okayama formulation of the excitation range (16):

$$R\,(\mu m) = (0.0276\,A/Z^{0.89}\,\rho)\,(E_0^{1.67} - E_c^{1.67}) \quad (2)$$

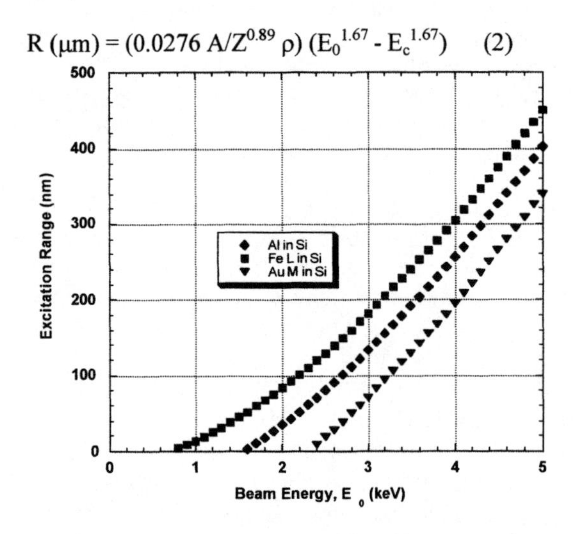

Figure 10. Range of excitation of various lines in a silicon matrix as a function of beam energy.

The range is less than 100 nm for these x-ray lines in silicon for an overvoltage of 2, and diminishes sharply as the overvoltage is reduced, so that for an overvoltage of 1.25, the excitation range is approximately 20 - 30 nm. The emergence of the high performance field emission gun SEM, which can still achieve nanometer-scale beam diameters at low energy, has made it possible to actually make use of this potential analytical resolution, greatly increasing interest in low energy x-ray microanalysis. The typical analytical FEG-SEM as currently fielded is equipped with a semiconductor (Si or Ge) energy dispersive x-ray spectrometer (EDS) with a limiting spectral resolution of 125 - 150 eV. While useful analyses can certainly be performed with the semiconductor EDS, the limited resolution of EDS (approximately 50 to 100 times the natural x-ray linewidth) and the aspects of the physics of x-ray generation impose some severe constraints. As the electron beam energy is reduced, x-ray microanalysis becomes insensitive to certain elements because of these factors.

Limits of Detection

At high overvoltage ($U > 3$), experience with semiconductor EDS x-ray spectrometry has shown that the concentration limit of detection for most elements in most matrices is approximately 0.001 mass fraction in the absence of peak interference (15). The limit of detection, C_{MDL}, depends strongly upon the characteristic peak-to-bremsstrahlung background (P/B) and upon the characteristic x-ray peak counting rate (P) (17):

$$C_{MDL} = 3.29\,a/[n\,\tau\,P\,(P/B)]^{1/2} \quad (3)$$

where a is the Ziebold-Ogilvie factor in the expression relating concentration to measured intensity ratio (unknown to pure element standard (18); "a" can be taken as approximately unity for general estimation purposes), n is the number of measurements, τ is the integration time per measurement, P is the peak counting rate, and B is the background counting rate. As the overvoltage is lowered, both the peak counting rate and the peak-to-background decrease. Figure (11) shows the behavior of P/B and C_{MDL} as a function of overvoltage as experimentally measured for Si Kα from pure silicon. Measurement conditions were 100, 200, and 1000 second accumulation times ($n{*}\tau$ product) with a beam current and detector solid angle chosen so that the EDS system deadtime was approximately 30% (total spectrum) for a 30 mm^2 Si detector with a resolution of 150 eV (Mn Kα). As can be seen in Figure 11 (a), operation at low overvoltage results in restrictions on the

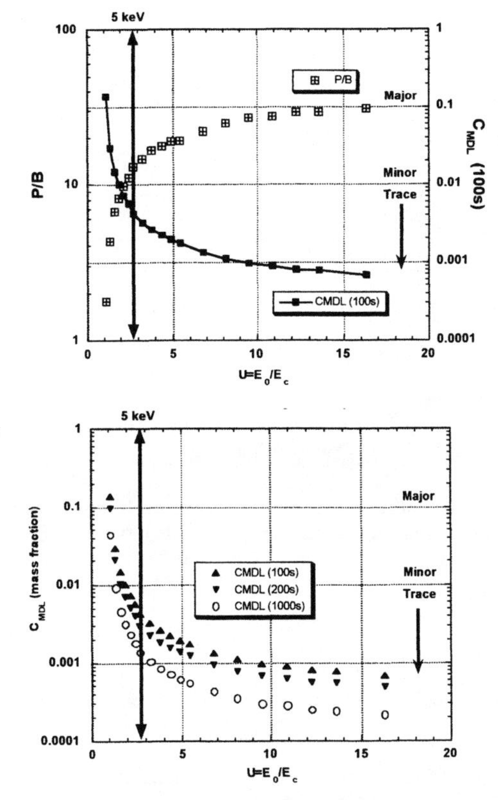

Figure 11 (a) Experimental measure of P/B for silicon and calculation of concentration limit for Si in a matrix of similar atomic number; (b) Calculated limit of detection derived from experimental P/B measurements

concentration limit of detection. Major (arbitrarily defined as C > 0.1 mass fraction) and minor (defined as 0.01≤ C≤ 0.1 mass fraction) constituents can still be detected with overvoltages as low as U = 2 for an accumulation time of 100 s. Trace constituents (defined as C < 0.01 mass fraction) above a level of 0.001 mass fraction cannot be reliably measured with U below 9.0. To reduce the overvoltage requirement to U = 3 for trace levels above 0.001 mass fraction requires an accumulation time of 1000 s, as seen in Figure 11(b). A higher resolution semiconductor EDS detector (e.g., 130 eV at Mn Kα) would improve the situation by a factor of 10 - 20%. Any peak interferences would serve to make the detection situation much more unfavorable.

Limitations Imposed by Fluorescence Yield

When an energetic electron scatters inelastically with a bound atomic electron resulting in the ejection of the atomic electron, the ionized atom is left in an excited state. The atom returns to the ground state through processes that involve electron transitions. The

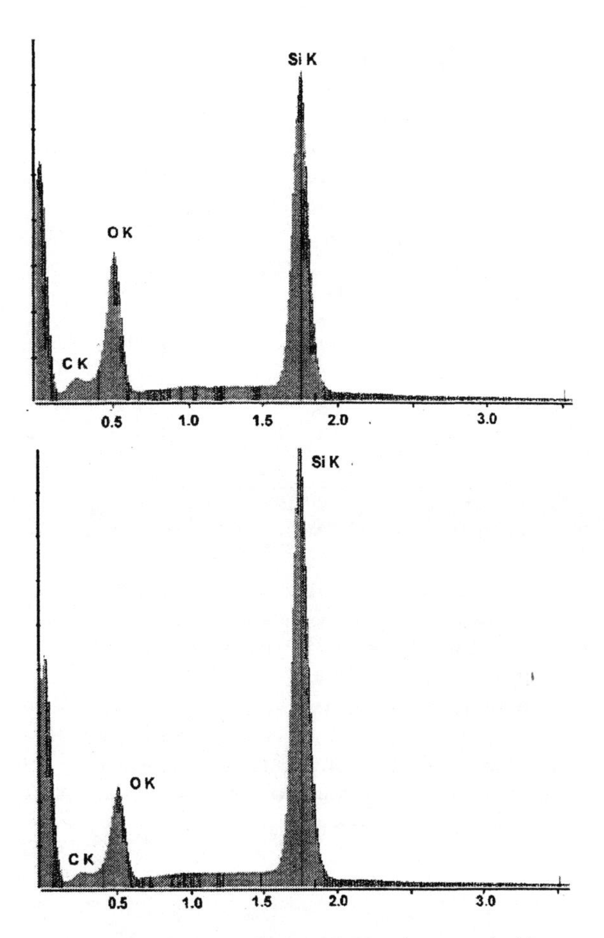

Figure 12. (a) SiO_2 excited at 5 keV and measured with a Si EDS (150 eV at Mn Kα) and (b) SiO excited at 5 keV and measured with a Si EDS (150 eV resolution)

difference in energy between the electron states can be released in the form of an x-ray photon, or the energy can be transferred to another atomic electron, which is ejected as an Auger electron. The fraction of ionizations that leads to photon emission is called the fluorescence yield, ω, which is related to the Auger yield, α:

$$\omega + \alpha = 1 \qquad (4)$$

For electron shells with low binding energy, the Auger process is strongly favored over x-ray emission. For example, for carbon K-shell ionization (E_K = 0.284 keV), the fluorescence yield is 0.00198. Table 2 lists some typical values of the fluorescence yield low ionization energy shells. Examination of this data reveals that an important consequence of having to select low energy L- and M- shells to analyze for intermediate and heavy atoms is that the fluorescence yields are substantially lower than that of similar energy K-shells, a factor of 3 to 5 poorer for L-shells and 10 - 30 poorer for M-shells. Thus, the x-ray spectra of these

Table 2 Fluorescence Yields

Element	Shell	Edge (keV)	ω
C	K	0.284	0.00198
Na	K	1.080	0.0192
Mg	K	1.303	0.0265
Si	K	1.848	0.0603
Zn	L	1.022	0.00736
Ga	L	1.117	0.00875
Ge	L	1.217	0.0103
Sm	M	1.080	0.00133
Eu	M	1.130	0.00137

elements with low beam energy excitation are likely to have characteristic peaks with low intensity relative to the background (x-ray bremsstrahlung), a problem that is further exacerbated by the poor resolution of the semiconductor EDS, which acts to spread the available peak information over a wide range of background.

The combination of low overvoltage and the low yield from L- and M- shell x-ray peaks can result in spectra which look extremely unfamiliar to analysts who are used to spectra excited in the conventional beam energy

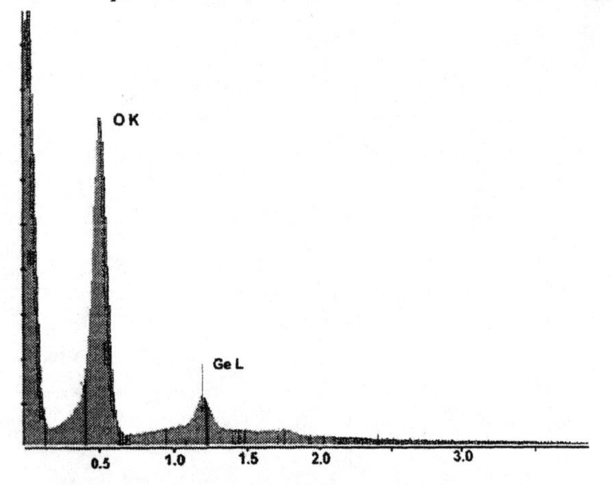

Figure 13. GeO_2 excited with a beam energy of 5keV and measured with a 150eV resolution Si detector.

range. As an example, Figure (12) shows the situation for (a) SiO_2 and (b) SiO excited with an incident beam energy of 5 keV and measured with a Si EDS (150 eV FWHM at Mn Kα). Both oxygen and silicon are measured with K-lines, and peaks for both are prominent in the spectrum, with the oxygen peak reduced relative to the silicon due to stronger absorption effects. By comparison, Figure (13) shows the spectrum for GeO_2. The Ge L-family of peaks (unresolved) is much reduced in intensity relative to the O and Si K-peaks due to the

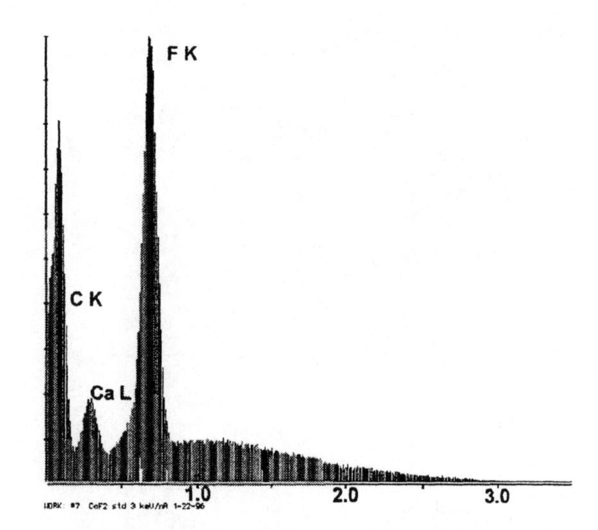

Figure 14. CaF_2 excited with an incident beam of 3 keV and measured with a 150 eV FWHM Si detector.

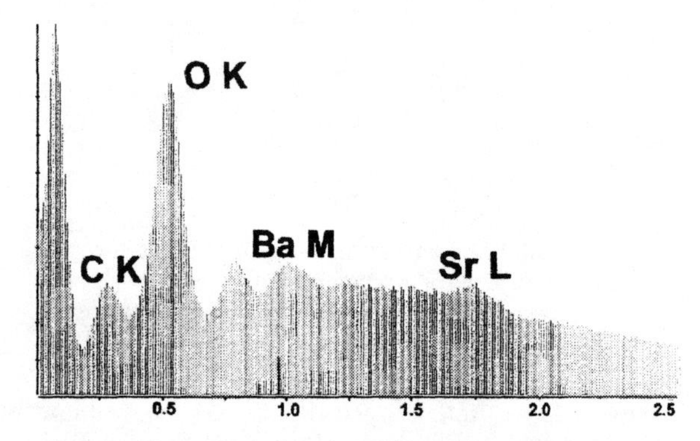

Figure 15. $BaCO_3$ with minor Sr (~0.05) excited at 5 keV and measured with a 150 eV Si EDS. The Ba L- and M- lines are marked. Note the low intensity for the Ba M series.

relative fluorescence yield differences, despite the fact that the overvoltage is actually greater for the Ge L-edge (E_{LIII} = 1.217 keV) than the Si K-edge (E_K = 1.848 keV). Similarly, Figure (14) shows the situation for CaF_2 excited with an electron beam energy of 3 keV, which is below the Ca K-edge (4.038 keV). Note that the F K peak is very prominent, but the Ca can only be detected through the Ca L peak, which is a factor of 10 lower relative to the F K, due to both the lower fluorescence yield from the L-edge, and the increased absorption of the lower energy Ca L-radiation.

An extreme example of the effects of excitation and fluorescence yield is shown in Figure (15), which presents the spectrum of $BaCO_3$ excited at 5 keV and measured with a Si EDS (150 eV FWHM at Mn Kα). The Ba M-family of peaks is just visible above background. In fact, the series of Ba-M "peak

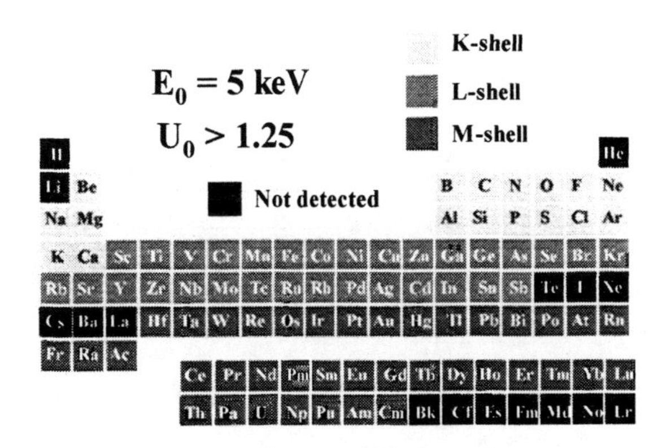

Figure 16. Selection of x-ray peaks available for analysis at 5 keV with conventional EDS.

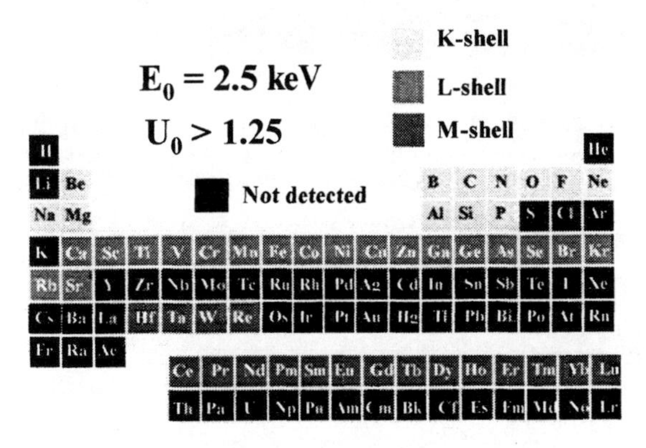

Figure 17. Selection of x-ray peaks available for analysis at 2.5 keV with conventional EDS

structures" is partially a result of the Ba-M absorption edges, which exist as discontinuities in the x-ray continuum (bremsstrahlung) and which are broadened into peak-like features as a result of the action of the detector broadening function.

An overall assessment of the impact of both low excitation and the fluorescence yield on elemental accessibility with a 5 keV incident beam is presented in Figure (16), while the situation with 2.5 keV is shown in Figure (17). At 5 keV, a few intermediate and heavy elements are inaccessible (black squares), while at 2.5 keV, the situation is much worse, with extensive sections of the Periodic Table effectively lost to SEM/semiconductor EDS measurement.

Future Developments

The physics of electron-excited x-ray production and propagation is determined by the choice of beam energy, the element(s) to be measured, and the matrix composition. Once the analyst's room to maneuver is restricted through the choice of low beam energy, the only possible route to improve the measurement process is to improve the x-ray spectrometry. From equation (3) and the spectra presented in Figures (12)-(15), it is clear that improving the spectral resolution will increase the peak-to-background, improving the visibility of low fluorescence yield peaks, reducing interferences, and lowering the limit of detection. Increasing the detector efficiency (i. e., solid angle) and reducing the pulse processing time will improve the peak counting rate, also contributing to improving the limit of detection.

Improvements to wavelength dispersive spectrometry

Wavelength dispersive x-ray spectrometry, based on Bragg diffraction of x-rays from a crystal is, in fact, the oldest form of x-ray spectrometry and is commonly incorporated in the instrument configuration referred to as the electron probe x-ray microanalyzer. The WDS is a high resolution (10 eV or less), "single channel", narrow bandpass spectrometer that must be mechanically scanned to cover a range of x-ray wavelengths (energy). The higher resolution of the WDS results in a higher P/B, yielding a lower limit of detection, than semiconductor EDS. Typically the WDS limit of detection is 10^{-4} to 10^{-5} mass fraction. The high spectral resolution has the added advantage of diminishing the likelihood of spectral interferences, although interferences may still occur for certain combinations of x-ray lines. However, the quantum efficiency of the WDS is significantly lower than the EDS due to the relative inefficiency of the diffraction process. Consequently, beam currents of the order of 100 nA or more are normally used to achieve adequate counting peak rates. Such current requirements are incompatible with small beam size, even with the thermally-assisted field emission gun. Additionally, when low beam energy excitation is considered, another problem becomes significant. Low energy x-rays originate from electron energy levels whose energy and population can be affected by chemical bonding which is manifested in the x-ray spectrum as shifts in the peak position and shape (19). Since the WDS only detects a narrow band of energy, for an accurate measurement of a peak that is subject to alteration due to chemical bonding effects, it is necessary to scan across the peak to accurately determine its position and the total intensity under the peak above background. This requirement for scanning further reduces the measurement efficiency and increases the need for beam current. Thus, WDS at low beam energy is normally performed under conditions that result in poorer spatial resolution than EDS.

Two developments may lead to improved WDS efficiency which can benefit low beam energy x-ray microanalysis. First, the WDS is a focusing device, wherein the x-rays diffracted at various points along a curved crystal are brought to a focus within the entrance slit of a gas proportional detector placed on the focusing circle of the spectrometer ("Rowland circle"). If the focused rays are allowed to diverge beyond the Rowland circle, they form a spatially-dispersed image of the peak (with a "single focusing" geometry WDS). By placing an imaging detector array off the Rowland circle in the dispersion plane, a wider energy window could be simultaneously measured, providing a direct image of the full peak and adjacent background regions (20). Second, the efficiency of the WDS can be improved by increasing its solid angle of collection. For low energy x-rays (0 - 1 keV), a parabolic reflective x-ray optic has been described that gathers a wide solid angle and presents a parallel beam to a flat crystal diffractor that scatters x-rays into a large window proportional gas detector. With this augmented WDS, limits of detection for light elements, e.g., C, as low as 10^{-5} mass fraction have been reported (21).

Microcalorimetry Energy Dispersive Spectrometry

The ideal x-ray spectrometer would have the resolution of the WDS combined with the energy dispersive character of EDS over the useful analytical range, 100 eV to 10 keV. This seemingly impossible requirement has actually been realized with the development of a practical "fast" microcalorimeter EDS at NIST (Boulder) (22). In the microcalorimeter spectrometer, the energy of an individual photon is measured as the temperature rise in an absorbing target maintained near 100 mK. The NIST microcalorimeter design incorporates critical features such as the use of an absorber maintained in the normal conducting state and a temperature sensor based upon a superconducting transition sensor which both contribute to a faster pulse response than earlier microcalorimeters. A novel superconducting quantum interference device is used as a stable, fast, high gain amplifier. With regard to analysis, the chief operating characteristics of the NIST microcalorimeter EDS are:

1. Energy resolution < 10 eV. As an example of the energy resolution applied in a difficult analytical situation, Figure (18) shows the separation of the Ba L-lines from the Ti K-lines in $BaTiO_3$ achieved with the NIST microcalorimeter EDS, and compared with the unresolved peaks in the spectrum obtained with a semiconductor EDS.
2. Limiting count rate (with beam blanking) ~ 1 kHz
3. Photon energy coverage: 250 eV - 10 keV. An example of this energy coverage is given in Figure (19),

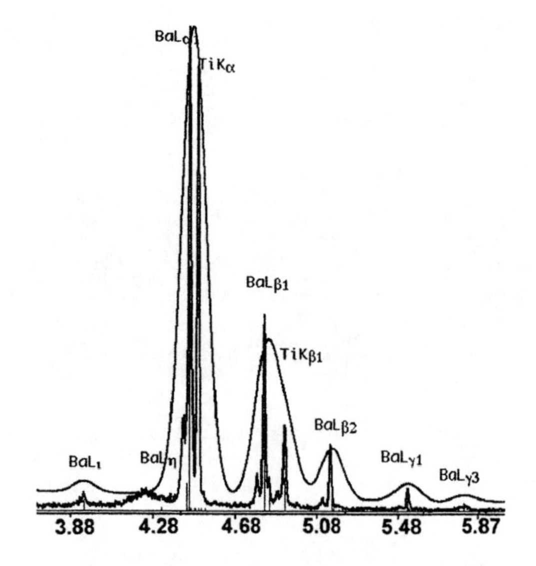

Figure 18. Comparison of spectra of $BaTiO_3$ obtained with the microcalorimeter EDS (FWHM=10 eV) and with the semiconductor EDS (FWHM=140 eV)

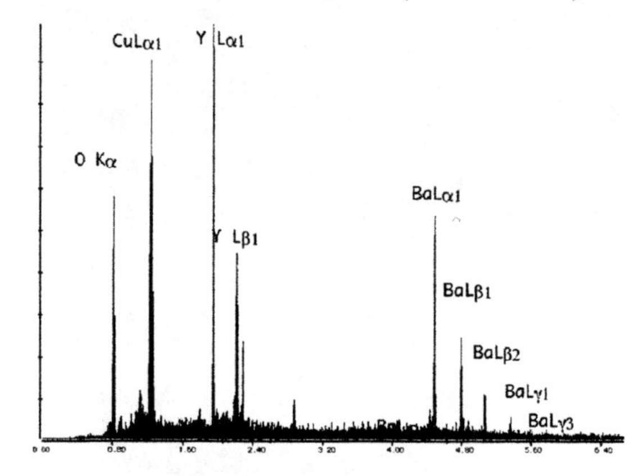

Figure 19. Microcalorimeter EDS spectrum of $YBa_2Cu_3O_7$ at a beam energy of 10 keV

which gives a spectrum for $YBa_2Cu_3O_7$ obtained with the microcalorimeter EDS, showing L-line detection for Cu, Y and Ba and K-line detection for O. Carbon K-shell x-rays at 0.282 keV have been successfully measured, as shown in Figure (20).

The NIST microcalorimeter EDS represents the leading edge of a revolution in x-ray spectrometry. In particular, microcalorimeter EDS technology should have a major and immediate impact on the application of the FEG-SEM to microanalysis at low beam energy, for example, in the analysis of particles during defect review in semiconductor processing. As an example, Figure (21) shows the spectrum obtained at 5 keV from WSi_2 with a 10 eV microcalorimeter EDS. The separation of the Si Kα line from the W M-family is virtually complete, thus eliminating the severe interference encountered with

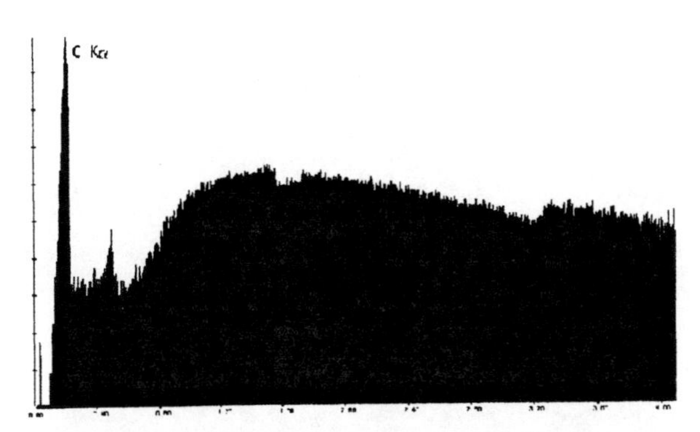

Figure 20. Microcalorimeter EDS spectrum of carbon on a logarithmic scale. Note the carbon coincidence peak and the discontinuities in the background due to absorption in the spectrometer materials.

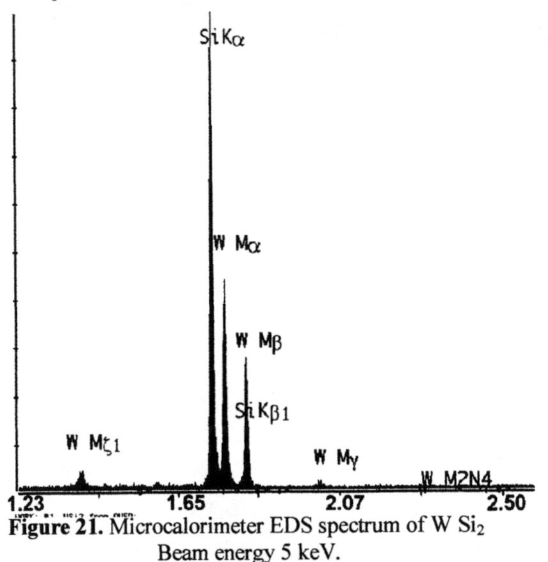

Figure 21. Microcalorimeter EDS spectrum of W Si_2 Beam energy 5 keV.

semiconductor EDS. The high P/B ratios evident in the spectra in Figures (18) – (21) will result in limits of detection similar to WDS. Finally, the resolution of the microcalorimeter EDS will make it possible to actually detect low energy L- and M- lines that have low fluorescence yield. As an example, Figure (22) demonstrates the detection of several of the low yield peaks of the titanium L-family and the barium M-family in $BaTiO_3$, which can be compared to Figure (15), obtained with semiconductor EDS. With conventional EDS, the Ti L-peaks are unresolved from the O K-peak, and the Ba M-peaks, while visible, are actually convolved with the Ba M-edge discontinuities in the background.

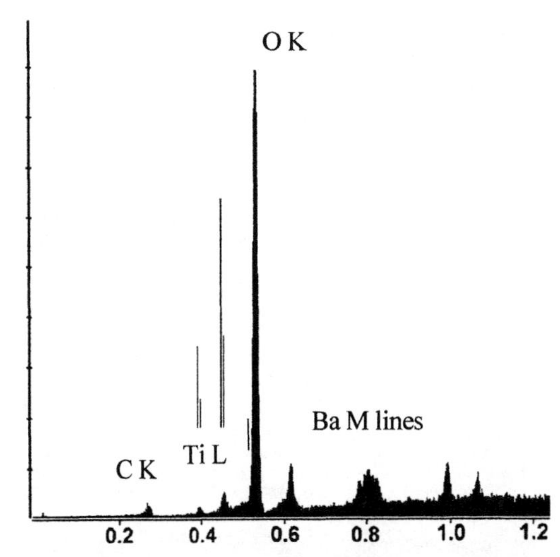

Figure 22. Microcalorimeter EDS spectrum of $BaTiO_3$, low energy region. Note detection of M-family members

ACKNOWLEDGEMENTS

Oak Ridge National Laboratory is managed for the U.S. Department of Energy by Lockheed Martin Research Corporation under contract DE-AC05-96OR22464. One of us (DCJ) acknowledges the Semiconductor Research Corporation for support for the work described here under contracts 96-LJ-419.001 and 96-LJ-419.002.

References

(1) *The National Road Map for Semiconductor Technology.* The Semiconductor Industry Association, 1997
(2) Joy D.C. Scanning **17**, 270-274 (1995)
(3) Joy D.C (1998), "Ultra-low energy imaging for metrology", Proc. S.P.I.E Meeting on Microlithography Santa Clara, CA. February 23-27th, 1998 (in press)
(4) Joy .D.C. Ultramicroscopy **37**, 216-222 (1991)
(5) Haider M, Zach J, and Wepf. R, "Physical Aspects of corrected LVSEM", Proc.ICEM-13, ed B.Jouffrey and C. Colliex, Paris, France 55-56 (1994)
(6) Haider M and Uhlemann S, Microscopy and Microanalysis **3**, Suppl.2, 1179-80 (1997)
(7) Muller H U, Volkel B, Hofmann M, Woll Ch, and Grunze M, Ultramicroscopy **50**, 57-64 (1993)
(8) Scheinfein M R, and Spence J C H, J.Appl.Phys. **73**, 2057-2060 (1993)
(9) Joy D.C., Joy C.S. and Bunn R.D. Scanning **18**, 533-538 (1996)
(10) Wells O.C. Appl.Phys.Lett., **16**, 151-153 (1970)
(11) Postek M.E. and Vladar A. E, Scanning **20**, 1-9 and 24-34 (1998)
(12) Erasmus S J, Holburn DM, Smith KCA "On line computation of diffractograms for analysis of SEM images", Inst. of Physics Conf. Ser. **52**, 73-76 (1980)
(13) Sato M and Orloff J Ultramicroscopy **41**, 181-192 (1992)
(14) Spence J C H, J.of Surf.Analysis **3**, 213-221 (1997)
(15).Goldstein, J. I., Newbury, D. E., Echlin, P., Joy, D.C., Romig, A. D., Jr., Lyman, C. E., Fiori, C., and Lifshin, E., *Scanning Electron Microscopy and X-ray Microanalysis,* 2nd edition Plenum Press: New York, 1992 .

(16) Kanaya, K. and Okayama, S. J Phys. D.: Appl. Phys., **5**, 43-48 (1972)
(17). Ziebold, T. O. Anal. Chem. **39**, 858-865 (1967).
(18) Ziebold, T. O. and Ogilvie, R.E. Anal. Chem. **36**, 322.-330 (1964)
(19). Bastin, G. F. and Heijligers, H. J. M.,"Quantitative Electron Probe Microanalysis of Ultralight Elements (Boron-Oxygen)" in *Electron Probe Quantitation*, eds. K. F. J. Heinrich and D. E. Newbury, Plenum :New York, 1991, 141-3
(20) Fiori, C. E., Wight, S. A., and Romig, A. D., Jr. *Microbeam Analysis 1991* (San Francisco Press) 327-8 (1991).
(21). Agnello, R., Howard, J., McCarthy, J., and O'Hara, D. Microscopy and Microanalysis, 3 (suppl 2) 889-890 (1997)
(22) Wollman, D. A., Irwin, K. D., Hilton, G. C., Dulcie, L. L., Newbury, D. E., and Martinis, J. M, J. Micros., **188,** 196-223 (1998)

Transmission Electron Microscopy: A Critical Analytical Tool for ULSI Technology

David Venables, David W. Susnitzky*, and A. John Mardinly*

Department of Materials Science and Engineering, North Carolina State University, Raleigh, NC 27695
**Materials Technology Department, Intel Corporation, 2200 Mission College Blvd., Santa Clara, CA 95052*

An overview of the capabilities and limitations of transmission electron microscopy (TEM) based analysis techniques in the context of the electronics industry is presented. The electron-beam/specimen interactions that enable morphological, crystallographic and compositional characterization with modern TEMs are briefly reviewed. Diffraction contrast, lattice and energy filtered imaging; energy dispersive x-ray spectrometry (EDS), and electron energy loss spectrometry (EELS) are reviewed and discussed. These techniques are illustrated through specific applications and case studies in the electronics industry. Particular emphasis is placed on sample preparation concerns, which represent a practical limitation to an expanded role for TEM as a critical analytical tool for ULSI technology.

INTRODUCTION

Transmission electron microscopy (TEM) provides for the characterization of materials at high spatial resolution through the use of a powerful suite of imaging, chemical analysis, diffraction and other techniques. These capabilities have led to an increasingly essential role for TEM-based analysis in process development, defect identification, yield improvement and root-cause analysis within the electronics industry. With continuing reductions in semiconductor device dimensions, the high spatial resolution of TEM-based techniques will be required to an even greater extent.

A transmission electron microscope consists of an electron source and a series of condenser, objective and imaging lenses placed within a vacuum column. The electrons emitted from the source are accelerated by a high potential (typically 200 kV) and the energetic electron beam is made incident on a thin (typically 100 nm), electron transparent sample placed within the objective lens. A phosphor screen on the exit side of the sample is used to view the images or diffraction patterns, which may be recorded on film or through a digital camera. Various detectors (e.g., x-ray detectors and electron spectrometers) may also be present for acquiring other kinds of information. The physical basis for obtaining information about the sample lies in the elastic (no electron energy loss) and inelastic (electron energy loss) scattering processes that occur when the electron beam interacts with the sample.

The most important elastic scattering process is electron diffraction from crystalline specimens. Diffraction in crystalline materials provides the basis for direct analysis of crystal structure and crystal orientation from an analysis of the diffraction patterns, diffraction contrast imaging of defects and discontinuities in the crystal [1-3] and lattice imaging [3,4], in which the periodic atomic arrangements may be directly observable. Thus, these techniques provide crystallographic, morphological and microstructural information about the sample.

Inelastic scattering processes provide complementary information about the sample through detection of characteristic x-rays emitted when electrons in the target atoms fill energy levels made vacant by the excitation of core electrons [3,5-6] and analysis of the energy lost by the transmitted electrons after undergoing such inelastic processes [3,7]. These techniques are known as energy dispersive x-ray spectrometry (EDS) and electron energy loss spectrometry (EELS), respectively. Since many inelastic processes are characteristic of the particular atoms in the target, they provide compositional information about the sample. Other signals (e.g., secondary, backscattered, and Auger electrons) are generated by the interaction of the electron beam with the specimen, however, these signals are not typically used in TEM-based analysis.

Exploiting these scattering processes for the needs of the electronics industry requires the production of thin sections of the feature of interest. Traditionally, the dimple/ion mill method, which consists of mechanical thinning followed by low energy (< 5 keV) argon ion sputtering, has been used to produce non-site-specific electron transparent specimens [8]. Mechanical thinning alone (the "wedge" technique) can also be used for this purpose, for sectioning of special test structures which are artificially elongated in one direction and for site-specific sectioning [9]. However, it is often difficult to apply the wedge technique consistently in the latter case. As a result, site specific sectioning usually requires the use of a focused ion beam (FIB) technique [10]. This sectioning method uses a high energy (25-50 keV), high current gallium ion beam to isolate a thin specimen from the region of interest. These systems may include scanning electron microscope capability to identify the region to be sectioned. A major limitation associated with this technique is the severe damage induced at the surface of the specimen by the gallium ions [11]. As a result, the utility of TEM for

CP449, *Characterization and Metrology for ULSI Technology: 1998 International Conference*
edited by D. G. Seiler, A. C. Diebold, W. M. Bullis, T. J. Shaffner, R. McDonald, and E. J. Walters
 1-56396-753-7/98/$15.00

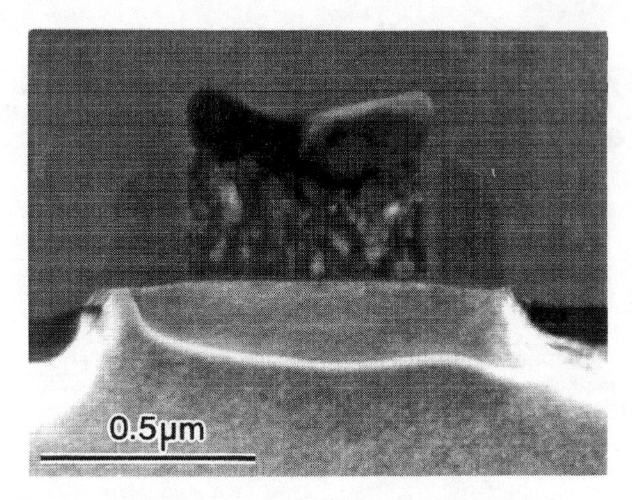

Figure 1. Weak-beam dark field micrograph showing a dislocation line connecting the source and drain of a transistor, which was the cause of an electrical short.

electronic device applications is limited primarily by specimen preparation concerns.

In the following sections, an overview of both the capabilities and the limitations of TEM-based techniques is presented in the context of the electronics industry. Diffraction contrast and lattice imaging modes of operation are reviewed in the first section, followed by a discussion of the techniques which yield compositional information in the second section. Examples illustrating the microstructural, morphological, crystallographic and compositional information obtained from these methods for process development, defect identification and root-cause analysis purposes are presented in each section. The role of sample preparation in enabling these analyses is specifically noted. The third section includes a more detailed discussion of sample preparation concerns, followed by a summary in the final section.

IMAGING TECHNIQUES

The diffraction of electrons by the planes of atoms in a crystal provides the basis for diffraction contrast imaging and lattice imaging, the two primary imaging modes in the TEM. Diffraction contrast imaging [1-3] is sensitive to long range lattice displacements associated with the presence of an extended defect, such as a dislocation. The specimen is purposely tilted to an orientation in which diffraction occurs primarily from only one plane of the crystal. This condition yields two strong electron beams -- the direct transmitted beam and a single diffracted beam. A physical aperture can be inserted on the electron beam exit side of the specimen to select either the direct or the diffracted beam to form an image. A bright field image is formed when the direct beam is selected and areas of the specimen which are strongly diffracting (e.g., the strained region around a defect core) thus appear dark in the image, while the rest of the field of view appears bright. The contrast is essentially reversed, i.e., the defective area appears bright on a dark background, when the diffracted beam is selected to form a dark field image. It is possible to further manipulate the diffracting conditions to produce weak-beam dark field images in which the contrast is enhanced and the resolution is improved.

The ability to detect and analyze extended defects in this manner is important, since the presence of such a defect in the active device region can degrade the electrical performance. A weak-beam dark field image of an extended defect is shown in Fig. 1. The bright horizontal line in the image is a dislocation that extends from the source to the drain of a transistor. Electrical measurements indicated that this particular device behaved as an electrical short circuit, and the TEM analysis clearly shows that the root-cause of the short circuit is the dislocation line. This analysis required sectioning of a specific transistor; hence, the focused ion beam method was employed for producing the specimen.

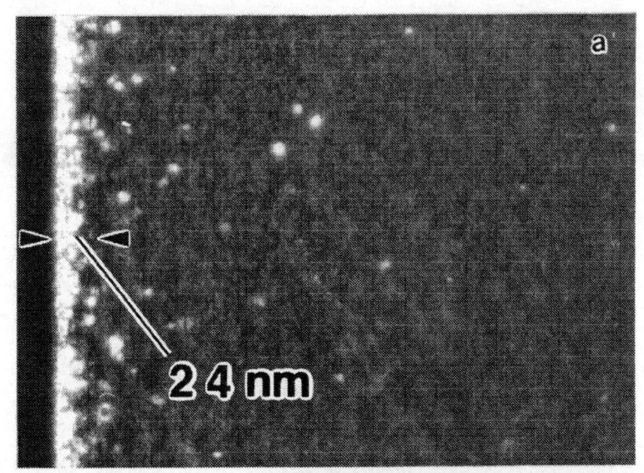

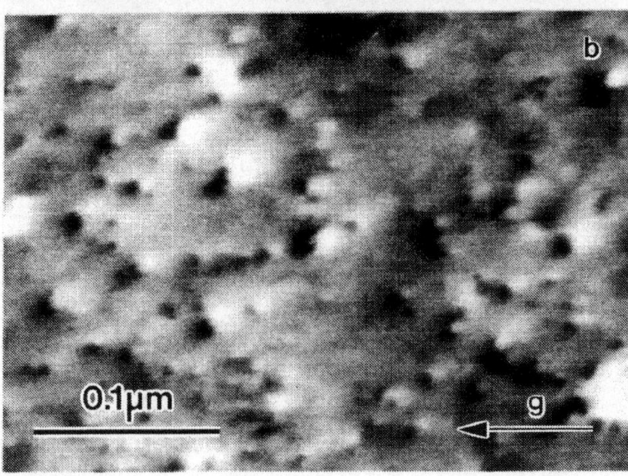

Figure 2. Hydrogen-induced defects in silicon resulting from low temperature hydrogen plasma exposure: a) cross-sectional weak-beam dark field image showing location of defects adjacent to the wafer surface and b) plan-view bright field image from which the strain field of the defects was determined to be compressive in nature. The vector **g** denotes the direction of the reflection used for imaging.

A more detailed analysis of such defects can be performed using diffraction contrast imaging. It is possible to determine the direction of the displacement vector associated with the defect, and to determine whether the displacement field is compressive or tensile in nature. This information may provide clues to the origin of the defects, although the time associated with performing the analysis may preclude its routine use. Specific procedures to perform such analyses are well established from both theoretical and experimental standpoints. An example of this kind of analysis is shown in Fig. 2. In this case, hydrogen plasma exposure was being investigated as a means of "dry" surface preparation of silicon wafers. For low substrate temperatures (150°C) defects near the surface of the wafer were observed, as shown in the weak-beam dark field cross-sectional image in Fig. 2a. Bright-field plan-view images of the defects, such as that shown in Fig. 2b, where analyzed according to established rules, and it was found that the strain field associated with the defects was compressive in nature. This information combined with quantitative correlations between the TEM data and secondary ion mass spectrometry depth profiles of the hydrogen distribution in the samples were consistent with the hypothesis that the defects consisted of platelets of hydrogen bound to silicon. Subsequent work showed that the defects did not form at temperatures above about 450°C, so that the plasma surface preparation was viable only at such temperatures [12, 13]. This research/developmental work was performed on unpatterned silicon wafers, and therefore conventional dimple/ion mill techniques were used to make the samples.

Significant limitations of diffraction contrast imaging are that the spatial resolution is relatively poor (~1 nm) and that the periodic nature of the crystal is not immediately obvious in the images. Lattice imaging [3,4] (also called phase contrast imaging, or high-resolution microscopy) overcomes these limitations. In this situation, the crystal is tilted so that many diffracted beams are formed. The diffracted beams and the direct transmitted beam then interfere with one another to form an image. The crystal acts like a phase grating (i.e., it introduces a phase change in the electron waves) so that the image reflects directly the periodicity of the lattice potential, within the resolution limits of the microscope. Provided the sample is very thin, the resolution is limited by the spherical aberration of the objective lens and the wavelength of the incident electrons. Higher beam energies improve the resolution by decreasing the electron wavelength; however, such energetic electron beams tend to cause significant damage to many specimens, including silicon. As a result of this restriction, many microscopes use 200 kV accelerating voltages. Modern commercial TEMs have point resolutions below 0.2 nm, while special purpose high voltage TEMs are approaching 0.1 nm point resolution.

This unparalleled spatial resolution provides a capability, unique among analytical methods, to characterize ultra-thin layers in terms of morphology, thickness and interface roughness. Three such applications of lattice imaging are shown in Figures 3-5. Lattice images of the near - surface region of the hydrogen plasma exposed

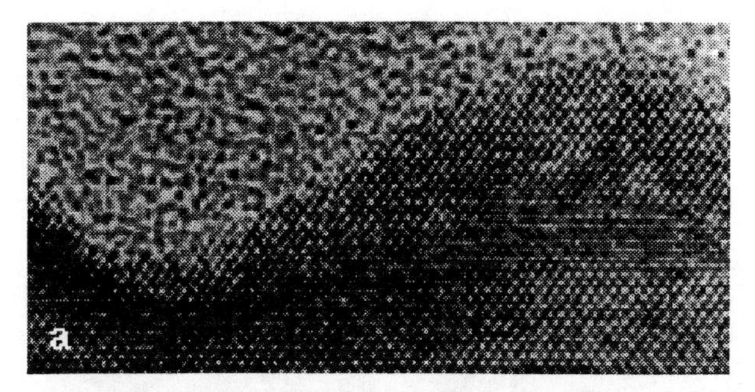

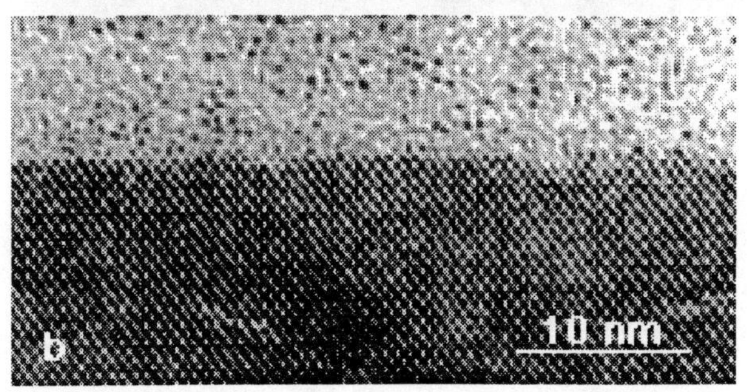

Figure 3. Cross-sectional lattice images of the silicon wafer surface a) after low temperature H plasma exposure, and b) of an untreated control sample.

wafers mentioned previously are shown in Fig. 3 [14, 15]. The silicon substrate shows lattice fringes from which the magnification was directly calibrated, since the spacing of the (111) planes in silicon is known to be 0.314 nm. A root mean square roughness value was then extracted from the images. This example compares a 300°C plasma exposure to an unexposed control sample. The 300°C exposure produced the roughest surfaces of any exposure temperature, however, 450°C exposures were similar to the control sample. A similar analysis can be performed at interfaces between materials which other techniques, such as atomic force microscopy, cannot access. However, such high resolution capability comes with the drawback of a severely restricted sampling area.

The high-resolution mode of imaging is useful also for the analysis of amorphous materials even though they exhibit no long range order. Fig. 4 shows the familiar example of a gate dielectric between the single crystal silicon substrate and the polycrystalline gate electrode. The amorphous gate dielectric shows a mottled contrast characteristic of amorphous materials. However, the thickness of the gate dielectric and the roughness of the top and bottom interfaces can still be assessed. Again, the lattice image in the silicon substrate can serve as a direct and accurate calibration of the magnification of the image. This example also illustrates quite vividly how device scaling is driving the microstructure of the devices further into a realm where the highest spatial resolution analytical tools, such as TEM, are required.

In the above examples, site-specific sample preparation techniques were not required to access the structures. However, single - bit failure analysis does require site-

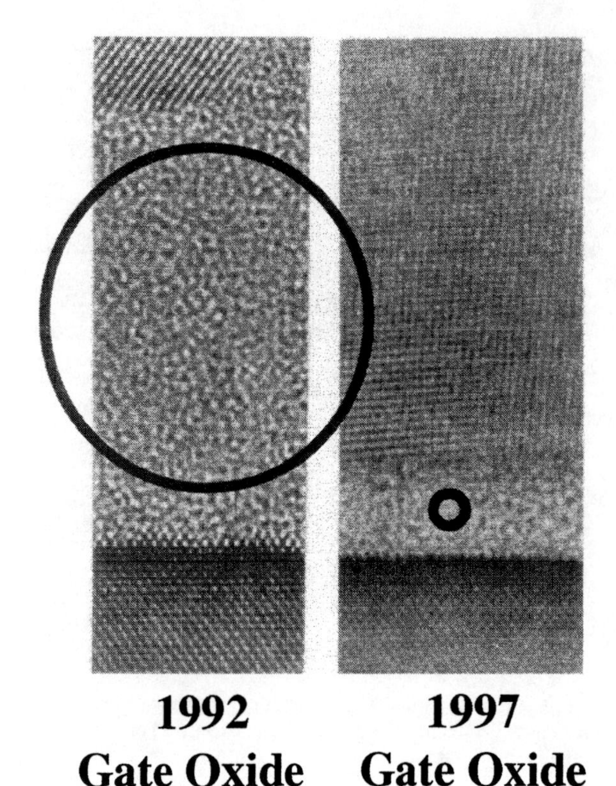

Figure 4. Lattice images of gate structures typical of 1992 and 1997 technologies. The large and small circles indicate the minimum electron beam size attainable with LaB_6 and field emission electron sources, respectively.

specific sectioning. A remarkable example of precision sectioning (combined with a healthy dose of good luck) is shown in Fig. 5. The high resolution image shows a defect at the edge of a 0.25 micron gate. The defect consists of a locally recrystallized area, only 5 nm in diameter, which was associated with a short through the gate dielectric. The defective gate was first located by electrical test methods, then isolated by voltage contrast and subsequently sectioned by the FIB method. This ability to locate, section and identify a 5 nm defect illustrates the power of TEM-based techniques.

ANALYTICAL TECHNIQUES

In addition to the microstructural, morphological and crystallographic information available from the primary imaging modes in TEM, compositional information about the specimen can be obtained at high spatial resolution by exploiting some of the signals produced by inelastic scattering processes. Inelastic interactions include phonon and plasmon excitation, inter- and intra-band transitions and inner-shell ionization. Of primary interest in TEM are the analysis of characteristic x-rays generated within the specimen and the analysis of the energy spectrum of electrons transmitted through the specimen. The incident electron beam can cause the ionization of target atoms by ejecting an inner-shell electron. When an electron from a higher energy level fills the hole created by this process, an x-ray may be emitted. The energy of the x-ray reflects the difference in energy between the two energy levels and is thus characteristic of the electronic structure of the target atom. Analysis of the energy spectrum of the emitted x-rays can thus be used to provide compositional information about the specimen [3,5-6].

Most TEMs are equipped with an energy dispersive x-ray detector, pulse processing electronics and computer analyzer for this purpose. In contrast to the imaging modes of operation, the energy dispersive x-ray spectrometry (EDS) data is usually acquired with the beam focused to a fine spot on the feature of interest. In bulk samples, the incident electrons undergo multiple scattering events and the size of the interaction volume (and x-ray generation volume) are generally quite large (~3 orders of magnitude) compared to the small probes that can be formed. For the thin-foil geometry of TEM specimens, the interaction volume is substantially smaller, and the spatial resolution is limited by the probe size. In this regard field emission electron sources are much preferred over other electron sources, since they provide probe sizes on the order of 1 nm.

Detection limits are determined by the relatively poor statistics involved in generating and detecting the emitted x-rays. The limited interaction volume produces few x-rays per incident electron. In addition, EDS detectors have a relatively small solid angle of collection, so that only about 1% of the emitted x-rays strike the detector. Finally, detection limits for light elements may be particularly poor because the low energy x-rays emitted from them tend to be absorbed by protective "windows" on the detectors, or even in the case of "windowless" detectors, by contamination layers that build up with time. Thus, there exists a trade-off between detection limit and spatial resolution : a large

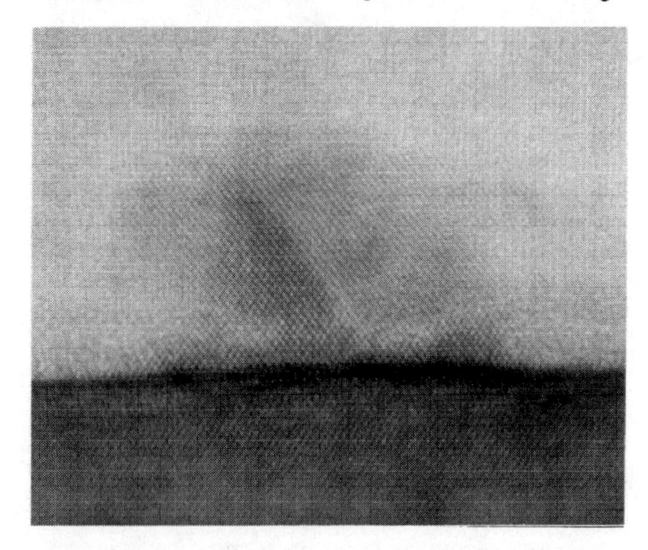

Figure 5. Lattice image of a defect at the edge of a 0.25 micron gate. The defect was isolated by voltage contrast and sectioned with the FIB technique.

TABLE 1. Characteristics of Analytical Techniques

Technique	Spatial resolution	Energy resolution	Detection limit
EDS	< 1.0 nm (w/ FEG)	~140 eV	0.1 - 1 wt % [a]
EELS	< 1.0 nm (w/ FEG)	0.3 eV (w/ FEG)	single atom sensitivity demonstrated [b]

[a] Detection limit worse for light elements

[b] Detection limit strongly dependendent on element / matrix combination and specimen thickness

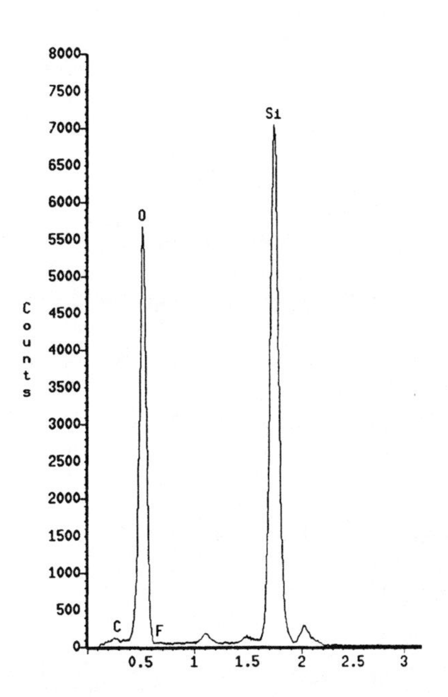

Figure 6. EDS spectrum taken from a silicon dioxide layer showing counts vs. x-ray energy.

probe size and a thick sample maximize the detection limit, while a small probe size and a thin sample produce the best spatial resolution. An additional limitation for EDS analysis is the poor energy resolution. As a result, some analyses are not possible with EDS because closely spaced x-ray peaks from different elements may not be resolvable. Table 1 shows typical detection limit, spatial resolution, and energy resolution estimates for EDS analysis in a TEM.

It is possible to quantify the data through either first principles analysis algorithms ("standardless" analysis) or through the use of standards. The standardless approach is characterized by a high degree of uncertainty, since not all of the relevant physical data is well known. The standards approach requires TEM samples of a standard material that contains the elements of interest in known quantities, which may be difficult to obtain. As a result, such spectra are usually used in a qualitative manner. Figure 6 shows an example of the x-ray spectrum measured from a silicon dioxide layer, illustrating the light element sensitivity achieved with EDS.

Compositional information about the specimen may also be obtained using electron energy-loss spectrometry (EELS), in which the energy spectrum of electrons transmitted through the specimen is analyzed [3,7]. These spectra contain information about nearly all of the inelastic scattering processes mentioned previously. The most commonly used features in EELS spectra are the ionization edges corresponding to the onset of inner-shell ionization for the particular elements within the specimen. This information is thus complementary to the x-ray data derived from EDS analysis. However, EELS analysis offers several significant advantages over EDS analysis. The superior energy resolution of electron spectrometers avoids the peak overlap difficulties that may be encountered in EDS spectra. In contrast to EDS spectra, chemical bonding information is available from the fine structure observed in the near and extended regions of the ionization edges. In addition, EELS is particularly useful for light element detection where EDS analysis may be problematic. Finally, since the electrons are strongly forward scattered, it is possible to detect most of the inelastically scattered electrons. When data is collected with parallel detection of the electrons and a field emission electron source, the detection limits may be much improved over EDS. In the extreme case of thorium atoms on a thin carbon support film, single atom detection has been demonstrated [16].

EELS data may be obtained with the beam focused to a fine spot on the region of interest as in EDS analysis, in which case the spatial resolution is again limited by the probe size. Table 1 summarizes detection limit, spatial resolution and energy resolution estimates for EELS

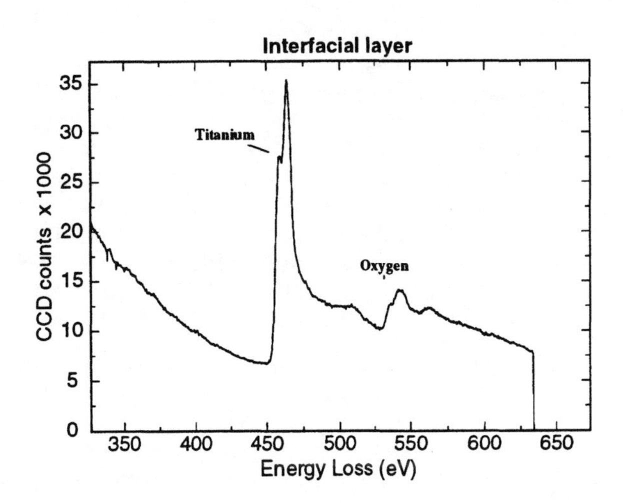

Figure 7. EELS spectrum confirming the presence of oxygen in a titanium layer.

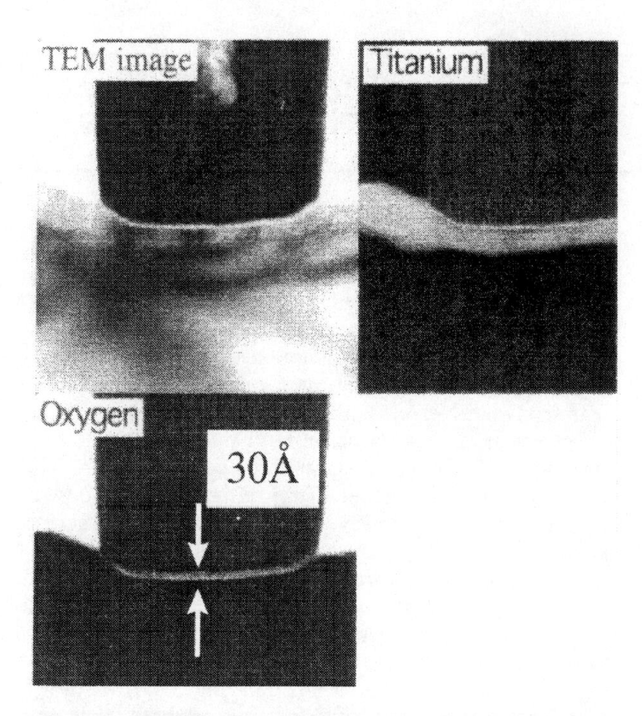

Figure 8. Conventional TEM image and energy filtered images showing a 3 nm thick oxygen-rich layer at the interface between a tungsten plug and the underlying titanium silicide which caused high contact resistance.

analysis. Generally, the data is used in a qualitative manner, although it is possible (but difficult) to quantify the data. An example of an EELS spectrum collected from a titanium layer in a device is shown in Fig. 7. The spectrum confirmed the presence of high oxygen levels in the interfacial region of the structure. This analysis could not be performed with EDS because the energy resolution of an EDS detector is insufficient to resolve the closely spaced Ti L and O K x-ray lines.

A relatively new, and extremely powerful use of the energy spectrum of transmitted electrons is energy-filtered imaging [17]. In this technique only a portion of the transmitted electron distribution is used for imaging. By selecting the portions of the spectrum corresponding to particular ionization edges (with the background removed), it is possible to produce a series of images which reflect the composition of the specimen. An example of this type of analysis is shown in Fig. 8. Electrical tests had identified this area as having a high contact resistance. The sample was sectioned by the FIB technique and energy-filtered imaging was used to provide composition maps of the contact region. The oxygen energy-filtered image shows the presence of an oxygen containing interfacial layer only 3 nm thick between the tungsten plug and the underlying titanium silicide, which was the source of the high resistance [18]. This example illustrates the power of TEM-based techniques, since there is essentially no other analytical technique with the spatial resolution and sensitivity to obtain such data.

One area in which these analytical techniques are generally unable to provide information is in the analysis of doping distributions in the semiconductor. The primary difficulty is that the dopant concentrations are usually below the detection limit of these techniques. To overcome these limitations, an indirect technique is used to visualize the two-dimensional distribution of the dopants. The etching rates of certain mixtures of hydrofluoric acid and nitric acid are sensitive to the resistivity of the silicon. By etching a TEM specimen in such an etchant, it is possible to induce a composition-related topography in the specimen [19-21]. Essentially, the more highly doped areas of the specimen are thinner than the more lightly doped areas as a result of the doping-dependent etching. When imaged in the TEM under appropriate conditions, a series of thickness fringes or contours can be seen in the differentially etched areas. (Thickness fringes arise from the interference of the electron waves excited in a specimen of varying thickness.) An example is shown in Fig. 9. This special test structure was patterned, implanted with arsenic and annealed to activate the arsenic. The sample was then sectioned using conventional dimple/ ion mill techniques and etched in a doping selective etchant. The highly doped source/drain region shows thickness fringes. By calibrating the one-dimensional profile of these fringes with secondary ion mass spectrometry data of the arsenic distribution, it is possible to assign concentration values to each fringe. Under the assumption that the etching rate is proportional to the doping concentration, each thickness fringe then

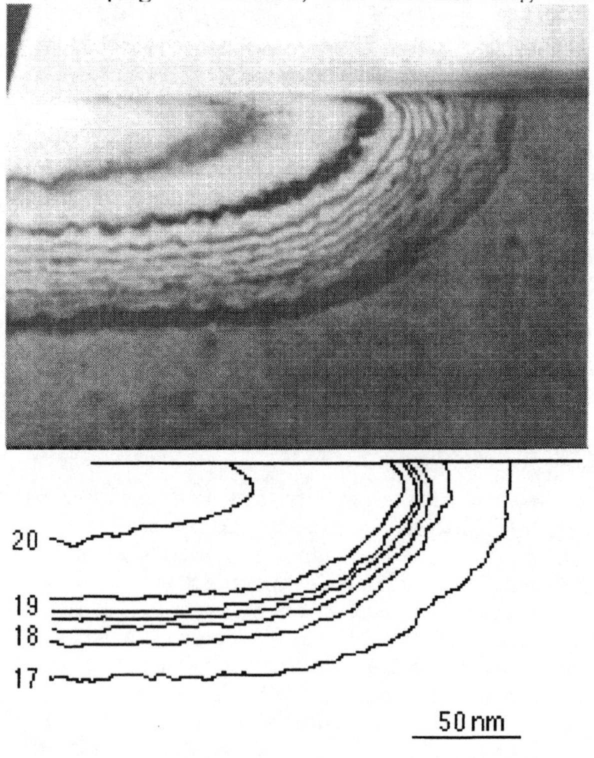

Figure 9. TEM image of selectively-etched arsenic-doped source/drain region, and two-dimensonal dopant contours derived from correlation of the TEM data and SIMS data.

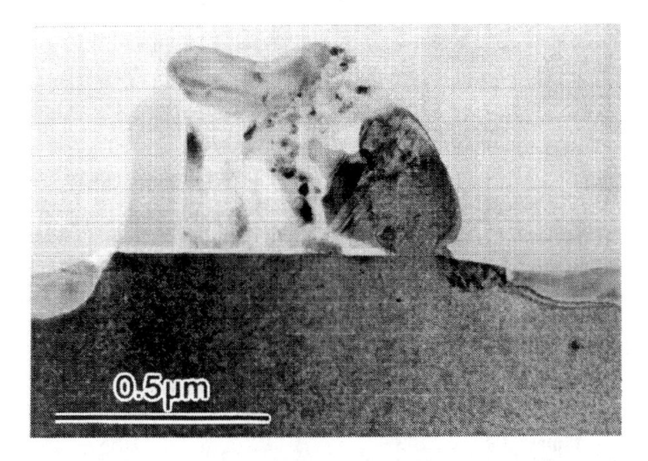

Figure 10. TEM image of a defective device sectioned by the FIB technique illustrating single-bit failure analysis.

corresponds to a contour of constant doping concentration. The detection limit of the technique is controlled by the etching conditions and is about $1x10^{17}$ cm^{-3}. The power of this indirect method for two-dimensional dopant profiling lies in the inherently high spatial resolution of TEM which permits the two-dimensional analysis of ultra-shallow junctions (including source/drain extensions) for ULSI applications.

SPECIMEN PREPARATION

TEM samples prepared by FIB techniques are required for most single bit failure analyses. As an example, a unit failed during a 1000 hr-long electrical stress test and liquid crystal analysis located a "hot spot" at a particular set of coordinates. A FIB cross section through this defect indicated that the silicide layer over the polysilicon layer was shorted to ground, although the exact nature of the short was not clear from SEM images. A thin cross-sectional TEM sample was then prepared through the same defect using the FIB technique. A TEM image recorded from the thin sample is shown in Fig. 10. It is apparent that the short is a silicon particle connecting the polysilicon electrode to the substrate, while nearly eliminating the nitride spacer. EDS analysis showed the presence of Ti-rich material in the polysilicon electrode above the silicon particle. Evidently, the silicon melted and recrystallized locally, thus displacing the silicide and nitride layers.

Despite the success of this type of analysis, there are physical limitations to the FIB technique of TEM sample preparation. These samples are typically prepared with 25-50 kV gallium ions which allow the TEM sample to be prepared with submicron precision. Gallium ion milling is known to modify the surfaces of TEM samples (and recall that both top and bottom surfaces are milled when preparing the samples) through a combination of amorphization and ion implantation [11].

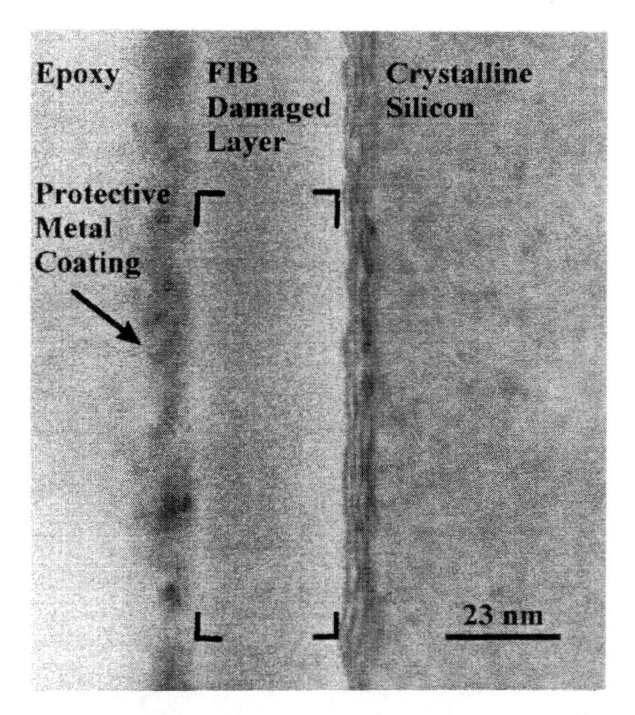

Figure 11. Cross-sectional TEM image of damaged layer created by gallium FIB milling.

Using typical FIB milling conditions the depth of 30 kV Ga^+ milling damage of silicon has been measured to be 28 nm, as shown in Fig. 11 . This thickness of damaged material corresponds closely to Monte Carlo simulations of 30 kV Ga^+ ion milling. The Monte Carlo calculations, plotted in Fig. 12, show that both silicon vacancies and

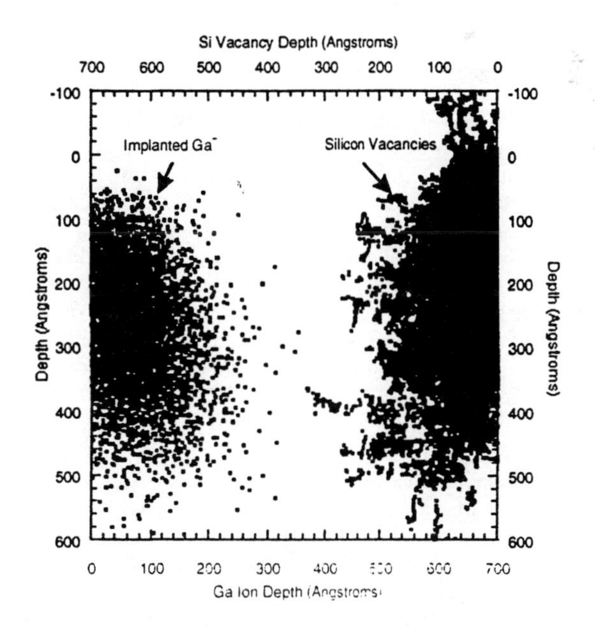

Figure 12. Monte-Carlo simulation of 30 kV Ga+ ion milling, of silicon, showing Ga+ and silicon vacancy distributions.

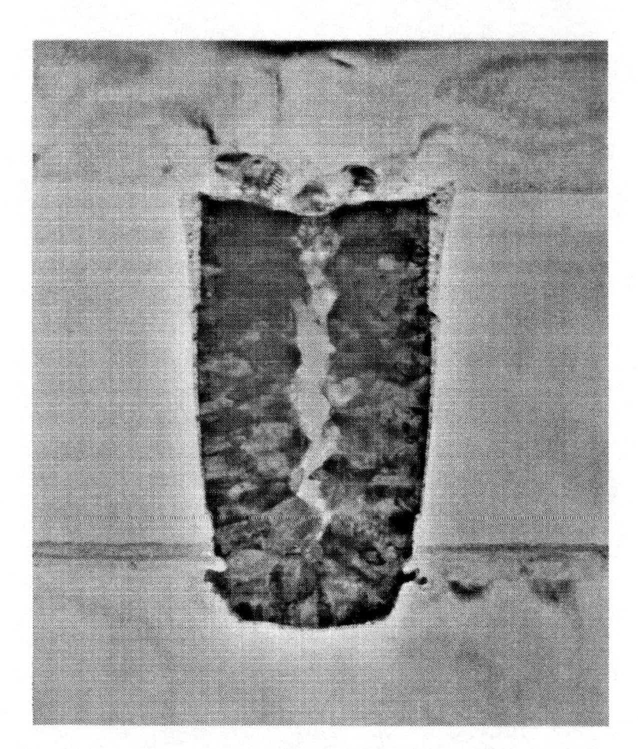

Figure 13. Cross-sectional TEM image of a tungsten interconnect that was prepared in the SEMAIM.

implanted Ga^+ ions are present at depths approaching 30 nm. This damaged material accounts for a large fraction (50% or more!) of the TEM sample, since the samples are only about 100 nm thick. The quality of a TEM lattice image is impaired by the presence of so much amorphous and damaged material in the sample. In addition, ion-beam induced compositional mixing of adjoining layers of material severely degrades analytical data obtained by EDS or EELS. Clearly, there is an emerging industry-wide need to develop site-specific approaches to TEM sample preparation that produce significantly less surface damage.

Argon ion milling is the only technique known now to provide thin specimens with significantly less surface amorphization than that provided by gallium FIB systems [22]. However, the endpoint control that can provide thin, centered cross-sections of small structures does not exist. Traditional argon ion mills are equipped with a low-power optical microscope that is totally inadequate for this task. The only alternative to the slow and tedious procedure of iterative milling and examination with electron optics is to have the ion mill and SEM integrated into a single tool. We have recently constructed such a tool, which we have named "SEMAIM", for SEM-Argon Ion Mill [23]. A TEM image of a tungsten interconnect that was milled in the SEMAIM is shown in Fig. 13. The sample is appropriately centered on the feature of interest and the microstructure of the tungsten is clearly discernible.

As the size of such features approaches that of traditional electron transparent TEM specimens, special requirements are placed on the preparation of cross-sectional

TABLE 2 Maximum specimen thickness (in nm) allowed to section interconnects at the given geometrical resolution for centering errors of 0, 10 and 40 nm.

0.25 micron technology

	Centering Error		
Resolution	0 nm	10 nm	40 nm
3 nm	69.5	60.5	15.1
2 nm	56.7	40.2	10.1
1 nm	40.0	20.1	5.0

0.18 micron technology

	Centering Error		
Resolution	0 nm	10 nm	40 nm
3 nm	58.3	42.5	10.6
2 nm	47.5	28.2	7.1
1 nm	33.5	14.1	3.5

0.13 micron technology

	Centering Error		
Resolution	0 nm	10 nm	40 nm
3 nm	49.4	29.9	7.5
2 nm	40.2	19.8	5.0
1 nm	28.4	9.9	2.5

samples. Since typical interconnects are cylindrical, there are simple geometrical effects that become significant as the radius of curvature scales with the shrinkage of devices. Geometrical "blurring" reduces resolution and becomes significant when the specimen is too thick, or off-center even though it may well be outstanding in its electron transparency and freedom from surface amorphization [24]. Table 2 provides examples of calculations for centering error and specimen thickness requirements for interconnects at future nodes on the SIA roadmap. These calculations show that at the 0.13 micron node, 1 nm resolution can only be obtained if the TEM samples are less than 28 nm thick if there is zero centering error, and 2.5 nm thick if the centering error is 40 nm. Relaxing the resolution requirement to 3 nm still requires samples less than 50 nm and 8 nm in thickness, respectively. It is not clear that any of the present specimen preparation techniques can routinely achieve such precision. Thus, specimen preparation may well be the limiting factor in deriving critical data from TEM-based analyses for ULSI technology.

SUMMARY

Transmission electron microscopy encompasses a range of high spatial resolution imaging and analytical techniques for the characterization of electronic materials and devices. Currently, TEM plays an essential role in process development, defect identification, yield improvement and root-cause analysis within the semiconductor industry. As semiconductor device dimensions continue to decrease, TEM will become a critical analytical tool for ULSI technology. Improvements in the design of field emission electron sources, lenses etc. and in the development of new, sophisticated analysis modes have led to the commercial availability of TEMs with an unprecedented combination of resolving power and analytical capability. However, a practical limitation to an expanded role for TEM-based techniques is the need for producing an extraordinarily thin, artifact-free specimen, precisely centered on a particular sub-quarter micron feature, in a timely and cost-effective manner. To meet this need, further improvement in existing techniques [25], and the development of novel specimen preparation techniques is urgently required.

ACKNOWLEDGMENTS

The authors would like to acknowledge the technical contributions of Sharon Darknell, Jian Duan, Robert Jamison, Kevin Johnson, Carmen Matos, Kian Sin Sim, Dennis Maher, Rose Zou, Joe Clark, Kim Christensen, and Ranju Datta. D. W. Susnitzky and A. J. Mardinly acknowledge the management support of Bob McDonald. D. Venables would like to acknowledge the financial support of the NSF Engineering Research Centers Program through the Center for Advanced Electronic Materials Processing at North Carolina State University.

REFERENCES

1. Hirsch, P. B., Howie, A., Nicholson, R. B., Pashley, D. W., and Whelan, M. J., *Electron Microscopy of Thin Crystals*, 2nd edition, Huntington, New York: Krieger, 1977.
2. Edington, J. W., *Practical Electron Microscopy in Materials Science*, New York: Van Nostrand Reinhold, 1976.
3. Williams, D. W., and Carter, C. B., *Transmission Electron Microscopy*, New York: Plenum Press, 1996.
4. Spence, J. C. H., *Experimental High-Resolution Electron Microscopy*, 2nd edition, New York: Oxford University Press, 1988.
5. Williams, D. W., *Practical Analytical Electron Microscopy in Materials Sceince*, 2nd edition, Mahawah, New Jersey: Phillips Electron Optics Publishing Group, 1987.
6. Joy, D. C., Romig, Jr., A. D., and Goldstein, J. I., *Principles of Analytical Electron Microscopy*, New York: Plenum Press, 1989.
7. Egerton, R. F., *Electron Energy Loss Spectroscopy in the Electron Microscope*, 2nd edition, New York: Plenum Press, 1996.
8. Bravman, J. C., and Sinclair, R., *Journal of Electron Microscopy Technique* **1**, 53-61 (1984).
9. Benedict, J. P., Anderson, R. A., and Klepeis, S. J., *Materials Research Society Symposium Proceedings* **254**, 121-140 (1992).
10. Basile, D. P., Boylan, R., Baker, R., Hayes, K., and Soza, D., *Materials Research Society Symposium Proceedings* **254**, 23-41 (1992).
11. Susnitzky, D. W., and Johnson, K. D., *Proceedings Microscopy Society of America*, (1998), in press.
12. Zou, Y., Venables, D., Christensen, K. M., and Maher, D. M., "An investigation of silicon (001) wafers after exposure to an atomic hydrogen plasma", presented at the Fall Materials Research Society Meeting, Boston, MA, Nov. 1994.
13. Zou, Y., "Surface and subsurface properties of Si (100) wafers after hydrogen plasma exposure", M. S. Thesis, North Carolina State University, 1997.
14. Maher, D. M., "The role of transmission electron microscopy in defect control and process variability", presented at the 42nd National AVS Symposium, Minneapolis, Minnesota, October 16-20, 1995.
15. Clark, J. C., "Physical and electrical characterization of hydrogen plasma exposed Si (100) wafers", M. S. Thesis, North Carolina State University, 1997.
16. Krivanek, O. L., Mory, C., Tence, M., and Colliex, C., *Microscopy, Microanalysis and Microstrucutre* **2**, 257 (1991).
17. Krivanek, O. L., Gubbens, A. J., Delby, N., and Meyer, D. E., *Microscopy, Microanalysis and Microstrucutre* **3**, 187 (1992).
18. McDonald, R. C., Mardinly, A. J. and Susnitzky, D. W., *Proceedings Microscopy Society of America* 449-50 (1997).
19. Sheng, T. T., and Marcus, R. B., *Journal of the Electrochemical Society* **128**, 881 (1981).
20. Maher, D. M., and Zhang, B., *Journal of Vacuum Science and Technology B* **12**, 347 (1994).
21. Neogi, S. S., Venables, D., Ma, Z., and Maher, D. M., *Journal of Vacuum Science and Technology B* , (1998), in press.
22. Schurke, T., Mandl, M., Zweck J., and Hoffman, H., *Ultramicroscopy* **41** 429-33 (1992).
23. Mardinly, A. J., and Jamison, R., Proceedings International Congress of Electron Microscopy (1998), in press.
24. Mardinly, A. J., and Susnitzky, D. W., *Materials Research Society Symposium Proceeding* (1998), in preparation.
25. Barna, A., Pecz, B, and Menyhard, M., *Ultramicroscopy* **70,** 161-171 (1998).

Microscopy and Spectroscopy Characterization of Small Defects on 200mm Wafers

C.R. Brundle,* Yuri Uritsky,*
Patrick Kinney,** Walter Huber,*** Andrew Green***

*Applied Materials, Inc., Santa Clara, CA 95054
**Microtherm, LLC, Minneapolis, MN 55413
***Sumitomo Sitix, Fremont, CA 94538

The standard method for the detection and mapping of small particles on 200mm unpatterned wafers is light scattering. The standard approach to determining what these particles are, and their origin, is to re-find them in a full wafer SEM and rely on SEM/EDX for analysis. At very small particle sizes (below 0.2 microns) the light scattering location inaccuracies make it hard to re-locate the particles in SEM and EDX can have difficulty because it is not a small volume technique. In addition EDX often does not have enough characterization power to lead to a root cause determination. A number of 200mm analytical alternatives have become available since the first conference in this series two years ago. Between them they provide better small volume and surface sensitivity, plus chemical and molecular information. These are: other modes of optical microscopy combined with wafer marking, AFM, SAM, FIB, and TOF-SIMS. We have evaluated all these using real defect situations on full wafers. A review of their capabilities is given here. In addition we discuss micro-calorimetry based EDX and micro-ESCA/XANES, techniques not yet available in 200mm form.

INTRODUCTION

As design rules shrink, specifications for allowable particle contamination on wafers during processing get more stringent. This trend will accelerate with the move from 200mm to 300mm. For processing equipment manufacturers, such as Applied Materials, the anticipated specifications (National Technology Roadmap for Semiconductors; Metrology supplement (1)) are a severe challenge, since equipment and thin film process developments occur about 2 years ahead of the roadmap. If we take the roadmap at face value, this means that we must have in place the means of eliminating 0.04µm particles by 2002. The industry, however, does not even have a way of detecting the presence of 0.04µm particles on monitor or blanket wafers, let alone determining their root cause and hence elimination! Current industry enforced specifications (as opposed to roadmap numbers) are around 0.12µm for 200mm and are being set at ~0.08µm for 300mm. The ability to reliably detect, accurately size (i.e. to know whether a spec is violated or not), and then to subsequently characterize well enough for root cause determination at the 0.08µm level for 200mm wafers is in its infancy, and for 300mm is non-existent for the simple reason that there is a lack of commercial analytical instrumentation to handle 300mm wafers.

The "standard" approach, when problems are encountered in meeting particle specifications for wafer process equipment, is to use light scattering for initial detection and location on monitor wafers (very fast; 30 seconds, run by anyone with minimal training), and then to re-find the features causing the light scattering (which may, or may not be particles) in a 200mm SEM/ EDX system, based on the light scattering X,Y coordinate file. One then gathers enough information (size, shape, morphology, qualitative elemental composition) so that the nature of the particle (if it is a particle) can be determined well enough to establish its origin (2). This is a very slow process; about 4 hours per wafer on average for a highly skilled analyst. A non-trivial point (but one which is often not appreciated) in this approach of marrying light scattering detection and SEM/ EDX analysis, is the required accuracy and transferability of the X,Y coordinate as determined by light scattering (3). The smaller the defect is, the smaller the field of view, FOV, (greater magnification) that must be used in SEM to find it and so the more accurate the X, Y input data must be to ensure the defect is located within that FOV. We will return to this later, but, even assuming one can find the defect/ particle in the SEM within an acceptable time, the analytical combination of SEM imaging and EDX elemental identification is not always sufficient. As the size decreases, EDX, which is not an ultra-small volume technique (typically a 1-4µm volume is sampled at 20kV beam energy and 0.4-1.5um at 10kV), has increasingly difficulty in separating the defect signal from that of the substrate (4,5). If the "particle" turns out to be a thin film patch, which is sometimes the case, EDX may essentially see no signal additional to the substrate background (6). EDX does not handle organics well, and does not provide

CP449, *Characterization and Metrology for ULSI Technology: 1998 International Conference*
edited by D. G. Seiler, A. C. Diebold, W. M. Bullis, T. J. Shaffner, R. McDonald, and E. J. Walters

the chemical state information often needed for a root cause determination. For example, knowing a particle contains aluminum when there are multiple sources of aluminum in the processing environment is insufficient. One needs to know if it is Al metal, Al oxide, or some other compound. If oxide, is it representative of ceramic (e.g. saphire), anodized Al, or Al metal oxidized somehow by the particle production process (6). If it is carbon based, is it graphite, oil, or what? Techniques are needed not only to handle smaller sizes, but which also have surface sensitivity, and provide chemical, molecular, and structural information. Such techniques have existed for some time, of course, though no one method can do all these things and they cannot all go down to the spatial resolution needed. Integration into commercial systems capable of taking full 200mm wafers and correctly handling the light scattering X, Y coordinate files has only come about in the last few years, however. AFM (Atomic Force Microscopy), SAM (Scanning Auger Microscopy), TOF-SIMS (Time of Flight Secondary Mass Spectroscopy), wafer marking/ optical bench combinations, and FIB (Focused Ion Beam) all exist now commercially, and it is our experience that all of them are needed given the variety of particle/defects encountered.

This review discusses the complementary use of the above techniques, together with the more standard SEM/ EDX approach for root cause determination on 200mm full wafers. It is based on on-going evaluations of their relative usefulness for real particle/ defect problems at Applied Materials and Sumitomo Sitix. In addition, because of their potential impact, we cover two analytical techniques not yet even on the commercial market, let alone available at 200mm. These are micro-ESCA/XANES (7,8)(defined here as having sub 1μm spatial resolution) and micro-calorimetery based X-ray emission (9).

INITIAL PARTICLE DETECTION; SIZE DETERMINATION; DETECTION LIMIT

We limit this review to discussion of particles on unpatterned wafers because the large majority of particle problems "owned" by the process tool manufacturer are detected using unpatterned wafers. These problems are either hardware degradation related, deposition or etching process related, or some interaction between these (6). The initial detection of the problem usually occurs while running Si monitor wafers routinely as a function of number of wafers processed, after cleaning steps, hardware changes, etc. Alternately, the problem may be detected on blanket film processed wafers. Within Applied Materials these problems would be connected either to new chamber/ process developments or marathon wafer runs for qualification. At customer sites on installed base equipment again the problem will usually surface using light scattering defect review tools on either monitor or blanket film wafers. At Sumitomo Sitix the product, using Applied Materials equipment, is epi, a blanket film product for which subsequent epi customers set extremely aggressive particle specifications.

Monitoring fully-processed patterned wafers using DRTs is a completely different arena. Here particle defects may be a result of the additional lithographic processes in addition to the tool/processes referred to above. Monitoring is often done by imaging (slower) rather than light scattering and the size of the defect that is routinely detectable is generally (but not always) larger, because of the difficulty in distinguishing defect from pattern. For this reason, and also because the tool and process specifications refer to either monitor or blanket film wafers, the problem still often reverts to doing the detective work on unpatterned wafers.

What are the detection limits of the existing light scattering tools? KLA-Tencor dominates the market and on polished bare Si monitor wafers their Surfscan tools can detect PSL (Polystyrene Latex) spheres down to 0.12μm routinely if the machine "recipe" is set correctly. Below this the false count rises dramatically because of the reduced signal to noise and consequent inability to distinguish particles from roughness and haze. Real particles, however, are not latex spheres and both the attempted sizing (from the light scattering intensity) and the limits of detection will vary, sometimes dramatically (i.e. factors of 10), with particle shape and material. On blanket films, the material of the film, the thickness, and the roughness will all affect the limits of detectablity and the sizing. Without expert calibration against standards these will be largely unknown (and therefore in error, since the typical user will assume the bare Si calibration), and with calibration the conclusion is always that detection limits are poorer, sometimes dramatically, than for polished monitor wafers. For example, on rough W films the detection limit may be as poor as 0.4μm in reality. Below this, particles may not be distinguishable from roughness. All this is important because it is no coincidence that the current industry particle specification, at 0.12μm, is the same as the light scattering detection limit on polished monitor wafers! The ultimate tool detection limit on polished Si wafers has become the *de-facto* specification of performance, irrespective of the fact that on real films the tool cannot usually perform close to this level (for the new 300mm light scattering DRTs the instrument specs may be as low as 0.08μm, which is why particle specs are being set there for 300mm). This means that, quite independent of the root cause analytical capabilities of SEM/ EDX, the SEM is an important tool for verification of particle size specification violations (and the reliability of the light-scattering DRT) from the 0.4μm down to sub 0.1μm range. For 0.25μm and above much faster imaging based optical tools such as the laser based Ultrapointe tool can (and are) also used for verification (10).

THE X,Y COORDINATE ISSUE; WAFER MARKING; BREAKING WAFERS

The wafer coordinate system is defined by the relative positions of the "V" notch (for 8" wafers) and the wafer center. The notch position is determined in the light scattering tool by scanning the notch "edge" and algorithm fitting to determine by point of the V. The center is determined by scanning the "edges" of the wafer and determining a diameter. Unfortunately this does not establish a very accurate or transferable origin (0, 0) because of (a) the difficulty of determining a true "edge" when that edge slopes, (b) an inadequate density of sampling points used (a data size issue), and (c) the irregularity and poor control specs on notch shapes and dimensions. This means that when re-establishing the same origin by the same procedure using an electron beam in the SEM, there will be at least a 100µm discrepancy, which means that the relocation error for the first particle in the SEM will be at least ±100µm. For a large (several micron), high contrast, particle, this will present no problem. The SEM field of view can be as large as 1000x1000µm, and the particles will be detectable against the background. In general detection of three such sized particles and their observed X, Y coordinates, compared to predicted coordinates, is sufficient to perform a transform for all other predicted coordinates in the file and eliminate the large 0,0 origin error. It should now be (and is) easier to re-find much smaller particles. Unfortunately, however, there are significant relative errors in the X, Y coordinates, in addition to the origin problem (11). These occur through a combination of issues to do both with stage accuracies and the actual measurement protocols. Some are systematic and can be reduced by adopting learning algorithms for a specific light scattering tool. Others are not systematic. For the Tencor 6200 series, however, the final limit is the grid size adopted by the laser scanner. The low throughput mode uses a 10x26µm grid size, which implies that the X, Y accuracy can never be better than ±13 micron. Our extensive experience with a variety of individual Tencor Surfscan™ 6200 tools is that casual use will result in inaccuracies still in the 100x100µm or greater range, and that the best that can be done is ±40µm. Unfortunately to guarantee detection of a 0.1µm particle of moderate contrast by SEM requires a field of view (FOV) reduction to 50x50µm (a magnification of 2000), which means the true X,Y value must be within 25µm of the predicted value. Here, then, is the nub of the problem - if one cannot find the particle within the 50x50µm FOV, it could be because (a) it is actually outside the FOV, (b) it is too low contrast to detect even at this magnification, (c) it was a false light scattering reading or (d) it has moved (possible, particularly for insulating particles under an e-beam). This is why it takes a skilled SEM operator and a long time to do a representative particle analysis on a full wafer. For 300mm wafers and specifications below 0.1µm this problem will get worse. In addition, the above scenario presumed there were at least 2 large particles to locate and perform the required transform to eliminate the origin error. This is becoming less and less likely as both the number and size of particles to be analyzed decreases.

The practical way out of these dilemmas is to provide accurate fiducalization of wafers and to be able to enter these into the light scattering files. The logical approach, at least for monitor wafers, would be to use laser pre-marked wafers for the monitoring. The fiducial marks would then be automatically detected and included in the X, Y light scattering file and would act as surrogate large initial particles for performing the transform. There is resistance to this on the basis of cost, convenience, and probably most importantly the mistaken belief that marking generates particles which will be confused with the particles we wish to detect. This is an erroneous belief when monitoring is done by making a "before" scan on the pristine wafer, an "after" scan once the wafer has been exercised through the tool, and then determining the "adders" from the difference. Since locating say, 100 particles on a 200mm wafer is literally similar to locating 100 grains of salt on an area the size of a football field, there is practically zero chance of confusing the positions of pre-existing defects with those of "adders".

The approach we adopt involves the use of the Micromark 5000™, a combination of a optical bench (taking 150, 200, and 300mm wafers) incorporating bright field, dark field, and Nomarski modes, cooled CCD camera detection, a laser based marking system, an accurate X, Y stage (±10µm) and sophisticated software for adding coordinates to existing Tencor files, updating files, and using the tool in a semi-automatic mode (12). In the wafer-marking mode laser marks of any size down to a micron can be placed rapidly anywhere on the wafer to within ±10µm accuracy. Typically, our procedure is to take the Tencor mapped wafer, drive to a predicted defect coordinate and examine for the defect using the dark field mode. Since this is similar to the Tencor light scattering approach it will usually show with high contrast (the dark field advantage) and can be easily detected at lower magnification (i.e. a wide FOV, ensuring it is in the FOV). Defects down to below 0.1µm are routinely and quickly found this way. Predicted X, Y coordinates (Tencor values) can then be updated in the file with an accurate value and./ or any number of laser marks can be made around the defect . There is then never any difficulty knowing whether you are in the correct place and exactly where a given defect is. An interesting extreme example of this application is shown in Figure 1. It shows the Nomarski image of a wafer marked defect, originally located by light scattering using a Tencor surfscan, on an epi wafer (13). The defect is clearly about 20µm long and a few microns wide, yet the Tencor sizing suggests a 200x200µm size! When examined by SEM and EDX, Figure 1(b), one can search all day and not find anything! There is no doubt we are in the correct location, because of the wafer marks, but there is no SEM contrast. The reason is because this particular "particle" is actually a crystalline silicon defect in the epi. Since it is of the same material as the substrate it has no atomic mass contrast in the SEM and no EDX signature. It has negligible topography contrast because it is actually only 0.05um high with very gently sloping edges (subsequently determined by AFM). In addition to solving the X,Y coordinate issue when examining full wafers by SEM, the marking procedure allows wafers to be broken into small pieces for instruments which cannot take full wafers and still guarantee to find the defects.

Finally if the particle is large enough (~0.3µm or bigger) the bright field and Nomarski modes of the Micromark 5000™ can be utilized to image the defect and sometimes provide useful defect classification information directly from this (e.g. size, shape), in a similar manner to a KLA imaging DRT.

EXAMPLES OF BOTH SUCCESS AND FAILURE OF ROOT CAUSE ANALYSIS BY RELIANCE ON SEM/EDX ALONE

Successful determination of particle origin using the SEM/ EDX approach is most likely to occur when (a) the particle consists of elements readily distinguishable from the substrate, (b) those elements carry a very unique signature of the particle, and (c) the particle is large enough to give a major contribution to the total EDX signal. A common example would be determination of a stainless steel mechanical-flake on a monitor Si wafer using the Ni/ Fe/ Cr ratio and the shape and morphology. Another often found particle is the product of O-ring degradation under plasma attack (2). Figure 2(a) shows a typical SEM of such a "particle". It is actually a conglomerate of much smaller particles. Its composition is non-uniform and EDX, Figure 2(b), shows that in some regions it is basically carbon (not much help for identification), but in others it contains Mg, and O, a strong signature of the filler material used in this particular type of O-ring. The F comes from the interaction with the plasma.

A typical, unsuccessful SEM/ EDX example is shown in Figure 3. The SEM shows a large area, "snow-flake" shape (14). The EDX shows virtually no distinction between on the "particle" and off it - just the Si substrate with a hint of carbon and oxygen in both cases. Experience tells us that this type of feature, which we have seen in numerous unrelated situations, is likely to be a very thin carbon-based patch, but the SEM/ EDX data alone cannot establish this (it could possibly be a water stain residue). Even if our customers were to trust our "experience", we could not establish the origin of the patch simply from the knowledge that it is carbon based. In fact this particular example represents a Krytox oil patch, as established through TOF-SIMS (see later).

COMPLIMENTARY APPROACHES FOR PARTICLE CHARACTERIZATION

What we are looking for are analytical techniques which have good spatial resolution, surface sensitivity, and provide elemental chemical and molecular information. The following in a list of available tools of the "Tencor compatible", 200mm variety, together with their main attributes and limitations.

Optical Techniques

Though light scattering is the standard tool for particle detection on unpatterned wafers, other optical approaches can provide characterization information. The white light optical bench/ wafer marking Micromark 5000™ has already been described. In addition to its ability to detect particles in dark-field mode below 0.1µm and its assistance on the X,Y coordinates accuracy issue, it, and other makes of optical benches, can provide shapes and sizes through imaging, if the defects are large enough (0.3µm), using a combination of dark-field, bright-field, and Nomarski modes. Confocal microscopes provide similar information and in addition can provide height information relative to the substrate (on, in, or below). The laser confocal Ultrapointe specializes in this (fast, easy to use, accurate X,Y, good down to 0.2µm) and in addition can provide crude materials information (e.g. metal or dielectric) through the optical properties affecting reflectivity (10).

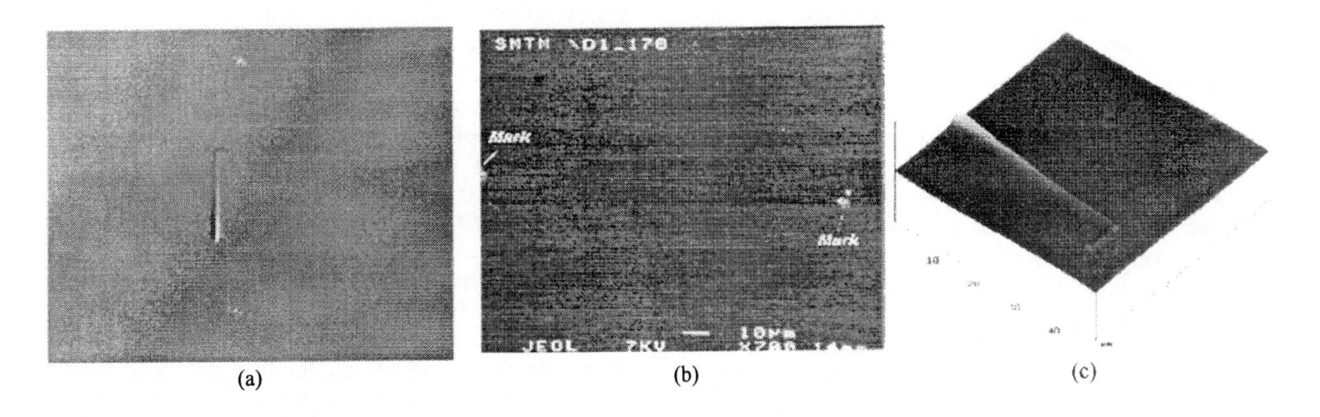

(a) (b) (c)

FIGURE 1. Crystalline Defect in Epi Wafer. a) Nomarski DIC image and wafer marks (MicroMark 5000); b) SEM image; c) 50 x 50 μm AFM scan.

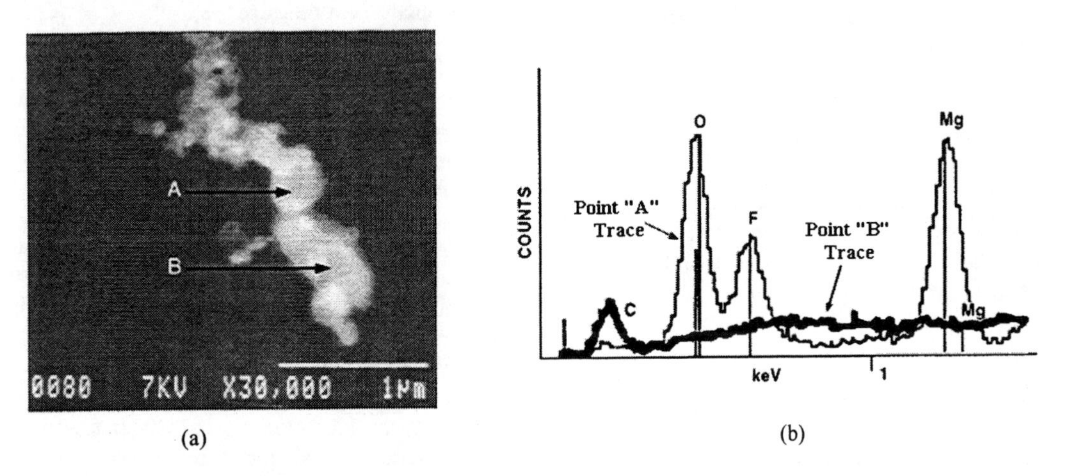

(a) (b)

FIGURE 2. a) SEM of multiple small particles (o-ring generated) on a Si monitor wafer; b) EDX showing o-ring signature at point "A" .

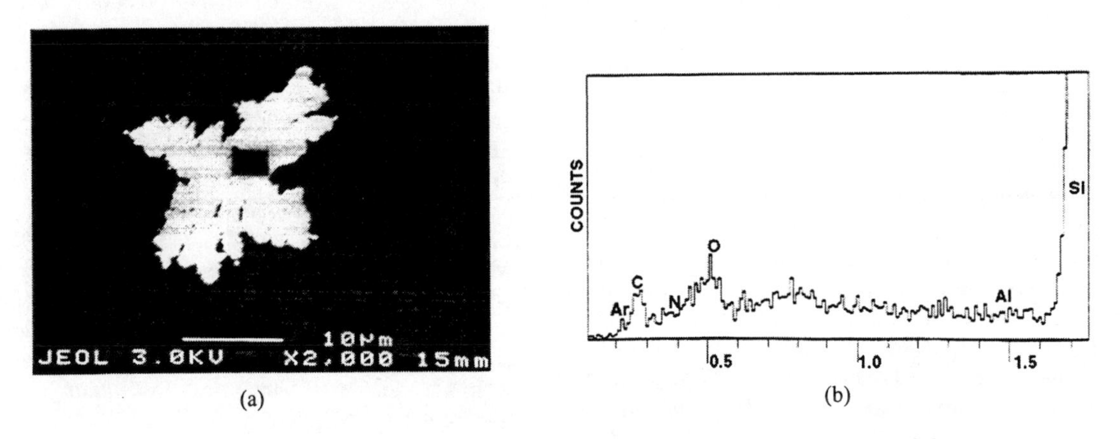

(a) (b)

FIGURE 3. a) SEM of a typical "snowflake particle" observed on a Si monitor wafer; b) EDX of the "snowflake".

Raman spectroscopy, providing much more chemical information through vibrational frequencies, is now available as an option on their 300mm version.

Atomic Force Microscopy, AFM

Once a defect is located, 200mm AFM systems can provide detailed profiles of features down to the nanometer scale. Since the AFM is basically a profilometer it does not matter if the defect is of identical material to the substrate, and so it is particularly good for distinguishing particles from bumps and pits. One disadvantage is that it is very slow to scan large areas at the pixel resolution necessary to locate small defects, i.e. another X, Y accuracy issue. This can be solved again by use of wafer marking, or more elegantly, Seiko (15) has installed Tencor-like light scattering initial detection and location in their AFM (unfortunately not available in the US). Another disadvantage is that no spectroscopy information is available (cf EDX with SEM; Raman with the Ultrapointe).

Focused Ion Beam, FIB

Long used in circuit design and repair applications, and as a tool for cutting sections for TEM analysis, FIB/ SEM combinations ("dual beam" systems) are increasingly being used for defect characterization. The FIB column micromachining capability allows sectioning into the centers of defects and into buried defects. The different contrast mechanisms of ion induced SEM, versus electron induced SEM, together with the materials delineating etch gases used in FIB, allow additional materials distinction in FIB'd features compared to ordinary SEM (16). The most powerful combination is to add EDX and quadrupole SIMS, available commercially as options, to provide spectroscopic characterization. FIB sectioning can be done on particles down to at least 0.2µm. The SIMS quadrupole mass spectrometer, however, requires high ion beam doses to achieve useful SIMS sensitivity, so would quickly destroy (typically in a few seconds) features this small.

Scanning Auger Microscopy, SAM

SAM has been a standard technique for small area surface analysis for many years. Though there is some variability (material dependence, kinetic energy of Auger electrons analyzed, angle of detection with respect to surface) for a flat vacuum/ material interface Auger typically gives information on approximately the top 10 atomic layers, with the signal decaying exponentially with depth. For particles on surfaces the situation is complex, because the high energy primary beam (typically 5-30Kev) can scatter laterally out of the particle and strike the adjacent substrate surface producing a component in the observed Auger spectrum which does not come from the particle surface. Thus, even a primary beam as small as 150A will not guarantee that the Auger signal comes entirely from the particle. Depending on particle size such artifacts can be minimized by going to either low primary beam energy for large particles (several µm), where the primary beam then does not escape the particle, or high primary energy for very small (thin) particles, where the primary beam passes right through with minimal scattering and is buried deep in the substrate (17).

High resolution imaging capability in the SEM mode is essential, since one is going to refind the defects from light scattering maps exactly as described for SEM/ EDX earlier. The PHI SMART 200™ system, with 150Å dark space resolution in the SEM mode has been able to locate particles down to 0.04µm in size and perform effective Auger analysis at this size (18).

A further advantage of Auger, not yet much explored for defect work, is to use the chemical state analysis capabilities in principle available from chemical shift information. There can be a problem with inadequate spectral resolution for chemical shift determination. The extent of this problem depends on instrument (analyzer) design, operating parameters, and exactly which shifts one is wanting to use. The PHI SMART 200™ uses a cylindrical mirror analyzer (CMA) which has limited spectral (energy) resolution, but has many other advantages. A hemispherical analyzer based system would significantly improve spectral resolution, but would have other drawbacks. Commercial 200mm wafer compatible hemispherical analyzer systems, however, do not yet exist. It is also feasible to add a low-cost, Auger attachment (a small commercially available system) to existing 200mm SEM/ EDX tools (19). It is not a scanning Auger system - i.e. spectra are acquired at one position, located by the SEM, and both sensitivity and spectral resolution are inferior to a stand-alone system, but, of course, it costs about one tenth of such a system! The major practical problem here is the poor vacuum of typical commercial SEMs. Since Auger is so surface sensitive it is necessary to keep the surface of the particle from being heavily carbon contaminated during analysis under the e-beam. Unfortunately this is very difficult in a non-UHV system, which means that the Auger data acquired will be seriously compromised. Even in UHV Auger systems it may be necessary to sputter clean particles before one can obtain a useful analysis, since the particle surfaces will have reacted with the air exposure prior to entering the Auger spectrometer. In general, if one needs analysis beyond the first 50Å or so of depth of the particle this will have to be obtained by sputter profiling. A good combination for covering the situation of both extremely small particles (or thin films) and large particles would be to add an EDX detector to the Auger system so that one could combine an analysis of the surface with one of the bulk of the particle without profiling.

Time-Of-Flight Secondary Ion Mass Spectrometry, TOF-SIMS

TOF-SIMS has extreme surface sensitivity. Used in the static mode the SMS signal will come from only the top couple of atomic layers. Although TOF SIMS has better trace sensitivity than many other methods it is very difficult to quantify the amount of a particular material from signal intensities. The real strength of static TOF-SIMS is its ability to give direct molecular information from the ion fragmentation patterns, particularly for organic and polymeric materials. When this information is obtained it is more powerful than the chemical shift information of Auger, since it can lead to a direct identification of the molecule present. One never knows in advance, however, if it is possible to obtain this level of information, so it is more of a hit and miss method than Auger. The ion beam columns used (similar to FIB columns) can focus down to sub 0.1μm, but because of the necessity to collect sufficient secondary ions for detection in the static mode (i.e. without destroying the information you are seeking) the limit on the particle sizes for molecular information is probably 0.25μm.

SELECTED EXAMPLES ILLUSTRATING THE PARTICLE CHARACTERIZATION CAPABILITIES

Confocal Laser Microscope, with Raman Spectroscopy

SEM, figure 4(a), showed massive numbers of large particles on a monitor wafer after being in a metal etch chamber where chlorine chemistry was used (the same particles were also observed on product wafers). EDX indicated the presence of chlorine. Figure 4(b) shows a low resolution Ultrapointe confocal image (20). Obviously such large particles are easy to detect. Using an experimental Raman attachment the Raman (vibrational) spectrum of one of these particles is shown in Figure 4 (b) (top trace). There is a one to one match with the spectrum of NH_4Cl shown in the bottom trace.

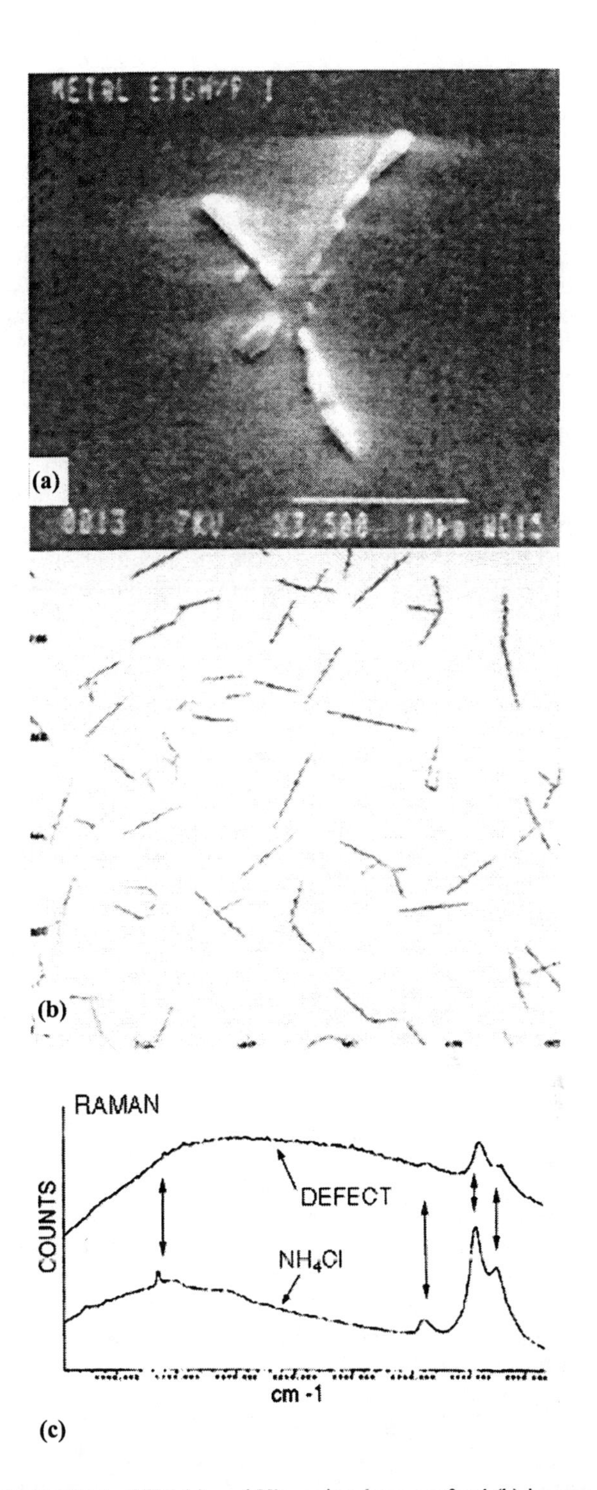

FIGURE 4. SEM (a), and Ultrapointe laser confocal (b) images of needle structure observed on an Si monitor wafer after being in a chamber involving Cl etch chemistry. c) Raman spectra of a needle particle (upper) and a NH_4Cl reference (lower).

AFM

In an evaluation of 200mm tools for characterizing defects on epi wafers, we looked at the same defects by SEM, AFM and FIB. Originally we tried doing this by relying just on Tencor 6200 maps for location of the defects and going to the different vendors for SEM and AFM instruments. This proved futile. The inaccuracies in the Tencor file locations, plus the inabilities of the SEM and AFM vendors' software to properly handle the X, Y coordinates issues resulted in none of the defects (about 20 per wafer) being located (this was in 1996). We therefore submitted the wafer to the Micromark 5000™, which easily located all the defects in dark field, took optical images (dark field, bright field and Nomarski), and wafer marked them. Particles could then be subsequently easily located in the AFM and SEM tools. Two things became immediately apparent from both SEM and AFM, (1) many of the defects thought to be particles turned out to be crystalline defects within the epi Si layer, and (2) for the latter type of defects Tencor light scattering sizing, using the default PSL on bare Si calibration, can be in error by up to a factor of 50! Of course, one should never use such a calibration for this situation, but, in fact, it is the automatic default and specifications based on it may be mandated by customers! An AFM example of a Si crystal defect is shown in Figure5(a) (13). It is a perfectly circular dome about 12µm in diameter and about 0.7µm high with a rough surface. Tencor 6200 light scattering, based on reference to PSL's on bare Si, suggests a 300µm size defect. Figure 5(b) shows the AFM of a genuine particle, a thin flake of Al_2O_3 about 1µm in diameter that is 0.1µm thick (13). Tencor 6200 sizes this at 0.1µm diameter, i.e. a factor of 10 too small. From a detailed study of many defects, real particles and crystalline defects, on epi wafers we concluded that the PSL-based particle sizing calibration roughly works (factor of 2) for fairly 3 dimensional, fairly smooth particles in the size range expected; ie. 0.1 to lum Outside these constraints the default sizing approach fails miserably, as would be expected from examination of the theoretical analyses that have been provided by Tencor. The two examples given above also show that the sizes can be both overestimated and underestimated, depending on the detailed intra-defect topography.

Figure 1(c) showed the AFM (Digital Instruments 7000™) of the defect detected by Tencor light scattering and imaged optically in the Micromark 5000™ in Figure 1(a) (13). As discussed earlier, SEM cannot detect this defect, Figure 1(b), because it is of the same material as the substrate and is so large laterally (many µm) but attains so little height, (50nm) that the angular change presented to the primary electron beam of the SEM, which controls the secondary electron emission yield (and hence contrast), is negligible. Thus in epi-layers very small "bump" Si defects will not be detectable by SEM in general. In CMP, where shallow pits produced by the polishing process, and detected by light scattering, are an issue, the same situation will occur: poor detection by SEM; easy detection and characterization by AFM.

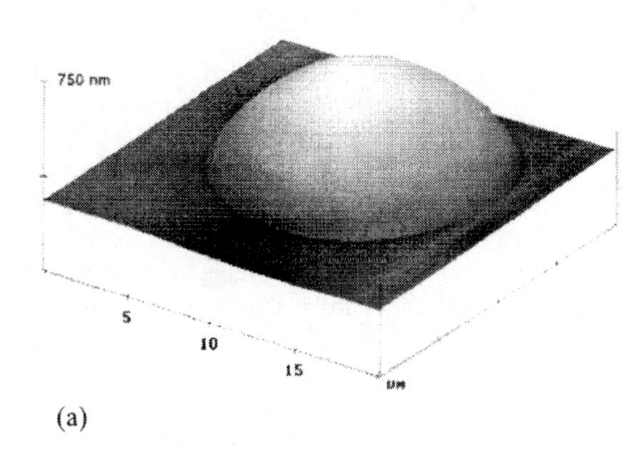

(a)

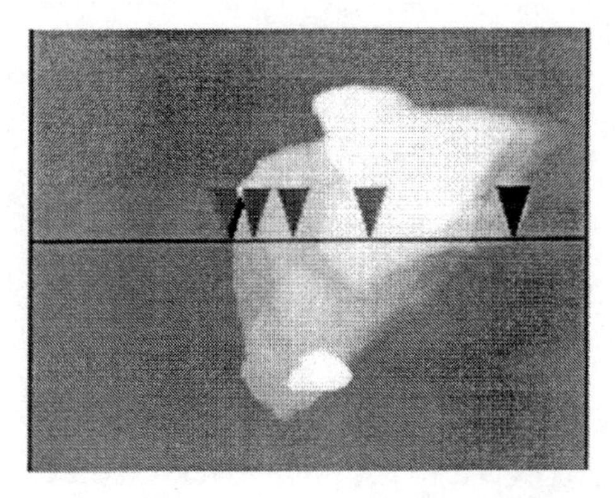

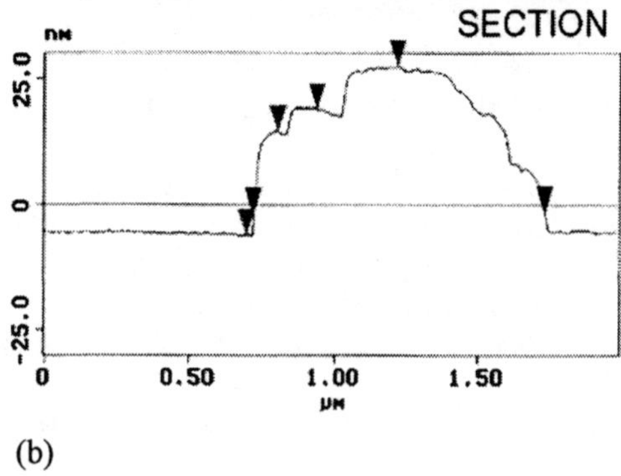

(b)

FIGURE 5. AFM (Digital 7000) of (a) a crystalline defect in an epi layer (Tencor greatly overestimates size); (b) an Al_2O_3 thin flake on an epi layer (Tencor greatly underestimates size).

FIB

Figure 6 shows an SEM image of a FIB cross-section of a 0.2µm Al_2O_3 particle sitting on a 1000Å of SiO_2 on a 200mm Si wafer. A 1000Å layer of Pt was pre-deposited on the defect prior to the FIB cut. This example is from a "real" particle defect problem, not a simulation, and the identification of it as Al2O3 came from a previous extensive study using SEM/ EDX/ Auger/ Raman in which we were also able to establish the origin (6). The particles were generated by plasma strikes on Al metal at the anodized/ metal interface within the gas distribution holes of a gas distribution plate. This interface was accessed by the plasma through cracks in the anodized coating. To definitely show that the particles came from the metal below the coating required eliminating ceramic chamber hardware and the anodized coating itself as sources, both of which are nominally Al_2O_3. The former was eliminated by demonstrating that the particle lacked any trace Cr (present in ceramic Al_2O_3 and easily detectable by fluorescence using a Raman microscope). The latter was eliminated because no S was found in the particles (by EDX, Auger). The anodization used at that time contained significant amounts of S. Other corroborating factors were also present and the interested reader is referred to the original article for this (6). The point of performing this later FIB cross-section was simply to see how easily the FIB tool in question, an FEI 820™, could find such real defect small particles from light scattering files and perform a FIB cut. The answer is quite easily. It had already been demonstrated earlier that the quadrupole SIMS attachment to FIB can identify and map Al from Al_2O_3 particles down to 0.1µm size on a Si substrate (22).

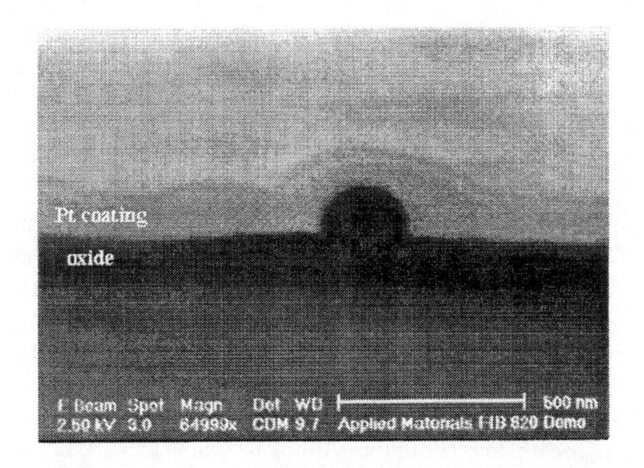

FIGURE 6. FIB cross-section (FEI 820) of a 0.2 µm hemispherical particle of Al_2O_3 on an oxide monitor wafer.

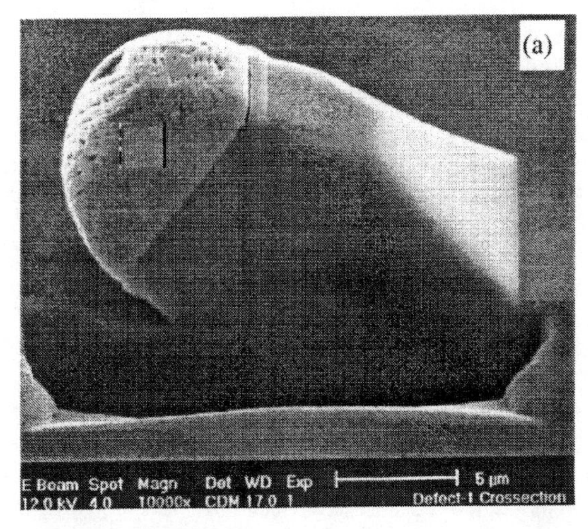

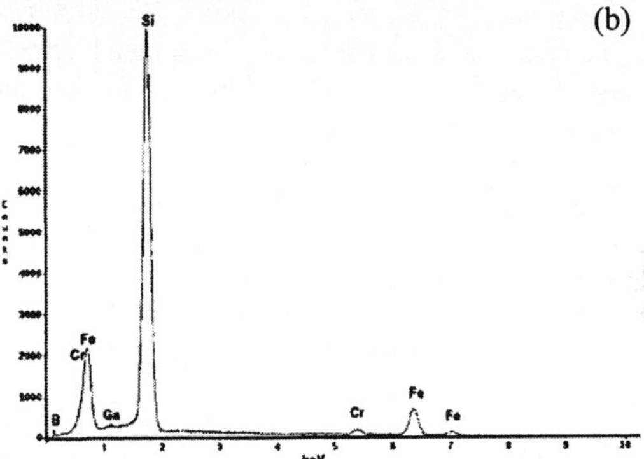

Figure 7. (a) FIB cross-section through a large crystalline defect found on an epi layer; (b) EDX from the indicated region of the FIB cross-section, showing the presence of Fe and Cr.

Figure 7(a) is a FIB cut through a large Si crystalline growth found in our epi-wafer defect studies. Figure7(b) is the EDX spectrum from the marked region of the spongy looking dome part of the defect. It clearly shows Fe and Cr (i.e. from steel), which may be the initiating cause of the defect growth (we found other defects giving the same results).

SAM

To date, only one company, Physical Electronics, makes a 200mm wafer SAM (the SMART 200). There are several already installed at fab sites, including AMD in Santa Clara, where Paul King has demonstrated the ability to locate, map and identify Fe particles from wet/ clean contamination down to as small as 400Å. We have used the SMART 200 for a number of situations where we have needed more detailed analytical information on smaller particles than could be effectively handled by EDX. Figure 8(a) represents such an example (23). It shows an SEM micrograph of a particle representative of a problem observed when running monitor wafers following a W etch-back process. The etch-back is performed using an SF_6 plasma, which generates F radicals, which in turn react with W, etching it away as gaseous WF_6. Unfortunately there are additional reactions owing to the presence of the TiN or Ti layer immediately below the W layer. The result is TiF_x products which can react with the chamber Al or Al_2O_3 hardware. The particle in figure 8(a) is indicative of this, since EDX indicates the presence of both Ti and Al. Figure 8(b) shows an Auger map of this same particle, clearly showing it to be a composite particle with a central Ti containing region and peripheral Al containing regions. Detailed Auger spectra from each of these regions indicate the Ti was present as either Ti oxide (or partially oxidized Ti), and the Al was metallic, oxidized at the surface. Thus the reaction of TiF_x must have occurred with bare Al inside the reaction chamber. A surprising finding is that the whole of the monitor wafer is covered with a thin layer (sub-monolayer because it is not observed by EDX) of Ti. Since this is a monitor wafer run long after the etch-back process, this means that TiF_x, or similar substance, remains in the process chamber and exhaust line with sufficient vapor pressure to deposit as Ti on the monitor wafer. This was the first example where surface contamination, as opposed to particle contamination issue was observed. Nothing like this had been seen before because of the lack of surface sensitive tools to detect such a phenomena. Since then we have observed several other examples by Auger, and also by ESCA (usable because spatial resolution is not needed) on wafer pieces.

(a)

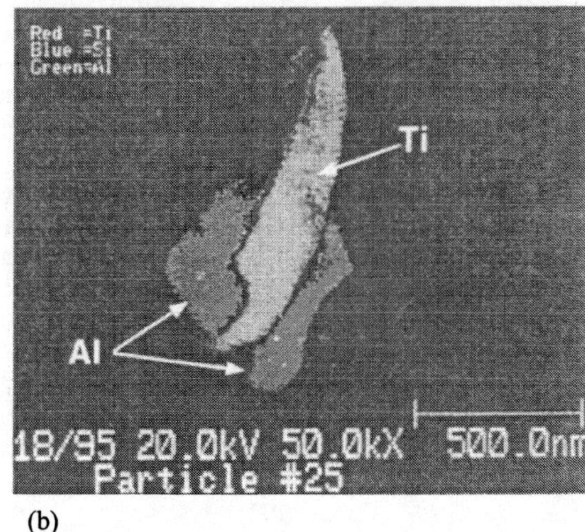

(b)

FIGURE 8. A Ti/Al composite particle found after W etchback process. (a) SEM image; (b) Composite SAM map of defect (reference 23).

TOF-SIMS

Figure 3 showed the SEM and EDX of a snow-flake like "particle" (13). As we discussed earlier, the EDX tells us nothing other than it is probably very thin and organic based (no difference was observed in the EDX "on" and "off" the particle). These defects were classified as a severe "particle" problem because they showed up in large numbers by Tencor light scattering (the de-facto particle specification). Even though they are large area, they are invisible, or almost invisible, under an optical microscope, again probably because that are so thin. Since we suspected an organic we thought TOF-SIMS might be able to identify which organic. At that time (1996) Charles Evans & Associates had just installed the prototype 200mm PHI-Evans system so we were able to run the full wafer, rather that to break it. Trying to find any of the snow-flake features using just the Tencor X, Y file again proved fruitless, so we wafer marked some regions containing clusters of the defects on the basis of their X, Y coordinates and then searched for these marks in the TOF-SIMS both optically and in the total ion-yield SIMS mode.

We easily found them this way and Figure 9(a) shows both a total negative ion yield and an F^- image of a cluster of 7 snowflakes. Clearly the snowflakes are strongly F containing. The full negative ion spectrum from mass to 200amu of one of the snowflakes is shown in Figure 9(b). The C, F, and O containing clusters marked are quite sufficient to identify the material as a fluoropolymer. On checking a library spectrum a match was found (including other peaks at higher mass) to Krytox, a well-known perfluoroether lubricant. The "particles" are droplets of Krytox oil, which was sufficient to immediately identify the source of the "particle failure" in this particular chamber. Again it should be noted that the material, Krytox, proved to be present not just in these droplets, which showed up by light scattering, but in fact was present as a thin layer over the whole wafer.

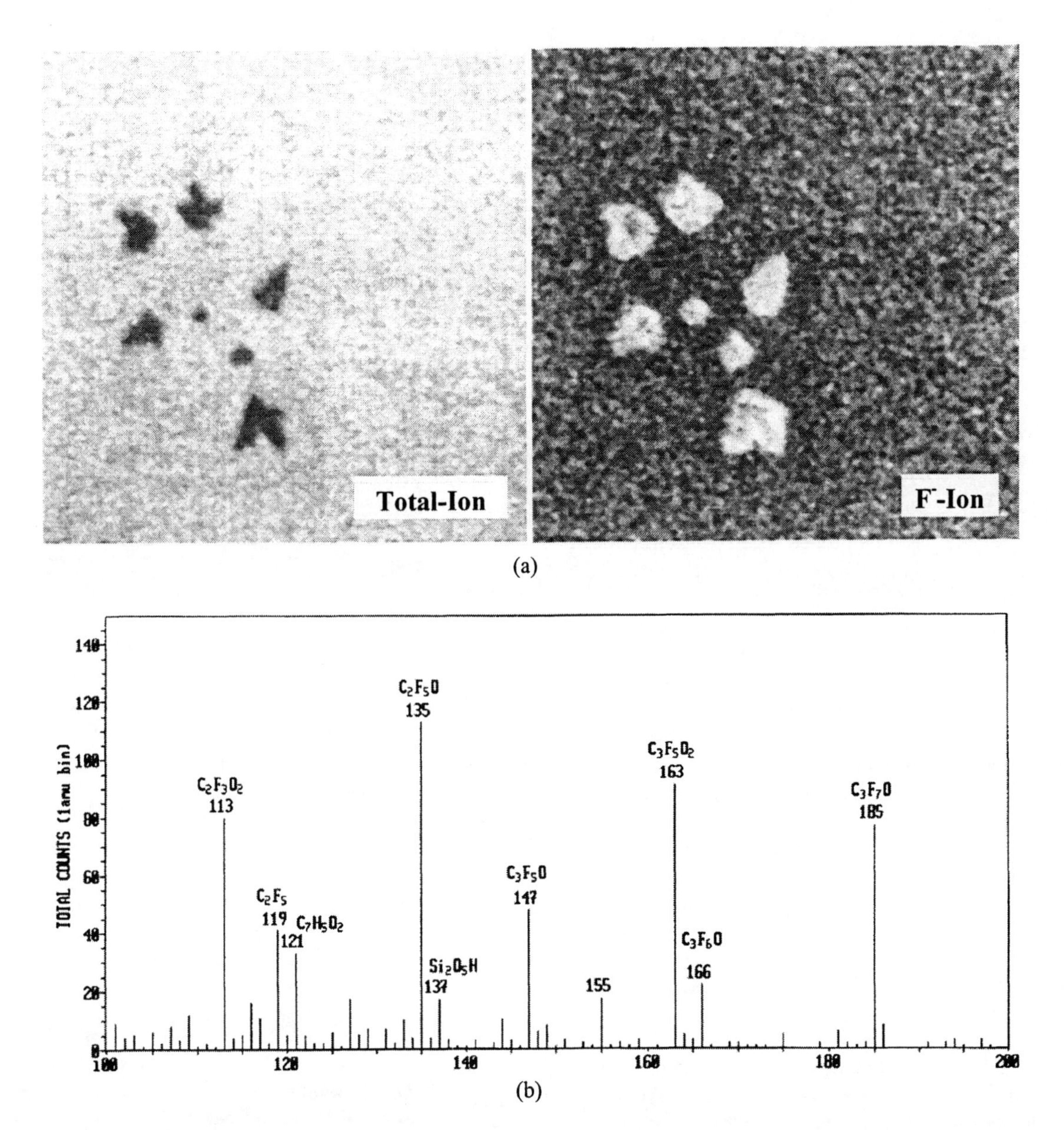

Figure 9. TOF-SIMS of "snowflake particles" found on a monitor wafer. (a) Total-ion image and F^- ion images; (b) SIMS spectrum from a snowflake particle in the 100-200 amu range.

Although its ability to identify organics is, perhaps, the most striking example of the characterization power of TOF-SIMS, it can also provide definitive molecular identification for inorganic contaminants also. One such example is ammonium phosphate. In a situation where particles grew to a few μm in size from nucleation sites probably smaller than 0.1μm on a TEOS layer, SIMS was able to identify both the positive ammonium and negative phosphate ions in the particles (24).

IMPORTANT TECHNIQUES NOT YET AVAILABLE FOR 200MM

Micro-calorimeter detector-based EDS

One of the main drawbacks of EDS was mentioned earlier; the inability to confine the EDS signal to a very small volume. A second drawback is the intrinsically poor spectral (i.e. energy) resolution of the Li drifted Si (or Ge) detectors used in EDS. Because of this many elemental X-ray emission spectral lines overlap, particularly in the crowded low energy region where many elements have X-ray lines, inhibiting their use for analysis purposes. A new detector, based on micro-calorimetry, where the energies of the impinging X-rays are distinguished by the amount of heat generated at the detector, has been developed and will soon be commercialized. It has been demonstrated to have over 10 times the spectral resolution of standard EDS detectors, making it comparable to WDX. This means that overlapping low energy lines no longer overlap and can be used for analysis. For example the EDS spectrum of Figure 7b, identifying a steel particle was taken at 12kV because it was necessary to use the high energy Fe and Cr lines to confirm their presence. The low energy lines overlap. With a micro-calorimeter detector the low energy Fe and Cr would be sharp and well-separated and the EDS spectrum could be taken at 3kV. Thus, in the majority of cases, it now becomes possible to confine the EDS volume analyzed to much smaller values by using only low accelerating voltages and still to get good analytical data. These two factors; vastly improved spectral resolution and confinement of the signal volume, will greatly extend the usefulness of the SEM/ EDX combination. It has been argued by Auger proponents that EDS runs out of steam below 0.2µm, and that Auger must be used (25). This is certainly not true for micro-calorimetry based EDS. Both techniques will be used; micro-calorimetry EDS for bulk analysis of sub-0.2µm particles and Auger for surface analysis and depth profiling.

Micro-ESCA/XANES

X-ray Photoelectron Spectroscopy, XPS (or Electron Spectroscopy for Chemical Analysis, ESCA, as it is sometimes known) is a powerful, general purpose technique for obtaining quantitative chemical state information at surfaces and interfaces (26). Existing commercial lab-based systems have been limited in the past for practical usage to about 20µm spatial resolution (usable count-rates). There has been an on-going effort at the Lawrence Berkeley Laboratory to use the soft X-ray Synchrotron Radiation from the Advanced Light Source, ALS, to implement a high flux sub-micron XPS beam line (27). This is supported by Intel, Applied Materials, and the ALS itself and the objective is to provide something close to a "commercial" analysis tool (i.e. user friendly in all aspects) for sub-micron applications on both small structures and for particles. To this end serious attention has been paid to the sample handling aspects and to the X, Y navigation procedure. The current version can handle only 2"x2" samples, but because of the availability of the Micromark 5000 we should be able to find and analyze small particles on cut up wafers. The beam line and end station have reached the stage where there is a need to benchmark its usefulness against the best lab-based systems. If successful there is some discussion on how to implement a 200mm capability and possibly a commercially run service.

Because the synchrotron source supplies X-rays over a wide energy range, it is possible also to run the same beam line in a different mode; New Edge X-ray Absorption Fine Structure, NEXAFS (also known as XANES, X-ray Absorption Near Edge Structure).

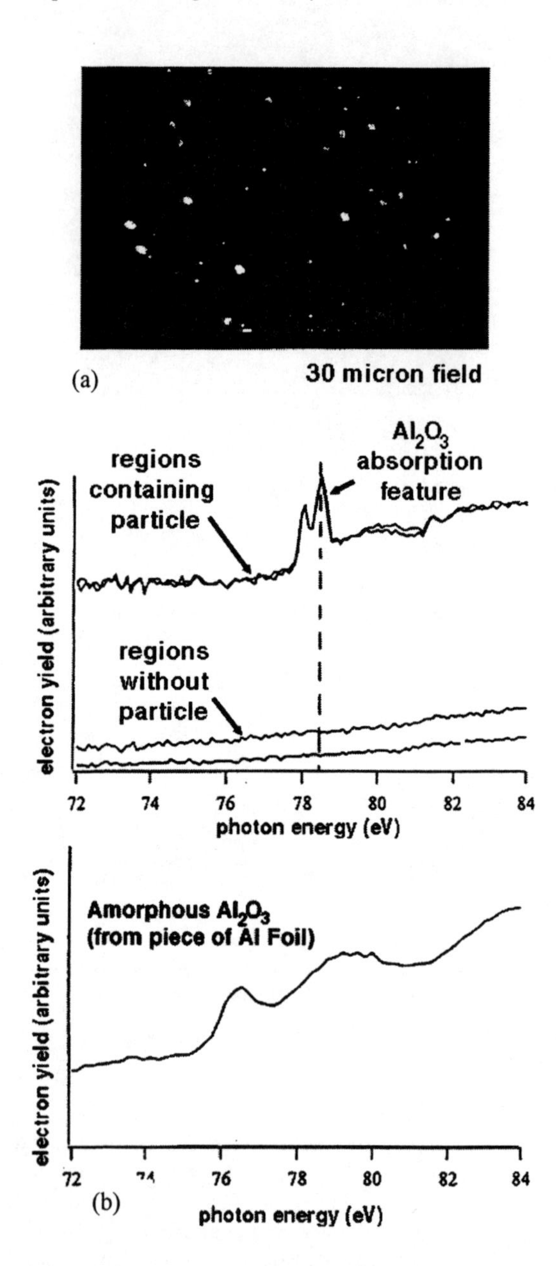

FIGURE 10. Synchrotron based Photo-microscopy of 0.5 mm Al_2O_3 particles on a 1" Si wafer. (a) secondary electron image at 78.5 eV photon energy; (b) Al(2p) absorption edge structure on the and particles and the Si substrate (top), on a piece of Al foil (bottom) (reference 8).

Basically this is a plot of absorption cross-section as a function of photon energy while crossing specific elemental absorption edges. The absorption is monitored by detecting the secondary electrons induced by the absorption process, which also makes it a surface sensitive technique. The edge identifies the element and shifts in the edges and related fine structure carry chemical state information, just like XPS. However, they are often larger and may carry structural information as well. Figure 10 represents a test example of the use of micro-XANES for particle detection and analysis (8). The sample consisted of a high density of 0.5µm Al_2O_3 particles deposited on a 1 inch Si substrate. Figure 10(a) is a 30µm FOV secondary electron image of the surface taken by sitting on the Al absorption edge peak at 78.5ev. The particles show up with high contrast. The absorption edge spectrum of one of the particles is shown in Figure 10(b), together with that of a piece of air oxidized Al foil taken at the same time. Both spectra, the Al_2O_3 on the foil and the Al_2O_3 particle are easily distinguishable from the absorption edge of Al metal, which is a featureless step occurring several ev lower in energy. The two oxide spectra are very different from each other, however. Comparison to literature spectra indicates that the foil Al spectrum is representative of amorphous Al_2O_3, whereas the particle spectrum has a match to crystalline α-Al_2O_3. So, in this case at least, it is possible not only to identify the particles as aluminum oxide (not metal), but also to say what crystalline phase it is. This is potentially extremely useful for particle root cause determination.

CONCLUSIONS AND FUTURE TRENDS IN FULL WAFER PARTICLE CHARACTERIZATION

The field appears to be bifurcating. On one side more and more effort is being put into rapid DRT, using both optical DRTs (Tencor, KLA, ADE, Orbot) and SEM based DRTs (KLA, JEOL, Opal) where the hope is to implement effective Automatic Defect Classification, ADC, schemes where defects are classified empirically by comparison to data base information on known defect types (28). Much (but not all) the effort here is on patterned product wafers. On the other side (of which we are representative), there is a demand for more sophistication in characterization on ever smaller particles so that root cause determination can be established for non-trivial situations. Most of this effort relates to unpatterned wafer analysis, with its concomitant much tougher X,Y accuracy problem, because that is where these problems surface. Both fields need, and are, rapidly transferring into the 300mm market. In the Defect and Thin Films Characterization Laboratory, DTCL, at Applied Materials we already have 300mm capability for light scattering (Tencor SP1); optical bench and wafer marking (the Micromark 5000); AFM (Digital Instruments 7000); and Ultrapointe Confocal Laser Microscopy (with a Raman system attachment). A 300mm version of the PHI SMART 200 Auger system is on order and a 300mm dual-beam SEM/ FIB is likely also. For those methods which are not either dark-field optical or SEM based, there remains a basic practical problem of the difficulty of first finding your defect (you can't analyze what you can't find!). AFM, SIMS, and micro-ESCA suffer from this. Integrating a high level dark-field optical system, or for the vacuum based techniques (SIMS, ESCA) a "dual beam" SEM column, plus paying serious attention to the X,Y accuracy problem would greatly enhance their practical full wafer particle analysis capabilities. For now these techniques require a wafer marking approach to be reliably successful.

We believe that as design rules get smaller and as the thin films used get thinner more and more emphasis will be placed on analysis in general at surfaces and interfaces (including, but not only, particles). For this reason we see both microcalorimetry based EDS and ESCA (both micro and larger area versions) as being important techniques for the future.

ACKNOWLEDGEMENTS

The authors would like to acknowledge helpful discussions with Paul King, Kent Childs, and Alain Diebold during the work that led to this paper. We are also grateful for the co-operation of Physical Electronics, FEI, and Digital Instruments during our comparative instrument evaluations.

REFERENCES:

1. Metrology Roadmap Supplement to the 1994 National Technology Roadmap for Semiconductors, A.C. Diebold, SEMATECH Technology Transfer Document #94102578A-TR, SEMATECH 2706, Montopolis Drive, Austin, TX78731

2. Uritsky, Y., Virendra, R., Ghanayem, S., and Wu, S., *Microcontamination*, **12, 5**, 25-29, May 1994.

3. Uritsky, Y., and Lee, H.Q., "Particle Analysis on 200 mm Notched Unpatterned Wafers", in Proceedings of the Symposium on Contamination Control and Defect Reduction in Semiconductor Manufacturing III, the Electrochemical Society, 1994, pp. 154-163.

4. Chernoff, D., Brundle, C.R., and Uritsky, Y. *Solid State Technology*, March 1996, 105.

5. Geiss, R. "Encyclopedia for Materials Characterization", Ed. Brundle, C.R. Evans, C.A., and Wilson, S., Chapter 3.1, Butterworth, Boston 1992.

6. Brundle, C.R., Uritsky, Y., and Tony Pan, J., *"Extending Particle Characterization Limits in Wafer Processing"*, Micro, **13, 7**, pp. 43, July/August, 1995.

7. Kiskinova, M.P., "ESCA Microscopy on Elettra: Chemical Characterization of Surfaces and Interfaces with Sub-micron Spatial Resolution", in Proceeding of the MRS 98 Spring Meeting, April 13-17, San Francisco, CA, 1998, 361.

8. Brundle, C.R., Uritsky, Y., Warwick, A., Hockett, R.S., A. Craig, C.Ayre, D.Dunham,T.Droubay, and B.P.Tonner, Proceedings of the Symposium on Contamination Control and Defect Reduction in Semiconductor Manufacturing IV, The Electrochemical Society, 1996

9. Wollman, D.A., Irwin, K.D., Hilton, G.C., Dulcie, L.L., Newbury, D.E., Martinis, J.M., "High-resolution, energy-dispersive microcalorimeter spectrometer for X-ray microanalysis", Journal of Microscopy, Vol. 188, Pt. 3, December 1997, 196-223.

10. Cai, C., Uritsky, Y., Francis, T., Xu, J.. Worster, B., "Defect Characterization by Using Laser Imaging Confocal Optical Microscopy (LICM)", in Proccedings of 191st Electrochemical Society Meeting, May 4-9, Montreal, 1997, 586.

11. Cai, M.-P, Uritsky, Y.S., and Kinney, P.D., in *Semiconductor Characterization: Current Status and Future Needs*, Eds. Bullis, W.M., Sailer, D.G., and Diebold, A.D., American Institute of Physics, Woodbury, NY, 1996.

12. Kinney, P.D., and Uritsky, Y., "Locating defects on wafers for analysis by SEM/EDX, AFM, and other microanalysis techniques" submitted for publication in the proceedings of the SPIE conference: *In-Line Characterization Techniques for Perofmrance and Yield Enhancement in Microelectronic Manufacturing II*, Santa Clara, CA, Sept. 23-24, 1998.

13. Brundle, C.R., Huber, W., Greene, A., unpublished results.

14. Brundle, C.R., Uritsky, Y., to be published

15. Fujino, N., Karino, I., Kobayashi, J., Kuramoto, K., Ohomori, M., Yasutake, M., Wakiama, S. "First Observation of 0.1 micron Size Particles on Si Wafers Using Atomic Force Microscopy and Optical Scattering", J. Electrochem. Soc., Vol. 143, No. 12, December 1996, 4125-4128.

16. Melngailis, J., *J. Vac. Sci. Technol.*, B **5, 2**, Mar/Apr 1987.

17. Childs, K.D., Narum, D., LaVanier, L.A., Lindley, P.M., Schuler, B.W., Mulholland, G., and Diebold, A.C. "A Comparison of Sub-Micron Particle Analysis by Auger Electron Spectroscopy, Time-of-Flight SIMS, and EDS," 42nd Nat'l Symp. of the AVS, Mpls., MN, October 16-25, 1995.

18. King, P., J.V.S.T (proceedings of 1997 symposium)

19. Chambers, S.A., Tran, T.T., and Hilerman, T.A. *J. Vac. Sci. Technol.*, A **15, 2**, Jan/Feb 1995.

20. Cai, C., Uritsky, Y., unpublished data

21. Stokowski, S., and Vaez-Iravani, M., these Proceedings

22. Diebold, A.C., Uritsky, Y., Brundle, C.R., Francis, T., Fatemi, H., Childs, K., and Clough, S., *Future Fab International,* **I, I**, 1996.

23. Uritsky, Y., Chen, L., Zhang, S., Wilson, S., Mak, A., Brundle, C.R., J. Vac.Sci.Tech. A **15**, 1319, 1997

24. Cai, C., Brundle, C.R., and Uritsky, Y., unpublished results

25. Childs, K., Watson, D.G., Paul, D.F., and Clough, S.P., these proceedings

26. Brundle, C.R., 'Encyclopedia for Materials Characterization", Ed. Brundle, C.R., Evans, C.A., and Wilson, S., Boston: Butterworth, 1992, pp 282.

27. Padmore, H.A., and Pianetta, P., these proceedings

28. Braun, A.E. "Defect Detection and Review Enter a New Era", Semiconductor International, May, 1998, p61.

X-ray Microscopy; an Emerging Technique for Semiconductor Microstructure Characterization

H. A. Padmore

Advanced Light Source, Lawrence Berkeley National Laboratory, Berkeley, CA 94720

The advent of third generation synchrotron radiation x-ray sources, such as the Advanced Light Source (ALS) at Berkeley have enabled the practical realization of a wide range of new techniques in which mature chemical or structural probes such as x-ray photoelectron spectroscopy (XPS) and x-ray diffraction are used in conjunction with microfocused x-ray beams. In this paper the characteristics of some of these new microscopes are described, particularly in reference to their applicability to the characterization of semiconductor microstructures.

INTRODUCTION

The latest generation of synchrotron radiation sources is characterized by having extremely high brightness. As the flux in a small focus is directly proportional to brightness, this type of source is optimized for microscopy, and at the ALS we have developed a range of such instruments with spatial resolutions from a few microns to less than 100 nm. For chemical identification and mapping, we have instruments based on traditional laboratory techniques such as infrared vibrational spectroscopy and x-ray photoelectron spectroscopy, as well as the complementary technique of x-ray absorption spectroscopy in which tunable synchrotron radiation has to be used. For structural characterization we have combined traditional single crystal and powder x-ray diffraction with a microfocused probe with micron dimensions. Although most of the systems developed so far can be classified as research tools, a system for micro-focused x-ray photoelectron spectroscopy (μ-XPS) of microstructured wafer samples has been developed in collaboration with Intel Components Research division and Applied Materials. This system has many of the tools found on commercial systems dedicated to the examination of wafers, including optical fiducialization, rapid sample introduction and high precision wafer motion systems. The intent of this paper is to give a brief overview of the capabilities of these systems.

MICROSCOPE CHARACTERISTICS

a) Microfocus XPS

A simple grating monochromator is used to produce tunable radiation from 250 - 1300 eV. The monochromator produces a vertical image of the synchrotron source, and a spherical mirror produces a horizontal stigmatic image at an adjustable pinhole aperture. This aperture of nominal size 20 x 40 μm provides the object for a pair of crossed elliptical mirrors to focus at 20 and 40:1 demagnification to a 1 μm focus. Experimental measurements have verified that a focus of approximately this size is formed. The mirror system is situated close to the sample, inside the measurement chamber, and is fully bakeable. Photoelectrons are energy analyzed using a commercial XPS analyzer (Phi Omni). A particular feature of the design is that the sample can be viewed using a high aperture optical microscope, and this is used to re-fiducialize the sample after transfer so that a coordinate reference system can be re-established. Together with an in-situ optical interferometer and an ultra-stable x-y sample stage, sub-micron absolute positioning and tracking can be achieved. The optical microscope can also read Tencor light scattering detection coordinate files, and move the sample stage to intercept selected particles. The sample transfer system and sample x-y stage are designed to work with sections of wafers up to 50 x 50 mm. The system was developed in collaboration with Intel Components Research Division, and Applied Materials. Technical aspects of the system are described in [1,2] and applications and quantification of performance is reported in [3,4].

CP449, *Characterization and Metrology for ULSI Technology: 1998 International Conference*
edited by D. G. Seiler, A. C. Diebold, W. M. Bullis, T. J. Shaffner, R. McDonald, and E. J. Walters

All aspects of the optical and x-ray microscopes are computer controlled. The system is used for the identification of surface chemical species, such as those produced after plasma processing, and for particle identification. One of the great advantages of synchrotron radiation over laboratory sources is that the photon energy can be tuned over wide ranges. In this way the cross section of a particular core level can be optimized, and the surface sensitivity can be 'tuned' by variation of the kinetic energy and hence the mean free path of electrons in the solid. For example, the cross section for carbon 1s photoemission increases by approximately a factor of 50 in going from Al K radiation (1486 eV) to 340 eV (approximately 50 eV above threshold). In addition, the kinetic energy of C 1s electrons at this energy would be in the minimum of the mean free path, and so surface features would be emphasized over the bulk. As many of the processes of interest relate to surface properties, for example adhesion, the ability to 'tune in' to the surface through an appropriate choice of photon energy is of key importance. Of equal importance for materials that are easily damaged such as polymers is the ability to increase the cross section of a required core level, and hence minimize the radiation dose. This effect can also be used to reduce the cross section of an unwanted component. One important feature of the μ-XPS system is that it is permanently installed on beamline 7.3.1.2, and dedicated to the characterization of microstructures. The synchrotron source runs 24 hrs/day, 5 days/week, and the monochromatized beam is shared equally between μ-XPS and another photoemission microscope (PEEM). This arrangement ensures that samples can be handled on relatively short notice.

b) Scanning Photo-Electron Microscopy (SPEM)

This system is similar to the μ-XPS apparatus described above, but uses an ultra-bright undulator source rather than a bending magnet radiation source, and uses zone plate focusing. The system is located on beamline 7.0.1 [5]. The detection system is the same as above, but the higher incident flux combined with the superb focusing characteristics of a zone plate allow a much better energy and spatial resolution. Zone plates are simply circular diffraction gratings whose period decreases towards the outer rings. They are made by electron beam lithography and patterned in gold, nickel or other metals depending on the photon energy range. If the zone plate is correctly made, the ultimate resolution is approximately given by the spacing of the outer most zone. In this case for reasons of flux we use a relatively coarse zone plate, and achieve a spatial resolution of around 0.2 μm. For transmission microscopy where lower signals can be tolerated, much finer zone plates can be used, and ones with 20 nm outer zones have been made, and a resolution of 30 nm has been reported. The SPEM produces spatially resolved maps by raster scanning the x-ray beam across a fixed sample, while recording a particular photoelectron peak intensity. This is done by moving the zone plate on a piezo driven stage in areas up to 100 by 100 microns, and allows the sample to have all the normal features of an ultra-high vacuum surface science system, such as sample heating and cooling. For examining larger areas, the sample stage itself can be moved over 20 x 20 mm. The zone plate system can be lowered out of the beam to allow direct observation of the sample using an optical microscope. This allows specific parts of the sample to be examined with reference to optical fiducial marks. Another characteristic of this system is that very high energy resolution can be achieved, better than 0.1 eV over the whole working range of photon energies (200 - 900 eV).

c) Photoemission microscopy (MAXIMUM)

The Multiple Application X-ray Imaging Undulator Microscope (MAXIMUM) is a scanning x-ray microscope like the μ-XPS and SPEM systems previously described, but achieves a microfocus using a Schwarzschild objective to demagnify a pinhole object at 20:1 demagnification [6]. This objective consists of a convex - concave mirror pair, with light entering at approximately normal incidence through a hole in the center of the concave mirror, with subsequent reflections from the convex and concave surfaces. The great advantage of this arrangement is that it can collect a large aperture while maintaining diffraction limited performance. In order to work at soft x-ray energies and so access a significant number of core levels, the optical surfaces have to be coated with a multilayer mirror. This consists of alternating ruthenium and boron carbide layers, and so the periodicity of this system defines the photon energy. In this case, an energy of 130 eV is used, and so means that the photoemission features seen are highly surface sensitive. This system was originally installed at the Synchrotron Radiation Center in Wisconsin and moved to beamline 12.0.1 in 1995. The 4 orders of magnitude higher brightness of the ALS translates directly into detected signal, and so the system can now work at its optimum performance. It achieves <100 nm spatial resolution, and has a total energy resolution of around 0.2 eV. The system is fully UHV compatible and has been used for a wide range of experimental studies on semiconductor materials [7,8, 9].

d) Photo-Emission Electron Microscopy (PEEM)

Photoemission electron microscopy using x-ray excitation (X-PEEM) combines two established techniques, x-ray absorption spectroscopy (XAS) and full field electron imaging to give a powerful tool for high-resolution chemical characterization of surfaces. In an X-PEEM system, monochromatic tunable radiation is condensed to illuminate a small field of view at the sample, typically 30 μm in diameter. Electrons photoemitted from this area are first accelerated in a high field region between the sample and the front of the first electron lens, and then subsequently imaged by an objective and several projector lenses at high magnification. The final image is then recorded on a phosphor screen - CCD system. This type of system has been used for many years using ultra-violet excitation, and by the use of appropriate filters, excellent work function contrast can be obtained. Such systems have achieved < 10 nm resolution (10). PEEM can also be combined with tunable synchrotron radiation however to give spectroscopic information [11]. If the energy of a photon beam is increased through the region of an x-ray absorption edge, the total yield of electrons from the sample will usually exhibit a series of sharp features within 10 eV of the edge. The Near Edge X-ray Absorption Fine Structure (NEXAFS) contains information on elemental composition, chemical bonding, bond orientation as well as magnetic state, and can be used as a unique fingerprinting tool. In order to record spectroscopic information with a PEEM, images are recorded at increments of increasing energy through an absorption edge. The integrated intensity in areas that show some feature of interest can be obtained by summing the intensities in individual pixels, and this intensity can then be plotted out through the series of images. The local NEXAFS spectrum can therefore be obtained. Because of the parallel nature of full field imaging, this type of microscopy can be fast, with whole volume data sets (a succession of images at increasing photon energy) recorded in a few minutes. Penetration depths for soft x-rays in solids are much greater than the range of low energy electrons, and this means that most primary photoelectrons are scattered to lower energies by the time they emerge from the surface. The energy spectrum imaged by the PEEM is therefore centered at a few eV and is around 10 eV wide. This combined with the chromatic nature of electron lenses gives a lower resolution limit of around 20 nm for a microscope operating at 10 KV/mm extraction potential. The mean free path of low energy electrons in solids is typically 2 - 5 nm and this sets the effective probing depth of the technique. Further advances in electron optics in which a chromatic aberration compensation system is used is expected to reduce the resolution limit to around 2 nm [12].

The ALS has two PEEM systems. The first PRISM [13] consists of an electrostatic objective lens typically working at an extraction potential of 10 KV, a movable aperture in the back focal plane, a projector lens and a channel plate-phosphor-CCD imaging system. The system has a spatial resolution of around 200 nm and is located on undulator beamline 8.0. This beamline is shared between several end stations, and gives a photon energy range of 70 - 1300 eV. A dedicated system, PEEM2, has been built on bending magnet beamline 7.3.1.1 [14]. This beamline was specifically optimized for PEEM for magnetic materials, and covers an energy range of 250 - 1300 eV [15]. The microscope operates at 30 KV, and has in addition to PRISM a stigmator-deflector at the objective back focal plane, an additional projector lens, and a set of adjustable back focal plane apertures. The system is currently being commissioned and is expected to give a resolution of 20 nm. The system is equipped with in-situ evaporation, and a sample preparation facility.

(e) Infrared microscopy

Synchrotron-based infrared (IR) beamlines provide a brightness advantages over conventional IR sources of 100 - 1000 in the near to mid IR region. This means a similar factor in intensity advantage for microscopy. At the ALS, bending magnet beamline 1.4 is dedicated to IR studies, and is equipped with a commercial IR interferometer and microscope for the 2 - 30 μm wavelength range [16]. The beamline extracts 40 x 10 mrads of light from the storage ring, passes it through a diamond isolation window, and uses parabolic collimating optics to produce a parallel beam for the interferometer and microscope. The 2 - 30 μm range is commonly used for the chemical fingerprinting due to the presence of many vibrational bands in this region. The system is being commissioned, but has so far been used to study a huge diversity of samples, from organic particles to bacteria used in bio-remediation. The whole system is located in a semi-clean room environment.

(f) Micro-diffraction

This system was originally developed to apply to the problems of thin film strain, particularly with reference to electromigration in integrated circuit interconnect structures. Although this problem has been widely studied using electron microscopy [17], and with focused x-ray probes [18], an understanding of stress at the spatial scale of single grains in Al-Cu lines is still lacking. The apparatus we have built for this task consists of a pair of elliptically bent mirrors to produce a 'white beam' micro-focus of 4 - 13 keV x-rays [19], a 4 crystal monochromator to monochromatize the incoming beam, a sample scanning stage to manipulate specific parts of a structure into the x-ray beam, and an x-ray CCD camera to record the diffraction patterns. A unique feature of the system is that the monochromator has co-linear input and output beams, and so by rotating to zero angle, the white x-ray beam can be directly focused on the sample. Rotating the crystals into the beam to provide monochromatic light causes no apparent beam motion at the micron level. White beam mode is used to record Laue diffraction patterns, from which we extract orientation information. Knowing orientation and the index of each diffracted beam, we also know the energy of each beam in the case of an unstrained grain. To measure strain then simply means that we have to insert the 4 crystal monochromator and measure the energy of a set of reflections. The energy differences to the unstrained case directly give strain information. We have obtained 0.8 x 1.5 μm resolution, and measured single grain orientation and strain in Al-Cu lines at this resolution. The ultimate strain resolution is simply set by the precision with which we can measure the energy of reflections, and this should be near 10^{-5}. Initial results on this system are reported in this conference [20]. A dedicated beamline for micro-diffraction, 7.3.3, has now been constructed and is undergoing commissioning. Commissioning of the end station will start in November this year following delivery of a new CCD x-ray detector and goniometer system [21].

(g) Micro-X-ray absorption spectroscopy

The system described above for micro-diffraction can also be used for micro-x-ray absorption spectroscopy (μ-XAS) in the 4 - 13 keV energy range. Using white beam, very rapid identification of the elements of interest can be made using fluorescence detection. Once these areas have been found, the monochromator can be scanned through an elemental absorption edge, and the NEXAFS features near the edge can be used as a unique chemical fingerprint. Most of the initial work using this technique has centered on analysis of environmental samples, one such example being the speciation of Zn as an oxalate form in a fungus [22]. It has also been employed to speciate the chemical form of Fe in an inclusion in solar cell silicon [23]. The ability to perform high spatial resolution chemical speciation in air, with samples in their native state make this technique particularly powerful.

Acknowledgements

This work was supported by the Director, office of Basic Energy Sciences, materials Sciences Division of the US Department of Energy, contract no. DE-AC03-76SF00098.

The work reported here represents that of many people working at the ALS and in university groups in the field of microscopy using x-rays.

Further information can be obtained through the ALS web site at www-als.lbl.gov, or by e-mail to contacts for each area:

a) Zahid Hussain; ZHussain@lbl.gov
b) Tony Warwick; Warwick@lbl.gov
c) Franco Cerrina; Cerrina@thor.xraylith.wisc.edu
d) Simone Anders; SAnders@lbl.gov
e) Michael Martin; MCMartin@lbl.gov
f) Alastair MacDowell; AAMacDowell@lbl.gov
g) Geraldine Lamble; GMLamble@lbl.gov

REFERENCES

(1) H. A. Padmore, G. Ackerman, R. Celestre, et al., *Synchrotron Radiation News*, **10(6)**, 18(1997)

(2) G. D. Ackerman, R. Duarte, K. Franck, et al., Proc. Materials Research Society Meeting, Symp. V. San Francisco April 1998

(3) F. Gozzo, B. Triplett, H. Fujimoto, et al., Proc. Materials Research Society Meeting, Symp. V. San Francisco April 1998

(4) Y. S. Uritsky, P. D. Kinney, E. L. Principe et al., Proc. Materials Research Society Meeting, Symp. V. San Francisco April 1998

(5) the beamline 7.0 spectromicroscopy program was initiated by Brian Tonner, Univ. Wisconsin, with continuing support for development from the ALS.

(6) F. Cerrina, G. Margaritondo, J. H. Underwood, M. Hettrick, M. A. Green, L. J. Brillson, A. Franciosi, H. Hochst, P. M. Deluca, and M. N. Gould, *Nucl. Instr. and Meth. in Phys. Res.* A **226**, 303 (1988)

(7) G. F. Lorusso, H. Solak, S. Singh, F Cerrina, J. H. Underwood and P Batson, this conference

(8) G. F. Lorusso, H. Solak, S. Singh, F. Cerrina, J. H. Underwood and P Batson, Proc. Materials Research Society Meeting, Symp. V. San Francisco April 1998

(9) F. Cerrina, A. K. Ray-Chaudhuri, W. Ng, S. H. Liang, S. Singh, J. T. Welnak, J. P. Wallace, C. Capasso, J. H. Underwood, J. B. Kortright, R. C. C. Perera, and G. Margaritondo, *Appl. Phys. Lett.* **63**, 63 (1993)

(10) G. Rempfer and O. H. Griffith, *Ultramicroscopy* **27**, 273-300 (1989)

(11) G. Harp and B. P. Tonner, *Rev. Sci. Instrum.*, **59**, 853 (1988)

(12) R. Fink, M. R. Weiss, E. Umbach, et al., *J. Elect. Spect. Relat. Phenom.* **84**, 231 (1997)

(13) B. P. Tonner, D. Dunham, T. Droubay and M. Pauli, . *Elect. Spect. Relat. Phenom.* **84**, 211 (1997)

(14) T. Stammler, S. Anders, H. Padmore, J. Stohr and M. Scheinfein, Proc. Materials Research Society Meeting, Symp. V. San Francisco April 1998

(15) H. A. Padmore, *AIP Conf. Proc. 389, 'X-ray and Inner-Shell Processes'*, Hamburg Sept. 1996, p 193

(16) Nicolet Magna 760 FTIR spectrometer combined with a Nic-Plan all reflecting IR microscope

(17) T. Marieb, P. Flinn, J. C. Bravman, D. Gardner et al., J. *Appl. Phys.* **78**, 1026 (1995)

(18) P. C. Wang, G. S. Cargill and I. C. Noyan, *Proc. Mat. Research Society,* **375**, 247 (1995)

(19) A. A. MacDowell, R. Celestre, C-H. Chang, K. Franck, M. R. Howells, S. Locklin, H. A. Padmore, J. R Patel and R. Sandler, *SPIE Proc.* **3152**, 126 (1998)

(20) C-H Chang, A. A. MacDowell, H. A. Padmore and J. R. Patel, this conference

(21) Bruker SMART CCD camera, computer and analysis codes, including an integrated Huber goniometer system

(22) G. M. Lamble, D. G. Nicholson, A. Moen and B. Bethelsen, submitted to XAFS X, Chicago, Aug. 1998

(23) S. McHugo, A. C. Thompson, G. Lamble et al., Proc. Materials Research Society Meeting, Symp. V. San Francisco April 1998

Analysis of Molecular Adsorbates on Si Surfaces with Thermal Desorption Spectroscopy

Norikuni Yabumoto
NTT Advanced Technology Corporation
Kangawa, JAPAN

Thermal desorption spectroscopy (TDS) is an effective tool for analysis of molecular adsorbates on a silicon wafer surface. Since the TDS uses a heat as a probe to desorb molecular absorbates from the wafer surface, it makes the molecular adsorbates themselves accessible for detection. Consequently, hydrogen, water molecule organic compounds and their fragments are detected with high sensitivity of less than 1×10^{12} atoms/cm^2 using a mass spectrometer. The change of the surface state due to heating is also found from the desorption temperatures. TDS is expected to be very useful to detect the change due to heat of even smaller amounts in the field of semiconductor manufacturing.

INTRODUCTION

The Si wafer is influenced by contact with the surrounding environment, including the equipment, clean room air, gas, water, chemicals, and human beings during LSI fabrication process. As result, in addition to particles, various impurities such as water, organic molecules, some molecules or ions, and metals attach to the wafer. Native oxide film is also grown at the same time. During the cleaning process, the microroughness of the surface may increase, and some defects existing in the crystals are enhanced.

For every process these issues must be analyzed beginning with the reception of the wafer, and continuing through cleaning, gate oxide film formation, junction formation, and ion implantation. Therefore, an optimum analytical method is demanded for each species of the contamination. Total reflection X-ray fluorescence (TXRF), secondary ion mass spectrometer (SIMS), inductively coupled plasma mass spectrometery (ICP-MS) and atomic absorption spectroscopy (AAS) can be applied to detect the metals to less than 10^{10} atoms/cm^2. For light elements, the analysis method is Auger electron spectroscopy (AES); for the native oxide film, X-ray photoelectron spectroscopy (XPS). However, it is necessary for particles to be detectable down to 35 nm, metals to less than 10^9 atoms/cm^2, and organic molecules to 2.5×10^{13} C atoms/cm^2 after ten years[1)]. These values are pretty close to the previous estimates for particles (50 nm) and metals (10^2 - 10^9 atoms/cm^2), but differ for organic molecules (10^{10} molecules/cm^2)[2)]. These requirements lead to the necessity to consider other analysis techniques.

The target of thermal desorption spectroscopy (TDS) is chiefly hydrogen, and molecular type adsorbates such as water and organic molecules. We applied TDS to investigate such adsorbates on the silicon surface after HF cleaning. We succeeded in detecting fragments of isopropyl alcohol that was used for the drying at the end of the cleaning step and some other organic adsorbates[3)]. Recently, TOF-SIMS[4)] and GC-MS[5)-7)] were applied to measure organic adsorbates as high-sensitive analysis methods of 10^{11} - 10^{12} molecules/cm^2. TDS also has the same high-sensitivity or more[8)].

This paper presents the results of a study of small amounts of molecular adsorbates on silicon wafers after cleaning. Oxidation of a hydrogen terminated silicon surface with organic contamination is also described.

THERMAL DESORPTION SPECTROSCOPY

Molecular adsorbates on a surface desorb from weak bonds in order of the bond strength as the temperature is increased. TDS has been used for over thirty years for the analysis of adsorbates. In general, it is a method by which the sample is heated in vacuum at a constant rate, and the desorbed adsorbates are analyzed with a mass spectrometer. From the temperature and the shape of the peak, the absorption state of the molecules can be deduced. When the signal can be distinguished from the background, highly sensitive analysis making use of the property of the mass spectrometer can be achieved. TDS uses heat as the analytical probe. As heat is a mild probe compared with high energy ions, adsorbates are desorbed from the surface as molecule states. However, it is necessary to understand that TDS is ineffective in analyzing materials that do not desorb due to heating.

The desorption rate N(t) can be expressed with

CP449, *Characterization and Metrology for ULSI Technology: 1998 International Conference*
edited by D. G. Seiler, A. C. Diebold, W. M. Bullis, T. J. Shaffner, R. McDonald, and E. J. Walters

Redhead equation[9].

$$N(t) = -d\sigma/dt = \nu_a\sigma_a\exp(-E/RT). \quad (1)$$

Here, n is the order of the desorption reaction, σ is the surface coverage of the adsorbate, v is the frequency factor, E is the activation energy for the desorption, R is the gas constant, and T is the absolute temperature for the desorption peak. From equation (1), the information obtained by the TDS is dominated by the order of the desorption, the activation energy, and the surface coverage which are important factors for the surface reaction. When TDS is applied to the evaluation of an LSI process, information regarding the kind of desorption and at what temperature it occurs is essential because it can determine the process temperature it is possible not only to study the desorption from the surface, but also contaminants in the film or the desorption via the thin film from the deeper layers[10)-11)].

Some kinds of DTS systems have been produced. Figures 1(a)-(c) show three representative systems. System (a) is the type where the heating and the detection part are in one chamber. The sample is heated near to the mass spectrometer. In order to protect the detector from the heat, only the back surface of the wafer is heated locally[12)-13)], or the quadruple mass spectrometer is protected by a liquid nitrogen shroud[14)]. All kinds of desorbed atoms and molecules can be analyzed in the mass spectrometer without being re-absorbed by the walls of the system. This system can be applied for the detection of the desorption of native oxide or aluminum chloride of plasma etching residue.

In system (b), the sample lies in the quartz tube and heated from all sides with an infrared furnace. Comparatively large samples can be examined and heated homogeneously. As the adsorbates desorb into the vacuum and are collected efficiently by the mass spectrometer, the sensitivity is high. The system (c) uses an atmospheric pressure ionization mass spectrometer (APIMS) or a gas-chromatograph (GC) as the detector. Since this system allows the measurement in the flowing gas, it is possible to follow both the desorption and the surface reaction at the same time. The following data were obtained by system (b) or system (c).

ANALYSIS OF ADSORBATES

Hydrogen

HF solution removes silicon oxide film and leads to a wafer surface terminated with hydrogen [15)]. Since no new native oxide films grows for a few hours, this cleaning is often used before gate oxide formation. Figures 2(a)-(b) show three-dimensional displays of TDS spectra after HF cleaning. The sample used was cut from a double-side mirror-polished silicon (100) wafer. The background mass spectra grow monotonically with the temperature as shown in (b), while various peaks at different mass number are clearly observed in (a) when the sample is present. Hydrogen desorbs due to the hydrogen-termination of silicon surface. Therefore, the desorption volume corresponds to around one mono-layer. The signals from water and oxygen fall below the background levels due to oxidation at about 500°C[16)]. Many other contaminant-related organic adsorbates besides hydrogen are present in small amounts.

After HF cleaning of the silicon surface, hydrogen is by far the biggest adsorbate, as much as 10 times more than the next molecules, water and ethylene. Considering the transmission efficiency of the quadruple mass spectrometer and the ionization rate of the ion source, more than 90% of the desorbed species is hydrogen. The TDS spectra of hydrogen for four different kinds of HF cleaning are shown in Fig. 3. The spectra were subtracted the backgrounds. Samples A and B are examples of wet HF solution cleaning, and samples C and D are examples of dry HF gas cleaning. Samples A and C were dried with inert gases, and samples B and D were exposed to IPA vapor during the drying process. The peak height at about 500°C is independent of the cleaning method, while the peak at about 400°C is higher for the samples washed in HF solution. No significant difference is observed depending on whether IPA vapor drying is used or not.

Froitzheim performed EELS and LEED on a 7x7 Si (111) surface after exposure to atomic hydrogen[17)]. At room temperature, both SiH_2 and SiH are created. When the surface is heated up to 380°C, only SiH remains. This SiH desorbs at 480°C, and disappears completely at 650°C, where the 7 x 7 structure appears again. Therefore, the hydrogen on the Si (100) surface cleaned with HF desorbs in the two reaction steps with heating as shown in Fig. 4. The surface changes from the di-hydride 1 x 1 structure to the mono-hydride 2 x 1 structure at about 400°C, and excess hydrogen desorbs. Then at about 500°C, the mono-hydride 2 x 1 structure changes to a non-hydrogen 2 x 1 structure and generates hydrogen.

Water

When water is present on the Si surface, it is not only the reason for silanol and the creation of a native

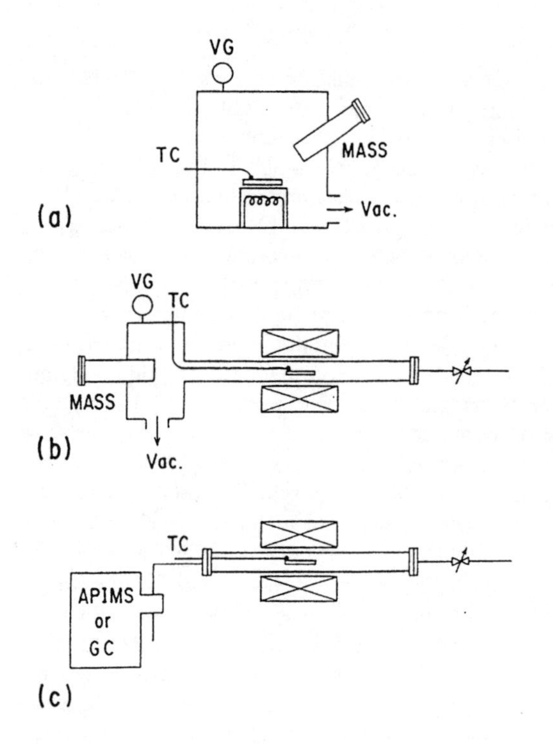

Figure 1. Representative thermal desorption systems.

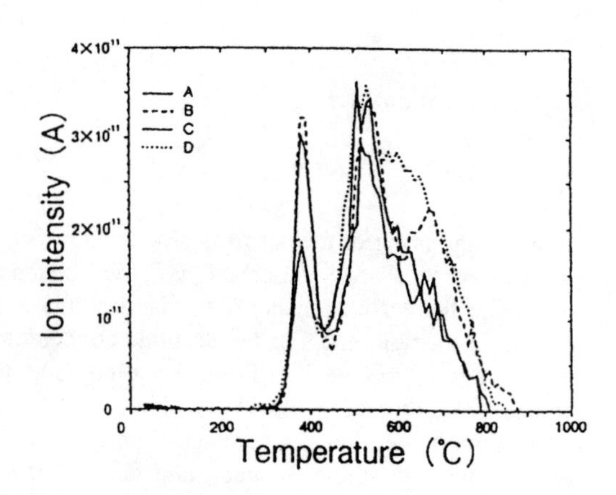

Figure 3. Thermal desorption spectra of hydrogen from the four different HF cleaning methods.

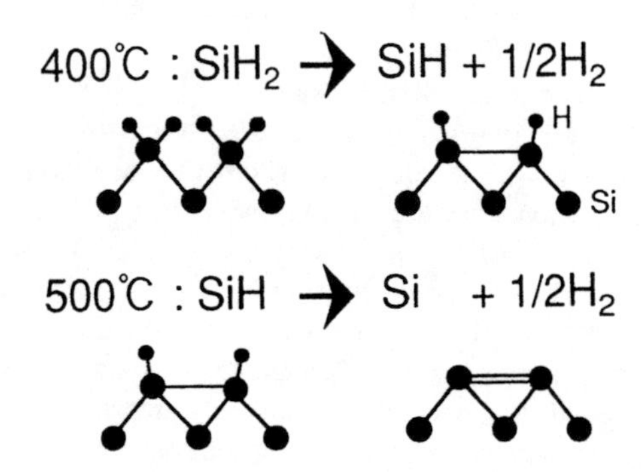

Figure 4. Identification of hydrogren desorbed at 400°C and 500°C.

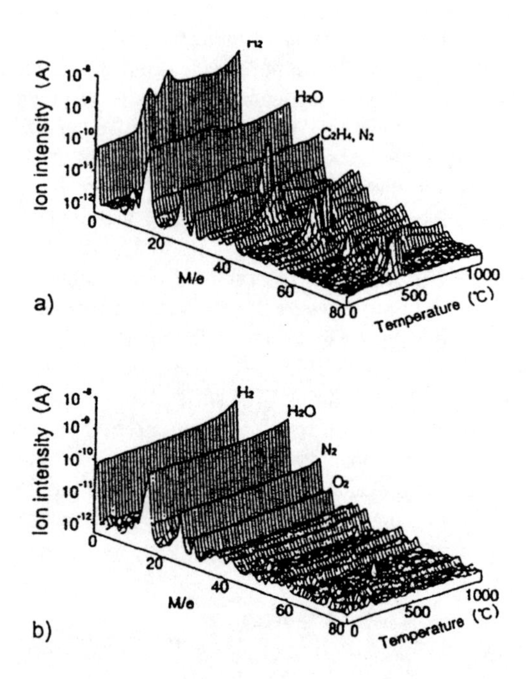

Figure 2. An example of three-dimensional display of TDS spectra of the HF-cleaned silicon wafer with ion intensity, temperature, and mass number.

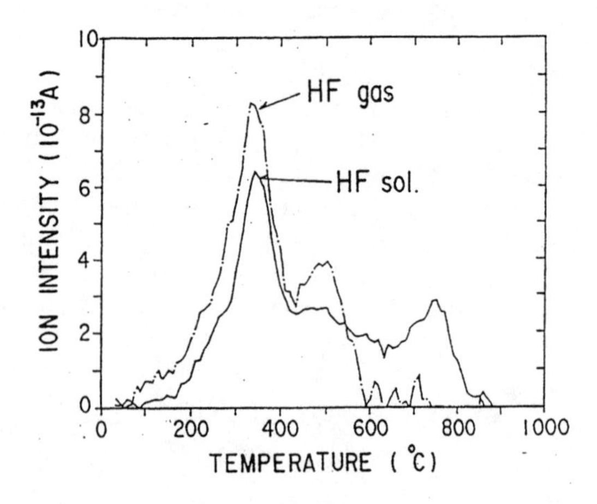

Figure 5. Thermal desorption spectra of H_2O after cleaning with an 1% HF solution and an 1% anhydrous HF gas.

oxide film, but it also dissolves some ions such as sulfate and ammonium or sodium, and becomes the source for new contamination. Therefore, water should be removed from the surface.

A water desorption spectrum with a peak at 340°C is often observed on the HF cleaned silicon surface for both HF solution cleaning and HF gas cleaning as shown in Fig. 5. However, as shown in the water spectrum in Fig. 2, there is a case that no desorbing water is observed after HF cleaning. On the other hand, this water peak is always observed for silicon surfaces with native oxide formed by either NH_4OH/H_2O_2 or HCl/H_2O_2 cleaning or for thermally oxidized silicon surfaces.

The amount of desorbing water in Fig. 5 is less than 0.1 mono-layer. Since the water desorbs at a high temperature compared with the boiling point (100°C), it is speculated that water chemically adheres to the surface due to hydrogen bonds when the native oxide remains on the surface. Figure 6 shows a model of the relationship between water molecules and silicon surfaces. Water molecules can easily form hydrogen bonds to the oxide surface, when they attach with (1) their positively charged hydrogen atoms (the electronegativity of O is 3.5, H is 2.2) to the surface oxygen with two lone pair electrons or (2) their negatively charged oxygen atom to the slightly positively charged hydrogen of the hydroxyl reduced which terminates the native oxide surface.[a] However, water molecules are repelled from the hydrogen-terminated silicon surface[b]. According to the difference of electronegativity of Si and H (Si:1.7, H:2.2), the surface hydrogen is slightly negatively charged. The oxygen atom with two lone pairs of water cannot approach to the negatively charged surface due to the electric repulsion. Therefore, the hydrogen-terminated silicon surface after HF cleaning is hydrophobic.

Organic compounds

Small amounts of many kinds of organic compounds are desorbed from the cleaned silicon surface as already shown in Fig. 2, which covers mass numbers from 1 to 80. Generally, the hydrocarbons most frequently detected after HF cleaning have two carbon atoms C_2H_x (x=1-6) followed by C_3H_y (y=1-8), C_4H_z (z=1-10), CH_w (w=1-4), and other ones with more than five carbon atoms as shown in Fig. 7. Since the hydrogen desorption corresponds to a mono-layer, the amount of the individual hydrocarbons corresponds to less than 1/100 mono-layer. Consequently, this figure also shows that the sensitivity of TDS for the organic adsorbates is less than 1×10^{12} molecules/cm^2.

When the wafer is contacted with a zig or is cleaned in a cleaning bath made of polyfluoroethylene, it is sometimes contaminated by the residue of polyfluroethylene. Figure 8 shows TDS spectra that appear after contact with polyfluoroethylene tweezers. The peaks of the mass number 31, 50, and 69 corresponding to CF, CF_2, and CF_3 appear together at the same temperature, 570°C.

When isopropyl alcohol (IPA) was used for drying at the end of the cleaning, a peak appeared of mass numbers 2/5. This peak corresponds to $CHCH_3OH$, which is a fragment of IPA molecule $CH(CH_3)OH$. However, the amount of IPA contamination was one magnitude smaller compared to other hydrocarbons that were observed on usual cleaned wafer.

Beside the adsorbates already discussed, molecules containing a benzene ring with mass number 77 and 78 are sometimes observed. It is speculated that the benzene ring is derived from dioctyl phthalate (DOP), dibutyl phthalate (DBP), or dibutyl hydroxy toluene (BHT), all of which are generally present in clean rooms. Some groups have reported that organic adsorbates adhere to the silicon surface from the environment, by using analysis methods based on an analyzer combined with thermal desorption. Using gas chromatograph[18] and using APIMS[19] as a detector, the Toshiba group detected dodecane and phthalic acid derived from DOP which came from clean room air. Hiroshima University also observed DOP using GC-MS[6]. The Sony group observed DBP and BHT on silicon wafers which had been stored in a wafer storage box[7]. These organic compounds are used as plasticizers or antioxidants. TDS, whose detector is a quadruple mass filter, hardly detects molecules with mass number more than 100. The detector should be selected for objective adsorbates.

OXIDATION OF ORGANIC ADSORBATES

Organic adsorbates react with silicon to form silicon carbide in addition to desorption at the heating process in inert gas or vacuum environment[20], and may cause secondary contamination. On the other hand, they might burn in oxygen to form carbon dioxide and water. Whether the silicon surface is terminated with hydrogen or covered with native oxide, hydrogen is present on the surface about 100 times more than organic adsorbates. This hydrogen also reacts with oxygen to form water. Therefore, if water is detected, it does not prove that the organic adsorbates on the silicon surface burn in the oxygen gas. However, if

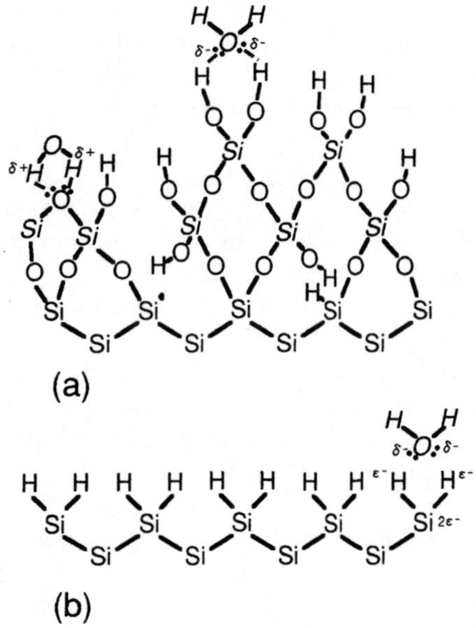

Figure 6. A model of relationship between H_2O molecule and silicon surface.

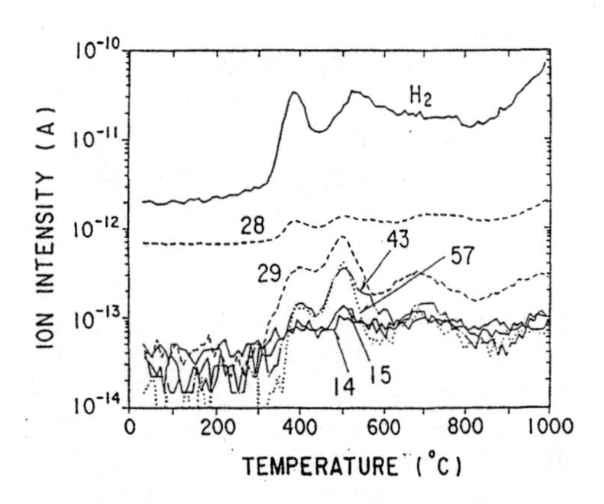

Figure 7. Thermal desorption spectra of hydrocarbons.

carbon dioxide is detected, this would constitute proof that organic adsorbates burn on the oxygen gas.

Carbon dioxide is detected by APIMS as a CO^{3-} ion in oxygen gas. Figure 9 shows the TD-APIMS spectra of CO^{3-} when oxygen is flowing. There are two peaks from 350°C to 650°C. These temperatures coincide with the ignition temperatures of various organic adsorbates. When the sample was cleaned with dilute HF, small amounts but many kinds of organic adsorbates attached to the surface. It should be noted that the CO^{3-} spectrum returns to the background level beyond 650°C, while some hydrocarbons continue to

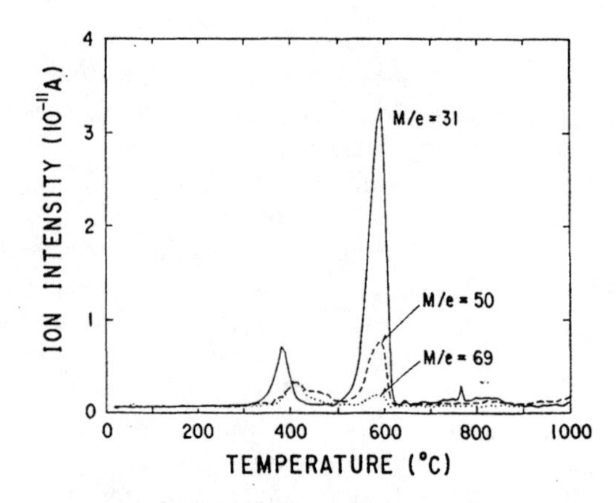

Figure 8. Thermal desorption spectra observed after contact with polyfluoroethylene tweezer.

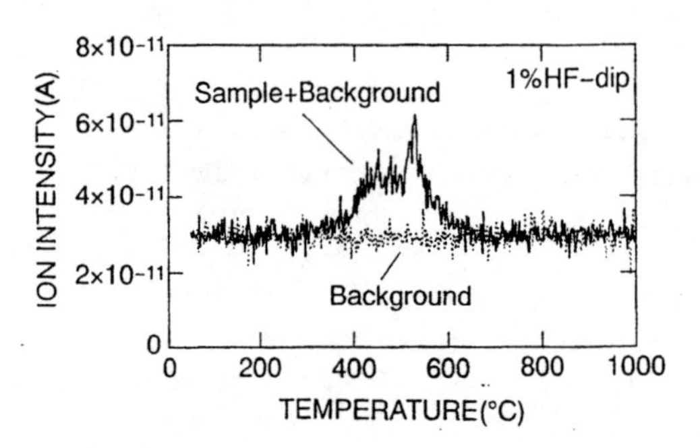

Figure 9. TD-APIMS spectra of CO^{3-} measured with flowing oxygen.

desorb up to 1000°C in vacuum as shown in Fig. 2 and Fig. 3. This suggests that organic adsorbates burn completely at 350-650°C.

SUMMARY

Thermal desorption spectroscopy (TDS) is an efficient technique to detect molecular adsorbates on a silicon wafer surface. Since the TDS uses heat as the probe to desorb molecular adsorbates from the wafer surface, it makes the molecule adsorbates themselves accessible for detection in contrast to a secondary ion mass spectrometer which destroys molecular adsorbates or creates new molecule-ions. The TDS system allows *in situ* observation of surface reactions during heating and cooling[21]. Therefore, TDS can also determine the temperature that adsorbates desorb from the wafer surface or that reactions occur. As differential thermal analysis or thermogravimetric analysis have contributed a lot in the evaluation of industrial thermal process,

TDS is expected to be very useful to detect the change due to heat of even smaller amounts in semiconductor manufacture.

REFERENCES

1. The National Technology Roadmap for Semiconductors: Technology Needs, 1997.
2. N. Yabumoto: UltraClean Technology, Vol. 4, 304 (1992).
3. N. Yabumoto, K. Minegishi, K. Saito, M. Morita and T. Ohmni: Semiconductor Cleaning Technology/1989, 90-9, Electrochem. Soc., ed. J. Ruzyllo and A. E. Novak, 265 (1990).
4. B. Schuler, P. Sander, and D. A. Reed: Proc. 7th Int. Conf. Secondary Ion Mass spectrometry (SIMS VII), Monterey, 851 (1989).
5. A. Shimazaki, M. Tamaoki, and Y. Sasaki: Ext. Abst. 188th electrochem. Soc., No. 476, CHICAGO, 755 (1995).
6. T. Takahagi, S. Shingubara, H. Sakaue, K. Hoshino, and H. Yashima: Jpn. J. Appl. Phys., 35, L818 (1996).
7. K. Saga and T. Hattori: J. Electrochem. Soc., Vol. 143, 3279 (1996).
8. N. Yabumoto, Y. Komine, Y. Kunni, T. Ohmi, S. Kojima, and K. Kubo: Proc. 22th Symp. ULSI Ultra Clean Technol., No. 6-8, Tokyo (1994).
9. P. Z. Redhead: Vacuum vol. 12, 203 (1962).
10. K. Murase, N. Yabumoto, and Y. Komine: J. Electrochem. Soc., Vol. 140, 1722 (1993).
11. K. Machida N. Shimoyama, J. Takahashi, N. Yabumoto, and E. Arai: IEEE Trans. Electron Device, Vol. 41, 709 (1994).
12. Y. Kobayashi, Y. Shinoda, and K. Sugii: J. Vac. Sci. Technol. A10, 2308 (1992).
13. N. Hirashita, M. Koayakawa, A. Aritmatsu, F. Yokoyama, and T. Ajioka: J. Electrochem. Soc., Vol. 139, 794 (1993).
14. H. Aoki, E. Ikawa, T. Kikkawa, Y. Teraoka and I. Nishiyama: Jpn. J. appl. Phys., 31 (1992) 2041.
15. T. Takahagi, I. Nagai, A. Ishida, H. Kuroda, and Y. Nagasawa:, J. Appl. Phys., Vol. 64, 3516 (1988).
16. N. Yabumoto, K. Saito, M. Morita, and T. Ohmi: Jpn. J. Appl. Phys. Vol. 30, L419 (1991).
17. H. Froitzheim, J. Kohler, and H. Lammering: surf. Sci. Vol. 149, 537 (1985).
18. A. Shimazaki: 1992 Int. Symp. Semiconductor Manufacturing Technology, Tokyo, 157 (1992).
19. M. Tamaoki, K. Takase, H. Sasaki, and a. Shimazaki: Ext. Adst. 41st Jpn. Soc. Appl. Phys., 28-a-ZQ-2, Kawasaki, 650 (1994).
20. N. Yabumoto, N. Kawamura, and Y. Komine: Ext. Abst. 184th Electrochem. Soc., No. 299, New Orleans, 490 (1993).
21. N. Yabumoto and Y. Komine: IEICE Trans. Electron., Vol. E75-C, 770 (1992).

Wet Chemical Analysis for the Semiconductor Industry
A Total View

Marjorie K. Balazs

Balazs Analytical Laboratory, Sunnyvale, California, 94089

The analysis of liquids to obtain information about semiconductor materials is known in the industry as "wet chemistry" and has been used since the beginning of the production of IC's. However, the analytical procedures never gained any significant attention until the mid 70's when the absolute measurement of phosphorus in PSG films by wet chemical analysis was incorporated by several industrial labs as the standard method of analysis.

Today, over 120 different procedures are used to gain specific information about incoming and processed materials used in the industry. These procedures cover ultra pure water, chemicals, thin films, and wafer cleanliness. Furthermore, they are used to evaluate the cleanliness of reactors, cleanrooms, and components of all kinds that are used in cleanrooms, wet benches and reactors.

This paper will cover a total look at the applications of wet chemical processes and the usefulness of the data obtained from these analytical techniques. The paper will cover not only those tests that one would expect to be done by wet processes such as the analysis of metals in chemicals, but will also cover many unusual applications of wet chemical analysis such as their usefulness in evaluating products from a variety of reactors.

Included in this part of the presentation will be a unique application to determine ion implantation contaminants and recent advances for analyzing 300mm wafers without breaking them and the analysis of contamination metals in copper thin films. Actual data will be provided for each of the analytical techniques presented.

INTRODUCTION

The analysis of liquids and solutions to obtain information about semiconductor materials, processes and the environment in which integrated circuits are produced is known in the semiconductor industry as wet chemistry. Even though wet methods have been used in the industry since the late 50's, there is still little understanding about these techniques. Yet, well over 120 procedures are used to obtain specific information in the industry's effort to maintain good quality control, improve yields and to help understand the sources of contamination.

It is easier to accept that wet analytical procedures can be used to measure materials used on the wet side of manufacturing such as ultra pure water, chemicals, and wet benches. However, as will be shown, wet procedures are also very useful in analyzing processed wafers, thin films, reactors, cleanrooms and components of all kinds.

There are many advantages to using wet chemical methods for measurements when low level specific information is needed for process evaluation and process control such as:

- Wet chemical measurements are primary, easily traceable to standards (i.e., NIST) and therefore absolutely quantitative.
- One can duplicate a matrix exactly and thus determine when and how the matrix causes interference and thus false data.
- Recovery studies can easily be done to verify the method.
- Since the procedures give accurate results, it is possible to compare data over time and to see baseline shifts or variations in processing.
- Wet chemical analysis is extraordinarily sensitive for measuring materials in the range of percentages down to parts-per-trillion (ppt) or atoms per square centimeter.

These facts not only make wet methods excellent for metrology, but also make them useful to standardize other methods of analysis such as Fourier Transform Infrared (FTIR) or complement "dry" instruments such as TXRF or SIMS.

The most obvious applications for wet chemical analysis are those where the samples are already in a liquid state. In fact for materials in this state, wet analysis is not only the right choice but generally the only choice. For example, tools like inductively coupled plasma-mass spectrometry (ICP-MS) are used to determine the concentration of metals in UPW (see **Table 1**). Furthermore, since water can easily be concentrated, the detection levels shown, in Table 1, can be lowered by a factor of 10 or 100. Therefore, parts-per-quadrillion (ppq) levels of metals in UPW can be measured today.

ICP-MS is also useful for measuring metals in chemicals. With chemicals, however, the analyst needs to do some preparation of the sample to be analyzed. Since the major constituent interferes with the accurate measurement of very low concentration of metals, it must be removed. Furthermore, during the preparation step, the sample must not become contaminated. To insure both accurate measurement and that no contamination has occurred, recovery studies are done. **Table 2** shows recoveries for HF after the removal of the matrix. The detection levels in this case are obtainable without concentrating the sample. In other words, after matrix removal the sample is brought back to the original volume.

CP449, *Characterization and Metrology for ULSI Technology: 1998 International Conference*
edited by D. G. Seiler, A. C. Diebold, W. M. Bullis, T. J. Shaffner, R. McDonald, and E. J. Walters

TABLE 1. Trace Metals Detection Limits in UPW by Inductively Coupled Plasma - Mass Spectrometry (ICP-MS).

Element		DL (ppb)
Aluminum	(Al)	0.003
Antimony	(Sb)	0.002
Arsenic	(As)	0.005
Barium	(Ba)	0.001
Beryllium	(Be)	0.003
Bismuth	(Bi)	0.001
Boron	(B)	0.05
Cadmium	(Cd)	0.003
Calcium	(Ca)	0.2
Cerium	(Ce)	0.001
Cesium	(Cs)	0.001
Chromium	(Cr)	0.004
Cobalt	(Co)	0.001
Copper	(Cu)	0.003
Dysprosium	(Dy)	0.001
Erbium	(Er)	0.001
Europium	(Eu)	0.001
Gadolinium	(Gd)	0.001
Gallium	(Ga)	0.001
Germanium	(Ge)	0.002
Gold	(Au)	0.003
Hafnium	(Hf)	0.006
Holmium	(Ho)	0.001
Indium	(In)	0.001
Iridium	(Ir)	0.002
Iron	(Fe)	0.02
Lanthanum	(La)	0.001
Lead	(Pb)	0.003
Lithium	(Li)	0.002
Lutetium	(Lu)	0.001
Magnesium	(Mg)	0.002
Manganese	(Mn)	0.002
Mercury	(Hg)	0.02
Molybdenum	(Mo)	0.004
Neodymium	(Nd)	0.001
Nickel	(Ni)	0.004
Niobium	(Nb)	0.001
Osmium	(Os)	0.002
Palladium	(Pd)	0.002
Platinum	(Pt)	0.009
Potassium	(K)	0.1
Praseodymium	(Pr)	0.001
Rhenium	(Re)	0.003
Rhodium	(Rh)	0.001
Rubidium	(Rb)	0.001
Ruthenium	(Ru)	0.002
Samarium	(Sm)	0.002
Scandium	(Sc)	0.01
Selenium	(Se)	0.02
Silicon	(Si)	0.5
Silver	(Ag)	0.001
Sodium	(Na)	0.007
Strontium	(Sr)	0.001
Tantalum	(Ta)	0.004
Tellurium	(Te)	0.005
Terbium	(Tb)	0.001
Thallium	(Tl)	0.006
Thorium	(Th)	0.003
Thulium	(Tm)	0.001
Tin	(Sn)	0.005
Titanium	(Ti)	0.002
Tungsten	(W)	0.005
Uranium	(U)	0.002
Vanadium	(V)	0.003
Ytterbium	(Yb)	0.001
Yttrium	(Y)	0.001
Zinc	(Zn)	0.005
Zirconium	(Zr)	0.005

10-97

TABLE 2. Detection Limits and Recoveries in 49% HF by ICP-MS.

Element		DL ppb	% Recovery
Aluminum	(Al)	0.01	114
Antimony	(Sb)	0.01	99
Barium	(Ba)	0.005	109
Beryllium	(Be)	0.01	102
Boron	(B)	0.2	82
Cadmium	(Cd)	0.01	90
Calcium	(Ca)	0.3	118
Chromium	(Cr)	0.01	89
Cobalt	(Co)	0.005	88
Copper	(Cu)	0.01	85
Gallium	(Ga)	0.005	94
Germanium	(Ge)	0.01	90
Iron	(Fe)	0.3	91
Lead	(Pb)	0.01	114
Lithium	(Li)	0.005	104
Magnesium	(Mg)	0.01	117
Manganese	(Mn)	0.01	94
Molybdenum	(Mo)	0.01	91
Nickel	(Ni)	0.01	91
Niobium	(Nb)	0.05	93
Potassium	(K)	0.3	112
Silver	(Ag)	0.01	107
Sodium	(Na)	0.01	98
Strontium	(Sr)	0.005	105
Tantalum	(Ta)	0.05	84
Thallium	(Tl)	0.05	97
Tin	(Sn)	0.01	96
Vanadium	(V)	0.01	88
Zinc	(Zn)	0.05	80
Zirconium	(Zr)	0.01	99

By using a variety of sample preparation procedures, chemicals of all types can be analyzed for metal content. **Table 3** contains a list of chemicals used in the semiconductor industry that are routinely analyzed. Furthermore, these chemicals can be measured to determine other aspects of their composition, such as concentration, anions and cations, organic contamination, (TOC in aqueous chemicals or specific compounds in organic solvents), moisture in organic materials and other specific tests to characterize the chemicals.

TABLE 3. List Of Chemicals Analyzed By Wet Chemical Procedures

Chemical	
1-Methyl-2-Pyrrolidone	2-Propanol
Acetic Acid	Acetone
Ammonium Hydroxide	Hexamethyldisilazane
Hydrochloric Acid	Hydrofluoric Acid
Hydrogen Peroxide	Methanol
Methyl Ethyl Ketone	Methyl Isobutyl Ketone
Mixed Acid Etchants	n-Butyl Acetate
Nitric Acid	SC1 Cleaning Solution
SC2 Cleaning Solution	Sulfuric Acid
TMAH	Ammonium Fluoride Solution
Buffered Oxide Etchants	Resist Strippers
Negative Photoresists	Positive Photoresists
Polyimide Solutions	Polyimide Resins
Photosensitizers	Photoresist Resins
Ethylene Glycol	Polymeric Materials
Polypropylene	Ployethylene
Polyfluorocarbons	Epoxy Resins
Non-Ionic Surfactants	Spin-on-Glass
Spin-on-Boron	Spin-on-Phosphorus
Tetraethyl Orthosilicate	Phosphorus Oxychloride
Germaniium Tetrachloride	Trimethylphosphite

Of these, TOC in etching and cleaning chemicals is frequently overlooked as a contaminant that could be harmful in cleaning chemicals. Generally they have not been observed to cause serious problems. However, considering that organic contaminants at ppmw levels are usually found in chemicals, they may be more deleterious in future processing (see **Table 4**). **Figures 1a and 1b** show the content of organic contaminants in hydrogen peroxide (H_2O_2). **Figure 2** shows the variation in concentration of TOC in sulfuric acid by vendor, lot and container type. Although these materials are oxidized under extreme conditions they could still cause problems under normal use, for example forming SiC on wafers.

TABLE 4. Summary of Current Detection Limits and Analytical Spike Recoveries for TOC in Selected Chemicals.

Chemical		**Average Recovery****	**TOC Detection Limits ppmw**
50-100%	Sulfuric Acid	93%	0.2*
85%	Phosphoric Acid	97%	0.1*
25-51%	Hydrofluoric Acid	97%	0.1*
35-71%	Nitric Acid	99%	0.5*
1-40%	Hydrogen Peroxide	103%	0.1
	Ammonium Hydroxide		
28-31%	NH_3	99%	0.2*
50%	Sodium Hydroxide	96%	0.4*
45%	Potassium Hydroxide	102%	0.4*
40%	Ammonium Fluoride	98%	0.1
BOE, Buffered Oxide Etchant		99%	0.1 - 0.2*

* Lower detection limits may be possible for more dilute solutions.

** Spike recoveries are for an organic standard (morpholine, C_4H_9NO) added after sample preparation. Calibration was performed using the same standard in deionized water.

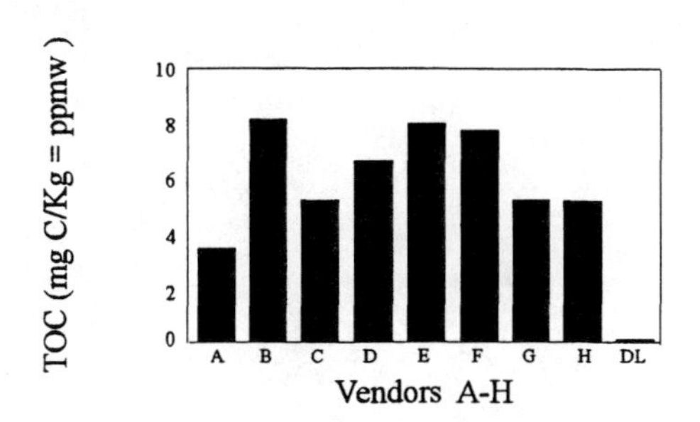

FIGURE 1a. TOC in 30% Hydrogen Peroxide.

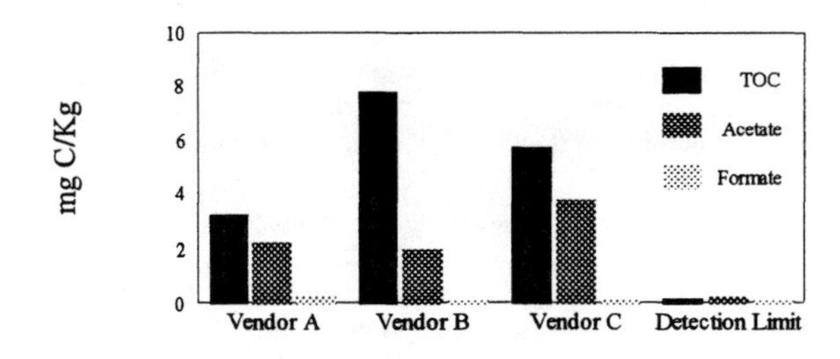

FIGURE 1b. TOC, Acetate And Formate Found In H_2O_2 After Catalytic Decomposition.

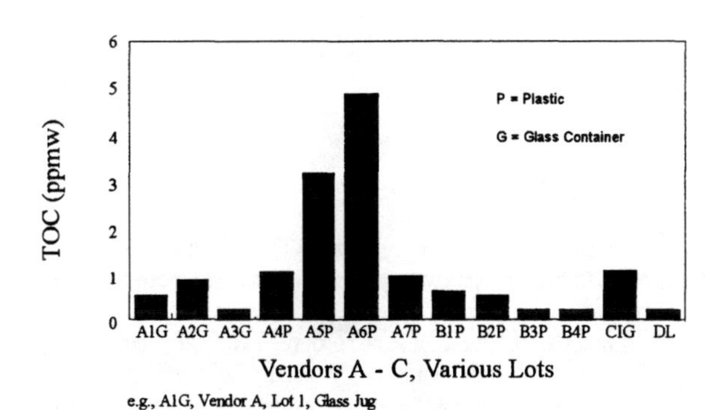

FIGURE 2. TOC in Concentrated H_2SO_4.

The most serious aspect in analyzing UPW and chemicals is preventing contamination while preparing the sample for measurement particular in the case of metals at the ppb, ppt and now sub-ppt levels. To reach the low levels sample collecting, vessel cleaning and concentration become extraordinarily important. While cleanroom conditions are adequate for sample collecting and vessel cleaning, they are not for sample concentration or measurement. Special equipment such as closed evaporators and automatic sampling units for ion coupled plasma-Mass Spectometer (ICP-MS) or graphite furnace atomic absorption spectrometer (GFAAS) where samples are in a holding pattern require filtered N_2 environments (see **Figure 3**). Although the final measurement can be a ten minute step, the extraordinary activity to preparing samples for measurement in absolutely clean and safe environments can take hours.

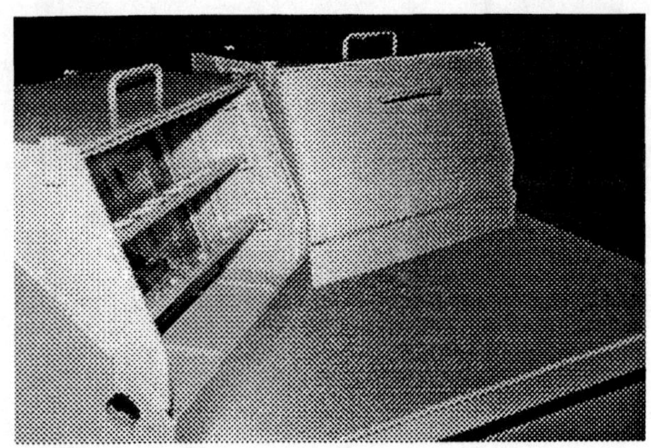

FIGURE 3. Closed Evaporator

Another important wet chemical application for the semiconductor industry is the measurement of metals on bare wafers, or in thin and thick oxides, nitrides or metallic films such as copper. To verify the accuracy of this technique, known quantities of metals from NIST standards are deposited directly on wafers declared clean by having gone through a Vapor Phase Decomposition - Inductively Coupled Plasma-Mass Spectrometry (VPD-ICP-MS) analysis just prior to being used. The solution is placed over the entire wafer, dried in a protective environment and processed (see **Table 5**).

In Table 5, it becomes very clear why recovery studies are so important. As noted, copper did not come off of the wafer using the standard VPD process. To remove copper efficiently another process known as drop scan ICP-MS is used. In this process any mixture of acids, bases, oxidizers or chelating materials can be made to remove specific metals from wafers. Using this procedure, copper can be removed efficiently, with 95% recovery.

Having methods to accurately measure metals in chemicals and in chemical baths, a correlation can now be made between metals in chemicals or UPW and those found to adhere to wafers (see **Table 6**). In this table two grades of hydrogen peroxide were evaluated after one proved to cause yield loss while the other did not. Although both chemicals were pure, especially considering the process where they were being used, the enhanced metals plating onto wafers in SC1 created a yield loss situation even though an SC2 clean followed. It has been observed through other failures involving metals on wafers that iron quantities over $15x10^{10}$ atoms/cm^2 generally cause failures.

Moving away from liquid and wafers into other areas, one where wet chemical analysis has proved to be extraordinarily useful is the measurement of cleanroom molecular contamination. A serious problem in production is hazy wafers. Although there are a wide variety of reasons for hazy wafers to occur such as particles, roughness, adsorbed organics or metal oxides such as Fe_4O_3, one cause is salt formations. These salts deposit on wafers directly when vapors such as HCl, HF or SO_3 react with NH_3 or organic amines. By scrubbing air through UPW or dilute chemical solutions, these contaminants known as "molecular contaminants" can be measured. Whether metallic, processing fumes, outgassed or leachable materials, these agents can be identified and usually related to both their source and effect on the wafer.

TABLE 5. Recovery Study of Selected Elements by Vapor Phase Decomposition - Inductively Coupled Plasma Mass Spectrometry (VPD-ICP-MS).

Wafer No.	Spike [ng]	Recovered [ng]	Blank [ng]	Percent Recovery
Aluminum				
1	1.26	1.02	0.076	89
2	2.35	2.30	0.421	90
3	4.47	4.36	0.038	94
			AV = 0.178	
Chromium				
1	0.654	0.644	<0.02	99
2	1.22	1.10	<0.02	90
3	2.32	2.28	<0.02	98
			AV = 0.02	
Copper (*)				
1	0.541	0.024	<0.01	4.4
2	1.01	<0.02	<0.01	<2.0
3	1.92	0.159	<0.01	8.3
			AV = 0.01	
Nickel				
1	0.567	0.607	<0.01	107
2	1.06	0.978	0.02	92
3	2.01	2.01	<0.01	100
			AV = 0.01	
Iron by GFAAS				
1	1.08	1.20	-	103
2	2.21	2.58	0.06	113
3	2.95	3.29	0.11	109
			AV = 0.085	
Sodium				
1	1.79	2.12	0.190	87
2	3.34	3.79	0.857	96
3	6.35	6.77	0.660	98
			AV = 0.569	

Remark (*): Another drop scanning method is used to get quantitative recovery for copper.

TABLE 6. Comparison of Trace Metal Concentration in Hydrogen Peroxide Solutions Used in SC1 Clean and on 6" bare Wafers

	Concentration (ppb) Hydrogen Peroxide		Surface Concentraion ($x10^{10}$ atoms/cm^2)	
Element	**Grade A**	**Grade B**	**Wafer A**	**Wafer B**
Aluminum (Al)	22	2.6	580	92
Chromium (Cr)	3.3	0.1	<1	<1
Iron (Fe)	6.7	<2	82	<5
Nickel (Ni)	2.6	<0.1	1.2	0.9
Sodium (Na)	10	2.3	12	14
Total	45	7	>600	>100

A company with just such a haze problem solicited help to find its cause and remove it. A wafer was evaluated by leaching it with UPW and running ion chromatography (IC) where results like those shown in **Table 7** are given. A wafer was also evaluated by VPD-ICP-MS. Finding that salt formation was the cause of the haze, air samples were run. Significant quantities of acid and base fumes were detected but no significant amount of metals.

TABLE 7. Surface concentration of Anions and Cations on 150mm Wafers by IC. (surface area: 364 cm^2, volume: 200 mL)

Ion	DL ions/cm^2	A ions/cm^2	B ions/cm^2	C ions/cm^2	Typical Range ions/cm^2
Anions					
Fluoride	1.7 E+12	*	*	*	<DL - 5 E+13
Chloride	1.9 E+11	1.7 E+12	2.0 E+12	3.3 E+12	<DL - 5 E+13
Nitrite	1.4 E+11	1.1 E+12	9.3 E+11	1.5 E+12	<DL - 1 E+13
Bromide	8.3 E+11	*	*	*	<DL
Nitrate	1.1 E+11	5.9 E+11	1.1 E+12	1.4 E+12	<DL - 1 E+13
Phosphate	7.0 E+11	*	*	*	<DL
Sulfate	1.7 E+11	4.8 E+11	1.8 E+12	7.9 E+11	<DL - 2 E+13
Cations					
Lithium	4.8 E+11	*	*	*	<DL
Sodium	1.4 E+11	1.9 E+12	1.3 E+13	3.2 E+12	<DL - 1 E+13
Ammonium	9.2 E+11	5.3 E+12	8.3 E+12	1.0 E+13	1E+12 - 2E+13
Potassium	1.7 E+11	*	*	1.2 E+12	<DL - 4 E+12
Magnesium	2.7 E+12	*	*	*	<DL - 2 E+12
Calcium	1.6 E+11	*	*	1.6 E+12	<DL - 4E+12

Changes in air flowrate, fresh air addition to cleanroom air and ventilation of the cleaning stations were made while air sampling continued. **Table 8** shows the improvement and the concentration of fumes finally obtained (2/94 → 5/94). These concentrations completely allowed for processing without haze formation. However, it must be stated, not seeing salts as haze on wafers does not mean that none are there. In routine evaluation of anions and cations on wafers after cleaning, we find small quantities of Cl^-, $SO_4^=$ and NO_x^-. The typical ranges that have been measured in a cleanroom of class 10 or better and regardless of the size of the IC manufacturer are shown in **Table 9**. Amine detection and identification are shown in **Figure 4**. **Figure 5** shows the Balazs Analytical Laboratory Air Sampler used in this study.

Another significant wet chemical application has been the measurement of metals in thin and thick films. A study of oxide films is shown in **Table 10.** Thermal, deposited and TEOS wafers were analyzed from several different sources. As is seen, the concentrations can be significant. Even though the situation has improved, metals in oxides, nitrides and other films are still quite high in films from various reactors today.

TABLE 8. Non-metallic Contamination at Various Sites in A Class 100 Cleanroom.

Element	Class 100 Cleanroom					
	Location 1		Location 2		Location 3	
ug/m^3	2/94	5/94	2/94	5/94	2/94	5/94
Ammonium	61	7.9	12	4.4	3.0	2.9
Fluoride	28	1.4	10	1.1	2.9	1.6
Chloride	0.14	<0.01	0.045	<0.01	0.14	0.16
Nitrite	33	3.3	9.0	1.4	6.9	3.7
Phosphate	<0.02	<0.02	<0.02	<0.02	<0.02	<0.02
Bromide	<0.02	<0.02	<0.02	<0.02	<0.02	<0.02
Nitrate	3.1	0.67	0.94	0.40	1.6	0.82
Sulfate	0.14	0.04	0.16	<0.02	0.51	0.25

TABLE 9. Cleanroom Air Analysis: Acids and Bases. Analysis using water scrubbers then IC.

Anhydrides:		Typical Range (ng/m^3)	pptv*
Ammonium	(NH_4^+)	1,000 - 10,000	1,400 - 14,000
Bromide	(Br^-)	<30	<9
Chloride	(Cl^-)	100-1,500	70 - 1,000
Fluoride	(F^-)	<600	<800
Nitrate	(NO_3^-)	500 - 5,000	200 - 2,000
Nitrite	(NO_2^-)	1,000 - 10,000	530 - 5,300
Phosphate	($HPO_4^=$)	<20	<5
Sulfate	($SO_4^=$)	<50 - 500	<13 - 130

* pptv = parts-per-trillion-volume

TABLE 10. Typical Concentration of Trace Metals In Thermal, TEOS and Grown or Deposited Oxide Layers From Different Sources. Surface Concentration in 10^{10} atoms/cm^2.

Sources		Thermal A	B	C	D	TEOS E	F	Deposited G	H	I
Aluminum	(Al)	440	180	2400	5400	20	3000	95	70	<25
Sodium	(Na)	-	-	-	-	<50	450	<30	<20	<20
Chromium	(Cr)	94	42	550	100	<1	-	<2	<0.7	<0.7
Iron	(Fe)	340	330	900	440	<10	150	40	<7	<7
Nickel	(Ni)	160	110	320	80	<1	-	<1	<0.7	1.4
Copper	(Cu)	160	150	20	-	-	-	-	-	-
Magnesium	(Mg)	-	-	-	-	-	30	-	-	-
Zinc	(Zn)	-	-	-	-	-	30	<5	<3	<3
Zirconium	(Zn)	-	-	-	-	-	2.5	-	-	-

1991 Study

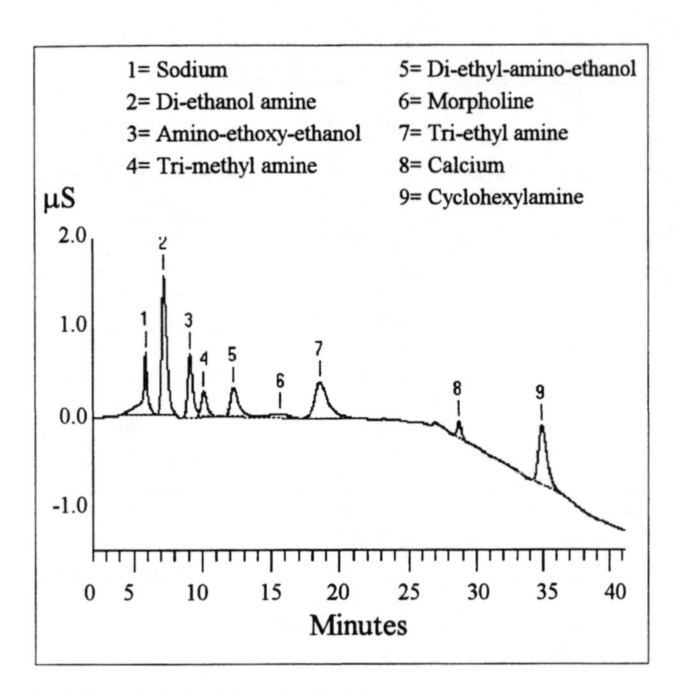

FIGURE 4. Amine Detection by IC.

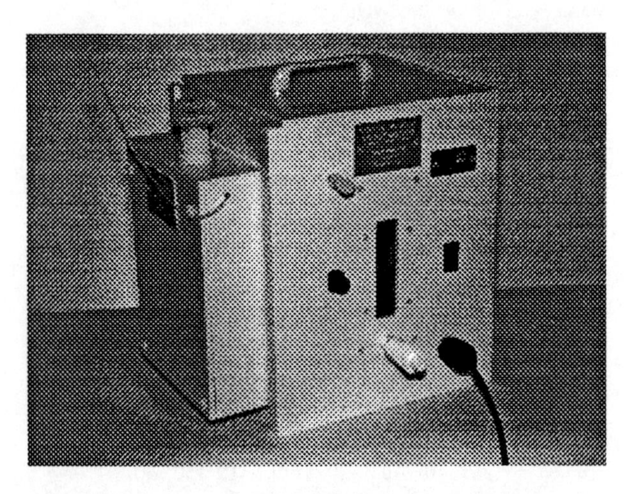

FIGURE 5. Balazs Analytical Laboratory Air Sampler

Chemicals, wet stations, reactors and air are not the only sources of ionic and metallic contamination in a cleanroom. Components of all kinds contain these contaminants which leach out into the air or directly onto wafers. **Table 11** lists both construction materials and components measured. To date, over 300 materials have been studied that go into cleanrooms, wet benches, reactors and construction materials of all kinds for UPW, chemical distribution, cleanroom construction, reactors and processing components. Numerous suppliers have requested studies for ions, metal, TOC, particle, deterioration or stability of materials and components in both liquid and vapor atmospheres. **Table 12** shows typical leachable ionic contaminants from cleanroom wipes and garments. **Table 13** shows the quantity of anions found when three different wafer carriers were rinsed with UPW.

TABLE 11. Materials

Wafers	Paints
O-Rings	Floor Tiles
Gloves	Tubing
Finger Cots	Potting Compounds
Metals	Motor Parts
Filters	Pumps
Gaskets	Safety Glasses
Hats	Polymers of all Kinds
Plastics	Carriers
Paper	Wafer Boxes
Valves	Cassettes
Pipes	Packing Materials
Garments	Photoresist
Resins	Cleanroom Construction Materials

TABLE 12. Extractable Ionics from Cleanroom Components

	Cleanroom Wipes		Cleanroom Garments	
Ion	**DL**	**ng/cm^2**	**DL**	**ng/cm^2**
Anions				
Fluoride	0.6	*	10.	*
Chloride	0.07	3.9	1.3	315
Nitrite	0.07	0.10	1.3	2
Phosphate	0.15	*	2.5	*
Bromide	0.15	*	2.5	8
Nitrate	0.15	0.77	2.5	2
Sulfate	0.15	0.54	2.5	1600
Cations				
Lithium	0.07	*	1.3	*
Sodium	0.07	1.7	1.3	844
Ammonium	0.15	2.4	2.5	*
Potassium	0.15	2.3	2.5	235
Magnesium	0.15	0.65	2.5	224
Calcium	0.07	2.4	1.3	2786

TABLE 13. Surface Concentration of Anions on Wafer Carrier by IC (surface area: 3600 cm^2, volume: 500 mL, leaching temp.: ambient)

Ion	DL ions/cm^2	A ions/cm^2	B ions/cm^2	C ions/cm^2	Typical Range ions/cm^2
Anions					
Fluoride	8.8 E+12	*	*	*	<DL
Chloride	4.7 E+10	6.1 E+11	4.0 E+11	4.5 E+11	<DL - 6 E+11
Nitrite	3.6 E+10	4.0 E+11	2.9 E+11	2.4 E+11	<DL - 5 E+11
Bromide	2.1 E+10	*	*	*	<DL
Nitrate	2.7 E+10	1.3 E+11	1.2 E+11	8.1 E+12	<DL - 3 E+11
Phosphate	1.8 E+10	*	*	*	<DL
Sulfate	4.4 E+10	1.1 E+11	6.1 E+10	*	<DL - 3 E+11

Contaminants that can cause n or p type doping are becoming an increasing problem in cleanrooms and on wafers. These materials can come from low level borates in UPW. However, UPW has generally not been the source of dopants that cause yield problems. The source is generally from the air or wafer containers. Wet chemical processes often can measure the boron (B) and phosphorus (P) but their sources are usually identified by using organic adsorbers and GC-MS analytical measuring procedures, because the dopants generally came from organic materials. **Tables 14** and **15** show the quantities of boron (B) and phosphorus found in various materials and from reactors.

TABLE 14. B and P in Various Components and Environments ng/cm^2.

Component or Environment	B	DL	P	DL
Garment	0.05	0.005	0.57	0.02
	0.09		2.0	
Plastic Container (clear)	0.39	0.02	0.34	0.02
	0.01			
(colored)	19.0			
Tape (Adhesive)	1.8	0.02	69.	0.03
Gasket (Rubber)	0.31	0.01	51	0.4
Wafer Box	35.		12.	
Mini Environment	0.04	0.02	ND	
Inspection Room	0.06		ND	
HEPA	0.06		ND	

TABLE 15. B and P on Bare Wafers 10^{10} atoms/cm^2

Ranges on Bare Wafers*	B	P
	39-1000	12-290
Reactor 1	4-90	12-19
Reactor 2	200-350	60-210
Reactor 3	20-32	ND-8

* From different reactors and wafer position in the reactor.

A serious case of n type contamination recently occurred in a fab where someone accidentally hit a water pipe with considerable force causing it to break. Water sprayed in high quantities across the fab. The fab was wiped down using large quantities of cleanroom wipes. After the incident n type doping contamination occurred uncontrollably. A study of air, surfaces, wipes and witness wafers identified the compound (an organophosphate), and its source (the cleanroom wipes). The fab had to be thoroughly wiped down and cleaned and considerable air exchanged before the problem disappeared.

Although not the subject of this paper, it is worth mentioning that the combination of wet chemistry, organics molecular measurements and contaminants found on wafers tied together very well for identifying contaminating materials and their sources. HEPA, ULPA, potting compounds, components of all kinds and wet benches fill the "cleanroom" with huge quantities of contaminants deleterious to wafers and product yield.

A novel application for measuring metals in thick films is to determine those that are implanted into a substrate during the doping process using an ion implanter. Since the different concentration profiles for implanting elements in a substrate during ion implantation is nearly the same as in an oxide film (see **Figure 6**), a wafer with about 2000 Å of oxide on it can become a test wafer. The ion flux implants the dopant and contamination metals into the oxide layer (see **Figure 7**). This layer is then analyzed by drop scan ICP-MS.

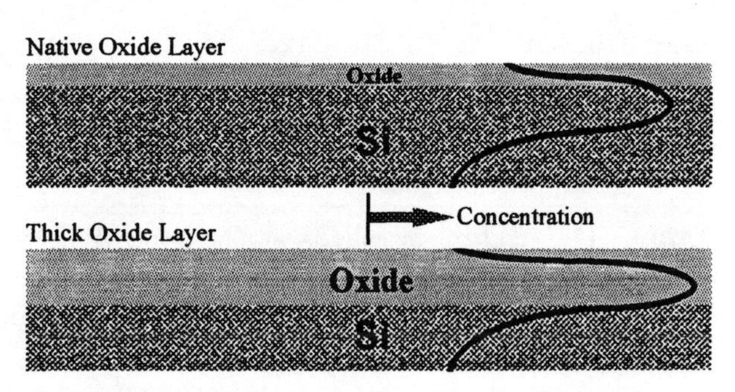

FIGURE 6. Concentration Profiles After Ion Implantation in Silicon and Silicon Dioxides.

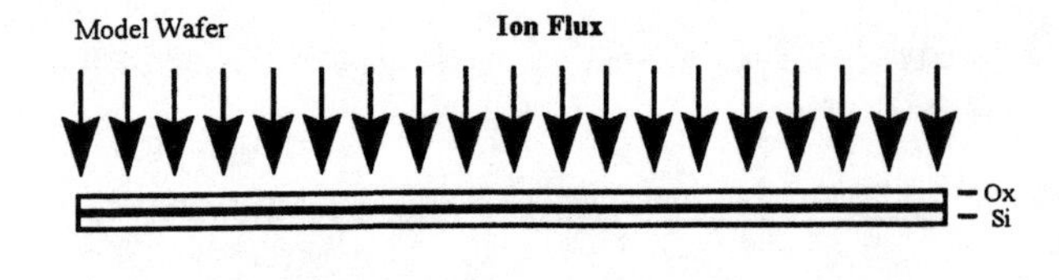

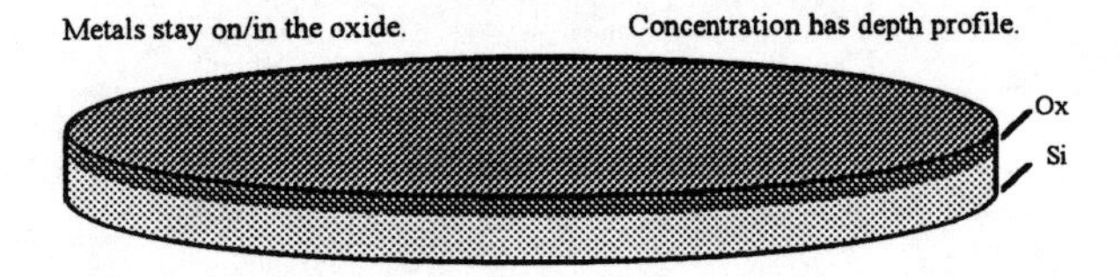

FIGURE 7. Measuring Trace Metals on Wafers from Ion Implantation Equipment.

The metals on the surface are removed and analyzed before the metals in the oxide are measured. A control wafer must be run to determine the metals in the oxide film which has not been exposed to the ion inplanter. **Table 16** reveals a fairly dirty ion implantation process. With this information, however, the engineers were able to greatly improve the situation.

Having measured numerous wafers for metallic contamination from ion implanters, the example above was not very unusual. The range for both surface metallic contamination and implanted metallic contamination is shown in **Table 17a and 17b.**

TABLE 16. High Energy Metallic Contamination (of Oxide) from an Ion Implantor. Trace Metals in Oxide Film on 150 mm Silicon Wafers 10^{10} atoms/cm^2.

Element		Detection Limit	Sample #1	#2	Control	Implanted Metal
Calcium	(Ca)	1.0	42	39	24	17
Potassium	(K)	0.2	18	12	8.5	7
Sodium	(Na)	0.5	22	14	16	2
Aluminum	(Al)	0.5	290	280	18	267
Iron	(Fe)	0.5	34	35	9.0	26
Chromium	(Cr)	0.5	4.7	3.0	0.8	3
Nickel	(Ni)	0.4	9.0	11	3.6	7
Zinc	(Zn)	0.2	18	17	19	0
Lithium	(Li)	0.1	*	*	*	*
Beryllium	(Be)	0.3	*	*	*	*
Magnesium	(Mg)	0.6	18	12	5.5	14
Titanium	(Ti)	0.8	236	45	*	141
Tungsten	(W)	0.01	35	27	*	31

* None detected above the detection limit.

TABLE 17a. Surface Metallic Contamination After Ion Implantation

Element	Concentration Range (1x10^{10} atoms/cm^2)
Aluminum	100 - 10,000
Iron	10 -100
Chromium	1 - 10
Nickel	1 - 10
Titanium	10 - 1,000

TABLE 17b. Range of Metal Contaminants Implanted Into the Substrate.

Element	Concentration Range (1x10^{10} atoms/cm^2)
Aluminum	10 - 10,000
Iron	10 -10,00
Chromium	10 - 5,000
Nickel	10 - 5,000
Titanium	0.1 - 50

The accurate measurement of thin film composition is made easy using wet chemical techniques. The film is simply stripped off of the wafer and the elements of interest measured by colorimetry, ICP-MS, GFAAS or inductively coupled plasma - optical emission spectometer (ICP-OES). PSG and BPSG are measured using colorimetry or ICP-OES. By getting absolute data traceable to NIST standards, an engineer can standardize secondary equipment such as FTIR or x-ray. It is important to understand, however, that the four methods mentioned (colorimetry, FTIR, ICP-OES, and x-ray) do not measure the same thing and thus often yield different answers. For example ICP-OES and x-ray measure only the total amount of phosphorus while FTIR measures the P-O and P-OH bonds and colorimetry measures only PO_4^{-3}. By changing the chemistry of the solution, however, colorimetry and only colorimetry can determine the concentration of P_2O_5 vs. P_2O_3 in the oxide film (see **Table 18**). Also, with colorimetry other compounds of phosphorus have been measured and the discovery made of occluded PH_3 from plasma reactors and the concentrations of this compound measured (see **Table 19**).

Table 18. Concentration on P_2O_5 vs. P_2O_3 in the Oxide Film

Sample	X-Ray %P	Colorimety Total P	P_2O_3	P_2O_5
1AD	7.6	7.5	2.7	4.8
2AD	5.0	4.9	1.5	3.4
3AD	2.2	1.8	0.5	1.3
1AN	8.0	5.9	1.1	4.8
2AN	5.1	4.3	0.4	3.9
3AN	2.3	1.7	<0.2	1.7

AD = as deposited, AN = annealed 800 °C in N_2

TABLE 19. Weight Percent Phosphorus in Plasma Doped Oxides.

No.	Total	P_2O_3	P_2O_5	PH_3
A-1	9.5	6.2 (65.3)*	2.9 (30.5)	0.4 (4.2)
A-2	5.6	4.0 (71.4)	1.1 (19.6)	0.5 (9.0)
A-3	5.2	3.9 (75.0)	0.9 (17.3)	0.4 (7.7)
A-4	5.0	4.1 (82.0)	0.4 (8.0)	0.5 (10.0)
A-5	5.4	3.9 (72.2)	1.0 (18.5)	0.5 (9.3)
B-1	4.1	3.3 (80.5)	0.6 (14.6)	0.2 (4.9)
B-2	4.1	3.4 (82.9)	0.5 (12.2)	0.2 (4.9)
C-1	5.0	3.8 (76.0)	1.1 (22.0)	<0.1 (2.0)
C-2	2.3	1.9 (82.6)	0.4 (17.4)	<0.05 (<1.0)
C-3	7.5	6.2 (82.7)	1.3 (17.3)	<0.05(<1.0)

* Numbers in parentheses are relative percents of weight percent phosphorus.

Today with the cost of 300 mm wafers, a change of the wet procedure has made it possible to measure phosphorus and boron in PSG and BPSG and return the intact wafer to the client for re-use. **Tables 20 and 21** compare the old method with the new and is a comparison of data using two completely different techniques. It is this kind of adaptability that illustrates the uniqueness, breadth and value of wet chemical testing.

Other measurements of thin films include Cu or Si in aluminum metalization, titanium/tungsten or silicide ratioing and metals in copper films (see **Table 22**). New methods are continuously being developed to deal with new films being used by the industry.

Studies such as the number of etches required to clean a quartz tool (see **Figure 7**), leachable metals and TOC from FRPP (see **Table 23**), the amount of fluoride contributed to hot UPW or peroxide solutions that go through fluoropolymer tubing, or a dynamic study of 100 foot of new tubing (see **Table 24**), illustrate the diversity of wet chemical analysis.

TABLE 20. Results for %P and %B in BPSG film on a 300 mm Wafer.

Method	%P (wt/wr) Colorimetry	%B (wt/wt) ICP-OES
Standard	4.85	3.36
New*	4.82	3.37

* Non-destructive analysis

TABLE 21. Non-destructive Analysis of 300 mm Wafers for %P and %B.

Wafer Sample	Colorimetry	ICP-OES
# 1	0.00	0.00
# 2	1.90	2.00
# 3	2.68	2.69
# 4	3.77	3.81
# 5	4.36	4.39
# 6	5.72	5.71
# 7	7.20	7.33
# 8	8.72	8.78

Table 22. Trace Impurities in Copper Films (ng/g).

Element		8" Wafer	Spike Recovery
Aluminum	(Al)	4	98
Arsenic	(As)	10	84
Antimony	(Sb)	3	95
Barium	(Ba)	1	94
Beryllium	(Be)	4	90
Boron	(B)	70	90
Cadmium	(Cd)	4	96
Calcium	(Ca)	300	95
Cerium	(Ce)	1	84
Chromium	(Cr)	5	93
Cobalt	(Co)	1	88
Gallium	(Ga)	3	88
Germanium	(Ge)	4	84
Lead	(Pb)	4	104
Lithium	(Li)	3	88
Iron	(Fe)	30	102
Magnesium	(Mg)	3	86
Manganese	(Mn)	3	92
Mercury	(Hg)	30	98
Molybdenum	(Mo)	6	81
Nickel	(Ni)	5	89
Potassium	(K)	150	113
Phosphorus	(P)	50	82
Silicon	(Si)	.5	
Sodium	(Na)	10	94
Strontium	(Sr)	1	114
Tin	(Sn)	10	94
Tungsten	(W)	10	89
Vanadium	(V)	4	94
Zinc	(Zn)	10	95
Zirconium	(Zr)	10	85

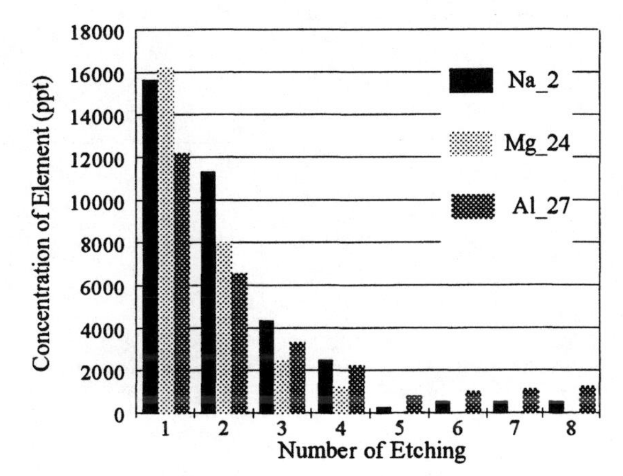

FIGURE 7. 0.5% HF Etching of Quartz Vessel #3.

TABLE 23. Leachable Trace Metals by ICP-MS.

Element		DL ppb (ng/l)	CP5 V-O Only FRPP
Aluminum	(Al)	0.05	0.15
Antimony	(Sb)	0.02	1.4
Arsenic	(As)	0.2	0.6
Barium	(Ba)	0.01	2.8
Boron	(B)	0.05	*
Calcium	(Ca)	0.03	*
Chromium	(Cr)	0.03	*
Cobalt	(Co)	0.02	*
Copper	(Cu)	0.05	*
Iron	(Fe)	0.1	0.2
Lead	(Pb)	0.05	*
Lithium	(Li)	0.03	*
Magnesium	(Mg)	0.02	0.09
Nickel	(Ni)	0.05	*

TABLE 24. Dynamic Rinse Study of 100 Foot Tubing.

Resistivity (megohm-cm)		TOC (ppb)		Particle Count (particles/ml)				
Minutes	**Filtrate**	**Minutes**	**Filtrate**	**Minutes**	**0.05µ**	**0.10µ**	**0.5µ**	**1.0µ**
				0	58.8	18.1	6.7	3.1
0	11.7	6	1.02	10	9.6	3.6	2.7	1.0
4	17.0	14	<0.50	20	14.3	3.9	1.4	0.0
6	17.6	22	<0.50	30	6.2	0.0	1.7	0.0
7	17.7	30	<0.50	40	6.9	1.9	3.1	1.7
8	17.8	46	<0.50	50	5.5	2.2	1.7	1.4
9	17.9	54	<0.50	60	3.2	0.0	0.0	0.0
10	18.0	69	<0.50	70	4.1	0.0	0.0	0.0
		77	<0.50	80	0.0	0.0	0.0	0.0
		93	<0.50	90	1.7	0.0	0.0	0.0
		100	<0.50	100	1.9	0.0	0.0	0.0
		123	<0.50	110	3.6	1.3	1.0	0.0
		HOURS		120	0.0	0.0	1.0	0.0
		3	<0.50	HOURS				
		4	<0.50	3	1.6	1.3	1.0	0.0
		5	<0.50	4	0.0	0.0	0.0	0.0
		6	<0.50	5	0.0	0.0	0.0	0.0
		7	<0.50	6	0.0	0.0	0.0	0.0
		8	<0.50	7	0.0	0.0	0.0	0.0
		9	<0.50	8	0.0	0.0	0.0	0.0
				9	0.0	0.0	0.0	0.0

	0 MIN	10 MIN	30 MIN	60 MIN	2 HRS	END
Fluoride	<0.05	<0.05	<0.05	<0.05	<0.05	<0.05
Chloride	0.08	0.06	<0.05	<0.05	<0.05	<0.05
Bromide	<0.05	<0.05	<0.05	<0.05	<0.05	<0.05
Nitrate	<0.05	<0.05	<0.05	<0.05	<0.05	<0.05
Phosphate	<0.05	<0.05	<0.05	<0.05	<0.05	<0.05
Sulfate	<0.05	<0.05	<0.05	<0.05	<0.05	<0.05
Lithium	<0.05	<0.05	<0.05	<0.05	<0.05	<0.05
Sodium	0.20	<0.05	0.10	0.09	<0.05	<0.05
Potassium	<0.05	<0.05	<0.05	<0.05	<0.05	<0.05
Silica	<3	<3	<3	<3	<3	<3
Calcium	<0.02	<0.02	<0.02	<0.02	<0.02	<0.02
Iron	<0.02	<0.02	<0.02	<0.02	<0.02	<0.02

No metals were detected above the DL's.

CONCLUSION

It is not possible to discuss the myriad of variations of wet chemical procedures that are requested by semiconductor engineers and suppliers. The applications discussed here, however, should illustrate the important role these procedures play for quality control and contamination-free manufacturing. In the future, there will be an even greater need to routinely obtain data from wet chemical procedures as specifications of materials and processes tighten.

MATERIALS CHARACTERIZATION - DOPANT PROFILING

Re-Examination of 2D Dopant Profiling Needs

Michael Duane

Advanced Micro Devices, Austin, Texas, 78741

The National Technology Roadmap for Semiconductors (NTRS) has established some stringent requirements on two-dimensional dopant profiling. This work explores the reasons behind these metrology requirements, and also presents the difficulty of achieving these goals from the metrologist's point of view. The goal of this paper is to present a more complete description of the metrology needs, including such issues as dopant statistics, surface profiling, active versus total concentrations, and physical regions of interest in a transistor in order to establish better metrology targets.

INTRODUCTION

Two-Dimensional (2D) dopant profiling has been a highly ranked need in both the process integration and TCAD (simulation) sections of National Technology Roadmap for Semiconductors (NTRS) since its inception. Although some progress has been made in addressing these needs, we are still far from the goals, and the targets continue to shrink. This lack of dopant metrology has *not* affected the rate of technology development. In light of this, the rationale behind the targets is re-examined so that compromises between the desired metrology and what is likely to be achieved can be explored.

METROLOGY TARGETS

The 1997 Roadmap lists the dopant spatial resolution requirements as 5 nm for 250 nm technology and 1 nm for the 70 nm generation (1). The entire row in the table is colored red, meaning *no known solutions*. For most generations, this target is about 1.5% of the transistor gate length (Lgate). This is a reasonable target considering the desired Lgate control of 3-sigma of <10% (2). This equates to 10% control in drain current. This 10% 3-sigma target implies 1-sigma control of roughly 3% in Lgate. Since a MOSFET is a symmetrical device, any movement in the lateral junction location is magnified by a factor of two when determining the source to drain spacing. This distance is often referred to as the effective channel length, Leff, although rigorously, the electrical channel length is not the same as the source to drain spacing. Comparing the two targets, we see that the desired resolution in the source to drain spacing (i.e., twice the lateral junction resolution) is approximately the same as the 1-sigma requirement on Lgate control. These targets were set by independent groups, but this comparison shows they are reasonably consistent. A simple statement of the need is that Leff is the most important MOSFET parameter, that we want to control this value to <10%, and therefore would like metrology that is significantly less than 10% of Leff.

The dopant *concentration* precision starts off at 5% for current technology and decreases to 2% at 70 nm (1). A footnote mentions the need for accurate reference materials (standards).

ANATOMY OF A JUNCTION

But what do these targets imply when we start considering the physical characteristics of a modern junction? The dopant concentration near the junction is approximately $10^{18}/cm^3$. A simple estimate of the average dopant spacing can be obtained from the reciprocal of the cube-root of the concentration, or 10 nm in this example. Thus, *the desired spatial resolution near the junction is already lower than the average dopant spacing!* There are two explanations for this paradox. One is the fact that the variation in carrier concentration (electrons or holes) is theoretically smoother than the dopant concentration (3). The carrier profile is more important to the device behavior, although the current metrology requirement is for the dopant profile. This distinction will be discussed more later. The second reason is that we typically perform electrical measurements on structures that are relatively wide, thereby averaging out the variations in source to drain spacing across the width of the structure. This latter point is underappreciated. Rather than the textbook picture of smooth dopant profiles with abrupt junctions, we are entering a regime where we need to consider the location of individual atoms, and how their random location and concentration affects devices (4-7).

For higher concentrations, the situation is improved, although the average dopant spacing only decreases by a factor of 2.2 as the concentration increases by 10x (e.g., at 10^{19} atoms/cm^3, the average dopant spacing is 4.6 nm, and decreases further to 2.2 nm at 10^{20} atoms/cm^3). The process technologist can gain a better understanding of the dopant metrologist's dilemma by converting the familiar units of atoms/cm^3 to atoms/um^3 or even atoms/nm^3: 10^{18} atoms/cm^3 = 10^6 atoms/um^3 = 10^{-3} atoms/nm^3. A cube that is 10 nm per side only has one dopant atom in it at 10^{18} atoms/cm^3. *Resolution finer than this requires*

CP449, *Characterization and Metrology for ULSI Technology: 1998 International Conference*
edited by D. G. Seiler, A. C. Diebold, W. M. Bullis, T. J. Shaffner, R. McDonald, and E. J. Walters

measuring fractions of an atom. There is an inherent tradeoff between concentration and resolution which is not reflected in the roadmap. There is some irony in the fact that the doping concentrations are lowest (and therefore the hardest to measure) in the regions of most interest. Similarly, changes made in some measurement methods in order to increase the lateral resolution also decrease the measurement sensitivity. For example, 10 nm SIMS spot size results in minute beam currents. Compounding this, the beam energies must be high, which affects the depth resolution (8).

What happens when we consider doping gradients? Assume a 100 nm junction depth, and the concentration dropping by 3 orders of magnitude in that distance. This is a somewhat conservative estimate, in that actual concentration drop will occur over a distance less than the junction depth. The concentration is changing by a decade every 30 nm. Near the junction, the dopant concentration is changing significantly over the average dopant spacing, as shown in the figure below.

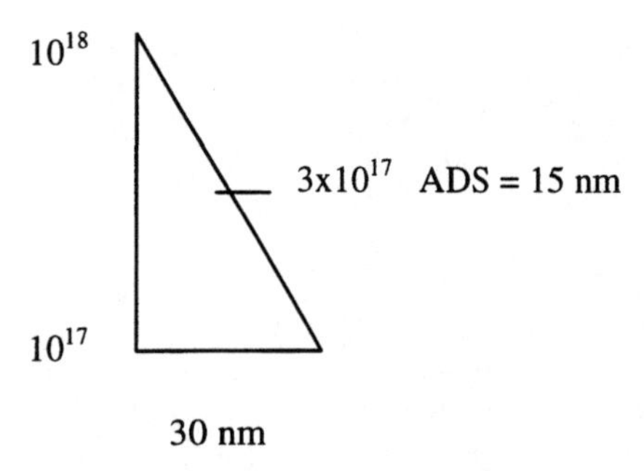

FIGURE 1. Assume the dopant concentration varies by 1 decade every 30 nm. For a background concentration of $3x10^{17}$ atoms/cm^3, the average dopant spacing (ADS) is 15 nm. Thus, *near the junction, the dopant concentration is changing by 3x over the ADS.*

This example leads to a second paradox. A 1D SIMS profile of the S/D dopant will show a smooth and continuous transition between 10^{18} and 10^{17} atoms/cm^3. But at a local level, the average dopant spacing at 10^{18} atoms/cm^3 is 10 nm, and 22 nm at 10^{17} atoms/cm^3. The sum of these two distances is equal to the 30 nm in the figure above. That is, we imagine the dopant as varying smoothly from 10^{18} to 10^{17} atoms/cm^3, yet a simple calculation of the average dopant spacing in this example shows there is no room over this distance for the intermediate concentrations ($4x10^{17}$, $5x10^{17}$, $6x10^{17}$, etc.) to "fit in". A smooth and continuous profile only occurs when we average over a large area, yet a transistor operates at a very local level.

Current devices may already be running into some of these statistical limits. Consider again the average dopant spacing of 10 nm near the junction. The minimum transistor width is often twice the minimum length. For 250 nm technology, this would mean that there would only be 50 dopant atoms along the width of the device near the junction. Several researchers have investigated the fluctuations that will result from the random variation in the number of dopant atoms, although these have been primarily focused on dopants in the channel and how that affects threshold voltage. The source to drain spacing is a key parameter in controlling the transistor, and the stochastic nature of this spacing may turn out to be a more important factor and hit us sooner than the random dopant variation in the channel.

The roadmap alludes to the need for 3D profiling as well, but it should be remembered that current 2D methods are already quasi-3D, in that they probe into the silicon a small, finite depth (10's of nm, depending on the concentration and technique). Most 2D methods require elaborate sample preparation, and it is difficult to imagine extending these techniques to 3D. An alternative approach may be to measure 2D profiles on two or more sides of a box, and to infer the 3D profile. More specifically, the conventional 2D cross-section coupled with a "top-down" view of the device might provide much of the desired information (9). In the top-down view, the poly gate would be removed, and local variations in the source to drain spacings would be measured.

Dopant statistics also has interesting implications for device simulation. Most simulations today are 2D, with a simulation grid that is finer than the average dopant spacing. Again, this is related to assumptions of carriers profiles varying more rapidly than dopant profiles, and of implicit averaging over wide structures. For 3D simulations, especially those statistical in nature, careful attention should be paid between the simulation grid and the average dopant spacing (10).

NEEDS AND REQUIREMENTS

This section will consider further what the dopant metrology needs are, particularly from a motivational viewpoint, and will also discuss important needs that are *not* in the current roadmap. The need for 2D profiling has been present for a long time, and is not new to deep sub-micron technologies. Indeed, it should be asked whether the need is any greater now than before. It is possible that the need was greatest at the time lightly doped drain (LDD) structures were introduced. Power supply voltages were still at 5 volts, and the resulting electric fields were so high that the transition region from high to low doping (i.e., the junction) had to be intentionally widened, which degraded the transistor performance (one of the key reliability-versus-performance tradeoffs the industry has faced). Since that time, supply voltages have better scaled with the channel length, and the "LDD" concentration has steadily increased to the point where it is more accurate to refer to this region between the channel and the deep source/drain (S/D) as the "S/D extension". Although this remains a key region for defining transistor behavior, the dynamic range (maximum - minimum concentration) has decreased substantially, and hot-electrons are less of a problem as the supply voltage continues to shrink. This has reduced somewhat the interest in detailed lateral profiling in this region of the transistor.

Implicit in the roadmap is profiling of the S/D. But other regions of the transistor have increased in importance. The reverse short channel effect (RSCE) is the unexpected maximum in threshold voltage as channel length is reduced (11). Damage from the S/D implant migrates to the channel region where it causes a pile-up of channel dopant at the surface. Although we have a reasonable understanding of this effect, this has increased the need for 2D profiling *in the channel*, where doping concentrations are lower than in the S/D extension (mid-10^{17} atoms/cm^3 to low-10^{18} atoms/cm^3). Because this effect is channel-length dependent, the need for profiling is on actual transistors, not large area test structures. Profiling of point-defects (vacancies and interstitials) in this region is almost as important as dopant profiling.

More recently, large amounts of "dose loss" have been observed at the silicon/oxide interface (12-13). It is not uncommon for 50% of a shallow implant to be trapped and electrically inactive at this interface. Note that this is primarily a 1D metrology problem. This effect creates a need for improved profiling near the silicon surface.

Dose loss leads into the issue of the different "types" of profiles. Although these definitions can be found in many textbooks, they are worth reviewing here. First, there is the difference between electrically active and the total concentration. Under the electrically active category, we can distinguish between electrically active dopant versus electrically active carrier concentration. There is also the need to distinguish between the "net" concentration (net dopant = donors - acceptors, and net carriers = electrons - holes) versus individual species. Different techniques measure different profiles. In many cases, the different types of profiles are similar. But it is important to distinguish which type of profile is being measured. It is the electrically active, net carrier profile that directly relates to device behavior. Of course, the carrier profiles are bias dependent (gate, source, drain and substrate biases). Techniques which allow biases to be applied to the device can be very instructional. However, there is not a strong need for this capability, as it is assumed that the carrier profiles can be determined from knowledge of the dopant profiles via a device simulator. For issues such as dose loss, it is important to be able to measure both the total and active concentrations. This may require two different measurement techniques.

A similar profiling problem arises at the silicon/silicide interface. It is important to have high dopant concentration at this interface to reduce contact resistance. During silicide growth, the dopants could segregate into the silicide. To date, dopant loss at this interface has not been reported as a significant problem, but it is mentioned here in case there are any special metrology issues arising from presence of a silicide. This interface is typically not very smooth, which leads to local variation in the distance between the silicide and the S/D junction. If this distance becomes too shallow, the diode leakage current can increase.

Another region of increased importance is in the poly gate. Historically, this region was considered to be a metal because it was so heavily doped. But as oxides get thinner, dopant penetration from the poly gate through the gate oxide becomes more of a problem, so the dopant concentration is reduced to help prevent this. This can lead to a side-effect known as "poly-depletion," whereby a small depletion layer forms at the poly/oxide interface under high gate bias. Poly-depletion is exacerbated by thinner gate oxides, and also contributes a more substantial fraction to the total effective gate capacitance as the oxide thickness is reduced. Technologists are trying to balance the tradeoffs between poly-depletion and dopant penetration (too little versus too much dopant at the poly/gate-oxide interface). The process technologist is interested in knowing the electrically active concentration at the bottom of the poly (tens of nm from the gate oxide). They are also interested in knowing whether any poly dopant penetrates through the gate oxide and into the channel, although electrical techniques (i.e., threshold voltage shifts) can satisfy this requirement. Actually, CV measurements can do a reasonable job of estimating the dopant concentration at the poly/oxide interface as well, although occasional direct measurement would be desirable.

We now return to the motivation for these stringent dopant profile requirements. This is largely driven by the need for "predictive" TCAD. Ideally, you would like to be able to accurately simulate a technology before you build it. However, there is great value in approximate results (14), which require less accuracy. Given the rapid pace of technology development, it is difficult, if not impossible, to achieve predictive TCAD. Although predictive TCAD remains a desirable goal, it may not be a realistic one, and it is this goal that imposes the strict dopant profiling requirements. We can also examine the dopant profiling needs from the other direction: given the current state of 2D profiling, *any* information is useful. As an example, simply determining the junction location may be as useful as the more complete dopant profile, and some measurement techniques are well suited for establishing junction locations (15-16). It is the lateral junction location *near the surface* that is of most interest. If one could correctly predict this lateral junction location and the 1D vertical S/D profile, you could have reasonable confidence in your ability to predict the whole profile (17).

As an interesting sidenote, *perfect* dopant metrology would not necessary result in predictive TCAD. Knowing the doping profile would allow you to predict the device performance with reasonable confidence, but it would not necessarily provide enough information to build a model for predicting the doping profile from a description of the processing conditions. Measuring the doping profile after each processing step would get you closer to the goal of predictive process simulation, but still would not guarantee that you would understand the physics well enough to build a model for predicting changes in the doping profile based on changes in the processing steps. In other words, perfect dopant metrology would allow predictive *device* simulation, but would only be a first step (albeit, an important one) in predictive *process* simulation. However, predictive device simulation alone is still valuable.

The source to drain spacing, or the lateral junction location, has been mentioned as a key requirement. It can be just as important to know this location *with respect to* the gate edge. The source to drain spacing controls the DC, or steady-state characteristics of a device. But the transient, or switching, characteristics of a device are heavily influenced by the amount of capacitance between the gate and the source/drain. The distance that the S/D diffuses under the gate controls this overlap capacitance. These dopant profiling requirements are summarized in the following table.

TABLE 1. Dopant Profiling Needs

Lateral junction location at silicon *surface*.
Distance from surface junction to gate edge (overlap).
2D *channel* profiles (top-down measurement?).
2D *point-defect* profiles.
1D surface profiling (dose loss).
1D active concentration at bottom of poly.

Rank ordered list of dopant profiling needs. In general, a distinction is not made between carrier or dopant profiles, under the assumption that dopant profiles can be converted to carrier profiles in a device simulator (the reverse process is not as straightforward).

ELECTRICAL ALTERNATIVES

It would be remiss to not mention the increased activity in inverse modeling to determine dopant profiles (18-19). In this approach, dopant profiles are inferred via iteration of electrical measurements and device simulation. Perhaps the greatest advantage to this technique is the possibility of measuring across-the-wafer variations. Most direct measurement techniques are too time-consuming to accomplish this. In fairness, 2D inverse modeling still takes several hours per profile, although this will improve as this approach matures. The question that always arises is whether a unique doping profile can be extracted. Theoretically, enough electrical measurements can be taken to result in a unique profile. When combined with conventional 1D profiling techniques, 2D inverse modeling could possibly satisfy most process integration and modeling needs.

SUMMARY

This paper has intended to build a bridge between process technologists and dopant metrology experts by more clearly stating the dopant profiling needs. Continued discussions between the two groups should lead to better compromises between what is desired and what is achievable. The current Roadmap goals are very aggressive, but that shouldn't imply that less aggressive results would not still be useful to technologists. TCAD and dopant metrology are not technology showstoppers, but small improvements in our predictive and diagnostic capabilities can have high payback.

ACKNOWLEDGMENTS

Discussions with Clayton Williams, of the University of Utah Physics Department, particularly on the issues of average dopant spacing and profile types, have been very helpful.

REFERENCES

1. The National Technology Roadmap for Semiconductors, Semiconductor Industry Association, Table 61, p. 181, 1997.
2. NTRS, Table 14, p. 46, 1997.
3. Xiang-Dong Wang, private communication.
4. H.-S. Wong and Y. Taur, IEDM Tech. Dig., p. 705, 1993.
5. K. Nishinohara, N. Shigyo, and T. Wada, IEEE Trans. Elec. Dev., Vol 39, p. 634, 1992.
6. J. T. Watt and J. D. Plummer, IEEE Trans. Elec. Dev., Vol. 41, p. 2222, 1994.
7. T. Mizuno, J. Okamura and A. Toriumi, VLSI Symposium Tech. Dig., p 41, 1993.
8. Mark Dowsett, private communication.

9. V. Ukraintsev, et al., Characterization and Metrology for ULSI Technology, TU-40, 1998.
10. P. A. Stolk, F. P. Widdershoven and D. B. M. Klaassen, SISPAD 97 Proceedings , p 153.
11. C. S. Rafferty, H.-H. Vuong, S. A. Eshraghi, M. D. Giles, M. R. Pinto, and S. J. Hillenius, IEDM Tech. Dig., p. 575,1993.
12. P. B. Griffin, S. W. Crowder and J. M. Knight, Applied Physics Letters, vol. 47, p. 482, 1995.
13. H.-H. Vuong, C. S. Rafferty, J. R. McCracken, J. Ning and S. Chaudhry, SISPAD 97 Proceedings, p. 85.
14. M. Duane, ChiPPS 97 Proceedings, also available at http://www.prismnet.com/~naomi/tcad/duane1.html
15. H. Edwards, et al., Appl. Phys. Lett., vol. 72, p. 698, 1998.
16. R. Kleinman, et al., IEDM 97, p. 691.
17. R. Goossens, private communication.
18. Z. K. Lee, M. B. McIlrath and D. Antoniadis, IEDM 97, p. 683.
19. A. Das, et al., IEDM 97, p. 687.

Two Dimensional Dopant Diffusion Study by Scanning Capacitance Microscopy and TSUPREM IV Process Simulation

J. Kim, J. S. McMurray, and C. C. Williams
Department of Physics, University of Utah, Salt Lake City, Utah 84112

J. Slinkman
IBM Microelectronics, Essex Jct., Vermont 05452

We report the results of a 2-step two-dimensional (2D) diffusion study by Scanning Capacitance Microscopy (SCM) and 2D TSUPREM IV process simulation. A quantitative 2D dopant profile of gate-like structures consisting heavily implanted n+ regions separated by a lighter doped n-type region underneath 0.56 μm gates is measured with the SCM. The SCM is operated in the constant-change-in-capacitance mode. The 2-D SCM data is converted to dopant density through a physical model of the SCM/silicon interaction. This profile has been directly compared with 2D TSUPREM IV process simulation and used to calibrate the simulation parameters. The sample is then further subjected to an additional diffusion in a furnace for 80 minutes at 1000C. The SCM measurement is repeated on the diffused sample. This final 2D dopant profile is compared with a TSUPREM IV process simulation tuned to fit the earlier profile with no change in the parameters except the temperature and time for the additional diffusion. Our results indicate that there is still a significant disagreement between the two profiles in the lateral direction. TSUPREM IV simulation considerably underestimates the diffusion under the gate region.

INTRODUCTION

Quantitative 2-dimensional (2D) dopant profiling is very important for the calibration of process simulators, which is identified in The National Technology Roadmap for Semiconductors as one of the significant needs of the semiconductor industry (1). Several promising techniques for high resolution 2D dopant profiling are currently under development, which include dopant sensitive chemical etch techniques, nano-SRP, and SCM. These techniques have been recently reviewed by A. C. Diebold *et al.* (2). The Scanning Capacitance Microscope (SCM) has been shown to be capable of quantitative 2D profiling (3,4,5,6). In a previous study, quantitative 2D dopant profiling of a gate like structure was obtained using SCM (7). Furthermore, we have made a direct comparison of the SCM results with the predictions of a TCAD process simulator TSUPREM4 (8). In this work, we perform a 2-step diffusion experiment and address the plausibility of calibrating the process simulator with the SCM measurements.

EXPERIMENTS

In the SCM technique, local capacitance is measured with the aid of a sensitive capacitance sensor electrically connected to the tip of an atomic force microscope (AFM). As the tip is scanned, both topographic and capacitance data are acquired simultaneously. For the measurements discussed, a NanoScope IIIa AFM/SCM, with a Dimension 3000 head manufactured by Digital Instruments, is used. Some of the measurements were performed with an RCA capacitance sensor (9). This sensor uses a 915 MHz probing voltage for capacitance detection. In this configuration, the input capacitance to the RCA sensor is about 1pF. Since most of this is stray capacitance, it is necessary to look for changes in capacitance to separate the small tip capacitance from the much larger stray capacitance. This is accomplished by applying a time varying bias voltage to the semiconducting sample. The bias voltage typically has a frequency between 5-50KHz. This bias voltage modulates the depletion capacitance in the semiconductor. This SCM configuration has been successfully used with both commercial and home-built AFM's (10,11).

While it is straightforward to measure the change in capacitance at each point with a lock-in amplifier, resolution degradation may occur in lightly doped areas since the resolution of the SCM is determined by the volume of silicon that is depleted. To overcome this problem, the SCM is operated in a constant-change-in-capacitance mode, where the change in capacitance measured by the sensor is small and is held constant by varying the amplitude of the bias applied to the sample with a feedback control (3). Figure 1 shows the schematic of our experimental setup. This leads to large bias voltages in heavily doped regions where the depletion is small

CP449, *Characterization and Metrology for ULSI Technology: 1998 International Conference*
edited by D. G. Seiler, A. C. Diebold, W. M. Bullis, T. J. Shaffner, R. McDonald, and E. J. Walters

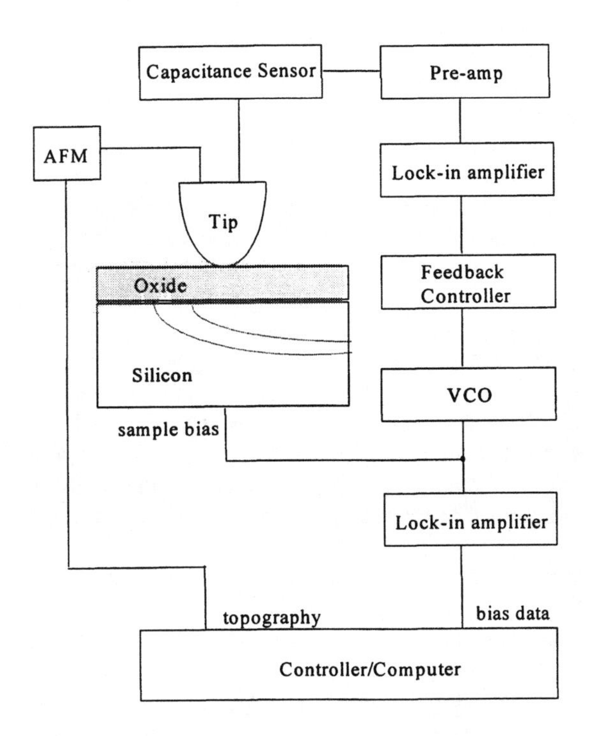

FIGURE 1. A schematic of experimental setup for constant-change-in-capacitance mode SCM measurement of cross-sectioned silicon wafer sample.

due to the large number of carriers and small bias voltages in lightly doped regions. Using this configuration, the SCM resolution is limited by the diameter of the tip.

The 2D profile reported here is obtained on an n-type sample (IBM XG33 sample, sample #1) that is implanted with phosphorus atoms. The substrate doping level is approximately 10^{15} cm^{-3} and the peak concentration is 8 X 10^{19} cm^{-3}, as measured by SIMS. The area of interest is a series of gate-like structures; heavily implanted regions separated by a lighter doped region underneath the gates. These structures are created by ion implantation, with 0.56 μm wide lines of oxide to mask the gate area. For the diffusion experiment, a piece of the wafer has been annealed in a nitrogen ambient furnace for 80 minutes at 1000C (sample #2). The samples are then cleaved and polished to the desired implanted area. The polishing is performed using a series of diamond embedded polishing pads. It is found that having a scratch-free surface is very important in order to avoid the modification of the capacitance signal by the topographic features. The smallest diamond particle size was 0.1 μm. At the final polishing step, a colloidal silica solution is used. Besides providing the final polish for the surface, this step also provides surface insulation and passivation. We have generated C-V curves after the polishing steps with the SCM and compared these measurements with results obtained on good thermal oxides, from which the thickness of the insulating layer between the tip and sample has been estimated to be ~3 nm. The curves also indicate that the insulating layer after polishing is of sufficient quality to be useful for SCM. We find that it is sometimes necessary to anneal the sample at ~200C to drive excess moisture off the surface. This additional annealing also seems to improve the quality of the insulating layer. A detailed study of sample preparation has been submitted separately (12).

The amplitude of the bias voltage measured in the constant change in capacitance mode is related to the dopant density through a conversion algorithm. The algorithm is based on a quasi-3D model of the tip sample capacitor (4). Using this method, high resolution vertical dopant profiles on silicon wafers have been achieved (3,4,5). These profiles are found to be in good agreement with vertical SIMS measurements.

The conversion algorithm requires several parameters: tip radius, peak dopant density and corresponding AC bias, oxide thickness, oxide dielectric constant, and size of the SCM sensor probing voltage. In this measurement a heavily doped silicon tip with a radius of 37nm (determined by imaging a Niobium film (13)) was used. The reference dopant value used in this conversion is 8 X 10^{19} atoms cm^{-3}, as determined by SIMS. Changing any of the other model parameters (oxide thickness, dielectric constant, or the amplitude of the sensor probing voltage) have a similar effect; they all affect the lowest dopant value generated with the conversion process (7). Currently none of these parameters are exactly known, but reasonable values can be chosen for two of the three, and the third varied as a free parameter within reasonable limits. The two values fixed in this conversion are the oxide thickness and the oxide dielectric constant. The thickness is chosen to be 3 nm, slightly larger than a native silicon oxide. The value of the dielectric constant used is 3.0. The value of the free parameter, the sensor probing voltage, is adjusted so that the dopant profile taken in the vertical direction in an implanted region most closely fits the SIMS profile. In this case, the probing voltages used were 1.15V peak for sample #1, and 1.7V peak for sample #2.

In our earlier report (7), we have shown that the average percentage difference of SCM and SIMS in the vertical profile is 15%. We have also shown that the spatial resolution is approximately the tip size (5). However, for the sample used in this experiment, the dopant profile is broad enough in both the vertical and lateral dimensions (a decade change in concentration over many tip diameters) that the accuracy in the lateral profile should be comparable to that in the vertical profile.

We used the commercially available process simulator, TSUPREM4 (14) for the prediction of the 2D dopant profiles, which can be cross-checked against SIMS in the 1D vertical direction. The full fabrication process of the XG33 sample was simulated with TSUPREM4 (8). We used "out-of-the-box" default model parameters for phosphorus diffusion with a minor adjustment to the best fit to the SIMS of the phosphorus dopant in the unmasked region of sample originally annealed in the oxidizing ambient at 900C.

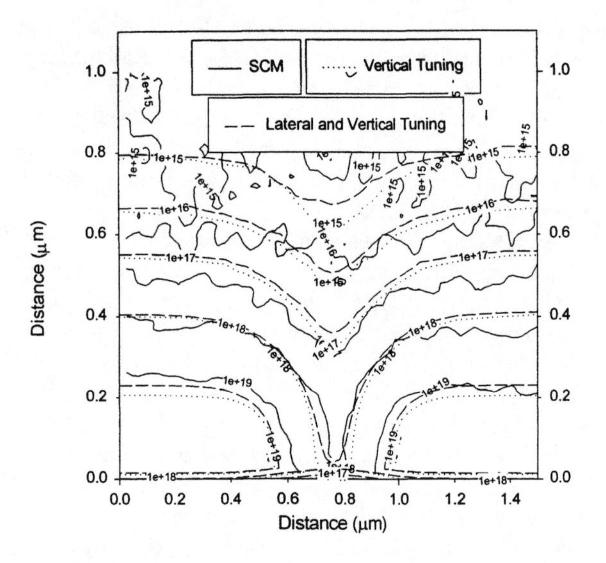

FIGURE 2. Comparison of TSUPREM4 simulation and SCM dopant profiles for the original XG33 sample (sample#1)

RESULTS AND DISCUSSION

In our previous study (8), we found that while the SIMS and SCM results show excellent agreement in the vertical profile, TSUPREM4 process simulation results using the "out-of-the-box" parameters underestimates the dopant concentrations between 0.1 μm and 0.4 μm ($8X10^{19}$ to $1X10^{18}$ cm^{-3}) under the surface. Furthermore the lateral diffusion underneath the gate is also underestimated by the TCAD simulation.

We have attempted further tuning of the TSUPREM4 process simulator to fit the full 2D SCM results by adjusting a few physically significant parameters. One of the results is shown in the contour plot of Fig. 2. For this comparison, a few adjustments in the parameters have been made for better fit in both vertical and lateral direction. To achieve the improved fit, the bulk interstitial recombination rate was increased by 10X and the surface interstitial recombination velocity was decreased by 0.1X. These adjustments improved the fit significantly, as seen in Fig. 2. While there are still some discrepancy close to the gate region, overall fit is much improved over the result obtained with "out-of-the-box" parameters.

With the simulator tuned both laterally as well as vertically, we now turn to the measurements on sample #2, which has been subjected to an additional diffusion in a furnace for 80 minute at 1000C. A vertical SIMS profile has been performed for this sample far from the masked region and used for cross-comparison. For the TSUPREM4 simulation, the parameters that provided best fit (vertically and laterally) for the original sample (sample #1) have been used, only the additional diffusion time and temperature has been changed. For this conversion, we have also explored the possibility of skipping the 1D SIMS measurement for this kind of diffusion experiments: that is, instead of using the peak concentration value obtained from the SIMS measurement, we have used the peak concentration predicted by the TS4 simulation and the lowest concentration (known substrate dopant concentration) to determine the proper sensor probing voltage value. The other parameters were kept the same as previous conversion for sample#1 (8). For the 2D SCM profiles no further optimization to SIMS profiles was used. Figure 3 shows the vertical profile of the diffused sample (sample #2) away from the gate. It can be seen that

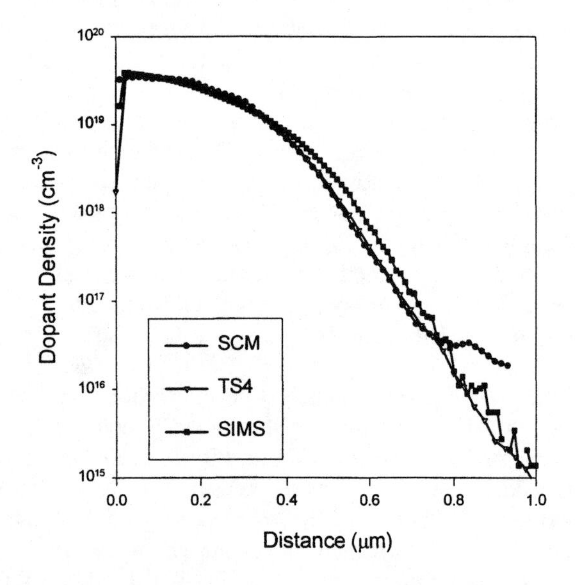

FIGURE 3. Vertical dopant profile of XG33 sample after an additional 80 minutes anneal (sample#2) at 1000C in the nitrogen ambient.

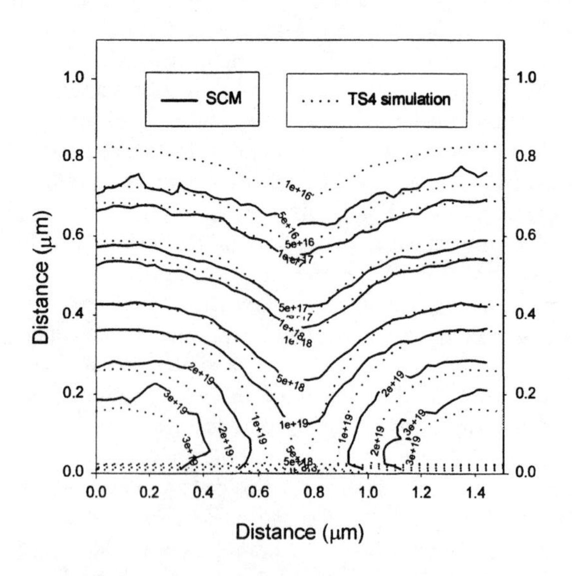

FIGURE 4. Contour plots of SCM and TSUPREM4 results on the diffused sample (sample#2).

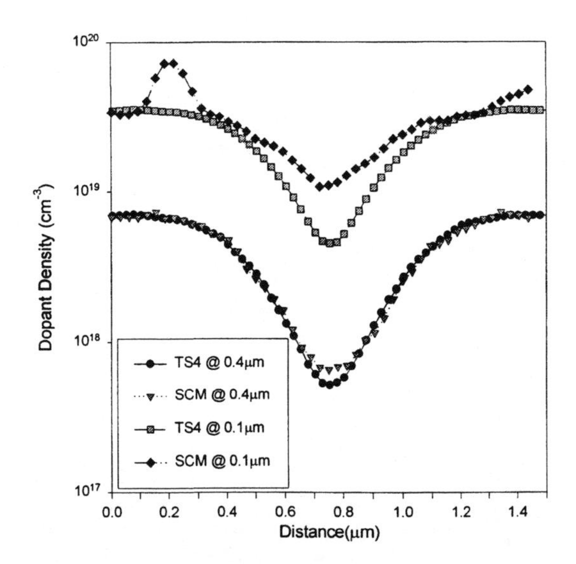

FIGURE 5. Lateral profile of the diffused sample at depths of 0.1 mm and 0.4 mm under the surface.

the agreement between SCM, TS4, and SIMS is good in the vertical direction down to $1X10^{17}cm^{-3}$. The TS4 simulation results have an excellent agreement with the SCM profile in the vertical direction.

A full 2D comparison of SCM and TS4 simulation is shown in Figure 4. As can be readily seen, the contour lines agree relatively well over most of the image. However, near and under the gate region, we can still see quite a good deal of disagreement between the two profiles. This is illustrated in Fig.5 where lateral cuts 0.1μm and 0.4 μm under the surface are plotted. While the two show excellent agreement at 0.4 μm, the TS4 result significantly underestimate the diffusion at 0.1 μm under the gate.

In addition to phosphorus diffusion coefficients for both interstitial- and vacancy-mediated diffusion species, TSUPREM4 also accounts for the diffusion and recombination of the interstitial and point defects themselves. The recombination and generation of point defects at the silicon surface and in the bulk are also accounted for. Specifically, when oxidation occurs, such as for the initially annealed XG33 sample (sample #1), the latter kinetics are dominant, since oxidation is known to inject interstitials into the silicon bulk from the surface. As a result, the interstitial concentration is far in excess of equilibrium during oxidation. During the final 80 minute anneal in the inert ambient, the point-defect will have relaxed to equilibrium. The TSUPREM4 coefficients for surface and bulk point-defect generation and recombination were tuned to fit the 900 C oxidation SIMS data.

For the simulation of the subsequent 80 minute inert anneal, it may well be the case that the point-defect generation model parameters are not accurate, causing TSUPREM4 to underestimate the lateral outdiffusion. It is most likely that the surface generation/recombination rate is incorrect, since the simulated, SIMS, and SCM vertical profiles at 80 minutes still match fairly well.

We also note that in actual practice, which is to say in the modeling of production level devices, which can require the modeling of upwards of order one hundred annealing steps (including furnace ramps), 2D simulations of actual MOSFET source-drain junctions more often than not underpredict the lateral outdiffusion for inert junction anneals. It is more the case for Boron (p+) source-drains, but it is also observed for Phosphorus (n+). Our result is consistent with the general trend. (15)

CONCLUSION

A two-step diffusion of dopants has been performed in two dimensions with SCM and TSUPREM4 process simulation. The process simulator has been tuned to best fit the SCM profile of the original sample both in vertical and lateral directions. This tuned simulator has been used for simulation of additional diffusion and compared with a SCM measurement on the diffused sample. Direct comparison of the 2D dopant profiles shows a moderate overall agreement. But underneath the gate, the process simulation underestimates the dopant diffusion. However, we believe that our results should be an impetus and a basis for improved model parameter tuning in 2D simulators such as TSUPREM4.

ACKNOWLEDGMENTS

The authors would like to thank Digital Instruments, SEMATECH, and SRC for their help and funding of this project and Wayne Wingert for sample preparation.

REFERENCES

1. The National Technology Roadmap for Semiconductors is published by the Semiconductor Industry Association, 4300 Stevens Creek Boulevard, Suite 271, San Jose, CA 95129
2. Diebold, A. C., Kump, M. R., Kopanski, J. J. and Seiler, D. G., *J. Vac. Sci. Technol. B* **14**(1) 196 (1996)
3. Huang, Y., Williams, C. C., and Slinkman, J., *Appl. Phys. Lett.* **66**(3) 344 (1995)
4. Huang, Y., Williams, C. C., Smith, H., *J. Vac. Sci. Technol. B* **14**(1) 433 (1996)
5. Huang, Y., Williams, C. C., and Wendman, M. A., *J. Vac. Sci. Technol. A* **14**(3) 1168 (1996)
6. Neubauer, G., Erickson, A., Adderton, D., *J. Vac. Sci. Technol. B* **14**(1) 426 (1996)
7. McMurray, J. S., Kim, J., Williams, C. C., *J. Vac. Sci. Technol. B* **15**(4), 1011 (1997)
8. McMurray, J. S., Kim, J., Williams, C. C., and Slinkman, J., *J. Vac. Sci. Technol. B* **16**(1) 344 (1998)

9. Palmer, R. C., Denlinger, E. J., and Kawamoto, H., RCA Rev. **43**, 194 (1982)
10. Williams, C. C., Slinkman, J., Hough, W. P., Wickramasinghe, H. K., *Appl. Phys. Lett.* **55**(16) 1662 (1989)
11. Kopanski, J. J., Marchiando, J. F., and Lowney, J. R., *J. Vac. Sci. Technol. B* **14**(1) 242 (1996)
12. Zavyalov, V., McMurray, J. S., and Williams, C. C., submitted for publication in the Proceedings of NIST
13. Westra, K. L., Mitchell, A. W. and Thompson, D. J., *J. Appl. Phys.* **74**(5) 3608 (1993)
14. TSUPREM4 User's Manual, Technology Modeling Associates, Palo Alto, CA, (1996)
15. For more discussions on the difficulties of calibrating process simulators and practical approaches to overcoming the difficulties, see S.S. Yu, H.W. Kennel, M.D. Giles, and P.P. Packan, 1998 International Electron Devices Meeting (IEDM) Technical Digest, p509 (1998). Also G. Le Carval, P. Schleiblin, D. Poncet, and P. Rivakin, Proceedings of the 1997 International Conference on Simulation of Semiconductor Processes and Devices, Cambridge, MA, p177 (1997)

Comparison of Measured and Modeled Scanning Capacitance Microscopy Images Across p-n Junctions

J. J. Kopanski, J. F. Marchiando, J. Albers, and B. G. Rennex

Semiconductor Electronics Division
National Institute of Standards and Technology
Gaithersburg, MD 20899-0001 USA

Scanning capacitance microscope (SCM) image contrast measured on ion implanted P^+/P and P^+/N junction structures with identical dopant profiles, as a function of SCM operating conditions, is compared to a theoretical model of the SCM based on a two-dimensional finite-element solution of Poisson's equation. Measured $\Delta C/\Delta V$ versus bias voltage curves were found to agree with the theoretical model. For the P^+/P structure, the lateral dopant profile is extracted using the model and compared to a Monte-Carlo simulation of the implanted profile. The image contrast of the P^+/N junction structure and the dependence of the apparent junction location on SCM bias voltage is compared to model predictions calculated using the known dopant profile. The SCM quiescent operating point where the apparent junction location in an SCM image and the actual metallurgical p-n junction coincide is defined.

INTRODUCTION

The scanning capacitance microscope (SCM), in its current configuration which utilizes a contact-mode atomic force microscope (AFM) for controlling the tip position and height, has been in existence since 1994 (1, 2). Because the contrast of SCM images of silicon depends on the local carrier concentration and because the measurement point-to-point spacing can be made less than 1 nm, SCM has been recognized since its earliest days as a very promising tool for measuring two-dimensional (2-D) dopant profiles (3). Indeed, the SCM can readily produce images of cross-sectioned transistors with state-of-the-art geometries that show contrast due to the dopant in the source and drain regions, S/D extensions, channel, and polysilicon gates (4, 5). Industrial need for 2-D dopant profile measurements and the introduction of commercial tools have resulted in the placement of many SCMs into the metrology and research laboratories of integrated circuit manufacturers.

Data interpretation schemes to translate 2-D SCM images of difference capacitance into 2-D dopant profiles are needed to make SCM images quantitative. First order interpretation of images of dopant profiles consisting of dopant gradients in like-type (P^+/P or N^+/N) substrates has proven relatively straightforward (4, 6). While SCM images of transistors containing p-n junctions are easily obtainable, images containing p-n junctions defy the simple interpretation scheme developed for dopant gradients in like-type substrates. A p-n junction causes a gradient driven charge redistribution and a resultant built-in depletion region which affects the SCM image. In addition, the difference in polarity of SCM signal between n-type and p-type material (4-5), prevents a meaningful application of the feedback loop commonly used to maintain a constant difference capacitance (4, 6). Experimentally, the apparent p-n junction location in SCM images changes with dc bias voltage. A recent comparison of junction depths of transistor test structures measured with SCM (7) produced varying estimates of the actual junction depth. Clearly, the SCM image contrast mechanisms need to be better defined.

In the proceedings of the previous workshop of this series (2), we published SCM images showing the dc bias voltage dependence of SCM images of simple lateral p-n junction test structures in silicon. Here, we revisit these structures and provide a detailed comparison between the experimentally measured SCM image contrast, and a theoretical model of SCM contrast based on a 2-D finite element solution of Poisson's equation. Data from two structures are presented, a P^+/P structure and a P^+/N junction. The structures were fabricated by an identical process with the only exception being the type of the starting substrate. The test structure pattern consists of 5 μm wide lines which were subjected to ion implantation alternating with 5 μm wide spaces which were masked from implantation. The implants were of 50 keV boron at a dose of 10^{14} cm^{-2} through 10 nm of silicon dioxide. The samples were annealed at 900 °C for 10 min in 10% hydrogen/90% nitrogen forming gas. This process produces a lateral dopant profile at the implant mask edge that varies from its peak concentration to the background level over about 400 nm. This relatively gradual transition reduces any steep dopant gradient effects in the SCM images. The existence of identical dopant profiles, in both a like-type and opposite-type substrate, allows us to clearly separate elements of SCM contrast due to a dopant gradient from effects due to the p-n junction.

For each structure, $\Delta C/\Delta V$ versus the bias offset voltage, V_{dc}, was measured at the center of the substrate region and at the center of the implanted region. These measured curves are compared to the theoretical dC/dV versus V_{dc} curves, which were calculated with the experimental model parameters. For the P^+/P sample, we extract the dopant profile from the SCM signal and compare it to a Monte-Carlo simulation (8) of the lateral dopant profile at the peak dopant density. For the P^+/N junction sample, we compare the measured SCM signal as a function of V_{dc} across the p-n junction to a theoretical simulation with a 2-D Poisson

CP449, *Characterization and Metrology for ULSI Technology: 1998 International Conference*
edited by D. G. Seiler, A. C. Diebold, W. M. Bullis, T. J. Shaffner, R. McDonald, and E. J. Walters
1998 The American Institute of Physics 1-56396-753-7/98/$15.00

model. Based on the observed behavior and model results, we suggest a technique to locate the "quiescent point" of an SCM image of an p-n junction, i.e., the bias conditions where the sign change in the SCM signal occurs at the metallurgical p-n junction.

MODEL OF THE SCM

The SCM has been modeled by two- and three-dimensional finite-element solutions of Poisson's equation which are described in detail elsewhere (9, 10). The finite-element method allows a more realistic simulation of the 3-D geometry of the SCM tip and its interaction with the underlying dopant gradient. A model that solves Poisson's equation is essential to include the interaction of the SCM with a p-n junction. The existence of a p-n junction requires consideration of both signs of charge carriers and their redistribution due to the dopant gradient. For uniformly doped material, the SCM tip is modeled by a spherical tip that smoothly terminates a conical shank. Due to the axial symmetry, a 3-D solution can be quickly found. To include the effect of a dopant gradient requires a true 3-D solution which is currently very time consuming to calculate. To simulate the SCM tip across a dopant gradient, we use a 2-D analog of the 3-D tip, i.e., a knife edge probe with a cross-section identical to the 3-D tip. While this is a 2-D solution, when scanned across the gradient in the steepest direction, this model preserves the interaction of the tip curvature with the dopant gradient.

The SCM model parameters consist of the SCM operating point, the SCM tip shape, and the sample conditions. The SCM operating point in constant dV mode consists of the applied ac voltage V_{ac} or ΔV, the dc bias voltage, V_{dc}, and the magnitude of the 915 MHz signal from the capacitance sensor, V_{HF}. The tip shape is reduced to just two parameters; the tip radius, r_{tip}, and the tip cone angle, CA. The sample parameters consist of the oxide thickness, t_{ox}, oxide dielectric constant, ϵ_{ox}, and the flatband voltage of the most lightly doped region, V_{fb}. The model zero-bias voltage is displaced from the applied zero-bias voltage by the flatband voltage of the sample. The flatband voltage of the sample is found from the voltage at which the peak SCM signal occurs, which is near but not exactly at the true flatband voltage (9). In practice, the output of the SCM capacitance sensor also depends critically on the tuning of the sensor and the coupling of the sensor to the sample and its surroundings. The coupling can vary depending on the proximity of the SCM tip and the sample surface to metallic parts of the sample stage and on the frequency of V_{ac}. As a check, it is advisable to confirm that n-type material is producing positive signal (sensor output in phase with V_{ac}), p-type material is producing negative signal (output 180° out of phase with V_{ac}), and insulating material producing negligible signal (less than the signal from the most highly doped regions).

The "FASTC2D" technique is used to extract dopant profiles from SCM images (4). A database of model differential capacitance (dC/dV) versus V_{dc} curves for a range of model parameters is calculated off-line. From these general solutions, the solution for a specific set of model parameters is found by interpolation. A conversion curve of modeled SCM signal versus dopant density is then determined from the interpolated dC/dV curves. After normalizing the SCM signal to the conversion curve, large numbers of SCM data points can be converted to dopant density in a few seconds on a personal computer.

P^+/P STRUCTURES

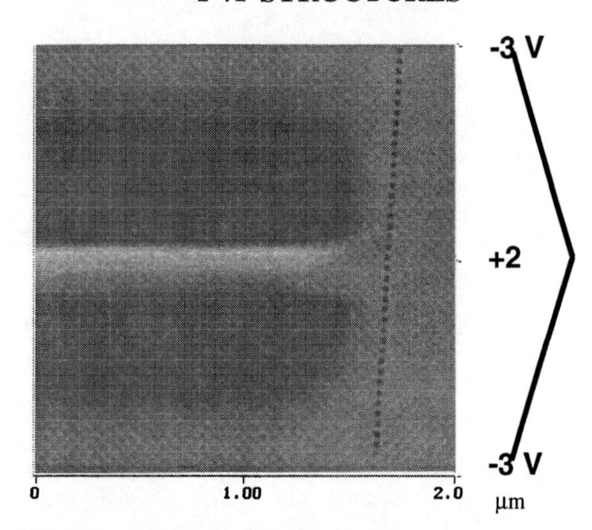

FIGURE 1. Constant dV image of P^+/P test structure. Contrast in the horizontal direction is due to the dopant gradient; contrast in the vertical direction is due to varying V_{dc}. Right side implanted.

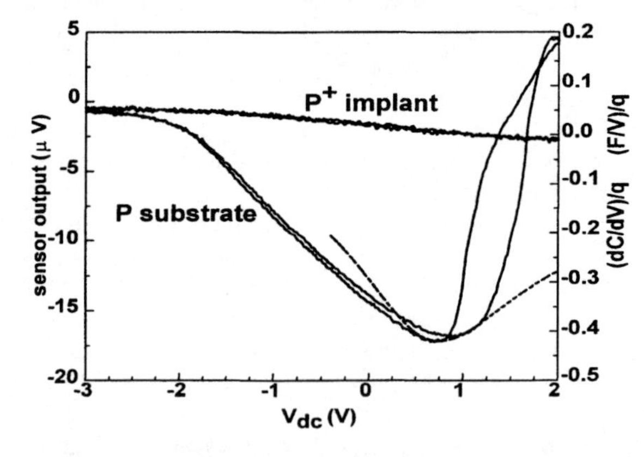

FIGURE 2. Experimentally measured $\Delta C/\Delta V$ versus V_{dc} curves for the a) P region and b) P^+ region. The theoretical dC/dV curve for the P region is also shown as the dashed line.

An SCM image of the P^+/P structure is shown as Figure 1. The image was acquired with $V_{ac} = 0.5$ V at 12 kHz and the sensor tuned to produce an output of less than 100 mV above its minimum output. The image contrast of Fig. 1 in the horizontal direction is due to the effect of the lateral variation in the implanted dopant, from the substrate (P) level at the left, across the lateral gradient, to the maximum implanted (P^+) level on the right side. The presumed location of the mask edge is indicated. The image contrast in the vertical direction is due to varying V_{dc} with the tip scan. V_{dc} was linearly increased from –3 V at the bottom of the image to +2 V at the image center, where V_{dc} then begins to decrease from +2 V to −3 V at the top of the image. Horizontal sections of this image are traces of SCM signal acquired across the dopant gradient at a single value of V_{dc}, while vertical sections correspond to $\Delta C/\Delta V$ versus V_{dc} curves acquired at a single value of dopant density. When measured in this manner, the signal of the $\Delta C/\Delta V$ curve is acquired in the same manner as for SCM images for dopant profile extraction. This measurement technique is proving useful to evaluate low-temperature oxides grown by chemical methods for cross-sectioned samples.

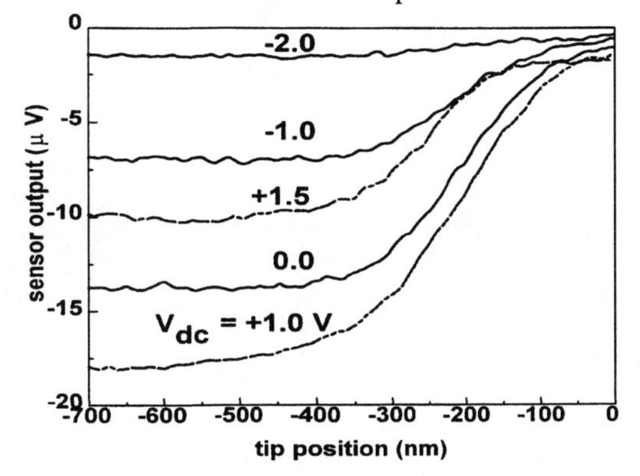

FIGURE 3. SCM signal measured across the P^+/P dopant gradient at different values of V_{dc}.

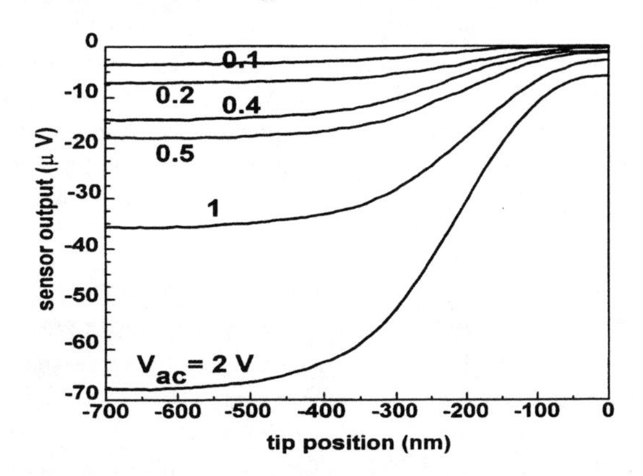

FIGURE 4. SCM signal measured across the P^+/P dopant gradient at different values of V_{ac}.

Figure 2 shows the $\Delta C/\Delta V$ versus V_{dc} curves, extracted from Fig. 1, from the substrate (P) and the implanted (P^+) regions. (V_{dc} is displayed relative to the traditional grounded substrate; however, since for SCM the tip is at virtual ground, the actual applied voltages are the opposite sign.) Both curves have a negative signal level, and the curve from the P region shows the characteristic peak in signal. A common non-ideality observed in $\Delta C/\Delta V$ curves measured with SCM is an increase in signal in the depletion region of the curve beyond the level expected theoretically. At high enough voltages, the signal will change sign. This behavior corresponds to a recovery from depletion condition towards the oxide capacitance. This may be due to mobile charges in the oxide of the sign which depletes the surface, or to conduction through the oxide.

Also shown in Fig. 2 is the theoretical dC/dV response calculated with the 3-D Poisson solver for an r_{tip} of 10 nm and the nominal sample parameters of t_{ox} of 10 nm, ϵ_{ox} of 3.9, and substrate doping of 10^{17} cm^{-3}. In general, the dC/dV response predicted with the Poisson solution is broader in V_{dc} than the response predicted by models based on quasi-2D models, i.e. sums of 1-D analytical solutions (4, 6). The agreement between the measured and model curves is good, though as discussed above, the experimental curve is distorted in depletion.

Figure 3 shows the effect of varying V_{dc} on the SCM signal measured across the dopant gradient. The maximum response is measured at the value of V_{dc} that produces the peak in the $\Delta C/\Delta V$ curve of the substrate. Interpretation is simpler and the dynamic range is greatest when the bias is adjusted to produce the peak response. Figure 4 shows the effect of varying V_{ac} on the SCM signal. Here, V_{dc} has been adjusted to the value which causes the SCM signal in the substrate to peak (about +800 mV). The SCM signal increases almost linearly with V_{ac} over the entire dopant profile. The constant dC mode of SCM (4, 6), where a feedback loop is utilized to automatically adjust V_{ac} to maintain a constant dC as the SCM tip is scanned across a dopant gradient, probably produces a more spatially localized measurement of dopant profile for gradients in like type substrates. However, since the change in the sign of SCM signal at a p-n junction may prevent a meaningful application of the constant dC mode, our purpose here is to determine the actual extent to which large values of V_{ac} cause variations in the determined dopant profile. Signal dynamic range increases with V_{ac} at the expense of spatial resolution in the measured dopant profile and possible loss of sensitivity to the highest dopant levels.

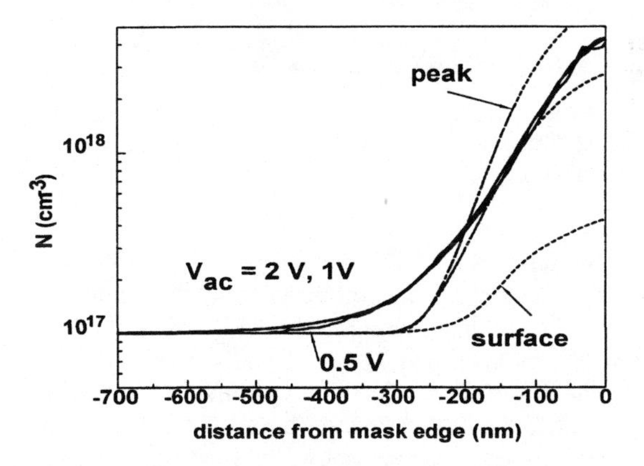

FIGURE 5. Dopant profiles extracted from the SCM line traces in Fig. 4 for V_{ac} = 2 V, 1 V, and 0.5 V. Monte-Carlo simulations of the lateral dopant profile at the surface, 100 nm beneath the surface, and at the depth of the peak concentration are shown as the dotted curves.

Figure 5 shows the dopant profiles extracted from some of the SCM line traces of Fig. 4 using the FASTC2D technique and a database of dC/dV curves calculated with the 2-D Poisson solver (projected to 3-D). For comparison, Monte-Carlo (UT-MARLOWE) simulations of the lateral dopant profile at the surface, 100 nm beneath the surface, and at the depth of the peak concentration are also shown. These profiles do not include the effect of the diffusion step; they are included because they were available. (Simulation of the actual diffused profile is underway.) The simulation was of 40000 histories of 50 keV boron, at a dose of 10^{14} cm^{-2}, through 10 nm of oxide, into (100) silicon with a substrate tilt of 7° and rotation of 30°. The lateral profile also varies with depth; SCM should measure some average profile between the minimum profile at the surface and the maximum profile at the peak concentration. Rather than a direct comparison, the simulated dopant profile is included to show the expected range and magnitude of the dopant profile extracted from the SCM signal. The extracted dopant profile changes very little as V_{ac} varies from 0.1 V to 2.0 V. Further study is needed, but constant dV mode SCM may produce dopant profiles above the mid 10^{17} cm^{-3} level that are well spatially localized, if V_{ac} is less than 2 V.

P^+/N JUNCTIONS

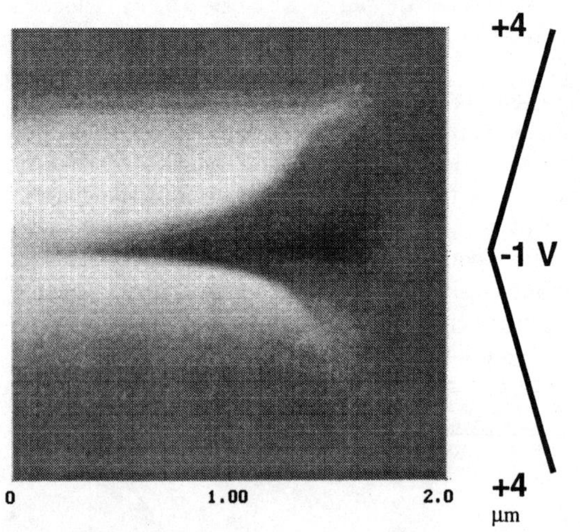

FIGURE 6. Constant dV image of the P^+/N test structure. The n-type substrate is on the left side, the P^+ implanted region is on the right side.

An SCM image of the P^+/N junction structure is shown as Fig. 6. The image was acquired under the same conditions and with the same tip as the image of the P^+/P sample, Fig. 1. However, in this image V_{dc} was linearly increased from +4 V at the top of the image to −1 V at the image center, where V_{dc} then begins to decrease from −1 V to +4 V at the bottom of the image. Figure 7 shows the $\Delta C/\Delta V$ versus V_{dc} curves, extracted from Fig. 6, from the substrate (N) and the maximum implant (P^+) regions. Also shown in Fig. 7 is the theoretical dC/dV response for the N substrate calculated with the 3-D Poisson solver.

While this structure differs from that imaged in Fig. 1 by only the type of the substrate, the SCM response measured across the dopant gradient is fundamentally different. The sign of the SCM signal changes from positive on the N substrate side of the image to negative on the P^+ implant side. The apparent junction location, i.e., the point where the signal changes sign, changes with V_{dc}. The more the value of V_{dc} biases the lightly doped side towards depletion, the more the location of the signal sign change (and the apparent junction) moves into the more lightly doped region. This region roughly corresponds to the built-in depletion region of the junction. Qualitatively, a voltage that would deplete carriers from the lightly doped side would also be of the correct sign to pull the opposite sign carriers from the other side of the junction. For this reason, SCM images of p-n junctions often appear to have "halos" around them.

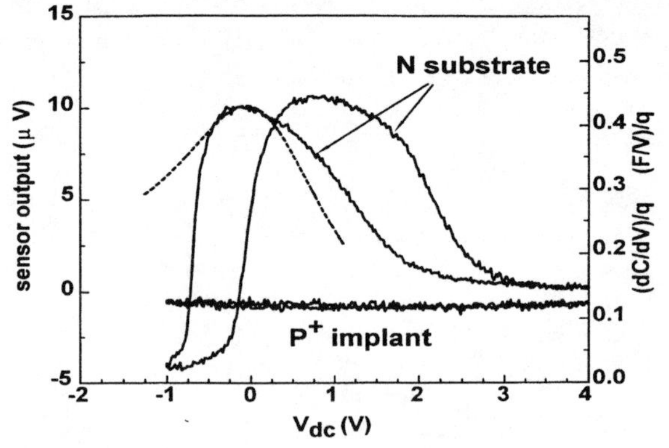

FIGURE 7. Experimentally measured $\Delta C/\Delta V$ versus V_{dc} curves for the a) N region and b) P^+ region. The theoretical dC/dV curve for the N region is also shown as the dashed line.

Figure 8 shows line traces of the SCM signal at values of V_{dc}-V_{fb} of +3.5 V, +2.5 V, +1.5 V, +0.5 V, and +0 V. The apparent junction location moves 600 nm between the extremes in V_{dc}. Obviously, an SCM "quiescent" operating point exists where the apparent junction location and the actual metallurgical junction coincide. An appealing first order estimate of the quiescent point is to 1) find the flatband voltage of the lightly doped side, and 2) add sufficient voltage to overcome the junction built-in voltage (about 0.9 V for this structure). For this structure, this quiescent point is at V_{dc} = 1.4 V, the flatband voltage, 0.5 V, plus the built-in potential, 0.9 V. This voltage will accumulate the N side, and forward bias the junction when the tip is in the P^+ side.

While it is possible to bias the p-n junction with the SCM, the actual voltage across the junction depends on the geometry of the sample and current paths, and the amount of voltage dropped across the MOS capacitor formed by the tip. In general, voltages that result in a forward bias condition may appear across the junction, while a reverse bias condition would likely result in a current path that did not include the junction. Thus, the actual quiescent point is likely to vary from the simple estimate and depend also on the actual dopant profile and sample geometry. Future simulation results will be used to refine the quiescent-point bias conditions. In actuality, the SCM signal is most directly related to the local carrier concentration profile, not dopant concentration. While this difference is neglected in the first order interpretation of dopant gradients in like-type substrates, image of p-n junctions emphasize the difference between carriers and dopant. Extraction of dopant profiles from SCM data around p-n junctions will require deconvolution from the carrier profile.

Figure 9 shows a 2-D Poisson solution simulation of the SCM signal across the p-n junction imaged in Fig. 6. The simulation is for a knife edge probe with an r_{tip} of 10 nm and CA of 10°, moving across the 2-D lateral dopant profile of the P^+/N test structure (obtained via UT-MARLOWE). The simulated profile is for the as-implanted condition and does not take into account the activation/diffusion step of the actual fabrication process. Consequently, the simulated profile was much steeper than the actual case and the simulated SCM signal makes its transition from peak to minimum over a shorter distance. In the experimental data, the extent of the region between p-type and n-type material is also being enhanced by a reservoir of mobile surface charges.

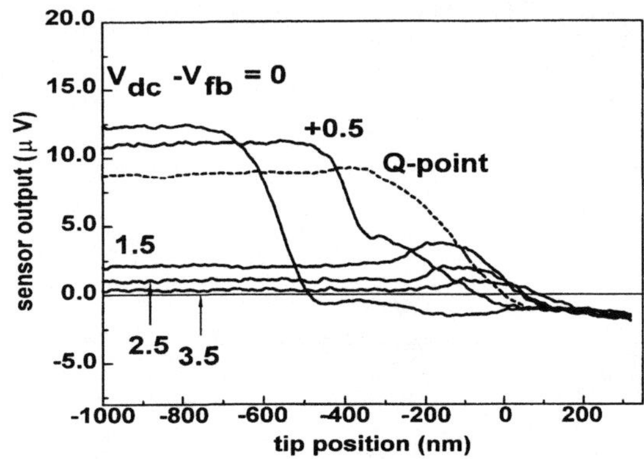

FIGURE 8. SCM signal measured across a p-n junction at different values of V_{dc}-V_{fb}.

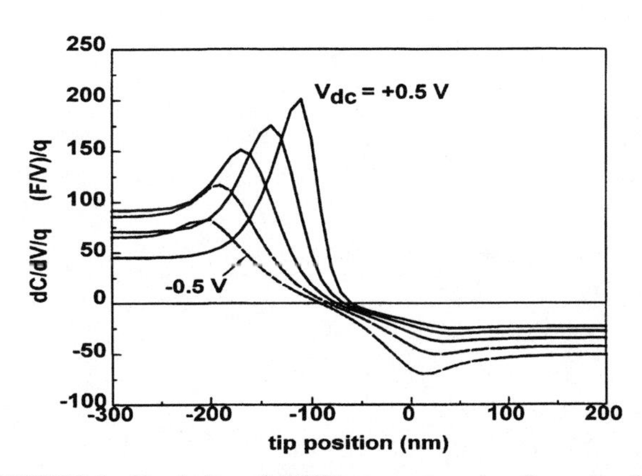

FIGURE 9. Simulation of dC/dV across a p-n junction using the SCM operating point and sample parameters of Fig. 6. V_{dc} varies from +0.5 V to –0.5 in 0.25 V steps.

In the simulation, the implant mask edge was located at 0 nm; the metallurgical p-n junction at the surface is at +35 nm. Apparent junction location shifts from 60 nm at +0.5 V_{dc} to 92 nm at –0.5 V_{dc}. The simulated SCM signal qualitatively reproduces the general features of the measured signal across p-n junctions, negative signal in the p-type region, positive signal in the n-type region, and a progress shift of the apparent junction location into the n-type region as the bias voltage is made more negative (in the direction to deplete the n-type region). The quiescent point of the simulation would occur at a bias voltage greater than +0.5 V. Significantly better agreement between modeled and measured signal is expected for simulations based on a better estimate of the dopant profile.

CONCLUSIONS

The contrast of SCM images of dopant gradients in like-type substrates and across p-n junctions agrees qualitatively with the predictions of a Poisson simulator which takes into account the interaction between the curved SCM tip and the sample. Measured $\Delta C/\Delta V$ versus V_{dc} agrees well with the theoretically predicted signal, though charge injection induces hysteresis into the measured curves. Apparent junction location in an SCM image changes in a predictable manner with the value of bias voltage. A first-order estimate of the bias necessary to produce an SCM image at its quiescent point is to add the junction built-in voltage (the voltage necessary to forward the junction if the tip is in the heavily doped side) to the flatband voltage of the side with lighter doping. SCM images of structures containing p-n junctions must be interpreted in terms of carrier concentration profiles. To obtain the dopant profile from the carrier profile will require additional interpretation.

ACKNOWLEDGMENTS

This work was supported by the National Semiconductor Metrology Program at the National Institute of Standards and Technology. Contribution of NIST, not subject to copyright.

REFERENCES

1. Huang, Y., and Williams, C. C., *J. Vac. Sci. Technol. B* **12**, 369-372 (1994).
2. Kopanski, J. J., Marchiando, J. F., and Lowney, J. R., in *Semiconductor Characterization: Present Status and Future Needs*, Bullis, W. M., Seiler, D. G., and Diebold, A. C., Eds., Woodbury, NY: AIP Press, 1996, pp. 308-312.
3. The National Technology Roadmap for Semiconductors, Semiconductor Industry Association, 4300 Stevens Creek Blvd., Suite 271, San Jose, CA 95129.
4. Kopanski, J. J., Marchiando, J. F., Berning, D. W., Alvis, R., and Smith, H. E., *J. Vac. Sci. Technol. B* **16**, 339-343 (1998).
5. Kopanski, J. J., Marchiando, J. F., and Alvis, R., "Practical Metrology Aspects of Scanning Capacitance Microscopy for Silicon 2-D Dopant Profiling," in *The Electrochemical Society Proceedings* 97-12, 1997, pp. 102-113.
6. Huang, Y., Williams, C. C., and Smith, H., J. Vac. Sci. Technol. B **14**, 433-439 (1996).
7. Ukraintsev, V. A., et al., *These Proceedings* (1998).
8. Obradovic, B., Wang, G., Snell, C., Balamurugan, G., Morris, M. F., Chen, Y., and Tasch, A. F., *UT-MARLOWE Version 4.1 User's Manual*, Austin, Texas: The University of Texas at Austin, 1998.
9. Marchiando, J. F., Lowney, J. R., and Kopanski, J. J., *Scanning Microscopy* **11**(2), (1997).
10. Marchiando, J. F., Kopanski, J. J., and Lowney, J. R., *J. Vac. Sci. Technol. B* **16**, 463-470 (1998).

INVERSE MODELING APPLIED TO SCANNING CAPACITANCE MICROSCOPY FOR IMPROVED SPATIAL RESOLUTION AND ACCURACY

J. S. McMurray and C. C. Williams

Department of Physics, University of Utah, Salt Lake City, Utah 84112

Scanning Capacitance Microscopy (SCM) is capable of providing two-dimensional information about dopant and carrier concentrations in semiconducting devices. This information can be used to calibrate models used in the simulation of these devices prior to manufacturing and to develop and optimize the manufacturing processes. To provide information for future generations of devices, ultra-high spatial accuracy (<10nm) will be required. One method, which potentially provides a means to obtain these goals, is inverse modeling of SCM data. Current semiconducting devices have large dopant gradients. As a consequence, the capacitance probe signal represents an average over the local dopant gradient. Conversion of the SCM signal to dopant density has previously been accomplished with a physical model which assumes that no dopant gradient exists in the sampling area of the tip. The conversion of data using this model produces results for abrupt profiles which do not have adequate resolution and accuracy. A new inverse model and iterative method has been developed to obtain higher resolution and accuracy from the same SCM data. This model has been used to simulate the capacitance signal obtained from one and two-dimensional ideal abrupt profiles. This simulated data has been input to a new iterative conversion algorithm, which has recovered the original profiles in both one and two dimensions.

In addition, it is found that the shape of the tip can significantly impact resolution. Currently SCM tips are found to degrade very rapidly. Initially the apex of the tip is approximately hemispherical, but quickly becomes flat. This flat region often has a radius of about the original hemispherical radius. This change in geometry causes the silicon directly under the disk to be sampled with approximately equal weight. In contrast, a hemispherical geometry samples most strongly the silicon centered under the SCM tip and falls off quickly with distance from the tip's apex. Simulation of the expected signal for each tip geometry shows significant differences in the expected resolution. This has also been explored experimentally with the SEMATECH #1 sample. This sample has a staircase dopant profile with 50 nm steps. Simulation of the expected signal for the SEMATECH #1 using a flat tip model shows good agreement with measured data. However, the flattened tip reduces the steps in the profile to inflections.

INTRODUCTION

Scanning Capacitance Microscopy (SCM) is capable of providing two-dimensional information about dopant and carrier concentrations in semiconducting devices. High quality quantitative two dimensional (2D) dopant profiles have been obtained on non-junction test structures.(1-2) This information can be used to calibrate models used in the simulation of devices prior to manufacturing and to develop and optimize the manufacturing processes. To provide information for future generations of devices, ultra-high spatial accuracy (<10 nm) will be required. In SCM measurements a sinusoidal voltage is applied between the tip and the semiconducting sample. This voltage produces and modulates a depletion region underneath the tip. The resolution of the SCM is approximately the radius of a tip which corresponds to the lateral extent of volume of silicon that is depleted by the applied bias. For this reason tip characteristics are very important.

For a tip to be useful for SCM it should be small, durable, and free from depletion itself. Currently three types of tips have been investigated. The smallest of these, heavily doped silicon, typically has a radius only slightly less than 10nm, about the target resolution. However, this tip suffers from depletion and is difficult to use for quantitative imaging. Magnetic Force Microscope (MFM) and silicided tips have a larger radius, ~35 and ~20 nm respectively. All tips that have been investigated are found to be susceptible to significant wear during measurement. This wear changes the shape of the tip from a hemispherical end to one that is flat. This geometric change significantly degrades the expected resolution of a tip. Given the challenges associated with current tips and the need for higher resolution, other options for improving resolution should be explored. One method, which may provide a means to obtain these goals, is inverse modeling of the SCM system.

INVERSE MODELING FOR SCM

Conversion of the SCM signal to dopant density has

CP449, *Characterization and Metrology for ULSI Technology: 1998 International Conference*
edited by D. G. Seiler, A. C. Diebold, W. M. Bullis, T. J. Shaffner, R. McDonald, and E. J. Walters

previously been accomplished with a physical model, which assumes that there is no dopant gradient in the sampling area of the tip. The conversion of data using this model produces results, for abrupt profiles, which do not have adequate resolution and accuracy for new technologies. This inaccuracy is due to dopant gradients in new technologies which are large enough that the dopant density changes significantly across the tip's sampling region. As a consequence, the capacitance probe signal represents an average over the local dopant gradient. This average will change slightly as the tip is moved a distance that is small compared to its radius. If the signal to noise ratio of the SCM is large enough to detect these small changes, then it is in principal possible to model the SCM/surface system providing a method for dopant profiles to be recovered with a resolution smaller than the tip diameter.

Physical Model of SCM/Si Interaction

The previous model consisted of the tip, placed on top of a thin oxide, covering uniformly doped silicon. The cylindrical symmetry inherent in this configuration allows the problem to be quickly reduced from three to two dimensions. Even so, it is too computational intensive to solve Poisson's equation numerically for this system to provide direct conversion of SCM data. Two approaches have been taken to deal with this problem. The first which has been implemented by those working at the National Institute of Standards and Technology (NIST) is to numerically solve Poisson's equation for a series of relevant cases and store the solutions in a data base (3). The solution from measured data is then found by interpolation on this data base. This solution is appealing in that it is a direct solution of Poisson's equation. However, the data base size grows as the product of the number of entries for each degree of freedom in the data base creating an enormous calculation requirements when a new degree of freedom is added or a change is necessary. A second approach, taken by the authors, is to look for reasonable approximations to reduce the effective dimension of the problem.

Accordingly the problem is broken into two parts; tip and oxide, and the silicon. Each of these sections are divided into annular regions centered under the tip. (See Figure 1.) The size of the annular regions is chosen such that the depletion depth across any one ring is approximately constant. The capacitive contribution of each section is then calculated and summed.

The shape of the tip effects the capacitance of the first section. Two tip shapes have been investigated. The first of these is for a tip with a hemispherical end. This is modeled by a sphere placed on top of the oxide. The interface between the oxide and the silicon is considered as a metal. The method of images is used to calculate the capacitance of each ring. (4)

The second shape is conical with the end truncated to a flat disk. This model approximates a worn tip. To calculate the capacitance in this configuration the oxide silicon interface is again assumed to be metallic. The capacitance for each ring under the disk is simply calculated as the parallel plate capacitance for a capacitor with plate separation equal to the oxide thickness and dielectric constant of the oxide and area of the ring. For rings not under the disk the capacitance is the series combination of the parallel plate oxide capacitance as above and the capacitance for a parallel plate capacitor with a dielectric constant of one and separation equal to the vertical distance from the center of the ring to the cone of the tip. This series combination is again multiplied by the area of the ring.

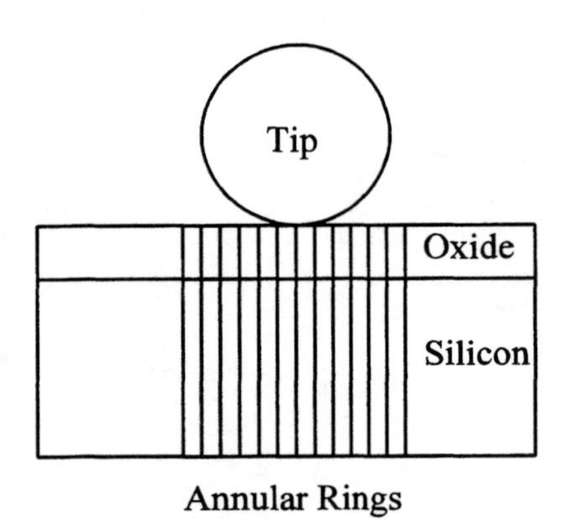

FIGURE 1. Cross-sectional view of physical model of SCM/Si system.

Each annular ring in the silicon is treated as a one dimensional MOS capacitor (5), with the insulator capacitance in the MOS equations case being the combination of the tip and oxide capacitances for the same annular ring. The silicon capacitance is calculated for the maximum and minimum voltages that are applied to the sample and the result is multiplied by the area of the ring. The series combination of this capacitance and the tip and oxide capacitance is calculated and the difference of the two resulting capacitances is taken yielding a change in capacitance for each ring.

The total change in capacitance is found by simply summing the contribution of each ring. The number of rings that are used is found by adding additional rings until the total change in capacitance is unaffected by the adding of an additional ring.

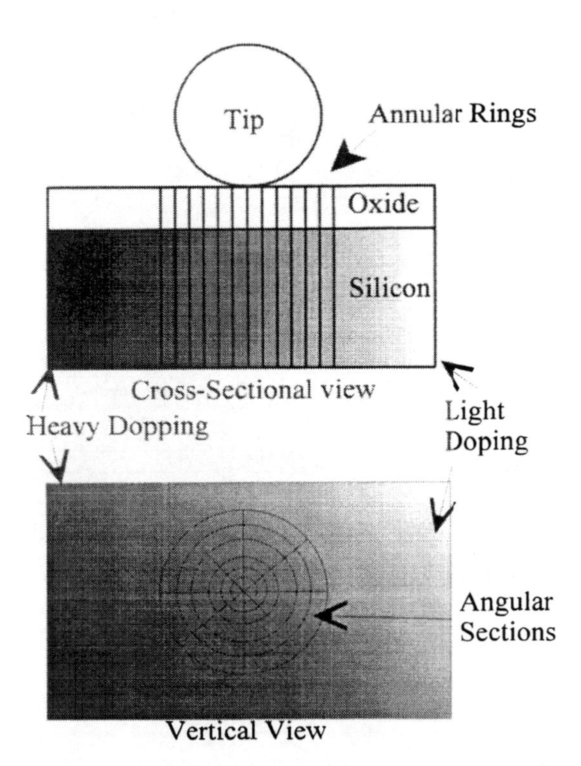

FIGURE 2. Cross-sectional and vertical view of physical model of SCM/Si system, showing both annular rings and angular sections. Darker shading represent heavier doping.

This model assumes a uniform dopant density throughout the tip's sampling region. As has been stated, this is not a good approximation for new technology. To allow for a non-uniform dopant density each annular ring is divided into n angular divisions, yielding small angular sections. (See Figure 2.) The dopant density is allowed to vary for each angular section of each ring. Each section of silicon is treated as a 1D MOS capacitor and the multiplication of the ring area is replaced by the multiplication of the angular section area. In this case the total change in capacitance is found by a sum over all angular sections for each annular ring and all rings.

For non-junction measurements the SCM is operated in a constant change in capacitance mode. This is done to keep the depletion region in the silicon small for the entire range of dopant densities. This is not only important for resolution but also for the approximation that the silicon can be treated as a series of 1D MOS capacitors. The condition for this to be valid is that the lateral components of the electric field in the silicon be small. This is accomplished with a feed back loop which compares the detected change in capacitance with a set change in capacitance. The AC bias voltage that is applied between the tip and the sample is varied such that the change in capacitance is held constant as the tip is scanned across a dopant profile. This has the effect of applying a small bias when the dopant density is light and the depletion region can become large and a large voltage when the doping is heavy and the depletion region is contained by the large number of carriers. When operated in this mode, the voltage that is applied between the tip and the sample is the experimental data that is recorded.

Simulation of Data using the Physical Model

With this model it is possible to explore the expected resolution of a dopant profile for a given tip size and shape. This is accomplished by picking a point of the dopant profile, assigning a desired AC voltage at this point and calculating the change in capacitance using the physical model that allows for dopant gradients. This change in capacitance is stored and will be referred to as the locked capacitance. This locked capacitance plays the same roll as the set capacitance in the feedback loop for obtaining experimental data. To calculate the voltage bias at every point an initial guess at the bias is made and the change in capacitance is calculated. If the change in capacitance is too large, compared to the locked capacitance, the voltage is decreased. If it is too small the voltage is increased. This is continued until the calculated change in capacitance is equal to the locked capacitance. In this way a bias profile that is related to the dopant density, tip size and tip shape is generated.

Figure 3 shows simulated data for flat and spherical tips with a 35 nm radius. The dopant profile used for this simulation was the SEMATECH #1 SIMS data (6). The figure also shows experimental data that was obtained with an MFM tip. For the spherical tip the 50 nm steps between 0.1 and 0.3 μm are clearly visible, while for the flat tip they appear as only small inflections. This example clearly shows the effect of tip shape on resolution. It also shows that the best fit to the experimental data comes by using a flat tip model. This is consistent with examination of tips after imaging with a scanning electron microscope (SEM). SEM images show that the tips often wear to a flat disk with the radius of the disk similar to that of the initial hemispherical radius.

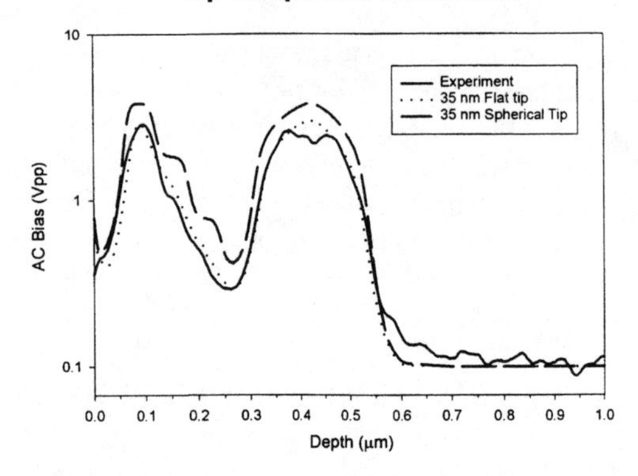

FIGURE 3. Simulation of bias data using the SEMATECH #1 dopant profile and two tip models. The simulated data is compared with experimental data.

Inverse Modeling Algorithm

Using this model it is possible to attempt to solve the inverse problem of finding the correct dopant profile from a given bias profile. Since the equations are not invertible, an iterative technique must be used. Our approach is to use the non-gradient model to invert the bias profile to a dopant profile yielding an initial approximation to the correct dopant profile. This approximate profile is then used to simulate a bias profile using the model that includes dopant gradients. This simulated profile's dopant density is then compared point by point with the measured profile, and the simulated profile is adjusted at each point. The simulated profile is increased at a point if the measured bias is greater than the simulated bias. If the measured bias is less than the simulated bias the dopant density is decreased. If they are the same, no change is made. The change in the dopant density is proportional to the percent difference in the measured and calculated bias. Using the adjusted dopant profile, a bias profile is again simulated and compared with the measured profile. This result is used to again adjust the dopant profile. This process is repeated until the difference between the measured and simulated bias profiles is small. The resulting dopant profile can have much better resolution than is seen in the original bias profile.

Inverse Modeling Results

Figure 4 shows results for a 1D simulation of a staircase structure. This simulation used a 60 nm diameter spherical tip on a profile with 20 nm steps. The dopant profile was used to simulate a bias profile with the model that includes dopant gradients. This bias profile was then converted back to a dopant profile using the model that assumes that the dopant density is uniform. As can be seen the steps have become inflections due to the large sampling area of the tip. After 25 steps in the iterative method, the dopant profile has regained much of its original shape. At step 99, of the algorithm has further improved the profile. The noise inherent in the data is amplified by this inverse method. The value of this noise was chosen to be 100 times smaller than the signal, which is

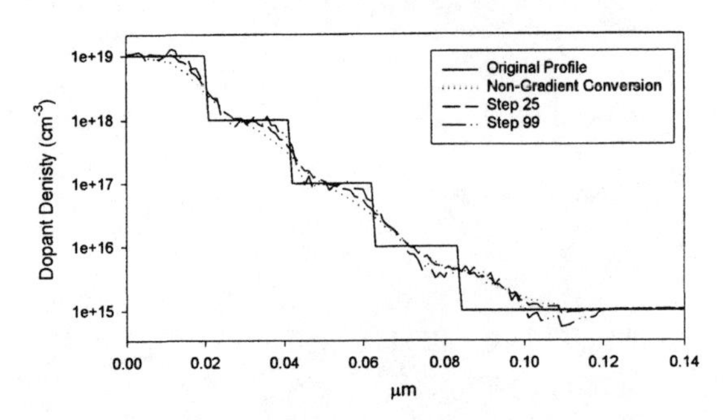

FIGURE 4. Demonstration of inverse modeling removing broadening of the profile due to a large tip.

approximately the calculated signal to noise in our current measurement system.

Figure 5 shows results for a 2D simulation. In this simulation the 35 nm radius spherical tip was used. In this simulation no noise was intentionally introduced. Figure 5a shows the original dopant profile in a ziggurat configuration with 50nm steps. Figure 5b shows the calculated bias profile. Note that broadening of the edges has occurred. Figure 5c show the converted dopant profile using the non-gradient model. Figure 5d shows a dopant profile after 100 steps of the algorithm, yielding a sharpened profile but one which still includes overshoot. This overshoot is found to be decreasing with each step.

The above results are encouraging but are not the solution to the entire problem. In new technologies the dopant gradients are sharp enough that carrier spilling can be a significant problem. The problem is very similar to a p-n junction. These effects are not currently included in the physical model and can have a significant impact on measured data. This work is currently underway. The understanding of how to model these effects will also greatly aid in the effort to provide quantitative data from p-n junction devices.

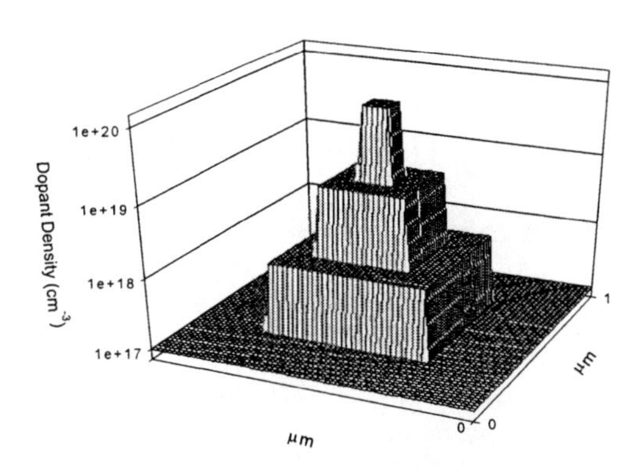

FIGURE 5a. 2D dopant test profile.

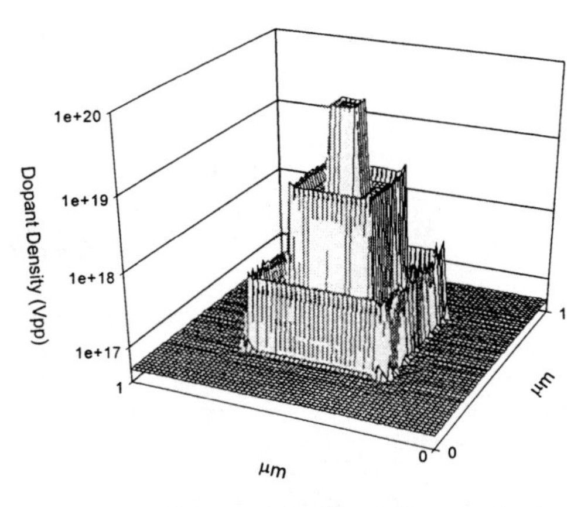

FIGURE 5d. Sharpened dopant profile using inverse modeling and gradient model. Profile is after 100 steps.

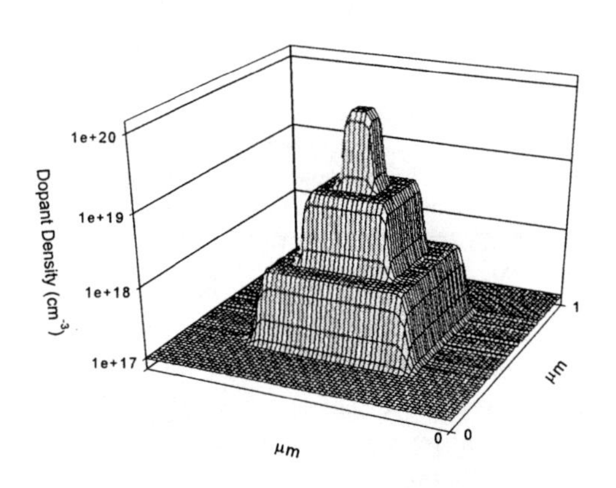

Figure 5c. Dopant profile converted from bias data using non-gradient model.

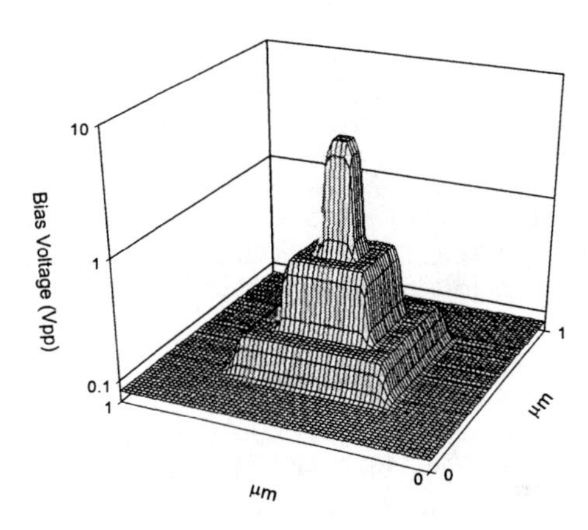

FIGURE 5b. Simulated bias data using gradient model.

CONCLUSIONS

An improved physical model has been developed which includes the effect of dopant gradients. This model has been used to simulated SCM bias data which is found to be in good agreement with measured data. An algorithm has been developed which shows improved resolution for 1 and 2D noiseless simulated data. Carrier spilling has not yet been included in this calculation.

REFERENCES

1. McMurray, J.S., Kim, J., Williams, C.C., *J. Vac. Sci. Technol* **B15**, 1011 (1997).
2. Kopanski, J.J., Marchiando, J.F., Smith, H.E., *J. Vac. Sci. Technol* **B16**, 339 (1998).
3. Marchiando, J.F., Kopanski, J.J.,Lowney, J.R., *J. Vac. Sci. Technol* **B16**, 463 (1998).
4. Huang Y., Ph.D. Thesis Physics Dept., University of Utah, 1995.
5. Sze, S. M. *Physics of Semiconducting Devices*, New York: John Wiley & Sons, 1981, pp. 362-375 .
6. Huang Y., Williams C.C., Wendman M.A., *J. Vac. Sci. Technol* **B14**, 1168 (1996).
7. A.C. Diebold, M.R. Kump J.J. Kopanski, D.G. Seiler , *J. Vac. Sci. Technol* **B14**, 196 (1996).

Silicon Surface Preparation for Two-Dimensional Dopant Characterization

V. A. Ukraintsev, F. R. Potts, R. M. Wallace, L. K. Magel, Hal Edwards, M.-C. Chang

Texas Instruments, Inc., Dallas, Texas, 75265

Most two-dimensional (2D) dopant characterization techniques deal with near surface carrier concentration or surface potential measurements using a device cross-section. Thus, a reproducible and well-characterized surface condition is essential for accurate measurements. We report on an X-ray photoelectron spectroscopy study of surface preparation techniques commonly used in 2D dopant characterization: (i) hydrogen termination, (ii) colloidal silica polishing, (iii) low temperature silicon oxidation. Although the presented results may be applied to any other technique, in this report we concentrate on scanning capacitance microscopy (SCM). For high quality SCM a thin, uniform, and charge-free insulating film has to be formed on top of the silicon sample. Two phenomena may have a critical impact on accuracy of SCM: (i) leakage through the insulating layer and (ii) surface charging which may cause unpredictable variations in flat-band voltage. We found that samples polished by colloidal silica have about one monolayer of oxidized silicon. If the polishing is followed by baking at 200°C in the air then 1 to 2 monolayers of silicon are oxidized and an ultra-thin insulating layer of (4-6) Å is formed. A significant leakage through the silicon oxide layer, which may alter SCM measurements, is expected in both cases. Silicon oxidation under UV irradiation in ozone ambient improves insulating property of silicon oxide film. Interface equivalent to ca. (8-15) Å of silicon dioxide is reproducibly formed by this technique. Importantly, all tested methods of silicon surface preparation are dopant independent. No noticeable surface charging has been observed for any studied method of silicon surface preparation. For silicon oxidized by UV-generated ozone the position of the Si $2p_{3/2}$ core-level relative to the Fermi level changes systematically with dopant type and concentration. The binding energy is higher for n-type silicon and lower for p-type doping. Such behavior is expected for silicon with a relatively low density of surface electronic states and hence unpinned Fermi level. We conclude that silicon oxidation by UV-generated ozone is a promising method of silicon surface preparation for high quality SCM.

INTRODUCTION

The high sensitivity of MOSFET characteristics to variations in dopant profiles leads to very strict requirements on doping metrology [1]. Over the last 10-15 years various analytical techniques have been employed in order to provide the desired accuracy of dopant characterization. However, only a few of these methods may be truly considered as bulk chemical analysis techniques [1]. The rest of the methods deal with near-surface carrier concentration or surface potential measurements using a device cross-section. To get a true bulk dopant distribution, all surface effects have to be minimized. Thus, for the majority of 2D dopant characterization techniques a reproducible and well-characterized surface condition is essential for accurate measurements.

In this paper we study the most common surface preparation techniques used in 2D dopant characterization: (i) hydrogen termination, (ii) colloidal silica polishing, and (iii) low temperature silicon oxidation. Although the presented results may be applied to any other technique, we this report we focus on scanning capacitance microscopy (SCM). SCM is a widely used dopant characterization technique and extensive background information for the method can be found elsewhere [2,3]. Scanning capacitance microscopy requires the presence of a thin insulating layer on top of the silicon sample to be studied. Naturally, silicon oxide is a very good candidate for such insulation. The silicon oxide layer plays a principal role in quantitative SCM. The sensitivity and dynamic range of the method strongly depend on silicon oxide thickness and dielectric constant. Moreover, since in many cases the depletion width determines the ultimate spatial resolution of SCM, the silicon oxide thickness should limit the spatial resolution as well. In other words, the silicon oxide layer must be thin enough (a few tens of angstroms or even less) to permit high spatial resolution and reliable detection for a wide dynamic range of dopant concentrations. However, an oxide that is too thin might have non-uniform thickness or dielectric constant or have high leakage, which would corrupt the capacitance measurements.

To operate the SCM properly, the tip-sample bias voltage should be around its flat-band value at every point across the area of interest [2,3]. Applied to SCM tip (metal electrode of MOS), the flat-band voltage serves to correct surface band bending caused by difference in tip and silicon work functions, surface electronic states, surface and oxide

CP449, *Characterization and Metrology for ULSI Technology: 1998 International Conference*
edited by D. G. Seiler, A. C. Diebold, W. M. Bullis, T. J. Shaffner, R. McDonald, and E. J. Walters

incorporated charged impurities. Considering the dependence of MOS flat-band voltage on semiconductor type and dopant concentration, this is not a trivial task. Oxide thickness may significantly affect the flat-band voltages. The thicker the oxide is, the higher voltage that should be applied to compensate the interface charges. The surface band bending related to interface charges can be neglected, however, if the oxide is thin enough [4].

It is evident that a surface insulating layer is a key element in SCM and hence must be thoughtfully characterized. X-ray photoelectron spectroscopy seems to be an appropriate analytical technique for such study since it can provide us with the chemical composition of the silicon top layer and the value of the surface potential [5,6].

RESULTS AND DISCUSSION

As-received, homogeneously doped natively oxidized silicon wafers were cleaved into ~ 1×1 cm^2 pieces and thoroughly cleaned in various boiled solvents. Several surface conditions with potential interest for SCM were studied: (a) as received natively oxidized silicon, (b) hydrogen-terminated silicon, (c) silicon polished with colloidal silica, (d) silicon polished with colloidal silica and baked in the air, (e) silicon polished with colloidal silica, baked in the air and oxidized by UV-generated ozone. Every surface preparation routine was applied to n- and p-type silicon of various dopant concentrations.

The hydrogen termination was performed by a 30 second dip of the samples in water-diluted 4.9% HF followed by ca. 3 minutes rinse in deionized (DI) water.

After solvent cleaning some samples were manually polished for 45 seconds, using Buehler's colloidal silica (siton), and thoroughly rinsed under tap water. An additional RCA silicon cleaning [7] was employed in some experiments on as-received and on siton-polished silicon but showed no significant effect on the measured surface potential.

To increase oxide film thickness two additional recipes were implemented:

(1) Solvent cleaned samples were manually polished for 45 seconds, using Buehler's colloidal silica, and rinsed under tap water. The samples were baked for 15 minutes in the air on hot plate set at 200°C.

(2) Some of these siton polished and baked samples were placed in the UVOCS UV ozone generator for 20 minutes.

Measurements were done on a Perkin Elmer PHI 5600 system using a monochromatic Al K_α radiation (1486.7 eV) and a pass energy of 23.5 eV.

The first set of experiments was performed on lightly doped n- and p-type silicon (ca. 2×10^{15} cm^{-3} phosphorous and boron, respectively). Chemical compositions of the top-most layer of silicon for various surface conditions are presented in Fig. 1. It is important to keep in mind, that in XPS, the photoelectrons are collected from a so called escape depth, which varies with the electron energy. For our experimental conditions, the escape depth may be estimated as (15-30) Å or (10-20) monolayers of silicon in the (100) crystallographic direction [8]. Thus, atomic concentrations presented in Fig. 1 correspond to multilayer silicon/air interface which may include adsorbates, silicon oxide and crystalline silicon. The actual surface (against bulk) contribution to the signal may be higher than one would expect from the simple estimations based on the value of escape depth. A non-flat interface can be responsible for this (surface roughness, silicon oxide 3D island formation, adsorbate agglomerates, *etc.*). Therefore, in the following discussion we do not rely on the absolute values of the atomic concentrations but rather focus on comparative analysis.

In all studied cases, the major detected elements were silicon, oxygen and carbon. Note that hydrogen can not be directly detected by XPS. Silicon was found in the form of bulk silicon and silicon oxide. Si $2p_{1/2}$ (ca. 100.0 eV) and Si $2p_{3/2}$ (ca. 99.4 eV) peaks were used to measure bulk silicon concentration. Peaks observed at ca. (102.5-103.5) eV were assigned to silicon bonded to various numbers of oxygen atoms from 1/2 for a Si-O-Si bridge to 4 for bulk SiO_2 [9].

The oxygen concentration detected using O1s peak at 532 eV was always higher than that calculated from the concentration of SiO_2. An additional oxygen may exist on the surface in form of physisorbed water or various organic adsorbates.

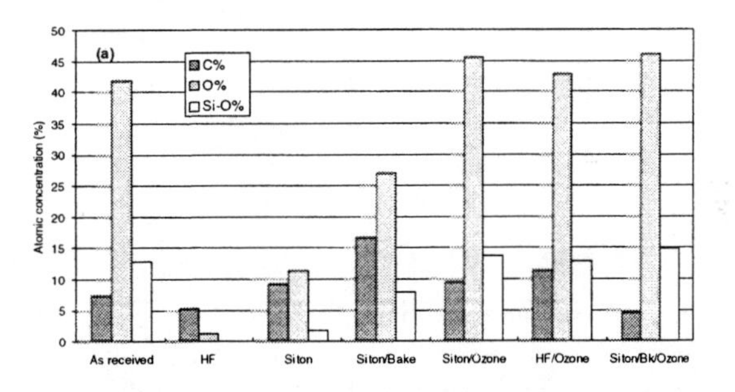

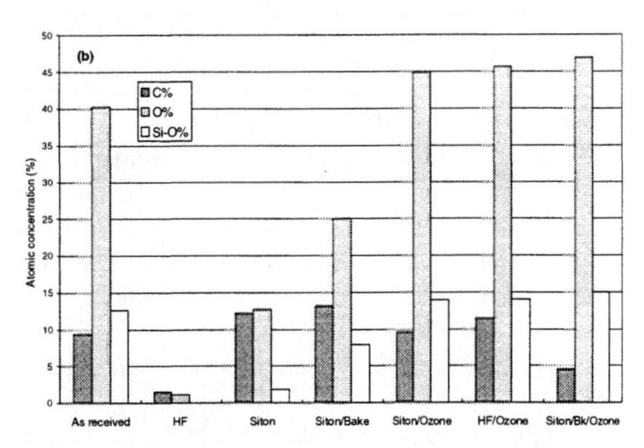

FIGURE 1. Chemical composition of top-most layer of silicon detected by XPS after employing various surface preparation techniques: (a) p-type 2×10^{15} cm^{-3} boron doped and (b) n-type 2×10^{15} cm^{-3} phosphorous doped silicon.

Carbon (C1s) was observed at an energy of 285 eV and probably corresponds to physisorbed hydrocarbons. No silicon carbide has been detected in the experiments since no signal was observed below 284 eV [10].

Several important conclusions can be derived from the data (Fig.1):

(1) Etching in diluted HF effectively removes silicon oxide and inhibits subsequent silicon surface oxidation as a result of hydrogen termination [11].
(2) The amount of silicon oxide for silicon polished by colloidal silica is about 6 times less than that for natively oxidized silicon.
(3) Baking of silicon in an air ambient significantly increases the amount of silicon oxide but keeps it below the level observed for natively oxidized silicon.
(4) Exposure to UV-generated ozone produces a silicon oxide layer with a thickness that slightly exceeds that of a native oxide.
(5) The data are very much similar for lightly doped n- and p-type silicon.

The second set of experiments was performed on wider variety of uniformly doped silicon samples: (i) n-type 5×10^{18} cm^{-3} antimony doped, (ii) n-type 2×10^{15} cm^{-3} phosphorous doped, (iii) p-type 1.3×10^{13} cm^{-3} boron doped, (iv) p-type 2×10^{15} cm^{-3} boron doped, and (v) p-type 9×10^{18} cm^{-3} boron doped. The (ii) and (iv) samples are identical to ones used in the first set of XPS experiments. All surface preparation procedures were kept unchanged.

Data obtained for silicon oxidized by UV-generated ozone are presented in Fig. 2. It is very important from the SCM stand point that none of tested surface preparation techniques revealed any definite dopant dependence. Therefore, no systematic error should be introduced to SCM by employing these methods of silicon preparation.

Thickness of Silicon Oxide Layer

According to XPS data available in a literature, the native oxide film formed at room temperature on Si(100) is equivalent to a ca. (8-12) Å thick layer of silicon dioxide [12,13,14,15]. Since the density of silicon atoms in silicon dioxide is about half of its density in crystalline silicon, the oxidation consumes (4-6) Å of silicon or 3-4 monolayers in the (100) crystallographic direction. Thus, the reduction of the oxidized silicon signal by a factor of 6 for the silicon polished by colloidal silica (Fig. 1) indicates that this sample has about one or even less than one monolayer of Si bonded to oxygen atoms.

This surprising (for the SCM community) result [16,17] is, however, in close agreement with a systematic study conducted by Pietsch *et al.* [18]. According to this study, a silicon surface polished by colloidal silica is predominately covered by dihydride species, a significant fraction of which is backbonded to oxygen. Such a chemical structure of the silicon top layer resembles an early stage of silicon native oxidation [11,19,20].

Siton polished and baked at 200°C silicon has ~ 3 times more silicon oxide; but, this amount is still considerably lower than for natively oxidized samples. We estimate that 1 to 2 monolayers of silicon are oxidized and form an ultra-thin and presumably an insulating layer of (4-6) Å.

The thickness of the silicon oxide layer formed by UV-generated ozone probably reaches a saturation similar to those observed for native oxidation [12,15]. Thus, 3 to 4 monolayers of silicon presumably oxidize, forming ca. (8-15) Å thick layer of silicon oxide.

Concluding, in agreement with Pietsch *et al.* [18] we found that colloidal silica polishing produces a passivated silicon surface where about one monolayer of silicon is bonded to oxygen. A significant leakage through the layer, which may corrupt SCM measurements, is anticipated [21]. Silicon oxidation in an ozone ambient "returns" silicon to its original or "as received" condition and forms an interface equivalent to ca. (8-15) Å thick layer of silicon dioxide. We speculate that during this process, silicon oxide thickness reaches its saturation value for room temperature oxidation [12].

Surface Impurities and Charging Effects

It is known [22,23] that high-resolution X-ray photoelectron spectroscopy is able to characterize silicon doping by accurate measurements of the Si 2p core-level position (E_{Si2p}) relative to the Fermi level. In the ideal case of an uncharged surface the binding energy should be higher for n- and lower for p-type silicon.

Unfortunately, in most cases this simple picture of the process does not work. The reason being surface band bending caused by various surface charging effects [4]. Surface layers of silicon differ in electronic structure from the bulk silicon. This alone may cause partial charging of surface atoms and hence surface band bending. Moreover,

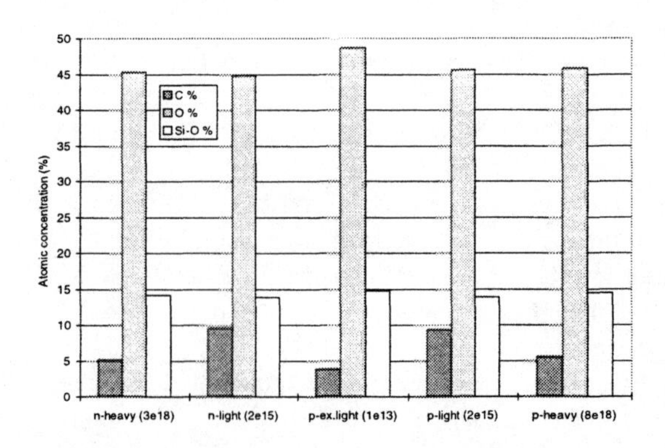

FIGURE 2. Chemical composition of top-most layer of silicon polished by colloidal silica, baked in the air and oxidized in UV-generated ozone.

various adsorbates, especially metal impurities, may affect the electronic structure of the silicon/air interface and change the Fermi level position near the surface.

As a result, the Fermi level may be pinned at the surface at a certain position with respect to the bulk silicon electronic structure. Since in XPS photoelectrons can be collected from a very thin surface layer of (15-30) Å, no doping dependent changes for the E_{Si2p} should be observed unless the Debye length or depth of surface charge screening is smaller or comparable to the photoelectron escape depth [24,25]. Thus, the higher the dopant concentration and the lower the surface charging, the more probable is observation of the dopant induced changes of the E_{Si2p}.

Binding energies for Si $2p_{3/2}$ core-level electrons obtained for various silicon samples treated by different surface preparation techniques are presented in Fig. 3. As expected the binding energy for heavily n-doped samples are higher than for heavily p-doped ones for all tested methods of surface preparation. Binding energies for lightly doped silicon reveal a similar tendency in general, but experience significant fluctuations in some cases.

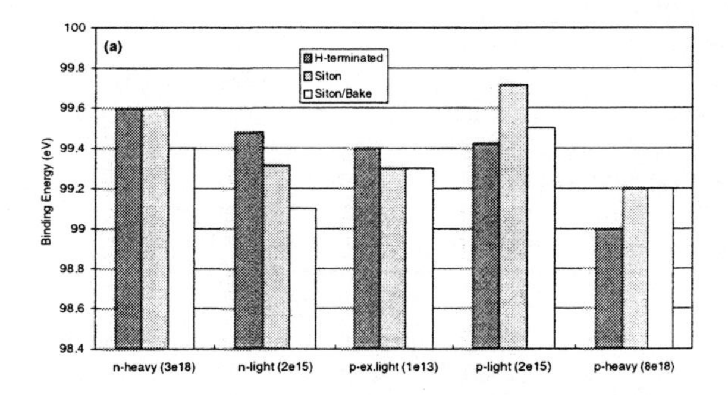

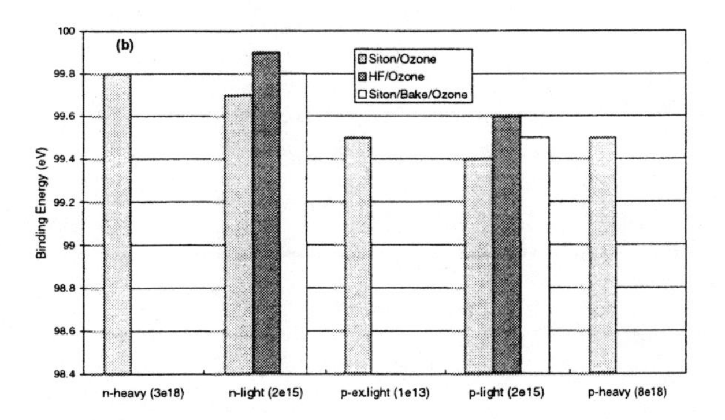

FIGURE 3. Changes of Si $2p_{3/2}$ core-level energy with dopant type and concentration for various surface preparation techniques: (a) hydrogen terminated silicon, siton polished silicon, baked in the air siton polished silicon; (b) baked in the air siton polished silicon oxidized in UV-generated ozone, siton polished silicon oxidized in UV-generated ozone, hydrogen terminated silicon oxidized in UV-generated ozone.

Hydrogen termination likely reduces density of the band gap surface states which appears in large, ca. (0.5-0.6) eV, variation of the E_{Si2p} with doping, Fig. 3(a). Since the Fermi level moves systematically through the band gap even at very low dopant concentrations it can be noticed again that surface concentration of impurities is probably low and causes no charging effects [22].

Polishing of silicon with colloidal silica changes surface condition significantly, Fig. 3(a). Though the E_{Si2p} for heavily n-is still higher than for heavily p-samples, no systematic variation of the binding energy with dopant is observed any more. The energy sporadic behavior suggests, however, no strong Fermi level pinning.

Oxidation of siton polished samples by UV-generated ozone improves the situation, Figure 3(b), suggesting that surface contamination by organic molecules and carbon may influence surface potential [22]. The dopant related shift of the E_{Si2p} for ozone oxidized samples is systematic, but the energy variation (ca. 0.3 eV) is less than the one observed for hydrogen terminated silicon, Fig. 3(a). On average slight increase of the binding energy, ca. (0.2-0.3) eV, is detected with respect to hydrogen terminated silicon. Such increase may be explained by minor charging of the silicon oxide film [6].

The following conclusions can be made:

(1) No strong surface charging was detected for any surface preparation method. In all cases the Si 2p binding energy was changing noticeably with dopant type and concentration.

(2) For hydrogen terminated silicon and silicon oxidized by UV-generated ozone, systematic changes of the E_{Si2p} with doping have been observed. The binding energy was higher for heavily n- and lower for heavily p-doped silicon.

(3) Hydrogen termination likely gives the lowest density of surface states since HF etched silicon shows systematic and the widest, ca. (0.5-0.6) eV, variation in the Si 2p binding energy. The other surface preparation methods probably create noticeable amount of surface electronic states which partially pin the Fermi level.

CONCLUSION

Silicon surface preparation techniques applicable for 2D dopant characterization were studied by X-ray photoelectron spectroscopy. It was found that high quality hydrogen termination can be achieved in a non-clean room environment by etching silicon in diluted HF (Fig. 1). This procedure leaves an uncharged and passivated silicon surface with an unpinned Fermi level. Large and systematic changes of Si 2p binding energy with dopant type and concentration indicate a low density of surface states and minor surface band bending. Formation of a silicon oxide layer likely increases the density of surface states, which may explain the observed reduction of the binding energy variation with doping. However, the Fermi level evidently remains unpinned in all tested cases including measurements on lightly doped samples (Fig. 3).

Silicon samples polished by colloidal silica have about one or even less than one monolayer of oxidized silicon (Fig. 1). The effective thickness of the insulating layer is estimated as (1-2) Å. According to a study conducted by Pietsch *et al.* [18], a silicon surface polished by colloidal silica is predominantly covered by dihydride species, a significant fraction of which is backbonded to oxygen. Such chemical structure of the silicon top layer resembles an early stage of native oxidation of hydrogen terminated silicon. If siton polishing is followed by baking in the air then 1 to 2 monolayers of silicon are oxidized and an ultra-thin insulating layer of (4-6) Å is formed. A significant leakage through the silicon oxide layer, which may alter SCM measurements, is expected in both cases [21].

Silicon oxidation under UV irradiation in an ozone ambient should improve the insulating properties of the silicon oxide film. It is found that an interface equivalent to ca. (8-15) Å of silicon dioxide can be reproducibly formed by this technique. We speculate that during this process, the thickness of the silicon dioxide layer approaches its highest possible value for room temperature oxidation.

Considering possible leakage through an ultra-thin insulating layer, silicon oxidized by UV-generated ozone is the best choice for today. According to XPS silicon oxide formed by ozone is similar to saturated native oxide and hence expected to be relatively uniform across the surface. This surface preparation technique consistently shows no surface Fermi-level pinning and a moderate density of surface electronic states. All these data suggest that silicon oxidation by UV-generated ozone is a promising method of silicon surface preparation for high quality scanning capacitance microscopy.

REFERENCES

1. R. Subrahmanyan, *J. Vac. Sci. Technol.* B10 358-368 (1992).
2. C. C. Williams, J. Slinkman, W. P. Hough, and H. K. Wickramasinghe, *J. Vac. Sci. Technol.* A8 895-898 (1990).
3. J. J. Kopanski, J. F. Marchiando, and J. R. Lowney, *J. Vac. Sci. Technol.* B14 242-247 (1996).
4. B. G. Streetman, *Solid State Electronic Devices*, Prentice-Hall, Englewood Cliffs, N. J., 1990, pp. 305-311.
5. *Practical Surface Analysis (Second Edition), Volume 1: Auger and X-ray Photoelectron Spectroscopy*, Edited by D. Briggs and M.P. Seah, John Wiley & Sons Ltd., 1990.
6. S. Iwata and A. Ishizaka, *J. Appl. Phys.* 79 (1996) 6653-6713.
7. W. Kern, D. A. Puotinen, *RCA Review* (1970) 187-206. We used the following four step cleaning procedure: (i) 4.9% HF 30 sec etching, (ii) NH_4OH:H_2O_2:H_2O=1:1:5 20 minutes etching at 80°C, (iii) H_2O:H_2O_2:HCl=6:1:1 15 minutes etching at 80C, (iv) 0.49% HF 120 sec etching.
8. R. Flitsch, S. I. Raider, *J. Vac. Sci. Technol.* 12 (1975) 305; M. F. Hochella, Jr., A. H. Carim, *Surf. Sci.* 197 (1988) L260.
9. M. M. Banaszak Holl, S. Lee, F. R. McFeely, *Appl. Phys. Lett.* 65 (1994) 1097-1099.
10. *Handbook of X-ray Photoelectron Spectroscopy*, Ed. Jill Chastain, Perkin Elmer Corporation, Eden Prairie, MN, USA
11. T. Takahagi, A. Ishitani, H. Kuroda, Y. Nagasawa, H. Ito, S. Wakao, *J. Appl. Phys.* 68 (1990) 2187-2191.
12. M. Morita, T. Ohmi, E. Hasegawa, M. Kawakami, and M. Ohwada, *J. Appl. Phys.* 68 (1990) 1272-1281.
13. D. Gräf, M. Grundner, R. Schulz, L. Mühlhoff, *J. Appl. Phys.* 68 (1990) 5155-5161.
14. M. L. W. van der Zwan, J. A. Bardwell. G. I. Sproule, and M. J. Graham, *Appl. Phys. Lett.* 64 (1994) 446-447.
15. G. F. Cerofolini, G. La Bruna, L. Meda, *Appl. Surf. Sci.* 93(1996) 255-266.
16. J. S. McMurray, J. Kim, C. C. Williams, J. Slinkman, *J. Vac. Sci. Technol.* B15 (1997) 1011-1014.
17. J. J. Kopanski, J. F. Marchiando, D. W. Berning, R. Alvis, H. E. Smith, "Scanning Capacitance Microscopy Measurement of 2D Dopant Profiles Across Junctions," *Proceedings of 4th International Workshop on Measurements, Characterization and Modeling of Ultra-Shallow Doping Profiles in Semiconductors*, M. Current, M. Kump and G. McGuire, Eds., Research Triangle Park, NC, USA, 1997, p. 53.1-9.
18. G. J. Pietsch, *Appl. Phys.* A 60 (1995) 347-363; G. J. Pietsch, G. S. Higashi, Y. J. Chabal, *Appl. Phys. Lett.* 64 (1994) 3115.
19. T. Takahagi, I. Nagai, A. Ishitani, H. Kuroda, Y. Nagasawa, *J. Appl. Phys.* 64 (1988) 3516-3521.
20. H. Ogawa, K. Ishikawa, C. Inomata, S. Fujimura, *J. Appl. Phys.* 79 (1995) 472-477.
21. B. Brar, G. D. Wilk, A. C. Seabaugh, *Appl. Phys. Lett.* 69 (1996) 2728-2730.
22. F. J. Himpsel, G. Hollinger, and R. A. Pollak, *Phys. Rev.* B28 (1983) 7014-7018.
23. S. Iwata and A. Ishizaka, *J. Appl. Phys.* 79 (1996) 6653-6713.
24. W. Eberhardt, G. Kalkoffen, C. Kunz, D. Aspens, M. Cardona, *Phys. Stat. Sol.* B88 (1978) 135-143.
25. F. J. Himpsel, P. Heimann, T.-C. Chiang, D. E. Eastman, *Phys. Rev. Lett.* 45 (1980) 1112-1115.

Dopant Characterization Round-Robin Study Performed on Two-dimensional Test Structures Fabricated at Texas Instruments

Vladimir A. Ukraintsev, R. Scott List and Mi-Chang Chang
Hal Edwards, Charles F. Machala, Richard San Martin

Texas Instruments, Inc., Dallas, Texas, 75265

Vladimir Zavyalov, Jeff S. McMurray and Clayton C. Williams

Department of Physics, University of Utah, Salt Lake City, Utah 84112

Peter De Wolf and Wilfried Vandervorst

IMEC, Kapeldreef 75, B-3001 Leuven, Belgium

David Venables, Suneeta S. Neogi, Diana L. Ottaviani

Department of Materials Science and Engineering, North Carolina State University, Raleigh, NC 27695

Joseph J. Kopanski, Jay F. Marchiando, Brian G. Rennex

National Institute of Standards and Technology, Gaithersburg, MD 20899

Jochonia N. Nxumalo, Yufei Li and Douglas J. Thomson

Department of Electrical and Computer Engineering, University of Manitoba, Winnipeg, Manitoba, R3T 5V6, Canada

The lack of a two-dimensional (2D) dopant standard and hence *a priori* knowledge of dopant distribution makes it impossible to unambiguously judge accuracy of any experimental or theoretical effort to characterize silicon doping in two-dimensions. Recently a strong progress has been made in quantitative scanning capacitance microscopy (SCM), scanning spreading resistance microscopy (SSRM), secondary electron (SE) and transmission electron microscopy (TEM) doping profiling. Several research groups have claimed an ability of quantitative 2D dopant characterization. A round-robin study involving various analytical techniques and comprehensive numerical simulations should help to evaluate an accuracy of available quantitative techniques and set some helpful standard for further development. We report on a world-wide round-robin study performed on CMOS 2D test structures fabricated at Texas Instruments (TI). Seven research groups, which represent an advanced SCM, SSRM and TEM dopant profiling, participated in the study. Related process information and 1D dopant profiles measured by secondary ion mass-spectroscopy (SIMS) were released to the sites. Process simulator TSUPREM4 tuned using SIMS and inverse electrical modeling was employed to simulate 2D dopant distributions for both PMOS and NMOS test structures. Method-to-method variations in pn-junction position are noticeably higher than 10 nm, the spatial resolution requested by TCAD. The statistical scatter in pn-junction position varies from 10 to 30%. Although interpretation of the data is still a challenging problem, controlled and reproducible surface preparation is an important step, which may help to reduce the observed diversity of dopant characterization results. A round-robin study on test structure capable of electrical characterization should link round-robin results to device performance and hence is necessary.

CP449, *Characterization and Metrology for ULSI Technology: 1998 International Conference*
edited by D. G. Seiler, A. C. Diebold, W. M. Bullis, T. J. Shaffner, R. McDonald, and E. J. Walters

INTRODUCTION

The lack of a 2D dopant standard and hence *a priori* knowledge of dopant distribution makes it impossible to unambiguously judge accuracy of any experimental or theoretical effort to characterize silicon doping in two-dimensions. Recently a significant progress has been made in quantitative doping profiling. Several research groups have claimed the ability of quantitative 2D dopant characterization [1]. A round-robin study involving various analytical techniques and comprehensive numerical simulation was undertaken in order to evaluate the accuracy of available quantitative techniques and to set some helpful standard for further development. We report on results of the first round of this worldwide round-robin study.

RESULTS

The round-robin study was performed on CMOS 2D test structures fabricated at TI (Fig. 1). Related process information and 1D dopant profiles measured by SIMS were released to the participating sites (Figs. 2, 3).

The following techniques are presented in the study: (1) numerical process simulation (TSUPREM4) by TI [2]; (2) dopant profiling and pn-junction delineation employing SCM by University of Utah [3] and National Institute of Standards and Technology [4]; (3) pn-junction delineation using scanning capacitance spectroscopy by TI [5]; (4) pn-junction delineation employing Schottky contact capacitance and resistance microscopy by University of Manitoba [6]; (5) dopant profiling and pn-junction delineation using SSRM by IMEC [7]; (6) dopant concentration profiling and pn-junction delineation applying wet etching followed by TEM by North Carolina State University [8].

The participants were asked to report the following data: (i) channel length, (ii) source, drain (S/D) and gate overlap, (iii) S/D junction depth, (iv) 2D dopant iso-concentration contours for S/D and channel regions, (v) lateral and vertical dopant profiles for these regions. The participants have agreed to not publish the results before reporting their results to TI.

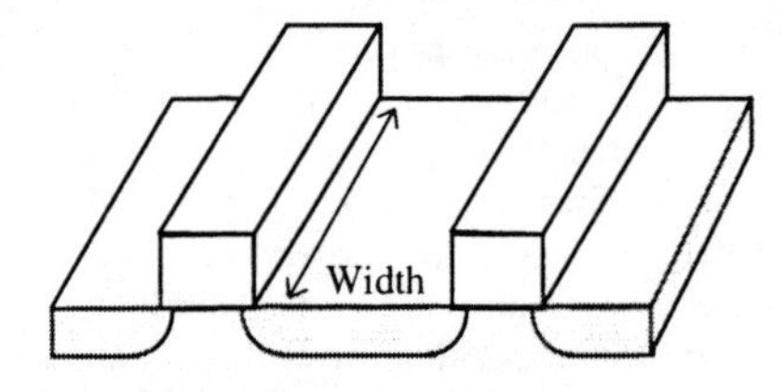

FIGURE 1. TI CMOS 2D test structure:
Pitch: 1.4 μm. Gate length: 0.4 μm. Gate width: 1 mm.
Doping (PMOS): P/1E17cm^{-3} N-well and B/1.5E15cm^{-2}/5KeV.
Annealing (PMOS): 20 min @ 900°C followed by RTA.
Doping (NMOS): B/1E17cm^{-3} P-well and As/1.5E15cm^{-2}/20KeV.
Annealing (NMOS): 140 min @ 800°C followed by RTA.

Metrology requirements imposed on 2D dopant profiling by TCAD were discussed in details elsewhere [9]. Dopant concentration sensitivity of (10^{17}-10^{20}) cm^{-3}, spatial resolution of 10 nm and accuracy of (20-50) % are necessary for proper characterization of sub-micron MOSFET and hence were requested from the participants.

Process simulator TSUPREM4 (TS4) tuned at TI using SIMS and inverse electrical modeling was employed to simulate 2D dopant distribution for both PMOS and NMOS test structures. Simulated dopant profiles for the NMOS test structure are in good agreement with 1D SIMS data (Fig. 2). At the same time, the boron profile simulated for PMOS structure disagrees with 1D SIMS data (Fig. 3). However, since pn-junction depth determined by TS4 agrees well with 1D SIMS prediction we assume that the

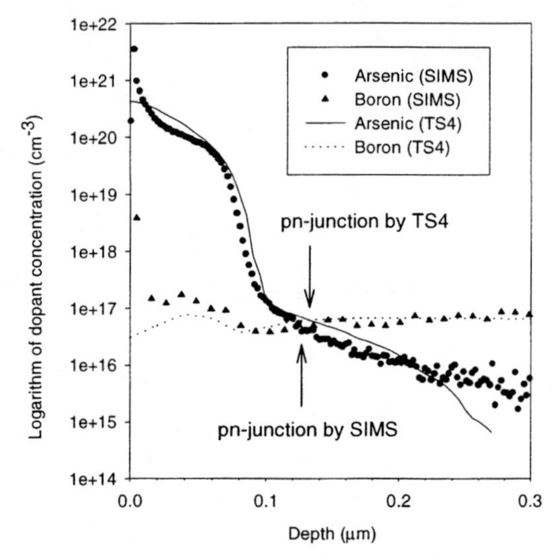

FIGURE 2. SIMS and TS4 dopant profilies for NMOS structure.

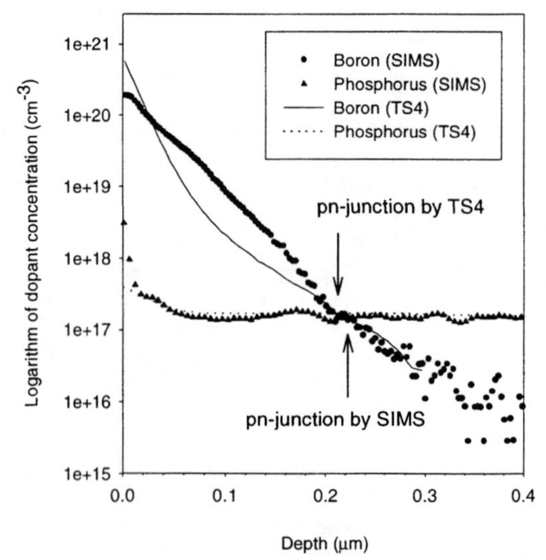

FIGURE 3. SIMS and TS4 dopant profilies for PMOS structure.

TABLE 1

PMOS

	SIMS	*TS4*	*TI SCM Cross-S*	*TI SCM Top-V*	*UM SCM*	*UM SRM*	*IMEC SSRM*	*UU SCM*	*NCSU TEM*	*NIST SCM*
S/D Depth (nm)	234†	216	182±7		180	205	165	250±30		235±14
Channel Length (nm)		(160)	186±36	209±31	200	380	145	125±40		81±13
S/D/G Overlap (nm)		120	(91±19)	(80±12)	80	7.5	127	138±40		161±11
Gate Length (nm)		(400)	368	369	360	395	399	400		403

NMOS

	SIMS	*TS4*	*TI SCM Cross-S*	*TI SCM Top-V*	*UM SCM*	*UM SRM*	*IMEC SSRM*	*UU SCM*	*NCSU TEM*	*NIST SCM*
S/D Depth (nm)	125†	135	198±13				135	100±15	100.5	131±33
Channel Length (nm)		(270)	121±8	131±12			175	320±30	228	295±25
S/D/G Overlap (nm)		65	(127±9)	(122±11)			110	37±10	56.5	50±13
Gate Length (nm)		(400)	375	375			395	394	353	395

† Gate oxide thickness is 20 nm for PMOS and 10 nm for NMOS. Numbers in brackets were not measured directly.

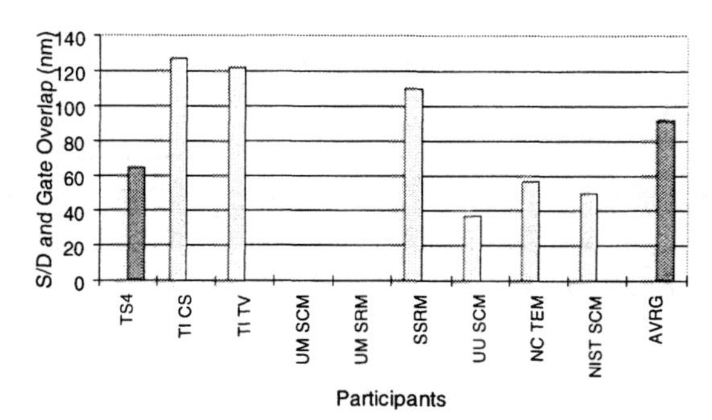

FIGURE 4. Source, drain and gate overlap for NMOS structure.

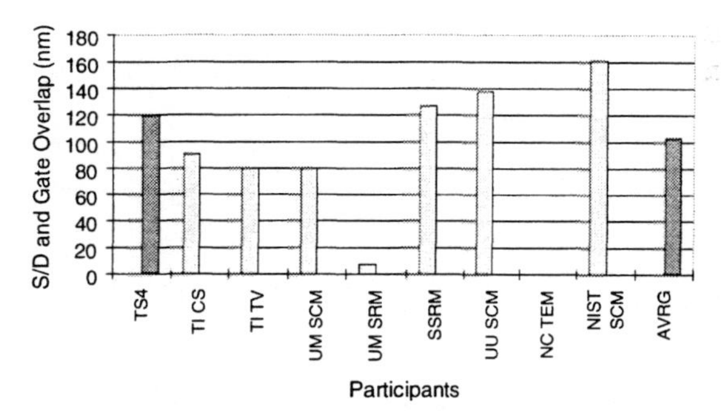

FIGURE 5. Source, drain and gate overlap for PMOS structure.

lateral pn-junction position is correctly simulated by TS4.

The round-robin results are summarized in Table 1 and Figs. 4-7. Method-to-method variation in pn-junction position are significantly higher than 10 nm, which is the spatial resolution requested by the TCAD community. Asymmetry between the left and the right sides of the device was usually observed.

Statistical scatter in pn-junction position (the source/drain and gate overlap) determined at a particular participating site varies from 10 to 30%. Shown in Table 1 standard deviations represent the statistical error but not a systematic error, which is unknown.

An average source, drain and gate overlap for PMOS is (103±27) nm with simulations showing 120 nm. An average source, drain and gate overlap for NMOS is (92±37) nm with a simulated value of 65 nm. Data reported for NMOS structure are clearly divided into two groups with significant, ~(50-60) nm, difference in the overlap value.

Among detected and potential problems which may cause the observed data divergence we want to notice the following: (i) dopant sensitive leakage through "insulating" layer of silicon oxide, (ii) inaccurate delineation of the gate edge, (iii) surface contamination during polishing, (iv) dopant lateral displacement during polishing, (v) selective dopant accumulation at surface during polishing.

The dopant sensitive leakage through "insulating" layer of silicon oxide is well illustrated by the data of the University of Manitoba group (Fig. 8). A nanoampere leakage current was detected for silicon sample polished by commonly used colloidal silica. The leakage may significantly influence SCM measurements. This problem as well as the problem of surface contamination during polishing have been addressed in recent study [10].

An accurate gate edge position determination is not a trivial task. During this study it was noticed that colloidal silica polishing and wet etching employed for dopant delineation may cause notable variation in apparent gate edge position (cf. gate lengths in Table 1). This problem can be resolved if a periodic test structure is used and gate size is accurately determined prior to sample cross-sectioning. In

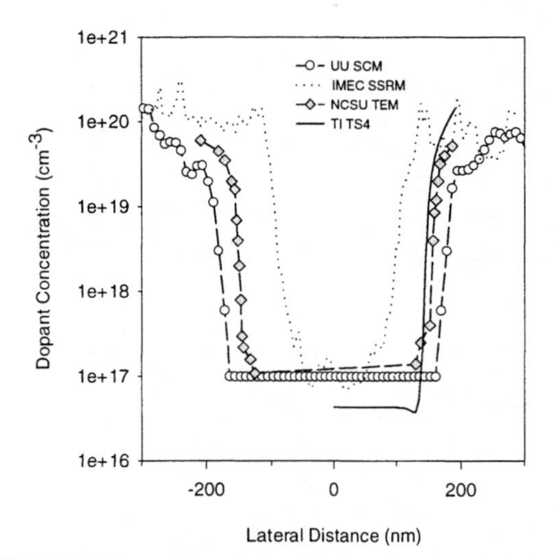

FIGURE 6. Lateral arsenic profiles for NMOS structure.

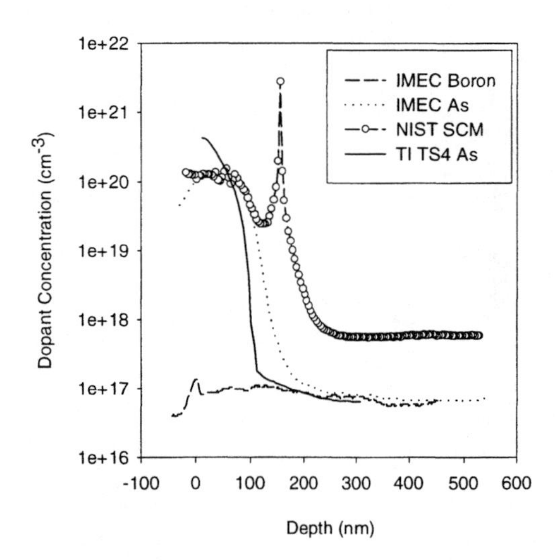

FIGURE 7. Vertical dopant profiles for NMOS structure.

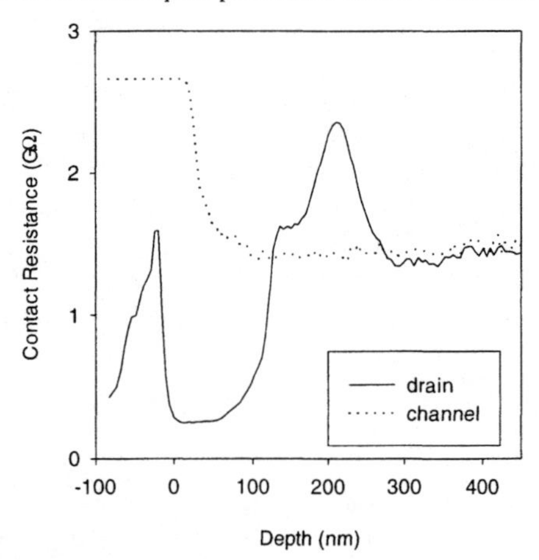

FIGURE 8. SRM data obtained for PMOS sample polished by colloidal silica. The tip is biased with respect to the sample and scanned while maintaining a small contact force ~ (1-10) μN.

this study TI has implemented scanning capacitance spectroscopy using both conventional cross-sectional analysis and a novel top-view approach. The latter helps to avoid mechanical polishing during sample cross-sectioning which may be a source of surface contamination and dopant redistribution.

The problem of dopant lateral displacement and dopant accumulation at silicon surface during polishing deserves careful consideration. To the best of our knowledge, this issue has not been addressed.

CONCLUSION

A round-robin study of dopant characterization was performed on 2D CMOS test structures fabricated at TI. Seven research groups which represent an advanced SCM, SSRM and TEM dopant profiling have participated in the study.

The first round has shown the following:

1) Site-to-site variation in pn-junction position are noticeably higher than spatial resolution requested by TCAD. PMOS overlap is (103±27) nm with simulated one of 120 nm. NMOS overlap is (92±37) nm with simulated one of 65 nm. Data reported for NMOS structure are divided in two groups with significant, ~(50-60) nm, difference in overlap.

2) Statistical scattering in pn-junction position determined at a particular participating laboratory varies from 10 to 30%. Therefore, none of the methods can guarantee the accuracy and resolution necessary for process simulators tuning at this time.

3) Although interpretation of SCM and SRM data is still a challenging problem, controlled and reproducible surface preparation is an important step, which may help to reduce the observed diversity of dopant characterization results.

4) A round-robin study on a test structure capable of electrical characterization should link round-robin results to device performance. Therefore we suggest that test structures with electrical characterization capabilities be used in the next round.

REFERENCES

1. *Proceedings of 4th International Workshop on Measurements, Characterization and Modeling of Ultra-Shallow Doping Profiles in Semiconductors*, M. Current, M. Kump and G. McGuire, Eds., Research Triangle Park, NC, USA, 1997.
2. TSUPREM4 User's Manual, TMA, 595 Lawrence Expressway, Sunnyvale, CA 94086.
3. J. S. McMurray, J. Kim, and C. C. Williams, *J. Vac. Sci. Technol.* B15 (1997) 1011-1014.
4. J. J. Kopanski, J. F. Marchiando, and D. W. Berning, R. Alvis, H. E. Smith, "Scanning capacitance microscopy measurement of 2-D dopant profiles across junctions," *Proceedings of 4th International Workshop on Measurements, Characterization and*

Modeling of Ultra-Shallow Doping Profiles in Semiconductors, M. Current, M. Kump and G. McGuire, Eds., Research Triangle Park, NC, USA, 1997, pp. 53.1-53.9.

5. H. Edwards, R. Mahaffy, R. McGlothlin, R. San Martin, E. U, M. Gribelyuk, R. S. List, C. K. Shih, and V. A. Ukraintsev, *Appl. Phys. Lett.*, accepted for publication.
6. J. N. Nxumalo, D. T. Shimizu, and D. J. Thomson, J. Vac. Sci. Technol. B14 (1996) 386; Y. Li, J. N. Nxumalo, and D. J. Thomson, "High resolution imaging of dopant profiles using small contact area metal-semiconductor C-V measurement," *Proceedings of 4th International Workshop on Measurements, Characterization and Modeling of Ultra-Shallow Doping Profiles in Semiconductors*, M. Current, M. Kump and G. McGuire, Eds., Research Triangle Park, NC, USA, 1997, pp. 63.1-63.7.
7. P. De Wolf, T. Clarysse, W. Vandervorst, J. Snauwaert, and L. Hellemans, *J. Vac. Sci. Technol.* B14 (1996) 380; P. De Wolf, T. Clarysse, and W. Vandervorst, and L. Hellemans and Ph. Niedermann, and W. Hänni, "Cross-sectional nano-SRP dopant profiling," *Proceedings of 4th International Workshop on Measurements, Characterization and Modeling of Ultra-Shallow Doping Profiles in Semiconductors*, M. Current, M. Kump and G. McGuire, Eds., Research Triangle Park, NC, USA, 1997, pp. 56.1-56.10.
8. S. S. Neogi, D. Venables, Z. Ma and D. M. Maher, "Factors affecting two-dimensional dopant profiles obtained by transmission electron microscopy of etched pn junctions in silicon," *Proceedings of 4th International Workshop on Measurements, Characterization and Modeling of Ultra-Shallow Doping Profiles in Semiconductors*, M. Current, M. Kump and G. McGuire, Eds., Research Triangle Park, NC, USA, 1997, pp. 66.1-66.12.
9. M. Duane, P. Nunan, M. ter Beek, and R. Subrahmanyan, *J. Vac. Sci. Technol.* B14 (1996) 218-223.
10. V. A. Ukraintsev, F. R. Potts, R. M. Wallace, L. K. Magel, Hal Edwards, M.-C. Chang, "Silicon surface preparation for two-dimensional dopant characterization," *these proceedings*.

Application of Scanning Probe Microscopy Nano-Indentation towards Nanomechanical Characterization of Polymer Films

Jin Xu*, Joe Hooker, Indira Adhihetty,
Paddy Padmanabhan, Tom Remmel, Wei Chen

Mail Drop 254, Motorola Inc.,2200 W. Broadway Rd., Mesa, AZ 85202

The extremely high measurement sensitivity and accuracy has made scanning probe microscopy (SPM) a valuable tool for detecting various kinds of tip surface interactions such as Van der Waals force for AFM, electron tunneling for STM, electric / magnetic forces for EFM/MFM, frictional force for LFM, etc. This paper presents a new technique, SPM nano-indentation technique, for monitoring the repulsive force between the sharp probe and material surface as a means to detect the mechanical properties of materials. The key advantages of SPM nano-indentation are its imaging capability which allows accurate measurement of indentation geometry and precise location of indentation probe for nanomechanical measurement. This particular study explores the areas of applicability of SPM for measuring mechanical properties such as Young's modulus of materials. Limit studies have been done in understanding the reproducibility of force curves, the effect of surface roughness on force curve, substrate on Young's modulus measurement, etc.. Examples of Young's modulus extraction for organic films will be presented.

INTRODUCTION

Over the past fifteen years, scanning probe microscopy (SPM) has evolved from an academic curiosity to a major characterization tool in the industry. Many SPM techniques have been invented and developed which take full advantage of the nano-meter 3D resolution offered by using a very sharp tip. A few examples are scanning capacitance, scanning resistance, scanning thermal, scanning Kelvin probe, magnetic/electric force etc..

This paper focuses on the development of SPM based nano-indentation technique for measuring mechanical properties of thin films. The SPM nano-indentation technique borrows the concepts from the traditional nano-indentation in that it utilizes the force curve obtained by driving a sharp probe into the sample surface to extract the mechanical properties of the materials (1-3). The main feature of SPM based nano-indentation is that the force applied to the surface is achieved by bending of a cantilever and therefore the force is usually orders of magnitude smaller than the force applied using the traditional nano-indenter. Forces down to pN range allow probing of very thin surface films and open the window of opportunities for examining mechanical properties of ultra soft and sensitive biological materials (4). Also, the nano-meter scale tip size permits characterization within small sample volume, which is particularly useful for measuring local property changes in multi-component and multi-phase polymer systems used in electronic devices and other applications. Furthermore, the high 3D resolution imaging capability of SPM allows accurate placement of indentation sites and measurement of indentation depth and material pile-up.

Mechanical properties such as hardness, Young's modulus and adhesion strength have a significant impact on ULSI device processing. This paper focuses on the development of SPM based nano-indentation technique and its application towards extraction of mechanical properties of polymer films. It is demonstrated that SPM nano-indentation provides reproducible indentation force curves and from which meaningful results of mechanical properties of polymer films can be extracted.

EXPERIMENTAL

The SPM system used for SPM nano-indentation work is a Digital Instruments D3000 scanning probe microscope. Figure 1 shows a schematic of the SPM set-up. Two types of probes, Si cantilever with integrated Si tip and stainless steel cantilever with diamond tip, were used for SPM nano-indentation study. The spring constant for Si probes was estimated to be ~ 50N/m and

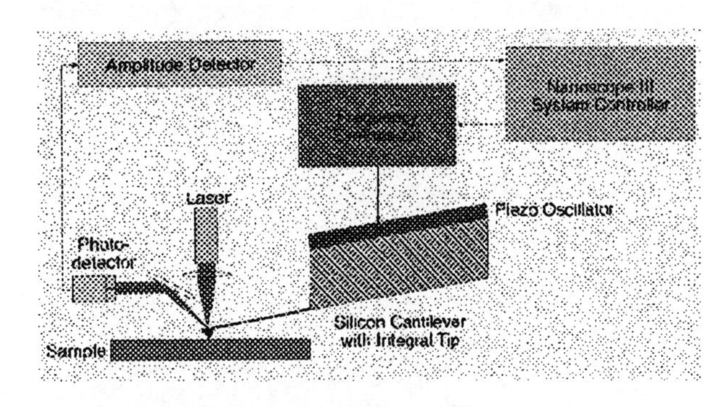

Fig.1: Block diagram of TappingMode^{TM} AFM set up (taken from an applications note by C. B. Prater and Y. E. Strausser from Digital Instruments).

CP449, *Characterization and Metrology for ULSI Technology: 1998 International Conference*
edited by D. G. Seiler, A. C. Diebold, W. M. Bullis, T. J. Shaffner, R. McDonald, and E. J. Walters

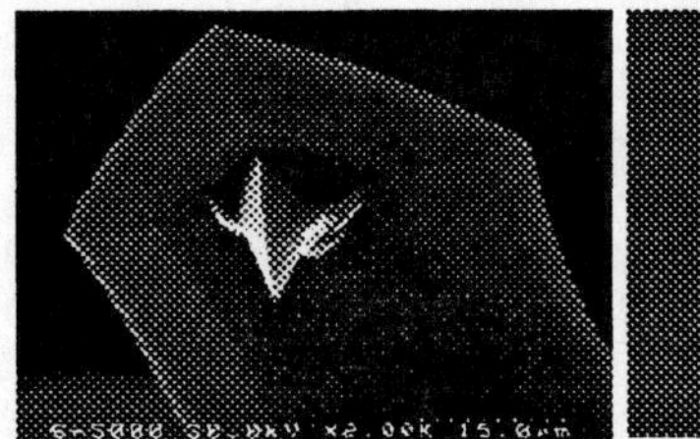

Figure 2: Secondary electron images a single-crystal Si probe. The tip radius can be readily measured from the high magnification SEM image.

the spring constant of stainless steel cantilever was measured using a micro-balance technique. Typical spring constant value of stainless steel cantilever is in the range of 100-300N/m, which is higher than that of standard Si cantilever. The different stiffness of the cantilever probe offers different applicable indentation force given the same typical cantilever deflection, which in term determines the measurable range of moduli (5). Figure 2 shows both low and high magnification field emission secondary electron (FESEM) images of a typical Si SPM probe. The shape and radius of the tip apex can be easily estimated from high magnification SEM image. The indentation is achieved by applying an appropriate voltage to extend the z-piezo tube which drives the probe assembly into the surface. The deflection of the cantilever as a result of the force between the tip and sample is measured via a laser beam which reflects off the back of the cantilever into a quadruple photo detector, therefore the change of the photo voltage V_t can be monitored as a function of the vertical piezo travel Δz_p.

PRINCIPLE AND METHODOLOGY

The principle and methodology used for SPM-based nano-indentation are similar to that for conventional nano-indentation technique. The specific methodology of SPM based nano-indentation was comprehensively summarized by VanLandingham et al (1). A few key steps from that methodology employed in our study towards extraction of Young's modulus are outlined below. Figure 3 is a simplified schematic of SPM force curve with photo detector voltage V_t as the Y-axis and z-piezo displacement Δz_p as the X-axis. Valuable information of the nano-meter scale mechanical response of the material can be extracted from the force curve. As shown in Fig. 3, at the "Repulsive Region" of the force curve, the cantilever bends and the amount of bending for a given amount of downward tip movement gives an indication of the material's elasticity. The slope of the force curve in the "repulsive region" is defined as the sensitivity parameter, which is photodiode voltage versus the piezo travel distance. This parameter depends on various factors, such as the cantilever geometry and the feedback laser alignment. Sensitivity is also material dependent. Under the same operation condition, material with higher Young's modulus gives steeper slope due to less indentation with the same applied force. When the sensitivity is calibrated on a material much stiffer than the cantilever, a conversion factor can be obtained which relates the photo detector voltage change with the actual cantilever bending.

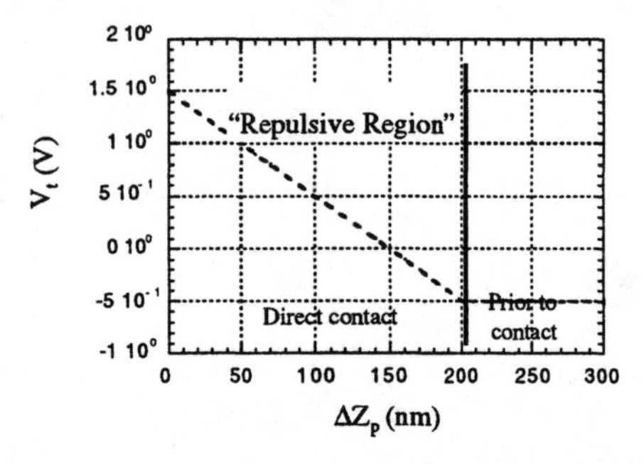

Figure 3: A force curve obtained directly from SPM system, shown on Y-axis is the photo detector voltage and X-axis is the z-piezo displacement.

The relationship between the z-piezo travel (Δz_p), the cantilever deflection (Δz_c) and the indentation depth (Δz_i) is,

$$\Delta z_p = \Delta z_c + \Delta z_i \qquad (1)$$

where Δz_p can be accurately converted through the voltage applied to the piezo tube, and Δz_c can be obtained once the photo detector voltage is "calibrated". In our experiments, single crystal Si sample (Young's modulus ~ 200GPa) was used as an "infinitely hard" material to correlate the cantilever deflection and the total z-piezo travel during the indentation. Once the z-piezo travel Δz_p and cantilever deflection Δz_c are obtained from the sample material and the single crystal Si, respectively, the indentation depth Δz_i can be easily calculated from Eq. (1). Given Δz_c, the applied force F can also be calculated if the cantilever spring constant k is known.

Figure 4 is a schematic plot of converted indentation force F vs. indentation depth Δz_i. The hysterisis between the loading and unloading portions of the force curve is largely due to the plastic deformation of the sample. At the start of the unloading curve, the force

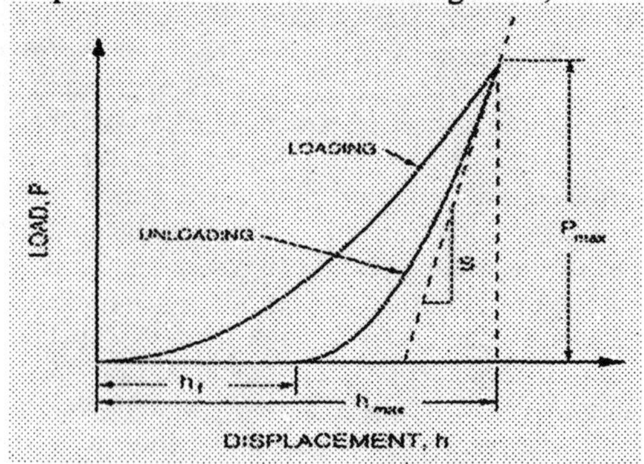

Figure 4: A schematic plot of converted load vs. indentation depth curve. The beginning portion of the unloading curve can be used to extract mechanical properties.

between the tip and sample is due to the elastic restoring force from the sample and therefore mechanical properties of the sample can be extracted. Sneddon developed relationship between indentation depth and load for several different indenter geometry (6), all of which can be represented in the form,

$$F = \frac{\xi E}{1-\upsilon^2}\Delta z_i^{\ m} = \xi E^* \Delta z_i^{\ m} \qquad (2)$$

where ξ is a constant which depends on the contact geometry, E and v are the sample elastic modulus and Poisson's ratio, respectively, and m is power law exponent characteristic of the indentation behavior. The parameter m can be determined from a curve fit of force P as a function of indentation depth Δz_i from a material with known E and v. However, without the availability of such material with known E and v, the approach we took is to use the fit from enough numbers of force vs. indentation curves with the power function $y=ax^m$. The variation of the exponent m is between 1.45 and 1.55 from all curve fittings with a correlation>0.95 using both Si and diamond tips , which suggested the best approximation for a parabolic tip. Therefore m was fixed at 1.5 for all data calculations. The FESEM picture of the Si tip also confirmed the parabolic shape. Furthermore, for a parabolic tip, ξ is equal to $4R^{0.5}/3$, where R is the tip radius (6). From our experiment, the R value of 10nm for Si tip and 100nm for diamond tip were used, which gave ξ at 1.33×10^{-4} $m^{0.5}$ for the Si tip and 4.22×10^{-4} $m^{0.5}$ for the diamond tip, respectively. This method has been used for extraction of Young's modulus for polymer films in this study.

RESULTS AND DISCUSSION

Figure 5(a) and (b) show two AFM images of a polymer surface before and after SPM nano-indentation by a diamond tip with load of 2, 4 and 6μN, respectively. The imaging was done using the same tip which performed the indentation. The three indents were inverted, as shown in Fig. 5(c), to reveal the sidewall of the indents. For systematic study of the SPM nano-indentation process, several effects, such as the surface roughness effect and substrate effect were investigated before performing nano-indentation on the polymer thin films.

1. Surface Roughness Effect

Surface roughness effect was investigated first to see if it affects the SPM nano-indentation force curve reproducibility. Seven materials, e.g. Epi-Si, Si, SOI, Pt, Poly-Si, Al and Cu with different surface roughness were selected. Si tip was used for this experiment. Each material was indented ten times and the sensitivity parameter was measured from the slope of the unloading curve each time to calculate the standard deviation from the ten experiments. No noticeable deformation was found in

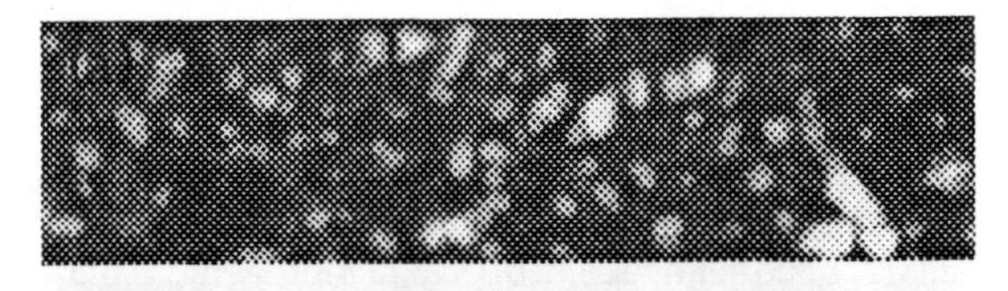

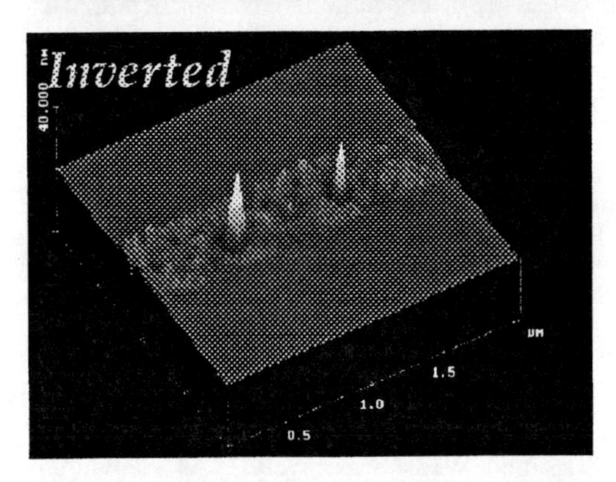

Figure 5: AFM images of a polymer surface (a) before indentation and (b) after three indentations with force of 2, 4, and 6μN. Panel (c) shows an inverted image of (b) showing the sidewall of the indents. Both images were taken with the diamond tip for the indentation.

any of the samples. In Fig. 6 the Y axis on the left is the averaged value of sensitivity for each material; and the Y axis on the right is the calculated standard deviation of the sensitivity values. The X axis represents the surface roughness. As shown in Fig. 6, the standard deviation of sensitivity went up as the surface roughness increased, which indicated that the rougher the surface, the less reproducible the indentation process. This trend implied that as the surface became less flat, there was possibly increased variation of the tip-sample contact geometry from each nano-indentation, which resulted less repeatability of the nano-indentation process.

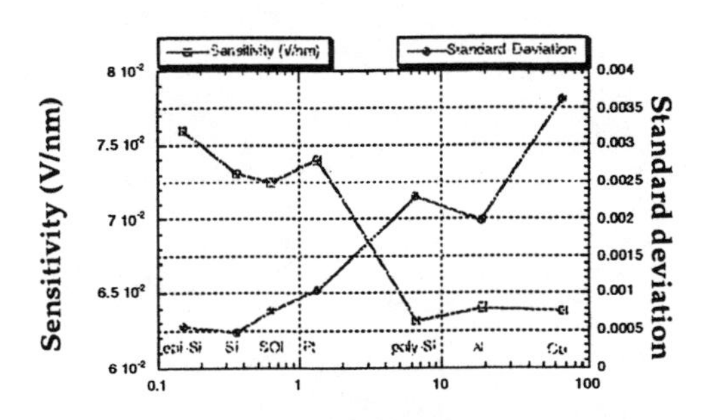

Figure 6: Effect of surface roughness on force curve reproducibility. As the roughness increased, the standard deviation of sensitivity also increased

2. Substrate Effect

Another effect which needs to be aware of is the so called "substrate effect". For traditional nano-indenter, the recommended maximum depth of indentation which the film properties alone are measured is thought to be about 10% of the film thickness (7). Nix et al also proposed a simple phenomenological relation using the ratio of indentation depth to the film thickness to describe the relative contributions of the film and substrate at the indentation process (8). For large indentation depth, the elastic properties of the substrate dominate the measurement; while for small indentation depth, the elastic properties of the film dominate. For SPM based nano-indentation, the similar substrate effect is also expected.

The substrate effect on SPM nano-indentation was studied using polymer films with Young's moduli of a few GPa, which are much smaller compared to the silicon substrate's 200GPa. Diamond tip was used to indent three polymer films (W19, W15 and W12) with different thickness each at 4kÅ, 10kÅ and 16kÅ. Same indentation force of 17μN was applied to each wafer to see if there is a difference on the extracted value of Young's modulus. The resulting indentation depth was at 140Å, 210Å and 520Å for W19, W15 and W12, respectively. As shown in Fig.7, W12 with the largest film thickness gave the lowest value of Young's modulus. As the film thickness decreased, the extracted Young's modulus values increased. The differences on the extracted Young's modulus values can be interpreted from the substrate effect. With the same indentation force, since the presence of the hard Si substrate caused less indentation depth for thinner films, therefore the Young's modulus value will increase from the extraction according to equation (2). In other words, the measured Young's modulus no longer truly represents the polymer film property alone, but a combined contribution from both film and substrate.

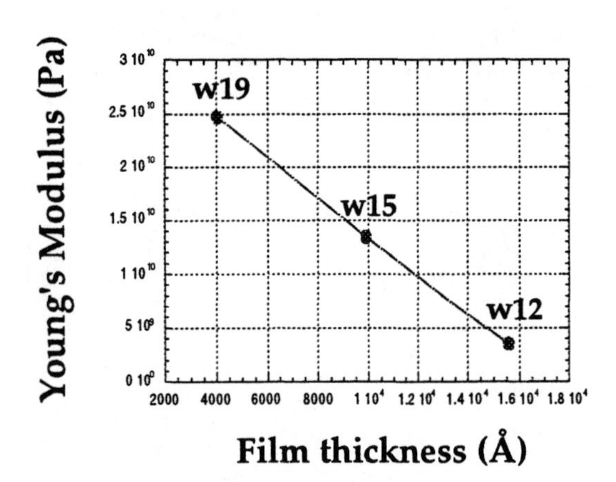

Figure 7: Substrate effect on extracted Young's Modulus. With the same indentation force, as the film thickness decreased from w12 to w19, the measured Young's Modulus value increased significantly.

3. SPM Nano-indentation on Polymer Thin Film

Nano-indentation was performed on several potential low dielectric polymer films. Figure 8(a) shows an example of a AFM force curve obtained from the polymer surface using a diamond tip. The curve was obtained at 1Hz loading frequency with the proper loading distance so that the retracting and the approaching curves in the non-contact regime were not separated by viscosity effect. A converted force curve of load as a function of indentation from polymer surface is shown in Fig. 8(b) as well. Using the model outlined above, E^* for one of the polymer films has been extracted and is shown in Fig. 9, where different indentation forces were used to extract Young's modulus from different indentation depth. From

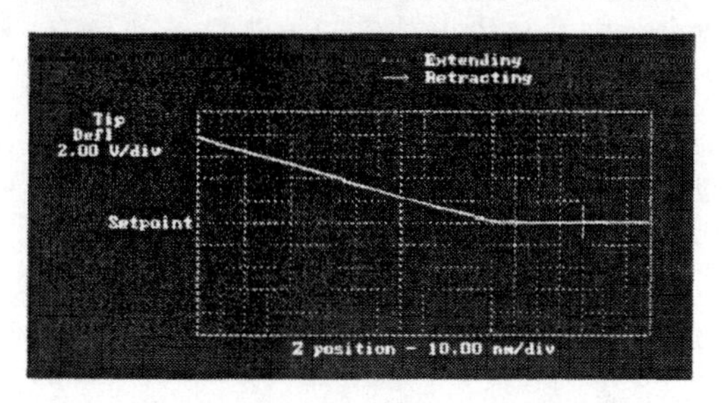

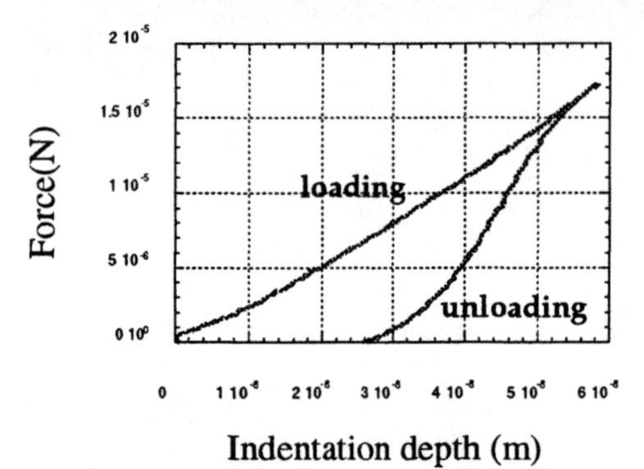

Figure 8: (a) A SPM force curve obtained from a polymer surface and (b) a converted load vs. indentation depth curve.

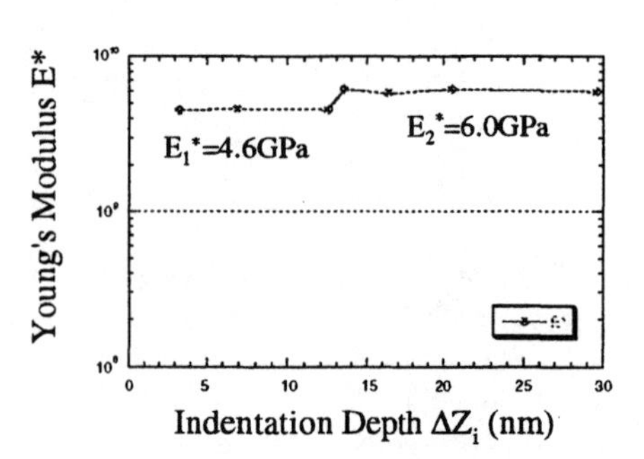

Figure 9: Extracted E* values for the polymer studied as a function of indentation depth. A softer surface layer is clearly detected.

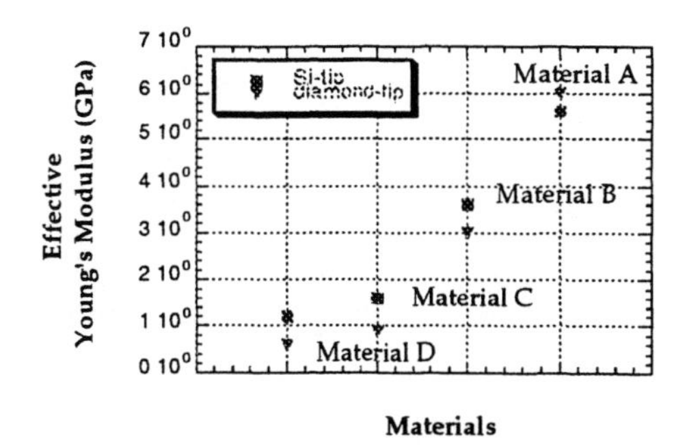

Figure 10: Extracted effective Young's moduli of four potential low dielectric polymer films using SPM nano-indentation with both Si and diamond tips.

the surface to about 20nm into the bulk, there are two different Young's moduli at 4.6GPa and 6.0GPa, respectively. The E^* for the bulk estimated at 6.0GPa is consistent with the value obtained by traditional nano-indentation technique. However, the smaller Young's modulus near the surface implied that there is a soft surface layer with a thickness of ~ 20nm. With the fine control of indentation force through cantilever deflection, SPM nano-indentation offers a promising feature for nano-meter scale depth profiling of mechanical properties of thin films. The measurements for all the four polymer films are plotted in Fig. 10 using both Si and diamond tips. The data appear to be in good agreement.

SUMMARY

SPM nano-indentation technique has been developed and applied to measure mechanical properties of polymer thin films which are of importance in future generation integrated circuits as inter-layer dielectric material. One of the polymer films measured showed bulk E^* to be ~ 6.0GPa, very close to the measured value from traditional nano-indenter. With applied force orders of magnitude smaller compared to that of traditional nano-indentation, the depth profiling capability of SPM nano-indentation is demonstrated through examining a 20nm soft "skin layer" of the polymer surface.

ACKNOWLEDGMENTS

Special thanks to Joy Watanabe, Jeff Wetzel and Betsy Weitzman for providing polymer samples for the study and their enthusiastic support for the project. Thanks to Jeff Elings for providing special probes and many useful discussions.

REFERENCES

1. M. R. VanLandingham, S.H. McKnight, G.R. Palmese, J.R. Elings, X. Huang, T.A. Bogetti, R.F. Eduljee and J.W., Jr.Gillespie, *Journal of Adhesion* v **64**, n 1-4, 31-59 (1997).

2. N.A. Burnham, R.J. Colton and H.M. Pollock, *Journal of Vacuum Science and Technology* **A9** (4), 2548-2556 (1991).

3. S.M. Hues, R.J. Colton, E. Meyer and H. -J. Guentherodt, *MRS Bulletin*, v **18**, n 1, 41-49 (1993).

4. A.L. Weisenhorn, M. Khorsand, S. Kasas, V. Gotzos and H.J. Butt, *Nanotechnology* **4**, 106-113 (1993).

5. M. R. VanLandingham, S.H. McKnight, G.R. Palmese, R.F. Eduljee, J.W. Gillespie, Jr. and R.L. McCullough, *Journal of Materials Science Letters* **16**, 117-119 (1997).

6. I.N. Sneddon, *International Journal of Engineering Science*, **3**, 47-57 (1965).

7. E.D. Nicholson and J.E. Field, *J. Hard Mater.* **5**, 89-132 (1994).

8. W.D. Nix, *Metallurgical Transactions*, v **20A**, 2217-2245 (1989).

Surface and Tip Characterization for Quantitative Two Dimensional Dopant Profiling by Scanning Capacitance Microscopy

V. V. Zavyalov, J. S. McMurray and C. C. Williams

Department of Physics, University of Utah, Salt Lake City, Utah 84112

In the present work, recent improvements in sample preparation and a tip quality evaluation are reported. A new method for cross-sectional surface preparation has been developed. The sample is heated at a temperature of 200-300C while being exposed to ultraviolet irradiation. This additional surface treatment improves dielectric layer uniformity, signal to noise ratio, and C-V curve flat band offset. The performance of three types of tips are also compared. It is shown that heavily doped silicon as well as worn tips are often depleted by the applied bias voltage causing errors in the Scanning Capacitance Microscope (SCM) measured dopant profile. When these effects are tested for and eliminated, SCM quantitative profiling over a 5-decade range of dopant density is demonstrated.

INTRODUCTION

Over the past few years Scanning Capacitance Microscope (SCM) is being increasingly involved in the measurement of two-dimensional (2-D) dopant profiles of sub-micron devices in the semiconductor industry (1-4). The possibility to fulfill the requirements (5) for 2-D mapping of a new device generation strongly depends on the experimental ability to control and maintain the state of the probe-sample interface in conjunction with the conversion algorithm based on a realistic physical model. Theoretically, the quasi-2-D model of hemispherical tip brought to the planar insulator-semiconductor surface has been developed and has proven a quick and convenient way to convert experimental SCM data (2-4). The experimental situation in SCM imaging remains not so clear because of the complexity of the 3D tip-air-insulator-semiconductor surface interaction. The major challenges in SCM sampling include the quality of an insulating layer between the tip and semiconductor and the tip quality itself. In this work, recent improvements in sample preparation and tip quality evaluation are reported.

SURFACE QUALITY

Several methods have been used to produce an insulating layer on the silicon surface for SCM imaging. It was found that most of the traditionally used dielectric layer could not fulfill the requirements for quantitative SCM dopant profiling (6). The surface produced by the final polishing step using a colloidal silica suspension solution has been found to be sufficient quality for SCM data acquisition (2,7). However, the quality of this surface is not always well controlled and the reproducibility of SCM results is often inadequate for quantitative analysis.

In Fig.1 a cross-sectional SCM 2-D image of the gate-like structure taken in ΔC mode (constant AC bias amplitude is applied to the sample and the change in capacitance is recorded) is shown. The sample was prepared by implanting 35 keV As ions with a dose of 10^{15} cm^{-2} into a p-type substrate with B-doping concentration of $10^{15} cm^{-3}$ and annealed at 950C for 80 min. P-n junctions are formed between 1.5μm gates with 1.5 μm spacing. This image was taken on a surface formed after polishing with a colloidal silica suspension. The poor quality of this dielectric layer can

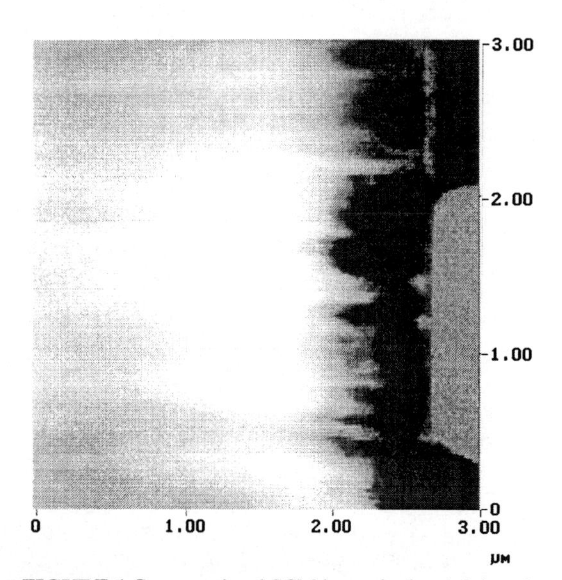

FIGURE 1. Cross-sectional SCM image in the vicinity of a p-n junction (image is centered between two gates) taken after polishing with colloidal silica suspension.

CP449, *Characterization and Metrology for ULSI Technology: 1998 International Conference*
edited by D. G. Seiler, A. C. Diebold, W. M. Bullis, T. J. Shaffner, R. McDonald, and E. J. Walters

be easily seen on samples with p-n junctions due to the strong contrast difference between p- and n-type sides of the junctions.

Several attempts were made to improve the quality of this surface. A good result has been achieved by sample heating at 300C for 30-40 min in an air environment. Nevertheless, the DC offset of the C-V curves remains very high even after heating. In Fig.2 the DC offset measured on the p- and n-type silicon substrates with doping concentration of 10^{15} cm^{-3} is shown. The offset value is measured as a DC bias corresponding to the maximum of the ΔC (dC/dV) signal with constant AC bias of 1V amplitude applied to the sample. This DC offset corresponds to the maximum slope of the C-V curve (dC\dV=max) and may be used to evaluate approximately the flat-band offset.

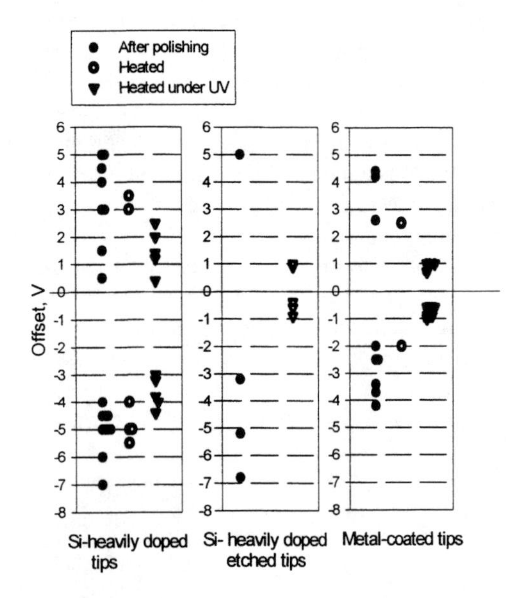

FIGURE 2. DC-offset of the maximum ΔC-signal measured on the silicon substrate with doping concentration of 10^{15} cm^{-3} (flat-band position of C-V curve).

The DC offset for the as-polished surface varies from sample to sample in a very wide range from +0.4V to +5V for p-type and from -2V to -7V for n-type substrates. After heating, it is still +(2-4)V for p- and -(2-5)V for n-type of semiconductor. The image in Fig.1 is a good illustration of SCM image taken on the surface with a large difference in flat-band position for the p- and n-type sides of the junction. The DC offset is chosen to compensate for the flat-band shift at the n-type region of the sample(dark region on the image). This DC offset pushes the p-type response (light region) far from the gate edge, so that there is no p-n junction contrast in vicinity of the gate edge.

Another disadvantage of the large DC offset value is the possibility that field-induced oxidation (FIO) will occur during SCM imaging. It has been observed that during AFM scanning along the silicon surface with a positive bias on the sample, additional oxide growth is induced by the tip-sample field (8). The rate of FIO depends on the scanning rate and on the difference in applied DC voltage and FIO threshold voltage found to be +(3-5)V. FIO leads to the decrease in SCM sensitivity, and when FIO occurs on consecutive scans, little useful information can be taken from this surface area.

The best result in our attempts to improve the dielectric layer quality has been achieved using ultra violet (UV) irradiation of the polished surface during the heat treatment. The UV exposure is performed using a "Minerallight" S89 standard UV lamp (from Ultra Violet Product, Inc). The combination of UV exposure with heating leads to the increase in signal/noise ratio and improves the uniformity of the ΔC signal along the polished surface.

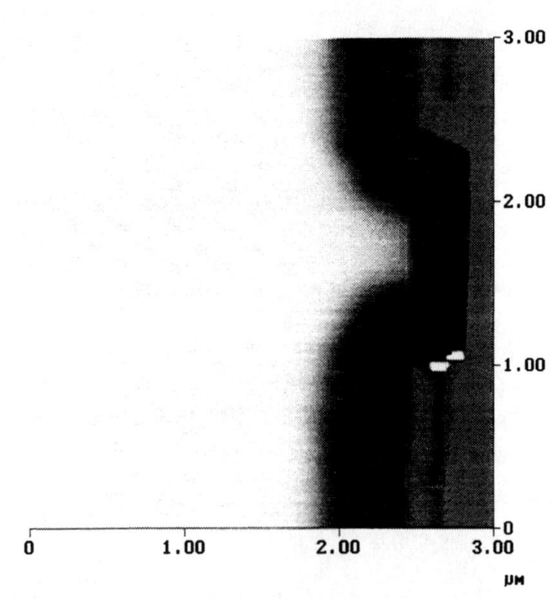

FFIGURE 3. Cross-sectional SCM image in the vicinity of p-n junction (single gate is shown) taken after heating under UV irradiation.

The most significant change is found in charging properties of the dielectric-semiconductor interface. First, the measured DC offset moves toward to 0V for all tips used (see Fig.2). It is found that for tips which can be treated as quasi-metallic, the DC offset slightly changes from sample to sample in a range from +0.6V to +1.0V for p-type and from -0.4V to -1.0V for n-type substrates. Fig.3 shows the SCM image of the same sample as in Fig.1. This image has been taken with 0V DC offset after heating with UV exposure of the polished surface. Due to the optimal offset for both sides of the p-n junction, the contrast between p- and n-type sides of junction is clearly seen in the vicinity of the gate. The 2D position of the built-in electrical junction (where the hole density is equal to the electron density) can be found (approximately) as the transition line between p- and n-type regions.

The combination of heating and UV irradiation also leads to the additional passivation and charge redistribution along the polished surface. We believe that streaky lines on the image in Fig.1 are due to the non-uniform charge distribution left after polishing (the direction of polishing is perpendicular to the sample edge). Note that the topographic image was taken simultaneously, and is very uniform and does not reveal any scan errors. As it can be seen in Fig.3, there are no such features after heating with UV irradiation.

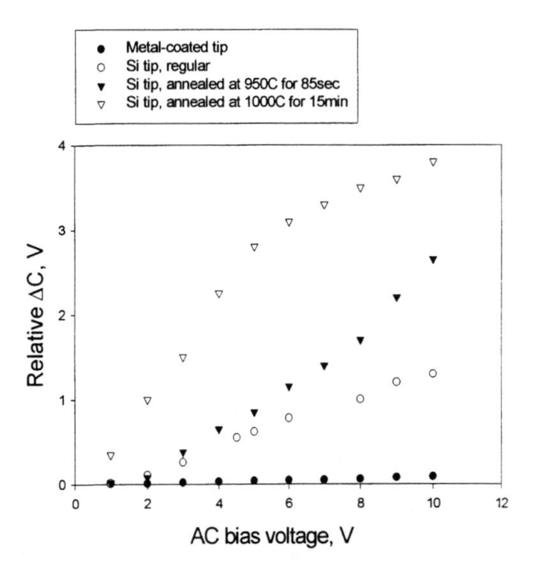

FIGURE 4. Tip depletion effect measured on the metal surface.

TIP QUALITY EVALUATION

In this work, three types of probing tips (provide by Digital Instruments, Inc.) are investigated for dopant profiling within the doping range of 10^{15}-10^{20}cm^{-3}. The heavily doped silicon tips were implanted by phosphorus atoms and annealed at 950C for 30s. The metal-coated tips are commercially available Co/Cr-coated silicon tips, such as those used for magnetic force microscopy. The silicided silicon tips are the third type characterized.

Two major disadvantages of heavily doped silicon tips are found. First, a native oxide on the tip apex exists and its quality varies from the tip to tip causing a variation in the measured DC offset. The DC offset measured with these tips is higher than those measured with the metal-coated tip even for the sample heated with UV exposure (see Fig.2). Field-induced oxidation of the heavily doped silicon tips is also observed during SCM imaging with either DC or AC bias voltage larger than the threshold FIO voltage. The additional oxidation of the tip apex significantly decreases the SCM sensitivity. To remove the oxide, the doped silicon tips were etched in 5% HF solution for 15-30s. After the oxide etch, the DC offset is practically the same as those measured by the metal-coated tips (Fig.2).

Second, a tip depletion (comparable with the depletion of heavily doped region of the sample) is observed. To measure depletion effect, the tip is brought to a metal surface and a ΔC signal is measured as a function of the AC bias voltage applied. In Fig.4 the effect measured by the heavily doped silicon tip is compared with those for the metal-coated tip. The ΔC signal dependence on the AC voltage measured with a metal-coated tip is found due to some instrumentation coupling and is two orders of magnitude less than the useful ΔC signal on a lightly doped silicon substrate. The effect measured with the regular heavily doped silicon tips is at least 10 times greater then this coupling. The heavily doped silicon tips were annealed to change the surface doping concentration of the tip apex. As it can be seen in Fig.4, the ΔC signal measured with annealed tips significantly increases compared with non-annealed tips.

The impact of the tip depletion on quantitative SCM dopant profiling is shown in Fig.5. The SEMATECH #1 sample with two peaks in the dopant profile and two steps (50nm width) on the trailing edge of the first peak is used. An abrupt vertical dopant profile was grown by a custom Si-Ge chemical vapor deposition system. In Fig.5 the vertical dopant profile measured by Secondary Ion Mass Spectroscopy (SIMS) is compared with SCM results. The SCM data is converted to dopant profile using the conversion algorithm described elsewhere (2). The SCM profile is pinned at one point in the dopant profile (10^{20}cm^{-3}). Using a single free parameter, the SCM data is converted to dopant concentration to fit the substrate level at 10^{15} cm^{-3}. Within this approach a good fit of the SIMS vertical dopant profile and the SCM data taken with the metal-coated tip is observed.

However, the results are different for data taken with regular heavily doped silicon tips. Since the SCM data is measured in ΔV mode (were the magnitude of AC bias is automatically changed to maintain the ΔC signal as constant), the depletion effect depends on the AC bias voltage (see Fig.4). Thus the contribution of the tip depletion to the total SCM signal depends on the doping concentration. As the result, the converted SCM data does not fit the dopant profile for three basic doping concentrations at 10^{20}, 10^{17} and 10^{15} cm^{-3} using the same approach as for the metal-coated tip (see Fig.5). In some cases such a SCM data set could be forced to fit the real dopant profile using a non-realistic

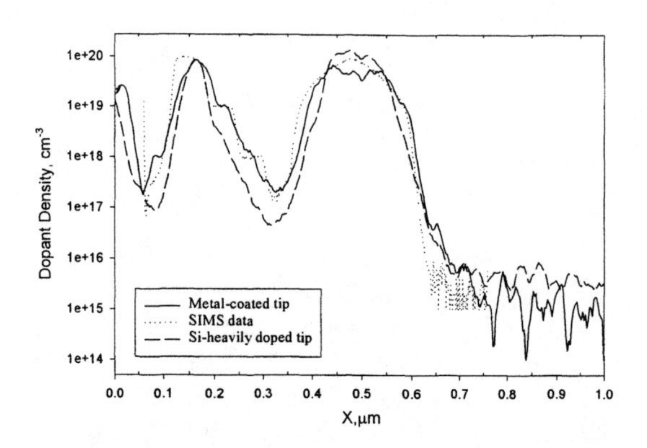

FIGURE 5. Comparison of converted SCM data with vertical SIMS dopant profile.

physical parameters (Boltzmann statistic, for example), but this corresponds only to curve fitting.

The SCM data measured with the metal-coated and silicided silicon tips is very consistent and the converted dopant profile in most cases corresponds to the data presented in Fig.5. It is found that even after a short time of SCM scanning (sometimes as short as several line scans), the very apex of the tip may be worn so that it is no longer hemispherical but rather flat. In some cases, the depletion of the worn tip is observed and the impact of the tip depletion on the quantitative SCM dopant profiling is basically the same as that shown for a heavily doped silicon tip.

CONCLUSION

The ability to control and maintain the state of the probe tip-dielectric layer-semiconductor interface for quantitative SCM dopant profiling is demonstrated. To improve the dielectric and charge properties of oxide-semiconductor interface, a new method of low temperature heating under ultra-violet exposure has been developed. This post polishing process allows us to produce the oxide layer with consistent quality and reproducibility. Three types of tips have been evaluated for quantitative SCM data acquisition in the doping range of 10^{15}-10^{20} cm^{-3}. It is shown that the heavily doped silicon tips can not be treated as quasi-metallic because of the tip apex oxidation and tip depletion. In some cases the worn metal-coated and silicided silicon tips may also be depleted by applied SCM bias voltage. Tip depletion distorts the SCM signal and causes the errors in the measured SCM dopant profile.

REFERENCES

1. Diebold A.C., Kump M.R, Kopanski J.J. and Seiler D.G., *Vac. Sci. Technol.* **B14**, 196(1996).
2. McMurray J.S., Kim J. and Williams C.C., *J. Vac. Sci. Technol.* **B15,** 1011 (1997).
3. McMurray J.S., Kim J. and Williams C.C., *J. Vac. Sci. Technol.,* **B16** 344 (1998).
4. Kopanski J.J., Marchiando J.E. and Lowney J.R., *Material Science and Engineering* **B44**, 46 (1997).
5. *The National Technology Roadmap for Semiconductors, Semiconductor Industry Association,* 4300 Stevens Creek Boulevard, San Jose, CA.
6. Huang Y. and Williams C.C., *J. Vac. Sci. Technol* **A14**, 1168 (1996).
7. Erickson A., Digital Instruments, Inc., private communication.
8. Ley L., Teuschler T., Mahr K., Miyazaki S. and Hunhausen M., *J. Vac. Sci. Technol* **B14**, 2845(1996).

Ultra-Shallow Junction Measurements: A Review of SIMS Approaches for Annealed and Processed Wafers

Gary R. Mount[1], Stephen P. Smith[1], Charles J. Hitzman[1]
Victor K.F. Chia[1] and Charles W. Magee[2]

1. *Charles Evans & Associates, 301 Chesapeake Drive, Redwood City, CA, USA, 94063*
2. *Evans East, 666 Plainsboro Road, Plainsboro, NJ, USA, 08536*

As device sizes continue to shrink, ion implant energies and implant depths have also decreased requiring improved characterization methods. SIMS continues to be an effective choice for characterizing both unannealed and annealed dopant distributions. New methodologies have been required, and are still under development in order to accurately and reproducibly measure very shallow dopant profiles. In this paper, we outline some of the fundamental issues, the methods in use today and show where more work is needed.

INTRODUCTION

Ultra-Shallow Junctions

Improved device performance has depended in part on decreased gate lengths. As gate lengths decrease, short channel effects begin to dominate device performance (1). Source/Drain depletion regions begin to interact with the channel and with each other and punchthrough can become problematic. Depletion region depth and lateral spread under the gate can be reduced using ultra-shallow implants in the S/D tip extensions. However, conductivity is lost with ultra-shallow S/D tip extensions since the conduction path cross-sectional area is reduced. Compensation for loss of conductivity can be accomplished by increasing the implant dose.

Much of the initial work on forming ultra-shallow junctions has focused on boron for three main reasons. First, PMOS devices made with implanted boron are particularly susceptible to punchthrough. Second, transient enhanced diffusion (TED) effects when annealing implanted boron make it more difficult to form shallow junctions than for NMOS dopants such as phosphorous and arsenic (2). Third, to achieve a shallow implant with a light element, very low acceleration energies are required and this has necessitated a new generation of low energy ion implanter (3).

NMOS shallow junctions are also needed. NMOS S/D tip extensions formed with arsenic or phosphorous implants are not subject to the same TED effects as boron and the implant energies can be higher. Ultra-shallow activated arsenic or phosphorous implants are easier to form than boron, but the characterization challenge is still very much the same.

Development of processes to form ultra-shallow junctions has created the need for improved characterization methods. Accurate and reproducible measurements of ultra-shallow implants both before and after annealing is critical to the assessment of USJ process development. Traditionally, Secondary Ion Mass Spectrometry (SIMS) has been shown to be highly effective providing high sensitivity with good depth resolution (4). High concentration ultra-shallow junctions have created the need to improve depth resolution and the near-surface quantification.

SIMS Analytical Requirements

Requirements for ultra-shallow dopant profile characterization can be generalized by saying that accurate quantification and good depth resolution are needed. However, the analysis needs to be tailored depending on the information required. Depth resolution can be emphasized at the expense of sensitivity or analysis time. Quantification accuracy through the surface oxide can be emphasized at the expense of detection limit and junction depth accuracy. For each situation we would prefer a single analytical solution where the most critical information is emphasized and the data is collected in a timely fashion. Process simulation, implanter evaluation, anneal treatments and process qualification, each have specific and not necessarily compatible requirements.

Process Simulation Development

Process simulation of device structures and device performance depends on accurately predicting the final activated dopant profile shapes and

CP449, *Characterization and Metrology for ULSI Technology: 1998 International Conference*
edited by D. G. Seiler, A. C. Diebold, W. M. Bullis, T. J. Shaffner, R. McDonald, and E. J. Walters
 1-56396-753-7/98/$15.00

concentrations. Process models are often refined and calibrated based on SIMS measurements (5). For modeling purposes, all aspects of the dopant profile shape may need to be known including accurate as-implanted dose, retained dose, concentration profile and junction depth. A single profile may not provide all of the required information.

Ion Implanter Evaluation

For the evaluation of ion implanter performance, dosimetry and profile shape must be assessed. Dosimetry from SIMS measurements requires accurate surface and near-surface quantification. Accurate standards and depth measurement are required for accurate dosimetry.

The profile tail shape at lower concentrations can provide information about channeling effects or energy contamination from the implanter if the substrate has been pre-amorphized (6). Profile tail shape can be affected by crater edge effects, primary beam scatter and by crater bottom roughening. Secondary ion extraction conditions, analysis chamber pressure, primary beam energy, and primary beam angle must all be chosen to avoid possible SIMS artifacts.

Anneal Process Evaluation

Dopant profile shape, the retained dose, and the junction depth are needed from profiles of annealed samples. Anneal process evaluations often look for the effects of fluorine from BF_2 and of pre-amorphization on the final junction depth. Comparison of profile shape before and after anneal is often needed and it is desirable to use the same analytical conditions for both measurements.

Process Qualification

New tools and production processes often need to reproduce results obtained by an existing process. High analytical precision needs to be emphasized in order to compare results so that processes can be accurately matched.

These requirements for SIMS evaluations of ultra-shallow dopants are pushing the envelope of SIMS capabilities. In this paper, we discuss each of these requirements in more detail.

SIMS Profile Optimization

Achievement of the analytical requirements depends on instrumentation with appropriate capabilities and on choosing the correct conditions. Consideration needs to be given to depth resolution, dynamic range and to accuracy.

Depth Resolution

Depth resolution is fundamentally limited by primary ion beam mixing (7). Cascade and recoil mixing effects can be reduced by lowering the primary ion beam energy. The component of beam energy that is normal to the sample surface is responsible for most mixing effects. By increasing the angle of incidence relative to the sample normal, the normal energy component is reduced and depth resolution is improved. Decreased primary beam energy decreases sputter yield and the primary ion column will deliver less beam current to the sample. Lower sputter rates result. Primary beam energy must therefore be chosen to achieve a balance between sufficient depth resolution and sufficient sputter rate in order to avoid extended analysis time.

Dynamic Range

High surface concentrations in the mid 10^{21} at/cm^3 range are typical of ultra-shallow dopant distributions and lead to high secondary ion flux and high detector count rates at the beginning of the profile. Positive secondary ion yields are increased by the presence of native oxide that further increases secondary ion flux. Secondary ion detector count rates must be kept down to avoid saturation in order to maintain linearity. At the same time we desire high ion yields and high count rates in order to measure the 1×10^{17} at/cm^3 junction depth and to achieve a 1×10^{16} at/cm^3 detection limit. To meet all of these criteria when oxide is on the surface, a dynamic range of more than six orders of magnitude is required. This dynamic range is not achievable with most detector systems. Many SIMS instruments would also suffer from memory effects in the profile tail due to analyte redeposition from instrument components.

Surface oxidizing conditions can be used to increase ion yields through the entire profile to the level of the surface oxide. The dynamic range requirements are reduced to about 5.3 orders of magnitude, which is within the capabilities of many detector systems. Surface oxidizing conditions can be achieved by bombarding at an angle normal to the sample surface (8) or by flooding the surface with oxygen while using oblique angle bombardment (9).

Memory effects can be reduced through instrument design. Lower extraction potentials and extraction lens geometry that minimizes the proximity of surfaces to the sample helps alleviate resputtering of material back on to the analyzed area.

Quantification Accuracy

SIMS quantification accuracy requires the use of accurate reference materials and requires that the ion yield conditions be known throughout the profile.

The first requirement can be met by using reference materials that are traceable to a qualifying agency such as NIST. Where such a material is not available, a round-robin traceable material can be used instead (10).

The requirement that ion yield conditions are known throughout the profile is not easily achieved. SIMS profiles often use oxygen or cesium primary ion beams that enhance electropositive and electronegative ion yields respectively. Concentrations of these beam elements in the sample are usually low before analysis except where there is surface oxide. The primary ion beam is implanted to a depth determined by the beam element, energy, angle and substrate material. Secondary ion yields do not reach equilibrium until the SIMS sputter etched crater reaches a depth of about 2.5 times the projected range (11). The region of non-equilibrium is called the SIMS transient region. The transient region can be further complicated by the presence of oxide on the sample surface. Oxide can cause large changes in secondary ion yield for both dopant and matrix elements. It might be possible to account for ion yield changes by normalizing the dopant intensity to the matrix intensity but the correction may not be accurate.

Secondary ion yield changes in the transient region can be minimized by using surface oxidizing conditions. Surface oxidizing conditions raise ion yields for dopant and matrix elements to the oxide level throughout the profile, including the transient region.

Secondary ion yields may also be affected by high concentrations of dopant at and near the surface. Ultra-shallow dopant levels can reach concentrations in the percent range. Concentrations this high are effectively changing the matrix material and may cause changes in secondary ion yields. Work still needs to be done to characterize possible ion yield changes while using surface oxidizing conditions.

Depth Accuracy

Profiles of ultra-shallow dopants result in craters that are 10nm to 100nm deep. The traditional method of depth scale determination in SIMS is to measure the crater depth with a stylus profilometer. With good equipment, good calibration references, and with careful measurement, this can still be done. Surface oxidizing conditions result in swelling of the crater bottom due to oxygen incorporation into the silicon lattice making the crater shallower than it would be without oxygen flood (12). Instead of measuring craters, it may be desirable to use a sputter rate reference such as a sample with a well-characterized delta doped layer. Careful monitoring of the primary beam current is necessary to ensure the sputter rate remains constant in between measurements on the sputter rate reference.

SIMS PROFILE EVALUATION

Profile Features

Figure 1 shows a SIMS profile of an ultra-shallow boron implant into silicon. It is a typical high dose; low energy unannealed ion implant into crystalline silicon showing a channeling tail. The profile is labeled showing features where special attention is required in order to generate accurate information.

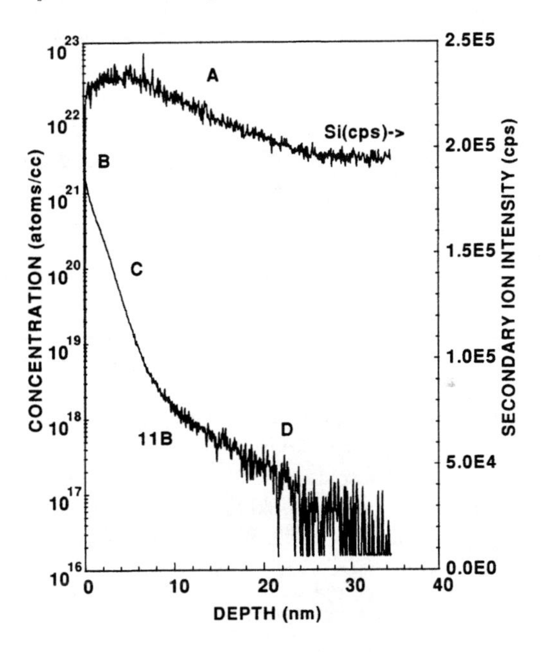

FIGURE 1: A SIMS ultra-shallow boron implant profile into silicon identifying areas that are important to consider in creating accurate SIMS profiles.

(A) The silicon matrix signal can change intensity by 20% over the first 25nm of the profile indicating a possible sputter rate change and/or an ion yield change.

(B) For high dose, low energy implants, a significant portion of the total dose can be contained in the surface oxide. Accurate quantification of dopant in the oxide is required. The sputter rate will also change between oxide and substrate.

(C) For low energy unannealed implants, SIMS depth resolution needs to be sufficient to resolve the steepest parts of the profile.

(D) The profile must have sufficient dynamic range to measure high near surface concentrations yet still provide close to 1 x 10^{16} at/cm^3 detection limits.

Crater edge rejection must be effective so that accurate tail shapes are reported enabling determinations of channeling, energy contamination and junction depths.

Recent results reported in the literature have shown that the change in matrix signal intensity (Feature A in Figure 1) is indeed due to sputter rate change, caused by crater bottom roughening (13,14). Crater bottom roughening happens when using a low energy oxygen primary beam at a 60° angle of incidence to the sample normal. This angle is the standard configuration for Phi 6600 series quadrupole SIMS. Cameca IMS series magnetic sector instruments using 4500 V secondary extraction with 2 keV net impact energy (for O_2) also have a similar angle of incidence. Atomika quadrupole instruments, which typically use primary beam bombardment normal to the sample, do not see the same change in matrix signal intensity and do not report crater bottom roughening (15).

(A) Investigation of Crater Bottom Roughening

The change in sputter rate during profiling caused by crater bottom roughening (area A in Fig. 1) could distort the depth scale reported for shallow dopant distributions in silicon. Roughening could also cause a loss in depth resolution. We have conducted a series of experiments looking at primary beam energy, angle to the sample normal, and the use of oxygen flooding.

Experimental

In order to investigate roughening further, we obtained an epitaxially grown silicon sample with five 5.4 nm thick layers and interface boron delta doping. XPS measurements reported 0.4 nm of surface oxide. Literature reports of crater bottom roughening used low energy O_2 primary beams and oxygen flooding at about 60° to the sample normal (13,14). Here we investigate 60°, 50° and 0° to observe the effect of incident angle and oxygen flooding on roughening and sputter rate. A Phi 6650 quadrupole was used for all measurements.

Results

Figure 2 shows a profile collected from the boron delta doped sample using a 500eV oxygen primary ion beam without oxygen flooding at 60° relative to the sample normal. Fig. 4 shows a profile collected from the same sample under identical conditions but at 50° to the normal. AFM images from the crater bottoms are shown in Figures 3 and 5.

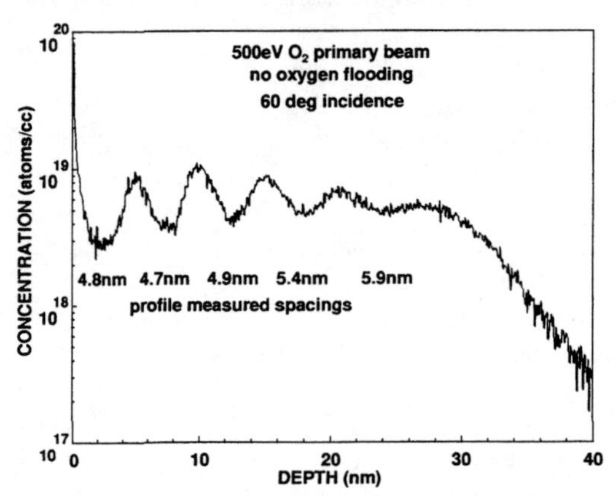

FIGURE 2. Silicon sample with 5.4 nm thick layers and interface boron delta doping profiled using a 500eV O_2 primary ion beam, no O_2 flooding and 60° incidence.

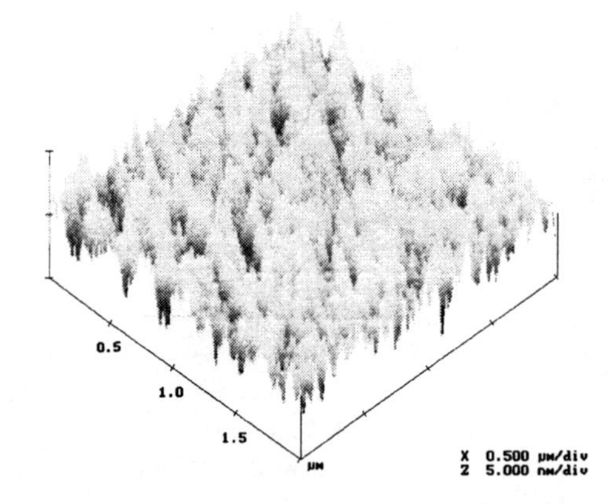

FIGURE 3: 2x2 μm AFM image of the SIMS crater bottom from Fig.2. RMS roughness = 1.15 nm. Average roughness = 0.92 nm. Rmax = 9.43 nm.

As expected from the degraded depth resolution, the 60° crater bottom is seen to be much rougher. The change of 60° to 50° reduced roughness by about a factor of four, which was surprising for such a small change in angle.

Figure 6 shows a profile collected from the same sample at normal incidence. The primary beam energy is 400 eV instead of 500 eV, which helps to reduce mixing effects and improve depth resolution. The effective mixing energy for the 500 eV / 50° case

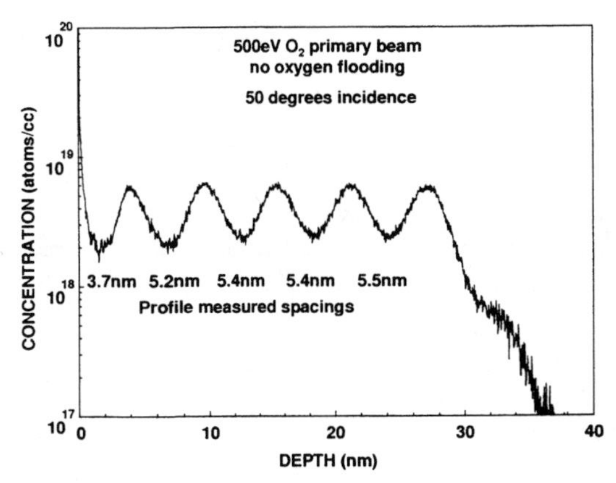

FIGURE 4: The incident angle was changed to 50° keeping all other conditions the same as in Fig.2. The delta spikes are more evenly spaced and depth resolution is improved.

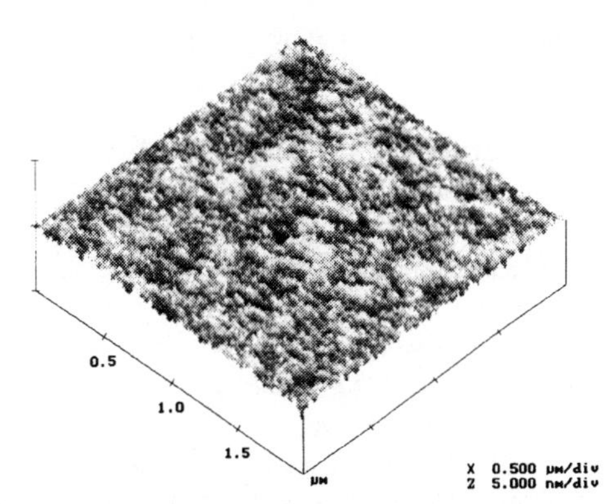

FIGURE 5: 2x2 μm AFM image of the Fig. 4 crater bottom. RMS roughness = 0.24 nm. Roughness average = 0.24 nm. Rmax = 2.4 nm.

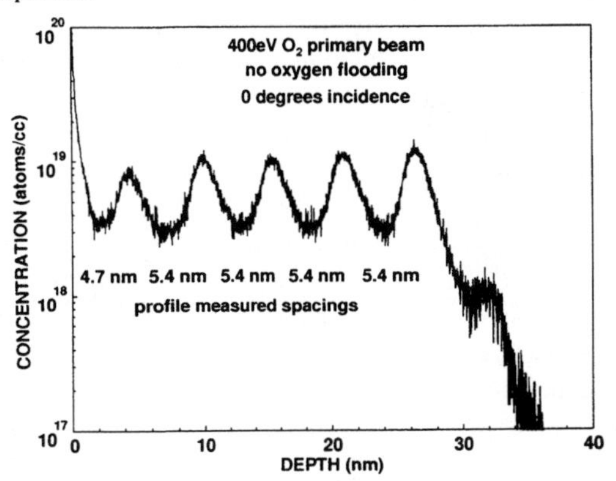

FIGURE 6: The incident angle was changed to 0° from the Fig.4 conditions and primary beam energy was lowered to 400eV. The delta spike spacing and depth resolution is improved.

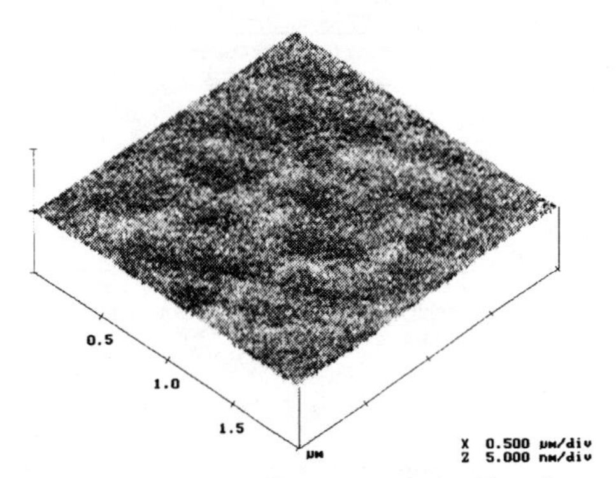

FIGURE 7: 2x2 μm AFM image of the Fig. 6 crater bottom. RMS roughness = 0.10 nm. Roughness average = 0.08 nm. Rmax = 1.0 nm.

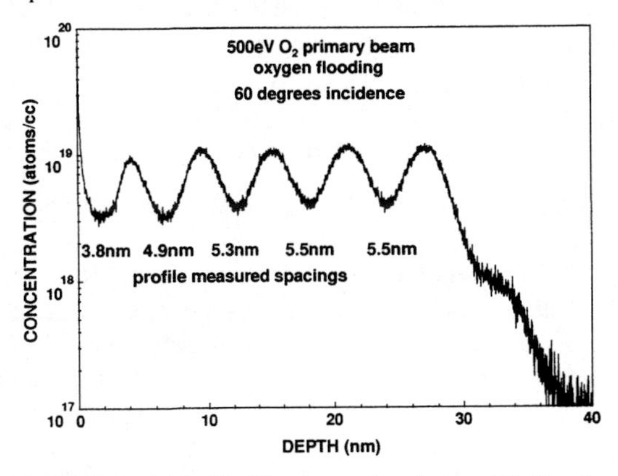

FIGURE 8: 500 eV O2 primary ion beam, O2 flooding and 60° incidence. Sputter rate change is reduced and depth resolution is improved comparing the same conditions with and without flooding.

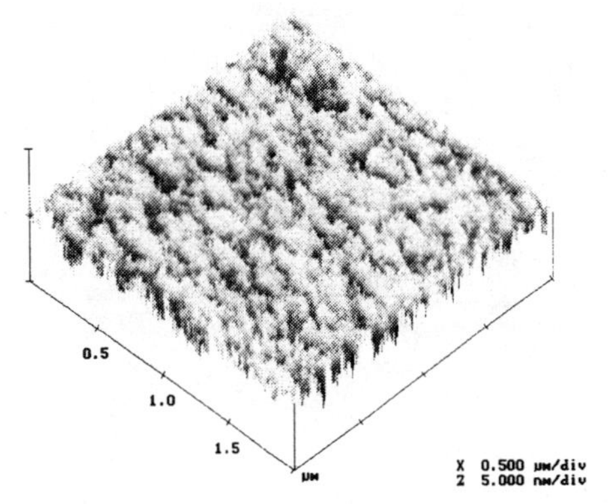

FIGURE 9: 2x2 μm AFM image of the Fig. 8 crater bottom. RMS roughness = 0.68 nm. Roughness average = 0.56 nm. Rmax = 7.3 nm.

was cos(50) x 500 = 321 eV. The normal incidence profile shows very good depth resolution with only a small increase in sputter rate in the top delta layer. Figure 7 shows the AFM image from the crater bottom reporting RMS roughness of 0.10 nm, average roughness (Ra) 0.08 nm and peak-to-valley range maximum (Rmax) of 1.0 nm. The virgin silicon surface shows RMS = 0.14 nm, Ra = 0.11 nm and Rmax = 1.3 nm.

Reports of change in sputter rate and crater bottom roughening have often been associated with the use of oxygen flooding (13,14,16). The 500 eV primary beam experiment was repeated using 60° and 50° incident angles, this time using oxygen flooding. Results are reported in Figures 8 through 11 and show that oxygen flooding improved depth resolution and reduced roughening compared to the non-flooded equivalent profile. Roughness results are summarized in Table 1.

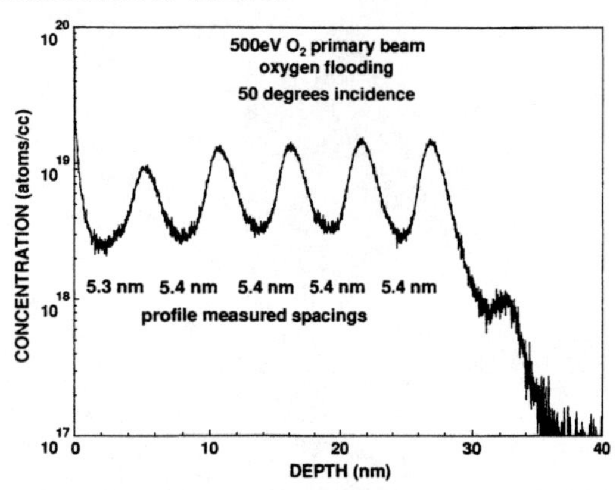

FIGURE 10: The incident angle was changed to 50° keeping all other conditions the same as in Fig.8. The delta spikes are the most evenly spaced yet and depth resolution is the best seen in these experiments.

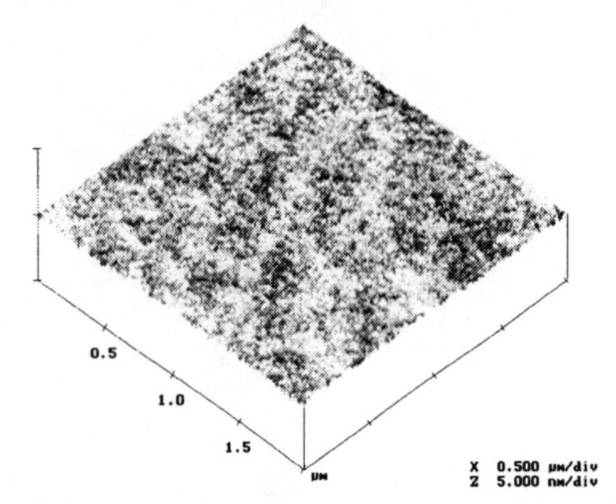

FIGURE 11: 2x2 μm AFM image of the Fig. 10 crater bottom. RMS roughness = 0.17 nm. Roughness average = 0.13 nm. Rmax = 1.7 nm.

Table 1: Roughness Summary

Energy (eV)	Angle (degrees)	O_2 Flood	RMS (nm)	Ra (nm)	Rmax (nm)
500	60	no	1.15	0.92	9.43
500	50	no	0.24	0.19	2.40
400	0	no	0.10	0.08	1.00
500	60	yes	0.68	0.56	7.25
500	50	yes	0.17	0.13	1.67

Even though the 50° oxygen flood result does not produce the smoothest crater bottom in this series of experiments, sputter rate consistency and depth resolution were the best (Fig. 10).

Discussion of Roughening

The severity of crater bottom roughening in silicon, using sub keV oxygen primary beams, appears to be largely a function of the incident angle rather than the oxygen flood condition. We have shown that oxygen flooding partially mitigates the roughening effect. While the use of a normally incident beam minimizes crater bottom roughening, improved depth resolution is not realized without reducing the primary beam energy. Our measurements showed a reduction in sputter rate of 5x when changing the beam angle from 50° incidence with flooding to 0° incidence without flooding keeping all other conditions the same. Sputter yields at normal incidence are already low (which allows stoichiometric oxide to form) and further reduction in primary beam energy slows the sputter rate yet again. The much higher sputter yields at oblique angles provide good sensitivity and timely analysis and continues to make this a very desirable analysis condition. These experiments have shown that the effects of roughening can be largely eliminated at oblique angles through the appropriate choice of conditions.

(B) Surface Oxide

Crater bottom roughening is not the only issue affecting accuracy and precision.

Unannealed shallow implants can have 10% or more of the total dose within the surface SiO_2 where the oxide thickness is typically 1 to 2 nm (area B in Figure 1). Profiling through surface oxide and into the substrate is a change of the matrix material and results in a change in sputter rate and ion yield. Surface oxidizing conditions, used to minimize SIMS equilibration (transient) effects, causes an increased sputter rate differential where oxide sputters between 2.1 and 2.4 times the rate in the silicon substrate (16). For boron, the ion yield relative to silicon also increases by about the same factor. Since the sputter rate and the ion yield both increase by about a factor of 2, the area under the curve is about the same compared with non-surface oxidizing conditions, if both are plotted using constant sputter rate and relative sensitivity factors. Comparing surface oxidizing conditions to non oxidizing conditions, the profile shows accurate dosimetry, but the shape is shifted towards the surface by about half the thickness of the oxide, and the concentration shown in the oxide, will be about 2x too high. The true profile shape when using any surface oxidizing conditions will require corrections.

Initial reports in the literature of unexplained shifts in profiles when using oxygen-flooding (17) are mostly explained by the increased oxide sputter rate. There have also been reports of energy dependence in the shift behavior (18) that cannot be explained by oxide sputter rate differentials. Further work is needed to elucidate the energy dependant behavior.

(C) Depth Resolution

The dopant distributions (area C in Figure 1), determines the depth resolution requirements of the SIMS profile. Generally, the lower the energy of the dopant implant, the higher the depth resolution requirement. Lower SIMS primary ion beam energy and/or higher incident tilt angle reduces primary ion beam cascade and recoil mixing effects, improving depth resolution (19). This general statement excludes crater bottom flatness and crater bottom roughening effects. The crater bottom will be flat assuming good instrumentation and alignment. Roughening effects were discussed above.

Dopant activation anneals cause diffusion that decreases distribution sharpness. Depth resolution requirements are reduced. However, it is desirable when comparing before and after anneal profiles to use the same analytical conditions. The maximum sputter rate achievable with high depth resolution for the unannealed profile then becomes even more important for timely analysis.

(D) Dynamic Range and Crater Edge Rejection

Surface concentrations in unannealed high dose profiles are often over 1×10^{21} at/cm^3. It is also desirable to achieve a 1×10^{16} at/cm^3 detection limit so that 1×10^{17} at/cm^3 junction depths can be more accurately measured. Over five orders of dynamic range are needed. Oxide on the sample surface enhances positive ion yields, further increasing dynamic range requirements for boron analysis to at least six orders of magnitude. Once count rates exceed 1M cps, detector saturation can become a problem.

Surface oxidizing conditions, besides minimizing the surface transient, have another important benefit in that dynamic range requirements are reduced to less than six orders of magnitude. Ion yields in the silicon substrate are increased close to the ion yields in the surface oxide and instrument transmission can be reduced to lower the maximum count rate.

The tail shape is also important to examine channeling, or to see if there is ion implanter energy contamination. For accurate tail shapes, it is vital that high surface dopant concentrations from around the SIMS crater edge are not collected. Crater edge rejection is made worse with oxygen flooding since higher instrumental operating pressures increase the probability of primary ion beam scatter. Scattering can be minimized by running the minimum pressure needed to saturate the sputtered surface.

DOPANT SPECIFIC PROFILING

Boron

Boron is well behaved in that it does not present challenges for mass resolution and does not appear to segregate during analysis. Ion yield increase through the surface oxide is balanced with the sputter rate increase and accurate quantification through the surface oxide is achieved. Shallow boron implants are often at high dose and the near-surface concentrations can exceed the solid solubility limit of 6×10^{20} at/cm^3. We have assumed that boron ion yields in silicon do not change at concentrations over the solid solubility limit. This assumption needs to be investigated.

Arsenic

Arsenic secondary ion yields are relatively high when a cesium primary ion beam is used and negative secondary ions are monitored. A 500eV cesium primary beam has sufficient depth resolution to report accurate arsenic tail shape and junction depth information for an unannealed 1keV arsenic implant. Fig. 12 shows an arsenic profile where the tail shape is accurately reported but it also shows an order of magnitude ion yield change through the surface oxide. As is the case for boron, arsenic ultra-shallow implants can have greater than 10% of the dose within the surface oxide. Potentially, corrections could be made for ion yield differences, but the SiO_2/Si transition is not instantaneous even with high depth resolution and a mathematical solution is not straightforward step function.

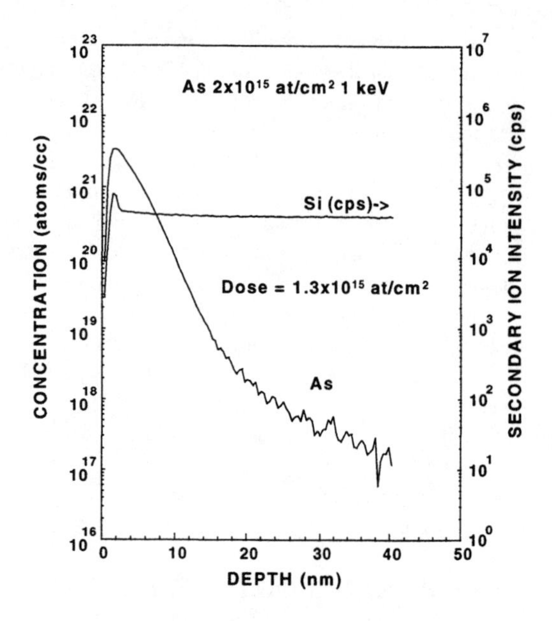

FIGURE 12. An arsenic profile collected with a 1keV Cs/60° incident beam. Quantification through the surface oxide is not accurate. Measured dosimetry is over 30% too low. The detection limit is about 1×10^{16} at/cm^3.

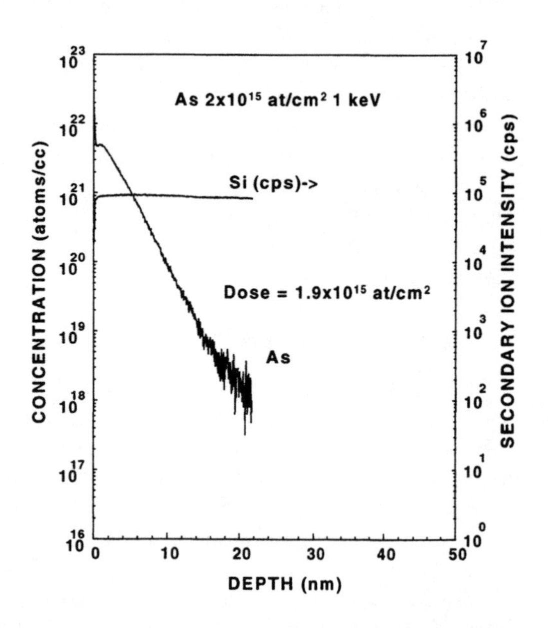

FIGURE 13. An oxygen primary ion beam (500eV/60°) and oxygen flooding were used to generate this profile. Dosimetry is within 5% of the expected value but detection is limited to 1×10^{17} at/cm^3.

Fig. 13 shows an arsenic shallow dopant profile collected using an oxygen primary ion beam with oxygen flooding while monitoring positive secondary ions. The measured dose is much closer to the expected value although it has not been shown that the relative arsenic sensitivity in oxide doubles to compensate for the increased sputter rate. The profile tail shape is the same as before. Dynamic range is reduced and a 1 x 10^{16} at/cm^3 detection limit is not achieved making measurement of the 1 x 10^{17} at/cm^3 junction depth difficult. The full arsenic profiling solution may require separate analyses for dosimetry and profile tail shape. If sensitivity to arsenic under oxygen flood conditions can be improved, a single solution is possible.

Phosphorous

Phosphorous profiles are subject to the same considerations as arsenic but with added analytical requirements. Silicon hydride at mass 31 interferes with phosphorous, also at mass 31. For quadrupole SIMS instruments, reduction of water vapor on the sample surface, reduction of analytical chamber water vapor partial pressure, and energy offset help to improve the phosphorous detection limit (20). We have not successfully suppressed the SiH interference on the sample surface.

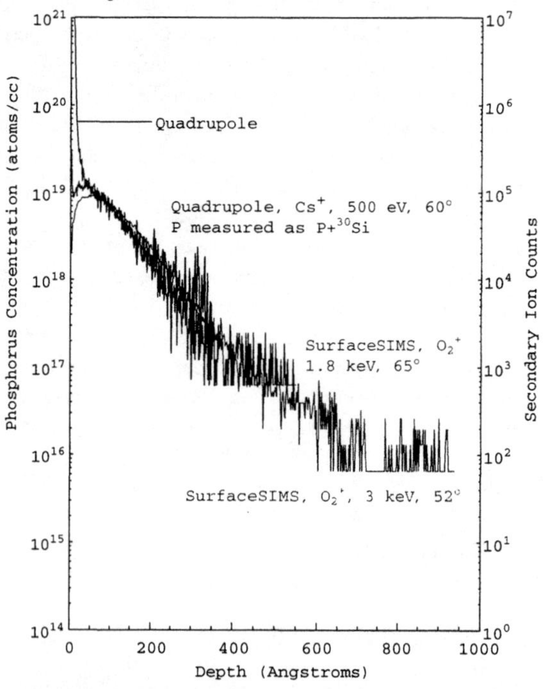

FIGURE 14: Overlay of ultra-shallow phosphorous profiles. Magnetic sector profiles (SurfaceSIMS) show primary beam mixing effects while the quadrupole results cannot suppress silicon hydride interference at the surface.

As in the arsenic case, low energy cesium can provide good depth resolution but phosphorous through the surface oxide is not accurately reported. Phosphorous can be measured through the surface oxide using an oxygen primary ion beam, surface oxidizing conditions, and high mass resolution, as provided by magnetic sector and time-of flight instruments. In Fig. 14, a magnetic sector instrument was used at high mass resolution with a 3 keV oxygen primary beam and oxygen flooding. Good dosimetry and profile shape were achieved with a 1 x 10^{16} at/cm^3 detection limit. Reducing the primary beam energy below 3 keV to improve depth resolution results in higher detection limits of 1 x 10^{17} at/cm^3. The full phosphorous profiling solution may require separate analysis for dosimetry and profile tail shape

CONCLUSIONS

1/ Accurate dosimetry can be achieved for boron ultra-shallow implants.

2/ Accurate profile shapes can be reported for boron, arsenic and phosphorous but corrections are needed through the surface oxide.

3/ Constant sputter rates through silicon using a low energy oxygen primary beam at oblique angle while using oxygen flooding is possible with an appropriate choice of beam energy and angle.

REFERENCES

1. Wolf, S., *Silicon Processing for the VLSI Era Volume 2 – Process Integration*, Sunset Beach, California: Lattice Press, ch. 5, pp.338-347, 1990.
2. Chason, E., Picrauz, S.T., Poate, J.M., Current, M.I., Borland, J.O., Diaz dela Rubia, T., Eaglesham, D.J., Holland, W.O., Law, M.E., Magee, C.W., Mayer, J.W. and Tasch, A.R., Mater. Res. Soc. Symp. Proc. 396, pp. 859-863, 1996.
3. Greenwell, D.W. and Brown, R.L., *Proceedings of the Eleventh International Conference on Ion Implantation Technology*, Piscataway NJ, IEEE Service Center, pp. 1-4, 1997.
4. Magee, C.W., Ultramicroscopy 14, pp.55-58, 1984.
5. Walk, H., Gluech, M. and Lerch, W., Process Simulation of Rapid Thermal Annealed Ultra Shallow Junction Formation in Inert and Oxidizing Ambient, presented at the MRS Spring Meeting, San Francisco, April 13-17, 1998.
6. Parab, K.B., Yang, S.-H., Morris, S.J., Tian, S., Rasch, A.F., Simonton, R. and Magee, C.W., J. Vac. Sci. Technol., B12, 1994.
7. Zalm, P.C., Proceedings of SIMS VIII, Chichester, West Sussex, John Wiley & Sons, pp. 307-314, 1992.
8. Clark, E.A., Dowsett, M.G., Spiller, G.D.T., Thomas, G.R., Augustus, P.D. and Sutherland, I., Vacuum 38, pp. 937-941, 1988.
9. Stingeder, G., Anal. Chem. 60, pp.1524, 1988.
10. ISO Guide 25, Geneve Switzerland, International Organization for Standardization, 1994.
11. Augustus, P.D., Spiller, G.D.T., Dowsett, M.G., Knightley, P., Thomas, G.R., Webb, R., and Clark, E.A., Proceedings of SIMS VI, Chichester, John Wiley & Sons, pp. 485-488, 1988.
12. Elst, K., Vandervorst, W., Bender, H. and Alay, J.L., Proceedings of SIMS IX, New York, John Wiley & Sons, pp. 617-620, 1993.
13. Ronsheim, P.A., Lee, K.L., Patel, S.B., Schuhmacher, M., Proceedings of SIMS XI, Chichester, West Sussex, John Wiley & Sons, pp. 301-304, 1998.
14. Alkemade, P.F.A., Jiang, Z.X., van den Berg, J.A., Badheka, R. and Armour, D.G., Proceedings of SIMS XI, Chichester, West Sussex, John Wiley & Sons, pp. 375-378, 1998.
15. Wittmaack, K., Surf. Interf. Anal., 24, pp.389-398, 1996.
16. Tian, C. and Vandervorst, W., J. Vac. Sci. Technol. A, 15(3), pp.452-459, 1997.
17. Wittmaack, K. and Corcoran, S.F., J. Vac. Sci. Technol. B, 16(1), pp. 272-279, 1998.
18. Dowsett, M.G., Cooke, G.A. and Chu, D.P., Proceedings of the 1998 International Conference on Characterization and Metrology for ULSI Technology, Poster TH-9, NIST, Gaithersburg MD, March 23-27, 1998.
19. Williams, P. and Baker, J.R., Nucl. Instrum. Meth., 182/183, pp. 15-20, 1981.
20. Magee, C.W. and Botnick, E.M., J. Vac. Sci. Technol. 19, pp. 47-51, 1981.

Two dimensional profiling of ultra-shallow implants using SIMS

G.A.Cooke, R.Gibbons and M.G.Dowsett

Department of Physics, University of Warwick, Coventry CV4 7AL, United Kingdom

The lateral spread of dopant under the implant mask edge and its behaviour during thermal processing is becoming increasingly important as device dimensions are reduced. Direct measurement of the distribution by high spatial resolution SIMS is not possible owing to the very few impurity atoms present in the analyte volume at junction concentrations. In this paper we describe a SIMS based technique, using a special sample structure, that may be used to access this information and discuss the instrumental requirements, resolution and detection limits, as well as presenting cross sectional dopant data.

INTRODUCTION

The continued reduction in device dimensions brings with it the requirement for more accurate modelling of the fabrication processes. Of major importance is the lateral spread of dopant under the mask edge, as this has a direct influence on the effective channel length (in MOS devices) and hence the electrical characteristics of the device. Indeed, for smaller technologies where the concentration varies rapidly, the lateral dopant profile at the source and drain must be deliberately engineered if the device is not to suffer unduly from carrier induced damage (i.e. hot electron effects). Although most TCAD packages include the provision for two (and even three) dimensional simulation, the algorithms on which they are based are often extensions of one dimensional (depth) observations and have not been verified in the lateral dimension. This approximation is probably valid in the bulk of the material which may be regarded as homogeneous, however, nearer the surface diffusion effects are likely to be more anisotropic, with effects such as dopant trapping at interfaces, defect enhanced diffusion and rapid changes in stress, all modifying the distribution during thermal processing. Monte Carlo calculations may be employed to model such effects but they are necessarily time consuming, especially in when complex structures are involved. Where interacting dopants, such as boron and phosphorous, are to be modelled their use is very limited.

The measurement of carrier profiles in cross section has been the subject of much research and a number of successful techniques have been developed (1,2,3). However, although the carrier profile is useful in determining the electrical properties of the distribution it may lead to inaccurate calibration of the simulator as inactive dopant, resulting from insufficient anneal or exceeding the solid solubility, may return to haunt the process engineer later in processing, as well as compromising device performance or lifetime. Despite the obvious need to model the true chemical profile, it is not surprising that there have been only a few attempts to measure it (4,5,6) when the required parameters and statistics are considered. If a spatially resolving analytical technique such SIMS is employed directly the expectation is pessimistic; assuming a 30nm spatial resolution a cubic voxel would contain about 1.4×10^6 atoms of which 14 000 would reach the detector (assuming a useful yield of 1%). For a modest statistical precision of 33% ten impurity ions must be counted. The voxel must therefore contain at least 1000 impurity atoms, representing an atomic concentration of 3.7×10^{19} atoms cm^{-3}. For modern devices 30nm is too big anyway!

The 2D SIMS Technique

Although direct application of SIMS is clearly not going to produce data of sufficient quality for dopant analysis, SIMS is perhaps best placed to fulfil the need for a two dimensional profiling tool as it is spatially resolving and has high sensitivity (especially for dopants, where fractional atomic concentrations in excess of 1 part in 10^9 may be achieved). If cross sectional profiles are required then the third dimension may be employed to overcome the limitations experienced by a direct measurement. The 2D SIMS technique (6,7,8), conceived by Prof. Hill of GEC-Marconi, ultilizes the third dimension to (i) increase the analyte volume and (ii) realise high lateral resolution while using a relatively large reactive probe to maximise the ion yield.

Figure 1 shows the detail of sample preparation. Initially a series of stripes are implanted into a substrate. Typically the stripes fill a 2 mm square area (i.e. 400 stripes each 2 mm long and 2 μm wide on a 5 μm pitch). The exact implant and mask conditions are chosen to reflect the process under investigation. It should be stressed that multiple dopants can be introduced and, with a little variation to the masking details, complete source-drain structures may be fabricated. Once implanted, the wafers are then subject to thermal processing, although the as-implanted profile may also be measured. Before the mask is removed a low energy ^{70}Ge marker is implanted at normal incidence (7). The stripes are next sectioned vertically by a series of anisotropically plasma etched trenches with a similar mark-space ration to the original

CP449, *Characterization and Metrology for ULSI Technology: 1998 International Conference*
edited by D. G. Seiler, A. C. Diebold, W. M. Bullis, T. J. Shaffner, R. McDonald, and E. J. Walters

mask. The angle between the implanted stripes and the trenches is typically 0.1°. To ensure planar erosion during the SIMS analysis, the trenches are next filled with polysilicon, with the oxide etch mask still intact. The test structure is then planarized by chemical mechanical polishing, leaving 5-30 nm of the mask remaining, this is then removed by wet etching in 2% HF solution, exposing the *original silicon surface*.

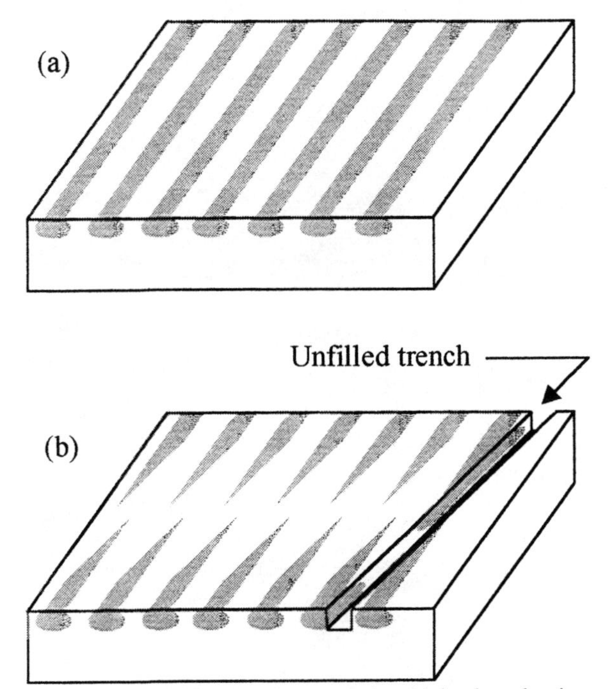

Figure 1 Detail of the sample preparation (a) implanted stripes are sectioned by a trench which is later filled as in (b)

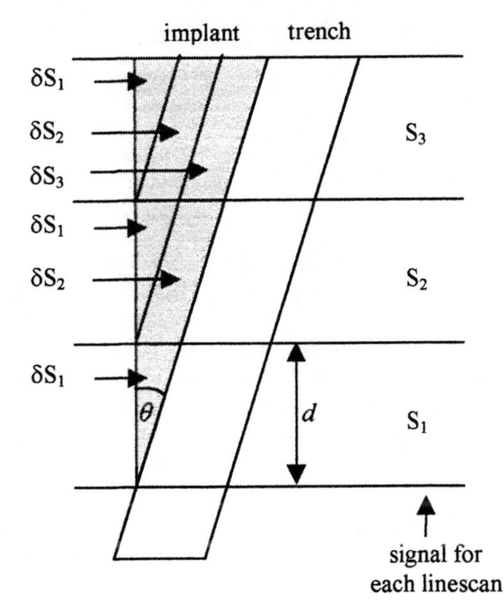

Figure 2 Analysis of the structure (only one stripe is shown for clarity)

The sample is analysed in a gated linescan mode, shown in figure 2. The primary beam is scanned orthogonally to the implant stripe direction and the signal for each *linescan* is recorded. This signal comprises the integrated signal from the same segments of all of the stripes and is proportional to the number of stripes, *N*, directly improving the sensitivity. It is clear from figure 2 that the small angle between the stripe and trench, θ, effectively magnifies the lateral dimension, so for a single linescan with a probe of

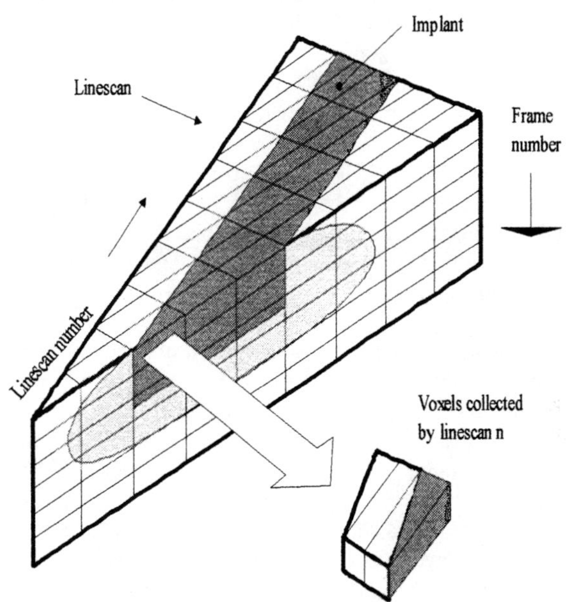

Figure 3 Section of one of the stripes showing the origin of the signal

diameter, *d*, the lateral dimension sampled is $d \tan \theta$, which may be regarded as the resolution of the technique. The difference between the signals from adjacent linescans is proportional to the lateral diffusion profile, and therefore differencing of a complete plane of linescans is representative of the lateral distribution at the depth of the plane. For example the signal arising from cell δS_n is,

$$\delta S_n = S_n - S_{(n-1)}$$

Successive scans expose deeper material, hence a full two dimensional profile may be accessed as in figure 3.

The fractional atomic concentration F_x, of an atomic species *X* in a measurement that consumes *N* sample atoms is given by,

$$F_x = n_x / N\tau_x,$$

where τ_x is the useful yield of species *X* and n_x is the number of ions of *X* detected. The elemental volume analysed by one pass is $\delta z m d^2 \tan\theta$ where δz is the depth of material removed and *m*, the total number of stripes traversed. Thus, assuming a matrix density, *D*, of atoms of average mass *M*,

$N = D\ \delta z m d^2 \tan\theta / M.$

The minimum absolute impurity detection level, $C_{x(min)}$, is therefore given by

$C_{x(min)} = n_{x(min)} / \delta z\ \tau_x\ m d^2 \tan\theta$

Where $n_{x(min)}$ is the minimum number of ions that must be detected for the required statistical precision (typically 10). Thus for a typical sample with, $\theta=0.1°$, $d=10\ \mu m$, $\delta z=5$ nm, $m=400$ and $\tau_x=0.01$, a detection limit of 2.8×10^{15} atoms cm^{-3} may in principle be achieved. This is coupled with a lateral resolution of 17 nm.

As the detection limit is dependent upon the size of the analyte, relaxing either the depth or lateral resolution will result in improved sensitivity. Thus adjacent cells (vertical or horizontal) may be integrated after data collection to improve the statistics for a particular investigation.

As the technique requires spatially separated signals to be compared, the collection efficiency over the entire sample area must be constant – to overcome this problem a planar implant is analysed using the same linescan technique and under the same conditions (including sample holder position). Data from this analysis are used to correct for efficiency variations across the analysis window, however, it should be stressed that the correction imposed by this method should be small (<30% across the entire 2 mm area) as errors due to statistically poor signals will be compounded.

Analytical Instrumentation

The 2D SIMS sample is typically 2 mm square, this size is restricted by the sensitivity requirements of the technique and any reduction in size will lead to loss of sensitivity. Therefore the instrument used for analysis must be capable of collecting ions efficiently from this entire area, this normally means that a quadrupole based instrument must be employed (although magnetic sector instruments with a large field of view, or with a scanning stage (9) could be used). Additionally, to obtain high depth resolution (and maintain it for the full depth of the profile without roughening or raster projection effects) low energy O_2^+ probes at normal incidence must be used. The primary beam scanning must be of sufficient precision to produce straight linescan, and flat bottomed craters to ensure that neither lateral or depth resolution is compromised.

Structures are currently analysed on the ASP group's two quadrupole instruments, both fitted with FLIG™ ultra-low energy ion guns, but with different secondary ion extraction optics.

The EVA 2000*FL* instrument has a modified parallel plate extraction and energy filtering system offering the full 2 mm field of view, whereas the EVA 3000 instrument has a very small field as shown in the upper plot of figure 4. Although this is useful for providing an element of optical gating for conventional depth profiling work, it hinders large area collection. To overcome this the extraction system is dynamically matched to the probe position (dynamic emittance matching, DEM), providing the much flatter extraction efficiency shown in the lower plot (10)

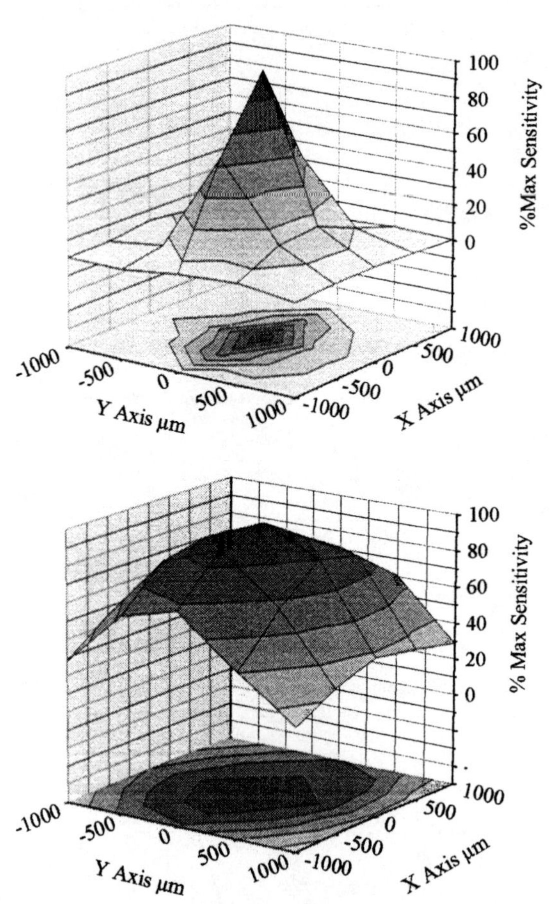

Figure 4 Area of high collection efficiency of an optically gated system (upper) is increased using DEM to provide suitable matching to the large sample area

The ultra low energy probes obtainable with these systems (as low as 100 eV O_2^+) enable both sub-nm depth resolution to be obtained and quantification from within 0.5 nm of the surface. This is a necessity when implants of 5 keV and below are to be analysed. Typical analysis conditions for a 5 keV boron implant are 200 nA of O_2^+, at 1keV,focussed into a 20-25 µm probe, at normal incidence.

It is vital for the ion beam to scan orthogonally with respect to the stripes. Any deviation will reduce the lateral resolution. We use two methods to achieve correct alignment. Alignment in-situ is available by mechanically rotating the scan assembly whilst imaging a group of parallel implanted lines near to the structure. The secondary ion signal is monitored as a function of beam displacement. When correct alignment is achieved the signal is constant for the length of the scan. If the structure is misaligned, the signal is modulated. Knowledge of the line spacing and length of scan permits the misalignment angle to be measured and a correction made. The other method involves accurately aligning the structure with a line previously sputtered on the sample holder. Using a precision stage assembly this method permits a number of samples to be aligned and thus reduces instrument time. In both cases, the alignment of the crater is checked after analysis. The measured misalignment, A, is used to calculate the effective probe diameter, d_{eff}, as

$$d_{eff} = d_p + l \tan A,$$

where, d_p is the probe diameter

Generally, however, the misalignment angle is better than 0.1°, leading to an effective increase in probe diameter of only 3.5 µm.

Data Processing and Calibration

Three groups of data are required for 2D analysis, the gated linescans from the structure (impurity and Ge marker), a set of data collected in a similar way from a planar implant, and a depth profile of the implant of interest, for which a test pad is sited next to the structure.

Initially, post equilibrium linescan data acquired from the planar implant are vertically integrated to improve the statistics. Next, the data are normalised (as there should be no variation in signal between different areas of the planar implant). The normalisation coefficients for each linescan are recorded and applied to the corresponding linescans in the data from the structure. Thus variation in efficiency of the extraction system can be compensated for. It is not sufficient to use a matrix channel to achieve this correction, as tests have shown that errors of over 50% may occur because of effects such as surface charging and topography changes in the polysilicon fill between the stripes.

Once corrected, the data from the special structure are differenced, permitting a two dimensional matrix to be filled. Calibration of the matrix is achieved by comparison of the depth profile measured on the sample with a depth profile extracted from the matrix. Although this method does not *require* the depth eroded from the patterned sample to be measured, such a measurement is made to ensure consistency.

Finally, the mask position is determined by knowledge of the true window width (measured by cross sectional TEM) and the symmetry of the Ge implant (which was made at normal incidence).

Results and Discussion

The results presented here are from a doubly implanted boron structure. The first implant was 5×10^{15} ions cm^{-2} at 7° tilt, into crystalline silicon. This was followed by a 5×10^{14} ions cm^{-2} implant at 45°. In both cases the implant energy was 5 keV. After implantation, the wafer was

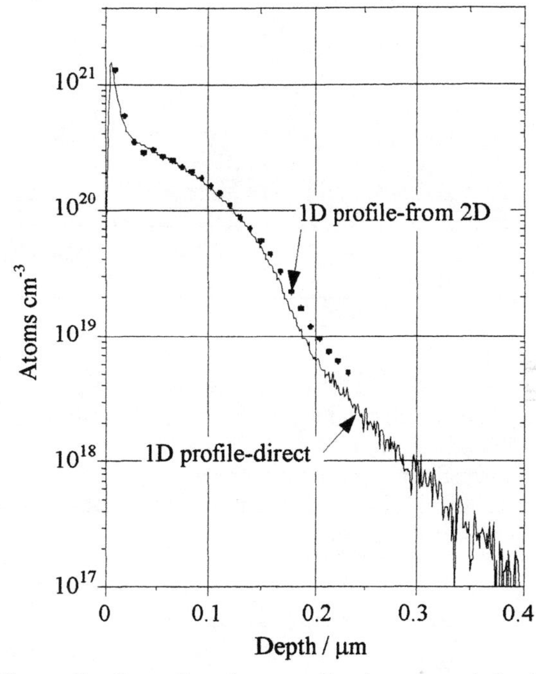

Figure 5 Comparison between directly measured depth profile and that extracted from the 2D SIMS technique

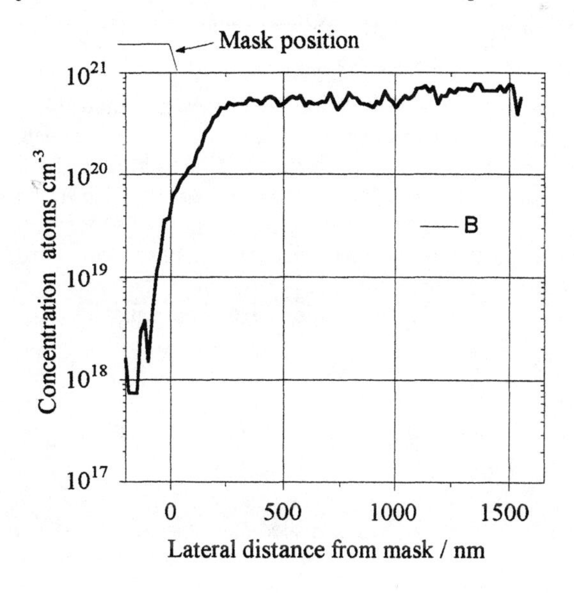

Figure 6 Measured lateral profile at 8 nm below the surface.

annealed at 1050°C for 20 seconds.

Figure 5 shows the comparison between the directly measured depth profile and that extracted from the two dimensional data set, with the depth scales being measured independently. There is good agreement between the two sets of data, with both showing the high concentration near surface feature indicative of precipitation of the implant. This is followed by the diffused implant and eventually a prominent tail due to channelling of the ions.

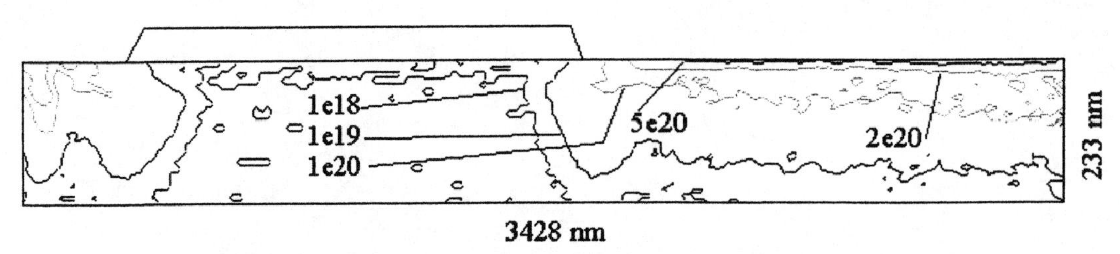

Figure 7 The full reconstructed two dimensional profile (the vertical scale has been magnified for clarity).

Figure 6 shows the reconstructed data at a depth of 8 nm from the surface. Again there is a transition between the region where precipitation may be expected to occur and the diffused region, though obviously the shape changes will be significantly affected by the mask edge. In this case the relatively high background precludes investigation of channelling at this depth.

Figure 7 shows the full cross sectional profile. The implant shape is roughly planar in the central region with side lobes extending under the mask edge. The variation of the dopant towards the edge of the window and under the mask is attributable in part to reflection of dopant ions by the mask, which, for this particular implant, was fabricated in 1 µm thick oxide.

There is a significant boron concentration immediately under the mask (approx. 10^{18} atoms cm^{-3}) which explains the poor detection limit of figure 6. The uniformity of the layer suggests that it is due to contamination of the surface during fabrication. This effect could of course be corrected for by simultaneous collection of both boron isotopes, the presence of B^{10} indicating the extent and concentration of contamination by environmental boron.

Conclusion

The 2D SIMS technique is a viable method for the study of cross sectional *dopant* profiles with high spatial resolution and with good sensitivity. As well as obtaining data from simple implants in both annealed and unannealed cases, it may also be used to study codiffusion effects. By increasing the number of masking layers it is feasible to construct complete source-drain structures, and as the technique is easily scaled, these may be at technologically important feature sizes.

Although the lateral magnification and enhancement in sensitivity is vested in the sample structure, the instrument used for analysis must fulfil stringent requirements. It must have a large field of view with a relatively flat collection efficiency, a means of sample alignment, and good mechanical and electrical stability, as a typical measurement takes up to 12 hours. Most importantly for shallow implants, sub-keV probes must be used to ensure that the profile can be quantified from very close to the surface. It is also necessary to maintain sub-nm depth resolution throughout the analysis which inevitably means that normal incidence O_2^+ must be used to prevent the onset of surface roughening (and raster projection effects) as rotation of the sample during analysis is obviously not possible. A strength of the technique is the simple data processing needed to extract the profiles.

References

1 DeWolf, P., Clarysse, T., Vandervorst, W.,Snauwaert, J. and Hellemans, L. *J.Vac.Sci.Technol.* **B14**(1996) 380

2 Venables, D. and Maher, D.M., *J.Vac.Sci.Technol.* **B14**(1996) 421

3 Many papers on this topic are contained within USJ-97, Eds. Kump. M., Current. M. and McGuire. G

4 von Criegern, R., Jahndel, F., Bianco, M. and Lange-Glieseler, R., *J.Vac.Sci.Technol.* **B12** (1994) 369

5 Goodwin-Johansson, S.H., Subrahmanyan, R., Floyd, C.E. and Massoud, H.Z., *J.Vac.Sci.Technol.* **B10** (1992) 369

6 Dowsett, M.G. and Cooke, G.A., *J.Vac.Sci.Technol.* **B10** (1992) 353

7 Cooke, G.A., Dowsett, M.G., Hill, C., Pearson, P., and Walton, A.J., *J.Vac.Sci.Technol.*, **B12** (1994) 243

8 Cooke, G.A., Pearson, P., Gibbons, R., Dowsett, M.G. and Hill, C., *J.Vac.Sci.Technol.* **B14**(1996) 348

9 Gibbons, R., Dowsett, M.G. and Richards, D.H., Secondary ion mass spectrometry SIMSX, eds.Benninghoven, A., Hagenhoff, B. and Werner, H.W., John Wiley and Sons, 1997, pp991-994.

10 Gibbons R, Dowsett M.G., Cooke G.A, Hill C., and Pearson P., *Secondary Ion Mass Spectrometry SIMSXI*, eds. Gillen G., Lareau R., Bennet J. and Stevie F., John Wiley and Sons, Chichester, 1998, pp313-316

Is Ultra Shallow Analysis Possible Using SIMS?

D. P. Chu, M. G. Dowsett, T. J. Ormsby, and G. A. Cooke

Department of Physics, University of Warwick, Coventry CV4 7AL, United Kingdom

The use of secondary ion mass spectrometry (SIMS) to analyse ultra shallow dopant profiles is now becoming routine. However, interpretation of the data is not straight forward, and the conventional method of effectively multiplying intensity and ion dose (time) axes by calibration constants to "quantify" the data is certain to produce serious inaccuracies. We demonstrate that for oxygen primary beams, analysis of silicon at normal incidence without oxygen flooding is currently the only analytical condition which leads to quantifiable, accurate profiles, and show that depth resolution better than 1 nm can be obtained from within 0.5 nm of the surface using sub-keV primary beams.

INTRODUCTION

Sub-micron process evaluation requires a depth profiling tool capable of quantitative analysis right from the surface, with a sub-nm depth resolution, concentration accuracy of 1%, and a depth scale accurate to within 1 nm. In principle, with the advent of ultra low energy secondary ion mass spectrometry (SIMS) depth profiling, the technique can meet such demands for a wide range of samples. However, the unquestioned use of reproducibility as a guide to data quality, rather than investigation to establish the accuracy of the technique leads to serious mistakes in quantitative analysis, just as it has throughout the history of the technique. Although recent publications illustrate the progress in ultra shallow profiling using sub-keV ion beams (1-5), the influence of analytical conditions and sample composition on the accuracy of the depth and concentration scales requires extensive investigation (3,6,7). Accurate ultra-shallow profiling requires the development of new quantification methods to complement new instrumentation.

A typical modern silicon sample consists of one or more wholly or partly processed dopant implants into a matrix which may be preamorphized with Ge or Si. In the most difficult cases, the implant energy may be as low as 100 eV, and the cumulative dose may be sufficiently high for the near surface region of the sample to be less than 70% silicon. There may be deliberate or inadvertent oxidation of the surface (3,8). Whilst valuable comparisons can be made between data from such samples, accurate quantification is currently extremely difficult because of matrix effects, changing erosion rates and other factors.

In the simplest case, however, the SIMS depth profile consists of two parts: a transient region starting at the surface, whose thickness is determined by primary beam species, energy and angle of incidence, and a steady-state region where the data are relatively simple to quantify. It is well known that both sputter yield and ionization rate vary in a nonlinear manner in the transient region. The analyst's task is therefore to choose conditions which minimize its width, and at least to find a way of correcting the overall depth scale for any effects due to the varying erosion rate (9). Portions of the dopant profile within the transient region are only quantifiable under special circumstances, for example if the sample surface is already coated with a thin oxide, and even here, the erosion rate is primary beam dose dependent and obtaining the true depth scale is difficult (10). It is therefore vital that the dopant profile lies beyond the transient region *and this is a prerequisite for accurate quantification*. Until recently it was believed that the surface transient region was due solely to the finite primary ion dose required to establish a steady-state chemistry in the sample surface. This was certainly the case for beam energies >2.5 keV. It was also known that steady state conditions were lost (for non-normal incidence) after 1-3 μm (11,12) with the onset of extreme roughening of the sample. In the last year it has been shown that the onset of roughening is within 10's nm of the surface at primary beam energies ~1 keV (6,13). Here, we show that at ≤500 eV roughening merges with the transient region at non-normal incidence, and quantitative profiling becomes impossible.

The inherent shape of an ultra shallow implant is not known and such samples give little warning when their profiles are seriously distorted. Conversely, profiles from closely spaced delta layers exhibit obvious changes when the surface roughens, e.g. depth resolution is lost, and apparent inter-peak spacing becomes dependent on measurement conditions. In this paper, we use such a sample to obtain an accurate relationship between signal intensity and depth, prior to obtaining a full concentration profile. The methods used here are extendable to more complex samples such as those with preamorphized and/or oxidised surfaces. We first discuss how the transient width can be minimised, and indeed whether its width can be measured from matrix transient signals. We then present data which reinforce the discovery of early surface roughening at non-normal incidence. Finally, we show that even at normal incidence new correction factors are required to obtain a depth scale accurate to 1 nm, and that in general, the combination of possible roughening and altered layer formation make the true depth eroded in silicon different from the surface profilometer measurement of crater depth.

CP449, *Characterization and Metrology for ULSI Technology: 1998 International Conference*
edited by D. G. Seiler, A. C. Diebold, W. M. Bullis, T. J. Shaffner, R. McDonald, and E. J. Walters

EXPERIMENTAL DETAILS

Experiments were carried out on two quadrupole SIMS instruments, EVA2000*FL* and EVA3000, equipped with floating low energy ion guns (FLIG™) designed by M.G.D. (14). The sample was grown by silicon MBE at a nominal temperature of 500°C and contained 10 boron delta doped layers with an approximate spacing of 18 nm. Immediately prior to analysis, the analyte surface was dipped in 5% HF solution in de-ionized water until the surface was fully hydrophobic. Therefore, all profiles started on a surface with similar conditions, and the measured parameters such as transient widths will be comparable. We have added other data from previous studies (7) on similarly treated float zone (FZ) silicon for comparison.

RESULTS AND DISCUSSION

Surface Transient Width and Sample Roughening

The silicon sputter yield varies continuously across the transient region, and is dependent on the initial surface condition of the sample. Although the relationship between erosion rate and primary beam dose is not yet established (15) it is possible to determine the average value from known parameters such as the transient shift Δz_{tr} (~0.9 nm keV^{-1} (9)) and the true bulk erosion rate $\dot{z}_{Si}$ provided the end point of the transient region can be detected. Note that if a profile is quantified using an erosion rate deduced directly from crater measurement, the transient region can appear more than four times narrower than its true width because the average transient erosion rate is higher than bulk (see Fig. 1). For normal incidence profiling with O_2^+ ions, we find that a wide range of matrix associated secondary ion signals stabilize at the same primary ion dose ϕ_{tr}. If it is *assumed* that both ionization rates *and* matrix sputter yield are constants after this point, ϕ_{tr} can be used to measure the absolute depth sputtered in silicon to achieve steady state z_{tr} (5,7):

$$z_{tr} = z_{app} - \dot{z}_{Si}(\phi_\delta - \phi_{tr}) + \Delta z_{tr} \qquad (1)$$

where z_{app} is the apparent depth of a marker layer fairly close to the surface (e.g. the centroid of a delta layer - see next section) and ϕ_δ is the total ion dose required to reach the marker. Our recent measurements (16) show that the above assumption is not quite correct, and the erosion rate continues to change slightly after the ion signals have stabilized. This results in a residual error which is ~0.5 nm for sub-keV profiling.

We show in Fig. 1 the apparent and corrected surface transient widths of an MBE grown silicon wafer for normal incidence O_2^+. Also included are similar data for FZ silicon. We can see that the true surface transient width reduces from over 11 nm to 0.5 nm as the primary beam energy decreases from 5 keV to 230 eV. In the sub-keV regime, the FZ result differs significantly from the MBE result, giving even narrower transients. The reason for this is not known - but it is likely to be associated with the high boron concentration on the surface of the MBE grown material due to segregation effects. It is important to note that based on this data one would expect the result for preamorphized silicon to be different again - especially if, as we suspect, the surface became loaded with atmospheric oxygen when removed from the implanter.

Nevertheless, it is clear that normal incidence O_2^+ profiling at sub-keV energies will provide quantitative data from 0.5 nm onwards.

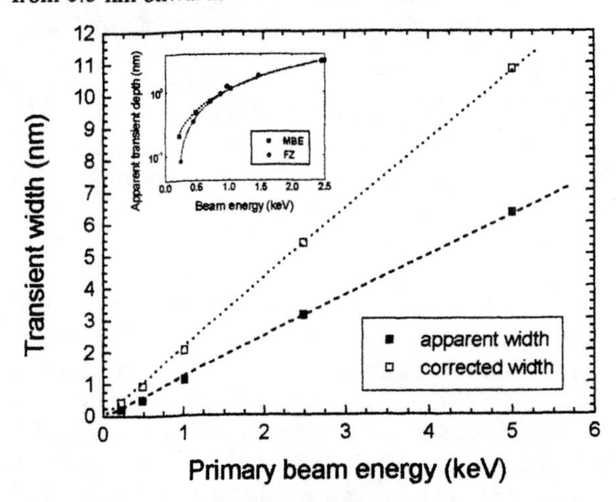

FIGURE 1. Apparent and corrected surface transient widths as a function of primary beam energy. The O_2^+ beam is at normal incidence.

It might be thought that increasing the primary beam incident angle θ would help to further reduce the surface transient width (and improve depth resolution), since the projected range of primary ions with a given energy will decrease $\propto \cos\theta$. However, this is incorrect as shown by Figs. 2 and 3. Firstly, it becomes difficult to identify the end of the transient region since matrix signals stabilize at different doses. One should perhaps take ϕ_{tr} as being given by the *last* signal to stabilize, but we do not at present know which this would be. Secondly, it is clear from this behaviour, and the loss of depth resolution on the MBE deltas that the surface is becoming rough almost immediately at non-normal incidence. Thus the chemical transient region has merged with the roughening region. (This would not be particularly obvious on an implant profile - but the recovered shape is nevertheless quite incorrect.) Fig. 2 shows that for a 500 eV O_2^+ primary beam the apparent transient width of the same sample increases from 0.47 nm at normal incidence to 25 nm or even 35 nm at 60° depending on which matrix channel is used as indicator (see insert of Fig. 2(b)).

Fig. 3 shows the complete behaviour for angles of bombardment from 0-60° using the SiO^+ and Si_2O^+ ions as the indicator. Data from other workers (6) demonstrates that using oxygen flooding makes these effects even worse.

Normal incidence bombardment in *vacuo* therefore becomes the only option at present for accurate depth

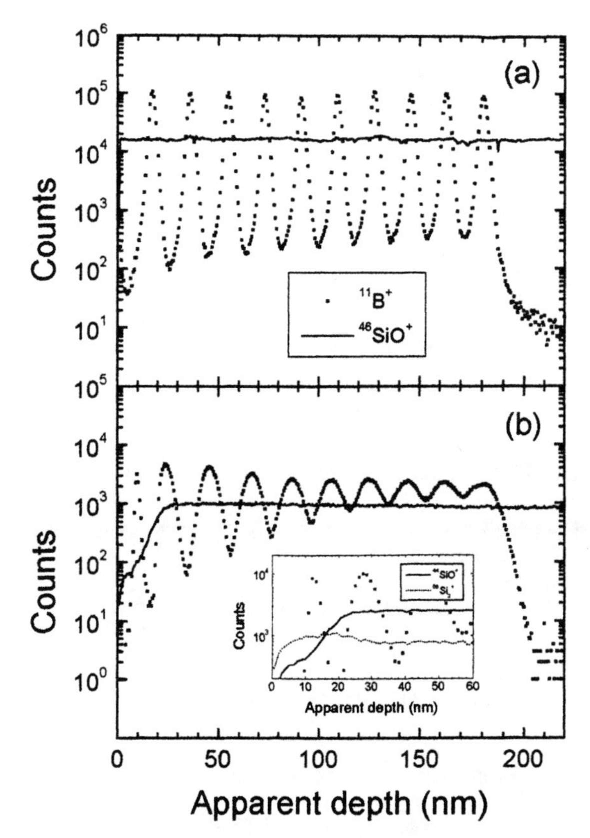

FIGURE 2. Depth profiles of ten boron δ layers in silicon using a 500 eV O_2^+ beam at (a) 0° and (b) 60°.

profile analysis using an oxygen beam at low energy for two reasons:

(i) The highest depth resolution is obtained; and it remains constant with increasing depth.

(ii) The narrowest transient region is obtained.

For angled bombardment, the inconvenience of zalar rotation might suppress the roughening (although this is not yet established for sub-keV beams) but normal incidence will still give the narrowest transients and the highest depth resolution.

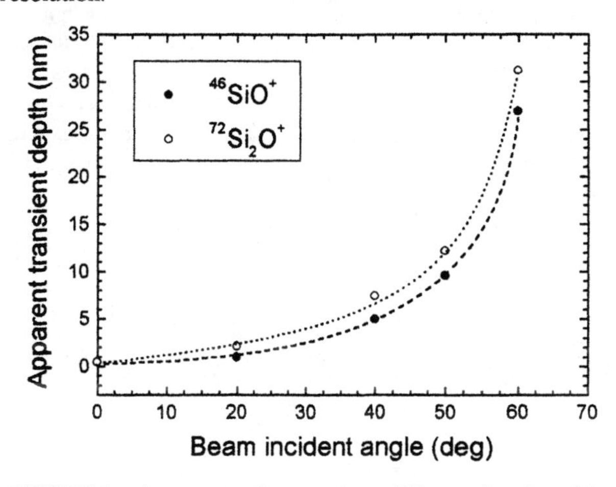

FIGURE 3. . Apparent surface transient width as a function of the incident angle of primary beam. The O_2^+ beam energy is 500 eV.

Systematic Errors in the Depth Scale

In a fully quantified SIMS profile the objective is to relate a particular ordinate in impurity concentration to its true depth within the (silicon) matrix. Leaving aside problems due to atomic mixing for the moment, we have already shown that two erosion rates (bulk and average transient) are required for this. These are almost universally obtained from surface profilometry of the SIMS crater. We now show that such measurements contain at least two possible sources of systematic error which become significant for shallow profiles.

Assuming that it is not significantly modified by exposure to air, the crater bottom is covered by a layer comprising a mixture of probe and matrix atoms. If roughening has occurred, this will also remain on the crater bottom. Assume that a profilometer measurement is made from the unmodified starting surface to the surface of this layer. It differs from the depth of matrix eroded for two reasons:

(i) The matrix density (matrix atoms cm^{-3}) is modified by the presence of the implanted probe atoms.

(ii) A profilometer stylus tip is typically 12 μm in diameter. This will ride over the crests of any roughening, resulting again in an underestimate of the true amount of matrix sputtered.

In the course of this work we quantified a third effect which is more subtle because it is impurity dependent: Suppose the primary beam induced mass transport is not solely due to random cascade mixing, but involves some preferential migration of the impurity into, or away from the altered layer (17). A delta layer would be observed to be shallower or deeper respectively than its true depth, as would the dopant intensity from any broader distribution.

It is important to determine whether or not these effects are significant. If we consider any impurity profile, it is clear that a depth calibration based on a constant erosion rate will leave the first ordinate after the transient out of position by the amount of the transient shift, and the last ordinate out of position by the terminal shift, the difference between the true and measured crater depths added to any migratory shift. In an early experiment to measure this latter effect at high beam energies (18) we used a method based on stopping a profile at the peak of a delta layer and assumed that the peak corresponded to the true depth of the feature. Because the asymmetry of the SIMS response is energy dependent, the peak position is itself displaced from the true position (the centroid of the response where random cascade mixing is the dominant mass transport mechanism (19)) by an energy dependent shift. In this work we used a different method: Assume that the true depth z_δ of a feature such as a delta layer is known by some means. Then if $z(E)$ is the apparent depth of (say) the centroid of any delta at a given energy E, the total energy dependent offset between the true and apparent positions is:

$$\Delta z_{tot}(E) = z_\delta - z(E). \tag{2}$$

This profile shift comprises dose dependent fractions of the transient shift Δz_{tr} due to changing erosion rate in the transient region and the terminal shift Δz_{te} due to the beam oxidation induced swelling of the crater base and directional relocation.

$$\Delta z_{tot}(E) = f_1(\phi)\Delta z_{tr}(E) + f_2(\phi)\Delta z_{te}(E) \quad (3)$$

where $f_1 \to 1$ and $f_2 \to 0$ as $\phi \to \phi_{tr}$, and $f_1 \to 0$ and $f_2 \to 1$ as $\phi \to \phi_{cr}$. The ϕ_{cr} is the total dose used to erode the SIMS crater. If Δz_{tot} is plotted vs. ϕ_δ for a particular energy, then

$$\Delta z_{tot}\Big|_{\phi=\phi_{tr}} = \Delta z_{tr} \quad (4)$$

and

$$\Delta z_{tot}\Big|_{\phi=\phi_{cr}} = \Delta z_{te} \quad (5)$$

Equation (5) gives the (beam energy dependent) correction required to convert the measured crater depth to the total thickness of silicon eroded plus the offset from directional relocation. In effect, if the impurity were to segregate *into* the altered layer, the impurity depth scale is extended beyond the depth of silicon eroded. The problem is to find a sample where z_δ is known accurately. It is known from this and other work (9) that the apparent centroid and apparent peak positions for boron deltas have essentially the same intercept z_0 if plotted against beam energy. Provided any systematic errors are beam energy dependent it can be assumed that $z_\delta \equiv z_0$. The results obtained by applying this strategy to the data from the boron delta layer sample are shown in Fig. 4. Further details of the calculation are given in Ref. (16).

Fig. 4 shows that the terminal shift correction can have serious impact on the depth calibration for low energy shallow profiles. It extents the total depth of a boron profile from a measured 20 nm crater by ~10% if a 500 eV beam is used, and would be of the same order as the measured depth when the beam energy reaches 5 keV. For boron, the magnitude is almost identical to the total thickness of beam synthesized oxide.

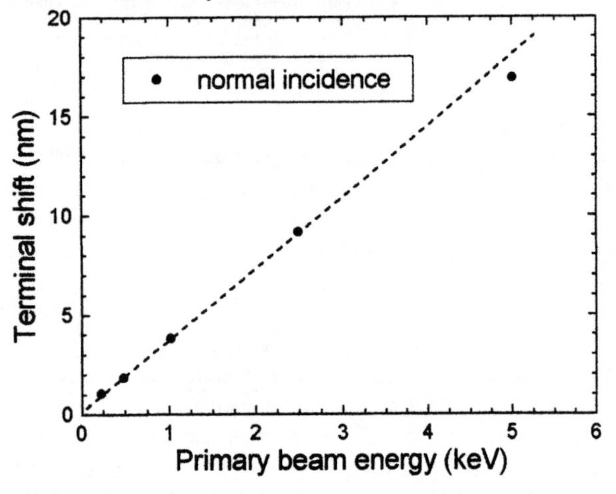

FIGURE 4. Terminal shift for boron in silicon as a function of primary beam energy. The O_2^+ beam is at normal incidence.

Based on the above arguments, we can now describe the overall correction procedure required to obtain an accurate depth scale for normal incidence profiling using a measured crater depth z_m (see Fig. 5):

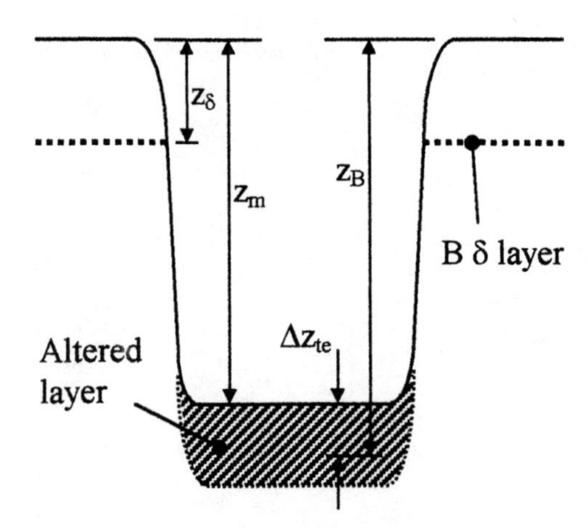

FIGURE 5. Schematic of the depth parameter notation.

The effective end of the depth scale z_B for the dopant is

$$z_B = z_m + \Delta z_{te}. \quad (6)$$

An apparent erosion rate $\dot{z}_{app}$ is calculated from

$$\dot{z}_{app} = z_m / \phi_{cr}. \quad (7)$$

The true width of the transient region is then:

$$z_{tr} = \dot{z}_{app}\phi_{tr} + \Delta z_{tr} \quad (8)$$

and the profile dopant related evolution rate is

$$\dot{z}_B = (z_B - z_{tr})/(\phi_{cr} - \phi_{tr}). \quad (9)$$

Finally, the true depth z of any ordinate at a beam dose ϕ where $\phi > \phi_{tr}$ is

$$z = z_{tr} + \dot{z}_B(\phi - \phi_{tr}). \quad (10)$$

Atomic Mixing in the Dilute Regime

The above procedure places a yield or signal intensity value at a depth which accurately corresponds to the dose at which it is observed. However, the intensity itself contains contributions from parts of the concentration profile originally at other depths because of atomic mixing. We have shown elsewhere that, for a dilute impurity level, a depth profile of a dopant feature can be described as a convolution of the feature and an empirically determined instrumental response function (20,21) The final stage in accurate profile quantification is therefore a deconvolution. The rules for this must be obeyed strictly, or the results will be rubbish concealed by a sophisticated mathematical process. They are:

(i) The impurity level must be dilute.

(ii) The response must be appropriate to the analytical conditions and the impurity/matrix combination.

(iii) The noise in the data must be accounted for on a rigorous statistical basis.

See Refs. (21) and (22) for more details.

The effectiveness of the process is demonstrated in Figs. 6(a) and (b) where the effects of both reduced beam energy and deconvolution are shown. The centroids of the first and second boron δ pairs are separated by 2 and 5 nm, respectively.

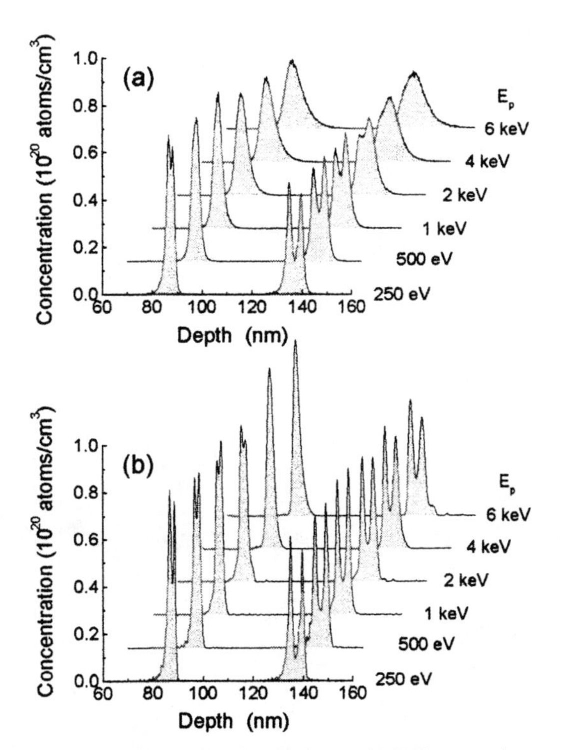

FIGURE 6. Concentration distributions of MBE grown boron δ pairs in silicon profiled with a normally incident O_2^+ primary beam of various energies E_p. (a) as measured and (b) deconvolved.

CONCLUSIONS

The data presented above together with that from other workers demonstrate that accurate ultra shallow profiles from silicon using oxygen primary ions can only be obtained using sub-keV beams at normal incidence. Under these conditions, sub-nm depth resolution can be achieved; and because the width of the transient region is minimized, the profile is quantitative from sub-nm depths. Bombardment at non-normal incidence with or without oxygen flooding leads to the onset of roughening almost immediately, accompanied by loss of depth resolution, changing erosion rate, and a wide transient region.

At normal incidence, a simple depth calibration procedure provides an undistorted profile fully corrected for systematic errors which have been ignored previously.

Deconvolution can further improve the depth resolution and alleviate profile distortion due to atomic mixing.

However, the work here was done on somewhat idealized samples, and specifically excludes effects due to preamorphization of silicon and non-dilute dopant concentrations. We are in the process of extending the work to cover such samples.

REFERENCES

1. Schuhmacher, M., Rasser, B., Renard, D., *Secondary Ion Mass Spectrometry SIMS XI,* eds. Gillen, G., Laureau, R., Bennett, J., and Stevie, F., John Wiley & Sons, Chichester, 1998, pp. 695-698.
2. Dowsett, M. G., *Secondary Ion Mass Spectrometry SIMS XI,* eds. Gillen, G., Laureau, R., Bennett, J., and Stevie, F., John Wiley & Sons, Chichester, 1998, pp. 259-264.
3. Wittmaack, K. and Corcoran, S.F., J. Vac. Sci. Technol. B **16**, 272-279 (1998).
4. Iltgen, K., Benninghoven, A. and Niehuis, E., *Secondary Ion Mass Spectrometry SIMS XI,* eds. Gillen, G., Laureau, R., Bennett, J., and Stevie, F., John Wiley & Sons, Chichester, 1998, pp. 367-370.
5. Dowsett, M.G., *Proc. ALC '97,* in press.
6. Wittmaack, K., submitted for publication.
7. Dowsett, M.G., Ormsby, T.J., Elliner, D.I., and Cooke, G.A., *Secondary Ion Mass Spectrometry SIMS XI,* eds. Gillen, G., Laureau, R., Bennett, J., and Stevie, F., John Wiley & Sons, Chichester, 1998, pp. 371-374.
8. Dowsett, M.G., Cooke, G.A., Elliner, D.I., Ormsby, T.J., and Murrell, A., *Secondary Ion Mass Spectrometry SIMS XI,* eds. Gillen, G., Laureau, R., Bennett, J., and Stevie, F., John Wiley & Sons, Chichester, 1998, pp. 285-288.
9. Wittmaack, K., Surf. Interface Anal. **24**, 389-398 (1996).
10. Wittmaack, K., Surf. Interface Anal., in press (1998).
11. Stevie, F.A., Kahora, P.M., Simons, D.S., and Chi, P., J. Vac. Sci. Technol. A **6**, 76-80 (1988).
12. Vajo, J.J., Doty, R.E., and Cirlin, E-H, J. Vac. Sci. Technol. A **14**, 2709-2720 (1996).
13. Alkemade, P.F.A., Jiang, Z.X., Van den Berg, J.A., Badheka, R., and Armour, D.G., *Secondary Ion Mass Spectrometry SIMS XI,* eds. Gillen, G., Laureau, R., Bennett, J., and Stevie, F., John Wiley & Sons, Chichester, 1998, pp. 375-378.
14. Dowsett, M.G., Smith, N.S., Bridgeland, R., Richards, D., Lovejoy, A.C., and Pedrick, P., *Secondary Ion Mass Spectrometry SIMS X,* eds. Benninghoven, A., Hagenhoff, B., and Werner, H.W., John Wiley & Sons, Chichester, 1997, pp. 367-370.
15. Wittmaack, K., Phil. Trans. Roy. Soc. Lond. A **354**, 2731-2764 (1996).
16. Chu, D.P., Ormsby, T.J., and Dowsett, M.G., in preparation.
17. Dowsett, M.G., Barlow, R.D., Fox, H.S., Kubiak, R.A.A., and Collins, R., J. Vac. Sci. Technol B **10**, 336-341 (1992).
18. Clark, E.A., Dowsett, M.G., Fox, H.S., and Newstead, S.M., *Secondary Ion Mass Spectrometry SIMS VII,* eds. Benninghoven, A., Evans, C.A., McKeegan, K.D., Storms, H.A., and Werner, H.W., John Wiley & Sons, Chichester, 1990, pp. 627-630.
19. Littmark, U. and Hofer, W.O., Nucl. Instrum. Meth. **168**, 329-342 (1980).
20. Dowsett, M.G., Rowlands, G., Allen, P.N., and Barlow, R.D., Surf. Interface. Anal. **21**, 310-315 (1994).
21. Chu, D.P. and Dowsett, M.G., Phys. Rev. B **56**, 15167-15170 (1997).
22. Dowsett,M.G. and Collins, R., Phil. Trans. Roy. Soc. Lond. A **354**, 2713-2729 (1996).

Ultra-shallow junction depth profile analysis using TOF-SIMS and TXRF

K. Iltgen[a], B. MacDonald[b], O. Brox[c], A. Benninghoven[c], C. Weiss[a], T. Hossain[b], E. Zschech[a]

[a] *AMD Saxony Manufacturing GmbH, Dresden/Germany*
[b] *AMD Inc., Austin/TX*
[c] *Physical Institute, University of Muenster/Germany*

The drive to fabricate ever shallower source/drain and channel junctions in DRAM and microprocessor production constitutes a great challenge for their analytical characterization. Ultra-shallow SIMS profiling requires high depth resolution and a small surface transient region. It also has to account for the interface between the native oxide and the Si bulk. We have investigated the shallow depth profiling capabilities of TOF-SIMS and TXRF.

TOF-SIMS is well established for high sensitivity microarea surface analysis. The technique features high mass resolution, high transmission and a parallel mass registration. Depth profiles are performed in the dual beam mode allowing an independent optimization of the analyzing and sputtering ion beam. The application of a low energy SF_5^+ sputtering ion beam provides depth resolution in the 0.5 nm range and minimizes the width of the near surface transient region. When depth profiling silicon, the oxidation state of the receding surface can be derived from the intensities of the positive secondary ion intensities $Si_2O_n^+$ (n=0....4). This allows the correction of changes in erosion rate and relative sensitivity factors which are caused by a varying surface oxidation state. For instance, such a variation of the surface oxidation state always occurs in the near surface region. The detection of the interface between the native oxide and the Si bulk and constant sputter and ion yields at this interface are accomplished by the incorporation of ^{18}O.

TXRF has proved to be a powerful technique for the analysis of metallic contaminations on wafer surfaces. One of the advanced applications of this technique is the characterization of ultra-shallow implant profiles. In contrast to the SIMS technique, TXRF provides only an integral value of implants for a near surface layer of some nm thickness. But TXRF is non-destructive, and, therefore, it can be used as a rapid in-fab characterization technique.

We will report on TOF-SIMS and TXRF depth profiles of shallow B and As implants in Si.

INTRODUCTION

Magnetic sector field SIMS (MS-SIMS) is very well suited for depth profiling deep implants because the high mass resolution and the high transmission of the analyzer result in good detection limits for the different dopants used in semiconductor processing. However, due to the high extraction field the application of low sputter ion energies, which for some implants need to be lower than 500 eV, is currently very difficult. Quadrupole SIMS (QP-SIMS) can be combined with low energy sputter ion sources but the analyzer features a lower transmission and only low mass resolution. As will be shown in this paper, very good shallow depth profiling results are achieved by TOF-SIMS.

The in-line characterization of the implant dose and its uniformity over the wafer is usually realized by analytical methods measuring either electrical surface properties or damage induced changes in optical or thermal properties. Here, TXRF is a promising technique because it also provides information on the in-depth distribution of the dopant and can be used for shallow implants.

FUNDAMENTALS

TOF-SIMS Depth Profiling

TOF-SIMS depth profiles are performed in the so-called dual beam mode [1]. For crater formation a low energy sputter ion beam is applied while analyzing the center of the sputter crater with a fine focussed, high energy ion beam. Both ion beams are pulsed with the TOF repetition rate of typically 10 kHz. Between two pulses of the analyzing ion beam the extraction field is turned off and the sputter ion beam is switched on, thus using the flight time of the analyzed secondary ions in the drift tube for sample erosion. Depth resolution and transient width

CP449, *Characterization and Metrology for ULSI Technology: 1998 International Conference*
edited by D. G. Seiler, A. C. Diebold, W. M. Bullis, T. J. Shaffner, R. McDonald, and E. J. Walters

are optimized by an appropriate choice of sputter conditions. Sensitivity is mainly determined by the current of the analyzing ion beam. For B in Si, for instance, a detection limit of $2E16/cm^3$ has been reported [2]. In contrast to QP-SIMS and MS-SIMS, where sensitivity decreases at lower sputter ion energies, the detection limits of TOF-SIMS are independent of the applied sputter conditions. Furthermore, the TOF analyzer features a high mass resolution and a parallel registration of all masses. For these reasons TOF-SIMS is especially powerful for shallow depth profiling and the analysis of a priori unknown components.

TXRF Depth Profiling

It is well known that, in the region of glancing incidence of X-rays, the penetration depth changes with the angle of incidence [3]. Above the critical angle of total reflection α_{crit} the penetration depth is in the range between 0.1 μm and 10 μm. Below the critical angle the penetration depth decreases drastically with the incident angle, and at very small incident angles the penetration depth reaches a constant value of only a few nm. The critical angle α_{crit} and the exact relation between penetration depth and incident angle depend on the X-ray energy and the investigated material. By recording angle dependent intensity profiles, TXRF can be used as a depth profiling technique.

INSTRUMENTATION

For all TOF-SIMS depth profiles a reflectron type mass spectrometer, developed at the University in Muenster, was applied. This spectrometer is equipped with four ion guns, two for analysing and two for sputtering. The analysis can be performed with a 10 keV EI-gun (Ar^+, Xe^+, O_2^+, SF_5^+ ...) or with a fine focussing 25 keV Ga-LMIG. The profiles shown in this paper were acquired with a 10 keV Ar^+ analysis ion beam. For sputtering, an additional EI-gun and a Cs-gun are available. The ion energies of the sputter ion sources can be varied between 300 eV and 10 keV. These ion guns are mounted at 52.5° to the surface normal and the impact angle of the sputter ion beam can be varied by a simple tilt. For depth profiling in positive SIMS mode, secondary ion yields can be enhanced by flooding the sample with O_2.

The TXRF measurements were performed at the Standford Synchrotron Radiation Lab [4] and with a commercial Rigaku 3700 TXRF system. The main advantages of the synchrotron radiation source are that it provides a higher intensity than conventional X-ray sources and a broad energy spectrum of highly collimated photons. Both the synchrotron setup and the Rigaku 3700 allow a stepwise increase of the angle of incidence. The angle scale is calibrated by a quick angle scan from which the critical angle is derived and compared to a theoretical known value.

RESULTS AND DISCUSSION

TOF-SIMS Depth Profiling

Depth Resolution and Transient Width

In SIMS depth profiling, depth resolution and the width of the surface transient region are mainly determined by the applied sputter conditions. Fig. 1 shows the main results of a systematic investigation optimizing depth resolution and the transient width [2].

The effect of different sputter conditions on depth resolution was studied by profiling δ-layers in Si. The measured signal shape of δ-layers has typically a sharp exponential rise and a relatively slow exponential decay. Depth resolution can be described by the characteristic length λ_d of the trailing edge. For Ar^+, O_2^+ and SF_5^+ sputtering Fig. 1.a depicts the decay length λ_d of B in Si as function of the sputter ion energy. The fact that depth resolution improves with decreasing sputter ion energy has driven the introduction of low energy floating ion columns providing relatively high current densities at low ion energies [5]. Another method to reduce the energy of the sputtering particle is the application of molecular sputtering. In a SF_5^+-molecule, for instance, each F-atom carries only 15% of the beam energy. At an ion beam energy of 600 eV the energy of a F-atom amounts 90 eV, which leads to a depth resolution of 0.53 nm (Fig. 1.a). Since the sputter yield of SF_5^+ is a factor of about 4 higher than the sputter yield of O_2^+, SF_5^+ sputtering also provides high erosion rates.

To minimize the width of the surface transient region, a HF-dipped silicon wafer was sputtered under different conditions. The HF-treatment removes the native surface oxide. Sputter equilibrium was supposed to be reached at the apparent transient depth z_{tr} at which all secondary ion intensities ($Si_mO_n^{\pm}$) are stabilized. As shown in Fig. 1.b, the apparent transient depth decreases with the sputter ion energy and is also reduced by the application of O_2-flooding. Moreover, the apparent transient depth is improved by molecular sputtering. The best apparent transient depth of z_{tr}=0.6 nm is achieved by 600 eV SF_5^+ sputtering.

As an example, Fig. 1.c shows a measurement of a 100 eV B implant acquired with the optimized sputter conditions of 600 eV SF_5^+ sputtering.

Detection of SiO_2/Si-Interface

The important influence of the surface oxide on ultra shallow SIMS profiles is demonstrated in Fig. 2. This figure depicts depth profiles of a 2 keV As implant which

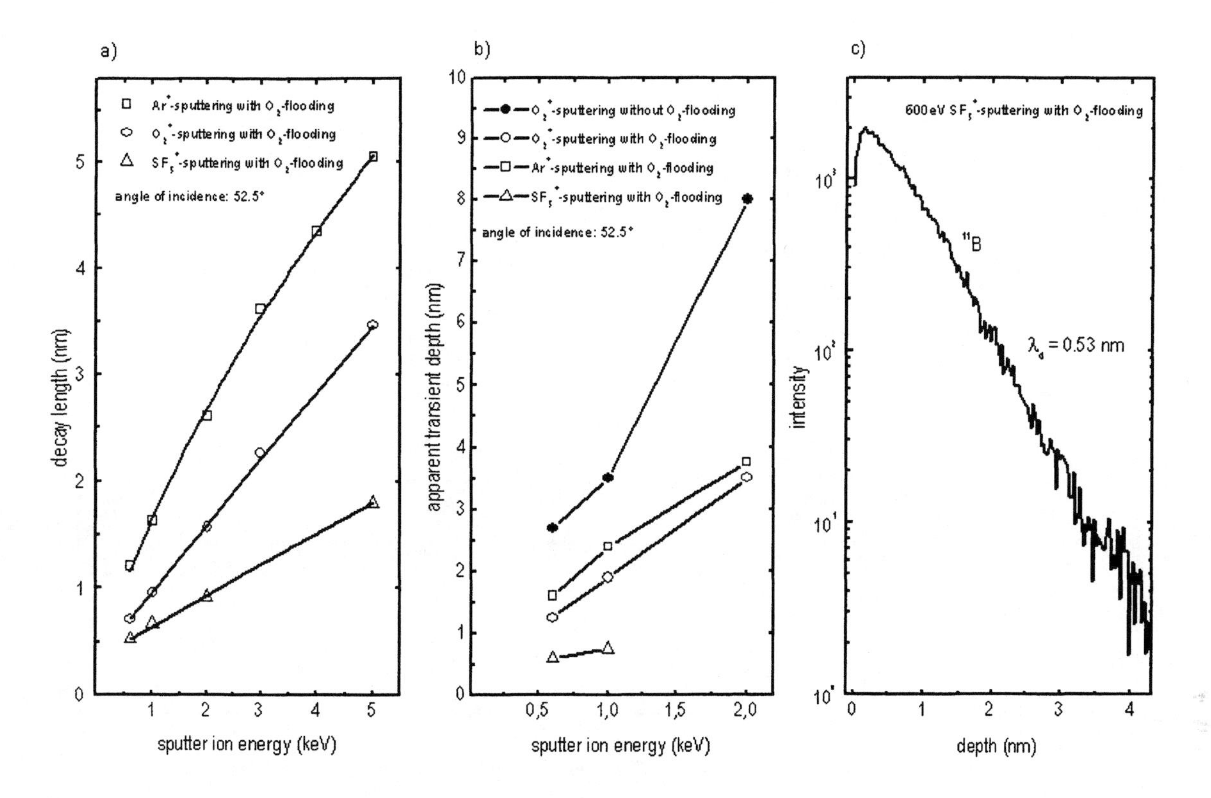

FIGURE 1. Influence of different sputter conditions on the decay length of B in Si (a) and the width of the surface transient region (b). TOF-SIMS depth profile of 100 eV B implant measured with 600 eV SF_5^+ sputtering (c).

was annealed for 10s at 1050 °C (nominal implant dose: 3e15/cm^2).

The profile in Fig. 2.a was acquired in the negative SIMS mode with 1 keV Cs^+ sputtering. These measurement conditions provide the best sensitivity for arsenic. As the profile shows, the annealing step drives a large amount of the implanted As into the bulk. The As concentration drops off at depth of about 40 nm. The thickness of the surface oxide can be derived by the SiO_2^--signal which reaches at a depth of 3.6 nm 50% of its maximum intensity. It has to be noted that the depth scale was established by assuming the same erosion rate for the surface oxide as for the Si bulk. However, for the applied sputter conditions the erosion rate of SiO_2 is found to be by a factor of about 1.2 smaller than the erosion of Si. Taking this difference into account, an oxide thickness of 4.2 nm is obtained. The relatively thick oxide was probably formed during or directly after the RTA process. At the SiO_2/Si-interface both the As^-- and the $AsSi^-$-signal exhibit a peak. However, none of them reflects the correct As distribution because they are distorted by severe matrix effects occuring at the SiO_2/Si-interface.

To measure the real As distribution, a constant matrix throughout the whole profile is crucial. This can be achieved either by sputtering with O_2^+ at normal incidence or by flooding the sample with O_2 and sputtering with O_2 or noble gas ions at any angle of incidence. For these conditions the sensitivity of As is about a factor of 10 worse than for Cs^+ sputtering. Figure 2.b shows the positive SIMS profile obtained with 2 keV Ar^+ sputtering and O_2-flooding. The little drop of the SiO^+-signal indicates that the SiO_2 formation in the Si bulk is not totally completed. Slightly more O_2 pressure would be necessary to form SiO_2 in the bulk. An intensive study of the effect of different oxidation states on various ion yields showed, however, that such little variations in silicon oxidation state hardly effect the secondary ion yields [2].

Although the measurement in Fig. 2.b provides the correct As distribution it contains no information on the position of the original SiO_2/Si-interface. Fig. 2.c shows a measurement acquired with 2 keV Ar^+ sputtering and flooding the sample with $^{18}O_2$. In this way the SiO_2/Si-interface can be detected by following the depth distribution of $Si^{16}O^+$. At the same time a constant matrix throughout this interface is achieved as indicated by the steady $^{30}Si^+$ and $Si^{18}O^+$-signals. The insert in Fig. 2.c shows the near surface region in more detail. The As-signal exhibits a maximum at the SiO_2/Si-interface. Nevertheless, the peak is not as strong as suggested by the Cs measurement in Fig. 2.a. The $Si^{16}O^+$-signal reaches at

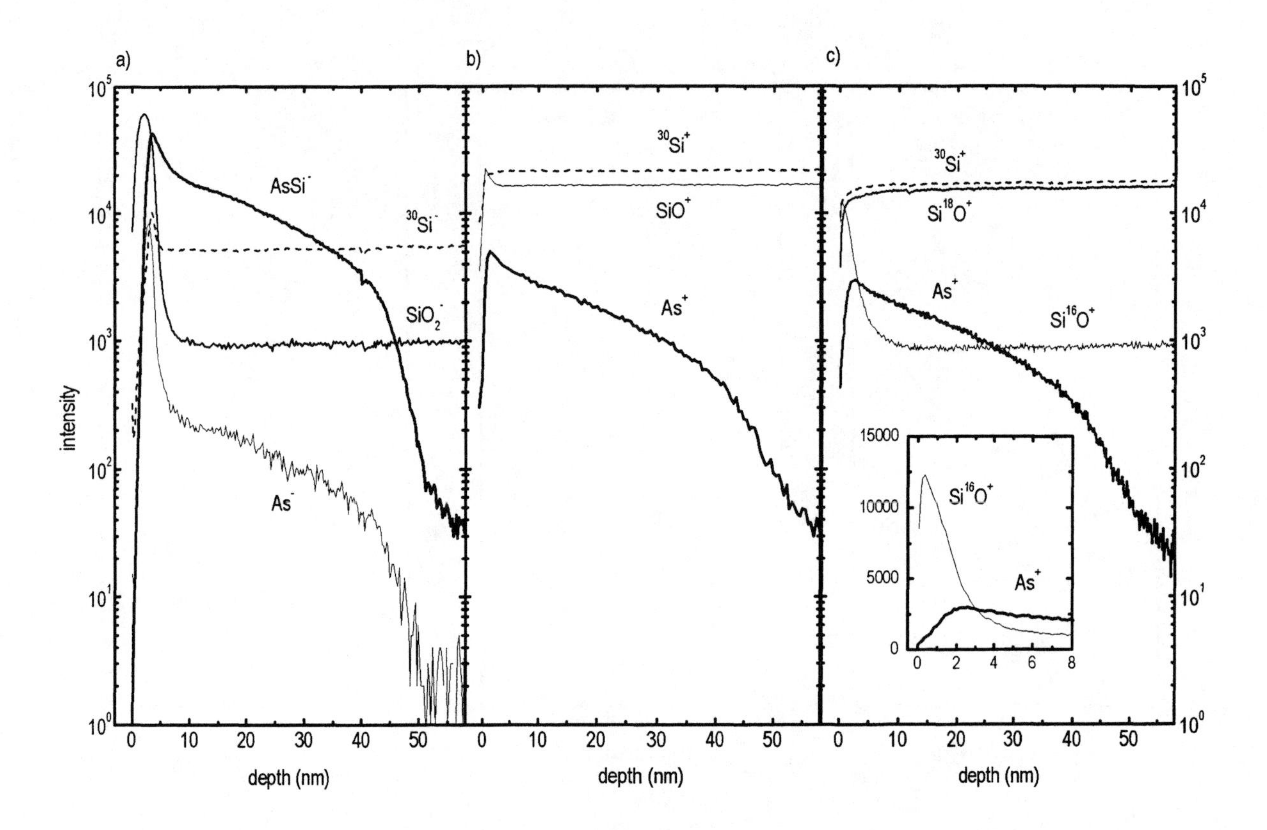

FIGURE 2. TOF-SIMS depth profiles of a 2 keV As implant which was annealed for 10 s at 1050 °C. The profiles were acquired with 1 keV Cs^+ sputtering (a), 2 keV Ar^+-sputtering with O_2-flooding (b), and 2 keV Ar^+ sputtering with $^{18}O_2$-flooding (c).

an apparent depth of 1.9 nm 50% of its maxium value. Note that despite the constant Si sputter yields the depth scale is not correct within the SiO_2-layer. Since the Si particle density is a factor of 2.2 lower in SiO_2 than in Si, the erosion rate in the SiO_2-layer is by the same factor higher than in the Si bulk. Taking the difference in erosion rate into account, an oxide thickness of 4.2 nm is obtained which is the same value obtained by the Cs measurement in Fig. 2.a.

TXRF Depth Profiling

The depth profiling capabilities of TXRF were studied at four wafers with As implants prepared at implant energies of 2 keV, 5 keV, 10 keV and 80 keV (nominal implant dose: 3E15/cm^2). The wafers with the 2 keV and 5 keV As implants were annealed for 10 s at 1050 °C. Since TXRF provides an integrated value of the As concentration in the near surface region, the TXRF results are compared to the cumulative dose measured by TOF-SIMS.

Fig. 3.a shows the TOF-SIMS results of these samples. For the reasons described in the previous section, the 2 keV and 5 keV implants were measured with Ar^+ sputtering while the 10 keV and 80 keV implants were analyzed with Cs^+ sputtering. Due to the annealing of the 2 keV and 5 keV implants, the As distribution of these implants is deeper than the As distribution of the non-annealed 10 keV implant.

Fig. 3.b shows TXRF results of the 2 keV and the 5 keV implants acquired at the SSRL Stanford. The incident angle was increased in steps of 0.01° between 0.01° and 0.1° and in steps of 0.005° between 0.1° and 0.13°. At each incident angle the As flourescence signal was recorded for 100 s. For the applied primary X-ray energy of 12 keV the critical angle of total reflection amounts about 0.15°. As the 2 keV implant and the 5 keV exhibit the same angle dependence of the As-signal, it can be concluded that in both samples the As depth distribution is very similar. This is confirmed by the TOF-SIMS measurements in Fig. 3.a.

Fig. 3.c depicts TXRF results of the 10 keV and 80 keV As implant acquired with the Rigaku 3700 system. The intensity of the selected Au Lβ primary radiation is a factor of about 100 lower than the beam intensity used for the measurements in Fig. 3.b. The acquisition time per incident angle was 100 s. Since the As distribution of the 10 keV implant is closer to the surface, this sample

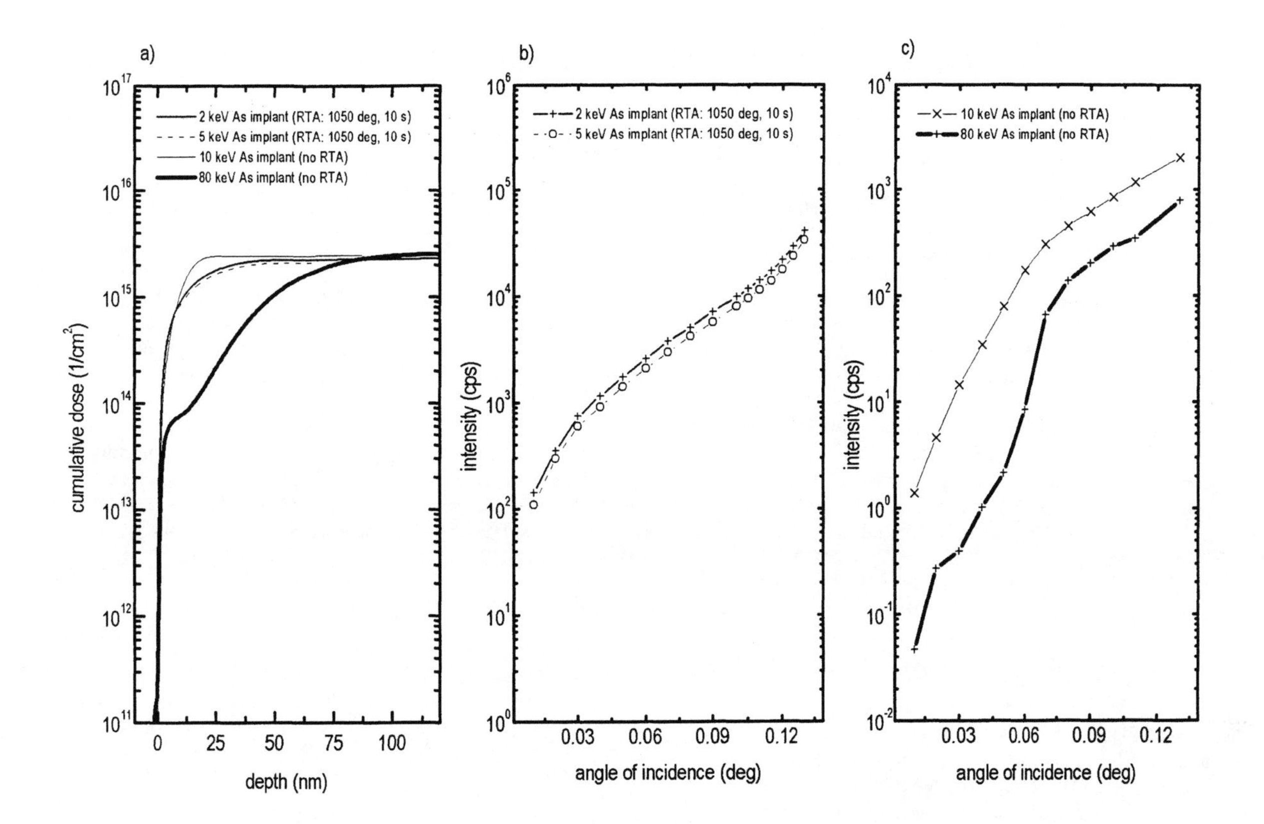

FIGURE 3. Cumulative dose of different As implants measured by TOF-SIMS (a). Angle resolved TXRF measurements of As implants acquired at the SSRL Stanford (b). Angle resolved TXRF measurements of As implants acquired by a Rigaku 3700 system (c).

provides a higher As flourescence signal than the 80 keV implant. The difference in signal intensity is especially high at small incident angles, where the penetration depth is small. The information depth increases with the angle of incidence and as a result the difference in signal intensity between the 10 keV implant and the 80 keV implant decreases.

TXRF has proved its capability as a non-destructive in-line depth profiling technique. Angle resolved TXRF measurements give at least qualitative information not only on the implant dose but also on the depth distribution of the dopant. Further improvement is expected by the experimental determination of the relation between angle of incidence and the penetration depth of the X-ray beam.

CONCLUSION

Ultra shallow SIMS profiling requires high depth resolution, a small width of the surface transient region and the detection of the interface between the surface oxide and the Si bulk. Depth resolution and the transient width are optimized by the application of molecular sputter ion species, such as SF_5^+. The detection of the SiO_2/Si-interface and constant sputter and ion yields at this interface are accomplished by the incorporation of ^{18}O. All this is achieved by TOF-SIMS depth profiling which provides a free choice of sputter conditions and parallel detection of all masses.

REFERENCES

[1] K. Iltgen, C. Bendel, E. Niehuis and A. Benninghoven, *JVST* **A15**, 460 (1997)

[2] K. Iltgen, *Thesis*, University of Muenster (1997)

[3] R. Klockenkaemper in *Total-Reflection X-Ray Fluorescence Analysis*, ed. by J.D. Winefordner, John Wiley & Sons (1997)

[4] P. Pianetta, N. Takaura, *Rev. Sci. Instrum.* **66**, 2 (1995)

[5] M.G. Dowsett, N.S. Smith, R. Bridgeland, D. Richards, A.C. Lovejoy and P. Pedrick, *SIMS X*, eds. A. Benninghoven et al., John Wiley & Sons (1996), p. 367

Fast Low Energy SIMS Depth Profiling for ULSI Applications

S.B.Patel and J.L.Maul

ATOMIKA Instruments GmbH, Bruckmannring 40, D-85764 Oberschleissheim, Munich, Germany

Accurate depth profiling of low energy implants using SIMS requires the use of low energy primary ion beams for two main reasons: minimise the surface transient so that steady-state sputtering conditions are achieved at shallower depths and improve the depth resolution by reducing the atomic-mixing effect. Recent SIMS tool development has reduced the analysis time at low beam energies, making it possible to routinely SIMS analyse low energy implants in the ULSI industry.

SIMS depth profiles of low energy boron and arsenic implants are presented. The data confirm previous findings that for accurately determining the junction depth of low energy boron implants in silicon, the energy of the O_2^+ primary beam must be no more than half the implant energy. In addition, low energy beams can reveal vital subtle detail in an implant profile which is distorted or lost at higher energies. Despite the very low primary beam energy capability of the SIMS tool used, extremely stable beam currents are obtained. The same tool is capable of carrying out fast measurements and obtain a high dynamic range.

INTRODUCTION

SIMS (Secondary Ion Mass Spectrometry) is an essential tool for depth profiling implantation profiles in semiconductors. Development of the ATOMIKA 4500 SIMS tool (1) has made it possible to use primary ion beams much lower in energy (sub-keV) than conventionally used, thereby improving the accuracy of SIMS depth profiles of low energy implants (2).

The aim of this paper is to demonstrate that this tool is essential for the analysis of low energy implants in silicon and show that the same tool is capable of fast depth profiling and produce profiles with sufficiently high dynamic range.

EXPERIMENTAL

All the analyses were carried out on standard 4500 SIMS tool. This instrument has oxygen and cesium primary ion guns, both of which are based on the FLIG (Floating Low Energy Ion Gun) (3) technology.

Depth calibration was done by measuring the crater depths using a Tencor Apha-Step 500 surface profiler. The concentration calibration was done by measuring the intensities from uniformly doped standards and then applying a calibration factor to the measured signal (impurity/matrix ratio) of the unknown sample.

A variety of different boron implanted samples were analyzed using normal incidence oxygen beam (O_2^+) with energies ranging from 250eV to 1keV. Low energy arsenic samples were analyzed using 500 eV and 1 keV cesium (Cs^+) beams at 60°.

RESULTS

Figure 1 shows the results of a 250 eV and 500 eV O_2^+ analysis of a 500 eV, $5x10^{14}$ atoms.cm^{-2} B^+ implanted silicon sample. The decay length, λ (depth in nm sputtered when the intensity drops by a factor of 1/e), is a measure of the depth resolution of the measurement and is stated for each profile. The dotted straight lines show the position at which the slope was taken in order to calculate λ.

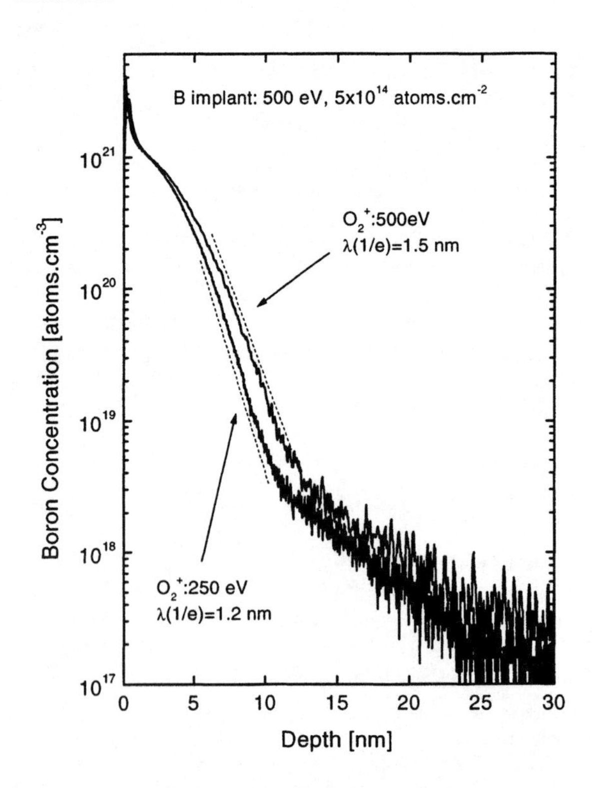

FIGURE 1 500 eV and 250 eV O_2^+ analysis of a 500eV B^+ implanted silicon sample.

CP449, *Characterization and Metrology for ULSI Technology: 1998 International Conference*
edited by D. G. Seiler, A. C. Diebold, W. M. Bullis, T. J. Shaffner, R. McDonald, and E. J. Walters

λ decreases from 1.5 nm to 1.2 nm as the primary beam energy is reduced from 500 eV to 250 eV. This implies that the profile measured with the 500 eV beam is artificially broadened to a greater extent than the profile measured with the 250 eV primary beam. Therefore, the lower 250 eV beam profile more accurately follows to the true boron distribution.

Figure 2 shows the results of a 2.5 keV, $5x10^{14}$ atoms.cm^{-2} BF_2^+ implanted sample with a similar boron penetration depth as the 500 eV B^+ sample. Comparison of Figs. 1 and 2 shows that the improvement in depth resolution with decreasing beam energy is similar for the two samples.

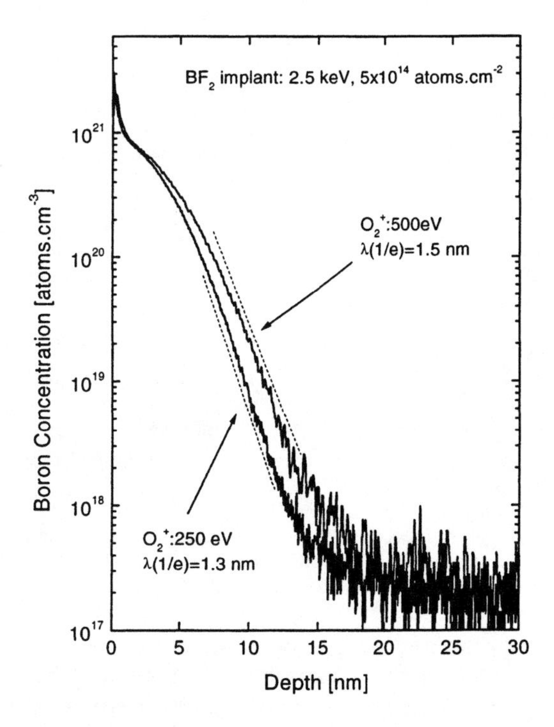

FIGURE 2 500 eV and 250 eV O_2^+ analysis of a 2.5 keV BF_2^+ implanted silicon sample.

Although the 250 eV measurements are more accurate than those carried out with a 500 eV O_2^+ beam, it is likely that energies lower than 250 eV will produce even more accurate profiles. The ultimate check would be to do measurements using successively lower beam energies, until the measured boron profile shows no improvement in depth resolution. Such a study is being carried out with various low energy implant samples and the results will be published shortly.

An important point is that the results presented here, support recent recommendations (4) that the energy of the O_2^+ analyzing beam should be no more than half that of the B^+ implantation energy. The results clearly demonstrate that the 500eV O_2^+ primary beam is far too high for the analysis of these types of boron implanted silicon samples.

Figure 3 shows a variety of different low energy boron implanted silicon samples analyzed using a 250 eV O_2^+ primary

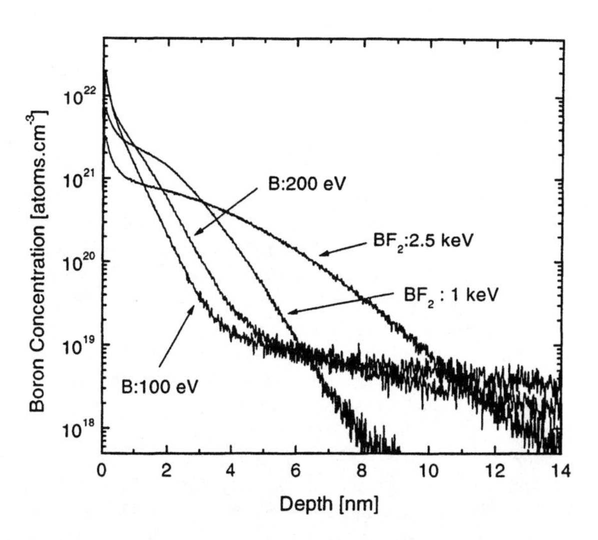

FIGURE 3 250 eV O_2^+ analysis of a several different boron implanted silicon sample.

beam. Differences in the boron profile shapes due to different implantation conditions can clearly be seen. The 200 eV and 100 eV B^+ implant samples have considerable energy contamination which is seen from about 4 nm and extends to several tens of nanometers.

Figure 4 shows a 3 keV BF_2^+ implanted sample analyzed using a 1000 eV, 500 eV and 250 eV O_2^+ primary beam. The sample

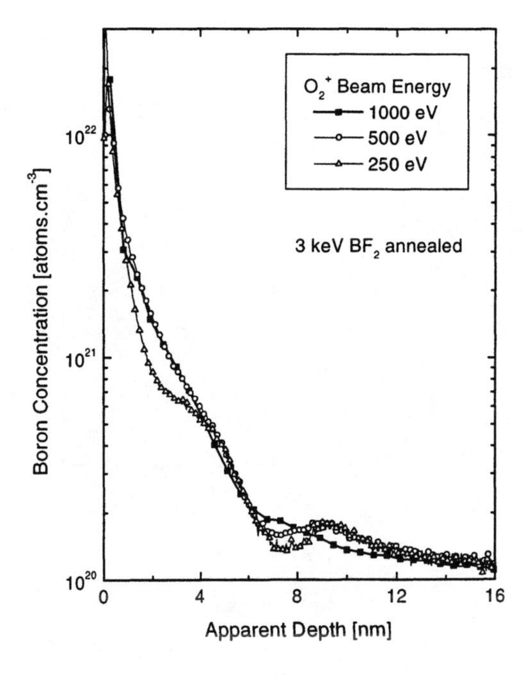

FIGURE 4 A 3 keV BF_2^+ annealed sample analyzed with different O_2^+ beam energies.

had been anneal after implantation and these results show how fine details of the boron distribution starts to appear as the energy of the O_2^+ beam is reduced. This type of information is important for understanding the annealing process involved.

It must be pointed out that all these low energy O_2^+ beam measurements have been carried out at normal incidence and without the use of an oxygen jet (also referred to as oxygen flood or oxygen leak). Several reports (5) have shown that the technique of using an oblique oxygen beam with the combination of oxygen flood causes inaccuracies in the measured profile due to artefacts associated with this technique (note that there may be matrices other than silicon where this technique is beneficial). In addition, recent finding (6,7) have shown that the influence of these artefacts to the measured profile is even more significant at low energies than previously reported. On the SIMS instrument used for this study, it is possible to vary the beam angle from normal incidence to angles greater than 60^0 independent of the primary beam energy, and in addition, it is even possible to use an oxygen jet. However, for all the boron data shown, the measurements were carried out at normal incidence without an oxygen jet in order to avoid the associated artefacts and thus obtain the most accurate SIMS depth profile.

With the ATOMIKA 4500 instrument high primary beam currents are possible at low energies whilst maintaining the beam spot size. Figure 5 shows a 1 keV O_2^+ analysis of a 5 keV, $1x10^{15}$ atoms.cm^{-2} B^+ implanted sample. The entire profile is accurately measured in only 8 minutes with an erosion rate of 23.1 nm/min.

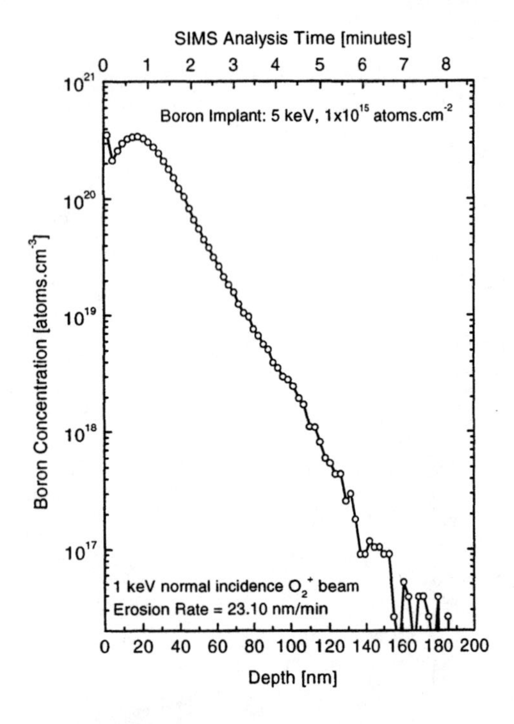

FIGURE 5 Fast depth profiling demonstrated with a 1 keV O_2^+ beam .

Figure 6 shows a 500 eV, $1x10^{15}$ atoms.cm^{-2} BF_2^+ implanted sample analyzed with a 500eV O_2^+ primary beam. Although the primary beam energy is too high for this sample in order to accurately measure the boron profile close to the surface, relatively higher beam energies can be used for simply determining the profile of the tail region of the boron distribution. Here the result shows that using a 500 eV beam energy, over five and a half order of magnitude dynamic range is possible with a detection limit around $3x10^{15}$ atoms.cm^{-3}.

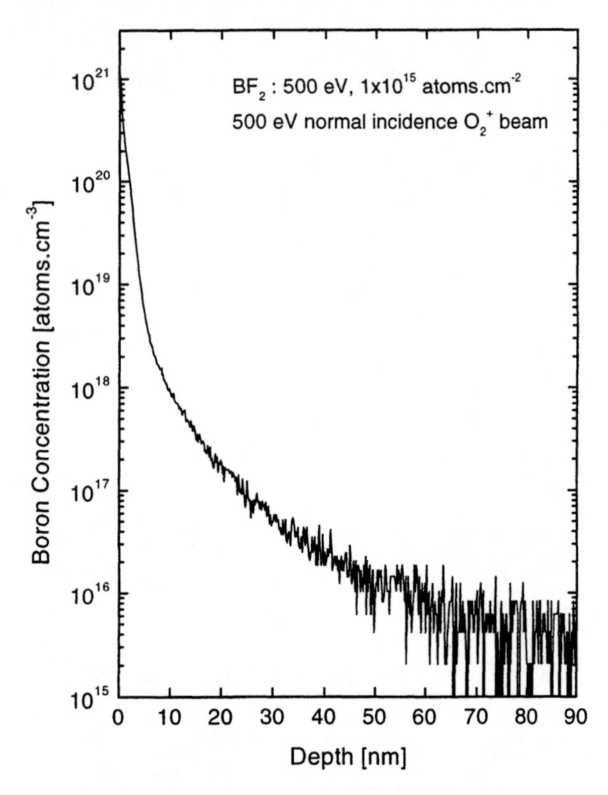

FIGURE 6 Low detection limit and large dynamic range possible with a 500eV O_2^+ beam.

Besides producing well-focussed low energy ion beams with high currents, exceptional current stability is achieved and is demonstrated in Fig. 7, which shows the current profile of a 170 eV O_2^+ beam over a long period of time.

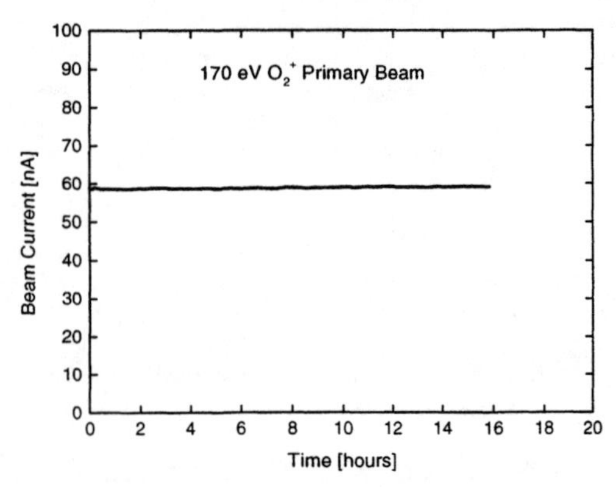

FIGURE 7 Current profile of a 170 eV O_2^+ beam measured over a long period of time.

Figure 8 shows two low energy arsenic implanted samples analyzed using a Cs^+ primary beam at 60° angle of incidence. Here, the low energy beam is clearly able to show differences in the arsenic distributions even at depth of a few nanometers. Figure 9 shows an 8 keV As^+ implanted sample analyzed using a 1 keV Cs^+ beam at 60°. The profile was recorded during a standard 1 keV Cs^+ set-up and shows how a detection limit of around $2x10^{15}$ atoms.cm^{-2} can routinely be achieved.

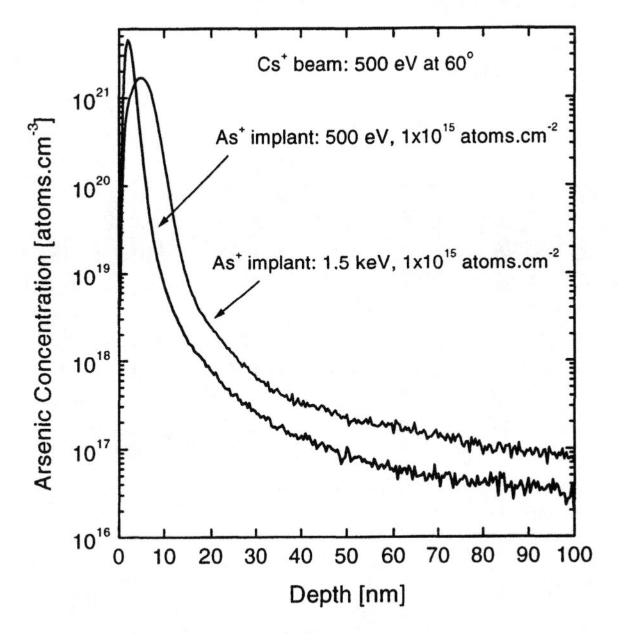

FIGURE 8 500 eV Cs^+ analysis of two low energy arsenic implanted silicon samples.

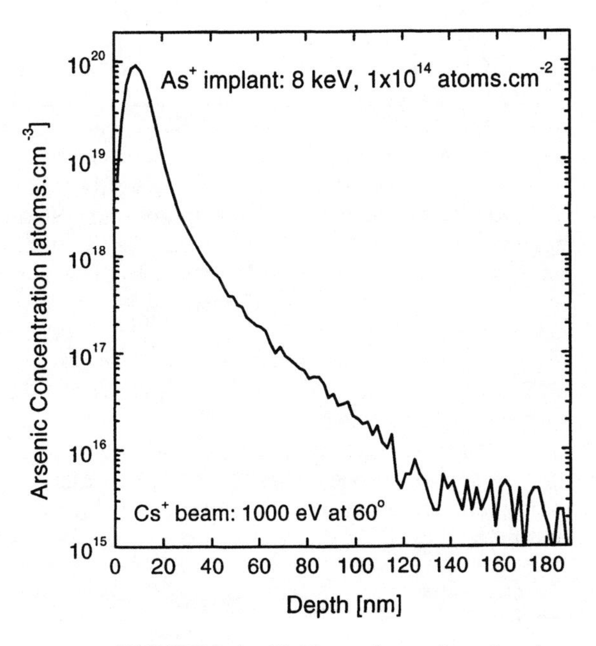

FIGURE 9 An 8 keV arsenic sample analyzed with a 1 keV Cs^+ beam.

CONCLUSIONS

It is necessary to use low energy primary beams in order to accurately measure low energy implants in silicon. For the analysis of low energy boron samples, a necessary requirement is that the energy of the O_2^+ beam is no more than half that of the implant energy.

With the ATOMIKA 4500 SIMS instrument, low energy implant profiles can be accurately measured and obtain high throughput, large dynamic range, low detection limits and exceptional beam stability even at very low energies.

REFERENCES

1. Maul, J. L., and Patel, S. B., Secondary Ion Mass Spectrometry, SIMS XI, Eds. Gillen, G., Lareau, R., Bennett, J., and Stevie, F., John Wiley and Sons, pp 707-710, 1997.
2. Wittmaack, K., Surf. Interface Anal. 21, pp 323, 1994
3. Dowsett, M. G., et al, J. Appl. Phys., 54, pp 6340-6345, 1983.
4. Wittmaack, K., Surf. Interface anal., 26, 1998.
5. Wittmaack, K., and Corcoran, S. F., J. Vac. Sci. Technol. B16, pp 272-279, 1998.
6. Wittmaack, K., to be published.
7. Wittmaack, K., Patel, S. B., and Corcoran, S. F., in these proceedings.

Resonance Ionization Mass Spectrometry - Applications to Surface Analyses and Depth Profiles of Semiconductors

T. J. Whitaker, K. F. Willey, W. R. Garrett, and H. F. Arlinghaus*

Atom Sciences, Inc., 114 Ridgeway Center, Oak Ridge, TN 37830, USA

Resonance Ionization (RI) uses wavelength-tunable lasers to selectively and efficiently ionize atoms of a specific element. Because pulsed lasers are typically used in RI, time-of-flight mass spectrometry is an inexpensive method of detecting the resulting ions and adding even more specificity to the process. We have combined the RI-mass spectrometry (RIMS) process with sputtering from a high-resolution ion gun to generate a uniquely sensitive surface and near-surface imaging technique, which has many advantages over conventional SIMS (secondary ion mass spectrometry). These advantages include much greater sensitivity, freedom from matrix effects, and almost interference-free data. The lack of interference results from the selectivity in the ionization process. This feature makes it possible to measure many analytes at concentrations that are not currently possible in SIMS. We present data showing how this technique can be used to measure depth profiles over a large dynamic range and produce high-resolution images of the concentration of several elements in various semiconducting substrates. We also briefly discuss the possibility of extending the technique to several elements that are important to the semiconductor industry, such as P, N, and C. These elements cannot be probed with one-photon RI steps using commercial laser systems because their first excited state is too high in energy.

INTRODUCTION

Secondary Ion Mass Spectrometry has become the standard for surface analysis and depth profiles. However, SIMS is inherently limited by the fact that it measures only ions produced during the sputtering process. Sputtering produces predominantly neutral atoms; the secondary ion yield being typically between 10^{-4} to 10^{-7} for most of the elements in the periodic table. This leads to a serious sensitivity limitation in SIMS when extremely small volumes must be probed, or when high lateral and depth resolution analyses are needed. Also, the secondary ion yield can vary by orders of magnitude as a function of surface contamination and matrix composition. While this change in the number of ions provides a challenge for the SIMS analyst desiring quantitative results, it makes almost no difference in the number of neutrals. For these reasons, several postionization methods have been developed that probe the neutral component. These techniques ionize the sputtered neutral atoms after emission from the sample surface. In secondary neutral mass spectrometry (SNMS) (1), the sputtered neutrals are ionized in a plasma or by electron-impact ionization. As a result, SNMS has much smaller matrix effects than SIMS. But even though SNMS is probing the more abundant neutral component, it is doing so with relatively inefficient ionization processes, and has therefore not improved sensitivity issues for small volumes. Surface analysis by laser ionization (SALI) (2) has been used to improve the ionization efficiency while retaining the advantages of probing the neutral component. In SALI, an intense laser beam is used to nonselectively ionize all the elements and molecules within the volume intersected by the laser beam. The laser beam is usually tightly focused to provide the intensity required for nonresonant multiphoton ionization, but this focussing drastically reduces the probed volume. SALI provides multi-element and molecular survey measurements with improved ionization efficiency over SNMS, but like SNMS, it suffers from isobaric interferences. Interferences from molecules, molecular fragments, and isotopes of other elements having the same mass as the analyte can make interpretation of the data difficult and quantitative analysis impossible. Very high mass resolution can reject such interferences but only at the expense of detection sensitivity.

* Present location: Westfälishche Wilhelms-Universität Münster, Germany

CP449, *Characterization and Metrology for ULSI Technology: 1998 International Conference*
edited by D. G. Seiler, A. C. Diebold, W. M. Bullis, T. J. Shaffner, R. McDonald, and E. J. Walters

We have developed and patented a technique that retains almost all the advantages of neutral postionization but improved detection efficiency and essentially eliminated interferences. In this technique, wavelength-tunable lasers selectively and efficiently ionize neutral atoms using unfocussed laser beams intersecting the sputtered plume. The technique, called Sputter-Initiated Resonance Ionization Spectroscopy (SIRIS), provides almost 100% ionization of the selected element in a relatively large volume. As the ionization is also selective for a specific element, mass interferences are greatly reduced or eliminated. However, the selectivity is also the source of the main compromise required by the technique, i.e. analysis of one element at a time.

EXPERIMENTAL

A sketch of the SIRIS experimental apparatus is shown Fig 1. A liquid metal (gallium) ion gun, capable of spot sizes below 50 nm in diameter, is used for high resolution sputtering. A higher current Atomika ion gun, which is not shown in the figure, is used for lower resolution measurements. During the time the pulsed primary ion beam from either of these guns is striking the target, voltages on the extraction electrodes are set so that electric fields retard positive secondary ions. After the primary ion pulse, but before the laser pulse, voltages are switched so that ions formed by laser resonance ionization have significantly different energy than secondary ions formed in the sputtering process. This allows the two electrostatic energy analyzers (ESA1 and ESA2) in the flight tube of the time-of-flight mass spectrometer to effectively discriminate against the secondary ions. Charge compensation for insulator analyses is possible using pulsed low energy electrons, which are introduced during the time interval between sputtering pulses.

The most common scheme for resonance ionization is shown in Fig. 2. The laser system that supplies all the necessary wavelengths is composed of a Nd:YAG laser and two dye lasers. The output of the Nd:YAG laser is frequency doubled or tripled to pump the dye lasers. Dye Laser 1 is usually frequency doubled or frequency mixed with the Nd:YAG fundamental to generate the appropriate ultraviolet (UV) wavelength. Dye Laser 2 typically does not require frequency upconversion. The final photoionization step is performed with the 1.06 μm fundamental from the Nd:YAG laser. Because the photoionization step is a nonresonant process (notwithstanding autoionizing structure), to saturate this step requires higher intensity laser light than do the resonant steps. The advantage in using an infrared wavelength for photoionization is that high intensities can be used without causing significant ionization from multiphoton absorption in molecules or atoms.

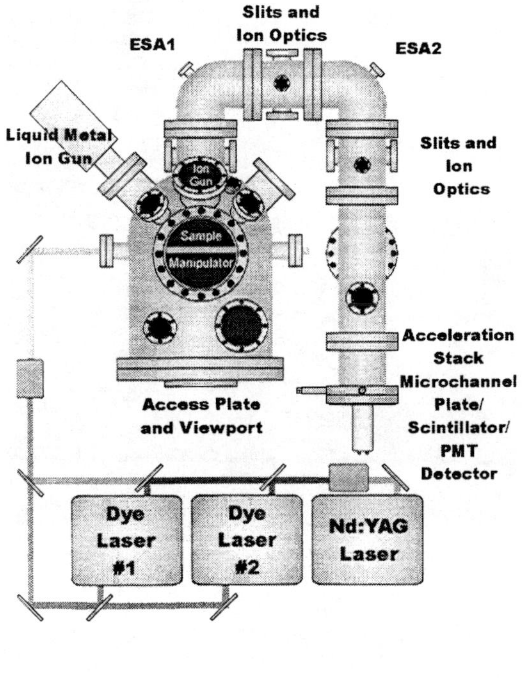

FIGURE 1. Sketch of the SIRIS apparatus. Several components have been omitted for clarity.

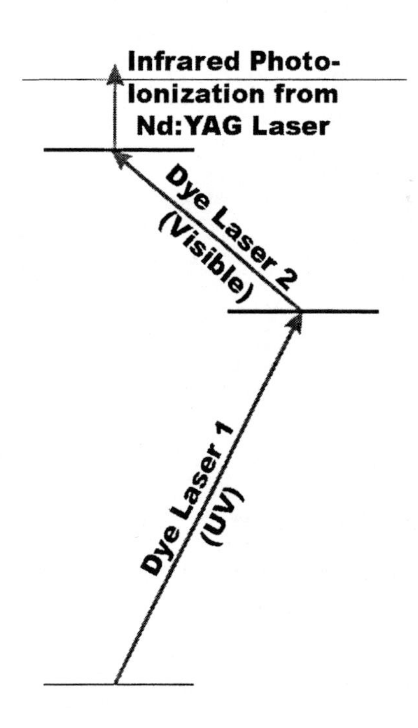

FIGURE 2. Typical dual resonance ionization scheme.

RESULTS

SIRIS depth profiles can be obtained by scanning the sample with a continuous ion beam in a raster pattern to etch away a rectangular crater and taking data with a pulsed ion beam in the center of the crater after each raster frame. Figure 3 shows a SIRIS depth profile of two wafers, one Si and the other SiO_2 on Si, which were simultaneously implanted with identical doses of Cu. Two features of SIRIS are exemplified in this data; the large dynamic range possible in SIRIS, and the lack of matrix effects. The large dynamic range (greater than six orders of magnitude in Fig. 3) is possible because of the lack of interferences. Lack of matrix effects is demonstrated by the fact that the integrated areas under the two peaks are within 3% of each other, even though one sample contained oxygen and therefore would have a dramatically enhanced secondary ion yield. However, the neutral fraction is essentially unaffected because it is so much larger than the ionic fraction. Other interesting effects observed in this data are the broadening of the depth profile in crystalline Si due to channeling effects and Cu pileup at the Si/SiO_2 interface in SiO_2.

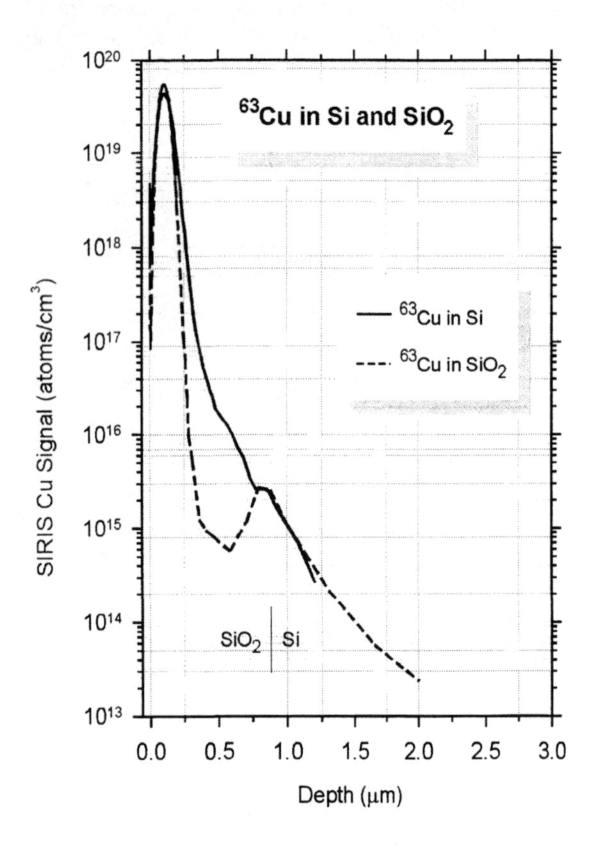

FIGURE 3. SIRIS depth profile of Cu implanted with 150 keV energy in Si and SiO_2 on Si. Dose was $5x10^{14}$ cm^{-3} for both samples.

In SIRIS, as in other microprobe techniques, the combination of 2-d imaging with depth profile capabilities allows visualization of three-dimensional distribution of elements. However, the extremely high overall efficiency of SIRIS (typically 3% to 8% useful yield for most elements) permits measurement of low concentrations of analytes in much smaller volumes, thus enabling higher lateral and depth resolution in these 3d visualizations.

Figure 4 shows a SIRIS image of the Cu concentration in and around a Cd inclusion in ZnCdTe film. This image was obtained by rastering the ion beam from the liquid Ga ion gun over the image area. Larger images require translating the sample under a fixed ion beam position. Figure 4 demonstrates the high resolution achievable with SIRIS even when measuring very low concentrations. The Cu "hot-spots", which are perhaps due to accumulation at grain boundaries, are reproducible and several are less than 0.5 μm. Furthermore, Cu in CdZnTe is an especially difficult analysis problem for other microprobe techniques because of interferences at the Cu masses. The elemental selectivity of SIRIS makes these measurements possible. This capability could greatly contribute to the study of dopant and contaminant distributions in semiconductor devices.

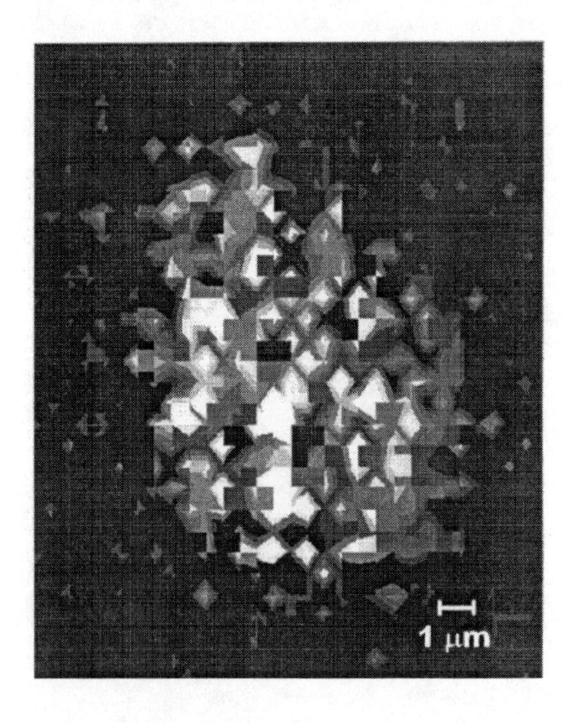

FIGURE 4. High-resolution image of the copper concentration in and around a Cd inclusion in CdZnTe film. The dark areas correspond to Cu concentrations less than 5 ppm (about $2x10^{17}$ cm^{-3}) and the light areas are over 25 ppm.

Applicability and Extension to Future Semiconductor Problems

Although we have only shown data here for Cu measurements, we have applied the SIRIS technology to a number of analytes in a wide variety of samples. A few examples include boron (3), plutonium (3), phosphorus, and arsenic in silicon, technetium in nickel, lead in zircon meteorites (for this measurement Pb isotope ratios were obtained to date the meteorites), and even tin attached as a label to DNA molecules (4). To measure phosphorus concentration, we were forced to use a laser scheme involving a simultaneous two-photon excitation. This technique requires high intensity UV light and can cause nonresonant ionization and interferences. In general, the main factor determining whether we can apply the preferred SIRIS laser scheme (shown in Fig. 2) is the wavelength of the first transition. Standard commercial laser systems using frequency upconversion in crystals limit us to wavelengths of about 195 nm and longer. Unfortunately, some elements that are becoming increasingly important to the semiconductor industry, such as nitrogen, have first transition wavelengths below this cutoff. Nitrogen requires 120 nm or shorter wavelengths for excitation from the ground state. For this reason, we have begun a study to assess the use of four-wave mixing to generate vacuum ultraviolet (VUV) wavelengths for use in SIRIS analyses.

In four-wave sum frequency mixing

$$\omega_{VUV} = \omega_1 + \omega_2 + \omega_3 \qquad (1)$$

where ω_{VUV} is the angular frequency of the VUV light and ω_i is the angular frequency of the ith laser. The power, P_{VUV}, generated at ω_{VUV} is proportional to

$$P_1P_2P_3[N\chi^{(3)}]^2F(b\Delta k) \qquad (2)$$

where P_i is the power of the ith laser beam, N is the density of the nonlinear vapor or gas, $\chi^{(3)}$ is the third-order nonlinear susceptibility, F is the phase-matching factor which is a function of the confocal-beam parameter (b) and the wavevector mismatch (Δk) between the generated VUV wave and the driving polarization. In mercury, several two-photon allowed states are available such that if one laser is tuned to ½ the energy of the two-photon resonance, not only is the experimental setup simplified by making $\omega_1 = \omega_2 = \omega_{UV}$ but $\chi^{(3)}$ is also greatly enhanced. In an unfocused beam, the condition $\Delta k = 0$ is required for constructive interference. With focussed beams (in sum frequency mixing), $b\Delta k = -2$ provides maximum conversion efficiency. To achieve this optimum phase matching at a given wavelength, a gas mixture is normally used and the concentrations adjusted.

Figure 5 shows a sketch of the heat pipe we have constructed for VUV generation. This heat pipe is considerably smaller than most in order to fit inside the SIRIS system. The total length from input window to output window is less than 18 cm. Hg is used as the nonlinear medium and Ar is used to phase match. We set λ_1 near

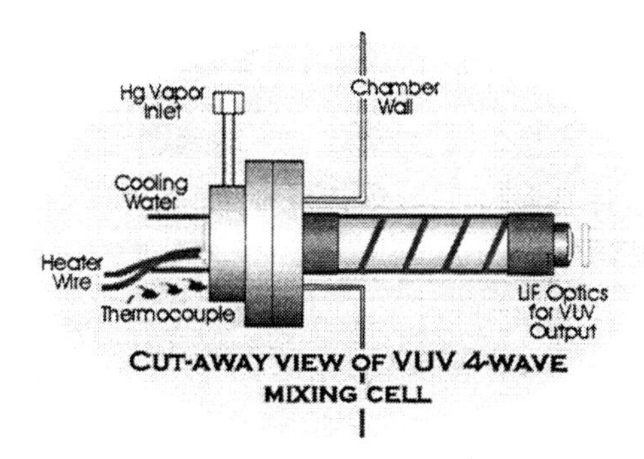

FIGURE 5. Sketch of heat pipe constructed to generate 120 nm light for SIRIS detection of N.

312.8 nm, which corresponds to ½ the energy of the two-photon allowed $7s\ ^1S_0 \leftarrow 6s^2\ ^1S_0$ transition in Hg. We then set λ_3 near 515.6 nm such that four wave mixing will generate the appropriate nitrogen wavelength at or below 120 nm.

At the present we have verified generation of VUV at 120 nm and we will soon begin experiments using this wavelength in SIRIS detection of atomic nitrogen. The main concern with such short wavelengths is that direct one-photon ionization of some low-ionization threshold molecules may cause interferences in the SIRIS data.

CONCLUSIONS

SIRIS provides specific capabilities that are not found in other analytical techniques, but it does so for only one element at a time. This will probably limit the role of SIRIS in semiconductor metrology to a rather narrow niche, but one of increasing importance as device dimensions continue to become smaller.

ACKNOWLEDGEMENTS

This work was funded in part by the Defense Advanced Research Projects Agency under contract DAAH01-97-C-R312.

REFERENCES

1. Benninghoven, A., Janssen, K.T.F., Tümpner, J., and Werner, H.W., eds., *Secondary Ion Mass Spectrometry - SIMS VIII* New York: John Wiley and Sons, (1992), articles therein.
2. Becker, C.H. and Gillen, K.T., *Anal. Chem.* **56**, 1671 (1984).
3. Arlinghaus, H.F., Spaar, M.T., Thonnard, N., McMahon, A.W., Tanigaki, T., Shichi, H., and Holloway, P.H., *J. Vac. Sci. Technol. A* **11(4)**, 2317-2323 (1993).
4. Arlinghaus, H.F., Kwoka, M.N., Guo, X.Q., and Jacobson, K.B., *Anal. Chem.* **69(8)**, 1510-1517 (1997).

Identification, simulation and avoidance of artifacts in ultra-shallow depth profiling by secondary ion mass spectrometry

K. Wittmaack,[a] S.B. Patel[b] and S.F. Corcoran[c]

[a] *GSF - National Research Center for Environment and Health, Institute of Radiation Protection, D-85758 Neuherberg, Germany,* [b] *ATOMIKA Instruments GmbH, Bruckmannring 40, D-85764 Oberschleißheim, Germany*
[c] *Intel Components TD Q&R, MS RA1-329, Hillsboro, OR 97212, USA*

Distortions and shifts of shallow implantation distributions of B in Si, recently observed in depth profiling by secondary ion mass spectrometry (SIMS) using 1.9 keV O_2^+ ions at oblique incidence (~ 60°) with oxygen flooding, have been reproduced at a beam energy of 1 keV. Measurements on samples containing a series of delta doping spikes revealed a pronounced initial drop in erosion rate followed by a more gradual decrease extending to at least 200 nm. These changes give rise to severe errors in depth calibration (shift up to 4 nm). The artifacts are due to bombardment-induced oxygen incorporation and surface roughening (ripple formation), the latter effect also causing a pronounced degradation in depth resolution. The essential features of the observed profile distortions can be simulated by a simple model. Ripple formation is not observed with normally incident oxygen beams in vacuum, in which case profiles with excellent resolution and minimum shift can be obtained, notably at ultra-low probe energy.

INTRODUCTION

Recently it has been reported that SIMS depth profiles of shallow implantation distributions of B in Si become heavily distorted if sputter erosion of the sample by means of a 1.9 keV O_2^+ beam at oblique incidence (~ 60°) is combined with oxygen flooding of the sample (1). The observed narrowing of the profiles was attributed to the transient oxygen incorporation of oxygen in combination with subsequent surface roughening (ripple formation), two phenomena which both reduce the sputtering yield of Si. Pronounced changes in erosion rate, rapid degradation in depth resolution and ripple formation have in fact been observed at a beam energy of 1 keV, using a Ge delta doped Si sample (2). Such samples are well suited for determining the changes in depth resolution and erosion rate as a function of depth, information that cannot be obtained using implanted samples. Very recent investigations on Si containing B deltas showed that the depth profiling artifacts occur over a rather wide range of impact angles (from ~39° to at least 62°, at 1 keV) and that they are significantly larger with than without oxygen flooding (3).

In this study we present a direct comparison of profile artifacts observed with implanted and delta doped samples subject to the same bombardment conditions. Particular attention has been devoted to identifying the onset of roughening by measuring matrix ion yield changes during sputter profiling. It was necessary to record the signal of at least two species which respond differently to the presence of oxygen at the surface, e.g., Si atomic and Si_2 dimer ions (4). With this approach the changes in erosion rate as well as the errors in concentration calibration can be assessed in some detail. By way of a model calculation we also show that, even though the retained implantation fluence may accidentally turn out correct, the profile measured with flooding is grossly incorrect.

EXPERIMENTAL

The SIMS depth profile measurements were performed using two quadrupole based Atomika Ion Microprobes, the Model 4100 at Intel and a Model 4500 Depth Profiler at Atomika Instruments, both featuring floating low-energy ion guns (5). In this work we used 1 keV O_2^+ beams, mostly at an impact angle of 62°±1° to the surface normal. Oxygen flooding was accomplished by means of a jet directed at the bombarded surface. Two types of samples were investigated, one containing a low-energy B implantation distribution and the other one a series of B delta spikes.

RESULTS AND DISCUSSION

Figure 1 shows a comparison of two B depth profiles of the implanted sample, obtained by SIMS analysis in vacuum as well as in combination with oxygen flooding (unless the oxygen flux or pressure is specifically quoted, the notion 'flooding' is meant to indicate that the oxygen flux was sufficiently high to achieve maximum secondary ion yield enhancement). Depth calibration of the two profiles was

CP449, *Characterization and Metrology for ULSI Technology: 1998 International Conference*
edited by D. G. Seiler, A. C. Diebold, W. M. Bullis, T. J. Shaffner, R. McDonald, and E. J. Walters
1998 The American Institute of Physics 1-56396-753-7/98/$15.00

based on individual surface-profilometer measurements of the sputtered crater depths and on the common assumption that the erosion rate was constant during the analysis. Concentration calibration was derived from B^+/Si^+ ratios and by setting the integral under the profile equal to the quoted implantation fluence. Clearly, the B profiles thus obtained differ significantly in all aspects, i.e. profile shape, location and height. Very similar differences have been reported for profiling at 1.9 keV (1). Using the arguments outlined in Refs. 1-3 we conclude that the profile obtained in vacuum is basically correct whereas the one recorded with flooding is heavily distorted.

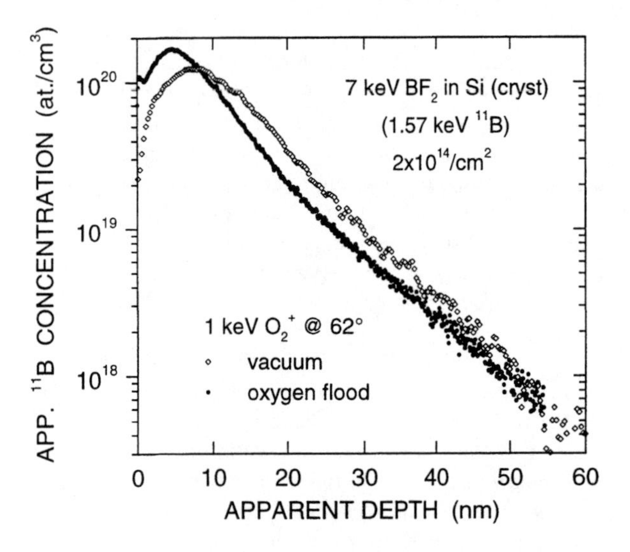

FIGURE 1. Comparison of SIMS depth profiles of B implanted in Si, measured with and without oxygen flooding. The total crater depth in the flooding experiment was 55 nm.

Depth profiles of the delta doped sample are depicted in Fig. 2. Two aspects are evident at first sight. Compared to the profile recorded in vacuum, all delta features measured with flooding are shifted towards the surface and become broader with increasing depth. This degradation in depth resolution is particularly clear for the last delta pair.

The peak shifts observed with oxygen flooding imply that the erosion rate decreased in course of sputter profiling. Knowing the true depth of delta location (3) one can derive mean erosion rates v averaged over the spacing of the deltas, as shown in Fig. 3. Within experimental accuracy v(vac) was constant during profile analysis. By contrast, v(flood) decreased continuously, quite rapidly at first and then more gradually. Owing to the fact that the first delta was located at a rather large distance of (41.5±1 nm) from the surface, details of the *initial* sputter rate changes could not be derived from the data for this sample. Note that, at the very beginning of sputtering, v must have been as large as in vacuum. But, as a result of oxygen incorporation, v soon started to decrease by an estimated factor of 2-3 (dashed curve). An additional decrease was subsequently brought about by the growth of ripples.

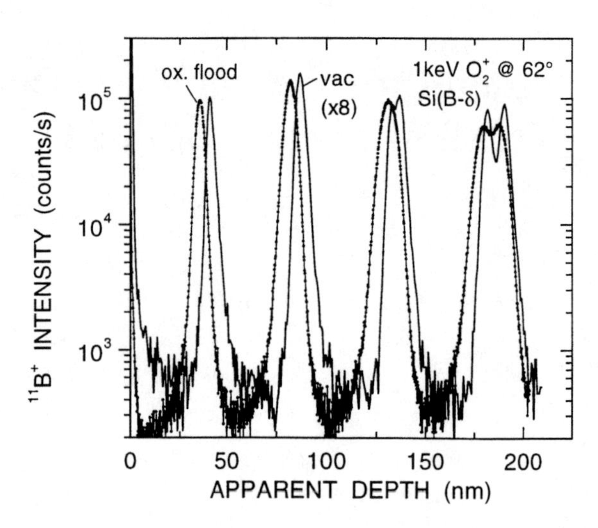

FIGURE 2. Comparison of SIMS depth profiles of B delta doping spikes in Si, measured with and without oxygen flooding. The first feature is a single delta, followed by three delta pairs with spacings of 1.7, 4.4 and 8.8 nm (± 0.2 nm). The end of the profiles corresponds to the total crater depth.

Further information about the changes in erosion rate may be derived from the depth dependence of the matrix reference signals shown in Fig. 4. In vacuum the Si^+ and Si_2^+ signals showed the well-known rapid initial decrease due to the removal of the native oxide. Thereafter the signals remained essentially constant, in accordance with the constant erosion rate shown in Fig. 3. With flooding the Si^+ intensity passed through a small minimum located at an apparent depth of about 1.5 nm and increased again to reach

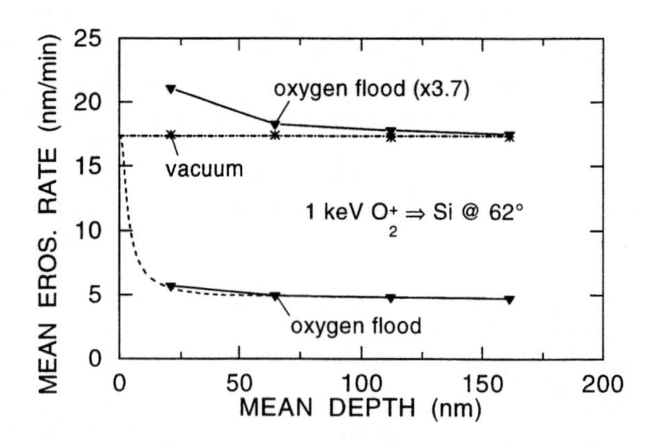

FIGURE 3. Mean erosion rates for sputter profiling at oblique beam incidence, in vacuum and with oxygen flooding. For better visibility the data obtained with flooding are also shown after multiplication by a suitable factor. The dashed curve is a rough estimate of the changes that must have occurred during the initial period of sputtering.

the maximum level at 3-4 nm. Upon further sputtering the Si^+ intensity started to decrease again, but with a rather small rate of change. Much larger changes in signal were observed with the Si_2^+ ions, in which case the intensity dropped by a factor of almost 3 during sputtering from the surface to 100 nm. Even beyond the depth of the Si^+ maximum, the decrease in Si_2^+ intensity still amounted to 30-40%. It is tempting, therefore, to assume that the Si_2^+ signal constitutes a semi-quantitative measure of the changes in erosion rate during sputter profiling with flooding. One consequence of this interpretation is that ripples may actually start to grow at an apparent depth of 3-4 nm, which will translate to real depth of about 8-12 nm.

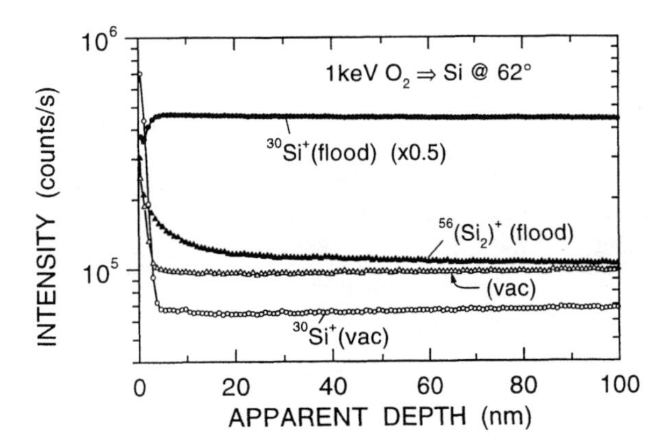

FIGURE 4. Variation of the Si monomer and dimer reference signals during sputter profiling in vacuum and with oxygen flooding.

Combining the results of Figs. 3 and 4 we arrive at the conclusion that it is misleading to interpret the stability of the Si^+ signal as reliable evidence for a constant erosion rate. Apparently the strong oxygen-induced secondary ion yield enhancement and the counteracting decrease in sputtering yield balance during the initial period of oxygen incorporation, so that only a small change in Si^+ signal remains (at 1.9 keV/57° (1) the change is even smaller).

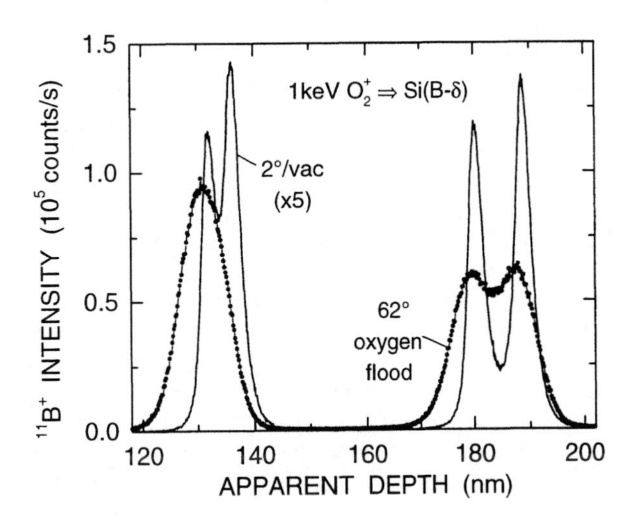

FIGURE 5. Comparison of the shape of two delta doping pairs profiled at normal incidence (2°) in vacuum and oblique incidence (62°) with flooding. The data are normalized to the same integral.

Another important aspect relates to the use of the Si^+ signal as a means of controlling the ionization probability of dopant ion signals (in this case B^+). Matrix ion signals commonly enter into calibration routines based on relative sensitivity factors (6) or impurity-to-matrix sensitivity ratios (1,7). The problem is that impurities often do not show the same dependence on the oxygen concentration as the matrix element (7,8). As a result, the sensitivity ratio may be different for the two major faces of the ripples which have very different orientations with respect to the beam (4). Both faces contribute to the impurity and matrix ion signals. This reasoning could explain the differences in areal density (retained dose) between profiles measured with and without flooding as well as anomalies observed during ripple growth on uniformly B doped Si samples (1). Problems of this kind have been avoided in Fig. 1 by deliberately using the quoted implantation fluence for concentration calibration. This procedure, however, cannot be applied in routine analyses, because one of the main purpose of SIMS profiling is to determine the retained dose rather than to consider it an input parameter.

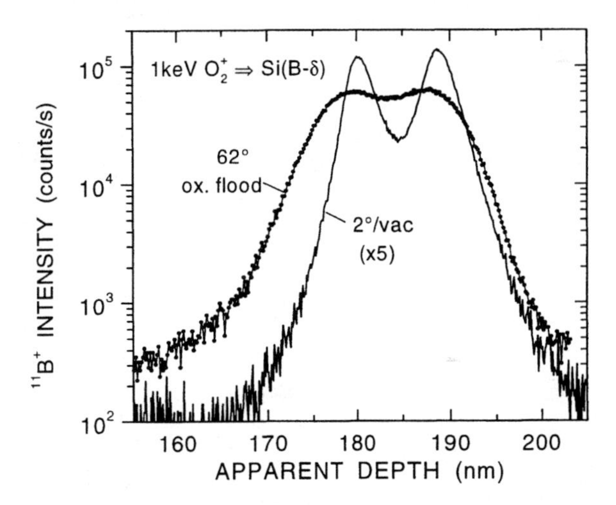

FIGURE 6. Details of the shape of profiles of the third delta pair measured at normal incidence in vacuum and oblique incidence with flooding.

Returning to the problem of depth resolution, Fig. 5 shows a comparison of SIMS profiles of the second and third delta pairs, measured at 62° with flooding and at essentially normal beam incidence (2°) in vacuum, i.e., at base pressure. The latter profiles are clearly much narrower than in the former.

An interesting detail of the profile shape becomes evident from Fig. 6 which shows an expanded section of the data of Fig. 5 on a semi-logarithmic scale. Although the profile width was found to be much narrower at normal beam incidence than at oblique incidence with flooding, the large depth tails and the section with the steepest exponential fall-off turned out to be almost the same under the two conditions. This result is in accordance with the observation that the decay length for B in Si using O_2^+ ion beams is essentially the same at 2° and ~ 60, in vacuum (9) as well as with flooding (10). The previous work was done under conditions where roughening did not (yet) occur. Hence the profile width was also largely independent of the impact angle (9). By contrast, the results of Fig. 6 reflect the pronounced increase in width produced by roughening. From the fact that, at the same time, the profile tails are not seriously affected, we can draw some conclusions concerning the shape of the ripples.

It is well known from basic SIMS studies that the profiles of delta distributions and rectangular distributions are very similar in shape at their leading in trailing edges (11). This result can easily be explained be realizing that the profile of a rectangular distribution with a width w may described as the convolution of the profile of a delta distribution over the depth interval w, with a constant weight factor (11,12). In the case of ripples passing through a delta doping distribution, a constant weight factor implies that the ripples have a saw-tooth shape. On the basis of this idea we have constructed the profile that one would get if in the measurement at normal incidence the surface had contained saw-tooth ripples with a peak-to-valley height of 6.24 nm. The results are shown in Fig. 7. The agreement between the measured and the simulated profile is surprisingly good. The remaining differences can be attributed to the simplifying assumption that peak and valley of all ripples were parallel to the sample surface. If one would allow for some statistical depth distribution of the ripples, the simulated profile would be smeared out at the leading and trailing edge, in accordance with the profile measured under conditions of ripple formation.

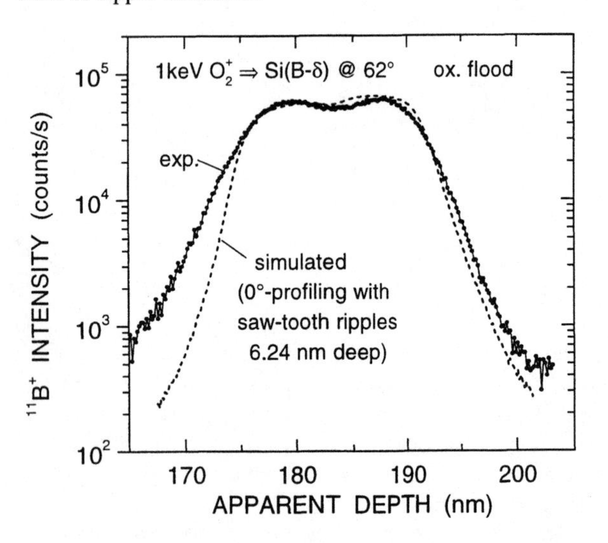

FIGURE 7. Comparison of the profile of a B delta pair measured at oblique beam incidence/flooding with a simulated profile that was constructed by convolution of the normal-incidence profile over a depth corresponding to the height of the assumed saw-tooth ripples.

One could argue that the roughening observed at oblique beam incidence may disappear at sufficiently high oxygen flux (or pressure). However, as Fig. 8 shows, the 62°-profiles are broader under all conditions. The best results were obtained in vacuum (pressure → 0 in Fig. 8), the worst at intermediate flux of the oxygen jet. An important aspect of the results is that even by application of very high pressures, i.e. >> 10^{-5} hPa (conditions outside the operating range of ion pumps), it will not be possible to achieve a depth resolution as good as by profiling at normal incidence in vacuum.

Additional experiments have shown that the depth resolution at angles up to 32° is almost as good as at 0° (3). There is, however, a difference in that the profile shift due to oxygen build-up (13) increases strongly with increasing impact angle (factor of 2.5 larger at 32° than at 0°). Hence the most reliable profiles of B in Si are obtained with low-energy oxygen beams at normal incidence. In fact, recent measurements have shown that B delta doping spikes with a spacing of only 2 nm can be well resolved using normally incident O_2^+ beams with energies of 250 eV (14) or even as low as 170 eV (15).

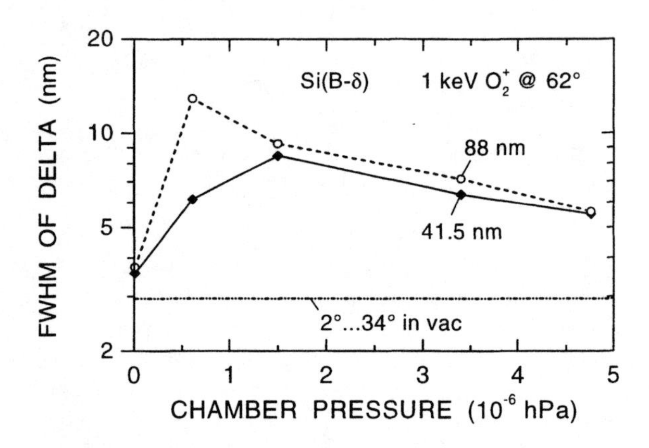

FIGURE 8. Full width at half maximum of single B delta doping spikes in Si vs the chamber pressure which varied due to jet-type oxygen flooding. The numbers denote the depth of delta location.

Even though it is satisfying to have technical means at hand which allow the profiling artifacts encountered at oblique incidence, notably with oxygen flooding, to be avoided, the question may be raised whether it is possible to simulate the observed profile degradation. This issue has been explored by means of a fairly simple model. The variation of the erosion rate v with time t was described by two slightly different analytical functions, labeled model 1 and 2, as shown in Fig. 9.

The real depth z_r was calculated by numerical integration of $v(t)$ up to the time that corresponds to a certain crater depth z_{cr}. For the purpose of comparing the two models we have chosen a fairly small depth, z_{cr} = 57 nm. The resulting dependence $z_r(z_{ap})$ and the profile shift $z_{sh}(z_{ap}) = z_r - z_{ap}$ are presented in Fig. 10 as dashed and solid lines (shift data multiplied by a factor of 10). Whereas a maximum shift similar to Fig. 1 is easy to reproduce, the depth dependence of the shift is seen to depend critically on the

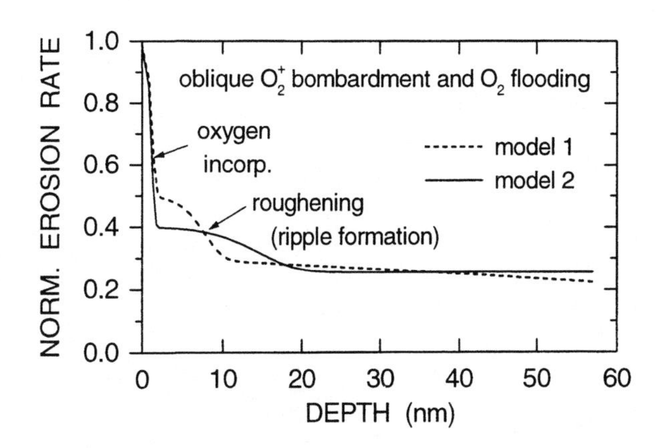

FIGURE 9. Assumed variation of the erosion rate during sputter profiling of Si with O_2^+ ions at oblique incidence with flooding.

details of the model. The effect of the crater depth is also shown (model 2 only). If we make the favorable assumption that the erosion rate becomes constant beyond some depth z^*, the extra amount of depth, δz, sputtered during the initial period of changing erosion rate can be calculated (δz = 5.58 nm for model 2). For $z_{ap} > z^*$ the apparent shift increases with increasing crater depth as $z_{sh} = \{1-(z_{ap}/z_{cr})\}\delta z$. This dependence is due to the fact that the *relative* deviation of the initial erosion rate from the mean increases with increasing crater depth.

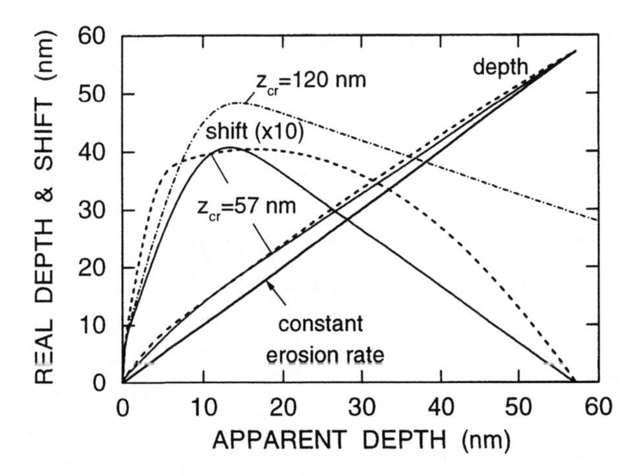

FIGURE 10. True (real) depth and profile shift calculated with the erosion rate models of Fig. 9, for two different crater depths. For z_{cr} = 120 nm only the shift is shown (model 2).

With this information at hand we can simulate distorted profiles under the simplifying assumption of a constant ionization probability of the impurity did not change during the measurement. The assumed (original) distribution is meant to reflect low-energy implantation with some channeling, see Fig. 11. The changing erosion rate determines not only the apparent depth scale but also the apparent concentration: the original distribution has to be multiplied

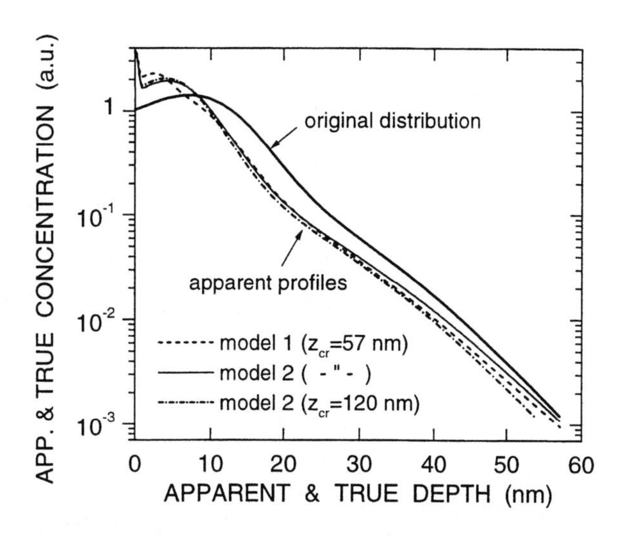

FIGURE 11. Comparison of the true doping distribution with three different simulated profiles which were distorted due to the depth dependent variations in erosion rate shown in Fig. 9.

by $v(z_{ap})/\langle v\rangle$, where $\langle v\rangle$ is the mean erosion rate averaged over the crater depth. This way mass is conserved, i.e., the *integrals* are the same for the original and the simulated profiles. The actual distributions of concentration versus depth, however, deviate strongly from each other, see Fig. 11. Hence we find, in agreement with Fig. 1, that even if one gets *'the dose right' in SIMS analysis with flooding, the measured implantation profile can be completely wrong*.

REFERENCES

1. Wittmaack, K., and Corcoran, S.F., *J. Vac. Sci. Technol.* **B16**, 272-279 (1998).
2. Jiang, Z.X., and Alkemade, P.F.A., in *Secondary Ion Mass Spectrometry SIMS XI*, Gillen, G., *et al.* (eds.), Chichester: Wiley, 1998, pp. 431-434.
3. Wittmaack, K., submitted for publication.
4. Wittmaack, K., *J. Vac. Sci. Technol.* **A8**, 2246-2250 (1990).
5. Dowsett, M.G., Smith, N.S., Bridgeland, R., Richards, D., Lovejoy, A.C., and Pedrick, P., in *Secondary Ion Mass Spectrometry SIMS X*, Benninghoven, A., *et al.* (eds.), Chichester: Wiley, 1997, pp. 367-370.
6. Wilson, R.G., *J. Appl. Phys.* **63**, 5121-5125 (1988).
7. Wittmaack, K., *Surf. Interface Anal.* **26**, 290-305 (1998).
8. Wittmaack, K., in ref. 2, p. 11-18.
9. Wittmaack, K., *J. Vac. Sci. Technol.* **B12**, 258-262 (1994).
10. Zalm, P.C., and Vriezema, C.J., *Nucl. Instrum. Meth. Phys. Res.* **B64**, 626-631 (1992).
11. Clegg, J.B., and Gale, I.G., *Surf. Interface Anal.* **17**, 190-196 (1991).
12. Wittmaack, K., in *Practical Surface Analysis*, Vol. 2: *Ion and Neutral Spectroscopy*, Briggs, D., and Seah, M.P. (eds.), Chichester: Wiley, 1992, pp.105-175.
13. Wittmaack, K., *Phil. Trans. R. Soc. London* **A354**, 2731-2764 (1996).
14. Dowsett, M.G., and Chu, D.P., *J. Vac. Sci. Technol.* **B16**, 377-381 (1998).
15. S.B. Patel, unpublished results (1998).

MATERIALS CHARACTERIZATION - GENERAL

High-Resolution Microcalorimeter Energy-Dispersive Spectrometer for X-ray Microanalysis and Particle Analysis*

D. A. Wollman,† G. C. Hilton,† K. D. Irwin,† L. L. Dulcie,† N. F. Bergren,†
Dale E. Newbury,‡ Keung-Shan Woo,§ Benjamin Y. H. Liu,§ Alain C. Diebold,§§
and John M. Martinis†

† National Institute of Standards and Technology (NIST), Boulder, Colorado 80303
‡ National Institute of Standards and Technology (NIST), Gaithersburg, Maryland 20899
§ University of Minnesota, Minneapolis, Minnesota 55455
§§ SEMATECH, 2706 Montopolis Drive, Austin, Texas 78731

We have developed a high-resolution microcalorimeter energy-dispersive spectrometer (EDS) at NIST that provides improved x-ray microanalysis of contaminant particles and defects important to the semiconductor industry. Using our microcalorimeter EDS mounted on a scanning electron microscope (SEM), we have analyzed a variety of specific sized particles on Si wafers, including 0.3 μm diameter W particles and 0.1 μm diameter Al_2O_3 particles. To compare the particle analysis capabilities of microcalorimeter EDS to that of semiconductor EDS and Auger electron spectroscopy (AES), we report measurements of the Al-Kα/Si-Kα x-ray peak intensity ratio for 0.3 μm diameter Al_2O_3 particles on Si as a function of electron beam energy. We also demonstrate the capability of microcalorimeter EDS for chemical shift measurements.

INTRODUCTION

Improved x-ray detector technology has been cited by SEMATECH's Analytical Laboratory Managers Working Group as one of the most important metrology needs for the semiconductor industry. In the Metrology Roadmap section of the 1997 National Technology Roadmap for Semiconductors (NTRS) (1), improved x-ray detector technology is listed as a key capability that addresses analysis requirements for small particles and defects. The transition-edge sensor (TES) microcalorimeter x-ray detector (2) developed at NIST has been identified as a primary means of realizing these detector advances, which will greatly improve in-line and off-line metrology tools that currently use semiconductor energy-dispersive spectrometers (EDS). At present, these metrology tools fail to provide fast and unambiguous analysis for particles less than approximately 0.1 μm to 0.3 μm in diameter. Improved EDS detectors such as the TES microcalorimeter are necessary to extend the capabilities of existing SEM-based instruments to meet the analytical requirements for future technology generations. With commercialization and continued rapid development, microcalorimeter EDS should be able to meet both the near-term NTRS goal of analyzing particles as small as 0.08 μm in diameter and the longer-term requirements of the semiconductor industry for improved particle analysis.

MICROCALORIMETER PERFORMANCE

Microcalorimeter EDS can already solve many materials analysis problems in the semiconductor industry. As illustrated in Table 1, the current performance of microcalorimeter EDS approaches that of high-resolution semiconductor EDS in terms of solid angle (4 msr using an polycapillary optic x-ray lens (3)) and maximum count rate (500 s^{-1}; over 1000 s^{-1} using a beam-blanker), while providing improved energy resolution comparable to that of a wavelength-dispersive spectrometer (WDS). The excellent energy resolution of our "general purpose" microcalorimeter EDS (~10 eV FWHM over the energy range 0 keV to ~10 keV) allows straightforward identification of closely spaced x-ray peaks in complicated spectra, including overlapping peaks in important materials (such as TiN and WSi_2) which cannot be resolved by semiconductor EDS. Recently, we have developed a TES microcalorimeter with an instrument-response energy resolution of 3.1 eV ± 0.1 eV FWHM (digital processing) and ~4 eV FWHM (analog processing) over the energy range 0 keV to ~2 keV (4). The ability to resolve severe peak overlaps using this detector is clearly observed in Fig. 1, in which we show an x-ray spectrum of TiN acquired in real time with our microcalorimeter EDS mounted on a SEM.

CP449, *Characterization and Metrology for ULSI Technology: 1998 International Conference*
edited by D. G. Seiler, A. C. Diebold, W. M. Bullis, T. J. Shaffner, R. McDonald, and E. J. Walters

TABLE 1. X-ray Spectrometer Comparison.[a]

Spectrometer Type	Energy Resolution (eV)	Maximum Count Rate (s^{-1})	Solid Angle (msr)	Collection Efficiency [b] (msr)
Semiconductor EDS (large area)	175 (at 6 keV)	30000	150	115
	145 (at 6 keV)	5000	150	115
Semiconductor EDS (high-resolution)	130 (at 6 keV)	3000	25	19
WDS (several diffracting crystals)	2 to 20	50000	8 to 25	0.8 to 2.5 [c]
Microcalorimeter EDS (NIST, with	7 (at 6 keV)	150	4	2
polycapillary x-ray optics)	3 (at 1.5 keV)	500	10	5

[a] A more complete table, including a comparison to other low temperature detectors, is presented in (2).
[b] Collection efficiency is defined as the product of the solid angle and the overall spectrometer efficiency at 1.7 keV.
[c] Because a WDS accepts only x-rays of a narrow energy band, its practical collection efficiency is further reduced (up to several orders of magnitude) when scanned over the entire energy range.

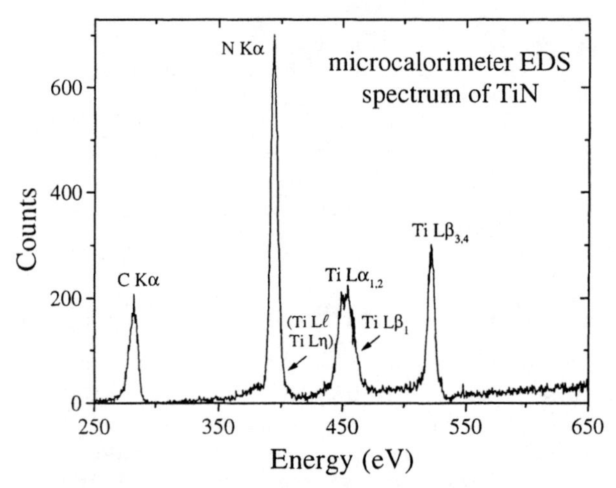

FIGURE 1. Microcalorimeter EDS x-ray spectrum of TiN, acquired under the following conditions: W filament, 2 keV beam energy, 320 s^{-1} input count rate, 230 s^{-1} output count rate, 27% dead time, 400 s live time, and a 45° x-ray takeoff angle. The original spectrum was corrected for energy nonlinearity, resulting in a constant energy binwidth of 0.45 eV per channel over the energy range presented. A significant C contamination peak developed during acquisition of all spectra reported in this work.

PARTICLE ANALYSIS

To demonstrate the usefulness of microcalorimeter EDS for particle analysis, we analyzed a variety of sub-micrometer particles produced at the University of Minnesota. This work was motivated by a SEMATECH project on detector development for particle analysis. As in previous particle analysis comparisons (5, 6), uniform particles of the desired size (0.3 μm diameter and 0.1 μm diameter) and material (Al_2O_3, W, and TiO_2) were selected using an electrostatic classification system and deposited on pieces of Si wafers under cleanroom conditions. These particles were then imaged and analyzed at NIST using SEM/microcalorimeter EDS.

In Fig. 2 we show a microcalorimeter EDS spectrum of a 0.3 μm diameter W particle on Si. Such particles cannot be analyzed using semiconductor EDS due to the severe peak overlaps between the Si-K and W-M x-ray lines. In Fig. 3 we present a microcalorimeter EDS spectrum of a 0.3 μm diameter TiO_2 particle, another particle type that is difficult to analyze using semiconductor EDS.

In order to compare the particle analysis capabilities of SEM/microcalorimeter EDS to that of other analytical methods, we analyzed 0.3 μm diameter Al_2O_3 particles on Si to determine the Al-Kα/Si-Kα peak intensity ratio as a function of electron beam energy. This particle-to-substrate

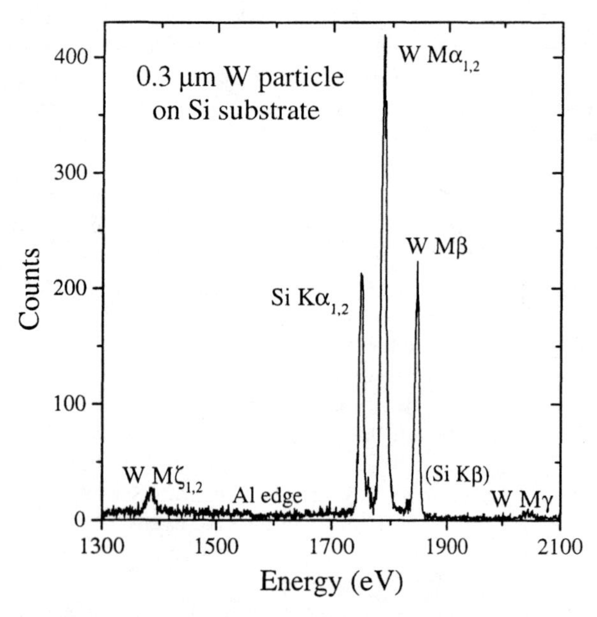

FIGURE 2. Microcalorimeter EDS x-ray spectrum of a 0.3 μm diameter W particle on a Si substrate, acquired under the following conditions: LaB_6 filament, 5 keV beam energy, 50 pA beam current, 63 s^{-1} input count rate, 60 s^{-1} output count rate, 5% dead time, 400 s live time, and a 45° x-ray takeoff angle. The original spectrum was corrected for energy nonlinearity, resulting in an energy binwidth per channel that increases from 0.7 eV to 1.1 eV over the energy range presented. The electron beam diameter was estimated to be less than the particle diameter.

ratio has been proposed as a figure of merit for particle analysis and has been useful in the semiconductor industry to determine optimum beam conditions for particle analysis. Particle-to-substrate ratios have been used previously to compare the particle analysis capabilities of field-emission SEM/semiconductor EDS (FE-SEM/EDS) and field-emission Auger electron spectroscopy (FE-AES) (5, 6).

The Al_2O_3 particles were imaged using a SEM and categorized by size and morphology. Compact (preferably round) Al_2O_3 particles with average diameters between 0.26 μm and 0.34 μm were selected for analysis. In Fig. 4 we show a microcalorimeter EDS spectrum of a typical 0.3 μm diameter Al_2O_3 particle acquired at a beam energy of 5 keV. All particle analyses were performed in spot mode with the electron beam at the approximate center of the particle. The smallest SEM electron probe sizes (corresponding to beam currents of 10 pA to 40 pA) were selected to keep the diameter of the electron beam less than that of the particle. This condition was difficult to satisfy at low beam energies (less than ~3 keV) using our LaB_6-filament SEM operating at a working distance of 39 mm. As a result, analyses of 0.3 μm diameter Al_2O_3 particles were

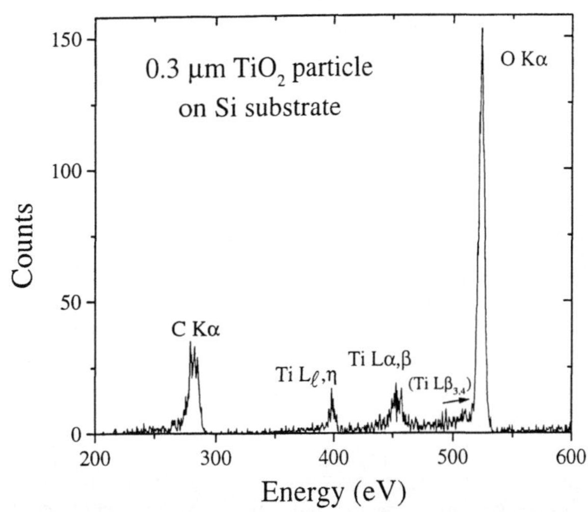

FIGURE 3. Microcalorimeter EDS x-ray spectrum of a 0.3 μm diameter TiO_2 particle on a Si substrate, acquired under the following conditions: LaB_6 filament, 1.8 keV beam energy, 0.53 nA beam current, 30 s^{-1} input count rate, 29 s^{-1} output count rate, 1% dead time, 400 s live time, and a 45° x-ray takeoff angle. The original spectrum was corrected for energy nonlinearity. The electron beam diameter was larger than the particle diameter.

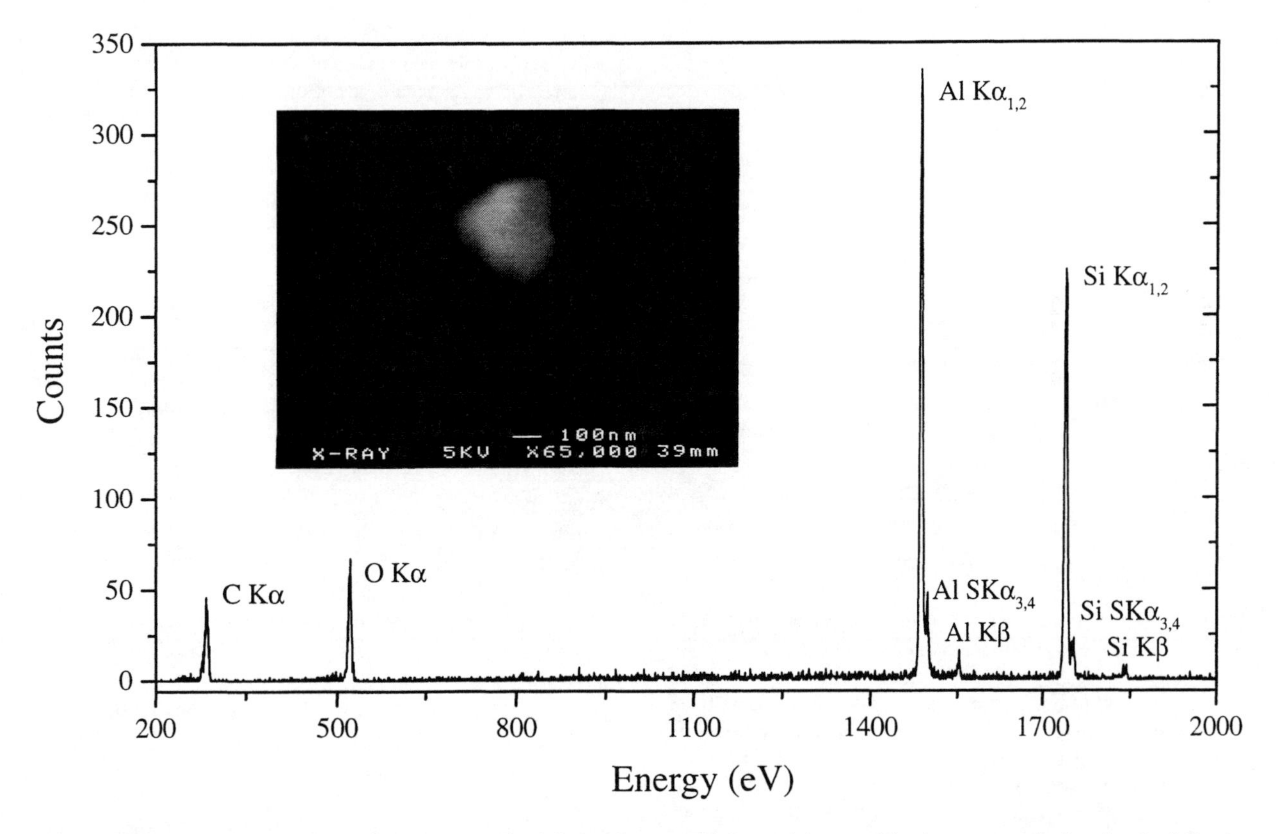

FIGURE 4. Microcalorimeter EDS x-ray spectrum of a 0.3 μm diameter Al_2O_3 particle on a Si substrate, acquired under the following conditions: LaB_6 filament, 5 keV beam energy, 40 pA beam current, spot mode, 200 s live time, 57 s^{-1} input count rate, 54 s^{-1} output count rate, 5% dead time, and a 45° x-ray takeoff angle. The spectrum was corrected for energy nonlinearity, resulting in a nonuniform energy binwidth per channel increasing from 0.4 eV to 0.9 eV over the energy range presented. Note the ability of the microcalorimeter EDS to resolve the Kα and satellite peaks of Al and Si. The electron beam diameter was estimated to be less than the particle diameter. A SEM micrograph of this particle obtained directly before analysis under the same SEM operating conditions is shown in the inset.

performed only at beam energies of 3 keV and higher.

In Fig. 5 we present background-subtracted Al-Kα/Si-Kα x-ray peak intensity ratios obtained from SEM/microcalorimeter EDS analyses of 0.3 μm diameter Al_2O_3 particles on Si. The Al/Si ratios from individual particle analyses at the selected electron beam energies were then averaged and are plotted on a logarithmic scale in Fig. 6.

Using the Al/Si ratio as a figure of merit (as suggested in (5, 6)), SEM/microcalorimeter EDS compares favorably with FE-SEM/EDS and FE-Auger for particle analysis in this analytical situation. For both SEM/microcalorimeter EDS and FE-SEM/EDS, the use of lower electron beam energies resulted in smaller x-ray generation volumes that were better matched to particle volumes, thus producing high particle-to-substrate ratios. In comparison, the Al/Si ratio for FE-Auger does not have such a strong dependence on beam energy. Unlike x-rays, Auger electrons have very limited range (on the order of a few nanometers) and are

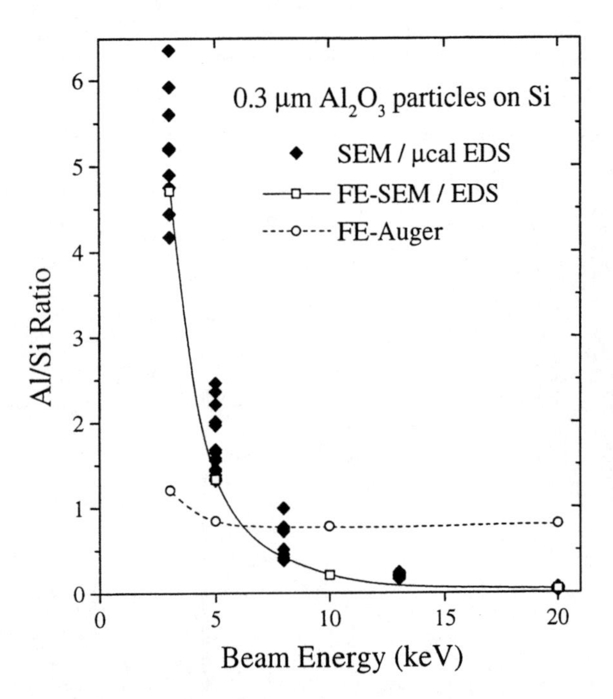

FIGURE 5. Comparison of SEM/microcalorimeter EDS (SEM/μcal EDS), FE-SEM/semiconductor EDS (FE-SEM/EDS), and FE-Auger analyses of 0.3 μm diameter Al_2O_3 particles on Si wafers. The SEM/microcalorimeter EDS data consist of background-subtracted Al-Kα/Si-Kα integrated peak intensity ratios from individual particle analyses, while the FE-SEM/EDS and FE-Auger data are averages over several particles (5, 6). The FE-Auger data were obtained using a whole-wafer field-emission scanning Auger system. The solid and dashed lines connecting the FE-SEM/EDS and FE-Auger data are provided as guides to the eye. The spread in the SEM/microcalorimeter EDS data is greater than that expected from counting statistics and is likely caused by differences in particle shape and electron beam position during analysis.

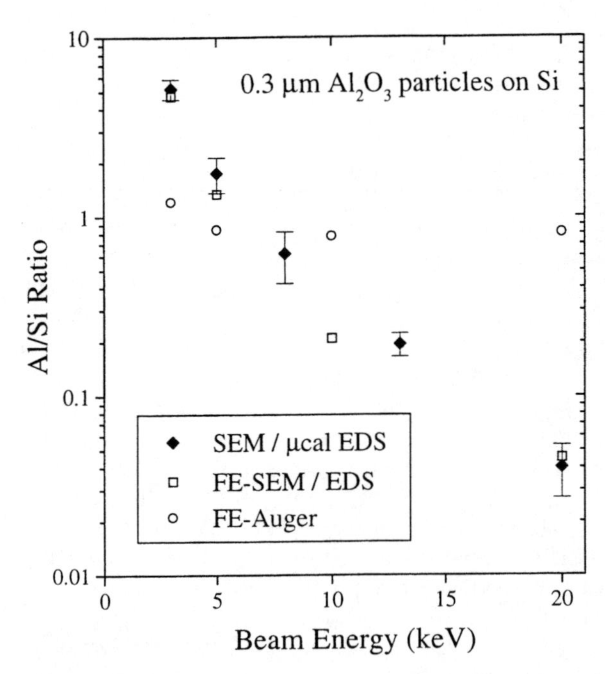

FIGURE 6. Comparison of average Al/Si ratios for SEM/microcalorimeter EDS (SEM/μcal EDS), FE-SEM/EDS, and FE-Auger analyses of 0.3 μm diameter Al_2O_3 particles on Si wafers. The SEM/microcalorimeter EDS data are averages of individual particle data presented in Fig. 5, with error bars equal to corresponding standard deviations. The average Al/Si ratio for SEM/microcalorimeter EDS is greater than that for FE-SEM/EDS by a factor of approximately 1.3. This factor is consistent with the difference in overall spectrometer efficiency (between microcalorimeter EDS and semiconductor EDS) caused by the presence of additional Al infrared-blocking x-ray windows in the microcalorimeter EDS.

collected only from the surface layer of the particle and substrate. The resulting particle-to-substrate ratio for FE-Auger is less dependent on the electron interaction volume (and thus beam energy) and can be greater than that of FE-SEM/EDS for smaller particles (5). A more complete understanding of the measured Al/Si ratios for x-ray and Auger particle analysis as a function of beam energy requires extensive modeling of the electron-beam/particle-substrate interaction (5) and is beyond the scope of this paper.

The ability of SEM/microcalorimeter EDS to analyze smaller Al_2O_3 particles was also investigated. In Fig. 7 we show microcalorimeter EDS spectra of Al_2O_3 particles as small as 0.1 μm in diameter. SEM/microcalorimeter EDS analysis of the 0.1 and 0.14 μm diameter Al_2O_3 particles was limited by the nonoptimal SEM performance at low beam energies, in which the diameter of the electron beam was larger than that of the particles. This limitation can be avoided in the future with the use of FE-SEM/microcalorimeter EDS.

By examining the role of x-ray and Auger particle analysis tools in microfabrication facilities, we may be able

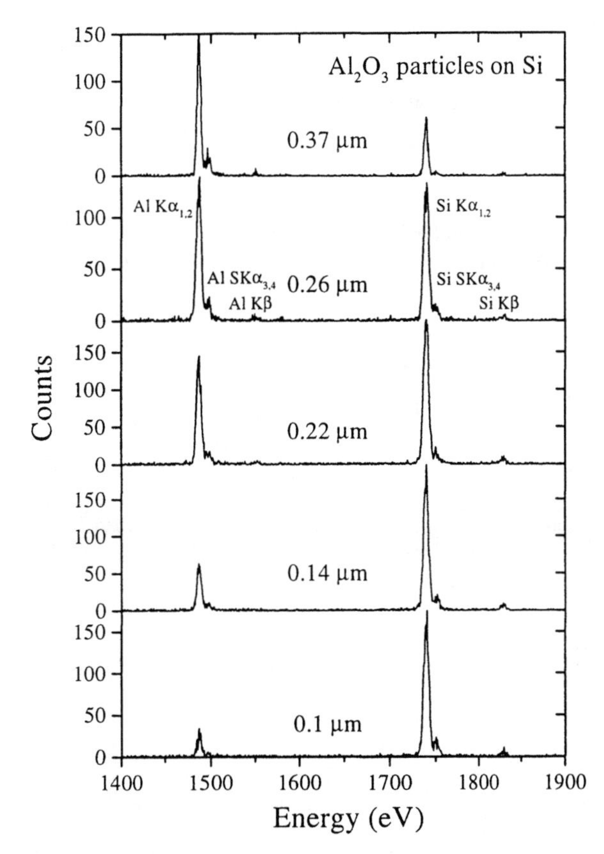

FIGURE 7. Microcalorimeter EDS x-ray spectra of several sized Al_2O_3 particles on a Si substrate, acquired under the following conditions: LaB_6 filament, 5 keV beam energy, 40 pA beam current, ~40 s^{-1} input count rate, ~40 s^{-1} output count rate, ~1% dead time, 150 s live time, and a 45° x-ray takeoff angle. The average diameter of each particle is displayed directly above its spectra. The original spectra were corrected for energy nonlinearity. The electron beam diameter was estimated to be larger than the diameters of the two smallest particles.

to project the future use of microcalorimeter EDS in the semiconductor industry. At present, after initial identification and mapping by light scattering, contaminant particles and defects are typically imaged and analyzed using whole-wafer FE-SEM/EDS defect review tools to characterize particle size, location, morphology, and composition (if possible, depending on particle size). Significant statistical information is gathered to characterize large numbers of contaminant particles, of which a representative subset is selected for exhaustive analysis by FE-SEM/EDS, FE-Auger (including depth profiling of particle composition in conjunction with ion sputtering), time-of-flight secondary ion mass spectrometry (TOF-SIMS), and other techniques. Often, particles/defects are observed at several process steps, and thus buried under the subsequent process layers, before a wafer is sent for detailed analysis. These samples are sent to whole-wafer focused ion beam (FIB) tools that prepare cross sections of particles for subsequent characterization by *in situ* SEM/EDS and *ex situ* TEM/EDS.

In the future, microcalorimeter EDS will provide significant benefits for all of the instruments described above (and others) that currently use semiconductor EDS. For example, commercialization of microcalorimeter EDS will allow the direct replacement of semiconductor EDS in existing defect review tools. At this initial stage of particle analysis, automation of analysis is critical to obtain as much statistical information as possible to characterize the contaminant particles and defects. The higher energy resolution of microcalorimeter EDS will provide easier qualitative identification of particle constituents, which in turn should allow improved autoclassification of particle x-ray spectra. In addition, the use of FE-SEM/microcalorimeter EDS during later stages of particle analysis will benefit from the new capability of microcalorimeter EDS to measure chemical shifts in x-ray spectra, as will be described in the following section.

CHEMICAL SHIFT MEASUREMENTS

Chemical shifts result from changes in electron binding energies with the chemical environment of atoms. Measurements of chemical shifts in analytical techniques such as x-ray photoemission spectroscopy (XPS) and Auger electron spectroscopy (AES) have been demonstrated to provide valuable chemical bonding state information (7).

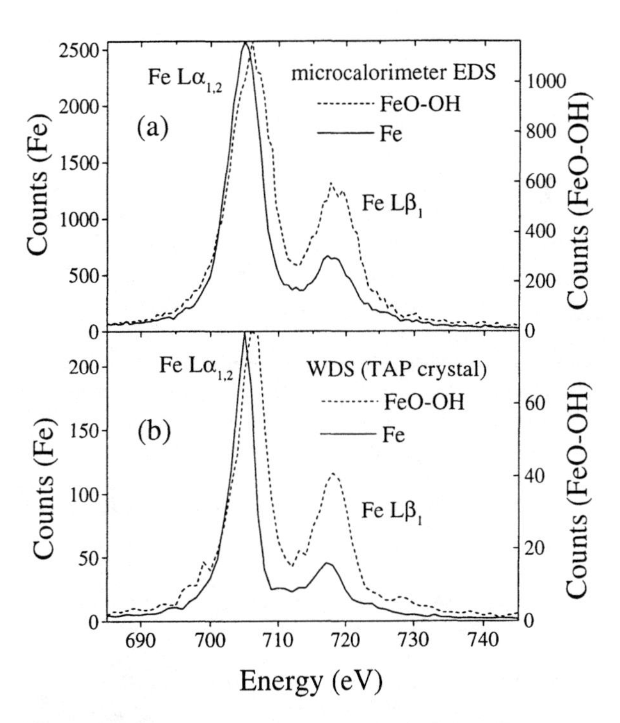

FIGURE 8. (a) Microcalorimeter EDS spectra and (b) WDS spectra of Fe (solid line) and FeO-OH (dashed line), from (11). The observed changes in the Fe-L peak positions and intensities result from chemical bonding effects. Good agreement is observed between the microcalorimeter EDS and WDS spectra.

For example, the ability to identify particle composition by distinguishing different oxidation states (for example, Al and Al_2O_3) using AES has been demonstrated to be useful in determining sources of contamination in semiconductor processing tools (8).

While chemical shift measurements are not as well established in x-ray spectroscopy, chemical shifts have been observed in WDS x-ray spectra as changes in x-ray peak positions, relative peak intensities, and peak shapes (9). These chemical shift effects can be significant (with x-ray peak shifts on the order of 1 eV), particularly for x-ray lines resulting from transitions involving valence electrons of light elements such as C (10). However, chemical shift measurements are not routinely performed in WDS analysis due to the extreme time penalty of scanning.

The improvement in energy resolution of our most TES microcalorimeter (4) now allows microcalorimeter EDS measurements of chemical shifts in x-ray spectra, as shown in Fig. 8 for Fe and FeO-OH (11) and in Ref. (4) for Al and Al_2O_3. The EDS operation of the microcalorimeter ensures that all peak shapes and integrated peak intensities are readily accessible. With further improvements in the energy resolution of microcalorimeter EDS, analysis using x-ray peak shapes and positions to provide may become practical and provide valuable chemical bonding state information for particle analysis.

CONCLUSION

The excellent energy resolution of microcalorimeter EDS provides improved capability for x-ray microanalysis of contaminant particles and defects, including the potential to provide valuable chemical bonding state information. With commercialization and further development, microcalorimeter EDS will extend the capabilities of SEM-based x-ray microanalysis instruments to help meet the analytical requirements for future technology generations.

ACKNOWLEDGMENTS

Particle specimens were produced at the University of Minnesota and analyzed using SEM/microcalorimeter EDS at NIST. The particle analysis work at NIST was supported in part by the NIST Office of Microelectronics Programs and SEMATECH.

REFERENCES

1. The National Technology Roadmap for Semiconductors (1997), The Semiconductor Industry Association, 4300 Stevens Creek Boulevard, Suite 271, San Jose, CA 95129.
2. Wollman, D. A., Irwin, K. D., Hilton, G. C., Dulcie, L. L., Newbury, D. E., and Martinis, J. M., *J. Microscopy* **188**, 196-223 (1997).
3. Wollman, D. A., Jezewski, C., Hilton, G. C., Xiao, Q.-F., Dulcie, L. L., and Martinis, J. M., *Microscopy and Microanalysis* **3** (Supplement 2), 1075-1076 (1997).
4. Irwin, K. D., Wollman, D. A., Hilton, G. C., Dulcie, L. L., Bergren, N. F., Martinis, J. M., and Huber, M. E., in preparation.
5. Childs, K. D., Narum, D., LaVanier, L. A., Lindley, P. M., Schueler, B. W., Mulholland, G., Diebold, and A. C., *J. Vac. Sci. Technol. A* **14**, 2392-2404 (1996).
6. Diebold, A. C., Childs, K., Lindley, P. M., Viteralli, J., Kingsley, J., Liu, B. Y. H., and Woo, K.-S., *J. Vac. Sci. Technol. A*, in press.
7. Briggs, D. and Riviere, J. C., *Practical Surface Analysis, Vol.1: Auger and X-ray Photoelectron Spectroscopy*, New York: John Wiley and Sons, 1990, pp. 85-141; Waddington, S. D., *ibid.*, pp. 587-594.
8. Brundle, D., Uritsky, Y., and Pan, J. T. *Micro* **13**, 43-56 (1995).
9. Fabian, D. F., Watson, L. M., and Marshall, C. A. W., *Rep. Prog. Phys.* **34**, 601-696 (1971).
10. Holliday, J. E., *J. Appl. Phys.* **38**, 4720-4730 (1967).
11. Wollman, D. A., Newbury, D. E., Hilton, G. C., Irwin, K. D., Dulcie, L. L., Bergren, N. F., and Martinis, J. M., *Proceedings of Microscopy and Microanalysis '98*, Altanta, GA (1998), in press.

Tungsten In-Film Defect Characterization

Y. Uritsky, S. Ghanayem, V. Rana, R. Savoy, S. Yang,

Applied Materials, Inc., Santa Clara, CA 95054

C.R. Brundle

Brundle & Associates, San Jose, CA 95125

Tungsten deposition and subsequent etch back are used in device manufacturing to make tungsten plug interconnects. This process utilizes an aggressive, fluorine radical rich, plasma chemistry to clean a deposition chamber and to remove the W film. This results in a variety of particle defects. Usually, the defects arise from two distinct sources: (a) hardware components degradation, and (b) inefficient evacuation of etch by-products. This paper describes new types of defects located in the W film (so-called "in-film" defects). The defects were analyzed by Scanning Electron Microscopy with Energy Dispersive X-ray Spectrometer (SEM/EDX) and Atomic Force Microscopy (AFM) on whole 200-mm wafers and then by Scanning Auger Microscopy (SAM), Electron Backscattered Diffraction (EBSD) and Transmission Electron Microscopy (TEM) on sectioned samples. Also, we have used a newly developed defect marking technique, which eliminates the problem of locating defects in analytical tools. Defects of interest fall into two classes: (a) W ring-like defects and (b) W hillocks. We show that the first class is formed due to contamination at the TiN glue layer (interface with the W). This inhibits W growth locally during deposition. The second class (hillocks) has the same composition as the rest of the film. These defects are large W single-crystal grains with no indication of foreign material or disruption at the TiN/W interface. The mechanism of formation of the W in-film hillocks is not understood.

INTRODUCTION

The yield of ULSI devices relies on the ability to tightly control a large number of process parameters. Defects, one of these parameters, are recognized as a root cause of greater than 50% of the losses and one that is difficult to control. This is especially true in case of blanket tungsten deposition and etch back processes that are used in the wafer processing industry to make via interconnects. In general, the W chemical vapor deposition (WCVD) is a two-step process: an initial nucleation step based on silane (SiH4) reduction of tungsten hexafluoride (WF6), is followed by a hydrogen reduction of WF6 to deposit the bulk of W film. After each wafer is processed, the chamber is cleaned with nitrogen trifluoride (NF3) plasma. In order to expose isolated W plugs the deposited W film is then removed using SF6 plasma etching. This complicated and aggressive chemistry (fluorine radicals) process is known to be "dirty". A variety of particle defects are formed due to a lack of control of the deposition and clean (W deposition chamber), inefficient by-product removal (W etch back chamber), and hardware components degradation (1,2). This paper describes the result of analyses of W-based defects formed in the W film during deposition, which partially remain on the surface after completion of the W etch back.

TEST PROCEDURE

Tungsten deposition and etch back were performed in an Applied Materials 200 mm wafer processing system. Three kinds of wafers were analyzed: (a) after WCVD (~ 400 nm) (b) after Physical Vapor Deposition (PVD) of the TiN adhesive layer prior to W deposition, and (c) the fully processed wafer (W deposition and etch back to an oxide film). Wafers of type (a) were scanned for defects using a Tencor Surfscan 6400, type (b) and (c) by Tencor 6200. Defects from the Tencor maps were then re-located (3) and analyzed in a JEOL 6600F field emission SEM equipped with a high resolution Oxford EDX system. After completion of SEM/EDX work, one W film wafer was then processed with the MicroMark 5000 (MicroTherm, LLC, Minneapolis, MN.) to eliminate problems in locating defects in subsequent analysis steps. The MicroMark 5000 is a unique instrument designed to locate defects on full-wafers (6, 8, or 12 inch) and then laser-mark the surface near selected defects. Once a defect has been marked, it is easily located in analytical tools, even after the wafer is cleaved. After marking, the wafer was analyzed in the scanning probe microscope (DI-7000) and sectioned to prepare samples for electron backscatter diffraction pattern (EBSD) analysis (4),

CP449, *Characterization and Metrology for ULSI Technology: 1998 International Conference*
edited by D. G. Seiler, A. C. Diebold, W. M. Bullis, T. J. Shaffner, R. McDonald, and E. J. Walters

SAM PHI 670 (CE&A Lab, Sunnyvale, CA), and for TEM (MAG Lab, Sunnyvale, CA).

RESULT AND DISCUSSION

1. Ring like defects

Type (a): After Deposition of the W film

Representative defects are shown in the Fig. 1. Commonly, they are approximately 30-50 μm in diameter, consisting of coaxial hole and adjacent annulus (Fig. 1a,c). The holes are about 10-15 μm in diameter, usually containing the low contrast, organic like (it was slightly melted and charged under e-beam) residue in the center (Fig. 1b). However, there were a few defects with inner hole contaminated by a residue of high contrast comparable with bulk W film (Fig. 1c). The surface of the annulus area looks recessed compared to the surface of the bulk W film, with different grain structure (Fig. 1a,c). In addition, we observed a few defects with holes of much smaller diameter (~2-3 μm), and one defect without the hole region at all.

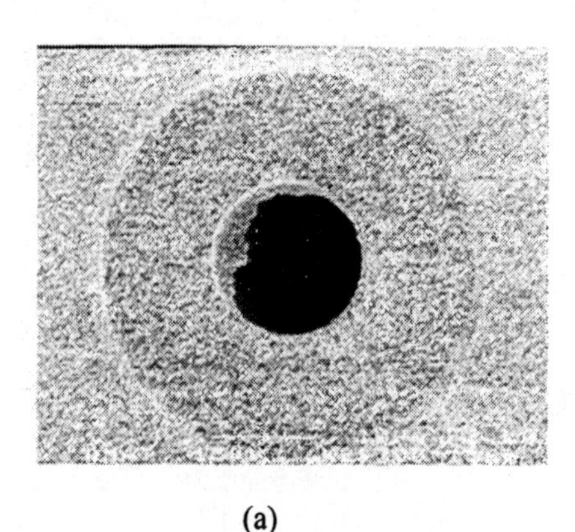

(a)

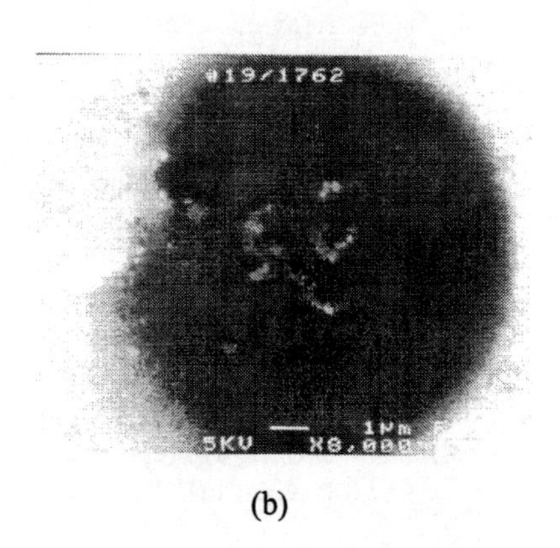

(b)

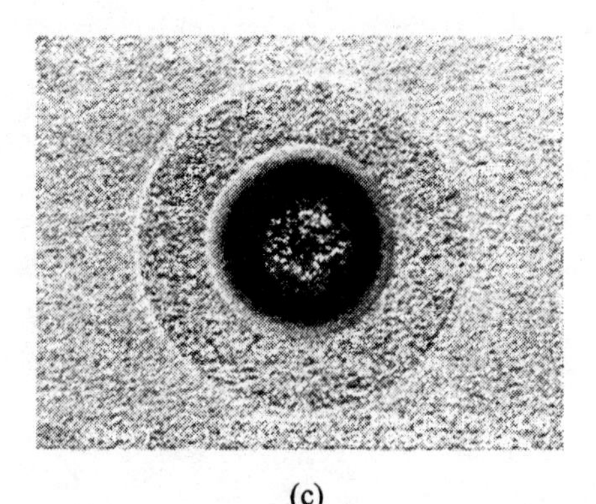

(c)

FIGURE 1a,b,c. SEM images of W-ring defects with "empty" inner hole with organic like contaminants in the hole region (a,b), and containing high contrast residue (c).

The SEM observations were corroborated and specified by the AFM analysis. Two W-rings with "empty" hole were scanned revealing similar to shown by SEM topographical features. Fig. 2a,b presents the surface plot and cross-sectional profile of defect shown in Fig. 1a,b. The cross-sectional profile was measured through bearing analysis (averaged Z value of each layer rather than a single line profile) of the area marked by an oblong box in Fig. 2b. It reveals the relationship between depth value and the area at that depth, offering good statistical measurements of surface steps. The magnification for the surface plot is 40X in the Z-axis. The plot shows that the inner hole is deeper and smaller than the annulus. The diameters of the hole and annulus are 15 and 27 μm. The average step height of the hole and adjacent annulus are approximately 400 and 100 nm, respectively. That means that the hole pierces through the W film (~ 400 nm thickness) to the TiN adhesive layer (compare with EDX and Auger data taken from "empty" hole), or even to the oxide, leaving TiN/Ti micro-islands in the central region.

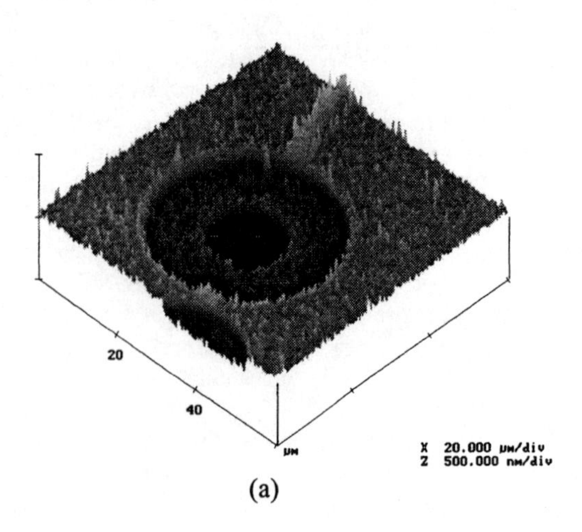

(a)

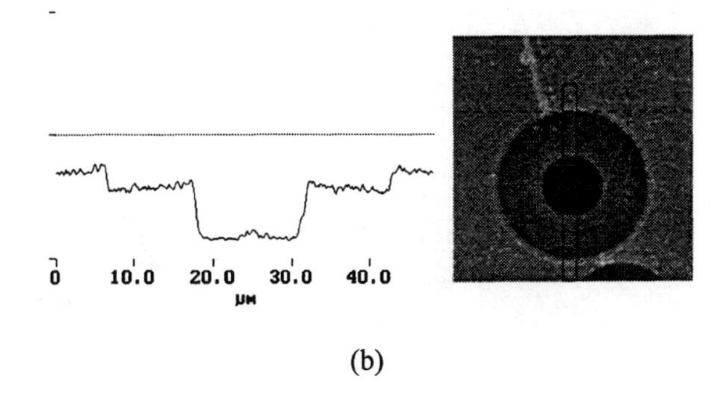

(b)

FIGURE 2a,b. AFM image (a) and cross sectional profile (b) of W-ring defect shown in the Fig. 1a

The described structural regions were analyzed for composition by EDS and AES. Fig. 3a presents the EDS spectrum obtained from the low contrast residue (e.g. Fig. 1b). The spectrum indicates the elements pertaining to the layers beneath the W (N, O, Si, and Ti), small amount of W, and C-F based constituents. The high contrast residue (e.g. Fig. 1c) is mainly W (Fig. 3b). Regarding the annulus area, the EDS analysis did not reveal any differences compared with the bulk W film.

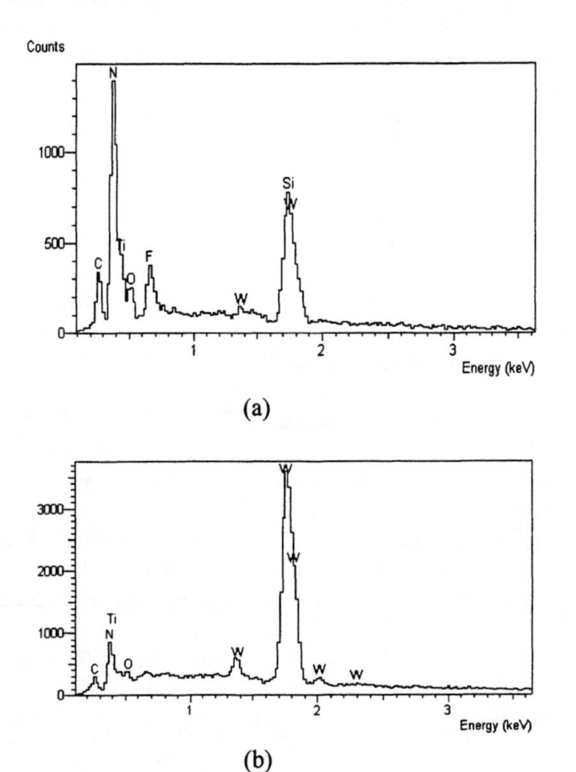

FIGURE 3a,b. EDS spectra obtained from the central region of (a) "empty" hole shown in the Fig. 1a,b and (b) from the high contrast residue (Fig. 1c).

The Auger results were in a good agreement with EDS, except for the lack of a F signal. The Fig. 4a shows the AES survey obtained from the low contrast residue (the same defect shown in the Fig. 1a,b) after sputtering off about 5 nm of surface to eliminate environmental hydrocarbon. The survey indicates the essential C along with W, Ti, and N, but no F. The lack of F is quite reasonable, because the high energy e-beam (20 kV) can easily desorb F from the surface (Kenton Childs, PHI, personal communication, and (5)). The surveys taken from the annulus and bulk film are identical and show the W, O, and trace of C which is probably due to previous SEM work (Fig. 4b).

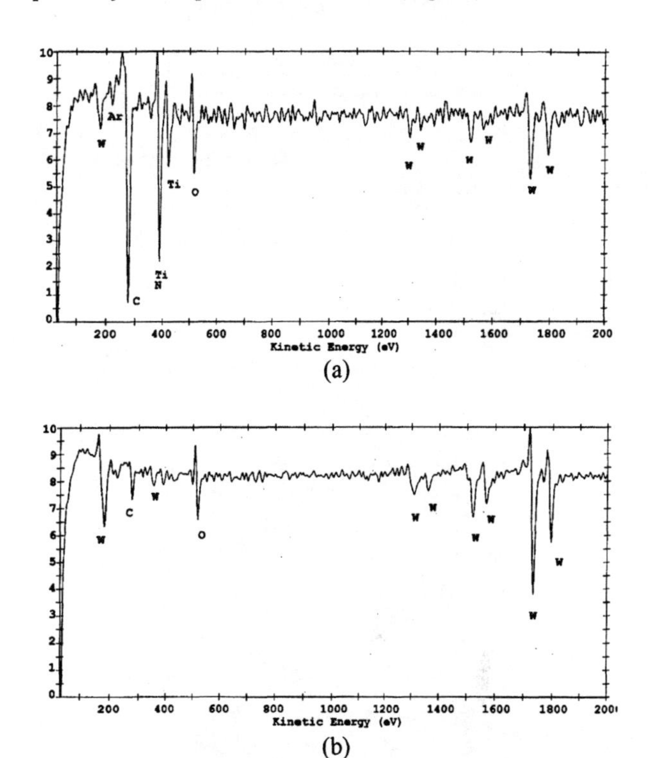

FIGURE 4a,b. The Auger surveys obtained from the (a) central region of the "empty" hole and (b) annulus area of defect shown in the Fig. 1a.

Type (b): The Initial TiN Film, Prior to W Deposition

The obtained results above suggest that the cause of formation of W rings is the foreign material (the major element found was the C) existing on a wafer prior to W deposition inhibiting the deposition of W in this location. The logical step to prove this concept was to locate them on an initial TiN film deposited by the PVD process (type (b) wafers). SEM/EDS analysis revealed three kinds of defects: (i) carbon-rich disk like spots about 5-6 μm in diameter with specific dendrite like "rainbow" in the perimeter (Fig. 5a, b); (ii) ruptures exposing an oxide film (Fig. 5c); (iii) occasional examples of Al-based particles (not shown here).

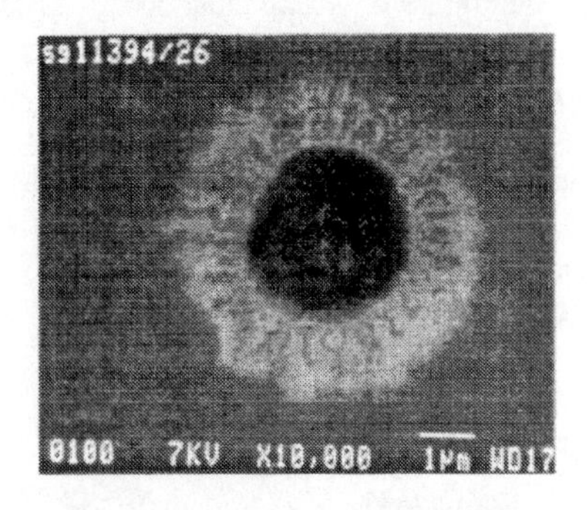

(a)

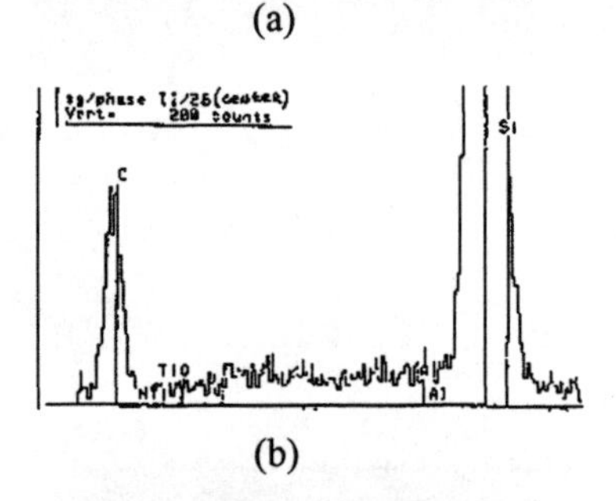

(b)

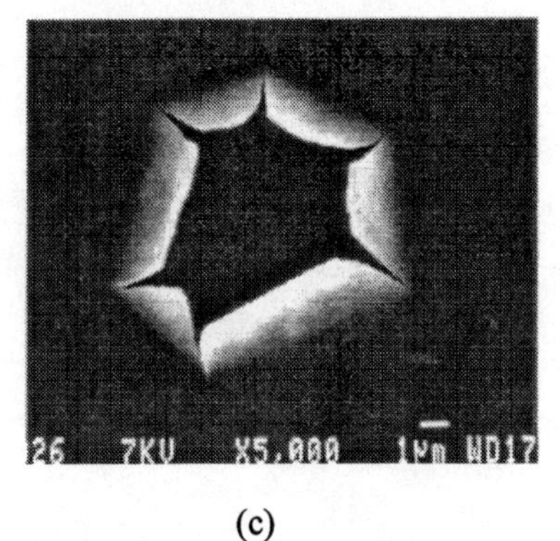

(c)

FIGURE 5 a,b,c. SEM images of typical defects found on the TiN wafer (a) quasi-disk with carbon rich central region (by EDS (b), there were no other compared to background elements) and dendrite perimeter, (c) film rupture exposing an oxide layer.

Type (c): The Fully Processed Wafer (W Deposition and Etch Back to an Oxide Layer)

After the final W etch back, the defects remain as W donut-shaped rings on a SiO_2 surface (the etch back removes the W layer and the TiN layer). These contain C and occasionally even Al at the donut center. An example of an "empty" and Al-containing donuts is shown in Fig. 6 a,b.

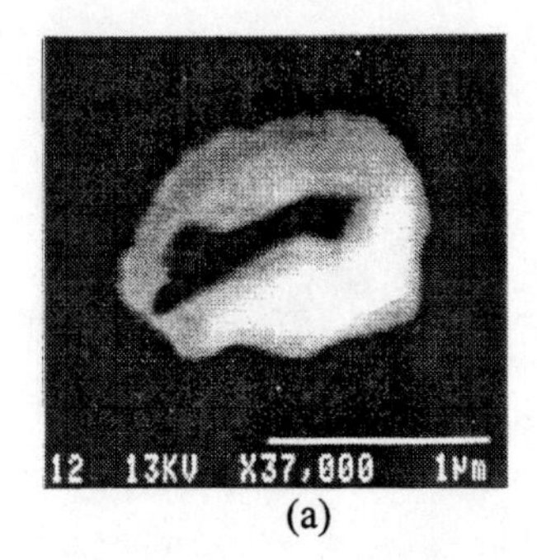

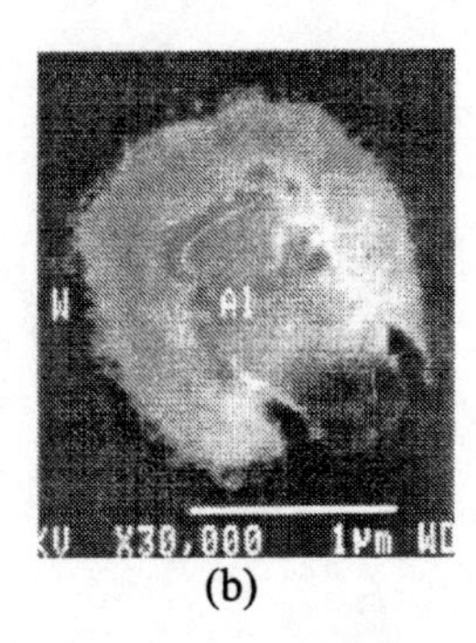

(a) (b)

FIGURE 6 a,b. SEM image of W donut like defects located on fully processed wafer (W deposition and etch back to an oxide layer). By EDX, the perimeter area is W, the central region is C, O, and Si for (a) and Al for (b).

2. Hillock defects

The hillocks or so-called "crystal" defects were observed only on W film wafers (type (a)). The majority of them were shaped as a semi-spheroid with a quasi oval base. A length-wise diameter ranged from 0.3-0.4 for the smallest to 2 μm for the largest one. Fig. 7a demonstrates a cluster of hillocks, but there were a few individual defects also. The SEM image shown in the Fig. 7a was taken on a small, approximately 10x20 mm, sectioned sample, tilted to 70 degree which is necessary for EBSD analysis (4). For this specific type of defect the high tilt appeared to be very useful to estimate their heights. Overall, the SEM showed heights varying within 0.2-0.7 μm.

The EDS results from any (small or large) defects under normal or tilted (70 degree) orientation was W only. This fact was confirmed by using the electron backscattered pattern (EBSP) technique (electron diffraction) in the SEM (4). The Kikuchi pattern obtained from the largest of Fig. 7a hillocks indicated a W single bcc alpha phase crystal (Fig. 7b), the same as expected from the bulk W film (strongly textured). The confidence index with which this phase was identified was 0.857, which is considered high enough to avoid confusion. Later we confirmed the EBSP analysis using TEM electron diffraction.

After determination of the chemistry (pure W) and structure (W bcc alpha phase single crystal) of the defects, the major interest was associated with observations of the W-TiN interface in order to try to understand the mechanism of defect growth. Combining the MicroMark 5000 defect marking approach with subsequent Focused Ion Beam milling we prepared a cross-section for TEM through a large hillock. The high resolution TEM image (Fig. 7c) together with TEM diffraction data (not shown) demonstrated that the hillock is essentially a large single crystal grain without indications of any foreign materials, or disruption at the W-TiN interface. Also, there was no evidence that the W interacts with the Ti under the TiN layer. There may be some activated sites where W nucleation begins earlier than in the other sites and growth proceeds rapidly. However, our experimental work did not revealed any evidence that such sites exist.

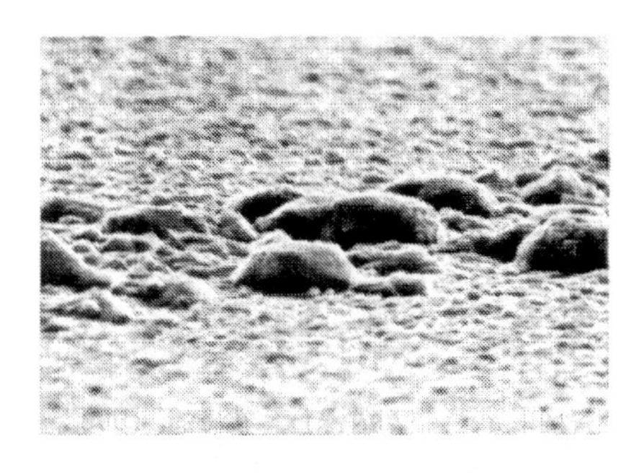

(a)

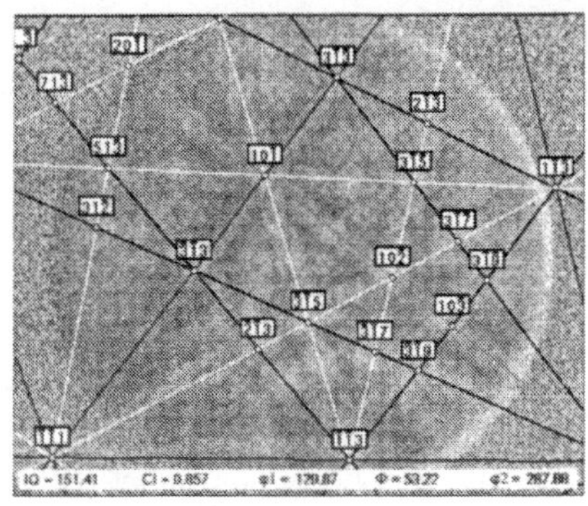

(b)

(c)

FIGURE 7a,b,c. (a) SEM image of hillocks located on W film wafer (70 degree tilt), (b) EBSP obtained from the largest hillock, and (c) high resolution TEM image of the hillock vicinity.

CONCLUSIONS

In analyzing defects on the three types of wafer (bare TiN films prior to WCVD; W films formed by WCVD on top of the TiN; fully processed wafers where the W and TiN has been etched-back to the SiO_2 substrate) a sophisticated set of analytical techniques had to be employed (SEM/EDX, SAM, AFM, TEM and EBSD). In addition the use of a recently developed defect marking approach (MicroMark 5000) allowed each technique used to easily find the same individual defect on either the full wafer, or on individual pieces.

On the bare TiN we found defects consisting of circular carbon-rich regions (probably originated from grease/oil droplets), ruptures through to the SiO_2 substrate, and Al-based particles.

On the CVD deposited W films we found circular defects which lacked any W in the center and had an annulus of reduced W thickness. We associate these with the circular defects found on the bare TiN surfaces. This type of defect, however, does not result in a defect on the etch-back product, because the etch-back process removes W and TiN down to the SiO_2 substrate anyway.

On etch-back wafers the defects are W "donuts" left on the bare SiO_2 surface which contain C flakes or Al particles in their centers. The formation mechanism we propose is that they originate from the presence of certain types of defect present on the TiN surface prior to W deposition. These are the TiN ruptures and Al particles we found on the bare TiN, and carbon flakes (not shown). These defects inhibit W growth locally, plus the W also piles up at their perimeter, resulting in thicker W at the perimeter. Etch-back then leaves some of the thicker W because of insufficient etch time to remove it. When the steps prior to the W deposition were made cleaner, the "W donuts" were eliminated in the final etch-back product.

The origin of the W hillocks, however, is still not understood (a similar phenomenon has been observed for Al-on-WCVD films, also without mechanistic explanation (6)). We demonstrated that the W hillock is a large W bcc single-crystal grain nucleated on the TiN glue layer. The reason for the different growth behavior of the hillocks is probably connected to local morphology differences in the TiN underlayer.

REFERENCES

1. Uritsky, Y., Rana, V., Ghanayem, S., Wu, S., *Microcontamination*, **V. 12 No. 5**, 25-29, May 1994.
2. Uritsky, Y., Chen, L., Zhang, S., Wilson, S., Mak, A., Brundle, C.R., *J Vac Sci Technol.,* **A15(3)**, 1319-1327, May/June 1997.
3. Uritsky, Y., Lee, H., **US patent # 5,381,004**, 1995.
4. Dingly, D.T., Alabaster, C., Coville, R., *Inst. Phys. Conf.* Ser. **98**, 451-454, (1989).
5. Uritsky, Y., Savoy, R., Kinney, P., Principe, E., Mowat, I., McCaig, L., "Evaluation of Single Semiconductor Defects Using Multiple Microanalysis Techniques", presented at the MRS Spring Meetings, San Francisco, CA, April 13-17, 1998.
6. Pico, C.A., Bonifield, T.D., *J. Mater. Res.*, **V. 8, No 5,** 1010-1018, May 1993.

Analysis of Submicron Defects Using an SEM-Auger Defect Review Tool

Kenton D. Childs, David G. Watson, Dennis F. Paul, and Stephen P. Clough

Physical Electronics, Inc., 6509 Flying Cloud Dr., Eden Prairie, MN, USA 55344

The challenges associated with analyzing semiconductor defects become greater as the device design rule decreases. According to the SIA National Technology Roadmap for Semiconductors, the current metrology requirement for particle analysis is 90 nm with the need to analyze 75 nm particles by the year 2001. These dimensional requirements are beyond the typical capabilities of current SEM/EDX defect review tools. Auger Electron Spectroscopy is a powerful method for measuring the surface composition of localized regions, and has been identified in the SIA roadmap as a primary technique for particle analysis. The ability of a state-of-the-art Auger defect review tool (DRT) to provide secondary electron and high spatial resolution elemental images is particularly effective in characterizing the often complex structure of semiconductor defects. Examples of Auger analysis from defects found at various process steps, on both unpatterned and patterned whole wafers, are shown. These examples highlight the ability of Auger to analyze both thin and laterally small or complex defects.

INTRODUCTION

The Semiconductor Industry Association (SIA) National Technology Roadmap for Semiconductors (1) (NTRS), 1997 Edition, forecasts production of 180 nm design rule devices during the year 1999 and 150 nm design rule devices two years later. The definition of a critical ("killer") defect, one which may cause device failure, is set nominally by the NTRS at 50% of the design rule - 90 nm and 75 nm, respectively. Therefore, it must be possible not only to detect defects of this size but to determine their origins in the process line. The latter challenge requires the ability to find and chemically analyze the defect on a 200 mm (or 300 mm) wafer.

The traditional method for "defect review" is to use a Defect Review Tool (DRT) based on a high spatial resolution Scanning Electron Microscope (SEM) equipped with an Energy Dispersive Xray Spectrometer (EDX or EDS). The EDX method has the advantage of fast spectral data acquisition and can be used in a mapping mode. However, current EDX instruments have some severe limitations when attempting to analyze particles in the sub-100 nm size range:

1. The high excitation beam energies (>10 keV) required to excite the x ray spectrum for heavier elements (e.g. first row transition metals) penetrate and produce signal from a large sample volume; typically (1 - 9) μm^3 which is much greater than the minimum dimensions of a critical defect. The spectrum of a small particle will, therefore, be dominated by signal contributions from the material surrounding the particle (2).

2. The x ray energy resolution of current EDX spectrometers is of the order of 140 eV which allows elemental but not chemical analysis of the defect.

Therefore, the EDX operator may face a serious data interpretation challenge in attempting to use EDX data to determine the source of a defect. For example, the spectrum of an Al particle on SiO_2 would look very similar to that of an Al_2O_3 particle on poly-Si.

Auger Electron Spectroscopy (AES) analyzes the electrons ejected from the sample by the competing atomic relaxation process during electron beam irradiation (3). Inelastic scattering of Auger electrons in the sample makes AES a very surface sensitive technique; typically the Auger signal arises from the top (0.5 - 3) nm of material because Auger electrons from deeper in the sample lose their characteristic kinetic energy before escaping the sample surface. This surface sensitivity, combined with small diameter (SEM size) probe beams and detection limits that vary no more than a factor of 10 across the periodic table, makes AES a very suitable technique for small particle or thin film defect analysis on semiconductor wafers. Examples of SEM-Auger analysis of structures and defects found on wafers after various stages of wafer processing are presented here.

CP449, *Characterization and Metrology for ULSI Technology: 1998 International Conference*
edited by D. G. Seiler, A. C. Diebold, W. M. Bullis, T. J. Shaffner, R. McDonald, and E. J. Walters

EXPERIMENTAL

The Auger defect review analyses were performed on a Physical Electronics, Inc. (PHI) *SMART-200* SEM-Auger DRT. This is a 200 mm full wafer instrument that accepts defect position information from optical or other defect detection systems (e.g. KLA-Tencor) and uses this information to navigate to the defects to be analyzed. The *SMART-200* is capable of high magnification SEM operation at primary beam energies of 500 eV to 25 keV. The Auger electrons are detected after passing through a high transmission cylindrical mirror analyzer (CMA). The analyzer and primary beam axes are coincident allowing analysis of surfaces and structures with large topographic features, including focused ion beam (FIB) cuts. In addition to small area spectral analysis, the *SMART-200* allows elemental and some chemical 2-D mapping as well as measurement of concentration versus depth profiles.

RESULTS AND DISCUSSION

50 nm Particle on Unpatterned Wafer

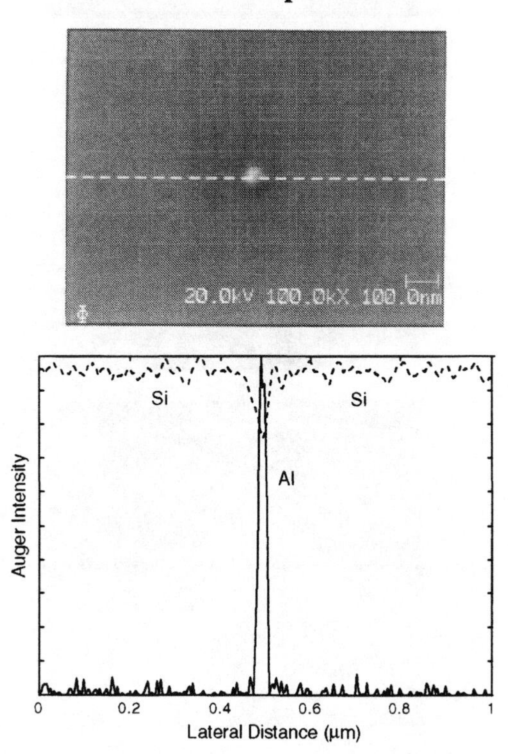

FIGURE 1. SEM image (upper) of a 50 nm diameter defect particle and Auger "linescans" for Al and Si recorded along the horizontal line (lower).

This example illustrates the current ability of Auger to meet the defect analysis requirements of 100 nm design technology. Fig 1 shows the SEM image (100 kX original magnification) of a 50 nm defect particle on an unpatterned 200 mm Si wafer. The horizontal line on the image in Fig. 1 indicates the position at which an Auger linescan (Auger intensity versus lateral distance) was collected and displayed as in Fig 1. Note that width of the peak in the Al linescan (at the half height) is also approximately 50 nm indicating that the spatial resolution

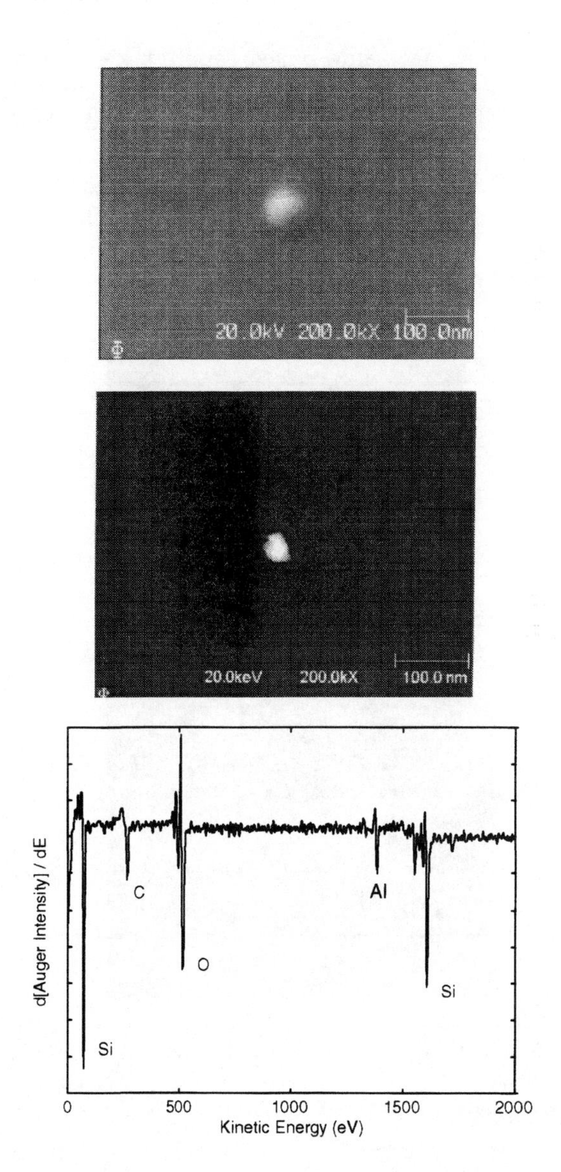

FIGURE 2. SEM image of 50 nm diameter defect particle on Si wafer (top), Auger map of Al (middle), and Auger spectrum

of the Auger signal is of the order of the primary beam diameter. The Si signal in the region of the particle arises from scattered primary electrons exciting the wafer surface adjacent to the particle.

Fig. 2 shows the same particle at 200 kX original magnification (top) and the Auger Al map (middle) and the spectrum of the particle, respectively. The lineshape of the Auger Al peak indicates directly that the particle is Al_2O_3; the Si signal arises from scattering of the primary beam.

17 nm Gate Oxide Cross Section

Fig. 3 shows the SEM image at 100 kX original magnification of a 17 nm gate oxide (SiO_2) in cross section and the overlaid Si (blue) and O (red) Auger electron images taken on the sample area from the SEM image Fig. 3. The red/blue region at the top of the Auger map is the surface of the poly-Si layer. The high surface specificity of Auger spectroscopy makes it possible to map a gate oxide layer of this thickness.

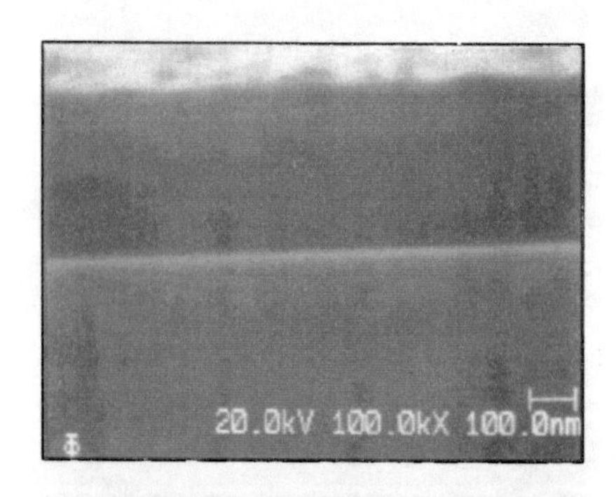

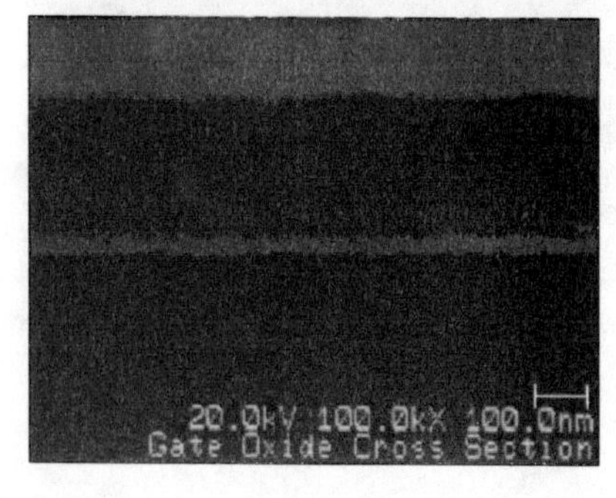

FIGURE 3. SEM image (upper) and corresponding overlaid Auger maps (lower) of O (red) and Si (blue) of a 17 nm thick gate oxide layer. The poly-Si surface oxide is also shown in red.

Particle Contamination after Poly-Si and W Deposition and Patterning

This example demonstrates the potential for complexity in defect review. SEM images of the FIB section of a contaminant particle at low (top) and a higher magnification (middle) are shown in Fig. 4 (4). The Auger electron intensity maps of O, Si, and W, taken at the higher magnification and overlaid in the bottom image in Fig. 4 indicate that the particle has a very complicated microstructure. From the distribution of the elements, it was determined that the particle is composed of SiO_2 and that it was introduced during the poly-Si deposition step.

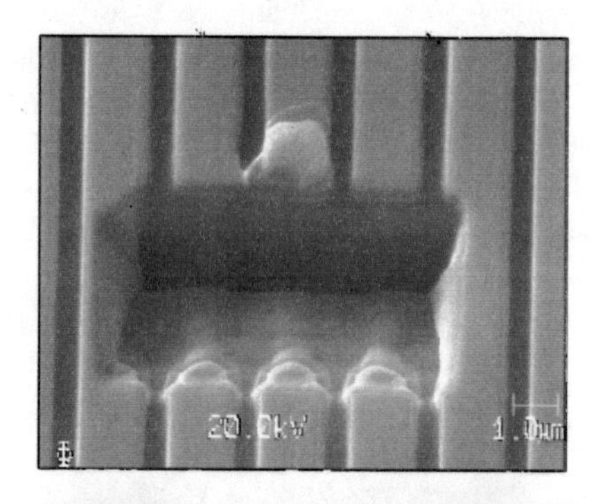

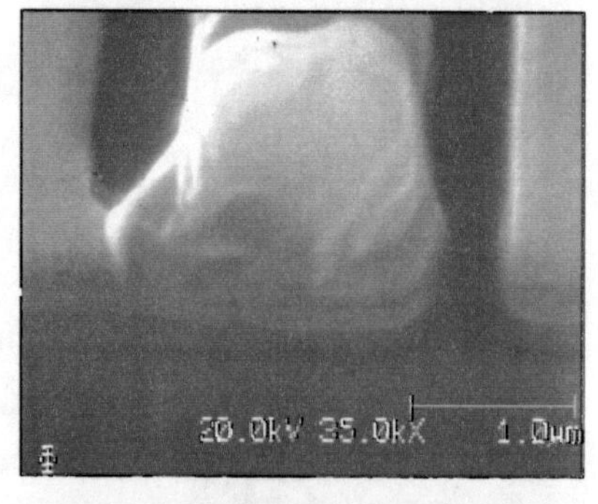

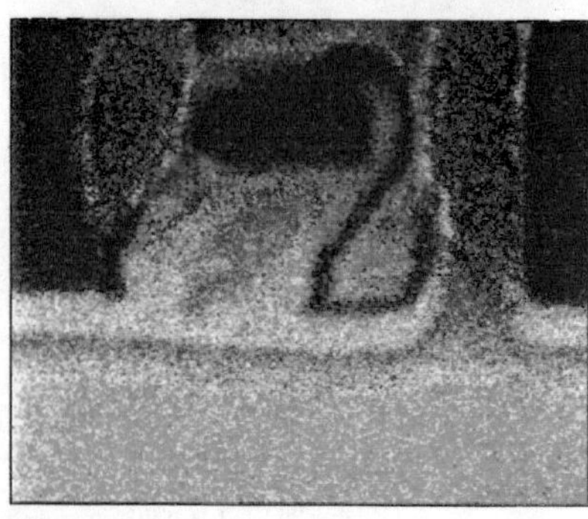

FIGURE 4. SEM image of FIB cross section of particle defect (top), the defect cross section at higher magnification (middle), and the overlaid Auger maps of O (red), Si (green), and W (blue) showing that the particle is SiO_2 and was introduced during poly-Si deposition. (Figure 4 images ©Institute of Environmental Sciences and Technology, (847) 255-1561)

Complex Defect After Poly-Si Etch Step

The defect shown in Fig. 5 was discovered on a poly-Si layer after plasma etching and photoresist stripping. Auger mapping analysis results are also shown in Fig. 5 where, C and Si are shown in red and blue respectively in each map and green represents V (upper right), S (lower left), and K (Lower right). Auger spectra (not shown) collected at various spots on the defect confirmed the presence of these elements; an Auger spectrum from a point outside the defect showed that the defect resides in the field oxide.

The circular shape of the defect suggests a drop of liquid that has dried on the wafer surface. The presence of S and
C on the defect suggests that it is residue from photoresist stripping step. The presence of V is unexpected and could be used as a fingerprint for determining the root cause; it may have originated from a wet chemical tank or plumbing.

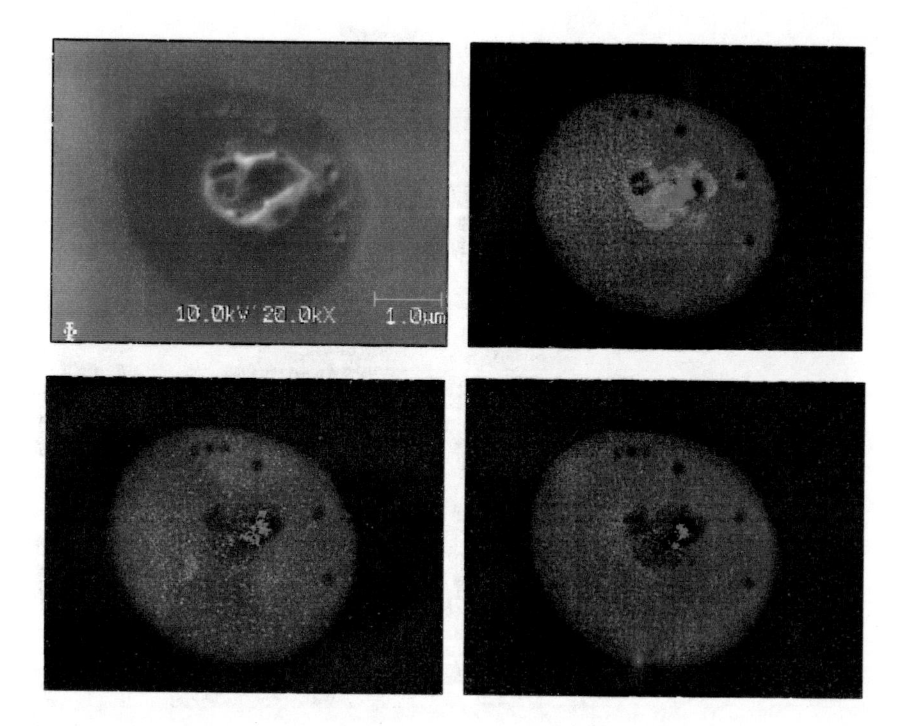

FIGURE 5. SEM image of residue from drop of photoresist stripping agent (upper left) and Auger maps of C (red) and Si (blue). The green area represents V (upper right), S (lower left), and K (lower right).

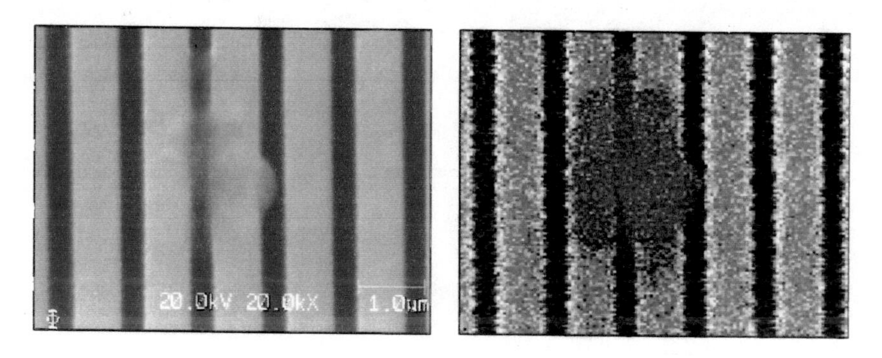

FIGURE 6. SEM image (left) of thin residue defect on TiN capped Al lines. Overlaid Auger maps of Ti (green) and Al (red) from area of thin residue defect (right).

Identification of Thin Residue on Metal Lines

The surface sensitivity of AES makes it appropriate for the analysis of thin film-like defects. Fig. 6 shows the SEM image (left) of TiN capped metal lines on SiO_2 with a nearly transparent thin film-like defect covering two of the lines and the substrate. The overlaid Auger maps of Ti and Al are also shown in Fig. 6.

Auger spectra taken from a pristine area on one of the lines and from the defect are shown in the upper and lower panels of Fig. 7, respectively. These spectra indicate that the defect contains metallic Al and F. The particle is a flake of Al metal and probably originates from the etch chamber. The source of the F is likely one of the periodic cleaning procedures used on the etch chamber.

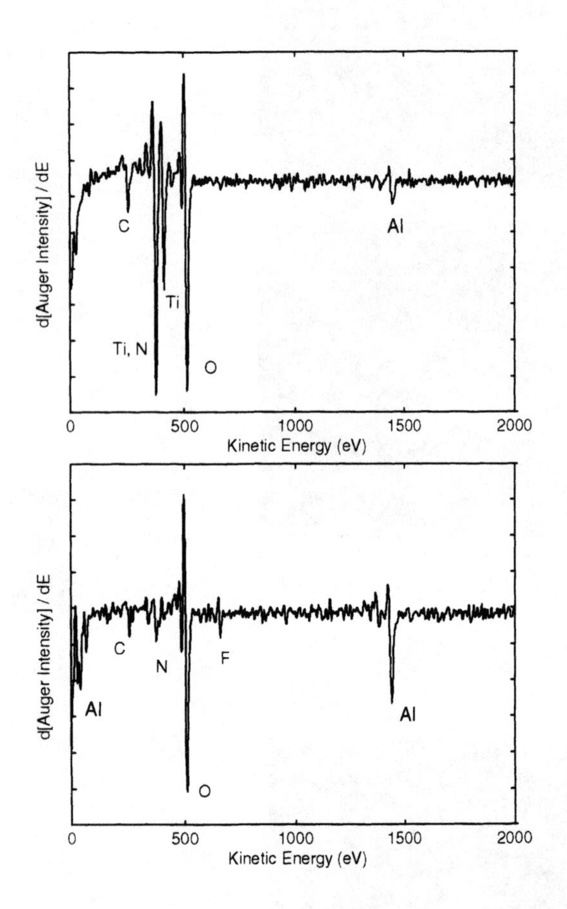

FIGURE 7. Auger spectrum from TiN capped Al line (upper) and from defect (lower) showing that the defect is fluorinated Al.

SUMMARY AND CONCLUSIONS

It is clear that an SEM combined with an integrated Auger Electron Spectroscopy capability provides elemental and chemical information about wafer defects that is not available when using current EDX technology. The limitations of EDX are due to the high energy required to excite the spectrum and the consequently large sampling volume of the technique. The surface sensitivity of AES restricts the sampled region to much smaller dimensions that will need to be analyzed on wafers in the immediate future.

REFERENCES

1. *The National Technology Roadmap for Semiconductors.* Semiconductor Industry Association (1997)

2. Childs, K. D. *et al. Journal of Vacuum Science and Technology B.* **14** (1996) 2392

3. Childs, K. D. *et al. The Handbook of Auger Electron Spectroscopy.* Physical Electronics, Inc. (1995)

4. Childs, K. D. *et al. 1996 Proceedings - Institute of Environmental Sciences of the 42nd Annual Technical Meeting on Contamination Control and Symposium on Minienvironments.* (The Institute of Environmental Sciences, Mount Prospect IL), page 147 (used with permission).

Polarized Light Scattering and its Application to Microroughness, Particle, and Defect Detection

Thomas A. Germer

Optical Technology Division
National Institute of Standards and Technology
Gaithersburg, MD 20899

In this paper, I will discuss the theory and application of polarized light scatter. Recent measurements have shown that some sources of scatter, including microroughness and subsurface defects, have well-defined polarizations for any specific pair of incident-scatter directions. By exploiting this knowledge, polarization techniques offer the possibility for large improvements in the sensitivity to defects and the discrimination of those defects from competing sources of scatter. The theoretical performance of a specific polarized light scattering instrument configuration will be analyzed to illustrate that a factor of 1.4 improvement in the minimum detectable particle or defect size can be readily attained.

INTRODUCTION

Optical scattering techniques have proven to be very useful for the detection of defects on silicon wafers in the ramp-up and high-yield stages of production. Since detectors for optical radiation can be extremely sensitive and fast, and since optical techniques lack vacuum requirements and, in many cases, stringent vibration tolerances, the techniques are ideal for tools with the high throughputs required for these stages of production. However, these techniques will inevitably run into barriers as feature sizes continue to be reduced. The particle sizes that will need to be detected are much smaller than the wavelength of the light used for the measurement, where the scattering cross section decreases rapidly. Therefore, small decreases in feature size translate into large increases required in the detection sensitivity. Furthermore, sources of delocalized scatter, such as microroughness, can dominate the signal, especially on blanket layers.

Since the scattered light intensity is affected by laser speckle noise, it is difficult to imagine that instruments relying upon the intensity alone will meet the new demands on optical inspection techniques. Since roughness, as spatial noise, has an associated randomness, fluctuations in the measured scattered light intensity prohibit detection of the smallest particles. Furthermore, since real-life particles are not perfect spheres, the directional distribution of light scattered by them will not necessarily replicate that of perfect spheres. Lastly, given the number of different types of defects that exist on wafers, an instrument designed with a small number of light collection directions that are optimized to enhance sensitivity to one class of defect in comparison to another will lack the discrimination capabilities to tell them all apart.

The polarization of scattered light can often yield significant information about the source of that scattered light[1,2]. Recent measurements have demonstrated that the light scattered by microroughness has a well-defined polarization that is independent of the microscopic details of the roughness[3]. Furthermore, other scattering sources, such as particles above the surface or defects below the surface give rise to polarizations that differ from those predicted by microroughness.

By using polarization-sensitive detectors, a system can be built which is insensitive to a particular scattering source[4]. Since these scattering sources yield well-defined polarizations in all scattering directions, the question of what optical geometry is most sensitive to different types of defects becomes moot since light over the entire scattering hemisphere can be utilized.

In this paper, I review the theory of polarized light scatter and present data and theoretical predictions to demonstrate the validity of the models and to illustrate the behavior of those theories. A specific design for a microroughness-blind instrument is presented, and the improvements in particle and defect detection limits are calculated.

THEORY

Theories for scattering from particulate contaminants and subsurface defects in the Rayleigh approximation and from microtopography have been developed elsewhere[1,5,6]. Each of these theories predicts a closed form expression for the Jones scattering matrix,

$$\begin{pmatrix} S_{ss} & S_{ps} \\ S_{sp} & S_{pp} \end{pmatrix} \qquad (1)$$

which relates the scattered electric field to the incident electric field. The p and s linear polarization states are defined such that the electric field is parallel and perpendicular, respectively, to the plane defined by the surface normal and the direction of propagation.

Within the Rayleigh approximation, where the size of the scatterer is much smaller than the wavelength of the light, a scatterer may be treated as a point polarizable dipole. The induced dipole moment is proportional to the local electric field, and it radiates locally in each direction with an amplitude and polarization determined by the projection of the dipole moment onto the plane perpendicular to the direction of propagation. The propagation of light to a detector must include the relevant reflections and

CP449, *Characterization and Metrology for ULSI Technology: 1998 International Conference*
edited by D. G. Seiler, A. C. Diebold, W. M. Bullis, T. J. Shaffner, R. McDonald, and E. J. Walters
1998 The American Institute of Physics 1-56396-753-7/98/$15.00

refractions that occur at each surface. Using this approximation, the polarization of light scattered by a particle above a surface, a defect below a surface, or a defect within a dielectric layer can be readily calculated[1,7].

The light scattered by small degrees of roughness has been calculated for a single interface[5], or by one or more interfaces in a stack of dielectric layers[6], using first-order vector perturbation theory. The solution to the roughness problem can be shown to be similar to that for defects with a small difference: the light scattered by the roughness of a particular interface is equivalent to dipoles generated by the electric field above that interface, which then radiate from below the interface (or vice versa)[8]. The polarization of the light scattered by a single microrough interface into a specific direction is only a function of the optical constant of the material and not of the power spectrum of the roughness. When the roughness of multiple interfaces contribute to scattered light, then the polarization also depends upon the correlation functions of the roughness between the different interfaces.

For scattering in the presence of a single interface, the largest contrast between particles, defects, and microroughness occurs when p-polarized light is incident on the sample at an oblique angle. In this case, the electric field inside the material has a different direction than that immediately outside the material. That is, for a high index material, such as silicon, the electric field just outside of the material is nearly perpendicular to the surface, while the electric field inside the material is nearly parallel to the surface. A detector viewing the polarization out of the plane of incidence effectively observes the direction of the dipole moment induced in the defect. For s-polarized incident light, the electric field direction does not depend upon location, resulting in a lack of sensitivity of the scattered light polarization to the location of the defect.

EXPERIMENT

Measurements of the polarization of light scattered by a sample given a fixed incident polarization, herein referred as bidirectional ellipsometry, have been measured using a goniometric optical scatter instrument[9]. Continuous wave p-polarized light of wavelength λ is allowed to be incident upon each sample at an angle θ_i. Light scattered into a polar angle θ_s and azimuthal angle ϕ_s (defined with respect to the plane of incidence) is collected with a polarization-analyzing detector. The results are presented in terms of a principle angle of polarization, η, and a degree of linear polarization, P_L, as functions of ϕ_s for fixed θ_i and θ_s. The angle η is measured with respect to s-polarization in a counterclockwise fashion looking into the beam. The parameter P_L lies in the range $0 \le P_L \le 1$, with $P_L = 1$ indicating linearly polarized light, and $P_L = 0$ indicating either circularly polarized light, or completely depolarized light.

The uncertainties in P_L tend to be dominated by random sources, including electronic noise and laser speckle. When $P_L = 1$, the value of η has a typical uncertainty determined by electronic noise and the alignment and quality of the optical elements. Although a complete discussion of these uncertainties is beyond the scope of this paper, the expanded uncertainties (with a coverage factor of $k = 2$) of η and P_L are not expected to exceed 5° and 0.05, respectively[9].

RESULTS AND DISCUSSION

Figure 1 shows bidirectional ellipsometry results for four silicon samples at 532 nm: two photolithographically generated microroughness standards (R1 and R2), the rough backside of a silicon wafer (R3), and a wafer exhibiting a high density of crystal originated particles (COP). The three rough samples (R1, R2, and R3) each scatter light at large angles by amounts differing by about two orders of magnitude from the next. The values of η for the rough samples agree very well with the microroughness model. That this agreement is so good even for the silicon wafer backside is surprising since the small-amplitude assumptions of the model are violated.

The polarization of the light scattered by the COP sample deviates from that of the microrough samples to a significant degree. In fact, the data seem to lie much closer to the subsurface defect model, in agreement with the current understanding that COPs are coalesced vacancies below the surface[10]. That there exist deviations of η from the subsurface defect model indicates that the model is incomplete, perhaps because the finite sizes and shapes of the defects have not been accounted for, or because a second source of scatter is interfering with the measurement.

The theory for scattering due to microroughness predicts the degree of linear polarization to be $P_L = 1$ for all of the

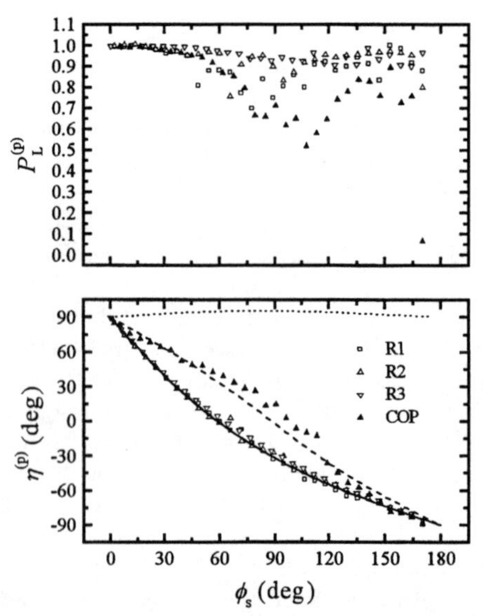

FIGURE 1 Bidirectional ellipsometry parameters for four different silicon samples for p-polarized incident light. The incident and scattering polar angles were $\theta_i = \theta_s = 45°$. The curves in the lower frame represent the predictions of the microroughness (solid), subsurface defect (dashed), and particle (dotted) models.

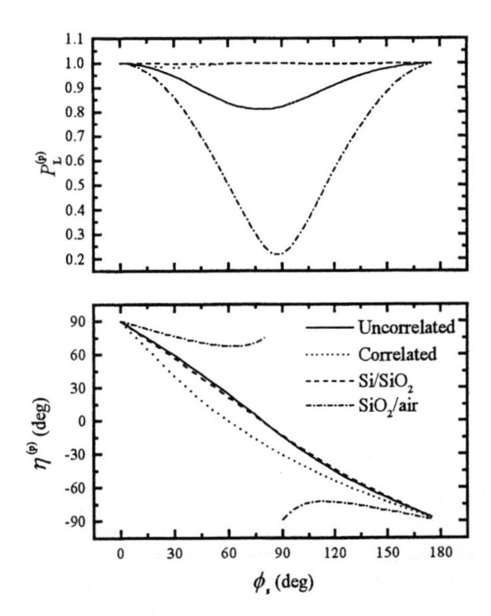

FIGURE 2 Predicted bidirectional ellipsometry parameters for scattering of p-polarized light from microroughness of a silicon wafer with a 5 nm thick oxide.

directions shown in Fig. 1. The data shows reasonable agreement with this prediction for most of the rough samples shown in Fig. 1, where most of the data has $P_L > 0.9$. The lowest scatter sample has randomly deviating data for directions near $\phi_s = 90°$, which may be an artifact associated with the low signals and scatter elsewhere in the laboratory. The values of P_L for the COP sample has a marked deviation from unity for directions $\phi_s > 90°$.

The particle scattering model, which is also shown in Fig. 1, yields a significantly different behavior from those of microroughness and subsurface defects. Since the electric field close to, but outside of, the surface is nearly perpendicular to the surface, the light is expected to scatter with nearly p-polarization ($\eta \sim 90°$).

The scattering from roughness in the presence of a dielectric layer becomes more complex due to the interference between the two interfaces. When more than one interface is rough, then one has effectively multiple sources of scatter. These sources can scatter coherently or incoherently, depending upon the correlation between the roughness of the interfaces. Figure 2 shows the effect that a 5 nm oxide layer can have on the bidirectional ellipsometry parameters. Shown are calculations for roughness from each interface alone and for correlated and uncorrelated roughness. There is a possibility that some of the measured deviations of P_L from unity observed for the microrough sample in Fig. 1 result from the existence of an uncorrelated native oxide.

NEW TOOLS

The finding that microroughness gives rise to scattering with a high degree of polarization in every direction, different from that of other scattering mechanisms, suggests that tools can be developed which are effectively blind to microroughness[4] These tools do not need to operate under the assumption that a specific detector position is ideal for a specific type of defect. In fact, light scattered into every direction can be collected and discriminated, allowing the elimination of specific scattering sources from the signal, without seriously sacrificing sensitivity to other defects.

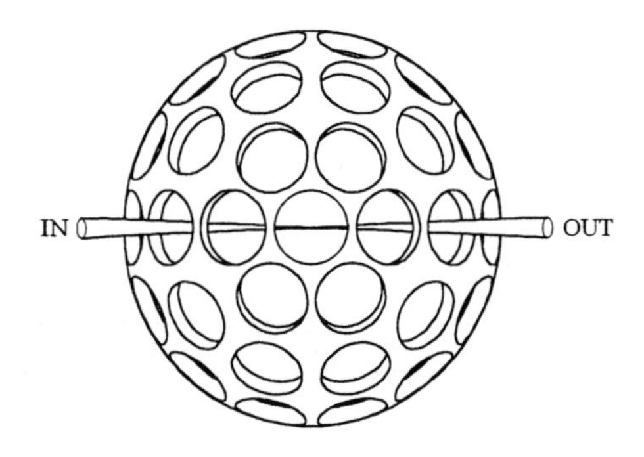

FIGURE 3 Example optical scatter instrument having twenty eight ports, each with a lens, polarizer, and a detector. Three extra ports are provided for the incident beam, the specular beam, and for support.

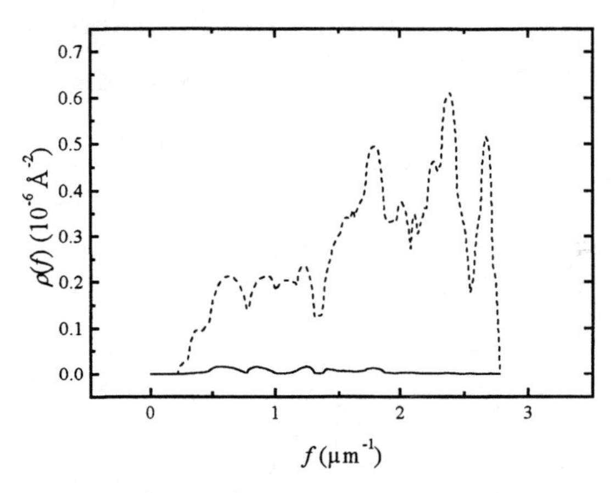

FIGURE 4 The microroughness response function for the instrument shown in Fig. 3. The solid line represents the response function for the system having polarizers aligned in a microroughness-blind configuration. The dashed line represents the response of the system when it does not have the polarization sensitive elements. The light has wavelength 633 nm, and is incident at an angle of 49° with p-polarization.

Such an instrument is illustrated in Fig. 3. A hemispherical shell contains 31 ports, with central polar angles of 0°, 24°, 49°, and 74°, each spanning a half-angle of 9.5°. Three ports are dedicated to the incident beam, the specular beam, and mechanical support (normal). Each of the other 28 ports holds a collection system with a lens, a polarizer, and a detector. All of the signals are assumed to be summed in this discussion. In this section, I will analyze the predicted performance of this system with and without polarizers.

A methodology has been developed to describe the sensitivity of an optical scatter instrument to micro-

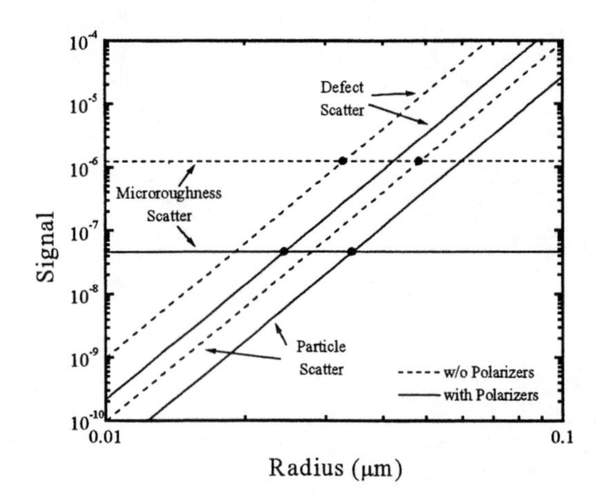

FIGURE 5 The calculated signal for particles as a function of radius for the system shown in Fig. 3 with and without polarization-sensitive detection. The horizontal lines correspond to the microroughness-induced scatter. The intersections of the microroughness/particle scattering functions are marked, indicating the decrease in particle size that can be detected with polarization-sensitive detection.

roughness[11]. A response function, $\rho(f)$, can be defined, so that the roughness-induced signal measured by an optical scatter instrument, H, is

$$H = \int_0^\infty df\ 2\pi f\ \rho(f)\, S(f), \qquad (2)$$

where $S(f)$ is the two-dimensional power spectral density (PSD) function of the surface height function. Figure 4 shows the function $\rho(f)$ for the scatter instrument illustrated in Fig. 3, with and without polarizers aligned to reduce microroughness-induced signal. Since each detector collects light over a finite solid angle, and thus the polarizer is only optimized for the center of the aperture, the extinction of microroughness-induced signal is not perfect.

It is possible to make an estimate of the minimum detectable particle size assuming the models for microroughness-induced and particle-induced scatter. For this calculation I will make an assumption that a typical roughness PSD function is given by

$$S(f) = 0.4\,(f \times \mu\text{m})^{-2.5}\,\text{Å}^2\ \ \mu\text{m}^2. \qquad (3)$$

The fractional standard deviation of the noise associated with laser speckle can be estimated to be

$$\sigma_H / H = (\Delta x\, \Delta f)^{-1/2} \approx 0.2, \qquad (4)$$

where Δx is the laser spot size and Δf is the instrument bandwidth from Fig. 4. The detection level for particles or defects is typically set about 5 times the standard deviation of the background noise to maintain the false count rate at a negligible level. Figure 5 shows the results of these calculations. I have assumed that the particle has an index of refraction of n = 1.5, a defect has an index of refraction of n = 1, and the laser spot illuminates a region of area A = 100 μm^2. The absorption depth of the substrate has been ignored, so the results only apply to a subsurface defect very close to the surface. Some of the signal from a defect or a particle is reduced by the addition of polarizers, however to a lesser degree than that for microroughness.

The intersection of the scatter curves from each mechanism yields an estimate of the particle or defect radius that can be detected above the noise of the microroughness-induced scatter. It can be seen from Fig. 4 that the use of the detection scheme shown in Fig. 3 should yield a factor of about 1.4 improvement in the detectable particle or defect radius. Further improvements can be made by making use of the different scattering sources' intensity distributions in addition to their polarizations, that is, by only collecting over select solid angles.

SUMMARY

It has been found that the polarization of light scattered by a surface can reveal the source of scattered light. Results of measurements of the polarization of light scattered by roughness were compared to those of theoretical models, and the agreement is very good. To demonstrate the utility of polarized light scatter techniques, the predicted performance of a particular device was calculated, yielding a factor of approximately 1.4 improvement in the radius of a detectable particle or defect. Although specific designs may yield different levels of improvement, it is expected that these techniques will extend the capabilities of light scattering tools for the inspection of silicon wafers.

ACKNOWLEDGMENTS

The author would like to thank Bradley Scheer of VLSI Standards, Inc., for supplying the microrough silicon samples, and Clara Asmail for many useful discussions.

REFERENCES

1. Germer, T. A., Appl. Opt. **36**, 8798–8805 (1997).
2. Germer, T. A., and Asmail, C. C., Proc. SPIE **3121**, 173–82 (1997).
3. Germer, T. A., Asmail, C. C., and Scheer, B.W., Opt. Lett. **22**, 1284–6 (1997).
4. Germer, T. A., and Asmail, C. C., Provisional U.S. Patent Application Filed (1997).
5. Barrick, D. E., *Radar Cross Section Handbook* (Plenum, New York, 1970).
6. Elson, J. M., J. Opt. Soc. Am. **66**, 682–94 (1976); Appl. Opt. **16**, 2873-81 (1977); J. Opt. Soc. Am. **69**, 48–54 (1979); J. Opt. Soc. Am. A **12**, 729–42 (1995).
7. Germer, T. A., Proc. SPIE **3275**, in press (1998).
8. Kröger, E. and Kretschmann, E., Z. Physik **237**, 1 (1970).
9. Germer, T. A., and Asmail, C. C., Proc. SPIE **3141**, 220–31 (1997).
10. Bender, H., Vanhellemont, J., Schmolke, R., Jpn. J. Appl. Phys. Part 2 **36**, L1217–20 (1997).
11. Germer, T. A., and Asmail, C. C., Proc. SPIE **2862**, 12–7 (1996).

Accurate Size Measurement of Monosize Calibration Spheres by Differential Mobility Analysis

George W. Mulholland and Marco Fernandez

Fire Science Division
National Institute of Standards and Technology
Gaithersburg, MD 20899-0001

A differential mobility analyzer was used to measure the mean particle size of three monosize suspensions of polystyrene spheres in water. Key features of the experiment to minimize the uncertainty in the results include developing a recirculating flow to ensure equal flows into and out of the classifier, an accurate divider circuit for calibrating the electrode voltage, and use of the 100.7 nm NIST SRM™ for calibrating the flow of the classifier. The measured average sizes and expanded uncertainties with a coverage factor of 2 are 92.4 nm ± 1.1 nm, 126.9 nm ± 1.4 nm, and 217.7 nm ± 3.4 nm. These calibration sizes were characterized by NIST to improve the calibration of scanning surface inspection systems.

INTRODUCTION

The 1997 National Technology Roadmap for Semiconductors (NTRS) discusses particles as small as 65 nm in diameter being a concern by year 2003. Polystyrene spheres have been widely used by the wafer scanner manufacturers for calibrating their instruments. There are three impediments to achieving accurate detection and measurement of the Roadmap targeted particle diameters. First, the scanner and test equipment manufacturers have made extensive use of non-certified, non-traceable particles as calibration standards. Second, some of the particles used were incorrectly sized by the particle suppliers (1). Third, improved methods of synthesizing monosize particles smaller than 100 nm are needed.

A joint project of National Institute of Standards and Technology (NIST), SEMATECH, and VLSI Standards, Inc., along with the collaboration of Duke Scientific Corp., has made strides toward developing accurate monosize standards covering the diameter range from 70 nm to 900 nm. In the first phase of this project, the following nominal sizes of polystyrene spheres have been targeted: 72 nm, 87 nm, 125 nm, 180 nm, and 216 nm. Five suppliers provided samples for each of these particle sizes. A differential mobility analyzer (DMA) and a scanning electron microscope (SEM) were used at NIST to select the best sample for each targeted size in terms of mean particle size and width of the size distribution. This study (1) showed that at least half of the size measurements by the suppliers differed by more than 5 % from the DMA size, with two differing by 20 % or more. This demonstrates that accurate measurement of particle size is an issue for particle sizes less than 250 nm.

As a first step it was decided to perform accurate sizing measurements on three of these five particle sizes. The preliminary particle size characterization based on the screening measurements for the peak in the size distribution and the half-width at half-height were as follows: 93 nm, 4.2 nm; 130 nm, 2.5 nm; and 224 nm, 4.0 nm. All three materials were made by Duke Scientific Corporation[1], which specified diameters of 89 nm ± 3 nm, 126 nm ± 3 nm, and 220 nm ± 6nm, using transmission electron microscopy with NIST Standard Reference Materials as internal standards (2).

These measurements were performed using a high ratio of the sheath flow to aerosol flow (40 to 1) in the DMA to enable the measurement of the width of the size distribution.. In the present study we focus on accurate measurement of the mean particle size. While the high flow ratio is best for high resolution, the accuracy is lower because of flow recirculation and electric field penetration near the inlet. Lower uncertainty in the mean particle size is obtained at lower flow ratios and a 10 to 1 flow ratio is used in this study.

This paper describes the experimental method used for accurately determining the mean particle size with the DMA and provides an uncertainty assessment for the peak particle size for each of the three sizes. Key features of the measurement are the calibration of the DMA using NIST SRM™ 1963 100.7 nm particle standard, the development of an accurate voltage calibration facility, and the experimental design for determining the contributions of repeatability and sample to sample variability on the uncertainty in the peak particle size.

EXPERIMENTAL PROCEDURE

The particle measurement system consists of an aerosol generation system, a differential mobility analyzer (DMA) for size selection, and a condensation nucleus counter for

---

[1] Certain commercial equipment or materials are identified in this paper to specify adequately the experimental procedure. Such identification does not imply recommendation or endorsement by the National Institute of Standards and Technology, nor does it imply that the materials or equipment identified are necessarily the best available for the purpose.

CP449, *Characterization and Metrology for ULSI Technology: 1998 International Conference*
edited by D. G. Seiler, A. C. Diebold, W. M. Bullis, T. J. Shaffner, R. McDonald, and E. J. Walters
1998 The American Institute of Physics 1-56396-753-7/98/$15.00

monitoring the aerosol concentration (see Fig. 1). A brief description of the instrumentation and methodology is given below; a detailed description is given by Kinney *et al.* (3) Several drops (1 drop per 200 cm^3 for 90 nm spheres, 2 drops per 200 cm^3 for 125 nm spheres, and six drops per 250 cm^3 for 220 nm spheres) of the 1 % by mass polystyrene spheres are diluted with de-ionized/filtered (0.2 µm pore size) water. The resulting suspensions are mixed by shaking and by placing the container in an ultrasonic bath for about 60 s. The suspension is nebulized at 107 kPa at gauge (15 psig) to form an aerosol with droplets containing the polystyrene spheres. The water evaporates as the aerosol flows at 83 cm^3/s (5 L/min) through a diffusion drier and then mixes with 40 cm^3/s of clean, dry air. The polystyrene aerosol is initially highly charged from the nebulization process and is "neutralized" with a bipolar charger so that the largest fraction of the particles have no charge, about 20 % have a +1 *e* and another 20 % have a -1 *e* charge, and much smaller fractions of multiply charged particles. It is the +1 *e* polystyrene spheres that are measured by the DMA.

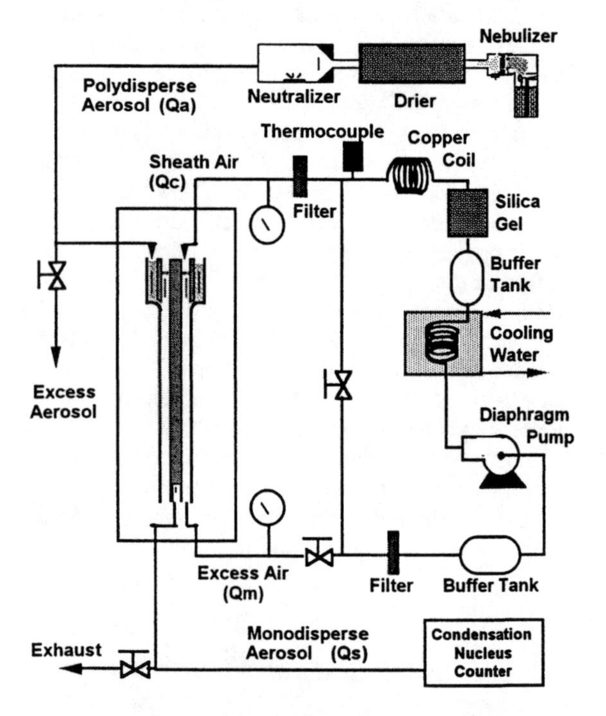

FIGURE 1. The particle measuring system includes an aerosol generation system (nebulizer), the DMA, and the condensation nucleus counter. The excess air is recirculated to match the sheath flow in this closed loop operation of the DMA.

The DMA consists of an inner cylinder rod connected to a variable (0 V to -11000 V) DC power supply and an outer annular tube connected to ground. Clean sheath air flows through the axial region while the charged aerosol enters through an axially symmetric opening along the outer cylinder. The positively charged polystyrene spheres move radially towards the center rod under the influence of the electric field. Near the bottom of the classifying region, a fraction of the flow consisting of near-monosize aerosol flows out of the DMA via a slit in the center rod. The particles next flow to a condensation nucleus counter, where the number concentration is measured. A typical measurement sequence is to measure the number concentration as a function of the voltage.

The quantity measured by the DMA is the electrical mobility, Z_p, defined as the velocity a particle attains under a unit electric field. Knutson and Whitby (4) derived an expression for the average value of Z_p for the particles entering the slit involving the electrode voltage, V, the sheath air flow, Q_c, the inner and outer radii of the cylinders, r_1 and r_2, and the length of the electrode down to the slit, L:

$$Z_p = \frac{Q_c}{2\pi VL} ln(r_2 / r_1) \tag{1}$$

This equation is valid provided the sheath air flow, Q_c, is equal to the excess flow, Q_m, leaving the classifier.

An expression for the electric mobility of a singly charged particle involving the particle diameter is obtained by equating the electric field force with the Stokes Drag force,

$$Z_p = \frac{eC(D_p)}{3\pi\mu D_p} \tag{2}$$

where μ is the air viscosity and e is the electron charge. The Cunningham slip correction $C(D_p)$, which corrects for the non-continuum gas behavior on the motion of small particles, is given by

$$C(D_p) = 1 + K_n[A_1 + A_2 \exp(-A_3 / K_n)] \tag{3}$$

where the Knudsen Number is twice the mean free path in air divided by the particle diameter ($K_n=2\lambda/D_p$) with A_1=1.142, A_2=0.558, and A_3=0.999 (5). For a measured value of Z_p, the particle diameter, D_p, is obtained iteratively from Eqs. (2) and (3).

In this study we operated the DMA with a flow ratio of 10 with a sheath air flow, Q_c, of 167 cm^3/s (10 L/min), and a aerosol inlet flow, Q_a, at a flow of 16.7 cm^3/s. For Eq. (1) to be valid, the sheath air and excess air flows, Q_c and Q_m, must be equal. As illustrated in Fig. 1, the excess air was circulated back into the inlet for the sheath air, assuring, in principle, that the flows were matched. The recirculating system includes the following components: two small pumps and buffer tanks before and after the pumps to minimize pulsation, cooling water and an ambient heat exchanger to cool the recirculated air, a filter to remove the particles, silica gel to remove water vapor, and a thermocouple. The measured leak rate was less than 0.017 cm^3/s or 0.01 % of the sheath air flow at a sheath air flow of 167 cm^3/s. This is to be compared with about a 1 % to 2 % difference in Q_c and Q_m using high resolution flow meters without flow recirculation. The nominal 167 cm^3/s flow of the sheath and excess air is set by adjusting the two valves indicated in Fig. 1 external to the classifier so that the classifier flow meter reads 167 cm^3/s. An automated soap-film flowmeter (Gilibtator-2) was plumbed into the system before the first valve to more accurately determine the flow. The temperature of the sheath air was monitored with a thermocouple to 0.1 °C, the ambient pressure to 13 Pa (0.1 mm Hg) using a Hg column barometer, and the pressure difference between ambient and the pressure in the classifier to the nearest 10 Pa (0.1 cm H_2O). The pressure and temperature in the classifier are needed for computing the air viscosity and the slip correction.

A typical experiment consists of starting the nebulizer, setting the voltage, collecting number concentration data for 45 seconds, then repeating the same process for a total of seven increasing voltages. Then the same sequence is repeated in reverse going from the highest to the lowest voltage. From a preliminary experiment, the voltage for the peak number concentration would be determined along with the voltages corresponding to about 80 %, 50 %, and 30 % of the peak concentration. For each voltage setting the concentration reading is obtained from the last 20 s of the interval to ensure particles classified at the previous voltage have exited the nucleus counter. The number concentration data is recorded with a PC.

A measurement sequence consists of measuring the voltage versus number concentration for all four particle sets at one time. A total of about 1.5 hour is required for this. This procedure is important, because the SRM particle is used to calibrate the other measurements as described below. Typical data sets for each of the four particle sizes are shown in Fig. 2. As the particle size increases from 90 nm to 220 nm, the peak voltage increases from about 1400 V to about 5900 V.

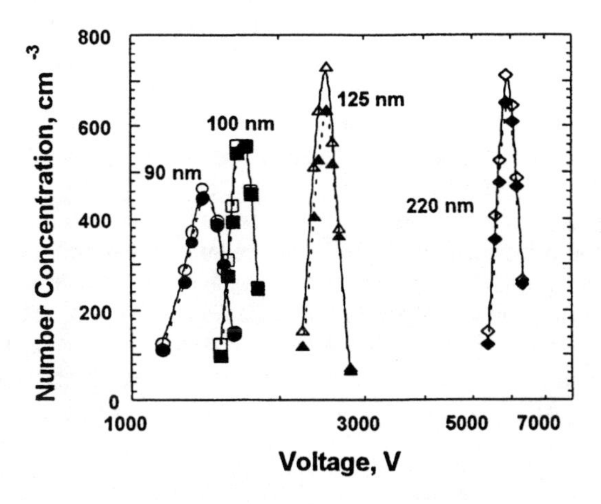

FIGURE 2. Plot of the number versus the voltage setting of the DMA for the three calibration particles and the 100 nm SRM.

A total of nine sets of data were collected for each particle size. For each particle size, the vendor supplied 3 samples from the batch of particles that were synthesized. In turn, for each of these three samples, three sets of measurements were made. By using this experimental design, both the sample to sample variability and the repeatability of the measurements could be assessed.

DATA ANALYSIS

The 100 nm NIST SRM1963 particles (actual size, 100.7 nm) are used to calibrate the DMA (6). First, the electrical mobility corresponding to the average voltage for the 100 nm SRM spheres is determined using Eq.(1) and the measured values for the sheath flow Q_c, the length of the classifier L, and the inner and outer radii r_1 and r_2. The electrical mobility is also computed for 100 nm SRM using Eqs. (2) and (3) based on the actual 100.7 nm diameter. Ideally these two values should be the same. If the measured value differs from the computed value, the value of the sheath flow in Eq. (1) is varied until the measured and computed mobilities are equal.

In some cases the average voltage is slightly different between the up-scan and the down-scan due to a slight drift in the aerosol generator. The particle concentration drifts downward on the order of 10 % over a 20 minute period. The difference in the peak voltage is typically 0.1 % or less with the largest difference being 0.3 %. In our analyses we use the average of the up-scan and down-scan results.

As described above, a corrected flow velocity is computed for each of the nine sets of experiments with each set consisting of the three unknown particle sizes and the 100 nm SRM. For each particle size, Eqs. (1) and (2) are used to compute the particle diameter corresponding to each of the seven voltages. Typical size distributions are plotted in Fig. 3 for the two smaller particle sizes.

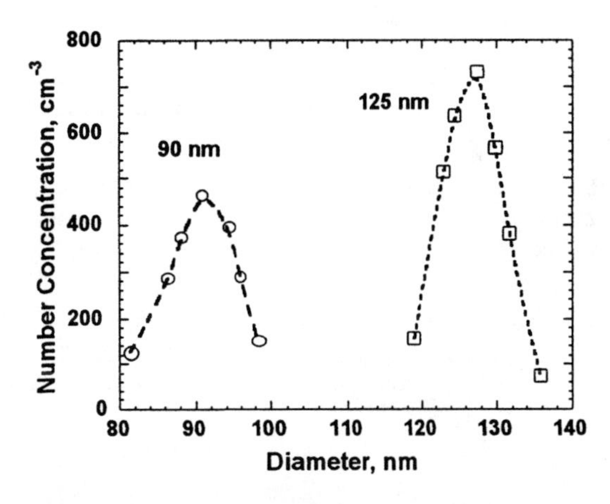

FIGURE 3. The number concentration versus voltage plots of Fig. 2 are expressed in terms of particle diameter for two smallest calibration particle sizes.

The peak diameter for the particle size distribution is computed for each distribution. The first three points and the last three points are fitted with cubic polynomials with the requirement that the 1st and 2nd derivatives be continuous at the middle point where the two polynomials meet We have also computed the number average diameter and obtained results in close agreement to the value given by the peak diameter.

UNCERTAINTY ANALYSIS

The components of uncertainty are divided into two categories: Type A are those evaluated by statistical methods and Type B are those evaluated by other means (7,8). These types correspond to random and systematic effects. For the particle sizing measurements, the Type A uncertainty is determined from the measurement repeatability and sample variability. The Type B uncertainty includes the uncertainty in the 100 nm SRM and the uncertainty in the various physical quantities appearing in Eqs. (1-3).

Type A Uncertainty

Two components contribute to the Type A uncertainty. One is the homogeneity of the sample: that is, sample-to-sample variability. The second is concerned with the measurement repeatability. Both of these components were obtained by making three repeat measurements on each of three different samples. The results are displayed in Fig. 4. It is seen that all the results are within about 1 % of the average of the peak diameters with the exception of one apparent outlier for the nominal 90 nm particle size. It is evident that the spread in the results are somewhat greater for the 90 nm particles relative to the larger sizes.

The analysis of variance shows no significant difference among diameters from the three different samples from each particle size. This finding translates into a between-sample deviation of zero.

The mean of the nine values of peak diameter, D_{peak}(avg), and the standard deviation of the mean, σ(A), are computed for each of the three particle sizes with the following results: **92.4 nm, 0.30 nm; 126.9 nm, 0.12 nm; 217.7 nm, 0.21 nm.** The standard deviation σ(A) is the total type A uncertainty.

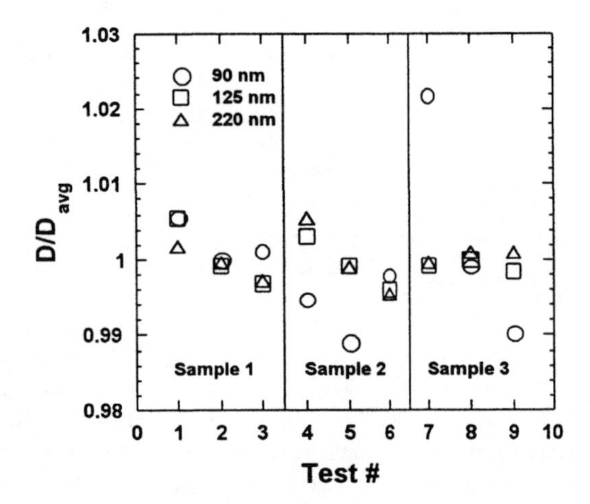

FIGURE 4. The ratio of the measured peaks in the size distribution to the average for each particle size are plotted to show the repeatability and sample to sample variability.

Type B Uncertainty

For five of the uncertainty components - voltage, the calibration particle size, pressure, temperature, and viscosity - the uncertainty analysis is straightforward. First an estimated standard deviation of the variable is obtained and then the resulting effect of changing the variable by one standard deviation on the particle diameter is obtained using Eqs. (1-3). Less direct analysis is required for slip correction, DMA resolution, and data analysis methodology. Here a brief account of the analysis is presented and the unique features of the study described. A more detailed description of the uncertainty analysis is contained in the paper on the 100 nm SRM (6).

Voltage Calibration

Because errors in the DMA voltage readings were observed in the range of 1 % to 3 % in a previous study (3), a high voltage (1000 V to 10000 V) calibration facility was set up. The facility consists of a high voltage divider and a digital voltmeter. The high voltage divider (Spellman High Voltage Electronics Corp., Model HVD-100-1) has a standard deviation equal 0.05 % of the nominal reading. The 10000 to 1 divider output was used resulting in output voltages to the digital voltmeter in the range of 0.1 V to 1.0 V. The digital voltmeter (Fluke, 8060A) also has a standard deviation of 0.05 % of the nominal reading over this range. Combining in quadrature the two standard deviations leads to a total uncertainty for the voltage calibration of 0.071 % of the reading. In computing the total uncertainty associated with the voltage measurement, the 1.0 V resolution of the DMA voltmeter is also included.

The effect of the change of voltage on particle size is determined via the particle mobility equations. As explained in more detail by Mulholland *et al.* (6), the voltage uncertainty affects the measurement of the unknown particle size directly but also indirectly through the calibration measurement of the 100 nm SRM. A change in the voltage for the calibration measurement will affect the corrected flow which will, in turn, affect the measured particle size. Table 1 contains the estimated voltage uncertainty and the resulting uncertainty in the particle size measurement.

Particle Standard

The 100 nm SRM has a combined uncertainty of 0.47 nm (6) and this uncertainty has the largest effect on the overall uncertainty in the calibration particles. Changing the diameter of the 100 nm calibration particle changes the corrected flow, which, in turn, affects all of the derived particle sizes. As seen in Table 1, the effect is about 0.5 % for each of the three sizes.

Pressure, Temperature, and Viscosity

The uncertainty in the pressure affects the mean free path λ, which in turn affects the slip correction. The 4×10^3 Pa uncertainty in the pressure results in changes of 0.13 % to 0.17 % in the three particle diameters. The temperature and viscosity uncertainties, listed for completeness, have a negligible effect on the overall uncertainty.

Slip Correction

The effect of the uncertainty in the slip correction on the particle size is subtle. As seen from Eqs. (2) and (3), the value of the slip correction affects the particle size, but the particle size also affects the slip correction. Two separate effects of uncertainty associated with the slip correction have been analyzed. The first, listed as Slip Correction A, is a result of the uncertainty in the constants A_1, A_2, and A_3 as determined by Allen and Raabe (5). The second, Slip Correction B, is the larger of the two and results from assessing the effect of using two different expressions for the slip correction (5,9). For the 200 nm particle size, the effect of the slip correction uncertainty (0.5 %) is as large as the effect of the 100 nm SRM.

DMA Resolution/Data Analysis

There are two issues regarding the data analysis that are discussed here even though their impact on the uncertainty analysis is negligible. First, the size distribution output of the DMA is broadened relative to the true size distribution. Secondly, there is a possibility that the peak size would be shifted by a change in the voltages selected for the analysis. Both of these effects were estimated using the DMA transfer function (4,6) and assuming Gaussian size distributions for the 3 calibration particles. In one set of calculations the voltages were also adjusted by 20 V to 50 V. As shown in Table 1, the largest effect was only 0.06 % of the particle size.

Computation of Total Uncertainty

The total Type B uncertainty, σ(B), is obtained as the root-sum-of-squares of the individual standard deviations. The total Type A and Type B uncertainty are also combined as a root-sum-of-squares to obtain the combined uncertainty, $u(D_{peak})$. The expanded uncertainty $U(D_{peak})$, defined such that there is an approximately 95 % level of confidence that the true average peak diameter is within ± $U(D_{peak})$ of the measured average, is calculated as $2u(D_{peak})$.

CONCLUSION

The values of the average peak diameters D_{peak}(avg) and the associated expanded uncertainty $U(D_{peak})$ are the following: **92.4 nm ± 1.1 nm, 126.9 nm ± 1.4 nm, and 217.7 nm ± 3.4 nm**. We believe that these three sizes together with the 100 nm SRM are the most accurately characterized particles in the size range less than 250 nm. The use of these materials together with other commercially available size standards based on these materials is expected to greatly improve the reliability of scanner calibrations.

Table 1. Uncertainties of Nominal 90 nm, 125 nm, and 220 nm Calibration Particles

Variable y, nominal value	**σ(y)**	**σ(90), nm**	**σ(125), nm**	**σ(220), nm**
Voltage				
1400 V	1.4 V	0.05		
2520 V	2.0 V		0.06	
5900 V	4.3 V			0.11
1650 V	1.5 V	0.05	0.06	0.11
Pressure				
1.0053×10^5 Pa	4000 Pa	0.16	0.21	0.29
Temperature				
22.0 °C	0.5 °C	0.02	0.01	0.04
Viscosity				
1.8277×10^{-5} Pa • s	7.3×10^{-9} Pa • s	0.02	0.03	0.06
100.7 nm SRM	0.47 nm	0.43	0.63	1.20
Slip Correction A		0.07	0.11	0.17
Slip Correction B		0.05	0.20	1.16
DMA Resolution		0.06	0.04	0.09
Spline Fit		0.01	0.06	0.05
σ(B), total class B uncertainty		0.47	0.71	1.71
σ(A), total class A uncertainty		0.30	0.12	0.21
$u(D_{peak})$, combined uncertainty		0.56	0.72	1.72
$U(D_{peak})$, expanded uncertainty		1.12	1.44	3.44

ACKNOWLEDGEMENTS

The authors thank SEMATECH for its financial support and for its technical support from Howard Huff and Randal Goodall.

REFERENCES

1. Mulholland, G.W., Bryner, N.P., Liggett, W., Scheer, B.W., Goodall, R.K., *Proceedings of the Society of Photo-Optical Instrumentation Engineers* **2862**, 104-118 (1996).
2. Duke, S.D. and Layendecker, E.B., "Internal Standard Method for Size Calibration of Sub-Micron Spherical Particles by Electron Microscopy," presented at Fine Particle Society (1988). Available from Duke Scientific Corp.
3. Kinney, P.D., Pui, D.Y.H., Mulholland, G.W., and Bryner, N.P., *J. Res. Natl. Inst. Stand. Techno* **96**, 147-176 (1991).
4. Knutson, E.O. and Whitby, K.T., *J. Aer. Sci.* **6**, 443-451 (1975).
5. Allen, M.D. and Raabe, O.G., *Aerosol Sci. Tech.*. 4, 269-286 (1985).
6. Mulholland, G.W., Bryner, N.P., and Croarkin, C., submitted to *Aerosol Sci. Tech.*.
7. Taylor, B.N. and Kuyatt, C.E., "Guidelines for Evaluating and Expressing the Uncertainty of NIST Measurement Results," NIST Technote 1297 (1994).
8. International Organization for Standardization, "Guide to the Expression of Uncertainty in Measurement," Geneva Switzerland (Corrected and reprinted 1995).
9. Hutchins, D.K., Harper, M.H., and Felder, R.L., *Aerosol Sci. Technol.* **22**, 202-218 (1995).

One Step Automated Unpatterned Wafer Defect Detection and Classification

Lie Dou, Daniel Kesler, William Bruno, Charles Monjak, and Jim Hunt

ADE Optical Systems, Charlotte, NC 28273, USA

Automated detection and classification of crystalline defects on micro-grade silicon wafers is extremely important for integrated circuit (IC) device yield. High training cost, limited capability of classifying defects, increasing possibility of contamination, and unexpected human mistakes necessitate the need to replace the human visual inspection with automated defect inspection. The Laser Scanning Surface Inspection Systems (SSISs) equipped with the Reconvergent Specular Detection (RSD) apparatus are widely used for final wafer inspection. RSD, more commonly known as light channel detection (LC), is capable of detecting and classifying material defects by analyzing information from two independent phenomena, light scattering and reflecting. This paper presents a new technique including a new type of light channel detector to detect and classify wafer surface defects such as slipline dislocation, Epi spikes, Pits, and dimples. The optical system to study this technique consists of a particle scanner to detect and quantify light scattering events from contaminants on the wafer surface and a RSD apparatus (silicon photo detector). Compared with the light channel detector presently used in the wafer fabs, this new light channel technique provides higher sensitivity for small defect detection and more defect scattering signatures for defect classification. Epi protrusions (mounds and spikes),slip dislocations, voids, dimples, and some other common defect features and contamination on silicon wafers are studied using this equipment. The results are compared quantitatively with that of human visual inspection and confirmed by microscope or AFM. This new light channel technology could provide the real future solution to the wafer manufacturing industry for fully automated wafer inspection and defect characterization.

INTRODUCTION

Automated defect inspection is becoming more and more necessary as the size of semiconductor devices continues to shrink and the wafer size continues to increase. Although laser-based Scanning Surface Inspection Systems (SSIS) are now being widely used for routine defect inspection of semiconductor wafers, visual inspection (VI) still plays an important role in the wafer manufacturing. This is due to the lack of defect classification capability of conventional SSISs. Conventional SSISs that collect scattered (dark channel) light from flaws on the wafer surface can not provide flaw image and classification information. An enhanced laser SSIS system equipped with a Quad Cell reconvergent specular detection (RSD) apparatus, normally known as light channel, has been developed and studied (1). Light channel monitors the movement and intensity of the reflected laser beam from the wafer surface (figure 1). This presentation describes the techniques and methodology to realize one step fully automated wafer defect inspection and classification using the enhanced SSIS.

THE ENHANCED LIGHT CHANNEL

Flaws such as pits, Epi stacking faults (ESF), dislocation and slips are "killer" defects. These types of defects can insipidly or catastrophically degrade the performance of a device. Identifying these defects and their principal source is very important because they affect the wafer yield and, hence, reduce the manufacture's profit. There are two major disadvantages when using a conventional SSIS without LC capability for defect inspection: (1) in some cases distinguishing different defects is as important as detecting them. For example, Epi mounds and surface particles may cause exactly same dark channel responses; however Epi mounds are yield killer defects while particles can be cleaned through reclean process. (2) Some killer defects do not scatter light due to smooth and shallow surface profiles. This type of defects will affect IC yield more and more as the device feature size become smaller and smaller. The enhanced SSIS using information from both light and dark channels can overcome the above difficulties and successfully detect and classify the killer defects (2).

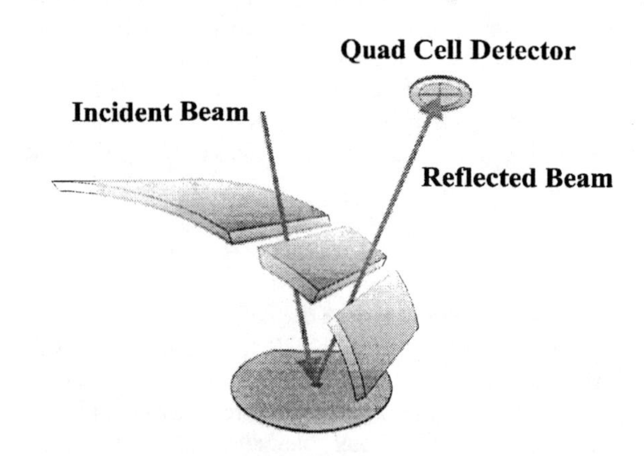

FIGURE 1. Reconvergent specular detection apparatus.

CP449, *Characterization and Metrology for ULSI Technology: 1998 International Conference*
edited by D. G. Seiler, A. C. Diebold, W. M. Bullis, T. J. Shaffner, R. McDonald, and E. J. Walters

The Quad Cell Light Channel Apparatus

The new enhanced SSIS Light Channel incorporates a multi channel silicon photo detector (Quad Cell, figure 2) which is used as the detection source.

The Quad Cell consists of four quadrants that are positioned parallel and perpendicular to the scan line.

FIGURE 2. Quad cell silicon photo detector used as LC detector.

The additional channels provided by the Quad Cell allows surface data to be collected that was not possible with conventional SSISs. Directional shifts of the specular beam can now be detected and quantified. The Quad Cell can detect very small deviations within the wafer surface plane. Small slope changes on the wafer surface cause a change of position of the reflected laser beam focal spot on the Quad Cell. Biasing between the four quadrants of the Quad Cell allows information to be obtained from one or any combination of the quadrants pertaining to these slope changes. Attenuation in the beam intensity can also be detected. Furthermore, the signal to noise ratio of the Quad Cell is much higher than traditional Light Channel detection devices. The result is greater sensitivity to light absorbing and large light scattering defects.

Defect Classification Architecture

Figure 3 shows the data collection and analysis architecture for the enhanced SSIS equipped with Quad Cell light channel. Detected event information from both light channel and dark channel will be sent to a data analysis and defect classification system. This system is based on statistical modeling using real defect samples and mathematical modeling. The classification algorithm will analyze all the DC and LC "signatures" of each event. Then the event will be identified as a certain type of defect. The quantified analysis results are displayed as a bin sorting results. Table 1 lists the characteristic "LC signatures" for typical yield killer defects.

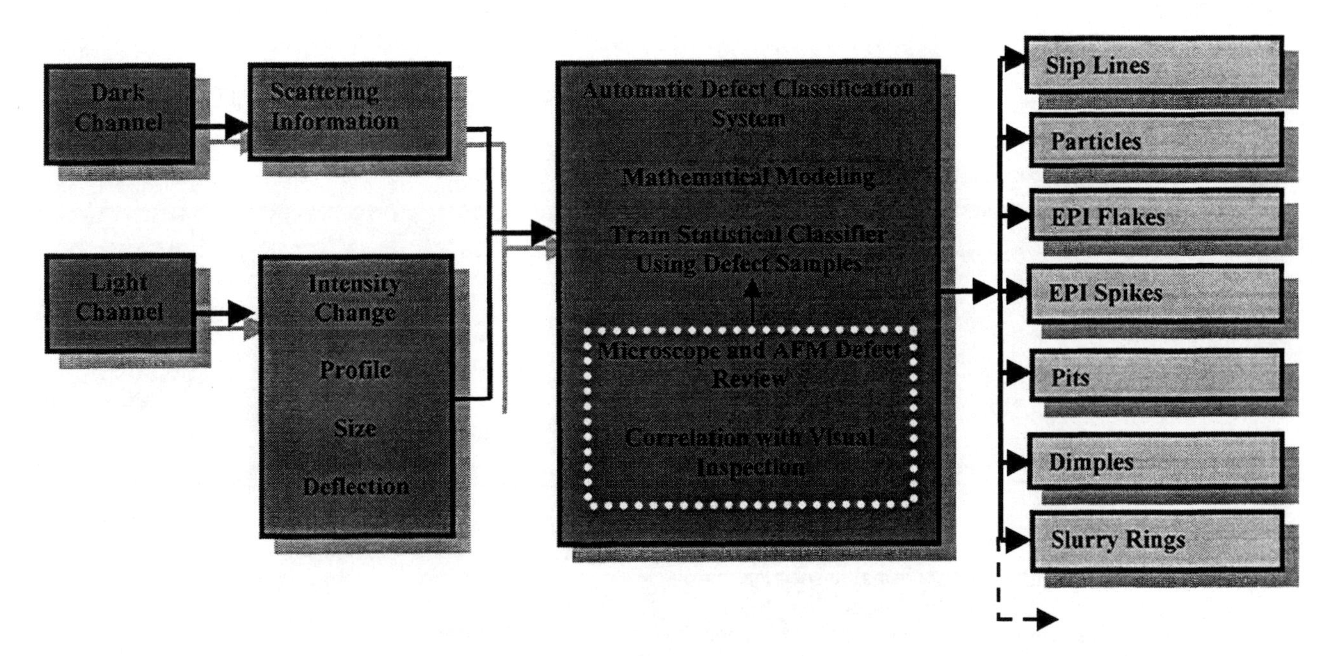

FIGURE 3. Automated defect classification architecture

TABLE 1. Defect Characteristic "Signatures" That Would Cause LC Response or Used For Defect Classification. *DC Responses Are PSL Scattering Equivalent.

Defect Type	Dc Response*	Defect "Signatures"
Slips	weak or no DC	slope, height, directional, start from edge, no light absorption
Big Particles	~0.3 um and up	absorb light, sharp slope change, physical size
Growth Hillocks	~0.7 um and up	regular slope change, defined physical sizes, weak or no light absorption, Protrusion
Pits	strong DC	absorb light, normally regular shape, slope change, caving inward
Dimples	No DC	Large physical size, slow slope change, no light absorption, curving inward.
Epi Stacking Faults (ESF)	Weak DC (depends on ESL's height)	Square in shape (100 surface), regular size (determined by Epi thickness), no absorption.

RESULTS AND DISCUSSIONS

To investigate the sensitivity of the Quad Cell, individual PSL standards ranging from 4-25 um have been measured. Shown in Figure-4 (a) and (b) are the maps and histograms of 4 um and 10 um PSLs.

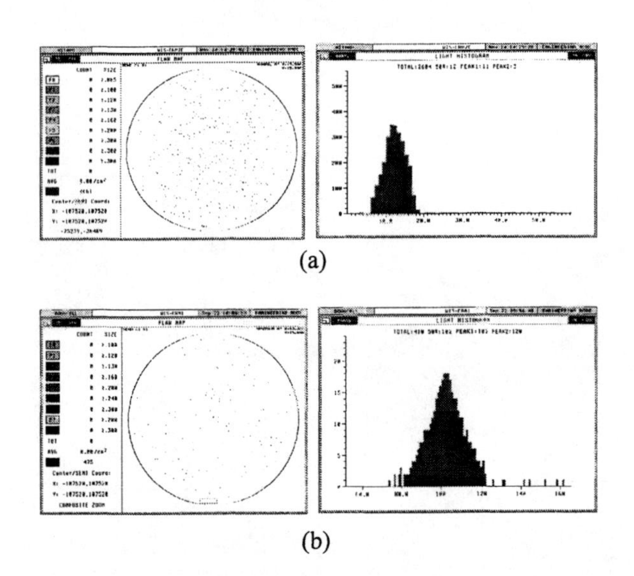

FIGURE 4. Quad Cell light channel maps and histograms of (a) 4 um and (b) 10 um PSLs.

In general, wafers should pass the final inspection by wafer manufacturers and incoming inspection by IC manufacturers before fabrication. However, some of the "killer" defects are extremely difficult to detect using the conventional particle scanner with only dark channel detection. In most cases, to detect this type of defect can also serve another function; Identification of the source of the defects within the process. Shown in figure 5 is a perfect example. These defects are small and very shallow pits. The possible source of the defects is that during the cleaning process the cleaning solution was not completely rinsed off. The solution residue then etched the wafer surface. The light channel map of the whole wafer clearly shows the spin pattern. When this wafer was measured by two laser surface scanners without light channel capability from different manufacturers, There are only 160~180 > 0.16 um particles detected with no pattern at all. The absence of a pattern is the result of no light scattered because these pits are too shallow and their edges are too smooth.

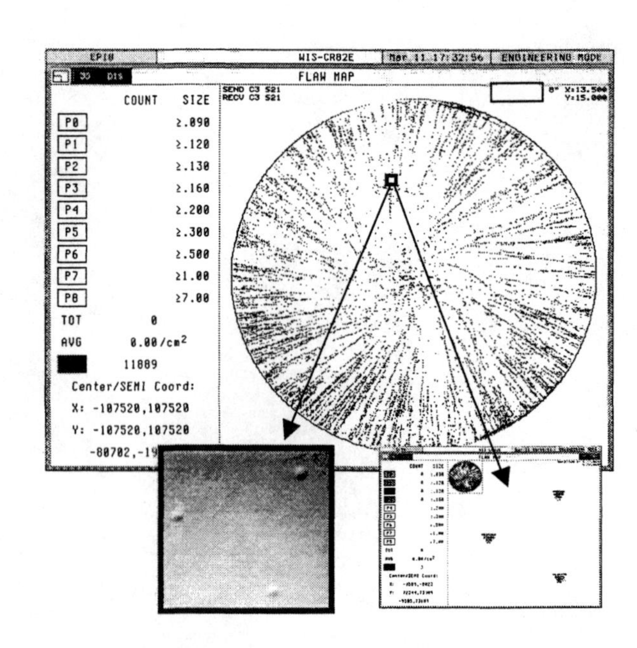

FIGURE 5. Light channel maps of small etch pits: (a) whole wafer, (b) enlarged area of 3 LC events, (c) microscopic picture of the same 3 pits.

Figure 6 shows a LC map of slurry rings. These defects scatter very little light so the dark channel that collects only scattering events was not capable of detecting them.

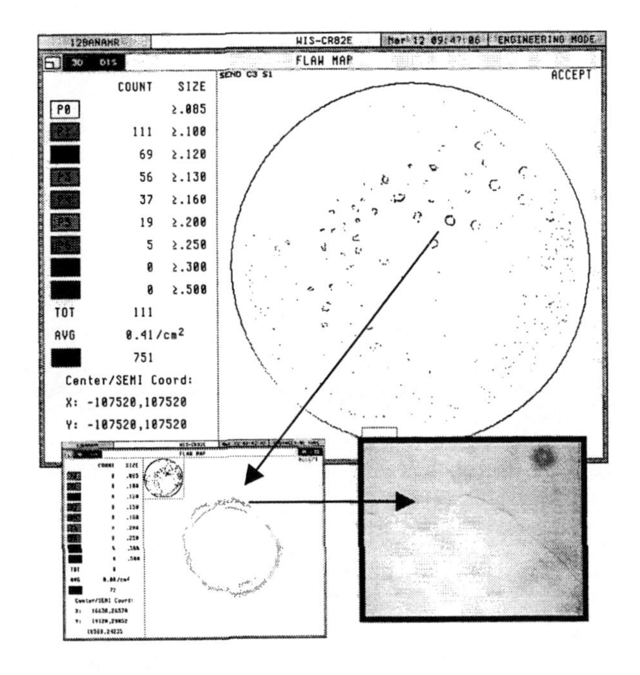

FIGURE 6. Quad Cell LC maps of slurry rings: (a) whole wafer map, (b) enlarged area of a ring, (c) microscopic image of a portion of the ring.

Figures 7-9 show additional defects that can be successfully detected and classified by enhanced SSIS with Quad Cell light channel.

Epi stacking faults are the most commonly observed defects resulting from surface contaminants of the substrates. Shown in figure 7 is an enhanced SSIS LC image of an Epi stacking fault defect on a (100) Epi wafer and the microscopic image of the same defect.

Figure 8a shows the map of slip lines detected by the Quad Cell light channel and 8b is the microscopic inspection result of the same wafer. One different phenomenon is that slip lines do not always trigger a dark channel response. Sometimes the surface steps are too small to scatter light with enough intensity to trigger the dark channel.

Figure 9 is the LC map and the microscopic picture of a pit that may be produced in the wafering process due to the excess hydrogen gas remaining in the molten silicon. Also shown in the figure is the microscopic image of the same pit.

The auto sorting results of the system correlated with that of visual inspection result very well.

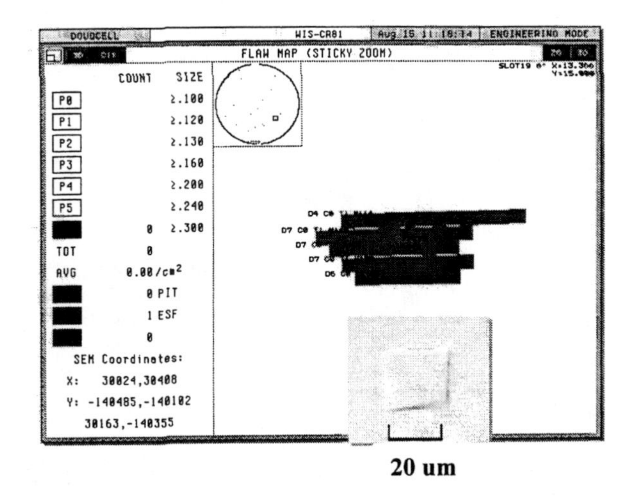

FIGURE 7. Light channel and microscopic images of a stacking fault defect.

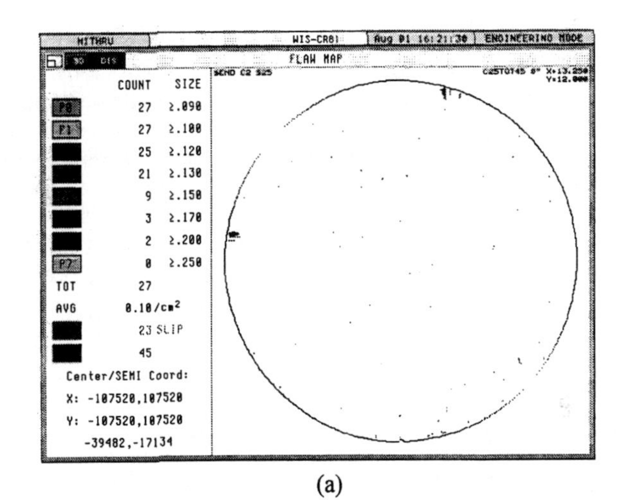

(a)

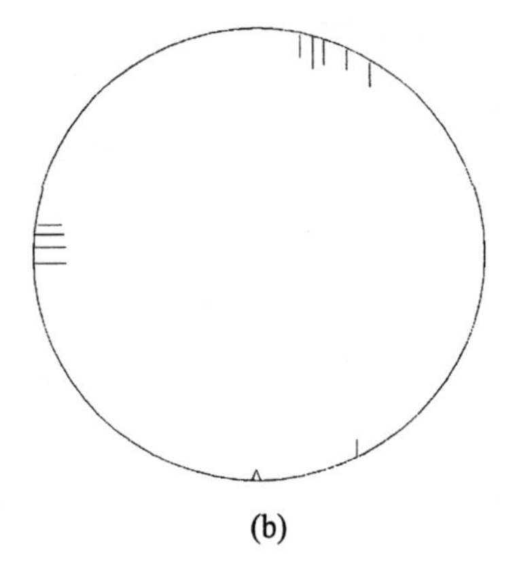

(b)

FIGURE 8. (a) Light channel map and (b) microscopic inspection result of a slipline wafer.

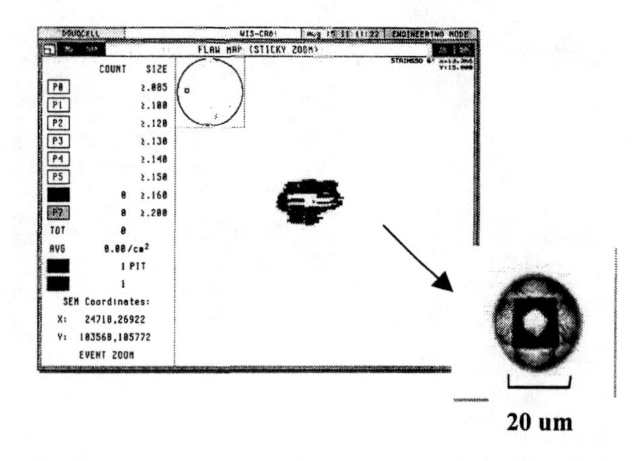

FIGURE 9. Light channel map and microscopic image of a pit.

CONCLUSION

Although there are other systems under development by different manufacturers to realize automatic defect classification, our study shows that the Quad Cell light channel apparatus can simultaneously detect particles, scratches, surface contamination and material defects with consistently precise repeatability while maintaining high throughput. The study results also reveal that the enhanced defect classification system can be used successfully to classify and quantify the surface "killer defects" and distinguish them from particles. It provides the solution for final wafer inspections, as well as the incoming inspection for IC manufacturers for a fully automated wafer inspection system.

REFERENCES

1. ADE Optical Systems, "CR81e Wafer Inspection System Specifications", September 1997.

2. Kesler, D., "CR81e Applied Technology Guide", Vol. 4 March 1998.

Mechanical Characterization of Thin Films*

David T. Read

Materials Reliability Division, National Institute of Standards and Technology, Boulder, CO 80303-3328

This paper describes a tensile testing technique that is applicable to the study of the mechanical properties of thin films. The approach has two main components: fabrication of silicon-framed tensile specimens; and testing these specimens in the microtensile testing apparatus. Speckle interferometry is used to obtain accurate displacements needed for measurement of Young's modulus. Typical results for electron-beam-evaporated films of copper and aluminum are compared to bulk values and to indentation results for thin films.

INTRODUCTION

Motivation

Thin films used in advanced electronic devices are formed by physical vapor deposition (PVD), chemical vapor deposition (CVD) or electrodeposition, their microstructure, and hence their mechanical properties, are often very different from those of bulk materials of the same chemical composition. Thus if one were to simply look up "aluminum" or "copper" in standard handbooks of materials, one would typically find data for the pure, annealed, bulk polycrystalline form, and these data would be wildly inaccurate for thin films. While materials used in electronics manufacturing have generally been carefully characterized electrically, they have not generally been characterized mechanically. Mechanical characterization, however, is becoming necessary because of the extreme mechanical and thermal material performance demanded in today's, and tomorrow's, designs.

Mechanical Testing of Thin Films

Tensile testing was selected because it is capable of producing unambiguous data on the yield strength, ultimate tensile strength, elongation, and Young's modulus. Other approaches, such as indentation, have been explored. Indentation testing, or hardness testing, is used for quality control of bulk production of commercial metals, but is not the method of choice for quantitative measurements. Some correlations between indentation and tensile results are included in this paper.

While the general principles of conventional tensile testing are applicable to thin films, conventional test equipment and techniques are not. Because vapor-deposited films are of the order of 1 μm thick, the failure loads are of the order of tens of milliNewtons, and the specimens cannot be handled directly. These conditions are contrary to traditional tensile testing, where the failure loads are thousands of Newtons and the specimens are handled directly in both fabrication and testing.

Several studies and reviews of the mechanical behavior of thin films have appeared in the literature (1–10). These show the variety of test techniques that have been developed, and results that have been obtained. Measurements of the Young's modulus of thin films have been plagued by controversy, especially the case of the "supermodulus" (11). The history of this controversy is beyond the scope of this paper. However, it shows that comparison of results obtained on various materials and by various techniques is much needed.

The specimens used in this technique have gauge sections that are less than 1 millimeter long. Therefore, the experimental challenge with these specimens is that the elastic displacements are typically 1 to 5 μm. Therefore, small displacements must be measured accurately. Other investigators have used larger specimens successfully (5), thus obtaining larger elastic displacements. However, the difficulties involved in handling and gripping centimeter-sized pieces of vapor-deposited thin films are significant.

An additional, and perhaps more important point about many of the previous techniques is that their specimen preparation involves processing steps incompatible with semiconductor fabrication facilities, such as coating the substrate with sodium chloride. The present techniques avoid the use of sodium chloride or any similar materials, and are intended to be compatible with commercial fabrication facilities.

TECHNIQUE

Specimen Preparation

The specimen design that we have used is shown in Fig. 1. It is a silicon-framed tensile structure that is formed using lithography and wet etching. A silicon wafer, polished on both sides, is oxidized. Rectangular windows are patterned in the SiO_2 on both sides; these will later serve as masks to define the through-wafer silicon etch. Then the front side is coated with the specimen film material of interest. Next the tensile coupons are patterned; the gauge section lies over

CP449, *Characterization and Metrology for ULSI Technology: 1998 International Conference*
edited by D. G. Seiler, A. C. Diebold, W. M. Bullis, T. J. Shaffner, R. McDonald, and E. J. Walters
1998 The American Institute of Physics 1-56396-753-7/98/$15.00

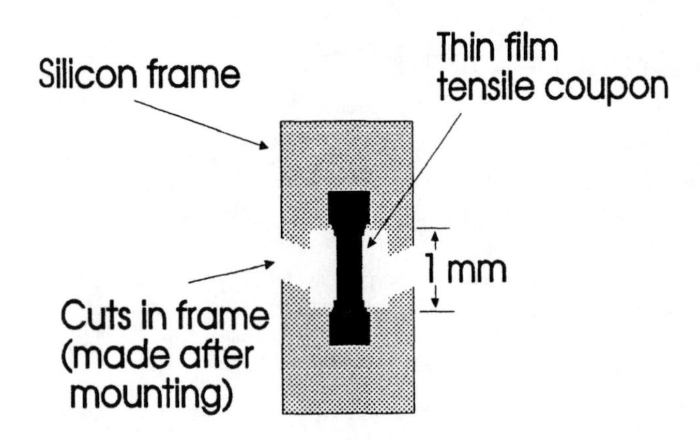

FIGURE 1. Silicon-frame tensile specimen, showing the tensile coupon and cuts made after specimen is mounted in the microtensile tester.

bare Si, and the ends lie over SiO_2 . To date subtractive processing of blanket films has been used, but additive processing would work as well. A bulk micromachining operation to remove the silicon from beneath the gauge sections of the tensile coupons is the final step in the process. The process places three requirements on the specimen material. It must:

1. Be capable of patterning in the shape of tensile coupons over Si and SiO_2;
2. Be resistant to the hydrazine hydrate etchant that is used to remove the silicon; and
3. Have excellent adhesion to SiO_2.

Several different metals, including aluminum, copper, titanium, and nickel have been shown capable of meeting these requirements. Epitaxial silicon specimens were fabricated by a more complicated process. Hydrazine hydrate was selected as the Si etchant because it does not attack Al and stops on SiO_2. Many commercial facilities avoid this material. However, such facilities may have alternative capabilities for etching Si, that would be compatible with this test procedure. One alternative, XeF_2, is being explored at NIST.

Test Apparatus

The test apparatus is shown in Fig. 2. Few piezo-driven microtensile devices have been described, but piezoelectric actuation is convenient for small specimens with limited elongation because the piezos are compact and have no sliding or rotating parts. The tester consists of less than a dozen simple mechanical components that were made from ordinary materials to ordinary tolerances in a conventional machine shop, and the commercially purchased piezoelectric stacks. These are driven with a commercial 500 V DC power supply. The displacement sensors were also purchased commercially. The piezoelectric stacks are 20 mm in diameter and 60 mm long. They are restrained by eight nylon rods that hold the moving head assembly to the base of the device. Stainless steel rods were tried, but the greater elastic compliance of nylon allows more displacement. The moving head assembly includes the

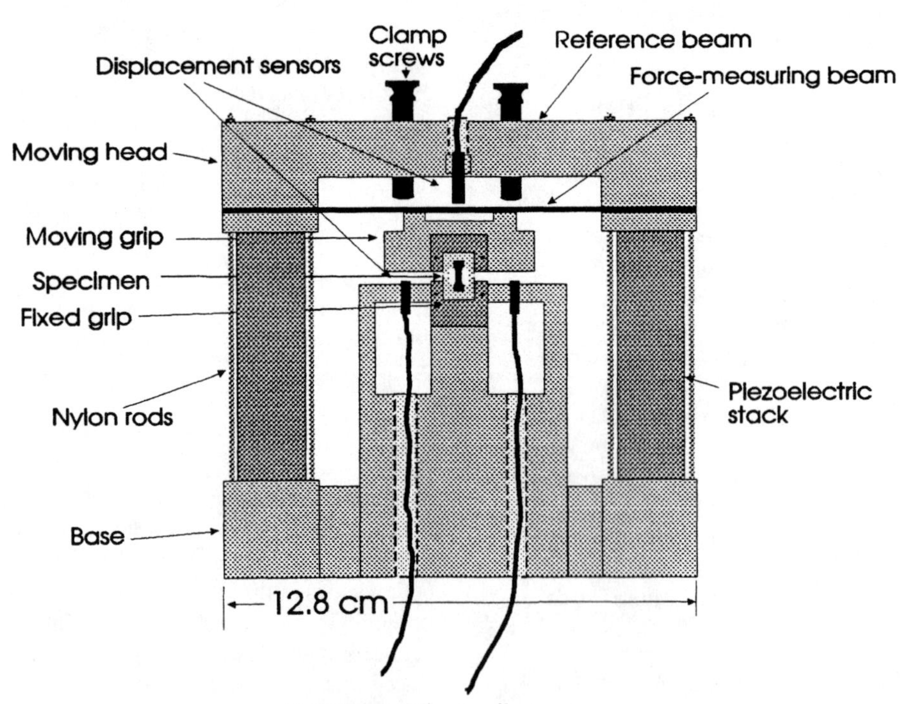

FIGURE 2. Microtensile tester.

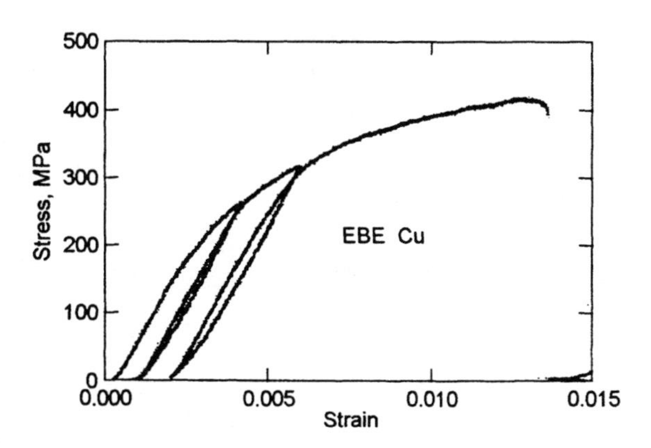

FIGURE 3 . Stress-strain curve for a copper thin-film specimen made by electron-beam-evaporation, including unloading and reloading to show hysteresis. This film has titanium cover layers.

reference beam, the force-measuring beam, and the moving grip; these are all cantilevered above the base plate to avoid friction.

A typical stress-strain curve for a copper film is shown in Fig. 3. Note that because the copper was deposited on bare silicon, which was subsequently etched away, the tensile strip consists of copper, optional titanium surface layers in this case, and possibly a thin layer of native SiO_2. Similar behavior is seen when the titanium on both sides of the coupon and the possible native oxide layer on the underside are etched away in dilute hydrofluoric acid. This record shows typical features for the copper and aluminum films tested by this procedure. Unloading followed by reloading produced hysteresis loops in the stress-strain plot. The yield strength is about 300 MPa, which is well above typical handbook values for pure polycrystalline copper. The ultimate tensile strength is about 400 MPa. The elongation to maximum load is about 1 percent strain, and failure follows soon thereafter. Aluminum behaves similarly, but with lower strength than copper. Epitaxial silicon is perfectly linear until failure.

The strain shown in Fig. 3 is obtained from the grip-to-grip displacement, as measured by the two noncontacting eddy-current displacement sensors. It is adequate for obtaining yield and ultimate strengths, but not for accurate measurement of Young's modulus; therefore, speckle interferometry was implemented.

Speckle Interferometry

Many variants of speckle interferometry have been reported in the literature, attesting to the power of the general technique. In contrast to the grid (12–22) and moiré (23–26) techniques, the speckle technique (27-30) does not require the preparation of a grating on the specimen surface. handle. In the present series of measurements, such a grating would have to survive the aqueous hydrazine etch used to remove the silicon substrate from underneath the specimen gauge length. This etch is very aggressive, so the creation of such a grating would be difficult.

A laser, beam splitter, and mirrors were used to create an optical field which is constant in the y- and z-directions, and has regularly spaced maxima and minima of optical intensity along the x-directions. (The z-direction is perpendicular to the specimen plane, and the x-direction is the direction of displacement.) This useful optical field occurs at the intersection of two coherent laser beams. It is conveniently obtained by extracting both beams from the output of the same laser and keeping the optical path lengths approximately equal. The distance between the intensity maxima is given by (27):

$$d = \lambda / 2 \sin \theta \quad , \qquad (1)$$

where λ is the wavelength of the light (632.8 nm in the present study) and θ is the deviation from vertical illumination. A digital camera was used to obtain many single-exposure images in each tensile test.

A novel method of analysis of a series of speckle images to obtain displacement information was used (31). Briefly, the light intensity record at each pixel is sinusoidal; each cycle indicates a displacement of d as given in (1). The records are analyzed individually using a Fourier method to count the cycles, to produce the displacement at each pixel. Figure 4 shows the displacement as a function of position along the tensile axis of a stressed TiAl multilayer tensile specimen. Each individual data point represents displacement at a pixel. The best-fit straight line is obscured by the data points. The slope of this line is the strain, which is then used along with the stress to get Young's modulus.

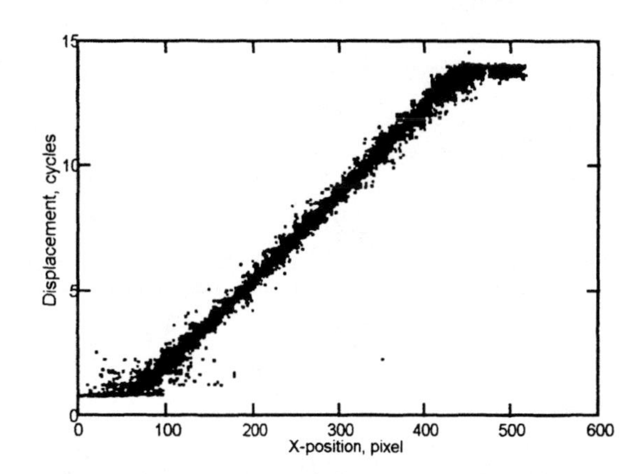

FIGURE 4. Displacement, measured in units of light intensity cycles, plotted against position along a tensile specimen under stress. Equation (1) gives the displacement corresponding to each intensity cycle, which is about 400 nm in the present case.

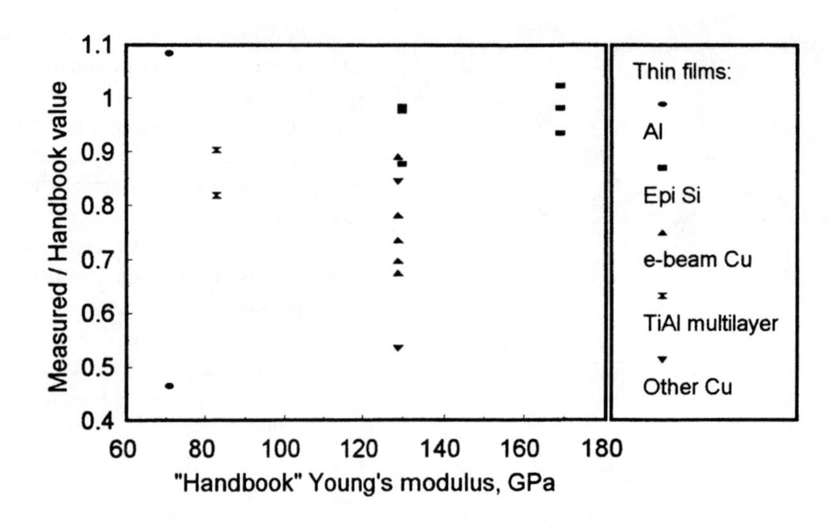

FIGURE 5. Ratio of measured values of Young's modulus to the handbook value plotted against the handbook value for several different specimens.

RESULTS

Measured values of Young's modulus for a variety of materials are plotted in Fig. 5. Consider copper, for which the most data are shown, including PVD copper by multiple methods and electrodeposited copper. The polycrystalline-average Young's modulus for copper at room temperature is 128 Gpa. (32). This value averages the different values in the high-symmetry crystal directions: 66.6 GPa along <100>, 130.3 along <110>, and 191.1 along <111>. Figure 5 shows that most results for metal films are well below the polycrystalline- average value. The experimental uncertainty is about 5 %, as derived from specimen-to-specimen variation. The statistical uncertainty associated with the strain measurement discussed above is around 1 %. The expected texture for PVD films is an excess population of grains with their <111> direction perpendicular to the substrate. X-ray studies showed the present *e*-beam evaporated Cu specimens to have a strong <111> texture. And it can be shown that in the face-centered-cubic crystal symmetry, all directions perpendicular to <111> have the same Young's modulus as occurs along the <110> direction, which is 130.3 GPa. Therefore, texture is not a likely explanation of the present results for PVD films. The cause of the unexpectedly low values has not been resolved, but porosity or some similar microstructural defect is suspected. Values of Young's modulus similar to the present result for electrodeposited copper were reported previously (33). Application of this technique to epitaxial Si specimens 15 μm thick gave values in agreement with the bulk values for single-crystal Si, as shown in Fig. 5.

Figure 6 shows a comparison of ultimate strength, inferred from indentation tests, compared to ultimate tensile strength

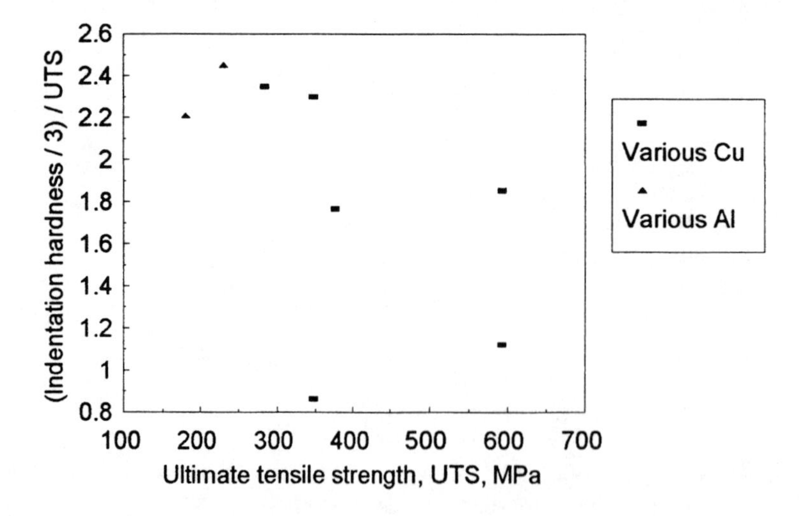

FIGURE 6. Ratio of estimated ultimate tensile strength from indentation measurements, given as indentation hardness over three, to tensile strength measured in tensile tests, plotted against the tensile ultimate strength.

Clearly, the estimates are highly scattered, and have a tendency to be greater than the tensile value. The present data do not represent a definitive study, but they do raise questions about the reliability of indentation tests for determination of Young's modulus or ultimate tensile strength. Indentation data may nevertheless be useful for quality control or other purposes.

A complete description is of the technique described here, and additional results, are compiled in (34).

ACKNOWLEDGEMENTS

Financial support of this work from the NIST Office of Microelectronics Programs is gratefully acknowledged.

REFERENCES

1. Hardwick, D. A., *Thin Solid Films* **154**, 109–124 (1987).
2. Alexopoulos, P. S. and O'Sullivan, T. C., *Annual Review of Materials Science* **20**, 391–420 (1990).
3. Brotzen, F. R., *International Materials Reviews* **39**, 24–45 (1994).
4. Schweitz, J.-A., *Materials Research Society Bulletin* **17**, 34–45 (1992).
5. Fartash, A., Schuller, I.K., and Grimsditch, M., *Review of Scientific Instruments* **62**, 494–501 (1991).
6 Ruud, J. A., Josell, D., Spaepen, F., Greer, A. L., *Journal of Materials Research* **8**, 112–117 (1993).
7. Small, M. K., Daniels, B. J., Clemens, B. B., and Nix, W. D., *Journal of Materials Research* **9** 25–30 (1994).
8. Kim, I. and Weil, R., *Thin Solid Films* **169**, 35–42 (1993).
9. Ding, X., Ko, W. H., and Mansour, J. M., *Sensors and Actuators* **A23**, 866–871 (1990).
10. Read, D.T. and Dally, J. W., *Journal of Materials Research* **8**, 1542–1549 (1993).
11. Davis, B. M., Seidman, D. N., Moreau, A., Ketterson, J. B., Mattson, J., and Grimsditch, M., *Physical Review* **B43**, 9304–9307 (1991).
12. Allais, L., Bornert, M., Bretheau, T., and Caldemaison, D., *Acta Metallurgical et Materialla* **42**, 3865–3880 (1994).
13. Parks, V. J., *Optical Engineering* **21**, 633–639 (1982).
14. Carolan, R., Egashira, M., Kishimoto, S., Shinya, N., *Acta Metallurgica and Materialla* **40**, 1629–1636 (1992).
15. Sadat, A. B. and Reddy, M. Y., *Experimental Mechanics* **29**, 346–349 (1989).
16. Morimoto, Y., Seguchi, Y., and Yamashita, M., *JSME International Journal, Series I* **34**, 37–43 (1991).
17. Duffy, J. and Chi, Y. C., *Materials Science and Engineering* **A157**, 195–210 (1992).
18. Haynes,A. R. and Coates, P. D., *Journal of Materials Science* **31**, 1843–55 (1996).
19. Nisitani, H. and Fukuda, T., *JSME International Journal, Series I* **35**, 354–360 (1992).
20. Mari, D., Marti, U. and Silva, P. C., *Materials Science and Engineering* **A158**, 203–206 (1992).
21. Goldrein, H. T., Palmer, S. J. P. and Huntley, J. M., *Optics and Lasers in Engineering* **23**, 305–318 (1995).
22. Kobayashi, A. S. (editor), *Handbook on Experimental Mechanics*, Society for Experimental Mechanics, Bethel, CT, 1993.
23. Weller, R., and Shepard, B.M., *Proc. Society for Experimental Stress Analysis* **6**, 35–38 (1948).
23. Parks, V.J., *Handbook on Experimental Mechanics*, Ed. A.S. Kobayashi, Prentice-Hall, Englewood Cliffs, N.J., 1987, pp. 282–313.
24. McDonach, A., McKelvie, J., and Walker, C.A., *Optics and Lasers in Engineering*, **1**, 85–105 (1980).
25. Post, D., *Experimental Mechanics* **31**, 276–280 (1991).
26. Jones, R., and Wykes, C., *Holographic and Speckle Interferometry*, Cambridge University Press, 1989.
28. Sirohi, R.S., *Speckle Metrology*, Marcel Dekker, New York, 1993.
29. Erf, R. K., *Speckle Metrology*, Academic Press, New York, 1978.
30. Dainty, J. C., *Laser Speckle and Related Phenomena*, Springer-Verlag, Berlin, 1984, pp. 203-254.
31. Read, D. T., *Measurement Science and Technology*, to be published, 1998.
32. Simmons, G., and Wang, H. *Single Crystal Elastic Constants and Calculated Aggregate Properties: A Handbook*, M. I. T. Press, Cambridge, 1971.
33. Lin, K., Kim, I. and Weil, R., *Plating and Surface Finishing*, July, 1988, pp. 52-56.
34. Read, D. T., *Tensile Testing of Thin Films: Techniques and Results*, National Institute of Standards and Technology, Gaithersburg, MD, Technical Note 1500-1, 1998.

High Sensitivity Technique for Measurement of Thin Film Out-of-Plane Expansion

Chad R. Snyder and Frederick I. Mopsik

National Institute of Standards and Technology, Polymers Division, Gaithersburg, MD 20899

An absolute, high sensitivity technique for the measurement of the out-of-plane expansion of thin films is presented. To demonstrate the ability of the technique to produce correct results for the thermal expansion of materials, measurements were performed on <0001> single crystal sapphire. A thin polymeric film designed for use as an interlayer dielectric (ILD) was measured to show the utility of the technique to resolve the measurement of displacement in thin films. Additionally, the ability of the technique to measure swelling (or hygrothermal expansion) was demonstrated on a typical epoxy molding compound.

INTRODUCTION

The National Technology Roadmap for Semiconductors targets "Measurement...of materials properties of packaging materials for the sizes, thicknesses, and temperatures of interest."(1) Due to the drive towards denser packages and the use of thin polymeric films as inner-layer dielectrics (ILDs), a technique is required that can make precision measurements of the coefficient of thermal expansion (CTE) of thin films. The current technique that is widely utilized by industry for determining the CTE of solid materials is thermomechanical analysis (TMA). However, the ASTM technique E 831-93 for CTE determination by TMA of solid materials calls for sample thicknesses between 2 mm and 10 mm.(2) These dimensions are far greater than those of interest as ILDs. A multi-user round-robin test compared the results obtained for the CTE of a 50 μm thick anisotropic polyimide film by TMA and a 1.1765 mm thick undoped, polished, <100> silicon. The results of this round-robin test clearly demonstrated that TMA is inadequate for the measurement of thin polymer films.(3) We have developed a technique utilizing a capacitance measurement for the accurate determination of the out-of-plane expansion of thin films on the order of 5 μm thick. This is important because the CTE is a key parameter for the design and reliability of polymer-containing products.(4)

Thermal expansion measurements have been performed on polymeric and non-polymeric materials, and hygrothermal expansion measurements have been performed on a bisphenol-A based epoxy with novolac hardener and fused silica filler.

EARLY ATTEMPTS AND CLAIMS

Due to the sensitivity of capacitance measurements, they have long been a technique for the measurement of small displacements. Resolution of capacitances to 1 nF/F are easily obtainable. Of the capacitance cells used for the measurement of polymeric films, most have problems that limit their reproducibility. Capacitance cell methods come in two general types: two-terminal and three-terminal methods. Of the two, the three terminal method has the least uncertainties.

The capacitance cell of Sao and Tiwary (5) was a two-terminal method that suffered from the problem of alignment of the electrodes plus expansion of the electrodes. The three-terminal method of Subrahmanyam and Subramanyam (6) suffers from several problems: electrode alignment, expansion and contraction of the entire apparatus, and the thermal spring constant which determines the loading conditions. Tong and coworkers (7) two-terminal technique suffers from electrode alignment problems, the requirement of specific specimen geometry, and stray capacitances and other fluctuations.

Our technique eliminates the problem of electrode alignment and uses the sensitivity of the three-terminal measurement. Previous measurements were reported on an earlier design (8) of our cell made from gold-coated Zerodur.(9)

EXPERIMENTS AND RESULTS

The absolute measurement technique that we have developed is based upon a parallel plate three-terminal guarded electrode. This technique derives from the well-known fact that the capacitance (C) of a parallel plate capacitor is given by:

$$C=\varepsilon\varepsilon_o A/d \qquad (1)$$

where ε is the dielectric constant of the material separating the plates (in the geometry of our experiment the material is air), ε_o is the permittivity constant (ε_o = 8.854 pF/m), A is the area of the plates, and d is the separation of the plates (which

CP449, *Characterization and Metrology for ULSI Technology: 1998 International Conference*
edited by D. G. Seiler, A. C. Diebold, W. M. Bullis, T. J. Shaffner, R. McDonald, and E. J. Walters
1998 The American Institute of Physics 1-56396-753-7/98/$15.00

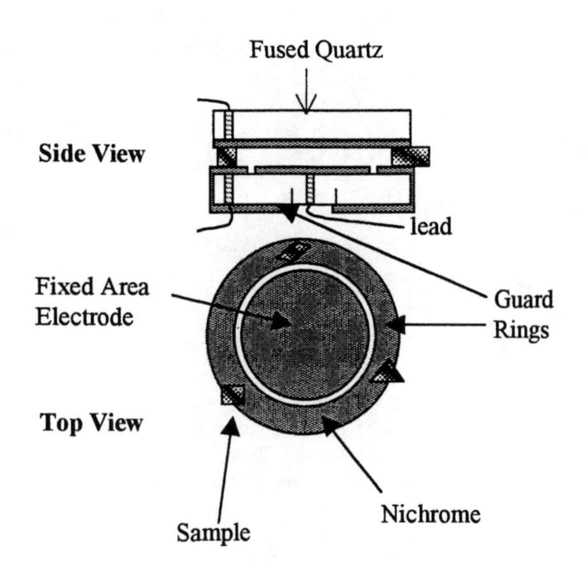

FIGURE 1. Schematic of capacitance cell described in the text.

corresponds to the thickness of the material being measured).It should be recalled that a good three-terminal guarded electrode, if properly constructed, allows the area, A, to be independent of the position of the opposite electrode since the area is defined solely by the guarded electrode, as long as the opposing electrode is much larger than the diameter defined by the guard gap. Furthermore, for any easily realized guard gap, the effective area of the electrode varies only very slowly with a change in plate separation.(10)

Fig. 1 is an illustration of the capacitance cell that we utilized in obtaining our data. It should be noted that the electrodes have been made from fused quartz coated with nichrome to obtain an optically flat surface. Calibration of the expansion of the electrode area (A) was obtained by performing measurements on 2 mm Zerodur (9,11) spacers. The spacers were measured using a specially designed caliper with a resolution of ±0.2 μm.(12) Capacitance measurements were obtained with an Andeen-Hagerling 2500 A 1 kHz Ultra-Precision Capacitance Bridge with a 1 nF/F resolution.

Measurements were performed inside an environmental chamber with a dry air purge so that either thermal expansion (zero relative humidity and temperatures ranging from -25 °C to 180 °C) or hygrothermal expansion (0 to 85 % relative humidity with temperatures ranging from 25 °C to 100 °C) could be measured. The temperature of the cell was calibrated with a resistance temperature device (RTD) mounted to the cell with thermally conducting paste. The RTD was calibrated against an ITS-90 standard reference thermometer. Calibration of the humidity was performed by the dew point method.(14)

The following protocol was followed prior to each run. In a laminar flow hood equipped with a high efficiency particle arresting (HEPA) filter, the electrode surfaces were cleaned with ethanol and distilled water and dried with lens tissue. High pressure filtered air was used to remove any remaining particles from the surface. All samples were cleaned in the same way. As is shown in Fig. 1, the samples were placed in the outside guard ring spaced $2\pi/3$ radians apart. The entire cell apparatus was then placed in a vacuum oven at 25 °C under 30 mm of Hg vacuum for 1 hour. (Evacuation of the cell was found necessary to simulate wringing, i.e., the removal of a trapped air layer caused by the contact of optically smooth surfaces.) The cell was then transferred to a Tenney (9) humidity/temperature chamber. The cell was equilibrated at the highest temperature of interest. Random temperatures were then sampled to eliminate any systematic errors induced by heating and cooling. The cell was allowed to equilibrate at any given temperature until the fluctuations in the capacitance were on the order of, or less than, 10^{-6} pF/pF. The pressure of the chamber was monitored during the run.

From the capacitance, the pressure, temperature, and relative humidity, the plate separation can be computed. Full details of the data analysis and reduction are given elsewhere.(13,14)

Thermal expansion measurements were performed on 0.5 mm <0001> single crystal sapphire and a 14 μm thick sample of a Dow Chemical Company's Cyclotene material. Swelling studies (hygrothermal expansion) were performed on a 1 mm thick epoxy molding compound - a bisphenol-A based epoxy with novolac hardener and fused silica filler. (This epoxy was obtained through National Semiconductor.)

RESULTS AND DISCUSSION

As discussed elsewhere (13), there are three levels in which the uncertainty in the thickness of the sample can be examined: [1] the isothermal stability as determined by measuring the sample's capacitance over long periods of time, *e.g.* 4 hours; [2] the reproducibility of the capacitance measurement determined by thermally (or hygrothermally) cycling the material and returning to the initial temperature; and [3] the reproducibility when measuring the capacitance after disassembling and reassembling the cell. The trend in magnitudes of the errors associated with each of the three above levels is [1]<[2]<<[3]. It is important to realize that the second level is the error that is associated with the measurement of the coefficient of thermal (or hygrothermal) expansion.

Aluminum Oxide (Sapphire)

The measurements on sapphire were performed to verify the ability of the technique to determine the CTE of a reference material. In Fig. 2, we have plotted $\Delta d/d_{25}$ for our experimentally determined data on <0001> sapphire. On the same graph, we have plotted the experimental data of Wachtman and coworkers (15,16) fitted to a cubic spline. Our data agrees well with the previous data. It should also be noted that Wachtman and coworkers used two different

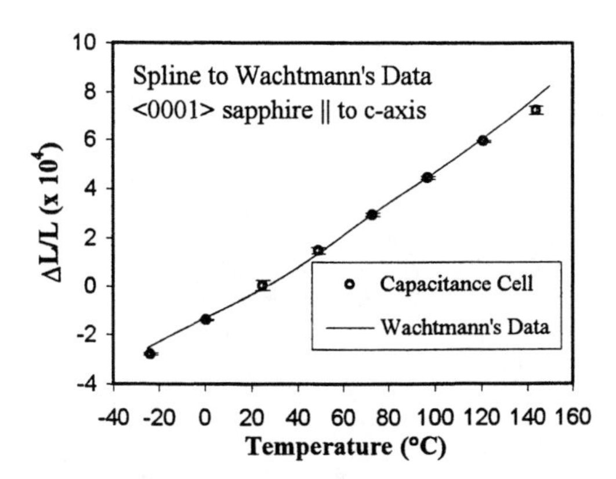

FIGURE 2. Expansion curve for single crystal <0001> oriented Al_2O_3 parallel to the principle axis. Note that the error bars correspond to the best estimate of two standard deviations in the experimental uncertainty.

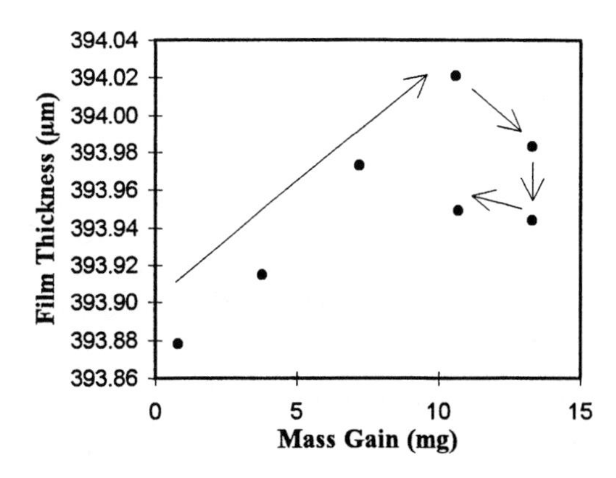

FIGURE 4. Epoxy thickness as a function of mass gain. Note the arrows indicate the direction in which the experiments were performed. The data points correspond to 2.2 %, 20 %, 40 %, 60 %, 60 %, and 40 %. The arrow pointing down at 13.3 mg mass gain corresponds to creep in the material with time. The associated errors bars are smaller than the size of the symbols. (12)

experimental setups and the data were then matched at 25 °C. Examination of the data shows that the curvature in our data is much smoother than that of Wachtman and coworkers indicating that our data are more self consistent. One other fact which makes our data set superior to that of Wachtman and coworkers is that there is a temperature uncertainty of ±1.4 K associated with their data, whereas our data has a temperature uncertainty of ±0.2 K. It should be recalled that excluding problems associated with asperities, our technique gives the same relative resolution in thickness regardless of the absolute thickness. Therefore, the results are extendable to films with thickness on the order of 5 µm. Furthermore, sapphire has an expansion coefficient on the order of 3 µm/(m·K). Polymer films have expansion coefficients an order of magnitude larger and therefore are far easier to resolve.

Cyclotene

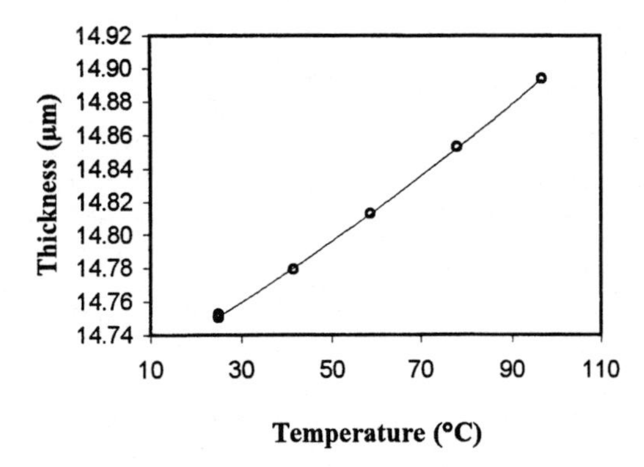

Figure 3. Expansion of the Dow Cyclotene as a function of temperature after subtraction of creep effects. The temperature profile followed was 100 °C, 25 °C, 60 °C, 40 °C, 80 °C, and 25 °C. (The manufacturer cites this Cyclotene's T_g as >350 °C.) The associated error bars are smaller than the size of the symbol.(12)

Because of the sensitivity of our technique, we were able to resolve creep of the sample at levels that ordinarily would not be observed. Therefore, the protocol that is given elsewhere was utilized to remove effects of creep for the Cyclotene.(13) To verify the consistency of the creep removal, the following temperature path was followed. Initially the sample was equilibrated at 150 °C for 20 hours. After the measured capacitance varied less than 1 µF/F the cell was equilibrated under the same circumstances for the following temperatures in the following order: 100 °C, 25 °C, 60 °C, 40 °C, 80 °C, and 25 °C. The data from each set was then corrected for creep and the curve in Fig. 3 was obtained. It is apparent from Fig. 3 that the data are self-consistent since all the data points fall on a single smooth curve. Additionally, the 25 °C point was reproduced within the experimental uncertainty.

Epoxy Molding Compound

The purpose of this paper has been to introduce the utility of our capacitance cell technique for various electronic packaging applications, therefore a full consideration of the behavior of the epoxy is beyond the scope of this work. (The reader is referred to Ref. 14 for further information.) A brief summary of the results and the implications are given here. In Fig. 4, we have plotted the film thickness as a function of the mass uptake of water. (It is important to note that mass uptake varied with humidity almost linearly.)

As can be seen from Fig. 4, the sample expands (swells) with increasing moisture uptake up to 40 %R.H. As the relative humidity was increased from 40 % to 60 %, two events were observed simultaneously. [1] The sample rapidly contracted, although the mass uptake increased, and [2] the sample began to "creep", *i.e.,* the thickness decreased as a function of time. To verify that the first event was not creep, the humidity was decreased to 40 %. After equilibration, it was apparent that the sample expanded upon decreasing the relative moisture content from 13.3 mg (at 60 % relative humidity) to 10.7 mg (at 40 % relative humidity). Therefore, events one and two were separate. Further examination of this anomalous behavior is the subject of a future publication.

What is important to note is that this technique has the capability to resolve changes in film thicknesses on the order of 2 nm/mm caused by swelling. Furthermore, creep is observable at an order of sensitivity higher, *i.e.*, 2 nm/cm. Therefore, this technique has been demonstrated to have utility over a wide range of environmental conditions and can therefore address many of the concerns of the packaging industry.

CONCLUSIONS

In summary, we have developed an absolute, high sensitivity technique for the measurement of the z-axis expansion of thin films and optimized the measurement procedures for use by materials users and producers. This technique can easily provide the relevant CTE data to a higher accuracy than is currently available.

REFERENCES

1. *The National Technology Roadmap for Semiconductors Technology Needs,* San Jose, CA: Semiconductor Industry Association, 1997, pp.148-150.
2. Storer, R.A (ed.), *Annual Book of ASTM Standards*, West Conshohocken, PA: American Society for Testing of Materials, 1997, Vol. 14.02, pp.548-552.
3. Schen, M., *et al.*, "An Industry/Government/University Partnership: Measuring Sub-Micrometer Strain in Polymer Films," Proceedings of the IPC Printed Circuits Expo., Boston, MA, April 24-27, 1994.
4. Neugebauer, C.A., in *Electronics Packaging Forum, Volume One*, New York: Van Nostrand Reinhold, 1990, ch. 1.
5. Sao, G.D.; Tiwary, H.V., *J. Appl. Phys.* **53**, 3040-3043 (1982).
6. Subrahmanyam, H.N.; Subramanyam, S.V., *J. Mat. Sci.* **22**, 2079-2082 (1987).
7. Tong, H.M.; Hsuen, H.K.D.; Saenger, K.L.; Su, G.W., *Rev. Sci. Instrum.* **62**, 422-430 (1991).
8. Schen, M.; Mopsik, F.I.; Wu, W.; Wallace, W.E.; Beck Tan, N.C.; Davis, G.T.; Guthrie, W., *Polym. Preprints* **37** (1), 180 (1996).
9. Certain commercial materials and equipment are identified in this paper in order to specify adequately the experimental procedure. In no case does such identification imply recommendation or endorsement by the National Institute of Standards and Technology, nor does it imply that the items identified are necessarily the best available for the purpose.
10. Lauritzen, J.I. Jr., *1963 Annual Report Conference on Electrical Insulation*, NAS-NRC Publ. 1141.
11. Schott material having a coefficient of thermal expansion of less than 0.05×10^{-6} $\mu m \cdot m^{-1} \cdot K^{-1}$.
12. Quoted uncertainties represent the best estimate of two standard deviations in the experimental uncertainty.
13. Snyder, C.R.; Mopsik, F.I., *To be submitted to J. Appl. Phys.*
14. Snyder, C.R.; Mopsik, F.I., *To be submitted to J. Appl. Phys.*
15. Wachtman, J.B. Jr.; Scuderi, T.G; Cleek, G.W., *J. Am. Chem. Soc.* **45**, 319-323 (1962).
16. Wachtman, J.B. Jr.; Scuderi, T.G.; Cleek, G.W., *ADI Auxilliary Publications Project, Photoduplication Service,* Document No. 7146, Washington, D.C.: Library of Congress, 1962.

The Study of Silicon Stepped Surfaces as Atomic Force Microscope Calibration Standards with a Calibrated AFM at NIST

V.W. Tsai[1], T. Vorburger[2], R. Dixson[2], J. Fu[2], R. Köning[2], R. Silver[2], E.D. Williams[3]

[1] *Dept. of Materials Engineering, University of Maryland, College Park, MD 20742*
[2] *National Institute of Standards and Technology, Gaithersburg, MD 20899*
[3] *Dept. of Physics, University of Maryland, College Park, MD 20742*

Due to the limitations of modern manufacturing technology, no commercial height artifact at the sub-nanometer scale is currently available. The single-atom steps on a cleaned silicon (111) surface with a height of 0.314 nm, derived from the lattice constant of silicon, have considerable potential as an atomic force microscope (AFM) calibration artifact at the sub-nanometer range. A metrology AFM developed at National Institute of Standards and Technology (NIST), called the calibrated AFM (C-AFM), is used to measure this type of surface. In this paper, the results of six sets of measurements made over a period of five months are presented. The calculation of the step algorithm and the uncertainty of the measurement are introduced and discussed briefly.

INTRODUCTION

The industrial applications of atomic force microscope (AFM) are important particularly in the semiconductor industry (1). Based on technology trends predicted by the National Technology Roadmap for Semiconductors (2), the thickness of the gate oxides of one gigabit dynamic random access memories (DRAMs) will be reduced to less than 4 nm by the year 2001. The behavior of devices can be affected by atomic-scale variations of the oxide layer (3,4). Therefore, the use of AFM as a metrology tool to characterize surface topography at nanometer and sub-nanometer scales becomes increasingly important. Although commercial AFM manufacturers provide calibration samples and procedures with their instruments, providing traceability for displacement measurements along three axes of motion is problematical. To address this need for AFM calibration standards, the National Institute of Standards and Technology (NIST) has developed a metrology AFM, called the calibrated AFM (C-AFM), that will be used to calibrate standards (5,6). These, in turn, can be used to calibrate commercial AFMs.

Due to the limitations of present fabrication technology, there is no commercial height calibration artifact available in the sub-nanometer range. The clean silicon (111) stepped surface, prepared in a ultra high vacuum (UHV) environment, appears to be stable in air for long periods after the native oxide has grown on the surface. This type of sample has been proposed as a calibration specimen for the z axis of AFMs in the nanometer range by Suzuki *et al.* (7) whose work was mainly focused on developing a standard algorithm of data analysis to tighten up the agreement between different measuring instruments. The calibration of the AFM relied on calibration artifacts (8) with step heights of 2 nm to 8 nm.

As an application of its capability to perform traceable measurements, the C-AFM is used to measure Si(111) samples. The preparation of these samples has been described elsewhere (9). Preliminary results of silicon step height measurements using the C-AFM have been previously reported (10). In this paper, the current status of C-AFM measurements on silicon steps and our data analysis algorithms are discussed in more detail.

EXPERIMENT

The C-AFM measures motion along the z axis with a calibrated capacitance gauge and motion along the x and y axis with laser interferometers (11). The measurement of a step height with the C-AFM uses the following procedure. The first step is the calibration of the capacitance gauge with a z-interferometer. This is done by driving the z-stage while recording the outputs of the capacitance gauge and interferometer. The calibration factor of the capacitance gauge is then determined by performing a second order least squares fit on this data. After calibrating the capacitance gauge, the z-interferometer assembly is replaced with the AFM beam-bounce head assembly for normal measurement (12). A scan is initiated with a 5 μm field of view, then zoomed on a clean area. One single atomic step is then centered in the image. Once the scan is initiated, the outputs of the capacitance gauge and x or y interferometer are recorded using customized software. The program is written to acquire data profiles synchronized with the AFM scan, so the capacitance gauge signal is composed as an image. Typically, one capacitance image contains 250 scan lines, each with 500 data points. Each set of measurements consists of twenty images, scanned repeatedly at the same position using the same settings.

The width between steps on the surface of the samples, which depends on the angle deviation of the surface plane from the Si(111) plane, generally ranges from 100 nm to 2 μm. In this work, two samples having terrace widths of about 1 μm and 500 nm (shown in Fig. 1) were used. These two samples were made and stored in the desiccator in air for more than one year. The image

CP449, *Characterization and Metrology for ULSI Technology: 1998 International Conference*
edited by D. G. Seiler, A. C. Diebold, W. M. Bullis, T. J. Shaffner, R. McDonald, and E. J. Walters
1998 The American Institute of Physics 1-56396-753-7/98/$15.00

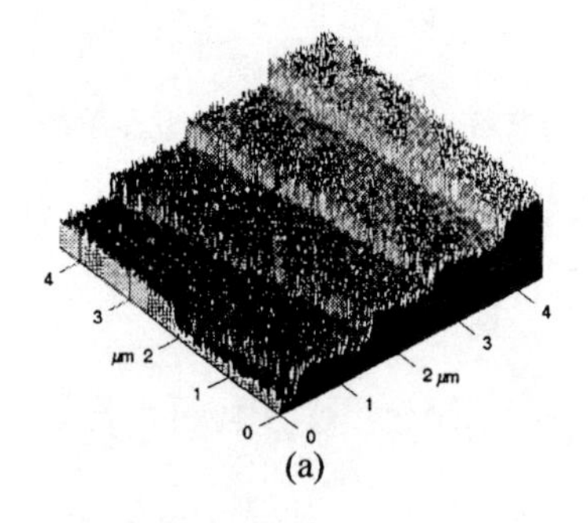

(a)

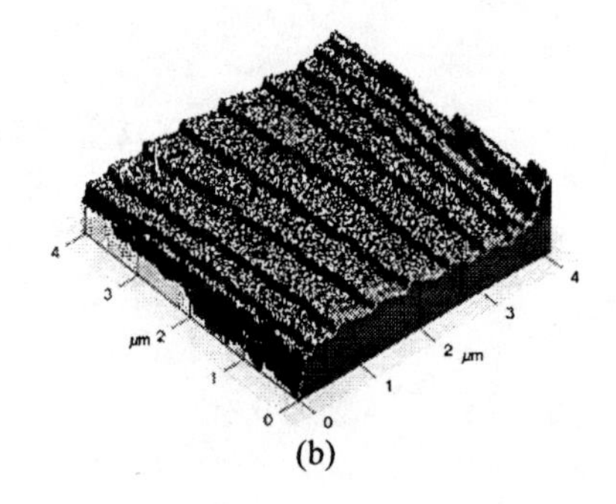

(b)

FIGURE 1. The Silicon (111) surfaces with (a) 1 μm wide terrace and (b) 500 nm wide terrace are measured by C-AFM.

TABLE 1. The scan parameters of the measurement

Measurement	No.1	No.2	No.3	No.4	No.5	No.6
Sample	a	a	a	a	a	b
AFM Head[a]	I	II	I	I	I	I
Operation[b]	C	C	C	C	INT-C	C
Relative Humidity(%)	30.4	25.2	26.3	13.2	12.7	28.9

[a] The head I,II represent two different AFM heads, using the beam-bounce surface sensing technique.
[b] "C" represents the contact mode and "INT-C" represents the intermittent contact mode.

data are acquired at a 0.5 Hz scan rate with a 1.6 μm to 1.9 μm scan size. During the scan, the stage is driven by a linear ramp waveform voltage with feedback loop controlled from the C-AFM controller. Therefore, the spacings between data points are essentially equal in time and distance.

Each time the system is set up for measuring the Si(111) step, a new tip is used. In the measurements for this work, the step edge is carefully aligned parallel with the *y*-axis so that the fast scan direction, which is the *x*-axis, is oriented normal to the step. In order to investigate the reproducibility of the silicon step measurement with the C-AFM, six sets of measurements, taken over a period of five months, were made with different setup conditions and scan parameters listed in Table 1.

RESULTS AND DISCUSSION

Although a standard step height algorithm has been developed primarily for stylus instruments (13), this method is not useful for all types of step geometries and instruments. A method for determining the step height specifically from AFM measurements has not yet been standardized. For the current C-AFM step height measurements, another approach for stylus step height measurements is adapted to calculate the single-sided step height. Due to the noise floor of the present instrumental setup, which is about 0.15 nm in a 1 Khz bandwidth, the reduction of noise by an averaging method is an important consideration while calculating the Si(111) atomic step. Another major consideration in analyzing the data is the definition of the step edge, or more explicitly, the location at which a height difference is to be computed. Before calculating the step, distortions in the capacitance gauge image, such as drift in the *z* axis and tilt in the *x-y* plane, are removed. The drift in the *z* axis and tilt along the *y* axis, the slow scan axis, are removed by subtracting the average value of each scan line from the scan line profile itself. The tilt in the *x*-direction is removed by fit-plane subtraction. Instead of using a whole image, only the bottom area of the step is selected as the reference plane to calculate the least squares fit coefficient to level the image. After correcting the distortion in the software, the step location is searched and determined from the entire image in the software. Once the step location is determined, the transition region around the step edge and the regions at the edges of the profiles are selected and excluded from the data analysis. These excluded (cut-off) regions are denoted as a percentage of the terrace width. Finally, the remaining data are analyzed and used to calculate the step height with an areal method and two profile methods.

The areal method is a plane fitting method to the upper terrace. The step height is calculated by extrapolating the fitted plane to the step location of the entire image. Because the center of the step location is insensitive to the relative tilt of the top and bottom terrace, the height of the fitted plane at this location is used to represent the step height value of the image.

In the first profile method each profile is aligned to have the same step location, then all the profiles are averaged. Lines are fit to the top and bottom sections of the average profile. The step height is then calculated by extrapolating these two lines to the step edge location. Because the step edge location determined from the entire image can be influenced by the noise signal, the step edge of the average profile, with improved signal-to-noise ratio, is defined again. Using a least squares fit to the entire profile, the intersection of the fitted line and data profile determines the step edge location.

The second profile method consists of calculating the step height from each profile then taking the average value from all the profiles. For each profile, the step height is calculated using linear fits to the top and bottom terraces and extrapolating these to the step edge. The average step height is calculated again by excluding some step values that differ from the original average step height by more than two standard deviations.

In Table 2, the average step height values and standard deviations resulting from these approaches using a cut-off region

TABLE 2. The result of the Si(111) step height measurements. Each data represents the average value with standard deviation of twenty repeated measurements.

units in nm	No.1	No.2	No.3	No.4	No.5	No.6
Areal method	0.294 ± 0.009	0.298 ± 0.009	0.290 ± 0.010	0.313 ± 0.004	0.303 ± 0.006	0.296 ± 0.009
Profile method I	0.295 ± 0.008	0.301 ± 0.006	0.302 ± 0.006	0.313 ± 0.003	0.305 ± 0.004	0.308 ± 0.007
Profile method II	0.295 ± 0.008	0.310 ± 0.005	0.299 ± 0.006	0.312 ± 0.004	0.305 ± 0.005	0.306 ± 0.007

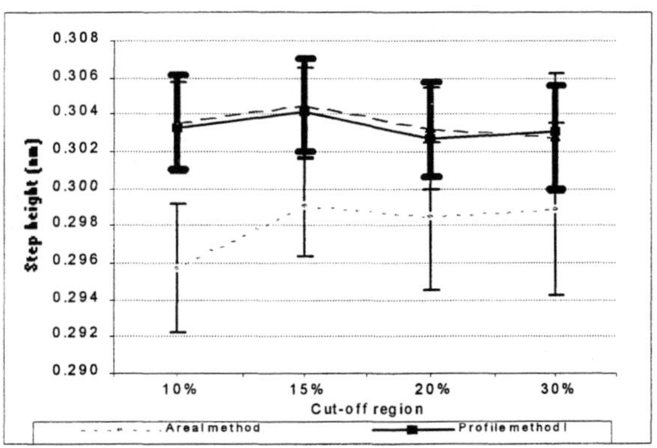

FIGURE 2. Changing the unanalyzed portion (cut-off region) of the data image, shows no significant effect on the average height of six sets of measurement. The error bar shows the standard deviation of twenty measurements.

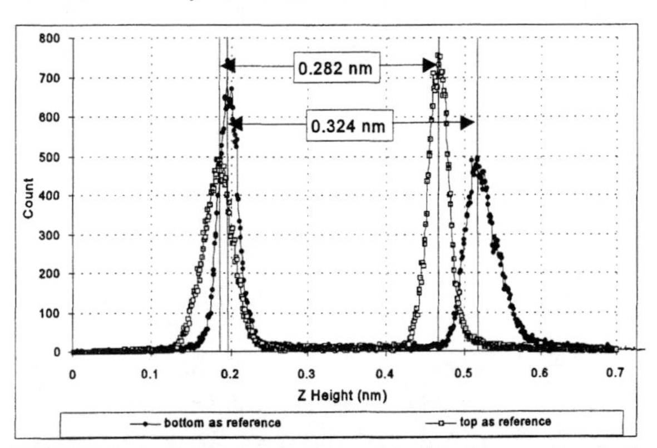

FIGURE 3. A discrepancy in the results of the histogram analysis arises from the tilt correction of the image, depending on the selection of reference plane.

of 15 percent are listed. The over all average heights of the six sets of measurements with the standard deviation of the means are (0.299 ± 0.003) nm, (0.304 ± 0.003) nm, and (0.305 ± 0.003) nm. After examining the differences among these approaches with computer simulated data, it was demonstrated that these three approaches are mathematically equivalent. In other words, when the same step location and selected region of the data are used, these three approaches yield the same result. The difference in the results given above is due only to implementation parameters of the three methods, primarily the selected portion of the image data and the step edge location. To explore this issue, we have done some parameter sensitivity studies. In Fig. 2 a plot of the average step height from the six data sets versus the excluded percentage of the terrace width is shown. The plot shows some variability in the standard deviation of the means. It also shows some change in the average values, particularly for the areal method, but indicates that the difference between the areal method and profile methods is only partially due to the amount of data included in the analysis. We believe the difference is also partially due to the location of the included data. Because the edge of the step in the image is not perfectly linear, the data selected in the areal method must be further away from the step edge than in the profile methods. This makes the areal method more sensitive to residual distortions in the image, particularly at the edges. The question of the most appropriate step height algorithm is under continuing investigation.

The histogram method (7) to determine the height of the step was also investigated. Because of the apparent non-parallelism of the upper and lower planes, believed to be primarily due to instrumental effects, the calculation of step height by this method is sensitive to the tilt correction of the image, as illustrated in the Fig.3. The data used here is from the No. 6 experiment. The calculated height of the step can be changed by choosing the upper or lower terrace as the reference plane for the tilt correction. The possible solution for this problem is averaging the step height value, calculated from using the top and bottom as reference planes respectively. This approach is currently under evaluation.

A general uncertainty budget for step height measurements with the calibrated AFM has been developed (14,15). The uncertainty budget for height measurements in the subnanometer range of the Si(111) step height is determined based on the above results. In the following discussion, the data analyzed by the average profile method, method I, are used. From the data, the overall reproducibility of the measurement is estimated by calculating the standard deviation of the mean from six average results. As mentioned above, the measurement reproducibility is 0.0025 nm. This estimation also includes measurement repeatability from six sets of the data is 0.0013 nm, calculated by the root-mean-square of the standard deviations of the means of each set. This uncertainty estimations compose the type A uncertainty. The type B uncertainty, estimated by non-statistical methods, is 0.95% (of the measured feature) which is determined from the general type B uncertainty of the C-AFM (14,15). In addition, this uncertainty component also includes an estimate of the possible dependence of the result on the analysis algorithm. Since the analysis methods used operate on slightly different subsets of the profile data. A conservative estimate of the difference between the areal method and profile methods of 0.005 nm is used here. By using a rectangular probability distribution model (16), the uncertainty due to the analysis algorithm can be calculated as 0.0029 nm. By using the root-sum-of-squares, the combined standard uncertainty of the measurement is 0.004 nm. Therefore, the final result of the silicon (111) single step measurement is (0.304 ± 0.008) nm, where an expanded uncertainty, using a coverage factor of k=2, is given. This value represents our best estimate of the silicon (111) step height as measured by the C-AFM.

SUMMARY

In summary, the current status of Si(111) atomic step height measurements with the C-AFM has been presented. Future effort will be directed toward better understanding of the sources of non-reproducibility, such as the influence of surface stability, the AFM operational parameters, and environmental effects. Efforts toward improved system stability and design are also ongoing. Additionally, questions concerning the most practical measurement protocol and meaningful definition of the step height for these samples need to be considered. Finally, the investigation of the utility of these samples in industrial applications is already under way (17).

ACKNOWLEDGMENTS

This work was supported in part by the National Advanced Manufacturing Testbed (NAMT), the National Semiconductor Metrology Program (NSMP), the NSF-MRSEC under grant # DMR 96-32521, and the NSF-FAW under grant # 90-23453 at the University of Maryland. We give special thanks to J. S. Villarrubia of NIST for helpful discussions about this work.

REFERENCES

1. T.V. Vorburger, J.A. Dagata, G. Wilkening, and K. Iizuka, Annals CIRP, **47(2)** (1997)
2. The National Technology Roadmap for Semiconductors 1997, Semiconductor Industry Association, San Jose (1997)
3. T. Ohmi, M. Miyashita, M. Itano, I Imaoka, and I. Kawanabe, IEEE Trans. Electron Devices, ED-**39**, 537 (1992)
4. T. Yamanaka, S.J. Fang, H.C. Lin, J.P. Snyder, and C.R. Helms, IEEE Electron Devices Lett. (to be published)
5. R. Dixson, V. W. Tsai, T. Vorburger, E. D. Williams, X. S. Wang, J. Fu, and R. Köning, Proc. ASPE, **15**, 70 (1997).
6. R. Dixson, T. V. Vorburger, P. J. Sullivan, V. W. Tsai, and T. H. McWaid, Proc. ASPE, **14**, 375 (1996).
7. M. Suzuki, S. Aoyama, T. Futstsuki, A.J. Kelly, T. Osada, A. Nakano, Y. Sakakibrara, Y. Suzuki, H. Takami, T. Takenoku, and M. Yasutake, J. Vac. Sci. Tech., **A14**, 1228 (1996)
8. T. Ohmi and S. Aoyama, Appl. Phys. Lett., **61(20)**, 2479 (1992)
9. V. Tsai, X.-S. Wang, E. D. Williams, J. Schneir, and R. Dixson, Appl. Phys. Lett., **71(11)**, 1495 (1997)
10. V. W. Tsai, T.V. Vorburger, P. Sullivan, R. Dixson, R. Silver, E.D. Williams, J. Schneir, Proc. ASPE, **14**, 212 (1996)
11. J. Schneir, T.H. McWaid and T.V. Vorburger, Proc. SPIE, **2196**, 166 (1994)
12. J. Schneir, T. McWaid, R. Dixson, V.W. Tsai, J.S. Villarrubia, E.D. Williams, and E. Fu., Proc. SPIE **2439**, 401(1995)
13. ISO/WD 5436-1995, Geometrical Product Specifications (GPS) - Surface Texture - Profile Method - Measurement standards
14. R. Dixson, R. Köning, T.V. Vorburger, J. Fu, and V. W. Tsai, Proc. SPIE, to be published.
15. R. Köning, R. Dixson, J. Fu, V.W. Tsai, T.V. Vorburger, E.D. Williams, and X.S. Wang, Proc. the 2nd Seminar on Quantitative Microscopy, Wien, Austria, 172 (1997)
16. "Guide to the Expression of Uncertainty in Measurement", ISO, Geneva 1993
17. T.V. Vorburger and E.C. Clayton, private communications.

Tip Characterization for Scanning Probe Microscope Width Metrology

S. Dongmo, J. S. Villarrubia, S. N. Jones, T. B. Renegar, M. T. Postek, and J. F. Song

National Institute of Standards and Technology, Gaithersburg, MD 20899, USA.

Determination of the tip shape is an important prerequisite for converting the various scanning probe microscopies from imaging tools into dimensional metrology tools with sufficient accuracy to meet the critical dimension measurement requirements of the semiconductor industry. Determination of the tip shape generally requires that the tip be used to image a "tip characterizer." Characterizing the characterizer has itself been an obstacle to progress, since most methods require the geometry of the characterizer to be known with uncertainty small compared to the size of the tip. We have recently developed a "blind" reconstruction method that allows the 3-dimensional tip shape to be estimated from an image of an *unknown* characterizer. This method has heretofore been tested only in simulations. We report here initial results of an experimental test of the technique. We compare the reconstructed profile of a tip to an independently measured profile from a scanning electron microscope (SEM). A relatively large (500 nm radius) diamond stylus profiler tip was used in order to test the principle of blind reconstruction in a size regime where SEM instrumental uncertainties are small compared to the tip size. The blind reconstruction and SEM profiles agree well, differing only by an amount comparable to the repeatability of the stylus profiler images (better than 2% of the sample corrugation in this instance).

INTRODUCTION

With resolution down to the atomic level, topographic scanning probe microscopy (SPM), chiefly scanning tunneling microscopy and atomic force microscopy (STM and AFM), offers the promise of dimensional measurements with unprecedented accuracy. This is particularly important for industries like semiconductor electronics and data storage that need accurate measurements of sub-micrometer structures. SPM is used more and more successfully for tasks that in the recent past were reserved exclusively for the more traditional methods of scanning electron microscopy (SEM), optical microscopy, and stylus profiling.

It is important, however, to understand that while resolution is a necessary condition for accuracy (what cannot be detected cannot be measured), it is not sufficient. A highly resolved image may still contain geometrical distortions. In SPM, one such distortion arises from scanner hysteresis, drift and other nonlinearities. Some SPMs that appear on the market or which are conceived and manufactured in laboratories, e.g., the National Institute of Standards and Technology's Molecular Measuring Machine (1) and Calibrated-AFM (2), have made substantial progress in removing scanner problems through the use of independent linear measurement of the scanner position (e.g., calibrated capacitance gauges, interferometry). A distortion not addressed by these developments is caused by the non-vanishing size of the tip. For example, when the tip is blunt, narrow holes in the surface may not appear in the image, while peaks and bumps will appear considerably broadened. This deformation, called dilation, is greater in the case of highly-corrugated surfaces, and is correlated to the size and the shape of the tip. When a linewidth pattern or other high aspect ratio feature on a surface is imaged with an unknown tip, it is not possible to say what part of the image is due to the tip and what part is due to the sample. Two illustrative scenarios are depicted in Fig. 1, one with a sharp tip and wide sample feature, the other with a blunt tip and sample feature of zero width at the top. The images produced are identical. Therefore, on the basis of the image alone, it is not possible to say which of the two sample widths (or any width in between) is the true one.

Consequently, it is essential to develop a method to determine tip geometry. To this end a number of researchers have proposed techniques of tip characterization. One can roughly classify them into 3 categories: first, those which image the tip using a microscopy other than SPM (3-5), for example, scanning or transmission electron microscopy (SEM or TEM), second, those which deduce the shape of the tip from its image of a sample that is known (a known "tip characterizer") (6-24), and third, blind methods, which use a tip characterizer but do not require it to be independently known (25-29).

Tips are difficult to measure by methods in the first category because these methods generally provide only a 2-d image of what is actually a 3-d object, because instrumental uncertainties may be significant at the size scale involved, and because they are time consuming, especially consider-

CP449, *Characterization and Metrology for ULSI Technology: 1998 International Conference*
edited by D. G. Seiler, A. C. Diebold, W. M. Bullis, T. J. Shaffner, R. McDonald, and E. J. Walters
1998 The American Institute of Physics 1-56396-753-7/98/$15.00

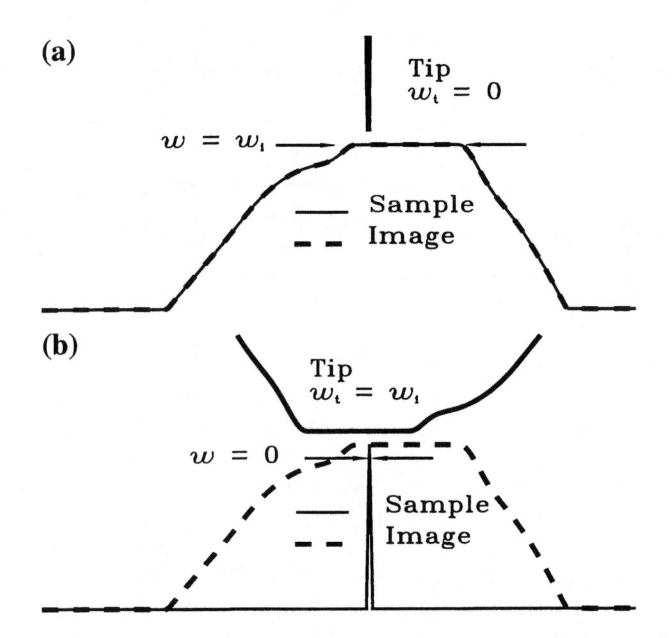

FIGURE 1. Two scenarios that produce the same image (dashed). In (a) a sharp tip ($w_t = 0$) faithfully images the sample. In (b) it is the sample which has zero width, and all of the observed width is due to the tip.

ing that tips require frequent characterization due to wear or damage suffered in collisions with the surface. Methods in the second category require that the characterizer be measured. This cannot generally be done with SPM since that would require a known tip. Instead the characterizer must be measured by an independent method like SEM or TEM with the attendant difficulties just mentioned.

We have developed a blind reconstruction method, which belongs to the third category. This method relies on the basic tools of mathematical morphology to derive an outer bound for the tip. This outer bound is a complex 3-d shape that has been shown in principle to well approximate parts of the tip that make contact with high curvature regions of the characterizer (25, 27).

If blind reconstruction can be shown to work and practical issues in its implementation overcome, the fact that it sidesteps the thorny issues associated with the first two categories of tip estimation methods could make it of great practical use for CD (critical dimension) measurements in the semiconductor industry. Until now, it has only been demonstrated in simulations. Simulations are useful to demonstrate that algorithms are derived without error from initial assumptions and the degree to which reconstructed tips might ideally be expected to agree with actual ones under typical conditions. However, no simulation can show that real instruments actually obey the assumed imaging model.

For that, one requires an experimental comparison. Here, we report initial results of such a comparison. The experiment compares a result obtained by blind reconstruction with the result provided by an independent measurement, in this case SEM. To avoid the more difficult SEM-related measurement issues alluded to above, we chose a stylus profiler as our SPM. Stylus profilers operate on a similar imaging principle to STM and AFM, producing images that may be modeled as dilations of the tip with the specimen. However, our stylus tip's radius was approximately 500 nm, an order of magnitude larger than typical AFM tips. With careful choice of characterizer, it is therefore possible in this way to test blind reconstruction in a size regime where SEM measurement uncertainties are small compared to the tip.

In what follows, we briefly review blind reconstruction, give experimental details concerning the acquisition and treatment of stylus profiler and SEM data to derive independent profiles of the stylus tip, and compare these profiles.

REVIEW OF BLIND RECONSTRUCTION

This section reviews blind reconstruction with the aim of providing a plausibility argument. The derivation of the algorithm will not be repeated here since this is available elsewhere (25).

In SPM, the measured image contains information concerning both the specimen and the tip. The key is to separate these. This is the main purpose of blind reconstruction.

Many SPM practitioners have observed tip "self-imaging" (similar to Fig. 1b) while scanning a blunt tip across a sharp protrusion on a surface (11, 30-32). This means that the inverted image of the protrusion can be considered as an estimate of the tip shape. Ideal self-imaging only occurs when the protrusion width, w, is 0. In any real case $w > 0$. Therefore the estimate is actually an outer envelope of the tip, and the dilation effect has to be taken into account.

More rigorously, one may regard all image protrusions as tip images, each broadened in different ways by the different underlying surface features. Figure 2(a) and (b) illustrate the idea. Figure 2(a) shows a surface, a tip, and the corresponding image. Two of the image maxima are labeled "1" and "2." Figure 2(b) shows two curves corresponding to the regions of the image in the neighborhood of these maxima. Regarding each of these as an independent outer bound on the tip, we can rule out any tip larger than the intersection of the two independently derived tips. This intersection is represented by the cross-hatched region in (b). The cross-hatched region is thus a better estimate than either curve taken alone. The extension to additional image maxima is obvious. What may be less obvious is that this method is not restricted to neighborhoods of maxima. For each point on the image, even those that are not local maxima, one can derive a corresponding outer bound on the tip.

This is evident from the iterative equation defining the full blind reconstruction result (25), which takes the form of an intersection over all points, x, within the image:

$$P_{i+1} = \bigcap_{x \in I} \{(I - x) \oplus [P_i \cap (x - I)]\} \cap P_i . \quad (1)$$

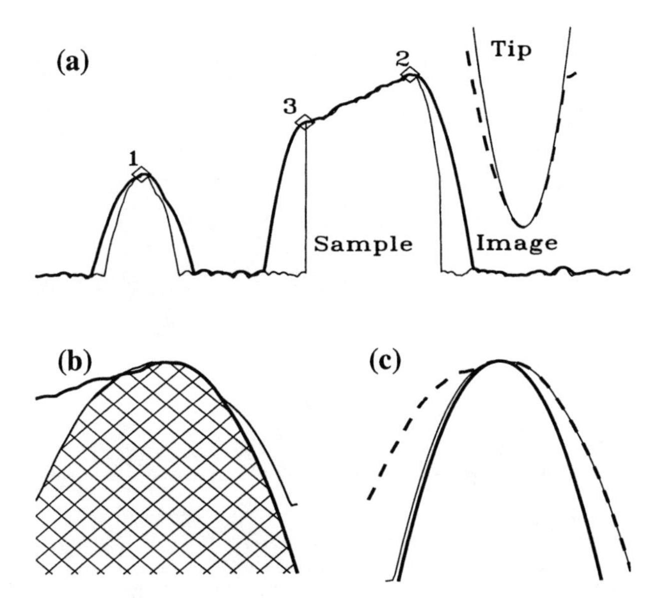

FIGURE 2. Illustration of the blind reconstruction method. (a) A surface (thin) and its image (thick) formed by dilation with the actual tip (continuous). Two image maxima and one non-maximum are labeled. The dashed tip is the fully converged blind reconstruction result. An example of blind reconstruction is given in (b) and (c) using only the three labeled image points. The first step, in (b), shows the overlap of the image maxima, 1 (thin) and 2 (thick), and their intersection (crosshatched area). The second step, in (c) shows the top of the crosshatched area (dashed) and the further improvement (thin continuous line) of the estimate obtained by including image point 3. The inverted actual tip (thick continuous line) is reproduced here for comparison.

Here, I is the set of all points contained in the image, $\oplus$ is the dilation operation of mathematical morphology, and the P_i describe tip shapes at various stages of convergence. For reasons related to the fact that self-imaging inverts the tip image, the P's are inverted descriptions of the tip—that is, they are reflected in the x, y, and z coordinates. Equation (1) is a recipe for constructing a sequence of P_i of decreasing size, eventually converging to a limit we call P_R. P_R is then the least outer bound on the tip shape available from the image alone.

As we have already remarked, Eq. (1) says that some tip information is obtained for points that are not local image maxima. This is illustrated in Fig. 2(c) where the dashed line represents our result from Fig. 2(b). The thin continuous line indicates the result after using the term in Eq. (1) for x equal to the point, a non-maximum, labeled "3" in the image. This result, while still an outer bound as it must be, is closer to the actual tip (thick continuous line), shown for comparison. This is the result after consideration of only three image points. The fully converged blind reconstruction result is compared to the actual tip in Fig. 2(a).

Equation (1) assumes the dilation model (8, 33) of imaging. That is, it assumes that for sample S

$$I = S \oplus P. \quad (2)$$

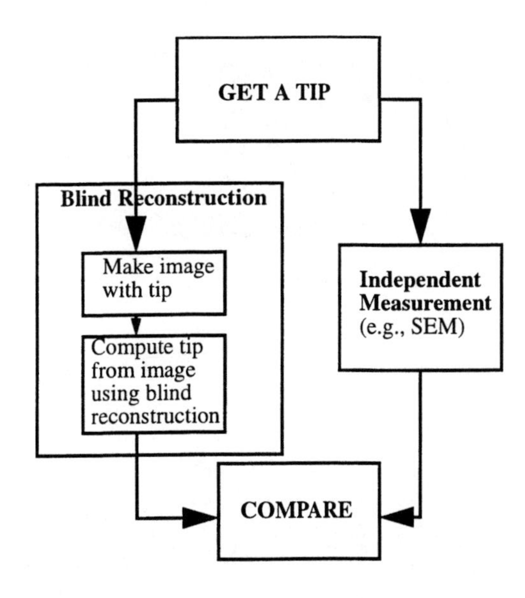

FIGURE 3. Design of the experiment.

This corresponds to the commonly held simple model of image formation, in which the tip height for a given lateral coordinate is that height at which the tip first makes contact with the specimen. (This is, incidentally, the same model assumed either explicitly or implicitly by the known-characterizer methods.)

In practice, of course, the dilation model is an approximation. The real image also reflects non-tip artifacts including noise, scanner nonlinearity, feedback loop response time and overshoot, piezo hysteresis and creep, etc. These instrumental artifacts produce some sharp spikes (upward-going and downward-going) on the image that cannot be expected to be removed totally by simple filtering.

To permit small deviations from the model, we employ a threshold parameter, which sets the maximum amount by which the image deviates from an ideal dilation (29). The reconstructed tip is permitted to be inconsistent with the image provided the magnitude of the inconsistency falls below this threshold.

EXPERIMENTAL

A flow diagram illustrating the main idea of the experimental comparison is displayed in Fig. 3. In the following sections, we explain the contents of the diagram's "Blind Reconstruction" and "Independent Measurement" boxes.

Obtaining a tip profile using blind reconstruction

The stylus profiler measurements were performed on a Rank Taylor-Hobson Talystep (34). The vertical and horizontal axes of this instrument are periodically calibrated

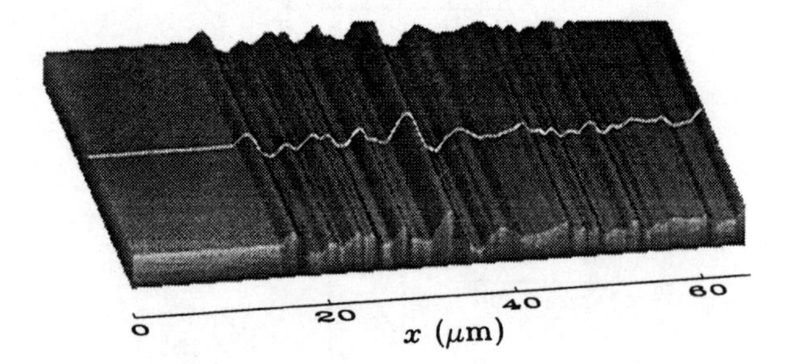

FIGURE 4. Schematic of the random roughness specimen illustrating the translational invariance along one direction. The z scale is exaggerated.

using known step height and sinusoidal pitch standards. The tip characterizer used in this experiment is the random roughness calibration specimen (Fig. 4) developed by Song (35). A smooth reference plane located at each side of the measuring area may be used as the start or the end of the random profile. This specimen's translational invariance along one axis is an important feature for our application because the stylus profiler measures only a profile rather than a full image. For an ordinary sample, it is not possible to infer a tip profile from an image profile—one needs to retain the full 3-d description. An exception occurs when the sample has translational invariance along an axis as does this one. In this case, profiles perpendicular to the ridges can be used to reconstruct a profile of the tip's "silhouette" (its projection onto the plane containing the profiler's scan).

We used a standard diamond stylus tip. The tip is held in an interchangeable holder with a flat on one side. When the tip is mounted on the Talystep, the stylus holder is clamped by a stylus adapter spring. The forward scan direction is perpendicular to the flat.

We recorded two 240 μm profiles consisting of 24,000 pixels at 10 nm intervals. The profiles had 230 nm rms roughness with a total corrugation (maximum to minimum point) of ~1.6 μm. Fig. 5 displays part of a trace.

The profiles became the I input in Eq. (1). The other required inputs are a threshold value and P_0, which is an initial crude outer bound. The threshold we used was 20 nm.

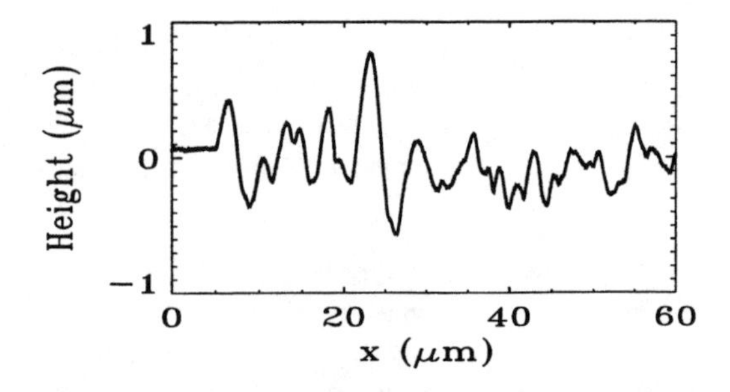

FIGURE 5. A section of the stylus profiler trace across the random roughness specimen. Part of the flat is visible at the left.

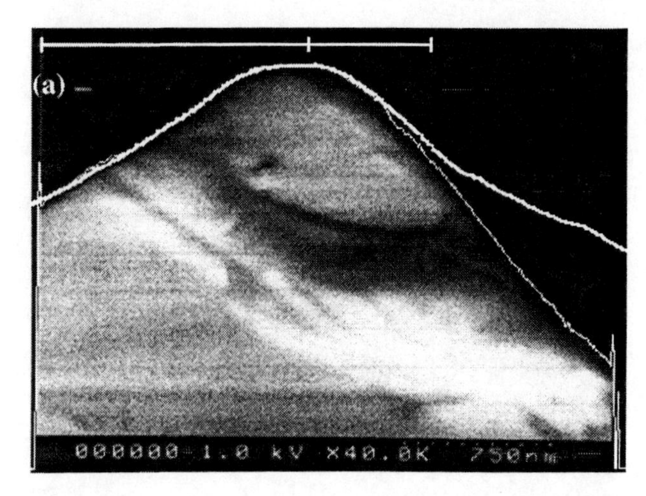

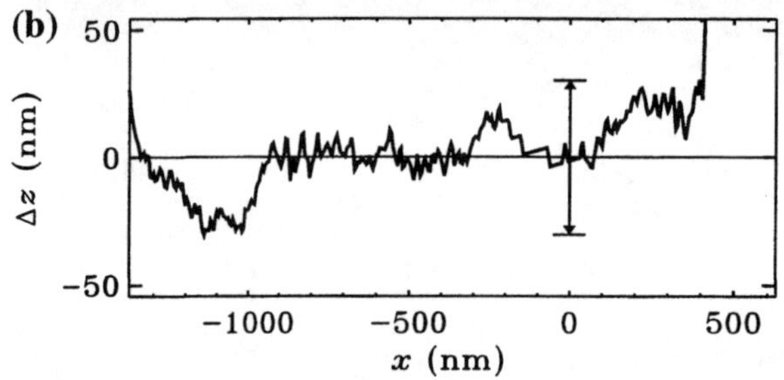

FIGURE 6. (a) Comparison of SEM (thin line) and blind reconstruction (thick line) profiles superimposed on the SEM image. The horizontal scale bar indicates the same 2000 nm wide region for which the difference between the curves is plotted in (b), where it is compared to stylus profiler repeatability (vertical bar).

This corresponds to just over 1% of the sample's corrugation. The choice was made independently of the SEM results (i.e., the threshold was not a fit parameter) by a method described elsewhere (29). For P_0 we used a tip that was completely flat for 2.5 μm on either side of the apex. That is,

$$P_0 = \begin{cases} 0 & \text{for } |x| \leq 2.5\ \mu\text{m} \\ -\infty & \text{everywhere else} \end{cases} \tag{3}$$

The resulting blind reconstruction profile is the thick curve in Fig. 6(a).

Obtaining a tip profile from the SEM

The tip was imaged in a high resolution scanning electron microscope, a cold field emission Hitachi S4500 (34), operated at 1 kV. We constructed a tip holder that allowed a small working distance during SEM imaging. The tip was cleaned with alcohol and coated by a thin (nominally < 20 nm) Pd-Au layer before being imaged to minimize surface charging. To calibrate SEM magnification, we used a NIST-certified standard reference material (SRM 484). The angle between the tip axis and the electron beam was set to 90° using the SEM's sample manipulator. The tip is therefore viewed directly from the side, and a profile can be determined from the image. The tip's rotational angle about its own axis was adjusted until the reference flat on the tip holder was parallel to the electron beam. This makes the desired profile perpendicular to the electron beam and insures that the SEM measures the same projection of the tip that was measured using the stylus profiler/blind reconstruction method.

When the electron beam strikes the stylus tip, secondary electrons are detected by the SEM's detector. When the beam misses the tip, other parts of the SEM chamber are far away and out of focus, so only background noise is measured at the detector. A profile curve was extracted from the resulting image by locating the points near the edge where the intensity first exceeds a threshold value high enough to discriminate against background counts. The SEM image and its profile (thin curve) are shown in Fig. 6(a).

COMPARISON OF BLIND RECONSTRUCTION AND SEM RESULTS

The comparison between the tip shape deduced by blind reconstruction and the image given by the SEM was done by overlaying the corresponding profiles using a least squares fit to determine the best relative positioning. The experiment was designed with a view to minimizing the number of free parameters required in this fit. For example, because we used instruments with calibrated scale (or magnification), the scale is not a free parameter. By careful positioning of the tip in the SEM as described above, we can forgo any parameters designed to correct for image distortion due to the SEM's angle of view.

The remaining parameters are the same that one would use to do this comparison manually, for example by laying one profile printed on a transparency on top of the other. One would move them to the same location and rotate them in the plane to point them in the same direction. This corresponds to two translational and one rotational parameter. By requiring the two profiles to have the same maximum value, we reduce the number of parameters that need to be fit by least squares to only two, the rotation and a horizontal translation. (One might think that the rotation could itself be fixed by careful alignment in the SEM. That is, however, not the case. The raster direction of the electron beam, and hence the image orientation, changes as the beam propagates down the column due to magnetic fields there. Consequently, the image orientation depends upon such instrument settings as the working distance and beam energy. The operator generally corrects this by rotating the image electronically. The result is a certain arbitrariness in this angle.)

The fit just described resulted in the overlay of profiles shown in Fig. 6(a). The residuals are shown in Fig. 6(b). For comparison, the vertical bar also shown there corresponds to $\pm\sigma$, where σ = 30 nm is the one standard deviation repeatability of the stylus traces (determined from differences in successively measured profiles at the same position on the sample). Far from the apex, the blind reconstruction result deviates to the high side. This is expected because only parts of the tip nearer to the apex than the corrugation of the characterizer have a chance of being touched by the characterizer. Blind reconstruction's outer bound may deviate from the true profile by a large amount in such inaccessible areas. However, the accessible areas within about 500 nm of the apex agree to within the repeatability of the stylus profiler.

A detailed uncertainty budget, taking into account all significant potential error sources is under development and will be published elsewhere along with application of the method to tips with more complicated profiles.

CONCLUSION

Improvement of SPM capabilities for nanometer-scale CD metrology depends upon estimating and correcting for the shape of the tip used to acquire the image. Blind reconstruction, which we proposed a few years ago, offers a potential resolution of this problem. Previous work showed its efficacy in simulations, but those studies could not address the extent to which noise and other instrumental artifacts present in real images would compromise the ability to recover the tip shape. The present study, in which blind reconstruction and SEM results for the same tip agree very well, demonstrates by example that such instrumental non-idealities need not be an insurmountable obstacle.

ACKNOWLEDGMENT

The authors gratefully acknowledge support from the National Semiconductor Metrology Program at NIST.

REFERENCES

1. E. C. Teague, L. W. Linholm, M. W. Cresswell, W. B. Penzes, J. A. Kramar, F. E. Scire, J. S. Villarrubia, and J. S. Jun, in Proceedings of the 1993 IEEE International Conference on Microelectronic Test Structures; Sitges, Barcelona, Spain; March 22-25, 1993, p 213.
2. J. Schneir, T. H. McWaid, J. Alexander, and B. P. Wilfley, J. Vac. Sci. Technol. B **12**(6), 3561-3566 (1994); J. Schneir, T. H. McWaid, R. Dixson, V. W. Tsai, J. S. Villarrubia, E. D. Williams, and E. Fu, SPIE Proceedings **2439**, 401-415 (1995).
3. B. Hacker, A. Hillebrand, T. Hartmann, and R. Guckenberger, Ultramicroscopy **42-44**, 1514-1518 (1992).
4. U. D. Schwarz, H. Haefke, P. Reimann, and H. J. Guntherodt, J. Microscopy **173**, 183-197 (1994).
5. J. A. DeRose and J. P. Revel, Microsc. Microanal. **3**, 203-213 (1997).

6. G. Reiss, J. Vancea, H. Wittmann, J. Zweck, and H. Hoffmann, J. Appl. Phys. **67**(3), 1156-1159 (1990); G. Reiss, F. Schneider, J. Vancea, and H. Hoffmann, Appl. Phys. Lett. **57**, 867-869 (1990).
7. D. J. Keller, Surf. Sci. **253**, 353-364 (1991).
8. H. Gallarda and R. Jain, SPIE Vol. **1464**, 459-473 (1991).
9. D. A. Grigg, P. E. Russell, J. E. Griffith, M. J. Vasile, and E. A. Fitzgerald, Ultramicroscopy **42-44**, 1616-1620 (1992).
10. D. J. Keller and F. S. Franke, Surf. Sci. **294**, 409-419 (1993).
11. L. Montelius and J. O. Tegenfeldt, Appl. Phys. Lett. **62**(21), 2628-2630 (1993).
12. J. E. Griffith and D. A. Grigg, J. Appl. Phys. **74**, R83-R109 (1993).
13. J. Vezenka, R. Miller, and E. Henderson, Rev. Sci. Instrum. **65**, 2249-2251 (1994).
14. C. Odin, J. P. Aime, Z. El. Kaadoum, and T. Bouhacina, Surf. Sci. **317**, 321-340 (1994).
15. S. Xu and M. F. Arnsdorf, J. Microsc. **173**, 199-210 (1994); J. Microsc. **187**, 43-53 (1997).
16. T. O. Glasbey, G. N. Batts, M. C. Davies, D. E. Jackson, C. V. Nicholas, M. D. Purbrick, C. J. Roberts, S. J. B. Tendler, and P. M. Williams, Surf. Sci. **318**, L1219-L1224 (1994).
17. N. Bonnet, S. Dongmo, P. Vautrot, and M. Troyon, Microsc. Microanal. Microstruct. **5**, 477-487 (1994).
18. P. Markiewicz and M. C. Goh, J. Vac. Sci. Technol. B **13**(3), 1115-1118 (1995).
19. D. L. Wilson, K. S. Kump, S. J. Eppell, and R. E. Marchant, Langmuir **11**, 265-272 (1995); D. L. Wilson, P. Dalal, K. S. Kump, W. Benard, P. Xue, R. E. Marchant, and S. J. Eppell, J. Vac. Sci. Technol. B **14**, 2407-2416 (1996).
20. K. F. Jarausch, T. J. Stark, and P. E. Russell, J. Vac. Sci. Technol. B **14**(6), 3425-3430 (1996).
21. M. Van Cleef, S. A. Holt, G. S. Watson, and S. Myhra, Journal of Microscopy **181**(1), 2-9 (1996).
22. K. I. Schiffmann, M. Fryda, G. Goerigk, R. Lauer, and P. Hinze, Ultramicroscopy **66**, 183-192 (1996).
23. V. J. Garcia, L. Martinez, J. M. Briceno-Valero, and C. H. Shilling, Probe Microscopy **1**, 107-116 (1997).
24. A. W. Marczewski and K. Higashitani, Computers Chem. **21**(3), 129-142 (1997).
25. J. S. Villarrubia, Surf. Sci. **321**, 287-300 (1994).
26. S. Dongmo, M. Troyon, P. Vautrot, E. Delain, and N. Bonnet, J. Vac. Sci. Technol. B **14**, 1552-1556 (1996).
27. J. S. Villarrubia, J. Vac. Sci. Technol. B **14**, 1518-1521 (1996).
28. P. M. Williams, K. M. Shakesheff, M. C. Davies, D. E. Jackson, C. J. Roberts, and S. J. B. Tendler, J. Vac. Sci. Technol. B **14**, 1557-1562 (1996).
29. J. S. Villarrubia, J. Res. Natl. Inst. Stand. & Technol. **102**(4), 425-453 (1997).
30. F. Jensen, Rev. Sci. Instrum. **64**(9), 2595-2597 (1993).
31. K. L. Westra, A. W. Mitchell, and D. J. Thomson, J. Appl. Phys. **74**(5), 3608-3610 (1993).
32. E. J. Van Loenen, D. Dijkkamp, A. J. Hoeven, J. M. Lenssinck, and J. Dieleman, Appl. Phys. Lett. **56**(18), 1755-1757 (1990).
33. G. S. Pingali and R. Jain, Proceedings IEEE Workshop on Applications of Computer Vision, (1992) pp. 282-289.
34. Certain commercial equipment is identified in this report in order to describe the experimental procedure adequately. Such identification does not imply recommendation or endorsement by the National Institute of Standards and Technology, nor does it imply that the equipment identified is necessarily the best available for the purpose.
35. J. F. Song, Surface Topography **1**, 303-314 (1988).

Crystallographic Characterization of Interconnects by Orientation Mapping in the SEM*

C. E. Kalnas[1], R. R. Keller[1], and D. P. Field[2]

[1]*National Institute of Standards and Technology, Materials Reliability Division, 325 Broadway, Boulder, CO 80303*
[2]*TexSEM Laboratories, 392 East 12300 South, Suite H, Draper, UT 84020*

We show in this paper how electron backscatter diffraction and orientation mapping within a scanning electron microscope can be used to measure local variations in crystallographic texture and grain boundary structure in interconnects. The reliability-limiting phenomena of stress voiding and electromigration are two examples of interconnect failure modes that depend strongly on local crystallographic structure. Several analysis examples are presented to show the utility of this technique for characterization of local microstructures in both copper- and aluminum-based lines. The advantages of a local measurement technique over a global texture method for orientation determination became immediately apparent in these investigations. This local approach to characterizing crystallography is expected to play an even larger role in technologies such as damascene or lift-off processing, where lines are deposited directly into precisely-defined geometries. Particularly in these more advanced processing technologies, one cannot extrapolate measurements of blanket film structure and properties to the case of narrow lines.

INTRODUCTION

The fact that metallizations for on-chip interconnects are fabricated with line dimensions of the order of 0.5 μm in thickness and 0.25 μm in width imposes stringent constraints on their reliability. Grain diameters that typically develop in physical-vapor-deposited metal films grown at relatively low temperatures are approximately as large as the film thickness (1). For interconnects with dimensions as given above, we expect the presence of single grains through the film thickness and across the line width, leading to a bamboo or near-bamboo structure. The behavior of such a material will be determined not simply by the global microstructure as inferred from blanket film characterizations, but rather by local variations in crystallographic features such as texture and grain boundary structures.

A notable example of the importance of local crystallography as opposed to global texture was demonstrated on aluminum-based interconnects by Rodbell *et al.* (2). They found that lines of stronger average texture showed shorter stress voiding lifetimes than lines of weaker average texture. This result, apparently contradictory to the conventional belief that stronger average texture results in better resistance to voiding, was clarified when measurements of local grain orientations were made. Grains adjacent to voids in the lines of stronger average texture were more randomly oriented than undamaged grains in the line. This suggested that a single off - $\langle 111 \rangle$ grain in the strongly textured material was more detrimental than such a grain in the weakly textured material. Such rapid changes in grain-to-grain misorientations necessarily lead to different grain boundary structures and sometimes detrimental properties in thin metal films for interconnects. A characterization method suitable for assessing interconnect microstructures at this scale is clearly necessary. Electron backscatter diffraction (EBSD) has been used successfully for making such characterizations on both aluminum and copper interconnect systems (2,3,4), revealing the significant effects of local variations in microstructure on voiding susceptibility.

EBSD in conjunction with scanning electron microscopy has been developed into a routine characterization tool over the past decade or so and has been applied to a variety of materials systems (5,6). Its very nature lends itself readily to the characterization of submicrometer metal structures such as on-chip interconnects. The scanning electron microscope (SEM) is usually operated to produce a practically monochromatic beam of electrons with energy in the range 10 to 40 keV, while the specimen is tilted to a large angle, typically 70°. The associated information volume contributing to the formation of the diffraction patterns is compatible with interconnect grain sizes and linewidths, as addressed in more detail in the next section.

We describe in this paper the application of an automated variant of EBSD, termed orientation mapping, as applied to interconnects. Orientation mapping systems consist of computer-controlled beam positioning, diffraction pattern capture and indexing, and software routines for analyzing crystallographic orientation relationships. Included here are examples of analyses carried out on aluminum alloy and copper lines subjected to stress- and electromigration-induced voiding. Emphasis is placed on analysis examples rather than details of the reliability physics, which is

*Partial contribution of the National Institute of Standards and Technology. Not subject to copyright in the U.S.

CP449, *Characterization and Metrology for ULSI Technology: 1998 International Conference*
edited by D. G. Seiler, A. C. Diebold, W. M. Bullis, T. J. Shaffner, R. McDonald, and E. J. Walters

beyond the scope of this paper. See references (2-4) for such descriptions.

EXPERIMENTAL CONSIDERATIONS

The basic configuration of an EBSD system as mounted in an SEM is shown in Fig. 1. A phosphor screen affixed to the end of a low-light video system allows the user to view in two dimensions the scattering of electrons from the specimen surface once the stationary beam is positioned in the desired location on the specimen. Typical SEM operating conditions for EBSD characterization include an accelerating voltage in the range 10 to 40 kV, a probe current of approximately 1 nA, and a short working distance. These conditions result in a sufficiently fine probe size while providing a signal-to-noise ratio large enough for accurate pattern indexing.

The information volume contributing to the formation of diffraction patterns is determined by two parameters: (a) the depth into the specimen from which electrons containing orientation information backscatter, and (b) the diameter of the electron probe as resolved on the specimen surface. An upper bound to the orientation-dependent information depth can be given by assuming normal beam incidence on the specimen surface. The Forward-Backward approximation (7), describing the formation of backscattered electron contrast, indicates that orientation-dependent contrast arises from a depth corresponding to approximately two absorption distances into the crystal. The resulting information depths using absorption data from Reimer (8) for 10 to 40 keV electrons in aluminum are approximately 0.06 to 0.12 μm. Analogous depths for copper are 0.03 to 0.05 μm. The penetration in a direction normal to the surface of specimens tilted 70° will be less. The electron probe diameter for a LaB_6 filament operated at 30 kV at a current of 1 nA is approximately 0.02 μm (9) and leads to a resolved spot size of approximately 0.06 μm on the specimen surface. We refer the reader to a recent study of measured versus calculated EBSD spatial resolution, given by Ren *et al.*

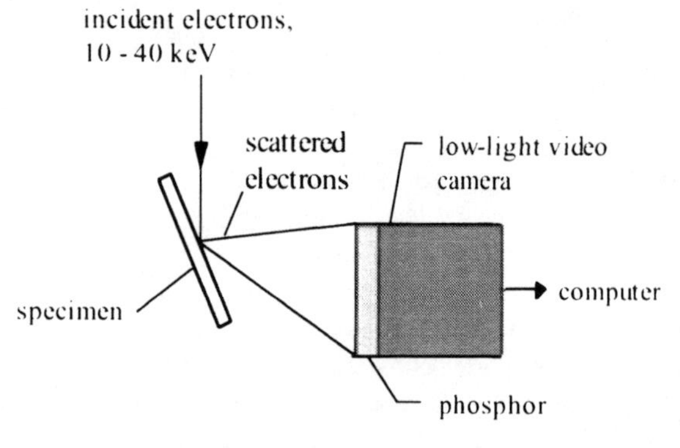

FIGURE 1. Schematic showing specimen and EBSD hardware in the SEM.

(10), for further details. The total EBSD analysis volume then has a diameter of approximately 0.1 μm or less and is therefore well-suited to the characterization of interconnect structures presently a couple of times larger than that. Much finer spatial resolution can be attained with the use of a field-emission electron source. Present-day instrumentation should be capable of characterizing interconnects with linewidths approaching 0.1 μm, forecasted for manufacture in 2006 (11). With further advances in SEM and EBSD hardware anticipated by that time, we may have a tool that is still applicable to interconnects targeted for manufacture 10 or more years from now.

The EBSD pattern contains an abundance of crystallographic information. Included are the three-dimensional orientation matrix and a semi-quantitative measure of the degree of lattice disorder. A single pattern suffices to assess the orientation matrix since scattering from the specimen occurs divergently and a very large solid angle of scattered electrons is detected. Therefore, all three primary lattice directions contribute to the symmetry of the resulting diffraction pattern. An example of an indexed EBSD pattern obtained from an aluminum alloy is shown in Fig. 2. The bright bands are Kikuchi bands and represent (simplistically) the traces of specific families of atomic planes as they would intersect the phosphor screen. As such, Kikuchi bands act as though rigidly attached to the specimen; hence, tilting the specimen results in translation of the diffraction pattern. Lower index crystallographic zone axes are observed where Kikuchi bands intersect. One possible orientation solution to this particular pattern is shown by the overlaid lines and zone axis indices. Note the large angle of detection, indicated by the fact that poles of

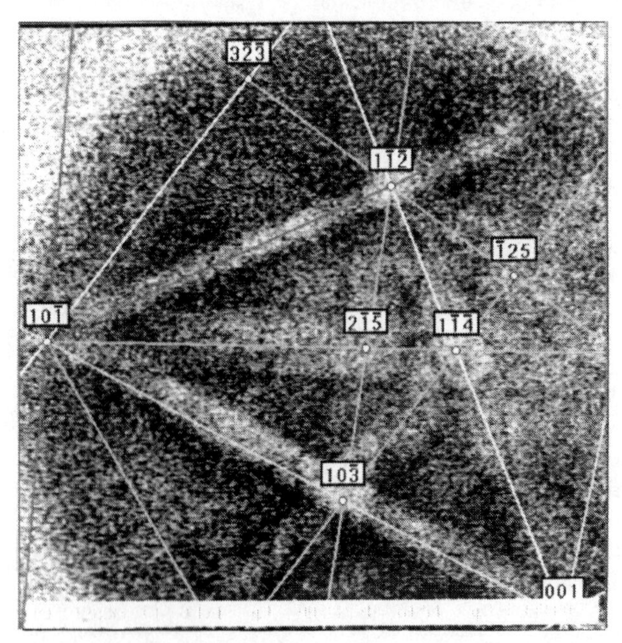

FIGURE 2. Example of indexed EBSD pattern obtained from an aluminum alloy interconnect.

both (100) and (110) types are visible. While not 100% exact, the solutions typically found through the pattern recognition and indexing algorithm are within approximately 1° of the absolute orientation, while relative misorientation determinations can be made to approximately 0.5° (12). The lattice disorder is measured by assessing the degree of diffuseness of the diffraction pattern. More diffuse patterns result upon scattering from a crystal where the lattice disorder is greater, due for instance to a high density of dislocations or point defects, or to disorder associated with a grain boundary.

A specific experimental requirement for EBSD characterization of interconnects is that the specimens must be either unpassivated or depassivated prior to imaging. An important consideration is to ensure that the underlying metal is left unaltered during depassivation. This is often done by selective chemical means, either by dissolving the passivating material in an appropriate acid solution or by reactive ion etching. Interconnects without passivating layers are likely to be subjected to stress states significantly different from those of a passivated line, implying that long-term relaxation processes may proceed differently from that expected for a passivated line. A further consequence of exposed interconnects is that the surface will likely react, leading to the thin, native passivating oxide for aluminum or to a more damaging corrosion oxide for copper.

Comprehensive microstructural characterizations are made by collecting a large number EBSD patterns and evaluating them in terms of crystallographic texture. Automated orientation mapping systems are capable of collecting and indexing many thousands of diffraction patterns over a period of several hours. In addition to the obvious spatial resolution advantages, the most significant advantage of orientation mapping over more conventional X-ray diffraction texture measurements centers on the fact that each piece of orientation data can be correlated directly back to the location on the specimen from which the measurement was taken.

ORIENTATION MAPPING OF INTERCONNECTS

During orientation mapping, the electron beam is moved by a computer system into an array of positions on the specimen surface, and the resulting EBSD pattern corresponding to each position is collected and automatically indexed by means of pattern recognition algorithms. A simple way to display the initial result is to create an image composed of contrast variations that depend on the variations in pattern diffuseness encountered while the beam is rastered across the specimen. Figure 3 shows such an image, obtained from a wide, pure aluminum line, with the beam stepped in 0.16 μm increments. Grain interiors appear brighter, corresponding to sharper EBSD patterns, whereas grain boundary regions appear darker. Overlaid on this image are bright lines delineating the grain boundaries. This interconnect segment has a polycrystalline grain structure, which is apparent in the image. Such information is normally unavailable through conventional secondary electron imaging or even by channeling contrast during backscattered electron imaging due to the excessive topographic contrast in those imaging modes. Although the spatial resolution available in the image quality map is limited by the EBSD information volume and the distance between beam positions, the user does not have to resort to the less common, and sometimes more tedious, analytical methods of transmission electron microscopy or focused ion beam imaging in order to see the grain structure. Similar analyses from aluminum alloy lines of width 0.5 μm and less have also been reported (13,14).

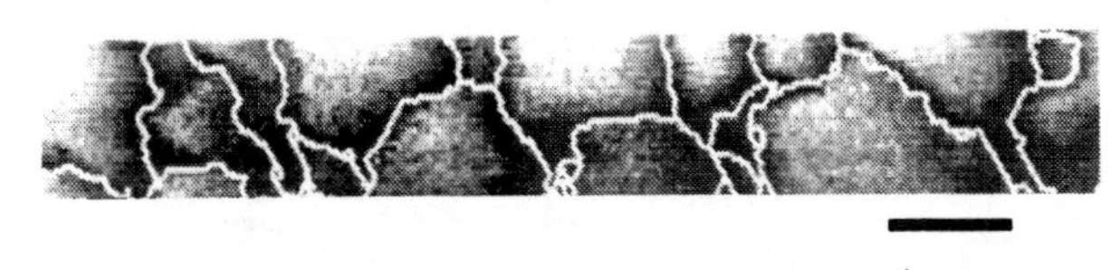

FIGURE 3. Orientation map showing grain structure and grain boundaries.

The same data set used to construct the image in Fig. 3 was used to calculate the crystallographic texture plots in Fig. 4. The orientations displayed in this figure correspond to data taken directly from the single interconnect segment shown above. The ⟨111⟩ fiber texture is apparent from these plots. The ability to denote the orientation of specific grains is evident by the orientation map in Fig. 5. Grains are highlighted according to how close their orientations are to

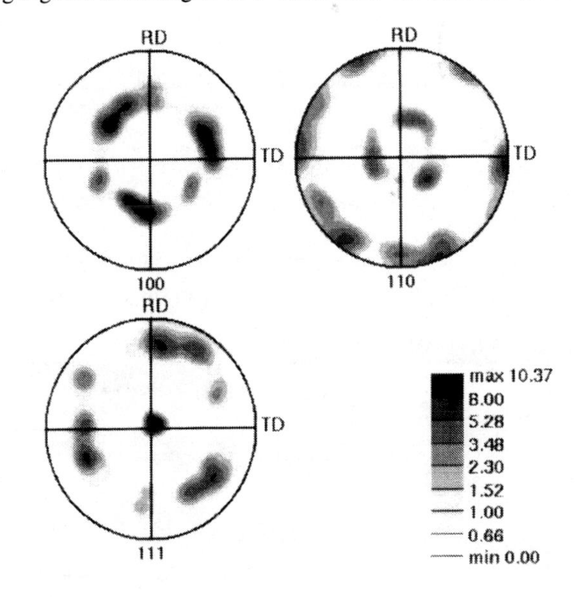

FIGURE 4. Pole figures showing predominant ⟨111⟩ fiber texture in aluminum interconnect segment. The intensity scale is given in terms of times random.

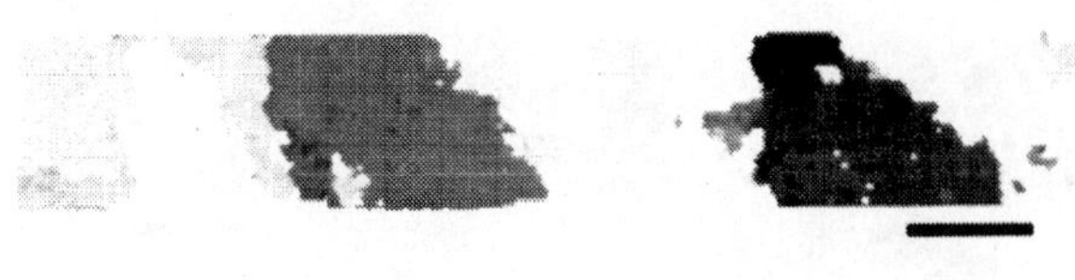

Figure 5. Orientation map indicating proximities of grain orientations within 10° of the ⟨111⟩ fiber texture. Darker regions are closer to ⟨111⟩.

the overall average ⟨111⟩ fiber texture of the line. Darker regions have surface normals most closely aligned with ⟨111⟩ directions.

The fact that the relationships between orientation measurements and microstructure are preserved means that we can also infer grain boundary structures to a certain extent. For instance, the distribution of minimum misorientation angles across boundaries is known. With the assumption of perfect columnar grain growth during film deposition, we can assess both the grain boundary plane crystallography and the tilt and twist character associated with any individual boundary. Such information can be useful in describing the diffusive properties of grain boundaries in interconnects. For instance, a local crystallographic characterization of the regions adjacent to stress voids in copper lines (3) revealed that voids formed at triple junctions where three grains intersected with orientations relatively far from the overall average ⟨111⟩ fiber texture. Further, the relatively higher misorientation angle boundaries forming voided triple junctions had a significant proportion of twist character to them as well as tilt character with rotation axes parallel to the film plane. Such grain boundary structures are associated with high in-plane diffusivities due to the dislocation structures comprising the boundaries. This suggested that local regions of the interconnect microstructure displaying high in-plane diffusivity were favored sites for significant growth of stress voids.

Orientation mapping has also been used to reveal the grains of damascene-processed lines. Bamboo structures are desired for interconnect structures due to their good resistance to electromigration voiding. Figure 6 shows an orientation map obtained from damascene-processed Al-

Figure 6. Orientation map showing the near-bamboo grain structure of damascene-processed Al-0.5% Cu lines.

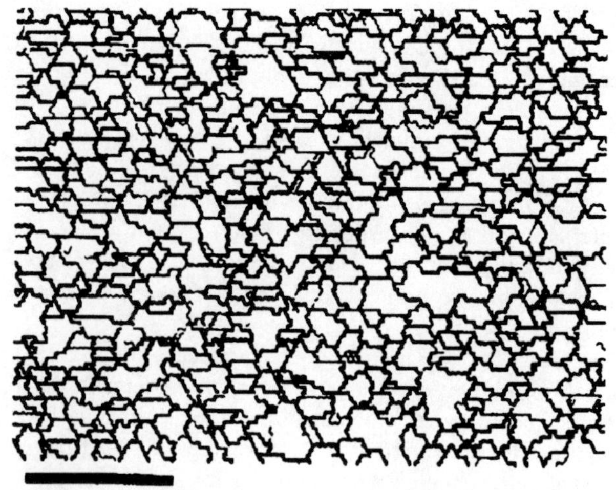

FIGURE 7. Orientation map obtained from field emission SEM data, showing grain boundaries in a blanket film of platinum with average grain diameter 0.075 μm. The beam was stepped in 0.02 μm increments. Thinner lines correspond to grain boundaries with misorientation angles less than 10°.

0.5% Cu lines. The contrast variations indicate how close the grain orientations are to the overall average ⟨111⟩ fiber texture. Note that the lines are actually near-bamboo, since several finer grains are visible along the line edges.

The advantages of the spatial resolution of a field emission microscope for orientation mapping are displayed in Fig. 7. Shown is a map displaying the grain boundaries in a blanket film of platinum with average grain diameter 0.075 μm. The data were taken on a SEM with a Schottky source running at an accelerating voltage of 20 kV and a probe current of 1 nA. The beam was stepped in 0.02 μm increments. This result demonstrates that crystallographic orientation mapping with the use of a field emission SEM is capable of revealing grain structure at the sub-0.1 μm level. This characterization method may also be useful for extremely narrow interconnects.

CONCLUSIONS

Orientation mapping is being used as a measurement tool to aid in the understanding and eventual control of stress voiding and electromigration. These failure mechanisms will become particularly critical as interconnect linewidths decrease and as new material systems such as Cu and low *k* dielectrics are introduced (11).

The effects of variables such as local texture and grain boundary misorientations on voiding can be assessed by orientation mapping. This characterization technique will be critical to understanding the interactions among the relevant microstructural variables and their effects on interconnect reliability. The orientation mapping method should find application to future interconnect systems.

ACKNOWLEDGMENTS

We thank T. N. D. Marieb for providing specimens. We also acknowledge J. A. Nucci for collaborating on the Cu interconnect research and the NIST Office of Microelectronics Programs for support.

REFERENCES

1. Sheppard, K. G. and Nakahara, S., *Process. Adv. Mater.* **1**, 27-39 (1991).
2. Rodbell, K. P., Hurd, J. L., and DeHaven, P. W., *Mater. Res. Soc. Symp. Proc.* **403**, Pittsburgh: Mater. Res. Soc., 1996, pp. 617-626.
3. Keller, R. R., Nucci, J. A., and Field, D. P., *J. Electron. Mater.* **26**, 996-1001 (1997).
4. Field, D. P. and Dingley, D. J., *J. Electron. Mater.* **25**, 1767-1771 (1996).
5. Field, D. P., *Ultramicros.* **67**, 1-10 (1997).
6. Juul Jensen, D., *Ultramicros.* **67**, 25-34 (1997).
7. Spencer, J. P., Humphreys, C. J., and Hirsch, P. B., *Philos. Mag.* **26**, 193-213 (1972).
8. Reimer, L., *Transmission Electron Microscopy*, New York: Springer-Verlag, 1984, ch. 7, p. 303.
9. Goldstein, J. I., Newbury, D. E., Echlin, P., Joy, D. C., Romig, Jr., A. D., Lyman, C. E., Fiori, C., and Lifshin, E., *Scanning Electron Microscopy and X-Ray Microanalysis*, 2nd ed., New York: Plenum Press, 1992, ch. 2, p. 62.
10. Ren, S. X., Kenik, E. A., Alexander, K. B., and Goyal, A. *Microsc. Microanal.* **4**, 15-22 (1998).
11. *The National Technology Roadmap for Semiconductors: Technology Needs*, 1997 edition, Semiconductor Industry Association, numerous references therein.
12. Dingley, D. J. and Randle, V., *J Mater. Sci.* **27**, 4545-4566 (1992).
13. Hurd, J. L., Rodbell, K. P., Gignac, L. M., Clevenger, L. A., Iggulden, R. C., Schnabel, R. F., Weber, S. J., and Schmidt, N. H., *Appl. Phys. Lett.* **72**, 326-328 (1998).
14. Field, D. P., Sanchez, Jr., J. E., Besser, P. R., and Dingley, D. J., *J. Appl. Phys.* **82**, 2383-2392 (1997).

HRTEM as a Metrology Tool in ULSI Processing

V.S. Kaushik, L. Prabhu, A. Anderson and J. Conner

Physical Analysis Laboratory, Advanced Product Research & Development Laboratory, Motorola Inc., 3501 Ed Bluestein Blvd., Austin, TX, 78723

Current ULSI processes involve the use of ultra-thin layers to achieve device performance. Proper control of the thickness of these layers is a manufacturing requirement. To this end accurate thickness measurement is an important metrology issue. High-resolution Transmission Electron Microscopy (HRTEM) provides a direct imaging method for materials and interfaces and can provide accurate thickness measurement. Crystal lattice images that are obtained can serve as internal calibration standards. There are a number of challenges to using this technique including careful sample preparation and choice of imaging conditions. The valid use of HRTEM as a metrology technique requires consideration of several aspects such as sample location and film properties. Optimal conditions for measuring amorphous layers are different from those for crystalline layers. We have used this technique to measure the thickness of amorphous dielectric layers (1-5 nm) and crystalline layers such as tantalum. HRTEM measurements have been used for the calibration of fab in-line metrology tools especially in those cases where the in-line tools do not already have necessary material parameters. Potential applications of this method include measurement of defect density and distribution for modeling and simulation.

INTRODUCTION

Shrinking device dimensions and low-voltage applications have resulted in the use of ultra-thin layers to achieve device performance in ULSI processing. For the fabrication of thin layers with consistent film properties, the process equipment has to be well-characterized. Film thickness is one of the critically controlled parameters for thin films. The accurate measurement of film thickness is therefore an important metrology issue. Transmission electron microscopy (TEM) has long been recognized as a valuable imaging and analytical tool in semiconductor processing. High-resolution (HR)TEM can provide images of materials and their interfaces enabling accurate thickness measurement. Although a less glamorous use of HRTEM, thickness measurement is a critical function in the metrology aspects of ULSI processing. While the TEM is currently not perceived as an in-line metrology tool in a manufacturing environment, the TEM serves as an accurate calibration standard for other in-line tools. TEM can also be used to examine other properties of the film such as crystallinity and homogeneity.

SAMPLE PREPARATION

The choice of sample preparation method will vary widely (1) depending on the nature of the sample or material being analyzed. Ultimately the sample should contain the area of interest in the electron transparent region which is typically 50-100 nm thick for HREM. Further the sample needs to be free of artifacts introduced during the sample prep operations. During equipment qualification a large number of samples may need to be examined in order to determine the process window for volume manufacturing operations. Traditional techniques of dimpling and ion milling can offer large thin areas with minimal artifacts. This is also a relatively rapid method suitable when there are many samples. We have successfully used this method for most samples of interest in ULSI processing such as thin dielectric films and metal or metal oxide films. For films with poor adhesion, the focused ion beam (FIB) can be used to prepare TEM samples. The popularity of the FIB as an FA tool may limit the volume of TEM samples thus prepared.

RESULTS

The principles and practice of HRTEM have been well described in the literature (2). Several precautions with regard to sample and microscope conditions need to be taken in order to obtain the most meaningful data. For most silicon based samples HRTEM images are obtained in the <110> zone axis of Si. We have used this imaging condition as a reference point in all our measurements.

Imageable Regions

For HRTEM imaging the choice of the region on the sample where imaging is performed is crucial. Since ion-milled edges are often amorphized, it is important to find an area on the sample that is thin enough for clear images and yet far enough from the edge to minimize ion-milling artifacts. The area chosen should also be representative of the sample. If local variations of thickness are observed on the sample, the scale of the observation should be

CP449, *Characterization and Metrology for ULSI Technology: 1998 International Conference*
edited by D. G. Seiler, A. C. Diebold, W. M. Bullis, T. J. Shaffner, R. McDonald, and E. J. Walters

taken into account. Often, there is a short-range variation of thickness over a few microns and a long-range variation as in center-to-edge of wafer. Even though the sample may contain thin regions 100's of microns long, the area that is practically examined in the TEM is very small (typically a few microns) in comparison to other in-line tools that typically provide averages over much larger areas (0.1-1.0 mm). This should be kept in mind when comparing data between different techniques.

Amorphous Films

To make accurate measurements on thin films, it is critical that the layer of interest be distinctly identified. This can often be achieved by capping the layer of interest by another layer of differing contrast without affecting the original microstructure.

For thin amorphous films, the layer of interest can be well defined with sufficiently distinct contrast when it is bounded on either side by crystalline layers. We have studied ultra-thin dielectric layers such as oxides and nitrides sub-1 nm to 5 nm using this technique. Figure 1 shows a thin dielectric layer deposited on a silicon substrate with a cap of polysilicon. Since this cap layer appears amorphous at the left end of Fig. 1, measurements were only made in the region where the boundaries on either side are crystalline and well-defined.

(Poly)-Crystalline Films

Crystalline or polycrystalline films may sometimes have rougher surfaces than amorphous films due to the grain structure. This may cause thickness measurements to vary. The thickness measurement can then be reported to include this variation. Unless the film has adhesion problems or is mechanically fragile, capping layers are less important for thin crystalline films since the epoxy used in sample prep can serve as a boundary.

FIGURE 1. HREM image of a thin 20Å (2.0 nm) dielectric layer delineated by crystalline layers on either side.

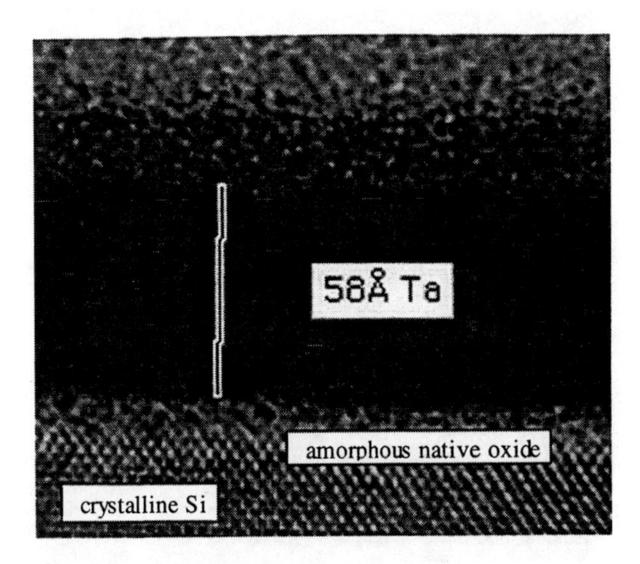

FIGURE 2. HREM image of a thin tantalum layer on an amorphous native oxide over a crystalline Si substrate.

Thin films of tantalum have been examined as shown in Fig. 2. These measurements were used for calibration of fab in-line metrology tools. The HRTEM technique is especially useful for those cases where the in-line tools do not already have necessary material parameters.

Rough Interfaces

For thin layers that are deposited or grown on pre-treated substrates, the interfacial roughness becomes a factor in choice of imageable region. Figure 3 shows a rough but well-defined Si surface. Figure 4 obtained at a thicker region from the same sample shows a dark band at the interface which is probably due to roughness along the sample thickness or z-direction. This roughness causes some uncertainty in the location of the interface. Since the scale of the roughness (~0.5 nm) is a significant fraction of the amorphous layer thickness in this case (~2.0 nm), this roughness is also a source of measurement errors and should be reported.

MICROSCOPE CALIBRATION

In most silicon-based processing, the silicon lattice at high resolution is readily available as an internal calibration standard. However, for thicker films which need to be imaged at lower magnifications, the microscope must be calibrated at those magnifications. Many TEM facilities currently use digitized image acquisition rather than traditional photographic negatives. If possible, it is better to set up the image acquisition procedure using the software to deliver calibrated images than to modify the images after acquisition. We have evaluated two different methods for calibration as described next.

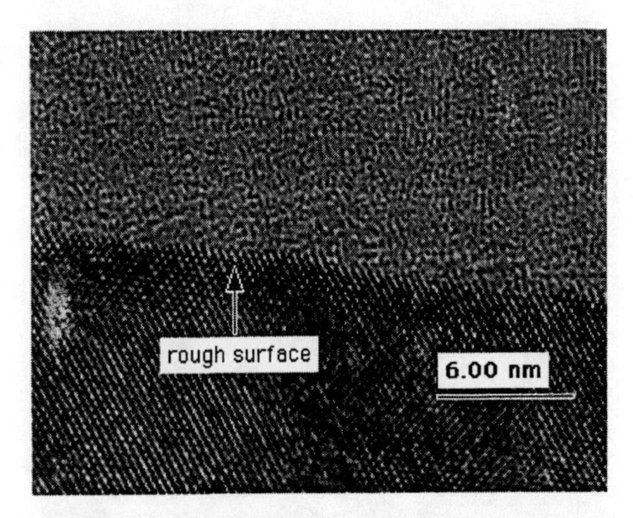

FIGURE 3. TEM image of a very thin region clearly showing measurable interfacial roughness on a well-defined surface.

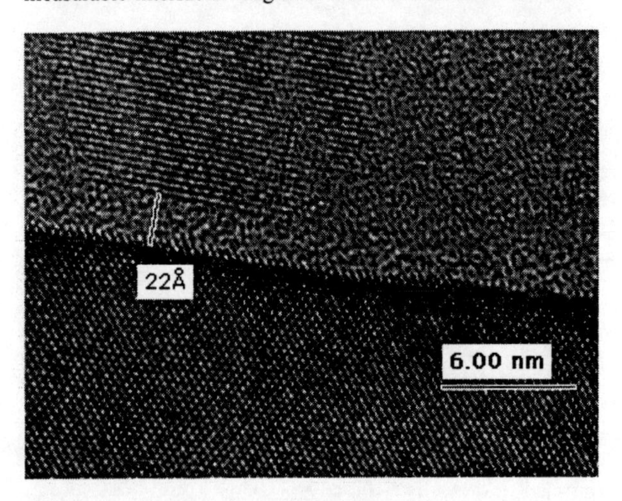

FIGURE 4. TEM image of a slightly thicker region of the same sample of Fig. 3 showing a dark line indicative of roughness averaged along the sample thickness.

Sub-micron Device Structures

Device structures fabricated in ULSI technologies typically provide features over a range of dimensions from 5-50 nm gate oxides to >500 nm gate electrodes and contacts. At HRTEM magnifications >250KX an image is obtained showing the gate dielectric and the Si lattice. The dielectric thickness can be measured using the Si lattice as a calibration standard. At intermediate magnifications ~100KX the gate oxide can be used as a standard to measure other features e.g. gate polysilicon. Note that for thicker gate oxides (>25 nm) the roughness is a smaller fraction of the thickness and this will lead to reduced error. This known dimension can in turn can be used at lower magnifications 10-25KX for poly thickness or contacts.

Calibration Standards

Another method is to obtain standards that have features or thin layers (3) of known dimensions such as multilayers deposited by molecular beam epitaxy (MBE). Although the sample may have hetero-epitaxial layers of differing contrast, the interfaces between the layers should be clearly defined at lattice-resolution. If this interface is not well-defined, it can introduce small errors in the measurement.

CONCLUSIONS

The use of HRTEM as a calibration technique for thickness measurement is rapidly increasing in our laboratory. TEM measurements are often used to calibrate other in-line measurement tools such as ellipsometers or critical dimension microscopes. This is useful for new materials or processes where the films fabricated have properties (e.g. refractive index or dielectric constant) that are not well-known.
In summary, the points to remember while using HRTEM for metrology are:
a) choice of location on the sample;
b) proper definition of the layer of interest;
c) effects and sources of thickness variations from roughness and crystallinity; and
d) calibration of the microscope.
Another potential area of application for this technique is the measurement of defect density distribution that might be of interest to modeling and simulations.

ACKNOWLEDGMENTS

The authors would like to gratefully acknowledge H.H. Tseng, R. Hegde, C. Hobbs and D. O'Meara for providing the processed silicon wafers. We also appreciate the management support of D. Sieloff and J. Mogab, and P. Tobin for helpful discussions.

REFERENCES

1. Bravman, J.C., Anderson,R., and McDonald, M.L., *Specimen Preparation for Transmission Electron Microscopy of Materials*, Materials Research Society Symposium Proceedings, Vol. 115, 1988.
2. Spence, J.C.H, *Experimental High-Resolution Electron Microscopy*. Oxford University Press, 1988.
3. McCaffrey, J.P., *Calibrating the TEM*, Microscopy Today, Vol. 97-9, November 1997, pp 16.

Transmission Electron Microscopy Investigation of Titanium Silicide Thin Films

A. F. Myers, E. B. Steel, and L. M. Struck

Surface and Microanalysis Science Division, Chemical Science and Technology Laboratory, NIST, Gaithersburg, MD 20899.

H. I. Liu and J. A. Burns

MIT Lincoln Laboratory, L-216, 224 Wood St., Lexington, MA 02173.

Titanium silicide films grown on silicon were analyzed by transmission electron microscopy (TEM), electron diffraction, scanning transmission electron microscopy (STEM), energy dispersive x-ray spectroscopy, and electron energy loss spectroscopy. The films were prepared by sequential rapid thermal annealing (RTA) at 675 °C and 850 °C of 16-nm-thick sputtered Ti on Si (001) wafers. In some cases, a 20-nm-thick TiN capping layer was deposited on the Ti film before the RTA procedure and was removed after annealing. HRTEM and STEM analysis showed that the silicide films were less than 0.1 µm thick. The capped film was more uniform, ranging in thickness from ~ 25 - 45 nm, while the uncapped film ranged in thickness from ~ 15 - 75 nm. Electron diffraction was used to investigate the titanium silicide phases present in the films; more than one phase was found to exist.

INTRODUCTION

Titanium disilicide ($TiSi_2$) has recently been investigated as an interconnect material for sub-0.25 µm CMOS devices, due to its low electrical resistivity and high thermal stability, as well as its self-aligning behavior (1-11). $TiSi_2$ forms in two different crystal structures. The metastable C49 structure is base-centered orthorhombic and forms at low temperatures; around 700 °C, C49 transforms to the stable C54 structure, which is face-centered orthorhombic. The C54 structure has a much lower resistivity (13-25 µΩ-cm) than the C49 structure (60-130 µΩ-cm) and is thus the desired variant for device interconnects (2, 3).

As the silicide film thickness decreases far into the sub-micrometer region or the line width decreases below a few micrometers, the annealing temperature required to convert the C49 phase to the C54 phase increases (4, 5). The higher temperature needed for C54 formation may lead to diffusion of dopants and interdiffusion between adjacent layers in the device, thereby degrading device performance.

Using a titanium nitride capping layer during silicide formation has been reported to improve the characteristics of the resulting silicide film, as determined primarily by measuring sheet resistivity (1, 7-11). The TiN capping layer lowers the resistivity, improves the film uniformity, suppresses line-width degradation effects and increases the thermal stability of the film. In the present study, blanket Ti silicide films grown on silicon were studied by several analytical electron microscopy techniques to investigate the role of the TiN capping layer in improving the quality of the silicide film. The following techniques were utilized: transmission electron microscopy (TEM), scanning transmission electron microscopy (STEM), selected area electron diffraction (SAED), energy dispersive x-ray (EDX) spectroscopy, and electron energy loss spectroscopy (EELS). The results of this microscopy study are presented.

EXPERIMENTS AND RESULTS

Titanium silicide films less than 0.1 µm in thickness were prepared by rapid thermal annealing (RTA) of 16-nm-thick Ti films, which were reactively sputtered on Si (001) wafers (6). The Si substrate was doped with arsenic to $3x10^{20}$ cm^{-3}. Prior to Ti deposition, the Si wafers were sputter cleaned with argon ions to remove the native oxide. In some cases, a 20-nm-thick TiN capping layer was deposited on the Ti prior to the first anneal. The first rapid thermal annealing (RTA) step was carried out at 675 °C for 80 s in a 100% N_2 ambient. Unreacted Ti and the TiN capping layer were removed by wet etching in 1:1:5 NH_4OH:H_2O_2:H_2O for 3.5 minutes at 50 °C. The final RTA step was carried out at 850 °C for 60 s, also in a N_2 ambient. The sheet resistivities of the resulting Ti silicide films were measured using the standard Van der Pauw method to be 6.0 Ω/sq and 60 Ω/sq for the capped and uncapped films, respectively (12).

The TEM, STEM, SAED, EDX, and EELS analyses were performed on a 300-keV field emission TEM, equipped with an energy dispersive x-ray spectrometer and a parallel acquisition EELS spectrometer. Specimens were prepared for TEM by first dimpling, then argon ion milling. TEM (Figures 1 and 2) and STEM (Figures 3 and 4) showed that neither the capped nor the uncapped titanium silicide films were uniformly thick, varying in thickness from approximately 25 - 45 nm for

CP449, *Characterization and Metrology for ULSI Technology: 1998 International Conference*
edited by D. G. Seiler, A. C. Diebold, W. M. Bullis, T. J. Shaffner, R. McDonald, and E. J. Walters

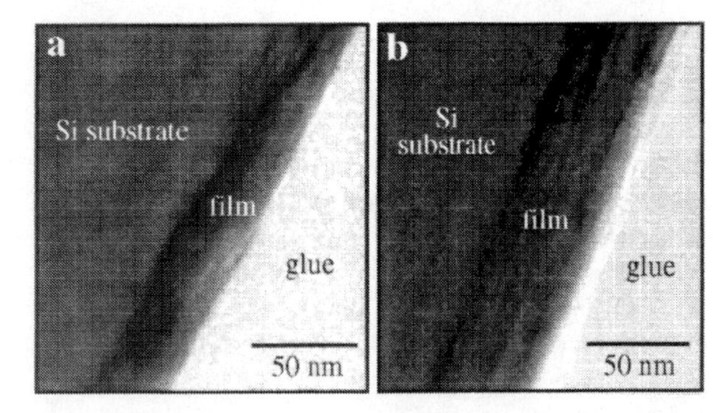

FIGURE 1. (a) and (b) show cross-sectional TEM images from two different regions of a Ti silicide film on a Si (001) substrate. This film, which was formed with a TiN capping layer, varied in thickness from approximately 25 nm to 45 nm.

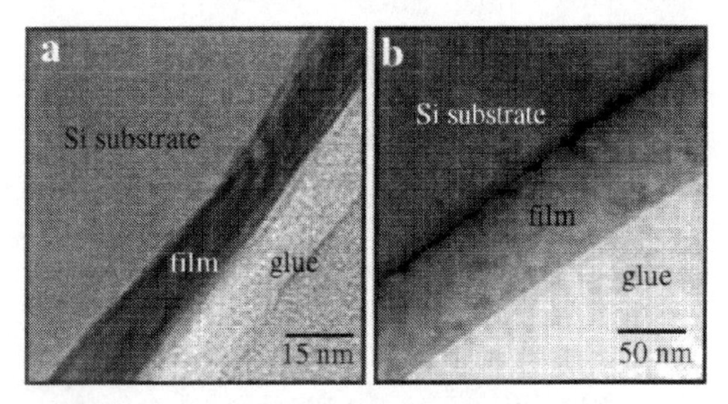

FIGURE 2. (a) and (b) show cross-sectional TEM images from two different regions of a Ti silicide film on a Si (001) substrate. This film, which was formed without a TiN capping layer, varied in thickness from ~ 15 nm to 75 nm.

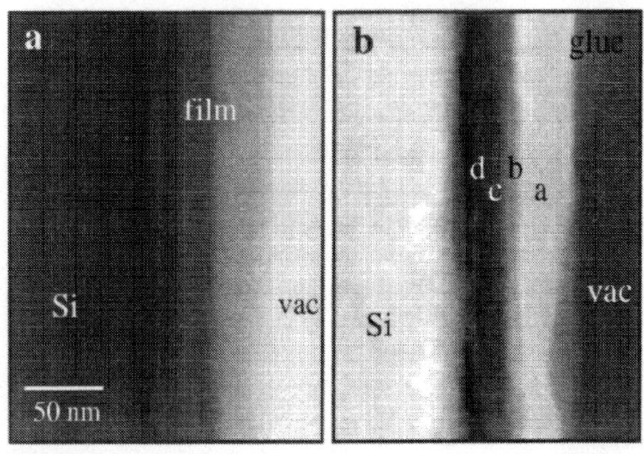

FIGURE 3. Bright field (a) and dark field (b) STEM images from the same region of the capped Ti silicide film. The letters a through d in the dark field image correspond to the EDX spectra in Figure 8.

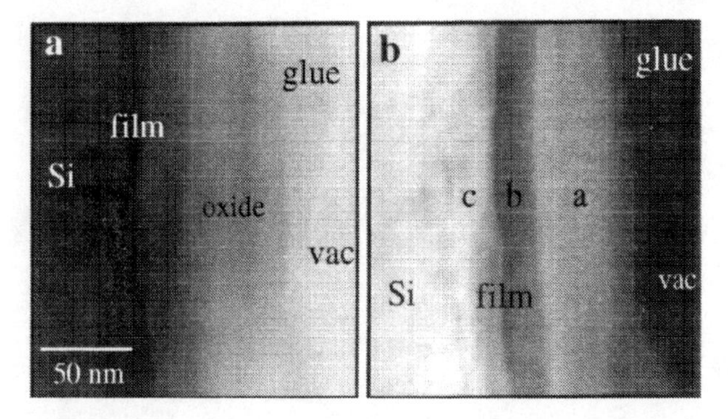

FIGURE 4. Bright field (a) and dark field (b) STEM images from the same region of the uncapped Ti silicide film. The letters a through c in the dark field image correspond to the EDX spectra in Figure 10.

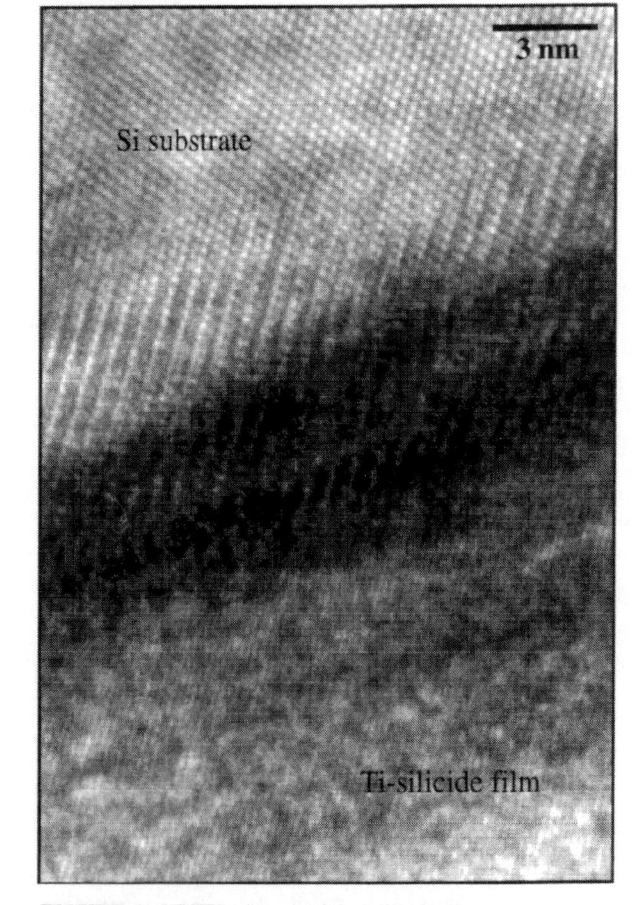

FIGURE 5. HRTEM image of the silicide/Si interface for the capped film. Si (111) and C54-$TiSi_2$ (022) fringes are visible.

the capped films and 15 - 75 nm for the uncapped films. Both films showed a banded contrast, produced by the presence of multiple grains in the films. The silicide/Si interface was highly defective, as evidenced by the dislocations clearly visible in Figure 2b and Figure 5. A high resolution TEM image of the Ti silicide/Si interface for the capped film is shown in Figure 5. The Si (111) interplanar spacing (3.136 Å) was used to calibrate the image magnification and to calculate the interplanar C54-$TiSi_2$ spacing (d-spacing) of the Ti silicide lattice fringes.

FIGURE 6. Electron diffraction pattern from the film shown in Figure 1. Reflections from more than one titanium silicide phase are superimposed on the Si [011] zone pattern in this image.

FIGURE 7. Electron diffraction pattern from the film shown in Figure 2. Reflections from more than one titanium silicide phase, as well as from unreacted Ti are superimposed on the Si [011] zone pattern in this image.

The measured d-spacing in the film is 2.09 Å; the (022) for is 2.093 Å. Thus, the film lattice fringes correspond to (022) planes in C54-$TiSi_2$.

SAED was used to identify the Ti silicide phases present in the films; more than one phase appears to exist. Figures 6 and 7 show typical electron diffraction patterns from the capped and uncapped titanium silicide films, respectively. The image analysis software MacLispix (13) was used to analyze the electron diffraction patterns and to aid in the identification of the phases present in the films. The Si (111) reflection was used to calibrate the camera constant of the diffraction patterns. Then the "measure" tool in MacLispix was used to determine the d-spacings of the titanium silicide reflections. The measure tool calculates the d-spacing of the selected reflection, labels the reflection, and gives the d-value on the image, as well as producing a table of the labels, d-values, and position of the reflections. For the capped film, electron diffraction showed the presence of the C49-$TiSi_2$, C54-$TiSi_2$, and Ti_5Si_3 phases, and the possible presence of the TiSi phase. For the uncapped film, electron diffraction showed the presence of C49-$TiSi_2$, TiSi and Ti_5Si_3, and the possible presence of C54-$TiSi_2$. A Ti_5Si_4 phase, which is not mentioned in the literature on electronic device applications of Ti silicide, is also listed in the JCPDS files (14) and on the Ti-Si phase diagram (15); since several reflections from both the capped and uncapped films match the d-spacings of this phase, its presence in these films cannot be ruled out. Unreacted titanium was also present in the uncapped film; the faint reflections close to the Si (111) reflections in Figure 7 correspond to Ti (001) planes. The faint spots near the central transmitted beam in Figure 7 arise from (020) planes of the C49-$TiSi_2$ phase.

To further investigate the composition of the films, EDX spectra were taken from various regions within the films (see Figures 3, 4, 8, and 10). This analysis was carried out in the STEM mode of the TEM, which allows x-ray spectra to be collected from localized regions in the specimen. Although quantitative EDX analysis has yet to be done, qualitatively the ratio of the Si peak to Ti peak changed throughout the film, indicating that the film composition is not uniform and that multiple silicide phases most likely exist. EDX measurement of Ti silicide standards for quantification of these film EDX spectra is in progress. Both the capped and the uncapped films showed the presence of oxygen near the film surface but not near the film/substrate interface. It is unclear whether the surface oxide formed during the film growth (the films were exposed to air between the first and second RTA processes, when transporting for etching) or later when the films were exposed to air after the final anneal. Because the titanium Lβ and oxygen Kα x-ray peaks overlap and absorption is an issue, EELS analysis was initiated to study the oxygen distribution in the films. Figure 9 shows background subtracted EELS spectra from the TiN-capped film. Near the film surface, oxygen was detected; near the Ti silicide/silicon interface, oxygen was not detected. This result is consistent with the EDX data obtained from this

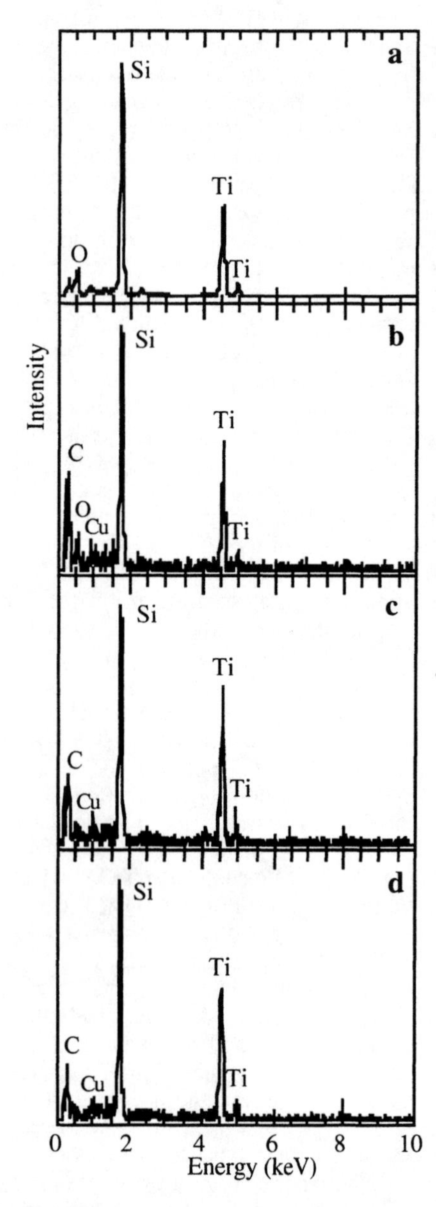

FIGURE 8. EDX spectra from the TiN-capped Ti silicide film shown in Figure 3. The oxygen K-line and Ti L-lines overlap in the low energy region (~ 0.5 keV) of the x-ray spectrum. The small peak between the Cu L-line and the Si K-line peaks corresponds to the As L-line.

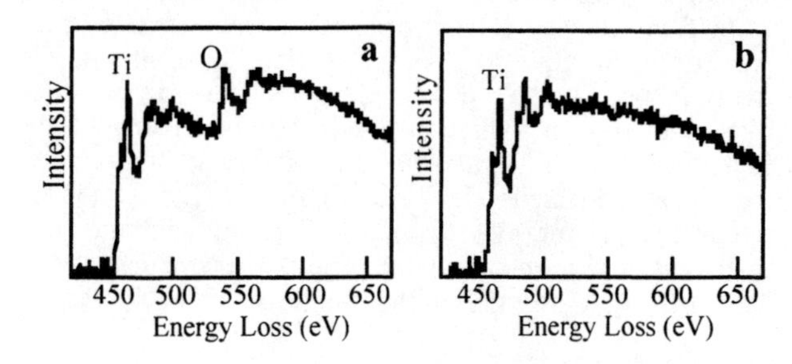

FIGURE 9. Background subtracted EELS spectra (a) from near the surface of the capped Ti silicide film and (b) from near the film/Si interface. The oxygen K-edge and Ti L-edge are clearly resolved in the EELS spectra, unlike in the x-ray spectra in Figure 8.

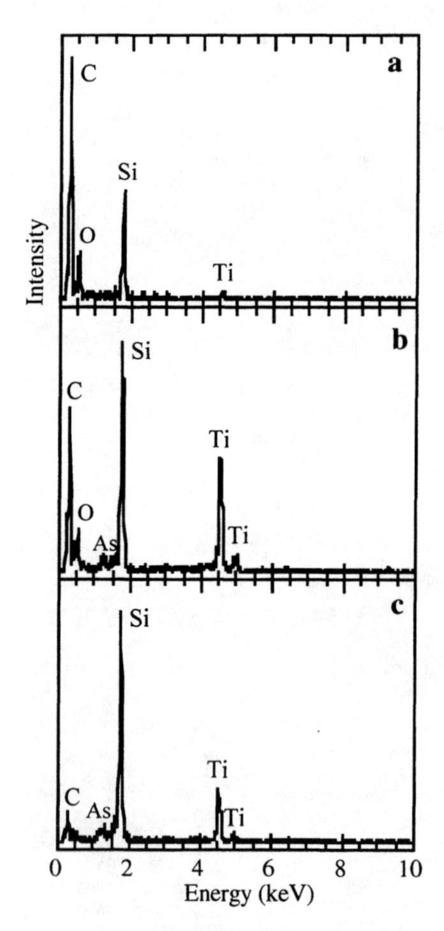

FIGURE 10. EDX spectra from the uncapped Ti silicide film shown in Figure 4. Spectrum (a) shows the presence of an oxide layer at the film surface; very little Ti exists in this surface oxide layer.

film. The arsenic dopant was also observed in the film; a small As L-line peak is visible in several of the EDX spectra in Figures 8 and 10.

DISCUSSIONS AND CONCLUSIONS

The quality of titanium silicide films with a 20-nm thick Ti nitride capping layer deposited prior to annealing was compared to Ti silicide films that did not have a TiN capping layer deposited. (Since the RTA was carried out in a nitrogen ambient, it is probable that the film specified as "uncapped" actually has a thin Ti nitride surface layer.) The 20-nm TiN cap was seen to improve the titanium disilicide film quality. The TiN-capped and the uncapped Ti silicide films were both grown from sputtered Ti initially 16 nm thick. However, the final silicide film was found to be thinner and more uniform for the capped film. Electron diffraction analysis also showed that the capped film contained more of the C54-$TiSi_2$ phase than the uncapped film. The C54-$TiSi_2$ phase, as well as the C49-$TiSi_2$ and Ti_5Si_3 phases, existed in the capped film, while the uncapped film contained mainly C49-$TiSi_2$, TiSi and Ti_5Si_3. The presence of the C54 phase in the uncapped film was not confirmed; only one measured d-spacing matched the C54 pattern, and that was an interplanar spacing which more closely matched the most intense Ti_5Si_3 reflection. The EDX spectra showed that the relative quantities of Si and Ti varied throughout both films, confirming the chemical nonuniformity of the film and that multiple silicide phases exist. The TiN capping layer thus appears to have enhanced the transformation from the C49 to the C54 phase.

Measurement of titanium silicide standards for quantification of the EDX and EELS results is in progress. Also, dark field imaging and detailed micro-diffraction studies are being initiated to determine the distribution of the various Ti silicide phases in these films.

The EDX and EELS spectra show oxygen near the film surface. One proposed reason the TiN improves silicide film quality is that it prevents oxidation (8). Since the samples have been exposed to air, we can neither rule out or confirm this possibility. More EDX and EELS analyses, including elemental mapping, will be performed to further probe the oxygen distribution in these Ti silicide films.

To summarize, a combination of analytical TEM techniques were used to investigate the morphology, crystallography (phases), and site-specific, qualitative elemental composition of uncapped and TiN capped titanium silicide films less than 100 nm in thickness. Several Ti silicide phases were found in both the uncapped and capped films. However, depositing a TiN capping layer prior to annealing was shown to increase the amount of the preferred, lower resistivity C54-$TiSi_2$ phase and to improve the uniformity of the films. As the dimensions of microelectronic devices become even smaller, reaching the sub-0.10 μm region and beyond, the use of such analytical electron microscopy techniques to study materials issues affecting electronic properties, such as sheet resistance, will become even more important.

REFERENCES

1. Tung, R. T., Maex, K., Pellegrini, P. W., and Allen, L. H., eds., *Silicide Thin Films — Fabrication, Properties, and Applications*, Mater. Res. Soc. Symp. Proc. **402**, (Materials Research Society, Pittsburgh, PA, 1996).
2. Murarka, S. P., *Silicides for VLSI Applications*, (Academic Press, New York, 1983).
3. Beyers, R., and Sinclair, R., *J. Appl. Phys.* **57**, 5240-5245 (1985).
4. Jeon, H., Sukow, C. A., Honeycutt, J. W., and Rozgonyi, G. A., and Nemanich, R. J., *J. Appl. Phys.* **71**, 4269-4276 (1992).
5. Roy, R. A. , Clevenger, L. A., Cabral Jr., C., Saenger, K. L., Brauer, S., Jordan-Sweet, J., Bucchignano, J., Stephenson, G. B., Morales, G., and Ludwig Jr., K. F., *Appl. Phys. Lett.* **66**, 1732-1734 (1995).
6. Liu, H. I., Burns, J. A., Wyatt, P. W., and Keast, C. L., *IEEE Transactions on Electron Devices*, in press (1997).
7. Matsubara, Y., Sakai, T., Ishigami, T., Ando, K., and Horiuchi, T., *Thin Solid Films* **270**, 537-543 (1995).
8. Apte, P. P., Paranjpe, A., and Pollack, G., *IEEE Electron Dev. Lett.* **17**, 506-508 (1996).
9. Shor, Y., and Pelleg, J., in *Silicide Thin Films — Fabrication, Properties, and Applications*, Mater. Res. Soc. Symp. Proc. **402**, Tung, R. T., Maex, K., Pellegrini, P. W., and Allen, L. H., eds., (Mater. Res. Soc., Pittsburgh, PA, 1996) pp. 107-112.
10. Nagabushnam, R. V., Sharan, S., Sandhu., G., Rakesh, V. R., Singh, R. K. and Tiwari, P., in *Silicide Thin Films — Fabrication, Properties, and Applications*, Mater. Res. Soc. Symp. Proc. **402**, Tung, R. T., Maex, K., Pellegrini, P. W., and Allen, L. H., eds., (Mater. Res. Soc., Pittsburgh, PA, 1996) pp. 113-118.
11. Ishigami, T., Matsubara, Y., Iguchi, M., and Horiuchi, T., in *Advanced Metallization for Future ULSI*, Mater. Res. Soc. Symp. Proc. **427**, Tu, K. N., Mayer, J. W., Poate, J. M., and Chen, L. J., eds. (Materials Research Society, Pittsburgh, PA, 1996) pp. 517-522.
12. The sheet resistance is commonly given in units of Ω/sq, where the "square" refers to one square unit of material. [Streetman, Ben G., *Solid State Electronic Devices*, 3rd edition, (Prentice Hall, Englewood Cliffs, NJ, 1990) p. 354.]
13. MacLispix is a freeware Macintosh image processing program written by David S. Bright at NIST. It is available at http://www-sims.nist.gov/division/microscopysoftware.html. This program includes tools for measuring d-spacings in spot patterns and ring diameters in ring patterns.
14. Joint Committee for Powder Diffraction Standards (JCPDS) files #27-0907 and #23-1079. (International Centre for Powder Diffraction Standards, 1997).
15. Massalski, T. B., Okamoto, H., Subramanian, P. R., and Kacprzak, L., *Binary Alloy Phase Diagrams*, Vol. 3, (American Society for Metals, Metals Park, OH,, 1996) pp. 3367, 3370-3371.

Focused Ion Beam Preparation for Cross-Sectional Transmission Electron Microscopy Investigation of the Top Surface of Unpassivated or Partially Processed ULSI Devices

H. Bender, P. Van Marcke, C. Drijbooms and P. Roussel

IMEC, Kapeldreef 75, B-3001 Leuven, Belgium, bender@imec.be

Alternative techniques for focused ion beam preparation of transmission electron microscopy specimens of partially processed or unpassivated ULSI devices are discussed and illustrated with case study examples. The loss of material due to milling with the Ga ion beam and the implantation damage in the top layers is estimated and compared with experimental observations. Different kinds of protective layers deposited in-situ in the focused ion beam (Pt), ex-situ (Au, glue remainders) or under standard device processing conditions (amorphous or polycrystalline Si) are examined.

INTRODUCTION

Transmission electron microscopy has a superb image resolution and is therefore essential for characterisation of state-of-the-art ULSI devices. Focused ion beam (FIB) is widely used nowadays for preparation of transmission electron microscopy (TEM) specimens in cases that site specific thinning on small features in the devices is necessary (1-5). In a development phase and for production control, non-passivated or only partially processed devices have to be analysed. In such cases often the top surface layers are of major importance. As one has to localise the region of interest first from a top view image, damage by the Ga^+ ion beam cannot be avoided even not by the use of as low currents as possible and by the fast deposition of a protective Pt-layer.

In this paper some possibilities to overcome this problem are discussed and illustrated by studies on advanced ULSI device structures. Possible solutions that are considered are : in-situ Pt deposition at low energy, ex-situ Au coverage, ex-situ glue coverage, and ex-situ deposition of amorphous or polycrystalline silicon under standard processing conditions before the wafers are transferred to TEM analysis.

EXPERIMENTAL

The procedures for TEM specimen preparation with FIB have already been discussed in detail in the literature (1-5). Thin strips are sawed from a silicon wafer at the region of interest with a thickness of directly ~50 μm so that no further polishing is necessary or with a thickness of approximately 250 μm requiring further mechanical polishing to ~50 μm before loading in the FIB. These strips are glued on a standard large slot copper TEM-grid and thinned stepwise by milling the top surface layers with decreasing beam currents until a thin slice remains at the selected spot. Typical thinned areas are 15-20 μm wide and 5-10 μm deep and can be prepared in 2-2.5 h milling time. Before the FIB thinning a protective metal layer (typically Pt) is deposited in-situ at the location of interest.

RESULTS AND DISCUSSION

Unprotected surfaces

During the localisation in the FIB of the region of interest, the Ga^+ ion beam damages and slightly sputters the surface of the sample. An estimate of these effects for bulk silicon is given in table 1 for some typical imaging conditions.

The sputtered depth d is calculated as

$$d = \frac{SIt}{A} \qquad (1)$$

with S the sputter yield, I the ion beam current, t the sputter time and A the scan area. The sputter yield for silicon with a 30 keV Ga ion beam is 0.15 $\mu m^3/nC$ (6). The localisation is typically done with beam currents of 4-11 pA at magnifications less than 10000 times. From the data in table 1 it follows that the sputtered depth is under these conditions less than 0.15 nm, i.e. less than the native oxide on the surface and hence for most applications negligible (unless for studies of e.g. thin gate oxides, or etch residues). The Pt deposition is normally done with a 150 pA beam current at a 10000 times magnification. Before the deposition normally one image is taken for positioning the Pt pattern. With a slow image grab (3.4s) the sputtered depth is again on the same order as for the localisation. Also with a slower acquired image the sputtered depth is only on the order of 1 nm. The actual sputtered depth varies with the material, but nevertheless it can be concluded that by careful control of the imaging conditions, the sputtering before the protective Pt layer is deposited, is

CP449, *Characterization and Metrology for ULSI Technology: 1998 International Conference*
edited by D. G. Seiler, A. C. Diebold, W. M. Bullis, T. J. Shaffner, R. McDonald, and E. J. Walters

TABLE 1. Sputtered depth d and implanted ion dose D in silicon during typical imaging conditions used before the Pt deposition

Beam current I	Imaging condition	Scan time t	**Sputtered depth d (nm)**			**Implanted ion dose D (ions/cm^2)**			
			Magnification			**Magnification**			
			1000	5000	10000	1000	5000	10000	-
pA		s	**Scan area A (μm^2)**			**Scan area A (μm^2)**			
			80000	3200	800	80000	3200	800	50
4	continuous	60	0.00	0.01	0.04	$1.9\ 10^{12}$	$4.7\ 10^{13}$	$1.9\ 10^{14}$	
11	continuous	60	0.00	0.03	0.12	$5.1\ 10^{12}$	$1.3\ 10^{14}$	$5.1\ 10^{14}$	
150	continuous	60	0.02	0.42	1.68	$7.0\ 10^{13}$	$1.7\ 10^{15}$	$7.0\ 10^{15}$	
150	image grab	3.4	0.00	0.02	0.09	$3.9\ 10^{12}$	$9.8\ 10^{13}$	$3.9\ 10^{14}$	
150	image grab	25.2	0.01	0.18	0.71	$2.9\ 10^{13}$	$7.3\ 10^{14}$	$2.9\ 10^{15}$	
150	image grab	41.9	0.01	0.29	1.18	$4.9\ 10^{13}$	$1.2\ 10^{15}$	$4.9\ 10^{15}$	
150	Pt deposition	1 (initial)							$1.9\ 10^{15}$

mostly not a severe limitation. Even with less stringent imaging conditions, the sputtered depth will not be more than a few nm.

The implanted dose D follows from

$$D = \frac{It}{A} \qquad (2)$$

The data for D in table 1 (neglecting the sputter effect) show that during the localisation step doses over 1.10^{14} ions/cm^2 are implanted in the top layer of the samples. Even a fast image grab before the Pt deposition with the 150 pA beam will add a dose of ~ 4.10^{14} ions/cm^2. The Pt deposition is done within a box of ~ 50 μm^2 with a current density of 3-5 pA/μm^2. During the initial Pt deposition, the growing Pt layer is still penetrated by the ion beam. It will take some 10 s before the Pt is thick enough to shield the substrate from implantation (the exact implantation depth in the Pt cannot be calculated as the materials properties are not well known). Already during the initial second, a dose over 1.10^{15} ions/cm^2 will be implanted. This dose cannot be lowered by varying the deposition conditions as deposition with lower current density would take longer time so that the total dose will not decrease.

TRIM simulations (7) have shown that in silicon with a 30 keV Ga ion beam, about 500 vacancies are created per incident ion. The distribution of these point defects has a tail reaching to 70 nm in the silicon. The critical point defect density for amorphization of silicon is 10^{22}/cm^3 (8). With a dose of 1.10^{15} ions/cm^2, this point defect density is exceeded up to 50 nm deep in the silicon according the TRIM simulations (7). It can be concluded that even with careful control of the imaging conditions the amorphization of free silicon surfaces cannot be avoided as the implanted dose during the initial Pt deposition is already high enough to create an amorphized top layer in the silicon. This problem also holds for SiGe layers, as well as for III-V materials (7). Other materials used in semiconductor devices which are less sensitive to amorphization (e.g. metals) or are anyhow amorphous (oxides, nitrides), will be damaged in the top layer by the ion implantation.

Figure 1 shows a XTEM image of the interface between the deposited Pt and an unprotected silicon surface at the bottom of an etched structure. The deposited Pt is a porous material consisting of small amorphous clusters with a diameter of ~ 5-10 nm. The silicon surface on the left of the picture is protected by the device structure and can be used as a reference surface (the bottom of the etched structure is slightly curved due to the etching, this shape is also observed on specimens prepared by means of conventional ion milling). The native oxide is seen as a bright layer on the surface of the undamaged silicon and in the transition region towards the damaged layer. The damaged layer consists of a bright grainy structured interface layer, a dark and a bright amorphous layer. Comparison of the interface position with the reference surface shows that no significant sputtering has occurred, in agreement with the above estimates. A slight swelling can be noticed of the damaged region due to the amorphization of the silicon. The amorphous silicon/silicon interface is situated at a depth of ~ 60 nm. An identical thickness of the amorphized silicon has also been observed in a specimen prepared by totally blind navigation to the position of the thinning, i.e. without any imaging before the Pt deposition. This shows that indeed the implanted dose during the initial Pt deposition exceeds the critical dose for amorphization and that this problem cannot be overcome without an ex-situ deposited protective layer.

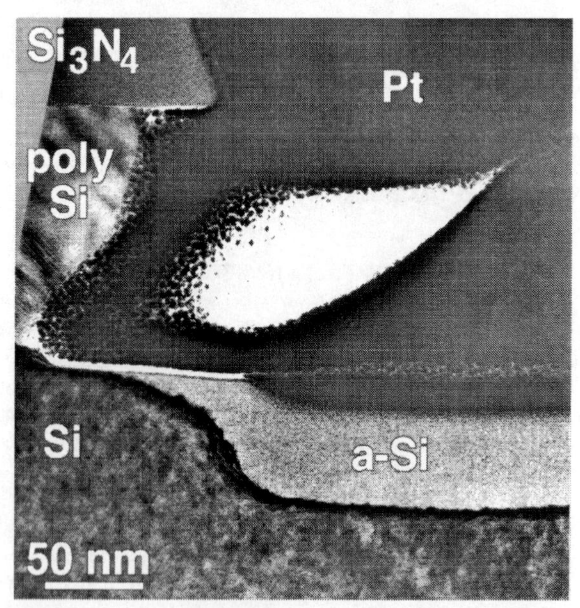

FIGURE 1. Cross-section TEM image of the damaged layer below the deposited Pt at the bottom of an etched structure.

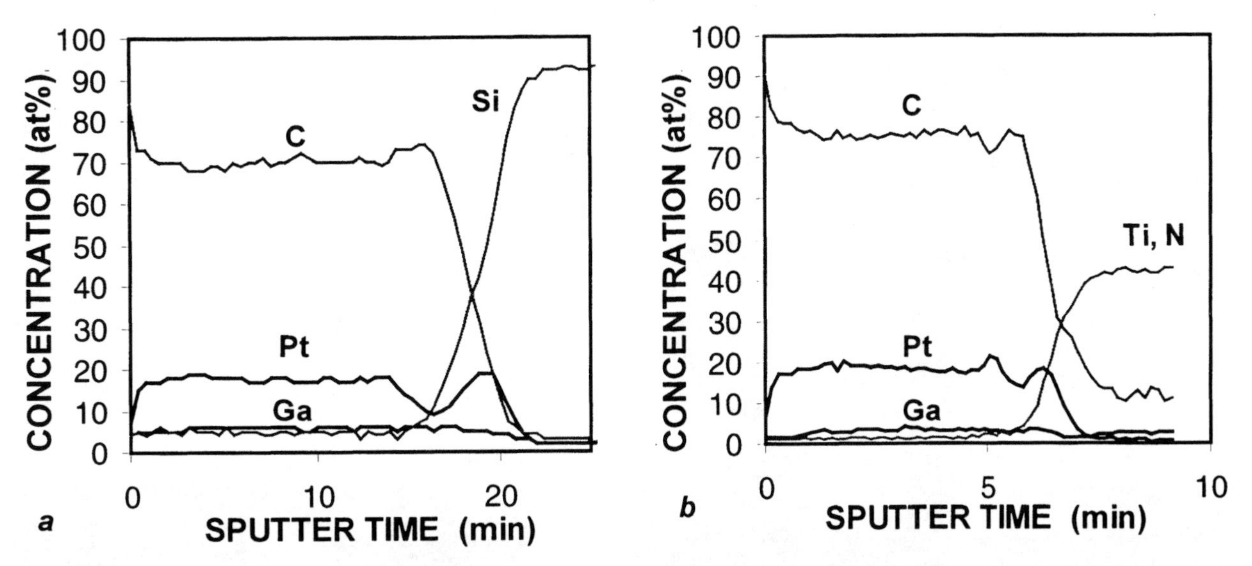

FIGURE 2. Auger depth profile through the Pt/substrate interface structure of a Pt pad deposited under standard conditions as used during TEM preparation on a) a blank silicon wafer and b) a TiN/Al/SiO_2/Si wafer.

The nature of the dark top layer is analysed by Auger electron spectroscopy depth profiling (Fig. 2a) through 70 nm thick Pt pads deposited on a blank silicon wafer under similar conditions as used during the TEM preparation. It can be noticed that the commonly called "Pt" layer actually consists of a C/Pt/Ga:70/20/5 mixture. This quantification is obtained with standard Auger sensitivity factors and without any correction for possible preferential sputtering effects which might be important in this porous material. At the interface with the silicon an increased C and decreased Pt concentration is present, while in the top silicon again a higher Pt concentration is present. These layers correspond with the bright interface layer and the dark amorphous layer on the TEM picture, respectively (Fig. 1). The Ga signal decreases at the same depth as the Pt. There is no O measurable with Auger spectroscopy at the interface. Hence except of the amorphization and the implantation, also a severe knock-in of Pt and C occurs during the initial deposition. The profiles are similar for depositions performed with much lower current density than standardly used and show also no difference when the silicon is strongly sputtered before the Pt deposition.

Figure 3 compares the interface structure of the Pt deposited on some common materials as used in micro-electronic devices. The interfacial layer consists in all cases of a brighter layer with fine granular structure and a

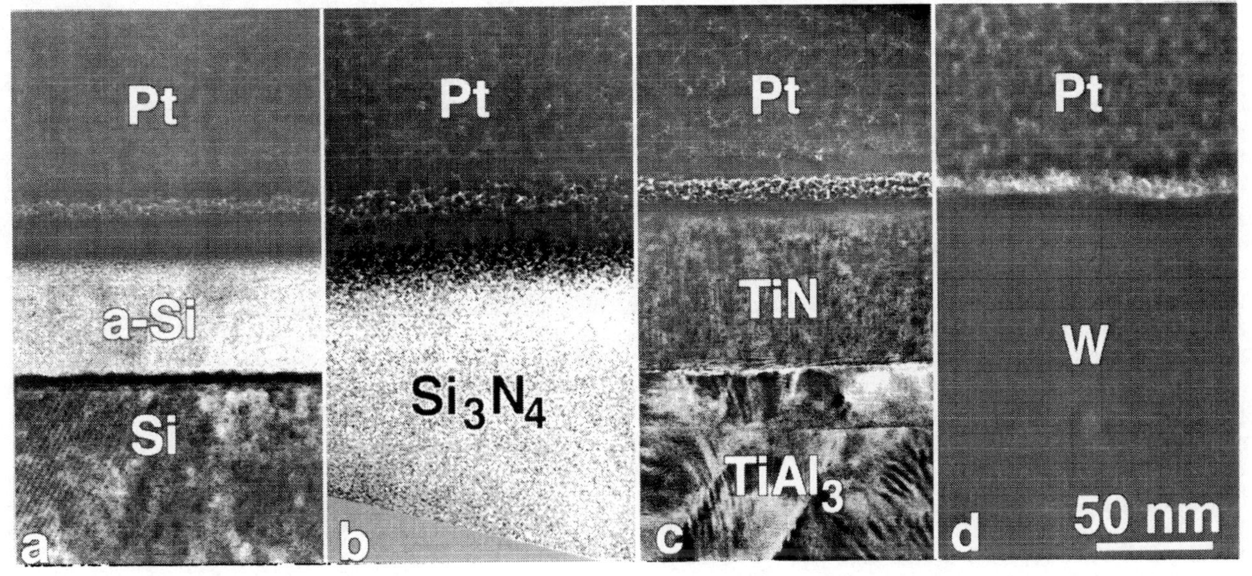

FIGURE 3. Cross-section TEM images of the interface of the deposited Pt with a) silicon, b) silicon nitride, c) TiN and d) W.

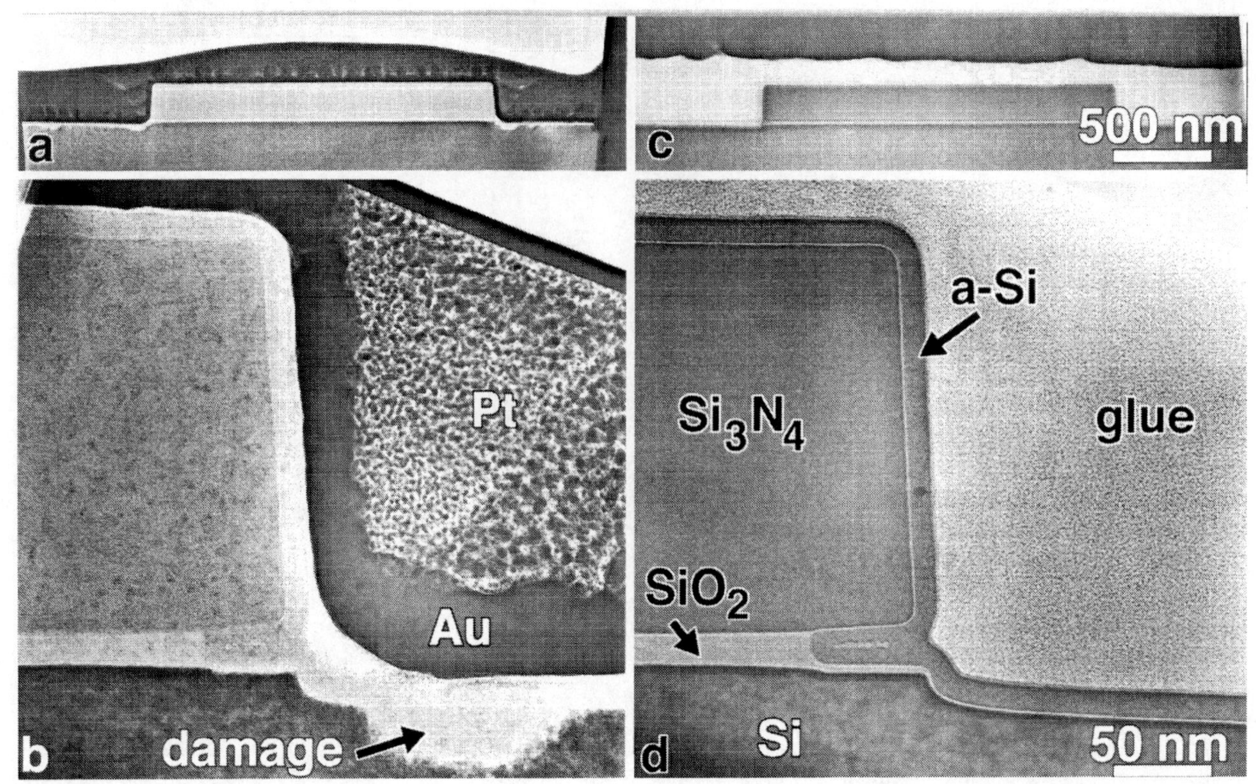

FIGURE 4. Cross-section TEM images of a PELOX isolation structure before the field oxidation : a) and b) with Au deposition as a protective layer, and c) and d) with a protective glue layer.

darker layer below it. For the TiN and W also a darker layer is observed on top of the bright interface layer. The Auger depth profile for Pt on TiN (Fig. 2b) shows two peaks in the Pt profile, one before the start of the substrate signal and a second one in the top of the TiN. The C profile has a small dip at the position of the first Pt peak. Hence in this case except of a knock-in of Pt and C, also an accumulation of Pt occurs above the interface. The detailed transient from a sputtering condition before the adsorption of the organic Pt molecules used for the metal deposition, towards the equilibrium between sputtering and deposition needs further investigation in order to fully understand the formation mechanism of the interface layer on different materials.

Decreasing the ion beam energy during the localisation and Pt deposition can reduce the depth of the amorphization of the silicon. The structure of the damaged layer is the same as when using the standard 30 keV ion beam, i.e. an interfacial layer, a Pt, C and Ga containing amorphous silicon layer, and an amorphous Si layer. The amorphous silicon/silicon interface is for a 10 keV ion beam at a depth of 35 nm, which is for many applications still too deep. The Auger depth profiles are similar as the one shown in Fig. 2a.

The amorphization depths reported here are larger than the values previously given by Walker and Broom (7) who found only 28 and 6 nm for the 30 and 10 keV ion beam, respectively. The difference is due to the fact that they measured the thickness of the amorphous layer starting from the bottom of the Pt knock-in layer. The comparison with the reference surface as done in the present paper (Fig. 1) shows that this is an incorrect evaluation of the damage layer.

Au protection

An ex-situ deposited protective layer is needed to avoid the damaged surface layer. Au sputtering as for scanning electron microscopy (SEM) is an obvious candidate for this purpose. It has however, several disadvantages. The thickness of the Au layer on the bottom of topography with a high aspect ratio is much less than on the planar surfaces, so that the bottoms of the structures are insufficiently or only partially covered. Also at the edges of steep topography the Au thickness is often insufficient and non-uniform as can be seen on figure 4a, b. Furthermore after some time, the Au layer starts growing in a porous dendrite-like structure (Fig. 4a) so that increased sputter time does not improve the coverage anymore. The correspondence between the pores in the Au and the amorphization of the silicon can clearly be seen on fig. 4a. This shows that it can be excluded that the amorphization is due to a faster milling rate of the silicon than of the Au so that below the Au, the specimen becomes so thin that the damaged sidewalls of the thinned strip start to overlap.

An additional disadvantage of the Au coverage is that the layer strongly absorbs the electron beam resulting in a strong contrast on the TEM images.

Glue protection

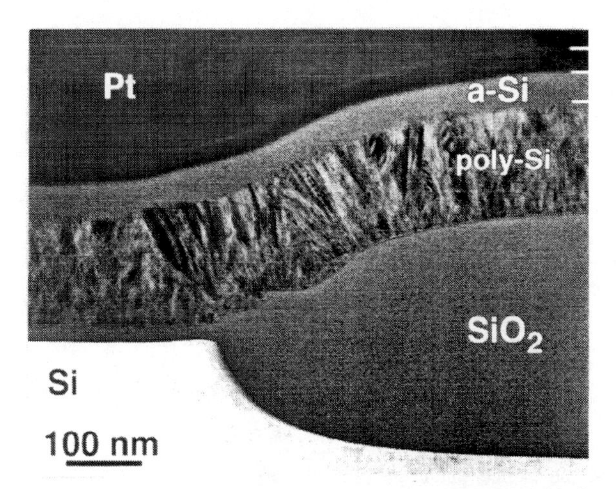

FIGURE 5. Cross-section TEM image of a PELOX isolation structure with a 100 nm poly-silicon capping layer.

An alternative and efficient protecting layer is found to be the remainders of the glue used during the polishing of the strips. An example is shown in Fig. 4 c,d. Unfortunately, this glue coverage is difficult to control so that sometimes a layer over the whole structure is present which smoothes all topography and which makes the localisation of the region of interest difficult (Fig. 4c).

Processing protective layers

Capping layers deposited under standard processing conditions on the wafers selected for TEM analysis give the best protection to the surface. Ideally such a layer should grow uniformly over all topography, so that structures with high aspect ratio are well protected while the topography remains visible hence facilitating the localisation of the region of interest in the FIB image. The deposition should require a low processing temperature and the layer should grow fast enough so that a thickness of 70-100 nm can be reached in a reasonable time. To have a good image contrast in the FIB a conducting layer is most preferable, while for a smooth contrast in the TEM a light and amorphous material is most suitable. Obvious candidates fulfilling these characteristics are amorphous or polycrystalline silicon and TEOS-oxide. Only in case of shallow topography Ti, TiN or Al can be considered.

Figure 5 shows an example of a poly-Si cap layer used to protect an isolation structure prepared by the PELOX process. The top of the poly layer is amorphized by the Ga ion beam and Pt is knocked-in as for the silicon substrates. The poly-Si layer is more than thick enough for protecting the structure of interest.

CONCLUSIONS

Using careful imaging conditions during the localisation step for FIB preparation, the removal of the top layer of the structure can be kept at a usually acceptable level on the order of 1-2 nm. However, damaging of the top layer during the initial stage of the Pt deposition cannot be avoided. For all materials a knock-in of Pt and C occurs in the uppermost layers and Ga is implanted. For materials as Si and SiGe also amorphization takes place. Lowering the ion energy results in more shallow implants and amorphization, however, generally still to an unacceptable depth. The only way to circumvent this ion damage in a FIB system, is the use of a dual beam FIB with which both the localisation and initial Pt deposition can be done with the electron beam instead of with the Ga ion beam. A drawback of this technique is the much slower deposition rate of the electron beam induced Pt deposition (and also the much higher price of a dual beam system).

A sputtered Au layer has several disadvantages, the major one is the bad coverage at topography with high aspect ratio so that damaging of the substrates cannot be fully avoided. On the other hand, an organic (i.e. glue remainders) layer is a very good protection, but would require a more controlled deposition process for practical applications.

A protective layer deposited during the device processing is to be preferred whenever possible from a processing point of view. Poly- or amorphous silicon, and TEOS-oxides are good candidates for such protective layers. A disadvantage of standard processing deposited protective layers on full wafers is that they will limit the possibilities to apply other analysis techniques to other parts of the same wafers.

The problems discussed in the present paper for protecting the surface during FIB preparation also hold for newly upcoming FIB preparation schemes extracting a thinned strip directly from the silicon wafer (9, 10).

ACKNOWLEDGEMENTS

The IMEC-processing groups are acknowledged for providing the processed wafers. The TEM observations are performed with the microscopes at EMAT, University of Antwerpen.

REFERENCES

1. Kirk, E.C.G., Williams, D.A., and Ahmed, H., *Inst. Phys. Conf. Ser.* **100**, 501-506 (1989).
2. Young, R.J., Kirk, E.C.G., Williams, D.A., and Ahmed, H., *Mat. Res. Soc. Symp. Proc.* **199**, 205-216 (1990).
3. Walker, J. F., Reiner, J. C., and Solenthaler, C., *Inst. Phys. Conf. Ser.* **146**, 629-634 (1995).
4. Pantel, R., Auvert, G., and Mascarin, G., Microelec. Engineering **37/38**, 49-57 (1997).
5. Bender, H., and Roussel, P., *Inst. Phys. Conf. Ser.* **157**, 465-468 (1997).
6. FEI Manual, Hilsboro OR, (1994).
7. Walker, J.F., and Broom, R.F., *Inst. Phys. Conf. Ser.* **157**, 473-478 (1997).
8. Cerva, H., and Hobler, G., *Inst. Phys. Conf. Ser.* **134**, 133-136 (1993).
9. Overwijk M.H.F., van den Heuvel, F.C., and Bulle-Lieuwma, C.W.T., *J. Vac. Sci. Technol. B* **11**, 2021-2024 (1993).
10. Sheng, T.T., Goh, G.P., Tung, C.H., and Wang, L.F., *J. Vac. Sci. Technol. B* **15**, 610-613 (1997).

Plan View TEM Sample Preparation Using The Focused Ion Beam Lift-Out Technique

F. A. Stevie, R. B. Irwin, T. L. Shofner[1], S. R. Brown, J. L. Drown[2], and L. A. Giannuzzi[2]

Lucent Technologies, 9333 S. John Young Parkway, Orlando, FL 32819
[1]Kirk Resources, 9333 S. John Young Parkway, Orlando, FL 32819
[2]University of Central Florida, Department of Mechanical, Materials, and Aerospace Engineering, 4000 Central Florida Boulevard, Orlando, FL 32816

Localized plan view TEM samples have been prepared from silicon semiconductor wafers using the focused ion beam lift-out technique. Two different methods of sample preparation before FIB machining were found to be successful: mounting cleaved samples sandwiched together or adding silver paint and cleaving through paint and samples. The plan view technique offers site specific TEM capability from a horizontal section rather than a vertical cross section. The sections can be taken from any layer and can be angled if desired. Results have been obtained from metal layers in a semiconductor device structure. TEM micrographs of tungsten plug arrays show non-uniform barrier layer coverage and tungsten grain size across the via. Hundreds of plugs have been cut through in one sample, thereby offering statistical as well as specific structural information. Metal and polysilicon lines have been examined for grain size and uniformity in a single micrograph. Plan view samples from continuous metal layers can also be made.

INTRODUCTION

The use of transmission electron microscopy (TEM) for analysis of semiconductor structures is expanding because of the need for higher magnification to examine ever smaller feature sizes. Also of importance is the capability to provide elemental information from an area as small as 5 nm in diameter using energy dispersive x-ray analysis of TEM specimens. Traditional methods of sample preparation for TEM have been successful for many applications, but have proven to be tedious and difficult or even impossible for use in site specific work (1-5). The application of focused ion beams (FIB) to preparation of samples for TEM has been rapid and extensive (6-27). Two significant advantages of FIB preparation are that it is site specific, with positional accuracy to about the diameter of the ion beam, and can provide a large area of uniform thickness. Areas of interest for FIB preparation have moved beyond semiconductor applications to metals, ceramics, powders, optical materials, and biology (26,27,31). However, the ability to provide plan view sections using FIB has just begun to be exploited (28).

Unlike the method first used for FIB sample preparation, where the sample was thinned to a few tens of micrometers before FIB sputtering, the development of the lift-out method of sample preparation has reduced or eliminated the need for sample treatment before FIB etching (29-31). This paper is concerned with the application of the lift-out approach to obtain plan view membranes from semiconductor structures, with an emphasis on the metal layers.

EXPERIMENTAL PROCEDURE

The focused ion beam system used in this work is the FEI Model 800, which provides a gallium ion beam as small as 7 nm diameter and can handle up to 200 mm diameter samples. A typical specification is 70 pA for a 25 nm diameter beam at 30 kV on the column. The lift-out method has been previously described in some detail (30) and is an adaptation of a method where the sample was removed by contact with a field emission tip (29). In the procedure used in this work, the TEM membrane is completely cut away from the substrate within the vacuum system. The specimen is allowed to rest in the trenches created in preparing the sample until the membrane can be micromanipulated in air using a drawn glass rod. The rod is brought against the TEM membrane which is removed with the aid of static electricity. The specimen is then deposited on a 3 mm diameter carbon coated TEM grid.

Most semiconductor plan view specimens are required to be taken from the top few micrometers of the wafer. Because the lift-out method requires some space at the very surface of the wafer to protect the membrane from falling off the bulk sample during lift-out, additional sample preparation had to be devised. Two approaches to plan view sample preparation were found to be successful. In the first method, the silicon wafer is cleaved and

CP449, *Characterization and Metrology for ULSI Technology: 1998 International Conference*
edited by D. G. Seiler, A. C. Diebold, W. M. Bullis, T. J. Shaffner, R. McDonald, and E. J. Walters

sample preparation had to be devised. Two approaches to plan view sample preparation were found to be successful. In the first method, the silicon wafer is cleaved and sandwiched together with another wafer, front to front, using epoxy or silver paint. The edge area of interest is mechanically polished using traditional metallurgical sample preparation methods, and then further polished, if necessary, using the FIB (Figure 1a). The sample is remounted at 90 degrees from the original polished cut, and the normal lift-out process is used from this point (Figure 1b). A sample holder was devised to mount the sample below the height restriction in the FIB entry chamber. The second approach is to add a small amount of silver paint to the sample at the area of interest, cleave the sample through the silver painted area, and proceed with FIB machining for lift out. Both methods of sample removal were successful after the initial technique development. If the layer of interest is not close to the surface, then a simple cleave of the sample followed by mechanical or FIB polishing will be sufficient.

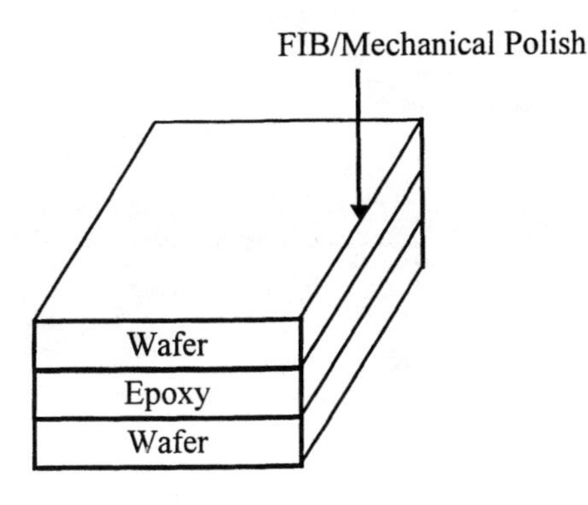

Figure 1a. Initial sample preparation is performed by polishing two wafer slices sandwiched with epoxy.

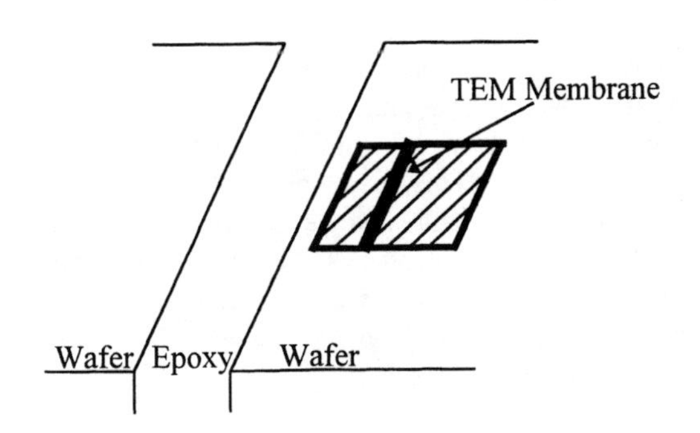

Figure 1b. TEM plan view specimen is cut out using the FIB on the edge of the structure shown in Figure 1a.

TEM analyses were performed using a Philips Model EM430 TEM operating at 300 kV.

RESULTS AND DISCUSSION

Figure 2 shows a low magnification TEM micrograph of an array of plugs, indicating that a large number of features can be examined in a single sample. One can obtain statistics that would normally be quite difficult to determine. It would be possible to prepare several specimens at different levels and provide a composite structure of the via, or to angle the section and observe the entire via structure with different levels of the via appearing at different points in the micrograph.

Figure 2. This TEM micrograph shows an array of over 200 tungsten plugs.

Figure 3 is a higher magnification TEM image of one tungsten plug with a titanium nitride barrier layer around the via. The grain size of tungsten in the plug is approximately 20 to 50 nm.

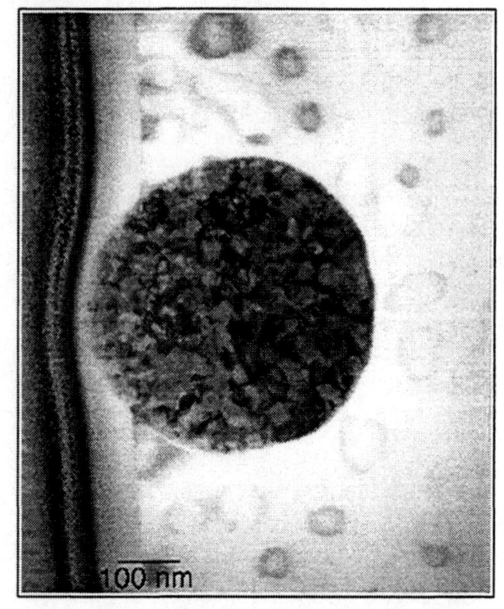

Figure 3. This figure shows a TEM micrograph of one tungsten plug from the array shown in Fig. 2.

Figure 4 shows both polysilicon and metal lines in the same image over a 20 μm x 5 μm area. Both lines are observed because the thickness of the TEM section includes the bottom of the metal lines and the top of the polysilicon lines. The metal lines are a TiN/Al/TiN/Ti structure from top down.

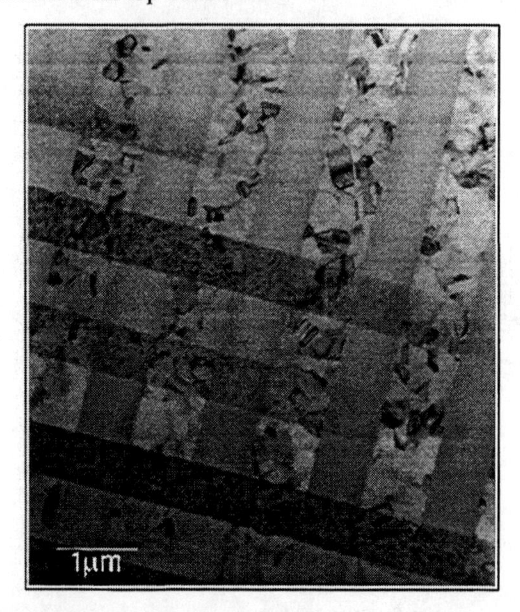

Figure 4. TEM micrograph that shows both polysilicon and metal lines.

It is possible to determine grain size of both lines as shown in the expanded view shown in Fig. 5, and to examine several lines at the same time. Previously, microstructural analysis was performed on blanket deposited thin films using conventional TEM plan view preparation. As the lines have been made smaller, the blanket deposited films may not be representative of the microstructure in the actual lines in the integrated circuit structure. Therefore, this FIB method of sample preparation is very important in the characterization of semiconductor structures.

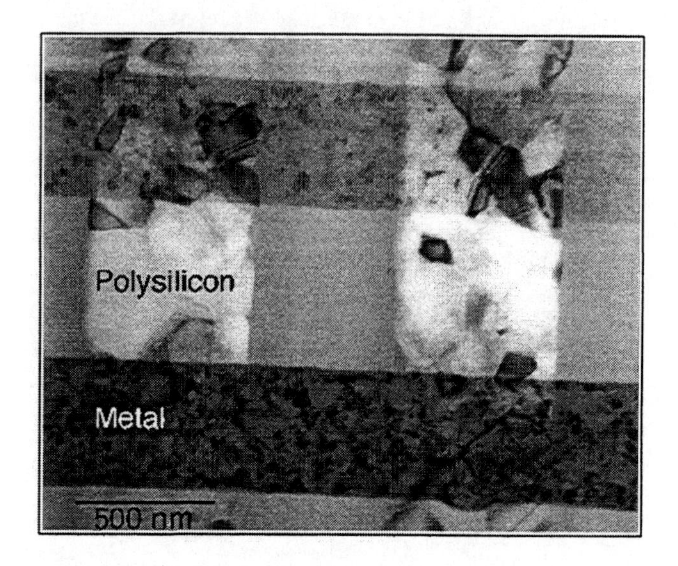

Figure 5. TEM micrograph showing higher magnification view of part of Fig. 4.

It is evident that this plan view is made at a slight angle which permits any feature, such as a metal runner, to be shown as a function of depth.

Plan view samples can also be obtained from blanket films using the lift-out technique. Figure 6 shows the plan view of an aluminum layer from the metal structure mentioned in Figure 4. This view can be used to obtain grain size measurements that can be compared with the measurements on individual lines as seen in Figures 4 and 5. Blanket film TEM plan view specimens can be compared with specimens prepared using other TEM sample preparation techniques.

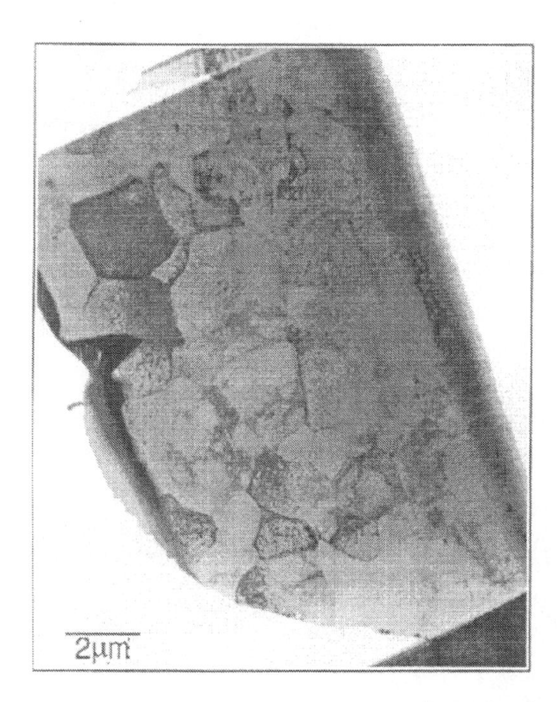

Figure 6. Plan view TEM micrograph of an aluminum layer.

SUMMARY

Plan view samples can be obtained using the lift-out method and provide a new and informative approach to examining metal layers in semiconductor structures. Future work will include changing the angle at which the plan view membrane will be cut out. Views at different angles may provide additional information on features such as the "bird's beak," and layer interfaces. A better way to mount and align the sample is being developed to achieve the desired angle of attack.

REFERENCES

1. Marcus, R. B., and Sheng, T. T., *Transmission Electron Microscopy of VLSI Circuits and Structures*, New York, NY, Wiley (1983).

2. Bravman, J. C., Anderson, R. M., and McDonald, M. L., Eds. *Specimen Preparation for Transmission Electron Microscopy of Materials*, Pittsburgh, PA, Materials Research Society Symposium Proceedings (1988) 115.

3. Anderson, R. M., Ed. *Specimen Preparation for Transmission Electron Microscopy of Materials I*, Pittsburgh, PA, Materials Research Society Symposium Proceedings (1989) **155**.

4. Anderson, R. M., Ed. *Specimen Preparation for Transmission Electron Microscopy of Materials II*, Pittsburgh, PA, Materials Research Society Symposium Proceedings (1990) **199**.

5. Anderson, R. M., Tracy, B., and Bravman, J. C., Eds. *Specimen Preparation for Transmission Electron Microscopy of Materials III*, Pittsburgh, PA, Materials Research Society Symposium Proceedings (1992) **254**.

6. Kirk, E. C., Williams, D. A., and Ahmed, A., *Inst. Phys. Conf. Series*, **100**, 501 (1989).

7. Young, R. J., Kirk, E. C. G., Williams, D. A., and Ahmed, H., *Materials Research Society Proceedings*. **199**, 205-209 (1990).

8. Park, K.-H., *Materials Research Society Proceedings* **199**, 271-275 (1990).

9. Lange, J. A., and Czapski, S., 17th Int. Symp. for Testing and Failure Analysis, ASM International, Materials Park, OH (1991) 397-400.

10. Dickson, N., Miller, J., Jackson, M., Kohn, S., Pyle, R., and Tatti, S., *SPIE Proceedings* **1802**, 155-166 (1992).

11. Mendez, H., Morris, S., Tatti, S., Dickson, N., Pyle, R. E., *SPIE Proceedings* **1802**, 126-132 (1992).

12. Black, E., Bridwell, J., and McConnell, R., *SPIE Proceedings* **1802**, 120-125 (1992).

13. Basile, D. P.,Boylan, R., Baker, B.,Hayes, K, and Soza, D. *Specimen Preparation for Transmission Electron Microscopy of Materials III*, New York, NY, Materials Research Society Symposium Proceedings (1992) **254**, 23.

14. Sanborn, C. E., and Meyers, S. A., *Mater. Res. Soc. Symp. On Specimen Preparation for Transmission Electron Microscopy of Materials-III*, Amer. Inst. Phys., NY (1992) 239-248.

15. Szot, J., Hornsey, R., Ohnishi, T., and Minagawa, S., *J. Vac. Sci. Technol* **B10**, 575-579 (1992).

16. Tarutani,M., Takai,Y., and Shimizu, R., *Jpn. J. Appl. Phys.* **31**, L1305-1308 (1992).

17. Yamaguchi, A., Shibata, M., and Hashinaga, T., *J. Vac. Sci. Technol.* **B11**, 2106-2020 (1993).

18. Nakajima, K., Sudo, S., Yakushiji, M., Ishii, T.,and Aoki, S., *J. Vac. Sci. Technol.* **B11**, 2127-2130 (1993).

19. Assayag, G. Ben, Vieu, C., Gierak, J., Sudraud, P., and Corbin, A., *J. Vac, Sci. Technol.*, **B11**, 2420-2426 (1993).

20. Tomikawa, T., and Shikata, S., *Jpn. J. Appl. Phys.* **32**, 3938-3942 (1993).

21. Assayag, C. Ben, Vieu, C., Gierak, J., Chaabane, H., Pepin, A., and Henoc, P., *J. Vac. Sci. Technol.* **B11**, 531-535 (1993).

22. Hull, R., Bahnck, D., Stevie, F. A., Koszi, L. A., and Chu, S. N. G. *Appl. Phys. Lett.* **62**, 3408-3410 (1993).

23. Stevie, F. A., Shane, T. C., Kahora, P. M., Hull, R., Bahnck, D., Kannan, V. C.,and David, E., *Surf. Int. Anal.* **23**, 61-68 (1995).

24. Kitano, Y., Fujikawa, Y,. Takeshita, H., Kamino, T., Yaguchi, T., Matsumoto, H., and Koike, H., *J. Electron Micros.* **44**, 376 (1995).

25. Hull, R., Stevie, F. A., and Bahnck, D., *Appl. Phys. Lett.* **66**, 341 (1995).

26. Muroga, A. and Saka, H. *Scripta Metallurgica et Materialia* **33**, 151-156 (1995).

27. Kitano, Y., Fujikawa, Y., Kamino, T., Yaguchi, T., and Saka, H., *J. Electron Micros.* **44**, 410-413 (1995).

28. Anderson, R., and Klepeis, S. J., *Specimen Preparation for Transmission Electron Microscopy of Materials IV*, Materials Research Society, Pittsburgh **480**, 187-192 (1997).

29. Overwijk, M. H. F., van den Heuvel, F. C., and Bulle-Lieuwma, C. W. T., *J. Vac. Sci. Technol.* **B11**, 2021 (1993).

30. Giannuzzi, L. A., Drown, J.L., Brown, S. R., Irwin, R. B., and Stevie, F. A., *Specimen Preparation in Materials for TEM Analysis IV*, Materials Research Society **480**, 19-27 (1997).

31. Giannuzzi, L. A., Drown, J. L., Brown, S. R., Irwin, R. B., and Stevie, F. A., *Microscopy Research and Techniques* (in press 1998)

X-ray photoemission electron microscopy for the study of semiconductor materials

Simone Anders[1], Thomas Stammler[1], Howard A. Padmore[1], Louis J. Terminello[2], Alan F. Jankowski[2], Joachim Stöhr[3], Javier Díaz[4], Aline Cossy-Favre[5], and Sangeet Singh[6]

[1]Lawrence Berkeley National Laboratory, 1 Cyclotron Road, Berkeley, CA 94720

[2]Lawrence Livermore National Laboratory, 7000 East Ave., Livermore, CA 94550

[3]IBM Almaden Research Center, 650 Harry Road, San Jose, CA 95120

[4]Departamento de Físic, Facultad de Ciencias, Universidad de Oviedo, Avda. Calvo Sotelo s/n, Oviedo, 33007, Spain

[5]EMPA, Dübendorf, Überlandstrasse 129, 8600 Dübendorf, Switzerland

[6]Center for X-ray Lithography, University of Wisconsin-Madison, Stoughton, WI 53589

Photoemission Electron Microscopy using X-rays (X-PEEM) is a novel combination of two established materials analysis techniques - PEEM using UV light, and Near Edge X-ray Absorption Fine Structure (NEXAFS) spectroscopy. This combination allows the study of elemental composition and bonding structure of the sample by NEXAFS spectroscopy with a high spatial resolution given by the microscope. A simple, two lens, 10 kV operation voltage PEEM has been used at the Stanford Synchrotron Radiation Laboratory and at the Advanced Light Source (ALS) in Berkeley to study various problems including materials of interest for the semiconductor industry. In the present paper we give a short overview over the method and the instrument which was used, and describe in detail a number of applications. These applications include the study of the different phases of titanium disilicide, various phases of boron nitride, and the analysis of small particles. A brief outlook is given on possible new fields of application of the PEEM technique, and the development of new PEEM instruments.

INTRODUCTION

NEXAFS spectroscopy is an established technique to study materials properties such as elemental composition, bonding structure, and molecular orientation (1). It is based on the availability of X-ray radiation of tunable wavelength which is produced by a synchrotron. PEEM using X-rays combines NEXAFS properties with high spatial resolution imaging. In addition to topological contrast which is present in a PEEM it is possible to acquire NEXAFS spectra which average over only sub-micron areas, therefore giving elemental and bonding information at with high spatial resolution.

PEEM PRINCIPLE

In an X-PEEM microscope the sample is illuminated by a monochromatized X-ray beam of a diameter larger than the field of view of the instrument. The X-rays are absorbed in the sample and cause photo electron emission which depends on the wavelength of the radiation, the elemental composition, and bonding structure of the elements in the sample. If there is a preferred orientation of molecules in the sample, the X-ray absorption and electron emission also depends on the relative orientation of the molecules and the electrical vector of the radiation for linearly polarized X-rays. A high electric field is applied between the sample and the electron optics to accelerate the electrons, and the electron optics is used to form an image of the emitted electrons. The topological contrast is achieved by the fact that the electric field is distorted at topological surface inhomogeneities which leads to deflection of the electron trajectories. PEEM is a full-field imaging technique and allows therefore real-time imaging of, for example, chemical reactions at surfaces or other time-dependent processes such as annealing phenomena or in-situ film growth. By imaging the sample at incrementally increased X-ray energy it is possible to obtain locally resolved NEXAFS spectra of the sample

CP449, *Characterization and Metrology for ULSI Technology: 1998 International Conference*
edited by D. G. Seiler, A. C. Diebold, W. M. Bullis, T. J. Shaffner, R. McDonald, and E. J. Walters

with an energy resolution given by the resolving power of the synchrotron beamline, and a spatial resolution given by the electron optics column. The highest spatial resolution for a UV PEEM described in the literature is 7 nm (2), and for a PEEM using X-rays 40 nm (3). The theoretical limit for a PEEM which is corrected for chromatic and spherical aberrations is about 2 nm (4).

EXPERIMENTAL

Figure 1 shows a schematic drawing of the PEEM used for the present experiments. It consist of a system of two electrostatic lenses operating at a nominal voltage of 10 kV. A movable aperture located in the backfocal plane of the objective lens limits the accepted pencil angle of the optics and determines the resolution of the microscope.

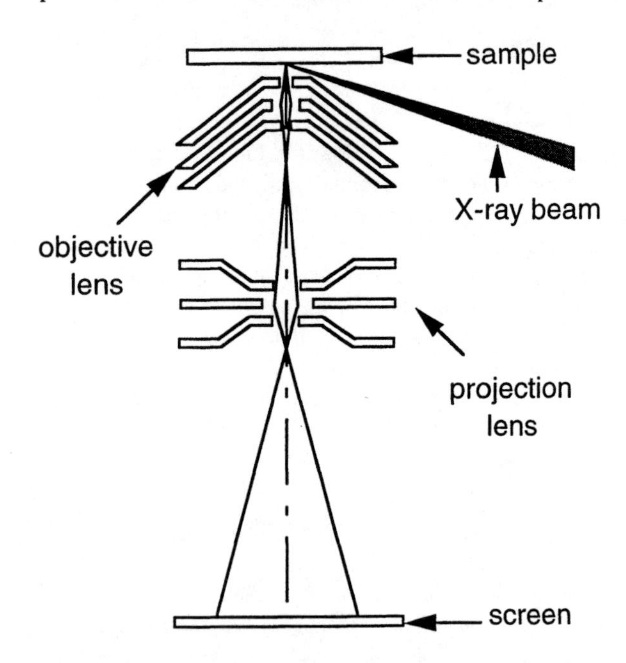

FIGURE 1: Schematic view of a PEEM with two lenses.

The PEEM was used at the Stanford Synchrotron Radiation Laboratory on beamline 10.1 and at the Advanced Light Source in Berkeley at beamline 8.0. The PEEM image is acquired by a microchannel plate, a phosphor screen, and a video camera. The spatial resolution of this microscope is 200 nm limited by misalignement, vibrations, and thermal drift. In a PEEM the sample needs to be conducting to some degree so that charging of the surface does not occur. In cases where this is problem such as, e.g., the study of diamond thin films or ceramic materials, a thin metal (gold or platinum) coating is applied to the surface to avoid surface charging. The probing depth of PEEM is given by the electron escape depth which is about 2-10 nm depending on the material. Due to the elemental specificity of the method, PEEM can study materials buried under a very thin film of another material.

APPLICATIONS

Study of Different Phases of Titanium Disilicide

Titanium disilicide exists in two different phases. One is a metastable C49 base-centered orthorhombic phase with a specific resistivity of 60-90 μΩ cm, the other one is a stable face-centered orthorhombic C54 phase with a specific resistivity of 12-15 μΩ cm (5). The low resistivity phase is desirable for ULSI interconnects, but it has been found that as dimensions shrink the high resistivity phase is formed preferentially (6, 7). We have studied a titanium disilicide pattern with structures of various dimensions which was deposited on a SiO_2 thin film on a Si wafer. The NEXAFS spectrum of both phases differs at the shoulder of the titanium L3 edge and is the basis for differentiation of the two phases in the PEEM microscope (8). Fig. 2 shows the titanium L3 edge absorption spectrum for the two phases.

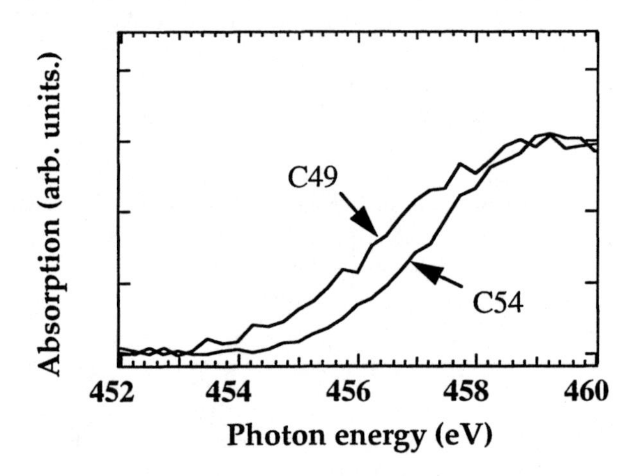

FIGURE 2: Titanium L3 edge absorption spectrum for the two phases of titanium disilicide.

Fig. 3 shows a detail of the patterned structure taken at an X-ray energy below (445.5eV), at the peak (459eV), and at the shoulder of the Ti L3 edge (455eV). Below the edge (Fig. 3a) the titanium pattern appears dark, at the peak (Fig. 3b) the pattern appears bright, whereas at the shoulder of the edge (Fig. 3c) bright edges and bright tips of the spokes are observed indicating a difference in the bonding structure - the formation of the low conductivity phase. As it can be seen in Fig. 2 the low conductivity phase shows stronger absorption at the shoulder of the titanium L_3 edge, therefore the secondary electron emission is higher, and areas of this phase appear brighter in the image. PEEM studies can identify the formation of the different phases of titanium disilicide with high spatial resolution.

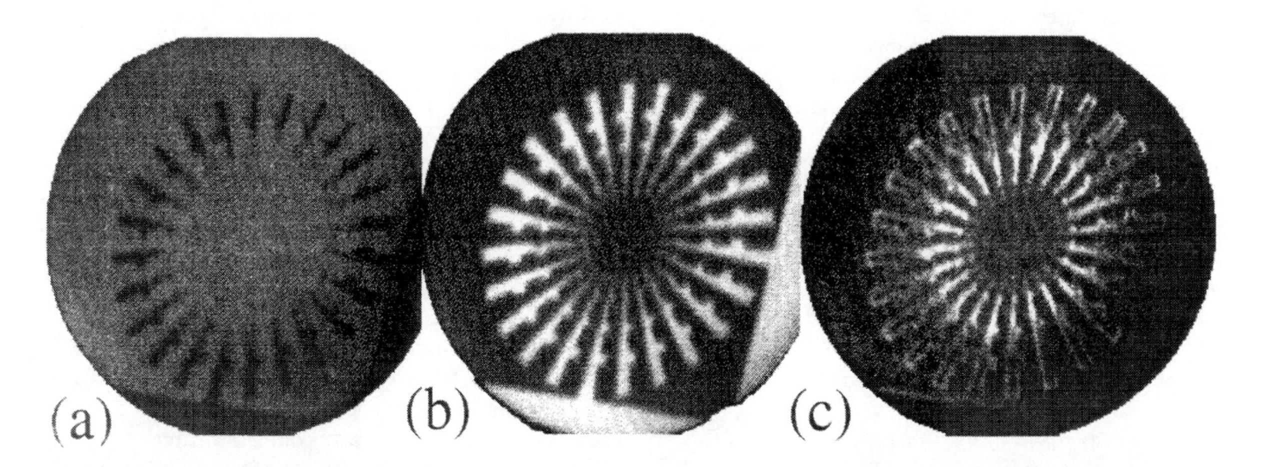

FIGURE 3: PEEM images of the patterned titanium disilicide structure, acquired at an X-ray energy (a) below (445.5 eV), (b) at the peak (459 eV), and (c) at the shoulder (455 eV) of the Ti L_3 edge. The entire spokes pattern is 80 μm in diameter.

Study of Different Phases of Boron Nitride

Boron nitride is an interesting material for technological applications including microelectronic devices, e.g., as gate insulator material on InP (9), and it has been discussed as a candidate for low-dielectric-constant applications (10). It exists in the sp^2-bonded hexagonal (h-BN), rhombohedral (r-BN), and turbostratic phases, and in the sp^3-bonded cubic (c-BN), and wurtzite (w-BN) phases. NEXAFS spectroscopy is very sensitive to variations in the local bonding structure and therefore is a suitable method to study the various phases of boron nitride (11).

We have investigated BN films which were grown by reactive sputter deposition from a pure boron target on silicon substrates. The samples were locally post-processed by ion bombardment with N_2^+ ions at energies of 90 and 180 keV with total doses of $1x10^{17}$ cm^{-2}, and with current densities of 80-100 μA cm^{-2}. Fig. 4 shows a PEEM image taken at the π* resonance of the boron K edge at 192eV. The small wedge on the lower part of the image is the implanted region.

Figure 5 shows the corresponding local NEXAFS spectra of the implanted and unimplanted region at the boron K edge which were taken by averaging over a region of about 50 μm diameter. Comparison with reference spectra from Jiménez et al. (11) shows that the unimplanted area exhibits a clear hexagonal structure with the boron π* resonance at 192eV (peak 1) and the σ* resonance at 197eV.

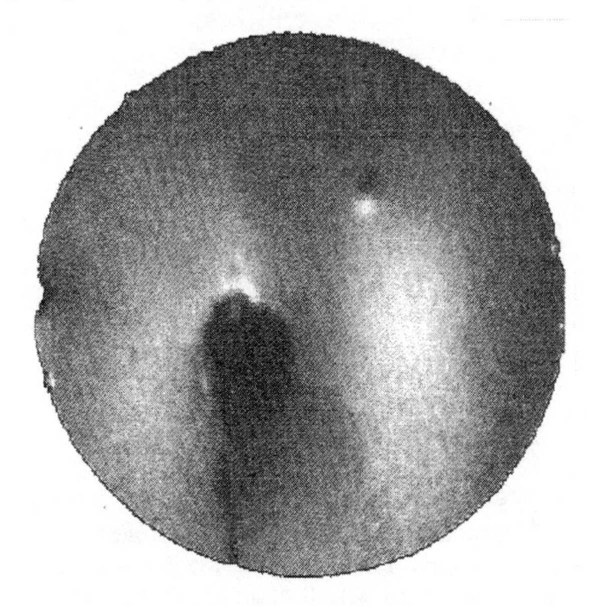

FIGURE 4: PEEM image taken at the π* resonance of the boron K edge at 192eV. The small wedge on the lower part of the image is the tip of the nitrogen implanted region. The field of view is 150 μm.

The spectrum is very similar to previously published spectra of films grown by either pulsed-laser or rf sputter deposition (11). The small additional features (peaks 2 to 4) just above the boron π* resonance are correlated with N-void effects in the hexagonal bonding, i.e. with the fact that boron atoms can be bound to three (peak 1), two (peak 2), one (peak 3), or no (peak 4) nitrogen atoms leading to four different peaks at the π* resonance. The spectrum in the implanted region exhibits stronger defect features, but no formation of the cubic phase which was observed when films were deposited by ion assisted laser-deposition at lower current densities of 10-70 μA cm^{-2} (11). The PEEM image shows that the unimplanted region is brighter than the implanted region due to the fact the boron π* resonance is stronger for the unimplanted h-BN as shown in Fig. 5. NEXAFS can easily distinguish between the different bonding states of the boron atoms in the thin film and PEEM can show the spatial distribution of the bonding states in the sample as Fig. 4 indicates. PEEM could also be a useful tool for the study of boron bonding states and boron distribution in boron implanted materials.

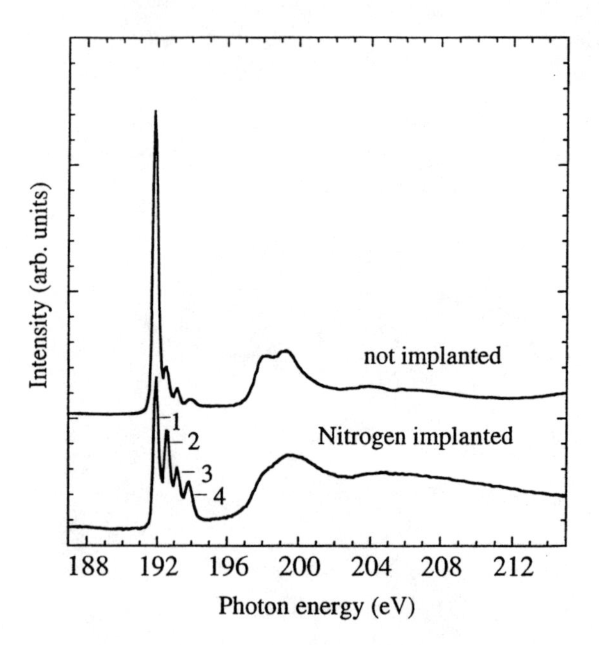

FIGURE 5: Local NEXAFS spectra at the boron K edge taken in the unimplanted and Nitrogen implanted region. The peaks correspond to: (1) - B bound to 3 N atoms, (2) - B bound to 2 N atoms, (3) - B bound to 1 N atom, (4) - B bound to no N atom.

Study of Small Particles

Particle contamination can be a severe problem for semiconductor manufacturing processes, and the identification of the elemental composition and chemical bonding state of small particles can help to identify sources of contamination. PEEM can be applied to the detection and analysis of small particles.

We have studied small particles on diamond-like carbon thin films which were formed by cathodic arc deposition (12). This material is a promising candidate for a cold electron emitter for flat panel displays. A number of small particles were found at the sample surface after deposition. These particles were several microns in diameter and showed an order of magnitude higher electron emission yield than the surrounding film. The particles were analyzed by taking local NEXAFS spectra on the particle and the surrounding film at the carbon and oxygen K edges. It was found that the particles contained a much higher concentration of oxygen than the surrounding film, and that the oxygen in the particles was bound mainly in carboxyl and carbonate groups. Enhanced field emission which is of crucial interest for flat panel display applications of diamond-like carbon seems to be connected to the enhanced oxygen content of these small particles.

In another experiment we investigated sliders of computer hard disks drives after a wear test which was performed to study the tribological behavior of the disks and sliders (13). Several small particles were observed at the slider surface after the wear test. Local NEXAFS spectra were taken of these particles and the surrounding slider surface area, and it was found that the particle composition and chemical bonding structure is identical to the amorphous carbon overcoat of the slider surface. This indicates that the particles originate from the slider surface and are not, e.g., dust particles from the environment.

CONCLUSIONS AND OUTLOOK

The combination of NEXAFS and PEEM gives a new tool for the study of materials problems and systems which are of interest to the semiconductor industry. It is possible to analyze the elemental composition and chemical bonding structure of samples with a high spatial resolution. The applications described in this paper are only a small number of possible materials systems which can be studied by this method.

Other materials which are of relevance to the semiconductor industry are polymers, e.g., for photoresist and low-dielectric-constant applications. Systems of polymers have been studied successfully by synchrotron-based X-ray microscopy techniques such as scanning X-ray transmission microscopy (STXM) (14) and PEEM (15). These methods are particularly sensitive to chemical bonding states in carbon-based materials such as polymers, and can yield comprehensive information about polymer bonding states, spatial distributions of various polymers in blends and multilayers, segregation and dewetting phenomena, polymer degradation, etc.

New PEEM instruments are being build around the world and installed at various synchrotrons. A new instrument (PEEM2) is just being commissioned at the ALS in Berkeley which will be much more user-friendly than existing research-type instruments. It is equipped with an automated, fast sample transfer system, numerous sample preparation and analysis capabilities, in-situ heating and deposition options, and computer control of the electron optics. The availability of new PEEM instruments which are more accessible for users from outside the synchrotron facilities will increase the number of experiments which can be performed and will broaden the fields of applications for these instruments.

ACKNOWLEDGMENTS

This work was supported by the Division of Materials Sciences, Office of Basic Energy Science, and performed under the auspices of the U.S. Department of Energy by the Advanced Light Source, LBNL, under Contract No. DE-AC03-76SF00098, and LLNL under contract No. W-7405-ENG-48. JD was supported by the Ministerio de Educación y Ciencia of Spain.

REFERENCES

1. Stöhr, J., *NEXAFS Spectroscopy*, New York: Springer, 1992.
2. Rempfer, G. F., and Griffith, O. H., *Ultramicroscopy* **27**, 273-300 (1989).
3. Bauer, E., Franz, T., Koziol, C., Lilienkamp, G., Schmidt, T., in: Rosei, R. (ed.), *Chemical, Structural, and Electronic Analysis of Heterogeneous Surfaces on the Nanometer Scale*, Kluwer Academic, Dortrecht, in press.
4. Fink, R., Weiss, M. R., Umbach, E., Preikszas, D., Rose, H., Spehr, R., Hartel, P., Engel, W., Degenhardt, R., Wichtendahl, R., Kuhlenbeck, H., Erlebach, W., Ihmann, K., Schlögl, R., Freund, H.-J., Bradshaw, A. M., Lilienkamp, G., Schmidt, Th., Bauer, E., and Brenner, G., *J. Electron Spectrosc. Relat. Phenom.* **84**, 231-250 (1997).
5. Jeon, H., Sukow, C. A., Honeycutt, J. W., Rozgony, G. A., and Nemanich, R. J., *J. Appl. Phys.* **71**, 4269-4276 (1992).
6. Roy, R. A., Clevenger, L. A., Cabral, Jr., C., Saenger, K. L., Brauer, S., Jordan-Sweet, J., Buccignano, J., Stephenson, G. B., Morales, G., and Ludwig, Jr., K. F., *Appl. Phys. Lett.* **66**, 1732-1734 (1995).
7. Saenger, K. L., Cabral, Jr., C., Clevenger, L. A., Roy, R. A., and Wind, S., *J. Appl. Phys.* **78**, 7040-7044 (1995).
8. Singh, S., Solak, H., Krasnoperov, N., Cerrina, F., Cossy, A., Díaz, J., Stöhr, J., and Samant, M., *Appl. Phys. Lett.* **71**, 55-57 (1997).
9. Paul, T. K., Bhattacharya, P., and Bose, D. N., *Appl. Phys. Lett.* **56**, 2648-2650 (1990).
10. Maeda, M., Jap. J. Appl. Phys. **29**, 1789-1794 (1990).
11. I. Jiménez, A. F. Jankowski, L. J. Terminello, D. G. J. Sutherland, J. A. Carlisle, G. L. Doll, Tong, W. M., Shuh, D. K., and Himpsel, F. J., *Phys. Rev. B* **55**, 12025-12037 (1997).
12. Díaz, J., Anders, S., Cossy-Favre, A., Samant, M., and Stöhr, J., *Appl. Phys. Lett.*, submitted for publication.
13. Anders, S., Stammler, T., Bhatia, C. S., Stöhr, J., Fong, W., and Bogy, D. B., *Spring Meeting of the Material Research Society*, San Francisco, 1998, to be published.
14. Winesett, D. A., Ade, H., Smith, A. P., Rafailovich, M., Sokolov, S., and Slep, D., *Microscopy and Microanalysis*, to be published (1998).
15. Anders, S., Stammler, T., Ade, H., Slep, D., Sokolov, J., Rafailovich, M., Heske, C., and Stöhr, J., *1997 Compedium of the Advanced Light Source*, Berkeley, to be published.

Neutron Reflectometry for Interfacial Materials Characterization

Eric K. Lin, Darrin J. Pochan, Rainer Kolb, Wen-li Wu, and Sushil K. Satija*

*Polymers Division and NIST Center for Neutron Research**
National Institute of Standards and Technology
Gaithersburg, MD 20899

Neutron reflectometry provides a powerful non-destructive analytic technique to measure physical properties of interfacial materials. The sample reflectivity provides information about composition, thickness, and roughness of films with 0.1 nm resolution. The use of neutrons has the additional advantage of being able to label selected atomic species by using different isotopes. Two examples are presented to demonstrate the use of neutron reflectometry in measuring the thermal expansion of a buried thin polymer film and measuring the change in polymer mobility near a solid substrate.

INTRODUCTION

The continued decrease in device size in the microelectronics industry has increased the importance of material properties in thin films and at interfaces. The interfacial material properties can be substantially different from those of the bulk material and can be the determining factor for the use of a particular material in a device. For example, a lack of adhesion of an interlayer dielectric material with the substrate can preclude its use in a device. With the increase in organic material components in microelectronics packaging and interconnects, the measurement of physical properties such as the coefficient of thermal expansion (CTE), adhesion kinetics, and the glass transition temperature in thin films and at interfaces is very important.

A powerful experimental tool to measure interfacial material properties is neutron reflectometry. The use of neutron reflectometry has enjoyed great success in studying organic materials at surfaces and interfaces (1). Neutron reflectometry is an in-situ, non-destructive measurement technique able to measure interfacial profiles and thin film thicknesses with very high resolution (~0.1 nm). Unlike other profiling techniques such as ion beam methods, the reflectivity measurement is averaged over macroscopic areas. The use of neutrons also provides the capability of selectively labeling an organic material component by deuteration. Thus, high scattering length contrast to neutrons can be generated between almost chemically identical species.

EXPERIMENTAL TECHNIQUE

The basic geometry of the experiment is schematically shown in Fig. 1 for the measurement of the reflectivity of a single component thin film. A highly collimated neutron beam is sent incident to the sample surface at small angles (typically less than 3°). The detector is placed at the same angle of reflection at the specular condition. In this configuration, the scattering vector, q [$q=(4\pi/\lambda)\sin\theta$, where λ is the wavelength and θ is the angle of incidence], is perpendicular to the sample surface. Below a critical angle, the neutron beam is totally externally reflected. At angles greater than the critical angle, the reflected intensity dramatically decreases as the neutron beam penetrates the top surface of the sample. Since the adsorption is generally low, the neutron beam has a penetration depth of several hundred nanometers. As the neutrons penetrate the film, the neutrons can be reflected from a second interface back towards the detector. The constructive and destructive interference with the neutrons reflected from the first interface gives rise to clear oscillations in the reflected intensity as a function of the incident angle.

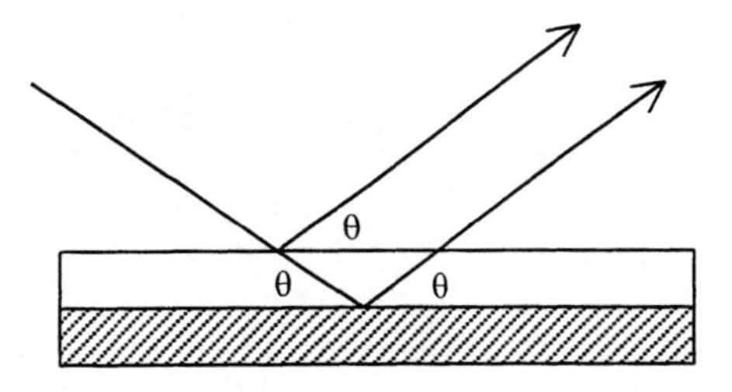

FIGURE 1. Schematic diagram of a reflectivity measurement of a single component thin film.

To determine the scattering length profile of the sample, a model profile is generated and then varied to best fit the reflectivity data. In general, the overall film thickness is determined from the spacing of the minima or maxima in

CP449, *Characterization and Metrology for ULSI Technology: 1998 International Conference*
edited by D. G. Seiler, A. C. Diebold, W. M. Bullis, T. J. Shaffner, R. McDonald, and E. J. Walters
1998 The American Institute of Physics 1-56396-753-7/98/$15.00

the reflectivity profile, the depth of the oscillations is controlled by the difference in the scattering length densities of each medium (between the vacuum and the film in this case), and any dampening of the oscillations in the reflectivity data is indicative of the interfacial roughness. The scattering length density is dependent upon nuclear interactions, which can greatly differ with different isotopes with similar atomic number. The reflectivity data are sufficient to determine layer thicknesses and interfacial roughnesses with a resolution of 0.1 nm.

There are some limitations for this experimental method. The phase information is lost and a model profile must be used to fit the data. The resulting parameters are model dependent and may limit the interpretation of the data. The use of independent measurements on the sample is invaluable for the reduction of the number of parameters needed to fit the reflectivity data. Neutron facilities are necessarily expensive and rare, so neutron reflectometry is not suitable for routine sample characterization. However, neutron reflectometry provides the opportunity to gain important information from samples that cannot be examined by other techniques.

To demonstrate the advantages of using neutron reflectometry, we present two examples of how it can be used to study important issues in electronics packaging: a) the measurement of the thermal expansion of a confined polymer film and b) the measurement of the mobility of polymer chains near a solid substrate. The experiments described here were performed at the NG-7 Reflectometer at the National Institute of Standards and Technology with a wavelength of 4.768 Å.

THERMAL EXPANSION OF A BURIED POLYMER LAYER

In this example, the ability of neutron reflectometry to accurately measure very small changes in thickness provides a method to determine changes in the thermal expansion behavior of confined thin polymer films. X-ray reflectivity and ellipsometry have been used to measure the thermal expansion (and the glass transition temperature from a change in the coefficient of thermal expansion) of films on silicon wafers (2,3). Polymers in the confined geometry of a thin film were found to have different thermal expansion and apparently different glass transition temperatures from those of the bulk state. These properties are further perturbed from bulk values when adhesive interactions are present between the polymer film and the substrate (4).

To deconvolute the effects of wall adhesion and chain confinement on the thermal properties of polymeric thin films, samples are prepared with a labeled polymer film confined between two non-adhesive polymeric layers. The

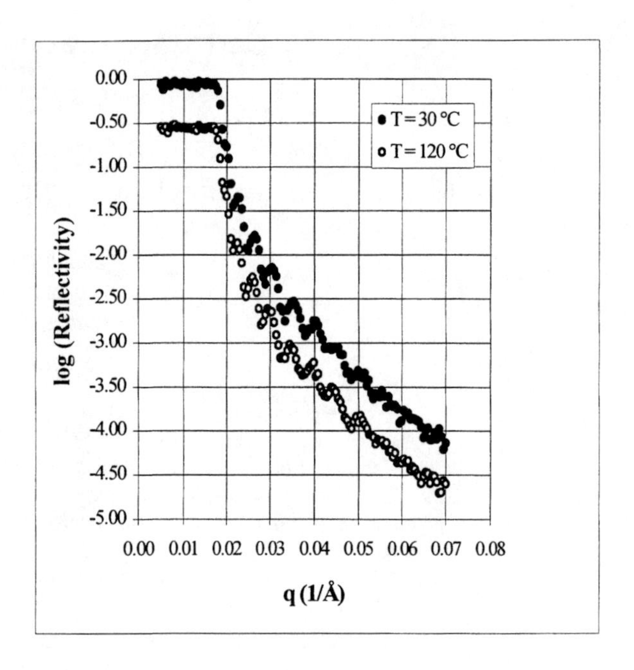

FIGURE 2. Neutron reflectivity data from a buried deuterated polystyrene layer 113 nm thick in between two polycarbonate layers 40.8 nm and 225 nm thick at two different temperatures. The curves are vertically offset for clarity.

labeled polymer is monodisperse deuterated polystyrene (d-PS) with a molecular mass, M_w (5), of 260,000 g/mol and the confining polymer is monodisperse polycarbonate with a molecular mass of 18,000 g/mol. To prepare the sample, the polycarbonate is initially spin-coated from a 1,2-dichloroethane solution onto a polished silicon wafer. After extensive annealing, the d-PS is spin-coated from a toluene solution onto a glass slide and then floated from water onto the lower polycarbonate layer. After annealing the resulting bilayer, the final polycarbonate layer is spin-coated from a 1,2-dichloroethane solution is floated onto the d-PS layer.

Figure 2 shows reflectivity data from one sample consisting of a lower polycarbonate layer 40.8 nm thick, a d-PS layer 113.0 nm thick, and an upper polycarbonate layer 225.0 nm thick at two different temperatures (6). The reflectivity is dominated by the characteristics of the d-PS layer. An increase in the d-PS thickness is evident from the shift of the oscillations in the data toward lower q values. The fitted d-PS thickness at 30 °C is 112.5 nm and the thickness at 120 °C is 114.9 nm. This small change in thickness is easily resolved from the reflectivity measurement. A full thermal expansion curve can then be determined and is shown in Fig. 3. This sample shows a glass transition temperature curve near the expected value of 100 °C from a change in the slope of the thermal expansion curve.

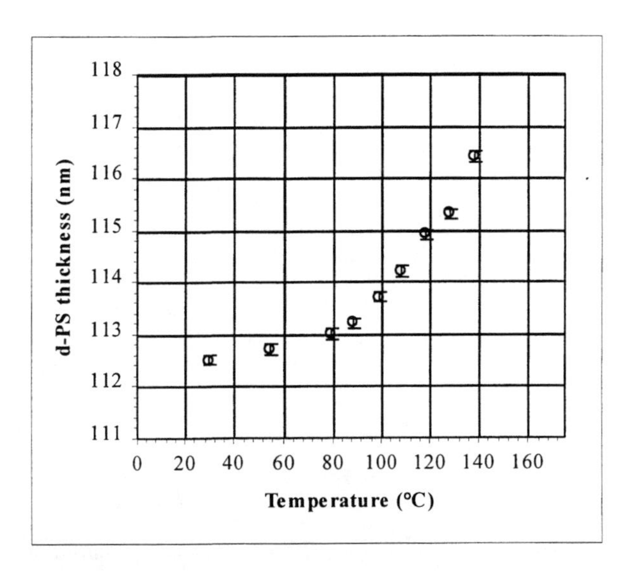

FIGURE 3. Thermal expansion curve for a deuterated polystyrene layer 112.5 nm thick in between two polycarbonate layers.

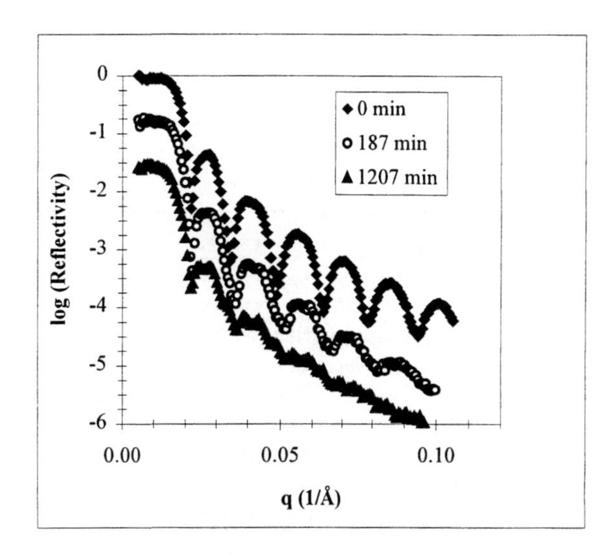

FIGURE 4. Reflectivity data from an interdiffusion experiment between deuterated PMMA and a hydrogenated PMMA layers after annealing for 0 min, 187 min, and 1207 min at 150 °C. The curves are vertically offset for clarity.

With other non-destructive experimental techniques such as X-ray reflectometry or ellipsometry, the contrast between the polycarbonate and the d-PS layer is very small because of the small difference in the electron density or the index of refraction. With neutrons, the scattering length density of d-PS is much greater than the polycarbonate, providing a strong signal from the component of interest.

POLYMER MOBILITY NEAR A SOLID SUBSTRATE

To understand the adhesion properties and kinetics of long chain molecules, a fundamental understanding of the mobility of chain molecules near interfaces is needed. In particular, information about the effect of chain connectivity and the chain-substrate interaction energy is required. Neutron reflectometry is an ideal technique to address this issue. The high spatial resolution of the technique enables one to measure chain movement over distances less than the characteristic length scale of the molecule. Deuteration provides strong contrast for one component between almost chemically identical species. Finally, the technique is non-destructive so that the same sample can be observed over a period of time for kinetic studies.

In this example, neutron reflectometry is used to study the interdiffusion kinetics between two thin (40.7 nm to 85.3 nm) poly(methyl methacrylate) (PMMA) layers on silicon oxide substrates. The samples are prepared with a deuterated PMMA, M_w = 260,000 g/mol, layer spin coated onto the silicon substrate. The second layer of hydrogenated PMMA, M_w = 262,000 g/mol, is spin-coated onto a glass slide and floated from water onto the d-PMMA layer. Initially, the interfacial width between the two layers is less than 1 nm wide. The mobility of the polymer chains in the deuterated layer is measured by observing the interfacial broadening between the deuterated and the hydrogenated layers as the sample is heated at 150 °C, 35 °C above the glass transition temperature of the polymers.

The broadening of the interface can be readily seen in the data shown in Fig. 4. The reflectivity data from the initially prepared sample have several sharp maxima and minima, indicating the presence of a sharply defined interface. The interfacial width from the best fit to this data is 0.9 nm. At the two longer annealing times, the oscillations in the reflectivity data begin to dampen at higher q values as a result of the broadening of the interface. The interfaces at 187 min and 1207 min are 6.92 nm and 14.6 nm wide, respectively. The mobility of the polymers can be determined from the rate at which the interface broadens.

The interdiffusion rate in these samples was examined as a function of the lower d-PMMA film thickness and was analyzed in terms of an effective diffusion constant. The effective diffusion coefficient of the polymer was found to be strongly dependent on distance from the silicon oxide surface (7). In recent studies, the effect of the polymer-surface interaction energy on the interdiffusion kinetics is determined through the use of silane coupling agents (8).

SUMMARY

In summary, neutron reflectometry is a powerful experimental technique to characterize the interfacial properties of materials near surfaces and interfaces. The use of neutrons precludes regular sample characterization because of the scarcity of neutron sources. However, the use of this high resolution, non-destructive technique provides the ability to address many important issues in interfacial materials characterization, particularly with the opportunity to label selectively a component of interest with isotopic substitution. The two examples presented above serve to illustrate some of the questions that can uniquely be addressed by neutron reflectometry.

ACKNOWLEDGMENTS

E. K. L. and D. J. P. thank the National Research Council and the National Institute of Standards and Technology Postdoctoral Research Associate Program for postdoctoral support.

REFERENCES

1. Russell, T. P., *Materials Science Reports*, **5**, 171-271, 1990.
2. Keddie, J. L., Jones, R. A. L., and Cory, R. A., *Europhyics Letters*, **27**, 59-64, 1994.
3. Orts, W. J., and van Zanten, J. H., Wu, W. L, and Satija, S. K., *Physical Review Letters*, **71**, 867-870, 1993.
4. Wallace, W. E., van Zanten, J. H., and Wu, W. L., *Physical Review E*, **52**, R3329-R3332, 1995.
5. According to ISO 31-8, the term "molecular weight" has been replaced by "relative molecular mass," symbol M_r. Thus, if this nomenclature were to be followed in this publication, one would write $M_{r,n}$ instead of the historically conventional M_n for the number average molecular weight, with similar changes for M_w, M_z, and M_v, and it would be called the "number average relative molecular mass." The conventional notation, rather than the ISO notation, has been employed for this publication.
6. The data throughout the manuscript and in the figures are presented along with the standard uncertainty (±) involved in the measurement.
7. Lin, E. K., Wu, W. L., and Satija, S. K., *Macromolecules*, **30**, 7224-7231, 1997.
8. Kolb, R., Lin, E. K., Wu, W. L., and Satija, S. K., in preparation.

Recent Developments in Neutron Depth Profiling at NIST

G.P. Lamaze, H.H. Chen-Mayer, and J.K. Langland

Analytical Chemistry Div., Chemical Science and Technology Laboratory, NIST, Gaithersburg, MD 20899

Neutron Depth Profiling [NDP] is a method of determining the concentration and location of certain isotopes in the near surface region of solids. While only a few isotopes are measurable by this technique, they happen to be isotopes of elements that are currently important to the semiconductor industry, namely boron, nitrogen, and lithium. NDP analysis is both quantitative and non-destructive; this makes it the reference method of choice for these elements. This paper discusses recent measurements at the National Institute of Standards and Technology (NIST) for each of these elements as well as recent improvements in the NDP facility. A brief explanation of the technique, including its advantages and limitations, is presented.

INTRODUCTION

Neutron depth profiling (NDP) is a technique for measuring the concentration and depth of certain light elements in the near surface region of solids. The elements most commonly analyzed by this technique are boron, nitrogen, and lithium. Boron is an important constituent in silicon semiconductors both as a dopant and in passivation layers. Many compounds of nitrogen are in use or proposed for semiconductor processes. Lithium compounds are important in many new battery technologies. Recent examples of measurements of these three elements are presented. Improvements in the facility and the ability for analyzing sub-millimeter samples are also discussed.

METHOD

Ziegler *et al* originally applied the NDP technique in 1972 (1) to the measurement of boron implanted in silicon. An NDP analysis begins with the capture of a "slow" neutron by a nucleus, which then emits a charged particle of well-known initial energy. As the charged particle exits the material, it loses energy in a predictable way and this energy loss is measured to determine the depth of the originating nucleus. For example, in the boron reaction, $^{10}B(n,\alpha)^{7}Li$, there are two outgoing alpha particles with energies of 1.472 (93.7% branch) and 1.776 MeV (6.3%), and two corresponding recoil ^{7}Li nuclei with energies of 0.840 and 1.014 MeV. All four particles carry the depth and concentration information.

FACILITY

A new liquid-hydrogen cold source has replaced the D_2O ice cold source used previously. Both the new and old cold sources have been discussed by Prask et al. (2) This new source reaches a lower effective temperature and more fully illuminates the neutron guides. The lower effective temperature gives two positive effects: improved neutron transport efficiency through the guides and increased neutron capture reaction rates. The Cold Neutron Depth Profiling (CNDP) instrument (3) is installed at the end of a curved super-mirror guide at the exit of Cold Tube West (CTW). The neutron guide acts as a lightpipe, with the longer wavelength neutrons preferentially guided to the chamber. Because the guide is curved, the NDP chamber does not directly view the reactor and its associated fast neutrons and gamma rays. The 13.5-cm thick sapphire filter used previously in the CTW beam is no longer necessary. Of all the instruments viewing the cold source, the CNDP instrument is the closest to the reactor and, consequently, the number of neutrons exiting the guide is the highest. The fluence rate at the entrance to the chamber is $2.45\pm0.05 \times 10^{9}$ n/cm^{2}s and the cadmium ratio is about 10^5. The gamma field at the sample position is about 200 mR/h.

While the absolute neutron fluence is not measured for each experiment, a run-to-run monitor is installed in the chamber to ensure that normalization between standards and unknowns can be performed correctly. We have replaced the fission chamber that was at the exit of the shutter with a boron deposit on silicon viewed by a surface barrier detector inside of the NDP vacuum chamber. This was done to reduce the gamma radiation in the region around the NDP chamber, which was caused by neutron capture in the aluminum of the fission chamber.

We are in the process of installing a monolithic lens for focusing the incoming neutron beam onto a small sample area (~200 μm). This will increase the number of neutrons per unit area and consequently the number of reactions within the sample. A small intense beam also permits sample scanning for lateral compositional mapping. One of the prototype monolithic lenses previously tested with a neutron beam (4) has been installed in the NDP chamber for testing. At the sample position, the lens gives a factor of eight gain in neutron fluence rate in an area of 0.21 mm diameter, full-width at half-maximum (FWHM) as measured with a neutron

CP449, *Characterization and Metrology for ULSI Technology: 1998 International Conference*
edited by D. G. Seiler, A. C. Diebold, W. M. Bullis, T. J. Shaffner, R. McDonald, and E. J. Walters
1998 The American Institute of Physics 1-56396-753-7/98/$15.00

camera. The first test concentrated on the lateral spatial resolution, i.e. moving a sample with composition variation across the focused beam and the charged particle response being measured. This test has demonstrated (5) that the spatial resolution for this particular sample is about 0.23 mm. As a part of developing the procedure for using a focused beam in the NDP measurements, we have also set up an automated scanning routine to search for the maximum charged particle response. In addition, we are developing a new alignment system that contains a charge-injection device camera with vacuum capability. It will provide a spatial resolution of 12 μm (pixel size) and an 8 bit dynamic range for the neutron intensity, and a possibility for charged particle imaging. Figure 1 shows the new vacuum-compatible motion stages, which will enable remote measurement of lens properties without breaking the vacuum. Further, the lens can be retracted out of the beam for measurements on the same sample with normal NDP geometry. The plans for using the lens include small samples, or a small area on a large sample. For example, it is interesting to study how the lithium distribution is altered within a sub-mm crater created by proton beam analysis on a $LiNbO_3$ sample. By scanning the sample laterally, we can also study samples of suspected boron non-uniformity caused by manufacturing processes.

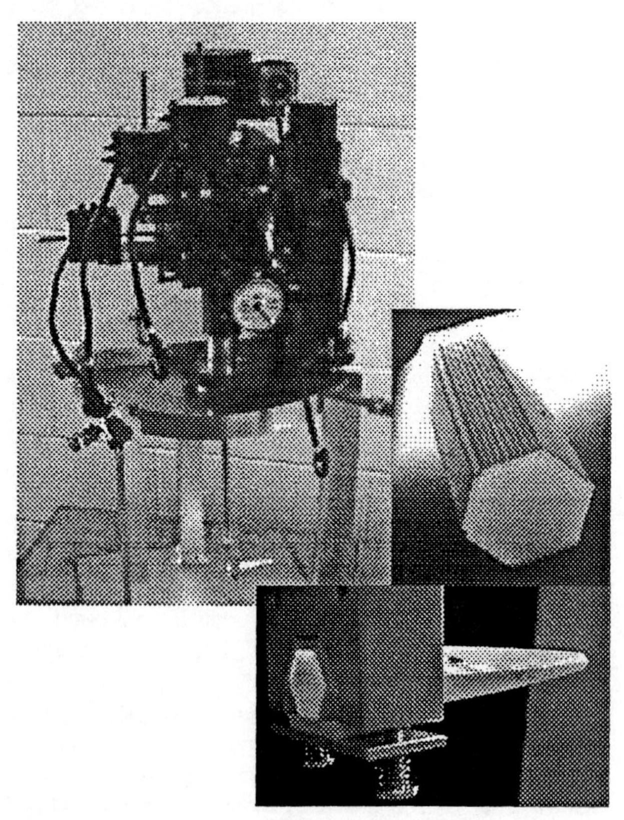

Figure 1. Apparatus for positioning and aligning lens. Stages are mounted on top of the vacuum chamber. Lens is shown in insets.

APPLICATIONS

Boron Measurements

Neutron depth profiling has been used to determine the elemental concentration, the film thickness, and the mass fraction of boron in boro-phosphosilicate glass coatings. The calibrated samples are then used as reference standards for other analytical techniques. We have previously used the alpha particle for this analysis. To improve the depth resolution, we have begun recently to use the outgoing ^{7}Li particle in our analysis method (6). This particle has a greater stopping power per unit depth and thus provides better depth resolution for thin films. We have made comparisons of the analysis of the lithium particle with the well-established routine analysis of the alpha particle. The data from a variety of matrices demonstrates the advantage of using the lithium ion in near surface profiling. These samples include: a thin (~2 nm) ^{10}B surface deposit, ^{10}B implanted in silicon, boron in diamond-like carbon, boron in epitaxially grown silicon, and BPSG layers on silicon. By making direct comparisons of the lithium ion and alpha particle stopping powers in the same matrix, the data can be used to validate stopping power codes such as TRIM (7), and to compare with other stopping power measurements.

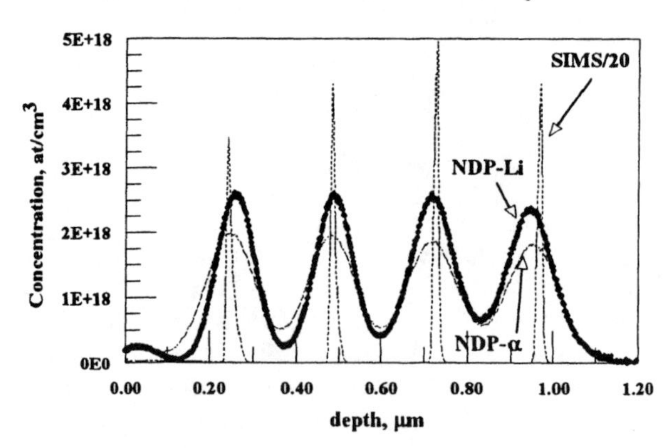

Figure 2. Comparison of boron distributions made with alpha particles, lithium particles and SIMS. SIMS concentrations are divided by twenty to be on same scale.

As one example of this type of analysis, we show in Fig. 2 the boron distribution obtained with both the alpha particle and the lithium recoil measured normal to the surface for a boron multilayer in silicon. The sample was epitaxially grown and the boron layers are thinner than the resolution of the neutron depth profiling system. This provides a means of comparing the resolution function of each particle. The boron concentration and location were

Table 1. Comparison of peak analysis with Helium and Lithium particles

	Helium particle Analysis		Lithium particle analysis	
Peak	Centroid, µm	Width, µm	Centroid, µm	Width, µm
1	0.247 ± .001	0.111 ± .002	0.262 ± .001	0.085 ± .001
2	0.482 ± .001	0.101 ± .002	0.492 ± .001	0.085 ± .001
3	0.719 ± .001	0.104 ± .002	0.723 ± .001	0.086 ± .001
4	0.951 ± .001	0.113 ± .005	0.949 ± .003	0.102 ± .008

also compared with secondary ion mass spectrometry (SIMS) measurements (8). Table I summarizes the depth and resolution for each of the peaks (together with their expanded uncertainties, 2σ) measured with both alpha particles and lithium recoils. The depth resolution for the lithium particle is always better than that for the alpha particle, even at a depth of almost one-micrometer. Unfortunately at that depth, the lithium recoil signal is overlapping a large background caused by photoelectron events from the sample (already subtracted in Fig. 2). Further efforts are necessary to reduce the background and find the appropriate subtraction procedures for the remaining background. As can also be noted in Fig. 2, the depth (peak centroid) determined from the two particles differs slightly, with the SIMS depth agreeing with the alpha for the first two peaks and being greater than the alpha and lithium for the last two peaks. This could be an indicator that the calculated stopping power is incorrect. We have continued our collaboration with Kevin Coakley of the Statistical Engineering Division of NIST on the modeling of detector response for NDP (9). Using the alpha particle results from this multilayer, he combines the known stopping-power data with straggling theory and multiple scattering effects to arrive at single model that describes the observed NDP spectra. Results of these calculations will be reported separately.

Nitrogen Measurements

Another element that can be studied by neutron depth profiling is nitrogen, which produces a proton from the reaction $^{14}N(n,p)^{14}C$. Because the incremental energy loss for protons is less than that for alpha particles, the depth resolution for nitrogen analysis is worse than that for boron. Typical resolution for analysis of nitrogen is 200 nm (2 σ), although resolution of 50 nm (2 σ) has been achieved by measuring the proton at angles of 70°. Figure 3 shows depth profiles for a sample of boron nitride grown on a layer of aluminum nitride. We are able to obtain simultaneously the profiles of the boron and nitrogen and thus measure the stoichiometry of the BN layer. A small amount of boron is also observed imbedded in the AlN layer. Other systems containing nitrogen that have been studied recently include titanium nitride, silicon nitride and nitrogen as a contaminant in layers of nickel aluminide. The detection limit for nitrogen measurements in thin films is 2 x 10^{14} at/cm^2. This corresponds to about 90 µg/g of nitrogen contaminant in a 1-µm thick layer of Ni_3Al.

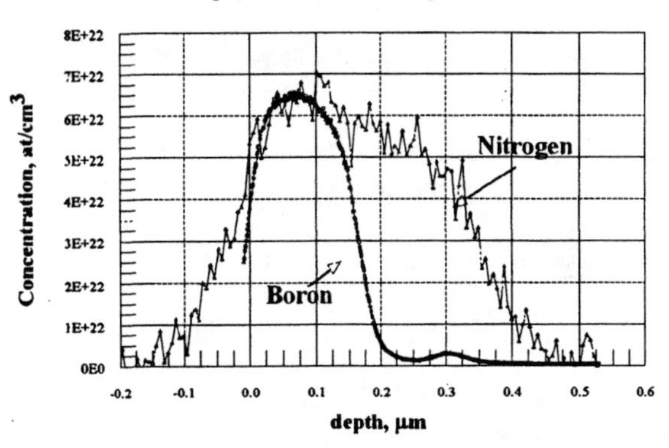

Figure 3. Boron and nitrogen profiles of boron nitride deposited on an aluminum nitride layer on silicon.

Lithium Measurements

We have made a variety of measurements of profiles of lithium in different matrices, e.g., $LiNbO_3$, lithium implants in diamond, and lithium in electrochromic multilayers. The latter are devices where the transport of lithium across conduction layers changes the optical transparency of the film from nearly opaque (coloring mode) to nearly transparent (bleaching mode). Such devices are intended for window and mirror coatings whose transmission could be easily controlled by the application of small voltages. Moreover, similar multilayers could be candidates for batteries for semiconductor applications. Neutron depth profiling measurements is a valuable tool in analyzing lithium transport across the boundary layers of such devices. Complete details of these measurements will be reported (10) separately, but a brief example is given here. The object was to measure the migration of lithium in an active device. The coated glass was placed in the sample position in the neutron beam and the electrodes were connected to a 3-volt supply through electrical feedthroughs in the vacuum chamber. Measurements were made in the coloring and bleaching modes, allowing sufficient time for the lithium to equilibrate between measurements. Figure 4 shows the results of these measurements. By taking the difference between the two spectra, the total amount of lithium transferred between the layers is obtained (positive on one side of the ion conductor and negative on the other). For this sample, 1.6 x 10^{17} atoms/cm^2 of lithium atoms were moved across the ion conduction layer. This is the equivalent of 25 mC of charge. Since the technique is non-destructive, repeated measurements of the lithium cycling from one state to the other could be made. Simultaneous measurements of lithium concentration and optical transmission demonstrated that the lithium transfer correlated with the change in film transparency.

SUMMARY

We have reported a new capability at the NIST NDP instrument to automatically position a monolithic neutron lens in front of a sample. This lens has a focal spot of 200 μm with a neutron intensity gain of about eight. A new method of using the lithium recoil for improved resolution measurements has been applied to a number of systems, with one specific example shown. We have described recent measurements on the elements of boron in silicon, boron and nitrogen in boron nitride and in aluminum nitride, and lithium in electrochromic multilayers. Further information on this instrument and other neutron instruments at the Center for Neutron Research is available at http://rrdjazz.nist.gov.

REFERENCES

1. J.F. Ziegler, G.W. Cole, and J.E.E. Baglin. J. Appl. Phys. **43** (1972) 3809.
2. H.J. Prask, J.M. Rowe, J.J. Rush, and I.G. Schroder. NIST J. Res. **98** (1993)1
3. R.G. Downing, G.P. Lamaze, J.K. Langland and S.T. Hwang. NIST J. Res. **98** (1993)109.
4. H. H. Chen-Mayer, D. F. R. Mildner, V. A. Sharov, J. B. Ullrich, I. Yu. Ponomarev, and R. G. Downing, J. Phys. Soc. Japan **65** Suppl. A (1996) 319-321.
5. H. H. Chen-Mayer, G. P. Lamaze, D. F. R. Mildner, and R. G. Downing, "Neutron Depth Profiling Using a Focused Neutron Beam", SPIE **2867** (1997) 140-143. 5th Intl. Conf. on Application of Nuclear Techniques, Neutrons in Research and Industry, Crete, Greece.
6. H. H. Chen-Mayer and G. P. Lamaze, NIMB, in press.
7. J.F. Ziegler. "The Stopping and Ranges of Ions in Matter" Pergamon Press Inc., New York, 1977.
8. D. Simons, private communication.
9. K.J.Coakley, R.G.Downing and G.P. Lamaze , H.C. Hofsass, J.Biegel, and C. Ronning. N.I.M. **A366** (1995) 137-144.
10. G.P. Lamaze, H. Chen-Mayer, M. Badding and L. Laby. *"In situ* measurement of lithium movement in thin film electrochromic coatings using cold neutron depth profiling". To be published.

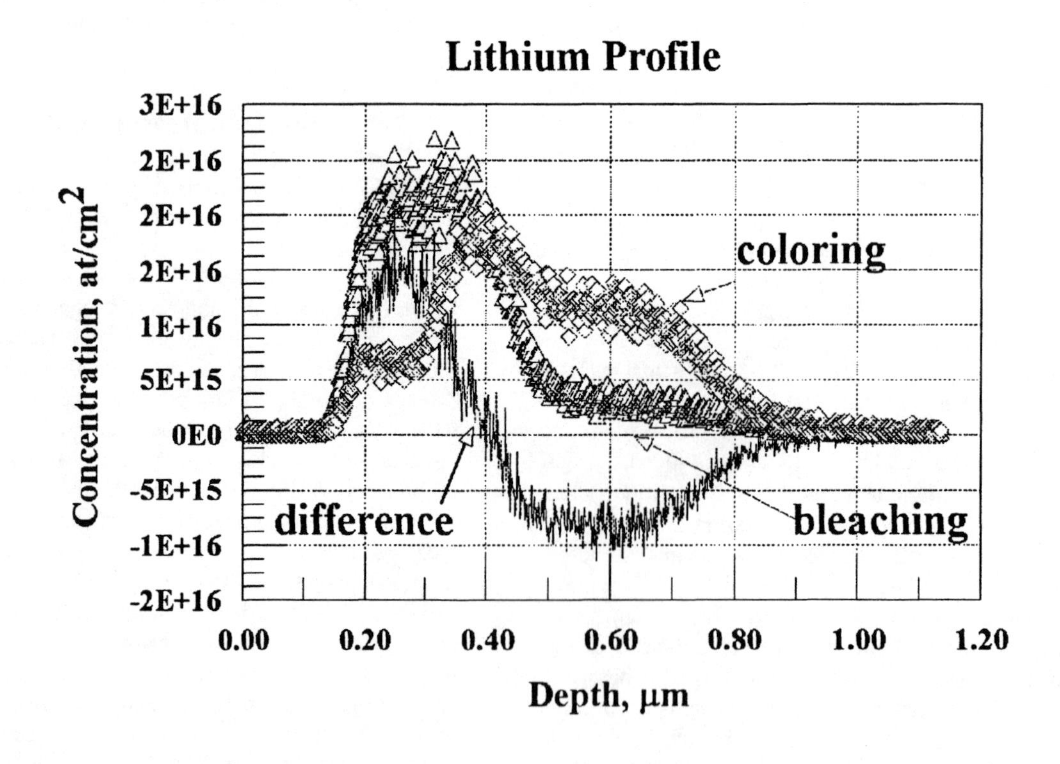

Figure 4. Lithium profiles of multilayer structure under different bias conditions.

The NIST Surface Analysis Data Center

C. J. Powell

Surface and Microanalysis Science Division, National Institute of Standards and Technology, Gaithersburg, MD 20899

J. R. Rumble, Jr., D. M. Blakeslee, and M. E. Dal-Favero

Standard Reference Data Program, National Institute of Standards and Technology, Gaithersburg, MD 20899

A. Jablonski

Institute of Physical Chemistry, Polish Academy of Sciences, ul. Kasprzaka 44/52, 01-224 Warsaw, Poland

S. Tougaard

Physics Department, Odense University, Campusvej 55, DK-5230 Odense M, Denmark

NIST has recently established a Surface Analysis Data Center to give greater visibility to its databases developed for applications in surface analysis. Two databases are currently available for this purpose, and a third will be released soon. These three databases are described briefly here.

Version 2.0 of the popular X-ray Photoelectron Spectroscopy Database (SRD 20) was released in 1997. This database has long been a valuable resource of binding-energy and related data for the surface analysis of a wide range of materials by x-ray photoelectron spectroscopy (XPS). Version 2.0 contains about 16,000 line positions, chemical shifts, and splittings. Pull-down menus are now available to initiate searches for the identification of unknown lines, to retrieve data on core-level binding energies, to retrieve data on chemical shifts, to initiate searches for compounds containing selected elements or for specified material classes, to perform searches based on compound names and other fields, and to display Wagner plots.

Version 1 of the Elastic-Electron-Scattering Cross-Section Database (SRD 64) was released in 1996. This database provides differential and total elastic-electron-scattering cross sections for simulations of signal-electron transport in XPS and Auger-electron spectroscopy (AES). Such simulations are important in the development of improved methods for correcting for elastic-electron-scattering effects in quantitative surface analyses by AES and XPS and for separating chemical information from effects due to the complex morphologies of semiconductor structures. Cross sections are given for all elements and for electron energies between 50 and 9,999 eV.

The Electron Inelastic-Mean-Free-Path Database (SRD 71) will be released in 1998. This database contains calculated and measured inelastic mean free paths (IMFPs) that are needed for quantitative surface analyses by AES and XPS. The IMFPs are available for elements, inorganic compounds, and organic compounds as a function of electron energy. IMFPs can also be obtained conveniently from various predictive formulae. Once a data source is selected, the user can choose to display the information graphically (that is, a plot of IMFP *versus* electron energy) or as values for one or more specified electron energies.

INTRODUCTION

Surface analysis is used extensively in the semiconductor industry for many purposes including the characterization of defects on wafers, the determination of composition-depth profiles of thin-film structures, and the measurement of depth distributions of implanted ions. The techniques in common use are Auger-electron spectroscopy (AES), x-ray photoelectron spectroscopy (XPS), and secondary-ion mass spectrometry (SIMS). We give brief descriptions here of three NIST databases that have been developed for use with AES and XPS. These databases are

CP449, *Characterization and Metrology for ULSI Technology: 1998 International Conference*
edited by D. G. Seiler, A. C. Diebold, W. M. Bullis, T. J. Shaffner, R. McDonald, and E. J. Walters

TABLE 1. Examples of line entries found for silicon using the chemical shift option of the NIST X-ray Photoelectron Spectroscopy Database. The suffix 'c' indicates a calculated chemical shift. The line notation is explained in the text.

Element	Line	Formula	Chemical Shift (eV)	Physical State
Si	δ2p	SiC	+1.1	powder
Si	δ2p	Si_3N_4	+2.5 c	chemical vapor deposition
Si	δ2p	SiN_x	+2.5	chemical vapor deposition
Si	δ2p	SiO	+3.4 c	vapor deposited
Si	δ2p	SiO_2	+4.3 c	sputter deposited
Si	δ2p	SiO_x	+4.1 c	reacted film
Si	$\delta KL_{23}L_{23}(^1D)$	SiC	–2.8	powder
Si	$\delta KL_{23}L_{23}(^1D)$	SiN_x	–5.1	chemical vapor deposition
Si	$\delta KL_{23}L_{23}(^1D)$	SiO_2	–7.7	powder
Si	$\delta\alpha 2p,KL_{23}L_{23}(^1D)$	SiC	–1.7	powder
Si	$\delta\alpha 2p,KL_{23}L_{23}(^1D)$	SiN_x	–2.6	chemical vapor deposition
Si	$\delta\alpha 2p,KL_{23}L_{23}(^1D)$	SiO_2	–3.9	passive oxide

products of the NIST Surface Analysis Data Center which was established recently to give greater visibility and focus for databases to be used in surface analysis.

NIST X-RAY PHOTOELECTRON SPECTROSCOPY DATABASE

Version 1.0 of the NIST X-ray Photoelectron Database (SRD 20) was released in 1989 (1). Version 2.0 of this database was released in 1997 with additional data (a total now of 16,000 line positions) and many additional features that make searches and displays both more efficient and more convenient.

Pulldown menus are available in version 2.0 to initiate searches for the identification of unknown lines in XPS ("Identify"), to retrieve data on core-level binding energies ("Element Search"), to retrieve data on chemical shifts ("Element Search"), to initiate searches for compounds containing selected elements or for specified material classes ("Formula Search"), to perform searches based on compound names and other fields ("Compound Name Search"), and to display Wagner plots ("Graph").

For the Identify option, the user first selects binding energy, Auger kinetic energy, or doublet separation. If binding energy is selected, for example, a screen appears on which the user can enter the unknown (i.e., measured) binding energy and an energy tolerance for up to four lines; a message at the bottom of the screen indicates the energy range of lines in the database. After initiating the search, the database displays information on the element, line, chemical formula, energy, and physical state for those stored line energies having energies within the defined energy range. This information can be sorted to fit the user's needs and can be printed. Additional details can be displayed (and printed) for a selected material: the measurement conditions (how the energy scale was calibrated and the method of charge reference for nonconductors), an evaluation of the data quality, and a literature reference. The same display capabilities exist for the other options.

For the Element Search option, a user selects binding energy data, Auger kinetic energy data, chemical shift data, or elemental data. If, for example, binding energy is selected, a screen appears on which the user can enter a symbol for an element. After the symbol is entered, the available lines for that element are displayed. For instance, if Si were entered, the 1s, 2s, 2p, and $2p_{3/2}$ photoelectron lines would be indicated (where $2p_{3/2}$ indicates that this subshell was resolved and measured). Up to four lines can be selected at a time for a search.

A new feature of version 2.0 is the ability to display chemical shifts of photoelectron lines or Auger lines. The user here has three options: review of previously measured chemical shifts, calculation of chemical shifts from measured data for a compound and elemental reference data (2), or both of the above options. Table 1 shows an example of line entries found following a search of chemical shifts for the Si 2p photoelectron line (indicated as δ2p), the Si main high-energy Auger line (indicated as $\delta KL_{23}L_{23}(^1D)$), and the modified Auger parameter (indicated as $\delta\alpha 2p,KL_{23}L_{23}(^1D)$. For this particular search, there were 151 matches, but Table 1 shows data for only 12 selected compounds. The suffix 'c' for some chemical shifts designates values calculated from elemental reference data; other values indicate measurements reported in a particular paper. The elemental reference data can be displayed.

For the Formula Search option, a user selects elements present or classes. In the former case, the user can find and display data for a particular compound by specifying up to nine elements present in the compound. In addition, data can be retrieved for compounds containing the selected elements plus any others that might be present. If classes is selected, the user can display data for a compound selected from a predefined set of classes, grouped under inorganic, organic, ligands and ligand centers, and other classes. If inorganic is selected, for example, the user must first specify cations or anions, and then specify subclasses. Examples of anion subclasses are hydroxide, nitride, oxide, and silicide among many others.

For the Compound Name Search option, a user seeks matches between a selected combination of individual data fields and the corresponding data in the database. The initial search is based on up to four search strings that are entered by the user, where each search string contains part of a compound name of interest. For example, a search could be made on 'silicide.' The results of any search can be further refined with up to five additional criteria.

For the Graph option, a user can display data for up to 12 compounds on a Wagner plot, another new feature of the database. Such plots display the modified Auger parameter, the kinetic energy of the principal Auger line, and the binding energy of the principal photoelectron line (3). These plots are very useful for identification of chemical state.

NIST ELASTIC-ELECTRON-SCATTERING CROSS-SECTION DATABASE

Version 1.0 of the NIST Elastic-Electron-Scattering Cross-Section Database (SRD 64) was released in 1996. This database provides relativistic and nonrelativistic elastic-electron-scattering cross sections for elements with atomic numbers between 1 and 96 and for electron energies between 50 and 9,999 eV. The differential and total cross sections have been calculated as described elsewhere (4).

Elastic-electron scattering is relevant to AES and XPS for two main reasons. First, experimental measurements and theoretical calculations have shown that elastic scattering can significantly modify the trajectories of the detected electrons of interest (i.e., the signal electrons). As a result, significant corrections (up to about 30%) may be necessary in the AES and XPS intensities used for quantitative analyses (5). The magnitude of the correction for a particular measurement depends on the element, the energy of the AES or XPS line of interest, the matrix, and the instrumental configuration (for XPS). Another related manifestation of elastic-electron scattering is modification of the surface sensitivity from that expected if elastic scattering is ignored (6). The mean electron escape depth in AES will be reduced by up to about 35% while it can be changed by up to ±30% in XPS for common measurement conditions. An important consequence of the latter result is that there is a restriction on the range of emission angles for which XPS data can be obtained and analyzed to obtain composition-versus-depth information (6).

A second reason for the relevance of elastic-electron scattering in AES concerns the role of topography on the Auger-electron signal intensity. Ideally, the signal intensity for a particular element would be proportional to elemental concentration, but there can be significant departures from proportionality (of up to about a factor of two) in the vicinity of the edges of semiconductor microstructures (7); similar deviations can occur near interfaces. If a focused electron beam in a scanning Auger instrument is scanned across aluminum or tungsten lines on a silicon substrate, for example, the signal intensities of the Auger lines may vary in a complicated way with beam position due to elastic-electron scattering in the vicinity of the line edges. In such cases, it can be difficult to separate chemical contrast from topographical contrast without a Monte-Carlo-type simulation for the microstructure of interest (7). Similar considerations also apply to the analysis of defects on wafers by AES (8).

A user of version 1.0 is prompted to select cross sections calculated with either relativistic or nonrelativistic models (although the relativistic cross sections are believed to be more reliable). From the main menu, the user will initially select the element of interest and the electron energy. A selection must then be made of one of three coordinate systems in which differential cross sections can be presented: $d\sigma/d(\sin\theta/2)$ versus $\sin(\theta/2)$, $d\sigma/d\Omega$ versus θ, or $d\sigma/d\theta$ versus θ where θ is the polar scattering angle and $d\Omega$ an element of solid angle. The differential cross section for the selected element, electron energy, and coordinate system is then displayed on the screen. The electron energy can be conveniently increased or decreased (using the keyboard or on-screen buttons) so that variations of the cross sections with energy can be readily visualized.

Figures 1 and 2 show examples of the differential elastic-scattering cross sections for Si at electron energies of 96 and 1621 eV (corresponding to the energies of the most intense Si LVV and KLL Auger transitions, respectively, in the differential Auger spectrum). The large changes in the shapes of the two cross sections can be easily seen.

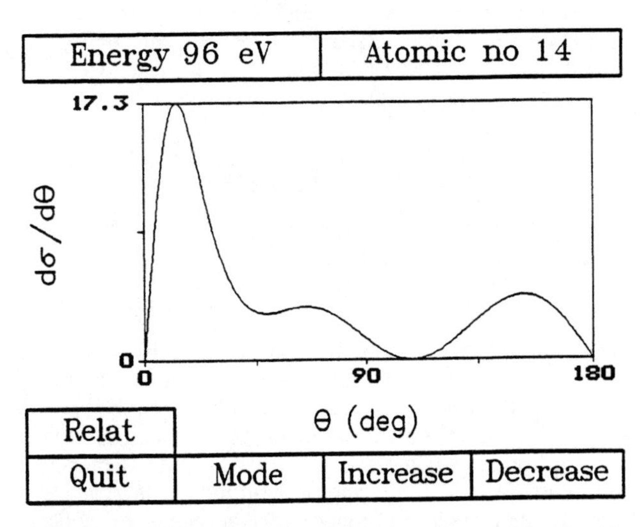

FIGURE 1. Differential elastic-scattering cross section for Si at 96 eV from the relativistic model plotted as a function of polar scattering angle θ. The ordinate units are a_0^2/radian where a_0 is the Bohr radius for hydrogen (0.529 Å).

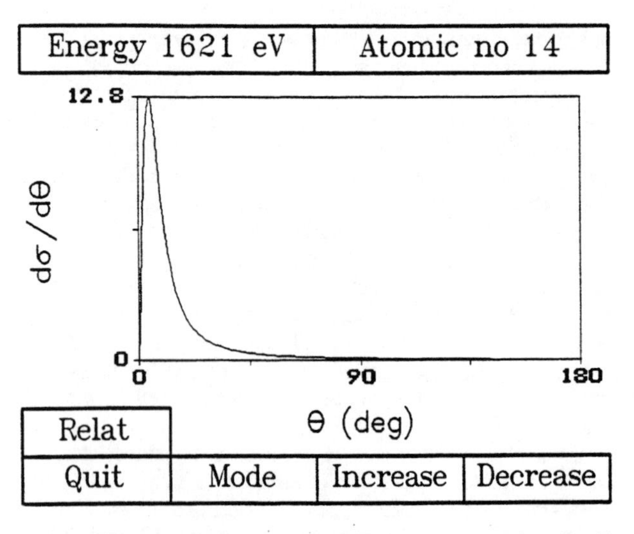

FIGURE 2. Differential elastic-scattering cross section for Si at 1621 eV from the relativistic model plotted as a function of polar scattering angle θ. The ordinate units are a_0^2/radian where a_0 is the Bohr radius for hydrogen (0.529 Å).

After displaying a differential cross section, the user can choose another coordinate system, create a file with the calculated cross section, and create a random number generator. The latter option creates parameters to activate the random-number generator which provides scattering angles for Monte Carlo simulations of electron transport; portable source codes that implement these generators are also available. A visual test of the quality of the random number generators can be made from an on-screen comparison of the histogram of generated scattering angles with the corresponding differential cross section.

It is possible to compare differential cross sections calculated using the relativistic and nonrelativistic models; such differences occur particularly for elements of high atomic number. Calculated cross sections can also be printed or saved for future use.

NIST ELECTRON INELASTIC–MEAN–FREE–PATH DATABASE

Version 1.0 of the NIST Electron Inelastic-Mean-Free-Path Database (SRD 71) will be released in 1998. This database provides values of electron inelastic mean free paths (IMFPs) for use in quantitative surface analyses by AES and XPS. In most commercial instruments, AES and XPS quantitative analyses are made with elemental sensitivity factors. This approach introduces systematic errors (of up to a factor of about two) that arise mainly from the neglect of IMFP variation with chemical state and from lineshape changes with chemical state (9).

A user often needs IMFP values at one or more specific electron energies or wishes to display IMFPs as a function of energy. The database can provide IMFP information from up to three types of sources: calculated IMFPs from experimental optical data for a limited number of materials, IMFPs measured by elastic-peak electron spectroscopy for some elemental solids, and IMFPs from predictive formulae for all materials. The calculated and measured IMFPs were generally reported in journal papers at specified electron energies, and it was convenient to fit these IMFPs with appropriate functions so that IMFPs could be found by interpolation at intermediate energies (10).

A user of version 1.0 can display or obtain IMFP values for elements, inorganic compounds, and organic compounds from the main menu. For a selected material in each class, the user can select a source of IMFPs (calculations, measurements, or predictive formulae) and then choose from a number of options for the electron energy range, IMFP units (nanometers, Angstroms, and milligrams per square meter), and the type of display (linear, logarithmic, or linear/logarithmic coordinates).

The user can display the IMFP results graphically, obtain an IMFP for a single electron energy, obtain IMFPs for multiple energies, and can create an IMFP Table for regularly spaced electron energies. The IMFPs from the latter two options can be stored in files for later processing (e.g., on-screen comparisons of IMFPs from different sources for the same material or comparisons of IMFPs for different materials, printing, or permanent storage).

Figures 3 and 4 show examples of the IMFP displays. Elemental IMFPs for Si, Al, Cu, and W are shown as a function of electron energy in Fig. 3 to illustrate the IMFP variations that can occur from material to material. Each curve in Fig. 3 was generated from the calculated IMFPs of Tanuma et al. (11) for each elemental solid. The calculated IMFPs for Al, Si, and Cu compare favorably with the corresponding experimental measurements (10).

Figure 4 shows IMFPs for silicon dioxide, silicon nitride, and titanium nitride to indicate the IMFP

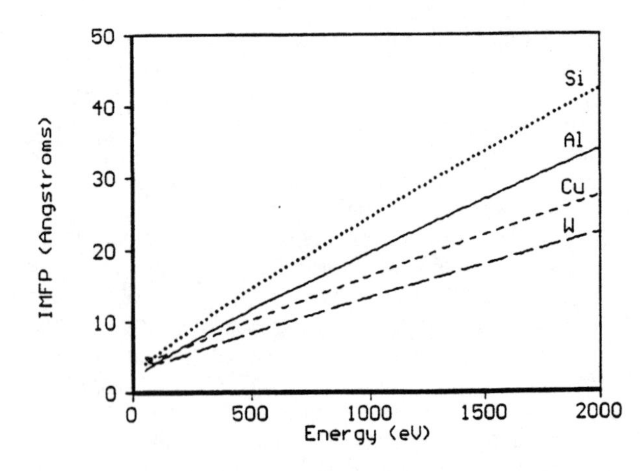

FIGURE 3. Inelastic mean free paths for Si, Al, Cu, and W as a function of electron energy. The elemental symbols have been added after the display was copied from the database screen.

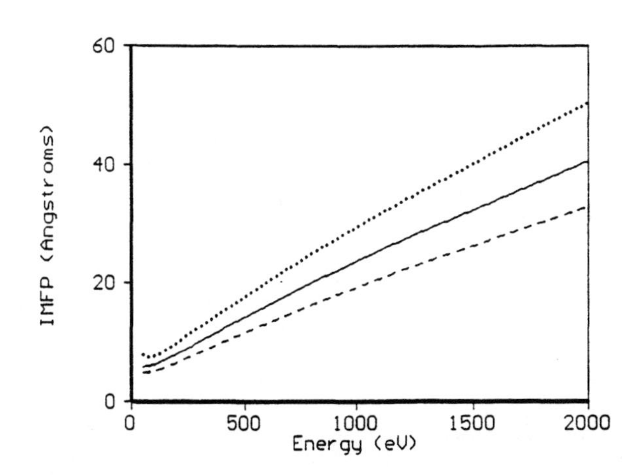

FIGURE 4. Inelastic mean free paths for silicon dioxide (dotted line), silicon nitride (solid line), and titanium nitride (dashed line) as a function of electron energy.

variations in materials of relevance to the semiconductor industry. In addition, a comparison of the IMFPs for silicon dioxide plotted in Fig. 4 with the IMFPs for silicon in Fig. 3 shows that the IMFP ratios vary from 1.46 at 100 eV to 1.29 at 200 eV, 1.21 at 500 eV, 1.20 at 1000 eV, and 1.19 at 2000 eV. The IMFPs for silicon dioxide in Fig. 4 were obtained by calculation from experimental optical data (12) while the IMFPs for the other two compounds were obtained from the IMFP predictive formula TPP-2M of Tanuma *et al.* (13).

SUMMARY

Brief descriptions have been given of version 2.0 of the NIST X-ray Photoelectron Spectroscopy Database (SRD 20), version 1.0 of the NIST Elastic-Electron-Scattering Cross-Section Database (SRD 64), and version 1.0 of the NIST Electron Inelastic-Mean-Free-Path Database (SRD 71) which are products of the NIST Surface Analysis Data Center. It is hoped to add other products later. Further information on these and other NIST databases can be obtained from the NIST Standard Reference Data Program's web site: www.nist.gov/srd.

REFERENCES

1. Powell, C. J., *Surf. Interface Anal.* **17**, 308-314 (1991).
2. Powell, C. J., *Appl. Surf. Science* **89**, 141-149 (1995).
3. Wagner, C. D., Gale, L. H., and Raymond, R. H., *Anal. Chem.* **51**, 466-482 (1979).
4. Jablonski, A., and Tougaard, S., *Surf. Interface Anal.* **22**, 129-133 (1993).
5. For example: Jablonski, A. and Powell, C. J., *J. Vac. Sci. Tech. A* **15**, 2095-2106 (1997) and references therein.
6. Jablonski, A., Tilinin, I. S., and Powell, C. J., *Phys. Rev.* B **54**, 10927-10937 ((1996); Powell, C. J., Jablonski, A., Tilinin, I. S., Tanuma, S., and Penn, D. R. (to be published).
7. El Gomati, M., Prutton, M., Lamb, B., and Tuppen, C. G., *Surf. Interface Anal.* **11**, 251-265 (1988).
8. Testoni, A. L., "New Models and Methods for Analysis of Particles and Defects," presented at the National Symposium of the American Vacuum Society, Philadelphia, Pennsylvania, October 14-18, 1996.
9. Powell, C. J., in *Quantitative Surface Analysis of Materials*, edited by N. S. McIntyre, Philadelphia: American Society for Testing and Materials Special Technical Publication 643, 1978, pp. 5-30.
10. Jablonski, A. and Powell, C. J. (to be published).
11. Tanuma, S, Powell, C. J., and Penn, D. R., *Surf. Interface Anal.* **17**, 911-926 (1991).
12. Tanuma, S, Powell, C. J., and Penn, D. R., *Surf. Interface Anal.* **17**, 927-939 (1991).
13. Tanuma, S, Powell, C. J., and Penn, D. R., *Surf. Interface Anal.* **21**, 165-176 (1994).

ULSI Technology and Materials: Quantitative Answers by Combined Mass Spectrometry Surface Techniques

M.Bersani[1], M.Fedrizzi[1], M.Sbetti[2], and M.Anderle[1]

1 ITC-IRST, 38050 Povo - Trento, Italy
2 SGS-Thomson, 20041 Agrate Brianza - Milano, Italy

The progressive microelectronics ULSI device shrinking towards improving the performances has driven the development of new materials and process technologies. A good example is given by oxynitride, an innovative material which is thought for the next generation of 0.25 µm MOS circuits. Oxynitrides have replaced thermal silicon oxides as gate insulator due to the properties of good masking against impurity diffusion, together with the excellent dielectric strength and the better resistance to dielectric breakdown. The strong request from microelectronics industries for a complete and accurate characterization of this new material and the technological processes concerned, has considerably stimulated the research, particularly in the field of analytical methodology. Secondary Ion Mass Spectrometry, linked since the beginning with microelectronics developement, shows again to be the most reliable and suitable microanalytical technique to give answers to this topics. In this work we present some examples of methodologies applied to an accurate quantitative characterization of this new material, together with its impact on the production processes. We show how the complementary employing of several mass spectrometry techniques, such as magnetic sector SIMS, SNMS and ToF-SIMS, can give a more complete overview both to process issues and to methodological developements of the techniques themselves.

INTRODUCTION

The developement of materials and processing in microelectronics needs an increasing parallel improvement of the analytical methodologies, required to give up-to-date answers to the state of the art in this field.

In this work we present the application of analytical methodologies based on mass spectrometry on materials of great interest in ULSI microelectronics, such as oxynitrides. In particular we show how we solved the analytical problems concerned with this topic: the change in RSF between two different matrices, and the analysis on thin films with thickness less than 100 Å.

MATERIALS

The new 0.25 µm technology needs a tunnel oxide thickness below 80 Å [1]. The quality of the oxide directly affects the performances of the memories in which it is utilized.

When the thickness is reduced under 100 Å, the necessary reliability and working life properties can not be guaranteed by the usual SiO_2.

A satisfactory solution, proposed since the second half of the 80's, consists in incorporating nitrogen in the oxide [1,2]. These oxides with nitrogen (oxynitrided), show several advantages in respect of the traditional oxides; in fact, on the same thickness, oxynitrided have a greater dielectric constant, an improved impurity diffusion resistance, a reduced degradation in presence of high electric fields [2]. These features are the reason why oxynitrided are an adfirmed presence in the production technology of the last generation devices In order to characterize this kind of materials, the analytical request that mass spectrometry has to answer is to determine, with the suitable accuracy and sensitivity, the exact content in nitrogen and the shape of its distribution in the SiO_2/Si systems. The analyses in the present work were performed on a set of representative oxynitride samples, obtained with N_2O and NO precursors [3]. Table 1 lists the main properties of the samples.

TABLE 1. Process data of the analyzed samples.

Samples	Nitridation time*	Precursors	Nitridation temperature *	Thickness [Å]
1	t_1	10% NO	T_1	70.9
2	t_2	10% NO	T_1	69.2
3	t_3	10% NO	T_1	70.3
4	t_3	50% NO	T_1	69.6
5	t_4	100% N_2O	T_2	70.5
6	t_2	50% NO	T_1	91.2
7	t_1	50% N_2O	T_3	240

* Confidential parameters.

CP449, *Characterization and Metrology for ULSI Technology: 1998 International Conference*
edited by D. G. Seiler, A. C. Diebold, W. M. Bullis, T. J. Shaffner, R. McDonald, and E. J. Walters

The thicknesses of the samples have been obtained by ellipsometry after the nitridation process.

EXPERIMENTAL

In order to develope a complete analytical methodology we have made dinamic SIMS analyses with a CAMECA IMS 4f, ToF-SIMS analyses with an ION-ToF IV, and SNMS analyses using a LEYBOLD INA-3.

Several measurement conditions have been investigated for the SIMS analyses. In particular we have varied the impact energy of the primary beam, in order to determine the best analytical conditions, that is a compromise between a good depth resolution, a reasonable beam uniformity, and a satisfactory dynamic range. In all the measurements we used a Cs primary beam, monitoring MCs^+ secondary ions (where M is the species of interest).

Table 2 shows the different impact energies used for SIMS analyses, together with the incidence angles in respect of the normal to the sample surface.

TABLE 2. Impact energies and angles used in SIMS analyses.

Impact energy [keV]	Incidence angle [deg]
2	70
3	52
4	46
5.5	42

In all the analytical conditions for SIMS measurements we have kept the sputtering rates below 0.5 Å/sec. In the configuration used, the sputtering rate doesn't vary significantly passing through SiO_2 and Si.

In ToF-SIMS measurements we used a 25 keV Ga beam. For sputtering we used a Cs beam at 1 keV and 5 keV impact energies.

As a sputtering agent, the INA-3 instrument uses an RF low-pressure Ar plasma. In this case the sample is polarized such in a way that the impact energy is 700 eV. The neutral sputtered particles are post-ionized in the plasma by electronic impact, with an 1% yield efficiency. This technique has no lateral resolution.

METHODOLOGY

The matrix effects represent the real problem in the quantification of SIMS analyses on multilayer samples; they can be reduced in part by monitoring MCs^+ ions [4,5]. However, we used suitable standards, obtained by low-energy N_2 implants in SiO_2 and Si. The ratio between the Relative Sensitivity Factors (RSF's) in the two matrices is ~ 3.5 (the lowest being that in SiO_2). In order to quantify the signal in the interface region between SiO_2 and Si, we used a so called layer-matching method. The signal of oxygen was taken as a marker for the change in matrix. The RSF of nitrogen in the interface zone was calculated step by step as a function of the variation of the oxygen profile, applying a relation elsewhere reported [6]. The RSF so obtained were applied to the experimental signal of nitrogen, following the formula

$$[N](x) = \frac{I_N(x)}{I_{Si}(x)} RSF(x) \qquad (1)$$

where [N] is the concentration of nitrogen
I_N is the signal of nitrogen
I_{Si} is the signal of silicon
and all these quantities are considered as a function of depth.

Figure 1 shows the procedure step: the signal of nitrogen as a function of depth (1.a) is used in (1) with the calculated RSF (1.b) to give the quantitative profile(1.c).

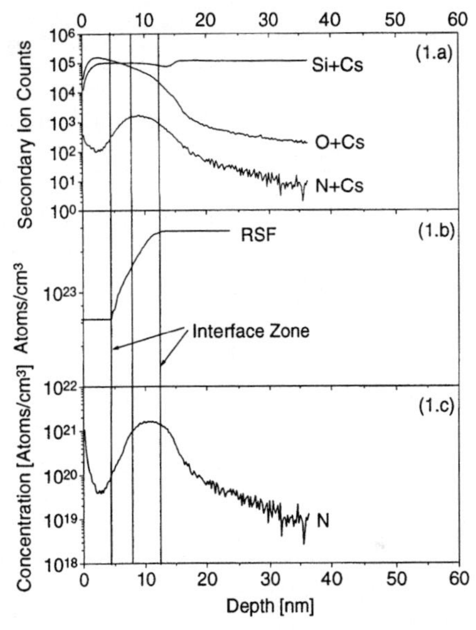

FIGURE 1. Application of the layer matching method to the quantification of SIMS analyses on sample 6 in Tab. 1.

RESULTS AND DISCUSSION

Figure 2 shows the qualitative profiles obtained on the same sample at different impact energies of the primary beam. It can be easily observed that reducing the energy we have a considerable improvement in depth resolution, keeping a good dynamic range. In particular we can notice the increasing of the peak level together with the decreasing of the Full Width at Half Maximum (FWHM).

The variation of the oxygen signal in the interface zone is sharper in the low energy analyses, and a plateau of the signal is more easily identified in the SiO_2 film. In this kind of analysis, the significant parameters we can obtain to characterize the nitridation process are: peak concentration (C_p), peak position in respect of the interface SiO_2/Si, Full Width at Half Maximum (FWHM) of nitrogen quantitative profile, and nitrogen dose.

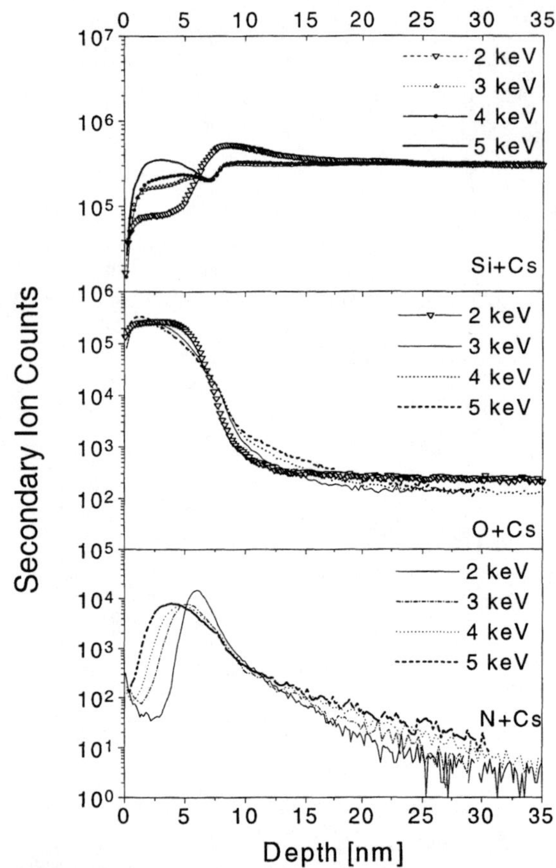

FIGURE 2. SIMS depth profiling at different impact energies: Si+Cs, O+Cs and N+Cs signals are presented.

Figure 3 shows C_p obtained with the 3 keV measurements as a function of the different treatments involving NO nitridation. It can be observed that there is a correlation between the nitridation process and the resulting C_p.

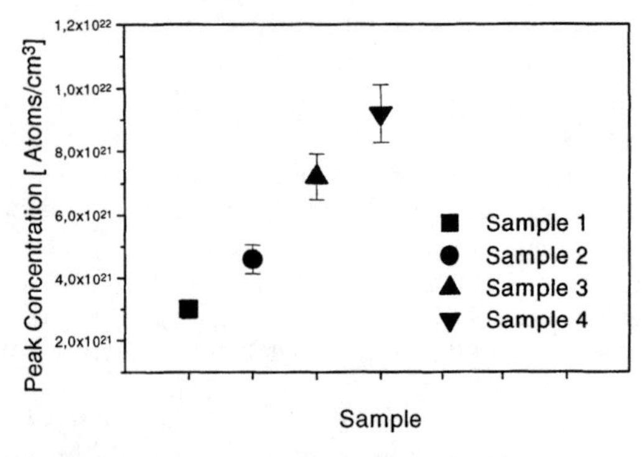

FIGURE 3. Peak concentration of nitrogen in different oxynitride.

Another example of the information that can be obtained from SIMS analyses is reported in Tab. 3, in which we compare the FWHM of two different samples with the same C_p. The results suggest that in the case of N_2O precursor the nitridation reaches a saturation, with a consequent spreading of the profile.

TABLE 3. FWHM comparison in different samples.

Sample Number	Precursor	Cp [atoms/cm^3]	FWHM [nm]
2	NO	$4.6x10^{21}$	34
5	N_2O	$4.2x10^{21}$	43

This kind of results cannot be obtained without a suitable analytical methodology, for example, without an adapting of the RSF's in the interface zone or in the case of a low depth resolution.

To verify the applicability of other analytical techniques on this kind of materials, ToF-SIMS analyses were performed on suitable samples, previously characterized by SIMS. Figure 4 shows the ToF-SIMS depth profile of sample 6 of Tab.1. In this case the analysis of the different species is parallel, and not sequential like in SIMS. In this way it is possible to follow an higher amount of interesting species.

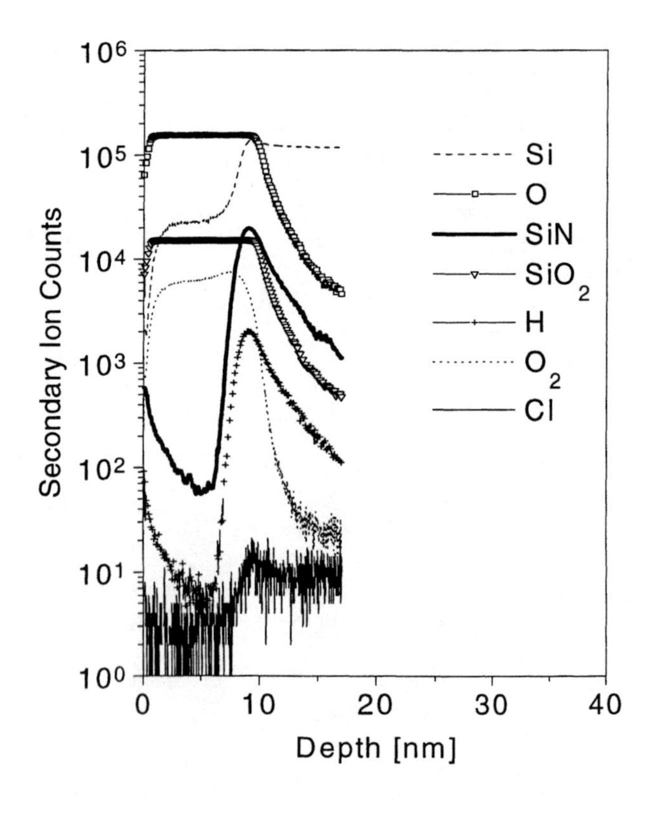

FIGURE 4. Sample 6 ToF-SIMS analyses with 1 keV impact energy.

The measurement in Fig. 4 has been performed with an 1 keV impact energy Cs primary beam. The depth resolution is improved in respect of SIMS. A reliable quantification methodology of the signals, in particular of the molecules containing nitrogen, is not immediately defined. Another difficulty is given by the correct evaluation of the role of the interface, which seems to be more significant in ToF-SIMS than in SIMS. On the other hand, a good quality of ToF-SIMS analyses is the easiness in drawing the profile of some contaminating elements, such as H and Cl, for which we can obtain a good depth resolution together with a good dynamic range.

Analyses taken at different impact energies show a dramatic dependence of the signals on the sputtering parameters. This fact must be more accurately investigated, in order to understand limits and advantages of a future ToF-SIMS quantitative approach. Figure 5 presents a profile carried out at a Cs impact energy of 5 keV.

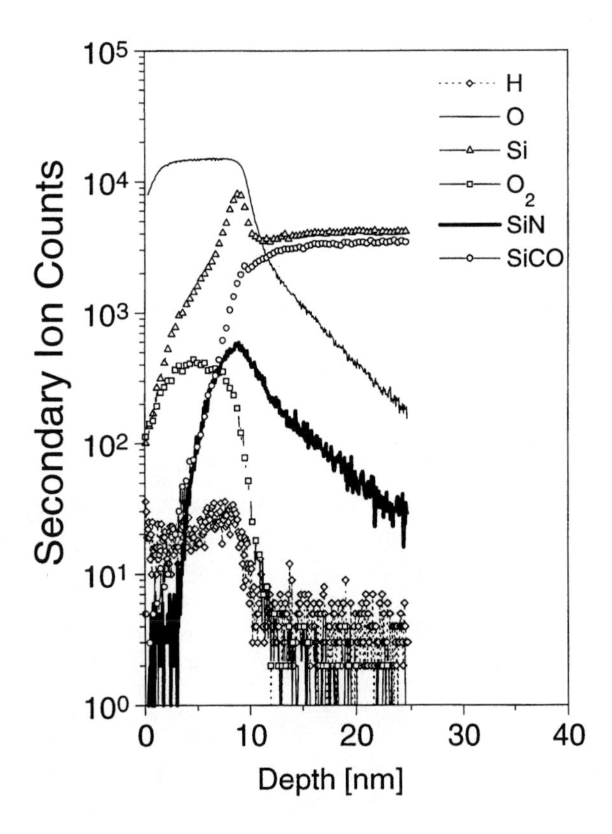

FIGURE 5. Sample 6 ToF-SIMS analyses with 5 keV impact energy.

Figure 6 shows the comparison between SNMS and SIMS results on the same sample with a high nitrogen concentration at the SiO_2/Si interface.

From a quantification point of view SNMS analysis is more straightforward than SIMS, due to the fact that sputtering and ionization are separated spatially and temporarly: this strongly reduces the matrix effects.

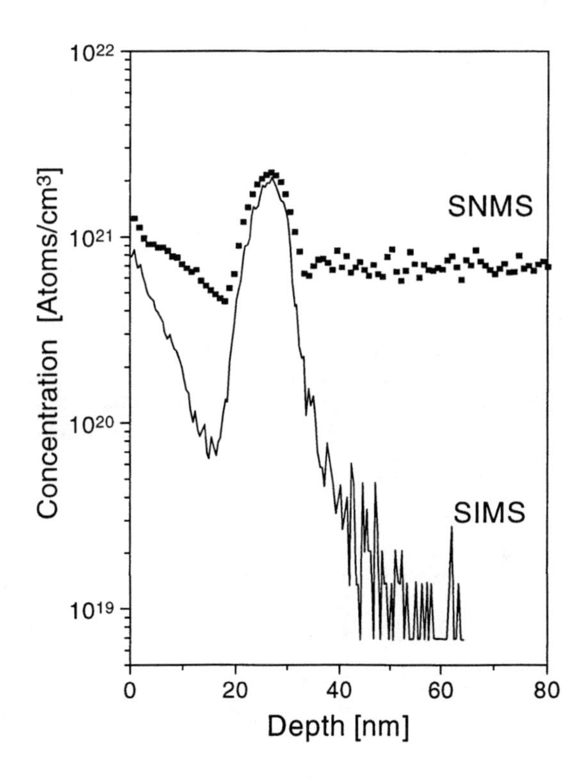

FIGURE 6. SIMS and SNMS comparison on sample 7 in Tab. 1.

Unfortunately the SNMS dynamic range is worse than SIMS one. The reason is that the quadrupole analyzer of SNMS instrument has not the necessary mass resolution to separate the ^{14}N and $^{28}Si_2^+$. In any case, the comparison between SNMS and SIMS quantitative results gives us a confirmation of our SIMS data quantification methodology.

CONCLUSIONS

In this work we show that a complementary use of mass spectrometry techniques can be proposed as an useful tool to define an analytical quantitative methodology. Our results obtained in the oxynitride film characterization by SIMS are in agreement with the literature [3,7]. The comparison with SNMS results confirms the reliability of our quantitative methodology, being anyway SNMS non suitable for systematic characterization of these films because of its reduced dynamic range. Finally, ToF-SIMS technique shows to be very promising in carrying out quantitative results from very thin films analyses. A complete investigation of matrix and interface effects in this case has to be done.

ACKNOWLEDGEMENTS

The authors would like to thank the collegues of the Physics Division at SGS-Thomson Microelectronics for the interesting discussions on these topics; the technicians of Physics and Chemistry of Surfaces and Interfaces Division of ITC-IRST for the qualified support in the analyses; Dr. R. Canteri for ToF-SIMS analyses.

REFERENCES

[1] Moslehi, M.M., Saraswat, K.C., *IEEE Transactions on Electronic Devices*, **32**, 106-123, (1983).
[2] Hori, T., Iwasaki, H., Tsuji K., *IEEE Transactions on Electronic Devices*, **36**, 340-349, (1989).
[3] Hegde, R. I., et al, *Applied Physics Letters*, **66**, 2882-2884, (1995).
[4] Gao, Y., *Journal of Applied Physics*, **64**, 3760, (1988).
[5] Moro, L., Canteri, R., Anderle, M., *Surface and Interface Analysis*, **18**, 765-772, (1992).
[6] Bersani, M., Fedrizzi, M., Ferroni, M., Savoia, G., Anderle, M., *Secondary Ion Mass Spectrometry SIMS XI*, New York, John Wiley & Sons, 1998, 1055-1058.
[7] Oakey, P. R., et al., *Secondary Ion Mass Spectrometry SIMS XI*, New York, John Wiley & Sons, 1998, 253-256.

Wafer and Bulk High-Purity Silicon Trace Element Analysis at the Texas A&M University Nuclear Science Center

Daniel James Van Dalsem

Nuclear Science Center, Building 1095, Nuclear Science Center Road, Texas A&M University, College Station, Texas 77843-3575

A trace element analysis program for wafer and bulk high-purity silicon (Si) samples has been operating at the Texas A&M University Nuclear Science Center (TAMU NSC) since 1996. Samples are irradiated in the NSC's 1-MW TRIGA research reactor at a thermal neutron fluence rate of 10^{13} n/cm^2/s for 14 hours. After an appropriate decay length, bulk samples are chemically etched to remove surface contamination while wafer surfaces are first rinsed with acid to determine surface contamination and then etched to obtain epitaxial layer contamination information. All samples, along with the appropriate etching solutions are analyzed using gamma-ray spectroscopy to quantitatively determine the various radioisotopes created during irradiation. Elements typically determined are antimony (Sb), arsenic (As), bromine (Br), cadmium (Cd), chromium (Cr), cobalt (Co), copper (Cu), gallium (Ga), gold (Au), iron (Fe), molybdenum (Mo), potassium (K), silver (Ag), sodium (Na) tungsten (W) and zinc (Zn). The potential exists to also determine cesium (Cs), iridium (Ir), lanthanum (La), mercury (Hg), rubidium (Rb), scandium (Sc), and zirconium (Zr). Detection limits range from 10^{14} down to 10^7 atoms/cm^2 in surface analysis and 10^{13} down to 10^8 atoms/cm^3 in bulk Si.

INTRODUCTION

The continuing trend of increasing single crystal Si wafer size necessitates the need for increasingly sensitive analytical techniques when characterizing high-purity Si. Among these analytical techniques is neutron activation analysis (NAA) for measuring the trace element impurities in the high-purity Si used for semiconductor devices. Since 1996, a trace element analysis program for wafer and bulk Si samples has been operating at the TAMU NSC.

Work is currently underway to develop the analysis capability for whole Si wafers up to 200 mm with an ultimate 300 mm goal. A whole wafer analysis facility has been designed, built and will be fully functional in 1998. Once characterized, it will provide trace element impurity information for the wafer's surface, epitaxial layer and bulk matrix. Furthermore, dedicated laboratory space has been completely renovated for the necessary post-irradiation chemical manipulations. Anticipated whole wafer analysis times are less than one month from wafer receipt, possibly two weeks depending on the required analysis complexity.

EXPERIMENTS AND RESULTS

Neutron Activation Analysis

The ideal trace element measurement technique would require no sample preparation, have no blank value to subtract from the measured value, allow all elements to be determined simultaneously and at ultratrace (<1 mg/kg) concentrations regardless of other elements in the sample. Neutron activation analysis of Si fulfills most of these requirements. In NAA, samples do not have to be dissolved prior to irradiation eliminating trace element contamination introduction from dissolution reagents such as acids. The sample only needs to be handled with clean instruments and enclosed in a clean encapsulation vessel prior to irradiation. Contamination introduced after irradiation is meaningless because it is not radioactive. Since there is no measurable radioactivity in a sample before irradiation, there is no blank value to subtract from the measured value. Furthermore, since NAA is a nuclear technique, an element's chemical form is irrelevant to its determination and thus another element's presence in the sample is also irrelevant. Except for nuclear interferences, NAA is virtually matrix independent.

The radioactivity induced in an element subjected to neutron activation using a nuclear reactor is:

$$A_{t(d)} = [N(\phi_{th}\sigma_0 + \phi_{epi}I_0)S]e^{-\lambda t(d)} \quad (1)$$

where

$A_{t(d)}$ = disintegrations per second at time t(d)
$t(d)$ = length of decay from end of irradiation
ϕ_{th} = thermal neutron fluence rate, n/cm^2/s
σ_0 = thermal neutron cross section, cm^2
ϕ_{epi} = epithermal neutron fluence rate, n/cm^2/s
I_0 = resonance integral cross section, cm^2
S = saturation factor $(1-e^{-\lambda t(i)})$
λ = decay factor, $\ln(2)/t_{1/2}$
$t(i)$ = length of irradiation
$t_{1/2}$ = radionuclide halflife

Neutron activation analysis is highly selective for various elements because of its many adjustable experimental parameters such as irradiation and decay

CP449, *Characterization and Metrology for ULSI Technology: 1998 International Conference*
edited by D. G. Seiler, A. C. Diebold, W. M. Bullis, T. J. Shaffner, R. McDonald, and E. J. Walters

length (1). Short irradiation lengths will enhance short-lived radionuclide measurements over longer-lived radionuclide measurements while longer irradiation lengths will do the opposite. Being able to adjust these parameters makes Si NAA a powerful analytical tool.

Bulk Silicon Analysis

Of the two types of high-purity Si trace element analysis, bulk Si analysis is the less complex and tedious. Upon sample receipt, loose surface contamination and Si dust are rinsed away. The samples are wrapped in aluminum (Al) foil and placed in an Al can, 3 cm deep and 6 cm in diameter. The can and its lid are mechanically sealed with epoxy between the can and lid to enhance the seal. The cans are loaded into a specifically designed stand that holds the cans in place during irradiation. This stand can accommodate as many as three cans simultaneously, thus allowing multiple sample analyses.

Samples are irradiated for 14 hours at a thermal neutron fluence rate of 10^{13} n/cm^2/s. After an appropriate decay length to allow the ^{31}Si ($T_{1/2}$=2.62 hours) generated during irradiation to decay away, the cans are removed from the stand and opened in a remote manipulation handling cell. The samples are transported to the chemical etching laboratory where they are unwrapped and chemically cleaned prior to gamma-ray spectroscopy.

Once unwrapped, the samples are immersed in an experimentally optimized mixture of concentrated hydrofluoric acid (HF), concentrated nitric acid (HNO_3) and water (H_2O). The HNO_3 oxidizes the surface Si and the HF dissolves the newly oxidized Si surface. Etch immersion times vary from sample to sample depending upon the sample's surface to volume ratio with fresh etch solution used for each sample. To prevent already removed contaminates from plating back on the Si surface, a Au and Cu carrier solution is added to the etching solution prior to etching. Previous work had shown contaminates, in particular Au and Cu, will plate back on the Si surface during chemical etching.

Upon removal from the etching solution, the samples are rinsed and dried prior to mounting in a counting holder using the same epoxy used to seal the Al cans. The samples are mounted to prevent movement in the holder between gamma-ray spectroscopy measurements. The mounting assures the samples are placed in the same identical position for each gamma-ray spectroscopy measurement.

Gamma-ray spectroscopy is performed using HPGe detectors capable of detecting gamma rays from tens of keV to several MeV in energy. Typically, the energy range measured is 140-2754 keV, well within the detectors' capabilities. The initial spectrum is acquired when the 1266 keV ^{31}Si gamma ray is still large enough to detect easily but not so large it dominates the spectrum. If the sample is counted too soon, it is too radioactive for the detector to accurately measure the various radionuclides' disintegration rates. If it is counted too late, there is insufficient activity to determine the sample's apparent volume using Si as an internal standard.

TABLE 1. Typical Trace Elemental Concentrations Measured (Atoms/cm^3) in Bulk Silicon

Element	Concentration Range	Form
Antimony	8E9 - 2E12	a,b,c
Arsenic	7E10 - 1E12	a,b,c,d
Bromine	3E10 – 7E11	a,b,c,d
Cadmium	9E11 – 2E12	a,b,d
Chromium	5E11 – 2E13	a,b,c,d
Cobalt	1E11 – 2E13	b,c,d
Copper	3E11 – 4E13	a,b,c,d
Gallium	5E10 – 1E11	c,d
Gold	4E8 – 3E11	a,b,c,d
Iron	2E14	b
Molybdenum	3E11 – 2E12	b,c,d
Potassium	2E13 – 2E14	a,c,d
Silver	1E11 – 4E12	a
Sodium	2E13	b
Tungsten	1E10	c,d
Zinc	3E12 – 4E13	a,b,c,d

[a]single crystal puck
[b]polycrystalline puck
[c]amorphous vapor-deposited polycrystalline piece
[d]polycrystalline beads

In the following days, each sample is counted once per day. A one-week decay period follows these counts and a final gamma-ray spectrum is obtained to measure any long-lived radionuclides whose activities were too low to determine earlier in the presence of the shorter-lived radionuclides. Table 1 lists the range of elemental concentrations seen in bulk Si NAA.

Silicon Wafer Analysis

Silicon wafer analysis differs from bulk Si analysis because not only are the Si matrix's trace element concentrations determined but the surface contamination and epitaxial layer trace element impurities are also determined. The surface contamination analysis is as much a measure of the contamination in getting the wafer from the fabrication line to the analysis facility as it is a measure of the contamination resulting from the fabrication line itself. Therefore, the measured surface contamination is a combination of both contamination from the fabrication line and from handling after being removed from the fabrication line.

The first difference between wafer and bulk Si analysis is the sample preparation. Because a bulk Si sample's surface is chemically cleaned after irradiation, surface contamination is not a concern in bulk analysis. However, a wafer's surface must remain as uncontaminated as possible until after irradiation. Thus, a different encapsulation method is employed for wafers, one that prevents a surface contamination increase at the time the wafer is prepared for irradiation.

TABLE 2. Quartz Dish Cleaning (2)

Procedure	Solution or Material	Duration
Soak	H_2SO_4:H_2O_2	10 minutes
Rinse	18 MΩ water	10 minutes
Soak	10% HF	1 minute
Rinse	18 MΩ water	10 minutes
Soak	HCl:H_2O_2:H_2O	2 minutes
Rinse	18 MΩ water	10 minutes
Dry	Class 100 Clean Room	until dry

Partial Wafer Analysis

Partial wafer analysis calls for encapsulating the wafer pieces in a specifically cleaned high-purity quartz dish with an identically cleaned quartz lid, Table 2. Before loading the wafer pieces into the quartz cup, their thickness must be measured with a micrometer. Once irradiated, they will be weighed and their surface area determined for surface contamination calculations. Since only a wafer's polished side is analyzed for surface contamination, the wafer pieces are packed in the cleaned quartz dish with the polished face of one wafer piece facing the polished face of another piece. Thus, partial wafer analysis is done on an even number of wafer pieces. The pieces must be kept stationary in the dish to prevent chipping the edges of the pieces. The room above and below the wafer pairs is filled with clean scrap wafer pieces. This packing scheme results in a "wafer sandwich" physically isolated from contamination sources. Since no Si is removed from the wafer's surface during surface contamination analysis and not enough Si is removed during epitaxial layer analysis to use as an internal standard as in bulk Si analysis, a neutron fluence rate monitor must be included in the quartz dish for irradiation (3). When analyzing wafer pieces, unpolished Si wafers are used as fluence rate monitors.

Like the bulk Si samples, the quartz dish is wrapped in Al foil and placed in an Al can for irradiation. The can is sealed, irradiated and opened after irradiation same as a can holding bulk Si samples except only one quartz dish fits into a can. The decay after irradiation is shorter for partial wafer analysis than for bulk Si samples because there is less total ^{31}Si radioactivity.

Upon removal from the Al can, the quartz dish is taken to the chemical etching laboratory. The Si wafer pieces are taken out of the dish for surface contamination analysis followed by epitaxial layer analysis. It is at this point the unpolished Si wafers included as neutron fluence rate monitors are also removed from the dish. They are weighed and placed in planar etch solution, a solution discussed below. The backs of the wafer pieces are covered with acid-resistant tape. This means only the polished side of the wafer piece is exposed to the chemical etching reagents.

The HF rinse is the first step in analyzing a wafer piece. Since HF does not dissolve Si itself, this first chemical etch step is a rinsing of the soluble components from the wafer surface. The native oxide layer is dissolved along with any materials soluble in HF. The radioactivity measured in this step is a measure of the surface contamination.

TABLE 3. Typical[a] Trace Elemental Concentrations Measured (Atoms/cm^2) in Wafer Silicon

Surface Contamination		Epitaxial Silicon Layer	
Element	**Concentration**	**Element**	**Concentration**
Arsenic	2E8-2E10	Arsenic	6E8-1E9
Antimony	3E10-4E11	Antimony	2E8-1E9
Bromine	6E8-9E9	Bromine	4E8-2E9
Copper	9E10-2E11	Copper	1E11-3E11
Gold	1E9-1E10	Gold	7E9-4E10
Manganese	2E9-9E10	Manganese	3E10-9E10
Sodium	3E11-6E12	Sodium	3E11-3E12
Tungsten	2E8-6E9	Tungsten	2E8-7E8

[a]Surface contamination and epitaxial layer concentrations from a single analysis of a customer's wafer pieces.

The next step is a planar etch using an experimentally optimized mixture of H_2O, concentrated acetic acid (CH_3COOH), concentrated HNO_3 and concentrated HF, plus the Au/Cu carrier solution. In this step, the pieces are soaked until the polished surface disappears. Once the pieces are soaked in the planar etch, they are rinsed and dried. The acid-resistant tape is removed and the pieces stored until the next day when they are subjected to the same bulk Si etching procedure described previously.

Once the acid-resistant tape is removed from the Si wafer pieces, the unpolished Si wafers are removed from the planar etch solution. They are rinsed, dried and weighed. The mass difference is used in determining the neutron fluence rate.

Once the surface chemical procedures are completed, the HF rinse, planar etch and fluence rate monitoring solutions are brought to a standard volume and counted on the same HPGe detectors used for bulk Si analysis. Since the detectors are efficiency calibrated with a National Institute of Standards and Technology (NIST) traceable standard, absolute disintegration rates are calculated for the various radionuclides in solution based on their measured count rates. These disintegration rates are used to calculate the mass of each trace element present and thus their concentration. Table 3 lists typical trace element concentrations for surface contamination and the epitaxial layer in wafer silicon.

DISCUSSIONS AND CONCLUSIONS

The bulk Si analysis procedure has been applied to both single crystal and polycrystalline Si samples. Shapes have included 50 mm diameter by 12.5 mm thick right cylinders ("hockey pucks"), assorted amorphous pieces produced by shattering vapor-deposited Si cylinders many times too large to fit into an irradiation can and beads ranging from a few mm to several mm in diameter. The analysis procedure has worked equally well on all bulk Si sample shapes with detection limits varying due to total sample size.

The partial Si wafer analysis procedure has been applied to a variety of sample wafers also with success. In particular, the HF rinse to measure surface contamination has proven successful compared to other surface contamination measurement techniques. It is quantitative because the wafer's entire polished surface is immersed in HF and the HF is quantitatively collected for gamma-ray spectroscopy. There are no correction factors for collection yield.

The Future: Whole Wafer Analysis

Because partial wafer analysis involves cleaving a wafer into small enough pieces to fit into a quartz dish for irradiation, it is more desirable to analyze a whole wafer at a time, uncleaved. There should be minimum handling between wafer removal from the fabrication line and being placed in an enclosure preventing contamination prior to irradiation. Ideally, the transfer process would add no surface contamination. It must be performed in as clean an environment as possible and the wafer enclosure as free of surface contamination as possible also.

The TAMU NSC is currently completing fabrication and testing of a whole wafer irradiation facility that will be capable of analyzing wafers up to 200 mm in diameter. The packaged wafers are enclosed in an Al holder held vertically in the irradiation facility. Depending upon the thickness of the package, multiple wafers can be irradiated simultaneously. To homogenize the fluence received by the wafer(s), the holder will be rotated on a central axle. Plans call for the TAMU NSC to eventually install an irradiation facility capable of handling 300 mm wafers. Preliminary measurements indicate an anticipated thermal neutron fluence rate of $6x10^{12}$ n/cm^2/s. This estimated fluence rate is based upon previous measurements made in similar irradiation core positions. Anticipated detection limits are equal to current partial wafer analysis limits. With future facility and technique refinements, elemental detection limits should drop by an order of magnitude.

A whole wafer irradiation facility fabrication involves more than making a device to hold an entire wafer. A facility must also be fabricated to handle the wafer irradiator after irradiation while the wafers are unloaded. Because anticipated radiation levels are expected to be high, >10 R/hr at one foot, long-handled tools are being fabricated to reduce the gamma-ray dose received by reactor personnel.

To assure a constant surface contamination measurement from wafer to wafer, an etching device is also being fabricated. It consists of two halves. The top half has a knife edge on the bottom which forms a water-tight seal between it and the wafer allowing solution to be poured directly on to the wafer surface. A *Plexiglass* plate fits over the opening in the top half to allow watching the wafer during etching and to serve as a shield from the ^{31}Si beta particles. Absorbed beta radiation dose to the extremities is thus reduced.

The bottom half of the etching device is a right cylinder with a shallow cavity for the wafer. When the two halves are connected, the top half's knife edge is pressed tight to the wafer which is held in place by the cavity in the bottom half. This configuration results not only in the same amount of surface area being etched each time but the same position on the wafer too. The etching device's reproducibility will be advantageous to comparing surface contamination results between wafers and trace element concentrations in the epitaxial layer of different wafers.

ACKNOWLEDGEMENTS

The author and the TAMU NSC wishes to thank Dr. Joseph A. Keenan and Texas Instruments, Incorporated for their invaluable contributions in establishing the wafer and bulk high-purity Si trace element analysis program at the TAMU NSC. Dr. Keenan's expertise in Si NAA has been utilized on countless occasions.

REFERENCES

1. Ehmann, W. D., Vance, D. E., *Radiochemistry and Nuclear Methods of Analysis*, New York, John Wiley & Sons, Inc., 1991, ch. 9, pp. 255-258.
2. Keenan, J. A., Gnade, B. E., and White, J. B., *Journal of the Electrochemical Society* **132**, 9, 2232-2236 (1985).
3. Larrabee, G. B., and Keenan, J. A., *Journal of the Electrochemical Society* **118**, 8, 1351-1355 (1971).

Prospects for Single Atom Sensitivity Measurements of Dopant Levels in Silicon

Richard R. Vanfleet, Mike Robertson*, Mike McKay**, John Silcox

School of Applied and Engineering Physics, Cornell University, Ithaca, NY 14853

The demands of the National Technology Roadmap for Semiconductors will necessitate measurement of dopant concentrations with greater spatial resolution than now possible. Current experimental and simulation experience indicate that Annular Dark Field (ADF) imaging in a Scanning Transmission Electron Microscope (STEM) should be able to determine dopant distributions with near atomic resolution. The ADF signal is derived from electrons diffusely scattered to high angles, resulting in contrast due to atomic number (Z-contrast) and defects. Atomic number scattering is proportional to Z^n (n is typically between 1.5 and 1.9), and is thus chemically sensitive. Similar to the bright field phase contrast techniques of Ourmazd (1), concentration profiles can be simply determined from microscope images. A simple model predicts approximate signal to noise ratios from 3.3 for one arsenic atom in a column of 100 silicon atoms to 18 for a single gold atom. Multislice simulations support this conclusion as does experimental work with silicon-germanium and compound semiconductor quantum wells. Boron atoms fill substitutional sites in silicon with a size misfit that locally strains the lattice. For high boron concentrations this strain has been seen by the ADF technique. Simulations predict that the strain field induced by a single boron atom will be visible at low temperatures. The current state of experiment and simulations is discussed with emphasis on imaging boron and gold in silicon.

INTRODUCTION

Current methods for measuring dopant concentrations do not yet have the spatial resolution that will be required for future generations of semiconductor technology. The ability to use the Annular Dark Field (ADF) imaging mode of a Scanning Transmission Electron Microscope (STEM)(2) to determine dopant concentrations has been pointed out earlier(3). Since then much progress has been made in determining the sensitivity of these methods with quantitative approaches including the determination(4, 5) of the indium concentrations within an InGaAs quantum well with close to single atom sensitivity. To carry out such observations at a quantitative level, it is necessary to acquire the images with linear detectors, preferably with a single electron sensitivity and to record those images digitally. A scanning system is ideally suited to such an endeavor and this has been carried out for the Cornell STEM(6,7).

An example of an ADF image is shown in Fig. 1. It is an image of a nominal 3 monolayer coverage of germanium on silicon. The image looks down the {110} axis and clearly images the atomic dumbells. The bright region on the right is the increase in scattering due to the germanium coverage. Quantification of the germanium concentration will be discussed shortly.

At room temperature the electron scattering that gives rise to the imaging characteristics will be predominantly the thermal diffuse scattering from the matrix, static elastic strain from defects and additional elastic scattering from impurities (dopants) and clusters of atoms in the matrix. This mode of imaging has been termed a 'Z-contrast' mode but while there is a substantial atomic number dependence, in practice any attempt to determine quantitative dopant profiles using this approach should be backed by careful simulation of the scattering intensities.

In a Scanning Transmission Electron Microscope (STEM), a very small (2Å) probe of electrons is formed at and transmitted through the sample. A signal is derived from the interaction of the probe with the sample at a point and then the probe is moved to a new point on the sample and the process repeated. The image is built up point by point as the probe is scanned over the sample. The Annular Dark Field detector is donut shaped with the inner hole centered on the optic axis of the microscope. The unscattered electrons pass

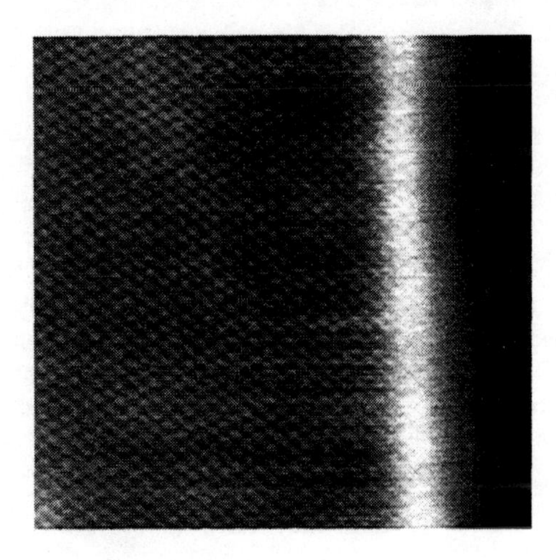

Figure 1 ADF image of germanium on silicon. The bright band on the right is the germanium covered surface.

* Currently at JDS Fitel, Inc., Ottawa, CA.
** Department of Physics, University of Texas El Paso.

CP449, *Characterization and Metrology for ULSI Technology: 1998 International Conference*
edited by D. G. Seiler, A. C. Diebold, W. M. Bullis, T. J. Shaffner, R. McDonald, and E. J. Walters

through the center of the inner hole and are used to form the bright field image or gather Electron Energy Loss Spectroscopy (EELS) data in parallel with the ADF signal. The contrast reversals with defocus and thickness variations inherent in normal TEM bright field diffraction contrast images do not occur. Thus direct measurements of the scattered intensity can be interpreted more readily with the ADF mode. Effects such as strain, and a thickness dependence of the probe in the sample add enough complications that simulations are necessary to extract the full information provided in the images. In this case the simulation is more extensive than in the TEM case but the resources are rapidly becoming available to carry this out at a reasonable cost.

In this paper we provide over simplified models and estimates to give an initial sense of the numbers involved. We also provide simulations of scattering from single dopant atoms in a silicon matrix. These estimates and simulations suggest that single atom sensitivity levels may be attainable for a number of interesting cases. This work builds on earlier experience (8,9,10,11,12) with simulations and experiment to assess the conditions and possible complications that might arise in determining dopant distributions to single atom sensitivity in silicon, on a scale commensurate with device scales targeted for the future.

Given that "TEM is becoming a required precision measurement tool for manufacturing and a necessary analytical tool for R&D and failure analysis support" (R.McDonald et al.(13)) and that high spatial resolution analytical methods using Electron Energy Loss Spectroscopy are now available to complement the ADF imaging, there is significant potential in developing this method to measure dopant concentrations on the nanoscale.

SIMPLE MODELS

A dopant atom replacing a silicon atom in the sample will alter the measured signal at that point. If the current density at a given dopant atom site i is J_i (t) where t is the depth in the column, then the change in signal, ΔS, can be expressed as J_i (t) (σ_{dop} - σ_{ref}) where σ_i is the cross section for scattering electrons into the ADF detector. This needs to be summed over all dopant atoms to obtain the resultant change. Whether this is detectable relative to a column of silicon atoms depends on how this signal compares to the noise in the pure silicon column. A crude estimate of the signal, S, from the column of N silicon atoms is N J_o σ_{ref} where J_o is the current density averaged down the column. The contrast, C, can then be defined as ΔS/S. For a small number of dopant atoms we can write C as:

$$C = \sum_i J_i(t)(\sigma_{dop} - \sigma_{ref}) / NJ_o\sigma_{ref}. \quad \textbf{(1)}$$

If we now use the result (14) that the ADF cross sections σ vary with atomic number as Z^n where $1.5 < n < 1.9$. then equation 1 becomes:

$$C = \sum_i J_i(t)(Z^n_{dop} - Z^n_{ref}) / NJ_oZ^n_{ref}. \quad \textbf{(2)}$$

Equation 2 indicates that the contribution to the contrast does depend on depth in the same way as the current density. If however, the concentration of dopants is relatively high, then J_i (t) is on average J_0 giving:

$$C = c_d(Z^n_{dop} - Z^n_{ref}) / Z^n_{ref}. \quad \textbf{(3)}$$

where c_d is the local concentration of dopant atoms (N_d/N).

Assuming noise in the measurement is due to collection statistics, and thus proportional to $S^{1/2}$, the signal to noise ratios can be estimated. ΔS is the signal of interest and the signal to noise becomes C $S^{1/2}$. For a single dopant atom of gold in 100 atoms of silicon, n=1.7, and typical signals on order of 10,000 counts, a signal to noise ratio of 18 is calculated (using the approximations of equation 3). Using the Rose criterion (15), even a single arsenic atom at a signal to noise value of 3.3 might be visible at these levels. In practice, using single electron detectors, careful imaging conditions, and carefully prepared samples, noise of order 0.2% is possible. Thus, sensitivities should be better than estimated here.

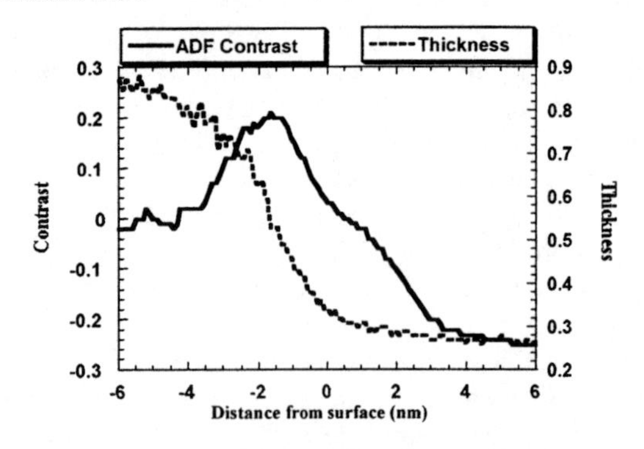

Figure 2 Uncorrected contrast across germanium layers (bulk silicon at left). Thickness from Electron Energy Loss Spectroscopy also shown.

As an example of the considerations outlined here, Fig. 2 shows a line scan (averaged perpendicular to the scan direction) across a nominally 3 monolayer coverage of germanium on silicon. This is the same sample as shown in Fig. 1. Also shown in Fig. 2 is the thickness for the same region. Using the Electron Energy Loss Spectrometer, images can be formed from electrons that have lost a specified energy. The ratio of the image of electrons having lost a plasmon energy to the image for zero energy loss is proportional to the thickness of the material creating that plasmon. The ratio of those images is shown as thickness in Fig. 2. Ignoring thickness effects in the contrast, a 6.5% concentration of germanium would be calculated. This is incorrect because of the rapid drop in thickness across the germanium region. With the thickness taken into account, the contrast is as shown in Fig. 3. The corresponding concentration is shown on the right hand scale. The vertical line marks the surface of the crystal. The contrast and thickness values above the surface are artifacts of the thickness correction and due to the glue used in the sample preparation. The glue plasmon is outside the energy window used to measure

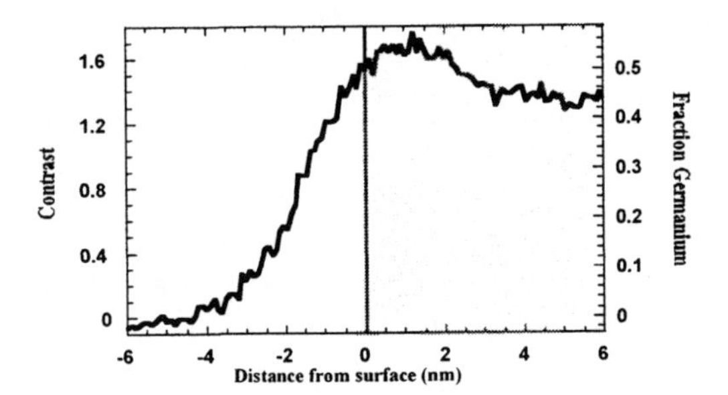

Figure 3 Thickness corrected germanium contrast and concentration. Shaded region to right indicates area where data is an artifact of the thickness correction in regions where there is no silicon or germanium (see text).

the thickness and thus any information in the processed image above the surface (indicated by shading in Fig. 3) is an artifact of the thickness correction. This type of artifact is only an issue when the plasmon energy shift is significant compared to the energy window used for the plasmon image, as in the case of the glue and silicon. In the germanium region where bonding, density, and therefore plasmon energy are similar to silicon, the thickness is mapped correctly.

This example indicates some of the issues and concerns that need to be addressed in this problem. First, it is essential to measure the overall thickness of the sample and as outlined above this can be accomplished by the use of plasmon imaging coupled with zero loss imaging. Second, other experimental considerations include minimization of surface contamination and damage. Both of these can generate excess noise that considerably reduces the sensitivity of the method. HF dips may accomplish this although a suitable protocol has not yet been established.

On the theoretical side the approximations incorporated in expressions 1 through 3 are extremely crude. The use of uniform electron densities and cross-sections is an extreme approximation. Indeed the only justification for this simple model is to generate an argument that accurate simulation for these cases would be worth the effort.

SIMULATIONS

Image simulation of the annular dark field imaging mode is carried out by a multislice method (9,16,17). The electron wave function for the focused probe is calculated from the microscope parameters. This wave function is then propagated numerically through the sample using established multislice techniques. The exit wave function is propagated to the detector and the signal is calculated based on the probability of finding an electron between the inner and outer angle of the dark field detector. Temperature effects are included using the Einstein model for phonons and the frozen phonon approximation(11). Each atom in the modeled crystal is given a random displacement. The simulated exit wave function is calculated and the process is repeated with new random displacements. The probability distribution is averaged at the detector for the different configurations. The RMS displacement is determined by and directly related to the temperature. Many different atomic configurations are calculated and averaged to achieve a converged simulated signal. This method has been shown to be in very good agreement with experimental measurements of convergent beam electron diffraction patterns (11) of silicon as well as ADF images of InP (12,18).

Simulations can also be used to track the shape and distribution of the electron beam as it interacts with the sample (19). As the electron wave function passes through the sample, it is compressed by the focusing effects of the atoms resulting in a channeling of the electron density on the atomic columns. The variation in channeling with depth is illustrated in Fig. 4 and 5. The size of the channeling peak changes with temperature. Lower temperatures increase the channeling peak and lower the ADF signal. Microscope and detector parameters used in all simulation in this manuscript correspond to those of the VG HB501 STEM at Cornell University.

SINGLE ATOMS

When dopant concentrations are low, the ADF signal and contrast should track with the channeling peak according to equations 1 and 2. Figure 4 shows the simulated ADF contrast for a single gold atom in 200 Å of silicon {110} as a function of depth. The maximum pixel in the silicon channeling peak is also shown. The channeling peak value has been normalized to the maximum contrast to better illustrate the correlation between the contrast and the channeling. The simulation was done for room temperature and indicates that imaging of single atoms of gold in silicon relative to a silicon column noise of 0.002 is feasible.

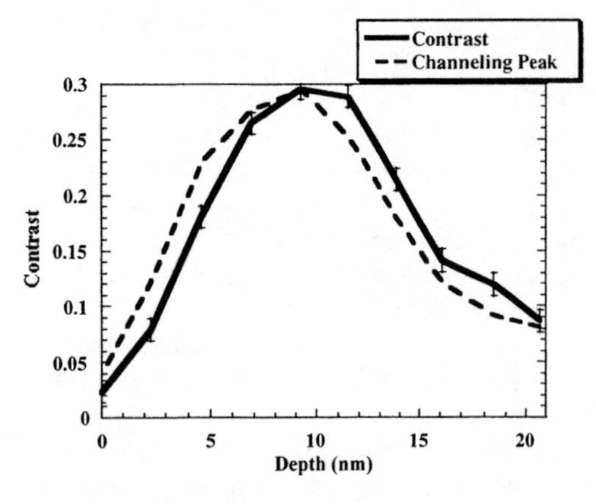

Figure 4 Simulated ADF contrast for a single gold atom and silicon channeling peak as a function of position in ~200 Å of silicon {110}. Channeling peak is scaled to the maximum of the contrast.

STRAIN CONTRAST

A second source of diffuse scattering is strain or disorder in the crystal lattice. Dislocations and grain boundaries show up bright in ADF images. Many dopants have a size misfit that produces significant strain in the lattice. Boron at a concentration of $4x10^{20}$ B/cm^3 has been observed to produce enough strain to be imaged (20). The static displacement field increases the diffuse scattering in the same manner as increasing the random displacements due to temperature. The static strain field can be masked by the phonon displacements, thus temperature is a vital parameter when imaging a strain field. There are currently no simple models that predict strain field contrast for ADF imaging. Therefore, multislice simulations have been used to predict image contrast.

A simple $c/r^2 \hat{r}$ strain field with c=-1.4 Å^2 is used for boron (21). The simulations predict no contrast at room temperature for a single boron atom(22). The contrast at low temperatures (liquid helium temperatures) for a single boron in ~220 Å of silicon {110} is shown in Fig. 5 as well as the peak pixel in the channeling peak (again normalized to the contrast peak value). Figure 5 is the contrast in the column of atoms adjacent to the boron atom where the strain field is strongest. Strain field imaging also tracks with the channeling peak and in the case of boron results in at most a 7% increase in signal.

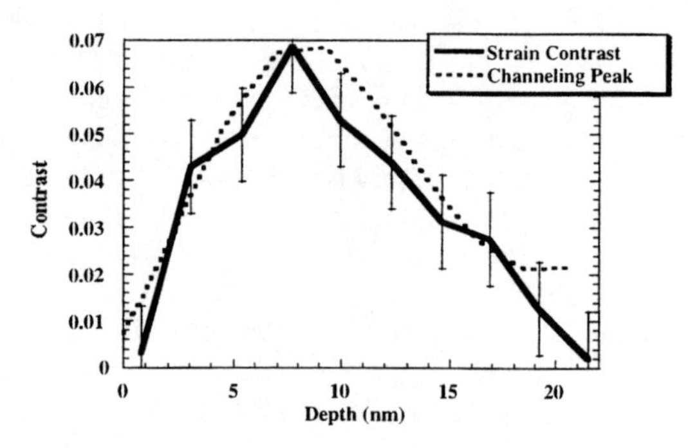

Figure 5 Simulated boron strain field contrast as function of position in ~220 Å of silicon {110}. Channeling peak maximum is scaled to peak contrast.

If boron concentrations are high and the sample is thick (1000's of Å), then the probe is able to interact with many boron atoms and their strain fields. This contrast is now visible at room temperature. The average atom in this region will have a displacement relative to its unstrained position

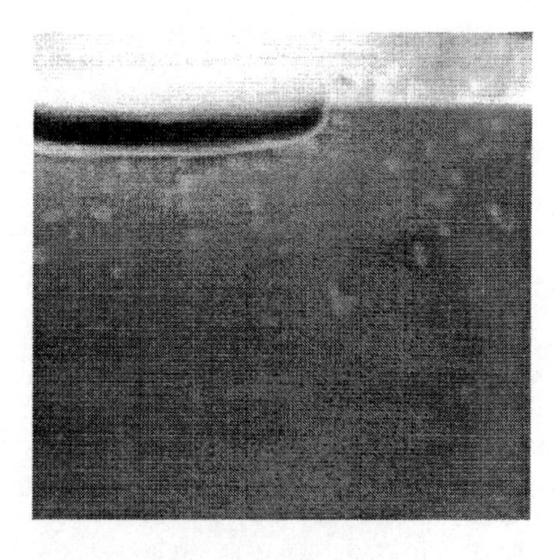

Figure 6 ADF image of the Source/Drain region of a PMOS Transistor. 10^{19} boron doping is in brighter region below the metal contact.

which will increase its diffuse scattering (as with low concentrations or thin samples), but without atomic resolution. Figure 6 shows a room temperature ADF image of a PMOS source/drain with a nominal boron doping of 10^{19} cm^{-3}. The brighter region under the contact is presumably due to the boron doping. Figure 7 shows contrast for a line scan downward from the contact.

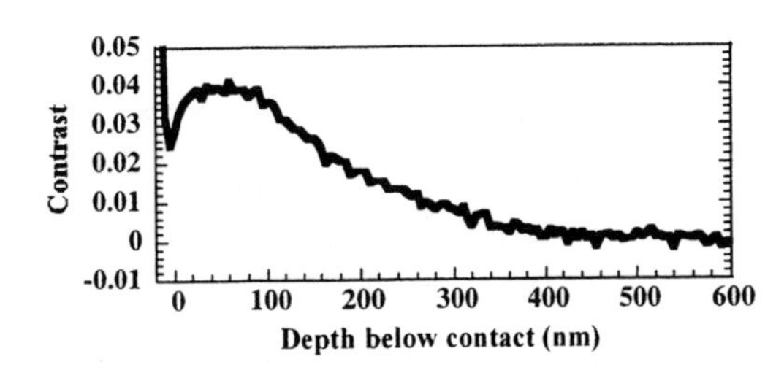

Figure 7 Experimental ADF contrast below the Source/Drain contact of Fig. 6.

SUMMARY

Annular Dark Field (ADF) imaging uses diffusely scattered electrons to form an image of the sample. This results in images that have atomic number and strain field contrast. To draw quantitative results from the images, corrections for sample thickness must be made. Electron Energy Loss Spectroscopy (EELS) in the Scanning Transmission Electron Microscope (STEM) can be used to measure thickness in parallel with the ADF image. Contrast in the corrected image can be directly related to the projected dopant concentration.

Imaging of single higher atomic number dopant atoms appears possible based upon image simulations. Image simulations for single low atomic number dopants also show sufficient strain field contrast, for imaging at low temperatures.

ACKNOWLEDGMENTS

This research was supported by an Air Force grant # F49620-95-1-0427. The Cornell STEM was acquired through the NSF (grant # DMR-8314255) and is operated by the Cornell Center for Materials Research (NSF grant # DMR-9632275). Mike McKay held an NSF REU summer assistantship through CCMR. Support and helpful discussions with Earl Kirkland and Mick Thomas are gratefully acknowledged. Samples of a device test structure were supplied by Sheldon Aronowitz and Lindor Hendrickson of LSI Logic, Santa Clara, CA.

REFERENCES

1. Ourmazd, A., Taylor, D.W., Cunningham, J., and Tu, C.W., *Phys. Rev. Lett.* **62**, 933 (1989).
2. Crewe, A.V., Langmore, J.P. and Isaacson, M.S., in *'Physical Aspects of Electron Microscopy and Microbeam Analysis'*, Ed. Siegel, B.M. and Beaman, D.R., (Wiley, New York 1975) p.47
3. S. J. Pennycook and J Narayan, *App.Phys. Lett.* **45**, 385 (1984).
4. Hillyard, S., PhD Thesis, Cornell University,1996.
5. Hillyard, S., Chen, Y.-P., Schaff, W.J., Eastman, L.F. and Silcox, J., *J. Mic.Soc. Am. Proc., Microscopy and Microanalysis 1995*, 294-5 (1995).
6. Kirkland, E.J., *Ultramicroscopy,* **32**, 349-364, (1990).
7. Kirkland, E.J. and Thomas, M.T., *Ultramicroscopy* **62**, 79-88 (1996).
8. Lee, T.C., PhD Thesis, Cornell University, 1995. Lee, T.C., Kottke, M., Watanabe, J., Fejes, P.L., Theodore, N.D., and Silcox, J., in preparation.
9. Kirkland, E.J. Loane, R.F. and Silcox, J., *Ultramicroscopy* **23**, 77-96 (1987).
10. Loane, R., Kirkland, E.J., Silcox, J., *Acta Cryst.* **A44,** 912 (1988).
11. Loane, R.F., Xu, P., and Silcox, J., *Acta Cryst.* A**47**, 267-278 (1991).
12. Hillyard, S., Loane, R.F., and Silcox, J., *Ultramicroscopy* **49**, 14-25 (1993).
13. McDonald, R.C.,Mardingly, A.J. and Susnitsky, D.W., *J. Mic. Soc. Am. Proc., Microscopy and Microanalysis*, 449-450 (1997).
14. Hartel, P., Rose, H. and Dinges,C., *Ultramicroscopy* **63**, 93-114 (1996).
15. Rose, A., *Adv. Elec.* **1**, 131 (1948).
16. The simulation code developed by E.J. Kirkland is currently being used.
17. Cowley, J.M., *Diffraction Physics*, Amsterdam: Elsevier, 1995.
18. Loane, R.F. , Xu, P., and Silcox, J., *Ultramicroscopy* **40**, 121-138 (1992).
19. Hillyard, S., and Silcox, J., *Ultramicroscopy* **52**, 325 (1993). Hillyard, S., and Silcox, J., *Mat.Res. Soc.Symp.Proc.* **332**, 361-372 (1994).
20. Perovic, D.D. and Paterson, J.H., *Proc. 49th Annual EMSA Meeting*, Ed. G.W. Bailey (San Francisco Press, San Francisco, 1991) p. 704. Perovic, D.D., Rossouw, C.J., and Howie, A., *Ultramicroscopy* **52**, 353, (1993).
21. Hall, C.R., Hirsch, P.B., Booker, G.R., *Phil. Mag.* **14**, 979 (1966).
22. Hillyard, S. and Silcox, J., *Ultramicroscopy* **58**, 6 (1993).

Novel Analytical Technique for On-Line Monitoring of Trace Heavy Metals in Corrosive Chemicals

Stan Tsai[1], Samantha H. Tan[1], Anthony F. Flannery[2], Christopher W. Storment[2], Gregory T.A. Kovacs[2]

[1]ChemTrace Corporation, 3563 Investment Blvd., Bldg. E6, Hayward, CA 94545-7306
[2]CISx202, Stanford University, Palo Alto, CA 94305-4075

A novel analytical technique is presented in this paper for trace metal monitoring in corrosive chemicals. The technique is an application of anodic stripping voltammetry with an array of micro-machined mercury ultramicroelectrodes as the probe electrode. The latter is resistant to hydrofluoric acid and has been proven to be accurate, reproducible and robust. In a case study, the probe was continuously used in a HF bath for twelve hours with no sign of failure. It demonstrated the strong potential of this technique to perform real-time on-line monitoring of trace metals in corrosive chemicals such as those used in semiconductor processing.

INTRODUCTION

In the semiconductor industry, contamination of trace heavy metals is well recognized as an important contributor to device failure (1). Process chemicals such as hydrochloric acid, hydrofluoric acid, sulfuric acid, ammonium hydroxide, nitric acid and their mixtures are frequently monitored to ensure that trace metal impurities stay within acceptable concentration ranges. As the integrated circuits (IC) are scaled to sub-micron linewidth, the monitoring practices increase in importance. At present, such analyses are conducted off-line (2,3), typically involve the sampling of the process chemical in a contamination free manner, transport of the sample container to an analytical lab, preparation of the sample, analysis by spectroscopic techniques and data calculations and interpretations. The spectroscopic techniques are typically inductively coupled plasma-mass spectrometry (ICP-MS) and graphite furnace atomic absorption spectroscopy (GFAAS). It is clear that the analytical process needs a team of experienced chemists, demands expensive lab set-up, and most importantly, often results in costly delays. Therefore, it is highly desirable to develop an analytical technique that is capable of on-line, real time detection of trace metals.

Anodic stripping voltammetry (ASV) has been known for a long time as an effective technique for the detection of heavy metal contamination at trace levels (4,5). The technique requires relatively simple instrumental setup and is extremely sensitive. The use of a mercury ultramicroelectrode as the working electrode adds several advantages to the ASV technique (6,7). With micron-sized electrode surfaces, it improves preconcentration efficiency, reduces background currents, increases sensitivities, and also eliminates the necessity to de-gas the test solution (8). When ultramicroelectrodes are grouped in an array, the detection is further improved by the multiplied analytical signal.

Kovacs et al (9,10) fabricated mercury ultramicroelectrode arrays (Hg-UMEA) using contemporary micro-machining techniques. In their approach, an array of iridium disk electrodes were "integrated" on a silicon chip and then electrochemically plated with mercury. The chip, except the electrode area, is encapsulated in a protective coating of PECVD silicon nitride. With this technique, the concerns of handling hazardous bulk mercury are eliminated. Micron-sized mercury electrodes can be mass-produced and can be precisely reproduced. The Hg-UMEA approach greatly simplified ASV and enhanced its performance. Detection of various heavy metals in natural water was demonstrated down to low parts per billion (ppb) levels (9,10).

However, the original work on the probe was not applicable to chemical matrices such as those used in semiconductor processing. In this study, a new probe was designed based on a similar Hg-UMEA approach. Silicon carbide was introduced as a protective coating to withstand the extremely corrosive nature of the process chemicals. This paper presents successful applications of this new probe for measurement of certain trace metals in some of the most corrosive chemicals used in semiconductor processing. It therefore suggests a new field of potential application for Hg-UMEA probes as a low-cost solution for on-line monitoring of heavy metals in IC processing.

EXPERIMENTAL

Reagents used in all the experiments were of analytical grade. Ultrapure water (UPW) from a RO/DI system was polished through a Millipore Milli-Q purification system before use and had a resisitivity of 18.2 MΩ. All standards and trace metal spiked samples were traceable to NIST. Their concentrations were verified using ICP-MS (Perkin

CP449, *Characterization and Metrology for ULSI Technology: 1998 International Conference*
edited by D. G. Seiler, A. C. Diebold, W. M. Bullis, T. J. Shaffner, R. McDonald, and E. J. Walters
 1-56396-753-7/98/$15.00

Elmer Sciex Elan 5000). All the components of the electrochemical cells and acid baths were made of plastics and pre-leached with acids to eliminate trace metal contamination.

Fabrication of Hg-UMEA

The iridium based UMEA was fabricated using micro-machining techniques. The fabrication process for the Ir UMEA was described in detail previously (10,11) and only a brief description is provided here. A 0.5 μm silicon oxide dielectric layer was first thermally grown on a silicon wafer to isolate the silicon substrate. On this insulating layer iridium was deposited and patterned using a sputter lift-off process to form arrays of ultramicroelectrodes. The entire probe, except for the Ir surfaces, was coated with a layer of PECVD silicon carbide. After fabrication, the Ir-UMEA probes were diced and soldered to copper leads. Each probe was inspected with an optical microscope for the integrity of the Ir disks. A probe that passed the visual test was plated with mercury. In the mercury plating step, the array of Ir disks was first electrochemically pre-conditioned by scanning it between -0.5 V and 0.5 V versus a Ag/AgCl reference electrode for at least 10 times at a scan rate of 1 V/sec. The array was then brought to a potential of -0.45 V vs. Ag/AgCl until the total desired deposition charge was accumulated. The deposition charge controlled the size of the mercury electrodes on the Ir disks. After the deposition step, the Hg-UMEA was ready for analytical use.

ASV Analysis

In this study, ASV was used for trace metal analysis of semiconductor process chemicals. For example, for the detection of copper (Cu^{++}), a deposition potential of -1.2 V vs. Ag/AgCl was applied to the sensor to preconcentrate copper ions into the mercury electrodes. The preconcentration time varied from 30 seconds to 30 minutes depending on the concentration range of interest. In general, the sensitivity of the technique increases proportionally when preconcentration time is increased. A square wave anodic potential sweep was then applied from -1.2 V to 0 V. The stripping current reported here was the net current (12) which was the current difference between the forward and the reverse currents. Figure 1 shows a typical stripping wave of 50 ppb copper in 50:1 HF with a deposition time of 2 minutes. The electrochemical tests were carried out with a PAR 273A potentiostat/galvanostat (EG&G, Princeton, NJ) interfaced to an IBM PC computer. Experiments were run on a Model 270 software (EG&G, Princeton, NJ). A three electrode system was employed with the Hg-UMEA as a working electrode, a Pt foil (about 0.5 cm^2) as the counter electrode, and a Ag/AgCl reference electrode.

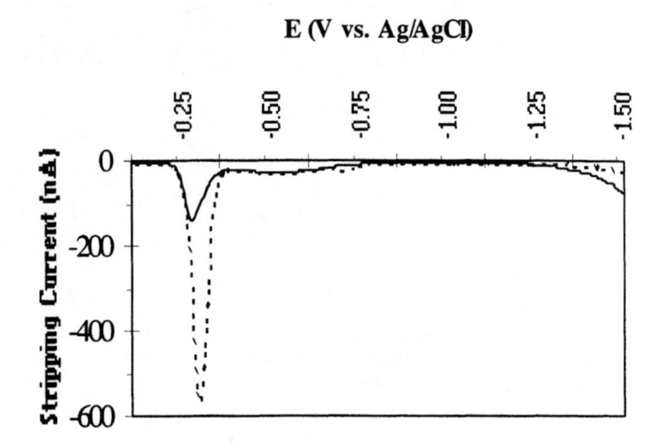

FIGURE 1. The stripping wave for the detection of copper in 50:1 HF solution. Conditions: 50 ppb Cu^{2+} in 50:1 HF, t_d = 2 minutes, E_d = -1.2 V vs. Ag/AgCl, E_{sw} = 25 mV, ΔE = 5 mV, and f = 60 Hz. Solid line indicates direct detection of Cu in 50:1 HF and the broken line indicates the detection of Cu using a medium exchange to 0.1 M sodium acetate after the deposition step.

RESULTS AND DISCUSSION

Mercury Ultramicroelectrode Array

For application in semiconductor process liquids, the Hg-UMEA probe needs to be resistant to corrosive chemicals. Since hydrofluoric acid (HF) solution is one of the most corrosive chemicals, it was used as a test case. The initial version of the probe was coated with PECVD silicon nitride. It was found that the film dissolved in 2% HF within five minutes. Various other thin films (i.e., alumina, diamond-like carbon, thin-film diamond and silicon carbide) were coated on silicon wafers and tested with a variety of chemicals including HF. Silicon carbide showed good compatibility with the sensor fabrication process and was inert to most HF containing chemicals (11). No sign of degradation was observed even after twelve hours of soaking in 49% HF. Therefore, an improved version of Hg-UMEA was fabricated with silicon carbide as the outer dielectric coating.

Figure 2 schematically shows a portion of the probe layout that was used in this work. The probe consisted of 20 iridium disks (working electrode) and rings (pseudo-reference electrode) and a large surrounding Ir band (counter electrode). Each Ir disk had a diameter of 10 μm. The disks were interconnected underneath with gold (Au) interconnects. To avoid inter-diffusion between the Ir disks, they were placed 500 μm apart from each other. The iridium rings were electrically isolated from the Ir disks. They were used as optional on-probe reference electrodes. The iridium band served as the optional on-probe counter electrode. The protective coating, silicon carbide, covered the entire chip except for the iridium disks, rings and band.

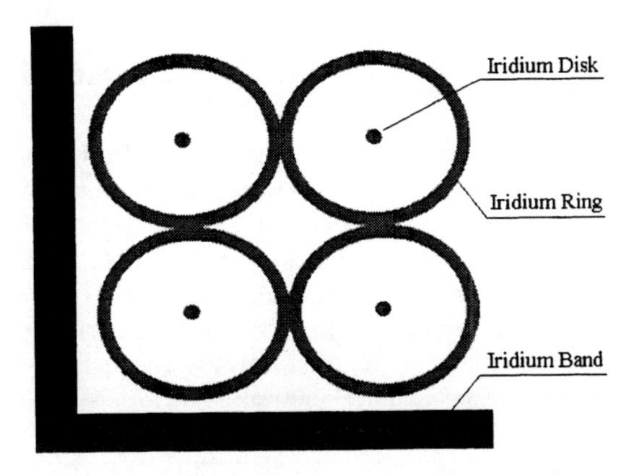

FIGURE 2. Schematic diagram of a section of an Ir-UMEA sensor. The entire 20 electrode array is laid out in an area of 2.5 mm x 4 mm.

An optical microscope was used to examine the physical appearance of the iridium/mercury array. Under optimum conditions, the iridium disks were smooth and flat. After the deposition step, the electroplated mercury electrodes were observed to have well defined hemispherical shapes. As mentioned before, the size of the mercury hemispheres was controlled by the amount of deposition charge. It was observed that these mercury hemispheres were mechanically stable.

Characteristics of the Hg-UMEA

In view of the application of the Hg-UMEA sensor to the semiconductor industry, a wide range of elements were tested in six important semiconductor process chemicals. It was found that the Hg-UMEA sensor was capable of measuring low ppb levels of heavy metals such as Bi, Cd, Cu, Pb, Sb, Tl and Zn in corrosive chemicals like HF, BOE, HNO_3, and diluted H_2SO_4, HCl, H_3PO_4. The analytical characteristics of the Hg-UMEA sensor is presented with respect to its linear dynamic range, detection limit, accuracy, reproducibility, and robustness. Sensor lifetime was demonstrated with the analysis of Cu in 50:1 HF. The analytical data obtained are summarized in Table 1.

TABLE 1. Summary of Analytical Characteristics of the Hg-UMEA.

Linear Concentration Range	~2 orders of magnitude, e.g. 0.5 - 20 ppb or 20 - 1,000 ppb depending on deposition time
Detection Limit	0.5 ppb
Accuracy	< 0.5%
Reproducibility	< 1.7 % RSD for 70 replicates within 12 hours
Robustness and Lifetime	No sign of degradation for 4 days with >500 analyses

Linear Dynamic Range

Anodic stripping voltammetry operates in two sequential steps. A cathodic potential is first applied for a period of time, whereby trace metals from the solution are deposited and accumulated into the mercury electrodes. An anodic potential scan is then applied to strip off the deposited metals, generating a stripping current. When a deposition time of one minute was used, copper showed a stripping current that was linear to the solution concentration from 20 to 500 ppb. The sensitivity obtained was 2.4 nA/ppb. At concentrations below 20 ppb, the stripping current was too small and failed to show good linearity (13). In contrast, stripping current for solutions above 500 ppb was too large and produced signal saturation. Thus, concentrations between 20-500 ppb represent the linear dynamic range for this technique. The Hg-UMEA showed excellent linearity within this range with a correlation coefficient of 0.999.

Since stripping current increases with deposition time, the lower working concentration range of the sensor was optimized by lengthening the deposition time. For example, the linear dynamic range could be shifted to cover 2-50 ppb when the deposition time was extended to 10 minutes. However, for practical purposes, the deposition time only can be varied from 30 seconds to 30 minutes. Therefore, the linear dynamic concentration range depending on the deposition time, is in range of 0.5 ppb to 1 ppm.

Similarly, stripping current is proportional to the number of Hg electrodes in an array. A wider linear dynamic range is potentially achievable with an array of more electrodes.

Detection Limit

In this study the detection limit was estimated by the lowest standard used for obtaining the linear dynamic range. With a 30 minute deposition time in a 0.5 ppb Cu solution, a good analytical signal was observed for the sensor. However, only concentrations above 0.5 ppb showed linear response. A lower detection limit may be achievable with extended deposition time, more electrodes, and/or a lower noise potentiostat.

Accuracy

The accuracy of this technique was evaluated by measuring solutions containing a known amount of Cu against an established ASV calibration curve. Both the calibration standards and the samples were prepared from a NIST Cu standard (SRM 3114) and verified by ICP-MS for their nominal concentrations. They were prepared in the same matrix and under identical test conditions. For example, to verify accuracy, a 20 ppb copper-spiked 0.1 M HNO_3 solution was prepared and analyzed. The stripping current obtained was compared to a calibration curve based on 10, 50 and 100 ppb standards of the same matrix. The spiked sample was measured to be 20.06 ppb with a variation of 0.3% from its nominal concentration.

Numerous experiments were performed with different concentration ranges (from 2 ppb to 500 ppb) in various matrices (e.g., 1.0M sodium acetate, 50:1 HF and 0.1M HNO_3) and also for various elements (e.g., Cu, Pb, and Zn.). Within the linear dynamic range, the measured concentrations deviated less than 1% from their nominal concentrations.

Reproducibility and Signal Drift

It is expected that different probes can generate different stripping currents even under identical test conditions. However, with current micromachining technology and an optimized mercury-plating procedure (14), this probe-to-probe variation was controlled within ± 10%. Indeed, since calibration is performed for each probe in real-life analysis, the effect of the variation is expected to be minimal. The measurement-to-measurement variation was characterized by performing repeated analysis using the same Hg-UMEA in the same solution. In one experiment, 70 plus analyses were repeated for a period of twelve hours in a 50:1 HF solution containing 50 ppb Cu. The stripping currents were extremely reproducible with a relative standard deviation (RSD) of only 1.7%. Comparing the currents at the beginning and at the end of the test period, the Cu signal drifted less than ± 4%

Robustness and Life Time

The Hg-UMEA probe proved to be very robust. The plated mercury hemispheres adhered well on the iridium bases and withstood shaking, rinsing and mild blowing under a N_2 gas stream. Monitored with an optical microscope, the physical appearance of the probe showed no change after the above described handling. The ASV stripping currents collected before and after probe handling were also the same. The probe was extremely durable in HF solution. In a lifetime test, one probe was continuously used for four days in 50:1 HF without failure. The mercury hemispheres of the probe were observed in to be in excellent condition after experiencing more than 500 analyses. Although some current drifts were observed, the final current was still within 75% of the initial value. The drift during such an extended time period can be easily compensated if the probe is periodically calibrated.

Continuous Monitoring of Copper in HF Bath

To fully demonstrate the performance of the Hg-UMEA, a HF bath was prepared and filled with 50:1 HF. The solution was continuously stirred to ensure the homogeneity of the solution. Copper impurities were spiked into the bath to simulate increasing levels of contamination. A Hg-UMEA probe was used to measure the concentration change. ASV was programmed to repeat analysis automatically at approximately ten minutes intervals. After initiation, the analyses were executed without manual interruption. During each cycle of analysis, the electrode potential was first set at -1.2 V vs. Ag/AgCl for 4 minutes for preconcentration. An anodic square wave potential sweep was then applied from -1.2 V to -0.1 V. The stripping current signal was recorded. The probe was then allowed to rest in the test solution for 5 minutes before the start of another cycle.

Figure 3 shows the behavior of the copper signal in the HF bath. Initially, the HF bath was cleaned and tested free from copper contamination. Only a background charging current of 17 nA was observed. After one hour, 50 ppb Cu^{++} was spiked into the HF bath. The stripping current showed an immediate increase to about 250 nA. The bath was then left undisturbed for the next 11 hours. A total of 70 analyses were performed during the period. At the end of the 11 hour period, an additional 10 ppb Cu^{++} was spiked into the solution. The stripping current showed another jump, increasing to 310 nA.

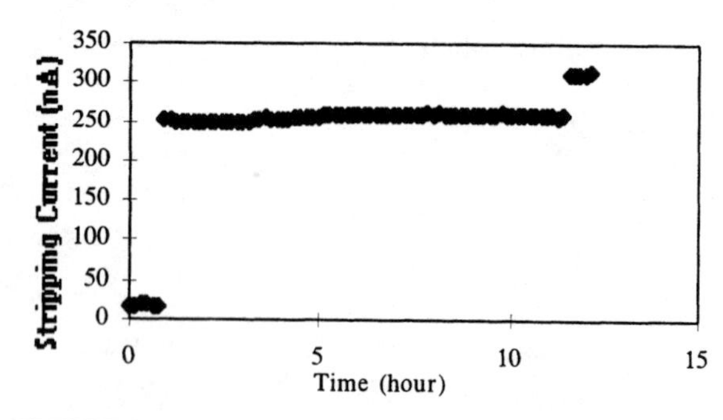

FIGURE 3. Continuous monitoring of copper impurity in a HF process path.

The stripping current was very stable with minimal signal drift. The 70 analyses carried out in the 50 ppb HF solution showed a 1.7% RSD. In other words, with the above experimental conditions, a concentration change of only 2.5

ppb in the HF process bath was detected at a 99.7% confidence.

The probe in this case showed ideal characteristics for a monitor. The probe was extremely durable in the HF solution. No human intervention was required for maintenance activity. For more than 12 hours, the Hg-UMEA probe generated consistent and accurate analytical signals in a HF bath and responded immediately to concentration changes.

Flow Cell and Medium Exchange

When immersed in harsh semiconductor processing liquids, the probe body was protected from corrosion by the silicon carbide coating. However, mercury as the working electrode was exposed and unprotected. Various problems arose when the Hg-UMEA was used directly in process chemicals during the current stripping step, including reduced sensitivities, high background currents, and mercury dissolution. To prevent this, the sensor was removed from the process chemical after the deposition step and placed in an inert electrolyte for the current stripping step. This process is referred to as medium exchange.

Medium exchange is known as an effective tool to minimize this type of problem (15). It involves a deposition of metal impurities from the sample solution, followed by stripping them in an inert electrolyte. A flow cell (see Figure 4) was constructed for medium exchange. In this study, the stripping wave was generated in an inert electrolyte (e.g., 0.1 M sodium acetate) rather than in the process chemical and certain advantages associated with medium exchange were realized. For example, after 5 minutes of pre-concentration at -1.2 V vs. Ag/AgCl, a stripping current of 140 nA was observed in a 50 ppb Cu^{++}/50:1 HF solution (Figure 1, solid line). With medium exchange to 0.1 M sodium acetate electrolyte, the same probe generated a stripping current of 590 nA (Figure 1, broken line). The signal sensitivity was increased by four times.

From previous experiments it was observed that a low ppb level of zinc (Zn) was not detectable by direct measurement in 50:1 HF due to the shielding effect of the strong hydrogen wave. With medium exchange to 0.1 M sodium acetate, the acidity of the electrolyte was reduced, resulting in a dramatically reduced hydrogen wave. Subsequently, detection of Zn in the low ppb range became possible. Similarly, direct analysis of metals in 1:1 HF and 6:1 BOE solution showed very high background currents and no meaningful analytical signal could be deduced. With the use of a flow cell and medium exchange, stripping currents were observed with the same sensitivities as those measured in an inert electrolyte.

More significantly, very poor performance was observed for some chemicals due to the instability of mercury in these corrosive mediums. In 1:1 HNO_3 the mercury hemispheres on Hg-UMEA were oxidized and dissolved, thus making ASV analysis impossible in nitric acid. To overcome this, the probe was first immersed in 0.1 M sodium acetate electrolyte in a flow cell. A potential of -1.2 V vs. Ag/AgCl was applied to the sensor before the introduction of 1:1 nitric acid. The sodium acetate electrolyte was removed, and 1:1 nitric acid was introduced. During this medium exchange, the sensor was held at -1.2 V vs. Ag/AgCl. After the required deposition time, the sodium acetate electrolyte was re-introduced. An ASV potential scan was then applied to the probe to generate a stripping current. Since the mercury electrodes were constantly biased under a potential of -1.2 V for the entire period of exposure, mercury oxidation was prevented. After the analysis, the mercury hemispheres were observed to be intact and shiny. Using this strategy, copper in 1:1 HNO_3 can be detected, although the sensitivity was only about 25% of that compared to an inert electrolyte. In summary, the application of a flow cell for medium exchange enhanced the performance of the Hg-UMEA and extended its application to a large group of corrosive semiconductor process chemicals.

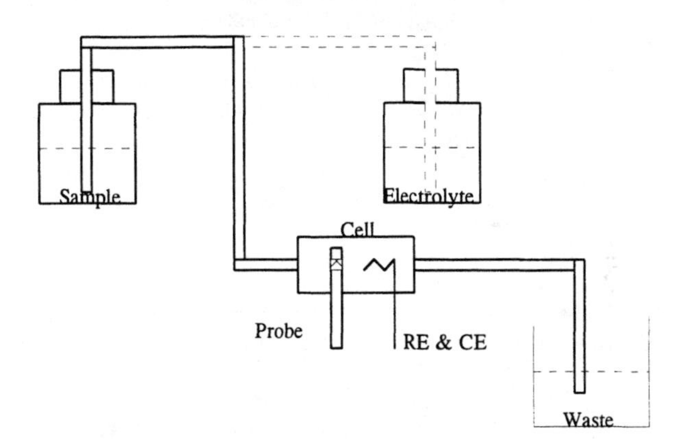

FIGURE 4. Schematic diagram of a flow cell for medium exchange

CONCLUSION

An HF resistant heavy metal sensor is successfully fabricated. The Hg-UMEA probe is sensitive, accurate, robust and reliable. The twelve-hour case study demonstrates that the Hg-UMEA probe is capable of performing continuous analyses. With the use of a flow cell for medium exchange, sub ppb level detection is possible for a large number of corrosive chemicals. The Hg-UMEA system has a strong potential to be used as an on-line sensor to monitor trace metal contamination in a semiconductor process bath.

ACKNOWLEDGMENTS

This research work has been supported by the National Science Foundation under award number 9461869. We like to thank Molly Raden Passer and Richard Bretz of ChemTrace for performing some initial probe characterization.

REFERENCES

1. Bullis W.M, Seiler D.G., and Diebold A.C., Editors, C.R. Helms, Silicon Surface Preparation and Wafer Cleaning: Role, Status and Needs for Advanced Characterization, *Semiconductor Characterization Present Status and Future Needs,* Woodbury, NY, American Institute of Physics, 1996, pp. 111-112.
2. Bullis W.M, Seiler D.G., and Diebold A.C., Editors, Critical Metrology and Analytical Technology Based on the Process and Materials Requirements of the 1994 National Technology Roadmap for Semiconductors, *Semiconductor Characterization Present Status and Future Needs,* Woodbury, NY, American Institute of Physics, 1996, pp. 31-36.
3. Gupta K., Tan S., Pourmotamed Z., Cristobal F., Oshiro N. and McDonald R., Contamination Control & Defect Reduction in Semiconductor Manufacturing III, *Electrochem.* *v* **94-9**, pp. 200-221, 1994.
4. Wang J., *Stripping Analysis Principles, Instrumentation, and Applications*, Deerfield Beach, FL, VCH, 1985, pp. 109-124.
5. Brainina K. and Neyman E., *Electroanalytical Stripping Methods,* John Wiley & Sons, 1993, pp. 104-119.
6. Wehmeyer D.R. and Wightman R.M., *Anal. Chem.* v **57**, pp. 1989-1993, 1985.
7. Tunon-Blanco P. and Costa-Garcia A., *Reviews on Analytical Chemistry Euroanalysis,* **VIII**, pp. 273-90, 1994.
8. Baranski A.S., *Anal. Chem.*, v **59**, pp. 662-6, 1987.
9. Kovacs G.T.A.; Storment C.W. and Kounaves S.P., *Sensors and Actuators B*, v **23,** pp. 41-7, 1995.
10. Kounaves S.P.; Deng W.; Hallok P.R.; Kovacs G.T.A. and Storment C.W., *Anal. Chem.*, v **66**, pp. 418-23, 1994.
11. Flannery A.F.; Mourlas N.J.; Storment C.W.; Tsai S.; Tan S.H. and Kovacs G.T.A., *"PECVD Silicon Carbide for Micromachined Transducers"* Proceedings of Transducer, 1997 International Conference on Solid-State Sensors and Actuators, Chicago, IL, June 19, 1997.
12. Osteryoung J.G. and Osteryoung R.A., *Anal. Chem.*, v **57**, pp. 101A-110A, 1985.
13. Kounaves S.P. and Deng W., *Anal. Chem.*, v **65**, pp. 375-9, 1993.
14. Tan S., Dougherty D., Tsai S., Bretz R., Kovacs G., Flannery A. and Storment C., Private Communication, NSF Report (SBIR#9461869), 1996.
15. Wang J., *Stripping Analysis Principles, Instrumentation, and Applications,* Deerfield Beach, FL, VCH, 1985, pp. 48-50 and pp. 87-90.

NSOMS Characterization of Semiconductors and Related Materials

Ran Liu, Nigel Cave*, Juan Carrejo, Wei Chen,
Tan-Chen Lee and Thomas Remmel

Semiconductor Products Sector, Motorola Inc.
*Mesa, Arizona, 85202 and Austin, Texas, 78721**

Near-field optical microscopy and spectroscopy have been developed and applied to optical characterization of semiconductor materials and devices. This paper will discuss the instrumentation of our NSOMS system and highlight its applications in three areas: direct near-field imaging, near-field magneto-optical imaging, and near-field spectroscopy and imaging.

Key words: NSOMS, magneto-optical, photoluminescence, electroluminescence, Raman.

INTRODUCTION

The device features in integrated circuits have been scaled down to the deep sub-micron region and optical characterizations of such small dimensions have been inevitably obstructed by the diffraction limit of far-field optics. The Near-field Scanning Optical Microscopy and Spectroscopy (NSOMS) has been developed to break the far-field diffraction limit and so permits the analysis of sub-micron areas (1). For example near-field techniques can easily achieve a resolution of 0.05 μm whereas conventional far-field systems are diffraction limited to ~0.8 μm at ~ 488 nm. NSOM has been widely applied to surface chemistry and biological molecular imaging (2). More recently there has been a surge of effort in using NSOM in solid state research and applications, such as near-field spectroscopy and imaging of quantum electronic structures (3,4), nanofabrication (5) and magneto-optical data storage [6]. However, there is much less effort in applying NSOM to practical microelectronic materials and devices (7) and turning this technique into powerful characterization tool in semiconductor industries.

In this work, we have explored applications of near-field optical microscopy and spectroscopy to a variety of microelectronic materials and devices. This paper will first present direct NSOM optical images of device structures and dislocation patterns of a partially relaxed SiGe film on Si substrate to demonstrate the NSOM spatial resolution and compare it with that of AFM. Then examples will be given in areas where the optical contrast mechanisms allow information to be gained that current electron and scanning probe microscopy can not provide. These areas include magneto-optical imaging using the Faraday and Kerr effects, fluorescence imaging, photo- and electro-luminescence spectroscopy and imaging, and near-field Raman spectroscopy.

INSTRUMENTATION

A commercially available Aurora near-field scanning optical microscope (TopoMetrix, Santa Clara, CA) is used for the direct optical imaging work. A tripod piezoelectric XYZ stage scans the sample relative to the tip while maintaining a constant distance from the tip to the sample surface. This distance is regulated by the optical shear force feedback system where a diode laser beam (670 nm) is focused onto the tip that is dithered at the resonance frequency by a piezoelectric tube and then reflected off sample surface to a detector. Laser beam from an Ar+ laser or a He-Ne laser is coupled into the NSOM tip. The reflected or transmitted light is collected with conventional far-field microscope objectives and detected with a photomultiplier tube (PMT). An interference bandpass filter at the exciting laser wavelength is inserted in the collecting optical path to filter out the feedback laser light and to ensure that only the reflected light from the exciting laser is detected.

For the polarization contrast imaging such as magneto-optical imaging, half wave and quarter wave plates are placed between the laser and the coupler and adjusted to obtain nearly linear polarization of the light out of the tip. A polarizer is inserted in the collecting optical path to probe different polarization states.

The selection of the spectrometer and detector system used for near-field spectroscopy was made with a view of optimizing the light throughput. A Kaiser Holospec 1.8i transmissional grating spectrometer was selected as it was significantly more efficient than a comparable reflection grating spectrometer (80% versus 10-20%). It has the disadvantage that the grating is optimized for a particular wavelength and dispersion. These disadvantages were mitigated by purchasing three gratings to allow high resolution Raman, low resolution Raman and a PL grating. A Princeton Instruments liquid nitrogen cooled, back-thinned CCD was selected to ensure exceptional sensitivity (80%). The CCD chip size was 1 inch wide consisting of 1752 pixels (pixel width of 15 μm) to allow enough data points being obtained for a narrow Raman

CP449, *Characterization and Metrology for ULSI Technology: 1998 International Conference*
edited by D. G. Seiler, A. C. Diebold, W. M. Bullis, T. J. Shaffner, R. McDonald, and E. J. Walters

peak. The spectrometer was calibrated using the emission lines from a neon lamp.

In case of photoluminescence and electroluminescence imaging, notch filters are used to block the exciting as well as feedback laser lines.

DIRECT NEAR-FIELD MICROSCOPY

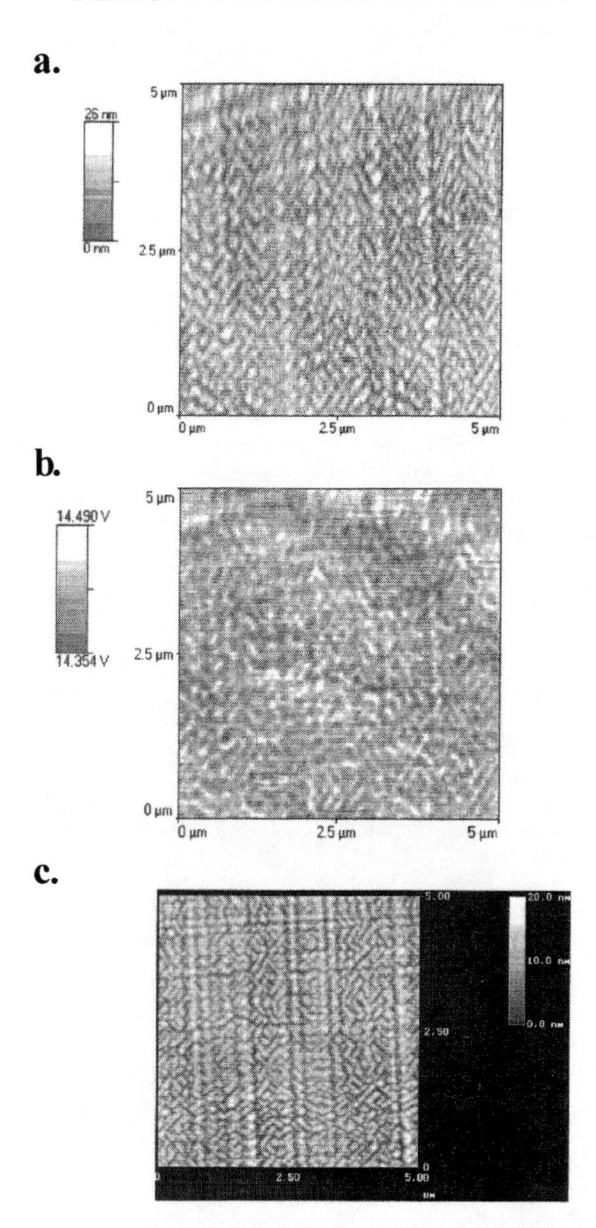

Figure 1. NSOM topographic (a), optical (b) images and AFM image (c) of dislocation lines in a relaxed $Si_{0.7}Ge_{.0.3}$ film on silicon.

The ability of NSOM to image semiconductor surfaces has been evaluated using a variety of semiconductor samples. The quality of the images that can be obtained using a NSOMS tip with 50 nm aperture is illustrated by the topographic and optical NSOMS images from a 140 nm thick $Si_{0.7}Ge_{0.3}$ film on silicon that had relaxed via the formation of misfit dislocations (Figure 1). Evidence of vertical dislocation lines along [110] axis is clearly visible in both images with island formation along [100] axis as well. This clearly demonstrates that NSOMS can resolve features down to 0.1 µm. The AFM image shown in Figure 2c, shows the same information as the topographic image, but with significantly improved resolution.

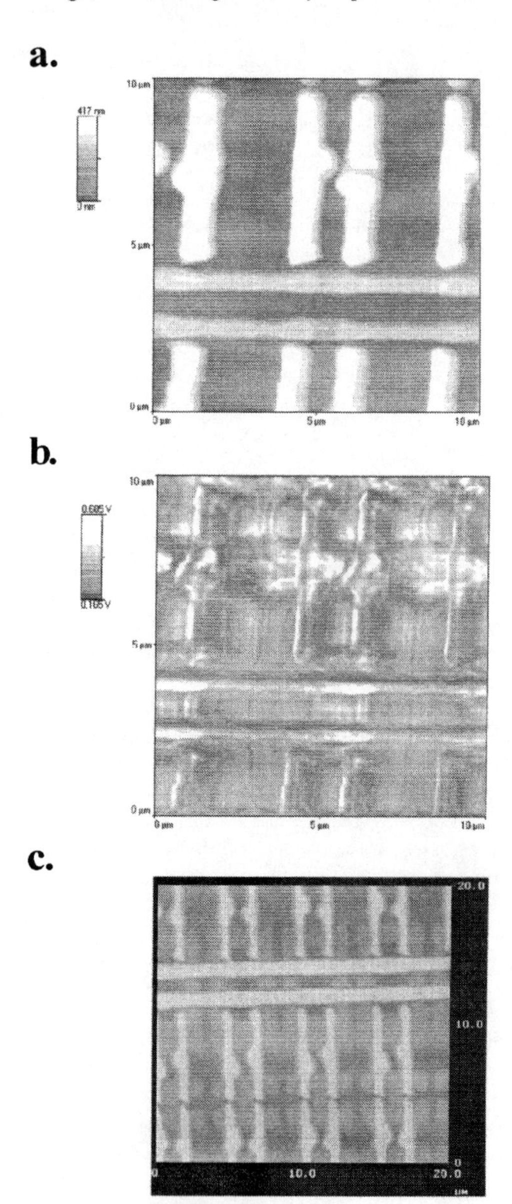

Figure 2. NSOM topographic (a), optical (b) images and AFM image (c) of a device structure.

The topographic and optical images of a device structure are presented in Figures 2a and 2b and AFM image is also shown in Figure 1c from this sample for comparison. The dominant topographic changes are caused by the doped polycrystalline Si world lines. It can be seen again that the AFM image exhibits superior resolution in surface topography. The optical image is noticeably contains more information but less well resolved than the topographic image due to complexity of the topographic shadowing effect and the interference effect of the 500 nm thick field silicon oxide.

In summary, in its current form, NSOMS for the simple optical imaging tasks works well and does yield information on the scale better than 0.1 μm. However, for general measurements of surface roughness, high resolution images of semiconductor devices and materials, AFM and SEM are the better choices. Near-field optical microscopy excels only when the optical contrast mechanisms can be exploited.

MAGNETO-OPTICAL IMAGING

The magneto-optical part of this work involves exploiting the Faraday (transmission) and Kerr (reflection) effects. When polarized light is transmitted or reflected from a magnetic material, the magnetic field changes the polarization direction of the light. Thus, by placing a polarizer in the transmitted or reflected beam path, the light reflected or transmitted from different domains can be contrasted.

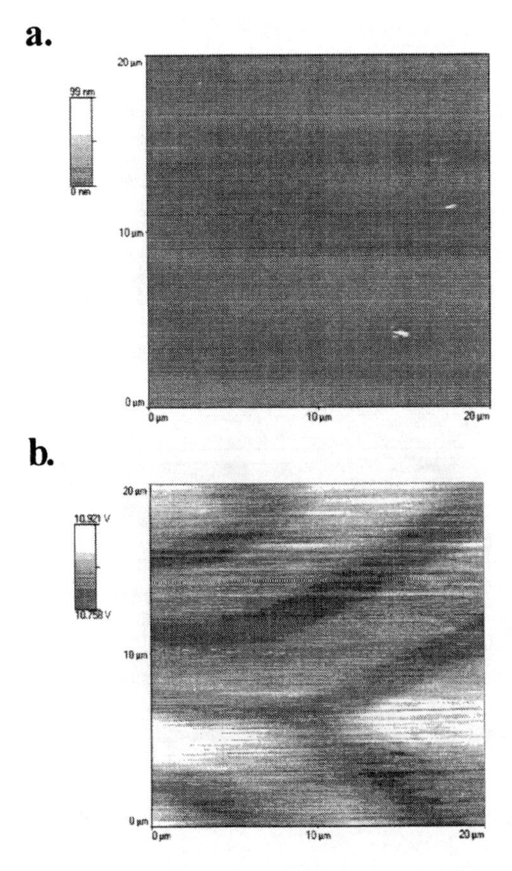

Figure 3. Topographic (a) and optical (b) images from a 2 μm yttrium-iron-garnet film on gadolinium-gallium-garnet substrate showing magnetic domains in the optical image.

Figure 3a shows the NSOMS topology image from a 2 μm thick film of yttrium-iron-garnet on gadolinium-gallium-garnet and Figure 3b shows the optical image acquired with polarization contrast. Significant contrast is detectable in the optical image that is not visible in the topographic image. The contrast is due to the light undergoing different amounts of Faraday rotation by the different magnetic domains in the magnetic layer.

Obtaining magnetic images in reflection has proved more difficult. Figures 4a and 4b present the topographic and optical images obtained from a patterned magnetic RAM (MRAM) structure consisting of sandwich NiFeCo, CoFe and Cu multilayered structure on a Si_3N_4 buffer layer. The topographic image just shows the patterned MRAM bitcell whereas significant contrast is detectable in the optical image. The light and dark shading are partly caused by the different magnetic domains rotating the polarization direction of the incident light to different degree (Kerr effect). In a cell of these dimensions, only one or two domains would normally be expected to be present. In the image, several areas of contrast can be seen, suggesting that some additional optical contrast is being detected which is confusing the image.

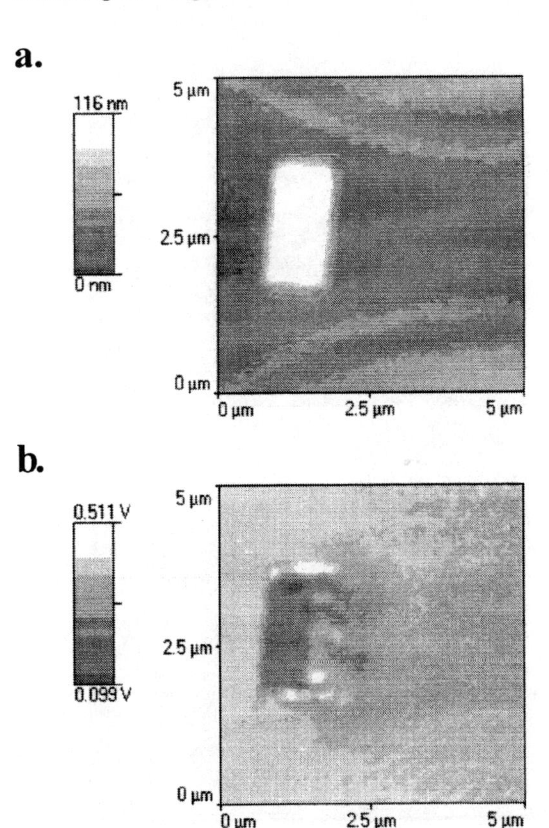

Figure 4. Topographic (a) and optical (b) images of an MRAM bitcell.

In order to isolate the magnetic effect, we have explored different modulation techniques. A modulation technique we successfully used so far involves adding a small coil to the commercial NSOM stage to provide an ac magnetic field perpendicular to the sample surface. By locking in the optical signal to this ac field, we succeeded in obtaining images of modulated intensity due to pure magnetic effect. Figures 5a and 5b are the topography and modulated magneto-optical images from an identical bitcell. Again the topographic image shows only the topography of the bitcell, but the optical image has huge contrast around the edge of the bitcell. The reason that there is little contrast in the middle of the cell is that the magnetic moments that are uniformly aligned to the long axis of the cell and it is hard to flip these moments into the vertical direction. However, near the edge, the magnetic moments are tilted and give rise to some vertical components that respond to the vertical ac field. This suggests that the edge effect might be a crucial factor affecting the switching behavior of small MRAM devices. Currently, we are also attempting other modulation configurations such as using horizontal field or combinations of horizontal and vertical fields. We also planned to investigate the dynamical switching behavior of actual MRAM circuits. In this case, unlike magnetic field microscopy (MFM) and scanning electron microscopy with polarization analysis (SEMPA), the contrast effects are not perturbed by electrically generated magnetic fields.

a.

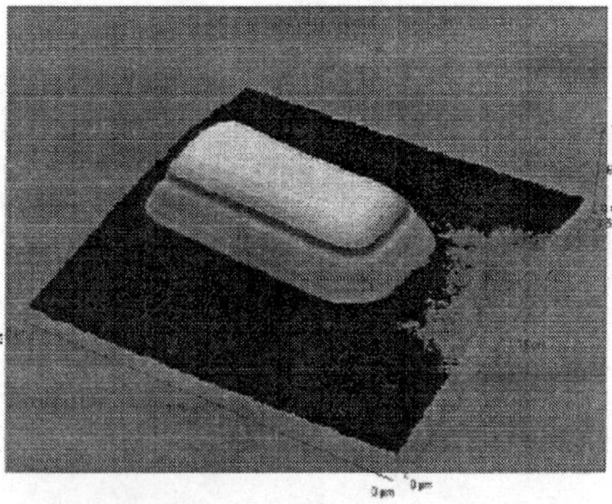

b.

Figure 5. Topographic (a) and modulated optical (b) images of a MRAM bitcell.

SPECTROSCOPY AND IMAGING

The other area where near-field optical techniques provide a major advantage is optical spectroscopy. Near-field spectroscopic techniques such as photoluminescence (PL) and Raman offer the ability to probe sub-micron areas and so provide information on the composition, crystallinity and stress. The difficulty with near-field spectroscopy techniques is the low light levels involved. Typically, for a 300-500Å tip aperture, there is a $\sim 10^6$ reduction in the output laser power as compared to the input power. While this is adequate for direct microscopy, spectroscopy becomes more difficult.

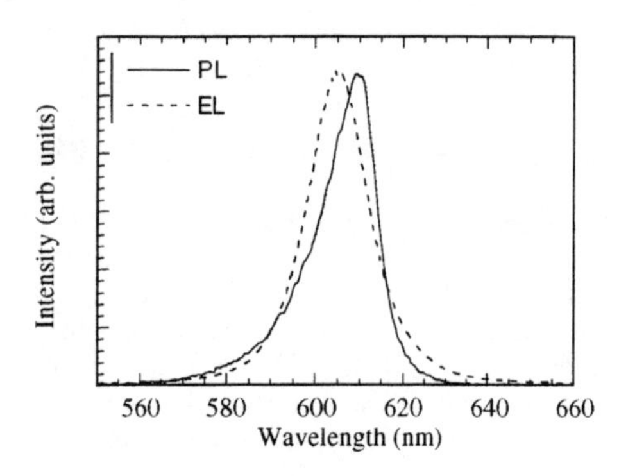

Figure 6. Near field photoluminescence spectrum (a) and EL spectrum (b) of InGaAlP/GaAs heterostructure LED.

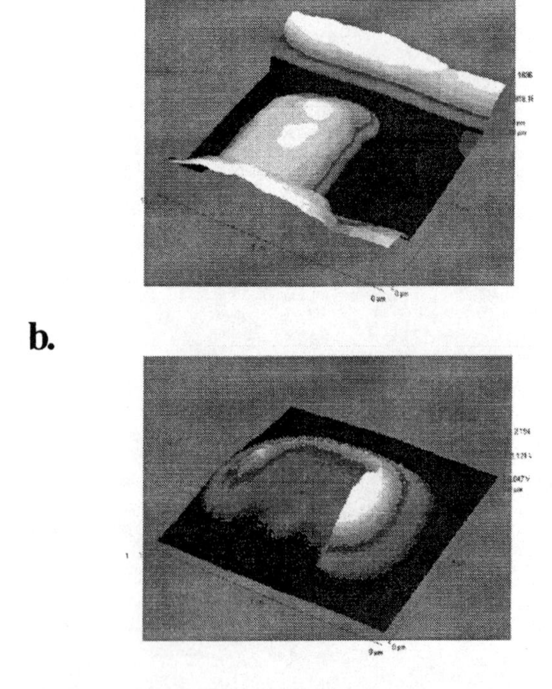

Figure 7. Topographic (a) and optical emission (b) images of a InGaP/InGaAlP LED pixel.

The photoluminescence (PL) work has concentrated on the visible region. Figure 6 shows a near-field PL spectrum from an MOCVD grown InGaP/InGaAlP heterostructure. For comparison, EL spectrum from an electrically powered pixel is also plotted. It can be seen that the EL peak is more symmetric than the PL spectrum. Using the tip collection mode, the emitted light was mapped. Figures 7a and 7b display the topography and emission images of the pixel. The EL emission is concentrated around the metal finger and much of the light is blocked by the metal contact.

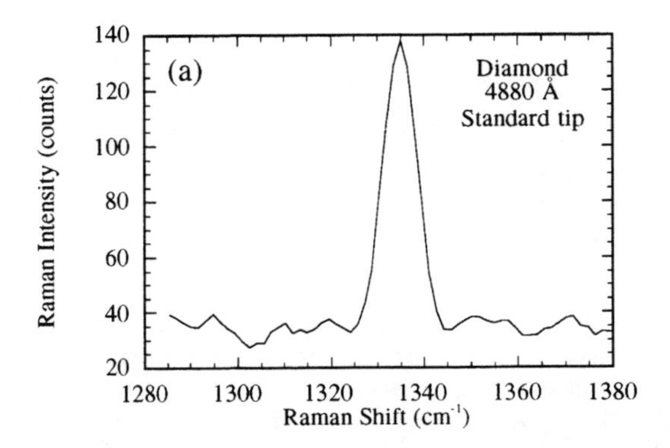

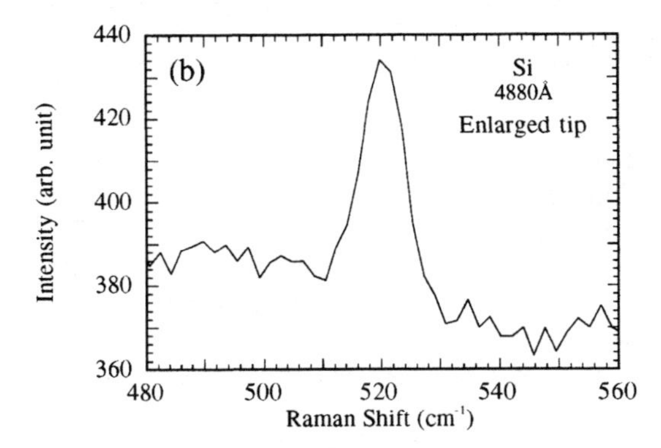

Figure 8. Near-field Raman spectra from diamond (a) and Si (b).

Raman spectroscopy using NSOM is more challenging due to the even lower signal levels. However, there is good reason to pursue it because Raman scattering can provide very rich information about materials properties such as stress and damage in Si-based devices. So far, little progress in the NSOM Raman spectroscopy has been made, especially on semiconductor materials. Most of the near-field spectra have been taken from transparent samples such as diamond (7) and $KTiOPO_4$ (8). In this case the near field condition is broken by the large optical penetration depth. Recently near-field Raman spectrum was obtained from Si LO phonon by Smith and co-workers (9) using He-Ne laser line at 633 nm. The penetration depth of light at this wavelength is as large as 2.4 µm, which relaxes the near field condition and gives rise to more Raman signal. However, near-field Raman spectroscopy using shorter wavelength laser lines is more desirable since the thickness of electrically active layers in most Si devices are less than 0.5 µm. This work is concentrated on obtaining near-field Raman spectra using the blue line (488 nm) from the Ar+ laser. Figure 8a plots the near-field Raman spectrum from diamond obtained using a high-resolution tip (500 Å). This spectrum is in sharp contrast to that published in (7), which showed an extremely broad near-field Raman peak. Our result demonstrates that it is possible to obtain as sharp LO phonon Raman peak in near-field as in far-field case. Raman signal from the LO phonon in Si is about an order of magnitude weaker than the one in diamond due to much smaller optical penetration depth. The Si Raman peak has also been obtained using the 488nm Ar+ laser line (Figure 8b), although a lower resolution (> 0.2 µm) tip had to be used. However, it is not yet practicable to perform Raman stress and damage mapping due to the very long accumulation time required for the Si Raman spectrum.

CONCLUSION

In conclusion, near-field optical techniques have been explored to allow the optical sub-micron analysis of semiconductor materials and devices. Near-field optical microscopy, magneto-optical and photoluminescence are currently available and the methodologies continue to be improved. The results to date using the NSOMS system appear extremely promising, particularly in the fields of magneto-optical imaging and spectroscopy. In the MO field, magnetic domains have been detected in both transmission and reflection. Magneto-modulation methods using magnetic fields have been developed to screen out some of the interfering optical contrast and so yield images containing just the magnetic domain information. The feasibility of the spectroscopy work has been proven by the detection of visible PL and EL spectra from InGaP/InGaAlP LED devices. In addition, the emission characteristics from LED pixels have also been imaged. For the first time the near-field Si Raman phonon spectra have been obtained using a blue laser line (488nm). Work is continuing to obtain the Raman spectrum of silicon with higher resolution tip.

REFERENCES

1. E. Betzig and J. K. Trautman, Science **257**, 189 (1992).
2. M. A. Paesler and P. J. Moyer, *Near-field Optics-Theory, Instrumentation and Applications*, John Wiley & Sons, (1996).
3. R. D. Grober, T. D. Harris, J. K. Trautman, E. Betzig, W. Wegscheider, L. N. Pfeiffer, and K. W. West, Appl. Phys. Lett. **64**, 1421 (1994).
4. H. F. Hess, E. Betzig, T. D. Harris, L. N. Pfeiffer, and K. W. West, Science **264**, 1740 (1994).
5. E. Betzig , J. K. Trautman, R. Wolfe, E. M. Gyorgy, and P. L. Finn, Appl. Phys. Lett. **61**, 142 (1993).
6. E. Betzig, J. K. Trautman, R. Wolfe, E. M. Gyorgy, P. L. Finn, M. H. Kryder, and C. H. Chang, Appl. Phys. Lett. **61**, 142 (1992).
7. W. M. Duncan, J. Vac. Sci. Technol. **A 14**, 1914 (1995).
8. C. L. Jahncke, M. A. Paesler, and H. D. Hallen, Appl. Phys. Lett. **67**, 2483 (1995).
9. D. Smith, S. Webster, M. Ayad, S. Evans, D. Fogherty, and D. Batcheler, Ultramicroscopy, May 1996.

A Microcontamination Model For Rotating Disk Chemical Vapor Deposition Reactors

R. W. Davis, E. F. Moore, D. R. Burgess, Jr. and M. R. Zachariah

Chemical Science and Technology Laboratory
National Institute of Standards and Technology
Gaithersburg, MD 20899

This paper presents preliminary results from a model currently under development for gas-phase generated submicron-size contaminant particles (i.e., microcontaminants) in rotating disk chemical vapor deposition reactors. These particles present a problem during semiconductor processing, and this model is intended as a useful tool for gaining a better understanding of this problem. A one-dimensional formulation is employed to model the central section of the reactor, a technique which allows the use of detailed chemical reaction sets. The existing Sandia SPIN code, which contains a solver for the reacting flow, is modified by the addition of an aerosol model for the particles. This model utilizes a moment transport formulation which accounts for convection, diffusion, gravity, thermophoresis, chemical production, coagulation and condensation. Results are presented primarily in terms of reactor performance maps which indicate film growth and contamination rates as functions of substrate temperature. The effects of variations in reactor operating parameters on these maps are discussed.

INTRODUCTION

Particle contamination is a major problem afflicting semiconductor processing. The bulk of this problem is now due to gas-phase generated particles as opposed to those particles which may enter with the process stream or flake off equipment surfaces. These gas-phase generated particles are generally small (i.e., submicron) compared with the other types of contaminants and can thus be referred to as "microcontaminants." It is the purpose of the present paper to describe a numerical model currently under development for the formation, growth and transport of microcontaminants in the commonly-used rotating disk chemical vapor deposition (CVD) reactor.

Rotating disk CVD reactors have been the subject of numerous modeling studies involving both nonreacting and reacting flows (1-4), as well as large (> 1μm) contaminant particle transport (5). The interest in this type of reactor stems from several factors. First, the flow is simple and well-characterized when care is taken in the selection of the operating regime. The simple flow patterns lead, in turn, to highly uniform film deposition. Finally, a modified form of the von Karman similarity transformation (1) can be employed to reduce the entire problem to a set of equations solely dependent on the axial coordinate over a large central portion of the reactor. This allows the use of much more detailed chemical reaction sets than would be possible in multidimensions. Using these ideas, a one-dimensional model for the flow and chemistry in the rotating disk reactor has been developed by Sandia National Laboratories and embedded in the SPIN code (6). This code computes velocity, temperature and species distributions, as well as surface deposition rates, in this reactor. The effort to be described here involves the addition of an aerosol model to the SPIN code in order to compute the formation, growth and transport of microcontaminants.

The study of microcontaminants in CVD reactors has been limited due to the great increase in computing resources required by the addition of aerosol dynamics to an already complicated multidimensional chemically reacting flow (7,8). Thus, the one-dimensional nature of the rotating disk formulation makes it an ideal candidate for the inclusion of an aerosol model. The model employed here is similar to that of Whitby and Hoshino (8) which utilized a moment transport formulation in conjunction with a lognormal size distribution function in a two-dimensional horizontal CVD reactor. The lesser dimensionality of the present model, however, allows the use of a much more detailed chemical reaction set.

The results to be presented here are preliminary in nature and serve to illustrate the type of information that will become available from this model during subsequent stages of development, particularly following planned validation experiments. These results will be presented primarily in terms of reactor performance maps which indicate film growth rate and contamination rate as functions of disk temperature. Reactor operating regimes that produce adequate growth rates within acceptable contamination levels can then be identified. The effects of parameter variations on these maps are assessed as an aid to the optimization of

CP449, *Characterization and Metrology for ULSI Technology: 1998 International Conference*
edited by D. G. Seiler, A. C. Diebold, W. M. Bullis, T. J. Shaffner, R. McDonald, and E. J. Walters

reactor operating conditions.

FLOW/CHEMISTRY SOLUTION

The reactor configuration under consideration here is an infinite radius rotating disk with axial coordinate z. The flow/chemistry solution algorithm is embedded in the SPIN code (6). Transport equations for radial and circumferential momentum, thermal energy and species are of the form

$$\rho\, u\, d\Phi/dz = d/dz\,(\Gamma\, d\Phi/dz) + S_\Phi\,, \quad (1)$$

where Φ is a general dependent variable, ρ is density, u is axial velocity, Γ is a diffusion coefficient and S_Φ is a source term. State and continuity equations complete the flowfield description. The source terms in the species transport equations are the chemical production rates via gas-phase reactions. These are derived from the law of mass action driven by a user-supplied reaction mechanism. The net flux of chemical species into the substrate results in a film growth rate. Gas-phase and surface chemical kinetics are handled by CHEMKIN (9) and SURFACE CHEMKIN (10), respectively, while the variable transport properties are determined from TRANSPORT (11). While boundary conditions are specified at the disk, there is no specification (at least for the results to be presented here) of the inflow velocity to the reactor. This velocity is allowed to assume a value determined from the natural suction induced by the spinning disk. This ensures a well-behaved flow with very uniform deposition.

AEROSOL MODEL

The aerosol model employed here is a moment transport formulation that accounts for particle formation, growth via condensation and coagulation, and transport via convection, diffusion, gravity and thermophoresis (8,12,13). A lognormal size distribution function is utilized because of its wide applicability in many physical systems (14) and its ease of mathematical manipulation (8,12). The moment transport equation for the present one-dimensional configuration is

$$\rho\,(u + c_k)\, d M_k / dz = d/dz\,(\Gamma_k\, d M_k / dz) + S_k\,, \quad (2)$$

where c_k is the transport velocity of M_k due to gravity and thermophoresis, Γ_k is the diffusion coefficient of M_k, and S_k is the moment source due to chemical production, coagulation and condensation. The kth moment, M_k, is defined as (14)

$$M_k = 1/\rho \int_0^\infty n(d_p) d_p^{\,k} d(d_p)\,, \quad (3)$$

where d_p is particle diameter. A lognormal number concentration frequency distribution function, n, is employed whereby (12)

$$n = N\,[\,(2\pi)^{1/2} \ln \sigma_g\,]^{-1} \exp\{\,-0.5 \ln^2 (d_p/D_{gn}) / \ln^2 \sigma_g\,\}. \quad (4)$$

Here N is the total number concentration of particles, σ_g is the geometric standard deviation and D_{gn} is the geometric mean size. Equation (2) is solved numerically for k = 0,3,6 which correspond to moments proportional to total number concentration (i.e., N), volume fraction and optical scattering power, respectively (15). All other moments can be determined from these three (8).

The modeling of each term in Eq. (2) is sufficiently complex that it will only be outlined here [see (12) for details]. The moment transport velocity due to gravity is derived by assuming that the gravitational force on a particle is

$$F_g = (\pi\, \rho_p / 6)\, d_p^{\,3}\, g\,, \quad (5)$$

where ρ_p is particle density and g is gravitational acceleration. The thermophoretic moment transport velocity in the free-molecular limit is based on the following force:

$$F_{th} = -P\, \lambda\, d_p^{\,2}\, \nabla T / T\,, \quad (6)$$

where P is the reactor pressure, λ is the mean free path of the gas molecules and T is temperature. The moment diffusion coefficient, Γ_k , is derived from the following particle diffusion coefficient:

$$D_p = k_B\, T\, (1 + 1.43\, Kn^{1.049}) / (3\pi\mu\, d_p)\,, \quad (7)$$

where k_B is the Boltzmann constant, Kn is the Knudsen number ($2\lambda / d_p$) and μ is gas viscosity.

The source term, S_k, in Eq. (2) presents the most difficult challenge to model. It consists of three parts:

$$S_k = S_{chem} + S_{coag} + S_{cond}\,, \quad (8)$$

where the three terms on the RHS represent chemical source, coagulation, and condensation, respectively. The chemical source term, S_{chem}, is the gas-phase formation rate of particle precursors and is an output from the chemical reaction model. The precursors can consist of monomers (single molecules) or clusters of monomers (i-mers) depending on the particular chemical reaction mechanism employed. Once formed, particles grow by means of particle-particle collisions (coagulation) as well as molecular condensation on particle surfaces. The determination of S_{coag} involves the evaluation of complex collision integrals [(8,12,13)] which, for a lognormal distribution, have been approximated by interpolation formulas (16,17). This leads to the following coagulation source terms:

$$M_0:\ S_{coag} = -\,4.90\, \rho^2 (k_B T/\rho_p)^{1/2}\, M_0^{31/16}\, M_6^{5/48} / M_3^{1/24}, \quad (9)$$

$$M_3: \quad S_{coag} = 0\,, \tag{10}$$

$$M_6: \quad S_{coag} = 2.69\, \rho^2 (k_B T/\rho_p)^{1/2}\; M_0^{5/48}\, M_3^{13/8}\, M_6^{13/48}. \tag{11}$$

Note that the volume fraction (~ M_3) does not change as a result of coagulation since particle volumes are conserved during this process.

The condensation source term, S_{cond}, is derived from the following equation for the volume growth rate of a particle (15):

$$dv_p/dt = \alpha\, P\, \pi\, d_p^2\, v_m / (2\,\pi\, m\, k_B\, T)^{1/2}\,, \tag{12}$$

where v_p is particle volume, t is time, α and m are sticking coefficient and mass of condensible species molecules (e.g., Si_2H_2) and v_m is molecular volume of the condensed phase (e.g., Si). Note that the condensation source term is of a different nature than the other terms in Eq. (2). This is because evaluation of this term, unlike the others, involves two-way coupling between the chemistry and the aerosol model. Without condensation, the flow/chemistry solution is not dependent on the aerosol solution, since the aerosols in question here are quite dilute. Thus, a single pass through the solution procedure is all that is necessary to obtain the aerosol solution once the flow/chemistry has been determined. When condensation is present, however, the particles act as a sink for gas-phase condensible species and thus affect the chemistry. An iterative procedure, which becomes more difficult as S_{cond} increases, is then required to obtain both the flow/chemistry and aerosol solutions. The importance of condensation depends on the value assigned to the sticking coefficient, α, in Eq. (12), where $0 \leq \alpha \leq 1$. Since a precise value for this parameter is highly system dependent, as well as a function of particle size and composition, it will be assumed for the purposes of this study that $\alpha = 0.1$. This is large enough to include the effects of condensation in the results without making the computations unduly burdensome.

The solution of the system of equations represented by Eq. (2) is accomplished by means of the LSODE package for the solution of ordinary differential equations (18). The iteration involving this solution coupled with the flow/chemistry solver is carried out on high performance workstations. The boundary conditions are that all moments are zero both at the disk surface and far above it. Particles impact, and thus contaminate, the substrate via a diffusive flux. It is the number and size distribution of this flux which is of primary concern in this modeling effort. In particular, it is of interest to determine the rate at which "killer" particles ($d_p \geq 0.12\mu m$) (19) impact the surface. Reactor operating conditions that maintain this large particle flux within acceptable limits are obviously important to delineate.

CHEMICAL REACTION MECHANISM

The gas-phase chemical reaction mechanism employed in this study consists of reactions involving silicon molecular chemistry that lead to silicon cluster formation (particle nucleation) and subsequent growth. The silicon molecular chemistry is adapted from the reaction set presented in Coltrin et al. (1). A number of reactions have been added to this set in order to adequately describe the chemistry over a range of temperatures and concentrations outside of those considered by these past researchers.

The silicon molecular chemistry can be divided into three general regimes. The first involves the decomposition of silane and the establishment of near steady-state concentrations of silane species, including SiH_4 and Si_2H_6. The second consists of reactions involving silylene species (e.g., SiH_2 and SiH_3SiH) and hydrogen elimination to form small silicon molecules. Finally, the third general regime consists of reactions involving these small silicon molecules and/or molecular clusters, including Si_2H_2, Si_2, Si and Si_3.

The reaction set describing silicon cluster formation essentially consists of enumerating all the possible combinations. Small clusters react to form larger and larger clusters which eventually become particle nucleation sources. Silicon hydride, Si_2H_2, is used as the condensible species.

RESULTS AND DISCUSSION

The baseline configuration employed for the initial tests of the microcontamination model consists of an atmospheric pressure reactor with a disk rotation rate of 1000 rpm and argon diluent containing 0.1% mole fraction of silane. The inlet flow velocity of approximately 3.35 cm/s is naturally induced by the disk rotation, and all inlet boundary conditions are applied at z = 2 cm. The effects of variations from the baseline conditions are presented by means of reactor performance maps over a disk temperature range of 1000 K to 1400 K, which is typical for silicon epitaxy (3).

Figure 1 depicts the total number concentration (M_0) and average particle size (M_1/M_0) as functions of height above the disk for the baseline case at disk temperatures of 1100 K and 1400 K. Both of these quantities are seen to reach their maximum values inside the thermal boundary layer just above the disk. While the average particle size is not a strong function of temperature, the total number concentration increases with temperature. As will be shown, this is reflected in greater substrate contamination levels at higher temperatures. The decrease in average particle size approaching the substrate is due to the fact that diffusion is the primary mechanism of particle transport toward the surface, and the diffusion coefficient, Eq. (7), varies inversely with d_p or d_p^2 depending on size (14). This inverse dependence of the diffusion coefficient on particle size is fortuitous in that it acts to reduce substrate contamination by large particles. Total number concentrations, M_0, near the surface are reduced due to the thermophoretic component of c_k which is independent of particle size (12,14).

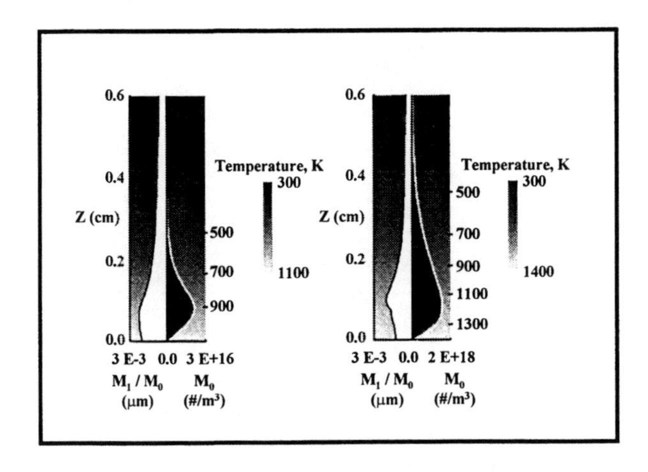

FIGURE 1. Total number concentration (M_0) and average particle size (M_1/M_0) with a thermal contour background for the baseline case with disk temperatures of 1100 K and 1400 K.

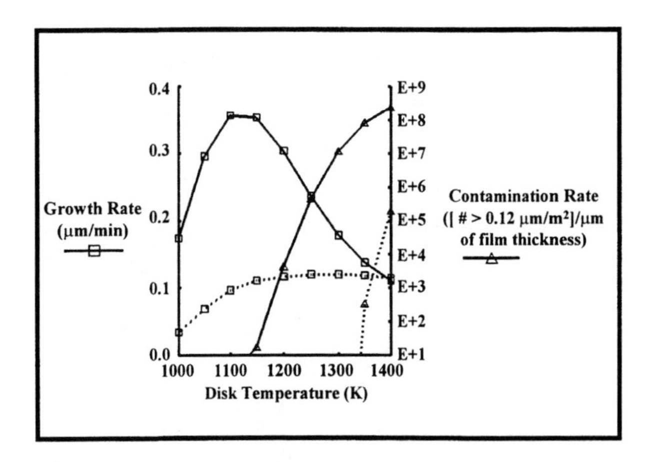

FIGURE 2. Reactor performance map for the baseline case and effect of pressure reduction (—— 1.0 atm; ------ 0.1 atm).

As noted previously, reactor performance maps are the primary output from this model. The performance map for the baseline case is presented in Fig. 2 as the solid plots. These two plots represent the temperature dependence of the surface film growth rate in μm / minute and the large "killer" particle surface contamination rate per m^2 per time required to deposit one μm of film thickness. This adjustment of the contamination rate to reflect variations in film growth rate provides a means of assessing the severity of the contamination problem independent of how much fabrication time is necessary. Values of the contamination rate above the horizontal axis (i.e., >10) would probably be considered unacceptable (19). However, it is noted that trends here are more reliable than absolute values since these results can be highly sensitive to modeling assumptions (such as sticking coefficient) that have yet to be experimentally validated. It can be seen from Fig. 2 that the surface growth rate peaks at about 1100 K and then decreases with temperature as the contamination rate rapidly rises. This drop in film growth rate is due to the presence of increasing numbers of gas-phase contaminant particles which act as an added surface for silicon deposition. This increase in gas-phase particles with temperature (along with a broad size distribution) also leads to the increasing surface contamination rate. Thus, it can be seen from the solid plots in Fig. 2 that the optimum disk temperature for the baseline case is approximately 1100 K.

The effects of variations in the reactor operating conditions from the baseline case will now be assessed. A set of plots (dashed lines) in Fig. 2 illustrates the difference between the baseline pressure of 1 atm and a reduced pressure of 0.1 atm. As can be seen, lowering the pressure results in both lower film growth rate and lower surface contamination, which are at least partially attributable to decreased reaction rates at the lower pressure. Therefore, if slower growth rates are acceptable, decreasing the reactor pressure is an effective method of contamination control. The effect of varying the disk rotation rate from 500 rpm to 1500 rpm is presented in Fig. 3. It is clear from this figure that increasing the rotation rate results in an increased film growth rate along with moderately decreased surface contamination. It is noted, however, that too large a disk rotation rate can result in undesirable recirculation zones adjacent to the walls in an actual reactor configuration (20). Finally, changes in performance due to increasing the inlet silane concentration from a mole fraction of 0.1% to 0.3% are depicted in Fig. 4. The growth rate, as expected, increases as does the contamination rate (albeit slightly) over most of the temperature regime. At high temperatures, however, increasing the silane concentration has almost no effect on both the growth and contamination rates. This is because the increased silane concentration is depleted at the high temperatures due to increased condensation on gas-phase particles. This reduces surface growth rate and results in larger particles which less readily diffuse to the substrate.

CONCLUDING REMARKS

Preliminary results have been presented from a microcontamination model for rotating disk CVD reactors which is currently under development. The model utilizes a one-dimensional formulation to model the cylindrical central section of the reactor. This reduction in dimensionality allows the use of detailed chemical reaction sets. Reactor performance maps have been presented that demonstrate decreased large-particle contamination rates with reduced pressure and increased disk rotation rate. It has also been noted that thermophoresis and a size-dependent particle diffusion coefficient act to ameliorate substrate contamination. Finally, because of the complexity of this model, the only method of quantitatively verifying the results is through comparisons with experimental data, an effort that is currently underway. An experimentally-validated microcontamination model should prove to be a powerful tool for both the design and operation of rotating disk reactors.

REFERENCES

1. Coltrin, M. E., Kee, R. J. and Evans, G. H., *J. Electrochem. Soc.* **136**, no. 3, 819-829 (1989).
2. Breiland, W. G. and Evans, G. H., *J. Electrochem. Soc.* **138**, no. 6, 1806-1816 (1991).
3. Jensen, K. F. and Einset, E. O., *Ann. Rev. Fluid Mech.* **23**, 197-232 (1991).
4. Ho, P., Coltrin, M. E. and Breiland, W. G., *J. Phys. Chem.* **98**, no. 40, 10138-10147 (1994).
5. Davis, R. W., Moore, E. F. and Zachariah, M. R., *J. Crystal Growth* **132**, 513-522 (1993).
6. Coltrin, M. E., Kee, R. J., Evans, G. H., Meeks, E., Rupley, F. M. and Grcar, J. F., "SPIN (Version 3.83): A FORTRAN Program for Modeling One-Dimensional Rotating - Disk/Stagnation-Flow Chemical Vapor Deposition Reactors," Sandia National Laboratories Report SAND91-8003, 1991.
7. Okuyama, K., Huang, D., Seinfeld, J. H., Tani, N. and Kousaka, Y., *Chem. Eng. Sci.* **46**, no. 7, 1545-1560 (1991).
8. Whitby, E. and Hoshino, M., *J. Electrochem. Soc.* **143**, no. 10, 3397-3404 (1996).
9. Kee, R. J., Rupley, F.M. and Miller, J. A., "CHEMKIN II: A FORTRAN Chemical Kinetics Package for the Analysis of Gas Phase Chemical Kinetics," Sandia National Laboratories Report SAND89-8009, 1989.
10. Coltrin, M. E., Kee, R. J. and Rupley, F. M., "SURFACE CHEMKIN (Version 4.0): A FORTRAN Package for Analyzing Heterogenous Chemical Kinetics at a Solid-Surface-Gas-Phase Interface," Sandia National Laboratories Report SAND90-8003B, 1991.
11. Kee, R. J., Dixon-Lewis, G., Warnatz, J., Coltrin, M. E. and Miller, J. A., "A Fortran Computer Code Package for the Evaluation of Gas-Phase Multicomponent Transport Properties," Sandia National Laboratories Report SAND86-8246, 1986.
12. Whitby, E. R., McMurry, P. H., Shankar, U. and Binkowski, F. S., "Modal Aerosol Dynamics Modeling," U.S. Environmental Protection Agency Report for Contract No. 68-01-7365, 1991.
13. Whitby, E. R. and McMurry, P. H., *Aerosol Sci and Tech.* **27**, No. 6, 673-688 (1997).
14. Hinds, W. C., *Aerosol Technology*, New York: John Wiley, 1982.
15. Friedlander, S. K., *Smoke, Dust and Haze*, New York: John Wiley, 1977.
16. Dobbins, R. S. and Mulholland, G. W., *Comb. Sci. And Tech.* **40**, 175-191 (1984).
17. Zachariah, M. R. and Semerjian, H. G., *AIChE Journal* **35**, no. 12, 2003-2012 (1989).
18. Radhakrishnan, K. and Hindmarsh, A. C., "Description and Use of LSODE, the Livermore Solver for Ordinary Differential Equations," Lawrence Livermore National Laboratory Report UCRL-ID-113855 (NASA Reference Publication 1327), 1993.
19. "The National Technology Roadmap for Semiconductors," Semiconductor Industry Association, 1994.
20. Biber, C. R., Wang, C. A. and Motakef, S., *J. Crystal Growth* **123**, 545-554 (1992).

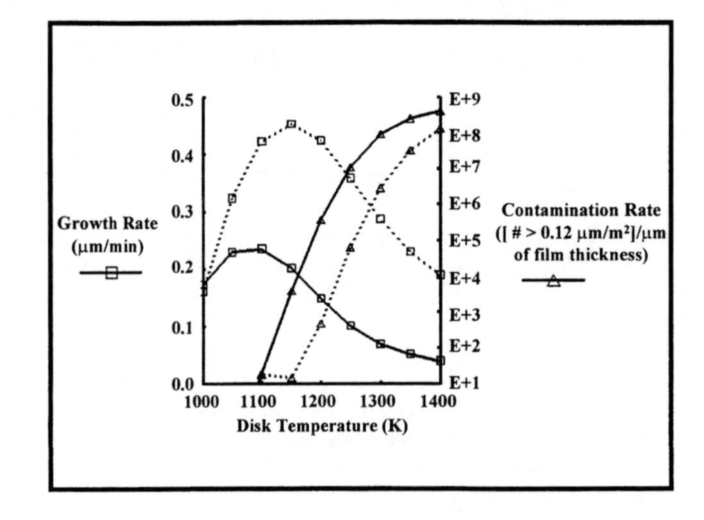

FIGURE 3. Effect of disk rotation rate on the reactor performance map (—— 500 rpm; ------ 1500 rpm).

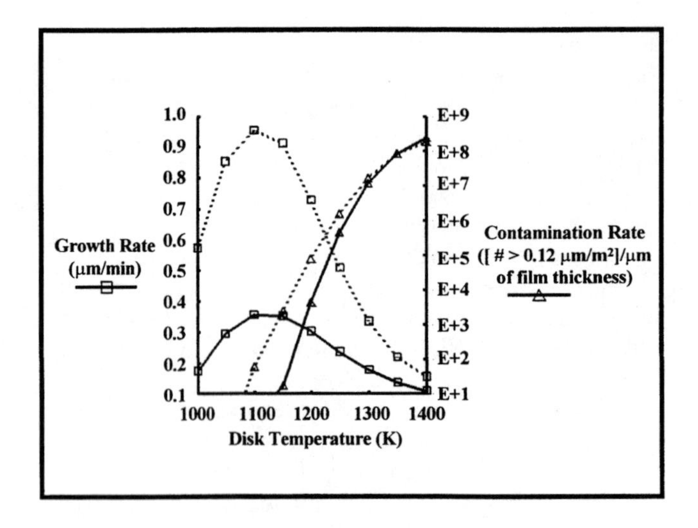

FIGURE 4. Effect of inlet silane mole fraction concentration on the reactor performance map (—— 0.1%; ------ 0.3%).

Properties of Process Gases Determined Accurately with Acoustic Techniques

John J. Hurly

Physical and Chemical Properties Division, Chemical Science and Technology Laboratory
National Institute of Standards and Technology, Gaithersburg, MD 20899

Abstract: Our laboratory has developed a highly-precise, automated apparatus for measuring sound speeds, viscosity, and thermal conductivity in hazardous and/or corrosive process gases throughout the temperature and pressure ranges of interest to the semiconductor processing community. The sound speeds are determined from resonances within a cylindrical cavity (or resonator) filled with the sample gas. The resonator is thermostated to within a few millikelvin between 200 K $\leq T \leq$ 480 K. The pressure range is 0.05 MPa $\leq P \leq$ 1.5 MPa, or 80% of the samples' vapor pressure, whichever is lower. The analysis of sound speeds provides the perfect gas heat-capacities, $C_p^o(T)$, and the non-idealities of these gases. The viscosity and thermal conductivity are determined in similar resonators with geometries optimized for measuring energy losses. These thermophysical properties are required by the semiconductor processing industry to model processes and to calibrate mass flow controllers (MFCs). The actual calibration of the MFCs is performed with nonhazardous surrogate gases, which we are also characterizing. We have obtained extensive sound speed data for the surrogate gases SF_6, CF_4, C_2F_6 and the semiconductor gas BCl_3. These results and their analysis are presented.

INTRODUCTION

The thermophysical properties of the hazardous gases used by the semiconductor processing industry are needed for process modeling and for the accurate calibration of mass flow controllers (MFCs). The actual calibration of the MFCs is performed with nonhazardous, surrogate gases, whose thermophysical properties are also required.

Our laboratory has developed a highly-precise, automated apparatus for measuring sound speeds, viscosity, and thermal conductivity in hazardous and/or corrosive process gases throughout the temperature and pressure ranges of interest for semiconductor processing (200 K $\leq T \leq$ 480 K and $P \leq$ 1.5 MPa). The analysis of measured sound speeds provides the perfect gas heat capacities, $C_p^o(T)$, and the non-idealities of these gases. We have obtained extensive data for the surrogate gases SF_6, CF_4, and C_2F_6, and have completed low pressure measurements in BCl_3. The viscosity and thermal conductivity are measured in similar resonators with geometries optimized for observing energy losses. The capabilities of the apparatuses for measuring viscosity and thermal conductivity have been demonstrated (1), and systems to perform these measurements in hazardous semiconductor process gases are being constructed.

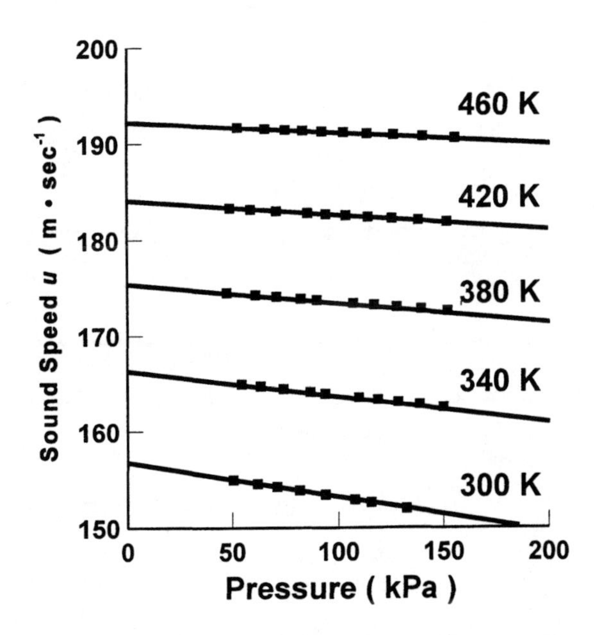

Figure 1 Measured sound speeds, $u(T,P)$, in boron trichloride along five isotherms. Isotherms at 320 K, 360 K, 400 K, and 440 K are not shown.

EXPERIMENTAL TECHNIQUE

The apparatus employed here is based on a similar device used for the successful study of more than 20 non-hazardous gases and gas mixtures (2,3). The precursor system and the acoustic model have been described in detail elsewhere (4,5). For the current work, the successful design of the previous apparatus has been modified to handle the hazardous, reactive and corrosive gases employed in the semiconductor processing industry. These modifications address the reactive nature of the subject gases, and their safe handling and disposal.

The entire apparatus is automated. All valving is air-operated from a remote process control panel and/or computer. The entire system is enclosed in a walk-in gas cabinet which conforms to Article 51 of the Uniform Fire Code for Semiconductor Fabrication Facilities. The cabinet encloses the resonator, gas manifold, pressure controller,

CP449, *Characterization and Metrology for ULSI Technology: 1998 International Conference*
edited by D. G. Seiler, A. C. Diebold, W. M. Bullis, T. J. Shaffner, R. McDonald, and E. J. Walters

Table 1. Summary of experimental T and P ranges

	T Range (K)	P Range (MPa)	No. Isotherms	No. Points
CF_4	$300 \leq T \leq 475$	$0.1 \leq P \leq 1.5$	8	114
C_2F_6	$175 \leq T \leq 475$	$0.1 \leq P \leq 1.5$	14	181
SF_6	$230 \leq T \leq 460$	$0.1 \leq P \leq 1.5$	15	280
BCl_3	$300 \leq T \leq 460^{\dagger}$	$0.05 \leq P \leq 0.15$	9	119

†only 65 points and isotherms ≤ 400K were used in the analysis.

TABLE 2. Coefficients to calculate $C_p^o(T)/R$ from eq. (2).

	CF_4	C_2F_6	SF_6	BCl_3
A_0	-0.34875	0.55648	-2.60970	3.88413
A_1	3.55103e-02	5.59322e-02	6.88968e-02	1.59872e-02
A_2	-3.97999e-05	-5.78139e-05	-7.90288e-05	-1.31660e-05
A_3	1.74882e-08	2.10239e-08	2.66873e-08	1.01223e-09
A_4	1.58249e+04	1.30662e+04	---	---

temperature bath, pumps, sample bottles and transducers. The cabinet provides secondary containment of the process gases, protecting personnel from a catastrophic failure. Passive and active interlocks are employed to prevent over pressurization or heating. The system as a whole has a maximum allowable working pressure (MAWP) of over 150% of our administrative limit of 1.5 MPa. All surfaces in contact with the sample gas are constructed of corrosion resistant alloys. Where practical, ASME Section IX certified welds are used for joints. Welds were internally borescoped to assure full penetration. Welding was performed in a Class 100 Clean room, using a high purity purge gas filtered to 0.02 micron, where moisture and oxygen levels were less than 1 ppm. Nickel gasket connections are utilized wherever mechanical connections are required. Where possible, commercially available vacuum pumps, pump oil, valves, and transducers which have been specifically designed for semiconductor processing service have been incorporated into the apparatus.

A brief description of the experimental set up follows. The heart of the apparatus is an acoustic resonator. The resonator is a heavy-walled, 14 cm long, cylindrical cavity that has been bored out of an Alloy 400 (~67% Ni, ~33% Cu) cylinder.

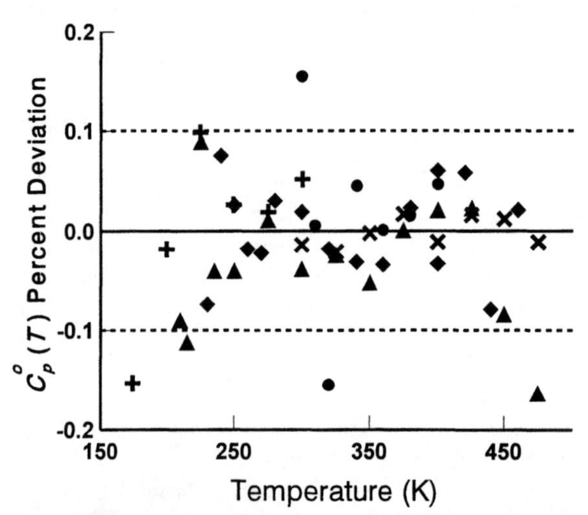

Figure 2 Percent deviation of experimental $C_p^o(T)/R$ from the fit to eq. (1) using the parameters in eq. (2) where: ✖ CF_4, ▲ C_2F_6, ◆ SF_6, the ✚'s denote the measurements of Ewing and Trusler (3) for CF_4, and ● BCl_3.

The cavity has an inner diameter of 6.5 cm and an outer diameter of 7.8 cm. Circular Alloy 400 plates (1.3 cm thick) are sealed to the ends of the cavity with bolts and gold o-rings. One end plate contains two thin metal diaphragms mounted flush with the interior surface of the cavity. These diaphragms isolate the sample gas, while coupling acoustic energy into and out of the cavity. The diaphragms are Alloy 400 disks (1 cm in diameter, 25 μm thick) that have been electron beam welded around their circumferences to small flanges mounted on the top end plate. Acoustic wave guides connect the diaphragms to two remote electro-acoustic transducers at ambient temperature. When sample gas is present in the resonator cavity, a pressure controller maintains an equal pressure of argon in the wave guides and transducer housings. Thus, the metal diaphragms are not stretched by a differential pressure. The wave guides are commercially purchased horn-shaped tubes with a length of 15 cm and a diameter that tapers exponentially from 0.12 cm to 0.33 cm. A 2.5 cm long, thin-walled, stainless-steel tube connects the horns to gas-tight transducer housings that are also maintained at the sample pressure. A metal screen located at

Table 3. Parameters for equations of state deduced from $u(T,P)$ measurements.

	b_o ($m^3 \cdot mol^{-1}$)	λ	ε / k_B (K)
	CF_4		
$B(T)$ ($cm^3 \cdot mol^{-1}$)	8.65202e-05	1.480236	191.8956
$C(T)$($cm^3 \cdot mol^{-1}$)^2	8.29090e-05	1.720815	124.2436
	C_2F_6		
$B(T)$($cm^3 \cdot mol^{-1}$)	1.39038e- 04	1.40462	286.2507
$C(T)$($cm^3 \cdot mol^{-1}$)^2	1.77713e-04	1.31904	307.2723
$D(T)$($cm^3 \cdot mol^{-1}$)^3	A_0 = 4.75509e-12	A_1 = -1.51714e-09	
	SF_6		
$B(T)$($cm^3 \cdot mol^{-1}$)	1.19796e-04	1.43660	294.4928
$C(T)$($cm^3 \cdot mol^{-1}$)^2	1.75705e-04	1.31771	313.0808
$D(T)$($cm^3 \cdot mol^{-1}$)^3	9.43273e-06	2.0	340.6818
	BCl_3		
$B(T)$($cm^3 \cdot mol^{-1}$)	1.73130e-04	1.51287	335.7466
$C(T)$($cm^3 \cdot mol^{-1}$)^2	1.65914e-04	1.13202	793.9712

the narrow end of the horn damps spurious resonances within the wave guides. When the wave guides are filled with argon, they strongly attenuate sound at frequencies above 8 kHz, limiting the frequencies which can be studied.

The sound generator is a commercially manufactured earphone speaker. It emits sound that is transmitted down a wave guide through one metal diaphragm into the resonator. The acoustic energy within the resonator is then coupled through the second metal diaphragm up the second wave guide to the sound detector which is a commercially manufactured microphone designed as a hearing aid. The frequency of the source is typically scanned through two longitudinal and one radial mode: (3,0,0), (4,0,0), and (0,0,1), for example, where the modes are labeled with the notation of Gillis (4). The frequency f and the width g of each resonance are measured using standard procedures (5) and instruments with a typical fractional uncertainty of less than 1×10^{-5}. The speeds of sound and their uncertainties are computed from weighted averages of the results for the three modes, and the uncertainties typically range from 10-100 ppm varying roughly as P^{-2}.

The resonator is suspended vertically in a well-stirred thermostated bath of either silicon oil or methanol. The bath is controlled within two millikelvin of the set-point. A 25 Ω capsule-type standard platinum resistance thermometer (SPRT), calibrated to ITS-90, is mounted in an aluminum block in thermal equilibrium with the resonator. Four-wire resistance measurements of the SPRT are performed by a high precision DC multi-meter.

A 13 kPa full-scale capacitance differential pressure transducer (DPT) is used to detect the differential pressure between the argon and the sample gas. The DPT is calibrated for pressure and temperature dependancies and is thermostated with a stability of ±0.1 K. Pressure measurements are made on the argon side of the DPT with a quartz-bourdon-tube differential pressure gage. The reference side of the gage is maintained below 2 Pa with a rotary pump. This bourdon tube gage has been calibrated with a dead weight gage to a standard uncertainty of $\sigma_P = 30\ \text{Pa} + 0.0001 \times P$.

Measurements are made along isotherms by initially loading the resonator to either 1.5 MPa or 80 % of the sample's vapor pressure, whichever is less. The temperature and pressure are allowed to equilibrate, and the frequencies and widths of the acoustic resonances are measured. The temperature is maintained and the pressure dropped in successive steps. The pressure is reduced by briefly opening a series of air-operated valves and collecting a portion of the sample gas in a cryotrap at 77 K. Once the pressure is reduced the apparatus is allowed to return to equilibrium and the frequencies and widths are measured at the new state point.

The effective radius and length of the cylindrical resonator are required in determining sound speed from the measured frequencies. These dimensions were accurately determined as functions of the temperature by measuring the resonance frequencies when the resonator was filled with argon, a gas for which the speed of sound is accurately known. Because the calibration and measurements were conducted in the same bath, there is a high degree of compensation for the effects of temperature gradients in the bath and for errors in the measurement of temperature.

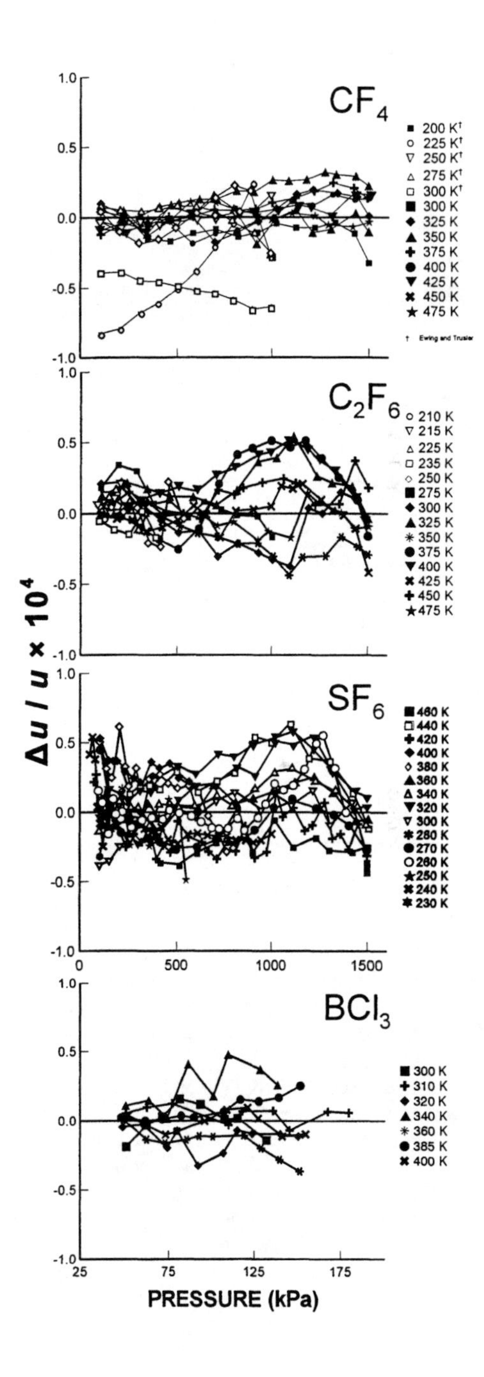

Figure 3. Deviations of measured sound speeds from those calculated from the deduced equation of state (eos) for the indicated systems, $\Delta u/u \times 10^4 = [\,(u_{meas.} - u_{eos})/\,u_{eos.}] \times 10^4$.

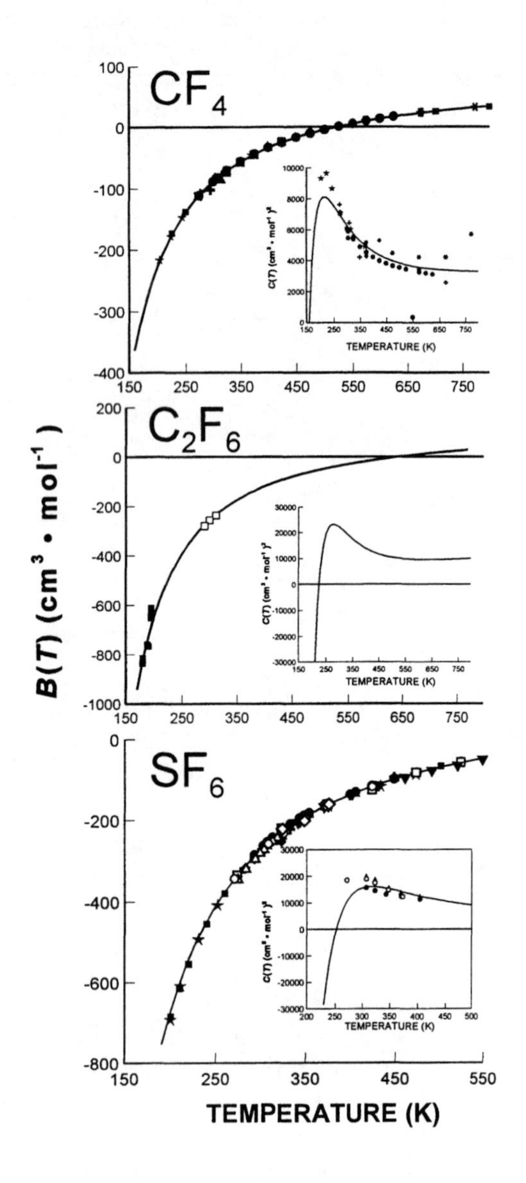

Figure 4 The second and third virial coefficients $B(T)$ and $C(T)$ as determined from the fit of the measured sound speeds to eq (3) shown as solid line. The previously published measurements, where available, are given as (references can be found in (11) unless otherwise noted) CF_4 :■ Dymond and Smith, ✚ Cawood and Patterson, ◆ MacCormack and Schneider, ▲ Hamman *et al.*, ● Douslin *et al.*, ✱ Kalfoglou and Miller, ★ Lange and Stein, ✖ Sigmond *et al.*, Bose *et al.*; C_2F_6: ■ Pace and Aston, □ Bell *et al.* (12); SF_6: ■ Dymond and Smith, □ MacCormack and Schneider, ✚ Hamann *et al.*, ▲ Clegg *et al.*, Δ Dymond and Smith, ● Mears *et al.*, ∇ Hajjar and MacWood, ✖ Nelson and Cole, ▼ Bellm *et al.*, O Sigmund *et al.*, ★ Hahn *et al.*, ◇ Hosticka and Bose, ◆ Santafe *et al.*.

RESULTS

Here we briefly review our $u(T,P)$ measurements in four systems: CF_4, C_2F_6, SF_6, and BCl_3. A detailed publication including the actual measured sound speeds is being prepared (6). The measurements discussed in the following section are summarized in Table 1. For CF_4 the highly precise sound speeds previously reported by Ewing and Trusler (7) are included in the present analysis. Ewing and Trusler's measurements cover the ranges 175 K $\leq T \leq$ 300 K and P < 1.0 MPa or 80% of the vapor pressure. The sound speed data for BCl_3 are shown in Fig. 1.

ANALYSIS

Each sound speed isotherm is fit to the acoustic virial equation of state:

$$u^2 = \frac{\gamma^o RT}{m}\left(1 + \frac{\beta_a P}{RT} + \frac{\gamma_a P^2}{RT} + \frac{\delta_a P^3}{RT} + \dots\right) \quad \textbf{(1)}$$

where u is the sound speed, m is mass, R is the gas constant, T is temperature, $\gamma^o(T)=C_p^o(T)/C_v^o(T)$, and β_a, γ_a, and δ_a, are the temperature-dependent acoustic virial coefficients. $C_p^o(T)$ is obtained from eq. (1) through the relation $C_p^o(T)/R = \gamma^0/(\gamma^0 - 1)$.

The measured $C_p^o(T)/R$ values were fit to the equation:

$$\frac{C_P^o(T)}{R} = A_0 + A_1\left(\frac{T}{\mathrm{K}}\right) + A_2\left(\frac{T}{\mathrm{K}}\right)^2 + A_3\left(\frac{T}{\mathrm{K}}\right)^3 + A_4\left(\frac{T}{\mathrm{K}}\right)^{-2} \quad \textbf{(2)}$$

and the coefficients presented in Table 2. Figure 2 shows the percent deviations of $C_p^o(T)/R$ determined from the individual isotherms from the fits to eq. (2).

The density virial coefficients, $B(T)$, $C(T)$, etc. can be directly related to the acoustical virial coefficients in eq. (1) through exact thermodynamic equations (8). It is therefore possible, given an algebraic expression for the temperature dependence of the density virial coefficients, to perform a non-linear fit of the virial equation of state

$$P = RT\rho\left[1 + B(T)\rho + C(T)\rho^2 + \cdots\right] \quad \textbf{(3)}$$

to the sound speed measurements, $u(T,P)$, thereby providing the density virial coefficients and their temperature dependence.

The algebraic expression for the temperature dependence of the density virial coefficients employed here is that determined for the square well potential. This form has the advantages of having realistic temperature dependencies and the ability (4) to extrapolate reasonable values beyond the experimental temperature range. In some cases an inverse-T polynomial is used in place of the square-well form. The equations expressing the first three density virials in terms of the square-well potential parameters are as follows:

$$B(T) = b_o\left[1 - (\lambda^3 - 1)\Delta\right] \quad \textbf{(4)}$$

$$C(T) = \frac{1}{8} b_o^2 \left(5 - c_1 \Delta - c_2 \Delta^2 - c_3 \Delta^3 \right)$$

$$c_1 = \lambda^6 - 18\lambda^4 + 32\lambda^3 - 15 \qquad \textbf{(5)}$$

$$c_2 = 2\lambda^6 - 36\lambda^4 + 32\lambda^3 + 18\lambda^2 - 16$$

$$c_3 = 6\lambda^6 - 18\lambda^4 + 18\lambda^2 - 6$$

$$D(T) = b_0^3 \Big(0.2869 + 1.634\Delta - 23.29\Delta^2 + 54.65\Delta^3 + 70.76\Delta^4 - 168.2\Delta^5 - 12.74\Delta^6 \Big) \qquad \textbf{(6)}$$

where $\Delta = e^{\varepsilon / k_B T} - 1$, and k_B is Boltzmann's constant. The adjustable parameters are: ε the well depth, λ the ratio of the width of the well to the diameter σ of the hard core, and the molar volume of the hard core $b_o = ⅔\pi N_a \sigma^3$, where N_a is Avogadro's constant. *D(T)* is known only for certain values of λ and eq. (6) is appropriate when $\lambda \equiv 2.0$.

The virial equation of state eq. (3) was fit to the *u(T,P)* measurements for each fluid, and the resulting parameters provided in Table 3. The $C_p^o(T)/R$ values were held fixed at the values given by eq. (2) with the parameters in Table 2. The measured sound speeds nominally have uncertainties of less than 0.01% that provide values of $C_p^o(T)$ with an uncertainty of less than 0.1%. The virial equation of state determined by this procedure typically reproduces measured gas densities to better than 0.1% (9,10).

The *u(T,P)* results for CF_4 were fit to a virial equation of state including *B(T)* and *C(T)* temperature dependent density virials given by eqs. (4) and (5). The measurements of Ewing and Trusler (3), with the exception of the 175 K isotherm, were included in the fit. The 175 K isotherm required higher virial terms than *C(T)*; hence it was omitted from the fit. The isotherms at 225 K and 300 K of Ewing and Trusler had to be corrected by multiplying the measured sound speeds by a constant (1.000065 and 1.000040 respectively) to have their zero pressure sound speeds correlate to that given by eqs. (1) and (2). Figure 3 shows the deviations of the measured sound speeds from those calculated from the deduced equation of state. In Fig. 3 the corrections made to the measurements of Ewing and Trusler are not included. The fit had 158 degrees of freedom, *v*, and χ^2/v was 0.74.

The *u(T,P)* measurements in C_2F_6 were fit to eq. (3) using a square-well *B(T)* and *C(T)*, but the higher densities required the inclusion of a *D(T)* term. The *u(T,P)* measurements did not provide enough detail to fit a square-well representation of *D(T)*; therefore, a two term polynomial was used, $D(T) = A_0 + A_1 T^{-1}$. The fit had $v = 169$ and a χ^2/v of 1.04. The deviation of the measured *u(T,P)* from that calculated from the determined equation of state can be seen in Fig. 3.

For SF_6 we covered enough of *P-T* space to use the square-well form of *D(T)*. The SF_6 fit had $v = 286$ and a χ^2/v of 0.25. The deviation of the measured sound speeds in SF_6 from those calculated from the deduced equation of state is given in Fig. 3.

Figure 4 shows the determined second and third virial coefficients, *B(T)* and *C(T)*, as represented by eqs. (4) and (5) using the parameters in Table 3. Previously published results are show for comparison. In almost all cases our virial coefficients reproduce the previously published measurements to within their experimental uncertainties. Also notice that Eqs. (4) and (5) extrapolate well outside of the experimental temperature ranges, and have the correct temperature dependencies.

Measurements of *u(T,P)* in boron trichloride, BCl_3, are shown in Fig. 1. Higher pressure measurements required heating the source bottle. Whenever heat was applied to the source bottle, the *u(T,P)* measurements were not reproducible. This indicated a change in sample composition. For this reason pressure measurement only up to the room temperature vapor pressure are considered. During the fitting process the results above 400 K appeared inconsistent with the rest of the data set. The resonator was loaded to and held at 500 kPa and 440 K, while *u(T,P)* measurements were taken with time. A linear change in the measured resonance frequencies with time was observed. There was no observable shift within experimental uncertainties at 400 K. These changes with time are an indication of changing composition. Hence the isotherms above 400 K are left out of the fit pending further analysis. The remaining isotherms between 300 and 400 K were fit as before and the results are provided in Tables 2 and 3. The fit had 59 degrees of freedom and a χ^2/v was 0.05. The deviations of measured *u(T,P)* from that calculated from the deduced equation of state are shown in Fig. 2. No previously published virial coefficients could be found in the literature to compare against. This demonstrates the uniqueness and value of our new facility for the characterization of semiconductor process gases.

REFERENCES

1. Gillis, K.A., Mehl, J.B., and Moldover, M.R., *Rev. Sci. Instrum.* **67**, 1850 (1996).
2. Goodwin, A. R. H., and Moldover, M. R., *J. Chem. Phys.* **95**, 5236 (1991).
3. Gillis, K.A., *Int. J. Thermophys.* **18**, 73 (1997).
4. Gillis, K.A., *Int. J. Thermophys.* **15**, 821 (1994).
5. Gillis, K.A., Goodwin, A.R.H., and Moldover, M.R., *Rev. Sci. Instrum.* **62**, 2213 (1991).
6. Hurly, J.J., To Be Published.
7. Ewing, M. B. and Trusler, J. P. M., *J. Chem. Phys.* **90**, 1106 (1989).
8. Gillis, K.A. and Moldover, M.R., *Int. J. Thermophys.* **17**, 1305 (1996).
9. Hurly, J. J., Schmidt, J. W., Boyes, S. J., and Moldover, M. R., *Int. J. Thermophys.* **18**, 579 (1997).
10. Hurly, J.J., Schmidt, J.W., and Gillis, K.A. , *Int. J. Thermophys.* **18**, 137 (1997).
11. Dymond, J.H. and Smith, E.B., *The Virial Coefficients of Pure Gases and Mixtures*, Oxford Press, New York 1980.
12. Bell, T.N., Bignell, C.M. and Dunlop, P.J., *Physica A* **181**, 221 (1992).

X-ray metrology for ULSI structures

D.K. Bowen, K.M. Matney and M. Wormington

Bede Scientific Incorporated, 14 Inverness Drive East, Suite G-104, Englewood, CO 80203, USA

Non-destructive X-ray metrological methods are discussed for application to both process development and process control of ULSI structures. X-ray methods can (a) detect the unacceptable levels of internal defects generated by RTA processes in large wafers, (b) accurately measure the thickness and roughness of layers between 1 and 1000 nm thick and (c) can monitor parameters such as crystallographic texture and the roughness of buried interfaces. In this paper we review transmission X-ray topography, thin film texture measurement, grazing-incidence X-ray reflectivity and high-resolution X-ray diffraction. We discuss in particular their suitability as on-line sensors for process control.

INTRODUCTION

The trend towards larger wafers makes destructive testing methods grossly uneconomic, and non-destructive X-ray metrological methods (1) are now very attractive for both process development and process control of ULSI structures. Besides being non-destructive, X-ray metrological methods have several other advantages. They can detect the unacceptable levels of internal defects generated by RTA processes in large wafers. As layers of dielectric and metal become thinner, accurate measurement of thickness and roughness by traditional methods becomes less and less accurate but can be performed by X-rays. Finally, the complexity of ULSI structures makes it necessary to monitor new parameters such as crystallographic texture and the roughness of buried interfaces. In this paper we review the following X-ray techniques for ULSI metrology: transmission X-ray topography, thin film texture measurement, grazing-incidence X-ray reflectivity and high-resolution X-ray diffraction. We discuss in particular their suitability as on-line sensors for process control.

X-RAY TRANSMISSION TOPOGRAPHY

This method allows the imaging of all dislocation structures above a few μm in length in complete silicon wafers. It uses diffraction contrast, in a manner similar to transmission electron microscopy but requires no specimen preparation. The geometry is shown in Figure 1a for section topography and in Figure 1b for projection (Lang) topography in which the wafer is scanned through the beam. For QC work, the source is a powerful rotating-anode generator. Current measurement times with such systems are a few seconds for section topography and below ten minutes for moderate resolution projection topography (enough to image individual dislocations) across an 8” wafer.

Section topography reveals defects in a slice through the wafer, such as the oxygen precipitates in Si shown in Figure 2. The image is formed by interference processes, and is fully calculable from first principles. However, simple visual inspection is sufficient to reveal defects such as oxygen-induced precipitation and stacking faults arising at a substrate-epilayer boundary, since the image of a perfect specimen is easily defined: a set of interference fringes parallel to the crystal surface but otherwise featureless. The interference fringes are easily destroyed by small amounts of crystal imperfection.

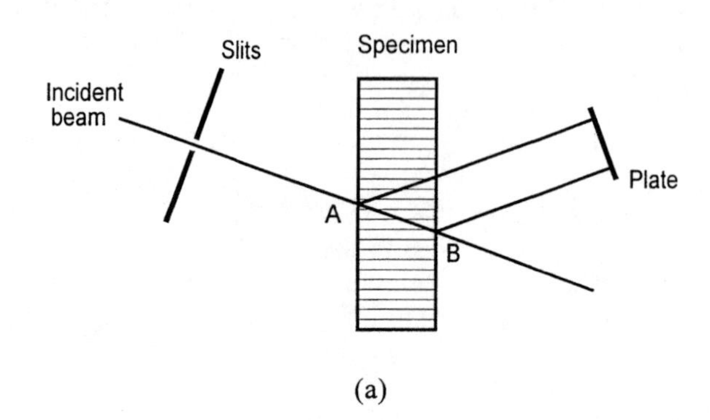

(a)

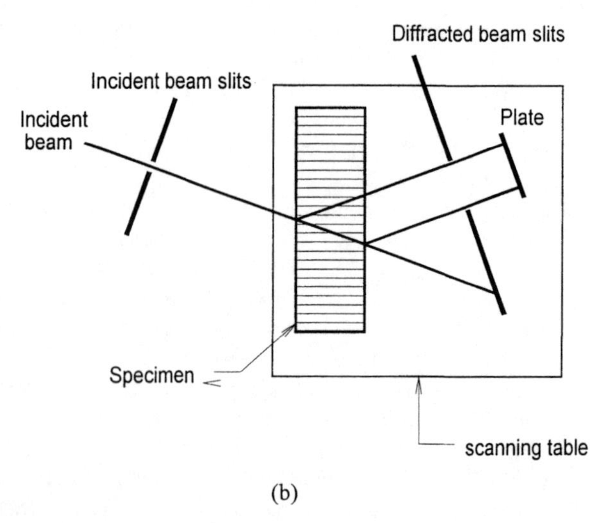

(b)

FIGURE 1. Geometry of (a) section topography, (b) Lang or projection topography.

CP449, *Characterization and Metrology for ULSI Technology: 1998 International Conference*
edited by D. G. Seiler, A. C. Diebold, W. M. Bullis, T. J. Shaffner, R. McDonald, and E. J. Walters

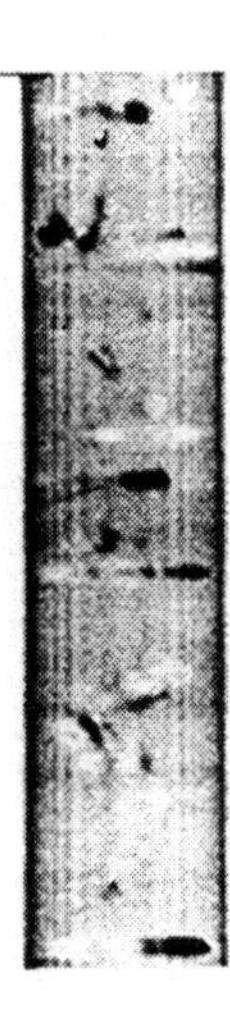

FIGURE 2. Section topograph of a Si wafer, showing oxygen precipitates. MoKα radiation, 220 reflection.

An example of dislocations and slip bands produced by slip during annealing (2) is shown in Figure 3. The contrast is strong in such images, and defective regions are easily recognized. The ideal perfect crystal in this method gives simply a featureless gray image. Slip bands may even be identified by color image segmentation methods and shown as a different color to the background.

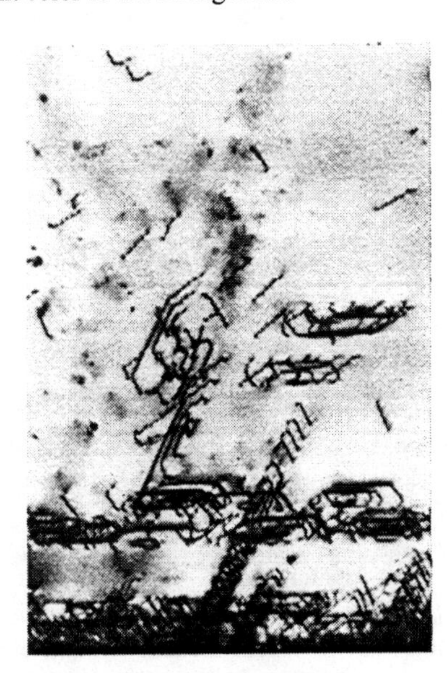

FIGURE 3. Transmission Lang topograph of Si wafer, showing dislocations and slip bands produced by slip on annealing. Field of view 1 mm wide. MoKα source, 220 reflection.

Lang topography is typically used in process development to develop annealing procedures that minimize dislocation generation and help to predict yield, and in process control to verify annealing procedures. Defects such as dislocations and slip bands generated near wafer mounting points in annealing processes, or those produced near stress-concentrating overlayers in device processing are easily seen. It is possible to construct a tool specialized for fabrication lines in which the measurement time would be less than two minutes, with digital image recording, processing and storage.

TEXTURE MEASUREMENT

Crystallographic texture assessment of metallic layers is already used in advanced fabrication facilities to control the growth of subsequent layers, etching behavior and electro-migration. In magnetic materials it also controls the shape of the magnetization curve. Texture measurement has traditionally been rather slow, but recent advances in X-ray optics, especially the use of polycapillary or Khumakhov lenses (3), have made some 30 – 100 x increase in speed and have brought the method into the QC domain. The instrumentation is shown in Figure 4, and texture maps made by the Schulz (4) method (using a slit) and the new method (5, 6) are shown in figure 5.

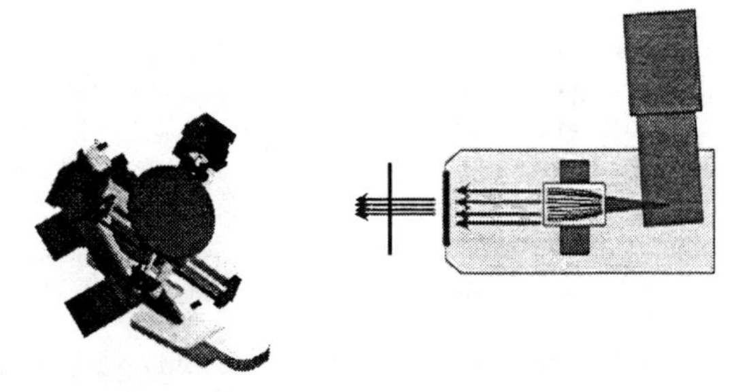

FIGURE 4. Geometry of the new method for texture measurement. A polycapillary optic is used to obtain a 20 × 20 mm beam collimated in both horizontal and vertical directions. A "texture stage" with two axes of rotation is used to mount the sample.

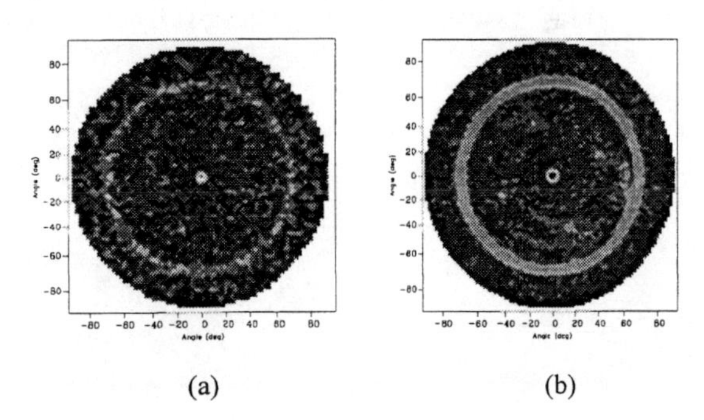

(a) (b)

FIGURE 5. (111) texture maps of 5 nm MnFe layer; (a) conventional Schulz method, (b) new polycapillary optic method with the same data collection time as (a). The fiber texture is shown much more clearly in (b).

Complete pole figure measurements may now be made at one point on layers less than 5 nm thick in less than five minutes with a standard detector and much less with an area detector. Partial pole figures or a deconvolved rocking curve measurement (7) to verify critical aspects of texture are possible in less than one minute.

GRAZING INCIDENCE X-RAY REFLECTIVITY

The measurement of interference fringes with X-rays incident at a grazing angle less than about four degrees provides measurement of film thickness for most materials over the range 1 – 1000 nm (8). The geometry is shown in figure 6.

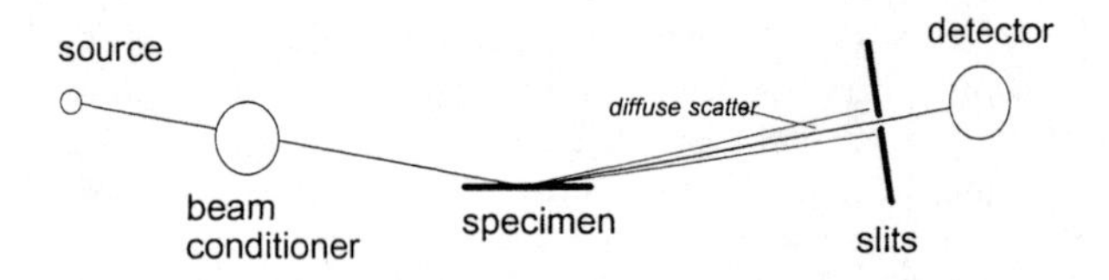

FIGURE 6. Geometry of X-ray reflectivity measurement. Specular reflectivity is measured by maintaining equal angles of incidence of the incident and scattered beams (θ-2θ scan). Diffuse or off-specular scatter is measured by scanning either the specimen or the detector axis with the other fixed, or by a θ-2θ scan with an offset of the specimen from the specular condition.

The accuracy at 10 nm thickness is about 0.2% and is about 2% even for 2 nm layers of native oxide on Si. Roughness and density are also determined, even for buried layers. Most importantly, *a priori* knowledge of optical constants is **not** required. This is because the only significant correction to the simple measurement of fringe spacing is the X-ray refractive index, and this is given accurately by the position of the critical angle at which the reflectivity drops sharply from near 100%.

Figure 7 shows excellent data obtained on an 8 Å native oxide layer on silicon, using synchrotron radiation. The simulated curve is modeled from first principles, and the thickness is determined to better than 1 Å. Data of this quality would not be obtained on a fabrication line, but one order lower dynamic range is feasible. Use of a shorter wavelength (e.g. Ag Kα at 0.51 Å) will show more fringes in a given angular range.

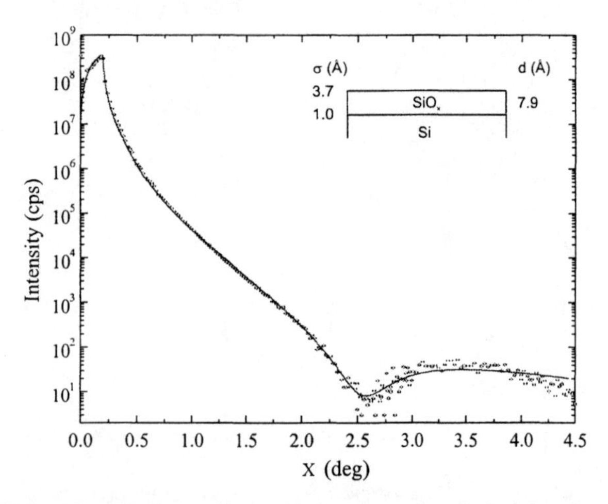

FIGURE 7. Native oxide on Si. Synchrotron radiation data from SRS, Daresbury Laboratory, wavelength 1.3926 Å. Modeling from first principles shows the thickness to be 8 Å.

Polymer films on Si are particularly easy to measure since the electron density differences are large. Figure 8 shows a PMMA layer spun onto Si, and measured in 2 mins with a conventional X-ray tube. With newer sources and optics, a ten-fold increase in speed could easily be obtained.

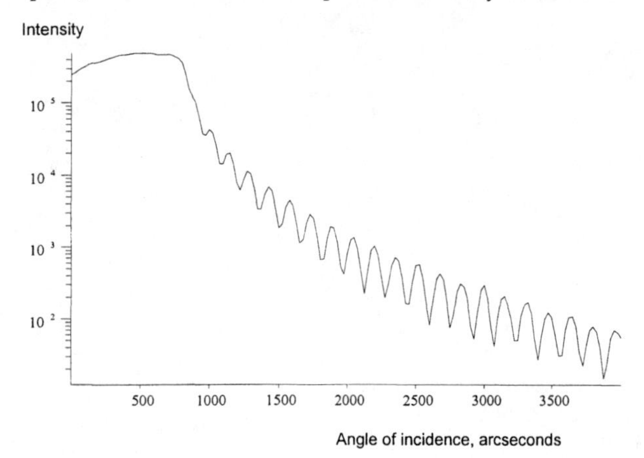

FIGURE 8. Grazing incidence specular reflectivity of a 97.8 nm PMMA layer on Si. 2 mins data collection time with a conventional X-ray tube running at 1 kW. Bede GXR1 reflectometer.

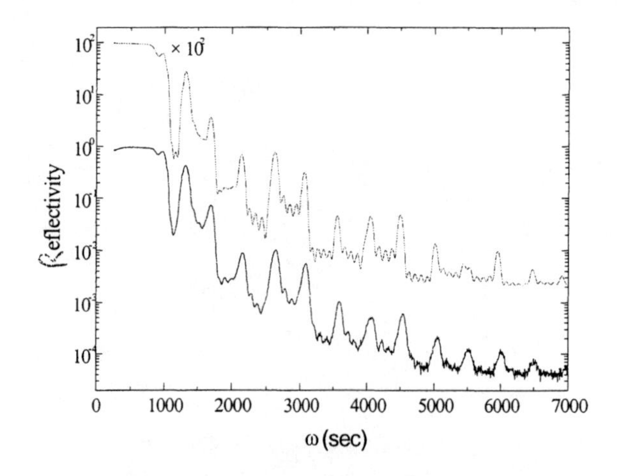

FIGURE 9. Specular reflectivity of a Si-Ge short period superlattice, comprising 5 × (229 Å Si + 94 Å $Si_{0.43}Ge_{0.57}$), an 86 Å Si cap and 25 Å native oxide layer. The Si layer has 5 Å roughness/width and the SiGe layers 10 Å roughness/width. The upper curve (displaced for clarity) is the theoretical simulation by which the above parameters were determined.

Figure 9 shows the specular reflectivity from a short-period superlattice of a heavier material, SiGe on Si. Very detailed modeling could be performed on this material, and these data were the first to show differential roughening between Si on SiGe and SiGe on Si, since confirmed by TEM and explained by lateral elastic relaxation of growth protuberances. Whilst this is epitaxial material, its dislocation density was far too high for structural detail to be revealed by high-resolution X-ray diffraction techniques.

3.1 Traceability of X-ray reflectivity methods

The traceability of measurements to NIST standards is an important issue, which involves both instrumentation and software verification. Instrumentally, this involves simple angular calibration of the specimen and detector goniometers. In a recent blind test, four separate vendors' measurements of the thickness of a thin film were within a scatter of 0.25% (0.5 Å in 200 Å) even on uncalibrated instruments. Measurement time for thickness measurements in this range can be less than one minute.

If interference fringes are observed beyond about twice the critical angle, no software modeling is necessary. The thickness is then given to < 1% by $d = \lambda/2\Delta\theta$, where d is the layer thickness, λ is the wavelength and $\Delta\theta$ the fringe separation. The refractive index is not required since for all materials it is only a few parts in 10^5 and at high scattering vectors it may be ignored. When less than two fringes are observed, however, modeling is required. The modeling software used is based upon non-controversial Fresnel and X-ray scattering theory with well-established X-ray optical constants. Differences between software programs are essentially differences in the numerical analysis methods used, and a system of code result verification against NIST codes would provide valuable traceability. The electron density is obtained from chemical formulae and bulk density as a close approximation, but the data contain their own density correction factors through the position of the critical angle. The case of thin oxide on silicon is especially insensitive to the modeling parameters. At very low thicknesses, the critical angle is determined by the density of the silicon substrate. This is the best known of all densities, since it is used in the determination of the Avogadro number (9). Errors in the density of the oxide have a negligible effect on the determination of the thickness.

Specular reflectivity alone does not distinguish between roughness and composition grading (electron density variation). However, diffuse scatter may be used to separate these parameters, with somewhat longer measurement times, and may also be used to determine lateral roughness correlation lengths. Vertical roughness correlations may also be determined from diffuse scatter, preferably using a detector-only scan. This allows one to tell whether successive layers are replicating the contours of previous layers, and thus, for example, whether the thickness of a dielectric layer is reasonably constant whatever the local undulations. This is illustrated in figure 10, which shows a diffuse scan for the SiGe specimen shown in figure 9. The fringe contrast in the diffuse scan, which matches that of the specular scan, shows that the layers are deposited with strong conformality or vertical correlation. The effect of the conformality on the diffuse scatter may be modeled with good accuracy, and it is quite possible to determine the average degree of vertical conformality of layers. This has implications in particular for dielectrics, where a uniform thickness is desirable whatever the roughness, to ensure predictable capacitance and to avoid dielectric breakdown.

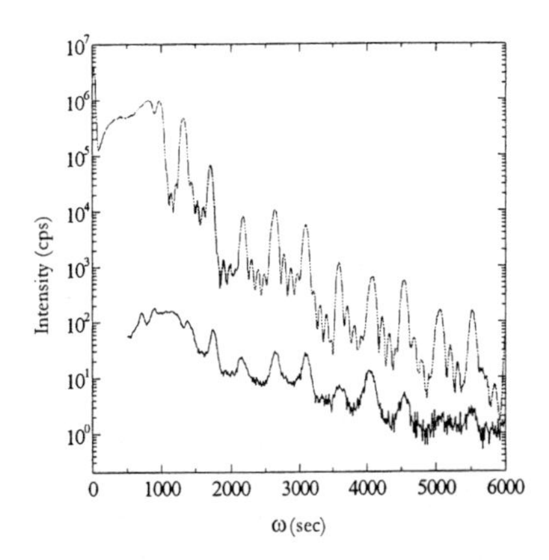

Figure 10. The specular reflectivity of the SiGe sample shown in figure 9, and the longitudinal scan (lower curve) taken using a θ-2θ scan with specimen offset from the specular position.

HIGH RESOLUTION X-RAY DIFFRACTION

The main uses of HRXRD are for measurement of doping levels through lattice parameter measurement, and monitoring of wafer quality. Multiple-reflection beam conditioners are used before the sample and a triple-axis analyzer after the sample. Since reflection ranges are only a few seconds in Si with typical X-radiation, very high resolution instruments are required, though these can still make rapid measurements.

Measurement of lattice parameter

This may be measured by means of the Bond method of absolute Bragg angle measurement (10), in which zero errors are removed by measurement on both sides of zero, or by the triple-axis method developed by Bowen and Tanner (11). The Bragg angle may be measured to one arc second relatively easily, and to 0.1 arc second with extreme care. This gives lattice parameters accurate to 1 in 10^6 or 10^7. (Methods do exist to make comparative measurements to about 1 in 10^8 but they are not practical for silicon wafers).

Figure 11 shows the effect of some dopants on the Si 004 (CuKα) Bragg angle, taken from published sources.

The X-ray Bragg angles, and hence the splitting of peaks between substrate and epilayer, depend upon the absolute difference in the strain, not on the ratio. From the above curves, the minimum doping level that may be directly measured from lattice parameter measurements is of the order of 10^{16} - 10^{17} cm^{-3} for boron.

The detectable difference between a substrate and an epilayer will depend on the base level of the material. For the common case of a p+ substrate doped at 10^{18} cm^{-3} and an epilayer doped at 10^{15} cm^{-3}, we will be able to monitor the substrate doping and its variation across the wafer quite accurately, but will not be able to determine the doping level of the epilayer.

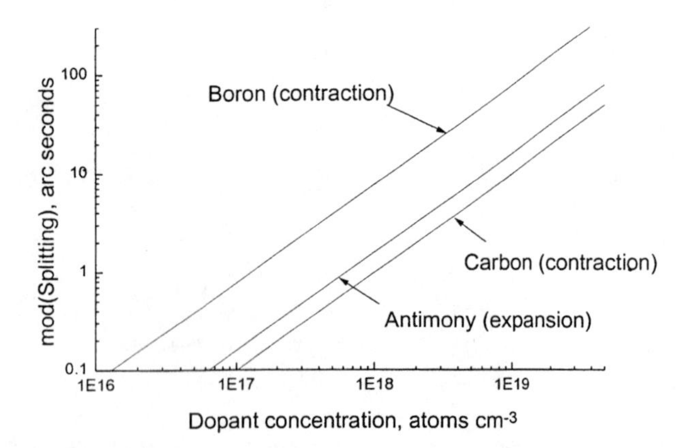

FIGURE 11. Effect of dopants on the Bragg angle, plotted on logarithmic axes.

Assessment of crystal perfection

Enormous efforts have gone into the growth and preparation of defect-free silicon, and so far as bulk dislocations in boules are concerned these have been very successful. However, the polishing process is not yet under complete control, as shown by the variation in perfection mapped in figure 12 for an as-supplied wafer. This shows the triple-axis FWHM, which for a perfect crystal is approximately 3 arc seconds, as a function of position on the wafer over a 100 mm diameter. The variation is from 3.5 to 4.6 arc seconds. Although this is very good material, imperfections are evident and could be a serious loss of yield in demanding applications such as image sensors.

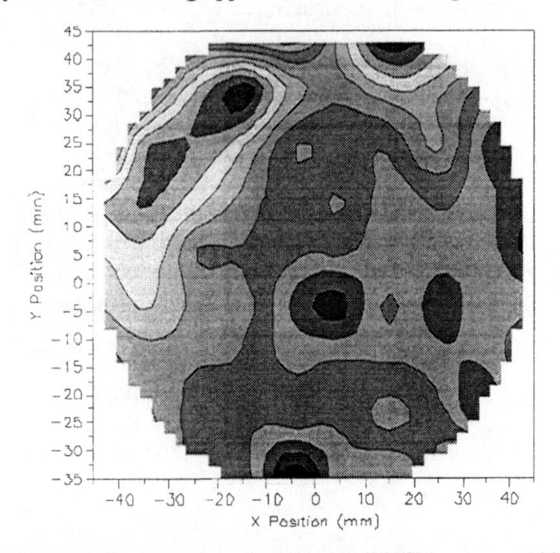

FIGURE 12. Variation of triple-axis FWHM across a 100 mm diameter of a Si wafer, as received (specimen courtesy Eastman Kodak Corporation). The contours range from 3.5 to 4.6 arc seconds.

We ascribe these imperfections to sub-surface damage caused by the mechanical or chemo-mechanical polishing process. In other wafers we have seen values up to 8 arc seconds for the triple-axis FWHM.

CONCLUSIONS

All the methods discussed require no specimen preparation, are non-destructive and may be performed in ambient clean-room conditions. The theory is well understood. In most cases, methods exist to provide automated interpretation with output of statistical control parameters and these are currently being developed commercially.

X-ray methods are averaging methods. The area sampled is of order 100 μm^2 – 100 mm^2, and they should be thought of as complementary to rather than competitive with probe methods such as SEM or AFM. However, it is our experience that the defects that they reveal vary in density relatively slowly across a wafer and it is quite possible to obtain excellent defect maps at an early stage of processing. With a 300 mm wafer costing about $1000, and with each successive major processing step multiplying the cost by a factor of about 10, the economic benefit of early structural characterization to this degree of detail is an evident manufacturing advantage.

REFERENCES

1 Bowen, D.K. and Tanner, B.K., *High Resolution X-ray diffractometry and topography*, Taylor & Francis, London 1998

2 Miltat, J.E.A. and Bowen, D.K., Physica Status Solidi, **(a) 3**, 431-445, (1970)

3 Kumakhov, M.A., Nucl. Instrum. Methods **48**, 283-286 (1990)

4 Schulz, L.G., J. Appl. Phys. **20**, 1030 1949

5 Matney, K.M., Wormington, M., Bowen D.K. and Xiao, Qi-Fan, *Two Dimensional Fiber Optic X-ray Collimators for Texture Mapping of Thin Films,* Denver X-ray Conference, Steamboat Springs, CO, August 1997

6 Kardiawarman, York, B.R., Qian, X.W., Xiao, Q.F. MacDonald, C.A. and Gibson, W.M., Proc. SPIE **2519**, 197-205 (1995)

7 Vaudin, Mark D., Rupich, Martin W., Jowett, Martha, Riley, G.N. Jr., and Bingert, John F., Journal of Materials Research, in press (1998)

8 Bowen, D.K. and Tanner, B.K., Nanotechnology **4** (1993) 175-182

9 Deslattes, R.D., Henins, A., Bowman, H.A., Schoonover, R.M., Carroll, C.L., Barnes, I.L., Machlan, L.A., Moore, L.J. and Shields, W.R., Phys. Rev. Lett. 33:463 1974

10 Bond, W.L., Acta Cryst., **13**, 814 1960

11 Bowen, D.K. and Tanner, B.K., Advances in X-ray Analysis, **37** (1994), 123 - 128

Flow Measurements in Semiconductor Processing; New Advances in Measurement Technology

S.A. Tison and A.M. Calabrese,

National Institute of Standards and Technology, Bldg 220 Rm. A55, Gaithersburg, MD 20899

Gas flow measurement, control, and distribution are an integral part in meeting present and future semiconductor processing requirements (1). Changes in processing and environmental concerns have put additional pressure not only on accurate measurement of the gas flow, but also in reducing flows. To address the need for more accurate metering of gas flows, NIST has developed primary flow standards which have uncertainties of 0.1% of reading or better over the flow range of 10^{-9} mol/s to 10^{-3} mol/s (0.001 sccm to 1000 sccm). These standards have been used to test NIST-designed high repeatability flow transfer standards (2) which can be used to document and improve flow measurements in the semiconductor industry (3). In particular two flowmeters have been developed at NIST; the first is a pressure-based flow sensor and the second a Doppler-shift flowmeter, both of which can be used for *in-situ* calibration of thermal mass flow controllers or for direct metering of process gases.

INTRODUCTION

Gas flow measurement, control, and distribution are important to many semiconductor processes (1). Changes in processing have in some cases resulted in reduced processing pressures and the need for accurate and reliable flow measurement below 10^{-4} mol/s (100 sccm). In addition to process changes, environmental concerns have provided additional incentives to reduce processing flows.

To address the need for more accurate metering of gas flows, NIST has developed primary flow standards which have uncertainties of 0.1% of reading or better over the flow range of 10^{-9} mol/s to 10^{-3} mol/s (0.001 sccm to 1000 sccm). These standards have been used along with NIST-developed high repeatability flow transfer standards (2) to document and improve flow measurements at calibration facilities which support the semiconductor industry (3). For flows larger than 1×10^{-4} mol/s (100 sccm) it was discovered that, while significant errors were being made in the calibration laboratories, these errors were typically less than 1% and were not the predominant cause of flow measurement errors being introduced into semiconductor processing. For flows lower than 1×10^{-4} mol/s (100 sccm) errors as large as 17% have been attributed to the calibration process.

Although significant errors can and have been introduced through improper calibration, many of the errors are artifacts of the operating principles of thermal mass flow controllers (TMFCs). TMFCs are typically calibrated with inert gases, and gas correction factors are applied when used with the process gases. The correction factors are typically not measured but predicted based upon gas properties and the specific TMFC design. This method of transferring flow measurement often results in large errors that are instrument dependent and are also sensitive to use conditions such as the ambient temperature (4,5). To achieve uncertainties of 1% with process gases it is necessary to calibrate TMFCs *in-situ* with the process gas or to develop a sensor that is more predictable or traceable to fundamental gas properties such as the gas speed of sound.

Two NIST-developed low-gas flowmeters are capable of being used for *in-situ* calibration of TMFCs or to directly meter gas flows for semiconductor processes. The first is a NIST developed pressure-based flow sensor (laminar flowmeter) that is able to measure flows over a range of 1×10^{-6} mol/s to 1×10^{-3} mol/s. This flowmeter has been evaluated over a four-year period and has shown a long-term stability of +/-0.15% of the indicated flow. The second flowmeter uses a Doppler-shift technique for measuring the gas flow. Both meters can be constructed of all stainless steel components and are useable with many semiconductor processing gases.

LAMINAR FLOWMETER

Laminar flowmeters measure the pressure loss of the gas as it passes through a physical restriction of some type. If the geometry is well known the pressure loss can be predicted using hydrodynamic theory. More typically, the pressure loss is measured and correlated to a mass flow which is determined by calibration with a primary flow standard. The hydrodynamic relation between the mass flow and the pressure loss for a particular geometry can be derived from hydrodynamics (6) as,

$$\mathrm{m} = \frac{\mathrm{K_1}}{\mu}\left[(2\rho\Delta P) - K_2\left(\frac{\rho\Delta P}{\mu}\right)^2\right] \tag{1}$$

where μ is the gas viscosity, ρ is the average gas density, ΔP is the differential pressure across the restriction, and K_1 and K_2 are geometrical constants. Traditionally, laminar flowmeters have been designed such that the pressure change is approximately 2.5 kPa. This small differential pressure is difficult to measure accurately, and results in

CP449, *Characterization and Metrology for ULSI Technology: 1998 International Conference*
edited by D. G. Seiler, A. C. Diebold, W. M. Bullis, T. J. Shaffner, R. McDonald, and E. J. Walters

the largest contributor to the uncertainty of the flow measurement. The NIST laminar flowmeter (LFM) design minimizes flow uncertainties by minimizing uncertainties in the pressure and temperature measurement.

LFM Design

The NIST design (2), shown schematically in Fig. 1, attempts to maximize the viscous pressure losses while maintaining laminar flow. The LFM achieves pressure drops as large as 1 MPa by using a rectangular channel with a high aspect ratio and helical flow path lengths of 300 cm or more. The pressure is measured upstream and downstream of the restriction with high-resolution (0.0001 kPa) pressure transducers. The temperature is measured with a 100 ohm platinum resistance thermometer which is mounted in the mid-section of the flowmeter. To minimize errors due to thermal gradients the LFM is mounted in an enclosure that is actively temperature controlled.

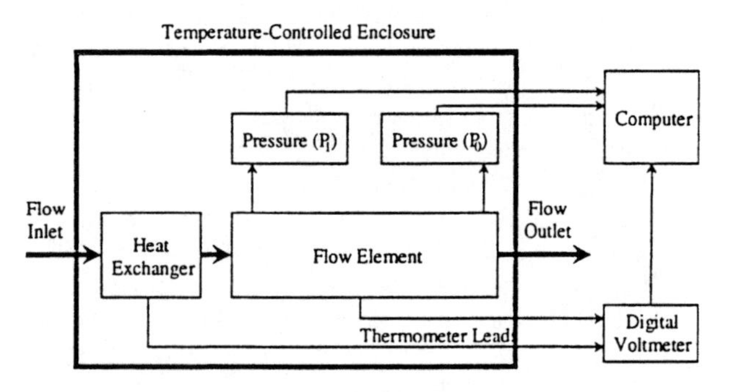

FIGURE 1. Laminar flowmeter schematic.

The LFM was characterized by direct comparison with a NIST constant volume flowmeter (2) over a range of 7 x 10^{-7} mol/s to 7 x 10^{-4} mol/s with nitrogen, argon, helium, and sulfur hexafluoride gases. The nitrogen flow is plotted as a function of the product of the average and differential pressures (which is approximately proportional to the gas flow) in Fig.2. The solid line is a fit of equation (1) to the data. For nitrogen flows larger than 0.15 g/s (5 x 10^{-4} mol/s), the measured flow deviates significantly from those predicted by equation (1). The additional pressure loss not predicted by equation (1) is due to secondary flows caused by the centrifugal forces of the gas as it traverse the helical path. The dotted line in Fig. 2 is a fit to the data using a correction (7) for the centrifugal force effects developed by Dean.

A significant amount of calibration data has been taken over the past four years to determine the stability of the LFM with time. The stability of the LFM is represented graphically in Fig. 3 as the difference between the true flow and the LFM flow as determined from a fit to the calibration data obtained in 1995. From Fig. 3, it can be seen that no significant change in the LFM calibration over

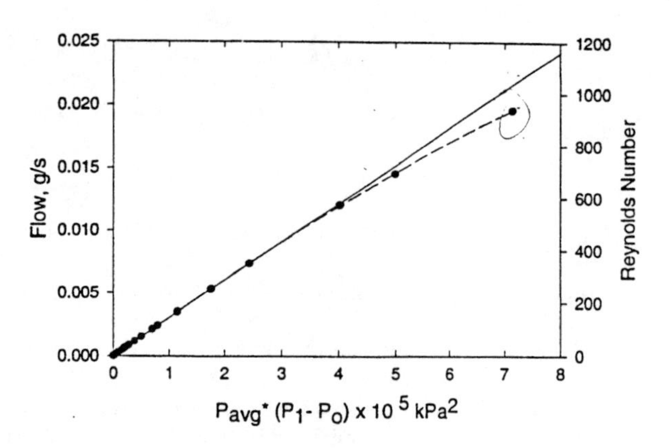

FIGURE 2. The solid line represents a fit of equation 1 to the nitrogen flow data. The dashed line is a fit to the data that incorporates corrections due to centrifugal forces.

the period of four years is detectable. The random scatter in the data is largely attributed to the primary constant volume flowmeter used to obtain the calibration data, which has a relative uncertainty of 0.05% (k=1).

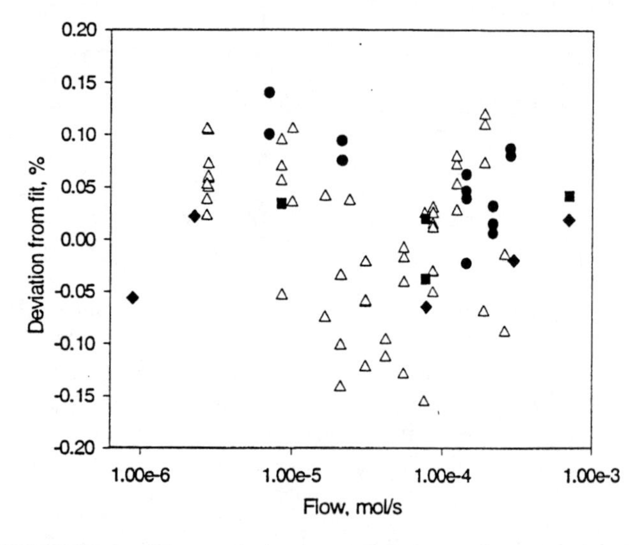

FIGURE 3. The symbols show deviations of the measured nitrogen flow from at fit to the 1995 data. The data taken at approximately one year intervals are represented by symbols; △ 1995, ● 1996, ◆1997, and ■1998.

DOPPLER-SHIFT FLOWMETER

An acoustic Doppler-shift flowmeter measures the flow-induced phase shift in an acoustic wave propagating along an unobstructed tube through which gas flows. Measurements of the upstream and downstream phase shifts enable the simultaneous determination of the mean gas velocity and the speed of sound in the gas, and measurements of the temperature and pressure enable computation of the mass flow rate. The goals for this flowmeter are (A) to measure gas flow in the range

0.74×10^{-6} to 740×10^{-6} mol/s, (B) to have a resolution of 7×10^{-8} mol/s over this range, (C) to work with multiple gases (including relevant process gases), (D), to have uncertainties of less than 1% over the range of the instrument, and (E) to be essentially a real-time instrument with a time resolution of 0.1 seconds.

Flowmeter Design

A prototype acoustic flowmeter (ACFM) has been developed which uses a stainless steel flowtube and brass couplers. Narrow-band piezoelectric transducers with stainless steel diaphragms are used as a source, and 0.318 cm condenser microphones are used as receivers. While not compatible with all gases because of their nickel diaphragms, these condenser microphones are useful in developing a laboratory instrument, and eventually piezoelectric transducers similar to the source will also be used as receivers. Because sound waves propagating in the tube walls would interfere with phase measurements in the gas, coupling directly into the gas is best achieved through small holes (~0.36 mm) in the tube. For a tube with rigid walls, the fundamental mode of acoustic propagation is a plane wave of phase velocity c (the speed of sound in free space), and for frequencies below the cutoff frequency $f_{cutoff} = c/(1.7d)$ (where d is the tube diameter), only this mode propagates without significant attenuation (8). Choosing d = 1.0 mm results in a cutoff frequency of about 200 kHz. This tube diameter choice is also prompted by flow considerations, since it is desirable to keep both Mach numbers and Reynolds numbers low ($M<0.07$ and $Re<2000$ for a maximum argon flow of 7×10^{-4} mol/s). Availability of appropriate transducers and the desire to stay well below 200 kHz led to a choice of 100 kHz for the source frequency. A schematic of the ACFM setup is given in Fig. 4. A dimension of 0.1 m between source and receiver was chosen to accommodate the transducer size and efficiency and the attenuation length (8) in the gas. Extending the tubes 0.2 m past each receiver sufficiently reduces any signal reflected off the end of the tube. This may also be accomplished via an impedance-matching horn.

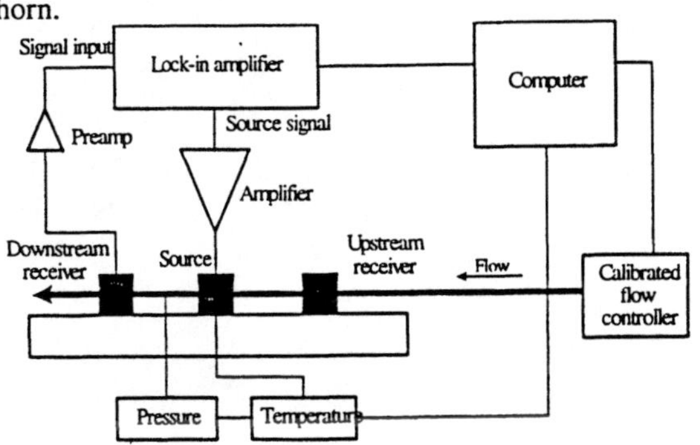

FIGURE 4. Schematic representation of ACFM

ACFM Results

The mean flow velocity (v) and mass flow rate ($\dot{m}$) for the ACFM are calculated with equation (2).

$$v = (\Delta\Theta)c^2/fL\,, \quad \dot{m} = \rho\, v(\pi r^2) \tag{2}$$

Here ($\Delta\Theta$) is the change in phase between a given flow and zero flow, c is the sound speed, f is the source frequency, L the distance between source and receiver, ρ is the gas density, and r is the tube radius. The mean velocity may be calculated in this manner because the acoustic wave number (k) is independent of the flow profile to first order(9).

To test the ACFM, the flow from a NIST-calibrated flow controller (2) is directed into the ACFM and simultaneous flow measurements are performed with each instrument. The phase is measured with a lock-in amplifier. Typically, the flow is increased from a minimum value to a maximum value, with 20 phase measurements made at each flow setting. Fig. 5 shows the argon flow measured with the ACFM as a function of the calibrated flow.

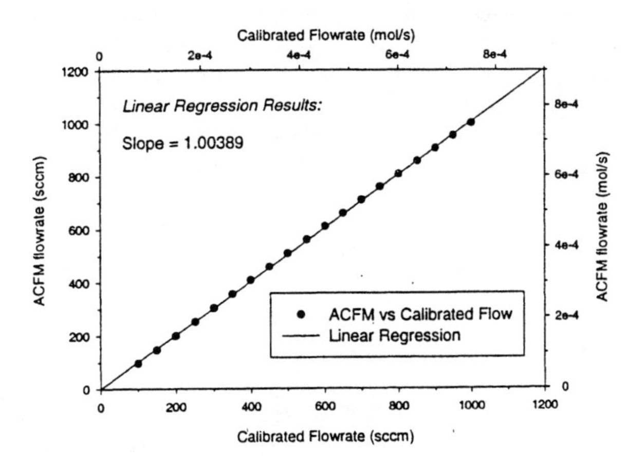

FIGURE 5. ACFM results for argon in a 1.0 mm inner diameter tube.

The slope is close to one, as expected, and the deviation from linearity is small. The phase changes by 11.7 radians over a range of 6.7×10^{-4} mol/s (900 sccm), giving a sensitivity of 9.72×10^{-9} radians/mol/s (0.013 radians/sccm). The standard deviation for a typical phase measurement is approximately 8×10^{-4} radians, indicating a flow resolution of 7.4×10^{-8} mol/s (0.1 sccm) is possible. The residuals for this data all fall within 1% of the maximum flowrate.

Similar experiments have been performed with a variety of gases (including N_2, He, and SF_6) and for different coupler and flowtube geometries. The results are consistent with those presented for argon in this paper. A more detailed account of the apparatus design, experimental procedure, and results analysis is currently being prepared for publication. There are several challenges remaining in order to develop a more robust instrument, including incorporation of piezoelectric transducers as receivers.

SUMMARY

Two types of flowmeters have been developed which operate in the range of 7 x 10^{-7}mol/s to 7 x 10^{-4} mol/s. The first is a pressure drop (laminar flowmeter) that has demonstrated a stability over a four year period of +/- 0.15% or better. This flowmeter is capable of being used with most gases and has been used to perform proficiency tests of primary flow standards maintained by thermal mass flow controller manufacturers and flowmeter users (3). While this flowmeter maintains excellent stability with time, it's usefulness is limited to gases with room-temperature vapor pressures greater than 1 MPa.

The second flowmeter measures the Doppler-shift induced by the flow to determine the gas velocity. Temperature and pressure measurements are made to calculate the gas density. The density is used along with the measured gas velocity to determine the mass flow of gas. This flowmeter has a very low pressure drop (20 kPa or less) and may be used with a wide variety of semiconductor process gases. The prototype acoustic flowmeter (ACFM) is capable of measuring flows with uncertainties of 1% of the instrument full scale value.

Either of these flowmeters can be used for *in-situ* calibration of other flowmeters such as thermal mass flow controllers or to measure process gas flows directly. Each flowmeter has unique capabilities and limitations that dictate their usefulness for particular applications.

ACKNOWLEDGEMENTS

The authors are grateful to Michael Moldover and Tony Goodwin for consultation regarding the ACFM. This work was funded in part by the National Semiconductor Metrology Program.

REFERENCES

1. National Technology Roadmap for Semiconductors, Sematech, Austin, Texas (1997)
2. Tison, S.A. and Berndt, L., "High-Differential-Pressure Laminar Flowmeter," Proc. of the 1997 ASME Fluids Engineering Summer Meeting, Vancouver, B.C., June 22-26, 1997.
3. Tison, S.A. and Doty, S.W., "Low-Gas-Flow Proficiency Testing," Proc. of the 1997 National Conference of Standards Laboratories (NCSL) annual meeting, NCSL, Atlanta Georgia, July 27-31, 1997
4 .Tison, S.A., *J.Vac. Sci. Technol.*, **14(4)**, 2582-2591, (1996).
5. Nelsen, D., *Semiconductor International*, October 1992, 72-74 (1992).
6. Schlichting, H., *Boundary Layer Theory*, New York, McGraw-Hill, 1960, pp. 68-69.
7. Berger, S.A., and Talbot, L., *Ann. Rev. Fluid Mech.*, **15**, 461-512 (1983).
8. Morse, P.and Ingard, K., *Theoretical Acoustics*, Princeton, NJ: Princeton University Press, 1968.
9. Robertson, B., J. Fluids Eng. **106**, 18 (1984).

A Self-Controlled Microcontrolled Microvalve

C. A. Rich and K. D. Wise

Center for Integrated Sensors and Circuits, Dept. of EECS, University of Michigan, Ann Arbor, MI 48109-2122

Integrated microvalves are needed for a broad range of semiconductor-processing-related applications. These include precision mass microflow controllers (μFCs) for dry etch systems, miniature gas chromatography systems for real-time monitoring, point-of-use semiconductor process reactant generators, and compact control systems for mini-environments. This paper reports a pneumatically actuated, integrated silicon microvalve, which was developed as a forerunner to an 8b μFC intended for the precision control of semiconductor process gases in the range from 0.1 to 10sccm. The structure was designed to be batch-fabricated and compatible with on-chip thermopneumatic actuation. Assembled single-bit μFC devices achieve the targeted flow rate of 5sccm (determined by an in-line flow channel) at 20psid (1034torr). The valve alone may achieve significantly higher flow rates. The leak rate is 0.08sccm under 26.1psig actuation pressure, and the valve can seal against pressures greater than 29psid (1500torr).

Keywords: microvalve, microflow, mass flow, gas chromatography, dry etch, RIE

INTRODUCTION

Despite the continued advance of MEMS-related technology, development of a high-performance, batch-fabricated, micromachined silicon microvalve for the precision control of fluids remains an elusive goal. Significant research has already progressed in this area, although no single valve structure has been popularly accepted by industry. A brief review of some potential applications demonstrates the relevance of this work to the semiconductor industry in particular.

Gas Metering Applications

Mass flow controllers (MFCs) fulfill the fundamental need of many fabrication processes to introduce precisely metered quantities of fluid reactants into controlled environments. As such, MFCs have a direct impact on process characteristics. For example, oxygen flow in a dry etch (such as RIE) has a substantial impact on process uniformity, anisotropy, selectivity, and aspect-ratio-dependent etching (ARDE) (1)-(3). Note that low-flow process gases, such as oxygen, often require flow rates on the order of 0.1-10sccm or less.

As a reflection of industry consensus, the 1997 National Technology Roadmap for Semiconductors (NTRS) discussion of "critical dimension" (CD) implies an acute need for better gas flow control in the future. The Interconnect Technology Working Group assumes a constant CD tolerance of ±8%, independent of absolute CD (4). Thus, if the etch characteristics of a process are highly dependent on the flow of a certain gas, a flow controlled to within 1-2% accuracy would still consume nearly a fourth of the total tolerance budget. Likewise, the NTRS anticipates that etch selectivity (5) and ARDE (6) will continue to be key issues for future technologies.

Current mass flow control technology, however, may not be adequate to meet future requirements. Commercial MFCs are poorly suited for low-flow semiconductor processing, with accuracies of ±0.1 or even ±1 sccm at low flow ranges. Conventional MFCs also suffer from slow time constants and inherent dead volumes that degrade transient response (7). *Dead volume* refers to the empty cavities in the downstream flow path of an MFC (or stand-alone valve) that must be pressurized with process gas before flow starts and must be depleted before flow will stop. This characteristic is similar to the parasitic capacitance of an electrical circuit and is likewise undesirable.

Embedded Fluid Control Applications

Applications for microvalves embedded in larger systems are also emerging. Of particular interest is the potential for a compact, integrated gas chromatography (GC) module. Micromachined GC (μGC) technology promises performance and portability well beyond present-day macro systems, but complete integration currently awaits a suitable valving system (8)-(11).

The Factory Integration section of the NTRS identifies metrology needs that could potentially be met by a μGC. For example, assessing the effect of factory performance specifications (which would include environmental contamination from gases and volatile organic compounds) on wafer yield would enable a reduction of factory overdesign and process material overspecification (12). Recent research has used GC methods to identify sources of yield-impacting contamination as obscure as HEPA filter potting compounds (13).

In situ process measurement and control, another pillar of the NTRS (14), would benefit greatly from μGC technology as well. GC technology has been pinpointed for the detection of airborne contaminants (15), the analysis of

CP449, *Characterization and Metrology for ULSI Technology: 1998 International Conference*
edited by D. G. Seiler, A. C. Diebold, W. M. Bullis, T. J. Shaffner, R. McDonald, and E. J. Walters

process gas impurities (16), and the evaluation of organic content in recycled and/or ultrapure water supplies (17). Some airborne contaminants commonly found in cleanrooms have been shown to substantially degrade deep-UV (DUV) photoresists; effective monitoring of such contaminants is readily achieved with GC technology (18).

Industry has already taken the first steps towards μGC technology. MTI Analytical Instruments, for example, offers a portable GC system that utilizes a silicon micromachined injector and solid-state detector (19). This completely self-contained system measures 15cm × 36cm × 36cm and can operate continuously for 6-8 hours. Detection of 1ppm takes place within 160s. Although not handheld size, the system nonetheless demonstrates commercial interest in the miniaturization of GC equipment.

The NTRS also anticipates advances in containment that would benefit from microvalves in general. For example, the Front End Processes section recommends point-of-use generation of process reactants to improve efficiency (20). Microvalves and microreaction chambers could facilitate localized generation of such reactants in precisely the quantities required (21), (22). Also of interest, the Factory Integration section foresees the rapid embrace of cluster tools and minienvironments to enhance flexibility, reduce chemical use, and improve contaminant control (23). Such equipment requires compact, inexpensive control devices to restrain the cost of redundancy and to keep machine footprints within reasonable limits.

DESIGN

Considering the breadth of possibilities, it is apparent that the requirements for microvalves may vary greatly; however, a general set of design metrics may be proposed. An ideal microvalve should be *robust, low-power, fast, simply fabricated, useful over a wide pressure and flow range,* and *low-leak.* Although compatibility with corrosive fluids is another concern, nearly all microvalves are micromachined out of silicon, and thus fluid compatibility is dependent on protective films rather than on the fundamental valve structure.

Industry and academia have both shown creativity in the development of stand-alone microvalves, although all of the devices reported so far (to the authors' knowledge) have significant limitations. In-depth reviews of the various technologies available are compiled elsewhere (24), (25). After evaluation of various actuation schemes, *pneumatic* actuation, as a precursor to thermopneumatic actuation, was chosen for development in this research. Pneumatic and thermopneumatic actuation exhibit the potential to perform acceptably by all six microvalve design metrics. Both of these methods rely on gas or vapor pressure to drive a valve diaphragm. (Thermopneumatic valves generate pressure by vaporizing a liquid.) The advantage of these two methods is an ability to generate large forces over a significant distance, since gas molecules will diffuse within a closed cavity to reach pressure equilibrium. The freedom to specify actuator size (by simply scaling the diaphragm), combined with the potential for considerable travel, allows microvalves to be constructed with comparatively high flow conductance. Substantial valve plate travel should also facilitate robust valves with a high tolerance for particulate contamination. Furthermore, actuation pressure is relative to ambient pressure; thus, nearly any inlet pressure may be valved if appropriate ambient and drive pressures are maintained.

Simple pneumatic actuation does require an off-chip compressed-air source, which may or may not be a concern depending on the application. Thermopneumatic actuation (discussed later) should provide a viable solution to this concern, however.

For the sake of simplicity in testing and characterization, the authors decided to initially construct devices consisting of only a single microvalve with an in-line flow channel. While not very useful as a variable flow controller, this simple 1-bit design nonetheless serves as proof-of-concept for a future 8-bit controller. Some design goals were also established for this device:

- *Robustness*: Employ a durable structure that is able to handle 10μm particulates without loss of performance
- *Fabrication*: Use only one silicon and two glass layers; primarily batch-fabricated
- *Leak rate*: Achieve a leak rate of < 0.01sccm at 20psid (1034torr) inlet–to–outlet and 26.1psig actuation pressure
- *Operating range*:
  Pressure: Be able to close against 29psid (1500torr)
  Flow: Conduct ~5sccm at 20psid for the entire μFC; achieve a flow for the valve alone of 1000sccm at 20psid

Figure 1 depicts the developed prototype structure. The corrugated diaphragm in the silicon layer functions as a valve plate and occludes the gas inlet when driven by pneumatic pressure. With no drive pressure applied, the valve is open and gas flows through the precision flow channel etched in the top glass cap, exiting through the gas outlet. The incoming pneumatic drive line is controlled by a 3-way solenoid valve fed from a compressed nitrogen source.

The flow channel and valve seat depths have been set at 20μm to accommodate the passage of large particulates. The remaining dimensions for the flow channel are subsequently sized to generate the anticipated flow rate of ~5sccm. A quasi-empirical model for gas flow rates through the valve structure and flow channel has been developed from the equations of Olsen, Berman, and Steckelmacher (26)-(28). Although their equations assume rectangular channels, an "uncurling" of the circular valve structure allows equivalent rectangular dimensions to be

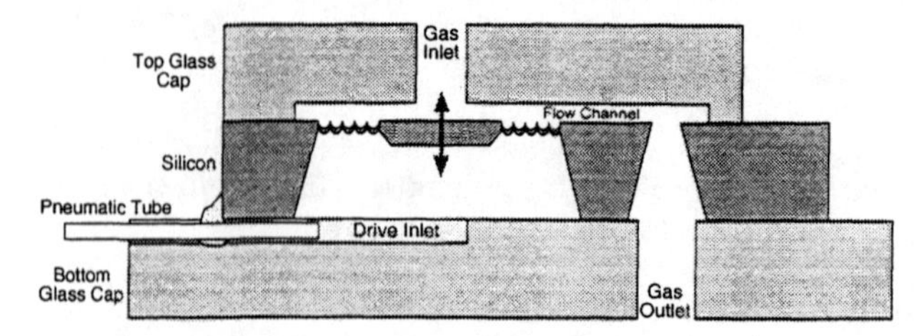

FIGURE 1. Schematic of a pneumatically actuated microvalve.

assigned for modeling purposes.

Design and fabrication of the valve diaphragm and pressure cavity draws upon previous research performed at the University of Michigan by Zhang (29). The overlaid labels of Figure 3 give major dimensions for the valve.

The silicon layer of Figure 1 requires four masks. Front and back deep-boron etch-stops define the diaphragm rim and boss (valve plate), the pressure cavity, and the outlet port edges. Next, an RIE etch defines the top side of the corrugations. Finally, a shallow boron etch-stop defines the corrugation thickness. An EDP etch is then used to bulk micromachine the diaphragm underside, pressure cavity, and gas port. This etch step is also used to separate the dies, since dicing might damage the released diaphragms.

The flow channel and valve seat in the top glass cap are defined by a single mask and etched using an HF-nitric acid etch. A groove is also saw-cut into the bottom glass cap for drive inlet access. After subsequent dicing, the gas inlet and outlet orifices are drilled in the top and bottom caps, respectively, with a diamond bit. Although individually drilled holes are not conducive to batch fabrication, future work may take advantage of batch-compatible ultrasonic glass drilling. Note that the flow channel dimensions, and thus gas flow rates, are completely determined by the separately processed glass cap. Next, the top glass cap is attached by electrostatic bonding, and the bottom glass cap is glued with epoxy (although electrostatic bonding could also be used). Finally, the pneumatic drive tube is inserted into the groove in the bottom glass cap and sealed with epoxy.

Flow and leak rates for the prototype device were expected to be less than 10sccm and as low as 10^{-3} sccm, respectively, depending on drive and inlet pressures. Since no commercial flowmeters were found that could measure such low rates accurately, an indirect approach has been developed: $\Delta P/\Delta t$ measurements for a known volume at the outlet of the device are taken as test data. Using a modified ideal gas law, $\Delta P/\Delta t \cdot V = \Delta n/\Delta t \cdot RT$, with appropriate substitutions, mass flow may then be found.

Zhang's work concluded the lifetime for corrugated diaphragms of similar construction to be at least several hundred thousand cycles without noticeable fatigue. No lifetime tests have yet been performed for the work reported here.

EXPERIMENTAL RESULTS

Final yield after assembly and processing was lower than anticipated, although most problems are understood and can be avoided in the future. The fabricated components were still sufficient, however, to allow the subsequent assembly and testing of several devices. There was also some initial concern that the valve boss would become electrostatically bonded to the top glass cap, thus sealing the valve shut permanently. Fortunately, such problems never occurred and all assembly efforts were successful in this regard. A photo and SEM shots of device structures are shown in Figures 2 and 3.

The circular markers of Figure 4 correspond to measured data for one of the devices tested. None of the valves were tested for maximum unregulated flow, due to difficulty in separating an intact valve from its channel.

FIGURE 2. Photograph of silicon and top glass cap. Bottom glass cap is not shown.

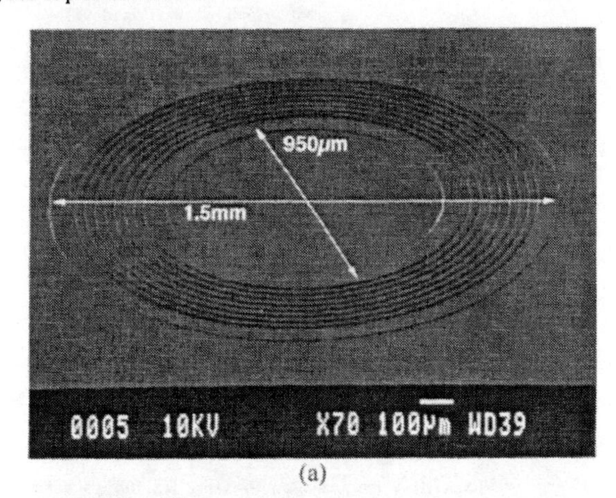

(a)

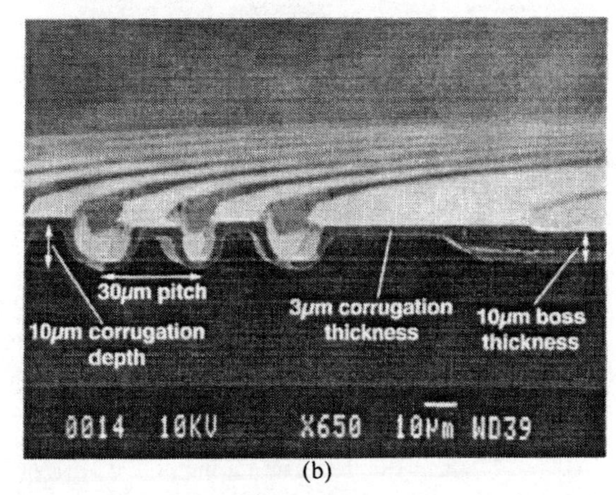

(b)

FIGURE 3. a) SEM shot of corrugated diaphragm/valve plate. b) Cross-section of corrugations.

During testing, it became apparent that thorough cleaning before the electrostatic bonding step is critical to a good valve seal. Any residue on the valve sealing surfaces degraded the leak rate considerably. It was also found, however, that particulates in the inlet gas stream were largely irrelevant. Valves survived the passage of "dirty" room air, water, alcohol, and other cleaning solvents with

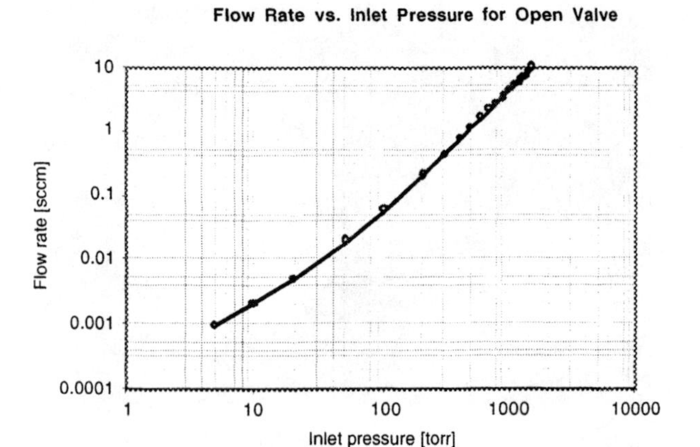

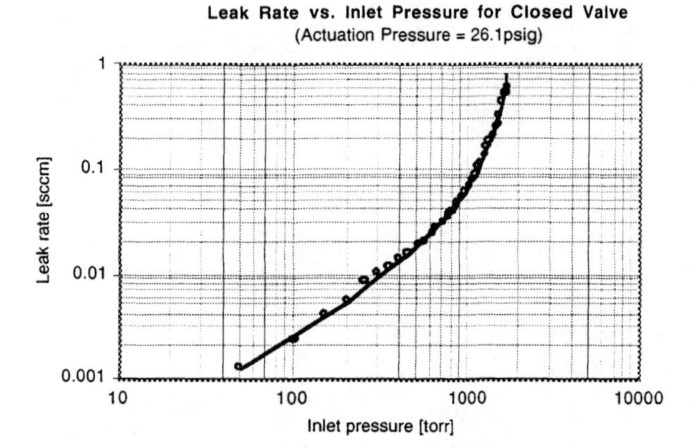

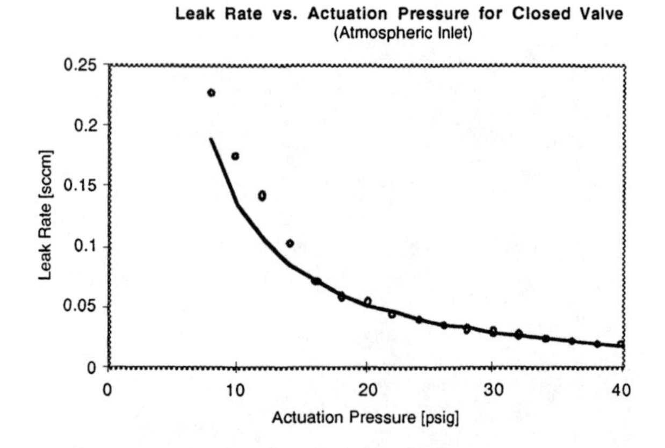

FIGURE 4. Flow and leak rate data for the prototype device. All valves were tested with outlets vented to vacuum (10^{-2} Torr). Circles correspond to data points and lines to performance predicted by the model.

little or no change in leak rates. In fact, the passage of liquids *improved* the leak rates slightly for several of the valves, presumably by washing out particulates.

Response time for the valves is largely determined by the relaxation time of drive pressure in the pneumatic drive line, which is dependent on tubing length and diameter. The drive lines were connected to a 3-way solenoid valve by 0.3m of 200μm ID PTFE tubing. The speed of the solenoid valve controlling the drive pressure also influenced the response time. A rough estimate by visual observation indicated a response time of a few tenths of a second.

DISCUSSION & CONCLUSIONS

The solid curves of Figure 4 correspond to the model predictions for the valve under test. Predicted and measured data values match to within a few percent on average. The range of accuracy required for a full μFC implementation would probably require flow channel dimensioning by means of Monte Carlo simulation. This would account for both viscous and molecular flow through the channel.

Additionally, the glass channels of the present device were etched using a wet-etch process, which, although sufficient for testing purposes, is not uniform enough for mass production. An ultrasonic or dry etch process would most likely be required to achieve the precision required for commercial applications.

Leak rates for the valve/μFC combination were less than 0.1sccm through 20psid, with 26.1psig drive pressure. This meets the design target of 0.01sccm only through 5.7psid (300torr); hence, further improvement in valve sealing ability is desirable. Leak rate is determined primarily by the geometry of the closed valve, with little contribution from the flow channel. Thus a larger maximum open flow yields a larger dynamic range. Adjustment of the model to remove the flow channel completely predicts a flow rate at 20psid on the order of 10^{4}sccm and leak rate of 10^{-1}sccm, giving a potential dynamic range of 100000:1 for a stand-alone valve.

The feasibility of the basic design and process flow developed for the prototype pneumatically actuated valve structure has thus been verified, as demonstrated by the realization of working microvalve/μFC structures. The present design exhibits many of the fluid-handling advantages of a good microvalve, such as low dead volume, low leak rate, high flow rate, and robustness, in spite of its requirement for an external air source.

FUTURE WORK

An 8-bit Microflow Controller

The pneumatically-actuated microvalve is only a proof-of-concept stepping-stone on the way to development of more useful devices. As a first attempt to demonstrate a pneumatic-microvalve-based system, efforts now seek to develop a pneumatically actuated μFC with 8b resolution. Such a flow controller entails routing eight binarily weighted, valved flow channels in parallel. This configuration takes advantage of the long-term stability of micromachined flow channels over proportional valving to develop a μFC with accuracy and repeatability exceeding that of conventional MFCs.

The targeted flow range for the device is 0.05-12.8sccm with 800torr inlet pressure and 26psig actuation pressure. This range slightly overlaps the expected leak rates to allow for future improvements in valve sealing.

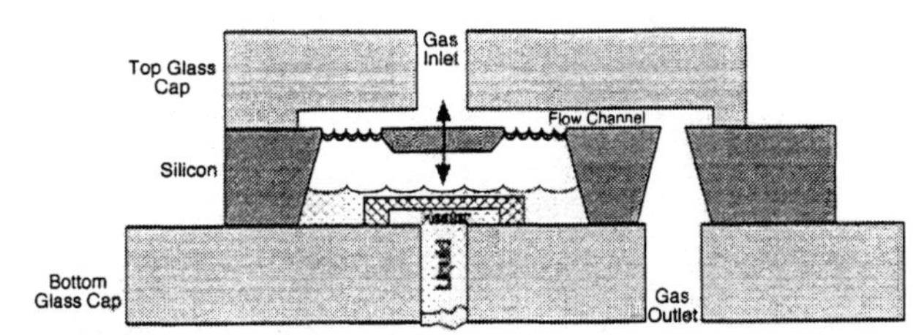

FIGURE 5. Schematic of a thermopneumatically actuated microvalve/μFC.

(One possibility is the deposition of parylene, a soft, conformal polymer, on the valve sealing surfaces.)

Thermopneumatically Actuated Microvalve

It is also desirable in the long term to implement a self-contained microvalve that may be completely regulated by electronic signals. To that end, a *thermo*pneumatically actuated microvalve is depicted in Figure 5.

Thermopneumatic actuation relies upon drive pressure developed by the vaporization of a heated fluid in a sealed cavity. The balance of the structure (diaphragm, flow channel, etc.) shown in Figure 5 is identical to that of the pneumatically actuated μFC. Thermopneumatic actuators previously fabricated at the University of Michigan by Bergstrom, et al. produced nearly 2atm in less than 100ms, with only 20mW of input power (30). Simulations predicted an optimized actuation time of less than 40ms for Bergstrom's elevated heater structure. Such response time and power consumption is competitive with other actuation schemes. Furthermore, the sustained drive pressure of such an actuator falls within a range appropriate for integration with the valve reported herein. Successful thermopneumatic valves of another design have already been demonstrated elsewhere by Zdeblick, et al. (31), (32).

Operating temperature range may be a concern, although a wide variety of actuation liquids with different boiling points are available (24). This may allow adequate freedom to select suitable actuation temperatures for many applications. The author anticipates completion of such a thermopneumatic microvalve as the final step in development of a robust, self-contained, useful microvalve assembly for integration into complete microsystems.

ACKNOWLEDGMENTS

The author would like to thank Dr. Janet Robertson for construction of an initial version of the microvalve flow model. Drs. Andrew Oliver and Yafan Zhang also graciously provided assistance with development of the fabrication process flow.

This work was supported by the Semiconductor Research Corporation, contract #97-FC-085.

REFERENCES

1. Legtenberg, R., Jansen, H., de Boer, M., and Elwenspoek, M., *J. Electrochem. Soc.* **142**, 2021-2028 (1995).
2. Morimoto, T., *Jpn. J. Appl. Phys.* **32**, 1253-1258 (1993).
3. Tuda, M., and Kouichi, O., *Jpn. J. Appl. Phys.* **36**, L518-L521 (1997).
4. 1997 *National Technology Roadmap for Semiconductors*, San Jose, CA: Semiconductor Industry Assoc., 1998, p. 94.
5. Ibid, p. 77.
6. Ibid, p. 105.
7. Esashi, M., *Sensors and Actuators A* **21-23**, 161-167 (1990).
8. Terry, S., Jerman, J., and Angell, J., *IEEE Trans. on Elect. Dev.* **ED-26**, 1880-1886 (1979).
9. Bruns, M., "Silicon micromachining and high speed gas chromatography," *Proc.*, Inter. Conf. Industrial Elect., Cntrl., Instr. and Automation, Power Elect. and Motion Cntrl., San Diego, CA, pp. 1640-1644, 1992.
10. Ocvirk, G., Verpoorte, E., Manz, A., and Widmer, H., "Integration of a micro liquid chromatograph onto a silicon chip," *Proc.*, 8th Inter. Conf. Solid-State Sensors and Actuators (Trans. '95), Stockholm, Sweden, pp. 756-759, 1995.
11. Reston, R., and Kolesar, S. Jr., *J. Micromech. Systems* **3**, 134-155 (1994).
12. 1997 *NTRS*, p. 120.
13. Lebens, J. A., McColgin, W. C., and Russell, J. B., *J. Electrochem. Soc.* **143**, 2906-2909 (1996).
14. 1997 *NTRS*, pp. 122-123.
15. Tamaoki, M., et al., "Effects of airborne contaminants in the cleanroom for ULSI manufacturing process," *Proc.*, IEEE/SEMI Adv. Semic. Manuf. Conf. and Wkshp. (IEEE 95CB35811), Cambridge, MA, pp. 322-326, 1995.
16. Carpio, R. A., and Lindt, E., *Semic. Int'l.* **12**, 164-167 (May 1989).
17. Frye, G. C., et al., *J. Inst. of Environ. Sci.* **39**, 30-37 (Sep-Oct 1996).
18. Dean, K. R., and Carpio, R. A., "Real-time detection of airborne contaminants in DUV lithographic processing environments," *Proc.*, Tech. Mtg., Inst. of Environ. Sci., Mt. Prospect, IL, pp. 9-16, 1995.
19. <http: //www.mtigc.com/products_and_solutions/p200.html>
20. 1997 *NTRS*, p. 78.
21. Caruana, C. M., *Chem. Engr. Prog.* **92**, 12-19 (Apr 1996).
22. Srinivasan, R. et al., "Chemical performance and high temperature characterization of micromachined chemical reactors," *Proc.*, 9th Inter. Conf. Solid-State Sensors and Actuators (Trans. '97), Chicago, IL, pp. 163-166, 1997.
23. 1997 *NTRS*, p. 130.
24. Barth, P., "Silicon microvalves for gas flow control," *Proc.*, 8th Inter. Conf. Solid-State Sensors and Actuators (Trans. '95), Stockholm, Sweden, pp. 276-279, 1995.
25. Krulevitch, P., et al., *J. Micromech. Systems* **5**, 271-281 (1996).
26. Berman, A. S., *J. Appl. Physics* **36**, 3356 (1965).
27. Olson, R. M., *Essentials of Engineering Fluid Mechanics*, 4th. New York: Harper & Row, 1980.
28. Steckelmacher, W., *Vacuum* **28**, 269-275 (1978).
29. Zhang, Y., "Non-planar diaphragm structures for high-performance silicon pressure sensors," Ph.D. thesis, University of Michigan, July 1994.
30. Bergstom, P., et al., *J. Microelect. Systems* **4**, 10-17 (1995).
31. Zdeblick, M. J. and Angell, J. B., "A microminiature electric-to-fluidic valve," *Proc.*, 4th Inter. Conf. on Solid-State Sensors and Actuators (Trans. '87), Tokyo, Japan, pp. 827-829, 1987.
32. Zdeblick, M. J., "Thermopneumatically-actuated microvalves and integrated electro-fluidic circuits," *Proc.*, IEEE Solid-State Sensor and Actuator Workshop, Hilton Head, SC, pp. 251-255, 1994.

X-ray photoelectron spectroscopy, depth profiling, and elemental imaging of metal/polyimide interfaces of high density interconnect packages subjected to temperature and humidity

David R. Jung,[1] Bola Ibidunni,[2] and Muhammad Ashraf[2]

[1]Science Applications International Corporation, Rockville, MD 20851,
[2]Sheldahl Corporation, Longmont, CO 80503

X-ray photoelectron spectroscopy (XPS) was used to analyze surfaces and buried interfaces of a tape ball grid array (TBGA) interconnect package that was exposed to temperature and humidity testing (the pressure cooker test or PCT). Two metallization structures, employing 3.5 and 7.5 nm Cr adhesion layers, respectively, showed dramatically different results in the PCT. For the metallization with 3.5 nm Cr, spontaneous failure occurred on the polymer side of the metal/polyimide interface. Copper and other metals were detected by XPS on and below this polymer surface. For the metallization with 7.5 nm Cr, which did not delaminate in the PCT, the metallization was manually peeled away and also showed failure at the polymer side of the interface. Conventional XPS taken from a 1 mm diameter area showed the presence of metals on and below this polymer surface. Detailed spatially-resolved analysis using small area XPS (0.1 mm diameter area) and imaging XPS (7 μm resolution) showed that this metal did not migrate through and below the metal/polymer interface, but around and outside of the metallized area.

INTRODUCTION

Semiconductor packaging is typically composed of multiple metal and polymer layers. Long-term adhesion of the metal/polymer interfaces is needed for the required lifetime of the products in which they are used. The surface analytical technique of x-ray photoelectron spectroscopy (XPS) has long been a primary method for assessing the chemical and physical reactions at metal/polymer interfaces and for diagnosing changes in those interfaces following temperature and humidity testing [1-3].

XPS is an extremely surface-sensitive analytical technique that provides information about the chemical composition (atomic percentages) and chemical states (e.g., oxidation states of metals) present in the near surface region of a material (the top 5 nm) [4]. This information is provided without the excessive tendency for sample degradation that is found in Auger electron spectroscopy (AES). AES, which otherwise has superior imaging resolution, also has a tendency to be plagued by charging problems, which frequently distort the chemical shifts and mask the corresponding chemical state information that is of interest.

A capability that has been available commercially only in the past few years (and that is increasingly finding use) is elemental imaging by XPS. One such instrument, the VG 220i XL, is capable of recording photoelectron images in a rapid "parallel" acquisition mode from sample regions 125, 250, and 1000 μm in width. The spectrometer employs a magnetic primary lens and performs optical Fourier transforms before and after the hemispherical energy analyzer. The imaging resolution of the system is dependent on an angular aperture setting. For the minimum setting of the aperture, an optimal resolution of 2 μm is specified. In practice, using a larger aperture setting yields images of good quality in 15 to 30 minutes with 7 μm spatial resolution.

EXPERIMENTAL

A tape ball grid array (TBGA) [5,6] interconnect package with approximately 100 to 200 μm -wide lines and spaces is composed of multilayer metallizations formed by an adhesive chromium layer, evaporated copper, and electroplated copper, nickel and gold on polyimide. The metallization is encapsulated with a solder mask, and solder balls are attached to the gold surface. Figure 1 schematically depicts the vertical and lateral structure (not to scale). The width of the solder ball is much greater than the width of the metallization lines.

X-ray photoelectron spectroscopy was performed on a VG 220i-XL instrument operated at a pressure of 5×10^{-10} torr or below and using a monochromatic Al K_α alpha source with a spot size of approximately 1 mm. Small area analysis (0.1 mm diameter) was defined by using the angular aperture and field of view aperture of the lens system of the instrument, rather than by using a small area x-ray beam. For imaging, the analyzer was operated in the CRR = 4 mode (constant retardation ratio defined as $E/\Delta E$, where E is the kinetic energy of the electron and ΔE is the pass energy of the analyzer) with an angular aperture setting of 3.5. These settings lower the energy and spatial

CP449, *Characterization and Metrology for ULSI Technology: 1998 International Conference*
edited by D. G. Seiler, A. C. Diebold, W. M. Bullis, T. J. Shaffner, R. McDonald, and E. J. Walters

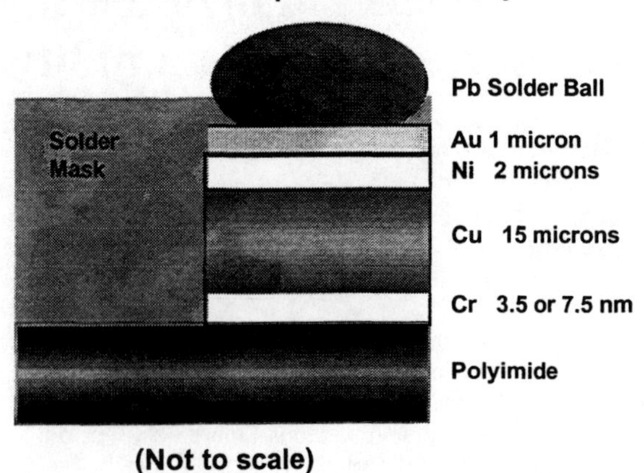

FIGURE 1. Structure of the TBGA with metallization thicknesses as marked. The lateral and vertical sizes are not drawn to scale.

resolution relative to their optimum values [viz., CRR = ∞ (or ΔE = 0) and AA = 0, respectively], but increase the count rate for imaging. Increasing the angular aperture value from 0 to 3.5 lowers the spatial resolution of the images from approximately 2 μm to 7 μm. Thus, for the roughly 100 μm wide metallizations of the TBGA, lower spatial resolution was traded for shorter image acquisition times.

FIGURE 2. TBGA (left) with 3.5 nm Cr after 96 h PCT showing spontaneous delamination. With 7.5 nm Cr (right) after 240 h PCT showing no delamination.

Elemental images were derived by subtracting a background image acquired at a binding energy 10 eV below that for the peak image. The acquisition times were 7 min. each for the Au 4f peak and background images, and 13 to 15 min. each for the C 1s, Ni 2p, and Cu 2p peak and background images of Fig. 7. Acquisition of an elemental image of a thick layer in 20 min. or less is typical for this instrument.

The samples were subjected to a pressure cooker test (PCT) consisting of exposure to 120 °C and 100% relative humidity (rh) for 48 to 240 hours. Following the PCT, the samples were allowed to cool and then stored in plastic bags until they were analyzed by XPS several weeks later. To analyze metal/polymer interfaces that did not spontaneously delaminate in the PCT, forceps were used to manually peel apart the metal from the polyimide just prior to insertion into the XPS instrument for analysis. Because even a few minutes in air is enough to oxidize Cr and Cu, the analysis focused on the location of the metals with respect to the metal/polymer interface, and not on their oxidation states. Furthermore, the chemical reactions of diffusing metal atoms or ions with the polyimide (that might have been analyzed in terms of the C 1s spectra) are obscured whenever Ar^+ ion sputtering is used, precluding a study of reactions below the immediate surface region.

RESULTS

After exposure to 100% rh and 125° C for 96 hours, the metallization delaminated from the polyimide for Cr adhesion layers of 3.5 nm but not for 7.5 nm (Figure 2).

In order to analyze the interface on the right side of Fig. 2 (7.5 nm Cr adhesion layer), the metallization was manually peeled using a forceps. The spontaneously

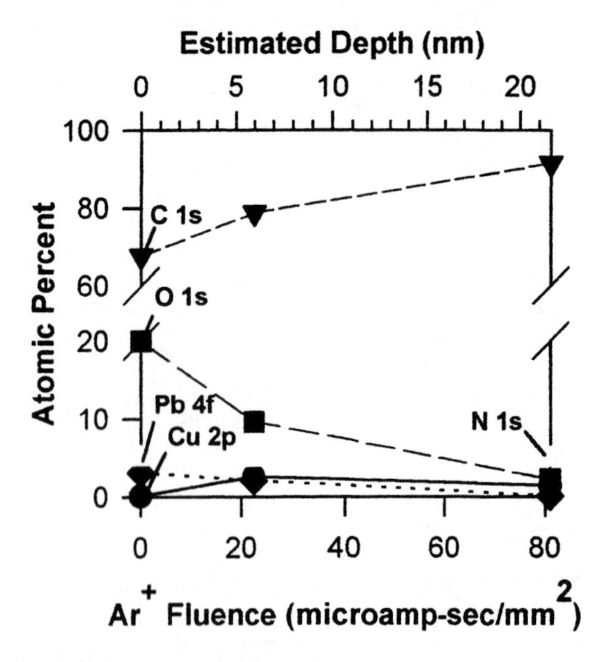

FIGURE 3. Sputter depth profile of polyimide side of the spontaneously-delaminated "S" interface opposite a metallization under a solder ball (see Fig. 2, left).

delaminated interface will be denoted "S" while the manually peeled interface will be denoted "P." XPS was used with either small (0.1 mm dia.) or large (1.0 mm dia.) area analysis and with and without sputter depth profiling using a rastered Ar^+ ion beam (3 keV).

XPS of the metal side of the "S" interface using a 0.06 mm analysis area showed significant C 1s and N 1s compositions at the surface (Table 1). This indicated that the failure of this interface occurred on the polymer side.

TABLE 1. Composition of metal side of the "S" interface without sputtering.

Element	Atomic Percent
Cr	56
C	22
O	17
N	3
Pb	2

Figure 3 shows a large area depth profile of the polyimide side of the "S" interface opposite a metallization under a solder ball. In this depth profile, the presence of Pb and Cu after exposure of the sample to a Ar^+ fluence of 80 microamp-s-mm^{-2} shows that the metals have migrated into the polyimide to a large degree. This evidence is consistent with the catastrophic failure of the metal/polymer interface through a mechanism of chemical catalytic degradation of the polyimide. The upper horizontal axis of Fig. 3 gives an estimate of the sample depth based on the sputtering rate that has been calibrated for a copper/nickel depth profiling standard (Geller Microanalytical). This is only intended as an order of magnitude estimate of depth because the sputter yield for polyimide is different from that for a metal.

For the "P" interface, XPS of the metal side showed large C 1s, N 1s, and O 1s peaks that indicated cohesive failure in the polymer and a strongly-adhered metal/polymer interface (Figure 4). Using a 1 mm diameter analysis area, Cu and Ni are observed also on the polymer side of that interface. Because these spectra were taken from the part of the TBGA where there were no solder balls, the possibility of finding Pb is excluded.

Figure 5 shows an optical micrograph of the polyimide side of the sample analyzed in Fig. 4. The lower part of the image is the polyimide surface revealed by peeling away the metallization. In the upper part of the image, two metallizations were initially present. In the upper left, the metallization has been peeled away, while in the upper right, the metallization is still present. Dark edges adjoin the metallized regions. To address the lateral distribution of the metal on the polyimide side of this "P" interface, the sample was positioned so that the strip beneath the metallization (lower part of Fig. 5) was centered in the XPS lens system. Figure 6 compares the survey spectra using 0.1 and 1.0 mm diameter analysis areas. The spectra show that Cu and Ni are not found within the strip beneath the metallization, but must be present outside this strip in regions that were not metallized.

Thus, the manually peeled ("P") interface that exhibited good adhesion in the pressure cooker test does not have metal below the peeled interface. Metal is present outside this interface, but that metal would have no effect on adhesion at and below the metal/polymer interface.

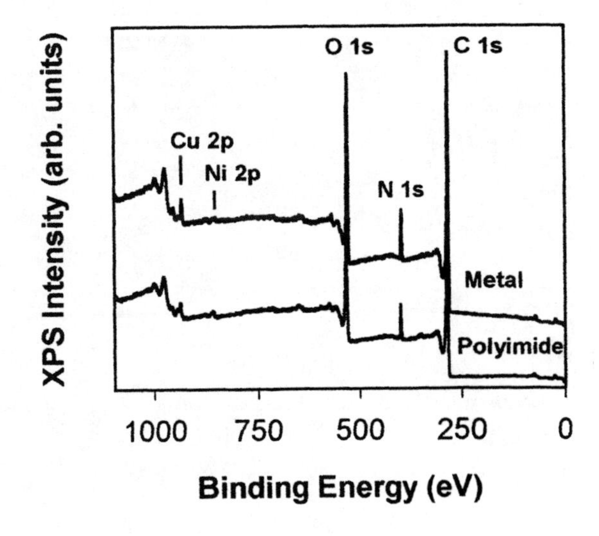

FIGURE 4. Comparison of XPS survey spectra using large area analysis (1 mm dia.) for the metal and polyimide sides of the "P" interface (without sputtering).

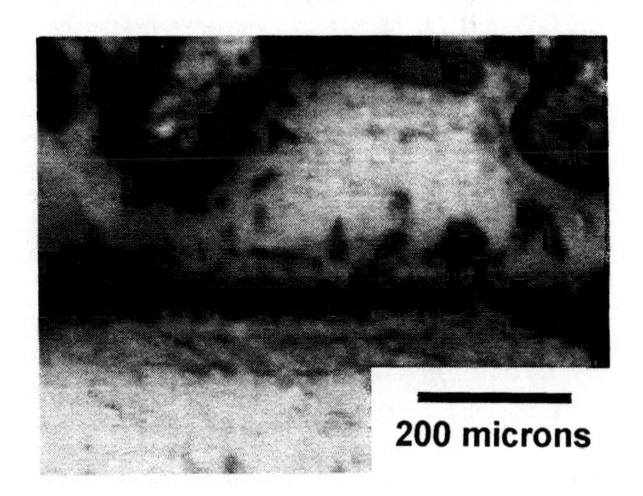

FIGURE 5. Optical micrograph of the polyimide side of the "P" interface, as discussed in the text.

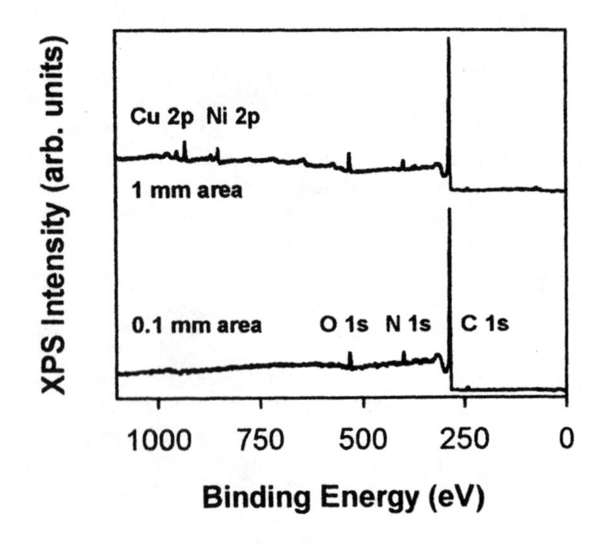

FIGURE 6. Comparison of XPS survey spectra using small area (0.1 mm dia.) and large area (1.0 mm dia.) analysis of the polyimide side of the "P" interface centered on the broad metallization region of the lower part of Fig. 5.

Figure 7 shows XPS imaging results for the same region of this sample shown in Fig. 5. Gold is present in the upper right corner in Fig. 7a because one of the metallizations was not manually peeled away, leaving the uppermost surface of the multilayer structure (Fig. 1) intact. In the image for C 1s (Fig. 7b), the dark lines outlining the edges of the metallization indicate the presence of material on top of the polyimide in these regions. This overlying material attenuates the C 1s photoemission intensity from below. This overlayer is clearly composed of Cu and Ni (Figs. 7c and 7d, respectively), as shown by the bright regions of the Cu 2p and Ni 2p images that match well the dark regions of the C 1s image (Fig. 7b). These imaging results are in accord with the presence of Cu and Ni only in the large area survey spectra of Fig. 6. The spatial resolution for these images has been measured to be 7 μm. At the edge of a step-like feature in the image, the spatial resolution is defined as the distance over which the image intensity drops from 80% to 20%. This spatial resolution is quite adequate for imaging features of the size typical in a microelectronics package (100 μm or greater).

CONCLUSIONS

For the thinner Cr layer, spontaneous failure occurred on the polymer side of the metal/polyimide interface. Massive copper migration into the polyimide was observed along with other elements. For the thicker Cr layer, Cu was not observed below the manually-peeled interface, but was observed on the non-metallized surface of the polyimide, especially near the edges of the metal lines and pads. For the "S" interface, the evidence is consistent with catastrophic failure of the metal/polymer interface through a

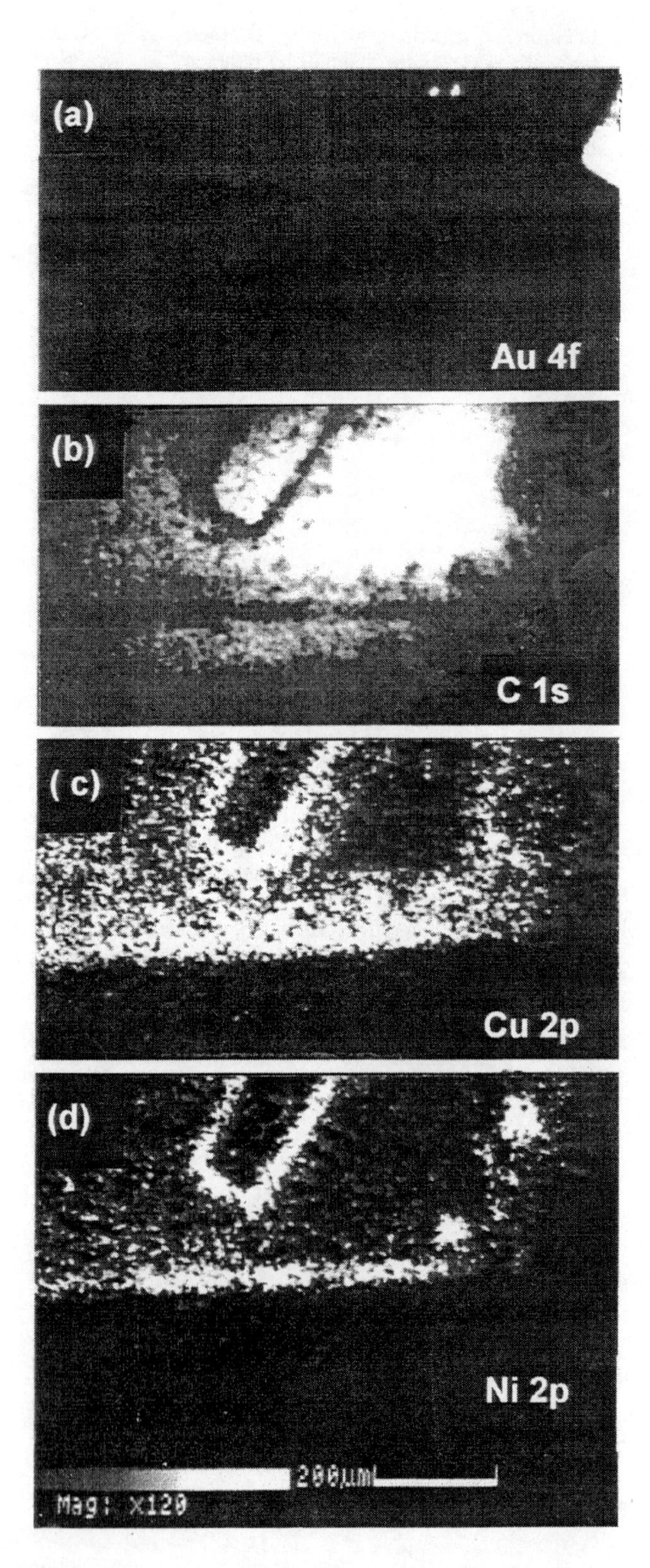

FIGURE 7. XPS imaging of C 1s, Au 4f, Cu 2p, and Ni 2p core levels showing the presence of metals only outside of the metallized areas.

mechanism of chemical catalytic degradation of the polyimide. For the "P" interface, the presence of metals on the surface of the polyimide but outside of the metallized regions cannot be the result of conventional diffusion. It may be caused by the migration of metal ions in solution in the water layer present on the sample during the pressure cooker test. Further work is needed to identify the mechanisms of metal diffusion and/or migration (both lateral and vertical) and the roles they may play in interface failure.

The results of this study demonstrate that imaging XPS can be applied quickly and with appropriate spatial resolution to characterize metallizations which form the high density interconnect packages that are of increasing importance in the microelectronics industry [5,6] Specifically, we have identified the presence of metals at interfaces of tape ball grid arrays that have been subjected to temperature and humidity. Imaging XPS and small area XPS provided information on the lateral distribution of copper and other metals on the surface of the array that would otherwise not have been available. This information was crucial in accurately distinguishing the effects of lateral and vertical metal migration at the spontaneously and manually-peeled metal/polyimide interfaces of the TBGAs.

REFERENCES

1. Jordan , J.L., Kovac, C.A., Morar, J.F., and Pollak, R.A., *Phys. Rev. B* **36,** 1369-1377 (1987).

2. Gerenser, L.J., *J. Vac. Sci. Technol. A* **6,** 2897-2903 (1988).

3. Furman, B.K., Childs, K.D., Clearfield, H., Davis, R., and Purushothaman, S., *J. Vac. Sci. Technol. A* **10,** 2913-2920 (1992).

4. Briggs, D., and Seah, M.P., Eds., *Practical Surface Analysis,* John Wiley & Sons, New York, 1990.

5. Kuzawinski, M.J., and Blackwell, K.J., *Electronic Packaging and Production* pp. 56-63, Sept. 1997.

6. Brathwaite, N., Huey, K., and Chang, P., *Electronic Packaging and Production* pp. 65-77, Nov. 1997.

Author Index

D

E

F

G

H

I

J

K

L

M

N

O

P

R

S

T

U

V

W

X

Y

Z

Key Words Index

NOTE - This index is intended to be used in concert with the Table of Contents which provides general guidance as to where various topics are discussed. Entries in this index identify only locations where there is significant discussion of the indexed term. Because the vast majority of the papers in the volume relate to silicon or metrology, these terms are not listed in the index except under special circumstances. Page numbers following the term are the numbers of the first page of the article in which the term appears.

T

U